# Deepwater Sedimentary Systems

# Deepwater Sedimentary Systems

## Science, Discovery, and Applications

Edited by

**Jon R. Rotzien**
Basin Dynamics, Conroe, TX, United States; University of Houston, Houston, TX, United States

**Cindy A. Yeilding**
bp America (retired), Houston, TX, United States

**Richard A. Sears**
Energy Resources Engineering, Stanford University, Stanford, CA, United States

**F. Javier Hernández-Molina**
Department of Earth Sciences, Royal Holloway, University of London, Egham, Surrey, United Kingdom

**Octavian Catuneanu**
Department of Earth and Atmospheric Sciences, University of Alberta, Edmonton, AB, Canada

ELSEVIER

Elsevier
Radarweg 29, PO Box 211, 1000 AE Amsterdam, Netherlands
The Boulevard, Langford Lane, Kidlington, Oxford OX5 1GB, United Kingdom
50 Hampshire Street, 5th Floor, Cambridge, MA 02139, United States



ISBN: 978-0-323-91918-0

For information on all Elsevier publications
visit our website at https://www.elsevier.com/books-and-journals

*Publisher:* Candice Janco
*Acquisitions Editor:* Amy Shapiro
*Editorial Project Manager:* Helena Beauchamp
*Production Project Manager:* Paul Prasad Chandramohan
*Cover Designer:* Greg Harris

Typeset by STRAIVE, India

# Contents

## 4. Deepwater passive margin foldbelts

*Md Nahidul Hasan and Paul Mann*



## 5. Salt tectonics in deepwater settings

*Michael R. Hudec and Martin P.A. Jackson*



## 6. Deepwater sedimentary processes

*Jon R. Rotzien, F. Javier Hernández-Molina, Marco Fonnesu and Antoine Thieblemont*

## 9. Contourites and mixed depositional systems: A paradigm for deepwater sedimentary environments

*F. Javier Hernández-Molina, Sandra de Castro, Wouter de Weger, Debora Duarte, Marco Fonnesu, Tatiana Glazkova, Adam Kirby, Estefanía Llave, Zhi Lin Ng, Oswaldo Mantilla Muñoz, Sara Rodrigues, Francisco J. Rodríguez-Tovar, Antoine Thieblemont, Adriano R. Viana and Shaoru Yin*



## 10. Mass transport processes, injectites and styles of sediment remobilization

*T.M. Alves, S. Cardona and M.C.N.L. Rodrigues*



## 11. Source-to-sink analysis of deepwater systems: Principles, applications and case studies

*E. Szymanski, L. Fielding and L. Davies*

## 16. Imaging and interpretation: Seismic, rock physics and image log analysis workflows for deepwater systems

*Sumit Verma, Shuvajit Bhattacharya, Tom Fett, Per Avseth and Ivan Lehocki*



## 17. Seismic rock physics and machine learning for deepwater stratigraphic intervals

*Anshuman Pradhan, Per Avseth and Tapan Mukerji*

## 18. Integrating forward stratigraphic modeling with basin and petroleum system modeling

*Allegra Hosford Scheirer, Keyu Liu, Jianliang Liu and Oliver Schenk*



## 19. Technical (engineering) advancements enabling deepwater exploration and production

*Lori Fremin, Richard A. Sears and Charlie Williams*



## 20. Economic considerations and market condition effects in deepwater

*Cheryl R. Collarini, Henry S. Pettingill and Joe L. Stires*

# Contributors

*Numbers in parentheses indicate the pages on which the authors' contributions begin.*

**Derek D. Adams** (1), EarthBridge Energy, Houston, TX, United States

**T.M. Alves** (361), 3D Seismic Laboratory, School of Earth and Environmental Sciences, Cardiff University, Cardiff, United Kingdom

**Per Avseth** (555, 593), Dig Science/NTNU, Oslo; Dig Science/NTNU, Trondheim, Norway

**Shuvajit Bhattacharya** (555), University of Texas at Austin, Austin, TX, United States

**L.M. Bonnell** (471), Geocosm LLC, Durango, CO, United States

**Rebecca L. Caldwell** (251), Chevron Technical Center, Chevron Corporation, Houston, TX, United States

**S. Cardona** (361), Sediment Mechanics Lab, Department of Geology & Geophysics, Texas A&M University, College Station, TX, United States

**Octavian Catuneanu** (443), Department of Earth and Atmospheric Sciences, University of Alberta, Edmonton, AB, Canada

**Cheryl R. Collarini** (693), Collarini Energy Experts, Covington, LA, United States

**L. Davies** (407), Petryx Ltd, Gaerwen, United Kingdom

**Sandra de Castro** (301), Department of Earth Sciences, Royal Holloway, University of London, Egham, Surrey, United Kingdom

**Wouter de Weger** (301), Department of Earth Sciences, Royal Holloway, University of London, Egham, Surrey, United Kingdom

**Zackary M. Donovan** (723), Episcopal High School, Bellaire, TX, United States

**Javier Dorador** (533), Departamento de Estratigrafía y Paleontología, Universidad de Granada, Granada, Spain

**Debora Duarte** (301), Department of Earth Sciences, Royal Holloway, University of London, Egham, Surrey, United Kingdom; Divisão de Geologia e Georrecursos Marinhos, Instituto Português do Mar e da Atmosfera—IPMA, Lisbon, Portugal

**I. Espejo** (471), TPS Consulting Services, Houston, TX, United States

**Tom Fett** (555), Consultant, San Antonio, TX, United States

**L. Fielding** (407), Petryx Ltd, Gaerwen, United Kingdom

**Marco Fonnesu** (179, 203, 301), ENI, Rome; Eni SpA, San Donato Milanese, Milan, Italy

**Lori Fremin** (673), Shell—Retired, Houston, TX, United States

**Yuqian Gan** (1), University of Texas, Austin, TX, United States

**Tatiana Glazkova** (301), Department of Earth Sciences, Royal Holloway, University of London, Egham, Surrey, United Kingdom

**Lisa R. Goggin** (251), Chevron Digital Products and Services, Chevron Corporation, Houston, TX, United States

**Md Nahidul Hasan** (119), Department of Earth and Atmospheric Sciences, University of Houston, Houston, TX, United States

**F. Javier Hernández-Molina** (179, 203, 301, 723), Department of Earth Sciences, Royal Holloway, University of London, Egham, Surrey, United Kingdom

**Allegra Hosford Scheirer** (625), Stanford University, Stanford, CA, United States

**Michael R. Hudec** (149), Bureau of Economic Geology, The University of Texas at Austin, Austin, TX, United States

**Martin P.A. Jackson** (149), Bureau of Economic Geology, The University of Texas at Austin, Austin, TX, United States

**David Jutson** (515), Paleo-Data Inc., New Orleans, LA, United States

**Adam Kirby** (301), Department of Earth Sciences, Royal Holloway, University of London, Egham, Surrey, United Kingdom

**Benjamin Kirkland** (51), CNOOC Energy USA LLC, Houston, TX, United States

**R.H. Lander** (471), Geocosm LLC, Durango, CO, United States

**Ivan Lehocki** (555), Dig Science/Lehocki Geospace, Oslo, Norway

**Jianliang Liu** (625), China University of Petroleum, Qingdao, China

**Keyu Liu** (625), China University of Petroleum, Qingdao, China

**Estefanía Llave** (301), Instituto Geológico y Minero de España (IGME, CSIC), Madrid, Spain

**Paul Mann** (83, 119), Department of Earth and Atmospheric Sciences, University of Houston, Houston, TX, United States

**Oswaldo Mantilla Muñoz** (301), Department of Earth Sciences, Royal Holloway, University of London, Egham, Surrey, United Kingdom; Ecopetrol S.A, Colombian Petroleum Co, Bogotá, Colombia

**Olmo Miguez-Salas** (533), Departamento de Estratigrafía y Paleontología, Universidad de Granada, Granada, Spain

**Jon Minken** (1), Santos, Adelaide, SA, Australia

**Tapan Mukerji** (593), Stanford University, Stanford, CA, United States

**Zhi Lin Ng** (301), Department of Earth Sciences, Royal Holloway, University of London, Egham, Surrey, United Kingdom

**Andrew Pepper** (51), This is Petroleum Systems LLC, Fredericksburg, TX, United States

**Henry S. Pettingill** (693), Rose and Associates, Houston, TX, United States; Geo Ventures International, Ainsa, Spain

**Anshuman Pradhan** (593), Stanford University, Stanford, CA, United States

**M.C.N.L. Rodrigues** (361), Laboratory on Basin Analysis (Laboratório de Análise de Bacias, LABAP), Departamento de Geologia—Setor Ciências da Terra, Universidade Federal do Paraná, Centro Politécnico—Jardim das Américas, Curitiba, Brazil

**Sara Rodrigues** (301), Department of Earth Sciences, Royal Holloway, University of London, Egham, Surrey, United Kingdom

**Francisco J. Rodríguez-Tovar** (301, 533), Departamento de Estratigrafía y Paleontología, Universidad de Granada, Granada, Spain

**Jon R. Rotzien** (1, 179, 203, 251), Basin Dynamics, Conroe; University of Houston, Houston, TX, United States

**Oliver Schenk** (625), Schlumberger, Aachen, Germany

**Richard A. Sears** (673, 723), Energy Resources Engineering, Stanford University, Stanford, CA, United States

**Joe L. Stires** (693), Noble Energy, Retired, Houston, TX, United States

**E. Szymanski** (407), Utah Geological Survey, Salt Lake City, UT, United States

**T.R. Taylor** (471), SedPetrology LLC, Tucson, AZ, United States

**Antoine Thieblemont** (179, 203, 301), TOTAL S.A., Exploration & Production, EXPLO/GTS/ISS/CLS, Pau Cedex; TotalEnergies, Courbevoie, France

**Sumit Verma** (555), University of Texas Permian Basin, Odessa, TX, United States

**Adriano R. Viana** (301), PETROBRAS, Rio de Janeiro, Brazil

**Ryan Weber** (515), Paleo-Data Inc., New Orleans, LA, United States

**Charlie Williams** (673), Center for Offshore Safety, Executive Director—Retired, Houston, TX, United States

**Diane Woodruff** (1), Occidental Petroleum, Houston, TX, United States

**Cindy A. Yeilding** (723), bp America (retired), Houston, TX, United States

**Shaoru Yin** (301), Key Laboratory of Submarine Geosciences, Second Institute of Oceanography, Ministry of Natural Resources, Hangzhou, China

# Preface

Jon R. Rotzien, F. Javier Hernández-Molina, Cindy A. Yeilding, Richard A. Sears and Octavian Catuneanu

## Why deepwater?

Deepwater surrounds every continent, and during geologic history, marine conditions were common and persisted on continental land masses. Much of Earth (~63%) consists of present-day deepwater settings as well as deepwater deposits that have accumulated to form the sedimentary record. The oceans serve as and harbor vast resources. Wind, solar, currents, petroleum, metal and other mineral resources can be found in deepwater, and growing forms of alternative energy are just beginning to be viable offshore with huge potential. However, deepwater sedimentary processes and products are, globally, not fully understood. This is due to their remote and often inhospitable locations for humans. The habitats, vast resources and significant unknowns are among some of the reasons encouraging further exploration into the deepwater environments that exist on Earth and perhaps other planets.

This book was originally created to serve academia and industry by providing a current treatise on the methodologies required for mapping and interpreting deepwater sedimentary systems. As the book evolved, it became clear to all contributors that with the growing interest and investment in offshore resources, the deepwater processes described and documented in this volume are integral to evaluation of many forms of offshore industry, not just oil and gas. There has been increasing interest in alternative energy resources in modern economies as documented in numerous white papers, conference proceedings and political reporting, and this too helped guide the book's purpose. While this text highlights the science and technology necessary for successful deepwater oil and gas exploration and production, it also recognizes opportunities in deepwater more generally for all forms of marine and energy resource management, ocean interactions and geohazard and climate-change mitigation.

This effort evolved due to the confluence of multiple events in the early 21st century. First, anecdotal evidence and commentary from technical leaders highlighted that the world does not know enough about deepwater sedimentary systems. Due to the immense scale of the oceans and several other barriers, both technical and geopolitical, enough instrumentation cannot be deployed at reasonable cost to adequately sample deep marine settings. Much of what we do know about deepwater systems is a result of oil and gas exploration and development over the last few decades, and while these efforts have taught us a lot about deepwater environments, much of that work has been local in nature and focused on the exploitation of hydrocarbon resources. The need for a better, more holistic way to explore and characterize the ocean floors and what lies below became apparent.

Second, it was observed that international peer-reviewed journals, such as the *AAPG Bulletin*, *Interpretation*, *Marine Geology*, *Deep-Sea Research* and *Marine and Petroleum Geology*, have been publishing deepwater sedimentary system papers at a high rate. Review papers, however, are not abundantly available to integrate new, complex information, and to provide clarity on research derived from the wide variety of deepwater stakeholders from industry, academia and government.

Third, the demand for information and technical short courses on deepwater turbidite, contourite, mass-transport deposit (MTD) and mixed systems by professional organizations has led to perhaps the greatest number of conferences on these topics in recent years since the 1990s. There are numerous conferences every year pertaining to deepwater sedimentary systems, and knowledge of deepwater sedimentary systems has increased exponentially. However, very few of these conferences have been developed into peer-reviewed books that would enable the broad distribution of deepwater knowledge to the global community. Deepwater systems remain a growing, important and technologically current topic.

Fourth, most large energy company resource portfolios today include ancient and present-day deepwater reservoirs. Industry technical meetings as well as corporate quarterly reports for supermajors and large independents highlight activities and resources focused on deepwater reservoirs. The topic of a reference book came up in conversation at one industry meeting, where an executive mentioned, "The knowledge of deepwater systems has increased so rapidly, we have a tough time keeping informed of what is relevant to the business." This comment reflects the recognition by many managers and researchers that there is a need for a reference book that employees and students can use to increase their knowledge in deepwater best practices—in not only geology, but also geophysics, engineering, economics and energy markets. This book is also intended for those who may be transitioning to deepwater assets after working many years in fluvial and shallow-marine environments, for example. *What do they need to know? What will make them valuable team players in their new deepwater roles?*

Fifth, recent reduction in student enrollment in the geosciences limits the number of specialists being trained and equipped to deal with energy resource management for global population growth. Information on deepwater systems needs to be taught in universities around the world as part of a background to understanding and managing our current and future sources of offshore energy and geological resources. Further, this volume strives to document much of the knowledge that risks being lost as researchers and industry professionals retire.

Sixth, travel hit an all-time high late in 2019, with many professionals meeting face-to-face to discuss the knowns, unknowns and *unknown unknowns* of deepwater sedimentary systems in various meetings. These meetings started friendships, business partnerships and collaborations, which undoubtedly will lead to advances in knowledge, shared skills and improved technology in deepwater systems. The COVID-19 pandemic in 2020 reduced this travel, allowing time to address technical challenges and encourage collaboration on meaningful projects despite the lack of face-to-face communication.

Through these realizations, and as attested by technical experts, it became apparent that a book on deepwater sedimentary systems is greatly needed at this time.

According to Kurt W. Rudolph, retired Chief Geologist of ExxonMobil, "In my ~40 years of trying to understand deepwater depositional systems, my appreciation for the variability between systems, within systems, and within reservoirs has grown tremendously. This heterogeneity puts a premium on applying multiple models, methods, and datasets; moreover, integration across disciplines and keeping humble in the face of uncertainty are two keys to sustained success. While this complexity is a challenge, it also offers gateways in terms of new concepts and exploration/development opportunities."

Susan Morrice, Founder and President of Belize Natural Energy, adds, "What a tremendous undertaking. This in-depth, applied and broad reaching volume will save thousands of human hours for future explorers. To have this collection from revered, experienced scientists at their fingertips will enable these explorers of the future to expand their minds further and take that leap of faith needed for extraordinary results that humanity needs today!"

Niall McCormack, PhD, Cofounder and CEO of Causeway GT and former Vice President of Exploration for BHP, says, "Deepwater sedimentary systems will play a major role in our ability to harvest resources from the earth in a sustainable manner for decades to come. This is the first volume of its kind that looks to integrate all elements of the system, the tool set used to describe them and the different uses that we have for these systems. We are at a point of transformation of our energy systems; how we study them and write about them needs to transform, too. This is the first book to do that."

Liz Schwarze, Vice President of Global Exploration for Chevron, says, "Characterizing rocks in the subsurface has never been more important as the industry looks for and develops new resources to provide affordable and reliable energy and as we apply our skills toward decarbonizing solutions such as carbon capture utilization and storage. Complex deepwater systems have all the elements—reservoir, traps and seals—and are a vital part of today's and tomorrow's energy ecosystem. Describing their architecture from seismic given outcrop analogs has long been our practice, but now with improved computing and data, we can harness the power of modeling to create that critical link between what we know from our data sets and the realities that those data sets represent. The opportunity to better describe uncertainty through finding not one but many models that fit our known data points drives smart data collection throughout the life of a project and gives us confidence to invest."

Mason Dykstra, Vice President at Enthought and former Geologic Advisor at Anadarko, mentions, "The depths of the oceans still hold riches and mysteries for those researchers intent on exploring in deepwater environments. Resources include oil and gas to fuel our current energy infrastructure, nodules of rare minerals crucial to the energy transition, as well as hosting the cables that connect our digital planet. In addition, the deep holds a record of the harm we are doing to our planet in the form of shipwrecks, drift nets, microplastics, and other detritus. The deep oceans are incredibly important to our planet, yet remain a relatively poorly understood environment, both in the modern, and in the ancient, all of which make a volume on deepwater sedimentary systems more relevant than ever."

Cindy Yeilding, editor and former Senior Vice President of BP, says, "Geoscience is one of the most important disciplines of the 21st century. Geoscientists have chosen to be, and been accredited as, experts in earth systems. We are therefore stewards for the environment. As humankind investigates opportunities in deepwater, we must apply knowledge of current and paleogeologic systems to assure that activities and infrastructure are safe and their environmental impacts are minimal. Knowledge of the processes and systems described in this volume are an essential starting point and we must continue to research offshore and share lessons learned as our knowledge of this vast frontier develops and matures."

F. Javier Hernández-Molina, editor, marine geologist and professor in Sedimentary Geology at the Royal Holloway University of London, says, "In recent decades, our knowledge and understanding of deepwater sedimentary systems has grown very rapidly and the interrelation between gravitational, hemipelagic and bottom current processes are proving to be more complex than previously thought. The reality is that there is still a lack of documented empirical data on the sedimentary processes on these sedimentary environments, especially in the bottom boundary layer—this is a great unknown! The monitoring and quantification of sedimentary processes, with the help of numerical models of gravity flows and bottom currents due to water masses circulation are essential in the future to fully understand and quantify the dynamics of sediment laden water and the associated secondary oceanographic processes (eddies, deep-sea storms, deep tides, internal waves, etc.). From this a new revolution in the knowledge of the deepwater sedimentary systems will be possible *to connect* hydrodynamics to process to deposit, enabling the development of new sedimentary models that better explain the formation of deepwater sedimentary intervals. A symbiosis of science, environment and industry is more needed than ever, promoting the investigation of deepwater sedimentary systems to adopt new perspectives that involve a greater knowledge of the processes, including the *differentiation of their* synchronous and asynchronous interactions, to understand the formation of deposits in deepwater sedimentary systems."

Richard Sears, editor, former VP of Shell and adjunct professor at Stanford University, says, "I remember the early days of offshore exploration, including deepwater, where the 'hey lookies' on seismic data attracted our attention and started us down the path of understanding what it was we were looking at. Our understanding of geologic processes and the geophysical techniques used to image them has brought us to where we are today. Because there has been, and is, much that we don't know, it is often an idea, a concept, that initiates the exploration process. As the idea matures, we look for evidence of it in our data sets. When we find it, it is often a case of 'I never would have seen it if I hadn't believed it'. This volume is the first attempt for deepwater to move beyond the wild concept stage, and organize and present what we know today so that when we develop new concepts, we are able to come down to Earth and properly understand and explain them to others."

Such a book that to date has not been tackled due to the technological newness of the subject and the resulting inconsistencies in terminology, interpretations and approach has been a considerable challenge. One way to address this challenge was to include a global, diverse group of contributors from different continents, schools of thought and occupations (industry vs. academic vs. government vs. entrepreneur). This book was not written from any one point of view. There are authors representing many backgrounds, with the aim of achieving a global perspective of the state of the science of deepwater sedimentary systems. A large and varied team has worked together to achieve the best results. The writing of this book has facilitated significant knowledge sharing across professions, industries and generations among the authors, and it is the intention that the readers and users of this book benefit from this broad perspective.

Refining the message of the book was difficult at times because there is a wide audience seeking information about deepwater sedimentary systems. Colleagues who reviewed the book proposal, some of whom represent heads of exploration, researchers or development geoscientists for energy companies, recommended the book remain steadfast in its message to inform the oil and gas industry. However, the intended audience for this book includes business professionals (science, engineering, technology, management), researchers (including professors, lecturers, pre- and post-docs and university students) and staff at government agencies around the world. A wide variety of professionals can use the information provided in this book, including those employed in wind and solar energy, carbon capture storage and use, geothermal energy, mining and metals, communications, investing and fund management, market analysis and insurance. This is intended to be a practical guide to characterizing and interpreting deepwater sedimentary systems, which it is hoped all audiences will appreciate. What will resonate with industry professionals is how the book is modeled after a typical supermajor energy company internal training program. During the first 15 years of employment, training is an important part of an employee's development. This book was crafted to reflect an internal training program in the geosciences, including skills and knowledge in geology, geophysics, engineering and economics, as well as some of the softer skills pertaining to creativity in the geosciences. Naturally, the book starts with big scales and fundamental topics of geology and then progresses to workflows and multidisciplinary, integrated topics. The book culminates with chapters on engineering capabilities and economics, demonstrating why subsurface geological and geophysical characterization is such an important foundation for successful upstream investment and resource development in deepwater. Many of the book authors teach their subject matter expertise internally at various organizations, externally at conferences and at meetings organized by professional organizations such as AAPG, EAGE, SEG and SPE. The authors have worked on every oil-producing continent and across the upstream value chain. Collectively, the authors have taught their expertise at hundreds of universities. This book will prepare the professional as well as the student very well for a career in mapping and interpreting deepwater systems to inform smart business decisions. The book's thesis is simple: it aims to provide huge value to those employed in the evaluation of deepwater opportunities.

Great volumes have been written on turbidite reservoirs (Weimer et al., 2000), deepwater systems (Pickering and Hiscott, 2015) and the sedimentology and stratigraphy of deepwater deposits (Shanmugam, 2006). There currently are no volumes that address all aspects and controls in a holistic fashion—and from large to small scale—on deepwater sedimentary systems. This comprehensive book will allow readers to:

- Understand the fundamental controls on deepwater sedimentary systems, their building blocks and their role in society (Chapter 1)
- Apply workflows and descriptions of how source rocks and fluid systems are modeled and interpreted in deepwater systems (Chapter 2)
- Characterize deep crustal architecture of sedimentary basins and their internal anatomy (Chapter 3)
- Analyze deepwater fold belts, which are difficult to image yet host huge accumulations of petroleum along passive margins (Chapter 4)
- Interpret salt and mobile substrates more generally for evaluating basin evolution, petroleum systems and trap analysis (Chapter 5)
- Observe and measure deepwater processes of sedimentation to determine the nature of the deposit and interpret likely depositional environment (Chapter 6)
- Identify sedimentary structures and common sedimentation units that link process to deposit and aid in stratigraphic correlation in sedimentary environments (Chapter 7)
- Describe the main depositional environments found in deepwater systems dominated by downslope currents and flows (Chapter 8)
- Differentiate turbidite systems from systems dominated by along-slope processes such as contour currents (Chapter 9)
- Characterize mass-transport deposits and complexes, and other more exotic forms of deepwater sedimentary processes and deposits (e.g., injectites), and note their role in the significant reshaping and constructing of deepwater sedimentary systems (Chapter 10)
- Apply methods in a holistic way in source-to-sink analysis by understanding the drivers and "reservoirs" for sediment along its path from mountain top to deep basin floor (Chapter 11)
- Use sequence stratigraphy to predict sedimentation patterns despite the sedimentological variability of deepwater systems (Chapter 12)
- Understand the range and variability of controls on reservoir quality of deepwater sandstone intervals (Chapter 13)
- Apply skills in micropaleontology to characterize and correlate deepwater sedimentary deposits (Chapter 14)
- Characterize deepwater ichnofacies and their likely sedimentary environments to infer their impact on reservoir quality of deepwater stratigraphic intervals (Chapter 15)
- Understand modern subsurface imaging and interpretation methods used to map and interpret deepwater sedimentary systems at various scales (Chapter 16)
- Use methodologies in seismic rock physics and machine learning to characterize deepwater reservoirs (Chapter 17)
- Visualize the intersection between forward stratigraphic modeling and deepwater sedimentation to enhance interpretation of deepwater stratigraphic intervals (Chapter 18)
- Understand the technological capabilities driving and enabling deepwater oil and gas exploration and development over the past half century (Chapter 19)
- Review the economic principles and global market factors that influence decisions to invest in deepwater exploration and development (Chapter 20) and, finally,
- Envision the uses of the deepwater areas—both subsurface and other environmental applications—over the next century (The Next 100 Years)

While it is exceedingly difficult to become an expert in all of these disciplines, it takes an integrated understanding of these skills and methods to achieve an unobstructed perspective of deepwater sedimentary systems.

For current deepwater practitioners, at all levels, this is a valuable compendium of current knowledge. For those looking at a future career involving deepwater energy systems, this volume represents their starting point: what we know about deepwater today and the directions that it is going. For those who do not intend to focus on deepwater per se, but see themselves as future thought leaders in business, academia or government, this text can provide them with invaluable knowledge of an environment, deepwater, that will undoubtedly be important in the affairs of mankind for decades to come.

Exploring and developing in the deepwater requires the skills of many individuals working toward a common goal—in the same manner this book was created. Thank you for your interest in deepwater sedimentary systems. Enjoy!

**Acknowledgments**
Understanding deepwater sedimentary systems is a collaborative effort. Deepwater nirvana probably cannot be achieved by one person or one team or one company. Its breadth involves the scientific and technology communities, energy companies, government organizations, deep-sea explorers and entrepreneurs. Through the course of our careers, we have been fortunate to meet, learn from and collaborate with many individuals and organizations, far too many to list or name. Some of the key individuals who helped guide the thinking and direction of this book are thanked in each chapter for their contributions, thoughtful conversations and interactions over many years. We owe the inspiration expressed in this book and much of the analysis to many colleagues, advisors, friends and teachers, above all to the distinguished group with whom we've been privileged to be associated at the University of Houston, University of Texas, South Dakota School of Mines and Technology, Stanford University, Royal Holloway University of London, University of Alberta, Colorado College, BP, Amoco, Shell, Devon and Hess. We have learned so much from them. And above all else, we have learned so much from the authors of this book who are thanked for their dedication to creating this modern treatise that brings a holistic perspective to deepwater sedimentary systems. Many of the concepts shared here originated from collaborations and conversations since ~1970. The editors wish to thank Cheryl Collarini for her help in developing and refining the message in this preface.

From the editors, and on behalf of the book team.

## References

Pickering, K.T., Hiscott, R.N., 2015. Deep Marine Systems: Processes, Deposits, Environments, Tectonics and Sedimentation: Wiley and American Geophysical Union. ISBN: 978-1-4051-2578-9.

Shanmugam, G., 2006. Deepwater Processes and Facies Models: Implications for Sandstone Petroleum Reservoirs. Elsevier, Amsterdam 476 pp.

Weimer, P., Slatt, R.M., Coleman, J., Rosen, N.C., Nelson, H., Bouma, A.H., Styzen, M.J., Lawrence, D.T. (Eds.), 2000. Deepwater Reservoirs of the World. GCSSEPM Foundation 20th Annual Bob F. Perkins Research Conference, 1105 p.

# Frontispiece

Deepwater sedimentary processes and deposits are among the most challenging aspects of Earth to observe and to measure. This book characterizes the geoscience of deepwater sedimentary systems—from the basin to the pore—and describes what makes them prolific sources of petroleum, among other natural resources. Deepwater sedimentary systems are underexplored, represent a frontier and host myriad opportunities for human-led endeavors. Understanding the geoscience of deepwater is one aspect of evaluating this potential. (*Top photo*) A classic turbidite composed of a complete Bouma sequence, Upper Miocene Mount Messenger Formation, Rapanui, Taranaki Basin, North Island, New Zealand. (*Bottom photo*) The Olympus Tension Leg Platform in the deepwater Gulf of Mexico, United States. Operated by Shell, Olympus, in approximately 3100 ft. (950 m) of water, is the third tension leg platform installed in the prolific Mars Basin. Top*: Jon R. Rotzien;* bottom*: Shell.*

Chapter 1

# Introduction to deepwater sedimentary systems

Jon R. Rotzien[a,b], Diane Woodruff[c], Derek D. Adams[d], Yuqian Gan[e] and Jon Minken[f]

[a]*Basin Dynamics, Conroe, TX, United States,* [b]*University of Houston, Houston, TX, United States,* [c]*Occidental Petroleum, Houston, TX, United States,* [d]*EarthBridge Energy, Houston, TX, United States,* [e]*University of Texas, Austin, TX, United States,* [f]*Santos, Adelaide, SA, Australia*

## Introduction

Deepwater sedimentary systems have been studied for more than a century. From the first oil production from Pliocene-aged turbidite reservoirs in California, USA, in the 1860s, to the first documentation of sediment gravity flows in deep Swiss Lakes in the late 19th century, deepwater sedimentary processes and deposits have garnered attention from scientists to entrepreneurs. Deepwater sedimentation occurs over a large part of the Earth today (~70% +), and ancient deepwater deposits are found onshore in sedimentary basins and rock outcrops. Although their sedimentary processes can only be observed indirectly or reconstructed at orders of magnitude smaller scale in research labs, deepwater processes include powerful currents, flows and cycles that balance Earth's chemical and sediment distribution. These processes also play a role in global and regional climate. Not only are deepwater stratigraphic intervals the sources of significant petroleum resources, developed at an increasingly rapid pace offshore since the late 19th and early 20th centuries, deepwater basins offer great potential for renewable energy sources such as wind, solar and geothermal. Deepwater stratigraphic intervals also host important elements and minerals used in electric devices, such as cell phones and car batteries. These are both finite and renewable energy resources. In particular, the discovery and development of hydrocarbon resources in present day deepwater basins has been extraordinary in its pace, global breadth and for the enormous economic value added through the utilization of these resources. Missing from the literature is a modern compilation of the world's understanding of deepwater sedimentary systems and the application of that knowledge to resource development and the impact on the rock and fluid envelopes of Earth—from the crust, ocean and lakes and atmosphere. Hence, this book attempts to fill that gap and serve as an academic reference to those studying and investigating ancient and modern deepwater systems, a comprehensive introduction for students and a review of the associated technology that forms the basis of oil and gas development in deepwater today.

This book is a collection of chapters written by leaders in their respective fields. The authors come from many different backgrounds—oil and gas, research, academia, government, entrepreneurs—and range from current and new PhD students researching deepwater systems never observed in detail before to business executives whose teams have pioneered technology to achieve exploration and production milestones for industry. The authors describe many different reasons to study deepwater systems based on their own research aims, experiences and perspectives. To some, it may seem like too broad a collection of topics and authors for one book. However, deepwater systems cannot be understood or taught by a few individuals or by one company alone. A global team effort is required to see the *whole playing field*. All disciplines—geology, geophysics, engineering, economics, the explorer's mindset—are called on to adequately assess and characterize deepwater sedimentary systems at various scales to inform scientific and business decisions. Approaches taken from each of these professions are interwoven throughout individual chapters and highlight the interdisciplinary and multi-disciplinary methods needed for problem-solving in deepwater sedimentary systems.

This introduction provides important context and foundational definitions for deepwater sedimentary systems. Use this chapter as a guide to key themes that can be read in-depth in each of the other 20 chapters, as well as the preface. Upon successful completion of this chapter, the reader will have been introduced to the following concepts and be fully prepared to explore them in more detail in the rest of this book:

1. The definition of deepwater sedimentary systems
2. The history of deepwater sedimentary systems studies

**Deepwater Sedimentary Systems.** https://doi.org/10.1016/B978-0-323-91918-0.00003-7

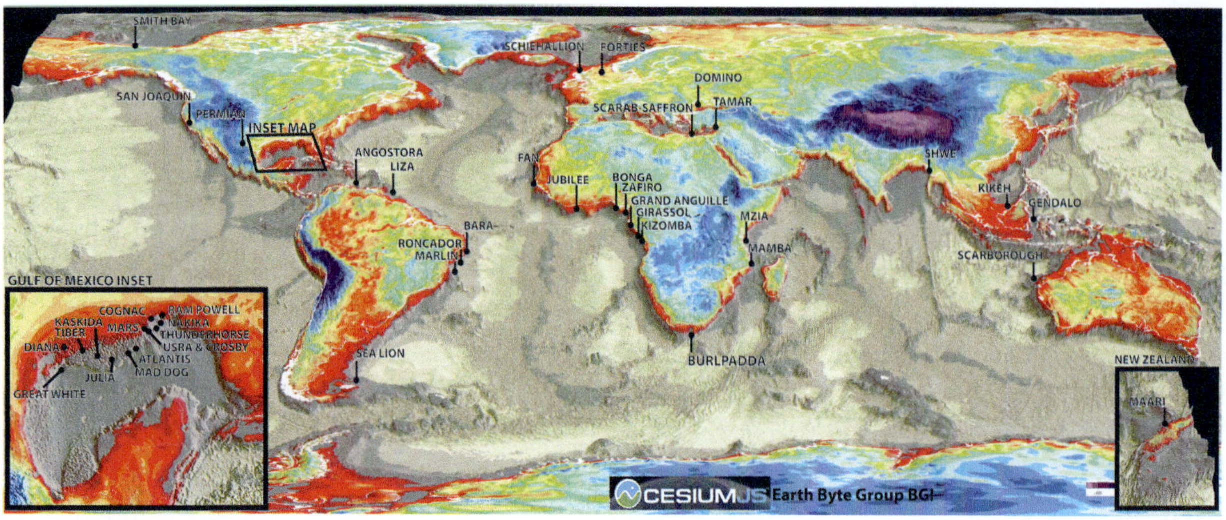


**FIG. 1** Global map of major oil and gas exploration successes in the last century. Note the concentration of large discoveries along passive margins in the US Gulf of Mexico and Atlantic Basin.

3. The drivers and regional controls of deepwater sedimentation
4. Common deepwater architectural elements
5. Deepwater terminology and its challenges in usage
6. Oil and gas exploration and development in deepwater
7. The exploration process and the explorer's mindset
8. Engineering principles of offshore and deepwater exploration and production
9. Economics of deepwater exploration and production
10. Key messages
11. Conclusions and future areas of investigation for understanding deepwater sedimentary systems

While *deepwater exploration* is commonly associated with petroleum resource exploration and extraction, exploration of all types is represented in ocean and deep lake frontiers. The above themes introduce and examine ocean and sedimentary processes, the frameworks and drivers of deepwater sedimentation, how deepwater systems are measured and quantified, why they are studied and how deepwater systems affect and are influenced by society and technology.

The global map of major oil and gas fields (Fig. 1) indicates that deepwater exploration and production covers surprisingly little footprint compared to the available regions for future exploration. Many more continental margins presumably host petroleum resources, yet have not been explored. Though a majority of the research and information to date on deepwater systems stems from the oil and gas industry, and therefore a weighted discussion in this book focuses on this industry, future applications extend beyond oil and gas for deepwater as the global energy mix changes. A primary objective of this book is to provide workflows for exploring and characterizing deepwater sedimentary systems for use in applied research, oil and gas exploration, future energy systems, climate and ocean modeling and basic scientific discovery. These aims are not only integral for scientific understanding, but also vital to industries and the communities they support.

To begin to understand deepwater sedimentary systems, it helps to start with the big picture. *What are they? How do they form? How are they measured or quantified? What is their significance?* Enjoy the dive from submarine canyon head to deepwater abyssal plain.

Note: this introduction—like all chapters in the book—contains comprehensive lists of references compiled by experts chronicling decades of history in the understanding of deepwater systems from many perspectives.

## Defining deepwater systems

Deepwater depositional systems are areas of net accumulation of sediment in water depths below wave base (generally greater than 200m (656 ft.) sensu Pickering and Hiscott, 2015), whereas deepwater *sedimentary systems* are not purely depositional (see Chapter 9). The 200m (656 ft.) depth commonly exceeds storm wave base, so that any deposits found at these depths are the products of sediment gravity flows or bottom currents. However, *deepwater* from the standpoint of water depth is defined differently by different operators and practitioners, and with the advent of new drilling technology and experience since the early 20th century (Weimer and Slatt, 2004; see Chapters 19 and 20), deepwater activity has become more commonplace. Deepwater systems can be found in lakes (lacustrine systems) or oceans (marine systems). Depositional products in deepwater systems range from siliciclastic gravel and mud to fine-grained carbonate and reef detritus. Petroleum systems elements, such as reservoir, seal and source rock are highly variable (Saller et al., 2004; see Chapter 2). For sediment to reach deepwater, it typically must first pass through terrestrial and shallow water environments. The transfer of sediment to deepwater can be a result of gravity acting on the sediment, such as rock fall, slide, slump, creep, debris flow or grain flow. Sediment can also be transported via liquefied or fluidized flow or turbidity flow processes (see Chapters 6 and 12). Other processes that affect deepwater sedimentary systems include currents operating in shelf and slope settings such as tides, waves, upwellings, internal waves, bottom currents and surface currents. Background sedimentation, or sedimentation occurring without the aid of sediment gravity flows or bottom currents, typically consists of pelagic sedimentation that is not necessarily derived from the coastal and shallow marine environments (see Chapters 6 and 7).

Engineering definitions for deepwater have been largely based on modern water depth, which requires different equipment and technology to extract, store and process petroleum. For example, a platform fixed to the sea floor can be used in shallower water depths but in a deeper water setting floating production solutions, such as a floating production storage and offloading facility (FPSO), with subsea tie backs may be required (see Chapter 19 for more production systems). Not surprisingly, technology has enabled the concept of deepwater to become deeper through time. In 1937, deepwater was defined as 14 ft. (~5 m) water depth offshore Creole, Louisiana, USA (Angelle, 2021). In the 1970s and 1980s companies ventured farther into the US Gulf of Mexico and deepwater operations moved from water depths more than 1000 ft. (~305 m) to greater than 5000 ft. (~1520 m). As discussed in Chapter 19, deepwater wells can be drilled in greater than 12,000 ft. (~3660 m) of water today.

*If sediment is deposited at deep paleo water depths, but then is uplifted in a mountain range, is the rock a deepwater deposit?* Yes, if the sediment accumulated in a deepwater environment, but now forms lithified rock in a sub-aerially exposed mountain belt, the rock is indeed deepwater in origin but is now located in a modern terrestrial environment. However, what if the rocks were deposited in a shallow lacustrine carbonate environment and are buried offshore today in 2 km (~ 1.25 mi.) water depth and 7 km (~ 4.35 mi.) below mudline? While the rocks are located in a modern deepwater environment, they were not formed in a deepwater sedimentary environment. In order to answer similar questions regarding the definition of deepwater deposits, the history of research and studies conducted on these systems must be understood to fully appreciate current thinking.

## History of study of deepwater sedimentary systems

The history of study of deepwater sedimentary systems started in the middle to late 19th century, grew steadily through the mid-20th century and has remained a focus of industry research studies from the 1990s to present day. This section and Fig. 2 highlight how deepwater sedimentary systems came to be so popular for industry, academia and government endeavors.

### Early days

Thick deepwater accumulations of relatively coarse-grained sediment are largely the product of sediment gravity flows, and great attempts have been made to study them even before the advent of modern reservoir characterization efforts. Forel (1885, 1887) made the first observations on density currents in deep lakes in Switzerland. Murray and Renard (1891) conducted the first sampling of deep-sea sedimentary deposits. Milne (1897) calculated sediment gravity flow speeds from submarine communication cable breaks around the world. Following that lead, scientists began to realize that some cable breaks were due to submarine landslides and ensuing flows following natural hazards, with the most notable being the 1929 Great Banks Earthquake offshore Newfoundland (Heezen and Ewing, 1952; Heezen and Drake, 1964; Piper et al., 1988). Because these natural events were rare, deep and difficult to predict and to monitor, lab experiments were designed to recreate sediment gravity flows in various tanks (Kuenen, 1950; Middleton, 1966), in which the scales of flow speeds, ambient water, flow density and submarine geomorphology were approximated to achieve realistic outcomes. Offshore examinations of natural habitats for sediment gravity flows, including submarine canyons, offered insight into turbidity currents (Johnson, 1938; and other studies by Daly, Stetson and Shepard), and eventually led to the term *turbidite* by Kuenen (1957). The intrigue and unknowns of the deep gained in popularity such that famous pioneers of deepwater processes (i.e., Kuenen, Miglorini and Shepard) organized the first major conference on sediment gravity flows and their deposits in 1948 at the 18th International Geological Congress in London, England.

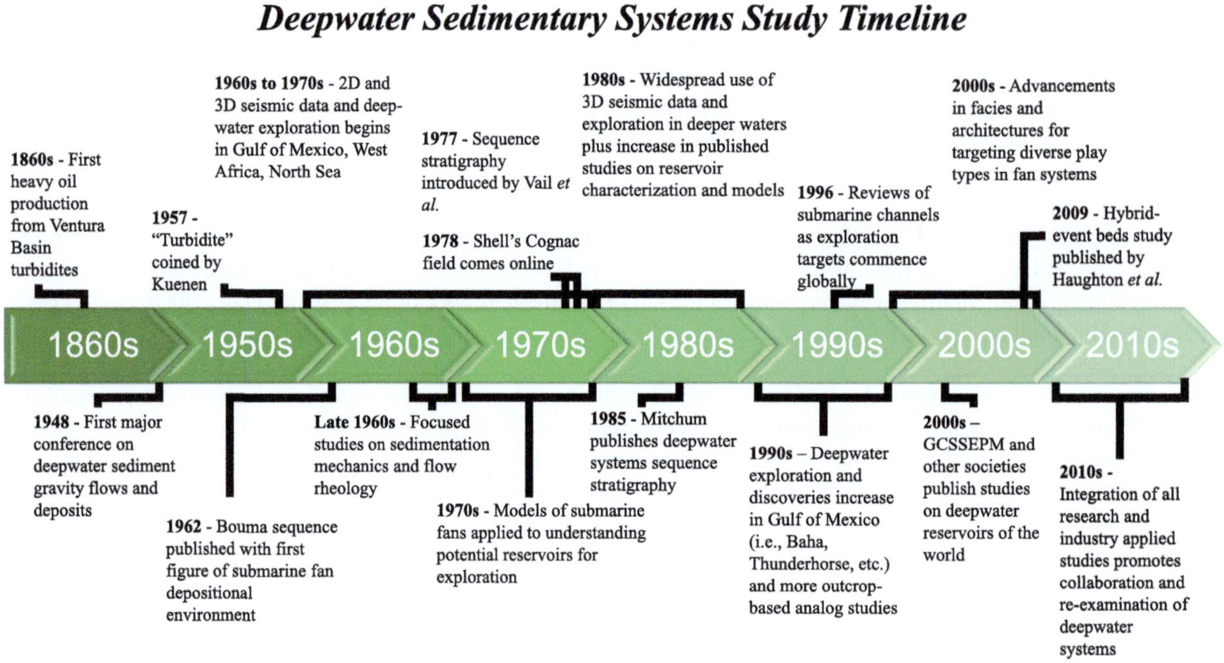


**FIG. 2** Timeline highlighting some of the key events in the history of deepwater sedimentary systems' studies. For additional detail, please see the accompanying text.

## Early 20th century

Knowingly and unknowingly, scientists began mapping and measuring ancient deepwater stratigraphy in the field in the early 20th century, identifying differentiating features of the deposits to contrast with those from more readily observable modern terrestrial and shallow marine depositional systems. These studies revealed characteristic sedimentary structures preserved in deepwater clastic deposits (Sheldon, 1928; Signorini, 1936) that could be used to understand the sedimentation processes associated with sediment gravity flows. Many of the beds, also termed *sedimentation units*, exhibited graded bedding (Bramlette and Bradley, 1940; Kuenen and Miglorini, 1950; Pettijohn, 1957). The graded beds of the Eocene-Oligocene Annot Sandstone, France with systematic arrangement of sedimentary structures in a vertical profile would later become part of the Bouma (1962) Sequence (see Chapter 7) and one of the first depictions of sediment distribution in a submarine fan environment (Fig. 3A). Also in the Annot Sandstone (Stanley, 1963) and in the USA (McBride, 1962), similar observations provided the foundation for new deepwater facies models. All of these deposits were interpreted to represent large-scale depositional systems, such as submarine fans that were first observed offshore California, USA, by Menard (1955). Additional examination by Crowell et al. (1966) elaborated the types of structures and the vertical sequencing of sedimentary structures, or first-order architectural elements, in the Ventura Basin, USA, where production of heavy oil from turbidite reservoirs first began in the 1860s by Josiah Stanford and others (Schwalbach et al., 2009). Sedimentation mechanics studies (Sanders, 1963) yielded new information on structures best known for sediment bypass, including the traction carpet (Dzulynski and Sanders, 1962) and a process that could enable long-lived and momentous turbulent suspensions of sediment in water, termed *auto-suspension* by Bagnold (1962). By the 1960s, flow rheology (Dott Jr., 1963) and along-slope (i.e., contour current) vs downslope processes and deposits had been observed and differentiated (Murphy and Schlanger, 1962). Hollister (1967) noted how contour current-sculpted sedimentary systems (see Chapter 9) differed substantially from downslope-oriented turbidite and mass transport depositional systems.

Prior to the 1970s, most deepwater sedimentation was interpreted to result from transport and deposition of fine-grained sand, silt- and clay-sized particles whereas larger gravel-sized conglomeratic formations were interpreted as everything *but* deepwater deposits (Tyler, 1972). Only later, high-density turbidity currents (Lowe, 1979, 1982) were identified as the transport mechanism for gravel-sized clasts. That invoked other physical processes (i.e., grain-to-grain interactions, hindered settling, buoyancy, etc.), not just fluid turbulence, to carry gravels in turbulent suspension and laminar traction (see Chapters 6 and 7).

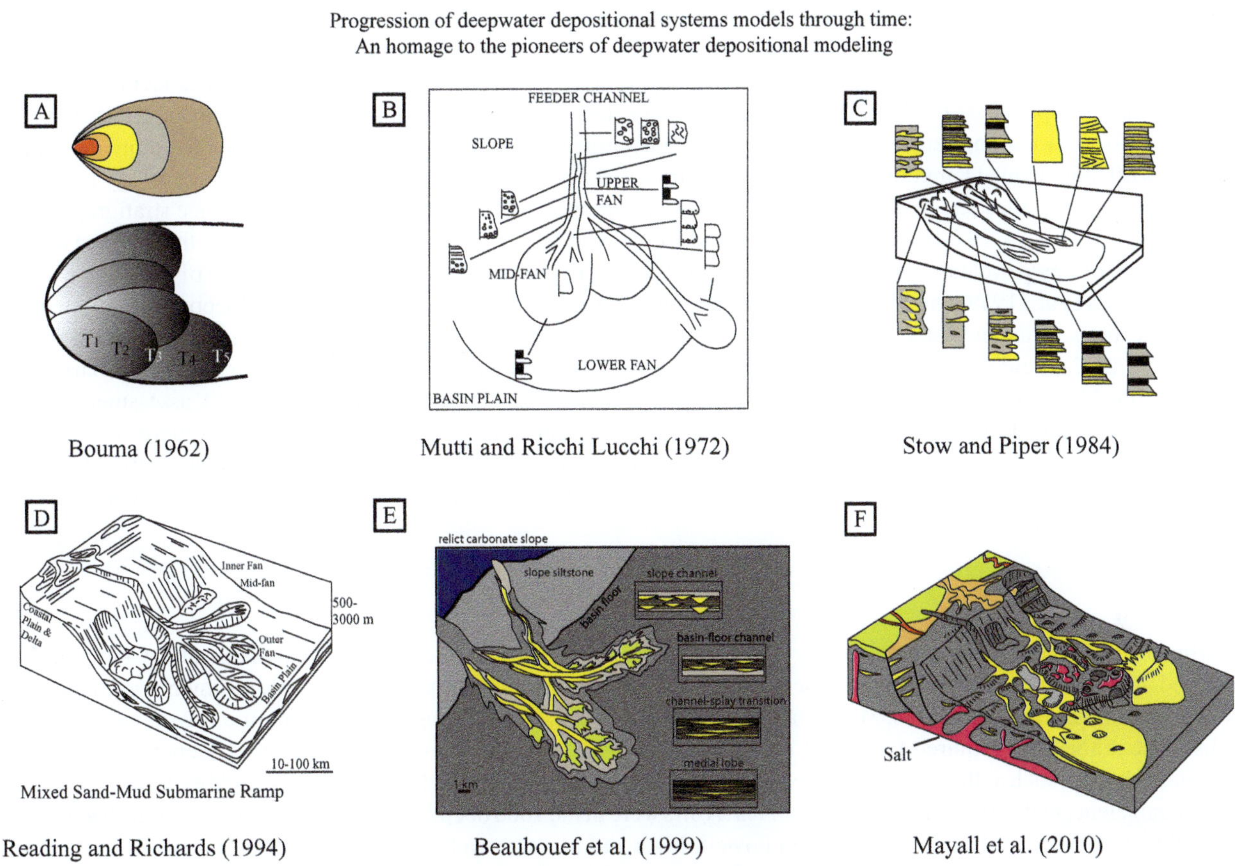


**FIG. 3** Deepwater sedimentary systems models from ~1960 to present. While the term submarine fan is used, note that the following diagrams show evidence of fans, fan valleys, aprons, mass transport complexes, contourite drifts and a variety of large-scale mass-movement deposits that could help construct a fan in the general sense of the term, but are not classically thought of as submarine fan deposits. In the 1960s (A), deepwater sedimentary models were used to explain the variation in Bouma (1962) sequences, with Ta intervals being deposited primarily in proximal positions, and Te intervals found in greater abundance in distal positions. In the 1970s (B), submarine fan models illustrated submarine channel, levee and splay (lobe) settings, each with a typical facies model incorporating gravelly to muddy sediment. The 1980s (C) showed canyon, channel, levee and splay and overbank environments in more detail with higher resolution on the bathymetric profile of the basin. The 1990s (D) witnessed a big increase in the diversity and settings for deepwater deposition, in part summarized by the excellent work of Reading and Richards (1994). An exponential increase in how to classify deepwater systems described muddy to gravelly depositional systems along point- and line-sourced continental margins. The expansion of deepwater outcrop knowledge illustrated key systems including the (E) Brushy Canyon Formation, USA, Tanqua-Karoo, South Africa, Jackfork Group, USA, and Great Valley Sequence, USA. Along with the reduction in oil prices throughout the 1990s, these publications by Beaubouef et al. (1999), Sullivan et al. (2000, 2004), Lowe (2004) and Sprague et al. (2005) highlighted the internal architecture of canyon, channel and splay environments. In Deptuck et al. (2007), (F) Mayall et al. (2010), Sylvester et al. (2011), McHargue et al. (2011), supermajor operators including Shell, BP and Chevron demonstrated their depositional system and channelized models for deepwater systems around the world. Publications in 2020 have shown how bottom currents affect the deposition of sediment brought downslope by gravity (De Castro et al., 2020; De Weger et al., 2020; Fonnesu et al., 2020; Miramontes et al., 2020).

## 1970s

Improvements in imaging and sampling techniques (see Chapters 16 and 19) in the 1970s helped refine submarine fan models to channelized and non-channelized splay environments, including the suprafan-splay (Normark, 1970, 1991), the depositional lobe (splay) (Mutti and Ghibaudo, 1972) and detached lobe (Mutti and Ricci Lucchi, 1972). These first realistic submarine fan renditions were drawn and used to explain reservoir and non-reservoir components and "decay" of turbidite reservoir quality away from high-energy feeder environments such as canyons and channels (Fig. 3B). However, mass transport-dominated environments showed very different, almost disorganized, morphology from turbidity current-dominated fans (Walker and Massingill, 1970). Systems with multiple smaller simultaneously active canyon-channel-gully environments, typical of active margins or high-latitude settings, modified by contour currents were shown by Stow and Piper (1984) (Fig. 3C). Enhancements in ocean data acquisition (see additional sections later in this chapter, as well as Chapters 16 and 19) yielded higher resolution images of the variety of seafloor features (Heezen and Hollister, 1971), offering inspiration to many play-based petroleum exploration cartoons. Submarine fan growth and sedimentation patterns were observed in the relatively unconfined fan environments such as the Bengal Fan offshore India (Curray and Moore, 1971) and Navy Fan offshore USA (Normark and Piper, 1972), whereas sedimentation patterns revealing flow confinement were observed using sidescan sonar in the Santa Barbara Channel, also offshore California, USA (Sanders, 1973).

With the widespread use of 2D and then 3D seismic data, deepwater exploration began in earnest in the 1960s and 1970s (Chapter 19). In addition, geopolitics (Yergin, 1990) encouraged Western companies to explore for more oil and gas resources offshore United Kingdom, USA and West Africa, yet new methods were required to predict lithology in offshore, frontier acreage. A milestone in deepwater sedimentary system research—Exxon's method of sequence stratigraphy—was born (Vail et al., 1977a, 1977b—see Chapter 12). Shell and several other companies began to employ this method to assess the presence of key petroleum risk elements (Chapter 2) along passive margins at a scale and order of magnitude smaller than how Sloss et al. (1949) and Sloss (1963) had previously defined the sequence. The science of deepwater sedimentation and applications of its understanding proliferated in process review papers and experiments on subaqueous debris flows (Hampton, 1972), field development case studies on stratigraphic architecture and facies geometry in billion-barrel fields including the Ventura-Rincon-San Miguelito-Santa Barbara trend, USA (Hsü, 1977; Hsü et al., 1988), assessments of stratigraphic traps in broad submarine fan environments and their geologic risks (Walker, 1978) and descriptions of wireline log signatures for deepwater reservoir and non-reservoir intervals (Selley, 1979). Deepwater reservoir characterization studies were essential to development planning for fields coming online, to name a few, Shell's Cognac and later Mensa and others in the US Gulf of Mexico.

## 1980s

Starting in the 1980s, the petroleum industry's widespread acceptance of 3D seismic data's immense value (Cranberg, 2020) enabled oil and gas operators to better characterize continental margin architecture and deepwater sedimentary fill. While not without its limitations and pitfalls, 3D seismic data was (and still is) the most useful and economic way of interpreting the subsurface at the scale required for the oil and gas industry (see Chapter 16). Operators ventured deeper into established petroleum basins, such as the US Gulf of Mexico, because production rates from deepwater wells could exceed 10,000 barrels of oil equivalent per day (boepd), whereas shelf reservoirs were in the 1000 boepd range (see Chapter 19). Large discoveries in the US Gulf of Mexico, beginning with Ram Powell, Mica and Mars, as well as the discoveries offshore West Africa and Brazil, established new deepwater play fairways. The nature of the prospect- and reservoir-level questions demanded higher specificity regarding the plumbing of deepwater sedimentary systems, their architecture, their stratal stacking patterns and their 3D heterogeneity. A varied and significant body of work was undertaken including: depositional models and vertical facies models of fine-grained, muddy turbidites (Stow and Shanmugam, 1980), grain size and structures and their specific hydraulic conditions of formation (Komar, 1985) and submarine fan variability and the range of submarine fan types (Nilsen, 1980). In addition, submarine canyons were discovered to play a significant role in deepwater systems (Shepard, 1981—see Chapter 8) as well as large-scale slides (Bugge, 1983—see Chapter 10), and debris flows (Takahashi, 1981; Johnson, 1984—see Chapter 6). The Deep Sea Drilling Program (DSDP, later known as International Ocean Drilling Program or IODP) (Warme et al., 1981) and other shipboard expeditions (Stow et al., 1985) made numerous discoveries about seafloor morphology and the near-subsurface around the world. Industry and academic personnel were constantly in need of the latest information and updates on deepwater sedimentary systems, and many primers, short courses (Nelson and Nilsen, 1984) and books (Pickering et al., 1986; Pickering et al., 1989) summarized the state of the scientific understanding. For years, sequence stratigraphy had been calibrated to and most useful in coastal to shallow marine systems, but now extended farther to deepwater areas with information from exploration wells and producing fields (Mitchum, 1985). During this time, subaerial and subaqueous gravity flows were observed for their analog value to turbidity flows (Simpson, 1982; Costa and Williams, 1983) including avalanches and dense-water outflows into standing bodies of water. These types of flows continue to aid the understanding of the anatomy of sediment gravity flows even at thousands of meters water depth.

## 1990s

With deepwater exploration growing in areas such as the Perdido Fold Belt of the US Gulf of Mexico (i.e., BAHA-1 wildcat well in 1996 led by a consortium of companies to drill one of the last remaining very large structures in North America at the time—see Chapter 4) and major discoveries including Auger, Mars and Ram Powell, the 1990s were a flurry of activities gathering information on the architecture and heterogeneity of deepwater reservoirs (Coleman et al., 2000). Additional modern submarine fan studies (Weimer, 1990) and reviews of turbidity currents (Middleton, 1993), turbidite facies and nomenclature assessments of select basins (sensu Ingersoll, 1990), and their global variability and applications (Mutti, 1992), allowed for differentiation of fan types, based on grain size, types of flows and whether the origin of the sediment was a point or line source (Reading and Richards, 1994) (Fig. 3D). Integration of these various scales of studies aided sedimentary system modeling efforts. Several universities (Anderson et al., 1999) and industry research labs sponsored expeditions to what became classic outcrops and led to the development of many deepwater conferences and publications. Reviews on submarine channels (Clark and Pickering, 1996) greatly benefited from petroleum exploration and discoveries in similar depositional environments in deepwater provinces globally (Sanders and Friedman, 1997). Research teams investigated outcrop analogs around the world including the Brushy Canyon Formation, USA (Beaubouef et al., 1999, Fig. 3E), Capistrano Formation, USA (Campion et al., 2000), Ross Formation, Ireland (Elliott, 2000), Mount Messenger Formation, New Zealand (Browne and Slatt, 2002; Browne et al., 2007), Hecho Group, Spain (Pickering et al., 1995), in addition to the classic Annot Sandstone, France (Pickering and Hilton, 1998), Tanqua-Karoo outcrop belt, South Africa (Sullivan et al., 2000) and Marnoso Arenacea Formation, Italy (Mutti and Ricci Lucchi, 1972; and many others), summarized in Weimer et al. (2000) and Nilsen et al. (2007) (see Chapter 8). Troublesome clay-rich reservoirs coined *slurry-flow deposits* shed light on the simultaneous interaction of turbidity current and debris flow processes (Lowe and Guy, 2000—see Chapters 6 and 7). Similar 'linked' debrites were found in the Ross Formation (Haughton et al., 2003), and were soon established as hybrid-event beds (Haughton et al., 2009), discussed more completely in Chapters 6 and 7. Stratigraphic trap plays offshore Angola urged additional outcrop studies on turbidite bed terminations against confining slopes (Haughton, 1994) and the evolution of turbidite deposition within confinement (Kneller and McCaffrey, 1999). Large conferences on turbidite reservoirs brought together the world's deepwater researchers to provide a current understanding of the science (Weimer et al., 2000) and produced tomes on deepwater outcrop analogs and subsurface producing oil and gas fields. The concepts presented at the end of the 20th century continue to be meaningful in that they allow the energy industry to bridge the gap between observation of sedimentary transport and depositional processes in order to build predictive models based on first principles of geoscience.

## Moving into the 21st century

Several key advancements from the late 1990s prompted integration in the early 2000s of sedimentary process response (Gardner et al., 2003), source-to-sink linkages (Mutti et al., 2003) and basin margin topographic influences (mobile substrates) on depositional models (Mayall et al., 2010) (Fig. 3F). Process advancements supplied a better understanding of flow evolution from debris flow to turbidity flow (Mohrig and Marr, 2003), flow transformation and how debris flows can transform into turbidity flows and vice versa (Waltham, 2004), flow surging (Lamb et al., 2004) and fluid mechanics of sediment gravity flows (Stickel and Powell, 2005) (see Chapters 6 and 7). Reviews of ancient submarine fans were used to compare and contrast the modern (Mattern, 2005). Characterization of alternative forms of deepwater sedimentation, including sandy injectites (Hurst and Cartwright, 2007), promoted new thinking as to how sediment can be emplaced in different sedimentary intervals after deposition (see Chapter 10). Elusive sandy submarine fans (Piper and Normark, 2001) to submarine braid plains (Hesse et al., 2001) and their geologic risks pertaining to reservoir presence and quality (Walker, 2004) added and refined diverse exploration play types to the operators' repertoire. Disquieting information of how the dogmatic principles of classic sequence stratigraphy may not apply to borderland fans detailed by Covault et al. (2007) led to a more complete understanding of the merits, limitations, pitfalls and uncertainties associated with the broad methodology of sequence stratigraphy (Catuneanu et al., 2009; Donovan, 2010—see Chapter 12). The first thorough reviews on deepwater channel architecture and hierarchy (Sprague et al., 2005; Mayall et al., 2006; Deptuck et al., 2007; McHargue et al., 2011; Sylvester et al., 2011) promoted the interrogation of the range and variability in deposits in these classic plays from slope to basin floor, and from muddy to gravelly systems. Reviews of deepwater sedimentation and stratigraphy provided useful frameworks to apply to the energy industry's workflows and internal exploration processes (Pickering et al., 1986, 1989; Shanmugam, 2006; Slatt, 2006; Talling et al., 2012; Pickering and Hiscott, 2015; Stow and Smillie, 2020). Major workshops led by the American Association of Petroleum Geologists (AAPG), Society for Sedimentary Geology (SEPM), the US Gulf Coast Association of Geological Societies (formerly GCAGS, now *GeoGulf*), European Association of Geoscientists and Engineers (EAGE) and British Sedimentological Research Group (BSRG) have also reinvigorated applied deepwater sedimentology and stratigraphy research. This rich collaboration of the deepwater sedimentary systems community has highlighted the strengths and weaknesses of the current understanding and has initiated a call-to-action to reexamine deepwater processes and deposits from a global perspective. As Donovan (2010) explains in a sequence stratigraphy review, it can be difficult to speak the same language regarding highly technical geoscience terms if there are multiple versions of the same concept. Bringing diverse groups together in conferences and through collaboration helps to break down these barriers in communication.

Observation of deepwater sedimentary processes in nature, in the lab and in numerical simulations has provided a robust physics-based framework for understanding process range and variability. Each of these decades has advanced the understanding of deepwater sedimentary systems and their practical applications, and has encouraged exploration into deeper waters through time (see Chapter 19). This brief history of the study of deepwater sedimentary systems aims to place this book into context. It is our goal to (i) integrate and build upon the work in these volumes, (ii) provide a total petroleum systems perspective and (iii) establish a framework for success for venturing into additional deepwater exploration frontiers to supply the world's growing energy and societal demands. Through these studies and achievements around the world over the past ~150 years, summarized in Fig. 2, scientists have observed similar drivers of sedimentation operating in different tectonic settings and basin margins, allowing a window into the controls on deepwater sedimentation.

## Regional controls on deepwater sedimentation

There exist numerous controls that determine the type, distribution and geometry of deepwater sedimentation. Three of them operate on a regional or basinal scale: tectonics and subsidence, relative sea level and sediment supply. Each in turn span a range of timescales, manifestation and effective extent. This section provides an overview of these three regional controls, and more information are referred to the citations and other chapters within this book.

### Tectonics and subsidence

Deepwater basin formation is underpinned by tectonic processes (see Chapters 3–5). These include faulting, folding and mantle subsidence (Ingersoll, 2011). For example, extensional or transform faulting can form a rift basin or type of strike-slip basins, such as a transtensional basin, respectively. Compressional forces and folding can create a foreland basin. Mantle subsidence controls the formation of a cratonic basin. Long-term tectonic activity associated with plate movements eventually forms large basins such as oceans. These basins are ultimately categorized based on their processes of formation and position relative to major geologic features such as volcanic corridors and plate boundaries (Fig. 4).

The driving force of sedimentary basin formation is tectonic subsidence, or the vertical displacement of a stratigraphic horizon downward into the Earth's crust. It is an important long-term ($10^6$ years) control on the creation of space available for sedimentation, or the *accommodation*. Landmark studies have demonstrated that the style and rate of tectonic subsidence are controlled by the tectonic setting (e.g., Dickinson, 1976; Angevine et al., 1990; Xie and Heller, 2009; Fig. 5). These studies summarize subsidence analysis in graphs that measure the vertical movement of a stratigraphic horizon at a specific location with respect to a datum (van Hinte, 1978), labeled on the $y$-axis and time labeled on the $x$-axis. To separate the contribution of tectonic load, sediment load and sea level change since the time of sediment deposition, subsidence analysis uses measurements and assumptions on stratigraphic thickness, lithology, paleobathymetry and chronostratigraphy depending on the availability and resolution of the data.

When a sedimentary layer subsides, porosity is reduced due to mechanical compaction, physio-chemical changes such as pressure solution and cementation. Understanding changes in a rock's porosity and related permeability is critical for assessing and predicting rates and volumes of fluid flow through and from layers of rock. Via direct measurement and modeling, reservoir quality can be predicted and integrated into the overall understanding of a sedimentary system (see Chapters 13, 16 and 18). Generalized mathematical equations are developed to calculate the amount of subsidence by the weight of the overlying sediment. After removing the local isostatic effects of sediment loading through a process known as backstripping (Steckler and Watts, 1978), or the quantitative removal of sedimentary layers going back in time, the remaining tectonic subsidence is determined. The result shows the idealized subsidence history of a basin that would have existed if only water, and *no sediment*, filled the subsiding basin.

Tectonic forces typically exert more influence on accommodation than the initial formation of a basin. This is especially true for local areas within basins, where tectonic movements can abruptly change the accommodation. One primary example is interpreted from the Messinian Salinity Crisis caused by the closure of the Mediterranean Sea due to the presence of the Gibraltar Arc (Garcia-Castellanos and Villaseñor, 2011). The tectonics associated with this event also affected the neighboring Black Sea (Bartol and Govers, 2009). The Black Sea is now in continued subsidence due to the ongoing collision between the Eurasian and African plates and the westward escape of the Anatolian block along the North and Anatolian faults in Turkey (Nikishin, 2003).

While tectonics does provide the prominent control on basin formation, some sedimentary basins continue to fill in the absence of tectonic activity. While passive margins form due to rifting, they can experience little to no active tectonic deformation in the drift phase of their development (see Chapters 3–5 for more detail on this topic). Passive margins are the most abundant type of continental margin (Fig. 6). Examples of current passive margins include the eastern and northern borders of North America, the US Gulf of Mexico, the West Coast of Africa and the northern coast of Australia. Passive margins exhibit kilometer-scale bathymetric changes from coast to basin floor. Basin floors along passive margins act as the final sink for sediment, especially a substantial volume of sand, forming Earth's largest depositional systems and many prolific petroleum fields (Fig. 1). Passive margins are a key theme throughout this book, and much of the data and many of the examples presented in other chapters also come from studies of passive margins.

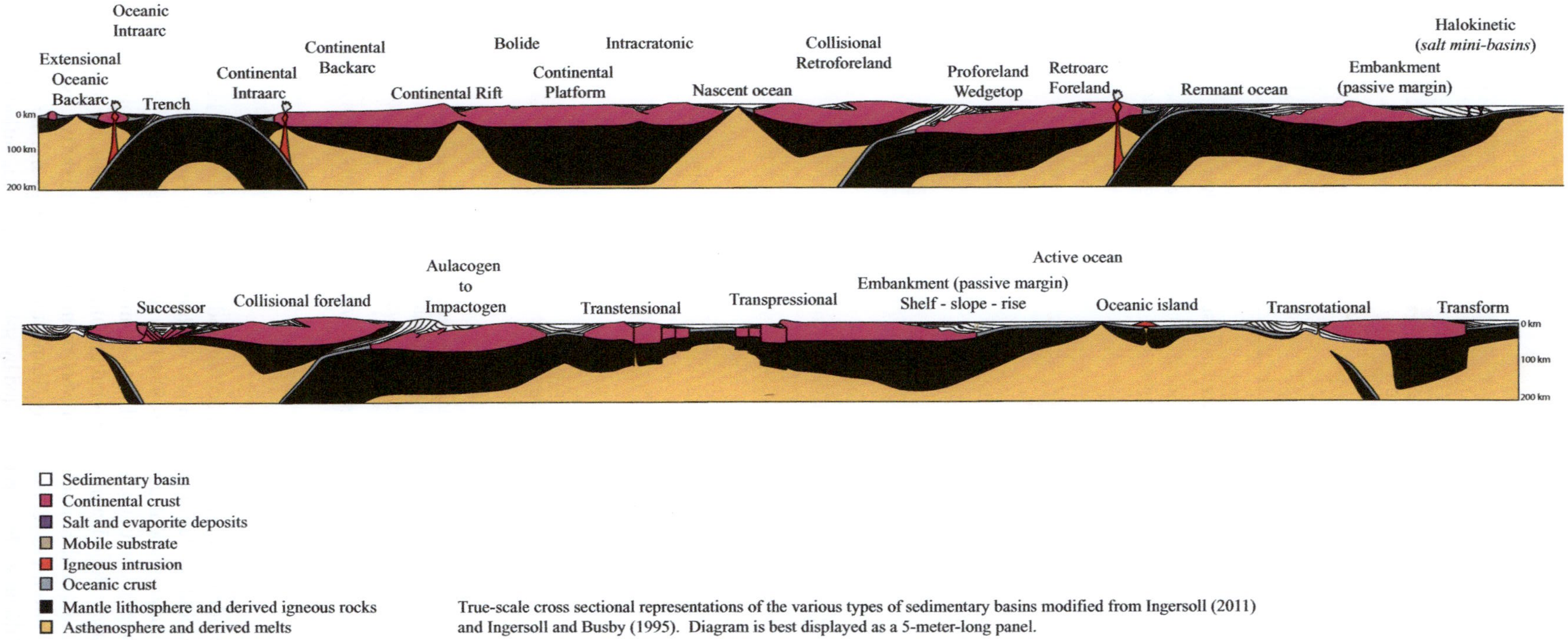


**FIG. 4** Deepwater tectonic settings are determined by subsidence type and what type of plate tectonic features are expressed in the vicinity of the basin. The major types of plate boundaries—convergent, divergent and strike-slip—produce a number of different basin types and are discussed further in Chapters 3 and 4. This 5-m-long panel developed and inspired based on the superb work by Ingersoll (2011) and references therein.

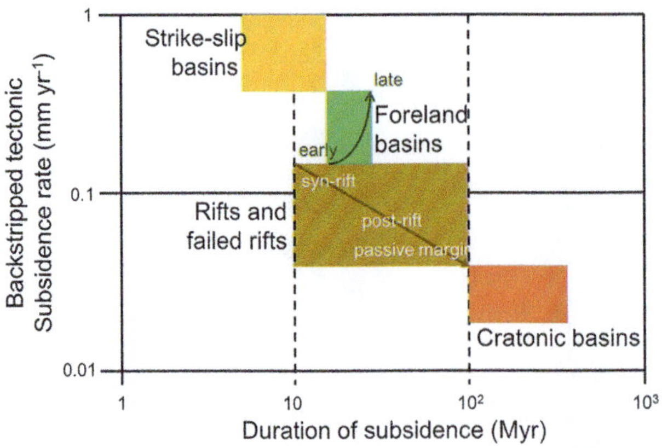


**FIG. 5** This plot shows the typical scales of subsidence rate and duration of subsidence for different types of sedimentary basins. *(Modified and redrafted from Allen, P.A., Allen, J.R., 2013. Basin Analysis: Principles and Application to Petroleum Play Assessment, third ed., Wiley-Blackwell, ISBN: 978-0-470-67377-5, pp. 632.)*

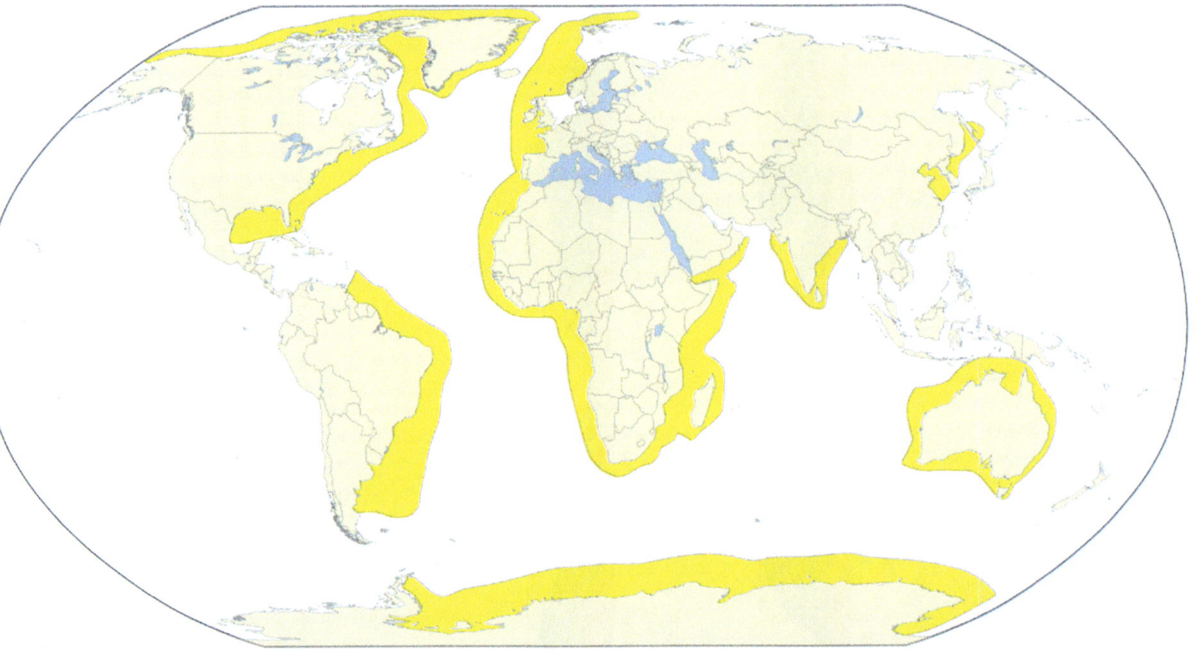

**FIG. 6** Global distribution of passive margins shown in yellow. The margins can be further defined by presence or lack of abundant seaward-dipping reflectors (SDR) during basin formation (volcanic vs non-volcanic passive margin). Interestingly, these passive margins are also the sites of the greatest thicknesses of sedimentary deposits in the world. *(Modified after Leroy, M., Gueydan, F., Dauteuil, O., 2008. Uplift and strength evolution of passive margins inferred from 2-D conductive modelling. Geophys. J. Int. 172 (1), 464–476. https://doi.org/10.1111/j.1365-246X.2007.03566.x and Geoffroym, L., 2005, Volcanic passive margins. Geoscience 337, 1395–1408. https://doi.org/10.1016/j.crte.2005.10.006 and references therein.)*

## Sea level

In most basins, sea level is the other major control on accommodation (lake level for lacustrine basins). Sea level can be divided into two main types: global (eustatic) sea level and relative sea level (see Chapter 12 for more information). The range in global sea level can vary by about 300m (~ 984 ft.), whereas relative sea level can vary internally in basins depending on local tectonic forces, climate (i.e., rainfall, ice volume) and sediment supply. At the Kimmeridgian-Tithonian stage boundary in the Late Jurassic Period, a time of global greenhouse conditions, the global sea level was 140m (~ 460 ft.) higher than today (Haq, 2018). At the Last Glacial Maximum (LGM) ~20,000 years ago when terrestrial ice volume peaked, sea level was 134m (~ 440 ft.) lower than today (Lambeck et al., 2014) and the coast was at or near the edge of the present-day continental shelf. Global sea level is controlled by solid Earth tectonic influences and climate and is moderated by changes in terrestrial ice volume. Sea level is cyclical over various time scales. One of the most common cycles, termed a

**TABLE 1** Sea level cycle order and generalized corresponding duration, thickness and aerial extent.

| Cycle order | Nomenclature | Thickness range (ft) | Aerial extent (sq miles) | Duration (Myr) | |
|---|---|---|---|---|---|
| | | | | Range | Mode |
| First | Megasequence | 1000+ | Global | 50–100+ | 80 |
| Second | Supersequence | 500–5000+ | Regional | 5–50 | 10 |
| Third | Sequence | 500–1500 | 50–50,000 | 0.5–5 | 1 |
| Fourth | Parasequence Set | 20–800 | 20–2000 | 0.1–0.5 | 0.45 |
| Fifth | Parasequence | 10–200 | 20–2000 | 0.01–0.1 | 0.04 |

Milankovitch Cycle, occurs on a ~40,000–100,000 year timescale. This cycle is caused by changes in the Earth's orbital parameters: eccentricity (how circular the orbit is), obliquity (the tilt of Earth's rotational axis) and precession (the "wobble" of the axis).

Sea level is reconstructed primarily through the study of micropaleontology and the analysis of biostratigraphic data (see Chapters 14 and 15), palaeogeographic mapping (Chapters 8–11), physical stratigraphy (Chapters 6–13) and stable isotope geochemical analysis (Chapter 11). Tectonic and climatic controls of sea level add up to produce different orders of sea level change (Mitchum Jr. and Van Wagoner, 1990; Table 1). While long-period, first-order ($10^6$–$10^7$) sea-level change exerts control on basin configuration, third- and fourth-order cycles of sea-level change are significant as they immediately alter the width of the continental shelf (Haq et al., 1988). If the shelf width is reduced during the lowering of sea level, this will theoretically promote the delivery of sediment to the shelf edge. The earliest theories of sequence stratigraphy proposed relative sea level as one of the primary controls on deepwater sediment delivery (Van Wagoner et al., 1987, 1988; Posamentier and Vail, 1988; Posamentier et al., 1988—see Chapter 12 for a more complete review of how sequence stratigraphy is applied to exploring deepwater sedimentary systems). These studies used seismic-scale examples from the US Gulf Coast, West Africa, onshore US (San Joaquin Basin, South Texas Basin, etc.) and many other basins to provide evidence for the method. Subsequent studies have refined how relative sea level affects deepwater sedimentation in the Pleistocene to Recent: the Danube fluvial valley of the Romanian Black Sea shelf and related fan deposits (Popescu et al., 2004); the Fuji-Einstein shelf-edge delta in the Gulf of Mexico and linked submarine channels (Sylvester et al., 2012) and the shelf-edge delta and stratigraphically equivalent submarine fan offshore Corsica (Deptuck et al., 2012). Sea level change can also alter coastal sediment transport, change littoral cell activity and facilitate the delivery of sand to deepwater (Covault et al., 2007; Covault and Graham, 2010).

## Sediment supply

Although sequence stratigraphy (Vail et al., 1977a, 1977b) stressed relative sea-level change, sediment supply is part of the triad of controls on deepwater sediment delivery (Carvajal and Steel, 2006; Carvajal et al., 2009; Catuneanu et al., 2009 and references therein—see Chapter 12). High-resolution well log studies in the upper Paleocene–lower Eocene Wilcox Group across South Texas, USA show that sediment supply can accentuate shoreline and delta migration across the shelf when sea level change is modest, especially when subjected to greenhouse conditions, or very warm periods in Earth's history (Zhang et al., 2016). If sediment supply is high, the effect of quickly rising sea level (transgression) can be reduced. After filling accommodation on the shelf, sediment tends to bypass the shelf and is sent to deepwater. This is a recurring theme in modern shelf-edge deltas (see Chapter 8), where sediment is delivered to the shelf edge and into deepwater even with rising sea level (e.g., Porebski and Steel, 2003) and a positive shelf-edge trajectory.

Additional shelf processes influence sediment supply to deepwater (Gan et al., 2021). Studies utilizing seismic data and modern analogs demonstrate the range of sediment transport processes along the shelf (Sweet, 2020 and references therein). When fluvial feeder systems reach the shelf margin, there is abundant evidence for the development of *organized*[a]

[a] Simply, the term organized is used to describe how channels, levees, splay and overbank environments are elegantly distributed in turbidite-dominated deepwater depositional systems, which tend to show a high level of architectural complexity and recognizable geometries that correspond to the various pathways and depositional patterns for turbidity flows. Disorganized refers to deepwater systems dominated by debris flows and mass-transport processes, which tend to show a low level of architectural complexity. These terms are used purely for simplication purposes.

deepwater sedimentary systems dominated by turbidity flow processes off the shelf edge. As Chapter 8 explains, turbidity flow-dominated systems display a much higher order of architectural organization of channels, levees, splays and other morphological elements than do systems dominated by mass transport processes and debris flows (*disorganized* systems according to some studies). These organized systems include submarine fans, fan valleys and aprons along a variety of basin margins (Fig. 4). These systems have a high level of architectural complexity. Away from the main fluvial feeder systems very little sediment is fed into deepwater (Sweet and Blum, 2016; Paumard et al., 2020). More than tens of kilometers away from the fluvial inputs, the sedimentation tends to be muddier, prone to mass transport (Prather et al., 2016) and overall poorer reservoir quality. Where there is low sediment input from the continent or if insufficient coarse-grained sediment reaches the shelf margin, the result is a mud-prone deepwater system, which will tend to be disorganized (Bourget et al., 2014).

Longshore drift can facilitate delivery of sediment from fluvial feeder systems to canyon heads and into deepwater. Modern studies of the Aquitaine shelf, Bay of Biscay, offshore France, show that during periods of intense erosion, longshore drift delivers large volumes of sediment south to the adjacent Capbreton shelf canyon creating a dynamic environment of erosion and deposition (Guiastrennec-Faugas et al., 2020). Large storms on the Swatch of No Ground shelf and canyon complex in the Bay of Bengal have liquefied shelf sediment and transported it into canyon heads (Rogers and Goodbred Jr., 2010). These studies indicate the importance of understanding the depositional and climatic processes from the coast to the shelf and how the upper limit of the marine depositional system influences sediment input into the deepwater basin. An understanding of the main conveyers of sediment into the basin enables better reservoir prediction (see Chapter 11 for more information). Carbonate sedimentary systems differ greatly from siliciclastic systems in their unique biological control in response to water depth changes, and their sensitivity to clastic sediment influxes (Chiarella et al., 2017). It is not uncommon for carbonate and clastic systems to be mixed in petroleum provinces, such as in the Delaware Basin, USA (Kvale et al., 2020).

An integrated research area, termed "Source to Sink" and abbreviated to "S2S" (see Chapter 11), seeks to unify the previously discussed principles of sediment transport and deposition from sediment origins to its final resting place. S2S highlights the importance of sediment routing patterns and supply in basins and the potential for reservoir-quality (high porosity, high permeability) sediment deposition in deepwater environments (e.g., Helland-Hansen et al., 2016; Nyberg et al., 2018). Qualitative methods introduced by Meade (1972, 1982), Dickinson (1997) and Hovius and Leeder (1998) have expanded significantly with onshore and offshore data to quantitatively analyze the link between the sediment provenance and deepwater basin fill (Sømme et al., 2009a,b,c). Case studies that quantify sediment transport from terrestrial catchments show how sediment is delivered into the deep basin through coastal, shelfal and continental margin processes (Walsh and Nittrouer, 2009; Septama and Bentley, 2017; Fielding et al., 2017).

## Substrate controls

When sediment arrives in deepwater, it tends to *conform to its container*. Sediment gravity flows in particular are bottom-hugging mixtures of particles and water that seek to find the lowest parts of the seafloor. These flows move through and fill the lowest areas of canyons, channels and recently abandoned areas in frontal splay environments, similar to how sedimentation events find their way to different parts of an alluvial fan. Seafloor bathymetry, however, is highly variable. The geometry of a basin is in constant dynamic balance among the previously described basin-controlling factors (see Chapters 11 and 12). The processes associated with and derivatives of salt diapirism (Chapter 5), mud diapirism, growth faults and fold belts (Chapter 4) and changing seafloor conditions can cause flows to experience relative confinement and unconfinement (see Chapters 6–10), allowing sedimentation patterns to constrict between bathymetric highs and spread out laterally when there are no topographic highs or barriers, forming complex channel-splay environments that morph from tributary, distributary and feeder system depending on the rugosity of the seafloor and its gradient (sensu Lomas and Joseph, 2004). In many of the world's most productive petroleum basins, including the US Gulf of Mexico, salt diapirism creates a ~30,000 mile$^2$ (77,700 km$^2$) area of the continental slope defined by ~5–10-mi (~8–16-km) diameter minibasins, or halokinetic basins (Ingersoll, 2011), all of which formed at slightly different times (see Chapter 5 for more detail). Much like salt's ductile nature, mud and unconsolidated sediment can also form diapirs and erupt onto the seafloor (see Chapter 10), forming mud volcanoes, or topographic highs that can buttress sediment gravity flows. Faults can break through the seafloor and cause submarine fan systems to adjust their sediment delivery rates and channel incision rates (Pizzi et al., 2021). While submarine fans are drawn as simple cartoons, in real life, these deepwater depositional systems can show high amounts of variability due to the uneven and unpredictable nature of the seafloors they inhabit.

### Summary of regional controls

Regional controls on deepwater sedimentation provide key information on all elements of geologic risk and uncertainty in petroleum exploration from source rock deposition to reservoir presence to trap creation. Tectonic-controlled subsidence, sea level and sediment supply provide primary controls on accommodation in deepwater sedimentary basins. Basin modeling based on the knowledge of these factors from outcrop, modern analog, seismic exploration, well log and core data and physics of sediment transport can help determine subsurface geologic risk and reduce the uncertainty in a variety of modeling parameters. Deepwater sedimentary systems as described here are quite variable depending on their input parameters, including basin type, grain size, feeder system and several other components (Reading and Richards, 1994). Many of these terms may be new to the reader, but deepwater systems do not lack in sophisticated and often misunderstood terminology.

## Challenges with terminology

Terminology for deepwater sedimentary systems can be confusing. Part of the reason why deepwater sedimentary systems have remained remote and inaccessible to the global community is the impenetrable lexicon used to convey geological observations and concepts. Different terms for the same feature, confusing naming conventions, inequality in citations depending on geographic region and school of thought have over the years constructed barriers to conversation (also discussed in Chapter 12). For example: What differentiates a *slope channel complex* vs a *base-of-slope channel complex*? Is this difference worth noting? There are significant differences in reservoir types between slope and base-of-slope channels (Beaubouef et al., 1999; Sullivan et al., 2000; Pyles et al., 2010, 2012; Hubbard et al., 2020 and references therein), and the two types of channel complexes may require different gross depositional environment (GDE) designations (discussed further in Chapter 8). Certainly there is overlap and gray area, too, in what is considered *confined* or *unconfined* deepwater sedimentation (see Chapter 6; Sinclair and Tomasso, 2002; Lomas and Joseph, 2004; Etienne et al., 2012). Other difficult conversations can arise from, for example:

- How are channelized or non-channelized environments differentiated?
- What designates a *channelized splay* vs a distributary channel environment?
- Where to draw the map boundaries between different depositional environments?
- When is an erosional surface deep enough or so pronounced that it becomes a scour and signifies confinement, such as in a channelized environment?
- Outside of reservoir modeling, is there a deposit that is truly a *sheet sand*?

Not only is this dense terminology difficult to read and to interpret in literature, but it also can create impasses at industry and academic meetings. Examples of terminology differences can be observed when examining models of submarine channel architecture and fill (see Chapter 8) among petroleum industry researchers (Sprague et al., 2005; Mayall et al., 2006; Deptuck et al., 2007; Sylvester et al., 2011; McHargue et al., 2011). Definitions for these models or terms may arise because a group works primarily in certain types of basins or depositional systems. Therefore, the terms and models are fit for their purpose. When the terms or models are extended to a new basin, they may not sufficiently describe the different sedimentary environments. A more objective approach is to use process, scale and deposit to make clear references and distinctions for widely variable terminology and cite a broad range of studies to achieve a global perspective.

Because terminology can be different from company to company and from institution to institution, a working set of *architectural elements*, or *building blocks* at various scales that construct a deepwater sedimentary system, is often used to clarify the concept of scale and type of deposit in deepwater systems. This is called architectural element analysis.

## Basic architectural elements and definitions

In deepwater environments, there are key stratigraphic surfaces that divide 3D intervals of rock, known as geobodies. There is a need to define a hierarchy in these geobodies for characterization, interpretation and modeling purposes.

### Scaling concepts

The concept of scale is important in the study of deepwater sedimentation (sensu Ghosh and Lowe, 1993; Reading and Richards, 1994; Coleman et al., 2000; Xu et al., 2017; Catuneanu, 2019a,b). In this chapter, architectural element analysis

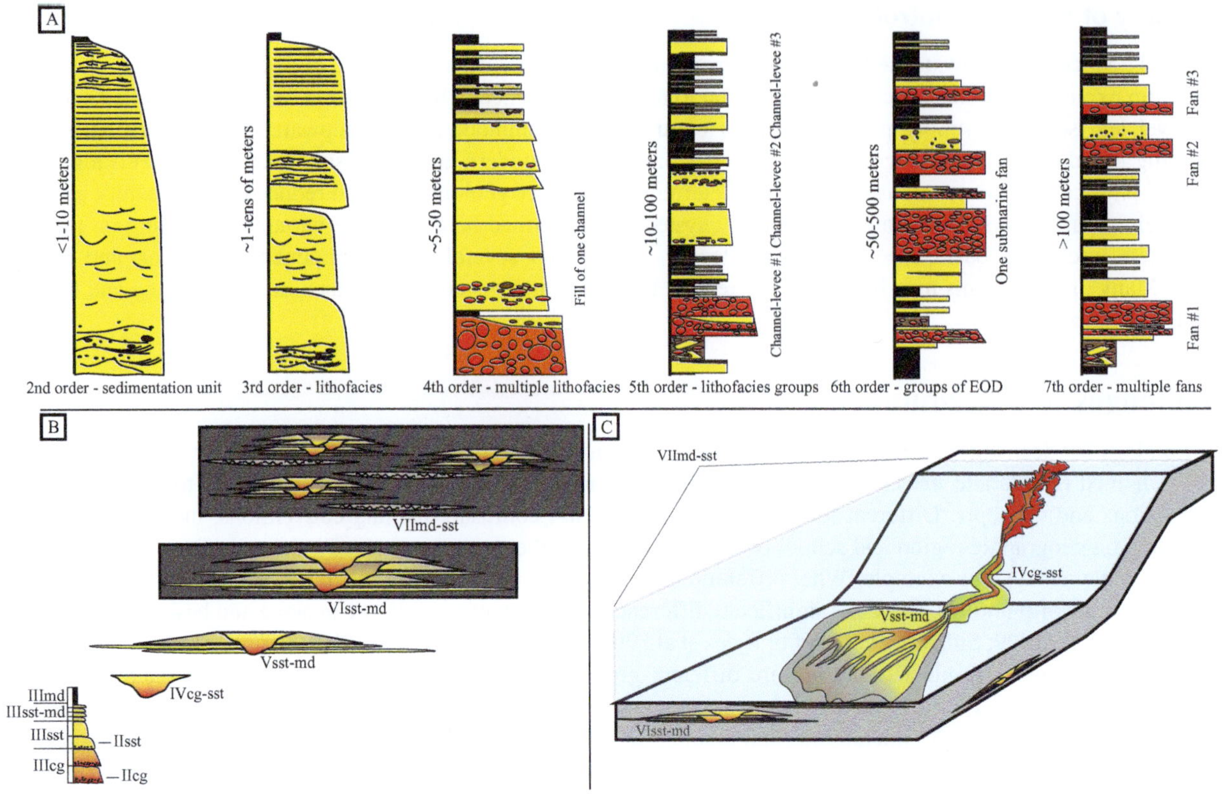


**FIG. 7** The hierarchy of a deepwater sedimentary system can be expressed in terms of architectural elements. Architectural elements range from the smallest scale (sub-bed-scale sedimentation divisions) to the largest scale (at the scale of a submarine fan and beyond). While there are several ways to define and determine architectural elements (Pickering et al., 1986, 1989, 1995), Ghosh and Lowe (1993) and Lowe and Ghosh (2004) propose a flexible and non-genetic method. This method can be used to define architectural elements in 1-, 2- and 3-dimensions, is purely descriptive and acknowledges that the greatest uncertainty resides at the upper end of the length scale. Beginning from upper left: (A) stratigraphic columns demonstrating architectural elements in one dimension—(from left to right)—first-order architectural elements are sub-bed-scale divisions, including individual Bouma divisions; second-order architectural elements are sedimentation units and represent the basis for many deepwater sedimentation models (i.e., Bouma, Lowe, Haughton, etc.); third-order architectural elements are lithofacies or mappable rock units; fourth-order architectural elements are lithofacies associations, often interpreted as genetic elements, such as a channel or levee deposit; fifth-order architectural elements are groups of lithofacies associations that often form channel-levee or splay complexes; sixth-order architectural elements are submarine fans or the fill of an entire canyon, often at formation scale and sub-divided by third- or fourth-order sequence boundaries; seventh-order architectural elements are multiple submarine fans, perhaps bounded by second-order sequence boundaries and so on. (B) This diagram shows the 2D expression of deepwater architectural elements, and (C) shows the 3D expression in a block diagram.

is employed to establish a hierarchical order in the range and variability of deepwater stratigraphy (Fig. 7; Table 2). Architectural element analysis defined by Ghosh and Lowe (1993) and Lowe and Ghosh (2004) is a powerful method that is purely descriptive, applies to sedimentary intervals in 1-, 2- and 3-dimensions, focuses on lithosomes or rock type instead of surfaces that are notoriously difficult to recognize in fine-grained intervals, and is open-ended and not limited, thus allowing for higher and higher scales of architecture not easily seen by humans. It is also conservative and different from other industry interpretation methods in that it avoids rote interpretation of the genesis of deposits. This type of analysis builds from other sedimentary classification schemes (Miall, 1985; Pickering et al., 1989, 1995) and is used extensively throughout the book.

To understand deepwater stratigraphy is to understand its basic building blocks. The sedimentation unit, or bed, is the basic building block of all deepwater petroleum reservoirs (see Chapter 7). Beds in deepwater successions are typically normally graded, and can have a wide range of surfaces and internal structures. Deepwater successions typically targeted as reservoir intervals are set apart in that they contain both sand and mud fractions, typically sorted with the coarser grains (gravels and sand) at the base and finer grains (silt and clay) at the top. Recognition of sedimentation units, or these individual building blocks, provides the basis for describing sedimentation and facies models in the following chapters of this book.

**TABLE 2** Architectural elements in deepwater settings.

| | Scale (thickness) | Definition | Deepwater example |
|---|---|---|---|
| First-order | Sub-bed (centimeter to meter) | Individual interval of a specific sedimentary structure | Planar laminated interval in a turbidite bed (Bouma $T_b$ interval) |
| Second-order | Bed (centimeters to multiple meters) | One sedimentation unit, deposited by a single sediment gravity flow | Bed composed of conglomerate and sandstone (Lowe $R_{23}S_3$ turbidite bed) |
| Third-order | A lithofacies (typically on the scale of meters to tens of meters) | Group of beds with similar traits, typically divided into lithofacies by lithology and bed thickness | Thick-bedded sandstone |
| Fourth-order | Groups of different lithofacies; "genetic" elements (meters to tens of meters) | A thick interval of one lithofacies, or two or more lithofacies that are adjacent, bounded by stratigraphically significant surfaces, that form a genetic element, such as the fill of a channel | Erosional surface overlain by mudstone, thin-bedded sandstone and mudstone, and thick-bedded sandstone overlain by mudstone (i.e., a typical fill of a channel) |
| Fifth-order | A group of fourth-order elements (tens of meters to hundreds of meters) | Two or more fourth-order elements | Deposits that form a channel-levee element |
| Sixth-order | A group of fifth-order elements (hundreds of meters) | Two or more fifth-order elements | Deposits that form a channel-levee complex |
| Seventh-order | A group of sixth-order elements (hundreds of meters to kilometers) | Two or more sixth-order elements | Submarine fan consisting of multiple channel-levee complexes and lobe complexes, or the fill of an entire canyon |
| Eighth-order | A group of seventh-order elements (typically kilometer scale) | Two or more seventh-order elements | Multiple submarine fans separated by second- to third-order sequence boundaries |

## Reasoning for scale-based modeling

Industry workflows are designed to take natural systems and import them into a model that is cognizant of geological concepts (Pyrcz et al., 2005; Hovadik and Larue, 2007). Modeling workflows are more geology and geophysics up front, then progress into more engineering, yet are iterative and interwoven in the upstream value chain (Fig. 8—see Chapters 2, 13, 14, 16–20). Key parts of deepwater sedimentary deposits include surfaces, the stratigraphy and architectural or depositional elements (geobodies). There is a need to employ geology-based interpretation of reservoirs, without the reliance on major surfaces (sensu Miall, 1985, 1988). Architectural element analysis ranks deposits in a hierarchy, from smallest scale elements at sub-bed scale (first order), to largest scale elements in an open-ended fashion (second order, third order and so on). Sediment gravity flows typically deposit one sedimentation unit, groups of similar sedimentation units form lithofacies, groups of lithofacies make a channel, splay, levee and so on up to fan scale and depositional sequence scale. This process of "scaling up" the basic building blocks of the total sedimentary system serves two primary purposes: it reveals the overall plumbing of the deepwater system and also yields predictive context.

Previous methods by Miall (1985, 1988, 1996) to examine fluvial systems primarily in outcrop form are insufficient for describing deepwater systems for a number of reasons. Other classification schemes for deepwater systems include those by Mutti and Normark (1987, 1991), Normark et al. (1993), Pickering et al. (1995), Gardner et al. (2003) and other architectural element analysis methods proposed by industry consultant groups, some that include genetic element linkage with particular deepwater deposits as part of the workflow.

## Basic deepwater architectural elements

For the purposes of architectural element analysis, first-order architectural elements are the smallest scale building blocks of a deepwater sedimentary system. These are individual sedimentary structure divisions in a bed (Chapter 6), such as a

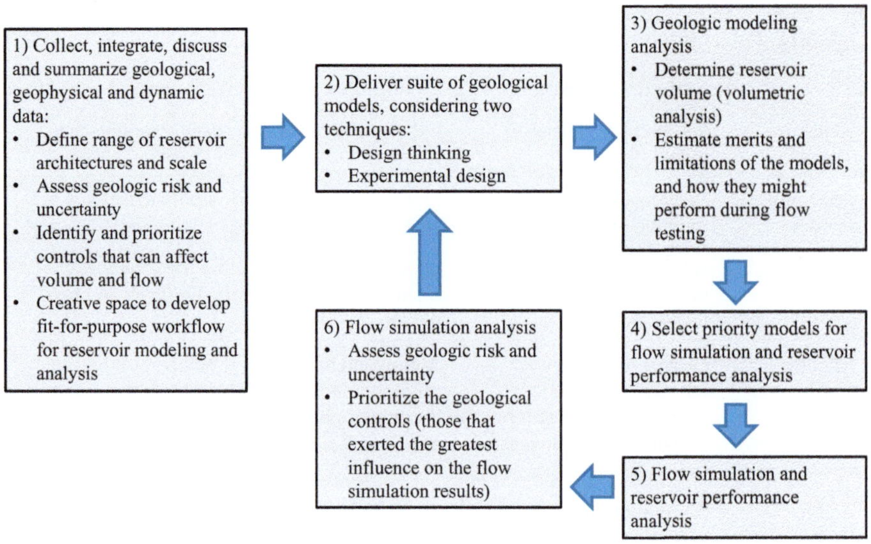


**FIG. 8** Industry workflow chart for modeling deepwater reservoirs. This is an iterative process and integrates a diverse set of geoscience and engineering disciplines. *(Modified from Hovadik, J.M., Larue, D.K., 2007. Static characterizations of reservoirs: refining the concepts of connectivity and continuity. Pet. Geosci. 13 (3), 195–211.)*

structureless division, commonly interpreted as a suspension deposit from a fully turbulent flow, or Bouma (1962) Ta division. Other individual Bouma (1962) divisions (Tb, Tc, Td, Te), or Lowe (1982) R1, R2, R3, S1, S2, S3 divisions (Tt, too), or Lowe and Guy (2000) slurry-flow units M1–7, or Haughton et al. (2009) hybrid event bed units H1, H2, H3, H4, H5 and many others, also fall within this category (Chapter 7).

Second-order architectural elements are individual beds, deposited by one event or sediment gravity flow. Second-order architectural elements can be derived from the range of sediment gravity flows, often noted as turbidite, debrite or hybrid event beds. Bed contacts serve as second-order architectural element boundaries.

Third-order architectural elements are packages of several similar sedimentation units, known as lithofacies (see Chapters 6–10, 17–18 for deepwater examples). These can be intervals of thick-bedded turbidites, or interbedded mudstone with thin sandstone turbidites and slurry-flow deposits. Third-order bounding surfaces divide groups of second-order sedimentation units. These can commonly refer to units known as depositional facies, such as the fill of portions of channels, levees, splays (Chapter 8) or drifts (Chapter 9).

Fourth-order architectural elements commonly consist of cyclic associations of contrasting third-order elements, separated by fourth-order surfaces. For example, (i) an erosional surface overlain by a thick interval of conglomerate, amalgamated sandstone, and thin-bedded sandstone and mudstone, representing the fill of a channel axis; (ii) onlap interval of thick-bedded sandstone, overlain by medium- and thin-bedded sandstone and mudstone, representing the fill of a channel margin or deposits of a lateral accretion package and (iii) 10–20-m-thick (~ 33–66 ft.) repeating thin-bedded sandstone, siltstone, and mudstone and thick-bedded mudstone, deposits commonly found in distal splay environments. Fourth-order architectural elements may show repetitive or interlayered third-order elements. These are referred to as lithofacies associations which correspond to specific depositional environments.

Fifth-order architectural elements commonly refer to the fill of significant features, such as canyons and channels, stacked multistory channel-levee complexes or individual small fans or aprons. Fifth-order architectural elements may represent shifting deposystems responding to a number of factors including longer term eustatic or tectonic cycles (Chapters 3 and 12).

Sixth-order architectural elements are often formation-level units representing, for example, multistory channel complexes, bounded by sixth-order surfaces. Sixth-order architectural elements can be submarine fans, like the Upper Mount Messenger fan apron (Rotzien et al., 2012) or the Amazon Fan (Pirmez et al., 1997). This is the largest unit defined by Lowe and Ghosh (2004), but this is an open-ended classification scheme and can account for fan complexes, which would be seventh-order architectural elements (Chapter 8) and so on.

## Oil and gas exploration and development

Many of the lessons learned about deepwater systems science and technology originate from the oil and gas industry, its geophysical and geological surveys, drilling campaigns and production histories (see Preface and Chapters 19 and 21). Exploration for oil and gas in deepwater clastic reservoirs has taken the energy industry to the corners of the globe from the deserts and tundra of onshore environments to the ultra-deepwater (definitions vary, but some have said in water depths >2000 m or ~ 6562 ft.) of remote continental margins (Fig. 1). Since the first oil discoveries of Brea-Olinda Field, onshore California, USA (Table 3 and Fig. 9) in the 1880s, success continued in California for the next century. Additional onshore deepwater reservoirs were discovered in 1911 at the Tabaquite Field, Trinidad, in 1925 at the Wheat Field in the Permian Basin, USA, and in 1939 at the Barataria Field—the first deepwater reservoir discovered in the US Gulf of Mexico (Cossey and Studlick, 2007). For the next several decades, incremental reserves were added in these established hydrocarbon

**TABLE 3** Key deepwater fields and their significance.

| Field | Year | Country | Basin | Water depth (m) | Resource | | Significance | Resource source |
|---|---|---|---|---|---|---|---|---|
| Brea-Olinda | 1880 | USA | Los Angeles | Onshore | 69 | mmbls | First deepwater clastic field? | http://www.laalmanac.com/energy/en14.php |
| Tabaquite | 1911 | Trinidad | | Onshore | – | mmbls | Trinidad deepwater clastics | Cossey and Studlick (2007) |
| Wilmington | 1921 | USA | Los Angeles | Onshore | 1200 | mmbls | Giant California field | Slatt et al. (1988) |
| Wheat | 1925 | USA | Permian | Onshore | 22 | mmbls | First deepwater clastic field in West Texas | Cossey and Studlick (2007) |
| Barataria | 1939 | USA | Onshore Gulf of Mexico | Onshore | – | mmbls | First deepwater reservoirs in the US Gulf of Mexico | Cossey and Studlick (2008) |
| Anguille Marine | 1962 | Gabon | Ogouue Delta | 25 | 345 | mmboe | Significant discovery offshore West Africa | |
| Forties | 1970 | UK | North Sea | 106 | 2750 | mmbls | First offshore giant field in UK North Sea | Whaley (2010) |
| Frigg | 1971 | Norway | North Sea | 100 | 4 | Tcf | First significant deepwater field offshore Norway | www.norskpetroleum.no; Heritier et al. (1980) |
| Chicontepec | 1978 | Mexico | Tampico-Misantla | Onshore | 12,300 | mmbls | Largest deepwater clastic field in the world | Busch (1992) |
| Scarborough | 1979 | Australia | Exmouth Plateau | 950 | 12 | Tcf | First deepwater discovery offshore Australia | https://www.woodside.com.au/what-we-do/australian-growth-projects/scarborough |
| Auger | 1984 | USA | Gulf of Mexico | 872 | 220 | mmbls | Significant US Gulf of Mexico minibasin play | |
| Marlim | 1985 | Brazil | Campos | 850 | 2700 | mmbls | Giant field offshore Brazil | ANP, Production to Oct. 2020 |

*Continued*

**TABLE 3 Key deepwater fields and their significance—cont'd**

| Field | Year | Country | Basin | Water depth (m) | Resource | | Significance | Resource source |
|---|---|---|---|---|---|---|---|---|
| Mars-Ursa | 1989 | USA | Gulf of Mexico | 896 | 1742 | mmbls | First giant field in the US deepwater Gulf of Mexico | Laherrere (2020) |
| Mica | 1990 | USA | Gulf of Mexico | 1328 | 61 | mmboe | First significant find of petroleum in the sub-salt of the US Gulf of Mexico | Weimer et al. (2017) |
| Schiehallion | 1993 | UK | West of Shetland | 400 | 425 | mmbls | One of the first deepwater fields in the UK Continental Shelf (see also Foinaven) | Leach et al. (1999) |
| Volund | 1994 | Norway | North Sea | 120 | 90 | mmboe | Deepwater clastic injectite field | Pernin et al. (2019); www.norskpetroleum.no |
| Zafiro | 1995 | Equatorial Guinea | Niger Delta | 660 | 1190 | mmbls | Deepwater exploration begins in Africa | |
| Bonga | 1995 | Nigeria | Niger Delta | 1000 | 1400 | mmbls | Giant field offshore Nigeria | Coajou and Insalaco (2018) |
| Girassol | 1996 | Angola | Lower Congo | 1400 | 1200 | mmbls | Start of successful exploration campaign offshore Angola | Coajou and Insalaco (2018) |
| Atlantis | 1998 | USA | Gulf of Mexico | 1600 | 575 | mmbls | First large US Gulf of Mexico fold belt discovery | |
| Scarab-Saffron | 1998 | Egypt | Nile | 550 | 5 | Tcf | Large gas field in the Nile Delta | Weimer and Slatt (2004) |
| Gendalo | 2000 | Indonesia | Kutei | 1425 | 1 | Tcf | Source rocks deposited in sediment gravity flow deposits | Saller et al. (2006) |
| Buzzard | 2001 | UK | North Sea | 96 | 550 | mmbls | Large stratigraphic trap in a mature basin | Doré and Robbins (2005) |
| Jack/St Malo | 2003 | USA | Gulf of Mexico | 2130–640 | 550 | mmboe | Commercial development of the US Gulf of Mexico Paleogene (deepwater Wilcox) reservoirs | https://www.chevron.com/projects/jack-stmalo |
| Jubilee | 2003 | Ghana | Tano | 1400 | 600 | mmbls | Large stratigraphic trap—that prompted significant exploration along the Atlantic Margins | Dailly et al. (2017) |

**TABLE 3** Key deepwater fields and their significance—cont'd

| Field | Year | Country | Basin | Water depth (m) | Resource | | Significance | Resource source |
|---|---|---|---|---|---|---|---|---|
| Tamar | 2009 | Israel | Levant | 1700 | 11 | Tcf | First deepwater field in the Levant Basin | H. Pettingill (Pers. Comm.) and Needham et al. (2017) |
| Prosperidade | 2010 | Mozambique | Rovuma | 1900 | 30 | Tcf | Giant East Africa discovery | Fletcher (2017) |
| Liza | 2015 | Guyana | Guyana-Suriname | 1700 | 1400 | mmbls | New and significant oil province | Chisholm (2020) |
| Sakarya | 2020 | Turkey | Black Sea | 2200 | 19 | Tcf | Largest discovery in the Black Sea | |

The reported fields may not be the largest within their basin but are considered to be significant to the geopolitical landscape and technical understanding of deepwater sedimentary systems.

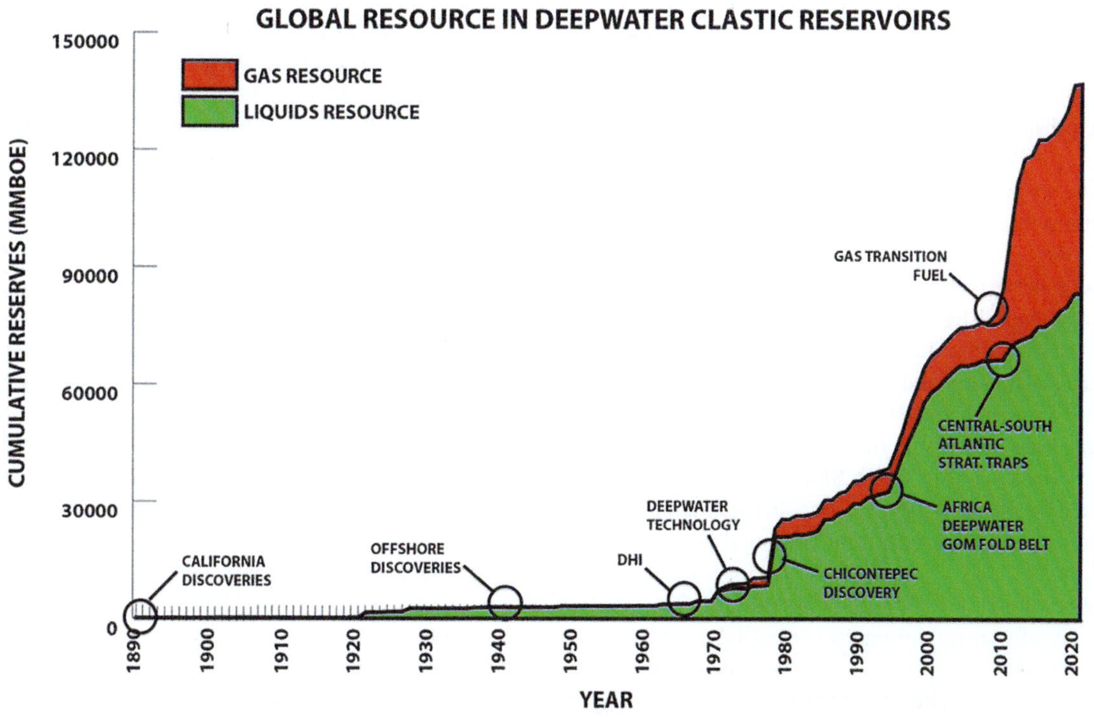


**FIG. 9** Cumulative size of hydrocarbon fields within deepwater sandstone intervals over time (Creaming curve). Not all fields are captured, but the ~275 fields captured here provide a useful dataset to recognize the main trends in the exploration for deepwater fields. Resource volumes are cumulative and interpreted at the end of field life as quoted by the operator. DHI is direct hydrocarbon indicator. This curve is typical of the exploration process in individual basins, which tend to show extended periods of flat to minimal growth in resource punctuated by the occasional sharp increase. Prior to these sharp upturns in resource, the discoveries have been made by the exploration teams, but the technology to produce from these frontier settings has not been developed at an economic scale. The first jump ~1920–1930 represents companies moving offshore into just a few to several feet of water. From 1970 to 1980, companies moved from the edge of the shelf into deepwater. In the 1990s, companies deployed a vast amount of deepwater production technology in offshore areas worldwide. Since the 1970s, there were several advances in deepwater technology including 3D seismic data, DHI analysis, drilling capabilities and offshore production (see Chapter 19). *(Data from ~275 fields compiled by the authors.)*

provinces. Exploration in shallow water began to emerge in the 1940s and 1950s as operators started to push offshore. It was not until the mid-1970s that reserves associated with deepwater sedimentary reservoirs began to contribute significantly to the global petroleum resource base. During this time, technology evolved to facilitate drilling in deepwater environments. Key technology included the recognition of direct hydrocarbon indicators (DHI) on seismic datasets followed by the development of deepwater drilling capabilities, and finally the deployment of production facilities (FPSO and Tension Leg Platform (TLP)). Later, more comprehensive 3D seismic datasets allowed for more detailed and accurate characterization and risk mitigation of prospects. With the ability to step past the modern continental margin with new drill rigs and produce hydrocarbons from new discoveries, passive margin explorers could now target deepwater slope and basin-floor plays. Recently, large discoveries have been made offshore Guyana (Chisholm, 2020) and the Black Sea in water depths exceeding 1700 and 2200 m ($\sim$ 5577 and 7218 ft.), respectively. As exploration has continued into deeper waters in more remote environments, petroleum reserves have continued to increase without a period of significant plateau, highlighting the remaining potential and importance of these reservoirs to help meet global future energy demand (Fig. 9).

Major changes in the creaming curves, such as those shown as sharp jumps in Fig. 9, can be related to the opening of new provinces for auction and leasing (geopolitics), technological advancements (Chapter 19) and improvements in business strategy and geological understanding. Several key factors have influenced the continued increase in reserve addition resulting in profound success for select deepwater reservoirs. In the 1990s, both Nigeria and Angola opened deepwater provinces to exploration resulting in significant discoveries in both areas. In 1995, the Neptune discovery in the sub-salt fold belts (see Chapter 4) of the US Gulf of Mexico prompted a significant period of exploration. In the 2010s, abundant gas was discovered onshore in the US, resulting in a recognition of natural gas as a potential bridge in the growth of alternative forms of energy. This reinterpretation also resulted in newfound value in gas reserves previously interpreted as stranded. This paradigm opened large gas plays in East Africa, the Black Sea and the Levant Basin (offshore Egypt, Israel, Lebanon and Syria) with discoveries by Eni, ExxonMobil, Kosmos and Noble (Chevron), for example. Industry had begun to test stratigraphic traps in remote basins along the Central-South Atlantic margin with notable success in the Tano Basin, Ghana, Sergipe-Alagoas Basin, Brazil and Guyana-Suriname Basin resulting in the continued rise in reserves to the present day.

These discoveries consisting of stratigraphic trap plays challenged the common exploration hubris that stratigraphic traps are riskier than structural traps (Chapter 2), even though stratigraphic traps have demonstrated giant reserves and are commonly the largest fields in their respective basin (Stirling et al., 2017; Merrill and Sternbach, 2017). Large stratigraphic prospects often remain untested until late in the exploration history of respective basins (i.e., Buzzard, UK). While large, undrilled structures in proven petroleum provinces are rare, future exploration may focus on subtle stratigraphic trap prospects, which can be interpreted in great abundance in deepwater sedimentary systems.

Stratigraphic trapping in deepwater reservoirs is relatively common as sandstone intervals are encased in sealing sedimentary deposits: hemipelagic, pelagic, muddy debrite or mass transport deposits (see Chapters 6–10). Driven by gravity, sand-prone deposits commonly fill accommodation along the slope and basin floor and are then draped by mud during a hiatus, or covered by debrites (see Chapter 12). Even in basins with a history of post-charge (see Chapter 2) tectonic activity, it is common for a stratigraphic component to be found in the petroleum trap (i.e., US Gulf of Mexico minibasin plays such as Lobster and Ram Powell described in Edman and Burk, 1998; Clemenceau et al., 2000). The base-of-slope position (Chapters 8 and 9) sets up a favorable geometry for stratigraphic traps. Isolation of sandstone intervals from slope feeder systems can result from bypass (forming a pinch out) or erosional truncation by mud-filled channels or mass transport deposits (Amy, 2019; Straccia and Prather, 1999). Bypass is a common occurrence at the base of slope in systems dominated by higher flow efficiency. In these systems, reservoir-quality sediment can be carried far out into the basin, away from the base-of-slope position (Mutti and Normark, 1987).

Deepwater sandstone intervals can form exceptionally high-quality reservoirs supporting high flow rates and high ultimate recovery (HRHU) of hydrocarbons. Commonly, the remote environments in which these fields are discovered necessitate a faster turnaround on initial investment, favoring reservoirs and fluids with properties to support production at high rates. An example of a high flow rate was the peak production plateau for the Forties Field in the North Sea, which saw production exceeding 500,000 barrels of oil per day (Hurst et al., 2005). Deepwater developments are sensitive to the number of required wells and the associated subsea equipment as those costs form a large part of the capital expenditure (see Chapters 19 and 20). Poor quality reservoirs or high viscosity, thick petroleum demand more wells, thus decreasing the economic viability of the project (see Chapter 20). The highest quality reservoirs in deepwater sandstone intervals are commonly associated with stacked channel and splay deposits (Chapter 8) as well as environments reworked by bottom currents to form hybrid or mixed turbidite-contourite systems (Chapter 9). These environments typically have sandstone reservoirs with high porosity and permeability (see Chapter 13) and lack strong vertical and horizontal partitioning (i.e., high reservoir connectivity—see Chapter 2). The rock quality and architecture permits sustained flow rates and high sweep efficiency, as well as the potential for pressure support, thus increasing overall petroleum recovery (Chapters 13 and 17).

As grains of sand sort and settle to the bed during the deposition of a classic turbidite (Bouma Sequence—see Chapters 6 and 7) what can be produced is a relatively high-quality reservoir with uniform flow units. These turbidites can be relatively unchanging along their lengths of 10s to 100s of km. The beds can be quite tabular with limited variability in bed thickness (Tokes and Patacci, 2018—see Chapters 8, 9 and 16). Similar uniform quality of reservoir flow units is rarely recognized in other siliciclastic or carbonate environments, which typically require a greater number of development wells to achieve a high ultimate recovery. For the northern Gulf of Mexico, development well spacing was closer in the shelfal reservoirs than in deepwater reservoirs, indicating more reservoir heterogeneity in shallow marine deposits (Weimer and Slatt, 2004—see Chapters 19 and 20). This common observation in many basins around the world reveals deepwater reservoirs can provide high recovery per well and contain higher net-to-gross ratios than their upslope equivalents.

Given the high cost of developing deepwater fields, careful and consistent reservoir characterization is needed prior to delivering final investment decisions. Primary sources of uncertainty include the architecture of the reservoir flow units and their inherent heterogeneity. Mass transport deposits, debrites (discussed more in Chapters 7 and 10) and hemipelagic intervals can form baffles or barriers, either predictively or more randomly distributed in and among sand-prone elements (Sullivan et al., 2000). Mudrock intervals can drape channel bases and margins (sensu Abreu et al., 2003). This may cause segmentation and isolation of petroleum pools or sub-optimal placement of both injector and producer wells, increasing development costs. Intraformational sealing geobodies, such as mudstone, can be present in confined and weakly confined slope channel complexes resulting in multiple petroleum pools with different reservoir features, such as the Okume Complex offshore Equatorial Guinea (Nasser et al., 2017). The mudstone can also serve to isolate sand-prone geobodies making it difficult to plan efficient water floods (i.e., Schiehallion, North Sea in Govan, 2006). The geometry and continuity of baffles and seals at field scale are typically sub-seismic and problematic to predict and therefore are routinely modeled stochastically. Where applicable, 4D seismic data workflows have been useful in recognizing unswept resources (Marler et al., 2014; Tura et al., 2009; Govan, 2006).

Deepwater petroleum reservoirs will remain important in the coming decades given the large volume of hydrocarbons in currently producing fields and discovered fields that are undeveloped. New fields will benefit from a variety of technology improvements, including high-quality seismic datasets which will derisk exploration prospects and result in better well placement during development, as well as complex well designs and optimized completions which will increase the ultimate recovery of the reserves. The large capital outlays demanded by deepwater exploration and development, particularly in frontier basins (Chapters 19 and 20), will make it challenging for all but the larger operators (i.e., supermajors and national energy companies), to pursue these types of plays. Basins with established production are primed for additional infrastructure-led exploration (ILX) where alternative plays that may have been previously deemed high risk (i.e., stratigraphic traps), novel reservoirs (i.e., injectites, Pernin et al., 2019) or intervals of bypassed pay or unswept oil are pursued. The close proximity to infrastructure may not necessitate high rate and high recovery given that the initial infrastructure investment has already been made (see Chapter 19 and 20). In addition, new projects are expected to have a better environmental footprint by reducing emissions through minimizing flaring and venting, however, this could require extra pipelines and injector wells, which will add to the development costs.

As discussed, petroleum discoveries have yielded immense amounts of data on deepwater sedimentary systems, their sedimentology, stratigraphy, rock properties, architectural elements and broader controls on deepwater processes. The applied research produced by joint partnerships among industry, academia and government over the past hundred years is astonishing in breadth and depth. One interesting outcome is the increase in accuracy and precision in measuring deepwater systems properties through time with modern technology and 4D seismic data acquisition.

## Mapping and interpreting deepwater sedimentary systems

While the sedimentary record (Bouma, 1962 and references therein) and submarine cable breaks provided the initial curiosity into deepwater processes, 21st century methods have enabled a much more detailed window into the diversity of deepwater sedimentary systems. The repertoire of available data is expansive and the scale of some of the key tools is shown in Fig. 10. In addition to the rock record, methods now include:

- High-resolution seafloor bathymetry and shallow penetrating acoustic methods (chirp sub-bottom profiles and multibeam echosounder data),
- Deep penetrating low-frequency seismic data, with multichannel seismic and airgun technology (Stock et al., 2007),
- 2D, 3D and 4D seismic datasets that permit interrogation of how petroleum moves through the reservoir and is renewed in fields over annual and decadal timescales,
- A suite of potential fields data including gravity and electro-magnetics,

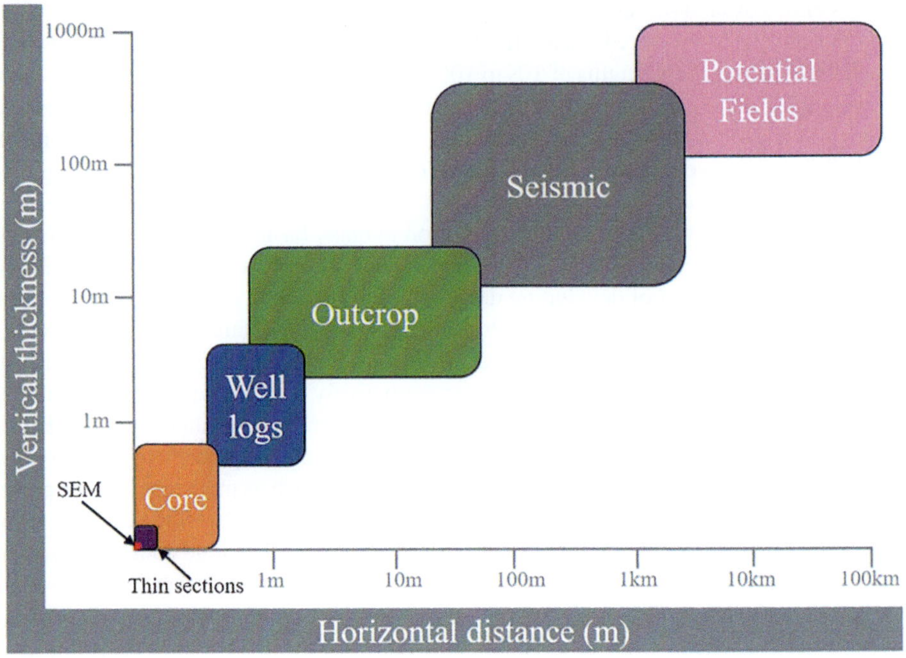


**FIG. 10** Length scale of investigation for key geologic tools, methods and datasets. Uncertainty is present in all subsurface datasets. All tools and methods have ranges of accuracy and precision. *(Modified after Nichols, G., Baker, S., 2015. Field-based training: luxury or necessity? Geo ExPro 12 (3), 28–31.)*

- Modern sampling of deepwater sedimentary accumulations using a variety of apparatuses, notably in submarine canyons including the Monterey Canyon offshore California, USA (Paull et al., 2018), and many others and
- Deployment of vast arrays of ocean and ocean bottom monitoring equipment (Clare et al., 2020).

While the rock record is still one of the most important datasets to link sedimentary processes and depositional products, new methods and data sources available for collaborative studies of deepwater systems are leaving less to the imagination about what goes on in the least sampled parts of Earth's surface.

## Seismic data

Table 4 explains the primary tools used by geoscientists to explore and interpret deepwater sedimentary systems, their scales of investigation, when they should be applied, their limitations, costs and timing. Acoustic seismic data remains one of the most important techniques used to understand the Earth (see Chapter 16). The continual development and application of seismic technology over the past 100+ years (reflection methods pioneered by J.C. Karcher) has enabled geoscientists to extract and interpret ever increasing detail from the subsurface, fueling operators' ability to explore further and deeper than ever thought possible.

At its most fundamental level, seismic data is an acoustic response to changes in rock density and seismic wave velocity within the Earth. While an extremely powerful tool, it does not provide an image of the subsurface but rather an approximation. Therefore, great care needs to be taken when attempting to interpret and build a geologic understanding based upon these data. Fig. 11A and B illustrate how changes in seismic frequency can make impactful differences in geologic interpretation. Relatively high frequencies (>50 Hz) are needed to resolve details at the outcrop or bed scale (Fig. 11A). This shows the sheer amount of geologic detail unseen by most typical deepwater seismic datasets (peak frequencies ranging from 10 to 20 Hz). Fig. 11B shows how changes in frequency (15 Hz vs 35 Hz) result in a more or less accurate interpretation of the subsurface and can sometimes result in entirely different geologic interpretations. A myriad of factors, assumptions and choices go into generating a seismic image with strong influence on their subsequent interpretation. Many of these parameters are controllable or calibrated, such as seismic acquisition parameters and design, types of technology employed and methods for processing the data, but many are beyond the interpreter's control, such as the actual physical parameters of the Earth (e.g., reservoir or target depth, lithology, compaction, bed thickness, etc.). Every seismic-related choice throughout the exploration and prospect maturation process has a cumulative effect on the final products and subsequent interpretation. Therefore the time, effort and costs associated with each choice must be considered in a larger business or scientific context. Cost–benefit analysis at almost every step is key to making sound commercial decisions in the near- and far-term. For example, an important decision can be made on whether to spend more money to acquire an ocean-bottom

**TABLE 4** **The primary tools used by geoscientists to explore and interpret deepwater sedimentary systems, their scales of investigation, when they should be applied, their limitations, costs and timing.**

| Geoscientists' tool | Scale of investigation | When to use | Primary uses/ products | Technical limitations | Costs and timing |
|---|---|---|---|---|---|
| Seismic data | 10s to 100s to 1000s of meters | Early exploration to production phases (all phases) | Basin analysis/ history, sequence stratigraphy, age correlation, trap mapping and evaluation: structural maps, thickness maps, facies maps, source presence maps, hydrocarbon drainage analysis/ migration, petroleum systems modeling, reservoir quality modeling | Resolving fine detail limited by acquisition, processing decisions/ assumptions and rock properties; need to integrate and calibrate with other data sources, primarily well-logs and core | Millions to 10s of millions USD (marine surveys); can take months to acquire and years to process or re-process |
| Potential fields (e.g., gravity and magnetic data) | kilometers to 100s of kilometers | Early exploration phase, new country entry decisions | Regional evaluation, basement mapping and evolution, presence of low density deposition (i.e., salt) | Low resolution, model assumptions, rarely a smoking gun | Fairly low cost, often acquired simultaneously with marine seismic; months. Aeromag/grav is inexpensive and can be acquired rapidly; weeks to months |
| Well logs (LWD +Wireline) | Centimeters (e.g., microresistivity formation images) to meters (most) | All phases | Lithology, rock properties, petrofacies models, sequence stratigraphy/age correlation, fluid samples, pressure data, reservoir connectivity | Operational and physical limitations, LWD availability/ quality of some tools | Hundreds of thousands to millions USD (offshore; rig time +service company cost, less for onshore); hours to days |
| Core, sidewall cores, cuttings | Centimeters to meters | Exploration drilling phase, development phase | Valuable in-situ rock sample, small-scale detail, petrofacies models, reservoir models | Small vertical sample of subsurface; lateral extrapolation of observed features | 10s of thousands to millions USD; Days to acquire, weeks to months to describe and integrate |
| Petrology, SCAL, SEM, fluid inclusions | Micrometers to millimeters | Exploration to development | Reservoir characterization, performance, connectivity, migration pathways | Time and cost, measurement error/ uncertainties | Thousands to millions USD; months |
| Geochemistry (PVT, fluid analyses, RockEval, etc.) | N/A | Early exploration to production | Commercial analysis, development planning, flow assurance | Sample contamination, operational risks | Thousands USD; months |

*Continued*

**TABLE 4 The primary tools used by geoscientists to explore and interpret deepwater sedimentary systems, their scales of investigation, when they should be applied, their limitations, costs and timing—cont'd**

| Geoscientists' tool | Scale of investigation | When to use | Primary uses/ products | Technical limitations | Costs and timing |
|---|---|---|---|---|---|
| Biostratigraphy/ chemostratigraphy | >10s of meters | Early exploration to development | Regional evaluation, age correlation | Age uncertainty depending on geologic time period/how diagnostic is biota? Sampling density, sample contamination | Thousands USD; weeks to months |
| Hydrocarbon systems models/ basin models/ reservoir quality models | 100s of meters to kms | Early exploration/ exploration | Regional/play to prospect analysis, source presence risk, source richness, source maturity, generation timing, porosity/ permeability estimates | Coarse model grid, many inherent assumptions, lack of calibration, non-uniqueness | Thousands to 10s of thousands USD; months |
| Geologic models | Meters to 10s of meters | Development planning to production phases | Simulating reservoir performance, production history matching, well planning | All models are wrong, some are useful | Thousands to 10s of thousands USD; weeks to months |
| Outcrop | Meters to 100s of meters | All phases | Analog data, visualizing scale and processes | Applicability, physical access | Thousands USD; months to years |
| Well tests/DSTs/ production data | 100s of meters to kilometers | Exploration drilling, development, production | Well rates, development planning | Length of test is proportional to lateral investigation distance, non-uniqueness/ uncertainty in interpretation | Millions to 10s of millions USD; days to months |

node (OBN) seismic survey instead of a conventional streamer survey earlier in the exploration phase in order to maximize the value of information in the near-term (e.g., data quality necessary for seismic inversion; see Chapters 16 and 17) and to provide development and production-stage project efficiencies in success case outcomes (e.g., 4D seismic data).

## Seismic acquisition and processing

Over time, seismic technology has advanced to literally include additional dimensions to the data (see Chapter 16). In many ways, each dimension is simply a higher density version of the previous. 2D data is a combination of many 1D seismic traces along a single vector. Initially, 3D data was essentially many 2D seismic lines acquired in a dense grid and processed together, however, modern acquisition systems with multiple cables and source arrays are acquiring some 3D information as the data are recorded. 4D simply represents a second or third 3D seismic survey acquired in the same location as the first but at a later time. 4D is primarily used to compare to the initial 3D survey in order to observe and interpret differences often caused by changes in fluid types, locations and their contacts with each other. 4D seismic data is used more routinely in long-lived oil and gas fields to determine how the fluids are moving through the reservoir to production. In some cases, the fields are replenished with newly sourced petroleum to replace the amount that has already been produced, and this too can be observed using 4D data.

Much of the innovation in deepwater seismic imaging was driven by operators' exploration efforts in the US Gulf of Mexico and other evaporite/salt basins around the world (e.g., West Africa sub-salt and pre-salt, North Sea, later in Brazil

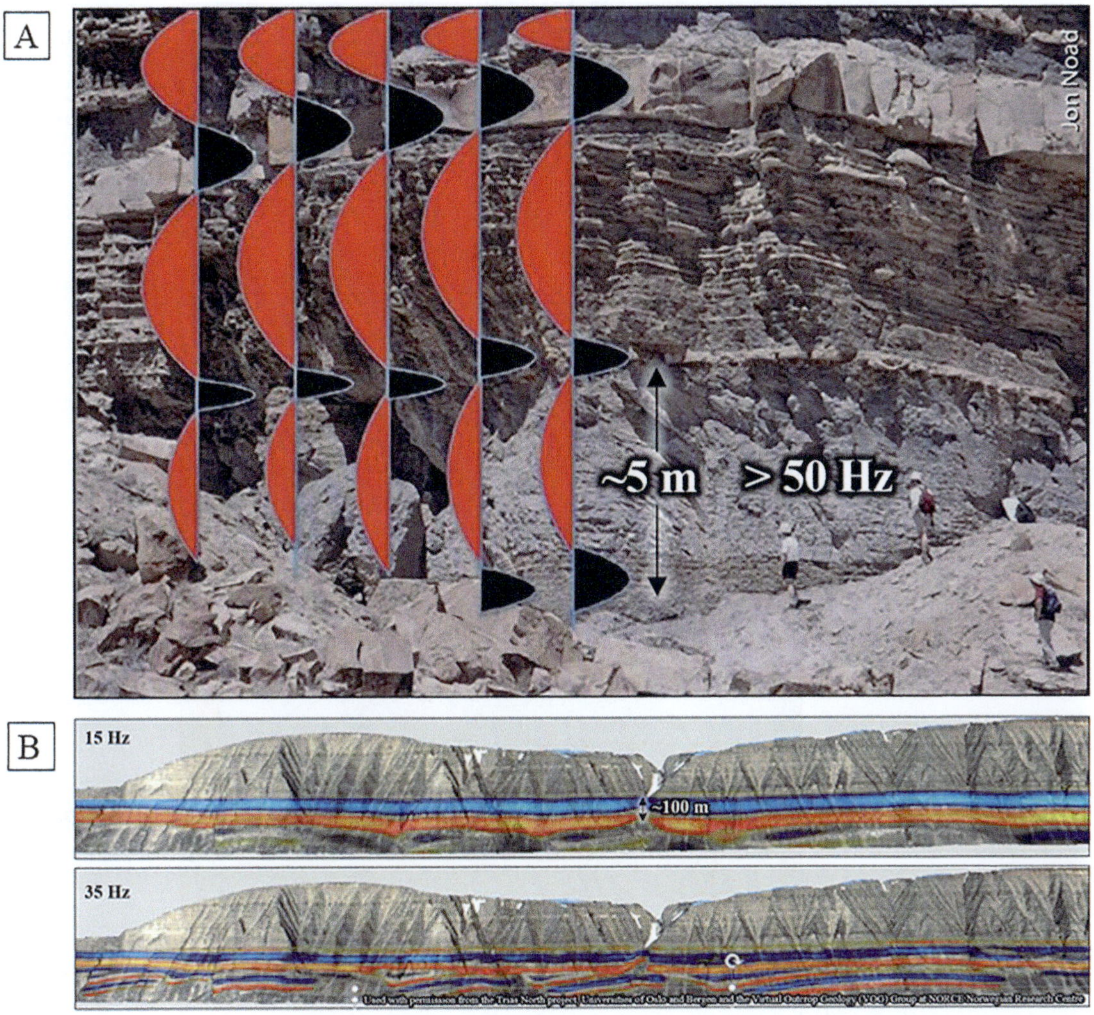


**FIG. 11** While seismic and well log interpretation are among the most abundantly employed methods in subsurface interpretation, outcrops of producing intervals and outcrop analogs provide significantly more detail on geologic heterogeneity, highlighting the necessity to "ground truth" and compare subsurface interpretations with direct measurements of available sedimentary intervals. (A) High resolution seismic "wiggle" traces overlaid on outcrop of parasequence-scale deposits in Woodside Canyon, Book Cliffs, Utah, USA. (B) Synthetic seismic image overlaid on a large-scale outcrop to highlight seismic resolution at varying frequencies. *(Panel A: Image used with Permission from Jon Noad.)*

pre-salt (Dellinger et al., 2017)). Salt geometries often reflect or disperse acoustic seismic energy, making processing difficult, hence limit its ability to provide an accurate subsurface image (Chapter 5). In order to overcome these challenges, continuous development of a whole host of technologies was and is still needed. Experiments in survey acquisition design showed that longer geophone streamers could provide the necessary seismic offsets—the distance between source and receiver—to image beneath these salt layers (standard vs long-offset data, Fig. 12A). Employing multiple airgun source boats (see Chapter 19) with shorter receiver streamer lengths (e.g., dual source acquisition) is another technique used to increase seismic offset.

As a seismic wave passes through the subsurface it encounters different rock layers and geometries depending on the compass direction (or azimuth) it travels to and from. Seismic surveys acquired in a single direction/azimuth can therefore be limited in their image accuracy (known as narrow-azimuth or NAZ data). By acquiring data from multiple azimuths, accuracy generally improves (multi-azimuth). Iteration from NAZ surveys to include as many azimuths as possible in a wide-azimuth configuration (WAZ) adds additional image detail and improved signal-to-noise ratios. This is achieved by sailing source and receivers in many linear directions or even in a circle, as with coil survey acquisitions. Representing the forefront and future of deepwater surveys is the most technologically advanced (and most costly) seismic acquisition OBN survey. OBN surveys involve lowering geophone receivers to the seafloor and therefore eliminate the effects of the water column (seismic multiples, noise, etc.) on the seismic signal. Also, if acquired early in the exploration timeline these

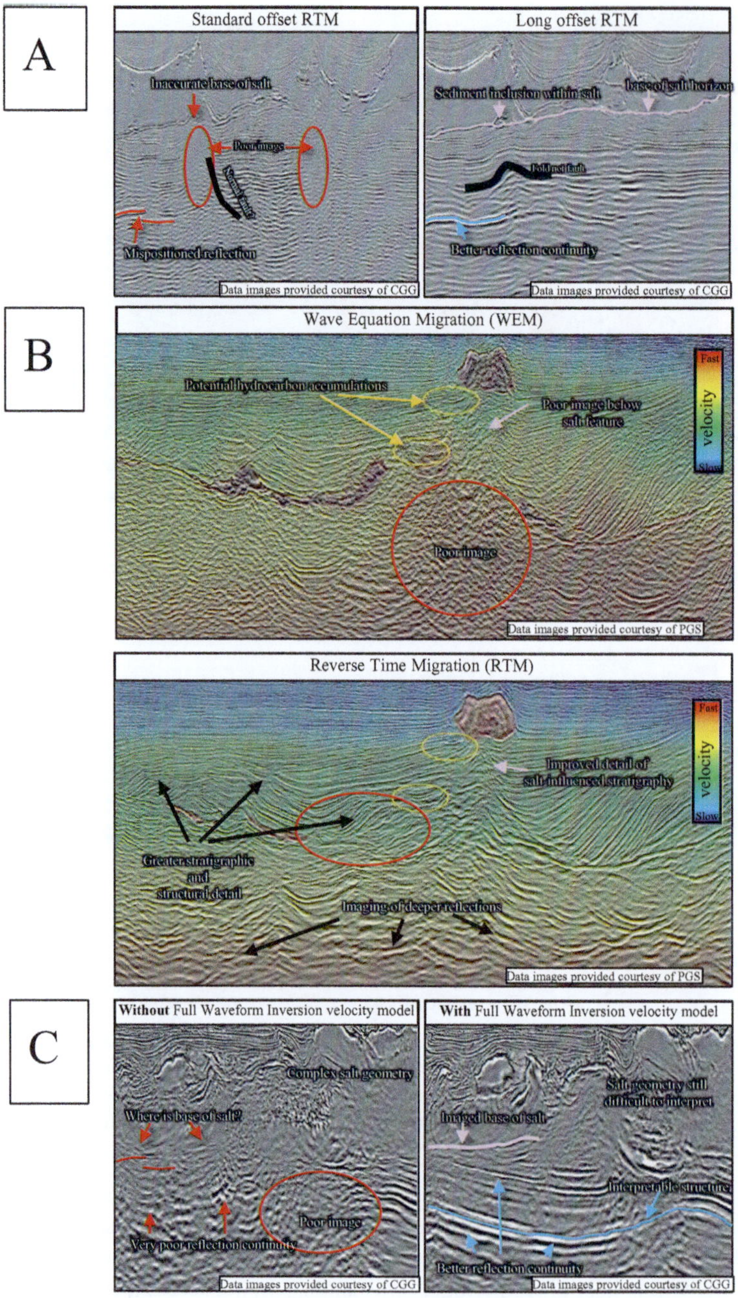


**FIG. 12** Examples of 2D seismic profiles that demonstrate the advances in acquisition and processing over the last few decades, from the top: (A) seismic image comparison between standard vs long-offset data from the US Gulf of Mexico. (B) Seismic image comparison between a legacy wave equation migration (WEM) and more modern reverse time migration (RTM). (C) Example of seismic imaging improvement by using full wave-form inversion (FWI) migrations to produce highly detailed and more accurate velocity models in shallow sections.

surveys are also ideal for 4D applications because the survey parameters (e.g., the locations of each of the nodes) are known and repeatable at a later date.

At the same time, information processing for geologically relevant interpretations occurs in step changes, especially due to the exponential increase in computational power over the past 20+ years. The emergence and widespread adoption of reverse time migrations (RTMs) and other computationally intensive methods by both seismic vendors and in-house processing teams has begun to unlock the secrets of difficult-to-image targets in sub-salt and pre-salt intervals (Fig. 12B). While generally lower in frequency, hence seismic resolution, hence geologic detail, migrations coupled with strong geologic interpretation, velocity model building skills (see Chapter 5) and deepwater sedimentary research, showed scientists an image of what *could* be (Dellinger et al., 2017), where previously was mostly noise (Fig. 12B). As computing power continues to increase, the use of full wave-form inversion (FWI) migrations to produce highly detailed velocity models in shallow sections is proving successful at further clarifying deeper targets in basins with heterogeneous velocity profiles (e.g., salt, overpressured shale intervals, densely faulted areas, structurally complex regions, etc.). The effect of more accurately defining the geometry and velocity of salt structures on the sub-salt image is illustrated in Fig. 12C.

While high-end migrations may be the vehicle for improved imaging of the subsurface, without proper interpretation, testing and iteration, they are but a technique without application. During processing or reprocessing of seismic data and even earlier at the acquisition stage if possible, continual reinforcement of project goals is critical to attaining real-world business success. Seismic processors need to understand the specific business needs of each project and the impact of their decisions on the bottom line while seismic interpreters need to have sufficient understanding of, and be heavily involved in, the processing decisions in order to accurately interpret final datasets. Some of the key decisions made by operators and service companies pertain to migration types and parameters, gather angle mutes, velocity and anisotropy assumptions, post-processing methods such as de-noise filters, and automatic gain control. It is possible that a standardized workflow of processing and post-processing steps can result in an improved image of the subsurface, but the end product can be so much more accurate and useful if constant communication, iteration, testing and mutual understanding is achieved between processors, interpreters and managers at all stages of a seismic project. With proper foresight during the acquisition and processing stages, higher-end, advanced techniques such as quantitative seismic interpretation products can be used to squeeze every last drop of useful information: 4D seismic data, seismic inversions, seismic facies models, pseudo-impedance models, etc. For a more detailed review of the applications of these products, please refer to Chapters 16 and 17.

## Well logs

Geoscientists' most common and important calibration points to the subsurface are geophysical logs (see Chapter 16 for more detail on certain logs). Geophysical logging or commonly "well bore" logging is a continuous recording of a parameter that can be obtained in real time (Logging While Drilling (LWD) or post drilling (Wireline)) along a well bore. Well logs provide information on bore hole diameter, geologic formation resistance to an electrical current, natural radioactivity of the sedimentary layers, velocity of sound propagation and numerous other parameters that have developed over the past 100 years (Rider and Kennedy, 2018) to provide information on the subsurface. These measurements at various scales have enabled geoscientists, petrophysicists and engineers to develop a more robust understanding of deepwater sedimentary systems, reservoir properties and fluids in the subsurface, allowing for calibration of large-scale seismic data to pore-scale, detailed reservoir characterization. Deepwater sedimentary intervals commonly contain units that have thin-bedded sandstone with mudstone, mixed clay and sand deposits (i.e., debrites) and mudstone clasts within a sandstone matrix (see Chapters 6–15). If insufficient types of well logs or well log data are acquired for these heterogeneous intervals, an incorrect petrophysical interpretation may be completed resulting in false valuation of resource in place and misunderstanding of the oil and gas asset. In particular, the interbedded and dispersed, clasts of clay can reduce the average resistivity across a reservoir interval resulting in hydrocarbon bearing zones that are underestimated or missed (Boyd et al., 1995).

Given the importance of this topic, comprehensive books of the various well logs and their measurements have been compiled (Rider and Kennedy, 2018). A short review of more of the recent advancements in several of the well logs is provided below (Fig. 13) with an emphasis on how they have increased the ability to characterize deepwater reservoirs. Over the next decades as more forms of alternative energy become available, it is expected that there will be a great need to maximize the efficiency of production of existing hydrocarbon reserves, map marginal reservoirs around existing infrastructure and delineate hydrocarbon-bearing intervals that were previously unrecognized by vintage logging techniques. An example of a marginal reservoir with comprehensive well logs and collocated core data from the North West Shelf of Australia is shown in Fig. 13 and highlights some of the well logs that are discussed.

Multi Array and Triaxial Induction Tools (i.e., AIT) are resistivity tools that are useful in evaluating thin-bedded intervals of pay because of their higher vertical resolution, allowing interpreters to view greater detail in the sedimentary layers. The higher resolution reduces "shoulder" effects that occur in muddy sandstone intervals. This can decrease the measured resistivity, and, in some instances, can decrease the calculated pay in the interval (Boyd et al., 1995; and others). Thin-bedded reservoirs ($<10$ cm or $\sim 3.9$ in. bed thickness—see Chapter 7) are found within most deepwater environments and are commonly found in greater abundance in submarine fan fringe (lobe fringe) and levees. The AIT tool may aid in accurately characterizing saturations and recognizing bypassed pay.

Nuclear magnetic resonance (NMR) can accurately define reservoir pore geometries, allowing for detailed understanding of saturation, porosity and permeability. It enables the petrophysicist to determine clay-bound,

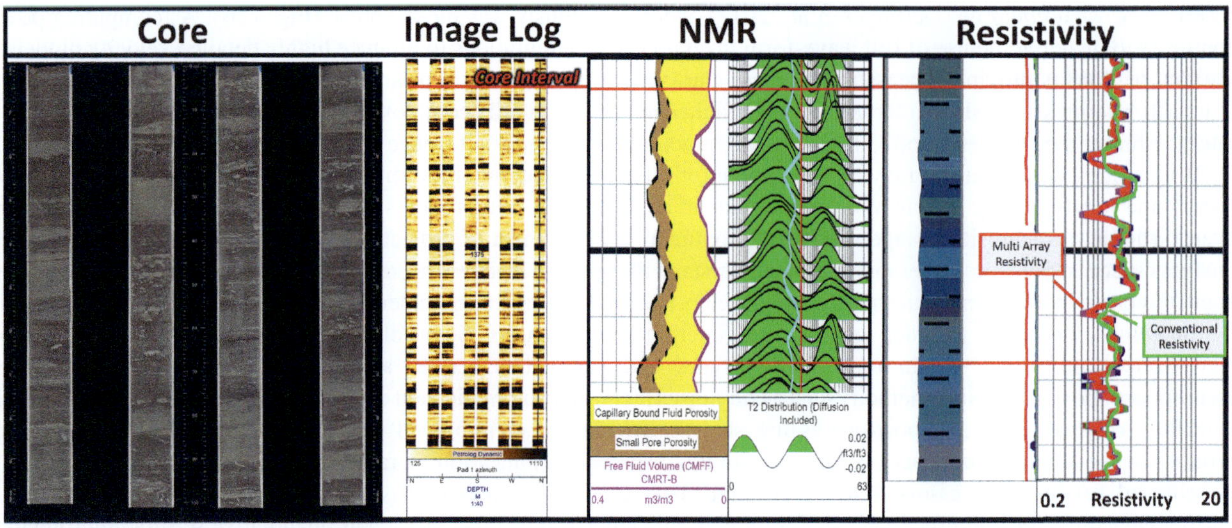


**FIG. 13** A suite of well log images and data used to interpret the subsurface. From left to right: core-calibrated example of log responses over a thin- to medium-bedded, water-bearing deepwater reservoir interval consisting of turbidites and transitional flow deposits with a variable amount of bioturbation. In the image log track, the sandstone intervals are resolved and sedimentary structures can be interpreted without whole core. The NMR log provides useful information on the presence of the free fluid volume, and the multi-array resistivity log (far right) provides detailed fluid saturation information.

capillary-bound and free-fluid volumes, which are important in debrites, thin-bedded intervals and mud-clast conglomerates (see Chapter 6). Structured and dispersed clays within these reservoir facies lead to an over interpretation of water saturation ($S_w$) in conventional logs as a result of bound clay. In these conditions, productive sandy units may be discounted in resource estimation, thus decreasing the value of the field or leaving these units undeveloped (Claverie et al., 2007). The derived NMR pore size and irreducible water saturation provides a more accurate understanding of the reservoir potential and hydrocarbon saturation. Additionally, impermeable units can more easily be differentiated (both visually and quantitatively) from permeable intervals, which permits a more accurate characterization of the most productive part of the reservoir (Potts et al., 2011).

Image log resolution has increased recently with higher well-bore coverage and a greater number of sensors (i.e., Formation MicroImager known as FMI and EARTH Imager) (Pöppelreiter et al., 2010). Millimeter-scale detail of sedimentary features can be resolved allowing the interpreter to recognize structures such as ripple cross lamination and fine mudstone clasts (Chapters 6 and 7) and zones of bioturbation (Chapter 15). This detail can provide information on the geometry of thin-bedded pay that is important to properly value an asset. Image logs are the closest dataset to conventional whole core for understanding detailed reservoir sedimentology in deepwater systems. Collocated core and image logs provide a powerful dataset used to develop geologic relationships that can aid in the development of depositional models (Slatt and Davis, 2010). In most petroleum provinces, whole core in all wells is cost prohibitive whereas image logs cost less, and therefore are more likely to be acquired. Additionally, dipmeter data provide useful information that may not always be apparent in the core (i.e., sediment transport direction and bed dips in debrites—see Chapter 16 for more information). Image logs have proven valuable in identifying mudstone intervals that drape channel margins and bases (i.e., mudstone drapes) in channel complexes, which can cause reservoir compartmentalization (Barton et al., 2010).

Formation testing tools with large inlets enable fluid samples and formation pressures to be obtained in challenging reservoirs that may have heavier fluids (i.e., Schlumberger Saturn Probe). These tools feature real-time downhole fluid analysis that can ensure sample quality by removing drilling fluid contamination and characterizing the fluids at reservoir conditions (i.e., $CO_2$ and C6+). When dealing with heavy fluids, deep reservoirs, low-quality or *marginal* reservoir lithofacies (i.e., slurry flow deposits, thin-bedded sandstone and mudstone—see Chapters 6 and 7) or the edges of the stratigraphic traps, being able to sample and confirm pressure connectivity and deliverability in these reservoirs may make the difference between a discovery or dry hole. Additionally, these tools capture samples for pressure, volume and temperature (PVT) analysis and flow assurance analysis in discoveries to improve the development planning process and ultimately enhance production volumes and life of the field. High asphaltic oils in the US Gulf of Mexico demonstrate the challenges of obtaining representative fluid samples (Nagarajan et al., 2012) and benefit from the larger inlets on modern formation testing tools. In thin-bedded intervals, it can be problematic to sample an individual sandstone reservoir within the smaller inlet/probes as it demands very accurate placement of the tool. With the larger inlet, it may be possible to sample several of the individual thin beds, providing information on the reservoir deliverability and fluid composition (see Chapter 2 for more information on these topics).

As oil and gas exploration for many operators from ~2015 to present focuses near existing infrastructure, the use of modern well logging technology helps operators find new hydrocarbon accumulations and identify previously overlooked or missed opportunities. Well bore logging technology has rapidly increased and developed over the past 50 years giving geoscientists, petrophysicists and engineers the tools they need for a more robust understanding of formations in the subsurface and their fluids. This short review of the recent advancements illustrates the importance to the characterization of deepwater reservoirs and reservoired fluids and will remain critical in the future exploration and analysis of deepwater sedimentary systems.

Mapping and interpreting deepwater sedimentary systems involves an accurate and precise measurement of their properties using the great variety of tools described above. New tools are invented and recalibrated each year, leading to more accurate and precise—and both direct and indirect—measurements of Earth's interior, fluid envelopes and deposits. The same tools are made better through each field assessment and each well drilled. Field techniques are also constantly refined to move further into the 21st century. All of these methods are useful to attain more information on deepwater systems and as such our knowledge of deepwater systems remains on an infinite frontier.

New regions in deepwater commonly described as "frontier" are areas of active exploration. How do explorers map and interpret deepwater systems with confidence? The exploration common process (ECP) was created for this purpose, and is described in the next section.

## The exploration common process

An intimate understanding of the deepwater sedimentary system is required for proper play fairway analysis and prospect evaluation. Maps of deepwater gross depositional environments (GDE), also known as environments of deposition (EOD), are a key product in technical evaluations and can drive exploration programs by underpinning commercial and business decisions (see Chapters 8–12). At their global extent, these maps for a given time period are known as the *Earth model*. In general, geoscientists use all available data (i.e., seismic, offset and analog wells, seeps and slicks, geochemical, etc.) to construct a model of past environments in order to predict the present-day hydrocarbon system. At the regional to play scale, GDE maps are the primary tool to synthesize multiple hydrocarbon play elements and their risks, notably reservoir and source presence, seal, reservoir quality (see Chapters 2 and 13 for more information) and trap. As investigation progresses toward the prospect scale, detail is added, often through additional or improved datasets, and the GDE map is refined to include estimates of additional play elements such as hydrocarbon migration pathways, reservoir quality and connectivity and seal integrity (Chapter 2).

Robust technical evaluation methods should take into account analysis and consideration of multiple geologic and depositional scenarios. These may be numerous during early exploration phases (play) and are subsequently distilled and refined as new data comes to light (prospect-level). It is important to keep an open mind to all possible outcomes when constructing a depositional model in order to prevent early bias toward one solution and to remain flexible when integrating new information. This approach can also help determine what types of data are most necessary to discriminate between favored models and can aid in a cost–benefit analysis of acquiring (i.e., buying/spending money) new data. Generating an extremely detailed and complete GDE map is not always necessary (or possible) to make an informed business decision and therefore following an exploration process has been shown to be the most efficient means at finding hydrocarbons.

While the exact origin of the ECP is difficult to pinpoint, certain works are credited for its establishment, including Vail et al. (1977a, 1977b) and Hubbard et al. (1985a,b, 1990). Now, this process is widely used by nearly every oil and gas company to explore for petroleum resources, with its own idiosyncrasies from group to group. The ECP is not limited to deepwater deposits, yet its proving grounds included offshore Gulf Coast, USA, West Africa and UK North Sea in the late 1960s and early 1970s. Today, the ECP is used around the world in the hunt for new conventional and unconventional plays.

### Sequence stratigraphy—AAPG memoir 26

A group of researchers at the Exxon Lab in Houston hailed from Northwestern University, where they adopted and translated the term 'depositional sequence' from their advisor L. L. Sloss to seismic scale in the exploration for oil and gas offshore Gulf Coast, USA, West Africa and UK North Sea. Sequence stratigraphy is discussed at length in Chapter 12 and the history of its inception and application is explained in detail in Donovan (2010).

The fundamental unit in sequence stratigrpahy is the depositional sequence, defined by Mitchum et al. (1977) as "a relatively conformable succession of genetically-related strata bounded at its top and base by their unconformities or their relative conformities." Therefore, sequence stratigraphy is the study of rock relationships within a chronostratigraphic framework of repetitive, genetically-related strata bounded by surfaces of erosion or non-deposition, or their correlative conformities (Van Wagoner et al., 1988) and is a method for establishing a chronostratigraphic correlation framework based primarily on geometric relationships. Used as a predictive tool for a whole host of play elements, this method has applications across a wide range of scales from basin, to regional and prospect, to field and finally reservoir to bed-scale. Born out of seismic data, this powerful prediction tool has enabled operators to successfully explore for petroleum resources for >60 yrs. and due to its refinement and ongoing testing, will persist for years to come and in new applications of finding and developing energy resources. One of the most promising and emerging frontier industries for the method of sequence stratigraphy is carbon dioxide sequestration (see Chapter 21).

While seismic is the dominant data type used to map out sequence stratigraphic surfaces and geometries in the subsurface, it is important to understand the scale, resolution and limitations of various data types (Fig. 10 and Table 4) and necessary to incorporate as much evidence as is reasonably possible to make sound geologic interpretations. These additional items include but are not limited to: sedimentary stacking patterns from offset and analog well logs (Fig. 13), conventional and side-wall cores, outcrop analogs, seismic attributes and fluid production data.

Prior to 1977, there were few examples that explained the formula of how exploration was performed. Exploration was formalized as a published method by Exxon in 1977 in AAPG Memoir 26. According to Vail et al. (1977a, 1977b), exploration mapping was a four-stage process:

1. Define the depositional sequence(s).
2. Define the internal character of the depositional sequences, using seismic facies analysis, to interpret the lateral extent of facies for a specified time or depth interval.
3. Transfer those facies to a basemap to create isochron or isopach maps (i.e., thickness maps).
4. Characterize the configuration, continuity, amplitude, frequency and interval velocity to interpret the subsurface geology, to make inferences on elements of the total petroleum system and its timing of critical play elements (i.e., source rock maturity, expulsion, trap timing and fill, etc.—see Chapters 2 and 18).

This method, while created decades in the past, is still used at companies both large and small worldwide, but with much faster computational and mapping hardware and software. Sometimes frontier exploration requires a wider lens, and evaluating the depositional sequence is insufficient for understanding the origins and potential of a basin. *Is there a scale above the depositional sequence scale?* Hubbard et al. (1985a,b) and Stewart (2020) look for higher order architecture of sequences, controlled by tectonic processes (tectonics discussed earlier in this chapter), known as *megasequences*.

## Megasequence mapping—Hubbard et al. (1985a,b) and Stewart (2020)

While Exxon and Shell developed the concepts presented in sequence stratigraphy about the same time, exploration in other frontier regions by BP prompted a new way of thinking. By identifying major tectono-stratigraphic intervals, Hubbard et al. (1985a,b, 1990) characterized what would become known as *megasequences*, facilitating another "top-down" approach to exploration (identifying the biggest features first; alternatively, architectural element analysis commonly identifies the smallest features first). Like classic Exxon sequence stratigraphy, the first aim was to identify the highest order stratigraphic surfaces that were not crosscut by other lower order stratigraphic surfaces. These surfaces defined the boundaries between megasequences, or where one megasequence ended, and a new tectonic phase had begun, ushering in potentially quite different styles of sedimentation and depositional architecture in a new petroleum system.

Hubbard et al. (1985a) took the outline in Vail et al. (1977a, 1977b) to greater detail, which came to be known as the ECP, also consisting of four steps. It was not that BP was uneasy of revealing the process; they understood it was in the analysis that differentiated competitors on the global stage of exploration. Megasequence mapping is at one scale greater than that defined by Vail et al. (1977a, 1977b):

- Identify the megasequence boundaries
  - Note geometric relationships including the combinations of truncation, onlap, toplap, downlap
  - Plot "quick-look" isopach maps to check the quality of the mapping progress
  - Identify the shelf position and begin delineation of key depositional environments
- Characterize internal attributes and traits
  - Interpret the seismic facies and perform seismic facies analysis

- Work with the understanding that specific lithologies or fluids cannot be determined directly using seismic data (Exxon understood this principle, too)
- Build a geologic model (or a few)
  - Demonstrate the validity of vertical succession of depositional environments in model
  - High-grade areas likely prone to reservoir, seal and source rock potential; discuss geologic risk and uncertainty
- Develop play concepts
  - Undertake detailed prospect mapping in small areas with favorable potential that satisfy all 6–7 key risk elements (see Chapter 2 for more information)
  - Complete play fairway analysis and discuss geologic risk and uncertainty as needed

## The modern exploration common process

The global synthesis of exploration workflows have produced a commonly accepted methodology for exploration and appraisal (Fig. 14 modified after Fraser, 2010). Any company wishing to standardize its exploration process has a workflow to enable portfolio analysis on a global level. While smaller companies can explore for hydrocarbons in a variety of creative

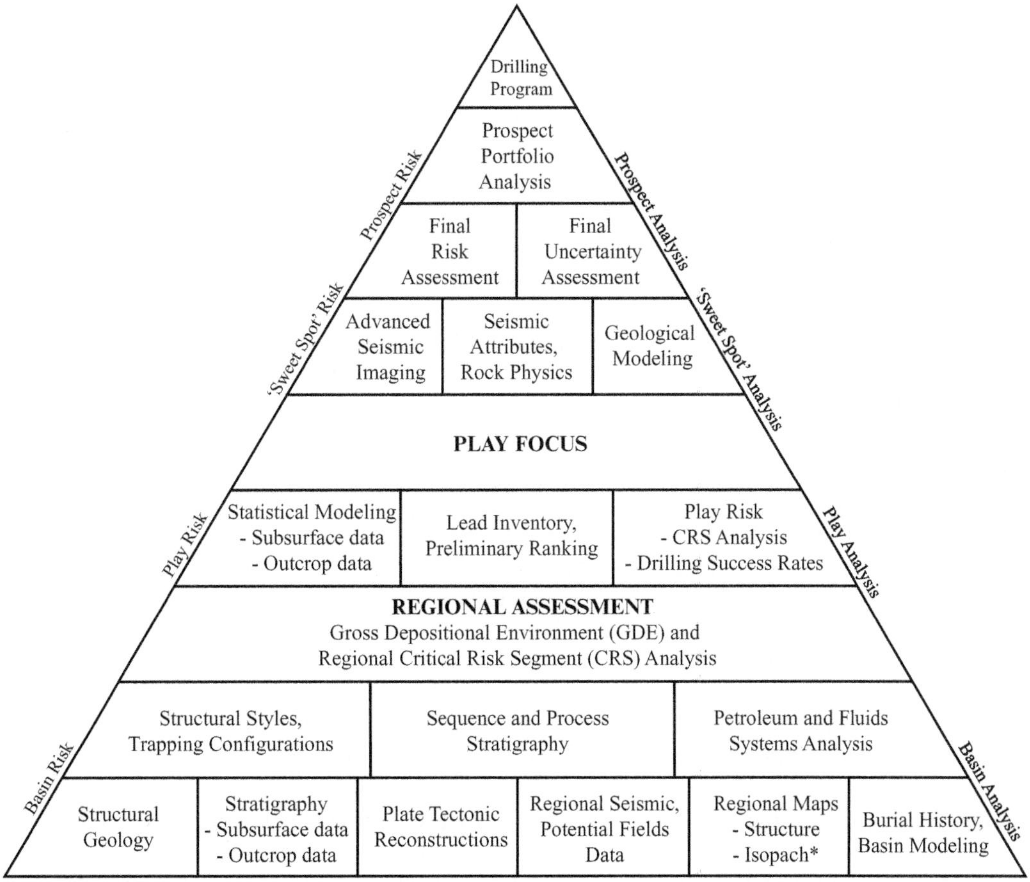


**FIG. 14** The exploration workflow or the "hierarchy of exploration" is part of the exploration common process and can be described using a *triangular diagram*, which is read from bottom to top. The first steps provide a geologic foundation in order to perform play fairway analysis, build a lead inventory and finally deliver a portfolio of ranked prospects. First, basin analysis provides a regional history of the area of interest: tectonic, burial history, paleogeography and paleoclimate. This stage illustrates exploration concepts or ideas. Second, play analysis illustrates where key play elements combine to form a viable play. It also serves as a stage to screen risk and acquire acreage. Risk parameters are different for each company. Third, fairway and lead identification reveal the most favorable areas for prospect development. Fourth, prospects are risked and ranked to provide "drill-ready" prospects. While this diagram was built from petroleum exploration workflows, the spirit of exploration can be used more generally for the assessment of natural resources found or produced in basins worldwide. *(Modified after Fraser, A.J., 2010, A Regional Overview of the Exploration Potential of the Middle East: A Case Study in the Application of Play Fairway Risk Mapping Techniques, in: Vining, B.A., Pickering, S.C. (Eds.), Petroleum Geology: From Mature Basins to New Frontiers—Proceedings of the 7th Petroleum Geology Conference: The Geological Society, London, 791–800; and many others including the Houston Explorers Club.)*

ways, at major and supermajor companies there is a need to unify this process for benchmarking purposes (Rudolph and Goulding, 2017), but foremost to achieve proper ranking of wildcats and exploration opportunities.

While the ECP is still used for some basins including recent frontier exploration in Guyana and Suriname (Oswald, 2019) and megasequence mapping remains a new ventures as well as retrospective process (Stewart, 2020), the emphasis has changed to (i) infrastructure-led exploration (ILX), (ii) bypassed pay and (iii) adjacent pay in currently producing assets. A process is needed to gain deep geological knowledge of a basin, especially super basins (Sternbach and Merrill, 2021) that continue to reveal new exploration plays as companies develop more accurate and sophisticated geological models. Deep geological knowledge primarily benefits companies that develop acreage in a basin for several decades, and is a driving theme at supermajors. These companies are able to observe the interaction between basin-scale processes (Chapter 2–5, 11, 12), sedimentation processes (Chapter 6 and 7) and depositional environment (Chapters 8–10), whereas smaller acreage teams need to seek data outside their core areas for basin context. With increased knowledge of the geology comes increased understanding of the new opportunities. Reduced investment in deepwater exploration from 2015 to present (later in this chapter, and in Chapters 19–21) portends an increase in ILX, seeking new fields close to production facilities as well as bypassed pay and nearby pay. Finding new petroleum near existing fields and within existing fields makes *big fields bigger*, a repeating refrain with many of the world's large oilfields (Fig. 9).

Thus, the new ECP still requires context provided by regional seismic and stratigraphic correlations, existing sequence stratigraphic interpretations for the adjacent mapped areas, a robust interpretation of seismic facies and geologic analog analysis. These steps are similar to the ECP of 1977, yet the tools used to evaluate the subsurface are much more powerful, with faster results and better imaging. The geology remains very complex. For example, whereas evaporite-related "turtle" structures in the 1990s were difficult to image and understand, now the pre-salt intervals offshore Brazil and Angola present challenging lithofacies to map and interpret. In the deepwater deposits of the Permian Basin region of the USA, downhole logging tools are incapable of determining oil saturation in mixed carbonate-clastic intervals such as the Wolfcamp Group. Still, exploration success from rote application of the process, or through serendipity, yields about the same results for large and small companies (Milkov and Navidi, 2020). While the petroleum industry has roughly the same access to the tools used for exploration, exploration success is remarkably lopsided, with large discoveries found by relatively few teams. This would indicate that creativity—something not described in exploration often—is one of the most important keys to success (Yeilding, 2019).

Following successful exploration, which is more in line with geological and geophysical wisdom and techniques, assessment and extraction of resources from deepwater areas calls for proper understanding and execution of the engineering principles described next.

## Engineering principles for deepwater petroleum exploration and production

Along with new technologies and advancements in geosciences, deepwater exploration and production has also benefited from advances in all forms of engineering. For the purpose of this book, this section will describe a few examples, but there are other resources for the reader to more thoroughly dive into the recent advancements in engineering technologies and tools (e.g., *Petroleum Engineering Handbook* Volumes I–VII, *Journal of Petroleum Technology magazine*, *Fundamentals of Petroleum Engineering SPE Course*, *Offshore Technology Conference*, etc.). The term *engineering* is used in this section in general to cover the full array of specialties that play a role in the oil and gas industry, including drilling, completions, production and reservoir, to name a few. The key advances in technology highlighted in this chapter are discussed in greater detail and illustrated using historical case studies in Chapter 19 of this book.

Advances in engineering have stemmed from the need to access previously difficult or impossible to reach deepwater reservoirs. Innovation is used to find new ways to drill and produce hydrocarbons more economically, efficiently and safely. Another key driver for new technology is the ability to tap into more reserves to keep pace with growing energy demand for a rapidly growing world. For example, drilling vertical wells in onshore unconventional plays simply was not efficient or economic. This limitation created an incentive to find another way to produce oil and gas from these types of resources more effectively. Advances in horizontal drilling and extended reach wells coupled with hydraulic fracturing were key to unlocking larger resources in onshore and even offshore basins. These technologies have truly reshaped the oil and gas industry by changing how explorers find and develop reservoirs that were previously assumed to be forever inaccessible.

As geoscientists continue to develop new play concepts and search for undiscovered resources, engineering solutions are needed to find new ways to reach these exploration targets and ultimately bring them to market. Companies have moved from onshore to offshore and further into deeper and deeper water in order to make new discoveries (see Chapter 19). Drilling in water depths exceeding 10,000 ft. ($\sim$ 3050 m) and in areas with potentially high temperature and pressure

(HPHT) reservoirs required state-of-the-art designs for drill ships, drilling techniques and subsea equipment. Deepwater dynamically positioned drill rigs with dual derrick and dual activity drilling systems are now utilized to not only drill these reservoirs but to drill them faster and cheaper than ever before. Subsea well-head technology like 15K or 20K BOPs (15,000–20,000psi; McCulley, 2013 and managed pressure drilling allow companies to safely drill HPHT reservoirs, ultimately opening the door to explore more play concepts. Without these innovations, opportunities like the deepwater Wilcox Group (see also Paleogene Trend or The Lower Tertiary) and Norphlet reservoir trends in the US Gulf of Mexico would not be drilled, discovered or produced.

Completions and facilities engineering teams have played a large role in improving the commerciality of deepwater deposits, especially during the development and production phases. Smart completions and single-trip multi-zone completions have been pivotal for both onshore and offshore deepwater developments. Single-trip multi-zone completions in lateral wells have driven a surge in production from the Permian Basin, USA. Smart completions have also been key to developing complex and extremely thick reservoirs, like the Wilcox. Facilities and production engineering teams continue to extract value and optimize production by advancements in subsea multi-phase pumping, downhole artificial gas lift for deep wells and other production enhancement technologies. Optionality for subsea developments, pipeline infrastructure and FPSOs enable operators to design the most economic and flexible plan to bring petroleum to downstream refineries and ultimately to market.

With all the technological advances in the engineering world, exploration teams continue to find new and creative concepts to test. If economics or technology do not support drilling new plays immediately, these opportunities are held in portfolios until the pieces are in place to finally test the geologic concepts. In mature basins, the straightforward and least expensive conventional deepwater prospects have largely been developed and produced. Later in an asset's life-cycle, as new technology becomes available, companies are able to return to the same fields and basins to target petroleum accumulations and reservoir facies that were previously thought to be second tier, lower quality or uneconomic. This natural progression of developing a basin's resources over time via continuous advances in engineering and geologic knowledge promotes a holistic understanding of deepwater deposits since all facies and stratigraphic elements could be potential exploration and development targets depending on the available technology. As companies continue to target more challenging, untested prospects, technology will need to keep pace in order to open up additional worldwide opportunities, and to ensure their commerciality.

Innovation is driven by interdisciplinary teams working together to integrate all data related to the exploration, appraisal and development of deepwater deposits. A highly collaborative team can more quickly recognize gaps in knowledge and limitations of current technology. When a need is identified, this opens the door for innovation and creativity to find a solution. Eventually, this work by a greater team leads to desired improvement and new technology to make a prospect or discovery viable and economic.

As outlined in an earlier section of this chapter (see Table 4 and Chapters 16–18), geoscientists have many tools available to investigate deepwater sedimentary systems. When a discovery is made, the first few wells (exploration and appraisal phase) are focused on the collection of geologic data to help the subsurface team improve their understanding of the resource in place and inform the potential development plan. Next, development wells collect additional valuable information to integrate into the subsurface model. Typically, geologic information is a secondary priority to critical engineering data during this phase of the overall project. Geoscientists use the data collected to build a geologic framework and to inform the team of a path forward to development and full-scale production. However, data is also required by the various engineering disciplines to help the greater project team move from a discovery to a producing field.

Engineering disciplines, such as reservoir, drilling, completions, production and facilities, can utilize geologic interpretations to improve development plans and designs within a project. Fig. 15 shows what geologic inputs are used to better constrain and inform various engineering aspects of a project. For example, seismic data can provide the location of potential baffles and barriers to fluid flow, such as faults or impermeable rock (e.g., an off-axis lithology that baffles connectivity between two channel axes—see Chapter 8–10 for examples), that reservoir engineers then incorporate in their reservoir models. This can then provide a more accurate estimate for well drainage radii and ultimately helps the reservoir engineer and team optimize well placement and well count. Such optimization results in a more efficient and economic development plan, providing a potentially larger profit margin for the project. During this integration and optimization phase, teams will find and enact the best options for a given project. Inadequate plans or designs can be the spark for innovation and often lead to the development of needed technologies. Additionally, when the greater team (from geoscientists to facilities engineers to commercial) focuses on integrating all data and makes a true interdisciplinary effort to optimize every aspect of a project, efficiency and safety improve, yet again leading to greater returns. As discussed in the next section, finding ways to make exploration projects more economic will continue to be critical in order for companies to justify developing and producing these discoveries.

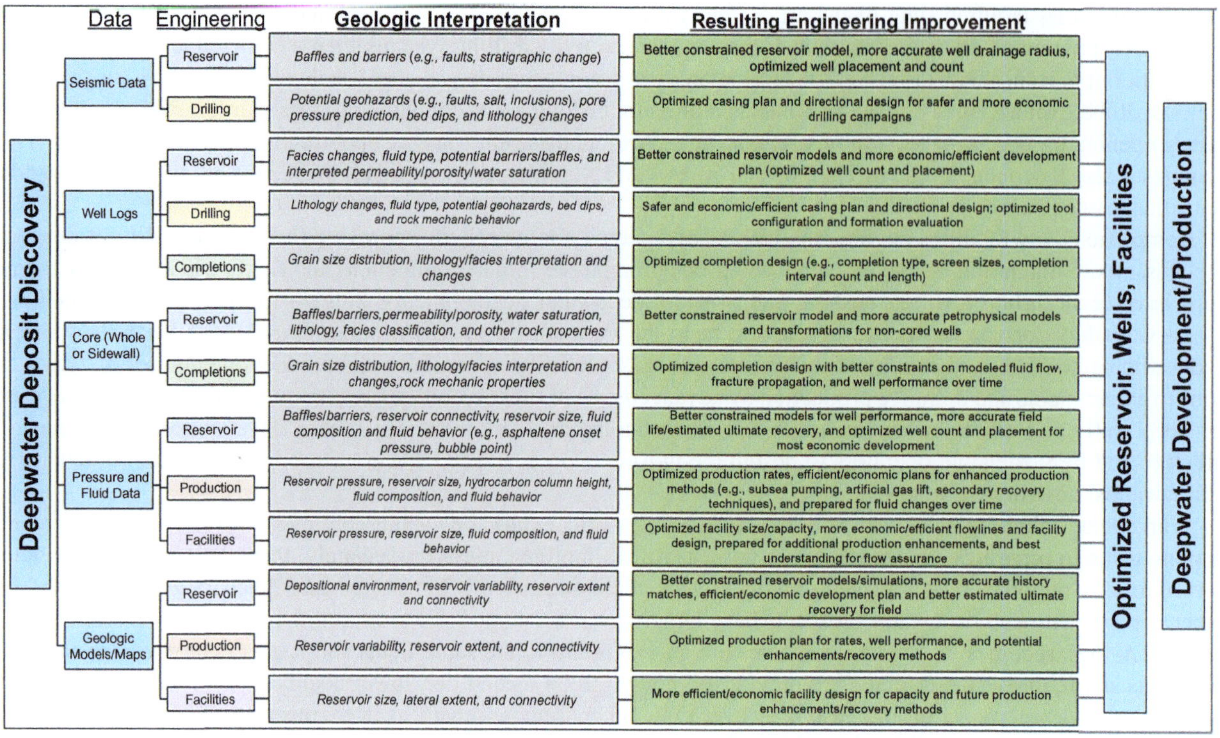


**FIG. 15** This flowchart provides specific examples of how data obtained for a deepwater discovery can be used by various engineering disciplines. Moving from left to right, the column of data (blue) obtained from the discovery includes seismic data to geologic models and maps. The next column (multicolored) indicates the engineering disciplines that may use that data, and the next column explains the geologic interpretation (gray) of the original data (first blue column). The far right column (green) highlights how the geologic interpretation can be used to improve the engineering results. All of the available data, learnings and improvements are utilized to optimize plans for the reservoir, wells and facility to bring the deepwater discovery to development and production. Note that this diagram is not an exhaustive list of data used in the engineering phase but is a representation of some of the data that is typically obtained and used in the initial discovery phase to add value to the project.

## Economics of deepwater exploration and production

In today's global economy of $90 trillion, the world is 80% powered by fossil fuels (Yergin, 2020). In the past 10–15 years alone, the USA has shifted from importing 60% of oil to now being the world's largest producer and a net exporter of oil and gas. Large economies in this global market still drive great demand for oil and gas. China alone imports 75% of its oil. Despite this shift in market distribution and the desire for advanced economies to decarbonize, it is clear that global demand for oil and gas will remain strong for at least the next few decades (BP Energy Outlook: 2020 Edition, n.d.; ExxonMobil, 2019). Therefore, it is important to understand deepwater systems as they contain vast undiscovered and undeveloped hydrocarbon-charged reservoirs. Exploration and development of these resources is critical for keeping pace with the world's continually growing energy needs. Chapters 19 and 20 provide more detail on the unique characteristics of deepwater resources that ultimately make these investments among the most economic greenfield developments available in the energy industry.

In the 2010s, the petroleum industry has seen significant volatility in commodity prices resulting in companies streamlining portfolios to focus spending on core areas, divest lower quality assets and adjust their oil to gas production ratios. Additionally, there are significantly fewer investment dollars available for companies to continue pursuing high risk opportunities. Particularly in the later 2010s, this has created a shift in the exploration process where companies of all sizes are more discriminating in the frontier plays and true wildcat exploration targets that they pursue. Many companies are also refocusing on "tie-back" opportunities around existing infrastructure or secondary recovery methods for more mature producing fields. This shift is documented in the International Energy Agency's (IEA) 2020 World Energy Investment report (IEA, 2020) in the global upstream oil and gas investment section. Based on data from 2010 to 2020, exploration spending has decreased from over 20% to now less than 10% of total upstream investment. With this significant drop in funds targeted for exploration, the annual global conventional resources discovered has dropped from ~48 billion barrels of oil equivalent (BBOE) in 2010 to ~15 BBOE in 2019 (Fryklund and Stark, 2016). Without new discoveries to replenish reserves, the industry is finding new ways to grow their portfolio and maintain production over time. Companies are focused on utilizing existing knowledge of produced reservoirs and basins to search for other targets. Originally deemed second tier opportunities, these more challenging and riskier reservoirs have now become "first tier" targets. This modification of the exploration process has driven companies to shift their attention or return to super basins or other prolific depositional settings in order to lower risk and increase returns in more challenging times (see Chapters 19 and 20 for more examples and discussion). Deepwater deposits continue to rise to the top of company portfolios due to their high volume potential and large areal extent compared to other plays. When oil demand outpaces current supply (see Chapters 19 and 20) as is often the case with commodity cycles, deepwater systems will once again be at the forefront of exploration portfolios including those that have not been extensively explored in the recent past, such as mixed deepwater sedimentary systems, combinations of turbidites, contourites and mass transport deposits (described in greater detail in Chapters 8–11).

Figs. 1 and 9 demonstrate that petroleum-bearing deepwater deposits have been found worldwide in great abundance and will continue to be a primary play for future exploration programs. However, despite the current nature of the world economy and global culture, strong geopolitical forces can challenge the development of discovered resources. Political instabilities, environmental concerns and agendas, regulation and policy changes and market volatility add to this complexity. Even if potential or discovered resources cannot be developed due to government mandates (for example: drilling moratoriums, environmentally protected lands and resources, etc.), explorers can still assess and prioritize the geologic, economic and sociogeopolitical risk of these opportunities as popular sentiment can change rapidly. As more developing nations strive to increase national wealth and as their citizens consume more energy, the energy industry will need to be poised to scale opportunities to deliver affordable energy that is accessible to all. In the meantime, there are currently many other nations that benefit from oil and gas exploration and production and therefore present less risk from a geopolitical perspective to target deepwater prospects. For example, Federal Lands, USA, earns royalties from oil and gas production and in the 2010s, the oil and gas industry was the second-largest contributor to the US Internal Revenue Service (federal tax service) due to these royalties. US State and Federal governments earn high annual revenues from petroleum production, especially in deepwater intervals found in the Permian Basin and Gulf of Mexico. Due to relatively stable regulations in some countries like the USA, companies have spent decades and tens of billions of dollars exploring and producing oil and gas in mature basins instead of pursuing opportunities in geopolitically riskier areas.

At this point in world history, many countries have begun to plan to transition away from fossil fuels as their primary source of energy to a balanced distribution that will include a significant portion of new and alternative energy resources. Over 120 countries have currently pledged to reach net-zero greenhouse gas emissions by the middle of the 21st century. This questions new petroleum exploration as achieving these goals would need substantially less upstream oil and gas investment. However, if the world should fall short of its commitments in the midst of an energy shift, a severe and potentially prolonged energy crisis may occur. Managing oil and gas supply in the process of a transition to alternative energy could arguably become more important and delicate with deepwater systems playing a key role in this complex and important task. Despite the progress in bringing alternative energy sources online, most yearly published energy reports (e.g., BP Energy Outlook (BP Energy Outlook: 2020 Edition, n.d.), Shell Scenarios (Shell, 2020), ExxonMobil Outlook for Energy (ExxonMobil, 2019), IEA World Energy Outlook (IEA, 2019), EIA Annual Energy Outlook U.S. Energy Information Administration, n.d.) still project high demand for oil and gas. Fig. 16 shows various projections for possible future global oil supply. The range for needed oil supply can vary from the base decline in supply needing no new investments (black line) to minimum supply asking for some additional investment to meet global oil demand (IEA SDS WEO 2020 case) derived from the Paris Agreement climate goals (blue line) to potentially the highest predicted supply needed if countries *cannot*[b] meet their targets or do not change policies (orange line). This large projected oil demand is predicted over the next few decades especially as developing nations still require sources of affordable energy (i.e., oil and natural gas) to meet the needs of growing populations and economies. Strong demand for fossil fuels has occurred over the past few decades in countries like China and India that have the largest and most rapidly growing economies in the world. Both countries rely on affordable and accessible energy sources like coal and oil to try to keep up with the rapidly growing energy demand of their populations. Even with new climate initiatives and goals, the need for better health and higher living standards in developing nations may outweigh the push for climate actions supported by developed countries.

In nations that are introducing new laws, regulations and tax treatments to achieve climate targets, many oil and gas companies are integrating these initiatives into their company goals and long-term business plans (see Chapters 20 and 21). Some companies have publicly stated their intention to become carbon neutral for certain business units (Table 5). Plans for carbon capture, utilization and storage (CCUS) are being announced, and forecasts calculate the related emissions

[b] It is a possibility that renewable energy resources may not be able to scale fast enough to meet demand.

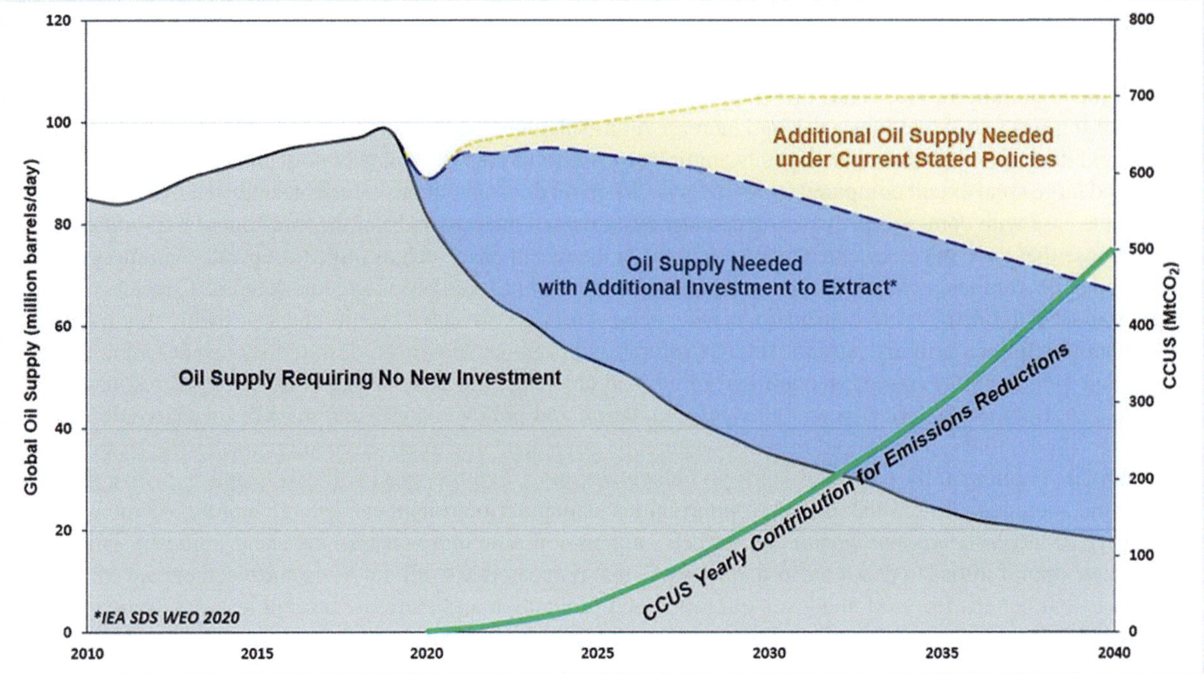


**FIG. 16** Global Oil Supply Chart based on current production. If there is no further investment in existing fields, the oil supply is projected to decline as shown by the *black curve*. The oil supply required to meet the forecasted demand based on the global emissions objective in accordance to the Paris Agreement is shown by the *blue dashed curve*. In order to achieve the targeted supply based on the goals set by governments and companies (IEA SDS WEO 2020 case), new investment in existing fields as well as discovery and development of new fields is required. This is in addition to the base decline. The highest oil supply projected based on the current commitments and trajectories under the stated policies is shown by the *dashed orange line*. In order to match the highest oil demand, new oil discoveries are required and will need to be brought to production. This is in addition to the anticipated discoveries and production included in the *blue curve*. The global emission goals result in a long-term plan to reduce oil demand over the next few decades. As demand is forecasted to decline, oil and gas companies are planning to increase their carbon capture, utilization and storage (CCUS) programs to contribute to the targeted global emissions reductions. By 2060, CCUS is projected to reduce cumulative emission by 27%, so as seen by the *green curve*, the CCUS contribution for these emissions reductions will increase yearly as more facilities and methods become available. *(Modified from figures in IEA Reports (IEA, 2020b, World Energy Outlook 2020, IEA, Paris. https://www.iea.org/reports/world-energy-outlook-2020; IEA, 2019. Transforming Industry through CCUS, IEA, Paris https://www.iea.org/reports/transforming-industry-through-ccus).)*

**TABLE 5** Carbon goals for some operating companies.[a]

| Company | Date of announcement | Stated climate goals | Target date | Method to achieve goals | References |
|---|---|---|---|---|---|
| BHP | September 2020 | Net zero emissions | 2050 | Operational efficiencies, low emissions technologies, product stewardship | https://www.bhp.com/investor-centre/climate-change-2020/ |
| BP | February 2020 | Net zero emissions | 2050 | Renewable energy, cut emissions from operations and upstream production | https://www.bp.com/en/global/corporate/news-and-insights/press-releases/bernard-looney-announces-new-ambition-for-bp.html |

**TABLE 5** Carbon goals for some operating companies—cont'd

| Company | Date of announcement | Stated climate goals | Target date | Method to achieve goals | References |
|---|---|---|---|---|---|
| Chevron | March 2021 | Net zero emissions | N/A | Renewable energy, carbon capture, hydrogen, eliminate flaring | https://www.chevron.com/sustainability/environment/lowering-carbon-intensity |
| China National Offshore Oil | June 2019 | Increase clean energy production | 2050 | Operational efficiencies, develop clean energy technologies, carbon sinks, increase green standardization | https://www.cnoocltd.com/attach/0/1f327d785d4b4fb4bb704eda72ae190a.pdf |
| China National Petroleum Corporation | May 2020 | Carbon neutrality | 2060 | Improved efficiencies, low carbon emissions technologies, carbon capture and storage, carbon sink forests | https://www.cnpc.com.cn/en/csr2020enhmshn/202105/031c7f9dff8140e9b403d26db8c9fbb6/files/bfe6a34a1b7f44d1a11c4b2ddaf7530e.pdf |
| ConocoPhillips | June 2021 | Net zero emissions | 2050 | Operational efficiencies, low carbon technologies, eliminate flaring | https://static.conocophillips.com/files/resources/conocophillips-2020-sustainability-report.pdf |
| Eni | March 2021 | Net zero emissions | 2050 | Operational efficiencies, carbon capture, reforestation, eliminate flaring | https://www.eni.com/en-IT/low-carbon/strategy-climate-change.html |
| Equinor | November 2020 | Net zero emissions | 2050 | Renewable energy growth, low carbon technologies, carbon capture | https://www.equinor.com/en/news/20201102-emissions.html |
| ExxonMobil | December 2020 | Net zero emissions | 2050 | Operational efficiencies, carbon capture, biofuels, eliminate flaring | https://corporate.exxonmobil.com/News/Newsroom/News-releases/2020/1214_ExxonMobil-announces-2025-emissions-reductions_expects-to-meet-2020-plan |
| Marathon | March 2020 | Reduce GHG emissions intensity | 2030 | Renewable energy, carbon capture, biofuels, low carbon technologies, operational efficiencies | https://sustainability.marathonpetroleum.com/reader/2 |

*Continued*

**TABLE 5** Carbon goals for some operating companies—cont'd

| Company | Date of announcement | Stated climate goals | Target date | Method to achieve goals | References |
|---|---|---|---|---|---|
| Occidental | November 2020 | Net zero emissions | 2050 | Carbon capture and storage, operational efficiencies, eliminate flaring, renewable energy | https://www.oxy.com/OurBusinesses/CarbonManagement/Pages/default.aspx |
| Petrobras | December 2020 | Net zero emissions | 2050 | Biofuel, operational efficiencies, renewable energy | https://petrobras.com.br/en/society-and-environment/environment/climate-changes/ |
| Repsol | December 2019 | Net zero emissions | 2050 | Carbon capture and storage, reforestation, natural climate sinks, renewable energy | https://www.repsol.com/en/press-room/press-releases/2019/repsol-will-be-a-net-zero-emissions-company-by-2050.cshtml |
| Saudi Aramco | July 2020 | Net zero emissions | 2050 | Carbon capture, operational efficiencies, renewable energy | https://www.aramco.com/en/news-media/news/2019/ocgi-new-measures-climate-challenges |
| Sinopec | March 2021 | Carbon neutrality | 2060 | Operational efficiencies, carbon capture, reduce flaring, carbon trading, carbon sinks, hydrogen | http://www.sinopecgroup.com/group/en/Resource/Pdf/SustainReport2020en.pdf |
| Royal Dutch Shell | March 2020 | Net zero emissions | 2050 | Renewable energy, nature-based solutions, hydrogen, carbon capture and storage, work with customers to mitigate individual emissions | https://www.shell.com/media/news-and-media-releases/2021/shell-accelerates-drive-for-net-zero-emissions-with-customer-first-strategy.html |
| Total SA | May 2020 | Net zero emissions | 2050 | Clean energy asset investment, renewable energy technologies, carbon sinks, energy efficiencies, eliminating flaring | https://www.total.com/media/news/total-adopts-new-climate-ambition-get-net-zero-2050 |

This table lists some oil and gas companies that have made public climate-related commitments and the details of the stated climate goals.
[a]*Note: This list is not the full list of companies pledging to reach net zero emissions or other climate-related targets. This is also a list as of August 2021 that is expected to change in future years.*

reductions from applying this method (Fig. 16; National Petroleum Council, 2019). CCUS requires large reservoirs and robust seals in order to permanently store $CO_2$, but this method can also be implemented earlier when companies target large developed reservoirs that could benefit from enhanced oil recovery utilizing $CO_2$ floods. As oil and gas companies begin to shift from exclusively producing petroleum to now finding ways to store carbon byproducts, targeting high porosity and permeability reservoirs with extensive lateral connectivity will be necessary for carbon dioxide storage just as it was desired for oil and gas exploration and development. The potential for deepwater facies to store large volumes of $CO_2$ ensures that continued research, understanding and exploration in these environments will play a key role for decades.

## Key Messages

1. Deepwater deposits are not only found in modern offshore basins, but also in onshore regions where tectonic processes have uplifted the sedimentary layers from original subaqueous deposition below wave base.
2. The history of study into deepwater sedimentary systems spans over 150 years and likely longer, yet society's knowledge of deepwater systems remains limited due to the relatively sparse sampling of oceans and deep lakes.
3. Climate, tectonics, sediment supply and derivatives of these controls (such as relative sea level) provide primary controls on the type, amount and abundance of sediment in deepwater systems. Some of these factors and resulting products are measurable; others are at such a scale (both time and space) that measurements are difficult and uncertain. There are myriad combinations that will be discussed in each chapter to provide a global appreciation for the scale and complexity of deepwater systems. Each deepwater sedimentary system is different.
4. As a way to provide clarity to the scale and type of deepwater deposit and compare depositional systems from around the world, architectural element analysis can be used to "speak the same language" in industry meetings and other discussions where time is of the essence.
5. Deepwater terminology is often misunderstood and can be routinely misused, as with any other frontier science. It helps to have robust operational definitions (see glossary for term definitions).
6. Global petroleum exploration maps and creaming curves suggest there are many more frontier deepwater successes on the horizon, augmented with new high-tech means of evaluating the subsurface to make informed business decisions. Other industries and societal needs can be addressed using energy resource management in deepwater areas; use of deepwater knowledge is not limited to the oil and gas industry, or just the energy industry (see Chapter 21).
7. While the workflows have remained relatively unchanged since the seminal texts of the 1970s and 1980s, the creative process in exploration coupled with high-resolution visualization of the Earth's crust and 21st century understanding of deepwater sedimentary systems has unleashed a new wave of promising exploration plays to fuel the growing demand for petroleum around the world.
8. Once found, deepwater resources require engineering expertise to provide not only a technical success—but one that is also economic and fits with societal norms.
9. The economics of deepwater exploration and production are, in principle, no different than that of onshore (see Chapters 19 and 20 for more information), but can be at a much larger scale due to the depth and size of resource that is required to trigger investment. Carbon dioxide capture, use and storage will be an important part of project economics today and likely into the future even as other technology becomes available.

## Conclusions

Deepwater sedimentary systems have been studied for more than a century. From early production in turbidite reservoirs to monitoring sediment gravity flows using equipment deployed in submarine canyons, deepwater sedimentary processes and deposits require creative, multi-disciplinary workflows and state-of-the-art technology to reach them and to understand them. Deepwater sedimentary deposits are not only found offshore today, but also in mountains that have uplifted basin-floor deposits due to the tectonic collisions. An understanding of deepwater sedimentary processes is required not only for established industries, such as oil and gas, but also for emerging industries including offshore wind and solar as well as geothermal projects that rely on porous and permeable deepwater rocks. Deepwater oil and gas development and technological achievement in established basins has encouraged frontier exploration success in basins with both similar and very different geology to satisfy petroleum demand on a global scale. While there are bits and pieces of the rich deepwater history told through reports, academic papers, presentations, books and conference proceedings, this chapter and the next 20 chapters aim to provide a comprehensive modern understanding of deepwater sedimentary systems and the application of that knowledge to resource exploration and development.

## Acknowledgments

The authors would like to acknowledge Richard Sears for the discussions that ultimately supplemented the Engineering Principles section; Daniel Wetzel (International Energy Agency) for the discussions and expertise that helped define the Economics Section; Cindy Yeilding for the discussions on the Gulf of Mexico, the exploration process at the Houston Explorers Club and throughout the drafting of this chapter; F. Javier Hernández-Molina for discussions on the broad scientific applications for deepwater research and Octavian Catuneanu for discussions on basin-scale processes and sequence stratigraphy. J.R.R. dedicates this chapter to Peter Lellis, whose geoscience, interpretation and leadership skills in the energy business were second to none. This work would not have been possible without conversations and guidance throughout the years from the following: D.R. Lowe, S.A. Graham, G.H. Browne, P.R. King, E. and S. Muller, J.R. Suter, A.D. Donovan, A. Hough, D. Nelson, The Drifters Research Group, P. Lellis, M. Brown, C. Collarini, S. Verma, M. Honarpour, C. Pirmez, Z. Sylvester, J. Pacht, W.D. McCaffrey, G. Apps, F.J. Peel, P.D.W. Haughton, M.D. Sullivan, T. McHargue, K.W. Rudolph, P. Myrow, C. Siddoway, J. Noblett, C. Bank, J.R. Gaither, R. M. Slatt, M.L. Sweet, Y. Gavillot, N. Uzunlar, A. Lisenbee, M. McKeown & family, E. Szymanski, B.E. Prather, C. VanDenburg, M. Fonnesu, P. Weimer, H. Pettingill, M.G. Rowan, B. Hudson, G. Fowler, L. Long, J. Stock, and organizations including AAPG, EAGE, HOT Engineering, Houston Explorers Club, Petroleum Investments Committee, GNS Science, Gulf Coast Section of SEPM, CERAWeek and many others.

## References

Abreu, V., Sullivan, M., Pirmez, C., Mohrig, D., 2003. Lateral accretion packages (LAPs): an important element in deep water sinuous channels. Mar. Pet. Geol. 20, 631–648.

Amy, L.A., 2019. A review of producing fields inferred to have upslope stratigraphically trapped turbidite reservoirs: trapping styles (pure and combined), pinch-out formation, and depositional setting. AAPG Bull. 103, 2861–2889. https://doi.org/10.1306/02251917408.

Anderson, K.S., Hickson, T.A., Crider, J.G., Graham, S.A., 1999. Integrating teaching with field research in the Wagon Rock project. J. Geosci. Educ. 47, 227–235. https://doi.org/10.5408/1089-9995-47.3.227.

Angelle, S.A., 2021. America's Energy Corridor: Louisiana Serving the Nation's Energy Needs: Louisiana Department of Natural Resources, Policy Paper. 37 pp.

Angevine, C.L., Heller, P.L., Paola, C., 1990. Quantitative sedimentary basin modeling. In: AAPG Continuing Education Course Notes Series. 32, p. 133, https://doi.org/10.1306/CE32509.

Bagnold, R.A., 1962. Auto-suspension of transported sediment; turbidity currents. Proc. R. Soc. Lond. A Math. Phys. Sci. 265, 315–331. https://doi.org/10.1098/rspa.1962.0012.

Bartol, Govers, 2009. Flexure Due to the Messinian-Pontian Sea Level Drop in the Black Sea., https://doi.org/10.1029/2009GC002672.

Barton, M., O'Byrne, C., Pirmez, C., Prather, B., van der Vlugt, F., Alpak, F.O., Sylvester, Z., 2010. In: Pöppelreiter, M., García-Carballido, C., Kraaijveld, M.A. (Eds.), Dipmeter and Borehole Image-Log Technology. AAPG Memoir, 92, pp. 195–211.

Beaubouef, R.T., Rossen, C.R., Zelt, F.B., Sullivan, M.D., Mohrig, D.C., Jennette, D.C., Bellian, J.A., Friedman, S.J., Lovell, R.W., Shannon, D.S., 1999. Deepwater sandstones, Brushy Canyon Formation, West Texas. In: American Association of Petroleum Geologists Continuing Education Course Note Series 40., https://doi.org/10.1306/CE40695. 48 pp.

Bouma, A.H., 1962. Sedimentology of Some Flysch Deposits: A Graphic Approach to Facies Interpretation. Elsevier, Amsterdam, p. 167.

Bourget, J., Ainsworth, R.B., Thompson, S., 2014. Seismic stratigraphy and geomorphology of a tide or wave dominated shelf-edge delta (NW Australia): process-based classification from 3D seismic attributes and implications for the prediction of deep-water sands. Mar. Pet. Geol. 57, 359–384. https://doi.org/10.1016/j.marpetgeo.2014.05.021.

Boyd, A., Darling, H., Tabanou, J., Davis, B., Lyon, B., Flaum, C., Klein, J., Sneider, R.M., Sibbit, A., Singer, J., 1995. The Lowdown on Low-Resistivity Pay. The Oilfield Review, Autumn Edition.

Anon., BP Energy Outlook: 2020 Edition. https://www.bp.com/content/dam/bp/business-sites/en/global/corporate/pdfs/energy-economics/energy-outlook/bp-energy-outlook-2020.pdf.

Bramlette, M.N., Bradley, W.H., 1940. Geology and Biology of North Atlantic Deep-Sea Cores: U. S. Geological Survey, Professional Paper 196-A. pp. 1–34. https://pubs.usgs.gov/pp/0196/report.pdf.

Browne, G.H., Slatt, R.M., 2002. Outcrop and behind-outcrop characterization of a Late Miocene slope fan (channel-levee complex), Mt Messenger Formation, New Zealand. AAPG Bull. 86, 841–862. https://doi.org/10.1306/61EEDBB6-173E-11D7-8645000102C1865D.

Browne, G.H., King, P.R., Arnot, M.J., Slatt, R.M., 2007. Architecture of base-of-slope fans, Mount Messenger Formation, Pukearuhe Beach, New Zealand. In: Nilsen, T.H., Shew, R.D., Steffens, G.S., Studlick, J.R.J. (Eds.), Atlas of Deepwater Outcrops. AAPG Studies in Geology 56, Tulsa, Okla, pp. 257–261, https://doi.org/10.1306/St561240 (Chapter 64).

Bugge, T., 1983. Submarine Slides on the Norwegian Continental Margin, With Special Emphasis on the Storegga Area. IKU Continental Shelf Publications 110, Norway. 152 pp.

Busch, D.A., 1992. Chicontepec Field, Mexico, Tampico-Misantla Basin. In: Foster, N.H., Beaumont, E.A. (Eds.), Stratigraphic Traps III: AAPG Treatise of Petroleum Geology. Atlas of Oil and Gas Fields, pp. 113–128.

Campion, K.M., Sprague, A.R., Mohrig, D., Lovell, P., Drzewicki, P.A., Sullivan, M.D., Ardill, J.A., Jensen, G.N., Sickafoose, D.K., 2000. Outcrop expression of confined channel complexes. In: Weimer, P., Slatt, R.M., Coleman, J., Rosen, N.C., Nelson, H., Bouma, A.H., Styzen, M.J., Lawrence, D.T. (Eds.), Deepwater Reservoirs of the World. Houston Gulf Coast Section SEPM Foundation, pp. 127–150, https://doi.org/10.5724/gcs.00.15.0127.

Carvajal, C.R., Steel, R.J., 2006. Thick turbidite successions from supply-dominated shelves during sea-level highstand. Geology 34, 665–668. https://doi.org/10.1130/G22505.1.

Carvajal, et al., 2009. Shelf-edge architecture and bypass of sand to deep water: influence of shelf-edge processes, sea level, and sediment supply. J. Sediment. Res. 79 (9), 652–672. https://doi.org/10.2110/jsr.2009.074.

Catuneanu, O., 2019a. Scale in sequence stratigraphy. Mar. Pet. Geol. 106, 128–159.

Catuneanu, O., 2019b. Stratigraphic cyclicity in deep-water settings. In: Rotzien, J.R. (Ed.), Basin Dynamics Bulletin. vol. 1(5). A Basin Dynamics Publication, pp. 1–25.

Catuneanu, O., Abreu, V., Bhattacharya, J.P., Blum, M.D., Dalrymple, R.W., Eriksson, P.G., Fielding, C.R., Fisher, W.L., Galloway, W.E., Gibling, M.R., Giles, K.A., Holbrook, J.M., Jordan, R., St, C.G., Kendall, C., Macurda, B., Martinsen, O.J., Miall, A.D., Neal, J.E., Nummedal, D., Pomar, L., Posamentier, H.W., Pratt, B.R., Sarg, J.F., Shanley, K.W., Steel, R.J., Strasser, A., Tucker, M.E., Winker, C., 2009. Towards the standardization of sequence stratigraphy. Earth Sci. Rev. 92 (1–2), 1–33. https://doi.org/10.1016/j.earscirev.2008.10.003.

Chiarella, D., Longhitano, S.G., Tropeano, M., 2017. Types of mixing and heterogeneities in siliciclastic-carbonate sediments. Mar. Pet. Geol. 88, 617–627.

Chisholm, T., 2020. Hess's journey into an emerging superbasin and ultra-high impact exploration for an independent E&P company. In: Houston Geological Society Annual University of Houston Sheriff Lecture, Nov. 9, Houston, Texas.

Claverie, M., Hansen, S.M., Daungkaew, S., Prickett, Z., Akinsanmi, O.B., Pillai, P.S., 2007. Applications of NMR logs and borehole images to the evaluation of laminated deepwater reservoirs. In: Paper Presented at the Asia Pacific Oil and Gas Conference and Exhibition, Jakarta, Indonesia, October. SPE-110223-MS. doi:10.2118/110223-MS.

Clare, M., Lintern, D.G., Rosenberger, K., Hughes Clark, J.E., Paull, C., Gwiazda, R., Cartigny, M.J.B., Talling, P.J., Perara, D., Xu, J., Parsons, D., Jacinto, R.S., Apprioual, R., 2020. Lessons learned from the monitoring of turbidity currents and guidance for future platform designs. Geol. Soc. Lond. Spec. Publ. 500, 605–634.

Clark, J.D., Pickering, K.T., 1996. Architectural elements and growth patterns of submarine channels: applications to hydrocarbon exploration. AAPG Bull. 80 (2), 194–220. https://doi.org/10.1306/64ED878C-1724-11D7-8645000102C1865D.

Clemenceau, G.R., Colbert, J., Edens, D., 2000. Production results from levee-overbank turbidite sands at Ram/Powell Field, deepwater Gulf of Mexico. In: Weimer, P. (Ed.), Deep-Water Reservoirs of the World, vol. 20. GCSSEPM Foundation, https://doi.org/10.5724/gcs.00.15.0241.

Coajou, A., Insalaco, E., 2018. Giant discoveries of the Gulf of Guinea (GoG): 1987–1997 and beyond. In: AAPG International Conference and Exhibition, Cape Town, South Africa, Nov. 4–7. Search and Discovery Article 30611.

Coleman, J.L., Browne, G.H., King, P.R., Slatt, R.M., Spang, R.J., Williams, E.T., Clemenceau, G.R., 2000. The inter-relationships of scales of heterogeneity in subsurface, deep water E and P projects—lessons learned from the Mount Messenger Formation (Miocene), Taranaki Basin, New Zealand. In: GCSSEPM Foundation 20th Annual Research Conference, Deepwater Reservoirs of the World, December 3–6.

Cossey, S.P.J., Studlick, J.R.J., 2007. A global overview of fields and discoveries in deepwater deposits. AAPG Stud. Geol. 56, 12–15.

Cossey, S.P.J., Studlick, J.R.J., 2008. Characteristics of worldwide hydrocarbon discoveries and fields in deep-water deposits. In: Schofield, K., Rosen, N. C., Pfeiffer, D., Johnson, S. (Eds.), Answering the Challenges of Production from Deep-Water Reservoirs: Analogues and Case Histories to Aid a New Generation, vol. 28. GCSSEPM Foundation.

Costa, J.E., Williams, G.P., 1983. Debris-Flow Dynamics (Videotape). U. S. Geological Survey Open File Report OF 84-0606., https://doi.org/10.3133/ofr84606.

Covault, Graham, 2010. Submarine fans at all sea-level stands: tectono-morphologic and climatic controls on terrigenous sediment delivery to the deep sea. Geology 38 (10), 939–942. https://doi.org/10.1130/G31081.1.

Covault, J.A., Normark, W.R., Romans, B.W., Graham, S.A., 2007. Highstand fans in the California borderland: the overlooked deep-water depositional systems. Geology 35, 783–786. https://doi.org/10.1130/G23800A.1.

Cranberg, A., 2020. Exploring for oil and gas in Hungary—an interview for basin dynamics. Bulletin 2 (8), 4.

Crowell, J.C., Hope, R.A., Kahle, J.E., Ovenshine, A.T., Sams, R.H., 1966. Deepwater Sedimentary Structures Pliocene Pico Formation Santa Paula Creek. California Division of Mines and Geology, Special Report 89, Ventura Basin, CA, https://doi.org/10.1306/BC743C47-16BE-11D7-8645000102C1865D. 40 pp.

Curray, J.R., Moore, D.G., 1971. Sedimentary and tectonic processes in the Bengal deep-sea fan and geosyncline. In: Burk, C.A., Drake, C.L. (Eds.), Continental Margins. Springer, New York, pp. 617–627.

Dailly, P., Henderson, T., Kanschat, K., Lowry, P., Sills, S., 2017. The Jubilee field, Ghana: opening the late Cretaceous play in the West African transform margin. In: Merrill, R.K., Sternbach, C.A. (Eds.), Giant Fields of the Decade 2000–2010. vol. 113. AAPG Memoir, pp. 257–272, https://doi.org/10.1306/13572010M1132997.

De Castro, S., Hernandez-Molina, F.J., Rodriguez-Tovar, F.J., Llave, E., Ng, Z.L., Nishida, N., Mena, A., 2020. Contourites and bottom current reworked sands: bed facies model and implications. Mar. Geol. 428. https://doi.org/10.1016/j.margeo.2020.106267, 106267.

De Weger, W., Hernandez-Molina, F.J., Flecker, R., Sierro, F.J., Chiarella, D., Krijgsman, W., Manar, A.M., 2020. Late Miocene contourite channel system reveals intermittent overflow behavior. Geology 48 (12), 1194–1199. https://doi.org/10.1130/G47944.1.

Dellinger, J., Brenders, A.J., Sandschaper, J.R., Regone, C., Etgen, J., Ahmed, I., Lee, K.J., 2017. The garden banks model experience. Lead. Edge 36, 151–158. https://doi.org/10.1190/tle36020151.1.

Deptuck, M.E., Sylvester, Z., Pirmez, C., O'Byrne, C., 2007. Migration–aggradation history and 3-D seismic geomorphology of submarine channels in the Pleistocene Benin-Major Canyon, western Niger Delta slope. Mar. Pet. Geol. 24, 406–433. https://doi.org/10.1016/j.marpetgeo.2007.01.005.

Deptuck, et al., 2012. Dimensions and architecture of late Pleistocene submarine lobes off the northern margin of East Corsica. Sedimentology. https://doi.org/10.1111/j.1365-3091.2007.00926.x.

Dickinson, W.R., 1976. Plate tectonic evolution of sedimentary basins. In: American Association of Petroleum Geologists Continuing Education Course Notes. vol. 1, pp. 1–56.

Dickinson, W., 1997. Panel throws down gauntlet. GSA Today 7, 25.

Donovan, A.D., 2010. The sequence stratigraphic family tree: understanding the portfolio of sequence methods. In: Application of Modern Stratigraphic Techniques: Theory and Case Histories. vol. 94. SEPM Special Publication, pp. 5–33, https://doi.org/10.2110/pec.55.04.0592.

Doré, G., Robbins, J., 2005. The Buzzard field. In: Petroleum Geology Conference Series. vol. 6. Geological Society, London, pp. 241–252, https://doi.org/10.1144/0060241.

Dott Jr., R.H., 1963. Dynamics of subaqueous gravity depositional processes. AAPG Bull. 47, 104–108. https://doi.org/10.1306/BC743973-16BE-11D7-8645000102C1865D.

Dzulynski, S., Sanders, J.E., 1962. Current Marks on Firm Mud Bottoms: Connecticut Academy of Arts and Science. vol. 42 Transactions, pp. 57–96 (ISBN/UPC: 9780208011077).

Edman, J.D., Burk, M.K., 1998. An integrated study of reservoir compartmentalization at Ewing Bank 873, offshore Gulf of Mexico. In: SPE. 49246. New Orleans, Louisiana, USA, pp. 653–668.

Elliott, T., 2000. Depositional architecture of a sand-rich, channelized turbidite system: the Upper Carboniferous Ross Sandstone Formation, western Ireland. In: Weimer, P., Slatt, R.M., Coleman, J., Rossen, N.C., Nelson, H., Bouma, A.H., Styzen, M.J., Lawrence, D.T. (Eds.), Deepwater Reservoirs of the World, Gulf Coast Section SEPM, 20th Annual Perkins Research Conference, pp. 342–373, https://doi.org/10.5724/gcs.00.15.0342.

Etienne, S., Mulder, T., Desaubliaux, M., Bez, G., Kwasniewski, A., Parize, O., Dujoncquoy, E., Salles, T., 2012. Multiple scale characterization of sand-rich distal lobe deposit variability: examples from the Annot Sandstones Formation. In: Eocene–Oligocene, SE France, Sedimentary Geology. vols. 273–274, pp. 1–18, https://doi.org/10.1016/j.sedgeo.2012.05.003.

ExxonMobil, 2019. Outlook for Energy: A Perspective to 2040. https://corporate.exxonmobil.com/-/media/Global/Files/outlook-for-energy/2019-Outlook-for-Energy_v4.pdf.

Fielding, L., Najman, Y., Millar, I., Butterworth, P., Ando, S., Padoan, M., Barfod, D., Kneller, B., 2017. A detrital record of the Nile River and its catchment. J. Geol. Soc. 174 (2), 301–317.

Fletcher, T., 2017. The Windjammer discovery: play opener for offshore Mozambique and East Africa. In: Merrill, R.K., Sternbach, C.A. (Eds.), Giant Fields of the Decade 2001–2010, 113. American Association of Petroleum Geologists, Memoir, pp. 273–304.

Fonnesu, M., Palermo, D., Galbiati, M., Marchesini, M., Bonamini, E., Bendias, D., 2020. A new world-class deepwater play type, deposited by the syn-depositional interaction of turbidity flows and bottom currents: the giant Eocene Coral Field in northern Mozambique. Mar. Pet. Geol. 111, 179–201. https://doi.org/10.1016/j.marpetgeo.2019.07.047.

Forel, F.A., 1885. Le ravin sous-lacustre des flueves glaciaires. C. r. hebd. Seanc. Acad. Sci. Paris 191, 725–728.

Forel, F.A., 1887. Le ravin sous-lacustre de Rhone dans le lac Leman. Bull. Soc. Vaud. Sci. Nat. 23, 85–107.

Fraser, A.J., 2010. In: Vining, B.A., Pickering, S.C. (Eds.), A Regional Overview of the Exploration Potential of the Middle East: A Case Study in the Application of Play Fairway Risk Mapping Techniques. Petroleum Geology: From Mature Basins to New Frontiers—Proceedings of the 7th Petroleum Geology Conference. The Geological Society, London, pp. 791–800.

Fryklund, B., Stark, P., 2016. Super Basins—The Basins that Keep on Giving: IHS Markit, September 2016. https://ihsmarkit.com/solutions/permian-basin.html.

Gan, Y., Almeida Júnior, F.N., Rossi, V.M., Steel, R.J., Olariu, C., 2021. Sediment transfer from shelf to deepwater slope: how does it happen? J. Sediment. Res., accepted.

Garcia-Castellanos, D., Villaseñor, A., 2011. Messinian salinity crisis regulated by competing tectonics and erosion at the Gibraltar Arc. Nature 480, 359–363. https://doi.org/10.1038/nature10651.

Gardner, M.H., Borer, J.M., Melick, J.J., Mavilla, N., Dechesne, M., Wagerle, R.N., 2003. Stratigraphic process–response model for submarine channels and related features from studies of Permian Brushy Canyon outcrops, West Texas. Mar. Pet. Geol. 20, 757–787. https://doi.org/10.1016/j.marpetgeo.2003.07.004.

Ghosh, B., Lowe, D.R., 1993. The architecture of deepwater channel complexes, Cretaceous Venado Sandstone Member, Sacramento Valley, California. In: Graham, S.A., Lowe, D.R. (Eds.), Advances in the Sedimentary Geology of the Great Valley Group. Sacramento Valley California, vol. 7. SEPM Pacific Section, pp. 51–65.

Govan, A., Primmer, T., Douglas, C., Moodie, N., Davies, M., Nieuwland, F., 2006. Reservoir management in a deepwater subsea field—the Schiehallion experience. SPE Res. Eval. Eng. 9 (04), 382–390. https://doi.org/10.2118/96610-PA. Paper Number: SPE-96610-PA.

Guiastrennec-Faugas, L., Gillet, H., Peakall, J., Dennielou, B., Gaillot, A., Jacinto, R.J., 2020. Initiation and evolution of knickpoints and their role in cut-and-fill processes in active submarine channels. Geology 49 (3), 314–319. https://doi.org/10.1130/G48369.1.

Hampton, M.A., 1972. The role of subaqueous debris flows in generating turbidity currents. J. Sediment. Petrol. 42, 775–793. https://doi.org/10.1306/74D7262B-2B21-11D7-8648000102C1865D.

Haq, 2018. Jurassic Sea-level variations: a reappraisal. GSA Today. https://doi.org/10.1130/GSATG359A.1.

Haq, B.U., Hardenbol, J., Vail, P.R., 1988. Mesozoic and Cenozoic chronostratigraphy and cycles of sea-level change. In: Wilgus, C.K., Hastings, B.S., Kendall, C.G.St.C., Ross, C.A., Van Wagoner, J.C. (Eds.), Sea-Level Changes: An Integrated Approach, vol. 42. SEPM Special Publication, https://doi.org/10.2110/pec.88.01.0071.

Haughton, P.D.W., 1994. Deposits of deflected and ponded turbidity currents, Sorbas Basin, Southeast Spain. J. Sediment. Res. 64, 233–246. https://doi.org/10.1306/D4267D6B-2B26-11D7-8648000102C1865D.

Haughton, P.D.W., Barker, S.P., McCaffrey, W.D., 2003. 'Linked' debrites—in sand-rich turbidite systems—origin and significance. Sedimentology 50, 459–482. https://doi.org/10.1046/j.1365-3091.2003.00560.x.
Haughton, P.D.W., Davis, C., McCaffrey, W., Barker, S., 2009. Hybrid sediment gravity flow deposits—classification, origin and significance. Mar. Pet. Geol. 26, 1900–1918. https://doi.org/10.1016/j.marpetgeo.2009.02.012.
Heezen, B.C., Drake, C.L., 1964. Grand Banks slump. AAPG Bull. 48, 221–225. https://doi.org/10.1306/BC743BE1-16BE-11D7-8645000102C1865D.
Heezen, B.C., Ewing, M., 1952. Turbidity currents and submarine slumps, and the 1929 Grand Banks earthquake. Am. J. Sci. 250, 849–873. https://doi.org/10.2475/ajs.250.12.849.
Heezen, B.C., Hollister, C.D., 1971. The Face of the Deep. Oxford University Press, New York, p. 659.
Helland-Hansen, W., Sømme, T.O., Martinsen, O.J., Lunt, I., Thurmond, J., 2016. Deciphering Earth's natural hourglasses: perspectives on source-to-sink analysis. J. Sediment. Res. 86 (9), 1008–1033. https://doi.org/10.2110/jsr.2016.56.
Heritier, F.E., Lossel, P., Wathne, E., 1980. Frigg field—large submarine-fan trap in Lower Eocene rocks of the Viking Graben, North Sea. In: Halbouty, M. (Ed.), Giant Oil and Gas Fields of the Decade: 1968–1978. AAPG Memoir 30. doi:10.1306/M30410C4.
Hesse, R., Klaucke, I., Khodabakhsh, S., Piper, D.J.W., Ryan, W.B.F., 2001. Sandy submarine braid plains: potential deepwater reservoirs. AAPG Bull. 85 (8), 1499–1521. https://doi.org/10.1306/8626CAEB-173B-11D7-8645000102C1865D.
Hollister, C.D., 1967. Sediment distribution and deep circulation in the western North Atlantic (Unpublished PhD dissertation). Columbia University, New York. 467 pp.
Hovadik, J.M., Larue, D.K., 2007. Static characterizations of reservoirs: refining the concepts of connectivity and continuity. Pet. Geosci. 13 (3), 195–211.
Hovius, N., Leeder, M.R., 1998. Clastic sediment supply to basins. Basin Res. 10, 1–5.
Hsü, K.J., 1977. Studies of Ventura field, California I: facies geometry and genesis of lower Pliocene turbidites. AAPG Bull. 61, 137–168.
Hsü, K.J., Kelts, K., Valentine, J.W., 1988. An analysis of sedimentary facies, and facies geometry: A model of longitudinal turbidity current transport: Ventura basin, California with special emphasis on the Ventura avenue field. In: 2009 Santa Barbara and Ventura Basins Tectonics, Structure, Sedimentation, and Oilfields along an East-West Transect. AAPG Pacific Section Guide, pp. 87–104.
Hubbard, R.J., Pape, J., Roberts, D.G., 1985a. Depositional sequence mapping as a technique to establish tectonic and stratigraphic framework and evaluate hydrocarbon potential on a passive continental margin. In: Berg, O.R., Woolverton, D.G. (Eds.), Seismic Stratigraphy II: An Integrated Approach to Hydrocarbon Exploration, 39. AAPG Memoir, pp. 79–91.
Hubbard, R.J., Pape, J., Roberts, D.G., 1985b. Depositional sequence mapping to illustrate the evolution of a passive continental margin. In: Berg, O.R., Woolverton, D.G. (Eds.), Seismic Stratigraphy II: An Integrated Approach to Hydrocarbon Exploration, 39. AAPG Memoir, pp. 93–115.
Hubbard, R.J., Edrich, S.P., Ratey, 1990. Geologic evolution and hydrocarbon habitat of the "Arctic Alaska Microplate". In: Brooks, J. (Ed.), Classic Petroleum Provinces. Geological Society of London, Special Publication 50. Geological Society of London, pp. 143–187.
Hubbard, S.M., Jobe, Z.R., Romans, B.W., Covault, J.A., Sylvester, Z., Fildani, A., 2020. The stratigraphic evolution of a submarine channel: linking seafloor dynamics to depositional products. J. Sediment. Res. 90, 673–686. https://doi.org/10.2110/jsr.2020.36.
Hurst, A., Cartwright, J., 2007. Relevance of sand injectites to hydrocarbon exploration and production. In: Hurst, A., Cartwright, J. (Eds.), Sand Injectites: Implications for Hydrocarbon Exploration and Production. vol. 87. AAPG Memoir, pp. 1–19, https://doi.org/10.1306/M871209 (book).
Hurst, A., Fraser, A.J., Fraser, S.I., Hadler-Jacobsen, F., 2005. Deep-water clastic reservoirs: a leading global play in terms of reserve replacement and technological challenges. Geol. Soc. Lond. Pet. Geol. Conf. Ser. 6, 1111–1120. https://doi.org/10.1144/0061111.
IEA, 2019. Transforming Industry through CCUS. IEA, Paris. https://www.iea.org/reports/transforming-industry-through-ccus.
IEA, 2020. World Energy Investment 2020. IEA, Paris. https://www.iea.org/reports/world-energy-investment-2020.
Ingersoll, R.V., 1990. Nomenclature of upper Mesozoic strata of the Sacramento Valley of California: review and recommendations. In: Ingersoll, R.V., Nilsen, T.H. (Eds.), Sacramento Valley Symposium and Guidebook. vol. 65. Pacific Section SEPM, pp. 1–3.
Ingersoll, R.V., 2011. In: Busby, C., Perez, A.A. (Eds.), Tectonics of Sedimentary Basins: Recent Advances, first ed. Blackwell Publishing Ltd. 664 pp.
Johnson, D., 1938. The origin of submarine canyons. J. Geomorphol. 1, 111–340.
Johnson, A.M., 1984. Debris flow. In: Brunsden, D., Prior, D.B. (Eds.), Slope Instability. John Wiley & Sons Ltd, Chichester, pp. 257–361.
Kneller, B.C., McCaffrey, W.D., 1999. Depositional effects of flow non-uniformity and stratification within turbidity currents approaching a bounding slope: deflection, reflection and facies variation. J. Sediment. Res. 69, 980–991. https://doi.org/10.2110/jsr.69.980.
Komar, P.D., 1985. The hydraulic interpretation of turbidites from their grain sizes and sedimentary structures. Sedimentology 32, 395–407. https://doi.org/10.1111/j.1365-3091.1985.tb00519.x.
Kuenen, P.H., 1950. Turbidity currents of high density. In: 18th International Geological Congress 1948, London, Reports, pt. 8, pp. 44–52.
Kuenen, P.H., 1957. Sole markings of graded greywacke beds. J. Geol. 65, 231–258. https://www.jstor.org/stable/30057070.
Kuenen, P.H., Miglorini, C.I., 1950. Turbidity currents as a cause of graded bedding. J. Geol. 58, 91–127. https://doi.org/10.1086/625710.
Kvale, E.P., Bowie, C.M., Flenthorpe, C., Mace, C., Parrish, J.M., Price, B., Anderson, S., DiMichele, W.A., 2020. Facies variability within a mixed carbonate-siliciclastic sea-floor fan (upper Wolfcamp Formation, Permian, Delaware Basin, New Mexico). AAPG Bull. 104 (3), 525–563.
Laherrere, J., 2020. US GOM OCS updated oil and gas production and reserves. ASPO France Special Report, 40 pp.
Lamb, M.P., Hickson, T., Marr, J.G., Sheets, B., Paola, C., Parker, G., 2004. Surging versus continuous turbidity currents: flow dynamics and deposits in an experimental intraslope minibasin. J. Sediment. Res. 74 (1), 148–155. https://doi.org/10.1306/062103740148.
Lambeck, et al., 2014. Sea level and global ice volumes from the last glacial maximum to the Holocene. PNAS. https://doi.org/10.1073/pnas.1411762111.
Leach, H.M., Herbert, N., Los, A., Smith, R.L., 1999. The Schiehallion development. In: Petroleum Geology Conference Series. vol. 5. Geological Society, London, pp. 683–692, https://doi.org/10.1144/0050683.
Lomas, S.A., Joseph, P., 2004. Confined turbidite systems. Geol. Soc. Lond., Spec. Publ. 222, 1. https://doi.org/10.1144/GSL.SP.2004.222.01.01.

Lowe, D.R., 1979. Sediment gravity flows: their classification and some problems of application to natural flows and deposits. SEPM Spec. Publ. 27, 75–82. https://doi.org/10.2110/pec.79.27.0075.

Lowe, D.R., 1982. Sediment gravity flows: II. Depositional models with special reference to the deposits of high-density turbidity currents. J. Sediment. Petrol. 52 (1), 279–297. https://doi.org/10.1306/212F7F31-2B24-11D7-8648000102C1865D.

Lowe, D.R., 2004. Deepwater sandstones: Submarine canyon to basin plain, western California. In: AAPG and Pacific Section; AAPG Guidebook and Lecture Notes. Artistic Printing, Salt Lake City. 79 p. (ISBN 1589550056, 9781589550056).

Lowe, D.R., Ghosh, B., 2004. A stratigraphic and architectural-element methodology for the subdivision and interpretation of deepwater clastic sequences: an example from the Cretaceous Venado Sandstone, Sacramento Valley, California. In: Appi, C.J., D'Avila, R., Viana, A. (Eds.), Deepwater Sedimentation: Technological Challenges for The Next Millennium. Brazilian Association of Petroleum Geologists Special Paper, pp. 42–56.

Lowe, D.R., Guy, M., 2000. Slurry-flow deposits in the Britannia Formation (Lower Cretaceous), North Sea: a new perspective on the turbidity current and debris flow problem. Sedimentology 47, 31–70. https://doi.org/10.1046/j.1365-3091.2000.00276.x.

Marler, S., Nicholls, M., Maguire, D., Hay, B., de Narvaez, J., 2014. Applying rock physics inversion for understanding depositional architecture in deepwater clastic reservoirs: Okume Field, Equatorial Guinea. In: Presented at the 84th SEG Annual International Meeting. https://doi.org/10.1190/segam2014-1241.1.

Mattern, F., 2005. Ancient sand-rich submarine fans: depositional systems, models, identification, and analysis. Earth Sci. Rev. 70, 167–202. https://doi.org/10.1016/j.earscirev.2004.12.001.

Mayall, M., Jones, E., Casey, M., 2006. Turbidite channel reservoirs—key elements in facies prediction and effective development. Mar. Pet. Geol. 23, 821–841. https://doi.org/10.1016/j.marpetgeo.2006.08.001.

Mayall, M., Lonergan, L., Bowman, A., James, S., Mills, K., Primmer, T., Pope, D., Rogers, L., Skeene, R., 2010. The response of turbidite slope channels to growth-induced seabed topography. AAPG Bull. 94 (7), 1011–1030. https://doi.org/10.1306/01051009117.

McBride, E.F., 1962. Flysch and associated beds of the Martinsburg Formation (Ordovician), central Appalachians. J. Sediment. Petrol. 32, 39–91. https://doi.org/10.1306/74D70C45-2B21-11D7-8648000102C1865D.

McCulley, R., 2013. Project 20K aims to unlock HP/HT deepwater reserves. Offshore Magazine. https://www.offshore-mag.com/deepwater/article/16761226/project-20k-aims-to-unlock-hpht-deepwater-reserves.

McHargue, T., Pyrcz, M.J., Sullivan, M.D., Clark, J.D., Fildani, A., Romans, B.W., Covault, J.A., Levy, M., Posamentier, H.W., Drinkwater, N.J., 2011. Architecture of turbidite channel systems on the continental slope: patterns and predictions. Mar. Pet. Geol. 28, 728–743. https://doi.org/10.1016/j.marpetgeo.2010.07.008.

Meade, R.H., 1972. Sources and sinks of suspended matter on continental shelves. In: Swift, D.J.P., Duane, D.B., Pilkey, O.H. (Eds.), Shelf Sediment Transport. Dowden, Hutchinson & Ross, Stroudsburg, pp. 249–262.

Meade, R.H., 1982. Sources, sinks, and storage of river sediment in the Atlantic drainage of the United States. J. Geol. 90, 235–252. https://www.jstor.org/stable/30066396.

Menard, H.W., 1955. Deep-sea channels, topography, and sedimentation. AAPG Bull. 39, 236–255. https://doi.org/10.1306/5CEAE136-16BB-11D7-8645000102C1865D.

Merrill, R.K., Sternbach, C.A. (Eds.), 2017. Giant Fields of the Decade 2000–2010, Memoir 113. AAPG. 322 pp.

Miall, A.D., 1985. Architectural-element analysis: a new method of facies analysis applied to fluvial deposits. Earth Sci. Rev. 22, 261–308.

Miall, A.D., 1988. Architectural elements and bounding surfaces in fluvial deposits: anatomy of the Kayenta Formation (Lower Jurassic), southwest Colorado. Sediment. Geol. 55, 233–262.

Miall, A.D., 1996. The Geology of Fluvial Deposits. Springer-Verlag, Berlin, p. 582.

Middleton, G.V., 1966. Experiments on density and turbidity currents: I. Motion of the head. Can. J. Earth Sci. 3, 523–546. https://doi.org/10.1139/e66-038.

Middleton, G.V., 1993. Sediment deposition from turbidity currents. Annu. Rev. Earth Planet. Sci. 21, 89–114. https://doi.org/10.1146/annurev.ea.21.050193.000513.

Milkov, A.V., Navidi, W.C., 2020. Randomness, serendipity, and luck in petroleum exploration. AAPG Bull. 104 (1), 145–176.

Milne, J., 1897. Suboceanic changes. Geogr. J. 10, 129–146 (259–289).

Miramontes, E., Eggenhuisen, J.T., Jacinto, R.S., Poneti, G., Pohl, F., Normandeau, A., Campbell, D.C., Hernández-Molina, F.J., 2020. Channel-levee evolution in combined contour current-turbidity current flows from flume-tank experiments. Geology 48 (4), 353–357. https://doi.org/10.1130/G47111.1.

Mitchum, R.M., 1985. Seismic stratigraphic expression of submarine fans. AAPG Mem. 39, 117–136. https://doi.org/10.1306/M39449C7.

Mitchum, R.M., Vail, P.R., Thompson, S., 1977. Seismic stratigraphy and global changes of sea level. Part 2: the depositional sequence as the basic unit for stratigraphic analysis. In: Payton, C.E. (Ed.), Seismic Stratigraphy—Applications to Hydrocarbon Exploration, 26. AAPG Memoir, pp. 53–62.

Mitchum Jr., R.M., Van Wagoner, J.C., 1990. High-frequency sequences and eustatic cycles in the Gulf of Mexico basin. In: Proceedings, Gulf Coast Section SEPM 11th Annual Research Conference, pp. 257–267, https://doi.org/10.5724/gcs.90.11.0257.

Mohrig, D., Marr, J.G., 2003. Constraining the efficiency of turbidity current generation from submarine debris flows and slides using laboratory experiments. Mar. Pet. Geol. 20, 883–899. https://doi.org/10.1016/j.marpetgeo.2003.03.002.

Murphy, M.A., Schlanger, S.O., 1962. Sedimentary structures in Ilhas sao Sebastiao Formations (Cretaceous), Reconcavo Basin, Brazil. AAPG Bull. 46, 457–477. https://doi.org/10.1306/BC743837-16BE-11D7-8645000102C1865D.

Murray, J., Renard, A.F., 1891. Report on Deep-Sea Deposits Based on Specimens Collected During the Voyage of H.M.S. Challenger in the Years 1872–1876. Government Printer, Challenger Reports, London. 525 pp.

Mutti, E., 1992. Turbidite sandstones: Agip-Instituto di Geologia. Università di Parma, Italy. 275 pp.

Mutti, E., Ghibaudo, G., 1972. Un esempio di torbiditi di coniode sottomarina esterina: le Arenarie di San delle Scienze di Torino, Classe di Scienze Fisiche, Mathematiche e Naturali, Serie 4, no. 16. 40 pp.

Mutti, E., Normark, W.R., 1987. Comparing examples of modern and ancient turbidite systems: problems and concepts. In: Leggett, J.K., Zuffa, G.G. (Eds.), Marine Clastic Sedimentology: Concepts and Case Studies. Springer, Dordrecht, The Netherlands, pp. 1–38, https://doi.org/10.1007/978-94-009-3241-8_1.

Mutti, E., Ricci Lucchi, F., 1972. Turbidites of the northern Apennines: introduction to facies analysis (English translation by Nilsen, T. H., 1978). Int. Geol. Rev. 20, 125–166. https://doi.org/10.1080/00206817809471524.

Mutti, E., Tinterri, R., Benvelli, G., di Biase, D., Gavanna, G., 2003. Deltaic, mixed and turbidite sedimentation of ancient foreland basins. Mar. Pet. Geol. 20, 733–755. https://doi.org/10.1016/j.marpetgeo.2003.09.001.

Nagarajan, N.R., Dong, C., Mullins, O.C., Honarpour, M.M., 2012. Challenges of heavy oil fluid sampling and characterization. In: SPE Annual Technical Conference and Exhibition, SPE 158450, San Antonio, Texas, 8–10 Oct.

Nasser, M., Cronister, C., Hay, B., Amos, C., 2017. 4D seismic close-the-loop impacts business decisions at the Okume Complex, Equatorial Guinea. Lead. Edge 36 (5), 394–400. https://doi.org/10.1190/tle36050394.1.

National Petroleum Council, 2019. Meeting the Dual Challenge—A Roadmap to At-Scale Deployment of Carbon Capture, Use, and Storage (CCUS) Report. https://dualchallenge.npc.org/downloads.php.

Needham, D.L., Pettingill, H.S., Christensen, C.J., French, and Karcz, Z., 2017. The Tamar giant gas field: opening the subsalt Miocene gas play in the Levant Basin. In: Merrill, R.K., Sternbach, C.A. (Eds.), Giant Fields of the Decade 2000-2010. AAPG Memoir 113. https://doi.org/10.1306/13572009M1133688.

Nelson, C.H., Nilsen, T.H., 1984. Modern and ancient deep-sea fan sedimentation. SEPM Spec. Publ. 14. https://doi.org/10.2110/scn.84.14. 404 pp.

Nikishin, A., 2003. The Black Sea basin: tectonic history and Neogene–Quaternary rapid subsidence modelling. Sediment. Geol. 156 (1–4), 149–168. https://doi.org/10.1016/S0037-0738(02)00286-5.

Nilsen, T.H., 1980. Modern and ancient submarine fans: discussion of papers by Walker, R. G., Normark, W. R.: SEPM spec. Publ. AAPG Bull. 64, 1094–1112. https://doi.org/10.1306/2F91943B-16CE-11D7-8645000102C1865D.

Nilsen, T.H., Shew, R.D., Steffens, G.S., Studlick, J.R.J. (Eds.), 2007. Atlas of Deepwater Outcrops. vol. 56. AAPG Studies in Geology, Tulsa, Okla, https://doi.org/10.1306/St561240.

Normark, W.R., 1970. Growth patterns of deep sea fans. AAPG Bull. 54, 2170–2195. https://doi.org/10.1306/5D25CC79-16C1-11D7-8645000102C1865D.

Normark, W.R., 1991. Turbidite elements and the obsolescence of the suprafan concept. Giorn. Geol. 53/2 (ser 3a), 1–10.

Normark, W.R., Piper, D.J.W., 1972. Sediment and growth pattern of navy deep-sea fan, San Clemente Basin, California borderland. J. Geol. 80, 198–223. https://www.jstor.org/stable/30059315.

Normark, W.R., Posamentier, H., Mutti, E., 1993. Turbidite systems: state of the art and future directions. Rev. Geophys. 31 (2), 91–116.

Nyberg, B., Helland-Hansen, W., Gawthorpe, R.L., Sandbakken, P., Eide, C.H., Sømme, T., Hadler-Jacobsen, F., Leiknes, S., 2018. Revisiting morphological relationships of modern source-to-sink segments as a first-order approach to scale ancient sedimentary systems. Sediment. Geol. 373, 111–133.

Oswald, E.J., 2019, Guyana Exploration: Presentation at Houston Explorers Club Leadership Luncheon, St Regis Hotel, Houston, 27 September.

Paull, C.K., Talling, P.J., Maier, K.L., Parsons, D., Xu, J., Caress, D.W., Gwiazda, R., Lundsten, E.M., Anderson, K., Barry, J.P., Chaffey, M., O'Reilly, T., Rosenberger, K.J., Gales, J.A., Kieft, B., McGann, M., Simmons, S.M., McCann, M., Sumner, E.J., Clare, M.A., Cartigny, J., 2018. Powerful turbidity currents driven by dense basal layers. Nat. Commun. 9, 1–9. https://doi.org/10.1038/s41467-018-06254-6.

Paumard, V., Bourget, J., Payenberg, T., George, A.D., Bruce Ainsworth, R., Lang, S., Posamentier, H.W., 2020. Controls on deep-water sand delivery beyond the shelf edge: accommodation, sediment supply, and deltaic process regime. J. Sediment. Res. 90, 104–130.

Pernin, N., Feuilleaubois, L., Bird, T., Reiser, C., 2019. Identifying and de-risking near-field opportunities through reliable pre-stack broadband attributes: examples from the Paleocene North Sea (UK–Norway) injectites play. Geol. Soc. Lond. Spec. Publ. 494, 445–459. https://doi.org/10.1144/SP494-2019-11.

Pettijohn, F.J., 1957. Sedimentary Rocks, second ed. Harper, New York. 718 pp., ISBN-10: 1114189804; ISBN-13: 978-1114189805.

Pickering, K.T., Hilton, 1998. Turbidite Systems of SE France: Application to Hydrocarbon Prospectivity. Vallis Press, ISBN: 0952731312. 229 pp., 9780952731313.

Pickering, K.T., Hiscott, R.N., 2015. Deep Marine Systems: Processes, Deposits, Environments, Tectonics and Sedimentation. Wiley and American Geophysical Union (ISBN: 978-1-4051-2578-9. ISBN: 978-1-405-12578-9).

Pickering, K.T., Stow, D.A.V., Watson, M., Hiscott, R., 1986. Deepwater facies, processes and models: a review and classification scheme for modern and ancient sediments. Earth Sci. Rev. 23, 75–174. https://doi.org/10.1016/0012-8252(86)90001-2.

Pickering, K.T., Hiscott, R.N., Hein, F.J., 1989. Deep-Marine Environments. Unwin Hyman, London, https://doi.org/10.1016/0040-1951(92)90284-D. 416 pp.

Pickering, K.T., Hiscott, R.N., Kenyon, N.H., Ricchi Lucchi, F., Smith, R.D.A., 1995. Atlas of Deep Water Environments: Architectural Style in Turbidite Systems. Chapman and Hall, London, ISBN: 978-94-011-1234-5, p. 333.

Piper, D.J.W., Normark, W.R., 2001. Sandy fans: from Amazon to Hueneme and beyond. AAPG Bull. 85, 1407–1438. https://doi.org/10.1306/8626CACD-173B-11D7-8645000102C1865D.

Piper, D.J.W., Shor, A.N., Hughes Clarke, J.E., 1988. The 1929 "grand banks" earthquake, slump, and turbidity current. In: Edward-Clifton, H. (Ed.), GSA Special Papers Sedimentologic Consequences of Convulsive Geologic Events. Geological Society of America, https://doi.org/10.1130/SPE229-p77.

Pirmez, C., Hiscott, R. N., Kronen Jr., J. D., 1997. Sandy turbidite successions at the base of channel-levee systems of the Amazon fan revealed by FMS logs and cores: Unraveling the lithofacies architecture of large submarine fans. In: Flood, R.D., Piper, D.J.W., Klaus, A., Peterson, L.C. (Eds.), Proceedings of the Ocean Drilling Program Leg 155: Amazon Fan, vol. 155, pp. 7–33. Compiled Electronic Index of the Proceedings of the Ocean Drilling Program, 612 pp.

Pizzi, M., Whittaker, A.C., Lonergan, L., Mayall, M., Mitchell, W.H., 2021. New statistical quantification of the impact of active deformation on the distribution of submarine channels. Geology 49 (8), 926–930. https://doi.org/10.1130/G48698.1.

Popescu, et al., 2004. The Danube submarine canyon (Black Sea): morphology and sedimentary processes. Mar. Geol. https://doi.org/10.1016/j.margeo.2004.03.003.

Pöppelreiter, M., Garcia-Carballido, C., Kraaijveld, M., 2010. Borehole image log technology: application across the exploration and production life cycle. In: Pöppelreiter, M., Garcia-Carballido, C., Kraaijveld, M. (Eds.), Dipmeter and Borehole Image Log Technology. 92. AAPG Memoir, https://doi.org/10.1306/13181274M923406.

Porebski, Steel, 2003. Shelf-margin deltas: their stratigraphic significance and relation to deepwater sands. Earth Sci. Rev. https://doi.org/10.1016/S0012-8252(02)00161-7.

Posamentier, H.W., Vail, P.R., 1988. In: Wilgus, C.K., Hastings, B.S., Kendall, C.G.S.C. (Eds.), Eustatic Controls on Clastic Deposition II—Sequence and Systems Tract Models. SEPM, https://doi.org/10.2110/pec.88.01.0125.

Posamentier, H.W., Jervey, M.T., Vail, P.R., 1988. In: Wilgus, C.K., Hastings, B.S., Kendall, C.G.S.C. (Eds.), Eustatic Controls on Clastic Deposition I—Conceptual Framework. SEPM, https://doi.org/10.2110/pec.88.01.0109.

Potts, R., Smith, C.H., Ramakrisha, S., 2011. Thin bed sand shale sequences decoded by NMR answers. In: Paper Presented at the SPE EUROPEC/EAGE Annual Conference and Exhibition, Vienna, Austria, May. Paper Number: SPE-142781-MS. doi:10.2118/142781-MS.

Prather, B.E., O'Byrne, C., Pirmez, C., Sylvester, Z., 2016. Sediment partitioning, continental slopes and base-of-slope systems. Basin Res. 29 (3), 394–416.

Pyles, D.R., Jennette, D.C., Tomasso, M., Beaubouef, R.T., Rossen, C., 2010. Concepts learned from a 3D outcrop of a sinuous slope channel complex: Beacon channel complex, Brushy Canyon Formation, West Texas, U.S.A. J. Sediment. Res. 80, 67–96.

Pyles, D.R., Tomasso, M., Jennette, D.C., 2012. Flow processes and sedimentation associated with erosion and filling of sinuous submarine channels. Geology 40 (2), 143–146.

Pyrcz, M.J., Catuneanu, O., Deutsch, C.V., 2005. Stochastic surface-based modeling of turbidite lobes. AAPG Bull. 89 (2), 177–191.

Reading, H.G., Richards, M., 1994. Turbidite systems in deepwater basin margins classified by grain size and feeder system. AAPG Bull. 78, 792–822. https://doi.org/10.1306/A25FE3BF-171B-11D7-8645000102C1865D.

Rider, M.H., Kennedy, M., 2018. The Geological Interpretation of Well Logs, third ed. Rider-French Consulting Limited, United Kingdom. ISBN 10: 0954190688, ISBN 13: 9780954190682.

Rogers, K.G., Goodbred Jr., S.L., 2010. Mass failures associated with the passage of a large tropical cyclone over the Swatch of No Ground submarine canyon (Bay of Bengal). Geology 38, 1051–1054. https://doi.org/10.1130/G31181.1.

Rotzien, J.R., Sharman, G., Masalimova, L.U., Burgreen, B.N., Lowe, D.R., Graham, S.A., 2012. Deepwater deposits of the Late Miocene Taranaki and East Coast Basins, New Zealand. In: Stanford Project on Deepwater Depositional Systems Field Trip & Guidebook, Feb. 14–22. New Plymouth & Gisborne, New Zealand, 165 pp.

Rudolph, K.W., Goulding, F.J., 2017. Benchmarking exploration predictions and performance using 20+ yr of drilling results: one company's experience. AAPG Bull. 101 (2), 161–176.

Saller, A., Werner, K., Sugiaman, F., Cebastiant, A., May, R., Glenn, D., Barker, C., 2004. Characteristics of Pleistocene deep-water fan lobes and their application to an upper Miocene reservoir model, offshore East Kalimantan, Indonesia. AAPG Bull. 92 (7), 919–949.

Saller, A.H., Lin, R., Dunham, J.B., 2006. Leaves in turbidite sands: the main source of oil and gas in the deep-water Kutei Basin, Indonesia. AAPG Bull. 90, 1585–1608.

Sanders, J.E., 1963. Concept of fluid mechanics provided by primary sedimentary structures. J. Sediment. Petrol. 33, 173–179. https://doi.org/10.1306/74D70DF3-2B21-11D7-8648000102C1865D.

Sanders, J.E., 1973. Map of parts of floor of Santa Barbara Channel, California, compiled from side-scanning sonar records. AAPG Bull. 57, 802–803. https://doi.org/10.1306/83D90C5A-16C7-11D7-8645000102C1865D.

Sanders, J.E., Friedman, G.M., 1997. History of petroleum exploration in turbidites and related deepwater deposits, 1997. prefacer. prefacer, In: Brice, W. R., Pees, S.T. (Eds.), History of Oil and Gas Exploration in North America: A Symposium Dedicated to Parke A. Dickey (1909–1994), Pioneer Petroleum Geologist in Pennsylvania. Northeastern Science Foundation, Troy, NY, United States, pp. 67–102. 19 (1–2) http://sedis.iodp.org/pub-catalogue/index.php?id=2001-003217.

Schwalbach, J.R., Miller, D.D., Rotzien, J.R., 2009. Pliocene turbidites of the Ventura Basin: Core workshop and field trip, complete with guidebook. In: Held in conjunction with the AAPG/SEPM Meeting Short Course, May 3–6, Ventura, CA. 28 pp.

Selley, R.C., 1979. Dipmeter and log motifs in North Sea submarine-fan sands. AAPG Bull. 63, 905–917. https://doi.org/10.1306/2F918496-16CE-11D7-8645000102C1865D.

Septama, E., Bentley, S.J., 2017. Source-to-sink sediment delivery in the Gulf of Papua from SEM-MLA-aided provenance analysis of deep-sea turbidite sands. AAPG Bull., 907–936.

Shanmugam, G., 2006. Deepwater Processes and Facies Models: Implications for Sandstone Petroleum Reservoirs. Elsevier, Amsterdam. 476 pp., eBook ISBN: 9780080458427, Hardcover ISBN: 9780444521613, Paperback ISBN: 9781493302352.

Sheldon, P.G., 1928. Some sedimentation conditions in Middle Portage rocks. Am. J. Sci. 15, 243–252. https://doi.org/10.2475/ajs.s5-15.87.243.

Shell, 2020. The Energy Transformation Scenarios. https://www.shell.com/promos/energy-and-innovation/download-full-report/_jcr_content.stream/1612814283728/d14d37b7dd060d78b65bfee3c7654520e1038 1aa/shell-energy-transformation-scenarios-report.pdf.

Shepard, F.P., 1981. Submarine canyons: multiple causes and long-time persistence. AAPG Bull. 65, 1062–1077. https://doi.org/10.1306/03B59459-16D1-11D7-8645000102C1865D.

Signorini, R., 1936. Determinazione del senso di sedimentazione degli strati nelle formazione arenacea dell' Appennino settentrionale. Boll. Soc. Geol. Ital. 55, 259–267.

Simpson, J.E., 1982. Gravity currents in the laboratory, atmosphere, and ocean. Annu. Rev. Fluid Mech. 14, 213–234. https://doi.org/10.1146/annurev.fl.14.010182.001241.

Sinclair, H.D., Tomasso, M., 2002. Depositional evolution of confined turbidite basins. J. Sediment. Res. 72, 451. https://doi.org/10.1306/111501720451.

Slatt, R.M. (Ed.), 2006. Stratigraphic Reservoir Characterization for Petroleum Geologists, Geophysicists, and Engineers. vol. 61. Elsevier. 492 pp., Hardcover ISBN: 9780444563651 eBook ISBN: 9780444563705.

Slatt, R.M., Davis, R.J., 2010. Calibrating borehole image and dipmeter logs with outcrops and behind-outcrop cores: case studies and applications to deepwater deposits. In: Pöppelreiter, M., García-Carballido, C., Kraaijveld, M. (Eds.), Dipmeter and Borehole Image Log Technology. AAPG Memoir, p. 92, https://doi.org/10.1306/13181282M92441.

Slatt, R.M., Boak, J.M., Goodrich, G.T., Lagoe, M.B., Vavra, C.L., Bishop, J.M., Zucker, S.M., 1988. Depositional facies, paleoenvironments, reservoir quality, and well log characteristics of Mio-Pliocene deep water sands, Long Beach Unit, Wilmington Field, California. In: Lomando, A.J., Harris, P. M. (Eds.), Giant Oil and Gas Fields: Soc. Econ. Paleon. Mineral., vol. 1. Core Workshop #12, pp. 31–88.

Sloss, L.L., 1963. Sequences in the cratonic interior of North America. Geol. Soc. Am. Bull. 74, 93–114.

Sloss, L.L., Krumbein, W.C., Dapples, E.C., 1949. Integrated facies analysis. In: Longwell, C.R. (Ed.), Sedimentary Facies in Geologic History. Memoir 39. Geological Society of America, pp. 91–123.

Sømme, T.O., Helland-Hansen, W., Martinsen, O.J., Thurmond, J.B., 2009a. Relationships between morphological and sedimentological parameters in source-to-sink systems: a basis for predicting semi-quantitative characteristics in subsurface systems. Brain Res. 21 (4), 361–387.

Sømme, T.O., Helland-Hansen, W., Granjeon, D., 2009b. Impact of eustatic amplitude variations on shelf morphology, sediment dispersal, and sequence stratigraphic interpretation: icehouse versus greenhouse systems. Geology 37 (7), 587–590.

Sømme, T.O., Martinsen, O.J., Thurmond, J.B., 2009c. Reconstructing morphological and depositional characteristics in subsurface sedimentary systems: an example from the Maastrichtian–Danian Ormen Lange system, Møre Basin. AAPG Bull. 93 (10), 1347–1377.

Sprague, A.R.G., Garfield, T.G., Goulding, F.J., Beaubouef, R.T., Sullivan, M.D., Rossen, C., Campion, K.M., Sickafoose, D.K., Abreu, V., Schellpeper, M. E., Jensen, G.N., Jennette, D.C., Pirmez, C., Dixon, B.T., Ying, D., Ardill, J., Mohrig, D.C., Porter, M.L., Farrell, M.E., Mellere, D., 2005. Integrated Slope Channel Depositional Models: The Key to Successful Prediction of Reservoir Presence and Quality in Offshore West Africa. CIPM. 13 pp. https://www.researchgate.net/profile/R-Beaubouef/publication/265807668_INTEGRATED_SLOPE_CHANNEL_DEPOSITIONAL_MODELS_THE_KEY_TO_SUCCESSFUL_PREDICTION_OF_RESERVOIR_PRESENCE_AND_QUALITY_IN_OFFSHORE_WEST_AFRICA/links/54e60c860cf277664ff20ec6/INTEGRATED-SLOPE-CHANNEL-DEPOSITIONAL-MODELS-THE-KEY-TO-SUCCESSFUL-PREDICTION-OF-RESERVOIR-PRESENCE-AND-QUALITY-IN-OFFSHORE-WEST-AFRICA.pdf.

Stanley, D.J., 1963. Vertical petrographic variability in Annot Sandstone turbidites. J. Sediment. Petrol. 33, 783–788. https://doi.org/10.1306/74D70F2E-2B21-11D7-8648000102C1865D.

Steckler, M.S., Watts, A.B., 1978. Subsidence of the Atlantic-type continental margin off New York. Earth Planet. Sci. Lett. 41, 1–13. https://doi.org/10.1016/0012-821X(78)90036-5.

Sternbach, C.A., Merrill, R.K., 2021. Introductory remarks: discovery thinking and giant fields. In: IMAGE 2021, Denver, 27-September.

Stewart, I., 2020. Tectonostratigraphic megasequences and chronostratigraphy. In: Scarselli, N., Adam, J. (Eds.), Regional Geology and Tectonics, second ed, pp. 113–117 (Chapter 7).

Stickel, J.J., Powell, R.L., 2005. Fluid mechanics and rheology of dense suspensions. Annu. Rev. Fluid Mech. 37, 129–149. https://doi.org/10.1146/annurev.fluid.36.050802.122132.

Stirling, E.J., Fugelli, E.M.G., Thompson, M., 2017. The edges of the wedges: a systematic approach to trap definition and risking for stratigraphic, combination and sub-unconformity traps. In: Bowman, M., Levell, B. (Eds.), Petroleum geology of NW Europe: 50 years of learning—Proceedings of the 8th Petroleum Geology Conference: Geological Society, London, Petroleum Geology Conference Series 2017. vol. 8, pp. 273–286, https://doi.org/10.1144/PGC8.19.

Stock, J.M., Clayton, R., Luyendyk, B., Davy, B., 2007. Scientific Report on Cruise NBP0607C 10 October to 23 October 2006 Lyttelton, New Zealand to Lyttelton, New Zealand. Seismological Laboratory, California Institute of Technology, Pasadena, CA, 25 pp.

Stow, D.A.V., Piper, D.J.W. (Eds.), 1984. Fine-Grained Sediments: Deepwater Processes and Facies. Geological Society Special Publication No. 15., https://doi.org/10.1144/GSL.SP.1984.015.01.38. 659 pp.

Stow, D.A.V., Shanmugam, G., 1980. Sequence of structures in fine-grained turbidites: comparison of recent deep-sea and ancient flysch sediments. Sediment. Geol. 25, 23–42. https://doi.org/10.1016/0037-0738(80)90052-4.

Stow, D.A.V., Smillie, Z., 2020. Distinguishing between deepwater sediment facies: turbidites, contourites and hemipelagites. Geosciences 10 (68). https://doi.org/10.3390/geosciences10020068.

Stow, D.A.V., Cremer, M., Droz, L., Normark, W.R., O'Connell, S., Pickering, K.T., Stelting, C.E., Meyer-Wright, A.A., DSDP Shipboard Scientists, 1985. Mississippi fan sedimentary facies, composition, and texture. In: Bouma, A.H., Normark, W.R., Barnes, N.E. (Eds.), Submarine Fans and Related Turbidite Systems. Springer-Verlag, New York, pp. 259–266, https://doi.org/10.1007/978-1-4612-5114-9_38.

Straccia, J.R., Prather, B.E., 1999. Stratigraphic traps in deepwater turbidite reservoirs at the base of depositional slope. In: Society of Petroleum Engineers Offshore Europe Conference, Aberdeen, Scotland, September 7–9, 1999, SPE-56894-MS. 14 pp https://www.researchgate.net/publication/265125419_Stratigraphic_traps_in_deepwater_reservoir_at_the_base_of_depositional_slope_SPE_56894.

Sullivan, M.D., Jensen, G.N., Goulding, F.J., Jennette, D.C., Foreman, J.L., Stern, D., 2000. Architectural analysis of deepwater outcrops: implications for exploration and development of the Diana sub-basins, western Gulf of Mexico. In: GCSSEPM Foundation 20th Annual Research Conference—Deepwater Reservoirs of the World, December 3–6., https://doi.org/10.5724/gcs.00.15.1010.

Sullivan, M.D., Foreman, J.L., Jennette, D.C., Stern, D., Jensen, G.N., Goulding, F.J., 2004. An integrated approach to characterization and modeling of deepwater reservoirs, Diana Field, western Gulf of Mexico, in integration of outcrop and modern analogs in reservoir modeling. AAPG Mem. 80, 215–234. https://doi.org/10.1306/M80924C11.

Sweet, M., 2020. Controls on the grain size of sediment routed to deep water: insights from quaternary systems. In: Rotzien, J.R. (Ed.), Basin Dynamics Bulletin, vol. 2(4). A Basin Dynamics Publication, pp. 1–13.

Sweet, M.L., Blum, M.D., 2016. Connections between fluvial to shallow marine environments and submarine canyons: implications for sediment transfer to deep water. J. Sediment. Res. 86, 1147–1162. https://doi.org/10.2110/jsr.2016.64.

Sylvester, Z., Pirmez, C., Cantelli, A., 2011. A model of submarine channel-levee evolution based on channel trajectories: implications for stratigraphic architecture. Mar. Pet. Geol. 28 (3), 716–727. https://doi.org/10.1016/j.marpetgeo.2010.05.012.

Sylvester, et al., 2012. Seismic Stratigraphy of a Shelf-Edge Delta and Linked Submarine Channels in the Northeastern Gulf of Mexico., https://doi.org/10.1016/j.marpetgeo.2010.05.012.

Takahashi, T., 1981. Debris flow. Annu. Rev. Fluid Mech. 13, 57–77. https://doi.org/10.1146/annurev.fl.13.010181.000421.

Talling, P.J., Masson, D.G., Sumner, E.J., Malgesini, G., 2012. Subaqueous sediment density flows: depositional processes and deposit types. Sedimentology 59, 1937–2003. https://doi.org/10.1111/j.1365-3091.2012.01353.x.

Tura, A., Dobbs, S., Davies, K., Zhang, J., 2009. Remaining oil thickness and well positioning using 4D at Alba field, North Sea. In: SEG Technical Program Expanded Abstracts. doi:10.1190/1.3255692.

Tokes, L., Patacci, M., 2018. Quantifying tabularity of turbidite beds: morphology and applications. In: AAPG Annual Convention and Exhibition, Article #90323, Salt Lake City, May 20–23.

Tyler, J.H., 1972. Pigeon Point Formation: an upper Cretaceous shoreline succession, Central California coast. J. Sediment. Petrol. 42 (3), 537–557. https://doi.org/10.1306/74D725B3-2B21-11D7-8648000102C1865D.

U.S. Energy Information Administration, Annual Energy Outlook 2021. AEO2021 https://www.eia.gov/outlooks/aeo/pdf/02%20AEO2021%20Petroleum.pdf.

Vail, P.E., Mitchum, R.M., Thompson III, S., 1977a. Relative changes of sea level from coastal onlap. In: Payton, C.E. (Ed.), Seismic Stratigraphy—Applications to Hydrocarbon Exploration. vol. 26. AAPG Memoir, pp. 63–82, https://doi.org/10.1306/M26490C5.

Vail, P.R., Todd, R.G., Sangree, J.B., 1977b. Seismic stratigraphy and global changes of sea level, part 5: chronostratigraphic significance of seismic reflections. In: Payton, C.E. (Ed.), Seismic Stratigraphy—Applications to Hydrocarbon Exploration. vol. 26. AAPG Memoir, pp. 99–116, https://doi.org/10.1306/M26490C7.

van Hinte, J.E., 1978. Geohistory analysis—application of micropaleontology in exploration geology. Am. Assoc. Pet. Geol. Bull. 62, 201–222.

Van Wagoner, J.C., Mitchum, R.M., Posamentier, H.W., Vail, P.R., 1987. An overview of sequence stratigraphy and key definitions. In: Bally, A.W. (Ed.), Atlas of Seismic Stratigraphy, Volume 1. Studies in Geology. vol. 27. American Association of Petroleum Geologists, pp. 11–14.

Van Wagoner, J.C., Posamentier, H.W., Mitchum, R.M., Vail, P.R., Sarg, J.F., Loutit, T.S., Hardenbol, J., 1988. An overview of sequence stratigraphy and key definitions. In: Wilgus, C.K., Hastings, B.S., Kendall, C.G.S.C., Posamentier, H.W., Ross, C.A., Van Wagoner, J.C. (Eds.), Sea Level Changes—An Integrated Approach. vol. 42. SEPM Special Publication, pp. 39–45.

Walker, R.G., 1978. Deepwater sandstone facies and ancient submarine fans: models for exploration for stratigraphic traps. AAPG Bull. 62, 932–966. https://doi.org/10.1306/2F9182E3-16CE-11D7-8645000102C1865Ds.

Walker, R.G., 2004. Facies models revisited. In: AAPG 2004 Annual Convention Abstracts Volume. Dallas, Texas, p. A144.

Walker, R.G., Massingill, J.V., 1970. Slump features on the Mississippi Fan, Northeastern Gulf of Mexico. GSA Bull. 81, 3101–3108. https://doi.org/10.1130/0016-7606(1970)81[3101:SFOTMF]2.0.CO;2.

Walsh, J.P., Nittrouer, C.A., 2009. Understanding fine-grained river-sediment dispersal on continental margins. Mar. Geol. 263, 34–45. https://doi.org/10.1016/j.margeo.2009.03.016.

Waltham, D., 2004. Flow transformations in particulate gravity currents. J. Sediment. Res. 74, 129–134. https://doi.org/10.1306/062303740129.

Warme, J.E., Douglas, R.G., Winterer, E.L. (Eds.), 1981. The Deep Sea Drilling Project: A Decade of Progress. vol. 32. SEPM Special Publication, https://doi.org/10.2110/pec.81.32. 564 pp.

Weimer, P., 1990. Sequence stratigraphy, facies geometries, and depositional history of the Mississippi Fan, Gulf of Mexico. AAPG Bull. 74, 425–453. https://doi.org/10.1306/0C9B2321-1710-11D7-8645000102C1865D.

Weimer, P., Slatt, R.M., 2004. Deepwater reservoir elements: channels and their sedimentary fill. In: Petroleum Systems of Deepwater Settings. Distinguished Instructor Series. 7. Society of Exploration Geophysicists and European Association of Geoscientists and Engineers, Tulsa. 490 pp.

Weimer, P., Slatt, R.M., Coleman, J., Rosen, N.C., Nelson, H., Bouma, A.H., Styzen, M.J., Lawrence, D.T. (Eds.), 2000. Deepwater reservoirs of the world. GCSSEPM Foundation 20th Annual Bob F. Perkins Research Conference. 1105 pp.

Weimer, P., Zimmerman, E., Hersh, H., Snyder, B., Leibovitz, M., Schwendeman, K., Hankins, B., et al., 2017. Atlas of fields and discoveries, central Mississippi Canyon, Atwater Valley, northwestern Lloyd Ridge, and western DeSoto Canyon protraction areas, northern deep-water Gulf of Mexico. AAPG Bull. 101 (7), 995–1002.

Whaley, J., 2010. The first UK giant oil field. Geo ExPro 7, 76–82.

Xie, X., Heller, P.L., 2009. Plate tectonics and basin subsidence history. Geol. Soc. Am. Bull. 121, 55–64. https://doi.org/10.1130/B26398.1.

Xu, C., Hamman, J., Brickell, A., Rotzien, J.R. (Eds.), 2017. Introduction to special section: integrating multiscale and multidiscipline data. Interpretation, 5 (2), pp. 1–2.

Yeilding, C.A., 2019. The explorers mindset. In: Presentation at Houston Explorers Club Quarterly Leadership Luncheon, St Regis Hotel, Houston, Texas, USA, 15 Feb.

Yergin, D., 1990. The Prize: The Epic Quest for Oil, Money & Power. Free Press (912 pp. ISBN-10: 1439110123, ISBN-13: 978-1439110126).

Yergin, D., 2020. The New Map: Energy, Climate, and the Clash of Nations/Daniel Yergin. Penguin Press, New York, p. 512 (Print, ISBN-10: 1594206430, ISBN-13: 978-1594206436).

Zhang, et al., 2016. Greenhouse shoreline migration: Wilcox deltas. AAPG Bull. 100 (12). https://doi.org/10.1306/04151615190.

Chapter 2

# Source rocks and petroleum systems in deepwater plays

**Benjamin Kirkland**[a] **and Andrew Pepper**[b]
[a]*CNOOC Energy USA LLC, Houston, TX, United States,* [b]*This is Petroleum Systems LLC, Fredericksburg, TX, United States*

## Introduction

In all petroleum plays, there is—somewhere in the system—a source rock that has generated petroleum. It is therefore an oft-cited strategy to "follow the source rock" in exploration. Although successful plays and prospects can ultimately have no traditionally productive reservoir or common traps, all play concepts require an active source rock. Of course, not all source rocks are similar in character, but it is truly ubiquitous that all petroleum accumulations begin with generation from the organic matter in a source bed.

Just as the source bed is essential to a play or prospect, so too is a subsurface petroleum system (or, fluid system) that delivers petroleum to a trap to form an accumulation. Whether petroleum is produced conventionally from a high permeability reservoir or unconventionally from a source bed or tight sand, the reservoir fluid phase is the net result of the entire generation, expulsion, migration and accumulation process through time. Fluid potential, controlled by variations in capillary entry pressure, fluid densities and pore pressure, ultimately controls the journey from source to trap, and the present-day accumulation can ultimately be viewed as a snapshot of a dynamic system evolving through geologic time.

In deepwater plays, the reservoir is generally a porous lithology deposited in a mud-dominated environment that is typically younger and shallower than the actively generating source rock. Production from the deepwater areas of the Niger Delta, Gulf of Mexico and Guyana basins (among many others) is principally from this type of play, and this is the type that is the focus of this chapter. Other reservoir play types are, however, exploited in present-day deepwater development areas including: reservoirs that underly source rocks (Norphlet play, Gulf of Mexico), source rocks that are coeval with up-dip reservoirs (pre-salt play, Brazil Atlantic Margin) and cool, shallow mudrocks that provide the feedstock for microbial generation of gas (common worldwide). For clarity, it is worth noting that the term *deepwater* can carry two very different connotations. The first is an operational definition referring to an offshore play in greater than 200 m (~656 ft) of water at present day, and the second is a geologic definition referring to deposition in a deepwater setting in the geologic past—such as the Bone Spring turbidite reservoirs of the present-day onshore Permian Basin (see Chapters 1 and 19 for further discussion on water depths). This chapter focuses on the many plays that fit both definitions as deepwater reservoir settings that are being exploited in present day deepwater environments.

This chapter uses as a framework the approach referred to as *petroleum systems analysis* as laid out by Pepper (2003). Envisioned as a holistic approach to conceptualizing the processes from petroleum generation through accumulation, the petroleum system concept was originally proposed by Magoon (1988) and detailed in AAPG Memoir No. 60 The Petroleum System—From Source to Trap (Magoon and Dow, 1994).

Magoon and Dow describe the petroleum system concept as an active source rock, its related oil and gas accumulations and the essential elements required for those accumulations to exist. This linkage requires a thorough understanding of the fundamental physical and chemical processes involved in delivery of petroleum from the source bed to a reservoir within a trap. Here, an updated petroleum system concept with three first-order petroleum systems elements is presented: charge, trap and reservoir Fig. 1 (Pepper, 2003). Each element has two sub-elements that aid in conceptualizing and quantifying the processes leading to evaluation of play or prospect size and chance of success. For oil and gas exploration to become a consistent business, it is critical for petroleum systems components to have clear definitions without gaps or overlap between workflow elements. The bulk of this chapter will be a stepwise summary of these petroleum systems elements and sub-elements, effectively creating a common conceptual workflow for all disciplines involved in prospecting.

Deepwater Sedimentary Systems. https://doi.org/10.1016/B978-0-323-91918-0.00012-8

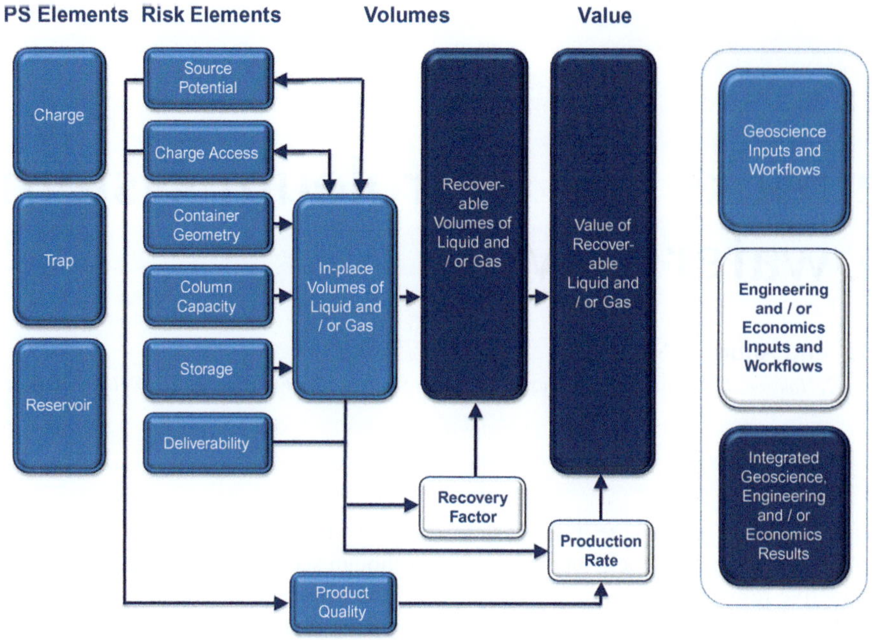


**FIG. 1** Petroleum systems summary chart including the role of petroleum systems analysis in a standard exploration workflow. The three petroleum systems elements (charge, trap, reservoir) are listed on the far left while their respective sub-elements (or, risk elements) are listed in the next column. The lines connecting the sub-elements to the in-place volumes step represent the subsurface assessment portion of petroleum systems analysis. Next, the engineering-focused steps—colored white—are combined to determine the volume of recoverable petroleum, its value and the rate at which it can be produced. Although some of steps in the workflow are led by petroleum systems analysis, others only represent involvement and integration with other subsurface disciplines.

As shown in Fig. 1, the first petroleum systems element is *charge*; this element encompasses the summary of processes that result in petroleum arriving to a trap. This, like the other two petroleum systems elements, comprises two sub-elements. For clarity, it is advisable to refer to all sub-elements in the following format: "*petroleum systems element: sub-element.*"

The first of these sub-elements is *charge: source potential*, which is a characterization of the source bed itself. A petroleum system can be viewed as a petroleum volume-creation engine in which the migrating and accumulating petroleum is introduced into the system by volume creation in the source bed. Because the petroleum system concept addresses the origin of a specific volume of petroleum, it is important to recognize that *charge: source potential* provides an ultimate upper limit on the petroleum volume of an entire petroleum system. It is not sufficient for a source bed to be present; rather, it must have sufficient generative potential so that, at its maximum level of maturity, it is able to overcome migration losses and deliver a volume to the play or prospect. Roller and Pepper (2018) and Pepper and Roller (2021) introduce a concept known as *ultimate expellable potential* for source rocks and the acronym UEP for ultimate expellable petroleum (as in a mass or volume). There are also specific terms UEO and UEG for ultimate expellable oil and ultimate expellable gas, respectively.

The companion sub-element *charge: access* encompasses the processes that are required to generate petroleum from a source bed and ultimately allow for that petroleum to migrate into a trap. First, in-situ labile organic matter is converted into smaller, simpler molecules of petroleum fluid that are expelled from the solid organic matter framework. Next, the expelled molecules collect in a pore system as a petroleum fluid phase which can then begin to migrate within and out of the source bed. The conversion process is driven by thermal stress and sorption controls when expulsion occurs from the organic matter. Finally, the expelled petroleum volumes migrate through pore space in a sedimentary section in the direction of lowest fluid potential—a result of capillary entry pressure, pore pressure and buoyancy.

The next petroleum systems element is *trap*; this can be viewed as a geologic container with capacity to contain a petroleum accumulation. Traps can be as simple as a dome-shaped feature with only one mappable edge sealing a petroleum accumulation. However, increasing numbers of edges can come from structural boundaries such as fault planes, unconformities or stratigraphic pinch outs. Purely stratigraphically defined containers in some deepwater depositional systems can require up to five sealing container edges including up-dip, lateral, top and bottom edges. Sealing edges of a trap can sometimes be difficult to define, but their presence is ultimately required as the *trap: container geometry* petroleum systems sub-element.

Not only must the trap have a geometry as defined by its edges, but each edge must also have the ability to oppose the buoyant force of a volume—or vertical column—of petroleum. This introduces the companion sub-element *trap:*

*column capacity*. A rock property known as *capillary entry pressure* must be at least as great as the buoyant force of the petroleum and any additional reservoir pressure along each container edge. Resistant capillary forces of the rock constituting the edge arise from the pressure required to allow a petroleum phase with a given interfacial tension to pass through the water-wet pore throats in a rock.

The final petroleum systems element is *reservoir*, and its first sub-element is *reservoir: storage*. Since it is the pore system that actually contains petroleum fluids, the presence of a specific lithology alone is not enough to constitute a reservoir (although complete absence of a high-porosity facies can be a common failure mechanism when testing new plays or individual prospects). A reservoir comprises a rock with a pore system *and* its occupying fluid.

When porosity is well-developed in a reservoir rock, that unit's thickness usually represents a key parameter of a discovery's assessment and can separate a commercial discovery from only a "technical success" situation. Because of its importance, common workflows have been developed to estimate thickness from seismic attributes in many well-imaged deepwater reservoir systems. The concepts to apply while using investigative geophysical workflows begin with porosity reduction in a clastic reservoir related to both mechanical and chemical compaction. Actively depositing deepwater systems can have high sedimentation rates which load the section with sediment and pore water faster than it can drain to hydrostatic equilibrium conditions. This can result in high pressure and a low compaction state resulting in reservoir porosities that bear little relationship to depth. In addition, thermal disequilibrium can keep reservoir temperatures far below a typical equilibrium temperature expected from simplistic geothermal gradients. The common result is that quartz cementation is delayed, preserving high intergranular volumes.

The companion sub-element *reservoir: deliverability*, while a step beyond traditional petroleum systems analyses, requires assessment of production rates from the static oil-in-place concept established in the *reservoir: storage* calculation. Commercial production in deepwater development projects requires high productivity from a relatively low number of producing wells. Integration with engineering concepts is crucial at this final stage of evaluation to estimate parameters for input into Darcy's equation for radial flow; this will ultimately govern deliverability. The variable concerning rock in this dynamic model is permeability and the variables for fluid are pressure, viscosity and formation volume factor. As mentioned, high overpressure characteristic of many deepwater plays can bolster deliverability and even compensate for scenarios in which viscosity may be relatively high.

All petroleum system elements are required to facilitate accumulation, but commerciality of those accumulations can also be heavily affected by in-reservoir processes acting on the reservoired petroleum, which is addressed in the final section of this chapter.

## Petroleum systems sub-element *Charge: Source potential*

A source rock is a rock containing sedimentary organic carbon with petroleum-generating potential: the capability to generate and expel oil and/or gas upon thermal maturation. There have been many efforts to try to place ranges and limits on organic richness or thickness to qualify source rock samples as "poor" or "good," but, in reality, the expellable potential of a source rock is an integral of thickness, organic richness and quality that is not effectively approximated by assessing any one single parameter (Pepper and Roller, 2021).

Deepwater reservoirs are often charged by shelf or slope source facies underlying the deepwater depositional section, but they can also be charged by source facies deposited in older stratigraphy from an earlier period of basin subsidence in basinal, paralic or even nonmarine environments. Organic geochemists have, for many decades, recognized that organic precursors, bacterial alteration, lithofacies and diagenesis determine the nature of the organic matter preserved in source rocks (Demaison, 1980; Peters and Cassa, 1994; Katz, 2005). The concept of *organofacies* presented in Pepper and Corvi (1995a) and shown in Fig. 2 and Table 1 is an effective source rock classification scheme that is recommended and employed in this chapter. In addition, a general review of source rock classifications as employed across the history of petroleum systems analysis was shared in PSA (2016) and facilitated discussion among the community. It is recommended to review this reference for comparison between the most common classification schemes.

### Source rock deposition

Although, in principle, source rocks can be coincidentally developed adjacent to deepwater reservoir plays, they are more commonly deposited significantly earlier than deepwater reservoirs in shallower water depths during early thermal subsidence in a *syn*-rift setting or even in a prerift megasequence. This opens the possibility that almost any underlying depositional environment and organofacies can charge deepwater structures provided the expelled volumes can overcome migration losses between source bed and reservoir. Some examples illustrating the wide range of organofacies in older, deeper source rocks that charge vertically into much younger deepwater reservoir plays include organofacies A, distal

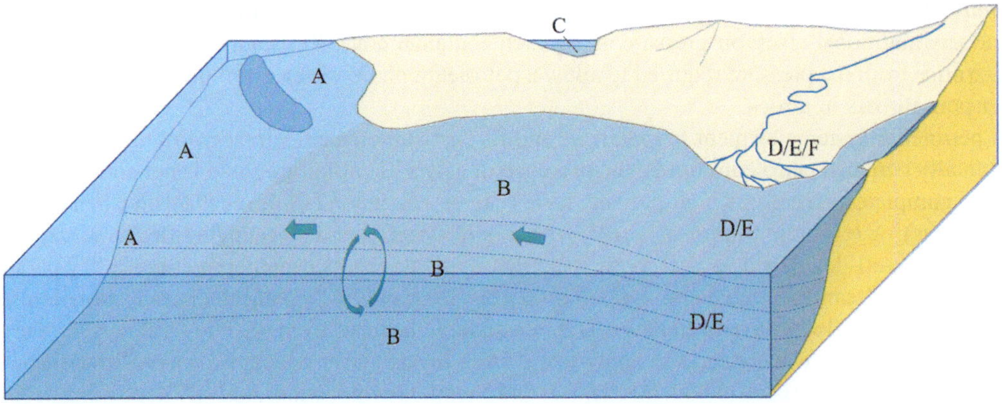


**FIG. 2** Generalized depositional environments for organofacies. Organofacies A, B and C are predominately derived from aquatic algal and bacterial precursors; organofacies D, E and F are predominately derived from terrigenous, higher-plant and bacterial precursors. Organofacies A and B are both made up of organic matter derived from marine algae and bacteria, but differences in the depositional environment and diagenesis result in different reaction kinetics: organofacies A organics are generally deposited in a clay-poor, carbonate depositional environment. This environment is low in detrital iron as compared with a clastic depositional environment, so sulfur generated by sulfate-reducing bacteria does not react with iron ions to form iron oxides. Instead, it becomes incorporated into kerogen molecules. A clastic depositional environment is more likely to result in lower-sulfur organofacies B. Both marine organofacies (A and B) are typically deposited in shelf or slope environments. These fairly shallow and proximal environments can be conducive to high organic production, preservation due to anoxia from the organic production, and slow sedimentation rates. Oceanic upwelling (indicated by the *circular green arrows)* is a process by which longshore littoral currents (indicated by *straight green arrows*) result in flow of deep, nutrient-rich ocean waters toward the surface; this can be a key accelerator of both production and preservation. Organofacies C precursors tend to be localized and site-specific freshwater algae and bacteria in large lacustrine environments. High production is possible due to runoff and smaller-scale marine-like processes, but anoxia is more likely to be a result of water column stratification. Thus, distribution of organic matter is often in the basin center rather than only along margins. Organofacies D and E are typically grouped to create a single terrigenous organofacies with petroleum-generating potential, and organofacies F represents the endmember of organic matter with little-to-no potential to generate. The combined organofacies D/E covers kerogens formed from all waxy, higher-plant derived organics such as wax, resin and lignin as well as bacterial precursors. Depositional environments for organofacies D/E can be in coastal, shelf, or slope settings due to transport of the terrigenous organic matter. Organofacies F is derived from woody, lignin-rich organic matter with very low hydrogen:carbon ratio. *(Modified from Pepper and Corvi (1995a).)*

**TABLE 1** Kerogen kinetic classification into organofacies A–F. In addition to the descriptions and features for each organofacies, a generalized relationship with the source rock "type" classification scheme from Tissot and Welte (1984) is included in the final column on the right.

| Organofacies | Description | Principal biomass | Sulfur incorporation | Environmental/age association | Possible IFP classification |
|---|---|---|---|---|---|
| A | Aquatic, marine, siliceous or carbonate/ evaporite | Marine algae, bacteria | High | Marine, upwelling zones, clastic-starved basins (any age) | Type IIS |
| B | Aquatic, marine, siliciclastic | Marine algae, bacteria | Moderate | Marine, upwelling zones, clastic basins (any age) | Type II |
| C | Aquatic, nonmarine, lacustrine | Freshwater algae, bacteria | Low | Tectonic nonmarine basins; minor on coastal plains (Phanerozoic) | Type I |
| D | Terrigenous, nonmarine, waxy | Higher plant cuticle, resin, lignin, bacteria | Low | Some (Mesozoic and younger) coastal plains; marine | Type III |
| E | | Higher plant cuticle, lignin, bacteria | | Some (Mesozoic and younger) coastal plains | |
| F | Terrigenous, nonmarine, wax-poor | Lignin | Low | Coastal plains (Late Paleozoic and younger) | Type III/IV |

Modified from Pepper, A., Corvi, P., 1995a. Simple kinetic models of petroleum formation. Part I: oil and gas generation from kerogen. Mar. Pet. Geol. 12(3), 291–319.

deepwater Gulf of Mexico, organofacies B proximal equivalents of the deepwater Gulf of Mexico, organofacies C, Neocomian–Barremian saline lacustrine *syn-* to postrift in Campos and Santos basins, Brazil, organofacies F, Triassic prerift Mungaroo Formation coals, Westralian Basin, Australia.

This chapter will focus on the marine source rocks that charge most deepwater petroleum systems and, in fact, most of the world's known oil in general. These are deposited typically as mudstones with the following conditions during deposition: high supply of organic matter from biologic productivity or lateral transport, anaerobic conditions for preservation and low sedimentation rates (Katz, 2005). All three processes play a role in defining *charge: source potential*, but it is reasonable that one or two efficient or elevated processes might compensate for other less favorable factors. Of course, there can also be negative compounding factors. For example, a very low sedimentation rate may be favorable for concentration but limiting for preservation since the organic matter will be subject to aerobic or anerobic bacterial degradation for a longer time during early burial.

Pepper (2016a,b,c), Pepper et al. (2017) and Pepper and Roller (2021) point to the global or mega-regional occurrence of marine source rocks during correlative, specific windows in geologic time termed *acmes*. This ongoing research employs the geologic age of peak source rock deposition in source beds as a correlation mechanism. In addition, local depositional conditions must concurrently allow widespread production and preservation of hydrogen-rich aquatic organic matter. Productivity may be a global or regional phenomenon, but all deposition is local.

It is a common misconception that deepwater, basinal environments are required for source rock deposition. In actuality, most of the world's prolific source rocks were deposited in relatively shallow water shelf to slope environments; development of a marine source rock stratigraphically adjacent to a deepwater turbidite reservoir is uncommon and requires unique conditions. This is because productivity occurs in the near-surface photic zone, and this zone is most productive where coastal upwelling or run-off introduce a high level of nutrients to the waters. So, deeper water results in higher settling time to the seabed during which, in an oxic water column, consumption by aerobes is extensive. Thus, net carbon flux to the seabed through the water column decays exponentially with increasing water depth. Fig. 3 shows an outcrop to basin center example via a regional UEP/UEO map for combined Cenomanian and Turonian acmes (late Cretaceous) in the Gulf of Mexico.

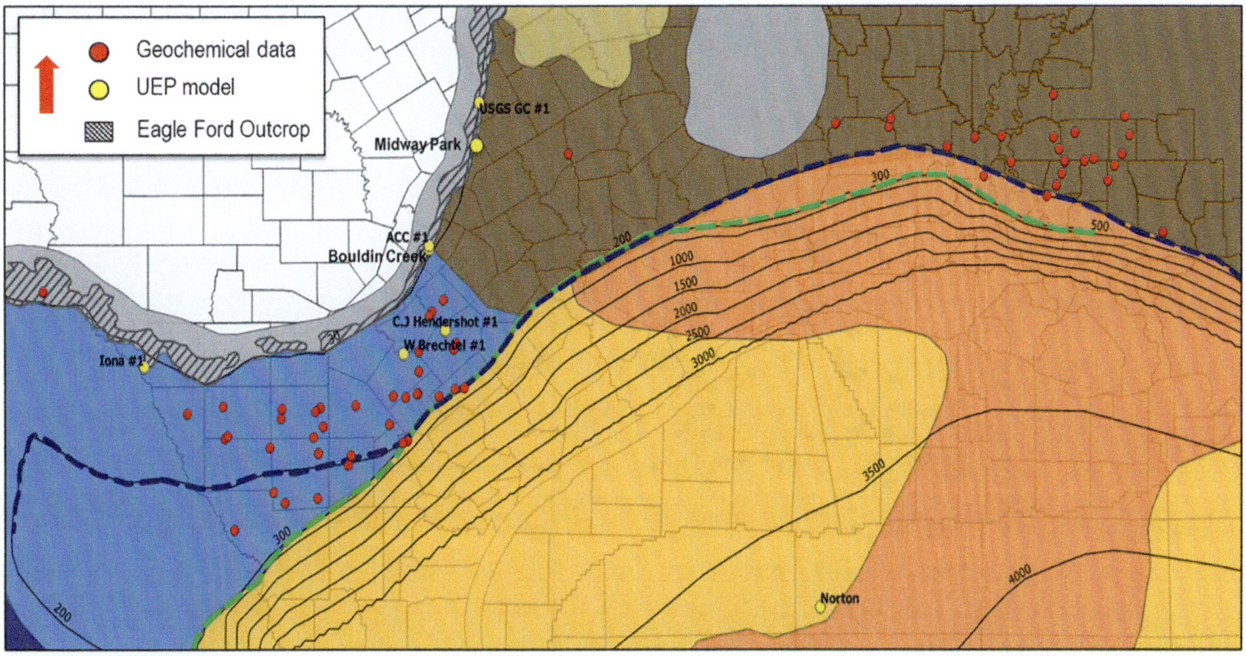


**FIG. 3** Map of UEP in northern Gulf of Mexico. Variation in ultimate expellable potential (UEP) from shelf to basin in the Eagle Ford source rock and equivalents in the northwestern Gulf of Mexico. Modified from Pepper et al. (2020a, b). Map shows: bathymetry (m); early Cenomanian carbonate shelf edge prior to Eagle Ford deposition as *thick dashed blue line*; Aptian shelf margin similarly in *green*; location of wells/outcrops with geochemical data (*red dots*) and UEP models (*yellow dots*). Lithofacies/organofacies colors are: clay-poor shelf carbonate organofacies A mudrocks in *blue*; clay-rich shelf organofacies B mudrocks in *brown*; clay-rich slope turbidite organofacies B mudrocks in *orange*; clay-rich pelagic organofacies B mudrocks in *gold*; *gray* areas missing by erosion. UEP is greatest in the Cenomanian and Turonian sections of the wells located up-dip of the early Cenomanian shelf edge; however, available biostratigraphic data show that the acmes are not actually associated with the oceanic anoxic event that is believed to have occurred at the Cenomanian-Turonian boundary. UEP at the equivalent down-dip slope facies encountered in the Norton well is low; the Eagle Ford is a prolific source rock deposited largely in a shelfal environment.

The most prolific expulsion potential is developed in carbonate rich/clay poor organofacies A in the southwest Texas shelf and Maverick intrashelf basin—the location of the Eagle Ford unconventional play. The clay-rich, condensed section in the deep basin center, at 3–4 km (1.9–2.5 mi) water depth during these acmes, is extremely thin resulting in much lower expulsion potential (Pepper et al., 2020a,b). Shelves with high productivity are optimum places to develop source rocks and are sometimes aided by upwelling to provide additional nutrients for massive support of productivity.

Source rock deposition in deepwater requires special circumstances. In normal conditions, open ocean circulation patterns bring cold, oxygenated waters from the poles into the deep oceans ensuring that bottom-feeding organisms can destroy what little organic matter rains to the deep ocean floor (Demaison, 1980). Bathymetric restriction is required to prevent this open circulation, which allows anoxic conditions to develop in the sediments, near-bottom waters and even far into the water column as seen in the present-day Black Sea.

Such conditions were developed periodically in the mid-Cretaceous, early opening history of the Atlantic Ocean where *oceanic anoxic events* (OAEs) were first documented in deepwater environments (Schlanger and Jenkyns, 1976). Ongoing research (Pepper et al., 2020a,b) in the Eagle Ford formation shows that the isotopic excursion used to date the acme at OAE 93.5 may not coincide with the timing of organic carbon deposition, which occurred at earlier and later times (95 and 93 Ma, respectively).

Fig. 4 shows a UEP log of DSDP Site 367, offshore Senegal, annotated with geologic time and illustrating acmes with high UEP; some of these acmes were originally identified as OAEs. Pepper et al. (2020a, b) showed subsidence profiles that

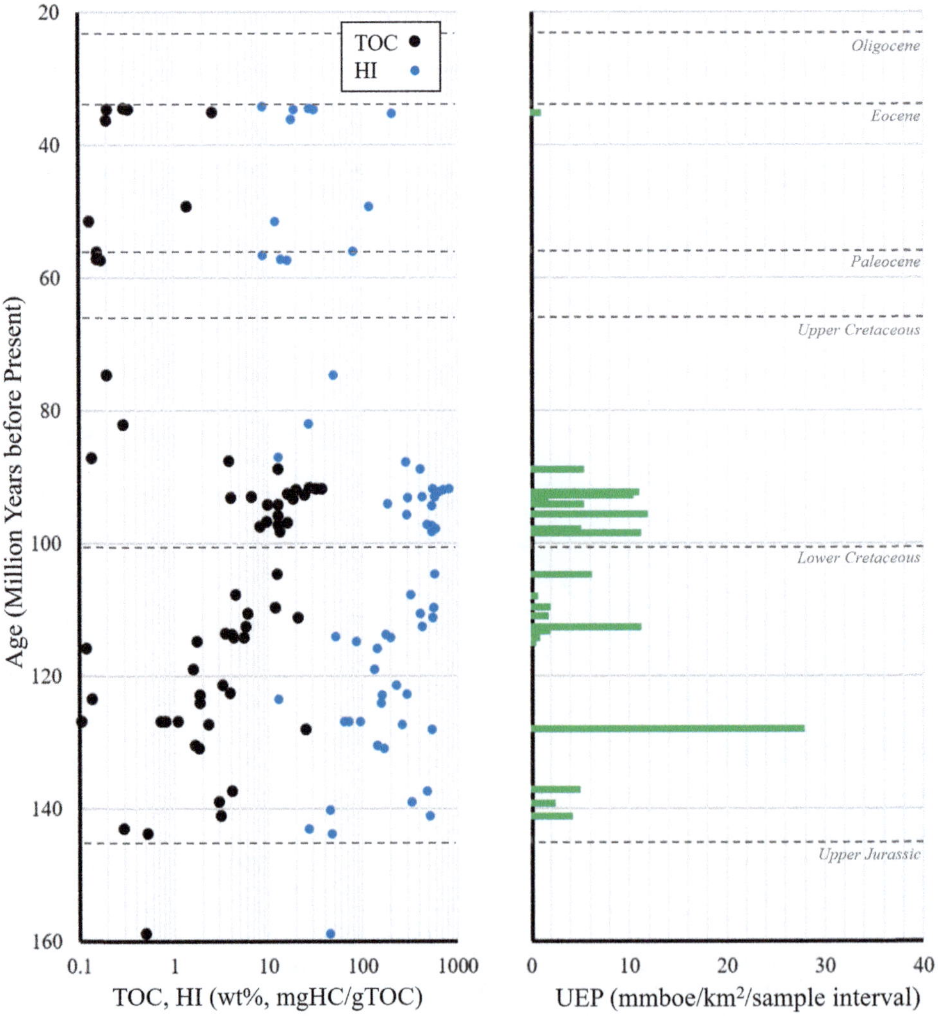


**FIG. 4** DSDP 367 UEP log. UEP has been calculated using the total organic carbon (TOC) and hydrogen index (HI) for the Deep Sea Drilling Program (DSDP) Site 367 well offshore Senegal. Because this well intersected rock deposited across a large geologic timeline, the source rock acmes in the Cenomanian–Turonian (Upper Cretaceous), Albian, and Aptian (Lower Cretaceous) sections are apparent. It should be noted that both a significant amount of organic carbon *and* a high hydrogen index are required to generate significant hydrocarbon volumes. A cutoff on TOC alone does not indicate a prolific source rock. *(Modified from Pepper et al. (2020a) and Pepper et al. (2020b).)*

indicate these Cretaceous source rocks were deposited in several kilometers of water depth on foundering Jurassic ocean crust. This is exceptional and reflects an unusual period of restricted deepwater circulation during the central Atlantic in the Cretaceous.

### Marine source rock diagenesis

Organic matter that avoids total aerobic degradation at or below the seabed polymerizes during the diagenetic process to form *kerogen* (Horsfield and Rullkotter, 1994). These shallowly buried sediments are in a zone of sulfate-reducing bacterial activity. At this point, marine source rocks with algae- and bacteria-derived biomass diverge toward two end-member organofacies: A or B. The resulting organofacies is dependent on the destination of sulfur in sulfate-derived hydrogen sulfide produced by sulfate-reducing bacteria (Fig. 5). Clays contain iron, so, in clay-rich facies, the hydrogen sulfide tends to react with this iron to form iron oxide minerals such as pyrite. In clay-poor environments—which are often, but not always, carbonate-rich—hydrogen sulfide reacts with kerogen and results in much more sulfur bound within the complex kerogen molecules (Horsfield and Rullkotter, 1994). This clay-poor vs clay-rich distinction forms the basis for the most fundamental division of marine organic matter: organofacies A and B, respectively (Pepper and Corvi, 1995a).

## Petroleum systems sub-element *Charge: Access*

In this petroleum system sub-element, the term *access* refers to the ability of a trap to access organic compounds bound in source rock. For that to occur, many individual physical and chemical processes must coalesce. Simply, kerogen in the source rock must transform into petroleum fluids and those fluids must flow through the subsurface into a given trap. When both of these processes occur, that trap is said to have access to migrating petroleum or become *charged*.

Kerogen breaks down to petroleum molecules in response to applied thermal stress—a process known as *generation*. Generation is a kinetic process driven by temperature and, to a lesser degree, geologic time. The time–temperature history

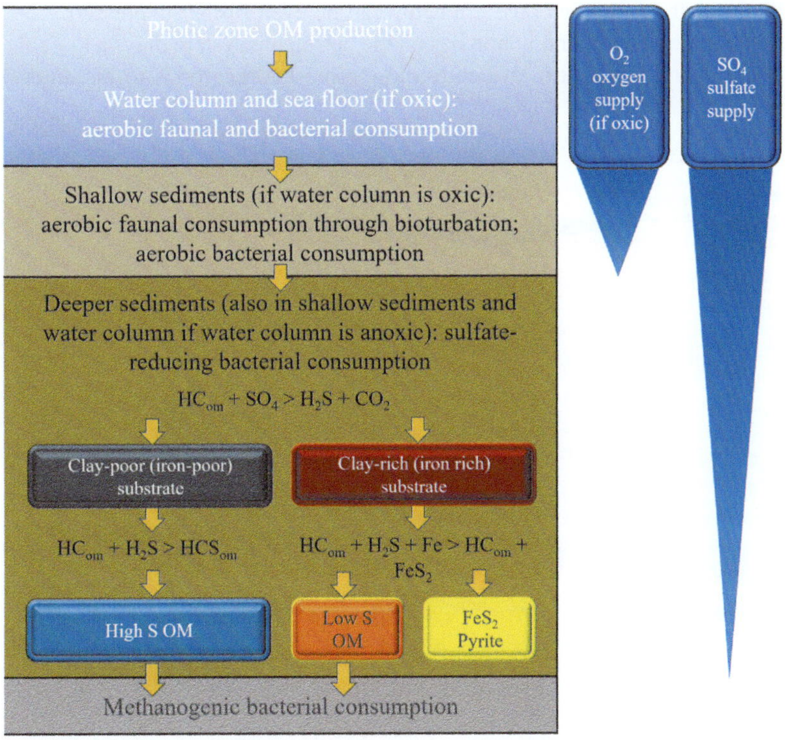


**FIG. 5** Role of sulfur in source rock deposition. The text box here is indicative of a 1D depth profile in a water column and into the sedimentary section of depositional environments for source rocks. Beginning at the top of the figure, organic matter (OM) is produced by photosynthesis and ultimately descends through the water column toward the seabed (base of blue rectangle). Within the sediments, sulfate-reducing bacteria generate hydrogen sulfide and carbon dioxide by metabolizing organic matter. As shown by the parallel downward diagenetic pathways, clay-poor or clay-rich sediments determine reactants for the hydrocarbons and hydrogen sulfide and form either high-sulfur organic matter (organofacies A) or low-sulfur organic matter and iron oxides (organofacies B).

of the source bed determines the cumulative thermal stress. In many deepwater basins, present-day burial depth is the same as maximum burial depth and the irreversible generation process is driven inexorably by increasing depth and temperature due to the heat engine within the earth.

## Thermal conditions of deepwater basins

Young, deepwater basins are often developed in oceanic passive margin settings overlying an area of the lithosphere that can range widely in age, composition and thickness. Under thermal equilibrium and slow sedimentation rate conditions, the heat flow at the base of the sedimentary section is controlled principally by two factors: the depth of the lithosphere-aesthenosphere boundary isotherm (Parson and McKenzie, 1978) and the heat generated through radiogenic mineral decay in the crust (Collaboration, 2011).

In oceanic basins like the Red Sea that are early in the seafloor-spreading stage, the lithosphere-aesthenosphere isotherm is close to the surface and total heat flow is dominated by this contribution. As the lithosphere ages and cools farther from the spreading center, the isotherm deepens and the thermal equilibrium depths settle to a more typical ~100 km after ~70 Ma (Pepper et al., 2021). Thus, in basins with Mesozoic source rocks, the heat flow from this source will be relatively uniform in a lateral sense with subdued variation as it is in the Gulf of Mexico (Fig. 6).

High heat flow conditions in passive-margin, deepwater play settings require some contribution from radiogenic mineral decay in granitic crust. The thickness and composition of the upper crust are therefore key controls on the lateral distribution of heat generated in addition to that from convection from the aesthenosphere.

These theoretical patterns in equilibrium heat flow are modified in many prolific petroleum basins with relatively old lithosphere and a thick, rapidly deposited sedimentary section. These basins do not have time to thermally equilibrate, particularly if the contribution from radiogenic mineral decay is low. Because cool sediments are very rapidly loaded into the depocenter, the whole lithosphere including the source beds enters a transient, depressed state. This phenomenon accounts for the extremely low geothermal gradients observed in some areas of the deepwater Gulf of Mexico. In salt basins, with an overlying, highly conductive salt canopy, average gradients can be even further depressed.

The effect this transient temperature phenomenon has on the petroleum system plays a key role in why many deepwater plays are so prolific. The cooler temperatures at depth allow more hydrocarbons to be generated though biogenic gas

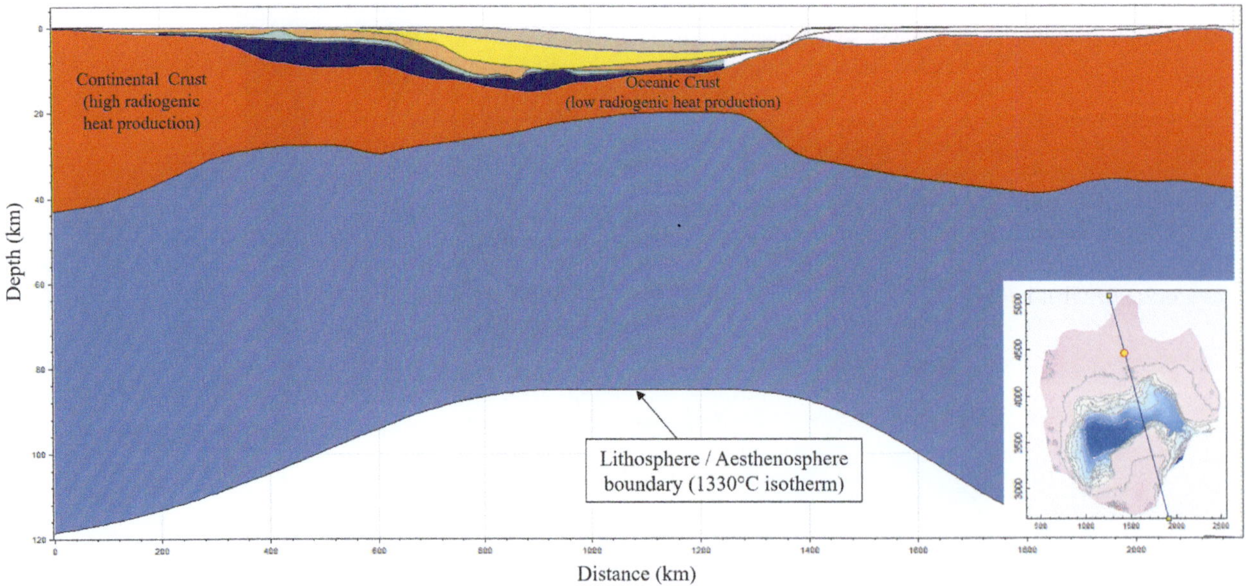


**FIG. 6** Gulf of Mexico lithosphere cross section. This cross section shows the crust in *orange* and uppermost mantle in *purple* (sometimes referred to as mantle lid) underlying the Gulf of Mexico. The two main sources of heat in sedimentary sections on Earth are shown here as radiogenic heat production (RHP) in the crust and depth proximity to the lithosphere-aesthenosphere boundary. Generally, continental crust contains a significantly higher concentration of the radioactive elements potassium, uranium and thorium than oceanic crust, so heat flow is typically higher when continental crust is underlying a sedimentary basin. However, mantle convection delivers high heat flow to a shallower depth in oceanic crust, so temperature prediction should be thought of as an inverse problem in basins without extensive calibration. *(Modified from Pepper and Roller (2021).)*

generation, prevent extensive quartz cementation in reservoirs, and prevent petroleum in both reservoirs and source rocks from cracking to gas.

## Kinetics of petroleum generation

Petroleum generation is modeled as a first order reaction and kinetic process governed by the Arrhenius equation:

$$k = Ae^{\frac{-E_a}{RT}} \tag{1}$$

This equation describes the rate $k$ of a chemical reaction via the pre-exponential constant $A$ and temperature $T$ as well as kerogen with an activation energy $E_a$. Time has a linear effect while temperature's effect is exponential (Pepper and Corvi, 1995a).

Aside from intracratonic basins, sedimentary basin cycles are rarely longer than ~200 million years (Woodcock, 2004); subsequently, source rocks younger than ~10 million years old are unlikely to be heated rapidly enough to reach petroleum-generating temperatures at or above 100°C. This time window indicates that there is a finite limit to the heating rates possible during sedimentary basin evolution. Using the above age limitations, the practical heating rate limits are 0.4–10°C per million years, and Pepper and Corvi (1995a) showed that an order of magnitude change in heating rate changes the required temperature for rapid oil generation by approximately 15°C. This observation imposes a practical limit on the time effect to approximately ±10°C, globally, compared with a standardized global heating rate of 2°C per million years. To put this in a perspective, time is one of a number of second order parameters in thermal modeling of deepwater basins. Another is the variation in sediment surface temperature, which can be as much as 25°C cooler in deepwater than at the shoreline.

Pepper and Corvi (1995a) also showed that, for six global organofacies from A to F, oil generation occurs in more-or-less alphabetic order at increasing thermal stress. For the two organofacies that charge most of the world's deepwater petroleum systems, organofacies A and B, the higher sulfur A typically generates ~10°C before lower sulfur B. This is a reflection of the previously-discussed lower activation energy required to break carbon-sulfur bonds vs carbon–carbon bonds.

It should be noted that, with the need to understand several temperature-controlled processes in deepwater petroleum systems, the first step in any kinetic modeling or analytical work should always be to verify the temperature data themselves. This poses a challenge given the increasing trend of reduction in data collection in expensive, modern deepwater wells. Wireline pressure and fluid sampling tools are often assumed to represent static formation temperatures, but, even if they sample formation fluid, it can take many days for a wellbore to equilibrate following the effects of relatively cool drilling mud circulating from the ocean riser to the deep wellbore (Luheshi, 1983).

## Petroleum expulsion into the pore space

Expulsion is the process by which smaller, simpler organic molecules cracked from very large, complex kerogen molecules leave the organic solid structure to enter the pore space of the source bed. During this process, the smaller molecules collect to form a petroleum fluid phase that is distinct from the aqueous phase. The expulsion efficiency and lag-time between generation of molecules and creation of a true fluid phase depends on the *hydrogen index* (HI) of the kerogen (Pepper and Corvi, 1995b); kerogens with high HI expel generated liquid petroleum molecules readily while those with low HI expel later in the generation window if they expel at all. Petroleum that does not expel remains sorbed in the organic kerogen framework as a feedstock for oil-to-gas cracking after which this carbon mass is readily expelled into the pore space as gas molecules (Pepper and Dodd, 1995).

The HI of any given organofacies can vary widely. As mentioned in the *charge: source potential* section, a layer-by-layer description of organofacies, total organic carbon (TOC) content, HI and thickness are the inputs to a complete description of the oil, gas, ultimate expellable potential of a source rock.

## Petroleum fluid phase

Once expelled molecules have collected in the source rock pore space, they begin to form a fluid *phase*. The density of the petroleum phase is different to that of the pore water, and this begins to generate buoyancy force. To understand petroleum fluid phases, it's necessary to understand a basic phase diagram as shown in Fig. 7. Phase diagrams in petroleum systems analysis serve a different purpose to the more familiar engineers' approach. Petroleum engineering work typically focuses on predicting the detailed behavior of a known petroleum fluid composition in response to predictable decreases in pressure

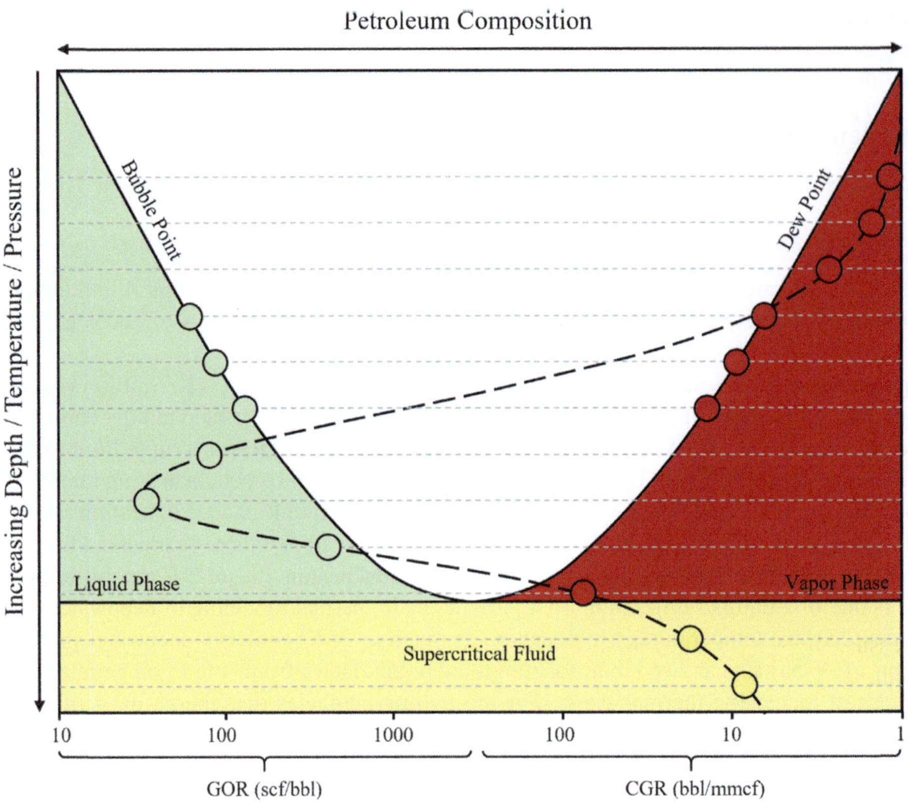


**FIG. 7** Phase diagram. This diagram shows petroleum composition on the X-axis (with heavier, liquid hydrocarbons with low gas-oil ratio (GOR) on the left and light, gas hydrocarbons with low condensate-gas ratio (CGR) on the right) vs depth, temperature and/or pressure on the Y-axis. The black dashed line is indicative of a typical hydrocarbon composition profile in a deepwater basin. Color-filled *circles* indicate an existing hydrocarbon phase at a given depth. As the composition curve crosses the dew point or bubble point, petroleum exists in either vapor phase (*red circles*), two phase (both *red* and *green circles* on edges of *white* space), or liquid phase (*green circles*). With very high temperature and pressure, petroleum can eventually cross the critical point to become a supercritical fluid (*yellow circles*) in which it takes on some properties of both liquid and vapor.

and temperature. In petroleum systems analysis, there is a need to understand the behavior of a wide range of compositions in a vast pressure and temperature space. This is done using a pressure–temperature function to create an equivalent depth axis against which composition can be plotted.

A reservoir, once deposited, can be viewed as moving down the *Y*-axis (depth increasing) through the phase diagram to meet migrating hydrocarbons expelled from more deeply buried source beds. Deepwater reservoirs in rapidly subsiding basins have a long window in which biogenic gas charge can be accessed from primary bacterial activity, so the shallowest reservoir fluid is a dry gas. When a migrating liquid petroleum phase meets this reservoir, larger molecules in the petroleum liquid will initially dissolve in the vapor phase as a condensate. When the rate of liquid addition becomes greater than the rate of biogenic gas addition, the condensate yield in the vapor will increase until it becomes limited by the dew-point curve. When this occurs, a saturated liquid phase known as the *oil leg* will form below the saturated vapor phase—called the *gas cap*. Further addition of liquid petroleum charge changes the proportions of the two phases, until a sufficient amount is supplied so that the gas cap is dissolved entirely into the oil leg. The liquid phase in this scenario becomes undersaturated. If higher GOR fluid from the maturing kitchen arrives, the gas fraction in the fluid will start to increase again, and the sequence may happen in reverse. Alternatively, the temperature and pressure may be high enough that this reversal occurs in the supercritical region where there is no distinction between liquid and vapor phases with changing composition. It is likely that fluid compositions in the pore systems of source beds also evolve through the supercritical region.

## Forces and directions of migration: Fluid potential

Buoyancy force is always toward the surface, but hydrodynamic forces and rock properties also control the direction of fluid flow through pore space in the subsurface. These forces can be quantified as directional vectors; however, it is conventional in subsurface evaluations to work instead with a scalar potential that is calculated for selected points or units.

This concept is covered extensively by England et al. (1987) who expand on earlier concepts by Hubbert (1953). Hantschel and Kauerauf (2009) also cover the topic in detail by including numerical simulation concepts.

*Fluid potential* $\Phi_F$ can be used as a concept that holistically allows for calculation of fluid flow in the subsurface. England et al. (1987) define $\Phi_F$ as "the work necessary to transfer a unit volume of fluid from reference conditions to the relevant (subsurface) conditions of interest." Further specification with the terms water potential $\Phi_W$ and petroleum potential $\Phi_P$ allows for a two-component concept that covers general petroleum migration concepts in subsurface evaluations.

To define $\Phi_F$ in clear, practical terms, the work done can first be shown as the expression:

$$\left[P - P^0\right]V - mg\left[z - z^0\right] + 2\gamma\left[\frac{1}{r_t} - \frac{1}{r_p}\right]\cos\beta \cdot V \tag{2}$$

This indicates the three key components of $\Phi_F$: work done against pressure (first term), work done against gravity (second term) and work done against capillary forces in porous rock units (third term). These three components are referred to as pore pressure, buoyancy force and capillary entry pressure, respectively, and all variables will be defined in the following section.

Since all porosity in the subsurface is first saturated with water, then $\Phi_W$ for a unit volume can be calculated with the equation:

$$\Phi_W = P - \rho_W g z \tag{3}$$

Here, $P$ is pore pressure, $\rho_W$ is the density of the formation water, $g$ is acceleration due to gravity ($9.8\,\mathrm{m \cdot s^{-2}}$) and $z$ is depth of the geologic unit. $\Phi_P$ can then calculated in reference to the existing $\Phi_W$:

$$\Phi_P = \Phi_W + (\rho_W - \rho_P)gz + \frac{2\gamma\cos\beta}{r_t - r_p} \tag{4}$$

This equation adds variables to the $\Phi_W$ equation; $\rho_P$ is the density of petroleum, $\gamma$ is the interfacial tension between the petroleum and water phases, $\beta$ is the contact angle between the petroleum-water meniscus and the pore throat, $r_p$ is the radius of the donating pore and $r_t$ is the radius of the receiving pore throat. The equation for $\Phi_P$ demonstrates that, in addition to pore pressure, both buoyancy and capillary entry pressure control whether a petroleum phase will displace the water phase in a porous rock. These principles are shown in Fig. 8, which illustrates how high capillary entry pressure of seal rock can lead to high fluid potential and offset the buoyancy of a petroleum accumulation in the underlying reservoir.

## Rates of petroleum phase migration

Early models of petroleum migration adopted from reservoir simulators treated petroleum flow as a viscous process governed by Darcy's Law; their algorithms moderated rates of vertical migration according to the permeability of mudrocks and viscosity of the petroleum phase. Such models controlled primarily by viscosity provided key early insights into the feasibility of vertical migration through mudstones and, even more importantly, led to an understanding that can be applied to differing degrees in all basins: that a time factor is involved in vertical charging. This means that a migration lag time needs to be considered in predicting charge access from deeply buried source rocks to shallow reservoirs. Petroleum charge may arrive at a trap millions to tens of millions of years after expulsion with results such as oil fields in the deepwater Gulf of Mexico positioned over the gas kitchen while uncharged and biogenic gas reservoirs are positioned above the oil window. Except where the source bed and reservoir are immediately adjacent, it is the timing of arrival at the trap that determines charge access risk, not necessarily the timing of expulsion in the source bed.

However, a more recent appreciation of petroleum expulsion rates offers another opportunity to understand the control on rates of vertical migration in a capillary-dominated—rather than viscous—view of the petroleum system (Fig. 9).

The rate of volume creation in this example (Fig. 9A) is perhaps the fastest of any petroleum system in the world: reaching a peak rate in the oil window of 3.6m of subsurface fluid per million years. However, when viewed on a commercial production timescale this corresponds to an incredibly low rate—approximately one petroleum molecule height per day. Given a slower heating rate more typical of old passive margins (Fig. 9B), the rate of volume creation is 10 times slower.

These almost unimaginably low rates make it likely that the geological system is governed by capillary forces, rather than concepts with such heavy control from viscosity. In this capillary-dominated school of thought, the expelled petroleum is injected into the pore volume of a carrier bed at such a low rate that it follows the lowest capillary entry pressure pathways

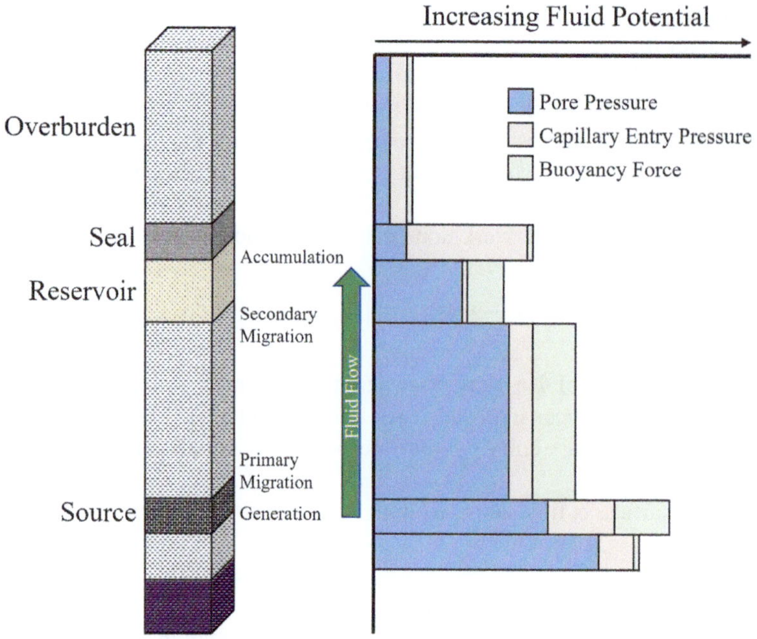


**FIG. 8** 1-D fluid potential diagram. A simplified 1-D petroleum system is illustrated by the block diagram on the left of this figure, and the corresponding fluid potential is shown on the plot to its right. The three components of fluid potential (pore pressure, capillary entry pressure and buoyancy force) are conceptually added for illustration of each of these by rock unit. Because fluids always flow from high fluid potential to low fluid potential, the *green arrow* is indicative of a petroleum fluid migrating toward the surface. However, the very high capillary entry pressure in the seal rock results in a barrier to upward flow and accumulation in the low fluid potential unit: the reservoir. Note that this diagram assigns only one value per variable for each layer or geologic unit (similar to a very simple 1D basin model), but, in reality, the values are gradational between units with possible significant variation.

through the medium. In large volumes of rock, this may only require a saturation loss of only a few percent of the pore space to span the rock (Mackenzie and Quigley, 1988). This opens the possibility that vertical migration can be modeled as a simple equation in which the rate of volume creation in the source bed governs the rate of ascent of vertical petroleum migration. In this case, the charge volume accessed by a trap is a simple subtraction of cumulative migration loss volumes from cumulative expelled volumes into the source bed. Fig. 10 shows a prototype of such a model using the Pompano section from Pepper and Yu (1995). This type of model runs in less than a minute, which offers the ability for real-time interactive migration modeling.

This migration concept predicts that the petroleum volumes arrive in the order in which they are expelled, with lower maturity, denser and more viscous fluids staying ahead of later expelled, more mature, less dense and more mobile fluid as is commonly observed in stacked pay systems in the deepwater Gulf of Mexico (Kirkland and Rushton, 2019). The higher the UEP and the greater its level of conversion, the higher the volume of charge reaching the trap. Since the volume expelled later during the gas window drives the earlier oil kitchen volume, it should not be surprising to see oil fields many kilometers above kitchens which are currently in the gas expulsion window.

## Pore pressure

Pore pressure is the pressure of fluids in the pore space of rock. For subsurface fluid flow and fluid potential, pore pressure is the most important variable, and pressure conditions can be a result of either local, regional or even basin-scale dynamic processes. There is also an interdependent relationship between pore pressure and porosity, and it is truly necessary to always consider both when discussing either variable.

To understand pressure or stress in the subsurface, it is first critical to frame the pore pressure range through depth in one vertical dimension (1D) with the fluid and sediment vertical stresses. These are referred to as the hydrostatic (or normal) pressure; and the lithostatic (or overburden) stress, respectively (Fig. 11). These curves represent the total weight of water and bulk rock (including pore water) at any given depth. In nearly all deepwater, subsiding basins, pore pressure will be a value greater than or equal to hydrostatic pressure but less than lithostatic stress.

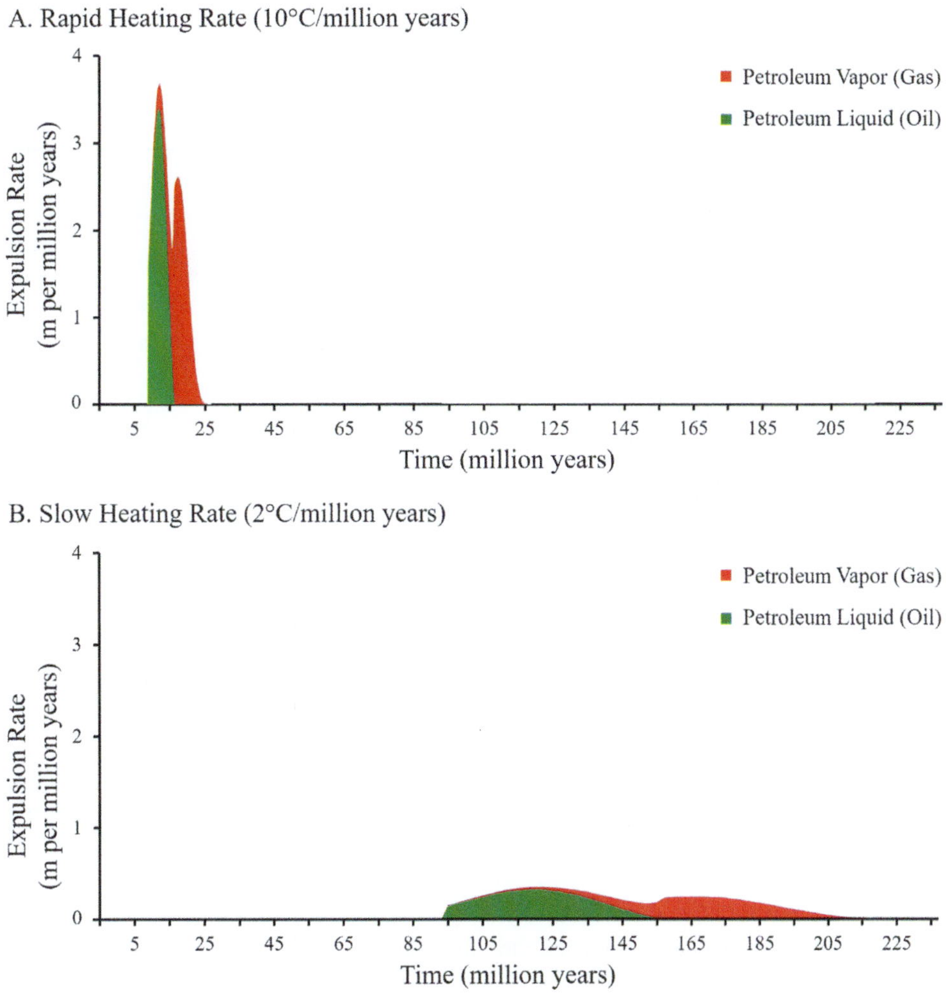


**FIG. 9** Expulsion rate as a function of heating rate. (A) This plot shows expulsion timing of a particular, modeled high-UEP source rock based on a rapid heating rate of 10°C per million years. To show a rate of migration as a function of volume creation, the volume units have been converted to height in meters above a source bed per million years. Note here that the petroleum has been expelled in its entirety within approximately 13 million years, however, this still only equates to the approximate height of one molecule per day of petroleum. (B) This plot shows results of the same model as the one above in (A) except that the heating rate has been lowered to 2°C per million years. In this case, it takes approximately 125 million years for all of the petroleum to be expelled.

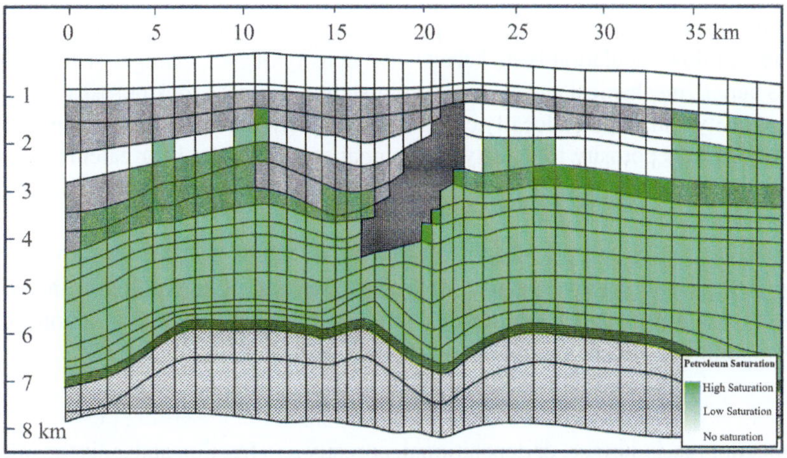


**FIG. 10** 2D migration model. This image is a representation of a prototype 2D migration model using the Pompano section of Pepper and Yu (1995) in which vertical migration is modeled as a simple overflow equation. The rate of volume creation in the source bed governs the rate of ascent of the migration front, and the charge volume accessed by a trap is a simple subtraction of cumulative migration loss volumes from cumulative expelled volumes into the source bed. This type of model runs in less than a minute which offers the ability for real-time interactive migration modeling. It is suggested here that this type of model may be faster, more useful and, in some ways, more physically accurate than some existing numerical migration simulations.

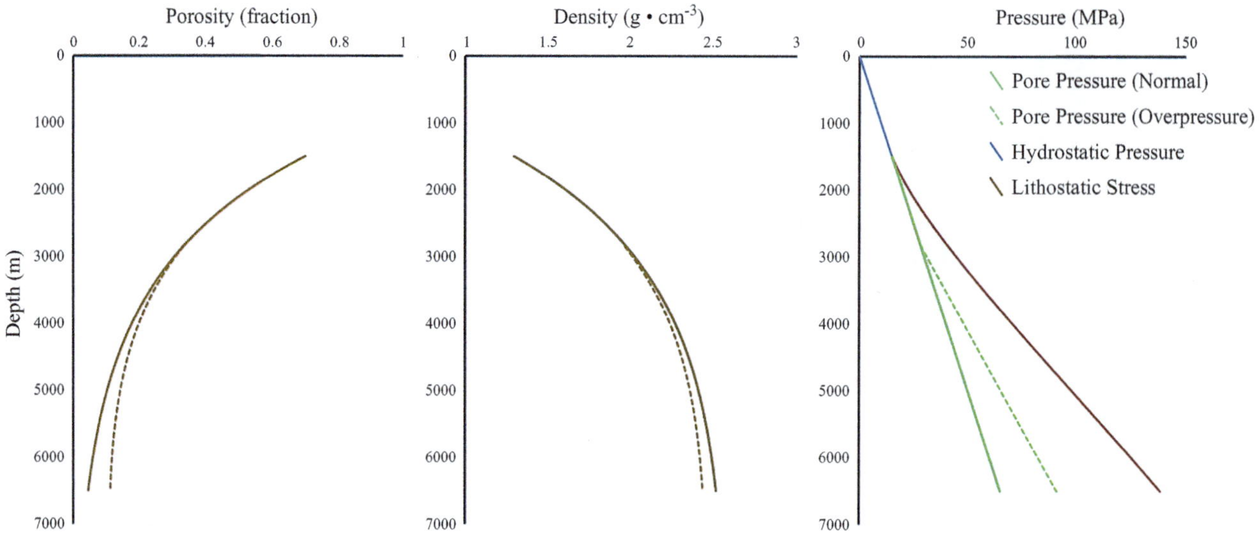


**FIG. 11** Pore pressure profiles. The three plots here show theoretical 1-D profiles through a normally pressured (*solid lines*) and an overpressured (*dashed lines*) sedimentary section. The blue line is the pressure of a continuous column of water (hydrostatic pressure), the brown line is the stress of the rock and its contained pore fluids (lithostatic stress) and the *green lines* are the pressure of the pore fluids. The *brown lines* within the porosity and density plots show the compaction trends with depth. The effects of overpressure (*dashed lines*) on the rock properties are sometimes referred to as *under-compaction* because the rock at depth has characteristics of less compacted rock that is typically found at shallower depths. In the overpressured case, the higher pore pressure increases the fractional porosity of the rock in a linear relationship, and this fractional porosity increase results in a linear decrease in bulk density since there is more low density fluid for a given volume. This figure shows an idealized pure mudstone sequence, but deepwater reservoirs in a similar under-compacted state can have excellent porosity that results in increased reservoir storage and deliverability.

Hydrostatic pressure ($\sigma_H$) can be calculated using the following equation:

$$\sigma_H = \rho_W g z \tag{5}$$

The density of the formation water is represented by $\rho_W$, $g$ is acceleration due to gravity (9.8 m s$^{-2}$) and $z$ is depth of the geologic unit. A geologic scenario in which pore pressure is the same as or very close to hydrostatic stress throughout the sedimentary section is indicative of high vertical permeability or a low rate of sediment loading.

Lithostatic stress ($\sigma_V$) calculation requires a different method as shown in the following equation:

$$\sigma_V = g \cdot \int_0^z \rho_R(z) dz \tag{6}$$

Here, $g$ is acceleration due to gravity (9.8 m·s$^{-2}$), $z$ is depth of the geologic unit and $\rho_R$ is the density of rock above depth $z$. The complicating variable in this calculation is $\rho_R$ because the density of rock changes with lithology, porosity, pressure and temperature. As it is buried, mechanical and chemical compaction decrease rock porosity and increase bulk density. To account for the variable density with depth, a locally calibrated compaction trend is used that either directly assigns density as a function of depth or converts porosity to density for comparison. A method for calculating the compaction trend of mudstones was first proposed by Athy (1930):

$$\rho_R = \rho_0 + (\rho_1 - \rho_0)e^{-kz} \tag{7}$$

This equation calculates rock density $\rho_R$ at a specific depth $z$ as a function of depth. The surface density is represented by $\rho_0$, the maximum possible (or, fully compacted) density is represented by $\rho_1$ and $k$ is a constant referred to as Athy's parameter. The same equation can also be modified to calculate rock porosity ($\varphi_R$):

$$\varphi_R = \varphi_0 + (\varphi_0 - \varphi_1)e^{-kz} \tag{8}$$

Similar to the density equation, $\varphi_0$ is surface, or depositional, porosity as a fraction and $\varphi_1$ is the minimum possible (or fully compacted) porosity. It is important to remember that these methods (Athy, 1930) represent a compaction trend for a single lithology. In real application, it is not uncommon that a single 1D lithostatic stress model may have multiple compaction trends or constant values due to lithologic variation, structural detachments or allochthonous geological bodies.

Once the hydrostatic and lithostatic stresses have been quantified, it is possible to place pore pressure in context. Still considering a 1D profile as in Fig. 11, the pore pressure will nearly always be at or above the pressure of a continuous, connected column of water—the hydrostatic pressure. As rocks compact, the pore volume decreases and water is expelled. When mudstones compact, the high depositional porosity (up to 70%) can relatively quickly be reduced to less than 10%, and the large fraction of water must be expelled into adjacent lower water potential ($\Phi_W$) lithologies that are generally much more porous. When the mudstones are compacted faster than the water can be expelled, the reduced pore volume leads to higher pore pressure in a process known as *disequilibrium compaction* (Dutta et al., 2021). In addition, if the porous lithologies are unable to drain the expelled pore water from compacting mudstones due to poor vertical or lateral hydraulic connectivity, pore pressure will continually build above the hydrostatic pressure with increasing depth.

The magnitude of the pore pressure above the hydrostatic pressure is referred to as *overpressure* ($P_{OP}$), or *excess pressure*, and is defined as:

$$P_{OP} = P - \sigma_H \tag{9}$$

In this equation, $P$ is the pore pressure and $\sigma_H$ is the corresponding hydrostatic pressure at the same depth. The magnitude of the lithostatic stress above pore pressure is referred to as effective stress ($\sigma'$). This represents the fraction of the overburden stress supported by sediment grain-to-grain contacts (rather than the fraction supported by overpressured pore fluid), and can be calculated according to Terzhagi's Principle (Terzaghi, 1925) with:

$$\sigma' = \sigma_V - P \tag{10}$$

Similar to Eq. (7), $P$ is pore pressure and $\sigma_V$ is lithostatic stress. Relating the concept of $\sigma'$ back to compaction, Athy's equation can be rewritten with $\sigma'$ replacing the depth variable:

$$\rho_R = \rho_0 + (\rho_1 - \rho_0)e^{-k\sigma\prime} \tag{11}$$

Compaction progresses with depth in a hydrostatic system only as effective stress increases with depth (Fig. 11).

In deepwater systems, the pressure regime can range from hydrostatic to highly overpressured depending on sedimentation rate, lithology and connectivity. Many prolific petroleum provinces occur in passive margin basins with geologically rapid sedimentation dominated by mudstones. These conditions are conducive to high overpressure. In the Gulf of Mexico Basin, it is not uncommon for wells to drill over 7500 m of mud-dominated sediment deposited entirely in the Neogene; overpressure in these plays can be very high. However, one lesson learned from the early exploration of the deepwater Gulf of Mexico was that deep wells in the deepwater basin did not suffer from the very high overpressures commonly encountered in early deepwater exploration of the present-day shelf break and upper slope. A key pore pressure concept to consider, though, is lateral connectivity. Permeable lithologies such as extensive turbidite deposits can allow fluids to easily flow up-dip (Fig. 12). Pore pressure, while first conceptualized in one dimension, is truly a 3D subsurface property.

## Buoyancy force

Buoyancy force is an upward (toward-the-surface) force acting on the lower density phase in a multiphase system. In petroleum systems analysis, there are at least two phases (petroleum liquid-water or petroleum vapor-water) and sometimes three phases (vapor–liquid-water). The density hierarchy is typically petroleum vapor < petroleum liquid < water. The buoyancy force ($F_B$) on a fluid can be calculated using Archimedes' principle:

$$F_b = -(\rho_A - \rho_B)g \tag{12}$$

Here, $\rho_A$ is the density of the higher-density fluid, $\rho_B$ is the density of the low-density fluid and $g$ is acceleration due to gravity ($9.8\,m \cdot s^{-2}$). Force is considered negative here due to direction toward surface.

## Capillary Entry Pressure

Capillary entry pressure ($P_{cap}$) is the pressure that is required for a nonwetting phase to pass through a pore throat into a rock and displace the wetting phase (Pittman, 1992). Early investigations were in terms of mercury injection (Washburn, 1921), but it has been applied over the years to common natural subsurface fluids with the following equation (using variables as defined in Eq. 4 above):

$$P_{cap} = \frac{-2\gamma\cos\beta}{r_t - r_p} \tag{13}$$

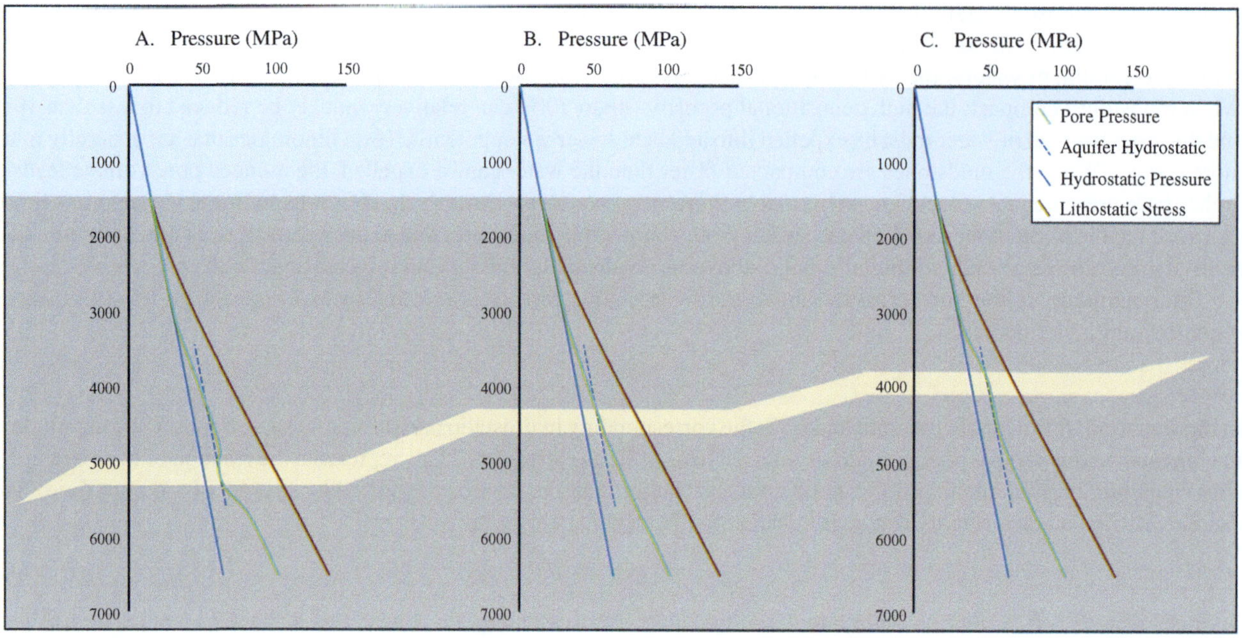


**FIG. 12** Pressure profiles along a dipping sand body. Three pressure profiles are shown along a monoclinal sand body that has one connected, overpressured water gradient (*dashed blue line*). (A) The deepest profile shows a pressure regression into the sand body from the surrounding shales. The shales have a similar compaction and pore pressure trend along strike, but the sand transmits overpressured fluid up-dip. This results in lower fluid potential and enhanced hydrodynamic sealing capacity. (B) The middle profile is a lateral point in the sand in which it is in general pressure equilibrium with the surrounding shales. This point is often referred to as the *centroid.* (C) The shallowest pressure profile shows that the sand body is higher pressure than the surrounding shales as fluid has been transmitted from down-dip. This can result in low sealing capacity due to higher fluid potential in the reservoir and drilling issues if the highly pressure sand was unexpected.

Capillary entry pressure is a concept that applies at all times to migrating petroleum, but because the denominator is the difference between the donating and receiving pores, it is important mostly at lithological interfaces with substantial porosity differences. For example, petroleum migrating through a high porosity carrier bed to a low porosity seal rock will encounter a high capillary entry pressure at that interface. However, petroleum migrating only through a high porosity carrier bed will only encounter low capillary entry pressures and migration direction will be influenced principally by buoyancy.

## Primary migration

Earlier in this section, *expulsion* was defined as the thermally driven process by which petroleum fluid molecules cracked from kerogen are delivered into the water-bearing pore space of the source bed. These expelled molecules collect to form a petroleum phase with a density difference relative to the water and, once that distinct phase has formed, the petroleum is capable of migrating.

In a source bed with a relatively large system of pores, primary migration first occurs as distribution of the petroleum phase within the bed until saturation builds in the pore system—an important process for unconventional plays. Once the petroleum phase has saturated connected pores within the source bed, the petroleum follows wetted pore throat networks into adjacent rock units above or below that have lower fluid potential than the remaining pores in the source bed pore system. At this point, further petroleum saturation of the source bed largely ceases, and additional expelled petroleum follows these established pathways into adjacent rock units. This process is termed *primary migration*. Because fluid potential generally increases with depth, primary migration is typically toward the surface. However, depending on capillary entry pressure and pore pressure, the direction of primary migration can also be downward against buoyancy force. As mentioned in the introduction to this chapter, this scenario is believed to be occur in the Norphlet play in the Gulf of Mexico (Godo, 2019; Kirkland and Brand, 2020).

## Secondary migration

Following primary migration, petroleum will continue to move through the subsurface and follow a path of lowest fluid potential ($\Phi_P$). This process of petroleum migration and displacement of existing pore water is referred to as *secondary migration*. When petroleum migrates into a porous and permeable zone such as a sand body, there is a balance between buoyancy force and capillary entry pressure that results in general, but not necessarily direct, up-dip migration. Geologic units with low capillary entry pressures and large lateral extents, such as extensive turbidite sand bodies, are most effective to facilitate secondary migration; these are referred to as *carrier beds*.

Consider a volume of petroleum that has displaced the pore water of an overburden rock just at the interface with the source rock during the primary migration process. As soon as that volume of petroleum is within the pore space of the overburden rock, the secondary migration process has begun (Fig. 13—Time Step 1). The petroleum volume will now continue to migrate upward to shallower depths within the overburden rock following a possibly tortuous path of lowest petroleum potential; local variation in capillary entry pressure can divert migrating petroleum from a direct path upward.

Once the petroleum migrates into a carrier bed (Fig. 13—Time Step 2), there is still a small force required to work against interfacial tension as the nonwetting fluid passes through a receiving pore throat that is significantly smaller in radius than the donating pore. Thus a connected petroleum column height is required to achieve sufficient buoyancy force to overcome capillary entry pressures even in secondary migration in sandstones. In addition, once pore space has become saturated with petroleum fluid, that pore space becomes an established migration pathway as long as the volume of

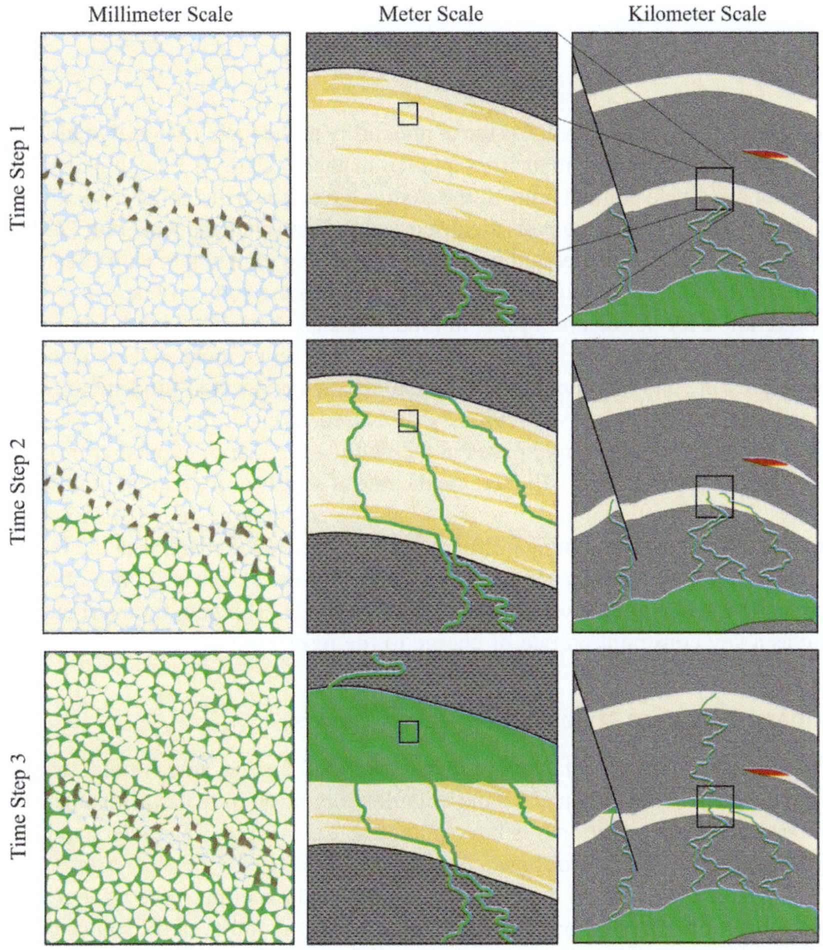


**FIG. 13** Secondary migration time steps. It has been established that migration of petroleum through a sedimentary section is driven by fluid potential and this concept is present on multiple scales from millimeters to meters to kilometers. Three time steps are shown here each with migrating petroleum following the path of lowest fluid potential at a different scale. Time Step 1 shows petroleum that has already undergone primary migration into an overlying carrier bed from the source rock. At this point, petroleum has not entered the reservoir but has displaced pore fluid in the underlying rock up to the interface. The reservoir shows heterogeneity within the sandy body and it is currently saturated with pore water. Time Step 2 shows migration into the reservoir from the underlying strata. At the meter scale, petroleum is largely migrating toward the surface due to buoyancy force but it is deflected laterally up-dip along the lower permeability zones in the reservoir. At the millimeter scale, petroleum is migrating up-dip along the lithic-rich streak toward lower fluid potential pores. However, some fluid streams are passing through the largest pore throats and migrating vertically. Time Step 3 shows petroleum beginning to accumulate within the reservoir. Petroleum has saturated most of the pore space in the reservoir with the exception of pores that are cut off by high potential throats and remain saturated with the original aqueous phase. A column has begun to build and increased fluid potential at the top of the reservoir, and it is beginning to leak vertically toward the next reservoir higher in stratigraphy.

petroleum is continuing to displace additional pore water. So, rather than individual droplets, liquid petroleum tends to migrate as connected fluid streams.

The reservoir-scale (or meter-scale) frames in Fig. 13 (middle column) not only illustrate the tendency of petroleum to migrate vertically in porous lithologies but it also highlights the effect of carrier bed and reservoir heterogeneity. In the case of an isotropic carrier bed, buoyancy force acts on petroleum while the pore pressure is equal for both aqueous and petroleum phases and the petroleum volume will migrate vertically upward to the shallowest strata of the carrier bed. If there are variations in porosity, permeability and pore throat diameter, then petroleum may migrate laterally to lower petroleum potential portions of the reservoir before breaking through vertically. Once petroleum has migrated to the sealing edge at the top of the reservoir, then the petroleum volume will accumulate in the carrier bed if petroleum potential of that overlying unit is greater. If petroleum potential is lower than the carrier bed, then the petroleum volume will continue to migrate vertically into that geologic unit. As shown in Fig. 13—Time Step 3, an accumulation of petroleum can lead to high enough buoyancy force that petroleum potential of the reservoir dynamically overtakes that of the overlying geologic unit, and petroleum will leak through the sealing edge.

Now that secondary migration has been defined in terms of the guiding fluid potential processes, the rate and magnitude of secondary migration can be considered. The most widely applied mathematical concept for fluid flow through porous media is Darcy's Law:

$$q = \frac{-k}{\mu} \nabla \Phi_P \tag{14}$$

In this application, $q$ is the flow rate per unit area, $k$ is the permeability of the rock, $\mu$ is the dynamic viscosity of the fluid and $\nabla \Phi_P$ is the petroleum fluid potential differentialacross a given distance. Darcy's Law, however, only considers the flow rate of an infinite volume or at a single point in time. Some studies have shown that other methods may better replicate rates across an actual sedimentary section; other methods are discussed in Hantschel and Kauerauf (2009) and Chapter 18 of this volume. In addition, a constant supply of generated petroleum is required to maintain a differential in petroleum potential. This supply is controlled by the rate of generation and expulsion from the source rock.

If rates of vertical migration for petroleum liquids in mudstones are controlled by the rate of supply from the source rock, then migration rates in highly permeable lithologies will be limited even more so. Upon reaching the critical height to overcome capillary entry pressure and migrate through sandstone, each connected fluid stream in Fig. 13 may travel very rapidly through the carrier bed to the trap. As previously discussed, even in the Gulf of Mexico—where petroleum is believed to be generated at one of the most rapid paces in the world—the rate of expulsion from the kitchen is incredibly slow. Structural focusing within the fetch area will increase the rate of supply as individual fluid streams coalesce into large-scale migration pathways, but a simple volume-time mass balance calculation (Fig. 10) shows that, even in such a fast system, an Upper Miocene reservoir containing a billion barrels of petroleum charged in the last 5 million years has only received fluid at an average rate of 0.5 barrels per day. The rate is slow because it will take a finite amount of time for the downdip reservoir above the kitchen to receive enough volume to create each new fluid stringer. This process again represents generation rate as a key constraint on rate of charge to the reservoir.

## Primary and secondary migration styles in deepwater basins

"Exploring from the source rock upward" and "bottom-up exploration" first became popular mantras in exploration in the late 1990s. This strategy helps to define and understand complex migration in real-world situations. The first step in this proven theory is to understand the distribution of source potential. This includes an understanding of source rock acmes, organofacies and UEP for a particular basin or play.

The next step is to delineate the extent of the thermally mature and expelling zone of source rock. Fig. 14 shows the extent of source rock in a theoretical passive margin play in black and the maturity of the source bed in green and red. In this schematic example, the shelf and the ultra-deepwater areas are thermally immature. The oil expulsion window (green) begins outboard of the present-day shelf break, and maximum thermal stress occurs beneath the sediment thick in the area of the present-day lower slope where gas is being expelled from the source bed (red).

The source distribution concept exploits the principle that migration is simply vertical in low permeability rocks and that the drive in a rapidly loaded shaley section will be upward until the petroleum volume expelled from the source bed has been exhausted. Unless alternate pathways exist (considered below), biogenic gas is the only charged fluid in reservoirs that are above the highest migrated thermogenic petroleum. While most of the theoretical wells overlying the mature source rock in Fig. 14 are successful, an example of a failure well that technically overlies the oil expulsion kitchen is represented by the theoretical "Well J."

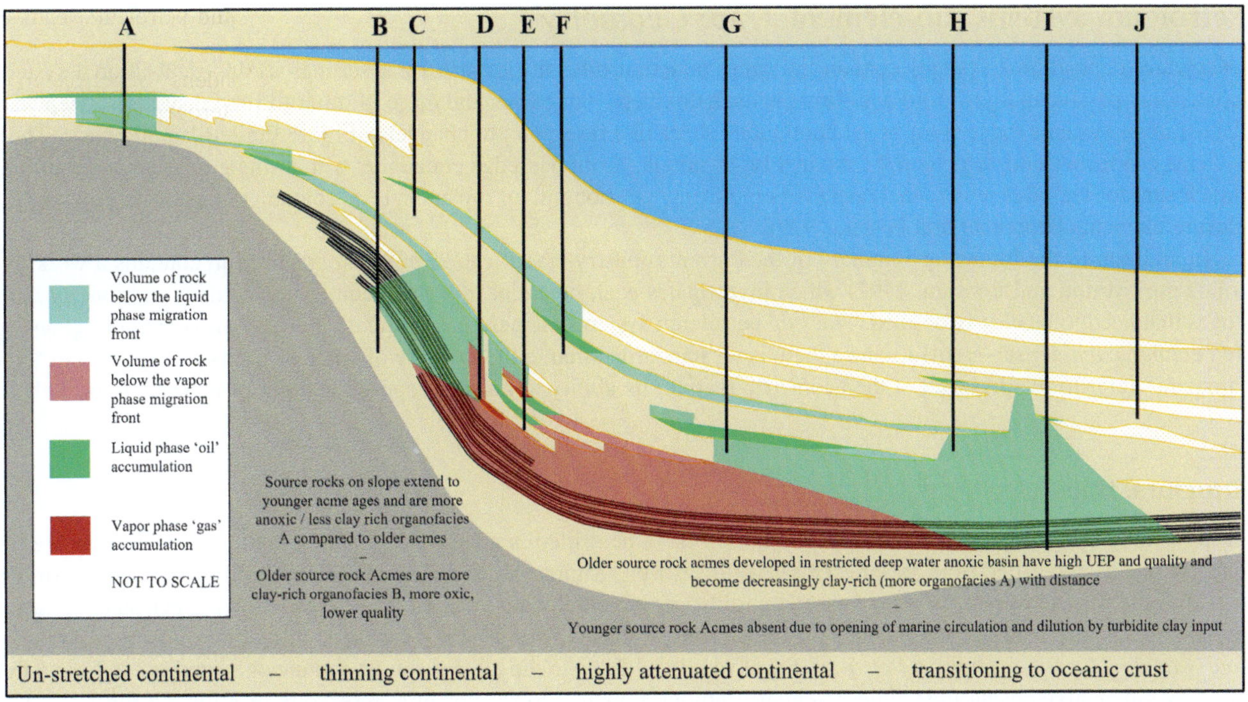


**FIG. 14** Schematic basin-scale migration diagram. Cartoon cross section—not to scale—of a passive margin with reservoirs deposited in shelf to deep basin settings. Sand bodies are shown with tan stipple, mudrocks with solid tan fill and source beds as *black lines* representing individual acmes with laterally varying distributions. Organofacies generally grade from clay-poor distal basin organofacies A to increasingly clay-rich, slope and proximal basin organofacies B. Patterns in geothermal profile follow the crustal architecture and are depressed by transient effects driven by recent sedimentation patterns. High thermal stress sufficient to reach the gas expulsion window occurs in the area of maximum sediment thickness beneath the present outer shelf/upper slope, decreasing inboard with thinning overburden and decreasing outboard due to decreasing radiogenic heat production from thinning continental crust, until the edge of the oil expulsion window is reached both inboard and outboard. Transparent green areas show primary migration of liquid petroleum fluid from the source beds via mudrocks, secondary migration in carrier beds and tertiary migration from traps; likewise vapor phase migration in transparent red. Liquid phase accumulations in green; vapor in red. See text for discussion of the outcomes at the different wells.

The third concept is the principle of "first carrier bed." This concept refers to the deepest carrier bed, or the first carrier bed that can influence migrating petroleum, and the direction in which migrating petroleum is directed by the first carrier bed's structural configuration. In the authors' experience, the first carrier bed in much of the west-central deepwater Gulf of Mexico, this is the Paleocene-Eocene Wilcox-equivalent basin-floor fan system. Absent one of these deep carrier systems, vertical migration will carry on into shallower depositional systems to the extent allowed by cumulative volumes created vs cumulative migration loss volumes.

Referring back to the conceptual basin in Fig. 14, Wells E and G have interesected the first carrier bed and found significant accumulations. Gas has been expelled from the red zone of source rock, but has migrated only to a certain height below the accumulation found by Well G. Migrating gas has filled the trap tested by Well E, saturated the petroleum liquid, created a two-phase reservoir and subsequently built a column that has overcome the column capacity of the seal and resulted in leakage to overlying carrier beds. Wells B, H and I also intersected first carrier beds, but they only found low-saturation petroleum shows indicated by the light green rock. Wells B and I were too close to the edge of the expelling zone of source rock and the interesected reservoirs did not have access to significant volumes. Well H simply did not drill the trap even though the exploration principals for migration were sound.

The accumulation at Well G is above the gas kitchen, but this reservoir is more vertically distant from the source bed as compared with the source bed underlying the accumulation at Well E. This oil field is an undersaturated liquid phase, despite sitting above the gas expulsion kitchen, because insufficient gas phase volume has been produced by the kitchen to overcome the vertical migration distance. The liquid column in the Well G accumulation is at column capacity and leaking, maintaining equilibrium with its seal. So the first carrier geometry of reservoir G defines not only the fetch area of the first carrier play system but also the additional petroleum charge budget available to all shallower accumulations charged by leakage: in this case from accumulations at Wells G to F, and from F to C. The fetch area for the accumulations at Wells D and E contributes the ultimate charge budget at the basin-edge heavy oil accumulation tested by well A. Collection and focusing of charge volumes at first carrier makes the remaining vertical charging process much more efficient.

## Petroleum systems sub-element *Trap: Geometry*

Deepwater depositional systems contain the same potential diversity of structural settings as any depositional system: extensional passive margins, fold-and-thrust belts, halokinetic salt basins and more. Similar to fluvial systems, deepwater depositional systems range from broad continuous sheet-like reservoirs to channels of more limited lateral extent. These systems can form multiple potential container types ranging from one-edge containers with a single top seal and four-way dip closure to five-edge monoclinal containers requiring top, bottom, up-dip and lateral container edges with only one dip closure element (Beaumont and Foster, 1999)

In addition to the following summary of the current industry classification of traps, this section contains a workflow following (Milton and Bertram, 1992) which investigates each potential trap as a geometric container with one or more edges that need to be clearly defined. The *trap* petroleum systems element is specifically divided into *container geometry* and *column capacity* sub-elements since there is no standardization in the industry of whether terms like "trap" and "containment" refer to the presence of the geometric feature, its ability to hold a proposed column of petroleum, or both.

### Structural traps

Traps formed by geological structures are often the first to be drilled in a new deepwater exploration area as they are frequently readily identified from seismic, offset wells or even features at the surface. Folding, faulting and unconformities can all form structural traps. In traps formed by folds, geologic units dip away from the crest allowing petroleum to accumulate; this is referred to as dip closure. If a folded trap dips away from the crest in all four directions, it is referred to as a *four-way dip closure* or simply a *four-way* (Fig. 15A). In addition to dip closure, faults and unconformities can form a trap edge by either juxtaposing high fluid potential lithologies against lower potential reservoirs or allowing for a mudrock "smear" to form in the plane (Fig. 15B). This mudrock smear is often referred to as *shale gouge* and is formed by grinding

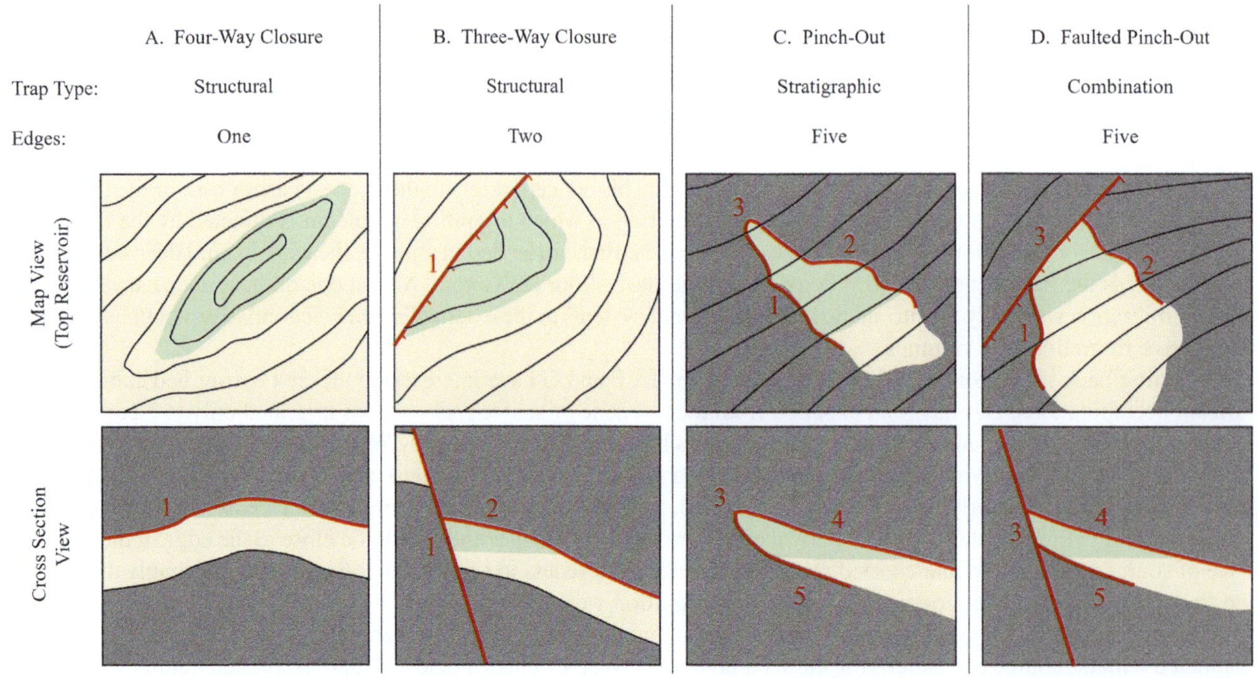


**FIG. 15** Examples of trap container geometries. (A) A four-way closure is the simplest structural trap. A reservoir is contained within a dome-shaped fold, and a single sealing edge above the reservoir forms the container. (B) A three-way closure is the other main type of structural trap. Two sealing edges, including an edge on top of the reservoir and a fault creating an up-dip edge, are required to create a container. (C) A stratigraphic trap has depositional or erosional edges. This trap example exists as a sand-rich, single turbidite lobe complex deposited in a mud-dominated deepwater environment. The limit of sand deposition creates three edges (left lateral, right lateral and base) while erosion or deposition create the top and up-dip edges. This type of trap requires five total edges to be effective. (D) A combination trap has both stratigraphic and structural edges containing the petroleum accumulation. In this trap example, most of the sealing edges are similar to the example in C, but the left lateral edge and up-dip edge are partially created by a structural feature: a fault.

of deformed rock the fault plane to a finer, poorly sorted and often well-cemented material with higher capillary entry pressure and lower permeability (Hippler, 1993). When dip closure forms a trap against a fault or unconformity, it is referred to as a *three-way* or even a *two-way* depending on the number of directions of dip closure.

### Stratigraphic traps

Stratigraphic traps are formed when a porous reservoir is encased in higher fluid potential lithologies—generally with at least one direction of dip closure (Fig. 15C). For this type of trap to contain a petroleum accumulation, it must be sealed above, below and on three sides. Deposition of individual sand bodies on a mud-dominated slope is a common way in which stratigraphic traps can be formed; these types of traps are often found in deepwater settings. In addition, up-dip pinch-outs of more laterally extensive sand fairways or even entire formations can form strike-trending stratigraphic traps. In some settings, unconformities can form what may be better described as a stratigraphic trap than a structural one—particularly when there is one direction of dip closure. Large-scale pinch-out-style stratigraphic traps may be identified from offsets wells, seismic data interpretation or surface geologic mapping, but smaller-scale traps often require analysis of geophysical anomalies on seismic data.

### Combination traps

Structural and stratigraphic traps in fact are terms for endmembers. In practice, it is possible that a three-way fault closure could have a stratigraphic edge midway through the oil leg or that a sand body contains an accumulation only where it is draped over a dipping structural ridge (Fig. 15D). There are also other types of trap that may not strictly fit into a definition for either stratigraphic or structural, such as a trap formed by a tilted oil–water contact due to hydrodynamic variations within a reservoir body.

### Defining the container edges of a trap in practice

These trap terminologies are widely used in the industry and are useful for communication, but properly defining a trap for prospect evaluation is best performed by identifying and mapping individual sealing edges of reservoirs (Milton and Bertram, 1992). For example, many geologists intuitively seek simple four-way traps as the first structures to be drilled in a new play because they can be reliably identified and mapped on a single container edge on the base of the seal (Fig. 15A). Perhaps less intuitive is a situation in which the reservoir is a channel draped across the same four-way structure. At first glance, this trap may appear to be more complex than the example in Fig. 15A, but, as long as the single base seal edge is present across the four-way closure, the channel still sits inside the same simple one edge container. As discussed in the next section, the required column capacity is the same in both cases.

Fig. 15C shows a much more complicated geometry that is typical of stratigraphic traps drilled along the Atlantic margin since the early 2000s. For this channel to form a container, it must have five edges: top, bottom, pinch-out of sand at the up-dip entry point and lateral edges on both margins of the channel. Since, unlike the four-way example, four channel edges are not contained inside a top-seal edge forming total dip closure, these five edges are required to prevent petroleum from migrating laterally out of the trap. Unfortunately, the presence or absence of the up-dip and lateral edges cannot always be determined from seismic data due to limitations in resolution or character definition; the difficulty in proving these edges now imparts risk on *trap: geometry* to the prospect. This is the reason that a geophysical petroleum fluid attribute—or direct hydrocarbon indicator (DHI)—is such an important subsurface evaluation tool in such plays.

## Petroleum systems sub-element *Trap: Column capacity*

If the petroleum systems sub-element *trap: geometry* concerns the presence or absence of edges required to define a trap, then *trap: column capacity* concerns the capacity of each edge to oppose buoyant force and pore pressure of a petroleum accumulation in the underlying reservoir. The vertical height of that accumulation is referred to as the *column*.

Fluid potential, which in seal rocks is controlled by capillary entry pressure but sometimes augmented by overpressure, determines the column capacity along each container edge. As covered in the *charge: access* section, migrating petroleum follows a path of lowest fluid potential both when migrating into a trap and when attempting to migrate out of it. For example, in traps that contain a fault as a sealing edge, petroleum will leak vertically if reservoir fluid potential reaches the overlying top seal's fluid potential before that of the fault plane edge. Regionally extensive turbidite sands can have significant relief with respect to the sediment surface, and column capacity in the deeper parts of the reservoir fairway can benefit from hydrodynamic sealing enhancement as shown in Fig. 12.

## Capillary-based column capacity

Column capacity as a function of capillary entry pressure in deepwater petroleum systems—as in any petroleum system—depends on the presence of sealing lithologies with very low permeabilities. Both capillary entry pressure and permeability depend on pore throat size. As covered in the *charge: access* section, these typically fine-grained rocks are often overpressured and therefore in a low state of compaction. This should be considered when evaluating column capacity in deepwater plays.

Before proceeding to describe the capillary pressure workflow, it's worth commenting on the role of country rock vs fault rock. Many geoscientists place high risk on fault seal but there are many lines of evidence that, especially in mud-rich deepwater depositional settings, mudstone country rock is the dominant medium for cross-stratal flow of both compaction water and petroleum. Even in peer-reviewed journals, it is possible to find mudstones being described as "impermeable," however, this is inconsistent with production from unconventional reservoirs that has changed the face of energy all over the world. Even the word *seal* itself can be misleading since it may appear to indicate a strict presence/absence property leading to the disproven notion of fine grained rocks being regarded as impermeable until they are fractured. Fine grained rocks have a pore structure enabling them to compact by exporting pore water and enabling parts of their pore volume to be penetrated by migrating petroleum during migration either in a primary or secondary sense.

Take for example a conceptual horizontal slice through any sedimentary basin. Fault planes will have orders of magnitude less area than with the vast area of sedimentary stratal surfaces. Furthermore, large lateral overpressure differences between fault blocks attest to the low permeability of fault plane materials. These concepts point towards petroleum migration through country rock and are consistent with laboratory measurements shown in Hippler (1993). This study showed that fault gouge exhibited up to 5000 times lower permeability and up to 20 times greater capillary entry pressure in fault plane rock vs original country rock.

## Capillary column capacity workflow

The equation governing the entry of a nonwetting fluid into a wetting fluid in was introduced as Eq. (13) in the *charge: access* section. As described there, an established workflow to evaluate this process is mercury injection porosimetry conducted on rock plugs gathered from wells. These data can provide useful insights, but it is unlikely that samples from a wellbore will sample the critical pathway migrating petroleum takes through the sealing lithology in the critical container edge. Instead, it is recommended to refer back to the concept of fluid potential.

As normally written, Eq. (13) contains both a rock and a fluid term, but a column capacity analysis seeks to understand a rock property that governs leakage. To do this, the equation can be rearranged so that only the rock term appears on the right-hand side:

$$\frac{P_{cap}}{2\gamma \cos \beta} \propto \frac{1}{r_t} - \frac{1}{r_p} \tag{15}$$

Recall that $P_{cap}$ is the capillary entry pressure, $r_p$ is the radius of the donating pore, $r_t$ is the radius of the receiving pore throat and $\gamma$ is the interfacial tension between the nonwetting petroleum and wetting aqueous phases. The capillary equation in its normal form is cumbersome because $P_{cap}$ is a measure of the rock property only for a given interfacial tension and must be accompanied by the assumed interfacial tension in order to be physically meaningful. But in this rearranged form, the right-hand side of the equation is now a reflection of the rock properties only since it reflects the contrast in radius between donating pore and receiving pore throat without qualification. Since, in typical high-quality deepwater sandstone reservoirs, $r_p$ will be very large compared to the adjacent mudrock $r_p$, the term is strongly dominated by the very small pore throat radius (very high $1/r_t$) of the mudrock. Capillary entry pressure is mostly a reflection of the sealing rock property.

A petroleum accumulation that is leaking at some rate through the seal is variably at capillary equilibrium with the sealing lithology, so the observed fluid properties of that accumulation on the left hand side of Eq. (15) can be used to invert the rock property that is balancing this force. The capillary entry pressure term ($P_{cap}$) is measured by the fluid potential of the petroleum at the height in the column where leakage is occurring—assumed to be the crest. It is often thought of as *pressure* that controls leakage from an accumulation because units of fluid potential are the same as pressure, but the concepts put forward for migration in the *charge: access* section apply to column capacity as well. The difference in fluid potential between the reservoir and seal in Fig. 8 is indeed the column capacity for that example. Conceptually, Fig. 13—Time Step 3 shows a petroleum accumulation in which the reservoir fluid potential is actively leaking as the fluid potential in the reservoir has become greater than the seal.

To understand origins of variations in the rock-specific term on the righthand side of Eq. (15) further, grain size and clay content—often expressed as the "less-than-2 μ" fraction of a mudstone—will control lithofacies-driven variations. Additionally, for any given sedimentary facies, pore and pore throat size are diminished by compaction. As stated throughout this chapter, in the relatively cool stratigraphy of many actively depositing deepwater systems, mechanical compaction is the dominant process rather than chemical compaction which requires greater thermal stress. Mechanical compaction is driven by effective stress, and, for a given mudrock type, the pore throat size will be heavily influenced by this value.

## Capillary column capacity prediction

Eq. (15) can be used to create a predictive tool calibrated for sealing lithologies of different facies in the same basin or even to compare mudstones and other fine grained rocks locally. Fig. 16 shows such a calibrated model for Neogene condensed section mudrocks in the deepwater Gulf of Mexico. Having calibrated the relationship between column capacity and effective stress, the relationship can be expanded to predict, for a nominal reservoir pressure and temperature, how the supportable column height varies for petroleum fluids as a function of GOR. The significant difference between liquid and vapor phase densities and its impact on column buoyancy is well appreciated in the industry, but the wide variation

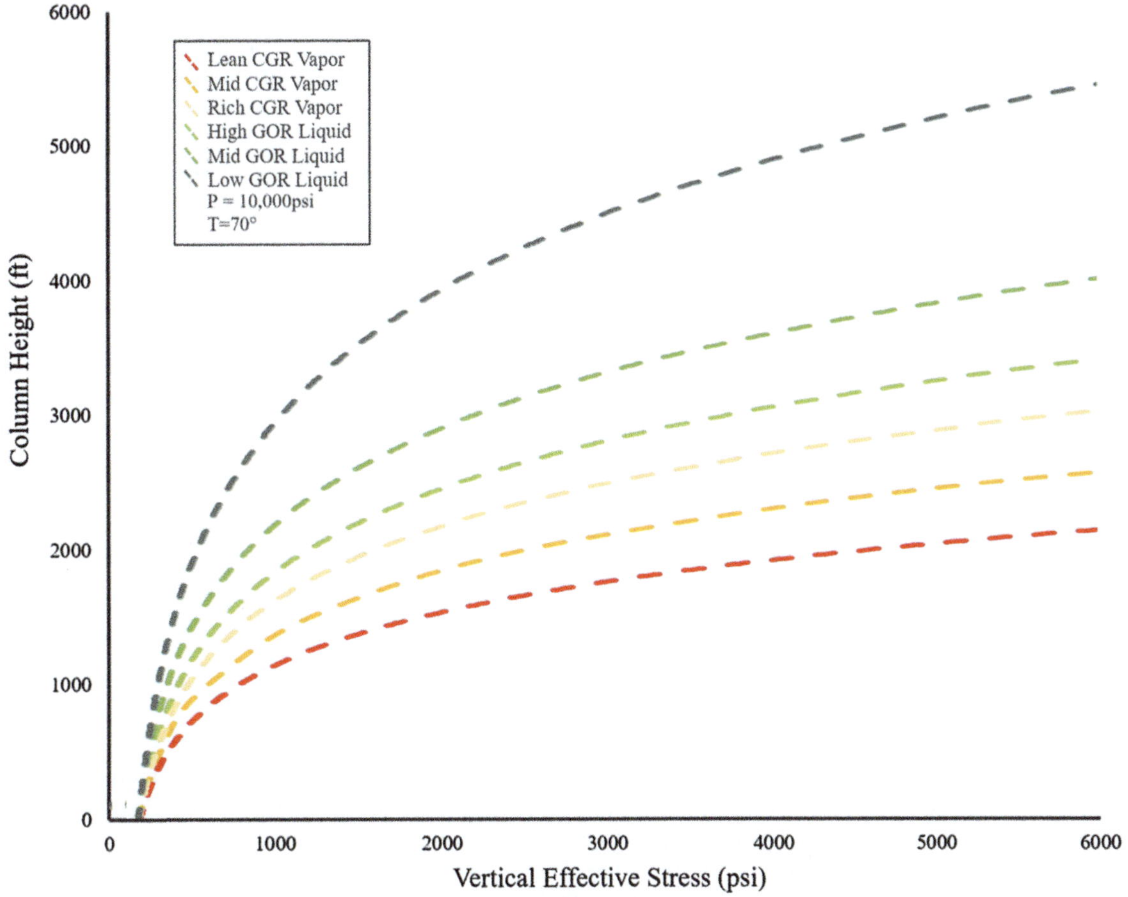


**FIG. 16** Column height (sealing capacity) vs vertical effective stress. This plot shows curves six petroleum fluid types including three liquid phases (300, 1000 and 2000 scf/bbl GOR) and three vapor phases (200, 33, 1 bbl, mmcf). Based on the concepts presented in Fig. 11, it seems intuitive that higher vertical effective stress and a greater state of compaction can result in larger column heights, but this plot illustrates—through Eq. (15) and the concepts discussed in *trap: column capacity*—that the lower buoyancy of heavier petroleum can have an extreme increasing effect on column capacity of a given seal rock.

in sealing capacity for petroleum liquids of varying GOR is also important. As the density difference between petroleum and water decreases, the model tends toward a "divide by zero" situation in the capillary equation. Many of the very large columns in the deepwater Gulf of Mexico are indeed low GOR, dense reservoir fluids.

### Hydrodynamic column capacity prediction

As in petroleum migration, hydrodynamics (or here, overpressure) can play a dominant role in trapping. Overpressure is a fluid phenomenon unlike capillary forces which act primarily at a rock interface where pore throat size changes (seal unit in Fig. 8). Capillary entry pressure variations tend to be in the hundreds of psi; overpressure differences between a reservoir and an overpressure maximum in an overlying mudrock can reach thousands of psi.

Column capacity controlled by hydrodynamic conditions results from a simple balancing of fluid potentials. A hydrodynamic seal is one in which the sealing lithology has a greater fluid potential than the reservoir due to greater pore pressure rather than simply capillary entry pressure. The scenario in Fig. 12 shows an example of how large, well-connected sand bodies can lead to hydrodynamic seals in down dip traps (Fig. 12A). It is believed that many large columns in plays around the world are facilitated by situations like this one.

## Petroleum systems sub-element *Reservoir: Storage*

A reservoir is a rock unit with properties that allow for petroleum accumulation and, ultimately, production to the surface. Globally, nearly any type of rock can function as a reservoir depending on the depositional, geomechanical and diagenetic history. In deepwater plays, it is most common that reservoirs are coarse-grained clastics with high porosity and permeability that have been preserved or developed throughout the diagenetic process. The process of clastic reservoir deposition and diagenesis is covered extensively in other chapters of this volume, so the focus here is on the role of the reservoir within a petroleum system. It is also worth noting that other reservoir facies can be economically important in deepwater developments even if they are significantly less common; a recent and relevant example is the microbial carbonate reservoirs in the Campos and Santos basins of Brazil (Muniz and Bosence, 2015).

As discussed in the *charge: access* section, fluid potential controls the direction and magnitude of subsurface fluid flow, and reservoirs represent unique, low-potential geologic units in a three-dimensional sense. Fluids flow into reservoirs and only flow outward if spilled out of the trap by inflow of additional petroleum fluids or if the sealing capacity is reached and they migrate vertically. The most common scenarios are that petroleum migrates into a reservoir from underlying or downdip carrier beds (secondary migration) or from underlying source rocks (primary migration). Once in the reservoir and within a container, petroleum is trapped by overlying, sealing rocks that have higher potential due to high capillary entry pressure of seal rock lithologies such as shale or marl. As petroleum continues to migrate into the reservoir, formation water is pushed downward by the accumulation of the more buoyant petroleum and either spills from the trap or remains as a higher-pressure water leg. This scenario is of course idealized, and, in nature, it is often found that reservoirs likely have small outflows through structural or stratigraphic breaches. As discussed in the *trap: column capacity* section, the lower the fluid potential difference between seal and reservoir, the greater the chance of vertical leakage.

The first consideration in evaluating the *reservoir* petroleum systems element is the likelihood that reservoir facies will be present with volume and properties that are sufficient for petroleum to saturate the pore space and accumulate. This is covered as the petroleum systems sub-element *reservoir: storage*.

Storage for a liquid petroleum phase reservoir is calculated by the volumetric equation:

$$OOIP = \frac{GRV \cdot c \cdot \varphi \cdot S_P}{FVF} \tag{16}$$

Where *OOIP* is stock tank barrels of oil originally in place in the reservoir, *GRV* stands for gross rock volume, $c$ represents the conversion factor from the respective volumetric unit to barrels, $\varphi$ represents fractional porosity, $S_P$ represents fractional petroleum saturation and *FVF* stands for formation volume factor—also called $B_o$. Determining gross rock volume is foundational and a subject in its own right (Tearpock et al., 2020), but, simply, it is the integral of area and thickness of the reservoir within the container. The focus of the following discussion will be on the porosity, saturation and formation volume factor terms. Together with the gross rock volume, these values provide the quantitative static reservoir description.

Of these three volumetric terms, porosity is the property solely of the rock. As shown in Eq. (11) in the section on *charge: access*, porosity decreases by mechanical compaction as a function of effective stress, but it can also decrease by mineral dissolution and precipitation during diagenesis as a function of thermal stress. This is often referred to as *chemical compaction*. For the simple case of quartz cementation—the dominant chemical compaction process in many deepwater reservoirs—reaction kinetics can be modeled as first order with a modified version of the Arrhenius equation (Eq. 1) as shown by (Waulderhaug, 1996).

Rapidly deposited and poorly drained deepwater fan systems—such as those in the Gulf of Mexico or Guyana-MSGBC conjugate basins—tend to build overpressure and thus potentially preserve high intergranular porosity at great depths due to under-compaction. Effective stress experienced by rapidly deposited, highly overpressured and fine-grained deepwater reservoirs can be 7 MPa (1000 psi) or less.

Short exposure times (high heating rates) experienced during rapid deposition significantly retard cementation at a given temperature compared with slowly heated basins. Also, as discussed in the *charge: access* section, deepwater basins can have especially cool thermal profiles.

The petroleum saturation term is a function of both rock and fluid properties. In terms of fluid properties, the density contrast and the interfacial tension between the petroleum and aqueous phases affect the ability of petroleum to displace water in the pore system (Eq. 4). Greater density contrast results in greater buoyancy and increases saturation as petroleum has greater force to overcome interfacial tension and capillary entry pressure. Capillary entry pressure is a property of the reservoir rock; poorer sorting and smaller pore throats result in reduced saturation.

The purely fluid term in the volumetric equation (Eq. (16)) is the formation volume factor (FVF). The FVF appears in the denominator and can significantly impact the oil in place; in volatile petroleum liquids, FVF can reach a value of up to 3. This means that only one-third of the reservoir fluid volume will be produced as oil at surface conditions and the rest of the reservoir fluid volume will be produced as gas. In many industry scenarios, determination of FVF for an undrilled prospect is often a job left solely for engineers, but, combined with that assessment, a holistic understanding of the petroleum system should always be integrated.

To reiterate the controls on GOR from the *charge: access* section: organofacies, thermal stress and degree of partial vs cumulative capture are important. It is also common that cool deepwater reservoirs receive additional increase in GOR from bacterial activity that has generated biogenic gas, and this biogenic gas contribution can also significantly increase FVF above what is expected solely from thermogenic petroleum.

## Petroleum systems sub-element *Reservoir: Deliverability*

The Darcy equation (Eq. 14) can be arranged to illustrate the factors in a deliverability calculation—an important part of the dynamic reservoir description in appraisal and development.

$$q = \frac{hk\nabla P}{\mu \cdot FVF} \tag{17}$$

In this application, $q$ is the flow rate per unit area, $h$ is the height of the well perforation, $k$ is the permeability of the rock, $\mu$ is the dynamic viscosity of the fluid, FVF is the formation volume factor of the fluid and $\nabla$P is the pressure differential across a given distance. It is important to keep in mind that deliverability of a reservoir is not explicitly required for a petroleum system to exist, but considering the application of petroleum systems analysis principles, it is discussed here as part of a workflow in industry assessments.

First, in terms of fluid properties, note that FVF appears again in the deliverability equation when considering the production rate at the surface. Second, pressure—in the form of the drawdown pressure between the formation and the wellbore—is a key driver in the production rate. The previously discussed overpressure in deepwater systems opens the window between bottom-hole pressure and saturation pressure and has a very positive effect on commerciality by allowing greater pressure drawdown and increased flow rate in expensive deepwater developments. Third, viscosity has a major impact on production rate since it appears in the denominator along with FVF. Viscosity of a reservoir fluid depends strongly on the reservoir fluid density, which, in addition to pressure and temperature, depends strongly on petroleum GOR and liquid density. For a given GOR, marine organofacies A fluids have a heavier stock-tank density than marine organofacies B, so prediction of important regional trends in density and viscosity begin with an understanding of organofacies.

In order to remove some effects of migration and alteration on petroleum fluids, an onshore unconventional play is useful as an analog for the effects of viscosity variation as shown in Pepper et al. (2020). This study introduces the case of the middle Permian Bone Spring and Spraberry-Dean turbidites that have been exploited in the Delaware and Midland subbasins, respectively, in the Permian Basin, USA. Development was initially with low-rate vertical conventional completions but has now been rejuvenated in the unconventional program to exploit associated lower permeability facies.

Although organic-rich themselves, these formations access charge from the underlying Wolfcamp Formation—itself now a prolific unconventional development target.

The organofacies A model predicts an extremely broad range of viscosities from highly viscous (ten or more centipoise) in low GOR oil at the expulsion onset to very low viscosities in the volatile oils in the Delaware Basin center. The organofacies B model predicts a much more limited range of viscosity due to the less dense associated stock tank liquid gravities generated from this organofacies.

These models show that viscosities of accessed petroleum and the resulting flow rates can range by four orders of magnitude from the expulsion kitchen edge to the dry gas window. Interpreted slow expulsion rates ensure that there is plenty of time for evolving petroleum fluids to migrate roughly in the order they are expelled; it is reasonable that more viscous petroleum is usually found in shallower, more permeable reservoirs. In deepwater development settings, this trade-off can counter the negative impact on well rate of a highly viscous, early-generated oil. Fluid viscosity should always be considered in context of all the factors in the deliverability equation.

## Key messages on in-reservoir petroleum fluid processes and properties

In addition to the charge history, processes including biodegradation, oil-to-gas cracking, mixing and fractionation can have significant impacts on fluid properties in present-day accumulations. Mixing of thermogenic charge with microbial gas indigenous to the formation can also raise gas-oil ratio and lower viscosity for improved production flow rates. In doing so, mixed microbial-thermogenic systems can add complexity and uncertainty to seismic attribute analysis and increase fluid compressibility beyond that expected for an expected low-GOR oil.

### Petroleum mixing

Many prolific petroleum basins have, in addition to their primary source rock, other secondary sources that can contribute to the petroleum system. In some cases, petroleum from these multiple sources can migrate into the same trap to form a mixture. These scenarios can be identified through petroleum geochemical analysis or simply by recognizing a geologically anomalous accumulation or a basin-scale trend in migration habit.

Often the first clue of multiple source rocks' contribution to a mixed petroleum fluid is recognized as a variation in bulk characteristics such as gravity or sulfur content. These variations can be attributed to differences in chemical composition; the resulting petroleum mixture will generally have properties proportional to the relative charge masses of the end-members. However, mixing can result in petroleum fluids with new properties due to a process known as *de-asphalting*. De-asphalting involves precipitation of asphaltenes from liquid petroleum due to mixing with light alkanes (Tissot and Welte, 1984). This can cause significant variation in petroleum density within an oil column or across a field. Organofacies A petroleum fluids have high asphaltenes, which are stabilized and distributed colloidally by large amounts of accompanying resins.

In addition to thermogenic petroleum products generated from maturation of source rocks, microbial gas—also referred to as biogenic gas—generated at low temperatures from detrital organics can contribute volumes of methane and very minor ethane to mixed petroleum accumulations. Milkov et al. (2007) and Kirkland and Rushton (2019) detailed examples of microbial-thermogenic mixing in Tertiary fields in the Gulf of Mexico. The results are shown in the idealized phase envelope in Fig. 7; microbial gas accumulations are usually in place prior to migration of oil into deepwater reservoirs. Oil that migrates into the gas accumulation dissolves first as a condensate, but, with sufficient addition of undersaturated oil, may eventually absorb the gas back into an undersaturated liquid phase. Mixed microbial-thermogenic fluids can have properties between lean gas-condensate and undersaturated oil.

### Biodegradation

Biodegradation is a common and natural process in relatively cool Cretaceous and Cenozoic deepwater reservoirs in which microbes present in subsurface pore water metabolize organic compounds in petroleum. The results of minor biodegradation may have limited effects on fluid properties, but extensive biodegradation can lead to extreme increases in density and viscosity of petroleum accumulations. This not only affects the production flow rate, but the degradation of desirable compounds also concentrates contaminants like sulfur, nickel, vanadium and acids.

Due to the production, transportation and refining challenges associated with heavily biodegraded oil, significant research has focused on the process definition and prediction of biodegradation. Both Wenger et al. (2002) and Larter et al. (2006) give a complete and clear overview of subsurface controls on biodegradation, and these studies build upon

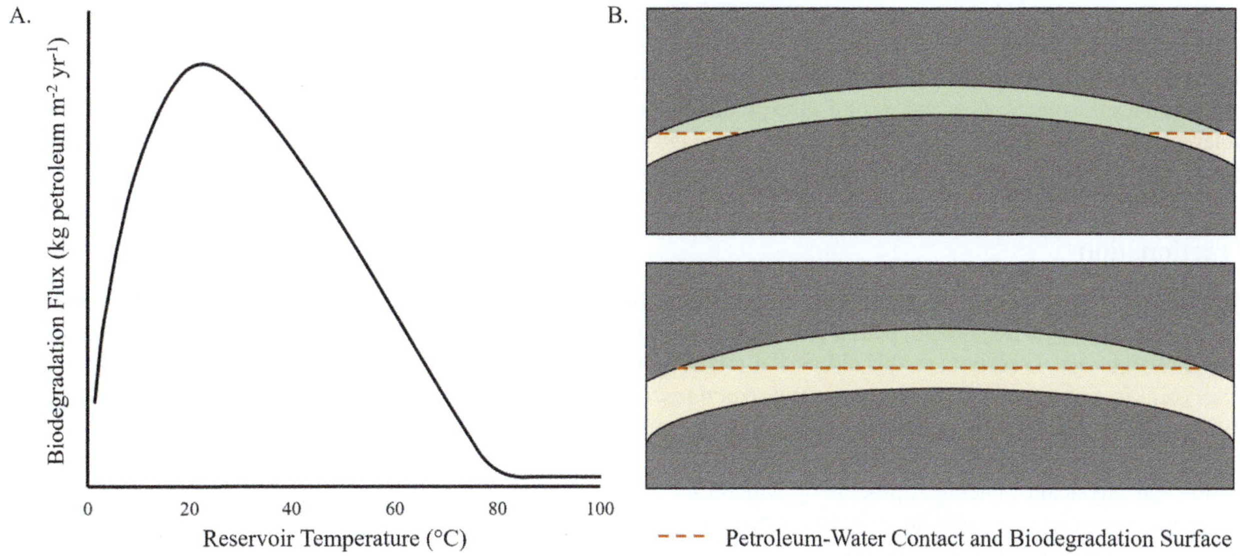


**FIG. 17** Biodegradation summary. Two of the key controls on biodegradation are summarized here. (A) is a summary of biodegradation flux as a function of temperature, which is generally the key control on predicting the risk of biodegraded petroleum. It should be noted that maximum microbial activity takes place in the range from 15°C to 40°C, so, even though microbes can exist and metabolize petroleum fluids up to 80°C, the risk of extensive biodegradation is substantially lower above 60°C and below 10°C. (B) Shows a simplified cross section of two petroleum accumulations. The oil–water contacts here are indicated by the red dashed lines, and the risk of significant biodegradation is substantially different between the two geometries. Looking back at A in this figure, the rate of biodegradation flux is in terms of area of the oil–water contact, so the upper reservoir will have significantly less biodegradation owing to lower contact area if all other conditions are held equal. *(Modified from Larter, S. et al., 2006. The controls on the composition of biodegraded oils in the deep subsurface: part II - geological controls on subsurface biodegradation fluxes and constraints on reservoir-fluid property prediction. AAPG Bull. 90(6), 921–938.)*

the observation by Pepper and Santiago (2001) that reservoir temperature is the primary controlling factor. Fig. 17 illustrates the typical microbial activity level from 0°C to 80°C at which the activity ceases and the reservoir is *pasteurized*. Biodegradation can occur at any temperature below 80°C unless the reservoir has cooled below that level since being pasteurized in the past, but this not typically a plausible thermal history in an actively depositing deepwater sedimentary section. In cases in which rapid burial induces thermal disequilibrium, cool temperature profiles allow for a very deep biodegradation window as well as biogenic gas generation window.

As illustrated in Fig. 17, microbial activity in biodegradation occurs at the oil–water contact, so some amount of water leg is required in the reservoir. Since this is an oxidative process, an oxygen supply is also essential. Because sedimentary basins are anoxic beginning at very shallow depths of burial, the source of molecular oxygen used by these anaerobes is recognized to be water (Larter et al., 2006). An alternate anaerobic mechanism is through reduction of dissolved sulfate dissolved into pore water from nearby evaporite sequences, but this is unlikely in most deepwater sequences (Pepper and Santiago, 2001). On the other hand, salinity significantly higher than that of seawater reduces bacterial activity dramatically (Larter et al., 2006).

Secondary factors on biodegradation rate and extent are volume and flow rate of the water leg, salinity, nutrient availability and petroleum charge rate. The composition of the initial, unaltered petroleum will also have a major impact on the properties of the biodegraded petroleum, and biodegradation is not mutually exclusive to other in-reservoir processes. Blumenstein-Weingartz (2011) detail a number of numeric models that attempt to integrate many of these factors to predict the rate of biodegradation.

Once biodegradation has been observed or modeled, there are multiple scales with which biodegradation extent can be described. These include the PM scale (Peters et al., 2004) and the Manco scale (Larter et al., 2021) with recent updates for extra heavy oil (Chang et al., 2021). The PM scale is the most widely applied, and it established a numeric ranking system for light (PM1–3), moderate (PM4–5), heavy (PM6–7), very heavy (PM8–9) and severe (PM10) biodegradation. These scales are all based on the relative abundance of compounds that are most easily, and first, metabolized. A generalized sequence of metabolization begins with normal alkanes and monocyclic aromatics before proceeding to branched and cycloalkanes and phenanthrenes in PM1–4. More complex compounds such as terpanes and steranes have been partially to completely metabolized in PM5–7, and, in extreme biodegradation cases PM8–10, even resistant compounds such as diasteranes and mono- and triaromatic steroids are affected. Each scale outlines an observed progression.

Despite ubiquity of bacteria and common occurrence thermal disequilibrium in sedimentary basins, nondegraded oils can persist to surprisingly shallow depths and low temperatures in deepwater plays. In the deepwater Gulf of Mexico, for example, controls other than biodegradation play an over-riding role in determining fluid properties. Low maturity organofacies A source rocks can result in heavy oils without biodegradation while organofacies B oils—with their characteristically light API gravity—maintain a higher gravity compared with organofacies A at any given level of biodegradation. This likely is a function of both organofacies and rapid, late charge timing.

## Fractionation

Several migration and containment processes can result in fractionation of petroleum into lighter and heavier molecular weight fractions. Fractionation of petroleum compounds during primary or secondary migration has already been addressed in the *charge: access* section, so this discussion focuses on fractionation of petroleum that has already accumulated in the reservoir. In some cases, the term *evaporative fractionation* has been proposed for in-reservoir fractionation (Thompson, 1988), but the process is likely related to saturation and phase behavior. While early conceptual models devised to explain fractionated petroleum proposed a process by which light hydrocarbons migrated out of a single undersaturated phase, there is no clear physical or chemical process by which this can happen in large volumes.

Since a petroleum accumulation exists at any point in time in an equilibrium phase state dependent on temperature, pressure, volume and chemical composition, any change to those boundary conditions can result in changes to the in-situ phase or phases of the petroleum. So, if the formation temperature or pressure of a reservoir with an accumulation decrease, it is possible that gas will begin to bubble out of a liquid phase. As discussed previously in this chapter, the gas will accumulate above the oil leg, and, in some containment cases, can migrate out of the reservoir towards a lower fluid potential geologic unit. The resulting two petroleum fluids include a migrated light mixture enriched in low-molecular weight hydrocarbons and a heavier mixture that remains in the reservoir that is enriched in aromatics and high-molecular weight saturates (Thompson, 2010).

There are many theoretical geologic scenarios which can facilitate fractionation; some of the easiest to visualize are structural changes to a trap after petroleum has accumulated within. Burial, uplift and erosion and salt tectonics are all processes that could affect temperature and pressure boundary conditions. However, it is much more common that changes in chemical composition of the accumulated petroleum lead to fractionated fluids. Thompson (2010) determined through geochemical analysis that fractionation is nearly universal—albeit to different degrees—based on a large global dataset. Interestingly, numerical simulations presented in Correa et al. (2012) replicated the interpreted process by Thompson, but they also indicated that late, high-maturity gas charge into an existing accumulation can facilitate fractionation. Persad et al. (1993) identified fractionation in the offshore Columbus basin southwest of Trinidad, and Gibson (1994) proposed that capillary entry pressure of fault gouge controls migration of gas or oil. This assumes that a separate free vapor phase must be present to migrate through fault gouge and leave the heavier liquid fraction behind. Considering these and other published examples, it seems likely that compositional change through the maturity window is the most common first step in fractionation of an accumulation in the reservoir.

## In-reservoir cracking (metagenesis)

In reservoirs with temperatures above 150°C, heavier, longer-chain components in petroleum fluid can be cracked to lighter compounds in a process known as *metagenesis* or *in-reservoir cracking*. Carbon–carbon, carbon-sulfur or other carbon-heteroatom covalent bonds can be broken in this process which, in extreme circumstances, can result in a liquid accumulation transforming entirely to a vapor phase with residual solids such as pyrobitumen and diamondoids (Tissot and Welte, 1984). However, this extent is uncommon in deepwater plays since rapid sedimentation often creates thermal disequilibrium and suppresses temperature with depth.

Analogous to the model described for kerogen-to-oil transformation (*catagenesis*) and oil-to-gas cracking of sorbed and pore fluid oil in the *charge: access* section, in-reservoir cracking of oil to gas in a reservoir is modeled as a chemical reaction with a rate constant $k$ described by the Arrhenius equation (Eq. 1). However, activation energies for in-reservoir cracking are higher with a narrower range than those for in-source cracking (Pepper and Dodd, 1995). The in-reservoir cracking process proceeds according to the strength of covalent bonds in the petroleum molecules with first breakage of weak carbon-sulfur bonds before proceeding to carbon–carbon and other carbon-heteroatom bonds (Horsfield and Rullkotter, 1994).

Just like gas generation within the source rock, the cracking process of petroleum in the reservoir is highly dependent on the composition of the hydrocarbon mixture. While the term "oil-to-gas cracking" is applicable in many circumstances, the metagenetic process is best viewed simply as degradation of longer-chain hydrocarbons into smaller molecules. Temperature and pressure conditions are key variables in controlling phase behavior, so the threshold for cracking oil to gas phase will vary between reservoirs and basins even if composition were to remain constant.

At the lowest thermal stress range of in-reservoir cracking (i.e., 150–170°C), the cumulative process may only be expressed in liquid phase accumulations as a slight decrease in average density and viscosity and a slight increase in gas-oil ratio (Kirkland and Brand, 2020). Above 170°C, models indicate that cracking of petroleum increases substantially, but evaluation of empirical datasets indicates that petroleum in reservoirs with temperatures as high as 185°C may not show evidence of extensive in-reservoir cracking. Empirical data and models agree that in-reservoir cracking is complete at a thermal stress of about 210°C forming dry gas with a small diamondoid condensate yield and solid pyrobitumen required to preserve hydrogen balance (Pepper and Dodd, 1995).

A relatively uncommon example of extensive in-reservoir cracking in present-day deepwater is in the Jurassic Norphlet Formation in the Mobile Bay area of the US Gulf of Mexico that can have formation temperatures over 215°C. Oil derived from the Smackover Formation source rock accumulated in the Norphlet reservoirs and underwent extensive in-reservoir cracking as well as thermochemical sulfate reduction due to the presence of sulfate minerals in the eolian sandstone (Mankiewicz et al., 2009). Oil from the Smackover likely had an unaltered cumulative gas-oil ratio of less than 1000 scf/bbl (Kirkland and Brand, 2020), but has, in some fields, been degraded to almost entirely methane, hydrogen sulfide and carbon dioxide.

## Conclusion

A complete petroleum systems assessment requires a holistic view of all the petroleum systems elements summarized in this chapter including their dependencies on each other. Only at that point can one assess the likelihood that a quantity of petroleum is delivered to a trap as well as the properties of that petroleum. An understanding of multiple fields is required for this assessment: petroleum geology, organic geochemistry of rock and petroleum, and the physics of heat flow, fluid flow and petroleum fluid phase behavior. One tool in the analytical workflow is numerical forward modeling of these physical processes; this is referred to as basin modeling or petroleum systems modeling.

These requirements are clearly technically demanding of any one individual, but with workflow experience, willingness to learn with multiple disciplines and exposure to data from real-world working petroleum systems, petroleum systems analysts can learn to deliver this integrated interpretation. Although the learning curve is steep and the work demanding, petroleum systems analysts can enjoy a role working closely with nearly every member of a project team. Another organizational model retains petroleum systems analysis—including basin modeling and organic geochemistry—as a distinct discipline delivered from a central, organization of specialists. This model requires excellent communication and close teamwork if the individual contributions are to be effectively integrated between the geochemist, basin modeler and the geologists in the project team—an integrated petroleum systems analysis is greater than the sum of its parts.

## Acknowledgments

The authors would like to thank the editors for the invitation to be a part of this book as well as their patience during the writing process. In addition, a hearty thanks is in order for their families; this sacrifice was not borne by the authors alone. Benjamin Kirkland would also like to extend thanks to Allegra Hosford Scheirer who helped ensure this chapter came together. For just a few of those who led, inspired and instructed him, he would like to thank Dave Moecher, Fausto Mosca, Kate Weissenberger, Lung-Chuan Kuo, Stephan Petmecky, Lei-Kuang Leu, John Pritchett, Art Leibold and his coauthor Andrew Pepper.

## References

Athy, L.F., 1930. Density, porosity, and compaction of sedimentary rocks. AAPG Bull. 14 (1), 1–24.

Beaumont, E., Foster, N., 1999. Chapter 2: classification of exploration traps. In: Treatise of Petroleum Geology/Handbook of Petroleum Geology: Exploring for Oil and Gas Traps. s.l.:s.n.

Blumenstein-Weingartz, I., 2011. Modelling of in-Reservoir Petroleum Biodegradation: Principles and Application. Library of the Aachen University of Technology, Aachen.

Chang, X.-C., et al., 2021. Investigation on the biodegradation levels of super heavy oils by parameter-striping method and refined Manco scale: a case study from the Chepaizi Uplift of Junggar Basin. Pet. Sci. 18 (1), 380–397.

Collaboration, T.K., 2011. Partial radiogenic heat model for Earth revealed by geoneutrino measurements. Nat. Geosci. 4, 647–651.

Correa, L.M., Penteado, H.L., Araujo, L.M., 2012. Compositional simulation of petroleum vertical migration and evaporative fractionation processes*. In: Nice, AAPG Hedberg Conference, Petroleum Systems: Modeling The Past, Planning The Future.

Demaison, G., 1980. Anoxic environments and oil source bed genesis. AAPG Bull. 64 (8), 1179–1209.
Dutta, N., Bachrach, R., Mukerji, T., 2021. Mechanisms of geopressue. In: Quantitative Analysis of Geopressure for Geoscientists and Engineers. Cambridge University Press, Cambridge, pp. 84–129.
England, W.A., Mackenzie, A.S., Mann, D.M., Quigley, T.M., 1987. The movement and entrapment of petroleum fluids in the subsurface. J. Geol. Soc. Lond. 144, 327–347.
Gibson, R.G., 1994. Fault-zone seals in siliciclastic strata of the Columbus Basin, offshore Trinidad. AAPG Bull. 78 (9), 1372–1385.
Godo, T., 2019. The Smackover–Norphlet petroleum system, deepwater Gulf of Mexico: oil fields, oil shows, and dry holes. Gulf Coast Assoc. Geol. Soc. J. 8, 104–152.
Hantschel, T., Kauerauf, A., 2009. Fundamentals of Basin and Petroleum Systems Modeling, 1 ed. Springer-Verlag Berlin Heidelberg, Berlin.
Hippler, S, 1993. Deformation microstructures and diagenesis in sandstone adjacent to an extensional fault: implications for the flow and entrapment of hydrocarbons. AAPG Bull. 77 (4), 625–637.
Horsfield, B., Rullkotter, J., 1994. Diagenesis,catagenesis, and metagenesis of organic matter. In: Magoon, L.S., Dow, W. (Eds.), AAPG Memoir 60 The Petroleum System - From Source to Trap. s.l.:s.n, pp. 189–199.
Hubbert, M.K., 1953. Entrapment of petroleum under hydrodynamic conditions. AAPG Bull. 37 (8), 1954–2026.
Katz, B., 2005. Controlling factors on source rock development—a review of productivity, preservation, and sedimentation rate. SEPM Spec. Publ. 82, 7–16.
Kirkland, B.T., Brand, J.H., 2020. A Characterization of the Smackover Source Rock in the Deepwater Gulf of Mexico and Derived Parameters for Petroleum Systems Modeling. Houston, GCAGS Transactions.
Kirkland, B., Rushton, S., 2019. Properties of Thermogenic and Microbial Hydrocarbon Mixtures at Gunflint Oil Field and Implications for Exploration in the Gulf of Mexico. Houston, GCAGS Transactions.
Larter, S., et al., 2006. The controls on the composition of biodegraded oils in the deep subsurface: part II - geological controls on subsurface biodegradation fluxes and constraints on reservoir-fluid property prediction. AAPG Bull. 90 (6), 921–938.
Larter, S., et al., 2021. A practical biodegradation scale for use in reservoir geochemical studies of biodegraded oils. Org. Geochem. 45, 66–76.
Luheshi, M.N., 1983. Estimation of formation temperature from borehole measurements. Geophys. J. R. Astron. Soc. 74, 747–776.
Mackenzie, A., Quigley, T., 1988. Principles of geochemical prospect appraisal. AAPG Bull. 72 (4), 399–415.
Magoon, L.B., 1988. Petroleum system - a classification scheme for research, resource assessment, and exploration. In: Magoon, L.B. (Ed.), Petroleum Systems of the United States. United States Government Printing Office, Washington, Washington, DC, pp. 2–15.
Magoon, L.S., Dow, W., 1994. The petroleum system. In: Magoon, L.S., Dow, W. (Eds.), AAPG Memoir 60 The Petroleum System - From Source to Trap. s.l.:s.n, pp. 3–24.
Mankiewicz, P.J., Pottorf, R.J., Kozar, M.G., Vrolijk, P., 2009. Gas geochemistry of the Mobile Bay Jurassic Norphlet formation: thermal controls and implications for reservoir connectivity. AAPG Bull. 93 (10), 1319–1346.
Milkov, A., et al., 2007. Compartmentalization and time-lapse geochemical reservoir surveillance of the Horn Mountain oil field, deep-water Gulf of Mexico. AAPG Bull. 91 (6), 847–876.
Milton, N.J., Bertram, G.T., 1992. Trap styles - an new classification based on sealing surfaces. AAPG Mem. 76 (7), 983–999.
Muniz, M., Bosence, D., 2015. Pre-salt microbialites from the Campos Basin (offshore Brazil): image log facies, facies model and cyclicity in lacustrine carbonates. Geol. Soc. Spec. Publ. 418.
Parson, B., McKenzie, D., 1978. Mantle convection and the thermal structure of the plates. J. Geophys. Res. 83 (89), 4485–4496.
Pepper, A., 2003. Petroleum Systems Analysis in the 21st Century – what should we be working on? In: Duppenbecker, S., Marzi, R. (Eds.), Discovery Series No. 7: Multidimensional Basin Modeling. AAPG/Datapages.
Pepper, A., 2016a. Back to the rocks: framework of depositional acmes in source rocks of the Gulf of Mexico Basin and North Caribbean margin. In: 2016 AAPG/Society of Exploration Geophysicists International Conference and Exhibition.
Pepper, A., 2016b. Back to the rocks: framework of depositional acmes in source rocks of the Gulf of Mexico Basin and North Caribbean margin. Gulf Coast Assoc. Geol. Soc. Trans. 66, 853–858.
Pepper, A., 2016c. Implications of early Gulf of Mexico tectonic history for distribution of Upper Jurassic to mid Cretaceous source rocks in deep water exploration areas of the US and Mexico. Gulf Coast Section SEPM Perkins-Rosen Conference. https://doi.org/10.5724/gcs.15.35.0036.
Pepper, A., Ahmad, N., Doebbert, A., 2017. Upper Paleocene–Lower Eocene 'Wilcox'-sourced petroleum systems of the Gulf of Mexico Basin: a re-evaluation and comparison with the coeval Patala Formation–sourced petroleum system in the Upper Indus Basin, Pakistan. Gulf Coast Assoc. Geol. Soc. Trans. 67, 481–488.
Pepper, A., Corvi, P., 1995a. Simple kinetic models of petroleum formation. Part I: oil and gas generation from kerogen. Mar. Pet. Geol. 12 (3), 291–319.
Pepper, A., Corvi, P., 1995b. Simple kinetic models of petroleum formation. Part III: modelling and open system. Mar. Pet. Geol. 12 (4), 417–452.
Pepper, A., Dodd, T., 1995. Simple kinetic models of petroleum formation. Part II: oil-gas cracking. Mar. Pet. Geol. 12 (3), 321–340.
Pepper, A., Roller, E., 2021. Ultimate expellable potentials of source rocks from selected super basins: what does "world class" look like? AAPG Bull. 105 (6), 1069–1097.
Pepper, A., Santiago, C., 2001. Impact of Biodegradation on Petroleum Exploration and Production, Observations and Outstanding Problems. Earth Systems Processes Conference, Geological Society of London, Geological Society of America, Edinburgh.
Pepper, A., Yu, Z., 1995. Influence of an inclined salt sheet om petroleum emplacement in the Pompano Field area, offshore Gulf of Mexico Basin. s.l. In: GCSSEPM Foundation 16th Annual Research Conference Salt, Sediment, and Hydrocarbons.

Pepper, A., Heister, L., Yarbrough, C., 2020a. Some observations on the Guiana basin petroleum system. In: Southeast Caribbean & Guiana Basins: Recent Studies and Advances in Understanding the Geology of Barbados, Guyana, Suriname, Trinidad and Tobago, and Venezuela, AAPG Virtual Research Symposium.

Pepper, A., Pradano, A., Heister, L., 2020b. Cenomanian-Turonian Organic Depositional Acmes 93 and 95 in the Gulf of Mexico: Paleobathymetry, Thickness, Lithofacies, Organofacies and Ultimate Expellable Potentials. s.l. GeoGulf Virtual Research Symposium.

Pepper, A., Doebbert, A., Laigle, J.M., Laigle, L., 2020c. Greater Permian Basin petroleum systems – what are we learning in the transition from conventional to unconventional? Houston Geol. Soc. Bull. 62 (7), 13.

Pepper, A., et al., 2021. Outcrop to Basin Center Thermal Evolution of the US Gulf of Mexico. AAPG-SEG IMAGE Conference, Denver.

Persad, K.M., Talukdar, S.C., Dow, W.G., 1993. Tectonic control in source rock maturation and oil migration in Trinidad and implications for petroleum exploration. In: Pindell, J.I., Perkins, R.F. (Eds.), Mesozoic and Early Cenozoic Development of the Gulf of Mexico and Caribbean Region–a Context for Hydrocarbon Exploration. Houston, GCSSEPM Foundation, pp. 237–249.

Peters, K., Cassa, M.R., 1994. Applied source-rock geochemistry. In: Magoon, L.S., Dow, W. (Eds.), AAPG Memoir 60 The Petroleum System - From Source to Trap. s.l.:s.n, pp. 93–120.

Peters, K., Waters, C., Moldowan, J.M., 2004. The Biomarker Guide, second ed. vol. 1 Cambridge University Press, Cambridge.

Pittman, E.D., 1992. Relationship of porosity and permeability to various parameters derived from mercury injection-capillary pressure curves for sandstone. AAPG Bull. 76 (2), 191–198.

PSA, 2016. https://www.youtube.com/watch?v=NbkHmNpUeEk. (Accessed 11 May 2020).

Roller, E., Pepper, A., 2018. Estimating the Ultimate Expellable Potential of Source Rocks: Defining "World Class" for Aquatic Organofacies with Examples from the Arabian, West Siberian, Bohai, and Williston Basins. AAPG Search and Discovery. http://www.searchanddiscovery.com/documents/2018/11055roller/ndx_roller.pdf. (Accessed 11 May 2020).

Schlanger, S., Jenkyns, H., 1976. Cretaceous oceanic anoxic events: causes and consequences. Geologie en Mijnbouw 55 (3–4), 179–184.

Tearpock, D.J., Bischke, R.E., Metzner, D.C., Brenneke, J., 2020. Applied Three-Dimensional Subsurface Geological Mapping: With Structural Methods, third ed. Pearson, s.l.

Terzaghi, K., 1925. Erdbaumechanik auf bodenphysikalischer Grundlage, First ed. F. Deuticke, s.l.

Thompson, K.F., 1988. Gas-condensate migration and oil fractionation in deltaic systems. Mar. Pet. Geol. 5 (3), 237–246.

Thompson, K.F., 2010. Aspects of petroleum basin evolution due to gas advection and evaporative fractionation. Org. Geochem. 41 (4), 370–385.

Tissot, B.P., Welte, D.H., 1984. Petroleum Formation and Occurence, second ed. Springer Verlag, Berlin Heidelberg.

Washburn, E.W., 1921. Note on a method of determining the distribution of pore sizes in a porous material. Proc. Natl. Acad. Sci. 7, 115–116.

Waulderhaug, O., 1996. Kinetic modeling of quartz cementation and porosity loss in deeply buried sandstone reservoirs. AAPG Bull. 80 (5), 731–745.

Wenger, L., Davis, C., Isaksen, G., 2002. Multiple controls on petroleum biodegradation and impact on oil quality. SPE Reserv. Eval. Eng. 5 (5), 375–383.

Woodcock, N., 2004. Life span and fate of basins. Geology 32 (8), 685–688.

# Chapter 3

# Crustal structure and tectonostratigraphy of rifted-passive margins with applications for hydrocarbon exploration

**Paul Mann**
*Department of Earth and Atmospheric Sciences, University of Houston, Houston, TX, United States*

## Introduction

The objectives of this chapter are to summarize the crustal structure and tectonostratigraphy of non-volcanic and volcanic, rifted-passive margins and to show how these controls influence deepwater sedimentation and the exploration of deepwater energy resources. Rifted-passive margins have an aggregate length of 105,000 km (~ 65,250 mi), which is greater than either spreading ridges (65,000 km or ~ 40,400 mi) or convergent plate boundaries (53,000 km or ~33,000 mi) (Bradley, 2008) (Fig. 1). Rifted-passive margins are largely known from geophysical mapping because they are largely covered by an average sedimentary thickness of 3–4 km (~ 1.9–2.5 mi) (Evenick, 2021). Because of their deep burial, rifted-passive margins are most directly known from deeply penetrating wells and seismic refraction and reflection subsurface mapping. Rifted-passive margins host the greatest accumulations of deepwater sedimentary deposits on Earth, and are therefore critical areas to study in order to gain a more complete understanding of deepwater systems discussed in following chapters.

The number of subsurface studies of rifted-passive margins worldwide has risen dramatically as improved geophysical imaging and academic and industry drilling of margins have increased. Rifts and rifted-passive margins provide the setting for about two-thirds of all giant oil and gas discoveries worldwide and for that reason are a major focus area for hydrocarbon exploration (Mann et al., 2003). Key exploration-related questions related to rifted-passive margins include: (1) Does a productive rifted-passive margin mean that its "look-alike" conjugate margin will also be productive? (2) Where should the deepwater limit of deepwater data acquisition and exploration be drawn at the deepwater, continent-ocean boundary or at the even deeper-water area underlain by oceanic crust? and (3) Should volcanic, rifted-passive margins be considered lower-priority exploration targets because magmatic activity has displaced or altered the normal deepwater sedimentary environments for organic-rich source rocks and highly porous and permeable reservoir rocks?

To accomplish these objectives, this chapter is divided into two parts. The first part summarizes the crustal structure and tectonostratigraphy of rifted passive margins with a focus on observations from marine geophysical surveys. Most of these observations are based on studies of the Gulf of Mexico and the circum-Atlantic margins which have been the main focus of the Conjugate Basins, Tectonics, and Hydrocarbons (CBTH) research group based at University of Houston. The compilation of observations was gathered from recent literature largely from the last 10 years and largely based on data from marine geophysical surveys that have been my central focus—rather than the area of experimental and analytical modeling. This compilation of observations from rifted-passive margins is by no means a complete summary—especially given the large number of recent papers on these topics from the world's rifted-passive margins. Results from the Gulf of Mexico (GOM) and Atlantic margins are emphasized. This review does not include the hydrocarbon-poor rift environments related to back-arc basins (Wallace et al., 2009) and strike-slip zones (Mann, 2007, 2012).

## Definition and significance of rifted-passive margin

"Rifts" are fault-bounded, elongate troughs beneath which the continental lithosphere has been extended, while "rifting" is the process by which the continental lithosphere stretches (Sengor, 2001) and is "necked" into a tapered, cross-sectional morphology (Sapin et al., 2021). Successful rifting leads to the formation of new ocean basins, whereas "failed" rifting

**Deepwater Sedimentary Systems.** https://doi.org/10.1016/B978-0-323-91918-0.00018-9

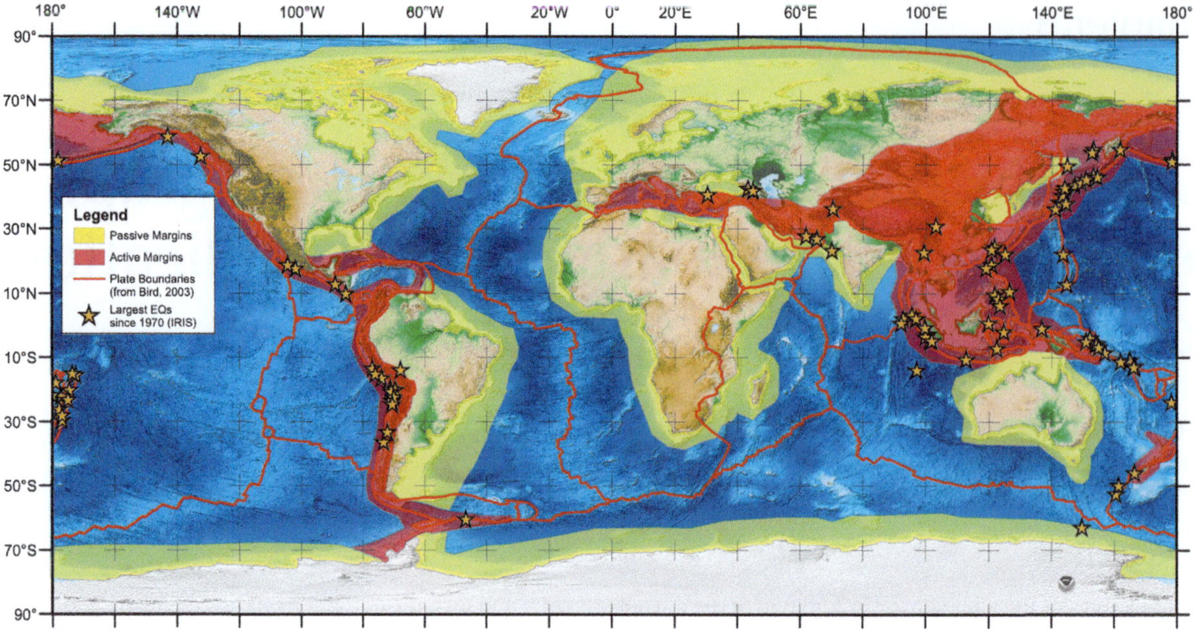


**FIG. 1** Map of rifted-passive (*green*) and active (*red*) plate margins of the world modified from passive margin compilation by Bradley (2008) and Mann (2015). The *stars* show largest earthquakes since 1970 that are confined to the active plate margins shown with *red* shading.

leads to an isolated zone of continental extension that is not associated with an adjacent area of oceanic crust with its thinner and well-defined crustal structure (Canales, 2014) (Fig. 1).

Small ocean basins in back-arc settings like the Black Sea and South China Sea can open to a width of a few hundred kilometers and then fail and then leave an isolated area of oceanic crust surrounded by thinned continental crust (Fig. 2). Continental rifting over a broad area generally ceases as oceanic spreading begins along a narrow and well-defined oceanic spreading ridge. Rifted-passive margins are one of the main stages in the Wilson cycle of ocean openings and closings through geologic time (Meschede, 2014). In general, rifted-passive margins are tectonically quiescent and therefore lack associated earthquake activity and active tectonic deformation that are characteristic of active plate boundaries, as shown in Fig. 1.

Rifted-passive margins exhibit a two-layer crustal structure with a lower level of rifted, continental crust that is overlain by a passive margin that is largely composed of undeformed, sedimentary rocks. For non-volcanic margins, the underlying, rifted margin consists of fault-controlled syn-rift, sedimentary wedges that form part of a subsiding weld that separates the rifted area of continental crust from newly formed and younger oceanic crust as seen in the Black Sea seismic reflection transect shown in Fig. 2. The overlying passive margin sedimentary rocks are deposited in the post-rift period and are controlled mainly by the thermal subsidence of the underlying rifted crust (McKenzie, 1978; Watts, 2012).

Faulting of the passive margin section can postdate the breakup unconformity and is commonly associated with gravitational sliding on the steepened slopes generated by crustal thinning along the underlying rifts. Regional areas of gravity-driven deformation characterized by downdip compression and updip normal faulting have been defined by Rowan et al. (2004) as "passive margin foldbelts" and are discussed in detail in Chapter 4 by Hasan and Mann (2022).

## Conceptual and numerical models for understanding the tectonic origin of rifted-passive margins

Brune (2016) and Chenin et al. (2021) have previously reviewed the conceptual development of models for rifted-passive margins that includes a comprehensive summary of analytical and numerical models for their evolution. A few key models for rifted-passive margin evolution are summarized in Table 1 that are important for understanding and placing the observations in this chapter into a broader, tectonic context. With the exception of geosynclines, these concepts have been widely adopted and modified over the years to explain the changing observations of rifted-passive margins.

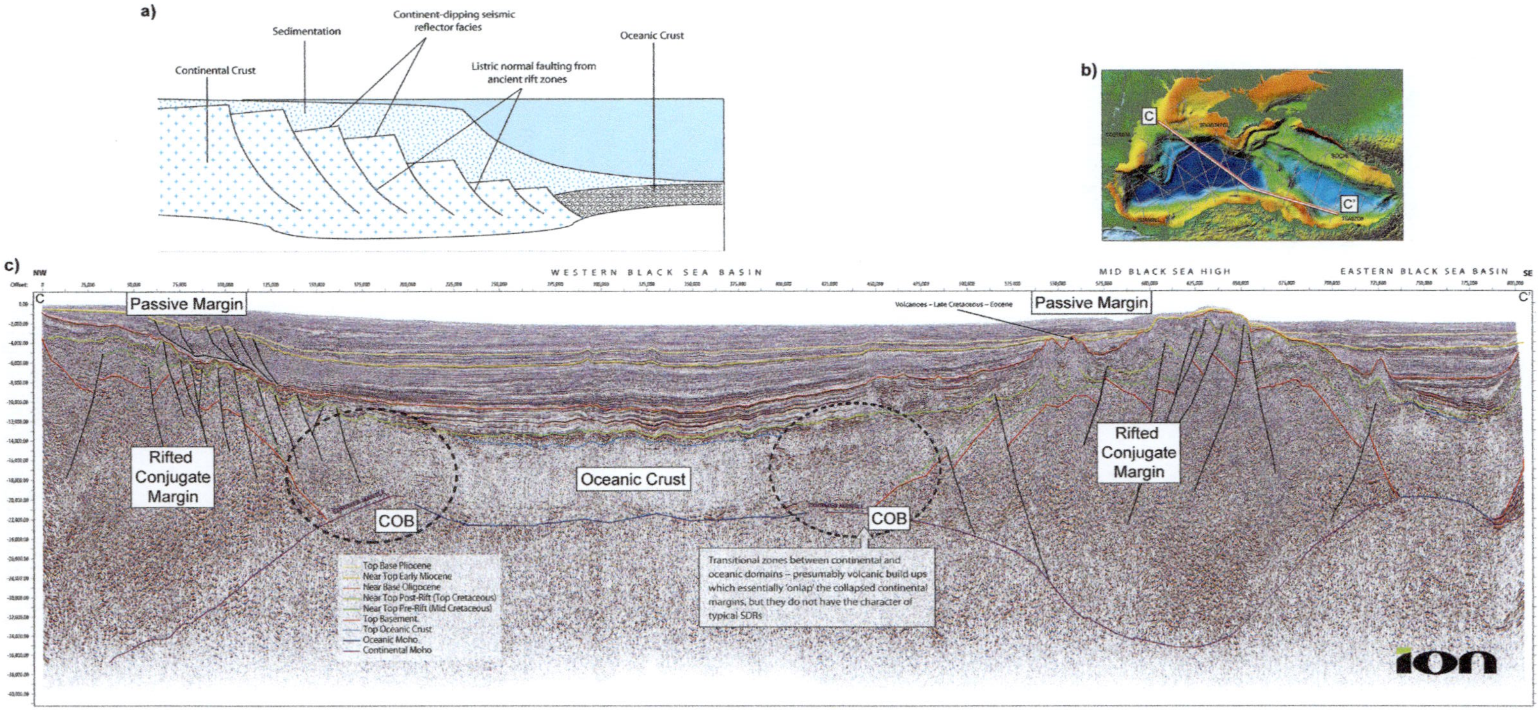


**FIG. 2** (A) Schematic section of a rifted continental margin overlain by a wedge of passive margin sedimentation extending seaward and tapering onto oceanic crust from https://en.wikipedia.org/wiki/Passive_margin. (B) Map of the Black Sea showing the location of the long seismic line shown in (C). (C) Deep-penetration seismic reflection image of conjugate, continental rifted margins of the Black Sea modified from Graham et al. (2013) that illustrate the main characteristics of non-volcanic passive margins and their relation to their underlying rifted margins. Complexities in the Black Sea conjugate passive margins include gravitational slumping in the western area and eruption of late Cretaceous-Eocene volcanoes in the eastern area.

**TABLE 1 Summary of six key models for understanding the tectonic origin of rifted-passive margins.**

| Source | Model | Observations |
|---|---|---|
| Dana (1873) | Proposed that geosynclines formed as deepwater areas downwarped between fixed continental masses | In deformed mountain belts, deepwater sedimentary units were observed within elongate, synclinal troughs that passively filled from the erosion of the adjacent, shallow-water areas. Passive margins were inferred to be the modern analogs of these ancient geosynclines |
| McKenzie (1978) | Proposed that pure shear thinning of the lithosphere occurred during the rifting of continental plates and was composed of two main stages that produced symmetrical conjugate margins. Stage 1 included rapid subsidence related to initial lithospheric thinning (the "rift phase") and Stage 2 included exponentially decreasing thermal subsidence (the "passive margin phase") | Two-phase subsidence history is a hallmark of rifted-passive margins. Early seismic reflection studies of continental margins show seaward-dipping normal fault systems along which crustal thinning occurred. This rift section is overlain by an undeformed passive margin related to thermal subsidence |
| Wernicke (1985) and Lemoine et al. (1987) | Simple shear thinning of the lithosphere that produces asymmetrical, conjugate margins. The proximal (footwall) side of the rift will have a greater proportion of crustal thinning (and thicker, syn-rift sedimentation) relative to mantle lithospheric thinning. The distal (hanging wall) side will exhibit the reverse effects | Low-angle normal fault systems (LANFs) are commonly observed in areas of continental extension including the Basin and Range province, USA, and the East African rift system. Asymmetry is also recognized from inverted Alpine rifted margins, where conjugate rift stratigraphy differs on the two margins |
| Dunbar and Sawyer (1989) | Finite element models were used to show that preexisting structure within the continental lithosphere formed by previous orogenic or rifting events control the structural architecture of rifted, continental margins. Where these basement trends parallel the rift direction, the rift zones are relatively narrow in width with smaller stretching factors because the lithosphere was weaker in this direction. Where these basement trends are orthogonal or oblique to the rift direction, the rift zone is wider and stretching factors are greater because the lithosphere was stronger in this direction. Narrow rift zones with parallel basement trends favor deeper mantle involvement and therefore are more prone to become volcanic margins. Wider rift zones with oblique basement trends favor less mantle involvement and therefore are more prone to become non-volcanic margins | The distribution of wide (fabric-orthogonal or -oblique) vs narrow (fabric-parallel) conjugate margins in the South Atlantic support predictions of this finite element model |
| Driscoll and Karner (1998) | Depth-dependent extension of the lithosphere was proposed to occur when the more ductile lower crust and mantle lithosphere accommodate more extension than the upper crust. Depth-dependent extension could occur within the framework of either the pure shear thinning model of McKenzie (1978) or the simple shear thinning model of Wernicke (1985) | Large discrepancies between measurements of upper crustal extension accommodated by brittle faulting and estimates of total crustal and lithospheric extension are observed at many rifted margins (Davis and Kusznir, 2004; Reston, 2009). Explanations of these discrepancies include: (1) limitations due to the resolution of seismic reflection data that prevent accurate quantitative extensional measurements and (2) complex, polyphase fault geometries have led to underestimates of total upper crustal extension. Deep-seismic profiling confirms the presence of mid-crustal detachment faults which separate more brittle horst and graben structures above the detachment from more ductile lower crustal boudinage-type structures below the detachment |
| Lavier and Manatschal (2006) | Numerical modeling showed a polyphase rift history that includes an initial and distributed lithospheric "stretching phase" of the more proximal continental area followed by a "thinning and mantle exhumation phase," where deformation becomes increasingly localized stretching in the more distal, oceanic area. The numerical model showed elements of both the pure shear model of McKenzie (1978) in its early stages and the more asymmetrical simple shear model of Wernicke (1985) in its later stages | Proximal sedimentation was observed by Chenin et al. (2021) to be older than distal settings on the Iberia-Newfoundland margin and on inverted Alpine margins and support thier predictions of this numerical modeling |

Dana (1873) **and the pre-plate tectonics geosyncline paradigm**. The 19th-century concept of static downwarps or mega-synclines on the Earth's surface coexisted with early plate tectonic concepts up through the 1970s. Progress was slow on understanding the tectonic controls of rifted-passive margins because geophysical imaging of the thickly-sedimented continental margins remained in its infancy during the 1960s and 1970s. As noted by McKenzie (1978), geosynclines correspond to passive margins that thermally subsided as a thermal response to lithospheric thinning.

McKenzie (1978) **model for rifting by pure shear of the lithosphere followed by thermal subsidence**. This elegantly simple model was consistent with the plate tectonics observations during the 1970s and 1980s. In this model, the first stage of the rifting process is the rise of a hot plume of "active mantle" material to the base of the lithosphere with a resulting dome formed at the Earth's surface that is deformed by horsts and grabens with accompanying volcanism. These brittle, crustal faults form as a consequence of pure shear and does not affect the lower continental lithosphere, which extends in a ductile manner (Meschede, 2015). The second stage of this model begins as the hot mantle plume eventually sinks with the cooler and denser lithosphere. This pulse of subsidence produces accommodation space for sedimentary rocks to be deposited in a broad and unfaulted saucer-shaped basin that initiates the passive margin phase. On a regional scale, this model predicts symmetrical, conjugate margins with a similar crustal structure.

Wernicke (1985) **model for simple shear thinning of the lithosphere followed by thermal subsidence**. This simple shear model invokes the existence and movement of a large-scale detachment fault that extends from the upper crust to the lower lithosphere and even into the asthenosphere. On the footwall side, deformation is intense in the shallow crust and broad in extent but does not extend at depth beneath the detachment. On the footwall side, the shallower area is undeformed, but the underlying area of the lithosphere is deformed by shearing along the detachment fault. Unlike the McKenzie (1978) model, the zone of the greatest fault displacement and thermal subsidence is not directly aligned with the rift—but is instead offset to one side. On a regional scale, the simple shear model would predict highly asymmetrical conjugate margins with a dissimilar crustal structure, heat flow, subsidence history and structural and basinal style. Unlike the McKenzie (1978) model with its symmetrical predictions, the Wernicke (1985) model provided an explanation for the various asymmetries that were observed on many conjugate margins (Davison, 1997).

Dunbar and Sawyer (1989) **used finite element modeling to propose that the preexisting crustal structure exerts variable resistance to rifting and that the crustal structure further controls the width of rift zones and their volcanic versus non-volcanic nature**. Where these basement trends parallel the rift direction, the rift zones are relatively narrow in width with smaller stretching factors because the lithosphere is weaker in this direction; where these basement trends are orthogonal or oblique to the rift direction, the rift zone is wider and stretching factors are greater because the lithosphere is stronger in this direction. Narrow rift zones with parallel basement trends favor deeper mantle involvement and therefore would be more prone to develop as volcanic margins. Wider rift zones with oblique basement trends favor less mantle involvement and therefore would be more prone to develop as non-volcanic margins (Reuber and Mann, 2019) (Fig. 3).

Driscoll and Karner (1998) **proposed depth-dependent stretching as a way for the more ductile lower crust and mantle lithosphere to accommodate more extension than the upper crust.** Depth-dependent stretching within the continental lithosphere can explain the subsidence of overlying sag basins without significant brittle faults affecting their sedimentary fill. Depth-dependent stretching could operate within the framework of either the pure shear model of McKenzie (1978) or the simple shear model of Wernicke (1985).

Lavier and Manatschal (2006) used finite element models to explain large areas of hyperextended continental crust that is asymmetrically distributed between the two rifted, conjugate margins. This model predicted a progressive shift in the locus of rifting and therefore combined elements of the pure shear and simple shear models (Chenin et al., 2021).

## Definition and significance of continent–ocean boundaries and continent–ocean transitions

The fundamental crustal weld or boundary for rifted continental margins is also called a "continent-ocean boundary" (COB) that separates continental lithosphere (Meschede, 2015; Eagles et al., 2015) from oceanic lithosphere (Canales, 2014). The term "continent-ocean transition" (COT) is widely used given the width and complexity of this transition between continental and oceanic lithosphere, as shown in the seismic transect of hte Black Sea in Fig. 2. Eagles et al. (2015) noted that criteria for picking a COT vary widely depending on the author and that the resulting variations in COB interpretations introduces a large degree of uncertainty into plate reconstructions.

Stretched continental crust or "transitional crust" is generally characterized by seaward-dipping normal faults, a prominent "breakup" unconformity that records the end of the syn-rift phase, and an overlying, post-rift sag basin formed by slower and gradually dissipating thermal subsidence that follows rifting.

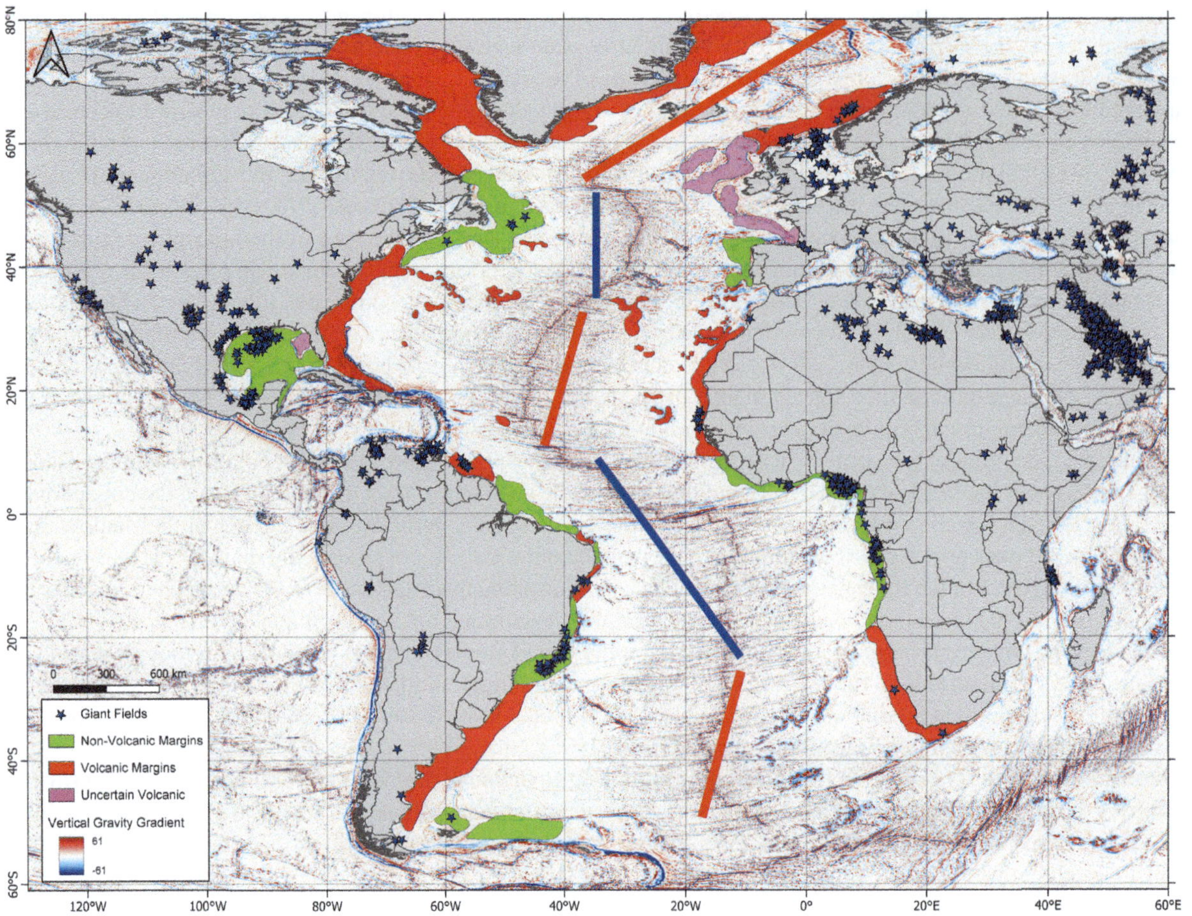


**FIG. 3** Compilation of volcanic (*red polygons*) and non-volcanic margins (*green polygons*) of the Mesozoic rifted margins of the Gulf of Mexico and circum-Atlantic margins. Rifted-passive margins with uncertain volcanic or non-volcanic origin are shown as *pink polygons*. Basemap is vertical gradient gravity (VGG) of Sandwell et al. (2014). The *red bars* indicate areas where the prerift orogenic grain was parallel to the Mesozoic rift direction, and the *blue bars* indicate areas where the prerift orogenic grain was orthogonal or at a high angle to the rift direction. The *blue stars* are giant oil and fields compiled from Dolson et al. (2021).

## Definition and significance of conjugate margins

A conjugate margin pair includes two rifted-passive margins now located on different tectonic plates and separated by an oceanic spreading ridge (Fig. 2). The conjugate margins were once conjoined along a shared rift system that preceded the breakup and formation of oceanic crust (Peace and Welford, 2020). As shown on the map of circum-Atlantic rifted margins in Fig. 3, conjugate margins that are now widely separated by the intervening oceanic crust of the Atlantic Ocean and the Gulf of Mexico can be rejoined by following the linear traces of fracture zones that connect points on one conjugate margin to the preoceanic spreading points on the other conjugate margin. The application of satellite gravity data from Sandwell et al. (2014) has been a critical technological innovation for more precise mapping of fracture zones that allowed for more precise reconstructions of conjugate margins.

Peace and Welford (2020) have noted that the fracture zone method for reconstruction may be complicated in some cases by the presence of post-breakup magmatic rocks, the preservation of relatively undeformed continental plateaus and broad zones of multiple failed rifts.

In recent years, marine surveys have systematically studied both widely separated margins with similar methods such as refraction and seismic reflection in order to improve the comparisons and stratigraphic correlations between the two conjugate margins. A selection of recent conjugate margin studies includes the following studies: conjugate basins of the Guinea Plateau of West Africa and the Demerara Plateau of northeastern South America by Olyphant et al. (2017), Museur et al. (2021) and Loncke et al. (2022); conjugate basins of the North Atlantic by Sandoval et al. (2019); conjugate basins of the Greenland-Norwegian Sea by Hopper et al. (2003) and Gernigon et al. (2020); conjugate basins of the South

China Sea by Song et al. (2019) and Bai et al. (2020); conjugate basins of the South Atlantic (Blaich et al., 2009; Beglinger et al., 2012; Reuber and Mann, 2019; Reuber et al., 2019; Wen et al., 2019); conjugate basins of Nova Scotia and Morocco (Louden et al., 2012; Song et al., 2020; Tari et al., 2012); conjugate basins of Iberia and Newfoundland (Causer et al., 2020); the conjugate basins of the Gulf of Aden (Leroy et al., 2010, 2012; Nonn et al., 2019; Mortimer et al., 2020); conjugate basins of the northern and southern Gulf of Mexico (Izquierdo-Llavall et al., 2022; Steier and Mann, 2019); conjugate basins of the Red Sea (Augustin et al., 2021; Bosworth et al., 2015) and conjugate basins surrounding Antarctica (Veevers, 2012; Ball et al., 2013; Nguyen et al., 2016; Gillard et al., 2015).

On a regional scale, these studies reconstructed either volcanic or non-volcanic conjugate margins as shown on the circum-Atlantic map in Fig. 3. In the Gulf of Mexico, the conjugate margins of the northeastern (US) portion can be restored quantitatively with the southern (Mexican) portion on the Yucatan Peninsula along a narrow, syn-rift zone in the central part of the basin, where thicker evaporite intervals are present (Liu, 2021; Rowan, 2022). This narrow, seam-like zone of thicker evaporites marking the site of the nascent or future oceanic spreading ridge is also observed in the Cenozoic Red Sea rift (Ligi et al., 2018; Mitchell et al., 2020; Augustin et al., 2021) and in the more obliquely-opening Gulf of California (Miller et al., 2021; Oskin and Stock, 2003; Umhoefer et al., 2018). In all of these different settings with variable evaporites, this central alignment of thicker evaporites tracks the initial influxes of seawater into a single, sub-sealevel, sediment-filled, proto-rift (Liu, 2021; Rowan, 2022). In the Gulf of Mexico, full closure of the conjugate margins is reached along this arcuate seam of thicker evaporites with realignment of gravity and magnetic highs along with similar bathymetric relief and styles of salt tectonics (Nguyen et al., 2016; Steier and Mann, 2019; Liu, 2021).

In the Gulf of Mexico, the late Jurassic widening of this central rift along a younger oceanic spreading center separates the evaporites on the two areas of rifted continental crust in the northern (Nguyen and Mann, 2016; Liu, 2021) and southern parts of the basin (Hasan and Mann, 2021). The overlying passive margin sections are not necessarily similar as by this time these margins were separated by a wider ocean and therefore were subject to variable paleogeography such as oceanic currents and proximity to adjacent land areas and sediment sources.

## Types and crustal structure of rifted passive margins

### Non-volcanic and volcanic conjugate margins defined and compared

Non-volcanic rifted-passive margins form with relatively small amounts of volcanic rocks and are characterized by "low-angle normal faults" (or LANFs—Morley, 2009) that bound half-graben structures and generally dip in the seaward direction (Fig. 4A). Franke (2013) pointed out that the term "non-volcanic rifted margin" is misleading because all margins generally exhibit some evidence for magmatism and intrusives, and for that reason, his preferred term is "magma-poor" rifted margin.

Areas of non-volcanic margins in around the Atlantic Ocean include the conjugate of Maritime Canada with the Iberian Peninsula in the North-Central Atlantic (Cawood et al., 2021) and the Equatorial Atlantic areas of northern Brazil and west-central Africa (Fig. 3). Based on the work by the CBTH research group (Steier and Mann, 2019; Liu, 2021), the rifted margins of the Gulf of Mexico are largely non-volcanic.

Preexisting, orogenic fabrics are either orthogonal to the trends of the rifted-passive margin as observed in the north-central Atlantic (Nance et al., 2015; Chenin et al., 2021) and Gulf of Mexico (Liu, 2021) or are orthogonal to oblique in the Equatorial Atlantic (Fig. 3). Also, apparent on the map in Fig. 3 is the clustering of giant oil and gas fields in areas of non-volcanic margins especially in the Equatorial Atlantic (Mann et al., 2003; Dolson et al., 2021). Other well-known areas of non-volcanic margins in areas outside of the Atlantic Ocean and Gulf of Mexico include northwest Australia (Bilal and McClay, 2022), the South China Sea (Zhou et al., 2018; Cameselle et al., 2020; Deng et al., 2020) and the Gulf of Corinth in Greece (Taylor et al., 2011; Nixon et al., 2016; Bell et al., 2018).

### Orthogonally-rifted, non-volcanic and hyperextended conjugate margins

The crustal structure of orthogonally-rifted, non-volcanic margins are traditionally divided into the proximal margin that includes the area of listric normal faults bounding syn-rift sedimentary wedges and the distal margin of extremely thinned continental crust that is potentially separated from the oceanic crust by a domain of exhumed, subcontinental mantle (Franke, 2013) (Figs. 4 and 5). The term "orthogonally-rifted" is used to describe rifting at a high angle to the trend of the margin and "obliquely rifted" for rifting at a low angle to the trend of the margin.

"Hyperextension" is defined as stretching of the crust such that the lower crust and upper crust become coupled and embrittled to the point that major faults can penetrate to the mantle and allow partial serpentinization of the uppermost

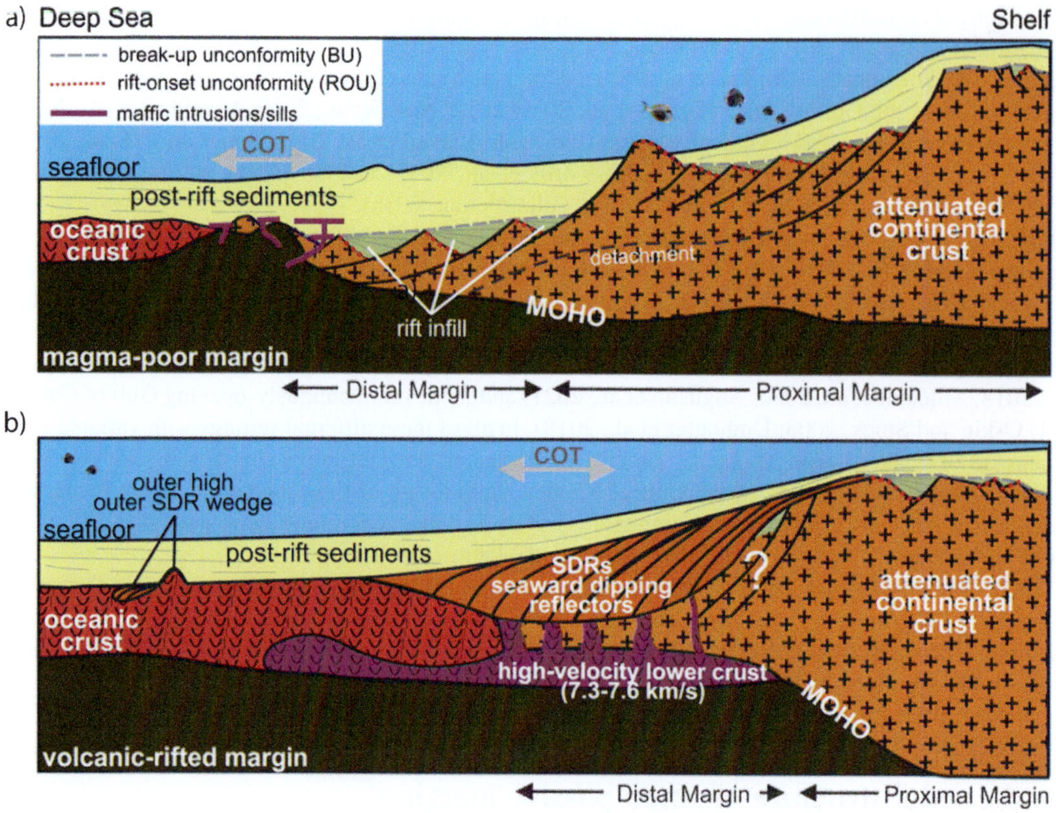


**FIG. 4** Comparisons from Franke (2013) of end-member examples of non-volcanic or magma-poor margins in A and volcanic rifted margins in B. (A) Common characteristics of a non-volcanic margin include: (1) a broad zone of rifted and necked continental crust of the proximal margin deformed by listric normal faults that generally dip seaward and sole out on a common, subhorizontal detachment surface. On the distal margin adjacent to the continent-ocean transition (COT), these faults may cut the entire crust and expose the upper mantle in a zone adjacent to oceanic crust. The rift-onset unconformity (ROU) forms the base of the half-grabens of the distal margin. The break-up conformity (BU) forms the top of the rifted section and marks the initiation of the passive margin. (B) Volcanic rifted margins show a more narrow zone of rifted and necked continental crust in the proximal margin that are partially buried by thick wedges of syn-rift volcanic flows in the distal margin that are expressed on seismic reflection data as seaward-dipping reflectors or "SDRs." These SDRs are underlain by a wide zone of high-velocity (Vp > 7.3 km/s) lower crust that is seaward of the continental necked zone. Oceanic crust adjacent to volcanic margins is commonly thicker that oceanic crust adjacent to non-volcanic margins and post-rift, passive margin sedimentation is generally thinner because SDRs fill much of the accommodation space created during the rifting phase.

mantle (Doré and Lundin, 2015; Gillard et al., 2019). Highly tapered, hyperextended margins are thought to result from slow extension rates, while more stubby-shaped volcanic margins are associated with faster extension rates (Doré and Lundin, 2015; Sapin et al., 2021).

The cross sections by Peron-Pinvidic et al. (2017) of the Congo margin of West Africa were proposed as the "archetypal, lower-plate rifted margin" based on the earlier Wernicke (1985) simple shear model for rifted-passive margins (Table 1). Deepwater exploration in Brazil, the Gulf of Mexico, Guyana-Suriname and West Africa are presently exploring the most distal, deepwater area that is underlain by the "external rift" and an associated zone of exhumed mantle with an overlying sag basin sealed by salt (Romito and Mann, 2022; Zalan, 2017).

Examples of semicontinuous belts of exhumed mantle localized at distal marginal rifts or outer marginal troughs along non-volcanic, rifted margins (Pindell et al., 2014) have been proposed based on seismic reflection data and documented by deep-sea drilling on the Newfoundland-Iberia conjugate margin (Sibuet et al., 2007; Péron-Pinvidic et al., 2013; Davy et al., 2016), the Brazil-Angola conjugate margin (Péron-Pinvidic et al., 2013), the Brazil-Gabon conjugate margin (Romito and Mann, 2022), the Norway-Greenland conjugate margin (Péron-Pinvidic et al., 2013), the margins of the South China Sea (Cullen et al., 2010; Savva et al., 2013; Li et al., 2016), the Canada Basin (Li et al., 2016); the Laptev Sea (Mazur et al., 2015; Lutz et al., 2018), the Norwegian margin (Osmundsen et al., 2016; Ferrand, 2020), the Red Sea (Ligi et al., 2018), the Somali margin (Mortimer et al., 2020), the conjugate margins of Australia and Antarctica (Gillard et al., 2015), the Porcupine Basin in Ireland (Whiting et al., 2021) and the conjugate margins of the Gulf of Mexico (Curry et al., 2018; Pindell et al., 2014).

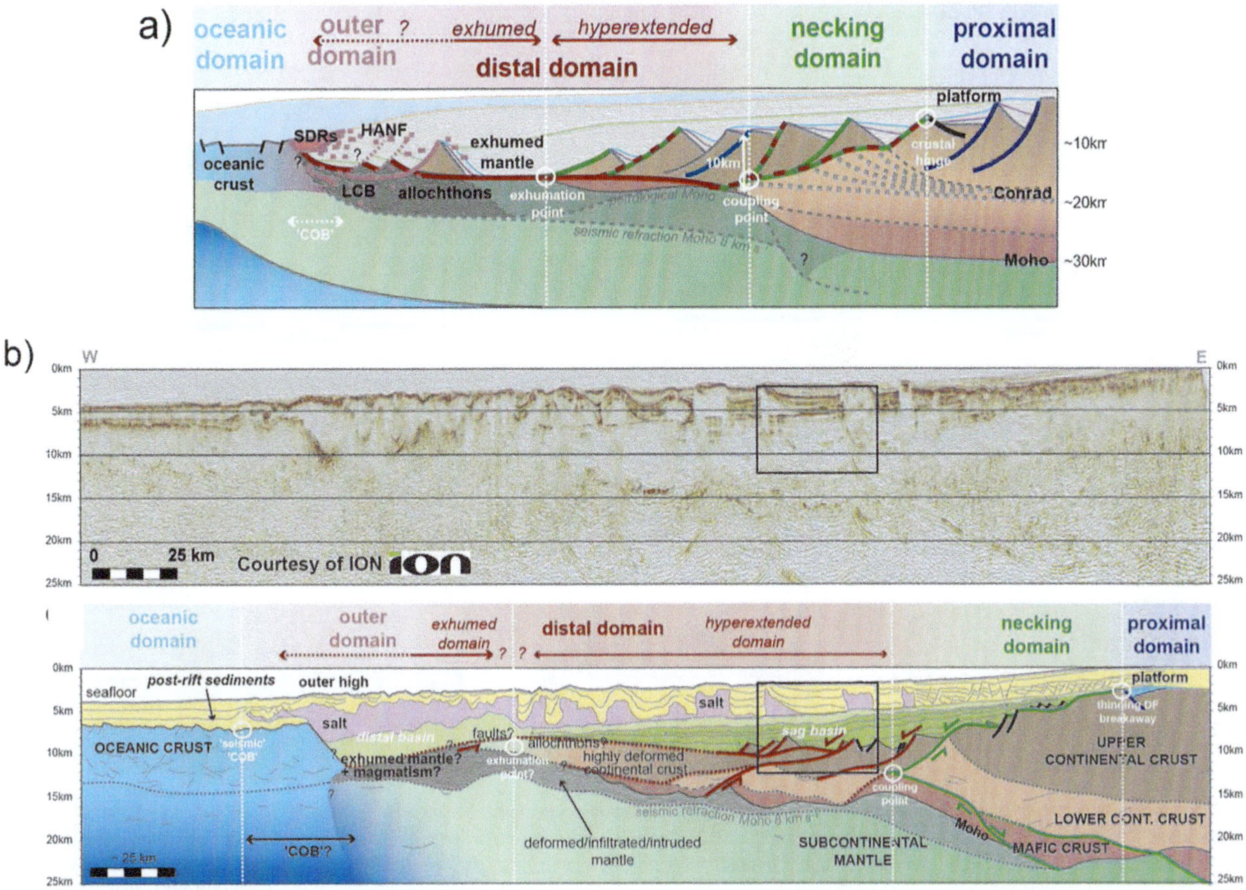


**FIG. 5** (A) Terminology proposed by Peron-Pinvidic et al. (2017) (modified from Unternehr et al. (2010), Masini et al. (2012) and Peron-Pinvidic et al. (2013)) for an orthogonally extended, non-volcanic rifted margin with exhumed mantle. (B) ION deep-penetration seismic line from Peron-Pinvidic et al. (2017) across the Congo margin of West Africa that is proposed by these authors as the "archetypal, lower-plate rifted margin" as predicted by the Wernicke (1985) simple shear model.

## Four major tectonostratigraphic megasequences on non-volcanic rifted margins and their associated unconformities

Rift sedimentary sequences on non-volcanic, rifted passive margins are divided into four main units: (1) the **prerift megasequence** predates the onset of rifting and generally displays tabular bedding units that are unaffected by faulting, or alternatively, highly folded and thrust-faulted orogenic belts as seen on deeply penetrating seismic reflection data from the Brazil-West African conjugate margins (Reuber and Mann, 2019); the top of the prerift megasequence and the base of the syn-rift megasequence is defined by the **rift onset unconformity** (Fig. 6); (2) the **syn-rift megasequence** accompanies the rift stage or stages and is characterized by generally seaward-dipping, normal faults bounding half-grabens and grabens containing syn-rift infill with wedge-like stratal geometries; the **breakup unconformity** truncates the wedge-shaped, syn-rift sedimentary units and separates them from the overlying sag basin and/or post-rift megasequence; (3) the **sag megasequence** is locally present and represents a transition from the rift stage to the passive margin stage and is characterized by saucer-shaped basins that overlie the rift section and onlap remnant highs of the underlying rift megasequence but generally remains unfaulted as seen for the presalt sag basins of Brazil (Karner and Gambôa, 2007) and (4) the **post-rift megasequence** forms the base of the passive margin or breakup phase and is characterized by tabular units (Fig. 6).

The origin of the breakup unconformity has been attributed to several different mechanisms: (1) a flexural rebound at the time of rifting; (2) convective upwelling of the asthenosphere and lithosphere with resulting heating and thermal expansion of the crust and (3) depth-dependent stretching of the lower crust that is not expressed by upper crustal normal faults (Franke, 2013; Chenin et al., 2021).

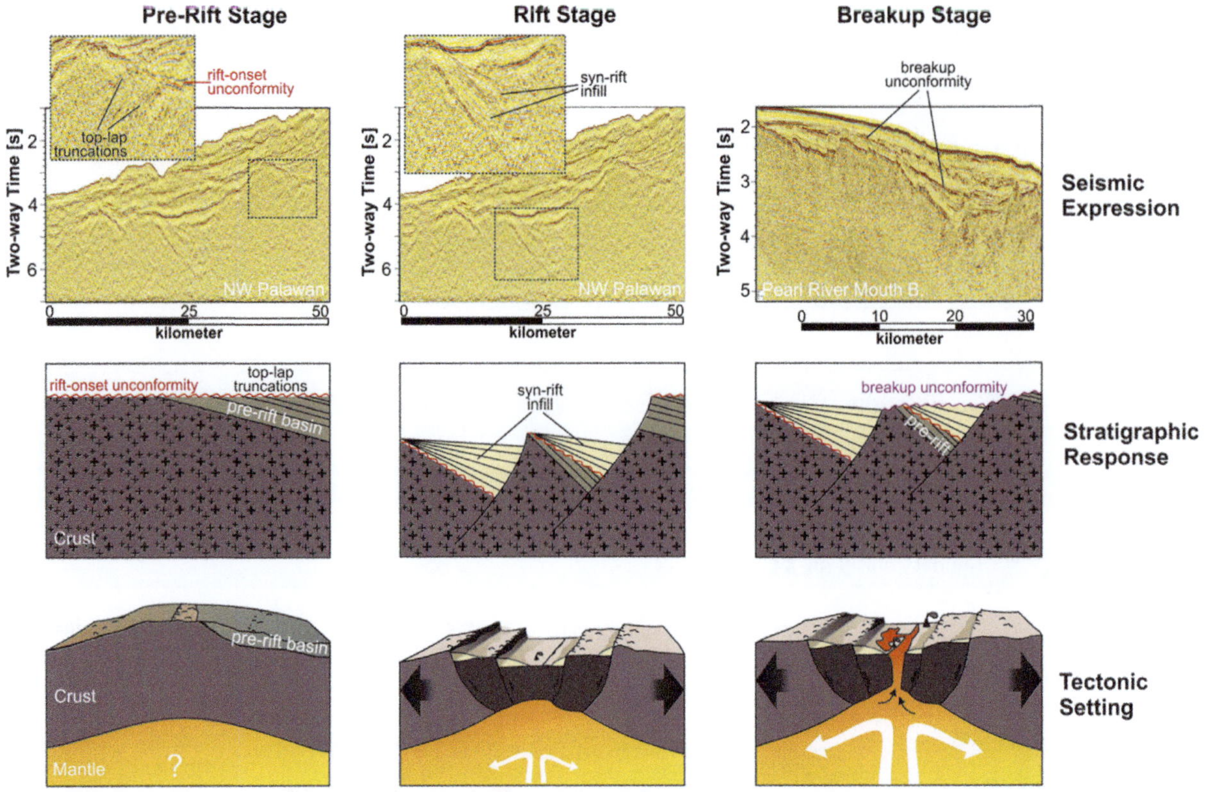


**FIG. 6** Schematic representation of the seismic expression, stratigraphic response and expression on seismic reflection data of the prerift, rift and breakup stages in the formation of rifted-passive margins from Franke (2013). Erosion resulting from pre or syn-rift doming will produce a rift-onset unconformity (ROU) which on seismic data can be seen as an angular unconformity with top-lap truncations of reflectors. Syn-rift infill shows wedge-shaped stratal geometries with the thinner end of the wedge on the dip slope of the hangingwall fault block and the thicker end of the wedge controlled by a normal fault scarp. Individual stratigraphic wedges commonly show discrete pulses of extension. At continental breakup, flexural rebound results in the uplift of the rift shoulders that is recorded by the breakup unconformity (BU) that separates the underlying wedge-shaped syn-rift units from the overlying, draped sedimentary rocks of the basal passive margin.

## Orthogonally-rifted, volcanic, conjugate margins

Volcanic rifted-passive margins form the majority of rifted-passive margins worldwide (Fig. 3) and are characterized by: (1) large amounts of volcanic and volcaniclastic sedimentary rocks that include 5–15-km-thick (~3–9 mi) wedges of volcanic flows manifested on seismic reflection data as "seaward-dipping reflectors," or SDRs (Reuber et al., 2019); (2) thinning of the continental crust is abrupt so the COT occurs over a short distance and in the vicinity of the SDRs and (3) the area seaward of the rifted continental crust is underlain by a high-velocity lower crust] (Vp > 7.3 km/s) (Menzies et al., 2002; Geoffroy, 2005; Franke, 2013; Tugend et al., 2018) (Fig. 4). The origin of the high-velocity crust is attributed to massive intrusions, mafic underplating and partially serpentinized mantle (Gillard et al., 2019).

Examples of volcanic margins in around the Atlantic Ocean include Greenland (Hopper et al., 2003), Angola (Empinotti et al., 2015; Fernandez et al., 2020), Namibia (Cartwright et al., 2012; McDermott et al., 2015; Paton et al., 2017; Norcliffe et al., 2018); the East Coast of North America (Bécel et al., 2020); the conjugate margins of the southern South Atlantic Ocean in South America and Africa (Reuber et al., 2019) and the Demerara Plateau-Guinea Plateau conjugate of the Central Atlantic Ocean (Museur et al., 2021; Loncke et al., 2022).

## Compilation of refraction transects from Atlantic non-volcanic and volcanic margins to illustrate margin asymmetry

Fig. 7A shows a compilation by Dias et al. (2015) of 11 seismic refraction profiles from the North and South Atlantic to illustrate the widespread occurrence of margin asymmetry as seen by width variations between conjugate margin pairs (Davison, 1997). The bracket labeled N represents their interpretation of the necked zone of continental lithosphere, while the bracket labeled H represents their interpretation of the hyperextended continental crust. The vertical lines labeled COB

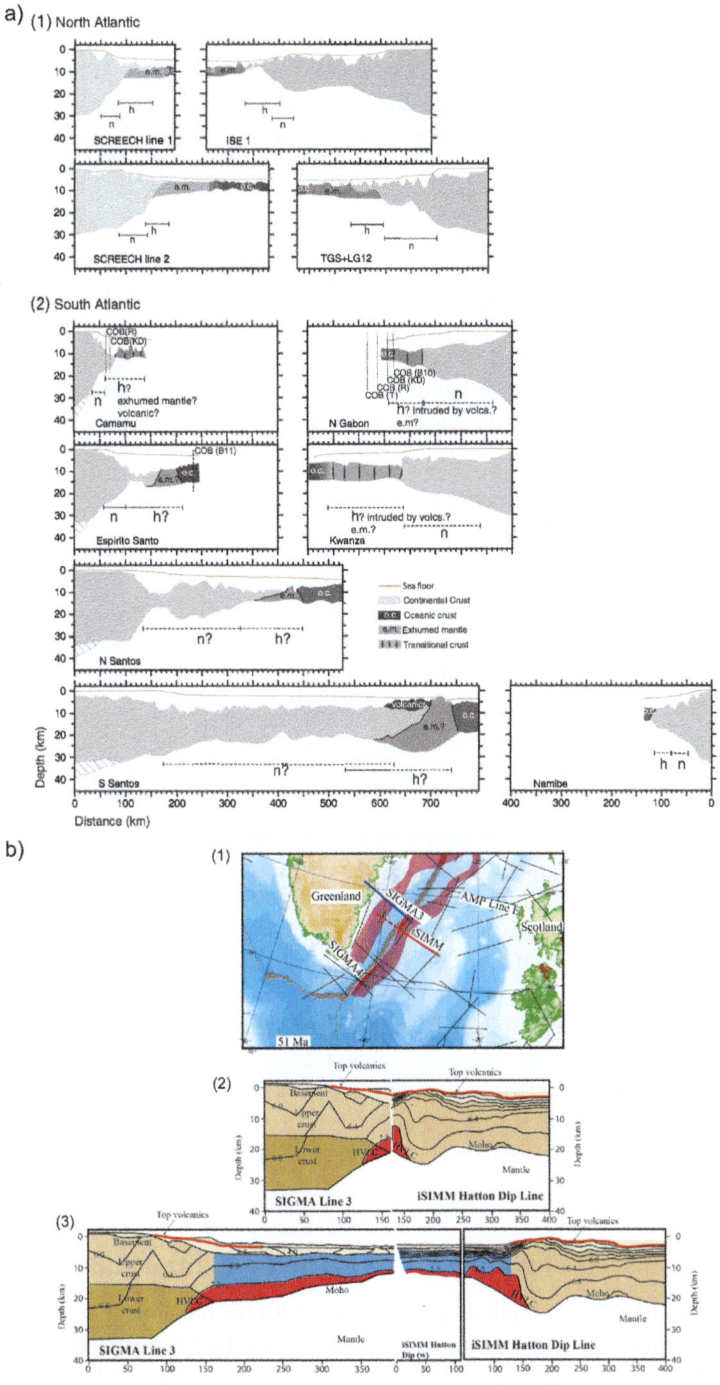

**FIG. 7** (A) Schematic cross sections based on seismic refraction data compiled from many sources by Dias et al. (2015) illustrating the rifted-passive margin crustal structure of some well-studied non-volcanic, conjugate margin pairs: (1) Newfoundland-Iberia conjugate margins of the North Atlantic. From top to bottom: Conjugate pair of Flemish Cap-Galicia Banks and conjugate pair of Flemish Pass-Southern Iberia Abyssal Plain. (2) South Atlantic Ocean. From top to bottom: Conjugate pair of Camamu-Gabon basins, conjugate pair of Espírito Santo-Kwanza basins, Northern Santos Basin and conjugate pair of Southern Santos-Namibe basins. Vertical *dotted lines* depict location of the continent-ocean boundary (COB). Horizontal *solid black lines* show the necking (n) and the hyperextended domains (h) interpreted at the western Iberia-Newfoundland margins and at the Espírito Santo margin. For the Camamu-Gabon, the COB locations by several different previous authors are indicated. (B) Deep structure of the Greenland margin and the Rockall Trough volcanic conjugate pair compiled by Biari et al. (2021) from several sources. (1) Reconstruction of the Greenland-Rockall Trough conjugate pair in the North Atlantic Ocean during the Early Eocene (51 Ma). Red polygons show the extent of the volcanic conjugate margins. Refraction lines shown below are SIGMA3 from the Greenland margin and iSIMM from the Rockall Trough. (2) Reconstruction of the margin at the initiation of seafloor spreading. (3) Present-day cross section showing full width of oceanic crust with its initial thickness at the COB of 16 km (~10 mi). Oceanic spreading is highly asymmetrical with the wider zone of oceanic crust on the Greenland side.

with a person's initials represent the COB interpretations by previous authors that demonstrate the point of Eagles et al. (2015) that there are few common criteria for picking a COB and that this leads to many different COB interpretations for the same area. EM stands for the exhumed mantle and OC stands for oceanic crust. What is striking is the highly-tapered areas of thinned continental crust in these orthogonal rifting settings and the large degree of asymmetry for the conjugate pairs as expressed by the variability in the widths of the zones of thinned, transitional continental crust and the presence or absence of a zone of exhumed mantle.

Fig. 7B is a taken from another compilation of circum-Atlantic data by Biari et al. (2021) showing a crustal transect across the volcanic-rifted margin between Greenland and the Rockall trough of the offshore area of the United Kingdom (red is high-velocity lower crust—HVLC—and the red line marks the top of SDR volcanic units). The HVLC is commonly inferred to be a magma chamber that underlies the areas of SDRs on both conjugates. The center diagram is a reconstruction of both margins at the time of initial seafloor spreading. These volcanic margins are more symmetrical in their prerift structure with a major difference being related to asymmetrical oceanic spreading that leads to the wider oceanic area on the Greenland side. A similar pattern of pronounced and asymmetrical oceanic spreading was described by Lin et al. (2019) for late Jurassic oceanic crust in the eastern Gulf of Mexico.

## Relation of volcanic rifted margins to mantle plumes

The link between mantle plumes, rifting and breakup of the supercontinent Pangea is a fundamental area of ongoing research. Early rift researchers like Turcotte and Emerman (1983) postulated two types of rifting: the first type is "active rifting" that occurs when mantle convection thins the lithosphere, causes crustal doming and results in lithospheric failure and rifting. The second type of rifting is called "passive rifting" that occurs when tensional stresses on the edge of the plate cause failure of the continental lithosphere; as a result, a mantle diapir or hotspot exploits this zone of induced weakness to the base of the crust and causes crustal thinning and rifting.

In Fig. 8A, regional doming above a rising mantle plume results in a characteristic, three-armed rift system superimposed on a large domed structure inferred to represent the area of an underlying, mantle hotspot. Two of the rift arms are successful in becoming oceanic spreading centers, while the third arm fails and remains a continental rift without oceanic spreading. Fig. 8B shows the Ethiopian Plateau as a modern analog to the idealized dome-above-hotspot model as shown in Fig. 8A. The subcircular Ethiopian Plateau is subdivided by the Main Ethiopian rift and two other fault systems, which have succeeded in creating oceanic crust in the Gulf of Aden and the Red Sea.

Volcanic margins are geographically associated with large igneous provinces associated with the heads of mantle plumes or hotspot tracks (Devey, 2015; Reuber et al., 2019). For that reason, plume head-related magmatism can cover a wide area but erupt volcanic rocks over a geologic time span as short as 2 Ma (Franke, 2013).

The Central Atlantic Magmatic Province (CAMP) is the most areally extensive event in Earth's history and formed during the late Triassic and earliest Jurassic between 173 and 201 Ma and was subsequently disrupted by the breakup of Pangea that dispersed CAMP magmatic rocks to southeastern North America, northwestern Africa and northeastern South America (Fig. 8C). Other studies discussing the effects of mantle plumes on early rifting of oceanic basins include the following: White (2014), Li et al. (2014), Whalen et al. (2015), Peace et al. (2020), Darros de Matos (2021), Dalziel and Lawver (2022) and Buchan and Ernst (2021).

## Comparison of regional transects of volcanic margins in the Central and South Atlantic

The cross-sectional geometries of SDRs compiled by Zinecker (2020) for five volcanic rifted-passive margins around the Atlantic Ocean area are shown in Fig. 9: (A) the Baltimore Trough of the East Coast of the USA (Biari et al., 2017); (B) the Demerara Rise, Suriname (Reuber et al., 2016); (C) Punta del Este Basin, Uruguay (McDermott et al., 2019; Reuber et al., 2019); (D) Orange Basin, Namibia (Bauer et al., 2000; Intawong et al., 2015) and (E) the Guinea Plateau (Zinecker, 2020).

The Guinea-Demerara examples from the Central Atlantic Ocean form conjugate margins (Zinecker, 2020) along with the two conjugate pairs from the South Atlantic Ocean (Reuber et al., 2019). Both inner, hinge-zone and outer SDRs as defined by McDermott et al. (2019) and Reuber et al. (2019) are common along volcanic rifted margins of the Atlantic margins.

## Four-stage model for the formation of SDRs along volcanic passive margins

A four-stage conceptual model for the evolution of volcanic-rifted margins from Reuber et al. (2019) is shown in Fig. 10. This model includes four tectonic stages based on the crustal structure derived from refraction data and deeply-penetrating seismic reflection data as described by Reuber et al. (2019).

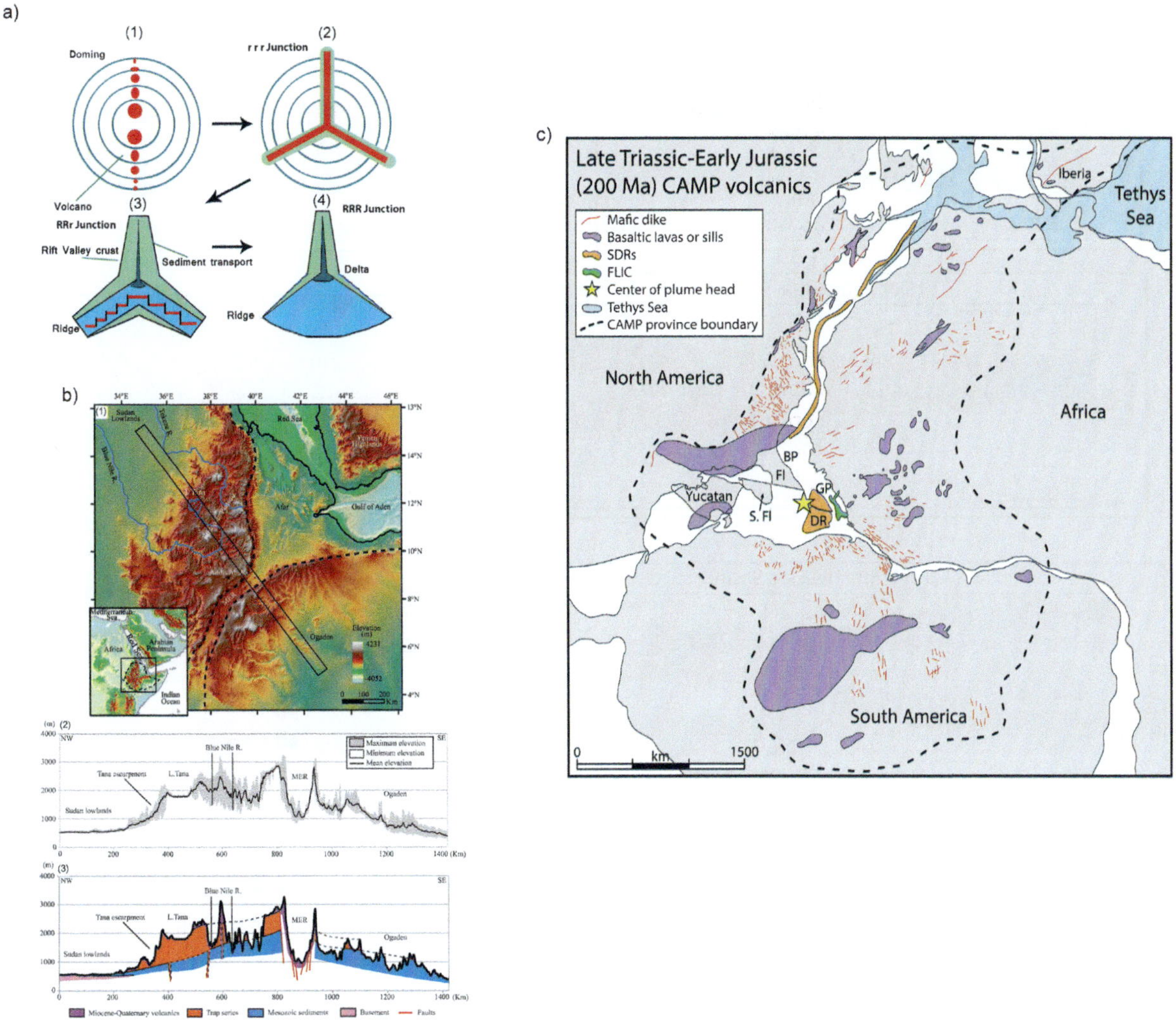


**FIG. 8** (A) Regional pattern triple-armed continental rift systems and oceanic spreading centers modified from https://commons.wikimedia.org/wiki/File:Aulacogendiagram.png. (1) Regional doming manifests the rise of a mantle plume with tensile stresses within the dome marked by a line of volcanoes indicated by the red circles. (2) Formation of a characteristic, triple-armed rift system overlying thinned continental crust in the domed region. (3) Transition of the triple-armed rift in continental crust to a zone of oceanic spreading along two of its arms; the third rift arm does not progress to oceanic spreading and instead forms a failed, continental rift (or aulacogen) that commonly becomes the pathway for a major river system. (4) Transition of the triple-armed rift to three oceanic spreading ridges that together form a ridge-ridge-ridge (RRR) plate boundary. (B) (1) Digital elevation model of Afar area of northeastern Africa from Sembroni et al. (2016) showing an example of a triple-armed rift in continental crust formed above a mantle plume and with oceanic spreading along two of its arms (Red Sea, Gulf of Aden) and continental rifting along its third arm (East African rift). (2) Topographic profile of the Main Ethiopian rift occupying a large dome related to a mantle plume. (3) Geologic profile of the Main Ethiopian rift. (C) Late Triassic-Early Jurassic plate reconstruction from Zinecker (2020) that was modified from Marzoli et al. (1999) and McHone and Puffer (2003) and shows the distribution of magmatism related to the Central Atlantic Magmatic Province (CAMP) of Latest Triassic-Early Jurassic (~204–193 Ma) age within the 7 million km$^2$ area within the *dashed lines* that includes areas of present-day North America, South America, Africa and Europe. Location of the Freetown Large Igneous Complex (FLIC) of Late Triassic (201.7 ± 0.7 Ma) age and late Jurassic SDRs (173 Ma) on the Guinea Plateau-Demerara Rise conjugate margin are also shown. The approximate center of the CAMP plume head as indicated by the radial mafic dike swarms compiled from Marzoli et al. (1999) is shown by the *yellow star*. **FL** = Florida block; **S. Fl** = South Florida block; **DR** = Demerara Rise; **GP** = Guinea Plateau.

The four evolutionary stages of the rifted-volcanic margin illustrate the characteristic features of a symmetrical volcanic margin. Fig. 10A shows the period of earliest igneous intrusions and volcanic activity on thinned continental crust that includes subaerial flood basalts emanating from flanking half-grabens. The rift axis is topographically elevated, and landward-dipping normal faults "shed" continental crust away from the rift axis. Plate divergence is minimal and is manifested by a localized area of rifting and magmatism as shown in Fig. 10.

During the second volcanic passive margin (VPM) stage shown in Fig. 10B, multiple basaltic flows cover the fully-ruptured continental crust (Pindell et al., 2014; Reuber et al., 2019). Half-graben structures are infilled with rift-related

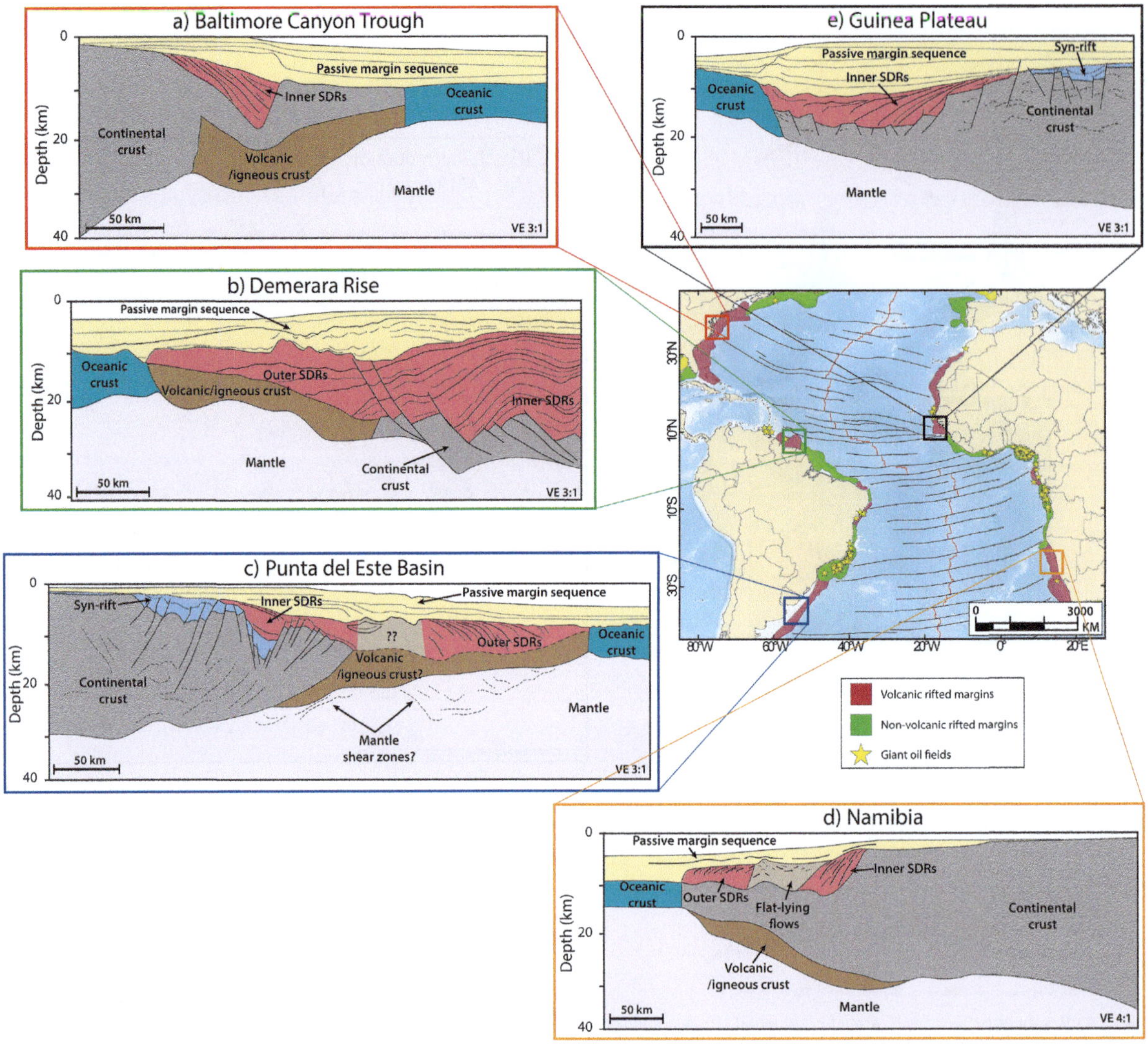


**FIG. 9** Compilation by Zinecker (2020) showing the geometries of SDRs in red from five volcanic-rifted margins from the circum-Atlantic Ocean. (A) Baltimore Trough, East Coast USA (Biari et al., 2017); (B) Demerara Rise, Suriname (Reuber et al., 2016); (C) Punta del Este Basin, Uruguay (McDermott et al., 2019; Reuber et al., 2019); (D) Orange Basin, Namibia (Intawong et al., 2015; Bauer et al., 2000) and (E) the Guinea Plateau (Zinecker, 2020). On the location map, the *red polygons* indicate volcanic rifted margins, the *green polygons* indicate non-volcanic rifted margins and the *yellow stars* show the locations of giant oil and gas fields around the circum-Atlantic. All five cross sections are shown at the same vertical scale.

sediments and extrusive volcanic rocks. A distance of about 20 km (~12.4 mi) separates the opposing continental plate edges at this stage and continued divergence results in margin subsidence and basinward rotation of the SDRs.

The third VPM stage (Fig. 10C) represents a period of prolonged volcanism during plate divergence with continuous symmetrical accretion of SDRs and underlying igneous crust along both conjugate margins. The rotated arrows indicate the directions of progressive basinward rotation of previously elevated to flat-lying flows to now, steeply dipping packages of extrusive basalts near the nascent, spreading center. Zones of SDRs/igneous crust can vary in width from tens to hundreds of kilometers as seen on the cross sections through five rifted-volcanic margins (Fig. 9).

In this third stage of the model (Fig. 10C), a typical 32-km-thick (~20 mi) continental crust is thinned and separated by a 160-km-wide (~100 mi) area of SDRs/igneous crust that is approximately 20 km (~12.4 mi) in thickness in its proximal area. Thinning of the continental crust and normal faulting is negligible at this stage because extension is accommodated along the oceanic spreading ridge.

The fourth and final VPM stage (Fig. 10D) is characterized by (1) cessation of voluminous volcanism; (2) active, oceanic spreading along the mid-ocean spreading ridge and (3) passive margin subsidence and sedimentation that includes marine drift-sequence stratigraphy that onlaps major, underlying and steeper dipping crustal boundaries and crustal

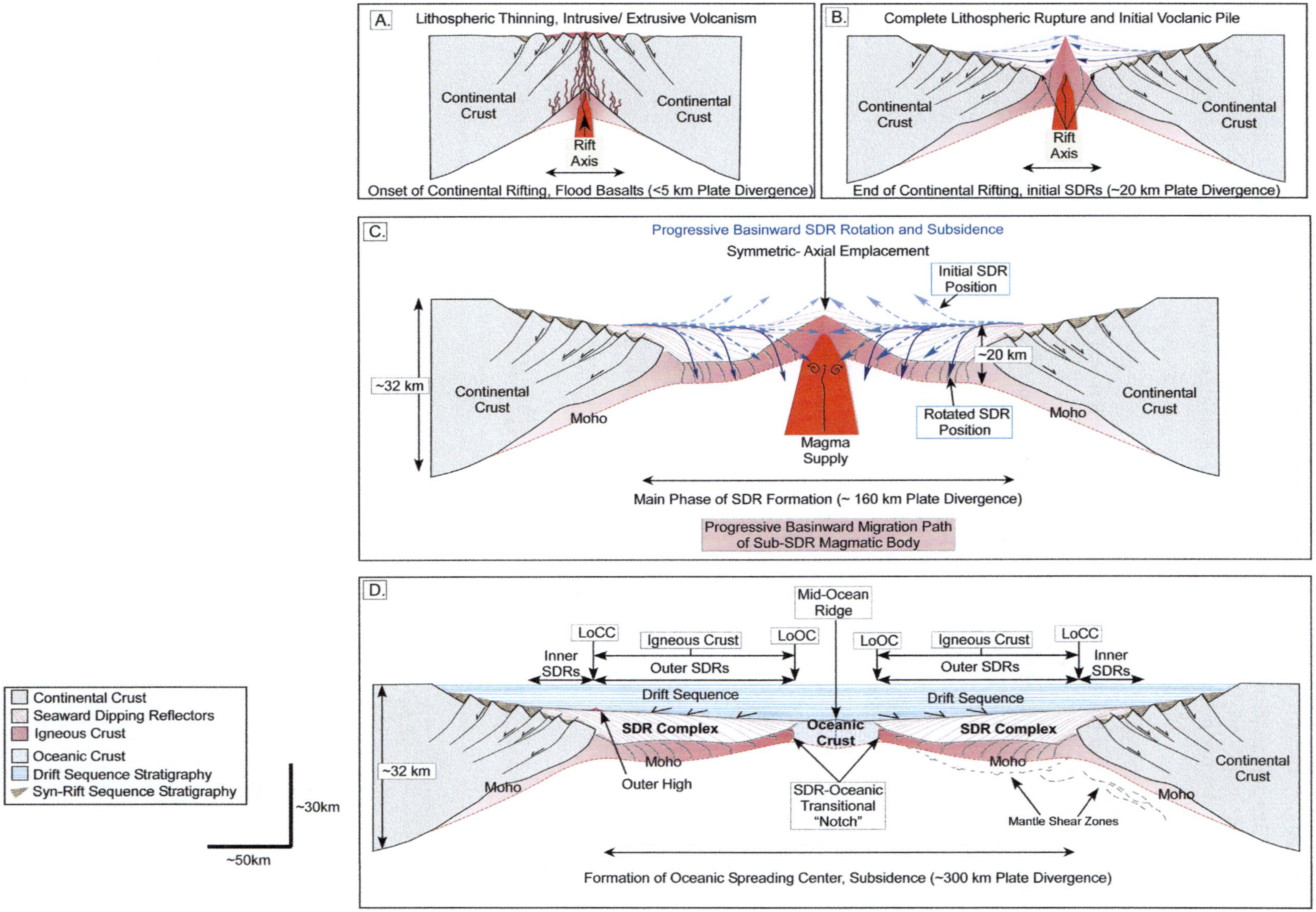


**FIG. 10** Four-stage conceptual model for the evolution of a volcanic rifted margin from Reuber et al. (2019). Stages (A–D) illustrate the primary features and components present on a typical symmetrical volcanic margin: (A) This volcanic rifted margin stage represents the period of earliest igneous intrusions and volcanic activity on thinned continental crust that includes subaerial flood basalts that emanate from flanking half-grabens. (B) During the second VPM stage, multiple basaltic flows cover fully ruptured continental crust (Pindell et al., 2014). (C) The third VPM stage represents a period of prolonged volcanism during plate divergence with continuous symmetrical accretion of SDRs and underlying igneous crust along both conjugate margins. The rotated arrows indicate the directions of progressive basinward rotation of previously elevated to flat-lying flows to now, steeply dipping packages of basalts near the nascent, spreading center. Thinning of the continental crust and normal faulting is negligible at this stage because the extension is accommodated along the oceanic spreading ridge. (D) The fourth and final VPM stage is characterized by (1) cessation of voluminous volcanism; (2) active, oceanic spreading along the mid-ocean spreading ridge and (3) passive margin subsidence and sedimentation that includes marine drift-sequence stratigraphy that onlaps major, underlying and steeper dipping crustal boundaries and crustal reflectors.

reflectors. This final stage includes conjugate VPMs with approximately 300 km (~ 190 mi) of intervening oceanic crust. The terms "limit of continental crust" (LoCC) and "limit of oceanic crust" (LoOC) are used to describe the major crustal boundaries.

The inner and outer SDRs indicate the relative positions of the SDRs to the continental margin. Outer highs are sometimes present at or near the LoCC, but Reuber et al. (2019) conclude that such outer highs are not common on the volcanic-rifted margins they describe in the South Atlantic. The SDR-oceanic boundary is commonly observed as a pronounced notch in the Moho (Reuber and Mann, 2019).

## Map showing orthogonally and obliquely-opening conjugate rifted margins of the Atlantic Ocean and Gulf of Mexico

The map of the conjugate, circum-Atlantic rifted-passive margins shows areas of orthogonal rifting of Triassic-Jurassic age and subsequent seafloor spreading with magnetic isochrons compiled by Seton et al. (2020) in the Central Atlantic Ocean between eastern North America and northwestern Africa (Fig. 11). Oblique rifting of Aptian-Albian age and subsequent seafloor spreading in the Equatorial Atlantic Ocean characterize the area between northeastern South America and central Africa, while orthogonal rifting of early Cretaceous age and subsequent seafloor spreading characterize the area between southwestern South America and southwestern Africa (Fig. 11).

The green lines represent magnetic isochrons of Cretaceous–Cenozoic oceanic crust, the dark blues represent Mesozoic oceanic crust and the single black line represents the COB. Radiating dikes are related to the prerift latest Triassic-early Jurassic (~201 Ma) Central Atlantic Magmatic Province (CAMP). Prerift shear zones of Precambrian age were compiled by Zinecker (2020) and are also shown on the map in Fig. 11.

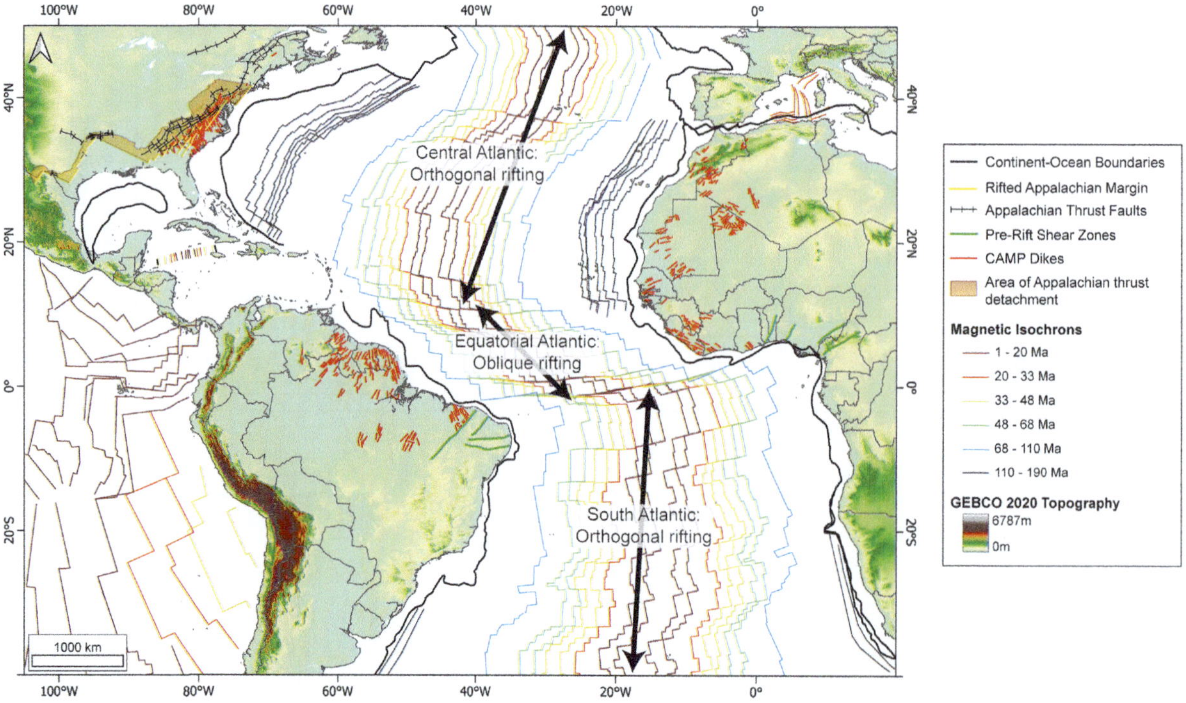


**FIG. 11** Map of the conjugate, circum-Atlantic rifted-passive margins showing areas of orthogonal Triassic-Jurassic rifting and subsequent seafloor spreading with magnetic isochrons from Seton et al. (2020) in the Central Atlantic Ocean between eastern North America and northwestern Africa; oblique Cretaceous rifting (Aptian-Albian) and subsequent Cretaceous-Cenozoic spreading in the Equatorial Atlantic Ocean between northeastern South America and central Africa; and orthogonal early Cretaceous rifting and subsequent seafloor spreading between southwestern South America and southwestern Africa. The *green lines* represent magnetic isochrons of Cretaceous-Cenozoic oceanic crust, the *dark blues* represent Mesozoic oceanic crust and the single *black line* represents the continent-ocean boundary (COB). Dikes related to the prerift latest Triassic-early Jurassic (~201 Ma) Central Atlantic magmatic province (CAMP) are shown.

## Cross-sectional differences between transform and orthogonal rifted margins

The crustal profiles of transform and orthogonally-rifted margins vary significantly, and their major differences are summarized in Fig. 12. On the orthogonally-rifted Iberian margin (Masini et al., 2011), the cross section of the zone of necking of the continental lithosphere is smoothly tapered and thinned by rifts, and this leads to a gently-sloping seafloor and an overlying, gently-sloping passive margin, sedimentary megasequence. In contrast, the cross section of the Ghana transform (Antobreh et al., 2009) exhibits a minor amount of strike-slip-controlled extension, an abrupt, subvertical truncation of the continental margin by the transform fault that underlies a more steeply-sloping seafloor and overlying, passive margin megasequence. The Ghana transform margin shown in Fig. 12 and the Guinea transform margin at the southern end of the Guinea Plateau shown in Fig. 13 from Zinecker (2020) show the same transform fault-related characteristics.

References that describe the structural and stratigraphic characteristics of obliquely-rifted and transform margins include the following: Morley et al. (1990), Faulds and Varga (1998), Basile and Allemand (2002), Lizarralde et al. (2007), Mounguengui and Guiraud (2009), Guiraud et al. (2010), Corti (2012), Nemčok et al. (2012), Brune et al. (2012), Bellahsen et al. (2013), Heine et al. (2013), Pagli et al. (2018), Karson (2021), Pérez-Gussinyé et al. (2020) and Kolawole et al. (2021)

## Key messages on the effects of preexisting crustal structure on rifting

As previously summarized in Table 1, Dunbar and Sawyer (1989) used finite element modeling to demonstrate that preexisting basement structural grain exerts variable resistance to rifting and that, in turn, controls the width of rift zones and their volcanic versus non-volcanic nature. Where these basement trends parallel the rift direction, the rift zones are relatively narrow in width with smaller stretching factors because the lithosphere was weaker in this direction; where these basement trends are oblique to the rift direction, the rift zone is wider and stretching factors are greater because the lithosphere is stronger in this direction. Narrow rift zones with parallel basement trends favor deeper mantle involvement and therefore are more prone to become volcanic margins. Wider rift zones with oblique basement trends favor less mantle involvement and therefore would be more prone to become non-volcanic margins.

Reuber and Mann (2019) tested these simple concepts related to basement structural grains using deeply-penetrating seismic profiles from the southern Brazilian rifted-passive margins that include the Pelotas Basin in the north (Fig. 14B) and the Punta del Este Basin 250 km (~ 155 mi) to the south (Fig. 14C). The interpreted profile from the Pelotas Basin illustrates the narrow (55 km or ~ 34 mi) zone of rifted continental crust within the rift-parallel fabric of the Dom Feliciano foldbelt, which exhibits a beta factor of 2.5. In this area, rifting was parallel to the structural grain so that the rift zones are relatively narrow and the beta factors are low because the lithosphere is weak in this direction (Reuber and Mann 2019).

The uninterpreted seismic reflection line across the Salado suture zone located in the Punta del Este Basin of offshore Uruguay is shown in Fig. 14C. This area of the margin shows a much wider (185 km or ~ 115 mi) zone of horsts and grabens that marks the necked zone of rifted continental crust. To the west, the interpreted continental crust is approximately 32 km

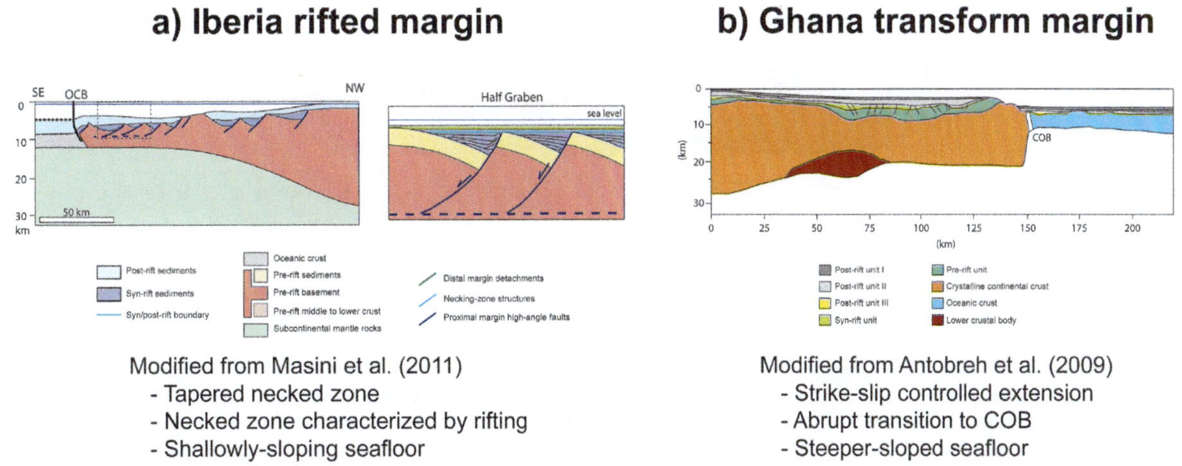


**FIG. 12** (A) The Iberia rifted margin in the Central Atlantic Ocean is an example of an orthogonally-rifted, low-taper, gently-sloping and non-volcanic rifted margin (Masini et al., 2011). (B) The Ghana transform margin in the Equatorial Atlantic is an example of an orthogonally-rifted, abruptly truncated, steeply-sloping and non-volcanic transform margin (Antobreh et al., 2009).

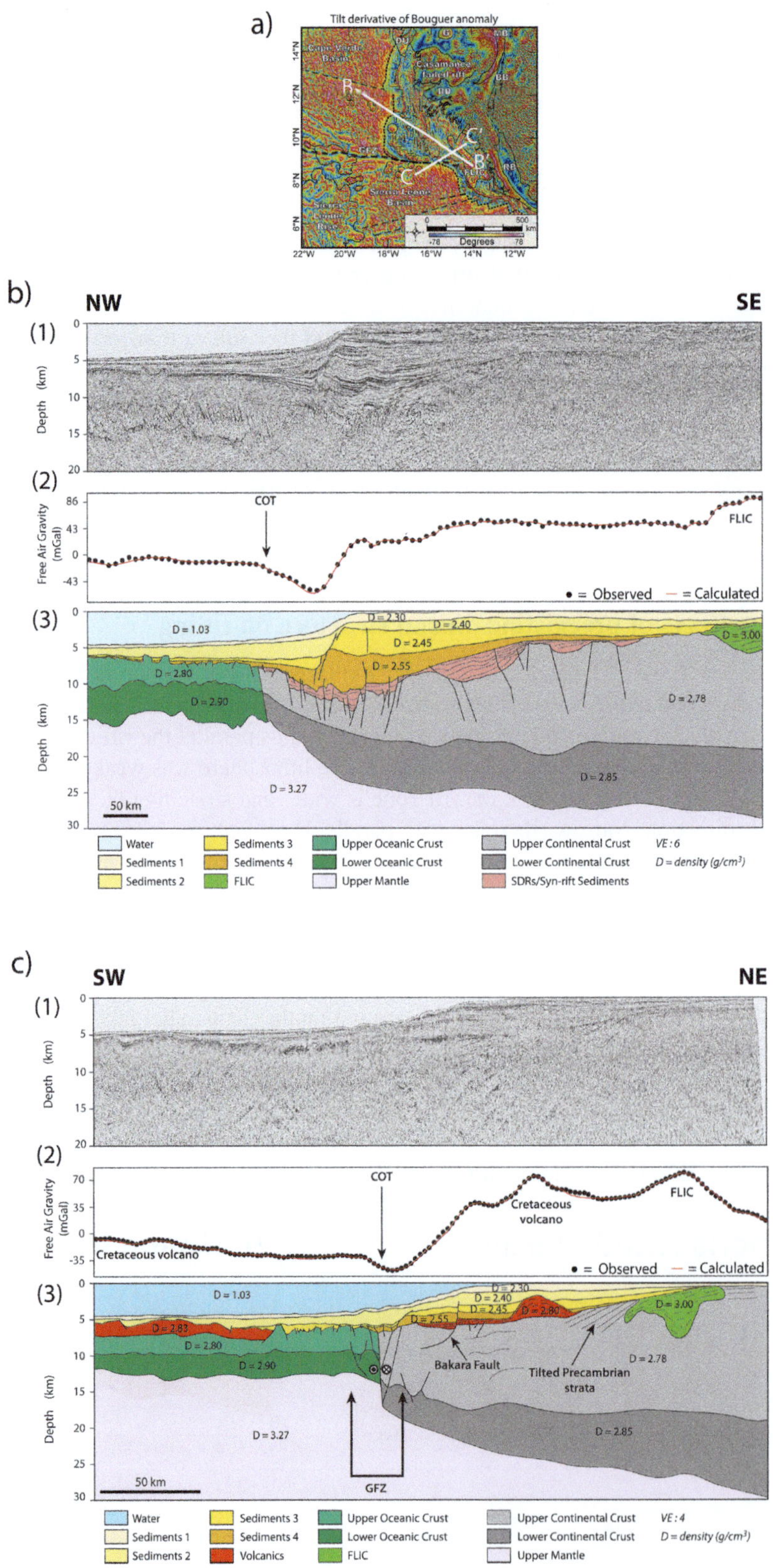


**FIG. 13** (A) Tilt derivative of the Bouguer gravity anomaly map of the Guinea Plateau compiled by Zinecker (2020) showing the Guinea Fracture Zone (**GFZ**), the Freetown layered igneous complex (**FLIC**), the Rokelide Belt (**RB**), the Bassaride Belt (**BB**), the Mauritanide Belt (**MB**), the Gasane volcanic plug (**G**), the Dakar Uplift (**DU**), the Bissau Depression (**BD**), the Casamance failed rift and the Sierra Leone Rise. (B) Forward-modeled 2D gravity profile in the northwest to southeast Central Atlantic dip direction across the Guinea Plateau. The location of the line is shown in (A) and vertical exaggeration is 6:1. (**1**) Uninterpreted 2D seismic profile used to constrain the upper 20 km (~12.4 mi) of the model. (**2**) Observed free-air gravity data (black circles) along the profile and the calculated gravity model (red line). (**3**) Multilayer crustal and stratigraphic cross section with model densities assigned to each block. (C) Forward-modeled 2D gravity profile in the northeast to southwest dip direction in the offshore Guinea area. The location of the line is shown in (A) and veritical exaggeration is 4:1. (**1**) Uninterpreted 2D seismic profile used to constrain the upper 20 km (~ 12.4 mi) of the model. (**2**) Observed free-air gravity data (black circles) along the profile and the calculated gravity model (red line). (**3**) Multilayer crustal and stratigraphic cross section with model densities assigned to each block. All figure parts modified from Zinecker (2020).

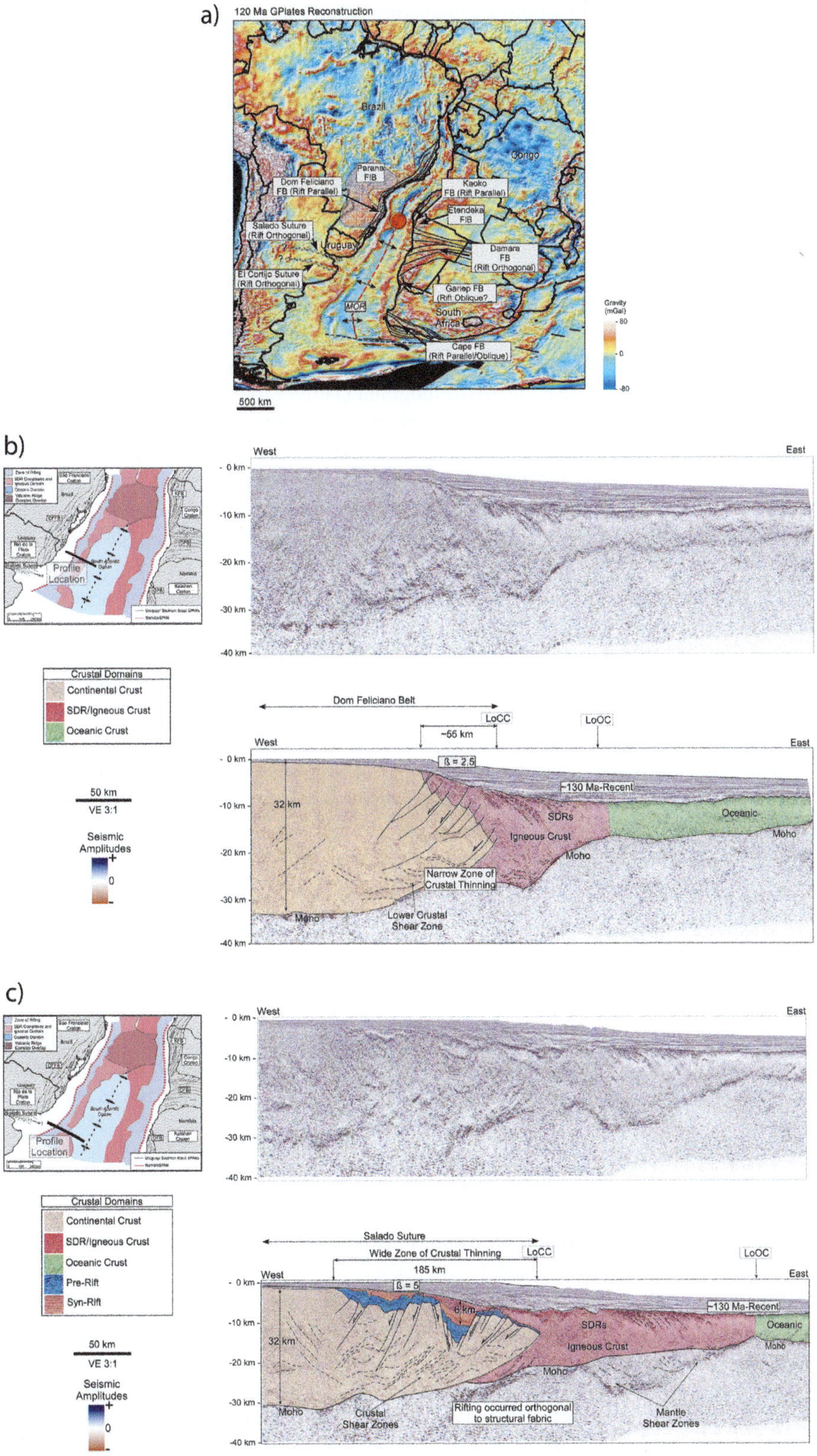


**FIG. 14** (A) Aptian (113–125 Ma) tectonic reconstruction from Reuber and Mann (2019) using a base map of free-air gravity data from Sandwell and Smith (2009). The South American margin has three tectonic features marking continental collision events: (1) the El Cortijo suture, (2) the Salado suture and (3) the Dom Feliciano foldbelt. The orientations of the two sutures relative to the paleo-rift axis of the South Atlantic are rift orthogonal, while the Dom Feliciano foldbelt is oriented rift parallel. The West African margin is flanked by four prominent foldbelts: (1) Kaoko, (2) Damara, (3) Gariep and (4) Cape. The orientations of these four foldbelts relative to the paleorift axis of the South Atlantic Ocean are rift parallel, rift orthogonal, rift oblique and rift parallel/oblique, respectively. (B) Uninterpreted west-to-east dip-oriented seismic reflection line across the rift-parallel Dom Feliciano fold belt of Paleozoic age located in the Pelotas Basin of offshore southern Brazil from Reuber and Mann (2019). Interpreted profile illustrating the narrow (55 km or ~ 34 mi) zone of rifted continental crust within the rift-parallel fabric of the Dom Feliciano fold belt. The beta factor is 2.5. Oceanic crust is present in the east of the SDRs/igneous crust. (C) Uninterpreted seismic reflection line across the Salado suture zone located in the Punta del Este Basin of offshore Uruguay from Reuber and Mann (2019). Seismic amplitudes are variable and range from steeply dipping, offlapping packages to laterally continuous and flat-lying events in the shallow section. Interpreted section in which a wide (185 km or ~ 115 mi) zone of horsts and grabens marks the rifted continental crust. To the west, the interpreted continental crust is approximately 32 km (~ 20 mi) in thickness, which is thinned in an eastward direction until its final continental rupture. The beta factor is 5.0. This LoCC is present in the west of the onset of SDRs/igneous crust and oceanic crust.

(~ 20 mi) in thickness, which is necked in an eastward direction until its final continental rupture adjacent to oceanic crust. In this area, rifting was parallel to the structural grain so that the rift zones are relatively wide and the beta factors are higher (5) because the lithosphere is stronger in this direction.

Profiles from other areas of southern Brazil were interpreted by Reuber and Mann (2019) in a similar way along with profiles from the west African conjugate margin and are shown as schematic cross sections in Fig. 15A. A graphical summary of the relationships include the following parameters for each South Atlantic transect area described by Reuber and Mann (2019): (1) the original, prerift crustal thicknesses, (2) the beta factor and (3) the post-rift, width of the rifted conjugate margins is shown in Fig. 15C. Lower beta factor values would indicate more narrow continental rift zones, whereas higher beta factors values will result in wider continental rift zones.

Profile 1 (Transect A-A′) in Fig. 15A-3 provides an example of the predicted relationship for pre-orogenic structural grains in the rift-parallel Dom Feliciano and Kaoko foldbelts that control lower beta factors and narrower zone of rifting on each conjugate margin. Profile 2 (Transect B-B′) in Fig. 15A-3 provides an example of the predicted relationship for pre-orogenic structural grains that are rift-orthogonal. The rift-parallel Dom Feliciano foldbelt experienced a brief period of rifting and resulting in a lower beta factor value of 2.5. In contrast, the rift-orthogonal Damara foldbelt was more resistant to rupture and underwent a more protracted rifting episode that resulted in a higher beta factor value of 4.3.

Profile 3 (Transect C-C′) in Fig. 15A-3 provides an example of the predicted relationship for pre-orogenic structural grains that are rift-orthogonal. The continental thinning of the more rift-resistant and rift-orthogonal Salado suture resulted in a wider zone of thinned continental crust with a higher beta factor value of 5. The arcuate and oblique orientation between the oblique Gariep foldbelt and the rift axis resulted in an intermediate beta factor value of 4 and an intermediate width zone of continental extension.

In Fig. 15B, Reuber and Mann (2019) modified the schematic model from Courtillot (1982) to illustrate how a propagating rift is affected by along-strike zones of wider, “locked” zones of continental extension between two plates. Their modification of the Courtillot (1982) model shows that the wider, locked zones correspond to areas of prerift, orogenic grains that are at right angles or oblique to the Cretaceous South Atlantic opening direction. Wider, locked zones in the South Atlantic like the rift-orthogonal Salado suture zone or Damara foldbelt exhibit higher beta factors. More narrow rift zones such as the rift-parallel Dom Feliciano and Kaoko foldbelts exhibit lower beta factors.

A crossplot of beta factor values versus angle of orientation, the compilation by Reuber and Mann (2019) shows the relationship of the beta factor to the structural orientations within orogenic belts that were crosscut by Early Cretaceous rift systems associated with volcanic and non-volcanic margins. In summary, their study supports the earlier numerical modeling of Dunbar and Sawyer (1989) showing the importance of basement fabrics on the beta factor and style of rifting.

References on the effects of preexisting crustal structures on rifting include the following: Dixon et al. (1987) on the control by Precambrian basement trends on Red Sea rifting; Morley (1999a, b) on the effects of basement fabrics on the East African rift system; Schumacher (2002) on the effect of preexisting structures on the Rhein graben; Krueger et al. (2018) on the control of basement features on oblique rifts of the Equatorial Atlantic; Henstra et al. (2019) on the controls of pre-existing normal faults on younger rifts; Heron et al. (2019) on the controls of basement structures on rifts in the Davis Strait; Jammes and Lavier (2019) on numerical experiments on the control of inherited crustal fabrics and rifting; Gouiza and Paton (2019) on the inherited controls on rifting of the Labrador Sea; Maestrelli et al. (2020) on experiments of rifts formed on inherited basement structures and Schiffer et al. (2020) on structural inheritance of rifts in the North Atlantic

## Tectonic significance of the marginal rift

Previous workers have identified “marginal rifts” or “outer marginal troughs” (Pindell et al., 2014; Peron-Pinvidic and Osmundsen, 2016) that are immediately adjacent to the area of late Jurassic oceanic crust that underlies the deep, central Gulf of Mexico (Hudec et al., 2013; Van Avendonk et al., 2015; Liu et al., 2019; Kenning and Mann, 2021; Rowan, 2022). A key element of the marginal rift concept is that an outer basement high exists seaward of the rift to confine continental or marine sediments along this seaward edge of the rift; this same outer high would form the landward edge of the late Jurassic oceanic crust. In the western Gulf of Mexico, Hudec et al. (2013) described this basement “inner ramp” or outer basement high which formed up to 4 km of bathymetric relief between the lower-lying, rifted continental crust and the higher-standing oceanic crust as shown schematically in the cross section in Figure 5 from Peron-Pinvidic et al. (2017).

For the non-volcanic rifted margins in the Gulf of Mexico, Pindell et al. (2014), Curry et al. (2018) and Minguez et al. (2020) all proposed that the marginal rift zone was underlain by a zone of serpentinized, subcontinental mantle that was exhumed during late Jurassic rifting (167–159 Ma) as the footwall of a low-angle and landward-dipping detachment fault. Pindell et al. (2014) proposed that the elongate, basement low originated as a mantle collapse feature that was partially filled

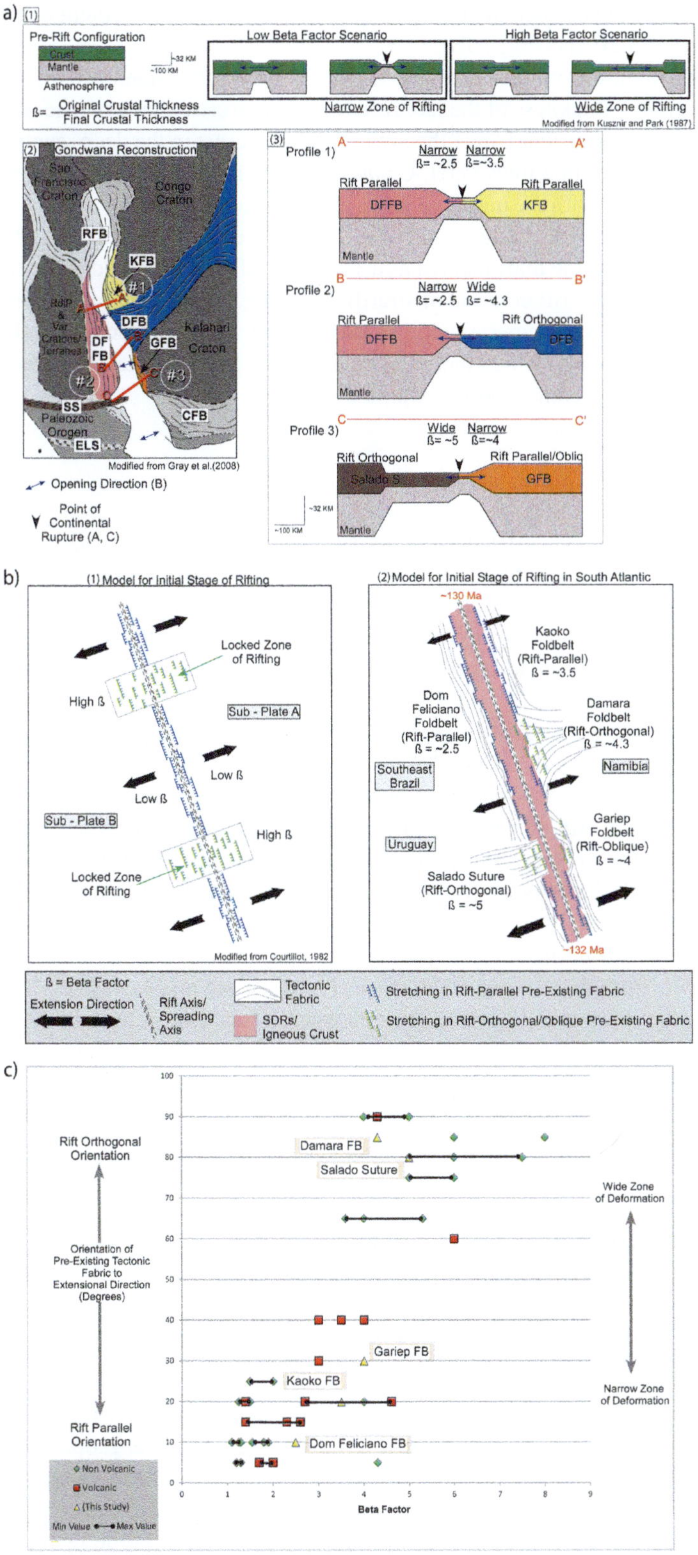


**FIG. 15** (A) (1) Summary of the relationships among the original, prerift crustal thicknesses, the beta factor and the post-rift, width of the rifted conjugate margins. Lower beta factors values will result in narrow continental rift zones, whereas higher beta factors values will result in wider continental rift zones (modified from Kusznir and Park (1987)). (**2**) Western Gondwana plate reconstruction showing the regional Brasiliano-Pan-African fold belts and sutures that intersect the South Atlantic conjugate rifted margins in our study area. (3) **Profile 1** (Transect A-A′) provides an example of the predicted relationship by Reuber and Mann (2019) for pre-orogenic structural grains in the rift-parallel Dom Feliciano and Kaoko fold belts that control lower beta factors and narrower zone of rifting on each conjugate margin. **Profile 2** (Transect B-B′) provides an example of the predicted relationship for pre-orogenic structural grains that are rift-orthogonal. The rift-parallel Dom Feliciano fold belt experienced a brief period of rifting and resulting in a lower beta factor value of 2.5. In contrast, the rift-orthogonal Damara fold belt experienced a longer, more resistant to rupture rifting episode and resulted in a higher beta factor value of 4.3. **Profile 3** (Transect C-C′) provides an example of the predicted relationship for pre-orogenic structural grains that are rift-orthogonal. The continental thinning of the rift-orthogonal Salado suture resulted in a wider zone of thinned continental crust with a higher beta factor value of 5. The arcuate and oblique orientation between the oblique Gariep fold belt and the rift axis resulted in an intermediate beta factor value of four and an intermediate width zone of continental extension. (B) (1) Modification of the schematic model from Courtillot (1982) by Reuber and Mann (2019), illustrating how a propagating rift is affected by along-strike zones of wider, "locked" zones of continental extension between two plates. (2) Modification of the Courtillot (1982) model by Reuber and Mann (2019) to show that the wider, locked zones correspond to areas of prerift, orogenic grains that are at right angles or oblique to the Cretaceous South Atlantic opening direction. Wider, locked zones in the South Atlantic like the rift-orthogonal Salado suture zone or Damara foldbelt exhibit higher beta factors. More narrow rift zones such as the rift-parallel Dom Feliciano and Kaoko fold belts exhibit lower beta factors. Regional trends of preexisting, orogenic basement fabrics are shown by the gray lines. (C) Crossplot of beta factor values versus angle of orientation from the South Atlantic compilation by Reuber and Mann (2019) that shows the relationship between the beta factor and the structural orientations within orogenic belts that were crosscut by Early Cretaceous rift systems. All figure parts modified from Reuber and Mann (2019).

with sediment and therefore was not a rift in the normal sense of continental crust with the same continental basement being extended and infilled with clastic sediment and evaporite intervals (Fig. 16B).

On the free-air gravity map in Fig. 16A, previous refraction stations as compiled from Nguyen and Mann (2016) are shown as white dots and constrain the generally thinner crustal thickness of the late Jurassic oceanic crust that underlies the deep, central Gulf of Mexico (Lin et al., 2019). The white polygons that fringe the area of oceanic crust in the deep, central Gulf of Mexico are the marginal rift system that is shown on seismic lines G, H, C and D in Fig. 16D. C and D, the interpreted seismic reflection line G-G′, the inferred sedimentary fill with the marginal rift is shown in purple and the pink color represents Jurassic salt deposits that were mainly concentrated in less faulted sag basin that overlies the marginal rift. The dip angles of both normal faults are 10–20°. Liu (2021) interprets the marginal rift as a sediment-filled rift formed at the tip of a highly-tapered, non-volcanic margin. The fill of the marginal rift consists of stratified, clastic, sedimentary rocks and evaporites and does not support this narrow feature as a zone of exhumed mantle as proposed in the cross section of the eastern Gulf of Mexico by Pindell et al. (2014). Liu (2021) also determined that the gravity across the marginal rift is consistent with a sediment-filled origin as opposed to a zone being composed of exhumed mantle.

### Tectonic origin of major salt basins

The three giant salt basins of the Gulf of Mexico, Brazil and West Africa all show similar map patterns and isopachs (Fig. 17A). In general, all three giant salt basins are asymmetrical with the thicker deposits and wider extent of the deposit to the west in the Gulf of Mexico and to the south in Brazil and Angola. In the Gulf of Mexico, the salt progressively thins and pinches out from west to east indicating that the source of seawater was from the Pacific Ocean and that the present-day area of Mexico formed a structural barrier between the Pacific Ocean and the low-lying rifted area of the Middle Jurassic Gulf of Mexico (Fig. 17).

For Brazil, hotspot activity during the early opening period led to the formation of the northwest-trending Rio Grande Rise that acted as a bathymetric barrier that separated seawater from the South Atlantic and the low-lying rifted area of the Campos-Santos salt basin (Fig. 17B). Similarly, the northeast-trending, conjugate hotspot track on the African margin acted as a bathymetric barrier that separated seawater from the South Atlantic and the lower-lying rifted area of the Angola salt basin. The influx of salt proved to be an excellent seal for the presalt reservoirs of both areas (Zalan, 2017) (Fig. 17B).

In the Central Atlantic the widest and thickest Jurassic salt deposits are found in Morocco and in eastern Canada, with smaller and less continuous salt bodies along the eastern margin of the US in the Baltimore Canyon, the Carolina trough and off the north-central coast of Cuba. Smaller and more discontinuous salt bodies are also present in Mauritania and Senegal. This indicates that the origin of the seawater was from the closing Tethys Ocean to the north and east. The Mediterranean salt giant is much younger and occupies a zone of subducting oceanic crust between Africa and Eurasia with the point of origin of seawater at the Straits of Gibraltar. Previous studies on these salt basins include the following: Davison (2007), Lentini et al. (2010), Mohriak (2014). Kukla et al. (2018) and Bellucci et al. (2021) on the Brazilian salt basins; Mitchell et al. (2020) and Augustin et al. (2021) on the Red Sea Basin; Cowie et al. (2016) on the salt basin in Angola and Tari et al. (2012) on the salt basins of the circum-Central Atlantic Ocean.

## Applications of knowledge of crustal structure and tectonostratigraphy of rifted-passive margins to hydrocarbon exploration

### Why do rifted-passive margins account for two-thirds of all giant oil and gas fields?

Early observations on the presence and controls of hydrocarbons in rifts focused on the active and largely subaerial East African rift system where the presence of post-rift sealing units were recognized as a critical element for preserving hydrocarbons formed in organic-rich, lake-filled rifts (Lambiase and Morley, 1999; Morley, 1999b) (Fig. 8B). As geophysical exploration and drilling progressed on the world's deepwater, rifted-passive margins, Mann et al. (2003) noted that about two-thirds of the 877 giant oil and gas fields discovered by 2000 were found in rift and sag basins and their overlying, marine, passive margin sequences (Fig. 3). These rifted-passive margin settings were highly productive as the result of abundant source rocks usually present in the rift or sag sequence and overlying reservoirs present in the syn-rift, sag or passive margin section (Fig. 6).

Mann et al. (2003) also noted the importance of the long-term tectonic stability and lack of recurrent deformation in passive margins as another key factor in their hydrocarbon productivity. Fraser et al. (2007) recognized the importance of various proposed rifting models on the origin and maturation of hydrocarbons formed in rifted-passive margins. These

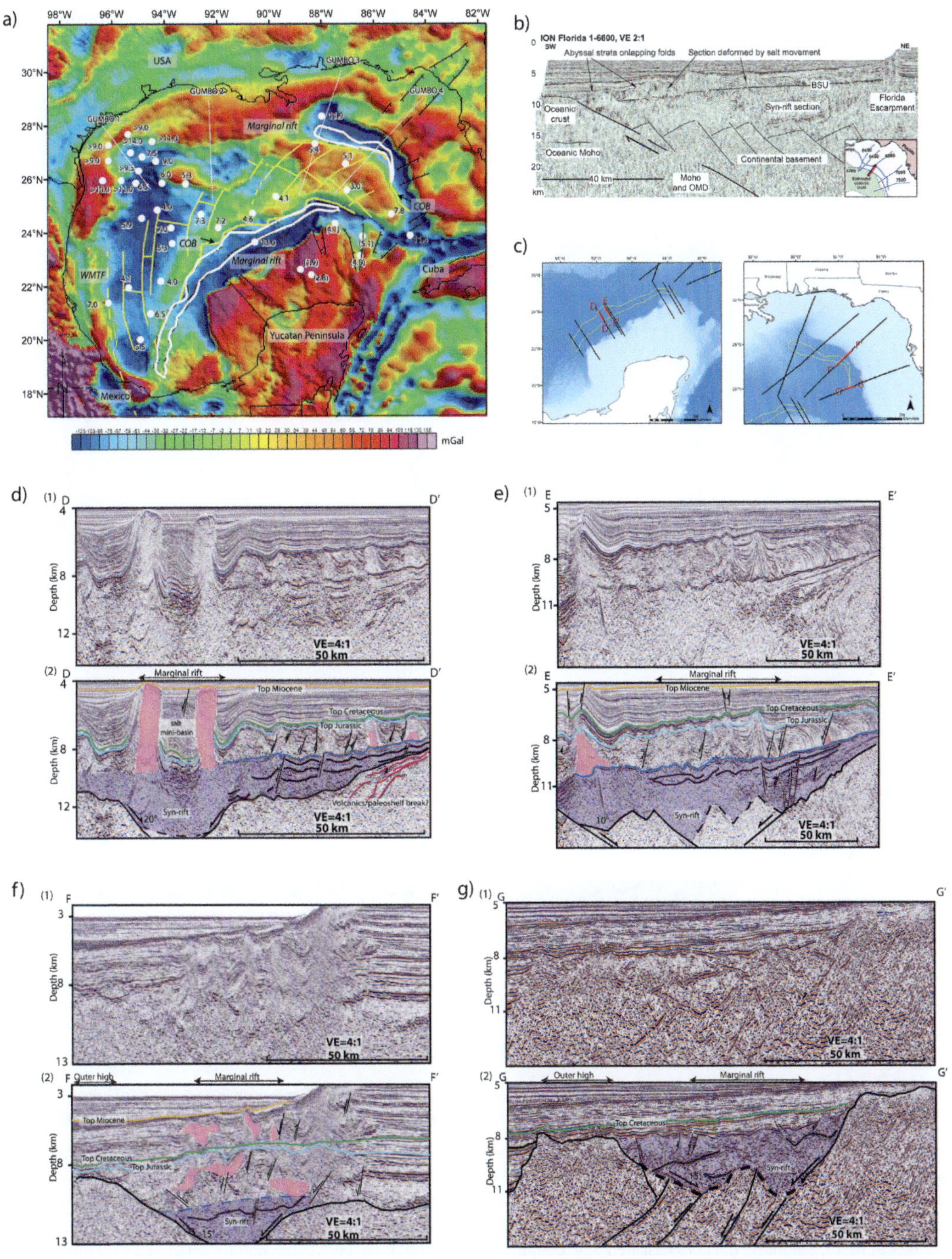

**FIG. 16** (A) Free-air gravity map (Sandwell et al., 2014) with previous refraction stations as compiled from Nguyen and Mann (2016) shown as the white dots. Numbers in parentheses next to the refraction station are the refraction-based crustal thicknesses in kilometers. The yellow line is the continent-ocean boundary (COB) as mapped by Sandwell et al. (2014) using satellite gravity. The white polygons in the southern and eastern Gulf of Mexico are the mapped extent of the circum-Gulf of Mexico marginal rift zone as show on the seismic profiles in D–G. (B) Interpretation of a deep-penetration ION line from the eastern Gulf of Mexico by Pindell et al. (2014), who proposed a low-angle detachment fault that dips landward and bounds a strip of exhumed mantle. An alternative interpretation by Liu (2021) is that this surface is the Moho underlying an unfaulted necked zone of thinned continental crust and that the marginal rift is sediment-filled and underlain by thinned continental crust rather than exhumed mantle. (C) Location maps showing the four seismic sections crossing the marginal rift as shown in D–G (D) (**1**) Uninterpreted seismic reflection line D-D′ across the marginal rift across the western Yucatan margin from Liu (2021). (**2**) Interpreted seismic reflection line D-D′. The purple color shows the inferred sedimentary fill with the marginal rift and pink color are Jurassic salt deposits that were mainly concentrated in less faulted sag basin that overlies the marginal rift. The dip angles of both normal faults are 20° and 10°. (E) (1) Uninterpreted seismic example of the marginal rift along E-E′ on the western Yucatan margin. (2) Interpreted seismic reflection line E-E′. (F) (**1**) Uninterpreted seismic reflection line F-F′ across the marginal rift in the northeastern Gulf of Mexico from Liu (2021). (**2**) Interpreted seismic reflection line F-F′. The true dip angles of both normal faults are 15°. (G) (1) Uninterpreted seismic example of the marginal rift along G-G′ in the northeastern Gulf of Mexico. (2) Interpreted seismic reflection line G-G′.

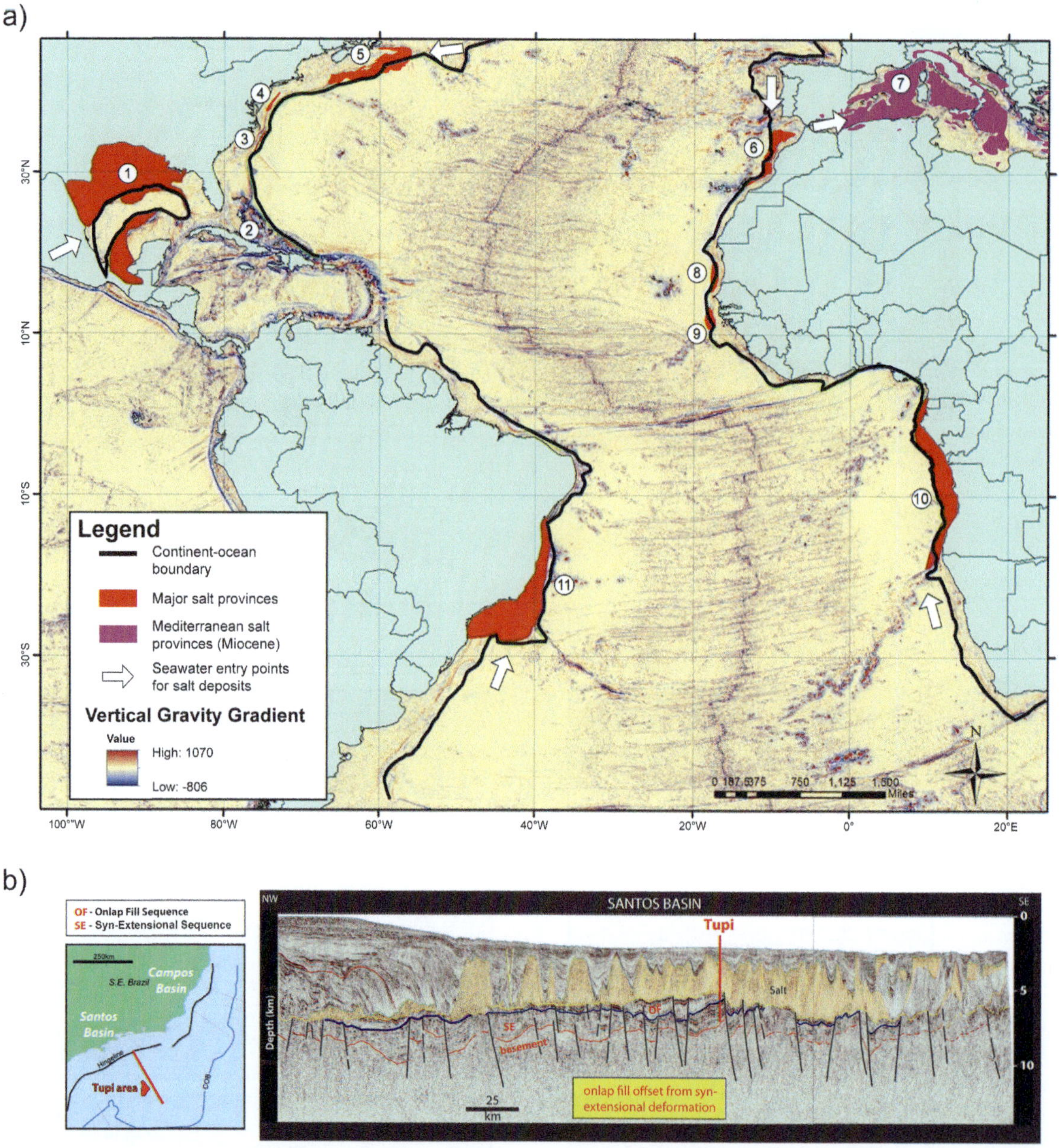


**FIG. 17** (A) Major salt basins of the Gulf of Mexico and the South Atlantic Ocean. In the Gulf of Mexico, the salt thins from west to east indicating that the source of seawater was from the Pacific Ocean and that the present-day area of Mexico formed a structural barrier between the Pacific Ocean and the low-lying rifted area of the Middle Jurassic Gulf of Mexico. For Brazil, hotspot activity during the early opening period lead to the formation of the northwest-trending Rio Grande Rise that acted as a structural barrier that separated seawater from the South Atlantic and the low-lying rifted area of the Campos-Santos salt basin. Similarly, the northeast-trending, conjugate hotspot track on the African margin acted as a structural barrier that separated seawater from the South Atlantic and the low-lying rifted area of the Angola salt basin. White arrows show the approximate point of origin for seawater filling the lower-lying rifted areas of the Gulf of Mexico and South Atlantic. Numbers represent the following areas with evaporitic deposits: (1) Gulf of Mexico; (2) Central Cuba; (3) Carolina trough; (4) Baltimore canyon; (5) Eastern Canada; (6) Morocco; (7) Mediterranean; (8) Mauritania; (9) Senegal; (10) Southwest Africa; (11) Southern Brazil. (B). Seismic section and location map modified from Lentini et al. (2010) showing the location of the Tupi giant oil field in the Santos Basin on a Cretaceous rift block sealed beneath a thick salt sequence of Aptian-Albian age (number 11 in A).

models included uniform stretching (McKenzie, 1978) and asymmetric stretching involving a major, low-angle shear zone cutting through to the mantle (Wernicke, 1985) (Table 1).

Beglinger et al. (2012) carried out a major regional study of west African rifted-passive margins by sub-dividing their stratigraphy and petroleum elements into prerift, syn-rift, sag and passive margin stages as shown schematically in Fig. 6. This tectonostratigraphy approach also allowed more precise stratigraphic correlations with areas on the South American conjugate margin. Davison and Underhill (2012) reviewed the sources, reservoirs and seals with emphasis on the North Sea and emphasized the importance of unconformity traps at the tops of beveled and tilted fault blocks.

Since 2000, 248 giant oil and gas fields have been added to the list of the giant fields of the world to bring their total number to over 1000—of which about two-thirds occupy rift or passive margin settings (Latham, 2021; Dolson et al., 2021). Improved 2D and 3D seismic imaging has led to many discoveries related to stratigraphic and combination traps related to thick, evaporite-sealed carbonate buildups and reefs and to submarine fan plays on passive margins (Dolson et al., 2021).

## How do hydrocarbon traps differ between orthogonal rifts and oblique rifts?

Nemčok et al. (2016) has reviewed the structure, stratigraphy and hydrocarbon trap types formed along largely submarine transforms during the early rift phase of the circum-Atlantic margins while Mann (2012) reviewed trap types for more subaerial strike-slip or shelfal strike-slip faults in California and Caribbean region. Refraction studies of oceanic transform faults and fracture zones along continental margins in areas of oblique opening in the Equatorial Atlantic have shown the major cross-sectional differences between the more gently-tapered, orthogonally-rifted margins and (Fig. 12A) and the abruptly truncated and "stubby" transform margins (Fig. 12B). The geometries of the tapered, orthogonally-rifted margins have been compiled globally by Sapin et al. (2021).

Grant (2020) has noted the importance of these two different margin types for the formation of stratigraphic traps along updip pinchouts of sandstone lobes of submarine fans. The lower relief, more tapered, orthogonally-rifted margins can host larger and more tightly sealed stratigraphic traps than more truncated, transform margins whose onlaps are along steeper surfaces and are more prone to updip leakage.

## Are non-volcanic rifted margins more hydrocarbon-rich than volcanic margins?

Giant oil and gas fields of the Atlantic rifted-passive margins show a close, spatial correlation with non-volcanic margins rather than along volcanic margins (Franke, 2013) (Fig. 3). Zalan (2017) explained this link between giants and non-volcanic margins as the result of accommodation space on non-volcanic margins that becomes filled by sedimentary rocks that potentially can serve as source, reservoir and seal rocks (Fig. 9). In contrast, the accommodation space on volcanic-rifted margins would largely be filled by magmatic rocks with limited potential to host hydrocarbons (Paton et al., 2017; Reuber et al., 2019) (Figs. 4, 9, 10).

Zalan (2017) also noted that volcanic margins exhibit greater amounts of post-rift and post-magmatic subsidence due to their greater heat loss and thermal subsidence (Fig. 9). For example, the passive margin section of the Pelotas basin of southern Brazil and northern Uruguay contains up to 9 km of mainly clastic sedimentary rocks. While the underlying rift stage is dominated my magmatic rocks, this overlying section contains source rocks deposited during late Cretaceous global anoxic events. These source rocks are overlain by slope and basin plan turbidite reservoirs and widespread stratigraphic traps.

For the conjugate volcanic margin Namibian margin in southwest Africa, Winter et al. (2022) summarized a 2021, light oil discovery that includes a widely-distributed, post-volcanic and overlying Barremian-Aptian source rock which charged stratigraphic traps in an onlapping Aptian sandstone unit. This promising light oil trend is currently being tested along strike both to the north and south. This recognition of complete and working hydrocarbon systems that functions independently of the underlying and "cooked" magmatic sections below is motivation for a fresh look at the potential of volcanic margins.

## Are "look-alike" hydrocarbon basins on conjugate margins the rule or the exception?

A tenet of rifted-conjugate margins is that their conjugate margin would be "look-alike" rift basins as the two margins were once contiguous in their prerift setting. In the Equatorial and South Atlantic during the late 1990s and early 2000s, a large discovery on one conjugate led to "conjugate chasing" of the supposed "look-alike" prospect on the other margin. For this reason, many regional studies were carried out in the Equatorial and South Atlantic (Lentini et al., 2010; Mello et al., 2012; Strozyk et al., 2017; Wen et al., 2019; Song et al., 2020; Plummer and Tilton, 2020) and the Gulf of Mexico (Steier and Mann, 2019; Snedden et al., 2020) to compare the stratigraphy, timing of events and known or inferred hydrocarbon systems of both conjugate margins.

For the Equatorial and South Atlantic, Zanella et al. (2018) compared the commercial volumes of oil and gas from five conjugate pairs to assess whether the two margins are truly "look-alike" from an exploration point of view: Santos-Namibe is much more productive on the Santos margin; Santos-Kwanza an approximate parity; Sergipe-Alagoas and Gabon are more productive on the Sergipe side; Tano-Barreirinhas are non-commercial on the Barreirinhas side;

and Sierre-Leone-Foz do Amazonas are non-commercial on both margins. The asymmetrical exploration performance of the two conjugates are related by Zanella et al. (2018) to several factors including: (1) asymmetrical widths of the basin margins such as seen for Santos (very wide fairway) and Namibe (more restricted fairway) (Fig. 7); (2) asymmetrical heat flows as inferred for the Kwanza (higher heat flow, more gas) and Santos (lower heat flow, more oil) and (3) asymmetrical burial histories for the Tano (commercial oil area relatively starved area removed from major rivers) and the Barreirinhas (noncommercial area in the gas window related to a Miocene to Recent clastic wedge related to the uplift of the Andes). In summary, there is a temptation to infer "look alike" basins but the reality is that major differences in their rifting or passive margin history may result in major differences in their hydrocarbon history.

## How does the crustal structure of the Santos basin control its rift and rift sag-based, presalt reservoirs?

In this section, I summarize the main hydrocarbon elements of the Campos-Santos basin of the Atlantic margin of southeastern Brazil which together contain 13 billion barrels of Brazil's proven oil. I relate these elements to the syn-rift, sag and passive margin tectonosequences of this very broad, orthogonally-rifted passive margin shown in Fig. 6. Exploration began in the more northern Campos basin during the latter half of the 20th century and focused on the passive margin-related sandstone and carbonate reservoirs that were part of large submarine fans of Late Cretaceous to Cenozoic age. The source rocks are presalt, late rift to sag-related Barremian calcareous shales with hydrocarbons passing along listric, normal faults and through a sag-related salt sequence (Mello et al., 2021a).

In 2005, drilling began to focus on the presalt interval of the Santos basin to the south of the Campos basin which was controlled by underlying rifts of Hauterivian to Late Barremian age and its overlying sag basin of Late Barremian to Late Aptian age. In 2007, the giant Tupi field was discovered with an estimated 5–8 billion barrels of oil equivalent in the presalt zone 18,000 feet below the ocean surface (Mello et al., 2021b). Subsequent drilling showed that productive carbonate reservoir in the sub-salt sag basin extended across the Campos, Santos and Espirito Santo basins. These discoveries highlight the importance of the syn-rift and sag sections for hosting the source rocks, the salt body for sealing the hydrocarbons and the sag basin carbonate rocks and overlying passive margin clastic fans for acting as reservoirs.

## How does the crustal structure of the Guyana margin control is passive-margin based sources, reservoirs and seals?

Discoveries of oil in stratigraphic traps within passive margin, submarine fan settings in Ghana, west Africa in 2007 spurred "conjugate hunting" in the years that followed for similar plays along the northeastern margin of South America (Guyana, Suriname, French Guiana) (Zanella et al., 2018). Although previously interpreted as a transform margin (Fig. 12A), the highly-tapered crustal structure of the Guyana margin indicates its formation as an orthogonally-rifted margin (Shipper and Mann, 2022) (Fig. 12B). This tapered crustal structure allows the formation of extensive stratigraphic traps to form in submarine fans of late Cretaceous age (Grant, 2020; Price et al., 2021). In 2015, the initial Liza discovery was followed by about 20 subsequent discoveries along a linear, margin-parallel trend in both Guyana and Suriname with an estimate size of 10 BBO and 30 TCF of gas (Price et al., 2021). The source is Cenomanian-Turonian shale, the reservoir is Late Cretaceous submarine fan and the seal is Paleogene shale, so the play is completely independent of the structure and stratigraphy of the underlying rift structure (Price et al., 2021; Trude et al., 2022).

## Does the continent-ocean boundary serve as an outer limit for deepwater exploration as the result of reduced heat flow in oceanic crust?

On either orthogonally-rifted or obliquely-rifted margins (Fig. 12), the continent-ocean boundary (COB) would mark an important boundary in the geothermal gradient between the relatively colder oceanic crust that lacks radiogenic heat production and thinned continental crust that is relatively warmer because of radiogenic heat production and therefore important for maturation of hydrocarbons (Khutorskoi and Polyak, 2016). The location of the continent-ocean boundary can vary widely as different workers use different criteria for its identification (Peace and Welford, 2020).

Dolson et al. (2021) propose several areas of giant oil and gas fields overlying oceanic crust that include the Niger delta, offshore India, East and West Africa and the Levant margin. The Niger delta has been displaced seaward as a passive margin foldbelt (Hasan and Mann, this volume) and differs from area of in situ marine sedimentation directly on Mesozoic oceanic crust as observed in the eastern Gulf of Mexico (Lin et al., 2019) and in southern Morocco (Galhom et al., 2022).

For in situ submarine fan sediments overlying Aptian age oceanic crust of the Congo-Angola basin, Anka et al. (2010) proposed a hydrocarbon system over that includes heat flow from Aptian oceanic crust, thick clastic sedimentation related to the paleo-Congo River and organic-rich source rocks within the distal part of the fan system. Petroleum systems on both passive margin foldbelts like the Niger data and areas of in situ marine deposition onto oceanic crust constitute an important area for future research.

## Conclusions

1. "Rifts" are fault-bounded, elongate troughs beneath which the continental lithosphere has been extended, while "rifting" is the process by which the continental lithosphere stretches or is "necked" into a tapered, cross-sectional morphology. Successful rifting leads to the formation of new ocean basins, whereas "failed" rifting leads to an isolated zone of continental extension that is not associated with an adjacent area of oceanic crust with its thinner and well-defined crustal structure.
2. Rifted-passive margins exhibit a two-layer crustal structure with a lower level of rifted, continental crust that is overlain by a largely undeformed passive margin of sedimentary rocks. For non-volcanic margins, the underlying, rifted margin consists of fault-controlled syn-rift, sedimentary wedges that form part of a subsiding weld that separate this rifted area of continental crust from newly formed and younger oceanic crust. The overlying passive margin sedimentary rocks are deposited in the post-rift period and are controlled mainly by the thermal subsidence of the underlying rifted crust.
3. Modern conceptual models for the formation of rifted-passive margins include: (1) pure shear thinning of the lithosphere with resulting margin symmetry; (2) simple shear thinning of the lithosphere with resulting margin asymmetry; (3) preferential extension along inherited orogenic grains that can control wider rifted margins when extension is fabric-oblique and to more narrow rifted margins when extension is fabric-parallel and (4) depth-dependent extension of the lithosphere occurs when the more ductile lower crust and mantle lithosphere accommodate more extension than the upper crust.
4. The fundamental crustal weld or boundary for rifted continental margins is also called a "continent-ocean boundary" (COB) that separates continental lithosphere from oceanic lithosphere. The term "continent-ocean transition" (COT) is widely used given the width and complexity of this transition between continental and oceanic lithosphere.
5. A conjugate margin pair includes two rifted-passive margins now located on different tectonic plates and separated by an oceanic spreading ridge. The conjugate margins were once conjoined along a shared rift system that preceded the breakup and formation of oceanic crust.
6. Non-volcanic rifted-passive margins form with relatively small amounts of volcanic rocks and are characterized by "low-angle normal faults" that bound half-graben structures and generally dip in the seaward direction.
7. The crustal structure of orthogonally-rifted, non-volcanic margins are traditionally divided into the proximal margin that includes the area of listric normal faults bounding syn-rift sedimentary wedges and the distal margin of extremely thinned continental crust that is potentially separated from the oceanic crust by a domain of exhumed, subcontinental mantle. I use the term "orthogonally-rifted" for rifting at a high angle to the trend of the margin and "obliquely-rifted" for rifting at a low angle to the trend of the margin.
8. Volcanic rifted-passive margins form the majority of rifted-passive margins worldwide and are characterized by: (1) large amounts of volcanic and volcaniclastic sedimentary rocks that include 5–15-km-thick wedges of volcanic flows manifested on seismic reflection data as "seaward-dipping reflectors," or SDRs; (2) thinning of the continental crust is abrupt so the COT occurs over a short distance and in the vicinity of the SDRs and (3) the area seaward of the rifted continental crust is underlain by a high-velocity lower crust ($Vp > 7.3$ km/s).
9. Doming above a rising mantle plume results in a characteristic, three-armed rift system superimposed on a large domed structure inferred to outline the active hotspot. Two of the arms are successful in becoming oceanic spreading centers, while the third arm fails and remains a continental rift without oceanic spreading.
10. The crustal profiles of transform and orthogonally-rifted margins are very different. On the orthogonally-rifted Iberian margin, the zone of necking of the continental lithosphere is smoothly tapered and thinned by rifts, and this leads to a gently-sloping seafloor and an overlying, gently-sloping passive margin megasequence. In contrast, transform margins exhibit a minor amount of strike-slip-controlled extension, an abrupt, sub-vertical truncation of the continental margin by the transform fault that leads to a more steeply sloping seafloor with a more steeply-sloping, overlying, passive margin megasequence.
11. Preexisting crustal structure exerts variable resistance to rifting and that, in turn, controls the width of rift zones and their volcanic versus non-volcanic nature. Where these basement trends parallel the rift direction, the rift zones are

relatively narrow in width with smaller stretching factors because the lithosphere was weaker in this direction; where these basement trends are oblique to the rift direction, the rift zone is wider and stretching factors are greater because the lithosphere is stronger in this direction. Narrow rift zones with parallel basement trends favor deeper mantle involvement and therefore are more prone to become volcanic margins. Wider rift zones with oblique basement trends favor less mantle involvement and therefore would be more prone to become non-volcanic margins.

12. Previous workers have identified "marginal rifts" or "outer marginal troughs" that are immediately adjacent to the area of late Jurassic oceanic crust that underlies the deep, central Gulf of Mexico. A key element of the marginal rift concept is that an outer basement high exists seaward of the rift to confine continental or marine sediments along this seaward edge of the rift. This same outer high would form the landward edge of the late Jurassic oceanic crust.
13. The three giant salt basins of the GOM and the South Atlantic Ocean in Brazil and West Africa all show similar map patterns and isopachs. In general, all three salt basins are asymmetrical with the thicker deposits and wider extent of the deposit to the west in the GOM and so the south in Brazil and Angola.
14. The above information on the crustal structure and tectonostratigraphy of non-volcanic and volcanic rifted-passive margins can be used to develop a better understanding of the following topics: (1) Why do rifted-passive margins account for two-thirds of all giant oil and gas fields?; (2) How do hydrocarbon traps differ between orthogonal rifts and oblique rifts?; (3) Are non-volcanic rifted margins more hydrocarbon-rich than volcanic margins?; (4) Are "look-alike" hydrocarbon basins on conjugate margins the rule or the exception?; (5) How does the crustal structure of the Santos basin control its rift and rift sag-based, presalt reservoirs?; (6) How does the crustal structure of the Guyana margin control is passive-margin-based sources, reservoirs and seals?; (7) Does the continent-ocean boundary serve as an outer limit for deepwater exploration as the result of reduced heat flow in oceanic crust?

## Acknowledgments

I thank Jon Rotzien and the co-editors of this volume for their invitation to contribute to this chapter and for their many helpful suggestions. I thank ION Geophysical and TGS for kindly providing seismic data used in this paper. I thank the industry sponsors of the Conjugate Basins, Tectonics, and Hydrocarbons (CBTH) Project at the University of Houston for their continued financial support. Special thanks to Mark Longacre and present and former CBTH students Mei Liu, Marcus Zinecker and Kenneth Shipper for many discussions that contributed to the ideas of this paper. I thank Jeff Storms for his expert assistance in editing and proofing the final figures and text.

## References

Anka, Z., Seranne, M., di Primio, R., 2010. Evidence of a large upper-Cretaceous depocenter across the continent-ocean boundary of the Congo-Angola basin: implications for paleo-drainage and potential ultra-deep source rocks. Mar. Pet. Geol., 601–611. https://doi.org/10.1016/j.marpetgeo.2009.08.015.

Antobreh, A.A., Faleide, J.I., Tsikalas, F., Planke, S., 2009. Rift– shear architecture and tectonic development of the Ghana margin deduced from multichannel seismic reflection and potential field data. Mar. Pet. Geol. 26, 345–368. https://doi.org/10.1016/j.marpetgeo.2008.04.005.

Augustin, N., van der Zwan, F.M., Devey, C.W., Brandsóttir, B., 2021. 13 million years of seafloor spreading throughout the Red Sea Basin. Nat. Commun. 12, 2427. https://doi.org/10.1038/s41467-021-22586-2.

Bai, Y., Wang, X., Dong, D., Brune, S., Wu, S., Wang, Z., 2020. Symmetry of the South China Sea conjugate margins in a rifting, drifting and collision context. Mar. Pet. Geol. 117. https://doi.org/10.1016/j.marpetgeo.2020.104397, 104397.

Ball, P., Eagles, G., Ebiner, C., McClay, K., Totterdell, J., 2013. The spatial and temporal evolution of strain during the separation of Australia and Antarctica. Geochem. Geophys. Geosyst. 14 (8), 2771–2799. https://doi.org/10.1002/ggge.20160.

Basile, C., Allemand, P., 2002. Erosion and flexural uplift along transform faults. Geophys. J. Int. 151 (2), 646–653. https://doi.org/10.1046/j.1365-246X.2002.01805.x.

Bauer, K., Neben, S., Schreckenberger, B., Emmermann, R., Hinz, K., Fechner, N., Gohl, K., Schulze, A., Trumbull, R.B., Weber, K., 2000. Deep structure of the Namibia continental margin as derived from integrated geophysical studies. J. Geophys. Res. Atmos. 105 (B11), 25829–25854. https://doi.org/10.1029/2000JB900227.

Bécel, A., Davis, J.K., Shuck, B.D., Van Avendonk, H.J.A., Gibson, J.C., 2020. Evidence for a prolonged continental breakup resulting from slow extension rates at the eastern North American volcanic rifted margin. J. Geophys. Res. Solid Earth 125. https://doi.org/10.1029/2020JB020093, e2020JB020093.

Beglinger, S.E., Doust, H., Cloetingh, S., 2012. Relating petroleum system and play development to basin evolution: West African South Atlantic basins. Mar. Pet. Geol. 30, 1–25. https://doi.org/10.1016/j.marpetgeo.2011.08.008.

Bell, R.E., Duclaux, G., Nixon, C.W., Gawthorpe, R.L., McNeil, L.C., 2018. High-angle, not low-angle, normal faults dominate early rift extension in the Corinth Rift, central Greece. Geology 46, 115–118. https://doi.org/10.1130/G39560.1.

Bellahsen, N., Leroy, S., Autin, J., Razin, P., d'Acremont, E., Sloan, H., Pik, R., Ahmed, A., Khanbari, K., 2013. Pre-existing oblique transfer zones and transfer/transform relationships in continental margins: new insights from the southeastern Gulf of Aden, Socotra Island, Yemen. Tectonophysics 607, 32–50. https://doi.org/10.1016/j.tecto.2013.07.036.

Bellucci, M., Aslanian, D., Moulin, M., Rabineau, M., Leroux, E., Pellen, R., Poort, J., Del Ben, A., Gorini, C., Camerlenghi, A., 2021. Salt morphologies and crustal segmentation relationship: new insights from the Western Mediterranean Sea. Earth Sci. Rev. 222. https://doi.org/10.1016/j.earscirev.2021.103818, 103818.

Biari, Y., Klingelhoefer, F., Franike, D., Funck, T., Loncke, L., Sibuet, J.-C., Basile, C., Austin, J.A., Rigoti, C.A., Sahabi, M., Benabdellouahed, M., Roest, W.R., 2021. Structure and evolution of the Atlantic passive margins: a review of existing rifting models from wide-angle seismic data and kinematic reconstruction. Mar. Pet. Geol. 126, 104898. https://doi.org/10.1016/j.marpetgeo.2021.104898.

Biari, Y., Klingelhoefer, F., Sahabi, M., Funck, T., Benabdellouahed, M., Schnabel, M., Reichert, C., Gutscher, M.-A., Bronner, A., Austin, J.A., 2017. Opening of the central Atlantic Ocean: implications for geometric rifting and asymmetric initial seafloor spreading after continental breakup. Tectonics 36 (6), 1129–1150. https://doi.org/10.1002/2017TC004596.

Bilal, A., McClay, K., 2022. Tectonic and stratigraphic evolution of the central Exmouth Plateau, NW Shelf of Australia. Mar. Pet. Geol. 136. https://doi.org/10.1016/j.marpetgeo.2021.105447, 105447.

Blaich, O.A., Faleide, J.I., Tsikalas, F., Franke, D., León, E., 2009. Crustal-scale architecture and segmentation of the Argentine margin and its conjugate off South Africa. Geophys. J. Int. 178 (1), 85–105. https://doi.org/10.1111/j.1365-246X.2009.04171.x.

Bosworth, W., Stockli, D.F., Helgeson, D.E., 2015. Integrated outcrop, 3D seismic, and geochronologic interpretation of Red Sea dike-related deformation in the Western Desert, Egypt – the role of the 23 Ma Cairo "mini-plume". J. Afr. Earth Sci. 109, 107–119. https://doi.org/10.1016/j.jafrearsci.2015.05.005.

Bradley, D.C., 2008. Passive margins through earth history. Earth Sci. Rev. 91 (1–4), 1–26. https://doi.org/10.1016/j.earscirev.2008.08.001.

Brune, S., 2016. Rifts and rifted margins: a review of geodynamic processes and natural hazards. In: Duarte, J.C., Schellart, W.P. (Eds.), Plate Boundaries and Natural Hazards. AGU Geophysical Monograph, vol. 219., https://doi.org/10.1002/9781119054146.ch2.

Brune, S., Popov, A.A., Sobolev, S.V., 2012. Modeling suggests that oblique extension facilitates rifting and continental break-up. J. Geophys. Res. Solid Earth 117 (B8). https://doi.org/10.1029/2011JB008860, B08402.

Buchan, K.L., Ernst, R.E., 2021. Plumbing systems of large igneous provinces (LIPs) on Earth and Venus: investigating the role of giant circumferential and radiating dyke swarms, coronae and novae, and mid-crustal intrusive complexes. Gondwana Res. 100, 25–43. https://doi.org/10.1016/j.gr.2021.02.014.

Cameselle, A.L., Ranero, C.R., Barckhausen, U., 2020. Understanding the 3D formation of a wide rift: the central South China Sea rift system. Tectonics 39. https://doi.org/10.1029/2019TC006040, e2019TC006040.

Canales, J.P., 2014. Layering of oceanic crust. In: Harff, J., Meschede, M., Petersen, S., Thiede, J. (Eds.), Encyclopedia of Marine Geosciences. Springer, Dordrecht, https://doi.org/10.1007/978-94-007-6644-0_23-1.

Cartwright, J., Swart, R., Corner, B., 2012. Conjugate margins of the South Atlantic: Namibia-Pelotas. In: Roberts, D.G., Bally, A.W. (Eds.), Regional Geology and Tectonics: Phanerozoic Passive Margins, Cratonic Basins and Global Tectonic Maps. Elsevier, pp. 202–221, https://doi.org/10.1016/B978-0-444-56357-6.00005-6.

Causer, A., Pérez-Díaz, L., Adam, J., Eagles, G., 2020. Uncertainties in break-up markers along the Iberia–Newfoundland margins illustrated by new seismic data. Solid Earth 11, 397–417. https://doi.org/10.5194/se-11-397-2020.

Cawood, A.J., Ferrill, D.A., Morris, A.P., Norris, D., McCallum, D., Gillis, E., Smart, K.J., 2021. Tectonostratigraphic evolution of the Orphan Basin and Flemish pass region—part 1: results from coupled kinematic restoration and crustal area balancing. Mar. Pet. Geol. 128, 105042. https://doi.org/10.1016/j.marpetgeo.2021.105042.

Chenin, P., Manatschal, G., Ghienne, J.-F., Chao, P., 2021. The syn-rift tectono-stratigraphic record of rifted margins (part II): a new model to break through the proximal/distal interpretation frontier. Basin Res., 1–44. https://doi.org/10.1111/bre.12628.

Corti, G., 2012. Evolution and characteristics of continental rifting: analog modeling-inspired view and comparison with examples from the East African Rift System. Tectonophysics 522, 1–33. https://doi.org/10.1016/j.tecto.2011.06.010.

Courtillot, V., 1982. Propagating rifts and continental breakup. Tectonics 1, 239–250. https://doi.org/10.1029/TC001i003p00239.

Cowie, L., Angelo, R.M., Kusznir, N., Manatschal, G., Horn, B., 2016. Structure of the ocean–continent transition, location of the continent–ocean boundary and magmatic type of the northern Angolan margin from integrated quantitative analysis of deep seismic reflection and gravity anomaly data. In: Sabato Ceraldi, T., Hodgkinson, R.A., Backe, G. (Eds.), Petroleum Geoscience of the West Africa Margin. vol. 438. Geological Society of London, Special Publications, pp. 159–176, https://doi.org/10.1144/SP438.6.

Cullen, A., Reemst, P., Henstra, G., Gozzard, S., Ray, A., 2010. Rifting of the South China Sea: new perspectives. Pet. Geosci. 16, 273–282. https://doi.org/10.1144/1354-079309-908.

Curry, M.A.E., Peel, F.J., Hudec, M.R., Norton, I.O., 2018. Extensional models for the development of passive-margin salt basins, with application to the Gulf of Mexico. Basin Res. 30 (6), 1180–1199. https://doi.org/10.1111/bre.12299.

Dalziel, I.W.D., Lawver, L.A., 2022. Does the British Isles Paleocene dike swarm reflect the former location of the Iceland hotspot? In: Foulger, G.R., Hamilton, L.C., Jurdy, D.M., Stein, C.A., Howard, K.A., Stein, S. (Eds.), In the Footsteps of Warren B. Hamilton: New Ideas in Earth Science. vol. 553. Geological Society of America Special Paper, https://doi.org/10.1130/2021.2553(31).

Dana, J.D., 1873. On some results of the earth's contraction from cooling including a discussion of the origin of mountains and the nature of the Earth's interior. Am. J. Sci. s3-5 (30), 423–443. https://doi.org/10.2475/ajs.s3-5.30.423.

Darros de Matos, R.M., 2021. Magmatism and hotspot trails during and after continental break-up in the South Atlantic. Mar. Pet. Geol. 129. https://doi.org/10.1016/j.marpetgeo.2021.105077, 105077.

Davis, M., Kusznir, N., 2004. Depth-dependent lithospheric stretching at rifted continental margins. In: Karner, G.D., Taylor, B., Driscoll, N.W., Kohlstedt, D.L. (Eds.), Rheology and Deformation of the Lithosphere at Continental Margins. Columbia University Press, pp. 92–137.

Davison, I., 1997. Wide and narrow margins of the Brazilian South Atlantic. J. Geol. Soc. London 154, 471–476.

Davison, I., 2007. Geology and tectonics of the South Atlantic Brazilian salt basins. In: Ries, A.C., Butler, R.W.H., Graham, R.H. (Eds.), Deformation of the Continental Crust: The Legacy of Mike Coward. vol. 272. Geological Society, London, Special Publications, pp. 345–359, https://doi.org/10.1144/GSL.SP.2007.272.01.18.

Davison, I., Underhill, J.R., 2012. Tectonics and sedimentation in extensional rifts: implications for petroleum systems. In: Gao, D. (Ed.), Tectonics and Sedimentation: Implications for Petroleum Systems. vol. 100. AAPG Memoir, pp. 15–42, https://doi.org/10.1306/13351547M1001556.

Davy, R.G., Minshull, T.A., Bayrakci, G., Bull, J.M., Klaeschen, D., Papenberg, C., Reston, T.J., Sawyer, D.S., Zelt, C.A., 2016. Continental hyperextension, mantle exhumation, and thin oceanic crust at the continent-ocean transition, West Iberia: new insights from wide-angle seismic. J. Geophys. Res. Solid Earth 121, 3177–3199. https://doi.org/10.1002/2016JB012825.

Deng, H., Ren, J., Pang, X., Rey, P.F., McClay, K.R., Watkinson, I.M., Zheng, J., Luo, P., 2020. South China Sea documents the transition from wide continental rift to continental break up. Nat. Commun. 11, 4583. https://doi.org/10.1038/s41467-020-18448-y.

Devey, C., 2015. Hotspot-ridge interaction. In: Harff, J., Meschede, M., Petersen, S., Thiede, J. (Eds.), Encyclopedia of Marine Geosciences. Springer Science.

Dias, A.E.S., Lavier, L.L., Hayman, N.W., 2015. Conjugate rifted margins width and asymmetry: the interplay between lithospheric strength and thermomechanical processes. J. Geophys. Res. Solid Earth 120, 8672–8700. https://doi.org/10.1002/2015JB012074.

Dixon, T.H., Stern, R.J., Hussein, I.M., 1987. Control of Red Sea rift geometry by Precambrian structures. Tectonics 6, 551–571. https://doi.org/10.1029/TC006i005p00551.

Dolson, J.C., He, Z., Horn, B.W., 2021. Two decades (2000–2020) and five paradigm shifts gleaned from AAPG's giant fields database. In: Sternbach, C., Merrill, R.K., Dolson, J.C. (Eds.), Giant Fields of the Decade: 2010–2020. vol. 125. AAPG Memoir, pp. 37–68.

Doré, T., Lundin, E., 2015. Hyperextended continental margins – knowns and unknowns. Geology 43, 95–96. https://doi.org/10.1130/focus012015.1.

Driscoll, N.W., Karner, G.D., 1998. Lower crustal extension across the Northern Carnarvon basin, Australia: evidence for an eastward dipping detachment. J. Geophys. Res. Solid Earth 103 (B3), 4975–4991. https://doi.org/10.1029/97JB03295.

Dunbar, J., Sawyer, D., 1989. How preexisting weaknesses control the style of continental breakup. J. Geophys. Res. B6, 7278–7292. https://doi.org/10.1029/JB094iB06p07278.

Eagles, G., Pérez-Díaz, L., Scarselli, N., 2015. Getting over continent ocean boundaries. Earth Sci. Rev. 151, 244–265. https://doi.org/10.1016/j.earscirev.2015.10.009.

Empinotti, L., Zalan, P., Heilbron, M., 2015. Arcabouco crustal profundo da parte centro-norte da margem de Angola. vol. 23 Brazilian Geociencias Petrobras, Rio de Janeiro. December, 2015.

Evenick, J.C., 2021. Glimpses into Earth's history using a revised global sedimentary basin map. Earth Sci. Rev. 215. https://doi.org/10.1016/j.earscirev.2021.103564, 103564.

Faulds, J.E., Varga, R.J., 1998. The role of accommodation zones and transfer zones in the regional segmentation of extended terranes. In: Faulds, J.E., Stewart, J.H. (Eds.), Accommodation Zones and Transfer Zones: The Regional Segmentation of the Basin and Range Province. vol. 323. Geological Society of America Special Paper, pp. 1–46, https://doi.org/10.1130/0-8137-2323-X.1.

Fernandez, O., Olaiz, A., Cascone, L., Hernandez, P., Pereira, A.D.F., Tritla, J., Ingles, M., Aida, B., Pinto, I., Rocca, R., Sanders, C., Herra, A., Tur, N., 2020. Geophysical evidence for breakup volcanism in the Angola and Gabon passive margins. Mar. Pet. Geol. 116, 104330. https://doi.org/10.1016/j.marpetgeo.2020.104330.

Ferrand, T.P., 2020. Transition from amagmatic to volcanic margin: mantle exhumation in the Vøring Basin before the Icelandic plume influence. Tectonophysics 776. https://doi.org/10.1016/j.tecto.2020.228319, 228319.

Franke, D., 2013. Rifting, lithosphere breakup and volcanism: comparison of magma-poor and volcanic rifted margins. Mar. Pet. Geol. 43, 63–87. https://doi.org/10.1016/j.marpetgeo.2012.11.003.

Fraser, S.I., Fraser, A.J., Lentini, M.R., Gawthorpe, R.L., 2007. Return to rifts—the next wave: fresh insights into the petroleum geology of global rift basins. Pet. Geosci. 13, 99–104. https://doi.org/10.1144/1354-079307-749.

Galhom, T., Mann, P., Rudolph, K., 2022. Jurassic-recent stratigraphy, structure, and hydrocarbon potential of the Mesozoic-Cenozoic rifted-passive margin of the Tarfaya-Dakhla basin of southern Morocco. Mar. Pet. Geol. 139. https://doi.org/10.1016/j.marpetgeo.2022.105626, 105626.

Geoffroy, L., 2005. Volcanic passive margins. C. R. Geosci. 337 (16), 1395–1408. https://doi.org/10.1016/j.crte.2005.10.006.

Gernigon, L., Franke, D., Geoffroy, L., Schiffer, C., Foulger, G.R., Stoker, M., 2020. Crustal fragmentation, magmatism, and the diachronous opening of the Norwegian-Greenland Sea. Earth Sci. Rev. 206, 102839. https://doi.org/10.1016/j.earscirev.2019.04.011.

Gillard, M., Autin, J., Manatschal, G., Sauter, D., Munschy, M., Schaming, M., 2015. Tectonomagmatic evolution of the final stages of rifting along the deep conjugate Australian-Antarctic magma-poor rifted margins: constraints from seismic observations. Tectonics 34, 753–783. https://doi.org/10.1002/2015TC003850.

Gillard, M., Tugend, J., Müntener, O., Manatschal, G., Karner, G.D., Autin, J., Sauter, D., Figueredo, P.H., Ulrich, M., 2019. The role of serpentinization and magmatism in the formation of decoupling interfaces at magma-poor rifted margins. Earth Sci. Rev. 196, 102882. https://doi.org/10.1016/j.earscirev.2019.102882.

Gouiza, M., Paton, D.A., 2019. The role of inherited lithospheric heterogeneities in defining the crustal architecture of rifted margins and the magmatic budget during continental breakup. Geochem. Geophys. Geosyst. 20, 1836–1853. https://doi.org/10.1029/2018GC007808.

Graham, R., Kaymakci, N., Horn, B., 2013. The Black Sea: something different? Geo ExPro (October), 58–62.

Grant, C., 2020. In: Future transform margin exploration: giving it a seal of approval. Transform Margins Workshop 2, Online Conference, December 14–16.

Guiraud, M., Buta-Neto, A., Quesne, D., 2010. Segmentation and differential post-rift uplift at the Angola margin as recorded by the transform-rifted Benguela and oblique-to-orthogonal-rifted Kwanza basins. Mar. Pet. Geol. 27, 1040–1068. https://doi.org/10.1016/j.marpetgeo.2010.01.017.

Hasan, M.N., Mann, P., 2021. Structural styles and evolution of the Campeche salt basin, southern Gulf of Mexico. Mar. Pet. Geol. https://doi.org/10.1016/j.marpetgeo.2021.105313.

Hasan, M.N., Mann, P., 2022. Deep water passive margin fold belts. In: Rotzien, J., Yeilding, C., Sears, R., Hernández-Molina, F.J., Catuneanu, O. (Eds.), Deepwater Sedimentary Systems. Elsevier. In this issue.

Heine, C., Zoethout, J., Müller, R.D., 2013. Kinematics of the South Atlantic rift. Solid Earth 4, 215–253. https://doi.org/10.5194/se-4-215-2013.

Henstra, G.A., Kristensen, T.B., Rotevatn, A., Gawthorpe, R.L., 2019. How do pre-existing normal faults influence rift geometry? A comparison of adjacent basins with contrasting underlying structure on the Lofoten Margin, Norway. Basin Res. 31, 1083–1097.

Heron, P.J., Peace, A.L., McCaffrey, K.J.W., Welford, J.K., Wilson, R., van Hunen, J., Pysklywec, R.N., 2019. Segmentation of rifts through structural inheritance: creation of the Davis Strait. Tectonics 38, 2411–2430. https://doi.org/10.1029/2019TC005578.

Hopper, J., Dahl-Jensen, T., Holbrook, S., Larsen, H., Lizarralde, D., Korennaga, J., Kent, G., Kelemen, P., 2003. Structure of the SE Greenland margin from seismic reflection and refraction data: implications for nascent spreading center subsidence and asymmetric crustal accretion during North Atlantic opening. J. Geophys. Res. 108 (B5). https://doi.org/10.1029/2002JB001996. 2003.

Hudec, M., Jackson, M.P.A., Peel, F.J., 2013. Influence of deep Louann structure on the evolution of the northern Gulf of Mexico. AAPG Bull. 97 (10), 1711–1735. https://doi.org/10.1306/04011312074.

Intawong, A., Huuse, M., Rodriguez, K., Hodgson, N., 2015. De-risking Source Rock Maturity in the Luderitz Basin using Near-surface Geotherm Estimates and Basin Modelling: Spectrum. https://www.spectrumgeo.com/wp-content/uploads/Intawong-Huuse-Rodriguez-Hodgson-De-Risking-Source-Rock-Maturity-in-the-Luderitz-Basin-using-Near-Surface-Geotherm-Estimates-and-Basin-Modelling.pdf. (Accessed 1 December 2021).

Izquierdo-Llavall, E., Ringenbach, R.C., Sapin, F., Rives, T., Callot, J.P., 2022. Crustal structure and lateral variations in the Gulf of Mexico conjugate margins: from rifting to break-up. Mar. Pet. Geol. 136. https://doi.org/10.1016/j.marpetgeo.2021.105484, 105484.

Jammes, S., Lavier, L.L., 2019. Effect of contrasting strength from inherited crustal fabrics on the development of rifting margins. Geosphere 15 (2), 407–422. https://doi.org/10.1130/GES01686.1.

Karner, G.D., Gambôa, L.A.P., 2007. Timing and origin of the South Atlantic pre-salt sag basins and their capping evaporites. In: Schrieber, B.C., Lugli, S., Bąbel, M. (Eds.), Evaporites Through Space and Time. vol. 285. Geological Society of London, Special Publications, pp. 15–35, https://doi.org/10.1144/SP285.2.

Karson, J.A., 2021. Oceanic transform faults. In: Alderton, D., Elias, S.A. (Eds.), Encyclopedia of Geology, second ed. vol. 3. Academic Press, United Kingdom, pp. 930–946.

Kenning, J., Mann, P., 2021. Cenozoic structural deformation between the Lamprea fold-belt and Salina del Bravo salt province by interacting salt and shale detachments, western Gulf of Mexico. J. Struct. Geol. 153. https://doi.org/10.1016/j.jsg.2021.104458.

Khutorskoi, M., Polyak, B., 2016. Role of radiogenic heat generation in surface heat flow formation. Geotectonics 50 (2), 179–195. https://doi.org/10.1134/S0016852116020047.

Kolawole, F., Firkins, M.C., Al Wahaibi, T.S., Atekwana, E.A., Soreghan, M.J., 2021. Rift interaction zones and the stages of rift linkage in active segmented continental rift systems. Basin Res. 33, 2984–3020. https://doi.org/10.1111/bre.12592.

Krueger, A., Norton, I., Casey, E., Matos, R.D., Murphy, M., 2018. Influence of Proterozoic heritage on development of rift segments in the Equatorial Atlantic. In: Search and Discovery Article #30577, Adapted from Extended Abstract Prepared for Oral Presentation given at 2018 AAPG Annual Convention & Exhibition, Salt Lake City, Utah, May 20-23, 2018.

Kukla, P.A., Strozyk, F., Mohriak, W.U., 2018. South Atlantic salt basins – witnesses of complex passive margin evolution. Gondwana Res. 53, 41–57. https://doi.org/10.1016/j.gr.2017.03.012.

Lambiase, J.J., Morley, C.K., 1999. Hydrocarbons in rift basins: the role of stratigraphy. Philos. Trans. R. Soc. Lond. A 357, 877–900. https://doi.org/10.1098/rsta.1999.0356.

Latham, A., 2021. Giant fields and exploration trends of the decade 2010–2019. In: Sternbach, C.A., Merrill, R.K., Dolson, J.C. (Eds.), Giant Fields of the Decade: 2010-2020. vol. 125. AAPG Memoir, pp. 11–20.

Lavier, L.L., Manatschal, G., 2006. A mechanism to thin the continental lithosphere at magma-poor margins. Nature 440, 324–328. https://doi.org/10.1038/nature04608.

Lemoine, M., Tricart, P., Boillot, G., 1987. Ultramafic and gabbroic ocean floor of the Ligurian Tethys (Alps, Corsica, Apennines): in search of a genetic imodel. Geology 15 (7), 622–625. https://doi.org/10.1130/0091-7613(1987)15<622:UAGOFO>2.0.CO;2.

Lentini, M.R., Fraser, S.I., Sumner, H.S., Davies, R.J., 2010. Geodynamics of the central South Atlantic conjugate margins: implications for hydrocarbon potential. Pet. Geosci. 16, 217–229. https://doi.org/10.1144/1354-079309-909.

Li, Z., Wan, L., Liu, J., Xue, L., Su, Y., 2014. Geodynamic characteristics and their effect on the petroleum geology conditions for passive rift basin in central-western African region. In: Search and Discovery Article #10644, Adapted from Extended Abstract Prepared in Conjunction with Oral Presentation Given at AAPG International Conference & Exhibition, Istanbul, Turkey, September 14–17, 2014.

Leroy, S., Lucazeau, F., d'Acremont, E., Watremez, L., Autin, J., Rouzo, S., Bellhahsen, N., Tiberi, C., Ebinger, C., Beslier, M.-O., Perrot, J., Razin, P., Rolandone, F., Sloan, H., Stuart, G., Lazki, A.A., Al-Toubi, K., Bache, F., Bonneville, A., Goutorbe, B., Huchon, P., Unternehr, P., Khanbari, K., 2010. Contrasted styles of rifting in the eastern Gulf of Aden: a combined wide-angle, multichannel seismic, and heat flow survey. Geochem. Geophys. Geosyst. 11 (7). https://doi.org/10.1029/2009GC002963, Q07004.

Leroy, S., Razin, P., Autin, J., Bache, F., d'Acremont, E., Watremez, L., Robinet, J., Baurion, C., Denèle, Y., Bellahsen, N., Lucazeau, F., Rolandone, F., Rouzo, S., Kiel, J.S., Robin, C., Guillocheau, F., Tiberi, C., Basuyau, C., Beslier, M.-O., Ebinger, C., Stuart, G., Ahmed, A., Khanbari, K., Al Ganad, I., de Clarens, P., Unternehr, P., Al Toubi, K., Al Lazki, A., 2012. From rifting to oceanic spreading in the Gulf of Aden: a synthesis. Arab. J. Geosci. 5, 859–901. https://doi.org/10.1007/s12517-011-0475-4.

Li, L., Stephenson, R., Clift, P.D., 2016. The Canada Basin compared to the southwest South China Sea: two marginal ocean basins with hyper-extended continent-ocean transitions. Tectonophysics 691, 171–184. https://doi.org/10.1016/j.tecto.2016.02.042.

Ligi, M., Bonatti, E., Bosworth, W., Cai, Y., Cipriani, A., Palmiotto, C., Ronca, S., Seyler, M., 2018. Birth of an ocean in the Red Sea: oceanic-type basaltic melt intrusions precede continental rupture. Gondw. Res. 54, 150–160. https://doi.org/10.1016/j.gr.2017.11.002.

Lin, P., Bird, D., Mann, P., 2019. Crustal structure of an extinct, late Jurassic-to-earliest Cretaceous spreading center and its adjacent oceanic crust in the eastern Gulf of Mexico. Mar. Geophys. Res. 40 (3), 395–418. https://doi.org/10.1007/s1100101909379-5.

Liu, M., 2021. Mesozoic Rift Evolution and Crustal Structure of the Gulf of Mexico Basin from Integration of Multiple Geological and Geophysical Datasets. PhD dissertation, University of Houston. 219 p.

Liu, M., Filina, I., Mann, P., 2019. Crustal structure of Mesozoic rifting in the northeastern Gulf of Mexico from the integration of seismic and potential fields data. Interpretation 7 (4), 1–33. https://doi.org/10.1190/INT-2018-0259.1.

Lizarralde, D., Axen, G.J., Brown, H.E., Fletcher, J.M., González-Fernández, A., Harding, A.J., Holbrook, W.S., Kent, G.M., Paramo, P., Sutherland, F., Umhoefer, P.J., 2007. Variation in styles of rifting in the Gulf of California. Nature 448, 466–469. https://doi.org/10.1038/nature06035.

Loncke, L., Mercier de Lepinay, M., Basile, C., Maillard, A., Roest, W., De Clarens, P., Patriat, M., Gaullier, V., Klingelhoefer, F., Graindorge, D., Sapin, F., 2022. Compared structure and evolution of the Demerara and Guinea transform margin plateaus. Tectonophysics, 822. https://doi.org/10.1016/j.tecto.2021.229112.

Louden, K., Wu, Y., Tari, G., 2012. Systematic variations in basement morphology and rifting geometry along the Nova Scotia and Morocco conjugate margins. In: Mohriak, W.U., Danforth, A., Post, P.J., Brown, D.E., Tari, G.C., Nemčok, M., Sinha, S.T. (Eds.), Conjugate Divergent Margins. vol. 369. Geological Society of London, Special Publications, pp. 267–287.

Lutz, R., Franke, D., Berglar, K., Heyde, I., Schreckenberger, B., Klitzke, P., Geissler, W.H., 2018. Evidence for mantle exhumation since the early evolution of the slow-spreading Gakkel Ridge, Arctic Ocean. J. Geodyn. 118, 154–165. https://doi.org/10.1016/j.jog.2018.01.014.

Maestrelli, D., Montanari, D., Corti, G., Del Ventisette, C., Moratti, G., Bonini, M., 2020. Exploring the interactions between rift propagation and inherited crustal fabrics through experimental modeling. Tectonics 39. https://doi.org/10.1029/2020TC006211, e2020TC006211.

Mann, P., 2007. Global catalogue, classification and tectonic origins of restraining- and releasing bends on active and ancient strike-slip fault systems. In: Cunningham, W.D., Mann, P. (Eds.), Tectonics of Strike-Slip Restraining and Releasing Bends. vol. 290. Geological Society of London, pp. 13–142.

Mann, P., 2012. Comparison of structural styles and giant hydrocarbon occurrences within four active strike-slip regions: California, Southern Caribbean, Sumatra, and East China. In: Gao, D. (Ed.), Tectonics and Sedimentation: Implications for Petroleum Systems. vol. 100. AAPG Memoir, pp. 43–93.

Mann, P., 2015. Passive plate margin. In: Harff, J., Meschede, M., Petersen, S., Thiede, J. (Eds.), Encyclopedia of Marine Geosciences. Springer, Dordrecht, pp. 1–8, https://doi.org/10.1007/978-94-007-6644-0_100-1.

Mann, P., Gahagan, L., Gordon, M., 2003. Tectonic setting of the world's giant oil and gas fields. In: Halbouty, M.T. (Ed.), Giant Oil and Gas Fields of the Decade 1990-1999. vol. 78. AAPG Memoir, pp. 15–105.

Marzoli, A., Renne, P.R., Piccirillo, E.M., Ernesto, M., Bellieni, G., De Min, A., 1999. Extensive 200-million-year-old continental flood basalts of the Central Atlantic Magmatic Province. Science 284 (5414), 616–618. https://doi.org/10.1126/science.284.5414.616.

Masini, E., Manatschal, G., Mohn, G., Ghienne, J.-F., 2011. The tectono-sedimentary evolution of a supra-detachment rift basin at a deep-water magma-poor rifted margin: the example of the Samedan Basin preserved in the Err nappe in SE Switzerland. Basin Res. 23, 652–677. https://doi.org/10.1111/j.1365-2117.2011.00509.x.

Masini, E., Manatschal, G., Mohn, G., Unternehr, P., 2012. Anatomy and tectono-sedimentary evolution of a rift-related detachment system: the example of the Err detachment (central Alps, SE Switzerland). GSA Bull. 123, 1535–1551.

McDermott, K., Gillbard, E., Clarke, N., 2015. From basalt to skeletons—the 200 million-year history of the Namibian margin uncovered by new seismic data. First Break 33, 77–85. https://doi.org/10.3997/1365-2397.33.12.83748.

Mazur, S., Campbell, S., Green, C., Bouatmani, R., 2015. Extension across the Laptev Sea continental rifts constrained by gravity modeling. Tectonics 34 (3), 435–448. https://doi.org/10.1002/2014TC003590.

McDermott, C., Collier, J., Lonergan, L., Fruehn, J., Bellingham, P., 2019. Seismic velocity structure of seaward-dipping reflectors on the South American continental margin. Earth Planet. Sci. Lett. 521, 14–24. https://doi.org/10.1016/j.epsl.2019.05.049.

McHone, J.G., Puffer, J.H., 2003. Flood basalt provinces of the Pangean Atlantic rift: regional extent and environmental significance. In: LeTourneau, P. M., Olsen, P.E. (Eds.), The Great Rift Valleys of Pangea in Eastern North America. Columbia University Press, pp. 141–154.

McKenzie, D., 1978. Some remarks on the development of sedimentary basins. Earth Planet. Sci. Lett. 40 (1), 25–32. https://doi.org/10.1016/0012-821X(78)90071-7.

Mello, M.R., De Azambuja Filho, N.C., Bender, A.A., Barbanti, S.M., Mohriak, W., Schmitt, P., De Jesus, C.L., 2012. The Namibian and Brazilian southern South Atlantic petroleum systems: are they comparable analogues. In: Mohriak, W., Danforth, A., Post, P., Brown, D., Tari, G., Nemcok, M., Sinha, S. (Eds.), Conjugate Divergent Margins. vol. 369. Geological Society of London, Special Publications, https://doi.org/10.1144/SP369.18.

Mello, M.R., Peres, W., Rostirolla, S., Pedrosa Jr., O., Piquet, A., Becker, S., Yilmaz, P., 2021a. The Santos Basin pre-salt super giant petroleum system: an incredible journey from failure to success. In: Mello, M.R., Yilmaz, P.O., Katz, B.J. (Eds.), The Supergiant Lower Cretaceous Pre-salt Petroleum Systems of the Santos Basin, Brazil. vol. 124. AAPG Memoir, pp. 1–34.

Mello, M.R., Rostirolla, S.P., Peres, W., Pedrosa Jr., O.A., Carvalho, M.D., Cardoso Netto, A., 2021b. Tupi Field: the largest oil producer in the South Atlantic realm, Santos Basin, Brazil. In: Mello, M.R., Yilmaz, P.O., Katz, B.J. (Eds.), The Supergiant Lower Cretaceous Pre-salt Petroleum Systems of the Santos Basin, Brazil. vol. 124. AAPG Memoir, pp. 395–430.

Menzies, M.A., Klemperer, S.L., Ebinger, C.J., Baker, J., 2002. Characteristics of volcanic rifted margins. In: Menzies, M.A., Klemperer, S.I., Ebinger, C. J., Baker, J. (Eds.), Volcanic Rifted Margins. vol. 362. Geological Society of America Special Paper, pp. 1–14.

Meschede, M., 2014. Wilson cycle: marine Geosciences. In: Harff, J., Meschede, M., Petersen, S., Thiede, J. (Eds.), Encyclopedia of Marine Geosciences. Springer, Dordrecht, https://doi.org/10.1007/978-94-007-6644-0_203-1.

Meschede, M., 2015. Lithosphere: structure and composition. In: Harff, J., Meschede, M., Petersen, S., Thiede, J. (Eds.), Encyclopedia of Marine Geosciences. Springer, Dordrecht, https://doi.org/10.1007/978-94-007-6644-0_86-1.

Miller, N.C., Lizarralde, D., Collins, J.A., Holbrook, W.S., Van Avendonk, H.J.A., 2021. Limited mantle hydration by bending faults at the Middle America Trench. J. Geophys. Res. Solid Earth 126 (1). https://doi.org/10.1029/2020JB020982.

Minguez, D., Hensel, E.G., Johnson, E.A.E., 2020. A fresh look at Gulf of Mexico tectonics: testing rotations and breakup mechanisms from the perspective of seismically constrained potential-fields modeling and plate kinematics. Interpretation 8 (4), SS31–SS45. https://doi.org/10.1190/INT-2019-0256.1.

Mitchell, N.C., Shi, W., Izzeldin, A.Y., Stewart, I.C.F., 2020. Reconstructing the level of the central Red Sea evaporites at the end of the Miocene. Basin Res. 33, 1266–1292. https://doi.org/10.1111/bre.12513.

Mohriak, W., 2014. In: Birth and development of continental margin basins: analogies from the South Atlantic, North Atlantic, and the Red Sea. AAPG Foundation Distinguished Lecture Search and Discovery Article 41502.

Morley, C.K., 1999a. Influence of preexisting fabrics on rift structure. In: Morley, C.K. (Ed.), Geoscience of Rift Systems— Evolution of East Africa. AAPG Studies in Geology No. 44, pp. 151–160, https://doi.org/10.1306/St44623C9.

Morley, C.K., 1999b. Comparison of hydrocarbon prospectivity in rift systems. In: Morley, C.K. (Ed.), Geoscience of Rift Systems— Evolution of East Africa. AAPG Studies in Geology No. 44, pp. 233–242, https://doi.org/10.1306/St44623C14.

Morley, C.K., 2009. Geometry and evolution of low-angle normal faults (LANFs) within a Cenozoic high-angle rift system, Thailand: implications for sedimentology and the mechanisms of LANF development. Tectonics 28, 1–30.

Morley, C.K., Nelson, R.A., Patton, T.L., Munn, S.G., 1990. Transfer zones in the East African rift system and their relevance to hydrocarbon exploration in rifts. AAPG Bull. 74, 1234–1253. https://doi.org/10.1306/0C9B2475-1710-11D7-8645000102C1865D.

Mortimer, E.J., Gouiza, M., Paton, D.A., Stanca, R., Rodriguez, K., Hodgson, N., Hussein, A.A., 2020. Architecture of a magma poor passive margin—insights from the Somali margin. Mar. Geol. 428, 106269. https://doi.org/10.1016/j.margeo.2020.106269.

Mounguengui, M.M., Guiraud, M., 2009. Neocomian to early Aptian syn-rift evolution of the normal to oblique-rifted North Gabon Margin (Interior and N'Komi Basins). Mar. Pet. Geol. 26 (6), 100–1017. https://doi.org/10.1016/j.marpetgeo.2008.11.001.

Museur, T., Graindorge, D., Klingelhoefer, F., Roest, W.R., Basile, C., Loncke, L., Sapin, F., 2021. Deep structure of the Demerara Plateau: from a volcanic margin to a transform marginal plateau. Tectonophysics 803, 228645. https://doi.org/10.1016/j.tecto.2020.228645.

Nance, R.D., Neace, E.R., Braid, J., Murphy, J.B., 2015. Does the Meguma Terrane extend into SW England? Geoscience Canada 4 (1), 61–76. https://doi.org/10.12789/geocanj.2014.41.056.

Nemčok, M., Henk, A., Allen, R., Sikora, P.J., Stuart, C., 2012. Continental break-up along strike-slip fault zones: observations from the Equatorial Atlantic. In: Mohriak, W.U., Danforth, A., Post, P.J., Brown, D.E., Tari, G.C., Nemčok, M., Sinha, S.T. (Eds.), Conjugate Divergent Margins. vol. 369. Geological Society of London, Special Publications, pp. 537–556, https://doi.org/10.1144/SP369.8.

Nemcok, M., Rybar, S., Sinha, S., Hermeston, S., Ledvenyiova, L. (Eds.), Transform Margins: Development, Controls, and Petroleum Systems, vol. 431. Geological Society of London Special Publication. https://doi.org/10.1144/SP431.15.

Nguyen, L.C., Hall, S.A., Bird, D.E., Ball, P.J., 2016. Reconstruction of the East Africa and Antarctica continental margins. J. Geophys. Res. Solid Earth 121 (6), 4156–4179. https://doi.org/10.1002/2015JB012776.

Nixon, C.W., McNeill, L.C., Bull, J.M., Bell, R.E., Gawthorpe, R.L., Henstock, T.J., Christodoulou, D., Ford, M., Taylor, B., Sakellariou, D., Ferentinos, G., Papatheodorou, G., Leeder, M.R., Collier, R.E.L.I., Goodliffe, A.M., Sachpazi, M., Kranis, H., 2016. Rapid spatiotemporal variations in rift structure during development of the Corinth Rift, central Greece. Tectonics 35, 1225–1248. https://doi.org/10.1002/2015TC004026.

Nonn, C., Leroy, S., Lescanne, M., Castilla, R., 2019. Central Gulf of Aden conjugate margins (Yemen-Somalia): tectono-sedimentary and magmatism evolution in hybrid-type margins. Mar. Pet. Geol. 105, 100–123. https://doi.org/10.1016/j.marpetgeo.2018.11.053.

Norcliffe, J.R., Paton, D.A., Mortimer, E.J., McCaig, A.M., Nicholls, H., Rodriguez, K., Hodgson, N., Van Der Spuy, D., 2018. Laterally confined volcanic successions (LCVS); recording rift-jumps during the formation of magma-rich margins. Earth Planet. Sci. Lett. 504, 53–63. https://doi.org/10.1016/j.epsl.2018.09.033.

Olyphant, J.R., Johnson, R.A., Hughes, A.N., 2017. Evolution of the Southern Guinea Plateau: implications on Guinea-Demerara Plateau formation using insights from seismic, subsidence, and gravity data. Tectonophysics 717, 358–371. https://doi.org/10.1016/j.tecto.2017.08.036.

Oskin, M., Stock, J., 2003. Cenozoic volcanism and tectonics of the continental margins of the upper Delfin Basin, northeastern Baja California and western Sonora. In: Johnson, S.E., Fletcher, J.M., Girty, G.H., Kimbrough, D.L., Martín-Barajas, A. (Eds.), Tectonic Evolution of Northwestern Mexico and the Southwestern USA. vol. 374. Geological Society of America, pp. 421–438.

Osmundsen, P.T., Péron-Pinvidic, G., Ebbing, J., Erratt, D., Fjellanger, E., Bergslien, D., Syvertsen, S.E., 2016. Extension, hyperextension and mantle exhumation offshore Norway: a discussion based on 6 crustal transects. Nor. J. Geol. 96, 343–372. https://doi.org/10.17850/njg96-4-05.

Pagli, C., Yun, S.-H., Ebinger, C., Keir, D., Wang, H., 2018. Strike-slip tectonics during rift linkage. Geology 47, 31–34. https://doi.org/10.1130/G45345.

Paton, D., Pindell, J., Bellingham, P., Horn, B., 2017. Evolution of seaward-dipping reflectors at the onset of oceanic crust formation at volcanic passive margins: insights from the South Atlantic. Geology 45, 439–442. https://doi.org/10.1130/G38706.1.

Peace, A.L., Phethean, J.J.J., Franke, D., Foulger, G.R., Schiffer, C., Welford, J.K., McHone, G., Rocchi, S., Schnabel, M., Doré, A.G., 2020. A review of Pangaea dispersal and large igneous provinces—in search of a causative mechanism. Earth Sci. Rev. 206, 102902. https://doi.org/10.1016/j.earscirev.2019.102902.

Peace, A.L., Welford, J.K., 2020. "Conjugate margins" – an oversimplification of the complex southern North Atlantic rift and spreading system? Interpretation 8 (2). https://doi.org/10.1190/int-2019-0087.1, SH33.

Pérez-Gussinyé, M., Andrés-Martínez, M., Araújo, M., Xin, Y., Armitage, J., Morgan, J.P., 2020. Lithospheric strength and rift migration controls on synrift stratigraphy and breakup unconformities at rifted margins: examples from numerical models, the Atlantic and South China Sea margins. Tectonics 39 (12). https://doi.org/10.1029/2020TC006255.

Peron-Pinvidic, G., Manatschal, G., Masini, E., Sutra, E., Flament, J.M., Haupert, I., Unternehr, P., 2017. Unravelling the along-strike variability of the Angola–Gabon rifted margin: a mapping approach. In: Sabato Ceraldi, T., Hodgkinson, R.A., Backe, G. (Eds.), Petroleum Geoscience of the West Africa Margin. vol. 438. Geological Society of London, Special Publications, pp. 49–76, https://doi.org/10.1144/SP438.1.

Péron-Pinvidic, G., Manatschal, G., Osmundsen, P.T., 2013. Structural comparison of archetypal Atlantic rifted margins: a review of observations and concepts. Mar. Pet. Geol. 43, 21–47. https://doi.org/10.1016/j.marpetgeo.2013.02.002.

Peron-Pinvidic, G., Osmundsen, P.T., 2016. Architecture of the distal and outer domains of the Mid-Norwegian rifted margin: insights from the Rån-Gjallar ridges system. Mar. Pet. Geol. 77, 280–299. https://doi.org/10.1016/j.marpetgeo.2016.06.014.

Pindell, J., Graham, R., Horn, B., 2014. Rapid outer marginal collapse at the rift to drift transition of passive margin evolution, with a Gulf of Mexico case study. Basin Res. 26, 1–25. https://doi.org/10.1111/bre.12059.

Plummer, M., Tilton, J., 2020. The Kwanza-Campos conjugate pair: mirror twins? Revisiting and exploring new frontiers. First Break 39, 63–70.

Price, L., Perkey, R., Oswald, E., Elkington, E., 2021. Liza Field, Guyana: the finding of a stratigraphic giant, from early exploration to production. In: Sternbach, C.A., Merrill, R.K., Dolson, J.C. (Eds.), Giant Fields of the Decade: 2010-2020. vol. 125. AAPG Memoir, pp. 433–456.

Reston, T.J., 2009. The structure, evolution and symmetry of the magma-poor rifted margins of the North and Central Atlantic: a synthesis. Tectonophysics 468 (1–4), 6–27. https://doi.org/10.1016/j.tecto.2008.09.002.

Reuber, K., Mann, P., 2019. Control of Precambrian-to-Paleozoic orogenic trends on along-strike variations in early Cretaceous continental rifts of the South Atlantic Ocean. Interpretation 7, 1–24. https://doi.org/10.1190/INT-2018-0257.1.

Reuber, K., Mann, P., Pindell, J., 2019. Hotspot origin for asymmetrical conjugate volcanic margins of the austral South Atlantic Ocean as imaged on deeply penetrating seismic reflection lines. Interpretation 7, SH71–SH97. https://doi.org/10.1190/INT-2018-0256.1.

Reuber, K., Pindell, J., Horn, B., 2016. Demerara Rise, offshore Suriname: magma-rich segment of the Central Atlantic Ocean, and conjugate to the Bahamas hotspot. Interpretation 4 (2), T31–T45. https://doi.org/10.1190/INT-2014-0246.1.

Romito, S., Mann, P., 2022. Crustal structure of the Camamu-Almada margin along the northeastern rift segment of Brazil from an integration of deep-penetration seismic reflection profiles, refraction, and gravity modeling. Tectonics. In press.

Rowan, M., 2022. The ocean-continent transition of late synrift salt basins: extension and evaporite deposition in the southern Gulf of Mexico and global analogs. In: Koeberl, C., Claeys, P., Montanari, A. (Eds.), From the Guajira Desert to the Apennines, and from Mediterranean Microplates to the Mexican Killer Asteroid: Honoring the Career of Walter Alvarez. vol. 557. Geological Society of America Special Papers, https://doi.org/10.1130/2022.2557(12).

Rowan, M., Peel, F.J., Vendeville, B.C., 2004. Gravity-driven fold belts on passive margins. In: McClay, K.R. (Ed.), Thrust Tectonics and Hydrocarbon Systems. vol. 82. AAPG Memoir, pp. 157–182.

Sandoval, L., Welford, J.K., MacMahon, H., Peace, A.L., 2019. Determining continuous basins across conjugate margins: the East Orphan, Porcupine, and Galicia interior basins of the southern North Atlantic Ocean. Mar. Pet. Geol. 110, 138–161. https://doi.org/10.1016/j.marpetgeo.2019.06.047.

Sandwell, D.T., Smith, W.H.F., 2009. Global marine gravity from retracked Geosat and ERS-1 altimetry: ridge segmentation versus spreading rate. J. Geophys. Res. 114, B01411. https://doi.org/10.1029/2008JB006008.

Sandwell, D.T., Müller, R.D., Smith, W.H.F., Garcia, E., Francis, R., 2014. New global marine gravity model from CryoSat-2 and Jason-1 reveals buried tectonic structure. Science 346, 65–67. https://doi.org/10.1126/science.1258213.

Sapin, F., Ringenbach, J.-C., Clerc, C., 2021. Rifted margins classification and forcing parameters. Sci. Rep. 11, 8199. https://doi.org/10.1038/s41598-021-87648-3.

Savva, D., Meresse, F., Pubellier, M., Chamot-Rooke, N., Lavier, L., Wong Po, K., Franke, D., Steuer, S., Sapin, F., Auxietre, J.L., Lamy, G., 2013. Seismic evidence of hyper-stretched crust and mantle exhumation offshore Vietnam. Tectonophysics 608, 72–83. https://doi.org/10.1016/j.tecto.2013.07.010.

Schiffer, C., Doré, A.G., Foulger, G.R., Franke, D., Geoffroy, L., Gernigon, L., Holdsworth, B., Kusznir, N., Lundin, E., McCaffrey, K., Peace, A.L., Petersen, K.D., Phillips, T.B., Stephenson, R., Stoker, M.S., Welford, J.K., 2020. Structural inheritance in the North Atlantic. Earth Sci. Rev. 206, 102975. https://doi.org/10.1016/j.earscirev.2019.102975.

Schumacher, M.E., 2002. Upper Rhine graben: role of preexisting structures during rift evolution. Tectonics 21, 1006. https://doi.org/10.1029/2001TC900022.

Sembroni, A., Faccenna, C., Becker, T.W., Molin, P., Abebe, B., 2016. Long-term, deep-mantle support of the Ethiopia-Yemen Plateau. Tectonics 35, 469–488. https://doi.org/10.1002/2015TC004000.

Sengor, A.M.C., 2001. Rifts of the world. In: Ernst, R., Buchan, K. (Eds.), Mantle Plumes: Their Identification through Time. vol. 352. Geological Society of America Special Paper, pp. 389–482.

Seton, M., Müller, R.D., Zahirovic, S., Williams, S., Wright, N.M., Cannon, J., Whittaker, J.M., Matthews, K.J., McGirr, R., 2020. A global data set of present-day oceanic crustal age and seafloor spreading parameters. Geochem. Geophys. Geosyst. 21 (10). https://doi.org/10.1029/2020GC009214.

Shipper, K., Mann, P., 2022. In: Testing rift versus transform models for the Guyana margin using 3D gravity model compared to seismic reflection data. SEG/AAPG IMAGE, The International Meeting for Applied Geoscience & Energy. In preparation.

Sibuet, J.-C., Srivastava, S., Manatschal, G., 2007. Exhumed mantle-forming transitional crust in the Newfoundland-Iberia rift and associated magnetic anomalies. J. Geophys. Res. 112. https://doi.org/10.1029/2005JB003856, B06105.

Snedden, J., Stockli, D., Norton, I., 2020. Paleogeographical reconstruction and provenance of Oxfordian aeolian sandstone reservoirs in Mexico offshore areas: comparison to the Norphlet aeolian system of the northern Gulf of Mexico. In: Davison, I., Hull, J., Pindell, J. (Eds.), The Basins, Orogens, and Evolution of the southern Gulf of Mexico and northern Caribbean. vol. 504. Geological Society of London, Special Publications, https://doi.org/10.1144/SP504-2019-219.

Song, T., Li, C.-F., Wu, S., Yao, Y., Gao, J., 2019. Extensional styles of the conjugate rifted margins of the South China Sea. J. Asian Earth Sci. 177, 117–128. https://doi.org/10.1016/j.jseaes.2019.03.008.

Song, C., Jiang, S., Wen, Z., Wang, Z., 2020. Comparison of the petroleum geology in the deep-water basins between the passive margin of Morocco and its conjugate margin of Canada. J. Earth Sci. 31 (5), 919–929. https://doi.org/10.1007/s12583-020-1340-8.

Steier, A., Mann, P., 2019. Late Mesozoic gravity sliding and Oxfordian hydrocarbon reservoir potential of the northern Yucatan margin. Mar. Pet. Geol. 103, 681–701. https://doi.org/10.1016/j.marpetgeo.2019.03.001.

Strozyk, F., Back, S., Kukla, P., 2017. Comparison of the rift and post-rift architecture of conjugated salt and salt-free basins offshore Brazil and Angola/Namibia, South Atlantic. Tectonophysics 716, 204–224. https://doi.org/10.1016/j.tecto.2016.12.012.

Tari, G., Brown, D., Jabour, H., Hafid, M., Louden, K., Zizi, M., 2012. The conjugate margins of Morocco and Nova Scotia. In: Roberts, D.G., Bally, A.W. (Eds.), Regional Geology and Tectonics: Phanerozoic Passive Margins, Cratonic Basins and Global Tectonic Maps. Elsevier, pp. 284–323.

Taylor, B., Weiss, J.R., Goodliffe, A.M., Sachpazi, M., Laigle, M., Hirn, A., 2011. The structures, stratigraphy and evolution of the Gulf of Corinth rift, Greece. Geophys. J. Int. 185, 1189–1219. https://doi.org/10.1111/j.1365-246X.2011.05014.x.

Trude, J., Kilsdonk, B., Grow, T., Ott, B., 2022. The structure and tectonics of the Guyana basin. In: Nemcˇok, M., Doran, H., Doré, A.G., Ledvényiová, L., Rybár, S. (Eds.), Tectonic Development, Thermal History and Hydrocarbon Habitat Models of Transform Margins: Their Differences from Rifted Margins. vol. 4. Geological Society of London, Special Publications, https://doi.org/10.1144/SP524-2021-117.

Tugend, J., Gillard, M., Manatschal, G., Nirrengarten, M., Harkin, C., Epin, M.-E., Sauter, D., Autin, J., Kusznir, N., McDermott, K., 2018. Reappraisal of the magma-rich versus magma-poor rifted margin archetypes. In: McClay, K., Hammerstein, J.A. (Eds.), Passive Margin: Tectonics, Sedimentation, and Magmatism. vol. 476. Geological Society of London, Special Publications, pp. 23–47.

Turcotte, D., Emerman, S., 1983. Mechanisms of active and passive rifting. Tectonophysics 94, 39–50.

Umhoefer, P.J., Darin, M.H., Bennett, S.E., Skinner, L.A., Dorsey, R.J., Oskin, M.E., 2018. Breaching of strike-slip faults and successive flooding of pull-apart basins to form the Gulf of California seaway from ca. 8-6 Ma. Geology 46 (8), 695–698. https://doi.org/10.1130/G40242.1.

Unternehr, P., Peron-Pinvidic, G., Manatschal, G., Sutra, E., 2010. Hyper-extended crust in the South Atlantic: in search of a model. Pet. Geosci. 16, 207–215.

Van Avendonk, H.J.A., Christeson, C.L., Norton, I.O., Eddy, D.R., 2015. Continental rifting and sediment infill in the northwestern Gulf of Mexico. Geology 43, 631–634. https://doi.org/10.1130/G36798.1.

Veevers, J.J., 2012. Reconstructions before rifting and drifting reveal the geological connections between Antarctica and its conjugates in Gondwanaland. Earth Sci. Rev. 111 (3–4), 249–318. https://doi.org/10.1016/j.earscirev.2011.11.009.

Wallace, L.M., Ellis, S., Mann, P., 2009. Collisional model for rapid fore-arc block rotations, arc curvature, and episodic back-arc rifting in subduction settings. Geochem. Geophys. Geosyst. 10 (5). https://doi.org/10.1029/2008GC002220, Q05001.

Watts, A.B., 2012. Models for the evolution of passive margins. In: Roberts, D.G., Bally, A.W. (Eds.), Regional Geology and Tectonics: Phanerozoic Rift Systems and Sedimentary Basins. Elsevier, pp. 32–57, https://doi.org/10.1016/B978-0-444-56356-9.00002-X.

Wen, Z., Shu, J., Song, C., Wang, Z., He, Z., 2019. Basin evolution, configuration styles, and hydrocarbon accumulation of the South Atlantic conjugate margins. Energy Explor. Exploit. 37 (3), 992–1008. https://doi.org/10.1177/0144598719840751.

Wernicke, B., 1985. Uniform-sense normal simple shear of the continental lithosphere. Can. J. Earth Sci. 22 (1), 108–125. https://doi.org/10.1139/e85-009.

Whalen, L., Gazel, E., Vidito, C., Puffer, J., Bizimis, M., Henika, W., Caddick, M.J., 2015. Supercontinental inheritance and its influence on supercontinental breakup: the Central Atlantic Magmatic Province and the breakup of Pangea. Geochem. Geophys. Geosyst. 16, 3532–3554. https://doi.org/10.1002/2015GC005885.

White, W.M., 2014. Hot spots and mantle plumes. In: Harff, J., Meschede, M., Petersen, S., Thiede, J. (Eds.), Encyclopedia of Marine Geosciences. Springer, The Netherlands, pp. 1–20, https://doi.org/10.1007/978-94-007-6644-0_14-1.

Whiting, L., Haughton, P.D.W., Shannon, P.M., 2021. From rifting to hyperextension: Upper Jurassic–Lower Cretaceous tectono-stratigraphy of the Porcupine Basin, Irish Atlantic Margin. Basin Res. 33 (2), 1662–1696. https://doi.org/10.1111/bre.12530.

Winter, F., Intawong, A., Robinson, J., 2022. The Orange basin – an underexplored oil giant? GeoExPro 19 (3), 45–46.

Zalan, P., 2017. Similarities and differences magma-poor and volcanic passive margins—Applications to Brazilian marginal basins. In: 14th Congress of the Brazilian Geophysical Society, August 3–6, 2015, Rio de Janeiro.

Zanella, E., Doran, H., Collard, J., 2018. Is a conjugate an analog? GeoExPro 15, 14–17.

Zhou, Z., Mei, L., Liu, J., Zheng, J., Chen, L., Hao, S., 2018. Continentward-dipping detachment fault system and asymmetric rift structure of the Baiyun Sag, northern South China Sea. Tectonophysics 726, 121–136. https://doi.org/10.1016/j.tecto.2018.02.002.

Zinecker, M., 2020. Structural and stratigraphic evolution of three Mesozoic, rifted-passive margins: Guinea Plateau, Demerara Rise, and the Southeastern Gulf of Mexico. Unpublished PhD dissertation, University of Houston. 302 p.

Chapter 4

# Deepwater passive margin foldbelts

Md Nahidul Hasan and Paul Mann
*Department of Earth and Atmospheric Sciences, University of Houston, Houston, TX, United States*

## Introduction

### What is a passive margin foldbelt?

Passive margin foldbelts (PMFB) are zones of gravitational failure along low-angle detachments that result in updip, extensional structures on the shelf and slope and downdip compressional structures in the deepwater basin (Rowan et al., 2004) (Fig. 1). Over the past few decades, interest has increased significantly in deepwater passive margin foldbelts primarily due to the presence of undrilled structural closures with reported natural oil seeps (Morley et al., 2011). Currently, deepwater PMFB have become a focus for deepwater exploration with several recent giant discoveries (Zhang et al., 2017), however, PMFB are challenging structural settings to interpret and model in the overall context of understanding deepwater sedimentary systems. This chapter advances some of the concepts presented in Chapter 3 by specifically addressing the traits and variability in passive margins foldbelts in select deepwater provinces. Further, this chapter serves two additional purposes: (i) it is intended to build a foundation for the themes of deepwater sedimentation presented in later chapters and (ii) it discusses the facets of exploration risk and uncertainty introduced in Chapters 1, 2 and throughout the book.

The term PMFB takes into account all three of these structural domains: an updip extensional domain marked by rift features, a middle transitional zone of sliding with an absence of either rifts or folds and a downdip compressional zone marked by thrusting, folding, diapir squeezing and salt nappe extrusion (Fig. 1). Counter-regional (landward-dipping) normal faults and associated rollovers can also form as local complexities produced by abrupt variations in the geometry of the detachment surface (Rowan, 2020). The variability in these structural domains plays a key role in all aspects of the upstream exploration process (Chapter 1) and evaluation of prospects (Chapter 2) in PMFB.

The width of these PMFB can range in scale from a few kilometers to hundreds of kilometers as shown in the many examples described in this chapter. The driver for detachment of PMFB is the potential energy stored in landward to seaward-sloping stratigraphic surfaces. Detachment in such settings commonly occurs along a low-friction evaporitic layer (see Chapter 5) or overpressured shale (for more information on mudrocks and their traits, see Chapters 2, 6, 7, 10, 12 and 14).

Common mechanisms that have been previously proposed to explain the locations of PMFB include: (1) rapid sedimentary loading related to a tectonic event from the hinterland and (2) the formation of a large delta that contributes to a localized and updip zone of anomalously thick clastic sedimentation (Morley et al., 2011). Both of these settings produce oversteepening of the passive margin that is necessary to generate sufficient shear stress for failure and resulting formation of the PMFB (Krueger and Gilbert, 2009).

## Objectives

The objective of this chapter is to explain the structural evolution, similarities, dissimilarities and controlling factors for the formation of shale- and salt-based, deepwater PMFB of the Gulf of Mexico, Atlantic Ocean and western Indian Ocean. In addition to the scientific relevance, these regions are selected for many reasons including data availability and diversity, petroleum prospectivity and for the quality of the ongoing research conducted by the Conjugate Basins, Tectonics, and Hydrocarbons industry-funded research consortium led by scientists at University of Houston (Fig. 2). Examples of PMFB from other passive margins along the eastern Indian Ocean and southeastern and eastern Asia are reviewed by Morley et al. (2011) and additional themes related to this topic are discussed in Chapters 3 and 5.

Studied PMFB in this chapter include (Table 1; Fig. 2): (1) the Atwater-Mississippi PMFB of the US Gulf of Mexico, (2) the Perdido PMFB of the US Gulf of Mexico, (3) the Mexican Ridges PMFB of the Mexican Gulf of Mexico, (4) the

**Deepwater Sedimentary Systems.** https://doi.org/10.1016/B978-0-323-91918-0.00016-5

FIG. 1 Schematic diagram showing different styles of gravitational failure detached on either salt or shale interval that result in differing structural types of passive margin foldbelts: (A) thrust-dominated shortening results in a fold-thrust belt at the base of the PMFB; (B) fold-dominated shortening with more upright folds at the base of the PMFB; (C) squeezing of pre-existing diapirs produces even higher relief, upright folds; (D) inflation of salt body resulting extrusion of a salt nappe at the base of the PMFB. *(Modified from Rowan, M.G., Peel, F.J., Vendeville, B.C., 2004. Gravity-driven fold belts on passive margins, in McClay, K.R., Ed., Thrust Tectonics and Hydrocarbon Systems, AAPG Memoir 82, 157–182. https://doi.org/10.1306/M82813C9.)*

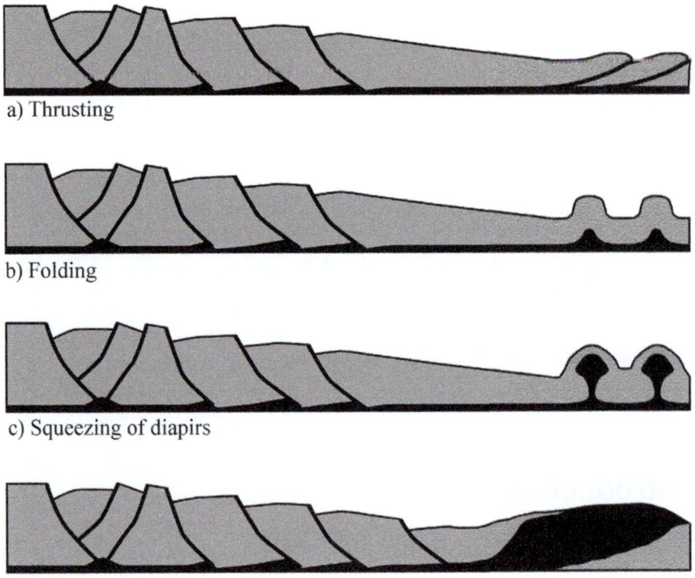


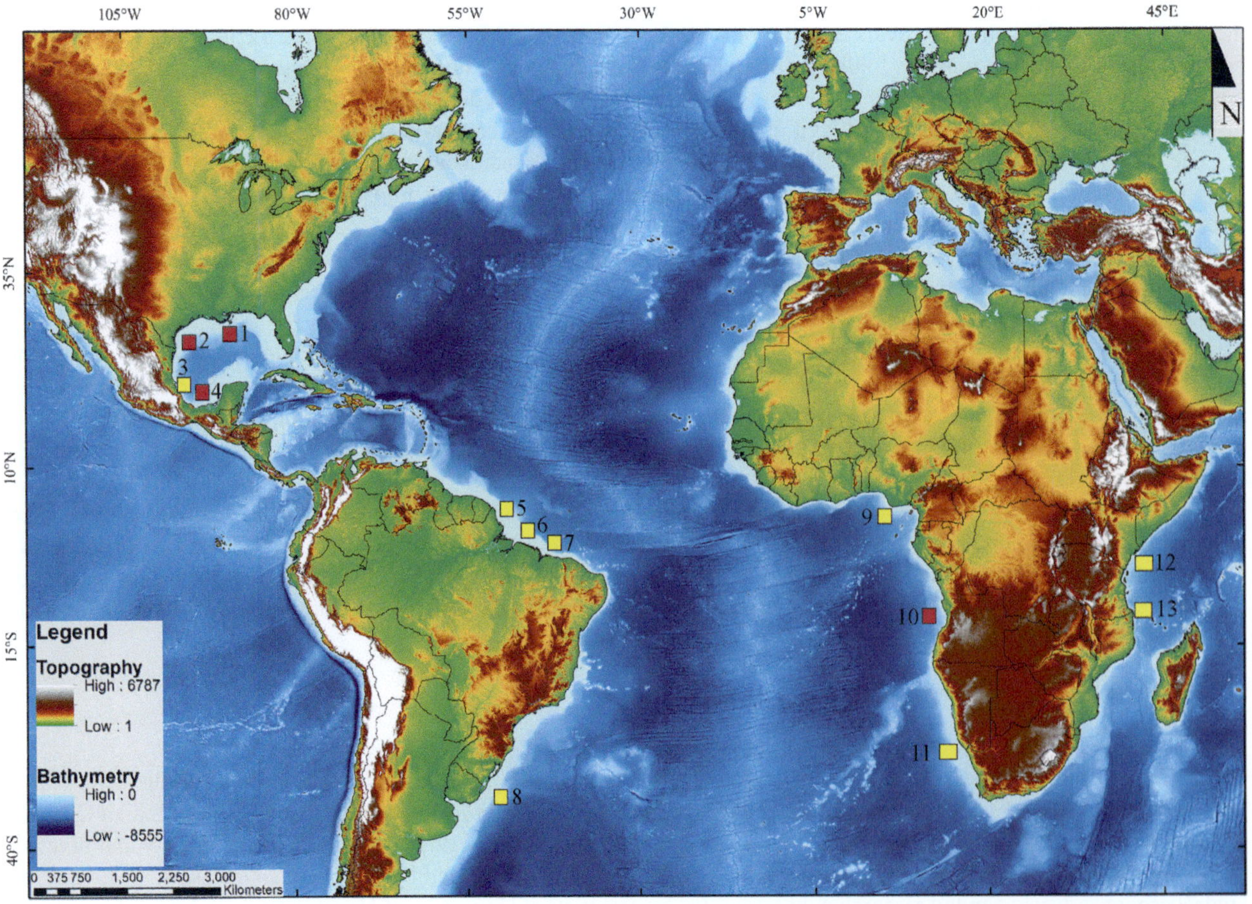


FIG. 2 Map of global topography (Olson et al., 2016) and bathymetry (GEBCO, 2020) showing the location and detachment type of the passive margin foldbelts of the Gulf of Mexico, Atlantic Ocean and western Indian Ocean. The colored rectangles represent the detachment type of the passive margin foldbelts (*red*=PMFB with gliding on a salt detachment and *yellow*=PMFB with gliding on a shale detachment). Thirteen PMFB described in this chapter include: (1) Atwater-Mississippi foldbelt of the US Gulf of Mexico (GoM), (2) Perdido foldbelt of the US Gulf of Mexico, (3) Mexican Ridges foldbelt of the Mexican Gulf of Mexico, (4) Campeche Salt Basin of the Mexican Gulf of Mexico, (5) Foz do Amazonas Basin related to the delta of the Amazon River of northern Brazil, (6) Pará-Maranhão Basin of northern Brazil, (7) Barreirinhas Basin of northern Brazil, (8) Pelotas Basin of southern Brazil and Uruguay, (9) Niger Delta of Nigeria, (10) Kwanza Basin of Angola, (11) Orange River Basin of South Africa, (12) Lamu Basin of Kenya and (13) Rovuma Basin of Mozambique.

**TABLE 1** Compilation of the main characteristics of the 13 passive margin foldbelts described in this chapter including their geographic locations, their basin name, the detachment type, the detachment surface age, the detachment slope angle and the bathymetric slope as measured from the seismic reflection lines and cross sections shown in this chapter.

| Geographic location | Basin name | Detachment type | Detachment surface age | Extensional domain (km) | Translational domain (km) | Compressional domain (km) | Detachment slope angle ($\beta$) | Bathymetric slope ($\alpha$) |
|---|---|---|---|---|---|---|---|---|
| Gulf of Mexico | Atwater-Mississippi foldbelt | Salt | Middle Jurassic | 260 | – | 150 | 3.5 | 0.9 |
| | Perdido foldbelt | Salt | Middle Jurassic | 300 | – | 215 | 5 | 1.5 |
| | Mexican Ridges foldbelt | Shale | Eocene–Oligocene | 65 | 15 | 120 | 7 | 2.3 |
| | Campeche Salt Basin | Salt | Middle Jurassic | 200 | – | 300 | 3 | 2.2 |
| Offshore South America | Foz do Amazonas Basin | Shale | Early Cretaceous; base Paleocene-Eocene; Early Pliocene | 85 | – | 120 | 3 | 1.1 |
| | Pará-Maranhão Basin | Shale and marls | Paleocene | 13 | 5 | 15 | 7 | 1.5 |
| | Barreirinhas Basin | Shale | Late Cretaceous | 40 | 13 | 13 | 4 | 2 |
| | Pelotas Basin | Shale | Paleocene-Eocene | 150 | 20 | 38 | 2 | 1.9 |
| Offshore West Africa | Niger Delta | Shale | Cretaceous | 190 | 22 | 85 | 1.5 | 1.4 |
| | Kwanza Basin | Salt | Early Cretaceous | 200 | | 160 | 2 | 2.2 |
| | Orange River Basin | Shale | Late Cretaceous | 140 | 20 | 95 | 1 | 1.1 |
| Offshore East Africa | Lamu Basin | Shale | Late Cretaceous, Late Oligocene | 13 | 10 | 20 | 3 | 1.9 |
| | Rovuma Basin | Shale | Late Eocene | 21 | 11 | 45 | 2 | 3.4 |

Campeche PMFB of the Mexican Gulf of Mexico, (5) the Foz do Amazonas PMFB associated with the Amazon Delta of northern Brazil, (6) the Pará-Maranhão PMFB of northern Brazil, (7) the Barreirinhas PMFB of northern Brazil, (8) the Pelotas PMFB of northern Brazil, (9) the Niger PMFB associated with the Niger Delta of Nigeria, (10) the Kwanza PMFB of Angola, (11) the Orange River PMFB of South Africa, (12) the Lamu PMFB of the Kenyan passive margin in the western Indian Ocean and (13) the Rovuma PMFB of the Mozambique passive margin in the western Indian Ocean.

Publicly-available and dip-oriented regional cross sections at similar vertical exaggerations are shown for each passive margin foldbelt system. These standardized maps and cross sections allow the scientific community to address the following six questions that are key for understanding the origin, structural evolution and deepwater hydrocarbon potential of PMFB:

(1) What are the controls on the variations in the lateral extent of the three characteristic structural domains of PMFB: the updip, extensional zone; the transitional zone and the downdip, compressional zone?
(2) What are the controls on the dip of the underlying detachment and the bathymetric slopes, and how do these variable dips affect the observed structural styles?
(3) Are there appreciable differences in the structural styles of PMFB based on their detaching on salt versus shale horizons?
(4) What are the causes of margin oversteepening that produce margin instability and result in the formation of PMFB?
(5) How do the taper angles of PMFB compare with the taper angles of accretionary prisms in subduction settings and subaerial fold and thrust belts?
(6) How do PMFB affect the components of deepwater petroleum systems that include source rocks, reservoirs, traps and hydrocarbon maturation?

## Previous studies of passive margin foldbelts: Main findings and implications

Passive margin foldbelts defined by updip extension and downdip shortening along either salt or shale detachments were first identified and systematically described by Rowan et al. (2004) (Fig. 1). This study noted that PMFB deformation includes elements of both gravity sliding along a basinward-dipping detachment surface and gravity spreading of a sedimentary wedge along a seaward-dipping bathymetric surface. Shortening at the base of the PMFB is driven by shelf and upper slope deposition that maintains the bathymetric slope and the gravitational potential with increased basinward tilting. The system can be driven by both thermal subsidence of the adjacent oceanic crust and increased basinward tilting related to the uplift of the continental block. The net shortening amounts and deformation rates of PMFB are much lower than those observed in either subaerial fold and thrust belts or subduction-related accretionary prisms because the driving, gravitational forces of thin-skinned PMFB are much smaller than the thicker-skinned forces related to the convergence of lithospheric plates.

Rowan et al. (2004) concluded that the main control on the variation in structural styles of PMFB was whether the detachment horizon beneath the PMFB was composed of shale with frictional strength—that produces basinward-verging thrust imbricates and associated folds because of the properties of plastic shale—or salt—that produces symmetrical fold above a viscous salt sheet with no internal frictional strength.

Morley et al. (2011) produced a global review of deepwater PMFB but widened the PMFB definition to include subduction-related accretionary prisms and deformed belts formed in arc-continent and continent-continent collisional zones. For the collisional types, these authors drew examples from southeast Asia and east Asia. Rowan et al. (2004), Morley et al. (2011) pointed out that the largest control on the structural styles of deepwater foldbelts *sensu lato* was shale vs. salt detachments—rather than their diverse tectonic settings.

Other than these two global reviews, most previous studies of PMFB have focused on specific examples of PMFB such as the Atwater-Mississippi fan foldbelt (Bouroullec et al., 2017; Bouroullec and Weimer, 2017; Peel et al., 1995), Perdido foldbelt (Fiduk et al., 1999; Trudgill et al., 1999), Mexican Ridges foldbelt (Salomón-Mora et al., 2009), Campeche Salt Basin (Hasan and Mann, 2021), Foz do Amazonas (Reis et al., 2016), Pará-Maranhão and Barreirinhas Basin (Oliveira et al., 2013), Pelotas Basin (Strozyk et al., 2017), Niger Delta (Bilotti and Shaw, 2005; Wu et al., 2015; Zhang et al., 2021), Kwanza Basin (Evans and Jackson, 2020; Hudec and Jackson, 2004), Orange River Basin (de Vera et al., 2010), Lamu Basin (Cruciani et al., 2017; Cruciani and Barchi, 2016) and Rovuma Basin (Cai et al., 2020). Academic and industry studies have produced a considerable amount of high-quality and deep-penetration seismic reflection data from PMFB, and several of those seismic profiles and cross sections are used in this analysis.

This chapter focusses only on the origins and evolution of PMFB from the Gulf of Mexico, south Atlantic Ocean and western Indian Ocean. This chapter does not include in its analysis the more broadly defined collision-related margins of the southeast and east Asia summarized by Morley et al. (2011).

# Examples of passive margin foldbelts in the Gulf of Mexico

## Regional setting of passive margin foldbelts in the Gulf of Mexico

The Gulf of Mexico (GoM) contains several well-described examples of gravity-driven PMFB that are detached on both salt and shale horizons (Fig. 3). The Gulf of Mexico evolved during a two-phase opening, with the first phase of rifting from the Triassic to the Middle Jurassic, and the second phase marked by seafloor spreading during the Late Jurassic (Marton and Buffler, 1999; Pindell and Kennan, 2009; Hudec et al., 2013; Eddy et al., 2014; Nguyen and Mann, 2015).

Following Phase 1 rifting (~210–163 Ma), seawater—probably from the Pacific Ocean—filled the broad sag basin overlaying the extinct rifts of the Bajocian-early Callovian and, through repeated episodes of evaporation, formed the Louann-Campeche salt deposit as part of a single, giant salt basin (Nguyen and Mann, 2015; Steier and Mann, 2019). The progressive, counterclockwise rotation of the Yucatan continental block formed an area of Late Jurassic, post-salt oceanic crust in the deep, central part of the Gulf of Mexico that separated the southern Campeche Salt Basin from the Louann Salt Basin.

In this section, the gravity-driven tectonics and hydrocarbon potential of the Gulf of Mexico are described in four key areas (Fig. 3): (a) Atwater-Mississippi fan PMFB, (b) Perdido PMFB, (c) Mexican Ridges PMFB and (d) Campeche PMFB. The basin margins of the Gulf of Mexico have undergone extensive gravitational gliding and spreading since the Middle Jurassic (Bouroullec and Weimer, 2017; Rowan et al., 2004). The major detachment layer for most of these PMFB is the Middle Jurassic salt—except in the case of the Mexican Ridges PMFB—where the detachment layer is Eocene–Oligocene shale (Peel et al., 1995; Salomón-Mora et al., 2011).

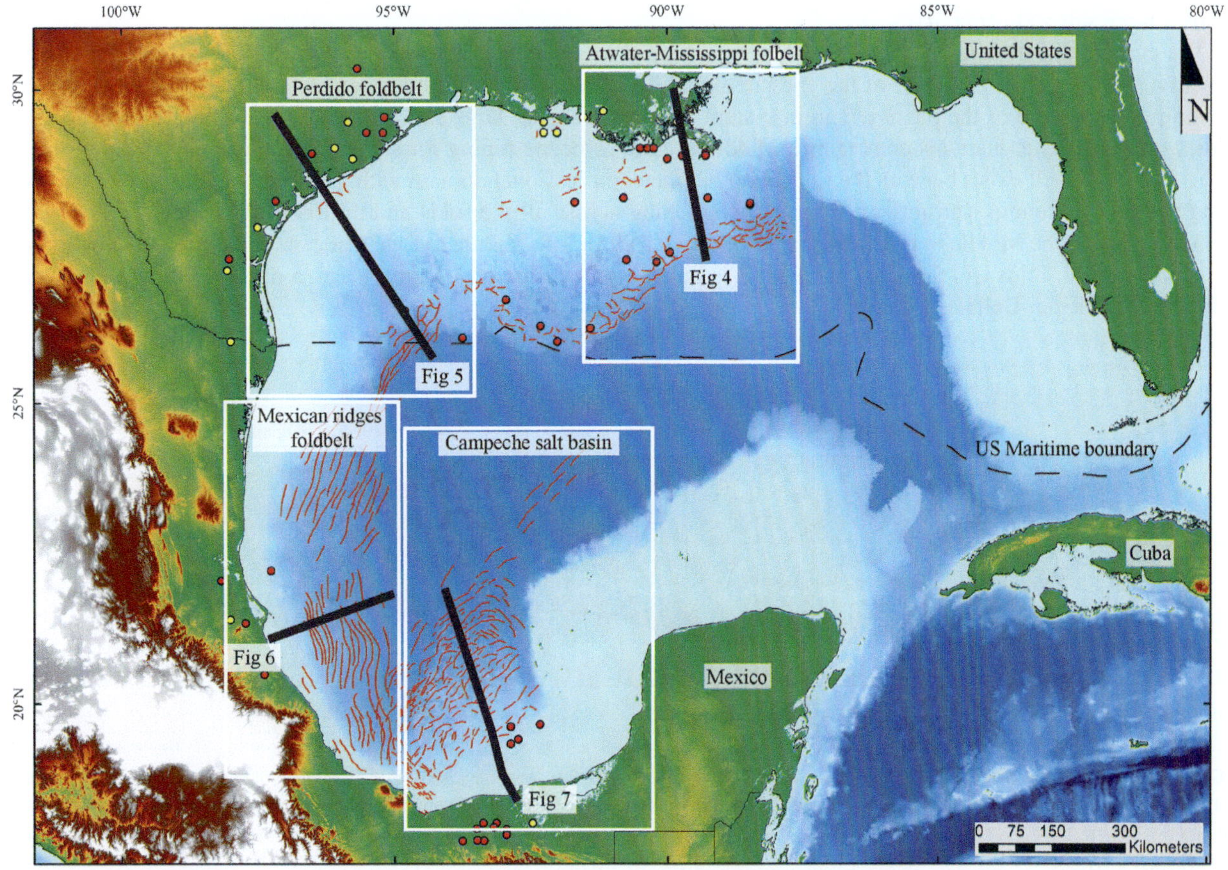


**FIG. 3** Bathymetric map of the Gulf of Mexico showing the distribution of passive margin foldbelts included in this study. PMFB include the salt-detached Atwater-Mississippi PMFB, salt-detached Perdido PMFB, salt-detached Campeche PMFB and shale-detached Mexican Ridges PMFB. US maritime boundary is shown by the *dashed black line* passive margin fold axes are shown by *red lines*; locations of the illustrated cross sections are shown by the *solid black lines* and the location of giant oil and gas fields are shown by the *red dots* (oil fields) and by the *yellow dots* (gas fields) (CBTH database compilation, 2020).

## Atwater-Mississippi fan passive margin foldbelt

### Structural provinces and age of formation

The Atwater-Mississippi fan and passive margin foldbelt in the northeastern US Gulf of Mexico is influenced by complex salt tectonics that controlled the formation of minibasins and salt-associated structures. The Mississippi Delta constitutes the primary sediment source to the slope and deepwater and produced a major influx of clastic deposition during the Miocene–Pliocene that shaped the present-day morphology of this PMFB (see Chapters 1, 4, 5, 8, 11 and 19 for more on this topic). The Atwater-Mississippi fan has a Neogene stratigraphic thickness of 3000–8500 m (~ 9843–27,887 ft.) (Bouroullec and Weimer, 2017).

Unlike other passive margins around the world where only the basal autochthonous salt level acts as the primary salt detachment, the thicker, remobilized salt in the Gulf of Mexico produces local areas where a high-level, secondary detachment and salt canopy is formed as seen for the Atwater-Mississippi fan PMFB (Bouroullec and Weimer, 2017). The extensive allochthonous salt-feeder system is likely controlled by thick Middle Jurassic salt deposition (Hasan and Mann, 2021). The enormous Neogene sediment deposition and complex, structural variability provides numerous listric normal faults, counter-regional faults, salt-withdrawal minibasins, salt canopy, folds and thrust structures (Fig. 4; Grando et al., 2009).

### Regional structural profile

The northwest-southeast regional, structural cross section of the Atwater-Mississippi PMFB (Fig. 4) shows a 260-km-wide (~ 162 mi.) extensional domain dominated by Paleogene-Early Miocene listric normal faults. The present-day shelf that affected by the extensional domain exhibits an extensive counter-regional salt-withdrawal basin filled with Miocene–Pliocene strata. The Middle Jurassic autochthonous salt is completely evacuated from the salt-withdrawal minibasin that forms a large salt weld (Fig. 4).

The compressional domain of the Atwater-Mississippi PMFB extends 150 km (~ 93 mi.) in the northwest-southeast direction (Fig. 4). The updip part of the compressional domain is dominated by relatively smaller salt bodies while the downdip to distal end of the passive margin system contains a single, large salt canopy. The injection of secondary salt bodies into the clastic intervals constrains a Middle to Late Miocene timing for salt emplacement.

A fold-thrust belt exists beneath the salt canopy in the compressional domain with the foldbelt detaching on an underlying and autochthonous salt horizon. The total shortening across all the folds and thrusts is 10 km (~ 6.2 mi.) with its formation during the Middle to Late Miocene (Peel et al., 1995). The Middle to Late Miocene timing of the allochthonous salt emplacement, folds and thrusts of the PMFB is linked to the influx of 4–8 km (~ 2.5–5 mi.) of clastic sedimentation from the Mississippi Delta.

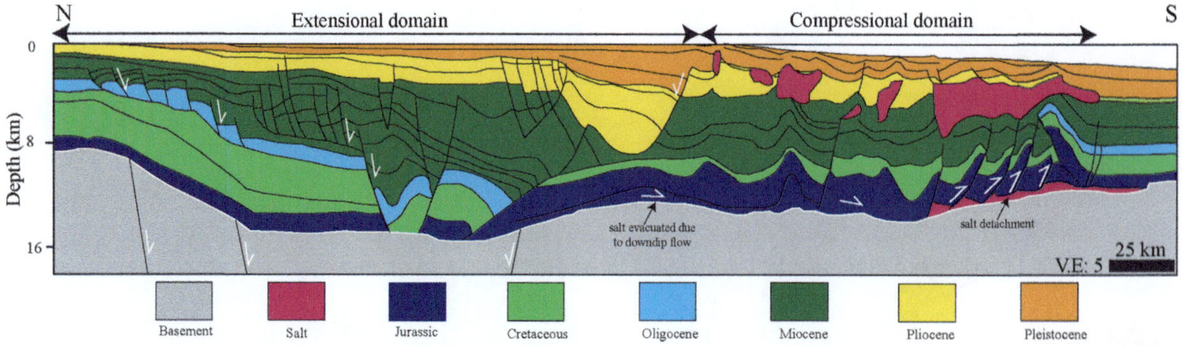


**FIG. 4** Regional northwest-to-southeast depth cross section through the Mississippi-Atwater passive margin foldbelt of the northern US Gulf of Mexico (location of cross section is shown on the map in Fig. 3). The lateral extents of the extensional and compressional domains are indicated. The compressional structures at the toe of the slope are cored by the Louann salt of Middle Jurassic age. The complete evacuation of Louann salt from the updip extensional domain inflates the downdip structures and, in some cases, results in an allochthonous sheet in the upper part of the section. *(The cross section is modified from Grando, G., Schleder, Z., Shackleton, R., Seed, G., Buddin, T., McClay, K., Borraccini, F., 2009. 3D Structural Evolution of the Salt Controlled Frampton Anticline, Atwater Valley Foldbelt, Deep Water Gulf of Mexico, AAPG search and discover article #50161, which had modified an earlier cross section by Peel, F.J., Travis, C.J., Hossack, J.R., 1995. Genetic structural provinces and salt tectonics of the Cenozoic offshore U.S. Gulf of Mexico: a preliminary analysis. In: Jackson, M.P.A., Roberts, D.G., Snelson, S., Salt Tectonics A Glob. Perspect. https://doi.org/10.1306/M65604C7.)*

### *Hydrocarbon potential*

The hydrocarbon potential of the Atwater-Mississippi PMFB is controlled by the interaction of salt tectonics and high sedimentation in the Neogene that provides the necessary structural traps and high-quality reservoir rocks. Extensive exploration and production with major discoveries are reported in the Atwater-Mississippi passive margin foldbelt and include Neptune, Mad Dog, Tahiti, Knotty Head, Genghis Khan, Big Foot and Thunder Horse (Morley et al., 2011).

Bouroullec et al. (2017) compiled 87 discoveries in the Atwater-Mississippi fan foldbelt and showed that 19 fields produce from structural traps with four-way dip closures (anticlines, salt pillows, turtle structures), 52 fields have combined structural-stratigraphic traps with three-way closures (salt-flank traps, foldbelt truncation traps) and 16 fields produce from stratigraphic traps (pinch outs, channels and valley fills). Most of the successful traps are combined structural-stratigraphic traps. Almost all of the fields produce from Oligocene to Pleistocene reservoirs (Bouroullec et al., 2017). In these fields, high-quality reservoirs are primarily turbidites deposited in basin-floor fan systems (see Chapters 6–11) originally fed by the Mississippi Delta.

## Perdido passive margin foldbelt

### *Structural provinces and age of formation*

The deepwater Perdido foldbelt, northwestern Gulf of Mexico (Fig. 3), includes an onshore extensional system, the Port Isabel foldbelt compressional zone and a downdip compressional zone defined by large detachment folds cored by autochthonous Middle Jurassic salt (Fig. 5). Updip extension of approximately 60 km (~ 37 mi.) suggests the downdip fold and thrust belt and salt canopy should collectively accommodate a similar amount of shortening (Trudgill et al., 1999).

Two orogenic events of the Cordilleran orogeny of the Pacific margin of North America controlled the evolution of the Perdido PMFB during the Cenozoic period (Fitz-Díaz et al., 2018): first, the Laramide orogeny during the Late Cretaceous to Eocene and; second, the post-orogenic uplift of North America from the Oligocene to present-day. Both events increased the seaward tilt of the western margin of the Gulf of Mexico and the resulting denudation led to large amounts of clastic sedimentation (Hudec et al., 2019, see Chapter 5). The combination of increased tilting and sedimentation destabilized the margin and gave rise to the Perdido PMFB system during the Oligocene, with gravity-related sliding continuing to the Miocene. The onset of deltaic deposition related to the Mississippi River in the Miocene–Pliocene resulted in the depositional abandonment in the Perdido PMFB and created a stable, gravitational equilibrium (Morley et al., 2011).

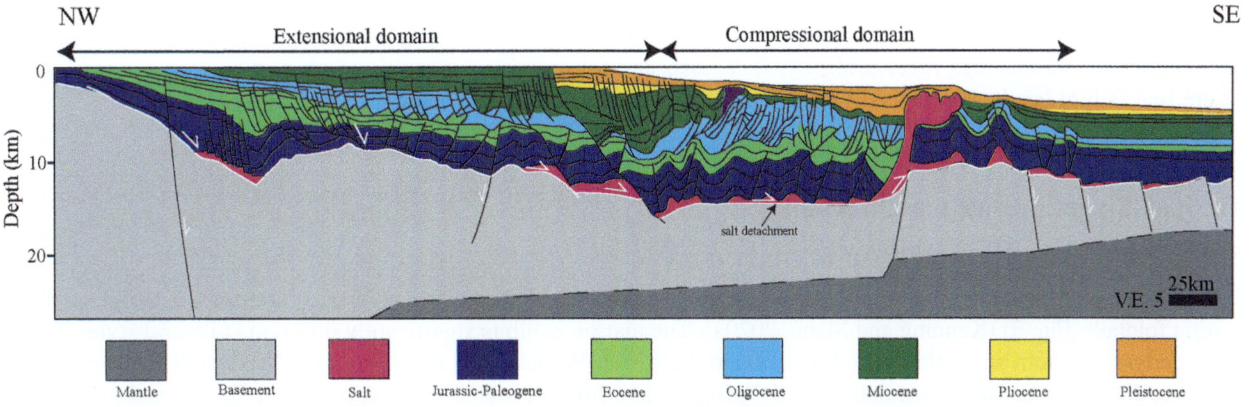


**FIG. 5** Regional, northwest-to-southeast, depth cross section through the Perdido passive margin foldbelt in the northwestern part of the US Gulf of Mexico (location of section is shown on the map in Fig. 3). The section shows the complex, stacked system of salt detachments. Numerous listric normal faults and zones of salt deflation characterize the updip extensional domain while thrusts, salt inflation and an upper salt canopy characterize the downdip compressional domain. Salt buttressing and emergent folds form along a basement ramp or "step-up fault" called the "Baha High" that also forms the approximate limit of the salt canopy. The Perdido foldbelt is the downdip and more external part of a more extensive belt of contraction that includes the Port Isabel foldbelt with its own salt canopy. *(The cross section is modified from Morley, C.K., King, R., Hillis, R., Tingay, M., Backe, G., 2011. Deepwater fold and thrust belt classification, tectonics, structure and hydrocarbon prospectivity: a review. Earth Sci. Rev. 104, 41–91. https://doi.org/10.1016/j.earscirev.2010.09.010, that is based on previous cross sections by Peel, F.J., Travis, C.J., Hossack, J.R., 1995. Genetic structural provinces and salt tectonics of the Cenozoic offshore U.S. Gulf of Mexico: a preliminary analysis. In: Jackson, M.P.A., Roberts, D.G., Snelson, S., Salt Tectonics A Glob. Perspect. https://doi.org/10.1306/M65604C7 and Trudgill, B.D., Rowan, M.G., Fiduk, J.C., Weimer, P., Gale, P.E., Korn, B.E., Phair, R.L., Gafford, W.T., Roberts, G.R., Dobbs, S.W., 1999. The Perdido Fold Belt, northwestern deep Gulf of Mexico, part 1: structural geometry, evolution and regional implications. Am. Assoc. Pet. Geol. Bull. 83, 1320–1336. https://doi.org/10.1306/00AA9A0C-1730-11D7-8645000102C1865D.)*

The primary detachment layer of the Perdido PMFB is the Middle Jurassic Louann salt. One unique feature of Perdido is the presence of a basement step-up called the "Baha High" by Hudec et al. (2019). Buttressing against the Baha High also forms the limit of the salt canopy in the Perdido PMFB.

Basinward of the salt canopy is a series of anticlines that vary in their structural profile from symmetrical to asymmetrical. These anticlines are associated with reverse faults and kink bands (Fig. 5).

### *Regional structural profile*

The northwest-southeast dip-oriented structural cross section of the Perdido PMFB traverses both the updip extensional and downdip compressional domains (Fig. 5). A 300-km-wide (~ 186 mi.) zone representing a thin-skinned extensional system detaches along Callovian-Bajocian salt. A secondary detachment with extensional growth faults can be observed along an Eocene horizon with a significant growth sequence during the Miocene.

The 215-km-wide (~134 mi.) downdip compressional domain is dominated by multi-level detachment structures, a salt canopy and anticlines with reverse faults on both limbs. Thrusts and folds form multiple detachment layers in the middle part of the cross-section. The structures detaching over salt are gentle detachment folds, whereas structures detaching along Eocene shale units represent strongly shortened fold and thrusts. On the lower slope of the Perdido PMFB, the basement step-up fault (Baha High) marks the distal edge of the salt canopy.

### *Hydrocarbon potential*

The cluster of large deepwater discoveries in the distal, downdip area of the Perdido PMFB includes BAHA, Trident, Great White, Silvertip and Tobago. All indicate a working petroleum system (Morley et al., 2011). Seafloor oil seeps have been reported in the downdip portion of Perdido and indicate the presence of a mature source rock, which is likely to Tithonian in age (Fiduk et al., 1999).

Maturity modeling by previous studies suggests that the Tithonian source rock passed into the peak oil window after the formation of the large structural closures. Trudgill et al. (1999) interprets some of the structures in the Perdido area were not covered by fine-grained sediment until 10.5 Ma, suggesting that seal integrity is the highest risk factor . The presence of numerous normal faults and reverse faults likely provide the migration pathways.

No significant discoveries have been reported in the middle slope in the basin, such as the Port Isabel exploration area (Morley et al., 2011). Buttressed by the Baha High in the downslope direction, the stress provided by updip extension is likely accommodated in the Port Isabel area. This observation suggests this zone is more structurally complex than the downdip compressional area (Peel et al., 1995). Thrusts and folds within the multi-detachment represent potential areas where seal failure is a key risk (see Chapter 2 for more information on this topic).

## Mexican Ridges passive margin foldbelt

### *Structural provinces and age of formation*

The Mexican Ridges extend over >400 km (~ 249 mi.) across the western Gulf of Mexico and form a Neogene passive margin foldbelt (Fig. 3) (Kenning and Mann, 2020a). The margin exhibits significantly different structural styles than the salt-filled northeastern and southern Gulf of Mexico. Callovian-Bajocian salt is absent in the Mexican Ridges because the area is underlain by oceanic crust that formed during the late Jurassic after the salt deposition (Nguyen and Mann, 2015; Kenning and Mann, 2020a).

The present-day area of the Mexican Ridges remained as a passive margin for much of the Mesozoic due to post-rift thermal subsidence. The Late Cretaceous to Middle Eocene Laramide orogeny shortened significant areas of eastern Mexico adjacent to the Mexican Ridges. From the Late Eocene to the Early Miocene, uplift and denudation contributed a large amount of clastic sediment to the deepwater (Salomón-Mora et al., 2009).

Progressive tilting and sediment loading resulted in gravity gliding along the Eocene–Oligocene shale detachment surface from the Middle Miocene (Morley et al., 2011; Salomón-Mora et al., 2009). The resultant passive margin foldbelt of the Mexican Ridges is characterized by a system of normal growth faults in the shelfal area known as the Quetzalcoatl extensional system, a neutral zone of shortening, and a train of folds along the continental slope (Yarbuh et al., 2018).

## *Regional structural profile*

The southeast-to-northwest regional, structural cross section of the Mexican Ridges PMFB traverses all elements of the PMFB that includes a 65-km-wide (~ 40 mi.) zone of extensional system, a 15-km-wide (~ 9 mi.) neutral zone of shortening and a 120-km-wide (~ 75 km.) zone of compression (Fig. 6).

The extensional domain is dominated by a listric normal fault system that detaches on Eocene shale. A group of synthetic-antithetic listric faults form grabens and create upper Miocene to Recent depocenters containing growth strata (Fig. 6). The translation zone is relatively narrow and characterized by the absence of any prominent structures related to either extension or compression.

The downdip compressional domain is dominated by series of detachment folds that are associated with reverse faulting. Most of the reverse faulting is associated with folds that are detached on a secondary layer of the Oligocene-Miocene age (Fig. 6). The observation of upper Miocene to Recent growth strata in the extensional domain indicates its synchronous timing with the deepwater foldbelts in the compressional domain. However, one key study proposes a time lag in the propagation of the folding in the downdip direction (Salomón-Mora et al., 2011).

## *Hydrocarbon potential*

In the Mexican Ridges, the main structural styles are detachment folds with local reverse faults that result in four-way dip closures. The key reservoir-seal pairs are interpreted as interlayered clastic deposits representing Cenozoic turbidite fan systems (Salomón-Mora et al., 2009).

The crustal composition of the Mexican Ridges passive margin foldbelt has important implications in terms of hydrocarbon maturity (Kenning and Mann, 2020a). Oceanic crust, having a lower heat flow, may cause the source rocks to remain immature at the time of foldbelt formation in Middle Miocene to Recent, especially in the proximal, updip area with less overburden.

The presence of bottom simulating reflectors (BSRs) has been described by previous authors and indicates a gas-mature system in the deepwater area (Kenning and Mann, 2020a). Oceanic spreading in the GoM is thought to have continued into the earliest Cretaceous (Nguyen and Mann, 2015), meaning that the most prolific Late Jurassic source rocks across the entire Gulf of Mexico may not exist above areas where oceanic crust is younger than Tithonian.

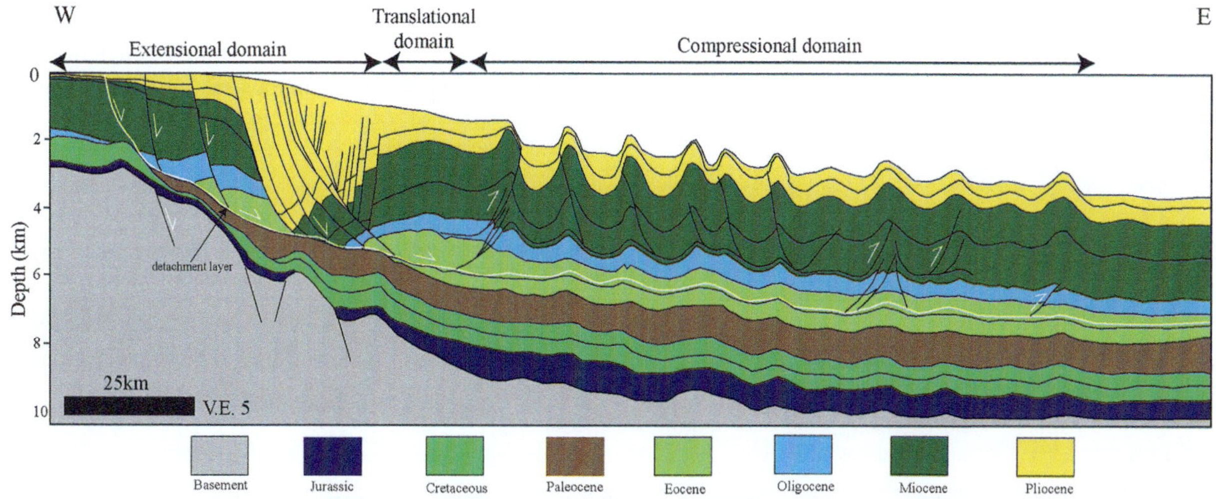


**FIG. 6** Regional, west-to-east depth cross section through the Mexican Ridges passive margin foldbelt along with western margin of the Mexican Gulf of Mexico (location of cross section shown on the map in Fig. 3). The structural domains include the updip Quetzalcoatl extensional domain characterized by listric growth faults; a narrow translational domain characterized by a neutral zone of overlying shortening structures and a downdip, compressional domain characterized by break thrusts and detachment folds. The slip on the updip listric normal faults and downdip folds and thrusts formed as a common detachment surface that followed a shale interval of Eocene–Oligocene age. *(The cross section is modified from a cross section by Salomón-Mora, L.E., Aranda-Garcia, M., Román-Ramos, J.R., 2009. Contractional growth faulting in the Mexican Ridges, Gulf of Mexico. AAPG Mem. 93–115. https://doi.org/10.1306/13191080M903332.)*

## Campeche passive margin foldbelt

### Structural provinces and age of formation

The Campeche passive margin foldbelt is defined by a 200-km-wide (~ 124 mi.) updip extensional zone and a coeval, 300-km-wide (~ 186 mi.) downdip compressional zone (Fig. 7). The extent of the translational zone is not obvious because numerous complex salt structures complicate the imaging, making it difficult to observe direct evidence of lateral translation.

Deformation due to passive diapirism and basinward tilting of the margin began during the Late Jurassic and Cretaceous with the formation of minor contractional structures. The Middle Miocene Chiapanecan orogeny (related to the shallow subduction of the Cocos plate) oversteepened and destabilized the margin with updip extension along the northeast-trending Macuspana and Comalcalco rift systems (Mitra et al., 2005; Pindell and Miranda, 2011). These extensional systems detach as listric normal faults during the Middle Miocene along the underlying autochthonous salt of Bajocian-early Callovian age. The updip extension of 20.5 km (~ 12.7 mi.) in the Comalcalco and Macuspana rifts contributed to the development of the deepwater compressional domain (Hasan and Mann, 2021).

Analysis of reflection continuity and stratal geometry in the compressional domain reveals the presence of Middle Miocene growth packages (Davison, 2020). Growth sequence and mass transport deposits (Chapter 10) in the Late Miocene to Recent indicate a major pulse of salt mobilization (Chapter 5) during this time (Hasan and Mann, 2021).

### Regional structural profile

The southeast-to-northwest regional structural cross section of the Campeche Salt Basin reveals three key observations: a part of the updip area is dominated by normal faults; the upslope extensional domain includes a zone of regional listric normal faults, detaching on both allochthonous salt and autochthonous salt; and several rollover structures and counter-regional normal faults have formed because of extensional deformation (Fig. 7). Rollover anticlines associated with the listric normal faults contain up to 4–6 km (~ 2.5–3.7 mi.) of Pliocene-Recent clasticdeposits.

The downdip area of compressional deformation is characterized by salt canopies, open salt-cored folds, salt diapirs and break thrusts. The allochthonous salt bodies have coalesced in the proximal part of this domain. Closely spaced upright diapirs concentrated in the outer marginal trough area and range from 6 to 8 km (~ 3.7–5 mi.) in relief. The bound mini-basins contain 8–10 km (~ 5–6 mi.) of clastic sedimentary infill (Fig. 7).

### Hydrocarbon potential

Hydrocarbon plays are most common in the shallow water extensional domain and the outer marginal trough due to the presence of Tithonian source rock, high-quality carbonate and clastic reservoirs and numerous salt structures (Davison, 2020; Hasan and Mann, 2021).

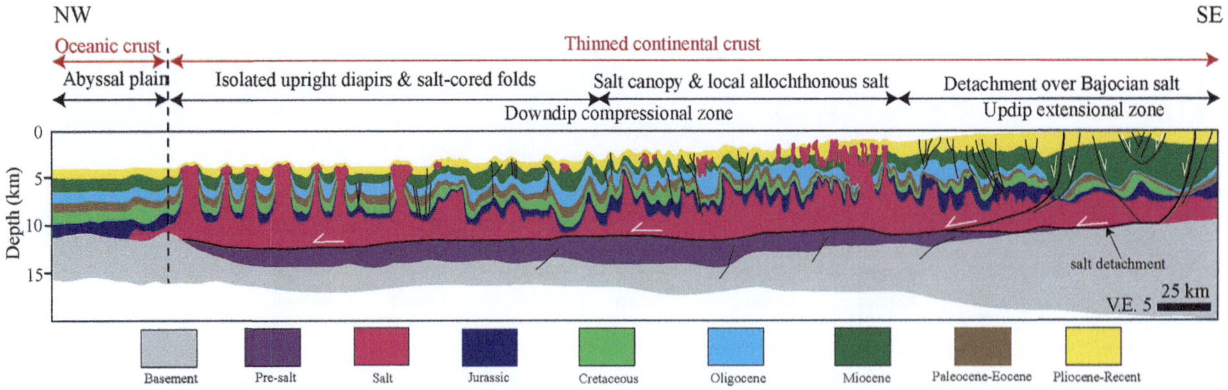


**FIG. 7** Regional, southeast-to-northwest structural cross section across the Campeche passive margin foldbelt (location of the cross section is shown on the map in Fig. 3). The lateral extent of the updip extension zone and downdip compressional zone is shown along with the across-strike variability of structural styles and the underlying types of crust, which generally dips seaward. The updip extensional zone consists of listric normal faults that detach on both allochthonous salt remobilized during the Late Miocene and autochthonous salt of the Bajocian-early Callovian age. Downdip contractional zone consists of salt-cored folds, thrusts, autochthonous salt diapirs, allochthonous salt canopy and detachment folds with kink bands. *(The cross section is modified from Hasan, M.N., Mann, P., 2021. Structural styles and evolution of the Campeche Salt Basin, southern Gulf of Mexico. Mar. Pet. Geol. 133, 105313. https://doi.org/10.1016/j.marpetgeo.2021.105313.)*

Updip salt rollers and rollover anticlines provide potential traps above the sloping salt detachment (Kenning and Mann, 2020b). Numerous salt-associated undrilled structures with diverse geometries are present in the downdip compressional domain and make trap potential the least risky element of the Campeche PMFB (Shann, 2020). Combined structural and stratigraphic traps are present along the upturned strata at the flanks of the salt diapirs (Kenning and Mann, 2020b).

Clusters of natural sea surface oil seeps from the Campeche Salt Basin have been described in studies of the updip extensional domain in the shallow offshore area and the downdip, deepwater compressional domain (Hasan and Mann, 2021; Kenning and Mann, 2020b). The abundance of seeps indicates a working petroleum system within the Campeche PMFB, although the presence of seeps also indicates likely seal failures related to active deformation.

The abundance of undrilled salt-associated structural traps, mature source rocks as shown by oil seeps, potential migration pathways and opening its leases to international oil companies has made the Campeche PMFB one of the most intensively explored margins over the past several years.

## Examples of passive margin foldbelts on the Atlantic margin of South America

### Regional setting of passive margin foldbelts along South America

The continental margin of South America forms areas of either rifted-passive margins (northeastern and southwestern Brazil, Uruguay and Argentina) or transform-rifted margins (northern Equatorial Brazil). All of these areas formed during the Late Jurassic breakup of Gondwana and resulted in the formation of the South Atlantic Ocean (Contreras et al., 2010). The conjugate African and Brazilian margin basins are separated by oceanic crust developed along the Mid-Atlantic Ridge (Strozyk et al., 2017).

In this section, the structural evolution of the passive margin foldbelts of the four salt-free basins of Brazil are described (Fig. 8): (a) Foz do Amazonas, (b) Pará-Maranhão, (c) Barreirinhas and (d) Pelotas. For all four, the essential elements and morphology of a passive margin foldbelt are present and include a detachment along overpressured shale and marl layers.

The Foz do Amazonas, Barreirinhas and Pará-Maranhão basins are located along the steep-sloped, transform-passive margin of northern equatorial Brazil (Oliveira et al., 2013). The more gently sloped Pelotas Basin formed along the volcanic-rifted-passive margin in southern Brazil and Uruguay.

### Foz do Amazonas passive margin foldbelt

#### *Structural provinces and age of formation*

The Foz do Amazonas, or "Mouth of the Amazon River", Basin is the largest offshore sedimentary basin of the Brazilian Equatorial Atlantic margin. Gravitational failure along this steep-sloped, transform-passive margin basin has controlled its present-day structural styles (Reis et al., 2010).

The Foz do Amazonas Basin was controlled by its transform-related slopes, steepening and inversion of the margin related to distant orogenic and erosional events in the Andean Mountains along the western margin of South America. Sediment supply to the Foz do Amazonas area dramatically increased during the Middle to Late Miocene due to the increased rate of Andean uplift, which deposited up to 10 km ($\sim$ 6 mi.) of clastics since the Middle Miocene (Reis et al., 2016). The period of rapid sedimentation oversteepened and destabilized the margin and resulted in gravitational failure and the formation of the passive margin foldbelt. The along-strike variation of sediment thickness due to this event controls the intensity and run-out distance of the gravity-driven deformation (Reis et al., 2010).

#### *Regional structural profile*

The southwest-northeast depth cross section through Foz do Amazonas basin shows a 85-km-wide ($\sim$ 53 mi.) extensional domain that is dominated by basinward-dipping listric normal faults and counter-regional, or landward-dipping, normal faults. Three detachment levels labeled as H1 (Early Cretaceous), H2 (base Paleocene-Eocene) and H3 (Early Pliocene) are indicated in the seismic profile (Fig. 9A). The main detachment level is the intermediate one (H2) located along the base of an overpressured, marine shale of Paleocene-Eocene age (Fig. 9).

A 120-km-wide ($\sim$ 75 mi.) compressional zone is composed of a zone of imbricated thrusts that dip landward. Each thrust fault detaches over the intermediate detachment horizon—H2 (base Paleocene-Eocene). The intensity of the landward-dipping and imbricated thrusting decreases in the landward direction (Fig. 9A and B).

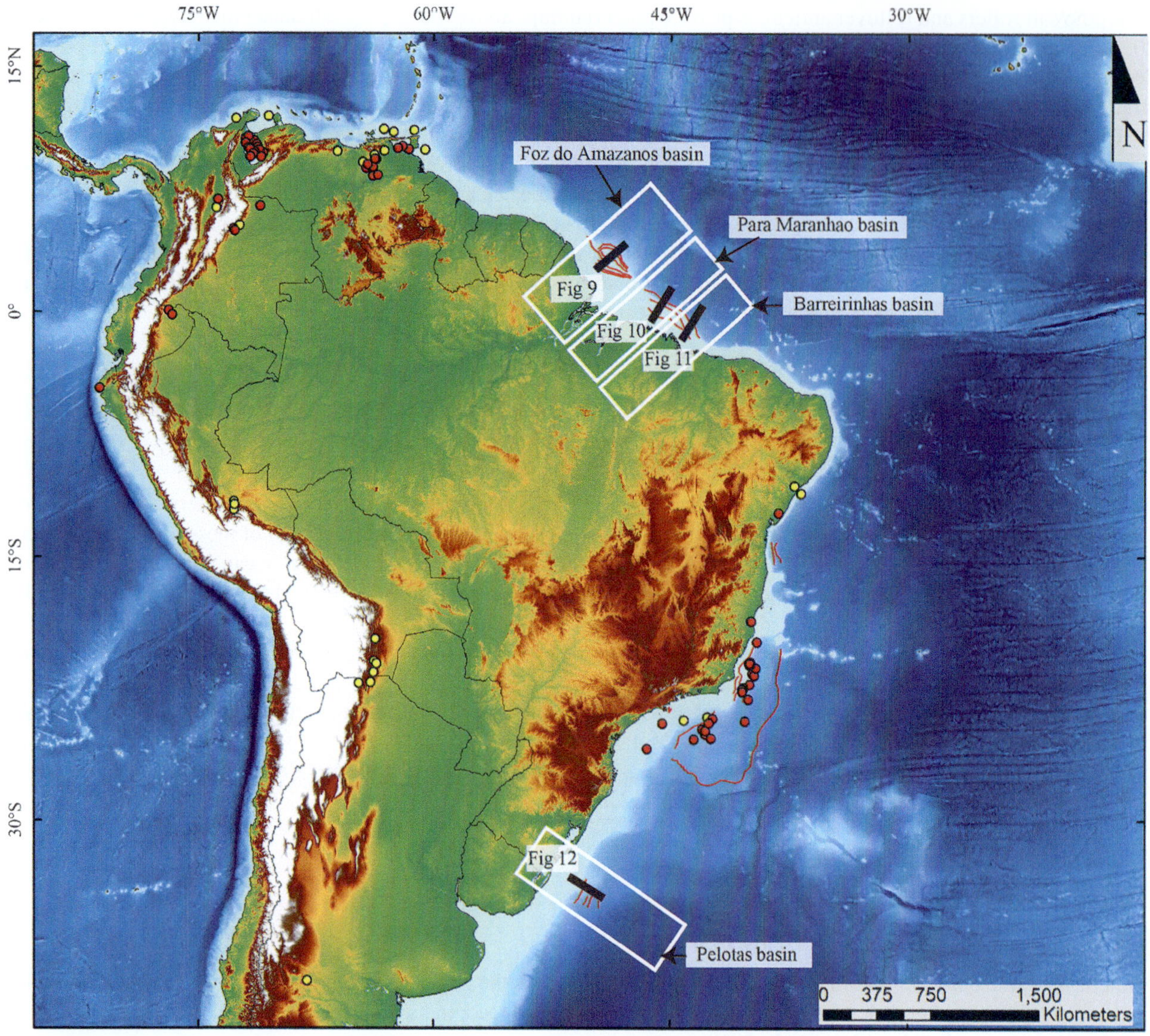


**FIG. 8** Bathymetric map of the South Atlantic passive margin of the South American continent showing the locations of four, shale-detached passive margin foldbelts: Foz do Amazonas Basin, Pará-Maranhão Basin, Barreirinhas Basin and Pelotas Basin. Passive margin fold axes are shown by *red lines*, locations of the illustrated cross sections are shown by the *solid black lines* and the location of the giant oil and gas fields are shown by the *red dots* (oil fields) and by the *yellow dots* (gas fields) (CBTH database compilation, 2020).

## Hydrocarbon potential

The Foz do Amazonas Basin is a frontier basin with only three deepwater wells—all of them penetrating sedimentary rocks no older than the Miocene (da Cruz et al., 2021). In the shelfal areas, Type-I source rocks of both Aptian and Cenomanian–Turonian ages have been identified along with the main reservoirs consisting of Upper Cretaceous and Cenozoic sandstone. A seismic facies analysis shows that a Late Cretaceous shale seals most of the basin (da Cruz et al., 2021). Using seismic reflection grids, direct indicators of working petroleum systems include bright spots, gas chimneys and bottom simulating reflectors (see Chapters 1, 2, 18–20). Two petroleum systems proposed for the Foz do Amazonas Basin include: (a) late Aptian source rocks paired with Late Cretaceous reservoir rocks (b) Cenomanian–Turonian source rocks paired with Late Cretaceous reservoirs (da Cruz et al., 2021).

There are three primary geologic risk factors in the Foz do Amazonas passive margin foldbelt. First, the downdip compressional domain is composed of active thrusts, indicating that timing and seal failure risk is high. Second, due to the lack of deepwater penetrations, the presence and quality of source rock in the basin remains speculative. Third, the abundant mass-transport complexes resulting from gravitational instabilities can act as barriers for vertical hydrocarbon migration (see Chapter 10).

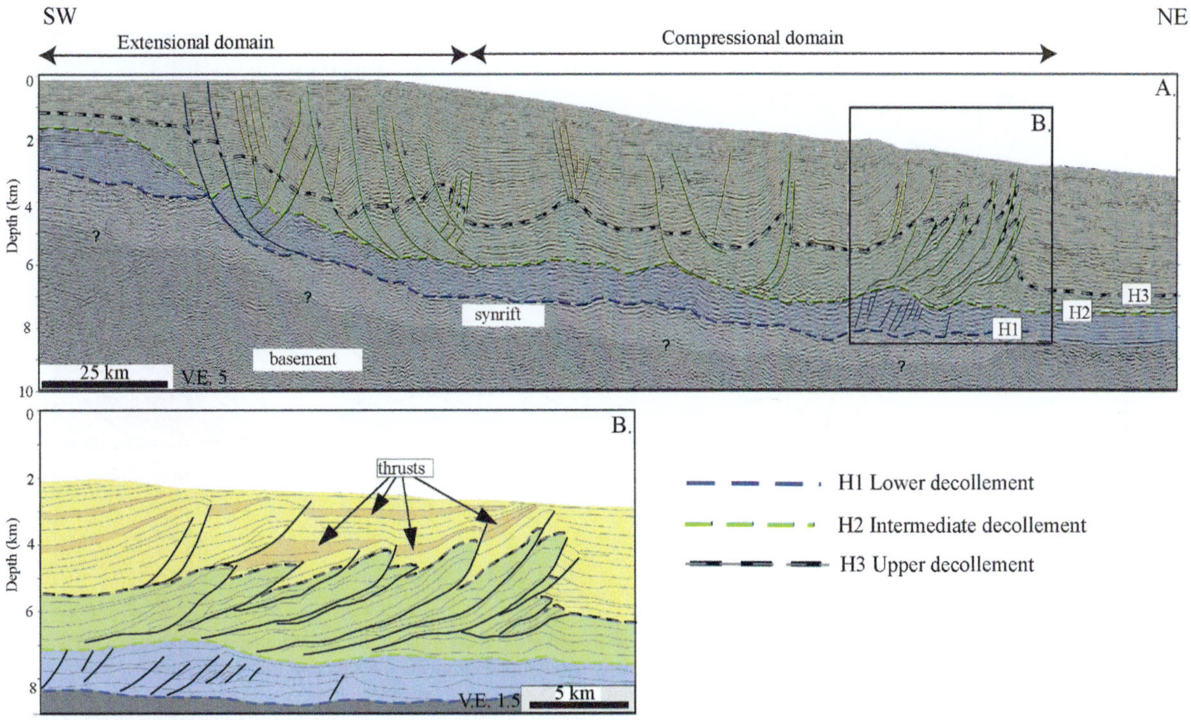


**FIG. 9** (A) Regional, southwest-to-northeast cross section in depth through Foz do Amazonas Basin (location on Fig. 8) showing three detachment levels labeled as H1 (detachment along an Early Cretaceous horizon), H2 (detachment along the base Paleocene-Eocene horizon) and H3 (detachment along an Early Pliocene horizon). The main detachment level is the intermediate one (H2) located at the base of a overpressured, marine shale of Paleocene-Eocene age. (B) Zoom of seismic section (location of zoomed area is boxed in Fig. 9A) showing the downdip fold and imbricated thrust belt of the Foz do Amazonas Basin. *(The cross-sections are modified from Reis, A.T., Araújo, E., Silva, C.G., Cruz, A.M., Gorini, C., Droz, L., Migeon, S., Perovano, R., King, I., Bache, F., 2016. Effects of a regional décollement level for gravity tectonics on late Neogene to recent large-scale slope instabilities in the Foz do Amazonas Basin, Brazil. Mar. Pet. Geol. 75, 29–52. https://doi.org/10.1016/j.marpetgeo.2016.04.011. Illustration used with permission from Elsevier.)*

## Pará-Maranhão passive margin foldbelt

### *Structural provinces and age of formation*

The Pará-Maranhão Basin in northeastern Brazil is located between the Foz do Amazonas Basin to the northwest and the Barreirinhas Basin to the southeast (Fig. 8). During formation of the Equatorial Atlantic in the Albian, a right-lateral shear across a broad zone shaped the steep margins of both the Pará-Maranhão and Barreirinhas basins (Oliveira et al., 2013). Following this oblique opening stage, the basin subsided as a steeply-sloped passive margin. The Pará-Maranhão Basin experienced significant gravity gliding to form its passive margin foldbelt (Fig. 10). Cenozoic gravity gliding occurred in response to depositional loading and thermal subsidence, thereby increasing the slope gradient and causing episodic reactivation of the bounding fracture zones (Oliveira et al., 2013).

The detachment surface occurs along the top of overpressured shale of Paleocene age. Overpressure is related to rapid sedimentation, burial and the development of this strong shale seal. It also signals the possible presence of underlying hydrocarbons (Oliveira et al., 2013).

### *Regional structural profile*

The southwest-northeast structural cross section of the Pará-Maranhão PMFB shows all three structural domains of the passive margin foldbelt: a 13-km-wide (~ 8 mi.) updip area of extension, a 5-km-wide (~ 3 mi.) zone of translation and a 15-km-wide (~ 9 mi.) zone of downdip compression (Fig. 10).

The extensional domain of the Pará-Maranhão PMFB is characterized by several large synthetic normal faults that are associated with prominent rollovers with growth strata. The thickness of growth strata can be up to 3 km (~ 1.9 mi.) (Fig. 10).

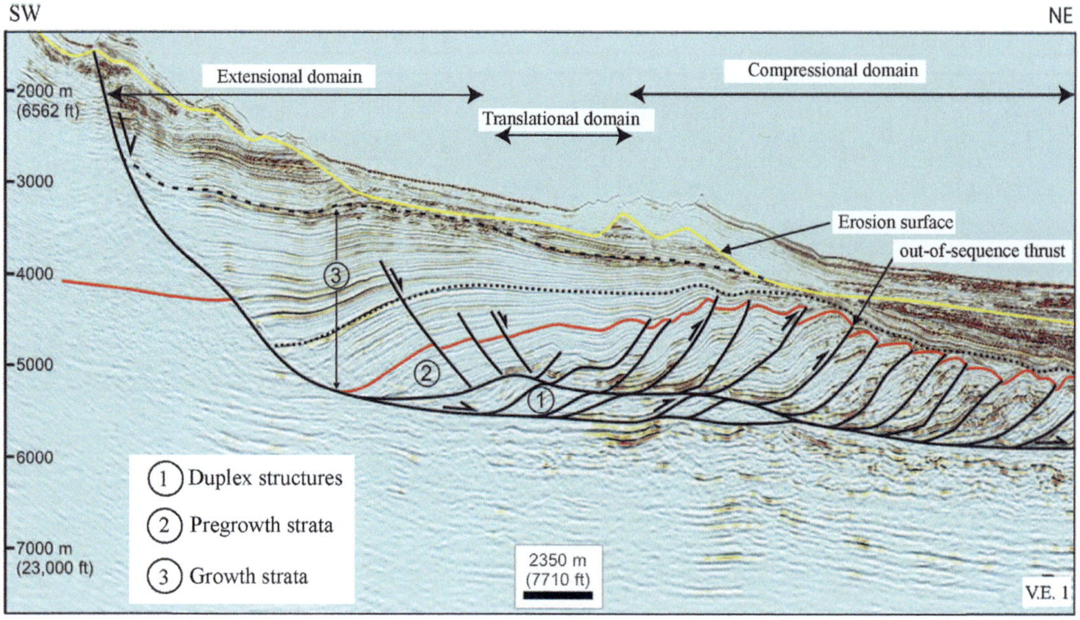


**FIG. 10** Regional, southwest-to-northeast, depth-converted, seismic reflection cross section through Pará-Maranhão passive margin foldbelt of the northern Brazilian passive margin (location of cross section shown on the map in Fig. 8) showing the lateral extent of the three, structural domains of the PMFB: an updip, extensional domain; a midslope translational domain and a downdip compressional domain. The main listric normal fault detaches along a horizon of shale and marl of Paleocene age. The downdip, compressional zone developed as a backstepping and landward-propagating sequence of younger thrusts developing in the hanging wall. *(The cross section is modified from Oliveira, M.J.R., Santarem, P., Moraes, A., Zalán, P.V., Caldeira, J.L., Tanaka, A., Trosdtorf, I., Jr., 2013. Linked extensional-compressional tectonics in gravitational systems in the Equatorial Margin of Brazil. In: Gao, D. (Ed.), Tectonics Sediment: Implications for Petroleum Systems. https://doi.org/10.1306/13351552M1003532.)*

The translational domain is relatively narrow with thrust duplexes and folding. The main detachment surface is Paleocene shale with minor marls. Imbricate thrusts dominate the downdip contractional domain without the presence of any large anticlines. The thrust structures backstep in age in the landward direction with the younger thrusts developing in the hanging wall (Fig. 10).

### *Hydrocarbon potential*

Sub-commercial occurrences of light oil are reported in Cenozoic carbonate reservoirs and Upper Cretaceous turbiditic reservoirs of the offshore Pará-Maranhão Basin (Oliveira et al., 2013). However, exploration interest has increased for this area following significant hydrocarbon discoveries in deepwater plays in the conjugate African equatorial margin and French Guiana (Jubilee and Zaedyus play) (Da Silva Pellegrini and Ribeiro, 2018). Three key plays associated with shallow water shelfal prograding clinoforms and deepwater turbidites have been interpreted by Da Silva Pellegrini and Ribeiro (2018). Potential source rocks of late Albian-early Cenomanian and Cenomanian–Turonian age have been identified in the oil maturity window from the conjugate West African margin (Da Silva Pellegrini and Ribeiro, 2018).

The rollover structures in the updip extensional domain and deepwater thrusts form traps of high exploration interest. A nominal total sediment thickness of 4 km (~ 2.5 mi.) in the updip extensional domain and 2 km (~ 1.9 mi.) in the downdip compressional domain makes source rock maturity the primary risk for the Pará-Maranhão Basin (Fig. 10). The imbricated thrusts in the downdip compressional domain also add the risk of breached reservoirs and few adjacent structural traps.

## Barreirinhas passive margin foldbelt

### *Structural provinces and age of formation*

The Barreirinhas Basin comprises an area of 46,000 $km^2$ (~ 17,760 sq. mi.) that is bounded to the northwest by the Foz do Amazonas Basin and to the southeast by an elevated high of crystalline basement (Oliveira et al., 2013). The tectonic and structural evolution of the Barreirinhas Basin is similar to the Pará-Maranhão Basin. The basin contains a basal pre-rift sequence composed of Paleozoic rocks that experienced rifting from late Aptian to late Albian. A 10-km-thick (~ 6 mi.) sequence of post-rift sediment was deposited in the deep and ultra-deepwaters (Fig. 11).

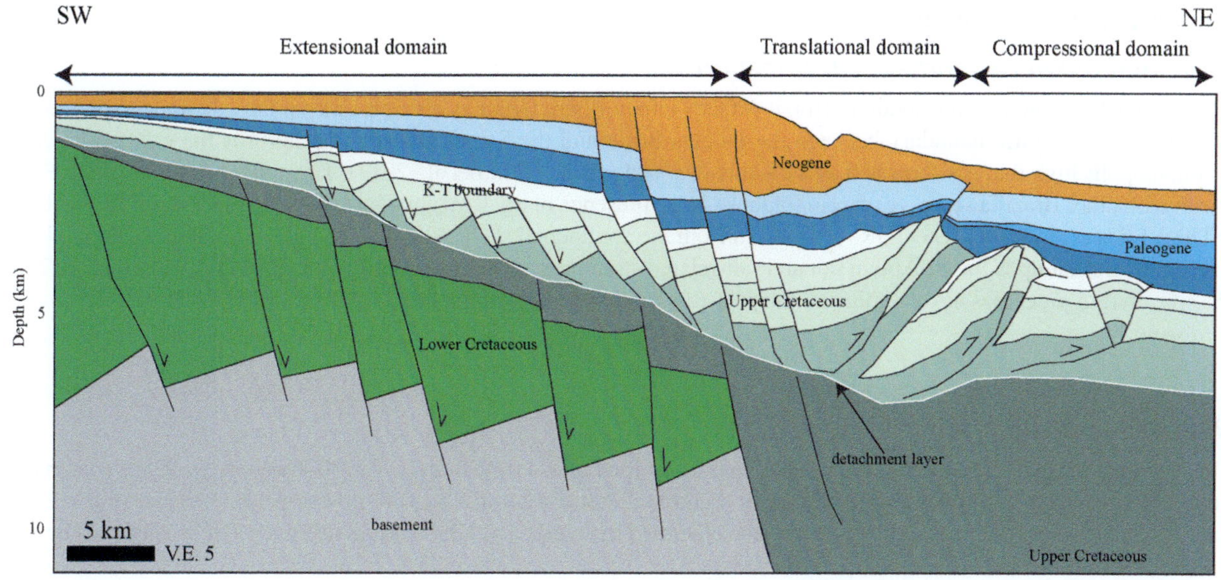


**FIG. 11** Regional, southwest-to-northeast depth cross section through the Barreirinhas Basin of northern Brazil (location of cross section shown on the map in Fig. 8). The lateral extent of the three structural domains are shown and include the zone of updip, listric normal faults of the extensional zone; the intermediate, midslope transitional area and the downslope fold and thrust belt of the compressional zone. The common basal detachment for the updip listric normal faults and downdip folds and thrusts developed along a marine shale unit of late Cretaceous age. *(The cross section is modified from Oliveira, M.J.R., Santarem, P., Moraes, A., Zalán, P.V., Caldeira, J.L., Tanaka, A., Trosdtorf, I., Jr., 2013. Linked extensional-compressional tectonics in gravitational systems in the Equatorial Margin of Brazil. In: Gao, D. (Ed.), Tectonics Sediment: Implications for Petroleum Systems. https://doi.org/10.1306/13351552M1003532.)*

The passive margin foldbelt system evolved as a single detachment layer of Late Cretaceous age and shows all three structural domains—extensional, translational and compressional. The timing of major translation occurred during the Paleogene. A previous study reported a total shortening of 5% based on sequential structural restoration (Oliveira et al., 2013).

## Regional structural profile

The southwest-northeast depth cross section through the Barreirinhas Basin shows 5–6 km (~ 3.1–3.7 mi.) total sediment thickness in the updip extensional domain (Fig. 11). The extensional domain is 40-km-wide (~ 25 mi.) with present-day bathymetric and detachment slopes of 2° and 4°, respectively (Table 1). Like other passive margins, the zone is dominated by listric normal faults that detach on Late Cretaceous overpressured shale.

The 13-km-wide (~ 8 mi.) translational domain lacks normal faults and thrust faults. Thrust structures dominate the downdip compressional domain. The pattern of growth strata indicates that the timing of deformation is Early Paleogene. Three large thrust structures are found in the area of downdip compression (Fig. 11).

## Hydrocarbon potential

The Barreirinhas passive margin foldbelt was explored in the 1960s through 1980s with only minor hydrocarbon production being established for a short period of time in some of the onshore fields (Oliveira et al., 2013).

The primary Cretaceous source rock in the Barreinhas PMFB is deeper (6–8 km, or ~ 3.7–5 mi.) than the equivalent located in the Pará-Maranhão Basin. Thicker overburden likely contributes to the source rock maturation (Da Silva Pellegrini and Ribeiro, 2018).

Exploration in the Barreirinhas PMFB includes four key points. First, Early Paleogene timing of deformation in the downdip compressional resulted in the presence of structural closures during peak expulsion of source rocks. Second, the detachment surface may act as an obstacle to hydrocarbon fluid migration from deeper source rocks. Third, possible periodic reactivation of faults may cause migration of fluids out of reservoirs. Fourth, the absence of drilled wells suggests that high-quality reservoir and source rock are two key geologic risks.

## Pelotas passive margin foldbelt

### Structural provinces and age of formation

The Pelotas Basin in offshore southern Brazil extends into northern Uruguay and remains a relatively underexplored basin (Fig. 8). The structural boundary between the Pelotas Basin and the Santos Basin to the north is the east-west-trending Florinanopolis high that is interpreted as a fracture or transfer zone (Conti et al., 2017). The main difference of the Pelotas Basin with other Brazilian passive margin foldbelts is the presence of thick magmatic wedges of seaward-dipping reflectors (SDRs)(Stica et al., 2014). The SDRs are present from the continental shelf to the lower slope of the Pelotas Basin. Unlike the Santos and Campos Basin, Aptian salt and salt-related structures are absent in the Pelotas Basin (Contreras et al., 2010). Passive margin processes have dominated the post-rift, passive margin of the Pelotas Basin during the late Cretaceous and Cenozoic.

### Regional structural profile

The regional northwest-southeast seismic section through the Pelotas PMFB reveals a 150-km-wide (~ 93 mi.) extensional zone, 20-km-wide (~ 12 mi.) zone of translation and a 38-km (~ 24 mi.) zone of downdip compression (Fig. 12). The extensional domain is significantly wider than the other two structural domains. The passive margin foldbelt detached along a Paleocene-Eocene interval of overpressured shale.

Antithetic faults accompany the thin-skinned, synthetic and listric normal faults of the extensional domain. Minor thrusting is observed in the translational domain, and thrust structures dominate the downdip compressional domain. The thrust structures form a backstepping sequence in the landward direction. The growth stratal patterns reveal that late Neogene deformation affected the PMFB (Fig. 12).

### Hydrocarbon potential

The Pelotas passive margin foldbelt is a frontier basin where significant oil and gas accumulations have yet to be discovered. Favorable source, reservoir and seal intervals have been proposed through seismic facies analysis in the pre-rift, syn-rift and post-rift megasequences by Conti et al. (2017). The source rocks interpreted in the Uruguayan portion of the Pelotas Basin are represented by a restricted marine source rock deposited during the Permian pre-rift phase, and two marine source rocks deposited during the Aptian-Albian and Cenomanian–Turonian post-rift phase (Conti et al., 2017). Up to 7 km (~ 4.3 mi.) thickness of Cretaceous and Cenozoic post-rift clastics were deposited in the basin, including potential Paleocene and Albian source rocks. In addition, numerous oil seeps have been reported in the basin, especially in slope areas at water depths of 500–800 meters (~ 1640–2624 ft.) (Geo Expro, 2013).

There are two potential risks for petroleum exploration in the Pelotas passive margin foldbelt. First, the presence of seaward dipping reflectors indicates volcanic activity, which causes higher transient heat flow and possible over-maturity of source rocks. Second, the present-day deformation of the passive margin foldbelt can lead to seal integrity risk.

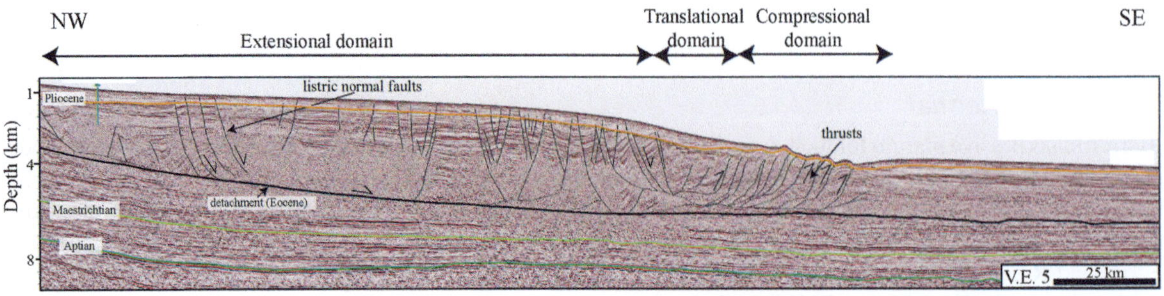


**FIG. 12** Regional, northwest-to-southeast depth cross section based on a deep-penetration seismic reflection line through Pelotas Basin of offshore Uruguay (location of cross section is shown on the map in Fig. 8). The lateral extent of the three structural domains are shown and include the updip extensional domain that is characterized by both synthetic and antithetic listric normal faults, the transitional zone in the midslope area and the downdip, compressional zone. The underlying detachment for the normal faults occurs on a shale interval of Paleocene-Eocene age. *(The seismic section is modified from Geo Expro, 2013. The Pelotas Basin Oil Province Revealed (September edition).)*

# Examples of passive margin foldbelts along West Africa

## Regional setting of passive margin foldbelts along West Africa

The continental margin of West Africa formed during the south-to-north progressive rifting of Gondwana and the separation of the South American and African continents. Rifting started initially in the south during the Late Jurassic and propagated northward (Ceraldi et al., 2017). During the syn-rift to passive margin transition, salt was deposited in the central area of the West African margin during Aptian time. The passive margin stage is dominated by regional uplift and seaward tilting of the African continent with related deposition of deepwater sedimentary intervals in submarine fans environments (Liu et al., 2008).

This section describes the structural evolution of three passive margin foldbelts of the offshore West African margin (Fig. 13): (1) Niger, (2) Kwanza and (3) Orange River. The gravity tectonics in the Niger and Orange River PMFB are controlled by overpressured shale, whereas in the Kwanza Basin deformation is controlled by salt remobilization (Morley et al., 2011; Rowan, 2020).

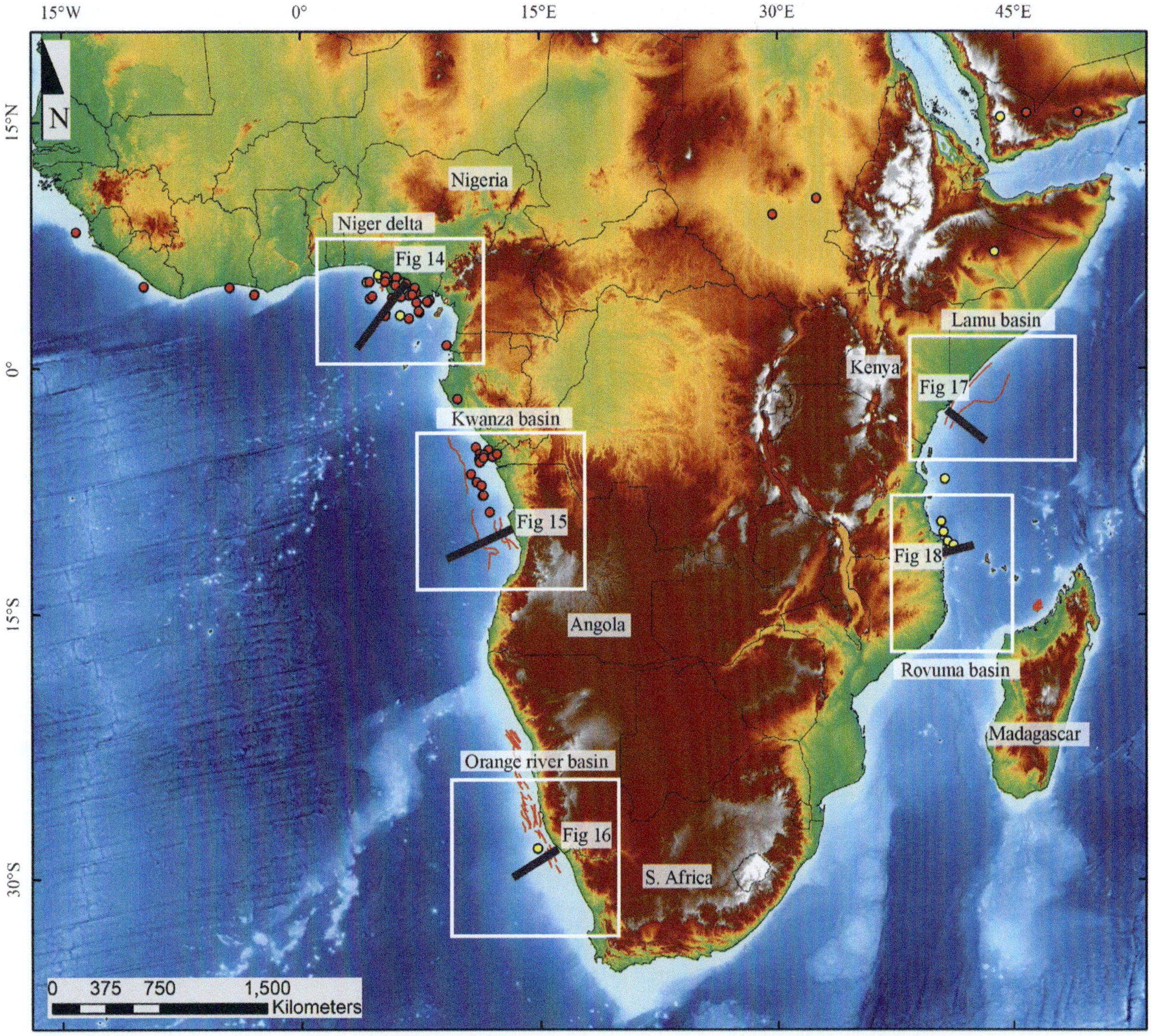


**FIG. 13** Bathymetric map of the offshore African margin showing the locations of five passive margin foldbelts formed along the passive margins of the African continent: the Niger PMFB along the Niger Delta of Nigeria; the Kwanza PMFB of Angola; the Orange River PMFB of South Africa; the Lamu Basin of Kenya and the Rovuma Basin of Mozambique. Passive margin fold axes are shown by *red lines*, locations of the illustrated cross sections are shown by the *solid black lines* and the location of the giant oil and gas fields are shown by the *red dots* (oil fields) and by the *yellow dots* (gas fields) (CBTH database compilation, 2020).

## Niger passive margin foldbelt

### Structural provinces and age of formation

The Niger passive margin foldbelt is a prolific hydrocarbon province located in the Gulf of Guinea (Zhang et al., 2021). The depositional history of the Niger Delta began in the Eocene with deposition of a clastic wedge that is 12 km (~ 7.5 mi.) thick. The Niger PMFB evolved above an overpressured shale detachment (Wu et al., 2015).

The gravity system of the Niger Delta PMFB is characterized as a headward delta-top zone of extension, a midslope zone of translation and a delta-toe zone of contraction (Steele et al., 2009; Wu et al., 2015). The distance from the updip extensional zone to the downdip contraction zone is approximately 300 km (~ 186 mi.). The overpressured shale detachment layer is the Akata Formation which has a slope of 1.4°.

### Regional structural profile

The regional cross section of the passive margin foldbelt developed on the Niger Delta shows the distribution of three main structural zones: (1) a 190-km-wide (~ 118 mi.) zone of updip extension; (2) a 22-km-wide (~ 14 mi.) zone of translation and (3) an 85-km-wide (~ 53 mi.) zone of downdip compression (Fig. 14). The updip extensional zone is characterized by both basinward and counter-regional growth faults and associated rollover depocenters. Gravity gliding along the listric normal faults creates additional accommodation in the updip extensional domain and increases the stress head in the downdip direction. The translational zone is characterized by large areas of little or no deformation interspersed with broad detachment folds above the Akata Formation (Corredor et al., 2005).

A zone of imbricate thrust structures dominates the downdip compressional domain. An older thrust system of Oligocene to Middle Miocene age has been reactivated by the deposition of Upper Miocene and Plio-Pleistocene sandstone splays (see Chapter 8 for more information on splays). This depositional reactivation resulted in various oversteepened thrust structures that were initially shale diapirs or mobile shale-cored structures (Corredor et al., 2005; Steele et al., 2009).

### Hydrocarbon potential

The Niger passive margin foldbelt ranks as one of the largest hydrocarbon provinces in the world with an estimated ultimate recovery of 40 billion barrels (Adegoke et al., 2017). Three different petroleum systems are present in the Niger: (1) Lower Cretaceous lacustrine system, (2) marine systems in Upper Cretaceous-Lower Paleocene and (3) a Cenozoic system.

The most prolific reservoirs are found in the delta growth faults of the extensional zone and in the deepwater elements in the midslope translational zone. Hydrocarbon accumulation in the deltaic reservoirs is likely to have migrated upward from downdip areas (Steele et al., 2009). Most of the giant fields of the Niger are in the translational domain (e.g., Bonga, Agbami, Akpo, N'nwa-Doro and Erha). The prospectivity in the translational zone is due to large structural closures containing excellent channelized deepwater reservoirs and mature source rocks in the oil window.

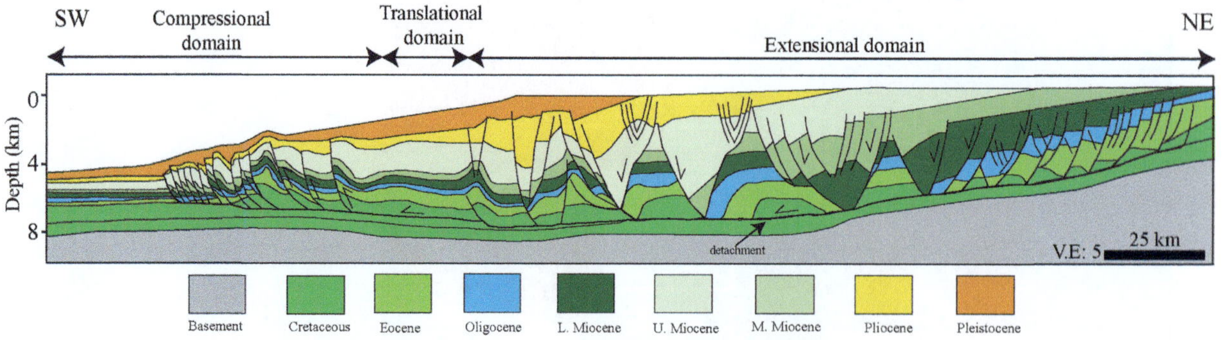


**FIG. 14** Regional, northeast-to-southwest depth cross section through Niger passive margin foldbelt (location the cross section is shown on the map in Fig. 13). The lateral extents of the three structural zones are shown and include the updip area of delta-top extensional faults, an intermediate translational domain and a zone of delta-toe of imbricate thrusts and folds. The underlying detachment for all three zones occurs within the Akata Formation, a prodelta marine shale of Cretaceous age. *(The cross section is modified from cross sections by Steele, D., Ejedawe, J., Adeogba, T., Grant, C., Filbrandt, J., Ganz, H., 2009. Geological framework of Nigeria linked shelf extension and deepwater thrustbelts, AAPG Hedberg Conference, October 4–9, Tirrenia, Italy. and Zhang, J., Wu, S., Hu, G., Yue, D., Xu, Z., Chen, C., Zhang, K., Wang, J., Wen, S., 2021. Role of shale deformation in the structural development of a deepwater gravitational system in the Niger Delta. Tectonics 40. https://doi.org/10.1029/2020TC006491.)*

The imbricate thrust belt in the contractional domain is not charged by hydrocarbons as the result of: (1) differing heat flow, as this domain overlies Cretaceous oceanic crust, (2) pulses of sediment loading that can cause thrust reactivation, a main root of seal failure, and (3) ineffective and unfocused hydrocarbon charge (Steele et al., 2009).

## Kwanza passive margin foldbelt

### Structural provinces and age of formation

The Kwanza Basin formed as a salt-influenced passive margin foldbelt offshore Angola (Fig. 13). Rifting of the Kwanza Basin initiated in the Early Cretaceous during the opening of the South Atlantic Ocean. A thick layer of Aptian salt was deposited shortly after rifting and its salt geometry reflects the syn-rift topography (Evans and Jackson, 2020; Hudec and Jackson, 2002). The salt accumulation is separated by a high, the Atlantic hinge zone, along which the salt is thin or absent. The basin is divided into two main parts: the western part known as the Outer Kwanza Basin that contains a 4-km-thick interval of salt and the eastern part known as the Inner Kwanza Basin (Morley et al., 2011).

The presence of salt, gradient of the slope, sediment loading and onland basement uplift events control the present-day structural configuration of the Kwanza Basin. Hudec and Jackson (2004) pointed out three phases of regional uplift: early Albian, Campanian, Miocene, that contributed to the evolution of the basin as a passive margin foldbelt.

### Regional structural profile

A northeast-southwest depth cross section shows a 200-km-wide (~ 124 mi.) zone of thin-skinned listric normal faults detaching over a relatively thin interval of Aptian salt. Prominent turtle structures and normal fault-bounded depocenters are located in the updip extensional domain (Fig. 15).

The downdip compressional domain is 160 km-wide (~ 100 mi.) and is dominated by a large-scale salt nappe, thickened salt plateau and squeezed diapirs. The salt nappe is approximately 100 kmwide (~ 62 mi.) and 2–4-km (~ 1.2–2.5 mi.) thick. The Atlantic hinge zone divides the extensional and compressional domains (Fig. 15). The translational domain is not easily discernible but has been interpreted by previous studies in the area of the Atlantic hinge zone (Evans and Jackson, 2020).

### Hydrocarbon potential

Pre-salt reservoir targets along the South Atlantic conjugate margins have been a major focus with significant discoveries concentrated in the Santos and Campos basins offshore Brazil. Using these Brazilian discoveries as analogs, the underexplored deepwater Kwanza Basin is now emerging as a prolific pre-salt hydrocarbon basin. The ultra-deepwater well Baleia-1 was the first commercial oil discovery in 1996 in the pre-salt section of the Inner Kwanza Basin. This well encountered oil shows in a 120-meter-thick (~ 394 ft.) sandstone reservoir. In 2011, due to recent deepwater discoveries in Brazil, Maersk drilled the Azul-1 pre-salt discovery, shortly following the Cameia-1 discovery well (Greenhalgh et al., 2012).

Although hydrocarbon discoveries in the Outer Kwanza Basin remain elusive, separate syn-rift basins divided by the Atlantic hinge zone formed the setting for pre-salt discoveries in the Inner Kwanza Basin (Hudec and Jackson, 2004). Restricted circulation during the Early Cretaceous led to the deposition of high-quality source rocks. During episodes

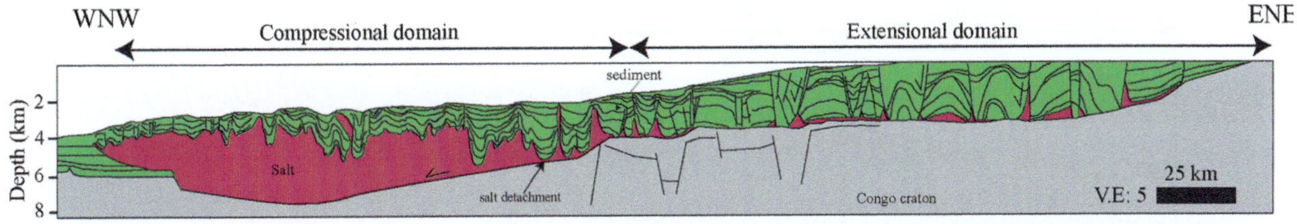


**FIG. 15** Regional, northeast-to-southwest depth cross section through the Kwanza passive margin foldbelt, offshore Angola (location of the cross section is shown on the map in Fig. 13). The two structural domains of the PMFB include the updip zone of listric normal faults detaching on Aptian salt and the downdip compressional area that includes large salt nappe of approximately 100 km (~ 62 mi.) in width and 2–4 km (~ 1.2–2.5 mi.) in thickness. *(The cross section is modified from previous cross sections by Evans, S.L., Jackson, C.A.L., 2020. Base-salt relief controls salt-related deformation in the outer Kwanza Basin, offshore Angola. Basin Res. 32, 668–687. https://doi.org/10.1111/bre.12390, Morley, C.K., King, R., Hillis, R., Tingay, M., Backe, G., 2011. Deepwater fold and thrust belt classification, tectonics, structure and hydrocarbon prospectivity: a review. Earth Sci. Rev. 104, 41–91. https://doi.org/10.1016/j.earscirev.2010.09.010, and Hudec, M.R., Jackson, M.P.A., 2004. Regional restoration across the Kwanza Basin, Angola: salt tectonics triggered by repeated uplift of a metastable passive margin. Am. Assoc. Pet. Geol. Bull. 88, 971–990. https://doi.org/10.1306/02050403061.)*

of relative uplift and subsidence of the west African continental margin, sediment overlying the Aptian evaporites have slowly migrated down the slope to create large structural closures in the basin (Greenhalgh et al., 2012).

## Orange River passive margin foldbelt

### *Structural provinces and age of formation*

The Orange River Basin is located in offshore Namibia and South Africa and records the development of a Late Jurassic to present-day volcanic-rifted margin (Granado et al., 2009). South Atlantic rifting is marked by a Late Jurassic to Late Cretaceous syn-rift megasequence, separated from the Late Cretaceous to present-day passive margin megasequence by the breakup unconformity. The post-rift evolution of the margin is characterized by gravity-driven failure that produced the Orange River passive margin foldbelt (de Vera et al., 2010).

Gravitational failure is recorded by its updip extensional domain linked to a downdip contractional domain detaching along a thin layer of overpressured shale of Turonian age. Sequential structural restoration shows a total extension of 24 km (~ 15 mi.) and a downdip shortening of 16 km (~ 10 mi.) (de Vera et al., 2010). The pattern of growth strata suggests that the gravity failure was short-lived, from the Coniacian to Santonian, and was likely the result of uplift of the West African craton and sediment loading by the Orange River Delta.

### *Regional structural profile*

The regional northeast-southwest depth cross section through the Orange River PMFB shows the main depositional megasequences and the late Cretaceous gravity-driven PMFB (Fig. 16). The extensional domain is 140km (~ 87 mi.) in length, the translational domain is 20km (~ 12 mi.) in length and the compressional domain is 95km (~ 59 mi.) in length.

The extensional domain is characterized by typical listric normal faults that detach on the Turonian interval. The average detachment slope is 1° with a bathymetric slope of 1.1°. Growth strata in the extensional domain show thickening along individual faults. The timing of slip is interpreted to be Coniacian to Santonian, although later reactivation of some extensional faults is likely (de Vera et al., 2010). The translational domain is relatively short, with a downdip length of 20km (~ 12 mi.). The 95-km-wide (~ 59 mi.) compressional domain is characterized by numerous high-angle landward-dipping imbricate thrust faults (Fig. 16).

### *Hydrocarbon potential*

Structural and stratigraphic traps in the Orange River PMFB include structural traps related to listric growth faults and stratigraphic traps related to turbidite-dominated submarine fans (Brownfield, 2016). There have been no discoveries to date in the Orange River PMFB. Hydrocarbons are likely generated from Barremian-Aptian and Cenomanian–Turonian marine Type II source rocks that are now buried by as much as 7 km (~ 4 mi.) of clastic, sedimentary rocks (Brownfield, 2016). Basin modeling studies have shown that the generation from Cenomanian–Turonian source rocks post-dates the timing of structural deformation, and therefore poses a significant risk for hydrocarbon accumulation.

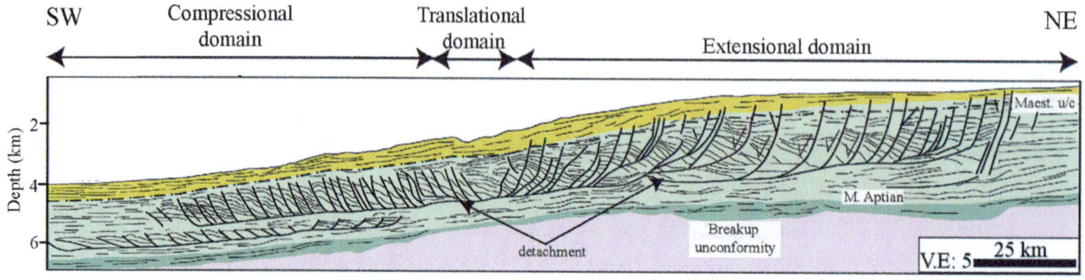


**FIG. 16** Regional, northeast-to-southwest depth cross section through the Orange River PMFB of late Cretaceous age along the south Atlantic passive margin of South Africa (location of the cross section is shown on the map in Fig. 13) The lateral extents of the three structural domains are shown with the updip extensional domain characterized by listric, normal faults; a narrow transitional zone and a downdip zone of imbricated thrust faults. *(The cross section is modified from previous cross sections by Grando, G., Schleder, Z., Shackleton, R., Seed, G., Buddin, T., McClay, K., Borraccini, F., 2009. 3D Structural Evolution of the Salt Controlled Frampton Anticline, Atwater Valley Foldbelt, Deep Water Gulf of Mexico, AAPG search and discover article #50161 and de Vera, J., Granado, P., McClay, K., 2010. Structural evolution of the Orange Basin gravity-driven system, offshore Namibia. Mar. Pet. Geol. 27, 223–237. https://doi.org/10.1016/j.marpetgeo.2009.02.003.)*

Furthermore, smaller gas discoveries (Kudu, Ibhubesi, K-J1) and gas seeps in shallow water indicate gas-prone source rock in the passive margin section that immediately overlies the Cretaceous seaward-dipping reflectors (Brownfield, 2016; de Vera et al., 2010).

Hydrocarbons can potentially migrate into Cretaceous and Cenozoic sandstone reservoirs that include deltaic and nearshore marine sandstone, turbidite sandstone, slope truncations along the present-day shelf and paleo-shelf edge and basin-floor fan reservoirs. Structural traps include growth-fault-related structures and rotated fault blocks within the continental shelf. Cretaceous and Cenozoic mudstone and shale rocks form the primary reservoir seals (Brownfield, 2016).

## Examples of passive margin foldbelts along East Africa

### Regional setting of passive margin foldbelts along East Africa

In this section, the structural evolution and hydrocarbon potential of two passive margin foldbelts of the offshore East African margin are described: (1) Lamu and (2) Rovuma.

The offshore Lamu and Rovuma basins are located along the passive continental margin of East Africa, stretching along the coast from Mozambique to Kenya (Fig. 13). The passive continental margin originated due to Karoo rifting and the Africa-Madagascar breakupduring the Middle Jurassic. Seafloor spreading ceased in the Early Cretaceous around 120–130 Ma. From the Early Cretaceous onward, both regions evolved as PMFB. The initiation of the East African Rift System took place during Eocene and is believed to have contributed to the passive margin foldbelt process (Cai et al., 2020).

### Lamu passive margin foldbelt

#### *Structural provinces and age of formation*

The passive margin foldbelt system in the Lamu Basin detached along two horizons, Late Cretaceous and Late Oligocene, that are both interpreted as overpressured shale. Although the presence of salt has been inferred, drilling of Kencan-1 and Garissa-1 and seismic facies analysis revealed that a siliciclastic unit forms the detachment level (Cai et al., 2020).

Analysis of growth strata indicates that the passive margin foldbelt remained active from Late Cretaceous to Early Miocene, but almost all PMFB-related deformation occurred before the late Paleocene. Locally, a series of post-rift-related seamounts forms a downslope barrier to the PMFB (Cruciani et al., 2017; Cruciani and Barchi, 2016).

#### *Regional structural profile*

The northwest-southeast depth cross section through the Lamu passive margin foldbelt shows the presence of two overlapping foldbelts driven by gravitational processes (Fig. 17). The older thrust system that detaches on the Late Cretaceous interval does not clearly exhibit the main elements of a PMFB. An overlying and younger detachment on a Late Oligocene shale unit does exhibit the main elements of a PMFB including: (1) a 13-km-wide ($\sim$ 8 mi.) zone of updip extension dominated by thin-skinned listric normal faults, (2) a 10-km-wide ($\sim$ 6 mi.) zone of translation characterized by absence of shortening and containing thick packages of Miocene-Recent sediment and (3) a 20-km-wide ($\sim$ 12 mi.) zone of downdip compression expressed as multiple imbricated thrust faults detached on a basinward dipping detachment surface.

#### *Hydrocarbon potential*

The structural cross section shows that the older and younger thrust belt both the older and younger PMFB in the Lamu Basin formed during the Late Paleogene to Recent time. Cai et al. (2020) predict gas discoveries in the Lamu PMFB because the primary Jurassic source rock was in the oil generation window during the Late Cretaceous and was in the gas window from Early Paleogene to Recent. Chances of gas discoveries are predicted because the gas generation window was in the Early Paleogene to Recent.

Cai et al. (2020) propose two main sources of risk for hydrocarbons in the Lamu PMFB: (1) Late Cenozoic seamounts are present and may create higher-than-expected heat flow and (2) source rock quality varies along the trend of the margin.

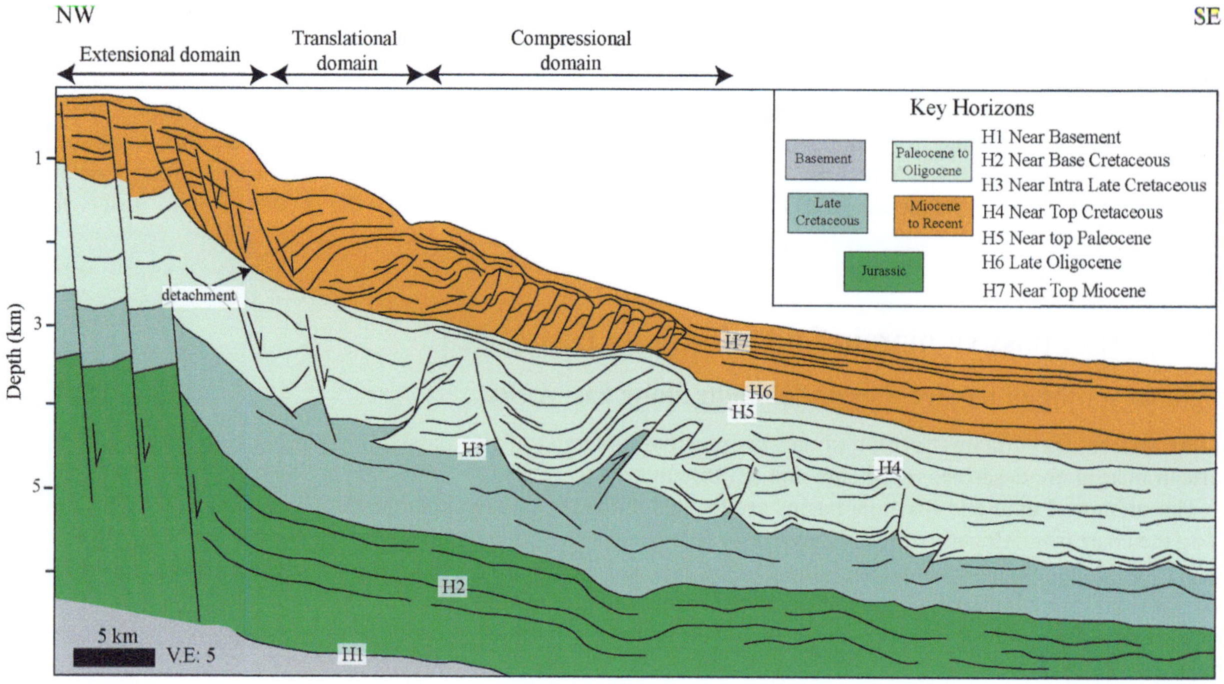


**FIG. 17** Regional, northwest-to-southeast depth cross section through the Lamu passive margin foldbelt on the Kenyan passive margin on the western Indian Ocean (location of the cross section is shown on the map in Fig. 13). The Lamu PMFB consists of a vertical stack of two thrust systems detached over basinward dipping detachment faults. The ages of the strata along the two detachment horizons are Late Cretaceous and Late Oligocene. *(The section is modified from a cross section by Cruciani, F., Barchi, M.R., 2016. The Lamu Basin deepwater fold-and-thrust belt: an example of a margin-scale, gravity-driven thrust belt along the continental passive margin of East Africa. Tectonics 35, 491–510. https://doi.org/10.1002/2015TC003856.)*

## Rovuma passive margin foldbelt

### Structural provinces and age of formation

The Rovuma passive margin foldbelt formed after the cessation of Mesozoic rifting and seafloor spreading. The downdip foldbelt developed mainly in the Oligocene and Miocene and detached along the top of a Paleocene shale horizon. An Oligo-Miocene deltaic system overlies the detachment horizon (Mahanjane and Franke, 2014).

Previous studies have proposed two principal causes for the formation of the Rovuma PMFB: (1) progradation of the Cenozoic Rovuma Delta and (2) uplift the adjacent land area in eastern Africa. Both mechamisms are linked to the early development of the East African Rift System (Cai et al., 2020; Mahanjane and Franke, 2014).

### Regional structural profile

The geometry of the passive margin foldbelt of the offshore Rovuma Basin comprises three domains: (a) a 21-km-wide (~ 13 mi.) updip extensional domain, (b) a 11-km-wide (~ 7 mi.) translational domain and (c) a 45-km-wide (~ 28 mi.) downdip compressional domain (Fig. 18). In the upslope extensional domain, listric normal faults detach on Eocene mudstones that are inferred to be undercompacted and overpressured.

The generation of accommodation creates a large rollover depocenter in the distal portion of the extensional and translation domains. The translation domain lacks prominentstructural features and basinward-dipping thrusts faults that characterize the downdip compressional domain. Based on the ages of growth strata, the timing of deformation is post-Oligocene.

### Hydrocarbon potential

Four source rock intervals have been reported in the Rovuma Basin: Permo-Triassic, Lower Jurassic, Lower Cretaceous and Upper Cretaceous. Geochemical analysis and seismic facies analysis show that a Late Jurassic source rock interval is widespread. Systematic basin modeling studies show that the timing of oil generation in the Rovuma Basin is Middle to Late Cretaceous, and that gas generation is Early Paleogene to Recent (Cai et al., 2020). These observations constrain a less

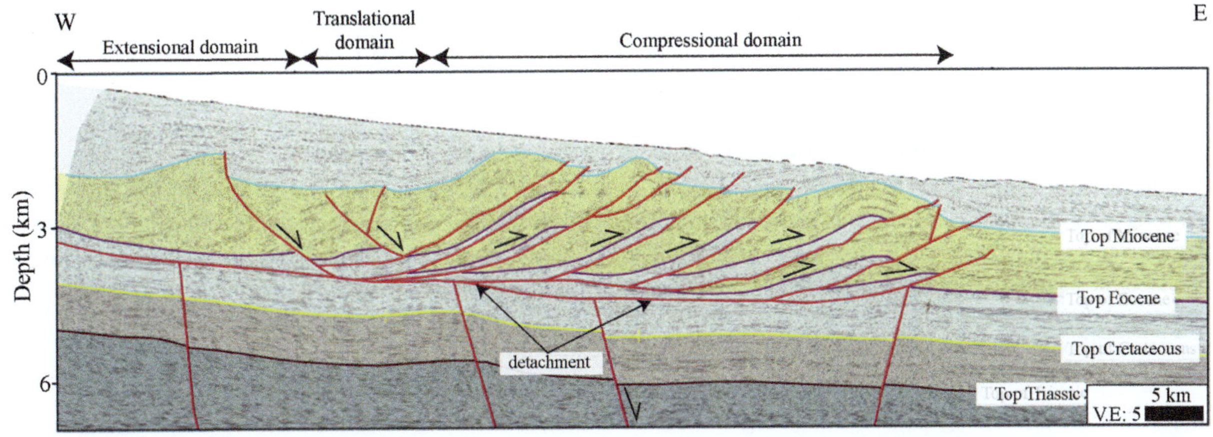


**FIG. 18** Regional, northwest-to-southeast depth cross section through the Rovuma Basin of Mozambique (the location of the cross section is shown on the map in Fig. 13). The lateral extents of the three structural zones are shown and include the updip extensional zone, a narrow transitional zone and the downdip compressional domain. The passive margin foldbelt system of the basin has been active from the late Eocene period and detached along the top of an Eocene shale horizon. *(The cross section is from a section by Cai, J., He, Y., Liang, J., Qiu, C., Zhang, C., 2020. Differential deformation of gravity-driven deepwater fold-and-thrust belts along the passive continental margin of East Africa and their impact on petroleum migration and accumulation. Mar. Pet. Geol. 112, 104053. https://doi.org/10.1016/j.marpetgeo.2019.104053. Figure used with permission from Elsevier.)*

favorable timing for oil generation and a more favorable timing for gas accumulation. The reservoirs are Late Cretaceous sandstone-rich channel-fills and splays that developed over wide areas of the East African margin.

Recent giant gas field discoveries such as Mamba, Prosperidade, Golfinho and Coral indicate a working petroleum system with gas mature source rock in the Rovuma Basin (Zhang et al., 2017). Reservoirs of the Coral and Mamba fields are composed of high-quality 100-meter-thick (~ 328 ft.), sandstone units formed by high-density gravity flows and bottom currents that have distributed the sands over tens of kms along the margin (see Chapters 7–9; Fonnesu et al., 2020).

# Discussion

## Settings of passive margin foldbelts

### *Transform-passive margins vs rifted-passive margins*

Most of the passive margin foldbelts of the Gulf of Mexico, Atlantic margins and Western Indian ocean margin are classified as rifted-passive margin settings. Rifted-passive margins include the following PMFB: (1) Atwater-Mississippi, Perdido and Campeche in the Gulf of Mexico; (2) Pelotas, Niger Delta and Orange River along the Atlantic margin and (3) Lamu and Rovuma in the Western Indian ocean.

In contrast, the Mexican Ridges, Foz do Amazonas, Pará-Maranhão and Barreirinhas PMFB are associated with transform-passive margins that characteristically have steeper basement and bathymetric gradients than rifted-passive margins.

The Mexican Ridges foldbelt evolved over oceanic crust. The Foz do Amazonas, Pará-Maranhão and Barreirinhas PMFB experienced initial transtension produced duringthe Aptian-Albian oblique opening of the Equatorial Atlantic Ocean. Differences in structural styles between PMFB in transform and rifted-passive margins have not yet been observed. For a more detailed description of tectonic and crustal models for rifted-passive and transform-passive margins, see Chapter 3.

### *Margin oversteepening related to orogenic events*

Orogenic events contribute to the passive margin foldbelt evolution by increasing the amount of seaward tilting of the passive margin (Rowan et al., 2004; Morley et al., 2011). Denudation resulting from this uplift contributes large amounts of clastic sediment to the basin, which acts to destabilize the margin even more. This combination and interplay of the increased slope due to gravity-gliding and sediment loading due to gravity spreading results in the structures demonstrated in the regional cross sections.

In the western Gulf of Mexico, the basement and surface slopes of the Perdido and Mexican Ridges foldbelt are controlled by the Laramide orogeny (Fitz-Díaz et al., 2018; Hudec et al., 2019). In the Campeche PMFB, the basement and surface slopes are controlled by the Middle Miocene Chiapanecan orogeny (Hasan and Mann, 2021). The northeastern South American PMFB including the Pará-Maranhão and Barreirinhas are similarly controlled by orogenic events related to the Andean orogeny (Reis et al., 2010).

### *Margin oversteepening related to cratonic uplift*

The influence of cratonic uplift has been proposed by previous studies for the evolution of the Kwanza PMFB. Cratonic uplift affects the PMFB in two ways: (1) by increasing basinward tilt and (2) by increasing sediment supply (Hudec and Jackson, 2004). The cratons in West Africa including the West African Craton Congo Craton and Kalahari Craton are considered the primary sediment sources of the large river systems that feed the offshore PMFB and have a direct influence on their evolution. Eastern African basins including the Lamu and Rovuma are assumed to be affected by the East African rift system, although the eastern branch of that rift system is located almost 500 km (~ 311 mi.) landward (Cai et al., 2020).

### *Margin oversteepening related to progradation of major deltas*

The Atwater-Mississippi, Foz do Amazonas and Niger PMFB are all strongly influenced by delta progradation. In the Atwater-Mississippi, 4–8 km (~ 2.5–5 mi.) of sediment has been deposited during the Middle to Late Miocene as the result of progradation of the Mississippi Delta (Peel et al., 1995). The Foz do Amazonas PMFB also accumulated 10 km (~ 6 mi.) of sediment in the Late Miocene (Reis et al., 2010). The Niger Delta is up to 12 km (~ 7.5 mi.) thick (Wu et al., 2015). The large amount of clastic sediment deposition in such a short period for each of these PMFB dominated their evolution through the gravity-spreading process. The timing of extensional faults, folds and thrusts in each basin coincides with the high amount of deposition resulting from delta progradation and indicates a causal relationship.

## Controlling factors for the formation of passive margin foldbelts

The controlling factors for passive margin foldbelts are divided into two end members: gravity gliding and gravity spreading (Rowan et al., 2004;). Gravity gliding is defined as the translation of a body along a detachment slope where gravitational energy is released due to this movement. In contrast, gravity spreading is defined as vertical collapse and lateral spreading of a body of rock under its weight. The direction and amount of detachment along the slope in a basin drive the gravity-gliding process, whereas depositional loading controls the gravity-spreading process.

The relative contribution of both controlling factors is still under debate (Brun and Fort, 2011; Rowan and Ratliff, 2012). Using forward modeling, Peel (2014) has shown that the contribution of gravity gliding has a positive linear relationship with the angle of the detachment surface. For example, with a low angle detachment surface (4°), gravity spreading contributes 65% of the total gravitational force, whereas, with a detachment angle of 8°, it is only 38% (Peel, 2014). Considering the range of detachment angles (2–7°) for the PMFB described in this chapter, it is likely that the passive margin foldbelt evolution is driven by similar combinations of slope-angle-controlled gliding and deposition-controlled spreading.

## Salt- vs shale-based detachments

Gravity-driven passive margin foldbelts are transported along either salt or shale detachments. The first-order structural style, which can range from a zone of updip extensional structures to a zone of lateral translation or even a zone of contractional structures, remains similar in all of this chapter's examples—regardless of whether salt or shale forms the detachment horizon.

In a more detailed analysis, it can be observed that some specific structural styles observed in PMFB depend on the salt vs shale detachment type. Symmetric detachment folds typically dominate gravity-driven systems associated with salt detachment. In contrast, shale-detached passive margin foldbelts are dominated by imbricate thrusts, fault-bend folds and fault propagation folds. The difference in structural style is likely the result of the differing rheologies of salt and shale (Oliveira et al., 2013).

In the subsurface, salt is mobile, viscous and readily deforms to fill the cores of folds, produce large diapirs and forms more extensive salt canopies than observed for shale. The reason for this is likely related to the more plastic behavior of shale that deforms only when the deviatoric stress overcomes its shear strength (Rowan et al., 2004).

The original distribution of autochthonous salt is also proposed as a control on fold geometry as described in PMFB including the Perdido and Campeche (Trudgill et al., 1999; Hasan and Mann, 2021). The presence of an originally thick salt layer can readily mobilize and fill the cores of detachment folds and produce gentle, symmetrical anticlines over large areas as seen in many PMFB in the Gulf of Mexico. In areas of a reduced and thinner salt supply, a more asymmetrical structural style is observed with tighter folds bounded by one or both sides by reverse faults (Rowan et al., 2004). For more information on salt tectonics in deepwater areas, see Chapter 5.

### Structural differences: Variations in updip extension and downdip compression

This chapter recognizes that the updip extensional domain of passive margin foldbelts is dominated by basinward-dipping listric normal faults associated with antithetic and landward-dipping normal faults (Figs. 4–6, 9A, 12 and 14). Evidence of expulsion rollovers associated with rapid sediment loading is observed in salt-controlled PMFB such as the Atwater-Mississippi, Perdido and Campeche (Figs. 4–5 and 7). The expulsion of the primary autochthonous salt results in allochthonous canopies that overlie the translational or compressional domains. These higher-level and extensive salt canopies are unique to salt-based PMFB and are not observed in shale-based PMFB. This observation is likely due to the higher mobility and lower viscosity of salt in comparison to shale (Wu and Bally, 2000).

The structural style of the downdip compressional domain depends on the salt vs shale nature of the underlying detachment. The compressional domain of shale-based systems is dominated by asymmetrical and imbricated thrust systems (e.g., Foz do Amazonas, Pará-Maranhão, Barreirinhas, Pelotas, Niger Delta, Orange River, Lamu and Rovuma). In contrast, salt basins such as the Atwater-Mississippi, Campeche, Perdido and Kwanza are dominated by symmetrical and upright detached and salt-filled anticlines with few thrust and reverse faults.

### Passive margin foldbelts compared within the framework of critical taper wedge theory

Bilotti and Shaw (2005), Mourgues et al. (2014) and Tesei et al. (2021) have all modeled passive margin foldbelts using critical taper wedge theory. The taper of a wedge is defined as the angle between its free surface and basal detachment. Critical taper wedge mechanics theory states that once a wedge reaches some critical taper angle, the wedge will grow self-similarly as material is either added to the wedge by deposition or tectonic accretion—or removed from the wedge by erosion or tectonic erosion. At the critical taper angle, the stress overcomes the strength of the detachment layer and the thrust wedge propagates basinward (Bilotti and Shaw, 2005; Davis et al., 1983).

The taper angle in the studied passive margin foldbelts ranges from 2.1 to 9.3° (Fig. 19). These taper angles are lower than the taper angles of accretionary wedges at active margins as reported by Davis et al. (1983). Evidence of propagation along a basal detachment results in a wide contractional domain in PMFB and shows that these broad PMFB attained critical taper at least for a limited time interval. The reason for attaining a critical taper stage at such a low taper angle may relate to the weak basal detachment consisting of either salt or overpressured shale (Bilotti and Shaw, 2005; Tesei et al., 2021).

Steeper basement slopes and higher detachment angles in the updip extensional domain commonly results in a narrower zone of listric normal faults observed in the Mexican Ridges, Pará-Maranhão and Lamu passive margin foldbelts (Figs. 6, 10 and 17). More gentle basement slopes and detachment angles result in wider zones of extensional deformation as observed in the Kwanza, Orange River, Pelotas and Campeche PMFB. However, the lateral and downdip limit of the passive margin foldbelt system is also controlled by the lateral extent of the underlying shale detachment or overpressured zone and by the buttressing effect of a downdip basement, step-up fault as found in the Perdido and Campeche PMFB (Hasan and Mann, 2021).

## Key messages

Exploration interest in deepwater passive margin foldbelts increased significantly over the past decade as more and more examples of PMFB have been described using seismic data grids. The growing interest is a result of numerous geological, engineering (Chapters 1 and 19) and economic (Chapter 20) factors including: the diminishing number of undrilled onshore and nearshore structural traps, thus urging companies to venture farther offshore; the abundance of attractive undrilled large traps in deepwater passive margin foldbelts, which are known to contain a variety of play types, reservoirs and seals; the presence of natural oil seeps indicating, in cases, multiple working petroleum systems; the availability of higher resolution wide-azimuth 2D and 3D seismic data in these structurally challenging deepwater provinces and advances in deepwater drilling technology (Morley et al., 2011).

There are many world-class examples of petroleum production from deepwater basins associated with mobile substrates formed by either salt or shale (see Chapter 19). Mobile salt or shale in a basin is attractive in the hydrocarbon exploration business as the remobilization of either can result in a variety of structural traps and seals. The presence of sandy turbidites,

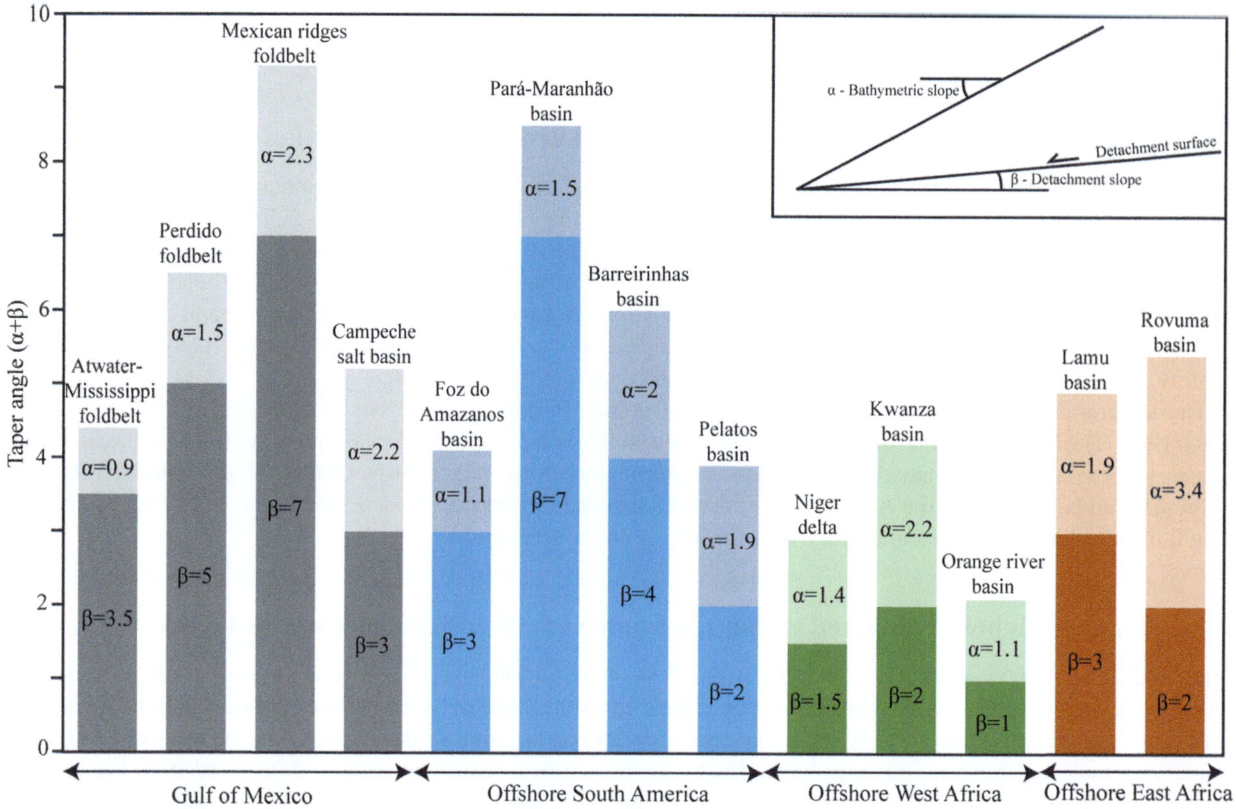


**FIG. 19** Bar chart summarizing the bathymetric slope ($\alpha$), detachment slope ($\beta$) and taper angle ($\alpha+\beta$) of each of 13 passive margin foldbelts described in this chapter. The translational zone and relative wide compressional domain indicate that the passive margin foldbelt system behaves as a critical taper wedge. The taper angles range from 2.1° to 9.3° and are relatively low compared with those of accretionary wedges at subduction zones with critical tapers in the range of >9° (Davis et al., 1983). The low taper angles of these PMFB compared to accretionary prisms may reflect their detachment along zones of elevated pore-fluid pressure as described for the Niger Delta PMFB by Bilotti and Shaw (2005).

widespread source rocks and abundant structural closures makes passive margin foldbelts an ideal area for offshore exploration.

Hydrocarbon exploration PMFB is much more extensive in salt-based systems compared with overpressured shale detachment systems. The salt-based PMFB, such as Atwater-Mississippi, Perdido, Campeche and Kwanza, provide the habitat for multiple, giant oil and gas discoveries (Figs. 3 and 13). Except for the productive Niger Delta and Rovuma examples, most shale-based PMFB lack significant hydrocarbon discoveries, although many of these deepwater, shale-dominated areas remain underexplored.

There are several key remaining questions for hydrocarbon exploration in passive margin foldbelts. First, how critical is the role of active deformation due to updip extension, which commonly poses a risk to seal failure? Seal failure drains reservoirs as the result of late-stage reactivation of faults (Hasan et al., 2021). Second, what is the stratigraphic location and extent of the detachment layer that can potentially act as a hydrocarbon fluid migration barrier? Third, how is reservoir-quality sediment routed through the sometimes tortuous pathways of PMFB to the deeper part of these basin? Structures that deform the seafloor can limit the transport of high-quality sediment. These themes resonate throughout many chapters in this book.

## Conclusion: Areas of future work for passive margin foldbelts

Deepwater passive margin foldbelts occur on a global scale and play an important role in the exploration and production business for many companies, countries and the economy. These complex structural provinces have been studied for decades, yet high-resolution seismic data and well penetrations have only recently become available so that researchers can compare and contrast their traits on a global scale. Based on the analysis presented in this chapter and the extensive results from prior studies listed in the references section, there exist several key unsolved challenges in passive margin foldbelt research. (1) What are the dominant controls on PMFB evolution? Three key controls include fluid pressure, overburden lithology and detachment dip, yet their relative impact on PMFB evolution remains an important research question. (2) What is the relative contribution of active and passive salt diapirism to not only the formation, but also the evolution of structures observed in PMFB? (3) What are the relative contributions of gravity-gliding and gravity-spreading in the development of PMFB, and how do these controls affect each of the main parts of the foldbelt? (4) Why do sequential restorations show that the amount of updip extension is typically higher than the amount of contraction? Additional geological and geophysical evidence from case studies as well as global synthesis studies are required to advance understanding in these principal research areas.

## Acknowledgments

We thank Jon Rotzien, Cindy Yeilding, Rich Sears, F. Javier Hernández-Molina and Octavian Catuneanu for the invitation to contribute this chapter and for their many constructive suggestions on the original manuscript. We also thank the industry sponsors of the Conjugate Basins, Tectonics, and Hydrocarbons (CBTH) consortium at the University of Houston for their continued financial support and technical discussions.

## References

Adegoke, O.S., Oyebamiji, A.S., Edet, J.J., Osterloff, P.L., Ulu, J.J., O.K.B.T.-C.F (Eds.), 2017. Chapter 2—Geology of the Niger Delta Basin. Elsevier, pp. 25–66, https://doi.org/10.1016/B978-0-12-812161-0.00002-8.

Bilotti, F., Shaw, J.H., 2005. Deep-water Niger Delta fold and thrust belt modeled as a critical-taper wedge: the influence of elevated basal fluid pressure on structural styles. Am. Assoc. Pet. Geol. Bull. 89, 1475–1491. https://doi.org/10.1306/06130505002.

Bouroullec, R., Weimer, P., 2017. Geometry and kinematics of Neogene allochthonous salt systems in the Mississippi Canyon, Atwater Valley, western Lloyd Ridge, and western DeSoto Canyon protraction areas, northern deepwater Gulf of Mexico. Am. Assoc. Pet. Geol. Bull. 101, 1003–1034. https://doi.org/10.1306/09011609186.

Bouroullec, R., Weimer, P., Serrano, O., 2017. Petroleum geology of the Mississippi Canyon, Atwater Valley, western DeSoto Canyon, and western Lloyd Ridge protraction areas, northern deepwater Gulf of Mexico: traps, reservoirs, and tectonostratigraphic evolution. Am. Assoc. Pet. Geol. Bull. 101, 1073–1108. https://doi.org/10.1306/09011610093.

Brownfield, M.E., 2016. Assessment of undiscovered oil and gas resources of the Orange River Coastal Province, Southwest Africa. compiler, In: Brownfield, M.E. (Ed.), Geologic Assessment of Undiscovered Hydrocarbon Resources of Sub-Saharan Africa. U.S. Geological Survey Digital Data Series 69–GG, p. 14, https://doi.org/10.3133/ds69GG. chap. 8.

Brun, J.-P., Fort, X., 2011. Salt tectonics at passive margins: geology versus models. Mar. Pet. Geol. 28, 1123–1145. https://doi.org/10.1016/j.marpetgeo.2011.03.004.

Cai, J., He, Y., Liang, J., Qiu, C., Zhang, C., 2020. Differential deformation of gravity-driven deepwater fold-and-thrust belts along the passive continental margin of East Africa and their impact on petroleum migration and accumulation. Mar. Pet. Geol. 112. https://doi.org/10.1016/j.marpetgeo.2019.104053, 104053.

CBTH Project, 2020. CBTH Phase V Atlas and Database Compilation.

Ceraldi, T.S., Hodgkinson, R., Backé, G., 2017. The petroleum geology of the West Africa margin: an introduction. Geol. Soc. Spec. Publ. 438, 1–6. https://doi.org/10.1144/SP438.11.

Conti, B., Perinotto, J.A.d.J., Veroslavsky, G., Castillo, M.G., de Santa Ana, H., Soto, M., Morales, E., 2017. Speculative petroleum systems of the southern Pelotas Basin, offshore Uruguay. Mar. Pet. Geol. 83, 1–25. https://doi.org/10.1016/j.marpetgeo.2017.02.022.

Contreras, J., Zühlke, R., Bowman, S., Bechstädt, T., 2010. Seismic stratigraphy and subsidence analysis of the southern Brazilian margin (Campos, Santos and Pelotas basins). Mar. Pet. Geol. 27, 1952–1980. https://doi.org/10.1016/j.marpetgeo.2010.06.007.

Corredor, F., Shaw, J.H., Bilotti, F., 2005. Structural styles in deepwater fold-and-thrust belts of the Niger Delta. Am. Assoc. Pet. Geol. Bull. 89, 753–780. https://doi.org/10.1306/02170504074.

Cruciani, F., Barchi, M.R., 2016. The Lamu Basin deepwater fold-and-thrust belt: an example of a margin-scale, gravity-driven thrust belt along the continental passive margin of East Africa. Tectonics 35, 491–510. https://doi.org/10.1002/2015TC003856.

Cruciani, F., Barchi, M.R., Koyi, H.A., Porreca, M., 2017. Kinematic evolution of a regional-scale gravity-driven deepwater fold-and-thrust belt: the Lamu Basin case-history (East Africa). Tectonophysics 712–713, 30–44. https://doi.org/10.1016/j.tecto.2017.05.002.

da Cruz, C.A., Severiano Ribeiro, H.J.P., da Silva, E.B., 2021. Exploratory plays of the Foz do Amazonas Basin, NW portion, in deep and ultra-deep waters, Brazilian equatorial margin. J. South Am. Earth Sci. 111. https://doi.org/10.1016/j.jsames.2021.103475, 103475.

Da Silva Pellegrini, B., Ribeiro, H.J.P.S., 2018. Exploratory plays of Pará-Maranhão and Barreirinhas basins in deep and ultra-deep waters, Brazilian Equatorial Margin. Braz. J. Geol. 48, 485–502. https://doi.org/10.1590/2317-4889201820180146.

Davis, D., Suppe, J., Dahlen, F.A., 1983. Mechanics of fold-and-thrust belts and accretionary wedges. J. Geophys. Res. Solid Earth 88, 1153–1172. https://doi.org/10.1029/JB088iB02p01153.

Davison, I., 2020. Salt tectonics in the Sureste basin, SE Mexico: some implications for hydrocarbon exploration. In: Davison, I., Hull, J.N.F., Pindell, J. (Eds.), The Basins, Orogens and Evolution of the Southern Gulf of Mexico and Northern Caribbean. Geol. Soc., London, Spec. Pub., vol. 504, pp. 345–359, https://doi.org/10.1144/SP.504-2019-227.

de Vera, J., Granado, P., McClay, K., 2010. Structural evolution of the Orange Basin gravity-driven system, offshore Namibia. Mar. Pet. Geol. 27, 223–237. https://doi.org/10.1016/j.marpetgeo.2009.02.003.

Eddy, D.R., Van Avendonk, H.J.A., Christeson, G.L., Norton, I.O., Karner, G.D., Johnson, C.A., Snedden, J.W., 2014. Deep crustal structure of the northeastern Gulf of Mexico: implications for rift evolution and seafloor spreading. J. Geophys. Res. Solid Earth 119, 6802–6822. https://doi.org/10.1002/2014JB011311.

Evans, S.L., Jackson, C.A.L., 2020. Base-salt relief controls salt-related deformation in the outer Kwanza Basin, offshore Angola. Basin Res. 32, 668–687. https://doi.org/10.1111/bre.12390.

Fiduk, J.C., Weimer, P., Trudgill, B.D., Rowan, M.G., Gale, P.E., Phair, R.L., Korn, B.E., Roberts, G.R., Gafford, W.T., Lowe, R.S., Queffelec, T.A., 1999. The Perdido fold belt, northwestern deep Gulf of Mexico, part 2: seismic stratigraphy and petroleum systems. Am. Assoc. Pet. Geol. Bull. 83, 578–612. https://doi.org/10.1306/00aa9c00-1730-11d7-8645000102c1865d.

Fitz-Díaz, E., Lawton, T.F., Juárez-Arriaga, E., Chávez-Cabello, G., 2018. The Cretaceous-Paleogene Mexican orogen: structure, basin development, magmatism and tectonics. Earth Sci. Rev. 183, 56–84. https://doi.org/10.1016/j.earscirev.2017.03.002.

Fonnesu, M., Palermo, D., Galbiati, M., Marchesini, M., Bonamini, E., Bendias, D., 2020. A new world-class deep-water play-type, deposited by the syndepositional interaction of turbidity flows and bottom currents: the giant Eocene Coral field in northern Mozambique. Mar. Pet. Geol. 111 (2020), 179–201. https://doi.org/10.1016/j.marpetgeo.2019.07.047.

GEBCO Compilation Group, 2020. GEBCO 2020 Bathymetric Grid. https://doi.org/10.5285/a29c5465-b138-234d-e053-6c86abc040b9.

Geo Expro, 2013. The Pelotas Basin Oil Province Revealed, vol. 10(4). https://www.geoexpro.com/articles/2013/12/the-pelotas-basin-oil-province-revealed. (Accessed 15 September 2021).

Granado, P., De Vera, J., McClay, K., 2009. Tectonostratigraphic evolution of the Orange Basin, SW Africa. Trab. Geol. 29, 321–328. https://doi.org/10.17811/tdg.29.2009.

Grando, G., Schleder, Z., Shackleton, R., Seed, G., Buddin, T., McClay, K., Borraccini, F., 2009. 3D Structural Evolution of the Salt Controlled Frampton Anticline, Atwater Valley Foldbelt, Deep Water Gulf of Mexico. AAPG. search and discover article #50161.

Greenhalgh, J., Borsato, R., Mathew, F., Duncan-Jones, G., Pimenta, I., Marques da Silva, J., Narciso da Silva, L., 2012. Petroleum plays and prospectivity in the Kwanza and Benguela Basins offshore Angola. Search and Discovery Article #10443, p. 7. accessed September 15, 2021 http://www.searchanddiscovery.com/documents/2012/10443greenhalgh/ndx_greenhalgh.pdf.

Hasan, M.N., Mann, P., 2021. Structural styles and evolution of the Campeche Salt Basin, southern Gulf of Mexico. Mar. Pet. Geol. 133. https://doi.org/10.1016/j.marpetgeo.2021.105313, 105313.

Hasan, M.N., Potter-McIntyre, S., Tedesco, S., 2021. Wrench faulting and trap breaching: a case study of the Kizler North Field, Lyon County, Kansas, USA. Midcont. Geosci. 2 (May 2021), 1–14. https://journals.ku.edu/mg/article/view/14892/13917.

Hudec, M.R., Jackson, M.P.A., 2002. Structural segmentation, inversion, and salt tectonics on a passive margin: evolution of the inner Kwanza Basin, Angola. GSA Bull. 114, 1222–1244. https://doi.org/10.1130/0016-7606(2002)114<1222:SSIAST>2.0.CO;2.

Hudec, M.R., Jackson, M.P.A., 2004. Regional restoration across the Kwanza Basin, Angola: salt tectonics triggered by repeated uplift of a metastable passive margin. Am. Assoc. Pet. Geol. Bull. 88, 971–990. https://doi.org/10.1306/02050403061.

Hudec, M.R., Norton, I.O., Jackson, M.P.A., Peel, F.J., 2013. Jurassic evolution of the Gulf of Mexico salt basin. Am. Assoc. Pet. Geol. Bull. 97, 1683–1710. https://doi.org/10.1306/04011312073.

Hudec, M.R., Dooley, T.P., Peel, F.J., Soto, J.I., 2019. Controls on the evolution of passive-margin salt basins: structure and evolution of the Salina del Bravo region, northeastern Mexico. GSA Bull., 1–16. https://doi.org/10.1130/b35283.1.

Kenning, J.J., Mann, P., 2020a. Control of structural style by large, Paleogene, mass transport deposits in the Mexican Ridges fold-belt and Salina del Bravo, western Gulf of Mexico. Mar. Pet. Geol. 115. https://doi.org/10.1016/j.marpetgeo.2020.104254, 104254.

Kenning, J.J., Mann, P., 2020b. Regional thermal maturity modelling of hydrocarbons along the deepwater Yucatan margin, southern Gulf of Mexico. In: Davison, I., Hull, J.N.L., Pindell, J. (Eds.), The Basins, Orogens and Evolution of the Southern Gulf of Mexico and Northern Caribbean. vol. 504. Geol. Soc. London, Spec. Publ., https://doi.org/10.1144/SP504-2019-252. SP504–2019–252.

Krueger, A., Gilbert, E., 2009. Deepwater Fold-Thrust Belts: Not all the Beasts are Equal. AAPG Search and Discovery, pp. 1–6. https://www.searchanddiscovery.com/documents/2009/30085krueger/images/krueger.

Liu, J.-P., Pan, X., Ma, J., Tian, Z., Chen, Y., Wan, L., 2008. Petroleum geology and resources in West Africa: an overview. Pet. Explor. Dev. 35, 378–384. https://doi.org/10.1016/S1876-3804(08)60086-5.

Mahanjane, E.S., Franke, D., 2014. The Rovuma Delta deepwater fold-and-thrust belt, offshore Mozambique. Tectonophysics 614, 91–99. https://doi.org/10.1016/j.tecto.2013.12.017.

Marton, G.L., Buffler, R.T., 1999. Jurassic-early Cretaceous tectono-paleogeographic evolution of the southeastern Gulf of Mexico basin. In: Mann, P. (Ed.), Sedmentary Basins of the World. vol. 4. Elsevier, pp. 63–91, https://doi.org/10.1016/S1874-5997(99)80037-9.

Mitra, S., Figueroa, G.C., Garcia, J.H., Alvarado, A.M., 2005. Three-dimensional structural model of the Cantarell and Sihil structures, Campeche Bay, Mexico. Am. Assoc. Pet. Geol. Bull. 89, 1–26. https://doi.org/10.1306/08310403108.

Morley, C.K., King, R., Hillis, R., Tingay, M., Backe, G., 2011. Deepwater fold and thrust belt classification, tectonics, structure and hydrocarbon prospectivity: a review. Earth Sci. Rev. 104, 41–91. https://doi.org/10.1016/j.earscirev.2010.09.010.

Mourgues, R., Lacoste, A., Garibaldi, C., 2014. The Coulomb critical taper theory applied to gravitational instabilities. J. Geophys. Res. Solid Earth 119, 754–765. https://doi.org/10.1002/2013JB010359.

Nguyen, L.C., Mann, P., 2015. Gravity and magnetic constraints on the Jurassic opening of the oceanic Gulf of Mexico and the location and tectonic history of the Western Main transform fault along the eastern continental margin of Mexico. Interpretation 4, SC23–SC33. https://doi.org/10.1190/INT-2015-0110.1.

Oliveira, M.J.R., Santarem, P., Moraes, A., Zalán, P.V., Caldeira, J.L., Tanaka, A., Trosdtorf Jr, I., 2013. Linked extensional-compressional tectonics in gravitational systems in the Equatorial Margin of Brazil. In: Gao, D. (Ed.), Tectonics Sediment: Implications for Petroleum Systems., https://doi.org/10.1306/13351552M1003532.

Olson, C.J., Becker, J.J., Sandwell, D.T., 2016. SRTM15_PLUS: Data Fusion of Shuttle Radar Topography Mission (SRTM) Land Topography with Measured and Estimated Seafloor Topography (NCEI Accession 0150537). NOAA National Centers for Environmental Information, Dataset. https://accession.nodc.noaa.gov/0150537.

Peel, F.J., 2014. The engines of gravity-driven movement on passive margins: quantifying the relative contribution of spreading vs. gravity sliding mechanisms. Tectonophysics 633, 126–142. https://doi.org/10.1016/j.tecto.2014.06.023.

Peel, F.J., Travis, C.J., Hossack, J.R., 1995. Genetic structural provinces and salt tectonics of the Cenozoic offshore U.S. Gulf of Mexico: a preliminary analysis. In: Jackson, M.P.A., Roberts, D.G., Snelson, S. (Eds.), Salt Tectonics A Glob. Perspect., https://doi.org/10.1306/M65604C7.

Pindell, J.L., Kennan, L., 2009. Tectonic evolution of the Gulf of Mexico, Caribbean and northern South America in the mantle reference frame: an update. In: James, K.H., Lorente, M.A., Pindell, J. (Eds.), The Origin and Evolution of the Caribbean Plate. vol. 328. Geol. Soc. London, Spec. Publ, pp. 1–55, https://doi.org/10.1144/SP328.1.

Pindell, J., Miranda, E., 2011. Linked kinematic histories of the Macuspana, Akal-Reforma, Comalcalco, and deepwater Campeche Basin tectonic elements, southern Gulf of Mexico. Gulf Coast Assoc. Geol. Soc. Trans. 61, 353–361.

Reis, A.T., Perovano, R., Silva, C.G., Vendeville, B.C., Araújo, E., Gorini, C., Oliveira, V., 2010. Two-scale gravitational collapse in the Amazon Fan: a coupled system of gravity tectonics and mass-transport processes. J. Geol. Soc. Lond. 167, 593–604. https://doi.org/10.1144/0016-76492009-035.

Reis, A.T., Araújo, E., Silva, C.G., Cruz, A.M., Gorini, C., Droz, L., Migeon, S., Perovano, R., King, I., Bache, F., 2016. Effects of a regional décollement level for gravity tectonics on late Neogene to recent large-scale slope instabilities in the Foz do Amazonas Basin, Brazil. Mar. Pet. Geol. 75, 29–52. https://doi.org/10.1016/j.marpetgeo.2016.04.011.

Rowan, M.G., 2020. Salt- and shale-detached gravity-driven failure of continental margins. In: Scarselli, N., Adam, J., Chiarella, D., Roberts, D.G., Bally, A.W. (Eds.), Regional Geology and Tectonics (Second Edition). Elsevier, pp. 205–234, https://doi.org/10.1016/B978-0-444-64134-2.00010-9.

Rowan, M.G., Ratliff, R.A., 2012. Cross-section restoration of salt-related deformation: best practices and potential pitfalls. J. Struct. Geol. 41, 24–37. https://doi.org/10.1016/j.jsg.2011.12.012.

Rowan, M.G., Peel, F.J., Vendeville, B.C., 2004. Gravity-driven fold belts on passive margins. In: McClay, K.R. (Ed.), Thrust Tectonics and Hydrocarbon Systems. vol. 82. AAPG Memoir, pp. 157–182, https://doi.org/10.1306/M82813C9.

Salomón-Mora, L.E., Aranda-Garcia, M., Román-Ramos, J.R., 2009. Contractional Growth Faulting in the Mexican Ridges. In: Bartolini, C., Román-Ramos, J.R. (Eds.), Petroleum Systems in the Southern Gulf of Mexico. AAPG Memoir 90, pp. 93–115, https://doi.org/10.1306/13191080M903332.

Salomón-Mora, L.E., Cruz-Mercado, M., Alsop, I., Archer, S., 2011. Tectonics, structure and hydrocarbon potential of the Mexican Ridges Fold Belt, western Gulf of Mexico. Gulf Coast Assoc. Geol. Soc. Trans. 61, 403–419.

Shann, M.V., 2020. The Sureste Basin of Mexico: its framework, future oil exploration opportunities and key challenges ahead. In: Davison, I., Hull, J.N. L., Pindell, J. (Eds.), The Basins, Orogens and Evolution of the Southern Gulf of Mexico and Northern Caribbean. Geol. Soc. London, Spec. Publ, vol. 504. p. SP504 https://doi.org/10.1144/SP504-2019-214.

Steele, D., Ejedawe, J., Adeogba, T., Grant, C., Filbrandt, J., Ganz, H., 2009. Geological framework of Nigeria linked shelf extension and deepwater thrustbelts. In: AAPG Hedberg Conference, October 4–9, Tirrenia, Italy.

Steier, A., Mann, P., 2019. Late Mesozoic gravity sliding and Oxfordian hydrocarbon reservoir potential of the northern Yucatan margin. Mar. Pet. Geol. 103, 681–701. https://doi.org/10.1016/j.marpetgeo.2019.03.001.

Stica, J.M., Zalán, P.V., Ferrari, A.L., 2014. The evolution of rifting on the volcanic margin of the Pelotas Basin and the contextualization of the Paraná-Etendeka LIP in the separation of Gondwana in the South Atlantic. Mar. Pet. Geol. 50, 1–21. https://doi.org/10.1016/j.marpetgeo.2013.10.015.

Strozyk, F., Back, S., Kukla, P.A., 2017. Comparison of the rift and post-rift architecture of conjugated salt and salt-free basins offshore Brazil and Angola/Namibia, South Atlantic. Tectonophysics 716, 204–224. https://doi.org/10.1016/j.tecto.2016.12.012.

Tesei, T., Cruciani, F., Barchi, M.R., 2021. Gravity-driven deepwater fold-and-thrust belts as critical Coulomb wedges: model limitations and the role of friction vs. fluid pressure. J. Struct. Geol. 153. https://doi.org/10.1016/j.jsg.2021.104451, 104451.

Trudgill, B.D., Rowan, M.G., Fiduk, J.C., Weimer, P., Gale, P.E., Korn, B.E., Phair, R.L., Gafford, W.T., Roberts, G.R., Dobbs, S.W., 1999. The Perdido Fold Belt, northwestern deep Gulf of Mexico, part 1: structural geometry, evolution and regional implications. Am. Assoc. Pet. Geol. Bull. 83, 1320–1336. https://doi.org/10.1306/00AA9A0C-1730-11D7-8645000102C1865D.

Wu, S., Bally, A.W., 2000. Slope tectonics: comparisons and contrasts of structural styles of salt and shale tectonics on the northern Gulf of Mexico with shale tectonics of offshore Nigeria in Gulf of Guinea. In: Mohriak, W., Talwani, M. (Eds.), Atlantic Rifts and Continental Margins. American Geophysical Union Geophysical Monograph, vol. 115, pp. 151–172, https://doi.org/10.1029/GM115p0151.

Wu, J.E., McClay, K., Frankowicz, E., 2015. Niger Delta gravity-driven deformation above the relict chain and Charcot oceanic fracture zones, Gulf of Guinea: insights from analogue models. Mar. Pet. Geol. 65, 43–62. https://doi.org/10.1016/j.marpetgeo.2015.03.008.

Yarbuh, I., González-Fernández, A., Spelz-Madero, R.M., Negrete-Aranda, R., Contreras, J., 2018. Development of detachment folds in the Mexican Ridges Foldbelt, Western Gulf of Mexico Basin. Tectonics 37, 2013–2028. https://doi.org/10.1029/2017TC004948.

Zhang, G.C., Qu, H.J., Zhao, C., Zhang, F.L., Zhao, Z., 2017. Giant discoveries of oil and gas exploration in global deepwaters in 40 years and the prospect of exploration. Nat. Gas Geosci. 28, 1447–1477. https://doi.org/10.11764/j.issn.1672-1926.2017.08.008.

Zhang, J., Wu, S., Hu, G., Yue, D., Xu, Z., Chen, C., Zhang, K., Wang, J., Wen, S., 2021. Role of shale deformation in the structural development of a deepwater gravitational system in the Niger Delta. Tectonics 40. https://doi.org/10.1029/2020TC006491.

Chapter 5

# Salt tectonics in deepwater settings

Michael R. Hudec and Martin P.A. Jackson✠
*Bureau of Economic Geology, The University of Texas at Austin, Austin, TX, United States*

## Introduction

The presence of salt changes almost everything about the geology of deepwater continental margins. To begin, salt tectonism deforms the seafloor, rerouting sediment-transport systems and thus controlling sediment distribution (see Chapters 6–12 in this volume). Furthermore, salt flow produces a unique suite of structures, creates new migration pathways for fluids and modifies heat flow (Jackson and Hudec, 2017). Thus, geologists exploring hydrocarbons in deepwater salt basins, or even just interpreting seismic data there, must have a working knowledge of salt tectonics.

Most deepwater salt basins are found on passive margins, although a few of these passive margins are now in convergent settings (Fig. 1). The reason for the dominance of passive margins in deepwater salt tectonics is that the early stages of continental rifting offer many opportunities for isolated conditions that can lead to salt deposition. The later evolution of these continental rifts into passive margins leaves the salt basins flanking the spreading ridge, on either side of the ocean (read more on this topic in Chapters 3 and 4).

The first hint that there might be something unusual about the geology of salt basin continental slopes came from bathymetric surveys conducted in the northern Gulf of Mexico during the 1940s and 1950s (e.g., Gealy, 1955). These surveys revealed a bizarre landscape of domes and basins unlike anything seen on continental shelves. Hydrocarbon exploration on the northern Gulf of Mexico slope was pioneered during the 1980s, with successes in that region leading to exploration efforts in many other deepwater basins. The primary technical obstacle to the advance of exploration into deeper water was drilling technology (Chapter 19). Beyond this, however, geologists had to learn to deal with a wide range of salt structures that were encountered for the first time: salt sheets, salt canopies and minibasins, to name a few. And so the struggle to understand deepwater salt tectonics began.

The impetus for the study of salt tectonics in deepwater comes largely from successes in exploring oil and gas in these basins, most notably in the northern Gulf of Mexico, Campos basin, Santos basin and Lower Congo basin (see also Chapters 1 and 19). In particular, the discovery of large volumes of oil below the Sigsbee salt canopy in the northern Gulf of Mexico forced geoscientists to study the genesis, deformation and internal structure of salt canopies in general. This research has been greatly aided by advances in the acquisition and processing of 3D seismic data, which have illuminated never-before-seen aspects of subsalt geology.

The goals of this chapter are first, to review the controls on salt distribution on deepwater passive margins; second, to describe the main types of salt structures that are common in these areas; third, to discuss ways in which the regional geology of deepwater passive margins can be modified through salt tectonism.

## Controls on salt deposition in deepwater

The two biggest influences on salt structural styles in deepwater are as follows: (1) the thickness of the salt and (2) the areal extent of the salt layer. In particular, a salt layer that is continuous from deepwater all the way up to the continental shelf allows a margin-scale linked kinematic system to form detached on the salt. The linked systems comprise extensional structures lying on the outer shelf and upper slope and contractional structures in deepwater. More localized salt basins in deepwater are much less likely to form such linked systems, and so have different structural styles.

The primary control on both thickness and areal extent of the salt layer is the timing of salt deposition with respect to rifting and continental breakup. Rowan (2014, 2015) defined four types of passive-margin salt basins, based on the rift stage

✠ Deseased.

**Deepwater Sedimentary Systems.** https://doi.org/10.1016/B978-0-323-91918-0.00020-7

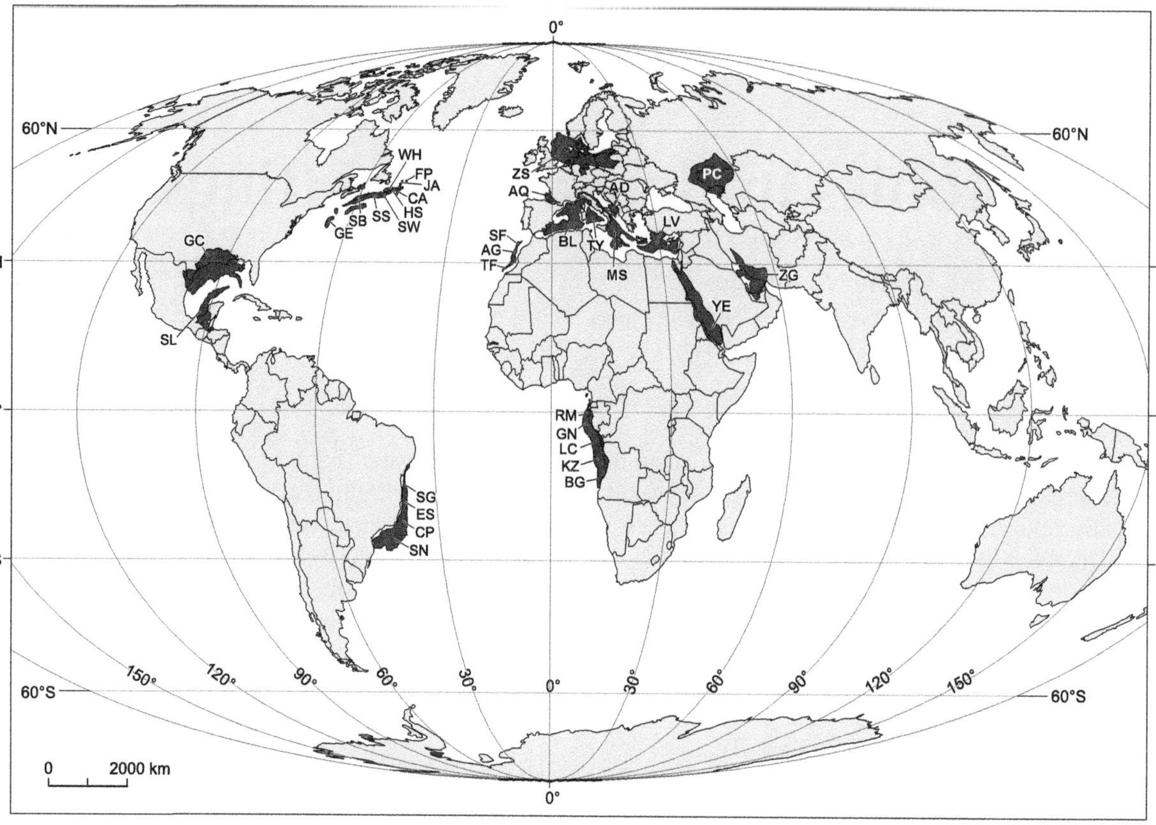


**FIG. 1** Large salt basins that are at least partly in deepwater. *AD*, South Adriatic; *AG*, Agadir; *AQ*, Aquitaine; *BG*, Benguela-Namibe; *BL*, Balearic; *CA*, Carson; *CP*, Campos; *ES*, Espirito Santo; *FP*, Flemish Pass; *GC*, Gulf Coast; *GE*, Georges Bank; *GN*, Gabon; *HS*, Horseshoe; *JA*, Jeanne d'Arc; *KZ*, Kwanza; *LC*, Lower Congo; *LV*, Levantine; *MS*, Messinian; *PC*, Pricaspian; *RM*, Rio Muni; *SB*, Sable; *SF*, Safi; *SG*, Sergipe-Alagoas; *SL*, Salina-Sigsbee; *SN*, Santos; *SS*, Scotian Slope; *SW*, South Whale; *TF*, Tarfaya; *TY*, Tyrrhenian; *WH*, Whale; *YE*, Yemeni; *ZG*, Zagros; *ZS*, Zechstein. *(Modified from Hudec and Jackson (2012)).*

at which salt deposition occurred: prerift, syn-stretching, syn-thinning and syn-exhumation. These models are based on the four-stage evolutionary model for magma-poor passive margins developed by Péron-Pinvidic and Manatschal (2009, Fig. 2). Each type gives rise to different salt distribution, and so, different structural styles in deepwater (Fig. 3).

In prerift salt basins (Fig. 3A), salt was deposited across the area prior to rifting, so shows no thickening or thinning across rift structures. The salt basin is dismembered during margin extension, with especially dramatic results in deepwater. The deepwater portions of the initially continuous salt basin are broken into small salt segments separated by broad expanses of rotated normal faults or exhumed crust and mantle. The salt is typically not very thick (although this depends on the setting in which the salt was originally deposited), and the salt layer is not continuous enough to host any significant linked kinematic systems. Diapirs, if they exist at all, are typically small and separated by broad nondiapiric regions. An example of the deepwater portions of a prerift salt basin can be found in the Bay of Biscay, offshore northern Spain.

Syn-stretching salt basins are deposited during the stretching phase of continental breakup when the upper crust is extended by steep, mostly planar faults (Fig. 2A). Surface deformation is decoupled from deeper levels by the ductile middle crust. Salt deposition is focused on the margins of the rift, where stretching-stage extensional faults are concentrated (Fig. 3B). Central parts of the rift (the future deepwater region) are relatively elevated at this stage so accumulate little salt. As a result, salt in the deepwater portions of syn-stretching salt basins is thin and disconnected, producing no significant salt structures. The best examples of this type of margin are the Iberia-Newfoundland conjugate margins in the North Atlantic Ocean.

The thinning (or hyperextension) stage of passive-margin evolution occurs when low-angle detachment faults form in the center of the rift (Fig. 2B). As in the stretching stage, the ductile middle crust detaches these faults from deformation in the lower crust and mantle. Because extension is now focused in the center of the rift, the deepwater portions of syn-thinning salt basins accumulate a thick, continuous salt layer (Fig. 3C). The base of salt is rugose, as the salt layer fills in the relief on the underlying, highly faulted succession. The continuity of the salt layer allows linked kinematic systems to develop, although this translation may be disrupted by underlying fault blocks once the salt begins to thin. Furthermore, the salt is typically thick

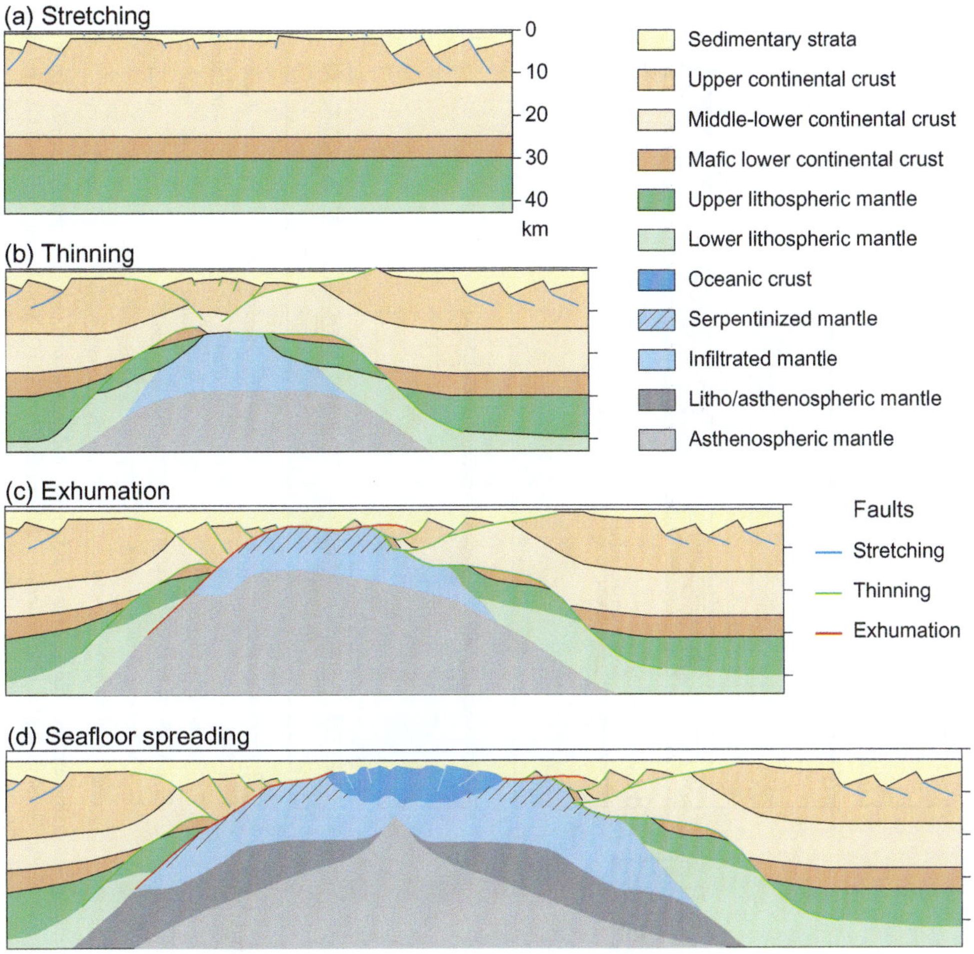


**FIG. 2** Conceptual sections showing a four-stage model for the evolution of magma-poor passive margins. (A) The stretching stage is characterized by high-angle faulting in the upper crust, decoupled from ductile stretching in the middle and lower crust. (B) In the thinning stage, lower-angle, inward-dipping conjugate normal fault thin the upper crust while outward-dipping faults thin the lower crust and upper lithospheric mantle. (C) Subcontinental mantle is exhumed along a low-angle fault that cuts through the entire lithosphere during the exhumation stage. (D) The spreading stage begins once continental crust completely separates and oceanic crust is accreted. *(Modified from Péron-Pinvidic and Manatschal (2009)).*

enough in these basins to allow growth of numerous diapirs, including in many cases allochthonous salt nappes at the distal end of the basin. Rowan (2014) suggests the northern Red Sea as an example of a syn-thinning salt basin.

The exhumation stage begins once a low-angle detachment fault cuts the entire lithosphere (Fig. 2C). As displacement increases the hanging wall slides off the footwall, eventually leading to exhumation of the lower crust and upper mantle. Extension during the exhumation stage is focused on the exhumation faults, so most of the other normal faults on the margin are inactive. Thus, salt deposited during exhumation is deposited continuously of the entire margin, with little or no fault offset on the base of salt (Fig. 3D). Furthermore, by the time exhumation occurs, the margin is typically several kilometers deep in the center, allowing in some cases 3–4 km of salt to accumulate. The resulting salt basin is ideally set up for a linked kinematic system to form, with updip extension compensated by downdip shortening. Such systems may accommodate many tens of kilometers of translation. The large thicknesses of salt in the basin center allow for formation of the full suite of salt structures, ranging from anticlines to stocks and walls, to salt sheets and canopies. Syn-exhumation salt basins are the largest and most hydrocarbon-prolific passive-margin salt basins, including the Gulf of Mexico and the conjugate margins of the South Atlantic (Fig. 1).

Thus, the structural style of salt on deepwater passive margins may vary dramatically based on the timing of salt deposition and on the geometry of any subsequent extensional structures. However, the most commonly encountered suite of structures, the most studied and the one of most interest to the petroleum industry is that found in syn-exhumation salt basins. The structural style of syn-exhumation basins will therefore be the focus of the rest of this chapter, with a caution to the reader that other styles occur.

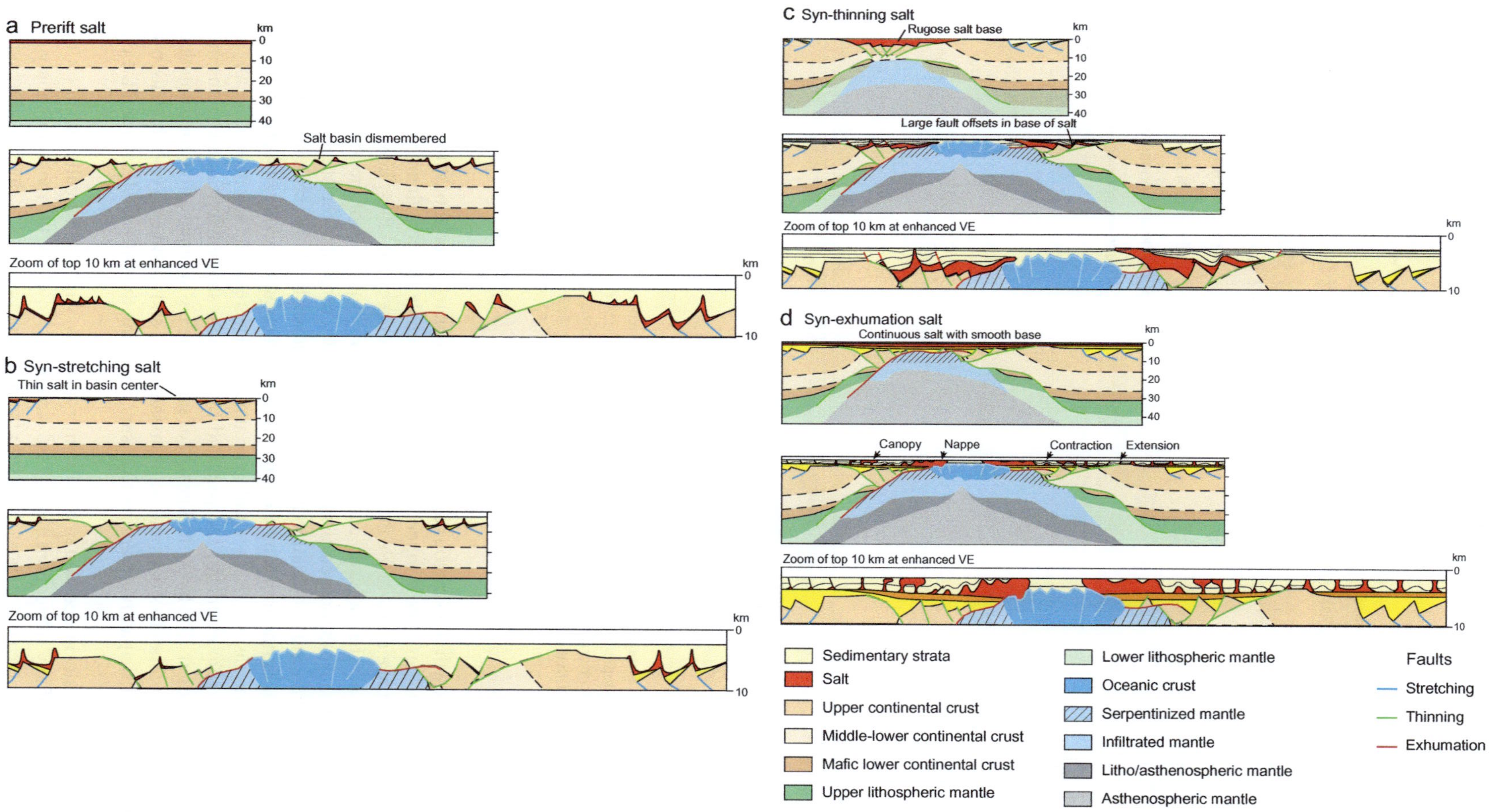


**FIG. 3** Evolution of the four types of passive-margin salt basins. (A) In prerift salt basins, the salt is dismembered during rifting, and so does not offer a continuous detachment surface. (B) In syn-stretching salt basins, salt in the future deepwater area is thin, most of it having been deposited over basement highs. Later extension also dismembered it. (C) Syn-thinning salt basins are relatively continuous and thick in deepwater, but typically feature a high-relief salt base, making for poor detachment surfaces. (D) Syn-exhumation basins tend to be continuous across the entire basin and very thick in deepwater. Furthermore, the salt base is relatively unfaulted, leading to an excellent detachment surface. *(Conceptual sections modified from Rowan (2014)).*

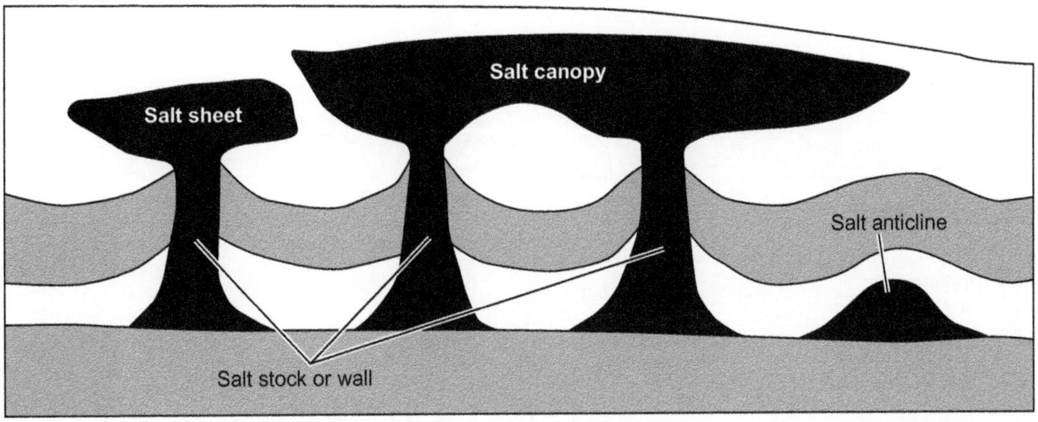


**FIG. 4** Schematic cross section showing salt structures characteristic of the deepwater regions of syn-exhumation salt basins. Salt stocks and salt walls are identical in cross section but differ along strike. Salt stocks are circular to elliptical in map view, in contrast to more elongate salt walls.

## Structural elements of deepwater salt systems

Two consequences of syn-exhumation salt basins are particularly important in contributing to the complexity of salt styles in deepwater. These are the large salt thicknesses and the superposition of toe-of-slope fold belts on earlier-formed structures. Thick salt means that salt stocks, walls, sheets and canopies are common elements in deepwater systems (Fig. 4). The fold belts deform these structures and create new ones.

All of the salt structures common to deepwater settings can be found in other environments, so in a sense there really is no such thing as "deepwater salt tectonics." However, the combination of thick salt and shortening produces some of the most spectacular salt geometries in the world, with a level of complexity found nowhere else. This section discusses the individual types of salt structures that form in this environment. The next section will consider the regional geology of these systems: how the evolution of these structures combines to accommodate the ongoing deformation of deepwater passive margins. Both the text and figures are based on the more extensive treatment of salt tectonics in Jackson and Hudec (2017), to which the reader is referred for more detail.

### Compressional folds

Compressional salt-cored folds are the simplest salt geometry found in deepwater settings. Jackson and Hudec (2017) listed five diagnostic features of these structures (Fig. 5). First, compressional folds form during shortening events, which are typically brief (a few million years) compared with the overall history of the basin. Thus, the most diagnostic feature of compressional folds is a long-lived prekinematic phase, followed by a short synkinematic period (the shortening event), followed by a postkinematic phase of quiescence. Although these synkinematic sedimentary intervals are typically deposited during a short time, they may be thick if folds have high relief and sediment supply is high. Second, compressional folds may form crestal thrusts or kink bands. Third, compressional thrusts build relief quickly, so they are commonly capped by erosional unconformities or high-relief onlap surfaces. Large erosional unconformities are unusual in deepwater settings, so their occurrence above salt-cored folds is quite diagnostic. Fourth, synkinematic sediment thinning onto contractional folds should be the same age on both sides of the fold crest. Fifth and finally, in many salt basins the shortening event causes folding but may also cause preexisting diapirs to expel salt sheets. Thus, salt sheets may be found emplaced at the level of the synkinematic section or, in many cases, lying on the crestal unconformity.

The most important variable in the geometry of compressional salt-cored folds is the existence of precursor structures. Fold geometries are simplest where there are no precursor structures so that a simple, layer-cake stratigraphy is deformed (Fig. 5). More complicated geometries form where preexisting salt anticlines are deformed. Two variants of preexisting folds are particularly interesting. In one, the early salt anticlines have shorter wavelengths than the later compressional folds, resulting in a complex, polyharmonic structure (Fig. 6). In the other variant, early salt pillows form at a variety of orientations with respect to the later shortening event (Fig. 7A). During shortening, pillows originally elongated normal to the shortening direction tighten and lengthen. Conversely, pillows originally elongated parallel to the shortening direction shorten in that direction and widen in a perpendicular direction (Fig. 7B). Regardless of their orientation, pillows being shortened tend to rise and their flanks to steepen. As the pillows fold, their crests can shift laterally if they comprise

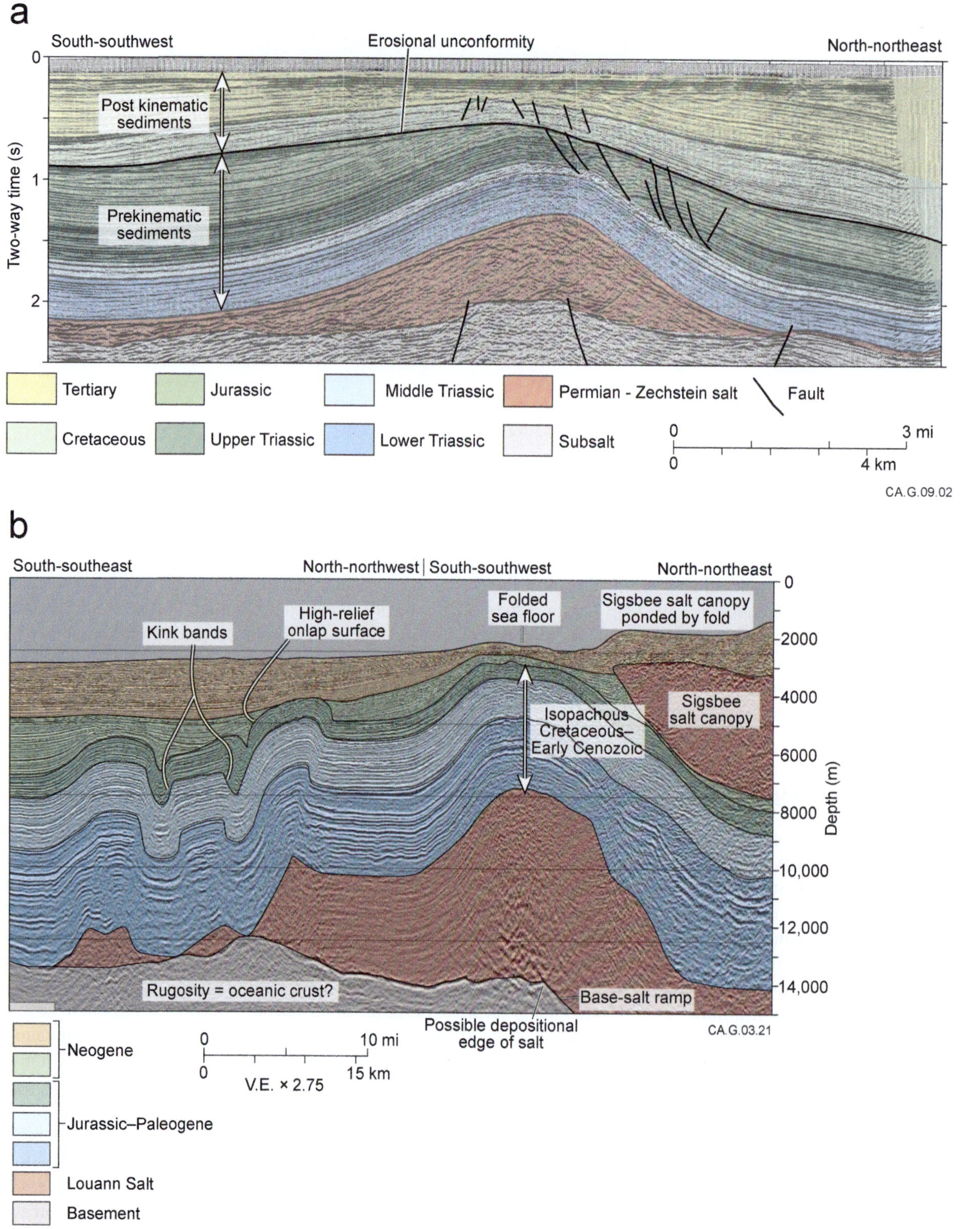


**FIG. 5** Diagnostic features of compressional folds. Folds show isopachous prekinematic section, comparatively brief synkinematic section and a much less-deformed postkinematic section. Erosional unconformities or high-relief onlap surfaces are common over fold crests. Salt sheets may be emplaced during shortening, and kink bands or thrust faults may form on fold flanks. (A) Seismic section from North Sea, (B) Seismic section from Perdido fold belt, northern Gulf of Mexico. *(Modified from Hudec and Jackson (2011)).*

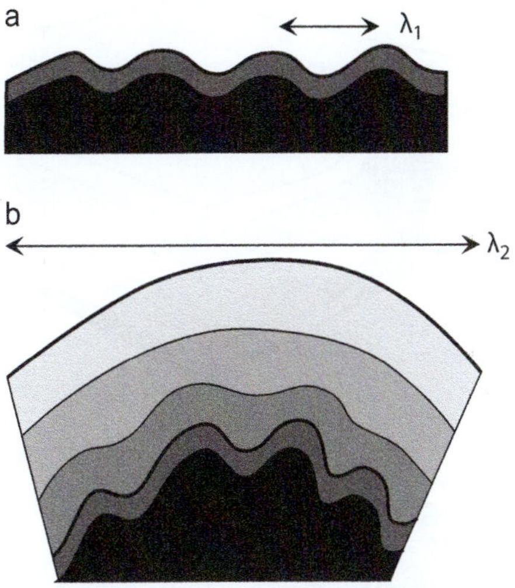


**FIG. 6** Multiwavelength growth folds form as they amplify and sediments thicken during shortening. (A) Dominant wavelength, $\lambda_1$, forms proportional to thickness of overburden *(shades of gray)* relative to salt *(black)*. (B) A larger wavelength, $\lambda_2$, forms after the overburden thickens and shortening resumes. *Modified from Jackson and Hudec (2017).*

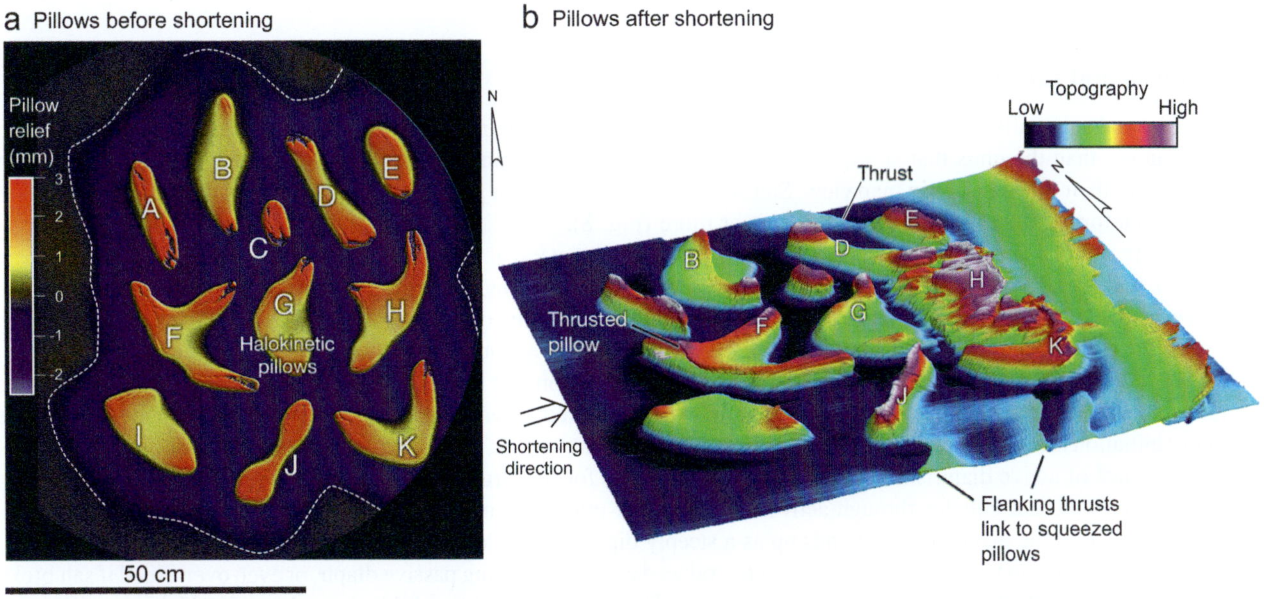


**FIG. 7** Images from a physical model showing evolution of a pillow fold belt. (A) Laser scan of model surface showing halokinetic rise of salt pillows prior to shortening. (B) Visualization of pillows after shortening. Pillows oriented perpendicular to shortening are more strongly shortened. The resulting array is highly irregular, very unlike the cylindrical structures typical of fold-and-thrust belts without precursor salt structures. *(Modified from Dooley et al. (2013)).*

nonparallel strata in ways that depend on the fold mechanism and prefolding geometry. Where pillows are laterally offset from one another the resulting buckle folds are also offset. An array of precursor pillows having various shapes and sizes creates an irregular template for any subsequent shortening. The resulting array of structures has been termed a pillow fold belt (Jackson et al., 2011; Dooley et al., 2013).

Regardless of the presence of precursor structures, most salt-cored anticlines have smoothly curved geometries in cross section. However, where there is a strong anisotropy imposed by interlayered beds of contrasting strength, more angular dip-domain folds form. Dip-domain folds have a kinked profile where each kink separates beds with more or less uniform dip. Typically the stronger layers are sandstone, chert or carbonate, whereas the weaker layers are shale, marl and especially evaporites.

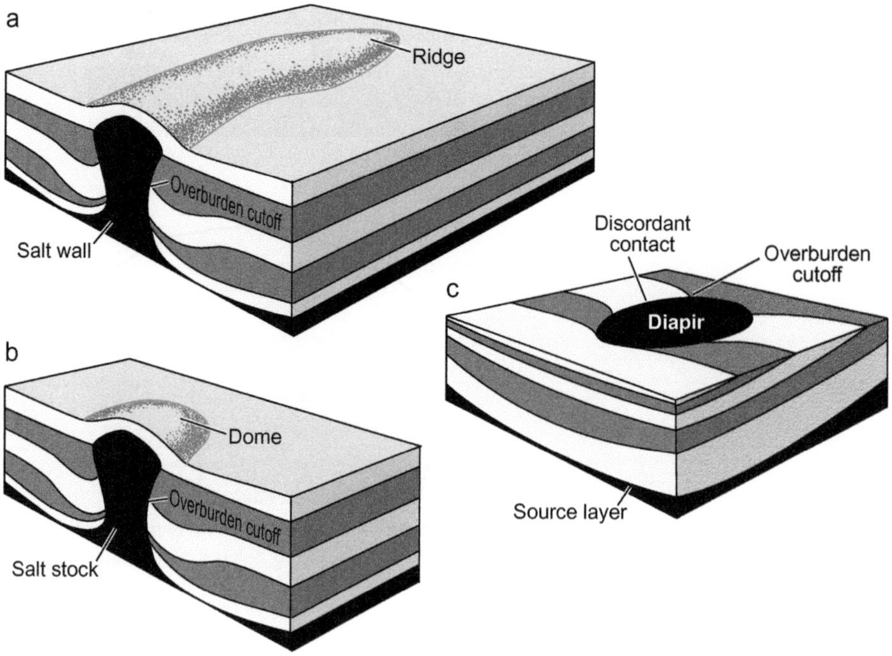


**FIG. 8** Comparison of salt stocks and salt walls. They differ in map view, but both are emplaced discordant to salt overburden. *(Modified from Jackson and Hudec (2017)).*

## Salt stocks and salt walls

Salt stocks and salt walls are both diapirs, that is, they pierce through their overburden and are in discordant contact with strata younger than the units that stratigraphically directly overlie the source layer. Stocks and walls appear identical in cross section; their difference is in map view. Salt stocks have a subcircular planform, whereas salt walls are elongate ridges having a planform axial ratio (ellipticity) of two or more (Fig. 8).

There are two main processes by which stocks and walls can initiate: reactive rise and active rise. Reactive diapirs form an extension, which can occur in deepwater settings during extension of the underlying basement. Thus, we might expect to see reactive diapirs in deepwater for salt deposited during the prerift, stretching or thinning stages of margin evolution. However, basement extension is largely over by the time salt is deposited in syn-exhumation salt basins, so these basins do not exhibit many (or any) reactive diapirs in deepwater. Instead, syn-exhumation salt basins feature a zone of extension and reactive diapirism on the shelf and upper slope, at the updip end of the linked kinematic system. In deepwater, the dominant mode of diapir initiation is active diapirism, in which pressurized salt rises to uplift, shoulder aside and break through its overburden. The dominance of active diapirism has important consequences for the structural styles of deepwater stocks and walls.

A stock or wall that initiates through active diapirism first uplifts, then ruptures its roof (Fig. 9). The ruptured roof is shouldered aside as the diapir rises and ends up as a steeply dipping flap along the diapir flank. Flaps typically dip parallel to the salt face. Flaps can be vertical if breakthrough produced a vertically rising passive diapir, or even overturned if salt broke out as an extruding salt sheet, flowing across the seafloor. The wavelength of flap folds depends on the thickness of the sediment package involved in the flap. A thick wedge of strata produces a wider zone of deformation and broader uplifted zone around the diapir (topographic halo) than a thin wedge folded to the same degree of rotation (Rowan et al., 2003).

Flap sediment may be intensely deformed by diapirism for three reasons. First, the sediment may be weakly consolidated, so rotating them to vertical or overturned dips should involve internal slumping, shearing and loss of cohesion. Flaps would probably be completely disaggregated if not for being buttressed against younger sediment deposited while the flap rotated. Second, outward movement of flaps shouldered aside by salt stocks circumferentially stretches the flaps, which are thin by a plastic extension (Nikolinakou et al., 2017). Third, flaps may be overpressured if composed of mudrocks and connected to deep, overpressured fluids.

A particularly common flap geometry in deepwater settings is the basal megaflap (Giles and Rowan, 2012; Graham et al., 2012). These thick successions of sediment are upturned at the base of a diapir and thin upward along the diapir flank (Fig. 10). The basal megaflap can pinch out or be eroded partway up the diapir flank, but in other instances, the megaflap can continue all the way up to the crest of the diapir, and even overturn beneath an overlying salt sheet. Basal megaflaps form as the salt layer inflates to form a pillow or anticline, which is then broken through by active diapirism. Sedimentary intervals in a flap are typically thin stratigraphically and by strain as the flap rotates.

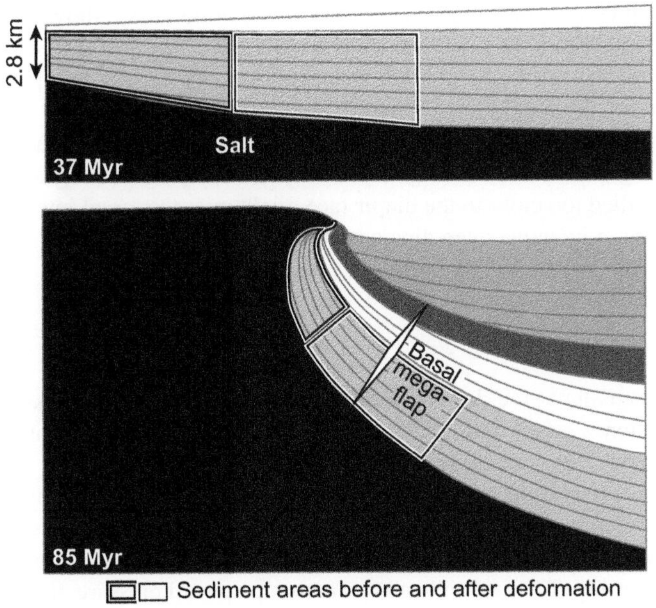


**FIG. 9** Finite-element model showing how active salt rise uplifts and ruptures its roof. The uplifted roof is shouldered aside to form a flap on the side of the diapir. *(Modified from Jackson and Hudec (2017)).*

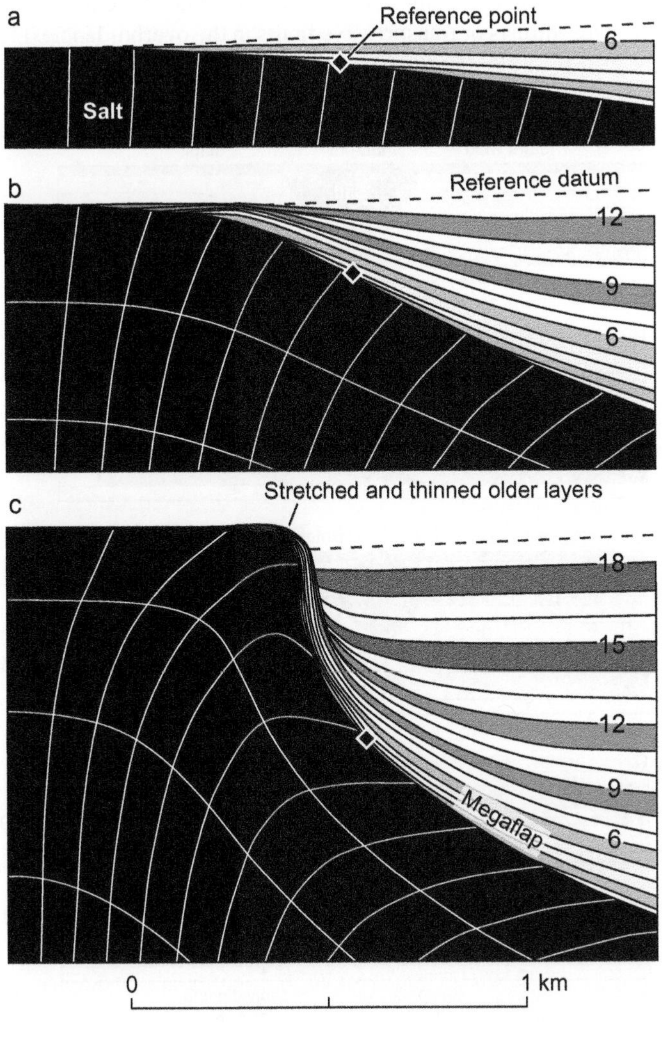


**FIG. 10** Finite-element model showing evolution of a basal megaflap that becomes onlapped by younger strata in the contact zone of a passive diapir. *(Modified from Jackson and Hudec (2017)).*

Basal megaflaps are most likely to originate where a salt contact dips moderately (a few tens of degrees), allowing a wedge of strata to onlap across the diapir crest and be folded upward by the rising salt (Schultz-Ela, 2003; Fig. 10). The style of folding depends on the depth, rock strength, anisotropy of the overburden, diapir shape and tilt and the molding ratio (balance of aggradation and salt-rise rates during folding).

The prevalence of basal megaflaps in many deepwater basins means that wells drilled on salt flanks often encounter elevated, condensed, shale-prone sections well above their regional elevations (Fig. 10). Furthermore, younger stratigraphy pinches out as it approaches the flap, meaning that a well drilled too close to the diapir face might miss the target interval altogether. These geometries, combined with the difficulties in imaging steep dips near the salt face (especially subsalt) make hydrocarbon traps against the sides of stocks and walls challenging to exploit in many deepwater settings.

## Squeezed diapirs

Because the deepwater portions of syn-exhumation salt basins lie at the downdip ends of long (up to hundreds of kilometers), continuous salt basins (Fig. 3D), they are common sites of salt-detached fold-and-thrust belts. The diapirs described earlier are caught up in this shortening so squeezed diapirs are a common structural style.

Because salt diapirs vary in size, profile, planform and orientation, each diapir responds differently to shortening. Despite this variation, squeezed stocks have several features in common (Figs. 11 and 12). First, a thick, arched roof; because tectonic forces are typically greater than buoyancy forces, a squeezed diapir can have a much thicker arched roof than that present above an active diapir formed without shortening. The roof arches in response to upward pressure of confined salt and to lateral compression of the roof. Second, a narrow or pinched-off diapir stem. Third, a salt pedestal at the base of this stem, which is the remnant of the downward-flaring diapir base. Fourth, continued rise of the diapir after the surrounding source layer is depleted. The rejuvenated diapir rises without receiving salt from the source layer if diapiric salt is forced upward by the converging flanks of the diapir. Fifth, folds, thrusts or transpressive faults in the overburden next to the diapir.

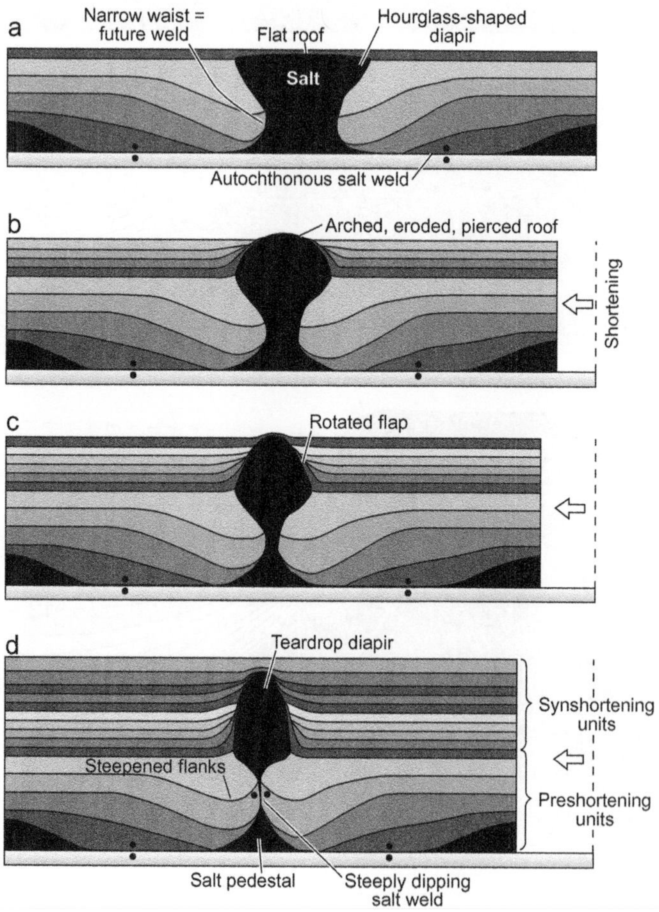


**FIG. 11** As a diapir is laterally squeezed, flanking strata steepen, the stem of the diapir narrows and can eventually pinch off, its bulb becomes taller and its roof arches. Passive diapirs stop growing halokinetically after exhausting their source layers but can be rejuvenated by lateral shortening as salt is displaced upward by the converging sides of the diapir. *(Modified from Jackson and Hudec (2017)).*

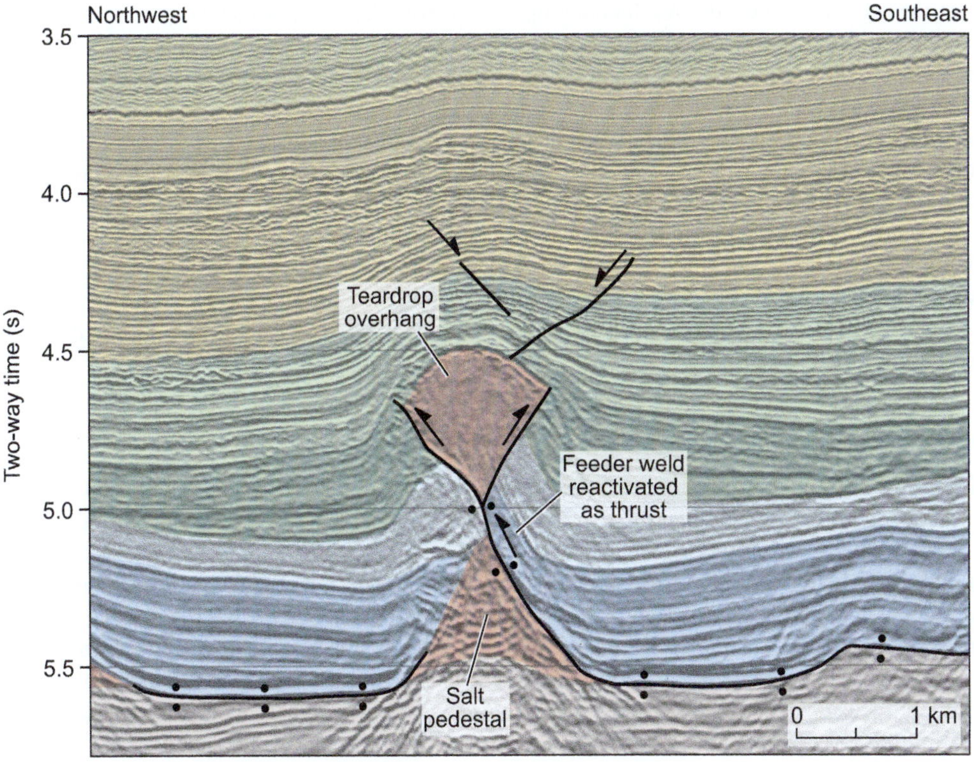


**FIG. 12** A pinched-off stem and northwest-directed thrusting of the hanging wall indicate shortening of the diapir. Most shortening occurred before the diapir's roof was deposited, but the roof continued to arch as shortening waned. Astrid fold belt, Lower Congo basin, southern Gabon. *(Modified from Jackson and Hudec (2017)).*

If the roof is thin and weak, rising salt disrupts the roof, allowing salt to break through to the surface and rise passively and extrude as a salt sheet. Elsewhere, ongoing sedimentation may keep the salt from spreading laterally. If so, salt in the shortening dome is forced upward as the diapir walls converge. In an hourglass-shaped diapir, the upper part of the diapir rises to form an inverted teardrop (sometimes called a *teardrop diapir*). With enough shortening, the waist of the hourglass can neck off as a weld, which separates the diapir's bulbous crest from a triangular pedestal below.

A salt stock shortens by expelling salt, whereas the adjacent country rocks shorten by folding and thrusting. This discontinuity creates complex fault patterns in physical models (Duffy et al., 2018; Fig. 13). Cross-faults allow a thrust salient to indent the soft stock like a piston and expel the salt. The stock becomes elliptical as one concave side of the

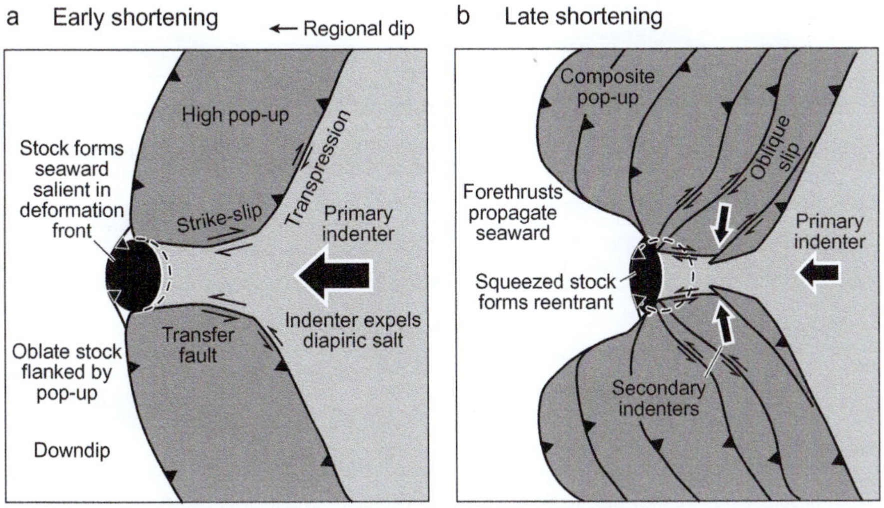


**FIG. 13** Contractional faulting around a buried squeezed stock evolves in two stages. (A) A primary indenter bounded by tear faults moves into a yielding stock. The stock is a salient in the deformation front. (B) As shortening continues, secondary indenters constrict the stock as the tear faults become transpressional. The stock changes from a salient to a reentrant in the deformation front. *(Modified from Jackson and Hudec (2017)).*

diapir overrides the other concave side. As the hanging-wall overburden rises, it makes space for diapiric salt to intrude downward into the source layer and diapiric pedestal. However, diapiric salt is more likely to flow upward than downward into the source layer if (1) the roof above the diapir is thin and can be pierced, (2) the salt is less dense than its overburden or (3) the source layer is thin.

Once a diapir has pinched shut, one flank may override the other as a thrust fault, offsetting the two halves of the diapir from each other. Most of the diapir is reduced to a continuous weld connecting its pedestal to a crestal teardrop or to an overlying salt sheet that typically flows seaward across the seafloor. Alternatively, a gently inclined thrust weld can offset the near-vertical welds marking the position of the former diapir. In general, the salt contacts of conical and low-amplitude diapirs act as favorably oriented thrust ramps, whereas taller diapirs are thrust off from their pedestals and incorporated within a crestal anticline (Callot et al., 2007). This tendency is especially pronounced if the diapir is asymmetric in cross section or leaning, in that the gentler flank can provide a favorable ramp to initiate a thrust. Thrusting can occur while the diapir is emergent or after it is buried.

## Salt sheets

A salt sheet is defined as a subhorizontal or moderately dipping, sheet-like salt diapir, sourced from a single-feeder diapir, whose breadth is several times greater than its maximum thickness. A salt canopy is a composite structure, made from the coalescence of two or more salt sheets. Salt sheets are typically tens of kilometers in lateral extent. Salt canopies range widely in scale from relatively modest canopies comprising only two sheets, up to the Sigsbee canopy in the northern Gulf of Mexico, which has hundreds of feeders and covers an area about the size of Switzerland.

Sheets and canopies are important in the deepwater portions of many salt-bearing passive margins for two reasons. First, the deepwater portions of syn-exhumation passive margins can host very thick salt (up to several kilometers thick), which is required to form salt sheets. Second, deepwater syn-exhumation salt margins are also sites of salt-detached shortening, which is critical in the formation of many salt sheets, as described in the following section.

Three types of salt sheets are common in deepwater settings: plug-fed extrusions, plug-fed thrusts and source-fed thrusts (Fig. 14; Hudec and Jackson, 2006). They are distinguished by the geometry of the feeder structure and by their

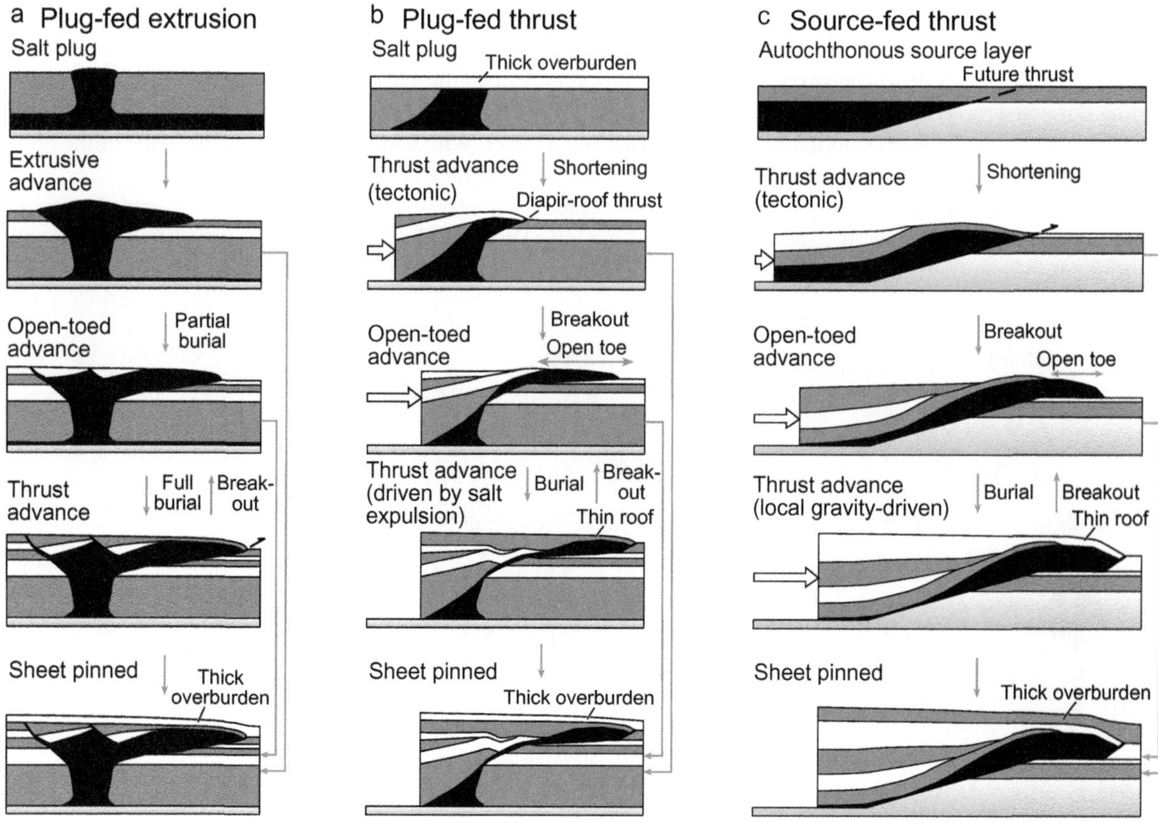


**FIG. 14** Most salt sheets evolve through a series of stages to reach their final form. The three most common evolutionary pathways, or lineages, are shown here. *Vertical arrows* between rows connect observed growth sequences. *(After Hudec and Jackson (2006)).*

initial mode of advance. They are quite distinct geometrically, although they can all have phases of extrusive advance, open-toed advance and thrust advance.

## Plug-fed extrusions

A plug-fed extrusion forms when a sheet extrudes from the top of a vertical or steeply dipping diapir (Fig. 14A). The term salt plug loosely to refers to both salt stocks and salt walls, although most plug-fed extrusions appears to be sourced from stocks. Plug-fed extrusions are probably the most common type of salt sheet.

Extrusive salt sheets spread whenever salt in a downbuilding diapir rises faster than dissolution, erosion and sediment aggradation can remove or contain it. Plug-fed extrusion may be initiated by decreases in dissolution, erosion, sediment aggradation rates or an increase in salt-rise rate.

However, even these factors may not be enough to initiate salt-sheet extrusion if the roof above the diapir is too thick and strong to be pierced easily. If so, the salt is encased and immobilized until something disrupts its overburden. For example, sedimentation adjacent to the diapir may pressurize salt in the source layer enough to overcome the overburden strength; extension can thin and weaken the overburden; shortening can uplift the overburden and subject it to erosional thinning, or erosion can thin the overburden in the absence of any lateral forces. Once the overburden is removed, salt may extrude rapidly if the load on the source layer is high. A sheet produced by breakout of overpressured salt most likely has a flat base because the sheet advances much faster than the sedimentary interval accumulates or aggrades. This effect is enhanced if salt breaks out during a sedimentary hiatus.

The lateral shortening of a salt diapir is especially effective in triggering plug-fed extrusions, because shortening simultaneously pressurizes the salt, expels salt from the diapir by inward convergence of the diapir walls and disrupts the roof. Plug-fed extrusions in orogenic belts are clearly related to shortening. Hudec and Jackson (2009) argued that most plug-fed extrusions on passive margins are also related to shortening.

The flow of an unbuttressed extrusive sheet is typically radially outward, with a preferred direction down the local slope. In most sedimentary environments, but especially in prograding margins, a roof does not accumulate uniformly over the entire sheet. Thus, open-toed advance typically follows extrusive advance. Open-toed advance continues until the sheet toe is buried, at which point the sheet may continue to advance by thrusting.

Once completely buried, a salt sheet may advance as a thrust as long as forces driving sheet advance can overcome the strength of the roof. Bigger sheets tend to advance by thrusting for a longer time. Thrust advance becomes more difficult as the roof thickens and strengthens, so even the largest sheet should eventually be pinned in place by its overburden.

## Plug-fed thrusts

Plug-fed thrusts form when a diapir's overburden is shortened by thrusting, carrying a sheet of salt from the diapir to the base of the thrust hanging wall (Fig. 14B). The style of deformation occurs during regional shortening if a thick, strong overburden prevents a laterally shortened salt plug from piercing the surface. Thrusting is facilitated if the plug is inclined because slipping along the more gently dipping diapir flank may nucleate the fault. Sheets of this type occur in orogenic belts and in deepwater fold-and-thrust belts (Grando and McClay, 2004; Callot et al., 2012; Chapter 4).

Many plug-fed thrusts never progress beyond initial thrusting. If the overburden is kilometers thick when shortening begins, salt is unlikely to pierce the surface without substantial erosion. Structural style typically changes abruptly along strike from a salt sheet near the plug to a salt-free thrust fault away from the plug (Fig. 13). Where diapirs predate shortening, transpressional thrust networks can form oblique to the regional shortening direction. Such thrusts tend to form parallel and adjacent to the flanks of salt walls even if this direction is strongly oblique to the regional maximum compressive stress.

Some plug-fed thrusts, however, have overburden thin enough to be eroded off during shortening. Erosion of the overburden above the thrust allows the salt to break out. If vigorous salt flow overcomes dissolution and sedimentation, open-toed sheet advance can begin (Fig. 14B). Extrusion can be rapid because salt is pressurized by tectonic compression or by a location at the toe of slope.

After breakout, an open-toed sheet extruding from a plug-fed thrust advances over the sediment surface until its feeder is completely pinched shut or the source layer is depleted of salt. The sheet then evolves through stages typical of open-toed advance: progressive burial, possible thrust advance (driven by gravity instead of by regional shortening) and eventual pinning (Fig. 14B). This postbreakout history can obscure an origin as a plug-fed thrust, especially if most of the prekinematic roof has been eroded.

## *Source-fed thrusts*

A source-fed thrust forms when a thrust imbricate detaches on autochthonous salt and carries a slice of salt in its hanging wall (Fig. 14C). In this terminology, source refers to the autochthonous salt layer, the source level for all the salt structures in the basin.

Source-fed thrusts range from thin slivers in the hanging walls of small thrusts to salt-cored nappes involving kilometer-thick salt sheets displaced by many kilometers. The largest thrust-fed breakouts form near the edges of salt basins, where the salt-lubricated detachment stiffens by changes in salt facies, thins or disappears. Here, translation, aided by a subhorizontal salt detachment, must be accommodated beyond the salt by thrusts or folds in the overburden. Thrust systems fringing salt basins can extend along strike for hundreds of kilometers (e.g., the Angola salt nappe that continues from the Benguela basin across the entire Kwanza Basin, and into the Lower Congo Basin).

The thickness of salt carried in a source-fed thrust depends on the level of the detachment within the salt (Fig. 15). Detachment is likely to follow the weakest layer within the evaporites—weak because of inherently low viscosity, shear-thinning or elevated pore pressure in the salt. In some cases, the salt does not flow appreciably within the thrust sheets and acts like just another layer (Fig. 16A). Elsewhere, the salt actively deforms, pinching and swelling within the thrust sheet to form thick salt accumulations (Fig. 16B).

In some areas, thrusts are localized at the crests of preexisting salt anticlines (e.g., Fig. 17) or salt plugs (Fig. 18). This localization occurs because sediment is thinner over the crest of the salt structure and thus weaker. In the case of salt plugs, physical models show that the diapirs start shortening first, as they are the weakest parts of the basin. Thrusts then propagate outward from the shortening diapirs, eventually linking to thrusts propagating outward from other diapirs to form complex 3D networks (Duffy et al., 2018). Eventually, after the diapir is completely welded shut, one flank of the diapir may overthrust the other (Fig. 18F).

## Canopies and sutures

The zone separating two coalesced salt sheets is a suture. Sutures can be subdivided into two types: autosutures, which form between two lobes of the same salt sheet, and allosutures, which form between salt sheets having separate feeders. Dooley et al. (2012) described suture types in detail and proposed terminology for their component parts.

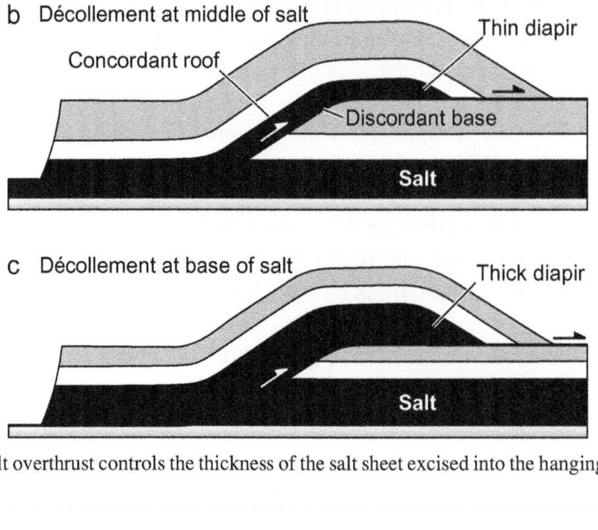


**FIG. 15** The décollement level of a salt overthrust controls the thickness of the salt sheet excised into the hanging wall of the thrust fault. *(Modified from Jackson and Hudec (2017)).*

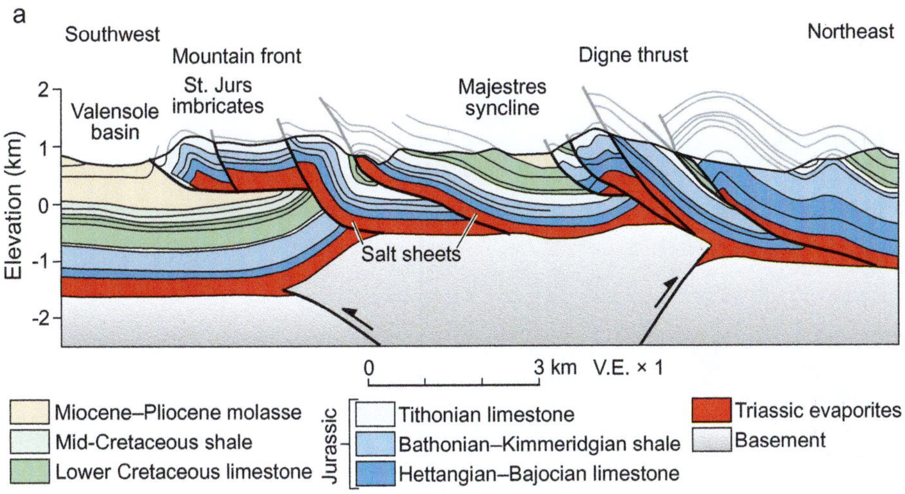


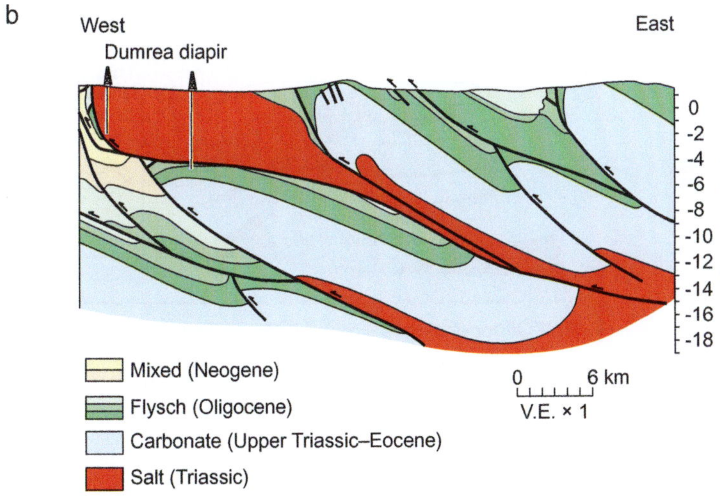


**FIG. 16** Examples of source-fed thrusts. (A) Thick salt forms source-fed thrusts in the subalpine Rhodanian basin (France). The salt sheet overrides equivalent stratigraphy in the Digne imbricates and overrides much younger Miocene-Pliocene molasse in the Valensole basin adjoining the mountain front. (B) Orogenic shortening involving a thick source layer can expel huge source-fed thrusts, such as the Dumrea diapir, a nearly horizontal allochthonous sheet 5700 m thick. The diapir is protected by brecciated cap rock. Cross section from the Albanides, Albania. *(After (A) Lickorish and Ford (1998); (B) Velaj et al. (1999)).*

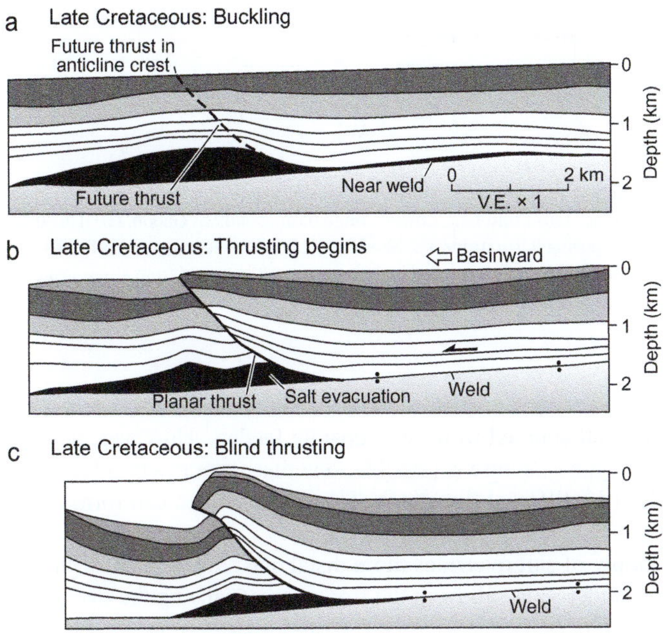


**FIG. 17** Thrust localized at the crest of a preexisting anticline. *(After Jackson et al. (2008)).*

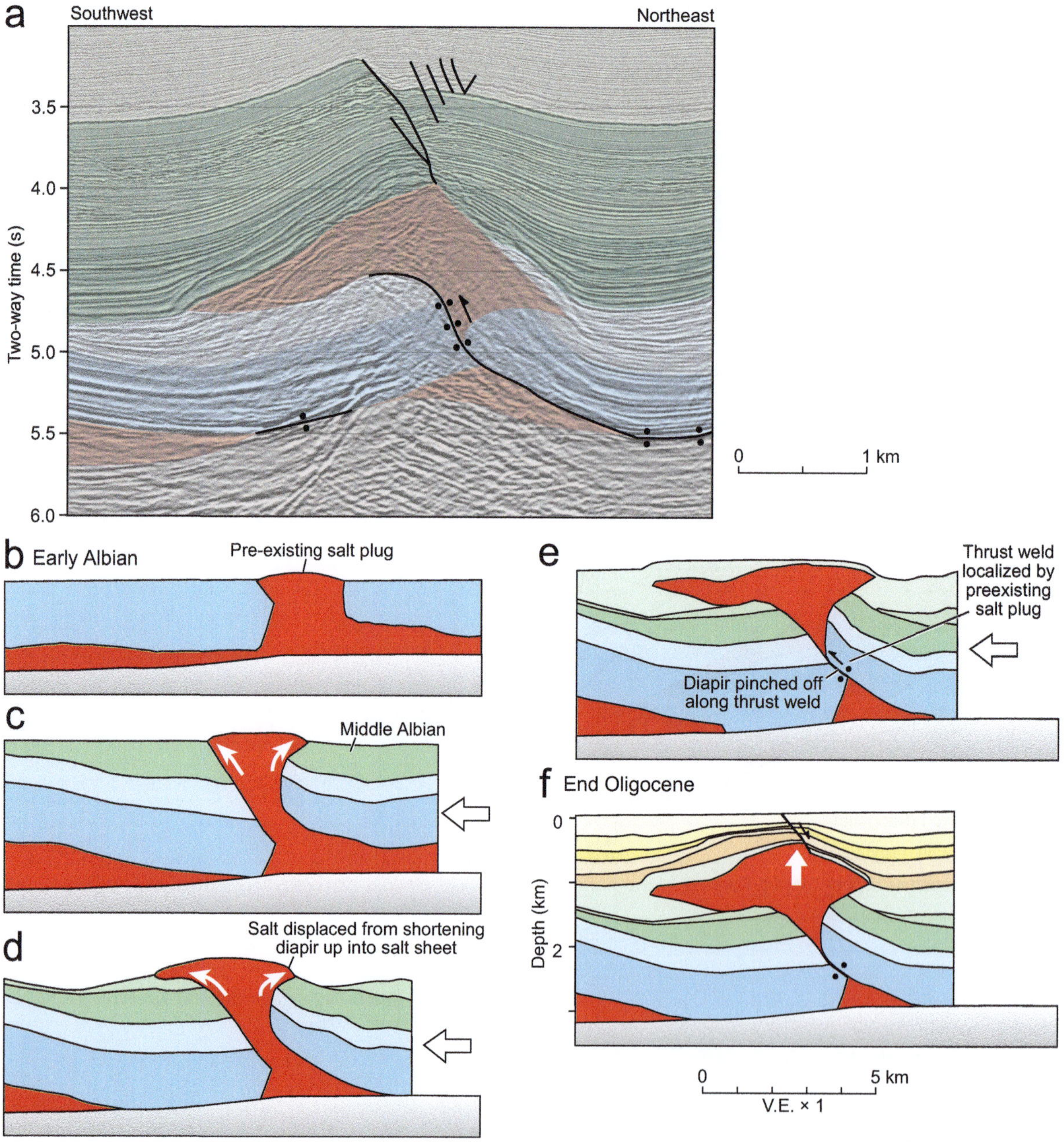


**FIG. 18** Thrust localized by preexisting salt plug. Lateral shortening can initiate or accelerate extrusion of allochthonous salt but can ultimately end extrusion when the waist of the diapir pinches off. (A) Seismic example from the Astrid fold belt, Lower Congo basin, southern Gabon. (B) This diapir grew passively until the mid-Albian. (C, D) Southward thrusting displaced salt, causing it to extrude. (E) Shortening closed the stem and cut off salt supply so that extrusion ended and the allochthonous sheet was buried. (F) As it was buried, the roof was arched by inflation of salt imported from along strike into the salt sheet. *(After Jackson et al. (2008)).*

## *Autosutures*

The traditional definition of a suture implies that it separates salt sourced from two separate feeders. However, physical models and 3D seismic data have demonstrated that a second type of suture is possible: one that separates two lobes of a single-feeder salt sheet that have moved at different speeds or in different directions. These autosutures can form in two ways: by overriding or by encircling.

Overriding autosutures form when salt near an active supply of salt overrides more distal and slower parts of the salt sheet (Fig. 19). The weight of the overriding salt lobe expels underlying salt from the frontal area, causing the overridden

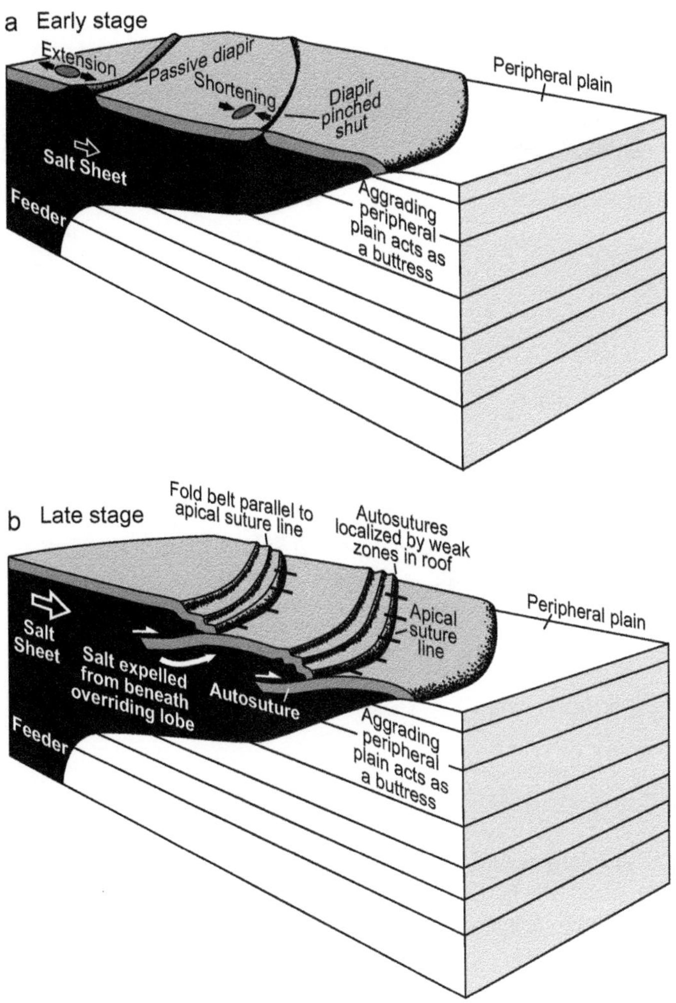


**FIG. 19** In an overriding autosuture, the roof of a salt sheet shortens where sheet advance is buttressed by the peripheral plain or by another sheet. Overriding autosutures may initiate at any weak part of the roof, especially reactive diapirs formed by previous stretching as the salt sheet spreads. The overridden part of the roof is depressed into the salt as it is loaded from the rear. *(After Dooley et al. (2012)).*

roof to subside into the salt. The overridden roof forms an upward-steepening intrasalt reflection on seismic data. The internal reflections may connect to the base of salt if subsidence is sufficient to weld the roof to the bottom of the sheet. However, overriding does not need this much subsidence, and most overriding autosutures flatten and terminate within the sheet.

Overriding autosutures can form whenever the rear of a salt sheet advances faster than its front. This phenomenon may occur if a thick, strong roof inhibits advance of the sheet toe, or if the toe is buttressed against an obstacle, such as peripheral-plain sediments or another salt sheet. Rapid advance of the rear of a sheet compresses its front. This shortening inflates the salt and may shorten the sheet roof in the form of small folds and thrusts. In the experiments of Dooley et al. (2012), autosutures formed primarily by exploiting preexisting zones of weakness in the roof, especially small-scale precursor diapiric walls that originally formed by stretching a thin roof and squeezed shut as the roof moved into a zone of shortening.

Encircling autosutures form when two lobes of a spreading salt sheet separate to bypass a topographic obstacle then rejoin on the downstream side of the obstacle (Fig. 20). This obstacle may be a thicker part of the salt-sheet roof, but the most effective obstacles are minibasins whose bases have nearly welded, slowing their transport to the front of the sheet. Depending on its strength, the obstacle may be deformed by passage of the enveloping salt.

In contrast to overriding autosutures, which form by imbrication of the top of salt, encircling autosutures involve the entire thickness of a spreading sheet and thus connect the top to the base of salt. If sediments aggrade as the salt lobes approach one another, the lobes may each climb slightly upsection toward one another, producing a subtle structural high along the basal suture line. This structural high may grow faster if roof sediment above the two lobes slumps into the gap between them, building a pile of mass-transport complexes that must be surmounted before the lobes can merge. This overridden pile of debris at the base of salt records the advance direction of the sheet.

**FIG. 20** In an encircling autosuture, the advancing salt sheet divided into two lobes to bypass a grounded minibasin or other topographic obstacles. The two lobes encircled this obstacle and rejoined on the downslope side. *(After Dooley et al. (2012)).*

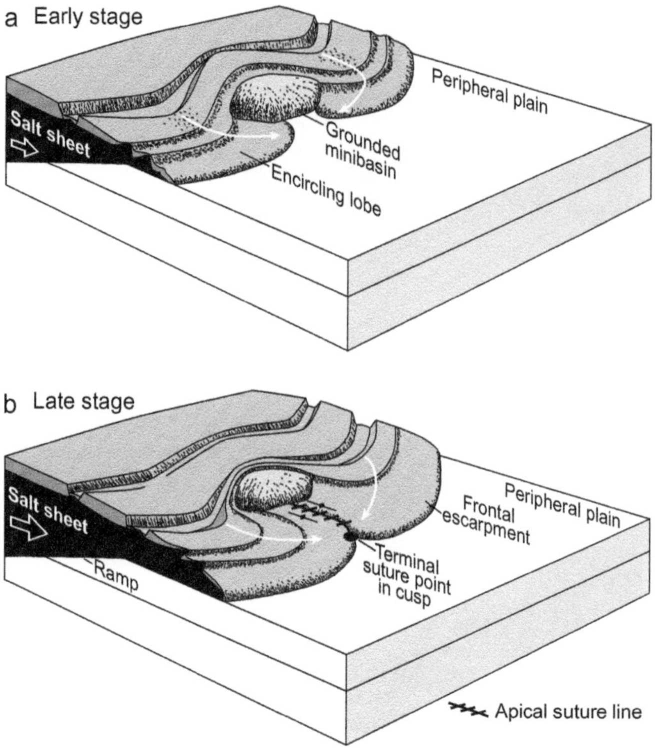


## *Allosutures*

A salt sheet typically climbs upsection away from its feeder, especially where the sedimentation rate is high relative to sheet advance. If subsalt strata are not deformed after a sheet is emplaced, the base of the salt sheet also climbs structurally away from its feeder and reaches its apex around the rim of the sheet. If two such sheets merge to form an allosuture, then the basal suture line should lie along a structural high in the base of salt (Fig. 21).

In most allosutures, one salt sheet overrides the other, generally because the overriding salt sheet has more salt or lies farther updip. Sheet override during suturing produces a strongly asymmetric, shallowly dipping allosuture. Overriding offsets the apical suture line in the direction of override, producing a suture that dips back toward the overriding sheet. In section view, asymmetric suture lines are commonly listric and connect the offset apical and basal suture points (Fig. 21).

As the overriding sheet advances, salt in the overridden sheet is expelled away from the suture, causing the suture to subside. This process deforms the suture, as suture lines that are initially steep become listric as sutures mature. Salt in the overridden sheet is displaced away from the suture, inflating the other parts of this sheet. Suture subsidence and inflation in front of the overriding sheet make it progressively more difficult for this sheet to continue advancing across the top of the overridden sheet. Instead, the overriding sheet begins to bulldoze its neighbor, pushing sediments trapped in the suture forward. This bulldozing stretches the suture, eventually tearing trapped sediments away from the basal suture line and obscuring the connection between the suture and the correlative basal suture line. Only a thin trail of sedimentary fragments connects the main body of suture sediments back to their basal point.

## Salt-evacuation geometries

The thick salt present in the deepwater portions of syn-exhumation salt basins allows for a wide range of large-amplitude salt-evacuation geometries, some of which are not found in any other settings. The four geometries focused on here are: turtle structures, minibasins, roho systems and counterregional systems.

## *Turtle structures*

A common, and very prolific, feature associated with stocks and walls in many deepwater regions is the turtle structure. A turtle forms as a result of shifting patterns of salt withdrawal—the initial depocenter forms in the center of the turtle, and then shifts to the two flanks. The flanks subside, leaving the original depocenter as a structural high

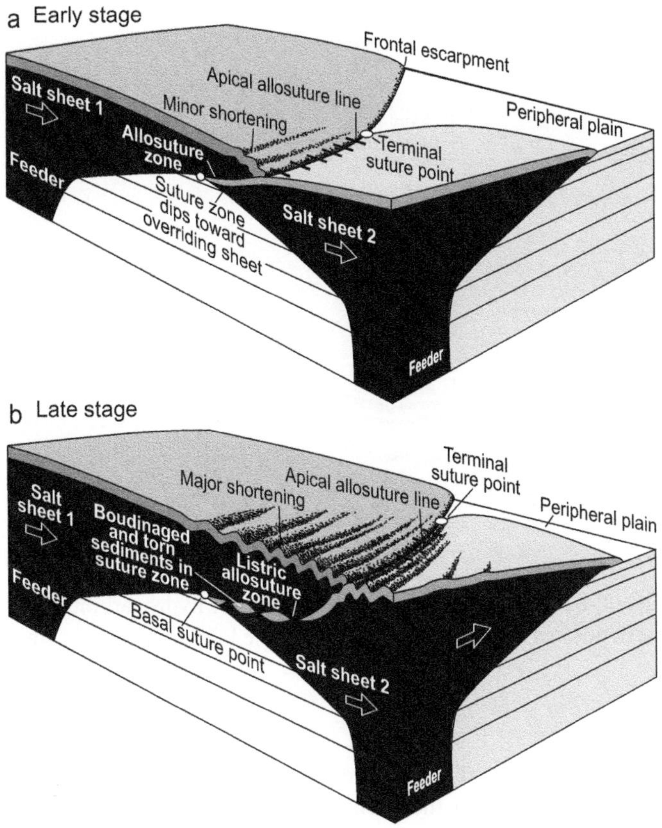


**FIG. 21** As one salt sheet overrides another in an asymmetric allosuture, sediments in the suture are depressed, stretched and dismembered to form boudins. These sediments eventually tear off from the basal suture line. Roof sediments near the apical suture line are shortened in both salt sheets, but especially in the overriding sheet. *(After Dooley et al. (2012)).*

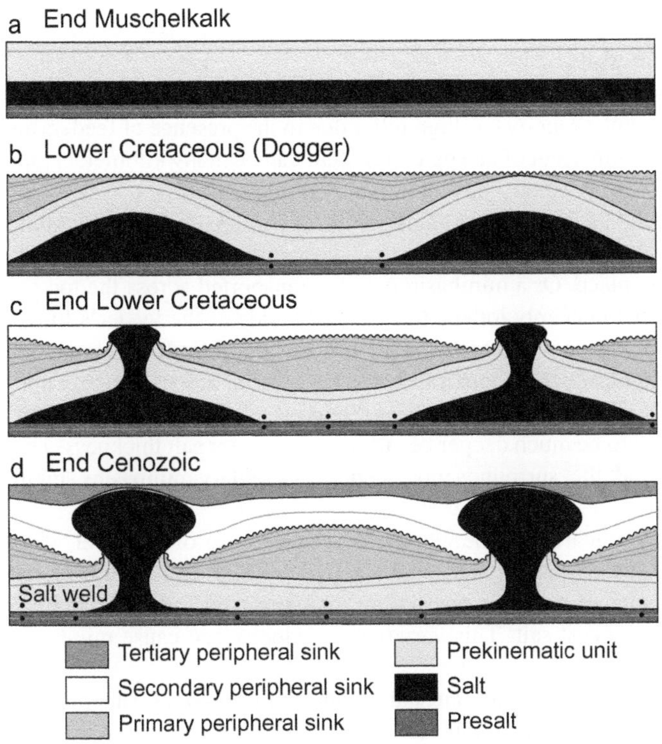


**FIG. 22** Trusheim's (1960) concept of evolving peripheral sinks during the growth of a salt pillow and diapir, based on examples from North Germany. His original restoration has been mirrored about the center of the line to give equal emphasis to the growth of diapirs and the intervening turtle-structure anticline.

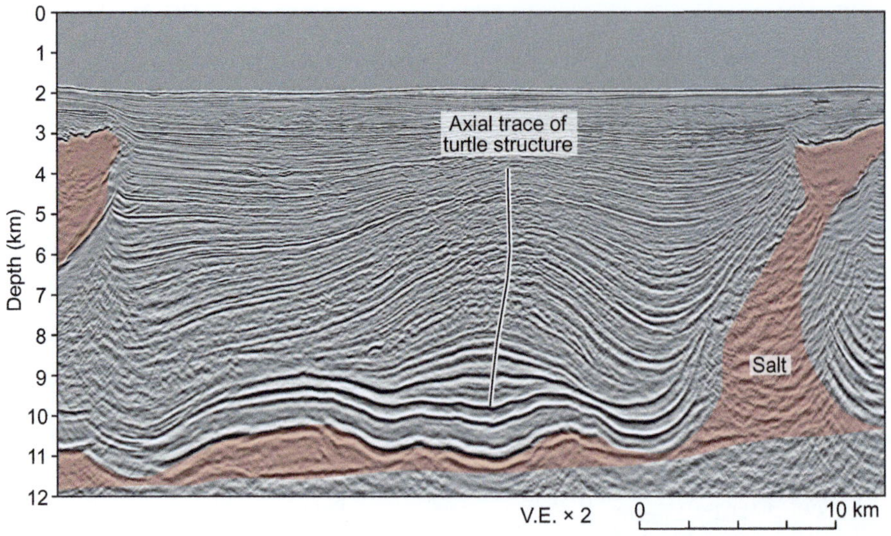


**FIG. 23** Seismic example of a Gulf of Mexico turtle-structure anticline produced by halokinesis. *(From Jackson and Hudec (2017)).*

(Figs. 22 and 23). Turtle structures can form in any water depth, but they are especially important for hydrocarbon exploration in deepwater, because a stacked series of sands may be deposited in the early depocenter. The later inversion by flank subsidence then places these sands in a structural high.

## *Minibasins and salt-stock canopies*

A minibasin is a small intrasalt basin largely surrounded by and subsiding into relatively thick autochthonous or allochthonous salt. Salt expelled from beneath a minibasin wells up around the minibasin margin, forming a network of salt walls, massifs or welds that surround the minibasin. A minibasin is thus a type of salt-withdrawal basin that can form only above relatively thick salt.

Minibasin structure depends in part on whether a minibasin forms above autochthonous salt (primary minibasin) or allochthonous salt (secondary minibasin; Fig. 24). By definition, the lateral edges of primary minibasins are salt structures. The type of salt structure is strongly controlled by the original salt thickness. Above moderately thick autochthonous salt, the lateral boundary may consist of isolated salt stocks connected by a polygonal network of salt anticlines. Where salt is thicker, minibasins may be surrounded by a network of salt walls.

The main difference between primary and secondary minibasins is that secondary minibasins typically subside into a much more complex salt geometry. The base of an allochthonous canopy has high relief due to the presence of feeders and basal sutures, which combine to form a base-salt geometry reminiscent of an egg carton. Salt can be many kilometers thick above feeders and much thinner above basal suture lines. This geometry has three main consequences.

First, secondary minibasins are likely to sink into salt feeders, because the thick salt there is more mobile than thinner salt above other parts of the canopy. Therefore, minibasins above feeders sink faster than those initiated in other parts of the canopy. A minibasin can initiate above a feeder and sink in place. Or a minibasin may be transported across the top of a feeder and subside so rapidly into the feeder that it becomes too deeply lodged to exit the far side of the feeder's funnel.

Second, the structural style of secondary minibasins depends on whether they sink into a feeder. Minibasins tend to be much simpler when they subside onto a subhorizontal base of allochthonous salt between feeders, a configuration analogous to subsidence of primary minibasins. Underlying salt is displaced sideways into flanking salt structures. Secondary minibasins that sink into feeders obviously have the potential to be much deeper because of the greater salt thickness. They can also rotate as they impinge on the funnel-shaped base of salt that surrounds most feeders. Secondary minibasins sinking into feeders are known as bowl or bucket minibasins.

Third, downslope transport of minibasins is greatly complicated by the egg-carton-shaped base of salt over which the system translates. Minibasins may be pinned in place if they cannot surmount a structural high in the base of salt, distorting the transport of all nearby units. Also, the velocity of downdip translation may vary from place to place in the canopy, with the most rapid transport being above the thickest salt. This variation in velocity can cause minibasins to move apart or collide.

The tendency of secondary minibasins to sink into feeders produces an unusual geometry in salt-stock canopies that have been loaded by thick sedimentary sequences. Minibasins tend to subside into the feeders, displacing salt into

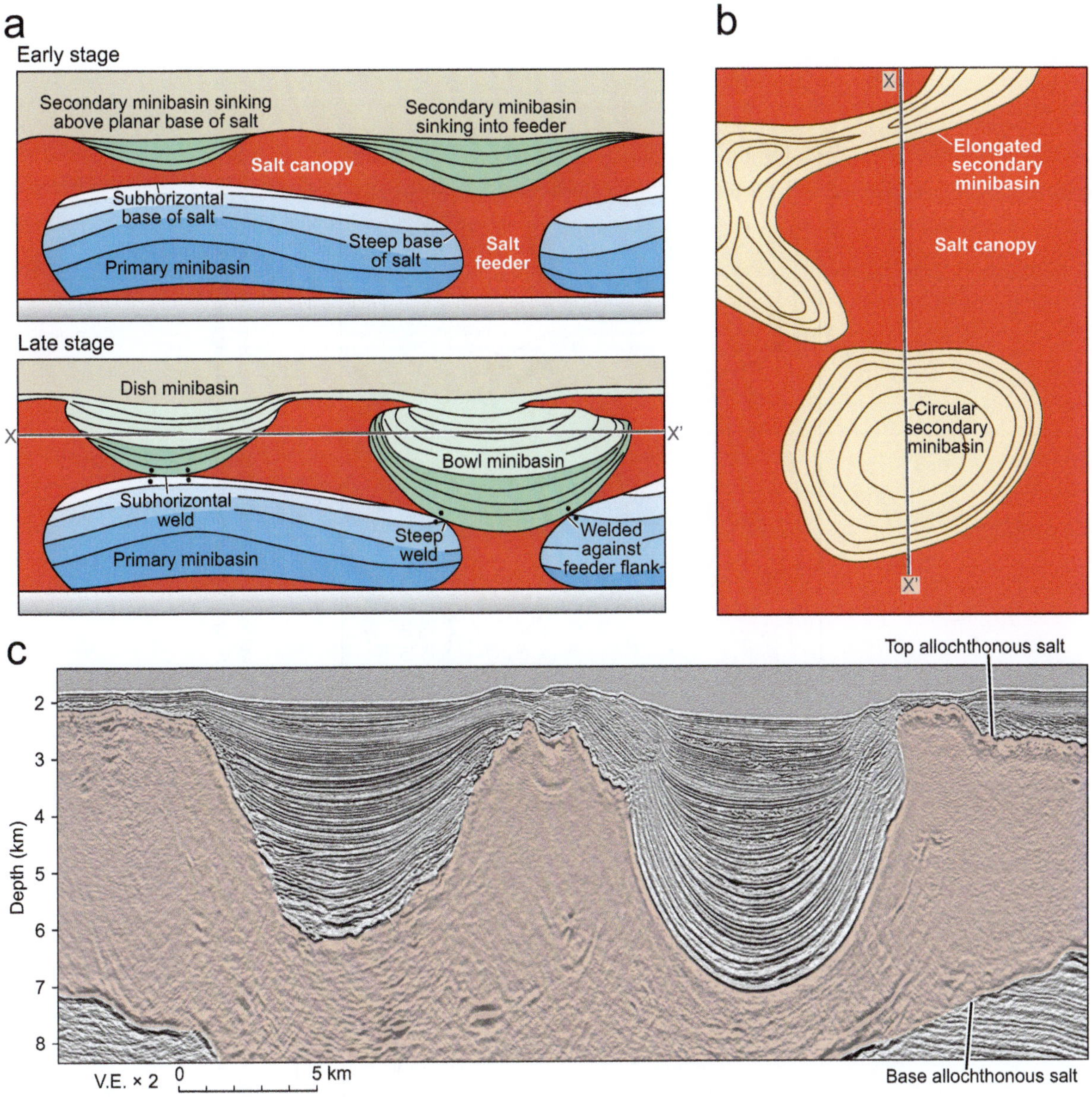


**FIG. 24** Minibasins are small basins, or depressions, that fill with sediment as they sink into thick salt. These growth synclines are laterally surrounded by salt or by salt welds. (A) Two-stage evolution of secondary minibasins sinking into allochthonous salt. An older, primary minibasin underlies the canopy. The second stage shows the level of the depth slice shown in (B). (B) Minibasin planforms are typically generalized as subcircular, but they can be any shape, including elongated. (C) Seismic example of dish-shaped minibasins on the continental slope of the northern Gulf of Mexico. *(After Jackson and Hudec (2017)).*

interfeeder areas (Hudec and Jackson, 2011; Fig. 25). Redistribution of salt inverts the initial salt distribution in map view. In some cases, salt expulsion from beneath minibasins may drive the final suturing of the component salt sheets in a salt-stock canopy.

## *Roho systems*

A roho system is a group of listric, basinward-dipping growth faults that sole out onto a salt sheet, salt canopy or allochthonous weld (Figs. 26 and 27). Sediment wedges in the fault blocks dip and thicken landward but become younger seaward. In map view, rohos are bounded by complex zones of strike-slip deformation, separating transported units above the roho from more stable blocks on either side.

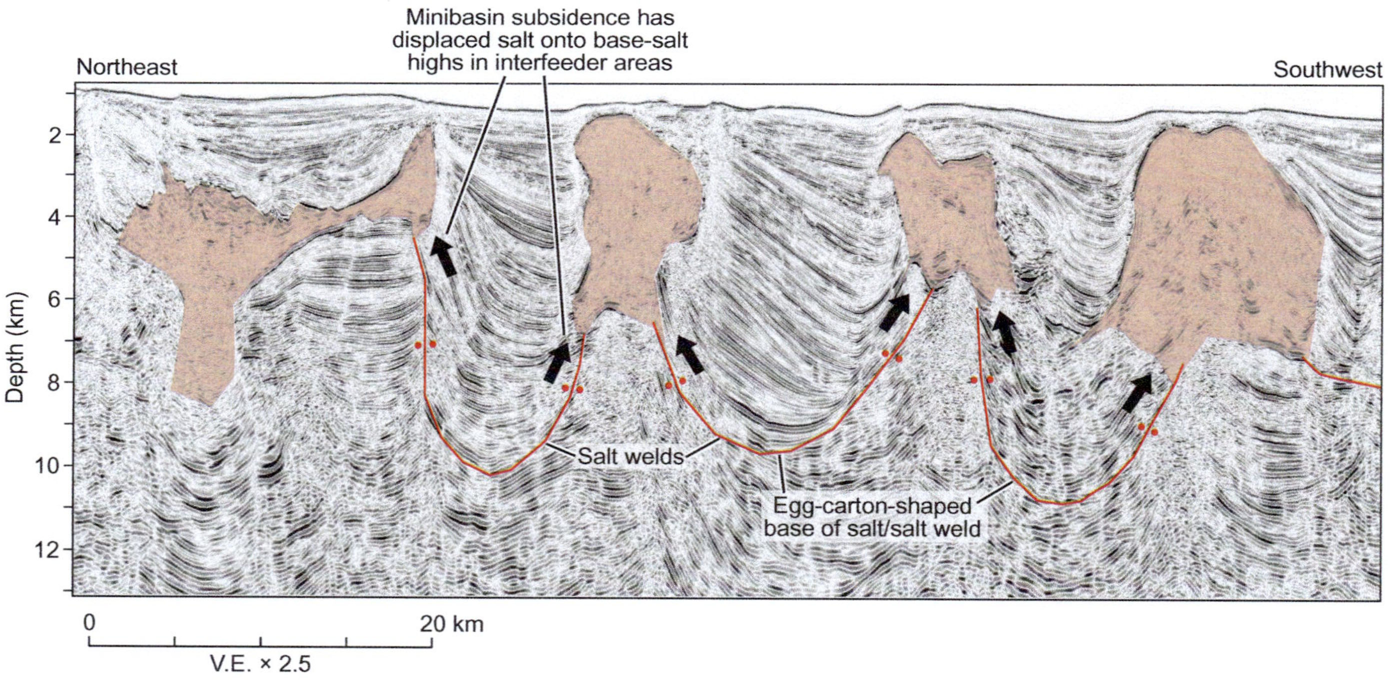


**FIG. 25** Amalgamation of a salt-stock canopy creates an egg-carton-shaped base of salt, parts of which may be welded by later minibasin subsidence. Here, minibasin subsidence has displaced salt from the base-salt structural lows near feeders up onto the interfeeder structural highs in the base of salt. High reflectivity in the section beneath the welds suggests that feeders lie out of the plane of this section. *(From Jackson and Hudec (2017))*.

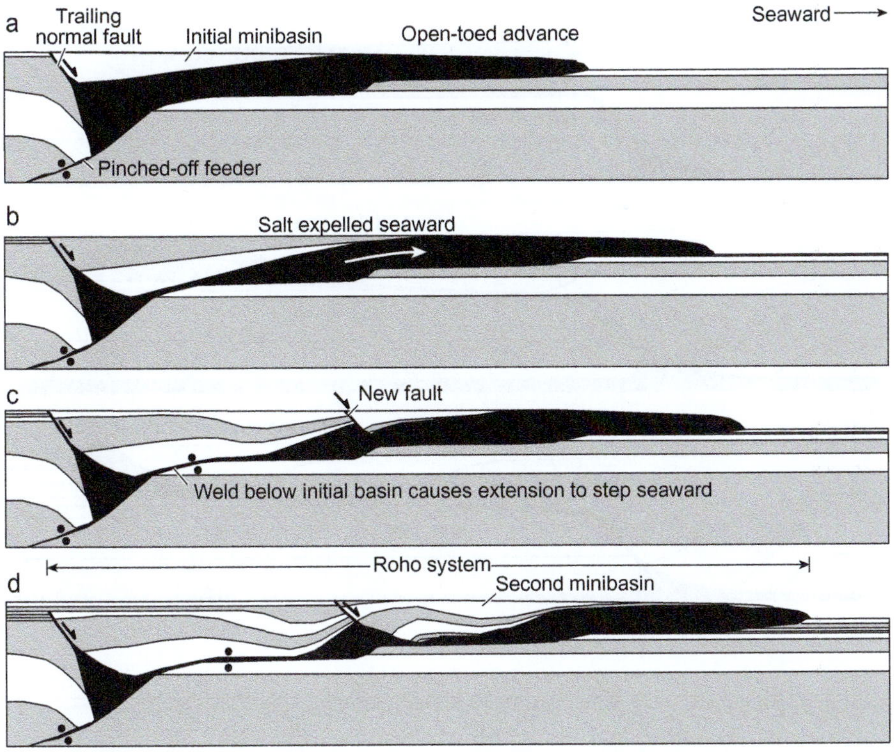


**FIG. 26** Schematic forward model of a roho shows sediment progradation driving seaward expulsion of salt from a salt sheet. In contrast to a counterregional system (Fig. 25), the thin roof is extended along a series of seaward-dipping normal faults. *(After Hudec and Jackson (2006)).*

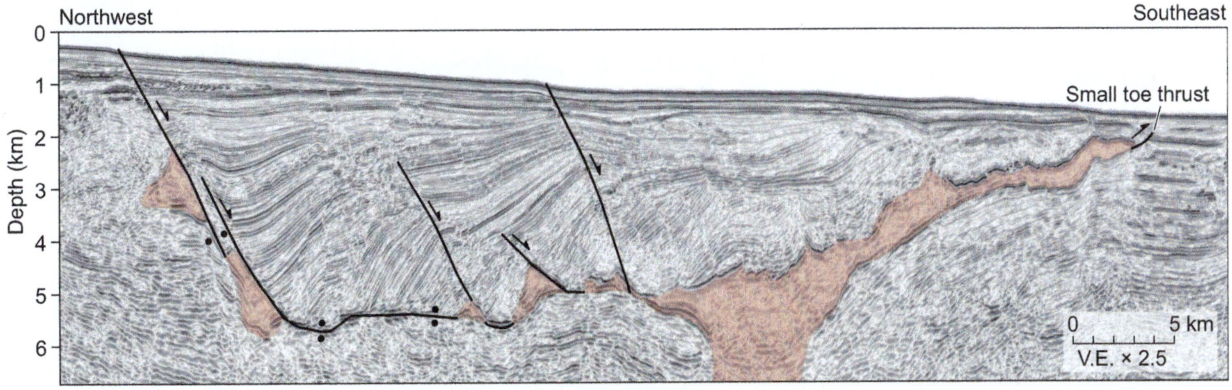


**FIG. 27** Extension above this still-active roho is driven by topographic slope. Judging from the comparatively small toe thrust at the seaward end of the roho, most of the shortening has been absorbed by narrowing of exposed salt before the toe was buried. *(Modified from Jackson and Hudec (2017)).*

Rohos form when the roof extends as sediment progrades across a salt sheet or canopy. The extension is driven by gravity, so rohos are favored by greater surface relief. This relief may be the product of a steep bathymetric slope or allochthonous salt that is long in the dip direction.

Rohos are linked kinematic systems in which extension near the updip end of the system must be balanced by shortening downdip. However, most rohos form during open-toed advance, so that the shortening occurs by narrowing of the exposed toe of the sheet or canopy (Fig. 26). This shortening, although real, is not recorded by roof structures. As a result, rohos are typically dominated by extensional structures, with only minor shortening in evidence near the toe of the system.

## Counterregional systems

A counterregional system is a landward-dipping, largely welded salt system whose upper part is a weld, remnant diapir or normal fault. Sediment wedges dip and thicken seaward, in contrast to rohos. The lower part of the system may ramp continuously down to the source layer or may include flat-lying segments. If the system has more than one ramp, it is a stepped counterregional system (Figs. 28 and 29). Counterregional refers to the landward dip of the master detachment, which is

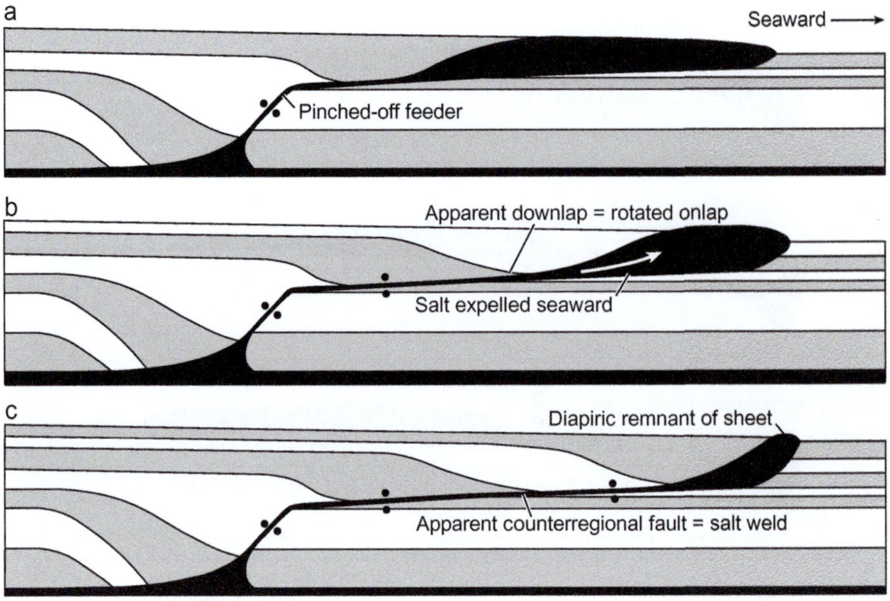


**FIG. 28** A schematic forward model of a stepped counterregional system shows sediment progradation driving seaward expulsion of salt from a salt sheet, as in a roho system (Fig. 24). *(Modified from Hudec and Jackson (2006)).*

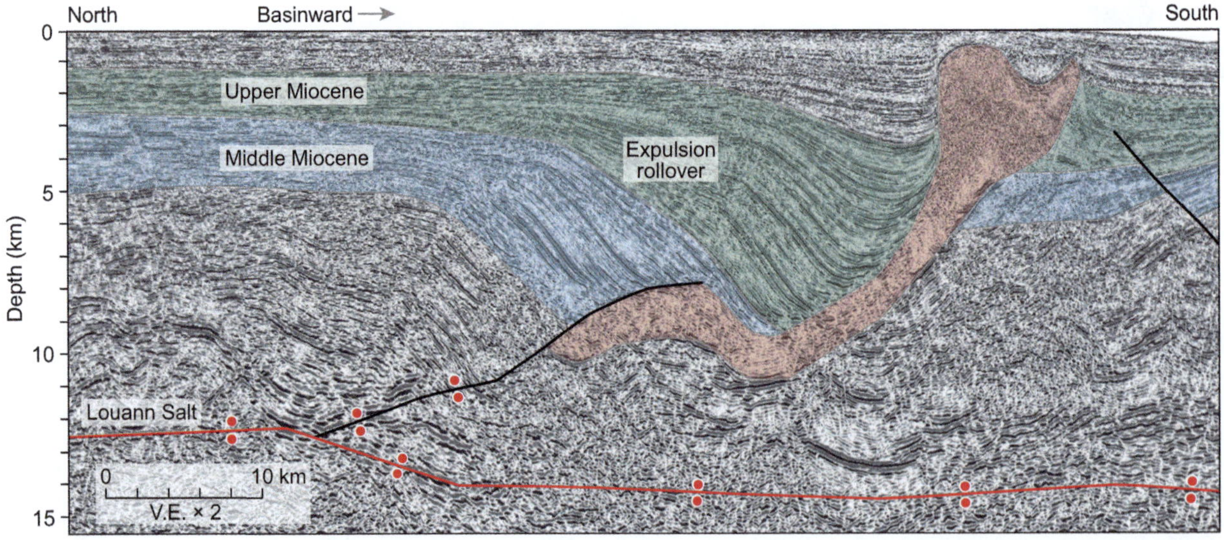


**FIG. 29** The expulsion rollover in this stepped counterregional system suggests that most of the salt was expelled in the middle-late Miocene. Salt escaped from the buried salt sheet through the basinward-leaning diapir at the seaward end of the counterregional structure. *(Modified from Jackson and Hudec (2017)).*

counter to the regional seaward dip of most normal faults on a passive margin. In map view, the seaward end of a counterregional system comprises a series of salt stocks joined by curved segments of welds or faults.

Although a counterregional system resembles a large landward-dipping normal fault, it forms primarily by salt expulsion during sediment progradation across the top of allochthonous salt and thus does not require much lateral translation (Fig. 28). The seaward-dipping sediment wedge is primarily an expulsion rollover, not a fault rollover. However, the upward propagation of normal faults beyond the tips of counterregional welds suggests that most counterregional systems probably involve at least minor amounts of slip. The lack of significant extension in most counterregional systems suggests that they are favored by settings having limited bathymetric relief, or areas where the roof was too thick and strong to fail in tension.

## Key messages on the regional geology of deepwater salt systems

There are gross similarities between all syn-exhumation deepwater salt margins. For example, they all involve relatively thick salt and lie at the downdip (compressional) end of linked kinematic systems. However, these generalizations conceal many complexities in deepwater-margin evolution, as well as considerable variation between margins. This section discusses some of the controls that regional geology can exert on the evolution of individual deepwater salt systems.

## Formation and propagation of deepwater compressional systems

Deepwater fold-and-thrust belts form where the downdip translation fed by updip extensional systems is impeded in some fashion. This impedance is most commonly caused by one of the following features (for additional information, see Chapters 3 and 4).

First, the downdip end of the salt detachment can act as a buttress to further seaward translation. In some cases, compressional structures form directly over the salt pinchout. In other situations, the seaward salt edge marks the root zone of a thrust fault that carries sediments further outboard. In either case, there is no translation on the surface beyond the edge of the weak salt layer.

Second, compressional systems may be located at any seaward flattening of the regional topographic slope. A particularly common type of fold belt occurs at the toe of the continental slope, although any regional change in topographic dip can localize shortening. Topographic slope is a major driver for salt-detached deformation, so changes in slope marks a change in stress state, and thus a potential locale for shortening. The likelihood that a slope change will localize shortening depends on both the magnitude of the dip change and the strength of the overburden, with large changes in topographic slope and thin, weak roofs both tending to localize strain.

Third, any buttress affecting the salt detachment disrupts seaward translation, and can thus localize shortening. Buttresses can include fault offsets of the base of salt, volcanic edifices or flattening (or reversal) of the detachment dip. Particularly complex geometries can form where buttresses are small compared with the length of the margin. In these situations, compressional structures form in small areas, whereas neighboring areas continue to translate seaward unimpeded. This creates fold-and-thrust belts with short strike lengths, separated from the rest of the margin by strike-slip systems.

Physical modeling (summarized by Brun and Fort, 2011) suggests that fold-and-thrust belts localized by structural buttresses (disruptions in the base of salt or the edge of the salt basin) propagate dominantly updip, away from the buttress. Fold-and-thrust belts localized by changes in bathymetry, on the other hand, shift with changing bathymetry. Thus, fold belts may shift downdip through time in a prograding margin, as the toe of slope moves seaward. These patterns of propagation can be complicated even further if the fold belts themselves are moved seawards by continuing translation.

Hudec and Jackson (2004) argued that shortening in the Kwanza Basin, Angola occurred through advance of a salt nappe at the seaward edge of the salt basin over most of the basin's history (Fig. 30). However, acceleration of translation due to continental uplift in the Miocene led to a much broader zone of shortening, initiating shortening 170 km (106 miles) updip of the salt nappe (Fig. 30C). Later, burial of the toe of the nappe in the Pliocene stopped nappe advance completely, allowing the broad zone of shortening to continue (Fig. 30D). The partitioning of shortening between the different potential loci was thus controlled by changes in regional slope and depositional patterns.

Like the Kwanza Basin, many passive margins have fold-and-thrust belts localized by combinations of features, forming complex patchworks of compressional systems. These systems can be difficult to unravel, especially if fold-and-thrust belts are formed at different times on different parts of the margin. It is quite true to say that deepwater salt systems tend to be dominated by compressional structures, but this generalization can conceal a great deal of complexity in time and space.

## Timing of shortening

Although deepwater salt systems lie at the downdip ends of passive margins throughout their history, observations from many basins indicate that they have not undergone continuous contractional deformation from their formation to the present day. Rather, they experience shortening only during those times when the forces driving seaward translation are sufficient to overcome the strength of the suprasalt sedimentary section.

The forces driving seaward translation are typically divided into two types: forces related to seaward dip of the salt detachment (gravity gliding), and forces related to seaward dip of the sediment surface (gravity spreading). There has been, and continues to be, considerable debate as to the relative importance of these two forces (Peel, 2014). However, this division serves as a convenient framework for discussion of timing of deepwater shortening. In general, shortening events in deepwater tend to be driven by some combination of tilting (increasing the tendency for gravity gliding; Fig. 30C) and sedimentation at the updip end of the system (increasing the tendency for gravity spreading).

In the northern Gulf of Mexico, Hudec et al. (2013) concluded that tilting and sedimentation were both required to drive deepwater fold-and-thrust belts. They suggested that the observed diachroneity in deepwater shortening arose as a result of shifting in depositional systems around the basin margin. In the northwestern Gulf of Mexico, major sediment influx

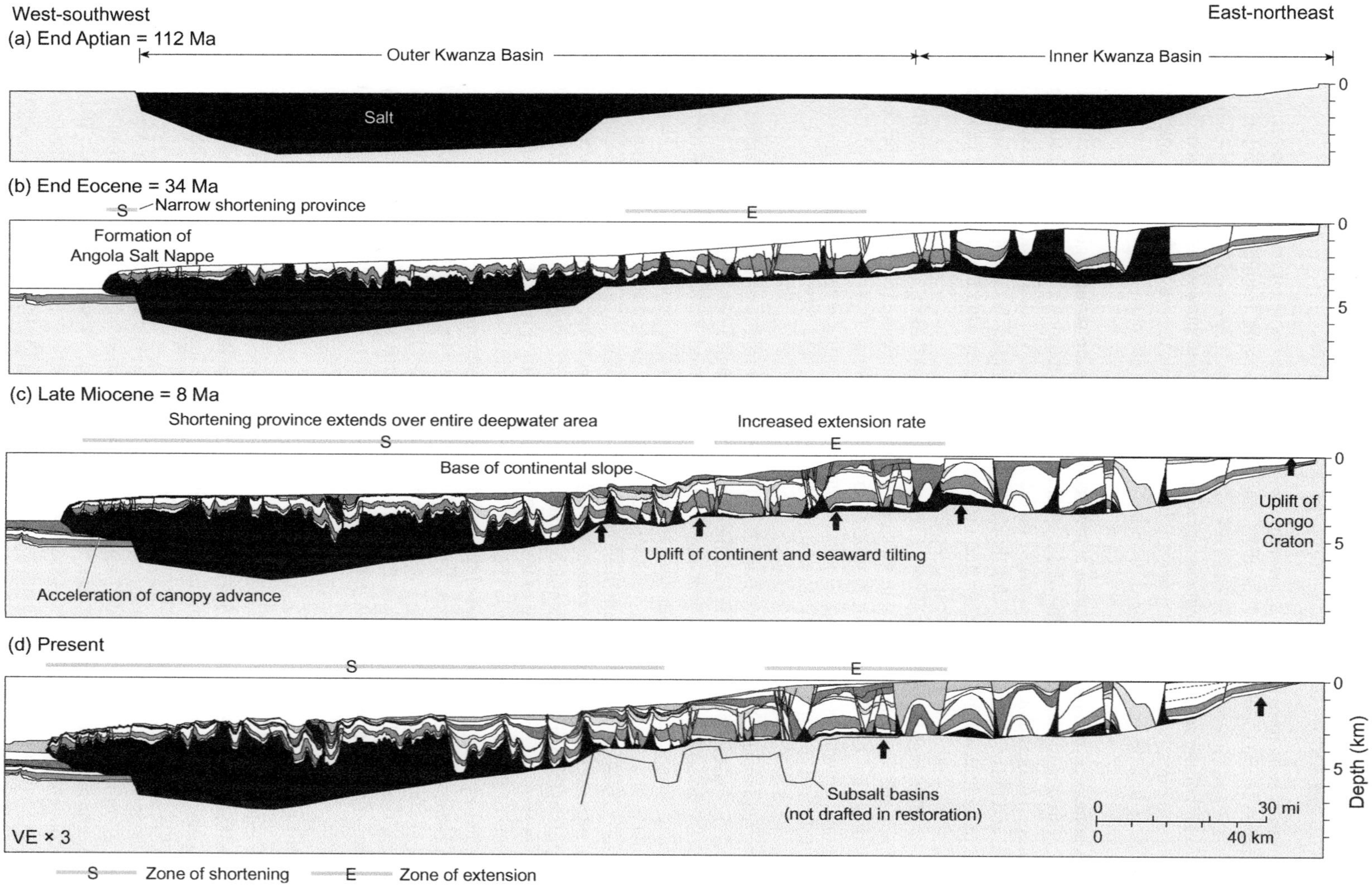


**FIG. 30** Sequential restoration of a section across the Kwanza Basin, Angola. (A) Aptian salt was deposited into two thick-salt basins, separated by a basement high. (B) Over the first 80 m.y. after salt deposition, slow extension on the upper slope led to slow overthrusting of the Angola salt nappe. (C) Continental uplift in the Miocene tilted the margin seaward, accelerating updip extension, translation and overthrusting of the Angola salt nappe. The zone of downdip shortening also widened dramatically. (D) A change in deepwater sediment distribution buried the toe of the Angola salt nappe, preventing it from advancing further. This enhanced the tendency for shortening to be accommodated further updip. *(Modified from Hudec and Jackson (2004)).*

occurred in updip areas during the Paleocene to early Oligocene. However, shortening in the downdip Perdido fold belt did not begin until the late Oligocene, when the margin tilted seaward. Further to the east, shortening did not occur during late Oligocene onset of tilting. Rather, large-scale shortening did not begin until major deltas were deposited during the Miocene. Fold-and-thrust belts are thus of different ages in the northeastern and northwestern parts of the Gulf of Mexico because the necessary driving forces were not in place at the same time.

Complex timing of shortening in deepwater fold-and-thrust belts is the rule, not the exception, especially when margins are long enough to have differences in geologic history along strike. Changes in depositional history, subsidence history, margin uplift or base-salt deformation along strike can all affect the timing of shortening in deepwater. Care should be taken in assessing the details of timing relationships along such margins: what is true along one part of the margin may not be true elsewhere.

## Canopy development and partitioning of translation between detachments

The deepwater regions of passive margins with syn-exhumation salt are likely sites for the development of salt sheets and salt canopies, due to the thick salt. However, sheets and canopies are not always developed, even on margins that have very thick salt. For example, the Kwanza basin, Angola, and Santos basin, Brazil, both have extremely thick salt but do not have widespread canopy development aside from the source-fed thrusts that have formed along the downdip edge of the salt basin. On the other hand, both the northern and southern Gulf of Mexico salt basins, plus the Lower Congo basin, West Africa have widespread canopy development throughout the deepwater regions. What governs the presence or absence of salt canopies in a deepwater passive margin?

To begin, one factor favoring canopy development is deposition of a large thickness of sediments on top of autochthonous salt. Thicker sediment piles compact and become denser, thus expelling more of the underlying salt into diapirs, and expelling this salt progressively more rapidly. When enough salt is expelled into diapirs at a high enough rate, they may rise fast enough to break out onto the surface to form salt sheets. When enough salt sheets form, they may amalgamate to form canopies. Salt expulsion under a thick sedimentary load can explain many of the canopies that lie beneath the present northern Gulf of Mexico shelf and slope, where suprasalt sediment thicknesses can locally exceed 15 km (9 mi).

However other margins, such as the southern Gulf of Mexico, have very well-developed canopy systems without such enormous thickness of sediment. What caused canopy formation in these areas? A common feature of thinner-overburden margins with widespread canopy development is widespread shortening. Shortening favors canopy development by squeezing diapirs and expelling their salt upward into salt sheets. Furthermore, diapir squeezing also breaks up the diapir roofs, creating free pathways toward the surface for any salt still trapped in the source layer at depth. Hudec and Jackson (2006) noted a worldwide correlation between deepwater shortening events and salt-sheet development. If enough salt sheets are expelled during shortening, canopies form. Thus, the canopies in the Lower Congo basin (Oligocene-Holocene) and southern Gulf of Mexico (Miocene-Holocene) both formed during major shortening events.

Once canopies have formed, the kinematics of passive margins may change dramatically. The canopies can form an alternative, shallower detachment level for seaward translation of sediments. In some cases, the shallower detachment may include several different canopies, connected by shale-based detachments in intercanopy areas. The deeper autochthonous-salt detachment may become inactive, because the thin cover over the shallower detachment is easier to deform than the full-thickness cover above the deeper detachment.

However, other considerations may keep the deeper detachment active in some situations. To begin, autochthonous-salt detachments tend to be much more widespread than the overlying canopies, so can accommodate translation over much wider areas. This can lead to complex situations in which the downdip, compressional toe of an autochthonous-salt detachment may be overlain by the updip, extensional breakaway zone of a shorter canopy-based detachment (Dooley et al., 2013). In other cases, the broader mechanics of the entire margin may favor activation of the deeper detachment (Gradmann et al., 2009). The development of canopies thus provides multiple options for seaward translation of sedimentary intervals, greatly complicating the regional picture.

## Zonation of deepwater salt structures

Salt thickness is one of the primary controls on salt flow, so systematic variations of salt thickness in deepwater regions can produce clear provinces in salt structural style. In general, salt thickens gradually toward the downdip end of syn-exhumation salt margins (Fig. 3D). This thickening is a product of both original depositional thickness of salt and post-depositional downdip flow of salt. However, there are several situations in which this pattern can be perturbed.

First, fault offsets in the base of salt during salt deposition can lead to abrupt changes in the salt thickness. This occurs in the deepwater Campos and Santos basins, Brazil, where depositional salt thickness is interpreted to have varied by 1–2 km across a series of basement faults (Davison et al., 2012). These steps in the base of salt influenced the seaward flow of salt, and have also played a role in sediment accommodation, with large minibasins forming above areas of thick salt. Another location where fault offsets play a major role in governing salt tectonics is in the eastern Gulf of Mexico (offshore Florida and Yucatán; Hudec and Norton, 2019). At the very downdip ends of these areas, salt thickened by 1–2 km (0.6–1.2 miles) into the outer troughs, producing diapir provinces in offshore Yucatán, in an area otherwise without diapirs.

Another situation in which salt thicknesses change abruptly in deepwater is where salt thins onto a source-fed thrust that carries salt out beyond the original edge of the salt basin. This scenario occurs in the northern Gulf of Mexico, where a Mesozoic-age source-fed thrust carried salt several tens of kilometers beyond the original edge of the basin. Salt is (as was) much thinner on this thrust than in the salt basin updip of the thrust. Hudec et al. (2013) argued that this thinning of salt at the downdip end of the basin produced zonation of salt structural styles, from large piercement diapirs and canopies in the thick-salt regions further updip to smaller salt cored folds near the downdip edge of salt. They argued that this zonation has influenced structural styles, seismic imaging and hydrocarbon prospectivity throughout the deepwater portions of the basin.

## Conclusions and future directions

Huge strides have been made in the past 30 years toward improving our understanding of deepwater salt tectonics. There are now have robust models for the evolution of salt sheets, salt canopies, salt stocks, salt anticlines, turtle structures and a host of other salt-related structures common in deepwater settings. However, there is still considerable uncertainty about how deepwater passive margins evolved during continental breakup, and how the salt structures were influenced by crustal separation. Recent research focuses on two aspects of the rifting process.

First, the location of crustal boundaries in deepwater is much debated (e.g., Rowan, 2020). The boundaries between thinned continental crust, exhumed lower crust, exhumed mantle and oceanic crust are in many cases uncertain, and difficult to resolve on currently available data. These boundaries have important implications for heat flow, and thus hydrocarbon maturation. Furthermore, depending on the timing of salt deposition with respect to crustal formation or exhumation, the creation of some types of crust may provide space for salt to flow into, causing updip extension without any obvious compressional structures at the downdip end of the margin.

Second, the original depositional configuration of the salt is poorly known (Hudec and Peel, 2019). Early models tended to assume that salt basins were deposited with a flat top, in many cases filling the basin all the way to sea level. If this were the case, reaching the present configuration in which the top of salt in deepwater is many kilometers below sea level would have required a rather spectacular collapse. This collapse would have had a major effect on the evolution of deepwater salt structures. However, more recent models have suggested that salt was deposited with a deposition slope on its upper surface, so that the updip end of the basin was near sea level while the downdip end was in many kilometers of water depth. If this is true, no major salt collapse occurred, and the earliest history of the basin was not nearly as dramatic as previously supposed.

As seen earlier, there are still many unknowns related to the earliest evolution of syn-exhumation, passive-margin salt basins. However, understanding of the subsequent evolution and present-day structure is quite mature. There are comprehensive models for the evolution of all major structural styles, and large numbers of analogs to assist in the seismic interpretation of present-day structures. There is also a strong understanding of how passive-margin evolution influences the development of fold-and-thrust belts, shallow detachments and structural zonation in deepwater settings. This knowledge of deepwater salt tectonics is an invaluable tool in understanding the history and evolution of deepwater sedimentary systems.

## References

Brun, J.-P., Fort, X., 2011. Salt tectonics at passive margins: geology versus models. Mar. Pet. Geol. 28, 1123–1145.

Callot, J.P., Jahani, S., Letouzey, J., 2007. The role of pre-existing diapirs in fold and thrust belt development. In: Thrust Belts and Foreland Basins, Frontiers in Earth Sciences, Part V. Springer, pp. 309–325.

Callot, J., Trocmé, V., Letouzey, J., Albouy, E., Jahani, S., Sherkati, S., 2012. Pre-existing salt structures and the folding of the Zagros Mountains. In: Alsop, G.I., Archer, S.G., Hartley, A.J., Grant, N.T., Hodgkinson, R. (Eds.), Salt Tectonics, Sediments and Prospectivity. Geological Society London, Special Publications, vol. 363, pp. 545–561.

Davison, I., Anderson, L., Nuttall, P., 2012. Salt Deposition, Loading and Gravity Drainage in the Campos and Santos Salt Basins. Geological Society London, Special Publications, vol. 363, pp. 159–174. no. 1.

Dooley, T.P., Hudec, M.R., Jackson, M.P.A., 2012. The structure and evolution of sutures in allochthonous salt. AAPG Bull. 96, 1045–1070.

Dooley, T.P., Jackson, M.P.A., Hudec, M.R., 2013. Coeval extension and shortening above and below salt canopies on an uplifted, continental margin: application to the northern Gulf of Mexico. AAPG Bull. 97, 1737–1764.

Duffy, O.B., Dooley, T.P., Hudec, M.R., Jackson, M.P.A., Fernandez, N., Jackson, C.A.-L., Soto, J.I., 2018. Structural evolution of salt-influenced fold-and-thrust belts: a synthesis and new insights from basins containing isolated salt diapirs. J. Struct. Geol. 114, 206–221.

Gealy, B.L., 1955. Topography of the continental slope in northwest Gulf of Mexico. GSA Bull. 66, 203–228.

Giles, K.A., Rowan, M.G., 2012. Concepts in Halokinetic-Sequence Deformation and Stratigraphy. Geological Society London, Special Publications, vol. 363, pp. 7–31. no. 1.

Gradmann, S., Beaumont, C., Albertz, M., 2009. Factors controlling the evolution of the Perdido Fold Belt, northwestern Gulf of Mexico, determined from numerical models. Tectonics 28, 1–28.

Graham, R., Jackson, M., Pilcher, R., Kilsdonk, B., 2012. Allochthonous salt in the sub-Alpine fold-and-thrust belt of Haute Provence, France. In: Alsop, G.I., Archer, S.G., Hartley, A.J., Grant, N.T., Hodgkinson, R. (Eds.), Salt Tectonics, Sediments and Prospectivity. Geological Society London, Special Publications, vol. 363, pp. 595–615.

Grando, G., McClay, K., 2004. Structural evolution of the Frampton growth fold system, Atwater Valley-Southern Green Canyon area, deep water Gulf of Mexico. Mar. Pet. Geol. 21, 889–910.

Hudec, M.R., Jackson, M.P.A., 2004. Regional restoration across the Kwanza Basin, Angola: salt tectonics triggered by repeated uplift of a metastable passive margin. Am. Assoc. Pet. Geol. Bull. 88, 971–990.

Hudec, M.R., Jackson, M.P.A., 2006. Advance of allochthonous salt sheets in passive margins and orogens. Am. Assoc. Pet. Geol. Bull. 90, 1535–1564.

Hudec, M.R., Jackson, M.P.A., 2009. Interaction between spreading salt canopies and their peripheral thrust systems. J. Struct. Geol. 31, 1114–1129.

Hudec, M.R., Jackson, M.P.A., 2011. The salt mine: a digital atlas of salt tectonics. In: The University of Texas at Austin, Bureau of Economic Geology. Udden Book Series No. 5; AAPG Memoir 99, 305 p.

Hudec, M.R., Norton, I.O., 2019. Upper Jurassic structure and evolution of the Yucatán and Campeche subbasins, southern Gulf of Mexico. AAPG Bull. 103 (5), 1133–1151.

Hudec, M.R., Peel, F.J., 2019. A critical review of models for deposition of the Louann salt, and implications for Gulf of Mexico evolution. In: Fiduk, J.C., Rosen, N.C. (Eds.), Salt Tectonics, Associated Processes, and Exploration Potential: Revisited: 1989-2019: 37th Annual GCSSEPM Foundation Perkins-Rosen Conference, pp. 222–238.

Hudec, M.R., Jackson, M.P.A., 2012. De Re Salica: fundamental principles of salt tectonics. In: Roberts, D.G., Bally, A.W. (Eds.), Regional Geology and Tectonics: Phanerozoic Passive Margins, Cratonic Basins, and Global Tectonic Maps. Elsevier, pp. 18–41. In press.

Hudec, M.R., Jackson, M.P.A., Peel, F.J., 2013. Influence of deep Louann structure on the evolution of the northern Gulf of Mexico. AAPG Bull. 97 (10), 1711–1735.

Jackson, M.P.A., Hudec, M.R., 2017. Salt Tectonics: Principles and Practice. Cambridge University Press. 498 p.

Jackson, M., Dooley, T., Hudec, M., McDonnell, A., 2011. The pillow fold belt: a key subsalt structural province in the northern Gulf of Mexico. AAPG Search and Discovery Article #10329, 23 p.

Jackson, M.P.A., Hudec, M.R., Jeanette, D.C., Kilby, R.E., 2008. Evolution of the Cretaceous Astrid thrust belt in the ultradeep-water Lower Congo Basin, Gabon. AAPG Bull. 92, 487–511.

Lickorish, W.H., Ford, M., 1998. Sequential restoration of the external Alpine Digne thrust system, SE France, constrained by kinematic data and synorogenic sediments. In: Mascle, A., Puigdefabregas, C., Luterbacker, H.P., Fernandez, M. (Eds.), Cenozoic Foreland Basins of Western Europe. vol. 134. Geological Society of London, London, pp. 189–211. In press.

Nikolinakou, M.A., Heidari, M., Hudec, M.R., Flemings, P.B., 2017. Initiation and growth of salt diapirs in tectonically stable settings: upbuilding and megaflaps. AAPG Bull. 101 (6), 887–905.

Peel, F.J., 2014. The engines of gravity-driven movement on passive margins: quantifying the relative contribution of spreading vs. gravity sliding mechanisms. Tectonophysics 633 (C), 126–142.

Péron-Pinvidic, G., Manatschal, G., 2009. The final rifting evolution at deep magma-poor passive margins from Iberia–Newfoundland: a new point of view. Int. J. Earth Sci. Eng. 98, 1581–1597.

Rowan, M.G., 2014. Passive-margin salt basins: hyperextension, evaporite deposition, and salt tectonics. Basin Res. 26, 154–182.

Rowan, M.G., 2015. Synexhumation salt basins: crustal thinning, subsidence, and accommodation for salt and presalt strata. In: Post, P.J., Coleman, J.L., Rosen, N.C., Brown, D.E., Roberts-Ashby, T., Kahn, P., Rowan, M.G. (Eds.), Petroleum Systems in Rift Basins, 31st Annual Gulf Coast Section SEPM Foundation Perkins-Rosen Research Conference, pp. 827–835.

Rowan, M.G., 2020. The South Atlantic and Gulf of Mexico salt basins: crustal thinning, subsidence and accommodation for salt and presalt strata. In: McClay, K.R., Hammerstein, J.A. (Eds.), Passive Margins: Tectonics, Sedimentation and Magmatism. Geological Society London, Special Publication, vol. 476, pp. 333–363.

Rowan, M.G., Lawton, T.F., Giles, K.A., Ratliff, R.A., 2003. Near-salt deformation in La Popa basin, Mexico, and the northern Gulf of Mexico: a general model for passive diapirism. AAPG Bull. 87, 733–756.

Schultz-Ela, D., 2003. Origin of drag folds bordering salt diapirs. AAPG Bull. 87 (5), 757–780.

Trusheim, F., 1960. Mechanism of salt migration in northern Germany. Am. Assoc. Pet. Geol. Bull. 44, 1519–1540.

Velaj, T., Davison, I., Serjani, A., Alsop, I., 1999. Thrust tectonics and the role of evaporites in the Ionian Zone of the Albanides. AAPG Bull. 83, 1408–1425.

Chapter 6

# Deepwater sedimentary processes

**Jon R. Rotzien[a,b], F. Javier Hernández-Molina[c], Marco Fonnesu[d] and Antoine Thieblemont[e]**

[a]*Basin Dynamics, Conroe, TX, United States,* [b]*University of Houston, Houston, TX, United States,* [c]*Department of Earth Sciences, Royal Holloway, University of London, Egham, Surrey, United Kingdom,* [d]*ENI, Rome, Italy,* [e]*TotalEnergies, Courbevoie, France*

## Introduction

This chapter explains the broad role of sedimentary processes and their linkage to the deposits found in deepwater sedimentary systems. Deepwater sedimentary processes are highly variable. Processes include pelagic and hemipelagic settling, mass movements and gravity flows as well as bottom currents, to biological and chemical forms of sedimentation. Pelagites are sedimentary intervals deposited by pelagic settling. Turbidites are sedimentary beds deposited by turbidity currents. Contourites are sedimentary deposits sculpted by bottom currents. These are just some of the end members along a continuum of deepwater sedimentary processes and their associated sedimentary deposits. The reason for understanding the diverse suite of sedimentary processes is simple: without a robust process-based understanding of deepwater sedimentation, facies associations and stratigraphic correlations cannot be interpreted with low uncertainty.

Sedimentary processes do not operate in a vacuum. In fact, sedimentary processes are not only controlled by sediment routing and oceanic water mass circulation, but also by large-scale continental drivers (Chapters 3–5) and many controls on a source-to-sink system (Chapters 11 and 12). Sedimentary processes provide a first-order control on resulting deepwater deposits and influence the ultimate success of a total petroleum system (Chapter 2).

From basin to basin, sedimentary facies models for each of the various depositional environments can be quite different (see Chapters 8 and 9 of this volume for more information on deepwater environments). This makes sense because every system is unique. These models are useful in that they show what is normal, and what is abnormal. A glimpse of sedimentary facies models from different basins reveals that some researchers apply modern entrenched terminology to describe sedimentation units (i.e., "this is a Bouma $T_b$ interval"), whereas others are fairly agnostic in the description of sub-bed-scale structures (i.e., "planar parallel lamination formed by upper stage flow regime," etc.). Both interpretations are equivalent because the observation and process match. Shanmugam (2020) highlights the various models enriched by frequent citation in the literature (described in Table 1 of Chapter 7; Bouma, 1962; Stow and Shanmugam, 1980; Lowe, 1982; Mutti, 1992; Haughton et al., 2009), yet emphasizes how observation should avoid interpretive terms implying the genesis of the deposit.

No model can completely explain the variety of deepwater sedimentary rocks. Furthermore, as additional data become available, scientists have realized that end member sedimentation units are rare. Scientists require and use many sedimentation models and will need to understand how many sedimentary facies are required to adequately model the stratigraphic interval of interest. Deepwater clastic rocks show a number of textures based on grain size distribution and flow processes (Lowe, 2004), which is the focus of this chapter. These textures have also been illustrated and summarized most recently by Pickering et al. (1989), Talling et al. (2012), Pickering and Hiscott (2015) and Stow and Smillie (2020), with an in-depth view of their ichnofacies fabrics provided in Chapter 15 of this volume.

This chapter is intended to serve four purposes: (i) to present the basic principles and physics of how the broad range of sedimentary deposits accumulate in a wide variety of deepwater environments; (ii) to provide context on the range and variability in sediment gravity flow processes and deposits, as well as bottom current processes, and why deepwater sediment transport and deposition is different from that of terrestrial and shallow marine environments; (iii) to provide information on foundational studies that have led to new understanding of sediment gravity flows and bottom currents; (iv) to illustrate that process is directly tied to deposit and the resulting porosity and permeability, and that these relationships are predictable.

**Deepwater Sedimentary Systems.** https://doi.org/10.1016/B978-0-323-91918-0.00021-9

# Sedimentary processes

## Hemipelagic and pelagic sedimentation

Pelagic sediment in the deep sea is largely produced by free-floating planktic and nektic organisms. Production of organic matter by plankton is due to the conversion of inorganic substances via oxygenic photosynthesis; see Fig. 1 and Chapter 2. The mineralized tests, or external shell, is tied to the euphotic environment with light, nutrients, temperature and population dynamics as the primary controlling factors. Phytoplankton, which are primary producers, consist of mineralized tests and include coccolithophorids constructed of calcium carbonate and siliceous diatoms constructed of silica. The major consumers are zooplankton. These represent planktic foraminifers and pteropods made of calcium carbonate and radiolarians made of silica (Hüneke and Henrich, 2011). All zooplankton and nekton also produce fecal pellets in the sedimentary record. Based on the composition, there are three main types of pelagic sediment: siliceous ooze, calcareous ooze and abyssal red clay (Chapter 7). Pelagic sediment distribution is primarily controlled by the distance from a continental landmass. This distance affects the dilution by detrital terrigenous components; water depth, which controls sediment preservation, and ocean fertility, which controls surface water productivity (Fig. 2; Thurman and Burton, 2001). During settling, pelagic particles may be dispersed laterally at various levels in the water column and along the seafloor by currents. Depending on size, density and shape, settling particles sink at various rates (from $<1$ to $>1000$ m/day ($\sim < 3$ to $> 3281$ ft.); Fowler and Knauer, 1986). At a water depth above the calcite compensation depth (CCD) or level (CCL), which is the depth at which the rate of supply of biogenic calcite equals its rate of dissolution (Fig. 3; Pälike et al., 2012), particles forming the sediment are composed largely of the calcareous or siliceous tests. At water depths below the CCL, only carbonate-free sediment accumulates, and pelagic sediment consists of siliceous tests or pelagic red clay (Hüneke and Henrich, 2011). Pelagic sediment accumulates at very low rates, typically less than 1 cm ($\sim$ 0.4 in.) per thousand years (Pickering and Hiscott, 2015) though it can be higher in seasonal "pulsed blooms" on highly productive basin margins (Shannon and Nelson, 1996) (Fig. 4). In areas of low surface productivity, very little biogenic material reaches the ocean

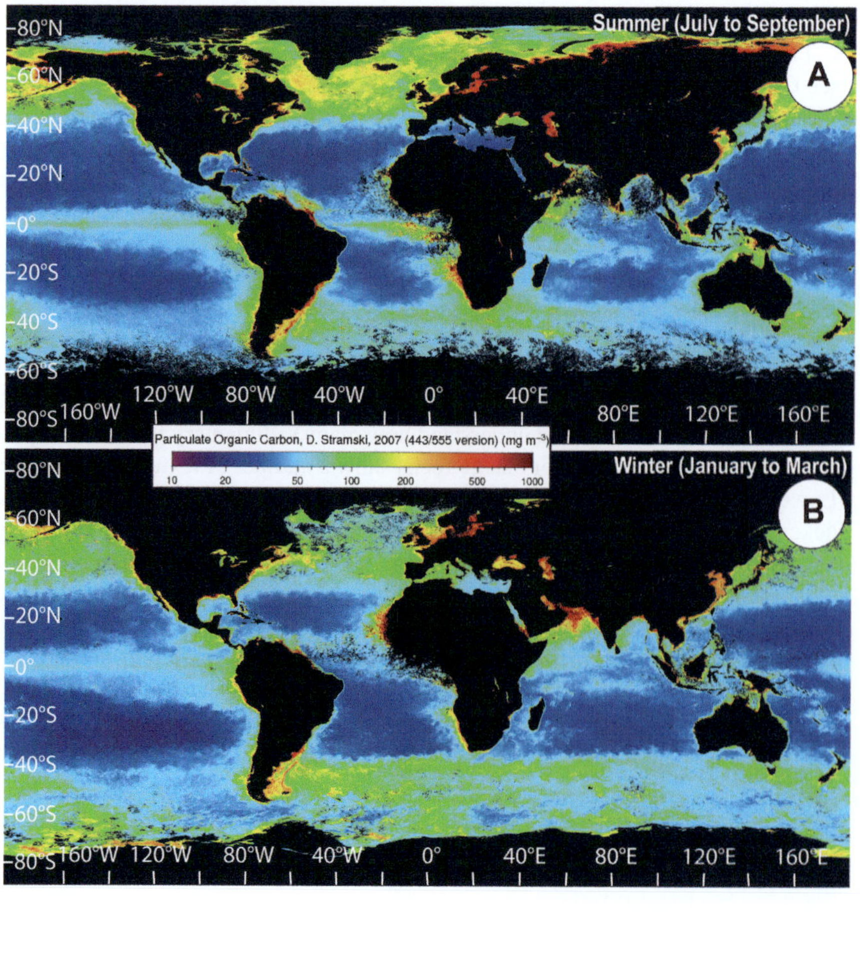


**FIG. 1** Concentration of particulate organic carbon (POC) in mg m$^{-3}$ for (A) Summer 2016 and (B) Winter 2016 calculated from Stramski et al. (2007). High POC values indicate areas with a higher sea-surface productivity.

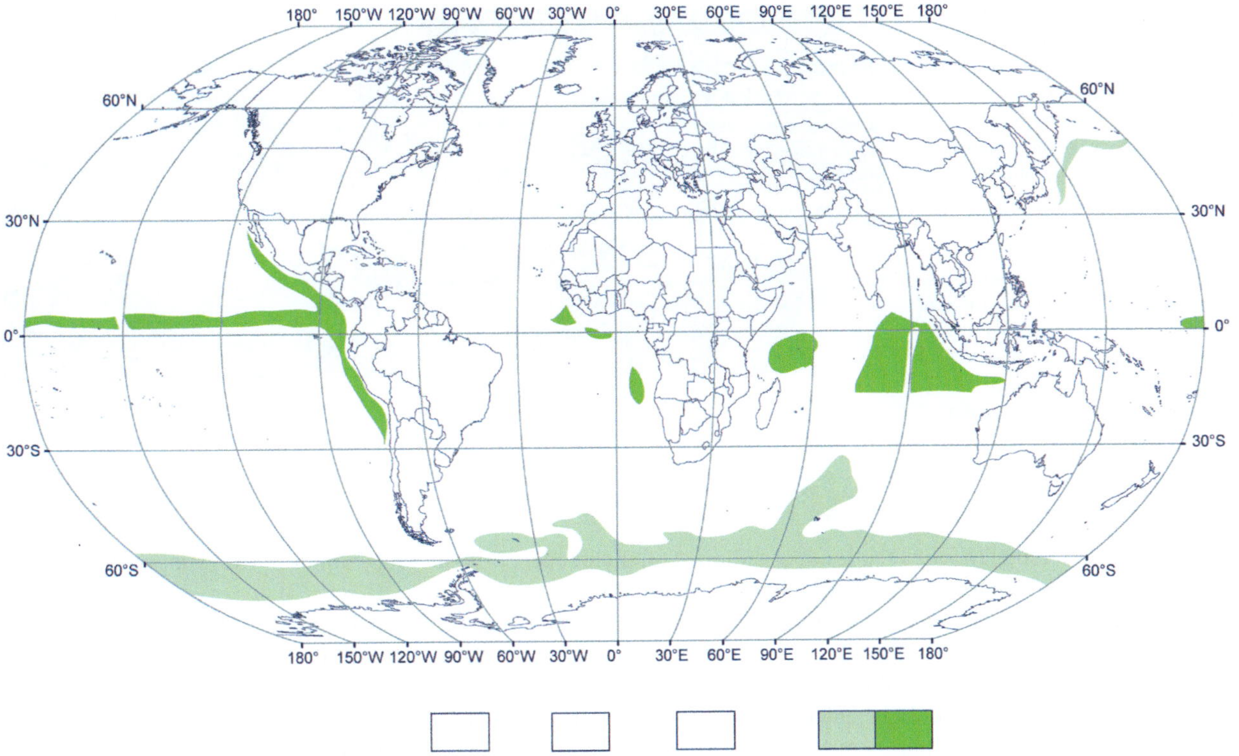


**FIG. 2** Global distribution of marine pelagic sediment types. *(Modified from Thurman, H.V., Burton, E.A., 2001. Introductory Oceanography, ninth ed. Prentice Hall, Englewood Cliffs.)*

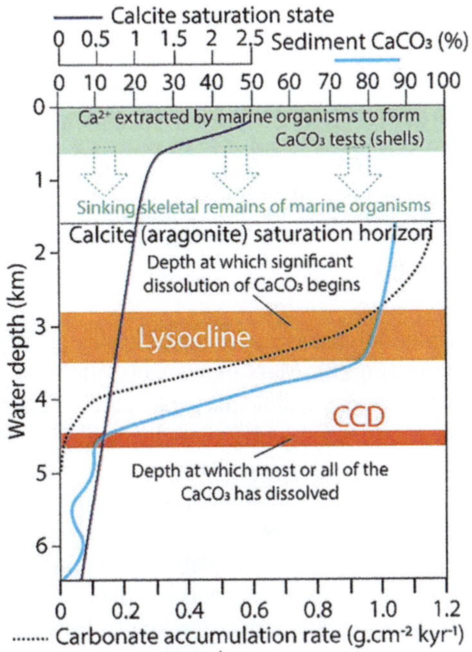


**FIG. 3** Illustration of the position of the calcite compensation depth (CCD) and lysocline, and their relationship to ocean bathymetry, carbonate accumulation rate (*dotted black line*), calcium saturation state (*dark blue line*) and $CaCO_3$ content (*light blue line*). Note that the depth of the lysocline, taken as the level below which there is rapid increase in calcite dissolution, and the CCD, which is the depth at which the rate of supply of biogenic calcite equals its rate of dissolution, vary within and between ocean basins. *(Modified after Pälike, H., et al., 2012. A Cenozoic record of the equatorial Pacific carbonate compensation depth. Nature, 488, 609–614. https://doi.org/10.1038/nature11360.)*

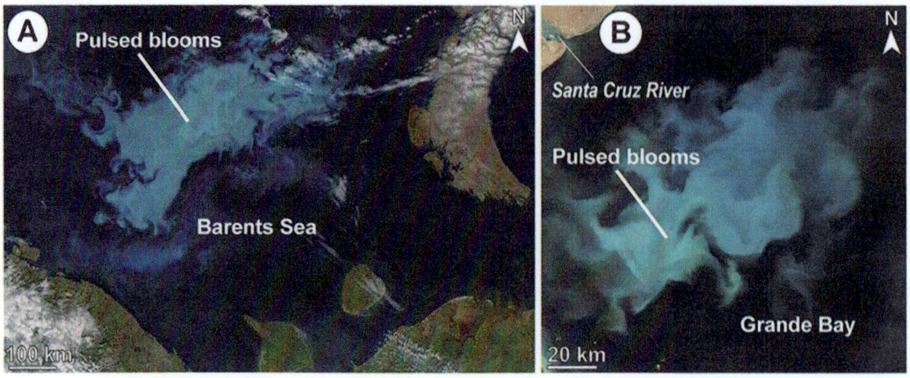


**FIG. 4** (A) Moderate Resolution Imaging Spectroradiometer (MODIS) image on NASA's Aqua satellite of blooms of coccolithophores in the Barents Sea, offshore Norway and Russia (acquired on July 6, 2016 by Jeff Schmaltz and Joshua Stevens). (B) Operational Land Imager (OLI) on Landsat 8 image acquired on January 2, 2021 of phytoplankton in Grande Bay, off Santa Cruz, Argentina. Bright green areas could be a mix of dinoflagellates, diatoms and coccolithophores; in the bluer areas, coccolithophores likely dominate. NASA Earth Observatory image courtesy of Lauren Dauphin.

floor, and abyssal red clay forms the residual sediment after carbonate dissolution (Chapter 7). Abyssal red clay typically exhibits the lowest sedimentation rates on Earth ($<0.5$ cm/ky or $<0.2$ in./ky).

Hemipelagic sediment is deposited along the continental margins (Chapter 3) and consists of biogenic material and a significant proportion ($>25\%$) of terrigenous or volcaniclastic material (Stow and Smillie, 2020). Because of the greater input of terrigenous siliciclastic sediment, hemipelagic settling has a high rate of sedimentation (5–15 cm/ky or 2–6 in./ky). This "normal" settling rate could be enhanced by physical mechanisms of flocculation, or particle aggregation, and biogenic processes such as the addition of fecal pellets (Stow et al., 2001). In many regions of the ocean, river supply is the most important contributor to terrigenous components. Terrigenous components are transported to the continental margins via surface plumes such as hypopycnal flows, or as a dilute suspension layer on the water column, or Nepheloid layers. These components can also be transported as a dense bottom layer or a sediment gravity flow. These are known as hyperpycnal plumes (Fig. 5; Walsh and Nittrouer, 2009). Although vertical settling is interpreted as the primary process for hemipelagic deposition, lateral advection by currents through the water column can also be responsible. In arid and semiarid tropical and subtropical ocean provinces, dust supply is the most important terrigenous contribution to hemipelagic sedimentation (Fig. 6; Middleton and Kang, 2017). In polar ocean provinces, sea ice, continental ice sheets and their outlet glaciers contribute to hemipelagic sedimentation. Sediment is incorporated into sea ice through (i) eolian transport; (ii) stirring up, resuspension and freeze-in of fine-grained terrigenous sediment in shallow shelf regions; (iii) river discharge onto sea ice and rock fall from cliffs in coastal areas (Fig. 7; Cuffey and Paterson, 2010; Domack and Powell, 2018) (Hüneke and Henrich, 2011).

## Introduction to mass movements

Mass movements transport sediment via gravitational forces acting on a body of sediment or rock. Mass movements include rock falls, creeps, slides, slumps and flows (Fig. 8). Mass movements often represent the precursor events to sediment gravity flows, and are mechanisms for long-distance sedimentary transport in deepwater environments. Most large-volume marine sedimentary deposits that are not hemipelagic, pelagic or chemical precipitates are originally derived from mass movements and gravity flows (Pickering et al., 1989). Since these mechanisms are driven by gravity, they transport sediment downslope and along the seafloor. If they have enough momentum, they can be extremely erosive, travel long distances over very shallow gradients—and even upslope.

Mass movements and gravity flows are implied as intergradational end members. Resulting deposits are often a combination of several processes active at different stages of transport and deposition (Posamentier and Martinsen, 2011). The products of the largest types of mass movements are submarine landslides known as *mass-transport deposits* or *mass transport deposits* (MTD) and *mass transport complexes* (MTC), discussed in detail in Chapter 10 of this volume. MTD mostly include sediment accumulations generated by mass movements (i.e., rock falls, creeps, slides and slumps) but also debris flows. MTC occur as the accumulation of one or more MTD and may also include deposits from turbidity currents (Bull et al., 2009). Often these types of deposits have sequence stratigraphic significance and form along sequence boundaries (see Chapter 12; for seismic examples on the North Slope, Alaska, USA see Chapter 16 of this volume).

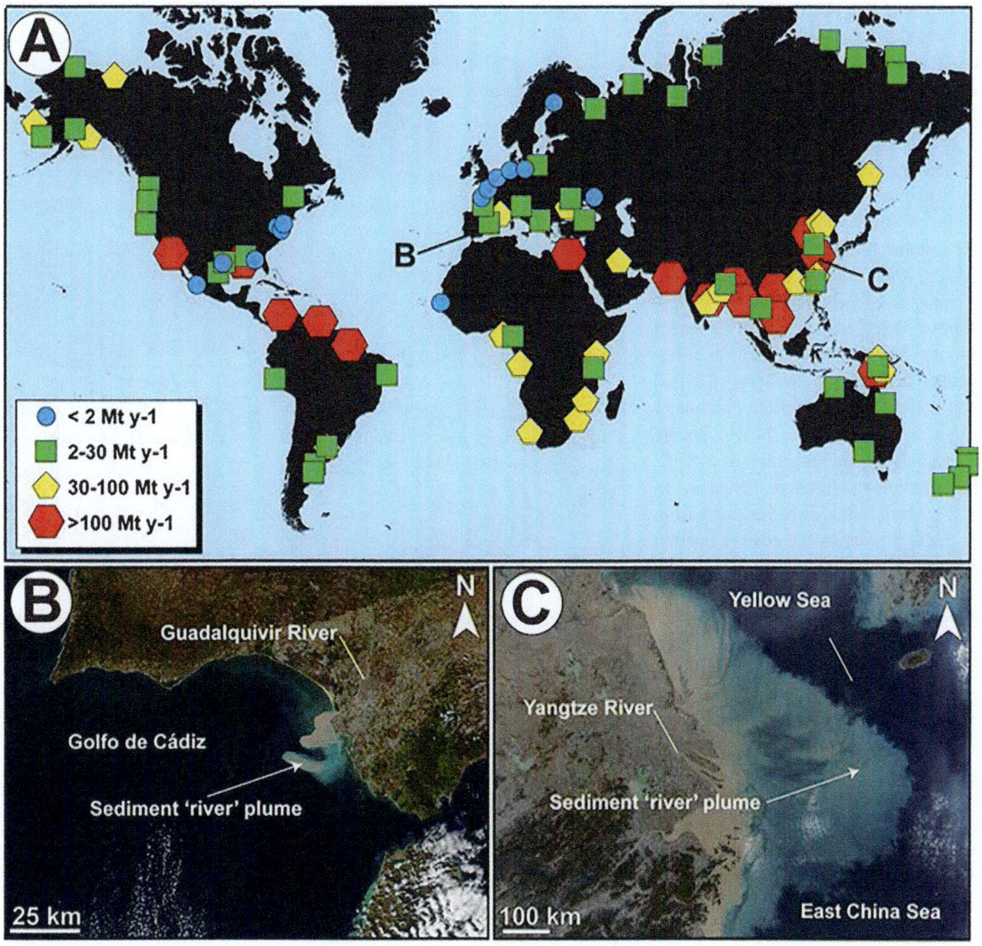


**FIG. 5** (A) Magnitude of sediment discharge at many large river mouths. Sediment discharge units are megatons per year (Mt $y^{-1}$). (B) Sediment "river" plume from the Guadalquivir River into the Golfo de Cádiz (Gulf of Cádiz) along Spain's southwestern coast. (C) Sediment "river" plume from the Yangtze River along China's Jiangsu province. Moderate Resolution Imaging Spectroradiometer (MODIS) on NASA's Aqua (B) and Terra (C) satellites captured these images on November 12, 2012 (B); November 9, 2017 (C). NASA Images courtesy of Jeff Schmaltz (B) and Jesse Allen (C). Locations of (B) and (C) are given in (A). *(Panel (A) Modified from Walsh, J.P., Nittrouer, C.A., 2009. Understanding fine-grained river-sediment dispersal on continental margins. Mar. Geol. 263, 34–45. https://doi.org/10.1016/j.margeo.2009.03.016.)*

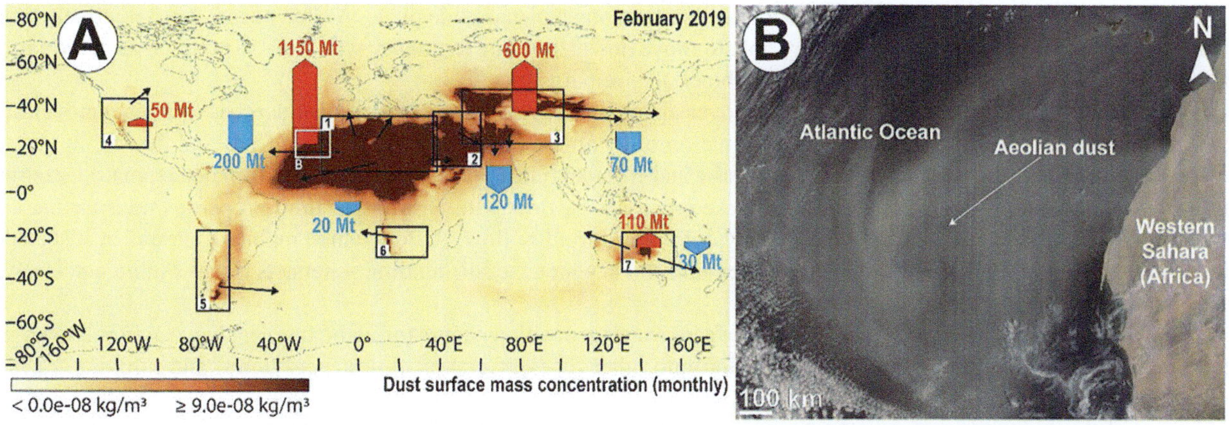


**FIG. 6** (A) Main routes of desert dust transport (*light blue arrows*) and locations of the major dust sources on the Dust Surface Mass Concentration (Monthly—February 2019) layer created from the M2TMNXAER (doi:https://doi.org/10.5067/FH9A0MLJPC7N). (1) Sahara; (2) Arabia; (3) Asia; (4) North America; (5) South America; (6) Southern Africa; (7) Australia. Dust emissions from different regions in Mt are indicated by *red arrows* and deposition to the oceans in Mt is indicated by *dark blue arrows*. (B) Dust storm swirled off the west coast of Africa (coastlines of Mauritania and Western Sahara, and the Canary Islands) and over the Atlantic Ocean. Moderate Resolution Imaging Spectroradiometer (MODIS) on NASA's Aqua satellite captured this image on January 15, 2012. NASA Image courtesy of Jeff Schmaltz. Location of (B) is given in (A). *(Panel (A) Adapted and modified from Middleton, N., Kang, U., 2017. Sand and dust storms: impact mitigation. Sustainability 9, 1053. https://doi.org/10.3390/su9061053.)*

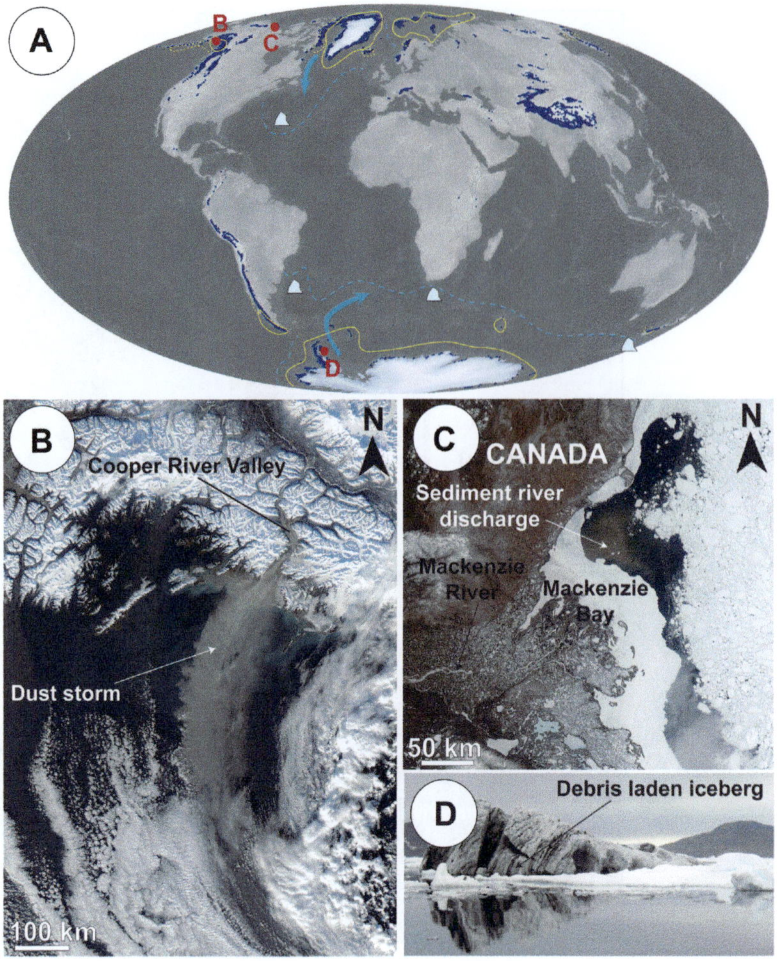


**FIG. 7** (A) Map of the global distribution of glaciers (within *yellow borders*). *Dark blue and white colors* highlight ice held in mountains regions and high latitude regional and continental ice sheets in Antarctica and Greenland. *Dashed light blue* line indicates the inferred generalized limit of iceberg drift, with the historical latitudinal limit of iceberg observations represented by *triangles*. Map by Robert Simmon. (B) Massive dust storm of glacial sediment off the coast of Alaska, USA, with Moderate Resolution Imaging Spectroradiometer (MODIS) on NASA's Aqua satellite on November 5, 2005. The plume comes from the Cooper River Valley. NASA Image courtesy of Jeff Schmaltz. (C) Sediment "river" plume from the Mackenzie River and its tributaries in the Canadian Arctic taken on May 30, 2020 with Moderate Resolution Imaging Spectroradiometer (MODIS) on NASA's Terra satellite. NASA Image courtesy of Lauren Dauphin. (D) Photograph of a debris laden iceberg observed along the Antarctic Peninsula. In general, release of fine glacial silt is dispersed away from the site of iceberg melting, whereas the coarser fraction (IRD) goes direct to the seafloor. *(Panel (A) From Domack, E.W., Powell, R., 2018. Chapter 7 - modern Glaciomarine Environments and sediments: an Antarctic perspective. In: Menzies J., van der Meer J.J.M. (2017) Past Glacial Environments (second edition), Elsevier, pp. 181-272. https://doi.org/10.1016/B978-0-08-100524-8.00030-0. Modified from Cuffey, K.M., Paterson, W.S.B., 2010. The Physics of Glaciers. Burlington, MA: Butterworth-Heinemann.)*

*Rock fall* is defined as the movement of loose debris that is removed from a steep slope and accumulates as talus along or near the base of the slope. Rock fall deposits consist of discrete blocks without internal deformation. Individual clasts or blocks can exceed hundreds of meters or even several kilometers in diameter.

*Creep* is a long-term, slow downslope movement of sediment without slip along a single detachment surface. Eventually creep leads to failure, and the process thus transitions into a slide or a slump.

*Slide* applies to a coherent mass of sediment that moves along a planar glide plane. A slide occurs when relatively undeformed strata of meter to kilometer scale are able to move downslope due to gravity with little to no internal deformation. Shearing occurs along a zone at the base of the block that remains intact.

A *slump* is a coherent mass of sediment that moves along a concave-up glide plane and undergoes rotational movements causing significant internal deformation and formation of compressional features including folds and thrusts.

A *flow* is a mixture of particles of sediment and water that can transport sediment via fluid turbulence, matrix strength, grain-to-grain collision—or a combination of these processes. A flow with Newtonian fluid properties transporting sediment via fluid turbulence is known as a turbidity current or turbidity flow. A flow which maintains grains in suspension due to grain collisions that generate dispersive pressure, and in which the surrounding water acts just as a lubricant, is a type of frictional debris flow named a grain flow.

A flow with plastic rheology and laminar state from which deposition occurs through "freezing" *en masse* is known as a cohesive debris flow. Debris flows often transition into other types of gravity flows (*mud flows*) and turbidity currents (Pickering et al., 1989; Mutti, 1992; Mutti et al., 2006) as more fluid is entrained in the flow.

In any of these mechanisms, initiation occurs when the shear at the base of the sedimentary interval exceeds the resistance of the detachment surface and is sustained by the development of a basal ductile shear zone. Triggers for mass movements include slope failure derived from sediment instability (arising from any number of factors including earthquakes and volcanic eruptions, or even storms), residual flows that follow debris flows and terrestrial river flooding events (forming

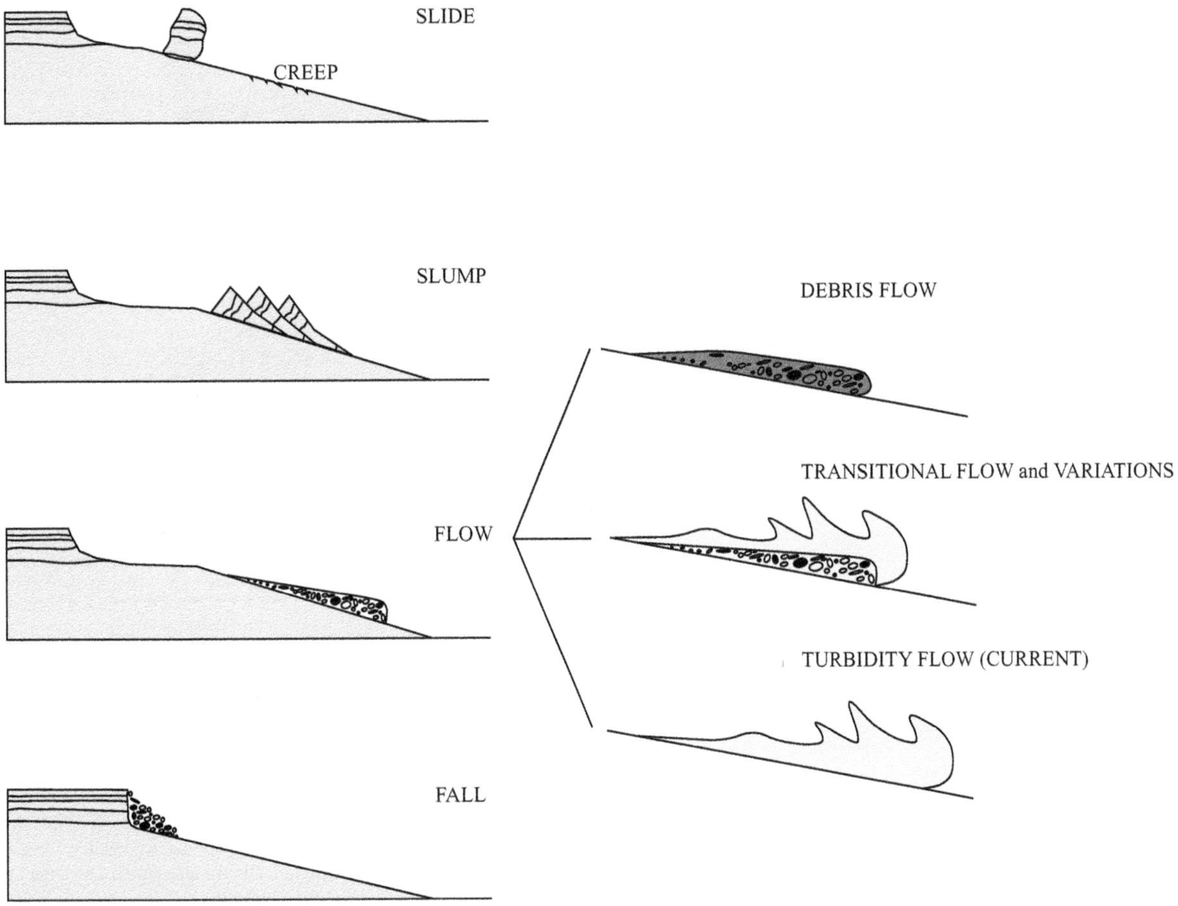


**FIG. 8** The various types of mass movements—slide, slump and falls—are precursors to sediment gravity flows, which can be represented by turbulent flow, debris flow or transitional flow. The types of flows create a certain type of deposit. Turbidity currents form turbidites; debris flows form debrites; transitional flows form transitional flow deposits. *(Diagram modified from Lowe, D.R., Graham, S.A., 2012. Turbidite Reservoirs and Depositional Systems. Short Course Lecture Notes. Stanford University, Approx 150 pp.)*

potential hyperpycnal flows—see Chapter 7). Rock fall, creep, slide and slump movements can serve as precursors for a sediment gravity flow, in which shearing penetrates the mass of actively deforming material. As the flow evolves, mud acquired by erosion or sourced from the partial disintegration of the sediment carried in the flow acts as a lubricant and creates an overpressure effect. The resulting reduction in friction among internal differential movements allows the mass to move quickly and chaotically in the form of a sediment gravity flow (Ogata et al., 2012).

## Sediment gravity flows

Sediment gravity flows are driven downslope by gravity acting on the solid particles, which are sustained by different supporting mechanisms. Middleton and Hampton (1973), classically subdivided the subaqueous sediment gravity flow types into *turbidity currents*, *fluidized sediment flows*, *grain flows* and *cohesive debris flows* as end-members types (see the following section). *Transitional flows* were later introduced to include flows in which cohesion is intermediate between frictional and laminar behavior (Baas et al., 2009).

One or more of these flow types may determine the structure of the flow when going downslope. The three principal parts of a typical sediment gravity flow, including the head, body and tail, vary in density, thickness and grain support mechanism along the length, width and height above the bed (Fig. 9). In an ideal setting along a smooth slope, the coarsest and densest fraction of the flow will form the nose or snout and is often a debris flow that runs out far ahead of the main body of the flow. Behind the nose, a much thicker head is formed by the billowing turbulent clouds, likely stratified with the high-density flow near the base and the lower density flow mixing with seawater above. Farther back, the tail of the flow is

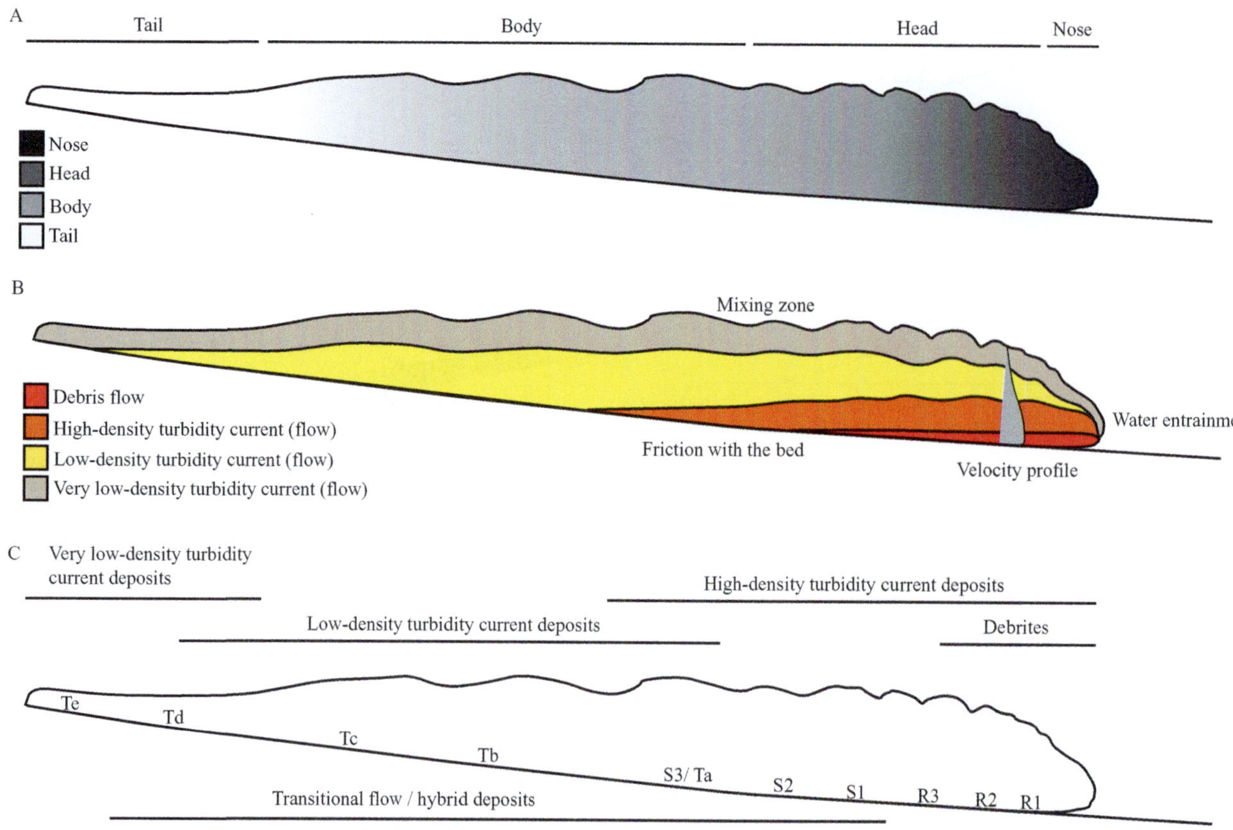


**FIG. 9** The anatomy of a sediment gravity flow indicates the type of transport and depositional mechanism, as well as resulting deposit. (A) Longitudinal anatomy of a sediment gravity flow shows the principal regions of the flow including the nose, head, body and tail. (B) All else equal, a sediment gravity flow can be partitioned into zones that favor transport by debris flow, high-density turbidity current, low-density turbidity current and very low-density turbidity current, which reveals (C) where in the flow typical sedimentation unit divisions are deposited. Note this is essentially a longitudinal mirror image of how sedimentation units and different facies are distributed in a deepwater environment (Lowe, 1982; Mutti, 1985, 1992).

thinner and has a lower sediment concentration. The tail decreases in thickness and sediment concentration until the end, or terminus, making the entire sediment gravity flow tens to hundreds of meters and even kilometers in length.

The partitioning or segregation of the rheology of the flow is not only longitudinal, but also vertical, revealing a classical internal stratification between a basal high-concentration flow and an upper low-concentration flow. These two parts of the flow are separated by a discontinuity with an abrupt gradient in velocity, density and viscosity (Postma et al., 1988). This is defined as bipartite turbidity current (Mutti, 1992), consisting of a relatively flat and smooth basal layer characterized by upward fluid escape, traction or cohesion and an overlying expanded and turbulent head (Manica, 2012) that forms due to the resistance encountered by the flow as it penetrates the ambient fluid (Fig. 9). The flow front's incorporation of ambient fluid and scour along the bed can cause substrate erosion and delamination leading to an increase in the size and density of the flow, also known as *flow bulking* (Fonnesu et al., 2016). The basal part of the flow is a dense zone of high pressure, causing fluid and sediment to move upward into the more turbulent part of the flow. The suppression of fluid pressure within the basal part of the flow causes deposition while the upper and rear parts of the flow, which are characterized by higher turbulence and large vortices generated by Kelvin–Helmholtz instabilities, may continue to bypass depositing sediment farther down depositional dip.

The model presented in Fig. 9 is a simplification of the structure of sediment gravity flow and best describes surge-like flows. When flows are prolonged to weeks instead of hours or minutes, the flow can stretch, which is typical for mud-rich turbidity currents including the Congo Canyon event measured by Azpiroz-Zabala et al. (2017).

It is important to understand the anatomy of a sediment gravity flow to be able to predict how and why it will deposit sediment preferentially in certain environments. This concept is expressed in how much sediment a flow can carry—its *capacity*—as well as what is the coarsest grain size a flow can carry—its *competency* (Kneller and Branney, 1995). Naturally, more robust flows are responsible for carrying and depositing the relatively coarse-grained sediment expressed as deepwater petroleum reservoirs. In turbidity currents, fluid turbulence is typically sufficient for suspending clay, silt and very fine-grained sand, yet additional mechanisms are required for coarser sand to gravel-sized grains. Additional inferences can be made on the physics of the flow including (i) whether it acted as a Newtonian fluid or Bingham plastic, (ii) its measure of turbulence or Reynolds Number (discussed in the following section) and (iii) the nature of flow steadiness and uniformity (Middleton, 1966, 1993; Lowe and Graham, 2012).

## Sediment gravity flow classification

Sediment gravity flows and flows originating from bottom currents show three types of dynamic behavior (Mulder and Alexander, 2001): fluid, mixed and plastic (Fig. 10). Fluid flows include sediment support mechanisms of turbulence, upward fluid flow and for the case of high-density gravelly flows, grain-to-grain interactions, and buoyancy effects. These mechanisms are relevant to all types of turbidity flows as well as liquefied or fluidized flows. *Turbidity currents* (also termed turbidity flows—the terminology is interchangeable) are characterized by Newtonian fluid rheology, in which sediment is supported by fluid turbulence and from which deposition occurs primarily through suspension settling and traction processes beneath a near-bed laminar sublayer. Upward fluid flow (fluidized flows) is not currently a mechanism that is invoked for long-distance transport of sediment. Debris flows include sediment support mechanisms of frictional grain collisions (grain flows) or plastic behavior due to the yield strength of matrix material (generally mud) to support cohesion between grains (cohesive debris flows). Mixed or transitional dynamic behavior—in which Newtonian fluid and Bingham plastic behaviors overlap—show both cohesive and grain-to-grain interactions as well as fluid turbulence during transport and deposition.

The flow classification scheme in Fig. 10 shows three primary dynamic behaviors for stand-alone flow types and associated particle support mechanisms. Though fluid turbulence provides sufficient particle support for sand, silt and clay grains in turbidity currents, a variety of sediment support mechanisms can operate simultaneously within the flow. It can therefore be difficult to deduce flow behavior from the sedimentary record. Considering the type of flows in which the cohesion is not the supporting mechanism, a key role is played by the flow density. Low-density flows are flows in which sediment concentration is $<10\%$ and are characterized by fully turbulent conditions while grain-to-grain support conditions dominate in high-density or *high-concentration* flows (sediment concentration $>25\%$) (Lowe, 1982; Talling et al., 2012). In between, moderate-density or *concentrated* flows ($\sim 10\%–25\%$ volume of sediment) combine grain-to-grain interactions, fluid-escape and fluid turbulence. Submarine flow classifications also need to differentiate between

<table>
<tr><th>Dynamic Behavior</th><th>Sediment Support Mechanism</th><th colspan="2">Flow Type</th></tr>
<tr><td rowspan="4">Fluid</td><td rowspan="3">Turbulence</td><td colspan="2">Very Low-Density Turbidity Currents</td></tr>
<tr><td colspan="2">Low-Density Turbidity Currents</td></tr>
<tr><td colspan="2">High-Density Turbidity Currents</td></tr>
<tr><td>Upward Fluid Flow</td><td colspan="2">Liquefied and Fluidized Flows</td></tr>
<tr><td>Mixed or Transitional</td><td>Mixed or Transitional</td><td colspan="2">Slurry Flows, Hybrid Event Flows, Argillaceous Flows</td></tr>
<tr><td rowspan="3">Plastic</td><td>Frictional Grain Collisions</td><td rowspan="3">Debris Flows</td><td>Frictional Debris Flows (Grain Flows)</td></tr>
<tr><td>Mixed or Transitional</td><td>Transitional Debris Flows</td></tr>
<tr><td>Cohesion Between Grains (Clays)</td><td>Cohesive Debris Flows</td></tr>
</table>

**FIG. 10** Sediment gravity flows can be described and divided into groups based on dynamic behavior, sediment support mechanism and flow type. Dynamic behaviors include fluid (Newtonian fluids), mixed or transitional and plastic (Bingham plastic). Sediment support mechanisms for fluid flows include turbulence and upward fluid flow or escape. Sediment support mechanisms for mixed or transitional flows include a combination of the previous mechanisms. Sediment support mechanisms for plastic flows include frictional grain collisions, cohesion between grains (clays) or mixed and can deposit sediment from frictional debris flows, transitional debris flows and cohesive debris flows. Sediment support mechanisms for mixed or transitional flows include a combination of the previous mechanisms and can deposit sediment from slurry-flows, hybrid flows and argillaceous flows, among flow variations that are currently being evaluated. *(Modified from Mulder, T., Alexander, T., 2001. The physical character of submarine density flows and their deposits. Sedimentology 48, 269–301. https://doi.org/10.1046/j.1365-3091.2001.00360.x; Lowe, D.R., Graham, S.A., 2012. Turbidite Reservoirs and Depositional Systems. Short Course Lecture Notes. Stanford University, Approx 150 pp.)*

fluidal and plastic flow behavior determined by the relative amount of dispersed clay. The latter has the matrix strength as a supporting mechanism exercised by the sticky proprieties of the clay particles present in the flow. Flows that have a fully cohesive behavior can be differentiated depending on the relative concentration and sorting of mud and other grain-size populations. These include mud flows, in which deposits have over 50% clay with the remaining parts split between silt and sand, and cohesive debris flows that contain >5% gravel with variable sand proportion (Mulder and Alexander, 2001). An intermediate class between fluidal and laminar flow behavior exists in mixed or transitional flow types. These include flows that are transient between these two regimes (Wang and Plate, 1996) and/or undergo flow transformation (Haughton et al., 2009). It is not possible to indicate an exact percentage of clay that can initiate the flow transformation because it also depends on flow velocity and grain concentration (Baas et al., 2009). However, Baas et al. (2009) suggests that clay concentration of at least 6% is required to induce transitional flow behavior.

### Bottom current processes

Bottom current processes are derived from water mass circulation and are important transport and sedimentary phenomena controlling deepwater (marine) sedimentation (see Chapter 9). Water masses are in constant movement in ocean basins (Rahmstorf, 2006; Kuhlbrodt et al., 2007) and any persistent current near the seafloor may be called a *bottom current*. Bottom currents are semipermanent water mass flows capable of eroding, transporting and depositing up to gravel-sized sediment on the seafloor, but most often clay, silt and fine-grained sand.

Primary bottom currents are related to thermohaline-induced geostrophic water mass circulation, but there are also a number of associated secondary processes such as benthic storms, overflows, interfaces between water masses, vertical eddies, horizontal vortices, tides and internal tides, internal waves and solitons, tsunami-related traction currents and rogue or cyclonic waves (Rebesco et al., 2014; Hernández-Molina et al., 2016; Yin et al., 2019). Some of these processes, including *internal waves*, are not well known and their consequences have not been fully considered though their associated energy and influence in shaping the seafloor is very important. Relatively dense water flows out into the ocean via narrow or shallow straits or over the continental margin and is steered to the right in the Northern Hemisphere and left in the Southern Hemisphere by Earth's rotation. When it is no longer constricted by the topography, it reshapes into a wider structure and adjusts under the forces of gravity, Earth's rotation and friction with the seafloor. The Earth's rotation tends to steer bottom currents to flow parallel to large-scale bathymetric contours, for example, along continental margins (see Chapters 3–5). Small-scale topography such as seamounts, ridges, straits, mounds, banks and canyons can disrupt a flow or cause it to accelerate. When the current velocity is sufficiently large, sediment is eroded and later deposited when velocity is reduced.

Currents can be either barotropic or baroclinic. *Barotropic flows* are currents with constant pressure surfaces parallel to surfaces of constant density, for example, tides, wind-driven currents and surface waves. In general, these flows are driven by pressure gradients caused by a sloping sea surface (Shanmugam, 2013). *Baroclinic flows* are currents with constant pressure surfaces not parallel to surfaces of constant density, such as internal waves and internal tides, that propagate on density surfaces known as *pycnoclines* in stratified waters. These flows are driven by pressure gradients caused by horizontal density differences and fronts. Baroclinic currents are found along the ocean floor of continental slopes and in submarine canyons.

Bottom currents are capable of generating depositional and erosional features at different scales (Heezen et al., 1959; Heezen et al., 1966; Heezen and Hollister, 1971; Faugères et al., 1999; Rebesco and Camerlenghi, 2008; Rebesco et al., 2014; Eberli and Betzler, 2019) and commonly are responsible for forming *Contourite Depositional Systems* (CDS) (Hernández-Molina et al., 2008). For more information about bottom current processes and their associated deposits see Chapter 9.

## Using processes to predict deposits

Deepwater flows and their associated processes generate a wide variety of deposits, and these processes can be simultaneously active, or partitioned, undergoing flow transformation (sensu Fischer, 1983; Waltham, 2004). Flow transformation causes a change in the rheology of the flow's support mechanism and depositional process. A normal chain of events can result in sediment gravity flow deposition as a debrite, high-density turbidity current deposit and low-density turbidity current deposit. But any combination can occur, as flows can go through multiple stages depending on how the flow evolves and its interaction with the substrate. For example, in the Miocene Marnoso-arenacea Formation, Italy, deposits within the same bed range from turbidites to debrites, capturing evidence for flow transformation (Amy and Talling, 2006). Alternatively, deposits found in other basins have documented the reverse order and various permutations (Mutti, 1992; Haughton, 2001; Yang et al., 2018; Talling, 2013). A flow's deposition ranges in character from grain flow, suspension, traction, cohesion or a mix of fluid turbulence and cohesion. Finally, the sediment gravity flow and its deposits can be influenced during transport or after deposition by other processes such as bottom currents.

## Facies tracts

Flow transformation and progressive bypass show how a flow can systematically deposit sediment along its length of transport (Fischer, 1983). Lowe (1982) and Mutti (1992) developed the concept of the *facies tract* as the deposits observed along the length of the same bed or bedset (sensu Campbell, 1967) on a flow's transport pathway (Fig. 11; Hussain et al., 2020). The vertical sequence of sub-bed-scale divisions (*first-order architectural elements* described in Chapters 1 and 7 of this volume) indicates the longitudinal fractionation of the flow through time (Mutti, 1992; Kneller and McCaffrey, 2003). Lateral distribution of first-order elements (see Chapter 1) records flow transformations that took place in 2D and 3D space (Mutti, 1992). Gravel-bearing flows tend to deposit first as a flow decelerates, with the clay depositing last due to Stokes' Law of settling velocity, forming facies including the deposits of cohesive debris flows and hyper-concentrated flows, gravelly high-density turbidity currents, sandy high-density turbidity currents and low-density turbidity currents. *Progressive bypass* is the concept that states coarse-grained sediment deposits first along the depositional pathway and is overlain by fine-grained deposits from the bypassing dilute part of the flow. During bypass, it is not uncommon for the overlying part of the flow to rework the previously deposited sediment.

Idealized facies tracts may not be preserved entirely if flows are starved of sediment or do not travel far into the basin (low flow efficiency), or if grain size variation is very low (no variety in sedimentary structures or facies tracts) (Fig. 12). The ability for a flow to carry sediment far into a basin (upward of a few thousand kilometers in the Amazon or Agadir basins offshore Brazil and Morocco, respectively), or *flow efficiency*, is primarily controlled by the amount of fine-grained sediment in the flow, which acts to enhance the buoyancy felt by grains in the flow. Additional fine material, such as mud, can be sourced from erosion of the bed in areas such as channels or the channel-lobe transition zone. The entrained clay may act to suppress turbulence near the bed and increase the cohesion of the flow (see Haughton et al., 2003, 2009), and/or the clay can be carried upward and incorporated into the overlying turbulent flow, thus sustaining the flow run-out (Tinterri et al., 2003). Highly efficient flows are often large flows carrying several km$^3$ of sediment with a substantial proportion of clay (>25%). These flows have relatively long intervals of sparse deposition or nondeposition until sediment is deposited,

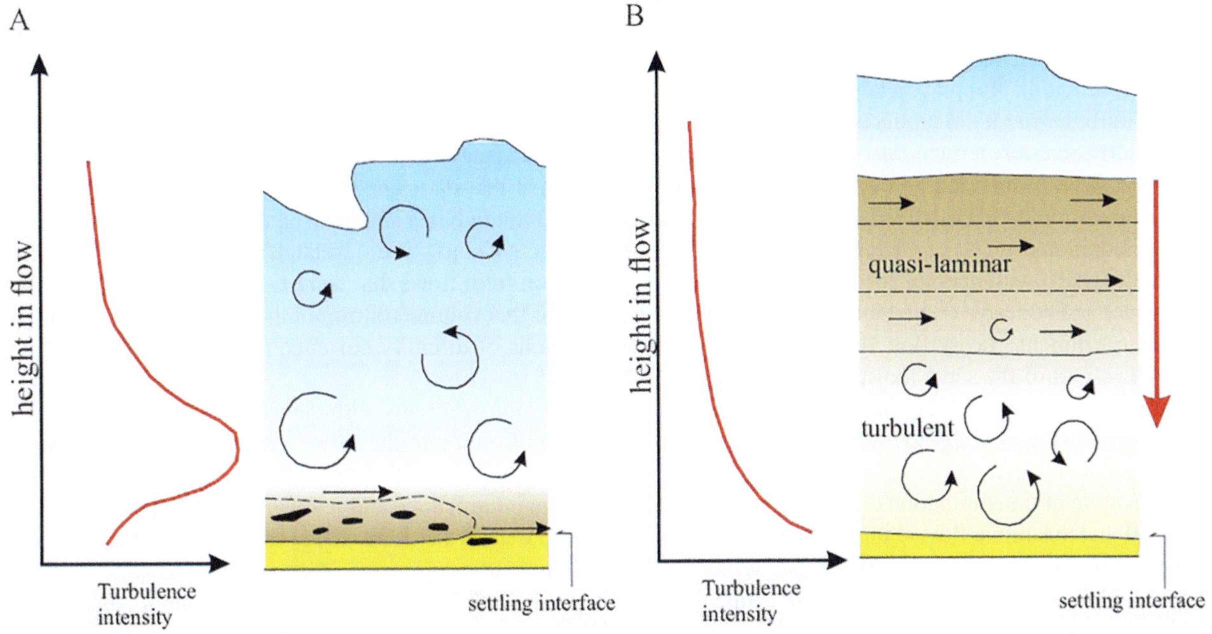


**FIG. 11** Schematic diagram representing alternative mechanisms of turbulence suppression in flows enriched in clay. (A) Longitudinal segregation model modified after Haughton et al. (2009) including a forerunning turbidity current followed by a debris flow; (B) vertical segregation including formation of quasi-laminar plug flow and basal turbulence enhancement modified after Baas et al. (2011). *(Modified from Haughton pers. comm; Hussain, A., Haughton, P.D., Shannon, P.M., Turner, J.N., Pierce, C.S., Obradors-Latre, A., Barker, S.P., Martinsen, O.J., 2020. High-resolution X-ray fluorescence profiling of hybrid event beds: Implications for sediment gravity flow behaviour and deposit structure. Sedimentology, 67(6), 2850–2882. https://doi.org/10.1111/sed.12722.)*

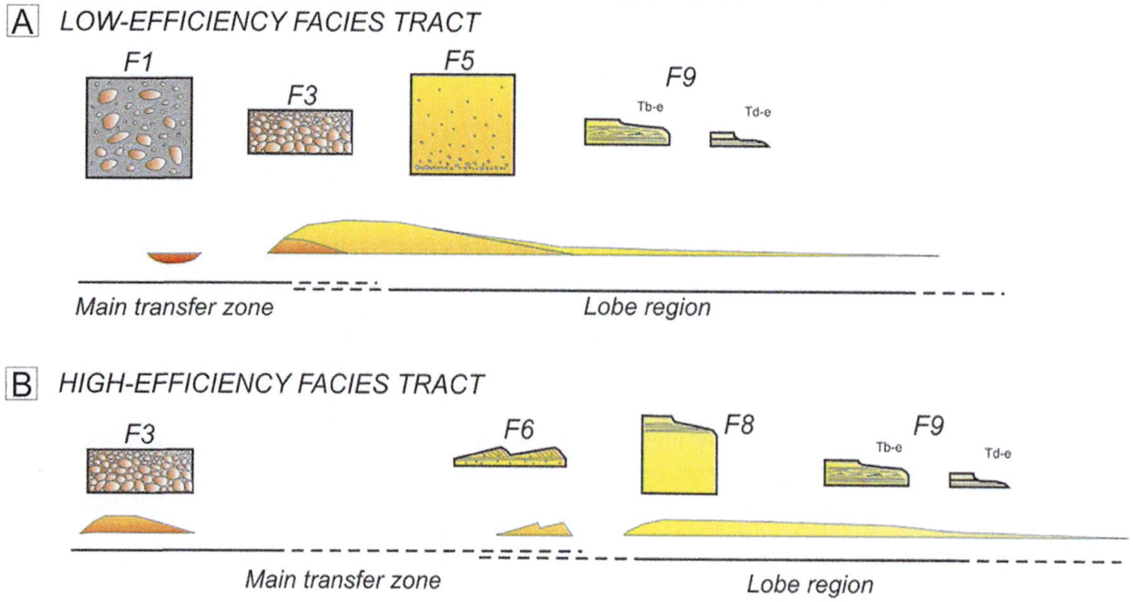


**FIG. 12** High- and low-efficiency flow and facies tracts based on Mutti (1992), which show the longitudinal segregation of facies based on the textural composition of the parent flow. High-efficiency facies tracts are formed by flows rich in silt and clay. These form detached facies tracts in which individual turbidite facies are efficiently segregated along the flow path. Sediment bypass is also a common trait. Low-efficiency facies tracts are formed by flows lacking significant amounts of clay and silt. These types of flows are typically unable to efficiently partition grain size. These flows deposit by freezing their sediment load soon after transformation without developing a full spectrum of facies. These flows form attached wedge-shaped beds.

known as *transfer zones*, and a facies tract that is discontinuous and not well connected, known as a *detached facies tract*. Low-efficiency flows are relatively small-volume flows containing coarser-grained sediment, with a much shorter transfer zone (i.e., sediment is deposited closer to the basin margin) and form an attached facies tract (Mutti et al., 1999).

## Flow transformation related to hybrid events

Flow transformation also plays a key role in defining the behavior of hybrid and transitional flows in which matrix support and fluid turbulence mix to produce a wide variety of muddy sandstone deposits. The transitional flow deposits studied in the UK North Sea were termed *slurry-flow* deposits due to their elevated clay content. Lowe and Guy (2000) interpreted that these beds were formed due to disaggregation of mud near the bed that dampened the turbulence and created a quasi-laminar flow from the elevated cohesion. Later, bipartite beds with sandy bases and debritic tops were termed *linked debrites* by Haughton et al. (2003) in the Western Irish Namurian Basin in County Clare, Ireland, and later described as hybrid event beds (HEB) with internal subdivisions representing deposition from flows that were fully turbulent, quasi-turbulent and laminar and cohesive (Haughton et al., 2009—see Chapter 7 of this volume for more information). HEB are a type of deepwater sediment gravity flow deposit emplaced by combinations of turbidity currents, transitional flows and debris flows, all as part of the same transport event (Fonnesu et al., 2015).

Two contrasting models for the dampening of turbulence in turbidity currents with elevated clay concentrations have been proposed (Fig. 11): (1) the incorporation and disaggregation of mud into the flow, increasing the surface area and distribution of clay that, in turn, increases cohesion of the flow (Lowe and Guy, 2000; Haughton et al., 2009). Flows undergoing this style of transformation often have a turbulent head and an intermediate and rearward part of the flow dominated by transitional to laminar flow regimes. (2) Sumner et al. (2009) and Baas et al. (2009, 2011) used open-channel flow experiments to demonstrate that turbulence damping is caused by the hydraulic separation of clay particles, which generates a sequence of *flow states*. Five flow states occur sequentially and include (Fig. 13): the turbulent flow, turbulence-enhanced transitional flow, lower transitional plug flow, upper transitional plug flow and a quasi-laminar plug flow. The damping of turbulence occurs from the top-down as the upper part of the flow has lower shear and is more easily dampened. The model also implies an early phase of turbulence enhancement determined by the onset of a plug flow (Baas and Best, 2002, 2007; Baas et al., 2009, 2011; Fig. 13B), which may occur in unison with phases of substrate erosion (Fonnesu et al., 2016). This second model highlights the role of the rate of deceleration of the flow based on the abundance of transitional flow deposits observed in or near channel mouth environments or topographic highs where flows would decelerate quickly.

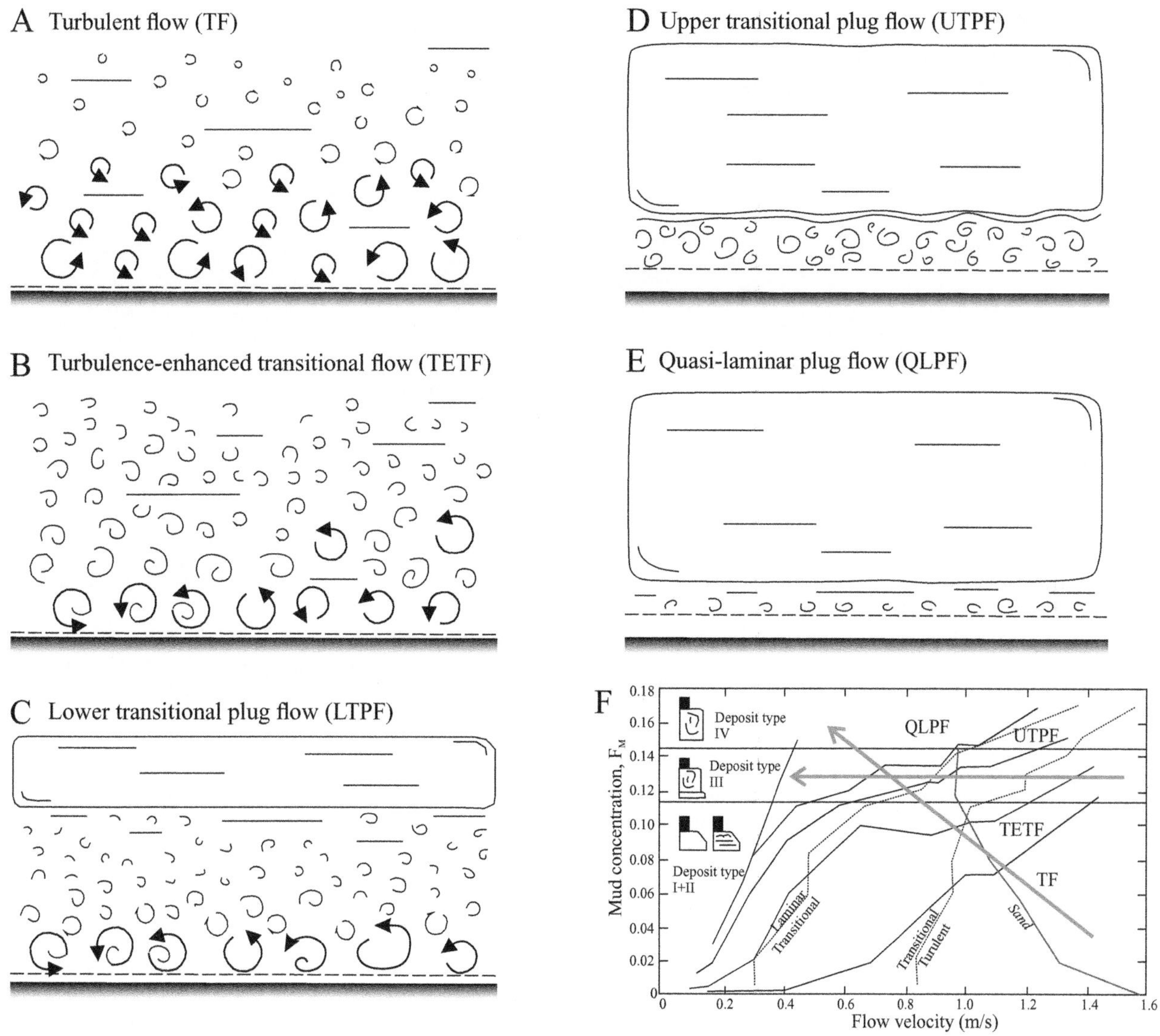


**FIG. 13** Transitional flow types as defined by Baas et al. (2009, 2011). (A) Turbulent flow (TF); (B) Turbulence-enhanced transitional flow (TETF); (C) Lower transitional plug flow (LTPF); (D) Upper transitional plug flow (UTPF); (E) Quasi-laminar plug flow (QLPF); (F) Phase-diagram showing the inferred ranges from different flow states in relation to the type of deposits observed (Baas et al., 2009). *Gray arrows* represent the trajectory through the phase diagram that would result in formation of hybrid event beds (Sumner et al., 2009). Deposits of type I and II comprise normally graded sandstone overlain by a mud cap (turbidite). Deposit type III comprises a clean sandstone overlain by an ungraded argillaceous sandstone and a mud cap. Deposit type IV comprises ungraded argillaceous sandstone and a mud cap.

## Examples of rock types

Sediment gravity flows generate a wide variety of deposits through various types of processes (Fig. 14) and will have different expressions in seismic data (see Chapters 16 and 17). Some of the primary intervals that function as subsurface porous reservoirs include deposits from a combination of turbidity currents, debris flows and flows that experience both fluid turbulence and matrix strength, such as slurry-flow deposits, hybrid event beds, transitional flow deposits or argillaceous sandstone beds (discussed in more detail in Chapter 7). Although these deposits do not always contribute to the total net, in terms of net sand in a gross interval, their presence controls reservoir heterogeneity, and reservoir porosity and permeability. In other cases, deepwater sediment gravity flow deposits enriched in argillaceous material may form internal seals or baffles, compartmentalizing the reservoir or contributing to different pressure regimes. Furthermore, recent discoveries found that the interaction of along-slope gravity flows with across-slope bottom currents processes may form mixed deposits (Shanmugam et al., 1993a, 1993b; Sansom, 2018; Rodrigues et al., 2022; see Chapter 9). In some cases,

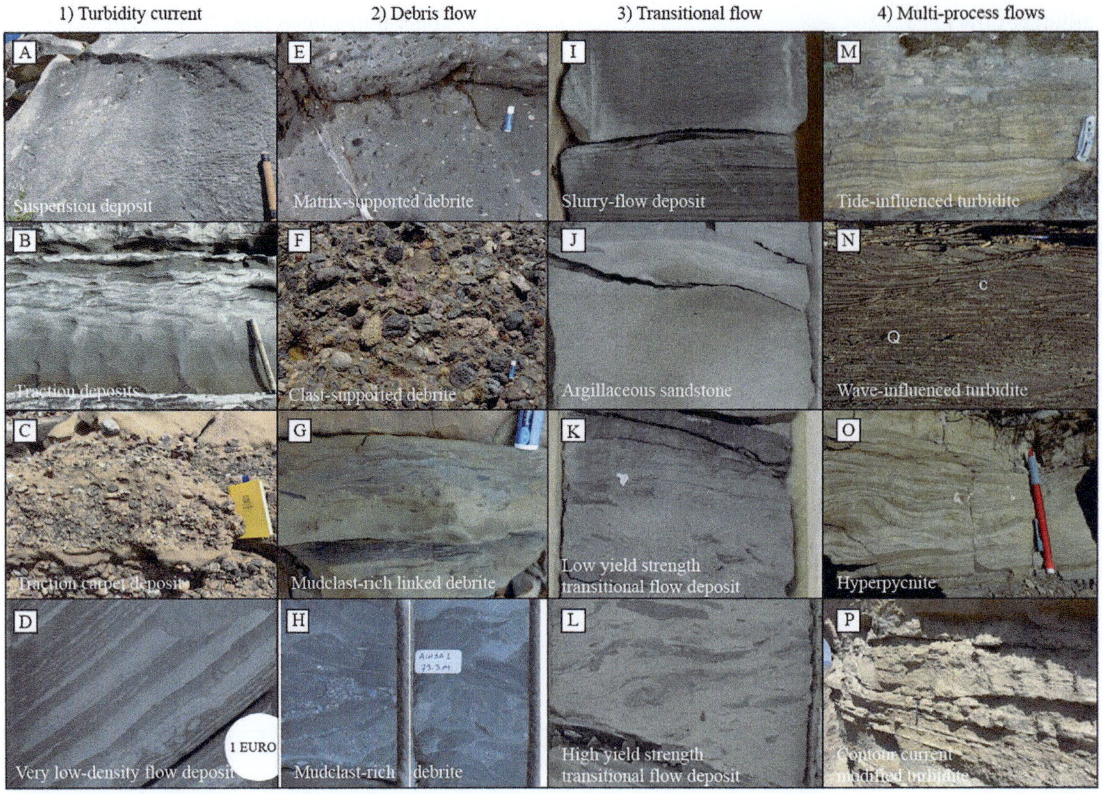


**FIG. 14** Rock fabric variability for principal classes of sediment gravity flow deposits including (1) turbidites, (2) debrites, (3) transitional flow deposits and (4) sediment gravity flow deposits that are influenced by terrestrial, shallow marine and bottom current processes. It shows different fabrics based on grain size and process of sedimentation. Wave-influenced deposit from Myrow et al. (2002). These fabrics reveal the sedimentation process and are strongly tied to the depositional environment. It is important to recognize these fabrics as a wholly separate (i) sedimentation unit or (ii) sub-bed-scale architectural element as part of a more organized sedimentation unit (example: a debrite as an H3 division, part of the Haughton et al. (2009) hybrid event bed sedimentation unit vs a debrite bed—the former tells potentially more about the depositional environment and grain size variation than the latter). This figure shows a more "granular" approach to classifying fabrics that are closely associated with main deposits, but may be part of any of the key deposit types (turbidite, debrite, transitional to hybrid event bed, other process). Turbidite intervals show a range of fabrics: (A) structureless or massive, indicating deposition from collapse of a fully turbulent flow from suspension, with dish and pillar structures resulting water elutriation following a grain flow, yet grain flow is not a robust mechanism for transporting large volumes of sediment over long distances; (B) structured sandstone including cross-stratified, planar laminated and ripple cross-laminated intervals formed by traction sedimentation; (C) inversely graded sandstone and conglomerate deposited as traction carpets via frictional freezing; (D) laminated and cross-laminated structures from traction and suspension from very low-density flows, which can be created by overbanking flows (and others in channel) that interact with other bottom currents (* large variability here from sediment waves to contour currents). Debris flow deposits show: (E) matrix-supported (with gravel clasts) and muddy (solely contorted mudstone) deposited via laminar, cohesive debris flow with matrix strength; (F) clast-supported debris flow (sometimes R1 deposit in Lowe, 1982) deposited via frictional freezing or as more loosely defined mass flow such as a rock fall/avalanche/flow in a high-gradient area, such as basinward of a steep slope; (G) mudclast-rich-linked debrite with relatively large mudclasts and a typically muddier matrix—this can also be anywhere in the sedimentation unit, including a mudclast-rich lag with contorted mudclasts at the base of the sedimentation unit, with enough mud to change the rheology of the flow (often classified as basal lag in many cases); (H) "mud on mud" disorganized muddy debrite, often with contorted and overturned intervals, each with varying proportions of mudstone, typically occurring from slumped hemipelagic intervals. Transitional flow textures include both (I) suspension and traction structures in slurry-flow deposits, enhanced with clay content 10%–35%; (J) mixed suspension and traction and matrix-strength processes to develop the hazy fabric of clay and sand-rich intervals, sometimes with outsized mudclasts and can be nongraded, graded or inversely graded; (K) low-yield strength transitional flow (potentially debris flow) textures often with small, wispy mudclasts, lacking significant lithification and therefore more easily disaggregated in a relatively low-concentration flow with fluid turbulence and matrix strength in operation (Amy et al., 2005); (L) high-yield strength transitional flows (potentially debris flow) with larger, more abundant mudclasts that appear to have maintained shape and have ductile, but more competent behavior as a result of higher lithification prior to transport. As lithification and mudclast density increases, the high-yield strength transitional flow deposit will approach that of the linked debrite mudclast. Other fabrics include those deposited by tides, waves, river flows and bottom currents. Tidal current propagation downsystem can create (M) rhythmic patterns in mud draping style and "on" vs "off" sedimentation, modifying the textures in the preceding categories. Wave-influenced sedimentation (N) can extend well below wave-base based on recent lab work, causing asymmetry in sedimentary structures, as well as creating wave-influenced turbidites (Myrow et al., 2002). Hyperpycnal deposits (O) often show thick intervals of structureless and structured sandstone with abundant carbonaceous matter. Bottom currents can modify the preceding deposits to rework the top or modify the entire bed (P), such as this sedimentation unit with an erosional base, yet is subtly reworked by along-slope currents.

these systems can be characterized by a very high net-to-gross due to the fact that the fine-grained part of the turbulent flow is pirated by the *syn*-depositional bottom current, allowing deposition of only the coarsest particles in the system axis (Fonnesu et al., 2020).

### Flow stratification and resulting rock properties

Turbulent flows stratify grain populations, sorting grains by size, density, surface texture and shape before and during deposition (Straub and Mohrig, 2008; Kane and Hodgson, 2011; Hansen et al., 2015), whereas cohesive debris flows create very poorly sorted deposits. Vertical grain-size gradients have been established in flume experiments (e.g., Straub and Mohrig, 2008; Straub et al., 2011) and from numerical simulations (e.g., Abd El-Gawad et al., 2012) but there are very few measurements from natural sediment gravity flows (e.g., Paull et al., 2018; Wang et al., 2020). The deposits from turbidity flows are typically normally graded, whereas cohesive debris flows result in ungraded, structureless and poorly sorted deposits. This is because relatively dense and large particles are concentrated at the base of the current while finer fractions are more homogeneously distributed (Garcia, 1994; Kneller and Buckee, 2000; Baas et al., 2005; Hansen et al., 2015). The intensity of turbulence controls the vertical variation in grain size in turbidity currents: highly turbulent flows will tend to homogenize sediment sizes. Therefore, the resulting sorting of grain populations from sediment gravity flows indicates the energy and *consistency of the energy* of the flows in any given deepwater environment. Since primary porosity in unconsolidated sediment is controlled by the grain size, sorting and packing of grains, the flow behavior has important implications for reservoir rock properties. As the sorting and grain size increase, both porosity and permeability will theoretically increase (e.g., Slatt, 2006; Fig. 15). Conversely, at the upper part of the flow, fine-grained sediment with abundant clay will reduce the permeability in the upper and more distal portions of the facies tract. In basins with more or less equal deposition of debrites and turbidites, the porosity and permeability of reservoir intervals can be highly complex due to the juxtaposition of both reservoir and non-reservoir intervals that can act as baffles to fluid flow.

## Sedimentation mechanics

Although the depositional record and submarine cable breaks provided the initial curiosity into deepwater deposits (Chapters 1 and 21), comparisons with sediment gravity flow analogs, as well as numerical simulation and tank experiments have enabled a much more detailed window into the diversity of sediment gravity flow processes. There are several concepts and equations that can help portray how sediment gravity flows operate and distribute sediment in a predictable manner.

### Reynolds number

While turbidity currents are as their name suggests, turbulent, deposits from these flows are formed via turbulent and near-bed laminar flow. In turbulent flow, there is extensive mixing expressed as vortices of fluid that move upward and downward within the flow. In laminar flow, there is no mixing within the flow. This can be explained physically by an understanding of Reynolds Number (R), which is a measure of inertial over viscous forces. When R is small, viscous forces dominate, and flow is laminar. When R is large, inertial forces dominate and mixing occurs, therefore flow is turbulent. In turbidity currents, velocities tend to be high and viscosities low, so flows are turbulent. Deposition occurs due to a decrease in velocity or depth of flow, which can happen in a number of scenarios in deepwater environments illustrated in Chapters 8–12 (Lowe and Graham, 2012).

### Froude number

Froude Number is a measure of inertial over gravitational forces and is expressed as the following equation:

$$\mathrm{Fr} = \frac{\mu}{\sqrt{gl}}$$

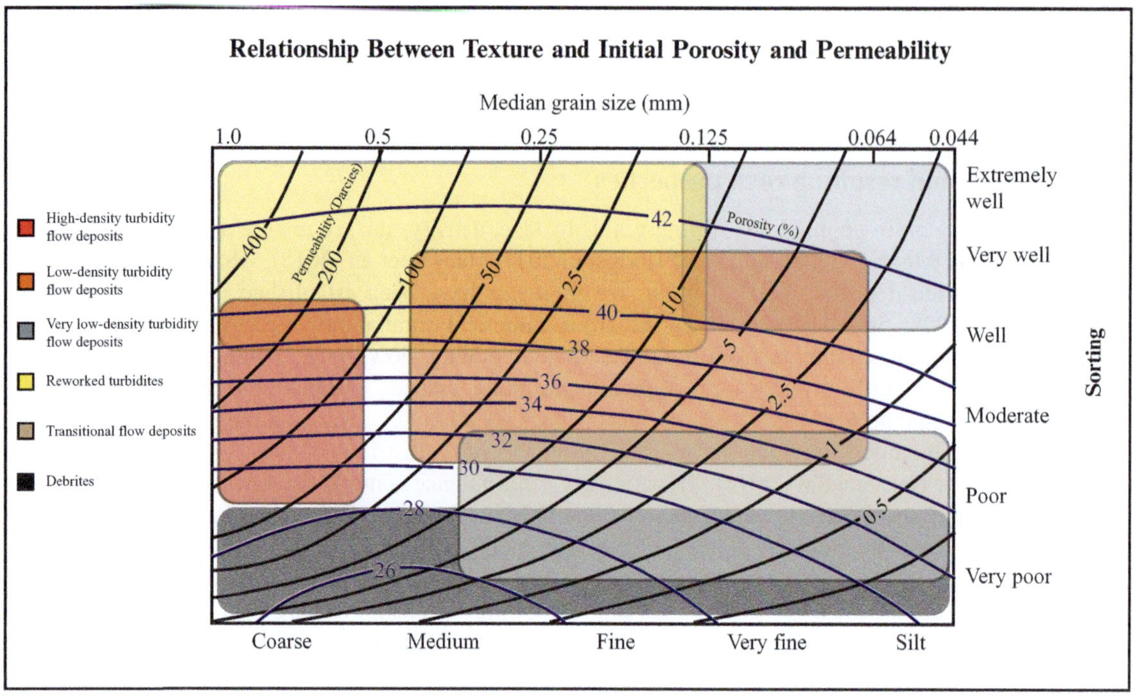


**FIG. 15** Relationship between rock type based on sedimentary process and sorting, grain size and resulting relative estimation of initial porosity and permeability. Estimates on sorting and grain size are for the principal rock types discussed in this paper that can function as key petroleum reservoir intervals. Key rock types include debrites, transitional flow deposits, low-density turbidity current deposits (classic turbidites), high-density turbidity current deposits, reworked turbidites and very low-density turbidity current deposits. Typical sorting and grain size ranges for the key rock types yield a distribution of porosity and permeability, and allow for ranking of deposits starting with high porosity and permeability: (i) high-density turbidity current deposits; (ii) reworked turbidites; (iii) low-density turbidity current deposits (classic turbidites); (iv) very low-density turbidity current deposits; (v) transitional flow deposits and finally; (vi) debrites. These estimates will be refined as more data become available from diverse rock types over the next few decades. *(Porosity and permeability diagram redrawn from Slatt, R.M. (Ed.), 2006. Stratigraphic Reservoir Characterization for Petroleum Geologists, Geophysicists, and Engineers, vol. 61, Elsevier, 492 p.)*

where $\mu$ is viscosity, $g$ is gravity or the external field and $\iota$ is length. If the Froude Number is less than 1, in which gravitational forces exceed the viscous forces, the flow is called a *subcritical flow*, otherwise known as *slow and tranquil* flow state. If the value is greater than 1, in which viscous forces exceed the gravitational forces, the flow is termed a *supercritical flow*, also known as a *fast and rapid* flow state (see Chapter 7 of this volume for additional explanation). If the value is equal to 1, it is known as *unity*. The Froude Number is often depth-averaged or applied as a densiometric function to compare experiments of sediment gravity flows in laboratory tanks, such as Euro Tank (sensu Pohl, 2019).

Lower stage plane beds and ripples are common bedforms in fine-grained deepwater environments that are formed by subcritical flows, developed beneath a near-bed laminar sublayer under a fully turbulent flow. Upper stage plane beds, antidunes and cyclic steps are types of bedforms formed by supercritical flows, in which the top of the flow is in phase with the bedform (Fig. 16). More recently, supercritical flows have been found to create a wider diversity of bedforms including a variety of scours, mounds, irregular lenses forming a diverse array of backsets and lamination configurations in the rock record (Ono et al., 2019). These sedimentary structures can form in any number of sedimentary environments, including deepwater, and the latter two are most commonly associated with abrupt breaks in gradient or morphological transition zones (i.e., base of the slope, submarine channel knickpoint and channel-lobe transition zone), or in the vicinity of hydraulic jumps. Deposits formed by supercritical flows have variable preservation potential, yet the depositional process via traction can enhance porosity and permeability leading to high reservoir quality.

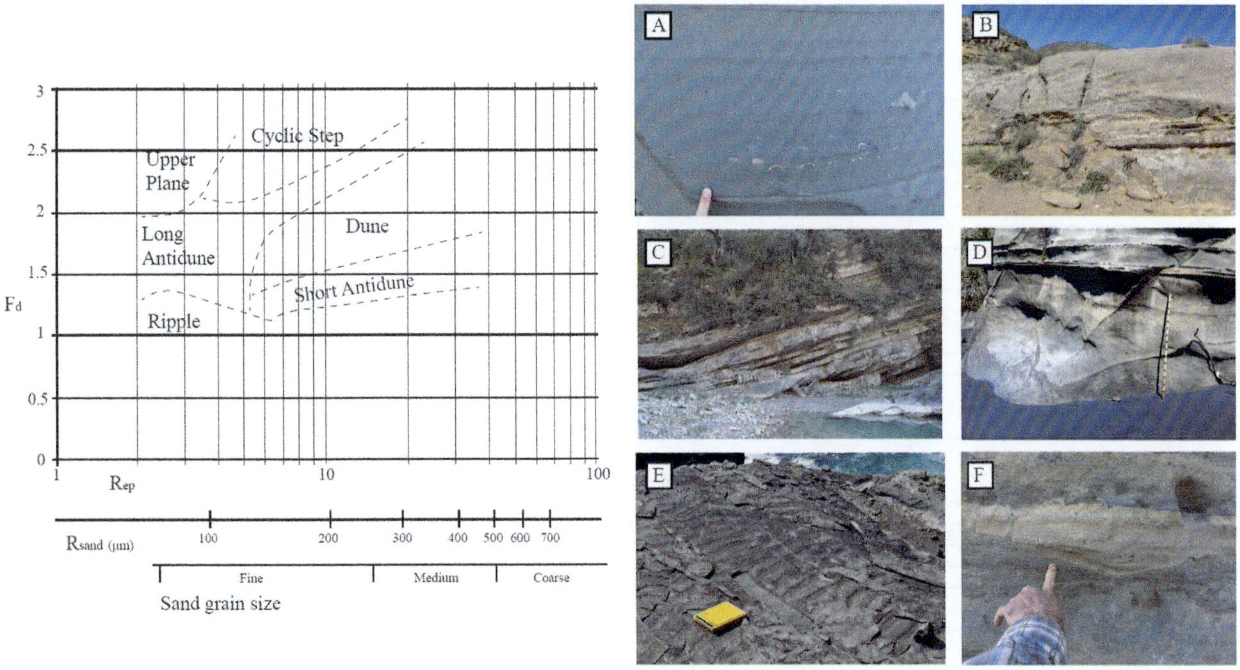


**FIG. 16** Sedimentary structure stability field for deepwater sediment gravity flows. Essentially, flow velocity and grain size determine the type of sedimentary structure. Note that in sand grain sizes, decreasing the velocity will generate an approximate Bouma Sequence. Field photographs illustrating the types of structures shown in the stability diagram: (A) Planar laminated sandstone with abundant shell hash, Upper Mount Messenger Formation (Pukearuhe Beach), Taranaki Basin, North Island, New Zealand; (B) cyclic step deposits indicating upslope accretion, Miocene, Sorbas Basin, Spain; (C) cross stratification produced by long antidunes, Ainsa Basin, Spain; (D) climbing dune-scale cross-stratification in the Lower Mount Messenger Formation, Tongaporutu, Taranaki Basin, New Zealand. (E) ripple cross-laminated sandstone, Carboniferous Ross Sandstone (near Cross, County Clare), Western Irish Namurian Basin, Ireland; (F) backsets infilling a shallow scour, thus preserving short antidunes in the rock record, Miocene, Tabernas Basin, Spain. *(Stability diagram modified and updated from Fedele, J.J., Hoyal, D., Barnaal, Z., Tulenko, J., Awalt, S., 2016. Bedforms created by gravity flows. In: Autogenic Dynamics and Self-Organization in Sedimentary Systems. SEPM Special Publication No. 106, pp. 95–121, Courtesy of J. Fedele. Field photographs from the various authors.)*

## Flow steadiness and uniformity

Whereas support mechanisms are governed in part by a measure of fluid turbulence, deposition can be predicted based on whether or not a flow is steady and/or uniform (Lowe and Graham, 2012). A steady flow is a flow in which there is no variation with time in the magnitude and direction of the velocity vector at each point in the flow field. A uniform flow is a flow in which the magnitude and direction of the velocity vectors are constant along the length of each streamline at any instant of time (all streamlines are parallel). The Bouma (1962) sequence of structures is interpreted *sensu stricto* as a measure of waning flow, or unsteadiness. This progression of sedimentary structures is shown in Fig. 16. However, flow nonuniformity can generate the same sequence of structures. Some examples of where flow nonuniformity would exist would be (i) decrease in gradient; (ii) flow out of conduit, such as a channel; (iii) flow into an area of unconfinement, from a confined space (Fig. 17). These are common play types in deepwater exploration and are linked directly to concepts of flow steadiness and uniformity.

## Newtonian vs plastic fluids

In turbidity currents that are neither decelerating nor accelerating—steady currents—the driving force of gravity balances the resisting forces of viscosity and friction (Fig. 18). Friction occurs between layers within the flow that are moving past one another due to differences in velocity with a specified thickness. This is known as shear stress and is calculated by the Newtonian equation of fluid flow. In ideal fluid flows, the force applied by gravity driving the fluid downslope is known as the applied shear stress. The applied shear stress is balanced by the friction between fluid layers or between the fluid and the surface over which it moves, known as the shear resistance. However, plastics do not begin to deform until the applied shear stress, or gravity force, exceeds a critical value. Whereas most debris flow deposits in deepwater environments are formed

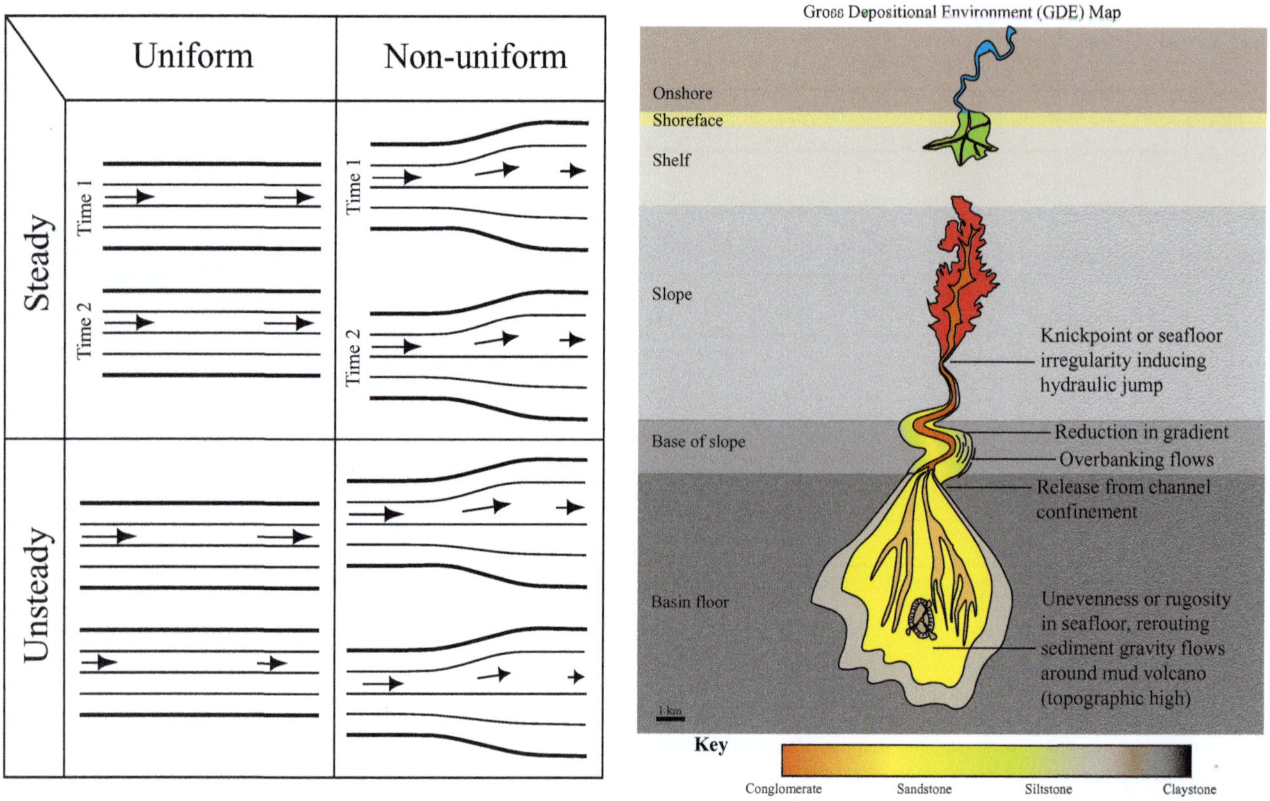


**FIG. 17** Steady and unsteady, uniform and nonuniform flows and environments of deposition where deepwater deposition takes place. In steady flows, there is no change between *time 1* and *time 2* in the velocity of the flow. In uniform flows, there is no change between *time 1* and *time 2* in the direction of the flow. Unsteady flows represent changes in velocity, and nonuniform flows represent a change in direction. While the Bouma Sequence records flow unsteadiness, a Bouma Sequence can also equally be deposited due to flow nonuniformity. The right part of the diagram shows where changes, or more specifically, reductions in confinement and gradient can cause simultaneous reductions in flow steadiness and uniformity, causing flows to become both highly unsteady and nonuniform. These are just a few examples. The processes linked to the depositional environment can create a first-order map of play types in any deepwater province. In other words, these physical observations lead to preferential deposition of sediment in certain areas of the deepwater realm. Furthermore, this is a positive feedback process *ceteris paribus*, such that flow steadiness and uniformity will become increasingly affected entering into certain depositional environments through the evolution of the deepwater depositional system. *(Diagram on left modified from Lowe, D.R., Graham, S. A., 2012. Turbidite Reservoirs and Depositional Systems. Short Course Lecture Notes. Stanford University, Approx 150 pp.)*

by cohesion and matrix strength, subaerial debris flows commonly found in alluvial fan environments are typically deposited through frictional grain interactions (Lowe and Graham, 2012).

## Linking process to deposit

Because the primary sediment delivery mechanism to deepwater areas is via sediment gravity flows, there are three principal classes of deepwater siliciclastic reservoirs based on depositional process: turbidites, debrites and transitional flow deposits (Fig. 10). *All of these may be modified during transport, during deposition, following deposition by fluvial, tidal, wave and bottom current processes.* Turbidites are typically split into two main classes: those that are deposited by low-density flows (Bouma, 1962), and those deposited by high-density flows (Lowe, 1982). The processes of sedimentation are interpreted from the suite of sedimentary structures in the bedding. Furthermore, low-density flows can be very low density (~1% sediment concentration; Stow and Shanmugam, 1980). But post-sediment gravity flow and *syn*-sediment gravity flow currents can also affect the resulting deposit, including tides (sensu Klein, 1971, 1998) often very similar to bottom-current-induced deposits, typhoon-triggered flows (Sequeiros et al., 2019), waves (Myrow et al., 2002), bottom currents (Stow, 1979; Shanmugam et al., 1993a, 1993b; De Castro et al., 2020) and bottom currents interacting with sediment gravity flows (Miramontes et al., 2020). At any time, these deposits may be involved in further mass transport, redepositing as an MTD, which incorporates faulting, folding, slumping and other forms of deformation and sediment transport (see Chapter 10). After that, the deposits can be altered again, most notably by injection or extrusion processes. These added

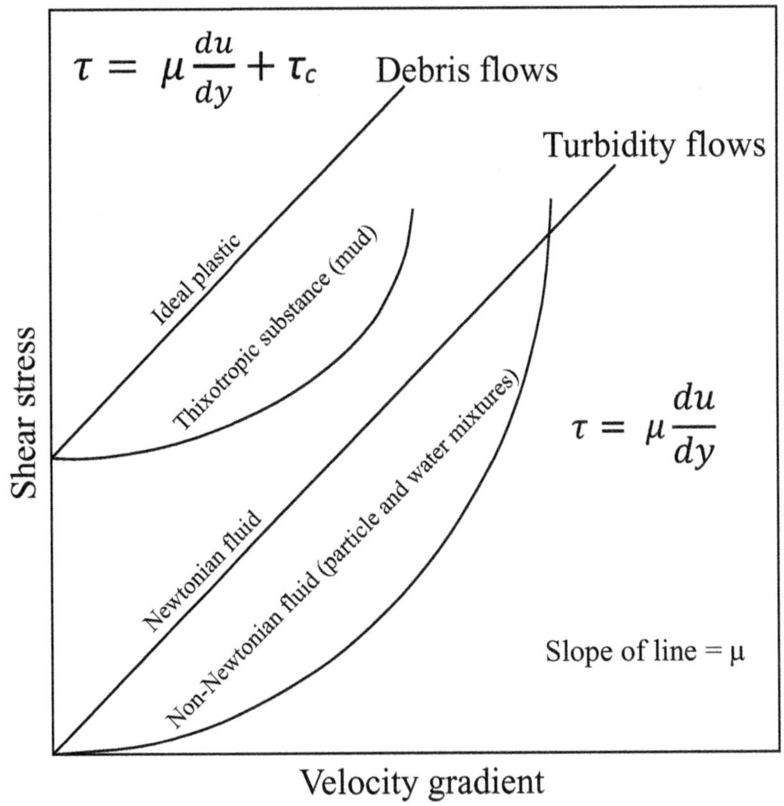


**FIG. 18** Shear stress vs velocity gradient diagram for Newtonian fluids, ideal plastics and other types of mixtures and substances. Ideal plastic substances require an initial shear stress value to be exceeded, known as the cohesive strength of the substance. *(Modified from Lowe, D. R., and S. A. Graham, 2012. Turbidite Reservoirs and Depositional Systems: Short Course Lecture Notes, Stanford University, approx. 150 p.)*

complexities can affect and modify these deposits left by mostly downslope flows, or they can cause wholesale redistribution of sediment in the deepwater realm. Therefore, sedimentation occurs along a continuum, or better yet an *iterative continuum*, in which sedimentary deposits of varying lithification can be modified to assume their ultimate sedimentary disposition. With these descriptions, it is assumed there is no post-depositional manipulation of the beds, which is a key assumption.

The following chapter details each type of model sedimentation unit, to showcase their range and variability. The rock types are the basic building blocks of petroleum reservoirs, and also present a first-order control on porosity and permeability in a petroleum reservoir. Each model is presented and is described in terms of sedimentation process with examples from specific basins as well as key uncertainties and controversies.

## Key messages

Most of deepwater is quite unremarkable, devoid of large-scale, high-energy features formed by sediment gravity flows, although is commonly affected by weak to very weak bottom currents due to the circulation of water masses. Most of the ocean floor actually consists of accumulations of hemipelagic to pelagic sediment, which eventually can be affected by bottom currents to some extent. *Where do relatively high-energy deposits derived from mass-transport, sediment gravity flows and bottom currents exist?* High-energy, active environments are found along the axis of a depositional system, such as a submarine fan. Fans typically consist of canyon, channel, levee, splay, overbank and drift environments (see Chapter 8) that feature high architectural complexity due to the processes associated with stratification in sediment gravity flows and potentially bottom currents. In the upper reaches of the submarine fan, gravelly deposits transported by high-density turbidity flows and debris flows tend to dominate along with MTD, their evacuation scars and record of sediment transport. The transition between the canyon and channel environment is generally marked by a change in the gradient of the slope, creating a zone of erosion and potentially a hydraulic jump, where supercritical flows dominate, forming zones of erosion and a variety of bedforms. Farther down in the channel-levee environment, coarser-grained and heterolithic deposits are found in the channel, and finer-grained to muddy deposits are typically found outside the channel. In the frontal splay environment, the dramatic loss of competency and capacity of the flows causes flow expansion and a vast accumulation of suspension deposits resulting from highly unsteady and nonuniform flows. Refined grain populations reach the frontal splay environment due to the overbanking of fine material into the levees, and coarser material is left upslope due to the change in Reynolds Number. In the splay environment, alternation of sandy and muddy sediment causes flows to contain more mud than in other environments, carrying the potential to cause repeated transitional flows and hybrid events. Far away from sediment input, distribution and remobilization points, hemipelagic and pelagic sediment accumulates in the overbank and may show subtle evidence of sediment waves caused by ambient currents or sediment gravity flows. This submarine fan is oriented perpendicular to the coast yet is located in a strait between two continental land masses, through which a contour current flows. Bottom currents are everywhere, but the most energetic current cores are at the base of the continental slope, terraces, along-slope channels, as well as the foot of sea-floor irregularities, as banks, highs, seamounts, gravitational scars and faults (see Chapter 9) (Rebesco et al., 2014). All of the submarine fan environments previously described are affected by bottom currents that reroute and transport sediment in the direction of the bottom current, also removing the finer-grained material and leaving the coarser-grained fraction of the original turbidite. So far, this is not unlike the Paleogene deepwater environments found in between modern-day Mozambique and Madagascar, described in Fonnesu et al. (2020). While the finer-grained sediment may be carried in the direction of the contour current, the channel-levee system may migrate *up current* due to the relatively weakly confined channel, summarized in Fig. 19. This example assumes that the full suite of grain sizes is provided by the coastal feeder system, the shelf-slope-basin floor profile is normal, sediment gravity flows dominate (uniformitarianism) and remobilization of deposits (catastrophism) is at a minimum. Further discussions on this topic can be found in Chapters 8, 9 and 18 of this volume.

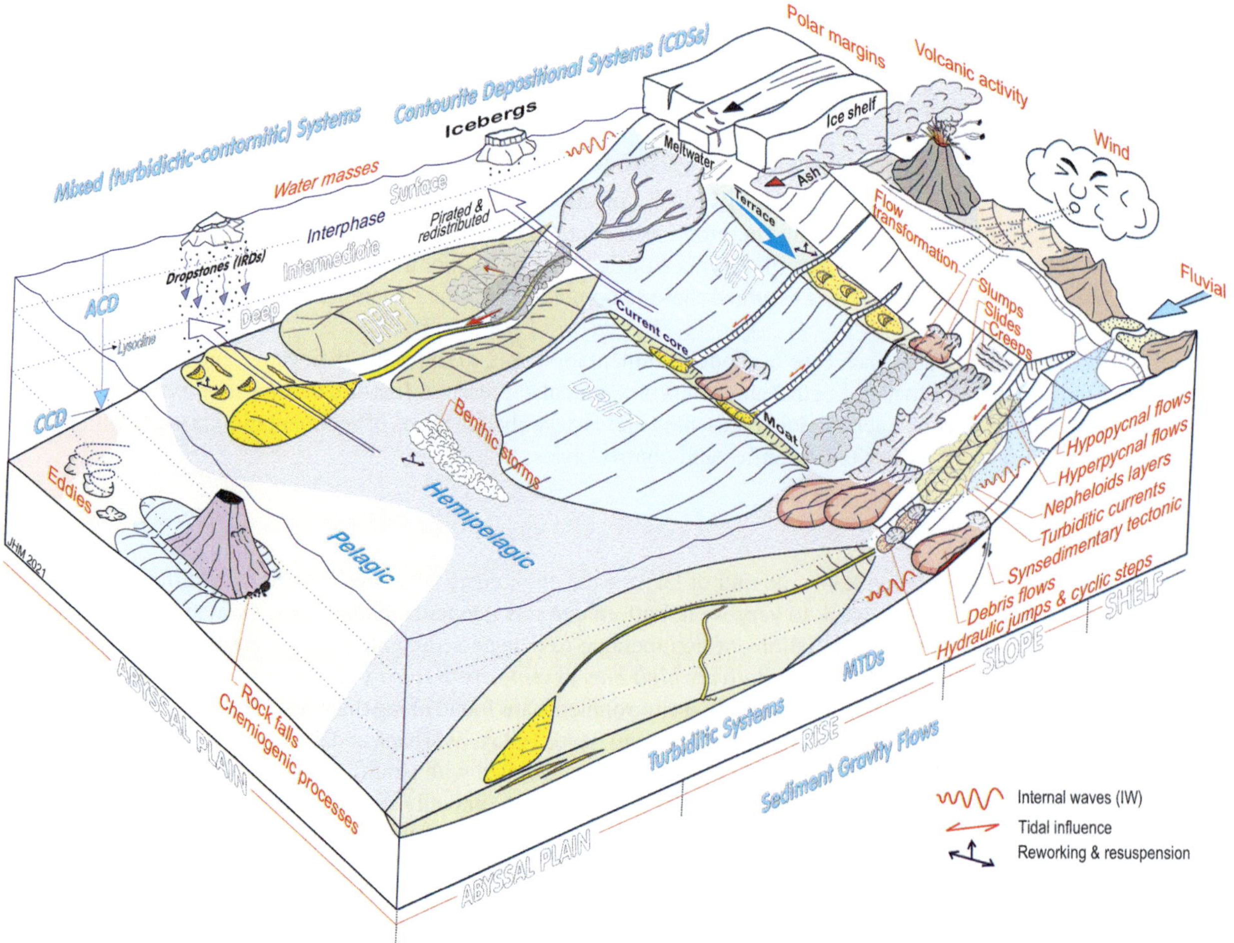


**FIG. 19** Conceptual sketch for sedimentary processes on deepwater (marine) environments. The processes interact laterally and vertically through time with synchronized or desynchronized interactions. In this way, a great variety of sedimentary facies and facies associations are found in deepwater sedimentary systems.

## Conclusions

Knowledge of sedimentary processes is critical to understand the broad range of deposits found in a wide variety of deepwater environments. Although quiet water conditions exist throughout most of the ocean causing hemipelagic and pelagic sedimentation to dominate, there are high-energy environments where abundant sediment gravity flows transport voluminous amounts of sediment from upslope feeder systems into deepwater. Perhaps the most recent change to this paradigm is the understanding of ocean water mass circulation, and the associated secondary processes that cause bottom currents along continental slopes to modify and redeposit sediment brought to deepwater by sediment gravity flows.

Sediment gravity flows exhibit three primary types of dynamic behavior: Newtonian fluid; Bingham plastic (to thixotropic); mixed to transitional. Turbidity currents exhibit Newtonian behavior and transport sediment to deepwater environments via fluid turbulence alone for sand-sized grains and finer, yet fluid turbulence is insufficient to transport coarse gravels. Grain-to-grain interactions, buoyancy effects, hindered settling are processes invoked to carry gravels to deepwater environments. Debris flows are supported by frictional or cohesive forces, yet there is often an overabundance of clay in deepwater, so debris flows have high matrix strength due to the cohesion provided by clay. There are also types of flows that exhibit both fluid turbulence and matrix strength during transport and deposition—these are broadly known as hybrid flows or transitional flows that form muddy sand or sandy mud deposits. These three primary types of dynamic behavior are responsible for forming a wide range of rock fabrics that can be modified by bottom current processes—and even wave, river or tidal processes. As a result, deepwater rock types are very different from those found in terrestrial or shallow marine environments based on these processes.

The sedimentation mechanics of sediment gravity flows and bottom currents permit the prediction of bedforms and rock types found in common deepwater environments of deposition. If a sediment gravity flow's competency and capacity are high due to gradient, sediment load or confinement, one could expect thick sandy deposits or erosional surfaces marking high-energy sediment transport downslope. Conversely, if the flow is incapable of carrying sediment, but it is just exiting an area of high energy, one should expect thick sandy deposits, similar to the apex of frontal splays, where flow expansion occurs and flows become highly unsteady and nonuniform. The concepts presented in this chapter illustrate that process is directly tied to deposit, and if the deposit can be predicted with some accuracy, other parameters such as porosity and permeability can be established, too, to gain an economic understanding of the resource in place (see Chapters 19 and 20 of this volume) or explore the broad range of seafloor sedimentary deposits for new resources.

## Acknowledgments

This chapter benefited greatly from constructive reviews from Octavian Catuneanu, Richard Sears and Cindy Yeilding. J. R. dedicates this chapter to Roger M. Slatt, whose decades of achievements inspired this book's vision and format. Professor Slatt intended to be an author of this work, yet he passed on too soon. This work would not have been possible without conversations and guidance throughout the years from the following: D. R. Lowe, S. A. Graham, E. Mutti, G. E. Hilley, G. H. Browne, P. R. King, E. and S. Muller, J. R. Suter, A. D. Donovan, D. Nelson, "The Drifters" research group (RHUL, UK), M. Honarpour, C. Pirmez, Z. Sylvester, W. D. McCaffrey, G. Apps, F. J. Peel, P. D. W. Haughton, D. D. Miller, J. R. Schwalbach, K. W. Rudolph, D. Adams, Y. Gavillot, N. Uzunlar, A. Lisenbee, M. McKeown & family, and organizations including AAPG, EAGE, HOT Engineering, GNS Science, Gulf Coast Section of SEPM, CERAWeek and many others.

## References

Abd El-Gawad, S.M., Pirmez, C., Cantelli, A., Minisini, D., Sylvester, Z., Imran, J., 2012. 3-D numerical simulation of turbidity currents in submarine canyons off the Niger Delta. Mar. Geol. 326–328, 55–66. https://doi.org/10.1016/j.margeo.2012.06.003.

Amy, L.A., Peakall, J., Talling, P.J., 2005. Density- and viscosity-stratified gravity currents: insight from laboratory experiments and implications for submarine flow deposits. Sediment. Geol. 179, 5–29. https://doi.org/10.1016/j.sedgeo.2005.04.009.

Amy, L.A., Talling, P.J., 2006. Anatomy of turbidites and linked debrites based on long distance (120 × 30 km) bed correlation, Marnoso Arenacea formation, northern Apennines, Italy. Sedimentology 53 (1), 161–212. https://doi.org/10.1111/j.1365-3091.2005.00756.x.

Azpiroz-Zabala, M., Cartigny, M.J., Talling, P.J., Parsons, D.R., Sumner, E.J., Clare, M.A., Pope, E.L., 2017. Newly recognized turbidity current structure can explain prolonged flushing of submarine canyons. Sci. Adv. 3 (10). https://doi.org/10.1126/sciadv.1700200.

Baas, J.H., Best, J.L., 2002. Turbulence modulation in clay-rich sediment-laden flows and some implications for sediment deposition. J. Sediment. Res. 72 (3), 336–340. https://doi.org/10.1306/120601720336.

Baas, J.H., Best, J.L., 2007. The dynamics of turbulent, transitional and laminar clay-laden flow over a fixed current ripple. Sedimentology 55 (3), 635–666.

Baas, J.H., McCaffrey, W.D., Haughton, P.D.W., Choux, C., 2005. Coupling between suspended sediment distribution and turbulence structure in a laboratory turbidity current. J. Geophys. Res. Oceans 110, 1–20. https://doi.org/10.1029/2004JC002668.

Baas, J.H., Best, J.L., Peakall, J., Wang, M., 2009. A phase diagram for turbulent, transitional, and laminar clay suspension flows. J. Sediment. Res. 79 (4), 162–183. https://doi.org/10.2110/jsr.2009.025.

Baas, J.H., Best, J.L., Peakall, J., 2011. Depositional processes, bedform development and hybrid bed formation in rapidly decelerated cohesive (mud-sand) sediment flows. Sedimentology 58, 1953–1987. https://doi.org/10.1111/j.1365-3091.2011.01247.x.

Bouma, A.H., 1962. Sedimentology of some Flysch Deposits: A Graphic Approach to Facies Interpretation. Elsevier, Amsterdam, p. 167.

Bull, S., Cartwright, J., Huuse, M., 2009. A review of kinematic indicators from mass-transport complexes using 3D seismic data. Mar. Pet. Geol. 26 (7), 1132–1151.

Campbell, C.V., 1967. Lamina, laminaset, bed and bedset. Sedimentology 8 (1), 7–26. https://doi.org/10.1111/j.1365-3091.1967.tb01301.x.

Cuffey, K.M., Paterson, W.S.B., 2010. The Physics of Glaciers. Butterworth-Heinemann, Burlington, MA.

De Castro, S., Hernandez-Molina, F.J., Rodriguez-Tovar, F.J., Llave, E., Ng, Z.L., Nishida, N., Mena, A., 2020. Contourites and bottom current reworked sands: bed facies model and implications. Mar. Geol. 428, 106267.

Domack, E.W., Powell, R., 2018. Chapter 7 – Modern Glaciomarine Environments and sediments: an Antarctic perspective. In: Menzies, J., van der Meer, J.J.M. (Eds.), Past Glacial Environments, second ed. Elsevier, pp. 181–272. https://doi.org/10.1016/B978-0-08-100524-8.00030-0.

Eberli, G.P., Betzler, C., 2019. Characteristics of modern carbonate contourite drifts. Sedimentology 66, 1163–1191. https://doi.org/10.1111/sed.12584.

Faugères, J.-C., Stow, D.A.V., Imbert, P., Viana, A.R., 1999. Seismic features diagnostic of contourite drifts. Mar. Geol. 162, 1–38. https://doi.org/10.1016/S0025-3227(99)00068-7.

Fischer, R.V., 1983. Flow transformations in sediment gravity flows. Geology 11, 273–274. https://doi.org/10.1130/0091-7613(1983)11<273:FTISGF>2.0.CO;2.

Fonnesu, M., Haughton, P., Felletti, F., McCaffrey, W., 2015. Short length-scale variability of hybrid event beds and its applied significance. Mar. Pet. Geol. 67, 583–603. https://doi.org/10.1016/j.marpetgeo.2015.03.028.

Fonnesu, M., Patacci, M., Haughton, P.D.W., Felletti, F., McCaffrey, W.D., 2016. Hybrid event beds generated by local substrate delamination on a Confined Basin floor. J. Sediment. Res. 86, 929–943. https://doi.org/10.2110/jsr.2016.58.

Fonnesu, M., Palermo, D., Galbiati, M., Marchesini, M., Bonamini, E., Bendias, D., 2020. A new world-class Deepwater play-type, deposited by the syndepositional interaction of turbidity flows and bottom currents: the giant Eocene Coral Field in northern Mozambique. Mar. Pet. Geol. 111, 179–201. https://doi.org/10.1016/j.marpetgeo.2019.07.047.

Fowler, S.W., Knauer, G.A., 1986. Role of large particles in the transport of elements and organic compounds through the oceanic water column. Prog. Oceanogr. 16, 147–194. https://doi.org/10.1016/0079-6611%2886%2990032-7.

Garcia, M.H., 1994. Depositional turbidity currents laden with poorly sorted sediment. J. Hydraul. Eng. 120, 1240–1263. https://doi.org/10.1061/(ASCE)0733-9429(1994)120:11(1240.

Hansen, L.A.S., Callow, R.H.T., Kane, I.A., Gamberi, F., Rovere, M., Cronin, B.T., 2015. Genesis and character of thin-bedded turbidites associated with submarine channels. Mar. Pet. Geol. 67, 852–879. https://doi.org/10.1016/j.marpetgeo.2015.06.007.

Haughton, P.D.W., 2001. Contained turbidites used to track sea bed deformation and basin migration, Sorbas Basin, south-East Spain. Basin Res. 13, 117–139. https://doi.org/10.1046/j.1365-2117.2001.00143.x.

Haughton, D.W., Barker, S., McCaffrey, W.D., 2003. "Linked" debrites in sand-rich turbidite systems—origin and significance. Sedimentology 50, 459–482. https://doi.org/10.1046/j.1365-3091.2003.00560.x.

Haughton, P.D.W., Davis, C., McCaffrey, W., Barker, S., 2009. Hybrid sediment gravity flow deposits—classification, origin and significance. Mar. Pet. Geol. 26, 1900–1918. https://doi.org/10.1016/j.marpetgeo.2009.02.012.

Heezen, B.C., Hollister, C.D., 1971. The Face of the Deep. Oxford University Press, New York.

Heezen, B.C., Hollister, C.D., Ruddiman, W.F., 1966. Shaping of the continental rise by deep geostrophic contour currents. Science 152, 502–508. https://doi.org/10.1126/science.152.3721.502.

Heezen, B., Tharp, M., Ewing, M., 1959. The floors of the oceans. I. The North Atlantic. In: Text to accompany the physiographic diagram of the North Atlantic. Special Paper 65. The Geological Society of America, New York.

Hernández-Molina, F.J., Stow, D.A.V., Llave, 2008. Continental slope contourites. In: Rebesco, M., Camerlenghi, A. (Eds.), Contourites. Developments in Sedimentology, vol. 60. Elsevier, Amsterdam, pp. 379–408. https://doi.org/10.1016/S0070-4571(08)10019-X.

Hernández-Molina, F.J., Wåhlin, A., Bruno, M., Ercilla, G., Llave, E., Serra, N., Roson, G., Puig, P., Rebesco, M., Van Rooij, D., Roque, C., González-Pola, C., Sánchez, F., Gómez, M., Preu, B., Schwenk, T., Hanebuth, T.J.J., Leal, S., Till, J.J., García-Lafuente, J., Brackenridge, R.E., Juan, C., Stow, D.A.V., Sánchez-González, J.M., 2016. Oceanographic processes and products around the Iberian margin: a new multidisciplinary approach. Mar. Geol. 378, 127–156. https://doi.org/10.1016/j.margeo.2015.12.00822.

Hüneke, H., Henrich, R., 2011. Pelagic sedimentation in modern and Ancient Oceans. Dev. Sedimentol. 63, 216–295. https://doi.org/10.1016/B978-0-444-53000-4.00004-4.

Hussain, A., Haughton, P.D., Shannon, P.M., Turner, J.N., Pierce, C.S., Obradors-Latre, A., Barker, S.P., Martinsen, O.J., 2020. High-resolution X-ray fluorescence profiling of hybrid event beds: implications for sediment gravity flow behaviour and deposit structure. Sedimentology 67 (6), 2850–2882. https://doi.org/10.1111/sed.12722.

Kane, I.A., Hodgson, D.M., 2011. Sedimentological criteria to differentiate submarine channel levee subenvironments: exhumed examples from the Rosario Fm. (upper cretaceous) of Baja California, Mexico, and the fort Brown Fm. (Permian), Karoo Basin, S. Africa. Mar. Pet. Geol. 28, 807–823. https://doi.org/10.1016/j.marpetgeo.2010.05.009.

Klein, G.D., 1971. A sedimentary model for determining paleotidal range. GSA Bull. 82, 2585–2592. https://doi.org/10.1130/0016-7606(1971)82[2585:ASMFDP]2.0.CO;2.

Klein, G.D., 1998. Classic tidalites—A partial retrospective view. In: Alexander, C.R., Davis, R.A., Henry, V.J. (Eds.), Tidalites: Processes and Products. vol. 61. SEPM Special Publication, pp. 5–14.

Kneller, B.C., Branney, M.J., 1995. Sustained high-density turbidity currents and the deposition of thick massive sands. Sedimentology 42, 607–616.

Kneller, B., Buckee, C., 2000. The structure and fluid mechanics of turbidity currents: a review of some recent studies and their geological implications. Sedimentology 47, 62–94. https://doi.org/10.1046/j.1365-3091.2000.047s1062.x.

Kneller, B.C., McCaffrey, W.D., 2003. The interpretation of vertical sequences in turbidite beds: the influence of longitudinal flow structure. J. Sediment. Res. 73, 706–713.

Kuhlbrodt, T., Griesel, A., Montoya, M., Levermann, A., Hofmann, M., Rahmstorf, S., 2007. On the driving processes of the Atlantic meridional overturning circulation. Rev. Geophys. 45, RG2001. https://doi.org/10.1029/2004RG000166.

Lowe, D.R., 1982. Sediment gravity flows: II. Depositional models with special reference to the deposits of high-density turbidity currents. J. Sediment. Petrol. 52 (1), 279–297. https://doi.org/10.1306/212F7F31-2B24-11D7-8648000102C1865D.

Lowe, D.R., 2004. Deepwater Sandstones: Submarine Canyon to Basin Plain, Western California: AAPG and Pacific Section AAPG Guidebook and Lecture Notes. Artistic Printing, Salt Lake City, p. 79.

Lowe, D.R., Graham, S.A., 2012. Turbidite Reservoirs and Depositional Systems: Short Course Lecture Notes. Stanford University, approx. 150 p.

Lowe, D.R., Guy, M., 2000. Slurry-flow deposits in the Britannia Formation (Lower Cretaceous), North Sea: a new perspective on the turbidity current and debris flow problem. Sedimentology 47, 31–70.

Manica, R., 2012. In: Schulz, H., Simoes, A., Lobosco, R. (Eds.), Sediment gravity flows: study based on experimental simulations. INTECH Open Access Publisher, pp. 263–286. https://doi.org/10.5772/28794.

Middleton, G.V., 1966. Experiments on density and turbidity currents: I. Motion of the head. Can. J. Earth Sci. 3, 523–546. https://doi.org/10.1139/e66-038.

Middleton, G.V., 1993. Sediment deposition from turbidity currents. Annu. Rev. Earth Planet. Sci. 21, 89–114. https://doi.org/10.1146/annurev.ea.21.050193.000513.

Middleton, G.V., Hampton, M.A., 1973. Sediment gravity flows: mechanics of flow and deposition. In: Middleton, G.V., Bouma, A.H. (Co-Chairmen), Turbidites and Deep Water Sedimentation. Soc. Econ. Paleontol. Mineral., Pac. Sect., Short Course, pp. 1–38.

Middleton, N., Kang, U., 2017. Sand and dust storms: impact mitigation. Sustainability 9, 1053. https://doi.org/10.3390/su9061053.

Miramontes, E., Eggenhuisen, J.T., Jacinto, R.S., Poneti, G., Pohl, F., Normandeau, A., Campbell, D.C., Hernández-Molina, F.J., 2020. Channel-levee evolution in combined contour current-turbidity current flows from flume-tank experiments. Geology 48 (4), 353–357. https://doi.org/10.1130/G47111.1.

Mulder, T., Alexander, T., 2001. The physical character of submarine density flows and their deposits. Sedimentology 48, 269–301. https://doi.org/10.1046/j.1365-3091.2001.00360.x.

Mutti, E., 1985. Turbidite systems and their relations to depositional sequences. In: Zuffa, G.G. (Ed.), Provenance of Arenites. D. Reidel Publishing Company, Dordrecht, pp. 65–93.

Mutti, E., 1992. Turbidite sandstones. Istituto di geologia, Università di Parma, Agip, p. 275.

Mutti, E., Tinterri, R., Remacha, E., Mavilla, N., Angella, S., Fava, L., 1999. An Introduction to the analysis of Ancient Turbidite Basins from an Outcrop Perspective: AAPG Continuing Education Course Note, No. 39., p. 61.

Mutti, E., Carminatti, M., Moreira, J.L.P., Grassi, A.A., 2006. Chaotic Deposits: Examples from the Brazilian Offshore and from Outcrop Studies in the Spanish Pyrenees and Northern Apennines, Italy. - A.A.P.G. Annual Meeting, April 9–12, Houston, Texas.

Myrow, P.M., Fischer, W., Goodge, J.W., 2002. Wave-modified turbidites: combined-flow shoreline and shelf deposits, Cambrian, Antarctica. J. Sediment. Res. 72 (5), 641–656. https://doi.org/10.1306/022102720641.

Ogata, K., Mutti, E., Pini, G.A., Tinterri, R., 2012. Mass transport-related stratal disruption within sedimentary mélanges: examples from the northern Apennines (Italy) and south-Central Pyrenees (Spain). Tectonophysics 568, 185–199. https://doi.org/10.1016/j.tecto.2011.08.021.

Ono, K., Plink-Bjorklund, P., Eggenhuisen, J.T., Cartigny, M.J.B., 2019. Froude supercritical flow processes and sedimentary structures: New insights from experiments with a wide range of grain sizes. Sedimentology. https://doi.org/10.1111/sed.12682.

Pälike, H., et al., 2012. A Cenozoic record of the equatorial Pacific carbonate compensation depth. Nature 488, 609–614. https://doi.org/10.1038/nature11360.

Paull, C.K., Talling, P.J., Maier, K.L., Parsons, D., Xu, J., Caress, D.W., Gwiazda, R., Lundsten, E.M., Anderson, K., Barry, J.P., Chaffey, M., O'Reilly, T., Rosenberger, K.J., Gales, J.A., Kieft, B., McGann, M., Simmons, S.M., McCann, M., Sumner, E.J., Clare, M.A., Cartigny, M.J., 2018. Powerful turbidity currents driven by dense basal layers. Nat. Commun. 9, 4114. https://doi.org/10.1038/s41467-018-06254-6.

Pickering, K.T., Hiscott, R.N., 2015. Deep Marine Systems: Processes, Deposits, Environments. Tectonics and Sedimentation. Wiley and American Geophysical Union. ISBN:978-1-4051-2578-9.

Pickering, K.T., Hiscott, R.N., Hein, F.J., 1989. Deep-Marine Environments. Unwin Hyman, London, p. 416.

Pohl, F., 2019. Turbidity Currents and their Deposits in Abrupt Morphological Transition Zones. PhD thesis, University of Utrecht. 210 p.

Posamentier, H.W., Martinsen, O.J., 2011. The character and genesis of submarine mass-transport deposits: insights from outcrop and 3D seismic data. In: Mass-Transport Deposits in Deepwater Settings. vol. 96. SEPM, Special Publication, Tulsa, pp. 7–38. https://doi.org/10.2110/sepmsp.096.007.

Postma, G., Nemec, W., Kleinspehn, K.L., 1988. Large floating clasts in turbidites; a mechanism for their emplacement. Sediment. Geol. 58, 47–61. https://doi.org/10.1016/0037-0738(88)90005-X.

Rahmstorf, S., 2006. Thermohaline Ocean circulation. In: Elias, S.A. (Ed.), Encyclopedia of Quaternary Science. Elsevier, Amsterdam, pp. 739–750.

Rebesco, M., Camerlenghi, A. (Eds.), 2008. Contourites. In: Developments in Sedimentology, vol. 60. Elsevier, p. 688.

Rebesco, M., Hernández-Molina, F.J., van Rooij, D., Wåhlin, A., 2014. Contourites and associated sediments controlled by deepwater circulation processes: state-of-the-art and future considerations. Mar. Geol. 352, 111–154. https://doi.org/10.1016/j.margeo.2014.03.011.

Rodrigues, S., Hernandez-Molina, F.J., Fonnesu, M., Miramontes, E., Rebesco, M., Campbell, D.C., 2022. A new classification system for mixed (turbidite-contourite) depositional systems: examples, conceptual models and diagnostic criteria for modern and ancient records. Earth Sci. Rev. https://doi.org/10.1016/j.earscirev.2022.104030.

Sansom, P., 2018. Hybrid turbidite-contourite systems of the Tanzanian margin. Pet. Geosci. 24, 258–276. https://doi.org/10.1144/petgeo2018-044.

Sequeiros, O., Pittaluga, M.B., Frascati, A., Pirmez, C., Masson, D.G., Weaver, P.P.E., Crosby, A.R., Lazzaro, G., Botter, G., Rimmer, J.G., 2019. How typhoons trigger turbidity currents in submarine canyons. Nature 9 (9220). https://doi.org/10.1038/s41598-019-45615-z.

Shanmugam, G., 2013. Modern internal waves and internal tides along oceanic pycnoclines: challenges and implications for ancient deep-marine baroclinic sands. AAPG Bull. 97, 767–811.

Shanmugam, G., 2020. The turbidite—contourite—tidalite—hybridite problem. In: Virtual Get Together "Virtual Geology Talk" And Video by the Drifters Research Group. Royal Holloway University of London. 2-July.

Shanmugam, G., Spalding, T.D., Rofheart, D.H., 1993a. Traction structures in deep-marine, bottom-current reworked sands in the Pliocene and Pleistocene, Gulf of Mexico. Geology 21, 929–932.

Shanmugam, G., Spalding, T.D., Rofheart, D.H., 1993b. Process sedimentology and reservoir quality of deepmarine bottom-current reworked sands (sandy contourites): an example from the Gulf of Mexico. AAPG Bull. 77, 1241–1259.

Shannon, L.V., Nelson, G., 1996. The Benguela: large scale features and processes and system variability. In: The South Atlantic. Springer Science and Business Media LLC, Berlin/Heidelberg, Germany, pp. 163–210. https://doi.org/10.1007/978-3-642-80353-6_9.

Slatt, R.M. (Ed.), 2006. Stratigraphic Reservoir Characterization for Petroleum Geologists, Geophysicists, and Engineers. vol. 61. Elsevier. 492 p.

Stow, D.A.V., 1979. Distinguishing between fine-grained turbidites and contourites off Nova Scotia. Sedimentology 26, 371–387. https://doi.org/10.1111/j.1365-3091.1979.tb00915.x.

Stow, D.A.V., Shanmugam, G., 1980. Sequence of structures in fine-grained turbidites: comparison of recent deep-sea and ancient flysch sediments. Sediment. Geol. 25, 23–42. https://doi.org/10.1016/0037-0738(80)90052-4.

Stow, D.A.V., Smillie, Z., 2020. Distinguishing between deepwater sediment facies: turbidites, contourites and hemipelagites. Geosciences 10 (68). https://doi.org/10.3390/geosciences10020068.

Stow, D.A.V., Huc, A.Y., Bertrand, P., 2001. Depositional processes of black shales in deep water. Mar. Pet. Geol. 18 (4), 491–498. https://doi.org/10.1016/S0264-8172(01)00012-5.

Stramski, D., Reynolds, R.A., Babin, M., Kaczmarek, S., Lewis, M.R., Röttgers, R., Claustre, H., 2007. Relationships between the surface concentration of particulate organic carbon and optical properties in the eastern South Pacific and eastern Atlantic oceans. Biogeosciences 5 (1), 171–201. https://doi.org/10.5194/bg-5-171-2008.

Straub, K.M., Mohrig, D., 2008. Quantifying the morphology and growth of levees in aggrading submarine channels. J. Geophys. Res. Earth Surf. 113, 1–20. https://doi.org/10.1029/2007JF000896.

Straub, K.M., Mohrig, D., Buttles, J., McElroy, B., Pirmez, C., 2011. Quantifying the influence of channel sinuosity on the depositional mechanics of channelized turbidity currents: a laboratory study. Mar. Pet. Geol. 28, 744–760. https://doi.org/10.1016/j.marpetgeo.2010.05.014.

Sumner, E.J., Talling, P.J., Amy, L.A., 2009. Deposits of flows transitional between turbidity current and debris flow. Geology 37 (11), 991–994. https://doi.org/10.1130/G30059A.1.

Talling, P.J., 2013. Hybrid submarine flows comprising turbidity current and cohesive debris flow: deposits, theoretical and experimental analyses, and generalized models. Geosphere 9, 460–488. https://doi.org/10.1130/GES00793.1.

Talling, P.J., Masson, D.G., Sumner, E.J., Malgesini, G., 2012. Subaqueous sediment density flows: depositional processes and deposit types. Sedimentology 59, 1937–2003. https://doi.org/10.1111/j.1365-3091.2012.01353.x.

Thurman, H.V., Burton, E.A., 2001. Introductory Oceanography, ninth ed. Prentice Hall, Englewood Cliffs.

Tinterri, R., Drago, M., Consonni, A., Davoli, G., Mutti, E., 2003. Modelling subaqueous bipartite sediment gravity flows on the basis of outcrop constraints: first results. Mar. Pet. Geol. 20 (6), 911–933. https://doi.org/10.1016/j.marpetgeo.2003.03.003.

Walsh, J.P., Nittrouer, C.A., 2009. Understanding fine-grained river-sediment dispersal on continental margins. Mar. Geol. 263, 34–45. https://doi.org/10.1016/j.margeo.2009.03.016.

Waltham, D., 2004. Flow transformations in particulate gravity currents. J. Sediment. Res. 74, 129–134. https://doi.org/10.1306/062303740129.

Wang, Z., Plate, E.J., 1996. A preliminary study on the turbulence structure of flows on non-Newtonian fluid. J. Hydraul. Res. 34, 345–361.

Wang, Z., Xu, J., Talling, P.J., Cartigny, M.J.B., Simmons, S.M., Gwiazda, R., Paull, C.K., Maier, K.L., Parsons, D.R., 2020. Direct evidence of a high-concentration basal layer in a submarine turbidity current. Deep-Sea Res. I 161. https://doi.org/10.1016/j.dsr.2020.103300, 103300.

Yang, T., Cao, Y., Friis, H., Liu, K., Wang, Y., 2018. Origin and evolution processes of hybrid event beds in the lower cretaceous of the Lingshan Island, Eastern China. Aust. J. Earth Sci. 65 (4), 517–534. https://doi.org/10.1080/08120099.2018.1433236.

Yin, S., Hernández-Molina, F.J., Zhang, W., Li, J., Wang, L., Ding, W., Ding, W., 2019. The influence of oceanographic processes on contourite features: a multidisciplinary study of the northern South China Sea. Mar. Geol. 415, 105967. https://doi.org/10.1016/j.margeo.2019.105967.

Chapter 7

# Deepwater sedimentation units

Jon R. Rotzien[a,b], F. Javier Hernández-Molina[c], Marco Fonnesu[d] and Antoine Thieblemont[e]

[a]*Basin Dynamics, Conroe, TX, United States,* [b]*University of Houston, Houston, TX, United States,* [c]*Department of Earth Sciences, Royal Holloway, University of London, Egham, Surrey, United Kingdom,* [d]*Eni SpA, San Donato Milanese, Milan, Italy,* [e]*TotalEnergies, Courbevoie, France*

## Introduction

This book characterizes deepwater sedimentary systems at all scales—from the tectonic and basin-scale drivers of deepwater sedimentation (Chapters 3–5), as well as the specific processes (Chapters 6 and 9) that deliver all types of grain sizes to deepwater environments. As mentioned in Chapter 1, the basic building block of any depositional system is the bed, or sedimentation unit. By recognizing, characterizing and being able to measure the wide range of sedimentation units in deepwater environments (Chapters 8–10), one can examine how the depositional system is built—bed by bed. Sometimes beds are removed due to erosion, and it is like pages being ripped out of a book. But with practice studying the wide range of sedimentation units in deepwater environments, one can reconstruct the history of the deepwater basin and determine what is upslope, downslope and to the sides—aiding in the prediction of sedimentary deposits away from control points such as wells, cores and select outcrops. Deepwater sedimentation units are often below the resolution of seismic data but can be roughly observed in well logs and especially full-bore Formation Micro Imager (FMI) data (Chapter 16) but are best observed in outcrop and in core. This chapter is designed to show the range and variability in endmember and hybrid types of sedimentation units found in deepwater environments prone to sediment gravity flows. While some of the sedimentation unit models presented are well established in peer reviewed research, others remain elusive until more data become available from the deep seas. This chapter is intended to describe a wide range of sedimentation units, yet emphasizes the greatest discoveries regarding deepwater sedimentation are yet to be found. The information is arranged in the following manner:

1. Introduction to deepwater sedimentation units
2. Turbidity current (flow) deposits, including coarse- to fine-grained deposits and specific cases of wave-modified turbidites, tide-modified turbidites and river-fed turbidites
3. Hybrid event bed (HEB) deposits
4. Debris flow deposits
5. Bottom current influenced deposits and reworked deposits (in brief, as Chapter 9 explains in more detail)
6. Pelagic and hemipelagic deposits
7. Mass transport deposits (MTD) (in brief, as Chapter 10 explains in more detail)
8. How to apply this information
9. Conclusion

Upon successful completion of this chapter, the reader will have a greater appreciation for:

- Characterizing fundamental sedimentation units found in deepwater environments,
- Interpreting the range and variability in 1-, 2- and 3-D sedimentation unit architecture and heterogeneity,
- Applying the merits, limitations, pitfalls in the nomenclature and how to interpret deepwater stratigraphic intervals constructed of multiple types of sedimentation units and
- Illustrating and characterizing the vertical and lateral relationships in sedimentary facies used to (i) correlate deepwater sedimentary deposits and (ii) build predictive models of the subsurface from bed-scale to submarine fan-scale.

**Deepwater Sedimentary Systems.** https://doi.org/10.1016/B978-0-323-91918-0.00009-8

## Sedimentation units

The fundamental building block of any deepwater depositional system is the sedimentation unit, or the deposit of one sediment gravity flow or event. Previous descriptions of the range of deepwater sedimentation units are provided in Ghibaudo (1992), Mulder and Alexander (2001), Bouma (2004), Pickering et al. (1989), Talling et al. (2012), Pickering and Hiscott (2015) and Stow and Smillie (2020). Deepwater sedimentation units are different from beds in shallow marine and terrestrial systems in that they typically have well-defined bed bases and tops. Herein are descriptions of turbidity current, transitional to hybrid flow, contour current and reworked deposits, as well as wave-, tide- and fluvial-modified turbidity current, debris flow and mass transport deposits, arranged in Table 1 roughly from high- to low-energy deposition. While these deposits can occur in any type of deepwater depositional system, certain deposits are found in higher abundance based on the controls of the depositional system including sediment type and supply, tectonic setting, relative sea level, climate and other source-to-sink factors (Chapters 1, 3–5, 11 and 12). Understanding the range and variability in sedimentation units is important for an accurate valuation of the subsurface, and there are myriad examples of how missteps in this process cost companies' significant write-downs. Applications of this skillset also extend to proper interpretation of the depositional environment, sampling of deepwater sediment for accurate climate and ocean calibration, mapping the distribution of contaminants in oceans and many others.

Most turbidites, which are the primary lithology in most deepwater petroleum reservoirs, are products of diverse sediment gravity flows. In other words, these flows transport sediment through a variety of mechanisms, simultaneously, via fluid turbulence, matrix strength and in between (Peakall et al., 2020) and can readily undergo flow transformation (sensu Waltham, 2004).

In general, researchers have been observing model sedimentation units with greater scrutiny, often finding differences within the same first-order architectural element. This has led to further diversification of sedimentation units and an increase in subdivisions. For example: the traditional description of the Bouma (1962) Sequence is of a single fining upward unit with five main divisions. Stow and Shanmugam (1980) divide the upper end of the Bouma Sequence into multiple subdivisions to describe very low-density turbidity current deposits. Another example is that Gutierrez (2019) and earlier studies divide the planar laminated and ripple cross-laminated intervals of the Bouma Sequence based on their geometry and angle of climb. Dodd et al. (2020) divide the H3 division (i.e., debritic division) of Haughton et al. (2009) into multiple divisions based on clay content and mudclast abundance. These three examples add significant complexity to the description of turbidite sedimentation units, but also add to the understanding of the range and variability of sedimentary structures formed from very dilute, fine-grained flows, low-density turbidity currents and HEB, respectively. These subdivisions are merited because they relate back to fundamental processes of sedimentation and are important because they have different reservoir quality values. Most importantly, the variations in sedimentary structures within sedimentation units can be related back to depositional environment, and in turn, bedding traits including porosity, permeability, reservoir extent, connectivity and deliverability (Walker, 2004). Other exotic deepwater deposits not formed by mass flows or sediment gravity flows including injectites, extrudites, mud volcanoes, sand volcanoes and fluid escape features are only mentioned briefly in this chapter and are discussed in detail in Chapter 10.

## Turbidity current deposits

Turbidity currents are responsible for depositing turbidites, a type of principal reservoir in deepwater stratigraphic successions (Tinterri et al., 2020). In turbidity currents, there are three main grain-size populations that form the basis for the three principal sedimentation units (sensu Lowe, 2004; Lowe and Graham, 2012): (i) grains $>10$ mm in diameter (pebble size particles and above) typically found at the base of flows forming laminar traction carpets (grain population one); (ii) grains from $\sim$0.25 to 10 mm in diameter (from medium sand to pebble size particles) can be found in suspension or as traction bedload (grain population two); and (iii) grains $<0.25$ mm (clay to fine sand particles) typically carried in suspension by fluid turbulence alone (grain population three). These grain populations roughly correspond to the main types of turbidity currents: low-density turbidity currents with sediment concentrations generally below 10% carrying mostly sand, or grain population two; high-density turbidity currents with sediment concentrations $>10\%$ carrying grain populations one and two; very low-density turbidity currents with low sediment concentrations typically carrying grain population three.

There are three main turbidite sedimentation unit models described in Fig. 1: (i) classic sandy turbidites represented by the Bouma (1962) Sequence; (ii) coarse-grained to gravelly turbidites represented by the Lowe (1982) Sequence; and (iii) fine- to very fine-grained turbidites represented by the Stow and Shanmugam (1980) model.

**TABLE 1** The fundamental building blocks of deepwater depositional systems include vast accumulations of pelagic and hemipelagic sedimentation, but in the interest of this paper, the relatively coarse-grained, high-energy flows are represented by several sedimentation unit models.

| | Model description | Principal divisions | Processes of sedimentation | Typical depositional environment | Depositional system type | Typical continental margin profile | Reservoir potential and quality | Bed thickness distribution |
|---|---|---|---|---|---|---|---|---|
| Bouma (1962) | Classic turbidite (low-density turbidity current deposit) | $T_a$, $T_b$, $T_c$, $T_d$, $T_e$ | Suspension, traction, hemipelagic settling | Canyon, channel, levee, lobe (splay), overbank | Apron, submarine fan, submarine fan valley | Active to passive | High | Highly variable; typically thin to very thick |
| Gonthier et al. (1984) | Classic contourite | Lower mudstone (F1), structured siltstone and very fine-grained sandstone (F4a), burrowed sandstone (F4b/F3), structured siltstone (F2), upper mudstone (F1) | Suspension, traction, hemipelagic settling | Overbank, levee | Drift, mixed turbidite and contourite system in all submarine fan types | Passive | Low | Typically thin- to medium-bedded deposits |
| Stow and Shanmugam (1980) | Fine-grained turbidite (very low-density turbidity current deposit) | T0, T1, T2, T3, T4, T5, T6, T7, T8 | Suspension, traction, hemipelagic, pelagic | Levee, overbank, distal lobe (splay) | Muddy apron | Passive | Low | Medium- to thin-bedded and less |
| Lowe (1982) | Coarse-grained turbidite (high-density turbidity current deposit) | $R_1$, $R_2$, $R_3$, $S_1$, $S_2$, $S_3$, $T_t$ | Traction, matrix strength, traction carpet, suspension | Canyon, channel, proximal lobe (splay) | Submarine fan valley, fan delta, Gilbert delta, talus slope, gravelly ramp | Active | Moderately high | Typically thick- to very thick-bedded |
| Mutti (1992) | Complete turbidite and debris flow spectrum defined along a facies tract | F1, F2, F3, F4, F5, F6, F7, F8, F9a, F9b, CgRF | Cohesive freezing, frictional freezing, traction carpets, en-masse deposition, traction-plus-fallout processes | Channel to splay (lobe) and sheet-like deposits, mostly in foreland basins | High to low efficiency turbidite system | Active | High | Highly variable; typically thick to very thin |

*Continued*

**TABLE 1** The fundamental building blocks of deepwater depositional systems include vast accumulations of pelagic and hemipelagic sedimentation, but in the interest of this paper, the relatively coarse-grained, high-energy flows are represented by several sedimentation unit models—cont'd

| | Model description | Principal divisions | Processes of sedimentation | Typical depositional environment | Depositional system type | Typical continental margin profile | Reservoir potential and quality | Bed thickness distribution |
|---|---|---|---|---|---|---|---|---|
| Lowe and Guy (2000) | Slurry flow deposit (transitional flow deposit) | $M_1$, $M_{2a}$, $M_{2b}$, $M_3$, $M_4$, $M_5$, $M_6$, $M_7$ | Suspension, mixed matrix strength and fluid turbulence, traction | Canyon, channel, lobe (splay) | Apron, submarine fan, submarine fan valley | Active to passive | Moderately high | Highly variable; typically thin to very thick |
| Haughton et al. (2003, 2009) | Linked debrite and hybrid event bed (transitional flow deposit) | H1, H2, H3, H4, H5 | Suspension, mixed matrix strength and fluid turbulence, traction, hemipelagic settling | Lobe (splay) | Apron, submarine fan, submarine fan valley | Active to passive | Moderate | Highly variable; typically thin to thick |
| Power et al. (2013) | Argillaceous sandstone (transitional flow deposit) | Lower sandstone interval (sensu Ta); upper muddy sandstone interval (argillaceous interval); mudstone top (sensu Te) | Suspension, mixed matrix strength and fluid turbulence | Channel, lobe (splay) | Submarine fan, submarine fan valley | Typically passive | Moderately high | Typically thick-bedded and less |
| De Castro et al. (2020) | Reworked contourite (turbidity current and contour current deposits) | Lower gravelly sandstone (F6), structured siltstone and very fine-grained sandstone (F4a), burrowed sandstone (F4b/F3), structured siltstone (F2), upper mudstone (F1) | Suspension, traction, hemipelagic settling | Channel, lobe (splay), levee, overbank | Drift, mixed turbidite and contourite system in all submarine fan types | Active to passive | Moderate | Typically thick-bedded and less |
| Rotzien (2019) and Gutierrez (2019) | Complete turbidite (high- and low-density turbidity current) | Lower antidune interval; $R_1$, $R_2$, $R_3$, $S_1$, $S_2$, $S_3/T_a$, $T_t$, $T_b$, $T_c$, $T_d$, $T_e$ | Traction, traction carpet, suspension, hemipelagic sedimentation | Canyon, channel, lobe (splay) | Submarine fan valley, submarine fan, apron, fan delta, Gilbert delta | Active | High | Typically thick- to very thick-bedded |

| | | | | | | | | |
|---|---|---|---|---|---|---|---|---|
| Myrow et al. (2002) | Wave-modified turbidite (modified low-density turbidity current deposit) | Lower cross-stratified interval, $T_a$, swaley and hummocky cross-stratified interval (SCS, HCS), turbidite divisions sensu $T_b$, $T_c$, $T_d$, $T_e$ | Suspension, traction, hemipelagic settling (wave-modified) | Shelf, canyon, channel and beyond | Storm-dominated shelf, wave-dominated delta, submarine fan valley | Passive | High | Typically medium- to thin-bedded |
| Klein (1971; 1978); Shanmugam (2006) | Tidally modified turbidite (modified low-density turbidity current deposit) | Lower structureless interval (sensu $T_a$), cross-stratified interval with reactivation surfaces, modified mud-draped laminated intervals (sensu $T_b$, $T_c$, $T_d$), $T_e$ | Suspension, traction, hemipelagic settling (tidally modified) | Shelf, canyon, channel | Tidally dominated shelf, tidally dominated delta, submarine fan, apron | Passive | Moderately high | Typically medium- to thin-bedded |
| Mulder et al. (2001, 2003) | Fluvially modified turbidite (modified low-density turbidity current deposit) or hyperpycnite | Lower structureless interval (sensu $T_a$), underlain by surge deposits (sensu Lowe, 1982), carbonaceous-rich turbidite divisions sensu $T_b$, $T_c$, $T_d$, $T_e$ | Suspension, traction, hemipelagic settling (fluvially modified) | Shelf, canyon, channel, lobe (splay), levee, overbank | Tidally-dominated shelf, tidally-dominated delta, submarine fan, submarine fan valley, apron | Active to passive | Moderately high | Highly variable; typically thin to thick |

The primary models describing most deepwater sedimentation units are designed and inspired based on the work of Bouma (1962), Klein (1971, 1998), Gonthier et al. (1984), Stow (1979), Stow and Shanmugam (1980), Lowe (1982), Lowe and Guy (2000), Mulder et al. (2001, 2003), Myrow et al. (2002), Haughton et al. (2003, 2009), Shanmugam (2006), Kane and Pontén (2012), Power et al. (2013), Capella et al. (2017), De Weger et al. (2020) and De Castro et al. (2020).

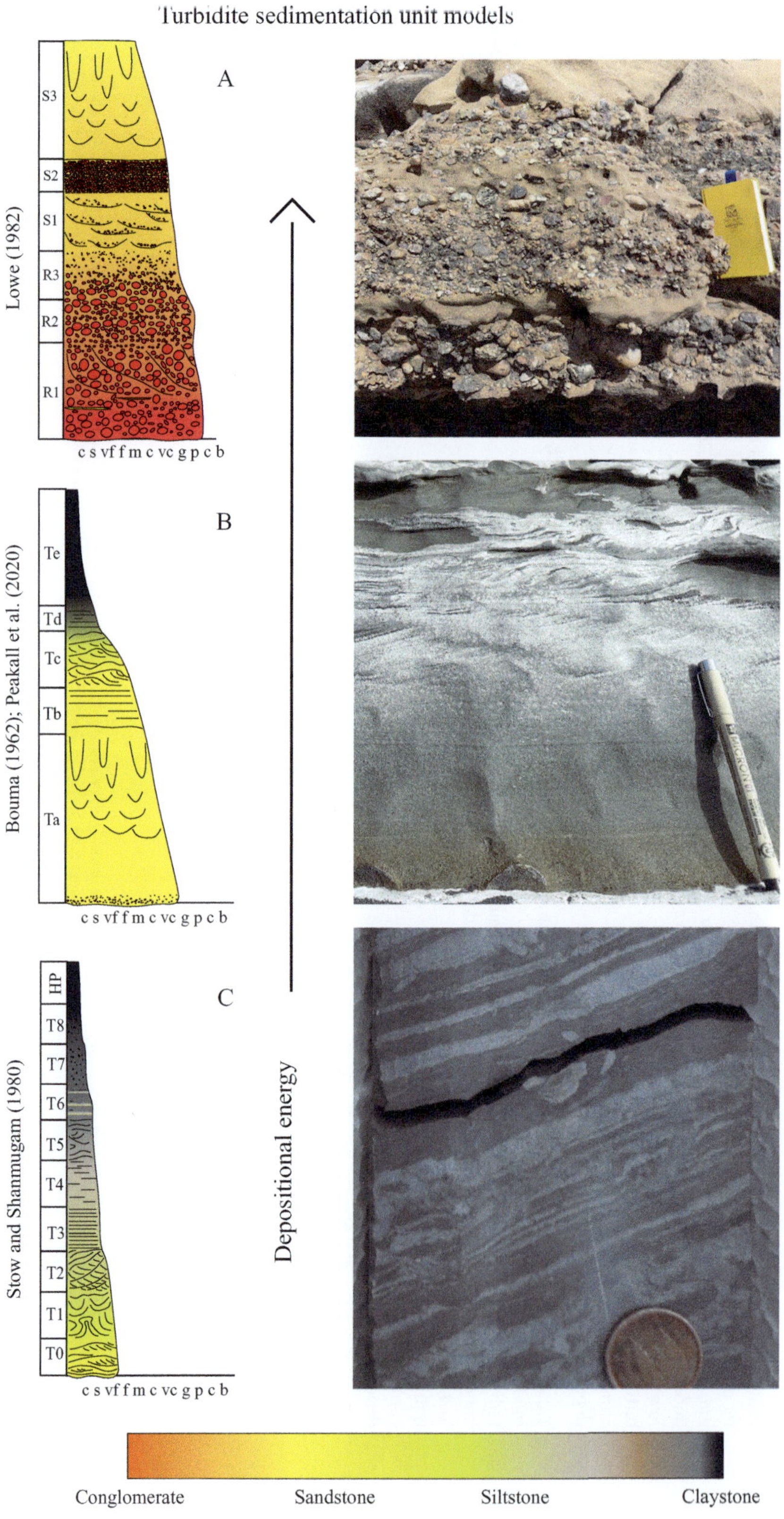


**FIG. 1** Deepwater sedimentation units formed by turbidity currents. Models include: (A) Lowe (1982) coarse-grained or high-density turbidity current sedimentation unit model with corresponding partial Lowe bed to the right, showing a well-developed R2 interval, Eocene Carmelo Formation, California, USA. (B) Bouma (1962) fine-grained, low-density or "classic" turbidity current sedimentation unit model modified by Peakall et al. (2020) with idealized bed to the right from the Upper Miocene Lower Mount Messenger Formation (Rapanui), Taranaki Basin, North Island, New Zealand; and (C) Stow and Shanmugam (1980) muddy or very low-density turbidity current sedimentation unit model with partial beds from the Eocene San Vicente Formation (Ainsa Quarry), Ainsa Basin, Spain.

## Low-density turbidity current deposits—"classic fine-grained turbidites"

The classic turbidite model described by Bouma (1962) has five main divisions, labeled $T_a$, $T_b$, $T_c$, $T_d$ and $T_e$ resulting from five stages of sedimentation, each division deposited by suspension ($T_a$, $T_d$, $T_e$) or traction ($T_b$, $T_c$) sedimentation (Fig. 1B). The five main stages of sedimentation represent a waning flow and result from flow unsteadiness, yet these stages can also occur due to flow nonuniformity. A sixth stage was added by Peakall et al. (2020) to show the transitional to debris flow stage preceding turbidite deposition leaving several bedding plane sole structures such as grooves, tool marks and flutes.

The $T_a$ division is structureless (unstructured) or "massive" sandstone formed by direct deposition from a fully turbulent flow, yet it can have a variety of dish and pillar structures (Lowe and Lo Piccolo, 1974). The overall appearance is created by instantaneous loss of energy in turbulent flow, leading to flow collapse and immediate suppression of grain movement along the bed, however, this is challenged by numerous articles citing auto-suspension (Bagnold, 1962), sustained high-density current (Kneller and Branney, 1995), a fine-grained "rain" (Cantero et al., 2012) or lofting (Steel et al., 2017) and others. Structureless sandstone is commonly found in deepwater environments, which are prone to episodic delivery of sediment, whereas almost any other environment is subject to continuous movement and reworking of the bed (i.e., rivers, shoreface, eolian environments), leaving a greater abundance of structured deposits.

Overlying the $T_a$ in a classic Bouma Sequence is the $T_b$ division. It is composed of planar parallel laminated sand, typically associated with high flow velocity. This sedimentary structure is associated with upper stage plane beds formed by traction beneath a near-bed laminar sublayer. These types of deposits have been misinterpreted in the past when they contain significant mud, and can represent transitional flow deposits, not true $T_b$ formed by traction (Stevenson et al., 2020). Further, the Bouma $T_b$ division is differentiated from crude stratification found in high-density turbidity current deposits (see Lowe, 1979 in next section) based on lamination thickness and sorting.

The Bouma $T_c$ interval consists of ripple cross-laminated sand and silt formed by traction beneath a near-bed laminar sublayer (Baas, 2003). The ripples that form the ripple cross-lamination are typically less than or equal to 3–4 cm (~ 1.2–1.6 in.) in height and less than 50 cm (~ 20 in.) in wavelength. The typical $T_c$ interval includes small foresets that form steepening-upward laminae, a record of the progradation of the ripple bedform in a unidirectional flow. There are five main types of ripple cross-lamination associated with deepwater deposits, and many variations within them: (i) starved or flaser ripple cross-lamination; (ii) single-layer ripple cross-lamination; (iii) low-angle climbing-ripple cross-lamination; (iv) high-angle ripple cross-lamination; (v) sinusoidal lamination, in which entire ripple form is preserved. The higher the angle of climb, the higher presumed sedimentation rate. As the angle of the cross-lamination climbs, the influence of suspension sedimentation increases as sediment is essentially raining out of suspension. Also, the flow expands while bedload traction transport is reduced, producing high-angle climbing-ripple cross-laminated sandstone ( Jobe et al., 2011). While these intervals are known as the $T_c$ division in many turbidite successions, care needs to be applied in that this type of sedimentary structure is not diagnostic of any single depositional environment. Climbing-ripple cross-laminated divisions can be identified in bottom current reworked turbidites and sandy contourites (Mutti et al., 2014; De Castro et al., 2020, 2021a,b; Fonnesu et al., 2020; Fuhrmann et al., 2020; De Weger et al., 2021).

The $T_d$ interval consists of laminated siltstone deposited primarily via suspension. Stow and Shanmugam (1980) later subdivided this interval and the underlying $T_c$ interval into nine divisions based on sedimentary structure variations, described later in this chapter.

The $T_e$ interval is the final interval of the Bouma sedimentation unit and is formed by gentle suspension settling of silt and clay from the low-density turbidity flow. There are two other mechanisms for forming this mudstone interval: (i) unrelated hemipelagic sedimentation, or (ii) pelagic sedimentation, also known as background sedimentation in deepwater environments. Overall, muddy bed tops containing $T_d$ and $T_e$ intervals are common in deepwater areas due to the potentially long pauses between relatively coarse-grained sedimentation events.

When lithified, the complete Bouma Sequence contains both porous reservoir and less permeable and finer grained rock. However, in nearly all sedimentary successions on Earth, there are variations in the typical Bouma Sequence that fall under the following categories: (i) incomplete or missing divisions; (ii) disproportionately thin or thick divisions; and (iii) alternation or reverse order of divisions.

The type section for the Bouma (1962) turbidite sedimentation unit is at Peira-Cava near the commune of Luceram in southeast France, in the Maritime Alps and surrounding areas including the entire Annot System. There are numerous examples of complete to incomplete Bouma Sequence locales around the world such as: (i) the Taranaki and East Coast basins, New Zealand; (ii) the Jackfork Group, Arkansas, USA; (iii) the Windermere Supergroup, British Columbia, Canada; (iv) the Tanqua-Karoo Basin, South Africa; (v) the Western Irish Namurian Basin, Ireland; (vi) the Ainsa Basin, Spain; (vii) the Marnoso-arenacea Formation (Northern Apennines), Italy; and many other locales. Note these examples are mostly active margin basins, but Bouma turbidites are often found in roughly equal abundance along passive margins.

## High-density turbidity current deposits—"coarse-grained to gravelly turbidites"

The coarse-grained to gravelly turbidite model by Lowe (1979, 1982) has six main divisions, labeled $R_1$, $R_2$, $R_3$ for the gravelly base and $S_1$, $S_2$, $S_3$ for the relatively finer grained sandy top. The bed results from six stages of sedimentation with each division deposited by traction ($R_1$, $S_1$—may be debris flow for $R_1$), traction carpet ($R_2$, $S_2$) or suspension ($R_3$, $S_3$) processes (Fig. 1A). As with the Bouma Sequence, the stages of sedimentation represent a waning flow and result from flow unsteadiness yet can also be formed due to flow nonuniformity.

The basal $R_1$ interval is characterized by the coarsest gravels in the sedimentation unit. Clast sizes can be in the boulder range, but more commonly cobble-grade. The $R_1$ interval is typically nongraded and can contain dune-scale cross-stratification, representing traction sedimentation. This type of cross-stratification can be several centimeters in thickness, such as in the Sites Sandstone, USA (Murray et al., 1996) to several meters in scale as documented in the Cerro Toro Formation, Chile (Hubbard et al., 2008). This interval may also represent a debrite, supported by cohesion and matrix strength, or by frictional forces (see Chapter 6).

The overlying $R_2$ interval consists of an inversely graded and gravelly traction carpet, deposited via frictional freezing. This is formed by shear of an overlying flow that agitates the grains, and the finer grains sieve through the coarser grains to the base of the flow. As the grains sort and settle, less and less space surrounds the grains until they have no room to move and freeze, thus preserving the inversely graded rock fabric (i.e., grains coarsen upward in the bed).

The $R_3$ interval consists of a normally graded (i.e., grains fine upward in the bed, also known as *graded*) gravelly deposit.

The S intervals consist mostly of sand and finer grained gravels including granules and pebbles and mirrors the depositional process from the underlying R intervals.

$S_1$ consists of dune-scale to low-angle sweeping cross-stratification formed by traction.

The $S_2$ interval is a traction carpet formed by frictional freezing.

The $S_3$ interval is structureless sand formed by suspension sedimentation (Lowe, 1988). It is deposited by the same sedimentary process and contains similar structures and grain sizes to the $T_a$ division, so they are considered equivalent. In some successions, the top of the $S_3$ is reworked by traction and can feature dune-scale cross-stratification.

Lowe (1982) was one of the first to acknowledge this oft missed division in deepwater successions, and labeled it the Lowe $T_t$ interval, representing dune-scale cross-stratification above the structureless Bouma $T_a$ interval, yet beneath the $T_b$ interval. A few studies, including Hoyal (2018) and Gutierrez (2019), place the $T_t$ below the $T_a$. In the $T_t$ interval, laminae are concave up and steepen upward typically >10 degrees. The relict dunes have relief of greater than 3 cm (~ 1.2 in.) and wavelengths that exceed 50 cm (~ 20 in.). Dunes interact with the entire depth of the flow, scale with it and require certain flow conditions to be met for an extended period of time (Arnott, 2012).

It is not uncommon in high-density turbidity current deposits to feature crude stratification consisting of thin, wavy, uneven to discontinuous intervals of well-sorted sand deposited via traction alternating with structureless sand deposited via suspension. These alternating layers are a few to several cm in thickness, depending on the grain size. This type of layering has been interpreted to represent the product of discrete surges or pulses in the flow. During sudden flow deceleration, suspension sedimentation ($S_3$/$T_a$) will occur. As the flow accelerates, it reworks the bed and forms a thin traction ($S_1$) or traction carpet ($S_2$) deposit.

While many have understood and worked deposits classically associated with deposition via low-density turbidity currents, forming the Bouma Sequence of sedimentary structures, fewer have had the chance to work high-density turbidity current deposits resulting from sediment concentrations greater than 10%. In these flows, coarser grains must be supported, at least in part, by concentration effects. The main concentration effects include grain-to-grain collisions, hindered settling, buoyancy effects and kinetic sieving (Lowe, 1982).

Coarse-grained units are typically created by three main stages of sedimentation (Lowe, 2004; Lowe and Graham, 2012). Sometimes a fourth stage can occur, which is the development of types of antidunes prior to the following three stages described here (Pohl et al., 2020 and references therein; Cornard and Pickering, 2019, 2020; Englert et al., 2020). First, if flows are steady and uniform, turbulence will reach the bed and form traction structures such as dunes or sweeping, low-angle cross-stratification. Second, if flows are waning, then the coarsest gravels will reach the bed and turbulence will be dampened, forming inversely graded traction carpets. Third, if flows are highly unsteady and nonuniform, then flows will collapse and bedload transport will be stopped, thus forming a structureless deposit.

The intervals R, S and T were chosen to represent first-order architectural elements of turbidite beds because the letters follow in the English alphabet. In some basins on Earth, nearly complete R-S-T turbidite sedimentation units can be found, consisting of $R_1$ through $T_e$ divisions (see Chapter 8). However, many basins do not provide the full suite of dynamic grain populations, or undergo the required changes in flow steadiness and uniformity to make this possible.

High-density turbidity current deposits often show less obvious structuring and a suite of Lowe divisions. Examples include: (1) the Pliocene Repetto Formation, Ventura Basin, California, USA; (2) the Eocene–Oligocene Annot Sandstone, France; (3) the Eocene Carmelo Formation, California, USA; (4) the Cretaceous Juniper Ridge Conglomerate and Venado Sandstone, California, USA; (5) the Cerro Toro Formation, Chile and associated deposits in the Neuquen Basin, Argentina; (6) the Rosario Formation, Baja California, Mexico; and many others.

### Very low-density turbidity current deposits—"Fine-grained turbidites"

Stow and Shanmugam (1980) proposed a model for fine-grained turbidites with nine divisions (T0–T8). The model follows this pattern, from base to top (Fig. 1C): (i) T0 consists of sandy to silty lenticular lamination, akin to starved ripple cross-lamination (corresponding to the Bouma $T_c$ interval) deposited by alternating traction and suspension sedimentation; (ii) T1 consists of convolute lamination, or lamination that has been perturbed from its original depositional fabric and has undergone a bit of soft-sediment deformation; (iii) T2 consists of low-amplitude climbing-ripple cross-lamination, typically small <2 cm (~ 0.8 in.) height preserved ripples; (iv) T3 consists of thin, alternating lamination with even spacing between laminae; (v) T4 consists of indistinct or poorly defined lamination and can have a wispy or faint appearance; (vi) T5 consists of wispy convolute lamination formed by soft-sediment deformation and not a primary depositional process (this interval represents the top of the Bouma $T_d$ equivalent interval comprised of the T1-T5); (vii) T6 consists of graded clay and silt lenses, darker in color due to abundance of clay; (viii) T7 consists of ungraded mud and pseudo-nodules; (ix) T8 consists of hemipelagic to pelagic sediment. While the Stow and Shanmugam (1980) model is attributed to turbidity currents, it shares many of the similarities with a waning contour current model (Mutti et al., 2014; De Castro et al., 2020, 2021a,b; Fonnesu et al., 2020; Fuhrmann et al., 2020; De Weger et al., 2021).

Mudstone in deepwater settings can be used to attribute its origins to certain depositional environments (Lazar et al., 2015). Mudstone can have multiple origins, ranging from dilute, low-density turbidity currents, hyperpycnal flows, sediment remobilization, debris flows or transitional flows (Boulesteix et al., 2019 and other associated studies). These observations challenge the idea that most mudstone intervals are merely the result of passive, hemipelagic to pelagic sedimentation. There is quite a bit of underappreciated detail in the silty to muddy intervals. These intervals can represent the distal remnants of a low-density flows and hold clues to their positions in a submarine fan environment (discussed in Chapter 8).

## Shallow-marine influenced turbidites

Simultaneous or postdepositional reworking by tides, waves and fluvial inputs can modify a turbidity current deposit. While these modifying processes can influence the deposition of any of the sediment gravity flow deposits mentioned in this section, the most described in the rock record are tidally modified turbidites, wave-modified turbidites and hyperpycnal and homopycnal flow deposits (Fig. 2).

### Wave processes

Wave-modified turbidites resemble tempestites or storm deposits. Wave-modified turbidites typically have five main divisions: (i) basal structureless division; (ii) swaley to hummocky cross-stratified sand (Mulder et al., 2009); (iii) planar to ripple cross-laminated sand; (iv) silt; and (v) clay top (Fig. 2A). The main differences include: (i) there are multidirectional to bidirectional current structures, not only unidirectional structures as observed in typical turbidites (sensu Donovan et al., 2015 and references therein); and (ii) beds are typically thinner (thin- to medium-bedded) and show less diversity in bed thickness (Myrow et al., 2002, 2008; Lamb et al., 2008). While wave-modified turbidites have been understood to form in epeiric shelf settings across the world, which are not common settings today, lab results suggest wave energy propagates into deeper water well below wave base by surging (Sequeiros et al., 2019).

### River-fed turbidites

Deltas are river-fed bodies of sediment that naturally prograde into a standing body of water. These river mouths can emit dense, sediment-laden flows during times of heavy rainfall and flooding, forming a flow so dense that it hugs the seafloor when entering the marine realm. These are known as hyperpycnal flows. Alternatively, the flows can be sediment laden, but not exceeding the density of seawater, and may float on top or at an intermediate level in the sea. These are known as hypopycnal flows. Hyperpycnal and hypopycnal flow deposits can accumulate on the delta front to the shelf in

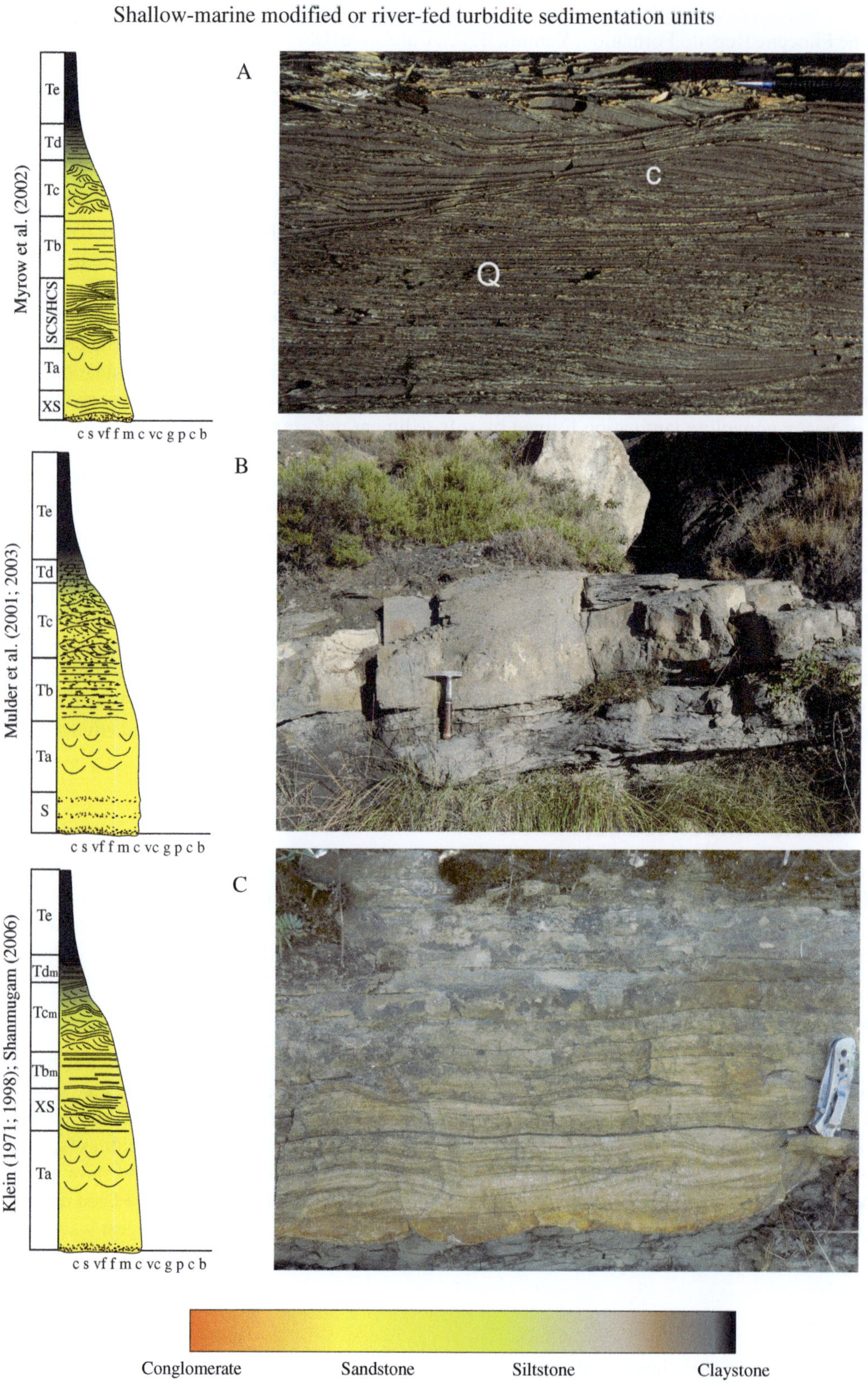


**FIG. 2** Deepwater sedimentation units formed by turbidity currents interacting with fluvial, wave and tide processes. Models include: (A) wave-modified turbidite, Antarctica (Myrow et al., 2002); (B) hyperpycnite, Lower Eocene Figols Group, Spain (Mutti et al., 1996; Mulder et al., 2001, 2003); and (C) tidally modified turbidite, Eocene-Oligocene Annot Sandstone, southeastern France (Klein, 1971, 1998; Shanmugam, 2006).

flood-dominated river-fed systems—especially during highstands or phases of denudation following episodes of tectonic uplift (Mutti et al., 1996, 2003; Tinterri, 2007) producing impressive accumulations of sheet-like sand beds containing hummocky cross-stratification (HCS), or, more rarely HEB ( Jin et al., 2021). These fluvio-turbidite systems (Mutti et al., 1996) promote both hyperpycnal and hypopycnal flows, each with the ability to carry abundant sediment, shallow-water microfossils and carbonaceous terrestrial material into deepwater.

These flows are typically prolonged, as they can create long-lived and well-"fed" fully turbulent flows with interstitial freshwater sustained by river discharge (Zavala and Arcuri, 2016). Hyperpycnal flow deposits typically show a sedimentation unit strongly resembling a low-density turbidity current deposit, yet with a fluvial signature (Mulder et al., 2001, 2003) including: (i) evidence of inversely graded first-order elements reflective of waxing flow; (ii) abundant carbonaceous matter originating from flooding onshore and incorporation of the plant matter into the flow; (iii) evidence of sustained flow including abundant traction structures; and (iv) alternation between various traction structures, showing evidence of surging and relaxation of current velocities at the sea bed (Fig. 2B). Hypopycnal flow deposits can accumulate in deepwater, especially in slope settings, and are considered ubiquitous in many deepwater subenvironments including channels, channel-lobe transition zones and in lobes (splays). These are characterized by thin-bedded sand-mud beds or couplets including structureless or parallel laminated sandstone and siltstone, current ripple cross-lamination, soft-sediment deformation or muddy sandstone intervals with pseudonodules. These beds are also termed *plumites* by Mutti (2019).

### Tidal processes

Tides can influence deposits well below wave base and deep into submarine canyons. It is unknown how deep these processes affect sedimentation in the deepwater realm, but the sedimentary signature can be found (Mulder et al., 2012). Tides can propagate in a flow down-system, much as wave action can, and provide a daily imprint on sedimentation in deepwater. Much as tidal sedimentation units in coastal and shallow marine areas show evidence of diurnal switches in current velocity, direction and slack water events, deepwater tidally modified turbidites show similarly distinct sedimentation units documented in the work of Klein (1971, 1998) and Shanmugam (2006) including: (i) mud drapes; (ii) primarily structured sandstone and siltstone sedimentation units; and (iii) rhythmic alternation of bed thicknesses (Fig. 2C).

## Hybrid event beds, slurry beds and transitional flow deposits

The previously described facies models set the stage for the observation of event beds (sensu Kneller and McCaffrey, 1999) characterized by a vertical association of matrix-poor and matrix-rich sandstone facies. Collectively these deposits are here referred to as hybrid event beds (HEB), following the terminology introduced by Haughton et al. (2009) which encompasses some of the deposits previously termed slurry beds (e.g., sandwich beds, linked-debrites, debrites sandwiched within co-genetic turbidites).

After the seminal paper by Lowe and Guy (2000) on a type of transitional flow deposit identified and named the "slurry-flow" deposit, the energy industry started to investigate these challenging reservoirs in major oil basins around the world, from the North Sea (UK), to the San Joaquin Basin (USA), to the Gulf of Mexico (USA), to the Norwegian Continental Shelf, among others. These deposits show evidence of both fluid turbulence and matrix strength during deposition in the form of readily recognizable structures derived from a turbidity current, structures that are attributable to debris flow—as well as sedimentary traits and structures unseen in pure turbidites or debrites, thus requiring their own classification. First noted in the UK North Sea by Chevron and categorized with M divisions—from $M_1$ to $M_7$ (Lowe and Guy, 2000) depending on characteristics, similar deposits were noted in the Western Irish Namurian Basin (Ireland) and classified using H divisions—H1 to H5 (Haughton et al., 2009). Other categorization schemes describe them as transitional flow deposits (Kane and Pontén, 2012) with a range of subdivisions (Koo et al., 2019), argillaceous sandstone beds in the US Gulf of Mexico and elsewhere (Power et al., 2013; Marchand et al., 2015), and even high- vs. low-yield strength debrites (Talling et al., 2012; Talling, 2013). These beds are characterized as nonerosional and typically have flat bed contacts. However, they may also show scoured bed bases (Fonnesu et al., 2016) and deformed tops due to soft-sediment deformation. The uneven bed contacts can even be a result of the relict topography of the debrite that is filled in by the next turbidite (Fonnesu et al., 2015).

Similar deposits have been known for a long time (Wood and Smith, 1959; Ricci Lucchi, 1965; Mutti et al., 1978; Van Vliet, 1978; Ricci Lucchi and Valmori, 1980). Recently, these types of beds received much attention from the research community and industry, encouraging a new round of outcrop studies (Haughton et al., 2003; Talling et al., 2004; Amy and Talling, 2006; Hodgson, 2009; Muzzi Magalhaes and Tinterri, 2010; Patacci et al., 2014; Terlaky and Arnott, 2014; Fonnesu et al., 2015, 2016, 2018; Southern et al., 2015; Spychala et al., 2017; Mueller et al., 2017, 2021; Pierce et al., 2018) and experimental research (Amy et al., 2006; Sumner et al., 2009; Baas et al., 2009, 2011) with the dual aim of understanding (i) how these flows operate and (ii) how the "unpredictable" poor quality reservoir facies might be distributed across deepwater depositional subenvironments more predictably.

Because of the proliferation of facies models developed from different basins, there is considerable overlap in like first-order architectural elements among the models. These deposits are variations on a theme of mixing and transformation of the turbulent and cohesive, plastic flow types.

## Slurry-flow deposits

Slurry-flow deposits were first recognized in the UK North Sea (Lowe and Guy, 2000) and are distinguished from turbidites based on anomalously high clay values in the matrix, ranging from about 5% to 35%. These high clay values change the rheology of the flow during deposition to quasiplastic. The clay-rich intervals contain many different types of sedimentary structures and structuring, and seven principal divisions have been identified—each possessing distinct sedimentary characteristics (Fig. 3A). These divisions were termed M divisions based on the type of sedimentary structure and amount of clay. The overall evolution of these flows changes from turbulent to transitional as clay is entrained in depositing sediment, and the viscosity, cohesion and resistance to shear increase. $M_1$ divisions are current structured to massive and represent deposition by both suspension, traction and matrix strength. Meso- and macro-banded intervals are known as $M_{2b}$, $M_{2c}$ and

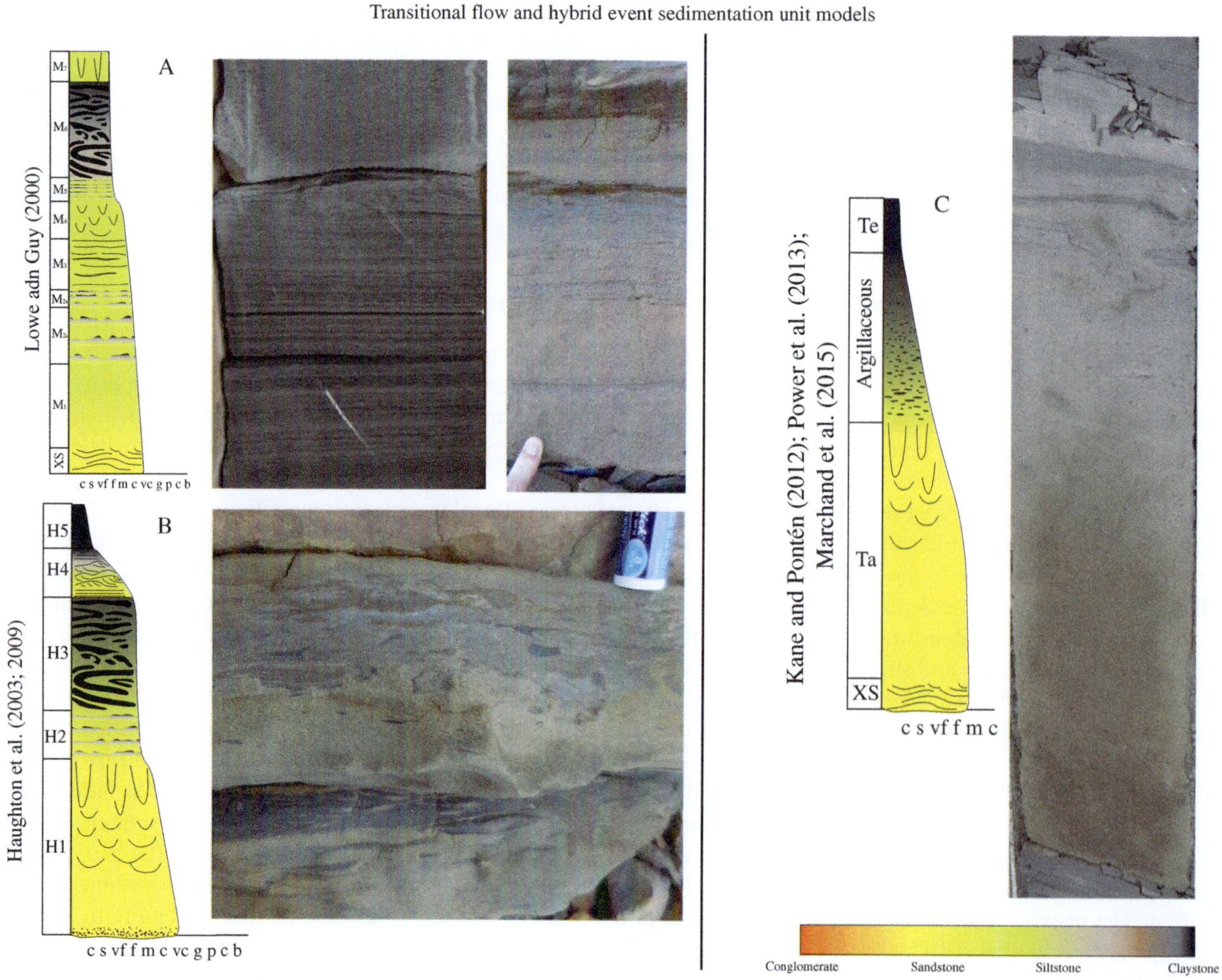


FIG. 3 Deepwater sedimentation units formed by transitional to hybrid flow behavior. Models include: (A) slurry-flow model of Lowe and Guy (2000) with examples of partial sedimentation units from the Permian Lower Spraberry Formation, Delaware Basin, USA and the Eocene Point Loma Formation, California, USA (from right to left); (B) hybrid event bed model of Haughton et al. (2003, 2009) with partial sedimentation units from the Carboniferous Ross Formation (Loop Head Peninsula, County Clare), Western Irish Namurian Basin, Ireland; and the (C) argillaceous sandstone bed of Power et al. (2013) with sedimentation unit from the Paleogene Wilcox Group, US Gulf of Mexico.

represent fluctuation between fully turbulent suspension sedimentation and that of transitional flow. Wispy laminated sandstone is classified as $M_3$ and is developed by traction processes and transitional flow. $M_4$ divisions are dish structured resulting from the deposition of grain flows or suspension sedimentation from a fully turbulent flow. $M_5$ divisions are micro-banded to flat-laminated and are formed by traction sedimentation. $M_6$ divisions are foundered and mixed layers that were originally laminated to micro-banded, having undergone soft-sediment deformation thus obscuring some of the original fabric. $M_7$ divisions are characterized by vertical water escape structures (sensu Lowe and Lo Piccolo, 1974). While these divisions are depicted in the model (Fig. 3A) as a complete succession, it is important to remember these divisions only commonly partner with 1–2 other divisions in most sedimentation units (Lowe and Guy, 2000). It is not currently understood why this is, and why more complete beds of this type have not yet been found.

## Linked debrites and hybrid event beds "sensu stricto"

Very commonly, sandstone intervals are regularly overlain by mudstone-clast-rich debritic intervals (Fig. 3B). These deposits may form two event beds, but very often are the result of one sediment gravity flow. Haughton et al. (2003) coined the term "linked" debrite to describe these beds in the Western Irish Namurian Basin. Later, the linked-debrites concept transmogrified to HEB (Haughton et al., 2009) and an idealized bed motif was proposed with up to five internal divisions.

The model includes from base to the top: a basal, generally graded and structureless sand (H1), sometimes including scattered mudclasts (H1b); a banded division comprising alternating sand and darker argillaceous (enriched in clay—*clayey sand*) bands (H2); a chaotic division of argillaceous sand with variable concentrations of mudclasts, sand patches, sand-injections or carbonaceous fragments that accumulated at the top of the division (H3); a fine- to very fine-grained sand that is usually graded and parallel or ripple cross-laminated (H4); and a silty to muddy capping division (H5). This is the model of Haughton et al. (2009) as produced by a flow changing rheology from turbulent to laminar and back to turbulent again. The transformations could be achieved by abundant eroded mud that can produce a cohesive flow through flow partitioning and transformation (Haughton et al., 2009; Muzzi Magalhaes and Tinterri, 2010; Fonnesu et al., 2016, 2018) and an increase in the fallout related to sudden decelerations, thus dampening the turbulence (Muzzi Magalhaes and Tinterri, 2010; Barker et al., 2008; Patacci et al., 2014).

The basal H1 division can be interpreted as a deposit created by a high-concentration (high-density) turbidity current (Haughton et al., 2009), or by the rapid aggradation and vertical segregation of turbulent and transitional flow. The latter is attributed to the supersaturation of sand due to insufficient turbulence or the addition of cohesive support (Baas et al., 2011). The H2, or banded division, implies a transitional frictional-cohesive flow (Lowe and Guy, 2000) with intermittent suppression of turbulence, or the modulation of turbulence with scouring and reworking of the underlying sand (Baas et al., 2009, 2011). The H3 division is interpreted to be deposited by a cohesive debris flow, which may range from low- to high-yield strength (sensu Talling et al., 2012; Talling, 2013), or may even be interpreted as a transitional flow deposit. The HEB concludes with the H4 and H5 divisions, which are interpreted to represent the deposits from a dilute wake and turbulent suspension cloud, similar to the Bouma $T_{bc(d)e}$ divisions.

The Haughton et al. (2009) model, like the other models, is often incomplete or highly variable in the real world. Not all divisions are necessarily present in any one bed, and the character of the internal divisions can change considerably (Haughton et al., 2010; Talling, 2013; Fonnesu et al., 2018). The H1 division can thin at the expense of the H3 division, pinching out distally and laterally until it completely disappears (Haughton et al., 2003; Amy and Talling, 2006). The H2 division may be absent, or subtle and easily overlooked. In other cases, it is greatly expanded and can dominate the deposit such as in the margins of the Lower Cretaceous Britannia Sandstone, UK North Sea (Lowe and Guy, 2000; Barker et al., 2008).

H4 may be absent or may wholly or partly have collapsed into the underlying H3 division (Fonnesu et al., 2015). Most importantly, the H3 division can have a variable character. The H3 can be dominated by mudclasts, or it can even contain large mudstone rafts (Haughton et al., 2003; Fonnesu et al., 2018). Other studies have documented H3 with swirly clast-poor muddy sand (Talling et al., 2004, 2012; Hodgson, 2009; Kane and Pontén, 2012; Kane et al., 2017) up to "starry night" texture with scattered sand grains dispersed in mud (Lowe and Guy, 2000).

Boundaries between different bed divisions can be sharp or transitional, and the proportion of different divisions can change considerably over short length scales (Talling, 2013; Terlaky and Arnott, 2014; Fonnesu et al., 2015). Nevertheless, HEB usually form laterally extensive deposits and lack substantial changes in bed thickness despite the change in internal character (Patacci et al., 2014; Fonnesu et al., 2015). Fonnesu et al. (2018) distinguished six different types of HEB in the Cretaceous-Paleocene Gottero Sandstone Formation (Italy) based on the texture of the H3 division and the size and shape of the clasts within it (Fig. 4). H3 division textures range from intact substrate blocks to well-mixed argillaceous sand, and from deformed slump-like, sand-injected mudstone to mudstone-clast-rich sandstone. Detailed field correlations document

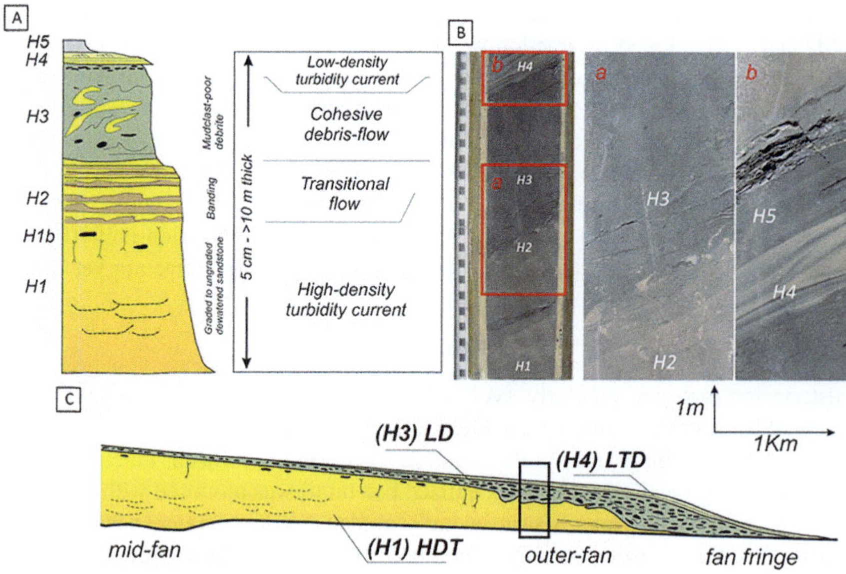


**FIG. 4** Vertical and longitudinal bed expression of a hybrid event bed according to Haughton et al. (2009) and followers. (A) Hybrid event bed in its ideal five-part internal structure and interpretation according to Haughton et al. (2009). (B) Example of hybrid event bed from the Paleocene Forties Fan, UK North Sea (Haughton et al., 2009). (C) Longitudinal bed expression passing from proximal to distal locations: mostly sandy beds to beds including a muddier division, reflecting the downslope flow evolution due to the transformation from frictional to cohesive flow. This general facies tract can be expressed at different scales and have many possible variations. Overall, this facies tract suggests that higher quality sandstone in HEB may be found in downslope, proximal positions.

the lateral transitions, which are arranged according to an increased intensity of soft-sediment deformation and mixing from the inferred area of mud acquisition of the individual flow to the margin of the event bed. Different types of H3 divisions can also be vertically superimposed within the same bed. Dodd et al. (2019) and Hussain et al. (2020) clearly document the occurrence of bi-partite H3 divisions consisting of basal sandier H3a and muddier H3b.

There are myriad models for HEB today, encompassing every possible variation of turbidite, transitional flow deposit and debrite constructing a sedimentation unit (e.g., Pierce et al., 2018). The term HEB is inclusive in that all the diverse types of beds with composite mixed rheology are a part of this geological family. It also became clear that turbulence damping may occur by more than one mechanism even in the same flow (Pierce et al., 2018). These observations compel scientists to think about what are the key traits that necessitate a HEB classification in all the areas they are found around the world including Cretaceous intervals in the North Falkland Basin (UK) (Dodd et al., 2020), the Neoproterozoic Windermere Supergroup of British Columbia (Terlaky and Arnott, 2014), as well as examples from different basins featuring hyperpycnal flows along delta fronts ( Jin et al., 2021).

Haughton et al. (2009) found that HEB typify over steepened slopes in the nascent stages of basin formation, especially in muddy basins in which flows can gather sufficient amounts of clay to change the flow rheology. However, more complex facies models link their development to the effect and impact of substrate erosion and delamination at more of a local scale (Fonnesu et al., 2016, 2018).

## Argillaceous sand beds and transitional flow deposits

Argillaceous sand beds (Fig. 3C) are bi- or tripartite sedimentation units consisting of: (i) a relatively silt- and clay-poor, generally structureless sand division; (ii) a strongly partitioned or gradational silt- and clay-rich division; and (iii) a mud top.

The bed make-up is like the one described by Haughton et al. (2009) for hybrid deposits, however it features gradational clay enrichment towards the top of the bed. This is similar to a mud-rich type of H3 division.

It most closely resembles a low yield strength debrite to transitional flow deposit because: (i) it generally lacks sizeable mud clasts greater than sand and granule size; (ii) there is little evidence of convolution or "swirled" textures (Talling, 2013) common in the higher matrix strength debritic intervals; and (iii) the sand, silt and clay grain size distribution is roughly equal. Enriched in silt, clay and ductile grains, the argillaceous and mud intervals are typically considered as non-reservoir, especially in highly compacted intervals.

These beds are interpreted to represent deposition via: (i) suspension collapse of a fully turbulent flow; (ii) transitional flow behavior governed by fluid turbulence and matrix strength, as clay-sized grains are concentrated near the bed and possibly disaggregated due to shear, thereby increasing the cohesion of the mixture, followed by; (iii) tail of the flow sedimentation including hemipelagic to pelagic sedimentation. The top of the unit can be complicated by deposition via low-density turbidity current processes. Beds grade from structureless $T_a/S_3$ divisions into silt- and clay-rich tops, taking up the place that would be low-density turbidity current divisions ($T_{bcd}$).

Thick intervals of these types of beds are found along sequence boundaries and typically in distal axial splay environments, first made somewhat famous in the deepwater Paleogene Wilcox Group Trend (US Gulf of Mexico) by scientists from companies including Shell, Chevron, BP, ExxonMobil and Statoil (Equinor), however, it was not until (i) Kane and Pontén (2012) described them, (ii) the Chevron group led by Power et al. (2013) coined the modern use of the term and (iii) BP's Marchand et al. (2015) characterized their reservoir quality in channel, splay and overbank environments. Other similar types of transitional flow deposits have been described by Kane et al. (2017) in the Permian Skoorstenberg Formation of the Karoo Basin (South Africa), and by Southern et al. (2017) in the Vøring Basin (Norwegian Sea) accounting for the detailed structure and fabric of the bed.

Examples of similar beds are also found in lacustrine deposits by Yang et al. (2018), Yang and Cao (2019) in Cretaceous intervals on Lingshandao, China, and Dodd et al. (2019) in the North Falkland Basin (UK). There are many forms of transitional flow deposits encompassing every possible variation of turbidite (sandy), transitional flow deposit (mixed sand and clay) and debrite (mudclast rip-up rich) in sedimentation units on a global scale (sensu Fonnesu, 2020). Out of context or rare occurrence of these types of beds in a sedimentary succession yields little information, therefore, observations of these are most helpful when specifically tied to a basin or reservoir interval, or if these occur as a repeating or abundant pattern in the stratigraphy.

## Hybrid event beds facies tracts and distribution across deepwater depositional systems

Hybrid event bed endmembers can be arranged in several different types of facies tracts. From proximal to distal, most of the documented cases show a typical longitudinal bed expression broadly passing from mostly sandy to including a muddier division. This facies tract reflects the downslope evolution of the flow derived from the transformation of a frictional flow to a cohesive flow. One remarkable example is from the Miocene Marnoso-arenacea Formation (Northern Apennines, Italy), where individual beds and their longitudinal facies transitions are correlated for over 120km (~ 75 mi.) (Amy and Talling, 2006).

Fonnesu et al. (2018) and (2020) describe the variability of the ideal facies tract in relation to the type of facies and processes involved, with a particular focus on the mechanisms of substrate entrainment, which is considered a key part in the prediction of the distribution and quality of the sand and mud deposits (Fig. 5).

Beds that are enriched in mudclasts or muddy rafts acquire their mud components essentially by intrusion and delamination of coherent pieces of substrate, with very limited bypass (Fig. 6A) (Eggenhuisen et al., 2010; Fonnesu et al., 2016). This mechanism produces tabular scours characterized by lateral injections and a tiered structure on the basin floor (Fonnesu et al., 2016) or against basin-floor topographic highs (Puigdefàbregas et al., 2004). Scours are often preserved along bed bases (Fonnesu et al., 2016; Kane et al., 2017; Patacci et al., 2020), but very often these features are missed due to their subtle nature. In the case that the rheology contrast between the flow and the substrate allows only the formation of shallow tabular scours, only fine-grained mudclasts would be entrained at the base of the flow. The clasts can be rapidly buried by the deposition of sand in suspension (Mutti and Nilsen, 1981; Postma et al., 1988), producing beds enriched in mudclasts in proximal locations. Alternatively, mudclasts can accumulate at the top of the aggrading bed, where they start to move as a shearing boundary layer due, in part, to the contemporaneous dewatering of the basal sand. In this case, the flow can produce deeper scours, and a larger quantity of substrate including not only mudstone clasts, but entire *substrate slabs*, can be entrained in the flow. This process will produce raft-bearing HEB. The rafts can be transported for short distances without suffering substantial deformation before being deformed by internal shear or invaded and broken apart by sand injection, thereby forming a chaotic texture. Both down current and lateral to the entrained rafts, a mud clast-rich flow, partially derived from the spalled raft fragments, will form clast-rich HEB similar to the HEB produced by direct smaller clast entrainment. In both of the facies tracts, the degree of mixing between the sand transported by the flow and the mud-rich components collected by the flow increase downsystem. It is reasonable to think that this process may permit the formation of argillaceous-type HEB if the flow had enough space to evolve and the process of particle comminution, hydraulic segregation and clay release remained efficient.

Mud entrainment can also occur due to turbulent erosion from unconsolidated substrate (Fig. 6B). In this case, mud is entrained by mechanical abrasion of the substrate, producing convex-downward scours developed as cut-and-fill features,

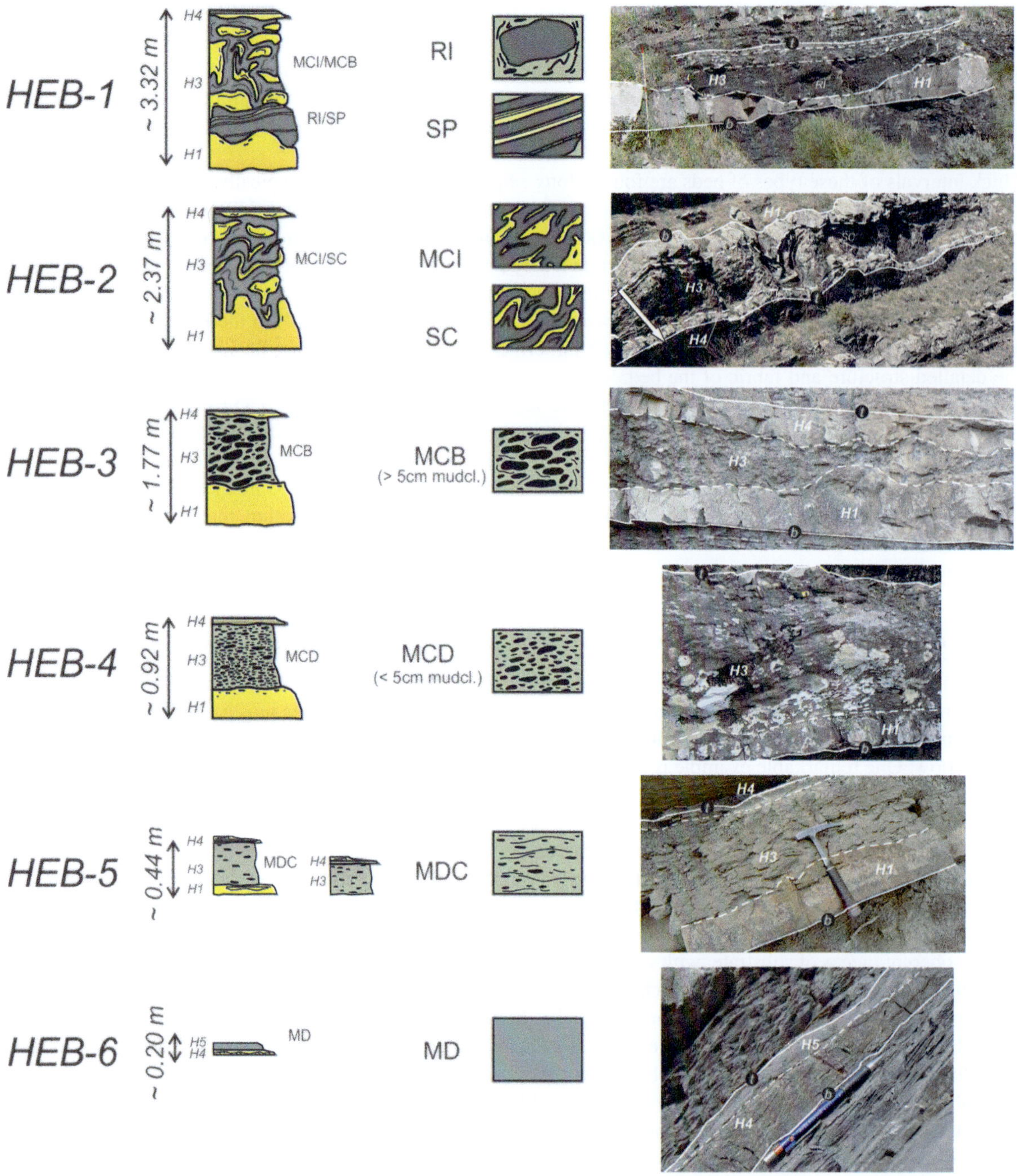


**FIG. 5** Hybrid event bed types from Fonnesu et al. (2018), showing the typical bed profile and photograph examples. Arrows indicate stratigraphic up when bedding is inverted ("b" base of the bed; "t" top of the bed). Typical H3 subfacies are indicated in the boxes.

megaflutes or mud-draped scours. Such features imply sediment bypass and are typically found in channels, channel-lobe transition zones and base-of-slope settings (Mutti and Normark, 1987; Kane et al., 2009; Peakall et al., 2020). This type of erosion usually provides fine-grained mudclasts and unconsolidated clay to the flow but not large coherent slabs. The resulting facies tract predicts a down-dip transition from sand (H1) to HEB in which the H3 division expands progressively down-dip at the expense of the H1 division. It culminates in the pinch out of the basal sand, followed by the distal pinch-out of the H3 division (Haughton et al., 2009). Because the two divisions can pinch-out in different places, it can result in HEB in which the H1 division is not preserved. Because of the nature of the entrainment, the texture of the H3 division is

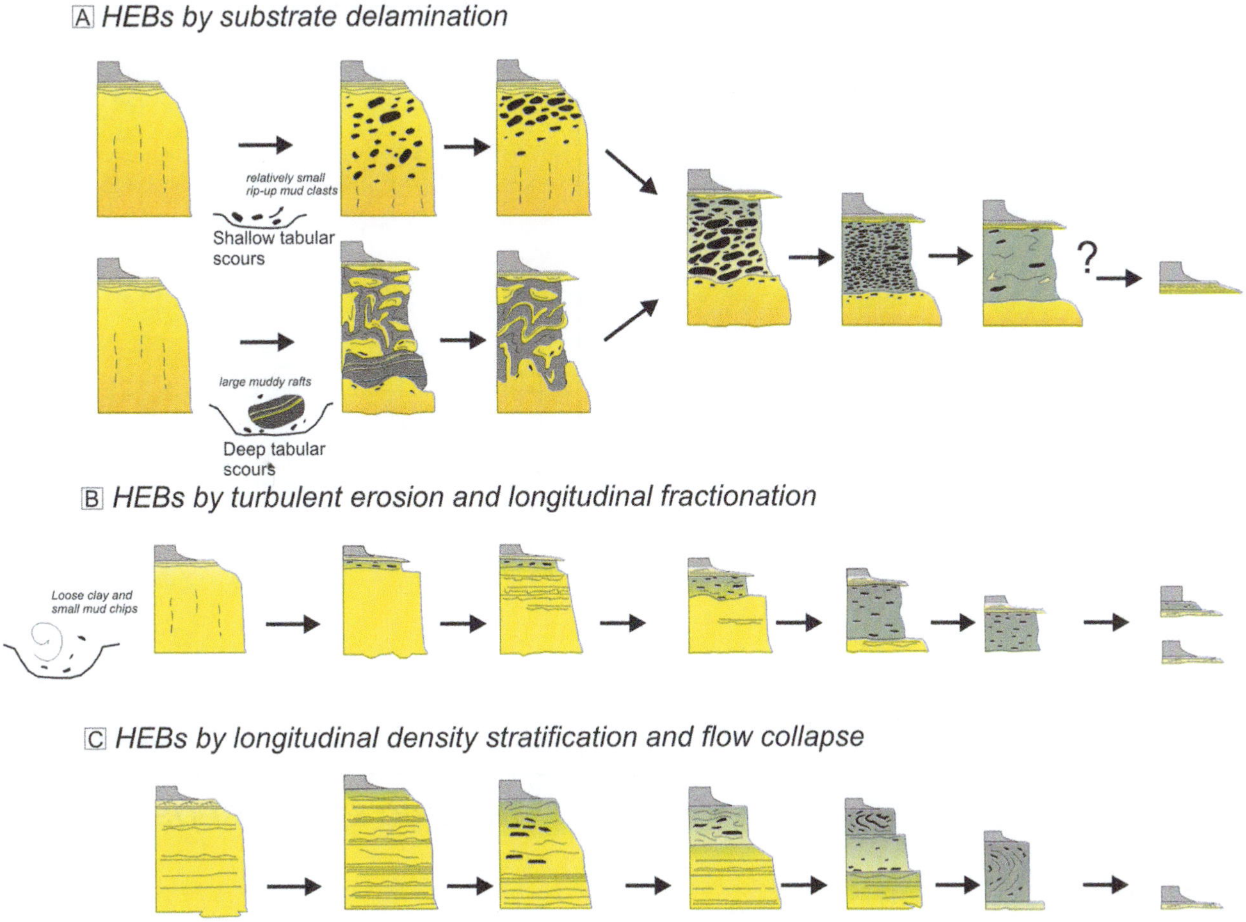


**FIG. 6** Facies tract types for hybrid event beds and slurry flow, argillaceous and transitional flow deposits. These are facies tracts in which the bed shows an enrichment of mud starting from a sandier bed. (A) Hybrid event beds forming from substrate delamination in high-concentration flow conditions (Fonnesu et al., 2016, 2018). (B) Hybrid event beds produced by turbulent erosion and longitudinal fractionation (Haughton et al., 2009). (C) Slurry, transitional and argillaceous sand deposits produced by longitudinal density stratification and flow collapse.

well-mixed and argillaceous with only scattered large mudstone clasts. These flows are known for efficient longitudinal and transverse segregation of components featuring lower settling velocities for the mica flakes, organic matter and clay floccules (Pyles et al., 2013). Further, the increase in clay concentration in the flow causes the flow to change from transitional to laminar. The basal sand is a "clean" sand and is interpreted as the deposit of a forerunning turbidity current that presages the deposition of the hybrid interval (Haughton et al., 2003, 2009).

Finally, a third mechanism accounts for the origin of event beds showing evidence of progressive turbulence dampening throughout the deposition of the beds (Fig. 6C), termed transitional flow deposits. In proximal positions, these beds show banding and fine-grained argillaceous intervals with progressive clay-enrichment towards the top. These beds typically lack the H4 division. In more distal positions, these beds have basal sandstone intervals with high amounts of dispersed clay, and where they pinch-out, are typically thin crudely laminated muddy sand beds (Kane and Pontén, 2012; Southern et al., 2017; Kane et al., 2017).

There are contrasting interpretations about the role of mud entrainment in the development of hybrid event beds. As discussed, erosional features thought to be a primary mechanism for mud entrainment have been observed in many outcrops and in cores from basins around the world for decades. The clay-enriched flow that forms a dense, low-yield strength layer progressively shuts off the transfer of turbulent kinetic energy from the base of the flow into the upper layers (Kane et al., 2017). Therefore, the flow incurs a downstream evolution from turbulent to transitional to laminar flow as it expands and decelerates (Kane and Pontén, 2012 and earlier studies), with the two divisions pinching out approximately in the same place (Kane et al., 2017). In this case, the basal sandstone is not retained as a forerunning turbidity current, rather it represents the product of distal flow collapse and transformation (Kane et al., 2017).

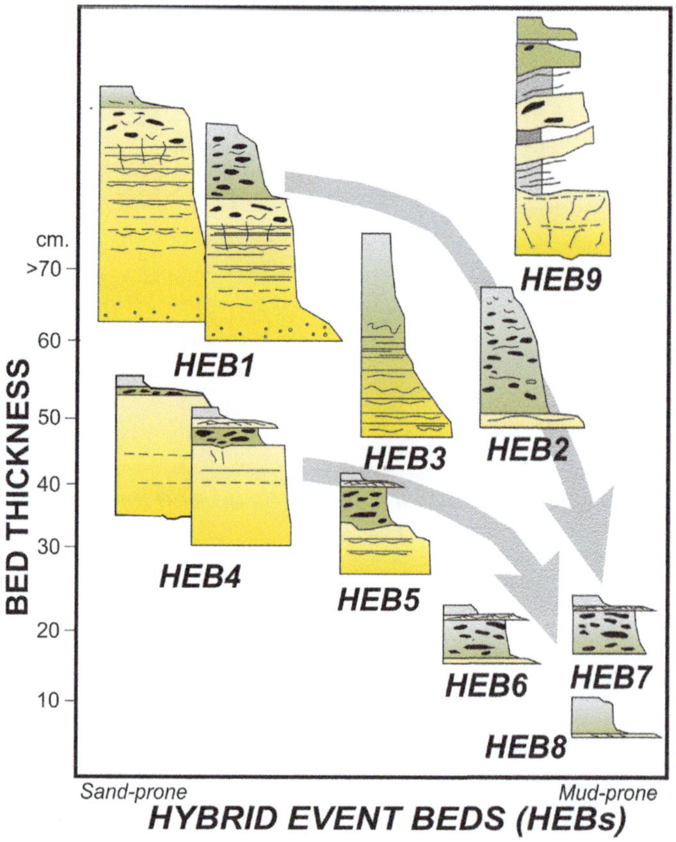


**FIG. 7** Hybrid event bed classification proposed by Pierce et al. (2018), based on mostly fine-grained hybrid event beds of the Carboniferous Ross Formation, Western Irish Namurian Basin, Ireland. The gray arrows group the beds into what are inferred to be separate facies tracts on the basis of their stratigraphic occurrence and stacking patterns.

## Summary

These facies tracts form numerous first-order architectural element configurations for the broad range of transitional flow deposits. The slurry-flow (Lowe and Guy, 2000), HEB (Haughton et al., 2009), transitional (Kane and Pontén, 2012) and argillaceous sand (Power et al., 2013) deposits show specific sedimentation units verifiable in basins and petroleum systems on a global scale. There exist numerous permutations of divisions and bedding styles making the overall class of transitional flow deposits among the most diverse types of deposits in the sedimentary record (Fig. 7).

## Debris flow deposits

### Matrix-supported debris flow deposits

Sediment gravity flows tend to have at least a small proportion of mud, and some are particularly enriched in mud which changes the flow rheology to a *debris flow*. For the major classes of debris flows, ranging from matrix-supported and cohesive to clast-supported and frictional (Fig. 8), this means that most deepwater debris flows are likely matrix-supported with a muddy matrix (Costa and Williams, 1983; Johnson, 1984; Talling, 2013). This is different from the subaerial realm in which frictional forces can dominate, leading to abundant framework gravel deposits in environments such as alluvial fans.

Debris flows and their deposits are characterized based on clast-size, distribution and degree of mixing and can be primarily muddy, sandy or gravelly. Debris flow deposits can be highly variable: (i) homogenized clay- and silt-rich with poor expression of original bedding; (ii) heterogeneous clay- and silt-rich with contorted and moderately mixed principal lithologies (high-yield strength); (iii) stratified clay- and silt-rich (low-yield strength); (iv) homogenized mud- and sand-rich with extensive mixing; (v) heterogeneous mud- and sand-rich without extensive mixing, but contorted and sheared bedding; (vi) stratified mud- and sand-rich; (vii) homogenized mud- and gravel-rich (Fig. 8B); (viii) moderately mixed and stratified mud- and gravel-rich (Fig. 8C); (ix) gravelly (frictional debris flow, potentially $R_1$ deposit in deepwater).

The muddy debris flows lacking conglomerate can be further characterized by the nature of the relative size of the mudclasts: fine (mm-scale), medium (cm-scale) and large (m-scale). Fine mudclasts can more efficiently distribute within the flow by abrasion and incorporation, changing the flow rheology to more of a slurry-flow deposit (see previous sections), in

**FIG. 8** Several types of debrites, ranging from debrites with a mud matrix to those that have a higher amount of sand in the matrix: (A) clast-supported debrite, Upper Cretaceous to Paleogene Pigeon Point Formation, near Pescadero, California, USA; (B) muddy matrix-supported debrite with floating pebble and granule clasts, Pigeon Point Formation; (C) variable muddy matrix-supported conglomerate, Pigeon Point Formation; (D) foundered and soft-sediment deformed pillow-structured debrite formed from flowage in a submarine channel-levee environment, with a muddy composition, Pigeon Point Formation; (E) linked debrite sedimentation units, Carboniferous Ross Formation, Loophead Peninsula locale, Western Irish Namurian Basin, Ireland, with a sandy to muddy sandstone matrix and medium-sized mudstone clasts; and (F) basal sedimentation unit debrite with sandy matrix and large, contorted mudstone clasts, in core, Eocene San Vicente Formation, Ainsa, Spain.

which cohesive and turbulent forces cause grains to move more freely. Medium and large mudclasts, with less equitable distribution in a flow, can show contortion and evidence of shearing in a high-yield strength flow. If they are well rounded and do not appear to have deformed to a great extent, they indicate deposition was likely more turbulent. Three common examples include: (i) large mudclasts in a stratified slurry flow to debris flow deposit showing inverse grading near the top of a bed; (ii) large mudclasts as part of a linked debrite at a bed top and (Fig. 8E) in the middle of a bed *sandwiched* between two intervals deposited by turbulent flow (Mutti and Nilsen, 1981); and (iii) mudclasts lining a scour as a lag deposited beneath a high-energy sediment gravity flow (Fig. 8F). In the latter two cases, these deposits were likely ripped up and redeposited, producing the rounded mudclasts. They are underlain and overlain by turbidite deposits, yet the final mode of deposition was likely governed by matrix strength due to the interaction of the muddy deposits. They show no soft-sediment or putty-like behavior that would resemble sticky, muddy deposits interacting together to form a cohesive bed layer.

There are multiple ways to interpret debrites, with the high-yield strength deposits showing competent bedding with folding, and the low-yield strength deposits (sensu Talling, 2013) showing more pervasive mixing with smaller clasts as well as some mud partitioning. These debrites have different levels of clay enrichment in the bed, typically increasing upward. These have higher matrix strength and are difficult to understand whether they are formed by transitional flow or debris flow processes.

## Clast-supported debris flows

Whereas most debris flow deposits in deepwater are formed by cohesion and matrix strength, subaerial debris flows commonly occurring in alluvial fan environments are typically deposited through frictional grain interactions, such as frictional debris flow deposits in Death Valley, California, USA. Clast-supported debrites (Fig. 8A) can form the basal division of the Lowe Sequence (Lowe $R_1$ interval).

In many cases it is difficult to discern whether a bed is in part a gravelly debris flow deposit, or a basal $R_{123}$ interval of a Lowe Sequence. There are at least three scenarios: (i) if the matrix is clay-rich, then that, in part, could provide matrix strength to facilitate movement as a debris flow; (ii) if the matrix is sand-rich and largely devoid of clay and the surrounding sedimentation units are turbidites, then the transport mechanism was likely a turbidity current; and (iii) if it is encased in other debrites and mudstone, it could be either option.

### Summary of transitional flows and debris flows

Debris flows often indicate sediment instability near steep gradients, but dilute flows can transport sediment long distances (sensu Mohrig et al., 1998; Talling, 2013 and references therein). These are typically mud-rich and can form barriers or baffles to vertical and lateral fluid flow, but typically form poor seals. Some reservoirs can be great gas producers (i.e., Britannia Field, UK North Sea, sensu Lowe and Guy, 2000; Lowe, 2004); others can be oil- and gas-producing as long as these deposits are not the primary reservoir lithofacies. These systems are fundamentally different in terms of architecture and planform geomorphology compared to turbidite-dominated depositional systems. Whereas downslope portions of debritic systems will be dominated by (i) HEB, (ii) transitional flow deposits, (iii) slurry-flow deposits and (iv) argillaceous sandstone beds, turbidite-dominated systems will have a strong tendency towards low-density turbidity current deposits, creating beds that include Bouma $T_{bcde}$ and $T_{cde}$ divisions. In some systems, like the Ballybunion locale of the Ross Sandstone, Ireland, the downslope or lower third to half of the depositional system will have much lower reservoir quality and would likely be excluded from an exploration program due to reservoir quality failure. Turbidite-dominated systems tend to show a high level of architectural complexity compared to debrite-dominated systems, which tend to lack morphological complexity (sensu Lowe and Graham, 2012).

## Bottom current influenced turbidites, contourites and reworked beds

### Contourites and reworked deposits

While sediment gravity flows provide a principal downslope transport mechanism to deepwater environments, along-slope contour currents can affect the sedimentation processes (see Chapter 9) and resulting deposits forming a variety of contourite drifts and associated geomorphic features (McCave and Tucholke, 1986; Faugères et al., 1999; Rebesco and Camerlenghi, 2008; Rebesco et al., 2014). Contour currents are also responsible for reworking sediment gravity flow deposits and skewing sediment deposition as evidenced by outcrop data (De Weger et al., 2020; Hüneke et al., 2021), core data (De Castro et al., 2020), seismic data (Bailey et al., 2020) and in tank models (Miramontes et al., 2020). Classically, contour currents were interpreted to be such low velocity and only capable of transporting very fine-grained sand, silt and clay (Stow and Shanmugam, 1980; McCave, 2008; Hüneke et al., 2021). New evidence shows that particularly robust currents can rework gravels and coarse-grained sand in areas of confinement and in oceanic gateways (Capella et al., 2017; De Weger et al., 2020). Contour current and reworked sediment gravity flow deposits, principally turbidites, have led to the discovery of large, multi-TCF deepwater fields offshore Africa (Fonnesu et al., 2020; Thieblemont et al., 2020), Canada (Donnelly et al., 2019) and in the US Gulf of Mexico (methane gas hydrate resource plays) (Meazell et al., 2020; Phillips et al., 2020).

Contour current deposits can be divided into three main classes based on principal grain populations (see Chapter 9): (i) classic contourites (Faugères et al., 1984; Gonthier et al., 1984); (ii) fine-grained reworked deposits (De Castro et al., 2020; Rodrigues et al., 2022) and (iii) medium-grained to gravelly reworked deposits (Capella et al., 2017; De Weger et al., 2020; Fonnesu et al., 2020).

### Classic contourites

Classic contourites (Fig. 9C) contain five principal divisions (Gonthier et al., 1984; Stow and Faugères, 2008): (i) a basal muddy division (Mutti F1); (ii) a silty sand division (F4a); (iii) a burrowed sandy division (F4b); (iv) a burrowed sandy silt division (F2); and (v) a capping mudstone division (F1). This upward coarsening and upward fining records a waxing and then waning contour current at the area of deposition. Sedimentary structures including ripple cross-lamination and planar lamination, representing both Froude subcritical and supercritical flow, are found if the contour current ever reached velocities high enough to deposit these divisions.

### Fine-grained reworked deposits

Fine-grained reworked deposits (Fig. 9B) contain the following divisions: (i) basal muddy division showing a coarsening up pattern, potentially underlain by an erosional surface or scour (representing bypass of a high-energy flow); (ii) thin structureless sand division; (iii) current structured sandy divisions including planar lamination and ripple-cross lamination; (iv) structured silt divisions; and (v) mudstone bed top. These divisions are interpreted to represent waxing flow conditions, then waning flow deposited originally by a primarily low-density turbidity current. Due to the prolonged and vacillating flow states, there can be multiple repeating structured divisions, like deposition in turbidity currents that are reflected and deflected off confining slopes summarized in Rotzien (2017).

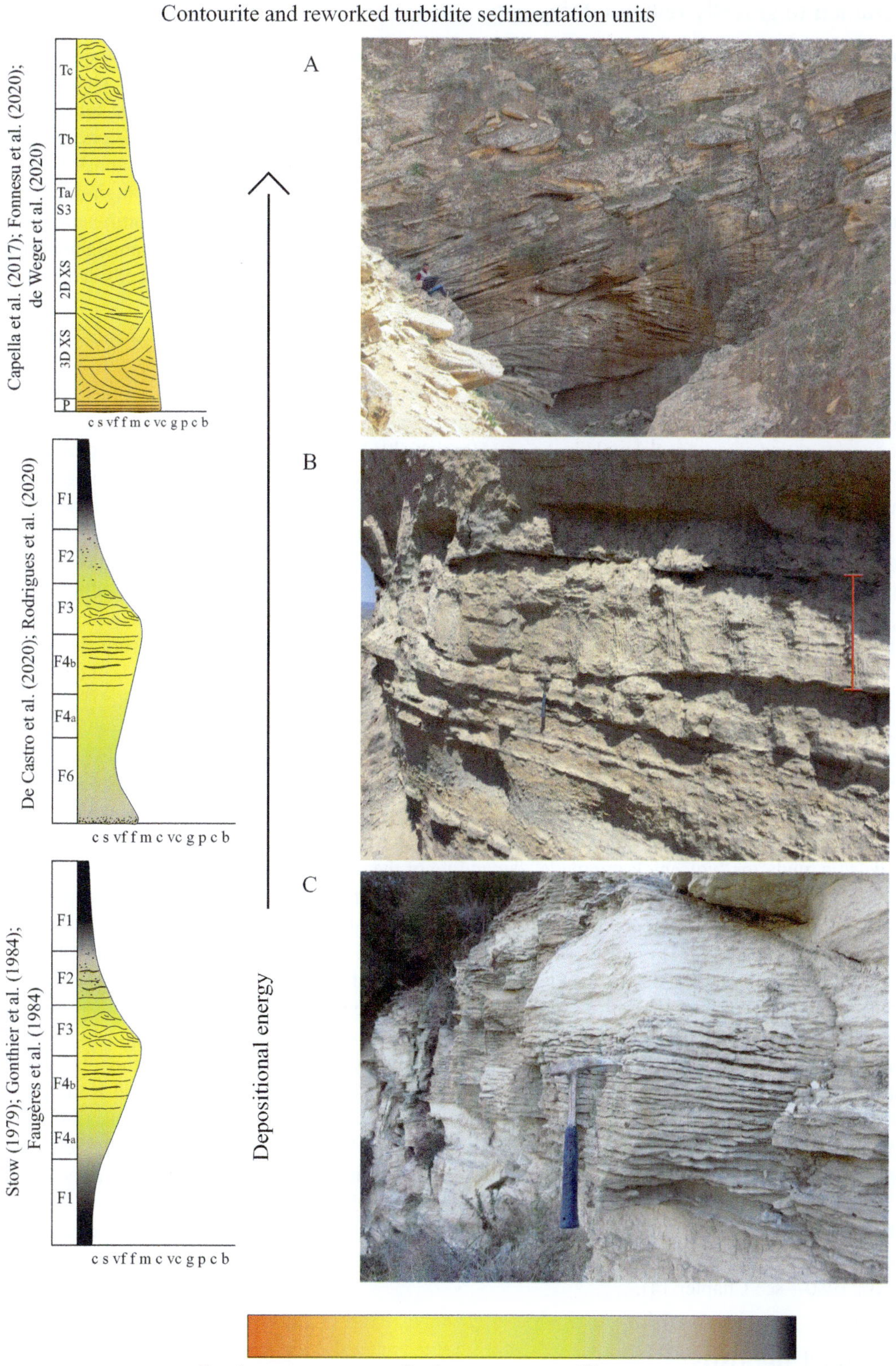


**FIG. 9** Deepwater sedimentation units formed by contour currents and the reworking of turbidity current deposits. Models include: (A) the coarse-grained and reworked model of Capella et al. (2017), De Weger et al. (2020) and Fonnesu et al. (2020) with partial beds from Morocco (Miocene); (B) the sandy contourite model of De Castro et al. (2020) with partial reworked bed, Miocene, Tabernas Basin, Spain; and (C) the classic contourite model of Gonthier et al. (1984) with sample bed from Cyprus (Miocene) (Kähler and Stow, 1998; Stow et al., 2002; Hüneke et al., 2021).

### Medium-grained to gravelly reworked deposits

Medium-grained to gravelly reworked deposits (Fig. 9A) show a variety of current structured divisions, including high-angle dune-scale cross-stratification and planar and trough cross-stratification (sensu Capella et al., 2017; De Weger et al., 2020; Fonnesu et al., 2020; De Weger et al., 2021). These deposits generally contain the following subdivisions: (i) basal planar laminated gravelly sandstone; (ii) 3D dune-scale cross-stratified gravelly sandstone; (iii) 2D dune-scale cross-stratified gravelly sandstone; (iv) structureless interval (rare); (v) planar laminated sandstone; and (vi) ripple cross-laminated sandstone. Medium-grained reworked deposits are principally deposited by high-density turbidity currents. Contour currents and sediment gravity flows are more robust and powerful in their combination here, evidenced by the lack of silt and clay at the top of the sedimentation unit. These deposits can feature a high amalgamation ratio, leading to high-quality reservoirs lacking local vertical compartmentalization. Studies suggest these beds can be pervasively and extensively reworked by more continual traction sedimentation than originally envisaged.

## Pelagic and hemipelagic deposits

Three main types of pelagic sediment (<25% detrital components) are recognized based on their composition: siliceous ooze, calcareous ooze and abyssal red clay (Hüneke and Henrich, 2011) (Figs. 10 and 11). Siliceous oozes are composed by at least 30% of the siliceous microscopic "shells" of plankton, such as diatoms and radiolaria, and, at lesser proportions, of either sponge spicules, silicoflagellates or both. Calcareous oozes are composed of at least 30% of the calcareous microscopic tests (external shells) of foraminifera, coccolithophores and pteropods (Fig. 10). Pelagic red clay (also known as simply red clay, brown clay or pelagic clay) contains less than 30% biogenic material. These types of sediment consist of eolian quartz, clay minerals, volcanic ash, subordinate residue of siliceous microfossils and authigenic minerals such as zeolites, limonite and manganese oxides. They may include other more exotic clast types including meteorite dust, fish bones and teeth, whale ear bones and manganese micro-nodules. These sediment types range from chocolate brown in the absence of reactive organic carbon due to manganese (Mn) and iron (Fe) minerals remaining in their oxidized states, to bright red in more deeply buried areas, due to the conversion of iron hydroxides to hematite (Hüneke and Henrich, 2011) (Fig. 11; D'Hondt et al., 2011).

By definition, hemipelagites are sedimentary deposits that comprise an admixture of biogenic pelagic material (generally >10%) and terrigenous or volcanogenic material (>25%), in which a significant proportion (over c.40%) of the terrigenous (volcanogenic) fraction is of silt size or greater (i.e., >4 μm), and the overall grainsize distribution is poorly sorted. When carbonate bioclast components are >30% they are cCalcareous muds" and if they have <30%, they are considered as "detritic muds" or "volcaniclastic muds" depending of the dominant composition. This type of sediment is generally considered as the "background" facies for many deepwater stratigraphic successions. Hemipelagic sedimentary deposits represent 15%–20% of the present-day seafloor with interval thicknesses exceeding several hundred meters along continental margins. Few attempts to define a facies model for these deposits have been proposed (Stow, 1985, 1986), where researchers try to define the detailed facies characteristics and represent the variety of these deposits (Stow and Tabrez, 1998). Stow and Smillie (2020) propose a classification of hemipelagites based on dominant source of sediment (e.g., glaciogenic, eolian, volcaniclastic, biogenic and organic-rich) (Fig. 12). Though hemipelagites are mostly structureless in oxygenated waters, where bioturbation is intense, in low oxygen and anoxic environments, where bioturbation is low or absent, parallel lamination is often preserved (Stow and Piper, 1984a, b; Brodie and Kemp, 1994; Stow and Tabrez, 1998). Studies also indicate low-angle lamination with a planar appearance after compaction from the bedload transport of silt grains or clay floccules (Schieber et al., 2007; Schieber and Southard, 2009; Schieber and Yawar, 2009; Yawar and Schieber, 2017). For nearly 50 years, there has been debate regarding some hemipelagic intervals and silty turbidites, causing researchers to introduce the term "hemiturbidite" (Stow and Wetzel, 1990; (Boulesteix, 2021)). Further, silty-muddy contourites (Stow, 1979; De Castro et al., 2020, 2021a, b) deposited by currents and hemipelagites deposited by suspension fallout can be very difficult to distinguish, especially when bioturbation is intense (Hesse, 1975; O'Brien et al., 1980—see Chapter 14).

## Mass transport deposits

Mass transport deposits (MTD) are formed by deposition of gravity driven dislodgements of poorly to partially lithified material, originating from an area upslope or on a topographically higher area, to an area of lower elevation (Shipp et al., 2011 and references therein) (see Chapter 10 in this book for more complete characterization of mass transport-dominated systems). MTD are often mixtures of the deposits previously described in this section that have been remobilized and

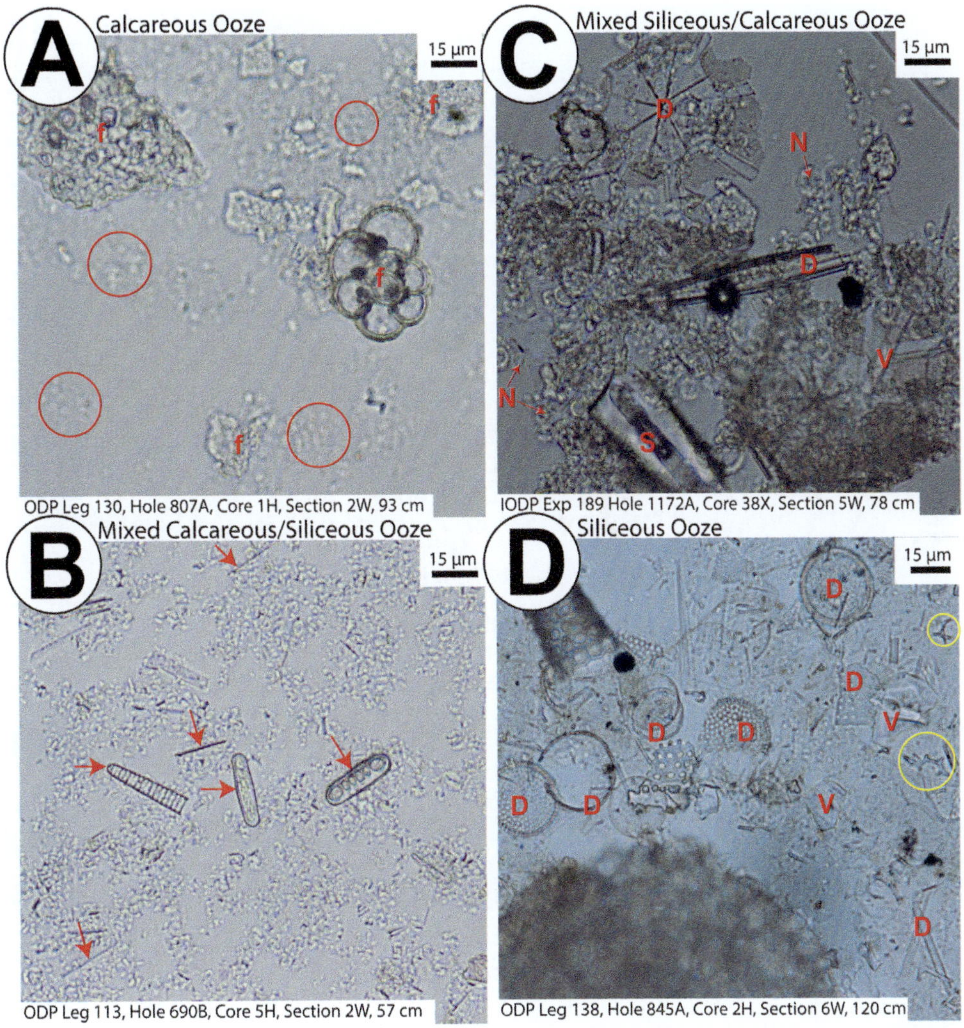


**FIG. 10** Different types of pelagic ooze. (A) Smear slide sample of a calcareous ooze. 60% filled with biogenic debris consisting of an approximately equal proportion of nannofossils (circled in red) and foraminifera (f). (B) Smear slide sample of a mixed ooze with approximately 75% covered by biogenic debris, mostly (92%) nannofossils and lesser pennate diatoms (8%; arrowed). (C) Smear slide sample of a mixed ooze with nannofossils (N; 60%), diatoms (D; 20%), sponge spicules (S; 5%) and volcanic debris (V; 3%). (D) Smear slide sample of a siliceous ooze that includes fragments of radiolarians (circled yellow), minor colorless volcanic glass (V) and common centric to pennate diatoms (30%; D). *(From Marsaglia, K., Milliken, K., Mark Leckie, R., Tentori, D., Doran, L., 2015. IODP Smear Slide Digital Reference for Sediment Analysis of Marine Mud. Part 2. Methodology and Atlas of Biogenic Components. International Ocean Discovery Program Technical Note 2, https://doi.org/10.14379/iodp.tn.2.2015.)*

undergone various degrees of folding, faulting and disaggregation during transport (Fig. 13). MTD are defined by their chaotic bedding, uneven top and generally flat base with overturned, faulted, folded and dismembered bedding (Cardona et al., 2020). In seismic data, they are described as chaotic to acoustically transparent with rugose tops and grooved bases (Posamentier and Kolla, 2003; Dykstra, 2005). MTD are commonly developed in slopes, splay and basin plain environments. While their internal lithological make up can be diverse, MTD tend to have average sand content of 8% and can exceed 5500 $km^3$ in volume (Kneller et al., 2016), making them the products of the largest gravity driven sedimentation events on Earth. Among the largest include the Storegga Slide and other events such as an 870-m-thick MTD with a volume 4100 $km^3$ and four additional 295-m-thick MTD with volumes about 1100 $km^3$. These MTD and other smaller deposits represent about ¼ of all sediment deposited along Barents Sea, offshore Russia and Norway (Safronova et al., 2017). MTD represent up to 50%–70% of deepwater deposits along particularly muddy continental margins (Reading and Richards, 1994) and comprise 24% of slope environments on average (Prather et al., 2016), though they can feature sand-rich intervals exceeding 50% net to gross (Sharman et al., 2015). MTD range in thickness from several meters to nearly a kilometer. MTD is the term for a single event deposit, whereas a mass transport complex (MTC) represents

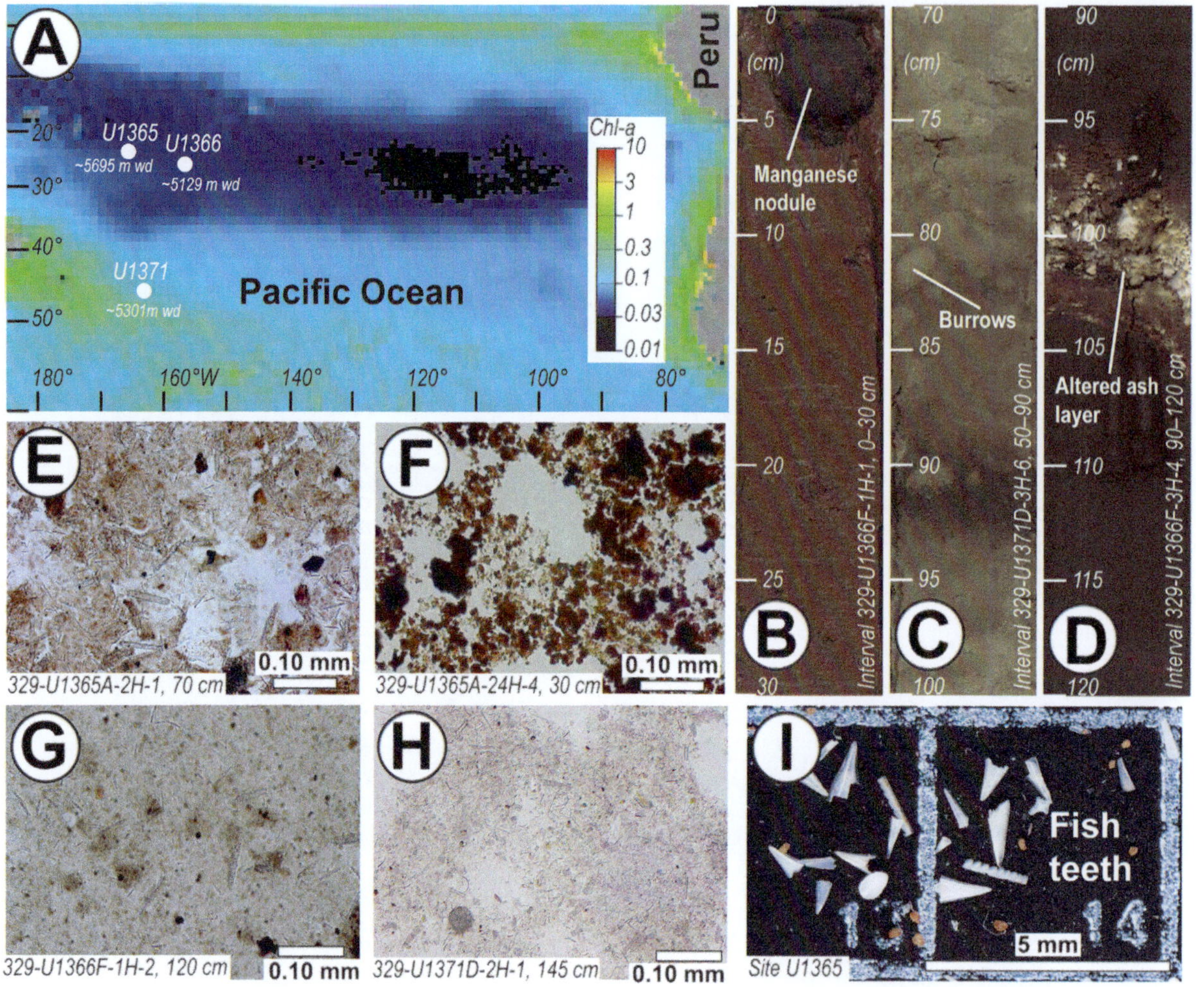


**FIG. 11** Examples of pelagic sediment. (A) Map of annual chlorophyll-a (Chl-a) concentrations. Expedition 329 sites located in (B) to (I) are given (white circles). (B–D) Core photographs, Site U1366 (B) and (D) and U1371 (C). (B) Dark brown zeolitic metalliferous pelagic clay. A manganese nodule is at the top of the core. (C) Olive-gray clay-bearing diatom ooze containing burrows. (D) Very dark brown metalliferous clay. Note white porcelaneous ash to altered ash layer. (E–H) Selected smear slide photomicrographs of clay sediment types, Site U1365 (E) and (F), Site U1366 (G) and Site U1371 (H). (E and F) Clay sediment with zeolite crystals. (G) Zeolitic metalliferous pelagic clay. (H) Clay-bearing diatom ooze. (I) Example of fish teeth, Site U1365. *(From D'Hondt, S., Inagaki, F., Alvarez Zarikian, C.A., et al., 2011. Proceedings of the Integrated Ocean Drilling Program Volume 329 Expedition Reports – South Pacific Gyre Subseafloor Life. Integrated Ocean Drilling Program Management International, Inc., Integrated Ocean Drilling Program. https://doi.org/10.2204/iodp.proc.329.2011.)*

multiple event deposits, like the hierarchical nature of how channels (Sprague et al., 2005) and splays (Prélat et al., 2009) are described. MTD can move at high speeds (Peel et al., 2019) and present a significant modern hazard to seafloor stability (Leon et al., 2017) and well planning through their particularly uneven arrangement of lithologies (Henry et al., 2018). MTD have also been successfully developed as hydraulically fractured (see the term "frac" or "fracking") reservoirs in the Permian Basin region, USA (Allen et al., 2013; Bhatnagar et al., 2019).

Most of the information on their extent, gross geometry, domains and nature of distribution is from seismic data (Dykstra, 2005; Moscardelli and Wood, 2008; Shipp et al., 2011; Alves, 2015; Omeru and Cartwright, 2019; Bhattacharya et al., 2020). In plan-view, MTD can take many forms, but are usually elongate, in the direction of dip and tend to be as architecturally complex as deposits formed by turbidity currents—in a chaotic sense. Further, they form snouts or pressure-ridged margins that can appear as frond-like in plan-view pattern, distinguishing them from other fan-shape patterns that more closely resemble deposition by turbidity current processes. Their makeup is described by Dykstra (2005) and is composed of four principal domains: (i) an updip scarp or failure plane, the site of the material evacuation known as the headwall (sensu Cardona et al., 2020); (ii) the updip portion of the MTD (upper translational); (iii) the middle

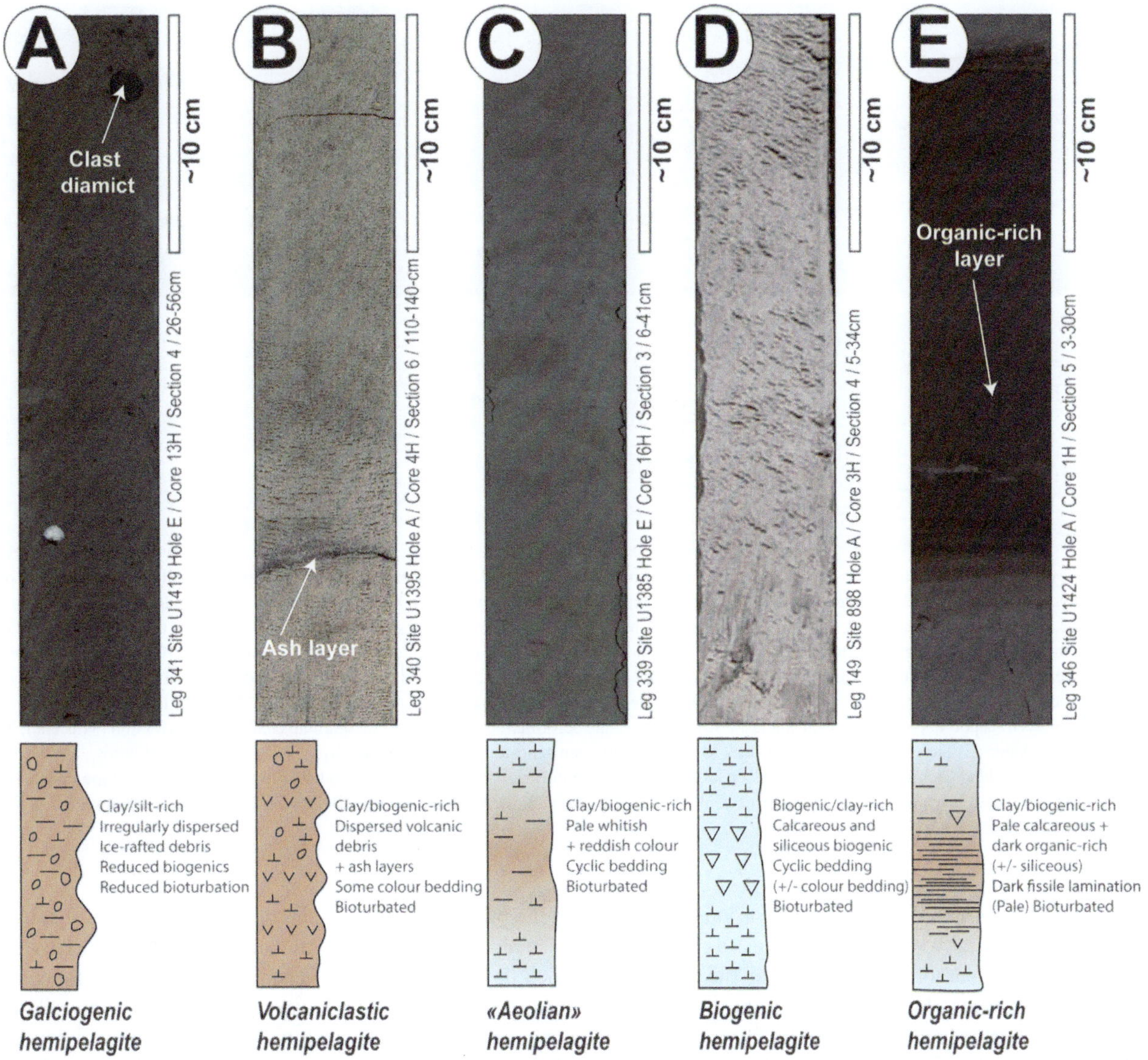


**FIG. 12** Core photographs of hemipelagic clay types: (A) Glaciogenic-dominated hemipelagite. Absence or low abundance of microfossils (few diatom-rich intervals). Composition of the mud is, on average, 70%–80% clay-size particles, 20%–30% silt-size particles and minor amounts (<2%) of volcanic glass. The primary minerals identified include quartz, plagioclase, K-feldspar, mica (muscovite to illite and biotite) and chlorite and/or kaolinite. Diamict clasts and lonestones are, in order of decreasing abundance, siltstone, argillite, sandstone, basalt, granitoid and chert. (Site U1419; Jaeger et al., 2014). (B) Volcaniclastic-dominated hemipelagite. Biogenic fraction contains calcareous nannofossils and planktonic and benthic foraminifers. Note the presence of ash fall layers (pumice) in the core section (Site U1395; Le Friant et al., 2013). (C) "Eolian"-dominated hemipelagite with terrigenous components including quartz, feldspar, clay minerals, detrital carbonate and biotite. The biogenic fraction is dominated by nannofossils with rare to common foraminifers and rare pteropods, ostracods, diatoms and sponge spicules (Site U1385; Stow et al., 2013). (D) Biogenic hemipelagite with terregineous materials >25%. Silty clay to clayey silt dominates this interval. The primary minerals identified include quartz, plagioclase, K-feldspar, mica (muscovite to illite and biotite) and clay minerals (illite and kaolinite, with a lesser proportion of smectite). Calcareous and siliceous marine skeletal debris (generally less than 10%) include foraminifers, sponge spicules, diatoms, radiolarians, dinoflagellates and various nannofossils, including coccoliths and discoasters (Site 898; Sawyer et al., 1994). (E) Organic-rich-dominated hemipelagite. The terrigenous components in this unit are dominated by clay and fine silty clay fractions. The primary minerals identified include quartz, plagioclase, K-feldspar and clay minerals (including smectite, illite and kaolinite and/or chlorite) as well as minor amounts of pyrite framboids in dark brown layers (organic-rich layers). Diatoms and siliceous sponge spicules dominate the biogenic fraction with some radiolarian-enriched levels. Calcareous nannofossils and foraminifers (mostly planktonic) can be abundant (up to 30%) in the organic-rich layers (Site U1424; Tada et al., 2015). *(Facies model from Stow, D., Smillie, Z., 2020. Distinguishing between deepwater sediment facies: turbidites, Contourites and Hemipelagites. Geosciences 10(2), 68, https://doi.org/10.3390/geosciences10020068.)*

**FIG. 13** Several types of mass transport deposits, ranging from disheveled to contorted and squeezed: (A) mass transport deposit located at the base of the fill of a canyon, lacking internally coherent bedding, pervasively internally deformed with large rafted sedimentary blocks up to 3 m (10 ft) in diameter. Grains range from cobble- to clay-sized, Venado Sandstone (Cortina Petrofacies), Monticello Dam Section, California, USA; (B) basin-scale El Gordo mass transport deposit consisting of overturned and gently folded and coherent turbidite beds, Miocene, Tabernas Basin, Spain; (C) mass transport deposit featuring three-dimensional isoclinal folding of thick- to very thick-bedded turbidite sandstone and bedded silstone and mudstone in the vicinity of Jamroll (Miocene), near Waikiki Beach and Te Kawau Pa, Mount Messenger Formation, Taranaki Basin, North Island, New Zealand; (D) 50- to 100-m-thick (~ 164–328 ft.) mass transport deposit rich in volcaniclastic strata, of variable deformation, documented in Sharman et al. (2015), Awakino (Miocene), North Island, New Zealand; (E) large slump or MTD of fine- to very fine-grained sandstone and mudstone, Bridges of Ross (Carboniferous), Western Irish Namurian Basin, Ireland and; (F) MTD emplaced on a slope between mudstone and interbedded sandstone and mudstone in slope channel- and gully-fill and associated levee deposits, Carboniferous Gull Island Formation, Western Irish Namurian Basin, Ireland. For more information on MTD, see Chapter 10.

(lower translational); and (iv) downdip portion or toe. Each of the principal domains consists of a certain set of traits and parts. The up-dip headwall or failure plane is a concave up surface representing the basal decollement or slide surface, often lined with motion-parallel grooves, tracing the route along which the MTD evacuated. The up-dip portion is thin where it pinches out against the headwall and thickens, typically downslope, to the upper and lower translational domains. The up-dip portion is typically characterized by fractures or zones of extension, or tensile fractures. These extensional zones can entirely separate the MTD into discrete blocks, known as slide blocks (Bhattacharya et al., 2020 and references therein). The middle of the MTD is a zone that is thicker than the up-dip portion and can be described as a zone of where extension in the up-dip portion—and contraction in the downdip portion—overlap. It should be noted that in some MTD, the extensional zone can comprise upwards of 70%–80% of the length of the MTD (Bhatnagar et al., 2019), so these are not simply divided into equal volume or distribution areas, they can be quite asymmetric. The downdip portion and distal end of the MTD is underlain by the failure plane, which eventually runs up through the stratigraphy to its terminus, where the tip of the MTD typically overthrusts younger stratigraphy. The downdip portion is a zone of contraction where thrusts, faults and overturned bedding is common and can form topography that later defines submarine channel routes (Ortiz-Karpf et al., 2015).

MTD are imaged along many continental margins, yet there are exceedingly few seismic-scale outcrop analogs such as the Awakino MTD (Sharman et al., 2015) and Rapanui MTD (Masalimova et al., 2016; Cardona et al., 2020) located in New Zealand. There are many others in different basins (Shipp et al., 2011). These are thick (>100 m or ~ 328 ft.), extensive (>2–10 km or ~ 1.2–6.2 mi.) and are largely interpreted to represent the remobilized deposits of what once were stable deepwater channel, splay and deltaic deposits. Internally, MTD can be pervasively deformed with no continuous bedding, or rather intact and competent very thick-bedded turbidites that are encased in masses of mildly to pervasively sheared and mixed original medium-bedded sandstone and mudstone. The orientation of the folds can be used to interpret the direction of transport, yet the internal deformation can be such that there are smaller-scale directional indicators pointing opposite ways, showing the complexity of MTD movement. If possible, it is best to observe the whole MTD in 3D instead of focusing on a small isoclinal fold that may give a spurious direction of transport.

MTD often occur due to the following reasons: (i) sediment instability due to pore pressure changes; (ii) sea level change, and they often mark sequence boundaries as part of the falling stage systems tract; (iii) tectonic activity including earthquakes; (iv) weather and climate change. MTD processes and deposits can offer several different explorations play types associated with their resulting topography, typically creating accommodation for later turbidite sedimentation ((Kneller et al., 2016); Henry et al., 2018).

## Summary of deepwater deposits

A key theme proposed in this chapter is to connect hydrodynamics to process to deposit. This enables researchers to build predictive models of the sedimentary landscape. The deposits described in the previous section can be described based on their sedimentation mechanics, process of sedimentation and genetically derived rock type designation. Mechanics range from fluid, transitional and plastic behavior. The type of flow, sedimentary process and the resulting rock type are summarized in Fig. 14 from Chapter 6 in this book and Figs. 1–10 from this chapter. Relative estimates of initial reservoir quality are shown in Fig. 15 in Chapter 6 in this book.

In deepwater sedimentation, the relative strength of sedimentary processes and grain sizes determines the overall fabric of the deposit, similar to how deltas are plotted on a 3D ternary diagram. While it is exceedingly rare to find to complete sedimentation units in any system, if flows wane gradually and are provided most grain sizes, there exist hypothetical *complete* models of sedimentation units that would form principal reservoirs (Fig. 14) in any deepwater system. However, most sedimentation units are missing one or more of the subdivisions. It is important to remember that all of these deposits can be remobilized in the form of MTD, injectites or extrudites that will obscure the original bedding fabric. These phenomena are discussed in Chapter 10 in this book.

### Nomenclature

Nomenclature is effective at placing beds within a framework, yet some beds show evidence of multiple flow rheologies and may be well-placed in different frameworks. For example, these frameworks may include the Lowe and Guy, Haughton and Bouma classifications. There must be cut-offs in place to determine in which category the bed will be placed. Each researcher will do this somewhat differently depending on the needs of the project. There is even an inveterate tendency to create new names for the same type of deposit, when it is merely subtle variation on an overarching theme, for example, using a completely new naming convention even though most of the first-order architectural elements are those described as H1–H5 divisions. This convention has resulted in very complicated characterization for deepwater deposits.

For example, some industry workflows divide stratigraphy into high-density flow deposits, low-density flow deposits, debrites and other types of rocks and plot them on pie diagrams. This method addresses process, yet it does not adequately address factors required for estimating initial fluid in place. Bed thickness-based classifications can be more effective at matching deposits with reservoir traits such as net to gross, porosity and permeability as discussed in the next section on *Lithofacies*.

## Lithofacies

In any given deepwater depositional system, there can be roughly 10 main deepwater lithofacies (*mappable* rock units = at sufficient spatial scale to be observed on a map) based on grain size, lithology and bed thickness (Campbell, 1967) including conglomerate (matrix- vs. clast-supported), very thick-bedded sandstone, thick-bedded sandstone, medium-bedded sandstone and mudstone, thin-bedded sandstone and mudstone, very thin-bedded sandstone and mudstone, muddy sandstone, chaotic deposits and mudstone, with their key traits listed in Table 2 and shown in Fig. 15. Incorrect interpretation of deepwater lithofacies and how they relate to each other is one of the main pitfalls in mapping deepwater depositional systems. These lithofacies tend to be deposited in broad erosional sheets, laterally continuous deposits or focused to marginal deposits based on availability of sediment and resulting deposition (Fig. 16). There can be many more lithofacies than 10, but this number is generally a good start for any basin on Earth. For example, each primary lithofacies can be subdivided further based on variability in sedimentary structures, sedimentation process, geometry and many other factors that define a sedimentary deposit, including petrofacies (Ingersoll, 1981) or ichnofacies (Rodríguez-Tovar et al., 2020 and references therein—see Chapter 15), palynofacies (McArthur et al., 2016) or biofacies (Weber, 2018—see Chapter 14) and many others. The number of lithofacies used in a study depends on the question being asked, but there tend to be about 10 common lithofacies themes in deepwater.

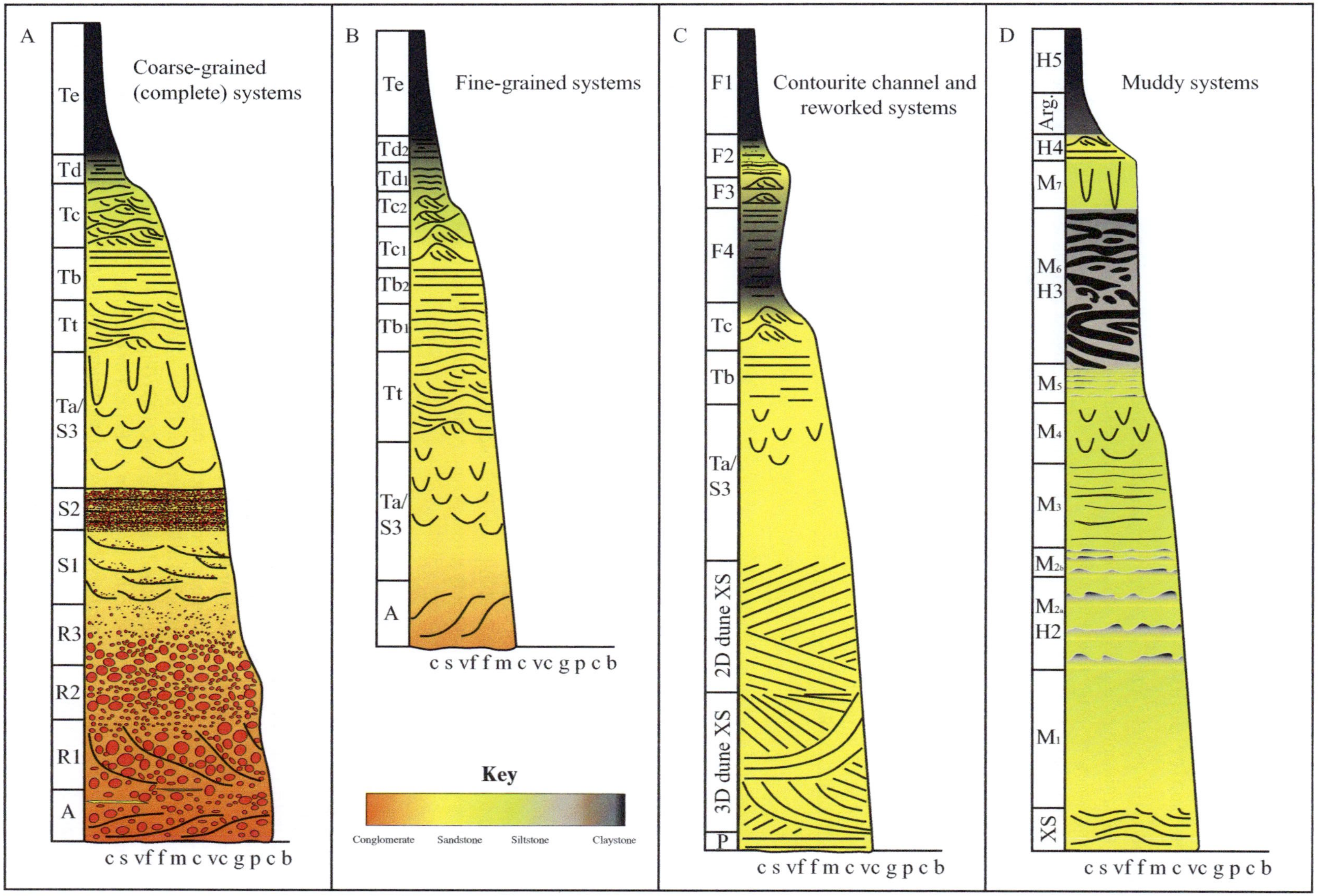


**FIG. 14** Idealized sedimentation units representing principal reservoir types in turbidite-dominated environments for both coarse-grained (Lowe, 1982) to complete (Bouma, 1962) and fine-grained (modified from Bouma, 1962; Gutierrez, 2019; Peakall et al., 2020) depositional systems, as well as mixed contourite and turbidite systems (modified from Capella et al., 2017; Fonnesu et al., 2020; De Castro et al., 2020; De Weger et al., 2020, 2021) and muddy systems (modified from Lowe and Guy, 2000; Haughton et al., 2003, 2009; Kane and Pontén, 2012; Power et al., 2013; Marchand et al., 2015; Fonnesu, 2020). If all grain sizes are provided, and the flows undergo each stage of sedimentation, the following sedimentation unit will form. In most cases, first-order architectural elements are missing, anomalously thick, repeated or are in a slightly different order. While this figure attempts to show what complete sedimentation units will look like if they go through every stage of sedimentation, it should be noted that all first-order architectural elements are not always present. In fact, that is typically the rule, not the exception.

**TABLE 2** Most deepwater depositional systems can be mapped using a 10-part lithofacies scheme.

| Lithology | Grain size | Grading | Bed thickness | Sedimentary structures | Secondary features | Sediment gravity flow divisions | Sedimentary process | Energy of setting | Typical depositional environment |
|---|---|---|---|---|---|---|---|---|---|
| Conglomerate (clast- and matrix-supported) | Gravel including pebbles, cobbles and boulders; can be well sorted to poorly sorted; matrix ranges from other gravels to sand to mud | Nongraded, poorly graded to normally graded, inversely graded traction carpets | Typically tens of centimeters to several meters | Cross-stratification, traction carpets, structureless, imbrication of gravels | Mudclast rip-ups, plowed bedding, mudstone blocks, armored mudclasts, flame structures, ball-and-pillow structures and other load structures, fractured shells, charcoal fragments | $R_1$, $R_2$, $R_3$, various debritic divisions | Debris flow, high-density turbidity current, modified grain flow, rock fall, mass transport | Very high to high energy | Canyon, channel |
| Very thick-bedded sandstone | Very coarse- to very fine-grained sand; can feature conglomerate including pebbles and granules near bed bases and along internal scour surfaces | Poorly graded to normally graded, inversely graded bed bases, inversely graded traction carpets | >1 m | Structureless sandstone, cross-stratification, crude stratification, planar lamination and ripple cross-lamination rarely at bed tops, convolute lamination, soft-sediment deformation, internal scour, dewatering structures | Bed amalgamation, fractured shells, mudclast rip-ups, load structures, concretions, charcoal fragments, very rare burrowing | $S_1$, $S_2$, $S_3$ ($T_a$/H1), $T_t$, $T_b$, $T_c$ and to a lesser extent $T_d$ and $T_e$; possible M and H divisions | High- and low-density turbidity current | High energy | Canyon, channel, lobe (splay) |
| Thick-bedded sandstone | Very coarse- to very fine-grained sand; can feature conglomerate including pebbles and granules near bed bases and along internal scour surfaces | Poorly graded to normally graded, inversely graded bed bases, inversely graded traction carpets | 30 cm to 1 m | Structureless sandstone, cross-stratification, crude stratification, planar lamination and ripple cross-lamination rarely at bed tops, convolute lamination, soft-sediment deformation, internal scour, dewatering structures | Bed amalgamation, fractured shells, mudclast rip-ups, load structures, concretions, charcoal fragments, very rare burrowing | $S_1$, $S_2$, $S_3$ ($T_a$/H1), $T_t$, $T_b$, $T_c$ and to a lesser extent $T_d$ and $T_e$; possible M and H divisions | High- and low-density turbidity current; transitional flow | High energy | Canyon, channel, lobe (splay) and to a lesser extent levee; rarely in overbank |

*Continued*

**TABLE 2** Most deepwater depositional systems can be mapped using a 10-part lithofacies scheme—cont'd

| Lithology | Grain size | Grading | Bed thickness | Sedimentary structures | Secondary features | Sediment gravity flow divisions | Sedimentary process | Energy of setting | Typical depositional environment |
|---|---|---|---|---|---|---|---|---|---|
| Medium-bedded sandstone and mudstone | Very coarse- to very fine-grained sand to silt and clay; can feature rare conglomerate including pebbles and granules near bed bases and along internal scour surfaces | Normal grading | 10–30 cm | Structureless sandstone, planar lamination and ripple cross-lamination, convolute lamination, soft-sediment deformation, internal scour, dewatering structures; rarely cross-stratification, crude stratification | Loading, fractured shelly material, sparsely burrowed caps (when present), small mudstone rip-ups | $S_3$ ($T_a$/H1), $T_b$, $T_c$, $T_d$, $T_e$; possible M and H divisions and to a lesser extent T0, T1, T2, T3, T4, T5, T6, T7, T8 | High- and low-density turbidity current; transitional flow | Moderate energy | Lobe (splay), levee and to a lesser extent canyon, channel and overbank |
| Thin-bedded sandstone, siltstone and mudstone | Medium- and fine- to very fine-grained sand, silt and clay | Normal grading | 3 to 10 cm | Planar lamination, ripple cross-lamination, flame structures; to a lesser extent structureless sandstone | Moderately burrowed, lack of mudstone rip-ups | $T_b$, $T_c$, $T_d$, $T_e$ and to a lesser extent $T_a$; possible M and H divisions and T0, T1, T2, T3, T4, T5, T6, T7, T8 | Low-density turbidity current; transitional flow; hemipelagic sedimentation | Low to very low energy | Levee, overbank, lobe (splay) |
| Very thin-bedded sandstone, siltstone and mudstone | Medium- and fine- to very fine-grained sand, silt and clay | Normal grading | <3 cm | Planar lamination, ripple cross-lamination, flame structures, mudstones show faint lamination | Moderately burrowed, lack of mudstone rip-ups | $T_b$, $T_c$, $T_d$, $T_e$; T0, T1, T2, T3, T4, T5, T6, T7, T9; possible M and H divisions and | Low-density turbidity current; hemipelagic sedimentation | Low to very low energy | Levee, overbank, lobe (splay) |

| | | | | | | | | | |
|---|---|---|---|---|---|---|---|---|---|
| Muddy sandstone to chaotic deposits | Medium- to very fine-grained sand, silt and clay, muddy matrix | Normal grading, poor grading to no grading | Up to several meters, but typically less than 1 meter to centimeters | Grain size and distribution banding, mudstone clasts, structureless sandstone, planar lamination, ripple cross-lamination, flame structures, convolute lamination, dewatering structures | Overturned bedding, injected sand and mud, disrupted layering, fluid escape structures | Highly variable and can include $M_1$, $M_2$, $M_3$, $M_4$, $M_5$, $M_6$, $M_7$, H1, H2, H3, H4, H5 | Debris flow (flow strength and cohesion); mixed turbulence and matrix strength (slurry flow, transitional flow, hybrid event, argillaceous sand) | Highly variable; high to low energy | Lobe (splay), overbank, levee, canyon, channel |
| Mudstone | Silt and clay | Normal grading to no grading | About 10cm to >1 m thick | Banding, mottling, lamination, grain aggregates | Burrowing, disrupted layering, injections, evidence of fluid escape | $T_d$, $T_e$; can be typically none; T6, T7, T8 and to a lesser extent T0, T1, T2, T3, T4, T5 | Dilute tails of low-density turbidity currents; debris flow; hemipelagic sedimentation; pelagic sedimentation | Typically very low to ultralow energy; can be high if remobilized (debrite or product of mass flow event) | Overbank, levee and to a lesser extent canyon, channel and lobe |

Some systems will have more; some less. Each lithofacies is deposited by a similar flow regime, has similar sedimentation structures, which translates to a typical range and variability in reservoir modeling traits including net to gross, porosity, permeability and reservoir distribution (lateral and vertical).

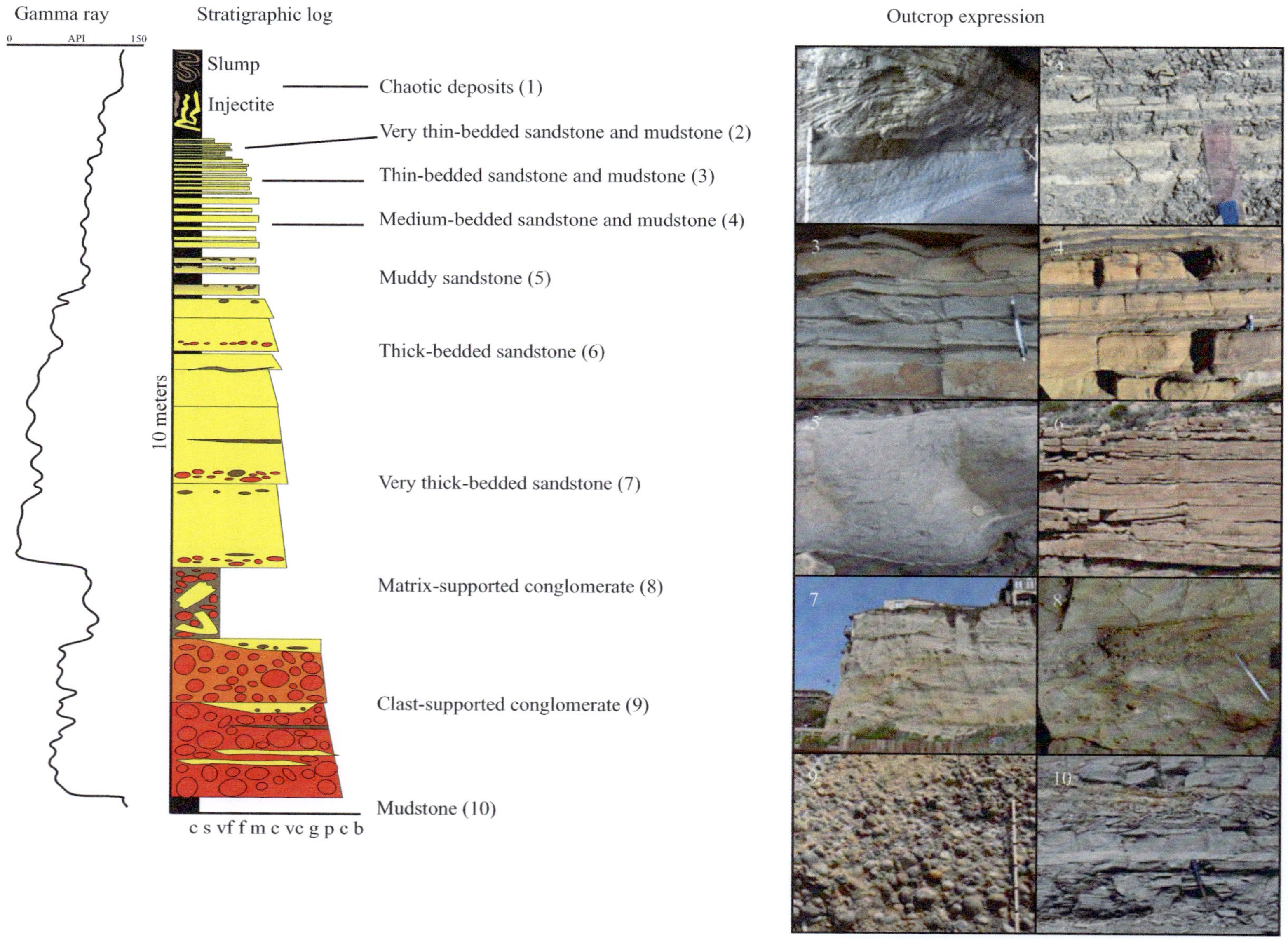


**FIG. 15** Key deepwater lithofacies, or mappable rock units that are distinguished using the Campbell (1967) bed thickness classification. Simply dividing rocks into lithofacies translates into meaningful differentiation of other metrics including porosity and permeability. In any given deepwater depositional system with the full range of grain sizes, and simply, there are about 10 main deepwater lithofacies including: conglomerate, including clast-supported (Eocene Carmelo Formation, California, USA) and matrix-supported (Late Cretaceous to Paleogene Pigeon Point Formation, California, USA) very thick-bedded sandstone (Miocene-Pliocene Capistrano Formation, California, USA), thick-bedded sandstone (Permian Brushy Canyon Formation, Texas, USA), medium-bedded sandstone and mudstone (Eocene Carmelo Formation, California, USA), thin-bedded sandstone and mudstone (Carboniferous Ross Formation, Ireland), very thin-bedded sandstone and mudstone (Upper Cretaceous Antelope Shale, California, USA), muddy sandstone (Carboniferous Ross Sandstone, Ireland), chaotic deposits (Miocene Mount Messenger Formation, New Zealand) and mudstone (Upper Cretaceous Eagle Ford Group, Texas, USA). Deposits tend to be distributed in a predictable fashion relating to sedimentation process, depositional environment and position along length of depositional system (Walker, 1967; Lowe, 1982; Kane and Hodgson, 2011; and others), but also along the off-axis zones and periphery of the depositional system.

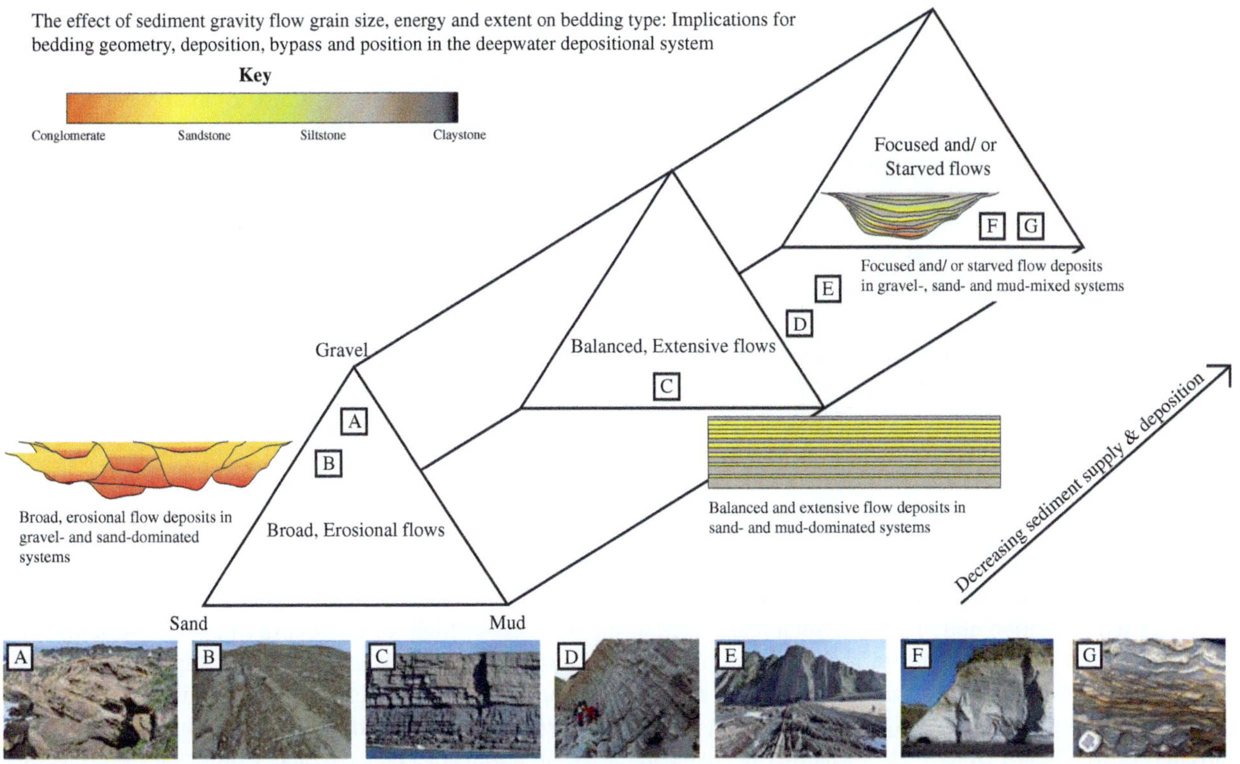


**FIG. 16** Three-dimensional bedding type diagram for gravelly, sandy and muddy types of deepwater sediment gravity flows. Key deepwater bedding types and geometries produced by (i) broad, erosional flows tend to create wide areas of lenticular bedding prone to bypass, typically in the upper zones of a deepwater depositional system; (ii) balanced, extensive flows tend to construct wide areas of tabular bedding with lower amounts of evidence for bypass, typically in the middle part of the depositional system; (iii) focused, starved flows tend to create areas of both bypass and terminal sedimentation throughout the depositional system. Erosional areas that are broad in areal extent tend to be complexly scoured and offset. Sheet-like deposits tend to be broad, sediment-rich areas are primarily depositional (not prone to bypass like the erosional areas). Starved areas tend to be sediment depleted and represent the last vestiges of that type of sediment along the basin profile—either in an upslope or downslope position. *Important note:* what is deposited is not always linked to what created the erosion.

Adhering to a simple lithofacies scheme is recommended for the following reasons: (i) it is bed thickness based, which is objective and focuses on the bed as the fundamental depositional unit of the reservoir; (ii) each lithofacies tends to have a statistically significant range of values including net to gross, porosity and permeability, such that these metrics can be input to both static and dynamic reservoir models; (iii) it is simple, accurate, effective and can be understood across disciplines; (iv) in each lithofacies "bin" there is room to describe the range and variability of process, deposit and reservoir characteristics. Lithofacies tend to be distributed in a predictable fashion relating to sedimentation process, depositional environment and position along length of depositional system (Walker, 1967; Lowe, 1982; Kane and Hodgson, 2011 and many others), but also along the off-axis zones and periphery of the depositional system. There is significant variation in these deposits, as the previous sections show just in the sand-rich deposits that tend to serve as reservoirs, carrier beds and part of the overburden.

A bed thickness classification does not account for all variation. For example, medium-bedded sandstone and mudstone deposits in the axial splay position can be different from the off-axis and peripheral zones in the same splay (Grecula et al., 2014; Brooks et al., 2018). These beds may have different net to gross, porosity and permeability profiles. Secondly, not all beds within the same "bin" will have the same properties. Very thick-bedded sandstone can be amalgamated or nonamalgamated. These two different bedding styles will have subtly different reservoir properties. Thirdly, the effects can be even greater if one bed, for instance, is planar laminated and well sorted whereas another bed of equal thickness is structureless and poorly sorted, and therefore has lower porosity and permeability values (highly variable from formation to formation). These observations yield information on beds that have wide, moderate and low variability in terms of sedimentary process, sedimentary structure and resulting rock properties.

## Conglomerate

Conglomerate in deepwater depositional environments (Walker, 1975; Cherry, 1993; Bozetti et al., 2018) can be supported by smaller gravel or sand clasts (clast-supported) or a muddy matrix (matrix-supported). Clast- or matrix-supported conglomerate can be interpreted as a debrite unless it shows evidence of current structures (R divisions). Conglomerate can be composed of two principal grain types: (i) lithic clasts, which can be from volcanic, sedimentary or metamorphic origin; or (ii) intrabasinal clasts, which are also known informally as *mudclast rip-ups*. Conglomerate typically accumulates in areas of confinement and along erosional surfaces in canyons (Cronin and Kidd, 1998; Lowe, 2004), channels (Hubbard et al., 2007) or in the proximal part of the channel mouth to channel-lobe (splay) transition zone (Hickson and Lowe, 2002).

Mudstone conglomerate, which can consist of variable sized clasts, provides evidence of eroded muddy substrate and incorporation into a relatively high-energy flow. While these mudclast-rich intervals can make up the base of a bed as part of a high-density flow deposit, these can be so abundant as to increase the cohesion of the flow, causing the resulting deposit to be interpreted as a debrite (Example: H3 division when the debrite is part of a HEB; sensu Haughton et al., 2009). Conglomerate, whether it is matrix- or clast-supported, or is composed of mudclast rip-ups, is interpreted to be deposited in the most high-energy environments capable of transporting such coarse material.

## Very thick-bedded sandstone

Very thick-bedded sandstone, or sandstone beds that exceed 1 m ($\sim$ 3.3 ft.) in thickness, can be composed of any previously discussed sedimentation unit. They are often interpreted as the products of high- to low-density turbidity currents. These beds can exhibit a wide, moderate and low variability in process and deposit. Wide variability deposits, or complete deposits (Fig. 14), include the gravelly sandstone beds of the Upper Cretaceous Pigeon Point Formation, USA (Tyler, 1972) composed of Lowe $R_{(1)23}S_{123}$ with overlying Bouma $T_{bcde}$ divisions, deposited via high- and low-density turbidity currents (Lowe, 2004). Other wide variability deposits are based on dominant grain size and sedimentary process (high-density turbidity current—14A; low-density turbidity current—14B; contour current—14C; transitional flow—14D). Moderate variability deposits, or partial deposits, include the sandstone beds of the Upper Miocene Upper Mount Messenger Formation, Tongaporutu, Taranaki, North Island, New Zealand, primarily composed of Lowe $S_1$ or Bouma $T_t$ divisions with thin $T_{bcde}$ bed tops, deposited via a low-density turbidity current (Masalimova et al., 2016). Low variability deposits, or incomplete deposits include, for example, the sandstone units of the Paleogene Merle Formation, USA, composed of Lowe $S_3$/Bouma $T_a$ divisions, with rare, structured bed tops (Anderson et al., 2006). An alternative high-energy flow deposit with low variability would be the gravelly sandstone of the Repetto Formation, Ventura Basin, USA, (Lowe $S_{23}$) in proximal splay settings (Schwalbach et al., 2009). Very thick-bedded successions alternating with thinner and finer-grained beds are also common in aggradational, ponded basins where the thick sandy beds also have thick muddy caps. In this case, thick and sheet-like HEB are common and laterally extensive. Examples of this observation include the Miocene Marnoso-arenacea Formation (Amy and Talling, 2006), the Oligocene-Miocene Castagnola Formation (Southern et al., 2015) or in the distal part of the Cretaceous Gottero Sandstone (Fonnesu et al., 2018), Italy. Very thick-bedded sandstone shows a wide variety of types of deposits based on process and depositional fabric, and each would need to be described separately depending on the studied interval.

## Thick-bedded sandstone

Thick-bedded sandstone, or sandstone beds that range from 30 cm to 1 m in thickness ($\sim$ 12–39 in.), is very similar to its bigger sibling *very thick-bedded sandstone*. While these beds can and do have mudstone components, it is generally a small proportion to the overall bed thickness and overall volume. They too can be composed of Lowe or Bouma type sedimentation units, or even the products of contour currents. Wide variability deposits include the gravelly sandstone beds of the Eocene-Oligocene Annot Sandstone at Chalufy, France or Upper Mount Messenger Formation sandstone, New Zealand composed of complete Bouma $T_{abcde}$ divisions, deposited primarily via a high- and low-density turbidity current. Moderate variability deposits, or partial deposits, include the sandstone beds of the Pliocene Pico Formation, Ventura Basin, USA primarily composed of Bouma $T_a$ divisions with thin $T_{(bcd)e}$ bed tops, deposited primarily via a high-density turbidity current. Low variability deposits, or incomplete deposits include the sandstone units of the Eocene San Vicente Formation near Ainsa, Spain, composed of Lowe $S_3$/Bouma $T_a$ divisions, with rare, structured bed tops. As with very thick-bedded sandstone and other lithofacies described in this section, thick-bedded sandstone shows a wide variety of types of deposits based on process and depositional fabric.

## Medium-bedded sandstone and mudstone

Medium-bedded sandstone and mudstone, or sandstone beds that range from 10 to 30 cm (~ 4–12 in.) tend to be composed of primarily low-density turbidity current deposits and reworked deposits. Mudstone bed tops become increasingly abundant, enough to necessitate the designation of mudstone when naming the lithofacies. Wide variability deposits include the sandstone of the Upper Cretaceous Panther Tongue Delta, USA or portions of the Eocene Carmelo Formation, USA composed of complete Bouma $T_{abcde}$ divisions, deposited primarily via low-density turbidity current. Moderate variability deposits include the sandstone beds of the Mapiri Formation, East Coast Basin, North Island, New Zealand composed of Bouma $T_{bd(d)e}$ divisions with nonexistent structureless bed bases deposited entirely via low-density turbidity current (Burgreen and Graham, 2014). Low variability deposits, or incomplete deposits include the sandstone units of the Sites Sandstone, Great Valley Group, Sacramento Basin, California, USA, with intervals composed of Lowe $S_3$/Bouma $T_{ae}$ divisions (Murray et al., 1996).

## Thin-bedded sandstone and mudstone

Thin-bedded sandstone and mudstone, or sandstone beds that range from 3 to 10 cm (~ 1.2–4 in.) tend to be composed of low-density turbidity current and reworked deposits. Mudstone can comprise the majority of these beds but is highly variable. Wide variability deposits include the sandstone beds and petroleum reservoir intervals of the Upper Mount Messenger Formation composed of nearly complete Bouma $T_{(a)bcde}$ divisions, deposited primarily via low-density turbidity current (Slatt et al., 1998; King and Browne, 2004). Moderate variability deposits, or partial deposits, include the sandstone beds of the Upper Cretaceous Juniper Ridge Conglomerate (USA) overbank deposits or background, composed of Bouma $T_{bc(d)e}$ divisions with nonexistent structureless bed bases deposited entirely via low-density turbidity current. Low variability deposits, or incomplete deposits include the sandstone units of the Cretaceous Fiske Creek Shale (USA), composed of Bouma $T_{(abcd)e}$ divisions. This lithofacies, along with medium-bedded sandstone and mudstone, is associated with more abundant transitional flow deposits, those of HEB, argillaceous sandstone beds and other hybrid deposits.

## Very thin-bedded sandstone and mudstone

Very thin-bedded sandstone and mudstone (< 3 cm or 1.2 in. bed thickness) shares the same traits as thin-bedded sandstone and mudstone, and is often grouped together, but there are a few exceptions. Excellent examples are found in levee deposits in the subsurface ( Janocko et al., 2013) and in outcrop (Morris et al., 2014). Variations include: (i) very little sandstone is present, and beds primarily consist of siltstone and claystone (Strachan et al., 2016); (ii) the process of sedimentation is different, these beds may be primarily transitional flow deposits, placing them in the *muddy sandstone* bin whereas the thin-bedded lithofacies includes typically turbidites; (iii) these deposits are somehow modified or reworked by other processes, and may represent the products of slumps or flowage (these intervals are more prone to slumping, as in the Rinevilla and Bridges of Ross locations in the Carboniferous Ross Formation, Ireland).

## Muddy sandstone

Muddy sandstone can typically represent the range and variability of slurry-flow, transitional flow, HEB and argillaceous sandstone bed deposits. Bed thickness variation can be high, ranging from very thick to very thin, so it is important to use a net to gross cutoff when determining how to apply this lithofacies in a modeling workflow. For example, some teams use 50% rule: if >50% of the bed is muddy sandstone, it goes in the muddy sandstone bin, otherwise the bed falls into one of the other sandstone or conglomerate categories. Additionally, if in a basin with high compaction, muddy deposits may represent 0% net to gross, so this point is moot. Slurry flow deposits include those in the UK North Sea described by Lowe and Guy (2000) and in the Sites Sandstone (USA) (Murray et al., 1996). HEB deposits include the type sections in the Western Irish Namurian Basin, such as Ballybunion (Pierce et al., 2018). Transitional flow and argillaceous sandstone beds are found in regions of petroleum exploration, but most famously the deepwater Wilcox Group (USA) (Kane and Pontén, 2012; Power et al., 2013).

## Mudstone

Mudstone or *mudrock*, previously and more commonly termed *shale*, can form thick intervals in deepwater settings, ranging from several cm to kilometers in thickness (Potter et al., 2005). Mudstone consists of a variety of types, and this heterogeneity is outlined in Boulesteix et al. (2020); principally by mudstone that is derived from a turbidity current or sediment gravity flow (deposited via suspension); that of remobilized mudstone (sediment gravity flow or mass movement) (Walker and Massingill, 1970); and that of a hemipelagic to pelagic origin. Further, while mudstone is a lumped term, it can include purer forms of siltstone, claystone and any variation of the two, even with sand mixed in Lazar et al. (2015). Mudstone intervals are important as they form the boundaries, seals and baffles for reservoirs (Worden, 2020).

## Chaotic deposits

While much of the stratigraphy derived from sediment gravity flows, contour currents, hemipelagic and pelagic sedimentation fits into the previous nine lithofacies, there exists a wide range of deposits that are more chaotic in nature due to mass transport processes and other forms of sediment remobilization discussed in detail in Chapter 10. These deposits may represent original bedding or bedding that has been modified by sediment remobilization that may have partially or completely obscured original bedding fabrics: (i) various types of debrites and transitional flow deposits if their reservoir traits are significantly different to exclude them from previous lithofacies listed; (ii) MTD (Lowe, 1972) and MTC (Shipp et al., 2011); (iii) slide blocks or remnant blocks (Alves, 2015; Bhattacharya et al., 2020) with variable degrees of slumping or overturned bedding; (iv) injected and extruded intervals (Hurst and Cartwright, 2007) such as those in the Panoche Hills Injectite Complex, USA, forming complex dykes and sills at an angle to the original stratigraphy; and (v) disrupted intervals due to fluid or semilithified sediment movement or upward escape typically found at the distal margins of submarine fan environments or along the slope (Chenrai and Huuse, 2017 and others) such as paramoudra. In conclusion, there is quite a broad range of chaotic deposits in deepwater, and they may or may not deserve their own lithofacies classification.

## Summary of downslope variation

Sedimentary deposits tend to be arranged predictively downslope in a deepwater depositional system based on overall grain size, flow type and depositional environment, with the coarsest grain sizes and highest energy deposits accumulating in the most proximal areas. Fig. 17 (modified from Lowe, 1982; Shanmugam, 2006 and others) shows how high-density, low-density, hybrid and very low-density flow deposits are organized on a typical continental margin profile. The themes of suspension, traction, traction carpet, hemipelagic and pelagic sedimentation are recurring themes from upslope to

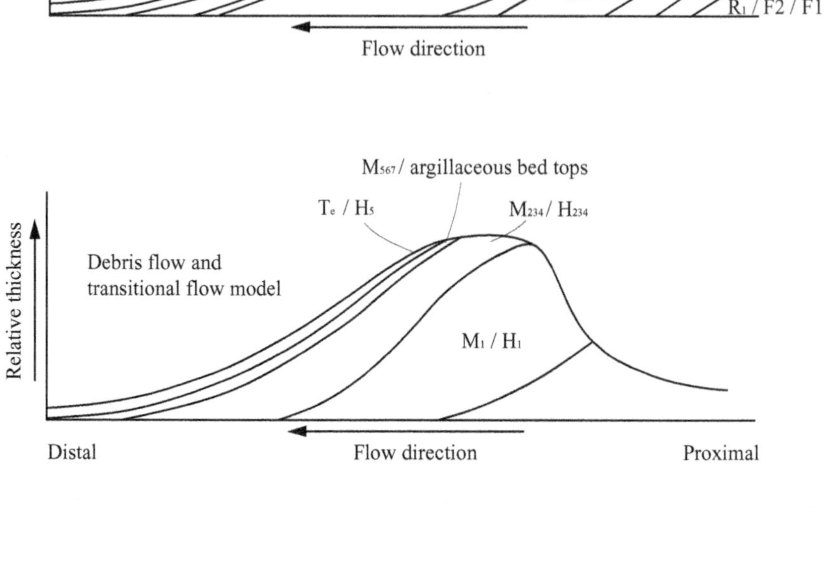


**FIG. 17** Downslope variation (proximal to distal) in first-order architectural elements and lateral variation in sedimentation style. Relative thickness of deposits along the axis of a deepwater depositional system. Original diagram from Lowe (1982) with modifications added from Stow and Shanmugam (1980), Mutti (1985, 1992), Lowe and Guy (2000), Shanmugam, 2006, Haughton et al. (2009), Kane and Pontén (2012), Power et al. (2013) and Marchand et al. (2015). T0–T8 divisions of low-density turbidity current deposits after Stow and Shanmugam (1980) and Shanmugam (2000); H1–H5 divisions of hybrid-event beds after Haughton et al. (2009) and others.

downslope and can be built into predictive models based on overall gradient of the continental margin. Further, the bedding shows variation from upslope to downslope, with upslope portions dominated by erosion and bypass (sensu Stevenson et al., 2015); medial portions dominated by broad lateral continuity of bedding; and distal portions or margins dominated by starvation of sediment. A typical deepwater depositional system shows that coarser grained deposits, requiring higher energy flows, are found in more proximal settings, whereas finer grained deposits are found in more distal, low-energy settings. There are exceptions (Giacomone et al., 2020) and even modifications for carbonate environments (Hairabian et al., 2019; Recouvreur et al., 2020) and mixed carbonate-siliciclastic environments (Chiarella et al., 2017 and references therein). For typical bedding and stacking patterns from an example deepwater depositional system for the Carboniferous Western Irish Namurian Basin, see Fig. 18. Further, key mappable lithofacies tend to be arranged predictively downslope and in 3D space in a typical deepwater depositional system (Fig. 19). Downslope sedimentation trends are based on a number of key factors related to the available sediment, sedimentation transport and deposition, gradient and accommodation. Simple models of sedimentation can be used to predict the type(s) of deposit and implications for the total petroleum system. Classic deepwater environments, as well as newly interpreted mixed environments (carbonate-clastic sensu Chiarella et al., 2017; contourite-turbidite sensu Hernández-Molina, 2021), open up a new playbook to deepwater research around the world—from the greater-than-a-century-old Permian Basin region, USA, to frontier offshore areas.

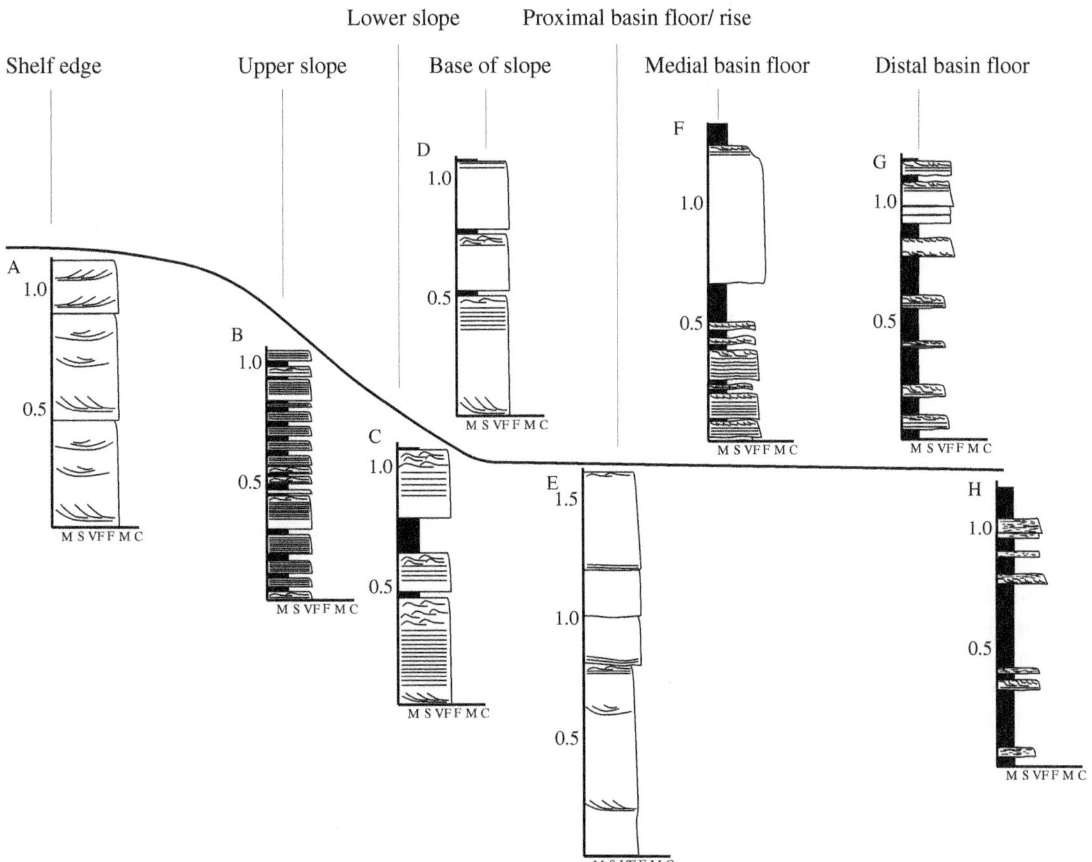


**FIG. 18** Example downslope change in sedimentation units in a fine-grained depositional system from shelf-edge delta to slope to basin floor, with an emphasis on turbidite sedimentation in deepwater, Western Irish Namurian Basin, Ireland (Rotzien, 2016). Sedimentation unit style from A to H (upslope and proximal to downslope and distal) varies predictably based on gradient, position relative to the slope and several other first principles including grain size, type of sediment gravity flow and processes relating to sediment transport and depositional environment. This diagram shows typical bed variation in the Tullig Cyclothem, Gull Island Formation, and Ross Sandstone Formation.

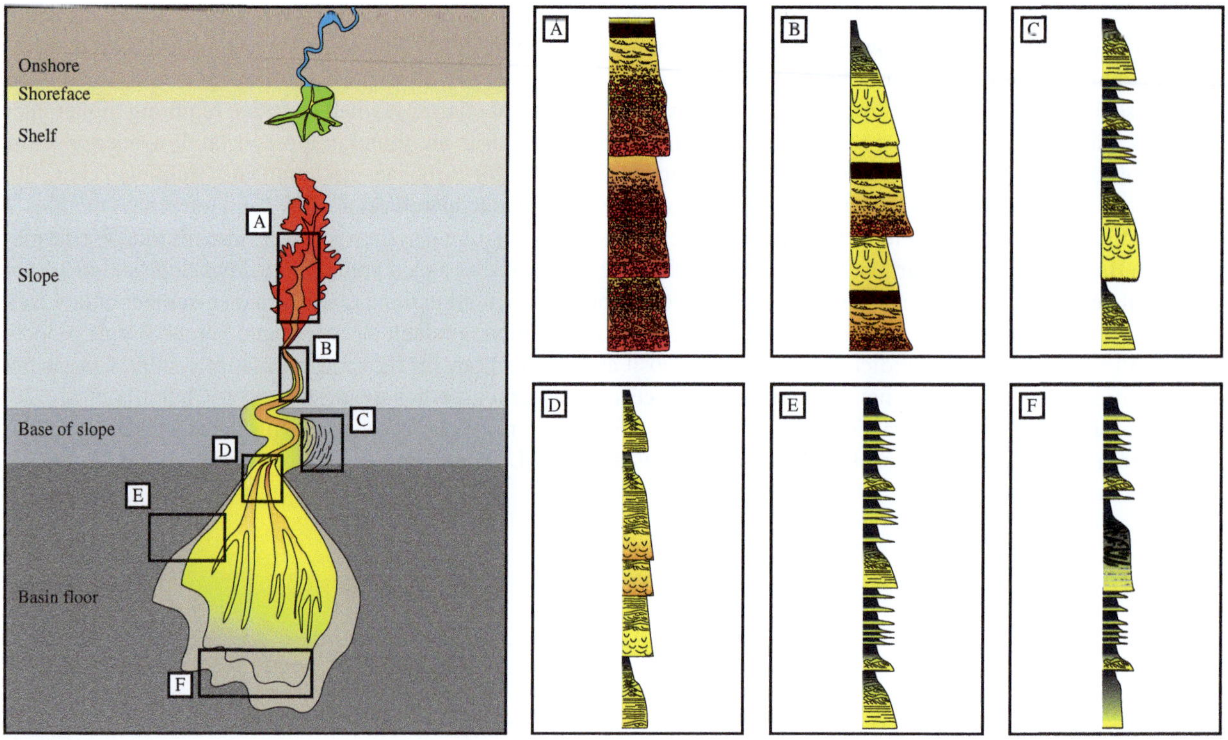


**FIG. 19** Sedimentary structure and sedimentation unit variability in a typical submarine fan depositional system. There are key differences from upslope to downslope and from axis to off-axis positions for typical reservoir intervals from (a) canyon, (b) channel, (c) levee, (d) channel-lobe (splay) transition zone, (e) off-axis splay environment and (f) distal, axial splay environment. The key differences include: (i) the presence of hybrid event beds along the axis and sedimentary structures in the off-axis and lateral positions; (ii) presence of a wide variety of structures in the overbank; (iii) similarity to fluvial systems and point bars within channels; (iv) a large structureless zone at hydraulic jumps and CLTZ. Note: this is a hypothetical diagram in response to a conceptual submarine fan model; your basin of interest may differ.

## Key messages

The range and variability of deepwater sedimentation unit models and styles can be daunting, but it is also very useful to assessing depositional environment and building information on the 3D architecture and evolution of the sedimentary basin. Each of the sedimentation unit models shows distinct first-order architectural element stacking patterns derived from a variety of sedimentary processes, including turbidity current, debris and transitional flows as well as mass transport, water masses circulation and bottom currents and associated oceanographic processes (fluvial influence, deep tides, internal waves, eddies). Their principal lithologies as well as the first-order architectural elements are distributed in a general fashion from upslope to downslope in a deepwater submarine fan depositional system, aiding in the predictability of deepwater sedimentation from basin to basin. In many cases, these disparate types of deposits can be interbedded, and recognition of their key similarities and differences can make the difference between effective exploration and unsuccessful dry holes. Turbidity current-generated deposits will show suites of Bouma (1962), Lowe (1982) and Stow and Shanmugam (1980) divisions, whereas transitional flow deposits will show evidence of Lowe and Guy (2000) and Haughton et al. (2003, 2009) divisions. Debrites will show masses of mixed and poorly sorted gravels, mud and sand. Sediment gravity flow deposits that are influenced by bottom currents or wave, tide and fluvial processes will have hybrid sedimentation units with evidence of each type of endmember sedimentation unit. The flow chart described in Fig. 20 can be helpful in linking the processes and grain size to expected reservoir potential.

One-dimensional stacking patterns, for instance as measured in a vertical core through flat-lying stratigraphy, reveal the Waltherian 3D relationships of the basin fill. In 3D, the distribution of these deposits indicate important time and spatial relationships in the basin, and will indicate where sedimentary processes such as debris flows vs. turbidity currents vs. bottom currents were most abundant and can be used to establish the distribution of petroleum systems elements (source rocks, reservoir rocks, seal rocks—Chapters 2, 12 and 18), how often and where submarine landslides occur (Chapter 10), source-to-sink sediment transport pathways (Chapter 11), how depositional environments are arranged in deepwater

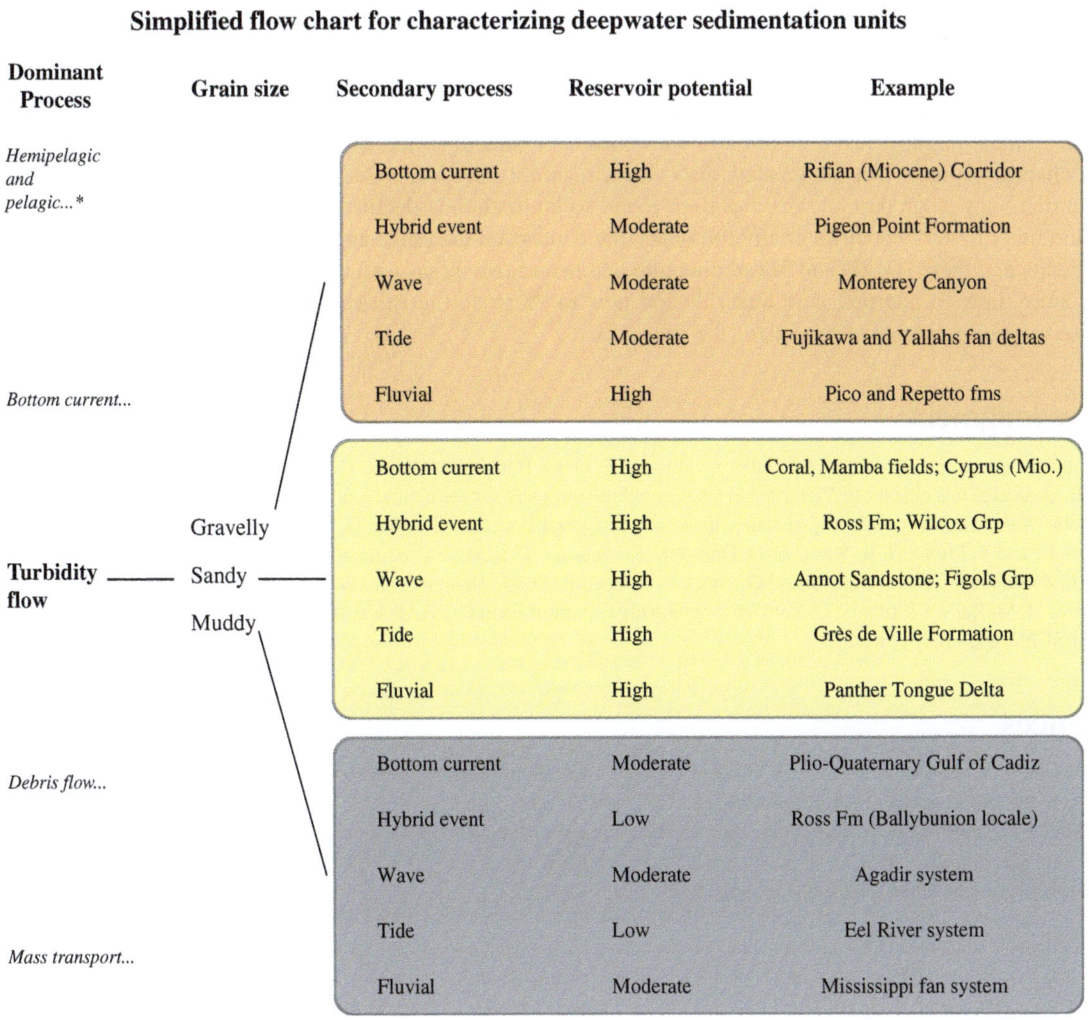


**FIG. 20** Simplified flow chart for characterizing deepwater sedimentation. This chart provides flow mechanism, typical grain size, secondary process, reservoir potential and examples of turbidity current deposits. Note that only turbidity flow (current) has been mapped out to include grain size, modification by secondary process, reservoir potential and example. This same workflow can be assigned to the other principal types of sedimentation, including hemipelagic and pelagic sedimentation, bottom current sedimentation, debris flow and mass transport sedimentation. *Note that hemipelagic and pelagic sedimentation is unlikely to include grain sizes coarser than fine silt.

settings (Chapters 8 and 9). While random types of sedimentation can occur, for instance large mass transport events, nature's ordering of these different types of deepwater deposits is often nonrandom and can allow for effective stratigraphic correlation of individual beds and bedding packages of similar lithofacies.

## Conclusions

This chapter described the basic building block of any depositional system—the bed, or sedimentation unit—by describing turbidity current models, including coarse-grained to fine-grained models and those modified by wave, tide, river and bottom current processes; hybrid event bed (HEB) models; debris flow models; bottom current models (in brief, as Chapter 9 explains in more detail); and how to apply this information.

Recognition, characterization and measurement of the variety of sedimentation units in deepwater environments reveal how a deepwater depositional system is built—bed by bed. Sometimes beds or parts of beds are removed due to erosion or reworked, creating amalgamation surfaces or refining the grain size population left by the reworking. But with practice studying the wide range of sedimentation units in deepwater environments, one can reconstruct the history of the deepwater basin and determine what is upslope, downslope and to the sides—aiding in the prediction of sedimentary deposits away from control points such as wells, cores and select outcrops. Deepwater sedimentation units are best observed in outcrop and in core, and relating bed stacking patterns to well log and seismic datasets requires specific knowledge regarding the resolution of the data to be integrated. This chapter showed the range and variability in endmember and hybrid types of sedimentation units found in deepwater environments prone to sediment gravity flows. While some of the sedimentation unit models presented are well established in peer reviewed research (i.e., Bouma, 1962; Lowe, 1982; Haughton et al., 2003, 2009, etc.), others remain elusive until more data become available on the deep sea.

This chapter explains that the greatest discoveries regarding deepwater sedimentation are yet to be found. However, there is a rich knowledge base of what has been found so far, including detailed process descriptions of fundamental sedimentation units found in deepwater environments, the context on the range and variability in primarily sediment gravity flow deposits and their 1D, 2D and 3D architecture and heterogeneity, as well as the merits, limitations and pitfalls in the nomenclature, how to interpret core intervals and how to illustrate the vertical and lateral relationships in sedimentary facies used to build predictive models of the subsurface.

## Acknowledgments

This chapter benefited greatly from constructive reviews by F. Javier Hernández-Molina, Octavian Catuneanu, Cindy Yeilding and Richard Sears. J. R. dedicates this chapter to Victoria and Charles, whose curiosity for the natural world inspired this work. This work would not have been possible without conversations and guidance throughout the years from the following: D. R. Lowe, S. A. Graham, E. Mutti, G. H. Browne, P. R. King, E. and S. Muller, J. R. Suter, A. D. Donovan, D. Nelson, "The Drifters" research group (RHUL, UK), M. Brown, M. Honarpour, C. Pirmez, Z. Sylvester, W. D. McCaffrey, G. Apps, F. J. Peel, P. D. W. Haughton, K. W. Rudolph, D. Adams, Y. Gavillot, N. Uzunlar, A. Lisenbee, M. McKeown & family, C. VanDenburg, and organizations including AAPG, EAGE, HOT Engineering, GNS Science, Gulf Coast Section of SEPM and many others.

## References

Allen, J., Schwartz, K., DeSantis, J., Koglin, D., Chen, F., 2013. Integration of structure and stratigraphy in bone spring tight oil sandstones using 3D seismic in the Delaware Basin. URTEC-1619830-MS, TX, pp. 2521–2527.

Alves, T.M., 2015. Submarine slide blocks and associated soft-sediment deformation in deep-water basins: a review. Mar. Pet. Geol. 67, 262–285.

Amy, L.A., Talling, P.J., 2006. Anatomy of turbidites and linked debrites based on long distance (120 × 30 km) bed correlation, Marnoso Arenacea formation, northern Apennines, Italy. Sedimentology 53 (1), 161–212.

Amy, L.A., Talling, P.J., Edmonds, V.O., Sumner, E.J., Lesueur, A., 2006. An experimental investigation of sand-mud suspension settling behaviour: implications for bimodal mud contents of submarine flow deposits. Sedimentology 53 (6), 1411–1434.

Anderson, K.S., Graham, S.A., Hubbard, S.M., 2006. Facies, architecture, and origin of a reservoir-scale sand-rich succession within submarine canyon fill: insights from wagon caves rock (Paleocene), Santa Lucia range, California, USA. J. Sediment. Res. 76, 819–838.

Arnott, R.W.C., 2012. Turbidites, and the case of the missing dunes. J. Sediment. Res. 82, 379–384.

Baas, J., 2003. Ripple, ripple mark, ripple structure. In: Middleton, G.V. (Ed.), Encyclopedia of Sediments and Sedimentary Rocks. Kluwer Academic Publishers, Dordrecht, Netherlands, pp. 565–568.

Baas, J.H., Best, J.L., Peakall, J., Wang, M., 2009. A phase diagram for turbulent, transitional, and laminar clay suspension flows. J. Sediment. Res. 79 (4), 162–183.

Baas, J.H., Best, J.L., Peakall, J., 2011. Depositional processes, bedform development and hybrid bed formation in rapidly decelerated cohesive (mud-sand) sediment flows. Sedimentology 58, 1953–1987.

Bagnold, R.A., 1962. Auto-suspension of transported sediment; turbidity currents. Proc. R. Soc. A 265, 315–331.

Bailey, W.S., McArthur, A.D., McCaffrey, W.D., 2020. Distribution of contourite drifts on convergent margins: examples from the Hikurangi subduction margin of New Zealand. Sedimentology. https://doi.org/10.1111/sed.12779.

Barker, S.P., Haughton, P.D.W., McCaffrey, W.D., Archer, S.G., Hakes, B., 2008. Development of rheological heterogeneity in clay-rich high-density turbidity currents: Aptian Britannia sandstone member, U.K. Continental shelf. J. Sediment. Res. 78, 45–68.

Bhatnagar, P., Verma, S., Bianco, R., 2019. Characterization of mass-transport deposits using seismic attributes: upper Leonard formation, Midland Basin. In: Rotzien, J.R. (Ed.), Basin Dynamics Bulletin, A Basin Dynamics Special Publication. vol. 1(2). Basin Dynamics, LLC, pp. 1–5.

Bhattacharya, S., Verma, S., Rotzien, J.R., 2020. Seismic Facies and 3D Attributes of Submarine Slides on the North Slope. Interpretation, Alaska, USA.

Boulesteix, K., 2021. Depositional processes and stratigraphic architecture of deep-water mudstones. The University of Manchester, Manchester, UK. Thesis, 223 pp.

Boulesteix, K., Poyatos-More, M., Flint, S.S., Taylor, K.G., Hodgson, D.M., Hasiotis, S.T., 2019. Transport and deposition of mud in deepwater environments: processes and stratigraphic implications. Sedimentology 66 (7), 2894–2925.

Boulesteix, K., Poyatos-Moré, M., Hodgson, D.M., Flint, S.S., Taylor, K.G., 2020. Fringe or backgrounApplicationSource characterizing deep-water mudstones beyond the basin-floor fan sandstone pinchout. J. Sediment. Res. 90 (12), 1678–1705. https://doi.org/10.2110/jsr.2020.048.

Bouma, A.H., 1962. Sedimentology of Some Flysch Deposits: A Graphic Approach to Facies Interpretation. Elsevier, Amsterdam. 167 pp.

Bouma, A.H., 2004. Key controls on the characteristics of turbidite systems. In: Geological Society of London Special Publication, 222, pp. 9–22.

Bozetti, G., Cronin, B.T., Kneller, B.C., Jones, A.M., 2018. Deepwater conglomeratic megabeds: analogues for event beds of the brae formation of the South Viking Graben, North Sea. In: Turner, C.C., Cronin, B.T. (Eds.), Rift-Related Coarse-Grained Submarine Fan Reservoirs: The Brae Play, South Viking Graben. vol. 115. AAPG Memoir, North Sea, pp. 114–118.

Brodie, I., Kemp, A.E.S., 1994. Variation in biogenic and detrital fluxes and formation of laminae in late quaternary sediments from the Peruvian coastal upwelling zone. Mar. Geol. 116 (3–4), 385–398.

Brooks, H.L., Hodgson, D.M., Brunt, R.L., Peakall, J., Poyatos-More, M., Flint, S.S., 2018. Disconnected submarine lobes as a record of stepped slope evolution over multiple sea-level cycles. Geosphere 14 (4), 1753–1779.

Burgreen, B., Graham, S.A., 2014. Evolution of a deepwater lobe system in the Neogene trench-slope setting of the East Coast Basin, New Zealand: lobe stratigraphy and architecture in a weakly confined basin configuration. Mar. Pet. Geol. 54, 1–22.

Campbell, C.V., 1967. Lamina, laminaset, bed and bedset. Sedimentology 8 (1), 7–26.

Cantero, M.I., Cantelli, A., Pirmez, C., Balachandar, S., Mohrig, D., Hickson, T.A., Yeh, T., Naruse, H., Parker, G., 2012. Emplacement of massive turbidites linked to extinction of turbulence in turbidity currents. Nat. Geosci. 5, 42–45.

Capella, W., Hernandez-Molina, F.J., Flecker, R., Hilgen, F.J., Hssain, M., Kouwen-Hoven, T.J., Van Oorschot, M., Sierro, F.J., Stow, D.A.V., Trabucho-Alexandre, J., Tulbure, M.A., De Weger, W., Yousfi, M.Z., Krijgsman, W., 2017. Sandy contourite drift in the late Miocene Rifian Corridor (Morocco): reconstruction of depositional environments in a foreland-basin seaway. Sediment. Geol. 355, 31–57. https://doi.org/10.1016/j.sedgeo.2017.04.004.

Cardona, S., Wood, L.J., Dugan, B., Jobe, Z., Strachan, L.J., 2020. Characterization of the Rapanui mass-transport deposit and the basal shear zone: mount messenger formation, Taranaki Basin, New Zealand. Sedimentology 67 (4), 2111–2148.

Chenrai, P., Huuse, M., 2017. Pockmark formation by porewater expulsion during rapid progradation in the offshore Taranaki Basin, New Zealand. Mar. Pet. Geol. 82, 399–413.

Cherry, S.T.J., 1993. The interaction of structure and sedimentary process controlling deposition of the upper Jurassic brae formation conglomerate, block 16/17, North Sea. In: Petroleum Geology Conference Series. vol. 4, pp. 387–400.

Chiarella, D., Longhitano, S.G., Tropeano, M., 2017. Types of mixing and heterogeneities in siliciclastic-carbonate sediments. Mar. Pet. Geol. 88, 617–627.

Cornard, P., Pickering, K.T., 2019. Supercritical-flow deposits and their distribution in a submarine channel system, middle Eocene, Ainsa Basin, Spanish Pyrenees. J. Sediment. Res. 89, 576–597.

Cornard, P., Pickering, K.T., 2020. Submarine topographic control on distribution of supercritical-flow deposits in lobe and related environments, middle Eocene, Jaca Basin, Spanish Pyrenees. J. Sediment. Res. 90, 1222–1243.

Costa, J.E., Williams, G.P., 1983. Debris-Flow Dynamics (videotape). U. S. Geological Survey Open File Report OF 84-0606.

Cronin, B.T., Kidd, R.B., 1998. Heterogeneity and lithotype distribution in ancient deep-sea canyons: point lobos deep-sea canyon as a reservoir analogue. Sediment. Geol. 115, 315–349.

De Castro, S., Hernandez-Molina, F.J., Rodriguez-Tovar, F.J., Llave, E., Ng, Z.L., Nishida, N., Mena, A., 2020. Contourites and bottom current reworked sands: bed facies model and implications. Mar. Geol. 428, 106267.

De Castro, S., Hernández-Molina, F.J., De Weger, W., Jiménez-Espejo, F.J., Rodríguez-Tovar, F.J., Mena, A., Llave, Estefanía, Sierro, F.J., 2021a. Contourite characterization and its discrimination from other deep-water deposits in the Gulf of Cadiz contourite depositional system. Sedimentology 68 (3), 987–1027. https://doi.org/10.1111/sed.12813.

De Castro, S., Miramontes, E., Dorador, J., Jouet, G., Cattaneo, A., Rodríguez-Tovar, F.J., Hernández-Molina, F.J., 2021b. Siliciclastic and bioclastic contouritic sands: textural and geochemical characterisation. Mar. Geol. 128, 105002. https://doi.org/10.1016/j.marpetgeo.2021.105002.

D'Hondt, S., Inagaki, F., Alvarez Zarikian, C.A., et al., 2011. Proceedings of the Integrated Ocean Drilling Program Volume 329 Expedition Reports – South Pacific Gyre Subseafloor Life. Integrated Ocean Drilling Program Management International, Inc., Integrated Ocean Drilling Program., https://doi.org/10.2204/iodp.proc.329.2011.

De Weger, W., Hernandez-Molina, F.J., Flecker, R., Sierro, F.J., Chiarella, D., Krijgsman, W., Manar, A.M., 2020. Late Miocene contourite channel system reveals intermittent overflow behavior. Geology 48 (12), 1194–1199.

De Weger, W., Hernández-Molina, F.J., Miguez-Salas, O., De Castro, S., Bruno, C., Chiarella, D., Sierro, F.J., Blackbourn, G., Amine Manar, M., 2021. Contourite depositional system after the exit of a strait: case study from the late Miocene South Rifian Corridor, Morocco. Sedimentology 68 (7), 2996–3032. https://doi.org/10.1111/sed.12882.

Dodd, T.J., McCarthy, D.J., Richards, P.C., 2019. A depositional model for deep-lacustrine, partially confined, turbidite fans: Early Cretaceous, North Falkland Basin. Sedimentology 66 (1), 53–80.

Dodd, T.J.H., Leslie, A.G., Gillespie, M.R., Dobbs, M.R., Bide, T.B., Kendall, R.S., Kearsey, T.I., Chiam, K., Goay, M., 2020. Deep to shallow marine sedimentology and impact of volcanism within the middle Triassic Palaeo-Tethyan Semantan Basin, Singapore. J. Asian Earth Sci. 196, 104371.

Donnelly, S., Beauboeuf, R., Bracht, R., Fitzsimmons, R., 2019. Regional Depositional Setting of the West Orphan Basin: The Importance of Contour Current Processes on Stratigraphic Trap Generation in a Hybrid Depositional System. AAPG Annual Convention, San Antonio.

Donovan, A.D., Staerker, T.S., Gardner, R.D., Pramudito, A., Wehner, M., Corbett, M.J., Lundquist, J.J., Romero, A.M., Henry, L., Rotzien, J.R., Boling, K.S., 2015. A sequence stratigraphic framework for establishing the chronostratigraphic relationships between the woodbine and eagle ford groups along the outcrop belt and shallow subsurface in Texas. J. Gulf Coast Assoc. Geol. Soc. 4, 67–87.

Dykstra, M., 2005. Dynamics of sediment mass-transport from the shelf to the deep sea. Unpublished PhD Thesis, UC Santa Barbara, 152 pp.

Eggenhuisen, J.T., McCaffrey, W.D., Haughton, P.D.W., Butler, R.W.H., 2010. Small-scale spatial variability in turbidity-current flow controlled by roughness resulting from substrate erosion: field evidence for a feedback mechanism. J. Sediment. Res. 80, 129–136.

Englert, R.G., Hubbard, S.M., Cartigny, M.J.B., Clare, M.A., Coutts, D.S., Hage, S., Clarke, J.H., Jobe, Z., Lintern, D.G., Stacey, C., Vendettuoli, D., 2020. Quantifying the three-dimensional stratigraphic expression of cyclic steps by integrating seafloor and deepwater outcrop observations. Sedimentology. https://doi.org/10.1111/sed/12772.

Faugères, J.C., Gonthier, E., Stow, D.A.V., 1984. Contourite drift moulded by the deep Mediterranean outflow. Geology 12 (5), 296–300.

Faugères, J.-C., Stow, D.A.V, Imbert, P., Viana, A., 1999. Seismic features diagnostic of contourite drifts. Mar. Geol. 162 (1), 1–38. https://doi.org/10.1016/S0025-3227(99)00068-7.

Fonnesu, M., 2020. Hybrid Event Beds: Facies, Processes and Distribution in Deepwater Turbidite Systems: Virtual Get Together "Virtual Geology Talk" and Video by the Drifters Research Group. Royal Holloway University of London. 16 June.

Fonnesu, M., Haughton, P.D.W., Felletti, F., McCaffrey, W.D., 2015. Short length-scale variability of hybrid event beds and its applied significance. Mar. Pet. Geol. 67, 583–603.

Fonnesu, M., Patacci, M., Haughton, P.D.W., Felletti, F., McCaffrey, W.D., 2016. Hybrid event beds generated by local substrate delamination on a Confined Basin floor. J. Sediment. Res. 86, 929–943.

Fonnesu, M., Felletti, F., Haughton, P.D., Patacci, M., McCaffrey, W.D., 2018. Hybrid event bed character and distribution linked to turbidite system sub-environments: the North Apennine Gottero Sandstone (north-west Italy). Sedimentology 65 (1), 151–190.

Fonnesu, M., Palermo, D., Galbiati, M., Marchesini, M., Bonamini, E., Bendias, D., 2020. A new world-class deepwater play type, deposited by the syn-depositional interaction of turbidity flows and bottom currents: the giant Eocene Coral Field in northern Mozambique. Mar. Pet. Geol. 111, 179–201.

Fuhrmann, A., Kane, I.A., Clare, M.A., Ferguson, R.A., Schomacker, E., Bonamini, E., Contreras, F.A., 2020. Hybrid turbidite-drift channel complexes: an integrated multiscale model. Geology 48, 562–568. https://doi.org/10.1130/G47179.1.

Ghibaudo, G., 1992. Subaqueous sediment gravity flow deposits: practical criteria for their field description and classification. Sedimentology 39 (3), 423–454.

Giacomone, G., Olariu, C., Steel, R., Shin, M., 2020. A coarse-grained basin floor turbidite system—the Jurassic Los Molles formation, Neuquen Basin, Argentina. Sedimentology 67 (7), 3809–3843.

Gonthier, E.G., Faugères, J.-C., Stow, D.A.V., 1984. Contourite facies of the Faro drift, Gulf of Cadiz. vol. 15 Geological Society of London, Special Publications, pp. 275–292.

Grecula, M., Hognestad, J., Price, S., Boya Ferrero, M., De Bruijn, G., Noraberg, K.T., Engenes, K., Mears, P., Van Ojik, K., McGarva, R., 2014. Interplay of fan-fringe reservoir deterioration and hydrodynamic aquifer: understanding the margins of gas development in the Ormen Lange field. In: McKie, T., Rose, P.T.S., Hartley, A.J., Jones, D.W., Armstrong, T.L. (Eds.), Tertiary Deep-Marine Reservoirs of the North Sea Region. vol. 403. Geological Society, London, Special Publications, pp. 157–183.

Gutierrez, M., 2019. ExxonMobil Process Stratigraphy: Connecting Facies and Architecture to Flow: British Sedimentological Research Group Annual General Meeting, Deepwater Depositional Systems Section, Egham, Surrey. 15 December.

Hairabian, A., Borgomano, J., Reimer, J.J.G., 2019. Models and parameters controlling carbonate slopes and associated gravity flow deposits. In: Rotzien, J.R., Cantrell, D., Dykstra, M., Gann, Y., Nicholson, T., Pradhan, A. (Eds.), Global Deepwater Depositional Systems: Geo Gulf Transactions for the 69th Annual GCAGS Convention and 66th Annual GCCSEPM Meeting, Houston, 23–25 October.

Haughton, P.D.W., Barker, S.P., McCaffrey, W.D., 2003. 'Linked' debrites—in sand-rich turbidite systems—origin and significance. Sedimentology 50, 459–482.

Haughton, P.D.W., Davis, C., McCaffrey, W., Barker, S., 2009. Hybrid sediment gravity flow deposits—classification, origin and significance. Mar. Pet. Geol. 26, 1900–1918.

Haughton, P.D.W., Davis, C.E., McCaffrey, W.D., 2010. Reply to comment by R. Higgs on "hybrid sediment gravity flow deposits—classification, origin and significance". Mar. Pet. Geol. 27, 2066–2069.

Henry, L.C., Wadsworth, J.A., Hansen, B., Hartman, K., 2018. Erosion and ponding of thunder horse Deepwater turbidites by mass transport complexes in Mississippi canyon based on image log sedimentology. Mar. Pet. Geol. 97, 639–658.

Hernández-Molina, F.J., 2021. Do you go with the flow? Observations on contourite and mixed systems: Virtual Get Together. Society of Sedimentary Geology. February 10.

Hesse, R., 1975. Turbiditic and non-turbiditic mudstone of cretaceous flysch sections of the East Alps and other basins. Sedimentology 22 (3), 387–416.

Hickson, T.A., Lowe, D.R., 2002. Facies architecture of a submarine fan channel-levee-overbank complex: the Juniper ridge conglomerate, Coalinga, California. Sedimentology 49 (2), 335–362.

Hodgson, D.M., 2009. Distribution and origin of hybrid beds in sand-rich submarine fans of the Tanqua depocentre, Karoo Basin, South Africa. Mar. Pet. Geol. 26, 1940–1956.

Hoyal, D., 2018. Deepwater Process Stratigraphy. Houston Geological Society Luncheon Talk, Houston, Texas. 24 October.

Hubbard, S.M., Romans, B.W., Erohina, T., Lowe, D.R., 2007. Facies and internal architecture of deepwater channel fill in the Cerro Toro formation (Campanian), northeast sierra del Toro, Chile. In: Nilsen, T., Shew, R., Steffens, G., Studlick, J. (Eds.), Deepwater Outcrops of the World Atlas: American Association of Petroleum Geologists Studies in Geology. vol. 56, pp. 140–142.

Hubbard, S.M., Romans, B., Graham, S.A., 2008. Deep-water foreland basin deposits of the Cerro Toro Formation, Magallanes basin, Chile: architectural elements of a sinuous basin axial channel belt. Sedimentology 55 (5), 1333–1359. https://doi.org/10.1111/j.1365-3091.2007.00948.x.

Hüneke, H., Henrich, R., 2011. Chapter 4 – Pelagic sedimentation in modern and ancient oceans. In: Deep-Sea Sediments, vol. 63. Elsevier, Amsterdam, pp. 215–351, https://doi.org/10.1016/B978-0-444-53000-4.00004-4.

Hüneke, H., Hernández-Molina, F.J., Rodríguez-Tovar, F.J., Llave, E., Chiarella, D., Mena, A., Stow, D.A.V., 2021. Diagnostic criteria using microfacies for calcareous contourites, turbidites and pelagites in the Eocene–Miocene slope succession, southern Cyprus. Sedimentology 68 (2), 557–592. https://doi.org/10.1111/sed.12792.

Hurst, A., Cartwright, J., 2007. Relevance of sand injectites to hydrocarbon exploration and production. In: Hurst, A., Cartwright, J. (Eds.), Sand Injectites: Implications for Hydrocarbon Exploration and Production. vol. 87. AAPG Memoir, pp. 1–19.

Hussain, A., Haughton, P.D., Shannon, P.M., Turner, J.N., Pierce, C.S., Obradors-Latre, A., Barker, S.P., Martinsen, O.J., 2020. High-resolution X-ray fluorescence profiling of hybrid event beds: Implications for sediment gravity flow behaviour and deposit structure. Sedimentology 67 (6), 2850–2882.

Ingersoll, R.V., 1981. Petrofacies, lithofacies, submarine-fan facies of the Great Valley group (sequence). In: Graham, S.A. (Ed.), Field Guide to the Mesozoic-Cenozoic Convergent Margin of Northern California: Pacific Section SEPM.

Jaeger, J.M., Gulick, S.P.S., LeVay, L.J., et al., 2014. Site U1419. In: Jaeger, J.M., Gulick, S.P.S., LeVay, L.J., et al. (Eds.), Proceedings of the Integrated Ocean Drilling Program. vol. 341., https://doi.org/10.2204/iodp.proc.341.105.2014. College Station, TX (Integrated Ocean Drilling Program).

Janocko, M., Nemec, W., Henriksen, S., Warchol, M., 2013. The diversity of deepwater sinuous channel belts and slope valley-fill complexes. Mar. Pet. Geol. 41, 7–34.

Jin, L., Shan, X., Shi, X., Fonnesu, M., Qiao, S., Kandasamy, S., Wang, H., Liu, S., Fang, X., Zou, 2021. Hybrid event beds generated by erosional bulking of modern hyperpycnal flows on the Choshui River delta front, Taiwan Strait. Sedimentology. https://doi.org/10.1111/SED.12862.

Jobe, Z.R., Lowe, D.R., Morris, W.R., 2011. Climbing-ripple successions in turbidite systems: depositional environments, sedimentation rates and accumulation times. Sedimentology 59, 867–898.

Johnson, A.M., 1984. Debris flow. In: Brunsden, D., Prior, D.B. (Eds.), Slope Instability. John Wiley & Sons Ltd, Chichester, pp. 257–361.

Kähler, G., Stow, D.A.V., 1998. Turbidites and contourites of the Paleogene Lefkara Formation, southern Cyprus. Sediment. Geol. 115, 215–231.

Kane, I.A., Hodgson, D.M., 2011. Sedimentological criteria to differentiate submarine channel levee subenvironments: exhumed examples from the Rosario formation (upper cretaceous) of Baja California, Mexico, and the fort Brown formation (Permian), Karoo Basin, South Africa. Mar. Pet. Geol. 28 (3), 807–823.

Kane, I.A., Pontén, A.S.M., 2012. Submarine transitional flow deposits in the Paleogene Gulf of Mexico. Geology 40 (12), 1119–1122.

Kane, I.A., McCaffrey, W.D., Martinsen, O.J., 2009. Allogenic vs. autogenic controls on megaflute formation. J. Sediment. Res. 79 (9), 643–651.

Kane, I.A., Pontén, A.S., Vangdal, B., Eggenhuisen, J.T., Hodgson, D.M., Spychala, Y.T., 2017. The stratigraphic record and processes of turbidity current transformation across deep-marine lobes. Sedimentology 64 (5), 1236–1273.

King, P.R., Browne, G.H., 2004. Unraveling the complexity of thin-bedded turbidite reservoirs: risk and reward in Taranaki Basin, New Zealand. Oil Gas Gazette 2 (2), 14–17.

Klein, G.D., 1971. A sedimentary model for determining paleotidal range. GSA Bull. 82, 2585–2592.

Klein, G.D., 1998. Classic tidalites—a partial retrospective view. In: Alexander, C.R., Davis, R.A., Henry, V.J. (Eds.), Tidalites: Processes and Products. vol. 61. SEPM Special Publication, pp. 5–14.

Kneller, B.C., Branney, M.J., 1995. Sustained high-density turbidity currents and the deposition of thick massive sands. Sedimentology 42, 607–616.

Kneller, B.C., McCaffrey, W.D., 1999. Depositional effects of flow non-uniformity and stratification within turbidity currents approaching a bounding slope: deflection, reflection, and facies variation. J. Sediment. Res. 69, 980–991.

Kneller, B., Dykstra, M., Fairweather, L., Milana, J.P., 2016. Mass-transport and slope accommodation: implications for turbidite sandstone reservoirs. AAPG Bull. 100 (2), 213–235.

Koo, W.M., Mohrig, D., Buttles, J., Sturmer, D., Pontén, A., Hess, T., 2019. Sand-mud couplets deposited by spontaneous remobilization of subaqueous transitional flows. Sedimentology 67 (1), 78–95.

Lamb, M.P., Myrow, P.M., Lukens, C.E., Houck, K., Strauss, J.V., 2008. Deposits of wave-influenced turbidity currents: Pennsylvanian Minturn Formation, Colorado, U.S.A. J. Sediment. Res. 78 (7–8). https://doi.org/10.2110/jsr.2008.052.

Lazar, O.R., Bohacs, K.M., Macquaker, J.H.S., Scheiber, J., Demko, T.M., 2015. Capturing key attributes of fine-grained sedimentary rocks in outcrops, cores, and thin sections: nomenclature and description guidelines. J. Sediment. Res. 85, 230–246.

Le Friant, A., Ishizuka, O., Stroncik, N.A., et al., 2013. Site U1395. In: Le Friant, A., Ishizuka, O., Stroncik, N.A., et al. (Eds.), Proceedings of the Integrated Ocean Drilling Program, volume 340: Tokyo (Integrated Ocean Drilling Program Management International, Inc.)., https://doi.org/10.2204/iodp.proc.340.105.2013.

Leon, R., Somoza, L., Urgeles, R., Medialdea, T., Ferrer, M., Biain, A., Garcia-Crespo, J., Mediato, J.F., Galindo, I., Yepes, J., Gonzalez, F.J., Gimenez-Moreno, J., 2017. Multi-event oceanic island landslides: new onshore-offshore insights from El Hierro Island, Canary Archipelago. Mar. Geol. 393, 156–175.

Lowe, D.R., 1972. Implications of three submarine mass-movement deposits, Cretaceous, Sacramento Valley, California. J. Sediment. Petrol. 42 (1), 89–101.

Lowe, D.R., 1979. Sediment gravity flows: their classification and some problems of application to natural flows and deposits. SEPM Spec. Publ. 27, 75–82.

Lowe, D.R., 1982. Sediment gravity flows: II. Depositional models with special reference to the deposits of high-density turbidity currents. J. Sediment. Petrol. 52 (1), 279–297.

Lowe, D.R., 1988. Suspended-load fallout rate as an independent variable in the analysis of current structures. Sedimentology 35 (5), 765–776.

Lowe, D.R., 2004. Deepwater Sandstones: Submarine Canyon to Basin Plain, Western California: AAPG and Pacific Section AAPG Guidebook and Lecture Notes. Artistic Printing, Salt Lake City, p. 79.

Lowe, D.R., Graham, S.A., 2012. Turbidite Reservoirs and Depositional Systems: Short Course Lecture Notes. Stanford University, p. 150. approx.

Lowe, D.R., Guy, M., 2000. Slurry-flow deposits in the Britannia formation (lower cretaceous), North Sea: a new perspective on the turbidity current and debris flow problem. Sedimentology 47, 31–70.

Lowe, D.R., Lo Piccolo, R.D., 1974. The characteristics and origins of dish and pillar structures. J. Sediment. Petrol. 44 (2), 484–501.

Marchand, A.M.E., Apps, G., Li, W., Rotzien, J.R., 2015. Depositional processes and impact on reservoir quality in Deepwater Paleogene reservoirs, US Gulf of Mexico. AAPG Bull. 99 (9), 1635–1648.

Masalimova, L.U., Lowe, D.R., Sharman, G.R., King, P.R., Arnot, M.J., 2016. Outcrop characterization of a submarine channel-lobe complex: the lower mount messenger formation, Taranaki Basin, New Zealand. Mar. Pet. Geol. 71, 360–390.

McArthur, A.D., Kneller, B., De Souza, P.A., Kuchle, J., 2016. Characterization of deep-marine channel-levee complex architecture with palynofacies: an outcrop example from the Rosario formation, Baja California, Mexico. Mar. Pet. Geol. 73. https://doi.org/10.1016/j.marpetgeo.2016.02.030.

McCave, I.N., 2008. Size sorting during transport and deposition of fine sediments: sortable silt and flow speed. Sedimentology 60, 121–142.

McCave, I.N., Tucholke, B.E., 1986. Deep current-controlled sedimentation in the western North Atlantic. The Western North Atlantic Region. Geological Society of America., https://doi.org/10.1130/DNAG-GNA-M.451.

Meazell, P.K., Flemings, P.B., Santra, M., Johnson, J.E., 2020. Sedimentology and stratigraphy of a deepwater gas hydrate reservoir in the northern Gulf of Mexico. AAPG Bull. 104 (9), 1945–1969.

Miramontes, E., Eggenhuisen, J.T., Jacinto, R.S., Poneti, G., Pohl, F., Normandeau, A., Campbell, D.C., Hernández-Molina, F.J., 2020. Channel-levee evolution in combined contour current-turbidity current flows from flume-tank experiments. Geology 48 (4), 353–357.

Mohrig, D., Ellis, C., Parker, G., Whipple, K.X., Hondzo, M., 1998. Hydroplaning of subaqueous debris flows. GSA Bull. 110 (3), 387–394.

Morris, E., Hodgson, D., Brunt, R., Flint, S., 2014. Origin, evolution and anatomy of silt-prone submarine external levees. Sedimentology 61, 1734–1763. https://doi.org/10.1111/sed.12114.

Moscardelli, L., Wood, L., 2008. New classification system for mass transport complexes in offshore Trinidad. Basin Res. 20 (1), 73–98.

Mueller, P., Patacci, M., Di Giulio, A., 2017. Hybrid event beds in the proximal to distal extensive lobe domain of the coarse-grained and sand-rich Bordighera turbidite system (NW Italy). Mar. Pet. Geol. 86, 908–931.

Mueller, P., Patacci, M., Di Giulio, A., 2021. Hybrid event bed distribution in a mixed siliciclastic-calcareous turbidite succession: a cross-current perspective from the Bordighera sandstone, Ligurian Alps, NW, Italy. Ital. J. Geosc. 140, 2.

Mulder, T., Alexander, J., 2001. The physical character of subaqueous sedimentary density flows an deposits. Sedimentology v. 48, 269–299.

Mulder, T., Migeon, S., Savoye, B., Faugères, J.-C., 2001. Inversely graded turbidite sequences in the deep Mediterranean. A record or deposits from flood-generated turbidity currents? Geo-Mar. Lett. 21, 86–93.

Mulder, T., Syvitski, J.P.M., Migeon, S., Faugères, J.-C., Savoye, B., 2003. Marine hyperpycnal flows: initiation, behavior and related deposits. A review. Mar. Pet. Geol. 20, 861–882.

Mulder, T., Razin, P., Faugères, J., 2009. Hummocky cross-stratification-like structures in deep sea turbidites: upper cretaceous Basque basins (Western Pyrenees, France). Sedimentology 56, 997–1015.

Mulder, T., Zaragosi, S., Garlan, T., Mavel, J., Cremer, M., Sottolichio, A., Sénéchal, N., Schmidt, S., 2012. Present deep-submarine canyons activity in the Bay of Biscay (NE Atlantic). Mar. Geol. 295–298, 113–127. https://doi.org/10.1016/j.margeo.2011.12.005.

Murray, C.J., Lowe, D.R., Graham, S.A., Martinez, P.A., Zeng, J., Carroll, A.R., Cox, R., Hendrix, M., Heubeck, C., Miller, D., Moxon, I.W., Sobel, E., Wendenbourg, J., Williams, T., 1996. Statistical analysis of bed-thickness patterns in a turbidite section from the Great Valley sequence, Cache Creek, Northern California. J. Sediment. Res. 66 (5), 900–908.

Mutti, E., 1985. Turbidite systems and their relations to depositional sequences. In: Provenance of Arenites. Springer, Dordrecht, pp. 65–93.

Mutti, E., 1992. Turbidite Sandstones. AGIP, Istituto di geologia, Università di Parma. 275 pp.

Mutti, E., 2019. Thin-bedded plumites: an overlooked Deepwater deposit. J. Mediterr. Earth Sci. 11, 1–20.

Mutti, E., Nilsen, T.H., 1981. Significance of intraformational rip-up clasts in deep-sea fan deposits. In: International Association of Sedimentologists—2nd European Regional Meeting, pp. 117–119.

Mutti, E., Normark, W.R., 1987. Comparing examples of modern and ancient turbidite systems; problems and concepts. In: Leggett, J.K., Zuffa, G.G. (Eds.), Marine Clastic Sedimentology; Concepts and Case Studies. Graham and Trotman, London, United Kingdom, pp. 1–38.

Mutti, E., Nilsen, T.H., Ricci Lucchi, F., 1978. Outer fan depositional lobes of the Laga formation (upper Miocene and lower Pliocene), East-Central Italy. In: Stanley, D.J., Kelling, G. (Eds.), Sedimentation in Submarine Canyons, Fans, and Trenches. Hutchinson & Ross, Inc., Stroudsburg, PA, United States, Dowden, pp. 210–222.

Mutti, E., Davoli, G., Tinterri, R., Zavala, C., 1996. The importance of ancient Fluvio-deltaic system dominated by catastrophic flooding in tectonically active basins. Mem. Sci. Geol. Padova 48, 233–291.

Mutti, E., Tinterri, R., Benevelli, G., di Biase, D., Cavanna, G., 2003. Deltaic, mixed and turbidite sedimentation of ancient foreland basins. Mar. Pet. Geol. 20 (6), 733–755.

Mutti, E., Cunha, R.S., Bulhoes, E.M., Arienti, L.M., Viana, A.R., 2014. Contourites and Turbidites of the Brazilian Marginal Basins. Search and Discovery Article (51069).

Muzzi Magalhaes, P., Tinterri, R., 2010. Stratigraphy and depositional setting of slurry and contained (reflected) beds in the Marnoso-arenacea formation (Langhian-Serravallian) northern Apennines, Italy. Sedimentology 57, 1685–1720.

Myrow, P.M., Fischer, W., Goodge, J.W., 2002. Wave-modified turbidites: combined-flow shoreline and shelf deposits, Cambrian, Antarctica. J. Sediment. Res. 72 (5), 641–656.

Myrow, P.M., Lukens, C., Lamb, M.P., Houck, K., Strauss, J., 2008. Dynamics of a transgressive prodeltaic system: implications for geography and climate within a Pennsylvanian intracratonic basin, Colorado, USA. J. Sediment. Res. 78. https://doi.org/10.2110/jsr.2008.061. no. 0–0.

O'Brien, N.R., Nakazawa, K., Tokuhashi, S., 1980. Use of clay fabric to distinguish turbidite and hemipelagic siltstones and silts. Sedimentology 27 (1), 47–61.

Omeru, T., Cartwright, J.A., 2019. The efficacy of kinematic indicators in a complexly deformed mass transport deposit: insights from the Deepwater Taranaki Basin, New Zealand. Mar. Pet. Geol. 106, 74–87.

Ortiz-Karpf, A., Hodgson, D.M., McCaffrey, W.D., 2015. The role of mass-transport complexes in controlling channel avulsion and the subsequent sediment dispersal patterns on an active margin: the Magdalena fan, offshore Colombia. Mar. Pet. Geol. 64, 58–75.

Patacci, M., Haughton, P.D.W., McCaffrey, W.D., 2014. Rheological complexity in sediment gravity flows forced to decelerate against a confining slope, Braux, SE France. J. Sediment. Res. 84, 270–277.

Patacci, M., Marini, M., Felletti, F., Di Giulio, A., Setti, M., McCaffrey, W., 2020. Origin of mud in turbidites and hybrid event beds: Insight from ponded mudstone caps of the Castagnola turbidite system (north-west Italy). Sedimentology 67 (5), 2625–2644.

Peakall, J., Best, J., Baas, J.H., Hodgson, D.M., Clare, M.A., Talling, P.J., Dorrell, R.M., Lee, D.R., 2020. An integrated process-based model of flutes and tool marks in deep-water environments: implications for palaeohydraulics, the Bouma Sequence and hybrid event beds. Sedimentology 67 (4), 1601–1666.

Peel, F., Apps, G., Duffy, O., Fernandez, N., Hudec, M., 2019. Mud Diapirism and Catastrophic Fluidization Triggered by MTC Collapse down a Salt Scarp: What Happens when a Hundred Billion Tonnes of Mud Falls down a 1-Km-High Slope? British Sedimentological Research Group AGM, Egham, Surrey. 14–15 December.

Phillips, S.C., Flemings, P.B., Holland, M.E., Schultheiss, P.J., Waite, W.F., Jang, J., Petrou, E.G., Hammon, H., 2020. High concentration methane hydrate in a silt reservoir in the northern Gulf of Mexico. AAPG Bull. 104 (9), 1971–1995.

Pickering, K.T., Hiscott, R.N., 2015. Deep Marine Systems: Processes, Deposits, Environments, Tectonics and Sedimentation. American Geophysical Union., ISBN: 978-1-405-12578-9. 688 pp.

Pickering, K.T., Hiscott, R.N., Hein, F.J., 1989. Deep-Marine Environments. Unwin Hyman, London. 416 pp.

Pierce, C.S., Haughton, P.D., Shannon, P.M., Pulham, A.J., Barker, S.P., Martinsen, O.J., 2018. Variable character and diverse origin of hybrid event beds in a sandy submarine fan system, Pennsylvanian Ross sandstone formation, western Ireland. Sedimentology 65 (3), 952–992.

Pohl, F., Eggenhuisen, J.T., Cartigny, M.J.B., Tilston, M.C., de Leeuw, J., Hermidas, N., 2020. The influence of a slope break on turbidite deposits: an experimental investigation. Mar. Geol. 424, 106160.

Posamentier, H.W., Kolla, V., 2003. Seismic geomorphology and stratigraphy of depositional elements in Deepwater settings. J. Sediment. Res. 73, 367–388.

Postma, G., Nemec, W., Kleinspehn, K.L., 1988. Large floating clasts in turbidites: a mechanism for their emplacement. Sediment. Geol. 58 (1), 47–61.

Potter, P.E., Maynard, J.B., Depetris, P.J., 2005. Mud and Mudstones. Springer. 297 pp.

Power, B., Covault, J., Fildani, A., Sullivan, M., Clark, J., Carson, B., Zarra, L., Romans, B., 2013. Facies analysis and interpretation of argillaceous sandstone beds in the Paleogene Wilcox formation, deepwater Gulf of Mexico. Trans. Gulf Coast Assoc. Geol. Soc. 63, 575–578.

Prather, B.E., O'Byrne, C., Pirmez, C., Sylvester, Z., 2016. Sediment partitioning, continental slopes and base-of-slope systems. Basin Res. 29 (3), 294–416.

Prélat, A., Hodgson, D.M., Flint, S.S., 2009. Evolution, architecture and hierarchy of distributary deepwater deposits: a high-resolution outcrop investigation from the Permian Karoo Basin, South Africa. Sedimentology 56, 2132–2154.

Puigdefàbregas, C., Gjelberg, J., Vaksdal, M., 2004. The Gres d'Annot in the Annot syncline: outer basin-margin onlap and associated soft-sediment deformation. In: Joseph, P., Lomas, S.A. (Eds.), Deepwater Sedimentation in the Alpine Basin of SE France: New Perspectives on the Gres D'Annot and Related Systems. vol. 221. Geological Society of London Special Publication, pp. 367–388.

Pyles, D.R., Straub, K.M., Stammer, J.G., 2013. Spatial variations in the composition of turbidites due to hydrodynamic fractionation. Geophys. Res. Lett. 40 (15), 3919–3923.

Reading, H.G., Richards, M., 1994. Turbidite systems in deepwater basin margins classified by grain size and feeder system. AAPG Bull. 78, 792–822.

Rebesco, M., Camerlenghi, A., 2008. Contourites. Developments in Sedimentology. vol. 60 Elsevier, Amsterdam.

Rebesco, M., Hernández-Molina, F.J., Van Rooij, D., Wåhlin, A., 2014. Contourites and associated sediments controlled by deep-water circulation processes: state-of-the-art and future considerations. Mar. Geol. 352, 111–154. https://doi.org/10.1016/j.margeo.2014.03.011.

Recouvreur, A., Fabregas, N., Mulder, T., Hanquiez, V., 2020. Geomorphology of a modern carbonate slope system and associated sedimentary processes: example of the giant great Abaco Canyon, Bahamas. Sedimentology 68 (1), 266–293.

Ricci Lucchi, F., 1965. Alcune strutture di risedimentazione nella formazione marnoso-arenacea romagnola. G. Geol. II 2, 265–292.

Ricci Lucchi, F., Valmori, E., 1980. Basin-wide turbidites in a Miocene, over-supplied deep-sea plain: a geometrical analysis. Sedimentology 27, 241–270.

Rodrigues, S., Hernández-Molina, F.J., Hillenbrand, C.-D., Lucchi, R.G., Rodríguez-Tovar, F.J., Rebesco, M., Larter, R.D., 2022. Recognizing key sedimentary facies and their distribution in mixed turbidite-contourite systems: the case of the Pacific margin of the Antarctic Peninsula. Sedimentology. https://doi.org/10.1111/sed.12978.

Rodríguez-Tovar, F.J., Miguez-Salas, O., Dorador, J., 2020. Image processing techniques to improve characterization of composite ichnofabrics. Ichnos 27 (3), 258–267.

Rotzien, J.R., 2016. Deltaic to deep-water depositional systems of the Western Irish Namurian Basin. In: Basin Dynamics Field Guide. 105 pp.

Rotzien, J.R., 2017. Stratigraphic terminations of the Eocene-Oligocene Annot sandstone, Southeast France. Trans. Gulf Coast Assoc. Geol. Soc. 67, 529–532.

Rotzien, J.R., 2019. The range and variability in sediment gravity flow deposits and petroleum reservoirs. In: Abstract and Presentation for the British Sedimentological Research Group Annual General Meeting, Egham, Surrey, 13–17 December.

Safronova, P.A., Laberg, J.S., Andreassen, K., Shlykova, V., Vorren, T.O., Chernikov, S., 2017. Late Pliocene-early Pleistocene deep-sea basin sedimentation at high-latitudes: mega-scale submarine slides of the North-Western Barents Sea margin prior to the shelf-edge glaciations. Basin Res. 29 (S1), 537–555.

Sawyer, D.S., Whitmarsh, R.B., Klaus, A., et al., 1994. Site 898. Proceedings of the Ocean Drilling Program, Initial Reports. vol. 149, https://doi.org/10.2973/odp.proc.ir.149.105.1994.

Schieber, J., Southard, J.B., 2009. Bedload transport of mud by floccule ripples—direct observation of ripple migration processes and their implications. Geology 37 (6), 483–486.
Schieber, J., Yawar, Z., 2009. A new twist on mud deposition: mud ripples in experiment and rock record. Sediment. Rec. 7 (2), 4–8.
Schieber, J., Southard, J.B., Thaisen, K., 2007. Accretion of mudstone beds from migrating floccule ripples. Science 318 (5857), 1760–1763.
Schwalbach, J.R., D.D. Miller, and J.R. Rotzien, 2009, Pliocene Turbidites of the Ventura Basin: Core Workshop and Field Trip, Complete With Guidebook, Held in Conjunction with the AAPG/SEPM Meeting Short Course, May 3–6, Ventura, CA, 28 pp.
Sequeiros, O., Pittaluga, M.B., Frascati, A., Pirmez, C., Masson, D.G., Weaver, P.P.E., Crosby, A.R., Lazzaro, G., Botter, G., Rimmer, J.G., 2019. How typhoons trigger turbidity currents in submarine canyons. Nature 9 (1), 9220. https://doi.org/10.1038/s41598-019-45615-z.
Shanmugam, G., 2000. 50 years of the turbidite paradigm (1950s-1990s): deep-water processes and facies models—a critical perspective. Mar. Pet. Geol. 17 (2), 285–342.
Shanmugam, G., 2006. Deepwater Processes and Facies Models: Implications for Sandstone Petroleum Reservoirs. Elsevier, Amsterdam, p. 476.
Sharman, G.R., Graham, S.A., Masalimova, L.U., Shumaker, L.E., King, P.R., 2015. Spatial patterns of deformation and paleoslope estimation within the marginal and central portions of a basin-floor mass-transport deposit, Taranaki Basin, New Zealand. Geosphere 11 (2), 266–306.
Shipp, R.C., Weimer, P., Posamentier, H.W., 2011. Mass-Transport Deposits in Deepwater Settings. vol. 96 SEPM Special Publication, Tulsa. 527 pp.
Slatt, R.M., Browne, G.H., Davis, R.J., Clemenceau, G.R., Colbert, J.R., Young, R.A., Anxionnaz, H., Spang, R.J., 1998. Outcrop-behind outcrop characterization of thin bedded turbidites for improved understanding of analog reservoirs: New Zealand and Gulf of Mexico. In: Society of Petroleum Engineers Annual Technical Conference and Exhibition, SPE 49563, pp. 845–853.
Southern, S.J., Patacci, M., Felletti, F., McCaffrey, W.D., 2015. Influence of flow containment and substrate entrainment upon sandy hybrid event beds containing a co-genetic mud-clast-rich division. Sediment. Geol. 321, 105–122.
Southern, S.J., Kane, I.A., Warchoł, M.J., Porten, K.W., McCaffrey, W.D., 2017. Hybrid event beds dominated by transitional-flow facies: character, distribution and significance in the Maastrichtian Springar Formation, north-west Vøring Basin, Norwegian Sea. Sedimentology 64 (3), 747–776.
Sprague, A.R.G., Garfield, T.G., Goulding, F.J., Beaubouef, R.T., Sullivan, M.D., Rossen, C., Campion, K.M., Sickafoose, D.K., Abreu, V., Schellpeper, M.E., Jensen, G.N., Jennette, D.C., Pirmez, C., Dixon, B.T., Ying, D., Ardill, J., Mohrig, D.C., Porter, M.L., Farrell, M.E., Mellere, D., 2005. Integrated slope channel depositional models: the key to successful prediction of reservoir presence and quality in offshore West Africa. In: CIPM. 13 pp.
Spychala, Y.T., Hodgson, D.M., Lee, D.R., 2017. Autogenic controls on hybrid bed distribution in submarine lobe complexes. Mar. Pet. Geol. 88, 1078–1093.
Steel, E., Buttles, J., Simms, A., Mohrig, D., 2017. The role of buoyancy reversal in turbidite deposition and submarine fan geometry. Geology 45 (1), 35–38.
Stevenson, C.J., Jackson, C.A.L., Hodgson, D.M., Hubbard, S.M., Eggenhuisen, J.T., 2015. Deepwater sediment bypass. J. Sediment. Res. 85 (9), 1058–1081.
Stevenson, C.J., Peakall, J., Hodgson, D.M., Privat, A., 2020. Tb or not Tb: banding in turbidite sandstones. J. Sediment. Res. 90 (8). https://doi.org/10.2110/jsr.2020.43.
Stow, D.A., Shanmugam, G., 1980. Sequence of structures in fine-grained turbidites: comparison of recent deep-sea and ancient flysch sediments. Sediment. Geol. 25 (1–2), 23–42.
Stow, D., Smillie, Z., 2020. Distinguishing between Deepwater sediment facies: turbidites, Contourites and Hemipelagites. Geosciences 10 (2), 68. https://doi.org/10.3390/geosciences10020068.
Stow, D.A.V., 1979. Distinguishing between fine-grained turbidites and contourites off Nova Scotia. Sedimentology 26, 371–387.
Stow, D.A.V., 1985. Fine-grained sediments in deep water: an overview of processes and facies models. Geo-Mar. Lett. 5, 17–23.
Stow, D.A.V., 1986. Deep clastic seas. In: Reading, H.G. (Ed.), Sedimentary Environments and Facies. Blackwell, Oxford, UK, pp. 339–444.
Stow, D.A.V., Kahler, G., Reeder, M., 2002. In: Stow, D.A.V., Pudsey, C.J., Howe, J.A., Faugères, J.-C., Viana, A.R. (Eds.), Deep-water Contourite Systems: Modern Drifts and Ancient Series, Seismic and Sedimentary Characteristics, vol. 22. Geological Society, London, Memoir, pp. 443–455.
Stow, D.A.V., Piper, D.J.W. (Eds.), 1984a. Fine-Grained Sediments: deepwater Processes and Facies. Geological Society Special Publication No. 15. 659 pp.
Stow, D.A.V., Piper, D.J.W., 1984b. Deepwater fine-grained sediments: facies models. Geol. Soc. Lond. Spec. Publ. 15 (1), 611–646.
Stow, D.A.V., Shanmugam, G., 1980. Sequence of structures in fine-grained turbidites: comparison of recent deep-sea and ancient flysch sediments. Sediment. Geol. 25, 23–42.
Stow, D.A.V., Tabrez, A.R., 1998. Hemipelagites: processes, facies and model. Geol. Soc. Lond. Spéc. Publ. 129, 317–337.
Stow, D.A.V., Wetzel, A., 1990. Hemiturbidite: a new type of deep water sediment. Proc. ODP Sci. Results 116, 25–34.
Stow, D.A.V., Faugères, J.C., 2008. In: Rebesco, M., Camerlenghi, A. (Eds.), Contourites. Developments in Sedimentology, vol. 60. Elsevier, Amsterdam, pp. 223–256.
Stow, D.A.V., Hernández-Molina, F.J., Alvarez Zarikian, C.A., et al., 2013. Site U1385. In: Stow, D.A.V., Hernández-Molina, F.J., Alvarez Zarikian, C.A., et al. (Eds.), Proceedings of the Integrated Ocean Drilling Program, Volume 339: Tokyo (Integrated Ocean Drilling Program Management International, Inc.)., https://doi.org/10.2204/iodp.proc.339.103.2013.
Strachan, L.J., Bostock, H.C., Barnes, P.M., Neil, H.L., Gosling, M., 2016. Non-cohesive silt turbidity current flow processes: insights from proximal sandy-silt and silty-sand turbidites, Fiordland, New Zealand. Sediment. Geol. 342. https://doi.org/10.1016/j.sedgeo.2016.06.017.
Sumner, E.J., Talling, P.J., Amy, L.A., 2009. Deposits of flows transitional between turbidity current and debris flow. Geology 37 (11), 991–994.
Tada, R., Murray, R.W., Alvarez Zarikian, C.A., et al., 2015. Site U1424. In: Tada, R., Murray, R.W., Alvarez Zarikian, C.A., et al. (Eds.), Proceedings of the Integrated Ocean Drilling Program, Volume 346: College Station, TX (Integrated Ocean Drilling Program)., https://doi.org/10.2204/iodp.proc.346.105.2015.

Talling, P.J., 2013. Hybrid submarine flows comprising turbidity current and cohesive debris flow: deposits, theoretical and experimental analyses, and generalized models. Geosphere 9, 460–488.

Talling, P.J., Amy, L.A., Wynn, R.B., Peakall, J., Robinson, M., 2004. Beds comprising debrite sandwiched within co-genetic turbidite: origin and widespread occurrence in distal depositional environments. Sedimentology 51, 163–194.

Talling, P.J., Masson, D.G., Sumner, E.J., Malgesini, G., 2012. Subaqueous sediment density flows: depositional processes and deposit types. Sedimentology 59 (7), 1937–2003.

Terlaky, V., Arnott, R.W.C., 2014. Matrix-rich and associated matrix-poor sandstones: avulsion splays in slope and basin-floor strata. Sedimentology 61, 1175–1197.

Thieblemont, A., Hernández-Molina, F.J., Ponte, J.P., Robin, C., Guillocheau, F., Cazzola, C., Raisson, F., 2020. Seismic stratigraphic framework and depositional history for cretaceous and Cenozoic contourite depositional systems of the Mozambique Channel, SW Indian Ocean. Mar. Geol. 425. https://doi.org/10.1016/j.margeo.2020.106192.

Tinterri, R., 2007. The lower Eocene Roda sandstone (south-central Pyrenees): an example of a flood-dominated river-delta system in a tectonically controller basin. Riv. Ital. Paleontol. Stratigr. 113, 223–255.

Tinterri, R., Civa, A., Laporta, M., Piazza, A., 2020. Chapter 17—Turbidites and turbidity currents. In: Scarselli, N., Adam, J. (Eds.), Regional Geology and Tectonics: Principles of Geologic Analysis, vol. 1, second ed. Elsevier, pp. 441–479.

Tyler, J.H., 1972. Pigeon point formation: an upper cretaceous shoreline succession, central California coast. J. Sediment. Petrol. 42 (3), 537–557.

Van Vliet, A., 1978. Early tertiary deepwater fans of Guipuzcoa, northern Spain. In: Stanley, D.L., Kelling, G. (Eds.), Sedimentation in Submarine Canyons, Fans, and Trenches. Dowen Hutchinson & Ross, Stroudsburg, pp. 190–209.

Walker, R.G., 1967. Turbidite sedimentary structures and their relationships to proximal and distal environments. J. Sediment. Petrol. 37, 25–42.

Walker, R.G., 1975. Generalized facies models for resedimented conglomerates of turbidite association. GSA Bull. 86, 737–748.

Walker, R.G., 2004. Facies models revisited. AAPG 2004 Annual Convention Abstracts Volume. Dallas, Texas., p. A144.

Walker, R.G., Massingill, J.V., 1970. Slump features on the Mississippi fan, northeastern Gulf of Mexico. GSA Bull. 81, 3101–3108.

Waltham, D., 2004. Flow transformations in particulate gravity currents. J. Sediment. Res. 74, 129–134.

Weber, R., 2018. Biostratigraphy: Paleo Data Inc. Short Course and Notes, Lafayette, Louisiana. 24 January, approx., 200 pp.

Wood, A., Smith, A.J., 1959. The sedimentation and sedimentary history of the Aberystwyth grits (upper Llandoverian). Q. J. Geol. Soc. Lond. 114, 163–195.

Worden, R., 2020, Lower Cretaceous Rodby and Paleocene Lista Shales: Characterisation and Comparison of Top-Seal Mudstones at Two Planned CCS sites, Offshore UK: ResTec 2.0 Conference by Petryx Ltd.

Yang, T., Cao, Y., 2019. Early cretaceous hybrid event beds on Lingshan Island, East China. In: Rotzien, J.R. (Ed.), Basin Dynamics Bulletin, A Basin Dynamics Special Publication. vol. 1, pp. 11–13. no. 4.

Yang, T., Cao, Y., Friis, H., Liu, K., Wang, Y., 2018. Origin and evolution processes of hybrid event beds in the lower cretaceous of the Lingshan Island, Eastern China. Aust. J. Earth Sci. 65 (4), 517–534.

Yawar, Z., Schieber, J., 2017. On the origin of silt laminae in laminated shales. Sediment. Geol. 360, 22–34.

Zavala, C., Arcuri, M., 2016. Intrabasinal and extrabasinal turbidites: origin and distinctive characteristics. Sediment. Geol. 337, 36–54.

Chapter 8

# Deepwater depositional environments

Jon R. Rotzien[a,b], Rebecca L. Caldwell[c] and Lisa R. Goggin[d]

[a]*Basin Dynamics, Conroe, TX, United States,* [b]*University of Houston, Houston, TX, United States,* [c]*Chevron Technical Center, Chevron Corporation, Houston, TX, United States,* [d]*Chevron Digital Products and Services, Chevron Corporation, Houston, TX, United States*

## Introduction

Deepwater depositional environments are areas of net sediment accumulation formed by the complex interaction of various sedimentary processes. Mapping deepwater depositional environments, often grouped in assemblages known as environments of deposition (*EOD*) or gross depositional environments (*GDE*) for simplicity, is a critical early step in the exploration common process (Chapter 1). Deepwater sedimentary processes (Chapter 6) include pelagic and hemipelagic sedimentation as well as deposits generated by sediment gravity flows and bottom currents (Chapter 7). In many cases, these sedimentary processes mix and overlap, creating complex stratigraphic successions of interbedded pelagites, turbidites and contourites. These types of deposits and their stratigraphic successions are found preferentially in certain depositional environments. The goal of this chapter is to describe the range and variability of deepwater environments, highlight how geoscientists apply this knowledge to map, interpret and model them, and discuss the advantages and uncertainties associated with these approaches. This chapter focuses primarily on environments constructed from sediment gravity flows. The following chapter, Chapter 9, characterizes environments featuring primarily along-slope sedimentary processes responsible for forming drifts and other morphological elements derived from bottom current activity. This chapter is divided into five main parts:

- an overview of the large-scale context and controls on deepwater depositional environments, including tectonic margin effects (Section "Introduction"), slope profiles (Section "Deepwater systems by tectonic setting") and the influence of shelf sediment delivery systems (Section "Slope profiles");
- descriptions of each of the main types of deepwater environments (primarily focusing on *depositional* environments—see *Chapter 9* for information on along-slope sedimentary environments featuring drifts and mixed environments), including canyons, channels, levees, splays (lobes), overbank (Section "Deepwater depositional environments"), contourite and mixed systems (Section "Contourite and mixed systems");
- further evaluations and considerations for understanding depositional environments, including typical grain-size distributions, methods for differentiating deepwater deposits, reservoir connectivity and compartmentalization, and common pitfalls leading to misinterpretation of deepwater environments (Section "Additional considerations");
- process-based modeling of deepwater depositional environments that incorporates the physical attributes of deepwater sedimentary processes at scale to create realistic predictions of stratigraphic architecture (Section "Advances in modeling depositional environments");
- how to put it all together as well as final conclusions and insights (Sections "Key messages", "Conclusion", "How are deltas different from submarine fans?").

This chapter is intended to provide a powerful visual impression of the scale and heterogeneity in deepwater depositional environments using a variety of outcrop examples. Application of the observations and methods presented in this chapter will be useful to professionals who map and interpret deepwater depositional systems and model them in 1-, 2- and 3-dimensional space. As described in Chapters 6 and 7, deepwater sedimentation is unique compared to shallow marine and terrestrial depositional systems. Individual sediment gravity flows tend to deposit one bed, or sedimentation unit, per event. Further, sediment gravity flows are not constrained by the confinement that flows through a river feel due to the reduction in density contrast between the deepwater flow and the ambient water, and therefore, sediment tends to distribute in broad patterns. However, as this chapter shows, that is not always the case. Deepwater environments share both similarities and differences with their upslope and terrestrial equivalents in a linked source-to-sink system (see Chapter 11 and section "How are deltas different from submarine fans?"). *Did you know that ancient channel systems on Mars have*

**Deepwater Sedimentary Systems.** https://doi.org/10.1016/B978-0-323-91918-0.00001-3

*similar sinuosity and scale to modern deepwater channel systems on ocean floors worldwide?* While the processes on Earth today may be entirely different to those operating on Mars in the past, this chapter aims to illustrate that process, deposit and environment are linked and can be modeled with accuracy for reservoir and seal prediction in exploration, appraisal and development using advanced computational methods. This chapter also invites the reader to think critically about the merits and limitations of commonly used subsurface interpretation methods in these environments (for further reading on this topic, see Chapter 16).

Deepwater depositional environments are investigated for diverse reasons by many types of researchers around the world. In any study, proper characterization of the depositional environment is required to yield meaningful analysis and conclusions. Simply, the foundational purpose of any study is to understand *what is the depositional environment(s) represented by the sedimentary deposit*. After arriving at possible solutions to this question, many other more impactful questions can be tackled. Questions driving the content in and inspiring the creation of this chapter include:

- What is the range and variability in types of sedimentation in different deepwater depositional environments?
- How do deepwater depositional environments fit together—like a puzzle—in a larger (regional) geologic context?
- How are deepwater depositional environments influenced by their settings and regional controls, such as tectonic setting?
- What are examples of mappable deepwater depositional environments and how do they form from the sedimentary processes operating in and around them?
- How do depositional environments scale in relation to sedimentation units (Chapter 7), and how does this fit into a hierarchical framework for deepwater depositional environments?
- What happens when turbidity currents and bottom currents mix? How do various sedimentary processes sculpt the deepwater depositional system?
- How are deepwater depositional environments observed and measured? What tools and analogs (e.g., from process-based computational modeling) are used to understand and predict their architectures?

## Deepwater systems by tectonic setting

Deepwater depositional systems occur in nearly all types of basins and tectonic settings, as described in the first part of this book (See Chapters 1–5 for more information). A compilation of the spectrum of deepwater basin types and tectonic settings along passive to active margins is shown in Fig. 1. While outcrop analogs are routinely used to show the

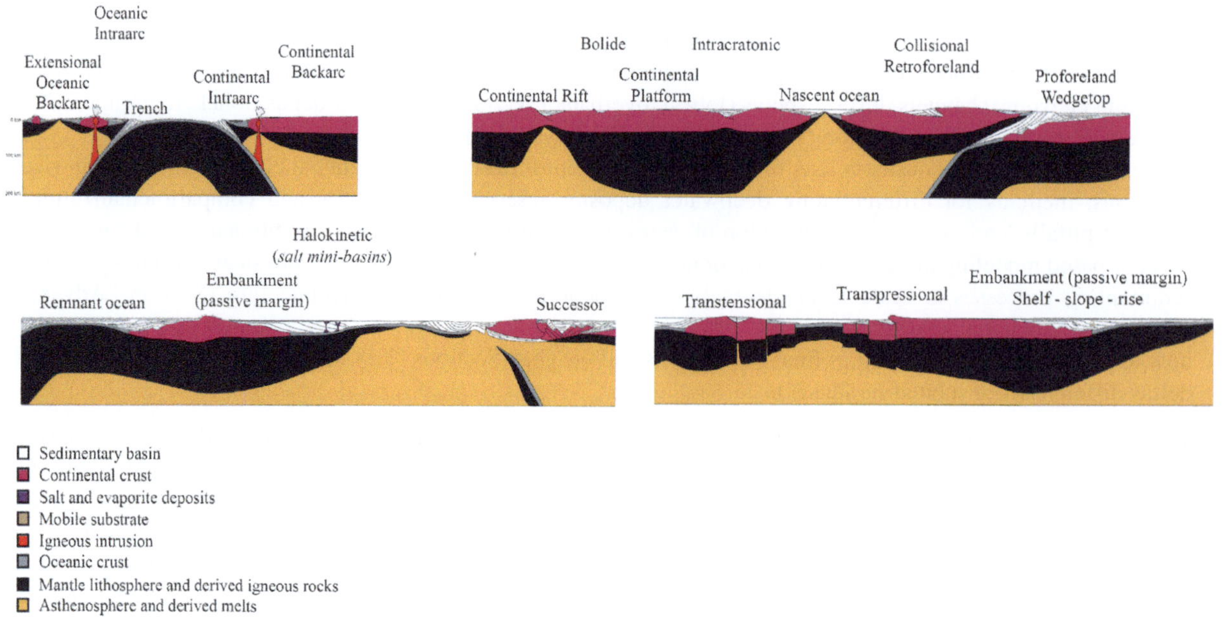


**FIG. 1** Selected tectonic regimes and deepwater basin types determined by subsidence type and what type of plate tectonic features are expressed in the vicinity of the basin. (A) Arc and trench type basins; (B) select continental, passive margin (embankment), transform and successor basins; (C) rift and collisional setting basins; (D) select hybrid transform basins. Developed, compiled and redrawn based on work by Ingersoll (2011) and references therein. For the complete 5-meter-long figure with the suite of tectonic settings, see Fig. 4 in Chapter 1.

reservoir-scale architecture of deepwater oil and gas fields, many of these analogs are found along active margins. However, many significant deepwater oil and gas fields are buried along passive margins. The scale of active vs passive margin submarine fan depositional systems is notably different with the one anomaly being the Bengal Fan, representing a mix of active and passive margin controls (Fig. 2). Active margin submarine fans are typically much smaller, in part due to basin confinement (~1/10th to 1/100th of the areal extent and volume), than fans found along passive margins. Passive margin fans achieve their great size and extent due to long-lived, continental-scale fluvial drainage systems that deliver sediment to deltas and shelves (see Section "Shelf systems: Supply to deepwater environments", and Chapters 1 and 11), especially during lowstands (Chapter 12) when shelf-edge deltas efficiently deliver typically finer-grained sediment (sand, silt, clay) to deepwater systems. The finer grain-size distributions fed to deltas on passive margins are also more easily carried by sediment gravity flows. Thus, there are multiple reasons for the size disparity between active and passive

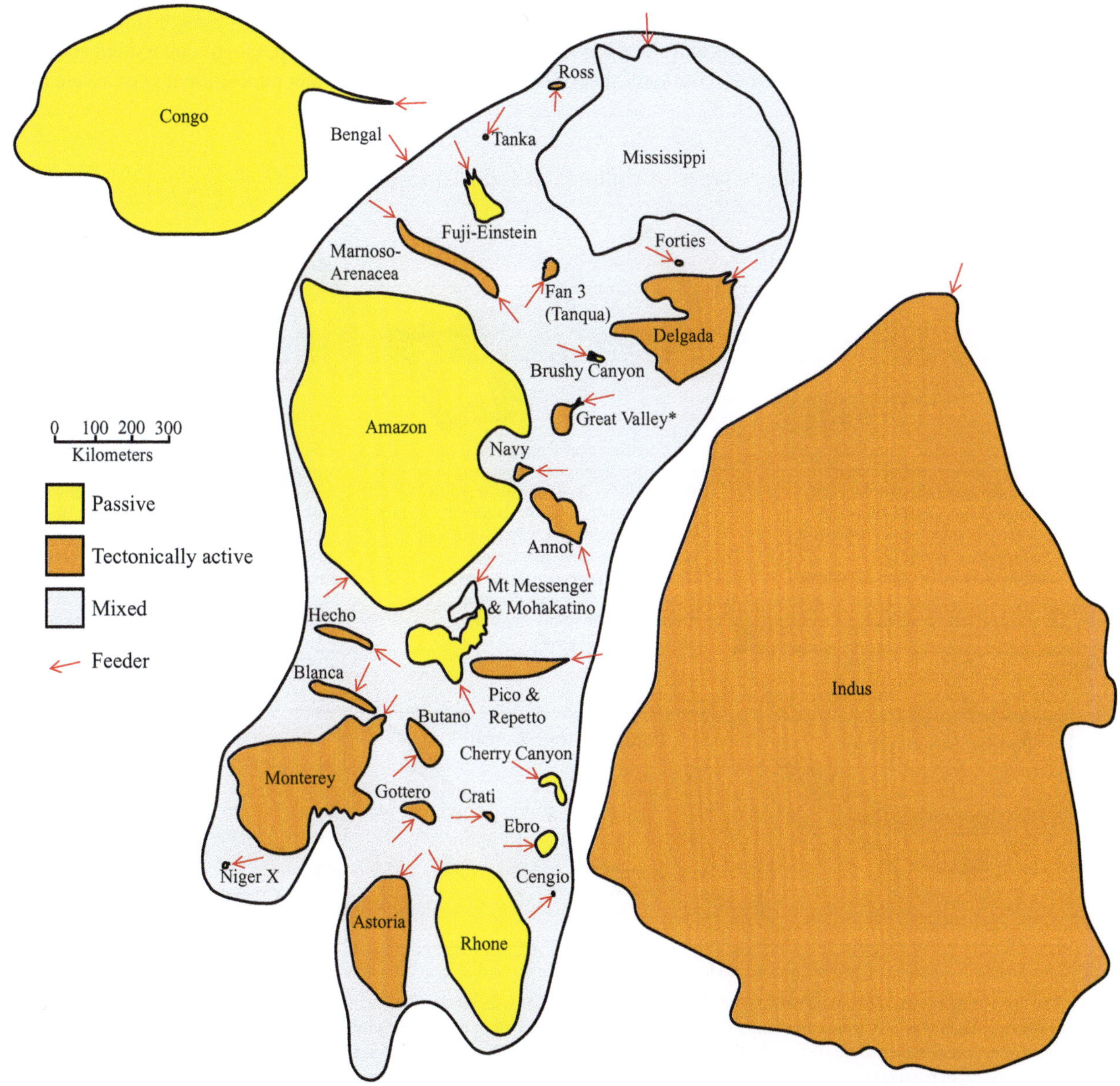


**FIG. 2** Submarine fan depositional systems from active and passive tectonic settings. Note that most of the deepwater depositional systems presented here have complex tectonic histories and have been simply categorized for the purposes of this diagram. *(Modified and redrawn from Barnes, N.E., Normark, W.R., 1985. Diagnostic parameters for comparing modern submarine fans and ancient turbidite systems. In: Bouma, A.H., Normark, W.R., Barnes, N.E. (Eds.), Submarine Fans and Related Turbidite Systems. Frontiers in Sedimentary Geology. Springer, New York, NY. https://doi.org/10.1007/978-1-4612-5114-9_3; Haughton, P., Kendall, C., 2006. Deepwater clastic systems. SEPM Sequence Stratigraphy Web. http://sepmstrata.org/Deep-waterClasIntro.html (Accessed 24 April 2015).)*

margin deepwater fans. The differences between these two classes of fans are good to keep in mind when comparing the stratigraphy from outcrops uplifted along active margins to that of deepwater systems buried tens of thousands of feet below mudline along passive margins.

## Slope profiles

Deepwater slope profiles, including basin margin or continental slope profiles, manifest in three principal types based on gradient and structural features: graded; above-grade stepped, ponded or perched; healed (Fig. 3). These three main types of slopes can be found along continental margins, or at a smaller scale along basin margins, and influence sediment transport, deposition and distribution in sediment gravity flow dominated depositional systems (summarized in Prather et al. (2016) and drawn in Fig. 3). Continental slopes are where deepwater sedimentation begins off of the continental shelf (Galloway, 1998), yet outer shelf and ramp profiles can accumulate sediment gravity flow deposits as well. Types of slope systems can be classified by delta position, on- vs off-axis and slope terrain—simple, chute, ponded, bypass—to understand and predict the nature of slope reservoir systems (leveed, confined, sheet) (Donovan, 2010). Deepwater depositional systems are further

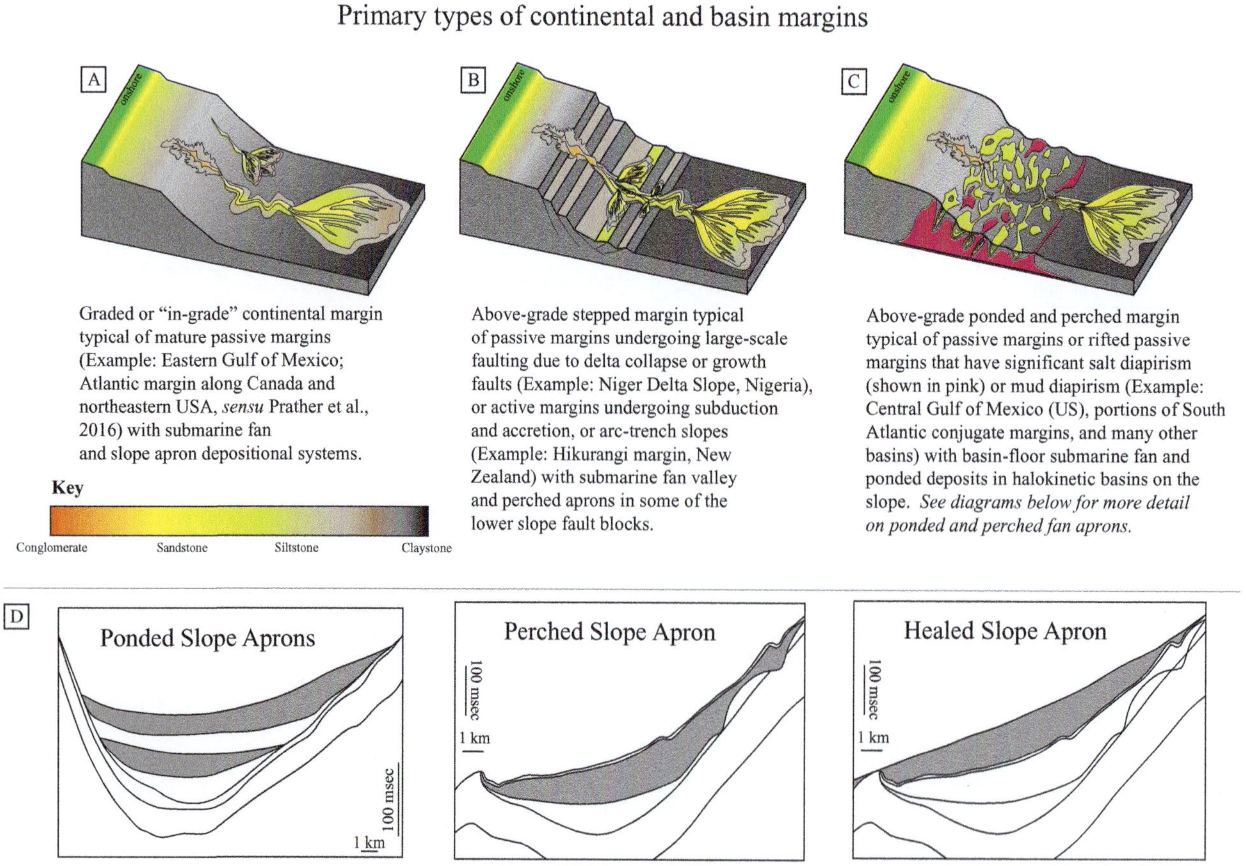


Cross-section line drawings of ponded, perched and healed slope aprons (in gray) derived from 2D seismic reflection data profiles (Prather et al., 2016).

**FIG. 3** The three primary types of continental margins illustrating submarine fan assemblages—including valleys and aprons—for turbidite- and debrite-dominated systems. Each type of system can be modified by mass transport and contour current activity. From left: (A) Graded or "in-grade" continental margin typical of mature passive margins (example: Eastern Gulf of Mexico, USA; Atlantic margin along Canada and northeastern USA); (B) Above-grade stepped margin typical of passive margins undergoing large-scale faulting due to delta collapse or growth faults (example: Niger Delta Slope, Nigeria), or active margins undergoing subduction and accretion, or arc-trench slopes (example: Hikurangi margin, New Zealand); (C) Above-grade ponded and perched margin typical of passive margins or rifted passive margins that have significant salt diapirism or mud diapirism (example: Central Gulf of Mexico, USA; portions of South Atlantic conjugate margins). (D) Ponded, perched and healed slope aprons in cross section. 1 km (~ 0.6 mi.) horizontal scale and 100 millisecond vertical scale. *(Modified from Prather, B.E., O'Byrne, C., Pirmez, C., Sylvester, Z., 2016. Sediment partitioning, continental slopes and base-of-slope systems. Basin Res. 29(3), 394–416 and references therein.)*

characterized by mappable GDE, often at a larger scale than individual environments: aprons, which include all line-sourced slope systems, defined as high- or low-relief as a proxy for sand content; valleys formed by incision or evacuation of sediment prior to filling by other processes; hemipelagic wedges of fine-grained sediment that typically thin downslope; contourite drifts formed by bottom currents; mass transport deposits (MTD), and associated top-fill. These main systems can be divided into the six primary deepwater environments (also read: *morphological element*): canyons, channels, splays, levees, overbank, drifts. Each depositional environment consists of its own set of traits depending on the type of GDE above. Some examples may read as follows: hybrid contourite-turbidite channel-levee complex as part of a slope apron; splay complex filling the bathymetric lows atop a mass transport complex; channel complex in a submarine valley and so on. These features can also be described with respect to position, such as proximity to the shelf-slope break. For example, there are several documented industry studies of deepwater slope channel-levee complexes offshore West Africa and in the US Gulf of Mexico.

Graded slopes are defined as having a smoothed concave-up theoretical reference profile and are efficient at transferring sediment to the basin floor (Fig. 3A). Graded slopes are typically sand-poor compared to aprons on stepped or ponded slopes since they are defined by smooth profiles that facilitate sediment bypass instead of deposition (see Chapter 1). These slopes result from long-term equilibrium of erosional and depositional processes and are often concave up, comprise thin aprons and are constructed of variably leveed channels and associated deposits. Examples include the Eastern US Gulf of Mexico and offshore Eastern Canada (Prather et al., 2016).

Above-grade slopes are steeper than equilibrium gradient and may be caused by rapid sea-level rise, tectonic over-steepening or alternating clastic and carbonate deposition (Recouvreur et al., 2020; Liu et al., 2021). Above-grade slopes (Prather, 2000, 2003; Prather et al., 2016) are synonymous with out-of-grade slopes (Pyles, 2008; Haughton et al., 2009) and are elevated above the level of a theoretical concave-up smoothed reference profile. There are three primary types of above-grade margins: stepped (Fig. 3B); ponded (defined as having no exit point—Fig. 3C) and perched (defined by an exit point—Fig. 3C); uneven (sensu De Weger et al., 2020). Uneven slopes are not stepped due to faulting or ponded due to the nature of the basin topography (Amy et al., 2007) or salt withdrawal (Chapter 5), but sculpted by other deepwater processes, such as bottom currents, to make an uneven above-grade profile.

Above-grade stepped slopes are characterized by faults that form alternating areas of lower and higher gradient (Fig. 3B). The faulting creates fault blocks that are typically rotated landward, therefore decreasing the local gradient and forming a variety of shelf-edge-parallel sub-basins. This area tends to form aprons, valleys or other small-scale features such as transient splay complexes (sensu Adeogba et al., 2005; Prélat et al., 2010) that take the shape of the space available for sedimentation. In summary, stepped slopes occur due to faulting that acts to raise the profile above the equilibrium concave-up profile. An example would be the Niger Delta Slope, offshore Nigeria (see Chapter 4) described in Adeogba et al. (2005).

Ponded and perched slopes occur on intraplate embankment margins typically due to salt diapirism (see Chapter 5), creating halokinetic basins (also known as *intraslope* basins or *minibasins* in the Gulf of Mexico) bounded on one or more sides by allochthonous salt (Prather et al., 1998). Small-scale ponded sub-basins also can form as a result of the movement of any type of mobile substrate through the stratigraphy. Above-grade ponded slopes can result from gravity-driven salt withdrawal that forms sub-basins, or sediment "ponds" (Fig. 3C). The stratigraphy that characterizes this basin-fill type is known as a *halokinetic sequence*, which is an unconformity-bound, thinned, and folded stratal package adjacent to a passive salt diapir(s) (Giles and Rowan, 2012). Stratigraphic analysis of halokinetic basin fill is described by Cumberpatch et al. (2020), Roca et al. (2020). These types of deposits share many of the traits with other modes of stratigraphic terminations, for example, basin-margin onlap of turbidite beds, Peira-Cava outlier and other parts of the Annot Sandstone depositional system, Southeast France summarized in Rotzien (2017, 2020). Salt withdrawal creates ponded areas in topographic lows, commonly ranging from ~15–50 km (~ 9–31 mi) in diameter. Ponded slopes can be healed, which is known as *healed-slope morphology*, by filling of the halokinetic basins and formation of a perched apron after the ponded minibasin is filled (bottom of Fig. 3). High-relief ponded aprons, in contrast, tend to have higher sand content, particularly in the proximal portions, where intense channelization and erosional processes promote relatively high connectivity within the various sand units. Since there are no significant through-going or bypass channels, abandonment of high-relief ponded aprons likely results from relative sea-level rise or avulsion of the shelf-to-slope delivery system (Prather et al., 2016 and references therein).

Filling and annealing of above-grade slopes, whether stepped, ponded or perched, can create healed slopes. These are slopes that have been readjusted over time (Ross et al., 1994) often feature erosional truncation, sediment bypass and marine onlap of apron complexes (Fig. 3A). Without seismic profiles that can illustrate a formerly stepped or ponded or perched slope, it can be difficult to distinguish a healed slope from a graded slope.

## Shelf systems: Supply to deepwater environments

Deepwater depositional environments are fed by updip coastal and shelf systems. Naturally, these upslope feeder systems have a profound effect on their basinward deposits. The transition within the source-to-sink (Chapter 11) sediment transport system—from coastal and shelf processes to deepwater processes—varies in character based on type of basin margin and location of the coast relative to the shelf edge. Deepwater depositional environments are thus intimately linked to the updip coastal and shelf systems that feed them. There are two widely recognized mechanisms for large-scale sediment transport from the coast to deepwater: (i) prograding margins characterized by the presence of shelf-edge delta(s); and (ii) margins dominated by cross-shelf and cross-slope bypass via submarine canyons (e.g., Fisher et al., 2021). This section describes how deepwater depositional systems vary and are influenced by these two different mechanisms depicted in Fig. 4. For further descriptions of these mechanisms (e.g., a review of shelf-edge deltas), see the studies referenced below.

Deepwater environments may be fed by updip shelf-edge deltas, linked with proximal deepwater deposits via slope channels and toe-of-slope fans or aprons. These systems are not commonly temporally linked to large basin-floor fan systems, though they can occur. Basinward of shelf-edge deltas it is common to find abundant sandy turbidite intervals interbedded with heterolithic, muddy deposits, often as minor chutes, gullies and lobes on the upper to middle reaches of the slope, and channels and laterally extensive "sheet-like" lobes on the lower slope (i.e., slope apron depiction in Plink-Björklund et al., 2001). These morphological elements are laterally continuous and organized in stable shelf-edge delta systems, which lack abundant growth faults, significant canyons and incised valley feeder systems. Slope-gravity collapse can result inlarge-scale slumps and slides (Alves, 2015) described in Chapter 6. Collapse features can also be abundant due to the high rate and delivery of sediment from proximal sources (i.e., high sediment supply). Muddier systems tend to have more of these collapse features than sandy systems, and can cause slope and basin-floor instability. In seismic data, these features are observed as discontinuous wedge-shaped geobodies characterized by chaotic and mounded reflectors (Porębski and Steel, 2003, also see Chapter 10 for more examples). Basin-floor fans tend to be not as well developed in these types of systems. Fans form more readily when linked to river-dominated shelf-edge deltas as opposed

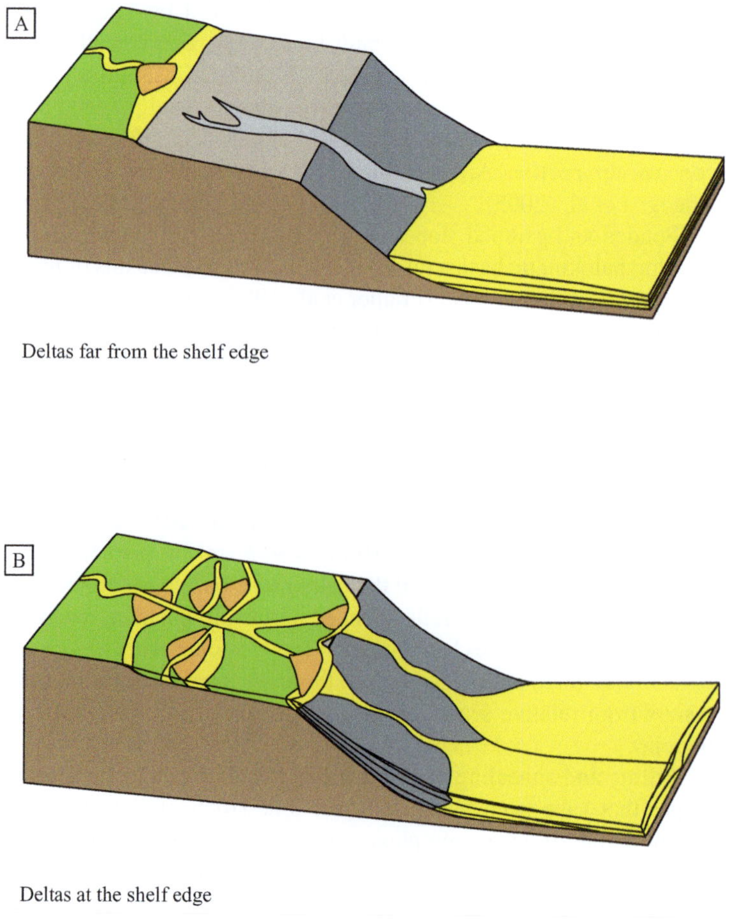


**FIG. 4** Examples of shelf-edge delta systems. (A) Deltas that are far from the shelf edge do not directly feed the continental slope. Rather, sediment is captured by a canyon that transports it to deepwater. Note the onlap of submarine fan deposits in this scenario. (B) Deltas that reach the shelf edge feed tributary channels and gullies at the top of the slope. This scenario provides a rather direct link of deltaic sedimentation to deepwater environments. Note the offlapping pattern of submarine fans along the base of slope. *(Modified from Fisher, W.L., Galloway, W.E., Steel, R.J., Olariu, C., Kerans, C., Mohrig, D., 2021. Deep-water depositional systems supplied by shelf-incising submarine canyons: recognition and significance in the geologic record. Earth Sci. Rev. 214, 103531.)*

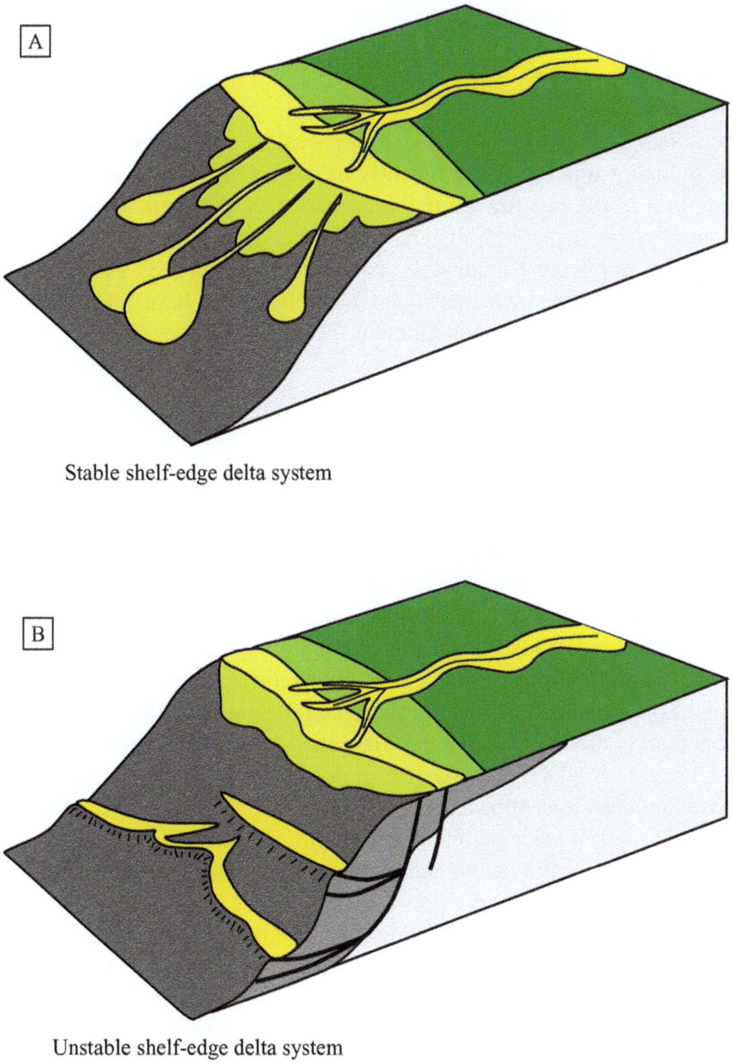


**FIG. 5** Shelf-edge delta systems: (A) stable systems consist of extensive mouth bar and perched mouth bar assemblages on the outer shelf and upper slope. The slope consists of abundant small, often coalescing, submarine fan aprons fed by slope channels and gullies. (B) Unstable shelf-edge delta systems are controlled by listric faults that tend to capture sediment along the slope in the form of minibasins oriented parallel to the coast, yet perpendicular to regional sediment transport direction. *(Modified from Porębski, S.J., Steel, R.J., 2003. Shelf-margin deltas: their stratigraphic significance and relation to Deepwater sands. Earth Sci. Rev. 62(3–4), 283–326.)*

to shelf-edge deltas dominated by wave or tidal processes (Dixon et al., 2012 and others). However, in unstable shelf-edge delta systems, factors such as high sediment supply, relative sea-level fall beyond the shelf edge and growth faulting along the shelf edge may induce canyon development or capture of antecedent canyons, thus causing sediment bypass to the base of slope and onto the basin floor forming a submarine fan (Fig. 5).

Finding the links between shelf-edge delta systems and deepwater deposits can be useful for identifying and predicting deepwater stratigraphy. Base-of-slope stratigraphic successions are often identified in seismic reflection data by offlap geometry, which indicates pinching out of the intervals onto the slope or transition into mud-prone slope deposits commonly termed in the literature as heterolithic intervals (e.g., Porębski and Steel, 2003; Fisher et al., 2021). In well logs and core, these base-of-slope to basin-floor deposits are characterized by stacked blocky sandstone alternating with heterolithic intervals. Overall, these can be remarkably sand-prone successions in stratigraphic trapping configuration (read more in Chapters 1 and 2).

The intimate connections among large-scale, source-to-sink processes, shelf-edge delta processes and their linked deepwater deposits allow the stratigrapher to make regional stratigraphic predictions. Shelf-edge delta systems and their equivalent deepwater deposits are recognized to routinely form from forced regression during relative sea-level fall (e.g. Posamentier and Kolla, 2003), or from systems with very high upstream sediment input and/or narrow shelves during stable to rising relative sea level (e.g., Carvajal and Steel, 2006; Koo et al., 2016; Harris et al., 2016; Chapter 12). These relationships permit correlation of stacking patterns on the shelf to the stacking of fans on the basin floor (e.g., Koo et al., 2016). There is also a link between the local shelf-edge delta process regime and related deepwater deposits. Generally, river-dominated shelf-edge deltas produce sand-rich, channelized deepwater systems with abundant basin-floor fans (e.g., Dixon et al., 2012; Laugier and Plink-Björklund, 2016). Tide- and wave-dominated shelf-edge delta regimes deliver less sand compared to their river-dominated counterparts, and, as a result, produce smaller or poorly defined channel and submarine fan systems. For further review on shelf-edge delta growth processes, geometries and facies, see Porębski and Steel (2003), Dixon et al. (2012), Laugier and Plink-Björklund (2016).

Deepwater systems headed by canyons deliver sediment to the basin floor via a process called *cross-shelf bypass*. The canyon essentially acts as a highway—transporting sediment quickly and efficiently to the basin floor. These systems deposit little to no sediment on the slope, and deposition is skewed toward the deep basin, as sediment sourced from the upstream coast and shelf is transported farther basinward in these shelf- and slope-incised systems. This example of basin-floor delivery creates an aggradational and onlap architecture against the continental margin (e.g. Fisher et al., 2021). Unlike prograding margins with shelf-edge delta systems, these systems do not generally record base-level changes or other characteristics of processes occurring on the shelf (e.g., Covault and Graham, 2010).

## Deepwater depositional environments

It is somewhat surprising when deepwater sedimentation is not indicative of the depositional environment. Because of this relationship, energy companies choose to explore for oil and gas resources by mapping and interpreting EOD, also known as GDE, which is a critical part of the exploration common process (Hubbard et al., 1985a, b, 1990—see Chapter 1 for more on the exploration process). Depositional environment mapping is also critical for basic scientific investigation and planning of drilling programs, for agencies such as the International Ocean Drilling Program (IODP), and for delineation of country boundaries in offshore regions. Processes, deposits and depositional environments are often interdependent and can create positive feedback cycles, which has led to the standardization of recognizable stratigraphic and seismic stratigraphic facies for certain depositional environments. For instance, channels provide confinement for flows with high Reynolds values (Chapter 6). If channels are kept open and "alive" by funneling high-energy flows, they may be filled by deposits with sedimentary structures showing evidence of high velocity (coarse-grained sediment, scours, bedforms, etc.). If channels are abandoned abruptly due to an avulsion, they may subsequently be filled by low-energy or highly variable deposits. Therefore, in order to successfully characterize deepwater sedimentary systems, it is important to understand the nature of deepwater environments, how they are formed, their 3D extent and the typical processes and deposits that sculpt them to be able to successfully evaluate deepwater exploration opportunities for any range of scientific and business endeavors. This is not an exhaustive description of deepwater environments, but rather a distillation or primer (as is much of this chapter) on the range and variability of several key environments found in submarine fan, fan valley and apron depositional systems in any type of tectonic setting (Fig. 1) and along any type of continental margin (Figs. 2 and 3).

When describing and characterizing deepwater depositional environments, it is important to provide the following information or evidence to build predictive models of the depositional system, which can vary greatly from basin to basin:

- definition and key traits;
- mode(s) of formation (i.e., why do they exist?);
- planform and cross-section architecture;
- processes of sediment transport and deposition, as well as events or processes that modify these environments;
- fill, including discussion on the range, variation and uncertainty;
- distinguishing features;
- predictive attributes and other systematic traits depending on key metrics including continental margin profile, grain size distribution and gradient.

These environments are not only limited to the deep sea, but can also be found in other non-marine regions such as deep lakes that extend below wave base. While there are myriad types of basins and continental margins, there tend to be six principal depositional environments including canyons, channels, levees, lobes (splays), overbank and drifts (Deptuck and Sylvester, 2018; this book), with gradational or sharp transitional zones between these types of environments including areas commonly known as the channel-lobe (splay) transition zone (CLTZ) (Wynn et al., 2002); canyon-channel transition zone (Anderson et al., 2006); inner and outer levee transition zone (Morris et al., 2014) and many more. As these environments collect sediment and are compacted, they are known as *morphological elements*, a subtle but key difference. In this way, the reservoirs of morphological elements are modeled in the subsurface to determine fluid flow characteristics in energy company workflows. Depositional environments (Fig. 6) exhibit architectural and lithological differences, and typically have important implications for energy company analysis of geologic risk pertaining to reservoir, quality, seal, trap and charge (Chapter 2) as well as seafloor hazards assessment.

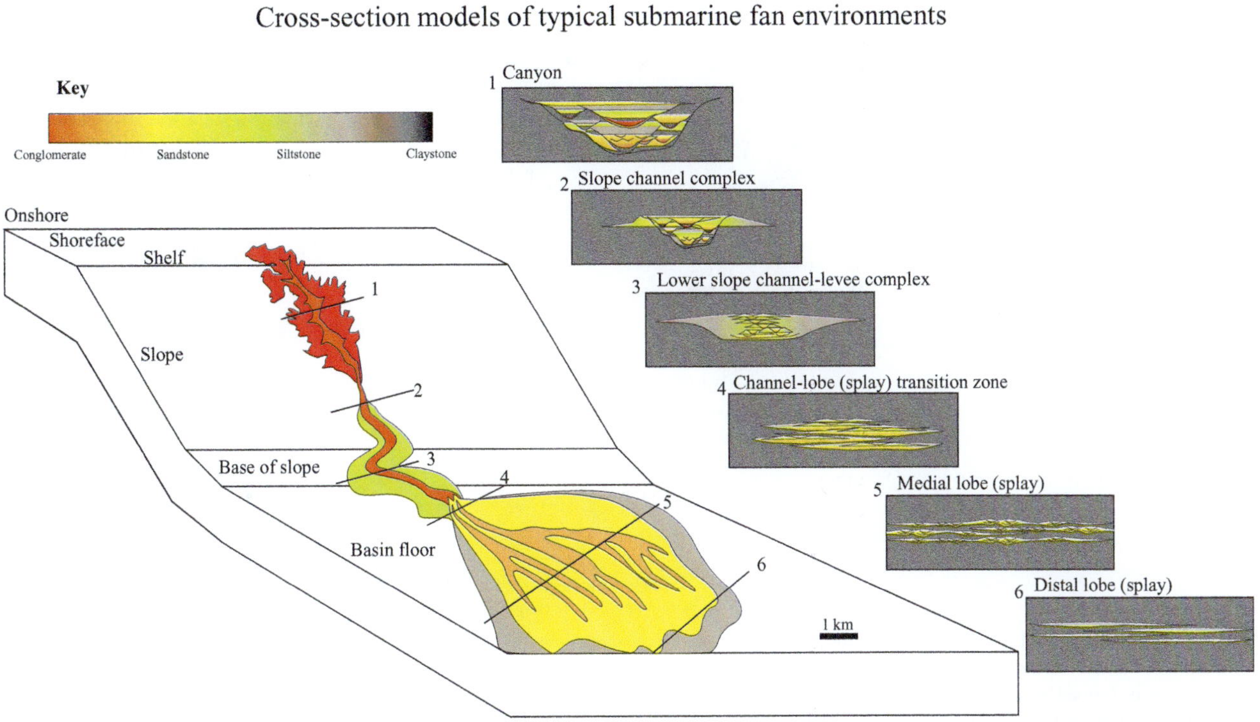


**FIG. 6** Profiles along the axis of a submarine fan depositional system. (1) Canyon environments can show multiple levels of mass transport deposits (MTD), gravelly braided channel environments creating "sheet-like" deposits, channel-levee systems and mass-flow deposits along canyon margins (Lowe, 2004). (2) Slope channel systems can show slump deposits, MTD and channel-levee systems flanked by poorly developed overbank levees (Hickson and Lowe, 2002; Campion et al., 2000; Abreu et al., 2003; Anderson et al., 1999; Beaubouef et al., 1999 and others). (3) Lower slope channel-levee complexes can show multiple stacked channel-levees in an overall outer-levee confined environment (Sylvester et al., 2011; Rotzien et al., 2014a). (4) The channel-lobe (splay) transition zone (CLTZ) can show multiple coalescing channel-levee, lobe and overbank complexes. (5) Medial lobe complexes show stacked deposits with areas of higher energy and erosion interpreted as distributary channel systems fed by the upslope channel-levee complex. (6) Distal lobe environments show these deposits encased in interlobe or hemipelagic mudstone.

While the following environments are described from the primacy of a turbidite-dominated system, there exists a plethora of other processes not directly associated with turbidity currents. Hemipelagic sedimentation can form mudstone "sheets" as a drape with thin-bedded (sensu Campbell, 1967) turbidites (also "mud belts" described by Sylvester et al. (2012) that tend to be thickest in the upper slope and thin downslope). Turbidites are generally celebrated for their high petroleum reservoir performance, yet many other factors that can operate coevally or post deposition include: bottom currents; sediment and fluid injection and escape such as injectites, pockmarks and mud volcanoes; sediment instability such as slides, mass transport deposits and falls; upslope migrating bedforms formed by overbanking currents such as sediment drifts, moats and waves. These are discussed in following chapters. Modern seafloor bathymetric datasets have been instrumental in observing the diverse nature of the seafloor (sensu Babboneau et al., 2010), in essence using modern seafloors and techniques to interpret sparse or incomplete datasets in systems of the past (see Chapters 1, 6, 7, 9–15).

## Canyons

Canyons typically represent the most landward portion of a significant deepwater depositional system (Fisher et al., 2021) along any type of continental margin (Cavazza and DeCelles, 1993; Laursen and Normark, 2002; Normark and Carlson, 2003; Gong et al., 2011). Submarine canyons are major capacitors and transporters of sediment to deepwater areas and commonly link river drainages (Klaucke and Hesse, 1996) and onshore incised valleys to the shelf and to deepwater (Sweet and Blum, 2016; Paull et al., 2018 and references therein). Canyons tend to be the deepest carved feature in the system and fully confine sediment gravity flows. Canyons can form within other sedimentary deposits, bedrock (Cronin and Kidd, 1998; Lowe, 2004; Anderson et al., 2006) or can be routed along a path defined by mobile substrate (Prather, 2000) such as salt (Fig. 7).

A submarine canyon is subdivided into at least four primary zones, from updip to downdip: (i) head; (ii) middle canyon; (iii) lower canyon; (iv) canyon-channel transition zone. Canyon heads are typically formed by headward erosion, providing

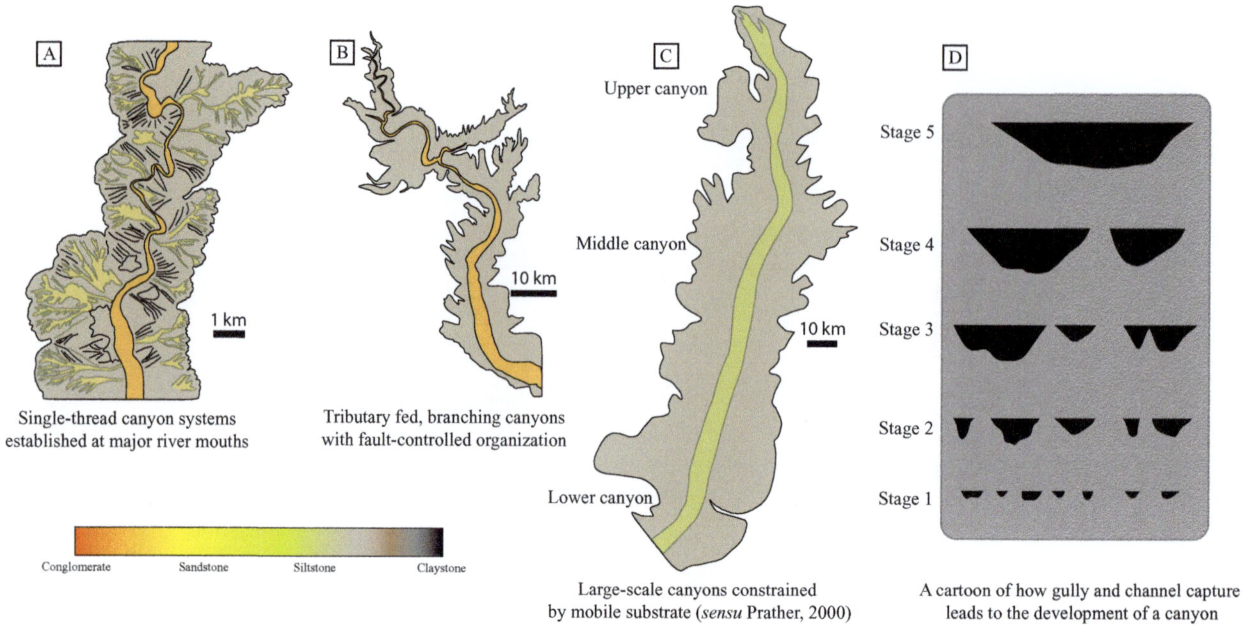


**FIG. 7** The geometry and fill of canyons is determined by several key factors including host rock, formation processes, sediment type and delivery mechanism ultimately controlled by tectonic setting, climate and sediment supply. Host rock can be bedrock, such as granite, or sedimentary deposits. Canyons can be formed by mass wasting processes, or excavation and erosion due to high-energy gravity flows. Sediment type can be highly variable and strongly depends on tectonic setting, upslope and terrestrial depositional environments and climate. Delivery mechanism ranges from mass wasting events to types of sediment gravity flows as well as river-, tide- and wave-related processes. Zones of mass wasting and failure shown in *brown*, with channels interpreted to have sediment gravity flows in *yellow* and *orange* to assist with depositional environment and lithology mapping and prediction. There are at least three principal canyon types described here with map-view interpretations provided of their architecture and extent: (A) single-thread canyon systems developed offshore and in connection with onshore fluvial-deltaic systems (example: Congo Canyon, offshore Democratic Republic of the Congo and Angola); (B) bedrock-confined canyon systems with branches and multiple tributaries, in part due to complex fault networks (example: Monterey Canyon, California, USA); (C) sedimentary deposit-confined canyons constrained and modified by mobile substrate (example: Mississippi Canyon, Gulf of Mexico, USA). (D) shows a cartoon depiction of how smaller gully and slope channel capture through time leads to the development of a canyon, similar to how Congo Canyon formed based on interpretation of 3D seismic data.

mass transport, slump, fall and flow deposits to the canyon margin and floor. Canyons are also excavated by powerful, erosional flows (Traer et al., 2015) that can originate from shelf systems such as deltas. Seismic study of the modern Congo Canyon for example, reveals this canyon system was preceded by the presence of numerous shelf-margin “rill-like” drainages which evolved to fewer but larger proto-canyon environments. These smaller canyons were ultimately replaced by the modern canyon system presumably through headward erosion and capture (example shown in Fig. 7D). The eventual migration of the head of the Congo Canyon shoreward of the coast allows this system to capture and funnel all transitory fluvial, coastal and shelf sediment directly to the deepwater environment.

Canyons can exceed 2.5 km (~ 1.5 mi) in depth and 10 km (~ 6.2 mi) in width, and lengths can reach hundreds of kilometers (Normark and Carlson, 2003; Covault et al., 2011; Fisher et al., 2021). Along dip, the longest canyons are concave up and drain long-lived and mature passive margins outboard of major deltas. Slightly concave canyon profiles indicate thermal cooling along passive margins, whereas, short, convex profiles drain actively deforming slopes and reflect seafloor uplift and deformation (Covault et al., 2011; Normark and Carlson, 2003). Typical planform profiles indicate that upper canyon margins are high-relief, jagged and scallop-shaped. Middle canyon environments show terraced margins that transition downslope into a smoother channel-levee environments (Fig. 17) with low relief (Babboneau et al., 2010). There are multiple classification schemes for canyons at different scales associated with feeder system, nature of shelf incision, continental margin and profile (concave to convex, Covault et al., 2011), shape of confinement (V- vs U-shaped, Babboneau et al., 2010) and climate summarized in Rotzien (2018); however, these existing classifications do not account for: (1) bed- to sub-bed-scale architectural elements and their process sedimentology, or (2) an upper depositional system-scale analysis of lithofacies, stacking patterns and reservoir distribution.

Landmark studies in mapping canyon environments and their stratigraphic evolution have led to a more complete understanding of their lithological variation from canyon to basin plain (Ghosh and Lowe, 1993; Lowe, 2004), their bathymetry and seafloor change through time (Paull et al., 2018), their role in sediment capture (Sweet and Blum, 2016), their ability to segment and partition sediment types along continental slopes (Clark and Steel, 2006; Prather et al., 2016), their typical timing of sedimentation events (Pirmez, 2011; Bain and Hubbard, 2016), their geometries, canyon-channel cross sections and longitudinal profiles (Normark and Carlson, 2003; Covault et al., 2011), their submarine geomorphology (Posamentier and Kolla, 2003; Jobe et al., 2011; McArthur and McCaffrey, 2019) and the sizes of submarine fans they create (Blum et al., 2017).

Submarine canyon-fill is highly variable and can feature deposits from slides, slumps, sediment gravity flows, falls and mass transport (Figs. 8–10 show a wide variability of canyon-fill from outcrop locales in California, USA). Farther downsystem and away from the canyon walls, deposits are generally less chaotic and finer grained. While the walls are highly unstable, the floor of the canyon where more flows course downsystem on a regular basis can feature more sediment gravity flow deposits (Greene and Surpless, 2017). Canyons may contain reservoir-quality sediment and can be ideal exploration targets because of their tendency to contain thick sediment fill, but the chaotic nature of these systems often may provide significant challenges to a drilling program (Allan et al., 2006; Gardiner, 2006; Stow and Mayall, 2000; Mayall et al., 2006). These diverse deposits in and around canyons may generate multiple play types: filled head scarps and or other nearby slide scars; gravity-driven deformation on the slope filled with out-of-canyon sediment; ponded lows atop mass transport deposits; base-of-slope deposits that are just upslope of mass transport deposits; re-worked downslope deposits (Kneller et al., 2016 and others).

Processes operating in canyon environments are diverse and can be derived from wave, tide, fluvial (hyper- to homopycnal), sediment gravity flow, as well as other forms of mass transport or mass wasting. High-energy processes dominate (Pohl, 2019) and typical deposits include the broad range of sediment delivered to the coastal and shallow marine environment and canyon head. Evidence of bypass including coarse-grained lag, as well as mudstone bypass surfaces, can be common. Canyons can feature abrupt changes in gradient along their length that are often classified as knickpoints (Sylvester and Covault, 2016), hydraulic jumps (Komar, 1971) and cyclic steps (Englert et al., 2020) over faults, escarpments, MTD surfaces, salt topography and locations at the base of slope, or over the shelf-slope break.

**FIG. 8** Outcrop expression of canyon deposits, the Eocene Carmelo Formation, California, USA. These rocks represent the uppermost fill of a canyon and consist largely of structured high-density turbidity current deposits: (A) canyon margin interpreted as a depositional contact, with various types of conglomerate adjacent to the Salinian Granite. Interpretation modified from Lowe (2004). (B) Second and third fining-upward cycles (sensu Cronin and Kidd, 1998) of the Carmelo Formation, south of Punta de los Lobos Marinos. View is to the south. Stratigraphic up is to the left in the photo. Note the conglomerate forms erosion-resistant ridges into the Pacific Ocean, whereas the mudstone intervals form recessive zones along the walkway and pocket beaches. One of the main uncertainties is whether or not the mudstone intervals are continuous across the canyon. (C) Lithofacies of the canyon-fill include thick-bedded, lenticular conglomerate with (D) evidence of basal traction sedimentation (dune-scale, cross-stratified conglomerate, typically Lowe $R_{12}$ intervals). Thick-bedded sandstone is generally structured and interpreted as Lowe $S_{12}$ intervals. *(Modified from South Dakota Mines Field Guide for Depositional Systems of Western California.)*

**Southern view of the Venado Sandstone at Monticello Dam**, Napa County, California. Stratigraphic up is to the left (east). Base of the canyon-fill is far right of picture, and top of canyon-fill transitions to the Antelope Shale to the left of the parking lot.

**Northern view of the Venado Sandstone at Monticello Dam**. Stratigraphic up is to the right (east). Typically medium-grained, thick-bedded sandstone, strongly resistant to erosion, abruptly transitions to mudstone of the Antelope Shale.

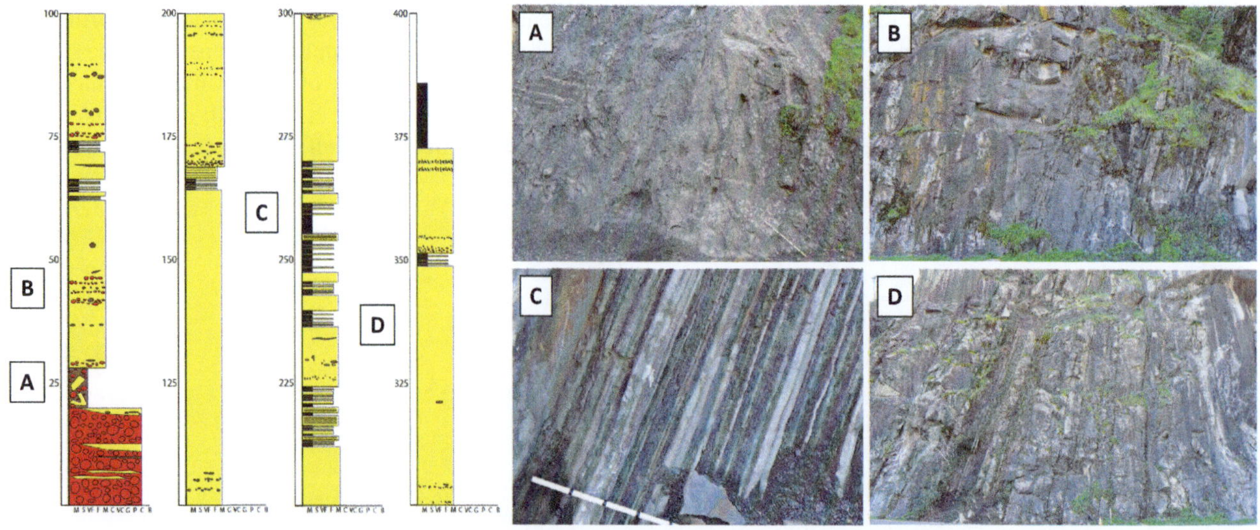


**Left: Canyon-fill stratigraphy of the Venado Sandstone, Monticello Dam section, Napa, California**. Section by Rotzien (2016). **Right:** A) 10-m + thick debrite at base of section, B) thick-bedded conglomeratic sandstone units likely representing a braid plain showing multiple intervals of traction sedimentation, C) thin- to medium-bedded turbidites and transitional flow deposits near a possible maximum flooding surface (MFS) and D) thick-bedded typically amalgamated sandstone turbidites with thin Bouma $T_{bcde}$ caps.

**FIG. 9** Outcrop expression of canyon-fill, the Late Cretaceous Venado Sandstone (Cortina Petrofacies), California, USA, representing the complete fill of a canyon consisting of mass transport deposits, and the coarse-grained deposits interpreted as both braided and leveed meandering channel deposits. Sandstone-rich intervals capped by the Antelope Shale. See the local extent of the outcrop in the top part of the figure. At the base of the figure, note the paired stratigraphic column with outcrop images, A, B, C and D. *(Modified from South Dakota Mines Field Guide for Depositional Systems of Western California.)*

Superb exposures of submarine canyon-fill—broadly organized from upper canyon (canyon head) to canyon-channel transition zone and each with its own distinct profile of reservoir quality, reservoir connectivity and sealing potential—crop out in central and western California, USA, and include the Carmelo Formation shown in Fig. 8 (Cronin and Kidd, 1998; Lowe, 2004), Panoche Formation (Greene and Surpless, 2017), Venado Sandstone shown in Fig. 9 (Cortina Petrofacies) (Ghosh and Lowe, 1993) and Merle Formation in Fig. 10 (Anderson et al., 1999, 2006). These Californian outcrops not only feature several hundreds of meters, if not kilometers, of the broad range of deposits including high- and low-density turbidity current, debris flow and mass transport deposits, sometimes remobilized, but also the evolution of canyon inception, growth and the multiple episodes of fill (Fig. 9) that can be used evaluate reservoir and seal potential as analogs for petroleum fields.

Using these outcrops, a generalized canyon-fill model can be divided into five evolutionary stages (Rotzien, 2018): (i) initial incision; (ii) bypass; (iii) high-energy deposition; (iv) lower energy deposition progressing to; (v) final

**Southwestern edge of the Merle Formation**, Monterey County, California. Note thick-bedded sandstone forms ~75-m-high cliffs and is surrounded by plains interpreted to be largely mudstone. Granite of the Gabilan Range is in the background, interpreted as the source of the feldspathic sandstone.

**Northwestern edge of the Merle Formation**, Monterey County, California. Note tabular nature of sandstone bedding. While bedding packages are laterally continuous, individual beds are not due to scours, lateral truncation and cut-and-fill architecture (Anderson ***et al.*** 1999; 2006).

**Lithofacies of the "Merle" Formation, Santa Lucia Basin, western California**: A) thick-bedded conglomeratic sandstone, with maximum grain size~1.5-m diameter; B) thick- to very thick-bedded sandstone. Note that no mudstone interval along any bed contact is continuous as a result of scouring flows; C) Bouma $T_b$ interval at the top of a bed. Sedimentary structures are rare in these intervals, by far the most common feature is structureless bedding (Bouma $T_a$/ Lowe $S_3$); D) overbank deposits consisting of thin- to very thin-bedded Bouma $T_{(bc)de}$ turbidites.

**FIG. 10** Outcrop expression of canyon- and channel-fill, Upper Cretaceous to Paleogene "Merle" Formation, Monterey County, California, USA, representing the canyon to channel transition zone consisting largely of thick- to very thick-bedded structureless sandstone (nearly 99% net:gross). *(Modified from South Dakota Mines Field Guide for* Depositional Systems of Western California.*)*

abandonment. Stacking patterns and spatial arrangement of lithofacies, large-scale bedforms, erosional surfaces and abundant disturbed strata indicate canyons host the highest energy sediment gravity flows within confinement in a deepwater depositional system.

In summary, canyons collect a variety of sediment depending on the nature of the coastal and shallow marine depositional environments that are in close proximity to the canyon. Canyon heads that are directly linked with a fluvial system (example: Congo Canyon), or those that incise narrow shelves along active margins tend to receive coarser sediment (example: Monterey Canyon), whereas canyon heads that are far from the coast along passive margins tend to collect fine-grained sediment (example: Amazon Canyon, Mississippi Canyon). Observations of ancient and modern canyons afford the opportunity to summarize major trends from upslope to downslope in these environments: (i) grain size fining as well as a decrease in range and variability in grain size; (ii) decrease in abundance of mass transport deposits and slumps, and increase in amount of sediment gravity flow deposits; (iii) decrease in abundance of frictional debris flows and increase in amount of cohesive debris flows; (iv) overall transition from R to S to T divisions (with decreasing evidence for bypass), then increase in M and H divisions, and finally T0–8 divisions (see Chapter 7); (v) a decrease in depth of scours and lenticular bedding (reduced amount of erosional surfaces), and increase in compensation stacking patterns; (vi) lower variability in bed thickness downsystem; (vii) an increase in flat bed contacts and more abundant and continuous mudstone intervals.

## Channels

Submarine channels are relatively deep and narrow conduits facilitating sediment transport through confinement. Confinement in submarine channels can be generated due to three main processes: (i) the construction of aggradational levees; (ii) erosion and incision from sediment transport and removal processes (Cronin et al., 2000; Wynn et al., 2002, 2007); (iii) mobile substrate or tectonic confinement (Anderson, 1998). Channels are often grouped by their planform architecture: tributary (confluence) vs distributary (Beaubouef et al., 1999); confined (Pyles et al., 2012; Pemberton et al., 2016) vs weakly confined (Moody et al., 2012) vs unconfined; leveed vs non-leveed (Mutti and Normark, 1987); straight vs meandering and sinuous (Wynn et al., 2007; Sylvester et al., 2013) vs braided (Belderson et al., 1984; Hesse, 1992; Wuellner and James, 1989; Rotzien et al., 2014b; Foreman et al., 2015). Channels are formed and sculpted by the range of deepwater sedimentary processes (Clark and Pickering, 1996), including turbidity currents, mass flows, MTD (Sylvester et al., 2012) and contour currents (De Weger et al., 2020) (Fig. 11).

Channels are typically formed and associated with preferentially focused, confined high-energy erosional and depositional events (Fildani and Normark, 2004; Di Celma et al., 2011; Hubbard et al., 2020). Channels often feature high-energy, coarser-grained flows capable of significant erosion (Crane and Lowe, 2007; Duerichen and Lowe, 2007), commonly identified by scours, mudstone rip-up clasts and lag deposits (Figs. 12 and 13). The sedimentary record of channels is more easily studied in outcrop compared to canyons due to their smaller scale (hundreds of meters to a few kilometers wide to tens of meters deep compared to multiple kilometers). Channels are generally larger than gullies (Surlyk, 1987; Shumaker et al., 2017) based on their more prominent and active role of sediment delivery to deepwater.

Successful modeling of the scale of channels requires an understanding of the diversity of outcrop and subsurface analogs (Nilsen et al., 2007). Establishment of the channel hierarchy (Sprague et al., 2005; *researchers now commonly employ hierarchy when describing deepwater systems, such as* Straub and Pyles, 2012 *or* Bell et al., 2020) permitted an analysis of the global range and variability in channel widths and thicknesses using outcrop and subsurface data sets by McHargue et al. (2011), revealing that most submarine channels are about 100–500 m (~ 328–1640 ft.) wide and about 7–12 m (~ 23–39 ft.) deep. Channel systems can range from individual storeys tens of meters in width and a few meters in thickness, or can form larger and more extensive channel belts (Campion et al., 2000; De Ruig and Hubbard, 2006) tens of meters in thickness and several kilometers in width. This hierarchical approach establishes a common interpretive descriptive terminology that links multiple scales of depositional patterns seen in core and outcrop to inferred depositional processes. However, simple application of this hierarchy is insufficient for modeling purposes, discussed later in this chapter.

Channels tend to migrate laterally early in their evolution, and then aggrade vertically (Deptuck et al., 2003, 2007) and are moderated by compensation (Straub et al., 2009). When levees are weakly developed, typically in the nascent stages of channel development, channels tend to migrate laterally. This is common when the overall depositional system is progradational and transitioning from a frontal splay complex, which is naturally sand-rich and mud-poor, to a channelized environment. Once a deep erosional conduit is established and levees are developed, vertical aggradation is common.

Large datasets (Pickering et al., 1995; Nilsen et al., 2007 and others) reveal trends in channel architecture from slope to basin floor (Stow and Mayall, 2000; Mayall and Stewart, 2000, 2001). Channel architecture is described from the canyon to channel transition zone, along the slope, to the base of slope, to the proximal part of the basin floor and finally the channel-lobe (-splay) transition zone by several studies (Porter et al., 2006; Straub and Pyles, 2012; Hubbard et al., 2020 and references therein). Overall, channels are narrow and deeply incised on the upper slope, and increase in width and decrease in depth toward the basin floor and onto the basin floor (Beaubouef et al., 1999). Planform geometry is a useful indicator of slope (gradient), internal channel architecture, sediment transport process, grain size and seafloor topography

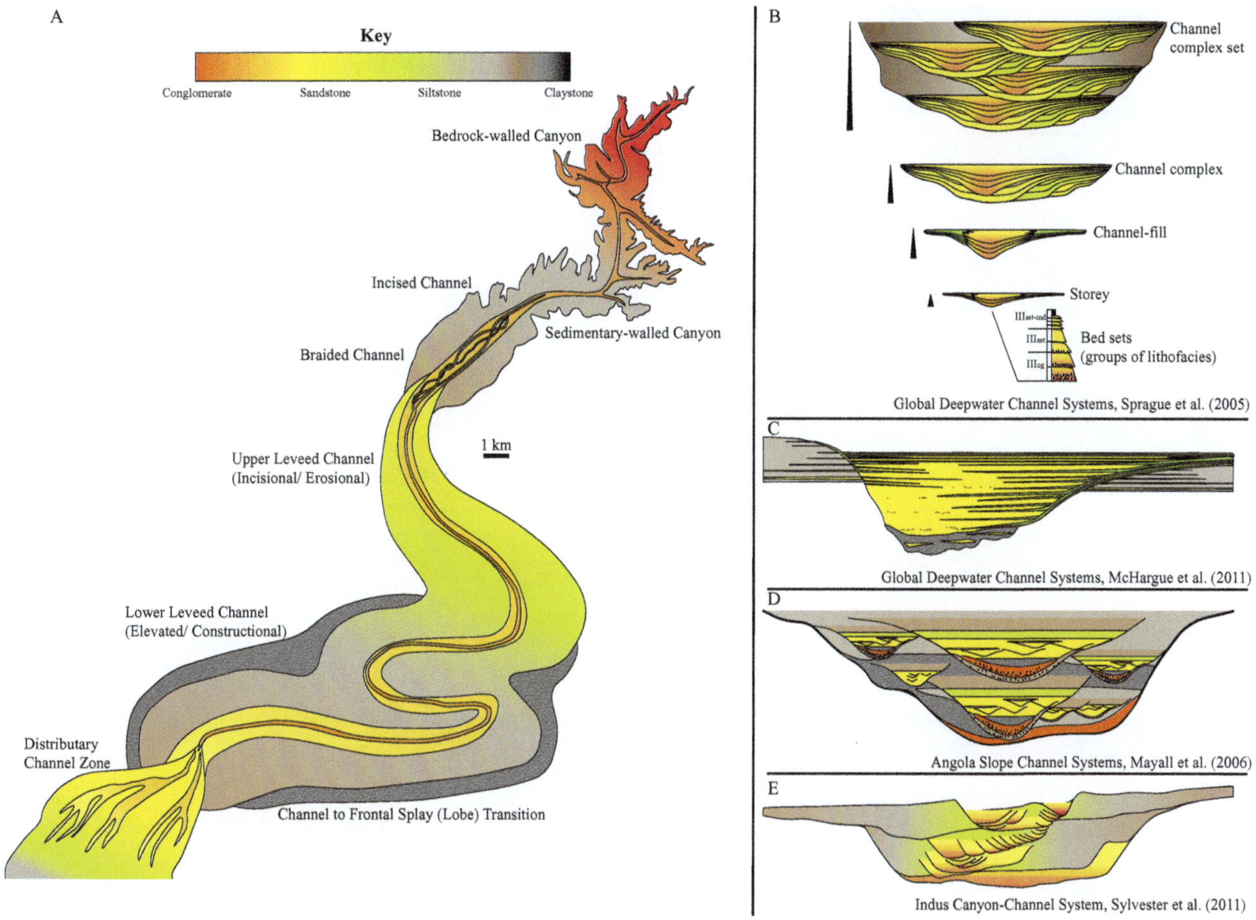


**FIG. 11** Generalized model for channelized environments of deposition (A). Typical environments include (i) incised channels, (ii) braided channels, (iii) upper leveed channel with narrower, sandier levee systems, (iv) lower leveed channel with wider, muddier levee systems, (v) distributary channels as part of the channel to frontal splay (lobe) transition zone and proximal portion of the splay complex. Most deepwater channel systems can be differentiated by their geomorphology and architecture, with common themes reflected in typical channel-fill architectural models by (B) Sprague et al. (2005) by Exxon-Mobil; (C) McHargue et al. (2011) by Chevron; (D) Mayall et al. (2006) by BP; (E) Sylvester et al. (2011) by Shell.

(Hesse and Rakofsky, 1992; Pyles et al., 2019; Pizzi et al., 2020; Mitchell et al., 2020, 2021), along with ultimate provenance, tectonic setting and other drivers. Sinuous channel systems tend to form on lower gradient, distribute reservoir about lateral accretion packages (LAP) (Abreu et al., 2003) that can appear to form "Grand Canyon-scale" incisional features, yet are formed by energy efficient smaller channels over time (Sylvester et al., 2011). These channel systems feature turbidity currents as a primary transport mechanism and are typically finer grained (Wynn et al., 2007). Straight or braided channel systems tend to form on steeper gradients, distribute reservoir-quality sediment about bars and terraces, feature a wide variety of sediment gravity flows and are typically coarser grained (Belderson et al., 1984; Foreman et al., 2015), similar to the observations of Leopold and Wolman (1957) on terrestrial channel systems. Additionally, turbidite-dominated channels show higher levels of predictable architectural complexity than debrite-dominated channels. Turbidity current-dominated channels have a variety of architectural elements, including LAP (Fig. 14) and a variety of bedforms (Lowe and Graham, 2012). This is due to the stratification of grain sizes in turbidity currents. The resulting architectural differences permit correct identification of turbidite-dominated channels from those transporting debris flows (Fig. 11B and D vs 11C).

Channelized reservoirs represent attractive petroleum exploration targets and have been the focus of many research studies. While some of the models represent straighter channels and feature symmetric channel-fill, asymmetric channel-fill often found in sinuous channels can show key differences in inner bend vs outer bend stratigraphy, similar to fluvial channels (Walker and Cant, 1984). Nearly all channels, regardless of scale, reveal a basic three-stage process from initiation to fill: *initiation* via scour; *inception of sedimentation and fill* by drape and finally *fill in earnest* by onlap (Fig. 12). Channelized reservoir connectivity is determined based on many factors at a variety of scales including rock type,

**FIG. 12** Top: Outcrop expression of channel and levee deposits of the Upper Cretaceous to Paleogene Pigeon Point Formation, California, USA. The left image shows the lower part of the formation at Bean Hollow State Beach, about 45 minutes west of Stanford. The recessively weathered units in the left of the photo are interpreted as levees, whereas the erosion-resistant sandstone and conglomerate to the right and up stratigraphic section are interpreted as channel deposits. The right photo captures the upper part of the formation at Pigeon Point Lighthouse. Here, the formation is dominated by matrix- and clast-supported pebble- to cobble-grade conglomerate interpreted as debrites. Bottom: Bed-scale photos of the formation at Bean Hollow State Beach. (A) Medium- to thin-bedded nearly complete turbidites with abundant soft-sediment deformation textures including disturbed mudstone representing flowage in the levee; (B) evidence of erosion and redeposition, as well as flame structures, indicating a near complete range of sedimentation processes in the levee; (C) load structures and ball-and-pillow indicating the soupy nature of the levee and lack of time for lithification in this active environment; (D) traction carpets in sandstone that filled the channel; (E) dish structures in thick-bedded channelized sandstone; (F) nearly complete R-S-T turbidites in the channel-fill. Overall, these deposits are interpreted to represent a channel-levee environment within a canyon (Lowe, 2004). *(Modified from Basin Dynamics Field Guide for Deepwater Depositional Systems of Western California.)*

amount and abundance of erosional surfaces, stratigraphic stacking, sinuosity, migration and aggradation patterns, and presence of levees and their construction and connectivity to the channel-fill. Reservoir heterogeneity can increase with the presence of lags, slumps (Fig. 11C) and other deposits not derived from more orderly turbidity currents (Fig. 11B, D). Modeling channels can be challenging at reservoir scale, so disparate outcrop examples can be used to refine an understanding of their geologic uncertainty (Nilsen et al., 2007). Industrial applications require an optimized understanding of geologic risk. Later in this chapter, modeling approaches address some of these challenges.

**FIG. 13** Outcrop expression of channel deposits of the Upper Miocene Upper Mount Messenger Formation, Pukearuhe Beach, Taranaki Basin, North Island, New Zealand: (A) channel deposits at the Pukearuhe Conglomerate locale, with multiple occurrences of truncation, onlap and downlap of a variety of rock types; (B) sedimentary and volcaniclastic deposits along an ancient channel margin at Waikaramarama, at the southern extent of Pukearuhe Beach. *(Modified from Basin Dynamics Field Guide for Deepwater Depositional Systems of the Taranaki Basin.)*

Channels are formed to carry and transmit flows, and not to accumulate deposits. The fill, preserving channelized environments in the rock record, can be genetically unrelated. A channel's fill (channel-fill) and architecture strongly relates to gradient and position on the slope or basin floor (Gardner et al., 2003). Models of channels indicate their shape, incisional patterns, fill and geometry, yet most models consist only of a cross section that may or may not have well control. Submarine channel variation is likely best displayed on a spectrum or continuum (Rotzien, 2019). Portions of this continuum from channel axis to margin and from shelf edge to basin floor are provided in Fig. 11.

Cross sectional variation in channels is described by identifying these parts of a channel's anatomy: axis (thalweg), off-axis, margin, and potentially inner levee, levee crest and outer levee to overbank (McHargue et al., 2011 and many others) shown in outcrop in Fig. 11. While there are many models, key models shown in Fig. 11 from Exxon and ExxonMobil (Beaubouef et al., 1999; Sprague et al., 2005), BP (Mayall et al., 2006), Shell (Deptuck et al., 2007; Sylvester et al., 2011), Chevron (McHargue et al., 2011) and others demonstrate the range and variability in channel-fill. While some channels are filled by high-energy turbidity current deposits (Campion et al., 2000), others contain significant intervals of mudstone and debrites (Porter et al., 2006; Mayall et al., 2006) and still others by mixed process flow deposits (Fonnesu et al., 2020). Most of the models show the difference between incised channels on the slope and constructional leveed channels near the base of slope and basin floor (Mayall and Stewart, 2000, 2001). While these are some of the most robust models for channel-fill, there is no single model that can explain channelized environments from upslope to downslope and in different depositional systems. Fig. 11 stresses the importance of understanding: (i) the hierarchy in channel-fill; (ii) the differences in channel architecture between sandy and muddy depositional systems; (iii) changes in channel

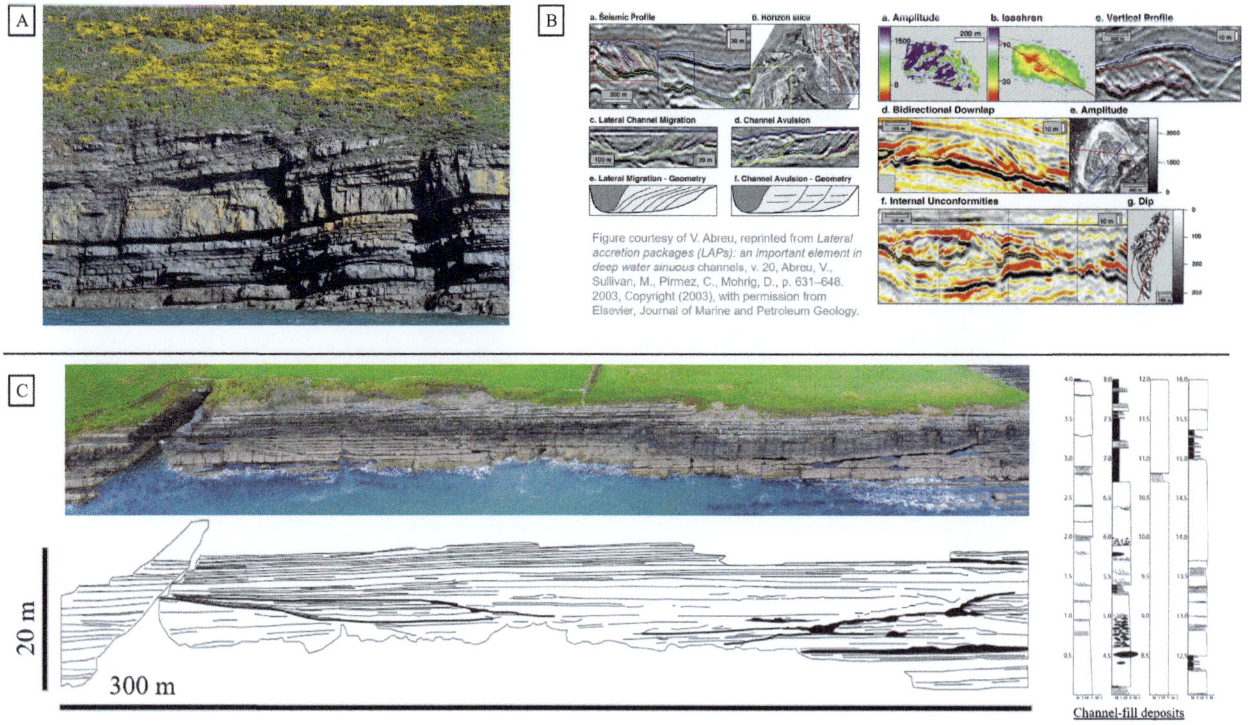


**FIG. 14** Outcrop expression of lateral accretion packages (LAP) at (A) Kilcloher Cliffs, Ireland, and (B) equivalent scale seismic data of LAP, offshore Africa (Abreu et al., 2003). (C) Channelized deposits and measured section (in meters) through the channel axis, Carboniferous Ross Formation, Rinevilla locale, County Clare, Ireland.

architecture and anatomy from upslope to downslope, and from overbank to axis; (iv) the differences between constructional and incisional channel environments; (v) the patterns (or lack thereof) in channel initiation, filling and abandonment.

Channel models provide key implications for reservoir presence and quality, distribution of shale drape and the heterogeneity challenging connectivity and net:gross from axis to margin. Generalized grain size profiles of channels shows coarser grained sediment in the axis, and finer grained sediment upwards and toward the margins if the flows that filled the channel felt confinement by the channel margin. If not, and flows are overfilling the channel, grain-size fining will be subtle. If flows are underfit, when flows do not feel confinement provided by the channel margins, bedding can sharply taper toward the channel margin and drape a margin with mudstone (Lowe et al., 2019), potentially forming channelized splays. In these cases, the full range of architectural elements described in Sprague et al. (2005) can be found within a channelized system. Or, if flows are overfit, sandstone will fill the channel from axis to margin. In these sandy systems, it is more common to have channel-levee connections, or a channel-fill that is connected to the levee-overbank (McHargue et al., 2011, and others).

## Levees

Levees are positive topographic features that form natural lateral boundaries to channel environments (Fig. 15). They are usually clay- and silt-rich, formed preferentially along turbidite channels due to overbanking flows. The deposition of cohesive muddy deposits at the expense of sandy deposits yields levees that are resistant to erosion. However, unconsolidated muddy deposits also make levees prone to slumping and flowage (Lowe, 2004). Levee deposits typically consist of interbedded sandstone and mudstone (sometimes described vaguely as "heterolithic" deposits), but the genesis of the deposits is highly variable (Cronin et al., 2000; King and Browne, 2004; Kane and Hodgson, 2011; Morris et al., 2014) and can be formed by sediment provided from the channel or from the overbank. Levees are generally a lower energy environment and the deposits tend to be thin-bedded, muddy, with a higher abundance of bioturbation. However, high-energy deposits can form levees, too, and there are numerous examples from the Pigeon Point Formation, USA (Lowe, 2004), Rosario Group, Mexico (Kane and Hodgson, 2011) and Juniper Ridge Conglomerate, USA (Hickson and Lowe, 2002). Overbanking flows can erode into levees, causing mudclast formation, plow-and-fill structures or bedding delamination. Levees can form petroleum reservoirs in important fields, such as Ram Powell (Browne and Slatt, 2002; Slatt, 2019).

Levees are interpreted to be built by three main mechanisms: (i) flow stripping, in which the less dense top of the sediment gravity flow separates from lower part and overtops channel; (ii) overflow of the entire flow over the channel margin into levee; (iii) partial overflow of the low-density fraction of the flow (Kane and Hodgson, 2011). However, levees can be constructed of overbank splays, including crevasse splays, sediment waves, contourites, mass transport deposits and a variety of slumps and hemipelagic to pelagic deposits (Morris et al., 2014) shown in cross section in Fig. 15. While many of the higher energy depositional environments also feature these processes and deposits, they often are more pronounced in the levee environment, which is lower energy and is not as susceptible to erosional flows compared to canyon, channel and some splay environments. Therefore, the subtle sedimentological signatures of the various processes can be better preserved than in high-energy axial environments.

One of the type section outcrop analogs for levees is the Paardekraal River Section, Permian Skoorsteenberg Formation, South Africa (Morris et al., 2014). Here, ~15-m-thick (~ 49 ft.) packages of levee deposits consist of fine-grained and pervasively ripple cross-laminated turbidites. The levees are adjacent to contemporaneous channel deposits. These, what are interpreted to be "classic levees" formed by the overbanking of channelized flows, are interbedded with crevasse splays and diverse other deposits. Paleocurrents in the levees diverge from the channel-based paleocurrents, similar to the Rosario Group, Mexico (Kane and Hodgson, 2011), and internal levees in the Mount Messenger Formation, New Zealand, shown in Fig. 16 (Rotzien et al., 2014a).

Levees are thickest near their crest where they interleave, merge and transition with channel margin deposits. Levees taper, or decrease in thickness to their terminus into the overbank realm, according to studies that quantified thickness decay rates (Pirmez et al., 1997; Straub and Mohrig, 2008). Levees can reach tens of kilometers in width along well-developed channel systems, such as the Amazon Fan, with overall height above the channel thalweg of hundreds of meters (Flood et al., 1991). Smaller levees can be less than 500m (~ 1640 ft.) in width and several meters in height above thalweg (Janocko et al., 2013a). Levees show both concave and convex geometry, often in the form of a wedge in cross section. In outcrop- or seismic-scale data, the wedge-shaped cross-section profile shows downlap onto out-of-channel strata, and can form high-amplitude reflection packages (HARP). Levees are generally larger and better developed along outer bends

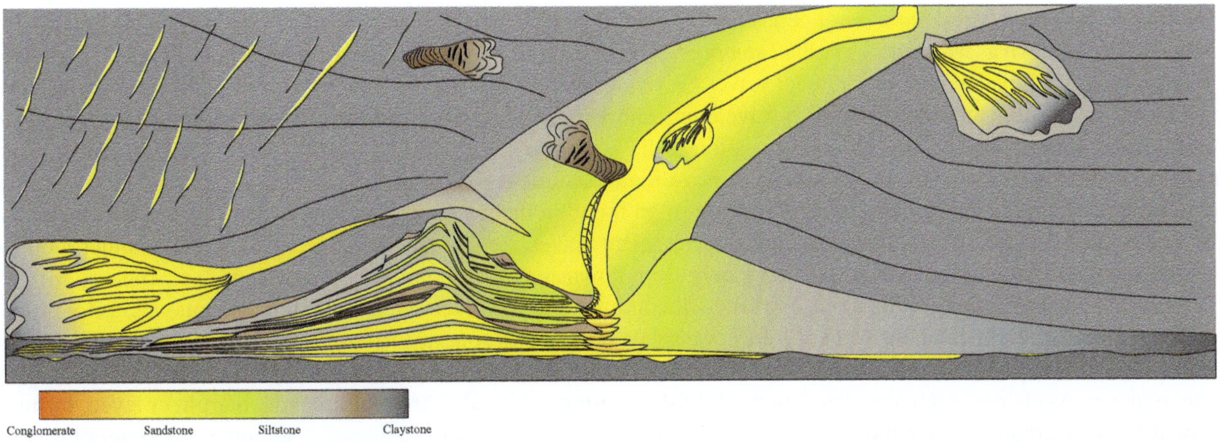


**FIG. 15** Deepwater levees are broadly divided into inner (internal) and outer (external) environments. Inner levees are constructed of several types of deposits that grade, merge and overlap with the channel margin, including those that form lobes and individual deposits of sediment gravity flows, slides, slumps and "flowage" (sensu Lowe, 2004) deposits. Outer levees are constructed of overbank flows generally formed by up to three distinct processes. Overbank and crevasse lobes, sediment waves and individual deposits of sediment gravity flows, slides, slumps and "flowage" deposits are considered normal levee deposits (Morris et al., 2014). Note the following: (i) how levees feature exponential decay in thickness away from the channel axis; (ii) average grain size decreases away from channel axis; (iii) the outer bend levee is better developed than the inner bend levee; (iv) levee deposits are sand-rich at channel inception, and sand-poor as the height-above-channel-thalweg (HACT) increases; (v) slumps and soft-sediment deformation are common in areas of higher gradient along the inner and outer levee, forming syndepositional faults. Various outcrop and subsurface data sets are used to constrain this model (Morris et al., 2014; Janocko et al., 2013a; (Kane and Hodgson, 2011)). However, many models fail to show the nature of instability in levees, which can be in internal and external zones. Levees play key roles in the development of avulsions and channel reestablishment. It is difficult to show the scale of sandy vs muddy bedding packages in the levees in this cartoon. Note that the thickness of these packages will vary along the length of the channel, making these packages lenticular in 3D, as well as shown in the cross section (sensu Waipingau locale, Taranaki Basin, New Zealand, in Browne et al., 2007).

**FIG. 16** Internal levee deposits of the Upper Miocene Upper Mount Messenger Formation, Taranaki Basin, North Island, New Zealand: (A) view north along the southern portion of Pukearuhe Beach; note how the inner levee deposits thin away from the channel margin to the south, toward the top of the outcrop. (B) and (C) show bed-scale images of medium- to thin-bedded structured turbidites in the levees.

in sinuous channel systems than inner bends (Khan and Arnott, 2011). Further levee asymmetry can be developed if contour currents are active (Miramontes et al., 2020).

Levees are divided into inner and outer domains, separated by a levee crest, or the highest topographic expression of the levee. Inner and outer levees can be formed for a single channel, or inner and outer levee complexes can be formed for channel complexes. Inner levee environments often mingle with several associated morphological elements including bars, terraces and lateral accretion packages. In many examples, it is difficult to tell where the levee begins and the other morphological element, such as a channel, ends (Fig. 15), and the stratigraphy can be difficult to tell apart (Fig. 17).

Levees can be a powerful recorder of depositional signatures, such as lateral accretion on the inner bend and erosion on the outer bend, known as *punctuated channel stepping*, which can lead to rhythmically bedded deposits. Levees are usually a prime site for the development of sediment waves (Wynn and Stow, 2002), or ridges that run perpendicular to the current that creates them, often up to a kilometer in wavelength and 10m ($\sim$ 33 ft.) in height from trough to crest that migrate and build upflow. Rarely observed in outcrop (Campion et al., 2011), new advances in seismic data resolution (Bhattacharya et al., 2020 and references therein) have made observation of sediment waves almost a normality in most levee-overbank environments (see Chapter 16). Sediment waves are often found in the levee-overbank area due to the supercritical flows that overbank from channelized turbidity currents. Sediment waves are deposited and sculpted in this manner alongside channelized environments and can be formed purely from downslope flows, not only along slope (i.e., contour currents).

When a levee breaks, a sediment gravity flow will continue into the levee-overbank environment, where it will likely expand and deposit sediment in the form of a crevasse splay. If the flows continue to emanate from the channel at that point, the downsystem portion of the channel can be cutoff, and a new route established (Ortiz-Karpf et al., 2015). This is known as an avulsion. Avulsions originate when a powerful flow overtops or breaks through a channel margin or levee, forming a seminal levee-overbank splay deposit. Avulsions are highly variable and can be forward-stepping, back-stepping, radial and single node (Kolla, 2007). Avulsions can be triggered by a change in discharge and bedload, increased channel sinuosity, limited space for deposition downsystem, the backfilling of channels by intrachannel splays (Prélat et al., 2010) or allogenic drivers (Kolla, 2007).

## Splays (lobes)

Splays, also known as lobes, are the default environment in high-energy deepwater settings due to the typically unconfined nature of deepwater basin physiography. The lack of strong density contrast between sediment gravity flows and ambient water column also allows flows to spread over wide areas, depositing sediment in lobate shapes that essentially conform to the shape of the container. A splay or lobe (the terms are typically interchangeable) is a body of sediment deposited in an area of unconfinement, originating from an area of relative confinement (Fig. 18) (sensu Normark, 1970, 1978; Lowe, 2004;

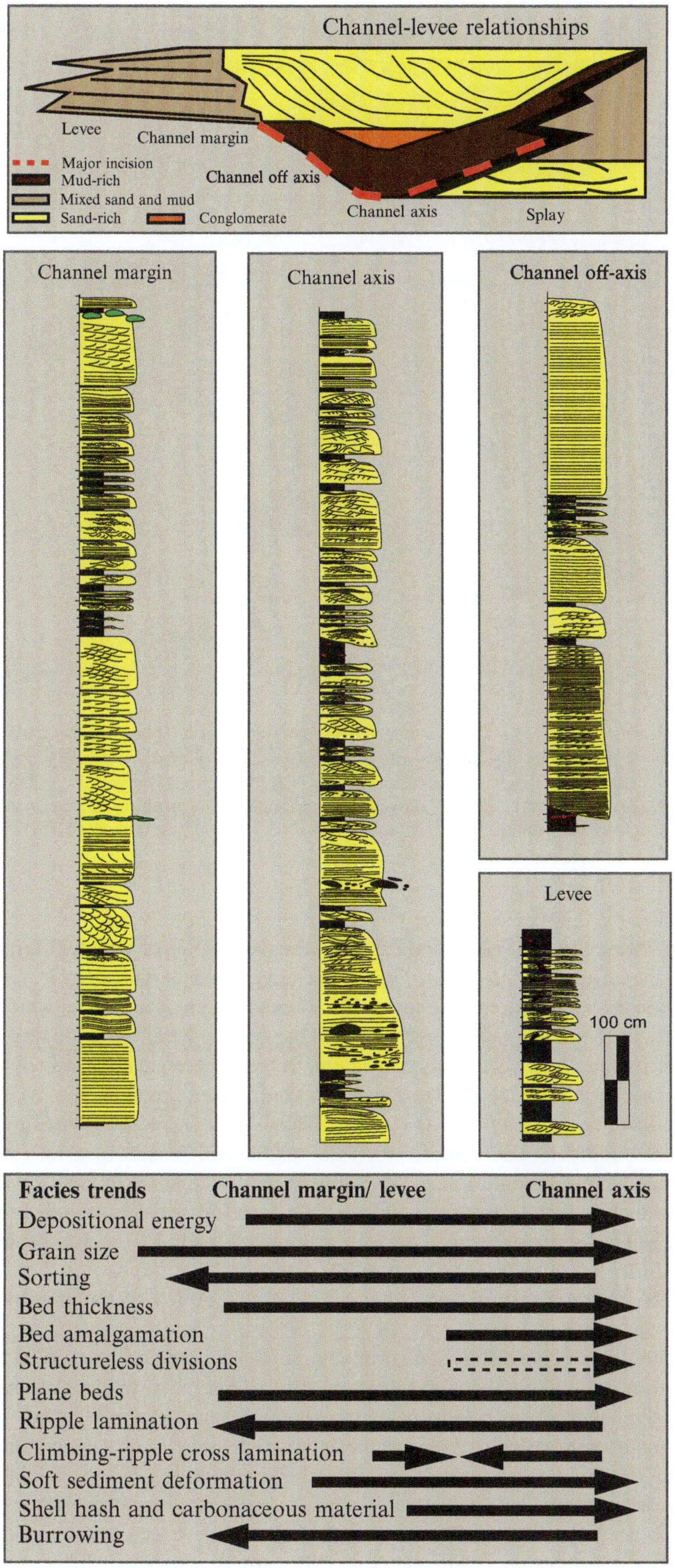


**FIG. 17** Comparison of channel and levee deposits in the Upper Miocene Mount Messenger Formation, Taranaki Basin, North Island, New Zealand. Bottom part of figure shows qualitative relationships of several sedimentological criteria used to differentiate deepwater deposits and environments of deposition.

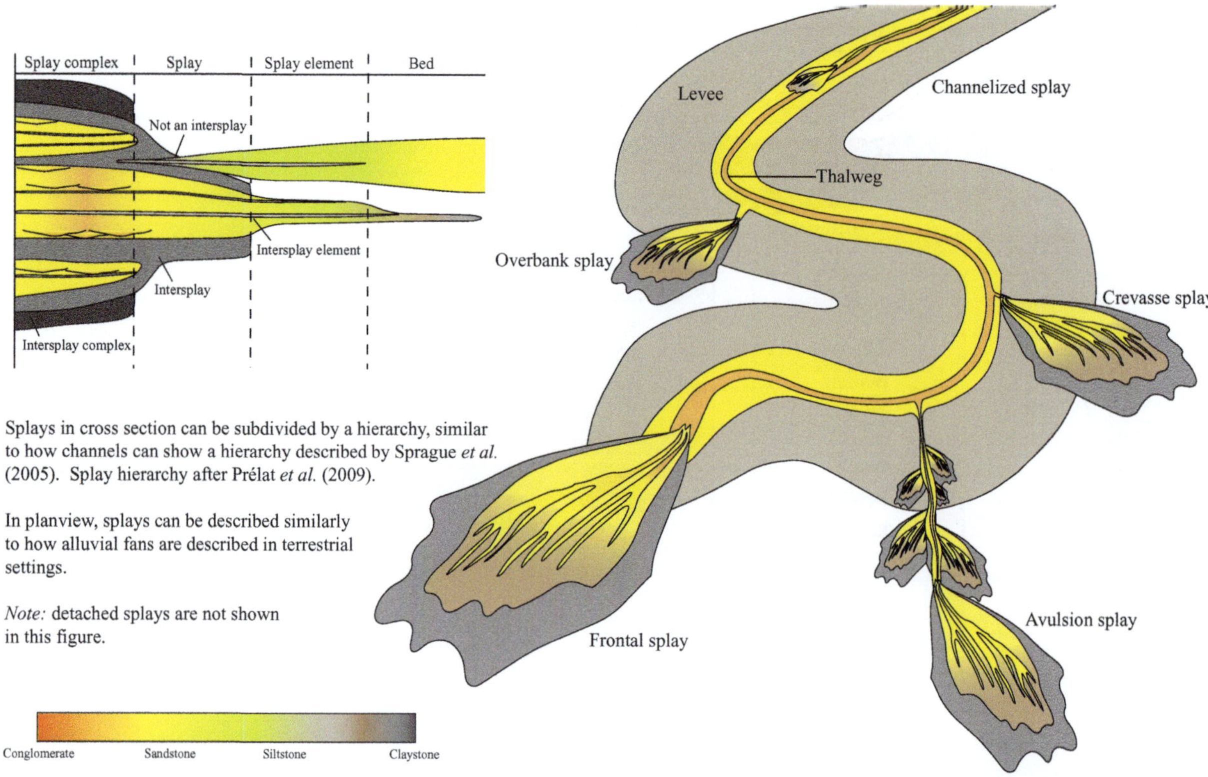


**FIG. 18** Splays (lobes) are terminal deposits located in areas of unconfinement, generally originating from areas of confinement. The only splays that are not terminal deposits are those that (i) are reworked and carried downsystem, such as the deposits originally in a channelized splay, or (ii) transient splays (see *transient fan* by Adeogba et al., 2005) that can be incorporated in subsequent flows and carried downsystem. The five principal types of splays shown here include channelized (Lowe et al., 2019), overbank (Puigdefàbregas et al., 2004), crevasse (Posamentier and Kolla, 2003), avulsion (Rotzien et al., 2014a; Ortiz-Karpf et al., 2015) and frontal types (Lowe, 2004; Hodgson, 2019). Splays in cross section can be subdivided by a hierarchy, similar to how channels can show a hierarchy described by Sprague et al. (2005).

Lowe et al., 2019; Hodgson, 2019; Cornard and Pickering, 2020). Please note that in some texts, *lobes* are referred to the terminal body of sediment located basinward of the channel mouth (i.e., frontal splay), whereas, bodies of sediment originating from the channel and deposited laterally to the channel are termed *splays*. These environments typically form in low-gradient areas and can readily show compensation stacking patterns due to the (i) low-gradient environment in which splays accumulate as well as (ii) the low degree of confinement. A splay environment will be present, therefore, if the flows do not feel confinement to a great degree. These environments are *distributary* in nature, as opposed to channelized environments which are *confined* (see Chapter 1 for additional discussion on distributary vs confined environments).

Splay deposits are constructed from repeated flows entering an area that suffer from a sudden reduction in flow steadiness and uniformity (a decrease in Reynolds' number—see Chapter 6). Flow expansion results in a loss of competency and capacity, and sedimentation occurs. The sedimentation can be rapid and thick in the proximal areas of the splay, and gradual and thin in the medial to distal areas of the splay. That is the case for turbidite-dominated splays, yet splays can be comprised of debrites, MTD (Sharman et al., 2015; Masalimova et al., 2016) and other diverse deposits. Splays have a wide range of sedimentation processes, stratigraphic architecture, depositional patterns and carry different challenges pertaining to their reservoir traits and connectivity compared to other submarine fan environments.

Compensation stacking of splay elements may be obvious in seismic data, but very subtle in core and outcrop data (Mutti and Sonnino, 1981). Compensation varies from longitudinal (up and downsystem) to lateral (side-to-side) and can be predictable (Straub et al., 2009). Because of the laterally continuous nature of splay deposits compared to canyon and channel deposits, at reservoir and outcrop scale, many of these deposits have been classified as "sheet" deposits through decades of studies. Every splay deposit must eventually terminate, so recently, *sheet* has been interpreted as a misleading term for these deposits. Traditionally, these features were interpreted as fairly uniform and were presented in models and sketches as broad, circular shapes to represent their distributary and unconfined nature. Today, splays are recognized to take many forms: the shape of their container (Sweet et al., 2019); elongate sediment bodies (Moscardelli and Wood, 2015), and even in fault-bounded topographic lows along arc-trench slopes (Burgreen and Graham, 2014; and others). The type of splay deposit varies depending on the available space provided for the splay to accumulate. Predictive arrangement and compensation style stacking patterns have been evaluated by a number of key researchers including Straub et al. (2009). In typical progradational deepwater systems, splay complexes precede channel development, leading to an overall upward fining profile (i.e., Christmas tree) in gamma ray, yet splays display various stacking patterns (Selley, 1979).

There are several different types of splays (*summarized* in Lowe et al., 2019), ranging from frontal, to overbank, to crevasse, to avulsion to channelized (Fig. 18)—very similar to terrestrial deposits associated with alluvial fan and fluvial environments. Frontal splays are downslope of the channel-lobe (splay) transition zone (CLTZ in Wynn et al., 2002) and form due to the loss of confinement of sediment laden flows at the terminus of the channel environment. Other types of splays occur in close proximity, but lateral to, channel environments. Overbank and crevasse splays often form downsystem and outside of an outer bend of a channel at an acute angle to the channel trajectory. Sometimes the splays form perpendicular to a channel (Posamentier and Kolla, 2003). An overbank splay is formed by a sediment-rich flow overtopping the levee and depositing sediment in the levee-overbank area. A crevasse splay is distinctly different in that it requires the flow to erode and cut into the levee or channel margin, carrying sediment through the gap into the levee-overbank. Naturally, crevasse splay deposits are thick and show abundant evidence of erosion whereas overbank splays typically lack these traits. If flows are repeatedly focused through a broken levee, then the splay will progress to a full avulsion splay and perhaps develop into a new route for the channel (Ortiz-Karpf et al., 2015). In this way, overbank and crevasse splays can be a precursor environment to a newly reoriented channel-levee system, part of the build-cut-fill-spill (BCFS) model of Gardner et al. (2003). In theory, types of overbank splays are more easily developed in sand-rich systems and less easily developed in mud-rich systems. In mud-rich systems, bank stability tends to keep flows channelized all the way to the frontal splay environment. Overbank splays also form HARP in levee-overbank areas (Flood et al., 1991; Pirmez and Flood, 1995). Splays can be connected to their deliverer of sediment, or detached (Van der Merwe et al., 2014); these are known as *detached splays*.

Splays are divided into sub-environments in a similar fashion to alluvial fans. Principal sub-environments include the splay apex, proximal splay, medial splay, distal splay and splay fringe. Main sediment distribution pathways in splay environments include distributary channels that widen and decrease in depth downsystem toward the splay fringe. The stratigraphic architecture and hierarchy of splays was first articulated in a hierarchical diagram by Prélat et al. (2009), modified in Fig. 18. Splay dimensions vary for size and type of splay, with most dimensions listed as minimum dimensions in outcrop studies, such as in one of the Permian type sections in South Africa: splay elements are typically 2 m (~ 6.6 ft.) thick, 5 km (~ 3 mi.) long and 3.5 km (~ 2.2 mi.) wide; splays are 5 m (~ 16 ft.) thick, 27 km (~ 17 mi.) long and 13 km (~ 8 mi.) wide; splay complexes are 50 m (~ 164 ft.) thick, 40 km (~ 25 mi.) long and 30 km (~ 19 mi.) wide (Prélat et al., 2009). One of the intervals, known as the intersplay or interlobe shown in Fig. 18, often thickens to a splay element away from the control point in this type section.

Splays are prone to a variety of sedimentation depending on the grain size and feeder system. Splays range from sand-rich braided splay complexes of the Pliocene Ventura Basin, California, USA, formed by an overabundance of coarse-grained sediment and vast subsidence, to mudstone-rich splays, consisting principally of hybrid event beds, such as in the Carboniferous Ross Sandstone, Ireland (Ballybunion locale). MTD-dominated splays (Sharman et al., 2015) can be sand-rich or sand-poor in specific intervals. While splay deposits are diverse in their sedimentology, stratigraphy and architecture, depending on the type of splay it is, a recurring theme is that splay deposits tend to be *less diverse*, and *less variable* than their upslope canyon and channel equivalents *within the same system*. That is, grain size, bedding type, bed thickness, depositional process trend toward lower statistical variance than their upslope counterparts (Rotzien, 2019). An example would be the Cretaceous Venado Sandstone (Cortina Formation), California, USA (Fig. 9), which is filled by conglomerate, debrites, thin-bedded sandstone and mudstone, medium-bedded sandstone and mudstone, and thick- to very thick-bedded sandstone and mudstone featuring a number of canyon-fill depositional environments (Ghosh and Lowe, 1993). The Sites Sandstone, interpreted as frontal splay deposits from the same depositional system, is filled by medium- to thick-bedded sandstone and mudstone and slurry-flow deposits (Murray et al., 1996). Observations like these will be tested as more data become available from oilfield calibration.

Turbidite-dominated splays generally show the following trends from apex to fringe: (i) grain-size fining; (ii) bed thinning; (iii) less erosion; (iv) increase in clay (Fig. 19). The proximal deposits are typically structureless sandy deposits due to flow non-uniformity, whereas the downsystem equivalents are structured and more prone to transitional flow. Bed contacts tend to be flat and lacking significant erosion compared to channels. As with many of these observations, the observations are relative to the other environments in the same depositional system. In the dive from proximal to distal splay deposits illustrated with outcrop examples in Fig. 19, key reservoir parameters change based on first principles, including net:gross, reservoir quality, connectivity and bed length, to name just four.

**FIG. 19** Deepwater splay (lobe) deposits generally ranked from proximal to distal: (A) channel-splay transition deposits of the Miocene Hikuwai Sandstone near Gisborne, New Zealand; (B) proximal axial splay deposits of the Miocene Lower Mount Messenger Formation, Tongaporutu, New Zealand; (C) medial splay deposits, Carboniferous Ross Sandstone near Bridges of Ross, Ireland; (D) juxtaposed proximal and distal splay deposits, Ross Sandstone near Kilcloher Cliffs; (E) medial splay deposits, Eocene-Oligocene Annot Sandstone near Peira-Cava, France; (F) medial to distal splay deposits, Hikuwai Sandstone; (G) distal splay deposits, Lower Mount Messenger Formation; (H) distal splay deposits of the Upper Cretaceous-Paleogene Zumaia Flysch, near Donostia-San Sebastian, Spain; (i) transitional deposits, representing distal splay environment, Upper Cretaceous Point Loma Formation, near San Diego, California, USA.

Vertical and lateral connectivity is affected by mudstone bed cap and mudstone interval preservation, which is a function of erosional strength and repetitiveness of flows, soft-sediment deformation, burrowing and abundance of distributary channels. Mudstone bed cap preservation is more common than in channels since fewer flows generally make it to the splay, allowing more time for muddy sediment to accumulate. However, this does not account for the types of flows that reach the splay environment. If erosional flows reach the splay environment, mudstone bed caps will be removed. In high-energy, axial areas of splays, bed amalgamation can be common due to abundant erosional flows, lack of mudcap deposition, loading and soft-sediment deformation. Soft-sediment deformation can be pronounced due to either the alacrity of flows delivered to the basin or buildup of fluid pressure in splay fringes, respectively. This can positively affect reservoir performance, but can also negatively impact seal capacity. Bioturbation has a similar effect. Enough bioturbation can reduce compartmentalization and improve connectivity. Too much bioturbation can reduce reservoir quality. Shallow channelization can also facilitate reservoir connectivity (Elliott, 2000). Shallow channels such as distributary channels, and other shallow scours filled by a few or several sedimentation units, can serve to connect the reservoir at multiple levels (Prélat et al., 2010). While splay environments may appear to have strong vertical partitioning due to the presence of mudstone bed caps, if traced along their length in outcrop, many of the bed surfaces are amalgamated due to the reasons listed above. In 3D, often there is a high degree of connectivity, making splays one of the more attractive environments for low compartmentalization and high connectivity.

Bed thicknesses and lengths vary depending on the position in the splay. Typical bed thicknesses range from very thick- to thick-bedded in proximal and axial positions, and thin- to very thin-bedded in distal and off-axis to fringe positions. Bed lengths, while shown in subsurface correlations to be greater than several kilometers, depends on the scale of observation. Bedding packages for splay elements and splays can extend for kilometers, yet individual beds typically are less than 1 km

(~ 0.6 mi.) in length, and in many cases, only several hundred meters (Browne and Slatt, 2002). In one giant field in the Ventura Basin, USA, individual beds showed large variation in thickness between wells spaced less than 100 ft. (~ 30 m) apart (Schwalbach et al., 2009). Splays typically show predictable stacking patterns, or thickening and thinning due to compensation stacking patterns.

Decay rate of reservoir facies occurs in different amounts in different directions, from along axis to lateral extent (Grecula et al., 2014). The key controls on reservoir quality and decay rate include (i) sediment transport and depositional processes and (ii) sediment volume, both of which control sedimentation unit type and bulk sand volume to distribute in a particular depositional environment. Another control is the bathymetry and volume of the container and input point(s) which will govern how sediment is distributed and in which directions (the grain of the system).

In conclusion, splays tend to take the shape of their container and are generally considered reservoir targets for their high degree of lateral and vertical connectivity. Splays can feature evenly distributed and laterally extensive bedding on the scale of kilometers to tens of kilometers. Splays offer the opportunity for thin-bedded pay zones, as they are more extensive than other environments in confinement and offer comparable net:gross and porosity and permeability values to channel deposits. However, thin-bedded pay zones may represent splay fringe environments connected to a larger reservoir, or they could represent sediment starvation to the basin. These two end-member cases illustrate some of the geologic uncertainty in exploring splay environments, and the scale of and difficulty in accessing well-exposed analogs of these systems has contributed to development of numerical analogs using computational stratigraphy (Section "Advances in modeling depositional environments").

## Overbank

The vast majority of deepwater environments includes seafloor outside and away from canyons, channels, levees and splays in areas prone to passive accumulation of sediment known as *background sedimentation* including dilute flows, and hemipelagic to pelagic sedimentation (Stow and Piper, 1984; Bankole, 2018). Overbank areas include basin plains (Sumner et al., 2012) and mudstone "sheets" as hemipelagic drape, with thin turbidites, or "mud belts" (Sylvester et al., 2012) on continental slopes, yet overbank areas abut and are transitional with the previously listed environments. One of the most argued zones includes the levee to overbank transition zone, which is often just referred to as the "levee-overbank" environment (Fig. 20). The overbank deserves special attention because it likely records more about Earth's geologic history than any other stratigraphic interval (see Preface, Chapters 9 and 21 for more information on the broad impacts of overbank research). There are generally rare flows or sediment gravity transport processes occurring here, so it is an interval featuring nearly continual accumulation of sediment with rare interruption. However, relatively high-energy events can occur in the overbank, including the zones proximal to channels, such as levees (Morris et al., 2014), making the overbank realm a diverse habitat. Overbank intervals can be subjected to mass transport, sediment instability, mobility including injection (Cobain et al., 2017), bottom currents and overbanking flows from channels carrying significant volumes of sediment. Gas, mud and other fluid escape features form mounds and pockmarks found in overbank areas (Chenrai and Huuse, 2017), but the genesis of these fluid escape structures is often related to and may occur above or adjacent to channelized environments as well ( Jobe et al., 2011). As a result, there are diverse processes not directly associated with turbidity currents that form main reservoirs in overbank settings. These settings are not always as muddy and moribund as traditionally envisioned.

The overbank is generally an area defined by relatively thick accumulations of mudstone, but can exceed 10–20% sandstone derived from sediment gravity flows, contour currents or other mass movements such as MTD (Fig. 21). Rare to abundant sediment gravity flow deposits including fine- to very fine-grained to silty turbidites typically with structures derived from low-density flows can be present (Chapter 6 and 7). Beds can show compensation stacking, though bedding typically lacks scours and other types of erosional and bypass features. Bedding is generally laterally continuous, tabular and features planar bed contacts (Fig. 21B). *How far do individual thin-bedded sedimentation units extend in overbank areas?* While bedding packages can extend for multiple kilometers, individual beds are typically much shorter. Samples from the North Island of New Zealand include <250 m (~ 820 ft.) (Fig. 21C and D) (Browne and Slatt, 2002) and <1 km (~ 0.6 mi.) (McArthur et al., 2016). In few basins on Earth such as the Basque-Cantabrian Basin, individual beds can be correlated for nearly 80 km (~ 50 mi.) (Mutti, 1977—Fig. 21E). The range of values spanning orders of magnitude demonstrate that it could prove quite challenging to understand and predict bed length. *So how far do the beds extend in hypothetical Basin A?* If flows have a high sediment volume and are reaching a flat ocean floor with little to no topography, the beds will likely be laterally continuous. However, if flows are reaching areas that are in confinement, bed lengths will likely be much shorter. Most researchers that are interested in the subject of bed length have constructed databases for different

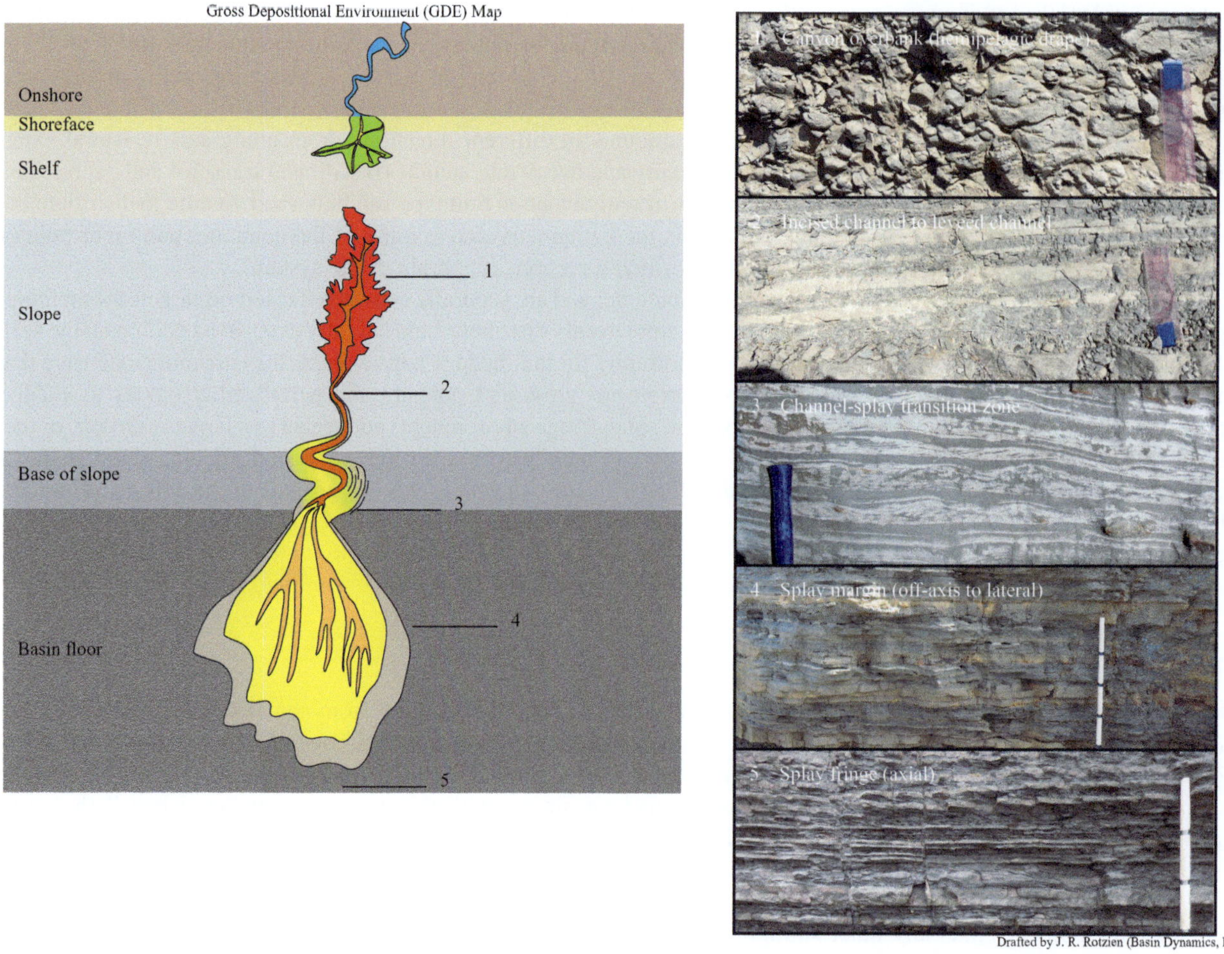


Drafted by J. R. Rotzien (Basin Dynamics, LLC)

**FIG. 20** Overbank areas are diverse and probably the least understood in the deepwater realm: (1) overbank area outside of the confinement of a canyon is typically hemipelagic in nature (Merle Formation, USA); (2) overbank near the incised channel tends to show thin- to very thin-bedded turbidites and slumped deposits (Juniper Ridge Conglomerate, USA); (3) strongly turbiditic sedimentation at the CLTZ and levee-overbank (Mount Messenger Formation, New Zealand); (4) interbedded transitional flow deposits and turbidites (Point Loma Formation, USA); (5) strongly transitional and hybrid bed sedimentation (Ross Formation, Ireland). Herein the focus is on five principal zones in the overbank. This is a minimum in terms of understanding the overbank area, and these areas can be much more diverse and require further scrutiny. The overbank represents an area of high potential for resource exploration and development into the future (see Chapters 9 and 21).

basins. These databases can be used for calibration purposes, and to answer questions regarding bed thickness decay rate (example: Grecula et al., 2014).

Overall, the overbank realm is generally lower net:gross than the levees, splays and channels of the same system. Paleocurrents tend to follow regional slope, and are usually at an angle to paleocurrents measured in the channels, levees and splays of the same system. Burrowing is generally more pronounced unless there are no burrowing organisms present. From proximal to distal (Fig. 20), bedding tends to change from hemipelagic (muddy) to sandy (turbiditic) to transitional flow (muddy) dominated (Brooks et al., 2018).

## Contourite and mixed systems

The previous sections described environments dominated by downslope sedimentation. However, it is important to consider both off-axis (lateral) sedimentation as well as along-slope currents, such as bottom currents or contour currents, that can re-work or wholly redistribute sediment from the shelf edge to the basin floor (Stow, 1979; Hernández-Molina, 2021). This is a brief overview, and for a more complete description, please refer to Chapter 9. Contourites are formed by currents that follow the same elevation along their journey of the slope, and can rework deposits that are delivered by sediment gravity flows. Contourite drifts and mounds vary in scale from tens of meters to tens of kilometers in length and width,

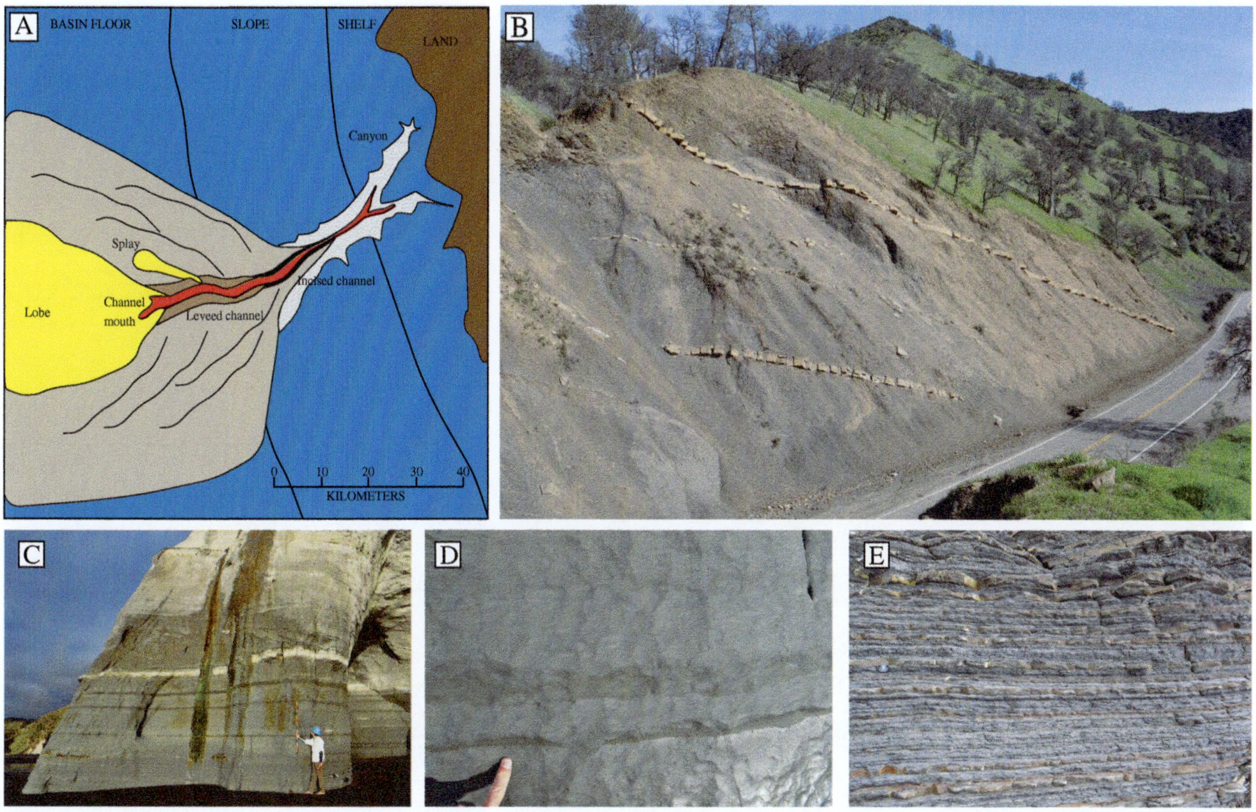


**FIG. 21** Typical overbank assemblages: (A) submarine fan valley depositional system modified from Lowe (2004), Lowe and Graham (2012) and Rotzien (2013) with overbank areas shown in tan and areas typically not greatly affected by sediment gravity flows shown in blue; (B) outcrop of the Cretaceous Fiske Creek Shale consisting of thin- to very thin-bedded, fine- to very fine-grained low-density turbidity current deposits, with two 1–2-m-thick (~ 3–7 ft.) Bouma Sequence beds near Guinda, California, USA; (C) coastal cliff of the Upper Miocene Mount Messenger Formation, Taranaki Basin, New Zealand, consisting of deposits from dilute, low-density turbidity currents, hemipelagic processes and volcanic ash; (D) heavily burrowed interval of very fine-grained sandstone turbidites and silty mudstone, Upper Miocene Mount Messenger Formation; (E) rhythmically bedded overbank to basin plain turbidites, Late Cretaceous to Paleogene, near Bilbao, Basque-Cantabrian Basin, Spain (lens cap for scale). Note the lateral continuity of many of the sandstone beds, as well as lack of erosional surfaces and variable amount of burrows.

to several hundred meters in height above the seafloor. There are entire deepwater systems formed purely from drifts and mounds from bottom currents, yet the most prospective areas for exploration are in mixed deepwater depositional systems with both turbidity current and contour current activity (Fig. 22). Drifts can take many shapes from detached, to plastered, to isolated, and tend to have four stages of development: initiate, grow, maintain; and migrate (see Chapter 9). They can modify any part of the sedimentary system previously described and form terrace drifts, channelized drifts, moats and architectures within these morphological elements as the distal parts of drift environments. While unaltered contourite deposits occurring in overbank areas are silt- and clay-prone, deposits delivered by sediment gravity flows and re-worked by robust contour currents—in *mixed systems*—can form high-quality sandy reservoirs in high-energy environments such as channels (Fonnesu et al., 2020; De Weger et al., 2020). However, overbank or non-channelized areas tend to be fine-grained and poor reservoir quality (De Castro et al., 2020). The term *mixed system* can also describe groups of environments in which processes operate simultaneously, not one after the other, potentially causing new sedimentary processes to be discovered in relatively uncharacterized deposits.

Bottom currents affect large areas of overbank areas along continental margins globally (Thran et al., 2018). These currents can form a number of bedforms (Rebesco et al., 2014) depending on current velocity, yet most currents are roughly tens of cm per second in velocity. Not only do bottom currents affect overbank areas, they also can change the entire character of the sedimentary system (Fig. 22) as modeled in tank experiments (Miramontes et al., 2020) and in exploration areas including West Orphan Basin, offshore Canada (Donnelly et al., 2019), offshore Mozambique (Thieblemont et al., 2020; Fonnesu et al., 2020) or offshore West Africa (Casson et al., 2020).

Since contourites tend to be poor reservoir quality overall, there is a need to develop mixed systems (with many dynamic components) models complete with reservoir properties and dimensional data for building models (see Chapter 9). Sizes

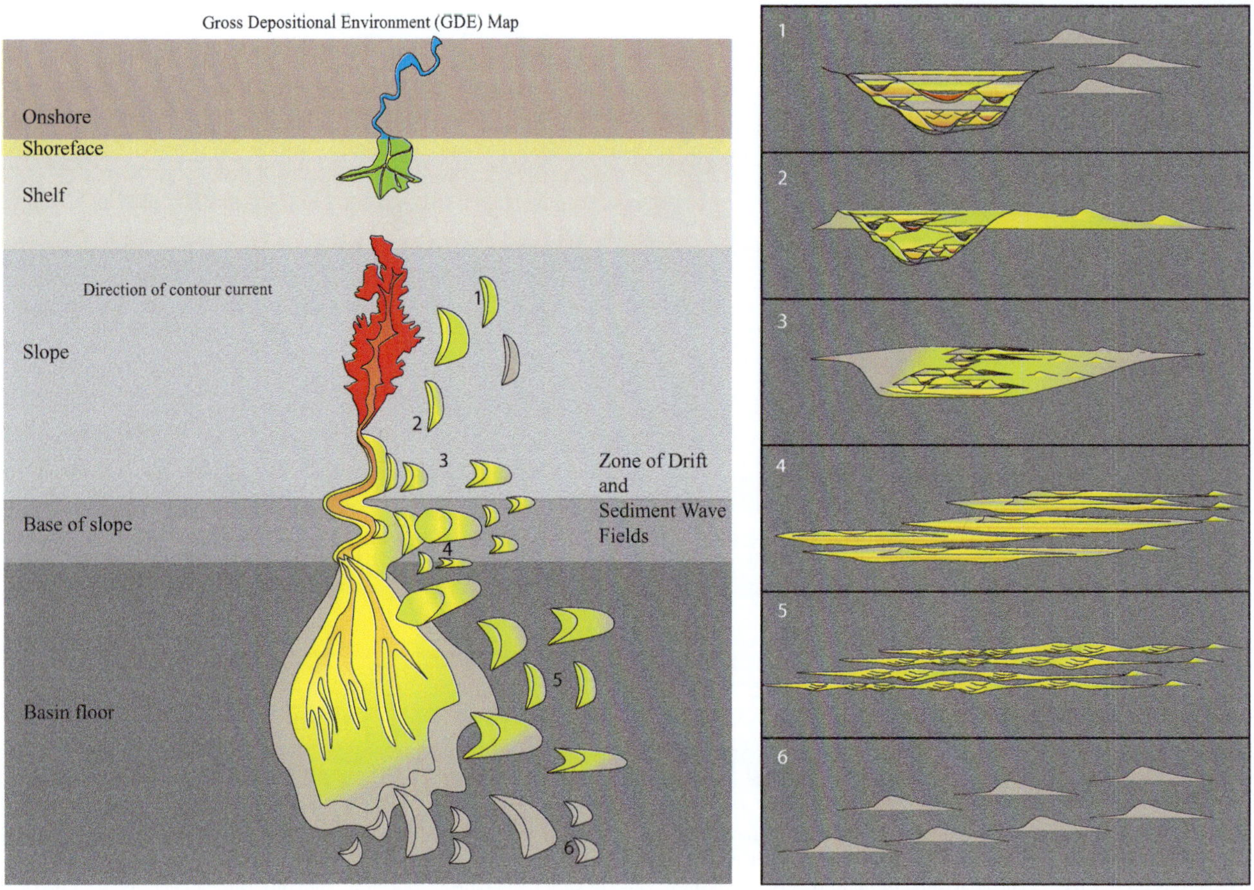


**FIG. 22** Simplified model GDE map for a mixed deepwater contourite and turbidite system. When sediment gravity flows, including turbidity currents and debris flows and all their variations, come together, and are combined with along-slope currents and mass transport processes, the system is termed a *mixed* deepwater sedimentary system (sensu Hernández-Molina, 2021). Note that while the system may migrate laterally in the direction of the bottom current, the system may also migrate *upflow* (sensu Fonnesu et al., 2020; and others). Example cartoon cross sections in the (1) canyon, (2) canyon-channel transition zone, (3) lower slope channel-levee complex, (4) channel-splay transition zone, (5) middle (medial) splay complex and (6) basin plain. See Chapter 9 for additional information.

and shapes of terraces, moats, mounds and contourite channels and their examples need to be cataloged and investigated to be applied in a predictive capacity using 3D seismic reflection data (i.e., where are likely areas for good reservoir, poor reservoir, no reservoir and seal). These systems also need to be understood from the standpoint of global circulation patterns and prospective time intervals. The challenges faced by industry include using observational data with prudence—seismic data mainly—to inform where is the high net:gross, best quality sandstone and requisite connectivity to serve as a reservoir. Furthermore, the outcrop analogs used could be very different from the subsurface (sensu Hernández-Molina, 2021 and references therein).

# Additional considerations

## Grain-size distribution

Grain-size distribution shows trends within depositional environments, as it does with lithology and first-order architectural element distribution downslope (see Fig. 17 in Chapter 7). Grain-size distribution and partitioning in deepwater environments is determined by multiple factors acting in unison: (i) provenance and caliber of the sediment provided to the system, including grain mineralogy; (ii) sediment transport and depositional processes; (iii) slope profile; (iv) scale of flows; (v) depositional environment.

If all the grain sizes are provided, and all else is kept equal, then a typical Lowe (1982), Stow and Shanmugam (1980) and Rotzien (2018) profile is sufficient (Fig. 17A in Chapter 7). However, if the system consists of fine-grained sand to silt and clay, only Bouma (1962) and finer-grained related sedimentation units will form. If only sand and silt is provided, then

Stow and Shanmugam (1980), Bouma (1962) divisions and hemipelagic divisions will be found. If all quartz is provided, then the depositional system will resemble quartz-rich systems such as the Permian Brushy Canyon Formation, USA (Beaubouef et al., 1999) or parts of the Carboniferous Jackfork Group, USA (Slatt et al., 1997) and some Cretaceous turbidite plays in the deepwater Gulf of Mexico. However, if the grains are varied in mineralogy, which can have a profound effect on grain shape and overall sedimentation process, then the system may resemble that of the deepwater Paleogene Wilcox Group, Gulf of Mexico (Kane and Pontén, 2012; Power et al., 2013; Marchand et al., 2015). In all of these situations, re-mobilized deposits including MTD can be abundant, but no more abundant than in mud-rich systems (Kneller et al., 2016; Prather et al., 2016). These are just a few examples that illustrate the spectrum of possibilities in depositional system type based on grain size, shape and sand:mud ratio (sensu Reading and Richards, 1994).

If the sedimentation transport processes are primarily via turbidity currents with a wide range of grain sizes, then there will be a high order of architectural complexity in the submarine fan. Turbidity currents are efficient size sorters of sediment apart from the "swash" effect and reverse helical flow along channel bends, therefore environments will be controlled by the sediment released above the channel margin and into the levee-overbank, and how much sediment reaches the end of the channel into the frontal splay environment. This order to the deepwater system is promoted through the following effects: (i) increase in sinuosity of channel systems with a number of possible bar types and channel margin elements such as terraces; (ii) focusing of the flows' power in different parts of the channel, leading to well-defined axis, off-axis and marginal areas of deposition with different bedding types and styles; (iii) higher levees increasing flow power and longer distance transport through channel systems farther into the basin; (iv) overall partitioning of depositional environments, as opposed to those formed by gravel-rich systems, which tend to form more alluvial fan-type or braided patterns (Walker, 1975; Rotzien et al., 2014b and references therein; Turner and Cronin, 2018). If sedimentation transport processes are primarily via debris flows, there will be a poor correlation to grain size partitioning in the system.

Transitional flow dominated systems show evidence of both types of grain distribution; grains will be coarser more proximal to the shelf, and transitional flows will mix the sand and clay grain populations leading to a more homogeneous distribution of reservoir and non-reservoir components in the deepwater system, be it a fan, apron or fan valley. In these systems, careful observation is required to find the high reservoir quality sections deposited by turbidity currents. This can be observed using the Haughton et al. (2009) modified downslope diagram (Fig. 17C in Chapter 7 of this volume).

Slope profile plays a key role in grain size partitioning. If the slope is concave up and typical of a cooled, mature passive margin, the slope will be generally an area of bypass and deposition of relatively muddy sediment, with well-sorted sand deposited at the base of slope with a majority on the basin floor (Covault et al., 2011; Prather et al., 2016). If the slope is convex up, or is associated with a number of faults or halokinetic minibasins, then coarser grains will be trapped in these areas of high sedimentation, and may never reach the base of slope unless *highways*, the playful term for "canyons" in Donovan (2010), are developed.

If flows are small and underfit for the size of the transport conduit, such as a canyon or channel, the flows and their deposits will never reach the terminus of the fan environment. In this case, most of the sand grains will be trapped in the upper to medial reaches of the deepwater depositional system (see channelized splays in Fig. 18). However, if flows are powerful and high in sediment yield, they will carry sediment far into the basin, coupled with the added effect of having more momentum from the suspended sediment. Flows like these can skew sediment richness farther downslope or have a more even theoretical sedimentation profile with book compensation stacking.

Depositional environment plays a key role in sediment distribution downslope, off-axis and to the periphery of the axis of sedimentation. Debrite-dominated environments show a very low degree of architectural differentiation (i.e., channelization and grain size distribution are somewhat moot points, because there is little difference between different points in the system), so exploration teams tend to focus on turbidite-dominated systems. But the drill bit will encounter all types of deepwater deposits, so the fewer surprises in terms of lithological estimation, the better. Because turbidity currents are stratified, the coarsest grains follow the low points of the topography in confinement, such as a canyon or channel, and the tall cloud of finer grains tends to spillover into areas not in confinement, such as a levee or overbank area. Therefore, the coarsest fraction will be in the canyons, channels, proximal and axial portions of the frontal splay complex, with finer grained and lower net:gross deposits in the levees and overbank areas.

Because of these five key principles, predictive stratigraphic models of where the best reservoirs should be in a deepwater depositional system can be built and modified as new data become available (example presented in Fig. 6). Turbidity currents stratify grain populations, and grains are sorted based on size, density and shape. Sorting and grain size are the key controls on primary porosity and permeability. As sorting and grain size increase based on the controlling factors described herein, porosity and permeability both increase (Slatt, 2006 and others). Mode of deposition control the sedimentary rock fabric and its resultant porosity and permeability. Understanding the sedimentary processes operating in various depositional environments improves prediction of poro-perm results and pore throat distribution estimates.

## Differentiating deepwater deposits

There needs to be a way to more quantitatively evaluate sedimentary process to determine depositional environment. The range and variability in deepwater deposits is useful in that it provides a way to readily distinguish one deposit from another to determine the depositional environment and estimate the lateral continuity of the reservoir. Overall, canyon-fill tends to show the highest variability in deposits, followed by channels, levees, splays and overbank in terms of bed thickness and lithofacies present. Previous studies have shown that depositional environments can be distinguished using bed statistics (Carlson and Grotzinger, 2001; Chen and Hiscott, 1999; Sylvester, 2007), yet ratios (Walker, 1967; Rotzien et al., 2014a) can also be useful in distinguishing the key deepwater depositional environments (Fig. 23). Ratios of first-order architectural elements may include individual Bouma (1962) $T_a$, $T_b$, $T_c$, $T_d$, $T_e$ divisions or individual Lowe (1982) $R_1$, $R_2$, $R_3$, $S_1$, $S_2$, $S_3$ divisions ($T_t$ too). The sum of the thicknesses of each first-order element are plotted on ternary diagrams. Using only these thicknesses, the data plot in certain areas on the diagram, revealing depositional environments in the same way Gazzi-Dickinson point counts are used to reveal tectonic settings. The concept of Turbidite Ratios remains in its nascent stages, and will require additional calibration using global data sets over the next decades.

In addition to the utility of ratios, observational data sets on stratigraphic terminations are used to estimate lateral bed continuity in core data. This assumes the "sheet sandstone" fallacy. All sediment gravity flow deposits must end somewhere. Lateral terminations are important because they can be helpful in the case of successful stratigraphic traps (Walker, 1978; Haughton, 1994; Allan et al., 2006; Gardiner, 2006; Hansen et al., 2019), or detrimental in that gross rock volume and connectivity is less than anticipated (Bakke et al., 2014). Stratigraphic terminations are notoriously difficult to estimate using core data or seismic reflection data (Vail et al., 1977a, b). Deepwater stratigraphic terminations occur in all settings, from salt basins (Giles and Rowan, 2012), to uneven basin floors (Apps, 1987), along basin margins (Sinclair and

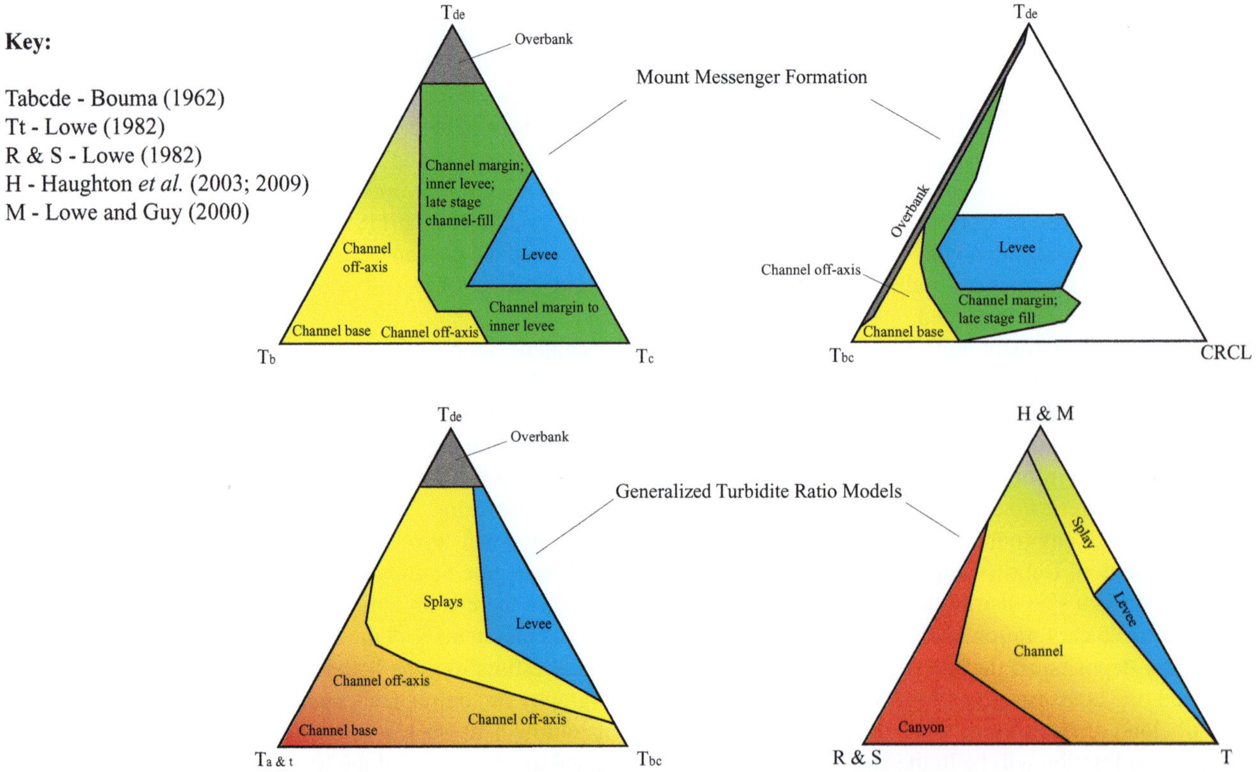


**FIG. 23** Turbidite Ratios applied to determining the depositional environment. The upper diagrams modified from Rotzien et al. (2014a) are results for subsets of data from outcrops of the Upper Miocene Mount Messenger Formation, Taranaki Basin, North Island, New Zealand. The lower diagrams present more generalized models for depositional environments using classic to broad coarse-grained, fine-grained and transitional flow first-order architectural elements. These diagrams show the differences in process and deposit in different depositional environments and can be used to compare reservoirs in different parts of the oilfield. The method fails in successions that (i) have thick intervals of hemipelagic to flow-derived mudstone; (ii) beds that are generally less than 10 cm (~ 4 in.) in thickness or (iii) in areas that have great diversity in sedimentary structures. There is considerable overlap in types of beds in different systems for the generalized diagrams. Walker (1967) was among the first to use ratios at the sub-bed-scale in turbidite depositional systems.

Tomasso, 2002), along the depositional system and within depositional environments. This concept of "how do sandstone bodies end" is critical in each phase of the upstream exploration and production process, and the specificity of the question requires higher resolution stratigraphic data.

There exist multiple classification schemes at different scales for deposits associated with confinement, bed termination, lapout, drape and pinchout: Type I, II and III (Haughton, 1994); conformable, transitional, simple, complex onlap (Stanbrook et al., 1999); stage 1, 2 and 3 (Kneller and McCaffrey, 1999); Type A and B (McCaffrey and Kneller, 2001); sequence stratigraphy terms defined in Vail et al. (1977a, b); abrupt onlap to feathering aggradational onlap (Smith and Joseph, 2004); type O and P (Bakke et al., 2014); pinch out and lateral terminations against salt diapirs (Collins et al., 2015). Overall, these studies describe several observations on stratigraphic terminations (Rotzien, 2017): thickening upward sandstone bedding patterns; evidence of flow reflection, deflection and decoupling (Kneller and McCaffrey, 1999); groove marks at odds with orientations of ripple cross-laminated sandstone, and delayed grain size fining (Lowe, 1982); overturned bedding (Puigdefàbregas et al., 2004) and fluid escape structures; a high amount of bed thickness variability; abundant mudclast rip-ups and gravel trains as evidence of high-energy and surging turbidity currents; wide range of amalgamation ratios (AR), which calculates amount of amalgamated bed contacts over total number of bed contacts; complex sediment gravity flow deposits in the lower section of the termination interval (M, H divisions); transition from finer grained, strongly structured to unstructured, and coarser grained deposits.

## Reservoir connectivity and compartmentalization

Reservoir connectivity is determined by porous and permeable intervals touching in 3D space and is influenced by rock type and its distribution, and can be identified by differences in fluid pressure and compositional differences (see Chapter 2). If reservoirs are not connected, they are interpreted as compartmentalized. Key depositional environments known for connectivity and compartmentalization challenges include: (i) channel complexes (channel to channel connectivity) (Barton et al., 2009); (ii) channel-levee complexes (channel to levee connectivity) (Sullivan et al., 2000, 2004; Browne and Slatt, 2002; Campion et al., 2000; Sylvester et al., 2011; McHargue et al., 2011); (iii) channel to splay connectivity (Van der Merwe et al., 2014); (iv) splay complexes (splay to splay connectivity) (Brooks et al., 2018) including in salt provinces (Franks and Uchytil, 2017; Henry et al., 2018; Weber, 2018). Comparisons in several publications show the differences between slope channel complex connectivity (Abreu et al., 2003) vs basin floor channel complex connectivity (Beaubouef et al., 1999). Other environments—including canyons—that are prone to deeply erosional flows, despite being mudstone rich in certain cases, tend to be well-connected in 3D space (sensu Lowe, 2004).

Compartmentalization can be due to stratigraphic, structural, diagenetic or syndepositional factors, either individually or in concert. Reservoir compartments can be formed by the following stratigraphic examples:

- basal, lateral and top seal (see Chapter 2) formed by mudstone or non-permeable rock (Barton et al., 2009), as well as erosional unconformities (Miall, 2016);
- injections and other forms of post-depositional sediment mobility including MTD (see Chapter 10) juxtaposing permeable and non-permeable rock (Hurst et al., 2011);
- compaction of fine-grained sedimentary rock reducing its porosity and permeability to exceedingly small values (i.e., transitional flow deposits in the UK North Sea (Lowe and Guy, 2000) vs the US Gulf of Mexico (Marchand et al., 2015)).

Further, compartmentalization can occur due to faulting and structural displacement of reservoir and non-reservoir units including the presence of deformation bands in sandstone-rich reservoirs (see Chapter 13) and salt diapirism (see Chapter 5). Diagenetic barriers can be formed by cement or authigenic clay (Higgs et al., 2015). Finally, syndepositional flowage, mass transport, slide or deposition of non-porous, non-permeable rock in between reservoir or carrier bed units can promote reservoir compartmentalization.

## Pitfalls leading to depositional environment misinterpretation

The goal of this chapter is to adequately describe deepwater depositional environments and their equivalent depositional elements (morphological elements; resulting 3D geobodies), how they are formed, where they are found in a deepwater depositional system as well as provide references and analogs. Misinterpreting rock types and depositional environments poses a risk to accurate subsurface interpretation, which can cause numerous challenges pertaining to reservoir assessment and final investment decisions. There are at least 11 major pitfalls that can lead to misinterpretation of rock types and overall deepwater depositional environment:

**(i)** believing that a depositional model in one system can work in another disparate system, often resulting from working the same basin or area of interest for too long (there is a need to keep multiple working models, sensu Yeilding, 2019);
**(ii)** building interpretations on interpretations, not the actual data or results of a study;
**(iii)** emphasizing certain data to the exclusion of other data that might have equal weight, such as the overreliance on seismic data for those unfamiliar with its merits and limitations (Long, 2019; Oswald, 2019; Chapter 1);
**(iv)** brief application of risk and uncertainty, usually performed at the end of the study, whereas it should be promoted throughout the study (Peel and Brooks, 2015; Chapter 2);
**(v)** not fully updating knowledge post well with new information to inform future business decisions (this is especially the case with dry holes);
**(vi)** performing an improper well tie, thus preventing accurate interpretation of the stratigraphy and how it correlates to geophysical logs and seismic reflection data;
**(vii)** not putting into writing specific pre-drill predictions, which can prevent an accurate benchmarking process for exploration and production (Rudolph and Goulding, 2017; Rudolph, 2020);
**(viii)** using analogy by proximity with basins that are not far from one another, however, these basins can feature entirely different stratigraphy and petroleum systems, such as halokinetic basins (Ingersoll, 2011) or segmented basins along active margins (example: California Borderland, USA);
**(ix)** not fully using stratigraphic data from nearby wells to perform accurate correlations and play out scenarios for what might be found in the next drilling opportunity (Furin, 2019);
**(x)** uncertainty predictions are too narrow leading to exploration surprises—both the good and the bad (Milkov and Navidi, 2020; Milkov and Samis, 2020);
**(xi)** conceptual, often oversimplified and averaged models of depositional systems tends to bias interpretations and can lead to missed opportunities. It is often best to maintain a strong awareness of the true range of complexity that is likely present in any studied or sampled system (this chapter).

## Advances in modeling depositional environments

One of the goals of this chapter is to illustrate the relationships found among deepwater environment, process, deposit and ultimate stratigraphy. The connections between these aspects of depositional systems can be utilized by geoscientists to predict characteristics of deepwater stratigraphic architecture. These so-called process-stratigraphic relationships rely on the community's conceptual and quantitative understanding of deepwater depositional systems, formed via observation of both modern and ancient systems (e.g., via wells, outcrop, seismic) and experimental studies (e.g., deepwater flume experiments, forward physics-based modeling algorithms). Proper application of these process-stratigraphic relationships, through data interpretation, theoretical concept application and/or computational modeling, can yield improved reservoir location, characterization and modeling predictions for exploration, appraisal and development.

This section outlines the role process-based computational modeling of deepwater depositional systems can play in utilizing process-stratigraphic relationships to characterize deepwater depositional systems, interpret observed geology and decrease uncertainty related to reservoir management practices. Process-based computational modeling represents one of the leading tools for geoscientists to expand knowledge of the process-stratigraphic relationships in deepwater settings. Direct observations of deepwater depositional processes are difficult, due to the challenges of data collection in extreme environments and the long timescales on which these processes act. Reproducing these systems through physical experimentation is both physically and financially impractical on large spatial scales and is limited by the inability to honor field-scale flow parameters in laboratory settings. Process-based computational modeling is also a valuable tool for direct use in exploration and reservoir management practices. In the absence of substantial direct observations and measurements of deeply buried target reservoirs, this technology provides the ability to develop extensive sets of physically realistic digital analogs.

This section includes a brief overview of deepwater process-based computational modeling and highlights why and how geoscientists apply this approach to scientific and business needs. The field of process-based modeling is expansive, and this chapter focuses only on the relationship to meso-scale deepwater depositional environments, with a focus on basin-floor fans. It does not include a detailed description of basin-scale or individual turbidity current-scale deepwater process-based sedimentary modeling, nor does it include process-based modeling of other depositional environments (e.g., coastal, shelf). It also does not include a description of other forward-stratigraphic modeling techniques (e.g., surface-, or event-based modeling) (e.g., Kaplan et al., 2016), additional reservoir modeling and simulation techniques, nor an in-depth explanation of how the output of process-based models can be used in reservoir or basin modeling (see Chapter 18).

## Process-based modeling of deepwater sedimentary systems

Process-based sedimentary computational modeling is the practice of simulating the physical (and sometimes chemical and/or biological) processes that compose sedimentary systems via mathematical representation and calculation using computers (e.g., Slingerland and Kump, 2011). Typically, process-based models will include some representation of fluid flow (e.g., Reynolds-averaged Navier–Stokes equations), related sediment erosion, transport and deposition, and the resulting changes to the bed surface (e.g., Exner equation). The exact sedimentation processes included will vary among models (e.g., the practitioner may choose to separately represent turbidity current, debris flow or pelagic sedimentation), and how the processes are mathematically represented can vary (e.g., rule-based, diffusion, advection–diffusion). Additional sedimentary system processes that may be included depend on various factors, such as scale of the system to be modeled, desired use of the model output and computational needs (e.g., less complex models typically require less computational power). For example: Do sedimentary-tectonic feedbacks need to be included in a basin-scale model? If the sedimentary system includes carbonate sediment, what additional chemical processes should be included?

A process-based sedimentary model that preserves some representation of the sediment that is deposited through time can be considered a process-based stratigraphic modeling approach, and results in the output of a stratigraphic model. The stratigraphic model typically records the three-dimensional spatial distribution of grain sizes and types. If the goal of the sedimentary modeling practice is to output a stratigraphic model, additional geologic processes may need to be included. For example, do diagenetic processes (see Chapter 13) need to be modeled if the ultimate goal is to predict porosity and permeability? Or might the practitioner consider converting to porosity and permeability via an empirically derived relationship from data gathered from the reservoir of interest?

The practitioner must consider the size of the system of interest, as well as the resolution needed to accurately represent the chosen processes, and settle on a size and resolution combination that is reasonable for the computational power available. For example, if spatial variation in tectonics are considered important, the model will need to be large enough to capture those signals. However, if the process of grain-scale sorting is considered important, the resolution of the model must be high enough to resolve that process. With current typical computational power, a model including such a large size and high resolution would be very difficult to run (and take a long time!). However, if both processes are deemed very important for the task at hand, practitioners may explore the viability of nested models. Similarly, temporal scale and resolution are important when determining the length of geologic time, processes and size of model that can be reasonably captured by a given computational setup. Temporal resolution of the processes refers to the length of time represented by a given "time step" in the modeling algorithm, and is related to spatial resolution and "speed" of the process being modeled (*for further details, see references on the Courant–Friedrichs–Lewy condition*). Temporal resolution of the output refers to how much geologic time has passed between each output "snapshot" that is recorded in the resulting model (e.g., a geologic surface). Both the need to represent a longer amount of total geologic time, or a higher temporal resolution in the output model, will require running the modeling algorithm for a longer amount of time. These considerations are important when planning modeling practices and feasibility of completing a study by prescribed deadlines.

An important aspect of applying process-based modeling results to exploration and appraisal or reservoir management workflows involves selecting the correct model, or set of models, to represent the reservoir of interest. Practitioners utilize a wide variety of methods for selecting proper "analog" models, and determining the ideal method relates closely to how one plans to apply the model. For example, if plans include developing a deterministic model to be used directly in reservoir simulation workflows and well planning activities, extensive conditioning to available well logs and production data is a best practice. This is an extremely difficult task, and developing best practices is at the forefront of modern research. However, if instead the practitioner plans to develop a suite of models that capture the fundamental stratigraphic architectural components (e.g., gross connectivity or heterogeneity estimates) of the target reservoir in order to reduce uncertainty related to reservoir characterization practices, then extensive conditioning may be less essential.

Deepwater process-based modeling techniques for meso- to large-scale systems have received increased attention over the last couple decades, though studies developing or applying them remain relatively sparse compared to those for other depositional environments, such as fluvial or shallow marine systems. Reza et al. (2006) proposed *ModDRE* which incorporates process-based techniques and stochasticity to model deepwater channels, accompanying lobes/sheets and intervening condensed hemipelagic shale sections, with a focus on the influence local gradient exerts on channel geometries combined with randomly chosen modifiers. *Cellular Automata for Turbidite Systems (CATS)* (Salles et al.,

2007, 2008a,b; Albertão, 2010; Albertão et al., 2015) is based on the non-stationary model of Salles (2006) and models turbulent, low-concentration (<9%) flows including the processes of flow distribution, water entrainment, and erosion and deposition. *FanBuilder* simulates turbidity currents, transport and deposition of multiple grain sizes and includes a high-resolution, shock-capturing solution of the discontinuous flow front (Groenenberg, 2007; Groenenberg et al., 2009, 2010; Athmer et al., 2010; Wang, 2015; Wang et al., 2017; Empinotti, 2020). *MassFLOW-3D* solves for fluid flow with flow turbulence simulated using the renormalization group (RNG) of equations, and models sediment bedload transport, entrainment of sediment and accounts for hindered settling as a function of sediment concentration (Heimsund, 2007; Aas et al., 2010a,b, 2014; Janocko et al., 2013b; Basani et al., 2014; Ge et al., 2017, 2018; Howlett et al., 2019). Waltham et al. (2008) present a process-based modeling solution utilizing gravity and turbulent Newtonian flow stress equations, and sediment deposition from suspension, which is assumed to be the dominant process. ExxonMobil developed deepwater process-based modeling approaches simulating turbidity current erosion as a function of flow velocity and grain-size distributions, and deposition as a function of settling velocity and concentration (Sun et al., 2008, 2010; Miller et al., 2008; Michael et al., 2010; Hoyal et al., 2018; Shashank et al., 2019; Wahab et al., 2019). Abd El-Gawad et al. (2012) simulate fluid flow, mass conservation for solute and sediment mixture and incorporate a velocity–pressure coupling over a deformable bed (Huang et al., 2005, 2007, 2008). Various studies have applied the modeling package *DionisosFlow* (Granjeon, 1996), utilizing diffusion-based equations to model large-scale deepwater sedimentary systems (Hawie et al., 2018, 2019; Nader et al., 2018). Chevron's high-fidelity computational stratigraphy program *CompStrat* (e.g., Willis and Sun, 2019) includes a deepwater module that simulates the erosion, transport and deposition of sediment associated with turbidity currents over a deformable bed, incorporating entrainment, suspended and bedload transport, and utilizing depth-parameterized flow shape functions (Fig. 24) (Caldwell et al., 2018; Sun et al., 2018; Laugier et al., 2019; Caldwell and Sun, 2019; Goggin, 2020). Burgess et al. (2019) presented *Lobyte3D*, a reduced-complexity model that simulates flow velocity and associated sediment transport and deposition as a function of bathymetric gradient and flow

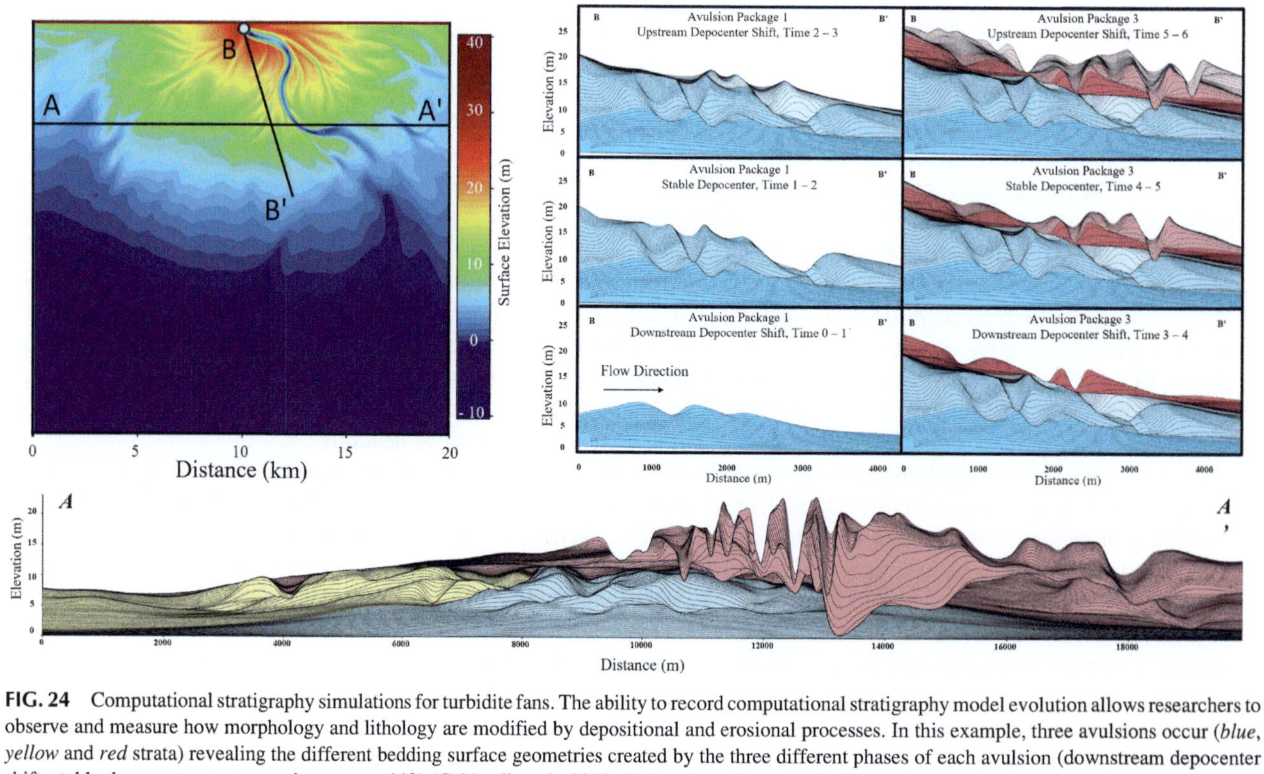


**FIG. 24** Computational stratigraphy simulations for turbidite fans. The ability to record computational stratigraphy model evolution allows researchers to observe and measure how morphology and lithology are modified by depositional and erosional processes. In this example, three avulsions occur (*blue*, *yellow* and *red* strata) revealing the different bedding surface geometries created by the three different phases of each avulsion (downstream depocenter shift, stable depocenter, upstream depocenter shift) (Caldwell et al., 2018; Caldwell and Sun, 2019). By running physics-based models, researchers can build multiple predictive analogs which can be compared to measured field data (e.g., wells) and utilized to better understand the depositional and stratigraphic system. *(Images from Caldwell, R.L., Sun, T., Willis, B., Baumgardner, S., Harris, A., Sullivan, M., 2018. Linking deepwater fan depositional processes and stratigraphic architecture using computational stratigraphy: GSA 2018 Annual Meeting, Indianapolis, Paper No. 46–16. In: Geological Society of America Abstracts with Programs. vol. 50, No. 6. and Caldwell, R.L., Sun, T., 2019. Deepwater distributary system avulsion processes and geomorphic signatures. In: AGU Fall Meeting Abstracts, vol. 2019, pp. EP13D-2171).)*

thickness over an evolving bed topography, and incorporates hemipelagic deposition. Descriptions of how these types of models have been applied is in the following Section "Goals and use of process-based modeling".

## Goals and use of process-based modeling

A fundamental goal of hydrocarbon exploration and reservoir characterization practices is to form accurate predictions of stratigraphic deposits that cannot be fully "seen". Remote imaging using seismic acquisition is typically limited by insufficient resolution to resolve important stratigraphic heterogeneity. Seismic signal attenuation due to depth and complexity of overlying sections degrades the ability to image deeply buried deposits at a high-enough resolution to accurately characterize the reservoirs and optimize well planning. High-resolution vertical well data can be gathered, but these methods are costly and the ultimate lateral resolution of this observation method is lacking. The difference in scale between these two observation methods—well and seismic data—can significantly influence the accuracy of subsurface predictions (Fig. 25). To fill the voids left by available observation methods, exploration and development geologists often predict the spatial distribution of reservoir characteristics based on knowledge gained from well-studied stratigraphic deposits (analog cases), such as well-exposed outcrops. However, stratigraphy is exceedingly variable, and relying on the relatively limited number of available, measurable "analog cases" to formulate production plans is insufficient. To limit exposure to significant uncertainty, companies desire access to expansive sets of system-scale analogs that match observed well data and fit within the geologic context interpreted from available seismic data. The use of system-scale analogs helps predict stratigraphic facies and plan appropriate well programs.

This data gap and need for accurate, high-resolution analogs of deepwater depositional systems has driven the development and application of high-resolution, deepwater process-based forward stratigraphic modeling technologies. Practitioners can develop extensive sets of realistic digital analogs that fill the resolution gaps left by observation methods (Fig. 26; Amaru, 2019). Advanced practices ensure the analogs are conditioned to available data, and honor the physics of depositional processes. By using observed, interpreted or reasonable values for parameters that set the depositional system (e.g., flow rate, sediment caliber and type paleotopography), and matching the output to observed stratigraphy (e.g., well data), practitioners constrain the creation of physics-adherent models that accurately predict the presence and architectural characteristics of reservoirs in data-limited scenarios.

The output process-based stratigraphic models can be used to uncover process-stratigraphic relationships within deepwater systems that aid the geoscientist in predicting stratigraphic architecture away from sparse well data control points and

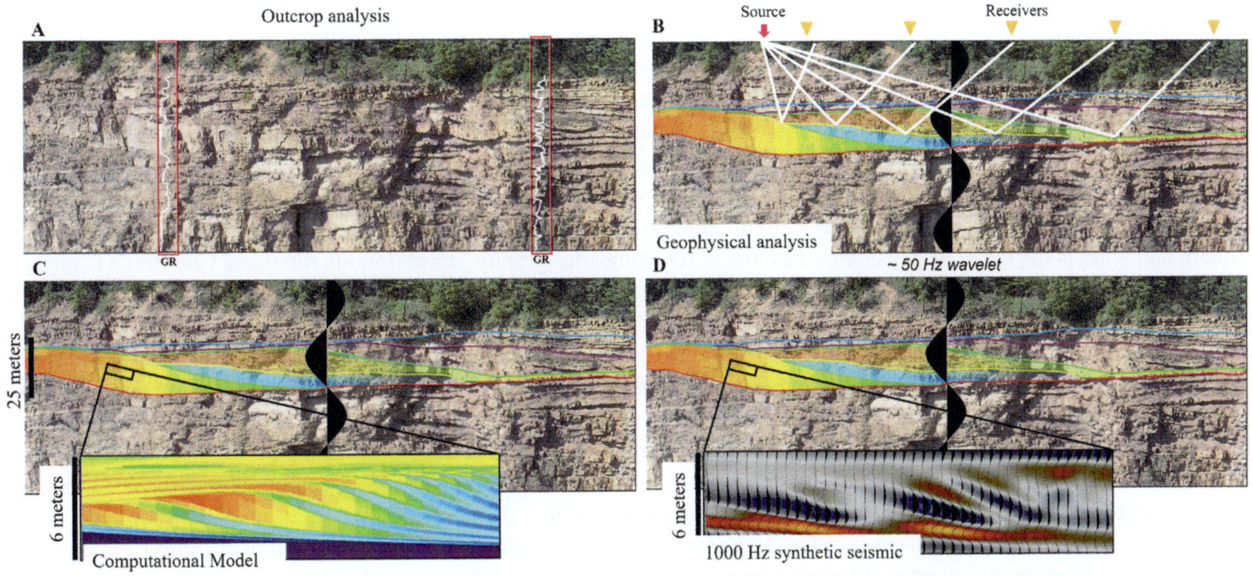


**FIG. 25** Data resolution influences interpretive understanding. Predictive models built using analogs from previous studies are a key tool used to optimize hydrocarbon production. The finest scale analogs are based on samples (thin-sections, cores and outcrops) and larger scale remote sensing (well logs and seismic) provide analog resolutions ranging from <0.5–1 m (~ 1.6–3.3 ft.) to tens to hundreds of meters, respectively. Data-conditioned computational stratigraphy models provide process-aligned predictive models that transcend previous analog scaling limitations. *(Carboniferous Jackfork Group, Arkansas, USA. Images from Goggin, L., 2020. Exploring the limits of seismic interpretation techniques through the use of computational stratigraphy models. In: AAPG Distinguished Lecture Series, 2019-20 Season.)*

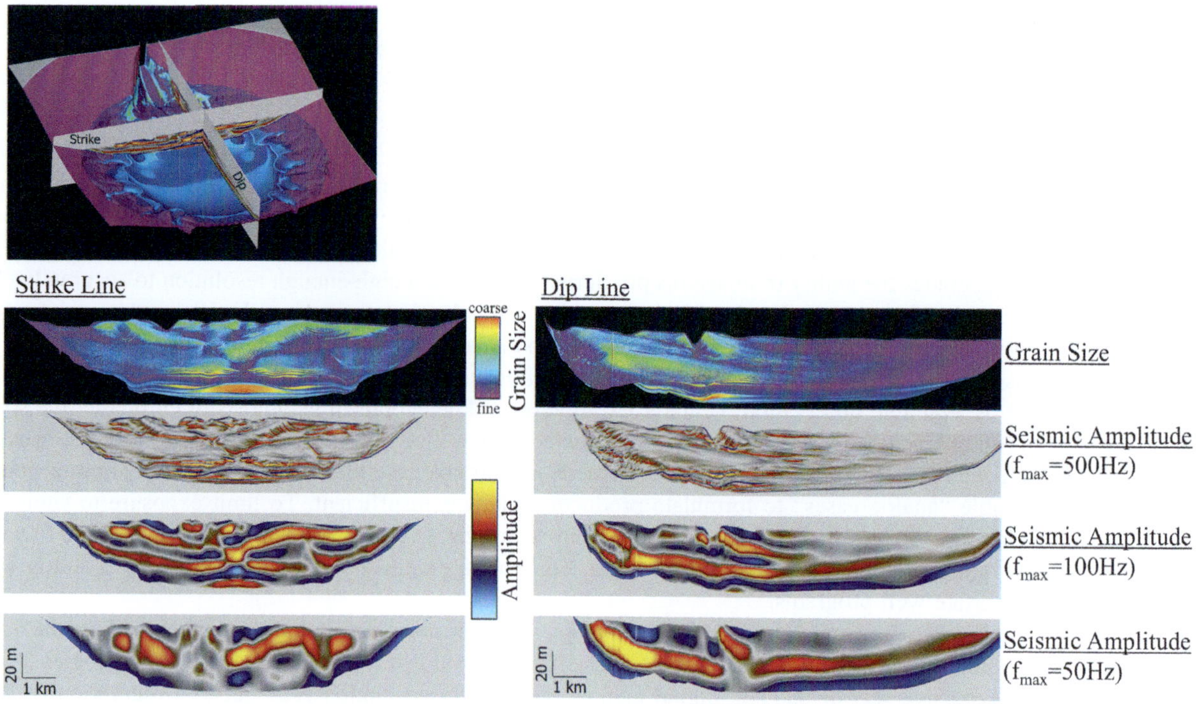


**FIG. 26** Deepwater ponded example of computational stratigraphy model, highlighting the importance of seismic resolution for interpreting stratigraphic character. Computational stratigraphy models can be converted to synthetic seismic models of selected frequencies (Amaru et al., 2017). In this example, ponded deepwater model has been converted to three synthetic seismic resolutions (500, 100 and 50 Hz) to investigate the relationship between frequency degradation and interpreted seismic geomorphology. Interpreted geomorphologies evident at the different frequencies are drastically different, influencing the uncertainty inherent in the reservoir. Vertical exaggeration = 50 ×.$f_{max}$ is maximum frequency. *(Images from Amaru, M., et al., 2019. Integrated computational stratigraphy reservoir characterization and seismic validation. In: AAPG International Conference & Exhibition, Buenos Aires.)*

below the resolution of seismic data. For example, deepwater deposits exhibit hierarchical trends in the three-dimensional organization and size of geobodies (see Chapter 1), as set by the hierarchical scales of their depositional processes. This creates a predictable framework of important stratigraphic features (e.g., connectivity, heterogeneity) that geoscientists use to build predictions. However, gaps remain due to the limited abundance of directly observable data and the relatively poor resolution of seismic imaging. The power of controlled experimentation using process-based computational modeling combined with observable data (e.g., outcrops) can further elucidate the stratigraphic characteristics of the geobodies and place them in their process-stratigraphic and hierarchical frameworks to improve predictions (e.g., Caldwell et al., 2018; Laugier et al., 2019; Caldwell and Sun, 2019). For example, it has been recognized that the avulsion process (discussed in previous sections) sculpts deepwater distributary system architecture through a hierarchy of scales, but what is the process mechanism, and what is the range of physically reasonable resulting stratigraphy? Experiments run using Chevron's *CompStrat* program capture avulsions within deepwater distributary systems (Fig. 24). Results suggest the avulsion process sets stratigraphic architecture, including predictable hierarchies and geobody characteristics (Caldwell et al., 2018; Caldwell and Sun, 2019). As another example, Athmer et al. (2010) used the *FanBuilder* program to elucidate the process-stratigraphic relationships between relay ramp topographies and the resulting distribution of turbidites. Burgess et al. (2019) use *Lobyte3D* to explore the signatures created by autogenic vs allogenic signals in deepwater fan stratigraphy, which are useful tools for the geoscientist to place the deposit within the broader regional stratigraphic context.

Process-stratigraphic relationships uncovered via computational modeling have been compared to observed deepwater deposits and further applied by geoscientists to basins and reservoirs of interest. Albertão et al. (2015) applied *CATS* to show that flow concentration and substratum grain size exert strong control on resulting deposit distribution and thickness, with results that compare well with observations in the Campos Basin, offshore Brazil. *CATS* even underwent a successful "blind test" via data gathered from a newly drilled well. Groenenberg et al. (2010) applied an iterative approach using *FanBuilder*, constraining model input parameters with interpretations from outcrop observations of the Tanqua depocenter, Karoo Basin, South Africa, by comparing the output to observations and subsequently updating the computational model. This study illuminated process-stratigraphic relationships between instability in distributive channel position, lobe growth and resulting stratigraphic complexity. Aas et al. (2014) used the *MassFLOW-3D* program to replicate observed thickness

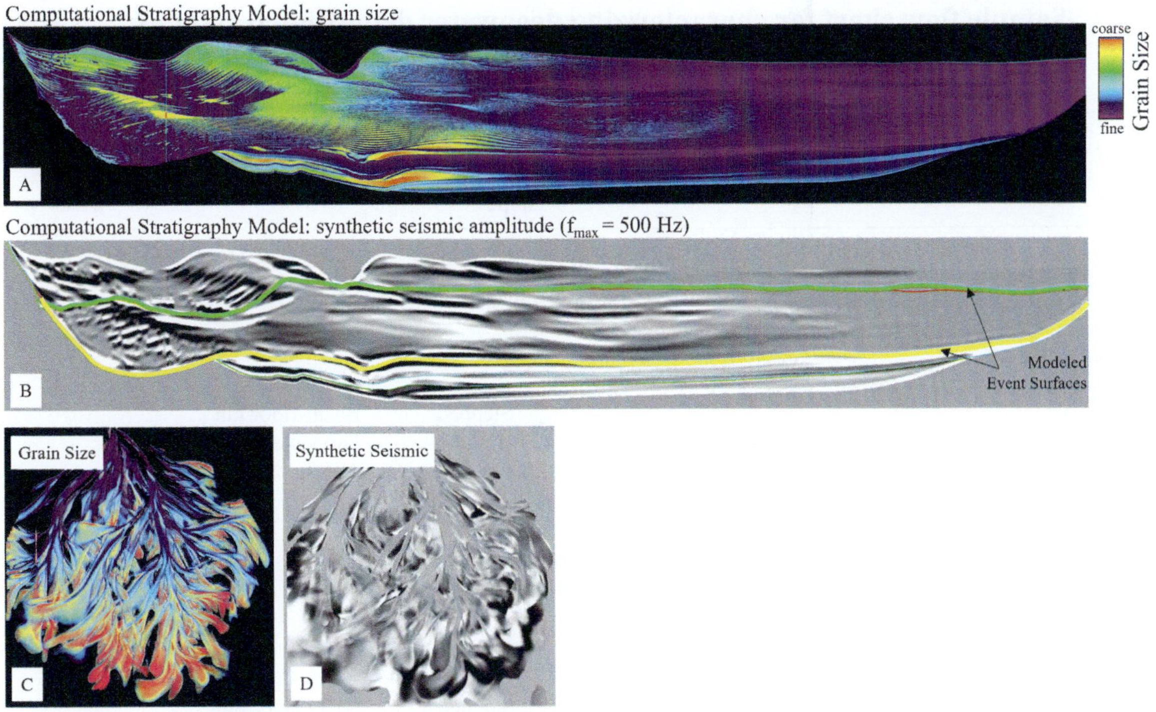


**FIG. 27** Computational stratigraphy grain-size models (A and C) are converted to seismic models (B and D). Reservoir properties in the models are then tied to the varying seismic response, delivering a scalable seismic analog. Field-acquired or well-constrained seismic signatures can then be compared to the models to assist in reservoir characterization and well planning. Example shown for CompStrat deepwater turbidite model. (A and B) cross-section views. (C and D) map views. *(Images from Amaru, M., et al., 2019. Integrated computational stratigraphy reservoir characterization and seismic validation. In: AAPG International Conference & Exhibition, Buenos Aires and Goggin, L., 2020. Exploring the limits of seismic interpretation techniques through the use of computational stratigraphy models. In: AAPG Distinguished Lecture Series, 2019-20 Season.)*

and grain-size distribution from an outcrop of the Eocene–Oligocene Annot Sandstone, Peïra Cava sub-basin, France, and found most accurate results occurred when flow scour was limited, highlighting the importance of cohesive beds on the process-stratigraphy of deepwater systems. Abd El-Gawad et al. (2012) applied process-based modeling to a sinuous channel in the modern, continental slope of the Niger Delta (see Chapter 4), accurately reproducing grain-size distribution as measured from piston cores and describing relationships between bathymetry and flow processes.

Post-processing algorithms may be further applied to the output of a process-based stratigraphic model for direct utilization in exploration and appraisal or reservoir management workflows. The practitioner can convert the output volume, typically represented by spatial variation in grain sizes, to porosity (e.g., Amaru et al., 2017) and/or permeability using various methods. The converted model can be used for flow diagnostics studies (e.g., to determine well flux and recovery efficiency) (e.g., Willis and Sun, 2019), as input to reservoir simulation methods (e.g., Willis et al., 2020), or in combination with other reservoir modeling methods, such as multi-point geostatistics (e.g. Willis et al., 2020; Mullins et al., 2021). The practitioner may further estimate model velocities and densities through general geophysical or empirical relationships, and apply forward seismic algorithms to estimate the seismic volume that would result from the represented reservoir (e.g., Amaru et al., 2017; Goggin et al., 2017). This allows the practitioner to compare the seismic signature of the analog model (Fig. 27) to actual seismic volumes (e.g., compare Fig. 27 to Miller et al., 2008). Process-based stratigraphic models may also be used in petroleum system basin modeling (Hosford Scheirer and Hitchcock, 2020 and references therein; Chapter 18).

## Key messages

To interpret depositional environments, one needs to see the big picture. An understanding of the basin or continental margin (Chapter 3–4), basin type and feeder system and other source-to-sink factors (Chapter 11) will determine the variety of depositional environments in any given deepwater depositional system. These controls on the system will show evidence of primary sediment delivery mechanism—whether it is via sediment gravity flow, mass transport or bottom current. It can be helpful to use a flow chart, such as the sample provided in Fig. 28.

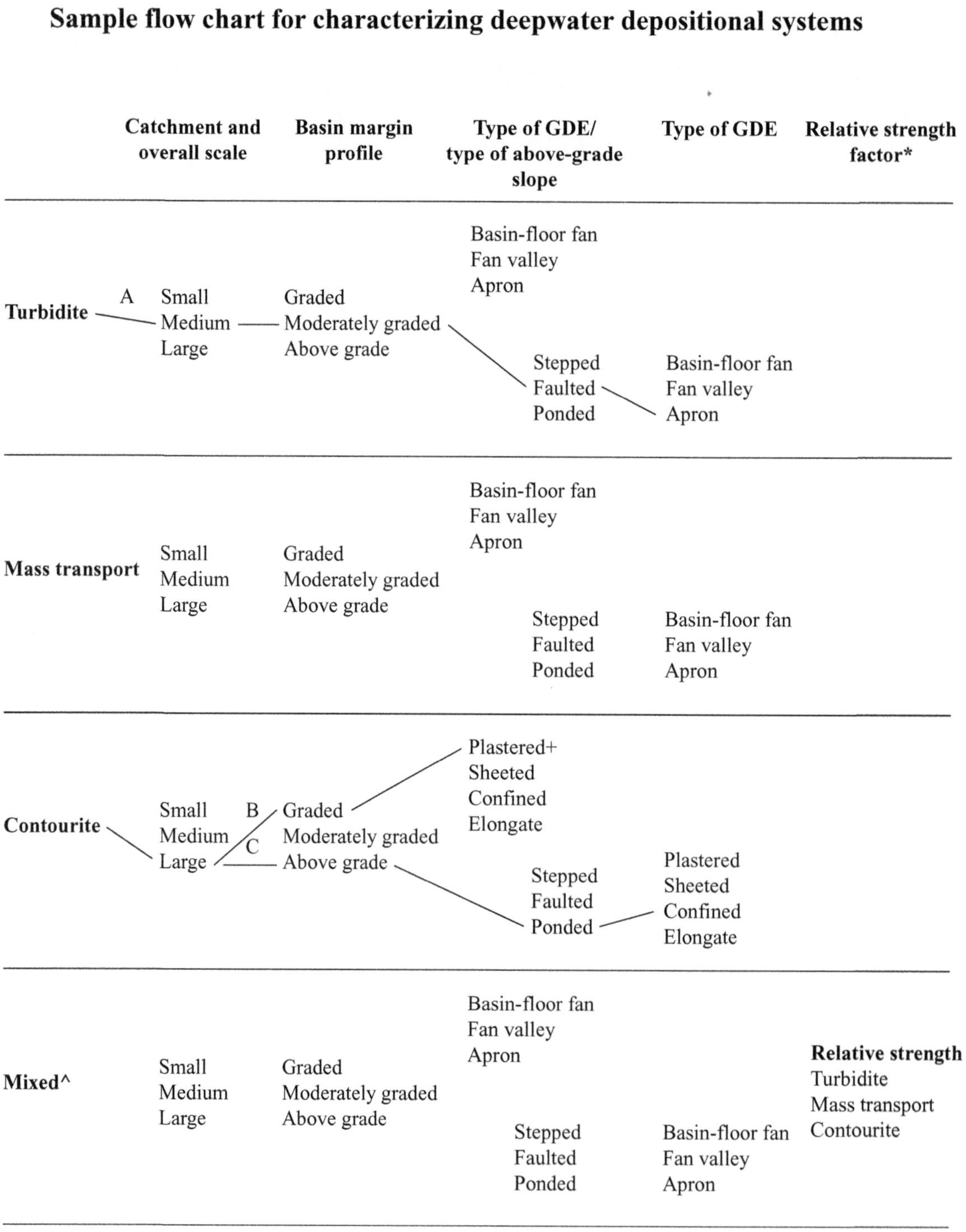


**FIG. 28** Simplified flow chart for characterizing deepwater sedimentary systems. Major factors in controlling type of deepwater gross depositional environment are overall process of sedimentation, catchment and size of system, basin margin type, position and amount of incision controlling extent of GDE and, if above grade, the classification of the type of above-grade slope. Hypothetical deepwater sedimentary system types labeled (A), (B) and (C).

It may only become apparent how a system is constructed after bed-scale observations can be made; or observations can be made at the seismic scale. It will likely take both (bottoms up and top down) to reduce the uncertainty in the interpretation. Methods for bottoms up using bed-scale observations are presented in this chapter as well as Chapters 6 and 7; methods for interpreting GDE at seismic scale are presented in this chapter as well as Chapters 3–5, 9–12 and 16. Following a robust mapping and interpretation of the area of interest, process-based modeling can be used for a sub-seismic interpretation of where the porous and non-porous rock intervals should be found.

Any broad generalized summary of a depositional system or environment is an oversimplification. The true complex nature of these systems is essentially infinite and exceptions are the rule. As cautioned above, oversimplification leads to the potential for mis-interpretation and missed opportunity. Keep in mind that every system described is but one example of what may accurately describe a subsurface environment that will never be fully "seen". In that spirit, most of deepwater is quite unremarkable, devoid of large-scale, high-energy features. Most of the ocean floor actually consists of accumulations of hemipelagic to pelagic sediment found in overbank environments. *Where do relatively high-energy deposits derived from mass transport, sediment gravity flows and bottom currents exist?* High-energy, active environments may be found along the axis of a depositional system, such as a submarine fan or in any location where potential energy is high (local or regional slopes). Fans typically consist of canyon, channel, levee, splay and overbank environments that can feature high architectural complexity due to the processes associated with sediment gravity flows. In the upper reaches of the submarine fan and in the confinement of canyon and channel systems, gravelly deposits transported by high-density turbidity currents and canyon-wall collapse or debris flows tend to dominate. Farther down in the channel-levee environment, coarser-grained and heterolithic deposits are found in the channel, and finer-grained to muddy deposits are found outside the channel due to stratification of the sediment gravity flows. In the frontal splay environment, the dramatic loss of competency and capacity of the flows causes the flows to expand and can trigger a thick accumulation of suspension deposits. Refined grain populations tend to reach the frontal splay environment due to the overbanking of fine material into the levees, and coarser material is left upslope. In the splay environment, alternation of sandy and muddy sediment causes flows to contain more mud than in other environments. As the distance from sediment input, distribution and remobilization points increases, hemipelagic and pelagic sedimentation dominates the overbank environment and, where bottom currents are present, may be re-worked into sediment waves. Suppose this submarine fan is oriented perpendicular to the coast, yet is located in a strait between two continental land masses, through which a contour current flows. Bottom currents are abundant, but the most energetic environments are at the base of the continental slope, terraces, along-slope channels, as well as at the foot of sea-floor irregularities, such as banks, highs, seamounts, gravitational scars and faults (Rebesco et al., 2014). All of the submarine fan environments previously described are affected by bottom currents that reroute and transport sediment in the direction of the bottom current, also removing the finer-grained material and leaving the coarser-grained fraction of the original turbidite. So far, this is not unlike the deepwater environments found in the Paleogene between modern day Mozambique and Madagascar, described in Fonnesu et al. (2020) and in Chapter 9. While the finer-grained sediment may be carried in the direction of the contour current, the channel-levee system may migrate *up current* due to the relatively weakly confined channel.

## Conclusions

Deepwater sedimentation can occur in any number of tectonic settings including compression, extension or transform regimes (Ingersoll, 2011). Typically, deepwater sedimentation styles are divided into two groups based on the basin margin: active vs passive margin. While submarine fans and their variations are the typical expression of a high-energy, sediment-rich deepwater depositional system, fans are typically much larger along passive margins than active margins. Any sediment gravity flow-fed deepwater depositional system will be constructed of many, if not all, of the six primary deepwater environments, including canyons, channels, levees, lobes (splays), overbank and drifts.

Canyons represent the main environment responsible for delivering vast amounts of sediment to deep parts of the basin. Canyons typically fully confine sediment gravity flows. Canyons can reach tens of kilometers in width and a few kilometers in depth. Canyons gain width and depth through mass wasting of their margins and evolution of rill-type conduits to larger and larger capacitors of sediment. Canyons are typically intimately linked to their feeder system, often a delta. In some cases, such as the Congo Canyon, the canyon head is incised into the mouth of the river delta, thereby providing a direct link for sediment delivery on its source-to-sink voyage downsystem.

Channels are the downslope equivalent of canyons and can confine flows due to erosion and incision into the seafloor, or by growth of the height of the levee above the thalweg, a modeling parameter often termed HACT (height above the channel thalweg).

Levees are a positive topographic feature that act to bound channels. Levees form through three often overlapping processes: (i) spillover of the fine-grained fraction of the sediment gravity flow along the channel margins; (ii) overbanking of sediment along channel bends due to run up of the flow along the channel margins; (iii) continuation of the upper fraction of the flow out of the channel due to momentum as the lower part of the flow remains in the channel. Levee growth can increase the sinuosity of the channel, and can increase a system's efficiency by allowing flows to transmit sediment farther into the basin (see Chapter 6). Levees preferentially form along channels at lower gradient and from the abundance of turbidity currents; slope channels do not have well-developed levees compared to basin-floor channels. Further, debrite-dominated systems do not show this level of architectural complexity of overlapping channel-levee complexes.

Splays (lobes) are bodies of sediment in an area of unconfinement originating from an area of confinement. Splays can take many forms, including those described as frontal, overbank, crevasse, avulsion, channelized or detached. Avulsion splays play an important role for channel avulsions and serve to establish new routes for sediment transport in deepwater environments. Naturally, splays are distributary environments compared to channels, which are interpreted as confined environments.

The overbank is the default area of deposition in deepwater since submarine fan features are relatively rare considering the scale of the oceans. Overbank settings are anywhere far from input from canyons or channels. Much of the ocean floor consists of sediment largely derived from pelagic and hemipelagic processes. However, the overbank area can have diverse types of deposits, including those originating from bottom currents, sediment gravity flows, other mass flows including mass transport deposits and slumps, as well as more exotic forms of sedimentation such as mud volcanoes and sandy injectites. Overbank is also often difficult to determine in some cases where levee and splay deposits are so thinly bedded. In these cases, descriptive combinations such as *levee-overbank* or *splay-overbank* are used. Overbank areas remain the least surveyed areas on the ocean floor, so may represent areas where future scientific discoveries will concentrate.

Perhaps one of the biggest changes to deepwater thinking inspired by high-resolution global seafloor mapping campaigns is the recognition of sedimentary features associated with along-slope currents or bottom currents. These processes create environments including drifts, mounds, moats and other hybrid environments such as channel-drifts. Depositional systems that have both active delivery of sediment gravity flows and bottom currents are termed *mixed* and can be further investigated in the next chapter, Chapter 9.

While the characterization of basins, continental margins and their deepwater environments has led to a standardized terminology and hierarchy (a way of talking the same language), process-based modeling efforts reveal how time lines (sequence stratigraphic surfaces) and idealized geobodies (3D shapes of canyon-fill, channel-fill, etc.) are often not in sync, generating a complex, heterogeneous distribution of porous and non-porous sedimentary deposits. Process-based modeling represents the culmination of concepts presented in this chapter and in Chapters 6 and 7 to demonstrate how petroleum resources can be more efficiently and effectively produced from porous reservoirs.

## How are deltas different from submarine fans?

When mapping frontier areas, many times the question is asked: *is this a delta or submarine fan?* While the question may seem simple, it can lead to a meaningful discussion. *Why not?* Deltas and submarine fans, especially frontal splay deposits, exhibit very similar patterns characterized by feeder channels, distributary channel systems, broad, fan-like shapes, grain-size fining from axis to off-axis and from proximal to distal, as well as similar scale (deltas and submarine fans can be tens to hundreds of kilometers in width and length). Up close, sedimentation units show a variety of key differences including characteristic traits of river-, wave- and tide-dominated sedimentation vs the sedimentation of sediment gravity flows and their processes related to fluid turbulence (turbidity currents) and matrix strength (debris flows). Context is necessary to distinguish between submarine fan environments and deltaic environments—the inquiring scientist needs to define upslope from downslope and landward from basinward and fit key environments along systems tracts. For additional discussion on how deltaic and deepwater sedimentation varies and has been confused in the past in both famous outcrops and deepwater oil and gas fields, see Chapters 6, 7 and 19.

## Acknowledgments

This chapter benefited greatly from constructive reviews by F. Javier Hernández-Molina, Octavian Catuneanu, Cindy Yeilding and Richard Sears as well as Chevron. The authors would like to thank Grant Wach for guiding this chapter during its early stages. J. R. dedicates this chapter to Sherrill (Sherry) Hossom Muller whose rare and exceptional gift for drawing and interpreting the natural world helped inspire and encourage many of the depositional models presented in several chapters in this book. This work would not have been possible without conversations and guidance throughout the years from the following: D. R. Lowe, S. A. Graham, G. H. Browne, P. R. King, M. Arnot, E. and S. Muller, J. R. Suter, A. D. Donovan, A. Marchand, A. Hough, D. Nelson, P. M. Myrow, J. A. Noblett, C. Siddoway, *The Drifters* Research Group, D. Chiarella, R. Weber, M. Brown, S. Verma, M. Honarpour, C. Pirmez, Z. Sylvester, J. Pacht, W. D. McCaffrey, G. Apps, F. J. Peel, P. D. W. Haughton, M. D. Sullivan, B. B. Chan, G. R. Sharman, T. McHargue, K. W. Rudolph, D. Adams, J. R. Gaither, M. L. Sweet, L. Fielding, F. Pohl, Y. Gavillot, N. Uzunlar, A. Lisenbee, M. McKeown & family, E. Szymanski, B. E. Prather, C. VanDenburg, M. Fonnesu, M. G. Rowan, B. Fisher, R. Steel, C. Olariu, T. Sun, M. Amaru, E. Meyers, A. Madof, and organizations including AAPG, EAGE, HOT Engineering, GNS Science, Houston Explorers Club, Gulf Coast Section of SEPM, CERAWeek and many others. The California and New Zealand studies conducted from 2011-2013 were made possible by the Siemon W. Muller Graduate Research Fellowship at Stanford University.

## References

Aas, T.E., Howell, J.A., Janocko, M., Jackson, C.L., 2010a. Control of Aptian palaeobathymetry on turbidite distribution in the Buchan graben, outer Moray firth, Central North Sea. Mar. Pet. Geol. 27 (2), 412–434.

Aas, T.E., Howell, J.A., Janocko, M., Midtkandal, I., 2010b. Re-created early Oligocene seabed bathymetry and process-based simulations of the Peïra cava turbidite system. J. Geol. Soc. 167 (5), 857–875.

Aas, T.E., Basani, R., Howell, J., Hansen, E., 2014. Forward modelling as a method for predicting the distribution of deep-marine sands: an example from the Peïra cava sub-basin. Geol. Soc. Lond. Spec. Publ. 387 (1), 247–269.

Abd El-Gawad, S., Cantelli, A., Pirmez, C., Minisini, D., Sylvester, Z., Imran, J., 2012. Three-dimensional numerical simulation of turbidity currents in a submarine channel on the seafloor of the Niger Delta slope. J. Geophys. Res. Oceans 117 (C5).

Abreu, V., Sullivan, M., Pirmez, C., Mohrig, D., 2003. Lateral accretion packages (LAPs): an important element in deep water sinuous channels. Mar. Pet. Geol. v. 20, 631–648.

Adeogba, A.A., McHargue, T.R., Graham, S.A., 2005. Transient fan architecture and depositional controls from near-surface 3-D seismic data, Niger Delta continental slope. AAPG Bull. 89 (5), 627–643.

Albertão, G., 2010. Control of the Submarine Palaeotopography on the Turbidite System Architecture: An Approach Combining Structural Restorations and Sedimentary Process-Based Numerical Modeling, Applied to a Brazilian Offshore Case Study. (Doctoral Dissertation Bordeaux 1).

Albertão, G.A., Eschard, R., Mulder, T., Teles, V., Chauveau, B., Joseph, P., 2015. Modeling the deposition of turbidite systems with cellular automata numerical simulations: a case study in the Brazilian offshore. Mar. Pet. Geol. 59, 166–186.

Allan, M.R., Goffey, G.P., Morgan, R.K., Walker, I.M., 2006. The deliberate search for the stratigraphic trap: an introduction. Geol. Soc. Lond. Spec. Publ. 254, 1–5.

Alves, T.M., 2015. Submarine slide blocks and associated soft-sediment deformation in deep-water basins: a review. Mar. Pet. Geol. 67, 262–285.

Amaru, M., et al., 2019. Integrated computational stratigraphy reservoir characterization and seismic validation. In: AAPG International Conference & Exhibition, Buenos Aires.

Amaru, M., Sun, T., Goggin, L., Harris, A., 2017. Integration of computational stratigraphy models and seismic data for subsurface characterization. Lead. Edge 36 (11), 947a1–947a6.

Amy, L., Kneller, B.C., McCaffrey, W.D., 2007. Facies architecture of the Grès de Peira Cava, SE France: landward stacking patterns in ponded turbiditic basins. Geol. Soc. Lond. 164, 143–162.

Anderson, K.S., 1998. Facies Architecture of Two Paleogene Structurally-Controlled Turbidite Systems, Central California. Stanford University Ph. D. thesis, p. 314.

Anderson, K.S., Hickson, T.A., Crider, J.G., Graham, S.A., 1999. Integrating teaching with field research in the wagon rock project. J. Geosci. Educ. 47, 227–235.

Anderson, K.S., Graham, S.A., Hubbard, S.M., 2006. Facies, architecture, and origin of a reservoir-scale sand-rich succession within submarine canyon fill: insights from wagon caves rock (Paleocene), Santa Lucia range, California, U.S.A. J. Sediment. Res. 76, 819–838.

Apps, G.M., 1987. Evolution of the Grès d'Annot Basin, SW Alps. Ph D thesis, University of Liverpool.

Athmer, W., Groenenberg, R.M., Luthi, S.M., Donselaar, M.E., Sokoutis, D., Willingshofer, E., 2010. Relay ramps as pathways for turbidity currents: a study combining analogue sandbox experiments and numerical flow simulations. Sedimentology 57 (3), 806–823.

Babboneau, N., Savoye, B., Cremer, M., Bez, M., 2010. Sedimentary architecture in meanders of a submarine channel: detailed study of the present Congo turbidite channel (Zaiango project). J. Sediment. Res. 80, 852–866.

Bain, H.A., Hubbard, S.M., 2016. Stratigraphic evolution of a long-lived submarine channel system in the late cretaceous Nanaimo group, British Columbia, Canada. Sediment. Geol. 337, 113–132.

Bakke, K., Kane, I., Martinsen, O.J., Petersen, S.A., Johansen, T.A., Hustoft, S., Jacobsen, F.H., Groth, A., 2014. Seismic modeling in the analysis of Deepwater sandstone termination styles. AAPG Bull. 97 (9), 1395–1419.

Bankole, S.A., 2018. Methods, Microstructure, and Mudrocks: Towards an Improved Understanding of Deepwater Mudrocks. PhD thesis, Heriot-Watt University, Edinburgh, UK.

Barton, M., O'Byrne, C., Pirmez, C., Prather, B.E., Van Der Vlugt, F., Alpak, F.O., 2009. Turbidite channel architecture: recognizing and quantifying the distribution of channel-base drapes using core and dipmeter data. In: Poppelreiter, M., Garcia-Carballido, C., Kraaijveld, M. (Eds.), Dipmeter and Borehole Image Log technoLogy. 92. AAPG Memoir, https://doi.org/10.1306/13181284M923289.

Basani, R., Janocko, M., Cartigny, M.J., Hansen, W.M., Eggenhuisen, J.T., 2014. MassFLOW-3D TM as a simulation tool for turbidity currents?: some preliminary results. IAS Spec. Publ. 46, 587–608.

Beaubouef, R.T., Rossen, C.R., Zelt, F.B., Sullivan, M.D., Mohrig, D.C., Jennette, D.C., Bellian, J.A., Friedman, S.J., Lovell, R.W., Shannon, D.S., 1999. Deepwater sandstones, Brushy Canyon Formation, West Texas. American Association of Petroleum Geologists Continuing Education Course Note Series 40, p. 48.

Belderson, R.H., Kenyon, N.H., Stride, A.H., Pelton, C.D., 1984. A braided distributary system on the Orinoco fan. Mar. Geol. 56, 196–206.

Bell, D., Hodgson, D.M., Pontén, A.S.M., Hansen, L.A.S., Flint, S.S., Kane, I.A., 2020. Stratigraphic hierarchy and three-dimensional evolution of an exhumed submarine slope channel system. Sedimentology 67 (6), 3259–3289.

Bhattacharya, S., Verma, S., Rotzien, J.R., 2020. Seismic Facies and 3D Attributes of Submarine Slides on the North Slope. Interpretation, Alaska, U.S.A.

Blum, M.D., Milliken, K.T., Pecha, M.A., Snedden, J.W., Frederick, B.C., Galloway, W.E., 2017. Detrital-zircon records of Cenomanian, Paleocene, and Oligocene Gulf of Mexico drainage integration and sediment routing: implications for scales of basin-floor fans. Geosphere 13 (6), 2169–2205.

Bouma, A.H., 1962. Sedimentology of some Flysch Deposits: A Graphic Approach to Facies Interpretation. Elsevier, Amsterdam, p. 167.

Brooks, H.L., Hodgson, D.M., Brunt, R.L., Peakall, J., Poyatos-More, M., Flint, S.S., 2018. Disconnected submarine lobes as a record of stepped slope evolution over multiple sea-level cycles. Geosphere 14 (4), 1753–1779.

Browne, G.H., Slatt, R.M., 2002. Outcrop and behind-outcrop characterization of a late Miocene slope fan (channel-levee complex), Mt messenger formation, New Zealand. Am. AAPG Bull. 86, 841–862.

Browne, G.H., King, P.R., Arnot, M.J., Slatt, R.M., 2007. Architecture of base-of-slope fans, Mount Messenger Formation, Pukearuhe Beach, New Zealand. In: Nilsen, T.H., Shew, R.D., Steffens, G.S., Studlick, J.R.J. (Eds.), Atlas of Deepwater Outcrops. AAPG Studies in Geology 56. AAPG, Tulsa. Okla, pp. 257–261 (Chapter 64).

Burgess, P.M., Masiero, I., Toby, S.C., Duller, R.A., 2019. A big fan of signals? Exploring autogenic and allogenic process and product in a numerical stratigraphic forward model of submarine-fan development. J. Sediment. Res. 89 (1), 1–12.

Burgreen, B., Graham, S.A., 2014. Evolution of a deepwater lobe system in the Neogene trench-slope setting of the East Coast Basin, New Zealand: lobe stratigraphy and architecture in a weakly confined basin configuration. Mar. Pet. Geol. 54, 1–22.

Caldwell, R.L., Sun, T., 2019. Deepwater distributary system avulsion processes and geomorphic signatures. In: AGU Fall Meeting Abstracts. vol. 2019, p. EP13D-2171.

Caldwell, R.L., Sun, T., Willis, B., Baumgardner, S., Harris, A., Sullivan, M., 2018. Linking deepwater fan depositional processes and stratigraphic architecture using computational stratigraphy: GSA 2018 Annual Meeting, Indianapolis, Paper No. 46–16. In: Geological Society of America Abstracts with Programs. vol. 50. No. 6.

Campbell, C.V., 1967. Lamina, laminaset, bed, and bedset. Sedimentology 8, 7–26.

Campion, K.M., Sprague, A.R., Mohrig, D., Lovell, P., Drzewicki, P.A., Sullivan, M.D., Ardill, J.A., Jensen, G.N., Sickafoose, D.K., 2000. Outcrop expression of confined channel complexes. In: Weimer, P., Slatt, R.M., Coleman, J., Rosen, N.C., Nelson, H., Bouma, A.H., Styzen, M.J., Lawrence, D.T. (Eds.), Deepwater Reservoirs of the World. Houston Gulf Coast Section SEPM Foundation, pp. 127–150.

Campion, K.M., Dixon, B.T., Scott, E.D., 2011. Sediment waves and depositional implications for fine-grained rocks in the Cerro Toro formation (upper cretaceous), Silla Syncline, Chile. Mar. Pet. Geol. 28, 761–784.

Carlson, J., Grotzinger, J.P., 2001. Submarine fan environment inferred from turbidite thickness distributions. Sedimentology 48, 1331–1351.

Carvajal, C.R., Steel, R.J., 2006. Thick turbidite successions from supply-dominated shelves during sea-level highstand. Geology 34 (8), 665–668.

Casson, M., Calves, G., Huuse, M., Sayers, B., Redfern, J., 2020. Cretaceous continental margin evolution revealed using quantitative seismic geomorphology, offshore northwest Africa. Basin Res. https://doi.org/10.1111/bre.12455.

Cavazza, W., DeCelles, P.G., 1993. Geometry of a Miocene submarine canyon and associated sedimentary facies in southeastern Calabria, southern Italy. GSA Bull. 105, 1297–1309.

Chen, C., Hiscott, R.N., 1999. Statistical analysis of facies clustering in submarine-fan turbidite successions. J. Sediment. Res. 69 (2), 505–517.

Chenrai, P., Huuse, M., 2017. Pockmark formation by porewater expulsion during rapid progradation in the offshore Taranaki Basin, New Zealand. Mar. Pet. Geol. 82, 399–413.

Clark, J.D., Pickering, K.T., 1996. Architectural elements and growth patterns of submarine channels: applications to hydrocarbon exploration. AAPG Bull. 80 (2), 194–220.

Clark, B.E., Steel, R.J., 2006. Eocene Turbidite-population statistics from shelf edge to basin floor,Spitsbergen, Svalbard. J. Sediment. Res. 76, 903–918.

Cobain, S.L., Hodgson, D.M., Peakall, J., Shiers, M.N., 2017. An integrated model of clastic injectites and basin floor lobe complexes: implications for stratigraphic trap plays. Basin Res. 29, 816–835.

Collins, J., Kenyon-Roberts, S., Cullen, B., White, J., Bordas-Le Floch, N., Downey, J., 2015. Arran Field: a complex heterolithic reservoir on the margins of the Forties Fan System. Geol. Soc. Lond. Spec. Publ. 403 (1). https://doi.org/10.1144/SP403.10.

Cornard, P., Pickering, K.T., 2020. Submarine topographic control on distribution of supercritical-flow deposits in lobe and related environments, middle Eocene, Jaca Basin, Spanish Pyrenees. J. Sediment. Res. 90, 1222–1243.

Covault, J.A., Graham, S.A., 2010. Submarine fans at all sea-level stands: tectono-morphologic and climatic controls on terrigenous sediment delivery to the deep sea. Geology 38 (10), 939–942.

Covault, J.A., Fildani, A., Romans, B.W., McHargue, T., 2011. The natural range of submarine canyon-and-channel longitudinal profiles. Geosphere 7, 313–332.

Crane, W.H., Lowe, D.R., 2007. Architecture and evolution of the paine channel complex, cerro toro formation (upper cretaceous), Silla syncline, Magallanes basin, Chile. Sedimentology 55, 979–1009.

Cronin, B.T., Kidd, R.B., 1998. Heterogeneity and lithotype distribution in ancient deep-sea canyons: point lobos deep-sea canyon as a reservoir analogue. Sediment. Geol. 115, 315–349.

Cronin, B.T., Hurst, A., Celik, H., Turkmen, I., 2000. Superb exposure of a channel, levee and overbank complex in an ancient deepwater slope environment. Sediment. Geol. 132, 205–216.

Cumberpatch, Z.A., Kane, I.A., Soutter, E.L., Hodgson, D.M., Jackson, C.A.-L., Kilhams, B.A., Poprawski, Y., 2020. Interactions of deepwater gravity flows and active salt tectonics. J. Sediment. Res. In press.

De Castro, S., Hernandez-Molina, F.J., Rodriguez-Tovar, F.J., Llave, E., Ng, Z.L., Nishida, N., Mena, A., 2020. Contourites and bottom current reworked sands: bed facies model and implications. Mar. Geol. v. 428, 106267.

De Ruig, M.J., Hubbard, S.M., 2006. Seismic facies and reservoir characteristics of a deep-marine channel belt in the Molasse foreland basin, Puchkirchen formation, Austria. AAPG Bull. 90 (5), 735–752.

De Weger, W., Hernandez-Molina, F.J., Flecker, R., Sierro, F.J., Chiarella, D., Krijgsman, W., Manar, A.M., 2020. Late Miocene contourite channel system reveals intermittent overflow behavior. Geology 48 (12), 1194–1199.

Deptuck, M.E., Sylvester, Z., 2018. Submarine fans and their channels, levees, and lobes. In: Micallef, A., Krastel, S., Savini, A. (Eds.), Submarine Geomorphology. Springer International Publishing, pp. 273–299.

Deptuck, M.E., Steffens, G.S., Barton, M., Pirmez, C., 2003. Architecture and evolution of upper fan channel-belts on the Niger Delta slope and in the Arabian Sea. Mar. Pet. Geol. 20, 649–676.

Deptuck, M.E., Sylvester, Z., Pirmez, C., O'Byrne, C., 2007. Migration–aggradation history and 3-D seismic geomorphology of submarine channels in the Pleistocene Benin-major canyon, western Niger Delta slope. Mar. Pet. Geol. 24, 406–433.

Di Celma, C.N., Brunt, R.L., Hodgson, D.M., Flint, S.S., Kavanaugh, J.P., 2011. Spatial and temporal evolution of a Permian submarine slope channel-levee system, Karoo basin, South Africa. Int. J. Sediment Res. 81, 579–599.

Dixon, J.F., Steel, R.J., Olariu, C., 2012. Shelf-edge delta regime as a predictor of deep-water deposition. J. Sediment. Res. 82 (9), 681–687.

Donnelly, S., Beauboeuf, R., Bracht, R., Fitzsimmons, R., 2019. Regional depositional setting of the West Orphan Basin: The importance of contour current processes on stratigraphic trap generation in a hybrid depositional system. In: AAPG Annual Convention, San Antonio.

Donovan, A.D., 2010. What are the shelf and slope breaks and why the rise and run of the intervening slope matters for deepwater plays and sequence models. Houst. Geol. Soc. Bull. 52 (9), 15. 17.

Duerichen, E.T., Lowe, D.R., 2007. Sedimentology and architecture of coarse-grained submarine channel fill, Lobitos Village Section, Talara Basin, Northwestern Peru. AAPG Stud. Geol. 56 (Atlas of Deepwater Outcrops, Chapter 140).

Elliott, T., 2000. Depositional architecture of a sand-rich, channelized turbidite system: The Upper Carboniferous Ross Sandstone Formation, western Ireland. In: Weimer, P., Slatt, R.M., Coleman, J., Rossen, N.C., Nelson, H., Bouma, A.H., Styzen, M.J., Lawrence, D.T. (Eds.), Deepwater Reservoirs of the World. In: 20th Annual Perkins Research Conference. Gulf Coast Section SEPM, pp. 342–373.

Empinotti, T., 2020. Geological Controls on Depositional Architecture of Turbiditic Deposits from Stepped Slope-to-Basin Profiles. Thesis.

Englert, R.G., Hubbard, S.M., Cartigny, M.J.B., Clare, M.A., Coutts, D.S., Hage, S., Clarke, J.H., Jobe, Z., Lintern, D.G., Stacey, C., Vendettuoli, D., 2020. Quantifying the three-dimensional stratigraphic expression of cyclic steps by integrating seafloor and Deepwater outcrop observations. Sedimentology. https://doi.org/10.1111/sed/12772.

Fildani, A., Normark, W.R., 2004. Late quaternary evolution of channel and lobe complexes of Monterey fan. Mar. Geol. 206, 199–223.

Fisher, W.L., Galloway, W.E., Steel, R.J., Olariu, C., Kerans, C., Mohrig, D., 2021. Deep-water depositional systems supplied by shelf-incising submarine canyons: recognition and significance in the geologic record. Earth Sci. Rev. 214, 103531.

Flood, R.D., Manley, P.L., Kowsmann, K.O., Appi, C.J., Pirmez, C., 1991. Seismic facies and late quaternary growth of Amazon submarine fan. In: Weimer, P., Link, M.H. (Eds.), Seismic Facies and Sedimentary Processes of Submarine Fans and Turbidite Systems. Springer-Verlag, New York, pp. 415–434.

Fonnesu, M., Palermo, D., Galbiati, M., Marchesini, M., Bonamini, E., Bendias, D., 2020. A new world-class Deepwater play type, deposited by the syndepositional interaction of turbidity flows and bottom currents, the giant eocene coral field in northern Mozambique. Mar. Pet. Geol. 111, 179–201.

Foreman, B., Lai, S., Komatsu, Y., Paola, C., 2015. Braiding of submarine channels controlled by aspect ratio similar to rivers. Nat. Geosci. 8, 700–703. https://doi.org/10.1038/ngeo2505.

Franks, S.G., Uchytil, S.J., 2017. Geochemistry of formation waters from the subsalt Tubular Bells Field, offshore Gulf of Mexico: implications for fluid movement and reservoir continuity. AAPG Bull. 100 (6), 943–967.

Furin, M., 2019. Alaska's recent giant Pikka discovery, wildcatting ain't dead yet: 'Yuge discovery, yuge potential'. In: Presentation at Houston Explorers Club Leadership Luncheon, St Regis Hotel, Houston, 19–Nov.

Galloway, W.E., 1998. Siliciclastic slope and base-of-slope depositional systems: component facies, stratigraphic architecture, and classification. AAPG Bull. 82 (4), 569–595.

Gardiner, A.R., 2006. The variability of turbidite sandbody pinchout and its impact on hydrocarbon recovery in stratigraphically trapped fields. In: Allen, M.R., Goffey, G.P., Morgan, R.K., Walker, I.M. (Eds.), The Deliberate Search for the Stratigraphic Trap, vol. 254. Geological Society, London, pp. 267–288. Special Publications.

Gardner, M.H., Borer, J.M., Melick, J.J., Mavilla, N., Dechesne, M., Wagerle, R.N., 2003. Stratigraphic process–response model for submarine channels and related features from studies of Permian brushy canyon outcrops,West Texas. Mar. Pet. Geol. 20, 757–787.

Ge, Z., Nemec, W., Gawthorpe, R.L., Hansen, E.W., 2017. Response of unconfined turbidity current to normal-fault topography. Sedimentology 64 (4), 932–959.

Ge, Z., Nemec, W., Gawthorpe, R.L., Rotevatn, A., Hansen, E.W., 2018. Response of unconfined turbidity current to relay-ramp topography: insights from process-based numerical modelling. Basin Res. 30 (2), 321–343.

Ghosh, B., Lowe, D.R., 1993. The architecture of deepwater channel complexes, Cretaceous Venado Sandstone Member, Sacramento Valley, California. In: Graham, S.A., Lowe, D.R. (Eds.), Advances in the Sedimentary Geology of the Great Valley Group, Sacramento Valley California. vol. 7. SEPM Pacific Section, pp. 51–65.

Giles, K.A., Rowan, M.G., 2012. Concepts in halokinetic-sequence deformation and stratigraphy. Geol. Soc. Lond. Spec. Publ. 363 (1), 7–31.

Goggin, L., 2020. Exploring the limits of seismic interpretation techniques through the use of computational stratigraphy models. In: AAPG Distinguished Lecture Series, 2019-20 Season.

Goggin, L., Sun, T., Amaru, M., Harris, A., Dutranois, A., Madof, A., 2017. 3-D volumetric interpretation with computational stratigraphy models. Houst. Geol. Soc. Bull. 60 (3), 21.

Gong, C.L., Wang, Y.M., Zhu, W.L., et al., 2011. The central submarine canyon in the Qiongdongnan Basin, northwestern South China Sea: architecture, sequence stratigraphy, and depositional processes. Mar. Pet. Geol. 28, 1690–1702.

Granjeon, D., 1996. Modélisation Stratigraphique Déterministe - Conception et Applications d'un Modèle Diffusif 3D Multi-Lithologique, Géosciences Rennes. Université de Rennes 1, Rennes.

Grecula, M., Hognestad, J., Price, S., Boya Ferrero, M., De Bruijn, G., Noraberg, K.T., Engenes, K., Mears, P., Van Ojik, K., McGarva, R., 2014. Interplay of fan-fringe reservoir deterioration and hydrodynamic aquifer: understanding the margins of gas development in the Ormen Lange field. In: McKie, T., Rose, P.T.S., Hartley, A.J., Jones, D.W., Armstrong, T.L. (Eds.), Tertiary deep-marine reservoirs of the North Sea Region. vol. 403. Geological Society, London, Special Publications, pp. 157–183.

Greene, T.J., Surpless, K.D., 2017. Facies architecture and provenance of a boulder-conglomerate submarine channel system, panoche formation, great valley group: a forearc basin response to middle cretaceous tectonism in the California convergent margin. Geosphere 13 (3), 32.

Groenenberg, R.M., 2007. Process-Based Modelling of Turbidity-Current Hydrodynamics and Sedimentation. Thesis.

Groenenberg, R.M., Sloff, K., Weltje, G.J., 2009. A high-resolution 2-DH numerical scheme for process-based modeling of 3-D turbidite fan stratigraphy. Comput. Geosci. 35 (8), 1686–1700.

Groenenberg, R.M., Hodgson, D.M., Prelat, A., Luthi, S.M., Flint, S.S., 2010. Flow–deposit interaction in submarine lobes: insights from outcrop observations and realizations of a process-based numerical model. J. Sediment. Res. 80 (3), 252–267.

Hansen, L.A.S., Hodgson, D.M., Pontén, A., Bell, D., Flint, S., 2019. Quantification of basin-floor fan pinchouts: examples from the Karoo Basin, South Africa. Front. Earth Sci. 7 (12), 20.

Harris, A.D., Covault, J.A., Madof, A.S., Sun, T., Sylvester, Z., Granjeon, D., 2016. Three-dimensional numerical modeling of eustatic control on continental-margin sand distribution. J. Sediment. Res. 86 (12), 1434–1443.

Haughton, P.D.W., 1994. Deposits of deflected and ponded turbidity currents, Sorbas Basin, Southeast Spain. J. Sediment. Res. 64, 233–246.

Haughton, P.D.W., Davis, C., McCaffrey, W., Barker, S., 2009. Hybrid sediment gravity flow deposits – classification, origin and significance. Mar. Pet. Geol. 26, 1900–1918.

Hawie, N., Covault, J.A., Dunlap, D., Sylvester, Z., 2018. Slope-fan depositional architecture from high-resolution forward stratigraphic models. Mar. Pet. Geol. 91, 576–585.

Hawie, N., Covault, J.A., Sylvester, Z., 2019. Grain-size and discharge controls on submarine-fan depositional patterns from forward stratigraphic models. Front. Earth Sci. 7, 334.

Heimsund, S., 2007. Numerical Simulation of Turbidity Currents: A New Perspective for Small-and Large-Scale Sedimentological Experiments (Master's thesis). The University of Bergen.

Henry, L.C., Wadsworth, J.A., Hansen, B., Hartman, K., 2018. Erosion and ponding of thunder horse Deepwater turbidites by mass transport complexes in Mississippi canyon based on image log sedimentology. Mar. Pet. Geol. 97, 639–658.

Hernández-Molina, F.J., 2021. Do you go with the flow? Observations on contourite and mixed systems. In: Virtual Get Together. Society of Sedimentary Geology. Feb. 10.

Hesse, R., 1992. Continental slope sedimentation adjacent to an ice margin. I. Seismic facies of labrador slope. Geo-Mar. Lett. 12, 189–199.

Hesse, R., Rakofsky, A., 1992. Deep-sea channel/submarine-Yazoo system of the Labrador Sea: a new deepwater facies model. Am. Assoc. Pet. Geol. Bull. 76 (5), 680–707.

Hickson, T.A., Lowe, D.R., 2002. Facies architecture of a submarine fan channel-levee- overbank complex: the Juniper ridge conglomerate, Coalinga, California. Sedimentology 49 (2), 335–362.

Higgs, K.E., Arnot, M.J., Brindle, S., 2015. Advances in grain-size, mineral, and pore-scale characterization of lithic and clay-rich reservoirs. AAPG Bull. 99 (7), 1315–1348.

Hodgson, D., 2019. LOBE or 'Laterally Offset Bulbous Elements': Keynote Lecture for British Sedimentological Research Group AGM, Royal Holloway. University of London.

Hosford Scheirer, A., Hitchcock, J. (Eds.), 2020. Basin and Petroleum System Modeling 13th Annual Industrial Affiliates Meeting. Stanford University. 10–11 Nov.

Howlett, D.M., Ge, Z., Nemec, W., Gawthorpe, R.L., Rotevatn, A., Jackson, C.A.L., 2019. Response of unconfined turbidity current to deep-water fold and thrust belt topography: orthogonal incidence on solitary and segmented folds. Sedimentology 66 (6), 2425–2454.

Hoyal, D.C.J.D., Sun, H., Vishnampet, R., Shringarpure, M., Prasad, B., Fedele, J.J., Denommee, K.C., 2018. Insights into controls on submarine fan evolution from depth-averaged model of turbidity currents. In: AGU Fall Meeting Abstracts. vol. 2018, p. EP51A-01.

Huang, H., Imran, J., Pirmez, C., 2005. Numerical model of turbidity currents with a deforming bottom boundary. J. Hydraul. Eng. 131 (4), 283–293.

Huang, H., Imran, J., Pirmez, C., 2007. Numerical modeling of poorly sorted depositional turbidity currents. J. Geophys. Res. Oceans 112 (C1).

Huang, H., Imran, J., Pirmez, C., 2008. Numerical study of turbidity currents with sudden-release and sustained-inflow mechanisms. J. Hydraul. Eng. 134 (9), 1199–1209.

Hubbard, R.J., Pape, J., Roberts, D.G., 1985a. Depositional sequence mapping as a technique to establish tectonic and stratigraphic framework and evaluate hydrocarbon potential on a passive continental margin. In: Berg, O.R., Woolverton, D.G. (Eds.), Seismic Stratigraphy II: An Integrated Approach to Hydrocarbon Exploration. vol. 39. AAPG Memoir, pp. 79–91.

Hubbard, R.J., Pape, J., Roberts, D.G., 1985b. Depositional sequence mapping to illustrate the evolution of a passive continental margin. In: Berg, O.R., Woolverton, D.G. (Eds.), Seismic Stratigraphy II: An Integrated Approach to Hydrocarbon Exploration. vol. 39. AAPG Memoir, pp. 93–115.

Hubbard, R.J., Edrich, S.P., Ratey, R.P., 1990. Geologic evolution and hydrocarbon habitat of the "Arctic Alaska microplate". In: Brooks, J. (Ed.), Classic Petroleum Provinces. vol. 50. Geological Society of London, Special Publication, pp. 143–187.

Hubbard, S.M., Jobe, Z.R., Romans, B.W., Covault, J.A., Sylvester, Z., Fildani, A., 2020. The stratigraphic evolution of a submarine channel: linking seafloor dynamics to depositional products. J. Sediment. Res. 90 (7), 673–686.

Hurst, A., Scott, A., Vigorito, M., 2011. Physical characteristics of sand injectites. Earth-Sci. Rev. 106 (3–4), 215–246.

Ingersoll, R.V., 2011. In: Busby, C., Perez, A.A. (Eds.), Tectonics of Sedimentary Basins: Recent Advances, first ed. Blackwell Publishing Ltd, p. 664.

Janocko, M., Nemec, W., Henriksen, S., Warchol, M., 2013a. The diversity of Deepwater sinuous channel belts and slope valley-fill complexes. Mar. Pet. Geol. 41, 7–34.

Janocko, M., Cartigny, M.B.J., Nemec, W., Hansen, E.W.M., 2013b. Turbidity current hydraulics and sediment deposition in erodible sinuous channels: laboratory experiments and numerical simulations. Mar. Pet. Geol. 41, 222–249.

Jobe, Z.R., Lowe, D.R., Uchytil, S.J., 2011. Two fundamentally different types of submarine canyons along the continental margin of Equatorial Guinea. Mar. Pet. Geol. 28 (3), 843–860.

Kane, I.A., Hodgson, D.M., 2011. Sedimentological criteria to differentiate submarine channel levee subenvironments: exhumed examples from the Rosario formation (Upper Cretaceous) of Baja California, Mexico, and the fort Brown formation (Permian), Karoo Basin, South Africa. Mar. Pet. Geol. 28 (3), 807–823.

Kane, I.A., Pontén, A.S.M., 2012. Submarine transitional flow deposits in the Paleogene Gulf of Mexico. Geology 40, 1119–1122.

Kaplan, R., Pyrcz, M.J., Strebelle, S., 2016. Deepwater reservoir connectivity reproduction from MPS and process-mimicking geostatistical methods. In: Geostatistics Valencia 2016. Springer, Cham, pp. 601–611.

Khan, Z.A., Arnott, R.W.C., 2011. Stratal attributes and evolution of asymmetric inner- and outer-bend levee deposits with an ancient deepwater channel-levee complex with the Isaac Formation, southern Canada. Marine and Petroleum Geology 28, 824–842.

King, P.R., Browne, G.H., 2004. Unraveling the complexity of thin-bedded turbidite reservoirs: risk and reward in Taranaki Basin, New Zealand. Oil Gas Gaz. 2 (2), 14–17.

Klaucke, I., Hesse, R., 1996. Fluvial features in the deep sea: new insights from the glacigenic submarine drainage system of the Northwest Atlantic Mid-Ocean channel in the Labrador Sea. Sediment. Geol. 106, 223–234.

Kneller, B.C., McCaffrey, W.D., 1999. Depositional effects of flow non-uniformity and stratification within turbidity currents approaching a bounding slope: deflection, reflection and facies variation. J. Sediment. Res. 69, 980–991.

Kneller, B., Dykstra, M., Fairweather, L., Milana, J.P., 2016. Mass-transport and slope accommodation: implications for turbidite sandstone reservoirs. AAPG Bull. 100 (2), 213–235.

Kolla, V., 2007. A review of sinuous channel avulsion patterns in some major deep-sea fans and factors controlling them. Mar. Pet. Geol. 24, 450–469.

Komar, P.D., 1971. Hydraulic jumps in turbidity currents. GSA Bull. 82, 1477–1481.

Koo, W.M., Olariu, C., Steel, R.J., Olariu, M.I., Carvajal, C.R., Kim, W., 2016. Coupling between shelf-edge architecture and submarine-fan growth style in a supply-dominated margin. J. Sediment. Res. 86 (6), 613–628.

Laugier, F.J., Plink-Björklund, P., 2016. Defining the shelf edge and the three-dimensional shelf edge to slope facies variability in shelf-edge deltas. Sedimentology 63 (5), 1280–1320.

Laugier, F.J., Sun, T., Sullivan, M., Harris, A., Madof, A., 2019. Heterogeneity in unconfined deepwater fans: insights from a quantitative evaluation of outcrop, subsurface, and computational stratigraphy. In: AAPG Annual Convention and Exhibition, San Antonio.

Laursen, J., Normark, W.R., 2002. Late Quaternary evolution of the San Antonio submarine canyon in the central Chile forearc (~33°S). Mar. Geol. 188 (3–4), 365–390.

Leopold, L.B., Wolman, M.G., 1957. River channel patterns: braided, meandering, and straight. In: USGS Professional Paper 282-B, no. iv, pp. 39–85, https://doi.org/10.3133/pp282B.

Liu, L., Ambrose, W.A., Lawton, T.F., Stockli, D.F., 2021. Tectonic controls on the evolution of mixed carbonate-siliciclastic systems: insights from the Late Paleozoic Ouachita-Marathon foreland, United States. Basin Res. https://doi.org/10.1111/BRE.12557.

Long, L., 2019. The progress of the energy reforms and exploration in Mexico. In: Presentation and Press Release, Houston Explorers Club Leadership Luncheon, St Regis Hotel, Houston, 3–May.

Lowe, D.R., 1982. Sediment gravity flows: II. Depositional models with special reference to the deposits of high-density turbidity currents. J. Sediment. Petrol. 52 (1), 279–297.

Lowe, D.R., 2004. Deepwater sandstones: submarine canyon to basin plain, western California. In: AAPG and Pacific Section AAPG Guidebook and Lecture Notes. Artistic Printing, Salt Lake City, p. 79.

Lowe, D.R., Graham, S.A., 2012. Turbidite reservoirs and depositional systems. In: Short Course Lecture Notes. Stanford University, p. 150. approx.

Lowe, D.R., Guy, M., 2000. Slurry-flow deposits in the Britannia formation (lower cretaceous), North Sea: a new perspective on the turbidity current and debris flow problem. Sedimentology 47, 31–70.

Lowe, D.R., Graham, S.A., Malkowski, M.A., Das, B., 2019. The role of avulsion and splay development in deepwater channel systems: sedimentology, architecture, and evolution of the deepwater Pliocene Godavari "A" channel complex, India. Mar. Pet. Geol. 105, 81–99.

Marchand, A.M.E., Apps, G., Li, W., Rotzien, J.R., 2015. Depositional processes and impact on reservoir quality in Deepwater Paleogene reservoirs, US Gulf of Mexico. AAPG Bull. 99 (9), 1635–1648.

Masalimova, L.U., Lowe, D.R., Sharman, G.R., King, P.R., Arnot, M.J., 2016. Outcrop characterization of a submarine channel-lobe complex: the lower mount messenger formation, Taranaki Basin, New Zealand. Mar. Pet. Geol. 71, 360–390.

Mayall, M., Stewart, I., 2000. The architecture of turbidite slope channels. In: Weimer, P., Slatt, R.M., Coleman, J., Rosen, N.C., Nelson, H., Bouma, A.H., Styzen, M.J., Lawrence, D.T. (Eds.), Deepwater Reservoirs of The World. Gulf Coast Section SEPM 20th Bob F. Perkins Research Conference, pp. 578–586.

Mayall, M., Stewart, I., 2001. The architecture of turbidite slope channels. In: Fraser, S.I., Fraser, A.J., Johnson, H.D., Evans, A.M. (Eds.), Petroleum Geology of Deepwater Depositional Systems, Advances in Understanding 3D Architecture. The Geological Society Conference, March 20–22, 2001 Proceedings, Burlington House, Picadilly, London. The Geological Society, unpaginated.

Mayall, M., Jones, E., Casey, M., 2006. Turbidite channel reservoirs – key elements in facies prediction and effective development. Mar. Pet. Geol. 23, 821–841.

McArthur, A.D., McCaffrey, W.D., 2019. Sedimentary architecture of detached deep-marine canyons: examples from the East Coast Basin of New Zealand. Sedimentology 66 (3), 1067–1101.

McArthur, A.D., Kneller, B., De Souza, P.A., Kuchle, J., 2016. Characterization of deep- marine channel-levee complex architecture with palynofacies: an outcrop example from the Rosario formation, Baja California, Mexico. Mar. Pet. Geol. 73. https://doi.org/10.1016/j.marpetgeo.2016.02.030.

McCaffrey, W.D., Kneller, B.C., 2001. Process controls on the development of stratigraphic trap potential on the margins of confined turbidite systems, and aids to reservoir evaluation. AAPG Bull. 20, 789–807.

McHargue, T., Pyrcz, M.J., Sullivan, M.D., Clark, J.D., Fildani, A., Romans, B.W., Covault, J.A., Levy, M., Posamentier, H.W., Drinkwater, N.J., 2011. Architecture of turbidite channel systems on the continental slope: patterns and predictions. Mar. Pet. Geol. 28, 728–743.

Miall, A.D., 2016. The valuation of unconformities. Earth Sci. Rev. 163, 22–71.

Michael, H.A., Li, H., Boucher, A., Sun, T., Caers, J., Gorelick, S.M., 2010. Combining geologic-process models and geostatistics for conditional simulation of 3-D subsurface heterogeneity. Water Resour. Res. 46 (5).

Milkov, A.V., Navidi, W.C., 2020. Randomness, serendipity, and luck in petroleum exploration. AAPG Bull. 104 (1), 145–176.

Milkov, A., Samis, J.M., 2020. Turning dry holes from disasters to exploration wisdom: decision tree to determine the key failure mode for segments in conventional petroleum prospects. AAPG Bull. 104 (2), 449–475.

Miller, J.K., Sun, T., Li, H., Stewart, J., Genty, C., Li, D., Lyttle, C., 2008. Direct modeling of reservoirs through forward process-based models: can we get there? In: International petroleum technology conference. OnePetro.

Miramontes, E., Eggenhuisen, J.T., Jacinto, R.S., Poneti, G., Pohl, F., Normandeau, A., Campbell, D.C., Hernández-Molina, F.J., 2020. Channel-levee evolution in combined contour current-turbidity current flows from flume-tank experiments. Geology 48 (4), 353–357.

Mitchell, W.H., Whittaker, A.C., Mayall, M., Lonergan, L., 2021. New models for submarine channel deposits on structurally complex slopes: examples from the Niger delta system. Mar. Pet. Geol. 129, 105040. https://doi.org/10.1016/j.marpetgeo.2021.105040.

Mitchell, W.H., Whittaker, A.C., Mayall, M., Lonergan, L., Pizzi, M., 2020. Quantifying the relationship between structural deformation and the morphology of submarine channels on the Niger Delta continental slope. Basin Res. 33 (1), 186–209.

Moody, J.D., Pyles, D.R., Clark, J., Bouroullec, R., 2012. Quantitative outcrop characterization of an analog to weakly confined submarine channel systems: morillo 1 member, Ainsa Basin, Spain. AAPG Bull. 96 (10), 1813–1841.

Morris, E., Hodgson, D., Brunt, R., Flint, S., 2014. Origin, evolution and anatomy of silt- prone submarine external levees. Sedimentology 61, 1734–1763. https://doi.org/10.1111/sed.12114.

Moscardelli, L., Wood, L., 2015. Morphometry of mass-transport deposits as a predictive tool. GSA Bull. 128 (1). https://doi.org/10.1130/B31221.1.

Mullins, J., van Der Vegt, H., Howell, J., 2021. Combining process-based models and multiple-point geostatistics for improved reservoir modelling. Pet. Geosci. 27 (3).

Murray, C.J., Lowe, D.R., Graham, S.A., Martinez, P.A., Zeng, J., Carroll, A.R., Cox, R., Hendrix, M., Heubeck, C., Miller, D., Moxon, I.W., Sobel, E., Wendenbourg, J., Williams, T., 1996. Statistical analysis of bed-thickness patterns in a turbidite section from the Great Valley sequence, Cache Creek, Northern California. J. Sediment. Res. 66 (5), 900–908.

Mutti, E., 1977. Distinctive thin-bedded turbidite facies and related depositional environments in the Eocene Hecho group (south-Central Pyrenees, Spain). Sedimentology 24, 107–131.

Mutti, E., Normark, W.R., 1987. Comparing examples of modern and ancient turbidite. In: Leggett, J.R., Zuffa, G.G. (Eds.), Marine Clastic Sedimentology: Concepts and Case Studies. Graham and Trotman, London, pp. 1–37.

Mutti, E., Sonnino, M., 1981. Compensation cycles: a diagnostic feature of sandstone lobes (abstract). In: International Association of Sedimentologists, 2nd European meeting, Bologna, pp. 120–123.

Nader, F.H., Souque, C., Lecomte, J.C., Deschamps, R., Chauveau, B., Granjeon, D., Frascati, A., 2018. Advanced 3-D Forward Stratigraphic Modeling of the East-Mediterranean Frontier Deepwater Basins: An Approach for Enhancing Reservoir Fairways Predictions. Search and Discovery Article, vol. 11133, pp. 20–23.

Nilsen, T.H., Shew, R.D., Steffens, G.S., Studlick, J.R.J. (Eds.), 2007. Atlas of Deepwater outcrops: Tulsa, Okla. AAPG Stud. Geol. 56.

Normark, W.R., 1970. Growth patterns of deep sea fans. AAPG Bull. 54, 2170–2195.

Normark, W.R., 1978. Fan valleys, channels, and depositional lobes on modern submarine fans, characters for recognition of sandy turbidite environments. AAPG Bull. 62, 912–931.

Normark, W.R., Carlson, P.R., 2003. Giant Submarine Canyons: Is Size any Clue to their Importance in the Rock Record? GSA Special Paper, vol. 370, pp. 175–190.

Ortiz-Karpf, A., Hodgson, D.M., McCaffrey, W.D., 2015. The role of mass-transport complexes in controlling channel avulsion and the subsequent sediment dispersal patterns on an active margin: the Magdalena fan, offshore Colombia. Mar. Pet. Geol. 64, 58–75.

Oswald, E.J., 2019. Guyana exploration. In: Presentation at Houston Explorers Club Leadership Luncheon, St Regis Hotel, Houston, 27–Sept.

Paull, C.K., Talling, P.J., Maier, K.L., Parsons, D., Xu, J., Caress, D.W., Gwiazda, R., Lundsten, E.M., Anderson, K., Barry, J.P., Chaffey, M., O'Reilly, T., Rosenberger, K.J., Gales, J.A., Kieft, B., McGann, M., Simmons, S.M., McCann, M., Sumner, E.J., Clare, M.A., Cartigny, J., 2018. Powerful turbidity currents driven by dense basal layers. Nat. Commun. 9, 1–9. https://doi.org/10.1038/s41467-018-06254-6.

Peel, F.J., Brooks, J.R.V., 2015. What to expect when you are prospecting: how new information changes our estimate of the chance of success of a prospect. AAPG Bull. 99 (12), 2159–2171.

Pemberton, E.A.L., Hubbard, S.M., Fildani, A., Romans, B., Stright, L., 2016. The stratigraphic expression of decreasing confinement along a deepwater routing system: outcrop example from southern Chile. Geosphere 12 (1), 114–134.

Pickering, K.T., Hiscott, R.N., Kenyon, N.H., Ricchi Lucchi, F., Smith, R.D.A., 1995. Atlas of Deep Water Environments: Architectural Style in Turbidite Systems. Chapman and Hall, London, p. 333.

Pirmez, C., 2011. sediment flux from source to sink in the brazos-trinity depositional system, western gulf of Mexico. In: AGU Chapman Conference for Source-to-Sink, Oxnard, CA. 27 January.

Pirmez, C., Flood, R.D., 1995. Morphology and structure of Amazon Channel. In: Flood, R.D., Piper, D.J.W., Klaus, A., et al. (Eds.), Proc. ODP, Init. Repts. 155: College Station, TX (Ocean Drilling Program), pp. 23–45.

Pirmez, C., Hiscott, R.N., Kronen Jr., J.D., 1997. Sandy turbidite successions at the base of channel-levee systems of the Amazon fan revealed by FMS logs and cores: Unraveling the lithofacies architecture of large submarine fans 155, p. 7–33. In: Flood, R.D., Piper, D.J.W., Klaus, A., Peterson, L.C. (Eds.), Proceedings of the Ocean Drilling Program Leg 155: Amazon Fan. Compiled Electronic Index of the Proceedings of the Ocean Drilling Program, p. 612.

Pizzi, M., Lonergan, L., Whittaker, A.C., Mayall, M., 2020. Growth of a thrust fault array in space and time: an example from the Deepwater Niger Delta. J. Struct. Geol. 137, 1–20.

Plink-Björklund, P., Mellere, D., Steel, R.J., 2001. Turbidite variability and architecture of sand-prone, deep-water slopes: eocene clinoforms in the Central Basin, Spitsbergen. J. Sediment. Res. 71 (6), 895–912.

Pohl, F., 2019. Turbidity Currents and their Deposits in Abrupt Morphological Transition Zones. PhD thesis, University of Utrecht, p. 210.

Porębski, S.J., Steel, R.J., 2003. Shelf-margin deltas: their stratigraphic significance and relation to Deepwater sands. Earth Sci. Rev. 62 (3–4), 283–326.

Porter, M.L., Sprague, A.R.G., Sullivan, M.D., Jennette, D.C., Beaubouef, R.T., Garfield, T.R., Rossen, C., Sickafoose, D.K., Jensen, G.N., Friedmann, S. J., Mohrig, D.C., 2006. Stratigraphic organization and predictability of mixed coarse- and fine-grained lithofacies successions in a lower Miocene deepwater slope-channel system, Angola Block 15. In: Harris, P.M., Weber, L.J. (Eds.), Giant Hydrocarbon Reservoirs of the world: From Rocks to Reservoir Characterization and Modeling. AAPG Memoir 88/SEPM Special Publication, pp. 281–305.

Posamentier, H.W., Kolla, V., 2003. Seismic geomorphology and stratigraphy of depositional elements in Deepwater settings. J. Sediment. Res. 73, 367–388.

Power, B., Covault, J., Fildani, A., Sullivan, M., Clark, J., Carson, B., Zarra, L., Romans, B., 2013. Facies analysis and interpretation of argillaceous sandstone beds in the Paleogene Wilcox formation, Deepwater Gulf of Mexico. In: Gulf Coast Association of Geological Societies Transactions. vol. 63, pp. 575–578.

Prather, B.E., 2000. Calibration and visualization of depositional process models for above- grade slopes: a case study from the Gulf of Mexico. Mar. Pet. Geol. 17, 619–638.

Prather, B.E., 2003. Controls on reservoir distribution, architecture and stratigraphic trapping in slope settings. Mar. Pet. Geol. 20, 529–545.

Prather, B.E., Booth, J.R., Steffens, G.S., Craig, P.A., 1998. Classification, lithologic calibration, and stratigraphyic succession of seismic facies of intra-slope basins, deep-water Gulf of Mexico. AAPG Bull. 82 (12).

Prather, B.E., O'Byrne, C., Pirmez, C., Sylvester, Z., 2016. Sediment partitioning, continental slopes and base-of-slope systems. Basin Res. 29 (3), 394–416.

Prélat, A., Hodgson, D.M., Flint, S.S., 2009. Evolution, architecture and hierarchy of distributary Deepwater deposits: a high-resolution outcrop investigation from the Permian Karoo Basin, South Africa. Sedimentology 56, 2132–2154.

Prélat, A., Covault, J.A., Hodgson, D.M., Fildani, A., Flint, S.S., 2010. Intrinsic controls on the range of volumes, morphologies, and dimensions of submarine lobes. Sediment. Geol. 232, 66–76.

Puigdefàbregas, Gjelberg, J., Vaksdal, M., 2004. The Grès d'Annot in the Annot syncline: outer basin-margin onlap and associated soft-sediment deformation. In: Joseph, P., Lomas, S.A. (Eds.), Deepwater Sedimentation in the Alpine Basin of SE France: New Perspectives on the Grès d'Annot and Related Systems. vol. 221. Geological Society, London, Special Publications, pp. 367–388.

Pyles, D.R., 2008. Multiscale stratigraphic analysis of a structurally confined submarine fan: Carboniferous Ross Sandstone, Ireland. AAPG Bull. 92, 557–587.

Pyles, D., Schwartz, D.E., Bouroullec, R., Jennette, D.J., Gordon, G.S., 2019. On the relationship among submarine-channel stacking patterns, environmental controls during deposition, and reservoir architecture: case studies of three Neogene outcrops in Southern California. J. Sediment. Res. 17 (1), 178–197. https://doi.org/10.2110/sepmsp.110.18.

Pyles, D.R., Tomasso, M., Jennette, D.C., 2012. Flow processes and sedimentation associated with erosion and filling of sinuous submarine channels. Geology 40 (2), 143–146.

Reading, H.G., Richards, M., 1994. Turbidite systems in Deepwater basin margins classified by grain size and feeder system. AAPG Bull. 78, 792–822.

Rebesco, M., Hernández-Molina, F.J., Van Rooij, D., Wåhlin, A., 2014. Contourites and associated sediments controlled by deepwater circulation processes: state-of-the-art and future considerations. Mar. Geol. 352, 111–154.

Recouvreur, A., Fabregas, N., Mulder, T., Hanquiez, V., 2020. Geomorphology of a modern carbonate slope system and associated sedimentary processes: example of the giant great Abaco Canyon, Bahamas. Sedimentology 68 (1), 266–293.

Reza, Z.A., Pranter, M.J., Weimer, P., 2006. ModDRE: a program to model Deepwater-reservoir elements using geomorphic and stratigraphic constraints. Comput. Geosci. 32 (8), 1205–1220.

Roca, E., Ferrer, O., Rowan, M.G., Munoz, J.A., Butille, M., Giles, K.A., Arbues, P., de Matteis, M., 2020. Salt tectonics and controls on halokinetic-sequence development of an exposed deepwater diapir: the Bakio Diapir, Basque-Cantabrian Basin, Pyrenees. Mar. Pet. Geol. 123, 104770.

Ross, W.C., Halliwell, B.A., May, J.A., Watts, D.E., Syvitski, J.P.M., 1994. Slope readjustment: a new model for the development of submarine fans and aprons. Geology 22, 511–514.

Rotzien, J.R., 2013. Processes of Sedimentation, Stratigraphic Architecture and Provenance of Deep-water Depositional Systems: Examples from the Upper Miocene Upper Mount Messenger Formation. Taranaki Basin, New Zealand and the Pliocene Repetto and Pico Formations, Ventura Basin, California (PhD thesis), Stanford University, Stanford, CA, USA. 218 pp.

Rotzien, J.R., 2017. Stratigraphic Terminations of the Eocene-Oligocene Annot Sandstone, Southeast France. vol. 67 Gulf Coast Association of Geological Societies Transactions, pp. 529–532.

Rotzien, J.R., 2018. Reservoir and seal trends and anomalies in submarine canyon environments. In: 21st century progress and limitations derived from outcrop studies of the Great Valley Group, western California, U.S.A. AAPG GTW Transactions, Beijing, China.

Rotzien, J.R., 2019. Practical Methods for Mapping and Interpreting Deepwater Reservoirs (GEO 53). Short Course Notes and Exercises. HOT Engineering, Stavanger, Norway. 636 pp.

Rotzien, J.R., 2020. The Eocene-Oligocene Annot Sandstone Depositional System: Relevance, Processes, Products and Models: Virtual Get Together. The Drifters Research Group, Royal Holloway, University of London, May 14.

Rotzien, J.R., Lowe, D.R., King, P.R., Browne, G.H., 2014a. Processes of sedimentation and stratigraphic evolution of the upper miocene upper mount messenger formation, Taranaki Basin, New Zealand. J. Mar. Pet. Geol. 52, 22–41.

Rotzien, J.R., Lowe, D.R., Schwalbach, J.R., 2014b. Processes of sedimentation and stratigraphic architecture of deepwater braided lobe complexes: the pliocene repetto and pico formations, Ventura Basin, U.S.A. J. Sediment. Res. 84, 910–934.

Rudolph, K.W., 2020, Interview for basin dynamics bulletin, v. 2, no. 9–10, 9 pp.

Rudolph, K.W., Goulding, F.J., 2017. Benchmarking exploration predictions and performance using 20+ yr of drilling results: one company's experience. AAPG Bull. 101 (2), 161–176.

Salles, T., 2006. Modélisation Numérique du Remplissage Sédimentaire des Canyons et Chenaux Sous-Marins par Approche Génétique. (Doctoral dissertation, Bordeaux 1).

Salles, T., Lopez, S., Cacas, M.C., Mulder, T., 2007. Cellular automata model of density currents. Geomorphology 88 (1–2), 1–20.

Salles, T., Lopez, S., Eschard, R., Lerat, O., Mulder, T., Cacas, M.C., 2008a. Turbidity current modelling on geological time scales. Mar. Geol. 248 (3–4), 127–150.

Salles, T., Mulder, T., Gaudin, M., Cacas, M.C., Lopez, S., Cirac, P., 2008b. Simulating the 1999 Capbreton canyon turbidity current with a cellular automata model. Geomorphology 97 (3–4), 516–537.

Schwalbach, J.R., Miller, D.D., Rotzien, J.R., 2009. Pliocene turbidites of the Ventura Basin. In: Core Workshop and Field Trip, Complete with Guidebook, Held in Conjunction with the AAPG/SEPM Meeting Short Course, May 3–6, Ventura, CA, p. 28.

Selley, R.C., 1979. Dipmeter and log motifs in North Sea submarine-fan sands. AAPG Bull. 63, 905–917.

Sharman, G.R., Graham, S.A., Masalimova, L.U., Shumaker, L.E., King, P.R., 2015. Spatial patterns of deformation and paleoslope estimation within the marginal and central portions of a basin-floor mass-transport deposit, Taranaki Basin, New Zealand. Geosphere 11 (2), 266–306.

Shashank, S., Sun, H., Hoyal, D.C., Shringarpure, M., Kushwaha, A., 2019. Supervised deep learning based approach for automated segmentation and tracking of lobes deposited by turbidity currents. In: AGU Fall Meeting Abstracts. 2019, p. EP54C-07.

Shumaker, L.E., Jobe, Z.R., Graham, S.A., 2017. Evolution of submarine gullies on a prograding slope: insights from 3D seismic reflection data. Mar. Geol. 393, 35–46.

Sinclair, H.D., Tomasso, M., 2002. Depositional evolution of confined turbidite systems. J. Sediment. Res. 72 (4), 451–456.

Slatt, R.M. (Ed.), 2006. Stratigraphic Reservoir Characterization for Petroleum Geologists, Geophysicists, and Engineers. vol. 61. Elsevier, p. 492.

Slatt, R.M., 2019. Interview for Basin Dynamics Bulletin. Basin Dynamics Bulletin, v. 1, no. 4–5, 20 pp.

Slatt, R.M., Weimer, P., Stone, C.G., 1997. Reinterpretation of depositional processes in a classic flysch sequence (Pennsylvanian Jackfork group), Ouachita Mountains, Arkansas and Oklahoma: discussion. AAPG Bull. 81 (3), 449–459.

Slingerland, R., Kump, L., 2011. Mathematical Modeling of Earth's Dynamical Systems. Princeton University Press.

Smith, R., Joseph, P., 2004. Onlap stratal architectures in the Grès d'Annot: geometric models and controlling factors. Geol. Soc. Lond. Spec. Publ. 221, 389–399. https://doi.org/10.1144/GSL.SP.2004.221.01.21.

Sprague, A.R.G., Garfield, T.G., Goulding, F.J., Beaubouef, R.T., Sullivan, M.D., Rossen, C., Campion, K.M., Sickafoose, D.K., Abreu, V., Schellpeper, M.E., Jensen, G.N., Jennette, D.C., Pirmez, C., Dixon, B.T., Ying, D., Ardill, J., Mohrig, D.C., Porter, M.L., Farrell, M.E., Mellere, D., 2005. Integrated Slope Channel Depositional Models: The Key to Successful Prediction of Reservoir Presence and Quality in Offshore West Africa. CIPM. 13 pp.

Stanbrook, D.A., Clark, J.D., Pringle, J.K., 1999. Turbidite onlap geometry in the grand coyer sub-basin, results from digital photogrammetry and recent fieldwork. In: 38th Annual Meeting of the British Sedimentological Research Group, Edinburgh, UK, 19–21 Dec.

Stow, D.A.V., 1979. Distinguishing between fine-grained turbidites and contourites off Nova Scotia. Sedimentology 26, 371–387.

Stow, D.A.V., Mayall, M., 2000. Deepwater sedimentary systems: new models for the 21st century. Mar. Pet. Geol. 17, 125–135.

Stow, D.A.V., Piper, D.J.W. (Eds.), 1984. Fine-Grained Sediments: Deepwater Processes and Facies, p. 659. Geological Society Special Publication No. 15.

Stow, D.A.V., Shanmugam, G., 1980. Sequence of structures in fine-grained turbidites: comparison of recent deep-sea and ancient flysch sediments. Sediment. Geol. 25, 23–42.

Straub, K.M., Mohrig, D., 2008. Quantifying the morphology and growth of levees in aggrading submarine channels. J. Geophys. Res. Earth Surf. 113. https://doi.org/10.1029/2007JF000896.

Straub, K.M., Pyles, D.R., 2012. Quantifying the hierarchical organization of compensation in submarine fans using surface statistics. J. Sediment. Res. 82, 889–898.

Straub, K.M., Paola, C., Mohrig, D., Wolinsky, M.A., George, T., 2009. Compensational stacking of channelized sedimentary deposits. J. Sediment. Res. 79, 673–688.

Sullivan, M.D., Jensen, G.N., Goulding, F.J., Jennette, D.C., Foreman, J.L., Stern, D., 2000. Architectural analysis of deepwater outcrops: implications for exploration and development of the Diana sub-basins, western Gulf of Mexico. In: GCSSEPM Foundation 20th Annual Research Conference – Deepwater Reservoirs of the World, Dec. 3–6.

Sullivan, M.D., Foreman, J.L., Jennette, D.C., Stern, D., Jensen, G.N., Goulding, F.J., 2004. An integrated approach to characterization and modeling of Deepwater reservoirs, Diana field, western Gulf of Mexico, in integration of outcrop and modern analogs in reservoir modeling. AAPG Mem. 80, 215–234.

Sumner, E.J., Talling, P.J., Amy, L.A., Wynn, R.B., Stevenson, C.J., Frenz, M., 2012. Facies architecture of individual basin-plain turbidites: comparison with existing models and implications for flow processes. Sedimentology 59 (6), 1850–1887.

Sun, T., Ghayour, K., Miller, J., 2008. Process-based modeling of deepwater depositional systems. In: GEO 2008, p. cp-246. European Association of Geoscientists & Engineers.

Sun, T., Ghayour, K., Hall, B., Miller, J., 2010. Process-based modeling of deep water depositional systems. In: Seismic Imaging of Depositional and Geomorphic Systems. vol. 30. Society for Sedimentary Geology.

Sun, T., Laugier, F.J., Caldwell, R., Harris, A., Sullivan, M., 2018. Computational stratigraphy of Deepwater sedimentary systems. In: AGU, Fall Meeting, Abstract OS12A-02.

Surlyk, F., 1987. Slope and deep shelf gully sandstones, upper Jurassic, East Greenland. AAPG Bull. 71 (4), 464–475.

Sweet, M.L., Blum, M.D., 2016. Connections between fluvial to shallow marine environments and submarine canyons: implications for sediment transfer to deep water. J. Sediment. Res. 86, 1147–1162.

Sweet, M.L., Gaillot, G.T., Jouet, G., Rittenour, T.M., Toucanne, S., Marsset, T., Blum, M.D., 2019. Sediment routing from shelf to basin floor in the Quaternary Golo System of Eastern Corsica, western Mediterranean Sea. GSA Bull. 132 (5–6), 1217–1234.

Sylvester, Z., 2007. Turbidite bed thickness distributions: methods and pitfalls of analysis and modelling. Sedimentology 54 (4), 1–24.

Sylvester, Z., Covault, J.A., 2016. Development of cutoff-related knickpoints during early evolution of submarine channels. Geology 44 (10), 835–838.

Sylvester, Z., Pirmez, C., Cantelli, A., 2011. A model of submarine channel-levee evolution based on channel trajectories: implications for stratigraphic architecture. Mar. Pet. Geol. 28 (3), 716–727.

Sylvester, Z., Deptuck, M.E., Prather, B.E., Pirmez, C., O'Byrne, C., 2012. Seismic stratigraphy of a shelf-edge delta and linked submarine channels in the northeastern Gulf of Mexico: SEPM Special Publication No. 99, Application of the Principles of Seismic Geomorphology to Continental-Slope and Base-of-Slope Systems. In: Case Studies from Seafloor and Near-Seafloor Analogues, pp. 31–59.

Sylvester, Z.N., Pirmez, C., Cantelli, A., Jobe, Z.R., 2013. Global (latitudinal) variation in submarine channel sinuosity: comment. Geology 41. https://doi.org/10.1130/g33548c.1, e287.

Thieblemont, A., Hernández-Molina, F.J., Ponte, J.P., Robin, C., Guillocheau, F., Cazzola, C., Raisson, F., 2020. Seismic stratigraphic framework and depositional history for cretaceous and Cenozoic contourite depositional systems of the Mozambique Channel, SW Indian Ocean. Mar. Geol. 425. https://doi.org/10.1016/j.margeo.2020.106192.

Thran, A.C., Dutkiewicz, A., Spence, P., Muller, R.D., 2018. Controls on the global distribution of contourite drifts: insights from an eddy-resolving ocean model. Earth Planet. Sci. Lett. 489, 228–240.

Traer, M.M., Fildani, A., McHargue, T., Hilley, G.E., 2015. Simulating depth-averaged, one- dimensional turbidity current dynamics using natural topographies. J. Geophys. Res. Earth Surf. 120 (8), 1485–1500.

Turner, C.C., Cronin, B.T., 2018. Rift-Related Coarse-Grained Submarine Fan Reservoirs: The Brae Play, South Viking Graben, North Sea. AAPG Memoir, p. 115, https://doi.org/10.1306/AAPG115.

Vail, P.E., Mitchum, R.M., Thompson III, S., 1977a. Relative changes of sea level from coastal onlap. In: Payton, C.E. (Ed.), Seismic Stratigraphy—Applications to Hydrocarbon Exploration. vol. 26. AAPG Memoir, pp. 63–82.

Vail, P.R., Todd, R.G., Sangree, J.B., 1977b. Seismic stratigraphy and global changes of sea level, part 5: chronostratigraphic significance of seismic reflections. In: Payton, C.E. (Ed.), Seismic Stratigraphy—Applications to Hydrocarbon Exploration. vol. 26. AAPG Memoir, pp. 99–116.

Van der Merwe, W.C., Hodgson, D.M., Brunt, R.L., Flint, S.S., 2014. Depositional architecture of sand-attached and sand-detached channel-lobe transition zones on an exhumed stepped slope mapped over a 2500 km2 area. Geosphere 10 (6), 1076–1093.

Wahab, A., Hoyal, D.C., Sun, H., Shringarpure, M., Straub, K.M., 2019. Statistical characterization of submarine fans over a wide range of hydrodynamic and sediment-transport parameters: a numerical campaign. In: AGU Fall Meeting Abstracts. 2019, p. OS33C-1800.

Walker, R.G., 1967. Turbidite sedimentary structures and their relationships to proximal and distal environments. J. Sediment. Petrol. 37, 25–42.

Walker, R.G., 1975. Generalized facies models for resedimented conglomerates of turbidite association. GSA Bull. 86, 737–748.

Walker, R.G., 1978. Deepwater sandstone facies and ancient submarine fans: models for exploration for stratigraphic traps. AAPG Bull. 62, 932–966.

Walker, R.G., Cant, D.J., 1984. Sandy fluvial systems. In: Walker, R.G. (Ed.), Facies Models, fourth ed. Pearson Prentice Hall, Saddle River, New Jersey, p. 662. Geoscience Canada Reprint Ser. 1, Fig. 1, p. 72, in Boggs Jr., S., ed., 1995, Principles of sedimentology and stratigraphy.

Waltham, D., Jaffey, N., MacLean, S., Zampetti, V., 2008. Stratigraphic modelling of turbidite prospects to reduce exploration risk. Pet. Geosci. 14 (3), 273–280.

Wang, X., 2015. Turbid Flows and Their Deposits on Slopes with Minibasins: A Modelling Approach. (Thesis).
Wang, X., Luthi, S.M., Hodgson, D.M., Sokoutis, D., Willingshofer, E., Groenenberg, R.M., 2017. Turbidite stacking patterns in salt-controlled minibasins: insights from integrated analogue models and numerical fluid flow simulations. Sedimentology 64 (2), 530–552.
Weber, R., 2018. Biostratigraphy. In: Paleo Data Inc. Short Course and Notes, Lafayette, Louisiana, 24–Jan, p. 200. approx.
Willis, B., Sun, T., 2019. Relating depositional processes of river-dominated deltas to reservoir behavior using computational stratigraphy. J. Sediment. Res. 89 (12), 1250–1276.
Willis, B.J., Kalla, S., Sun, T., 2020. Defining lithic patterns within river-dominated delta deposits for geostatistical simulation. Pet. Geosci. 27 (1). petgeo2019-105.
Wuellner, D.E., James, W.C., 1989. Braided and meandering submarine fan channel deposits, Tesnus formation, Marathon basin, West Texas. Sediment. Geol. 62, 27–45.
Wynn, R.B., Stow, D.A.V., 2002. Classification and characterization of Deepwater sediment waves. Mar. Geol. 192, 7–22.
Wynn, R.B., Kenyon, N.H., Masson, D.G., Stow, D.A.V., Weaver, P.P.E., 2002. Characterization and recognition of Deepwater channel-lobe transition zones. AAPG Bull. 86 (8), 1441–1462.
Wynn, R.B., Cronin, B.T., Peakall, J., 2007. Sinuous deepwater channels: genesis, geometry and architecture. Mar. Pet. Geol. 24 (6–9), 341–387.
Yeilding, C.A., 2019. The explorers mindset. In: Presentation at Houston Explorers Club Leadership Luncheon, St Regis Hotel, Houston, 15–Nov.

Chapter 9

# Contourites and mixed depositional systems: A paradigm for deepwater sedimentary environments

F. Javier Hernández-Molina[a], Sandra de Castro[a], Wouter de Weger[a], Debora Duarte[a,b], Marco Fonnesu[c], Tatiana Glazkova[a], Adam Kirby[a], Estefanía Llave[d], Zhi Lin Ng[a], Oswaldo Mantilla Muñoz[a,e], Sara Rodrigues[a], Francisco J. Rodríguez-Tovar[f], Antoine Thieblemont[g], Adriano R. Viana[h] and Shaoru Yin[i]

[a]*Department of Earth Sciences, Royal Holloway, University of London, Egham, Surrey, United Kingdom,* [b]*Divisão de Geologia e Georrecursos Marinhos, Instituto Português do Mar e da Atmosfera—IPMA, Lisbon, Portugal,* [c]*Eni SpA, San Donato Milanese, Milan, Italy,* [d]*Instituto Geológico y Minero de España (IGME, CSIC), Madrid, Spain,* [e]*Ecopetrol S.A, Colombian Petroleum Co, Bogotá, Colombia,* [f]*Departamento de Estratigrafía y Paleontología, Universidad de Granada, Granada, Spain,* [g]*TOTAL S.A., Exploration & Production, EXPLO/GTS/ISS/CLS, Pau Cedex, France,* [h]*PETROBRAS, Rio de Janeiro, Brazil,* [i]*Key Laboratory of Submarine Geosciences, Second Institute of Oceanography, Ministry of Natural Resources, Hangzhou, China*

## Introduction

The prevailing understanding of deepwater sedimentary systems has traditionally focused on gravitational processes, wherein turbiditic currents or mass-transport processes transfer sediments from shallow settings to deepwater dominated by pelagic sedimentation (see Chapter 6). But, over the last few decades, many authors have helped document and elucidate the role of the ocean circulation plays, in shaping the seafloor and controlling deepwater sedimentation. Deep-marine basins do not exactly represent the weak, low-energy, sediments-starved environments traditionally assumed. Bottom currents associated with water masses actively interact to shape the seafloor along continental margins, abyssal plains or in areas of bathymetric irregularities (scarps, highs, banks or seamounts). Bottom currents can mingle with other pelagic or gravity flow processes either synchronously or asynchronously to form new deposits. This chapter summarizes key concepts, processes and deposits associated with deepwater currents and sedimentary environments. With this chapter, readers can obtain a broader vision of the sedimentary processes that take place in deepwater environments, and their relationship with new types of deposits that have importance in the evolution of sedimentary basins and implications for many different scientific conceptual frameworks, environmental integrity, ocean policy, energy geosciences and mineral resources

The current consensus on contourites defines such deposits as "sediments deposited or substantially reworked by the persistent action of bottom currents" (Stow et al., 2002; Rebesco, 2005; Rebesco et al., 2008, 2014). Water masses move continually through ocean basins (e.g., Rahmstorf, 2006; Kuhlbrodt et al., 2007) and any persistent current near the seafloor is considered as bottom current (Rebesco et al., 2008). Bottom currents are semipermanent water-mass flows capable of eroding, transporting and depositing up to gravel-sized sediments from or in seafloor environments. Most bottom currents erode, transport and deposit clay, silt and fine-grained sand. Bottom currents generate depositional and erosional features at different scales (Fig. 1) and the combination of these features form a Contourite Depositional System (CDS; Hernández-Molina et al., 2003, 2008b). Bottom current features sometimes develop during depositional and erosional phases. These are referred to as mixed features (Fig. 1). Depositional, erosional or mixed features result from differences in the hydrodynamic characteristics of the bottom current. Similarly, differences in hydrodynamic characteristics can produce a wide range of depositional sedimentary facies.

At sedimentary facies scale in outcrops, in core material or well data, or from images of seafloor settings (videos, submarine photographs, etc.), the deposits discussed herein are often referred to as "contourites" or "contouritic facies." Large to very large deposits are identified based on their morphology and long-term sedimentary architecture in 2D or 3D seismic data (or very occasionally in outcrops). Here, the term "*contouritic drifts*" or just "*drifts*" is used (Figs. 1–3).

The original research that sought to define drifts linked these features to the depositional action of bottom currents (Heezen and Hollister, 1964, 1971). Recent studies on contourite drifts from both modern settings and the ancient record

**Deepwater Sedimentary Systems**. https://doi.org/10.1016/B978-0-323-91918-0.00004-9

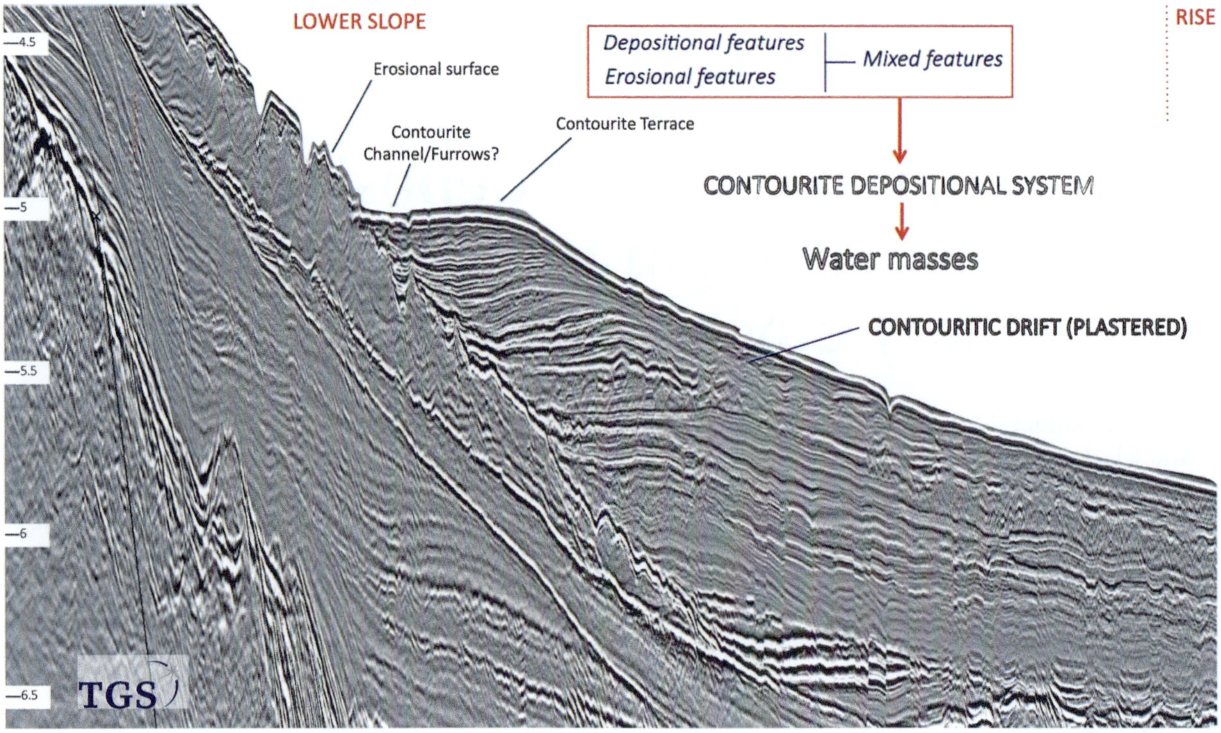


**FIG. 1** Example of a contourite depositional system (CDS) along the lower slope and continental rise of the Argentine margin, where large, associated depositional and erosional features show both lateral and vertical trends. The large drift represents a plastered drift. *(*Seismic line courtesy from *TGS, vertical scale in seconds).*

have used a variety of names for these features including "*sedimentary mounds*," "*sedimentary ridges*," "*sedimentary drifts*," or "*outer ridges*" (Rebesco et al., 2008). Thus far, literature sources have identified nine large drifts that share common morphologies and appear along most continental margins and abyssal plains (McCave and Tucholke, 1986; Faugères et al., 1999; Rebesco, 2005; Rebesco and Camerlenghi, 2008; Rebesco et al., 2014; Eberli and Betzler, 2019).

All drifts show a predominance of muddy sediments in areas with low ($<20\,\text{cm s}^{-1}$, $\sim 7.8\,\text{in s}^{-1}$) bottom current velocity (Davies and Laughton, 1972; Tucholke and Ewing, 1974; McCave, 1982; McCave and Tucholke, 1986; Tucholke, 2002; Miramontes et al., 2019; Thieblemont et al., 2019; Rajappa et al., 2020). The current's core shapes erosional elements (channels, moats, terraces, etc.) and determines the coarser-grained (sandier) deposits in both lateral and vertical dimensions (Fig. 2). As illustrated in Fig. 2, the large drifts are interpreted to represent a first-order depositional feature at basin scale associated with water masses (Hernández-Molina et al., 2003, 2008a,b, 2016a; Kirby et al., 2021a). Nevertheless, drifts are not always observed to be actively forming in deepwater environments and are thus interpreted to represent relict features from a shift in a water mass' velocity, depth or thickness (Fig. 2). For example, weakened currents during interglacial stages condition inactive drifts compared to glacial stages (Hernández-Molina et al., 2016b; Lofi et al., 2016). Smaller secondary depositional features such as sedimentary waves can appear in response to interactions of water mass cores or branches with irregularities along the seafloor (Miramontes et al., 2019; Thieblemont et al., 2019; Yin et al., 2019).

When bottom currents interact with other depositional processes such as pelagic or gravity flows, a "*Mixed Depositional System*" forms (Mulder et al., 2008; Rebesco et al., 2014). These express morphological elements and sedimentary facies that differ from those found in pure pelagic, gravitational or contourite domains (Rodrigues et al., 2021).

## Along-slope oceanographic processes

### Water masses versus ocean currents

A water mass is an identifiable body of water with a common formational history, whose physical properties differ from that of the surrounding water and which originated in a particular region of the ocean (Britannica, 2015; Emery and Meincke, 1986). Oceanographers usually divide water masses into simplified surface, intermediate and deepwater masses. In practice, water masses usually exhibit more complexity and researchers are able to identify additional water masses within these three categories for both modern and ancient oceans. An ocean current is a continuous, directional movement of sea water

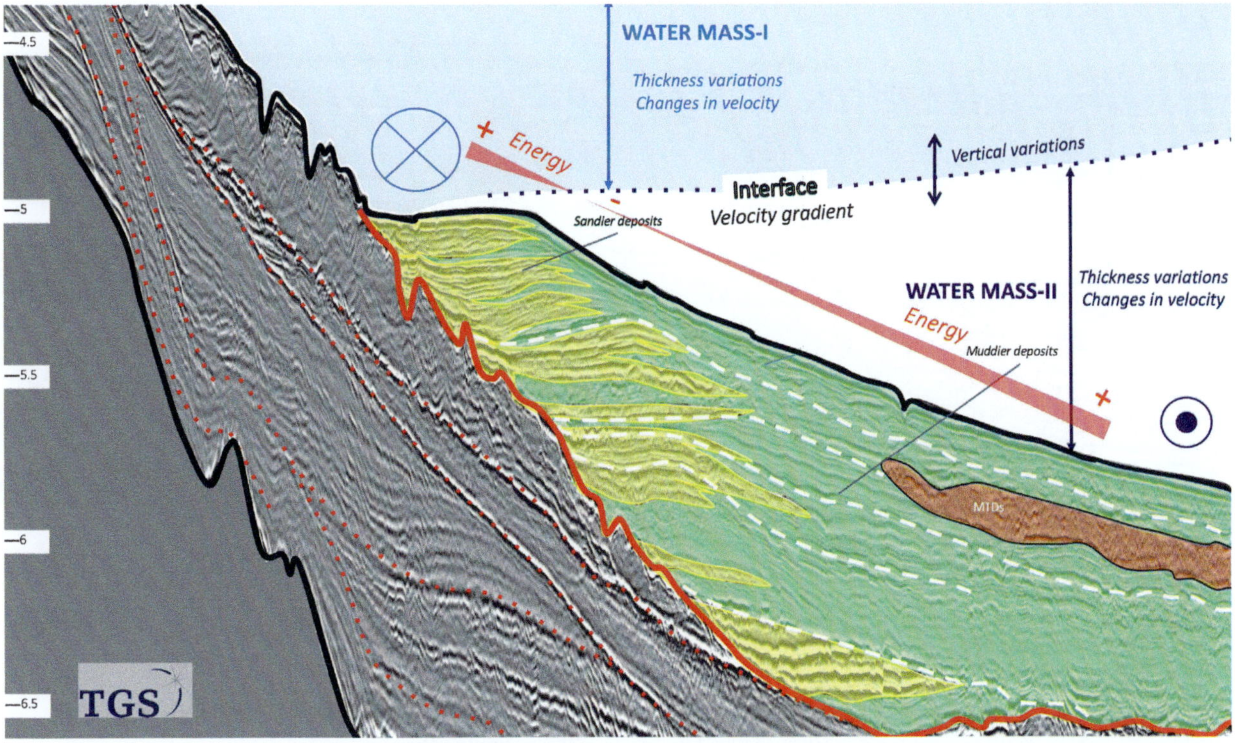


**FIG. 2** Example of the contourite depositional system (CDS) from Fig. 1 showing how the position of the large drift relates to the associated water mass-II, with its proximal top as a sub-horizontal surface (a contouritic terrace) related to the long-term influence of the interface (pycnocline) between two water masses (I and II). See the text for further explanation. The greater sedimentary thickness of the drift records the minimum energy of the bottom current. The red triangles indicate the relative energy related to the water masses but not to the processes associated to interfaces. A very basic correlation between seismic facies and sedimentary facies can be seen from muddier deposits *(light green)* and potentially coarser, sandier sediments *(yellow)*. *Light brown* deposit represents mass-transport deposits, MTD. *(*Seismic line courtesy from *TGS, vertical scale in seconds)*.

generated by forces acting upon the water including wind, the Coriolis Effect, waves, cabbeling and other temperature and/or salinity differences (Tomczak, 1999; Britannica, 2015). Given their independent definitions, water masses and ocean currents may not coincide but, over longer periods of weeks, months, years, etc., do show some correlation. Long-term average data tend to show each as similar to the other in terms of their major properties (Cochran et al., 2019).

## Oceanic currents and associate secondary processes

Fig. 4 shows ocean bottom currents related to the global thermohaline-induced geostrophic water-mass circulation (Wunsch, 2002; Rahmstorf, 2006; Kuhlbrodt et al., 2007; Cochran et al., 2019). Relatively dense water flows out into the ocean via narrow or shallow straits, or over the continental margin, and veers to the right or left (right in the Northern Hemisphere, left in the Southern Hemisphere) due to their travel between different inertial frames of Earth's rotation (Fig. 4). Once bathymetry no longer constrains a water mass, it spreads into a wider structure in response to gravity, Earth's rotation and friction with the seafloor. The Earth's rotation tends to steer bottom currents in a direction parallel to large-scale bathymetric contours. For example, water masses tend to flow along continental margins. Small-scale bathymetric features such as seamounts, ridges, straits, mounds, banks and canyons can disrupt a flow or cause it to accelerate. A sufficiently large current velocity can erode sediments and deposit it later once the velocity declines (Kennett, 1982; Stow et al., 2002; Rebesco and Camerlenghi, 2008).

Currents can be either barotropic or baroclinic. Barotropic flows are currents with constant pressure surfaces parallel to surfaces of constant density. These include tides, wind-driven currents and surface waves. In general, these flows are driven by pressure gradients caused by a sloping sea surface (Shanmugam, 2013). Baroclinic flows are currents whose surfaces of constant pressure and constant density do not strike parallel to each other and which propagate on density surfaces known as pycnoclines in stratified waters. Internal waves and internal tides represent examples of baroclinic flows. Pressure gradients caused by horizontal density differences and fronts drive baroclinic flows. These currents appear along the ocean floor of continental slopes and in submarine canyons.

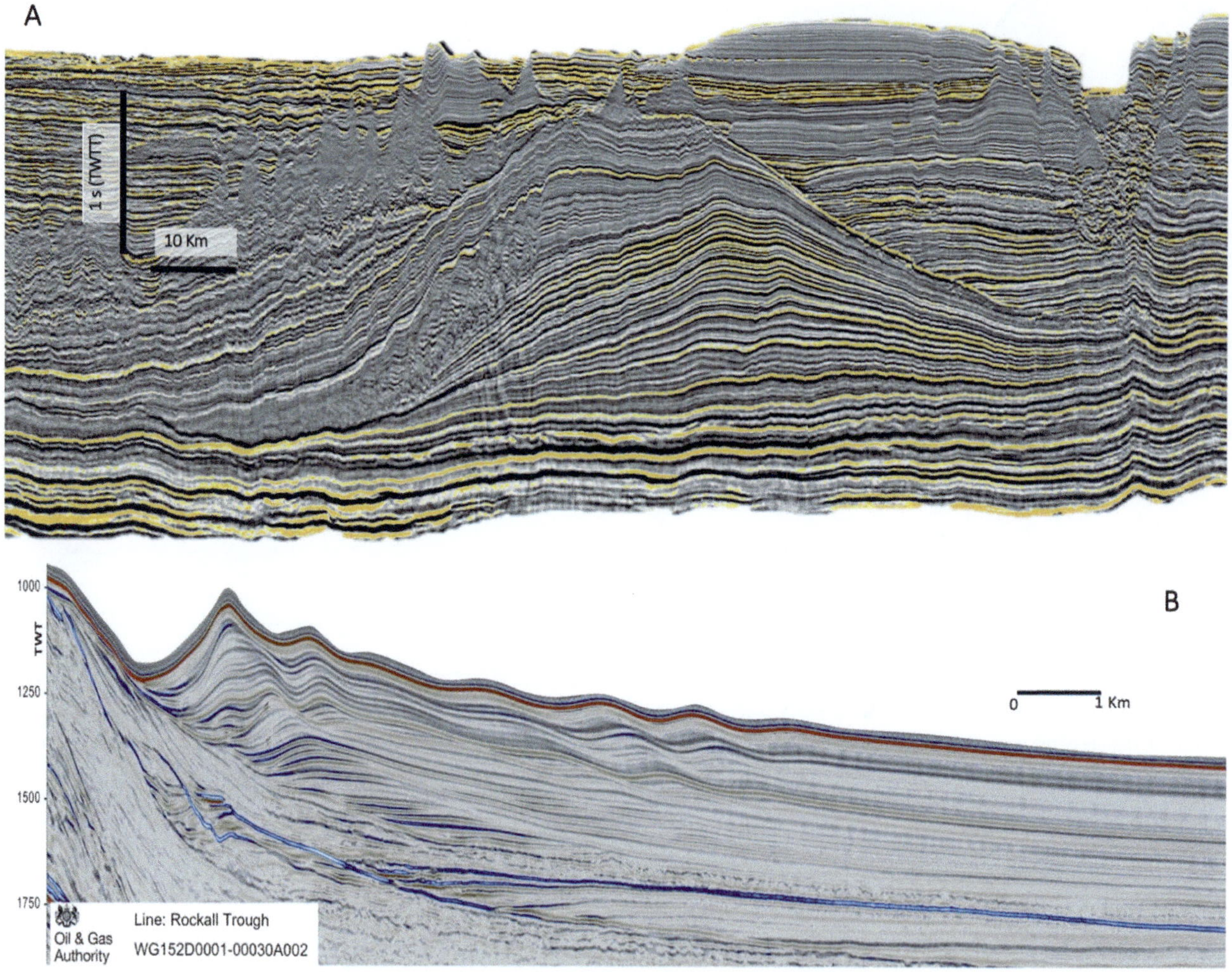


**FIG. 3** (A) Example of a large, buried mounded drift along the Mozambique Channel. (B) Example of a modern day large elongated, mounded and separated drift along the Rockall Trough. Vertical scale in time (milliseconds). *((A) From Thiéblemont, A., Hernández-Molina, F.J., Ponte, J.-P., Robin, C., Guillocheau, F., Cazzola, C., Raisson, F., 2020. Seismic stratigraphic framework and depositional history for Cretaceous and Cenozoic contourite depositional systems of the Mozambique Channel, SW Indian Ocean. Mar. Geol. 425, 106192. doi:10.1016/j.margeo.2020.106192. Data image courtesy of INP and WesternGeco. (B) Data courtesy from Oil and Gas Authority, UK).*

Global water-mass circulation carries a number of important implications for deepwater sedimentation as follows:

1. Deepwater formation. Modern-day thermohaline-induced geostrophic water-mass circulation models hold that deepwater formation takes place in a few localized areas (Fig. 4), the most important being those localized in the polar regions (Rahmstorf, 2006; Kuhlbrodt et al., 2007). This circulation model became fully established at the end of the Miocene up to the base of the Quaternary and has functioned throughout the Quaternary despite some fluctuation during glacial and interglacial stages. This oceanographic model does not apply to the geological past. Interpretations of climate and sedimentation prior to the Miocene must refer to earlier global or regional paleocirculation models.
2. Spreading of deepwater masses occurs mainly along the western portion of the marine basin (Fig. 4) as deep western boundary currents (DWBC) influenced by the Coriolis Effect (McCave et al., 1980; McCave and Tucholke, 1986). In ancient oceanic basins, similar bottom currents would flow along the western portions of marine basins when past forcing conditions resembled those of the present.
3. Fig. 5 shows the global distribution of the average annual current speed (in $cm^{-1}$) for geostrophic water-mass circulation based on calculations given in Thran et al. (2018). Bottom currents associated with geostrophic water-mass circulation exhibit low average speeds ($\sim$ 1–3 $cm^{-1}$, $\sim$ 0.3–1.1 in $s^{-1}$; Rebesco et al., 2014) but eastern boundary currents exhibit higher speeds (15–100 $cm^{-1}$, $\sim$ 5.9–39.3 in $s^{-1}$). In addition to that, the $>$ 15 $cm^{-1}$ ($> \sim$ 5.9 in $s^{-1}$) speeds occur around trenches, seamounts, ridges, banks, channels, straits and gateways ($\sim$300 $cm^{-1}$, $\sim$ 118.1 in $s^{-1}$ in the Strait of

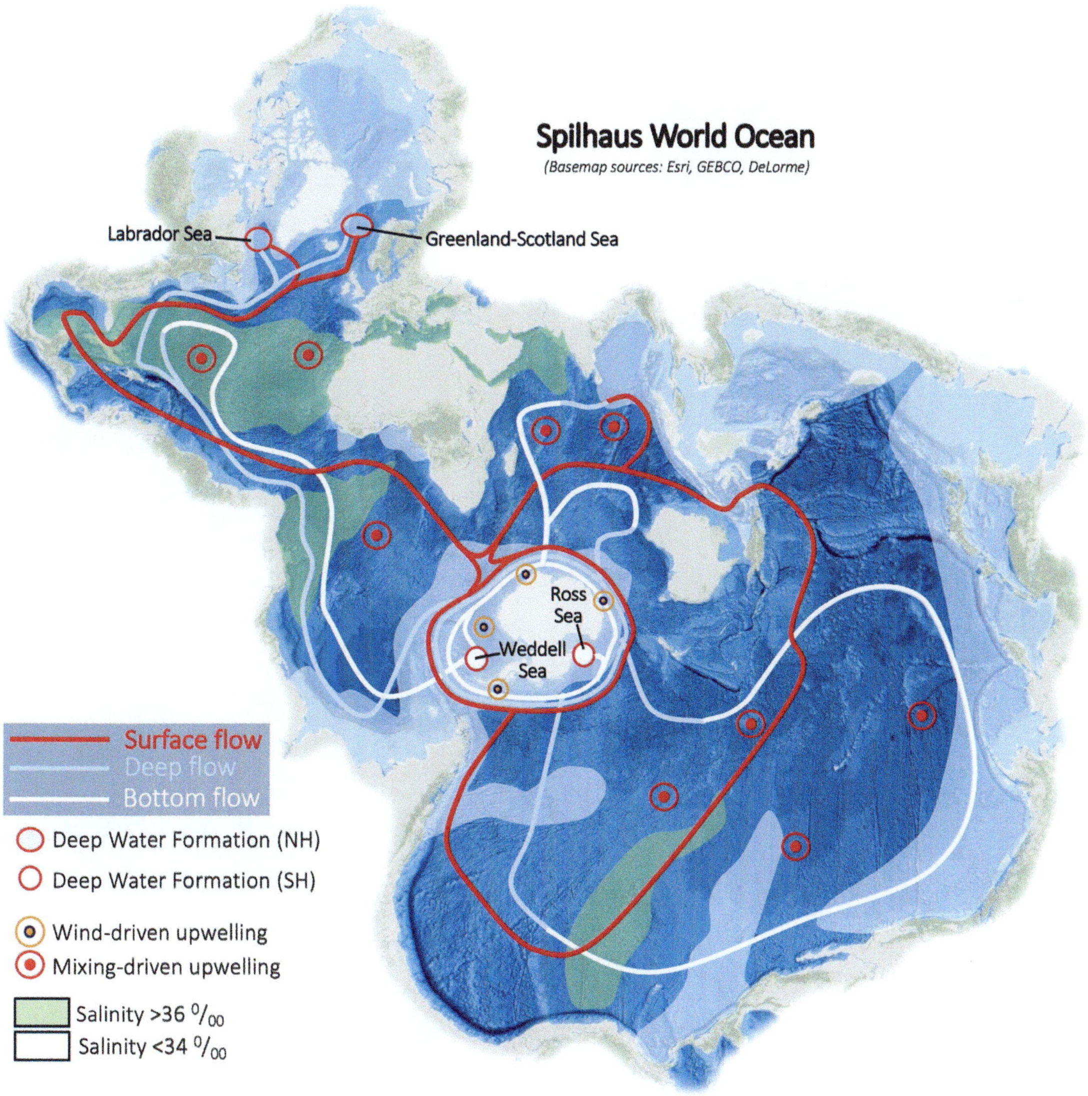


**FIG. 4** Global thermohaline circulation (Spilhaus World Ocean projection from Esri, GEBCO, DeLorme). *Red*: surface currents; *Light blue*: deepwater; *White*: bottom water; *White circles with a red line*: main sites of deepwater formation. In the Atlantic, warm, saline waters flow northward from the Southern Ocean into the Labrador and Nordic seas. By contrast, deepwater formation occurs in the North Pacific, whose surface waters are consequently fresher. Deepwaters formed in the Southern Ocean become denser than waters from the North Atlantic and therefore the former spread at deeper levels. Note the small, localized deepwater formation areas in comparison to the widespread zones of mixing-driven upwelling. Wind-driven upwelling occurs along the Antarctic Circumpolar Current. *(Oceanographic data from Rahmstorf, S., 2006. Encyclopedia of Quaternary Science, Illustration by E. Paul Oberlander, 2009, WHOI. http://oceanworld.tamu.edu/resources/oceanography-book/abruptclimatechange.htm; Kuhlbrodt, T., Griesel, A., Montoya, M., Levermann, A., Hofmann, M., Rahmstorf, S., 2007. On the driving processes of the Atlantic meridional overturning circulation. Rev. Geophys. 45, RG2001).*

Gibraltar, Mulder et al., 2003). Geostrophic water-mass circulation depends on the local seafloor physiography and obstacles. The lower areas of bathymetric features can amplify bottom current flow by a factor of two and enable local turbulent flow.

4. Bottom current velocity controls the concentration and thickness of fine-grained sediments in suspension (nepheloid layers). The highest sediment concentrations occur along western boundaries and more generally within oceanic gyres (Hollister and McCave, 1984). Nepheloid layers also depend on interfaces between water masses (McCave, 1972) because these can capture basinward sediment transport by gravity flows (Wells and Dorrell, 2021). Oceanic basins commonly host several water masses, some of which circulate in opposite directions and progressively and effectively transport fine-grained sediments from gravity flows.
5. Interfaces (pycnoclines) between water masses influence deepwater sedimentation through secondary oceanographic processes.

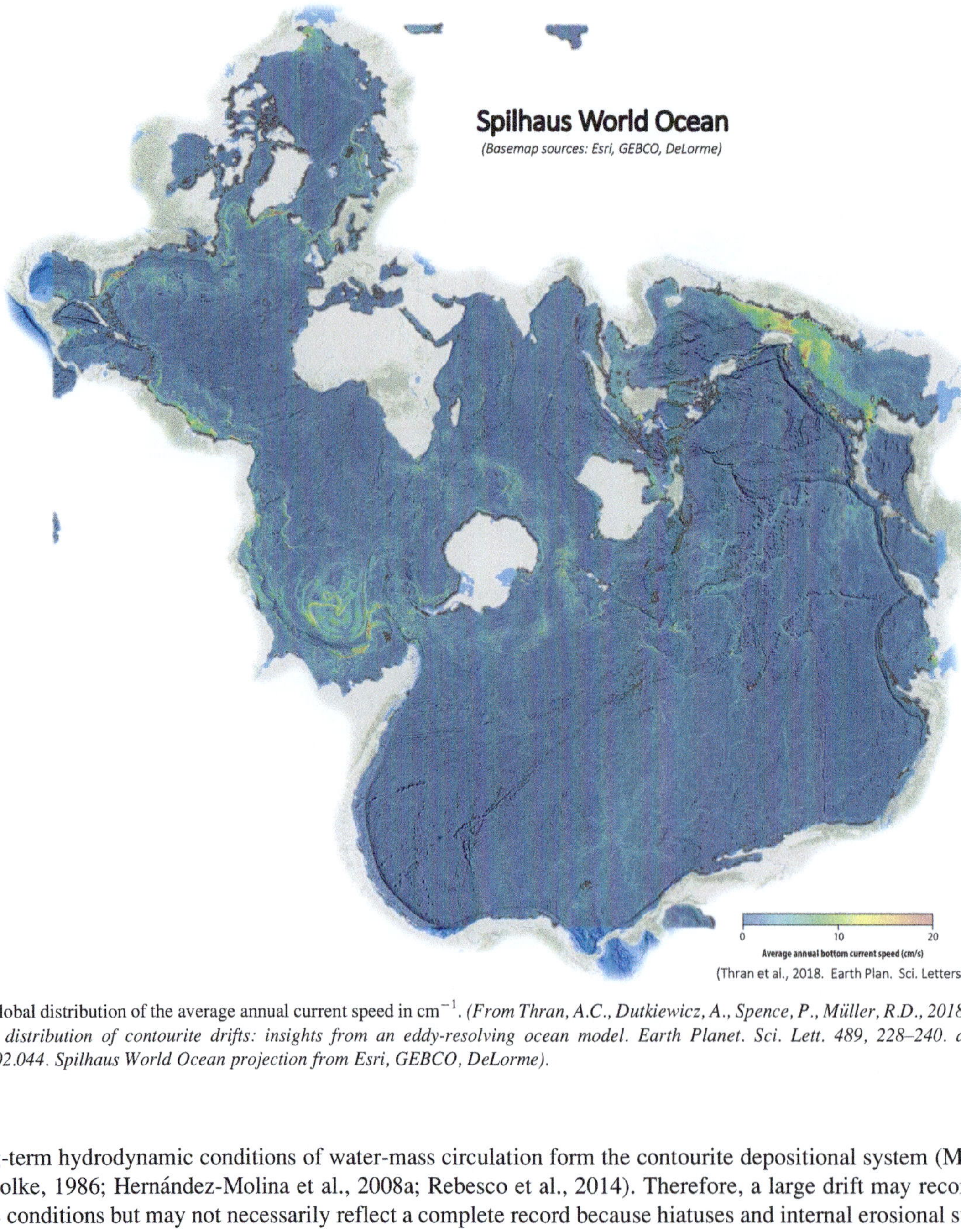


**FIG. 5** Global distribution of the average annual current speed in $cm^{-1}$. *(From Thran, A.C., Dutkiewicz, A., Spence, P., Müller, R.D., 2018. Controls on the global distribution of contourite drifts: insights from an eddy-resolving ocean model. Earth Planet. Sci. Lett. 489, 228–240. doi:10.1016/j.epsl.2018.02.044. Spilhaus World Ocean projection from Esri, GEBCO, DeLorme).*

6. Long-term hydrodynamic conditions of water-mass circulation form the contourite depositional system (McCave and Tucholke, 1986; Hernández-Molina et al., 2008a; Rebesco et al., 2014). Therefore, a large drift may record some of those conditions but may not necessarily reflect a complete record because hiatuses and internal erosional surfaces are always present (Faugères et al., 1999; Llave et al., 2001; Hernández-Molina et al., 2008b, 2016a). Fig. 6 shows the global distribution for both contourite depositional systems and mixed systems developed during the Mesozoic and Cenozoic over the durations of millions of years (Rebesco et al., 2014; Thran et al., 2018).

The secondary processes of benthic storms, overflows, interfaces between water masses, vertical eddies, horizontal vortices, tides and internal tides, internal waves and solitons, tsunami-related traction currents and rogue or cyclonic waves all influence water-mass circulation (Shanmugam, 2006, 2013, 2021; Rebesco et al., 2014; Hernández-Molina et al., 2016c; Yin et al., 2019). Some of these processes are not well understood and their influence on deepwater sedimentation dynamics and seafloor morphology remains uncertain (e.g., Maier et al., 2019; Miramontes et al., 2019). Shanmugam (2006, 2013, 2017, 2021), Rebesco et al. (2014) and Hernández-Molina et al. (2016c) describe general aspects of how secondary processes influence deep-marine sedimentation. Specific details about these processes can also be found in reports on benthic storms (Hollister, 1993; Gardner et al., 2017; Miguez-Salas et al., 2020), overflows (Legg et al., 2009; Glazkova et al., 2021; Yin et al., 2021), interfaces between water masses (Hernández-Molina et al., 2009; Preu et al., 2013), vertical eddies

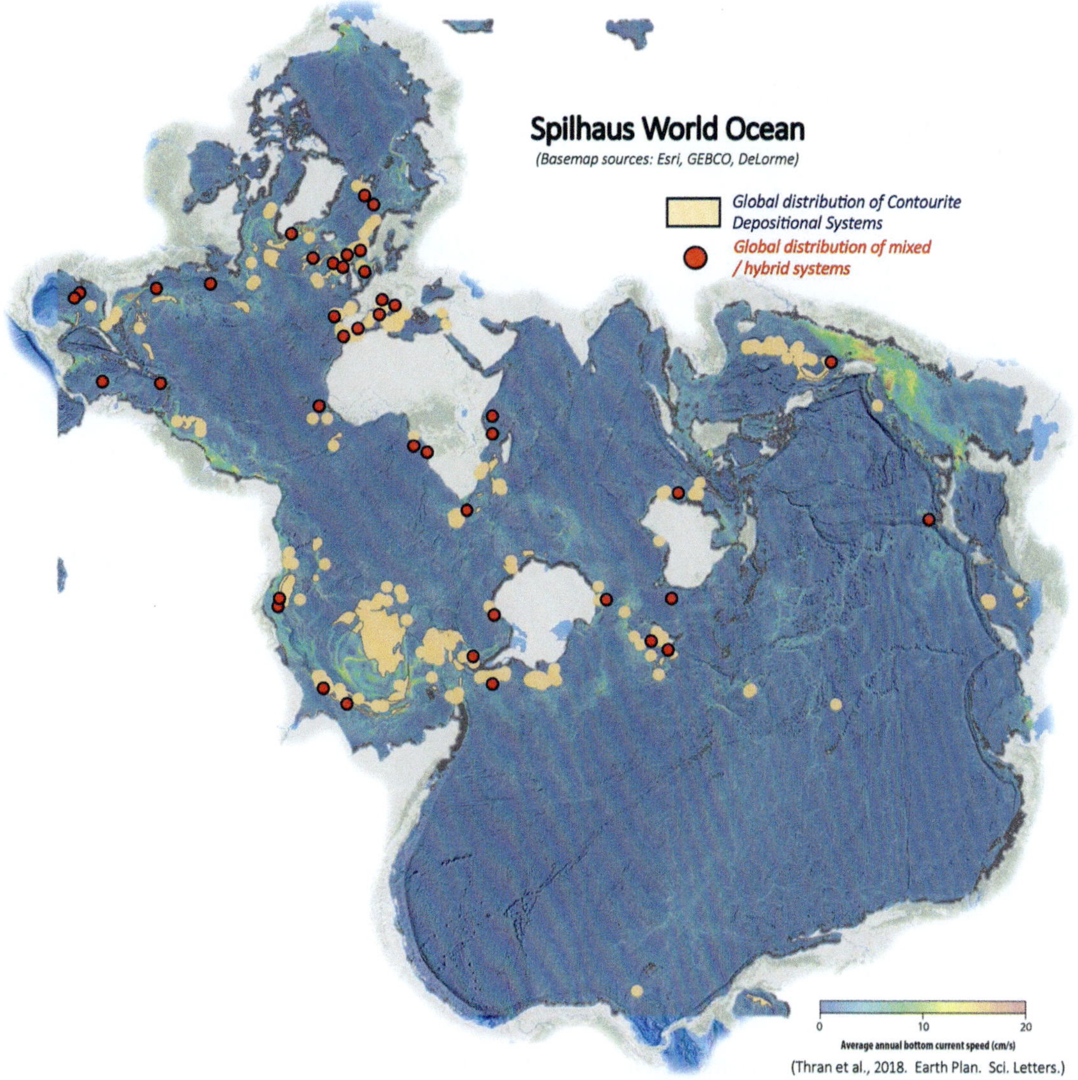


**FIG. 6** Global distribution for both contourite depositional systems and mixed systems (Spilhaus World Ocean projection from Esri, GEBCO, DeLorme). Background with global average annual bottom current speed in $cm^{-1}$ from Thran et al. (2018). The compilation of contourite systems is partially from Rebesco et al. (2014) and Thran et al. (2018). The compilation for the mixed systems has been elaborated for this chapter based on the following authors in alphabetic order: Amblas and Canals, 2016; Amblas et al., 2006; Badalini et al., 2016; Bailey et al., 2021a,b; Batchelor et al., 2021; Bouma and Treadwell, 1975; Brackenridge et al., 2013; Camerlenghi et al., 1997; Campbell and Deptuck, 2012; Carter and McCave, 2002; Casson et al., 2020; Cauxeiro et al., 2020; Chen et al., 2013, 2016, 2020; Cowan et al., 2008; Creaser et al., 2017; de Castro et al., 2020, 2021a,b; Deptuck and Kendell, 2020; Ding et al., 2021; Elliot and Parson, 2008; Enjorlas et al., 1986; Ercilla et al., 2011, 2019; Escutia et al., 2002; Esmerode et al., 2008; Fonnesu, 2013; Fonnesu et al., 2020; Forwick et al., 2015; Fuhrmann et al., 2020; Georgiopoulou et al., 2021; Gong et al., 2013, 2016, 2018; Hanquiez et al., 2010; Heritier et al., 1979; Hernández-Molina et al., 2010, 2017; Hillenbrand et al., 2008; Howe et al., 2002; Huang and Jokat, 2016; Ito, 1996, 2002; Knutz et al., 2002; Kuvaas et al., 2005; Larter et al., 2019; Locker and Laine, 1992; Lu et al., 2003; Lucchi and Rebesco, 2007; Lucchi et al., 2002; Marchès et al., 2007, 2010; Maselli and Kneller, 2018; McMaster et al., 1989; Mencaroni et al., 2021; Michels et al., 2001, 2002; Millot and Monaco, 1984; Miramontes et al., 2021; Moraes et al., 2007; Mourlot et al., 2018; Mutti, 1992; Mutti et al., 1980, 2014; Normandeau et al., 2019; Nugraha et al., 2018; Owens, 2017; Palamenghi et al., 2015; Palermo et al., 2014; Pandolpho et al., 2021; Preu et al., 2013; Rasmussen et al., 2003; Rebesco et al., 1996, 1997, 1998, 2002, 2007, 2016; Rodrigues et al., 2021, 2022a,b,c; Salabarnada et al., 2018; Sansom, 2017, 2018; Scheuer et al., 2006; (Shanmugam et al., 1993a,b); Solli et al., 2008; Stanley, 1987, 1988, 1993; Takashimizu et al., 2016; Trincardi et al., 2007; Tucholke, 2002; Tucholke and Mountain, 1986; Uenzelmann-Neben, 2001, 2006; van Li et al., 2020; van Raaphorst et al., 2001; Viana et al., 1998, 1999; Zhong et al., 2015).

(Hollister and McCave, 1984; Liu et al., 2017; López-Quirós et al., 2020), horizontal vortices (Preu et al., 2013; Miramontes et al., 2019), deep tides and internal tides, and internal waves and solitons (Apel, 2000, 2004; Jackson, 2004; Butman et al., 2006a; Shanmugam, 2006, 2012, 2014; Miramontes et al., 2020a).

Fig. 7 shows how secondary processes operate in a marine basin. They specifically function to (A) locally modulate the average current speed and direction of the water mass with overflows, deep-sea storms, eddies and internal waves acting as

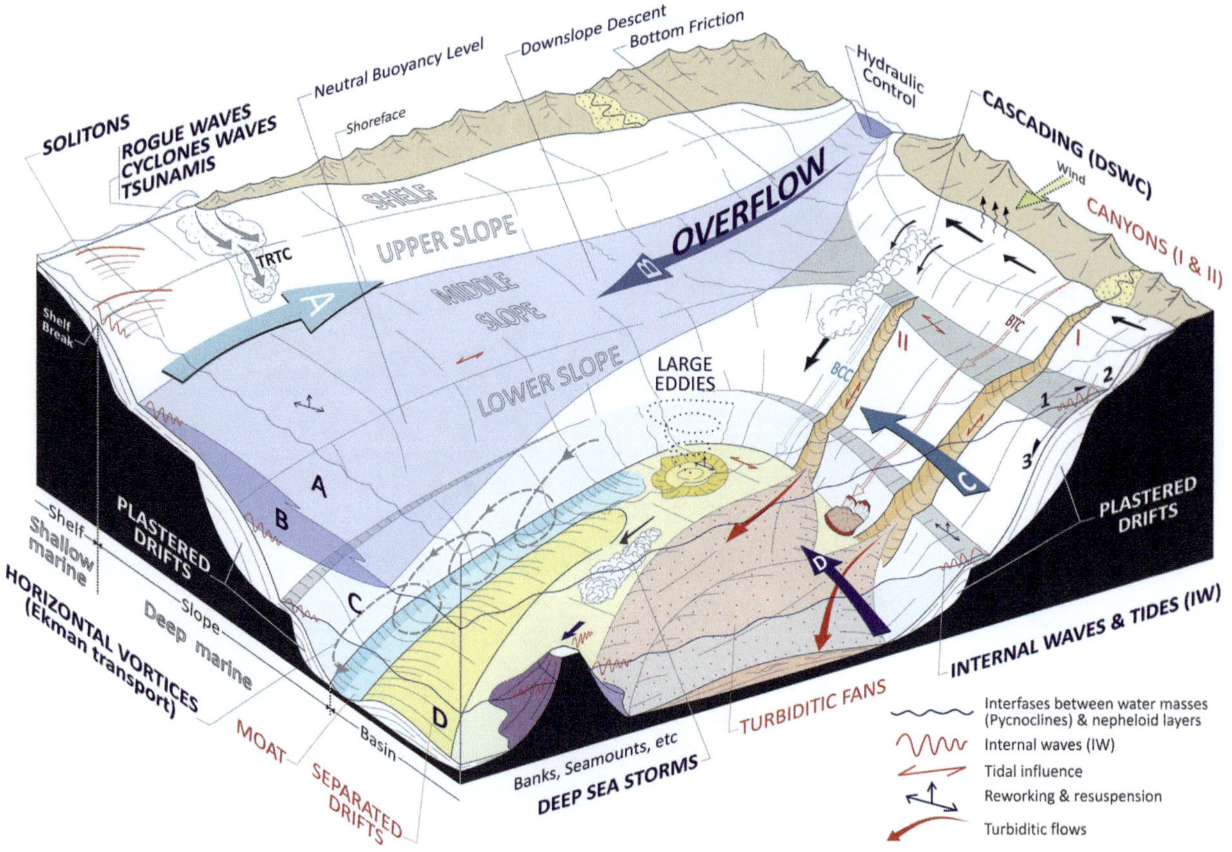


**FIG. 7** 3D sketch depicting possible oceanographic processes in deepwater environments. Velocity along the seafloor depends on density currents and overflows as well as on barotropic currents and intermittent processes such as cascading, giant eddies, deepsea storms, vortices, internal waves, internal tides, tsunamis, cyclone waves and rogue waves. *(From Rebesco, M., Hernández-Molina, F.J., Van Rooij, D., Wåhlin, A., 2014. Contourites and associated sediments controlled by deep-water circulation processes: state of the art and future considerations. Mar. Geol. 352, 111–154. doi:10.1016/j.margeo.2014.03.011, with permission from Elsevier).*

higher energy processes, (B) influence benthic activity along the seafloor and within the sediments to determine ichnofacies and (C) control local sedimentary facies associations.

## Present-day versus ancient global oceanographic circulation models

The modern-day global thermohaline-induced geostrophic water-mass circulation model has been developed only very recently and differs significantly from past circulation models. Plate tectonic factors determining the closing and opening of ocean gateways (Fig. 8) have induced pronounced changes in deepwater circulation and poleward temperature gradients on geological time scales. Kennett (1982), Knutz (2008), Seibold and Berger (2010) and Uenzelmann-Neben et al. (2017) provide useful summaries of past circulation models. Literature sources also provide detailed descriptions of Mesozoic and Cenozoic circulation models (e.g., Hay, 2009; Potter and Szatmari, 2009; Hamon et al., 2013; Pérez-Díaz and Eagles, 2017; Straume et al., 2020).

Plate tectonic configurations carry three major long-term implications for oceanic circulation: (1) general westward equatorial circulation through the Tethys Ocean ceases to exist after the Mesozoic. The disappearance of the Tethys leads to progressive replacement by circulation from the poles to the equator; (2) the opening of the Tasmania Seaway and Drake Passage increases deepwater formation in polar regions. The Antarctic Circumpolar Current (ACC) becomes fully established at the end of the Eocene and beginning of the Oligocene. This current strongly influences global circulation and climatic conditions; (3) closing and opening of ocean gateways change deepwater circulation. The modern circulation system became established at around 3–2.6 Ma with the final closure of the Central America Seaway (CAS). This closure ended the transition from the warm greenhouse climate of the late Cretaceous and early Cenozoic to the predominance

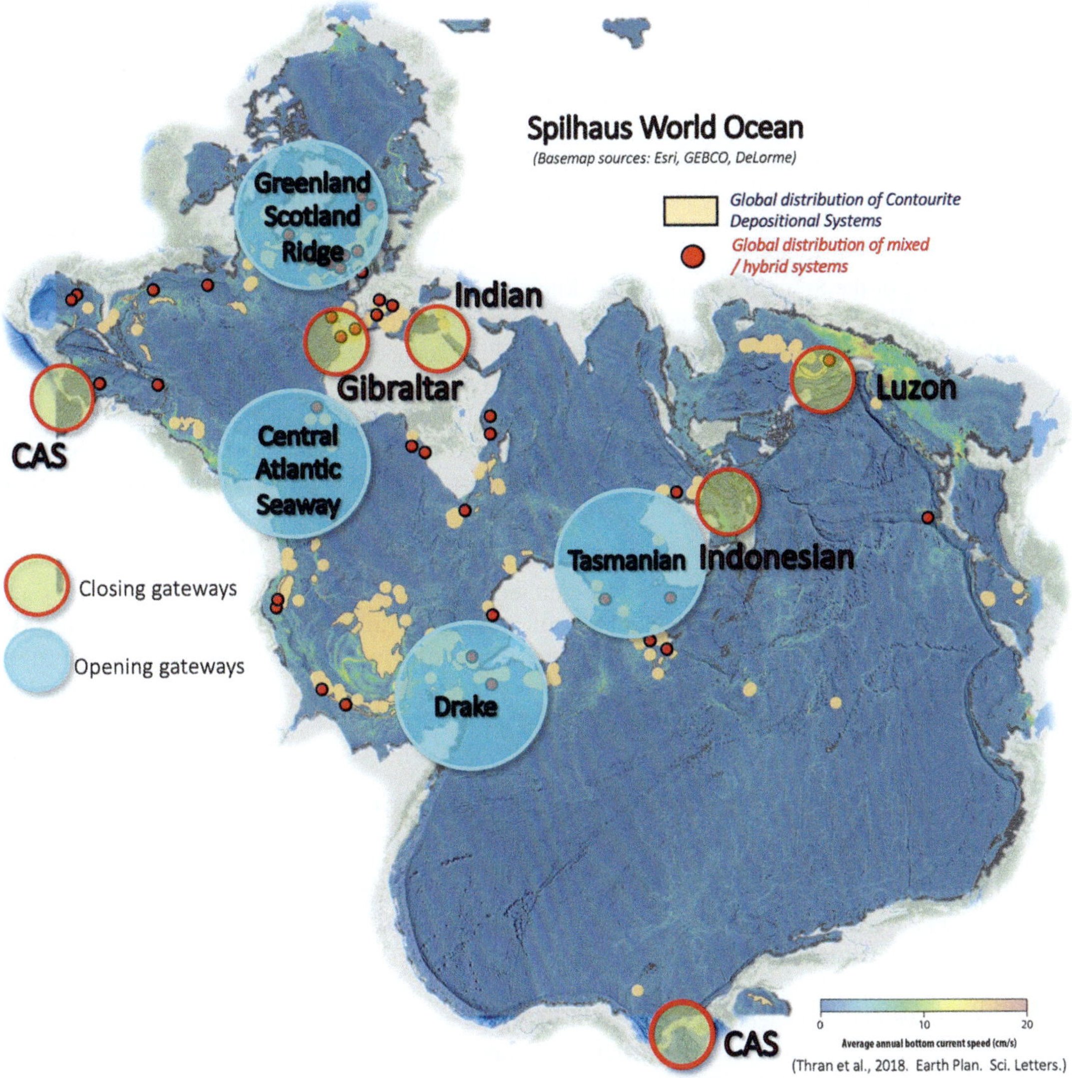


**FIG. 8** Distribution of important gateways during the Cenozoic. Closing gateways indicated in *yellow* and opening gateways indicated in *blue* (Spilhaus World Ocean projection from Esri, GEBCO, DeLorme). Closing and opening of gateways can induce drastic changes in deepwater circulation, determining the onset of the modern global thermohaline circulation from Fig. 4. References for the distribution of contourites and mixed depositional systems in Fig. 6. *(Based on Kennett, J., 1982. Marine Geology. Prentice-Hall, 813 pp.; Knutz, P.C., 2008. Palaeoceanographic significance of contourite drifts. In: Rebesco, M., Camerlenghi, A. (Eds.), Contourites Developments in Sedimentology, vol. 60. Elsevier, Amsterdam, pp. 511–535; Potter, P.E., Szatmari, P., 2009. Global Miocene tectonics and the modern world. Earth Sci. Rev. 96, 279–295; Seibold, E., Berger, W.H., 2010. The Sea Floor. An Introduction to Marine Geology, third ed. Springer Verlag, 369 pp.; Hamon, N., Sepulchre, P., Lefebvre, V., Ramstein, G., 2013. The role of eastern Tethys seaway closure in the Middle Miocene Climatic Transition (ca. 14 Ma). Clim. Past. 9, 2687–2702. doi:10.5194/cp-9-2687-2013; Uenzelmann-Neben, G., Weber, T., Grützner, J., Thomas, M., 2017. Transition from the cretaceous ocean to Cenozoic circulation in the western South Atlantic—a twofold reconstruction. Tectonophysics 716, 225–240. doi:10.1016/j.tecto.2016.05.036; Straume, E.O., Gaina, C., Medvedev, S., Nisancioglu, K.H., 2020. Global cenozoic paleobathymetry with a focus on the northern hemisphere oceanic gateways. Gondwana Res. 86, 126–143. doi:10.1016/j.gr.2020.05.011; Yin, S., Hernández-Molina, F.J., Hobbs, R., Gao, J., Ding, W., Yang, C., Lin, L., Zhang, H., Wu, Z., Li, J., 2021. Contourite processes associated with the overflow of Pacific Deep Water within the Luzon Trough: conceptual and regional implications. Deep-Sea Res. I: Oceanogr. Res. Pap. 103459).*

of northern hemisphere glaciations during the Quaternary (last 2.6 Ma) and the formation of the present-day global thermohaline-induced geostrophic water-mass circulation. Additional Cenozoic plate tectonic events influencing this shift include:

- Opening of the Tasman Seaway (late Eocene)—initial development of the Antarctic Circumpolar Current (ACC) around part of Antarctica.

- Opening of the Drake Passage (late Eocene/early Oligocene)—full development of the Antarctic Circumpolar Current (ACC), which contributes to the climatic isolation of the continent, the onset of ice accumulation in Eastern Antarctica and the final initiation of the Antarctic Bottom Water (AABW).
- Opening of a deepwater connection between the Arctic and the North Atlantic (early to middle Miocene)—full initiation of the North Atlantic Deep Water (NADW).
- Closure of the Indian Gateway (middle Miocene) that represented the end of the global equatorial circulation and the Mediterranean Sea contribution to the Atlantic Meridional Overturning Circulation (AMOC). Closure coincided with a period of major global climatic and oceanographic changes during the Middle Miocene Climatic Transition (MMCT) from approximately 14.2 to 13.8 Ma.
- Closure and opening of the Strait of Gibraltar (late Miocene and early Pliocene)—initiation of the present-day Mediterranean Outflow Water (MOW) with enhanced Meridional Overturning Circulation (MOC).
- Closure of the Indonesian seaway (late Miocene) with the final connection between the Pacific and Indian oceans.
- Closure of the Central American seaway (late Pliocene/early Quaternary)—final establishment of the Gulf current and onset of ice accumulation in the Arctic, an event that amplifies climatic variations.
- The modern-day global thermohaline-induced geostrophic water-mass circulation became fully stablished during the Quaternary. Both glacial stages and terminations of glacial stages bring higher production and spreading of saltier, more voluminous AABW. By contrast, the NADW experienced spreading and enhanced circulation during interglacial stages (e.g., Boyle and Keigwin, 1987; Duplessy et al., 1988; Rahmstorf, 2006; Lund et al., 2011).

Along with the northward drift of continents toward the northern hemisphere and the increase of the Earth's albedo, these oceanographic events contributed to the general global cooling of the Earth (Kennett, 1982; Zachos et al., 2001; Hamon et al., 2013; Straume et al., 2020). Most of these events occurred during periods of active bottom currents and the formation of contourite and mixed systems. For example, the initiation of many bottom current deposits began around the Eocene/Oligocene transition and was reactivated in the middle Miocene.

## Identification and characterization of contourite depositional systems in the seismic record: Morphological elements and associated seismic facies

Contourite drifts include a wide range of morphological features, for which several classification schemes have been proposed (for details see Faugères et al., 1999; Stow et al., 2002; Rebesco and Camerlenghi, 2008; Rebesco et al., 2014). Classifications are not always mutually exclusive due to gradations and overlap among drift features.

Contourites generally include unique depositional, erosional and mixed elements. As depositional elements, drifts (Fig. 9) are classified as: (1) elongated and mounded drifts (separated and detached), (2) sheeted drifts (plastered and abyssal sheet), (3) channel-related drifts (sheeted, patches and fans), (4) confined drifts, (5) fault-controlled drifts, (6) patch drifts (sheeted and mounded), (7) infill drifts, (8) mixed drifts and (9) delta drifts (McCave and Tucholke, 1986; Faugères et al., 1999; Rebesco, 2005; Rebesco and Camerlenghi, 2008; Rebesco et al., 2014; Eberli and Betzler, 2019). All drifts show varying degrees of mounding and gradation between drift types. Other minor depositional features include various bedforms (e.g., barchan dunes, sand ribbons) and sediments wave fields, which can be generated along the larger morphological features by secondary oceanographic processes.

Erosional elements subdivide into linear or areal features (Fig. 10). Linear features include moats, contourite channels, marginal valleys (or scours) and furrows. Areal features include abraded surfaces (find initial classifications in Hernández-Molina et al., 2008b; García et al., 2009).

Mixed elements are features formed by a combination of depositional and erosional processes (Fig. 11) such as contourite terraces and the subcircular depressions defined as "*honeycomb structures*" (García et al., 2016; Sun et al., 2017; Yin et al., 2019). Terraces are subhorizontal features that appear within both the present-day seafloor and the sedimentary record, and which have developed over time in constructional and erosive phases (Hernández-Molina et al., 2009; Preu et al., 2013).

Large features appear clearly in 2D seismic reflection profiles (Fig. 3) and bathymetric data due to their distinctive morphologies, seismic characteristics and their overall architecture relative to the basin (Faugères et al., 1999; Nielsen et al., 2008). Few studies have described contourites from 3D seismic datasets (e.g., Campbell and Mosher, 2016; Gong et al., 2017; Kirby et al., 2021a; Mantilla et al., 2022). The drifts appear clearly in seismic data and from seismic stratigraphic analysis because their seismic facies, seismic attributes and sedimentary architecture differ from those of other deepwater systems. Differences in seismic sources (different vertical and horizontal scales), the timing of data collection and seismic methods can generate differing images for the same sedimentary record. In spite of these methodological

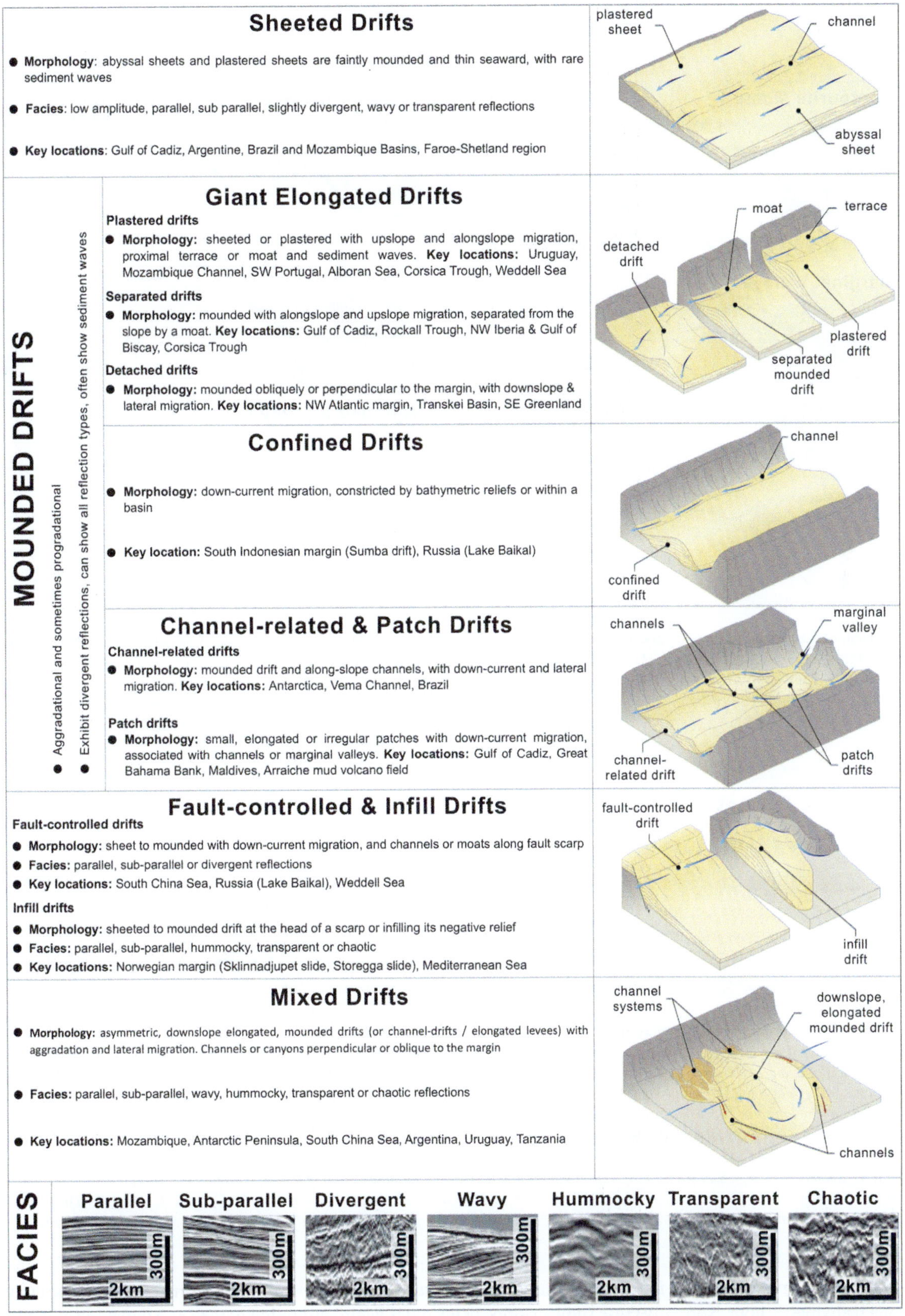


**FIG. 9** Main types of contourite drifts and associated erosional features with inferred bottom current pathways. *(Based on Faugères, J.-C., Stow, D.A.V., Imbert, P., Viana, A.R., 1999. Seismic features diagnostic of contourite drifts. Mar. Geol. 162, 1–38. doi:10.1016/S0025-3227(99)00068-7; Rebesco, M., Hernández-Molina, F.J., Van Rooij, D., Wåhlin, A., 2014. Contourites and associated sediments controlled by deep-water circulation processes: state of the art and future considerations. Mar. Geol. 352, 111–154. doi:10.1016/j.margeo.2014.03.01. Key seismic facies of contourite drifts from Thiéblemont, A., Hernández-Molina, F.J., Ponte, J.-P., Robin, C., Guillocheau, F., Cazzola, C., Raisson, F., 2020. Seismic stratigraphic framework and depositional history for Cretaceous and Cenozoic contourite depositional systems of the Mozambique Channel, SW Indian Ocean. Mar. Geol. 425, 106192. doi:10.1016/j.margeo.2020.106192).*

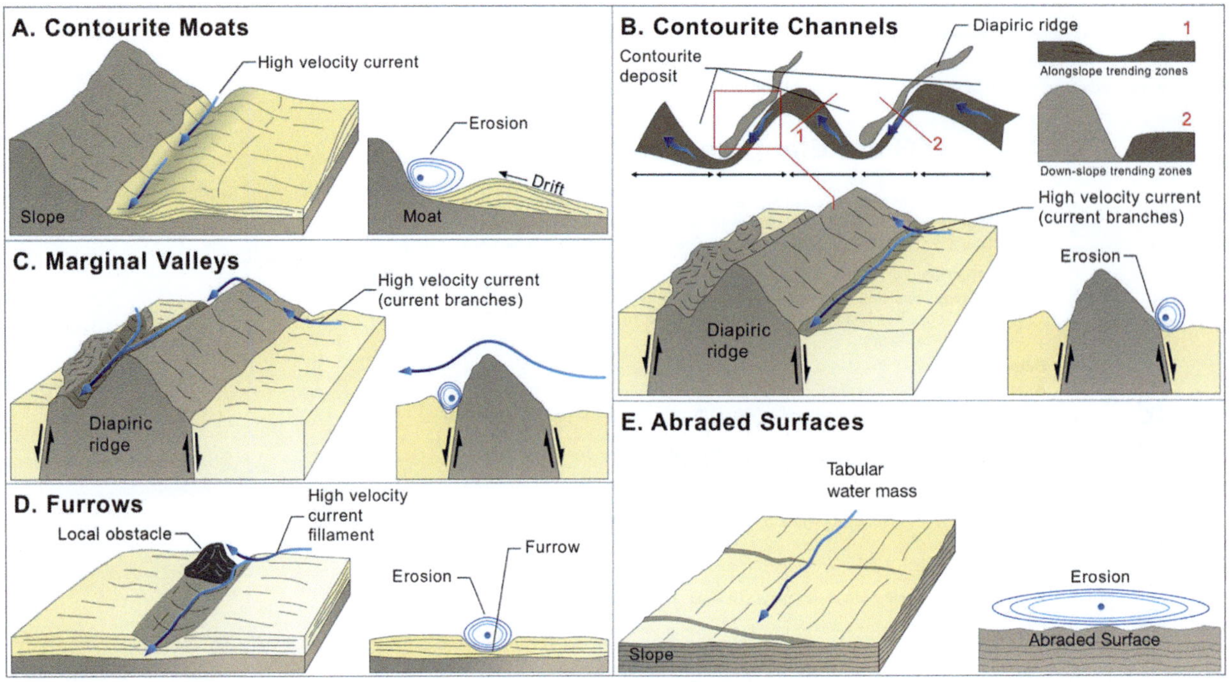


**FIG. 10** Dominant erosional features generated by bottom currents. *(From Hernández-Molina, F.J., Llave, E., Stow, D.A.V., 2008b. Continental slope contourites. In: M. Rebesco, A. Camerlenghi (Eds.), Contourites. In: Developments in Sedimentology, vol. 60, Elsevier, pp. 379–408. doi:10.1016/S0070-4571(08)10019-X; García, M., Hernández-Molina, F.J., Llave, E., Stow, D.A.V., León, R., Fernández-Puga, M.C., Díaz del Río, V., Somoza, L., 2009. Contourite erosive features caused by the Mediterranean outflow water in the Gulf of Cadiz: quaternary tectonic and oceanographic implications. Mar. Geol. 257, 24–40).*

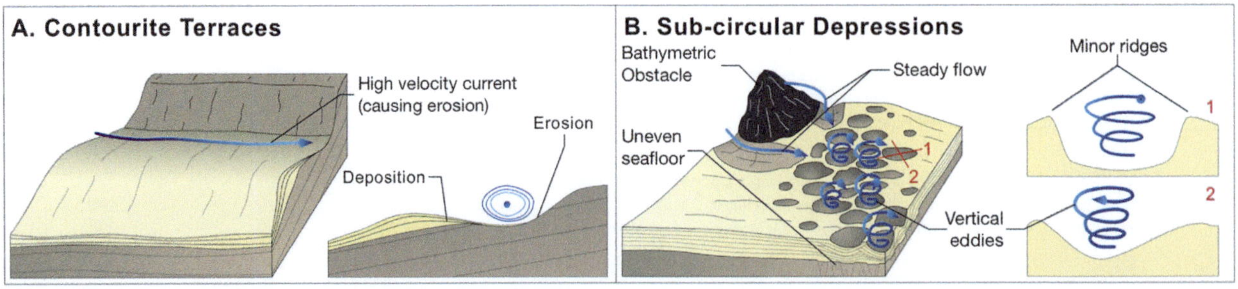


**FIG. 11** Mixed elements features (contourite terraces and the sub-circular depressions) formed by the combination of depositional and erosional phases. *(Based on Hernández-Molina, F.J., Paterlini, M., Violante, R., Marshall, P., de Isasi, M., Somoza, L., Rebesco, M., 2009. Contourite depositional system on the Argentine slope: an exceptional record of the influence of Antarctic water masses. Geology 37, 507–510; Preu, B., Hernández-Molina, F.J., Violante, R., Piola, A.R., Paterlini, C.M., Schwenk, T., Voigt, I., Krastel, S., Spiess, V., 2013. Morphosedimentary and hydrographic features of the northern Argentine margin: the interplay between erosive, depositional and gravitational processes and its conceptual implications. Deep-Sea Res. I: Oceanogr. Res. Pap. 75, 157–174; García, M., Hernández-Molina, F.J., Alonso, B., Vázquez, J.T., Ercilla, G., Llave, E., Casas, D., 2016. Erosive sub-circular depressions on the Guadalquivir Bank (Gulf of Cadiz): interaction between bottom current, mass-wasting and tectonic processes. Mar. Geol. 378, 5–19; Sun, Q., Cartwright, J., Lüdmann, T., Wu, S., Yao, G., 2017. Three-dimensional seismic characterization of a complex sediment drift in the South China Sea: evidence for unsteady flow regime. Sedimentology 64(3), 832–853; Yin, S., Hernández-Molina, F.J., Zhang, W., Li, J., Wang, L., Ding, W., Ding, W., 2019. The influence of oceanographic processes on contourite features: a multidisciplinary study of the northern South China Sea. Mar. Geol. 415, 105967).*

challenges, six criteria can serve to identify drifts in seismic records (Fig. 12): (1) large drifts form parallel to the continental slope or a bathymetric high. Normally these show upslope or up-relief lateral migration as well as a down-current migration, which can be difficult to determine given the large scale of the deposits; (2) they exhibit a basal surface, which constitutes a prominent truncation surface and marks the onset of drift formation; (3) the internal reflections (layers) and the top surface do not show stratigraphic conformity; (4) drifts exhibit widespread internal erosive discontinuities; (5) drifts exhibit a hierarchical range of high- to low-energy but large-scale deposits; (6) drifts are associated with large-scale bedforms. Given

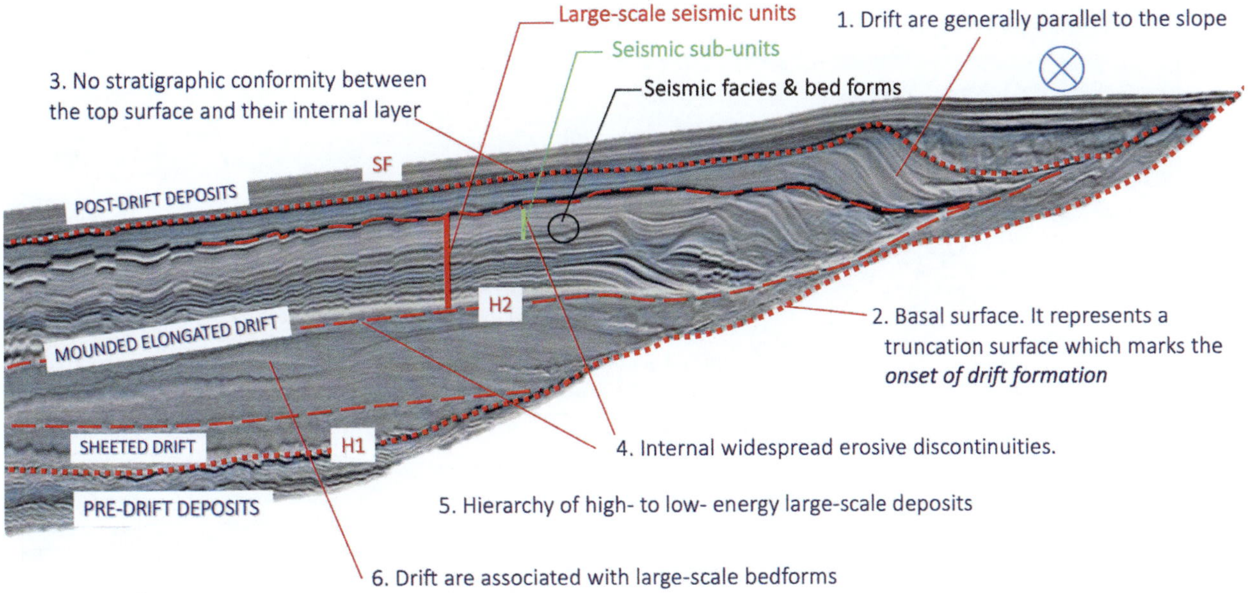


**FIG. 12** Example of a large contouritic drift in the Flemish Pass area, offshore Newfoundland (Laake et al., 2017). The three spatial scales and criteria for identifying contourite deposits in seismic data are included. Note the upslope migration of the long-term evolution of the drift with an onset stage (sheeted unit), a growing stage (mounded deposits) and a buried stage (post-drift deposits). The current direction is indicated in *blue*.

these factors, cores, well data or other sampling should be used to check and improve seismic interpretations wherever possible.

Contourite deposits are identified in seismic data at three different spatial scales (Faugères et al., 1999; Hernández-Molina et al., 2008b; Nielsen et al., 2008). The first-order scale analyzes the deposits and main seismic units on the largest spatial scale by tracing features' overall geometry and major regional discontinuities. The second-order scale covers intermediate spatial scales in which smaller seismic units (or subunits), subtler variations and minor discontinuities appear. The third-order scale covers the highest resolution and smallest spatial scale in which seismic facies, seismic attributes, bedforms and other smaller features (furrows, etc.) arise. This is a top-down methodology similar to classic sequence stratigraphy (Chapter 12) and opposite that of architectural element analysis (a bottom-up approach described in Chapter 1). Other aspects of sedimentary, tectonic and oceanographic context can affect drift evolution. Discontinuities and seismic facies record key changes at a basin-wide and even at inter-basin scale (Rebesco and Camerlenghi, 2008). Due to their complex evolution, contourite features can exhibit many different seismic facies (Faugères et al., 1999; Thiéblemont et al., 2020). These range from low-amplitude, discontinuous or transparent reflections to high-amplitude, and laterally continuous facies in various configurations (Fig. 9). 2D and 3D seismic (including seismic attributes) data and well log imaging technology are quickly evolving (Chapters 16 and 17) and will enhance the contourite and mixed (turbidite-contourite) systems identification at the three scales, but especially at the third-order scale.

Contourite depositional systems tend to show three long-term evolutionary stages in seismic data, including an onset stage, a drift-growth stage and a burial stage. Drifts of various ages identified along different margins consistently show these three stages (e.g., Rebesco et al., 2002; Hernández-Molina et al., 2014b; Creaser et al., 2017; Kirby et al., 2021a; Ng et al., 2021; Rodrigues et al., 2021a). The commonality of features and architecture described above demonstrates that drifts share mechanistic and evolutionary relationships, especially with the water masses that form them.

## Mixed (turbidite-contourite) depositional systems: Their recognition in the seismic record

Most continental margins host interactions between along- and down-slope processes that create complex features and deposits in deepwater settings (also see Chapter 8). Frequent interactions may form mixed (turbidite-contourite) systems (Fig. 13) composed of bottom current reworked sands (BCRS), pure turbidites and contourites or other deepwater deposits (Viana et al., 1998; Mulder et al., 2008; de Castro et al., 2020; Rodrigues, 2021—also see Chapters 6 and 7). Some studies

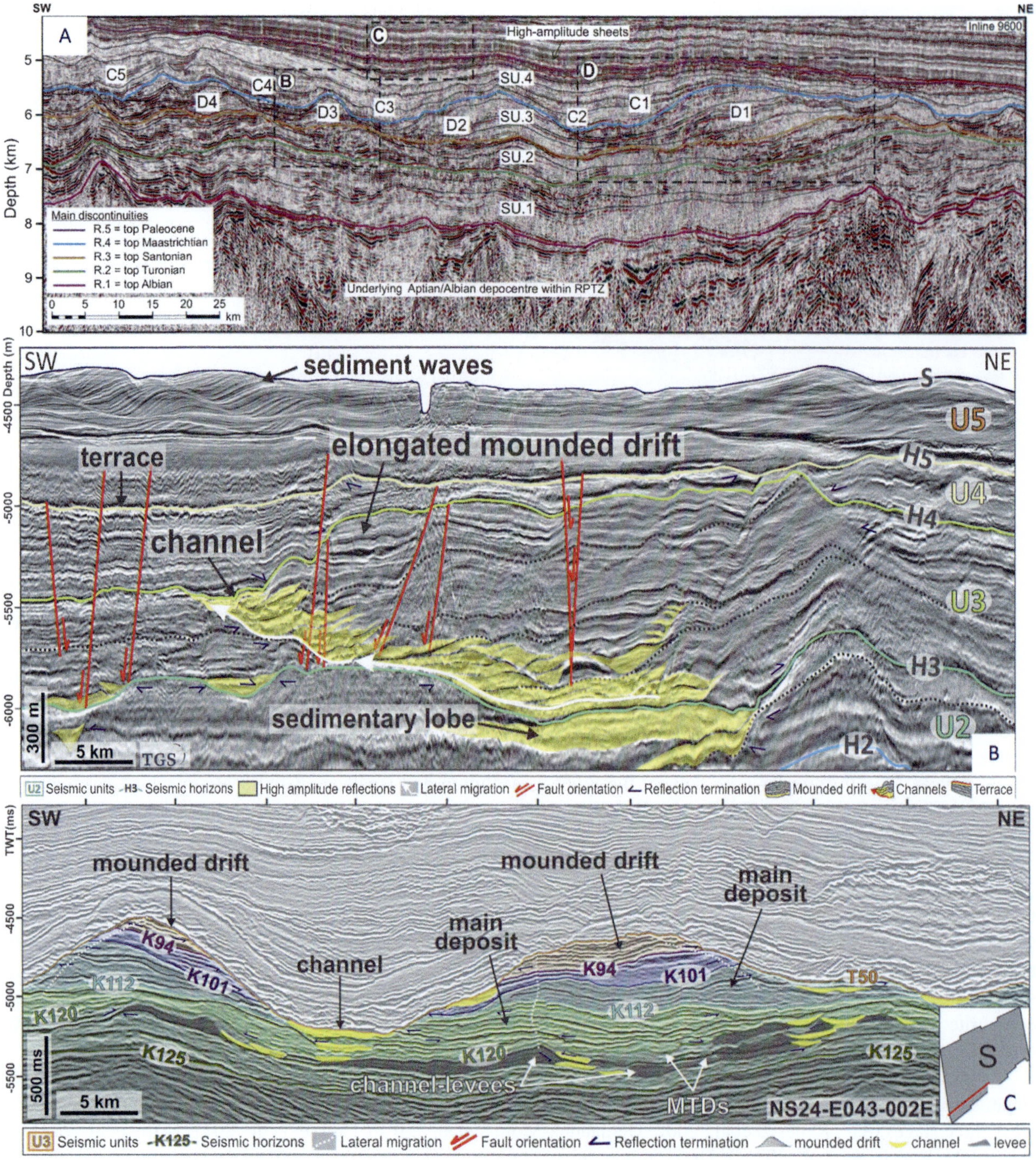


**FIG. 13** Examples of mixed (turbidite-contourite) depositional systems from (A) the Uruguayan Margin, (B) the Argentinian Margin and (C) the Canadian Margin (Deptuck and Kendell, 2020; Rodrigues et al., 2021c). These three examples include predrift deposits, sheeted drift deposits at the base, large, and asymmetric mounded drifts and postdrift deposits. *((A) From Creaser, A., Hernández-Molina, F., Badalini, G., Thompson, P., Walker, R., Soto, M., Conti, B., 2017. A Late Cretaceous mixed (turbidite-contourite) system along the Uruguayan Margin: sedimentary and paleoceanographic implications. Mar. Geol. 390, 234–253. doi:10.1016/j.margeo.2017.07.004, with permission from Elsevier; (B) From Rodrigues, S., Hernández-Molina, F.J., Kirby, A., 2021. A Late Cretaceous mixed (turbidite-contourite) system along the Argentine Margin: paleoceanographic and conceptual implications. Mar. Pet. Geol. 123, 104768. doi:10.1016/j.marpetgeo.2020.104768, with permission from Elsevier. (C) From Rodrigues, S., Deptuck, M., Kendell, K, Campbell, C., Hernández-Molina, F.J., 2022b. Cretaceous to Eocene mixed turbidite-contourite systems offshore Nova Scotia (Canada): spatial and temporal variability of down- and along-slope processes. Mar. Petrol. Geol. 138, 105572. https://doi.org/10.1016/j.marpetgeo.2022.105572).*

have drawn a distinction between mixed and hybrid systems. These arguments tend to assert that mixed systems deal with a wider spectrum of features and deposits formed under synchronous and asynchronous interactions (Mulder et al., 2008; Rodrigues, 2021; Rodrigues et al., 2022c), whereas hybrid systems only form under synchronous interactions (Sansom, 2018; Fonnesu et al., 2020; Fuhrmann et al., 2020; Rodrigues et al., 2021).

Mixed systems can be identified and characterized according to criteria of (1) orientation, (2) external morphology, (3) lateral migration, (4) stacking patterns and (5) seismic facies. Mixed systems can consist of up to four distinct depositional elements including down-slope elongated mounded drifts, channel-drifts, elongated levees and reworked submarine fans or splays (lobes) (Rodrigues et al., 2021—see Chapter 8). Depositional features exhibit asymmetric cross-sections, wherein a concave crest at the top connects a steep, rough flank and a smooth, depositional flank (Fig. 14). External morphology appears almond- to lens-shaped and flattens seaward (Fig. 14). Submarine channels or canyons also represent depositional features. These contour high relief areas to form V- to U-shaped features with downslope orientations (Fig. 14). Both mounded drifts and submarine channels migrate uniformly down- or up-current according to the bottom current direction (Fig. 14) and their stratigraphic evolution (Fig. 15) varies depending on the interaction between down- and along-slope processes (Georgiopoulou et al., 2021). In terms of seismic facies, mounded drifts generally appear as low-amplitude reflections (Figs. 13 and 14), whereas submarine channels appear as high-amplitude reflections that extend laterally toward the adjacent drifts (Fonnesu et al., 2020; Rodrigues et al., 2021).

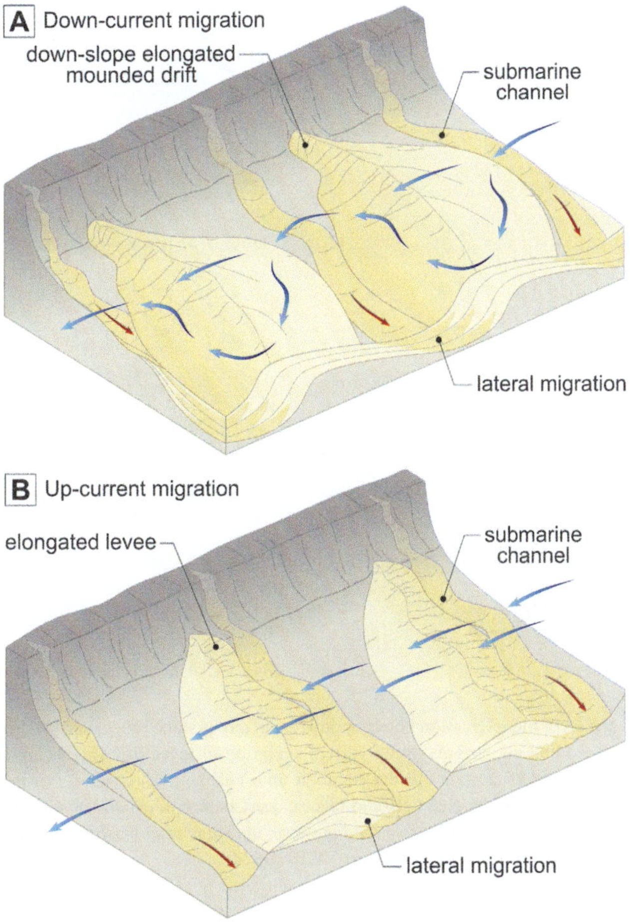


**FIG. 14** Conceptual model(s) for mixed depositional systems, characterized as (A) down-current migrating submarine channels and down-slope elongated mounded drifts or (B) up-current migrating submarine channels and elongated levees. *Blue arrows* represent an along-slope bottom current, while *red arrows* denote down-slope turbidity currents.

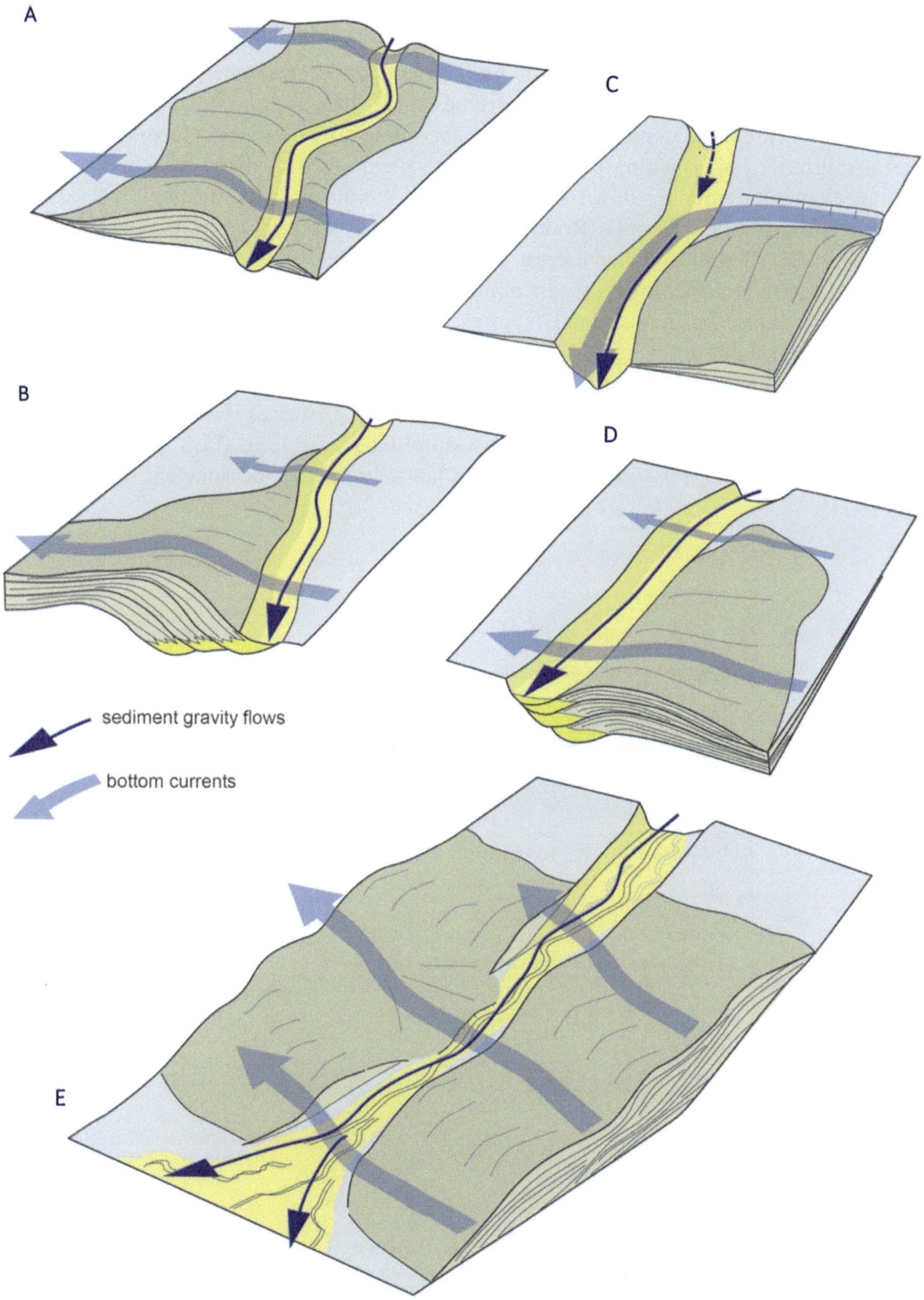


**FIG. 15** Contrasting stratigraphic outcomes of the interaction between down- and along-slope processes: (A) pirating of flow tops and heightened levee asymmetry; (B) forced migration counter to bottom currents by up-current accretion of levee drifts; (C) capture and diversion of bottom currents by channels; (D) lateral stepping forced by spilling of contour current sediment into channel; (E) channel downslope extension on account of accretion and base of slope re-profiling. For more information on channels and levees, see Chapter 8. *(Based on Georgiopoulou, A., Owens, M., Haughton, P. D.W., 2021. Channel and inter-channel morphology resulting from the long-term interplay of alongslope and downslope processes, NE Rockall Trough, NE Atlantic. Mar. Geol. 441, 106624. https://doi.org/10.1016/j.margeo.2021.106624).*

Key examples of mixed (turbidite-contourite) depositional systems have been identified offshore of Argentina (Rodrigues et al., 2021), Uruguay (Creaser et al., 2017), Mozambique (Fonnesu et al., 2020; Miramontes et al., 2021), Tanzania (Sansom, 2018; Fuhrmann et al., 2020), The Antarctic Peninsula (Rebesco et al., 1998, 2002; Hernández-Molina et al., 2017; Rodrigues et al., 2022b), southeast Greenland (Rasmussen et al., 2003), Canada (Deptuck and Kendell, 2020; Rodrigues et al., 2022b) as well as along the southwestern Iberian Margin (Mencaroni et al., 2021) and in the South China Sea (Gong et al., 2013). Innovative experiments by Miramontes et al. (2020b) have demonstrated how mixed (turbidite-contourite) depositional systems form in response to across- and along-slope sedimentary processes and that when along- and across-slope currents are contemporaneous, channels migrate in up-current direction.

## Deposits and sedimentary facies model

Within a contourite depositional system, transport capacity can vary significantly as a function of bottom current velocity. The bottom shear stress ($\tau = \rho u_*^2$, where $\rho$ is seawater density and $u_*$ is friction velocity) determines the real capacity for erosion, transport or deposition depending on the local behavior of the bottom currents. Associations between relative bottom current velocities and dominant contourite facies are addressed in McCave and Hall (2006), Stow et al. (2009), McCave et al. (2017), Culp et al. (2021), de Castro et al. (2021a) and de Weger et al. (2021).

The highest bottom current velocities are generally identified at the core of the bottom current (e.g., McCave and Tucholke, 1986; Faugères et al., 1999; de Castro et al., 2021a; de Weger et al., 2021). Bottom current velocity decreases regionally in a down-current direction because of turbulent mixing with ambient water masses, the decreasing density of the water mass and subsequent decrease in acceleration due to gravity. It can also decrease due to topographic gradients in which bottom currents are generated by geostrophic water-mass circulation and overflows. Therefore, the core-lateral decrease in bottom current velocity primarily relates to Coriolis effects, which drive the core toward the continental margin (e.g., Rebesco et al., 2014). This generates a core-perpendicular flow-velocity profile with diminished flow-velocity in a direction opposite of that induced by Coriolis deflection (Fig. 16). This core-perpendicular velocity profile forms sedimentary subenvironments including channels (associated with high flow-velocity of the core), the channel-drift transitional domain (decreased flow-velocity) and the drift (weak flow-velocity). As shown in Fig. 16, the energy associated with bottom currents in every subenvironment controls the dominant sedimentary facies, benthic activity, ichnofacies (Chapter 15) and microfacies. As noted above, local topographic features such as seamounts, ridges, straits, mounds, banks and canyons can disrupt the flow pattern and lead to local acceleration of the flow, especially along the foot of their flanks.

Another important process that has to be considered is how bottom currents are able to interact with sediment gravity flows simultaneously sourced from the continental margin. Flume experiments performed by Miramontes et al. (2020b) demonstrate how an along-slope slower bottom current can substantially deflect a much quicker across-slope turbidity current cloud, and modify the channel-levee system architecture (Fig. 17). The experiments show how an increase of contour-current velocity determines an increase of channel-levee asymmetry suggesting that the direction of migration of the channel should go against the direction of the bottom current (Fig. 14B). The result was completely in agreement with previously proposed depositional models based on seismic and well core facies analysis from the Mozambique margin (Fonnesu, 2013; Fonnesu et al., 2020) that postulate how syndepositional bottom currents may be responsible for important deviation in the gravity flow turbulent clouds, preventing the deposition of fine-grained sediments in the axis of the systems and deflecting them into the adjacent sedimentary drifts (Fig. 18).

### Contouritic sedimentary facies: Diagnostic criteria and classifications

Specific contourite sedimentary facies have been described by a number of studies (e.g., Stow and Lovell, 1979; Stow, 1982; Stow and Holbrook, 1984; Stow and Piper, 1984; Pickering et al., 1989; Faugères and Stow, 1993; Stow et al., 1996, 2002, 2013a,b; Gao et al., 1998; Faugères et al., 1999; Rebesco, 2005; Llave et al., 2006; Øvrebø et al., 2006; Shanmugam, 2006, 2012, 2013; Stow and Faugères, 2008; Rebesco et al., 2014; Hüneke et al., 2020; de Castro et al., 2020, 2021a,b; Hüneke et al., 2020; Stow and Smillie, 2020; de Weger et al., 2021).

Two primary conceptual frameworks have emerged for the characterization and identification of contourites. The first focuses on the presence and abundance of bioturbation (Stow, 1982; Gonthier et al., 1984), and the second asserts that primary sedimentary structures represent the basic criterion for contourite recognition (Shanmugam et al., 1993a,b; Shanmugam, 2000; Martín-Chivelet et al., 2003, 2008). The two different approaches (Stow vs Shanmugam) arose because the former researchers were addressing finer-grained contourite deposits within drifts, while the latter researchers addressed sandier parts of contourite and mixed depositional systems. A pragmatic approach to each of these frameworks would apply them to different systems depending on research objectives. However, criteria based on primary sedimentary processes are preferable to those based on secondary processes such as biological activity. The presence and abundance of diagnostic ichnological features along with key microfacies can confirm contouritic processes (Fig. 16). Further information concerning ichnology and contourite deposit systematics can be found in Chapter 15, as well as Wetzel et al. (2008), Rodríguez-Tovar and Hernández-Molina (2018), Dorador et al. (2019, 2021a,b), Miguez-Salas and

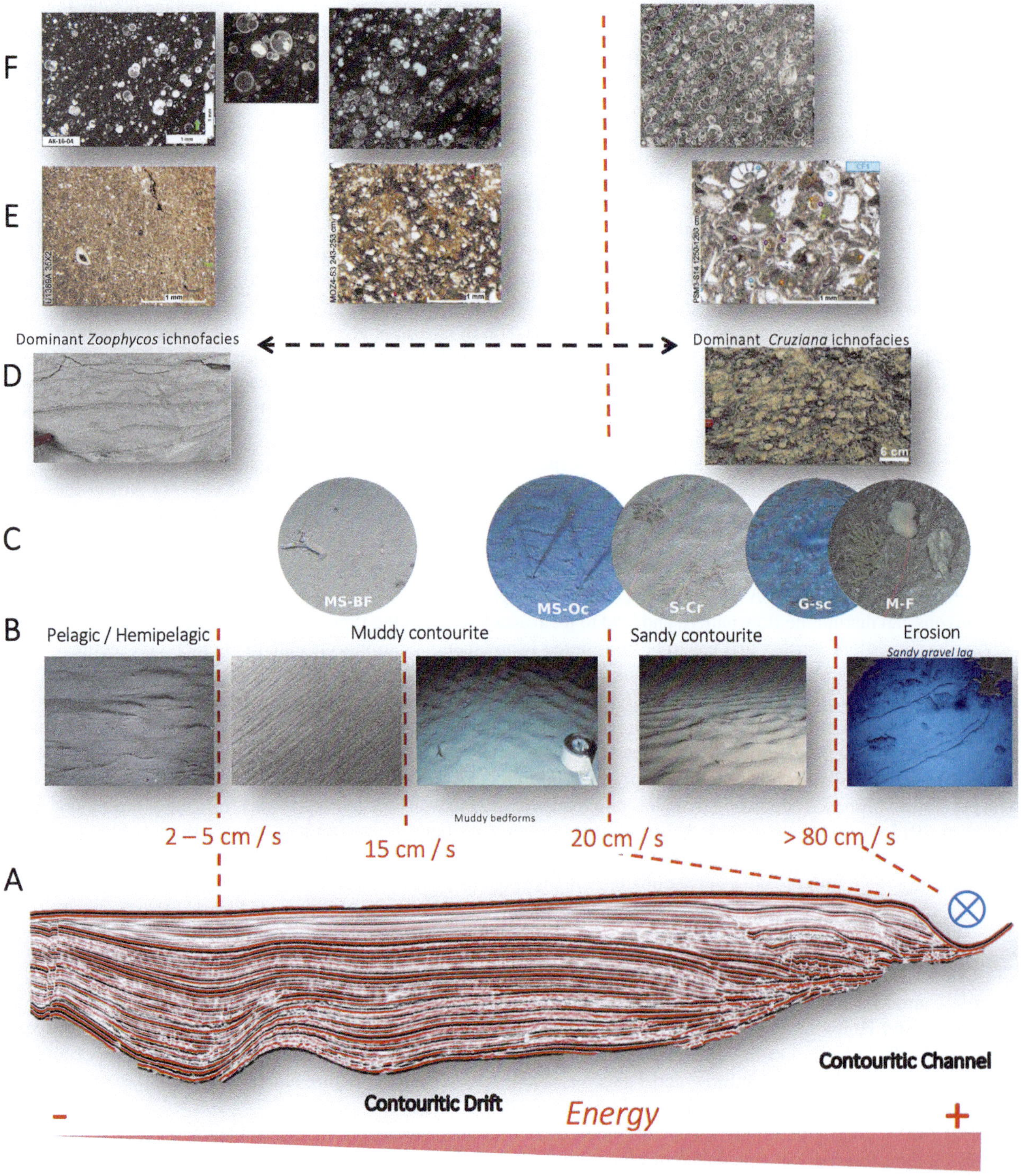


**FIG. 16** (A, B) Sketch summarizing relations between sedimentary deposits and bottom currents in a hypothetical profile perpendicular to the main core of a current associated with a water mass. Flow-velocity diminishes in the direction opposite of Coriolis deflection of the core. Sketches show erosional (contouritic channel) and adjacent depositional (contouritic drift) features. Relative bottom current velocities and their association with dominant facies follow interpretations given in McCave and Hall (2006), Stow et al. (2009), McCave et al. (2017) and Culp et al. (2021). Sedimentary deposits for contour currents can be found in Stow and Faugères (2008), Brackenridge et al. (2018), de Castro et al. (2020, 2021a,b), Hüneke et al. (2020), Stow and Smillie (2020), Yu et al. (2020) and de Weger et al. (2021). (C) Benthic communities' activity varies based on local bottom current velocity and type of substrate (see further details in Lozano et al., 2020), which determine different biotopes, for example: mud and sandy mud dominated by burrowing fauna (MS-BF); mud and sandy mud dominated by sea pens (MS-Oc); sand and gravelly sand with ripples and crinoids (S-Cr); rocky substrate comprising slab and chimney with large suspension feeders (M-F); gravelly substrate with solitary corals (G-sc). (D) Ichnofacies also change based on velocity conditions and type of substratum or deposits (Rodríguez-Tovar and Hernández-Molina, 2018; Dorador et al., 2019, 2021a,b; Miguez-Salas and Rodríguez-Tovar, 2019; Rodríguez-Tovar et al., 2019a; Miguez-Salas et al., 2020). Examples of microfacies include (E) late Quaternary and (F) ancient (Eocene to Miocene) record that exhibits key diagnostic features for contourite determination. These differ from sandier to muddier deposits (details from microfacies can be found in de Castro et al., 2020, 2021a,b; Hüneke et al., 2020; de Weger et al., 2021).

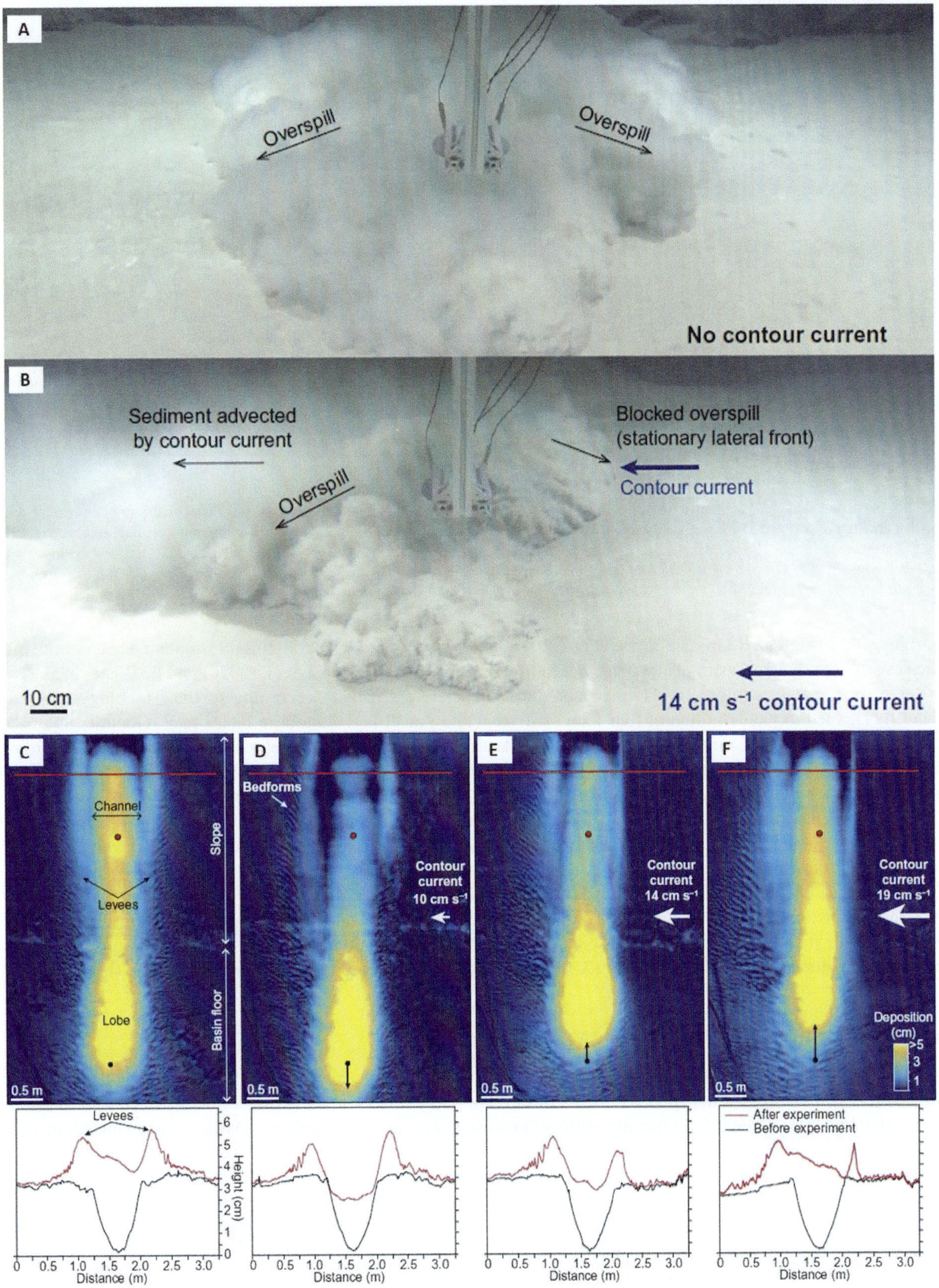


**FIG. 17** Frontal video frames and deposition maps and cross section of flume tank experiments. In (A) and (C) a 17% concentration turbidity current flowing at the in standing water at time-averaged downslope velocity maximum of 87 cm/s ( ~ 34.2 in $s^{-1}$) without bottom current interaction. (B), (D), (E) and (F) show the effects of the same flow with an along slope bottom current with 10, 14 and 19 cm/s (3.9, 5.5 and 7.4 in $s^{-1}$) velocity, respectively. *(From Miramontes, E., Jouet, G., Thereau, E., Bruno, M., Penven, P., Guerin, C., Le Roy, P., Droz, L., Jorry, S.J., Hernández-Molina, F.J., Thiéblemont, A., Silva Jacinto, R., Cattaneo, A., 2020a. The impact of internal waves on upper continental slopes: insights from the Mozambican margin (southwest Indian Ocean). Earth Surf. Process. Landforms 45, 1469–1482. https://doi.org/10.1002/esp.4818; Miramontes, E., Eggenhuisen, J.T., Jacinto, R.S., Poneti, G., Pohl, F., Normndeau, A., Campbell, D.C., Hernandez-Molina, F.J., 2020b. Channel-levee evolution in combined contour current–turbidity current flows from flume-tank experiments. Geology 48(4), 353–357. https://doi.org/10.1130/G47111.1, with permission from Geological Society of America).*

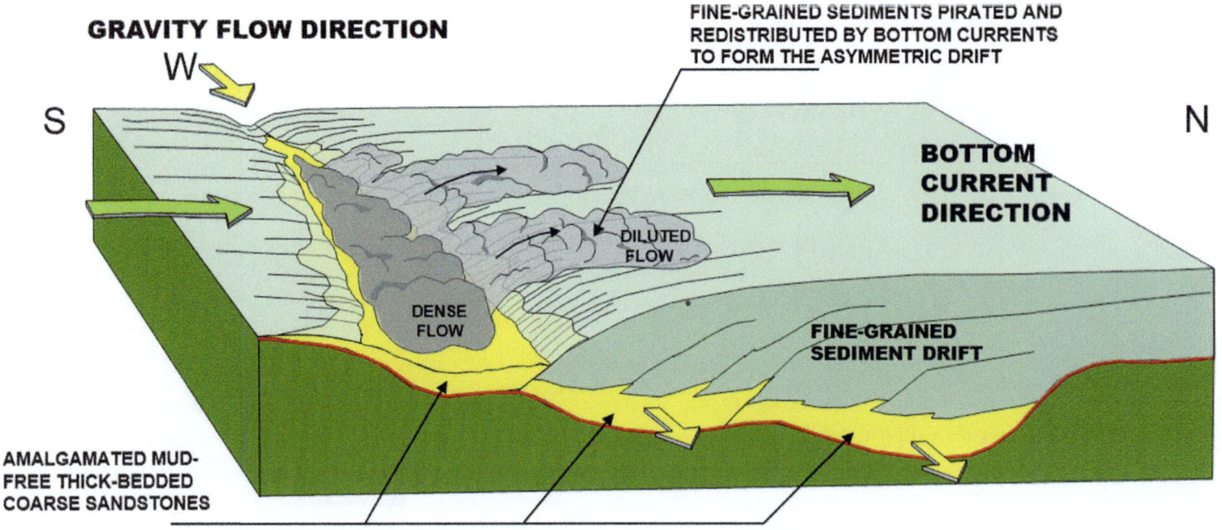


**FIG. 18** Depositional model describing the Mamba-Coral systems of Mozambique. This model shows the depositional processes of downslope oriented sand-rich gravity flows affected by syndepositional northward directed bottom currents resulting in asymmetric-migrating channel architecture and efficient sediments partitioning. *(From Fonnesu, M., Palermo, D., Galbiati, M., Marchesini, M., Bonamini, E., Bendias, D., 2020. A new world-class deep-water play-type, deposited by the syndepositional interaction of turbidity flows and bottom currents: the giant Eocene Coral Field in northern Mozambique. Mar. Pet. Geol. 111, 179–201. https://doi.org/10.1016/j.marpetgeo.2019.07.047, with permission from Elsevier).*

Rodríguez-Tovar (2019), Rodríguez-Tovar et al. (2019a) and Miguez-Salas et al. (2020). Information on contourite microfacies can be found in de Castro et al. (2020, 2021a,b), Hüneke et al. (2020), Yu et al. (2020) and de Weger et al. (2021).

Researchers have traditionally classified contourite deposits exclusively according to their lithology and texture (Fig. 19). This led to the definition of four (clay, silt, sand and gravel lag) contourite facies (e.g., Gonthier et al., 1984; Brackenridge et al., 2018; de Castro et al., 2021a). Clastic contourite deposits are generally categorized according to grain-size as muddy-, silty-, sandy- and gravel-contourites (Stow and Faugères, 2008). This classification can be subsequently divided based on lithology (Shanmugam, 2006, 2012, 2013; Stow and Faugères, 2008) as volcaniclastic-, shale-clast or shale-chip layers, calcareous-, siliceous bioclastic- and chemogenic-contourites. As such, contourite facies and their distribution depend on sediments source and supply (primarily from gravitational processes) and on the transport capacity of bottom currents.

Recent research has proposed new diagnostic criteria for contouritic sedimentary facies (Alonso et al., 2016; Brackenridge et al., 2018; de Castro et al., 2020, 2021a,b; Hüneke et al., 2020; Stow and Smillie, 2020; Yu et al., 2020; de Weger et al., 2021). Statistical analysis including Principal Component Analysis (PCA) of textural, microfacies, ichnological features and geochemical data (X-ray fluorescence, XRF) can help differentiate deposits (Fig. 20).

## Sandy contourites and bottom-current reworked sands

Sandy contourites (fine- and coarse-grained) are sand deposits that are fully reworked (redeposited) by bottom currents. In other words, bottom currents transport sediments from point *a* to point *b* where it is subsequently deposited. Sandy contourites show no clear evidence of other depositional processes. Fine-grained sandy contourites usually appear in areas or features defined by relatively weak bottom currents such as contourite drifts or at the channel-drift transition (Figs. 21 and 22). In these subenvironments, bioturbation may overprint primary depositional structures (Chapter 15). In these cases, bioturbation usually consists of *Planolites* and *Thalassinoides* ichnofabrics that can be indicative of increasing substrate cohesion from soft (*Planolites*) to firm ground (*Thalassinoides*) conditions. The presence of *Thalassinoides* could indicate breaks in sedimentation or a decline in sedimentation rate that has led to early compaction (firm ground) of initial soft ground (Rodríguez-Tovar et al., 2019a,b; see Chapter 15).

Coarse-grained sandy contourites with clast sizes ranging from medium-grained sand to gravel usually indicate a contouritic subenvironment characterized by energetic bottom currents. Examples include contourite channels or contourite terraces. In these depositional settings, sandy contourites usually consist of beds up to a meter in thickness and appear related to dunes in 2D and 3D seismic records (de Weger et al., 2020, 2021). Sandstones from these depositional

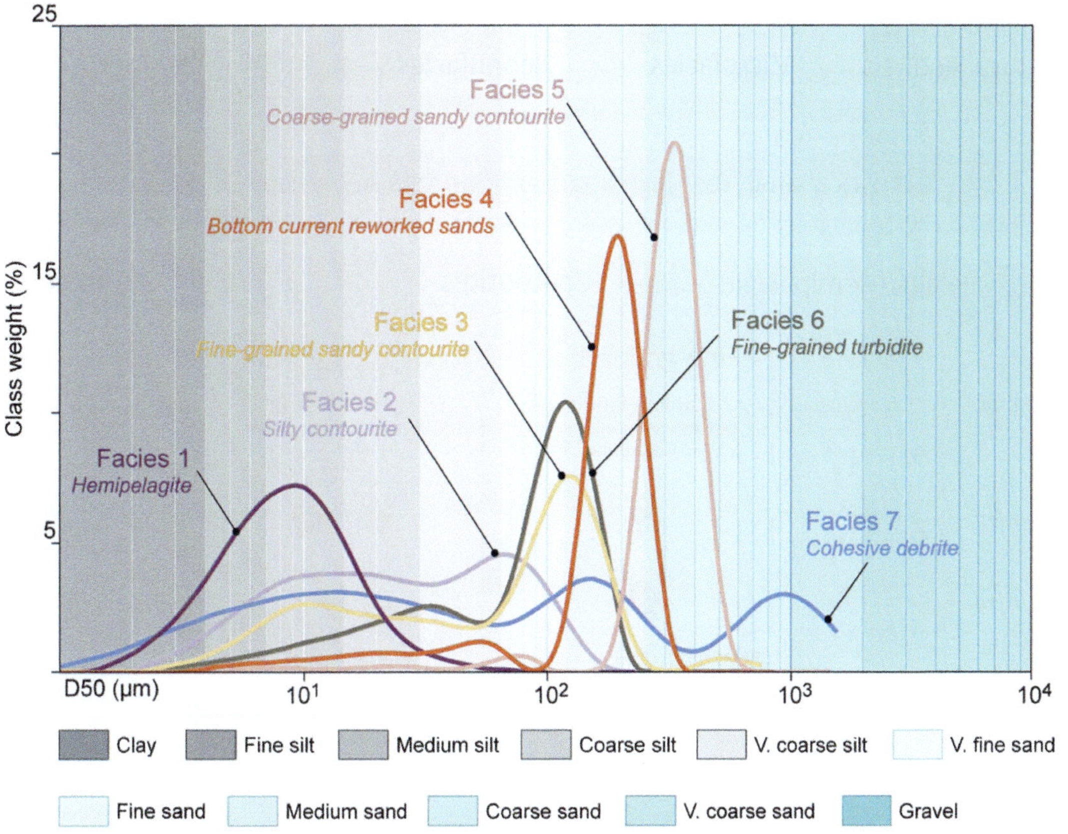


**FIG. 19** Grain-size density curves for dominant contourite deposits compared with reworked sands, debrites, hemipelagic and fine-grained turbidites. *(From de Castro, S., Hernández-Molina, F.J., de Weger, W., Jimenez-Espejo, F.J., Rodríguez-Tovar, F.J., Mena, A., Llave, E., Sierro, F.J., 2021a. Contourite characterization and its discrimination from other deep-water deposits in the Gulf of Cadiz contourite depositional system. Sedimentology, 68(3), 987–1027. https://doi.org/10.1111/sed.12813, with permission from John Wiley & Sons Ltd).*

subenvironments are generally moderately sorted making it difficult to identify trace fossils. With higher concentrations of fine-grained sediments, bioturbation becomes easier to identify. The abundance and location of bioturbation generally resembles that observed in fine-grained contourites and consists of the ichnofabrics of *Planolites* and *Thalassinoides*. As with fine-grained contourites, these ichnofabrics are related to the intermittent behavior of bottom current velocities and sediments input.

Turbidites reworked by bottom currents were defined by Stanley (1987, 1988). Later, the recognition criteria proposed by Shanmugam et al. (1993a,b) was derived from a turbidite-dominated, mixed system in the Gulf of Mexico, where contouritic reworking of sediments mainly occurred between turbidite channels on the slope. The sandy contourites consist of discretely rippled and well-sorted, very fine- to fine-grained sand layers. These are referred to as BCRS and they form because of reworking by bottom currents. Shanmugam et al. (1993a,b) asserted that the term BCRS is preferable to that of sandy contourite (Stow and Lovell, 1979; Gonthier et al., 1984) because it distinguishes these from more conventional sandy contourites, as bioturbation is not considered a diagnostic criterion for BCRS.

Recent studies have further defined BCRS as formed by partial, near-source reworking of a noncontouritic deposit by a bottom current (de Castro et al., 2020, 2021a,b). The partial reworking of a turbidite division (usually the top) with only minor sediments transport would qualify as BCRS. BCRS may or may not include a component of lateral displacement. They usually appear in a contourite drift environment (Figs. 21 and 22), for example, at the channel-drift transition (de Castro et al., 2020, 2021a; de Weger et al., 2021) or in mixed levee-drift systems (Shanmugam et al., 1993a,b; Fonnesu et al., 2020; Fuhrmann et al., 2020; Rodrigues et al., 2021b). Mutti et al. (2014) also defined a facies association similar to BCRS in which genetically distinct processes (turbidity flows and bottom currents) produce vertically and laterally associated facies. Most of these reworked sands show only minor bioturbation, but *Planolites* is usually present. *Planolites* is produced by a softground tracemaker that commonly develops bioturbation horizons throughout reworked beds. *Planolites* ichnofabric consists of simple, horizontal, actively filled burrow systems of deposit-feeding or predaceous

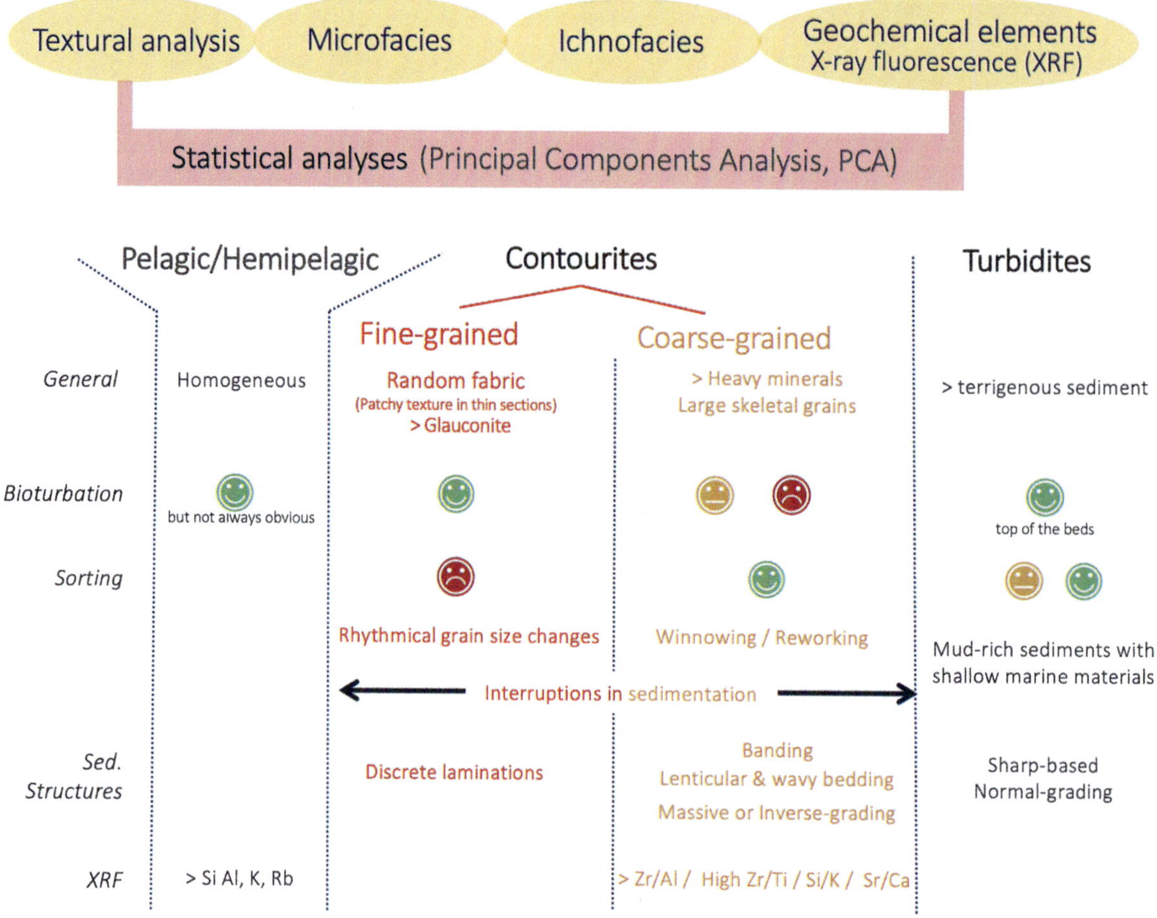


**FIG. 20** Statistical analyses of textural, microfacies, ichnological features and geochemical data (X-ray fluorescence, XRF) can differentiate muddy and sandy contourites from other deepwater deposits such as hemipelagic and fine-grained turbidites. Evidence of reworking, winnowing, rhythmic grain size variations, laminations, banding, lenticular to wavy bedding and massive or inverse-grading can differentiate muddy and sandy contourites. *(From Brackenridge, R.E., Stow, D.A.V., Hernández-Molina, F.J., Jones, C., Mena, A., Alejo, I., Ducassou, E., Llave, E., Ercilla, G., Nombela, M.A., Pérez-Arlucea, M., Frances, G., 2018. Textural characteristics and facies of sand-rich contourite depositional systems. Sedimentology, 65, 2223–2252; Hüneke, H., Hernández-Molina, F., Rodrıguez-Tovar, F., Llave, E., Chiarella, D., Mena, A., Stow, D., 2020. Diagnostic criteria using microfacies for calcareous contourites, turbidites and pelagites in the Eocene-Miocene slope succession, southern Cyprus. Sedimentology. doi:10.1111/sed.12792; de Castro, S., Hernández-Molina, F.J., de Weger, W., Jimenez-Espejo, F.J., Rodríguez-Tovar, F.J., Mena, A., Llave, E., Sierro, F.J., 2021a. Contourite characterization and its discrimination from other deep-water deposits in the Gulf of Cadiz contourite depositional system. Sedimentology 68(3), 987–1027. https://doi.org/10.1111/sed.12813).*

organisms (Pemberton and Frey, 1982). They tend to be associated with low-sedimentation rates or discontinuous sedimentation (Rodríguez-Tovar et al., 2019a,b).

The exploration of mixed systems also highlights a further class of bottom-current influenced deposits (Rodrigues et al., 2022c). The axial part of Mozambique Mamba and Coral systems are dominated by amalgamated thick-bedded sedimentation units interpreted to be deposited by high-concentration turbidity flows (Fonnesu et al., 2020—see Chapters 6 and 7). However, these beds are characterized by a noticeable lack of associated fine-grained facies conventionally deposited by the dilute and turbulent part of the turbidity flow. Beds are well graded from coarse- to medium-grained sandstone with moderate sorting. They display an atypical inverse trend of petrophysical properties with the top of the bed characterized by higher porosity and permeability values with respect to the bed base (Fonnesu, 2013). These atypical beds are interpreted to represent the deposits of high-concentration turbidity flows in which the fine-grained turbulent cloud has been stripped away by the syndepositional bottom current during their deposition, resulting in high-quality reservoirs (Fonnesu et al., 2020). This class of events, despite apparently classifiable as high-density turbidity flows (see Chapter 7), can be also archivable in the BCRS spectrum *as bottom-current influenced turbidites.*

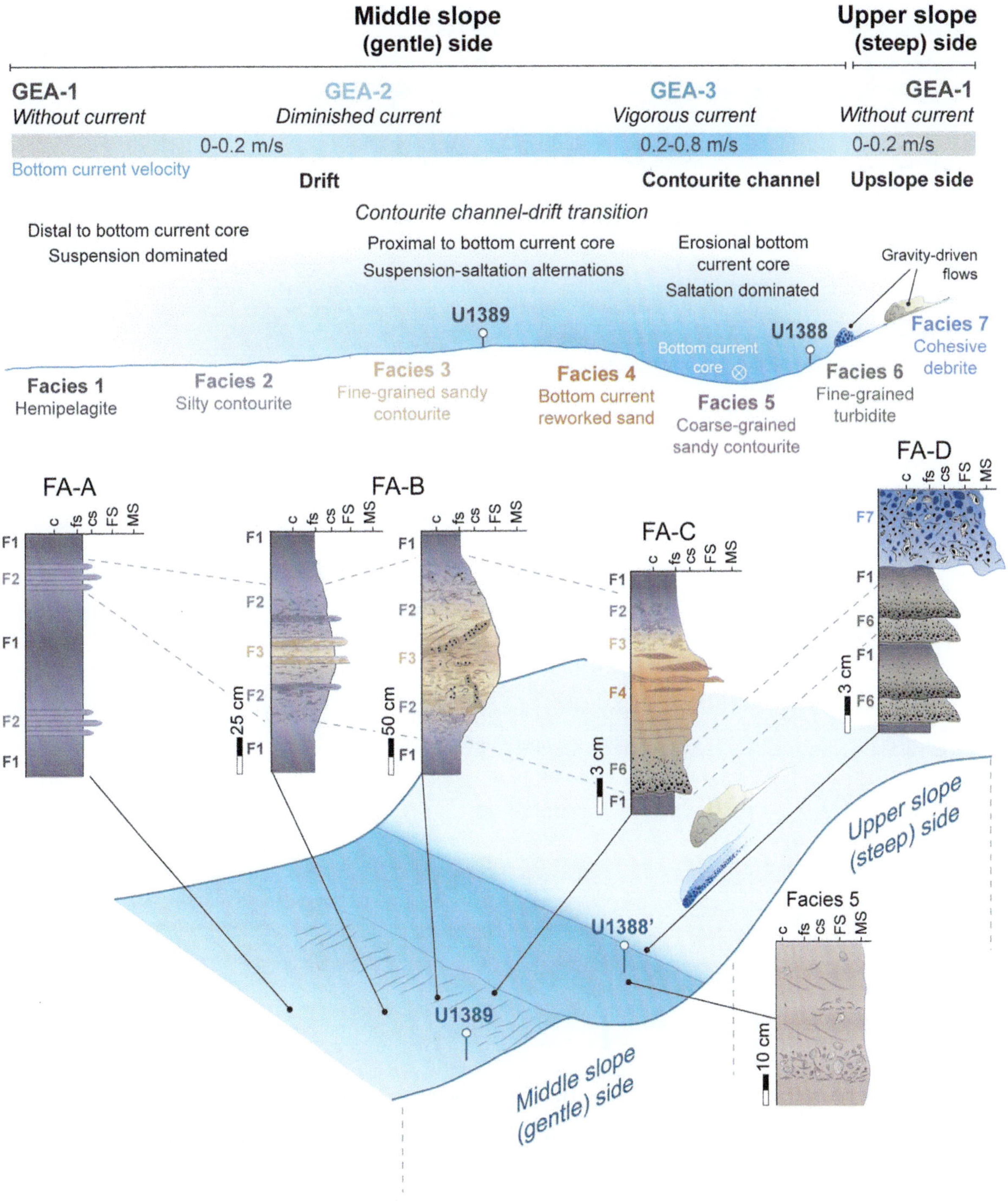


**FIG. 21** Conceptual depositional model for the continuous development of late Quaternary facies associations, geochemical element associations (GEA) and depositional processes for different contouritic physiographic domains. *(From de Castro, S., Hernández-Molina, F.J., de Weger, W., Jimenez-Espejo, F.J., Rodríguez-Tovar, F.J., Mena, A., Llave, E., Sierro, F.J., 2021a. Contourite characterization and its discrimination from other deep-water deposits in the Gulf of Cadiz contourite depositional system. Sedimentology 68(3), 987–1027. https://doi.org/10.1111/sed.12813, with permission from John Wiley & Sons Ltd).*

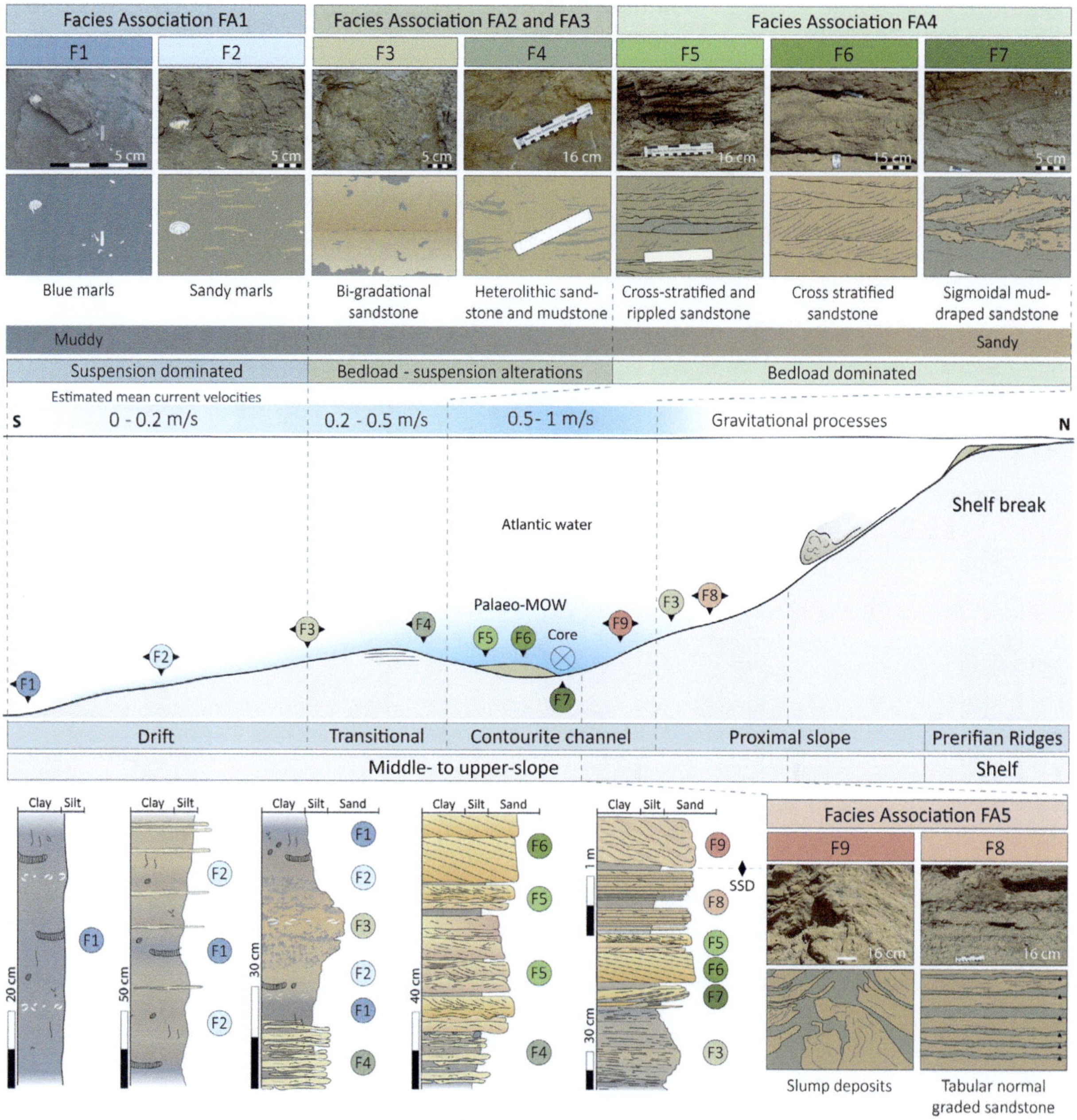


**FIG. 22** Sketch showing relations between the sedimentary facies, facies associations and different depositional and erosional elements from the proximal continental slope to the contourite drift in the ancient record. The facies associations are related to their dominant depositional process and associated current velocities. The panels in the bottom left corner show examples of observed facies stacking patterns. *(Late Miocene, from de Weger, W., Hernández-Molina, F.J., Miguez-Salas, O., de Castro, S., Bruno, M., Chiarella, D., Sierro, F.J., Blackbourn, G., Manar, M.A., 2021. Contourite depositional system after the exit of a strait: case study from the late Miocene South Rifian Corridor, Morocco. Sedimentology. https://doi.org/10.1111/sed.12882, with permission from John Wiley & Sons Ltd).*

## Facies model and facies associations: Integrating modern systems and ancient record systems

The standard bi-gradational sequence defined by Gonthier et al. (1984) and later updated by Stow and Faugères (2008) has been widely used as a diagnostic framework (Figs. 23–25) for modern and ancient bottom current and contourite mounded drift deposits (Duan et al., 1993; Balaky and Tamar-Agha, 2017; Capella et al., 2017; Rodríguez-Tovar et al., 2019a,b; Hüneke et al., 2020; van Li et al., 2020; Zhang et al., 2020; de Castro et al., 2021b). Stow and Faugères (2008) proposed the base-cut-out and top-cut-out contourite sequences, whereas Brackenridge et al. (2018) described a variety of contouritic

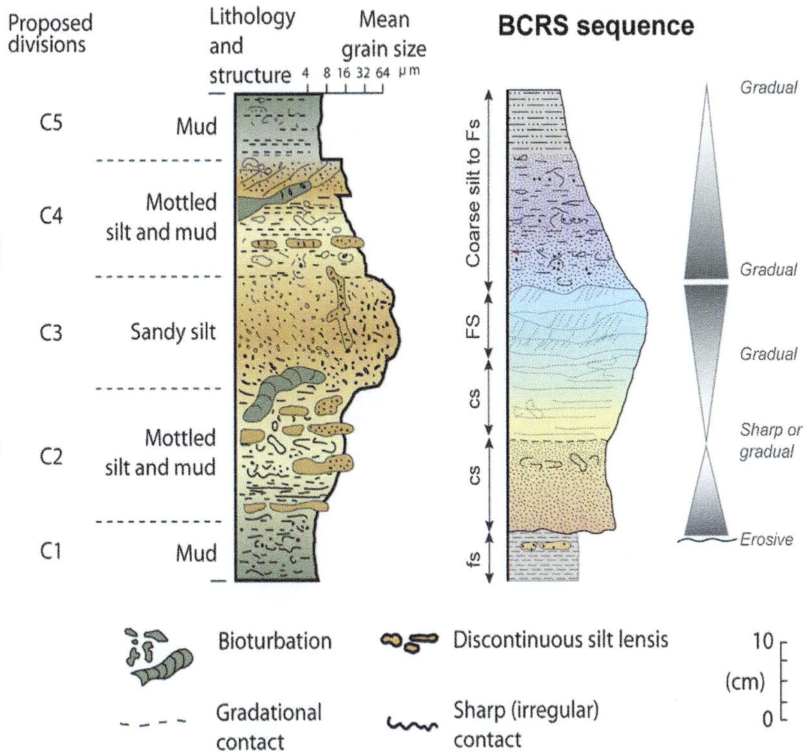


**FIG. 23** Sketch on the left shows the standard bi-gradational sequence defined by Gonthier et al. (1984) and later divided into five divisions by Stow and Faugères (2008). The sketch on the right shows bottom-current reworked sands (BCRS). *(From de Castro, S., Hernández-Molina, F.J., Rodriguez-Tovar, F.J., Llave, E., Ng, Z.L., Nishida, N., Mena, A., 2020. Contourites and bottom current reworked sands: bed facies model and implications. Mar. Geol. 428, 106267. https://doi.org/10.1016/j.margeo.2020.106267; de Castro, S., Hernández-Molina, F.J., de Weger, W., Jimenez-Espejo, F.J., Rodríguez-Tovar, F.J., Mena, A., Llave, E., Sierro, F.J., 2021a. Contourite characterization and its discrimination from other deep-water deposits in the Gulf of Cadiz contourite depositional system. Sedimentology 68(3), 987–1027. https://doi.org/10.1111/sed.12813, with permission from Geological Society, London and John Wiley & Sons Ltd).*

sand sequences (C3 intervals). However, none of these studies addressed spatio-temporal associations among deepwater deposits in different contouritic and physiographic domains.

The novel facies models proposed by de Castro et al. (2021a) and de Weger et al. (2021) advanced contourite research by characterizing the interrelated processes that form contourite drifts (Figs. 21 and 22). Sedimentary characteristics (sedimentary facies) and sedimentary facies associations can be assigned to the full range of contourite depositional and erosional domains including the drift, drift-channel transition and contourite channel (Figs. 21 and 22). In both modern and ancient records, facies changes primarily relate to variations in the intensity of the Mediterranean Outflow Water (MOW) and paleo-MOW (respectively), which decrease away from its core. Coarser-grained facies (Figs. 21 and 22) are usually dominant in proximal areas (with respect to the bottom current core) and reflect the migration of large-scale dunes in the contourite channel (Capella et al., 2017; de Weger et al., 2020). These deposits grade into sandy, finer-grained facies in the channel-drift transition and distal areas of the drift farther away from the main core of the bottom current, where fine-grained sedimentation prevails (Figs. 21 and 22). In the ancient record (Fig. 22), vertical sedimentary facies stacking patterns record tectonically induced, southward migration of the CDS and the intermittent behavior of the paleo-MOW. As indicated by the modern record, precessional- and millennial-scale climate variations drove these changes (Sierro et al., 2020; de Castro et al., 2021a; de Weger et al., 2021). Bi-gradational sequences have also been reported in mixed systems, specifically in parts of asymmetric mounded drifts, being associated with other deepwater deposits (Rodrigues et al., 2022a,c).

## Identifying contourites and mixed systems along active continental margins and within active tectonic settings

Contourites and mixed systems commonly appear along active margins or in deepwater settings subject to active tectonics (Bailey et al., 2021a,b; Liu et al., 2021). These settings have not enjoyed detailed analysis by contourite researchers. Studies have documented analogs for drifts along the active margins of the Pliocene-Quaternary Eastern Sunda Arc (Indonesia) (Reed et al., 1987), the Pliocene-Quaternary Southwest Adriatic Margin (Pellegrini et al., 2016), the Pliocene-Quaternary Hikurangi Margin (Bailey et al., 2021b), the Gulf of Cadiz (Spain) in the late Miocene (Ng et al., 2021) and Pliocene-Quaternary (García et al., 2009; Hernández-Molina et al., 2016b) and the late Miocene Rifian Corridor (Morocco) (Capella et al., 2017; de Weger et al., 2021).

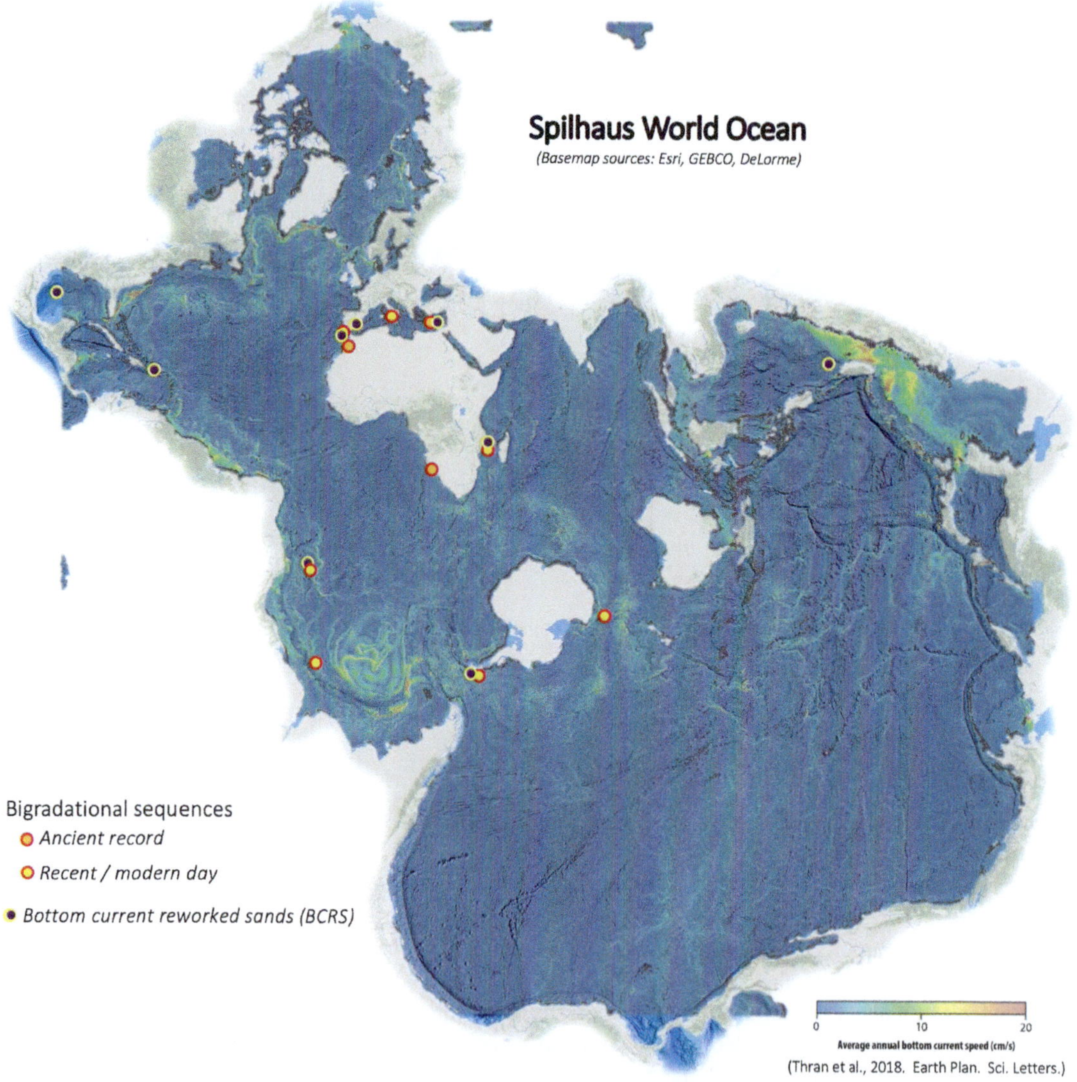


**FIG. 24** Examples where bigradational sequences in contourites and bottom current reworked sands (BCRS) have been identified in recent to modern-day deposits and ancient sedimentary records (Spilhaus World Ocean projection from Esri, GEBCO, DeLorme). Compilation for contourites based on (by alphabetic order): Alonso et al., 2016; Balaky and Tamar-Agha, 2017; Brackenridge et al., 2018; Capella et al., 2017; de Castro et al., 2020, 2021a,b; de Weger et al., 2020; Duan et al., 1993; Gonthier et al., 1984; Hernández-Molina et al., 2014a, 2016a; Hüneke et al., 2020; Mutti et al., 2014; Rodrigues et al., 2022a; Rodríguez-Tovar et al., 2019a,b; Sierro et al., 2020; Stow et al., 2013b; Viana and Faugères, 1998; Viana et al., 2007; Zhang et al., 2020. Compilation for bottom current reworked sands, BCRS (by alphabetic order): de Castro et al., 2020, 2021a,b; Fonnesu et al., 2020; Fuhrmann et al., 2020; Gong et al., 2016; Mutti, 1990; Rodrigues et al., 2021b; Sansom, 2018; Shanmugam et al., 1993a,b; Stanley, 1987, 1988, 1993.

Deepwater sedimentation depends on tectonics, climate and sea level fluctuations (e.g., Artoni et al., 2007; Leeder, 2011—see Chapters 1–8, 10–15, 18). At a regional scale, tectonic activity can control the evolution of contourite and mixed systems through subsidence and uplift and by the closing and opening of oceanic gateways, all of which in turn regulate sediments input and slope stability (Faugères and Stow, 2008; de Weger et al., 2020; Kirby et al., 2021a). The development of local deformational structures, such as folds, faults or diapirs (Chapters 4 and 5), determine seafloor topography. These structures create geomorphologic highs or depressions, which can obstruct or favor down- and along-slope sedimentary processes (Clark and Cartwright, 2009; Hernández-Molina et al., 2016a,b,c). The Gulf of Cadiz provides a clear example of structural control on contourite development. Diapiric structures determine bottom-current circulation by dividing it into several pathways, while fault-related seafloor depressions host contourite channels (García et al., 2009; Hernández-Molina et al., 2016a,b,c; Ng et al., 2021; Duarte et al., 2022).

Lithospheric uplift or local slope instabilities can increase sediment supply as demonstrated by examples like the Betic-Rif Orogeny of the western Mediterranean (Capella et al., 2017; de Weger et al., 2020) or the Andean Orogeny along the

**FIG. 25** Example of a contourite bi-gradational sequence from the ancient record, in the early Miocene deposits from Pakhna Formations, Cyprus. Main lithologies and microfacies for each division are included. *W*, wackestones; *P*, packstones; *G*, grainstones; *Ch*, *Chondrites*; *Pl*, *Planolites*.

South American margin (Morales et al., 2017; Kirby et al., 2021a). Tectonic activity can thus lead to a predominance of down-slope processes in which reworked sediments become an important component of drift construction (e.g., Liu et al., 2021).

Drifts developing in active settings resemble those observed from passive margins, although deposit dimensions depend on tectonically controlled accommodation and sediment supply. Contourite drifts developed along active margins generally show smaller dimensions, less continuity and appear less frequently than their passive margin counterparts, especially in areas of increased accommodation segmented by uplift and deformational structures (Bailey et al., 2021b; Duarte et al., 2022). Larger, continuous deposits appear in areas of regional tectonically induced subsidence (Duarte et al., 2022). In active tectonic settings, deformation plays an important role in controlling the morphology and stacking pattern of contourite deposits (e.g., Reed et al., 1987; Bailey et al., 2021b; Ng et al., 2021). Tectonic structure or other bathymetric features can also influence basin shape and connectivity among basins, which in turn affect bottom current circulation (Verdicchio and Trincardi, 2008; Pellegrini et al., 2016). Tectonic activity along active margins confines the slope and restricts circulation to narrow passages, which serve as corridors for bottom currents (Reed et al., 1987). Bathymetric restrictions can further accelerate currents (Hollister and McCave, 1984; Reed et al., 1987).

Most drift case studies have focused on recent examples because these tend to have experienced only limited deformation. Older records, and especially those from more active margins, would show modification of the configuration and other critical characteristics of contourite drifts (Ng et al., 2021). Drift type and evolution vary according to the slope gradient and presence of irregular bathymetric features (Bailey et al., 2021a; Ng et al., 2021). Continuous uplift along convergent margins leads to erosion and tends to prevent the preservation of contourite features (Ng et al., 2021). The development of multiple, interchangeable bottom current pathways can disrupt the formation and preservation of features (Pellegrini et al., 2016). The role of contourites in reconstructing paleogeography and tectonic history, and also the degree to which deformation can obscure contourite records, lends critical imperative to efforts of defining robust criteria for identifying ancient contourites along active margins (Hüneke and Stow, 2008).

Recent work addressing the question of how large contourite drifts behave as part of subduction processes (e.g., the Nankai Trough) has suggested that these features indeed influence fault-slip behaviors ranging from large tsunamigenic earthquakes (>8 Mw) to slow-slip and very low-frequency earthquakes (VLFE) (Tilley et al., 2021). Large mounded drift deposits seaward of the trench introduce porosity and permeability heterogeneities into otherwise homogenous, mud-dominated sediments. Normal faults within contourite deposits, more extensive than previously documented, may also enhance along-strike fluid transport. The locations of the very low-frequency earthquakes clusters within subduction margins can depend on the cooccurrence of these deposits and anomalously high pore pressure (Tilley et al., 2021).

## Control factors

Bottom currents (intensity and variation) and sediment supply control contourite sedimentation (e.g., Stow et al., 2002; Rebesco et al., 2014 and references therein). Additional factors determine details of contourite development and preservation. These include changes in the intensity and lateral to vertical paths of bottom currents and variation in the quantity and composition of sediment supply.

### General margin configuration

Tectonic effects exert long-term influences on deepwater sedimentation by determining slope gradient of the seafloor, subsidence, remnant marginal platforms and the opening and closing of oceanic gateways, which can accelerate or hinder bottom currents. The large-scale shape and configuration of ocean basins also determines the distribution and variability of bottom-current intensities in tandem with topographic and bathymetric boundaries and the Coriolis effect. When bottom currents interact with complex continental slope morphologies established by tectonic activity (diapirs, uplifted fault blocks, banks, ridges, seamounts, etc.), multiple current pathways emerge, as do more complex dynamics. These include the formation of several different branches of the current along with secondary flows, small fluxes (filaments), internal waves, local turbulence associated with eddies, overflows, helicoidal flows and upwelling and downwelling flows (e.g., Hernández-Molina et al., 2008b). Both passive and active margin settings display similar contourite sedimentation but vary in terms of scale with much smaller contourites developing along convergent margins (by an order of magnitude) relative to those developed along passive margins (Bailey et al., 2021a).

### Sediment supply

Tectonics, sea level fluctuations and oceanographic factors determine the amount and type of sediment supplied to the different continental margin domains. Active slope tectonics may trigger more effective gravitational processes delivering sediment from shallow to deepwater environments along submarine canyons, or by turbidity flows, slides and debris flows (see Chapters 6 and 7). These provide material that can be redistributed by bottom currents through continued erosion, transport and deposition, but these processes can obscure contourite history (e.g., Bender et al., 2012; Mulder et al., 2013; de Castro et al., 2020; Fonnesu et al., 2020). The influence of sea level on bottom current generation and intensity is complex (Chapter 12) and depends on geological and oceanographic context. However, its influence on sediment influx into a basin can be explained in conventional sequence stratigraphic frameworks for siliciclastic systems. Deepwater clastic sedimentary systems experience enhanced sediment transport during sea level fall and lowstand periods (RST and LST) and diminished sediment supplies during times of high relative sea level, when clastic sediment remains on the shelf and upper slope (Brackenridge et al., 2011; see Chapter 12).

### Pronounced density contrasts (pycnoclines)

Different erosional and depositional contourite features can form due to the interaction of water mass interfaces and the seafloor. The development of large contourite terraces relates to the long-term position of the interface (Hernández-Molina et al., 2009, 2011; Preu et al., 2013; Rebesco et al., 2014; Thieblemont et al., 2019; Kirby et al., 2021a). The strength of the interface depends on spatial and seasonal factors as well as other oceanographic processes, including internal tides, internal waves, eddies, solitons, etc. Internal wave and pulse-like density fronts created at these interfaces influence contourite feature development, for example, by resuspending sediments or eroding the seafloor (Hernandez-Molina et al., 2011; Preu et al., 2013; Shanmugam, 2013; Hanebuth et al., 2015; Droghei et al., 2016; Wienberg and Titschack, 2017; Wienberg et al., 2020).

## Eustatic changes

Many studies have emphasized the significance of glacioeustatic changes in the formation of large depositional and erosional features at continental margin and basin scales. These have identified vertical contourite stacking patterns, sequences and facies that vary systematically throughout the Quaternary. Examples include contourites in the Gulf of Cadiz (Hernández-Molina et al., 2002, 2016a,b,c; Llave et al., 2006, 2007; Bender et al., 2012; Roque et al., 2012; Brackenridge et al., 2018; Lebreiro et al., 2018; Mestdagh et al., 2019; de Castro et al., 2020; García et al., 2020), the western and northwestern Iberian margin (Bender et al., 2012; Dorador and Rodríguez-Tovar, 2016; Van Rooij et al., 2016; Mena et al., 2018; Liu et al., 2020; Teixeira et al., 2020; Rodrigues et al., 2020; Teixeira et al., 2019), the Alboran Basin between Spain and Morocco (Ercilla et al., 2016; Juan et al., 2016, 2020; Alonso et al., 2021), the North Sea (Hohbein and Cartwright, 2006; Batchelor et al., 2021), Antarctica (Maldonado et al., 2005; Lucchi and Rebesco, 2007; Escutia et al., 2009; García et al., 2016; Hillenbrand et al., 2021; López-Quirós et al., 2020), the Tyrrhenian Sea (Miramontes et al., 2016), the Argentina and Uruguay margin (Hernández-Molina et al., 2017), the Mozambican margin (Thieblemont et al., 2019), the Bahama Bank (Mulder et al., 2019) and the Capo Vaticano upper slope, Italy (Martorelli et al., 2016, 2021).

Sea level fluctuations exert a stronger influence on sediment supply than does vertical or other spatial changes in water mass flow (Chapter 12). Climatic variation strongly influences intermediate to deepwater circulation (e.g., Rohling et al., 2015). Rapid cooling (e.g., D/O stadials) and warming events (e.g., D/O interstadials) associated with Milankovitch forcing can enhance and reduce the circulation during colder and warmer periods, respectively, depending on the water masses. The vertical and lateral spatial distribution of contourite facies associations can vary with precession-driven, millennial-scale climate variability (Toucanne et al., 2012; Angue Minto'o et al., 2015; Miramontes et al., 2016; de Castro et al., 2020; Sierro et al., 2020; Martorelli et al., 2021). For example, intermediate and deep Mediterranean water masses became enhanced during cold periods (e.g., Cacho et al., 2000, 2006; Rogerson, 2002; Sierro et al., 2005, 2020; Hernandez-Molina et al., 2011; Rogerson et al., 2012; Toucanne et al., 2012; Ercilla et al., 2016; Juan et al., 2016, 2020; Alonso et al., 2021). During cold intervals (i.e., glacial periods), the Mediterranean Outflow Water (MOW) plume along the Atlantic Iberian continental margin shows signs of enhancement after its exit through the Strait of Gibraltar, where it descended to deeper depths relative to those indicated for warm intervals (e.g., Schönfeld and Zahn, 2000; Schönfeld et al., 2003; Hernández-Molina et al., 2006, 2014a,b; Llave et al., 2006; Voelker et al., 2006; Van Rooij et al., 2010; Roque et al., 2012; Brackenridge et al., 2013; Lofi et al., 2016; van Dijk et al., 2018; García et al., 2020; Mestdagh et al., 2020). Recently, Martorelli et al. (2021) have described the possibility of short-term climate variation (possibly multi-centennial scale) during the middle-late Holocene. Effects of this variation may also propagate in water masses through vertical and lateral positional changes in water mass core and/or interfaces and changes in the intensity of bottom current cores (Preu et al., 2013).

# Implications

Contourite and mixed (turbidite-contourite) systems carry implications for many different scientific conceptual frameworks and social spheres. These include (a) basin analysis, (b) accommodation in deepwater basins, (c) paleoclimatology and paleoceanography, (d) links between oceanographic and sedimentary processes, (e) geohazards, (f) environmental integrity, (g) ocean policy, (h) energy geosciences and (i) mineral resources.

## (a) Basin analysis

Water-mass circulation and bottom currents shape continental margins and adjacent abyssal plains (Hernández-Molina et al., 2008a,b). They also determine sedimentary stacking patterns (e.g., Rebesco et al., 2002; Creaser et al., 2017; Thiéblemont et al., 2020; Kirby et al., 2021a; Ng et al., 2021; Rodrigues et al., 2021a,b). Some contourite features such as drifts form at basin-scale. The detection and interpretation, therefore, require regional basin-scale analysis and should be considered in future basin modeling to maximize the understanding of a sedimentary basin (Chapter 18). Detailed 3D seismic surveys cannot often detect these features due to their dimensions. The contouritic drifts store sediments at different depths along the continental slopes, rises or abyssal plains and have to be seriously considered in future "Source-to-Sink" analysis (Chapter 11).

Plastered drifts along continental slopes and contourite terraces represent major features that define the physiography of the slope (Hernández-Molina et al., 2009; Preu et al., 2013; Mosher et al., 2017). These form at any depth along the upper, middle and lower slope. At shallower depths along the upper slope (Fig. 26), contourite terraces and drifts can be incorrectly identified as part of the continental shelf, especially in ancient records lacking clear paleodepth indicators.

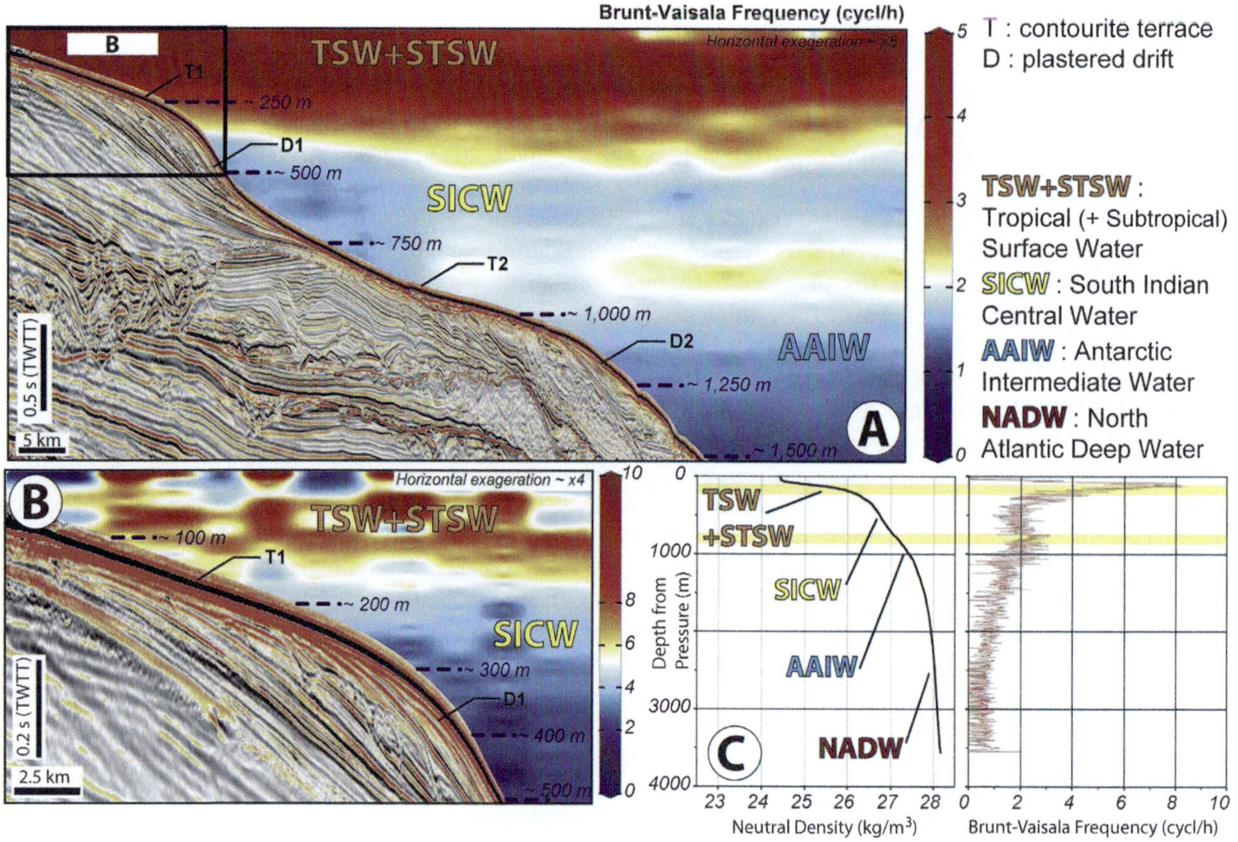


**FIG. 26** Example of contourite terraces and plastered drifts along the upper and middle continental slope of the Mozambican continental margin. (A) Seismic and hydrographic vertical section, (B) inset illustrating links between terraces and interfaces belonging to the SICW and TSW+SRSW, (C) plots showing pressure (m) vs neutral densities (kg/m$^3$) and buoyancy (cycl/h) for regional water masses. The *yellow stripes* indicate areas potentially correlated with internal wave generation at interfaces. *(From Thieblemont, A., Hernández-Molina, F.J., Miramontes, E., Raisson, F., Penven, P., 2019. Contourite depositional systems along the Mozambique Channel: the interplay between bottom currents and sedimentary processes. Deep-Sea Res. I 147, 79–99. https://doi.org/10.1016/j.dsr.2019.03.012, with permission from Elsevier).*

Large mounded or plastered contourite drifts commonly appear at the transition between the lower slope and continental rise (Faugères et al., 1999; Hernández-Molina et al., 2008a,b). The depth of these features depends on the depth of the deepest water mass (Fig. 27). Basin-scale drifts along the foot of the slope (FOS, or *base of slope*) demonstrate this relationship. Examples include (a) the South Atlantic (Hernández-Molina et al., 2009, 2016c; Preu et al., 2013) under the influence of the Antarctic Bottom Water (AABW), (b) the North Atlantic (Mosher et al., 2017) under the influence of

**FIG. 27** Examples of contourite terraces and plastered drifts along the lower continental slope of the northwest Atlantic continental margin. The position of these large bottom current features changes the position and extension of the foot of the slope (FOS) as well as the overall margin morphology. *(From Mosher, D.C., Campbell, D.C., Gardner, J.V., Piper, D.J.W., Chaytor, J.D., Rebesco, M., 2017. The role of deep-water sedimentary processes in shaping a continental margin: the Northwest Atlantic. Mar. Geol. 393, 245–259. https://doi.org/10.1016/j.margeo.2017.08.018, with permission from Elsevier).*

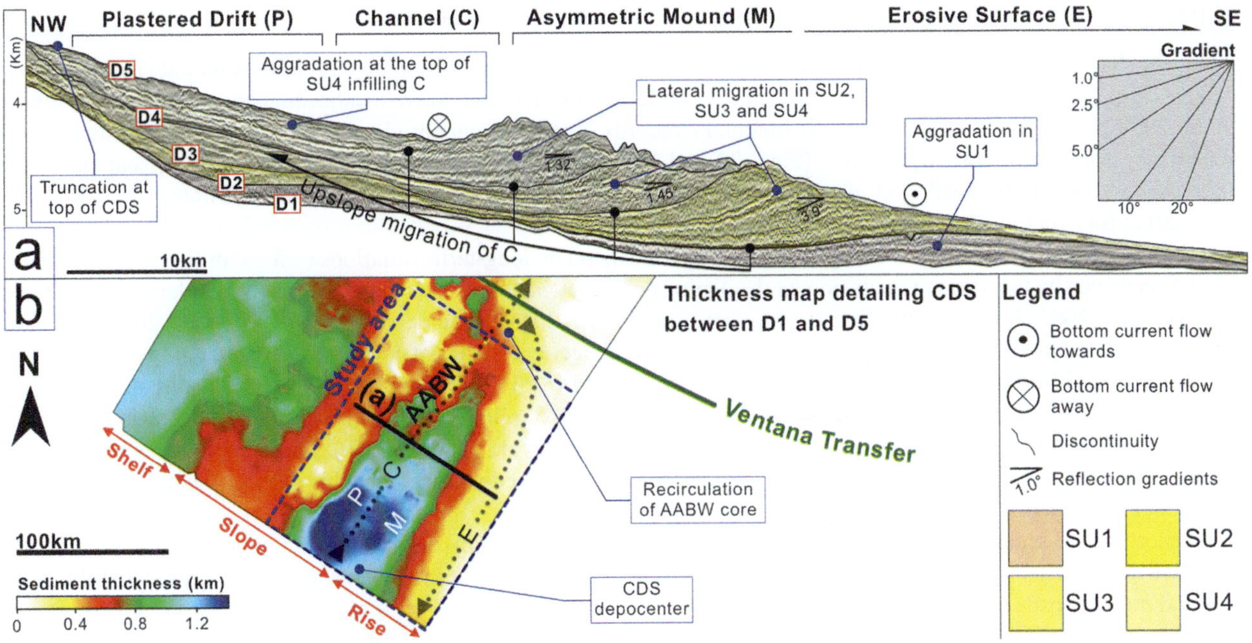


**FIG. 28** Example of large, Oligocene- to Mid-Miocene-aged sedimentary bodies showing prominent upslope lateral migration along the base of the Argentine slope. (A) Seismic profile detailing morphological features, sedimentary stacking pattern, reflection gradients and the AABW cores. (B) Map detailing the depocenter distribution and sedimentary thickness of this contourite depositional system. *(From Kirby, A., Hernández-Molina, F. J., Rodrigues, S., 2021b. Lateral migration of large sedimentary bodies in deep-marine systems: a case study offshore of Argentina. Sci. Rep. (Accepted), with permission from Springer Nature Limited).*

the North Atlantic Deep Waters (NADW) and Western Boundary Currents (DWBC), (c) the Canada Basin, Arctic Ocean (Mosher and Boggild, 2021) under the influence of the Arctic Ocean Bottom Current (AOBC) and (d) the South China Sea (Yin et al., 2019) under the influence of bottom water circulation (BW). Large mixed systems also form along with the FOS (e.g., Tucholke and Mountain, 1986; Locker and Laine, 1992; McGinnis and Hayes, 1995; McGinnis et al., 1997; Escutia et al., 2002; Rebesco et al., 2002; Hernández-Molina et al., 2017; Rodrigues et al., 2021, 2022a,c). The systematic position of contourite and mixed depositional features demonstrates the important role of water mass depth in shaping FOS morphology along continental margins.

Flatter terraces along continental slopes can create unique dynamics and gravitational processes. The abrupt change in seafloor physiography can trigger hydraulic jumps in turbidity flows. Coarse-grained sand is transported offshore from the adjacent shelf deposit along the terraces and can then enjoy subsequent reworking by bottom currents (Hernández-Molina et al., 2017, 2018). Abrupt changes in seafloor morphology can also influence submarine canyons and gullies that cut across the slope (e.g., Butman et al., 2006b).

On somewhat smaller spatial scales, discontinuities make it possible to identify particular sedimentary geometries that resemble clinothems or other signs of lateral sedimentary migration in deepwater environments (Fig. 28). These geometries have appeared offshore in Argentina (Hernández-Molina et al., 2010; Kirby et al., 2021b), Mozambique (Thiéblemont et al., 2020), South Africa (Niemi et al., 2000), northwestern Australia (Mantilla et al., 2022) and in the western equatorial Atlantic (Tucholke, 2002). A clearer understanding of diagnostic criteria and mechanisms for lateral sedimentary migration at different depths will help integrate the overall understanding of continental margin evolution.

## (b) Accommodation concepts for deepwater basins

As noted at length above, water masses strongly influence the formation of contourite depositional systems. The thickness of water masses and their boundaries (interfaces) can particularly control accommodation available for lateral and vertical sedimentary accretion of contouritic drifts (Hernández-Molina et al., 2009; Preu et al., 2013; Kirby et al., 2021b; Rodrigues et al., 2021, 2022a,b). This mechanism can explain the distribution of mud-dominated deposits along the slope (Figs. 2, 26–28) at different depths but does require additional analysis and verification, given its general nature and broad implications.

## (c) Paleoclimatology and paleoceanography

Contourite depositional systems represent a sedimentary record of paleoceanographic circulation and paleoclimatological changes (Fig. 27). These records offer expanded but contingent information relative to adjacent deposits (see discussions in Knutz, 2008 and Rebesco et al., 2014). Sedimentary records clearly document changes in the energy and direction of modern-day and paleo-current pathways (Figs. 29 and 30) on multicentennial, millennial and million-year timescales (e.g., Llave et al., 2006; Voelker et al., 2006; Knutz et al., 2007; Toucanne et al., 2007, 2021; Knutz, 2008; Rogerson et al., 2012; Sierro et al., 2020; Martorelli et al., 2021).

The onset and burial of contourite depositional systems depend on long-term variations (>5–10 million years) in paleo-circulation driven by the opening or closing, or deepening or shallowing of oceanic gateways and other plate tectonic events (e.g., Wunsch, 2002; Knutz, 2008; Zhang et al., 2011; Sijp et al., 2014; Straume et al., 2020; Ng et al., 2021; Yin et al., 2021). Punctuated pulses in mantle convective activity can force tectonic cycles of compression and sagging on shorter time scales of a few million to >0.4 million years (Rudge et al., 2008; Lovell, 2010; Ng et al., 2022). Ocean circulation changes of approximately 0.8–0.9 million-year duration overprinted by 2–2.5 million-year shifts have been attributed to such tectonic pulses (Cloetingh et al., 1990; Knutz, 2008; Hernández-Molina et al., 2014b, 2016b). These in turn have been invoked by Hernández-Molina et al. (2014b, 2016b) to explain variations in the sedimentary stacking pattern of contourite drifts and the formation of internal discontinuities and hiatuses (Fig. 30).

At shorter time scales (<0.4 Ma), climatic and eustatic changes driven by orbital cycles and oceanographic processes can cause deepening or shoaling of water masses and bottom currents. These in turn influence the vertical distribution of sand and mud deposits as documented in acoustic facies, stratigraphic stacking patterns and thicknesses of contourite features (Llave et al., 2001, 2006; Rogerson et al., 2012; Hernández-Molina et al., 2016b; Liu, 2019; Mestdagh, 2020; Sierro et al., 2020; de Castro et al., 2021b; Martorelli et al., 2021).

## (d) Links between oceanographic and sedimentary processes

Bottom currents develop bedforms within muddy and sandy deposits in deepwater environments (Fig. 31). The position of sandy bedforms reflects periods of intensified bottom currents associated with boundaries or internal discontinuities within drifts (see further details in Mutti et al., 2014; Campbell and Mosher, 2016; Hernández-Molina et al., 2017).

Identification and interpretation of these features can thus help decipher oceanographic and sedimentary processes in both modern day and ancient records. "Along-slope oceanographic processes" section summarizes water-mass circulation and associated secondary processes. Settings dominated by along-slope processes with vigorous bottom currents ($>0.4\,\mathrm{ms}^{-1}$) show redistribution of sandy deposits delivered by downslope channel systems. The sandy deposits located within channels, moats and furrows relate to sand ribbons and barchan dunes (Mutti et al., 2014; Hernández-Molina et al., 2018). In settings with weaker bottom currents ($<0.4\,\mathrm{ms}^{-1}$), 2D dune-like bedforms commonly appear with crests striking oblique to the main current flow. Contour currents remove fractions of the sand deposits delivered by the downslope channels and redistribute them along the slope where sediments accumulates in troughs associated with the 2D and 3D dune-like bedforms (Hernández-Molina et al., 2018; de Weger et al., 2020, 2021).

## (e) Geohazards

Bottom currents influence the distribution, composition and physical properties of contourites, and these in turn influence submarine slope stability (Bryn et al., 2005; Laberg et al., 2005; Laberg and Camerlenghi, 2008; Rebesco et al., 2014; Neves et al., 2016; Miramontes et al., 2018; Teixeira et al., 2019; Lobo et al., 2021; Rodrigues et al., 2021b). Gravitational stability depends on spatial and temporal variations in the clastic sedimentary processes of erosion, transport and deposition. As shown in Fig. 32, muddy contourites characterized by fine-grained sediments distributions, low-permeability and high pore-water content tend to form over-pressurized gliding planes that appear clearly in plastered drifts (Hernández-Molina et al., 2017; Miramontes et al., 2018).

As outlined in Rebesco et al. (2014), contourite deposits fail due to a range of factors including: (a) the geometry and location of the deposit, (b) low shear strength (high water content from relatively high sedimentation rates and grain-size homogeneity), (c) under-consolidation and excess pore pressure (in low-permeability, fine-grained contourites and high-porosity siliceous ooze layers) and (d) loading and gas charging. These topics are addressed in Laberg and Camerlenghi (2008), Rebesco et al. (2014), Miramontes et al. (2018), Brackenridge et al. (2020), Gauchery et al. (2021) and Yenes et al. (2021).

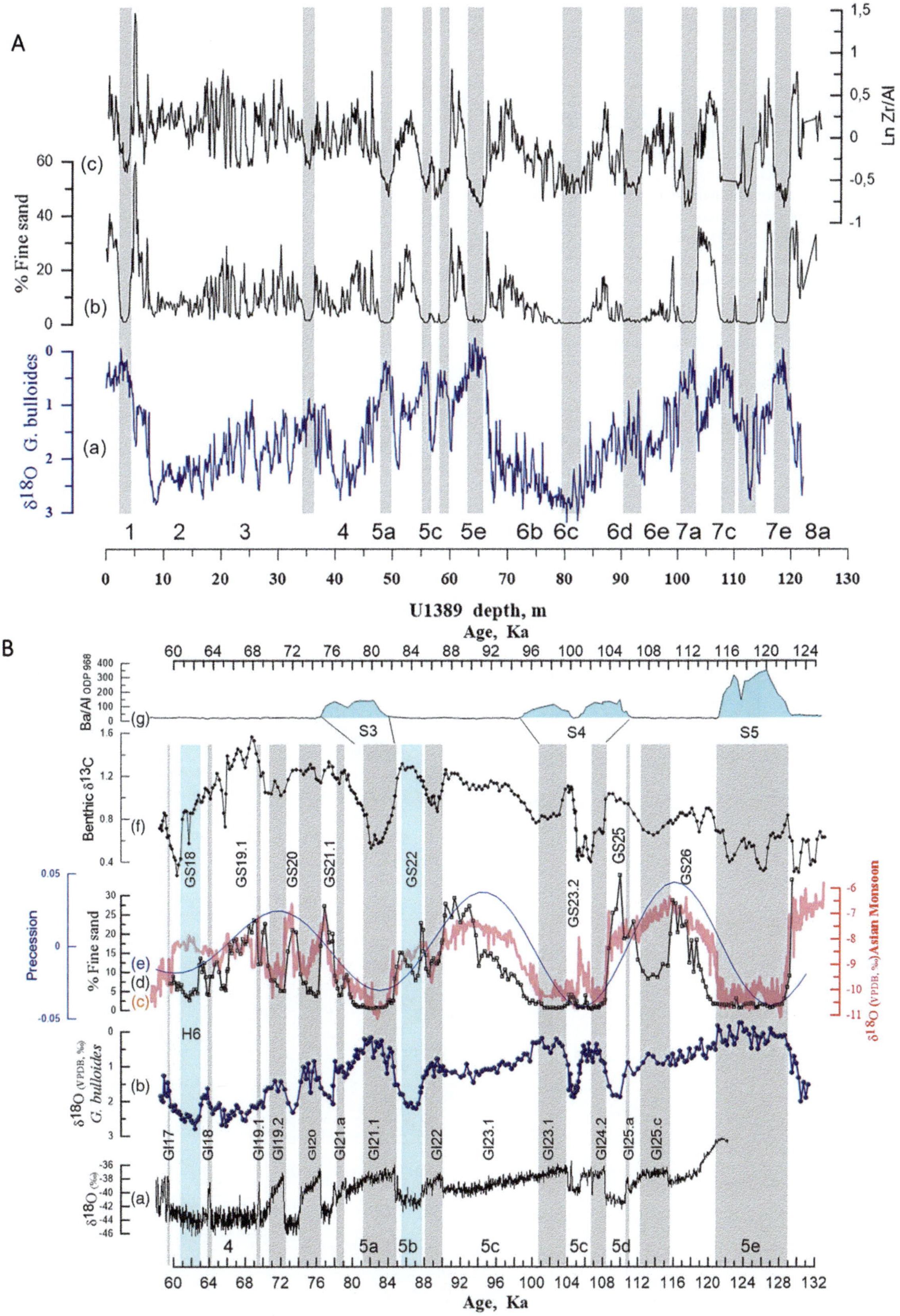


**FIG. 29** Proxy records of paleoceanographic and paleoclimatological changes. (A) Change in strength of the Mediterranean outflow water (MOW) over the last 250 kyr from IODP Site U1389 according to (a) IODP Site U1389 *Globigerina bulloides* δ18O record, (b) percent fine sand (62–150 μm fraction), (c) Zr/Al record. *Gray vertical bars* show the intervals with very weak MOW (minimum percent of fine sand). (B) Example of a millennial-scale changes in MOW strength and Mediterranean overturning from IODP Site U1389 during MIS 3. (a) δ18O NGRIP2 (‰ V-SMOW) with the GICC05 timescale, (b) *Globigerina bulloides* δ18O from IODP Site U1389, (c) percentage fine sand (62–150 μm fraction), (d) δ18O from Asian speleothems *(red)*, (e) *Cibicidoides pachyderma* δ13C from IODP Site U1389, (f) Mean SS along the Corsican margin. *Gray bands* show Greenland interstadials. Blue bands show HSs. GI2 to GI17 denote Greenland interstadials. GS3 to GS13 denote Greenland stadials. S2 denotes Sapropel 2. *(Further detail and explanation for A and B in Sierro, F.J., Hodell, D.A., Andersen, N., Azibeiro, L.A., Jiménez-Espejo, F.J., Bahr, A., Flores, J.A., Ausin, B., Rogerson, M., Lozano-Luz, R., Lebreiro, S., Hernández-Molina, F.J., 2020. Mediterranean Overflow over the last 250 ky. Freshwater forcing from the tropics to the ice sheets. Paleoceanogr. Paleoclimatol. 35 e2020PA003931, with permission from Wiley Online Library).*

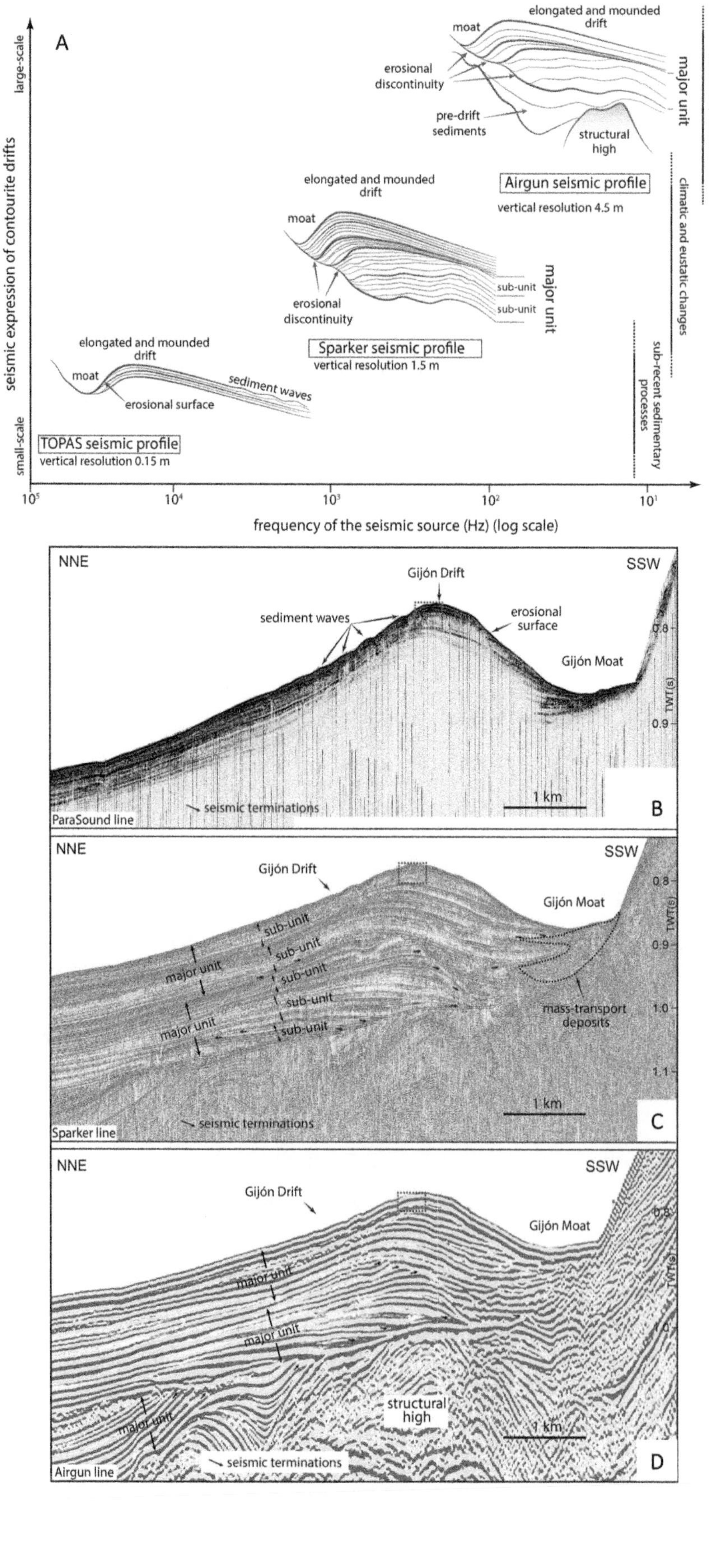


**FIG. 30** Example seismic records of sedimentary cycles for contouritic drifts at different geologic time scales. (A) A sketch showing the relationship between seismic data, seismic imagery of contourite drifts and long/short-term tectonic, climatic, eustatic, sedimentary variations. Interpreted ParaSound (B), sparker (C) and airgun (D) seismic profiles showing morphology and internal features of the Gijon drift, Cantabrian Sea. *(From (A, D) Liu, S., 2019. Contourite Depositional System in the Le Danois Bank Region, Southern Bay of Biscay: Sedimentary, Tectonic and Paleoceanographic Implications (PhD thesis). Ghent University, Ghent, Belgium).*

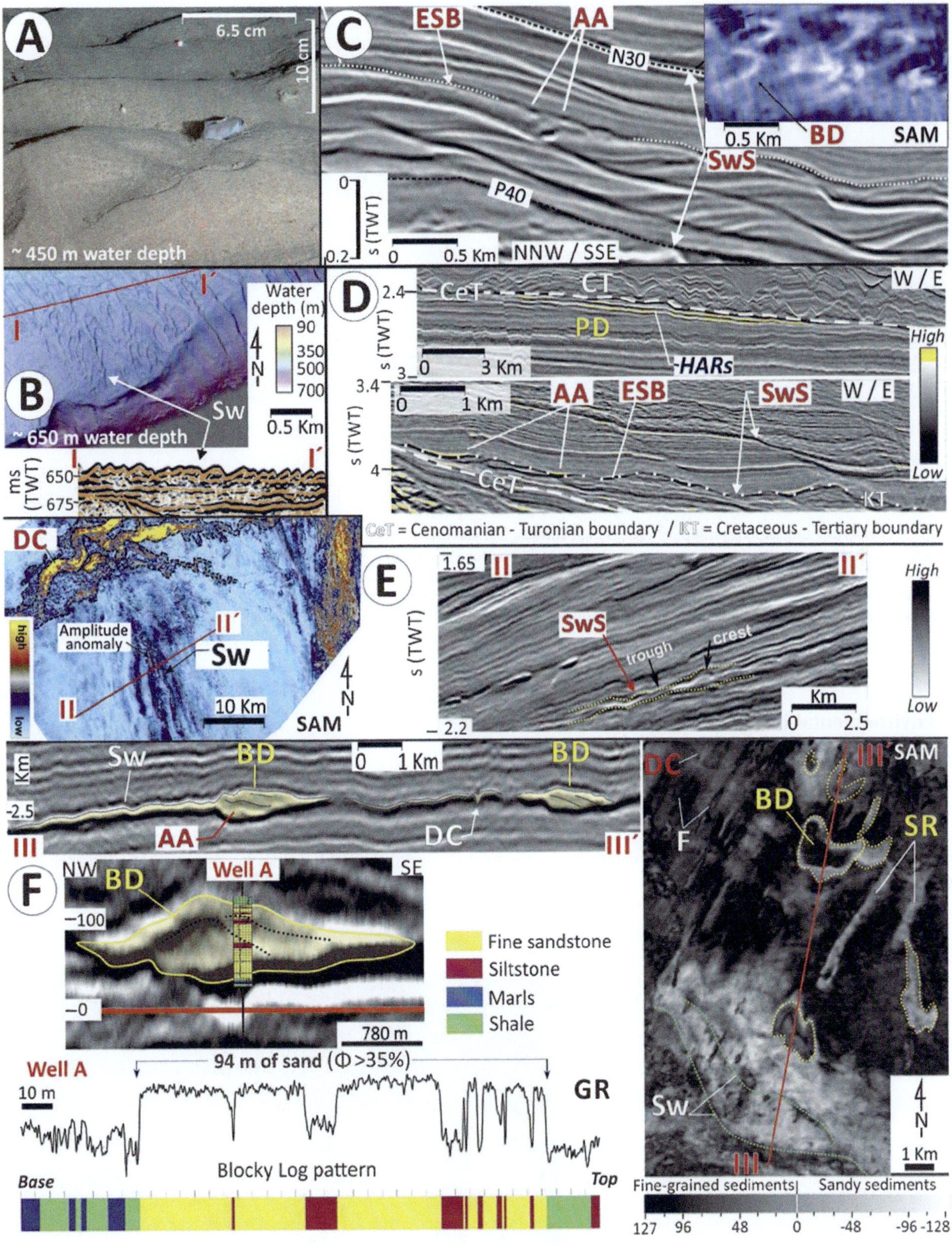


**FIG. 31** Examples of bedforms along contourite terraces of modern sandy deposits (A) Porcupine Seabight, Irish slope. (B) Modern sandy deposits in the Mozambique Channel. (C) Middle Miocene sequence, Canada, illustrating two-dimensional (2D) dune-like bedforms and barchan dunes with inset from AA surface. (D) Mozambique Channel sequence showing high-amplitude reflections (HARs) (late Cretaceous) and large bedforms (Paleocene to Miocene). (E) 2D dunes, Late Cretaceous, Namibian margin. (F) Oligocene sequence, Brazil, indicating thick sand bodies topped by barchan dunes accessed by a well (see further details in Mutti et al., 2014; Campbell and Mosher, 2016; Hernández-Molina et al., 2018, with permission from Geological Society of America). Abbreviations: *AA*, amplitude anomaly barchan "horns"; *BD*, Barchan dunes; *CT*, contourite terrace; *DC*, downslope channels; *ESB*, erosional surface with barchan bedforms; *F*, furrows; *HARDs*, high-amplitude reflections; *PD*, plastered drifts; *SAM*, seismic amplitude map; *SR*, sand ribbons; *SW*, 2D wave-like bedforms; *SwS*, layered sediments wave sequence.

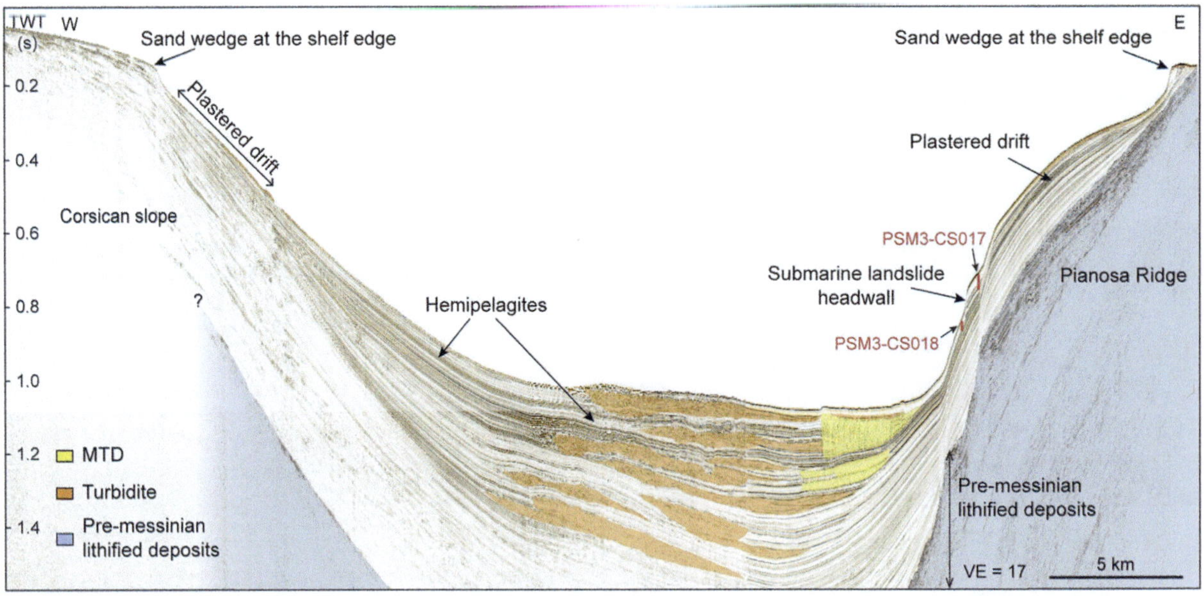


**FIG. 32** Example of genetic relationship between bottom currents and submarine slope stability within the Corsica Trough (northern Tyrrhenian Sea, offshore France and Italy). Slide features appear along the lower slope and distal part of a plastered drift. These contribute to mass transport deposits (MTD) appearing in the adjacent basin. *(From Miramontes, E., Garziglia, S., Sultan, N., Jouet, G., Cattaneo, A., 2018. Morphological control of slope instability in contourites: a geotechnical approach. Landslides 15(6), 1085–1095. doi:10.1007/s10346-018-0956-6, with permission from Springer).*

## (f) Environmental integrity

Intuitively, contourite and mixed depositional systems represent the physical environment of deepwater, benthic ecosystems (e.g., Lozano et al., 2020; Miguez-Salas et al., 2020). For example, they host cold-water corals (CWC) (Van Rooij et al., 2003, 2007a,b; Huvenne et al., 2009; Hebbeln et al., 2016, 2019; Lüdmann et al., 2016). Many examples of CWC mounds from around the world's ocean (Fig. 33) are associated with drifts due to their shared interdependencies with bottom currents, climate and food supply (e.g., De Mol et al., 2005, 2011; Lüdmann et al., 2016; Titschack et al., 2009, 2016).

**FIG. 33** Example of associations between bottom currents and deep, cold-water corals found in the Porcupine Seabight. *(Courtesy of MARUM, Univ. of Bremen, Germany).*

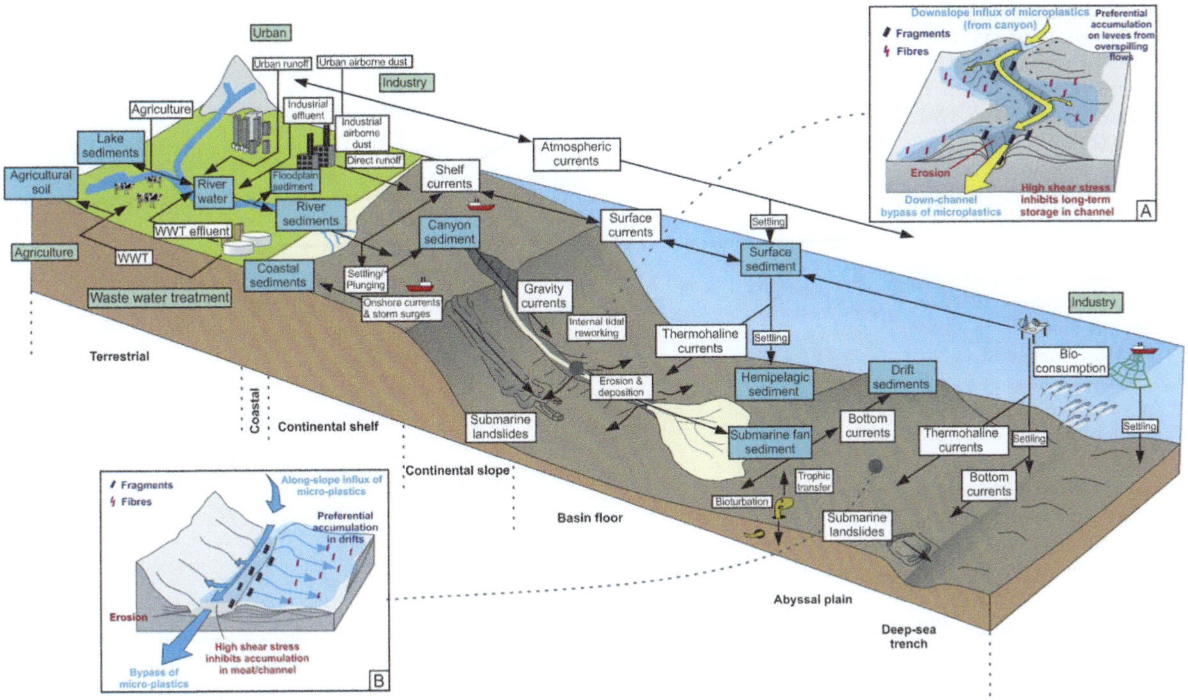


**FIG. 34** Microplastic input, transport vectors and sinks. *Green boxes* represent primary input, *blue boxes* represent temporary and permanent sinks, *white boxes* represent transport mechanisms and *arrows* show transport vectors. Insets show the potential distribution and transport of microplastics in (A) a channel-levee system and (B) a bottom current moat and drift system. *(From Kane, I.A., Clare, M.A., 2019. Dispersion, accumulation, and the ultimate fate of microplastics in deep-marine environments: a review and future directions. Front. Earth Sci. 7, 80. doi: 10.3389/feart.2019.00080, with permission from Frontiers).*

Moreover, contourite and mixed depositional systems represent favorable settings for the accumulation of microplastics (Kane and Clare, 2019; Kane et al., 2020). The marine plastic reservoir including pelagic debris and microplastics (<5 mm) is estimated to have reached ~5.25 trillion particles (van Sebille et al., 2015). These damage marine and coastal ecosystems throughout the world. The adverse impacts propagate into economic and public health spheres. Similar to natural sediments, microplastics are deposited in deepwater settings and are thus influenced by water masses and associated bottom currents (Fig. 34). This topic is further addressed in Horton and Dixon (2017), Kane and Clare (2019), Kane et al. (2020), Chiarella and Hernández-Molina (2021) and Chapter 21 of this volune.

## (g) Ocean policy

Research on contourite and mixed depositional systems (e.g., Hernández-Molina et al., 2009, 2010; Mosher et al., 2017; Mosher and Yanez-Carrizo, 2021) provides critical, objective constraints on determining the exact position of the foot of the slope (Fig. 35). These systems can specifically mark the outer limits of the continental shelf beyond 200 nautical-miles, a legal spatial designation in article 76 of the United Nations Convention on the Law of the Sea (UNCLOS; ABLOS, 2006). The current UNCLOS *Commission on the Limits of the Continental Shelf* (CLCS) also cites this morphological and spatial boundary. These policies define responsibilities and rights of states to respectively protect and extract resources from the natural marine environment. As such, precise and accurate identification of contourite and mixed depositional systems can have economic, social and environmental consequences (http://www.un.org/Depts/los/convention_agreements/convention_overview_convention.htm). Laws grant coastal states rights to explore for and extract natural resources. States are also expected to conserve and effectively manage the natural marine environment, which includes marine life, the water column and the seafloor. States must balance conservation objectives with energy production initiatives, which can now include alternative resources (as the production of energy from the water, currents and winds) in addition to oil and gas (see Chapter 21).

Research on modern deep-sea processes can also inform risk management in the telecommunications industry. Currently, >1,500,000 km (~ 932.050 mi.) of cable runs along the seafloor (https://submarine-cable-map-2020.telegeography.com) such that safe, stable seafloor for this infrastructure has become scarce. Oceanographic processes,

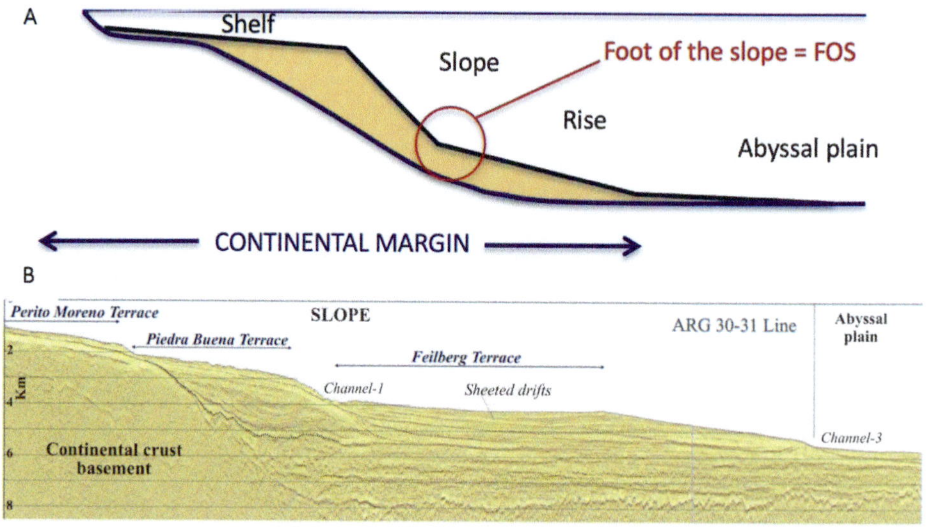


**FIG. 35** (A) Idealized sketch of a passive continental margin from the Manual on Technical Aspects of the United Nations Convention on the Law of the Sea—1982 (ABLOS, 2006) compared with a real profile (B) of an area of the Patagonian continental margin (Hernández-Molina et al., 2009) strongly influenced by circulation of different water masses and their associated bottom currents. Policies state that the foot of the slope (FOS) shall be determined as the point of maximum change in the gradient at its base (Article 76, UNCLOS) but areas hosting contourite terraces and plastered drifts can show remarkable variation in slope morphological profiles thus introducing considerable uncertainty to the assumed models shown in A.

including internal waves, eddies and storms can damage submarine cables and the interests that maintain them in the deepwater environment (such as Alcatel-Lucent). The cost of cable damage tends to be on the order millions of dollars. Cable users such as financial institutions can also suffer enormous losses due to interrupted communication. Detailed understanding of bottom currents, seafloor morphology and contourite systems has helped identify risk zones, including passages through straits, channels and other seafloor features generating water currents that can damage cables (see Chapter 21).

## (h) Energy geosciences

### *Deepwater petroleum plays*

Deepwater petroleum exploration has tended to focus on gravity flow deposits (see Chapters 1 and 6). Technological advances have shifted exploration toward deeper waters (>2000 m., see Chapter 21), where bottom currents, contourites and mixed systems play a primary role in determining the distribution of oil and gas reservoirs (Viana et al., 2007; Fonnesu et al., 2020). Understanding how these phenomena operate along continental margins can assist hydrocarbon exploration by identifying bottom currents that have affected or reworked sand with high reservoir potential (Viana, 2008).

Sandy contourites can exhibit considerable thickness, consisting of sorted sand with wide lateral extent (100 m or 328 ft.) and considerable vertical thickness (10 m/32.8 ft.), all of which contribute to deepwater reservoir and play potential (Viana and Rebesco, 2007; Viana, 2008; Mutti and Carminatti, 2012; Mutti et al., 2014; Shanmugam, 2012, 2016, 2017). Sandy contourites commonly appear in (i) contourite channels, (ii) contourite terraces and (iii) at the exit of straits or gateways (Fig. 34). Contourite terraces with sand-rich deposits have been described at different water depths along continental slopes and rises (Viana et al., 2002; Hernández-Molina et al., 2017; Miramontes et al., 2020b; de Castro et al., 2021b). These deposits often exhibit petrophysical characteristics of high porosity, high permeability (Yu et al., 2020) and high lateral and vertical transmissibility of fluids (Viana, 2008; Shanmugam, 2012). The deepwater setting of some sandy contourite systems can pose challenges during the drilling stages of exploration but their economic potential incentivizes the development of technologies to overcome challenges (see Chapters 19 and 20).

Mud-rich deposits identified in contourite and mixed depositional systems (Figs. 36 and 37) can act as effective caprock seals and, where they are relatively rich in organic matter, may become potential hydrocarbon source rocks or potential shale-gas reservoirs (Viana, 2008). Similarly, to the turbidite systems, mass-transport deposits (MTD, Chapter 10) are conspicuous components in the stratigraphic record in contourite channels, deeply modifying the shape of channels, contributing to their sediments infill and occasionally acting as stratigraphic seals for hydrocarbon reservoirs (Moscardelli et al., 2006; Alves et al., 2014; Cardona et al., 2016).

Interest in contourite depositional systems has grown with the recognition that reservoirs formerly interpreted as turbidites exhibit anomalous reservoir qualities that can be attributed to mixed system processes. In essence, mixed systems appear more commonly along continental margins than previously thought (Rebesco et al., 2014; Fonnesu et al., 2020; Rodrigues et al., 2021, 2022c). Exploration paradigms often ignore along-slope processes due to the many unknowns

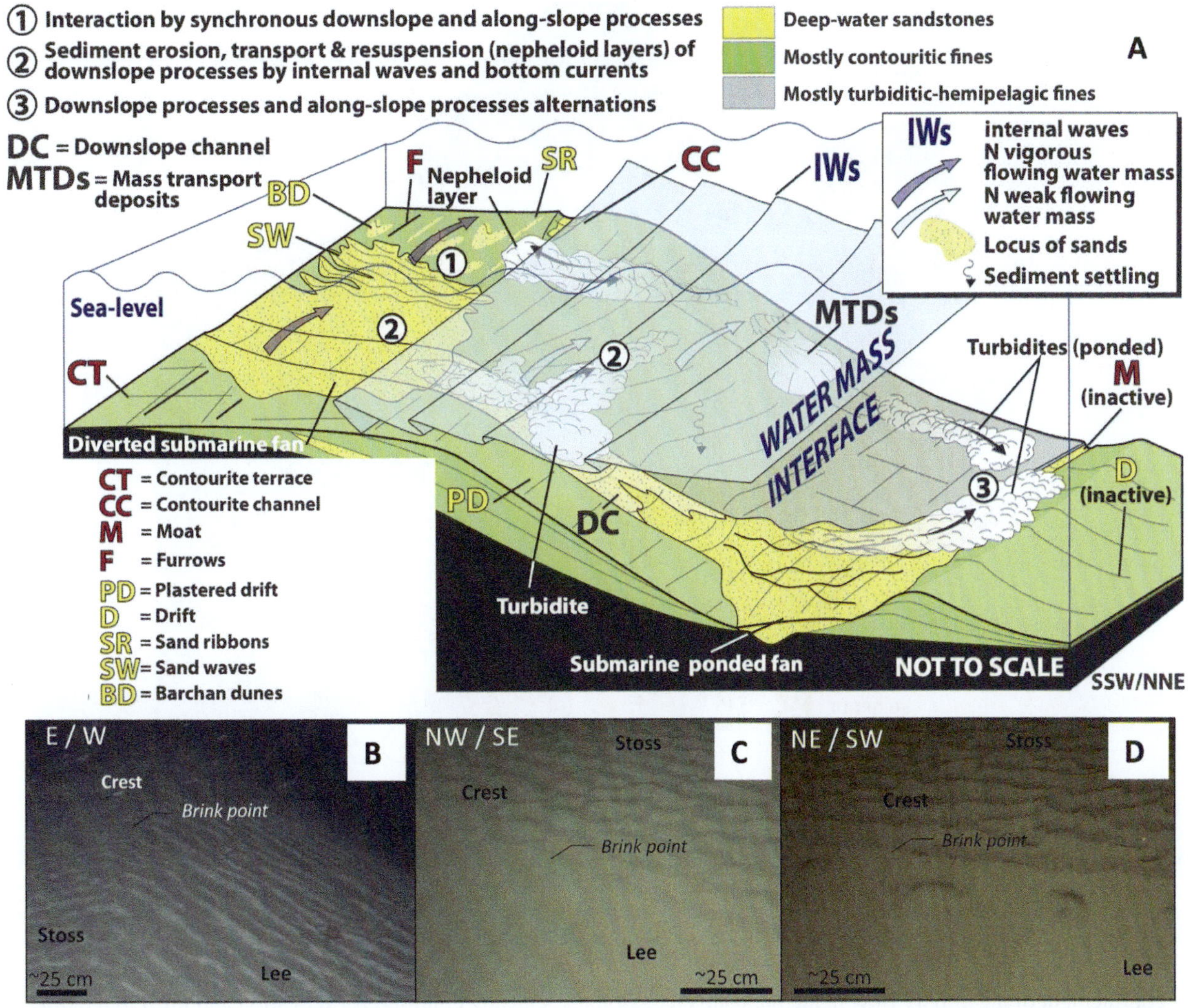


**FIG. 36** (A) 3D sketch showing the different settings in which along- and down-slope processes interact at different temporal scales indicating the location of sandy deposits. At bottom (B–D), examples of well sorted, deepwater sandy deposits forming dunes and ripples along contourite channels in the Gulf of Cadiz based on photographs and videos taken during the MOWER cruise (2014). *(From Thieblemont, A., Hernández-Molina, F.J., Miramontes, E., Raisson, F., Penven, P., 2019. Contourite depositional systems along the Mozambique Channel: the interplay between bottom currents and sedimentary processes. Deep-Sea Res. I 147, 79–99. https://doi.org/10.1016/j.dsr.2019.03.012; Based on Hernández-Molina, F.J., Larter, R., Maldonado, A., 2017. Neogene to quaternary stratigraphic evolution of the Antarctic Peninsula, Pacific Margin offshore of Adelaide Island: transitions from a non-glacial, through glacially-influenced to a fully glacial stage. Glob. Planet. Chang. 156, 80–111. doi:10.1016/j.gloplacha.2017.07.002; Fonnesu, M., Palermo, D., Galbiati, M., Marchesini, M., Bonamini, E., Bendias, D., 2020. A new world-class deep-water play-type, deposited by the syndepositional interaction of turbidity flows and bottom currents: the giant Eocene Coral Field in northern Mozambique. Mar. Pet. Geol. 111, 179–201. https://doi.org/10.1016/j.marpetgeo.2019.07.047).*

concerning mixed systems (Creaser et al., 2017). This can lead to significant differences between reservoir models and sedimentary information derived from core. In the context of hydrocarbon exploration, bottom currents can winnow and rework fine- and coarse-grained sediment delivered by gravity flows to channel-levee systems (Figs. 37–39). The "drift" or levee part of this system consists primarily of fine-grained sediment (silts and muds), which can serve as an effective seal for trapping hydrocarbons (see Bailey et al., 2021a). Mamba and Coral from northern Mozambique demonstrate how the syndepositional interaction between downslope-oriented gravity flows and across-slope bottom currents is able to accumulate large quantities of mud-free sandstone in channel axis with exceptional net to gross, extension and thickness (Fonnesu et al., 2020). The flow stripping process produces a peculiar architecture in which the system-axis migrates upflow, against the direction of the bottom current. This forms an asymmetric contourite levee due to the deviated part of the turbulent suspension cloud invading one flank of the channel axis (Figs. 14 and 38). This levee invasion induces

**FIG. 37** Examples of late Cretaceous mixed depositional systems along the Uruguayan continental margin including seismic amplitude anomalies, which may indicate sandier deposits within submarine canyons and overbank deposits. These could represent reservoirs. Between canyons, large asymmetric mounded drifts include prominent mud-dominated deposits, which could act as seals. *(Courtesy of BG Group, now Shell; Creaser, A., Hernández-Molina, F., Badalini, G., Thompson, P., Walker, R., Soto, M., Conti, B., 2017. A late cretaceous mixed (turbidite-contourite) system along the Uruguayan margin: sedimentary and paleoceanographic implications. Mar. Geol. 390, 234–253. https://doi.org/10.1016/j.margeo.2017.07.004).*

the confinement of the gravity flow in the opposite channel flank, causing continuous excavation and infill in the direction opposite of the bottom current. This unusual architecture challenges routine seismic interpretation and volumetric estimation because the top of the reservoir cannot be consistently picked, however, the top reflector tends to decrease in amplitude outside the channel axis into contemporaneous shaly levee-drift deposits (Fig. 38). Only in the far-stack volume are interpreters able to more easily identify the top reservoir upon which gas-in-place volumes can be correctly calculated. A deeper discussion of the implications of mixed systems as reservoirs can be seen in Rodrigues et al. (2022c).

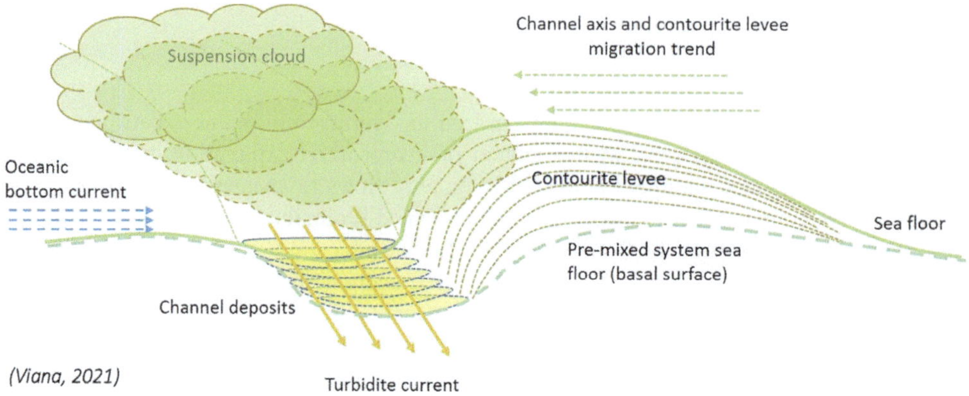


**FIG. 38** Schematic representation of the characteristics of a mixed contourite-turbidite system where the gravity-flow infill of the channel axis migrates against the trend of bottom currents due to the building of the asymmetric contourite levee that invades one flank of the channel axis. This levee invasion induces the confinement of the gravity flow in the opposite channel flank and its continuous excavation and infill in that trend in opposition of the bottom current flow trend.

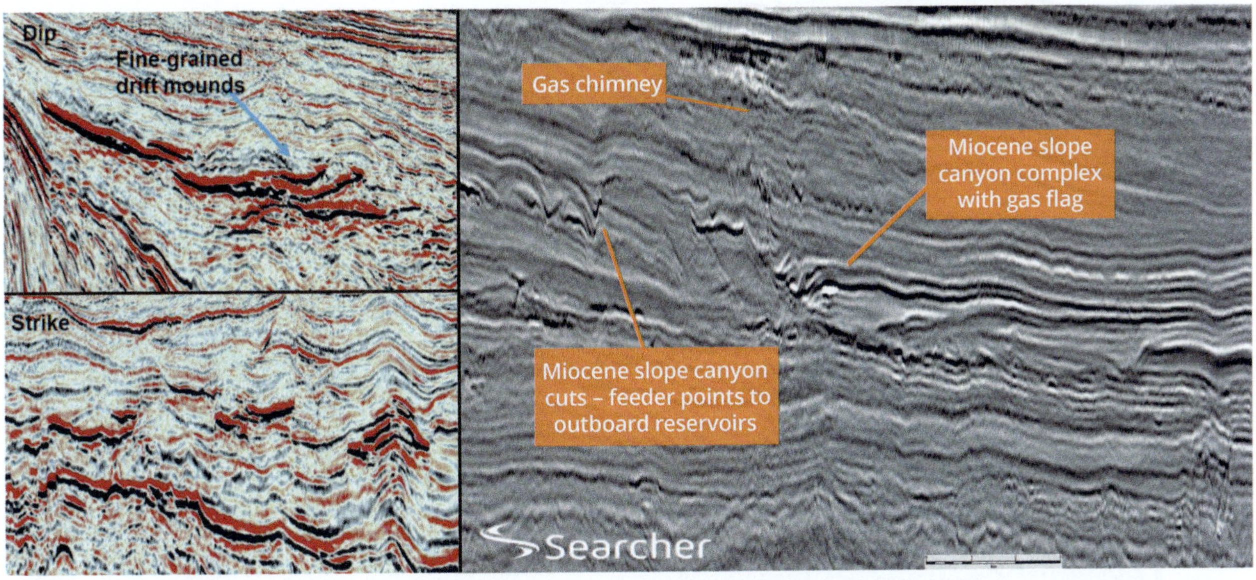


**FIG. 39** Examples of mixed depositional systems along the Argentine continental margin with evidence of sandier deposits (reservoirs?) from seismic amplitude anomalies. Both the overlying fine-grained deposits of the mounded drifts or hemipelagic deposits on their tops can serve as seals. *(Courtesy of Searchers; Davies, D., Hoiles, P., Rodriguez, K., Hodgson, N., 2020. GEO ExPro-offshore Argentina: tertiary play potential in the Malvinas Basin. GeoExpro, 17 (4). https://www.geoexpro.com/articles/2020/10/offshore-argentina-tertiary-play-potential-in-the-malvinas-basin; Rodriguez, K., Hodgson, N., 2019. Hydrocarbon potential of hybrid depositional systems. GeoExpro, 16(4). https://www.geoexpro.com/articles/2019/10/hydrocarbon-potential-of-hybrid-depositional-systems, with permission from GeoPublishing Ltd.)*

Recent research also highlights the role of bottom currents on stratigraphic trap development (Amy, 2019; Counts et al., 2021). The formation of moat erosions trending along-slope affects the continuity of downslope-oriented deepwater channels formed on the underlying sequence, interrupts the sand continuity and favors the formation of updip stratigraphic traps. Also, the winnowing effect of bottom currents on deepwater sandy deposits is potentially able to form detached or isolated sandstone bodies, with implicit trapping potential.

Many oil and gas companies have reported interest in exploring contourite and mixed systems as potential hydrocarbon reservoirs (Figs. 37, 39–41) in petroleum systems (Chapter 2), such as Petrobras interests offshore Brazil (e.g., Mutti et al., 1980, 2012, 2014; Viana et al., 2007; Viana, 2008), BG Group (now Shell) interests offshore Uruguay and Tanzania (e.g., Thompson et al., 2015; Sansom, 2017, 2018), BP interests also offshore Tanzania (Decal et al., 2016), ENI interests offshore Mozambique (e.g., Palermo et al., 2014; Fonnesu et al., 2020), Cairn Energy interests offshore Senegal (Warner, 2017) and Searcher's interests offshore Argentina (Rodriguez and Hodgson, 2019; Davies et al., 2020).

In order to boost the understanding of contourites and mixed systems, oil and gas companies are increasing their participation in Joint Industry Projects (JIP), such as the one at Royal Holloway, University of London (RHUL, United Kingdom) from 2018 to 2022 and supported by BP (United Kingdom), ENI (Italy), TotalEnergies (France), ExxonMobil (United States), Wintershall Dea (Germany) and Spectrum/TGS (United Kingdom) have supported efforts focused on the "*Seismic recognition and sediments characterisation of deepwater contourite depositional systems and their implications for the oil industry.*" The JIP has sought to integrate the relevant academic understanding of contourite and mixed-drift depositional systems into the industrial domain and support further scientific advances on such depositional systems with industry expertise, data and funding.

As energy exploration continuously moves toward frontier environments (Chapters 19–21), such as ultra-deepwater settings, it highlights the importance of furthering the understanding of contourites and mixed systems in order to better forecast the occurrence of high-quality hydrocarbon bearing reservoirs (Chapter 13) related to those systems. The continuous increase of geological and geophysical data quality and processing solutions coupled with the industry move toward deeper water settings and recent important discoveries in contourite and mixed systems strengthen the relevance of this research topic, and, as a consequence, more and more of these reservoirs are expected to be discovered in the future (Chapters 8 and 21).

## *Capture and storage of $CO_2$ (CCS)*

Carbon dioxide ($CO_2$) represents the most abundant anthropogenic greenhouse gas in the atmosphere (GCCSI, 2020). While the need for hydrocarbons continues to persist, the energy industry can alleviate global $CO_2$ emissions by capturing

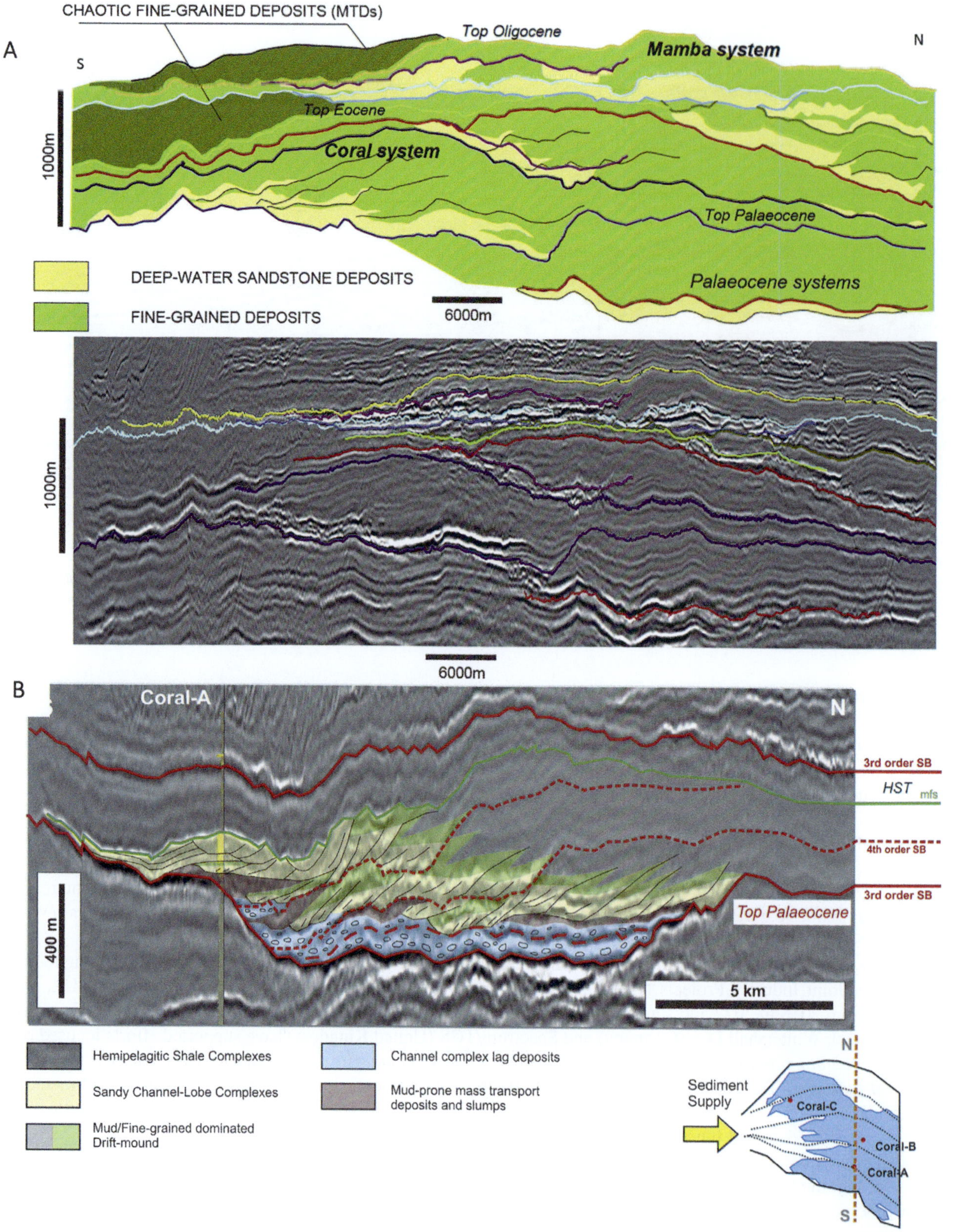


**FIG. 40** (A) Example of a world-class, deepwater petroleum play in a mixed system from northern Mozambique highlighting the stratigraphic distribution of the Mamba and Coral systems (Oligocene-Eocene-Eocene in age). Within each system, individual sand-bodies show a diachronous seismic geometry expressed by a unidirectional (southward) lateral stacking of component units and consistent lateral transitions from sand (high-amplitude responses) to shale facies (transparent seismic features) along gently southward dipping strata. This geometry characterizes most of the Mamba and Coral systems. Each sequence is capped by extensive fine-grained deposits that serve as regional seals. (B) Example of the Coral system showing the main stratigraphic surfaces and seismic facies interpretation. The overall southward migration and offset stacking pattern of the Coral channel complex appears as migrating-channels flanked by prominent asymmetric drift-mounds along northern flanks. The reservoir shows a systematic internal organization linked to 4th and 5th order base-level cycles (*SB*, sequence boundary; *HST*, highstand system tract; *mfs*, maximum flooding surface—see Chapter 12). *(From Fonnesu, M., Palermo, D., Galbiati, M., Marchesini, M., Bonamini, E., Bendias, D., 2020. A new world-class deep-water play-type, deposited by the syndepositional interaction of turbidity flows and bottom currents: the giant Eocene Coral Field in northern Mozambique. Mar. Pet. Geol. 111, 179–201. https://doi.org/10.1016/j.marpetgeo.2019.07.047, with permission from Elsevier).*

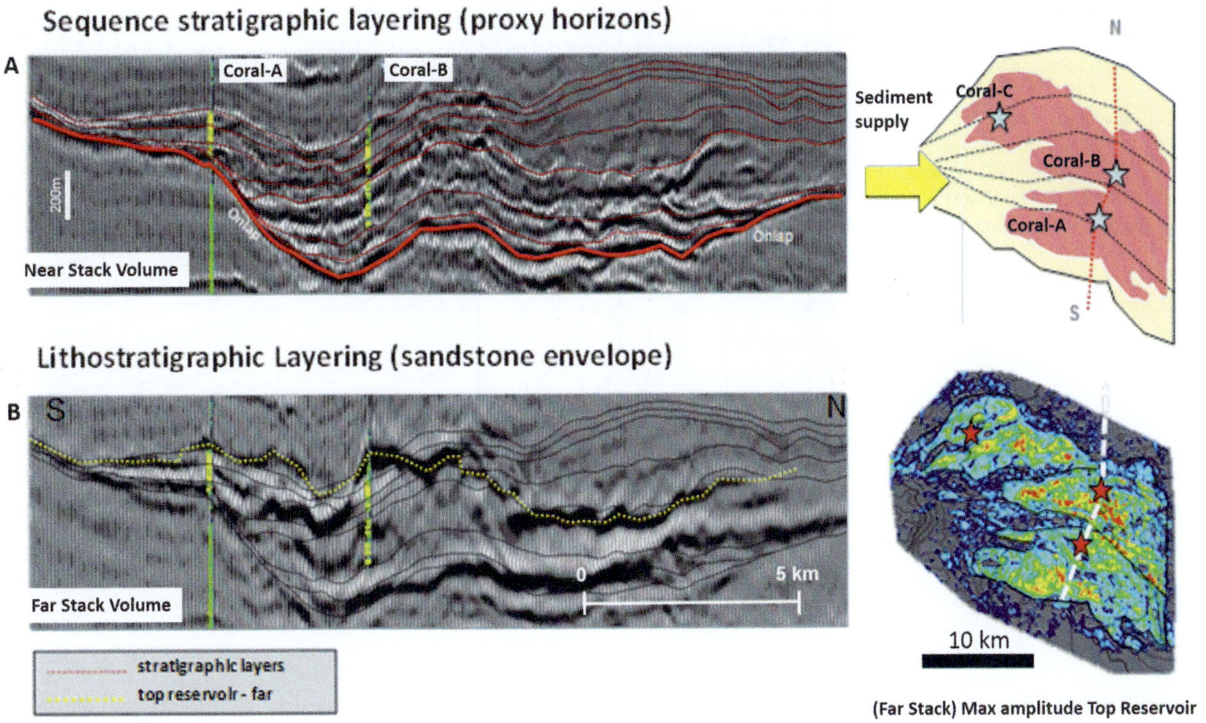


**FIG. 41** Comparison between near- and far-stack seismic data from the Coral system. (A) On the near-stack coeval stratigraphic surfaces can be picked from the system axis toward the external levee section; (B) on the Far-stack the *yellow horizon* represents the top of the gas-bearing sands formed by laterally stacked fan elements. The maps on the right represent the extension of the interpreted sedimentary bodies and seismic anomaly represented by the warmer and brighter colors. *(From Fonnesu, M., Palermo, D., Galbiati, M., Marchesini, M., Bonamini, E., Bendias, D., 2020. A new world-class deep-water play-type, deposited by the syndepositional interaction of turbidity flows and bottom currents: the giant Eocene Coral Field in northern Mozambique. Mar. Pet. Geol. 111, 179–201. https://doi.org/10.1016/j.marpetgeo.2019.07.047, with permission from Elsevier).*

and storing $CO_2$ below the subsurface in natural reservoirs before it ever reaches the atmosphere (CCS; Carbon Capture and Storage). This method aims to reduce atmospheric $CO_2$ levels and climate impacts, and deepwater sedimentary basins are potentially favorable for that (House et al., 2006). The International Energy Agency (IEA, 2017) has proposed that CCS could account for at least 14% of target reductions in $CO_2$ emissions by 2060 in order to stay within the 2 °C warming by 2100 scenario (Pereira et al., 2021). The Global CCS Institute (GCCSI, 2020) states that only 26 CCS facilities currently operate in the world and identifies the need for more subsurface storage facilities (see Chapter 21).

$CO_2$ capture and storage are increasingly recognized as a strategy for mitigating global climate change by both the EU (European Commission, 2019) and the Paris Climate Agreement (UNFCCC, 2015). Several CCS facilities consist of abandoned offshore oil and gas fields. Examples include the Sleipner gas field in block 15/9 of the North Sea, about 250 km (~ 155 mi.) west of Stavanger, Norway (Equinor and partners). The reservoir characteristics and subsurface architecture required for CCS strongly resemble those required for hydrocarbon reservoirs (e.g., grain-size heterogeneity, connectivity, porosity and permeability,Chapters 1, 2, 6–8, 10, 11–18). Petroleum companies have furthermore collected detailed information on these systems for decades. The continuing demand for petroleum and scope and intensity of climate risk will incentivize the development of deepwater CCS sites as part of mitigation strategies. Future exploration of offshore deepwater settings (Chapters 19–21) that include contourites and mixed systems can theoretically address both exploration and mitigation strategy needs (Chapter 8). Safe storage of $CO_2$ depends on several additional key factors. The $CO_2$ must be injected below a depth of ≥800 m (2624 ft.) (Fig. 42) and as a supercritical fluid (800–900 kg/m3) to make storage economically viable. A strong effective seal, ideally with a 4-way dip closure, must cap the reservoir. Pereira et al. (2021) and CO2CRC (2008) discuss additional challenges and constraints.

## (i) Mineral resources

Mineral resources from deepwater system are currently under quick expansion (Maestro-González, 2021; https://www.isa.org.jm/es, see Chapter 21). Erosion, transport and deposition by bottom currents may influence the accumulation of

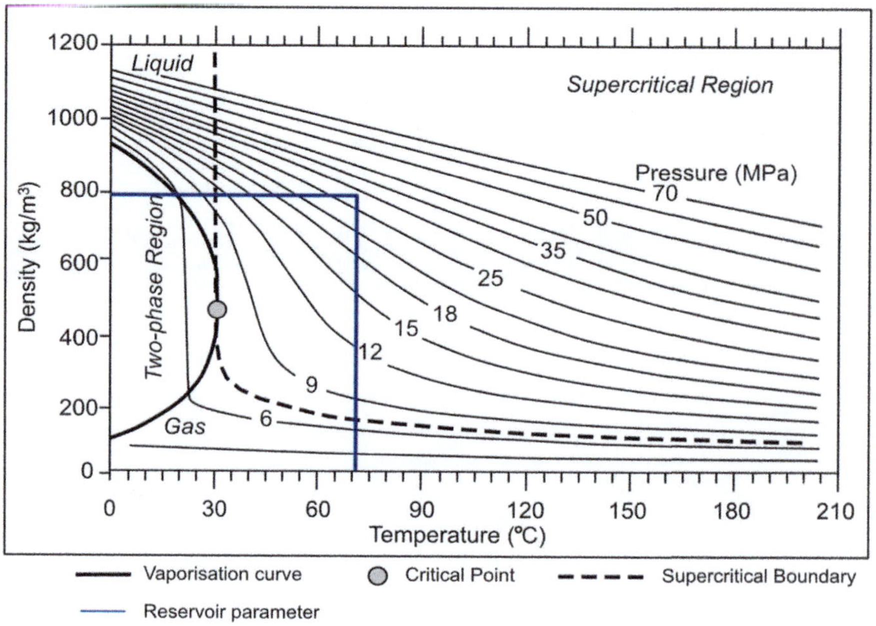


**FIG. 42** Geophysical conditions required for supercritical $CO_2$ storage. These include critical temperatures and pressures of 31.1 °C and 7.39 MPa, which generally occur at subsurface depths of 800 m (~ 2625 ft.). *(From Onyebuchi, V.E., Kolios, A., Hanak, D.P., Biliyok, C., Manovic, V., 2018. A systematic review of key challenges of CO2 transport via pipelines. Renew. Sust. Energ. Rev. 81, 2563–2583. https://doi.org/10.1016/j.rser.2017.06.064, with permission from Elsevier).*

economic mineral deposits, for example, the ferromanganese (Fe-Mn) crusts and nodules distribution (e.g., Charles et al., 2020; Zhong et al., 2021) or the distribution of deepwater mud highly enriched in rare-earth elements (Ohta et al., 2021). A mechanistic understanding of how bottom currents and other secondary oceanographic processes including internal waves and tides, benthic storms, eddies and vortices generate and otherwise influence depositional and erosional features in deepwater environments can resolve questions of where and why Fe-Mn nodules and crusts accumulate (Aleynik et al., 2017; Juan et al., 2018; Spearman et al., 2020). This understanding is also critical for deciphering paleoceanographic reconstructions (e.g., González et al., 2010).

## Key messages

From the first works on contourite deposits (Heezen and Hollister, 1964, 1971; Heezen et al., 1966) to the present, there has been a great advance in the knowledge of erosional and depositional features related to water masses and bottom currents, which has been conditioned on the one hand by an increase in the exploration of deepwater environments, as well as a greater ability to observe these environments with higher resolution. Despite this, contourite and mixed (turbidite-contourite) depositional systems are still considered a minority or exotic despite the ubiquitous occurrence and their common large sizes.

Contourite and mixed (turbidite-contourite) depositional systems are frequent in the modern and ancient deepwater environments being typically interrelated between gravitational or hemipelagic and pelagic deposits. The onset and evolution of these systems in the geologic record have a close, and intriguing, link between tectonic, gravitational processes and bottom currents. Nowadays, exist enough basic distinguishing criteria for identifying these deepwater systems on large (seismic) or small (sedimentary facies and facies association) scales. These criteria, and how they are to be applied, have been summarized in this chapter. Contourite and mixed (turbidite-contourite) systems carry implications for many different scientific conceptual, environmental and economic frameworks (Chapter 21) and they should be seriously considered in future basin analysis (Chapter 18), source-to-sink analysis (Chapter 11) and on the deepwater sequence stratigraphic framework (Chapter 12).

Research into contourite and mixed (turbidite-contourite) depositional systems is impacting decisions by policymakers in determining the boundaries of sea frontiers (United Nations) and informs policymakers and industry (in relation to exploration and telecommunication) leading to better exploration of economic resources and management of the deepwater environment (see Chapter 21). As hydrocarbon exploration moves into deeper water, research into these systems becomes more relevant because such deposits are more common in these deep settings and they could include sandy deposits with interest as new reservoirs. In addition, this research will have great research and exploratory potential for the next decades in supporting the industrial development of new models for hydrocarbon exploration and $CO_2$ capture and storage, which in turn has led to the discovery of energy resources during the energy transition, improved environmental conditions and a reduction in the effects of climate change.

## Conclusions and final considerations

Contourite and mixed depositional systems are critical features whose influence extends into many different spheres in deep oceans. Critical uncertainties remain as to how to identify and interpret these systems in deepwater environments for different scientific, economic and social applications. Marine basins commonly host large primary features deposited and modified by long-term circulation of water masses. A wide range of smaller secondary features demonstrate complex interactions of secondary processes that also influence these systems. Research thus far has found that processes interact and evolve in ways that are much more complex than previously assumed. New survey and monitoring tools, experiments and numerical models can help advance understanding. The bottom boundary layer just above the seafloor, for example, continues to elude systematic understanding. Multidisciplinary research and collaboration between marine geologists, benthic biologists and oceanographers can build on and advance deepwater sedimentological paradigms.

No evidence has emerged that contourite and mixed depositional systems were more or less common in the past, compared with their present-day distribution, but consensus has not been reached regarding which criteria are necessary and sufficient for their systematic identification in the ancient record. A few examples of contourite and mixed depositional systems appear in Paleozoic, Mesozoic and Cenozoic sections affirming assumptions that they adhere to basic uniformitarian principles.

The common occurrence of these systems in has significantly shifted the understanding of deepwater processes and sedimentary environments. Paradigms now clearly acknowledge the role of water masses and bottom currents in shaping the seafloor, modifying continental margin morphology and conditioning sedimentary stacking patterns through time. Questions regarding how bottom currents have shaped the classic deepwater stratigraphic record and past ocean basins could and should be used to guide a vast body of future research.

Bottom currents have been interpreted as permanent or semipermanent deepwater processes but recent research has documented systematic shifts in water mass depth, intensity and spatial position. Bottom currents exhibit dynamic and intermittent behavior at different scales. Periods of enhanced velocity may or may not synchronize with assumed external drivers such as gravitational sedimentary processes. Uncertainties also persist regarding secondary processes and their sedimentary records. All of these areas represent potential targets and objectives for future research (Chapters 21).

Factors known to influence contourite and mixed depositional systems interact at different scales. Tectonic activity, for example, influences these systems through downslope sediments transport and the onset and long-term evolution of water masses. In the longer term, remnant marginal platforms, structural highs and the opening of gateways all represent tectonically dependent influences that appear in contourite deposits. On shorter time scales, (glacio-)eustatic changes driven by orbital cycles and oceanographic processes can drive the deepening or shoaling of water masses and affect bottom current velocities. These in turn determine the vertical and lateral distribution of sandier and muddier deposits, their associated acoustic facies and the sedimentary stacking patterns and thicknesses of contourite and mixed depositional systems.

Contourite and mixed depositional system research is best conducted as an integrated effort that combines scientific, environmental and economic objectives. This approach optimizes the sharing of basic understanding, technological advances, funding and public good. Deepwater depositional environments are complex systems with global implications.

## Acknowledgments

This work was funded through the Joint Industry Project (JIP) supported by BP (United Kingdom), ENI (Italy), TOTAL (France), ExxonMobil (United States), Wintershall Dea (Germany) and TGS (United Kingdom) within the framework of "*The Drifters*" *Research Group* at Royal Holloway, University of London (RHUL). D.D. thanks the FCT (Fundação para a Ciência e a Tecnologia)—the Portuguese Science Foundation through the PhD grant SFRH/BD/115962/2016. We thank the editorial team, and in particular *J. R. Rotzien* (Basin Dynamics), for their revisions and suggestions that helped us to improve this chapter.

THE DRIFTERS RESEARCH GROUP
BOTTOM CURRENTS & DEEP-WATER SEDIMENTATION

## References

ABLOS, 2006. A Manual on Technical Aspects of the United Nations Convention on the Law of the Sea—1982. Prepared by the IHO, IAG, IOC Advisory Board on Law of the Sea (ABLOS), Special Publication, No. 51. International Hydrographic Bureau MONACO (March), fourth ed.

Aleynik, D., Inall, M.E., Dale, A., Vink, A., 2017. Impact of remotely generated eddies on plume dispersion at abyssal mining sites in the Pacific. Sci. Rep. 7 (1), 1–14. https://doi.org/10.1038/s41598-017-16912-2.

Alonso, B., Ercilla, G., Casas, D., Stow, D.A.V., Rodríguez-Tovar, F.J., Dorador, J., Hernández-Molina, F.J., 2016. Contourite vs gravity-flow deposits of the Pleistocene Faro Drift (Gulf of Cadiz): sedimentological and mineralogical approaches. Mar. Geol. 377, 77–94. https://doi.org/10.1016/j.margeo.2015.12.016.

Alonso, B., Juan, C., Ercilla, G., Cacho, I., López-González, N., Rodríquez-Tovar, F.J., Dorador, J., Francés, G., Casas, D., Vandorpe, T., Váquez, J.T., 2021. Paleoceanographic and paleoclimatic variability in the Western Mediterranean during the last 25 cal. kyr BP. New insights from contourite drifts. Mar. Geol. 437. https://doi.org/10.1016/j.margeo.2021.106488.

Alves, T.M., Kurtev, K., Moore, G.F., Strasser, M., 2014. Assessing the internal character, reservoir potential, and seal competence of mass-transport deposits using seismic texture: a geophysical and petrophysical approach Mass-Transport Deposits and Seismic Texture. AAPG Bull. 98, 793–824.

Amblas, D., Canals, M., 2016. Contourite drifts and canyon-channel systems on the Northern Antarctic Peninsula Pacific margin. In: Dowdeswell, J.A., Canals, M., Jakobsson, M., Todd, B.J., Dowdeswell, E.K., Hogan, K.A. (Eds.), Atlas of Submarine Glacial Landforms: Modern, Quaternary and Ancient. Geol. Soc. London Mem., vol. 46, pp. 393–394, https://doi.org/10.1144/M46.17.

Amblas, D., Urgeles, R., Canals, M., Calafat, A.M., Rebesco, M., Camerlenghi, A., Estrada, F., De-Batist, M., Hughes-Clarke, J.E., 2006. Relationship between continental rise development and palaeo-ice sheet dynamics, northern Antarctic Peninsula Pacific margin. Quat. Sci. Rev. 25, 933–944. https://doi.org/10.1016/j.quascirev.2005.07.012.

Amy, L.A., 2019. A review of producing fields inferred to have upslope stratigraphically trapped turbidite reservoirs: trapping styles (pure and combined), pinch-out formation, and depositional setting. Am. Assoc. Pet. Geol. Bull. 103 (12), 2861–2889.

Angue Minto'o, C.A., Bassetti, M.A., Morigi, C., Ducassou, E., Toucanne, S., Jouet, G., Mulder, T., 2015. Levantine intermediate water hydrodynamic and bottom water ventilation in the northern Tyrrhenian Sea over the past 56,000 years: new insights from benthic foraminifera and ostracods. Quat. Int. 357, 295–313.

Apel, J.R., 2000. Solitons near Gibraltar: views from the European remote sensing satellites. Report GOA 2000-1Global Ocean Association, Silver Spring, MD.

Apel, J.R., 2004. Oceanic internal waves and solitons. In: Jackson, C.R., Apel, J.R. (Eds.), Synthetic Aperture Radar Marine User's Manual. US Department of Commerce, National Oceanic and Atmospheric Administration, Silver Spring, pp. 189–206.

Artoni, A., Rizzini, F., Roveri, M., Gennari, R., Manzi, V., Papani, G., Bernini, M., 2007. Tectonic and climatic controls on sedimentation in late miocene cortemaggiore wedge-top basin (Northwestern Apennines, Italy). In: Lacombe, O., Lavé, J., Roure, F., Vergés, J. (Eds.), Thrust Belts and Foreland Basins: From Fold Kinematics to Hydrocarbon Systems. Springer, Berlin, pp. 431–456.

Badalini, G., Thompson, P., Wrigley, S., Walker, R., Argent, J., Hernández-Molina, F.J., Soto, M., de Santa Ana, H., 2016. Giant cretaceous mixed contouritic-turbiditic systems, offshore uruguay: the interaction between rift-related basin morphology, contour currents and downslope sedimentation. Search and Discovery Article #30482.

Bailey, W.S., McArthur, A.D., McCaffrey, W.D., 2021a. Distribution of contourite drifts on convergent margins: examples from the Hikurangi subduction margin of New Zealand. Sedimentology 68, 294–323. https://doi.org/10.1111/sed.12779.

Bailey, W., McArthur, A., McCaffrey, W., 2021b. Sealing potential of contourite drifts in deep-water fold and thrust belts: examples from the Hikurangi Margin, New Zealand. Mar. Pet. Geol. 123, 104776. https://doi.org/10.1016/j.marpetgeo.2020.104776.

Balaky, S.M., Tamar-Agha, M.Y., 2017. Sedimentology and stratigraphy of Geli Khana Formation (Anisian–Ladinian), a contourite depositional system in the northeastern passive margin of Arabian plate, northern Iraq-Kurdistan region. Arab. J. Geosci. 10 (5), 118.

Batchelor, C.L., Bellwald, B., Planke, S., Ottesen, D., Henriksen, S., Myklebust, R., Johansen, S.E., Dowdeswell, J.A., 2021. Glacial, fluvial and contour-current-derived sedimentation along the northern North Sea margin through the quaternary. Earth Planet. Sci. 566, 116966. https://doi.org/10.1016/j.epsl.2021.116966.

Bender, V.B., Hanebuth, T.J.J., Mena, A., Baumann, K.-H., Francés, G., von Dobeneck, T., 2012. Control of sediment supply, palaeoceanography and morphology on late Quaternary sediment dynamics at the Galician continental slope. Geo-Mar. Lett. 32, 313–335.

Bouma, A.H., Treadwell, T.K., 1975. Deep-sea dune-like features. Mar. Geol. 19, M53–M59. https://doi.org/10.1016/0025-3227(75)90078-X.

Boyle, E.A., Keigwin, L., 1987. North Atlantic thermohaline circulation during the past 20,000 years linked to high-latitude surface temperature. Nature 330, 35–40.

Brackenridge, R., Stow, D.A.V., Hernández-Molina, F.J., 2011. Contourites within a deepwater sequence stratigraphic framework. Geo-Mar. Lett. 31, 343–360.

Brackenridge, R.A., Hernández-Molina, F.J., Stow, D.A.V., Llave, R., 2013. A Pliocene mixed contourite-turbidite system offshore the Algarve Margin, Gulf of Cadiz: seismic response, margin evolution and reservoir implications. Mar. Pet. Geol. 46, 36–50. https://doi.org/10.1016/j.marpetgeo.2013.05.015.

Brackenridge, R.E., Stow, D.A.V., Hernández-Molina, F.J., Jones, C., Mena, A., Alejo, I., Ducassou, E., Llave, E., Ercilla, G., Nombela, M.A., Pérez-Arlucea, M., Frances, G., 2018. Textural characteristics and facies of sand-rich contourite depositional systems. Sedimentology 65, 2223–2252.

Brackenridge, R.E., Nicholson, U., Sapiie, B., Stow, D.A.V., Tappin, D.R., 2020. Indonesian Throughflow as a Preconditioning Mechanism for Submarine Landslides in the Makassar Strait. Geological Society, London, Special Publications 500, pp. 195–217. 1 April 2020 https://doi.org/10.1144/SP500-2019-171.

Britannica, The Editors of Encyclopedia, 28 October 2015. Water Mass. Encyclopedia Britannica. https://www.britannica.com/science/water-mass. (Accessed 27 September 2021).

Bryn, P., Berg, K., Stoker, M.S., Haflidason, H., Solheim, A., 2005. Contourites and their relevance for mass wasting along the Mid-Norwegian Margin. Mar. Pet. Geol. 22, 85–96. https://doi.org/10.1016/j.marpetgeo.2004.10.012.

Butman, B., Alexander, P.S., Scotti, A., Beardsley, R.C., Anderson, S.P., 2006a. Large internal waves in Massachusetts Bay transport sediments offshore. Cont. Shelf Res. 26, 2029–2049.

Butman, B., Twichell, D.C., Rona, P.A., Tucholke, B.E., Middleton, T.J., Robb, J.M., 2006b. Sea floor topography and backscatter intensity of the hudson canyon region offshore of New York and New Jersey. U.S. Geological Survey Open-File Report 2004-1441, CD-ROM. Also available online at http://pubs.usgs.gov/of/2004/1441/.

Cacho, I., Grimalt, J.O., Sierro, F.J., Shackleton, N., Canals, M., 2000. Evidence for enhanced Mediterranean thermohaline circulation during rapid climatic coolings. Earth Planet. Sci. Lett. 183, 417–429.

Cacho, I., Shackleton, N., Elderfield, H., Sierro, F.J., Grimalt, J., 2006. Glacial rapid variability in deep-water temperature and $\delta^{18}O$ from the Western Mediterranean Sea. Quat. Sci. Rev. 25 (23–24), 3294–3311.

Camerlenghi, A., Crise, A., Pudsey, C.J., Accerboni, E., Laterza, R., Rebesco, M., 1997. Ten-month observation of the bottom current regime across a sediment drift of the Pacific margin of the Antarctic Peninsula. Antarct. Sci. 9, 426–433. https://doi.org/10.1017/S0954102097000552.

Campbell, D.C., Deptuck, M.E., 2012. Alternating bottom-current-dominated and gravity-flow-dominated deposition in a lower slope and rise setting—Insights from the seismic geomorphology of the Western Scotian Margin, Eastern Canada. Application of the Principles of Seismic Geomorphology to Continental-Slope and Base-of-Slope Systems: Case Studies from Seafloor and Near-Seafloor Analogues. Soc. Sed. Geol. Tulsa, Spec. Publ. 99, pp. 329–346.

Campbell, C., Mosher, D.C., 2016. Geophysical evidence for widespread Cenozoic bottom current activity from the continental margin of Nova Scotia, Canada. Mar. Geol. 378, 237–260. https://doi.org/10.1016/j.margeo.2015.10.005.

Capella, W., Hernández-Molina, F.J., Flecker, R., Hilgen, F.J., Hssain, M., Kouwenhoven, T.J., van Oorschot, M., Sierro, F.J., Stow, D.A.V., Trabucho-Alexandre, J., Tulbure, M.A., Yousfi, M.Z., Krijgsman, W., 2017. Sandy contourite drift in the late Miocene Rifian Corridor (Morocco): reconstruction of depositional environments in a foreland-basin seaway. Dev. Sedimentol. 355, 31–57.

Cardona, S., Wood, L.J., Day-Stirrat, R.J., Moscardelli, L., 2016. Fabric development and pore-throat reduction in a mass-transport deposit in the Jubilee Gas Field, Eastern Gulf of Mexico: consequences for the sealing capacity of MTDs. In: Submarine Mass Movements and Their Consequences. Springer, Cham, pp. 27–37.

Carter, L., McCave, I.N., 2002. Eastern New Zealand drifts, miocene—recent. In: Stow, D.A.V., Howe, J.A., Faugères, J.-C., Viana, A.R. (Eds.), Deep-Water Contourite Systems: Modern Drifts and Ancient Series, Seismic and Sedimentary Characteristics. Geol. Soc. London Mem., vol. 22, pp. 385–407, https://doi.org/10.1144/GSL.MEM.2002.022.01.27.

Casson, M., Calvès, G., Huuse, M., Sayers, B., Redfern, J., 2020. Cretaceous continental margin evolution revealed using quantitative seismic geomorphology, offshore northwest Africa. Basin Res. 33 (1), 66–90. https://doi.org/10.1111/bre.12455.

Cauxeiro, C., Lopez, M., Hernández-Molina, F.J., Miguel, A., Cauxeiro, G., Caetano, V., 2020. Contourite vs turbidite outcrop and seismic architectures. Angolan Min. Oil Gas J. 1, 20–26. https://doi.org/10.47444/amogj.v1i1.9.

Charles, C., Pelleter, E., Révillon, S., Nonnotte, P., Jorry, S.J., Kluska, J.-M., 2020. Intermediate and deep ocean current circulation in the Mozambique Channel: new insights from ferromanganese crust Nd isotopes. Mar. Geol. 430, 106356. https://doi.org/10.1016/j.margeo.2020.106356.

Chen, H., Xie, X., Van Rooij, D., Vandorpe, T., Huang, L., Guo, L., Su, M., 2013. Depositional characteristics and spatial distribution of deep-water sedimentary systems on the northwestern middle-lower slope of the Northwest Sub-Basin, South China Sea. Mar. Geophys. Res. 34, 229–257. https://doi.org/10.1007/s11001-013-9191-7.

Chen, H., Xie, X., Zhang, W., Shu, Y., Wang, D., Vandorpe, T., Van Rooij, D., 2016. Deep-water sedimentary systems and their relationship with bottom currents at the intersection of Xisha Trough and Northwest Sub-Basin, South China Sea. Mar. Geol. 378, 101–113. https://doi.org/10.1016/j.margeo.2015.11.002.

Chen, Y., Yao, G., Wang, X., Lv, F., Shao, D., Lu, Y., Cao, Q., Tang, P., 2020. Flow processes of the interaction between turbidity flows and bottom currents in sinuous unidirectionally migrating channels: an example from the Oligocene channels in the Rovuma Basin, offshore Mozambique. Sed. Geol. 404, 105680. https://doi.org/10.1016/j.sedgeo.2020.105680.

Chiarella, D., Hernández-Molina, F.J., 2021. Go with the flow: the role of gateway and straits on plastic distribution. Geol. Today 37, 66–69. https://doi.org/10.1111/gto.12345.

Clark, I.R., Cartwright, J.A., 2009. Interactions between submarine channel systems and deformation in deepwater fold belts: examples from the Levant Basin, Eastern Mediterranean Sea. Mar. Pet. Geol. 26, 1465–1482. https://doi.org/10.1016/j.marpetgeo.2009.05.004.

Cloetingh, S., Gradstein, F.M., Kooi, H., Grant, A.C., Kaminski, M., 1990. Plate reorganization: a cause of rapid late Neogene subsidence and sedimentation around the north Atlantic? J. Geol. Soc. Lond. 147, 495–506.

CO2CRC, 2008. Storage capacity estimation, site selection and characterisation for CO2 storage projects. CO2CRC Report RPT08-1001, Cooperative Research Centre for Greenhouse Gas Technologies, Canberra.

Cochran, J.K., Bokuniewicz, H.J., Yager, P.L. (Eds.), 2019. Encyclopedia of Ocean Sciences, third ed. Academic Press. 4306 pp.

Counts, J.W., Amy, L., Georgiopoulou, A., Haughton, P., 2021. A review of sand detachment in modern deep marine environments: analogues for upslope stratigraphic traps. Mar. Pet. Geol. 132, 105184. https://doi.org/10.1016/j.marpetgeo.2021.105184.

Cowan, E.A., Hillenbrand, C.-D., Hassler, L.E., Ake, M.T., 2008. Coarse-grained terrigenous sediment deposition on continental rise drifts: a record of Plio-Pleistocene glaciation on the Antarctic Peninsula. Palaeogeogr. Palaeoclimatol. Palaeoecol. 265, 275–291. https://doi.org/10.1016/j.palaeo.2008.03.010.

Creaser, A., Hernández-Molina, F., Badalini, G., Thompson, P., Walker, R., Soto, M., Conti, B., 2017. A Late Cretaceous mixed (turbidite-contourite) system along the Uruguayan margin: sedimentary and paleoceanographic implications. Mar. Geol. 390, 234–253. https://doi.org/10.1016/j.margeo.2017.07.004.

Culp, J., Parent, A.M., Abolfazli, E., Strom, K., Romans, B.W., 2021. Advective sorting of silt by currents: a laboratory study. Sedimentology. https://doi.org/10.1111/sed.12889.

Davies, T.A., Laughton, A.S., 1972. Sedimentary processes in the North Atlantic. In: Laughton, A.S., Berggren, W.A. (Eds.), Initial Reports of Deep Sea Drilling Project, 12. U.S. Government Printing Office, Washington, DC, pp. 905–934.

Davies, D., Hoiles, P., Rodriguez, K., Hodgson, N., 2020. GEO ExPro-offshore Argentina: tertiary play potential in the Malvinas Basin. GeoExpro 17 (4). https://www.geoexpro.com/articles/2020/10/offshore-argentina-tertiary-play-potential-in-the-malvinas-basin.

de Castro, S., Hernández-Molina, F.J., Rodriguez-Tovar, F.J., Llave, E., Ng, Z.L., Nishida, N., Mena, A., 2020. Contourites and bottom current reworked sands: bed facies model and implications. Mar. Geol. 428, 106267. https://doi.org/10.1016/j.margeo.2020.106267.

de Castro, S., Hernández-Molina, F.J., de Weger, W., Jimenez-Espejo, F.J., Rodríguez-Tovar, F.J., Mena, A., Llave, E., Sierro, F.J., 2021a. Contourite characterization and its discrimination from other deep-water deposits in the Gulf of Cadiz contourite depositional system. Sedimentology 68 (3), 987–1027. https://doi.org/10.1111/sed.12813.

de Castro, S., Miramontes, E., Dorador, J., Jouet, G., Cattaneo, A., Rodriguez-Tovar, F.J., Hernandez-Molina, F.J., 2021b. Siliciclastic and bioclastic contouritic sands: textural and geochemical characterisation. Mar. Pet. Geol. 128, 105002. https://doi.org/10.1016/j.marpetgeo.2021.105002.

De Mol, B., Henriet, J.-P., Canals, M., 2005. Development of coral banks in Porcupine Seabight: do they have Mediterranean ancestors? In: Freiwald, A., Murray Roberts, J. (Eds.), Cold-Water Corals and Ecosystems. Springer, Berlin, Heidelberg, pp. 515–533.

De Mol, L., Van Rooij, D., Pirlet, H., Greinert, J., Frank, N., Quemmerais, F., Henriet, J.P., 2011. Cold-water coral habitats in the Penmarc'h and Guilvinec Canyons (Bay of Biscay): deep-water versus shallow-water settings. Mar. Geol. 282 (1–2), 40–52.

de Weger, W., Hernández-Molina, F.J., Flecker, R., Sierro, F.J., Chiarella, D., Krijgsman, W., Manar, M.A., 2020. Late Miocene contourite channel system reveals intermittent overflow behavior. Geology 48 (12), 1194–1199.

de Weger, W., Hernández-Molina, F.J., Miguez-Salas, O., de Castro, S., Bruno, M., Chiarella, D., Sierro, F.J., Blackbourn, G., Manar, M.A., 2021. Contourite depositional system after the exit of a strait: case study from the late Miocene South Rifian Corridor, Morocco. Sedimentology. https://doi.org/10.1111/sed.12882.

Decal, C., Stiring, E.J., Hossack, J., 2016. Contourite deposition over time along the offshore Tanzania Margin: controls, architectures, and effects on hydrocrabon prospectivity. In: East Africa: From Research to Reserves. 13-15 April. The Geological Society, UK. Abstract.

Deptuck, M.E., Kendell, K.L., 2020. Atlas of 3D seismic surfaces and thickness maps, central and southwestern Scotian Slope. CNSOPB Geosci. Open File Rep. 2020-002MF – 2020-006MF. 51 pp, https://doi.org/10.13140/RG.2.2.36421.37604.

Ding, W., Li, J., Fang, Y., Tang, Y., 2021. Morphotectonics and evolutionary controls on the Pearl River Canyon system, South China Sea. Mar. Geophys. Res. 34, 221–238. https://doi.org/10.1007/s11001-013-9173-9.

Dorador, J., Rodríguez-Tovar, F.J., 2016. Stratigraphic variation in ichnofabrics at the "Shackleton Site" (IODP Site U1385) on the Iberian Margin: paleoenvironmental implications. Mar. Geol. 377, 118–126.

Dorador, J., Rodríguez-Tovar, F.J., Mena, A., Francés, G., 2019. Lateral variability of ichnological content in muddy contourites: weak bottom currents affecting organisms' behavior. Sci. Rep. 9 (1), 1–7.

Dorador, J., Rodríguez-Tovar, F.J., Mena, A., Francés, G., 2021a. Deep-sea bottom currents influencing tracemaker community: an ichnological study from the NW Iberian margin. Mar. Geol. 437, 106503. https://doi.org/10.1016/j.margeo.2021.106503.

Dorador, J., Rodríguez-Tovar, F.J., Miguez-Salas, O., 2021b. The complex case of Macaronichnus trace fossil affecting rock porosity. Sci. Rep. 11, 1975.

Droghei, R., Falcini, F., Casalbore, D., Martorelli, E., Mosetti, R., Sannino, G., Santo-leri, R., Chiocci, F.L., 2016. The role of internal solitary waves on deep-water sedimentary processes: the case of upslope migrating sediment waves off the Messina Strait. Sci. Rep. 6, 36376.

Duan, T., Gao, Z., Zeng, Y., Stow, D.A.V., 1993. A fossil carbonate contourite drift on the Lower Ordovician palaeocontinental margin of the middle Yangtze Terrane, Jiuxi, northern Hunan, southern China. Sediment. Geol. 82, 271–284.

Duarte, D., et al., 2022. Structural Control and Tectono-Sedimentary Evolution of the Gulf of Cadiz Since the Late Miocene. Implications for the Contourite Depositional System, SW Iberia (in prep.).

Duplessy, J.C., Shackleton, N.J., Fairbanks, R.G., Labeyrie, L., Oppo, D., Kallel, N., 1988. Deepwater 895 source variations during the last climatic cycle and their impact on the global deepwater circulation. 896. Paleoceanography 3, 343–360. https://doi.org/10.1029/PA003i003p00343.

Eberli, G.P., Betzler, C., 2019. Characteristics of modern carbonate contourite drifts. Sedimentology 66, 1163–1191. https://doi.org/10.1111/sed.12584.

Elliot, G.M., Parson, L.M., 2008. Influence of sediment drift accumulation on the passage of gravity-driven sediment flows in the Iceland Basin, NE Atlantic. Mar. Petrol. Geol. 25 (3), 219–233. https://doi.org/10.1016/j.marpetgeo.2007.07.002.

Emery, W.J., Meincke, J., 1986. Global water masses-summary and review. Oceanol. Acta 9 (4), 383–391.

Enjorlas, J.M., Gouadain, J., Mutti, E., Pizon, J., 1986. New turbiditic model for the Lower Tertiary sands in the South Viking Graben. In: Spencer, A.M. (Ed.), Habitat of Hydrocarbons on the Norwegian Continental Shelf. Norwegian Petroleum Society, Oslo, pp. 171–178.

Ercilla, G., Casas, D., Vázquez, J.T., Iglesias, J., Somoza, L., Juan, C., Medialdea, T., León, R., Estrada, F., García-Gil, S., Farran, M., Bohoyo, F., García, M., Maestro, A., ERGAP Project and Cruise Teams, 2011. Imaging the recent sediment dynamics of the Galicia Bank region (Atlantic, NW Iberian Peninsula). Mar. Geophys. Res. 32 (1–2), 99–126. https://doi.org/10.1007/s11001-011-9129-x.

Ercilla, G., Juan, C., Hernández-Molina, F.J., Bruno, M., Estrada, F., Alonso, B., Casas, D., Farran, M., Llave, E., García, M., Vázquez, J.T., d'Acremont, E., Gorini, C., Palomino, D., Valencia, J., El Moumni, B., Ammar, B., 2016. Significance of bottom currents in deep-sea morphodynamics: an example from the Alboran Sea. Mar. Geol. 378, 157–170.

Ercilla, G., Juan, C., Periánez, R., Alonso, B., Abril, J.M., Estrada, F., Casas, D., Vázquez, J.T., d'Acremont, E., Gorini, C., Moumni, B.E., Do Couto, D., Valencia, J., 2019. Influence of alongslope processes on modern turbidite systems and canyons in the Alboran Sea (southwestern Mediterranean). Deep-Sea Res. I: Oceanogr. Res. Pap. 144, 1–16. https://doi.org/10.1016/j.dsr.2018.12.002.

Escutia, C., Nelson, C.H., Acton, G.D., Eittreim, S.L., Cooper, A.K., Warnke, D.A., Jaramillo, J.M., 2002. Current controlled deposition on the Wilkes Land continental rise, Antarctica. Geol. Soc. London Mem. 22, 373–384. https://doi.org/10.1144/GSL.MEM.2002.022.01.26.

Escutia, C., Bárcena, M.A., Lucchi, R.G., Romero, O., Ballegeer, M., Gonzalez, J.J., Harwood, D., 2009. Circum-Antarctic warming events between 4 and 3.5 Ma recorded in sediments from the Prydz Bay (ODP Leg 188) and the Antarctic Peninsula (ODP Leg 178) margins. Glob. Planet. Chang. 69, 170–184. https://doi.org/10.1016/j.gloplacha.2009.09.003.

Esmerode, E.V., Lykke-Andersen, H., Surlyk, F., 2008. Interaction between bottom currents and slope failure in the Late Cretaceous of the southern Danish Central Graben, North Sea. J. Geol. Soc. London 165, 55–72. https://doi.org/10.1144/0016-76492006-138.

European Commission, 2019. The European Green Deal. COM 2019 640 final, 11 December https://ec.europa.eu/info/sites/info/files/europeangreen-deal-communication_en.pdf.

Faugères, J.-C., Stow, D.A.V., 1993. Bottom-current-controlled sedimentation: a synthesis of the contourite problem. Sediment. Geol. 82, 287–297.

Faugères, J.-C., Stow, D.A.V., 2008. Contourite drifts: nature, evolution and controls. In: Rebesco, M., Camerlenghi, A. (Eds.), Contourites. Developments in Sedimentology, vol. 60. Elsevier, Amsterdam, pp. 257–288.

Faugères, J.-C., Stow, D.A.V., Imbert, P., Viana, A.R., 1999. Seismic features diagnostic of contourite drifts. Mar. Geol. 162, 1–38. https://doi.org/10.1016/S0025-3227(99)00068-7.

Fonnesu, F., 2013. The mamba complex supergiant gas discovery (Mozambique): an example of turbidite fans modified by deepwater tractive bottom currents. The 12th PESGB/HGS Conference on African E&P, London.

Fonnesu, M., Palermo, D., Galbiati, M., Marchesini, M., Bonamini, E., Bendias, D., 2020. A new world-class deep-water play-type, deposited by the syndepositional interaction of turbidity flows and bottom currents: the giant Eocene Coral Field in northern Mozambique. Mar. Pet. Geol. 111, 179–201. https://doi.org/10.1016/j.marpetgeo.2019.07.047.

Forwick, M., Laberg, J.S., Hass, H.C., Osti, G., 2015. The Kongsfjorden Channel System offshore NW Svalbard: downslope sedimentary processes in a contour-current-dominated setting. Arktos 1 (17). https://doi.org/10.1007/s41063-015-0018-4.

Fuhrmann, A., Kane, I.A., Ferguson, R.A., Schomacker, E., Bonamini, E., Contreras, F.A., 2020. Hybrid turbidite-drift channel complexes: an integrated multiscale model. Geology 48. https://doi.org/10.1130/G47179.1.

Gao, Z.Z., Eriksson, K.A., He, Y.B., Luo, S.S., Guo, J.H., 1998. Deep-Water Traction Current Deposits. A Study of Internal Tides, Internal Waves, Contour Currents and Their Deposits. Science Press, Beijing.

García, M., Hernández-Molina, F.J., Llave, E., Stow, D.A.V., León, R., Fernández-Puga, M.C., Díaz del Río, V., Somoza, L., 2009. Contourite erosive features caused by the Mediterranean outflow water in the Gulf of Cadiz: quaternary tectonic and oceanographic implications. Mar. Geol. 257, 24–40.

García, M., Hernández-Molina, F.J., Alonso, B., Vázquez, J.T., Ercilla, G., Llave, E., Casas, D., 2016. Erosive sub-circular depressions on the Guadalquivir Bank (Gulf of Cadiz): interaction between bottom current, mass-wasting and tectonic processes. Mar. Geol. 378, 5–19.

García, M., Llave, E., Hernández-Molina, F.J., Ercilla, G., Alonso, A., Lobo, F.J., Casas, D., Fernández-Salas, L.M., Mena, A., 2020. Interplay between tectonics, gravitational and bottom current processes in the upper and middle continental slope of the Gulf of Cadiz. Mar. Pet. Geol. 121, 104595.

Gardner, W.D., Tucholke, B.E., Richardson, M.J., Biscaye, P.E., 2017. Benthic storms, nepheloid layers, and linkage with upper ocean dynamics in the western North Atlantic. Mar. Geol. 385, 304–327. https://doi.org/10.1016/j.margeo.2016.12.012.

Gauchery, T., Rovere, M., Pellegrini, C., Cattaneo, A., Campiani, E., Trincardi, F., 2021. Factors controlling margin instability during the Plio-Quaternary in the Gela Basin (Strait of Sicily, Mediterranean Sea). Mar. Pet. Geol. 123, 104767. https://doi.org/10.1016/j.marpetgeo.2020.104767.

GCCSI, Global Carbon Capture and Storage Institute, 2020. The Global Status of CCS: 2020. Australia https://www.globalccsinstitute.com/wp-content/uploads/2020/12/Global-Status-of-CCS-Report-2020_FINAL_December11.pdf.

Georgiopoulou, A., Owens, M., Haughton, P.D.W., 2021. Channel and inter-channel morphology resulting from the long-term interplay of alongslope and downslope processes, NE Rockall Trough, NE Atlantic. Mar. Geol. 441, 106624. https://doi.org/10.1016/j.margeo.2021.106624.

Glazkova, T., Hernández-Molina, F.J., Dorokhova, E., Mena, A., Roque, C., Rodríguez-Tovar, F.J., Krechik, V., Llave, E., Kuleshova, L., 2021. Sedimentary processes in the Discovery Gap (Central-NE Atlantic): an example of a deep marine gateway. Deep-Sea Res. I: Oceanogr. Res. Pap. Submitted.

Gong, C., Wang, Y., Zhu, W., Li, W., Xu, Q., 2013. Upper miocene to quaternary unidirectionally migrating deep-water channels in the Pearl River Mouth Basin, northern South China Sea. AAPG Bull. 97 (2), 285–308. https://doi.org/10.1306/07121211159.

Gong, C., Wang, Y., Zheng, R., Hernández-Molina, F.J., Li, Y., Stow, D.A.V., Xu, Q., Brackenridge, R.E., 2016. Middle Miocene reworked turbidites in the Baiyun Sag of the Pearl River Mouth Basin, northern South China Sea margin: processes, genesis, and implications. J. Asian Earth Sci. 128, 116–129. https://doi.org/10.1016/j.jseaes.2016.06.025.

Gong, C., Peakall, J., Wang, Y., Wells, M.G., Xu, J., 2017. Flow processes and sedimentation in contourite channels on the northwestern South China Sea margin: a joint 3D seismic and oceanographic perspective. Mar. Geol. 393, 176–193.

Gong, C., Wang, Y., Wang, Y., Rebesco, M., Salon, S., Steel, R.J., 2018. How do turbidity flows interact with contour currents in unidirectionally migrating deep-water channels? Geology 46 (6), 551–554. https://doi.org/10.1130/G40204.1.

Gonthier, E., Faugères, J.-C., Stow, D.A.V., 1984. Contourite facies of the Faro Drift, Gulf of Cadiz. In: Stow, D.A.V., Piper, D.J.W. (Eds.), Fine Grained Sediments, Deepwater Processes and Facies. Geological Society, London, Special Publication 15, pp. 275–291.

González, F.J., Somoza, L., Lunar, R., Martínez-Frías, J., Martín Rubí, J.A., Torres, T., Ortiz, J.E., Díaz del Río, V., 2010. Internal features, mineralogy and geochemistry of Fe-Mn nodules from the Gulf of Cadiz: the role of the Mediterranean outflow water undercurrent. J. Mar. Syst. 80, 203–218.

Hamon, N., Sepulchre, P., Lefebvre, V., Ramstein, G., 2013. The role of eastern Tethys seaway closure in the Middle Miocene Climatic Transition (ca. 14 Ma). Clim. Past 9, 2687–2702. https://doi.org/10.5194/cp-9-2687-2013.

Hanebuth, T.J., Zhang, W., Hofmann, A.L., Löwemark, L.A., Schwenk, T., 2015. Oceanic density fronts steering bottom-current induced sedimentation deduced from a 50 ka contourite-drift record and numerical modeling (off NW Spain). Quat. Sci. Rev. 112, 207–225.

Hanquiez, V., Mulder, T., Toucanne, S., Lecroart, P., Bonnel, C., Marchès, E., Gonthier, E., 2010. The sandy channel-lobe depositional systems in the Gulf of Cadiz: gravity processes forced by contour current processes. Sed. Geol. 229, 110–123. https://doi.org/10.1016/j.sedgeo.2009.05.008.

Hay, W.W., 2009. Cretaceous oceans and ocean modeling. In: Hu, X., Wang, C., Scott, R.W., Wagreich, M., Jansa, L. (Eds.), Cretaceous Oceanic Red Beds: Stratigraphy, Composition, Origins, and Paleoceanographic and Paleoclimatic Significance. SEPM Special Publication No. 91, pp. 243–271.

Hebbeln, D., Van Rooij, D., Wienberg, C., 2016. Good neighbours shaped by vigorous currents: cold-water coral mounds and contourites in the North Atlantic. Mar. Geol. 378, 171–185.

Hebbeln, D., Bender, M., Gaide, S., Titschack, J., Vandorpe, T., Van Rooij, D., Wintersteller, Wienberg, C., 2019. Thousands of cold-water coral mounds along the Moroccan Atlantic continental margin: distribution and morphometry. Mar. Geol. 411, 51–61. https://doi.org/10.1016/j.margeo.2019.02.001.

Heezen, B.C., Hollister, C.D., 1964. Deep sea current evidence from abyssal sediments. Mar. Geol. 1, 141–174.

Heezen, B.C., Hollister, C.D., 1971. The Face of the Deep. Oxford University Press, New York. 650 pp.

Heezen, B.C., Hollister, C.D., Ruddiman, W.F., 1966. Shapings of the continental rise by deep geostrophic contour currents. Science 152, 502–508.

Heritier, F.E., Lossel, P., Wathne, E., 1979. Frigg Field—large submarine fan trap in Lower Eocene rocks of North Sea Viking Graben. AAPG Bull. 63 (11), 1999–2020. https://doi.org/10.1306/2F918856-16CE-11D7-8645000102C1865D.

Hernández-Molina, F.J., Somoza, L., Vázquez, J.T., Lobo, F., Fernández-Puga, M.C., Llave, E., Díaz-del Río, V., 2002. Quaternary stratigraphic stacking patterns on the continental shelves of the southern Iberian Peninsula: their relationship with global climate and paleooceanographic changes. Quat. Int. 92, 5–23.

Hernández-Molina, F.J., Llave, E., Somoza, L., Fernández-Puga, M.C., Maestro, A., León, R., Barnolas, A., Medialdea, T., García, M., Vázquez, J.T., Díaz del Río, V., Fernández-Salas, L.M., Lobo, F., Alveirinho Dias, J.M., Rodero, J., Gardner, J., 2003. Looking for clues to paleoceanographic imprints: a diagnosis of the gulf of Cadiz contourite depositional systems. Geology 31, 19–22.

Hernández-Molina, F.J., Llave, E., Stow, D.A.V., García, M., Somoza, L., Vázquez, J.T., Lobo, F., Maestro, A., Díaz del Río, V., León, R., Medialdea, T., Gardner, J., 2006. The contourite depositional system of the Gulf of Cadiz: a sedimentary model related to the bottom current activity of the Mediterranean Outflow Water and the continental characteristics. Deep-Sea Res. II 53, 1420–1463.

Hernández-Molina, F.J., Maldonado, A., Stow, D.A.V., 2008a. Abyssal plain contourites. In: Rebesco, M., Camerlenghi, A. (Eds.), Contourites. Developments in Sedimentology. vol. 60. Elsevier, pp. 345–378, https://doi.org/10.1016/S0070-4571(08)10018-8.

Hernández-Molina, F.J., Llave, E., Stow, D.A.V., 2008b. Continental slope contourites. In: Rebesco, M., Camerlenghi, A. (Eds.), Contourites. Developments in Sedimentology. vol. 60. Elsevier, pp. 379–408, https://doi.org/10.1016/S0070-4571(08)10019-X.

Hernández-Molina, F.J., Paterlini, M., Violante, R., Marshall, P., de Isasi, M., Somoza, L., Rebesco, M., 2009. Contourite depositional system on the Argentine slope: an exceptional record of the influence of Antarctic water masses. Geology 37, 507–510.

Hernández-Molina, F.J., Paterlini, M., Somoza, L., Violante, R., Arecco, M.A., de Isasi, M., Rebesco, M., Uenzelmann-Neben, G., Neben, S., Marshall, P., 2010. Giant mounded drifts in the Argentine Continental Margin: origins, and global implications for the history of thermohaline circulation. Mar. Pet. Geol. 27 (7), 1508–1530. https://doi.org/10.1016/j.marpetgeo.2010.04.003.

Hernández-Molina, F.J., Serra, N., Stow, D.A.V., Llave, E., Ercilla, G., Van Rooij, D., 2011. Along-slope oceanographic processes and sedimentary products around the Iberian margin. Geo-Mar. Lett. 31, 315–341.

Hernández-Molina, F.J., Llave, E., Preu, B., Ercilla, G., Fontan, A., Bruno, M., Serra, N., Gomiz, J.J., Brackenridge, R.E., Sierro, F.J., Stow, D.A.V., García, M., Juan, C., Sandoval, N., Arnaiz, A., 2014a. Contourite processes associated with the Mediterranean outflow water after its exit from the Strait of Gibraltar: global and conceptual implications. Geology 42, 227–230.

Hernández-Molina, F.J., Stow, D.A.V., Alvarez-Zarikian, C.A., Acton, G., Bahr, A., Balestra, B., Ducassou, E., Flood, R., Flores, J.A., Furota, S., Grunert, P., Hodell, D., Jimenez-Espejo, F., Kim, J.K., Krissek, L., Kuroda, J., Li, B., Llave, E., Lofi, J., Lourens, L., Miller, M., Nanayama, F., Nishida, N., Richter, C., Roque, C., Pereira, H., Sanchez Goñi, M.F., Sierro, F.J., Singh, A.D., Sloss, C., Takashimizu, Y., Tzanova, A., Voelker, A.H.L., Williams, T., Xuan, C., 2014b. Onset of Mediterranean outflow into the North Atlantic. Science 344, 1244–1250.

Hernández-Molina, F.J., Soto, M., Piola, A.R., Tomasini, J., Preu, B., Thompson, P., Badalini, G., Creaser, A., Violante, R.A., Morales, E., Paterlini, M., de Santa Ana, H., 2016a. A contourite depositional system along the Uruguayan continental margin: sedimentary, oceanographic and paleoceanographic implication. Mar. Geol. 378, 333–349. https://doi.org/10.1016/j.margeo.2015.10.008.

Hernández-Molina, F.J., Sierro, F.J., Llave, E., Roque, C., Stow, D.A.V., Williams, T., Lofi, J., Van der Schee, M., Arnáiz, A., Ledesma, S., Rosales, C., 2016b. Evolution of the gulf of Cadiz margin and Southwest Portugal contourite depositional system: tectonic, sedimentary and paleoceanographic implications from IODP expedition 339. Mar. Geol. 377, 7–39.

Hernández-Molina, F.J., Wåhlin, A., Bruno, M., Ercilla, G., Llave, E., Serra, N., Roson, G., Puig, P., Rebesco, M., Van Rooij, D., Roque, C., González-Pola, C., Sánchez, F., Gómez, M., Preu, B., Schwenk, T., Hanebuth, T.J.J., Leal, S., Till, J.J., García-Lafuente, J., Brackenridge, R.E., Juan, C., Stow, D.A.V., Sánchez-González, J.M., 2016c. Oceanographic processes and products around the Iberian margin: a new multidisciplinary approach. Mar. Geol. 378, 127–156. https://doi.org/10.1016/j.margeo.2015.12.00822.

Hernández-Molina, F.J., Larter, R., Maldonado, A., 2017. Neogene to quaternary stratigraphic evolution of the Antarctic Peninsula, Pacific Margin offshore of Adelaide Island: transitions from a non-glacial, through glacially-influenced to a fully glacial stage. Glob. Planet. Chang. 156, 80–111. https://doi.org/10.1016/j.gloplacha.2017.07.002.

Hernández-Molina, F.J., Campbell, S., Badalini, G., Thompson, P., Walker, R., Soto, M., Conti, B., Preu, B., Thiéblemont, A., Hyslop, L., Miramontes, E., Morales, E., 2018. Large bedforms on contourite terraces: sedimentary and conceptual implications. Geology 46 (1), 27–30. https://doi.org/10.1130/G39655.1.

Hillenbrand, C.-D., Camerlenghi, A., Cowan, E.A., Hernández-Molina, F.J., Lucchi, R.G., Rebesco, M., Uenzelmann-Neben, G., 2008. The present and past bottom-current flow regime around the sediment drifts on the continental rise west of the Antarctic Peninsula. Mar. Geol. 255, 55–63. https://doi.org/10.1016/j.margeo.2008.07.004.

Hillenbrand, C.-D., Crowhurst, S.J., Williams, M., Hodell, D.A., McCave, I.N., Ehrmann, W., Xuan, C., Piotrowski, A.M., Hernández-Molina, F.J., Graham, A.G.C., Grobe, H., Williams, T.J., Horrocks, J.R., Allen, C.S., Larter, R.D., 2021. New insights from multi-proxy data from the West Antarctic continental rise: implications for dating and interpreting Late Quaternary palaeoenvironmental records. Quat. Sci. Rev. 257, 106842. https://doi.org/10.1016/j.quascirev.2021.106842.

Hohbein, M., Cartwright, J., 2006. 3D seismic analysis of the West Shetland Drift system: implications for Late Neogene palae-oceanography of the NE Atlantic. Mar. Geol. 230, 1–20.

Hollister, C.D., 1993. The concept of deep-sea contourites. Sediment. Geol. 82, 5–11.

Hollister, C., McCave, I., 1984. Sedimentation under deep-sea storms. Nature 309, 220–225. https://doi.org/10.1038/309220a0.

Horton, A.A., Dixon, S.J., 2017. Microplastics: an introduction to environmental transport processes. WIREs Water 2018, e1268. https://doi.org/10.1002/wat2.1268.

House, K.Z., Schrag, D.P., Harvey, C.F., Lackner, K.S., 2006. Permanent carbon dioxide storage in deep-sea sediments. Proc. Natl. Acad. Sci. USA 103 (33), 12291–12295. https://doi.org/10.1073/pnas.0605318103.

Howe, J.A., Stoker, M.S., Stow, D.A.V., Akhurst, M.C., 2002. Sediment drifts and contourite sedimentation in the northeastern Rockall Trough and Faroe-Shetland Channel, North Atlantic Ocean. In: Stow, D.A.V., Howe, J.A., Faugères, J.-C., Viana, A.R. (Eds.), Deep-Water Contourite Systems: Modern Drifts and Ancient Series, Seismic and Sedimentary Characteristics. Geol. Soc. London Mem., vol. 22, pp. 65–72, https://doi.org/10.1144/GSL.MEM.2002.022.01.06.

Huang, X., Jokat, W., 2016. Middle Miocene to present sediment transport and deposits in the Southeastern Weddell Sea, Antarctica. Glob. Planet. Chang. 139, 211–255. https://doi.org/10.1016/j.gloplacha.2016.03.002.

Hüneke, H., Stow, D.A.V., 2008. Identification of ancient contourites: problems and palaeoceanographic significance. In: Rebesco, M., Camerlenghi, A. (Eds.), Contourites. Developments in Sedimentology. vol. 60. Elsevier, Amsterdam, pp. 323–344.

Hüneke, H., Hernández-Molina, F., Rodrıguez-Tovar, F., Llave, E., Chiarella, D., Mena, A., Stow, D., 2020. Diagnostic criteria using microfacies for calcareous contourites, turbidites and pelagites in the Eocene-Miocene slope succession, southern Cyprus. Sedimentology. https://doi.org/10.1111/sed.12792.

Huvenne, V.A.I., Van Rooij, D., De Mol, B., Thierens, M., O'Donnell, R., Foubert, A., 2009. Sediment dynamics and palaeo-environmental context at key stages in the challenger cold-water coral mound formation: clues from sediment deposits at the mound base. Deep-Sea Res. I Oceanogr. Res. Pap. 56, 2263–2280.

IEA, International Energy Agency, 2017. Energy Technology Perspectives 2017: Catalysing Energy Technology Transformations. OECD, https://doi.org/10.1787/energy_tech-2017-en.

Ito, M., 1996. Sandy contourites of the Lower Kazusa Group in the Boso Peninsula, Japan: Kuroshio-current-influenced deep-sea sedimentation in a Plio-Pleistocene forearc basin. J. Sed. Res. 66 (3), 587–598. https://doi.org/10.1306/D42683B5-2B26-11D7-8648000102C1865D.

Ito, M., 2002. Kuroshio current-influenced sandy contourites from the Plio-Pleistocene Kazusa forearc basin, Boso Peninsula, Japan. Geol. Soc. London Mem. 22 (1), 421–432. https://doi.org/10.1144/GSL.MEM.2002.022.01.29.

Jackson, C.R., 2004. An Atlas of Oceanic Internal Solitary Waves. Global Ocean Associates, Rockville, MD (prepared for Office of Naval Research).

Juan, G., Ercilla, G., Hernández-Molina, F.J., Estrada, F., Alonso, B., Casas, D., García, M., Farran, M., Llave, E., Palomino, D., Vázquez, J.T., Medialdea, T., Gorini, C., d'Acremont, E., El Moumni, B., Ammar, A., 2016. Seismic evidence of current-controlled sedimentation in the Alboran Sea during the Pliocene and Quaternary: palaeoceanographic implications. Mar. Geol. 378, 292–311.

Juan, C., Van Rooij, D., De Bruycker, W., 2018. An assessment of bottom current controlled sedimentation in Pacific Ocean abyssal environments. Mar. Geol. 403, 20–33. https://hdl.handle.net/10.1016/j.margeo.2018.05.001.

Juan, C., Ercilla, G., Estrada, F., Alonso, B., Casas, D., Vázquez, J.T., d'Acremont, E., Medialdea, T., Hernández-Molina, F.J., Gorini, C., Moumni, B., Valencia, J., 2020. Multiple factors controlling the deep marine sedimentation of the Alboran Sea (SW Mediterranean) after the Zanclean Atlantic Mega-flood. Mar. Geol. 423, 106–138.

Kane, I.A., Clare, M.A., 2019. Dispersion, accumulation, and the ultimate fate of microplastics in deep-marine environments: a review and future directions. Front. Earth Sci. 7, 80. https://doi.org/10.3389/feart.2019.00080.

Kane, I.A., Clare, M.A., Miramontes, E., Wogelius, R., Rothwell, J.J., Garreau, P., Pohl, F., 2020. Seafloor microplastic hotspots controlled by deep-sea circulation. Science 368, 1140–1145.

Kennett, J., 1982. Marine Geology. Prentice-Hall. 813 pp.

Kirby, A., Hernández-Molina, F.J., Rodriguez, P., Conti, B., 2021a. Sedimentary tacking pattern of plastered drifts: an example from the Cenozoic on the Uruguayan continental slope. Mar. Geol. 440, 106567. https://doi.org/10.1016/j.margeo.2021.106567.

Kirby, A., Hernández-Molina, F.J., Rodrigues, S., 2021b. Lateral migration of large sedimentary bodies in deep-marine systems: a case study offshore of Argentina. Sci. Rep. (Accepted).

Knutz, P.C., 2008. Palaeoceanographic significance of contourite drifts. In: Rebesco, M., Camerlenghi, A. (Eds.), Contourites Developments in Sedimentology. vol. 60. Elsevier, Amsterdam, pp. 511–535.

Knutz, P.C., Jones, E.J.W., Howe, J.A., van Weering, T.J.C., Stow, D.A.V., 2002. Wave-form sheeted contourite drift on the Barra Fan, NW UK continental margin. In: Stow, D.A.V., Pudsey, C.J., Howe, J.A., Faugères, J.-C., Viana, A.R. (Eds.), Deep-Water Contourite Systems: Modern Drifts and Ancient Series, Seismic and Sedimentary Characteristics. Developments in Sedimentology, vol. 22, pp. 85–98, https://doi.org/10.1144/GSL.MEM.2002.022.01.08.

Knutz, P.C., Zahn, R., Hall, I.R., 2007. Centennial-scale variability of the British Ice Sheet: implications for climate forcing and Atlantic meridional overturning circulation during the last deglaciation. Paleoceanography 22, PA1207. https://doi.org/10.1029/2006PA001298.

Kuhlbrodt, T., Griesel, A., Montoya, M., Levermann, A., Hofmann, M., Rahmstorf, S., 2007. On the driving processes of the Atlantic meridional overturning circulation. Rev. Geophys. 45, RG2001.

Kuvaas, B., Kristoffersen, Y., Guseva, J., Leitchenkov, G., Gandjukhin, V., Løvås, O., Sand, M., Brekke, H., 2005. Interplay of turbidite and contourite deposition along the Cosmonaut Sea/Enderby Land margin, East Antarctica. Mar. Geol. 217 (1–2), 143–159. https://doi.org/10.1016/j.margeo.2005.02.025.

Laake, A., Perdomo, J., Seymour, N., Imamshah, A., 2017. The importance of omnidirectional sampling for geologic interpretation of seismic data, Flemish Pass area, offshore Newfoundland. In: SEG International Exposition and 87th Annual Meeting, 1872-1876.

Laberg, J.S., Camerlenghi, A., 2008. The significance of contourites for submarine slope stability. In: Rebesco, M., Camerlenghi, A. (Eds.), Contourites. Developments in Sedimentology, vol. 60. Elsevier, Amsterdam, pp. 537–556.

Laberg, J.S., Stoker, M.S., Dahlgren, K.I.T., de Haas, H., Haflidason, H., Hjelstuen, B.O., Nielsen, T., Shannon, P.M., Vorren, T.O., Van Weering, T.C.E., Ceramicola, S., 2005. Cenozoic alongslope processes and sedimentation on the NW European Atlantic margin. Mar. Pet. Geol. 22, 1069–1088.

Larter, R.D., Hogan, K.A., Hillenbrand, C.-D., Smith, J.A., Batchelor, C.L., Cartigny, M., Tate, A.J., Kirkham, J.D., Roseby, Z.A., Kuhn, G., Graham, A., Dowdeswell, J.A., 2019. Subglacial hydrological control on flow of an Antarctic Peninsula palaeo-ice stream. Cryosphere 13, 1583–1596. https://doi.org/10.5194/tc-13-1583-2019.

Lebreiro, S.M., Antón, L., Reguera, M.I., Marzocchi, A., 2018. Paleoceanographic and climatic implications of a new Mediterranean Outflow branch in the southern Gulf of Cadiz. Quat. Sci. Rev. 197, 92–111.

Leeder, M.R., 2011. Tectonic sedimentology: sediment systems deciphering 848 global to local tectonics. Sedimentology 58, 2–56. https://doi.org/10.1111/j.1365-3091.2010.01207.x.

Legg, S., Briegleb, B., Chang, Y., Chassignet, E.P., Danabasoglu, G., Ezer, T., Gordon, A.L., Griffies, S., Hallberg, R., Jackson, L., Large, W., Özgükmen, T.M., Peters, H., Price, J., Riemenschneider, U., Wu, W., Xu, X., Yang, J., 2009. Improving oceanic overflow representation in climate models: the Gravity Current Entrainment Climate Process Team. Bull. Am. Meteorol. Soc. 90, 657–670.

Liu, L., Silver, D., Bemis, K., Kang, D., Curchitser, E., 2017. Illustrative Visualization of Mesoscale Ocean Eddies, Computer Graphics Forum. Wiley Online Library, pp. 447–458.

Liu, W., Fedorov, A., Xie, S.-P., Hu, S., 2020. Climate impacts of a weakened Atlantic Meridional Overturning Circulation in a warming climate. Sci. Adv. 6, eaaz4876. https://doi.org/10.1126/sciadv.aaz48 76.

Liu, S., Hernández-Molina, F.J., Lei, Z., Duarte, D., Chen, H., Wang, C., Lei, Y., Zhuo, H., Huang, S., Zhang, L., Su, M., 2021. Fault-controlled contourite drifts in the southern South China Sea: tectonic, oceanographic, and conceptual implications. Mar. Geol. 433, 106420. https://doi.org/10.1016/j.margeo.2021.106420.

Llave, E., Hernández-Molina, F.J., Somoza, L., Díaz-del-Río, V., Stow, D.A.V., Maestro, A., Alveirinho Dias, J.M., 2001. Seismic stacking pattern of the faro-albufeira contourite system (gulf of Cadiz): a quaternary record of paleoceanographic and tectonic influences. Mar. Geophys. Res. 22, 487–508.

Llave, E., Schonfeld, J., Hernández-Molina, F.J., Mulder, T., Somoza, L., del Rio, V.D., Sanchez-Almazo, I., 2006. High-resolution stratigraphy of the Mediterranean outflow contourite system in the Gulf of Cadiz during the late Pleistocene: the impact of Heinrich event. Mar. Geol. 227, 241–262.

Liu, S., 2019. Contourite Depositional System in the Le Danois Bank Region, Southern Bay of Biscay: Sedimentary, Tectonic and Paleoceanographic Implications (PhD thesis). Ghent University, Ghent, Belgium.

Llave, E., Hernández-Molina, F.J., Somoza, L., DAV, S., Díaz del Río, V., 2007. Quaternary evolution of the contourite depositional system in the Gulf of Cadiz. In: Viana, A., Rebesco, M. (Eds.), Economic and Palaeoceanographic Significance of Contourite Deposits. Geol. Soc. Lond. Spec. Publ. 276, pp. 49–80.

Lobo, F.J., López-Quirós, A., Hernández-Molina, F.J., Pérez, L.F., García, M., Evangelinos, D., Bohoyo, F., Rodríguez-Fernández, J., Salabarnada, A., Maldonado, A., 2021. Recent morpho-sedimentary processes in Dove Basin, southern Scotia Sea, Antarctica: a basin-scale case of interaction between bottom currents and mass movements. Mar. Geol. 441, 106598. https://doi.org/10.1016/j.margeo.2021.106598.

Locker, S.D., Laine, E.P., 1992. Paleogene-Neogene depositional history of the middle U.S. Atlantic continental rise: mixed turbidite and contourite depositional systems. Mar. Geol. 103, 137–164. https://doi.org/10.1016/0025-3227(92)90013-8.

Lofi, J., Voelker, A.H.L., Ducassou, E., Hernández-Molina, F.J., Sierro, F.J., Bahr, A., Galvani, A., Lourens, L.J., Pardo-Igúzquiza, E., Pezard, P., Rodríguez-Tovar, F.J., 2016. Quaternary chronostratigraphic framework and sedimentary processes for the Gulf of Cadiz and Portuguese Contourite Depositional Systems derived from Natural Gamma Ray records. Mar. Geol. 377, 40–57.

López-Quirós, A., Lobo, F.J., Escutia, C., García, M., Hernández-Molina, F.J., Pérez, L.F., Bohoyo, F., Evangelinos, D., Salabarnada, A., Maldonado, A., Naveira Garabato, A.C., 2020. Geomorphology of Ona Basin, southwestern Scotia Sea (Antarctica): decoding the spatial variability of bottom-current pathways. Mar. Geol. 422, 106113. https://doi.org/10.1016/j.margeo.2020.106113.

Lovell, B., 2010. A pulse in the planet: regional control of high-frequency changes in relative sea level by mantle convection. J. Geol. Soc. 167, 637–648.

Lozano, P., Fernández-Salas, L.M., Hernández-Molina, F.J., Sanchez Leal, R.F., Sánchez-Guillamón, O., Palomino, D., Farias, C., Lopez-Gonzalez, N., Garcia, M., Vazquez, J.T., Vila, Y., Rueda, J.L., 2020. Multiprocesses interaction in shaping the seafloor and controlling substrate types, habitats and benthic communities of the Gulf of Cádiz. Mar. Geol. 423, 106139. https://doi.org/10.1016/j.margeo.2020.106139.

Lu, H., Fulthorpe, C., Mann, P., 2003. Three-dimensional architecture of shelf-building sediment drifts in the offshore Canterbury Basin, New Zealand. Mar. Geol. 193, 19–47. https://doi.org/10.1016/S0025-3227(02)00612-6.

Lucchi, R.G., Rebesco, M., 2007. Glacial contourites on the Antarctic Peninsula margin: insight for palaeoenvironmental and palaeoclimatic conditions. In: Viana, A.R., Rebesco, M. (Eds.), Economic and Palaeoceanographic Significance of Contourite Deposits. Geol. Soc. London Spec. Publ. 276, pp. 111–127, https://doi.org/10.1144/GSL.SP.2007.276.01.06.

Lucchi, R.G., Rebesco, M., Camerlenghi, A., Busetti, M., Tomadin, L., Villa, G., Persico, D., Morigi, C., Bonci, M.C., Giorgetti, G., 2002. Mid-late Pleistocene glacimarine sedimentary processes of a high-latitude, deep-sea sediment drift (Antarctic Peninsula Pacific margin). Mar. Geol. 189, 343–370. https://doi.org/10.1016/S0025-3227(02)00470-X.

Lüdmann, T., Paulat, M., Betzler, C., Möbius, J., Lindhorst, S., Wunsch, M., Eberli, G.P., 2016. Carbonate mounds in the Santaren Channel, Bahamas: a current-dominated periplatform depositional regime. Mar. Geol. 376, 69–85.

Lund, D.C., Adkins, J.F., Ferrari, R., 2011. Abyssal Atlantic circulation during the Last Glacial Maximum: 1009 Constraining the ratio between transport and vertical mixing. Paleoceanography 26 (PA1213), 1010. https://doi.org/10.1029/2010PA001938.

Maestro-González, A., 2021. The future of mining: the exploitation of marine geological resources as global commons. In: Martín-Ramírez, J., Bauzá-Abril, B. (Eds.), Security in the Global Commons and Beyond. Springer Nature Switzerland AG, pp. 51–68.

Maier, K.L., Gales, J.A., Paull, C.K., Rosenberger, K., Talling, P.J., Simmons, S.M., Gwiazda, R., McGann, M., Cartigny, M.J.B., Lundsten, E., Anderson, K., Clare, M.A., Xu, J., Parsons, D., Barry, J.P., Wolfson-Schwehr, M., Nieminski, N.M., Sumner, E.J., 2019. Linking direct measurements of turbidity currents to submarine canyon-floor deposits. Front. Earth Sci. 7, 144. https://doi.org/10.3389/feart.2019.00144.

Maldonado, A., Barnolas, A., Bohoyo, F., Escutia, C., Galindo-Zaldívar, J., Hernández-Molina, J., Jabaloy, A., Lobo, F.J., Nelson, C.H., Rodríguez-Fernández, J., Somoza, L., Vázquez, J.-T., 2005. Miocene to recent contourite drifts development in the northern Weddell Sea. Glob. Planet. Chang. 45, 99–129. https://doi.org/10.1016/j.gloplacha.2004.09.013.

Mantilla, O., Hernández-Molina, F.J., Scarselli, N., 2022. Can deepwater bottom currents generate clinothems? An example of a large, asymmetric mounded drift in Upper Jurassic to Lower Cretaceous sediments from northwestern Australia. Geology. https://doi.org/10.1130/G50068.1.

Marchès, E., Mulder, T., Cremer, M., Bonnel, C., Hanquiez, V., Gonthier, E., Lecroart, P., 2007. Contourite drift construction influenced by capture of Mediterranean outflow water deep-sea current by the Portimão submarine canyon (Gulf of Cadiz, South Portugal). Mar. Geol. 242, 247–260. https://doi.org/10.1016/j.margeo.2007.03.013.

Marchès, E., Mulder, T., Gonthier, E., Cremer, M., Hanquiez, V., Garlan, T., Lecroart, R., 2010. Perched lobe formation in the gulf of Cadiz: interactions between gravity processes and contour currents (Algarve margin, southern Portugal). Sed. Geol. 229, 81–94. https://doi.org/10.1016/j.sedgeo.2009.03.008.

Martín-Chivelet, J., Fregenal-Martínez, M.A., Chacón, B., 2003. Mid-depth calcareous contourites in the latest Cretaceous of Caravaca (Subbetic Zone, SE Spain). Origin and paleohydrological significance. Sediment. Geol. 163, 131–146.

Martín-Chivelet, J., Fregenal Martínez, M.A., Chacón, B., 2008. Traction structures in contourites. In: Rebesco, M., Camerlenghi, A. (Eds.), Contourites. Developments in Sedimentology, vol. 60. Elsevier, Amsterdam, pp. 159–181.

Martorelli, E., Bosman, A., Casalbore, D., Falcini, F., 2016. Interaction of down-slope and along-slope processes off Capo Vaticano (southern Tyrrhenian Sea, Italy), with particular reference to contourite-related landslides. Mar. Geol. 378, 43–55.

Martorelli, E., Bosman, A., Casalbore, D., Chiocci, F., Conte, A.M., Bella, L.D., Ercilla, G., Falcini, F., Falco, P., Frezza, V., Gaglianone, G., Giaccio, B., Mancini, M., 2021. Mid-to-late Holocene upper slope contourite deposits off Capo Vaticano (Mediterranean Sea): high-resolution record of contourite cyclicity, bottom current variability and sandy facies. Mar. Geol. 431, 106372. https://doi.org/10.1016/j.margeo.2020.106372.

Maselli, V., Kneller, B., 2018. Bottom currents, submarine mass failures and halokinesis at the toe of the Sigsbee Escarpment (Gulf of Mexico): contrasting regimes during lowstand and highstand conditions? Mar. Geol. 401, 36–65. https://doi.org/10.1016/j.margeo.2018.04.001.

McCave, I.N., 1972. Transport and escape of fine-grained sediment from shelf areas. In: Swift, D.J.P., Duane, D.B., Pilkey, O.H. (Eds.), Shelf Sediment Transport. Dowden, Hutchinson & Ross, Stroudsburg, pp. 225–248.

McCave, I.N., 1982. Erosion and deposition by currents on submarine slopes. Bull. Inst. Géol. Bassin d'Aquitaine 31, 47–55.

McCave, I.N., Hall, I.R., 2006. Size sorting in marine muds: processes, pitfalls, and prospects for paleo-flow speed proxies. Geochem. Geophys. Geosyst. 7, Q10N05. https://doi.org/10.1029/2006GC001284.

McCave, I.N., Tucholke, B.E., 1986. Deep Current Controlled Sedimentation in the Western North Atlantic. Geological Society of America, Boulder, CO, pp. 451–468.

McCave, I.N., Lonsdale, P.F., Hollister, C.D., Gardner, W.D., 1980. Sediment transport over the Hatton and Gadar contourite drifts. J. Sediment. Petrol. 50, 1049–1062.

McCave, I.N., Thornalley, D.J.R., Hall, I.R., 2017. Relation of sortable silt grain-size to deep-sea current speeds: calibration of the 'Mud Current Meter'. Deep-Sea Res. 127, 1–12. https://doi.org/10.1016/j.dsr.2017.07.003.

McGinnis, J.P., Hayes, D.E., 1995. The roles of down-slope and alongslope depositional processes: southern Antarctic Peninsula continental rise. In: Cooper, A.K., Barker, P.F., Brancolini, G. (Eds.), Geology and Seismic Stratigraphy of the Antarctic Margin. Antarct. Res. Ser., vol. 68. American Geophysical Union, Washington, DC, pp. 141–156.

McGinnis, J.P., Hayes, D.E., Driscoll, N.W., 1997. Sedimentary processes across the continental rise of the southern Antarctic Peninsula. Mar. Geol. 141, 91–109.

McMaster, R.L., Locker, S.D., Laine, E.P., 1989. The early Neogene continental rise off the eastern United States. Mar. Geol. 87, 137–163. https://doi.org/10.1016/0025-3227(89)90058-3.

Mena, A., Francés, G., Pérez-Arlucea, M., Hanebuth, T.J.J., Bender, V.B., Nombela, M.A., 2018. Evolution of the Galicia Interior Basin over the last 60 ka: sedimentary processes and palaeoceanographic implications. J. Quat. Sci. 33, 536–549. https://doi.org/10.1002/jqs.3032.

Mencaroni, D., Urgeles, R., Camerlenghi, A., Llopart, J., Ford, J., Serra, C.S., Meservy, W., Gràcia, E., Rebesco, M., Zitellini, N., 2021. A mixed turbidite—contourite system related to a major submarine canyon: the Marquês de Pombal Drift (south-west Iberian margin). Sedimentology. https://doi.org/10.1111/sed.12844 (in press).

Mestdagh, T., 2020. Late Quaternary Stratigraphic and Sedimentary Evolution of the Northern Gulf of Cadiz Continental Margin: New Perspectives From Shelf-to-Slope Correlation and Borehole-Seismic Integration (PhD thesis). Ghent University, Ghent, Belgium.

Mestdagh, T., Lobo, F.J., Llave, E., Hernández-Molina, F.J., van Rooij, D., 2019. Review of the late Quaternary stratigraphy of the northern Gulf of Cadiz continental margin: new insights into controlling factors and global implications. Earth Sci. Rev. 198, 102944.

Mestdagh, T., Lobo, F.J., Llave, E., Hernández-Molina, F.J., García Ledesma, A., Puga-Bernabéu, A., Fernández-Salas, L.M., Van Rooij, D., 2020. Late quaternary multi-genetic processes and products on the northern Gulf of Cadiz upper continental slope (SW Iberian Peninsula). Mar. Geol. 427. https://doi.org/10.1016/j.margeo.2020.106214.

Michels, K.H., Rogenhagen, J., Kuhn, G., 2001. Recognition of contour-current influence in mixed contourite-turbidite sequences of the western Weddell Sea, Antarctica. Mar. Geophys. Res. 22, 465–485. https://doi.org/10.1023/A:1016303817273.

Michels, K.H., Kuhn, G., Hillenbrand, C.-D., Diekmann, B., Futterer, D.K., Grobe, H., Uenzelmann-Neben, G., 2002. The southern Weddell Sea: Combined contourite-turbidite sedimentation at the southeastern margin of the Weddell Gyre. In: Stow, D.A.V., Pudsey, C.J., Howe, J.A., Faugères, J.-C., Viana, A.R. (Eds.), Deep-Water Contourite Systems: Modern Drifts and Ancient Series, Seismic and Sedimentary Characteristics. Geol. Soc. London Mem., vol. 22, pp. 305–324, https://doi.org/10.1144/GSL.MEM.2002.022.01.22.

Miguez-Salas, O., Rodríguez-Tovar, F.J., 2019. Ichnofacies distribution in the Eocene-Early Miocene Petra Tou Romiou outcrop, Cyprus: sea level dynamics and palaeoenvironmental implications in a contourite environment. Int. J. Earth Sci. 108, 2531–2544. https://doi.org/10.1007/s00531-019-01775-x.

Miguez-Salas, O., Huffard, C.L., Smith, K.L., McGill, P.R., Rodríguez-Tovar, F.J., 2020. Faunal assemblage changes, bioturbation and benthic storms at an abyssal station in the northeastern Pacific. Deep-Sea Res. I Oceanogr. Res. Pap. 160, 103277. https://doi.org/10.1016/j.dsr.2020.103277.

Millot, C., Monaco, A., 1984. Deep strong currents and sediment transport in the northwestern Mediterranean Sea. Geo-Mar. Lett. 4, 13–17. https://doi.org/10.1007/BF02237968.

Miramontes, E., Cattaneo, A., Jouet, G., Théreau, E., Thomas, Y., Rovere, M., Cauquil, E., Trincardi, F., 2016. The Pianosa contourite depositional system (northern Tyrrhenian Sea): drift morphology and plio-quaternary stratigraphic evolution. Mar. Geol. 378, 20–42.

Miramontes, E., Garziglia, S., Sultan, N., Jouet, G., Cattaneo, A., 2018. Morphological control of slope instability in contourites: a geotechnical approach. Landslides 15 (6), 1085–1095. https://doi.org/10.1007/s10346-018-0956-6.

Miramontes, E., Garreau, P., Cattaneo, A., Caillaud, M., Jouet, G., Pellen, R., Hernández-Molina, F.J., Clare, M.A., 2019. Contourite distribution in the NW Mediterranean Sea: new insights from hydrodynamic modelling. Geomorphology 333, 43–60.

Miramontes, E., Jouet, G., Thereau, E., Bruno, M., Penven, P., Guerin, C., Le Roy, P., Droz, L., Jorry, S.J., Hernández-Molina, F.J., Thiéblemont, A., Silva Jacinto, R., Cattaneo, A., 2020a. The impact of internal waves on upper continental slopes: insights from the Mozambican margin (southwest Indian Ocean). Earth Surf. Process. Landforms 45, 1469–1482. https://doi.org/10.1002/esp.4818.

Miramontes, E., Eggenhuisen, J.T., Jacinto, R.S., Poneti, G., Pohl, F., Normndeau, A., Campbell, D.C., Hernandez-Molina, F.J., 2020b. Channel-levee evolution in combined contour current–turbidity current flows from flume-tank experiments. Geology 48 (4), 353–357. https://doi.org/10.1130/G47111.1.

Miramontes, E., Thiéblemont, A., Babonneau, N., Penven, P., Raisson, F., Droz, L., Jorry, S.J., Fierens, R., Counts, J.W., Wilckens, H., Cattaneo, A., Jouet, G., 2021. Contourite and mixed turbidite-contourite systems in the Mozambique Channel (SW Indian Ocean): link between geometry, sediment characteristics and modelled bottom currents. Mar. Geol. 437, 106502. https://doi.org/10.1016/j.margeo.2021.106502.

Moraes, M., Maciel, W.B., Braga, M.S., Viana, A.R., 2007. Bottom-current reworked Palaeocene-Eocene deep-water reservoirs of the Campos Basin, Brazil. In: Viana, A.R., Rebesco, M. (Eds.), Economic and Palaeoceanographic Significance of Contourite Deposits. Geol. Soc. London Spec. Publ. 276, pp. 81–94, https://doi.org/10.1144/GSL.SP.2007.276.01.04.

Morales, E., Chang, H.K., Soto, M., Corr ̂ea, F.S., Veroslavsky, G., de Santa Ana, H., Conti, B., Daners, G., 2017. Tectonic and stratigraphic evolution of the Punta del Este and Pelotas basins (offshore Uruguay). Pet. Geosci. 23, 415–426. https://doi.org/10.1144/petgeo2016-059.

Moscardelli, L., Wood, L., Mann, P., 2006. Mass-transport complexes and associated processes in the offshore area of Trinidad and Venezuela. AAPG Bull. 90, 1059–1088.

Mosher, D.C., Boggild, K., 2021. Impact of bottom currents on deep water sedimentary processes of Canada Basin, Arctic Ocean. Earth Planet. Sci. Lett. 569, 117067. https://doi.org/10.1016/j.epsl.2021.117067.

Mosher, D.C., Yanez-Carrizo, G., 2021. The elusive continental rise: insights from residual bathymetry analysis of the Northwest Atlantic margin. Earth Sci. Rev. 217, 103608. https://doi.org/10.1016/j.earscirev.2021.103608.

Mosher, D.C., Campbell, D.C., Gardner, J.V., Piper, D.J.W., Chaytor, J.D., Rebesco, M., 2017. The role of deep-water sedimentary processes in shaping a continental margin: the Northwest Atlantic. Mar. Geol. 393, 245–259. https://doi.org/10.1016/j.margeo.2017.08.018.

Mourlot, Y., Calvès, G., Clift, P.D., Baby, G., Chaboureau, A.-C., Raisson, F., 2018. Seismic stratigraphy of Cretaceous eastern Central Atlantic Ocean: basin evolution and palaeoceanographic implications. Earth Planet. Sci. Lett. 499, 107–121. https://doi.org/10.1016/j.epsl.2018.07.023.

Mulder, T., Voisset, M., Lecroart, P., Le Drezen, E., Gonthier, E., Hanquiez, V., Faugères, J.-C., Habgood, E., Hernández-Molina, F.J., Estrada, F., Llave, E., Poirier, D., Gorini, C., Fuchey, Y., Volker, A., Freitas, P., Lobo Sanchez, F., Fernandez, L.M., Morel, J., 2003. The gulf of Cadiz: an unstable giant contouritic levee. Geo-Mar. Lett. 23 (1), 7–18.

Mulder, T., Faugères, J.-C., Gonthier, E., 2008. Mixed turbidite-contourite systems. In: Rebesco, M., Camerlenghi, A. (Eds.), Contourites. Developments in Sedimentology, vol. 60. Elsevier, Amsterdam, pp. 435–456, https://doi.org/10.1016/S0070-4571(08)10021-8.

Mulder, T., Hassan, R., Ducassou, E., Zaragosi, S., Gonthier, E., Hanquiez, V., Marchès, E., Toucanne, S., 2013. Contourites in the Gulf of Cadiz: a cautionary note on potentially ambiguous indicators of bottom current velocity. Geo-Mar. Lett. 33, 357–367.

Mulder, T., Ducassou, E., Hanquiez, V., Principaud, M., Fauquembergue, K., Tournadour, E., Chabaud, L., Reijmer, J., Recouvreur, A., Gillet, H., Borgomano, J., Schmitt, A., Moal, P., 2019. Contour current imprints and contourite drifts in the Bahamian archipelago. Sedimentology 66, 1192–1221.

Mutti, E., 1990. Relazioni tra stratigrafia sequenziale e tettonica. In: Atti del 75simo congresso nazionale della Societa' Geologica Italiana 'La Geologia Italiana degli Anni 90'. Mem. della Soc. Geol. Ital. Rome, Italy. Soc. Geol. Ital. 45, 627–655.

Mutti, E., 1992. Turbidite Sandstones. Agip, Istituto di geologia, Università di Parma. 275 pp.

Mutti, E., Carminatti, M., 2012. Deep-water sands of the Brazilian offshore basins: search and Discovery. In: AAPG Annual Convention & Exhibition.

Mutti, E., Barros, M., Possato, S., Rumenos, L., 1980. Deep sea fan turbidite sediments winnowed by bottom currents in the Eocene of the Campos Basin, Brazilian offshore. In: Int. Assoc. of Sedimentol., 1st Eur. Reg. Meet. Abstracts. 114 pp.

Mutti, E., Cunha, R., Bulhões, E., Arienti, L., Viana, A., 2014. Contourites and turbidites of the Brazilian Marginal Basins. Search and Discovery Article #90189. 43 pp.

Neves, M., Roque, C., Luttrel, K., Vazquez, J.-T., Alonso, B., 2016. Impact of sea-level rise on earthquake and landslide triggering offshore the Alentejo margin (SW Iberia). Geo-Mar. Lett. 36 (6), 415–424.

Ng, Z.L., Hernández-Molina, F.J., Duarte, D., Roque, C., Sierro, F.J., Llave, E., Manar, M.A., 2021. Late Miocene contourite depositional system of the Gulf of Cádiz: the sedimentary signature of the paleo-Mediterranean Outflow Water. Mar. Geol., 106605. https://doi.org/10.1016/j.margeo.2021.106605.

Ng, Z.L., Hernández-Molina, F.J., Ledesma, S., Sierro, F.J., Duarte, D., Llave, E., Roque, C., Arnáiz, A., 2022. Late Miocene evolution of the eastern Deep Algarve basin: interaction of bottom currents and gravitational processes in a foredeep setting. Mar. Pet. Geol. 141, 105695. https://doi.org/10.1016/j.marpetgeo.2022.105695.

Nielsen, T., Knutz, P.C., Kuijpers, A., 2008. Seismic expression of contourite depositional systems. In: Rebesco, M., Camerlenghi, A. (Eds.), Contourites. Developments in Sedimentology, vol. 60. Elsevier, pp. 301–321.

Niemi, T.M., Ben-Avraham, Z., Hartnady, C.J.H., Reznikov, M., 2000. Post-Eocene seismic stratigraphy of the deep ocean basin adjacent to the southeast African continental margin: a record of geostrophic bottom current systems. Mar. Geol. 162, 237–258.

Normandeau, A., Campbell, D.C., Cartigny, M.J.B., 2019. The influence of turbidity currents and contour currents on the distribution of deep-water sediment waves offshore eastern Canada. Sedimentology 66, 1746–1767. https://doi.org/10.1111/sed.12557.

Nugraha, H.D., Jackson, C., Johnson, H.D., Hodgson, D.M., Reeve, M.T., 2018. Tectonic and oceanographic process interactions archived in Late Cretaceous to Present deep-marine stratigraphy on the Exmouth Plateau, offshore NW Australia. Basin Res. 31, 405–430. https://doi.org/10.1111/bre.12328.

Ohta, J., Yasukawa, K., Nakamura, K., Fujinaga, K., Iijima, K., Kato, Y., 2021. Geological features and resource potential of deep-sea mud highly enriched in rare-earth elements in the Central Pacific Basin and the Penrhyn Basin. Ore Geol. Rev., 104440. https://doi.org/10.1016/j.oregeorev.2021.104440.

Øvrebø, L.K., Haughton, P.D.W., Shannon, P.M., 2006. A record of fluctuating bottom currents on the slopes west of the Porcupine Bank, offshore Ireland—implications for Late Quaternary climate forcing. Mar. Geol. 225, 279–309.

Owens, M., 2017. The Role of Alongslope and Downslope Sedimentary Processes in the Construction of Cenozoic Large-Scale Mounded Features: A 3D Seismic Study From the NE Rockall Trough (MSc thesis). National University of Ireland, Dublin, Ireland, https://doi.org/10.1016/j.marpetgeo.2005.01.008.

Palamenghi, L., Keil, H., Spiess, V., 2015. Sequence stratigraphic framework of a mixed turbidite-contourite depositional system along the NW slope of the South China Sea. Geo-Mar. Lett. 35, 1–21. https://doi.org/10.1007/s00367-014-0385-z.

Palermo, D., Galbiati, M., Famiglietti, M., Marchesini, M., Mezzapesa, D., Fonnesu, F., 2014. Insights into a new super-giant gas field-sedimentology and reservoir modelling of the coral reservoir complex, offshore northern Mozambique. In: Offshore Technology Conference-Asia, Offshore Technology Conference., https://doi.org/10.4043/24907-MS.

Pandolpho, B.T., Klein, A.H., Dutra, I., Mahiques, M.M., Viana, A.R., Bueno, G.V., Machado, A.A., Camargo, Y.L., Hercos, C.M., Lima, Y., Filho, A.F., Theodoro, C.E., 2021. Seismic record of a cyclic turbidite-contourite system in the Northern Campos Basin, SE Brazil. Mar. Geol. 434, 106422. https://doi.org/10.1016/j.margeo.2021.106422.

Pellegrini, C., Maselli, V., Trincardi, F., 2016. Pliocene–Quaternary contourite depositional system along the south-western Adriatic margin: changes in sedimentary stacking pattern and associated bottom currents. Geo-Mar. Lett. 36, 67–79. https://doi.org/10.1007/s00367-015-0424-4.

Pemberton, S.G., Frey, R.W., 1982. Trace fossil nomenclature and the Planolites-Palaeophycus dilemma. J. Paleontol. 56, 843–881.

Pereira, P., Ribeiro, C., Carneiro, J., 2021. Identification and characterization of geological formations with CO2 storage potential in Portugal. Pet. Geosci. 27 (3). https://doi.org/10.1144/petgeo2020-123. petgeo2020-123.

Pérez-Díaz, L., Eagles, G., 2017. South Atlantic paleobathymetry since early cretaceous. Sci. Rep. 7, 11819. 10.1038/s41598-017-11959-7.

Pickering, K.T., Hiscott, R.N., Hein, F.J., 1989. Deep Marine Environments. Clastic Sedimentation and Tectonics Unwin Hyman Ltd, London.

Potter, P.E., Szatmari, P., 2009. Global Miocene tectonics and the modern world. Earth Sci. Rev. 96, 279–295.

Preu, B., Hernández Molina, F.J., Violante, R., Piola, A.R., Paterlini, C.M., Schwenk, T., Voigt, I., Krastel, S., Spiess, V., 2013. Morphosedimentary and hydrographic features of the northern Argentine margin: the interplay between erosive, depositional and gravitational processes and its conceptual implications. Deep-Sea Res. I: Oceanogr. Res. Pap. 75, 157–174.

Rahmstorf, S., 2006. Encyclopedia of Quaternary Science. Illustration by E. Paul Oberlander, 2009, WHOI http://oceanworld.tamu.edu/resources/oceanography-book/abruptclimatechange.htm.

Rajappa, S., Kurian, J.P., Tyagi, A., Murukesh, N., 2020. Interaction between seabed morphology and water mass in the South Eastern Arabian Sea: deductions of water mass flow. Geo-Mar. Lett. https://doi.org/10.1007/s00367-020-00663-7.

Rasmussen, S., Lykke-Andersen, H., Kuijpers, A., Troelstra, S.R., 2003. Post-Miocene sedimentation at the continental rise of Southeast Greenland: the interplay between turbidity and contour currents. Mar. Geol. 196, 37–52. https://doi.org/10.1016/S0025-3227(03)00043-4.

Rebesco, M., 2005. Contourites. In: Selley, R.C., Cocks, L.R.M., Plimer, I.R. (Eds.), Encyclopedia of Geology. Elsevier, Oxford, pp. 513–527.

Rebesco, M., Camerlenghi, A., 2008. Contourites. Developments in Sedimentology, vol. 60 Elsevier, Amsterdam, p. 688.

Rebesco, M., Larter, R.D., Camerlenghi, A., Barker, P.F., 1996. Giant sediment drifts on the continental rise west of the Antarctic Peninsula. Geo-Mar. Lett. 16, 65–75. https://doi.org/10.1007/BF02202600.

Rebesco, M., Larter, R.D., Barker, P.F., Camerlenghi, A., Vanneste, L.E., 1997. The history of sedimentation on the continental rise west of the Antarctic peninsula. In: Cooper, A.K., Barker, P.F. (Eds.), Geology and Seismic Stratigraphy of the Antarctic Margin II. AGU Ant. Res. Series, pp. 28–49, https://doi.org/10.1029/AR071p0029. 0066–4634, 71, 0-87590-884-5.

Rebesco, M., Camerlenghi, A., Zanolla, C., 1998. Bathymetry and morphogenesis of the continental margin west of the Antarctic Peninsula. Terra Antart. 5, 715–725.

Rebesco, M., Pudsey, C., Canals, M., Camerlenghi, A., Barker, P., Estrada, F., Giorgetti, A., 2002. Sediment drift and deep-sea channel systems, Antarctic Peninsula Pacific Margin. In: Stow, D.A.V., Pudsey, C.J., Howe, J.A., Faugères, J.-C., Viana, A.R. (Eds.), Deep-Water Contourite Systems: Modern Drifts and Ancient Series, Seismic and Sedimentary Characteristics. Geol. Soc. London Mem., vol. 22, pp. 353–371, https://doi.org/10.1144/GSL.MEM.2002.022.01.25.

Rebesco, M., Camerlenghi, A., Volpi, V., Neagu, C., Accettella, D., Lindberg, B., Cova, A., Zgur, F., The Magico Party, 2007. Interaction of processes and importance of contourites: Insights from the detailed morphology of sediment Drift 7, Antarctica. In: Viana, A.R., Rebesco, M. (Eds.), Economic and Palaeoceanographic Significance of Contourite Deposits. Geol. Soc. London Spec. Publ. 276, pp. 95–110, https://doi.org/10.1144/GSL.SP.2007.276.01.05.

Rebesco, M., Camerlenghi, A., Van Loon, A.J., 2008. Contourite research: a field in full development. In: Rebesco, M., Camerlenghi, A. (Eds.), Contourites. Developments in Sedimentology, vol. 60. Elsevier, pp. 3–10.

Rebesco, M., Hernández-Molina, F.J., Van Rooij, D., Wåhlin, A., 2014. Contourites and associated sediments controlled by deep-water circulation processes: state of the art and future considerations. Mar. Geol. 352, 111–154. https://doi.org/10.1016/j.margeo.2014.03.011.

Rebesco, M., Özmaral, A., Urgeles, R., Accettella, D., Lucchi, R.G., Rüther, D., Winsborrow, M., Llopart, J., Caburlotto, A., Lantzsch, H., Hanebuth, T., 2016. Evolution of a high-latitude sediment drift inside a glacially-carved trough based on high-resolution seismic stratigraphy. Quat. Sci. Rev. 147, 178–193. https://doi.org/10.1016/j.quascirev.2016.02.007.

Reed, D.L., Meyer, A.W., Silver, E.A., Prasetyo, H., 1987. Contourite sedimentation in an intracratonic forearc system: eastern Sunda Arc, Indonesia. Mar. Geol. 76, 223–241. https://doi.org/10.1016/0025-3227(87)90031-4.

Rodrigues, S., 2021. Sedimentary Model for Mixed Depositional Systems: Conceptual and Economic Implications (PhD thesis). Royal Holloway, University of London, UK. 331 pp.

Rodrigues, S., Roque, C., Hernández-Molina, F.J., Llave, E., Terrinha, P., 2020. The sines contourite depositional system along the SW Portuguese margin: onset, evolution and conceptual implications. Mar. Geol. 4301 (06357), 12.

Rodrigues, S., Hernández-Molina, F.J., Kirby, A., 2021. A Late Cretaceous mixed (turbidite-contourite) system along the Argentine Margin: Paleoceanographic and conceptual implications. Mar. Petrol. Geol. 123, 104768. https://doi.org/10.1016/j.marpetgeo.2020.104768.

Rodrigues, S., Hernández-Molina, F.J., Fonnesu, M., Miramontes, E., Rebesco, M., Campbell, D.C., 2022c. A new classification system for mixed (turbidite-contourite) depositional systems: examples, conceptual models and diagnostic criteria for modern and ancient records. Earth Sci. Rev. 104030. https://doi.org/10.1016/j.earscirev.2022.104030.

Rodrigues, S., Lucchi, R.G., Hillenbrand, C.-D., Rodríguez-Tovar, F.J., Hernández-Molina, F.J., Larter, R.D., Rebesco, M., 2022a. Recognizing key sedimentary facies and their distribution in mixed turbidite-contourite systems: The case of the Pacific margin of the Antarctic Peninsula. Sedimentology 69 (4), 1953–1991. https://doi.org/10.1111/sed.12978.

Rodrigues, S., Deptuck, M., Kendell, K., Campbell, C., Hernández-Molina, F.J., 2022b. Cretaceous to Eocene mixed turbidite-contourite systems offshore Nova Scotia (Canada): spatial and temporal variability of down- and along-slope processes. Mar. Petrol. Geol. 138, 105572. https://doi.org/10.1016/j.marpetgeo.2022.105572.

Rodrigues, S., Hernández-Molina, F.J., Fonnesu, M., Miramontes, E., Rebesco, M., Campbell, D.C., 2022c. A new classification system for mixed (turbidite-contourite) depositional systems: Examples, conceptual models and diagnostic criteria for modern and ancient records. Earth-Science Reviews 230, 104030. https://doi.org/10.1016/j.earscirev.2022.104030.

Rodriguez, K., Hodgson, N., 2019. Hydrocarbon potential of hybrid depositional systems. GeoExpro 16 (4). https://www.geoexpro.com/articles/2019/10/hydrocarbon-potential-of-hybrid-depositional-systems.

Rodríguez-Tovar, F.J., Hernández-Molina, F.J., 2018. Ichnological analysis of contourites: past, present and future. Earth-Sci. Rev. 182, 28–41.

Rodríguez-Tovar, F.J., Hernández-Molina, F.J., Hüneke, H., Llave, E., Stow, D., 2019a. Contourite facies model: improving contourite characterization based on the ichnological analysis. Sediment. Geol. 384, 60–69. https://doi.org/10.1016/j.sedgeo.2019.03.010.

Rodríguez-Tovar, F.J., Hernández-Molina, F.J., Hüneke, H., Chiarella, D., Llave, E., Mena, A., Miguez-Salas, O., Dorador, J., de Castro, S., Stow, D.A.V., 2019b. Key evidence for distal turbiditic- and bottom-current interactions from tubular turbidite infills. Palaeogeogr. Palaeoclimatol. Palaeoecol. 533, 109233. https://doi.org/10.1016/j.palaeo.2019.109233.

Rogerson, M., 2002. Climatic influence on sediment transport in the Mediterranean outflow current (Gulf of Cadiz, Spain). Newsl. Micropalaeontol. 66, 16–17.

Rogerson, M., Rohling, E.J., Bigg, G.R., Ramirez, J., 2012. Paleoceanography of the Atlantic-Mediterranean exchange: overview and first quantitative assessment of climatic forcing. Rev. Geophys. 50, RG2003.

Rohling, E.J., Marino, G., Grant, K.M., 2015. Mediterranean climate and oceanography, and the periodic development of anoxic events (sapropels). Earth Sci. Rev. 143, 62–97.

Roque, C., Duarte, H., Terrinha, P., Valadares, V., Noiva, J., Cachão, M., Ferreira, J., Legoinha, P., Zitellini, N., 2012. Pliocene and quaternary depositional model of the Algarve margin contourite drifts (Gulf of Cadiz, SW Iberia): seismic architecture, tectonic control and paleoceanographic insights. Mar. Geol. 303, 42–62.

van Li, H., Loon, A.J., He, Y., 2020. Cannibalism of contourites by gravity flows: explanation of the facies distribution of the Ordovician Pingliang Formation along the southern margin of the Ordos Basin, China. Canad. J. Earth Sci. 57 (3), 331–347. https://doi.org/10.1139/cjes-2018-0225.

Rudge, J.F., Shaw Champion, M.E., White, N., McKenzie, D., Lovell, B., 2008. A plume model of transient diachronous uplift at the earth's surface. Earth Planet. Sci. Lett. 267, 146–160.

Salabarnada, A., Escutia, C., Röhl, U., Hans Nelson, C., McKay, R., Jiménez-Espejo, F.J., Bijl, P.K., Hartman, J.D., Strother, S.L., Salzmann, U., Evangelinos, D., López-Quirós, A., Flores, J.A., Sangiorgi, F., Ikehara, M., Brinkhuis, H., 2018. Paleoceanography and ice sheet variability offshore Wilkes Land, Antarctica—part 1: insights from late Oligocene astronomically paced contourite sedimentation. Clim. Past 14 (7), 991–1014. https://doi.org/10.5194/cp-14-991-2018.

Sansom, P., 2017. Turbidites v Contourites: hybrid systems of the Tanzanian margin. In: Extended Abstract, 16th PESGB/HGS African E&P Conference, London, 31 August-1 September 2017.

Sansom, P., 2018. Hybrid turbidite-contourite systems of the Tanzanian margin. Geol. Soc. London Petrol. Geosci. 24, 258–276. https://doi.org/10.1144/petgeo2018-044.

Scheuer, C., Gohl, K., Larter, R.D., Rebesco, M., Udintsev, G., 2006. Variability in Cenozoic sedimentation along the continental rise of the Bellingshausen Sea, West Antarctica. Mar. Geol. 277, 279–298. https://doi.org/10.1016/j.margeo.2005.12.007.

Schönfeld, J., Zahn, R., de Abreu, L., 2003. Surface and deep-water response to rapid climate changes at the Western Iberian Margin. Global Planet. Change 36, 237–264.

Schönfeld, J., Zahn, R., 2000. Late Glacial to Holocene history of the Mediterranean outflow. Evidence from benthic foraminiferal assemblages and stable isotopes at the Portuguese margin. Palaeogeogr. Palaeoclimatol. Palaeoecol. 159, 85–111.

Seibold, E., Berger, W.H., 2010. The Sea Floor. An Introduction to Marine Geology, third ed. Springer Verlag. 369 pp.

Shanmugam, G., 2012. New perspectives on deep-water sandstones: origin, recognition, initiation, and reservoir quality. In: Handbook of Petroleum Exploration and Production. vol. 9. Elsevier, Amsterdam, p. 524.

Shanmugam, G., 2000. 49 years of the turbidite paradigm (1950s-1990s): deep-water processes and facies models a critical perspective. Mar. Pet. Geol. 17, 174–231.

Shanmugam, G., 2006. Deep-Water Processes and Facies Models: Implications for Sandstone Petroleum Reservoirs: 5 (Handbook of Petroleum Exploration and Production). Elsevier Science. 496 pp.

Shanmugam, G., 2013. Modern internal waves and internal tides along oceanic pycnoclines: challenges and implications for ancient deep-marine baroclinic sands. AAPG Bull. 97 (5), 767–811.

Shanmugam, G., 2014. Modern internal waves and internal tides along oceanic pycnoclines: challenges and implications for ancient deep- marine baroclinic sands: reply. AAPG Bull. 98, 858–879.

Shanmugam, G., 2016. Slides, Slumps, Debris Flows, Turbidity Currents, and Bottom Currents. Reference Module in Earth Systems and Environmental Sciences. Elsevier. 87 p. (Online).

Shanmugam, G., 2017. The contourite problem. In: Mazumder, R. (Ed.), Sediment Provenance. Elsevier, pp. 183–254.

Shanmugam, G., 2021. Mass Transport, Gravity Flows, and Bottom Currents. Downslope and Alongslope Processes and Deposits. Elsevier, https://doi.org/10.1016/C2019-0-03665-5. 608 pp.

Shanmugam, G., Spalding, T.D., Rofheart, D.H., 1993a. Process sedimentology and reservoir quality of deep-marine bottom-current reworked sands (sandy contourites): an example from the Gulf of Mexico. Am. Assoc. Pet. Geol. Bull. 77, 1241–1259.

Shanmugam, G., Spalding, T.D., Rofheart, D.H., 1993b. Traction structures in deep-marine bottom-current reworked sands in the Pliocene and Pleistocene, Gulf of Mexico. Geology 21, 929–932.

Sierro, F.J., Hodel, D.A., Curti, J.H., Flores, J.A., Reguer, I., Colmenero-Hidalgo, E., Barcena, M.A., Grimalt, J.O., Cacho, I., Frigola, F., Canals, M., 2005. Impact of iceberg melting in the Mediterranean thermohaline circulation during Heinrich events. Paleooceanography 20, PA2019.

Sierro, F.J., Hodell, D.A., Andersen, N., Azibeiro, L.A., Jiménez-Espejo, F.J., Bahr, A., Flores, J.A., Ausin, B., Rogerson, M., Lozano-Luz, R., Lebreiro, S., Hernández-Molina, F.J., 2020. Mediterranean Overflow over the last 250 ky. Freshwater forcing from the tropics to the ice sheets. Paleoceanogr. Paleoclimatol. 35. e2020PA003931.

Sijp, W.P., von der Heydt, A.S., Dijkstra, H.A., Flögel, S., Douglas, P.M.J., Bijl, P.K., 2014. The role of ocean gateways on cooling climate on long time scales. Glob. Planet. Chang. 119, 1–22.

Solli, K., Kuvaas, B., Kristoffersen, Y., Leitchenkov, G., Guseva, J., Gandjukhin, V., 2008. The cosmonaut sea wedge. Mar. Geophys. Res. 29, 51–69. https://doi.org/10.1007/s11001-008-9045-x.

Spearman, J., Taylor, J., Crossouard, N., Cooper, A., Turnbull, M., Manning, A., Lee, M., Murton, B., 2020. Measurement and modelling of deep sea sediment plumes and implications for deep sea mining. Sci. Rep. 10, 5075. https://doi.org/10.1038/s41598-020-61837-y.

Stanley, D.J., 1987. Turbidite to current-reworked sand continuum in Upper Cretaceous rocks, U.S Virgin Islands. Mar. Geol. 78, 143–151. https://doi.org/10.1016/0025-3227(87)90073-9.

Stanley, D.J., 1988. Turbidites reworked by bottom currents: upper cretaceous examples from St. Croix, U.S. Virgin Islands. Smithson. Contrib. Mar. Sci. 33. https://doi.org/10.5479/si.01960768.33. 79 pp.

Stanley, D.J., 1993. Model for turbidite-to-contourite continuum and multiple process transport in deep marine settings: examples in the rock record. Sed. Geol. 82, 241–255.

Stow, D.A.V., Smillie, Z., 2020. Distinguishing between deep-water sediment facies: turbidites, contourites and hemipelagites. Geoscience 10, 68.

Stow, D.A.V., 1982. Bottom currents and contourites in the North Atlantic. Bull. Inst. Géol. Bassin d'Aquitaine 31, 151–166.

Stow, D.A.V., Faugères, J.C., 2008. Contourite facies and the facies model. In: Rebesco, M., Camerlenghi, A. (Eds.), Contourites. Developments in Sedimentology, vol. 60. Elsevier, Amsterdam, pp. 223–256.

Stow, D.A.V., Faugéres, J.C., Howe, J.A., Pudsey, C.J., Viana, A.R., 2002. Bottom currents, contourites and deepsea sediment drifts: current state-of-the-art. Geo. Soc. London Mem. 22, 7–20.

Stow, D.A.V., Hernández-Molina, F.J., Llave, E., Bruno, M., García, M., Díaz del Río, V., Somoza, L., Brackenridge, R.E., 2013a. The Cadiz Contourite Channel: sandy contourites, bedforms and dynamic current interaction. Mar. Geol. 343, 99–114.

Stow, D.A.V., Hernández-Molina, F.J., Alvarez Zarikian, C.A., The Expedition 339 Scientists, 2013b. Proceedings IODP, 339. Integrated Ocean Drilling Program Management International, Tokyo, https://doi.org/10.2204/iodp.proc.339.2013.

Stow, D.A.V., Hernández-Molina, F.J., Llave, E., Sayago-Gil, M., Díaz del Rio, V., Branson, A., 2009. Bedform-velocity matrix: the estimation of bottom current velocity from bedform observations. Geology 37, 327–330. https://doi.org/10.1130/G25259A.1.

Stow, D.A.V., Holbrook, J.A., 1984. North Atlantic contourites: an overview. In: Stow, D.A.V., Piper, D.J.W. (Eds.), Fine Grained Sediments, Deep-Water Processes and Facies. Geological Society, London, Special Publication 15, pp. 245–256.

Stow, D.A.V., Lovell, J.P.B., 1979. Contourites: their recognition in modern and ancient sediments. Earth-Sci. Rev. 14, 251–291.

Stow, D.A.V., Piper, D.J.W., 1984. Deep-water fine-grained sediments: facies models. In: Stow, D.A.V., Piper, D.J.W. (Eds.), Fine Grained Sediments, Deep-Water Processes and Facies. Geological Society, London, Special Publication 15, pp. 611–646.

Stow, D.A.V., Reading, H.G., Collinson, J., 1996. In: Seas, D., In: Reading, H.G. (Eds.), Sedimentary Environments, third ed, pp. 395–453.

Straume, E.O., Gaina, C., Medvedev, S., Nisancioglu, K.H., 2020. Global cenozoic paleobathymetry with a focus on the northern hemisphere oceanic gateways. Gondwana Res. 86, 126–143. https://doi.org/10.1016/j.gr.2020.05.011.

Sun, Q., Cartwright, J., Lüdmann, T., Wu, S., Yao, G., 2017. Three-dimensional seismic characterization of a complex sediment drift in the South China Sea: evidence for unsteady flow regime. Sedimentology 64 (3), 832–853.

Takashimizu, Y., Kawamura, R., Rodríguez-Tovar, F.J., Dorador, J., Ducassou, E., Hernández-Molina, F.J., Stow, D.A.V., Alvarez-Zarikian, C.A., 2016. Reworked tsunami deposits by bottom currents: circumstantial evidences from late pleistocene to early holocene in the Gulf of Cádiz. Mar. Geol. 377, 95–109. https://doi.org/10.1016/j.margeo.2015.09.009.

Teixeira, M., Terrinha, P., Roque, C., Rosa, M., Ercilla, G., Casas, D., 2019. Interaction of alongslope and downslope processes in the Alentejo Margin (SW Iberia)—implications on slope stability. Mar. Geol. 410, 88–108.

Teixeira, M., Terrinha, P., Roque, C., Voelker, A., Silva, P., Salgueiro, E., Abrantes, F., Naughton, F., Mena, A., Ercilla, G., Casas, D., 2020. The Late Pleistocene-Holocene sedimentary evolution of the Sines 1 Contourite Drift (SW Portuguese Margin): a multiproxy approach. Sed. Geol. 407 (105737), 1–17.

Thieblemont, A., Hernández-Molina, F.J., Miramontes, E., Raisson, F., Penven, P., 2019. Contourite depositional systems along the Mozambique Channel: the interplay between bottom currents and sedimentary processes. Deep-Sea Res. I 147, 79–99. https://doi.org/10.1016/j.dsr.2019.03.012.

Thiéblemont, A., Hernández-Molina, F.J., Ponte, J.-P., Robin, C., Guillocheau, F., Cazzola, C., Raisson, F., 2020. Seismic stratigraphic framework and depositional history for Cretaceous and Cenozoic contourite depositional systems of the Mozambique Channel, SW Indian Ocean. Mar. Geol. 425, 106192. https://doi.org/10.1016/j.margeo.2020.106192.

Thompson, P., Badalini, G., Wrigley, S., Hendy, J., Walker, R., Argent, J., Soto, M., de Santa Ana, H., Gristo, P., Hernández-Molina, F.J., 2015. Travelling without moving: ever-changing seabed morphology and the influence of contour currents, from Cretaceous to Present-day, within the Pelotas Basin, offshore Uruguay. In: IAS 31st Meeting of Sedimentology, Kraków, 22nd-25th June 2015. Abstract volume.

Thran, A.C., Dutkiewicz, A., Spence, P., Müller, R.D., 2018. Controls on the global distribution of contourite drifts: insights from an eddy-resolving ocean model. Earth Planet. Sci. Lett. 489, 228–240. https://doi.org/10.1016/j.epsl.2018.02.044.

Tilley, H., Moore, G.F., Underwood, M.B., Hernández-Molina, F.J., Yamashita, M., Kodaira, S., Nakanishi, A., 2021. Heterogeneous sediment input at the Nankai Trough subduction zone: implications for shallow slow earthquake localization. Geochem. Geophys. Geosyst. 22. https://doi.org/10.1029/2021GC009965. e2021GC009965.

Titschack, J., Fink, H.G., Baum, D., Wienberg, C., Hebbeln, D., Freiwald, A., 2016. Mediterranean cold-water corals—an important regional carbonate factory? Depos. Rec. 2, 74–96.

Titschack, J., Thierens, M., Dorschel, B., Schulbert, C., Freiwald, A., Kano, A., Takashima, C., Kawagoe, N., Li, X., 2009. Carbonate budget of a cold-water coral mound (Challenger Mound, IODP Exp. 307). Mar. Geol. 259, 36–46.

Tomczak, M., 1999. Some historical, theoretical and applied aspects of quantitative water mass analysis. J. Mar. Res. 57 (2), 275–303.

Toucanne, S., Jouet, G., Ducassou, E., Bassetti, M.A., Dennielou, B., Minto'o, C.M.A., Lahmi, M., Touyet, N., Charlier, K., Lericolais, G., Mulder, T., 2012. A 130,000-year record of Levantine Intermediate Water flow variability in the Corsica Trough, western Mediterranean Sea. Quat. Sci. Rev. 33, 55–73.

Toucanne, S., Mulder, T., Schönfeld, J., Hanquiez, V., Gonthier, E., Duprat, J., Cremer, M., Zaragosi, S., 2007. Contourites of the Gulf of Cadiz: a high-resolution record of the paleocirculation of the Mediterranean outflow water during the last 50,000 years. Palaeogeogr. Palaeoclimatol. Palaeoecol. 246, 354–366.

Toucanne, S., Soulet, G., Riveiros, N.V., Boswell, S.M., Dennielou, B., Waelbroeck, C., Bayon, G., Mojtahid, M., Bosq, M., Sabine, M., Zaragosi, S., Bourillet, J.F., Mercier, H., 2021. The North Atlantic Glacial Eastern Boundary Current as a key driver for ice-sheet—AMOC interactions and climate instability. Paleoceanogr. Paleoclimatol. 36. https://doi.org/10.1029/2020PA004068. e2020PA004068.

Trincardi, F., Verdicchio, G., Miserocchi, S., 2007. Seafloor evidence for the interaction between cascading along-slope bottom water masses. J. Geophys. Res. 112, F03011. https://doi.org/10.1029/2006JF000620.

Tucholke, B.E., 2002. The Greater Antilles Outer Ridge: development of a distal sedimentary drift by deposition of fine-grained contourites. In: Stow, D.A.V., Pudsey, C.J., Howe, J.A., Faugères, J.-C., Viana, A.R. (Eds.), Deep-Water Contourite Systems: Modern Drifts and Ancient Series, Seismic and Sedimentary Characteristics. Geol. Soc. London Mem., vol. 22, pp. 39–55, https://doi.org/10.1144/GSL.MEM.2002.022.01.04.

Tucholke, B.E., Ewing, J.I., 1974. Bathymetry and sediment geometry of the Greater Antilles Outer Ridge and vicinity. Geol. Soc. Am. Bull. 85, 1789–1802.

Tucholke, B.E., Mountain, G.S., 1986. Tertiary paleoceanography of the western North Atlantic Ocean. In: Vogt, P.R., Tucholke, B.E. (Eds.), The Geology of North America, The Western North Atlantic Region. vol. M. Geol. Soc. Am., Boulder, pp. 631–650, https://doi.org/10.1130/DNAG-GNA-M.631.

Uenzelmann-Neben, G., 2001. Seismic characteristics of sediment drifts: an example from the Agulhas Plateau, southwest Indian Ocean. Mar. Geophys. Res. 22, 323–343. https://doi.org/10.1023/A:1016391314547.

Uenzelmann-Neben, G., 2006. Depositional patterns at Drift 7, Antarctic Peninsula: along-slope versus down-slope sediment transport as indicators for oceanic currents and climatic conditions. Mar. Geol. 233, 49–62. https://doi.org/10.1016/j.margeo.2006.08.008.

Uenzelmann-Neben, G., Weber, T., Grützner, J., Thomas, M., 2017. Transition from the cretaceous ocean to Cenozoic circulation in the western South Atlantic—a twofold reconstruction. Tectonophysics 716, 225–240. https://doi.org/10.1016/j.tecto.2016.05.036.

UNFCCC, 2015. Adoption of the Paris Agreement—Proposal by the President. UNFCCC–United Nations Framework Convention on Climate Change, Paris, France. https://unfccc.int/sites/default/files/resource/docs/2015/cop21/eng/l09r01.pdf.

van Dijk, J., Ziegler, M., de Nooijer, L.J., Reichart, G.J., Xuan, C., Ducassou, E., Bernasconi, S.M., Lourens, L.J., 2018. A saltier Glacial Mediterranean outflow. Paleoceanogr. Paleoclimantol. 33, 179–197.

van Raaphorst, W., Malschaert, H., van Haren, H., Boer, W., Brummer, G.-J., 2001. Cross-slope zonation of erosion and deposition in the Faeroe-Shetland Channel, North Atlantic Channel. Deep-Sea Res. I 48, 567–591. https://doi.org/10.1016/S0967-0637(00)00052-2.

Van Rooij, D., Campbell, C., Rueggeberg, A., Wahlin, A., 2016. The contourite log-book: significance for palaeoceanography, ecosystems and slope instability. Mar. Geol., 1–4.

Van Rooij, D., Blamart, D., Kozachenko, M., Henriet, J.-P., 2007a. Small mounded contourite drifts associated with deep-water coral banks, Porcupine Seabight, NE Atlantic Ocean. In: Viana, A.R., Rebesco, M. (Eds.), Economic and Palaeoceanographic Importance of Contourite Deposits. Geological Society, London, Special Publication 276, pp. 225–244.

Van Rooij, D., Blamart, D., Richter, T., Wheeler, A., Kozachenko, M., Henriet, J.P., 2007b. Quaternary sediment dynamics in the Belgica mound province, Porcupine Seabight: ice-rafting events and contour current processes. Int. J. Earth Sci. 96, 121–140.

Van Rooij, D., De Mol, B., Huvenne, V., Ivanov, M.K., Henriet, J.-P., 2003. Seismic evidence of current-controlled sedimentation in the Belgica mound province, upper Porcupine slope, southwest of Ireland. Mar. Geol. 195, 31–53.

Van Rooij, D., Iglesia, J., Hernández-Molina, F.J., Ercilla, G., Gómez-Ballesteros, M., Casas, D., Llave, E., De Hauwere, A., García-Gil, S., Acosta, J., Henriet, J.-P., 2010. The Le Danois Contourite Depositional System: interactions between the Mediterranean Outflow Water and the upper Cantabrian slope (North Iberian margin). Mar. Geol. 274, 1–20.

van Sebille, E., Wilcox, C., Lebreton, L., Maximenko, N., Hardesty, B.D., van Franeker, J.A., Eriksen, M., Siege, D., Galgani, F., Lavender Law, K., 2015. A global inventory of small floating plastic debris. Environ. Res. Lett. 10 (12), 124006. https://doi.org/10.1088/1748-9326/10/12/124006.

Verdicchio, G., Trincardi, F., 2008. Shallow-water Contourites. In: Rebesco, M., Camerlenghi, A. (Eds.), Contourites. Developments in Sedimentology, vol. 60. Elsevier, Amsterdam, pp. 409–433.

Viana, A., 2008. Economic relevance of contourites. In: Rebesco, M., Camerlenghi, A. (Eds.), Contourites. Developments in Sedimentology, vol. 60. Elsevier, Amsterdam, pp. 493–510, https://doi.org/10.1016/S0070-4571(08)10023-1.

Viana, A.R., Rebesco, M., 2007. Economic and palaeoceanographic significance of contourite deposits. Geol. Soc. Lond. Spec. Publ. 276. https://doi.org/10.1144/GSL.SP.2007.276.

Viana, A.R., Almeida Jr., W., Almeida, C.F.W., 2002. Upper slope sands—the late quaternary shallow-water sandy contourites of Campos Basin SW Atlantic margin. In: Stow, D.A.V., Pudsey, C.J., Howe, J.A., Faugères, J.C., Viana, A.R. (Eds.), Deep-Water Contourite Systems: Modern Drifts and Ancient Series, Seismic and Sedimentary Characteristics. Geological Society, London, Memoirs, vol. 22, pp. 261–270.

Viana, A.R., Almeida Jr., W., Nunes, M., Bulhoes, E.M., 2007. The economic importance of contourites. In: Viana, A.R., Rebesco, M. (Eds.), Economic and Palaeoceanographic Significance of Contourite Deposits. Geol. Soc. London Spec. Publ., vol. 276, pp. 1–23, https://doi.org/10.1144/GSL.SP.2007.276.01.01.

Viana, A.R., Almeida, J.W., Machado, L.W., 1999. Different Styles of canyon infill related to gravity and bottom current processes: Example from the upper slope of the SE Brazilian Margin. In: Sexto Congresso International da Sociedade Brasileira de Geofisica, Rio de Janeiro, SBGF014., https://doi.org/10.3997/2214-4609-pdb.215.sbgf014. 4 pp.

Viana, A.R., Faugères, J.-C., Stow, D.A.V., 1998. Bottom current controlled sand deposits—a review of modern shallow- to deep-water environments. Sediment. Geol. 115, 53–80. https://doi.org/10.1016/S0037-0738(97)00087-0.

Viana, A.R., Faugères, J.C., 1998. Upper slope sand deposits: the example of Campos Basin, a latest Pleistocene-Holocene record of the interaction between along slope and down slope currents. In: Stoker, M.S., Evans, D., Cramp, A. (Eds.), Geological Processes on Continental Margins: Sedimentation, Mass-Wasting and Stability. Geol. Soc. Spec. Publ., vol. 129, pp. 287–316.

Voelker, A.H.L., Lebreiro, S.M., Schonfeld, J., Cacho, I., Erlenkeuser, H., Abrantes, F., 2006. Mediterranean outflow strengthening during northern hemisphere coolings: a salt source for the glacial Atlantic? Earth Planet. Sci. Lett. 245, 39–55.

Warner, M., 2017. Cairn in Senegal. Creating an Emerging Oil Basin on the Frontier. https://www.cairnenergy.com/media/2148/af_en_cairn-in-senegal-dakar-final-1.pdf.

Wells, M.G., Dorrell, R.M., 2021. Turbulence Processes Within Turbidity Currents. Annu. Rev. Fluid Mech. 53 (1), 59–83. https://doi.org/10.1146/annurev-fluid-010719-060309.

Wetzel, A., Werner, F., Stow, D.A.V., 2008. Bioturbation and biogenic sedimentary structures in contourites. In: Rebesco, M., Camerlenghi, A. (Eds.), Contourites. Developments in Sedimentology, vol. 60, pp. 183–202.

Wienberg, C., Titschack, J., 2017. Framework forming scleractinian cold-water corals through space and time: a late Quaternary North Atlantic perspective. In: Rossi, S., Bramanti, L., Gori, A., Saco, O., del Valle, C. (Eds.), Marine Animal Forests: The Ecology of Benthic Biodiversity Hotspots. Springer, Cham, pp. 699–732.

Wienberg, C., Titschack, J., Frank, N., de Pol-Holz, R., Fietzke, J., Eisele, M., Kremer, A., Hebbeln, D., 2020. Deglacial upslope shift of NE Atlantic intermediate waters controlled slope erosion and cold-water coral mound formation (Porcupine Seabight, Irish margin). Quat. Sci. Rev. 237, 106310.

Wunsch, C., 2002. What is the thermohaline circulation? Science 298 (5596), 1179–1181. https://doi.org/10.1126/science.1079329, PMID 12424356.

Yenes, M., Casas, D., Nespereira, J., López-González, N., Casalbore, D., Monterrubio, S., Alonso, B., Ercilla, E., Juan, C., Bárcenas, P., Palomino, D., Mata, P., Martínez-Díaz, P., Pérez, N., JVázquez, J.T., Estrada, F., Azpiroz-Zabala, M., Teixeira, M., 2021. The Guadiaro-Baños contourite drifts (SW Mediterranean). A geotechnical approach to stability analysis. Mar. Geol. 437 (2021), 106505. https://doi.org/10.1016/j.margeo.2021.106505.

Yin, S., Hernández-Molina, F.J., Hobbs, R., Gao, J., Ding, W., Yang, C., Lin, L., Zhang, H., Wu, Z., Li, J., 2021. Contourite processes associated with the overflow of Pacific Deep Water within the Luzon Trough: conceptual and regional implications. Deep-Sea Res. I: Oceanogr. Res. Pap., 103459.

Yin, S., Hernández-Molina, F.J., Zhang, W., Li, J., Wang, L., Ding, W., Ding, W., 2019. The influence of oceanographic processes on contourite features: a multidisciplinary study of the northern South China Sea. Mar. Geol. 415, 105967.

Yu, X., Stow, D.A.V., Smillie, Z., Esentia, I., Brackenridge, R.E., Xie, X., Bankole, S., Ducassou, E., Llave, E., 2020. Contourite porosity, grain size and reservoir characteristics. Mar. Pet. Geol. 117, 104392.

Zachos, J., Pagani, H., Sloan, L., Thomas, E., Billups, K., 2001. Trends, rhythms, and aberrations in global climate 65 Ma to present. Science 292, 686–693.

Zhang, X., Zhang, T., Lei, B., Zhang, J., Yong, J., 2020. A giant sandy sediment drift in early Silurian (Telychian) and its multiple sedimentological process. Mar. Pet. Geol. 113, 1040.

Zhang, Z., Nisancioglu, K., Flatøy, F., Bentsen, M., Bethke, I., Wang, H., 2011. Tropical seaways played a more important role than high latitude seaways in Cenozoic cooling. Clim. Past 7, 801–813.

Zhong, G., Cartigny, M.J.B., Kuang, Z., Wang, L., 2015. Cyclic steps along the South Taiwan Shoal and West Penghu submarine canyons on the northeastern continental slope of the South China Sea. GSA Bull. 127 (5–6), 804–824. https://doi.org/10.1130/B31003.1.

Zhong, Y., Chen, Z., Gonzales, F.J., Ortiz, J.E., Blanco, L., Marino, E., Liu, Q., 2021. Insights into the origin of ferromanganese-rich deposits associated with South China Sea contourite depositional systems. Mar. Pet. Geol. 133, 105257. https://doi.org/10.1016/j.marpetgeo.2021.105257.

Chapter 10

# Mass transport processes, injectites and styles of sediment remobilization

T.M. Alves[a], S. Cardona[b] and M.C.N.L. Rodrigues[c]

[a]*3D Seismic Laboratory, School of Earth and Environmental Sciences, Cardiff University, Cardiff, United Kingdom,* [b]*Sediment Mechanics Lab, Department of Geology & Geophysics, Texas A&M University, College Station, TX, United States,* [c]*Laboratory on Basin Analysis (Laboratório de Análise de Bacias, LABAP), Departamento de Geologia—Setor Ciências da Terra, Universidade Federal do Paraná, Centro Politécnico—Jardim das Américas, Curitiba, Brazil*

## Introduction

Sediment remobilization is a common process associated with the early stages of sediment burial, diagenesis and fluid migration in sedimentary basins. It can be divided into two main types depending on relative timing and depth of occurrence: (a) mass transport deposits, or complexes, materializing erosion and an overall redistribution of near-seafloor strata and (b) relatively deeper sediment injections and fluid-flow features capable of extruding sediment onto the seafloor and surface. This chapter will focus on mass transport as a process and review associated deformation styles. The aim is to understand the diagnostic features that identify these processes in outcrop and seismic data of hydrocarbon-rich basins. The styles of sediment remobilization resulting from fluid flow and the build-up of overpressure in sediment will also be summarized in this work and identified as a key process occurring below the surface in many a sedimentary basin.

Mass transport deposits (MTDs) and mass transport complexes (MTCs) are often used interchangeably in the scientific community despite representing distinct scales and degrees of instability on continental slopes. The various types of mass movements occurring in nature generate a wide spectrum of deposits, which are better referred to as MTCs when they are clearly associated in space and time (Pickering and Hiscott, 2016). Such a broad definition considers all types of gravitational flows, including noncohesive turbidites and grainflows, as comprising MTCs. However, this chapter considers sediment remobilization as reworking of previously deposited sediments; therefore, these sediments tend to retain a degree of cohesion during their transport and hence exhibit laminar behavior. This is a key characteristic that distinguishes remobilized deposits (e.g., debrites, slumps and slides) from turbidites (Middleton and Hampton, 1976). In parallel, the most recent literature discriminates between MTDs and MTCs based on their geometry and character on seismic, borehole and outcrop data. When a single landslide event is believed to have generated a single deposit this is called a MTD. When several stacked MTDs are identified on seismic and sedimentological data, often based on the recognition of distinct basal glide zones separating successive, but discrete MTDs, the term most often used is MTC (Pickering and Hiscott, 2016). Significantly, while MTDs and MTCs typically occur in marine or lacustrine (i.e., subaqueous) environments, sediment injection and associated fluid flow can occur in both onshore and offshore sedimentary basins (Hurst et al., 2011; Andresen, 2021).

The notion that mass transport complexes (MTCs) and deposits (MTDs) may be associated with hydrocarbon accumulations is not novel among explorationists (Fairbridge, 1946). Such deposits have historically been considered as depositional facies to avoid from a viewpoint of hydrocarbon exploration (Posamentier and Kolla, 2003; Weimer and Slatt, 2004), but it is now understood that MTCs and MTDs can contain hydrocarbon source intervals (Tanavsuu-Milkeviciene and Sarg, 2012; Johnson et al., 2015), reservoirs (Shanmugam et al., 2009; Meckel, 2011; Bhatnagar et al., 2019) or constitute competent seal units (Godo, 2006; Alves, 2010a; Algar et al., 2011; Cardona et al., 2016, 2020b; Kessler and Jong, 2018; Amy, 2019)—see Chapter 7 in this book. In addition, the relief created by the emplacement of MTCs and MTDs on the seafloor can influence the pathways of post-emplacement turbidite flows (Armitage et al., 2009; Jackson and Johnson, 2009; Kneller et al., 2016; Ward et al., 2018; Henry et al., 2018), creating space on the seafloor for "healing phase" top-fill reservoir targets (Wood et al., 2015). Examples of such a control on hydrocarbon trapping and accumulation are documented in the Ubit field of Nigeria with ~2 billion barrels of oil (BBO) (Clayton et al., 1998), the Tarn field in the North Slope Borough of Alaska with ~100 MMBO, and the Meltwater field also in the North Slope of Alaska with ~50 MMBO (Houseknecht and Schenk, 2007; Houseknecht, 2019), to name three examples.

**Deepwater Sedimentary Systems.** https://doi.org/10.1016/B978-0-323-91918-0.00013-X

Sediment remobilization occurs in multiple geological settings and can markedly impact the depositional architecture of sedimentary basins (Roy et al., 2019; Palan et al., 2020; Wenau et al., 2021). Furthermore, it is capable of redistributing subsurface stresses, fluid and heat, in quasi-instantaneous episodes at the geological time scale (Ho et al., 2018; Roelofse et al., 2020). An example is the loss of effective pressure below the paleo-seafloor when large landslides occur via the sudden escape of fluid to the surface, as documented in large, blocky MTDs (Alves, 2010b, 2015). Vast areas with sand injections—themselves representing fluid migration paths and overpressure markers—are also strong enough to redistribute fluid and sediment through previously competent seal units (Sun et al., 2017). Finally, the largest of MTCs on continental margins, namely those associated with the mobilization of slide blocks (also named megaclasts) and subsequent deposition of megabreccias, have been systematically shown to form key markers of tectonism in sedimentary basins. This tectonism can occur at a regional scale (Alves, 2015; Festa et al., 2016; Naranjo-Vesga et al., 2020), or locally in association with tectonically controlled topography (Alves and Cupkovic, 2018).

As summarized in Chapter 6, MTDs and MTCs comprise a wide spectrum of deposits and represent distinct styles of sediment movement or remobilization. MTDs and MTCs include rock falls, creeps, slides, slumps and flows presenting some degree of internal cohesion—therefore excluding turbidites and noncohesive grainflows from their definition. They can comprise more than 50% of the sedimentary record of a single continental margin and are known to respond to varied tectonically and climatically driven triggers (Masson et al., 2006). Hence, modern sediment remobilization processes fall into the realm of Geohazards in economic and physical terms.

This chapter is divided into specific sections addressing the aspects of sediment remobilization in sedimentary basins (Fig. 1). After this introduction, a review of the current knowledge (state-of-the-art) about fluid remobilization is developed to inform readers of recent developments under this theme. The regional distribution of field analogs and recognized areas with sediment remobilization are summarized in the following sections and complemented with relevant examples in seismic data. The chapter will conclude with an analysis of how MTDs or complexes at different scales may help defining the potential impact of such sediment remobilization in petroleum plays, highlighting the importance of studying sediment remobilization as a potential geohazard.

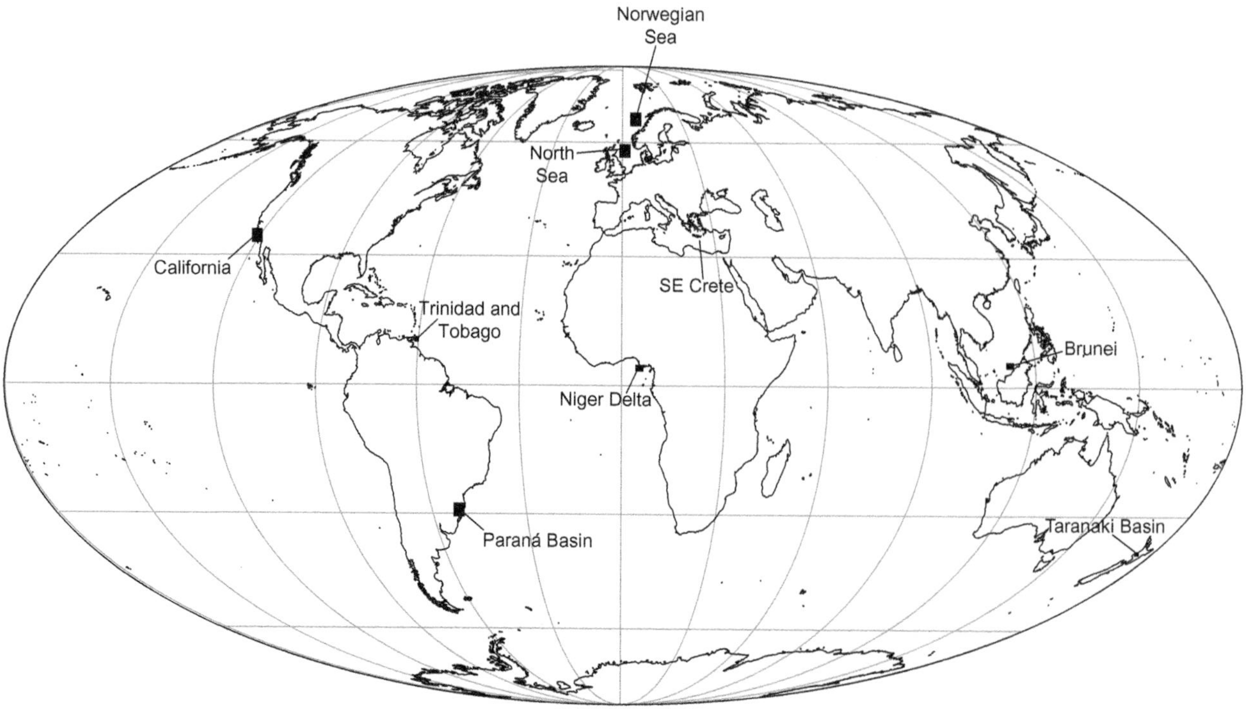


**FIG. 1** World map indicating the location of the case studies documented in this chapter. The distribution of the sediment remobilization features referred to in this chapter is associated with the regional tectonic setting and includes: (A) mass-wasting on tectonically active continental margins, (B) sand injection features in discrete overpressured basins, (C) mass-flows (nonglacial) occurring in proglacial marine deposits in relation to ice-margin retreat in an intracratonic basin. Main triggers of sediment remobilization are varied and have a clear relationship to local physiography and the stress history of sediments below these same remobilization features.

## Data and methods

This work uses three-dimensional (3D) seismic and outcrop data from distinct regions to illustrate common aspects of MTCs and MTDs. In particular, our work will present seismic examples from Trinidad and Tobago, Niger Delta and the North Sea, and outcrop data from Crete (Eastern Mediterranean), Taranaki Basin (New Zealand) and Paraná Basin (South Brazil) (Fig. 1).

### Seismic interpretation

Seismic data in this work include vertical seismic profiles of high resolution and quality complemented by seismic-attribute maps from the published literature. To characterize MTCs, seismic attributes of interest include root-mean-square (RMS) amplitude, maximum magnitude and coherence (or variance), which are all capable of highlighting structural and depositional features of particular areas of continental slopes affected by slope instability (Alves et al., 2014). In detail, RMS amplitude maps depict the average squared amplitude from individual samples within a defined interval (Brown, 2011). Maximum magnitude undertakes a similar process to RMS amplitude, but the seismic amplitudes computed are all positive, thus stressing the presence of the higher amplitude features against low-amplitude background strata. Coherence attributes convert a seismic volume of continuity (normal reflections) into a volume of discontinuity, allowing features such as faults, folds and ridges and stratigraphic boundaries, to be emphasized on time slices (Marfurt and Alves, 2015). The locations of the seismic surveys and outcrops used in this study are shown in Fig. 1.

### Outcropping mass transport deposits in SE Crete (Eastern Mediterranean)

In SE Crete a series of Miocene-Quaternary units were deposited on the tectonically active continental slopes that bordered what were, at the time, multiple islets and transtensional/extensional basins in the South Aegean Sea (Alves and Cupkovic, 2018; Sakellariou and Tsampouraki-Kraounaki, 2019) (Fig. 1). This area has been used as an analog of a marine rift basin, distinct from the Gulf of Corinth and Sperchios Basin in Central Greece insomuch as slope depositional processes dominate over the shallow-marine to continental facies predominating in Central Greece. Furthermore, the erosion of distinct and spatially variable basement units allows a more complete understanding of the effect of basement lithology and regolith types on sediment distribution.

Mass transport and turbidite deposits predominated around Crete during a Serravalian-Tortonian "syn-rift" phase that led to a widespread deepening of transtensional/extensional basins. At the end of the Miocene at least four (4) individual islands composed what is now the elongated E-W island of Crete, and the slopes of these islands bordered transtensional basins roughly striking N20 and N70 (Alves and Lourenço, 2010). Marine deposits predominated in this overall tectonic setting, leading to the concomitant deposition of proximal, coarser deposits near the four highs (islands) that existed at the time with slope turbidites and MTDs on adjacent slopes and deep-marine clays in the main basin depocenters. In SE Crete, it is possible to map and document the juxtaposition of such facies and to record the geometry of mass-wasting deposits sourced from distinct basement lithologies.

### Outcropping mass transport deposits in the Taranaki Basin (New Zealand)

In the North Island, New Zealand, Late Miocene (Tortonian) outcrops of MTDs are exposed in the Lower Mount Messenger Formation (LMMF). Following its initial development within a synrift setting the Taranaki Basin evolved as a passive margin in Late Cretaceous to Paleogene times (King and Thrasher, 1996; Strogen et al., 2017). Structural, sedimentological and paleogeographic evidence shows that the Taranaki Basin remained a passive margin throughout much of the Paleogene (King and Thrasher, 1992; Baur et al., 2014; Strogen et al., 2014).

The case study used in this chapter is the Rapanui MTD (RMTD). The strata analyzed in this study were deposited in middle to lower bathyal water depths during the Late Miocene (King et al., 1993) (Fig. 1). The shelf was relatively narrow (c. 10–15 km, or 6.2–9.3 mi, wide) and located immediately east of the study area (King et al., 2011). In general, the regional Late Miocene paleoflow direction in the Taranaki basin was to the northwest (King et al., 2011; Masalimova et al., 2016). These clastic sediments were mixed with coeval volcaniclastic material derived from the submarine andesitic Mohakatino volcanic arc in the northern offshore part of the basin (Giba et al., 2013; Shumaker et al., 2018) and were punctuated by several MTDs (King et al., 2011; Rotzien et al., 2014; Sharman et al., 2015; Masalimova et al., 2016).

## Outcrop examples from Paraná Basin (S Brazil)

The outcrop examples from South Brazil correspond to MTDs from the Itararé Group, exposed in three regions along the east border of Paraná Basin, as studied by Rodrigues et al. (2020). The Paraná Basin forms a large elongated (NNE-SSW) intracratonic basin (up to 1,600,000 $km^2$, or 617,763 $mi^2$) situated in southeast South America (Fig. 1). The stratigraphy and structural evolution of the Paraná Basin were controlled by tectonic trends inherited from a heterogeneous basement, which comprises cratonic terrains and orogenic belts agglutinated during the Brazilian Orogeny (Zalán et al., 1990). Several eustatic–tectonic cycles reactivated the basement structural trends and controlled sedimentation from the Ordovician to the Early Cretaceous (Milani et al., 1994).

The Itararé Group (Late Mississippian to Cisuralian) forms the lower half of a Permian-Carboniferous supersequence that is up to 2.5 km (1.6 mi) thick (França and Potter, 1991; Holz et al., 2010; Schneider et al., 1974; Zalán et al., 1990). Sediments were mostly accumulated in marginal to relatively deep-marine environments during multiple deglaciation episodes associated with the late Paleozoic ice age that affected southwestern Gondwana (Vesely and Assine, 2006; Valdez-Buso et al., 2019; Fallgatter, 2015). França and Potter (1991) subdivided the Itararé Group into three basin-wide lithostratigraphic intervals that correspond broadly to the formations previously defined by Schneider et al. (1974) and can be correlated with the three palynozones defined by Souza (2006). The examples in this study include MTDs from the Lagoa Azul, Campo Mourão and, mainly, the Taciba Formation, at the best outcrop exposures of the southernmost region of Rio do Sul (Fig. 1).

After the deposition of the Itararé Group, the Paraná Basin was affected by tectonic deformation related to the reactivation of ancient basement faults. This resulted from tectonic convergence at the active margin of the South American Plate and opening of the Atlantic Ocean (Milani, 2004, 1997; Soares, 1991). Yet, this post-sedimentary deformation is easy to differentiate from mass transport deformation—brittle structures formed during the youngest deformation episodes crosscut several layers and deposits and are associated with regional structures (Rostirolla et al., 2000, 2002, 2003; Trzaskos et al., 2006).

# Classifying sediment remobilization using seismic and outcrop data

Fundamentally, the processes behind sediment remobilization (i.e., mass transport) occur when the shear component of gravity as an underlying force surpasses the shear strength of a deposit due to allogenic or autogenic factors (Dott, 1963; De Blasio, 2011). Such phenomena are more likely to occur in poorly consolidated sediments (i.e., sediments near the surface or seafloor) than in deeply buried (> ~100 m, ~ 328 ft.) units, although evidence for large-scale remobilization of slide blocks is recorded in areas such as offshore Morocco (Dunlap et al., 2010), SE Brazil (Gamboa et al., 2010), Alaska's North Slope (Bhattacharya et al., 2020) and off Greenland (Cox et al., 2020). Therefore, sediment remobilization is not an uncommon phenomenon and can occur in subaerial and subaqueous environments. Consequently, deposits resulting from sediment remobilization have been documented across different geologic periods in both siliciclastic (Ogiesoba and Hammes, 2012; Cardona et al., 2016; Sun and Alves, 2020; Gutierrez and Snedden, 2021; Steventon et al., 2021) and carbonate settings (Eyles and Eyles, 2001; Jablonská et al., 2018; Le Goff et al., 2020). More recently, they have also been documented in mixed-composition settings (Moscardelli et al., 2019; Cumberpatch et al., 2020; Walker et al., 2021).

Following the definition of Pickering and Corregidor (2005), deposits resulting from sediment remobilization were routinely termed by the energy industry as MTCs. This term was first introduced by Weimer (1989) and its original definition had a clear sequence stratigraphic meaning. Later, Weimer and Shipp (2004) considered the term MTC as applying to "... features at a scale that can only be completely imaged on volumetrically large seismic surveys." Recent studies have demonstrated that deposits identified as MTCs are often composed of multiple individual MTDs typically below the resolution of conventional seismic data (Alves and Lourenço, 2010; Dykstra et al., 2011; Sobiesiak et al., 2017; Cardona et al., 2020b; Jablonska et al., 2021). Although numerous geoscientists use the terms MTDs and MTCs interchangeably, we prefer the definition of MTCs as depositional architectural complexes comprised of several coeval or closely chronologically and genetically related MTDs (Alves and Lourenço, 2010; Pickering and Hiscott, 2016; Cardona et al., 2020a). This hierarchical classification is also helpful when distinguishing between MTCs on the seismic scale (~50–200 ms two-way-traveltime (TWT) time thickness or >10 m, 33 ft., thick) from those identified at the scale of the outcrop, core and borehole data, i.e. the so-called MTDs (<10 m, 33 ft., thick). In addition, submarine MTCs can be further classified into attached or detached MTCs based on their relative source areas. Attached-MTCs are those sourced either from shelf or slope settings (i.e., shelf-attached or slope-attached MTCs), whereas detached-MTCs are sourced from isolated bathymetric highs (Moscardelli and Wood, 2008; Ortiz-Karpf et al., 2018) (Fig. 2). In general, attached-MTCs have areas over

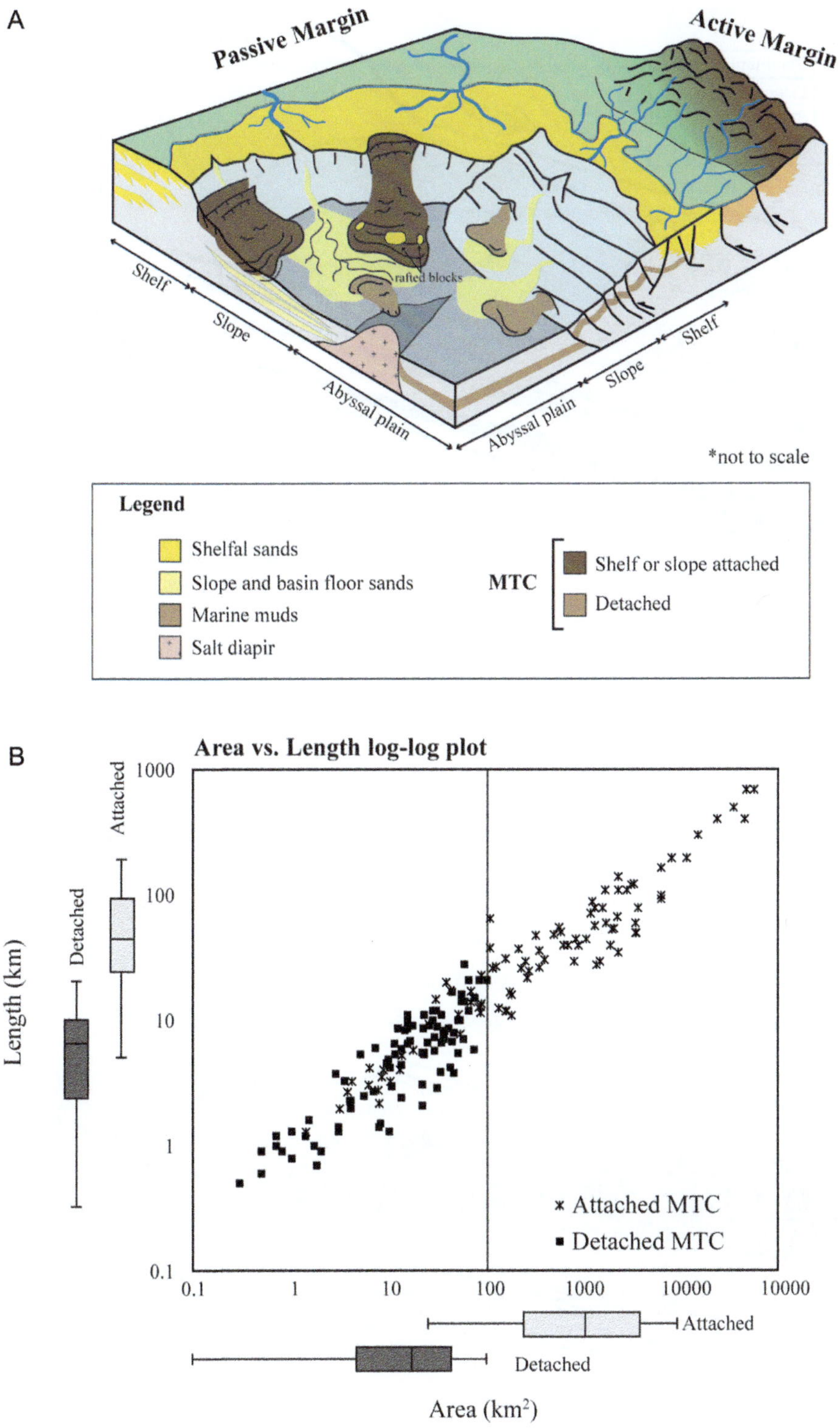


**FIG. 2** (A) Schematic model of attached and detached mass transport complexes (MTCs) on active and passive margins. (B) Log-log scatter plot of length (km) versus area ($km^2$) for several detached and attached MTCs. Note that attached-MTCs (either slope or shelf attached) typically cover larger areas when compared to detached MTCs. *(Data for plot found in Moscardelli, L., 2007. Mass Transport Processes and Deposits in Offshore Trinidad and Venezuela, and their role in Continental Margin Development. PhD Dissertation Thesis, The University of Texas at Austin, 185 pp.)*

100 $km^2$ (38.6 $mi^2$) and are approximately an order of magnitude broader than most detached-MTCs (Moscardelli and Wood, 2016).

The planform and cross-section architectures of MTCs, which are typically imaged in seismic data, can be divided into three strain-dominated domains, from proximal to distal along dip-direction with respect to the sediment source; the headwall, the translational and the toe domains (Lewis, 1971; Bull et al., 2009; Steventon et al., 2019) (Fig. 3). Each domain

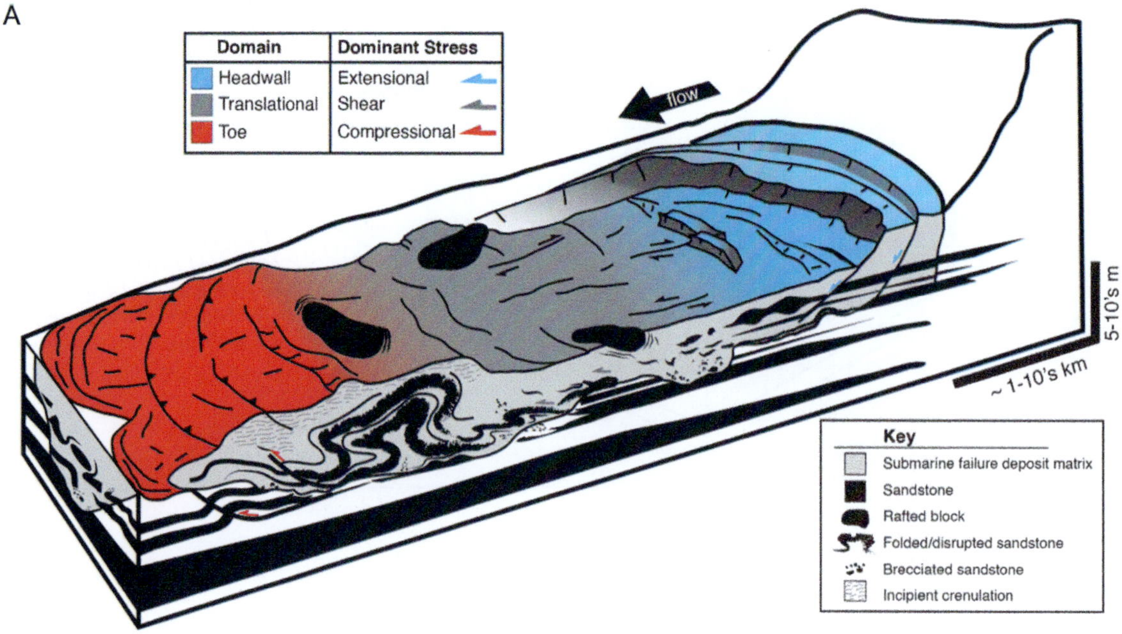


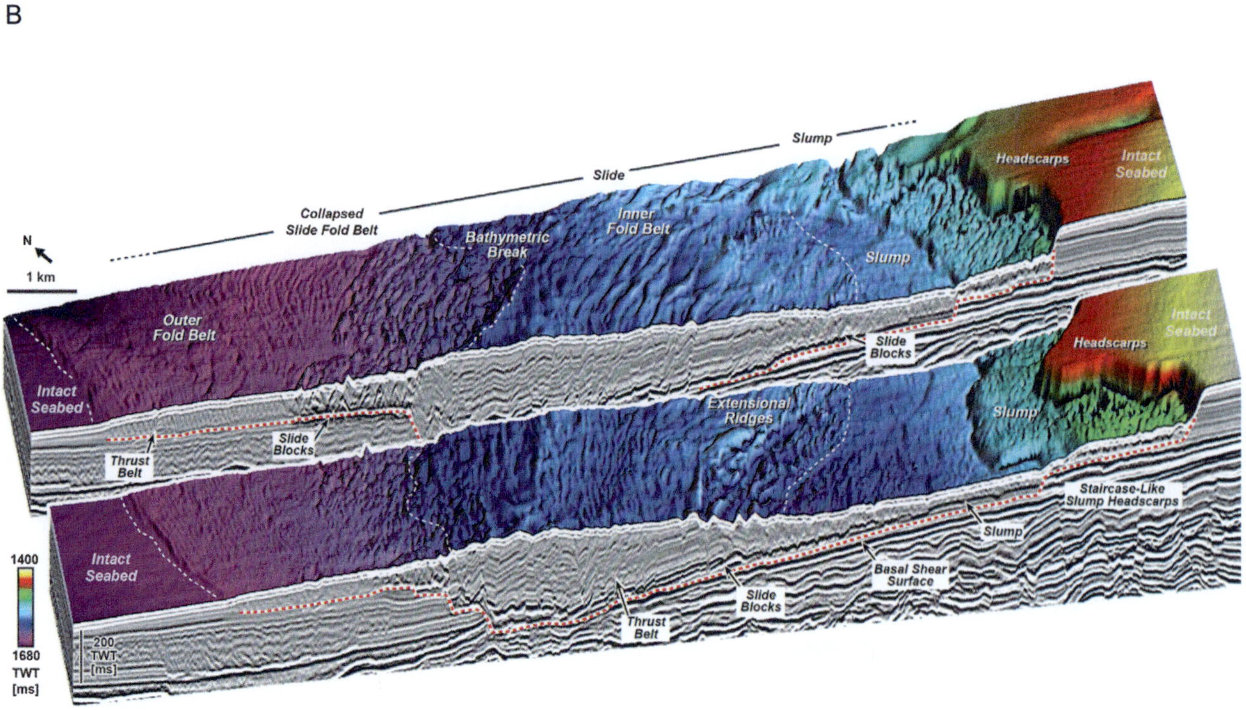


**FIG. 3** (A) Idealized submarine failure deposit with the three typical MTD domains and corresponding dominant stress. (B) Seismic example of a submarine mass transport complex (MTC) at the seafloor in Exmouth Plateau, NW shelf of Australia, whereby stress-dominant domains are characterized by the identification of strain features. *(Figure is modified from Scarselli, N., 2020. Submarine landslides—architecture, controlling factors and environments. A summary. In: Scarselli, N., Adam, J., Chiarella, D., Roberts, D.G., Bally, A.W. (Eds.), Regional Geology and Tectonics—Second Edition. Vol. 1: Principles of Geological Analysis, Elsevier, pp. 417–439, https://doi.org/10.1016/B978-0-444-64134-2.00015-8 (chapter 16).)*

is characterized by a locally dominant stress regime manifested by distinct syn-depositional deformation structures (e.g., boudinage, folds, faults, pressure ridges, flow fabrics, etc.). In general, the headwall domain of MTCs is dominated by extensional stress, manifested by the omnipresence of normal faults and extensional ridges (Martinsen and Bakken, 1990; Bull et al., 2009; Gamboa and Alves, 2016; Doughty-Jones et al., 2019). The translational domain is characterized by structures associated with longitudinal shear stress (Bull et al., 2009; Cardona et al., 2016; Safadi et al., 2017; Steventon et al., 2019). The toe domain is dominated by compressional stress caused by the arrest of the mass flow front, which creates a buttressing effect (Farrell, 1984; Frey-Martínez et al., 2006; Eng and Tsuji, 2019; Nugraha et al., 2020). When outcrops of MTDs are sufficiently exposed, it is possible to identify these key domains and their associated syn-depositional deformation structures (Alves and Lourenço, 2010; Alves, 2015; Sharman et al., 2015; Le Goff et al., 2020; Cardona et al., 2020a, b).

## Sediment remobilization as an overarching physical process

Sediment remobilization reflects a broad spectrum of phenomena, from soft-sediment instability near the sea floor to deeper fluid-driven injection of sand and fluid usually related with the build-up of overpressures below the seafloor. Sediment remobilization also records different scales and shows distinct areal distributions in sedimentary basins, from discrete features such as fluid pipes and chimneys to widespread fields of mud volcanoes and sand injection features (Milkov, 2000; Hurst et al., 2011). Close to the seafloor, MTCs and MTDs can be of different scales and potentially affect continental slopes for long periods of time. Sediment remobilization occurs on most continental margins and is recorded at outcrop in multiple regions of the world.

Several studies have focused on the geomorphological characterization of lithological and structural variations within MTDs using remote data (e.g., seismic, side-scan sonar), core and outcrop information with the purpose of understanding the process of mass failure and its controlling factors (Moscardelli et al., 2006; Frey-Martínez, 2010; Alves and Lourenço, 2010; Dykstra et al., 2011; Ogata et al., 2014; Moscardelli and Wood, 2016; Sobiesiak et al., 2017; Cardona et al., 2020a). The distribution, frequency and internal character of submarine landslides have revealed some key insights. According to Hühnerbach and Masson (2004), a series of landslide headscarps occur at a water depth of 1000–1300m (3281–4265 ft.) and the influence of slope gradient on landslide distribution is seemingly limited. However, Hühnerbach and Masson (2004) also recognized that the largest landslides occur in the gentler (lower angle) continental slopes, where strata involved in the mass transport of sediment can affect broader areas. In parallel, Urgeles and Camerlenghi (2013) identified the presence of many small failures on active margins, proving at the same time that passive margins show fewer but relatively larger landslides. More recently, Moscardelli and Wood (2016) have analyzed the morphometry of several MTCs from different settings around the world. Morphometric parameters such as length, area, volume and thickness of MTCs suggest that the geometry, geological setting and causal mechanisms are to a variable degree linked, and that predictive models can be applied in areas with incomplete or low-qualitydata.

MTDs consist of chaotic, convoluted strata resulting from meters to hundreds of meters of sediments being translated downslope, sometimes for hundreds of kilometers. In seismic data, these deposits are commonly composite bodies of blocky to highly disrupted, chaotic seismic facies with variable amplitude (Moscardelli and Wood, 2008; Bull et al., 2009; Sawyer et al., 2009; Posamentier and Martinsen, 2011; Gamboa and Alves, 2015; Scarselli et al., 2016). They are common geomorphic features and structures, most of which were generated during mass flows, that may allow an understanding of MTDs' development. Many of these features can also be used as kinematic indicators (Woodcock, 1979; Farrell, 1984; Strachan, 2008; Bull et al., 2009). Ultimately, the identification of geomorphic features and structures within MTDs allows the definition of three main regions within these deposits: (1) a headwall region dominated by extension, which comprises a headwall scarp, extensional ridges, blocks, normal faults and boudins, (2) a translational region dominated by basal shearing and sediment downslope movement where a transition from extensional deformation (or stretching) to compression occurs and (3) a toe region dominated by compression and marked by the development of folds and thrust systems in frontally confined landslides or pressure ridges resulting from thrusts in frontally emergent/unconfined landslides (Martinsen and Bakken, 1990; Frey-Martínez et al., 2006) (Fig. 3).

The basal shear zones of MTDs document different styles of interaction between the material being remobilized and the substrate (Alves and Lourenço, 2010; Alves, 2015; Sobiesiak et al., 2018; Cardona et al., 2020a, b). MTDs may also show some distinctive geomorphic features and structures such as compressional and extensional features not restricted to their toe and headwall regions respectively as a result of the presence of blocks, substrate topography or complex soft-sediment deformation processes (Strachan, 2008; Alsop and Marco, 2011, 2014). Finally, some blocks may travel beyond the edge of the toe region, the so-called outrunner blocks, to form basal shear surface striations (Nissen et al., 1999; Moscardelli et al., 2006; Kumar et al., 2021).

## Sediment remobilization as a geohazard in submarine environments

Mass movements are commonly attributed to different triggering mechanisms, usually a combination of local and regional factors such as slope geometry and changes in slope gradient, evolving tectonic settings, local and far-field seismicity, high sedimentation rates and seafloor deformation associated with salt tectonics, pore-fluid overpressure, destabilization of gas hydrates, isostatic rebound, glacial-eustatic low-stands, climate change and rapid variations in sea level (Nisbet and Piper, 1998; McAdoo et al., 2000; Posamentier and Kolla, 2003; Sultan et al., 2004; Lee et al., 2007; Alves et al., 2009; Bertoni et al., 2013; Berton and Vesely, 2016). In addition, mass flow initiation, and the distribution of associated deposits, depend on changes in rheological properties, liquefaction, diagenesis, differential compaction and fluid expulsion (Kvalstad et al., 2005; Bryn et al., 2005; Locat et al., 2014). For the well-studied Storegga Slide, offshore Norway, the failure is considered to have been preconditioned by high pore-fluid pressure related to high sedimentation rates that followed glacial retreat and was later triggered by an earthquake (and subsequent slope undercutting) on the lower continental slope in association with isostatic rebound (Haflidason et al., 2003; Kvalstad et al., 2005; Bellwald et al., 2019). A local control of strike-slip faults is also suggested for its northern flank, near the Modgunn Arch (Song et al., 2020).

Mass movement on continental slopes, volcanic islands or seamounts is often damaging to infrastructure and life, with the added danger of being able to trigger tsunamis that may affect coastal populations (Nisbet and Piper, 1998; Masson et al., 2006; Paris et al., 2020). In the particular case of volcanic islands, their evolution is a result of periods of rapid volcanic-edifice growth, followed by the large-scale collapse of their flanks (Boulesteix et al., 2013; Hunt et al., 2013), phenomena that often involve the remobilization of volumes of rock greater than 100 $km^3$ (24 $mi^3$) and comprising some of the largest mass movements in the world (Masson et al., 2002, 2008; Oehler et al., 2008). In the Canary Islands, NW Africa, the last 2 million years saw the onset of 11 different landslides that remobilized strata and rocks from different eruptive complexes (Masson et al., 2002; Boulesteix et al., 2013). These 11 landslides triggered tsunamis such as that associated with flank failure on the eastern coast of Tenerife (Güìmar and La Orotava mega-landslides) as proven by the accumulation of tsunami deposits on the Agaete Valley of Gran Canaria (Pérez-Torrado et al., 2006; Giachetti et al., 2011; Paris et al., 2018). Modeling studies have predicted the future collapse of the Cumbre Vieja volcano of La Palma as a single block with a volume of 500 $km^3$ (120 $mi^3$) and a thickness of 1400 m (4593 ft.), but recording different amplitudes and impact in coastal areas. Models have shown that the ensuing tsunami would mostly affect near-field areas of the Canary Islands and Northwest Africa, potentially inundating the Atlantic coastline of the United States of America and Western Europe in a worst-case scenario (Ward and Day, 2001; Mader, 2001; Gisler et al., 2006; Lovholt et al., 2008; Abadi et al., 2012; Tehranirad et al., 2015). As of the final compilation of this chapter (October 2021), the Cumbra Vieja volcano is erupting and feeding large volumes of lava to the SW flank of La Palma. Large-volume landslides resulting from flank collapses of ocean island volcanoes are thus considered one of the main causes of mega-tsunamis (Paris et al., 2018, 2020), but catastrophic tsunami waves can also be trigged by relatively small subaqueous landslides such as that of Anak Krakatau in Indonesia (September 2018). This submarine landslide remobilized a volume of 0.22–0.3 $km^3$ (Grilli et al., 2019; Zengaffinen et al., 2020) and generated a tsunami with a height approaching 13 m (43 ft.), resulting in the loss of more than 400 lives (Muhari et al., 2019).

## Sediment remobilization due to fluid flow and subsurface overpressure

Sediment remobilization is also considered a geohazard when associated with subsurface fluid flow, as the sudden release of overpressured fluid may jeopardize offshore drilling sites, their associated infrastructure, new $CO_2$ sequestration storage sites and even structures for geothermal, wind and solar energy production (Roelofse et al., 2020 and Chapter 21 of this book). In parallel, characterizing and dating fluid flow and structures associated with mass wasting can provide information concerning past events of fluid migration in a sedimentary basin. It also contributes to a better characterization of the magmatic processes leading to the escape of heat and hydrothermal fluids from deeper parts of the crust.

Mud volcanoes are a category of fluid vent structures (Fig. 4A) formed in sedimentary basins as the surface expression of hydrocarbon migration. This results in the extrusion of a three-phase mixture of solids and fluids (fine-grained solid-phase derived from underlying sediments and at least two fluid phases, water and gas—mainly methane, and occasional liquid hydrocarbons (Dimitrov, 2002; Etiope, 2015; Judd and Hovland, 2009; Kopf et al., 2001; Levin, 2005; Mazzini and Etiope, 2017). They are the surface expression of subsurface processes marked by the movement of large masses of sediments and fluids, such as diapirs, diatremes, domes, dewatering pipes, mud intrusions, mud mounds, chimneys and pipes (Kopf, 2002; Mazzini and Etiope, 2017; Skinner and Mazzini, 2009). The mixture of solids and fluids is commonly sourced from depths of several kilometers (Blouin et al., 2019; Kirkham, 2015; Stewart and Davies, 2006).

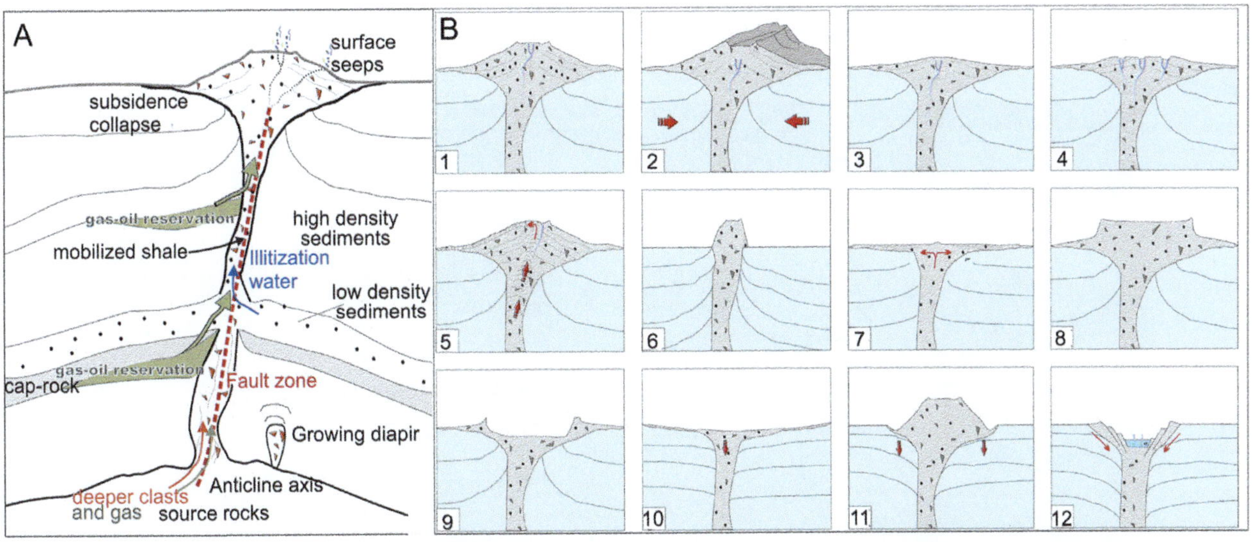


**FIG. 4** (A) Conceptual drawing indicating the main elements that compose most mud volcanoes and corresponding sources of fluids (from Mazzini and Etiope, 2017). (B) Mud volcanoes morphologies: (1) conical, (2) elongated, (3) pie-shaped, (4) multicrater, (5) growing diapir-like, (6) stiff neck, (7) -swamp-like, (8) plateau-like, (9) impact craterlike, (10) subsiding structure, (11) subsiding flanks and (12) sink-hole type.

Mud volcanoes occur on tectonically active margins (e.g., compressional zones of accretionary complexes, fold-and-thrust belts), passive margins, deep sedimentary basins related to active plate boundaries as well as delta regions and areas with important salt diapirism (Milkov, 2000; Kopf, 2002; Etiope, 2015; Mazzini and Etiope, 2017). They are located in petroliferous basins, both offshore (Black Sea, Gulf of Cadiz, Caspian Sea, Mediterranean Sea, Gulf of Mexico, Indian Ocean, Caribbean Sea, Norwegian Sea, Atlantic Ocean, Pacific Ocean, China Sea) and onshore in countries as Colombia, Azerbaijan, Papua New Guinea, Japan, Romania (see Mazzini and Etiope, 2017 and references herein), along anticline axes, strike slip faults, normal faults and fault-related folds.

Fluid vents vary significantly in size and geometry depending on the fluid rheology and source, related eruption processes and subsequent erosion (Kopf, 2002; Murton and Biggs, 2003; Mazzini and Etiope, 2017). They may display various morphologies such as conical, elongated, pie-shaped, multicrater, growing diapir-like, stiff neck, swamp-like, plateau-like, impact crater-like, subsiding structure, subsiding flanks and sink-hole type (Roelofse et al., 2020) (Fig. 4B). One common morphology observed near fluid vents is that of discrete, isolated "hills," which can be classified as mud cones (with flank gradients $> 5°$) or mud pies (with slope gradient $< 5°$) (Kopf, 2002).

Mud volcanoes are usually composed of a group of cones and crater systems (Mazzini and Etiope, 2017). Mud volcanoes may display areal extension ranging from the order of a square meter up to several square kilometers, a width of 4 km (2.5 mi) onshore and up to 12 km (7.5 mi) offshore and a height of a few centimeters to hundreds of meters (Orange et al., 2009; Yusifov and Rabinowitz, 2004; Mazzini and Etiope, 2017) (Fig. 5). The migration of fluid and mud is mainly driven by a combination of gravity-driven instability of shales and fluid overpressure build-up in shales, reservoir rocks or fractures, followed by the hydrofracturing of impermeable barriers (Kopf, 2002; Mazzini and Etiope, 2017; Revil, 2002). Fluid overpressure may be caused by fast rates of sediment deposition (Judd and Hovland, 2007; Wu et al., 2019), tectonic subduction and compression (Conrad et al., 2018), seismic-induced shock and gas buoyancy (Kopf et al., 2001).

Overpressure in shales may result in volumetric expansion due to the generation of hydrocarbons and other additional mechanisms such as the thermal effect in pore fluids (as temperature gradient increases), dehydration reactions (e.g., illitization of clay minerals), disequilibrium compaction, related to lithostatic loading or compressive tectonic stresses and pressure dissipation by fluid flow (Mazzini and Etiope, 2017; Revil, 2002). In contrast, gravity driven instability generally occurs due to rapid sedimentation in subsiding basins and blanketing of low density clay-bearing strata, which can be buoyant in surrounding units (Mazzini and Etiope, 2017). Intragranular overpressure also tends to increase by mechanical compaction during the gradual burial of clays or during sudden depositional events (slides, slumps, thick turbidite deposits) associated with high rates of sedimentation and subsidence. Rapid subsidence of porous sediments rich in water and organic matter and the subsequent thermal maturation of hydrocarbons during early diagenesis also produce fluid-mud mixtures that migrate upward under high pressures (Dimitrov, 2002; Kopf, 2002; Mazzini and Etiope, 2017; Milkov, 2000).

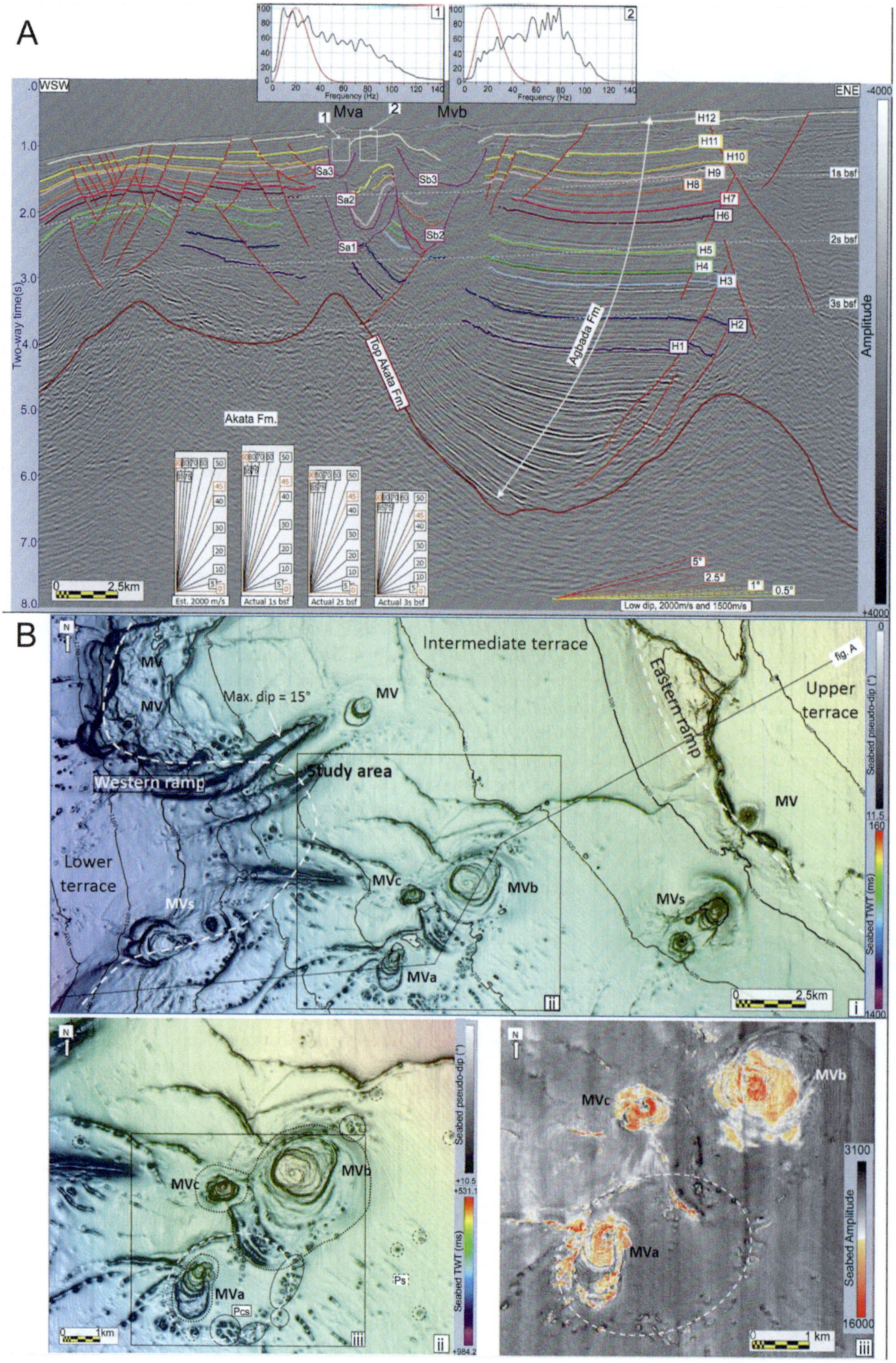


FIG. 5 Example of seismic section (A) and seafloor maps from 3D seismic data (B) from mud volcanoes in the western lobe of Niger Delta system, offshore Nigeria, studied by Dupuis et al. (2019). The studied features of the mud volcanoes MVa and MVb are highlighted by the magenta surfaces Sa1 to Sb3. The insets 1–2 show the frequency spectra (black curve) within the areas (white boxes) below and on the flanks of MVa, respectively. The maps include: (i) and (ii) two-way time map of the seafloor (in color) with gradient overlain on it and indicating the location of mud volcanoes based on Dupuis et al. (2019). Mud volcanoes MVa, MVb and MVc are three of the mud volcanoes studied in Dupuis et al. (2019), and (iii) map of seafloor amplitudes.

Hydrofracturing may result from this process and also by increases in fluid pressure, tectonic stresses, fault reactivation and seismicity (Mazzini and Etiope, 2017).

Sand injections can occur as dykes, sills, conical and saucer-shaped intrusions with wings (Fig. 6), and are usually identified as discordant, high-amplitude anomalies on seismic profiles (Andresen and Clausen, 2014; Cartwright et al., 2008; Hurst et al., 2005, 2011; Huuse et al., 2007) (Fig. 7). According to Hurst et al. (2011), the injection of sand into overburden units requires fluid overpressure as a precondition, initiating hydrofractures and driving fluid flow. The development of sand injectites results from fluid overpressures (usually the generation of excess water and, sometimes, hydrocarbons), the hydrofracturing of sealing strata, liquefaction, fluidization and the injection of sand (Duranti and Hurst, 2004; Hurst et al., 2011; Jolly and Lonergan, 2002; Vigorito and Hurst, 2010). Different triggering mechanisms are associated with the formation of sand injectites: (i) local earthquakes resulting in sand liquefaction (Obermeier, 1996, 1998; Rosseti, 1999; Boehm and Moore, 2002; Obermeier et al., 2005); (ii) overpressure caused by rapid loading and good seal integrity (Allen, 1985; Hildebrandt and Egenhoff, 2007; Strachan, 2002; Truswell, 1972); (iii) thermal pressurization (Ujiie et al., 2010); (iv) fluid migration causing increased overpressure (Davies et al., 2006; Lonergan and Cartwright, 1999).

Seismicity and rapid loading (by submarine landslides or sediment derived from storm waves) are considered the most typical triggering mechanisms for sand injection and extrusion (Truswell, 1972; Strachan, 2002; Hildebrandt and Egenhoff, 2007; Boehm and Moore, 2002; Jonk, 2010; Jonk et al., 2007; Obermeier, 1996; Obermeier et al., 2005), though seismicity is unlikely to act alone considering the energy required to fluidize and inject the 10s of $km^3$ of sand (Vigorito and Hurst, 2010). The overpressure caused by the rapid migration of fluid into depositional sand bodies can be related to: (i) the formation of polygonal faults in mudstones (Cartwright et al., 2003; Cartwright and Dewhurst, 1998; Wattrus et al., 2003); (ii) mineralogical phase changes (Davies et al., 2006); (iii) the rapid migration of hydrocarbon gas (Brooke et al., 1995); (iv) the decomposition of gas hydrates during periods of eustatic sealevel change or ocean warming (Hurst et al., 2011). Importantly, the migration of fluids from deep sedimentary sources is one of the factors responsible for the development of large-scale sand injectites—some capable of crosscutting more than 200 m (656 ft.) of fine-grained strata (Hurst et al., 2003b, 2011; Huuse et al., 2005).

The North Sea was the first province where subsurface sand injectites, up to several 100 of meters high and 1000 of meters wide, where recognized in the literature and proven in wells and seismic data (Lonergan and Cartwright, 1999; Huuse et al., 2005; Hurst and Cartwright, 2007; Andresen and Clausen, 2014; Huuse et al., 2007; Huuse and Mickelson, 2004; Shoulders et al., 2007). The recent study of Andresen and Clausen, (2014) described injectites from this province as comprising geometries ranging from basal sills with wings to V-shaped and conical injectites 300–3700m (984–12,139 ft.) in width and up to 150m (492 ft.) in height. The formation of these injectites was due to overpressure caused by rapid differential loading during the Oligocene, combined with a possible influx of fluids from underlying Paleozoic half-grabens. The authors suggested the Upper Paleocene sand within the Lista Formation as the source of the injectites, which are connected with the top of these formations by potential feeder conduits.

For the sand injectites in the Tertiary petroleum reservoirs of the northern North Sea suggested trigger mechanisms include seismicity, rapid loading and fluid migration (Lonergan and Cartwright, 1999; Jolly and Lonergan, 2002; Duranti and Hurst, 2004; Huuse et al., 2004). In the south Viking Graben, injectites consist of dikes (discordant to bedding) and sills (concordant to bedding), with thicknesses ranging from subcentimeter to meter scale, and commonly associated with injection breccias (up to 10-m thick), sand-supported with mud-clasts ( Jonk et al., 2005). Sand injection was possibly triggered by earthquake activity and may have been facilitated by petroleum fluids (Huuse et al., 2004; Jonk et al., 2005).

Sand injectites that comprise significant hydrocarbon reservoirs also occur in the Paleogene section of the northern North Sea (Hurst et al., 2003a; Dixon et al., 1995; Lonergan et al., 2000). Large-scale tabular dykes associated with extrusions (subhorizontal extensions) occur along the margins of the Nauchlan Member (Late Eocene Alba Formation), which comprises a deepwater channel fill extensively modified by post-depositional sand remobilization and injection that terminates at the Eocene-Oligocene unconformity (Duranti and Hurst, 2004) (Fig. 7D). These injectites resulted from a sudden increase in the ratio of overpressure to confining pressure, possibly caused by static liquefaction and enhanced by hydrocarbon gas (Duranti and Hurst, 2004). Fluid overpressure and associated hydraulic fracturing resulted in the severe disruption of adjacent mudstones, as proven by the fragments of mudstone that were incorporated into the injectites. According to Duranti and Hurst (2004), two main phases of sand injection occurred at different burial depths: the first was a shallow burial phase (below 100m, 328 ft.) that produced thin folded dykes and sills, while the second phase—with injectites at the boundary between the Eocene and Oligocene—was deeper (reflecting a burial depth of about 300m, or 984 ft.), more voluminous and formed large-scale tabular wing-like dykes that project from the edges of the channel fill. Numerous sharp-sided, thick dykes and sills were also formed (Duranti and Hurst, 2004).

In central California, at least three sand injection complexes have been the subject of several recent studies: (a) the Panoche Giant Injection Complex (Fig. 7A; Vigorito and Hurst, 2010; Scott et al., 2013; Vigorito et al., 2008;

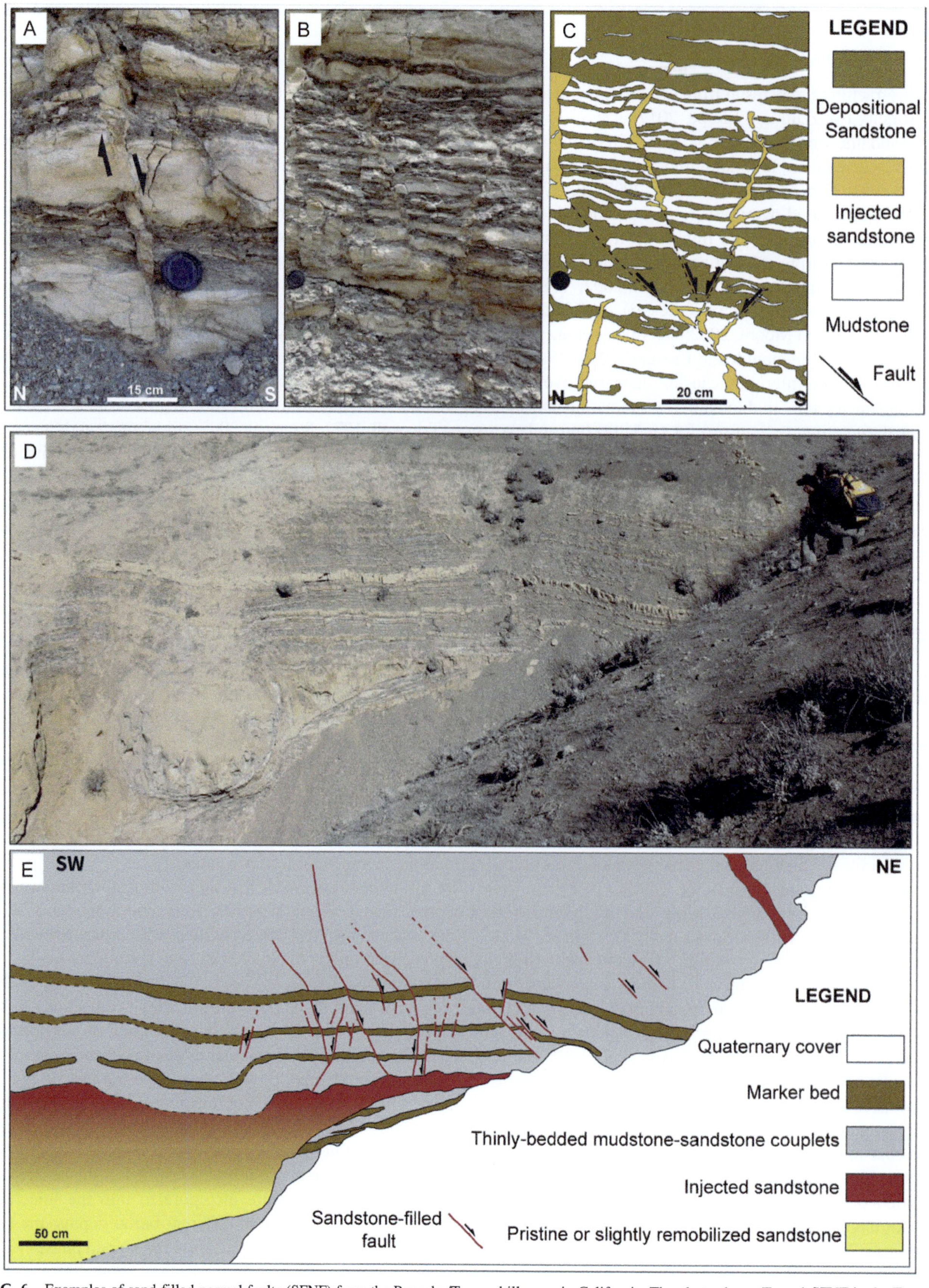


**FIG. 6** Examples of sand-filled normal faults (SFNF) from the Panoche/Tumey hills area in California. The photo shows Type 1 SFNF in the Tumey Gulch area, namely: (A) SFNF with centimeter-scale offsets and openings of a few centimeters. (B) Field photograph and (C) associated line drawing of conjugate sets of Type 1 SFNF with centimeter-scale offsets. In the photos, faults die out upwards over tens of centimeters, and the overlying layers are undeformed. The anomalous curvatures observed in some of the fault planes are related to "post-emplacement" compaction. (D) Panoramic photo of Type 1 SFNF from in the Tumey Gulch area (Outcrop O2). (E) Corresponding line drawing of a SFNF array overlying a remobilized turbidite sandstone body. Once again, the anomalous curvature shown by the fault planes is mainly due to "post-emplacement" deformation processes. *(Figures are modified from Palladino, G., Alsop, G.I., Grippa, A., Zvirtes, G., Phillip, R.P., Hurst, A., 2018. Sandstone-filled normal faults: A case study from central California. J. Struct. Geol. 110, 86–101, https://doi.org/10.1016/j.jsg.2018.02.013.)*

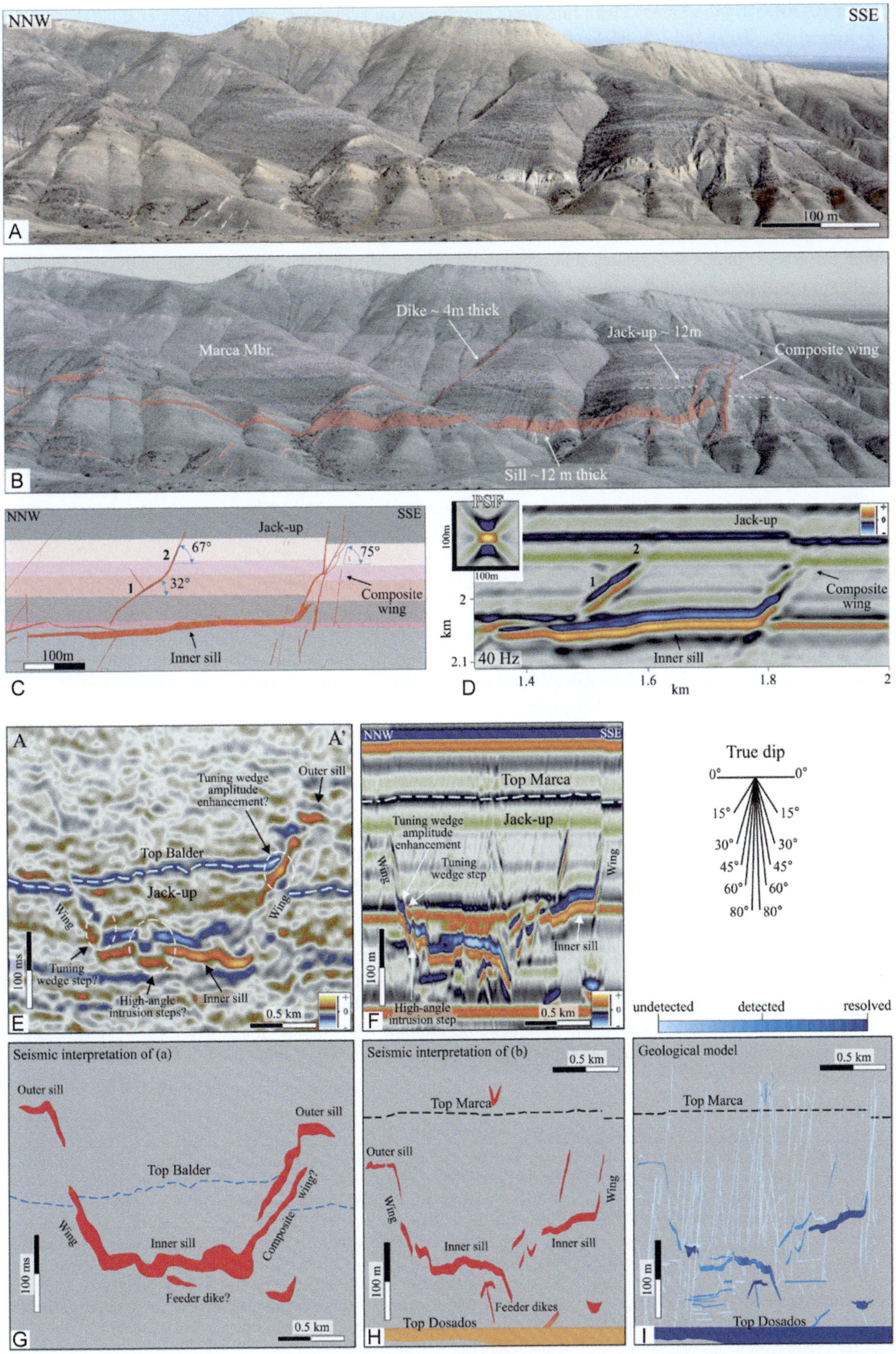


FIG. 7 Complex geometries and imaging of sandstone intrusions as detailed in Grippa et al. (2019) based on information from Central California, USA. The figure highlights the Right Angle Canyon in the Panoche hills area. (A) Detail of the southern wing of the Right Angle Canyon (RAC) and portion of *(Continued)*

Palladino et al., 2018), (b) the Tumcy Giant Injection Complex (TGIC) (Figs. 6A and 7B; Huuse et al., 2007; Zvirtes et al., 2020; Zvirtes et al., 2019) and (c) the Santa Cruz Injection Complex (SCIC) (Fig. 6B–E; Boehm and Moore, 2002; Palladino et al., 2020; Scott et al., 2009; Thompson et al., 2007). The Panoche Giant Injection Complex (PGIC) comprises a well-defined system of sand dikes and sills intruded within the Upper Cretaceous to Paleocene Moreno Formation (Vigorito and Hurst, 2010). Different architectural elements are present in the PGIC, from its base to its top, such as parent units, intrusive bodies and extrudites (Vigorito et al., 2008; Hurst et al., 2011; Scott et al., 2013). In the PGIC, parent units consist of turbiditic channel-complexes and isolated sandstone channels, which occur in the lower part of the Moreno Formation (Vigorito et al., 2008). Injectites correspond to interconnected single or multilayered sills and dikes with a thickness ranging from centimeters to meters, while extrudites are mound-like sand bodies located in the upper part of the Moreno Formation (Vigorito and Hurst, 2010). These extrudites are composed of fine- to medium-grained sands and link to the underlying intrusive complex via isolated dikes. According to Vigorito and Hurst (2010), the PGIC was the result of a large-scale overpressure event that occurred in the Lower Paleocene, involving an area of at least 1500 $km^2$, or 579 $mi^2$. At the base of the injection complex, the estimated pore-fluid pressures likely reached 22.26–25.08 MPa, or 0.81 and 0.95 of lithostatic pressure (Vigorito and Hurst, 2010). A diachronous timing for sand injection, occurring together with fluid pressure in excess of the lithostatic pressure within the source units, was also suggested by Vétel and Cartwright (2010) based on the analysis of cross-cutting relationships in the field. The authors recognized that the opening of the sand intrusions did not obey a systematic sense of movement, a character resulting from the presence of short-range mechanical interactions between adjacent sills and dikes.

Two different families of sandstone-filled normal faults occur in the lower portion of the Moreno Formation and, particularly, at the top of PGIC (Palladino et al., 2018). The first normal-fault family resulted from the rapid increase of pore-fluid pressure in poorly consolidated sandstones, with associated hydraulic failure of the overlying host strata, i.e., pressure build-up and overburden collapse extension. The second family resulted from regional extensional and related draining of fluidized sand toward active tectonic structures.

The TGIC is composed of a network of dikes and sills emplaced in an interval that is c. 450 m (1476 ft.) thick in the so-called Kreyenhagen Shales (Middle to Upper Eocene; Huuse et al., 2007; Zvirtes et al., 2019). According to Zvirtes et al. (2020), the parent units of the TGIC are the turbiditic channels of the Kreyenhagen Formation, lacking a contribution from the underlying Lodo and Domengine formations. When considering its stratigraphic and structural relationships, the TGIC consists of lower and upper intrusive intervals (Zvirtes et al., 2019). The lower intrusive interval comprises a sill complex with a stepped, staggered and multilayered geometry connected by narrow, short low- and high-angle dykes with planar and irregular geometry. The upper intrusive interval is a network of interconnected sills, dykes, irregular intrusive bodies and injection breccias, which show several intrusive shapes. These range from sheet-like intrusions with planar margins to highly irregular, bulbous and curved margins to asymmetric saucer-shaped intrusions with large wings emanating from the channelized turbidites (parent units; Zvirtes et al., 2019) (Figs. 6A and 7E–G).

As the TGIC intrudes the Lower-Middle Eocene Kreyenhagen Shale, but does not affect the overlaying Miocene Temblor Formation, the timing of sand injection must have occurred between the Middle Eocene and the Miocene (Palladino et al., 2018). In The Kreyenhagen Shale were identified sandstone-filled normal faults associated with pressure build-up with overburden collapse extension (Palladino et al., 2018), and sandstone-filled contractional structures related to regional contractional tectonics (Palladino et al., 2016).

The SCIC corresponds to a network of partially tar-saturated and unsaturated injectites emplaced in the Santa Cruz Mudstone (Scott et al., 2009; Thompson et al., 1999). Four different architectural elements occur in the SCIC, including parent units, injectites, extrudites and sandstone-filled faults (Palladino et al., 2020). The parent unit is commonly identified as the Santa Margarita Sandstone (Boehm and Moore, 2002; Thompson et al., 2007); however, some studies indicate multiple parent sandstones based on mineralogical data (Scott et al., 2009; Clark, 1981). The injectites consist of a

**FIG. 7, CONT'D** the southern inner sill. (B) Geological interpretation of the photo shown in a) highlighting how discordant is the sandstone intrusion and the overall jack-up of the overlying mudstone. (C) Geological model of the southern RAC wing. (D) Synthetic seismic model of the southern wing using a zero-phase Ricker wavelet with a peak frequency of 40 Hz to convolve the reflectivity model. (E) Seismic section through a saucer-shaped sandstone intrusion in the Volund field, Norwegian North Sea (Huuse et al., 2004; Schwab et al., 2014). Note that emplacement of the inner sill causes the jack-up of the overlying strata, as also shown in the outcrop photos above. Dashed white ovals highlight the occurrence of steps within the inner sill reflections and at the base of the wing reflection, as well as amplitude enhancement effect. (F) The 40 Hz synthetic seismic section from the RAC shown above displaying a similar geometry to the Volund section. (G) Seismic interpretation of (E) showing the main sandstone intrusion features on seismic data. (H) Seismic interpretation of (F) showing the geometry of the saucer-shaped intrusion as a seismic interpreter would see them. (I) Geological model of the RAC showing its high number of high-angle sandstone intrusions, which are unlikely to be detected by a seismic survey.

well-developed intrusive network that includes single or swarms of dikes, water and tar saturated, occasional sills and locally preserved saucer-shaped intrusions (Palladino et al., 2020) (Fig. 6B–E). The thicknesses of individual injectites generally range from a few centimeters to a decimeter. Yet, some isolated dikes or sills are up to a few meters wide and the Yellowbank/Panther beach sill is, at least, 15 m (49 ft.) thick (Thompson et al., 1999; Scott et al., 2009). Sandstone-filled faults are sand injections emplaced directly along tectonic structures, predominantly normal faults and less common strike-slip and compressional fault planes (Palladino et al., 2020). Extruded sands occur as multiple, laterally discontinuous tar-saturated mounds emplaced in different stratigraphic levels of the Santa Cruz Mudstone (Boehm and Moore, 2002; Hurst et al., 2005; Palladino et al., 2020). They span hundreds of meters, are several meters thick and, locally, the original structure of sand volcanoes shows multiple conduits and laminated flanks (Palladino et al., 2020).

The age of emplacement of the SCIC is constrained to the Late Miocene (7–9 Ma), as extrudites occur within the Santa Cruz Mudstone (Palladino et al., 2020). In detail, two distinct sand injection phases are identified (Palladino et al., 2020). The first phase (Late Miocene) resulted in large volumes of sand emplaced within the top-seal units and controlled by compaction and compressional tectonic processes. This phase was followed by hydrocarbon accumulation within newly injected sandstones. The second phase is related to a series of brittle tectonic events associated with the San Andreas/San Gregorio Fault System, which promoted the remobilization and accumulation of sand along newly formed fault planes, mostly high-angle extensional faults. As a result of this brittle deformation, the top seal was breached and previously accumulated hydrocarbons were leaked. The age of this phase is still uncertain but ranges from the Late Miocene to Quaternary (Palladino et al., 2020).

## Economic and societal relevance of sediment-remobilization processes

Processes of sediment-remobilization in deepwater settings, such as submarine landslides, turbidity currents, and subsurface fluid flow and seafloor expulsion (Clare et al., 2017), can have societal, economic and ecological impacts (Talling et al., 2014). Submarine landslides may affect marine biological communities due to the disturbance and modification of seafloor ecology during their emplacement, or by acting as habitat hotspots on their scars and remobilized elements (De Mol et al., 2009). This process is also considered a mechanism for the dispersion of species between isolated islands, thus influencing local evolution (Caujapé-Castells et al., 2017).

The economic impact of submarine landslides is related to damage of strategically important seafloor infrastructure including telecommunication cables, production platforms and hydrocarbon pipelines caused by these processes (Carter et al., 2014; Forsberg et al., 2016; Mosher et al., 2010; Piper et al., 1999; Pope et al., 2017; Shipp et al., 2004; Thomas et al., 2010). The repair of cables and subsea infrastructure networks may cost hundreds of millions of dollars, while the subsequent interruption of global communication, global financial trading and global supply chains can potentially be immense as trillions of dollars are traded per day on this network (Carter et al., 2014; Carter et al., 2009). In parallel, damage to seafloor hydrocarbon infrastructure (platforms and pipelines) can result in production delays and marine environment contamination due to uncontrolled losses of hydrocarbons (Skogdalen and Vinnem, 2012; Kaiser et al., 2009). The cost of damage to hydrocarbon pipelines due to submarine instability processes is estimated in more than $400 million per year (Mosher et al., 2010).

Submarine landslides may also cause tsunamis, such as the 3000 $km^3$ (720 $mi^3$) Storegga Slide that occurred offshore Norway at ~8.2 ka b.p. (Talling et al., 2014). Tsunamis comprise a threat to many a coastal community and may result in large numbers of fatalities (Harbitz et al., 2014; Tappin et al., 2001; Ward, 2001). Even small submarine landslides (i.e., less than 0.1 $km^3$, 0.02 $mi^3$) can be dangerous, causing tsunami waves as high as tens of meters (Von Huene et al., 2004; Bohannon and Gardner, 2004) and landward flooding (retrogression) leading to loss of life (Vardy et al., 2012). The destruction potential of such events has been associated with the volume, vertical displacement, water depth and velocity of the remobilized sediments (Álvarez-Gómez et al., 2011; McAdoo and Watts, 2004; Schnyder et al., 2016).

Hazard assessments for possible future tsunamis triggered by landslides have been a concern among researchers and nonscientists throughout the world (Paris et al., 2020). Based on the identification of landslide populations and their morphology, numerical simulations can help estimate the impact of past-tsunamis resulting from landslides and areas affected by such events (Lovholt et al., 2015; Yavari-Ramshe and Ataie-Ashtiani, 2016). Future large-scale landslides and ensuing tsunamis, such as the tsunami predicted for future flank collapse of the Cumbre Vieja volcano in the Canary Islands (Tehranirad et al., 2015; Ward and Day, 2001), can only be characterized using a combined geophysical, geological and modeling approach. Climate change is another issue that makes risk assessments for landslide-tsunamis relevant in the short to long term; paleoclimatic changes and eustatic variations have been identified as major factors triggering instability factors on some continental margins (McMurtry et al., 2004; Quidelleur et al., 2008; Boulesteix et al., 2013; Berton and Vesely, 2016).

Economically speaking, submarine landslides can generate thick, extensive MTDs, playing an important role as seal or reservoir intervals in deepwater petroleum systems (Moscardelli and Wood, 2008; Gamberi et al., 2011; Posamentier and Martinsen, 2011; Alves et al., 2014). MTDs associated with turbidites have been identified as some of the world's largest oil and gas fields (Barley, 1999). In terms of their sedimentological character, MTDs cover a wide range of depositional facies, from MTDs with a highly mud content (i.e., remobilized slope muds) to complex deposits with interconnected and interfingering sand intervals, i.e., slumped levees or channel bodies. Consequently, MTDs may act as competent barriers to fluid flow (top seals or intra-reservoir barriers and baffles) or, instead, reservoirs and conduits to fluid flow (Shultz, 2004; Dykstra et al., 2011). The impact of submarine landslides in petroleum systems is also related to the distribution of stratigraphic traps on the slope as a result of seafloor remobilization, the geometry of facies developed within MTDs and the morphology of these deposits in which local lows are usually controlled by their internal structure (Kneller et al., 2016). Reservoir geometry can be influenced by the different slumps, slides and debris flows, domains identified in submarine landslides, as these flow processes usually lead to discontinuous and compartmentalized sand reservoirs (Shanmugam et al., 1995). In turn, turbidity currents generally form continuous sheet-like sand bodies, with the caveat that, where MTDs are present, or in areas of slide scars, turbidite systems location, nature and geometry may vary widely (Brami et al., 2000; Kneller et al., 2016).

Fluid flowing in buried MTDs, and its eventual expulsion onto the seafloor, may constitute a hazard in multiple settings. Sediment remobilization resulting in mud volcanoes and sand extrusion due to fluid flow and subsurface overpressure may also impact subsea infrastructures such as telecommunication cables, production platforms and hydrocarbon pipelines (Kilb, 2008; Lupi et al., 2013). Sediment injection and/or free gas in the subsurface may result in slope instability once they modify sediment shear strength, compressibility and effective stress (Chillarige et al., 1997; Evans, 2010; Riboulot et al., 2013). Mud volcanoes represent geohazards due to the potentially violent release of large amounts of hydrocarbons and mud, the degradation of sediments at seafloor and quicksand effect and episodic dissociation of submarine gas hydrates (Mazzini and Etiope, 2017). They are also one of the geological sources of methane that are currently considered a major contributor to the atmospheric methane budget (Etiope, 2015). At shallow depths, the presence of gas may result in gas kicks and blowouts while drilling, subsidence and leaks outside casing, and issues with cementing wells (Nimblett et al., 2005). Natural seepage of fluids at the seafloor may be associated with sediment remobilization at depth, such as eruptive mud volcanoes with associated caldera collapses (Gray et al., 2013), and sand extrusions related to sand liquefaction, fluidization and injection (Hurst et al., 2011). The expulsion of fluids on the seafloor can create corrosive pore fluids and locally modify seafloor geotechnical properties (Thomas et al., 2010); hence, it can lead to problems for pipeline design, particularly where seeps occur in high spatial densities (Gafeira et al., 2012; Moss et al., 2012).

Sand injectites can influence traps, reservoirs, seals and fluid migration in sedimentary basins (Duranti et al., 2002; Hurst and Cartwright, 2007; MacLeod et al., 1999). They may affect the distribution and geometry of hydrocarbon reservoirs, connect otherwise isolated sand bodies to form possible hydrocarbon migration routes and thief zones (Hurst and Cartwright, 2007; Satur and Hurst, 2007). Therefore, sand injectites may impact hydrocarbon plumbing systems and represent potential drilling hazards (Hurst and Cartwright, 2007; Andresen and Clausen, 2014). Mud volcanoes are common in several petroleum provinces worldwide, including major hydrocarbon exploration and production regions (e.g., the North Sea, the Caspian Sea, the Gulf of Mexico, the Black Sea, the Sea of Okhotsk, the Sea of Japan) and often related to modern (Mazzini and Etiope, 2017; Cortes et al., 2018). These geological features consist of ideal targets for hydrocarbon exploration, as they indicate the existence of relevant subsurface reservoirs.

## Key messages: Outcrop examples from Crete, New Zealand and Paraná

### Fault-bounded slopes in SE Crete

Slope deformation styles in SE Crete include the presence of MTDs with distinct lithologies reflecting a marked contrast between remobilized mass-wasting deposits and "background" slope deposits. Hence, it is relatively straightforward to distinguish not only the source areas of MTDs on Crete, but also their exact geometries when exposed in the field (Fig. 8). Specific structures are associated with the largest MTDs: (a) recumbent folds of ductile material are common in areas where the slopes were tectonically oversteepened, or where sliding strata were buttressed in a locally emergent configuration sensu Frey-Martínez et al. (2006), (b) bed-parallel shearing occurs sporadically as small, localized slope-instability features, (c) rolling of larger blocks under breccia-conglomerate carpets, (d) sliding of large carbonate blocks occurs over siliciclastic slope strata and (e) moderate sliding of carbonate blocks and siliciclastic volumes of rock (slumps) is observed over detachment faults.

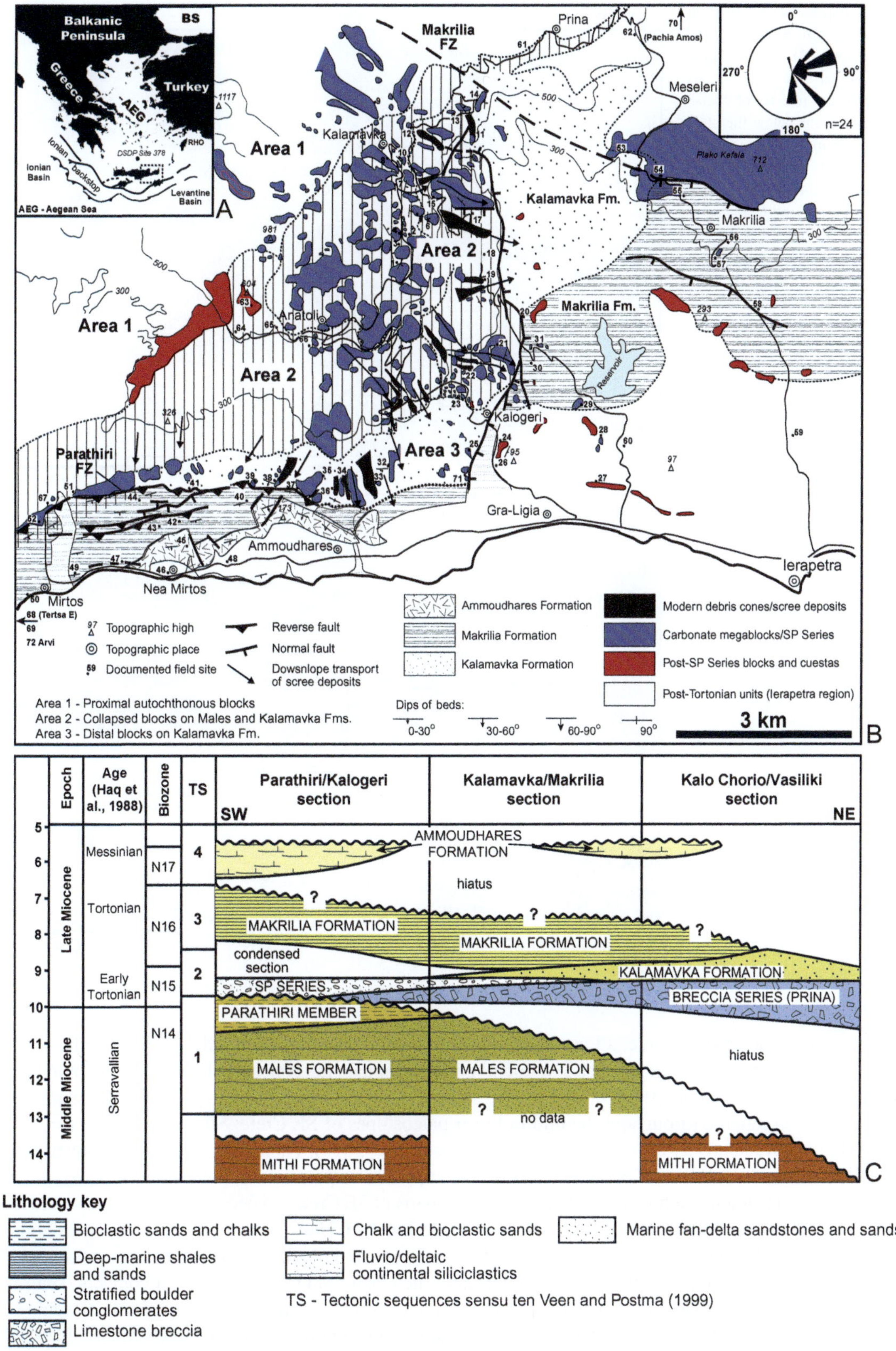


**FIG. 8** General map of SE Crete's slide blocks modified from Alves (2015). The small map in (A) shows the location of the investigated slope successions in SE Crete. In the geological map of SE Crete's paleoslope shown in (B), Area 1 comprises autochthonous carbonate blocks and breccia-conglomerates showing limited gravitational collapse. Area 2 comprises disrupted deepwater (carbonate) fan cones, carbonate megabreccias and boulder conglomerates. Area 3 includes carbonate fan cones, collapsed blocks and minor debris-flow deposits (boulder conglomerates) deposited in distal regions of the palaeoslope. (C) Stratigraphic panel for SE Crete, where a tectonic trough (Ierapetra Basin) was subject to extensional and transtensional movements since, at least, the end of the Serravalian. The stratigraphic panel is modified from Postma and Drinia (1993) and van Hisnbergen and Meleunkamp (2006). SP-Stratified Prina Series.

**FIG. 9** (A) Collapsed slide block at Location 33 in Fig. 8A, showing associated basal features. (B) Note the dual character of the basal shear zone at this location 33—up to 2m (6.6 ft.) of coarse immature breccias occur below the slide block and change abruptly into in a sandier deformed area with slope siliciclatic (sandy) material. The thickness of the basal glide zone reaches more than 4m (13 ft.) in its thickest zone.

Fig. 9 shows, as an outcrop example, a Late Miocene slide block transported over siliciclastic turbidites on the paleoslope of SE Crete. The area in question comprises autochthonous carbonate fan cones and boulder conglomerates, with slide blocks proving the widespread collapse of fault-bounded slopes during the Miocene (Fig. 8). Slide blocks reflecting large-distance transport are scattered throughout the mid to lower paleoslopes of SE Crete. Slide blocks occur mainly over the Parathiri Member and Kalamavka Formation (see Alves and Lourenço, 2010). Their size varies from 10 to 100m (33 to 328 ft.), with individual blocks >100m (328 ft.) long occurring in particular parts of the paleoslope (Fig. 8).

Other silicilastic MTDs are common toward the most distal parts of SE Crete's paleoslope, where carbonate blocks were sparser and debris-flows dominated instability processes (Figs. 8 and 9). Here, scarce polymictic breccia-conglomerates give rise to siliciclastic debrites with a minor presence of centimeter-sized blocks. Basal contacts are commonly erosional below individual blocks and slope strata, and also in debrites/siliciclastic MTDs. In fact, basal contacts comprise distinct lithologies and structures. Alves and Lourenço (2010) have shown that polymictic breccia–conglomerates and siliciclastic (sandy to conglomeratic) blocks are often embedded within a debris-flow matrix below the largest blocks. In Fig. 9 is shown an example of one of such limestone slide blocks under which a basal polymictic breccia changes into an area of siliciclastic slope sediment that was also deformed by the movement of the block. Below the first 1–2m (3.3 to 6.6 ft.) of basal strata, in which the polymictic breccia is observed, the basal slip plane is formed by contorted fragments of strata (sandy to silty)

**FIG. 10** (A) Detail of a sandy MTD in the Ammoudhares Formation near Location 50 in Fig. 8A. (B) Note the complex arrangement of siliciclastic material and small blocks within the imaged MTD, in which internal folding (slumps) alternate with lithified blocks of sediment. The basal glide zone is sharp and very thin (<30 cm) in this location.

embedded in a clayey matrix (Fig. 8B). Below other blocks, sheath folds are visible and often intercalated with more chaotic strata comprising sandy and silty material injected during the basal deformation event (Fig. 10).

A series of other structural features appear on SE Crete's paleoslope, from shear structures at the basal shear zone of MTDs and blocks, to bookshelf sliding and microstructural features indicating sediment shearing and local slope collapse (Figs. 11 and 12). Fig. 12 shows S-C fabrics, foliation and pinch-and-swell structures, which are observed below a 40-m tall slide block. In particular, S-C fabrics and pinch-and-swell structures are diagnostic of the direction of movement of MTDs and slide blocks (see Alves, 2015).

Comparisons between statistical data from slide blocks (and MTDs) in SE Crete and high-resolution seismic data from 60 published case studies have been presented in Alves and Lourenço (2010) as an attempt to understand the scale relationship between the thickness of failed megablocks and the corresponding thickness of deformed basal strata. In Fig. 13 are shown key graphs illustrating the thickness of failed strata vs. thickness of basal shear surfaces, as published in Alves and Lourenço (2010). Of importance in these graphs is the apparent scale discrepancy among the seismic data and outcrop examples studied by the authors—the difference between the average thickness of basal shear surfaces recorded in geophysical data (12.8 m, 42 ft., on average, reaching a maximum of 65 m, or 213 ft.), and the thickness observed in SE Crete (3.1 m, 10.2 ft., on average) is clear on the graphs. Nevertheless, both the submerged and outcropping examples show failed strata to be 2.0–20 times thicker than the underlying basal shear surface, a value further investigated in Alves (2015), who demonstrated that the ratio between the thickness of failed deposits and the corresponding thickness of deformed strata at the base of the blocks ($R$) is usually in the order of $5 < R < 10$. Average ratios between failed strata and their corresponding

**FIG. 11** (A) Example of bookshelf sliding of slope strata at Location 12 in Fig. 8A. This type of sliding may have occurred post-depositionally by the readjustment of the paleo-slope to tectonic oversteepening or local overpressure increase. (B) Collapse feature near Location 68 in Fig. 8A. Here, tectonic oversteepening and withdrawal of soft slope strata may have caused the local collapse feature imaged in the photograph. The depositional facies suggest the strata in this photo belong to the Makrilia Formation (see Fig. 8B).

basal shear surfaces reach 5:1 for the outcropping examples from SE Crete and just above 9:1 for documented examples from offshore landslides (Fig. 13). Such threshold values are important as the thickness of seismically imaged basal shear zones is often not resolved on industry seismic data; the ratio $R$ can thus be used to predict the thickness of such deformed basal shear zones.

## The Rapanui MTD outcrop, Miocene Mount Messenger Formation, Taranaki Basin (New Zealand)

The outcrops of the Rapanui MTD (RMTD) are dominated by soft-sediment deformation associated with submarine mass flows (Fig. 14). The RMTD is sandwiched between two sandstone units (Lobe 1 and Lobe 2) interpreted as submarine fan lobes by Masalimova et al. (2016). Based on sedimentary structures, slide blocks, matrix disruption and deformation features, three domains in the RMTD were identified: headwall (Fig. 15), translational (Fig. 16) and toe domains (Fig. 17). The thickness of the RMTD varies spatially from 4 m (13 ft.) near the headwall, to at least c. 15 m (49 ft.) in the toe, reflecting a thickness inflation of about 27%. The average thickness is 7.4 m (24 ft.) and the RMTD maintains this value

**FIG. 12** Detailed examples of soft-sediment deformation in the basal shear zones of blocks, Location 11 (Fig. 8A). (A) Local carbonate megabreccias below a 20 m thick block (66 ft.). (B) Foliated slope strata with local S-C fabrics indicating that the transport direction of the block above is to the right (east).

through the translational domain. Although the true dimensions of outcropping MTDs are generally unknown (Martinsen, 1989), the length of the RMTD can be estimated as approximately 10 km (6.2 mi) using the average deposit thickness (see Moscardelli and Wood, 2016, for methodology).

The headwall domain of the RMTD is not fully exposed but does make up approximately 15% of the outcrop extent. Where this domain is exposed, the RMTD has an average thickness of 4 m (13 ft.) and is dominated by structures reflective of extensional strain, for example, boudinage, fractures and clastic injectites (Fig. 15). Slide blocks are scarce in the headwall domain. The original bedding of the RMTD protolith can still be recognized and is comprised of thinly-interbedded sandstones with siltstones and mudstones (Fig. 15B and D). The syndepositional intrafolial folds in the headwall ($n = 11$) have orientations from upright to recumbent and their interlimb angles are predominantly close (30–70°). Furthermore, the headwall is devoid of isoclinal-shaped and ptygmatic-shaped folds, and open-shaped folds (interlimb angle between 70° and 120°) were only documented in this domain.

The deformation structures described in the headwall are confined within the RMTD and the interaction between the RMTD and the underlying deposits is minor. The translational domain in the RMTD has an average thickness of

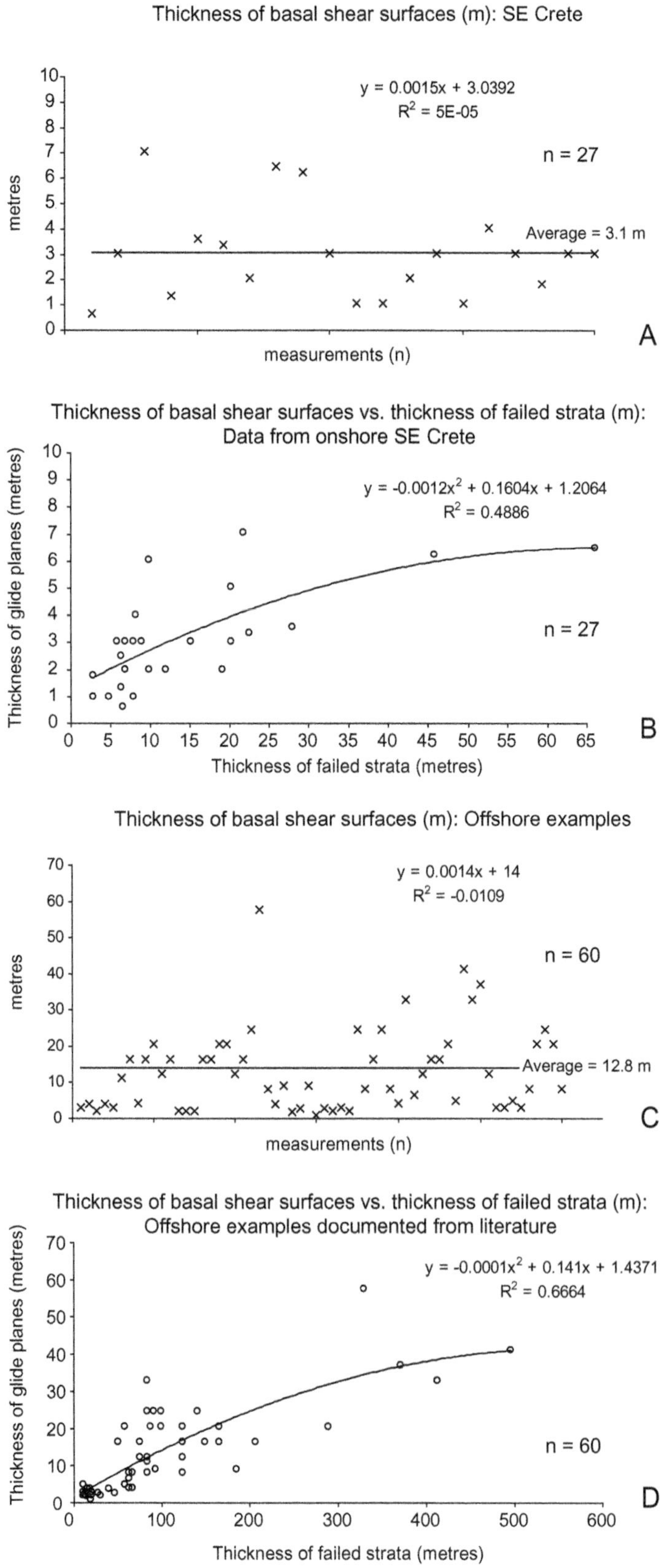


**FIG. 13** Graphs showing the scale relationships between the thicknesses of failed strata above vs. the thickness of basal shear zones (*R* ratio). Measurements were taken from several authors as described in Alves and Lourenço (2010) and Alves (2015)—Hampton et al. (1996), Gardner et al. (1999), Gee et al. (1999, 2006, 2007), DePlus et al. (2001), Bohannon and Gardner (2004), Haflidason et al. (2004), Lee (2005), Frey-Martínez et al. (2006), Greene et al. (2006), Lee et al. (2006), Vanneste et al. (2006), Hjelstuen et al. (2007), Minisini et al. (2007), Normarck et al. (2007), Sultan et al. (2007), Bull and Cartwright (2009), Moscardelli and Wood (2008), Alves and Cartwright (2009), Alves et al. (2009).

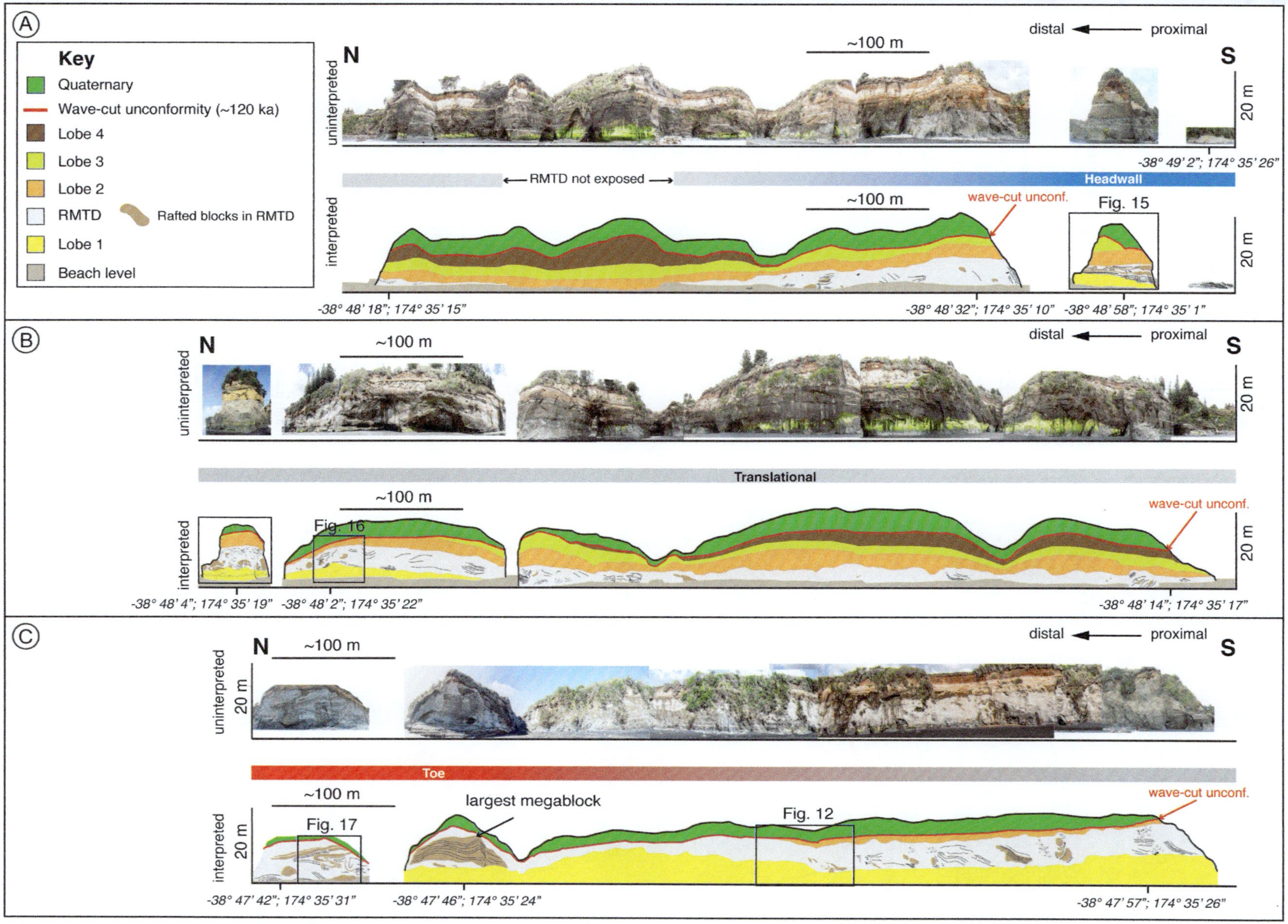


**FIG. 14** Photomosaic panels of the Rapanui mass transport deposit (RMTD) with approximate coordinates. Outcrops are exposed semi-obliquely to the northwest depositional dip. See Fig. 4D for the approximate map location for each panel. The RMTD is sandwiched between two sandstone units interpreted as submarine fan lobes and here informally labelled as Lobe 1 and Lobe 2. Slide blocks inside the RMTD have been traced to scale. (A) Uninterpreted (above) and interpreted (below) panel covering part of the headwall into the translational domain. (B) Uninterpreted (above) and interpreted (below) panel in the translational domain. Note the increase in slide blocks inside the RMTD matrix. (C) Uninterpreted (above) and interpreted (below) panel covering part of the translational into the toe domain. Here, the upper contact of the RMTD is sharply truncated by the Rapanui wave cut unconformity. *(Panels are modified from Cardona, S., Wood, L.J., Dugan, B., Jobe, Z., Strachan, L.J., 2020a. Characterization of the Rapanui mass-transport deposit and the basal shear zone: Mount Messenger Formation, Taranaki Basin, New Zealand (J. Baas, Ed.). Sedimentology 67, 2111–2148. https://doi.org/10.1111/sed.12697.)*

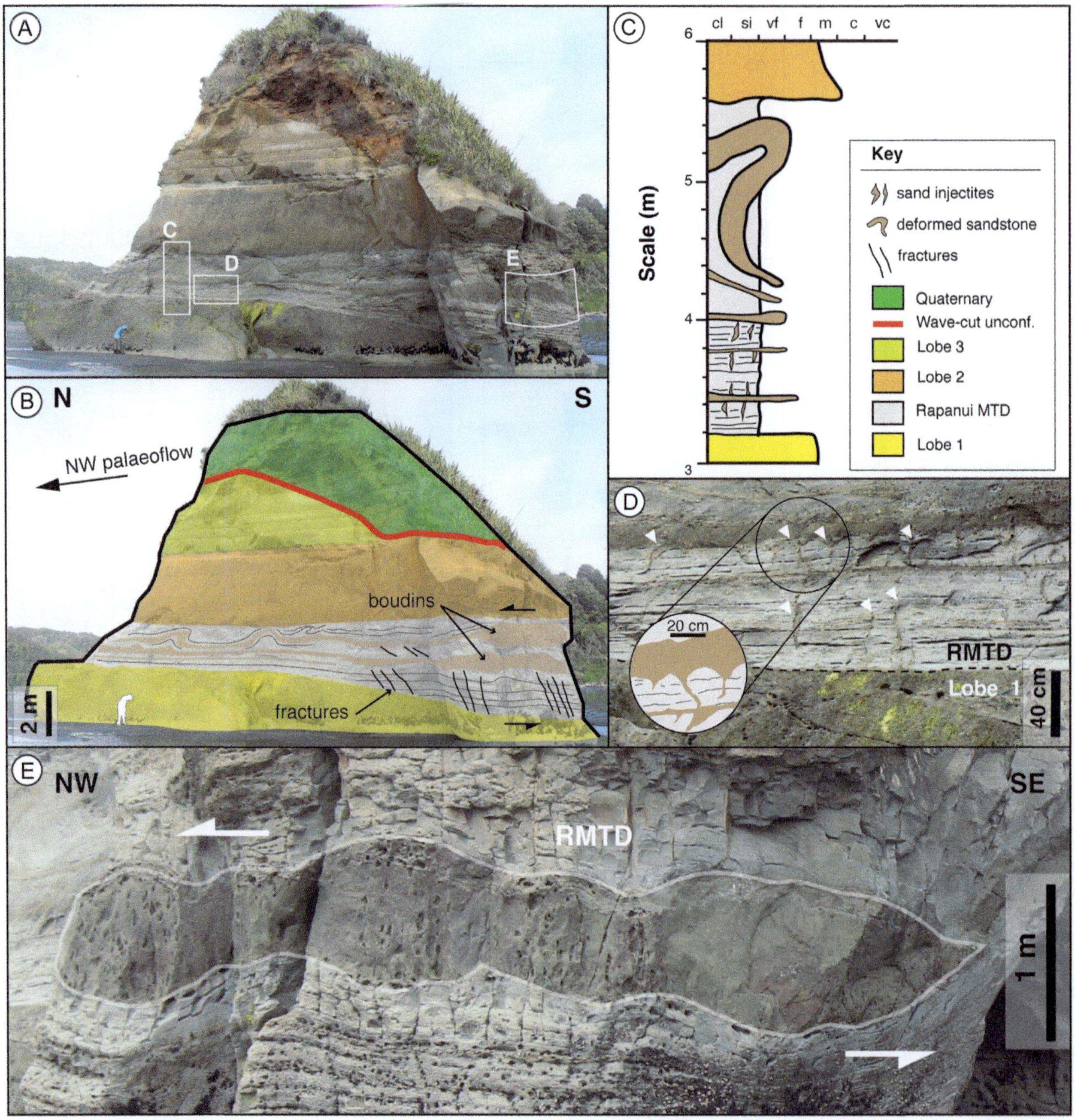


**FIG. 15** (A) Uninterpreted and (B) interpreted key outcrop in the headwall domain of the Rapanui mass transport deposit (RMTD)—see Figs. 4D and 5A for location. Note that deformation structures within the RMTD are associated with extensional stress and comprise fractures, boudinage and clastic injectites. However, deformation did not completely obliterate the original stratification of the RMTD "protolith." Fractures abruptly stop against thick sandstone layers and substratum. (C) Representative stratigraphic measured section in the headwall of the RMTD. Note that sandstone beds deformed plastically (presenting, for example, folds and boudinage) but disaggregation is minor. (D) Close up of the sand injectites in the RMTD. These cuspidate injectites (white triangles) have lengths around 10–20 cm and formed during sudden dilation of brittle beds (i.e., sandstone). The contact between the RMTD and the underlying deposits (Lobe 1), marked with the dashed line, is sharp and deformation is limited within the RMTD. (E) Close-up of metre-scale, shear-band boudin in a thick sandstone bed. This type of boudinage is associated with ductile deformation and large lateral displacement. Boudins only formed in the sandstone beds and can be used to estimate shear sense (see white arrows). *(Figure modified from Cardona, S., Wood, L. J., Dugan, B., Jobe, Z., Strachan, L.J., 2020a. Characterization of the Rapanui mass-transport deposit and the basal shear zone: Mount Messenger Formation, Taranaki Basin, New Zealand (J. Baas, Ed.). Sedimentology 67, 2111–2148. https://doi.org/10.1111/sed.12697.)*

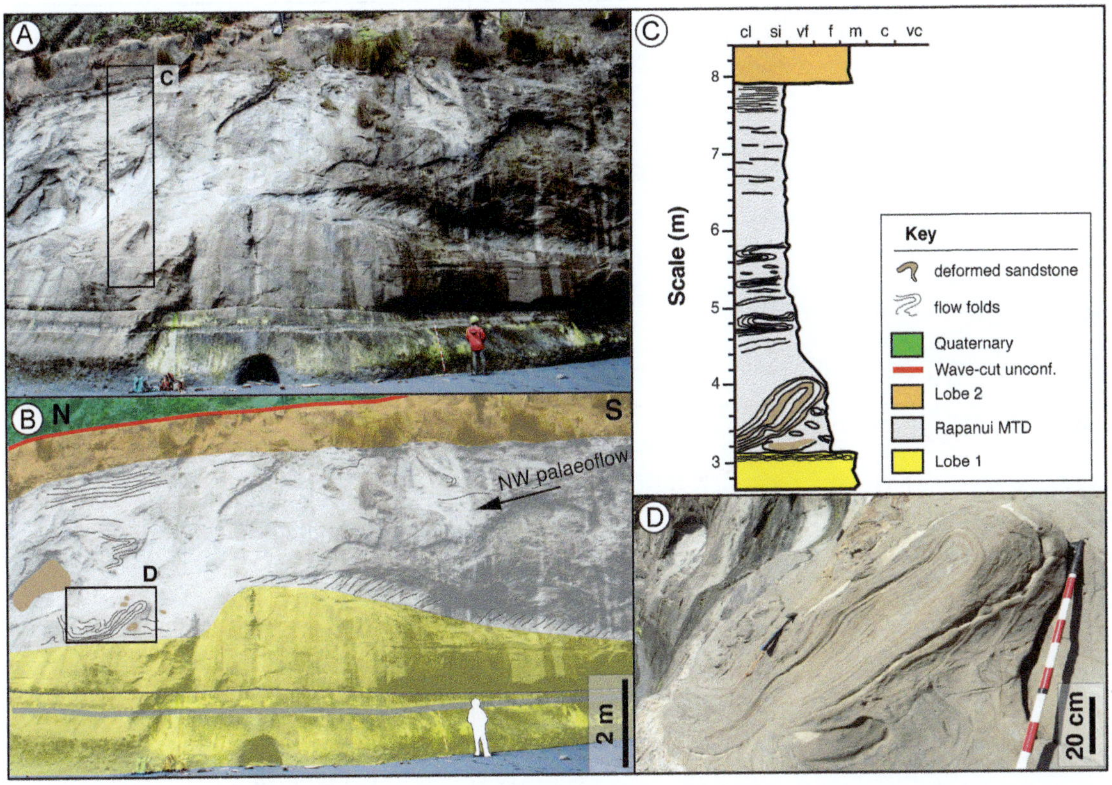


**FIG. 16** (A) Uninterpreted and (B) interpreted key outcrop in the translational domain of the Rapanui mass transport deposit (RMTD)—see Figs. 4D and 5B for location. Note the homogenization/stratal disruption of the RMTD matrix; nearly all primary bedding/stratification has been completely obliterated. Slide sandstone blocks are present in the matrix and some blocks show disaggregation (see Fig. 10). The basal shear zone here is irregular and shows evidence of entrainment and fluidization (see partially fluidized and folded sandstone from substratum in the close-up in (D). Flow folds or flowage structures are identified in the matrix resulting from partial to complete fluidization conditions (viscoplastic nature) during mass transport. (C) Representative stratigraphic measured section in the translational domain. (D) Close-up of fluidized sandstone from the substratum in the basal shear zone. *(Figure modified from Cardona, S., Wood, L.J., Dugan, B., Jobe, Z., Strachan, L.J., 2020a. Characterization of the Rapanui mass-transport deposit and the basal shear zone: Mount Messenger Formation, Taranaki Basin, New Zealand (J. Baas, Ed.). Sedimentology 67, 2111–2148. https://doi.org/10.1111/sed.12697.)*

7.4m (24 ft.) and comprises 60% of the outcrop extent. The transition into this domain can be identified by conspicuous stratal disruption and homogenization of the RMTD matrix (Ogata et al., 2012). In this domain, nearly all the original bedding/stratification from the RMTD protolith has been obliterated (Fig. 16). Flow folds or flowage fabric are also pervasive in this domain. Brodzikowski and Van Loon (1985) defined flow structures as those resulting from partial or complete deformation via fluidization (viscoplastic nature) without tectonic influence. Syndepositional folds ($n = 35$) are common in this domain and are primarily moderately inclined to recumbent tight-shaped folds with an average interlimb angle of 16°. No open-shaped folds were identified in this domain. Folds become rootless up-section in the RMTD and decrease in number. The described structures are characteristic of low-viscosity Bingham flow deposits (Tripsanas et al., 2008).

The toe domain exhibits the greatest thickness in the RMTD, reaching at least 15 m (49 ft.) even though its upper contact is truncated here by the Early Holocene Rapanui wave cut surface. The extent of this domain is estimated to be around 25% of the RMTD exposure. The largest megablock (c. 35m, or 115 ft., wide) in the RMTD is located in the toe (Fig. 14C). The RMTD matrix in the toe is substantially enriched in fine to medium-sand fraction in comparison with the other domains (samples S 8, S 9 and S 10 in Fig. 2B). Similar to the translational domain, syndepositional folds are abundant in the toe ($n =$ 33) and open shaped folds are absent. Most folds in this domain are moderately inclined to recumbent tight-shaped folds with an average interlimb angle of 17°. The dominant deformation structures in the toe are associated with a compressional stress regime such as meter-scale sheath-like folds of sandstone beds (Fig. 17B and D). These meter-scale folds are associated with pressure ridges (Frey-Martínez et al., 2006; Bull et al., 2009) and high shear strains (Farrell, 1984; Bradley and

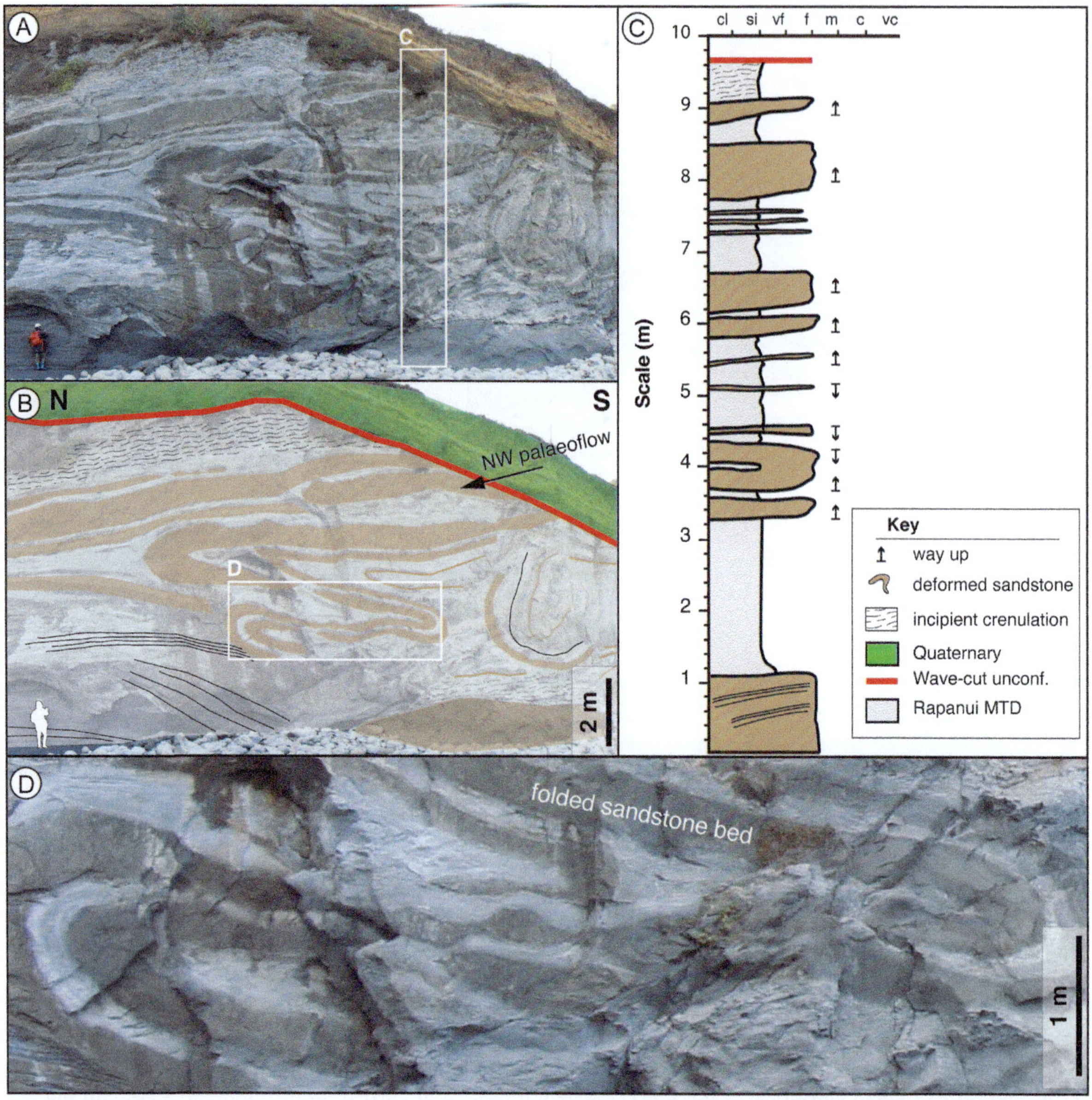


**FIG. 17** (A) Uninterpreted and (B) interpreted key outcrop in the toe domain of the Rapanui mass transport deposit (RMTD)—see Figs. 4D and 5C for location. The dominant deformation structures shown here are meter-scale compressional folds. These types of folds are probably associated with pressure ridges formed in the toe domain. Some folds display a sheath-like geometry such as the one in the close up in (D). Matrix near hinge of folds displays an incipient crenulation. (C) Representative stratigraphic measured section in the toe of the RMTD. (D) Close-up of sheath fold. *(Figure modified from Cardona, S., Wood, L.J., Dugan, B., Jobe, Z., Strachan, L.J., 2020a. Characterization of the Rapanui mass-transport deposit and the basal shear zone: Mount Messenger Formation, Taranaki Basin, New Zealand (J. Baas, Ed.). Sedimentology 67, 2111–2148. https://doi.org/10.1111/sed.12697.)*

Hanson, 1998). Additionally, the mud-rich portions surrounding some folds show a weak crenulation texture formed in response to folding (Fossen, 2016) (Fig. 17B).

## Outcrop analogs from the Itararé group, Paraná Basin (Southern Brazil)

MTDs have been identified along all the stratigraphic section and area of exposure of the Itararé Group (Carneiro and Costa, 2006; Carvalho and Vesely, 2017; Eyles et al., 1993; Gama et al., 1992; Mottin et al., 2018; Rodrigues et al., 2020, 2021; Schemiko et al., 2019; Suss et al., 2014; Vesely et al., 2018; Vesely and Assine, 2006; Valdez-Buso et al., 2019; D'Ávila, 2009). The full extent and thickness of a single MTD is usually difficult to assess due to the limited exposure of the Itararé

Group. The thickness of some MTDs generally ranges from about 5 m (16 ft.) to 10s of meters (Carvalho and Vesely, 2017; Mottin et al., 2018; Schemiko et al., 2019). When the boundaries between MTD and non-MTD strata are exposed, they show sharp aspects, with the base of MTDs being commonly erosive and irregular. The top surface of MTDs is usually flat but low amplitude relief has been described locally, on which fine-grained facies may be ponded (Vesely et al., 2018).

Although the Itararé Group presents penecontemporaneous deformation associated with glaciotectonics (Aquino et al., 2016; Fedorchuk et al., 2019; Rosa et al., 2019), most deformation structures described in this group resulted from subaqueous mass movements associated with turbidites, with or without a definite record of glacial influence (Salamuni et al., 1966; Schneider et al., 1974; França and Potter, 1991; Gama et al., 1992; Eyles et al., 1993; Vesely et al., 2005; Carneiro and Costa, 2006; Vesely and Assine, 2006; D'Ávila, 2009; Suss et al., 2014; Carvalho and Vesely, 2017; Valdez-Buso et al., 2019; Mottin et al., 2018; Vesely et al., 2018; Schemiko et al., 2019; Rodrigues et al., 2020). Several structures have been identified within these MTDs such as folds, faults, boudins, shear features (as quarter-like structures, sheared lamination, etc.), plus deformation at the borders of intrabasinal clasts (ICs) in diamictites, injectites and others (Vesely and Assine, 2006; Suss et al., 2014; Carvalho and Vesely, 2017; Valdez-Buso et al., 2019); Mottin et al., 2018; Vesely et al., 2018; Schemiko et al., 2019; Rodrigues et al., 2020, 2021).

The MTDs of the Itararé Group consist mostly of: (1) large allochthonous ICs of sandstones, rhythmites and mudstones; (2) deformed sandstone, rhythmite and shale; (3) heterogeneous (banded matrix) to homogeneous (massive matrix) sandy-muddy diamictites with dispersed granules to boulders of ICs (sandstones, shale, rhythmites and plant fragments) and extra-basinal clasts (granites and metamorphics), with some striated and faceted. Heterogeneous diamictite corresponds to pebbly-sandy-mudstone containing disrupted fragments of deformed strata in which the original bedding is still preserved (Vesely and Assine, 2006; Suss et al., 2014; Carvalho and Vesely, 2017; Valdez-Buso et al., 2019; Mottin et al., 2018; Vesely et al., 2018; Schemiko et al., 2019; Rodrigues et al., 2020, 2021). Based on the degree of deformation, disaggregation and mixing of sediments of the remobilized layers, MTDs have been described as slides, slumps and debris flows, or a transition between all these mass-movement deposits, in the three stratigraphic units comprising the Itararé Group (Vesely and Assine, 2006; Suss et al., 2014; Carvalho and Vesely, 2017; Valdez-Buso et al., 2019; Mottin et al., 2018; Vesely et al., 2018; Schemiko et al., 2019; Rodrigues et al., 2020, 2021).

Large ICs consist mainly of deltaic sandstone and were identified within the Taciba Formation in the southernmost and northernmost areas (Fig. 18; outcrops Ib-1, RS-2, RS-4 and RS-5). These clasts extend from tens of meters to 100 m (328 ft.) and are a few tens of meters thick (Figs. 19A–E and 23F). Well preserved bedding, sometimes tilted, occurs within these clasts, usually accompanied by sedimentary structures with little or no deformation (Rodrigues et al., 2020). Some of these clasts were described resting on internally deformed diamictite (Figs. 19D and 23F, Ib-1 and RS-2, respectively). Although large ICs have shown little internal deformation, folds are identified at the base of sandstone ICs; these may be related to the clasts' downslope movement and emplacement. Normal faults and sand injections near to, and in the contact with diamictite, may also relate to the clasts' translation (Fig. 19D and E).

MTDs with slide to slump features were also described in the Taciba Formation in the Rio do Sul region (e.g., RS-1 and RS-2) (Fig. 18). These MTDs display well preserved bedding with sedimentary structures, and partially preserved bedding with up to 5% of matrix, a result of sediment mixing. Structures consist mainly of normal and reverse faults and folds (symmetric to asymmetric, upright to recumbent folds), as well as symmetric and asymmetric boudins. Other shear-related features, such as quarter-like structures and intrastratal detachment surfaces, are also observed (Fig. 20A–G). When present, the matrix is massive and formed of partially mixed zones with little or no bedding/lamination. These mixed zones occur in the vicinity of disrupted layers with internal bedding/lamination preserved and deformed (Fig. 20F and G).

Outcrop RS-1 exemplifies MTDs with deformed layers and a relative lack of matrix. This MTD consists of rhythmite deformed by recumbent folds and, locally, symmetric boudins (Fig. 20A–C). In turn, outcrop RS-3 presents at least two intervals of deformed rhythmite and sandstone that are separated by an interval of rhythmite with no clear deformation (Fig. 20D–H). In the lower MTD, exposed in the southernmost part of outcrop RS-3, normal faults and associated thin shear zones, subparallel to the lamination, correspond to the extensional domain of the MTD. The upper MTD is better exposed and easily accessed in the central to northernmost part of the outcrop, where it shows partially to highly folded layers locally disrupted in symmetric boudins (Fig. 20E–G). Folds are asymmetric, inclined to recumbent and may show complex patterns with local refolding and thrust faults (Fig. 20E). Normal faults can deform the folded layers, relating to the main deformation stage or, instead, to a later stage of the mass movement marked by stress relaxation in the compressional domain of the upper MTD. A flow direction toward the NW is reinforced by the location of extensional (lower MTD) and compressional domains (upper MTD) to the south and north, respectively.

MTDs described as debris flows were identified in the three regions (Fig. 18) within the Lagoa Azul (CTM-1), Campo Mourão (outcrop CTM-2) and Taciba formations (RS-2, Ib-1, Ib-2 e CTM-3). These debris flows consist of sandy-muddy diamictites with a banded to mostly massive matrix that, usually, forms more than 75% of the volume of the deposit

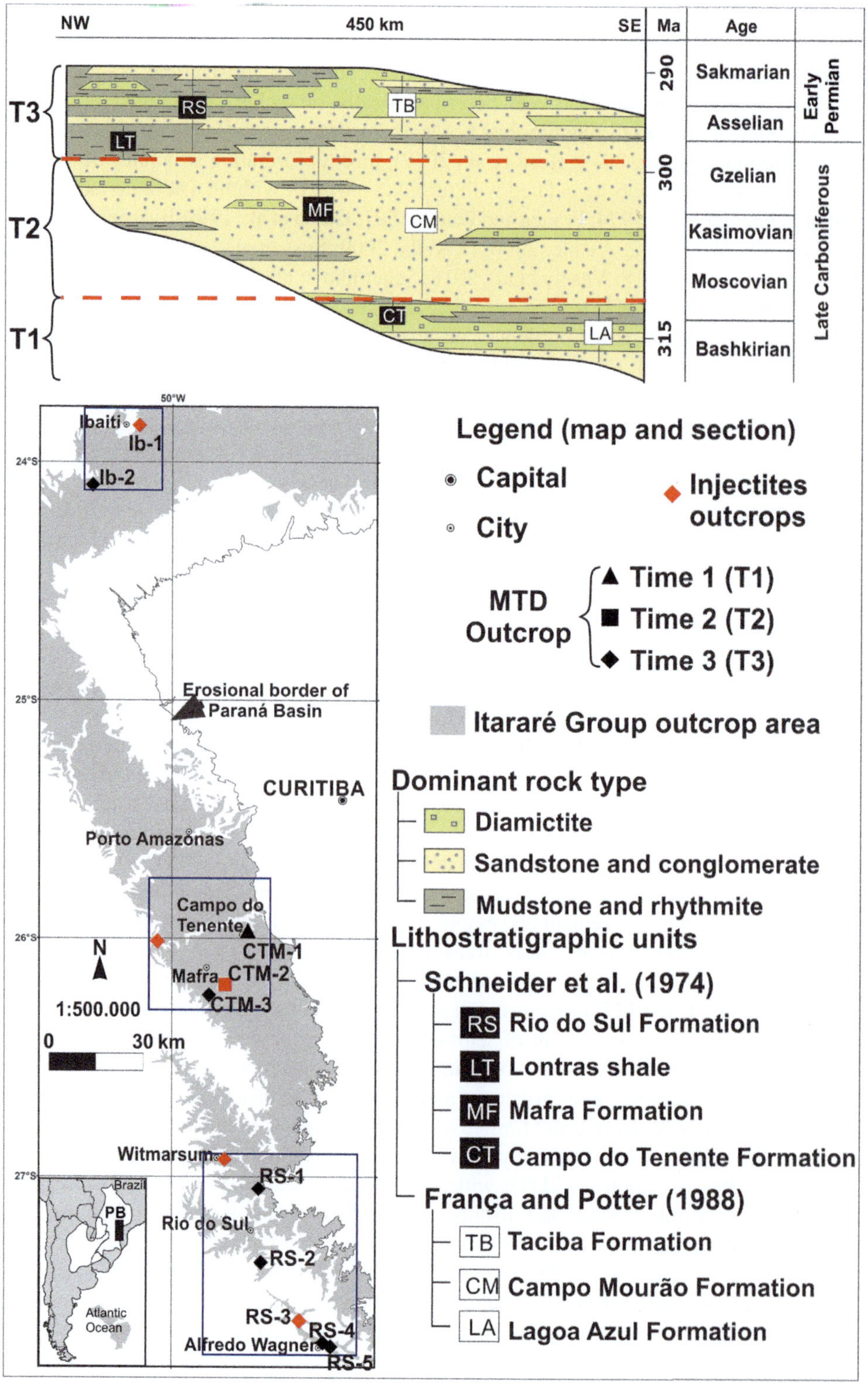


**FIG. 18** Location map and stratigraphic setting of MTDs in the Itararé Group (Paraná Basin in South Brazil) in the three areas of Ibaiti (Ib), Campo do Tenente-Mafra (CTM) and Rio do Sul (RS). The geographic location and stratigraphic positions of each mass transport deposit (outcrop locality) are indicated by codes and symbols, respectively. Sand injectites were documented in the outcrops indicated in red. *(Modified from Rodrigues, M.C.N.L., Trzaskos, B., Alsop, G. I., Vesely, F.F., 2020. Making a homogenite: an outcrop perspective into the evolution of deformation within mass-transport deposits. Mar. Petrol. Geol. 112, 104033. https://doi.org/10.1016/j.marpetgeo.2019.104033.)*

(Rodrigues et al., 2020). The matrix tends to show little (banded diamictites) or no bedding/lamination (massive diamictites), a character that subtly changes on a meter scale within the same outcrop (Rodrigues et al., 2021). In contrast, ICs dispersed within the matrix are disrupted and highly deformed with poorly to well-preserved bedding.

Deformation structures include folds, normal and inverse faults (with or without a sand/clay smear), symmetric and asymmetric boudins and other shear-related features, such as sheared laminations and/or fragments (Figs. 21–23). These

**FIG. 19** Large intrabasinal clasts (ICs) of sandstone: (A) without internal deformation (outcrop RS-5; Rodrigues et al., 2020); (B) with tilted bedding (adapted from Rodrigues et al., 2020); (C) with localized folding (outcrop RS-4); (D) resting on banded diamictite (Dm; outcrop Ib-1) and (E) internally deformed by normal faults.

structures are visible in deformed remnants of beds or banded matrix, as well as within sedimentary clasts formed by the rupture or boudinage of larger sandstone blocks. Internal deformation structures in ICs do not affect the surrounding matrix and have been interpreted to predate the clasts rupture (Rodrigues et al., 2020). Deformation styles in ICs thus include faulting, mutual injection between clasts and matrix sediments, and, more commonly, shearing and "grooves" that affect their borders (Figs. 21B, C, G, H and 23A, C, D, G). Disrupted lenses described in the matrix result from shearing of ICs and remnants of lamination. Other records of deformation described in some diamictites correspond to the preferential orientation of extra- and ICs long axes that tend to be roughly parallel to the flow direction (Rodrigues et al., 2020).

Debris flows from Lagoa Azul Formation were identified in the Campo do Tenente-Mafra region (outcrop CTM-1; Figs. 18 and 21A–E). The diamictite in CTM-1 is rich in ICs consisting of sandstone and occasionally rhythmites that are deformed internally by faulting and shearing. These ICs tend to occur in the same level with rounded/eye-shaped features interpreted as symmetric boudins. Some clasts are deformed by normal faults and tend to show sheared features (mostly at the top and borders of clasts) and grooves/"scratch" marks at their borders. Sheared sand fragments and remnant lamina in the matrix, together with the faults and shear bands in ICs, indicate shearing during mass flow (Rodrigues et al., 2020).

Mass-transported diamictite from Campo Mourão Formation was described in Campo do Tenente-Mafra region (outcrop CTM-2; Figs. 18 and 21F–M). This diamictite reveals a higher content of sand compared to other diamictites. ICs consist of mudstone and rhythmites with internal shear deformation of lamina. Cobble- to boulder-size clasts display sheared borders on which sediments or matrix fragments are incorporated, while intrabasinal granules tend to show a preferential orientation. Subhorizontal to inclined shear planes with mostly normal kinematics and clay smears were described in the matrix and in fractures crossing ICs. The matrix varies from massive to banded along outcrop CTM-2. Banded matrix display folds, asymmetric boudins and shear features.

Mass transport diamictites from the Taciba Formation were described in Ibaiti, Campo do Tenente-Mafra e Rio do Sul regions (outcrops Ib-1, Ib-2, CTM-3 and RS-2) (Figs. 18, 22 and 23). Diamictites in outcrop Ib-1 and Ib-2 record a heterogeneous matrix (Fig. 22A–D and F). At outcrop Ib-1, the diamictite is characterized by banded matrix deformed by reverse faults and folds that are locally disrupted or show fluidization and injection at their hinges (Fig. 22A–C). Diamictite at outcrop Ib-2 shows sandy lamina that are sheared and with other evidence for folded beds (Fig. 22F).

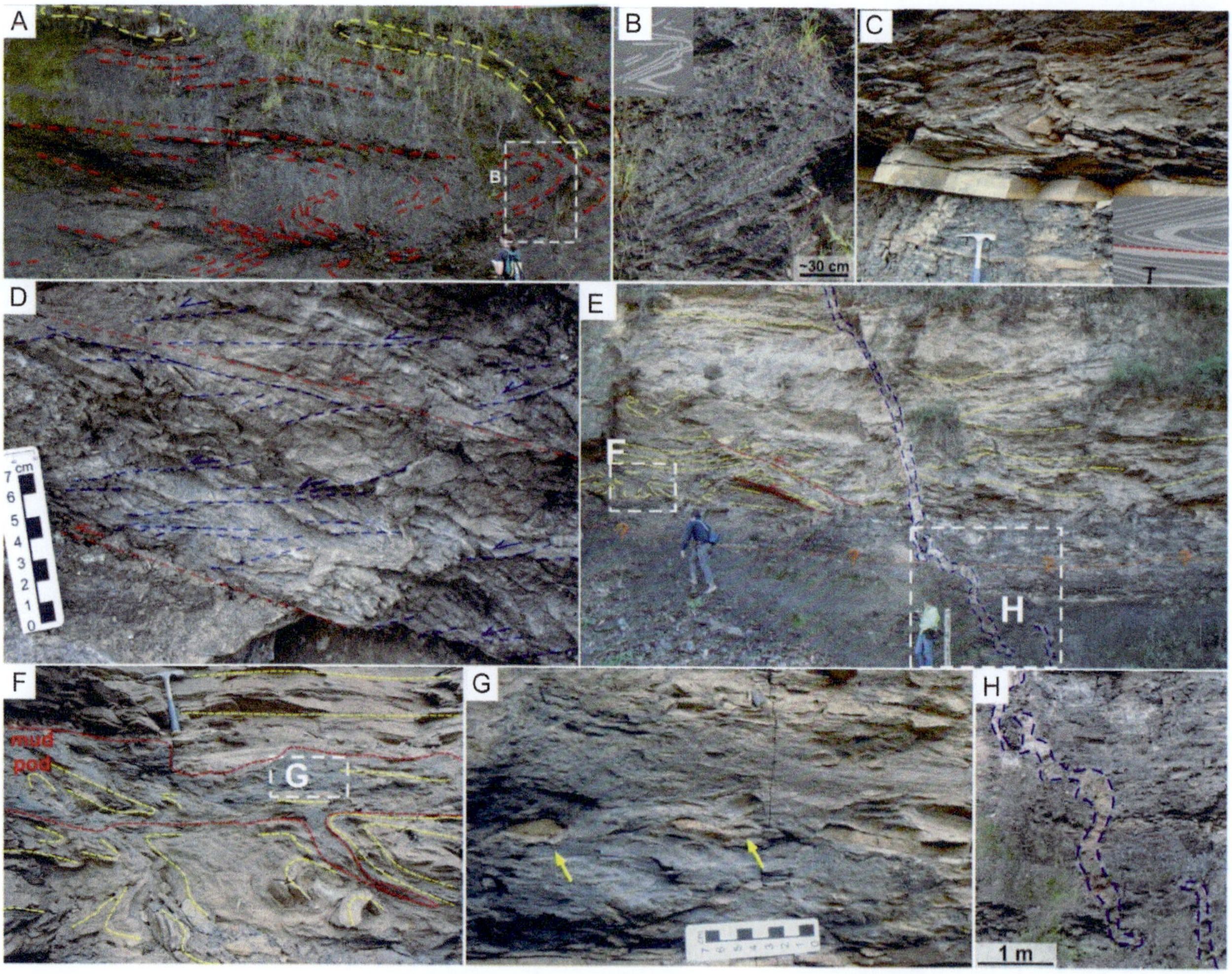


FIG. 20 MTDs at outcrops RS-1 (A–C) and RS-3 (D–K) revealing slides/slumps in the Itararé Group. MTD outcrop RS-1: (A) Interval of rhythmite folded with symmetric boudin at the limb of a thicker sandstone layer that is folded (adapted from Rodrigues et al., 2020). (B) Detail of recumbent folds and (C) recumbent folds with hinge thickening in thicker sandstone layers, next to the bottom of the MTD (possible detachment surface indicated by the red dashed line) (Rodrigues et al., 2020). Lower MTD in outcrop RS-3: (D) normal faults associated to subhorizontal or low-angle inverse faults in rhythmite. Upper MTD in outcrop RS-3: (E) symmetric to asymmetric folds and thrust faults in a sandstone and rhythmite interval, crosscut by sand injectites with "en echelon" pattern; (F) asymmetric to recumbent folds in rhythmites with mud-layers with no preserved lamination, sand-layers folded partially disrupted and pods of mud (dark gray material broadly highlighted by red dashed line; E and F adapted from Rodrigues et al., 2020); (G) symmetric boudins (indicated by yellow arrows) in limbs of folded sand layers (Rodrigues et al., 2020); (H) detail of sand injectites, with a partially folded geometry that possibly results from compaction that affects the upper MTD and the underlying rhythmite. Possible base limit of the upper MTD indicated by the orange dashed lines (F).

MTDs at outcrops CTM-3 and RS-2 tend to have a homogeneous matrix (i.e., they are massive) with portions showing a heterogeneous matrix with disrupted lamina and preserved remnants of the original bedding (Fig. 23A–E and F–I). Shear planes and zones characterized by clay and sand smears occur in the matrix at both outcrops. At CTM-3, diamictite in the shear planes/zones show both normal and reverse kinematics, whereas shear planes/zones at RS-2 show mostly reverse faults. Shear features such as "S-C" fabrics and asymmetric boudins were formed in these shear zones within diamictite intervals. ICs consist of sandstones and rhythmites that, usually, display internal deformation by folds and faults and border deformation. Reverse faults that affect both matrix and ICs were also identified in these diamictites.

Recent studies in the Campo do Tenente-Mafra and Rio do Sul regions have interpreted MTDs as the result of delta slope collapse caused by high sedimentation rates associated with intervals of maximum ice retreat (Suss et al., 2014; Fallgatter, 2015; Carvalho and Vesely, 2017; Valdez-Buso et al., 2019; Schemiko et al., 2019). These MTDs have been related to progradational or progradational-aggradational stacking patterns of a clinoform system with fluvio-deltaic to marine deposits, which sourced turbidites and noncohesive density flows more distally on the delta slope (Carvalho and Vesely, 2017; Schemiko et al., 2019). According to Carvalho and Vesely (2017). A rapid progradation of fluvio-deltaic

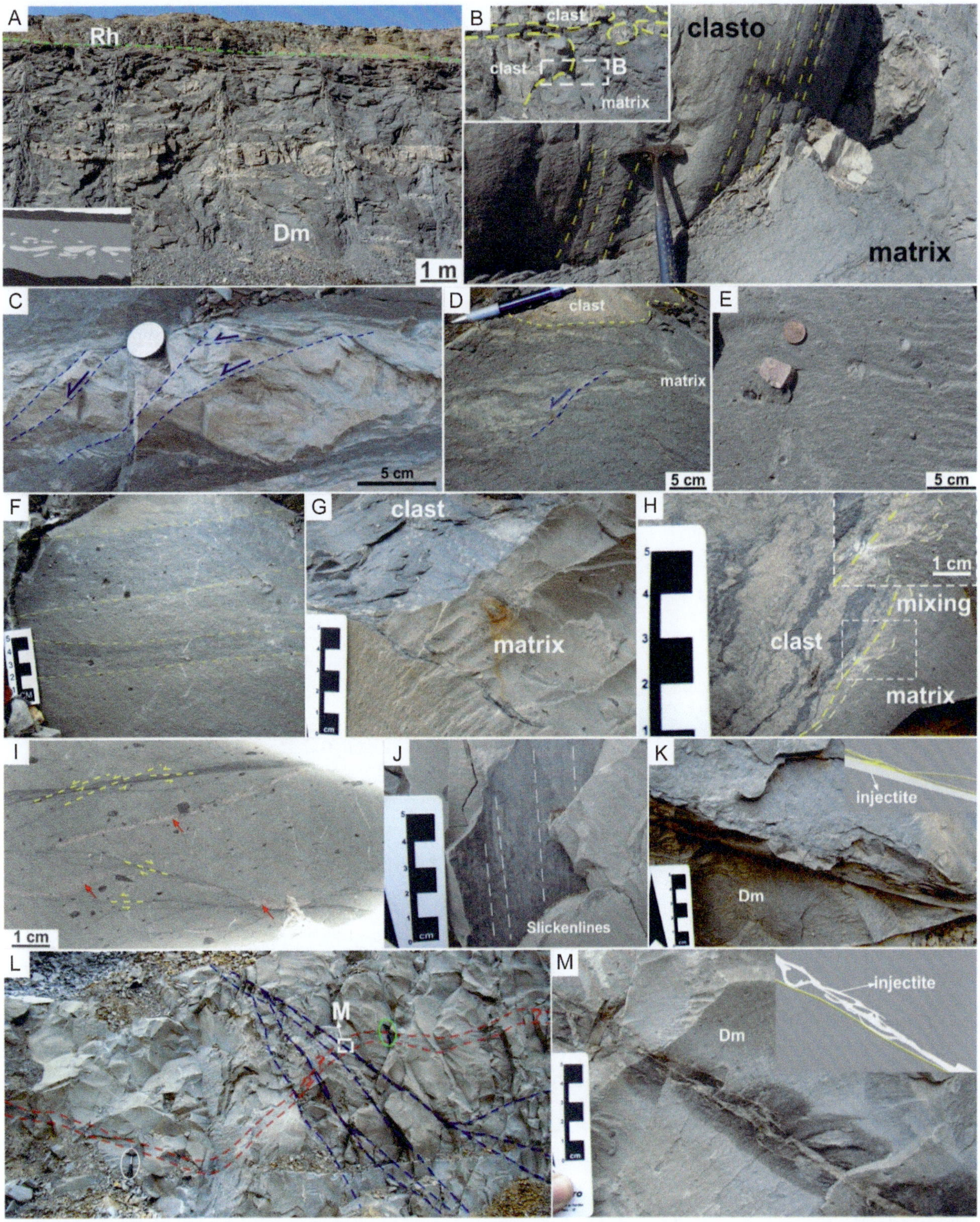


**FIG. 21** MTDs at outcrops CTM-1 (A–E) and CTM-2 (F–M) showing debris flows in the Lagoa Azul and Campo Mourão formations, respectively. MTD outcrop CTM-1: Diamictite (Dm) with (A) sandstone clasts deformed as symmetric boudins, with rhythmite deposited on the top (contact approximate indicated by green dashed line) (modified from Rodrigues et al., 2020); (B) these sandstone clasts formed by boudinage were subsequently deformed by grooves/stretch marks at the borders (A and B modified from Rodrigues et al., 2020); (C) some sandstone clasts were deformed by normal faults and shearing at the borders forming films of sediments incorporated by the matrix, which can be heterogeneous with remnant sand laminations sheared and deformed by normal faults (D; Rodrigues et al., 2020) or massive (E; Vesely et al., 2018, Rodrigues et al., 2020). MTD outcrop CTM-2: Diamictite (Dm) with massive (homogenous) to heterogenous matrix with discrete textural/compositional banding (F; highlighted by yellow dashed lines) and mud granules dispersed (Rodrigues et al., 2020). This diamictite also shows larger deformed clasts such as mud clasts with sheared borders (G) and rhythmite clasts with internal laminations sheared and folded and with sheared borders, with mixed zones resulting from the partial incorporation of sediment clasts by the matrix (H) (Rodrigues et al., 2020). (I) Clay smear faults (planes and zones) with anastomosed patterns that deform some of the sand injectites (Rodrigues et al., 2015, 2021). (J) Displacement surface (dark gray surface; looking down on the plane) with slickenlines in clay smear fault (highlighted by white lines; Rodrigues et al., 2020). (K) Tabular injectite parallel to clay smear fault (yellow lines in the drawing) (adapted from Rodrigues et al., 2020). (L) Sand injectites subparallel to clay smear faults (highlighted by blue dashed lines) that crosscut open folds in banded matrix (highlighted by red dashed lines). (M) Irregular anastomosed injectite subparallel to clay smear fault (yellow lines in the drawing).

**FIG. 22** MTDs at outcrops Ib-1 (A–E) and Ib-2 (F) showing debris flows in the Taciba Formation, in the Ibaiti region. MTD outcrop Ib-1: (A) Diamictite with matrix characterized by well-defined textural/compositional banding partially folded (highlighted by yellow dashed lines) and sand injections (highlighted by red dashed lines). (B) Asymmetric fold in banded matrix (Rodrigues et al., 2020). (C) Banded matrix folded and locally disrupted crosscut by thin later sand injection. (D) Injectites in the form of sills and dikes associated to the first stage of injection (Rodrigues et al., 2020). (E) Folded sandy sill (highlighted by red dashed line) (modified from Rodrigues et al., 2020). MTD outcrop Ib-2: (F) Photomosaic and interpreted sketch of mass transported diamictite with metric-scale symmentric and asymmetric folds (modified from Mottin et al., 2018).

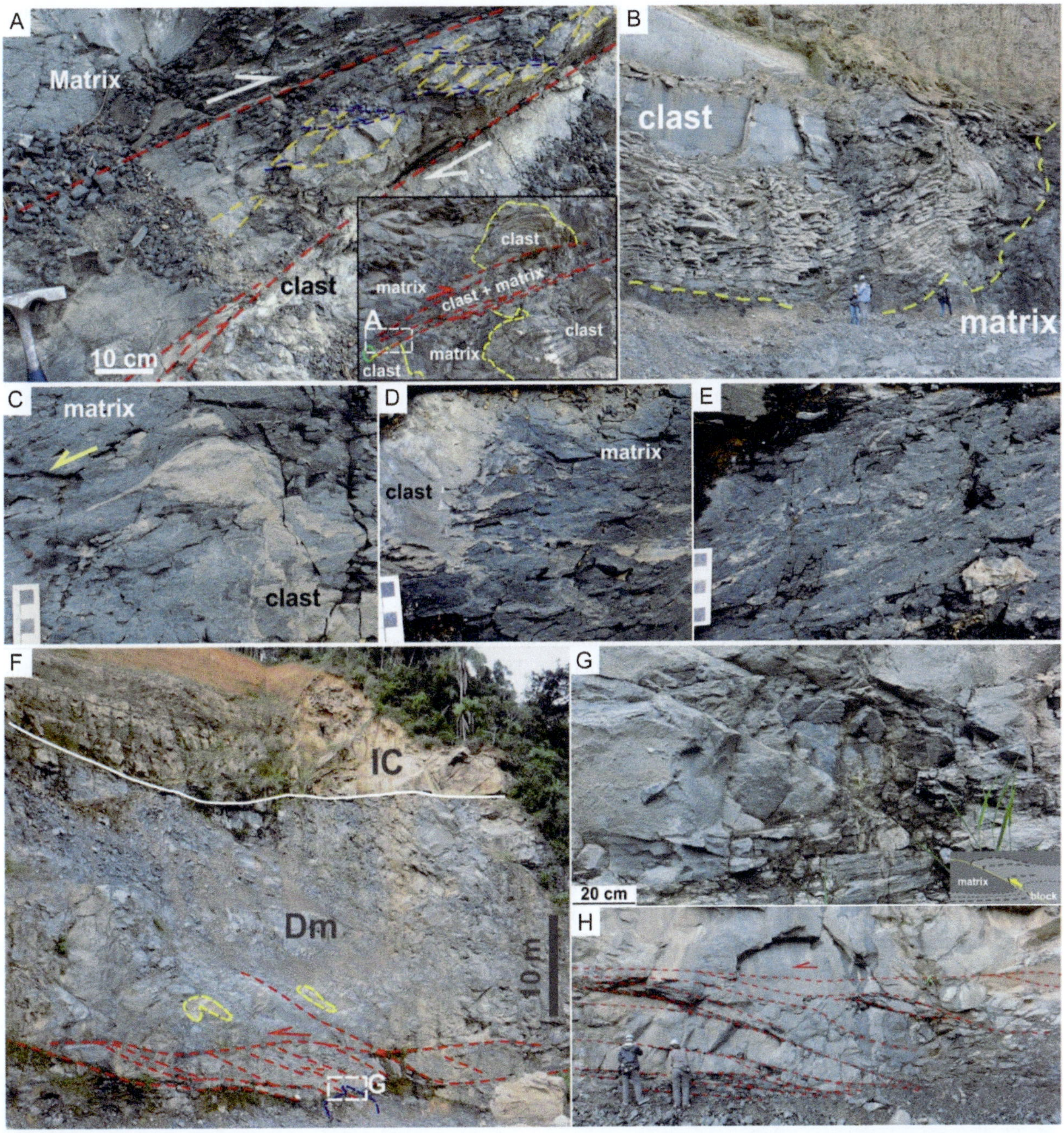


**FIG. 23** MTDs at outcrops CTM-3 (A–E) and RS-2 (F–J) revealing debris flows in the Taciba Formation, in the regions of Campo do Tenente-Mafra and Rio do Sul, respectively. MTD outcrop CTM-3: (A) Reverse fault affecting the diamictite matrix and intrabasinal clasts of rhythmite, which show internal deformation by folds not related to the external deformation; within the fault zone occur matrix and clasts fragments deformed, commonly with sigma shapes (highlighted by blue dashed lines). (B) Intrabasinal clast of rhythmite with internally deformed decimetric to decametric folds (Rodrigues et al., 2020). (C) Shearing at the borders of sand-rich rhythmite clast and resulting sandy films and fragments incorporated by the matrix. (D) Sandstone clast with border disintegration, mutual injection (matrix injection in clast and clast sediments injection in matrix) and incorporation of clast sediment by matrix (Rodrigues et al., 2020). (E) Diamicite matrix with sheared and disrupted sand lamination (Rodrigues et al., 2020). MTD outcrop RS-2: (F) Diamictite (Dm) with inverse shear zones (red-dashed lines) and deformed intrabasinal clasts of rhythmite (blue-dashed lines) and sandstone (yellow-dashed lines), and large intrabasinal clast (IC) resting on the diamictite (adapted from Rodrigues et al., 2021). (G) Detail of inverse fault and associated drag fold that affect both matrix and rhythmite clast (Rodrigues et al., 2020). (H) Inverse shear zones associated with clay smear (adapted from Rodrigues et al., 2020).

systems with recurrent collapse of their sand-rich portion is suggested by the occurrence of allochthonous fluvial and deltaic blocks in the MTDs. Nevertheless, MTDs in the Ibaiti region were interpreted as resulting from isostatic rebound and base-level fall due to ice-margin retreat, which caused slope instability and the resedimentation of previously accumulated glaciomarine sediments (Mottin et al., 2018).

## Sand injections in outcropping MTDs

Sand injections have been described at outcrop in the Campo Mourão and Taciba formations by Rodrigues et al. (2020) (Figs. 18, 20E, H, 21I, K–M, 22A, C–E and 24A–C). The injectites are composed of very fine to very coarse sand and may show silt, granules (including mud fragments) and fragments of the host rock. These structures seem to be pre-, syn- to post-MTD. According to Rodrigues et al. (2020), no parental rock was identified in most cases. Where the parental rock was identified, it corresponds to sand layers within the MTD (Fig. 24A and C). Injectites with breccia features were described in one of these cases, sometimes with sheared host-rock fragments (Figs. 24A and B). In other cases, the injections seem to be syn-MTD once they deform and are deformed by MTD structures (Fig. 24C). Such observations indicate sand injection within the MTD were due to increases in pore-fluid pressure of the sand layers during the mass flow. The trigger mechanism for most sand injection cases and the relationship with MTDs is still not clear and requires further investigation; nevertheless, the recurrence of injectites at outcrops suggests mass flows as a trigger mechanism for sand injection (Strachan, 2002).

At outcrop Ib-1, injectites crossing the diamictite consist of first-phase dikes and sills, in places folded with the banded matrix (Fig. 22A, D and E). This suggests that the injection of sand predates the mass flow, which naturally occurred when the host rock and injectites were still unconsolidated. In a second phase, thin injectites crosscut the first-stage injectites and folds in the banded matrix (Fig. 22C). This second phase of injectites occurred in the last phases of the mass flow or even later. Thin irregular to tabular injectites were also described in diamictite intervals at outcrop CTM-2 (Rodrigues et al., 2020) (Fig. 21I and K–M). These injectites crosscut the deformed banded matrix and clay smear shear planes or zones. They also occur parallel to subparallel to the shear structures. The temporal relationship between injectites and MTD structures suggests intrusion syn- and later to post-mass flow, but no possible parental rock was identified. The slumped MTDs at outcrop RS-3 also present injectites, which crosscut the deformed layers as dikes with an "en échelon" pattern (Fig. 20E and H). These injectites seem to have occurred after the cessation of mass flow; they were identified locally and without a clear relationship with tectonic structures. They may have been injected during the relaxation stage of the mass flow (Fig. 24).

# Concluding remarks

Several studies have demonstrated the direct and diverse relationship between different sediment remobilization and petroleum plays, including MTCs and MTDs, mud volcanoes and sand injections. Mud volcanoes and sand injectites are related to overpressured fluid flow occurring in different basins, onshore or offshore, and in different tectonic contexts. Mud volcanoes result from the flow of pressurized fluids with mud at depth, particularly hydrocarbons that ascend to the surface and commonly occur in association with petroleum systems. Therefore, mud volcanoes are considered ideal indicators of hydrocarbon plays that can be targeted for further exploration. Sand injections (or injectites) can influence

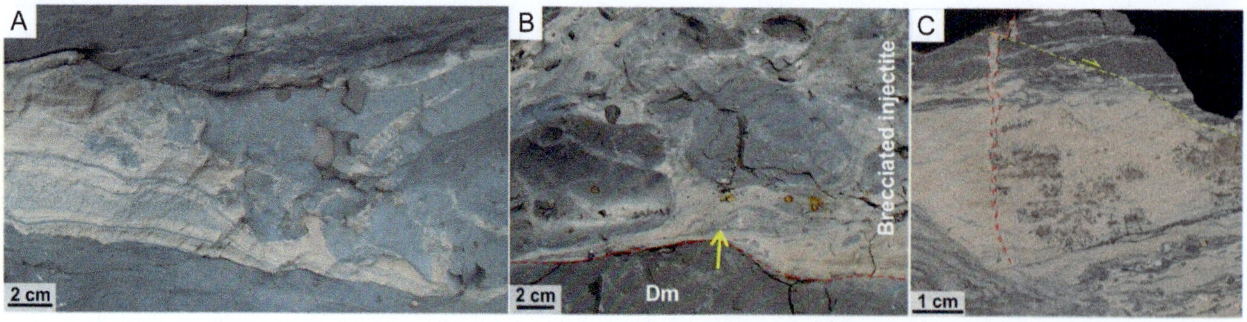


**FIG. 24** Examples of sand injections identified in MTD of Itararé Group in the Campo do Tenente-Mafra (A and B) and Rio do Sul regions (C) (see location in Fig. 1). (A) Sandstone layer partially fluidized and injected in diamictite matrix; note the irregular portions of the host rock partially to totally enveloped by the injections. (B) Breccia resulting from sand injection in diamictite; note that some host rock fragments show shearing (indicated by yellow arrow) (adapted from Rodrigues et al., 2020). (C) Sand injectite that crosscuts deformed laminations of rhythmite and is also ductily sheared with the host layers. Although the parental layer is not present in the sample, it is a sand layer within the MTD.

different aspects of petroleum plays, including reservoir, seal, trap and fluid migration. The distribution and geometry of injectites may affect the distribution of hydrocarbons reservoirs, connect isolated sand bodies, break through rocks that would otherwise be potential seals and indicate fluid migration routes and thief zones. Thus, the occurrence of sand injectites in petroleum systems has consequences for hydrocarbon plumbing systems.

Deepwater traps continue to be important targets for future hydrocarbon exploration and are likely to deliver future volumes in both mature and frontier basins (Amy, 2019; Collard, 2020). The several examples discussed here show that MTCs and MTDs, particularly the thicker and more extensive, may act as seals (top seals or intra-reservoir barriers) or reservoirs in deepwater petroleum systems. In addition, the irregular morphology of these deposits and of the seafloor affected by submarine landslides can influence the deposition of turbidites, which are potential deepwater reservoirs.

Submarine landslides can generate deposits with widely varying characteristics, geometries, morphologies, structures and depositional facies, as exemplified in this chapter. Although MTDs show domains with preferential styles of deformation, a wide variety of structures at seismic to subseismic scales and deformation styles, can occur along with these deposits. Furthermore, submarine landslides can be triggered by several coexisting mechanisms in different types of basins and tectonic contexts. Future studies of these deposits at different scales (at seismic to outcrop scale) will help address their different aspects and relationship with other deposits, thus identifying potential impacts on petroleum systems. Integration of seismic data with core data, and comparisons with analogous outcrops are probably the best approaches to understanding the formation of these deposits and their impacts on petroleum plays. Seismic data resolution is still not adequate enough to define the potential of MTCs to act as a source, reservoir or seal, and therefore, integration of different datasets is paramount to address the subsurface uncertainty. Moreover, not all MTCs/MTDs are made equal and explorationists need to assess them in a case-by-case scenario.

The phenomena resulting in sediment remobilization generate geohazards and need to be considered in offshore and subsurface projects. Because sediment remobilization can be directly or indirectly related to climate change, understanding how different sediment remobilization processes occur and their roles as geohazards is highly relevant. For instance, changes in marine dynamics, relative sea level and sediment input can favor the destabilization of submarine slopes and trigger large landslides with the potential to affect infrastructure on the seafloor and generate tsunamis, depending on the volume of remobilized sediments.

The remobilization of sediments associated with fluid flow, including mud volcanoes and sand extrusion (associated with the formation of sand injections), can be triggered by several mechanisms such as seismicity, high deposition rate and fluid migration. Large sediment remobilization related to submarine landslide may also result in fluid overpressure in underlying sediments and trigger fluid migration, mud volcanism and sand extrusions. Studies also indicate that these sediment remobilizations can affect subsea infrastructures as well as expulsing fluid and hydrocarbons to the surface to contaminate marine and lacustrine environments. In addition, sediment remobilization associated with fluid flow, particularly mud volcanoes, is can be a significant source of methane. Therefore, these sediment remobilization processes can contribute to climate change. Lastly, as offshore infrastructure keeps growing in the energy transition landscape, understanding such phenomena and associated deposits will be important for the responsible development of offshore infrastructure.

## Acknowledgments

Tiago M. Alves thanks V. Lykousis (HCMR), I. Zananiri, N. Carras and K. Perissoratis (IGME–Athens) for permits to the field campaigns in Crete (2008, 2009, 2013 and 2015–2018). His work on slide blocks in SE Crete was partly sponsored by Royal Society grant RSSV 2008/R3, TEPA's (Total-Angola) RAFTS consortium, Petrobras and Husky Energy. The COST Action ES1301: FLOWS—Impact of Fluid circulation in old oceanic Lithosphere on the seismicity of transform-type plate boundaries: new solutions for early seismic monitoring of major European Seismogenic zones is acknowledged for a meeting and field trip in 2016. M.C.N.L. Rodrigues thanks the Brazilian National Council for Scientific and Technological Development for funding the research on the Paraná Basin (CNPq, grant 461650/2014–2, PQ 302842/2017-9 and PQ 306780/2019-4) and the Coordination for the Improvement of Higher Education Personnel (CAPES) for providing a PhD scholarship. In addition to the Federal University of Paraná, colleagues from the Laboratory on Basin Analysis (LABAP) and Professor Ian Alsop (University of Aberdeen) contributed to the research undertaken in Southern Brazil. S. Cardona thanks the Sediment Mechanics Lab at Texas A&M University for funding while writing this chapter. Malcolm Arnot, Greg Browne, Mimi Do and Estefania Lopez are thanked for assistance with fieldwork in New Zealand, and Lesli Wood at the Sedimentary Analogs Database (SAnD) research program and Zane Jobe at the Chevron Center of Research Excellence (CoRE) at Colorado School of Mines for financial, technical and logistical support for fieldwork. Finally, we thank the comments provided by Editors Jon R. Rotzien, F. Javier Hernández Molina, Octavian Catuneanu, Cindy A. Yeilding and Richard A. Sears, which have significantly improved this chapter.

# References


Abadi, S.M., Harris, J.C., Grilli, S.T., Fabre, R., 2012. Numerical modeling of tsunami waves generated by the flank collapse of the Cumbre Vieja volcano (La Palma, Canary Islands): Tsunami source and near field effects. J. Geophys. Res. 117, C05030. https://doi.org/10.1029/2011JC007646.

Algar, S., Milton, C., Upshall, H., Roestenburg, J., Crevello, P., 2011. Mass-transport deposits of the deepwater northwestern Borneo margin—characterization from seismic-reflection, borehole, and core data with implications for hydrocarbon exploration and exploitation. In: Shipp, R.C., Weimer, P., Posamentier, H.W. (Eds.), Mass-Transport Deposits in Deepwater Settings. SEPM (Society for Sedimentary Geology), pp. 351–366, https://doi.org/10.2110/sepmsp.096.351.

Allen, J.R.L., 1985. Principles of Physical Sedimentology. Blackburn Press, Caldwell. 272 pp.

Alsop, G.I., Marco, S., 2011. Soft-sediment deformation within seismogenic slumps of the Dead Sea Basin. J. Struct. Geol. 33, 433–457. https://doi.org/10.1016/j.jsg.2011.02.003.

Alsop, G.I., Marco, S., 2014. Fold and fabric relationships in temporally and spatially evolving slump systems: a multi-cell flow model. J. Struct. Geol. 63, 27–49. https://doi.org/10.1016/j.jsg.2014.02.007.

Álvarez-Gómez, J.A., Aniel-Quiroga, Í., González, M., Olabarrieta, M., Carreño, E., 2011. Scenarios for earthquake-generated tsunamis on a complex tectonic area of diffuse deformation and low velocity: the Alboran Sea. Western Mediterranean. Mar. Geol. 284, 55–73. https://doi.org/10.1016/j.margeo.2011.03.008.

Alves, T.M., 2010a. 3D Seismic examples of differential compaction in mass-transport deposits and their effect on post-failure strata. Mar. Geol. 271, 212–224. https://doi.org/10.1016/j.margeo.2010.02.014.

Alves, T.M., 2010b. A 3-D morphometric analysis of erosional features in a contourite drift from offshore SE Brazil. Geophys. J. Int. 183, 1151–1164. https://doi.org/10.1111/j.1365-246X.2010.04827.x.

Alves, T.M., 2015. Submarine slide blocks and associated soft-sediment deformation in deep-water basins: a review. Mar. Pet. Geol. 67, 262–285. https://doi.org/10.1016/j.marpetgeo.2015.05.010.

Alves, T.M., Cartwright, J.A., 2009. Volume balance of a submarine landslide in the Espírito Santo Basin, offshore Brazil: quantifying seafloor erosion, sediment accumulation and depletion. Earth Planet. Sci. Lett. 288, 572–580.

Alves, T.M., Cupkovic, T., 2018. Footwall degradation styles and associated sedimentary facies distribution in SE Crete: insights into tilt-block extensional basins on continental margins. Sediment. Geol. 367, 1–19. https://doi.org/10.1016/j.sedgeo.2018.02.001.

Alves, T.M., Lourenço, S.D.N., 2010. Geomorphologic features related to gravitational collapse: submarine landsliding to lateral spreading on a Late Miocene-Quaternary slope (SE Crete, eastern Mediterranean). Geomorphology 123, 13–33. https://doi.org/10.1016/j.geomorph.2010.04.030.

Alves, T.M., Kurtev, K., Moore, G.F., Strasser, M., 2014. Assessing the internal character, reservoir potential, and seal competence of mass-transport deposits using seismic texture: a geophysical and petrophysical approach. AAPG Bull. 98 (4), 793–824.

Alves, T.M., Cartwright, J., Davies, R.J., 2009. Faulting of salt-withdrawal basins during early halokinesis: effects on the Paleogene Rio Doce Canyon system (Espírito Santo Basin, Brazil). AAPG Bull. 93, 617–652. https://doi.org/10.1306/02030908105.

Amy, L.A., 2019. A review of producing fields inferred to have upslope stratigraphically trapped turbidite reservoirs: trapping styles (pure and combined), pinch-out formation, and depositional setting. AAPG Bull. 103, 2861–2889. https://doi.org/10.1306/02251917408.

Andresen, K.J., 2021. The Norwegian-Danish Basin: A Dynamic Setting for Subsurface Sand Remobilization—Established Concepts on Distribution and Controlling Factors. Geological Society, London, pp. 47–68, https://doi.org/10.1144/SP493-2018-026. Special Publications, 493.

Andresen, K.J., Clausen, O.R., 2014. An integrated subsurface analysis of clastic remobilization and injection; a case study from the Oligocene succession of the eastern North Sea. Basin Res. 26, 641–674. https://doi.org/10.1111/bre.12060.

Aquino, C.D., Valdez, B.V., Faccini, U.F., Milana, J.P., Paim, P.S.G., 2016. Facies and depositional architecture according to a jet efflux model of a late Paleozoic tidewater grounding line system from the Itararé Group (Paraná Basin), southern Brazil. J. South Am. Earth Sci. 67, 180–200.

Armitage, D.A., Romans, B.W., Covault, J.A., Graham, S.A., 2009. The influence of mass-transport-deposit surface topography on the evolution of turbidite architecture: the Sierra Contreras, Tres Pasos formation (Cretaceous), Southern Chile. J. Sediment. Res. 79, 287–301. https://doi.org/10.2110/jsr.2009.035.

Barley, B., 1999. Deepwater problems around the world. Lead. Edge 18, 488–494. https://doi.org/10.1190/1.1438319.

Baur, J., Sutherland, R., Stern, T., 2014. Anomalous passive subsidence of deep-water sedimentary basins: a prearc basin example, southern New Caledonia Trough and Taranaki Basin, New Zealand. Basin Res. 26, 242–268. https://doi.org/10.1111/bre.12030.

Bellwald, B., Hjelstuen, B.O., Sejrup, H.P., Stokowy, T., Kuvås, J., 2019. Holocene mass movements in west and mid-Norwegian fjords and lakes. Mar. Geol. 407, 192–212. https://doi.org/10.1016/J.MARGEO.2018.11.007.

Berton, F., Vesely, F.F., 2016. Stratigraphic evolution of Eocene clinoforms from northern Santos Basin, offshore Brazil: evaluating controlling factors on shelf-margin growth and deep-water sedimentation. Mar. Pet. Geol. 78, 356–372. https://doi.org/10.1016/j.marpetgeo.2016.09.007.

Bertoni, C., Cartwright, J., Hermanrud, C., 2013. Evidence for large-scale methane venting due to rapid drawdown of sea level during the Messinian Salinity Crisis. Geology 41, 371–374. https://doi.org/10.1130/G33987.1.

Bhatnagar, P., Verma, S., Bianco, R., 2019. Characterization of mass transport deposits using seismic attributes: Upper Leonard Formation, Permian Basin. Interpretation 7, SK19–SK32. https://doi.org/10.1190/INT-2019-0036.1.

Bhattacharya, S., Verma, S., Rotzien, J.R., 2020. 3D seismic imaging of the submarine slide blocks on the North Slope, Alaska. Interpretation 8, SR37–SR44. https://doi.org/10.1190/INT-2020-0038.1.

Blouin, A., Imbert, P., Sultan, N., Callot, J.P., 2019. Evolution model for the Absheron mud volcano: from in-situ observations to numerical modeling. J. Geophys. Res. Earth Surf. 124 (3), 766–794. https://doi.org/10.1029/2018JF004872.

Boehm, A., Moore, J.C., 2002. Fluidized sandstone intrusions as an indicator of paleostress orientation, Santa Cruz, California. Geofluids 2, 147–161.

Bohannon, R.G., Gardner, J.V., 2004. Submarine landslides of San Pedro Escarpment, southwest of Long Beach, California. Mar. Geol. 203, 261–268. 1016/S0025-3227(03)00309-8.

Boulesteix, T., Hildenbrand, A., Soler, V., Quidelleur, X., Gillot, P.-Y., 2013. Coeval giant landslides in the Canary Islands: implications for global, regional and local triggers of giant flank collapses on oceanic volcanoes. J. Volcanol. Geotherm. Res. 257, 90–98. https://doi.org/10.1016/J.JVOLGEORES.2013.03.008.

Bradley, D., Hanson, L., 1998. Paleoslope analysis of slump folds in the Devonian Flysch of Maine. J. Geol. 106, 305–318.

Brami, T.R., Pirmez, C., Archie, C., Heeralal, S., Holman, K.L., 2000. Late Pleistocene deep-water stratigraphy and depositional processes, offshore Trinidad and Tobago. In: Weimer, P., Slatt, R.M., Coleman, J., Rosen, N.C., Nelson, H., Bouma, A.H., Styzen, M.J., Lawrence, D.T. (Eds.), Deep-Water Reservoirs of the World: Gulf Coast Section SEPM 20th Annual Research Conference. GCSSEPM, Houston, TX, pp. 104–115, https://doi.org/10.5724/gcs.00.15.0104.

Brodzikowski, K., Van Loon, A.J., 1985. Inventory of deformational structures as a tool for unravelling the Quaternary geology of glaciated areas. Boreas 14, 175–188.

Brooke, C.M., Trimble, T.J., Mackay, T.A., 1995. In: Hartley, A.J., Prosser, D.J. (Eds.), Characterisation of Deep Marine Clastic Systems: Special Publication. 94. Geological Society, London, pp. 95–101.

Brown, A.R., 2011. Interpretation of Three-Dimensional Seismic Data, seventh ed. American Association of Petroleum Geologists, Tulsa, https://doi.org/10.1190/1.9781560802884.

Bryn, P., Berg, K., Forsberg, C.F., Solheim, A., Kvalstad, T.J., 2005. Explaining the Storegga slide. Mar. Petrol. Geol. 22, 11–19.

Bull, S., Cartwright, J., Huuse, M., 2009. Small-scale insights into seismic-scale slumps: a comparison of slump features from the Waitemata Basin, New Zealand, and the Møre Basin, Off-shore Norway. In: Mosher, D., Shipp, R.C., Moscardelli, L., Chaytor, J.D., Baxter, C.D.P., Lee, H.J., Urgeles, R. (Eds.), Submarine Mass Movements and Their Consequences, pp. 257–266.

Cardona, S., Wood, L.J., Day-Stirrat, R.J., Moscardelli, L., 2016. Fabric development and pore-throat reduction in a mass-transport deposit in the Jubilee Gas Field, Eastern Gulf of Mexico: consequences for the Sealing Capacity of MTDs. In: Lamarche, G. et al. eds., Submarine Mass Movements and their Consequences, Cham, Springer International Publishing. Advances in Natural and Technological Hazards Research 41, 27–37. https://doi.org/10.1007/978-3-319-20979-1_3.

Cardona, S., Wood, L.J., Dugan, B., Jobe, Z., Strachan, L.J., 2020a. Characterization of the Rapanui mass-transport deposit and the basal shear zone: Mount Messenger Formation, Taranaki Basin, New Zealand (J. Baas, Ed.). Sedimentology 67, 2111–2148. https://doi.org/10.1111/sed.12697.

Cardona, S., Wood, L., Moscardelli, L., Dunlap, D., 2020b. Cannibalization and sealing of deepwater reservoirs by mass-transport complexes—the Jubilee field, Gulf of Mexico. Interpretation 8, SV17–SV30. https://doi.org/10.1190/INT-2019-0274.1.

Carneiro, C.D.R., Costa, F.G.D., 2006. Estruturas atectônicas da Bacia do Paraná em Campinas (SP): deformação sin-sedimentar no Subgrupo Itararé. Terrae Didática 2 (1), 34–43.

Carter, L., Burnett, D., Drew, S., Hagadorn, L., Marle, G., Bartlett-McNeil, D., Irvine, N., 2009. Submarine cables and the oceans—connecting the world. In: UNEP-WCMC Biodiversity Series. vol. 31. ICPC/UNEP/UNEPWCMC. 64 pp.

Carter, L., Gavey, R., Talling, P., Liu, J., 2014. Insights into submarine geohazards from breaks in subsea telecommunication cables. Oceanography 27 (2), 58–67.

Cartwright, J.A., Dewhurst, D., 1998. Layer-bound compaction faults in fine grained sediments. Geol. Soc. Am. Bull. 110, 1242–1257.

Cartwright, J.A., James, D.M.D., Bolton, A.J., 2003. In: Van Rensbergen, P., Hillis, R., Maltman, A., Morley, C. (Eds.), Subsurface Sediment Mobilization. 216. Special Publication, London, pp. 223–242.

Cartwright, J., James, D., Huuse, M., Vetel, W., Hurst, A., 2008. The geometry and emplacement of conical sandstone intrusions. J. Struct. Geol. 30, 854–867.

Carvalho, A.H., Vesely, F.F., 2017. Facies relationships recorded in a Late Paleozoic fluvio-deltaic system (Paraná Basin, Brazil): insights into the timing and triggers of subaqueous sediment gravity flows. Sediment. Geol. 352, 45–62.

Caujapé-Castells, J., García-Verdugo, C., Marrerorodríguez, Á., Fernández-Palacios, J.M., Crawford, D.J., Mort, M.E., 2017. Island ontogenies, syngameons, and the origins and evolution of genetic diversity in the Canarian endemic flora. Perspect. Plant Ecol. Evol. Syst. 27, 9–22.

Chillarige, A.V., Morgenstern, N.R., Robertson, P.K., Christian, H.A., 1997. Seabed instability due to flow liquefaction in the Fraser River delta. Can. Geotech. J. 34 (4), 520–533.

Clare, M.A., Vardy, M.E., Cartigny, M.J.B., Talling, P.J., Himsworth, M.D., Dix, J.K., Harris, J.M., Whitehouse, R.J.S., Belal, M., 2017. Direct monitoring of active geohazards: emerging geophysical tools for deep-water assessments. Near Surf. Geophys. 15, 427–444. https://doi.org/10.3997/1873-0604.2017033.

Clark, J.C., 1981. Stratigraphy, Palaeontology, and Geology of the Central Santa Cruz Mountains, California Coast Range. U.S. Geol. Surv. 51 pp. Professional Paper 1168.

Clayton, C.A., et al., 1998. Ubit Field Rejuvenation: a case history of reservoir management of a Giant Oil Field, Offshore Nigeria. SPE Annual Technical Conference and Exhibition. Society of Petroleum Engineers, https://doi.org/10.2118/49165-MS.

Collard, J., 2020. Searching for future super-basins. In: Global Super Basins, Sugar Land. AAPG.

Conrad, J.E., Prouty, N.G., Walton, M.A.L., Kluesner, J.W., Maier, K.L., McGann, M., Brothers, D.S., Roland, E.C., Dartnell, P., 2018. Seafloor fluid seeps on Kimki Ridge, offshore southern California: links to active strike-slip faulting. Deep-Sea Res. II 150, 82–91. https://doi.org/10.1016/j.dsr2.2017.11.001.

Cortes, J.E., Aguilera, R., Wilches, O., Osorno, J.F., Cortes, S.I., 2018. Organic geochemical insights from oil seeps, tars, rocks, and mud volcanoes on the petroleum systems of the Sinú-San Jacinto basin, Northwestern, Colombia. J. South Am. Earth Sci. 86, 318–341.

Cox, D.R., Huuse, M., Newton, A.M.W., Gannon, P., Clayburn, J., 2020. Slip sliding away: enigma of large sandy blocks within a gas-bearing mass transport deposit, offshore northwestern Greenland. AAPG Bull. 104, 1011–1043. https://doi.org/10.1306/10031919011.

Cumberpatch, Z., Soutter, E., Kane, I.A., Casson, M., 2020. Evolution of a mixed siliciclastic-carbonate deep-marine system on an unstable margin: the Cretaceous of the Eastern Greater Caucasus, Azerbaijan. Basin Res. https://doi.org/10.31223/OSF.IO/WSVU6.

Davies, R.J., Huuse, M., Hirst, P., Cartwright, J., Yang, Y., 2006. Giant clastic intrusions primed by silica diagenesis. Geology 34, 917–920.

De Blasio, F.V., 2011. Physical models of giant subaqueous rock avalanches. EPL (Europhys. Lett.) 96, 59004. https://doi.org/10.1209/0295-5075/96/59004.

De Mol, B., Huvenne, V., Canals, M., 2009. Cold-water coral banks and submarine landslides: a review. Int. J. Earth Sci. 98, 885–899.

Deplus, C., Le Friant, A., Boudon, G., Komorowski, J.-C., Villemant, B., Harford, C., Ségoufin, J., Cheminée, J.-L., 2001. Submarine evidence for large-scale debris avalanches in the Lesser Antilles arc. Earth Planet. Sci. Lett. 192 (2), 145–157. https://doi.org/10.1016/S0012-821X(01)00444-7.

Dimitrov, L.I., 2002. Mud volcanoes—the most important pathway for degassing deeply buried sediments. Earth Sci. Rev. 59, 49–76.

Dixon, R.J., Schofield, K., Anderton, R., Reynolds, A.D., Alexander, R.W.S., Williams, M.C., Davies, K.G., 1995. Sandstone diapirism and clastic intrusion in the Tertiary submarine fans of the Bruce-Beryl Embayment, Quadrant 9, UKCS. In: Hartley, A.J., Prosser, D.J. (Eds.), Characterisation of Deep-Marine Clastic Systems: Special Publication, vol. 94. Geological Society, London, pp. 77–94.

Dott, R.H., 1963. Dynamics of subaqueous gravity depositional processes. AAPG Bull. 47, 104–128. https://doi.org/10.1306/BC743973-16BE-11D7-8645000102C1865D.

Doughty-Jones, G., Lonergan, L., Mayall, M., Dee, S., 2019. The role of structural growth in controlling the facies and distribution of mass transport deposits in a deep-water salt minibasin. Mar. Pet. Geol. 104, 106–124. https://doi.org/10.1016/J.MARPETGEO.2019.03.015.

D'Ávila, R.S.F., 2009. Sequências Deposicionais do Grupo Itararé (Carbonífero e Eopermiano), Bacia do Paraná, na Área de Dr. Pedrinho e Cercanias, Santa Catarina, Brasil. Porto Alegre. Ph.D. thesis, Universidade do Vale do Rio dos Sinos, UNISINOS. 192 pp.

Dunlap, D.B., Wood, L.J., Weisenberger, C., Jabour, H., 2010. Seismic geomorphohgy of offshore Morocco's east margin, Safi Haute Mer area. AAPG Bull. 94, 615–642. https://doi.org/10.1306/10270909055.

Dupuis, M., Imbert, P., Odonne, F., Vendeville, B., 2019. Mud volcanism by repeated roof collapse: 3D architecture and evolution of a mud volcano cluster offshore Nigeria. Mar. Petrol. Geol. 110, 368–387. https://doi.org/10.1016/j.marpetgeo.2019.07.033.

Duranti, D., Hurst, A., 2004. Fluidisation and injection in the deep-water sandstones of the Eocene Alba Formation (UK North Sea). Sedimentology 51, 503–531.

Duranti, D., Hurst, A., Bell, C., Groves, S., Hanson, R., 2002. Injected and remobilised Eocene sandstones from the Alba Field, UKCS: core and wireline characteristics. Pet. Geosci. 8, 99–107.

Dykstra, M., Garyfalou, K., Kertznus, V., Kneller, B., Milana, J.P., Molinaro, M., Szuman, M., Thompson, P., 2011. Mass-transport deposits: combining outcrop studies and seismic forward modeling to understand lithofacies distributions, deformation, and their seismic stratigraphic expression. In: Mass-Transport Deposits in Deepwater Settings. SEPM (Society for Sedimentary Geology), pp. 293–310, https://doi.org/10.2110/sepmsp.096.293.

Eng, C., Tsuji, T., 2019. Influence of faults and slumping on hydrocarbon migration inferred from 3D seismic attributes: Sanriku-Oki forearc basin, northeast Japan. Mar. Pet. Geol. 99, 175–189. https://doi.org/10.1016/J.MARPETGEO.2018.10.013.

Etiope, G., 2015. Natural Gas Seepage. The Earth's Hydrocarbon Degassing. Springer International Publishing, Switzerland, https://doi.org/10.1007/978-3-319-14601-0. 199 pp.

Evans, T.G., 2010. A systematic approach to offshore engineering for multiple-project developments in geohazardous areas. In: Frontiers in Offshore Geotechnics II. CRC Press, London, UK, pp. 3–32.

Eyles and Eyles, 2001. Subaqueous mass flow origin for Lower Permian diamictites and associated facies of the Grant Group, Barbwire Terrace, Canning Basin, Western Australia. Sedimentology 47, 343–356. https://doi.org/10.1046/j.1365-3091.2000.00295.x.

Eyles, C.H., Eyles, N., Franca, A.B., 1993. Glaciation and tectonics in an active intracratonic basin: the late Palaeozoic Itarare Group, Parana Basin, Brazil. Sedimentology 40, 1–25.

Fairbridge, R.W., 1946. Submarine slumping and location of oil bodies. AAPG Bull. 30, 84–92. https://doi.org/10.1306/3D9337C8-16B1-11D7-8645000102C1865D.

Fallgatter, C., 2015. Confined to unconfined deep-water systems of the Paraná (Brazil) and Paganzo (Argentina) basins. Ph.D. thesis, Unisinos. 191 pp.

Farrell, S.G., 1984. A dislocation model applied to slump structures, Ainsa Basin, South Central Pyrenees. J. Struct. Geol. 6, 727–736. https://doi.org/10.1016/0191-8141(84)90012-9.

Fedorchuk, N.D., Isbell, J.L., Griffis, N., Vesely, F.F., Rosa, E.L.M., Montanez, I., Mundil, R., Yin, Q.Z., Iannuzzi, R., Roesler, G., Pauls, K.N., 2019. Carboniferous glaciotectonized sediments in the southernmost Paraná Basin, Brazil: ice marginal dynamics and paleoclimate indicators. Sediment. Geol. 389, 54–72. https://doi.org/10.1016/j.sedgeo.2019.05.006.

Festa, A., Ogata, K., Pini, G.A., Dilek, Y., Alonso, J.L., 2016. Origin and significance of olistostromes in the evolution of orogenic belts: a global synthesis. Gondwana Res. 39, 180–203. https://doi.org/10.1016/j.gr.2016.08.002.

Forsberg, C.F., Heyerdahl, H., Solheim, A., 2016. Underwater mass movements in lake Mjøsa, Norway. In: Submarine Mass Movements and Their Consequences. Springer International Publishing, pp. 191–199.

Fossen, H., 2016. Structural Geology. Cambridge University Press, Cambridge, UK. 524 pp.

França, A.B., Potter, P.E., 1991. Stratigraphy and reservoir potential of glacial deposits of the Itararé Group (Carboniferous-Permian), Paraná Basin, Brazil. AAPG Bull. 75, 62–85.

Frey-Martínez, J., 2010. 3D seismic interpretation of mass transport deposits: implications for basin analysis and geohazard evaluation. In: Submarine Mass Movements and Their Consequences. Springer Netherlands, Dordrecht, pp. 553–568, https://doi.org/10.1007/978-90-481-3071-9_45.

Frey-Martínez, J., Cartwright, J., James, D., 2006. Frontally confined versus frontally emergent submarine landslides: a 3D seismic characterisation. Mar. Pet. Geol. 23, 585–604. https://doi.org/10.1016/j.marpetgeo.2006.04.002.

Gafeira, J., Long, D., Diaz-Doce, D., 2012. Semi-automated characterisation of seabed pockmarks in the central North Sea. Near Surf. Geophys. 10 (4), 303–314.

Gama Jr., E.G., Perinotto, J.A.J., Ribeiro, H.J.P.S., Padula, E.K., 1992. Contribuição ao estudo da ressedimentação no Grupo Itararé: tratos de fácies e hidrodinâmica deposicional. Rev. Bras. Geociênc. 22, 228–236.

Gamberi, F., Rovere, M., Marani, M., 2011. Mass-transport complex evolution in a tectonically active margin (Gioia Basin, Southeastern Tyrrhenian Sea). Mar. Geol. 279, 98–110.

Gamboa, D., Alves, T.M., 2015. Spatial and dimensional relationships of submarine slope architectural elements: a seismic-scale analysis from the EspIrito Santo Basin (SE Brazil). Mar. Pet. Geol. 64, 43–57. https://doi.org/10.1016/j.marpetgeo.2015.02.035.

Gamboa, D., Alves, T.M., 2016. Bi-modal deformation styles in con fi ned mass-transport deposits: examples from a salt minibasin in SE Brazil. Mar. Geol. 379, 176–193. https://doi.org/10.1016/j.margeo.2016.06.003.

Gamboa, D., Alves, T., Cartwright, J., Terrinha, P., 2010. MTD distribution on a "passive" continental margin: the Espírito Santo Basin (SE Brazil) during the Palaeogene. Mar. Pet. Geol. 27, 1311–1324. https://doi.org/10.1016/j.marpetgeo.2010.05.008.

Gee, M.J.R., Masson, D.G., Watts, A.B., Allen, P.A., 1999. The Saharan debris flow: an insight into the mechanics of long runout submarine debris flows. Sedimentology 46, 317–335. https://doi.org/10.1046/j.1365-3091.1999.00215.x.

Gee, M.J.R., Gawthorpe, R.L., Friedmann, S.J., 2006. Triggering and evolution of a giant submarine landslide, offshore Angola, revealed by 3D seismic stratigraphy and geomorphology. J. Sediment. Res. 76, 9–19. https://doi.org/10.2110/jsr.2006.02.

Gee, M.J.R., Uy, H.S., Warren, J., Morley, C.K., Lambiase, J.J., 2007. The Brunei slide: a giant submarine landslide on the North West Borneo Margin revealed by 3D seismic data. Mar. Geol. 246, 9–23. https://doi.org/10.1016/j.margeo.2007.07.009.

Giachetti, T., Paris, R., Kelfoun, K., Pérez-Torrado, F.J., 2011. Numerical modelling of the tsunami triggered by the Güimar debris avalanche, Tenerife (Canary Islands): comparison with field-based data. Mar. Geol. 284, 189–202. https://doi.org/10.1016/J.MARGEO.2011.03.018.

Giba, M., Walsh, J.J., Nicol, A., Mouslopoulou, V., Seebeck, H., 2013. Investigation of the spatio-temporal relationship between normal faulting and arc volcanism on million-year time scales. J. Geol. Soc. 170, 951–962. https://doi.org/10.1144/jgs2012-121.

Gisler, G., Weaver, R.P., Gittings, M.L., 2006. SAGE calculations of the tsunami threat from La Palma. Sci. Tsunami Hazards 24 (4), 288–301.

Godo, T.J., 2006. Identification of stratigraphic traps with subtle seismic amplitude effects in Miocene channel/levee sand systems, NE Gulf of Mexico. Geol. Soc., 127–151. https://doi.org/10.1144/GSL.SP.2006.254.01.07. Special Publication 254.

Gray, T., Dingler, J., Wood, G., 2013. Multiscale seabed seepage manifestations in the South Caspian Basin, Azerbaijan. Offshore Technology Conference.

Greene, H.G., Murai, L.Y., Watts, P., Maher, N.A., Fisher, M.A., Paull, C.A., Eichhubl, P., 2006. Submarine landslides in the Santa Barbara Channel as potential tsunami sources. Nat. Hazards Earth Syst. Sci. 6, 63–88.

Grilli, S.T., et al., 2019. Modelling of the tsunami from the December 22, 2018 lateral collapse of Anak Krakatau volcano in the Sunda Straits, Indonesia. Scientific Rep. 9, 11946. https://doi.org/10.1038/s41598-019-48327-6.

Grippa, A., Hurst, A., Palladino, G., Iacopini, D., Lecomte, I., Huuse, M., 2019. Seismic imaging of complex geometry: forward modeling of sandstone intrusions. Earth Planet. Sci. Lett. 513, 51–63. https://doi.org/10.1016/j.epsl.2019.02.011.

Gutierrez, M.A., Snedden, J.W., 2021. Integrated characterization and failure mechanism for a mid-Pleistocene mass transport complex-dominant interval in the Mars Ursa Basin, northern Gulf of Mexico, USA. Interpretation 9, T253–T274. https://doi.org/10.1190/INT-2019-0293.1.

Haflidason, H., Sejrup, H.P., Berstad, I.M., Nygård, A., Richter, T., Bryn, P., Lien, R., Berg, K., 2003. A weak layer feature on the Northern Storegga Slide Escarpment. In: European Margin Sediment Dynamics. Springer Berlin Heidelberg, Berlin, Heidelberg, pp. 55–62, https://doi.org/10.1007/978-3-642-55846-7_7.

Haflidason, H., Sejrup, H.P., Nygard, A., Mienert, J., Bryn, P., Lien, R., Forsberg, C.F., Berg, K., Masson, D., 2004. The Storegga Slide: architecture, geometry and slide development. Mar. Geol. 213 (1–4), 201–234. https://doi.org/10.1016/j.margeo.2004.10.007.

Hampton, M.A., Lee, H.J., Locat, J., 1996. Submarine landslides. Rev. Geophys. 34, 33–59. https://doi.org/10.1029/95RG03287.

Harbitz, C.B., Løvholt, F., Bungum, H., 2014. Submarine landslide tsunamis: how extreme and how likely? Nat. Hazards 72, 1341–1374.

Henry, L.C., Wadsworth, J.A., Hansen, B., Hartman, K., 2018. Erosion and ponding of Thunder Horse deepwater turbidites by mass transport complexes in Mississippi Canyon based on image log sedimentology. Mar. Pet. Geol. 97, 639–658. https://doi.org/10.1016/J.MARPETGEO.2018.08.006.

Hildebrandt, C., Egenhoff, S., 2007. Shallow-marine massive sandstone sheets as indicators of palaeoseismic liquefaction—an example from the Ordovician shelf of Central Bolivia. Sediment. Geol. 202, 581–595.

Hjelstuen, B.O., Eldholm, O., Faleide, J.I., 2007. Recurrent Pleistocene mega-failures on the SW Barents Sea margin. Earth Planet. Sci. Lett. 258, 605–618.

Ho, S., Hovland, M., Blouet, J.-P., Wetzel, A., Imbert, P., Carruthers, D., 2018. Formation of linear planform chimneys controlled by preferential hydrocarbon leakage and anisotropic stresses in faulted fine-grained sediments, offshore Angola. Solid Earth 9, 1437–1468. https://doi.org/10.5194/se-9-1437-2018.

Holz, M., França, A.B., Souza, P.A., Iannuzzi, R., Rohn, R., 2010. A stratigraphic chart of the Late Carboniferous/Permian succession of the eastern border of the Paraná Basin, Brazil, South America. J. South Am. Earth Sci. 29, 381–399. https://doi.org/10.1016/j.jsames.2009.04.004.

Houseknecht, D.W., 2019. Petroleum systems framework of significant new oil discoveries in a giant Cretaceous (Aptian-Cenomanian) clinothem in Arctic Alaska. AAPG Bull. 103, 619–652. https://doi.org/10.1306/08151817281.

Houseknecht, D.W., Schenk, C.J., 2007. Outcrops of turbidite channel facies in the Torok Formation: reservoir analogs for the Alaska North Slope. In: Nilsen, T., Shew, R.D., Steffens, G.S., Studlick, J.R.J. (Eds.), An Atlas of Deepwater Outcrops—Models and Analogs. Am. Assoc. Petrol. Geologists Geol. Stud, pp. 373–377, https://doi.org/10.1306/1240980St5685.

Hühnerbach, V., Masson, D.G., 2004. Landslides in the North Atlantic and its adjacent seas: an analysis of their morphology, setting and behaviour. Mar. Geol. 213, 343–362. https://doi.org/10.1016/J.MARGEO.2004.10.013.

Hunt, J.E., Wynn, R.B., Talling, P.J., Masson, D.G., 2013. Multistage collapse of eight western Canary Island landslides in the last 1.5 Ma: sedimentological and geochemical evidence from subunits in submarine flow deposits. Geochem. Geophys. Geosyst. 14, 2159–2181. https://doi.org/10.1002/ggge.20138.

Hurst, A., Cartwright, J., 2007. In: Hurst, A., Cartwright, J. (Eds.), Sand Injectites: Implications for Hydrocarbon Exploration and Production. AAPG Memoir, 87, pp. 1–19.

Hurst, A., Cartwright, J.A., Duranti, D., 2003a. In: Van Rensbergen, P., Hillis, R.R., Maltman, A.J., Morley, C.K. (Eds.), Subsurface Sediment Mobilization. vol. 216. Special Publication, London, pp. 123–137.

Hurst, A., Cartwright, J.A., Huuse, M., Jonk, R., Schwab, D., Duranti, D., Cronin, B., 2003b. Significance of large-scale sand injectites as long-term fluid conduits: evidence from seismic data. Geofluids 3, 263–274.

Hurst, A., Cartwright, J.A., Duranti, D., Huuse, M., Nelson, M., 2005. Sand injectites: an emerging global play in deep-water clastic environments. In: Doré, A., Vining, B. (Eds.), Petroleum Geology: North-west Europe and Global Perspectives. Proceedings of the 6th Petroleum Geology Conference. Geological Society, London, pp. 133–144.

Hurst, A., Scott, A., Vigorito, M., 2011. Physical characteristics of sand injectites. Earth-Sci. Rev. 106, 215–246. https://doi.org/10.1016/J.EARSCIREV.2011.02.004.

Huuse, M., Mickelson, M., 2004. Eocene sandstone intrusions in the Tampen Spur area (Norwegian North Sea Quad 34) imaged by 3D seismic data. Mar. Pet. Geol. 21, 141–155.

Huuse, M., Duranti, D., Steinsland, N., Guargena, C.G., Prat, P., Holm, K., Cartwright, J.A., Hurst, A., 2004. Seismic characteristics of largescale sandstone intrusions in the Paleogene of the South Viking Graben, UK and Norwegian North Sea. In: Davies, R.J., Cartwright, J., Stewart, S.A., Underhill, J.R., Lappin, M. (Eds.), 3D Seismic Technology: Application to the Exploration of Sedimentary Basins. vol. 29. Mem. Geol. Soc. Lond, pp. 257–271.

Huuse, M., Cartwright, J.A., Gras, R., Hurst, A., 2005. Kilometre-scale sandstone intrusions. In: Doré, A.G., Vining, B.A. (Eds.), Petroleum Geology: North-West Europe and Global Perspectives—Proceedings of the 6th Petroleum Geology Conference. Geological Society of London, pp. 1577–1594.

Huuse, M., Cartwright, J., Hurst, A., Steinsland, N., 2007. In: Hurst, A., Cartwright, J. (Eds.), Sand Injectites: Implications for Hydrocarbon Exploration and Production. AAPG Memoir, 87, pp. 21–35.

Jablonská, D., Di Celma, C.N., Alsop, G.I., Tondi, E., 2018. Internal architecture of mass-transport deposits in basinal carbonates: a case study from southern Italy (C. Betzler, Ed.). Sedimentology 65, 1246–1276. https://doi.org/10.1111/sed.12420.

Jablonska, D., Pitts, A., Di Celma, C., Volatili, T., Alsop, G.I., Tondi, E., 2021. 3D outcrop modelling of large discordant breccia bodies in basinal carbonates of the Apulian margin, Italy. Mar. Pet. Geol. 123, 104732. https://doi.org/10.1016/J.MARPETGEO.2020.104732.

Jackson, C.A.-L., Johnson, H.D., 2009. Sustained turbidity currents and their interaction with debrite-related topography; Labuan Island, offshore NW Borneo, Malaysia. Sediment. Geol. 219, 77–96. https://doi.org/10.1016/J.SEDGEO.2009.04.008.

Johnson, R.C., Birdwell, J.E., Brownfield, M.E., Mercier, T.J., 2015. Mass-Movement Deposits in the Lacustrine Eocene Green River Formation, Piceance Basin, Western Colorado. AAPG, https://doi.org/10.3133/ofr20151044.

Jolly, R.J.H., Lonergan, L., 2002. Mechanisms and control on the formation of sand intrusions. Geol. Soc. Lond. 159, 605–617.

Jonk, R., 2010. Sand-rich injectites in the context of short-lived and long-lived fluid flow. Basin Res. 22, 603–621. https://doi.org/10.1111/j.1365-2117.2010.00471.x.

Jonk, R., Cronin, B.T., Hurst, A., 2007. Variations in sediment extrusion in basin-floor, slope, and delta-front settings: sand volcanoes and extruded sand sheets from the Namurian of County Clare, Ireland. In: Hurst, A., Cartwright, J. (Eds.), Sand Injectites: Implications for Hydrocarbon Exploration and Production. American Association of Petroleum Geologists Memoir, Tulsa, pp. 221–226.

Jonk, R., Hurst, A., Duranti, D., Mazzini, A., Fallick, A.E., Parnell, J., 2005. Origin and timing of sand injection, petroleum migration and diagenesis: the Tertiary petroleum system of the south Viking Graben, North Sea. Am. Assoc. Pet. Geol. Bull. 89, 329–357.

Judd, A., Hovland, M., 2007. Seabed Fluid Flow. Cambridge University Press, Cambridge. 475 pp.

Judd, A., Hovland, M., 2009. Seabed Fluid Flow: The Impact on Geology, Biology and the Marine Environment. Cambridge University Press, Cambridge, UK.

Kaiser, M.J., Yu, Y., Jablonowski, C.J., 2009. Modeling lost production from destroyed platforms in the 2004–2005 Gulf of Mexico hurricane seasons. Energy 34 (9), 1156–1171.

Kessler, F., Jong, J., 2018. Hydrocarbon retention in clastic reservoirs of NW Borneo—examples of hydrocarbon trap, reservoir, seal and implications on hydrocarbon column length. Berita Sedimentol. 40, 6–44.

Kilb, D., 2008. Throwing mud. Nat. Geosci. 1 (9), 572–573.

King, P.R., Thrasher, G.P., 1992. Post-Eocene development of the Taranaki Basin, New Zealand: convergent overprint of a passive margin. In: Watkins, J. S., Zhinqiang, F., McMillen, K.J. (Eds.), Geology and Geophysics of Continental Margins. AAPG Memoir 53. https://doi.org/10.1306/M53552.

King, P.R., Thrasher, G.P., 1996. Cretaceous-Cenozoic geology and petroleum systems of the Taranaki Basin, New Zealand. Inst. Geol. Nucl. Sci. v. Monogr. 13, 244.

King, P.R., Scott, G.H., Robinson, P.H., 1993. Description, correlation and depositional history of Miocene sediments outcropping along North Taranaki coast. Inst. Geol. Nucl. Sci. v. Monogr. 5, 199.

King, P.R., Ilg, B.R., Arnot, M., Browne, G.H., Strachan, L.J., Crundwell, M., Helle, K., 2011. Outcrop and seismic examples of mass-transport deposits from a late Miocene deep-water succession, Taranaki Basin, New Zealand. In: Mass-Transport Deposits in Deepwater Settings. SEPM (Society for Sedimentary Geology), pp. 311–348, https://doi.org/10.2110/sepmsp.096.311.

Kirkham, C.B., 2015. A 3D Seismic Interpretation of Mud Volcanoes Within the Western Slope of the Nile Cone. Cardiff University.

Kneller, B., Dykstra, M., Fairweather, L., Milana, J.P., 2016. Mass-transport and slope accommodation: implications for turbidite sandstone reservoirs. AAPG Bull. 100, 213–235. https://doi.org/10.1306/09011514210.

Kopf, A.J., 2002. Significance of mud volcanism. Rev. Geophys. 40, 1–52.

Kopf, A., Klaeschen, D., Mascle, J., 2001. Extreme efficiency of mud volcanism in dewatering accretionary prisms. Earth Planet. Sci. Lett. 189 (3–4), 295–313.

Kumar, P.C., Omosanya, K.O., Eruteya, O.E., Sain, K., 2021. Geomorphological characterization of basal flow markers during recurrent mass movement: a case study from the Taranaki Basin, offshore New Zealand. Basin Res. 33, 2358–2382. https://doi.org/10.1111/bre.12560.

Kvalstad, T.J., Andresen, L., Forsberg, C.F., Berg, K., Bryn, P., Wangen, M., 2005. The Storegga slide: evaluation of triggering sources and slide mechanics. Mar. Pet. Geol. 22, 245–256. https://doi.org/10.1016/J.MARPETGEO.2004.10.019.

Lee, H.J., 2005. Undersea landslides: extent and significance in the Pacific Ocean, an update. Nat. Hazards Earth Syst. Sci. 5, 877–892.

Lee, H.J., Locat, J., Desgagns, P., Parsons, J.D., McAdoo, B.G., Orange, D.L., Puig, P., Wong, F.L., Dartnell, P., Boulanger, E., 2007. Submarine mass movements on continental margins. In: Continental Margin Sedimentation. Blackwell Publishing Ltd., Oxford, UK, pp. 213–274, https://doi.org/10.1002/9781444304398.ch5.

Lee, H., Ryan, H., Kayen, R.E., Haeussler, P.J., Dartnell, P., Hampton, M.A., 2006. Varieties of submarine failure morphologies of seismically-induced landslides in Alaskan fjords. Norwegian J. Geol. 86, 221–230.

Lonergan, L., Lee, N., Johnson, H.D., Cartwright, J.A., Jolly, R.J.H., 2000. Remobilization and injection in deepwater depositional systems: implications for reservoir architecture and prediction. In: Wiemer, P., Slatt, R.M., Coleman, J., et al. (Eds.), Deep-water Reservoirs of the World. GCSSEPM Foundation 20th Annual Bob F. Perkins Research Conference, pp. 515–532.

Le Goff, J., Slootman, A., Mulder, T., Cavailhes, T., Ducassou, E., Hanquiez, V., Jaballah, J., Reijmer, J.J.G., 2020. On the architecture of intra-formational Mass-Transport Deposits: insights from the carbonate slopes of Great Bahama Bank and the Apulian Carbonate Platform. Mar. Geol. 427, 106205. https://doi.org/10.1016/j.margeo.2020.106205.

Levin, L.A., 2005. Ecology of cold seep sediments: interactions of fauna with flow, chemistry and microbes. Oceanogr. Mar. Biol. Annu. Rev. 43, 11–56.

Lewis, K., 1971. Slumping on a continental slope inclined at 1–4. Sedimentology 16, 97–110. http://onlinelibrary.wiley.com/doi/10.1111/j.1365-3091.1971.tb00221.x/abstract. (Accessed October 2014).

Locat, J., Leroueil, S., Locat, A., Lee, H., 2014. In: Krastel, S., et al. (Eds.), Submarine Mass Movements and Their Consequences. Advances in Natural and Technological Hazards Research, 37. Springer, Cham, https://doi.org/10.1007/978-3-319-00972-8_1.

Lonergan, L., Cartwright, J.A., 1999. Polygonal faults and their influence on deep-water sandstone reservoir geometries, Alba Field, United Kingdom Central North Sea. Am. Assoc. Pet. Geol. Bull. 83 (3), 410–432.

Lovholt, F., Pedersen, G., Gisler, G., 2008. Oceanic propagation of a potential tsunami from the La Palma Island. J. Geophys. Res. 113, C09026. https://doi.org/10.1029/2007JC004603.

Lovholt, F., Pedersen, G., Harbitz, C.B., Glimsdal, S., Kim, J., 2015. On the characteristics of landslide tsunamis. Phil. Trans. R. Soc. A 373, 20140376.

Lupi, M., Saenger, E.H., Fuchs, F., Miller, S., 2013. Lusi mud eruption triggered by geometric focusing of seismic waves. Nat. Geosci. 6 (8), 642–646.

MacLeod, M.K., Hanson, R.A., Bell, C.R., McHugo, S., 1999. The Alba field ocean bottom cable seismic survey: impact on development. Lead. Edge 18, 1306–1312.

Mader, C.L., 2001. Modelling the La Palma landslide tsunami. Sci. Tsunami Hazards 19, 150–170.

Marfurt, K.J., Alves, T.M., 2015. Pitfalls and limitations in seismic attribute interpretation of tectonic features. Interpretation 3, 1F–T41. https://doi.org/10.1190/INT-2014-0122.1.

Martinsen, O.J., 1989. Styles of soft-sediment deformation on a Namurian (Carboniferous) delta slope, Western Irish Namurian Basin. Ireland. Geol. Soc. London Spec. Publ. 41, 167–177.

Martinsen, O.J., Bakken, B., 1990. Extensional and compressional zones in slumps and slides in the Namurian of County Clare, Ireland. J. Geol. Soc. 147, 153–164. https://doi.org/10.1144/gsjgs.147.1.0153.

Masalimova, L.U., Lowe, D.R., Sharman, G.R., King, P.R., Arnot, M.J., 2016. Outcrop characterization of a submarine channel-lobe complex: the Lower Mount Messenger Formation, Taranaki Basin, New Zealand. Mar. Pet. Geol. 71, 360–390. https://doi.org/10.1016/j.marpetgeo.2016.01.004.

Masson, D.G., Watts, A.B., Gee, M.J.R., Urgeles, R., Mitchell, N.C., Le Bas, T.P., Canals, M., 2002. Slope failures on the flanks of the western Canary Islands. Earth Sci. Rev. 57, 1–35. https://doi.org/10.1016/S0012-8252(01)00069-1.

Masson, D.G., Harbitz, C.B., Wynn, R.B., Pedersen, G., Lovholt, F., 2006. Submarine landslides: processes, triggers and hazard prediction. Philos. Trans. R. Soc. A: Math. Phys. Eng. Sci. 364, 2009–2039. https://doi.org/10.1098/rsta.2006.1810.

Masson, D.G., Le Bas, T.P., Grevemeyer, I., Weinrebe, W., 2008. Flank collapse and large-scale landsliding in the Cape Verde Islands, off West Africa. Geochem. Geophys. Geosyst. 9. https://doi.org/10.1029/2008GC001983.

Mazzini, A., Etiope, G., 2017. Mud volcanism: an updated review. Earth Sci. Rev. 168, 81–112.

McAdoo, B.G., Watts, P., 2004. Tsunami hazard from submarine landslides on the Oregon continental slope. Mar. Geol. 203, 235–245. https://doi.org/10.1016/S0025-3227(03)00307-4.

McAdoo, B., Pratson, L., Orange, D., 2000. Submarine landslide geomorphology, US continental slope. Mar. Geol. 169, 103–136. https://doi.org/10.1016/S0025-3227(00)00050-5.

McMurtry, G.M., Watts, P., Fryer, G.J., Smith, J.R., Imamura, F., 2004. Giant landslides, mega-tsunamis, and paleo-sea level in the Hawaiian Islands. Mar. Geol. 203 (3–4), 219–233.

Meckel, L.D., 2011. Reservoir characteristics and classification of sand-prone submarine mass-transport deposits. In: Ship, R.C., Weimer, P., Posamentier, H.W. (Eds.), Mass-Transport Deposits in Deepwater Settings. Society of Sedimentary Geology (SEPM), pp. 423–452. http://sepm.org/CM_Files/sp96TOCandabstracts.pdf.

Middleton, G.V., Hampton, M.A., 1976. Subaqueous sediment transport and deposition by sediment gravity flows. In: Stanley, D.J., Swift, D.J.P. (Eds.), Marine Sediment Transport and Environmental Management. Wiley, New York, pp. 197–218.

Milani, E.J., 1997. Evolução Tectono-estratigráfica da Bacia do Paraná e seu Relacionamento com a Geodinâmica Fanerozóica do Gondwana Sulocidental. Ph.D. Thesis, Instituto de Geociências da Universidade Federal do Rio Grande do Sul, Porto Alegre, 255 pp.

Milani, E.J., 2004. Comentários sobre a origem e evolução tectônica da Bacia do Paraná. In: Bartorelli, A., Carneiro, C.D.R., Brito Neves, B.B. (Eds.), Geologia do Continente Sul-Americano: Evolução da Obra de Fernando Flávio Marques de Almeida. São Paulo, Beca Editora, pp. 265–279.

Milani, E.J., França, A.B., Schneider, R.L., 1994. Bacia do Paraná. Rev. Bras. Geociênc. 8, 69–82.

Milkov, A., 2000. Worldwide distribution of submarine mud volcanoes and associated gas hydrates. Mar. Geol. 167, 29–42. https://doi.org/10.1016/S0025-3227(00)00022-0.

Minisini, D., Trincardi, F., Asioli, A., Canu, M., Foglini, F., 2007. Morphologic variability of exposed mass-transport deposits on the eastern slope of Gela Basin (Sicily channel). Basin Res. 19, 217–240.

Moscardelli, L., Wood, L., 2008. New classification system for mass transport complexes in offshore Trinidad. Basin Res. 20, 73–98. https://doi.org/10.1111/j.1365-2117.2007.00340.x.

Moscardelli, L., Wood, L., 2016. Morphometry of mass-transport deposits as a predictive tool. Geol. Soc. Am. Bull. 128, 47–80. https://doi.org/10.1130/B31221.1.

Moscardelli, L., Wood, L., Mann, P., 2006. Mass-transport complexes and associated processes in the offshore area of Trinidad and Venezuela. AAPG Bull. 90, 1059–1088. https://doi.org/10.1306/02210605052.

Moscardelli, L., Ochoa, J., Hunt, I., Zahm, L., 2019. Mixed siliciclastic–carbonate systems and their impact for the development of deep-water turbidites in continental margins: a case study from the Late Jurassic to Early Cretaceous Shelburne subbasin in offshore Nova Scotia. AAPG Bull. 103, 2487–2520. https://doi.org/10.1306/02151917318.

Mosher, D.C., Moscardelli, L., Shipp, R.C., Chaytor, J.D., Baxter, C.D., Lee, H.J., Urgeles, R., 2010. Submarine Mass Movements and Their Consequences. Springer, Netherlands, pp. 1–8.

Moss, J.L., Cartwright, J., Cartwright, A., Moore, R., 2012. The spatial pattern and drainage cell characteristics of a pockmark field, Nile Deep Sea Fan. Mar. Pet. Geol. 35 (1), 321–336.

Mottin, T.E., Vesely, F.F., de Lima Rodrigues, M.C.N., Kipper, F., de Souza, P.A., 2018. The paths and timing of late Paleozoic ice revisited: new stratigraphic and paleo-ice flow interpretations from a glacial succession in the upper Itararé Group (Paraná Basin, Brazil). Palaeogeogr. Palaeoclimatol. Palaeoecol. 490, 488–504. https://doi.org/10.1016/j.palaeo.2017.11.031.

Muhari, A., Heidarzadeh, M., Susmoro, H., Nugroho, H.D., Kriswati, E., Supartoyo, Wijanarto, A.B., Imamura, F., Arikawa, T., 2019. The December 2018 Anak Krakatau Volcano Tsunami as inferred from post-Tsunami field surveys and spectral analysis. Pure Appl. Geophys. 176, 5219–5233. https://doi.org/10.1007/s00024-019-02358-2.

Murton, B.J., Biggs, J., 2003. Numerical modelling of mud volcanoes and their flows using constraints from the Gulf of Cadiz. Mar. Geol. 195, 223–236.

Naranjo-Vesga, J., Ortiz-Karpf, A., Wood, L., Jobe, Z., Paniagua-Arroyave, J.F., Shumaker, L., Mateus-Tarazona, D., Galindo, P., 2020. Regional controls in the distribution and morphometry of deep-water gravitational deposits along a convergent tectonic margin. Southern Caribbean of Colombia. Mar. Pet. Geol. 121, 104639. https://doi.org/10.1016/J.MARPETGEO.2020.104639.

Nimblett, J.N., Shipp, R.C., Strijbos, F., 2005. Gas hydrate as a drilling hazard: examples from global deepwater settings. Offshore Technology Conference.

Nisbet, E.G., Piper, D.J.W., 1998. Giant submarine landslides. Nature 392, 329–330. https://doi.org/10.1038/32765.

Nissen, S.E., Haskell, N.L., Steiner, C.T., Coterill, K.L., 1999. Debris flow outrunner blocks, glide tracks, and pressure ridges identified on the Nigerian continental slope using 3-D seismic coherency. Lead. Edge 18, 595. https://doi.org/10.1190/1.1438343.

Normarck, W.R., MacGann, M., Sliter, R., 2007. Age of Palos Verdes submarine debris avalanche. southern California. Mar. Geol. 203, 247–259.

Nugraha, H.D., Jackson, C.A.-L., Johnson, H.D., Hodgson, D.M., 2020. Lateral variability in strain along the toewall of a mass transport deposit: a case study from the Makassar Strait, offshore Indonesia. J. Geol. Soc. 177, 1261–1279. https://doi.org/10.1144/jgs2020-071.

Obermeier, S.F., 1996. Use of liquefaction-induced features for palaeoseismic analysis—an overview of how seismic liquefaction features can be distinguished from other features and how their regional distribution and properties of source sediment can be used to infer the location and strength of Holocene paleo-earthquakes. Eng. Geol. 44, 1–76.

Obermeier, S.F., 1998. Liquefaction evidence for strong earthquakes of Holocene and latest Pleistocene ages in the states of Indiana and Illinois, USA. Eng. Geol. 50, 227–254.

Obermeier, S.F., Olson, S.M., Green, R.A., 2005. Field occurrences of liquefaction-induced features: a primer for engineering geologic analysis of paleoseismic shaking. Eng. Geol. 76, 206–234.

Oehler, J.-F., Lénat, J.-F., Labazuy, P., 2008. Growth and collapse of the Reunion Island volcanoes. Bull. Volcanol. 70, 717–742. https://doi.org/10.1007/s00445-007-0163-0.

Ogata, K., Mutti, E., Pini, G.A., Tinterri, R., 2012. Mass transport-related stratal disruption within sedimentary mèlanges: examples from the northern Apennines (Italy) and south-central Pyrenees (Spain). Tectonophysics 568–569, 185–199.

Ogata, K., Mountjoy, J.J., Pini, G.A., Festa, A., Tinterri, R., 2014. Shear zone liquefaction in mass transport deposit emplacement: a multi-scale integration of seismic reflection and outcrop data. Mar. Geol. 356, 50–64. https://doi.org/10.1016/j.margeo.2014.05.001.

Ogiesoba, O., Hammes, U., 2012. Seismic interpretation of mass-transport deposits within the upper Oligocene Frio Formation, south Texas Gulf Coast. AAPG Bull. 96, 845–868. https://doi.org/10.1306/09191110205.

Orange, D.L., Teas, P.A.D.J., Baillie, P., Johnstone, T., 2009. Using seaseep surveys to identify and sample natural hydrocarbon seeps in offshore frontier basins. In: Proceedings of The Annual Convention, vol. 33. Indonesian Petroleum Association, pp. 363–384.

Ortiz-Karpf, A., Hodgson, D.M., Jackson, C.A.-L., McCaffrey, W.D., 2018. Mass-transport complexes as markers of deep-water fold-and-thrust belt evolution: insights from the southern Magdalena fan, offshore Colombia. Basin Res. 30, 65–88. https://doi.org/10.1111/bre.12208.

Palan, K., Green, A.N., Andrews, B., Sink, K., Wiles, E.A., 2020. A morphometric analysis of the fluid flow features of the southern Orange Basin, South Africa. Mar. Geol. 423, 106145. https://doi.org/10.1016/J.MARGEO.2020.106145.

Palladino, G., Grippa, A., Bureau, D., Ian, G., Alsop, Hurst, A., 2016. Emplacement of sandstone intrusions during contractional tectonics. J. Struct. Geol. 89, 230–249.

Palladino, G., Alsop, G.I., Grippa, A., Zvirtes, G., Phillip, R.P., Hurst, A., 2018. Sandstone-filled normal faults: a case study from central California. J. Struct. Geol. 110, 86–101. https://doi.org/10.1016/j.jsg.2018.02.013.

Palladino, G., Rizzo, R.E., Zvirtes, G., Grippa, A., Philipp, R.P., Healy, D., Alsop, G.I., 2020. Multiple episodes of sand injection leading to accumulation and leakage of hydrocarbons along the San Andreas/San Gregorio fault system, California. Mar. Pet. Geol. 118, 104431. https://doi.org/10.1016/j.marpetgeo.2020.104431.

Paris, R., et al., 2018. Mega-tsunami conglomerates and flank collapses of ocean island volcanoes. Mar. Geol. 395, 168–187. https://doi.org/10.1016/J.MARGEO.2017.10.004.

Paris, A., Heinrich, P., Paris, R., Abadie, S., 2020. The December 22, 2018 Anak Krakatau, Indonesia, Landslide and Tsunami: preliminary modeling results. Pure Appl. Geophys. 177, 571–590. https://doi.org/10.1007/s00024-019-02394-y.

Pérez-Torrado, F.J., Paris, R., Cabrera, M.C., Schneider, J.-L., Wassmer, P., Carracedo, J.-C., Rodríguez-Santana, Á., Santana, F., 2006. Tsunami deposits related to flank collapse in oceanic volcanoes: the Agaete Valley evidence, Gran Canaria, Canary Islands. Mar. Geol. 227, 135–149. https://doi.org/10.1016/J.MARGEO.2005.11.008.

Pickering, K.T., Corregidor, J., 2005. Mass-transport complexes (MTCs) and tectonic control on basin-floor submarine fans, middle Eocene, south Spanish Pyrenees. J. Sediment. Res. 75 (5), 761–783.

Pickering, K.T., Hiscott, R.N., 2016. Deep Marine Systems: Processes, Deposits, Environments, Tectonics and Sedimentation. Wiley & American Geophysical Union. 424 pp https://www.wiley.com/en-us/Deep+Marine+Systems%3A+Processes%2C+Deposits%2C+Environments%2C+Tectonics+and+Sedimentation-p-9781118865422. (Accessed April 2020). ISBN 978-1-4051-2578-9.

Piper, D.J.W., Cochonat, P., Morrison, M., 1999. The sequence of events around the epicentre of the 1929 Grand Banks earthquake: initiation of debris flows and turbidity currents inferred from sidescan sonar. Sedimentology 46, 79–97. https://doi.org/10.1046/j.1365-3091.1999.00204.x.

Pope, E., Talling, P.J., Carter, L., Clare, M.A., Hunt, J.E., 2017. Damaging sediment density flows triggered by tropical cyclones. Earth Planet. Sci. Lett.

Posamentier, H.W., Kolla, V., 2003. Seismic geomorphology and stratigraphy of depositional elements in deep-water settings. J. Sediment. Res. 73, 367–388. https://doi.org/10.1306/111302730367.

Posamentier, H.W., Martinsen, O.J., 2011. The character and genesis of submarine mass-transport deposits: insights from outcrop and 3D seismic data. In: Shipp, R.C., Weimer, P., Posamentier, H.W. (Eds.), Mass-Transport Deposits in Deepwater Settings. SEPM Tulsa, pp. 7–38.

Postma, G., Drinia, H., 1993. Architecture and sedimentary facies evolution of a marine, expanding outer-arc half-graben (Crete, late Miocene). Basin Res. 5, 103–124.

Quidelleur, X., Hildenbrand, A., Samper, A., 2008. Causal link between Quaternary paleoclimatic changes and volcanic islands evolution. Geophys. Res. Lett. 35 (2), 5.

Revil, A., 2002. Genesis of mud volcanoes in sedimentary basins: a solitary wave-based mechanism. Geophys. Res. Lett. 29 (12). https://doi.org/10.1029/2001GL014465.

Riboulot, V., Cattaneo, A., Sultan, N., Garziglia, S., Ker, S., Imbert, P., Voisset, M., 2013. Sea-level change and free gas occurrence influencing a submarine landslide and pockmark formation and distribution in deepwater Nigeria. Earth Planet. Sci. Lett. 375, 78–91. https://doi.org/10.1016/j.epsl.2013.05.013.

Rodrigues, M.C.N.L., Folmann, M., Trzaskos, B., Vesely, F.F., 2015. Registro estrutural de depósitos de transporte em massa do Permiano Inferior da Bacia do Paraná. XV Simpósio Nacional de Estudos Tectônicos, Vitória. Brasil, Anais.

Rodrigues, M.C.N.L., Trzaskos, B., Alsop, G.I., Vesely, F.F., 2020. Making a homogenite: an outcrop perspective into the evolution of deformation within mass-transport deposits. Mar. Petrol. Geol. 112, 104033. https://doi.org/10.1016/j.marpetgeo.2019.104033.

Rodrigues, M.C.N.L., Trzaskos, B., Alsop, G.I., Vesely, F.F., Mottin, T.E., 2021. Diversidade de estilos estruturais em depósitos de transporte em massa. Bol. Parana. de Geosci. 78, 83–109.

Roelofse, C., Alves, T.M., Gafeira, J., 2020. Structural controls on shallow fluid flow and associated pockmark fields in the East Breaks area, northern Gulf of Mexico. Mar. Pet. Geol. 112, 104074. https://doi.org/10.1016/J.MARPETGEO.2019.104074.

Rosa, E.L.M., Vesely, F.F., Isbell, J., Kipper, F., Fedorchuk, N., Souza, P.A., 2019. Constraining the timing, kinematics and cyclicity of Mississippian-early Pennsylvanian glaciations in the Paraná Basin, Brazil. Sediment. Geol. 384, 29–49.

Rosseti, D.D.F., 1999. Soft-sediment deformation structures in late Albian to Cenomanian deposits, São Luís Basin, northern Brazil: evidence for palaeoseismicity. Sedimentology 46, 1065–1081.

Rostirolla, S.P., Assine, M.L., Fernandes, L.A., Artur, P.C., 2000. Reativação de paleolineamentos durante a evolução da Bacia do Paraná – o exemplo do alto estrutural de Quatiguá. Rev. Bras. Geociênc. 29, 639–648.

Rostirolla, S.P., Appi, C.J., Mancini, F., Kraft, R.P., Bartoszeck, M.K., 2002. Fraturas controladoras da migração de fluidos nos arenitos Vila Velha, Grupo Iraraté, Bacia do Paraná. Rev. Bras. Geociênc. 31, 349–356. https://doi.org/10.1590/S0102-36162010000200004.

Rostirolla, S.P., Mancini, F., Rigoti, A., Kraft, R.P., 2003. Structural styles of the intracratonic reactivation of the Perimbó fault zone, Paraná basin, Brazil. J. South Am. Earth. Sci. 16, 287–300. https://doi.org/10.1016/S0895-9811(03)00065-8.

Rotzien, J.R., Lowe, D.R., King, P.R., Browne, G.H., 2014. Stratigraphic architecture and evolution of a deep-water slope channel-levee and overbank apron: the Upper Miocene Upper Mount Messenger Formation, Taranaki Basin. Mar. Pet. Geol. 52, 22–41. https://doi.org/10.1016/j.marpetgeo.2014.01.006.

Roy, S., Senger, K., Hovland, M., Römer, M., Braathen, A., 2019. Geological controls on shallow gas distribution and seafloor seepage in an Arctic fjord of Spitsbergen, Norway. Mar. Pet. Geol. 107, 237–254. https://doi.org/10.1016/j.marpetgeo.2019.05.021.

Safadi, M., Meilijson, A., Makovsky, Y., 2017. Internal deformation of the southeast Levant margin through continued activity of buried mass transport deposits. Tectonics 36, 559–581. https://doi.org/10.1002/2016TC004342.

Sakellariou, D., Tsampouraki-Kraounaki, K., 2019. Plio-quaternary extension and strike-slip tectonics in the aegean. In: Transform Plate Boundaries and Fracture Zones. Elsevier, pp. 339–374, https://doi.org/10.1016/B978-0-12-812064-4.00014-1.

Salamuni, R., Marques Filho, P.L., Sobanski, A.C., 1966. Considerações sobre turbiditos da Formação Itararé (Carbonífero Superior), Rio Negro-PR e Mafra-SC. Boletim da Sociedade Brasileira de Geologia 15, 1–19.

Satur, N., Hurst, A., 2007. Sand-injection structures in deep-water sandstones from the Ty formation (Paleocene), Sleipner Øst Field, Norwegian North Sea. In: Hurst, A., Cartwright, J. (Eds.), Sand Injectites: Implications for Hydrocarbon Exploration and Production. American Association of Petroleum Geologists Memoir, Tulsa, pp. 113–117.

Sawyer, D.E., Flemings, P.B., Dugan, B., Germaine, J.T., 2009. Retrogressive failures recorded in mass transport deposits in the Ursa Basin, Northern Gulf of Mexico. J. Geophys. Res. 114, B10102. https://doi.org/10.1029/2008JB006159.

Scarselli, N., McClay, K., Elders, C., 2016. Seismic geomorphology of cretaceous megaslides offshore Namibia (Orange Basin): insights into segmentation and degradation of gravity-driven linked systems. Mar. Pet. Geol. 75, 151–180. https://doi.org/10.1016/J.MARPETGEO.2016.03.012.

Schemiko, D.C.B., Vesely, F.F., Rodrigues, M.C.N.L., 2019. Deepwater to fluvio-deltaic stratigraphic evolution of a deglaciated depocenter: the early Permian Rio do Sul and Rio Bonito formations, southern Brazil. J. South Amer. Earth Sci. 95, 102260. https://doi.org/10.1016/j.jsames.2019.102260.

Schneider, R.L., Muhlmann, H., Tommasi, E., Medeiros, R.A., Daemon, R.A., Nogueira, A.A., 1974. Revisão estratigráfica da Bacia do Paraná. SBG, 28 Congresso Brasileiro de Geologia, Porto Alegre 1. pp. 41–65.

Schnyder, J.S.D., Eberli, G.P., Kirby, J.T., Shi, F., Tehranirad, B., Mulder, T., Ducassou, E., Hebbeln, D., Wintersteller, P., 2016. Tsunamis caused by submarine slope failures along western Great Bahama Bank. Sci. Rep. 6, 35925. https://doi.org/10.1038/srep35925.

Schwab, A.M., Jameson, E.W., Townsley, A., 2014. Volund Field: development of an Eocene sandstone injection complex, offshore Norway. In: McKie, T., Rose, P.T.S., Hartley, A.J., Jones, D.W., Armstrong, T.L. (Eds.), Tertiary Deep-Marine Reservoirs of the North Sea Region. vol. 403. Geol. Soc. (Lond.), pp. 247–260, https://doi.org/10.1144/SP403.4. Spec. Publ. 2015.

Scott, A., Vigorito, M., Hurst, A., 2009. The process of sand injection: internal structures and relationships with host strata (Yellowbank Creek Injectite Complex, California, U.S.A.). J. Sediment. Res. 79, 568–583.

Scott, A., Hurst, A., Vigorito, M., 2013. Outcrop-based reservoir characterization of a kilometer-scale sand-injectite complex. AAPG (Am. Assoc. Pet. Geol.) Bull. 97, 309–343.

Shanmugam, G., Bloch, R.B., Mitchell, S.M., Beamish, G.W.J., Hodgkinson, R.J., Damuth, G.E., Straume, T., Syversten, S.E., Shields, K.E., 1995. Basin-floor fans in the North Sea: sequence stratigraphic models vs. sedimentary facies. Am. Assoc. Pet. Geol. Bull. 79, 477–511.

Shanmugam, G., Shrivastava, S.K., Das, B., 2009. Sandy debrites and tidalites of pliocene reservoir sands in upper-slope canyon environments, offshore Krishna-Godavari Basin (India): implications. J. Sediment. Res. 79, 736–756. https://doi.org/10.2110/jsr.2009.076.

Sharman, G.R., Graham, S.A., Masalimova, L.U., Shumaker, L.E., King, P.R., 2015. Spatial patterns of deformation and paleoslope estimation within the marginal and central portions of a basin-floor mass-transport deposit, Taranaki Basin, New Zealand. Geosphere 11, 266–306. https://doi.org/10.1130/GES01126.1.

Shipp, C., Nott, J.A., Newlin, J.A., 2004. Physical characteristics and impact of mass transport complexes on deepwater jetted conductors and suction anchor piles. In: Shipp, C., Nott, J.A., Newlin, J.A. (Eds.), Offshore Technology Conference, Houston, Texas. Offshore Technology Conference, Houston, Texas. Physical characteristics and impact of mass transport complexes on deepwater jetted conductors and suction anchor piles.

Shoulders, S.J., Cartwright, J., Huuse, M., 2007. Large-scale conical sandstone intrusions and polygonal fault systems in. Tranche 6, Faroe-Shetland Basin. Mar. Petrol. Geol. 24, 173–188. https://doi.org/10.1016/j.marpetgeo.2006.12.001.

Shultz, A.W., 2004. Facies and environments of the Lobo Trend, Webb and Zapata Counties, Texas: examples from core. South Texas Geol. Soc. Bull. 45, 13–28.

Shumaker, L.E., Sharman, G.R., King, P.R., Graham, S.A., 2018. The source is in the sink: deep-water deposition by a submarine volcanic arc, Taranaki Basin, New Zealand. In: Kane, I. (Ed.), Sedimentology, https://doi.org/10.1111/sed.12475.

Skinner, J.A., Mazzini, A., 2009. Martian mud volcanism: terrestrial analogs and implications for formational scenarios. Mar. Pet. Geol. 26, 1866–1878. https://doi.org/10.1016/j.marpetgeo.2009.02.006.

Skogdalen, J.E., Vinnem, J.E., 2012. Quantitative risk analysis of oil and gas drilling, using Deepwater Horizon as case study. Reliab. Eng. Syst. Saf. 100, 58–66. https://doi.org/10.1016/j.ress.2011.12.002.

Soares, P.C., 1991. Tectônica sinsedimentar cíclica na Bacia do Paraná: controles. Tese para provimento de vaga de Professor Titular. Universidade Federal do Paraná, Curitiba. 131 pp.

Sobiesiak, M.S., Alsop, G.I., Kneller, B., Milana, J.P., 2017. Sub-seismic scale folding and thrusting within an exposed mass transport deposit: a case study from NW Argentina. J. Struct. Geol. 96, 176–191. https://doi.org/10.1016/j.jsg.2017.01.006.

Sobiesiak, M.S., Kneller, B., Alsop, G.I., Milana, J.P., 2018. Styles of basal interaction beneath mass transport deposits. Mar. Petrol. Geol. 98, 629–639.

Song, J., Alves, T.M., Omosanya, K.O., Hales, T.C., Ze, T., 2020. Tectonic evolution of strike-slip zones on continental margins and their impact on the development of submarine landslides (Storegga Slide, northeast Atlantic). GSA Bull. 132, 2397–2414.

Souza, P.A., 2006. Late Carboniferous palynostratigraphy of the Itararé Subgroup, northeastern Paraná Basin, Brazil. Rev. Palaeobot. Palynol. 138, 9–29.

Steventon, M.J., Jackson, C.A.-L., Hodgson, D.M., Johnson, H.D., 2019. Strain analysis of a seismically imaged mass-transport complex, offshore Uruguay. Basin Res. https://doi.org/10.1111/bre.12337.

Steventon, M.J., Jackson, C.A.-L., Johnson, H.D., Hodgson, D.M., Kelly, S., Omma, J., Gopon, C., Stevenson, C., Fitch, P., 2021. Evolution of a sand-rich submarine channel-lobe system and impact of mass-transport and transitional flow deposits on reservoir heterogeneity: Magnus Field, northern North Sea. Petrol. Geosci. https://doi.org/10.1144/petgeo2020-095. p. petgeo2020-095.

Stewart, S.A., Davies, R.J., 2006. Structure and emplacement of mud volcano systems in the South Caspian Basin. AAPG Bull. 90, 771–786.

Strachan, L., 2002. Slump initiated and controlled syndepositional sandstone remobilisation: an example from the Namuriam of County Clare, Ireland. Sedimentology 49, 25–41.

Strachan, L.J., 2008. Flow transformations in slumps: a case study from the Waitemata Basin, New Zealand. Sedimentology 55, 1311–1332. https://doi.org/10.1111/j.1365-3091.2007.00947.x.

Strogen, D., Bland, K.J., Nicol, A., King, P., 2014. Paleogeography of the Taranaki Basin region during the latest Eocene–Early Miocene and implications for the 'total drowning' of Zealandia. New Zeal. J. Geol. Geophys. 57, 110–127. https://doi.org/10.1080/00288306.2014.901231.

Strogen, D.P., Seebeck, H., Nicol, A., King, P.R., 2017. Two-phase Cretaceous–Paleocene rifting in the Taranaki Basin region, New Zealand; implications for Gondwana break-up. J. Geol. Soc. 174, 929–946. https://doi.org/10.1144/jgs2016-160.

Sultan, N., et al., 2004. Triggering mechanisms of slope instability processes and sediment failures on continental margins: a geotechnical approach. Mar. Geol. 213, 291–321. https://doi.org/10.1016/j.margeo.2004.10.011.

Sultan, N., Voisset, M., Marsset, B., Marsset, T., Cauquil, E., Colliat, J.-L., 2007. Potential role of compressional structures in generating submarine slope failures in the Niger Delta. Mar. Geol. 237, 169–190.

Sun, Q., Alves, T., 2020. Petrophysics of fine-grained mass-transport deposits: a critical review. J. Asian Earth Sci. 192, 104291. https://doi.org/10.1016/J.JSEAES.2020.104291.

Sun, Q., Alves, T., Xie, X., He, J., Li, W., Ni, X., 2017. Free gas accumulations in basal shear zones of mass-transport deposits (Pearl River Mouth Basin, South China Sea): an important geohazard on continental slope basins. Mar. Pet. Geol. 81, 17–32. https://doi.org/10.1016/j.marpetgeo.2016.12.029.

Suss, J.F., Vesely, F.F., Santa Catharina, A., Assine, M.L., Paim, P.S.G., 2014. O grupo itararé (Neocarbonífero-eopermiano) entre Porto Amazonas (PR) e Mafra (SC): Sedimentação gravitacional em contexto marinho deltaico com influência glacial. Geociencias 33, 701–719.

Talling, P., Clare, M., Urlaub, M., Pope, E., Hunt, J., Watt, S., 2014. Large submarine landslides on continental slopes: geohazards, methane release, and climate change. Oceanography 27 (2), 32–45.

Tanavsuu-Milkeviciene, K., Sarg, J.F., 2012. Evolution of an organic-rich lake basin—stratigraphy, climate and tectonics: Piceance Creek basin, Eocene Green River Formation. Sedimentology 59, 1735–1768. https://doi.org/10.1111/j.1365-3091.2012.01324.x.

Tappin, D.R., Watts, P., Mcmurtry, G.M., Lafoy, Y., Matsumoto, T., 2001. The Sissano, Papua New Guinea tsunami of July 1998 – offshore evidence on the source mechanism. Mar. Geol. 175, 1–23.

Tehranirad, B., Harris, J.C., Grilli, A.R., Grilli, S.T., Abadie, S., Kirby, J.T., Shi, F., 2015. Far-field tsunami threat in the North Atlantic basin from large scale flank collapses of the Cumbre Vieja volcano, La Palma. Pure Appl. Geophys. 172, 3589–3616.

Thomas, S., Hooper, J., Clare, M., 2010. Constraining geohazards to the past: impact assessment of submarine mass movements on seabed developments. Submarine Mass Movements and Their Consequences. Springer, Netherlands, pp. 387–398.

Thompson, B.J., Garrison, R.E., Moore, C.J., 1999. A late Cenozoic sandstone intrusion west of S. Cruz, California. Fluidized flow of water and hydrocarbon-saturated sediments. In: Garrison, R.E., Aiello, I.W., Moore, J.C. (Eds.), Late Cenozoic Fluid Seeps and Tectonics along the San Gregorio Fault Zone in the Monterey Bay Region, California GB-76. Annual Meeting of the Pacific Section. American Association of Petroleum Geologists, Monterey, CA, pp. 53–74.

Thompson, B.J., Garrison, R.E., Moore, C.J., 2007. In: Hurst, A., Cartwright, J. (Eds.), Sand Injectites: Implications for Hydrocarbon Exploration and Production, vol. 87. American Association of Petroleum Geologists, Memoir, pp. 151–162.

Tripsanas, E.K., Piper, D.J.W., Jenner, K.A., Bryant, W.R., 2008. Submarine mass-transport facies: new perspectives on flow processes from cores on the eastern North American margin. Sedimentology 55, 97–136.

Truswell, J.F., 1972. Sandstone sheets and related intrusions from Coffee bay, Transkei, South Africa. J. Sediment. Petrol. 42, 578–583.

Trzaskos, B., Vesely, F.P., Rostirolla, S.P., 2006. Eventos tectônicos recurrentes impressos no arcabouço estratigráfico do Grupo Itararé na região de Vila Velha, Estado do Paraná. Bol. Paran. Geoc., 89–104.

Ujiie, K., Kameyama, M., Yamaguchi, A., 2010. Geological record of thermal pressurization and earthquake instability of subduction thrusts. Tectonophysics 485, 260–268. https://doi.org/10.1016/j.tecto.2010.01.002.

Urgeles, R., Camerlenghi, A., 2013. Submarine landslides of the Mediterranean Sea: trigger mechanisms, dynamics, and frequency-magnitude distribution. J. Geophys. Res. Earth Surf. 118, 2600–2618. https://doi.org/10.1002/2013JF002720.

Valdez-Buso, V., Aquino, C.D., Paim, P.S.G., Souza, P.A., Mori, A.L., Fallgatter, C., Milana, J.P., Kneller, B., 2019. Late Palaeozoic glacial cycles and subcycles in western Gondwana: correlation of surface and subsurface data of the Paraná Basin, Brazil. Palaeogeogr. Palaeoclimatol. Palaeoecol. 531 (B), 108435. https://doi.org/10.1016/j.palaeo.2017.09.004.

van Hisnbergen, D.J.J., Meleunkamp, J.E., 2006. Neogene supradetachment basin development on Crete (Greece) during exhumation of the South Aegean core complex. Basin Res. 18, 103–124.

Vanneste, M., Mienert, J., Bünz, S., 2006. The Hinlopen Slide: a giant, submarine slope failure on the northern Svalbard margin. Arctic Ocean. Earth Planet. Sci. Lett. 245, 373–388.

Vardy, M.E., L'Heureux, J.S., Vanneste, M., Longva, O., Steiner, A., Forsberg, C.F., Haflidason, H., Brendryen, J., 2012. Multidisciplinary investigation of a shallow near-shore landslide, Finneidfjord, Norway. Near Surf. Geophys. 10 (4), 267–277.

Vesely, F.F., Assine, M.L., 2006. Deglaciation sequences in the Permo-Carboniferous Itararé Group, Paraná Basin, southern Brazil. J. South Amer. Earth Sci. 22, 156–168. https://doi.org/10.1016/j.jsames.2006.09.006.

Vesely, F.F., Assine, M.L., Rostirolla, S.P., 2005. Deformação penecontemporânea no Grupo Itararé, Bacia do Paraná: estilos estruturais, dinâmica sedimentar e implicações paleogeográficas. Simpósio Nacional de Estudos Tectônicos. Curitiba, Anais. SBG núcleo PR, Curitiba (CD-ROM), vol. 10.

Vesely, F.F., Rodrigues, M.C.N.L., Da Rosa, E.L.M., Amato, J.A., Trzaskos, B., Isbell, J.L., Fedorchuk, N.D., 2018. Recurrent emplacement of non-glacial diamictite during the late Paleozoic ice age. Geology 46 (7), 615–618. https://doi.org/10.1130/G45011.1.

Vétel, W., Cartwright, J., 2010. Emplacement mechanics of sandstone intrusions: insights from the Panoche Giant Injection Complex, California. Basin Res. 22, 783–807. https://doi.org/10.1111/j.1365-2117.2009.00439.x.

Vigorito, M., Hurst, A., 2010. Regional sand injectite architecture as a record of pore pressure evolution and sand redistribution in the shallow crust: insights from the panoche giant injection complex, California. J. Geol. Soc. Lond. 167, 889–904.

Vigorito, M., Hurst, A., Cartwright, J., Scott, A., 2008. Regional-scale subsurface sand remobilization: geometry and architecture. Geol. Soc. Lond. 165, 609–612.

Von Huene, R., Ranero, C.R., Watts, P., 2004. Tsunamigenic slope failure along the Middle America Trench in two tectonic settings. Mar. Geol. 203, 303–317.

Walker, W., Jobe, Z.R., Sarg, J.F., Wood, L., 2021. Progradational slope architecture and sediment distribution in outcrops of the mixed carbonate-siliciclastic Bone Spring Formation, Permian Basin, west Texas. Geosphere 17, 1268–1293. https://doi.org/10.1130/GES02355.1.

Ward, S.N., 2001. Landslide tsunami. J. Geophys. Res. Solid Earth 106, 11201–11215.

Ward, S.N., Day, S.J., 2001. Cumbre Vieja volcano: potential collapse and tsunami at La Palma, Canary Islands. Geophys. Res. Lett. 28, 3397–3400. https://doi.org/10.1029/2001GL013110.

Ward, N.I.P., Alves, T.M., Blenkinsop, T.G., 2018. Submarine sediment routing over a blocky mass-transport deposit in the Espírito Santo Basin, SE Brazil. Basin Res. 30, 816–834. https://doi.org/10.1111/bre.12282.

Wattrus, N.J., Rausch, D.E., Cartwright, J., 2003. In: Van Rensbergen, P., Hillis, R., Maltman, A., Morley, C. (Eds.), Subsurface Sediment Mobilization: Special Publications, vol. 216. Geological Society, London, pp. 323–334.

Weimer, P., 1989. Sequence stratigraphy of the Mississippi fan (Plio-Pleistocene), Gulf of Mexico. Geo-Mar. Lett. 9, 185–272. https://doi.org/10.1007/BF02431072.

Weimer, P., Shipp, C., 2004. Mass transport complex: musing on past uses and suggestions for future directions. In: Offshore Technology Conference, Offshore Technology Conference, pp. 1–10, https://doi.org/10.4043/16752-MS.

Weimer, P., Slatt, R.M., 2004. Deepwater reservoir elements: mass-transport complexes and slides. In: Petroleum Systems of Deepwater Settings. 7. Society of Exploration Geophysicists and European Association of Geoscientists and Engineers, pp. 7-1–7-32, https://doi.org/10.1190/1.9781560801955.ch7.

Wenau, S., Spiess, V., Zabel, M., 2021. Giant seafloor depressions caused by slope failures and bottom currents on the Namibia Continental Margin. Geochem. Geophys. Geosyst. 22. https://doi.org/10.1029/2020GC009548. p. e2020GC009548.

Wood, L.J., Moscardelli, L., Dunlap, D.B., Cardona, S., 2015. Healing-phase, top-trapped fills associated with mass transport complexes: controls on turbidite deposition in chaotic margin settings. In: AAPG Geoscience Technology Workshop (GTW) Sixth Annual Deepwater and Shelf Reservoir, Houston.

Woodcock, N., 1979. Sizes of submarine slides and their significance. J. Struct. Geol. 1, 137–142. https://doi.org/10.1016/0191-8141(79)90050-6.

Wu, T., Sahling, H., Feseker, T., Rendle-Bühring, R., Wei, J., Wintersteller, P., Marcon, Y., Pape, T., Römer, M., Bohrmann, G., 2019. Morphology and activity of the Helgoland Mud Volcano in the Sorokin Trough, northern Black Sea. Mar. Pet. Geol. 99, 227–236.

Yavari-Ramshe, S., Ataie-Ashtiani, B., 2016. Numerical modeling of subaerial and submarine landslide-generated tsunami waves – recent advances and future challenges. Landslides 13 (6), 1325–1368.

Yusifov, M., Rabinowitz, P.D., 2004. Classification of mud volcanoes in the South Caspian Basin, offshore Azerbaijan. Mar. Pet. Geol. 21, 965–975.

Zalán, P.V., Wolff, S., Astolfi, M.A.M., Vieira, I.S., Conceição, J.C.J., Appi, V.T., Neto, E.V.S., Cerqueira, J.R., Marques, A., 1990. The Paraná Basin, Brazil. AAPG Memoir 51, Tulsa. pp. 681–708.

Zengaffinen, T., Løvholt, F., Pedersen, G.K., Muhari, A., 2020. Modelling 2018 Anak Krakatoa Flank Collapse and Tsunami: effect of landslide failure mechanism and dynamics on Tsunami generation. Pure Appl. Geophys. 177, 2493–2516. https://doi.org/10.1007/s00024-020-02489-x.

Zvirtes, G., Hurst, A., Philipp, R.P., Palladino, G., Grippa, A., 2019. The Tumey Giant Injection Complex, Tumey Hill, California (USA). In: Silcock, S., Huuse, M., Bowman, M., Hurst, A., Cobain, S. (Eds.), Subsurface Sand Remobilization and Injection, vol. 493. Geological Society, London, Special Publications. 34 pp. https://doi.org/10.1144/ SP493-2019-3.

Zvirtes, G., Philipp, R.P., Hurst, A., Palladinoa, G., De Ros, L.F., Grippa, A., 2020. Petrofacies of Eocene sand injectites of the Tumey Giant Injection Complex, California (USA). Sediment. Geol. 400, 105617.

Chapter 11

# Source-to-sink analysis of deepwater systems: Principles, applications and case studies

E. Szymanski[a], L. Fielding[b] and L. Davies[b]
[a]*Utah Geological Survey, Salt Lake City, UT, United States,* [b]*Petryx Ltd, Gaerwen, United Kingdom*

## Introduction

Source-to-sink (S2S) analysis is a wide-ranging, multidisciplinary and dynamic research field that evolves continuously with the development of novel ideas about the nature of interconnected earth systems and the introduction of improved technologies. This chapter presents the subject from the perspective of a geoscientist tasked with characterizing the modern and ancient deposits of deepwater depositional systems for resource assessment. S2S is valuable for its power to reduce geologic uncertainty, and guide business decisions, in the sectors that exploit deepwater deposits from petroleum and critical mineral exploration to carbon capture and sequestration.

This chapter begins with an overview of fundamental definitions and business-case rationale regarding the utility of the science in economic resource assessment and extraction, then explores the ways in which quantitative techniques—geo-/thermochronology, petrography and chemostratigraphy, for instance—are used to build accurate geologic models of subsurface stratigraphic frameworks for improved exploration, appraisal and development activities in deepwater depositional settings. Drawing from published scientific articles for methodology, case studies and examples of practical applications of S2S analysis, readers will gain a fundamental understanding of techniques and workflows important to this field. This chapter also introduces practical advice for scaling and framing business-relevant projects and highlights valuable technical resources—some open access—that are available to geoscientists interested in conducting S2S work.

## What is source-to-sink analysis?

The phrase "source-to-sink analysis" is most often used to describe either any or all parts of a suite of analytical techniques and workflows that examine the effects of catchment (source) and basin (sink) dynamics on the generation, transport, composition, distribution and depositional timing of clastic sediments in modern and ancient sedimentary systems. S2S analysis, commonly abbreviated as "S2S", provides a holistic perspective of depositional systems by integrating onshore to offshore datasets of diverse size, classification and scale and providing a framework for cross-validation of analytical results from interrelated geologic disciplines—tectonics, climatology, stratigraphy and petrography, for example.

S2S work is multidisciplinary by nature, using field, laboratory, experimental and/or numerical modeling methods to deliver products that quantify sediment composition, volume and organization within the entire depositional system. As such, S2S analysis is used alongside other conventional techniques to reduce exploration uncertainty associated with clastic reservoir presence, abundance and quality at margin, play and prospect scales by constraining sediment budget dynamics, basin time-temperature histories, drainage system evolution and chronostratigraphy within sedimentary packages. S2S is used to reduce exploration risk uncertainty associated with most every geologic risk element (see chapter 2) and it can be scaled to address fit-for-purpose business decisions at any point in the exploration review process, whether the focus is frontier basin characterization, play extension and/or prospect generation. S2S is often valued after discovery as well, and the same methods may be used to refine appraisal estimates, define the reservoir petrological characteristics and stratigraphic architecture of a resource-bearing play for accurate development and play extension scenarios. Some examples of practical applications of S2S include:

**Deepwater Sedimentary Systems.** https://doi.org/10.1016/B978-0-323-91918-0.00014-1

1. Studying the size, scale and longevity of modern analogs to accurately model the depositional extent, stratigraphic thickness and constituent grain composition of ancient deepwater deposits, which are used to predict reservoir presence and quality and inform diagenesis models.
2. Conducting heavy mineral provenance analysis to identify sediment source terranes (i.e., the origination point of sediment) and determine sediment flux (i.e., the volume of sediment delivered to an outfall point) over time to forecast derivative sediment composition, improve reservoir presence and quality prediction and reduce reservoir uncertainty along trend, especially when steering exploration in frontier regions.
3. Using isotope geochronology and chemostratigraphy to calculate the maximum depositional age (MDA) and depositional rate for subsurface intervals-of-interest and building chronostratigraphic frameworks for well correlation, especially when lacking biostratigraphic data.
4. Applying thermochronologic techniques to detrital heavy minerals extracted from well materials (e.g., sidewall core) to derive the postdepositional thermal history of source and reservoir intervals, which constrains petroleum system timing (e.g., source rock maturation/charge initiation, structural trap formation) and provides important time-temperature history input to basin models.

## Anatomy of a source-to-sink system

S2S systems are constructed of sets of unique segments, system elements and environmental units that lie in distinct relative positions to one another along the system profile (Fig. 1). The boundaries of these segments, elements and environmental units, and the extent of geologic processes operating within them, are gradational and often shift up and down the system, depending on the relative controls of tectonics and climate, among other factors. Climate, and climate change, affects the processes operating in all system elements while mountain building (i.e., creation of relief), sediment storage and release mechanisms and sea-level change are the dominant controls in the erosion, sediment transfer and deposition segments, respectively. Each segment, and the elements therein, are interconnected and understanding the feedback mechanisms between individual elements and segments is a key competency that every practitioner should develop.

Elements within S2S systems often exhibit symbiotic behaviors, conveniently allowing the analysis of any individual or set of elements, from anywhere within the system, to provide a greater understanding about whole-system behavior.

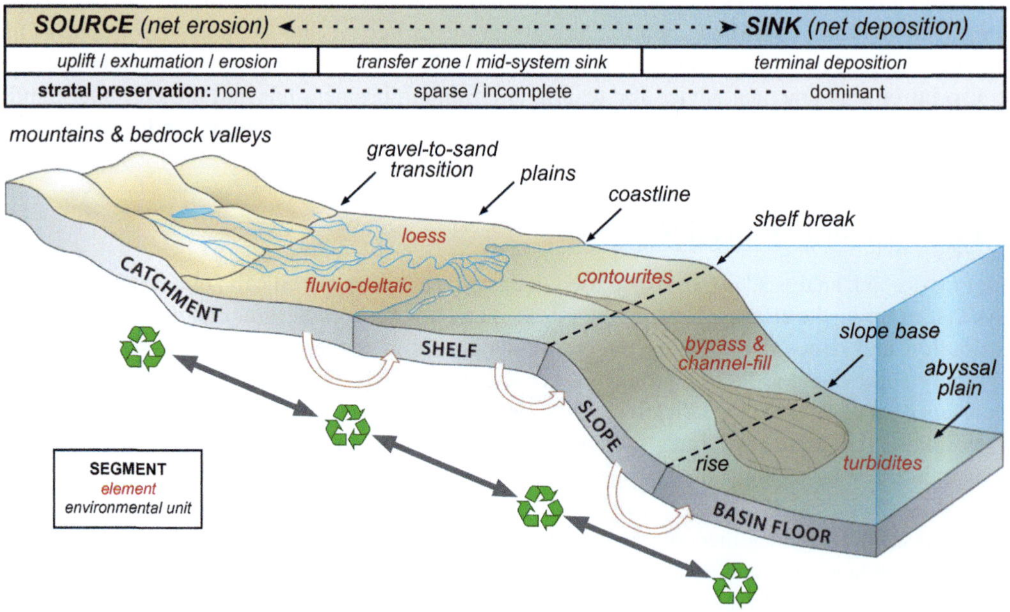


**FIG. 1** The source-to-sink system. Relative hypsometric configuration of the various Source-to-Sink System Profile Segments, Elements, Environmental Units and their boundaries. Defining these segments and environmental units is important to accurate S2S analysis. S2S examines the entire depositional system. The extent of processes operating within the system are gradational and often shift up and down the system, depending on the relative controls of tectonics and climate, among other factors. White arrows illustrate the direction of sediment transfer in the system through time. Ancient terminal sinks deposits (e.g., basin floor) may be subsequently displaced tectonically and subsequently form up-dip system segments (e.g., catchment), inverting the ancient system, allowing sediment to be recycled through the new system and introducing the concept of "sink-to-source" analysis. *(Modified from Blum and Hattier-Womack (2009), Graham et al. (2011), Helland-Hansen et al. (2016), National Science Foundation MARGINS (2004), and Sømme et al. (2009).)*

This means that systems can be studied both "forwards" and "backwards"! Inverse S2S modeling, sometimes referred to as sink-to-source analysis, typically relies on data and products that lie basinward of the coastline (e.g., sediment isopach maps, heavy mineral abundances and sedimentary petrology) to develop models and insights about mid- and upper-system elements. Such studies analyze mid-system and terminal depositional system elements to constrain system dynamics in elements further toward the hinterland; for example, submarine channel and fan dimensions mapped in age-constrained 3D seismic data may yield proportional estimates of catchment extent and sediment flux calculations at key time intervals, or a mix of geo-/thermochronologic, petrologic and stratigraphic data derived from deepwater well samples at different depth intervals can aid the construction of paleo-topographic, paleo-drainage and sediment source terrane maps that depict the surface of Earth at different times in the geologic past. The creation, use and refinement of these products are explored throughout this chapter.

Generally, sediment transfer in the system occurs in a basinward direction during the lifetime of the system though practitioners need to be aware that system elements may be displaced entirely with the passage of geologic time—through either tectonic inversion processes or icehouse/greenhouse cycles, for example—and sedimentary strata from prior depositional lows may be uplifted, eroded and "recycled" through a subsequent system. Although the concept of sediment recycling is straightforward, the study of it can be quite complex since elements from some systems undergo multiple periods of sediment recycling that can span supercontinent cycles. Analytical tools such as detrital mineral geochronology are useful to discern sediment recycling patterns and timescales (see "Common analytical techniques").

The size, longevity and influence of modern S2S systems around the world are as varied as the tectonic and climatic environments in which they organize (Fig. 2 and Table 1). In the business of petroleum exploration, system elements within

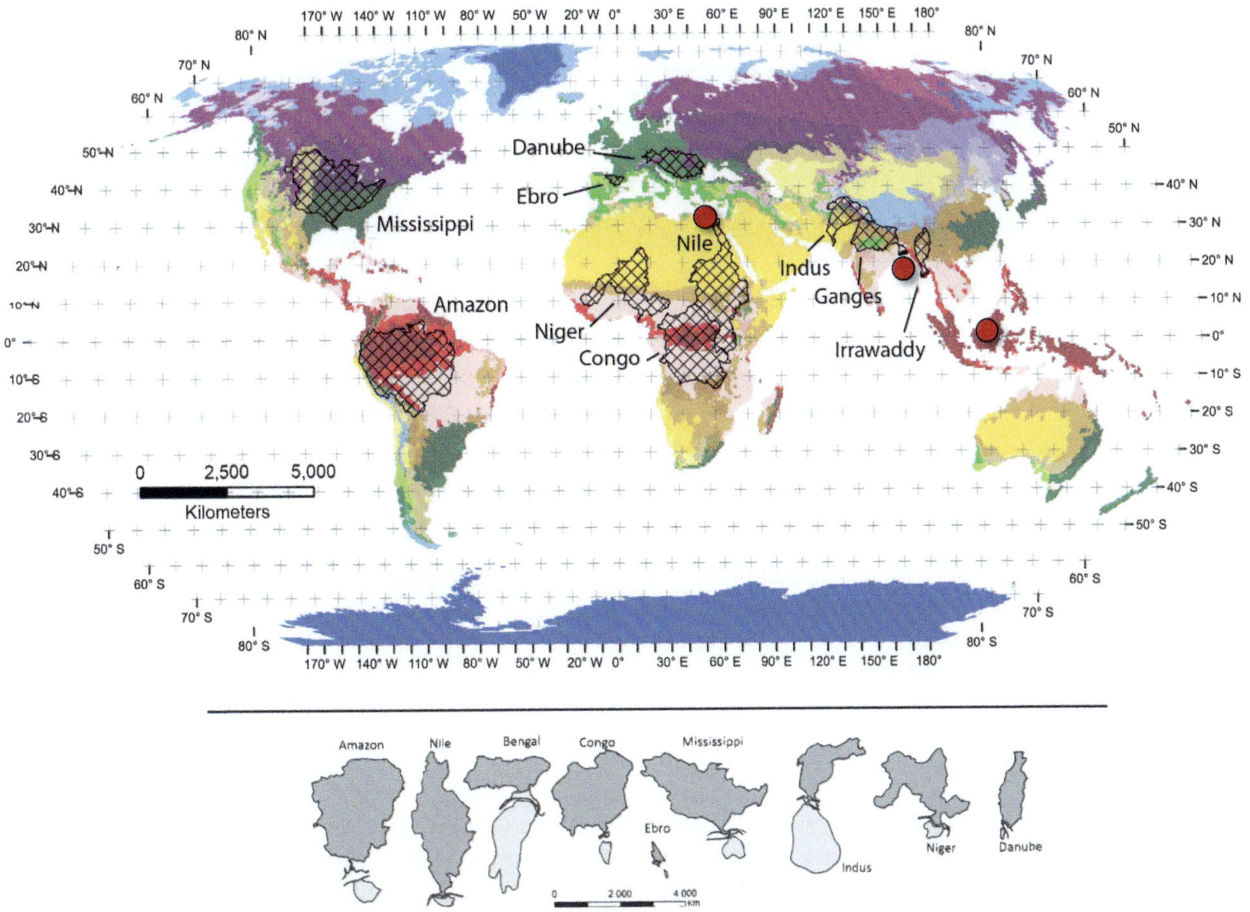


**FIG. 2** Source-to-sink systems of the modern world. S2S systems exist at many different scales around the globe and they lie within (and across) each type of climate zone. World map of the Köppen-Geiger climate classification observed using CRU TS 2.1 temperature and GPCC Full v4 precipitation data, period 1976–2000 (http://koeppen-geiger.vu-wien.ac.at/shifts.htm). Red circles show the approximate geographic location of case studies described in this chapter: Egypt; Myanmar; Borneo. Climate zones from Kottek et al. (2006). Areal extents and geometries of modern source-to-sink systems modified from Helland-Hansen et al. (2016). Catchment/watershed extents derived from the Global Runoff Data Centre (https://www.bafg.de/GRDC/EN/01_GRDC/grdc_node.html). See Table 1 for further information. Figure concept inspired by Sømme et al. (2009) and Nyberg et al. (2018).

**TABLE 1 The scale of S2S systems and elements.**

| Catchment/ Primary River | Continent | Geographic area | Primary climate zone(s) | Terminal sink | Basin area (sq. km.\|sq. mi.) | Submarine Fan area (sq. km.\|sq. mi.) |
|---|---|---|---|---|---|---|
| Amazon | South America | Brazilian Margin | Af, Am, Aw | North Atlantic Ocean | 5,965,897 I 2,303,446 | 330,000 I 127,414 |
| Congo | Africa | Sub-Saharan Africa | Af, Am, Aw | South Atlantic Ocean | 3,705,226 I 1,430,596 | ~300,000 I ~115,831 |
| Nile | Africa | Egyptian Margin | Bwh, Bsh, Aw | Mediterranean Sea | 3,353,206 I 1,294,680 | 700,000 I 270,272 |
| Mississippi-Missouri | North America | Midwestern United States | Dfa, Cfa, BSk | Gulf of Mexico | 3,240,619 I 1,251,210 | >300,000 I >115,831 |
| Niger | Africa | West Africa | Bwh, Bsh, Aw | Gulf of Guinea | 2,122,994 I 819,693 | *n/a* |
| Meghna-Padma-Ganges–Brahmaputra | Asia | NE Indian Ocean | Aw, Cwa, Csa | – | 1,621,000 I 625,872 | 3,000,000 I 1,158,306 (Bengal Fan) |
| Indus | Asia | NW Indian Ocean | Bwh, Bsh (and numerous) | Arabian Sea | 865,013 I 333,983 | 1,100,000 I 424,712 |
| Danube | Europe | Southest Europe | Cfb, Dfb | Black Sea | 795,316 I 307,073 | *relict* |
| Irrawaddy | Asia | Southeast Asia | Cwa, Aw, Bsh | Andaman Sea | 422,917 I 163,289 | *n/a* |
| Ebro | Europe | Eastern Spain | Cfb, Bsk, Csa | Mediterranean Sea | 85,460 I 32,996 | 5000 I 1931 |

Submarine Fan information partly from Barnes and Normark (1985) and Shanmugam (2016). Basin Area from *Global Runoff Data Center* (https://www.bafg.de/GRDC/EN/01_GRDC/grdc_node.html). Climate zones from (Kottek et al., 2006).

ancient rift margins (e.g., the North Sea or Gulf of Mexico) are often studied because, during their lifetime, those settings typically organize all four exploration risk elements (reservoir, seal, trap and source) beside one another, making exploration in those regions historically favorable, whether the targets lie within syn-rift lacustrine or pre-salt environments or post-rift deepwater fans, for example.

## Timescales and toolboxes

Each S2S system has a distinct lifetime, typically determined by the tectonic and climatic setting that it lies within—short-run systems in active orogenic zones within tropical climates may be relatively short-lived (thousands-millions of years) due to shifting drainage divides, high denudation rates and rapid basin filling processes while continental-scale drainages within stable cratons may last for millions to tens-of-millions of years (e.g., the greater Mississippi River catchment that feeds the northern Gulf of Mexico through the Mississippi Delta). Additionally, internal S2S system dynamics are not static during their lifetime. They are ever-changing on both the geological, and even human, timescale because the geologic processes that operate within them work on a range of temporal scales from historical (thousands of years) to intermediate (tens of thousands to millions) to deep time (millions to billions).

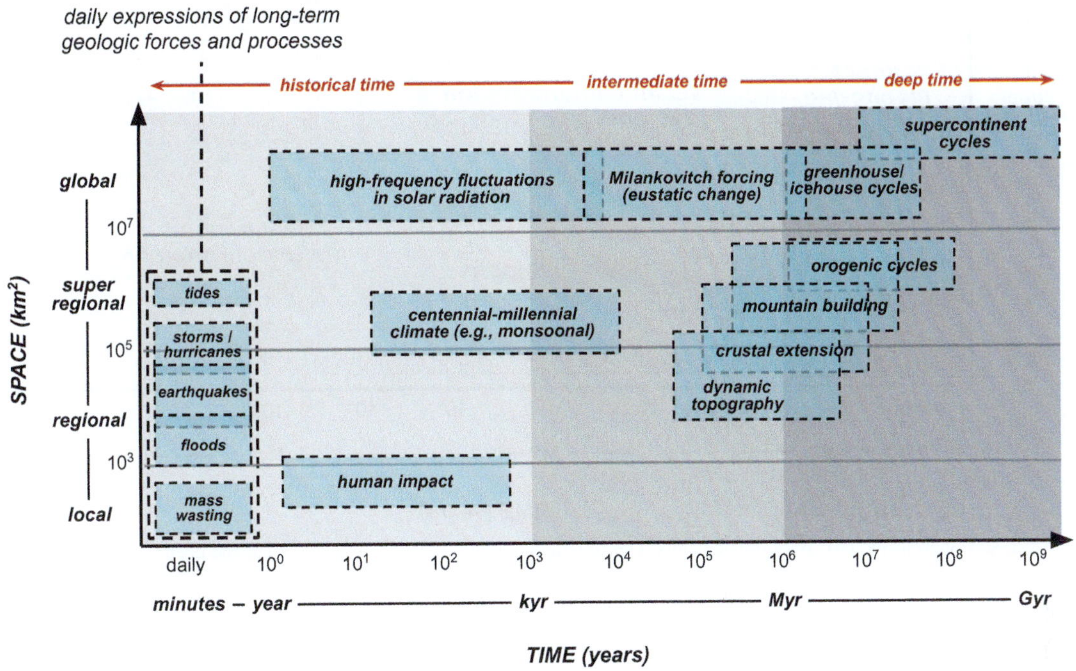


**FIG. 3** Source-to-sink timescales. The period and spatial extent of primary S2S forcing factors, and the effective range of different chronometric tools that may be applied to study them. These factors may be used to scale S2S projects according to the spatial extent and timescale of investigation. Background colors refer to temporal scales (historical, intermediate and deep time) discussed in the text. *(Modified from Helland-Hansen et al. (2016); and Romans et al., (2016).)*

S2S system behavior is primarily a response to geological and other forcing factors and each of those factors can influence a system at different spatial and temporal scales (Fig. 3). Dam building along a river significantly alters the river hydraulics which, in turn, affects the inherent ability of that river to transport sediment to the sink whether it be through mid-system sediment retention (behind dams) or a reduction in the erosive power of the stream itself because of the rise in local base level. Changes in sea level caused by Milankovitch forcing and climatic shifts, both operating on intermediate timescales, may disrupt S2S dynamics across an entire system because of their super-regional to global scale. These extensive and powerful natural forces create lasting effects in S2S systems that are often preserved in the rock record. As an example, consider that the architecture of stacked deltaic clinoforms, often easily observed in seismic data, resulted from gradual and predictable changes in the up-dip fluvial responses to climate and sea-level change (e.g., Blum and Hattier-Womack, 2009; Blum and Törnqvist, 2000). See chapter 8 for additional information.

S2S practitioners must also keep these spatio-temporal boundaries in mind when considering the appropriate approach for system characterization, which often depends on project scale (see "Project framing considerations"). Each technique and tool, just like the geologic processes they intend to study, has a sweet spot for the greatest application (Fig. 4). Because S2S system elements are inextricably linked, S2S practitioners have many ways in which they may characterize the entire system by analyzing discrete elements; besides the added confidence in an interpretation reached through multidisciplinary analyses, this convenience also may allow practitioners to circumvent potential access limitations to certain datasets due to either availability, quality and/or cost. For example, a practitioner may reduce the uncertainty associated with a targeted deepwater clastic reservoir by applying geomorphological scaling relationships to paleo-topographic models at syn-depositional time and use catchment and canyon metrics to estimate the extent of a deepwater fan system—this approach may be faster and lower cost than a prohibitively expensive purchase of new or reprocessed 3D seismic data in an exploration area of interest.

The spatio-temporal dynamism of system behavior highlights the importance of using multiple tools to accurately characterize a system and even prioritizes estimation techniques (e.g., system scale modeling) over precision techniques to characterize a system at any snapshot in time. There are distinct, and sometimes nuanced, advantages to multiple approaches, even when using the same high-level method. Brewer et al. (2020) took this approach when comparing the accuracy of three different methods, based on different input parameters (Table 2), to calculate estimates of sediment volume and flux:

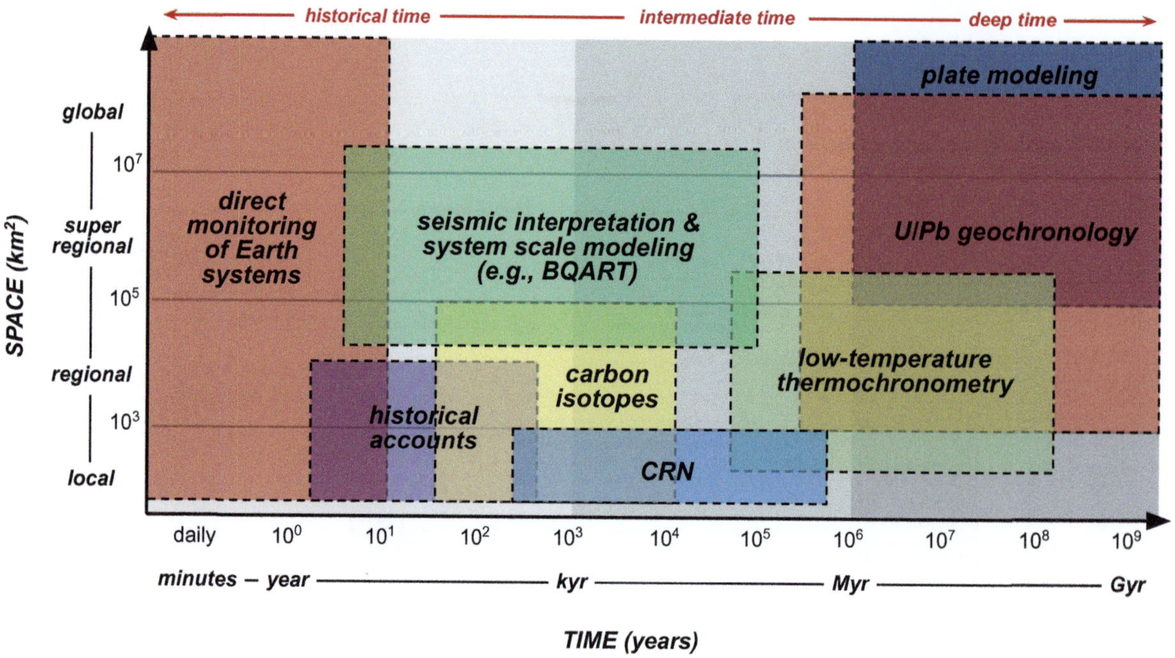


**FIG. 4** Source-to-sink analysis toolbox. Background colors refer to temporal scales (historical, intermediate and deep time) discussed in the text.

**TABLE 2** Methods for calculating sediment flux.

| Parameter | BQART model | Geomorphological scaling relationships | "Fulcrum" model |
|---|---|---|---|
| Catchment drainage area (A) | X | X | |
| Maximum catchment relief (R) | X | | |
| Catchment-averaged temperature (T) | X | | |
| Catchment-averaged lithology (L) | X | | |
| Glacial erosion factor (I) | X | | |
| Trapping efficiency of lakes and reservoirs (TE) | X | | |
| Human-influenced soil erosion factor (EH) | X | | |
| Water discharge (Qw) | X | | |
| Fluvial-palaeochannel length | | X | |
| Slope length | | X | |
| Basin-floor fan volume | | X | |
| Duration of deposition, derived from age model | | X | |
| Fluvial-palaeochannel averaged bankfull depth (Hbf) | | | X |
| Fluvial-palaeochannel bankfull width (Bbf) | | | X |
| Palaeochannel hydraulic radius (Rbf) | | | X |
| Median bedload grain size (D50) | | | X |
| Submerged density of bedload grains (ρs-w) | | | X |
| Bankfull Shields number for dimensionless shear stress ($\tau^{*}$bf50) | | | X |
| Fluvial-palaeochannel slope (S) | | X | X |
| Year-averaged bankfull duration (tbd) | | | X |

Parameters for which input data are potentially required for each sediment flux estimation method. Different combinations of geomorphological scaling relationships, which thus require different input parameters, can be used to estimate sediment flux. Fluvial-paleochannel bankfull width (Bbf) and averaged bankfull depth (Hbf) are used to estimate hydraulic radius (Rbf). Some combinations of Bbf and Hbf are inconsistent with input data; Rbf is used as a filtering parameter to remove such combinations. Table information summarized from Holbrook and Wanas (2014), Nyberg et al. (2018, 2021), Sømme et al. (2009), Syvitski and Milliman (2007) and references therein.
Modified from Brewer et al. (2020).

1. the BQART empirical model that relies heavily on paleoclimate and paleo-catchment relief input parameters (Syvitski and Milliman, 2007; Nyberg et al., 2021)
2. empirical geomorphological scaling relationships that leverage the dimensions of system segments (e.g., catchment extent, shelf width and slope; Sømme et al., 2009; Nyberg et al., 2018);
3. the "fulcrum" model that uses the paleo-hydrological parameters of trunk river channels to estimate down-system sediment flux (Holbrook and Wanas, 2014).

Brewer et al. (2020) conclude that, while each method generated plausible ranges of estimated sediment flux that are, importantly, within at least one order of magnitude of the reference values, the ability of each method to accurately estimate sediment flux varies greatly with its intended application. BQART models produce well-constrained results for submodern systems but highly uncertain results in deep-time systems. Estimates derived from geomorphological scaling relationships yield results with a broad distribution of values over one-to-three orders of magnitude for both submodern and deep-time sediment routing systems which may be good enough for frontier exploration but are too inaccurate for focused prospect risking. And sediment flux estimates generated using the "fulcrum" model are most sensitive to the duration of bankfull discharge, which is poorly constrained in deep-time systems which, again, renders its application to ancient systems ineffective for detailed exploration risking.

As with most S2S projects and the tools used to characterize them, work should be conducted iteratively and first-pass results, along with their inherent uncertainty, should be used as guidance to select more appropriate and, in most cases, multiple accurate characterization methods that may be employed in parallel.

## Project framing considerations

S2S uses myriad tools and techniques to characterize each or any number of discrete system elements and practitioners should consider a number of project framing decision criteria when planning an approach to system characterization, from deadline-dependent, business-driven guidelines to the viability of each technique to accurately measure the system elements of interest. Fig. 5 outlines the interconnectedness of S2S-specific tools, techniques and products and highlights their relevance relative to exploration business strategies, whether they be short-cycle deadlines for prospect generation and evaluation where useful products like stratigraphic models need to be generated within weeks to months or more ambitious goals like play extension and frontier exploration, where a more complete understanding of the entire S2S system is needed

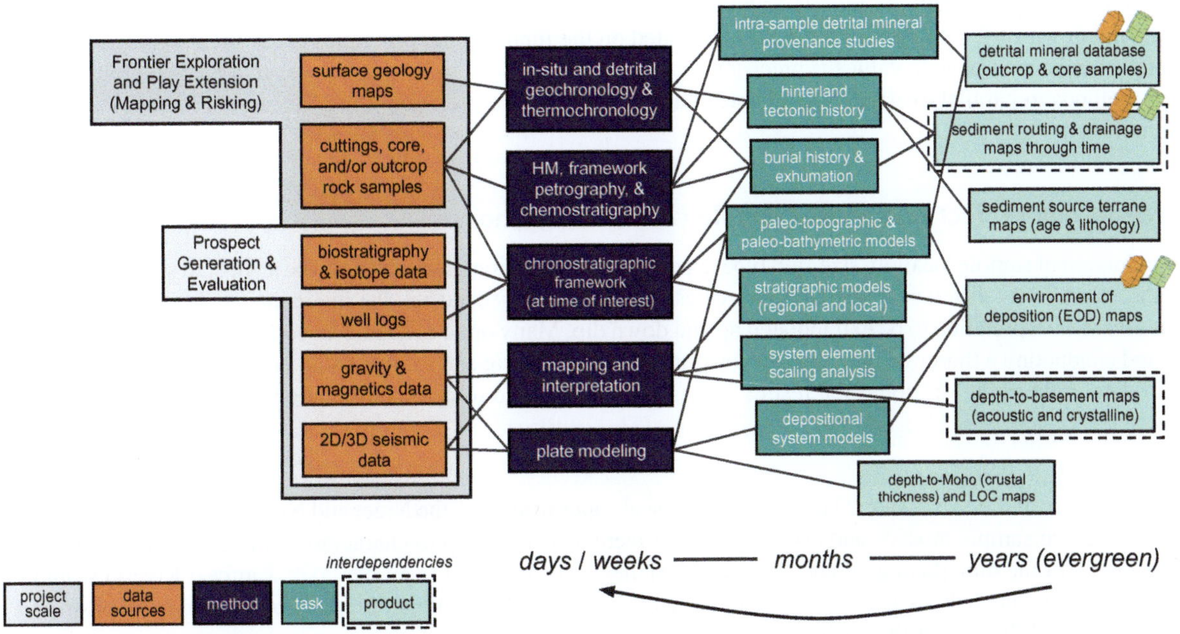


**FIG. 5** S2S project framing workflow. This workflow illustrates the techniques and approaches to use when making considerations for both data and time availability when framing the project (weeks vs. months vs. evergreen). This list focuses on S2S-related tools and techniques. Other viable risk assessment techniques, such as reservoir characterization and basin modeling, should be used in combination for a unique and accurate evaluation. Evergreen data should be updated periodically for use in shorter-cycle evaluations. The apatite and zircon crystal icons label S2S products that benefit especially from geochronological and thermochronological data inputs.

to generate supra-regional environment of deposition (EOD) maps and depth-to-Moho values for input into basin models. While S2S products themselves may be used to guide exploration decisions, many of them may also serve as input to other viable risk assessment techniques—framework grain composition values may be input to sandstone diagenesis modeling for reservoir quality and paleo-bathymetric maps may be used for basin models that predict expulsion, hydrocarbon phase and migration timing.

While it is possible to derive computational forward models based purely on theoretical inputs, the S2S products that influence business decisions are always grounded in data. For instance, reliable EOD maps (chapters 1 and 8) require information from the sink. Biostratigraphic age control (chapter 14) and paleo-bathymetry information from wells are two foundational inputs for a reliable chronostratigraphic framework which, in turn, ensures accurate correlation and interpretation of mappable horizons in seismic data to produce stratigraphic interpretations and, eventually, local and regional EOD maps. The products that take years to develop should be updated periodically so they may be deployed for use in shorter-cycle evaluations.

Practitioners should begin S2S work with high-quality metadata because the reliability and accuracy of derivative models and products are entirely reliant on the spatial and temporal resolution of input data—the classic phrase from the early days of computer science is "garbage in, garbage out". Reliable geologic maps and quantitative mapping tools are essential for developing accurate and reproducible models of both modern and ancient systems. At minimum, S2S analysis requires maps of Surface Geology—the rocks that crop out at the modern surface—and Basement Terranes—the cratonic bedrock underlying sedimentary strata—with self-consistent age and lithology data, standardized terms and global coverage, to create derivative products with consistent metadata that are useful in modern and ancient S2S analysis (e.g., sediment routing maps; Fig. 5). The use of high-quality input data ensures reliable outputs and allows rapid iteration of derivative models for streamlined global play and prospect risking. Many useful resources have been published in downloadable digital formats ready for integration into a S2S workflow (e.g., the Global Lithological Map database; Hartmann and Moosdorf, 2012).

## Fundamental techniques, tools and data

This section highlights the importance of reliable geologic data for accurate sediment source terrane characterization, provenance determinations, and deepwater reservoir quality prediction. It provides an overview of common analytical techniques, tools and methods including a discussion of primary scientific principles, analytical timelines, how data are reported and interpreted and the pitfalls or blind spots associated with each technique. Practitioners are guided on how to scale their own fit-for-purpose S2S projects, educated on the fundamental principles of the most widely used analytical techniques and encouraged to value the precision introduced by multidisciplinary studies that use several independent analytical methods in parallel to arrive at a reliable and reproducible interpretation. (See Chapter 14 for a review of Applied paleontology.)

### Source-to-sink project scaling and sample collection methods

Determination of sample location, size, and type depends on the question that is trying to be answered since S2S can address hypotheses on vastly different spatiotemporal scales. And, as discussed earlier, the interconnectedness of S2S system elements allows interpretation to be extrapolated up and down dip. Many options should be considered before heading into the field and conducting a thorough literature review prior to field planning and sample collection is considered a best practice. Useful data in the right location often results in less bias and a more robust interpretation.

#### *Where to sample?*

What is the scale of the S2S system? For a continental-scale appraisal (e.g., the Niger and Nile Rivers or the Bengal Fan) it is most practical to sample modern and ancient river sediment as it will help to characterize a larger area and aggregate the heterogeneity of the hinterland geology. Whereas a smaller scale appraisal (e.g., the Po or Zambezi Rivers) would benefit more from individual bedrock ages within its hinterland in order to get the kind of resolution needed to reconstruct a more useful source-sink history of the basin.

Are there parts of the hinterland that no longer exist that will need a proxy from elsewhere? Occasionally it is necessary to characterize a paleo-hinterland by using data from an area that may not currently be draining into that system but has the same provenance characteristics. Are there unconventional sediment sources that also need to be considered, such as contourites, loess or eolian deposits?

### *What to sample?*

In addition to determining *where* to collect samples, it is also important to consider *what kind* of sample to collect. Grain size and depositional facies are extremely important to take into consideration as sampling or comparing data from different grain sizes or facies will start to introduce bias from effects such as hydraulic sorting (chapters 6–10, 13). Sampling a levee sediment in one hinterland or core and comparing it to a channel sand from another hinterland or core would not be a true comparison, largely due to hydraulic sorting altering the concentration of heavy minerals, biasing zircon populations and postdepositional diagenesis.

Mud samples are often overlooked in S2S studies because the focus is usually to determine the provenance of sandy reservoir intervals. However, mud-rich cores in the sink can be analyzed at a higher frequency than sand-rich intervals, which may give a more continuous picture of the evolution of a system over time.

## Common analytical techniques

### *U-Pb-Hf geochronology*

The dating of individual zircons taken from either bedrock, core samples or detrital sediment samples allows a study to get a better understanding of the timing of tectonic events that may have affected all or part of your source catchment area. Knowing either the crystallization or maximum depositional age (MDA) of the geology of a catchment area can reveal which regions may or may not have been supplying sediment to the sink at any given time. First-order volcanic zircons are important for MDA calculations and assumptions regarding the sediment retention factor within a system. In craton-sourced siliciclastic sequences, the presence of first-order volcanic zircon correlates strongly to increased lithic content and, therefore, to degraded hydrocarbon reservoir quality as pore-clogging clay often develops from weathered feldspar within volcanic lithic grains.

Geochronology is most powerful when samples are analyzed from both the source and the sink. In this fashion, the set of samples can be used to directly tie the sediment back to its source area. However, even without samples from the sink, having a well-characterized hinterland will help to predict what sediment might have reached the source area, and when. Zircon is the most common mineral used in geochronology studies due to its abundance and resilience. Zircon can be found in igneous, metamorphic and sedimentary rocks. It is also highly resistant to weathering and alteration which means its age signature is relatively well preserved.

#### Dating

Dating zircons requires measuring the concentration of U and Pb radiogenic isotopes within a single grain to give the U/Pb isotope ratio. U exists in relatively high concentrations within zircon, but any Pb present is a result of radioactive decay and this is used to determine the age of the rock. Two different isotopes ratios are most commonly measured:

$$^{238}\mathrm{U}/^{206}\mathrm{Pb}\ (\text{Half Life} = 4.47\,\text{Ma})$$

$$^{235}\mathrm{U}/^{207}\mathrm{Pb}\ (\text{Half Life} = 704\,\text{Ma})$$

There are a number of different ways to extract isotopic data; the most commonly used techniques are Laser Ablation-Inductively Coupled Plasma-Mass Spectrometry (LA-ICP-MS), and Sensitive High Resolution Ion Microprobe (SHRIMP). LA-ICP-MS uses a laser to generate fine particles from the sample for analysis, while SHRIMP uses an ion beam. Data provided by both techniques can be used to generate two different U/Pb ratios—$^{238}$U/$^{206}$Pb and $^{235}$U/$^{207}$Pb. These ratios are used to create a Concordia diagram (Ludwig, 2000, Fig. 6), which is then used to calculate the age of the zircon. The age of the zircon is determined by the point of intersection of the two different series.

Once zircon ages have been calculated from the Concordia diagram, they may be plotted in several ways: Kernel Density Estimation (KDE); Probability Density Plot (PDP); Multidimensional Scaling (MDS); Cumulative Frequency plots. Most often used are either KDE or PDP combined with a histogram and the number of grains analyzed (Fig. 6). KDE plots use Gaussian distributions but do not take into consideration the analytical uncertainty associated with each analysis. This is important to note when working with considerably older samples that are prone to Pb-loss. PDP plots use Gaussian distributions but also take into account analytical uncertainty. This sometimes has the effect of over-representing younger grains, which suffer less from Pb-loss and associated uncertainties, but this is resolved when the PDP is plotted alongside a histogram showing the frequency at which the grains occur.

The way zircons are selected for analysis varies greatly and can often introduce bias at an early stage. It is important that samples selected in the field avoid bias from processes such as hydraulic sorting. This can be done by sampling rocks that

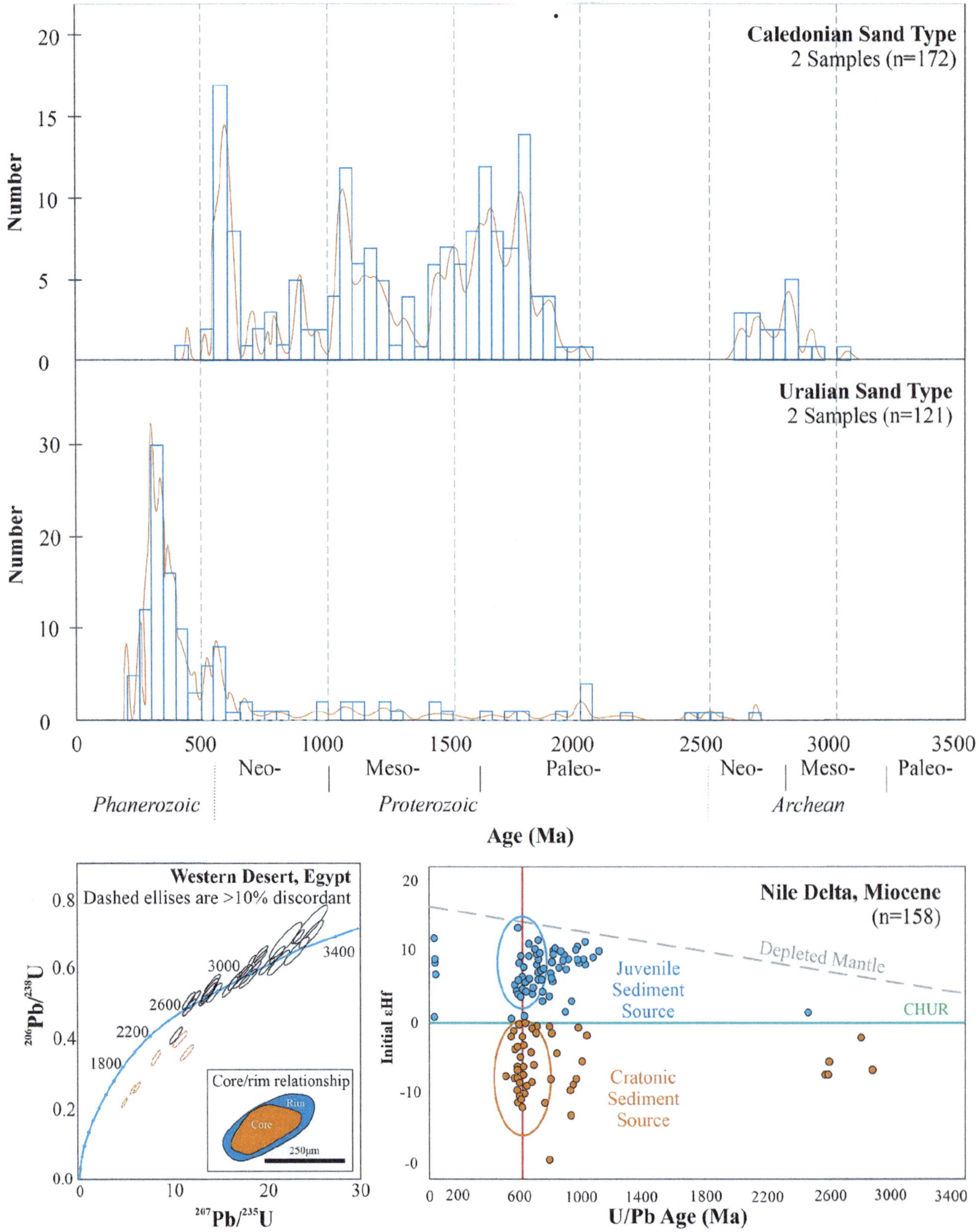


**FIG. 6** Detrital zircon (DZ) data plotting and interpretation. (A) Probability Density Plot (PDP) (red line) to show the accuracy of results, plotted alongside a histogram (blue bars) to illustrate the number of grains per grain population (Fleming et al., 2016). (B) A Concordia diagram showing the age data from a single zircon grain. The size of the ellipse relates to the discordance of the analyses. Inset shows a cathodoluminescence (CL) image of zonation within a zircon and a potential core and rim which could have been formed during two separate events. (C) When U/Pb zircon ages are plotted against εHf of the same grain, previously homogenous grain populations appear to be two different grain populations, one derived from a mafic rich hinterland, and one from a more cratonic hinterland ($n$ = grain numbers analyzed from Miocene Nile Delta core samples. *(Modified from Fielding et al. (2018).)*

are of a similar grain size and depositional facies. For example, comparing data taken from a channel to a levee will affect the quality of the results as smaller zircons (potentially from a particular source area) will be represented disproportionately within the levee deposits compared to the channel deposits.

The zonation within zircons can also lead to complications during analysis. It is good practice to obtain cathodoluminescence (CL) images of zircons prior to analysis so that when sampling using laser ablation, grain boundaries, cores, rims, cracks and other traces of radiation damage can be avoided.

### Lu/Hf

Lu/Hf isotope ratios are also often measured along with U/Pb ratios in zircon. The data is extracted from the zircon grain in much the same way, and ideally in exactly the same location as the original U/Pb analysis (i.e., the same laser spot). The Lu/Hf ratio of zircon is an excellent indicator of whether a sample is derived from a juvenile or cratonic source. This is particularly useful if source area sediment is suspected to have been repeatedly recycled. The sediment samples may share the same age signature across a large hinterland, yet their Lu/Hf ratios may differ, thus allowing the study to pinpoint the true origins of the sediment (Fig. 6).

## *Thermochronology*

Modern thermochronology is a theoretically complex and methodologically diverse field of study that has evolved rapidly in the past couple of decades with the advent of new thermochronometers, the improved understanding of their intrinsic behaviors, the refinement of established techniques and their creative application to solve geological problems associated with landscape evolution, tectono-structural deformation and history of the Earth. Many publications address the in-depth technical and methodological details of thermochronological systems and their application (e.g., Wolf et al., 1996; Stockli et al., 2000; Ehlers and Farley, 2003; Reiners, 2005; Reiners et al., 2005; Schoene, 2014; Malusà and Fitzgerald, 2019; Reiners and Brandon, 2006). Specific to basin analysis and resource exploration, Armstrong (2005), Peyton and Carrapa (2013) and Schneider and Issler (2019) present thorough overviews of practical applications of low-temperature thermochronology, including special considerations for sampling strategies and computational modeling. This section focuses on the products and applications of the most common, low-temperature thermochronology techniques that are routinely applied to S2S analysis (*see* Fig. 5), specifically in situ and detrital apatite and zircon (U-Th)/He and fission-track thermochronology and hybrid geo-/thermochronology methods including detrital zircon U-Pb and (U-Th)/He "double-dating". This is the context in which thermochronology is addressed throughout the rest of this chapter.

Thermochronology is the quantitative study of the thermal histories of rocks using temperature-sensitive radiometric dating methods. $^{40}Ar/^{39}Ar$, K-Ar, fission-track and (U-Th)/He are examples of common thermochronology methods. The term thermochronometer most often refers to a pairing of the dating method with a host mineral or bulk rock. Apatite and zircon are commonly used thermochronometers in S2S studies because they are also common accessory minerals in the types of rock that yield sand-prone siliciclastic sediments. Apatite and zircon also withstand chemical and physical abrasion quite well so they can be transported long distances, found in strata within most S2S system segments and can be recycled through multiple episodes of deposition and erosion (Fig. 1). Other thermochronometers like hematite, magnetite, rutile and titanite (e.g., Reiners and Farley, 1999; Wolfe and Stockli, 2010; Cooperdock and Ault, 2020) are less common in detrital S2S studies because they are typically sourced from relatively rare sediment source terranes; however, their sometimes rare presence in sedimentary systems may act as unique discriminant minerals in sediment provenance studies and, as thermochronometers, they have robust applications in other disciplines that may constrain up-dip system dynamics that control sediment generation such as the onset and evolution of orogenic processes and faulting.

### Fission-track and (U-Th)/He system fundamentals

Apatite and zircon fission-track (FT) and (U-Th)/He thermochronology are related in that they analyze the parent and daughter nuclides of the Uranium decay series (mostly). Their behavior affects the crystalline lattices of the host mineral. FT involves analyzing damage accumulated within the crystal lattice of a uranium-rich mineral (a.k.a. "tracks") caused by the natural spontaneous nuclear fission decay of $^{238}U$. The (U-Th)/He technique requires accurate measurement of parent-daughter radionuclides: $^{4}He$ accumulates volumetrically in the thermochronometer crystal lattice during the α-disintegration of $^{238}U$, $^{235}U$, $^{232}Th$, their daughter products and $^{147}Sm$.

As opposed to geochronometers (*see* section "U-Pb-Hf geochronology"), thermochronometers are open systems in the sense that the accrual and retention of their daughter products is controlled by the unique behavior of their crystal domains within thermal conditions. In (U-Th)/He systems, $^{4}He$ accumulates and is retained at temperatures between ~30–90°C (~ 85–195°F) for apatite (Farley, 2000, 2002; Ehlers and Farley, 2003) and ~130–200°C (~ 265–390°F) for zircon (Reiners and Brandon, 2006). In fission-track systems, tracks are partially annealed (i.e., erased) at temperatures between ~60–120°C (~ 140–250°F) for apatite (e.g., Gleadow et al., 1986; Laslett et al., 1987; Green et al., 1989a; Fitzgerald and Gleadow, 1990; Ketcham et al., 1999) and ~180–310°C (~ 355–590°F) for zircon (Yamada et al., 1995, 2007; Tagami, 2005; Tagami and Dumitru, 1996; Ketcham, 2019)—above and below those respective ranges, tracks are either developed and retained or nearly completely annealed. The apt terms for these temperature ranges are partial retention zone (PRZ) for (U-Th)/He age dating and partial annealing zone (PAZ) for fission-track age dating. The time when a particular thermochronometric mineral passed through these temperature windows, called the cooling age, may be calculated by measuring the relative amounts of parent nuclides and daughter product within a crystal. Occurring within the PRZ/PAZ range, the closure temperature ($T_C$) refers to the temperature of a thermochronometric system at the time determined by its cooling age. At this point, the system has cooled sufficiently to the point where there is no longer significant diffusive loss of the parent or daughter isotopes from the system (Braun et al., 2006).

The aforementioned information is intended as a guideline for framing the applicability of thermochronometers to S2S problems. In practice, there is considerable variability in $T_C$ and other system behaviors depending on real-world factors such as cooling rate and magnitude, grain size, radiation damage and grain chemistry to name a few. Reiners and Ehlers (2018) provide a solid basis for developing a better understanding of these techniques.

## Common applications

Thermochronometers cool within their host rocks as they approach the surface through geologic processes such as block faulting, upwarping, rapid denudation and geothermal gradient relaxation (Ehlers and Farley, 2003) and, conversely, they may heat up during postdepositional burial, crustal thickening from structural overthrusts, geothermal gradient intensification, hydrothermal fluid circulation and, in some rare cases, exposure to transient, high-intensity thermal fluxes at the surface of the Earth (e.g., forest fires; Mitchell and Reiners, 2003). Considering a mean annual surface temperature of 10°C (50°F) and the fact that subsurface temperatures generally increase with depth at a rate of ~20–30°C/km (~ 110–138° F/mi), common low-temperature thermochronometers are sensitive to temperature changes between the earth's surface and the top several kilometers of the crust, which is ideal for investigating the tectonic-sedimentary processes that influence, and occur within, S2S systems over a wide range of space and time (Fig. 4). Specific to petroleum resource assessment, the coincidence between the apatite and zircon (U-Th)/He PRZs and fission-track PAZs with the peak thermal maturity of liquid and gas-phase hydrocarbon generation between ~60–180°C (~ 140–355°F) means that these thermochronometers can resolve the timing, rate and magnitude of subsurface thermal processes important to hydrocarbon system development. Especially relevant to the characterization of deepwater sedimentary environments, thermochronology is often used in S2S analysis to create products that delineate the generation, transport and deposition of sediments from orogenic onset and evolution to basin formation and fill processes (e.g., regional and local stratigraphic models and EOD maps; Fig. 5).

Thermochronology may be applied in situ, meaning "in the original place", by collecting samples from sediment source terranes to analyze hinterland dynamics and/or anywhere in the detrital spectrum of the sediment routing system (e.g., Reiners and Brandon, 2006; Malusà and Fitzgerald, 2019; Stockli and Najman, 2020). It is a best practice to draw data from two or more S2S system segments to characterize entire system behavior. However, the utility of the thermochronometric suite ensures that even single-point datasets, like those obtained from offshore wells and piston cores or modern river outfalls and delta systems, act as aggregators of up-dip data and provide insight to S2S system dynamics, especially when two or more thermochronometers are used together to determine different parts of a sample's thermal history.

The juxtaposition of an idealized footwall and hanging wall pair (Fig. 7) demonstrates key S2S system concepts on a small scale—geology is fractal after all. In this situation, exhumation and denudation of the footwall rock create cooling of the hosted thermochronometers, sometimes through their closure temperature, during sequential stages labeled A, B and C. The thermochronometers then enter the sedimentary system where they are transported and then buried in an adjacent basin. This creates an inverted record of exhumation since the detritus that was first generated from the footwall (A) reaches the depositional sink ahead of detritus generated during subsequent footwall denudation stages (B and C).

Thermochronometers collected in situ from footwall rocks at the surface of the Earth, prior to erosion and transport, may yield the cooling history of the host rock and allow geologists to calculate the timing, rate and magnitude of orogenic exhumation. Samples collected in the sink, especially when taken from boreholes, may yield either their predepositional cooling history or some version of the sedimentary host rock burial history, depending on whether they experienced sufficient postdepositional thermal resetting at a depth illustrated by the hachured "$T_C$" line in Fig. 7. If not thermally reset by

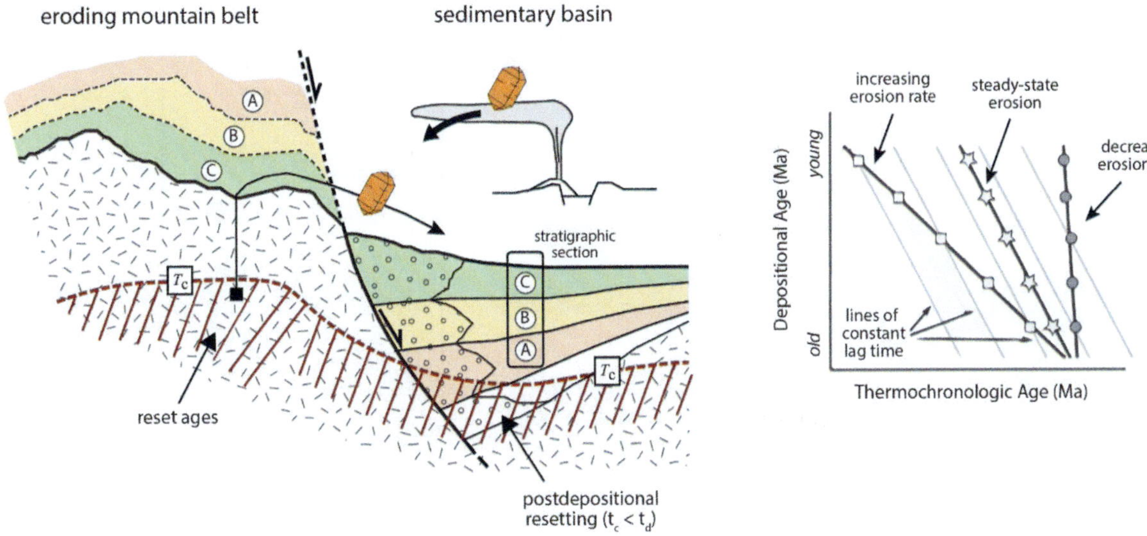


**FIG. 7** Thermochronology for basin analysis. Calculating orogenic exhumation, erosion rates and depositional lag from detrital mineral ages. *(Modified from Rahl et al. (2007) and Stockli and Najman (2020).)*

burial and heating, the thermochronometers could provide important information about the sediment source terranes in the hinterland from which they eroded. Their cooling ages may also be used to calculate depositional lag-time, which is defined as the difference in time between the closure age of a thermochronometer in its source region and the depositional age of its sedimentary host strata (*see* Zeitler et al., 1986; Rahl et al., 2007; Saylor et al., 2012).

Lag-time analysis of sedimentary basin fill, illustrated on the right side of Fig. 7, provides insight into the exhumation history of adjacent eroded orogens relative to steady-state erosion—an increased erosion rate is indicated by a relatively thin stratigraphic interval with a wide range of successive cooling ages while a relatively thick stratigraphic interval with a narrow band of cooling ages indicates a decreased erosion rate. When volcanic material enters the system, either directly through ash fall or flow, or rapid transport and deposition, lag-time is essentially zero. This scenario provides a unique opportunity to date the maximum depositional age of the strata directly. Determining the so-called MDA is not always straightforward and one of several methods may be best depending on the conditions (Coutts et al., 2019) but this technique is essential for building chronostratigraphic frameworks that are used to derive basin fill history and myriad S2S products (Fig. 5).

Detrital S2S work becomes more sophisticated and powerful with "double-dating" which typically determines the U-Pb geochronologic and (U-Th)/He thermochronometric age of either a single detrital mineral phase—often the same individual mineral is used for both analyses—or a set of complementary systems. S2S studies that combine U-Pb crystallization ages (*see* section "U-Pb-Hf geochronology") with thermochronometer cooling ages may gain detailed insight on landscape evolution, sediment provenance, chronostratigraphy and basin history. Fig. 8 illustrates how the zircon geo-/thermochronometer may be used to differentiate sediment derived from a set of unique sediment source terranes in a basin where the sampled sediments have not reached temperatures sufficient to reset the thermochronometer ages.

Sediment source A is volcanic ash derived from a local eruptive center. Zircon from this source yields overlapping U-Pb and (U-Th)/He ages because they crystallized and cooled in a geologic instant. Sediment source terranes B, C and D are a homogenous plutonic stock with the same crystallization age that was determined from dating xenocrystic zircon with U-Pb geochronology. However, their cooling histories are quite different: "B" experienced slow postmagmatic cooling and exhumation; "C" developed into a thrust sheet during a period of regional structural contraction; "D" was exhumed within a footwall during a subsequent period of extension—each of these unique histories is captured in their significantly different zircon (U-Th)/He cooling ages. Combined, the age spectra from each individual sediment source terrane (A, B, C and D) define the detrital zircon record found in the half-graben strata (E) so, if the in situ geo-/thermochronologic signal of each terrane is known, the provenance of the sediment within the basin may be determined by matching the detrital record to its in situ counterpart. For confident discrimination between sediment source terranes with different U-Pb ages, >100 grains should be dated from a detrital sample and large "*n*" datasets are preferred (Vermeesch, 2004; Pullen et al., 2014).

This straightforward sediment provenance exercise may be applied to S2S systems of all scales, from the local to the continental, to derive important information about any system element of interest. Applications of detrital geo-/thermochronology include deriving catchment tectono-climatic and erosive processes (e.g., Ruiz et al., 2004; Stock

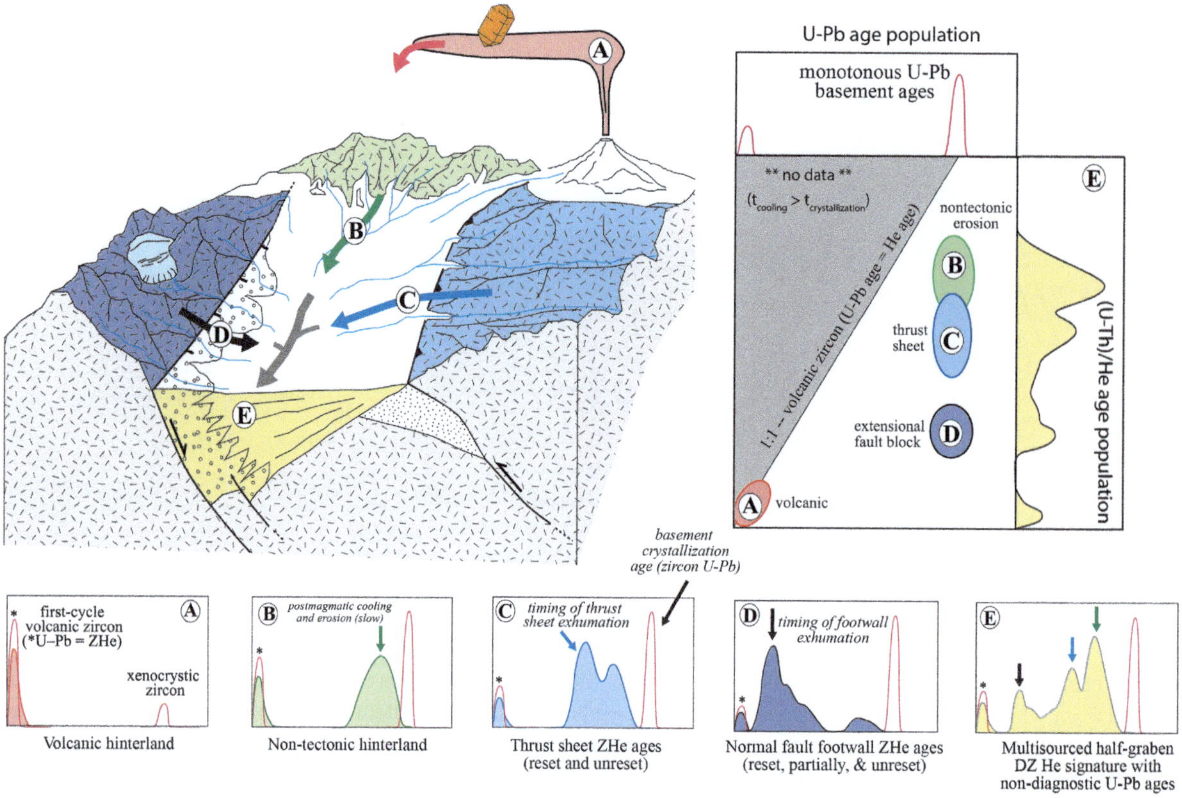


**FIG. 8** The utility of double-dating. U-Pb & U-Th/He double-dating for S2S provides several insights on landscape and basin evolution. This diagram illustrates how the zircon geo –/thermochronometer may be used to aid analysis of basin history, sediment provenance, chronostratigraphy. *(Modified from Reiners et al. (2005) and Stockli and Najman (2020).)*

et al., 2006; Carrapa, 2009; Mason et al., 2019), charting continental-scale drainage evolution (e.g., Blum and Pecha, 2014; Sharman et al., 2017; Zhang et al., 2019 *see* "S2S for frontier exploration and play extension"), unraveling the tectonic history of accreted terranes (e.g., Gehrels et al., 2003; Zapata et al., 2020) and even exploring inaccessible terranes (e.g., Tochilin et al., 2012; Thomson et al., 2013).

In a broad sense, thermochronology remains a relatively underutilized tool in subsurface characterization and basin analysis throughout the private sector. However, this is slowly evolving with the publishing of exploration-relevant studies, greater understanding and acceptance of the inherent uncertainties of different techniques and development of niche methods that address specific aspects of subsurface characterization. Novel applications include resolution of subsurface reservoir thermal histories by analyzing the formation timing and thermal evolution of pore-filling cements, further exploring the utility of thermochronometric minerals derived from fossilized biogenic material (e.g., Peppe and Reiners, 2007; Slabic and Lapen, 2017) and leveraging thermochronology data from fracture-filling cements and fault gouge (e.g., Ault, 2020; Cooperdock and Ault, 2020; Ault et al., 2016) to derive a chronological history of kinematic structural deformation in basins where these features are important to resource assessment.

## A word of caution

A single measured thermochronometer age does not inherently provide enough data for robust interpretation. In fact, very different thermal conditions and geologic processes may produce the same cooling age. As stated succinctly by Gautheron and Zeitler (2020), "Context is required for even qualitative interpretation of data, usually involving dates from other systems as well as geological information and constraints". Geo –/thermochronology subject matter experts should be consulted during data reduction and interpretation because misinterpretation of thermochronologic data can lead to wildly inaccurate assumptions about the target of study. To properly employ proprietary thermochronological data, S2S practitioners require accurate and reproducible measured thermal histories and some framework of geologic context to interpret what those thermal histories might mean so it is essential to establish a close working relationship with a reputable laboratory that welcomes iterative discussion about data reduction and geologic interpretation. Finally, beware of off-the-shelf geo-/

thermochronology products, including geo-/thermochronological age databases, that do not offer professional and expert consultation as well.

## Heavy mineral analysis and petrography

Traditional heavy mineral analysis and petrography data give valuable insight into the make-up of a hinterland and have been heavily relied on in the past as a standalone technique (e.g., Lewin et al., 2020; Morton et al., 2012; Shukri, 1950). Recently, these datasets are being used to supplement studies and are treated more cautiously due to the effects that weathering and diagenesis can have on a sample during transport, recycling and burial. The petrographic signature of a hinterland may look considerably different once it reaches the depocenter and undergoes postburial diagenesis. This, in turn, can lead to incorrect interpretations about the paleo-drainage of a fluvial system.

Both heavy mineral and petrographic analyses require samples to be point counted according to techniques such as the Gazzi-Dickinson method (Ingersoll et al., 1984), or the Galehouse (1969) "Area Method", which uses 200 grain counts per sample. A 2 φ-wide grain size-window is used in order not to lose a significant part of the heavy mineral population, and to obtain a faithful quantitative description of the detrital assemblage.

Heavy mineral assemblages can be used to calculate the following indices:

- MMI (metasedimentary minerals index): relative abundance of chloritoid, staurolite, kyanite, andalusite and sillimanite (Garzanti and Andò, 2007). MMI is used to estimate the average metamorphic grade of metaigneous and metasedimentary source rocks.
- ZTR (zircon, tourmaline, rutile) indicates the percentage of stable zircon, tourmaline, rutile and other Ti oxides over total transparent heavy minerals. ZTR varies from 0 to 100 (Hubert, 1962).

Ternary plots using quartz (Q), feldspar (F) and lithic (L) grain abundances are commonly used to help determine the tectonic provenance of samples (e.g., Dickinson, 1985; Fig. 9).

One of the benefits of manual petrographic framework studies over automated methods such as QEMSCAN—an abbreviation of "*quantitative evaluation of minerals by electron microscopy*"—is that lithic grains can also be broken down into indices (Garzanti and Vezzoli, 2003). For example, the metamorphic index (MI) varies from 0 in sediment derived from sedimentary and volcanic cover rocks to 500 in sediment derived from high-grade basement rocks. Very low- to low-rank metamorphic lithics can be subdivided into metasedimentary (Lms) and metavolcanic (Lmv) categories. Medium- to high-rank metamorphic lithics can be subdivided instead into felsic (metapelite, metasammite, metafelsite; Lmf) and mafic (metabasite; Lmb) categories. Knowing these individual parameters can sometimes help to distinguish one catchment area from another when studying particularly homogenous catchment areas.

## Whole-rock radiogenic isotope geochemistry (Sr/Nd) (see Fig. 10)

$^{87}$Sr occurs naturally in Earth as a result of the decay of $^{87}$Rb, which has a half-life of 48.8 billion years (Nebel, 2014). Sr/Rb isotope ratios have been widely used in the dating of rocks as they are relatively stable when subjected to processes like weathering and diagenesis. The Sm/Nd isotope system is similarly stable and is also useful for sediment fingerprinting and characterizing hinterlands, whereby $^{147}$Sm decays to $^{144}$Nd with a half-life of 106 billion years. Using the two isotope systems in conjunction creates a much more robust proxy and can give valuable insights as to whether a sink sediment has received sediment from continental or mantle-derived sources—or a combination of the two. This technique is particularly useful for studies where sink data may not be available (e.g., in frontier basins) as data from the hinterland can be used to model the mixing of sediments from different hinterlands over time to give an idea of what may be found in the depocenter.

This method traditionally uses fine-grained mud samples ($<63\,\mu m$/0.0025 inch. fraction) from both the source and the sink and can be useful when only mud-rich core samples are available. Because of its effectiveness, it can also prove useful for determining the provenance of reservoir seals (Aplin and Macquaker, 2011). Samples are weighed, leached, dissolved and separated in ion exchange columns to extract radiogenic elements such as Sr and Rb, and the bulk rare earth element (REE) fraction, which includes Sm and Nd. They are then analyzed in a mass spectrometer along with standards to monitor analytical uncertainty. Sm/Nd ratios are often reported using the epsilon notation (εNd), which normalizes the samples to the Chondritic Uniform Reservoir (CHUR) standard. Negative values imply cratonic origins and positive values indicate juvenile mafic sources (CHUR = 0).

Whole-rock radiogenic isotope geochemistry provides one of the only techniques that still works in mud-rich intervals.

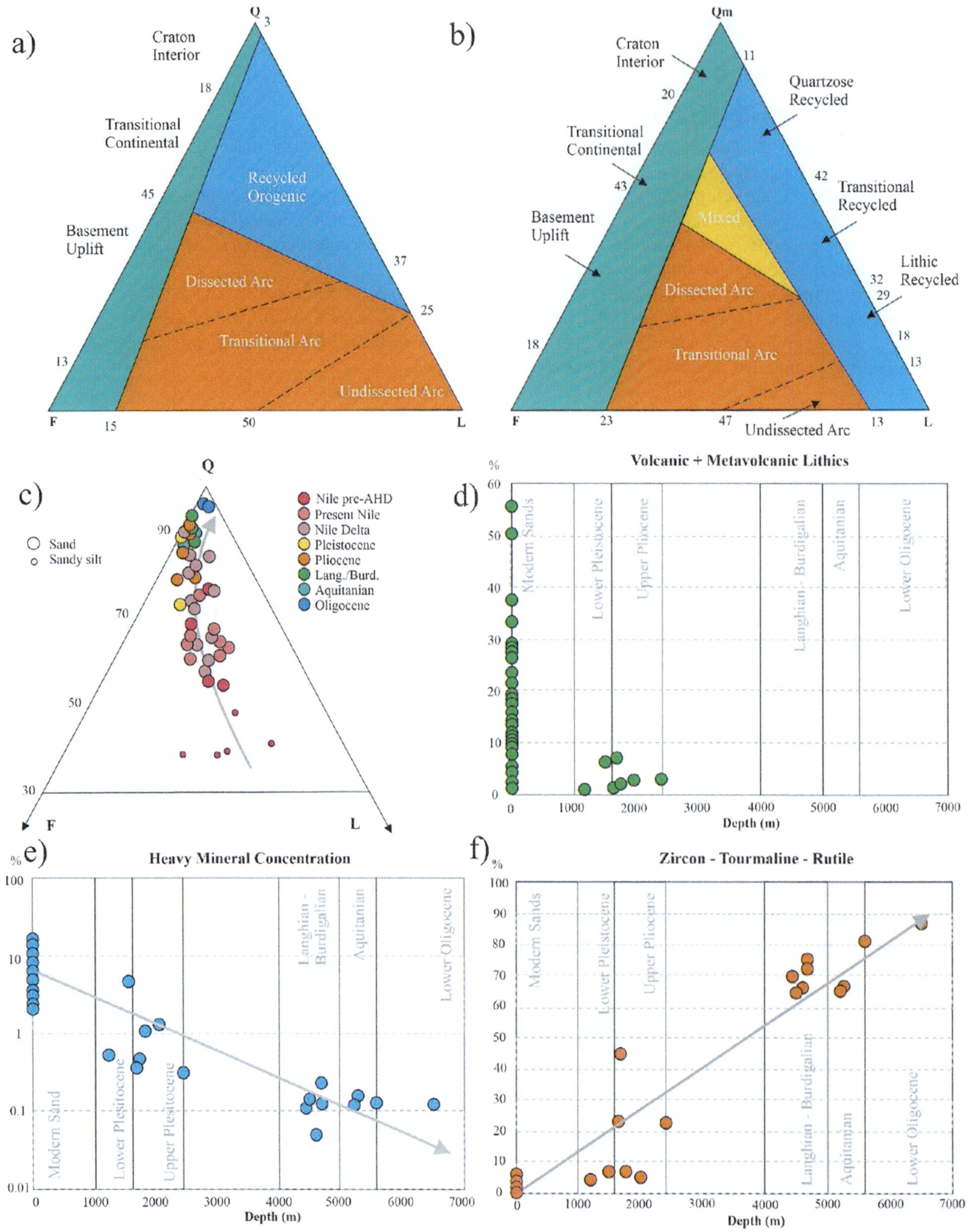


**FIG. 9** Ternary plots. Top: (A) and (B) Original Dickinson (1985) tectonic provenance ternary diagrams illustrating the different zones that sandstones fall into depending on their QFL content. Middle: (C) QFL shows how the proportion of quartz within a sample increases with time due to over representation as a result of diagenesis and the removal of feldspars and lithics (Fielding et al., 2018). (D) The volcanic and metavolcanic component is also seen to decrease with depth/time. Bottom: (E) Heavy mineral concentration plot highlights the overprinting effects that diagenesis can have on a sample once it has been buried as it shows heavy mineral concentrations decreasing dramatically with depth/time. (F) There is also an inverse relationship of the ZTR maturity index with depth/time.

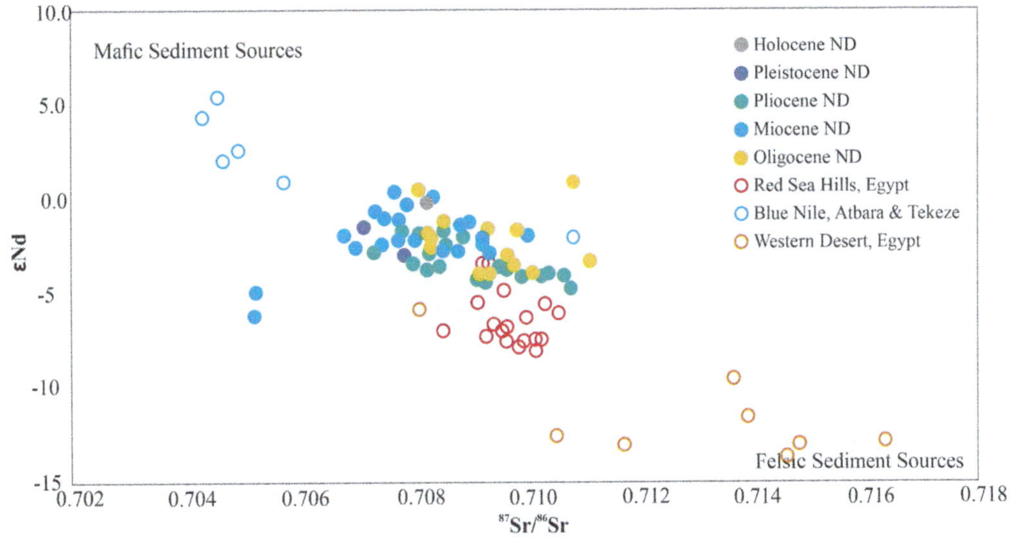


**FIG. 10** Sr/Nd data. Source Sr/Nd data (circles) and sink data from the Nile Delta core (dots). Better quality reservoir sediments will plot toward the bottom left of the chart, with negative epsilon Nd values and higher Sr ratios, while mafic rich and therefore potential poorer quality sediments will plot up toward the top left with higher epsilon Nd values and lower Sr ratios. This chart illustrates the mixing of cratonic and felsic hinterlands; hence, the delta samples plot on a mixing curve between the end members. Delta core sediments also show a fluctuation in the amount of mafic sediment that they receive over time.

### *Major and trace element geochemistry* (see *Fig. 11*)

Major and trace element geochemistry datasets can be used to better understand and even predict the composite mineralogy of a bedrock sample or to help shed light on sediment mixing from multiple source areas when studying detrital samples.

Ten major and minor elements are usually measured as percentage oxides: $SiO_2$, $TiO_2$, $Al_2O_3$, $Fe_2O_3$, MnO, MgO, CaO, $Na_2O$, $K_2O$ and $P_2O_5$, and 20 trace elements are measured at ppm levels: Rb, Sr, Y, Zr, Nb, Ba, Pb, Th, U, Sc, V, Cr, Co, Ni, Cu, Zn, Ga, Mo, As and S. These concentrations are normalized to standard trace element concentrations such as Post Archean Australian Shale (PAAS) compiled by McLennan and Taylor (1985) and for context and to allow comparisons of different datasets. V, Cr, Co, Ni and Cu concentrations from the trace analysis of detrital samples are of particular interest in S2S studies as their absence or presence indicates the quantity and location of mafic material within a catchment area. X-ray fluorescence (XRF) geochemistry analysis can also be used in sediment modeling, where the composite mineralogy of a sink sediment can be predicted using hinterland datasets.

## Building, testing and calibrating S2S models

The optimal application for geological models and maps that result from S2S analysis (see *products* in Fig. 5) is their integration into risk assessment workflows that guide exploration and basin characterization. In this context, the models produced will take all the available data, and with the help of a set of logical rules and processes, provide a set of predictions about the subsurface. This may be a prediction of the size of a submarine fan system in a yet un-surveyed area of the subsurface, or a set of predictions about the probable mineralogical assemblages expected in a given sediment package. In the latter case, mineral assemblage predictions may be used by petrophysicists or reservoir quality experts to assess probable porosity and permeability of reservoir intervals, giving a set of uncertainty parameters prior to drilling. This section focuses on the process of S2S modeling using physical system parameters and scaling those models into products useful for exploration.

### From the source to the shore

The Bengal Fan is one of the largest modern sediment deposits in the world, at over 1400 km (745 mi.) wide and up to 16.5 km (10.3 mi.) thick. The clastic material contained in the fan is derived from an area > 2 million square kilometers (> 772,200 square miles), crossing India, Bangladesh, Bhutan, Nepal and China. It is undoubtedly one of the most

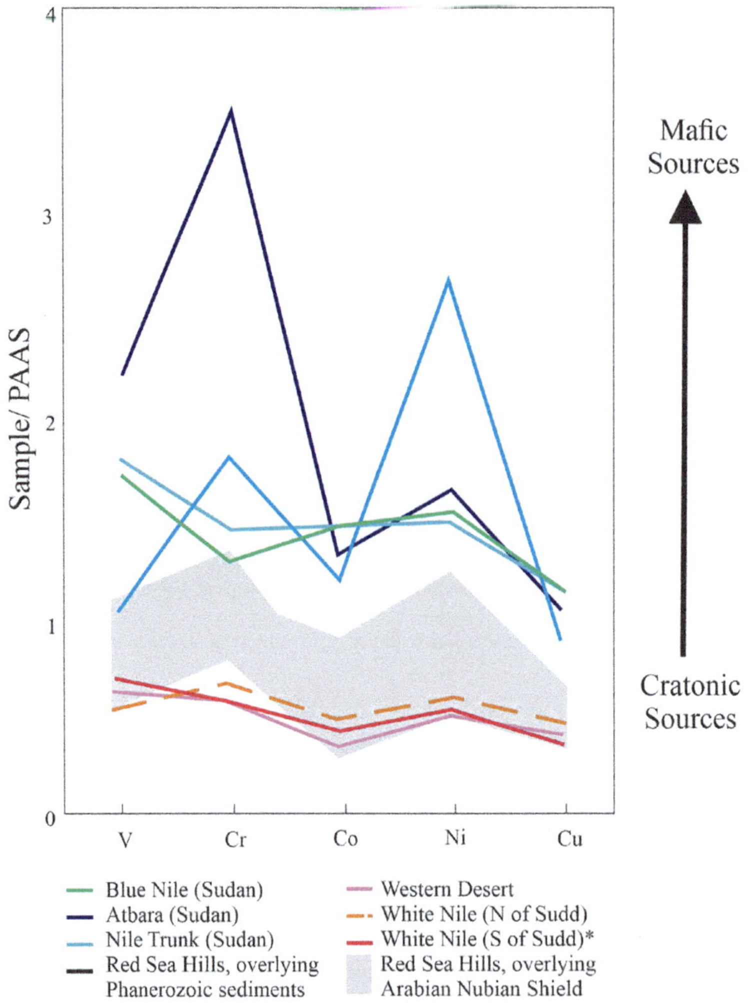


**FIG. 11** XRF data. An example of XRF data from modern river muds of the Nile River catchment, normalized to Post Archaean Australian Shale (PAAS: McLennan and Taylor, 1985). V, Cr, Co, Ni and Cu are shown here as increased amounts of these trace elements usually indicates a mafic provenance. This plot highlights the characterization of the cratonic (felsic) hinterland sources (i.e., Western Desert and White Nile) from the juvenile (mafic) hinterland sources (i.e., Blue Nile, Atbara). This plot also shows how underlying sediments can strongly control the provenance signature of a sediment if multiple episodes of recycling occur in an area. Here we see the Red Sea Hills provenance signature varying depending on whether it was overlying Phanerozoic or Arabian Nubian Shield sediments. *(Data from Fielding et al. (2017).)*

significant S2S systems available for study, but in many ways, it is an aberration. At 3000km (1158 mi.) long, the Bengal Fan is far longer than scaling relationships would classically predict. It contains a far larger proportion of the world's modern sediment generation than measurement of its drainage area alone would predict. It is however a brilliant illustration of why detailed and nuanced understanding of hinterland processes is key to predicting sedimentary intervals in the subsurface.

The Bengal Fan is the product of rivers draining the Himalaya mountain range, itself the product of one of the most violent tectonic collisions of the last hundred or so million years. Estimates of sediment generation indicate that 70% of the sediment in the Bengal Fan is derived from an area of possibly as low as 5% of the total drainage network of the Ganges and Brahmaputra rivers. The Bengal Fan is therefore a cautionary tale for any researcher studying ancient depositional systems. Forget to adequately model just one critical driver, be it climate, or tectonics, or geography, and the model will not just be wrong, but it will be *out of this world incorrect*. It also serves as a reminder that Earth can and does produce extraordinary phenomena. Modelers like to reduce uncertainty and take conservative guesses where data do not provide absolute inputs; the Bengal Fan demonstrates that sometimes a good model permits an extraordinary set of results without tending toward average outcomes.

## Drainage network analyses

One of the principal mechanisms for transporting sediment from mountain to shore is via rivers. Observations of streams from high-energy uplands to slow and languorous rivers meandering across floodplains follow a set of predictable patterns governed by basic hydrodynamic and physical principles. Strahler was one of the first to document river-length patterns, after whom is named the Strahler (1957) order, which orders streams according to their position in a drainage network. Fresh emerging streams are indexed as "1", and where two "1 s" meet a trunk river, a "2" is denoted. Two "2" rivers make a "3", but a "1" draining into a "3" does not increase its order (Fig. 12).

Other such patterns include bifurcation ratio, which places a value on the degree to which a basin is balanced in map view; longitudinal profiles, which show how far a river has incised into Earth's surface as it tends toward the most efficient profile and drainage density, which is a value of the total length of river channels per unit area across a basin.

These indices and values are important because they reveal how far a drainage system has moved toward equilibrium. Once a drainage system has reached full equilibrium, the amount of insight that can be gathered becomes limited. Measurements of the system can be used to build models back in time, but what is known about the system is that now *there is no signal remaining*, which reveals how the rivers may have looked in the past. Practitioners can identify discrepancies from equilibrium, which can be applied to understand how the drainage network has evolved through time.

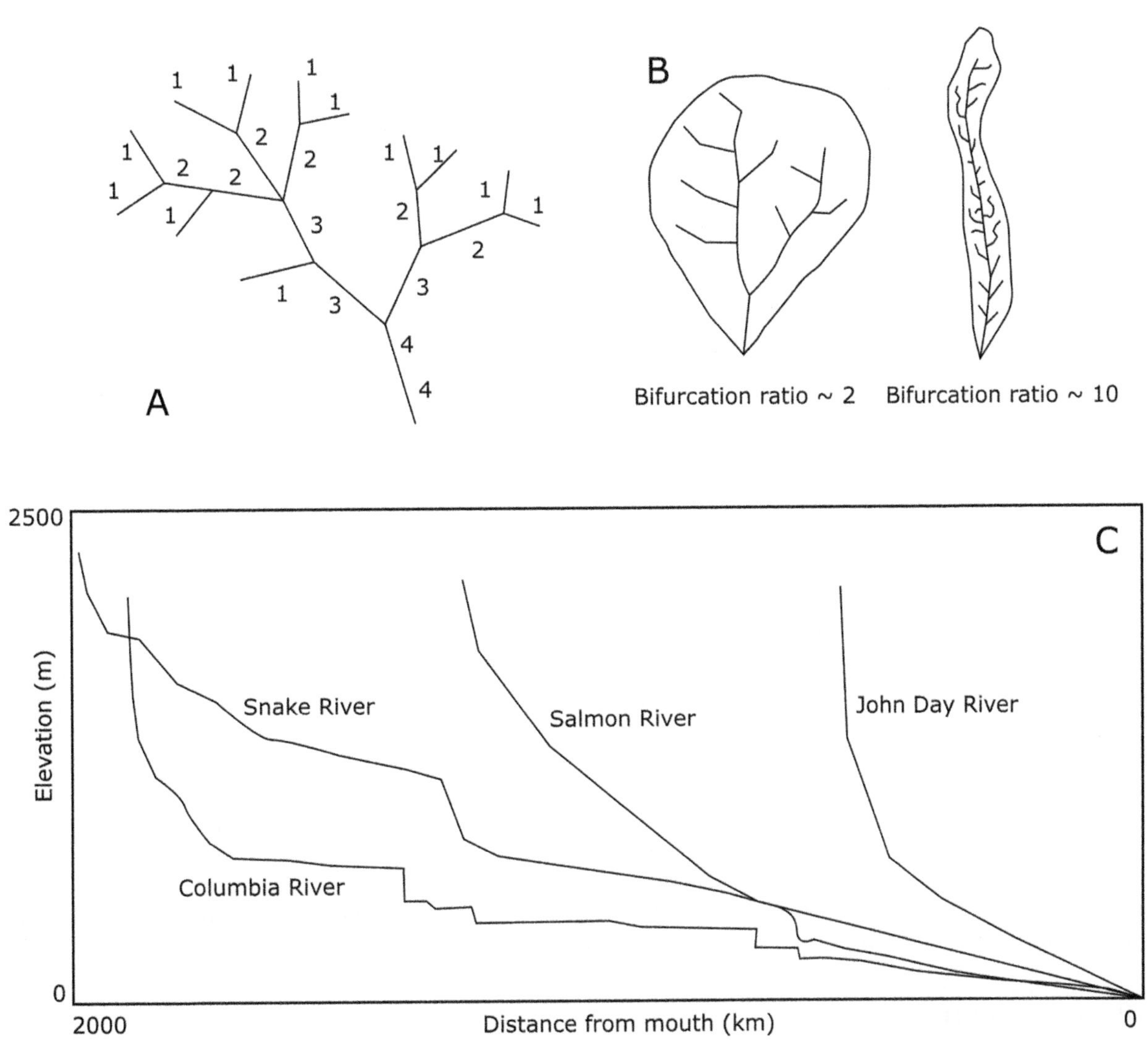


**FIG. 12** Drainage network analysis. (A) Strahler ordering; (B) illustration of different bifurcation ratios as a metric for parameterizing basin shape and tectonic influence; (C) different longitudinal profiles of rivers, note the mainstream Columbia River which resembles a power law trace whereas the John Day River shows a much steeper fall.

The goal of drainage network analysis is to create a conceptual model of how a drainage network looked in the past. If the topic is a Cretaceous submarine fan system, results may indicate the volume of the fan and the character of the sediment in the fan. Additional results would suggest the nature of the continental drainage system, and the presence of tectonic activity, both of which would exert control on sediment volume and caliber.

## *Longitudinal profiles and knickpoints*

When the longest continuous stream in a drainage basin is plotted with elevation on the *y*-axis and distance downstream on the *x*-axis, the longitudinal profile of how a river falls from its source all the way to its sink can be assessed. In this manner, scientists have been able to study how river systems evolve over time by comparing the profiles of streams on newly emerged volcanic islands with those in ancient cratonic environments like South Africa, Brazil or Australia. Over time, rivers cut down into bedrock pursuing the most efficient path toward a low energy state. In Fig. 12, this equilibrated state closely resembles the power law graph. Geology can conspire to upset this balance and the river system which has met or is approaching equilibrium. The profile may reveal kinks and drops in the river system. These are known as *knickpoints* and are signals of tectonic or climatic events, geologic boundaries or a mix of these and other factors.

A real-life example of a knickpoint is a waterfall. Imagine plotting a river sports enthusiast going over a waterfall. There will be a sudden drop in elevation for that person! Over time a waterfall will continue to erode into the country rock and cliff collapse will occur. This will result in the waterfall stepping back upstream and the knickpoint migrating as well. This means that in order to understand why a knickpoint like a waterfall has occurred, scientists need to look downstream for a cause. Sometimes, however, knickpoints are subtler and the change in elevation is more gradual and only becomes obvious when plotted in this way (*see* Fig. 12).

In any given depositional system, typically multiple longitudinal profiles can be plotted and interrogated to understand the evolution of a drainage network. The objective is to (i) highlight areas where there are knickpoints in the profiles and (ii) use geologic maps and an understanding of the tectonic regime to determine the cause of a given knickpoint. In one case, a waterfall may coincide with a change in lithology downstream. As a river passed from a harder lithology into softer rock, the river began to incise more deeply causing a change in elevation. Over time, this became a waterfall which then started to migrate upstream. Alternatively, a river gradually incising into the country rock may have been subject to localized uplift from compression in a foreland basin. The river kept pace with the uplift, but the resulting profile shows an artificial flattening of the river profile as it passes through this domain. In this latter example, it is particularly significant to the S2S model as areas with accelerated incision will be relatively over-represented in terms of the contribution of sediment from those country rocks to the sink. Longitudinal profiles therefore reveal an invaluable history of which parts of a drainage basin contribute the most sediment to the sink. Longitudinal profiles for all of the world's rivers can be used to estimate the relative contributions of different hinterland domains to the sediment observed in corresponding basins. This is the basic method researchers used to calculate that 70% of sediment in the Bengal Fan came from an area amounting to just 5% of the drainage basin.

### Strahler order and bifurcation ratio

As briefly mentioned earlier, the Strahler order is an index, which permits the organization of streams according to maturity and hierarchy in a river system. This is a useful tool as *lower order rivers* (Strahler 1–2) tend to incise more deeply into hinterlands, and are therefore more likely to be net contributors of sediment than *higher order streams*, which may be removing large volumes of sediment into floodplains and intermediate storage domains.

Strahler indices for streams can be used to calculate the bifurcation ratio, which is an attribute of the "balance" of a Strahler tree. While it has been subject to some criticism over the years due to the inability of bifurcation ratio to adequately account for tectonic or climatic drivers, it is in its simplest form a useful metric for representing the geographic characteristics of a basin. A basin with extreme bifurcation values will be skewed heavily toward either the incising or depositing stages of a fluvial system, and an understanding of this ratio as a first-look is a valuable means for determining how far a drainage network has moved toward a state of hydrodynamic equilibrium (Fig. 12).

### Wind gaps and water gaps

The drainage network is a representation of all of the streams and rivers draining a geographic area into a sediment sink. The edge of a drainage network, where it abuts neighboring networks is called a *drainage divide* and often follows tectonic and physiographic features like hills and mountain ranges as the streams try to move from higher to lower elevations. The shape of a drainage boundary is therefore often a product of the complex interplay of geology and climate. Artifacts in this relationship are wind gaps and water gaps.

A *wind gap* is so named because it is an area in a river system or fluvial valley in which air is found instead of water. Wind gaps occur when a river system, which has followed a path for a period of time, is suddenly changed, causing the path of the river to avulse. In a mature river system, an oxbow lake is an example of a wind gap after it has dried out. Previously, the river flowed along its path, but river avulsion took place causing the river to move to a new path, often leaving a river path, which eventually dries out.

Higher up in a fluvial system a wind gap can occur when a period of localized uplift causes a river to change its path. Where previously a stream flowed straight down a hillside toward the coast, a rapidly uplifting zone redirects the path toward a new downhill trajectory. The result is a wind gap observable in the hillside, which preserves the information that a river used to flow there. This information is significant as it can reveal the arrangement of drainage networks in the past. If researchers are investigating sand packages generated prior to and during this reorganization, then it indicates that the river system passed through this area. Further, for a period prior to the river avulsion, an intense period of incision occurred, likely resulting in over-sampling sediment generated from the country rock in this particular location.

### Drainage density

Drainage density ($D_d$) is defined as the total length of surface streams in a drainage basin divided by the area. Drainage density is a product of surface run-off and will tend toward a central value for all basins as equilibrium is reached. The central value is determined by (i) the amount of water in the system falling as precipitation and (ii) the permeability of the country rock or soils. In a limestone terrane, the drainage density will be artificially lowered as rivers cut into the substrate forming subterranean rivers, conversely, very wet basins with high degree of precipitation will have higher drainage densities. Drainage density is associated with "flashiness" or the abruptness of high flow rates or discharge juxtaposed with very low rates of flow. This is a measurement of the variability in surface run-off and a key driver for flooding events. Large water throughput and flooding are both drivers for increased sediment transport. When modeling a drainage system, higher drainage density is associated with a greater volume of sediment deposition in the sink.

When uncovering information about the history of a drainage basin, evidence of previous drainage density is useful if it can be found. Examples are ancient dry riverbeds and evidence in landscapes of incision surfaces or peneplains.

When creating models for sediment generation, it is therefore important to account for drainage density to predict sediment throughput. In the case of the Bengal Fan model, huge precipitation variability creates a high drainage density. As a result, large sediment volumes beyond those expected in a conventional scaling model are produced. Estimating Drainage density in a model scenario is therefore critical to making more accurate predictions.

### A workflow for conducting drainage network analysis

1. Derive modern drainage network from modern maps and programmatically using hydrology tools such as Hydrology toolset in ESRI ArcGIS mapping tools. The rivers visible today in maps and via satellite imagery are the product of the entire history of their development and the geology; programmatically derived drainage is the result of a model that takes the shortest paths from elevation data. Contrasting these drainage morphologies provides a fast filter for wind gaps and other artifacts, which reflect paleo-drainage systems.
2. Define key river systems and calculate Strahler orders for the system. Compare Strahler orders and bifurcation ratios to determine the degree to which a drainage system has achieved a state of balance. Systems that have not reached equilibrium record prior drainage areas and give clues as to paleo-drainage outlines.
3. Plot longitudinal profiles. Highlight knickpoints and departures from baseline onto geologic maps to determine lithological derived features such as changing hardness (and therefore erodibility) across rock types, and tectonic knickpoints, which may indicate a previous watershed boundary or river capture.
4. Using wind gaps and knickpoints of known ages (cross-referenced with structural or exhumation histories) modify modern drainage networks. Bring together structural data and drainage observations to build a paleo-drainage outline for a particular period in Earth history.

## *Paleo-Earth Systems Models as a means for predicting first-order drainage areas and their application to scaling relationships*

### What is a Paleo-Earth Systems Model (PESM)

A Paleo-Earth Systems Model (PESM) is for many the end product of a S2S workflow, a model that represents all terrestrial and hydrosphere interactions at a given point in time, and from that allows modeling of sediment build-up. PESM in their

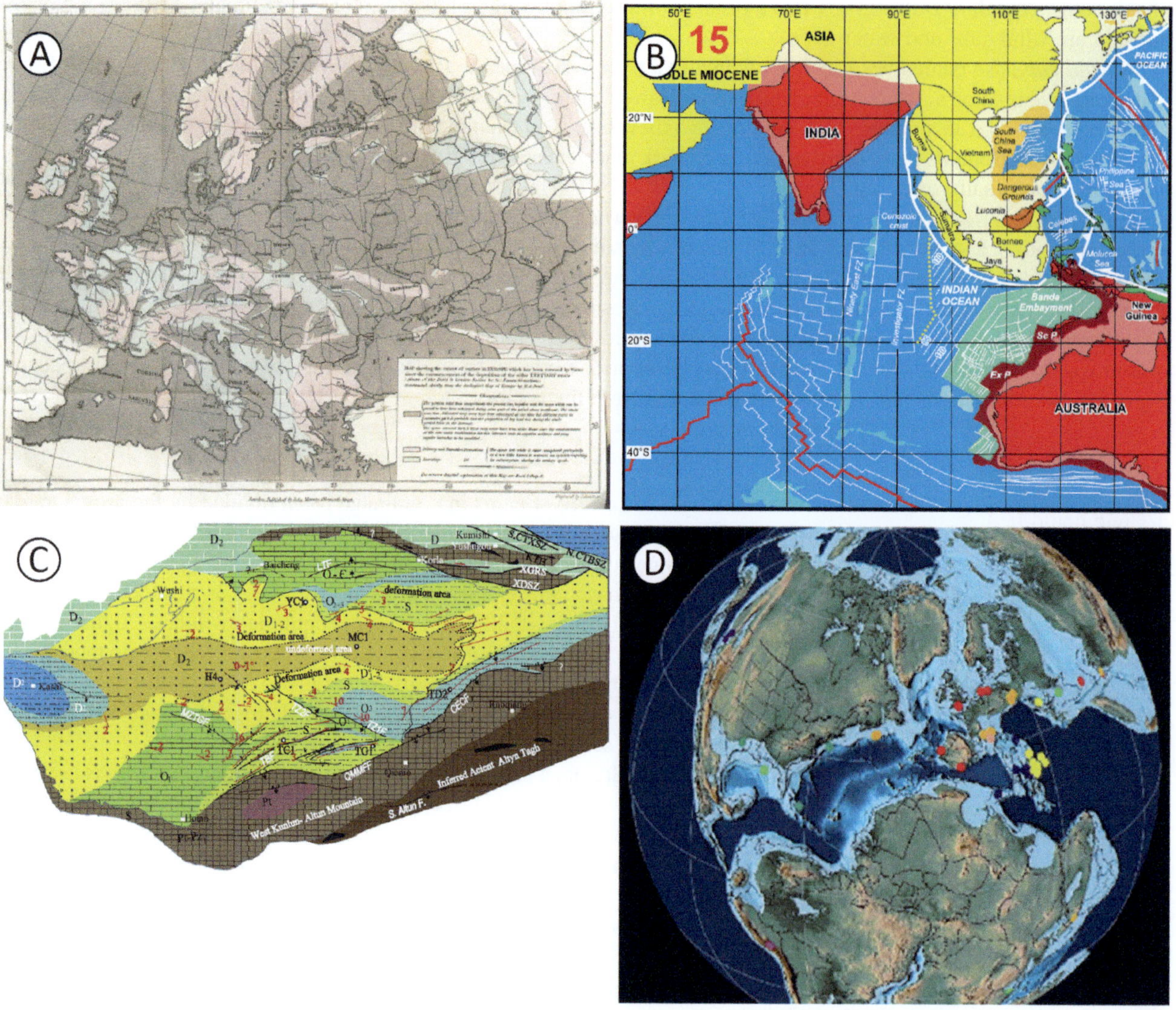


**FIG. 13** PESM through time, including simple paleo-geographic reconstructions to modern coupled Earth-atmosphere-ocean models. (A) Charles Lyell's paleogeographic map of NW Europe. (B) Reconstruction of the Indonesian and Indian Ocean regions at 15 Ma ((Hall, 2012). (C) Paleogeographic reconstruction from the Tarim basin prior to Late Devonian sedimentation (He et al., 2016). (D) Screenshot of PALEOMAP from the EarthByte group.

current form were created in order to model source rock facies for hydrocarbon exploration but have been utilized in various forms for decades by geologists interested in reconstructing Earth's surface as a means for exploration and understanding (Fig. 13).

By virtue of the amount of work which goes into paleo-Earth systems modeling, these tools are often offered to the hydrocarbon industry by specialist research groups and consultants who produce global models and sell nonexclusive licenses to results either on a consortium basis or as subscription. For a generalized and global picture, global models can be the most robust, but for specific regions like southeastern Asia, Mexico or northern Africa, the most robust and thorough models are often produced by specialist research groups who have spent many decades collecting data and improving regional models without the need to make their models "edge-fit" with surrounding regions models. While the lack of fit with surrounding models may seem a drawback to regional models, it often results in pictures that honor all of the data with less geologic information ignored in order to fit a globally consistent plate model. The result is a model that probably comes closer to the truth, as can be evidenced by scrutinizing a global model in parts of the world where various artifacts of "edge-fit" models have to accommodate inconsistencies and prevent the entire model from breaking. Regions that are an excellent bellwether for the overall veracity of the global model include the Russian far east (Okhotsk Sea), Southeast Asia (Sulu Sea, West Papua, Sulawesi and Timor), Antarctica (Palmer Land and Weddell Sea), Arctic (specifically the Kara-Laptev Sea region) and Mexico-Guatemala-Honduras. These regions often have less primary geologic data, can be at high latitudes or are on an Euler pole to a region with significant deformation (plate kinematic theory is beyond the scope of this textbook, but for general purposes, an Euler pole represents a theoretical line about which a deforming plate is rotating and it defines two points where it intersects with the earth's surface where deformation stops; these points are often hard to define accurately and in many plate models are where a lot of skeletons are hidden) and lots of data, and/or are subject to less scrutiny due to their being comparatively less attractive for hydrocarbon exploration by key clients or sponsors. Plate kinematic theory is beyond the scope of this book, but for general purposes, an Euler pole represents a theoretical line about which a deforming plate is rotating. It defines two points where it intersects with Earth's surface and where deformation stops; these points are often hard to define accurately. As a result, "skeletons" of imperfect models are hidden away in these regions, invoking improbable strike-slip systems with many hundreds of kilometers throw (e.g., Sorong Fault Zone), contentious plate fragments, which appear with no geologically meaningful explanation for their origin or demise (e.g., Sulu Sea), spinning or backstepping plate motion vectors (when plate motions are animated and viewed as a movie stepping forward in time) and/or models that completely disconnect with local geology and the rocks (e.g., Sea of Okhotsk).

Many of these paleo-geographic inconsistencies are ignored and brushed aside by users in order to not let perfection be the enemy of performance, but their presence should always be noted. In Asia, the presence of one of the largest such plate model problems was the the Sorong Fault Zone, invoked by some tectonics experts as *the bacon slicer*—a strike-slip system with throw of upwards of a thousand kilometers, which displaced tiny fragments of continental crust westwards. Inevitably, this interpretation has been proven incorrect. While it may now seem farcical to believe that continental fragments the size of Crete could be translated many hundreds of kilometers with no apparent damage to the country rock nor to the surrounding oceanic crust, the study of this system was responsible for solving not just the mystery of the Sorong Fault Zone, but also shed new light on the principal mechanisms of deformation across much of the Southeast Asian archipelago and the role of extensional tectonics in a region historically interpreted by compression from east to west (*see* Fig. 13).

## Building a PSEM

***Plate models and tectonic systems visualization*** Plate models are regional or often global structural models, which delineate contiguous crustal fragments and plot their deformation and relative movement in geographic space through time. They are useful because a working plate model will reveal the spatial relationships between an area of interest and other crustal fragments, and the climatic zone or zones.

The basic process for developing a new plate model is to fix a reference frame based on today's model for the plate configuration of an area. Then, based on data such as structural and tectonic notions, paleo-magnetic reversal stripes in oceanic crust, and geochronological data, construct a picture of what Earth looked like in a time period of interest. It is of primary significance to construct a map that reflects Earth's configuration during the time period of interest. Therefore, rather than define a map in today's plate configuration, researchers want to rewind time to the relevant period or stage and show how the rocks related at that point in "paleo-space", or when the continents, plate, mountains and basins were in their previous arrangement. As a general rule of thumb, working from a blank canvas is not advised as the volume of data missed is likely to be large. GPlates from the EarthByte group (https://www.earthbyte.org/) remains one of the world-leading free resources for starting to build a plate model. GPlates is standalone software and is also accessible via a web portal. The EarthByte team have been producing global plate models for decades and are constantly improving the published models. This can be layered with geologic data in the form of rock data from wells and/or outcrops with age and lithological information. Errors or inconsistencies in the area of interest can be solved and the model can be adapted in the software or by exporting a "timeslice" to a GIS platform for amendment.

Once into paleo-space, and once happy that the plate model honors the data (and for basic or first attempt models, it is advised to work with the EarthByte model until the practitioner is positive it needs altering), the next task is to start to reconstruct paleo-geology.

***Geology and paleo-geology*** Today, researchers can access near contiguous geologic maps for most of Earth. If a researcher goes back to the Miocene, rotates the geologic map and deletes any rocks which postdate the Miocene, the researcher can create a partial geologic map of the world. A Cretaceous rock for example, which can be seen on a modern geologic map, would have existed in the Miocene (though it may not have been in the form of an outcrop), and so the researcher can start to try and fill in the gaps based on what is known about the intervening time. For example, it is known that in the last 5 million years, the Nazca Plate has been subducting under the South America Plate, and that the Australian Plate has continued to collide with the Eurasian Plate. In both of these cases, it can be explained where the missing rocks have gone, and infer their compositions (in the former case, largely homogenous oceanic crust with pelagic sediment, in the latter, a mix of crust types based on the location in the collision). Making some educated guesses about geological development, the researcher can start to build a picture of the paleo-geology and the basic structure of what Earth looked like in the time period of interest.

A key stage when reconstructing paleo-geology is considering thermochronological data, which provide information about the cooling history of rocks and therefore which parts of the hinterland were emerging at a given time (see section "Thermochronology"). Detailed (U-Th)/He or fission-track thermochronology can provide absolute exhumation rates for areas of the hinterland that were uplifting, weathering, and eroding, and therefore contributing the most sediment. In the Himalayas today, nearly 50% of the sediment delivered is being generated by an area accounting for less than 2% of the geographic area. Understanding where the rocks are moving reveals where sediment is being formed due to erosion.

***Paleo-geography*** The next step is to move from paleo-geological understanding and transpose this into a representation of what the climatic and depositional environments looked like at this time, or the paleo-geography. By taking all the rock data and plotting them onto a map in paleo-space, a physiographic map can be developed, which delineates areas of high topography (perhaps indicated by i-type arc-volcano igneous deposits of the corresponding age) where rivers were incising, terrestrial depositional environments (for example lacustrine sediments), shelf and slope environments and relative climatic variations across the area of interest. Once all of the data have been input, an at least partial picture should emerge of what the area looked like as if viewed from space. Building a paleo-geographic map is undoubtedly the most complex, time-consuming and important phase of arriving at a PESM, and is invaluable in itself as an exercise. If the distribution of any petroleum play element is being considered, then a review of the paleo-geography is a fundamental (though all too commonly overlooked) process which any petroleum geologist should undertake.

Once paleo-geography is created, knowledge of modern systems can be applied to create a rudimentary topographic map for the timeslice. Using rules of thumb such as the mean height above sea level most arc-volcano systems emerge, or the mean water depth of carbonate reefs, it is possible to build a topographic map for the area of interest, which honors the geologic data and inferences made about paleo-geography.

***Climate and drainage*** If paleo-topography and paleo-bathymetry are available for the entire planet during the time period of interest, then it is possible to run a climate model such as the Hadley Center Coupled Ocean Atmosphere General Circulation Model (HadCM3; Collins et al., 2001). The HadCM3 provides a grid of data for every day of a "model" year with a graticule of 1.25 × 1.25 degree grids split into 20 elevation levels. The resulting outputs can be used to predict seasonality of climate, storminess and the propensity for climatic features such as monsoon, as well as mean temperatures and monthly rainfall estimates. The data can be used by the petroleum geologist to provide boundary conditions for subsequent models or interpretations.

Alternatively, climate information can be gathered directly from the geologic record from, for example, biostratigraphic data from known occurrences of climate-specific species, and from the occurrence of sedimentary products in the rock record. These data reveal less specific information than the data of mean monthly or annualized rainfall, but they have the added veracity of being based on primary observations and therefore not the product of a model that itself is the product of prior models.

With a paleo-topography and paleoclimate model, it is possible to use modeling software such as the Hydrology toolset in ESRI ArcGIS to build a model for an onshore river system. The drainage system built will give a prediction for which geographic areas were draining into the sink at the relevant time period and effectively define the geography of the source area. It is therefore a very useful output and can be further refined by integrating the observations and revelations made while conducting drainage network analysis.

***Summary*** Drainage network analysis, plate modeling, paleo-geologic and paleo-geographic reconstructions and paleo-climate modeling can provide a broad array of data, which help to understand the world when clastic sedimentary systems were eroding, transporting and depositing sediment. Each of these processes are models in of themselves and their outputs can be useful for S2S modeling, which uses assumptions made about Earth in the past to predict volume, form and composition of sediment bodies in the subsurface.

- □ **Plate model and paleo-geologic model**: Representation in map view of where crustal fragments were during the time period of interest. geologic information such as emergent hinterland areas and the composition of those rocks.
- □ **Paleo-geographic model**: Description of the erosional and depositional environments in paleo-space.
- □ **Paleo-drainage reconstruction**: A description of fluvial systems, which were present over the area of interest during target time interval.
- □ **Paleoclimate**: Either a model output or climatic conditions derived from known living habits of flora and fauna found in the rock record.

## Building landward sediment models

### Nonnumeric modeling and by-the-eye approximation

Once paleo-geography and paleo-geology have been produced, S2S practitioners often build conceptual models for sediment generation based upon rules of thumb and their experience as geologists. While these interpretations are often referred to as "models", they are approximate at best and do not constitute modeling in any meaningful way due to the preponderance of guesstimates, generalizations and analog-guided thinking that go into them. The workflow is as follows: 1. Paleo-geography and paleo-drainage definition. 2. Paleo-geology definition. 3. Prediction of sand magnitude and direction based on rules such as "Basaltic hinterlands create muddy feldspathic sands", and "Granite hinterlands produce moderate volumes of clean quartzose sands", and traffic-light style estimates for good-, intermediate- and poor-quality clastic units. In better instances, an experienced geologist may invoke their experience of similar systems and data from wells in those systems; while this always produces geologically meaningful results, the complexity of sediment routing systems makes the application of analogs dubious at best.

As a rule, if a reasonable quantitative estimate (complete with error margins) cannot be put on a prediction about the subsurface, the "prediction" is operationally of little more value than an educated guess (Tetlock and Gardner, 2016). Unpublished experiments on the effectiveness of un-calibrated guesses in subsurface workflows studied the success rate of historic predicted petroleum prospects, comparing numerically risked versus qualitative hydrocarbon prospects. Results indicated that prospects on which quantitative estimates were placed showed a better chance of being correct than randomly assigned results; conversely, prospects which had been assigned but with no uncertainty associated performed no better than random model. In these experiments, prospects generated with no uncertainties assigned showed only a relatively trivial benefit between guesses of highly experienced geologists and those of relatively new petroleum geologists (postgraduate career $<5$ years).

## Scaling relationships

Sømme et al. (2009) published the first set of observations, which related drainage networks in the hinterland with submarine fan geometries offshore. This linear regression has been built upon by various other authors and has in some cases been paired with paleo-drainage reconstructions to predict submarine fans in the subsurface. The general equation for modeling most commonly used is as follows:

$$Fl = 2.4199 Rl^{0.6713}$$

where $Fl$ is Fan length and $Rl$ is the length of the longest river in the drainage.

While the application of this equation has been used widely (in no small part due to the simplicity of the model and the ease with which anyone can apply it and acquire model "solutions" to a system), extreme care should be taken. The equation is a linear best fit to data, which represent a small part of Earth's history and are by their very existence in the geologic record anomalies and hugely impacted by preservation bias. This equation will incorrectly predict the sizes of the Bengal and Nile River fans systems by many orders of magnitude (two of the most significant modern fan systems on Earth) and was not designed by the authors as a means for calculating fan sizes in paleo-systems so much as a series of observations about the systems under study. A notable feature of this model is that the addition of new data to the database used to generate this equation by Davies et al. (2020) found that the linear model became less robust with reduced statistical significance.

## Mass balance studies

Mass balance studies are a novel and robust means for making quantitative predictions about the subsurface and take inspiration from the practice of double-entry accounting in the financial industry. In accounting, every transaction is entered into two accounts which must always equal; this allows for simple auditing as a correctly balanced account should always be equal, with credits and debits equaling. Similarly, in mass balancing, any sediment seen in the subsurface (sink) must be accountable from a corresponding hinterland (source). Practitioners will produce a numerical model for sediment volume over a given period, and then using seismic volumes compare with the volume of sediment actually present. This method has benefits as it allows practitioners to iterate and improve models over and over in order to arrive at an acceptable account on the source side. Once this answer has been arrived at, the other outputs can be trusted as relatively sensible due to the confidence granted by double-accounting. A general workflow for mass balance studies is as follows:

1. (Paleo-) geologic maps grouped and coded for mineralogical domains, e.g., granite types in an igneous hinterland domain are grouped together under "granitic hinterland zone 1", etc.
2. Paleoclimate models studied to derive maximum, minimum and median rainfall and temperature data.
3. Paleo-drainage analysis utilized to define source regions.

4. Thermochronology data studied to account for exhumation rates across hinterland and therefore volume of crust converted into sediment.
5. geologic domains intersected with exhumation rates in order to calculate volume of "granitic sand" versus "recycled sand" or "basaltic sand".
6. Total sediment mass calculated from hinterland domains.
7. Total sediment calculated is compared with sediment estimates from depth-migrated seismic data.
8. Revise and iterate until the result of step 7 is acceptable.
9. Review sediment composition data from step 5 utilized for subsurface predictions.

### Numerical modeling of clastic bodies

A new and exciting form of numerical modeling has been devised recently, which utilizes modern large datasets available freely from published and universally accessible datasets collected worldwide. This method reduces the reliance on generalizations and places greater significance on making use of data and minimizing assumptions, which have to be made by the S2S practitioner.

Step 1: Paleo-geography and paleo-drainage. Building a paleo-geographic map of the area of interest allows for the extrapolation between data points presented by rock and core data preserved to the modern day. Utilizing this data as a starting grid and using known relationships from present day geography, such as the knowledge that between a deepwater sandstone and a terrestrial paleosol, there must have been coastal depositional environments, allows a robust and meaningful model to be generated upon which other suppositions may be made, such as paleo-drainage models, which will provide a network from which the practitioner can begin to characterize the catchment areas of sediments.

Step 2: Hinterland characterization. Intersect paleo-drainage with overlapping inorganic geochemical data for area of interest. Filter and remove all rock data which postdates the time period of interest. Using a normative calculation such as CIPW norm calculation for igneous rocks, calculate the bulk mineralogical composition of the hinterland domain.

Step 3: Paleo-geography and paleoclimate inputs. Using paleo-drainage outputs, estimate the longest river length for onshore clastic systems. Intersect global biostratigraphic species occurrences (e.g., www.paleobiodb.org) to determine climatic conditions.

Step 4: Sediment modeling. Using the generalized equation below, it is possible to estimate the mineralogy of sediment generated in a given basin or region.

$$Ms \simeq Mh^*(C\,lim\,\Delta^*Tec\Delta^*Geog\Delta) \pm x$$

where *Ms* is mineralogical composition of the hinterland (source), *Mh* is mineralogical composition of hinterland, $Clim\Delta$ is chemical weathering rate expressed as a function of climatic variance, $Tec\Delta$ is the volume of sediment generated by tectonic exhumation and/or dynamic topography, $Geog\Delta$ is the relative size of the subaerial fluvial system and $x$ is geologic factors which may serve to speed up or alter normal weathering processes, for example climatic conditions during the Great Oxygenation Event at 2.4–2.0 Ga, which resulted in move from a mildly reducing environment to strongly oxidizing, drastically altering chemical weathering pathways before and after.

Many of the factors in this equation must be estimated, although work by Davies and Fielding demonstrates that relatively little specificity in tectonic and geographic factors (e.g., high/medium/low exhumation rates) is required to produce robust and precise predictions in modern systems and therefore a repeatable mathematical model which can be used in paleosystems. This is of particular benefit as metaanalysis of thermochronology data indicates that data may be over-interpreted by inexpert analysts, to the degree of deriving precise but wildly inaccurate estimates for exhumed crust.

The role of physical weathering is considered to have relatively little impact on the chemical (and therefore mineralogical) composition of sediments in bulk calculations. Physical processes have a large impact on mineral sorting at the scale of outcrops or members, with heavy minerals for example concentrated in depositional facies of fluvial sediment based on water velocity.

## Key messages on case studies and practical applications

This section introduces a series of case studies that demonstrate resource evaluation in basins of different exploration maturity—from mature basin characterization to frontier exploration and play extension to white space exploration—and each section draws from published articles and data resources that are framed within the theme. S2S in each of these settings requires unique problem-solving techniques to overcome data gaps and significant exploration risk uncertainties. Case studies include: (1) a sediment provenance study of Oligocene-to-recent-aged strata in the Nile S2S system using single-grain zircon U-Pb-Hf geochronology and whole-rock radiogenic isotope (Sr/Nd) techniques (among others) to demonstrate the control that basement lithology, sedimentary recycling and modern geomorphological features have in determining the composition of sediment reaching a continental-scale depocenter; (2) a multidisciplinary study in modern-day Myanmar that uses several geochronometers in tandem with fission-track thermochronology and petrography to explore the effects of orogenic onset and continental-scale drainage network reorganization on deepwater deposits in a region with a complex tectonic framework; (3) an evaluation of paleo-catchment evolution, sediment routing and clastic sediment delivery from watersheds along the Borneo margin in search of valuable elements and minerals.

## S2S in mature basins

One part of the world that has benefitted from many years of sediment provenance studies is the Oligocene to recent sedimentation, offshore Nile Delta cone (Fig. 14). Single grain (e.g., zircon) and bulk rock analyses of both the Nile source and sink demonstrate the role that basement lithology, sedimentary recycling and modern geomorphological features can play on the composition of sediment reaching a continental-scale depocenter. The vast number of wells drilled in the area means that there is an abundance of sink data, which can be used to ground truth S2S sediment dispersal predictions. Recent studies such as Fielding et al. (2017, 2018) have been able to use some of this data to shed light on the evolution of one of the longest rivers in the world and its influence on reservoir quality in the Nile Delta cone over time. One of the main considerations in this particular S2S study was to determine if there has been a connection between the Blue Nile and the main Nile since the Oligocene, and if there was, how has the contribution of volcanic sediment eroded from the Ethiopian continental flood basalts affected the reservoir quality of the Nile Delta.

The study by Fielding et al. (2018) used a multitechnique approach and carried out U/Pb zircon and Hf analyzes, whole-rock radiogenic isotope analysis (Sr/Nd), XRF mineral geochemistry and heavy mineral and petrographic analysis. Some of these techniques played a larger part in addressing the Blue Nile connection with the Nile Cone sands than others. Historically, most provenance studies carried out on the Nile Delta up until recently have relied primarily on heavy mineral, petrography, landform and radar analyses (e.g., Burke and Wells, 1989; McCauley et al., 1986). In these instances, heavy

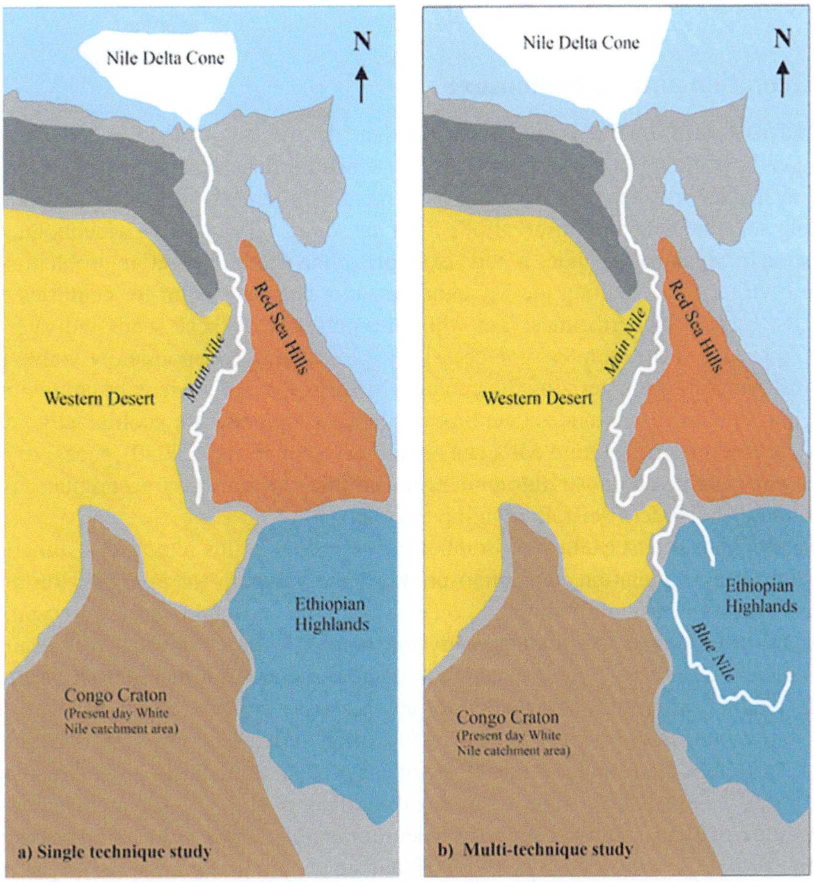


**FIG. 14** Nile River case study. Two different paleo-geographic reconstructions for the Nile drainage basin. Study (A) shows possible paleo-drainage outcomes when using a single technique. Study (B) shows a different drainage configuration as a result of a multitechnique study. The difference between the two systems means that areas of high reservoir potential could be overlooked if only a single provenance technique was used to reconstruct the drainage history.

mineral analysis of Nile Delta sediment showed no evidence of a connection with the Ethiopian Highlands, and landform and radar analyses have only been carried out on a small area of Egypt and not further afield.

U/Pb zircon analyses were key in Fielding et al. (2018) in that it showed a small zircon population that could have only been derived from the Ethiopian Highlands was present in all the Nile Delta core samples analyzed as far back as the Oligocene (Fig. 6). Despite being able to match both the zircon U/Pb age, and Hf content of the delta core zircons to rhyolitic zircons proximal to Lake Tana, the statistically small number of grains found in the core samples meant that more data were needed to make a more robust conclusion about the connection of the Blue Nile to the Nile Delta cone sediment in the Oligocene and onwards. When paired with additional datasets, the argument for an Ethiopian sediment source became much more apparent. Sr/Nd analysis of Nile Delta muddy intervals strongly showed that a highly mafic sediment source has been present in the Nile Delta system since the Oligocene (Fig. 10). After a thorough characterization of the hinterland and comparing the delta core samples to multiple possible end members, it was shown that the mafic source could have only come from the Ethiopian Highlands and a paleo-Blue Nile. Not only did this relationship agree with the U/Pb zircon dataset and show that the Ethiopian continental flood basalts had been contributing volcanic sediment to the delta since the Oligocene, but it was also able to show how sediment delivery from this area fluctuated over time. Further, these types of datasets can be used to help guide which parts of the stratigraphy may be more problematic for reservoir quality due to their volcanic lithic content than others. Sr/Nd data (Fig. 10) show how the Oligocene and Pliocene have a lower component of mafic detritus than the Miocene and Pleistocene. A recent Oligocene-age gas discovery in the West Nile Delta (Hoadar Discovery) shows how understanding the fluctuating contribution of different parts of the hinterland helps to better predict reservoir quality.

Fielding et al. (2018) highlight the importance of a multidisciplinary approach. If zircon analysis had been used as a standalone technique, then a more tentative interpretation would have been placed on the final conclusion. When used in conjunction with another technique that is also unaffected by weathering and postburial diagenesis, this interpretation became solid and actionable. Using heavy minerals alone would have led to a completely different interpretation due to the effects of diagenetic overprinting (Garzanti et al., 2018; *see* Fig. 9) masking any sediment contributions from the Ethiopian Highlands.

## S2S for frontier exploration and play extension

Frontier exploration and play extension often occur in basins where data are available from both the source and sink regions, albeit sometimes to greatly differing degrees. It is therefore essential that S2S studies in these regions use data from whatever sources are available, whether it be direct measurement or inferences drawn from modeling other system elements, because yielding accurate data and observations from any or all S2S system elements may be very challenging.

Petroleum exploration in Myanmar provides a good example of the need for creative problem-solving in S2S analysis. Petroleum seeps have been recognized along the Myanmar onshore coastal margin for centuries and are often exploited through hand-dug wells, even in modern times. Yet while the presence of these seeps indicates working hydrocarbon systems in the region, there are few examples of successful exploration for economically viable petroleum resources in offshore areas of interest. Further, characterizing the geologic history of the offshore is fraught with challenges in this tectonically dynamic region. Challenges include accounting for the development of a geologically young mountain belt, the Indo-Burman Range, that emerged at the Oligo-Miocene boundary (Najman et al., 2020), access to robust geologic outcrop and borehole samples with reliable chronostratigraphic ages and thick sedimentary accumulations in the modern Bay of Bengal that often obscure even modern seismic imaging techniques.

Zhang et al. (2019) provide a robust example of applied S2S research in this region and, although their research is not driven expressly by petroleum exploration, their map products are valuable for inferring the most-likely depositional timing, location and extent, and petrological composition of potentially favorable clastic reservoirs, the source rocks that may charge them, and the sealing units that are necessary for trapping hydrocarbons in the subsurface. Their multidisciplinary study in modern-day Myanmar uses several geochronometers (detrital zircon U-Pb-Hf, rutile U-Pb and mica Ar-Ar) in tandem with detrital zircon fission-track thermochronology, bulk rock Sr-Nd isotope geochemistry and petrography to deduce the paleo-drainage evolution of several large rivers that drain varied sediment source terranes in Myanmar and the eastern India-Asia collision zone. By using multiproxy sediment provenance control and extrapolating their observations from sample sites to other time-correlative strata in the region, Zhang et al. (2019) created a series of regional-scale restorations for three key time steps that show changes in paleo-drainage network configurations through time (Fig. 15).

One area of frontier petroleum exploration offshore Myanmar occurs in the Bay of Bengal, southwest of the Indo-Burman Range (IBR). Potential conventional reservoir targets lie at many different stratigraphic intervals due to the long history of clastic sedimentation in this region and this discussion will focus on the time steps illustrated Fig. 15.

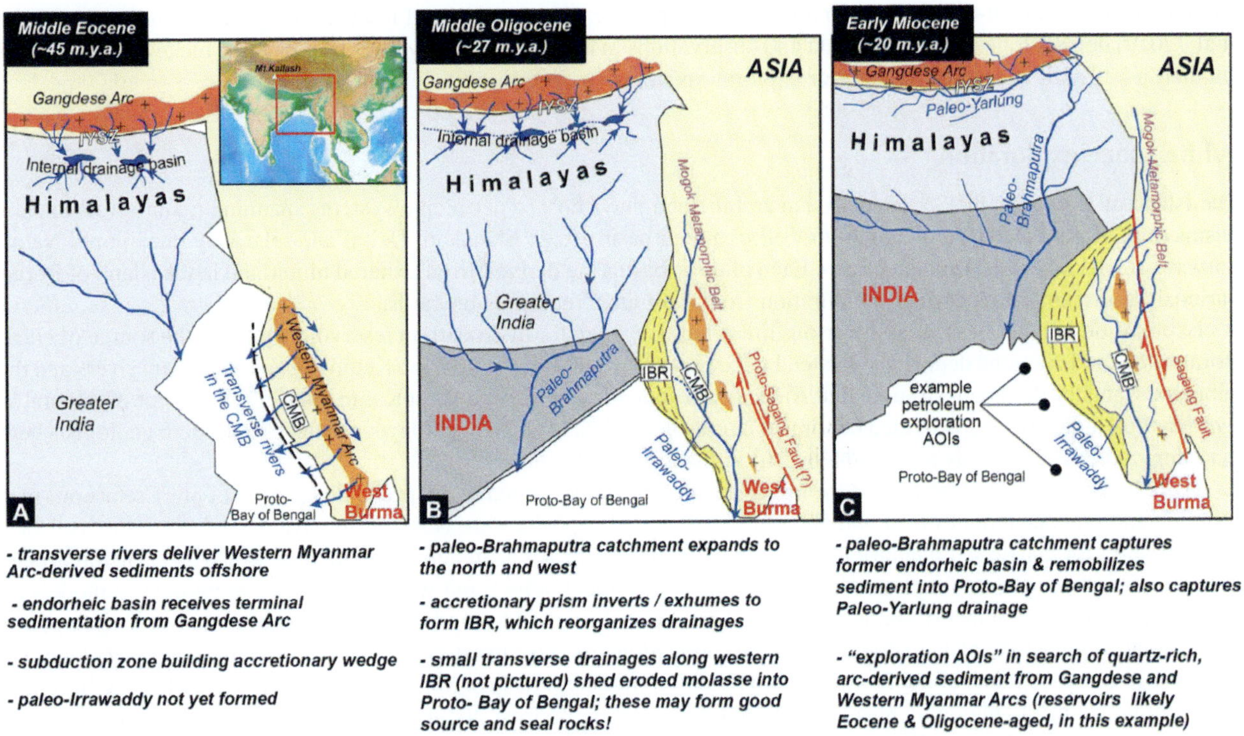


**FIG. 15** Myanmar case study. A series of schematic, paleo-geographic restorations of the eastern Himalayan-Tibet and modern-day Myanmar regions illustrates the development of this margin from the Middle Eocene to Early Miocene. *(Modified from Zhang et al. (2019).)*

Explorationists in this region may search for siliciclastic reservoirs that formed when quartz-rich, arc-derived sediment from the Gangdese and Western Myanmar Arcs was deposited offshore. Because of the complex and dynamic tectonic setting here, it is essential to understand when and where those drainages developed and these schematic reconstructions of paleo-drainage scenarios—modeled entirely on empirical data—show how this happened.

During the mid-Eocene (Fig. 15A), transverse rivers in the Central Myanmar Basin (CMB) delivered Western Myanmar Arc-derived sediment offshore and partially into the accretionary prism that will become the IBR. The Gangdese Arc, formed from the collision of proto-India and Asia, shed sediment southward into an endorheic basin south of the Indus-Yarlung Suture Zone (IYSZ) where it remains sequestered for some time. The paleo-Irrawaddy was not yet formed. This time step is important to understand since it sets the stage for later events but the deepwater strata developed at this time are eventually consumed into the active subduction zone. In the mid-Oligocene (Fig. 15B), the paleo-Brahmaputra catchment expands to the north and west and, due to its increased size, it likely begins developing a defined submarine canyon that feeds sediment into the proto-Bay of Bengal. Small transverse drainages along the western flanks of the nascent IBR (not pictured) shed eroded molasses, which may form high-quality source and seal rocks, into the proto-Bay of Bengal. By the Oligo-Miocene boundary, the accretionary wedge is partially inverted with the exhumation of the IBR (Najman et al., 2020), which reorganizes drainages in modern-day Myanmar and cuts off the potentially quartz-rich sediment sources of the former Western Myanmar Arc. Transitioning into the early Miocene (Fig. 15C), the paleo-Brahmaputra catchment incorporates the paleo-Yarlung drainage (and its arc-derived sediment) and remobilizes Eocene-Oligocene-aged sediment from the former endorheic basin. This is likely a significant time of clastic reservoir development given the quartz-rich composition of the sediment source terranes in the mid-and-upper-system regions and the development of a canyon system that delivers it southward into the proto-Bay of Bengal.

This case study is presented for demonstration purposes only. Explorationists working in the region would integrate quantitative methods to develop accurate models for input into exploration risk assessment. At minimum, the S2S-related maps and products (*see* Fig. 5) that are needed for confident exploration in this region are: (1) an accurate tectonic plate model that can accommodate the complex kinematics of this region; (2) EOD maps at desired source, reservoir and seal time intervals; (3) sediment routing maps with underlying sediment source terrane maps for the same periods; (4) a series of S2S system-scaling exercises to determine the size, duration and depositional extent of canyon feeder systems and submarine fans in the AOI; (5) a reliable detrital mineral database that captures the range of sediment compositions draining into the exploration AOI—this is essential for accurate petrographic modeling of reservoir diagenesis. The work of Zhang et al. (2019) demonstrates nicely how data and observations in a known area may be extrapolated to make inferences about another, a key skill necessary for frontier exploration and play extension.

## White space exploration

The island of Borneo in the Southeast Asian archipelago plays host to petroleum systems spanning mature well-explored basins (e.g., Baram Delta), moderately to well-explored basins (e.g., Mahakam Delta) and relatively unexplored Natuna, Sarwawak, Makassar and Tarakan basins. Each of these basins are derived from a shared hinterland in the island of Borneo, variously sourcing sediment over the duration from the Late Cretaceous onwards.

Exploration of these basins has at various times not only been for hydrocarbon reservoirs, but also the source of elusive Borneo gold and diamond deposits indicated by the presence of alluvial gold in the Ketungau and Mahakam rivers and their tributaries, and the diamonds of suspected Archean age found in Central and West Kalimantan. The presence of mineral and hydrocarbon indicators has provided continued interest in the island and its geology since the first modern geologists began mapping and describing this remarkable island in the early 20th century.

Work by Davies (2014) focused on the hinterland domain widely cited as the probable location of pre-Cretaceous metamorphic and igneous terranes from which gold and diamond deposits were believed to have been derived, as well as the source region for clastic sedimentary intervals bearing zircons of Archean age found in adjacent basins. The presence or absence of cratonic crustal fragments had been subject to critical debate and was an important ingredient to models predicting the supply of mature ortho-quartzites into surrounding basins and a model favored by exploration geologists seeking a high-quality sandstone play analogous to those found along the northwestern Australian margin. Igneous rocks of Cretaceous age made up much of the Schwaner Mountains, a mountain range exhumed after ~90 Ma, which reconfigured the fluvial drainage of the region, depositing sediment into the neighboring Natuna, Barito and Ketungau basins.

Extensive fieldwork across West and Central Kalimantan provided both hinterland samples (Granitoids ranging from K-granite to tonalite compositions, pelitic- and psammitic- metamorphic rocks), as well as modern sediment. U-Pb SHRIMP and LA-ICPMS dating on zircon grains from all sample types revealed a hinterland entirely absent of Archean grains. The absence of any zircon grains across the hinterland, including in any modern sediment (which were collected as a means for sense-checking the coverage of hard-rock sampling strategies), and in contrast to sediment from neighboring Sarawak and Java where small numbers of ancient zircon grains have been identified in hinterland rocks (and often attributed to long-distance deposits from SW Borneo), changed the understanding of this hinterland from that of an ancient cratonic fragment, capable of delivering large quantities of ortho-quartzite sand during the Eocene and Oligocene, to the locus of an I-type subduction arc system producing large volumes of intermediate to acidic volcanogenic sediment from approximately 130–85 Ma (Breitfeld et al., 2020).

The presence of mature hydrocarbon basins with world-class clastic reservoirs derived at least in part from this volcanic hinterland directly contradicts the often-cited "rule" that volcanic archipelago are incapable of producing high-quality reservoir sand. Until recently the excellent petroleum resources across Java, Sumatra and other petroliferous basins had been variously attributed to luck from presubduction sedimentation, anomalous concentration or preservation of clastic units, or altogether ignored by proponents of such rules. The presence of such excellent clastic deposits in relatively close proximity to hinterlands from which they were derived in this case highlights the extreme chemical weathering potential associated with tropical environments and the significance of climate to the successful modeling of sandstone composition.

The absence of metamorphic or igneous deposits capable of producing diamonds in West Kalimantan makes the provenance of reported finds a continued mystery. Primary gold observed during analyses of intrusive rocks provides compelling evidence of a working hydrothermal system in the Schwaner highlands, capable of producing the large volumes of alluvial gold which are removed annually by artisan workers panning in the rivers. The checkered history of gold exploration in Indonesian Borneo (as dramatized in the film Gold, by Stephen Gaghan) may offer an explanation for the continued reticence of international investors to explore deep into the Borneo jungle looking for gold, offering a perhaps unlikely protection to the exquisite and beautiful fauna and flora which call this magnificent island their home.

# Conclusion

S2S analysis is an impactful way for explorationists to characterize modern and ancient depositional systems and, depending on the reliability of available data, it may be used to reduce risk uncertainty in deepwater resource exploration and assessment at many scales from mature basins to white space exploration. While many trusted S2S tools and algorithms closely approximate the myriad geologic factors that regulate the erosion, transport and deposition of clastic material in sedimentary systems, there always exist geologic complexities and unknown variables that impact even the most well-constrained studies.

Key challenges include reducing the knowledge gaps that currently prevent the creation of predictive S2S frameworks. Current priorities for improving applied S2S research include: detailed petrological characterization of diverse sediment source terranes, their erodibility and mineralogical fertility in different tectonic and climatic regions, specifically zircon and other useful geo-/thermochronometers (e.g., Augustsson et al., 2018; Moecher and Samson, 2006; Chew et al., 2020), and biases inherent to popular analytical and modelling techniques that modify constituent grain composition and amplify signal-to-noise ratios in empirical datasets (e.g., Tolosana-Delgado, 2012; Sláma and Košler, 2012; Lukens et al., 2020; Markwitz et al., 2020); modeling of complex intra-system processes such as the role of sediment storage mechanisms in S2S systems (e.g., Huffman et al., 2022; Pizzuto et al., 2017) and the improved use of multiple sedimentary signals of distinct origin to facilitate stratigraphic record inversion (e.g., Sharman et al., 2019). Additionally, robust S2S analysis requires a strong understanding of crystalline basement structural controls on syn-rift siliciclastic and evaporite deposition which, once rift margins subside and the basin deepens and floods, is essential for delimiting the potential space for deepwater sedimentation. Even if deepwater strata are not the exploration target, much can be learned about early basin structural and stratigraphic architecture in well-imaged and data-rich regions with large petroleum reserves like the Barents Sea, Brazilian Cretaceous rift margin and the North American Gulf of Mexico (e.g., Hassaan et al., 2021; do Amarante et al., 2021; Frederick et al., 2020).

Fortunately, the science surrounding S2S analysis evolves rapidly, and today's technical uncertainties are often reduced in short time as innovative technical approaches tackle difficult problems. Relatively new and exciting techniques in computational modeling and machine learning examine S2S systems in the context of "big data" and aim to improve model accuracy. Novel approaches include mining vast catalogues of vintage data for amassing statistically viable datasets, developing means for overcoming hurdles related to digitizing (formerly) analog data and building data models and management tools that address challenges related to legacy data classification schemes which often reflect the unique ways that individual analysts categorized rock data. Impressive examples of new approaches in the sediment provenance realm specifically include quantitative analysis of sediment source terranes and "unmixing" of complex detrital mineral data signals (e.g., Sharman and Johnstone, 2017; Sundell and Saylor, 2017; Saylor et al., 2019; Saylor and Sundell, 2016), machine learning algorithms that use sand modal composition to characterize the relative control of tectonic and environmental factors that generate sediments (Johnson et al., 2022) and the use of sedimentology principles to obtain more detailed provenance information (Malusà and Garzanti, 2019).

## Acknowledgments

The authors wish to thank the editors for the invitation to write this chapter. Constructive reviews by Jon Rotzien, Cindy Yeilding, and Richard A. Sears improved the content and helped focus the narrative. We stand on the shoulders of giants and gratefully acknowledge the many geoscientists whose hard work and scientific expertise we highlight in this chapter. The source-to-sink community is fortunate to have many talented researchers working in this multidisciplinary and dynamic scientific space.

## References

Aplin, A.C., Macquaker, J.H., 2011. Mudstone diversity: origin and implications for source, seal, and reservoir properties in petroleum systems. AAPG Bull. 95 (12), 2031–2059.

Armstrong, P.A., 2005. Thermochronometers in sedimentary basins. Rev. Mineral. Geochem. 58 (1), 499–525.

Augustsson, C., Voigt, T., Bernhart, K., Kreißler, M., Gaupp, R., Gärtner, A., Hofmann, M., Linnemann, U., 2018. Zircon size-age sorting and source-area effect: the German Triassic Buntsandstein Group. Sediment. Geol. 375, 218–231.

Ault, A.K., 2020. Hematite fault rock thermochronometry and textures inform fault zone processes. J. Struct. Geol. 133, 104002.

Ault, A.K., Frenzel, M., Reiners, P.W., Woodcock, N.H., Thomson, S.N., 2016. Record of paleofluid circulation in faults revealed by hematite (U-Th)/He and apatite fission-track dating: an example from Gower Peninsula fault fissures, Wales. Lithosphere 8 (4), 379–385.

Barnes, N.E., Normark, W.R., 1985. Diagnostic parameters for comparing modern submarine fans and ancient turbidite systems. In: Submarine Fans and Related Turbidite Systems. Springer, New York, NY, pp. 13–14.

Blum, M.D., Hattier-Womack, J., 2009. Climate change, sea-level change, and fluvial sediment supply to deepwater depositional systems. External Controls on Deep-Water Depositional Systems. vol. 92 SEPM Society for Sedimentary Geology.

Blum, M., Pecha, M., 2014. Mid-Cretaceous to Paleocene North American drainage reorganization from detrital zircons. Geology 42 (7), 607–610.

Blum, M.D., Törnqvist, T.E., 2000. Fluvial responses to climate and sea-level change: a review and look forward. Sedimentology 47, 2–48.

Braun, J., Van der Beek, P., Batt, G., 2006. Quantitative Thermochronology: Numerical Methods for the Interpretation of Thermochronological Data. Cambridge University Press.

Breitfeld, H.T., Davies, L.B., Hall, R., Armstrong, R., Foster, M., Lister, G., Thirlwall, M., Grassineau, N., Hennig-Breitfeld, J., Van Hattum, M.W.A., 2020. Mesozoic PaleoPacific subduction beneath SW Borneo: U-Pb geochronology of the Schwaner Granitoids and the Pinoh Metamorphic Group. Front. Earth Sci. 8, 536.

Brewer, C.J., Hampson, G.J., Whittaker, A.C., Roberts, G.G., Watkins, S.E., 2020. Comparison of methods to estimate sediment flux in ancient sediment routing systems. Earth Sci. Rev. 207, 103217.

Burke, K., Wells, G.L., 1989. Trans-African drainage system of the Sahara: was it the Nile? Geology 17 (8), 743–747.

Carrapa, B., 2009. Tracing exhumation and orogenic wedge dynamics in the European Alps with detrital thermochronology. Geology 37 (12), 1127–1130.

Chew, D., O'Sullivan, G., Caracciolo, L., Mark, C., Tyrrell, S., 2020. Sourcing the sand: accessory mineral fertility, analytical and other biases in detrital U-Pb provenance analysis. Earth Sci. Rev. 202, 103093.

Collins, M., Tett, S., Cooper, C., 2001. The internal climate variability of HadCM3, a version of the Hadley Centre coupled model without flux adjustments. Climate Dynam. 17, 61–81. https://doi.org/10.1007/s003820000094.

Cooperdock, E.H., Ault, A.K., 2020. Iron oxide (U-Th)/He thermochronology: new perspectives on faults, fluids, and heat. Elements 16 (5), 319–324.

Coutts, D.S., Matthews, W.A., Hubbard, S.M., 2019. Assessment of widely used methods to derive depositional ages from detrital zircon populations. Geosci. Front. 10 (4), 1421–1435.

Davies, L.B., 2014. Cretaceous crust in SW Borneo: Petrological, geochemical, and geochronological constraints from the Schwaner Mountains. In: Proceedings of the Indonesian Petroleum Association 38th Annual Convention and Symposium.

Davies, L.B., Fielding, L., Fielding, S., 2020. A Better Appreciation of Reservoir Quality Through Data-Driven and Algorithmically-Generated Framework Mineralogy Prediction. ResTech 2.0.

Dickinson, W.R., 1985. Interpreting provenance relations from detrital modes of sandstones. In: Provenance of Arenites. Springer, Dordrecht, pp. 333–361.

do Amarante, F.B., Jackson, C.A.L., Pichel, L.M., Scherer, C.M.D.S., Kuchle, J., 2021. Pre-salt rift morphology controls salt tectonics in the Campos Basin, offshore SE Brazil. Basin Res. 33 (5), 2837–2861.

Ehlers, T.A., Farley, K.A., 2003. Apatite (U-Th)/He thermochronometry: methods and applications to problems in tectonic and surface processes. Earth Planet. Sci. Lett. 206 (1–2), 1–14.

Farley, K.A., 2000. Helium diffusion from apatite: General behavior as illustrated by Durango fluorapatite. J. Geophys. Res. Solid Earth 105 (B2), 2903–2914.

Farley, K.A., 2002. (U-Th)/He dating: techniques, calibrations, and applications. Rev. Mineral. Geochem. 47 (1), 819–844.

Fielding, L., Najman, Y., Millar, I., Butterworth, P., Ando, S., Padoan, M., et al., 2017. A detrital record of the Nile River and its catchment. J. Geol. Soc. London 174 (2), 301–317.

Fielding, L., Najman, Y., Millar, I., Butterworth, P., Garzanti, E., Vezzoli, G., Barfod, D., Kneller, B., 2018. The initiation and evolution of the River Nile. Earth Planet. Sci. Lett. 489, 166–178.

Fitzgerald, P.G., Gleadow, A.J.W., 1990. New approaches in fission track geochronology as a tectonic tool: examples from the Transantarctic Mountains. Int. J. Radiat. Appl. Instr. D Nucl. Tracks Radiat. Meas. 17 (3), 351–357.

Fleming, E.J., Flowerdew, M.J., Smyth, H.R., Scott, R.A., Morton, A.C., Omma, J.E., et al., 2016. Provenance of Triassic sandstones on the southwest Barents Shelf and the implication for sediment dispersal patterns in northwest Pangaea. Mar. Pet. Geol. 78, 516–535.

Frederick, B.C., Blum, M.D., Snedden, J.W., Fillon, R.H., 2020. Early Mesozoic synrift Eagle Mills Formation and coeval siliciclastic sources, sinks, and sediment routing, northern Gulf of Mexico basin. GSA Bull. 132 (11–12), 2631–2650.

Galehouse, J.S., 1969. Counting grain mounts; number percentage vs. number frequency. J. Sediment. Res. 39 (2), 812–815.

Garzanti, E., Andò, S., Limonta, M., Fielding, L., Najman, Y., 2018. Diagenetic control on mineralogical suites in sand, silt, and mud (Cenozoic Nile Delta): implications for provenance reconstructions. Earth Sci. Rev. 185, 122–139.

Garzanti, E., Andò, S., 2007. Heavy mineral concentration in modern sands: implications for provenance interpretation. Dev. Sedimentol. 58, 517–545.

Garzanti, E., Vezzoli, G., 2003. A classification of metamorphic grains in sands based on their composition and grade. J. Sediment. Res. 73 (5), 830–837.

Gautheron, C., Zeitler, P.K., 2020. Noble gases deliver cool dates from hot rocks. Elements 16 (5), 303–309.

Gehrels, G.E., Yin, A., Wang, X.F., 2003. Detrital-zircon geochronology of the northeastern Tibetan plateau. Geol. Soc. Am. Bull. 115 (7), 881–896.

Gleadow, A.J.W., Duddy, I.R., Green, P.F., Lovering, J.F., 1986. Confined fission track lengths in apatite: a diagnostic tool for thermal history analysis. Contrib. Mineral. Petrol. 94 (4), 405–415.

Graham, S.A., Romans, B.W., Covault, J.A., 2011. Source-to-sink in the stratigraphic record: capturing the long-term, deep-time evolution of sedimentary systems. AGU Chapman Conference Program. pp. 32–33.

Green, P.F., Duddy, I.R., Laslett, G.M., Hegarty, K.A., Gleadow, A.J.W., Lovering, J.F., 1989. Thermal annealing of fission tracks in apatite 4. Quantitative modelling techniques and extension to geological timescales. Chem. Geol. Isot. Geosci. Section 79 (2), 155–182.

Hall, R., 2012. Late Jurassic–Cenozoic reconstructions of the Indonesian region and the Indian Ocean. Tectonophysics 570, 1–41.

Hartmann, J., Moosdorf, N., 2012. The new global lithological map database GLiM: a representation of rock properties at the Earth surface. Geochem. Geophys. Geosyst. 13 (12).

Hassaan, M., Faleide, J.I., Gabrielsen, R.H., Tsikalas, F., 2021. Effects of basement structures and Carboniferous basin configuration on evaporite distribution and the development of salt structures in Nordkapp Basin, Barents Sea—part I. Basin Res. 33 (4), 2474–2499.

He, B., Jiao, C., Xu, Z., Cai, Z., Zhang, J., Liu, S., Li, H., Chen, W., Yu, Z., 2016. The paleotectonic and paleogeography reconstructions of the Tarim Basin and its adjacent areas (NW China) during the late Early and Middle Paleozoic. Gondwana Res. 30, 191–206.

Helland-Hansen, W., Sømme, T.O., Martinsen, O.J., Lunt, I., Thurmond, J., 2016. Deciphering Earth's natural hourglasses: perspectives on source-to-sink analysis. J. Sediment. Res. 86 (9), 1008–1033.

Holbrook, J., Wanas, H., 2014. A fulcrum approach to assessing source-to-sink mass balance using channel paleohydrologic parameters derivable from common fluvial data sets with an example from the Cretaceous of Egypt. J. Sediment. Res. 84 (5), 349–372.

Hubert, J.F., 1962. A zircon-tourmaline-rutile maturity index and the interdependence of the composition of heavy mineral assemblages with the gross composition and texture of sandstones. J. Sediment. Res. 32 (3), 440–450.

Huffman, M.E., Pizzuto, J.E., Trampush, S.M., Moody, J.A., Schook, D.M., Gray, H.J., Mahan, S.A., 2022. Floodplain sediment storage timescales of the laterally confined meandering Powder River, USA. J. Geophys. Res. Earth Surf. 127 (1).

Ingersoll, R.V., Bullard, T.F., Ford, R.L., Grimm, J.P., Pickle, J.D., Sares, S.W., 1984. The effect of grain size on detrital modes: a test of the Gazzi-Dickinson point-counting method. J. Sediment. Res. 54 (1), 103–116.

Johnson, J.I., Sharman, G.R., Szymanski, E., & Huang, X. (2022). Machine learning applied to a modern-Pleistocene petrographic data set: The global prediction of sand modal composition (GloPrSM) model. Journal of Geophysical Research: Earth Surface, 127, e2022JF006595. https://doi.org/10.1029/2022JF006595.

Ketcham, R.A., 2019. Fission-track annealing: from geologic observations to thermal history modeling. In: Fission-Track Thermochronology and its Application to Geology, pp. 49–75.

Ketcham, R.A., Donelick, R.A., Carlson, W.D., 1999. Variability of apatite fission-track annealing kinetics: III. Extrapolation to geological time scales. Am. Mineral. 84 (9), 1235–1255.

Kottek, M., Grieser, J., Beck, C., Rudolf, B., Rubel, F., 2006. World Map of the Köppen-Geiger climate classification updated. Meteorol. Z. 15, 259–263. https://doi.org/10.1127/0941-2948/2006/0130.

Laslett, G.M., Green, P.F., Duddy, I.R., Gleadow, A.J.W., 1987. Thermal annealing of fission tracks in apatite 2. A quantitative analysis. Chem. Geol. Isot. Geosci. Section 65 (1), 1–13.

Lewin, A., Meinhold, G., Hinderer, M., Dawit, E.L., Bussert, R., Lünsdorf, N.K., 2020. Heavy minerals as provenance indicator in glaciogenic successions: an example from the Palaeozoic of Ethiopia. J. Afr. Earth Sci. 165, 103813.

Ludwig, K.R., 2000. Decay constant errors in U-Pb concordia-intercept ages. Chem. Geol. 166 (3–4), 315–318.

Lukens, C.E., Riebe, C.S., Sklar, L.S., Shuster, D.L., 2020. Sediment size and abrasion biases in detrital thermochronology. Earth Planet. Sci. Lett. 531.

Malusà, M.G., Fitzgerald, P.G., 2019. Application of thermochronology to geologic problems: bedrock and detrital approaches. In: Fission-Track Thermochronology and its Application to Geology. Springer, Cham, pp. 191–209.

Malusà, M.G., Garzanti, E., 2019. The sedimentology of detrital thermochronology. Fission-Track Thermochronology and its Application to Geology. Springer, Cham, pp. 123–143.

Markwitz, V., Kirkland, C.L., Gessner, K., 2020. Provenance bias between detrital zircons from sandstones and river sands: a quantification approach using 3-D grain shape, composition and age. Geosci. Front. 11 (3), 835–842.

Mason, C.C., Romans, B.W., Stockli, D.F., Mapes, R.W., Fildani, A., 2019. Detrital zircons reveal sea-level and hydroclimate controls on Amazon River to deep-sea fan sediment transfer. Geology 47 (6), 563–567.

McCauley, J.F., Breed, C.S., Schaber, G.G., McHugh, W.P., Issawi, B., Haynes, C.V., Grolier, M.J., El Kilani, A., 1986. Paleodrainages of the eastern Sahara—the radar rivers revisited (Sir-A/B implications for a mid-tertiary Trans-African drainage system). IEEE Trans. Geosci. Remote Sens. 4, 624–648.

McLennan, S., Taylor, S., 1985. The Continental Crust: Its Composition and Evolution: An Examination of the Geochemical Record Preserved in Sedimentary Rocks. Blackwell Scientific, Oxford.

Mitchell, S.G., Reiners, P.W., 2003. Influence of wildfires on apatite and zircon (U-Th)/He ages. Geology 31 (12), 1025–1028.

Moecher, D.P., Samson, S.D., 2006. Differential zircon fertility of source terranes and natural bias in the detrital zircon record: implications for sedimentary provenance analysis. Earth Planet. Sci. Lett. 247, 252–266.

Morton, A., Mundy, D., Bingham, G., 2012. High-frequency fluctuations in heavy mineral assemblages from Upper Jurassic sandstones of the Piper Formation, UK North Sea: relationships with sea-level change and floodplain residence. Geol. Soc. Am. Spec. Pap. 487, 163–176.

Najman, Y., Sobel, E.R., Millar, I., Stockli, D.F., Govin, G., Lisker, F., et al., 2020. The exhumation of the Indo-Burman Ranges, Myanmar. Earth Planet. Sci. Lett. 530, 115948.

National Science Foundation MARGINS, 2004. Science Plans. https://www.nsf-margins.org/Publications/SciencePlans/MARGINS_SciencePlans2004.pdf.

Nebel, O., 2014. Rb–Sr dating. In: Rink, W., Thompson, J. (Eds.), Encyclopedia of Scientific Dating Methods. Springer, Dordrecht, https://doi.org/10.1007/978-94-007-6326-5_116-1.

Nyberg, B., Helland-Hansen, W., Gawthorpe, R.L., Sandbakken, P., Eide, C.H., Sømme, T., 2018. Revisiting morphological relationships of modern source-to-sink segments as a first-order approach to scale ancient sedimentary systems. Sediment. Geol. 373, 111–133.

Nyberg, B., Helland-Hansen, W., Gawthorpe, R., Tillmans, F., Sandbakken, P., 2021. Assessing first-order BQART estimates for ancient source-to-sink mass budget calculations. Basin Res. 33 (4), 2435–2452.

Peppe, D.J., Reiners, P.W., 2007. Conodont (U-Th)/He thermochronology: initial results, potential, and problems. Earth Planet. Sci. Lett. 258 (3-4), 569–580.

Peyton, S.L., Carrapa, B., 2013. An introduction to low-temperature thermochronologic techniques, methodology, and applications. In: Knight, C., Cuzella, J. (Eds.), Application of Structural Methods to Rocky Mountain Hydrocarbon Exploration and Development. vol. 65. AAPG Studies in Geology, pp. 15–36.

Pizzuto, J., Keeler, J., Skalak, K., Karwan, D., 2017. Storage filters upland suspended sediment signals delivered from watersheds. Geology 45 (2), 151–154.

Pullen, A., Ibáñez-Mejía, M., Gehrels, G.E., Ibáñez-Mejía, J.C., Pecha, M., 2014. What happens when n = 1000? Creating large-n geochronological datasets with LA-ICP-MS for geologic investigations. J. Anal. At. Spectrom. 29 (6), 971–980.

Rahl, J.M., Ehlers, T.A., van der Pluijm, B.A., 2007. Quantifying transient erosion of orogens with detrital thermochronology from syntectonic basin deposits. Earth Planet. Sci. Lett. 256 (1–2), 147–161.

Reiners, P.W., Farley, K.A., 1999. Helium diffusion and (U-Th)/He thermochronometry of titanite. Geochim. Cosmochim. Acta 63 (22), 3845–3859.

Reiners, P., 2005. Zircon (U-Th)/He thermochronometry. Rev. Mineral. Geochem. 58 (1), 151–179.

Reiners, P.W., Brandon, M.T., 2006. Using thermochronology to understand orogenic erosion. Annu. Rev. Earth Planet. Sci. 34, 419–466.

Reiners, P.W., Campbell, I.H., Nicolescu, S., Allen, C.M., Hourigan, J.K., Garver, J.I., et al., 2005. (U-Th)/(He-Pb) double dating of detrital zircons. Am. J. Sci. 305 (4), 259–311.

Reiners, P.W., Ehlers, T.A. (Eds.), 2018. Low-Temperature Thermochronology: Techniques, Interpretations, and Applications. vol. 58. Walter de Gruyter GmbH & Co KG.

Romans, B.W., Castelltort, S., Covault, J.A., Fildani, A., Walsh, J.P., 2016. Environmental signal propagation in sedimentary systems across timescales. Earth Sci. Rev. 153, 7–29.

Ruiz, G.M.H., Seward, D., Winkler, W., 2004. Detrital thermochronology—a new perspective on hinterland tectonics, an example from the Andean Amazon Basin, Ecuador. Basin Res. 16 (3), 413–430.

Saylor, J.E., Sundell, K.E., 2016. Quantifying comparison of large detrital geochronology data sets. Geosphere 12 (1), 203–220.

Saylor, J.E., Stockli, D.F., Horton, B.K., Nie, J., Mora, A., 2012. Discriminating rapid exhumation from syn-depositional volcanism using detrital zircon double dating: implications for the tectonic history of the Eastern Cordillera, Colombia. Bulletin 124 (5–6), 762–779.

Saylor, J.E., Sundell, K.E., Sharman, G.R., 2019. Characterizing sediment sources by non-negative matrix factorization of detrital geochronological data. Earth Planet. Sci. Lett. 512, 46–58.

Schneider, D.A., Issler, D.R., 2019. Application of low-temperature thermochronology to hydrocarbon exploration. In: Fission-Track Thermochronology and its Application to Geology. Springer, Cham, pp. 315–333.

Schoene, B., 2014. U-Th-Pb geochronology. In: Treatise on Geochemistry. vol. 4. Elsevier, pp. 341–378.

Shanmugam, G., 2016. Submarine fans: a critical retrospective (1950–2015). J. Palaeogeogr. 5 (2), 110–184.

Sharman, G.R., Johnstone, S.A., 2017. Sediment unmixing using detrital geochronology. Earth Planet. Sci. Lett. 477, 183–194.

Sharman, G.R., Covault, J.A., Stockli, D.F., Wroblewski, A.F.J., Bush, M.A., 2017. Early Cenozoic drainage reorganization of the United States Western Interior-Gulf of Mexico sediment routing system. Geology 45 (2), 187–190.

Sharman, G.R., Sylvester, Z., Covault, J.A., 2019. Conversion of tectonic and climatic forcings into records of sediment supply and provenance. Sci. Rep. 9 (1), 1–7.

Shukri, N.M., 1950. The mineralogy of some Nile sediments. Quart. J. Geol. Soc. 106 (1–4), 466–467.

Slabic, A., Lapen, T.J., 2017. U-Th-Pb chronology of fish bone in petroleum source rocks: an example from the Eagle Ford Group, Texas. Geol. Soc. Am. 49 (1). https://doi.org/10.1130/abs/2017SC288948.

Sláma, J., Košler, J., 2012. Effects of sampling and mineral separation on accuracy of detrital zircon studies. Geochem. Geophys. Geosyst. 13 (5).

Sømme, T.O., Helland-Hansen, W., Martinsen, O.J., Thurmond, J.B., 2009. Relationships between morphological and sedimentological parameters in source-to-sink systems: a basis for predicting semi-quantitative characteristics in subsurface systems. Basin Res. 21 (4), 361–387.

Stock, G.M., Ehlers, T.A., Farley, K.A., 2006. Where does sediment come from? Quantifying catchment erosion with detrital apatite (U-Th)/He thermochronometry. Geology 34 (9), 725–728.

Stockli, D.F., Najman, Y.M., 2020. Earth's dynamic past revealed by detrital thermochronometry. Elements 16 (5), 311–317.

Stockli, D.F., Farley, K.A., Dumitru, T.A., 2000. Calibration of the apatite (U-Th)/He thermochronometer on an exhumed fault block, White Mountains, California. Geology 28 (11), 983–986.

Strahler, A.N., 1957. Quantitative analysis of watershed geomorphology. Eos. Trans. AGU 38 (6), 913–920.

Sundell, K.E., Saylor, J.E., 2017. Unmixing detrital geochronology age distributions. Geochem. Geophys. Geosyst. 18 (8), 2872–2886.

Syvitski, J.P., Milliman, J.D., 2007. Geology, geography, and humans battle for dominance over the delivery of fluvial sediment to the coastal ocean. J. Geol. 115 (1), 1–19.

Tagami, T., 2005. Zircon fission-track thermochronology and applications to fault studies. Rev. Mineral. Geochem. 58 (1), 95–122.

Tagami, T., Dumitru, T.A., 1996. Provenance and thermal history of the Franciscan accretionary complex: constraints from zircon fission track thermochronology. J. Geophys. Res. Solid Earth 101 (B5), 11353–11364.

Tetlock, P.E., Gardner, D., 2016. Superforecasting: The Art and Science of Prediction. Random House.

Thomson, S.N., Reiners, P.W., Hemming, S.R., Gehrels, G.E., 2013. The contribution of glacial erosion to shaping the hidden landscape of East Antarctica. Nat. Geosci. 6 (3), 203–207.

Tochilin, C.J., Reiners, P.W., Thomson, S.N., Gehrels, G.E., Hemming, S.R., Pierce, E.L., 2012. Erosional history of the Prydz Bay sector of East Antarctica from detrital apatite and zircon geo-and thermochronology multidating. Geochem. Geophys. Geosyst. 13 (11).

Tolosana-Delgado, R., 2012. Uses and misuses of compositional data in sedimentology. Sediment. Geol. 280, 60–79.

Vermeesch, P., 2004. How many grains are needed for a provenance study? Earth Planet. Sci. Lett. 224 (3–4), 441–451.

Wolf, R.A., Farley, K.A., Silver, L.T., 1996. Helium diffusion and low-temperature thermochronometry of apatite. Geochim. Cosmochim. Acta 60 (21), 4231–4240.

Wolfe, M.R., Stockli, D.F., 2010. Zircon (U-Th)/He thermochronometry in the KTB drill hole, Germany, and its implications for bulk He diffusion kinetics in zircon. Earth Planet. Sci. Lett. 295 (1–2), 69–82.

Yamada, R., Tagami, T., Nishimura, S., Ito, H., 1995. Annealing kinetics of fission tracks in zircon: an experimental study. Chem. Geol. 122 (1–4), 249–258.

Yamada, R., Murakami, M., Tagami, T., 2007. Statistical modelling of annealing kinetics of fission tracks in zircon; reassessment of laboratory experiments. Chem. Geol. 236 (1–2), 75–91.

Zapata, S., Patiño, A., Cardona, A., Parra, M., Valencia, V., Reiners, P., Oboh-Ikuenobe, F., Genezini, F., 2020. Bedrock and detrital zircon thermochronology to unravel exhumation histories of accreted tectonic blocks: an example from the Western Colombian Andes. J. South Am. Earth Sci. 103, 102715.

Zeitler, P.K., Johnson, N.M., Briggs, N.D., Naeser, C.W., 1986. Uplift history of the NW Himalaya as recorded by fission-track ages on detrital Siwalik zircons. In: Symposium on Mesozoic and Cenozoic Geology, pp. 481–494.

Zhang, P., Najman, Y., Mei, L., Millar, I., Sobel, E.R., Carter, A., et al., 2019. Palaeodrainage evolution of the large rivers of East Asia, and Himalayan-Tibet tectonics. Earth Sci. Rev. 192, 601–630.

# Chapter 12

# Sequence stratigraphy of deepwater systems

**Octavian Catuneanu**
*Department of Earth and Atmospheric Sciences, University of Alberta, Edmonton, AB, Canada*

## Introduction

Sequence stratigraphy is a type of stratigraphy that relies on stacking patterns for the definition, nomenclature, classification and correlation of stratal units and bounding surfaces. The method examines the stratigraphic cyclicity and the related changes in sedimentation regimes that can be observed at the scales afforded by the resolution of the data available. Seismic stratigraphy, which is the precursor of modern sequence stratigraphy, set the early standards for the scales and applications of the methodology (Payton, 1977). The scope of sequence stratigraphy was subsequently expanded to all scales and datasets by integrating seismic with borehole and outcrop data. Beyond the research of the basin-fill architecture, sequence stratigraphy provides the genetic framework to rationalize and predict the distribution of economic deposits that relate to sedimentary processes, from petroleum systems to coal beds, mineral placers and aquifers.

Following the emergence of seismic stratigraphy in the 1970s, the development of sequence stratigraphy took place informally for more than three decades. The result was the proliferation of an unnecessarily complex terminology (e.g., several synonymous terms for the same concept), the misusage of terms (e.g., the same terms used with different meanings by different authors) and the proposal of different approaches to the delineation of sequences (Fig. 1). These confusions were addressed and clarified with the publication of formal guidelines by the International Commission on Stratigraphy (Catuneanu et al., 2011). The model-independent application of sequence stratigraphy is enabled by the observation of stratal stacking patterns at all scales permitted by the data available. On practical grounds, the identification of systems tracts is more important than the delineation of sequences. A systems tract is a linkage of contemporaneous depositional systems, forming the subdivision of a sequence (Brown and Fisher, 1977). Each type of systems tract is defined by a specific stratal stacking pattern, which results from the interplay of accommodation and sedimentation (Catuneanu, 2022). Systems tracts afford insights into the sediment dispersal patterns across a basin at different stages of sequence development, which helps to rationalize and predict the distribution of facies and natural resources in the sedimentary record. The construction of a framework of systems tracts and bounding surfaces fulfills the practical purpose of sequence stratigraphy. Beyond this, the delineation of sequences becomes a matter of model-dependent organization of systems tracts into stratigraphic cycles (Fig. 1). The different models in Fig. 1 reflect alternatives to (1) the selection of the sequence boundary, and (2) the nomenclature of systems tracts that are used to subdivide sequences. To a large extent, these proposals stem from the observation of sedimentary-basin fills placed in different geological settings. Therefore, none of these models is universally applicable. At the core of the methodology is the identification of stratal stacking patterns that are present within a study area, without any preconceived model assumptions (see Catuneanu, 2022, for a detailed discussion of sequence stratigraphic methodology and nomenclature).

The stratigraphic architecture and sedimentological makeup of deepwater systems are complex, due to the interplay of several allogenic and autogenic processes (Fig. 2), and the co-existence of multiple sediment sources and modes of sediment transport. Notably, sedimentation in deepwater settings relates to gravity-driven processes and pelagic/hemipelagic fallout in both enclosed and oceanic basins, as well as to contour currents and associated oceanographic processes in basins connected to the global ocean (see further details in Chapter 9). Despite this complexity, reproducible field criteria can be defined to identify the elements of the deepwater sequence stratigraphic framework, in a manner that is independent of any basin-specific (nondiagnostic) variability (Catuneanu, 2020). Accommodation and sedimentation on the shelf, which depend on relative sea-level changes, the production of extrabasinal and intrabasinal sediment and the physiography of the basin, define the "dual control" on shoreline trajectories and sediment supply to the shelf edge. Both elements of this

**Deepwater Sedimentary Systems**. https://doi.org/10.1016/B978-0-323-91918-0.00010-4

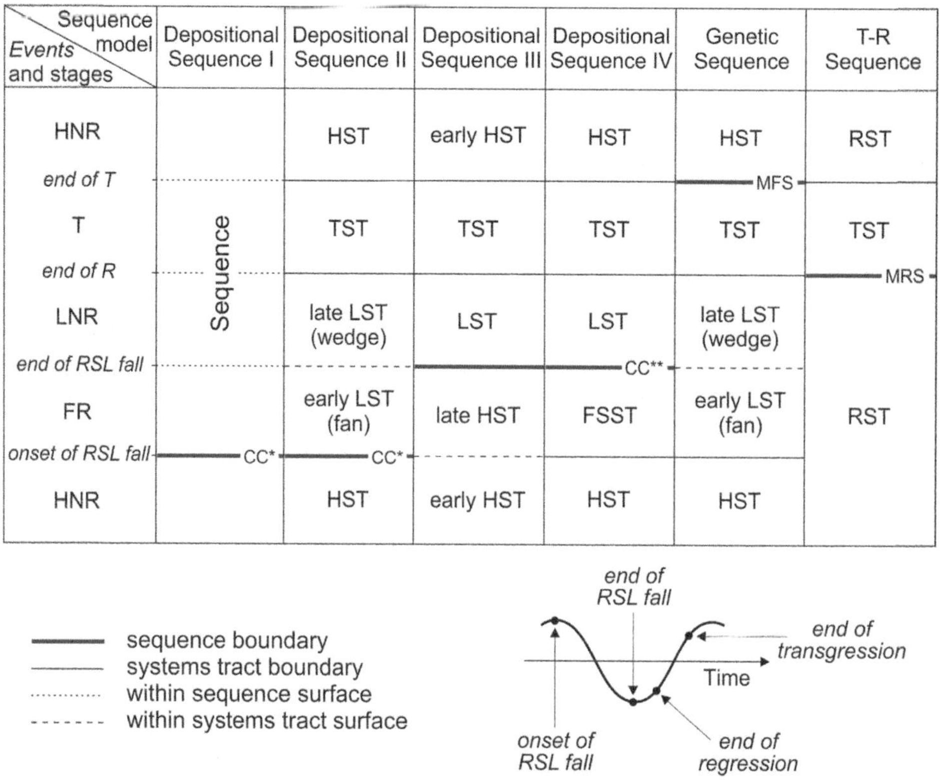


**FIG. 1** Approaches to sequence stratigraphy: nomenclature of systems tracts and timing of sequence boundaries. While the concept of "systems tract" was introduced in the 1970s (Brown and Fisher, 1977), its usage in sequence stratigraphy only started in the 1980s (Posamentier and Vail, 1985). Depositional sequence I: Mitchum et al., 1977; depositional sequence II: Posamentier et al., 1988; depositional sequence III: Van Wagoner et al., 1988; depositional sequence IV: Hunt and Tucker, 1992; genetic sequence: Galloway, 1989; T-R sequence: Johnson and Murphy, 1984. *Abbreviations*: CC*, correlative conformity in the sense of Posamentier et al. (1988), herein referred to as the "basal surface of forced regression"; CC**, correlative conformity in the sense of Van Wagoner et al. (1988), herein referred to as the "correlative conformity"; FR, forced regression; FSST, falling-stage systems tract; HNR, highstand normal regression; HST, highstand systems tract; LNR, lowstand normal regression; LST, lowstand systems tract; MFS, maximum flooding surface; MRS, maximum regressive surface; R, regression; RSL, relative sea level; RST, regressive systems tract; T, transgression; T-R, transgressive-regressive; TST, transgressive systems tract. *(From Catuneanu, O., Galloway, W.E., Kendall, C.G.S.C., Miall, A.D., Posamentier, H.W., Strasser, A., Tucker, M.E., 2011. Sequence stratigraphy: methodology and nomenclature. Newsl. Stratigr. 44(3), 173–245.)*

"dual control" contribute in discernable ways to the architecture and makeup of the stratigraphic record and can be separated based on stratigraphic and sedimentological criteria (Catuneanu, 2019a).

The sequence stratigraphy of deepwater systems is underlain by the genetic relationship between the stratigraphic shoreline transit cycles on the shelf and the corresponding cycles of change in gravity flows in the deepwater setting (Posamentier and Kolla, 2003; De Gasperi and Catuneanu, 2014; Catuneanu, 2020; Fig. 3). The correlation of shelf and deepwater sequences may be hampered by processes of erosion or nondeposition at the shelf edge or on the slope, but can be demonstrated where age data are available (e.g., by means of biostratigraphy—see Chapter 14; Gutierrez Paredes et al., 2017). However, the preservation of a continuous sedimentation record from the shelf to the deepwater setting is not required for the construction of a deepwater sequence stratigraphic framework (van der Merwe et al., 2010). The distinction between "shelf" and "deepwater" systems applies to shelf-slope settings, which include a shelf edge. However, this distinction can also be extrapolated to ramp settings, which are devoid of a shelf edge, whereby the shoreline transit area is equivalent to the shelf, and the distal area dominated by sediment-gravity flows defines the deepwater setting (Fig. 4). Therefore, in a general sense, the limit between shallow- and deepwater environments is defined by the dominant modes of sediment transport, such as the extent to which gravity controls sedimentary processes, rather than bathymetric or physiographic criteria.

Sedimentation cycles in the deepwater environment can have different origins (e.g., changes in the location of sediment entry points along the shelf edge, changes in the location of depositional elements on the sea floor, changes in relative sea level and/or changes in sediment supply), and not all have stratigraphic significance. The meaning of sedimentary cyclicity

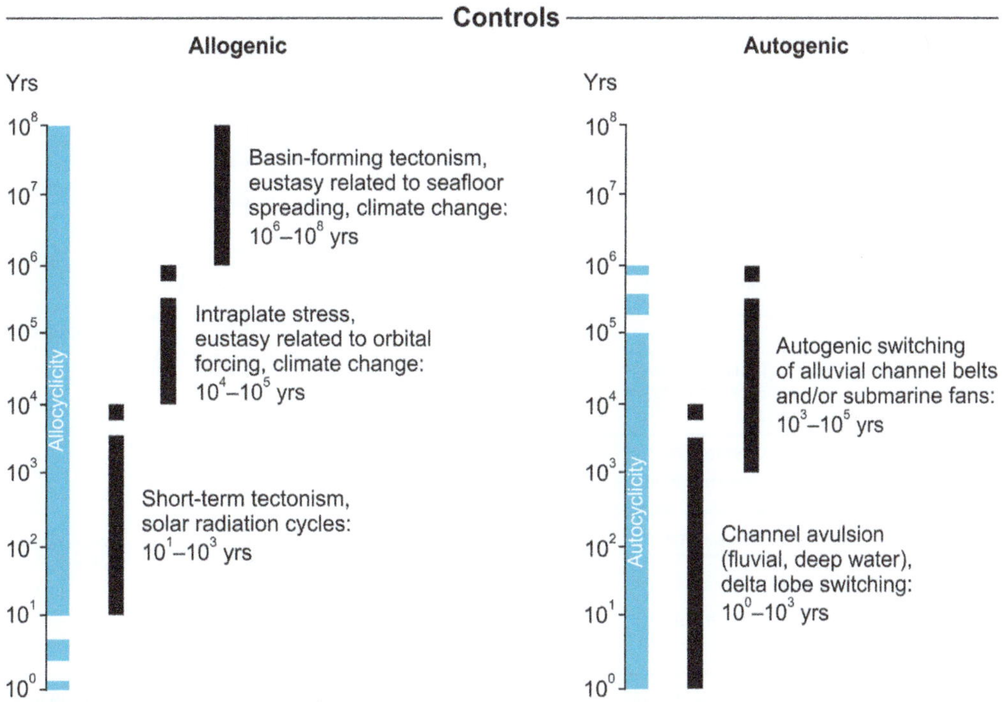


**FIG. 2** Controls on stratigraphic cyclicity. Note the overlap between the timescales of allogenic and autogenic processes. Long-term climate changes on timescales of $10^6$–$10^7$ years relate to greenhouse-icehouse cycles; shorter term climate changes on timescales of $10^4$–$10^5$ years define glacial-interglacial cycles, which are most evident during icehouse regimes. Solar radiation cycles also induce climatic fluctuations associated with a complex interplay of ice-sheet dynamics, atmospheric circulation and thermohaline circulation (Csato et al., 2014). Short-term tectonism includes stages of fault reactivation in fault-bounded sedimentary basins. In addition to tectonism and glacio-isostasy, the rates of subsidence are also affected by sediment loading and compaction during the entire evolution of sedimentary basins. The distinction between the allogenic and autogenic controls on stratigraphic cyclicity is often challenging and has no bearing on the sequence stratigraphic methodology. The observation of stratal stacking patterns, which the methodology is based upon, is decoupled from the interpretation of the underlying controls. *(From Catuneanu, O., 2019a. Model-independent sequence stratigraphy. Earth Sci. Rev. 188, 312–388.)*

in the deepwater setting is difficult to clarify solely from the analysis of vertical profiles (e.g., individual outcrops or wells). Some of the sedimentary cycles observed at specific locations are stratigraphic (i.e., delineated by the recurrence of sequence stratigraphic surfaces), whereas others reflect lateral shifts of depositional elements without a stratigraphic significance (i.e., sedimentological cycles). The separation between sedimentological and stratigraphic cycles requires the reconstruction of regional depositional trends within a 3D stratigraphic framework. Seismic data are critical for this purpose, as they provide the means to clarify the relative chronology of different types of gravity flows at regional scales (Fig. 5). In addition to the grids of 2D seismic lines, 3D seismic data also afford the plan-view visualization of depositional elements (i.e., seismic geomorphology; Fig. 6), providing further constraints to discriminate between the various types of gravity-driven processes (Fig. 7).

The deepwater setting is unique in the sense that the scales of sedimentological and stratigraphic cycles at any location are not mutually exclusive. In all other settings, sedimentological cycles are typically nested within the smallest scale stratigraphic cycles (Fig. 8; Catuneanu, 2019b). The difference relates to the patterns of sediment dispersal and deposition in the deepwater environment, whereby areas away from the paths of gravity-driven transport can accumulate lithologically monotonous pelagites or contourites for periods of time encompassing multiple stratigraphic cycles. At such locations, the facies that contribute to the definition of sequences and systems tracts may be missing and the scale of sedimentological cycles defined by the recurrence of the same types of depositional elements can exceed the scale of stratigraphic cycles. This can deepen the confusion with respect to the sedimentological versus stratigraphic nature of sedimentary cycles observed at specific locations. In this case, clarifying the origin of sedimentary cycles requires correlation with other sections or integration with geochemical data that can reveal stratigraphic patterns within apparently monotonous fine-grained successions (Harris et al., 2013, 2018; LaGrange et al., 2020).

Another source of confusion relates to the relative sea-level changes that are relevant to the formation of stratigraphic cycles in the deepwater setting. As changes in relative sea level are variable across a sedimentary basin, due to variations in

| Shoreline | Shelf edge | | |
|---|---|---|---|
| | Bathymetry | Processes | Trajectory |
| Highstand normal regression | > SWB | Hemipelagic | Aggradation |
| **MFS** | | | |
| Transgression | > SWB<br>0 - SWB | Hemipelagic<br>Mudflows, slumps<br>Finer turbidites | Aggradation or starvation<br>Retrogradation (collapse)<br>Aggradation |
| **MRS** | | | |
| Lowstand normal regression | 0 | Finer turbidites<br>Finer grainflows | Progradation to aggradation |
| **CC** | | | |
| Forced regression | 0<br>SWB - 0<br>> SWB | Coarser grainflows<br>Coarser turbidites<br>Mudflows, slumps<br>Hemipelagic | Progradation with downstepping<br>Aggradation to progradation<br>Retrogradation (collapse)<br>Aggradation |
| **BSFR** | | | |
| Highstand normal regression | > SWB | Hemipelagic | Aggradation |

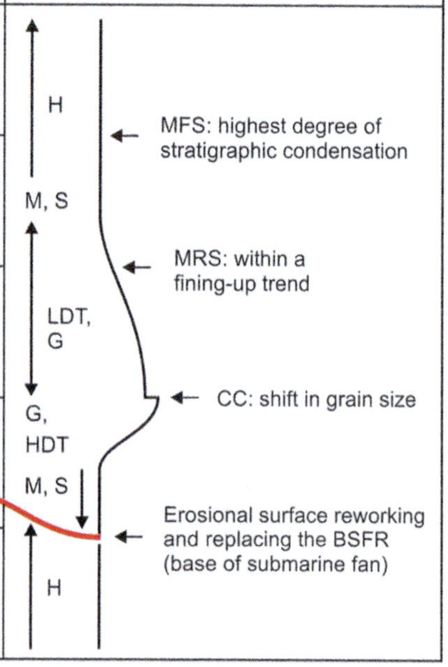


**FIG. 3** Shoreline versus shelf-edge trajectories, and corresponding deepwater processes, in the case of siliciclastic settings that are subject to high-magnitude changes in relative sea level (i.e., relative sea level below the elevation of the shelf edge at the end of forced regression). The shelf edge may or may not backstep during forced regression and transgression, depending on its stability. The delivery of riverborne sediment to the deepwater setting is most effective when the bathymetry at the shelf edge is less than the SWB. Unfilled incised valleys across a submerged shelf may further enhance the transfer of sediment from river mouths to the shelf edge even during stages of highstand in relative sea level. In such cases, as well as in the case of atypically narrow shelves, riverborne sediment supply (both volume and grain size) to the deepwater setting may increase overall. This diagram illustrates common trends observed in the rock record. Any variability in the sedimentological makeup of systems tracts needs to be assessed in the context of each sedimentary basin. *Abbreviations*: BSFR, basal surface of forced regression; CC, correlative conformity; MRS, maximum regressive surface; MFS, maximum flooding surface; SWB, storm wave base; H, hemipelagic sediment; M, mudflows; S, slumps; HDT, high-density turbidity flows; G, grainflows; LDT, low-density turbidity flows.

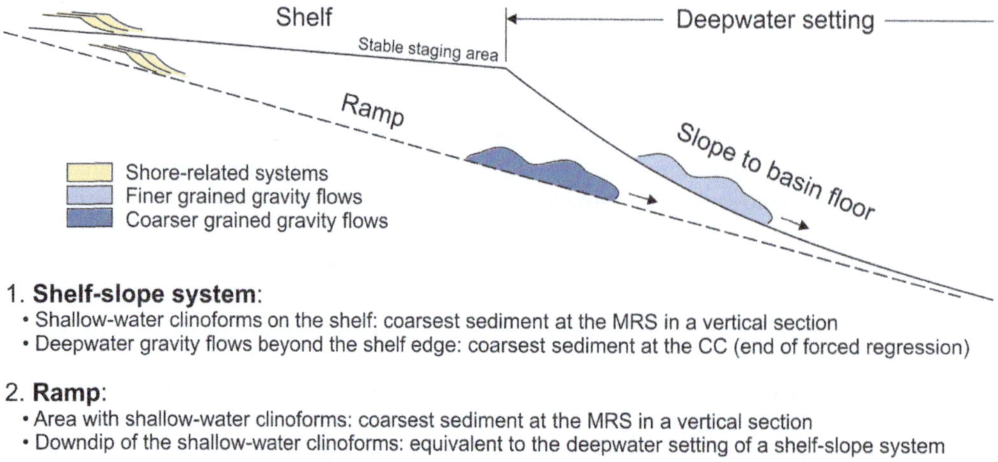


**FIG. 4** Shelf-slope versus ramp systems in siliciclastic settings. Shelf-slope systems may develop in a variety of tectonic settings with low-gradient basin margins (e.g., divergent continental margins, but also some convergent margins, back-arc basins and forelands). Ramps are more typical of fault-bounded basins with steeper gradient basin margins, which promote a coarser clastic sediment supply. Downdip of the shore-related clinoforms, the ramp setting is similar to the deepwater setting of a shelf-slope system in terms of the dominant processes of sediment transport and deposition, except that the gravity flows tend to include coarser sediment. In the absence of gravity flows (e.g., not enough gradient or sediment supply), the ramp beyond the shallow-water clinoforms becomes equivalent to a shelf in terms of processes of sediment transport and deposition (e.g., storm currents, sedimentation from suspension). *Abbreviations*: MRS, maximum regressive surface; CC, correlative conformity.

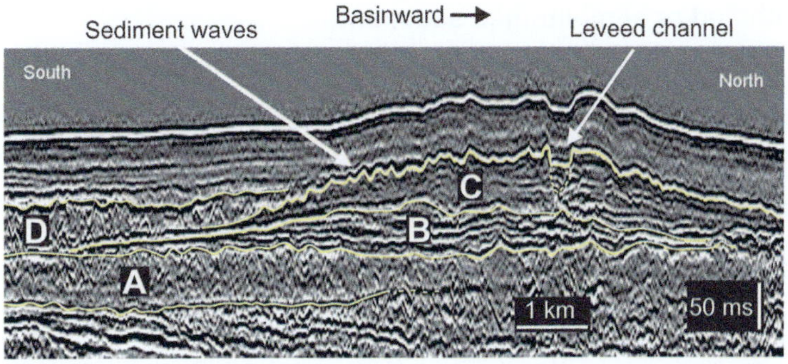


**FIG. 5** Seismic transect showing the common elements of a submarine fan complex in a basin-floor setting. A—mudflow deposits (chaotic internal facies) of the early falling-stage systems tract; B—frontal-splay turbidites (parallel reflections, convex-up top) of the late falling-stage systems tract; C—leveed channel of the lowstand systems tract or early transgressive systems tract; D—mudflow deposits of the late transgressive systems tract. Note the progradation of gravity-flow deposits into the basin from A to C, and the retrogradation from C to D. *(Pleistocene, Gulf of Mexico; modified after Posamentier, H.W., Kolla, V., 2003. Seismic geomorphology and stratigraphy of depositional elements in deep-water settings. J. Sediment. Res. 73(3), 367–388.)*

**FIG. 6** Spectral decomposition images from 3D seismic timeslices, illustrating the morphology of deepwater turbidite channels and splays in a basin-floor setting. *(Mahanadi Basin, offshore India; from Catuneanu, O., 2022. Principles of Sequence Stratigraphy. second ed., Elsevier, Amsterdam; images courtesy of Reliance Industries, India.)*

subsidence rates along dip and strike directions, it becomes important to specify that only the changes in relative sea level *at the shoreline* are relevant to the timing of deepwater stratigraphic sequences. This work addresses the common sources of confusion in the interpretation of deepwater sedimentary cycles and provides guidelines for the construction of sequence stratigraphic frameworks in deepwater settings.

**1. Lithified or semi-lithified sediment:**

| Type of transport | Diagnostic features |
|---|---|
| Rock falls: lithified sediment, small scale<br>Slides: lithified sediment, large scale | Original stratification preserved and undeformed |
| Slumps: semi-lithified sediment, small or large scale | Original stratification preserved and deformed; commonly muddy |

**2. Gravity flows (loose sediment):**

| Type of flow | | Clast-support mechanism |
|---|---|---|
| Fluidal flows | Turbidity flow<br>*Liquefied flow* [1]<br>*Fluidized flow* [1] | Water turbulence<br>Pore-water pressure<br>Water escape |
| Debris flows | Grainflow [2] (non-cohesive debris flow) | Grain collision (sands) |
| | Mudflow (cohesive debris flow) | Grain cohesion (clays) |

[1] secondary flows, within the 'traction carpet' of turbidity flows
[2] stand-alone flows, or within the 'traction carpet' of turbidity flows

**FIG. 7** Classification of gravity-driven processes in deepwater settings. *(Modified after Reading, H.G., 1996. Sedimentary Environments: Processes, Facies and Stratigraphy. third ed., Blackwell Science, 687 pp.)*

| Sedimentary cycles | Defining features | Origin of cycles | Subdivisions | Bounding surfaces |
|---|---|---|---|---|
| STRATIGRAPHY: sequences | Stratigraphic stacking patterns | Allogenic [1] or autogenic [2] | Systems tracts | Sequence stratigraphic |
| SEDIMENTOLOGY: bedsets | Sedimentological stacking patterns | Autogenic [3] or allogenic [4] | Beds, bedsets | Facies contacts |

[1] Interplay of tectonism, climate, and eustasy; [2] Autogenic changes in sediment supply (channels, channel belts, deltas);
[3] Migration of channels, macroforms, bedforms; [4] Tidal or fairweather-storm cycles, seasonal changes in energy/supply.

**FIG. 8** Classification of sedimentary cycles. The distinction between stratigraphic and sedimentological cycles is most challenging in deepwater settings, whereby areas away from the paths of gravity-driven transport can accumulate lithologically monotonous pelagites or contourites for periods of time that span multiple stratigraphic cycles. At such locations, the scale of sedimentological cycles defined by the recurrence of same-type depositional elements can be either smaller or larger than the scale of stratigraphic cycles. The definition of stratigraphic and sedimentological cycles is independent of the interpretation of the underlying controls, as both types of sedimentary cycles can be produced by a combination of allogenic and autogenic processes.

## Controls on stratigraphic cyclicity

Stratigraphic cyclicity in the deepwater setting is controlled by the interplay of accommodation and sedimentation on the shelf, which modifies the sediment supply to the shelf edge. The timing of systems tracts and bounding surfaces depends on the balance between relative sea-level changes and sedimentation in coastal environments, which controls shoreline trajectories and the delivery of extrabasinal and intrabasinal sediment to the deepwater environment (Figs. 3 and 9).

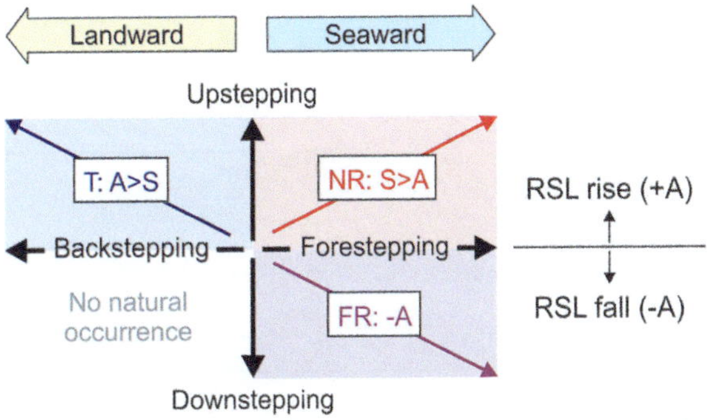


**FIG. 9** Shoreline trajectories, as defined by combinations of lateral (forestepping or backstepping) and vertical (upstepping or downstepping) shoreline shifts. All combinations are common in nature, except for transgression during RSL fall. The stratal stacking patterns that define systems tracts in downstream-controlled settings are linked to shoreline trajectories: normal regression (forestepping and upstepping), forced regression (forestepping and downstepping) and transgression (backstepping and upstepping). The amounts of shoreline upstepping and downstepping quantify the magnitudes of RSL rise and fall, whereas the extent of lateral shoreline shifts depends on the magnitude of RSL changes, sediment supply and topographic gradients. *Abbreviations*: NR, normal regression; FR, forced regression; T, transgression; A, accommodation; S, sedimentation; RSL, relative sea level; +A, positive accommodation; −A, negative accommodation.

Deposition or erosion in the deepwater setting is constrained by several factors, including the mechanisms of sediment transport (e.g., types of gravity-driven processes; Fig. 7) and the physiography of the basin (e.g., seafloor gradients, or barriers to flow). In basins connected to the global ocean, contour currents may also scour the seafloor or accumulate sediment in relation to or independently of relative sea-level changes and shoreline trajectories. In the latter case, the deepwater unconformities and contourites are unrelated to the sequence stratigraphic framework (i.e., they do not provide criteria for the definition of systems tracts and bounding surfaces; Fig. 10). However, if changes in relative sea level modify the paths, energy and sediment load of contour currents, the different products of contour current activity can be rationalized within the sequence stratigraphic framework (Fig. 11; Llave et al., 2001, 2006; Brackenridge et al., 2011; Mestdagh et al., 2019). Contourites can also be integrated in the sequence stratigraphic framework by correlation with areas of accumulation of gravity-flow deposits, where the delineation of sequences is more evident (e.g., Fonnesu et al., 2020).

The distinction between accommodation (i.e., relative sea-level changes) and sedimentation (i.e., base-level changes) is important in coastal environments, where their interplay generates the three types of shoreline trajectory that define the

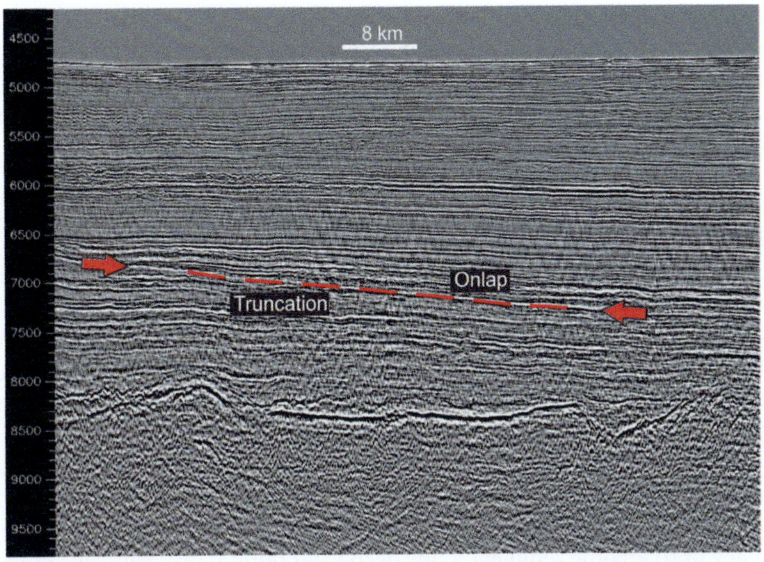


**FIG. 10** Regional unconformity cut by deepwater contour currents in a basin-floor setting (Bay of Bengal, offshore India; image courtesy of Reliance Industries, India). Where erosion is triggered by tilting of continental margins, contourite-bounding unconformities may coincide with sequence stratigraphic surfaces (e.g., a basal surface of forced regression at the onset of continental uplift and forced regression). Where erosion relates only to changes in the pattern of deepwater circulation, the unconformity that marks the new seafloor equilibrium profile may be unrelated to accommodation and shoreline trajectories.

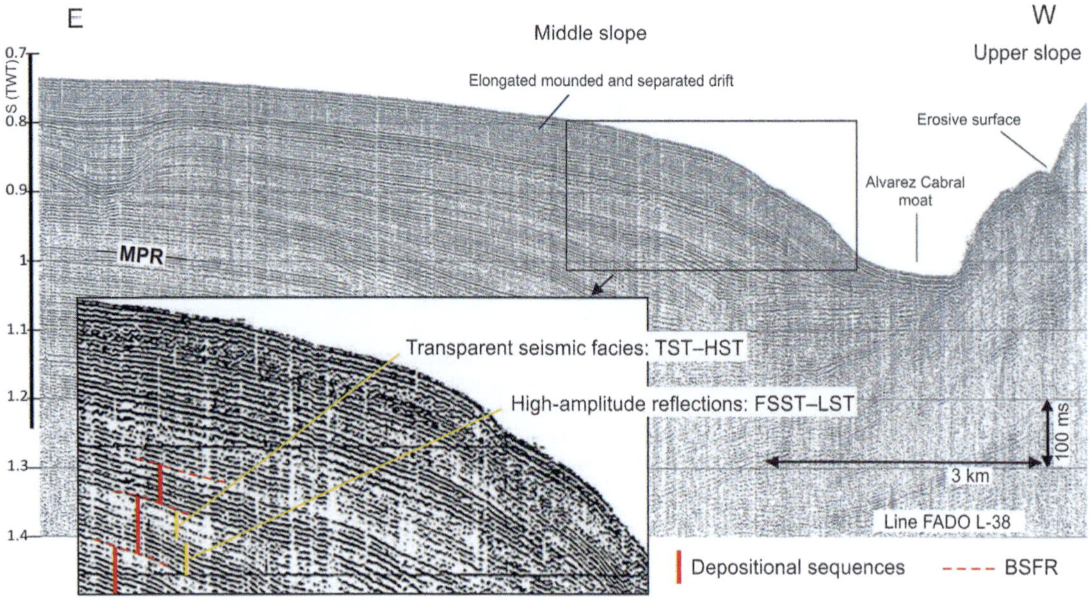


**FIG. 11** Stratigraphic cycles in Quaternary contourites on the continental slope of the Gulf of Cadiz. Changes in seismic facies are attributed to fluctuations in the strength of the Mediterranean outflow water, which in turn are linked to relative sea-level changes. Systems tracts are interpreted based on the correlation of the slope contourites with the sequences identified on the shelf, as afforded by high-resolution seismic data calibrated with IODP (International Ocean Discovery Program) well data. The MPR marks a change from an obliquity-controlled periodicity prior to c.900–920ka (41 kyr glacial-interglacial cycles during the Upper Pliocene-Middle Pleistocene) to a dominant eccentricity-driven periodicity (100 kyr cycles) during the Middle-Late Pleistocene (Llave et al., 2001). Siliciclastic sediment supply to contourite systems is typically higher during forced regressions and lowstands and lower during transgressions and highstands (Mestdagh et al., 2019). *Abbreviations*: MPR, Middle Pleistocene Revolution; TST, transgressive systems tract; HST, highstand systems tract; FSST, falling-stage systems tract; LST, lowstand systems tract; BSFR, basal surface of forced regression. *(Modified after Llave, E., Hernandez-Molina, F.J., Somoza, L., Diaz-del-Rio, V., Stow, D.A.V., Maestro, A., Alveirinho Dias, J.M., 2001. Seismic stacking pattern of the Faro-Albufeira contourite system (gulf of Cadiz): a quaternary record of paleoceanographic and tectonic influences. Mar. Geophys. Res. 22, 487–508, and Mestdagh, T., Lobo, F.J., Llave, E., Hernandez-Molina, F.J., Van Rooij, D., 2019. Review of the late quaternary stratigraphy of the northern Gulf of Cadiz continental margin: new insights into controlling factors and global implications. Earth Sci. Rev. 198, 102944.)*

systems tracts and bounding surfaces (Fig. 9; Catuneanu, 2019a). This distinction becomes less meaningful away from the coastline, where the rates of accommodation and sedimentation no longer influence shoreline trajectories. Both accommodation and sedimentation record a 3D variability across the basin, with rates that change along dip and strike directions; however, only the rates of accommodation and sedimentation at the shoreline are relevant to the timing of systems tracts and bounding surfaces, including those in deepwater settings. For example, the Plio-Pleistocene stratigraphic sequences of the Mississippi fan complex include falling-stage systems tracts that formed during stages of glaciation and forced regression on the shelf (i.e., relative sea-level fall), while the deepwater setting, where deposition occurred, experienced higher rates of subsidence and consequent relative sea-level rise (Weimer, 1990; Weimer and Dixon, 1994; Madof et al., 2019; Catuneanu, 2020).

Sediment supply to the shoreline depends on the production and distribution of extrabasinal and/or intrabasinal sediment. The production of extrabasinal sediment is typically unrelated to changes in relative sea level, reflecting the rates of weathering and erosion of extrabasinal source areas. The production of intrabasinal sediment depends in part on the relative sea level, as well as on other factors such as water temperature and chemistry. Variations in sediment supply to the shoreline can modify the stacking patterns that define systems tracts, but only in terms of the manifestation of normal regressions versus transgressions during stages of relative sea-level rise (Fig. 9). Forced regressions remain controlled solely by stages of relative sea-level fall and play a key role in the definition of stratigraphic cyclicity in deepwater settings (Fig. 3; Posamentier and Kolla, 2003; van der Merwe et al., 2010; De Gasperi and Catuneanu, 2014; Catuneanu, 2020).

While the relative sea level is only one element of the "dual control" on the stratigraphic architecture, changes thereof affect the transfer efficiency of riverborne sediment to the deepwater setting, the shelf capacity to retain extrabasinal sediment and the production of intrabasinal sediment. The transfer efficiency of riverborne sediment to the deepwater setting is higher during relative sea-level fall, as forced regressions typically expose the shelf, bringing the river mouths close to or even at the shelf edge. At the same time, stages of relative sea-level fall decrease the amount of accommodation on the shelf, lowering the shelf capacity to retain extrabasinal sediment. In carbonate settings, falls in relative sea level expose and shut down carbonate factories, lowering the production of carbonate sediment. Therefore, relative sea-level changes, and the distinction between lowstands and highstands in relative sea level, remain critically important for understanding the patterns of sediment distribution across a basin, including sediment supply to the deepwater setting. Any basin-specific variability in the type and amount of sediment supply to the shelf edge (e.g., siliciclastic vs. carbonate settings; the increase in riverborne sediment supply to deepwater in the case of narrow shelves or shelves dissected by unfilled incised valleys; Sinclair and Tomasso, 2002; Clift, 2006; Covault et al., 2007; Covault and Graham, 2010; Sweet and Blum, 2016; Catuneanu, 2020; Sweet, 2020) is not diagnostic to the definition of systems tracts, and needs to be rationalized on a case-by-case basis.

## Stratigraphic versus sedimentological cycles

The nested depositional cycles that build the sedimentary record can be observed from sedimentological to stratigraphic scales (Fig. 8). Within the transit area of a shoreline (i.e., typically on a shelf), where changes in depositional environment are most frequent, sedimentological cycles (beds, bedsets) are always nested within the lowest rank systems tracts and component depositional systems. In this case, the scales of stratigraphic and sedimentological cycles at any location are mutually exclusive: sedimentological cycles form without changes in depositional system and systems tract, and their stacking patterns define the lowest rank systems tracts; in contrast, stratigraphic cycles (i.e., sequences) involve changes in depositional systems and systems tracts and their stacking patterns define systems tracts of higher hierarchical ranks (Catuneanu, 2019b).

Outside of the shoreline transit area (i.e., in fully continental settings and in deepwater settings), stratigraphic cyclicity may develop without changes in depositional environment. In this case, establishing the relationship between sedimentological and stratigraphic cycles depends on the ability to recognize sequences and component systems tracts. In fully continental settings, the development of subaerial unconformities affords the delineation of depositional sequences. In this case, the distinction between sedimentological and stratigraphic cycles is straightforward and bedsets are always nested within the lowest rank sequences. The two types of sedimentary cycles become more difficult to separate in deepwater settings, where both sedimentological and stratigraphic units may develop without significant interruptions in sedimentation (Fig. 12). In this case, sedimentological cycles defined by the recurrence of same-type depositional elements can be either smaller or larger than the stratigraphic cycles that define sequences; i.e., the scales of sedimentological and stratigraphic cycles at specific locations are no longer mutually exclusive.

The sedimentological makeup of deepwater systems depends on sediment supply (extrabasinal vs. intrabasinal), as well as on the interaction and relative contributions of several independent processes of sediment transport and deposition, including gravity-driven mass failures, contour currents and suspension fallout (Fig. 13; details in Chapters 6–10). While all types of deposit are important to evaluate in hydrocarbon reservoir studies, not all are equally relevant to the construction of the sequence stratigraphic framework. Specifically, the sequence stratigraphic framework describes the cyclicity recorded by gravity-driven processes, which is linked to shoreline trajectories and can be used for the designation of systems tracts (Fig. 3). For this reason, sedimentological work must precede the stratigraphic analysis in order to separate the depositional elements that originate from processes with different degrees of relevance to sequence stratigraphy.

The distinction between stratigraphic cycles and sedimentological cycles in deepwater settings is difficult based only on vertical profiles (e.g., well data). The identification of stratigraphic trends requires an integration of seismic and sedimentological data from multiple locations in order to constrain the relative chronology of the different types of gravity-driven processes at regional scales (Figs. 5 and 12). Without this regional perspective, the sedimentological versus stratigraphic meaning of sedimentary cycles observed at specific locations can be equivocal. In areas with frequent gravity flows, sedimentological cycles are usually nested within stratigraphic cycles (Fig. 12). However, in areas of continuous sedimentation dominated by nondiagnostic processes (i.e., hemipelagic fallout, contour currents), sedimentological cycles defined by the recurrence of same-type depositional elements (e.g., mudflow deposits) can also exceed the scale of stratigraphic cycles.

## Sequence stratigraphic framework

The same types of systems tracts and bounding surfaces can form in all downstream-controlled tectonic and depositional settings and at all stratigraphic scales (Catuneanu, 2019a, b). The deepwater setting is no exception, with a stratigraphic architecture that includes falling-stage, lowstand, transgressive and highstand systems tracts (Fig. 14). The formation of these systems tracts and their bounding sequence stratigraphic surfaces is linked to shoreline trajectories and changes thereof on the shelf, which modify the sediment supply to the shelf edge during the shoreline shelf-transit cycles (Fig. 3). The dominant and subordinate types of gravity flows that contribute to the development of deepwater sequences

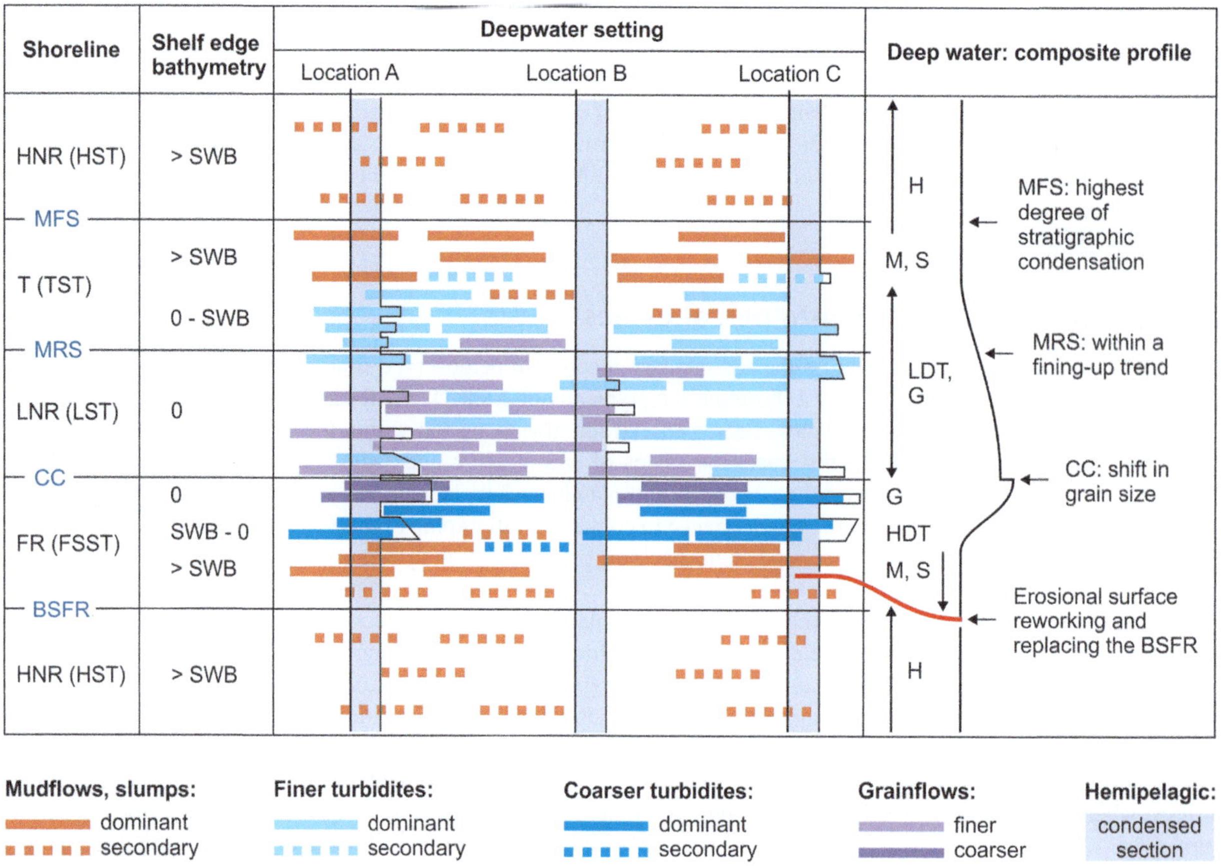


**FIG. 12** Stratigraphic cyclicity in the deepwater setting during a shoreline shelf-transit cycle. The composite profile describes the relative chronology of the dominant gravity-driven processes in the case of wide and low-gradient siliciclastic shelves that are subject to high-magnitude changes in relative sea level (i.e., shelves exposed at times of lowstand, and submerged at times of highstand). The sedimentological makeup of the composite profile (e.g., grain size and the types of gravity flows) may be modified by basin-specific factors. The stratigraphic criteria that define the deepwater sequence stratigraphic framework relate to the *trends of change* in gravity flows during the shoreline transit cycles on the shelf and are independent of the sedimentological variability of the composite profile. Note the difference in terms of timing, frequency and grading trends between the stratigraphic cycle defined by the composite profile and the sedimentological cycles at specific locations. The latter reflect the shifts in the place of accumulation of depositional elements during a stratigraphic cycle. *Abbreviations*: HNR, highstand normal regression; FR, forced regression; LNR, lowstand normal regression; T, transgression; HST, highstand systems tract; FSST, falling-stage systems tract; LST, lowstand systems tract; TST, transgressive systems tract; BSFR, basal surface of forced regression; CC, correlative conformity; MRS, maximum regressive surface; MFS, maximum flooding surface; SWB, storm wave base; H, hemipelagic sediment; M, mudflows; S, slumps; HDT, high-density turbidity flows; G, grainflows; LDT, low-density turbidity flows. *(Modified from Catuneanu, O., 2019b. Scale in sequence stratigraphy. Mar. Pet. Geol. 106, 128–159.)*

during the various stages of a relative sea-level cycle are summarized in Fig. 15, based on outcrop and subsurface studies (Posamentier and Kolla, 2003; van der Merwe et al., 2010; De Gasperi and Catuneanu, 2014; Gutierrez Paredes et al., 2017; Sweet et al., 2019; Sweet, 2020). However, departures from the common trends may be recorded as a function of basin-specific variations in physiography and sediment supply (e.g., narrower and steeper shelves enable the delivery of coarser siliciclastic sediment to the shelf edge, thus increasing the contribution of high-density turbidity flows and grainflows to the deepwater system; Sinclair and Tomasso, 2002; Clift, 2006; Covault and Graham, 2010; Sweet and Blum, 2016; Catuneanu, 2020; Sweet, 2020).

The composition and the relative development of systems tracts in deepwater depend to a large extent on the depositional setting on the shelf. In siliciclastic settings, the highest sediment supply to the shelf edge is commonly delivered at the end of forced regression (i.e., "lowstand shedding"; Posamentier and Allen, 1999; Catuneanu, 2006; Sweet et al., 2019), while the lowest sediment supply is typically associated with the maximum flooding surface (e.g., Gutierrez Paredes et al., 2017; Fig. 3). In carbonate settings, sediment delivery to the shelf edge follows different trends, with the highest supply during times of highstand, when the carbonate factory is usually most productive (i.e., "highstand shedding"; Schlager et al., 1994). These "reciprocal sedimentation" patterns, which involve the "lowstand shedding" of clastics and the "highstand shedding" of carbonates, have been observed before the emergence of sequence stratigraphy (Van Siclen, 1958;

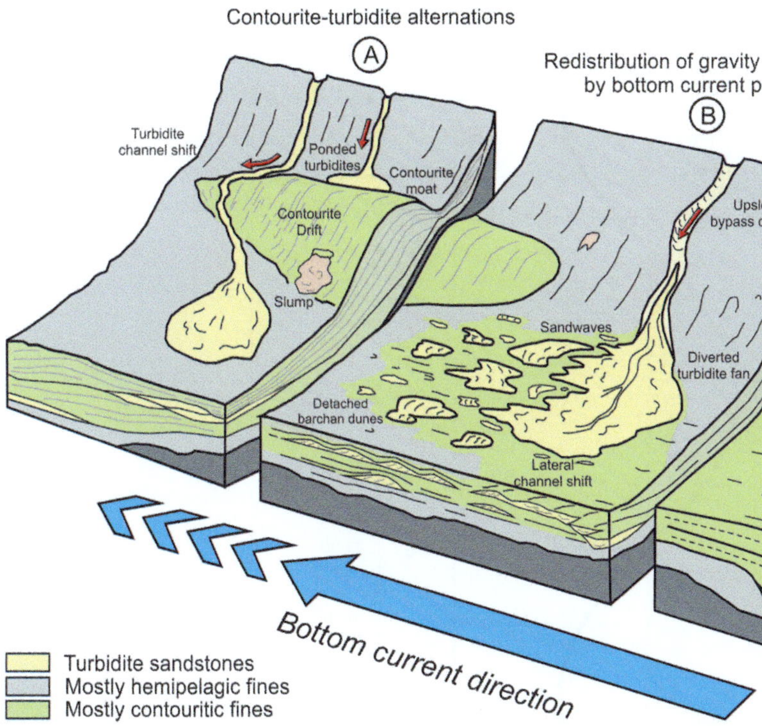


**FIG. 13** Interplay of gravity flows and contour currents in mixed deepwater systems. Contour currents may alternate with gravity flows (A), redistribute previously accumulated gravity-flow deposits (B) or interact with gravity flows at *syn*-depositional time (C). While all types of deposits are important to evaluate in hydrocarbon reservoir studies, only gravity flows provide the criteria for the identification of systems tracts (see text for details). *(Modified after Fonnesu, M., Palermo, D., Galbiati, M., Marchesini, M., Bonamini, E., Bendias, D., 2020. A new world-class deep-water play type, deposited by the syndepositional interaction of turbidity flows and bottom currents: the giant Eocene coral field in northern Mozambique. Mar. Pet. Geol. 111, 179–201.)*

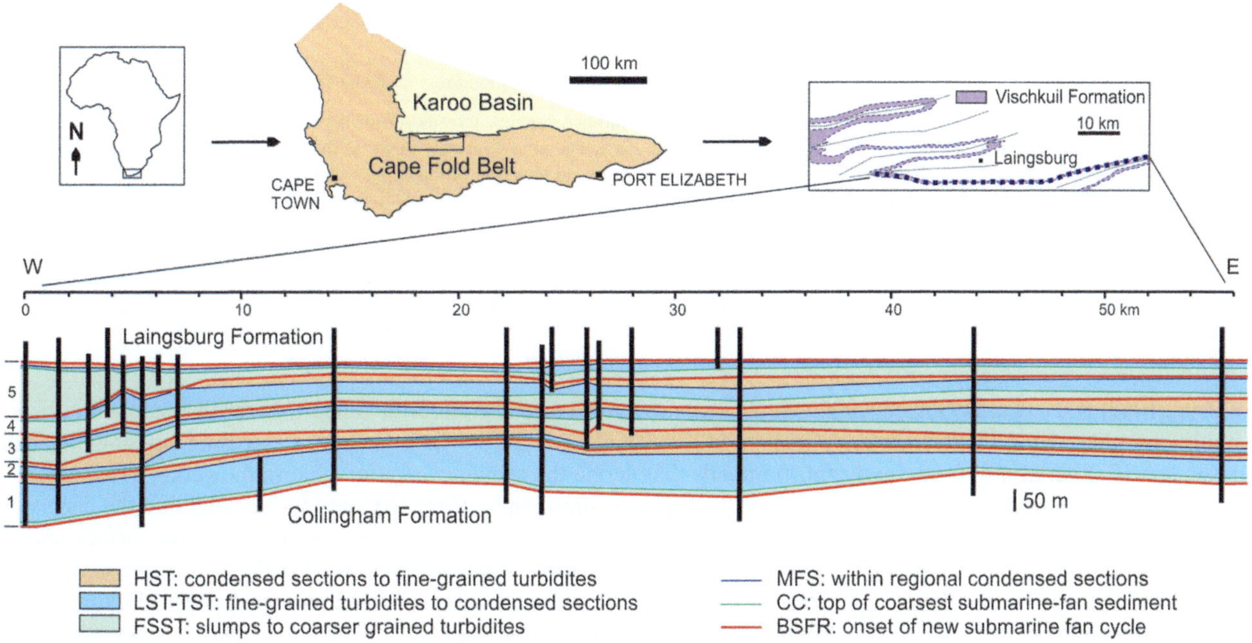


**FIG. 14** Sequence stratigraphy of the Permian Vischkuil Formation in a deepwater setting. The Vischkuil Formation consists of five depositional sequences (1–5), which correspond to five submarine fan complexes. Stratigraphic cyclicity is controlled by glacio-eustatic fluctuations during an icehouse climatic regime, likely on timescales of $10^5$ years. Note the lateral continuity of the stratigraphic architecture across the low-relief basin floor. Vertical bars represent measured outcrop sections. *Abbreviations*: FSST, falling-stage systems tract; LST, lowstand systems tract; TST, transgressive systems tract; HST, highstand systems tract; BSFR, basal surface of forced regression; CC, correlative conformity; MFS, maximum flooding surface. *(Karoo Basin, South Africa; modified after van der Merwe, W.C., Flint, S.S., Hodgson, D.M., 2010. Sequence stratigraphy of an argillaceous, deepwater basin-plain succession: Vischkuil formation (Permian), Karoo Basin, South Africa. Mar. Pet. Geol. 27, 321–333.)*

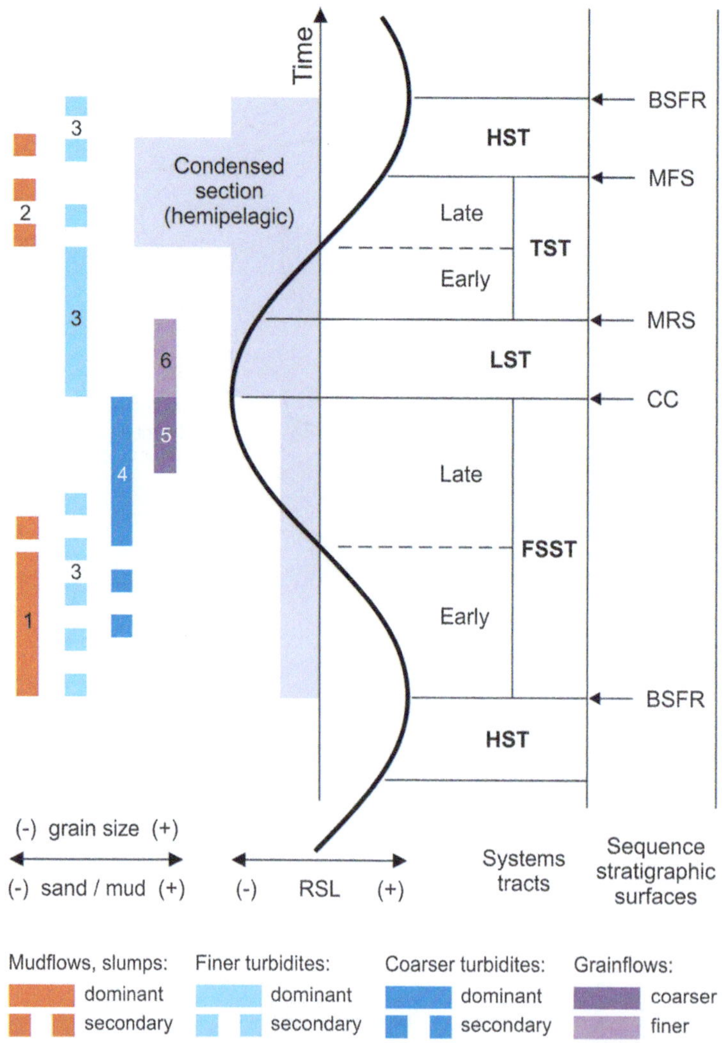


**FIG. 15** Commonly observed trends of change in deepwater gravity flows during a shoreline transit cycle on the shelf. Diagnostic for the identification of systems tracts and bounding surfaces are the trends of change rather than the actual types of gravity flows, as the latter depend on several basin-specific variables (see text for details). Conducive factors: 1—muds on the outer shelf, and lowering of the storm wave base; 2—muds on the outer shelf, and instability at the shelf edge; 3—mixed mud and sand at the shelf edge, lower sand-to-mud ratio; 4—mixed sand and mud at the shelf edge, higher sand-to-mud ratio; 5—coastal systems at the shelf edge, coarser sediment; 6—coastal systems at the shelf edge, finer sediment. *Abbreviations*: FSST, falling-stage systems tract; LST, lowstand systems tract; TST, transgressive systems tract; HST, highstand systems tract; BSFR, basal surface of forced regression; CC, correlative conformity; MRS, maximum regressive surface; MFS, maximum flooding surface; RSL, relative sea level. *(Modified from De Gasperi, A., Catuneanu, O., 2014. Sequence stratigraphy of the Eocene turbidite reservoirs in Albacora field, Campos Basin, offshore Brazil. AAPG Bull. Am. Assoc. Pet. Geol. 98(2), 279–313.)*

Wilson, 1967) and can affect the development of both gravity-driven and contourite systems. Notwithstanding these commonly observed trends, the deepwater systems need to be analyzed case-by-case, in order to separate any nondiagnostic (basin specific) variability from the diagnostic trends that define the sequence stratigraphic framework.

Among the seven surfaces of the "conventional" sequence stratigraphy (Fig. 16), only four extend into the deepwater environment: the basal surface of forced regression, the correlative conformity, the maximum regressive surface and the maximum flooding surface (Figs. 3, 12 and 15). The basal surface of forced regression and the correlative conformity, which mark changes in the direction of shift of the relative sea-level, tend to have the strongest physical expression in the deepwater setting (Figs. 3 and 12). Field data indicate that the basal surface of forced regression marks an increase in siliciclastic sediment supply and/or a decrease in carbonate sediment supply to the shelf edge (Fig. 17). In contrast, the correlative conformity marks a decrease in siliciclastic sediment supply and/or an increase in carbonate sediment supply to the shelf edge (Fig. 18). These trends relate to the shelf capacity to retain riverborne sediment and to produce carbonate sediment, both lowered during relative sea-level fall. Seismic data also reveal that the basal surface of forced regression is

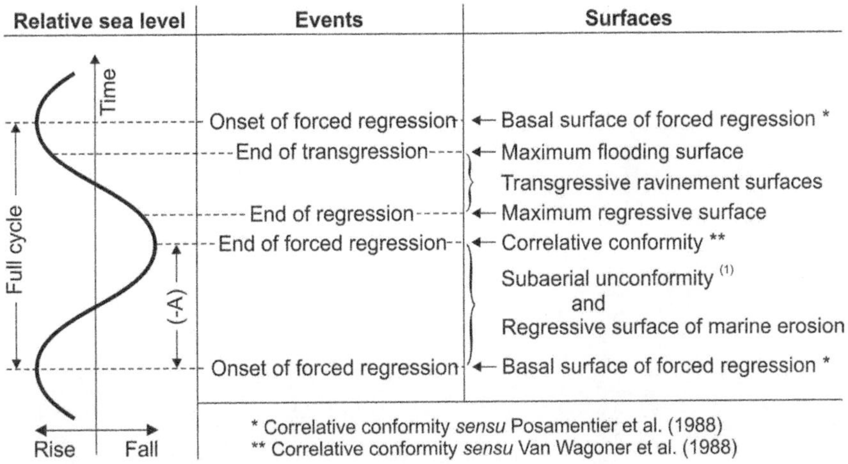


**FIG. 16** Timing of sequence stratigraphic surfaces relative to the main events of a relative sea-level cycle. *Notes*: [(1)] where the shoreline trajectory during forced regression is steeper than the landscape gradient; subaerial unconformities may also form during transgression where the landscape gradient is steeper than the shoreline trajectory (Catuneanu and Zecchin, 2016); (−A), negative accommodation. *(Modified after Catuneanu, O., Willis, A.J. and Miall, A.D., 1998. Temporal significance of sequence boundaries. Sediment. Geol. 121, 157–178; Embry, A.F., Catuneanu, O., 2002. Practical Sequence Stratigraphy: Concepts and Applications. Canadian Society of Petroleum Geologists, Short Course Notes, 147 pp.)*

commonly scoured by the passage of early falling-stage mudflows (Fig. 19A), whereas the correlative conformity marks a change from coarser to finer-grained turbidites (Fig. 19B).

The maximum regressive surface and the maximum flooding surface tend to be lithologically cryptic, as they develop during relative sea-level rise within fining-upward and condensed intervals, respectively (Figs. 3 and 12). Seismic data indicate no discernable difference between the lowstand and early transgressive turbidity flows (Fig. 5), and a paucity or absence of mudflows within the lowstand—early transgressive interval that encompasses the maximum regressive surface (Fig. 19C). Similarly, the plan-view expression of the maximum flooding surface is often undistinguishable from that of other paleoseafloors within condensed sections that span the underlying late transgressive and the overlying highstand deposits (Fig. 19D). For these reasons, the identification of sequence stratigraphic surfaces in the subsurface is best approached starting with the analysis of 2D seismic lines, which afford the observation of stratal terminations and stacking patterns, particularly along dip directions (Fig. 20). Once the systems tracts are identified based on 2D geometrical relationships, the bounding sequence stratigraphic surfaces can be imaged in plan view where 3D seismic data are available (Fig. 19).

Despite the cryptic expression of maximum regressive and maximum flooding surfaces in deepwater settings, sedimentological, biostratigraphic and geochemical criteria may still be defined to identify these surfaces within lithologically monotonous successions (Donovan et al., 2015; Turner et al., 2015, 2016; Gutierrez Paredes et al., 2017; Dong et al., 2018; Harris et al., 2018; Catuneanu, 2019a; LaGrange et al., 2020). In sedimentological terms, the maximum regressive surface approximates the end of grainflow deposition (e.g., collapse of beach sands at the shelf edge; Figs. 12 and 15), where siliciclastic shelves become entirely continental by the end of progradation. In carbonate settings, the maximum regressive surface may coincide with a drowning surface where the subsequent transgression and water deepening are rapid (Fig. 21). The maximum flooding surface typically marks the time of lowest sediment supply to deepwater in both siliciclastic and carbonate settings (Figs. 3, 12, 15 and 22), which often results in the highest abundance of plankton within the condensed sections (Fillon, 2007; Gutierrez Paredes et al., 2017).

The timing of the basal surface of forced regression and the correlative conformity is independent of sediment supply; i.e., these surfaces mark the onset and the end of relative sea-level fall, respectively (Fig. 15). In contrast, the maximum regressive surface and the maximum flooding surface form during relative sea-level rise, with a timing controlled in part by the sediment supply to the shoreline (Fig. 15). Relative sea-level changes modify the sediment supply to the shelf edge in both siliciclastic and carbonate settings, and remain an important component of the "dual control" on the stratigraphic architecture. The relationship between the relative sea level and the timing of sequence stratigraphic surfaces is evident in all depositional settings, from carbonate or mixed carbonate-siliciclastic (Figs. 17 and 18) to fully siliciclastic (Figs. 14 and 23). This relationship, as well as the identification of deepwater systems tracts and bounding surfaces, can be further validated by the correlation of deepwater systems with the shallow-water systems on the shelf, wherever enabled by stratal preservation and the availability of data (Figs. 17 and 20).

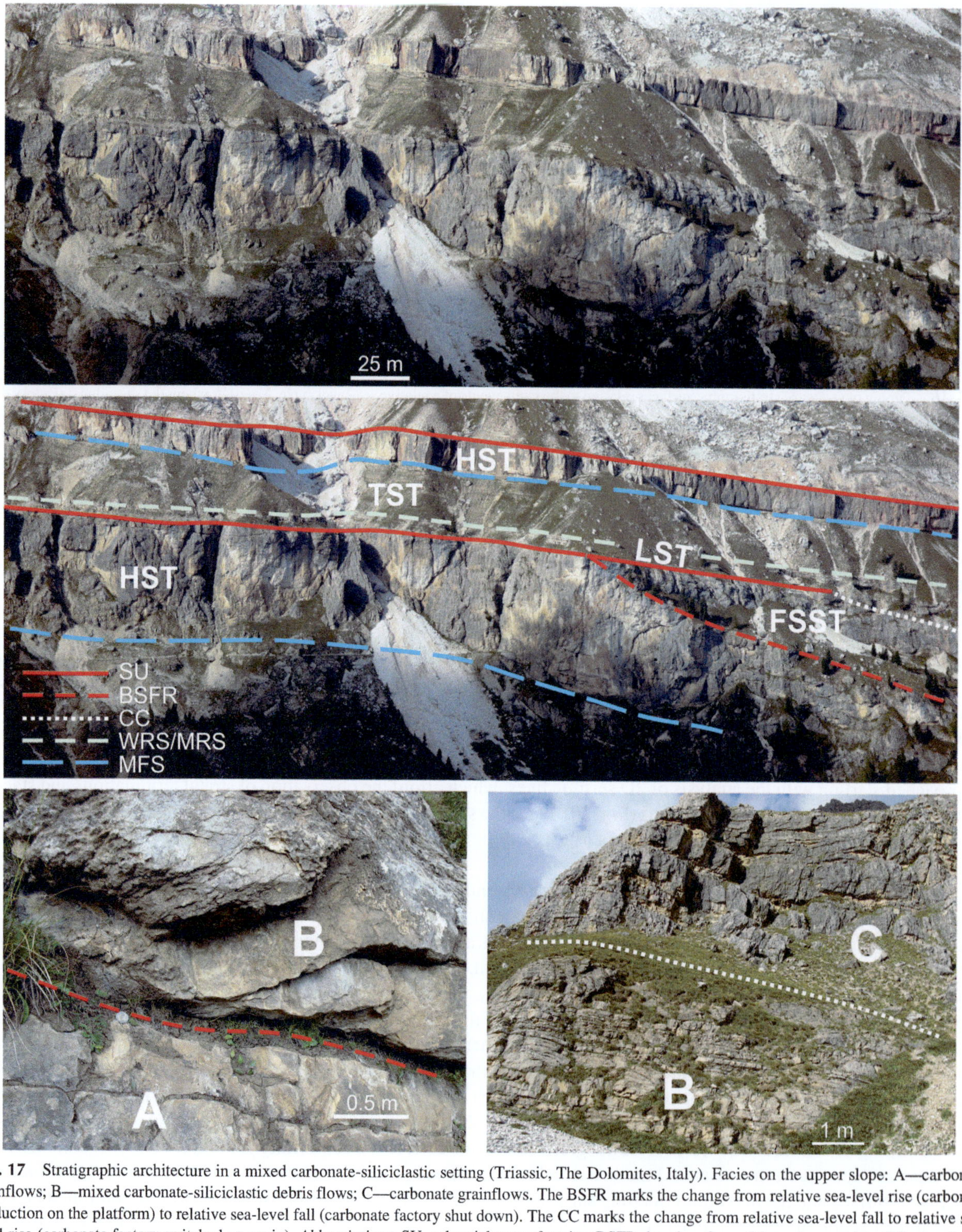


**FIG. 17** Stratigraphic architecture in a mixed carbonate-siliciclastic setting (Triassic, The Dolomites, Italy). Facies on the upper slope: A—carbonate grainflows; B—mixed carbonate-siliciclastic debris flows; C—carbonate grainflows. The BSFR marks the change from relative sea-level rise (carbonate production on the platform) to relative sea-level fall (carbonate factory shut down). The CC marks the change from relative sea-level fall to relative sea-level rise (carbonate factory switched on again). *Abbreviations*: SU, subaerial unconformity; BSFR, basal surface of forced regression; CC, correlative conformity; WRS/MRS, wave-ravinement surface reworking a maximum regressive surface; MFS, maximum flooding surface; HST, highstand systems tract; FSST, falling-stage systems tract; LST, lowstand systems tract; TST, transgressive systems tract.

**FIG. 18** Outcrop example of a correlative conformity (CC) in a mixed carbonate-siliciclastic deepwater setting (Triassic, The Dolomites, Italy; note person for scale). A—Siliciclastic turbidites of high-density flows (coarser grained, with a high sand-to-mud ratio, dominated by the divisions A and B of the Bouma sequence); B—carbonate turbidites of low-density flows (finer grained, with a low sand-to-mud ratio, dominated by the divisions C, D and E of the Bouma sequence). The CC marks a decrease in grain size and a change from siliciclastic riverborne sediment (relative sea-level fall: carbonate factory shut down) to carbonate sediment (relative sea-level rise: carbonate factory switched on).

## Sequences in fine-grained successions

The application of geochemical methods to constrain the sequence stratigraphic framework is generally underutilized, but is emerging as a powerful tool with great potential, particularly in fine-grained successions where the conventional methods of facies analysis are less effective (e.g., Harris et al., 2013, 2018; Playter et al., 2018; LaGrange et al., 2020). Both organic and inorganic geochemistry have relevance to sequence stratigraphy and can be used to identify systems tracts and bounding surfaces in mud-dominated successions and to reconstruct the history of relative sea-level changes during the evolution of a sedimentary basin (e.g., Slingerland et al., 1996; Harris et al., 2013, 2018; Dong et al., 2018).

### Organic geochemistry

Early work on the variability of total organic carbon (TOC) in fine-grained successions established a general relationship between paleobathymetry and the amount of TOC in the sediment, leading to the conclusion that the higher TOC amounts correlate with stages of transgression and water deepening (Creaney and Passey, 1993). According to this model, the highest TOC is expected at maximum flooding surfaces, which mark the end of transgressions, based on the idea that oxygen levels at the seafloor are controlled by water depth (i.e., as the water depth increases, so do the anoxia and the preservation potential of the organic matter that accumulates on the seafloor). Subsequent work demonstrated that in addition to *syn*-depositional bathymetric conditions and oxygen levels, other factors such as nutrient supply to the water column and sediment supply to the seafloor can also influence significantly the amounts of TOC in the sedimentary record (e.g., Hay, 1995; Tyson, 2001; Arthur and Sageman, 2005; Katz, 2005). However, because all these factors can be affected by one underlying control (i.e., changes in relative sea level), the general conclusions of Creaney and Passey (1993) still stand, as they often match the observations in many case studies (e.g., Slingerland et al., 1996; Fleck et al., 2002; Arthur and Sageman, 2005; Bohacs et al., 2005; Turner et al., 2016; Dong et al., 2018; Harris et al., 2018).

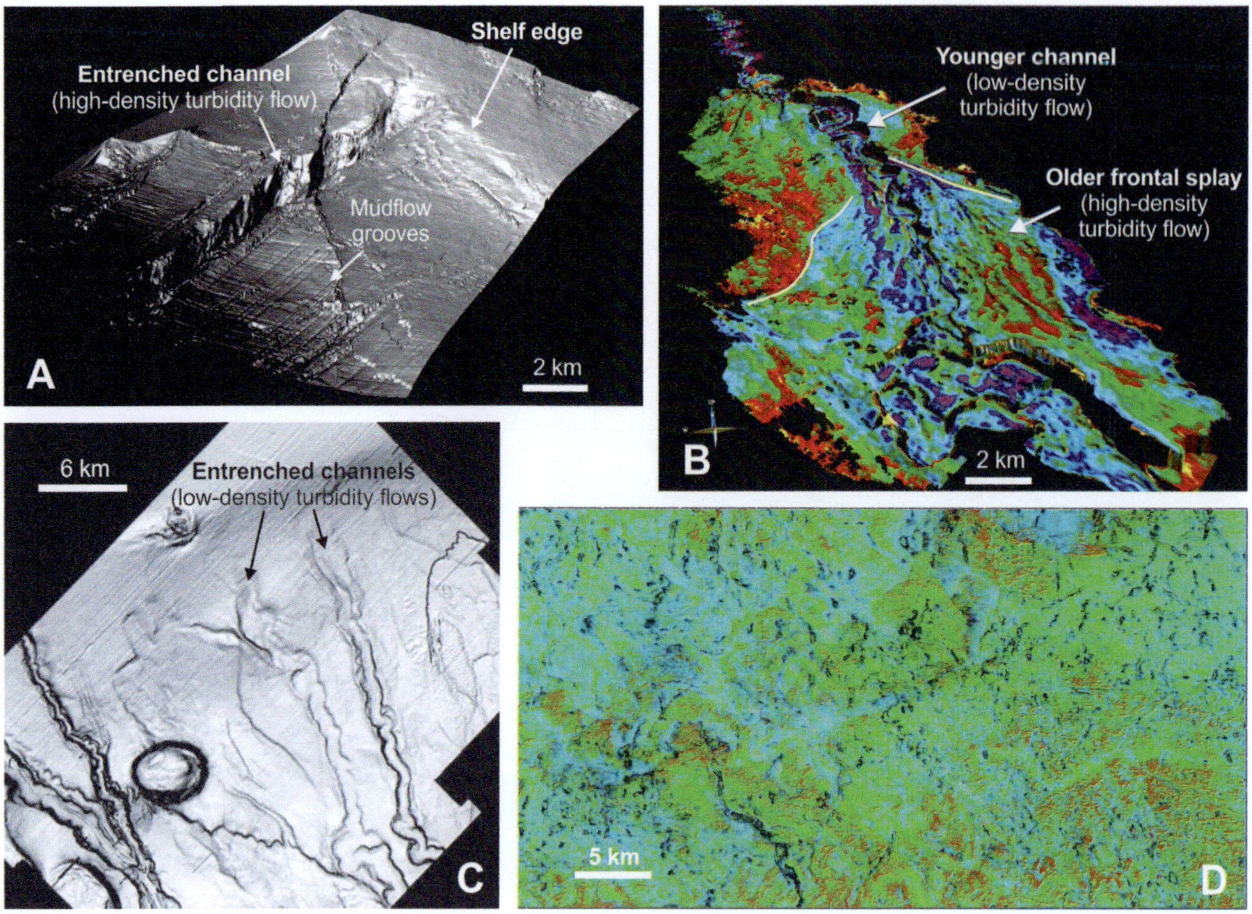


**FIG. 19** Plan-view expression of stratigraphic surfaces in deepwater settings from 3D seismic horizon slices. A—basal surface of forced regression (slope setting, Gulf of Mexico); B—correlative conformity (change from high- to low-density turbidity flows; slope setting, Gulf of Mexico); C—maximum regressive surface (slope setting, Gulf of Mexico); D—maximum flooding surface (within a condensed section, with fluid-escape faults; basin-floor setting, Bay of Bengal). The stratigraphic position of surfaces A, B and C is indicated in Fig. 20. *(From Catuneanu, O., 2022. Principles of Sequence Stratigraphy. second ed., Elsevier, Amsterdam; images A, B, and C courtesy of Henry Posamentier; image D courtesy of Reliance Industries, India.)*

In general terms, the accumulation and preservation of organic matter in organic-rich sediments require three main conditions (Dong et al., 2018):

1. Enhanced organic productivity (Pedersen and Calvert, 1990; Sageman et al., 2003; Tyson, 2001, 2005; Wei et al., 2012), which generally requires a high nutrient supply to the water column (Hay, 1995). In turn, nutrient supply can be enhanced by several factors, including upwelling of bottom waters, river runoff and nutrient cycling (Pedersen and Calvert, 1990; Arthur and Sageman, 2005; Algeo and Ingall, 2007), acting independently or in combination.
2. Enhanced preservation of organic matter (Demaison and Moore, 1980; Arthur and Sageman, 1994; Mort et al., 2007), which requires reducing conditions (i.e., oxygen deficiency) in the bottom waters.
3. Low sedimentation rate (Creaney and Passey, 1993; Tyson, 2005), which ensures a minimal dilution of the organic matter with detrital sediment. The effect of sediment supply on the accumulation and preservation of organic matter is, however, complex, as it can not only dilute the concentration of organic carbon (i.e., negative effect on the TOC amount) but also protect the organic matter on the seafloor from an oxygenated water column (i.e., positive effect on the TOC amount) (Tyson, 2001; Katz, 2005).

All three factors described above (i.e., productivity, preservation and sediment supply) interact to constrain the TOC amounts found in organic-rich petroleum source rocks or unconventional petroleum reservoirs (Tyson and Pearson, 1991; Harris et al., 2004; Rimmer et al., 2004; Bohacs et al., 2005; Katz, 2005). Despite the complexity of the variables

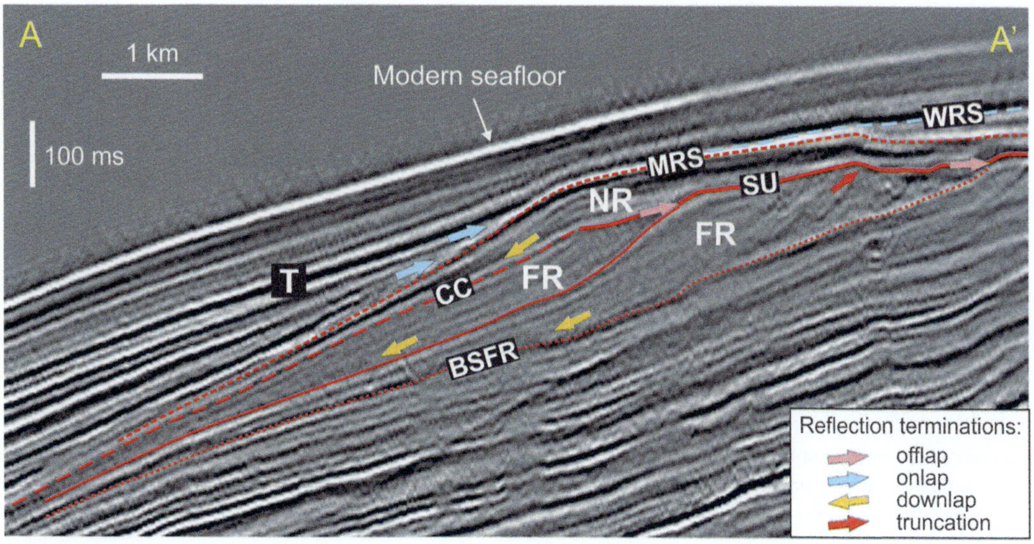


**FIG. 20** Stratal stacking patterns at seismic scale, in a shelf to slope setting (Pleistocene-Holocene, Gulf of Mexico; data courtesy of Henry Posamentier). The forced regression relates to the Late Pleistocene glaciation (sea-level fall and exposure of the continental shelf accompanied by fluvial erosion). The location of the shoreline at the shelf edge during the late stages of forced regression and lowstand normal regression enabled the progradation of the shelf-slope system, as seen on the 2D seismic line. The topset of the lowstand systems tract (NR) is a fluvial system at the seismic scale. However, this topset includes lower rank depositional systems at subseismic scales, from fluvial to shallow marine, related to higher frequency shoreline shelf-transit cycles. The clinoform surfaces within the seismic-scale systems tracts indicate episodic progradation during the higher frequency cycles. *Abbreviations*: FR, forced regression (progradation with downstepping); NR, normal regression (progradation with upstepping); T, transgression (backstepping); BSFR, basal surface of forced regression; SU, subaerial unconformity; CC, correlative conformity; MRS, maximum regressive surface; WRS, wave-ravinement surface.

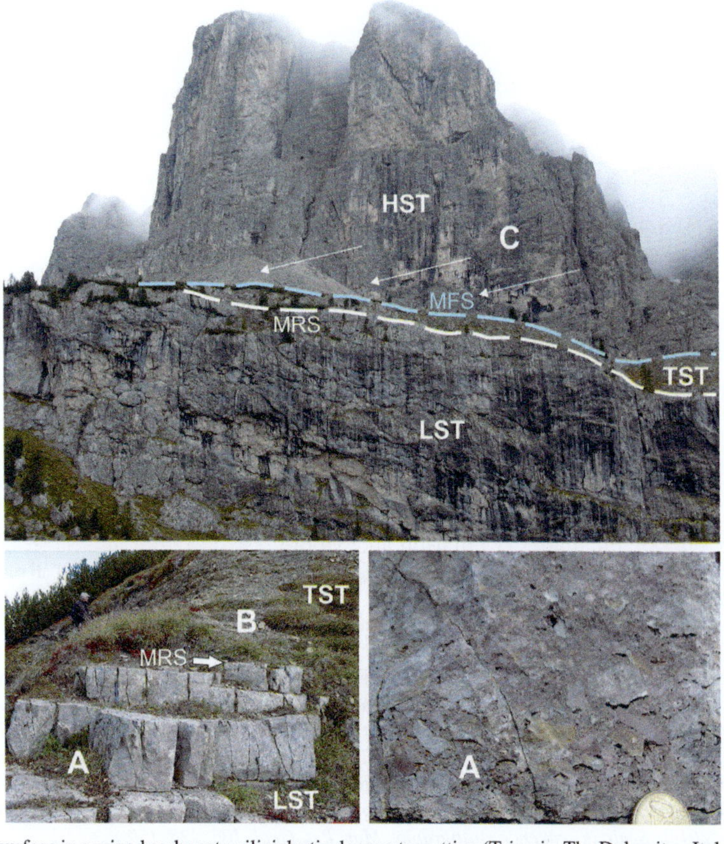


**FIG. 21** Maximum regressive surface in a mixed carbonate-siliciclastic deepwater setting (Triassic, The Dolomites, Italy). In this example, the maximum regressive surface coincides with a drowning surface generated by rapid transgression and water deepening. Facies: A—slope calcirudites; B—transgressive shales; C—slope calcarenites. *Abbreviations*: LST, lowstand systems tract; TST, transgressive systems tract; HST, highstand systems tract; MRS, maximum regressive surface; MFS, maximum flooding surface.

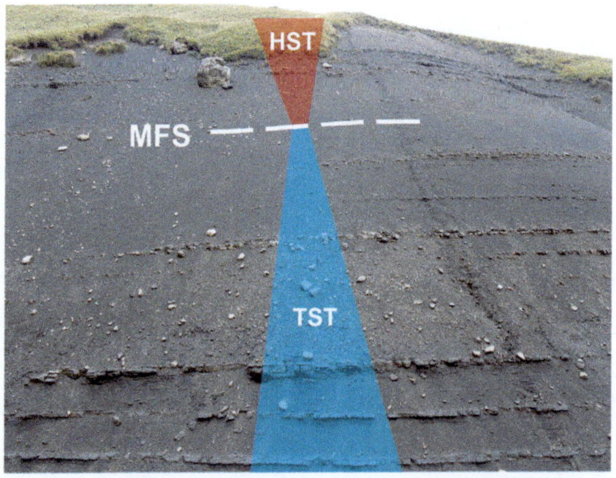


**FIG. 22** Maximum flooding surface (MFS) in a mixed carbonate-siliciclastic deepwater setting (Triassic, The Dolomites, Italy). The MFS corresponds to a time of minimum influx of platform sediment to the deepwater setting, leading to the highest degree of stratigraphic condensation. The frequency of turbidity flows decreases during transgression and increases during subsequent highstand normal regression. *Abbreviations*: TST, transgressive systems tract; HST, highstand systems tract.

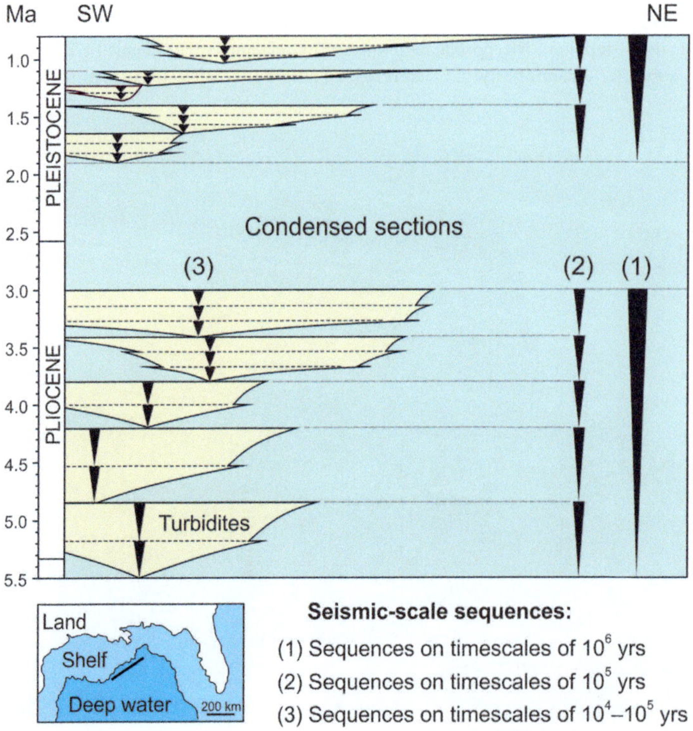


**FIG. 23** Nested architecture of stratigraphic cycles in the deepwater setting of the Gulf of Mexico. Sequences of all scales record an increase in sediment supply during deposition, driven by forced regressions of corresponding magnitudes that dominated the Plio-Pleistocene icehouse. In this example, sequences of all hierarchical ranks are dominated by the falling-stage systems tract, which is a typical signature of icehouse regimes. Coeval with the forced regressions on the shelf (i.e., stages of relative sea-level fall), which controlled the timing of sequences, the deepwater setting was subject to higher rates of subsidence and consequent relative sea-level rise. Only the relative sea-level changes *at the shoreline* are relevant to the timing of stratigraphic cycles (see discussion in Catuneanu, 2019a). *(See map for location; modified from Weimer, P., 1990. Sequence stratigraphy, facies geometries, and depositional history of the Mississippi Fan, Gulf of Mexico. Bull. Am. Assoc. Pet. Geol. 74, 425–453; Weimer, P., Dixon, B.T., 1994. Regional sequence stratigraphic setting of the Mississippi Fan complex, northern deep Gulf of Mexico: implications for evolution of the northern Gulf basin margin. In: Weimer, P., et al. (Eds.), Submarine Fans and Production Characteristics. Society for Sedimentary Geology (SEPM), Gulf Coast Section, 15th Annual Research Conference, Houston, Texas, pp. 373–381; Madof, A.S., Harris, A.D., Baumgardner, S.E., Sadler, P.M., Laugier, F.J., Christie-Blick, N., 2019. Geology 47(6), 545–549; Catuneanu, O., 2019b. Scale in sequence stratigraphy. Mar. Pet. Geol. 106, 128–159.)*

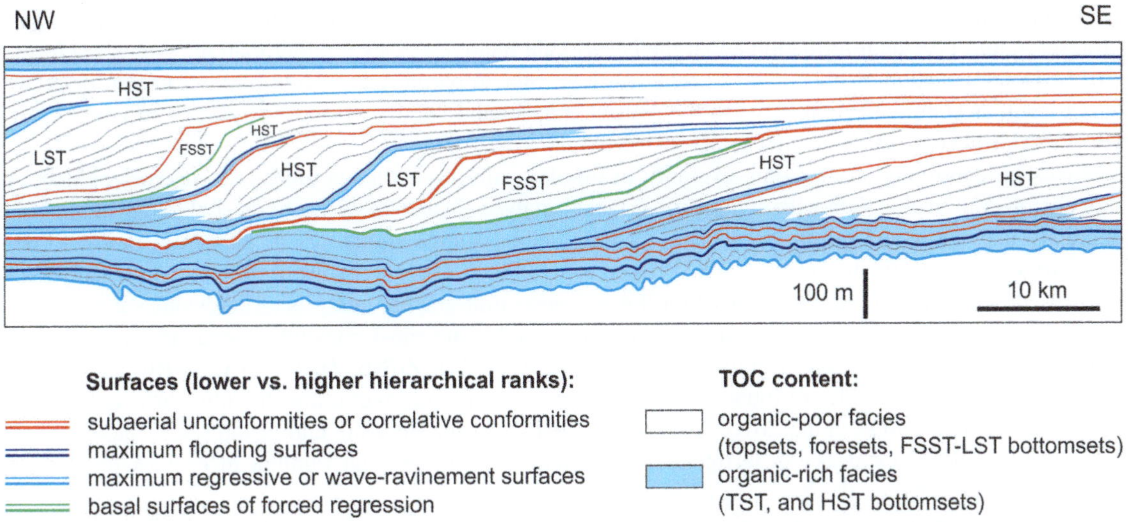


**FIG. 24** Distribution of total organic carbon (TOC) within a sequence stratigraphic framework, based on seismic acoustic impedance data calibrated with well data (Quintuco–Vaca Muerta system, Tithonian–Lower Valanginian, Neuquén Basin, Argentina; modified after Dominguez et al., 2016, 2020, and Reijenstein et al., 2020). The highest TOC amounts are found within transgressive systems tracts and the bottomsets of highstand systems tracts (minimum dilution with detrital sediment). Sequence stratigraphic surfaces are shown at two hierarchical levels; thicker lines indicate surfaces of higher hierarchical rank. *Abbreviations*: FSST, falling-stage systems tract; LST, lowstand systems tract; TST, transgressive systems tract; HST, highstand systems tract.

involved in the accumulation and preservation of organic matter, the stratigraphic analysis of the TOC in the sedimentary record is somewhat simplified by the fact that all three main factors have a common underlying control, namely, the relative sea level. Case studies in different sedimentary basins ranging in age from Paleozoic to Holocene showed that relative sea-level changes can exert a direct control on bioproductivity, redox conditions, detrital sediment supply and ultimately TOC concentrations (Fleck et al., 2002; Arthur and Sageman, 2005; Bohacs et al., 2005; Dong et al., 2018; Harris et al., 2018). For example, relative sea-level rise and transgression promote influxes of nutrient-rich upwelled water, which increase surface-water productivity; enhance water-column stratification, leading to the development of bottom-water anoxia; result in the trapping of terrigenous sediment in coastal and nearshore depositional environments, which favors the development of organic-rich condensed sections offshore. Consequently, the TOC content is typically enriched in transgressive systems tracts, and is depleted in systems tracts that accumulate during regression (Dong et al., 2018; Harris et al., 2018).

Due to their common link to the relative sea level, the proxies for restriction of water masses (Mo/TOC), redox conditions (Mo/Al and S/Fe) and productivity of biogenic silica converge to the conclusion that TOC values are typically the highest at or around maximum flooding surfaces (Harris et al., 2018). High-resolution geochemical data coupled with sedimentological data from core further document that the highest TOC values may in fact be recorded just above the highest rank maximum flooding surface within a studied section (Harris et al., 2018). This may be due to the fact that the horizon which corresponds to the deepest water at *syn*-depositional time is often *above* the maximum flooding surface at any location (Catuneanu et al., 1998; Zecchin et al., 2021). At the other end of the spectrum, the lowest TOC values are most often recorded at times of lowstand in relative sea level (Fleck et al., 2002; Dong et al., 2018; Harris et al., 2018; Dominguez et al., 2020). Recent examples of TOC distribution within a sequence stratigraphic framework have been documented by Dominguez et al. (2020) and Reijenstein et al. (2020), who found the highest TOC values within transgressive systems tracts and the bottomsets of highstand systems tracts (Fig. 24).

## Inorganic geochemistry

Besides TOC, inorganic geochemical proxies based on major, minor and trace elements can also be used to build the sequence stratigraphic framework and to reconstruct changes in relative sea level (Harris et al., 2013, 2018; Dong et al., 2018; Playter et al., 2018; LaGrange et al., 2020). Basic observations that relate major element geochemistry to relative sea level include the aluminum enrichment in the deepwater setting at times of lowstand (i.e., due to the increase in clay delivery during "lowstand shedding"), and the increase in carbonate delivery to the deepwater during "highstand shedding" (Harris et al., 2018; Dong et al., 2018). These observations can be further constrained and calibrated with paired studies of detrital and clay proxies which, when used in conjunction, can demonstrate changes in relative sea level (Li et al., 2000; Playter et al., 2018). Within the context of each sedimentary basin, the proxies used for detrital input (e.g., Sc/Zn and

Lu/Hf, higher during relative sea-level fall) and clay input (e.g., Zr/La, higher during relative sea-level rise and particularly during transgression) provide information about the basin-scale relative sea level rather than the global sea level (Playter et al., 2018). Documentation of trace element renewal (e.g., Mo) in partially restricted basins can also be related to the relative sea level, as communication between the restricted basin and the open ocean increases during stages of relative sea-level rise (Harris et al., 2013; Turner et al., 2016).

Challenges in identifying stratigraphic cyclicity in lithologically monotonous mudstone successions make the integration of chemostratigraphic data with conventional facies analyses increasingly common (LaGrange et al., 2020). Chemostratigraphy uses inorganic geochemical data, such as stable isotope signatures or elemental ratios, to define and correlate stratigraphic units (Pearce et al., 1999; Weissert et al., 2008). A recent review of the elements and elemental ratios that can be used as chemostratigraphic proxies for the sequence stratigraphic analysis of mudstone successions was provided by LaGrange et al. (2020). Among the elemental proxies that are most relevant to sequence stratigraphy, lows in terrigenous proxies (Ti/Al), minima in grain size proxies (Si/Al, Zr/Al, Zr/Nb, Zr/Rb, Y/Al, Y/Rb) and the increase in proxies for biogenic silica (Si/Al) are found to be closely associated with maximum flooding surfaces (Ratcliffe et al., 2012; Sano et al., 2013; Turner et al., 2016; Harris et al., 2018). At the opposite end of the scale, maximum regressive surfaces are characterized by peaks in terrigenous and grain size proxies and minima in proxies for biogenic silica (Ratcliffe et al., 2012; Sano et al., 2013; Harris et al., 2018).

The correct interpretation of elemental proxies requires that elemental data are calibrated with mineralogical data to ensure the proper usage of the former for the purposes of sequence stratigraphy (e.g., to discern between the depositional vs. diagenetic relevance of elemental data; LaGrange et al., 2020). So far, the most reliable application of geochemical proxies is for the identification of maximum flooding and maximum regressive surfaces which delineate transgressive-regressive cycles (LaGrange et al., 2020). Other surfaces that form in deepwater settings (i.e., the basal surface of forced regression and the correlative conformity) are not as easily identifiable with geochemical data, but can be identified based on different criteria (e.g., the petrography of framework grains; Figs. 17 and 18). Therefore, the integration of independent data sets always provides the best and most reliable approach to the identification of all elements of the sequence stratigraphic framework. For example, the position of a maximum flooding surface is best constrained where geochemical proxies are calibrated with biostratigraphic data (e.g., the highest abundance of microfossils and the highest ratio between distal and proximal benthic foraminifera within a condensed section; Gutierrez Paredes et al., 2017; Zecchin et al., 2021) and seismic stratigraphy (e.g., tracing downlap surfaces into the depocenters).

Much work is still needed to refine the geochemical criteria that afford the construction of a sequence stratigraphic framework, and integration with other methods such as sedimentary petrography, ichnology and biostratigraphy can improve the reliability of interpretations (see Chapters 14 and 15). Furthermore, the geochemical proxies in the toolbox of sequence stratigraphy need to be adjusted to the *stratigraphic age* of the basin under analysis, as the chemistry of oceans and atmosphere, with impact on all geochemical parameters, changed over time (Eriksson et al., 2004, 2005, 2010, 2013). Additional variability is introduced by the *tectonic setting*, as water circulation and chemistry in restricted basins differ from the conditions in the open ocean and by the *climatic regime* which modifies surfaces processes and associated detrital and clay proxies (Ruffell et al., 2002a, b—see Chapter 11).

## Stratigraphic scales in the deepwater setting

The sequence stratigraphic framework of deepwater systems records a nested architecture of stratigraphic cycles (i.e., sequences) that can be observed at different scales (Weimer, 1990; Weimer and Dixon, 1994; Gardner et al., 2003, 2009; Catuneanu, 2019b; Fig. 23). There are no temporal or physical standards for the scale of sequences that develop at different hierarchical levels. Scales are basin specific, reflecting the interplay of local and global controls on accommodation and sedimentation (Fig. 2). With the exception of orbital forcing, the periodicity of all other controls on stratigraphic cyclicity can vary unpredictably. Even in the case of orbital cycles, whose periodicities are most regular, the process-response correlation with stratigraphic sequences is still tentative because the orbital signal may be distorted or overprinted by local or regional processes (Strasser, 2018). Despite the nested architecture of stratigraphic cycles, the sequence stratigraphic framework is not truly fractal as sequences of different scales may differ in terms of underlying controls and internal composition of systems tracts (Catuneanu, 2019a, b).

At the smallest stratigraphic scales, systems tracts consist solely of beds and bedsets (i.e., sedimentological cycles; Fig. 8). At any larger scales, systems tracts consist of higher frequency (lower rank) sequences (i.e., stratigraphic cycles; Fig. 23). Seismic stratigraphy introduced by default a minimum scale for systems tracts and sequences, which had to exceed the vertical seismic resolution (i.e., typically in a range of $10^1$ m). As a result, the building blocks of the seismic stratigraphic framework are commonly observed at scales of $10^1$–$10^2$ m. The perception that sequences and their component systems tracts develop typically at scales of $10^1$–$10^2$ m is an artifact of seismic resolution, but it dominated stratigraphic thinking for decades. The reality of sequences and systems tracts at subseismic scales has become evident with the advances in high-resolution sequence stratigraphy (see discussion in Catuneanu, 2019b). It is now clear that most commonly, the building blocks of a seismic stratigraphic framework consist of higher frequency sequences that develop at subseismic scales; e.g., seismic-scale systems tracts ($10^1$–$10^2$ m) consist typically of sequences of $10^0$–$10^1$ m scales, which are different from and should not be confused with parasequences (Hernandez-Molina et al., 2000; Zecchin, 2007; Catuneanu and Zecchin, 2020).

The timescales of stratigraphic cycles in the deepwater setting reflect the cyclicity of shoreline shifts on the shelf within transit areas located updip of the shoreline trajectories of immediately higher hierarchical ranks (Catuneanu, 2019a, b). At each scale of observation, the width of the shoreline transit area depends on the gradient of the shelf, the magnitude of relative sea-level changes and the rates of sedimentation at the shoreline. Shoreline transit cycles can be observed at different scales (i.e., hierarchical levels), from $10^2$ to $10^3$ years (e.g., short-term variations in the balance between accommodation and sedimentation at the shoreline; Amorosi et al., 2017) to $10^4$–$10^5$ years (e.g., glacial-interglacial cycles related to orbital forcing, which are most evident during icehouse regimes; Burgess and Hovius, 1998; Porebski and Steel, 2006; Ainsworth et al., 2017, 2018) and $10^6$–$10^7$ years (e.g., in the case of long-term climate changes associated with icehouse-greenhouse cycles; Fig. 2). The Mississippi fan complex in the Gulf of Mexico provides an example of nested stratigraphic cycles related to forced regressions of different magnitudes during the Plio-Pleistocene icehouse (Fig. 23). In this example, sequences of all hierarchical ranks are dominated by the falling-stage systems tract, which is a typical signature of icehouse regimes (Weimer, 1990; Weimer and Dixon, 1994; Tesson et al., 2000; Fielding et al., 2006, 2008; Isbell et al., 2008; Zecchin et al., 2015; Sweet et al., 2019).

At each scale of observation, the growth and the progradation of slope clinoforms take place in incremental steps, reflecting the episodic delivery of sediment to the shelf edge (e.g., "highstand shedding" in the case of carbonates or "lowstand shedding" in the case of siliciclastic systems; Schlager et al., 1994; Posamentier and Allen, 1999; Sweet et al., 2019; Fisher et al., 2021). These cyclic fluctuations in sediment supply to the shelf edge are linked to the shoreline shelf-transit cycles, which are controlled by all factors that modify accommodation and sedimentation on the shelf. The location of the shoreline transit areas that develop at different stratigraphic scales also affects the bathymetry at the shelf edge during the shoreline transit cycles, influencing further the type of sediment that can reach the shelf edge (Fig. 3). At the "first-order" scale of shelf-slope systems, changes in sediment supply to the shelf edge relate to the second-order and lower rank stratigraphic cycles of shoreline shifts across the shelf (Catuneanu, 2019a, b). The scales of deepwater sequences that develop in relation to the shoreline shelf-transit cycles may range from subseismic ($10^0$–$10^1$ m) to seismic ($10^1$–$10^2$ m), and their identification depends on the resolution of the data available.

Scale is a key topic in sequence stratigraphy, with implications for methodology, nomenclature and the classification of stratigraphic cycles that develop at different scales. Stratigraphic cyclicity is basin specific in terms of timing and scales, reflecting the importance of local controls on accommodation and sedimentation. The scale of the smallest stratigraphic cycle that can be identified at any location depends on the resolution of the data available. In the context of seismic stratigraphy, the smallest identifiable "sequence" is typically not the smallest stratigraphic cycle within the study area, but the smallest stratigraphic cycle that is above the resolution of the seismic data. The inability to identify the smallest sequence in every study indicates that the classification of stratigraphic cycles is best approached starting with the "first-order" basin fill as the anchor for the definition of hierarchical ranks (Catuneanu, 2019b). The scale of the lowest rank systems tracts at any location defines the highest resolution that can be achieved with a stratigraphic study. A scale-independent approach to methodology and nomenclature is key to a consistent application of sequence stratigraphy across the entire range of geological settings, stratigraphic scales and types of data available.

## Discussion and key messages

The sequence stratigraphic methodology is independent of any basin-specific variables, such as the tectonic setting (e.g., active vs. passive margins—see Chapters 1, 3–5), the depositional setting (e.g., siliciclastic vs. carbonate) or the climatic regime at the time of deposition (see Chapter 11). However, the geometry of sequences, as well as their internal systems-tract composition and sedimentological makeup are highly variable as a function of the *syn*-depositional conditions of accommodation and sedimentation. Differences in the types of subsidence, depositional elements and basin physiography modify the physical attributes of the stratigraphic framework, and require versatile sedimentological skills to identify sequence stratigraphic units and bounding surfaces in various settings. Irrespective of this architectural and sedimentological variability, the sequence stratigraphic methodology always starts with the observation of stratal stacking patterns that afford the identification of systems tracts and bounding surfaces. Beyond the construction of the framework of systems

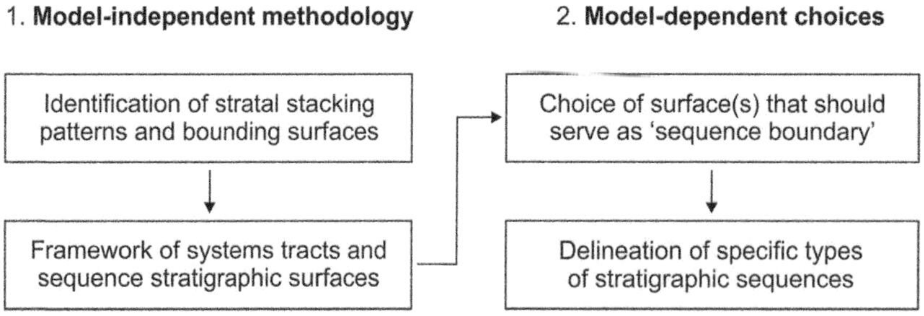


**FIG. 25** Workflow of sequence stratigraphy. Stratal stacking patterns provide a basis for the definition of all units and surfaces of sequence stratigraphy, at scales defined by the purpose of study or constrained by the resolution of the data available. Beyond the framework of systems tracts and bounding surfaces, the selection of the "sequence boundary" becomes a model-dependent choice, often guided by the mappability of the different types of sequence stratigraphic surfaces that are present within the study area. *(From Catuneanu, O., 2022. Principles of Sequence Stratigraphy. second ed., Elsevier, Amsterdam.)*

tracts and sequence stratigraphic surfaces, the selection of the "sequence boundary" becomes a matter of model-dependent delineation of stratigraphic cycles (Figs. 1 and 25; Catuneanu et al., 2011).

Each type of tectonic setting is unique in terms of subsidence mechanisms (e.g., extensional, thermal or flexural), sediment supply and dispersal patterns, location of depocenters and basin physiography. The variability imposed by the tectonic setting to the stratigraphic architecture is one of the reasons for the proposal of different models (Fig. 1), as each model is based on conclusions drawn from data collected in different types of sedimentary basins (e.g., "passive" margins, forelands or rifts). Contrasts between the stratigraphic frameworks that develop in different tectonic settings include the scales, the architecture and the systems-tract and sedimentological composition of sequences. All types of sedimentary basins are tectonically active because all require a tectonic mechanism to form, although they are generally classified into "passive" versus "active" with reference to the rates of subsidence and the variability in the rates of subsidence over time. For example, divergent continental margins are often referred to as "passive" margins because they are dominated by thermal subsidence, which operates with relatively low and constant rates, although they still record tectonic activity and the majority of accommodation is owed to subsidence.

For the purpose of stratigraphic analysis, the fact that the methodology is independent of the local controls on accommodation and sedimentation makes it unnecessary to customize the workflow to the tectonic setting, even though the mechanisms and the patterns of subsidence have an evident impact on the architecture of the stratigraphic framework. In this context, the broad classification of sedimentary basins into "passive" versus "active" is sufficient to describe tectonic settings in terms of the relative contributions of tectonism and sea-level changes to accommodation and stratigraphic cyclicity; i.e., tectonically "passive" settings are more susceptible to the influence of sea-level changes on the development of sequences (Fig. 23), whereas "active" settings such as convergent continental margins or rift basins are dominated by tectonic controls (Gawthorpe et al., 2003; Martins-Neto and Catuneanu, 2010; details in Chapter 3). These underlying conditions affect the areal extent, the composition and the degree of 3D temporal and physical variability of stratigraphic sequences.

Tectonism during the evolution of a sedimentary basin can also influence the development of ramps versus shelf-slope systems (Fig. 4). Sedimentary basins with a high tectonic activity (and hence, instability) are prone to the formation of ramps or poorly developed shelf-slope systems with narrower shelves. In contrast, tectonically "passive" basins can sustain the long-term progradation of slope clinoforms, given sufficient sediment supply, resulting in the development of well-defined shelf-slope systems with wider shelves (Fig. 24). Ramps and shelf-slope systems may also coexist within a sedimentary basin, due to variations along strike in tectonic activity and sediment supply (Dominguez et al., 2020), with ramps typically preceding the development of shelf-slope systems at any location (Fig. 24). The type of basin, and the degree of connectivity to the global ocean, also affects the modes of sediment transport and the types of depositional elements that dominate the stratigraphic framework. For example, gravity-driven processes may occur in all types of basins, albeit more frequently in tectonically active ones, whereas bottom currents are typically restricted to basins connected to the global ocean.

Facies predictions in areas away from data control points such as wells are enabled by the identification of systems tracts, which establish genetic relationships between same-age depositional systems (e.g., the link between shelf processes and the sediment supply to the deepwater setting). The modes of sediment transport also play an important role in the prediction of conventional petroleum reservoirs, as they control the sediment dispersal patterns and the location and orientation of depositional elements in the deepwater setting (Fig. 13—see Chapters 6–11, 16–18). Deepwater petroleum systems have

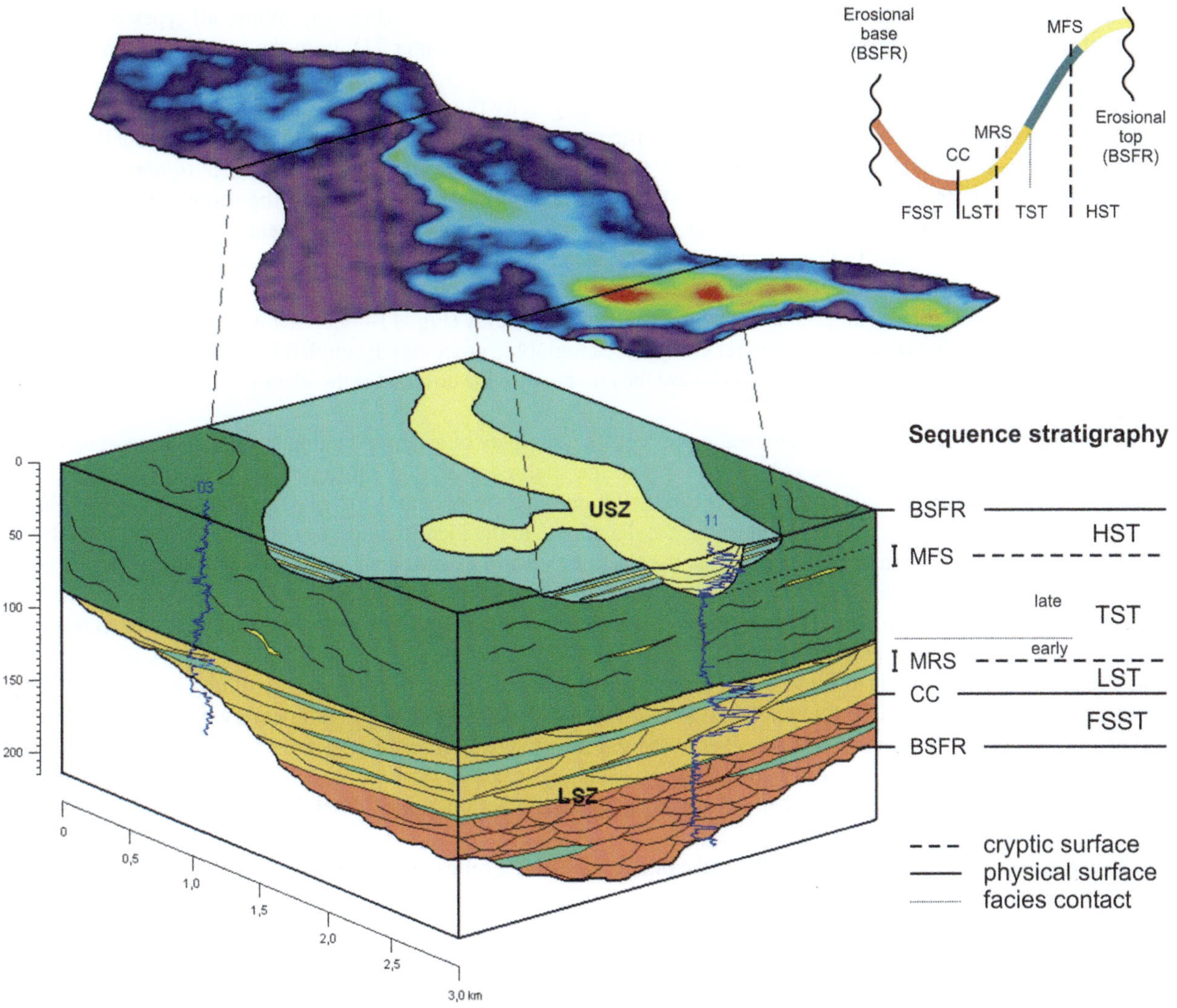


**FIG. 26** Stratigraphic framework of the Eocene petroleum reservoirs in the Albacora Field. The LSZ is subdivided into a coarser lower portion of vertically and laterally amalgamated channel complex deposits, and a finer-grained upper portion of laterally (but not vertically) amalgamated channel-fill deposits interbedded with laterally extensive shale units. The two portions of the LSZ are separated by a CC, which marks a sharp decrease in the volume and caliber of the sediment delivered to the deepwater setting. The pelitic unit between the LSZ and the USZ consists of cohesive debris flow and slump deposits. The USZ marks renewed progradation and sediment supply to the deepwater system following the MFS. *Abbreviations*: LSZ, lower sandstone zone; USZ, upper sandstone zone; BSFR, basal surface of forced regression; CC, correlative conformity; MRS, maximum regressive surface; MFS, maximum flooding surface; FSST, falling-stage systems tract; LST, lowstand systems tract; TST, transgressive systems tract; HST, highstand systems tract. *(Campos Basin, offshore Brazil; from De Gasperi, A., Catuneanu, O., 2014. Sequence stratigraphy of the Eocene turbidite reservoirs in Albacora field, Campos Basin, offshore Brazil. AAPG Bull. Am. Assoc. Pet. Geol. 98(2), 279–313.)*

been documented worldwide (see Chapters 2, 18–20), and can be rationalized within the sequence stratigraphic framework. The Campos Basin in offshore Brazil provides such an example, whereby the deposition of reservoirs and seal facies followed predictable patterns that correlate with the shoreline transit cycles on the shelf. In turn, the position of the shoreline relative to the shelf edge during the various stages of sequence development modified the volume and the caliber of the riverborne sediment delivered to the deepwater setting, and hence the types and the quality of petroleum reservoirs (Fig. 26). The latter aspect reflects both depositional and diagenetic processes, as discussed in detail in Chapter 13.

## Conclusions

Sedimentation in deepwater settings is the result of several independent processes of sediment transport and deposition, including gravity flows and suspension fallout in both enclosed and oceanic basins, as well as water masses circulation in basins connected to the global ocean where contour currents may operate (Fig. 13). While all types of deposits are important to the sedimentological makeup of deepwater systems (see Chapter 21), only the trends of change in gravity flows during the shoreline shelf-transit cycles provide dependable criteria for the construction of the sequence stratigraphic framework in all settings (Figs. 3 and 15). The actual types of gravity flows, which contribute to the formation of systems tracts, are nondiagnostic, as they depend on basin-specific parameters that control sediment supply to the shelf edge during the shoreline transit cycles: the physiography of the shelf (width, seafloor gradients, the presence or absence of unfilled incised valleys across the shelf); the magnitude and rates of relative sea-level changes at the shoreline; the rates of sedimentation at the shoreline and the extrabasinal versus intrabasinal origin of the sediment (see Chapter 11). Bottom (contour) currents further modify the sedimentological makeup of deepwater sequences in oceanic basins (Fig. 13) and contribute to the identification of systems tracts and bounding surfaces where variations in their sediment composition are tied to relative sea-level changes and shoreline shelf-transit cycles (Fig. 11—see Chapter 9). However, contourites and related unconformities may also form in relation to oceanographic processes decoupled from relative sea-level changes and shoreline transit cycles on the shelf, in which case they do not provide criteria for the identification of systems tracts and bounding surfaces (Fig. 10).

Stratigraphic cyclicity in the deepwater setting reflects the interplay of accommodation and sedimentation on the shelf, which controls shoreline trajectories, sediment supply to the shelf edge and the timing of deepwater sequences, systems tracts and bounding surfaces. Changes in the types, volume and composition of gravity flows during the shoreline shelf-transit cycles define stratigraphic trends that are diagnostic to the identification of systems tracts and bounding surfaces. Forced regressions invariably lower the shelf capacity to retain extrabasinal sediment and to generate intrabasinal carbonate sediment, thus modifying the sediment supply to the shelf edge, and are fundamental to the delineation of deepwater depositional sequences (Fig. 23). Notably, forced regressions are always driven by the fall in relative sea level on the shelf, independently of sedimentation (Figs. 14, 17, 18 and 23). Therefore, the relative sea level plays a key role in the stratigraphic architecture of deepwater systems, and the distinction between lowstands and highstands remains critical to understanding the distribution of extrabasinal and intrabasinal sediment across a sedimentary basin. Within the stratigraphic framework, the nondiagnostic variability in the sedimentological makeup of systems tracts reflects the unique tectonic and depositional settings of each sedimentary basin and needs to be rationalized on a case-by-case basis. The distinction between stratigraphic trends and the nondiagnostic sedimentological variability is key to the construction of the sequence stratigraphic framework.

The application of sequence stratigraphy to the deepwater setting relies on the construction of composite profiles that describe the relative chronology of the different types of gravity flows at regional scales (Fig. 12). This is typically accomplished with seismic data (Figs. 5 and 19B), which can be calibrated with well data (e.g., biostratigraphy—see Chapter 14; Gutierrez Paredes et al., 2017). Stratigraphic cyclicity relevant to the delineation of sequences is defined by the composite rather than local profiles. The place of accumulation of gravity-flow deposits depends on the position of the sediment entry points along the shelf edge, the types of gravity flows and the seafloor morphology (e.g., the location of depocenters controlled by structural elements, salt tectonics or mud diapirism—see Chapters 3–5). The allocyclic and autocyclic shifts of deepwater depositional elements (e.g., the lateral shifts of leveed channels) further enhance the offset between local trends and the regional composite profile in terms of timing and frequency of cycles, timing of coarsening- and fining-upward trends and timing of coarsest sediment (Fig. 12). The sedimentological cycles defined by local trends must not be confused with the stratigraphic cycles defined by regional trends. Notwithstanding these caveats, the stratigraphic architecture of basin floors that lack any significant relief can be remarkably consistent (van der Merwe et al., 2010; Fig. 14).

## Acknowledgments

I thank F. Javier Hernández-Molina, Jon Rotzien, Richard Sears and Cindy Yeilding for their constructive comments during the peer-review process, which helped shape the final version of this chapter.

## References

Ainsworth, R.B., Vakarelov, B.K., MacEachern, J.A., Rarity, F., Lane, T.I., Nanson, R.A., 2017. Anatomy of a shoreline regression: implications for the high-resolution stratigraphic architecture of deltas. J. Sediment. Res. 87, 1–35.

Ainsworth, R.B., McArthur, J.B., Lang, S.C., Vonk, A.J., 2018. Quantitative sequence stratigraphy. Am. Assoc. Pet. Geol. Bull. 102, 1913–1939.

Algeo, T.J., Ingall, E., 2007. Sedimentary $C_{org}$:P ratios, paleocean ventilation, and phanerozoic atmospheric $pO_2$. Palaeogeogr. Palaeoclimatol. Palaeoecol. 256, 130–155.

Amorosi, A., Bruno, L., Campo, B., Morelli, A., Rossi, V., Scarponi, D., Hong, W., Bohacs, K.M., Drexler, T.M., 2017. Global sea-level control on local parasequence architecture from the Holocene record of the Po plain, Italy. Mar. Pet. Geol. 87, 99–111.

Arthur, M.A., Sageman, B.B., 1994. Marine black shales: depositional mechanisms and environments of ancient deposits. Annu. Rev. Earth Planet. Sci. 22, 499–551.

Arthur, M.A., Sageman, B.B., 2005. Sea-level control on source-rock development: Perspectives from the Holocene Black Sea, the mid-cretaceous Western Interior Basin of North America, and the late Devonian Appalachian Basin. In: Harris, N.B. (Ed.), The Deposition of Organic-Carbon-Rich Sediments: Models, Mechanisms, and Consequences. vol. 82. Society for Sedimentary Geology (SEPM) Special Publication, pp. 35–59.

Bohacs, K.M., Grabowski, G.J., Carroll, A.R., Mankiewicz, P.J., Gerhardt, K.J., Schwalbach, J.R., Wegner, M.B., Simo, J.A., 2005. Production, destruction, and dilution—the many paths to source rock development. In: Harris, N.B. (Ed.), The Deposition of Organic-Carbon-Rich Sediments: Models, Mechanisms, and Consequences. vol. 82. Society for Sedimentary Geology (SEPM) Special Publication, pp. 61–101.

Brackenridge, R., Stow, D.A.V., Hernández-Molina, F.J., 2011. Contourites within a deepwater sequence stratigraphic framework. Geo-Mar. Lett. 31, 343–360.

Brown Jr., L.F., Fisher, W.L., 1977. Seismic stratigraphic interpretation of depositional systems: Examples from Brazilian rift and pull apart basins. In: Payton, C.E. (Ed.), Seismic Stratigraphy—Applications to Hydrocarbon Exploration. vol. 26. American Association of Petroleum Geologists Memoir, pp. 213–248.

Burgess, P.M., Hovius, N., 1998. Rates of delta progradation during highstands; consequences for timing of deposition in deep-marine systems. J. Geol. Soc. Lond. 155, 217–222.

Catuneanu, O., 2006. Principles of Sequence Stratigraphy. Elsevier, Amsterdam. 375 pp.

Catuneanu, O., 2019a. Model-independent sequence stratigraphy. Earth Sci. Rev. 188, 312–388.

Catuneanu, O., 2019b. Scale in sequence stratigraphy. Mar. Pet. Geol. 106, 128–159.

Catuneanu, O., 2020. Sequence stratigraphy of deep-water systems. Mar. Pet. Geol. 114, 1–13. https://doi.org/10.1016/j.marpetgeo.2020.104238.

Catuneanu, O., Zecchin, M., 2016. Unique vs. non-unique stratal geometries: relevance to sequence stratigraphy. Mar. Pet. Geol. 78, 184–195.

Catuneanu, O., Zecchin, M., 2020. Parasequences: allostratigraphic misfits in sequence stratigraphy. Earth Sci. Rev. 208 (103289), 1–17.

Catuneanu, O., Willis, A.J., Miall, A.D., 1998. Temporal significance of sequence boundaries. Sediment. Geol. 121, 157–178.

Catuneanu, O., 2022. Principles of Sequence Stratigraphy, second ed. Elsevier, Amsterdam, p. 454.

Catuneanu, O., Galloway, W.E., Kendall, C.G.S.C., Miall, A.D., Posamentier, H.W., Strasser, A., Tucker, M.E., 2011. Sequence stratigraphy: methodology and nomenclature. Newsl. Stratigr. 44 (3), 173–245.

Clift, P.D., 2006. Controls on the erosion of Cenozoic Asia and the flux of clastic sediment to the ocean. Earth Planet. Sci. Lett. 241 (3–4), 571–580.

Covault, J.A., Graham, S.A., 2010. Submarine fans at all sea-level stands: tectonomorphologic and climatic controls on terrigenous sediment delivery to the deep sea. Geology 38, 939–942.

Covault, J.A., Normark, W.R., Romans, B.W., Graham, S.A., 2007. Highstand fans in the California borderland: the overlooked deep-water depositional systems. Geology 35, 783–786.

Creaney, S., Passey, Q.R., 1993. Recurring patterns of total organic carbon and source rock quality within a sequence stratigraphic framework. AAPG Bull. 77, 386–401.

Csato, I., Catuneanu, O., Granjeon, D., 2014. Millennial-scale sequence stratigraphy: numerical simulation with Dionisos. J. Sediment. Res. 84, 394–406.

De Gasperi, A., Catuneanu, O., 2014. Sequence stratigraphy of the Eocene turbidite reservoirs in Albacora field, Campos Basin, offshore Brazil. AAPG Bull. Am. Assoc. Pet. Geol. 98 (2), 279–313.

Demaison, G.J., Moore, G.T., 1980. Anoxic environments and oil source bed genesis. Am. Assoc. Pet. Geol. Bull. 64, 1179–1209.

Dominguez, R.F., Continanzia, M.J., Mykietiuk, K., Ponce, C., Pérez, G., Guerello, R., Lanusse Noguera, I., Caneva, M., Di Benedetto, M., Catuneanu, O., Cristallini, E., 2016. Organic-rich stratigraphic units in the Vaca Muerta formation, and their distribution and characterization in the Neuquén Basin (Argentina). In: Unconventional Resources Technology Conference (URTeC) 2016, pp. 1–12. https://doi.org/10.15530-urtec-2016-2456851.

Dominguez, R.F., Catuneanu, O., Reijenstein, H., Notta, R., Posamentier, H.W., 2020. Sequence stratigraphy and the three-dimensional distribution of organic-rich units. In: Minisini, D., Fantin, M., Lanusse Noguera, I., Leanza, H.A. (Eds.), Integrated Geology of Unconventionals: The Case of the Vaca Muerta Play. vol. 121. AAPG Memoir, Argentina, pp. 163–200.

Dong, T., Harris, N.B., Ayranci, K., 2018. Relative sea-level changes and organic matter accumulation in shales of the middle and upper Devonian Horn River group, northeastern British Columbia, Canada: insights into sediment flux, redox conditions, and bioproductivity. Geol. Soc. Am. Bull. 130 (5/6), 859–880.

Donovan, A.D., Gardner, R.D., Pramudito, A., Staerker, T.S., Wehner, M., Corbett, M.J., Lundquist, J.J., Romero, A.M., Henry, L.C., Rotzien, J.R., Boling, K.S., 2015. Chronostratigraphic relationships of the woodbine and eagle ford groups across Texas. GCAGS J. 4, 67–87.

Eriksson, P.G., Altermann, W., Nelson, D.R., Mueller, W.U., Catuneanu, O. (Eds.), 2004. The Precambrian Earth: Tempos and Events. Developments in Precambrian Geology. 12. Elsevier, Amsterdam. 941 pp.

Eriksson, P.G., Catuneanu, O., Nelson, D., Popa, M., 2005. Controls on Precambrian sea-level change and sedimentary cyclicity. Sediment. Geol. 176 (1–2), 43–65.

Eriksson, P.G., Sarkar, S., Samanta, P., Banerjee, S., Porada, H., Catuneanu, O., 2010. Paleoenvironmental context of microbial mat-related structures in siliciclastic rocks. In: Seckbach, J., Oren, A. (Eds.), Microbial Mats: Modern and Ancient Microorganisms in Stratified Systems. Springer, Berlin, pp. 73–108.

Eriksson, P.G., Banerjee, S., Catuneanu, O., Corcoran, P.L., Eriksson, K.A., Hiatt, E.E., Laflamme, M., Lenhardt, N., Long, D.G.F., Miall, A.D., Mints, M. V., Pufahl, P.K., Sarkar, S., Simpson, E.L., Williams, G.E., 2013. Secular changes in sedimentation systems and sequence stratigraphy. Gondwana Res. 24, 468–489.

Fielding, C.R., Bann, K.L., MacEachern, J.A., Tye, S.C., Jones, B.G., 2006. Cyclicity in the nearshore marine to coastal, lower Permian, Pebbley Beach formation, southern Sydney Basin, Australia: a record of relative sea-level fluctuations at the close of the late Palaeozoic Gondwanan ice age. Sedimentology 53, 435–463.

Fielding, C.R., Frank, T.D., Birgenheier, L.P., Rygel, M.C., Jones, A.T., Roberts, J., 2008. Stratigraphic record and facies associations of the late Paleozoic ice age in eastern Australia (New South Wales and Queensland). In: Fielding, C.R., Frank, T.D., Isbell, J.L. (Eds.), Resolving the Late Paleozoic Ice Age in Time and Space. vol. 441. Geological Society of America Special Paper, pp. 41–57.

Fillon, R.H., 2007. Biostratigraphy and condensed sections in deepwater settings. In: Weimer, P., Slatt, R. (Eds.), Introduction to the Petroleum Geology of Deepwater Settings. AAPG Studies in Geology 57 AAPG/Datapages Discovery Series, vol. 8. American Association of Petroleum Geologists, pp. 499–577.

Fisher, W.L., Galloway, W.E., Steel, R.J., Olariu, C., Kerans, C., Mohrig, D., 2021. Deep-water depositional systems supplied by shelf-incising submarine canyons: recognition and significance in the geological record. Earth Sci. Rev. 214 (103531), 1–62.

Fleck, S., Michels, R., Ferry, S., Malartre, F., Elion, P., Landais, P., 2002. Organic geochemistry in a sequence stratigraphic framework: the siliciclastic shelf environment of cretaceous series, SE France. Org. Geochem. 33, 1533–1557.

Fonnesu, M., Palermo, D., Galbiati, M., Marchesini, M., Bonamini, E., Bendias, D., 2020. A new world-class deep-water play type, deposited by the syndepositional interaction of turbidity flows and bottom currents: the giant Eocene coral field in northern Mozambique. Mar. Pet. Geol. 111, 179–201.

Galloway, W.E., 1989. Genetic stratigraphic sequences in basin analysis. I. Architecture and genesis of flooding-surface bounded depositional units. Am. Assoc. Pet. Geol. Bull. 73, 125–142.

Gardner, M.H., Borer, J.M., Melick, J.J., Mavilla, N., Dechesne, M., Wagerle, R.N., 2003. Stratigraphic process-response model for submarine channels and related features from studies of Permian brushy canyon outcrops, West Texas. Mar. Pet. Geol. 20, 757–787.

Gardner, M., Borer, J.M., Romans, B., Baptista, N., Kling, E.K., Hanggoro, D., Melick, J., Wagerle, R., Dechesne, M., Carr, M., Amerman, R., Atan, S., 2009. Stratigraphic models for deep-water sedimentary systems. In: Schofield, K., Rosen, N.C., Pfeiffer, D., Johnson, S. (Eds.), Answering the Challenges of Production from Deep-Water Reservoirs: Analogues and Case Histories to Aid a New Generation. 28th Annual Gulf Coast Section SEPM Foundation, Bob F. Perkins Research Conference, 2008, pp. 77–175.

Gawthorpe, R.L., Hardy, S., Ritchie, B., 2003. Numerical modeling of depositional sequences in half-graben rift basins. Sedimentology 50, 169–185.

Gutierrez Paredes, H.C., Catuneanu, O., Romano, U.H., 2017. Sequence stratigraphy of the Miocene section, southern Gulf of Mexico. Mar. Pet. Geol. 86, 711–732.

Harris, N.B., Freeman, K.H., Pancost, R.D., White, T.S., Mitchell, G.D., 2004. The character and origin of lacustrine source rocks in the lower cretaceous synrift section, Congo Basin, West Africa. Am. Assoc. Pet. Geol. Bull. 88, 1163–1184.

Harris, N.B., Mnich, C.A., Selby, D., Korn, D., 2013. Minor and trace element and Re-Os chemistry of the upper Devonian Woodford shale, Permian Basin, West Texas: insights into metal abundance and basin processes. Chem. Geol. 356, 76–93.

Harris, N.B., McMillan, J.M., Knapp, L.J., Mastalerz, M., 2018. Organic matter accumulation in the upper Devonian Duvernay formation, Western Canada Sedimentary Basin, from sequence stratigraphic analysis and geochemical proxies. Sediment. Geol. 376, 185–203.

Hay, W.W., 1995. Paleoceanography of marine organic-carbon-rich sediments. In: Huc, A.Y. (Ed.), Paleogeography, Paleoclimate, and Source Rocks. Studies in Geology. vol. 40. American Association of Petroleum Geologists, pp. 21–59.

Hernandez-Molina, F.J., Somoza, L., Lobo, F.J., 2000. Seismic stratigraphy of the Gulf of Cadiz continental shelf: a model for late quaternary very high-resolution sequence stratigraphy and response to sea-level fall. In: Hunt, D., Gawthorpe, R.L.G. (Eds.), Sedimentary Responses to Forced Regressions. 172. Geological Society of London, Special Publication, pp. 329–361.

Hunt, D., Tucker, M.E., 1992. Stranded parasequences and the forced regressive wedge systems tract: deposition during base-level fall. Sediment. Geol. 81, 1–9.

Isbell, J.L., Cole, D.I., Catuneanu, O., 2008. Carboniferous-Permian glaciation in the main Karoo Basin, South Africa: stratigraphy, depositional controls, and glacial dynamics. In: Fielding, C.R., Frank, T.D., Isbell, J.L. (Eds.), Resolving the Late Paleozoic Ice Age in Time and Space. vol. 441. Geological Society of America Special Paper, pp. 71–82.

Johnson, J.G., Murphy, M.A., 1984. Time-rock model for Siluro-Devonian continental shelf, western United States. Geol. Soc. Am. Bull. 95, 1349–1359.

Katz, B., 2005. Controlling factors on source rock development—a review of productivity, preservation, and sedimentation rate. In: Harris, N.B. (Ed.), The Deposition of Organic-Carbon-Rich Sediments: Models, Mechanisms, and Consequences. vol. 82. Society for Sedimentary Geology (SEPM) Special Publication, pp. 7–16.

LaGrange, M.T., Konhauser, K.O., Catuneanu, O., Harris, B.S., Playter, T.L., Gingras, M.K., 2020. Earth Sci. Rev. 203, 103137.

Li, L., Keller, G., Adatte, T., Stinnesbeck, W., 2000. Late cretaceous sea-level changes in Tunisia: a multidisciplinary approach. J. Geol. Soc. Lond. 157, 447–458.

Llave, E., Hernandez-Molina, F.J., Somoza, L., Diaz-del-Rio, V., Stow, D.A.V., Maestro, A., Alveirinho Dias, J.M., 2001. Seismic stacking pattern of the Faro-Albufeira contourite system (gulf of Cadiz): a quaternary record of paleoceanographic and tectonic influences. Mar. Geophys. Res. 22, 487–508.

Llave, E., Schonfeld, J., Hernández-Molina, F.J., Mulder, T., Somoza, L., del Rio, V.D., Sanchez-Almazo, I., 2006. High-resolution stratigraphy of the Mediterranean outflow contourite system in the Gulf of Cadiz during the late Pleistocene: the impact of Heinrich event. Mar. Geol. 227, 241–262.

Madof, A.S., Harris, A.D., Baumgardner, S.E., Sadler, P.M., Laugier, F.J., Christie-Blick, N., 2019. Stratigraphic aliasing and the transient nature of deep-water depositional sequences: revisiting the Mississippi Fan. Geology 47 (6), 545–549.

Martins-Neto, M.A., Catuneanu, O., 2010. Rift sequence stratigraphy. Mar. Pet. Geol. 27, 247–253.

Mestdagh, T., Lobo, F.J., Llave, E., Hernandez-Molina, F.J., Van Rooij, D., 2019. Review of the late quaternary stratigraphy of the northern Gulf of Cadiz continental margin: new insights into controlling factors and global implications. Earth Sci. Rev. 198, 102944.

Mitchum Jr., R.M., Vail, P.R., Thompson III, S., 1977. Seismic stratigraphy and global changes of sea level. Part 2. The depositional sequence as a basic unit for stratigraphic analysis. In: Payton, C.E. (Ed.), Seismic Stratigraphy - Applications to Hydrocarbon Exploration. vol. 26. American Association of Petroleum Geologists, Memoir, pp. 53–62.

Mort, H., Jacquat, O., Adatte, T., Steinmann, P., Follmi, K., Matera, V., Berner, Z., Stuben, D., 2007. The Cenomanian/Turonian anoxic event at the Bonarelli level in Italy and Spain: enhanced productivity and/or better preservation? Cretac. Res. 28, 597–612.

Payton, C.E. (Ed.), 1977. Seismic stratigraphy—applications to hydrocarbon exploration. AAPG Mem. 26, 516.

Pearce, T.J., Besly, B.M., Wray, D.S., Wright, D.K., 1999. Chemostratigraphy: a method to improve inter-well correlation in barren sequences—a case study using onshore Duckmantian/Stephanian sequences (West Midlands, U.K.). Sediment. Geol. 124, 197–220.

Pedersen, T.F., Calvert, S.E., 1990. Anoxia vs. productivity: what controls the formation of organic-carbon-rich sediments and sedimentary rocks? Am. Assoc. Pet. Geol. Bull. 74, 454–466.

Playter, T., Corlett, H., Konhauser, K., Robbins, L., Rohais, S., Crombez, V., MacCormack, K., Rokosh, D., Prenoslo, D., Furlong, C.M., Pawlowicz, J., Gingras, M., Lalonde, S., Lyster, S., Zonneveld, J.-P., 2018. Clinoform identification and correlation in fine-grained sediments: a case study using the Triassic Montney formation. Sedimentology 65, 263–302.

Porebski, S.J., Steel, R.J., 2006. Deltas and sea-level change. J. Sediment. Res. 76, 390–403.

Posamentier, H.W., Allen, G.P., 1999. Siliciclastic Sequence Stratigraphy: Concepts and Applications. vol. 7 SEPM Concepts in Sedimentology and Paleontology. 210 pp.

Posamentier, H.W., Kolla, V., 2003. Seismic geomorphology and stratigraphy of depositional elements in deep-water settings. J. Sediment. Res. 73 (3), 367–388.

Posamentier, H.W., Vail, P.R., 1985. Eustatic controls on depositional stratal patterns. In: Soc. Econ. Paleont. Mineral. (SEPM) Research Conference No. 6, Sea Level Changes—An Integrated Approach. Houston, October 20–23, 1985. (abstract and poster).

Posamentier, H.W., Jervey, M.T., Vail, P.R., 1988. Eustatic controls on clastic deposition. I. Conceptual framework. In: Wilgus, C.K., Hastings, B.S., Kendall, C.G.S.C., Posamentier, H.W., Ross, C.A., Van Wagoner, J.C. (Eds.), Sea Level Changes—An Integrated Approach. vol. 42. SEPM Special Publication, pp. 110–124.

Ratcliffe, K.T., Wright, A.M., Schmidt, K., 2012. Application of inorganic whole-rock geochemistry to shale resource plays: an example from the eagle ford shale formation, Texas. Sediment. Rec. 10, 4–9.

Reijenstein, H.M., Posamentier, H.W., Bande, A., Lozano, F.A., Dominguez, R.F., Wilson, R., Catuneanu, O., Galeazzi, S., 2020. Seismic geomorphology, depositional elements, and clinoform sedimentary processes: impact on unconventional reservoir prediction. In: Minisini, D., Fantin, M., Lanusse Noguera, I., Leanza, H.A. (Eds.), Integrated Geology of Unconventionals: The Case of the Vaca Muerta Play, Argentina. vol. 121. AAPG Memoir, pp. 237–266.

Rimmer, S.M., Thompson, J.A., Goodnight, S.A., Robl, T.L., 2004. Multiple controls on the preservation of organic matter in Devonian-Mississippian marine black shales: geochemical and petrographic evidence. Palaeogeogr. Palaeoclimatol. Palaeoecol. 215, 125–154.

Ruffell, A.H., Price, G.D., Mutterlose, J., Kessels, K., Baraboshkin, E., Grocke, D.R., 2002a. Palaeoclimate indicators (clay minerals, calcareous nannofossils, stable isotopes) compared from two successions in the late Jurassic of the Volga Basin (SE Russia). Geol. J. 37, 17–33.

Ruffell, A.H., McKinley, J.M., Worden, R.H., 2002b. Comparison of clay mineral stratigraphy to other proxy palaeoclimate indicators in the Mesozoic of NW Europe. Phil. Trans. R. Soc. London 360, 675–693.

Sageman, B.B., Murphy, A.E., Werne, J.P., Ver Straeten, C.A., Hollander, D.J., Lyons, T.W., 2003. A tale of shales: the relative roles of production, decomposition, and dilution in the accumulation of organic-rich strata, middle-upper Devonian, Appalachian Basin. Chem. Geol. 195, 229–273.

Sano, J.L., Ratcliffe, K.T., Spain, D.R., 2013. Chemostratigraphy of the Haynesville shale. In: Hammes, U., Gales, J. (Eds.), Geology of the Haynesville Gas Shale in East Texas and West Louisiana, USA. 105. AAPG Memoir, pp. 137–154.

Schlager, W., Reijmer, J.J.G., Droxler, A., 1994. Highstand shedding of carbonate platforms. J. Sediment. Res. B64, 270–281.

Sinclair, H.D., Tomasso, M., 2002. Depositional evolution of confined turbidite basins. J. Sediment. Res. 72, 451–456.

Slingerland, R., Kump, L.R., Arthur, M.A., Fawcett, P.J., Sageman, B.B., Barron, E.J., 1996. Estuarine circulation in the Turonian western interior seaway of North America. Geol. Soc. Am. Bull. 108, 941–952.

Strasser, A., 2018. Cyclostratigraphy of shallow-marine carbonates—limitations and opportunities. In: Montenari, M. (Ed.), Stratigraphy & Timescales. vol. 3. Academic Press, UK, pp. 151–187.

Sweet, M.L., 2020. Controls on the grain size of sediment routed to deep water: insights from quaternary systems. Res. Insights 2 (4), 1–13.

Sweet, M.L., Blum, M.D., 2016. Connections between fluvial to shallow marine environments and submarine canyons: implications for sediment transfer to deep water. J. Sediment. Res. 86, 1147–1162.

Sweet, M.L., Gaillot, G.T., Jouet, G., Rittenour, T.M., Toucanne, S., Marsset, T., Blum, M.D., 2019. Sediment routing from shelf to basin floor in the quaternary Golo system of eastern Corsica, France, western Mediterranean Sea. Geol. Soc. Am. Bull. https://doi.org/10.1130/B35181.1.

Tesson, M., Posamentier, H.W., Gensous, B., 2000. Stratigraphic organization of late Pleistocene deposits of the western part of the Rhone shelf (Languedoc shelf) from high resolution seismic and core data. Am. Assoc. Pet. Geol. Bull. 84, 119–150.

Turner, B.W., Molinares-Blanco, C.E., Slatt, R.M., 2015. Chemostratigraphic, palynostratigraphic, and sequence stratigraphic analysis of the Woodford shale, Wyche farm quarry, Pontotoc County, Oklahoma. Interpretation 3 (1), SH1–SH9.

Turner, B.W., Tréanton, J.A., Slatt, R.M., 2016. The use of chemostratigraphy to refine ambiguous sequence stratigraphic correlations in marine mudrocks. An example from the Woodford Shale, Oklahoma, USA. J. Geol. Soc. 173, 854–868.

Tyson, R.V., 2001. Sedimentation rate, dilution, preservation, and total organic carbon: some results of a modelling study. Org. Geochem. 32, 333–339.

Tyson, R.V., 2005. The "productivity versus preservation" controversy: cause, flaws, and resolution. In: Harris, N.B. (Ed.), The Deposition of Organic-Carbon-Rich Sediments: Models, Mechanisms, and Consequences. vol. 82. Society for Sedimentary Geology (SEPM) Special Publication, pp. 17–33.

Tyson, R.V., Pearson, T.H., 1991. Modern and ancient continental shelf anoxia: an overview. In: Tyson, R.V., Pearson, T.H. (Eds.), Modern and Ancient Continental Shelf Anoxia. Geological Society, London, Special Publication, vol. 58, pp. 1–24.

van der Merwe, W.C., Flint, S.S., Hodgson, D.M., 2010. Sequence stratigraphy of an argillaceous, deepwater basin-plain succession: Vischkuil formation (Permian), Karoo Basin, South Africa. Mar. Pet. Geol. 27, 321–333.

Van Siclen, D.C., 1958. Depositional topography—examples (Louisiana and Texas) and theory. Am. Assoc. Pet. Geol. Bull. 42, 1896–1913.
Van Wagoner, J.C., Posamentier, H.W., Mitchum, R.M., Vail, P.R., Sarg, J.F., Loutit, T.S., Hardenbol, J., 1988. An overview of sequence stratigraphy and key definitions. In: Wilgus, C.K., Hastings, B.S., Kendall, C.G.S.C., Posamentier, H.W., Ross, C.A., Van Wagoner, J.C. (Eds.), Sea Level Changes—An Integrated Approach. vol. 42. SEPM Special Publication, pp. 39–45.
Wei, H.Y., Chen, D.Z., Wang, J.G., Yu, H., Tucker, M.E., 2012. Organic accumulation in the lower Chihsia formation (middle Permian) of South China: constraints from pyrite morphology and multiple geochemical proxies. Palaeogeogr. Palaeoclimatol. Palaeoecol. 353, 73–86.
Weimer, P., 1990. Sequence stratigraphy, facies geometries, and depositional history of the Mississippi Fan, Gulf of Mexico. Bull. Am. Assoc. Pet. Geol. 74, 425–453.
Weimer, P., Dixon, B.T., 1994. Regional sequence stratigraphic setting of the Mississippi Fan complex, northern deep Gulf of Mexico: implications for evolution of the northern Gulf basin margin. In: Weimer, P., et al. (Eds.), Submarine Fans and Production Characteristics. Society for Sedimentary Geology (SEPM), Gulf Coast Section, 15th Annual Research Conference, Houston, Texas, pp. 373–381.
Weissert, H., Joachimski, M., Sarnthein, M., 2008. Chemostratigraphy. Newsl. Stratigr. 42 (3), 145–179.
Wilson, J.L., 1967. Cyclic and reciprocal sedimentation in Virgilian strata of southern New Mexico. Geol. Soc. Am. Bull. 78, 805–818.
Zecchin, M., 2007. The architectural variability of small-scale cycles in shelf and ramp clastic systems: the controlling factors. Earth Sci. Rev. 84, 21–55.
Zecchin, M., Caffau, M., Catuneanu, O., 2021. Recognizing maximum flooding surfaces in shallow-water deposits: an integrated sedimentological and micropaleontological approach (Crotone Basin, southern Italy). Mar. Pet. Geol. 133 (105225), 1–30.
Zecchin, M., Catuneanu, O., Rebesco, M., 2015. High-resolution sequence stratigraphy of clastic shelves IV: high-latitude settings. Mar. Pet. Geol. 68, 427–437.

Chapter 13

# Reservoir quality and diagenesis of deepwater sandstones

R.H. Lander[a], L.M. Bonnell[a], T.R. Taylor[b] and I. Espejo[c]

[a]*Geocosm LLC, Durango, CO, United States,* [b]*SedPetrology LLC, Tucson, AZ, United States,* [c]*TPS Consulting Services, Houston, TX, United States*

## Introduction

Sandstone reservoirs are sources of hydrocarbons, geothermal heat, lithium and water while also providing storage for captured carbon, generated hydrogen, compressed air and contaminants. Reservoir quality describes the ability of the rock pore system to store and produce significant volumes of fluids over engineering timescales. Reservoir quality is strongly linked to permeability, which is a measure of the ability of a rock to transmit fluids. While total porosity commonly is used as a proxy for reservoir storage capacity, in most deepwater sandstones it is a poor index of reservoir quality given that micropores (pores with diameters $<10\,\mu m$) may make up a significant fraction of the total porosity while contributing little to the extractable fluid volume or permeability. For engineering purposes, fluid storage in conventional reservoirs is generally limited to macroscopic pores (diameters $>10\,\mu m$). Reservoir quality studies rely on the analysis of core plug sized samples (i.e., volumes on the order of tens of $cm^3$) and may involve predictive modeling. Such studies are done in collaboration with sedimentological, stratigraphical, petrophysical and geophysical studies that focus on the broader spatial characteristics of the reservoir, as described in other chapters in this volume.

Diagenetic alteration may substantially reduce both permeability and fluid storage capacity in deepwater sandstones. The nature and magnitude of diagenesis (the suite of processes that transform unconsolidated sediment to rock) is largely controlled by the depositional composition and texture of the reservoir of interest and its burial history. Virtually all sandstones have lost significant porosity and permeability compared to their depositional states in response to compaction (the reduction in bulk volume). Compaction results mainly from grain rearrangement and deformation under burial and tectonic loads but in some cases also may result from dissolution at grain contacts. In addition, most sandstone reservoirs have experienced at least some extent of cementation. Cementation involves the precipitation of solid phases within intergranular pore spaces. While cementation arises from a wide range of processes that occur under varying conditions and rates, it tends to progressively degrade reservoir quality with increasing time and temperature exposure.

There is nothing unique about the nature of diagenesis or reservoir quality of deepwater sandstones compared to sandstones from other geologic settings. However, because the diagenetic pathways of sandstones depend on depositional characteristics as well as burial and fluid flux histories, an area of interest may display characteristic patterns in reservoir quality that reflect the processes that dictate the sand depositional composition and texture (see Chapter 11; Heins and Kairo, 2007) as well as the filling, heatflow and tectonic history of the basin (see Chapters 3–10 and 12).

## Characterizing sandstone reservoir quality

Before describing the main controls on reservoir quality in deepwater sandstones, a few words about reservoir quality characterization methods are in order. While petrophysical core analysis is an essential source of porosity and permeability data for sandstone reservoirs, such data in isolation do not provide enough information to understand why the data have the values they do. Sandstone petrology, the study of the composition and texture of sandstones and how they evolve through geologic time, however, provides a basis for understanding and predicting porosity, permeability and many other aspects of sand and sandstone. A fundamental tool in petrology is petrography, which involves the detailed description and analysis of rock samples.

**Deepwater Sedimentary Systems.** https://doi.org/10.1016/B978-0-323-91918-0.00011-6

Essential petrographic data types for sandstone reservoirs include:

- *Modal analysis*: the volume abundances of solid rock components and macroscopic porosity. In addition to quantifying the present-day composition, modal data also provide essential constraints on the compaction state of a sample and its composition at the time of deposition. Modal classification schemes are hierarchical with the highest level being represented by macroporosity (pores >30 μm in diameter), grains, matrix (clay-sized material that was present at the time of deposition), grain replacements (where the grain material that was present at deposition now is made up of a different solid phase), matrix replacements (where the matrix composition has changed after deposition) and cements (where solid material has precipitated in an intergranular pore). Subcategories provide additional critical information such as distinguishing between intergranular macropores and macropores that form in response to grain dissolution. (See Taylor et al., 2022, Chapter 7 for recommended measurement procedures.)
- *Grain size analysis*: the grain size distribution, which is most commonly described by the mean grain size and sorting metrics such as the standard deviation in phi units (the negative of the log base two of the grain diameter in mm). The grain size distribution is a fundamental constraint on permeability in addition to other sediment and rock properties. Note that in sedimentary petrology sorting is determined based on silt- and coarser-sized grains and does not include detrital clay matrix. (See Taylor et al., 2022, Chapter 6 for recommended measurement procedures.)
- *Grain coating analysis*: the extent to which quartz grain surfaces have been covered by materials that inhibit the formation of quartz overgrowth cements. Grain coatings, when well-developed, can have an enormous impact on the extent of quartz cementation in sandstones, although other factors such as thermal history, grain size distribution, compaction state and framework grain composition also are important. (See Taylor et al., 2022, Chapter 8 for recommended measurement procedures.)
- *Paragenesis analysis*: study of the relative time history of diagenetic processes in a rock. (See Taylor et al., 2022, Chapter 5 for recommended analysis procedures.)

Standard analysis of thin sections using polarized-light microscopy is the most accurate and cost-effective method for collecting quantitative petrographic data for use in reservoir quality studies. While image analysis of thin sections and polished surfaces using automated mineralogy techniques (AM) (e.g., QEMSCAN™, Nanomin™, Mineralogic™, AMICS™, TIMA™, INCAMineral™, etc.) provides useful information about the spatial distributions of minerals and pores, in their current form they are an insufficient basis for collection of petrographic data. A fundamental shortcoming of such methods is that they do not generate a modal analysis. To illustrate the gap that exists between AM results and a modal analysis, consider the occurrence of quartz and illite in a sandstone. Quartz may be present in the form of detrital (i.e., was present at deposition) monocrystalline, polycrystalline and chert grains as well as small domains within detrital rock fragments such as those made up of schist. Additionally, quartz can occur as authigenic (i.e., formed after deposition) overgrowth cement or pore-lining microcrystalline cement. Distinguishing between these modes of occurrence of quartz is essential for reservoir quality studies given that the modal categories above affect rock properties in profoundly different ways.

Illite may occur in sandstone as detrital clay with pore-lining or pore-filling textures, clay laminae, rims that adhere on grain surfaces and within grains such as shale rock fragments. Illite formed during diagenesis commonly occurs as fine, pore-bridging crystallites. Here again it is critical to classify each of these modes of illite separately due to their differing influences on permeability and chemical compaction.

Grain size measurements derived from AM methods can be incorrect because they do not distinguish overgrowth cement from the host grain and are unable to recognize the dimensions of rock fragments that incorporate multiple minerals. A final shortcoming of AM techniques is that the spatial resolution generally is too low to provide an accurate basis for quantification of the volumes and spatial distributions of critically important grain coating materials where they occur.

## Depositional texture and composition

Sand depositional properties constrain the nature of diagenetic changes that can occur with burial. Depositional characteristics of particular significance include abundance of matrix and the mean size, sorting and composition of framework grains. The deepwater lithology scheme presented in Chapter 7 considers each of these characteristics except for framework grain composition (see overview in Table 2 in Chapter 7). Table 1 in this chapter builds on this classification scheme by emphasizing textural aspects of these lithologies that affect reservoir quality. The column that lists "reservoir quality potential associated with depositional texture" provides a rough guide to those lithology types that could have adequate reservoir quality based purely on depositional texture. Importantly, however, there is no guarantee that even the most favorable of these lithology types will have adequate reservoir quality: the suite of post-depositional processes that constitute diagenesis has the potential to reduce reservoir quality to subeconomic levels for any lithology type.

**TABLE 1** Reservoir quality related aspects of common lithology types in deepwater sandstones (as described in Table 2 from Chapter 7).

| | Lithology | Typical depositional environment | Mean grain size | Grain sorting (coarse silt and sand grains only) | Matrix abundance | Reservoir qualty potential associated with depositional texture | Potential diagenetic reservoir quality risk factors |
|---|---|---|---|---|---|---|---|
| 1 | Conglomerate (clast and matrix supported) | Canyon, channel | Gravel including pebbles, cobbles, and boulders; sandy or muddy matrix | Poorly to moderately sorted (well sorted possible but rare) | Moderate to high | Low to medium | Compaction, cementation |
| 2 | Very thick-bedded sandstone | Canyon, channel, lobe (splay) | Fine to very coarse sand (very fine possible but rare) | Moderate to well sorted | Low | High | Compaction, cementation |
| 3 | Thick-bedded sandstone | Canyon, channel, lobe (splay) and to a lesser extent levee; rarely in overbank | Very fine to very coarse sand, can include pebbles and granules near base | Moderate | Low to Moderate | Medium | Compaction, cementation |
| 4 | Medium-bedded sandstone | Lobe (splay), levee and to a lesser extent canyon, channel and overbank | Very fine to coarse sand (very coarse sand possible but rare) | Moderate | Moderate to high | Medium | Compaction, cementation |
| 5 | Thin-bedded sandstone, siltstone, and mudstone | Levee, overbank, lobe (splay) | Medium- and fine- to very fine-grained sand, silt, and clay | Well sorted to very poorly sorted | High | Medium to low | Compaction, cementation |
| 6 | Very thin-bedded sandstone, siltstone, and mudstone | Levee, overbank, lobe (splay) | Medium- and fine- to very fine-grained sand to medium, silt, and clay | Well to poorly sorted | High | Low to medium | Compaction, cementation |
| 7 | Muddy sandstone to chaotic deposits | Lobe (splay), overbank, levee, canyon, channel | Medium- to very fine-grained sand, silt, and clay | Well to poorly sorted | High | Low | Compaction |
| 8 | Mudstone | Overbank, levee and to a lesser extent canyon, channel and lobe | Silt and clay | (Not applicable) | Very high | Very low | Compaction |

Note that even the most favorable lithologies may have inadequate reservoir quality due to diagenetic alteration.

Deepwater sandstones display broad variations in mean grain size and framework grain composition from one basin to another as illustrated in Fig. 1. Within a stratigraphic unit, these depositional characteristics also can show significant variations. For example, a deepwater sandstone dataset that we have analyzed petrographically from the Cretaceous of offshore West Africa, displays decreasing feldspar content and poorer sorting with increasing mean grain size (Fig. 2). Substantial changes in reservoir quality can also occur over lateral scales of meters in response to textural variations associated with depositional processes as illustrated in an example near the end of the chapter.

Grain size exerts a fundamental control on permeability that tends to be maintained throughout the diagenetic evolution of sandstones as illustrated in Fig. 3. In this example involving diagenetically altered Mesozoic sandstones with similar framework grain compositions, the permeability of very fine-grained sandstones is roughly two orders of magnitude lower compared to coarse-grained samples with similar porosities—at least for porosities greater than ~15 vol%. At lower porosities where more extensive diagenetic alteration has occurred, however, the permeability contrast with grain size diminishes.

Mean grain size and grain sorting also affect depositional porosity as well as the surface area in contact with pore fluids, which influences permeability and the net rate of some chemical reactions. The thin section photomicrographs in Fig. 4 all have the same mean grain size but illustrate the considerable variations in grain sizes, shapes and compositions that can occur depending on grain sorting and origin. These laboratory sand packs have precisely defined grain size distributions based on mixes of sieve separates (Parker et al., 2016). The upper row of images shows well-rounded forms derived from unconsolidated sand from the St. Peter Sandstone (Ordovician) from the Upper Mississippi Valley, USA. Most of the grains from this source are made up of quartz. The bottom row in the figure includes images of modern sand from the Animas River in Durango, Colorado, USA that includes grains deposited in an alpine to high desert setting that are derived from a

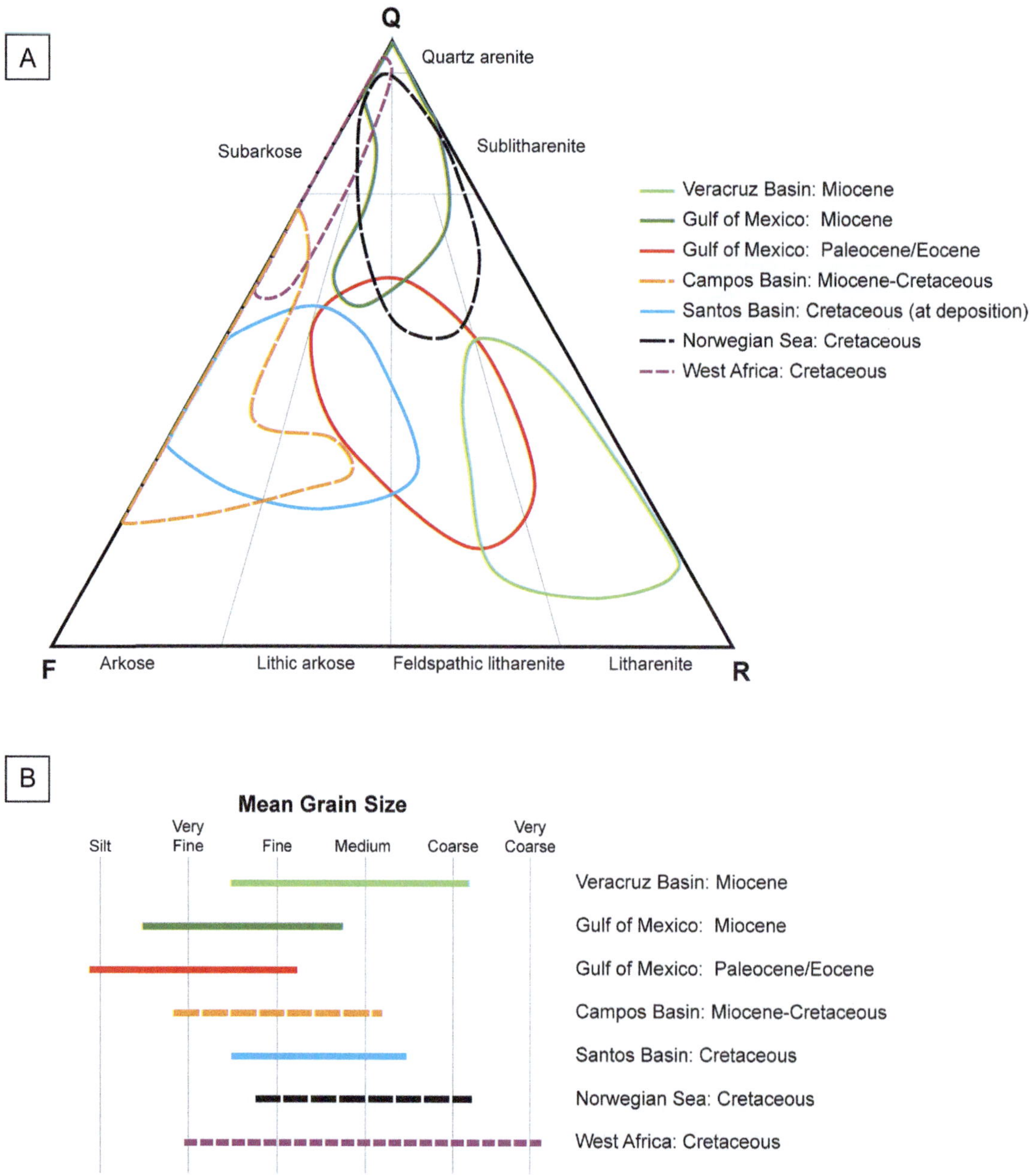


**FIG. 1** Example compositions and mean grain sizes for deepwater sandstones. (A) Framework grain compositions using the classification scheme of Folk (1980). (B) Variations in mean grain size. Compilation of data from Anjos et al. (2003), Fetter et al. (2009), Taylor et al. (2010), Dutton et al. (2012), Bahlis and De Ros (2013), Tobin and Schwarzer (2014), Marchand et al. (2015), Taylor et al. (2022) and unpublished datasets. The plots are intended to illustrate the broad range in composition and texture shown by deepwater sandstones and probably do not fully represent the variations that occur within the named basins and stratigraphic ages.

mélange of nearby rock types. In both sand sources, it is apparent that the intergranular porosity decreases systematically as sorting becomes poorer. In sandstones with abundant rigid grains, porosity variations associated with sorting differences can persist after compaction (Rossi and Alaminos, 2014).

An additional critically important textural characteristic of deepwater sandstones is the abundance and spatial distribution of "matrix". Matrix in sandstone reservoir quality studies refers to mud-sized material that is present at the time of deposition. Matrix is not included in the mean grain size or grain sorting calculations, which are limited to clasts that are at least upper silt in size. Matrix may occur as a pore-fill as illustrated in Fig. 5A, as laminae as shown in Fig. 5B or in patches that are related to bioturbation (see Chapter 15). Although matrix is microporous, the associated pores have tiny dimensions and high wetted surface area compared to intergranular pores. Consequently, matrix-rich sandstones tend to have low permeabilities as illustrated in Fig. 5C. Furthermore, matrix laminae produce large permeability anisotropies

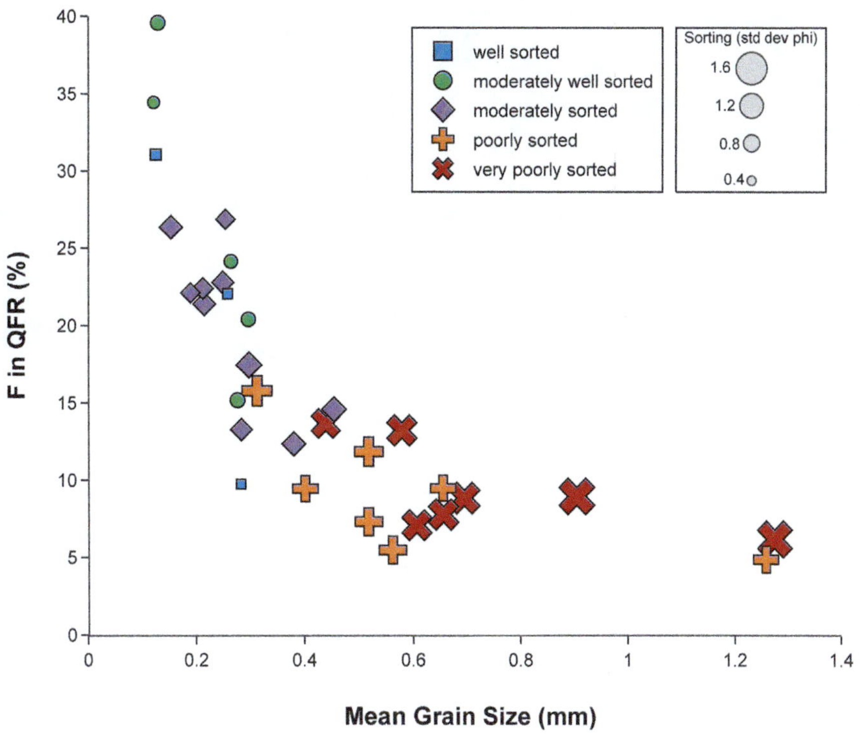


**FIG. 2** Variations in the percentage of grains made up of feldspar (QFR scheme of Folk, 1980) with mean grain size and sorting for the West Africa, Cretaceous dataset shown in Fig. 1. Note the decrease in both feldspar abundance and grain sorting (shown by symbol shape and size) with increased mean grain size.

where values are orders of magnitude lower perpendicular to bedding compared to parallel with bedding. Variations in the volumes and textures of matrix in deepwater sandstones reflect depositional processes and the nature of sediment input (see Chapters 6–11) that may be influenced by bioturbation (see Chapter 15).

Depositional framework grain compositions determine mechanical compaction response to load in addition to the nature of geochemical reactions that may occur. As an example, consider the three compositions shown in the thin section photomicrographs of Fig. 6, which depict a quartz arenite, an arkose and a litharenite with abundant volcanic rock fragments. The compaction significance of the associated variations in quartz, feldspar and rock fragment abundances is shown using a QFR diagram (Folk, 1980) in Fig. 7. Because quartz grains are the most resistant to deformation among these grain types, sandstones that are rich in quartz generally experience the least extent of mechanical compaction in response to a given maximum effective stress. However, in some cases, quartz-rich sandstones can experience significant "chemical compaction" that results from dissolution of quartz at grain contacts that are covered by illite. Controls on mechanical and chemical compaction are discussed at greater length in the following section.

Depositional framework grain composition also has a profound effect on the nature of chemical processes that occur during diagenetic alteration. As an example, the nearly pure quartz arenite shown in Fig. 6A is much less likely to be associated with the formation of clay minerals or zeolites compared to the arkose in Fig. 6B or the volcanic rock fragment-rich litharenite in Fig. 6C. The link between framework grain composition and formation of quartz, clay and zeolite cements is shown schematically in Fig. 8. Here, it is apparent that quartz overgrowth cements have the potential to be a major control on reservoir quality only when roughly half or more of the framework grains are made up of quartz. This compositional control reflects the fact that quartz overgrowths need quartz grains for nucleation. The linkage between framework grain composition and cement types and volumes is reviewed later in the chapter.

From this discussion, it is clear that one of the most important tasks in reservoir quality prediction for undrilled locations is to constrain the depositional composition and texture of potential reservoir intervals. Many factors influence these depositional characteristics including sediment sources, tectonic setting, climate, transport distance and residence time, and depositional processes as discussed in Chapter 11 and other chapters in this book.

## Compaction

The bulk volume of a sandstone decreases as a result of compaction, leading to a reduction in porosity. Porosity, however, also may be reduced by cementation without a change in bulk volume. Consequently, the term "compaction curve" should be avoided when discussing sandstone porosity trends.

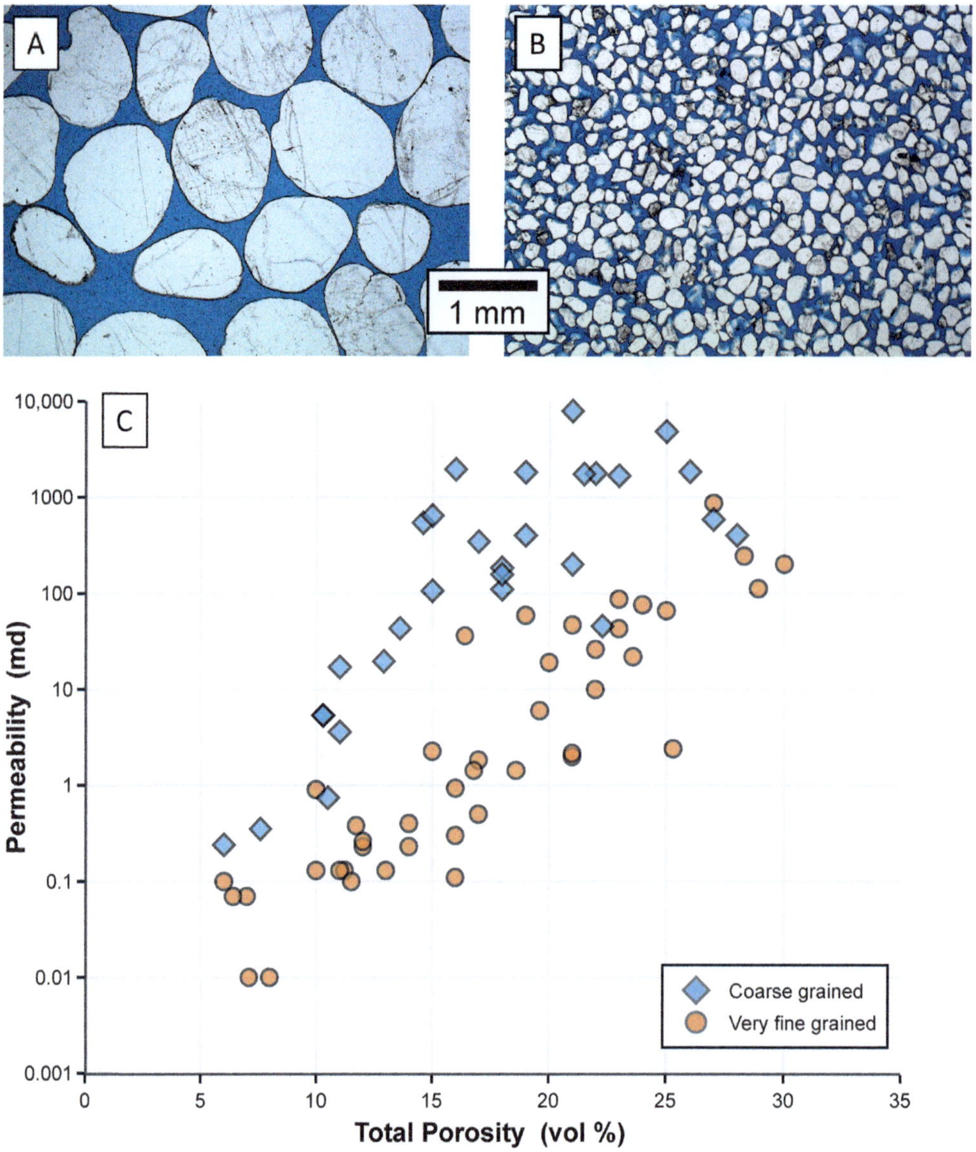


**FIG. 3** Impact of mean grain size on permeability. (A) Thin section photomicrograph of unconsolidated coarse-grained sand, (B) thin section image of unconsolidated fine-grained sand with the same magnification as A, (C) core plug porosity and permeability data from Jurassic aged sandstones. *Thin section photomicrographs use plane-polarized light.*

Compaction occurs in response to the load transmitted through the solid portion of the rock. This load is referred to as effective stress (Terzaghi, 1923; Biot, 1941). Where tectonic forces are negligible, the effective stress can be estimated by subtracting the fluid pressure from the lithostatic pressure (the pressure induced by the combined weight of the fluids and solids above the point of interest). Effective stresses therefore are reduced in rocks that have fluid pressures that exceed the hydrostatic gradient (overpressure) compared to rocks with hydrostatic pressure gradients. Consequently, overpressured sandstones may experience reduced extents of compaction compared to hydrostatic conditions, although the magnitude of this reduction depends on a number of other factors discussed at greater length later in this section.

The most widely used measurement for estimating the magnitude of sandstone compaction is intergranular volume (IGV) (e.g., Houseknecht, 1987a,b; Ehrenberg, 1995; Paxton et al., 2002). IGV is the volume fraction of the rock that is not made up of framework grains or grains that have been replaced or dissolved. Consequently, IGV represents the sum of the intergranular porosity, cement and matrix volume fractions. The magnitude of compaction can be estimated by comparing the measured IGV with an estimate of the depositional IGV. Depositional IGVs can be estimated using natural or laboratory experiment analogs. The dependency of depositional IGV on mean grain size and sorting is illustrated for experimental grain packs consisting of well-rounded quartz grains in Fig. 9. These results of Parker et al. (2016) echo

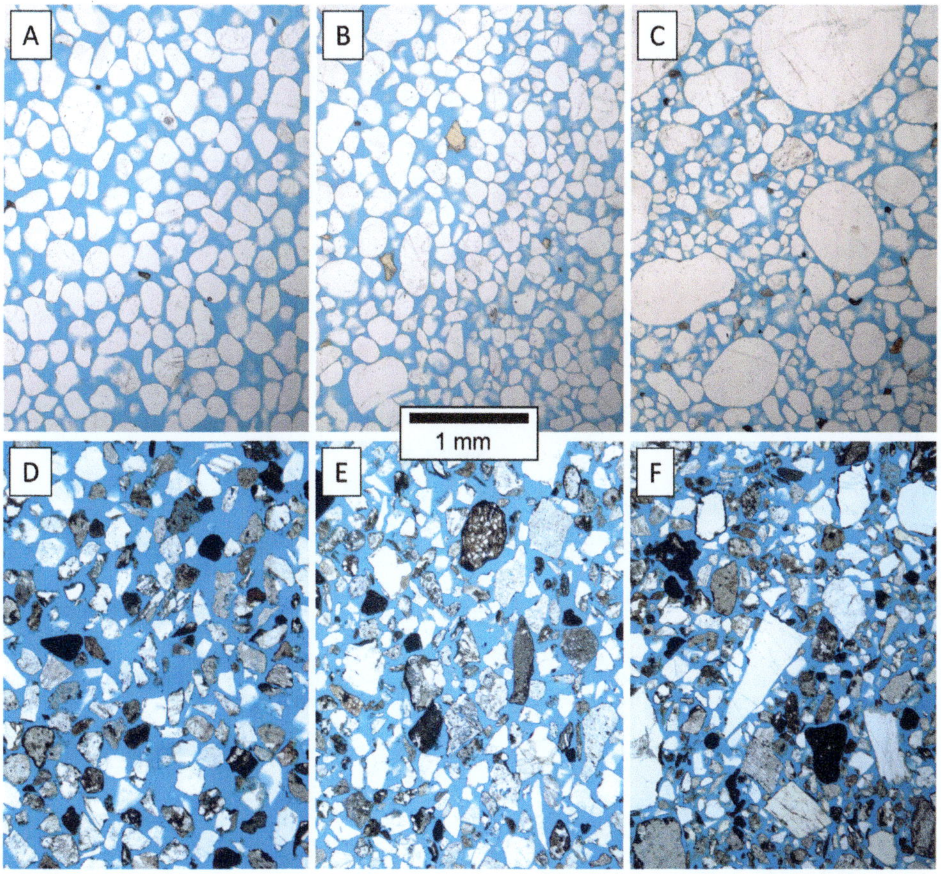


**FIG. 4** Thin section photomicrographs (*plane-polarized light*) with the same magnification showing fine-grained unconsolidated sand with the same sieve-based measured mean grain size of 0.23 mm. (A), (B) and (C) are quartz arenite composition, whereas (D), (E) and (F) have a feldspathic litharenite composition based on the classification scheme of Folk (1980). The grain size distribution is very well sorted in (A) and (D), moderately well sorted in (B) and (E), and poorly sorted in (C) and (F) (standard deviation in phi values of 0.18, 0.61 and 1.4, respectively).

those of the classic study of Beard and Weyl (1973) in that they show a decline in depositional IGV with poorer sorting. Although the IGV values for samples with medium- and coarse-grained mean grain sizes are similar, values for a given sorting tend to be greater in finer-grained samples. Some deepwater sandstone lithologies also have high IGV values due to abundant detrital clay matrix. Typically, however, the intergranular porosity of such samples is low given that the matrix occludes much or all of the intergranular volume. Other factors that may influence depositional IGV include depositional process, grain frictional properties and grain shape.

IGV is a useful metric in reservoir quality studies because it represents the maximum potential intergranular porosity in the absence of matrix or cement. The utility of distinguishing between IGV and porosity is illustrated by cross sections from 3D volumes in Fig. 10 that we created using the Cyberstone™ digital sedimentary petrology model (Lander and Bonnell, 2018; Guilkey et al., 2021). The figure includes two scenarios, both of which have the same depositional IGV value of 36 vol% and end up with the same total porosity of 15 vol% after burial. These scenarios, however, have significantly different final IGV values. In the scenario shown in Fig. 10C and D, decline in intergranular porosity is due exclusively to compaction so the final IGV and intergranular porosity both are 9 vol%. The large bulk volume reduction (shown by the area in gray in Fig. 10D) occurs in response to extensive ductile deformation of abundant shale rock fragments. The scenario in Fig. 10A and B results in much less compaction even though it was subjected to identical load history. In this case, the higher IGV value of 26 vol% occurs due to a greater proportion of rigid quartz grains compared to the scenario in Fig. 10C and D. The reduction in intergranular porosity from 36 to 14 vol%, in this more rigid grain-rich scenario, resulted from compaction as well as precipitation of 12 vol% of cement (shown in orange).

An additional significant difference between the two scenarios is in the volumes of constituent pore types. In the Fig. 10B scenario most of the porosity consists of intergranular pores (14 vol%), which are highly effective for fluid flow and storage with an additional small volume of micropores (1 vol%) that contribute little to reservoir quality. By contrast, the Fig. 10D scenario has more microporosity (6 vol%) and less intergranular porosity (9 vol%), leading to poorer reservoir quality despite having an identical total porosity.

These scenarios demonstrate the utility of modal data from petrographic analyses for assessing reservoir quality controls compared to relying on porosity values in isolation. From the IGV, intergranular porosity and cement modal abundances, it

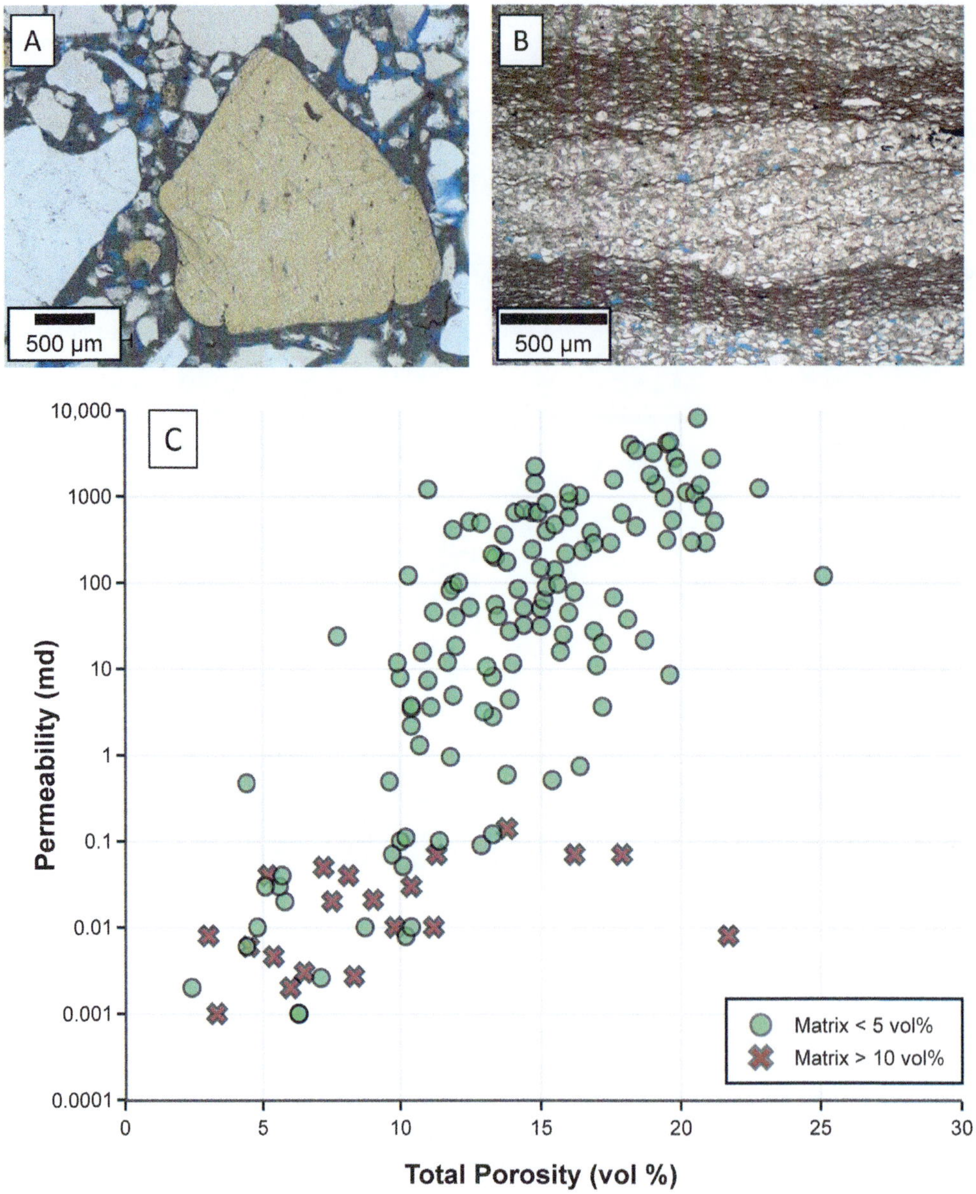


**FIG. 5** The mode of occurrence and abundance of matrix impacts sandstone porosity and permeability: (A) thin section photomicrograph showing pore-filling matrix, (B) thin section photomicrograph of matrix laminae and (C) porosity and permeability of Cretaceous sandstones with variable matrix abundances. *Thin section photomicrographs use plane-polarized light.*

is clear that reservoir quality reduction from the Fig. 10D scenario results exclusively from compaction, whereas the reduction in the Fig. 10B scenario occurs in response to both compaction and cementation. Such data are of fundamental importance when assessing the likely reservoir quality in undrilled regions that may have differing burial histories or depositional textures and compositions compared to available analog samples.

## Mechanical compaction

Mechanical compaction is an important cause of porosity loss in virtually every reservoir sandstone (Lundegard, 1991) and arises chiefly from grain rearrangement and deformation. The magnitude of mechanical compaction is related to sandstone composition and loading history.

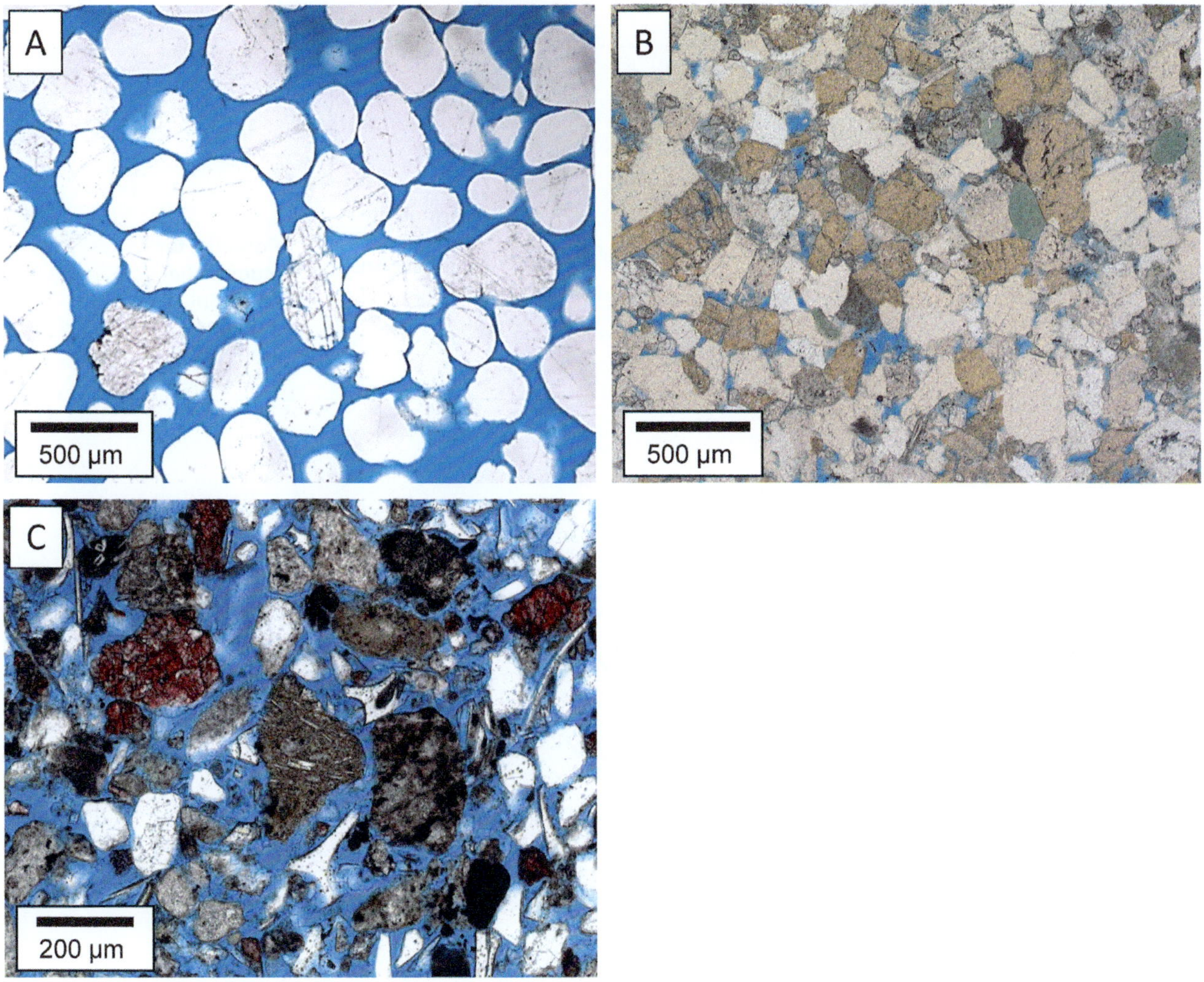


**FIG. 6** Thin section photomicrographs of sandstones with varying compositions: (A) quartz arenite, (B) arkose (feldspar-rich) and (C) litharenite (abundant rock fragments and, in this example, volcanic glass shards). *Thin section photomicrographs use plane-polarized light.*

## *Impact of framework grain composition*

As mentioned earlier, framework grain composition can vary widely in deepwater sandstones (Fig. 1). These compositional variations have important consequences for the compaction behavior of sandstones.

Paxton et al. (2002) demonstrated that the IGV values for rigid-grain rich sandstones from individual well locations typically are ~26vol% for samples with maximum burial depths greater than about 2000m (6562 feet). Data from two rigid grain rich datasets of Mesozoic age that we have studied are consistent with this result as illustrated in Fig. 11. Samples in this figure all experienced maximum effective stresses in excess of 25MPa (3626 psi), which is comparable to a burial depth of 2000m (6552 feet) under hydrostatic fluid pressures. Notably, these datasets show no systematic change in IGV with any of the following factors: effective stress over the range of ~25 to 65MPa (3626–9427 psi), thermal exposure with maximum temperatures ranging from ~80 to 170°C (176–338°F) and quartz cement abundances that range from ~1 to 15vol%. While it is possible that such sandstones may have experienced minor chemical compaction, the extent of any such compaction apparently is insensitive to the substantial effective stress, temperature and quartz cement variations within the datasets. Quartzose datasets from other regions, however, sometimes do show significantly lower IGV values that result from chemical compaction, as discussed in greater detail in a later section.

Much of our understanding of the mechanical behavior of sand with ductile grains comes from the seminal experimental study of Pittman and Larese (1991). These researchers created binary mixtures of rigid quartz grains with fourteen other grain types representing a broad range in ductile behavior ranging from very weak materials such as glauconite to stronger

FIG. 7 An illustration of how compaction may be influenced by framework grain composition (QFR classification scheme of Folk, 1980).

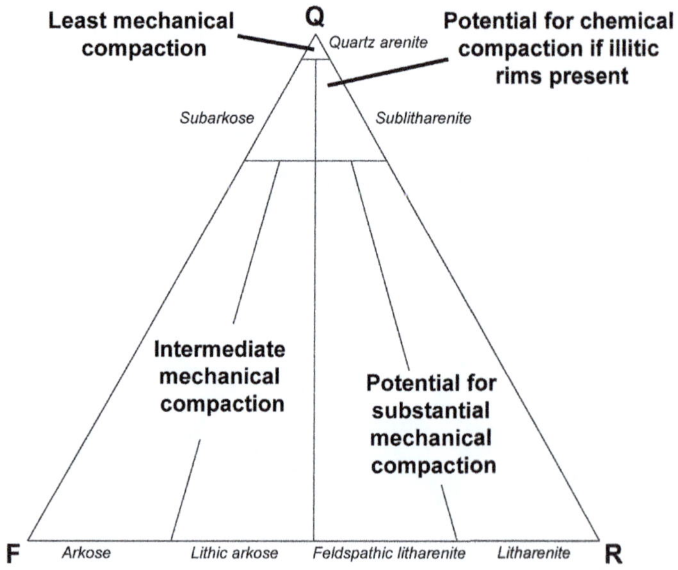


FIG. 8 Schematic illustration of the influence of framework grain composition on the potential occurrence of abundant diagenetic cements (QFR classification scheme of Folk, 1980).

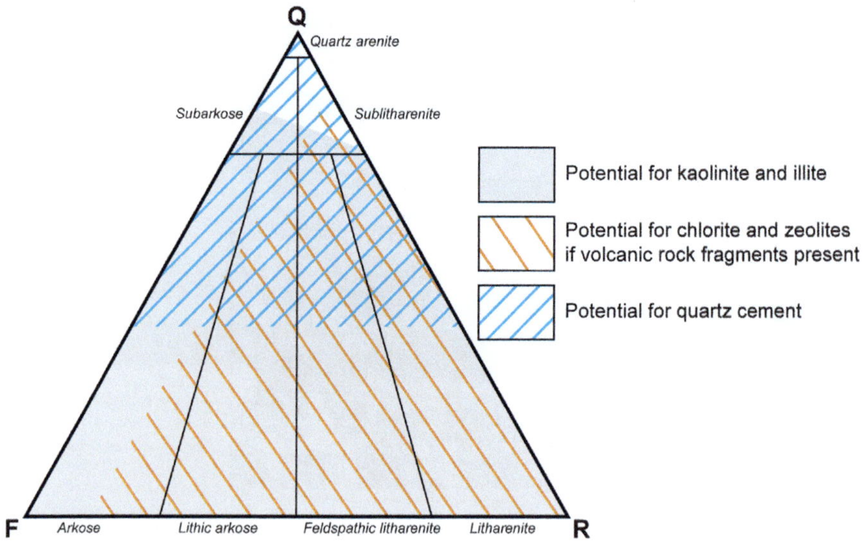


FIG. 9 Depositional intergranular volume in laboratory sand grain packs made up of mainly of rounded quartz grains as a function of mean grain size and sorting. Sorting becomes poorer with increasing $x$-axis value. Note that depositional intergranular volume and depositional porosity are identical for these experiments given that the sand packs do not include detrital matrix or microporous grains. Data of Parker et al. (2016).

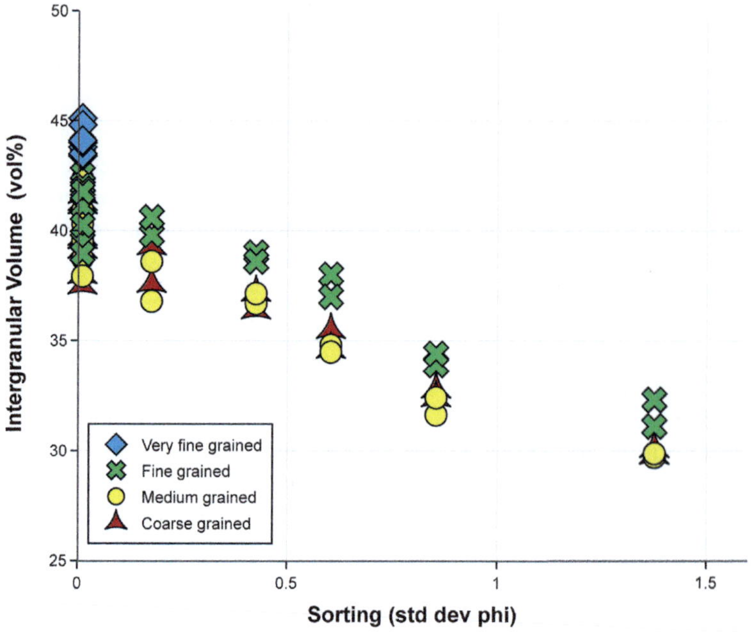

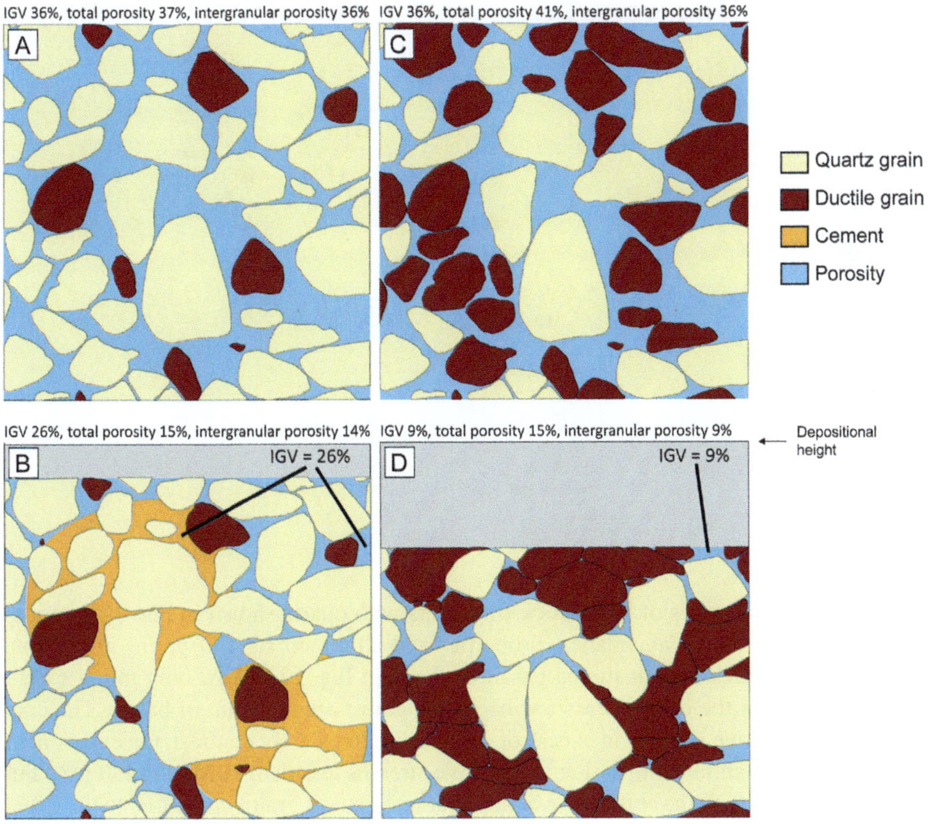


**FIG. 10** Vertical cross sections from Cyberstone 3D simulations for two sandstone scenarios that illustrate the differences between total porosity, intergranular porosity and intergranular volume (IGV). Both intergranular volume and intergranular porosity are 36 vol% for the two depositional scenarios in (A) and (C) despite differences in the framework grain compositions. The nature of diagenetic alteration for the scenarios is quite different after burial through time, however, as shown in (B) and (D). In (B), which has abundant rigid quartz grains, IGV has been reduced to 26 vol% by mechanical compaction. The intergranular porosity of 14 vol% is reduced in this sample not only by compaction but also by cementation. The total porosity of 15 vol% includes about 1 vol% of microporosity associated with the ductile grains. In (D), which has abundant microporous ductile grains, the total porosity likewise is 15 vol%. In this scenario, all of the porosity loss is due to compaction. The IGV and intergranular porosity values in this case both are identical at 9 vol% and the microporosity is 6 vol%.

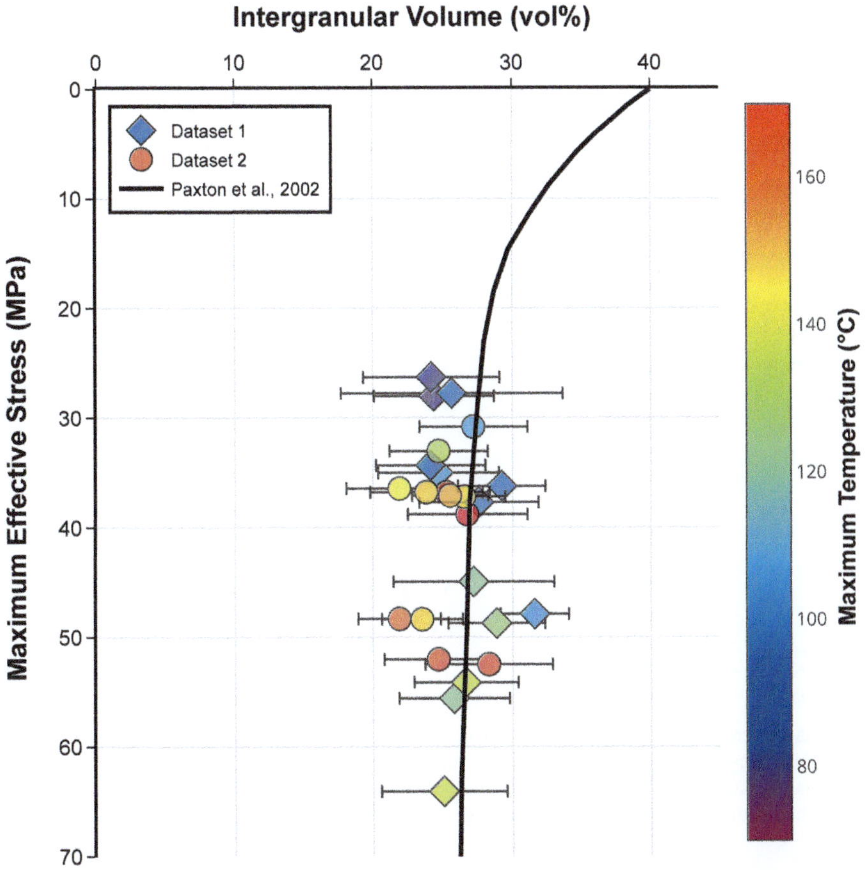


**FIG. 11** Intergranular volumes in quartzose sandstones as a function of maximum effective stress and maximum temperature from two datasets that were not among those studied by Paxton et al. (2002).

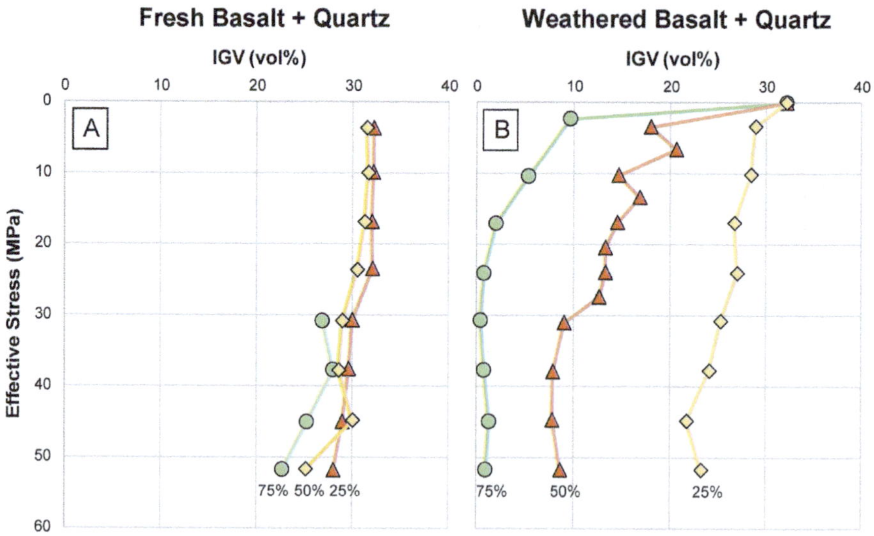


**FIG. 12** Experimental results of Pittman and Larese (1991) recast in terms of IGV as a function of effective stress and the percentage of a ductile grain type. (A) Fresh basalt grains mixed with quartz grains and (B) weathered basalt grains mixed with quartz grains.

metamorphic rock fragments. The results from mixtures of basalt rock fragments with quartz shown in Fig. 12 provide a particularly striking demonstration of how the volume abundance, properties and alteration state of grains affect compaction as a function of effective stress. Sand packs with unaltered basalt depicted in Fig. 12A show only minor extents of mechanical compaction with loading, even for the case where 75 vol% of the grains are made up of basalt. The compaction that does occur is mainly limited to effective stresses in excess of 25 MPa (3626 psi). By contrast, the experiments using weathered basalt in Fig. 12B indicate substantial compaction at effective stresses less than 10 MPa (1450 psi). Furthermore, these results show a far greater sensitivity to the abundance of the weathered basalt grains compared to the experiments involving unaltered basalt grains. The range in behavior for all admixed grain types is shown in Fig. 13A for experiments that were all conducted with 52 MPa effective stress (7542 psi), which is roughly equivalent to the effective stress expected at 4200 m (13,780 feet) burial depth under hydrostatic conditions. Of particular note in the plot is the wide range in IGV values where the grain mixtures have low quartz abundance.

Despite the broad range in the extent of compaction resulting from the variations in the abundance and mechanical properties of ductile materials, Lander and Walderhaug (1999) demonstrated that the decline in IGV with effective stress shown in the experimental data of Pittman and Larese (1991) and in natural sandstones is generally well approximated using an exponential function where the high effective stress asymptote is related to the framework grain composition. Lander

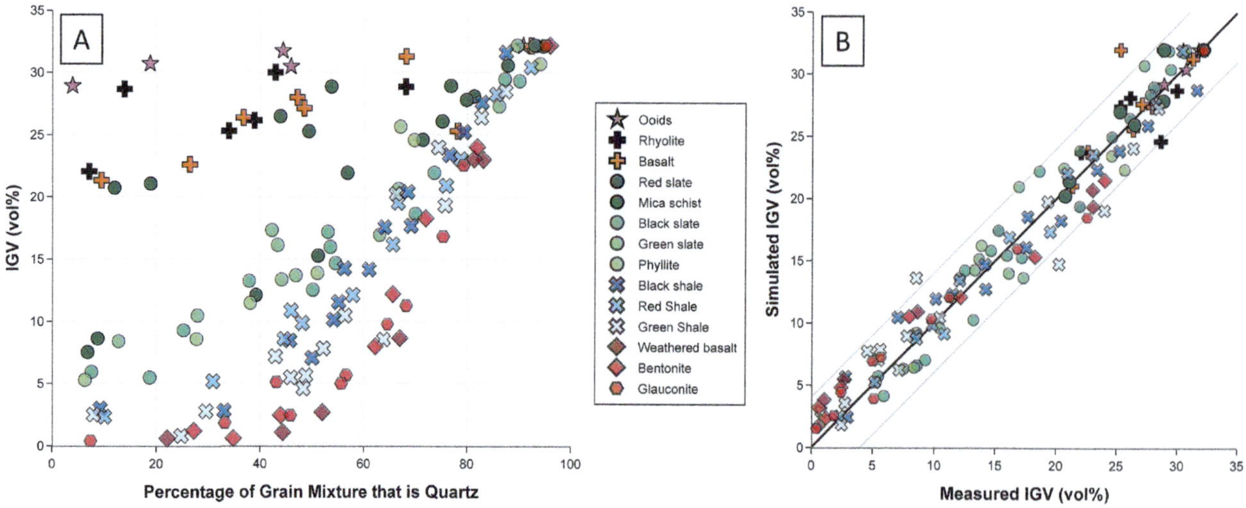


**FIG. 13** (A) Plot showing variations in IGV value with the percentage of grain mixtures made up of quartz based on data from Pittman and Larese (1991) experiments conducted using 52 MPa (7542 psi) effective stress. Note the large range in IGV values among admixed grain types that have low quartz percentages. (B) Comparison between IGV derived from the Pittman and Larese (1991) experiments depicted in (A) with values obtained from Touchstone simulations. The *thick black line* is a 1:1 reference, whereas the *thin gray lines* indicate a range of four volume percent around this reference.

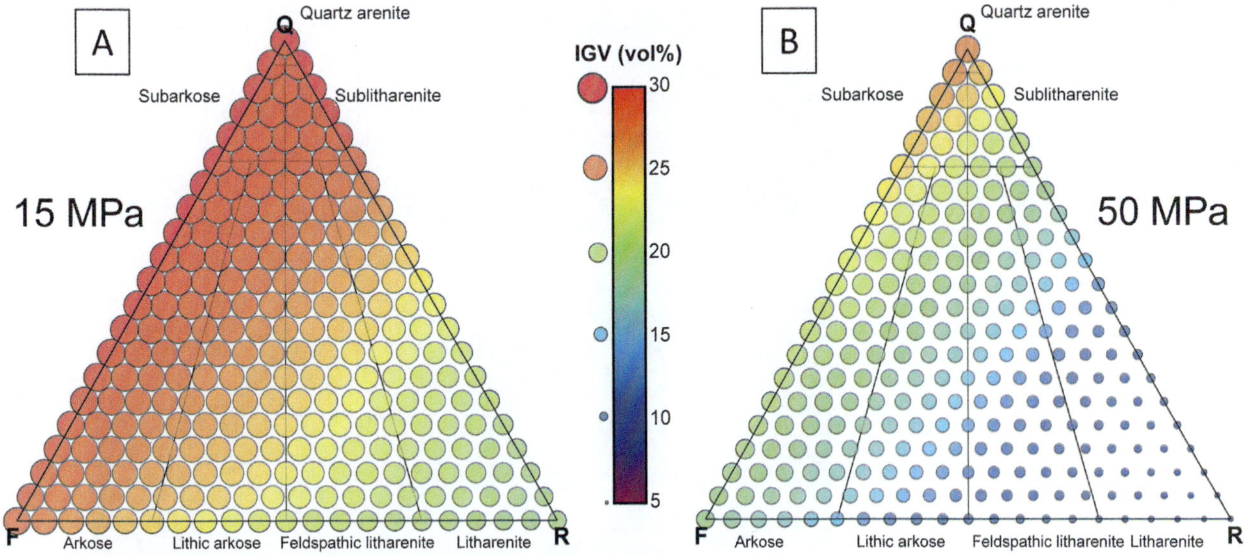


**FIG. 14** Simulated intergranular volume (IGV) as a function of framework grain composition for the simple scenario described in the text for points on a QFR diagram (Folk, 1980). The IGV values are indicated both by symbol size and color. Note that IGV values are likely to deviate substantially from this simple scenario in actual sandstone reservoirs. (A) Results for 15 MPa (2175 psi) maximum effective stress and (B) 50 MPa (7252 psi) maximum effective stress.

and Bonnell (2020) have subsequently extended this model to predict IGV behavior for mixtures of grain types with varying mechanical properties. The performance of this model as implemented in the Touchstone™ diagenesis and reservoir quality simulation system (developed by Geocosm LLC) is shown in Fig. 13B. Touchstone's compaction model is able to reproduce the observed variations in IGVs at 52 MPa (7542 psi) effective stress at an accuracy approaching the measurement uncertainty. In this analysis the mechanical strength parameters for each framework grain type were optimized to fit the data. In addition to simulating compaction, Touchstone considers the effects of replacement and cementation through geologic time and uses these results to simulate rock properties such as porosity, permeability and bulk density, mineralogy and chemistry (for more information about Touchstone see Chapter 12 of Taylor et al., 2022).

To illustrate how both effective stress and mixtures of framework grain types can affect IGV, we created a suite of simulation scenarios shown in Fig. 14 using Touchstone. These scenarios mix ductile rock fragments with similar behavior to the slate rock fragments used by Pittman and Larese (1991) (see Fig. 13A), with more rigid feldspars, and even more rigid quartz grains. At a maximum effective stress of 15 MPa (2175 psi) (Fig. 14A), which corresponds to ~1200 m (3937 feet) of burial under hydrostatic pressure, the variation in simulated IGV with framework grain composition is relatively limited with values ranging from ~20 to 32 vol% (as shown by both symbol size and color in the figure). Increasing the effective stress to 50 MPa (7252 psi) (Fig. 14B), however, leads to much larger IGV variations with composition with values ranging from ~7 vol% for the rock fragment apex compared to 27 vol% for the quartz apex. Thus, in the absence of cements and matrix, all compositions could have acceptable reservoir quality at 15 MPa (2176 psi), whereas, nearly 40% of the compositional range would have maximum potential intergranular porosity values of less than 15 vol% at effective stress of 50 MPa (7252 psi). Our intention with this example is to illustrate the broad compositional and burial history controls on reservoir quality: these results should not be used as a guide for assessing IGV values of reservoir sandstones given the simplifying assumptions that they employ.

## *Impact of cements and matrix*

An additional factor influencing mechanical compaction is the abundance and mechanical properties of matrix and cement. Such materials may act to reduce compaction magnitudes while also changing geomechanical and rock physics properties as discussed in Chapter 17. Regrettably, at present, there are no published systematic laboratory studies on the effects of abundances, spatial distributions, morphologies and mechanical properties of common cement types and textures on sandstone compaction. However, in collaboration with R. E. Larese (an independent consultant), we conducted a pair of laboratory experiments that illustrate the impact that cements can have on compaction behavior and porosity preservation. In the experiments, we compared the compaction behavior for a grain pack with 4 vol% calcite with a pore-lining geometry depicted in Fig. 15 with a control experiment lacking the cement. These packs are medium grained and well sorted with ~30

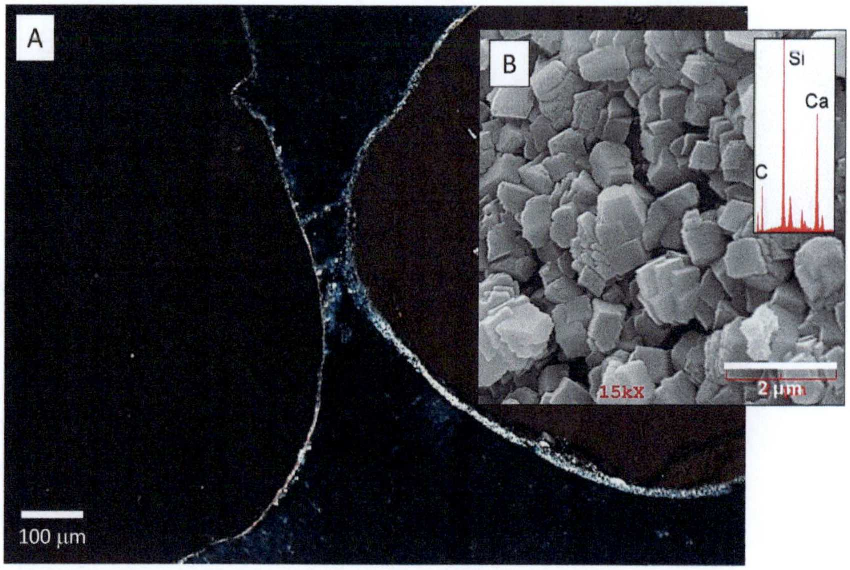


**FIG. 15** Synthetic calcite cement in a grain pack created in collaboration with R. E. Larese. (A) Thin section photomicrograph in cross-polarized light where calcite cement appears as birefringent material on the outer boundaries of quartz grains, and (B) SEM photomicrograph of calcite rhombs that precipitated on a quartz grain surface together with peaks for carbon ("*C*"), calcium ("*Ca*") and silicon ("*Si*") from an energy dispersive X-ray spectroscopy analysis of the material.

vol% of the grains being made up of shale rock fragments and the remainder being quartz. The control experiment lacking the carbonate cement compacted under 35 MPa (5076 psi) effective stress to an IGV of 18 vol%, whereas the cemented sample reached an IGV of 28 vol%. The strengthening effect of the cement in this case resulted in net porosity preservation given that the cemented sample after compaction has 24 vol% intergranular porosity compared to 18 vol% intergranular porosity for the companion sample lacking cement.

From this experiment, it is clear that cement, even in small volumes, can have a significant impact on mechanical compaction. The magnitude of this impact is a function of a number of factors such as (1) the compaction state and maximum effective stress over the time interval of cement precipitation, (2) the mechanical properties and contact bonding characteristics of the cements, (3) the texture and morphology of the cement and (4) the maximum effective stress reached after cementation. In general, however, cements that form early in the burial history have the largest potential to reduce compaction. Additionally, cements that line pores and consist of rigid, blocky, interlocking crystals that are tightly bound to grain surfaces are more likely to affect compaction compared to smaller, flexible crystallites that have weak attachments to their growth substrates.

### *Impact of fluid overpressure*

As discussed previously, fluid overpressures act to reduce the load transmitted through the solid portion of sediments and rocks and therefore cause reduced extents of compaction compared to hydrostatic conditions (Biot, 1941). The mere occurrence of fluid overpressure, however, does not guarantee a reservoir quality benefit. The magnitude of any such benefit depends on (1) the effective stress history (which, among other things, reflects the history of overpressure generation), (2) the mechanical behavior of the sandstone of interest (which reflects the depositional fabric, grain size distribution and the composition and properties of constituent grains, replacements, cements and matrix) and (3) whether other diagenetic processes such as quartz cementation are the dominant control on reservoir quality. In their assessment of the potential for reservoir quality preservation associated with fluid overpressures, Bloch et al. (2002) found that the largest potential for reservoir quality preservation from fluid overpressure development is in Tertiary or younger reservoirs that develop overpressures at shallow depths in response to compaction disequilibrium. In their assessment, older rocks have reduced potential for porosity preservation due to a greater likelihood that cementation is a primary control on reservoir quality and because of the potential that overpressures formed at least in part at greater depths of burial in response to processes such as oil to gas cracking, quartz cementation, smectite-illitization or other temperature-dependent processes.

To illustrate the influence that the timing of fluid overpressure development has on porosity preservation consider the scenario shown in Fig. 16A for a Miocene deepwater sandstone that is presently at 5000 m (16,404 feet) burial depth. The sandstone in the scenario is fine grained and has a subarkosic composition. We used a Touchstone model that is calibrated to sandstone data from the area of interest and considered two alternative scenarios for the development of fluid overpressures of 50 MPa (7252 psi), each of which results in a present-day effective stress of 12.5 MPa (1813 psi). The present-day fluid overpressure in these scenarios is close to a maximum likely value for its depth given that it is associated with fluid pressure that is near the fracture gradient. The "early overpressure" scenario in the figure assumes that overpressures developed at

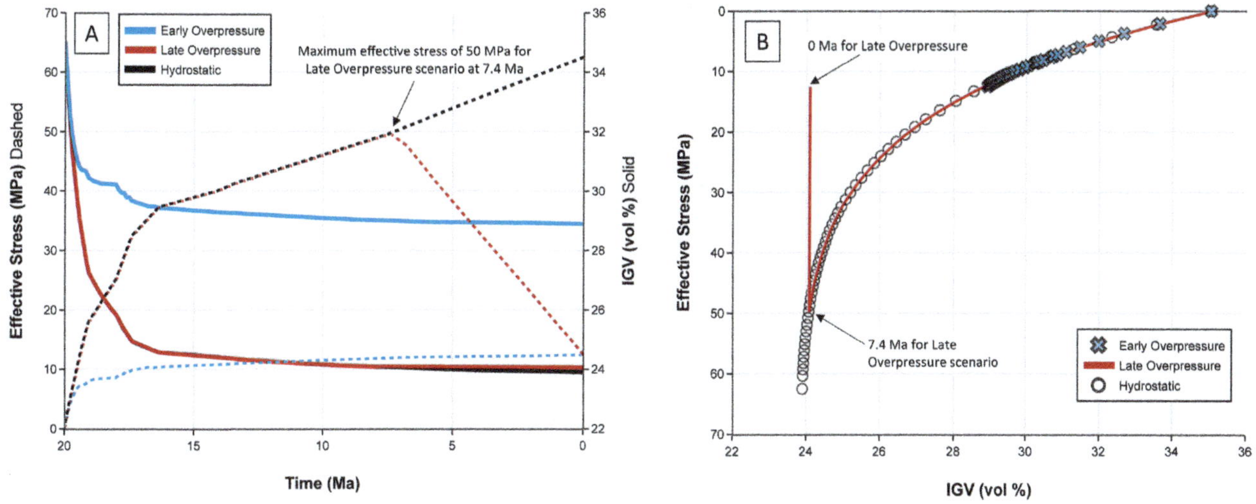


**FIG. 16** Touchstone simulations of compaction behavior for a deepwater fine grained, subarkosic sandstone for three alternative effective stress scenarios: "*Early Overpressure*" where the maximum effective stress occurs at the present-day, "*Late Overpressure*" where the maximum effective stress occurred ~7.4 Ma and "*Hydrostatic*" where no overpressure was present at any time in the burial history. (A) IGV (*solid lines*) and effective stress (*dashed lines*) through geologic time. (B) IGV as a function of effective stress history.

shallow depths from compaction disequilibrium that resulted from rapid initial rates of burial and the encasement of the sand body in low permeability mudrocks. In the "late overpressure" scenario, on the other hand, the overpressure is assumed to have started to develop after 4000 m (13,123 feet) of burial. In this scenario, a maximum effective stress of 50 MPa (7252 psi) was reached at 7.4 Ma as indicated by the arrow in Fig. 16A. We also consider a benchmark that assumes hydrostatic fluid pressures throughout the burial history of the sandstone. The present-day IGV values from the simulations are ~24 vol% for the late overpressure scenario and ~29 vol% for the early overpressure scenario. The late overpressure result is virtually identical to the result from the hydrostatic fluid pressure benchmark. Note that in reservoir sandstones the intergranular porosities for these scenarios will be lower than these simulated IGV values due to the presence of cements and matrix.

To understand the compaction behavior in these simulations, consider the decline in IGV with effective stress shown in Fig. 16B. Here, it is apparent that all of the IGV values from the early overpressure scenario fall on the IGV/effective stress curve defined by the hydrostatic benchmark. The lower limit of the IGV values for this scenario, however, reflects the associated maximum effective stress of 12.5 MPa (1813 psi) (blue dotted line on Fig. 16A). The IGV values for the late overpressure simulation also follow the hydrostatic curve in Fig. 16B until it reaches the maximum effective stress value of 50 MPa (7252 psi) at 7.4 Ma. From this point onward effective stress values decreased until they reached the present-day value of 12.5 MPa (1813 psi). Sandstone compaction during diagenesis is largely irreversible. Consequently, the IGV value from the late overpressure scenario reflects the extent of compaction resulting from a 50 MPa load despite the substantial decrease in effective stress that occurred from 7.4 Ma to the present.

## *Impact of tectonic stress*

A final factor that has the potential to significantly reduce reservoir quality through mechanical compaction is the occurrence of tectonic stresses (Chapter 3) associated with compressional regimes (Chapter 4) and diapirs (Chapter 5). Until recently most basin and petroleum systems modeling studies considered only vertical effective stress (Chapter 18). To illustrate the potential magnitude of reservoir quality degradation associated with tectonic stresses in a deepwater geologic setting, consider the much higher present-day values for mean effective stress derived from a model of the deepwater Perdido Foldbelt compared to the vertical effective stress as shown in Fig. 17 (Gradmann and Beaumont, 2012). These two metrics have similar values only at the base of the vertical section. Above this level, the mean tectonic stress exceeds the vertical effective stress by as much as 13 MPa (1885 psi). The complex pattern in effective stress through the profile reflects the interactions among gravitational and tectonic loads as influenced by the occurrence of fluid overpressures.

To illustrate the impact on reservoir quality of the tectonic component of the stress, we used Touchstone to simulate IGV values for a single sandstone type throughout the vertical section. This simple simulation assumes that the present-day effective stresses represent the maximum historical values and that cement strengthening is not significant. The results for this example indicate that IGV values may be as much as 5 vol% lower when tectonic stresses are accounted for compared to the case where they are ignored. These results show both the sensitivity of reservoir quality to tectonic stresses and

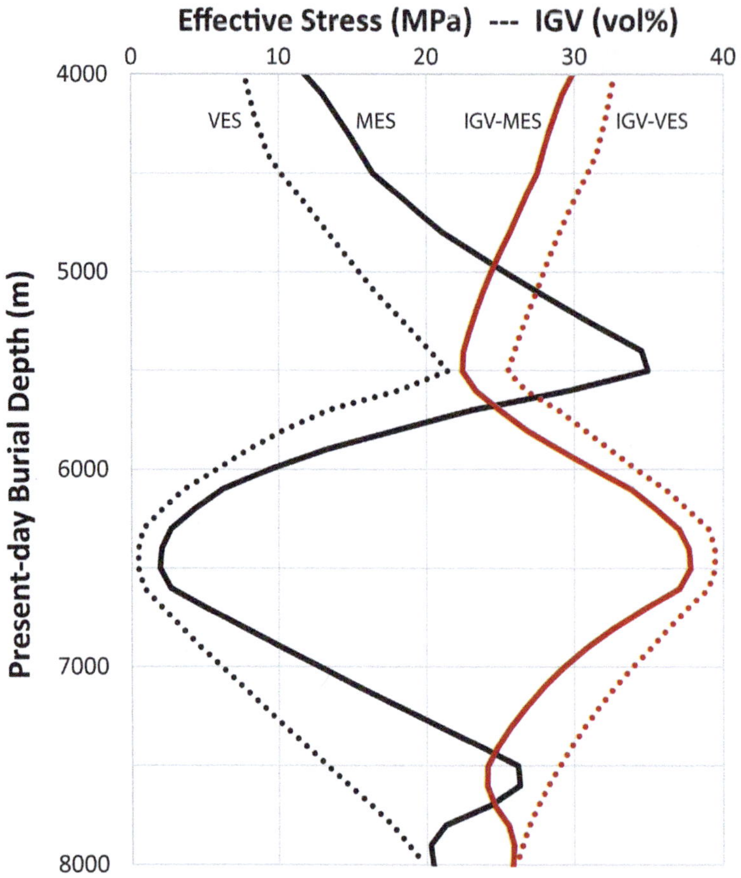


**FIG. 17** Tectonic stresses have the potential to significantly reduce reservoir quality of sandstones. The figure shows a vertical profile from the deepwater Perdido Foldbelt in the Gulf of Mexico at 250 km along model MQ-30 with the following results from the simulation by Gradmann and Beaumont (2012): "*MES*" is the mean effective stress and "*VES*" is the vertical effective stress at this position. Superimposed on the figure are our Touchstone simulation results: "*IGV-MES*" intergranular volume assuming that the *MES* values represent the maximum effective stresses, and "*IGV-VES*" intergranular volumes simulated with the same Touchstone model parameters but using the *VES* values.

the potential danger of making reservoir quality predictions using models that rely on inputs from basin and petroleum systems models that do not account for tectonic forces in regions where such forces are significant.

## Chemical compaction

Chemical compaction in sedimentary petrology refers to bulk volume reduction that occurs when mass dissolves from load-bearing contacts among grains and other solid materials (Houseknecht and Hathon, 1987). While "pressure solution" (e.g., Sorby, 1908; Weyl, 1959; Tada and Siever, 1989) commonly is considered to be synonymous with chemical compaction, we avoid this term given that there is considerable debate about the role of pressure in controlling chemical compaction in sandstones (Bjørkum et al., 1998). Where chemical compaction has a significant impact on reservoir quality, quartz and chert are the primary materials affected by the process, although contact dissolution also can occur in solid materials made up of other minerals.

Petrographic evidence for chemical compaction indicates that it takes place in sandstones with a wide range in depositional ages and geologic settings. In most cases, however, chemical compaction is not the primary control on the economic viability of deepwater sandstone reservoirs. Where chemical compaction is significant, it generally is associated with the presence of muscovite mica or illitic clay at quartz grain contacts (e.g., Heald, 1956; Sibley and Blatt, 1976; Houseknecht, 1988; Bjørkum et al., 1998) such as illustrated in Fig. 18A and B. In deepwater sandstones, illitic clay at grain contacts has the potential to occur when rims are present on the grains at the time of deposition (e.g., Wooldridge et al., 2017, 2018) or in sand subjected to dewatering (Porten et al., 2019). Stylolites in sandstones (e.g., Fig. 18C and D) may nucleate on illitic clay

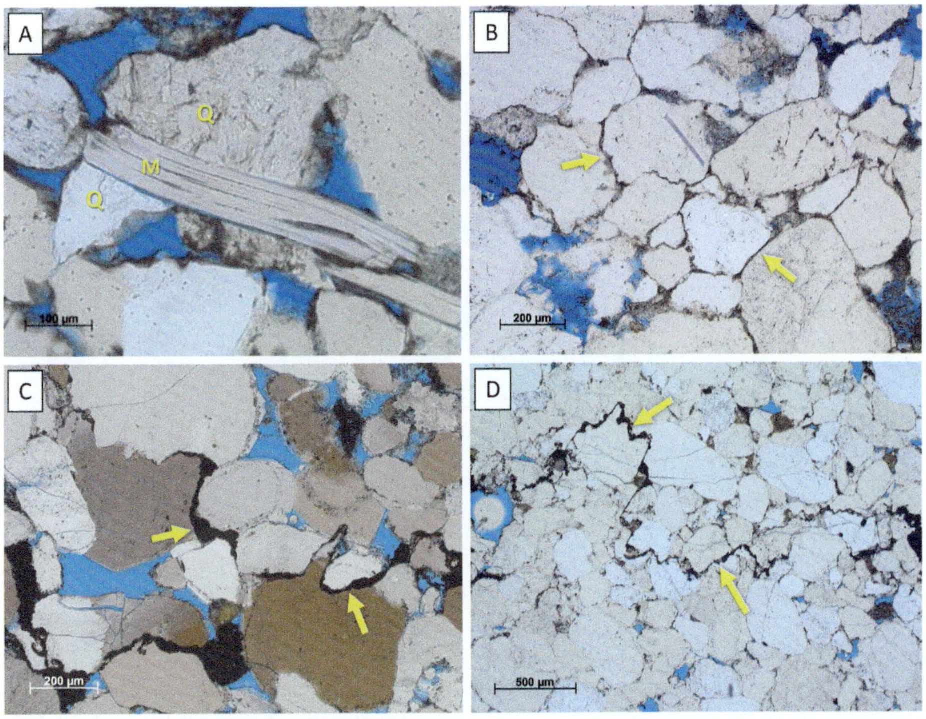


**FIG. 18** Examples of chemical compaction in sandstones. (A) Quartz grains "*Q*" have dissolved where they are in contact with a detrital muscovite grain "*M*". (B) Long grain contacts have developed for quartz grains that have illitic clay rims as indicated by the *yellow arrows*. In (C) and (D) stylolites are indicated by *yellow arrows*. These stylolites may have nucleated on illitic clay laminae or grain coatings. *All photomicrographs show plane-polarized light.*

laminae and reduce permeability in flow directions perpendicular to stylolite orientation (Walderhaug and Bjørkum, 2003; Walderhaug et al., 2006; Toussaint et al., 2018).

In addition to chemical compaction being promoted where quartz is in contact with muscovite and illite, other factors that may affect the process include temperature, contact stresses and areas, time, quartz cement abundance and fluid chemistry (e.g., Tada and Siever, 1989; Gratier and Guiguet, 1986; Kristiansen et al., 2011). The relative significance of contact stress, temperature and time in natural sandstones remains difficult to constrain despite extensive study of chemical compaction in the laboratory (e.g., De Boer, 1977; Gratier and Guiguet, 1986; Elias and Hajash, 1992; Niemeijer et al., 2002). The classical view of the process considers chemical compaction to result from increased solubility at load-bearing contacts compared to surfaces bordering intergranular pores (e.g., De Boer, 1977; Tada and Siever, 1989). Bjørkum (1996), however, provides striking petrographic examples that indicate significant contact dissolution occurs with negligible contact stress and argues that an essential aspect of the process is the presence of illite or muscovite in contact with quartz. Such observations led Alcantar et al. (2003) and Walderhaug et al. (2004) to suggest that the process is controlled by geochemical reactions that occur within a thin water film at contacts. Results from a suite of novel experiments support the notion of rapid quartz dissolution in response to a chemical corrosion process in the water film at the contact between quartz and muscovite (e.g., Alcantar et al., 2003; Greene et al., 2009; Kristiansen et al., 2011). This experimental approach also indicates far lower rates of dissolution for quartz-on-quartz contacts. A theoretical model for the contact dissolution process suggests that dissolution rates should be largely insensitive to contact stress (Kristiansen et al., 2011). Nonetheless, other experimental studies clearly demonstrate that increasing contact stress and temperature lead to increase rates of chemical compaction of quartz grain packs lacking illite coatings or muscovite grains (e.g., Elias and Hajash, 1992). To date no a priori quantitative models have been developed that are capable of accurately predicting the magnitudes of chemical compaction in sandstones away from sample control.

## Cementation, replacement and dissolution

Most deepwater sandstone reservoirs of current and future interest have been subject to reservoir quality degradation in response to geochemical alteration after deposition (see review by Morad et al., 2010). Geochemical reactions in sandstones may result in the formation of dissolution pores and precipitation of cements and replacements as shown for an example in Fig. 19. Although cements and replacements may form simultaneously in response to the same geochemical driving forces,

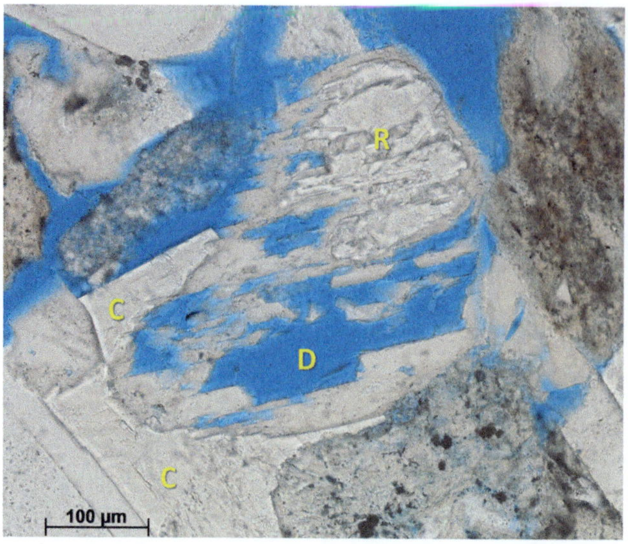


**FIG. 19** Thin section photomicrograph in plane light showing porosity from feldspar grain dissolution "*D*", calcite replacing a feldspar grain "*R*" and calcite cement "*C*". Although these three modal categories may be related in origin and timing, it is critical to distinguish between them for reservoir quality characterization and understanding.

they differ texturally in that cements precipitate in intergranular pores, whereas replacement occurs in volumes formerly occupied by solid materials such as grains, matrix or earlier cement generations. Distinguishing between cement and replacement is essential for reservoir quality studies because a given volume of cement will cause a larger reduction in permeability compared to the same volume of material replacing grains or matrix. Additionally, this distinction is needed to accurately assess intergranular volume (IGV), which, as discussed above, is an important index of the compaction state of sandstones.

The formation of grain dissolution porosity has been called upon as a mechanism for improving reservoir quality at depth. However, the formation of grain dissolution porosity often is associated with a net reduction reservoir quality due to the associated precipitation of cements. If the associated cements are highly microporous clay minerals (e.g., Chuhan et al., 2001), the permeability will tend to decline substantially even if total porosity is largely unchanged given that dissolution pores and micropores are substituted for larger and better-connected intergranular pores. Even when there appears to be a net increase in porosity due to grain dissolution, the volumes typically are small (<3 vol%) (Taylor et al., 2010). Thus, the occurrence of grain dissolution porosity is unlikely to be the key control on the economic viability of conventional deepwater sandstone reservoirs.

The sections below review some of the more volumetrically significant alteration products in deepwater sandstones. The controls on quartz cementation will be described first given the importance of this cement type for reservoir quality in many deepwater sandstones. The sections that follow provide overviews of the occurrence of authigenic clay, zeolite and carbonate minerals, each of which has the potential to significantly affect reservoir quality, depending on sand depositional composition and burial history.

## Quartz

Quartz is among the most common cement types found in quartzose deepwater sandstones, where it typically forms as overgrowths that nucleate on detrital quartz grains (e.g., Link and Welton, 1982; Moraes, 1991; Houseknecht and Ross, 1992; Helmold and Van De Kamp, 1984; Macaulay et al., 1992; Ryu and Niem, 1999; Sullivan et al., 1999; Anjos et al., 2000; Anjos et al., 2003; Worden and Barclay, 2000; Mansurbeg et al., 2008; Dutton and Loucks, 2010; Dutton et al., 2012; Bahlis and De Ros, 2013; Marchand et al., 2015; Zhang et al., 2015; Fu et al., 2016; Oliveira and De Ros, 2018; Higgs et al., 2019; Porten et al., 2019; Huang et al., 2020, Bello et al., 2021; Hansen et al., 2021; Liu et al., 2021; Liu and Xiong, 2021).

The novel hypothesis regarding the control on quartz overgrowth cementation proposed by Walderhaug (1994a, b, 1996, 2000) is among the most important breakthroughs in sedimentary petrology. This work led to the development of models with unprecedented accuracy in reproducing quartz cement volumes in sandstones that have wide variations in thermal exposure and geologic settings (e.g., Walderhaug, 1994b, 1996, 2000; Lander and Walderhaug, 1999; Walderhaug et al., 2000; Taylor et al., 2004, 2010, 2015; Lander et al., 2008; Tobin et al., 2010; Tobin and Schwarzer, 2014).

The quartz cementation process can be broken down into three main steps: (1) release of dissolved silica into solution, (2) transport of this silica and (3) precipitation of the silica as overgrowths that nucleate on preexisting quartz surfaces. Prior to Walderhaug's work academic and industry studies focused on the first two of these steps. Walderhaug, however, proposed that it in fact is the third step that controls the extent of quartz cementation in most reservoir sandstones. The notion that silica supply is not the rate limiting control on quartz cementation in most sandstones is consistent with fluid chemistry from sandstone reservoirs that almost invariably show silica concentrations that are near or above equilibrium with quartz (e.g., Bazin et al., 1997a, b; Land, 1997). Silica sources are likely to be local given that fluid fluxes through reservoir sandstones during the time of active quartz precipitation and the concentration of dissolved silica are both low (Bjørklykke and Egeberg, 1993; Giles et al., 2000). Silica sources may occur within the sandstone body or interbedded lithologies and include silica that forms (1) as a by-product of reactions such as those involving feldspar alteration and the formation of illite, (2) from dissolution of clasts made up of materials such as chert and biogenic silica and (3) from dissolution of quartz at grain contacts and stylolites.

Walderhaug (1994b) proposed that the rate of quartz growth per unit surface area is a function of temperature and that the net rate of growth for the sandstone as a whole also depends on the surface area for overgrowth nucleation, which is related to size and abundance of quartz grains, grain coating coverage and intergranular porosity (nucleation area diminishes as intergranular pores are filled). The temperature-dependent term describing the rate of quartz growth per unit surface area changed from a logarithmic form in the first version of the model (Walderhaug, 1994b; Lander and Walderhaug, 1999) to an Arrhenius formulation (Walderhaug, 2000) that has a similar overall temperature dependence, but a more rigorous foundation in chemistry.

Despite the monumental improvement that this model brought to the state-of-the-art in understanding and predicting quartz cement distributions in sandstones, it does show systematic errors with respect to the estimated nucleation surface area (Walderhaug, 2000). This error has to do with much faster rates of growth on noneuhedral growth surfaces compared to crystal facets and the fact that overgrowths that form on smaller grains or crystal domains develop fully euhedral forms with less growth compared to those that form on larger grains (Lander et al., 2008). This nucleation domain effect is apparent in Fig. 20 where an overgrowth on a single large quartz domain grew much further into an intergranular pore compared to one that formed on a small domain of a polycrystalline grain.

An important additional finding by Lander et al. (2008) is that the kinetic parameters reported by Walderhaug (1994b, 2000) are not universally applicable to all sandstone reservoirs. Instead, for accurate quartz cement predictions, it is essential that the kinetic parameters are constrained based on high-quality, internally consistent petrographic measurements and thermal history reconstructions from the area of interest. To illustrate the importance of applying appropriate kinetic parameters, consider a dataset consisting of Jurassic sandstones in offshore Norway that have been exposed to a wide range in thermal histories (Fig. 21A–C) with a dataset that is made up of Miocene sandstones from offshore southeast Asia where all samples have been exposed to temperatures in excess of 150°C (302°F) (Fig. 21D–F). The procedure we use in this

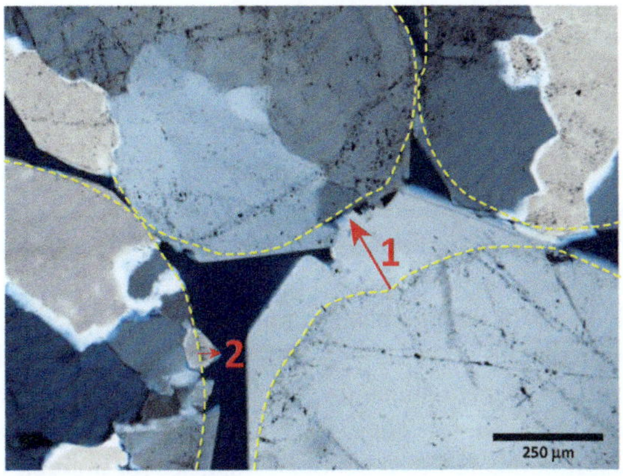


**FIG. 20** Cross-polarized light photomicrograph of quartz grains (outlined by *yellow dashed lines*) and associated overgrowths. Notice that the overgrowth on the grain with the single crystal domain is thicker (*arrow 1*) compared to an overgrowth that formed on the small domain of a polycrystalline grain (*arrow 2*) that abuts the same pore.

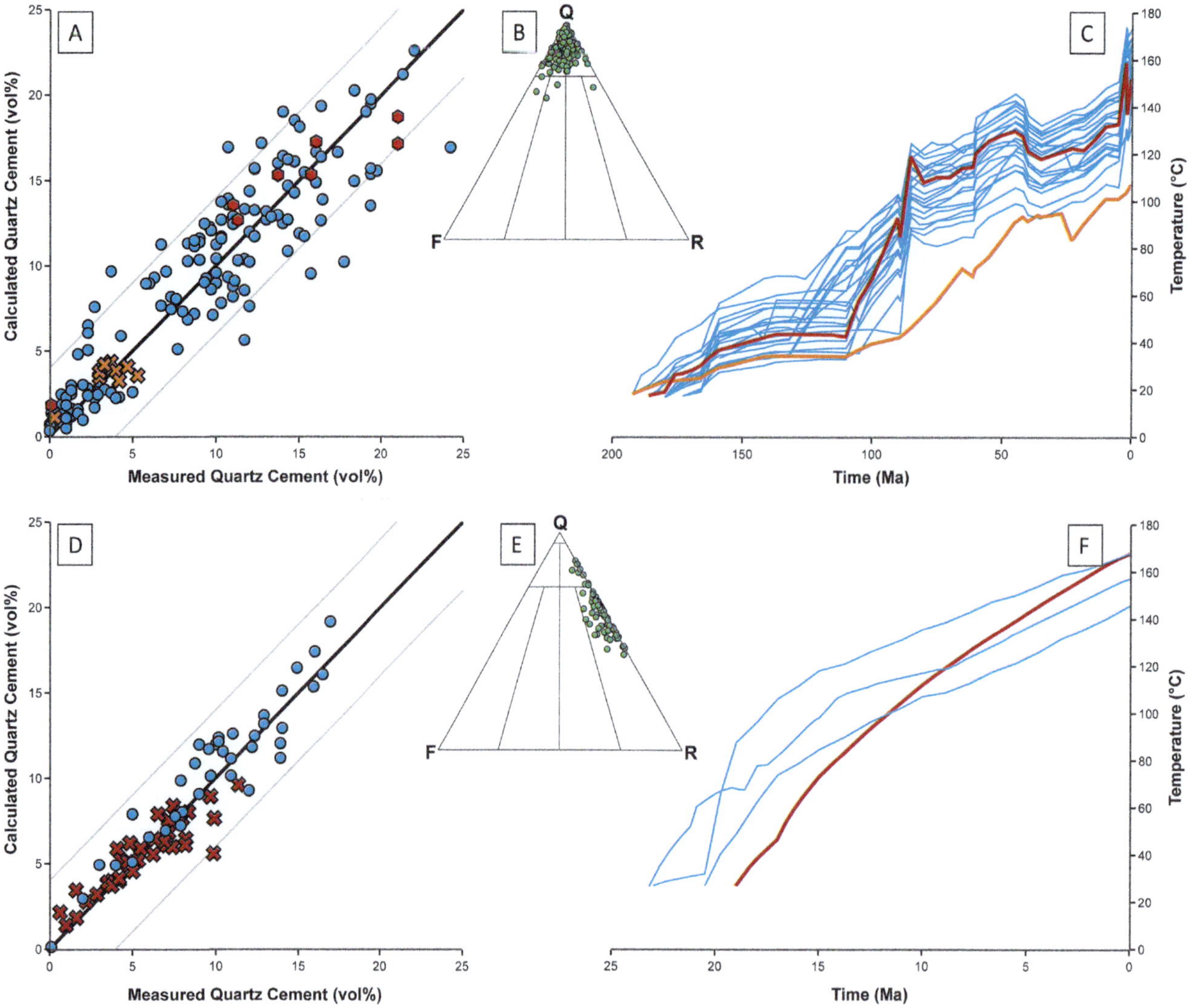


**FIG. 21** Examples of optimized quartz cement model performance in Touchstone. (A) Quartz cement abundance from petrographic measurements against simulated present-day values for Jurassic sandstones from offshore Norway. Samples with the *blue circle* symbols were used to optimize kinetic parameters but those with *red hexagon* and *orange* "*x*" symbols were not. (B) QFR diagram for samples in A. (C) Temperature histories for the samples in A, using the same sample color codes. (D) Miocene sandstones in offshore southeast Asia where simulated present-day quartz cement values are compared with petrographic measurements. Samples with *blue circle* symbols were used to constrain kinetic parameters and samples with *red* "*x*" symbols were not. (E) QFR diagram for samples in D. (F) Temperature histories for the samples in D, using the same sample color codes.

analysis involves adjusting kinetic parameters to achieve the best fit possible between the simulation results and the measured quartz cement values for the samples shown in blue in Fig. 21A and D. The samples shown in orange and red in the figure are withheld from the optimization procedure and thus represent an independent test of the kinetic values. As shown in the figure, quartz cement abundances are reproduced at a degree of accuracy that is comparable to the measurement uncertainty. Importantly, however, the kinetic parameters for each of the two datasets were optimized independently of each other and are not the same.

When the kinetic parameters obtained for the two basins are switched (i.e., the optimized parameters from the Norway dataset are used on the southeast Asia dataset and vice versa), the quartz cement model performance is poor as shown in Fig. 22. In particular, the simulated quartz cement abundances are significantly overpredicted for the Norway dataset while being underpredicted for the southeast Asia dataset. These results imply that the rates of quartz growth per unit surface area are faster in the southeast Asia dataset compared to the Norway dataset for a given temperature.

The causes of the differences in apparent quartz overgrowth kinetics in these and many other datasets that we have worked with are unclear but may be related to the following factors: (1) methodologies used to simulate thermal histories, (2) methodologies for petrographic data collection, (3) differing extents of quartz crystal defects and strain, which vary depending on the grain source (Ernst and Blatt, 1964; Seyedolali et al., 2002), (4) differences in small scale nucleation

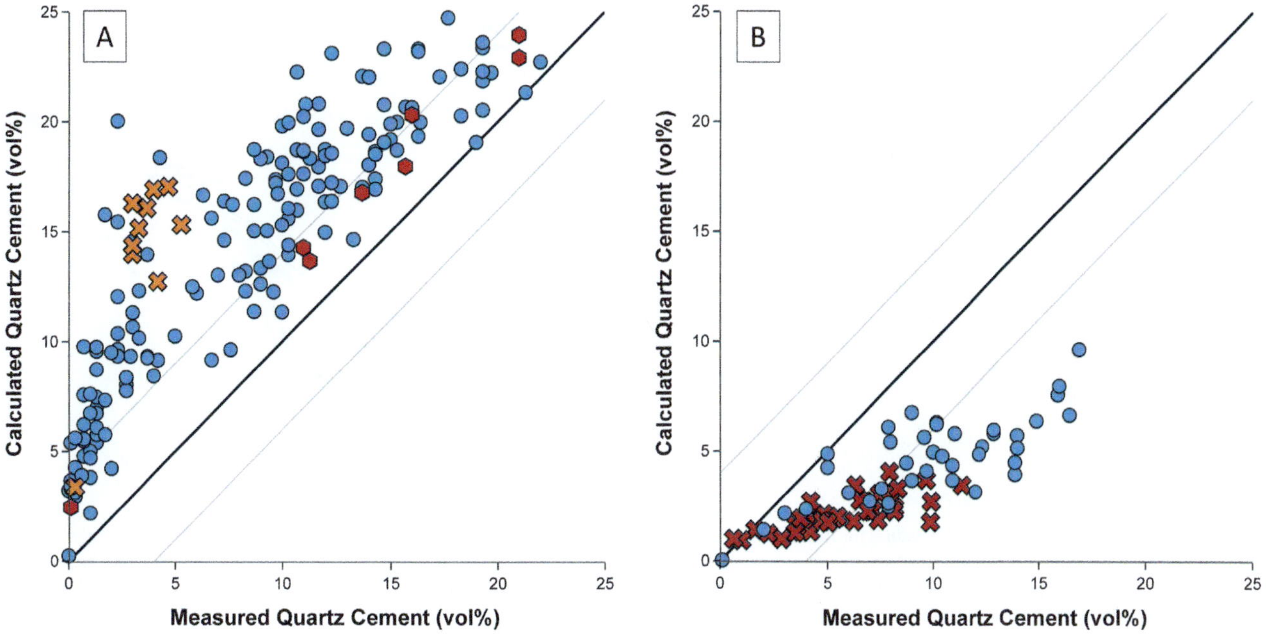


**FIG. 22** Present-day quartz cement simulation results compared with petrographic measurements when optimized cementation kinetics for the datasets in Fig. 21 are switched. (A) Results for Jurassic sandstones from offshore Norway when using the optimized parameters from the southeast Asia example shown in Fig. 21D–F. (B) Results for southeast Asia example when using the optimized parameters from Jurassic sandstones from offshore Norway shown in Fig. 21A–C.

discontinuities on the surfaces of quartz grains that could reflect differences in sand provenance and transport history (Lander et al., 2008) and (5) fluid composition and fluxes through time.

## *Grain coatings*

Grain coatings can significantly limit the extent of quartz cementation in quartz grain-rich sandstones with high thermal exposures (Heald and Larese, 1974; Pittman et al., 1992) as demonstrated in Fig. 23. Grain coatings in the context of quartz overgrowth analysis represent thin coatings of material on quartz grain surfaces (typically <30 μm) that impede the ability of overgrowths to occlude intergranular pores. Although clay minerals such as chlorite and its associated mixed-layer phases commonly occur as coatings, coatings are not limited to chlorite or to clay minerals in general. Examples of authigenic coatings made up of other minerals include pore-lining siderite, iron oxy-hydroxides, apatite, clinoptilolite and microcrystalline quartz. Grain rims of materials such as clay minerals or iron oxy-hydroxides that are present at the time of deposition also may act as coatings (Wilson, 1992; Bloch et al., 2002; Dowey et al., 2017; Wooldridge et al., 2017, 2018). Additionally, coatings in deepwater sandstones can develop from infiltration associated with dewatering (Houseknecht and Ross, 1992; Porten et al., 2019). The development of infiltrated coatings in deepwater settings is strongly affected by primary sediment permeability and promoted by rapid deceleration during deposition as well as high net rates of deposition such as in proximal lobe and channel-lobe transition areas of submarine fans (Chapters 7 and 8 and Porten et al., 2019). In Cretaceous deepwater sandstones in the Vøring Basin, infiltrated clay coatings can cover 80–97 area% of quartz grain surfaces where they are associated with quartz cement volumes of 1–5 vol% compared to estimated volumes of well over 20 vol% had they not been present (Porten et al., 2019). Such well-developed coatings, however, are not always present in deepwater sandstones.

Coating measurements consider the areas of grain surfaces. These data therefore cannot be directly determined from the volume abundance of the coating material as derived from a standard modal analysis. Methods for coating analysis are described in Chapter 6 of Taylor et al. (2022).

The impacts of grain coatings on quartz overgrowth development are illustrated in Fig. 24 for a sandstone with well-developed siderite coatings. In the image, it is clear that overgrowths formed only where discontinuities in the coatings exposed the outer surface of quartz grains to pore fluids. Where coatings are continuous, on the other hand, overgrowths were unable to grow into intergranular pores, resulting in a significant net preservation in porosity compared to what would be expected if the coatings were not present. The significance of coatings for porosity preservation are illustrated by the simulation results shown in Fig. 25 that use an imaged chlorite coating distribution from a reservoir sandstone as input for

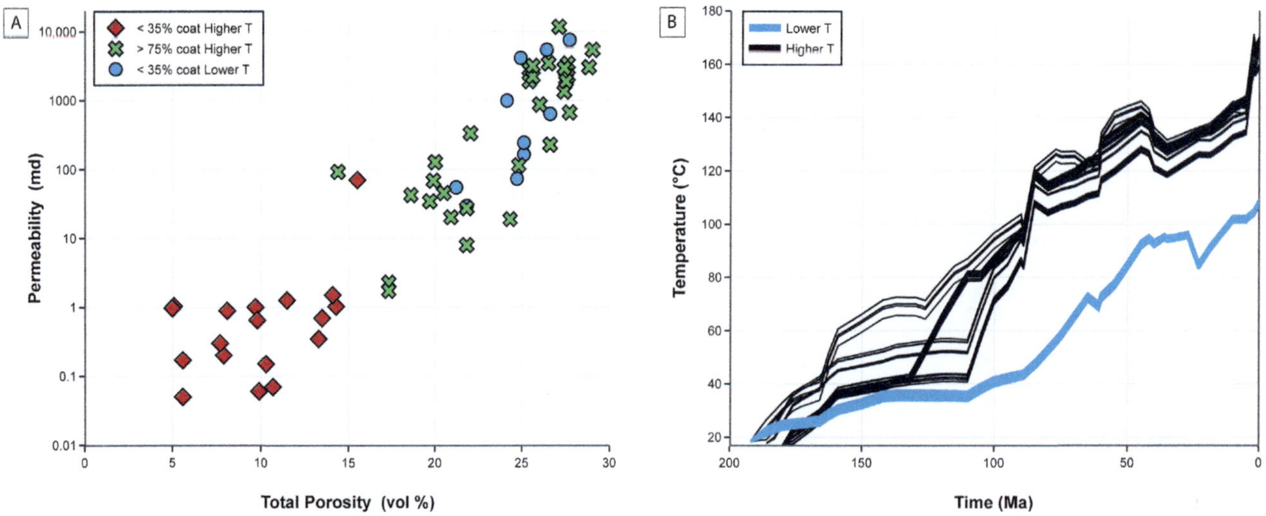


**FIG. 23** (A) Porosity and permeability values for Jurassic age sandstones. Symbols are linked to the area % of the quartz grains affected by grain coatings as well as the temperatures histories in (B). Note that samples with low grain coatings and high thermal exposures (*<35% coat Higher T*) have low porosity and permeability but samples with the same thermal exposure and high coating values (*>75% coat Higher T*) have porosities and permeabilities that are comparable to low coating samples with low thermal exposure (*<35% coat Lower T*).

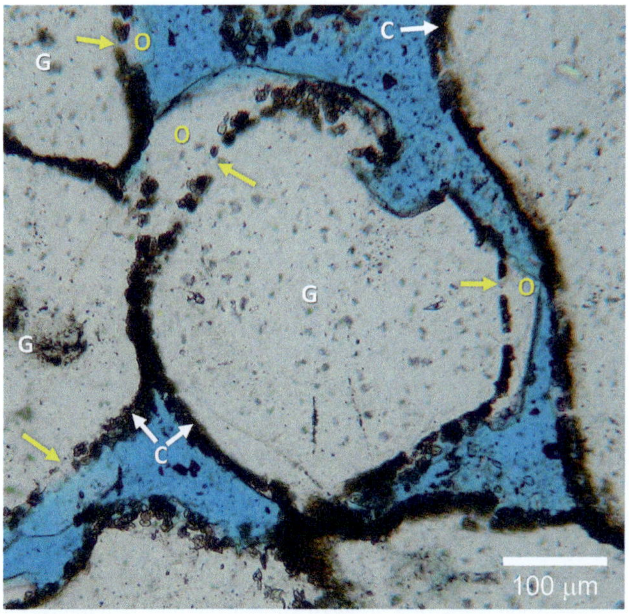


**FIG. 24** Plane-light thin section photomicrograph illustrating siderite coatings (*white arrows* with "*C*") on quartz grains ("*G*") where coating breaks such as those indicated by the *yellow arrows* allowed quartz overgrowths ("*O*") to grow into intergranular pores.

the set of quartz cementation results shown on the left column but omit the coating in the results shown in the right column. These simulations, which we made using the Prism2D™ code (Lander et al., 2008), assume a constant temperature of 135°C (275°F) and span a time interval of 20 m.y. Reducing the grain coating from the ~80 area% coverage on the left to no coverage on the right resulted in an increase in quartz cement abundance of as much as ~15 vol%.

It is important to note that quartz cementation does occur in the simulation with the chlorite coating given that the coating coverage is not perfect (Fig. 25). Moreover, it is clear that given sufficient additional temperature exposure, the coated simulation eventually would lose most of the remaining intergranular pore space due to quartz cementation. This dependency of coating effectiveness with thermal exposure is illustrated by Jurassic sandstones with varying thermal exposures in Fig. 26. Sandstones with high thermal exposure (larger yellow triangles) must have >95 area% grain coating

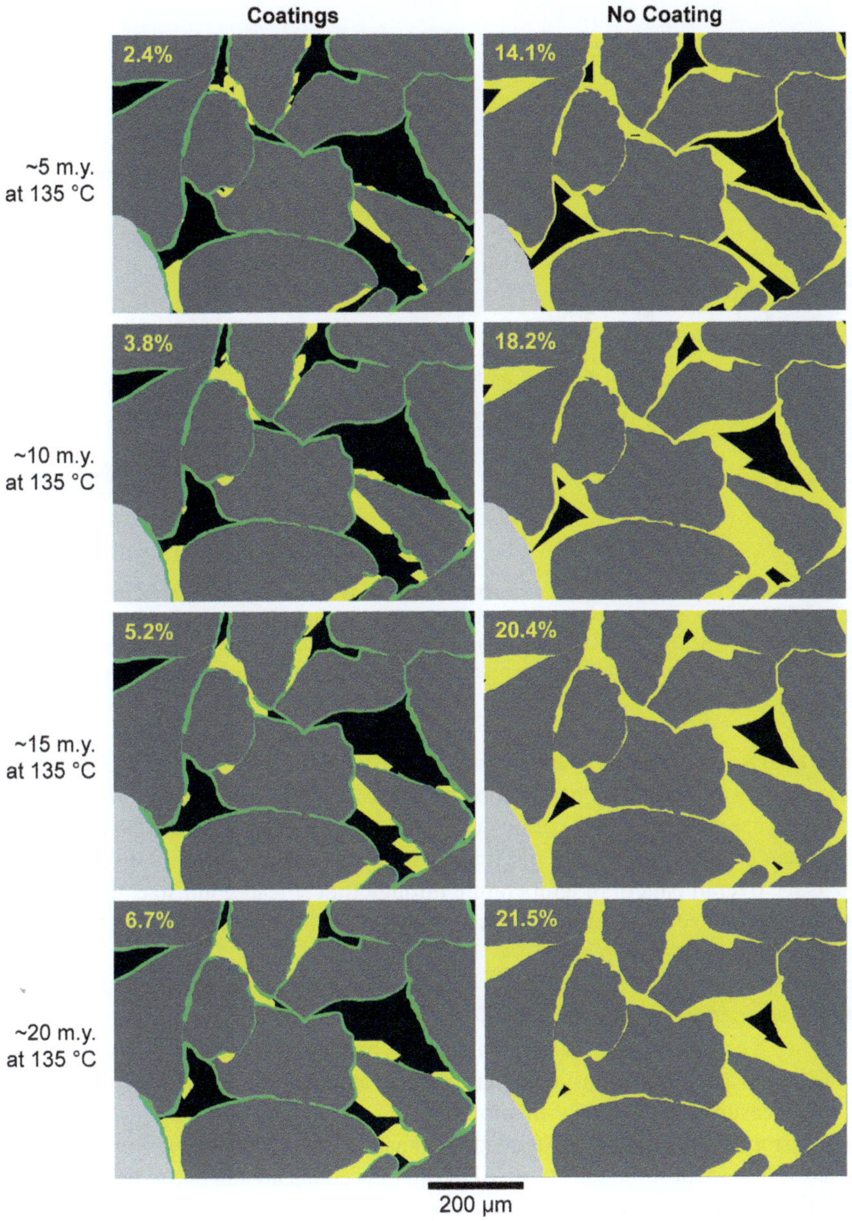


**FIG. 25** Prism2D simulations of quartz cement growth where the input geometry is derived from an interpreted thin section photomicrograph of a chlorite coated sandstone where the coating material is considered in the left set of results but omitted from the results in the right column. Color codes are *black* for porosity, *yellow* for quartz overgrowth cement, *dark gray* for quartz grains, *light gray* for nonquartz grains and *green* for chlorite. The values shown with the *yellow text* indicate the simulated quartz cement abundance. The cumulative thermal exposure is indicated by the temperature and time duration along the left side of the simulation results.

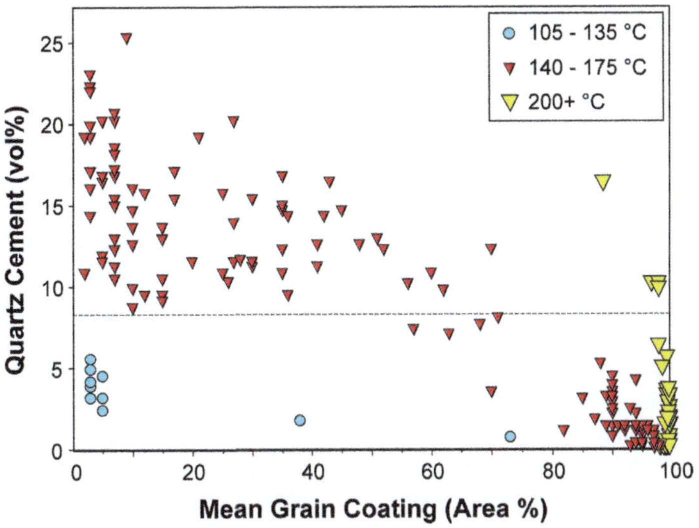


**FIG. 26** Quartz cement abundance with grain coating coverage for Jurassic sandstones with varying thermal exposures, as exemplified by the maximum temperatures indicated in the symbol legend. Sandstone samples all have <10 vol% matrix and <10 vol% cements other than quartz. The *dashed horizontal line* shows 8 vol% quartz cement. As thermal exposure increases, sandstones with a given quartz cement volume tend to be associated with increased grain coat coverage.

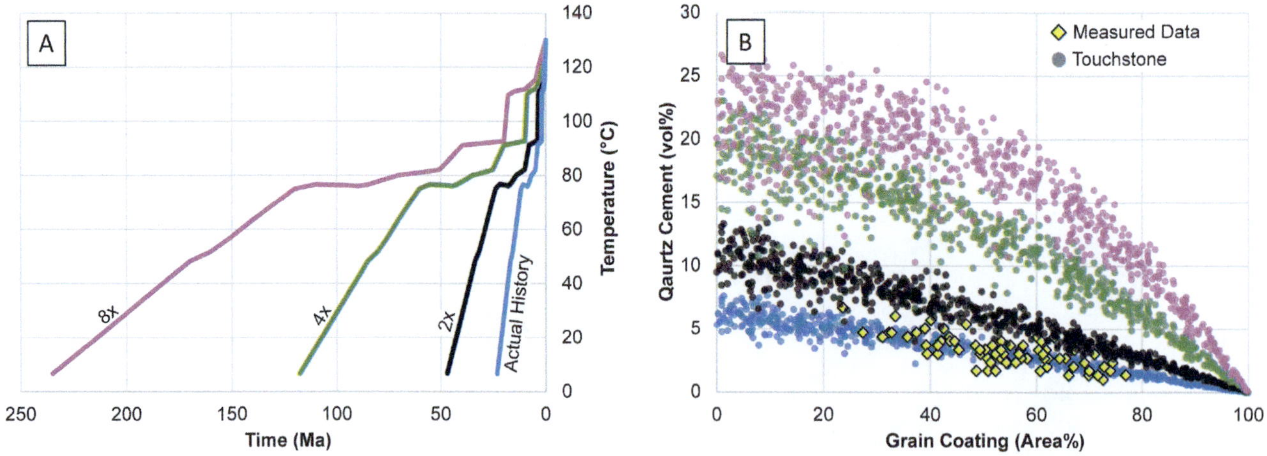


**FIG. 27** Sensitivity study of the effect of time and grain coating coverage on quartz cement abundances. (A) Temperature history scenarios that all use the same temperature values but have twice the time for the *black line*, four times the time for the *green line* and eight times the time in the *magenta line* compared to the actual dataset temperature history shown in *blue*. (B) The *yellow diamond* symbols show measured values from the dataset that was used to optimize Touchstone model parameters. The *circle* symbols show Touchstone simulation results. The colors of these symbols correspond to the temperature history scenarios in (A).

to limit the amount of quartz cement to less than 8 vol% (shown by the dashed line). By contrast, samples with low thermal exposure (blue circles) have little quartz cement even when they have almost no grain coating. Samples with intermediate thermal exposure (smaller red triangles) in this example must have grain coat coverage in excess of 50 area% to limit the amount of quartz cement that forms to less than 8 vol%. The amount of grain coat coverage needed to limit quartz cementation also depends on other factors such as the overall pattern of temperature through time, mean grain size, the extent of compaction, the quartz grain abundance, the abundance and texture of detrital matrix and the abundances and textures of cements that form prior to, or concurrently with, quartz precipitation.

The scenarios depicted in Fig. 27 are designed to give insights into the role of time in controlling the effectiveness of grain coating coverage. In these simulations, we used the same sequence of temperature values in a Touchstone model but increased the elapsed time by factors of 2, 4 and 8 compared to the actual temperature history for a deepwater reservoir. The Touchstone simulation accurately reproduces the variation in measured quartz cement abundance observed in the dataset as shown by the correspondence between the measurements (yellow diamonds in Fig. 27B) and simulation results (blue circles) when using the actual temperature history (Fig. 27A). We extended the range in coating considered in the Touchstone model, however, beyond the measured values to show the sensitivity of the quartz cement abundance to the theoretical limits in coating coverage. With each doubling of the time exposure, the simulated quartz cement for a given grain coating coverage increases until intergranular pores are completely occluded in some of the longer time interval scenarios. As the thermal exposure increases, the correspondence between quartz cement and grain coating coverage becomes increasingly nonlinear. Similar patterns emerge in cases where the thermal exposure is increased by raising temperature while holding the elapsed time constant. This simulated pattern in decreasing coating effectiveness with increased thermal exposure is consistent with data from sandstone reservoirs (Fig. 26).

## Clays and zeolites

Authigenic clay minerals and zeolites in reservoir sandstones typically form in response to chemical reactions involving detrital material. This localization of the associated reactions reflects the low concentrations of aqueous species containing aluminum and silicon in formation fluids, which limit the potential for mass transport of these essential elements. An additional factor that can influence the nature and extent of reaction is the flux of externally derived brines through deepwater sandstones, particularly when it comes to minerals incorporating K, Na, Ca and Mg. Predictive models have been developed for some but not all reactions involving clay minerals and zeolites. In particular, kinetic models have been developed for the formation of illite from smectite (Pytte, 1982; Altaner, 1986; Pytte and Reynolds, 1989; Velde and Vasseur, 1992; Huang et al., 1993; Elliott and Matisoff, 1996), illite from kaolinite (Franks and Zwingmann, 2010; Lander and Bonnell, 2010), chlorite from mafic minerals and glass (Bonnell et al., 2014) and albitization of calcic plagioclase (Perez and Boles, 2005), which may be associated with the formation of kaolinite, laumontite and carbonates. Quantitative models are lacking, however, for predicting the abundance of smectite, most zeolites and most modes of occurrence of chlorite and kaolin in sandstones.

## Smectite

Smectite (Fig. 28A) and mixed layer phases incorporating smectite are common in deepwater sandstones (e.g., Ramseyer and Boles, 1986; Noh and Boles, 1993; Ryu and Niem, 1999; Mansurbeg et al., 2006, 2008; Fu et al., 2016; Dutton et al., 2012; Stammer, 2014; Zhang et al., 2015; Higgs et al., 2019; Baiyegunhi et al., 2020; Huang et al., 2020; Jackson et al., 2020; Liu et al., 2021). Smectite is a compositionally and structurally variable clay mineral that forms at relatively low thermal exposures and is characterized by having an expandable interlayer region. Trioctahedral smectites are rich in Fe or Mg and often result from alteration of mafic volcaniclastic glass or associated labile minerals such as pyroxene and may be interlayered with, or react to form, chlorite (Beaufort et al., 2015). Dioctahedral smectite tends to be richer in Si and forms from the alteration of intermediate to silicic volcanic materials. With increasing thermal exposure this smectite type forms mixed-layer phases with illite and the smectite component diminishes (e.g., Hower et al., 1976). See McKinley et al. (2003) for a review of the occurrence of smectite in sandstones.

## Kaolin

Kaolinite (Fig. 28B) and dickite are aluminous clay mineral polymorphs that are common alteration products in deepwater sandstones (e.g., Helmold and Van De Kamp, 1984; Tieh et al., 1986; Hayes and Boles, 1992; Houseknecht and Ross, 1992; Anjos et al., 2000; Worden and Barclay, 2000; Mansurbeg et al., 2006, 2008; Dutton and Loucks, 2010; Dutton et al., 2012; Zhang et al., 2015; Fu et al., 2016; Higgs et al., 2019; Porten et al., 2019; Baiyegunhi et al., 2020; Huang et al., 2020; Bello et al., 2021). Kaolinite generally forms at low thermal exposures ($< \sim 60°C$ (140°F) in many cases) as a byproduct of the alteration of various aluminosilicate minerals such as feldspars and micas. Its occurrence has been linked in some reservoir sandstones to significant fluxes of meteoric water associated with subaerial recharge (Kantorowicz, 1984; Bjørlykke and Aagaard, 1992; De Ros, 1998; Marfil et al., 1998; Ketzer et al., 2003; Oliveira and De Ros, 2018) as well as to acidic fluids associated with organic matter and $CO_2$ (Ehrenberg, 1991; Gaupp et al., 1993; Lanson et al., 1996). Kaolinite in deepwater sandstones in the Shetland-Faroes Basin has been attributed to meteoric water incursion during a sea-level lowstand in association with hyperpycnal flows (Mansurbeg et al., 2006). The common occurrence of kaolinite in deepwater sandstones makes it seem unlikely, however, that meteoric water incursion is a requirement for its formation. For example, authigenic kaolinite formed in Cretaceous submarine fan deposits of the Vøring Basin, which are thought to have been hundreds of kilometers from the paleoshoreline (Porten et al., 2019). Kaolinite in such deepwater environments likely forms as a product of the alteration of thermodynamically unstable phases such as plagioclase, amphibole and other heavy minerals (Walderhaug and Porten, 2007; Taylor et al., 2010).

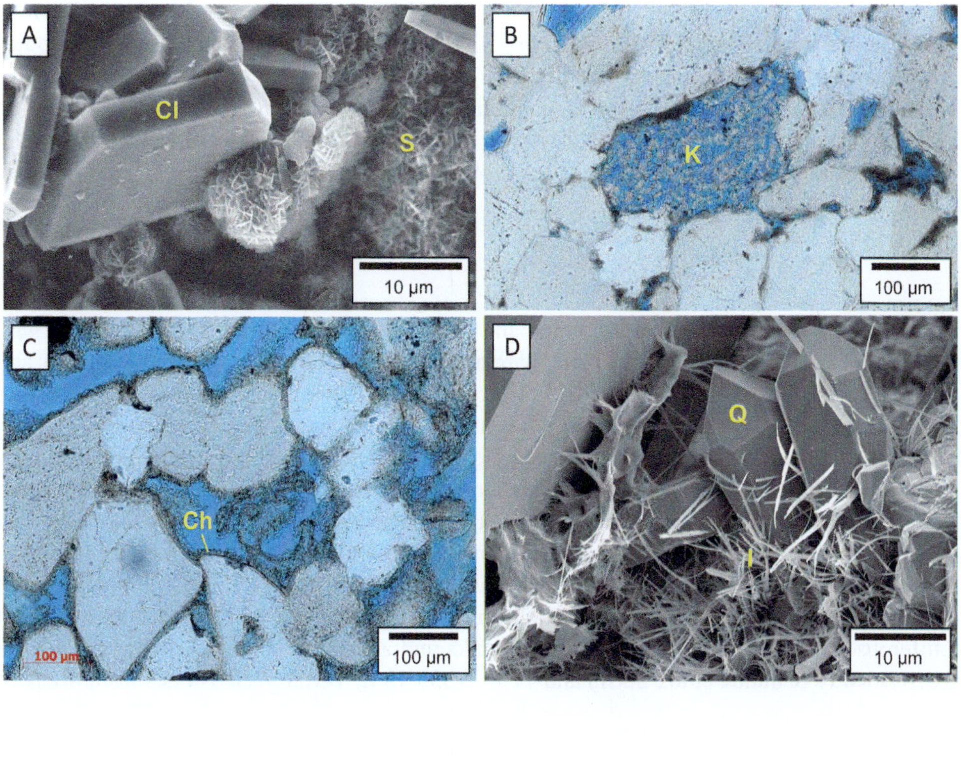


**FIG. 28** Examples of authigenic clay and zeolite in sandstones. (A) Scanning electron microscope image showing smectite ("*S*") and the zeolite clinoptilolite ("*Cl*"). (B). Thin section with kaolinite ("*K*") replacing a feldspar (plane-polarized light). (C) Thin section with pore-lining chlorite ("*Ch*") cement (plane-polarized light). (D) Scanning electron microscope image with fibrous illite ("*I*") cement and authigenic quartz ("*Q*") cement.

To get a sense of the volumes of kaolinite that can form from reactions involving feldspar, our calculations indicate that ~3.3 vol% of kaolinite could form by Al mass balance from the albitization of 5 vol% plagioclase with An 67 mol% assuming that the kaolinite has 50% microporosity (albitization is the process where detrital calcic plagioclase is replaced by nearly pure albite). Wholesale dissolution of that same volume of plagioclase could result in ~ 8.2 vol% microporous kaolinite, whereas dissolution of 5 vol% of pure albite and K-feldspar would produce ~ 5.0 and ~ 4.6 vol% microporous kaolinite, respectively.

At high thermal exposure kaolinite may react to form illite if a potassium source is available. This reaction is kinetically controlled (Franks and Zwingmann, 2010; Lander and Bonnell, 2010) so the temperatures at which kaolinite illitization occurs are variable as discussed in a later section. If potassium is not available to form illite, kaolinite may react to form dickite at temperatures that range from roughly 70 to 140°C (158–284°F) (Tilley and Longstaffe, 1989; Ehrenberg, 1993; Macaulay et al., 1993; Beaufort et al., 1998; Mansurbeg et al., 2006, 2008). This reaction also is likely kinetically controlled. Sandstone texture and composition and fluid compositions and fluxes also affect the nature and magnitude of reactions involving kaolinite.

## *Chlorite*

Chlorite (Fig. 28C) is an iron and magnesium bearing clay mineral that commonly occurs in deepwater sandstones (e.g., Link and Welton, 1982; Helmold and Van De Kamp, 1984; Hayes and Boles, 1992; Houseknecht and Ross, 1992; Noh and Boles, 1993; Ryu and Niem, 1999; Sullivan et al., 1999; Anjos et al., 2000; Anjos et al., 2003; Dutton, 2008; Mansurbeg et al., 2008; Dutton and Loucks, 2010; Dutton et al., 2012; Bahlis and De Ros, 2013; Fu et al., 2016; Higgs et al., 2019; Baiyegunhi et al., 2020; Huang et al., 2020, Jackson et al., 2020; Bello et al., 2021; Hansen et al., 2021; Liu et al., 2021; Liu and Xiong, 2021). Chlorite is of particular interest in reservoir quality studies because it can form thin coatings on grain surfaces. When sufficiently well developed, such coatings have the potential to substantially reduce the extent of quartz cementation in reservoirs with high thermal exposures (Heald and Larese, 1974). The impacts of grain coatings including chlorite are discussed at greater length in the section on quartz cementation.

A common mechanism for chlorite formation in deepwater sandstones involves the alteration of intermediate to mafic volcaniclastic material. Examples of the deepwater occurrence of such chlorite include Cretaceous sandstones from the Santos Basin of Brazil (Anjos et al., 2003; Bahlis and De Ros, 2013) and Paleocene sandstones from the West of Shetlands area (Sullivan et al., 1999). Chlorite or precursor phases also may form where iron-bearing fresh waters mix with seawater (e.g., Ehrenberg, 1993; Griffiths et al., 2021). In such cases, ooids and pelletal grains that include chloritic material may form syndepositionally by accretion of shallow water clays. Examples of chloritic clasts are found in shallow marine Cretaceous sandstones from Pakistan (Berger et al., 2009) and the US Gulf Coast (Ryan and Reynolds, 1996), Jurassic sandstones from the Norwegian Continental Shelf (Ehrenberg, 1993; Bloch et al., 2002; Griffiths et al., 2021) and Permian sandstones from the Carnarvon Basin in Western Australia (Baker et al., 2000). Chlorite cement as well as chloritic ooids and pellets in turbidites of the Cretaceous Agat Formation from the northern North Sea are interpreted as having formed in a shallow marine setting prior to being redeposited in a deepwater setting (Hansen et al., 2021). Interactions with evaporitic brines also may affect the occurrence and properties of chlorite (Taylor et al., 2004; Ajdukiewicz et al., 2010). For more information regarding chlorite origin, occurrence and properties in reservoir sandstones, see the reviews of Worden et al. (2020) and Dowey et al. (2017).

## *Illite*

Illite (Fig. 28D) often forms in deepwater sandstones (e.g., Link and Welton, 1982; Ramseyer and Boles, 1986; Houseknecht and Ross, 1992; Macaulay et al., 1992; Worden and Barclay, 2000; Mansurbeg et al., 2008; Dutton and Loucks, 2010; Dutton et al., 2012; Stammer, 2014; Fu et al., 2016; Higgs et al., 2019; Baiyegunhi et al., 2020; Bello et al., 2021; Liu and Xiong, 2021; Liu et al., 2021) from kaolinite or smectitic clay as discussed earlier. These reactants may be authigenic or detrital in origin. A potassium source also is needed for illite to form. This source could be in situ K-feldspar dissolution (e.g., Chuhan et al., 2001) or potassic brine that is externally derived (e.g., Gaupp et al., 1993). Illite reaction pathways for both clay reactant types are kinetically controlled, although the reaction involving kaolinite appears to require greater thermal exposure compared to smectitic reactants (Lander et al., 1990; Elliott and Matisoff, 1996; Franks and Zwingmann, 2010; Lander and Bonnell, 2010). Because of the kinetic control on reaction rates, the temperatures at which illite forms are variable. For instance, the reaction of kaolinite to form illite is associated with temperatures of at least 100°C (212°F) in Permocarboniferous sandstones from Saudi Arabia (Franks and Zwingmann, 2010), 140°C (284°F) in Jurassic sandstones from offshore Norway (Chuhan et al., 2001) and 170°C (338°F) in Tertiary sandstones from offshore southeast Asia (Lander and Bonnell, 2010). Additionally, the reaction to form illite may cease in the presence of remaining reactant clay if insufficient potassium is available (Środoń and Eberl, 1984; Chuhan et al., 2001).

Authigenic illite that forms from kaolinite often is characterized by fibrous forms (Fig. 28D) that may significantly reduce permeability by increasing wetted surface area, flow path tortuosity and irreducible water saturation (McHardy et al., 1982; Kantorowicz, 1984, 1990; Bjørlykke and Aagaard, 1992; Howard, 1992; Gaupp and Okkerman, 2011). Illite that forms from smectitic precursors, on the other hand, tends to inherit the "corn flake" morphology of the precursor clay (Fig. 28A) (e.g., Lander et al., 1990; Taylor et al., 2022). Consequently, the impact of illitization on sandstone permeability is likely less significant for smectite-derived illite compared to illite that forms from kaolinite.

### *Zeolite*

Zeolites are aluminosilicate minerals that in deepwater sandstones form mainly from the alteration of volcaniclastic material (Hay and Sheppard, 2001). A generalized alteration succession from silicic glass to clinoptilolite/heulandite (Fig. 28A) to analcime to albite is observed in a variety of lithologies with increasing exposure to temperature or highly saline/alkaline brines (Hay and Sheppard, 2001). Macaulay et al. (2012) found that Oligocene deepwater sandstones from the Perdido Fold Belt in the Gulf of Mexico experience dramatic reductions in permeability as abundant silicic glass shards within them react to form clinoptilolite. These researchers also found that the transition of glass-rich to clinoptilolite-rich compositions occurs abruptly within wells and appears to be related to incursions of brines that may have migrated along faults but is insensitive to observed present-day temperature variations. Clinoptilolite also has formed in volcaniclastic deepwater sandstones from Western Oregon, USA (Chan, 1985; Ryu and Niem, 1999) and the zeolite analcime occurs in volcaniclastic deepwater sandstones from the Irish Rockall Basin ( Jackson et al., 2020) together with chlorite. The zeolite laumontite commonly is associated with the albitization of calcic plagioclase. It has been found in deepwater sandstone reservoirs of Tertiary age with estimated maximum temperatures of 90°C (194°F) and above (Helmold, 1980; Link and Welton, 1982; Helmold and Van De Kamp, 1984; Noh and Boles, 1993; Ryu and Niem, 1999). Our calculations indicate that ~3.4 vol% of laumontite could form from albitization of 5 vol% calcic plagioclase with a detrital An content of 67 mol%, assuming Al and Ca mass balance constraints.

## Carbonates

Common authigenic carbonate minerals in deepwater sandstones include calcite (Fig. 29A) and siderite (Fig. 24) as well as dolomite and its iron-rich counterpart, ankerite (Fig. 29B) (e.g., Link and Welton, 1982; Chan, 1985; Tieh et al., 1986; Boles and Ramseyer, 1987; Cavazza and Gandolfi, 1992; Noh and Boles, 1993; Houseknecht and Ross, 1992; Macaulay et al., 1992; Carvalho et al., 1995; Hendry et al., 1996; Anjos et al., 2000; Worden and Barclay, 2000; Browne and Slatt, 2002; Mansurbeg et al., 2006, 2008; Dutton, 2008; Fetter et al., 2009; Dutton and Loucks, 2010; Dutton et al., 2012; Fu et al., 2016; Bell et al., 2018; Oliveira and De Ros, 2018; Higgs et al., 2019; Al-Mufti and Arnott, 2020; Baiyegunhi et al., 2020; Huang et al., 2020; Liu and Xiong, 2021; Liu et al., 2021; Bello et al., 2021;

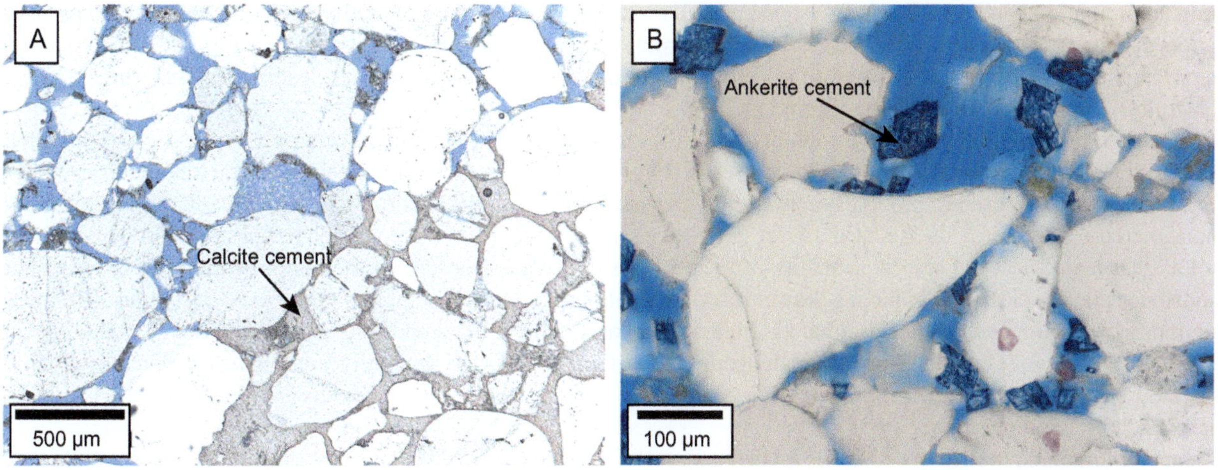


**FIG. 29** Examples of carbonate cements in sandstones as illustrated by plane-light photomicrographs of thin sections. (A) Concretionary calcite cement. (B) Ankerite cement.

Hansen et al., 2021). Among carbonate cements in sandstones, calcite generally is the most volumetrically abundant type (Primmer et al., 1997; Taylor et al., 2022). This cement type may be derived from detrital sources within the sand (e.g., bioclasts or carbonate rock fragments), interbedded carbonate lithologies (De Souza and De Assis Silva, 1998) or moderate to long scale migration of carbonate saturated fluids (Taylor, 1990; Taylor and Land, 1996). Calcite frequently forms as concretions. In deepwater sandstones, these concretions often are isolated (e.g., Browne and Slatt, 2002 Figs. 8 and 9) and therefore do not represent significant barriers for fluid-flow. Carbonate cemented layers, however, also have been described in deepwater sandstones (Mansurbeg et al., 2006), where they may create localized baffles or barriers to vertical flow. Continuous layers of carbonate cements that extend over lateral distances of hundreds of meters such as those found in shallow marine settings (e.g., Bjørkum and Walderhaug, 1990; Walderhaug and Bjørkum, 1998), however, may be less common in deepwater settings given that they typically are associated with large, laterally continuous volumes of shelly material.

An additional potential mechanism for calcite cementation in deepwater arkosic sandstones is the albitization of detrital calcium-rich plagioclase grains (e.g., Boles and Ramseyer, 1987; Boles, 1998; Oliveira and De Ros, 2018). As an example of the potential of this reaction for generating calcite, our calculations indicate that if a sandstone containing 5 vol% plagioclase with 67 An mol% is fully albitized, $\sim$ 1.2 vol% calcite could form from the resulting Ca released by the reaction. Calcite of this origin formed at temperatures on the order of 80–140°C (176–284°F) in Miocene deepwater sandstones in the San Juan Basin, where it also is associated with significant mass transfer (Boles, 1998).

Siderite, dolomite and ankerite rarely are the dominant controls on reservoir quality at the field scale in deepwater sandstones (e.g., Boles and Ramseyer, 1987; Noh and Boles, 1993; Browne and Slatt, 2002; Mansurbeg et al., 2008; Dutton and Loucks, 2010; Dutton et al., 2012; Fu et al., 2016; Bell et al., 2018; Higgs et al., 2019; Baiyegunhi et al., 2020; Huang et al., 2020; Liu and Xiong, 2021; Liu et al., 2021; Bello et al., 2021; Hansen et al., 2021). Siderite and dolomite cements in deepwater sandstones often appear to form near the depositional surface where they may or may not occur as small concretions (e.g., Boles and Ramseyer, 1987; Boles, 1998; Porten et al., 2019). Ankerite and Fe-rich dolomite, by contrast, are generally associated with temperatures in excess of 100°C (212°F) (e.g., Boles and Ramseyer, 1987).

Accurate prediction of carbonate cement volumes remains elusive because carbonate minerals tend to be much more soluble in aqueous solutions compared to common aluminosilicate minerals in sandstones and this allows for greater potential for transport into or out of reservoirs. Reconstructing carbonate solubility through geologic time is complicated by dependencies on biogenic activity, formation water chemistry, $pCO_2$ and temperature, factors that collectively can be difficult to constrain in the geologic past (see review by Morad, 1998). For more information about carbonate cementation in sandstones see the reviews of Morad (1998), Morad et al. (2010) and Worden et al. (2018).

## Assessing reservoir quality risk factors

From the discussion earlier, it is clear that there are numerous depositional and diagenetic factors that influence sandstone reservoir quality. The flow chart depicted in Fig. 30 attempts to identify how these factors influence reservoir quality risk associated with oil exploration and production from deepwater sandstones. The factors and thresholds for other applications will differ depending on the fluid type; targeted flow rates; reservoir thickness; occurrence of structural features such as faults, deformation bands and fracture systems; extent of aquifer support and production methods, among other economic factors discussed in Chapter 20. For instance, lower permeabilities may be more acceptable for gas production or storage of compressed air or hydrogen compared to oil production from conventional reservoirs, leading to less stringent decision constraints compared to those in the figure.

It is impossible to create a reservoir quality risk flow chart that provides a comprehensive consideration of all relevant factors while also being simple enough to be useful. Rigorously accounting for interactions of factors requires computer models such as those used in previous sections and discussed further in the section that follows. The purpose of the flow chart, therefore, is to provide the basis for an initial assessment of textural, compositional, burial history and diagenesis factors that may affect the reservoir quality of deepwater sandstones. Because of the simplifications, it is possible that sandstones that are classified as having "lower reservoir quality risk" may nonetheless prove to be uneconomic for oil production, whereas others that are classified as "higher reservoir quality risk" could be economically viable. As an example, deepwater turbidite reservoirs from the Paleogene of the Gulf of Mexico generally have lower permeabilities than is expected for conventional oil production (Marchand et al., 2015) but are economically viable due to thick producing intervals, among other factors.

Input information needed to use the flow chart consists of (1) expected deepwater sandstone lithology types at the area of interest (see Table 1), (2) petrographic data from applicable analog samples and (3) temperature and effective stress reconstructions of the area of interest as generated by basin/petroleum systems models (see Chapter 18). Nodes 1–4 assess the

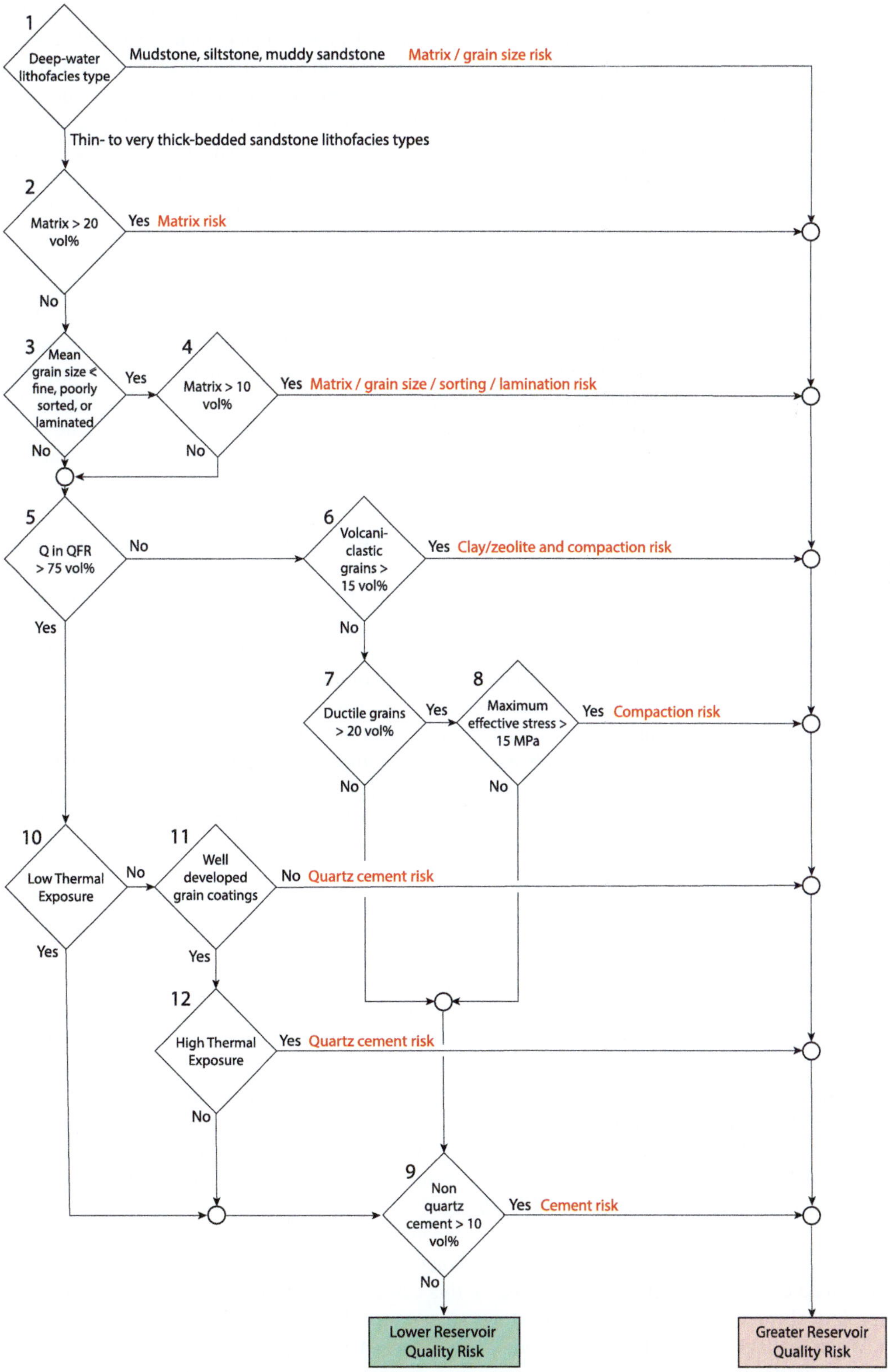


**FIG. 30** Flow chart designed to provide an initial assessment of risk factors associated with reservoir quality for oil production from deepwater sandstones. The numbers above and to the left of the diamond symbols correspond to the node numbers discussed in the text. Please see the text for a discussion of the assumptions and limitations associated with this flow chart.

impact of depositional texture on reservoir quality risk factors. Node 1 considers the deepwater lithology types listed in Table 1. Those lithologies that are characterized as incorporating abundant detrital clay matrix generally are considered non-reservoir for conventional oil production, irrespective of any other considerations and therefore are classified as having greater reservoir quality risk. Node 2 considers the remaining lithology types, all of which include significant volumes of sand-sized grains. In some cases, these lithologies have high reservoir quality risk due to the presence of significant volumes of detrital clay (Fig. 5). Node 3 assesses the influence of the grain size distribution for favorable outcomes from Node 2. Sandstones with fine or finer mean grain sizes, poor sorting or well-developed laminations are likely to have lower permeabilities compared to coarser, better sorted and more homogeneous counterparts (Fig. 3). Such samples, however, still have the potential to have adequate reservoir quality if they contain relatively low volumes of detrital matrix as tested in Node 4.

Nodes 5–8 consider the influence of framework grain composition for those sandstones with the most favorable outcomes from Nodes 3 and 4. Node 5 assesses the fraction of framework grains made up of quartz in analog samples based on the Q value from the QFR scheme of Folk (1980), which may be determined using petrographic data. Samples that are comparatively poor in quartz grains are further assessed by Node 6, which considers the volume abundance of volcaniclastic grains. Such grains tend to be reactive, producing clays (often chlorite) for mafic materials and illite/smectite or zeolites for intermediate to silicic materials. Furthermore, altered volcanic material can become quite ductile, leading to increased compaction (Fig. 12). The 15 vol% cutoff for this node is arbitrary. In reality, the volume fraction threshold will vary depending on the nature of the volcanic material, the fluid compositions and fluxes that the sandstone has experienced and the effective stress history.

Cases in Node 6 that do not contain abundant volcaniclastic materials are assessed in Node 7, which considers whether there are abundant ductile grains of nonvolcanic origin. Such grains could include shale rock fragments, intrabasinal rip-up clasts, phyllosilicate-rich metamorphic fragments, glauconite clasts or other mechanically weak grain types. It is worth noting that shale rock fragments and rip-up clasts are sometimes referred to as "pseudomatrix", given that in their compacted state they may be misidentified as detrital clay matrix. Here again, the 20 vol% threshold in Node 7 is only a rough approximation. As shown in Fig. 13A, the magnitude of compaction associated with a given volume fraction of ductile materials varies significantly depending on the mechanical behavior of the grain type. For those sandstones with abundant ductile grain volumes, the next consideration in Node 8 is whether they have been exposed to effective stresses in excess of 15 MPa (2176 psi), which is equivalent to ~1200 m (3937 feet) of burial under hydrostatic fluid conditions. This effective stress cutoff is arbitrary but consistent both with experimental compaction studies that show rapid compaction with effective stress for samples that contain abundance ductile grains (Fig. 12) and the simulation results from Fig. 14.

The favorable outcomes from Nodes 7 and 8 are assessed in Node 9, which considers a nonquartz cement cutoff of 10 vol %. Alteration of nonquartz grains in these sandstones could reduce reservoir quality such as when feldspar alteration generates kaolinite, illite, calcite or laumontite cement or when carbonate bioclasts or rock fragments react to form carbonate cement. Additionally, carbonate or other cements may form in sandstones in association with fault-related fluid fluxes or from material derived from interbedded lithologies. When the cement volume exceeds the 10 vol% threshold the reservoir quality risk is deemed to be greater. The favorable outcomes from Node 9 are classified as having lower reservoir quality risk.

The remaining decision nodes (10–12) have to do with broad controls on quartz cementation in sandstones with abundant detrital quartz grains. Models of quartz cementation show roughly linear increases in cement volumes with time and exponential increases with temperature, while also being highly sensitive to composition and texture. Therefore, it is not possible to provide a simple definition for "Low thermal exposure" in Node 10 without the aid of quartz cementation models that consider the entire thermal history of the location of interest in addition to important compositional and textural characteristics. For Mesozoic and younger aged sandstones, however, an example of low thermal exposure could be considered temperatures less than ~110°C (230°F) for ~10 m.y. The temperature threshold, however, drops with increased depositional age. In Paleozoic aged rocks, for instance, increased time can make up for lower temperature exposure (e.g., Lander and Walderhaug, 1999; Lander et al., 2008).

Node 11 considers whether grain coatings are well-developed for cases where the area of interest has a thermal exposure that exceeds the threshold in Node 10. Grain coating assessments should be based on measurements conducted on analog samples as described earlier. Well-developed coverage in this context could be considered to be values in excess of ~65 area%. Reservoir quality risk will be great where coatings are not well-developed due to the potential for significant quartz cementation. When coatings are well-developed, Node 12 considers whether a high thermal exposure is expected at the area of interest. As discussed in the section above on grain coatings, progressively greater coating coverage is needed to preserve reservoir quality as thermal exposure increases (Figs. 26 and 27). Thus, there is no simple time or temperature threshold associated with Node 12, although in Mesozoic and younger sandstones, temperature in excess of ~130°C (266°F) for ~20 m.y. could be considered high thermal exposure. For cases with well-developed coatings that have intermediate thermal exposure, a final consideration in Node 9 is whether there could be significant volumes of nonquartz cements.

## Key messages

While the flow chart from the section above may be a useful means for an initial assessment of the factors that could impact reservoir quality, quantitative numeric models offer a more rigorous way to account for the complexity of interacting depositional and diagenetic factors while also explicitly simulating permeability and the volume fractions of intergranular, grain dissolution and microscopic pores. As with most things in geology, sandstone reservoir quality and diagenesis are almost never controlled by only one factor but instead reflect the cumulative effect over time of multiple interacting processes where the outcome is strongly influenced by the starting point. In this section, we use diagenesis simulations to illustrate scenarios of how reservoir quality changes through geologic time in deepwater sandstones in response to variations in burial history, framework grain composition and depositional facies. The purpose of these simulations is to illustrate a few scenarios of different factors affecting diagenesis and reservoir quality of deepwater sandstones. Thus, these results should not be taken as a quantitative or semiquantitative guide with universal applicability. The best approach for assessing deepwater reservoir quality is to conduct a reservoir quality analysis and prediction study in concert with source to sink (Chapter 11), depositional environment (Chapter 8) and basin modeling (Chapter 18) studies. To learn more about the concepts and workflows associated with predrill reservoir quality predictions, see the discussion of process-oriented diagenesis and reservoir quality models in Chapter 12 of Taylor et al. (2022).

In the first portion of the analysis, we consider the impact that the broad range in framework grain composition has on reservoir quality variations for the best depositional lithology type in deepwater sandstones. In the second part of the analysis we assess reservoir quality variability over lateral distances of meters to kilometers within a submarine fan complex.

## Sensitivity to framework grain composition

In this portion of the analysis, we assume that the sandstones of interest all have an ideal depositional texture: the "very thick-bedded" lithology type (Table 1 in this chapter and Table 2 in Chapter 7) with a well sorted, unimodal texture, medium mean grain size, no detrital clay matrix, no laminations and poorly developed grain coatings (30 area%). We consider, however, the impact that the broad range in framework grain compositions that are found in deepwater sandstones can exert on reservoir quality by selecting the quartzose, lithic-rich and feldspathic compositions indicated by 1, 2 and 3, respectively, on Fig. 31A. Note that the selected compositions are within the range observed in deepwater sandstones. The lithic-rich scenario contains abundant volcanic rock fragments with mafic compositions. This scenario is among the worst potential lithic-rich cases given the highly reactive nature of volcanic rock fragments. Where lithic fragments are made up of other less reactive materials the associated cement volumes could be far lower than the example described as follows.

In the simulations, kaolinite and illite formation from the alteration of feldspars and chlorite formation from the alteration of mafic volcanic fragments are considered along with the nucleation and growth of quartz cement. We subjected

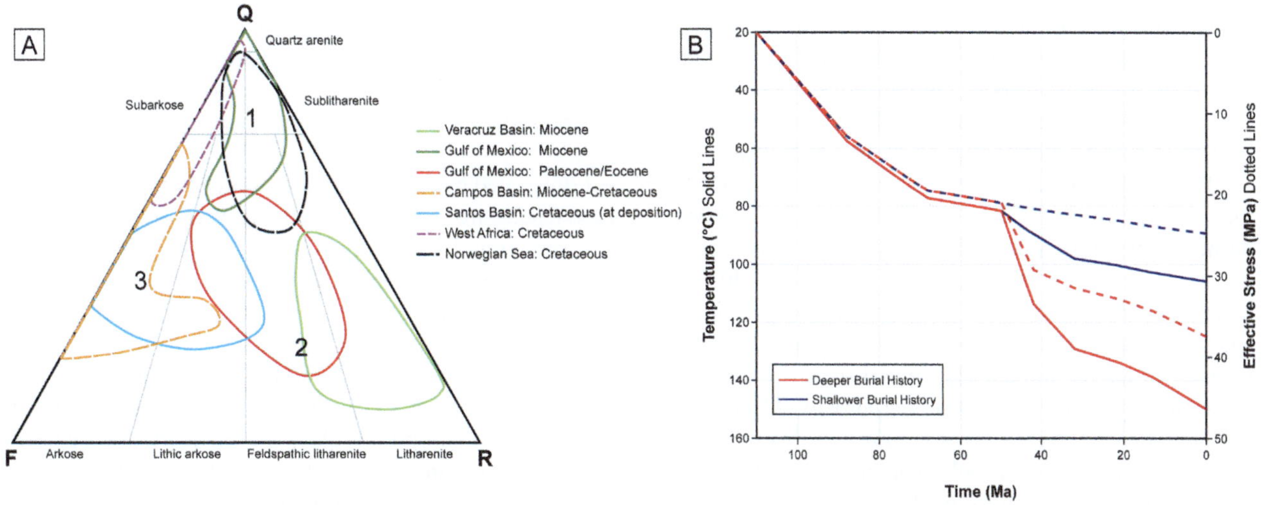


**FIG. 31** Input for diagenesis and reservoir quality simulation scenarios. (A) Framework grain compositions for *1—quartzose*, *2—lithic-rich* and *3—feldspathic* scenarios. (B) Effective stress and temperature through time for the *Shallower Burial History* and *Deeper Burial History* cases.

these synthetic sandstones to the two alternative burial histories depicted in Fig. 31B. These burial histories are identical from the time of deposition of 110 Ma through 50 Ma. From this point onward, however, the "shallower" scenario sees only moderate increases in temperature and effective stress, whereas the "deeper" scenario experiences significant increases in these factors due to greater depths of burial. Combined, these compositional and burial history scenarios result in enormous ranges in simulated reservoir quality through geologic time and at the present-day as shown in Fig. 32.

The permeability simulations in Fig. 32A and B show drops in values ranging from two to seven orders of magnitude from the time of deposition to the present day. The permeabilities for the quartzose scenario (green lines) at any given time always exceed that of the other compositions, whereas the lithic scenario (black lines) almost always has the lowest values. The total porosity results (Fig. 32C and D), on the other hand, do not show the same pattern. Here, the lithic sandstone initially has the highest porosity due to the presence of microporous grains at the time of deposition. During the initial stages of burial, however, this compositional scenario loses porosity at a faster rate compared to the other two compositions. Additionally, the feldspathic scenario (orange lines) generally has the highest total porosity even where it has permeability values that are orders of magnitude lower than the quartzose scenario due to differences in the nature of the pore types for these two scenarios, as discussed later.

Compaction is the dominant cause of the reduction in porosity and permeability in the earliest part of the burial histories as indicated by the simulated intergranular volume (IGV) values with time shown in the dashed lines in Fig. 31E and F. All three compositions experience the majority of the IGV decline from 110 to 70 Ma during the initial phase of burial, which is identical for the two burial history cases. The porosity loss in the quartzose scenarios is almost entirely due to compaction over this timeframe. The lithic-rich and feldspathic scenarios, however, lose considerable additional intergranular porosity from chlorite and kaolinite cementation, respectively, as shown in Fig. 31G and H. This cementation occurs in response to the alteration of mafic volcanic rock fragments for the lithic-rich scenario and feldspar for the feldspathic scenario. In the simulations, kaolinite and chlorite form at relatively low thermal exposures prior to 50 Ma, when the two burial history cases diverge. Thus, the timing of formation of these clays is insensitive to the overall thermal exposure difference between the two burial history cases.

The next major cause of reservoir quality degradation in the simulations is quartz overgrowth precipitation (Fig. 31G and H). Simulations from both burial histories have significant differences in the quartz cement abundance depending on the framework grain composition. In particular, quartz overgrowth cement is most abundant for the quartzose compositions due to the greater nucleation area from more abundant detrital quartz grains and the reduced extent of intergranular porosity loss from compaction and clay cementation. There also is a significant difference in results between the two alternative burial histories with substantially greater quartz cement in the deeper burial case. Note that the lithic scenario has just trace amounts of quartz cement even in the deeper burial case due to the occurrence of grain coating chlorite (modeled to have formed from the alteration of the abundant mafic volcanic rock fragments) and because there is little or no remaining intergranular porosity by the time the rock is hot enough for significant quartz cementation to occur.

The final diagenetic process that is displayed in the simulations is the reaction of kaolinite and K-feldspar to form illite. This reaction is important only in the deeper burial case where illite began to form at ~30 Ma in the quartzose and feldspathic scenarios. Little illite forms in the lithic scenario due to the paucity of a kaolinite reactant (kaolinite typically is associated with neutral to acidic pore fluid pH but in this scenario, the pH likely would have been high due to hydrolysis of volcanic glass). The volume of illite that formed is substantially greater in the feldspathic scenario compared to the quartzose scenario. In both compositions, however, the reaction eventually stopped. In the quartzose simulation, the reaction stopped when all available K-feldspar was consumed, whereas in the feldspathic composition, the reaction stopped because all intergranular pore space was occluded.

The combined effects of compaction and cementation are best illustrated by the intergranular porosity results shown by the solid lines in Fig. 31E and F. Early in the burial histories, these results are comparable to the IGV values shown by the dashed lines on the same plots. After IGV decline mostly ceases, intergranular porosity continues to drop in response to cementation. For the lithic-rich scenarios, most of this porosity loss occurs by 60 Ma. In the quartzose scenarios, by contrast, most of the cementation porosity loss occurs from 40 Ma to the present due mainly to quartz cementation. For the feldspathic scenario, little porosity loss occurs after 60 Ma for the shallower burial history but substantial porosity loss occurs from 50 to 20 Ma in the deeper burial scenario in response to the combination of quartz and illite cementation.

Returning to the analysis of the porosity differences among the scenarios, although the feldspathic scenario tends to have the greatest overall porosity, much of this porosity is associated with microporous clays. In the quartzose scenarios, on the other hand, the overall volumes of clay are much lower and the intergranular porosities therefore are greater compared to the other compositional scenarios exposed to the same burial conditions. Intergranular porosity is much more effective for fluid flow in sandstones compared to micropores or grain dissolution pores. Consequently, the quartzose scenarios have higher permeabilities in response to their greater intergranular porosities compared to the other lithology types that are exposed to the same burial conditions.

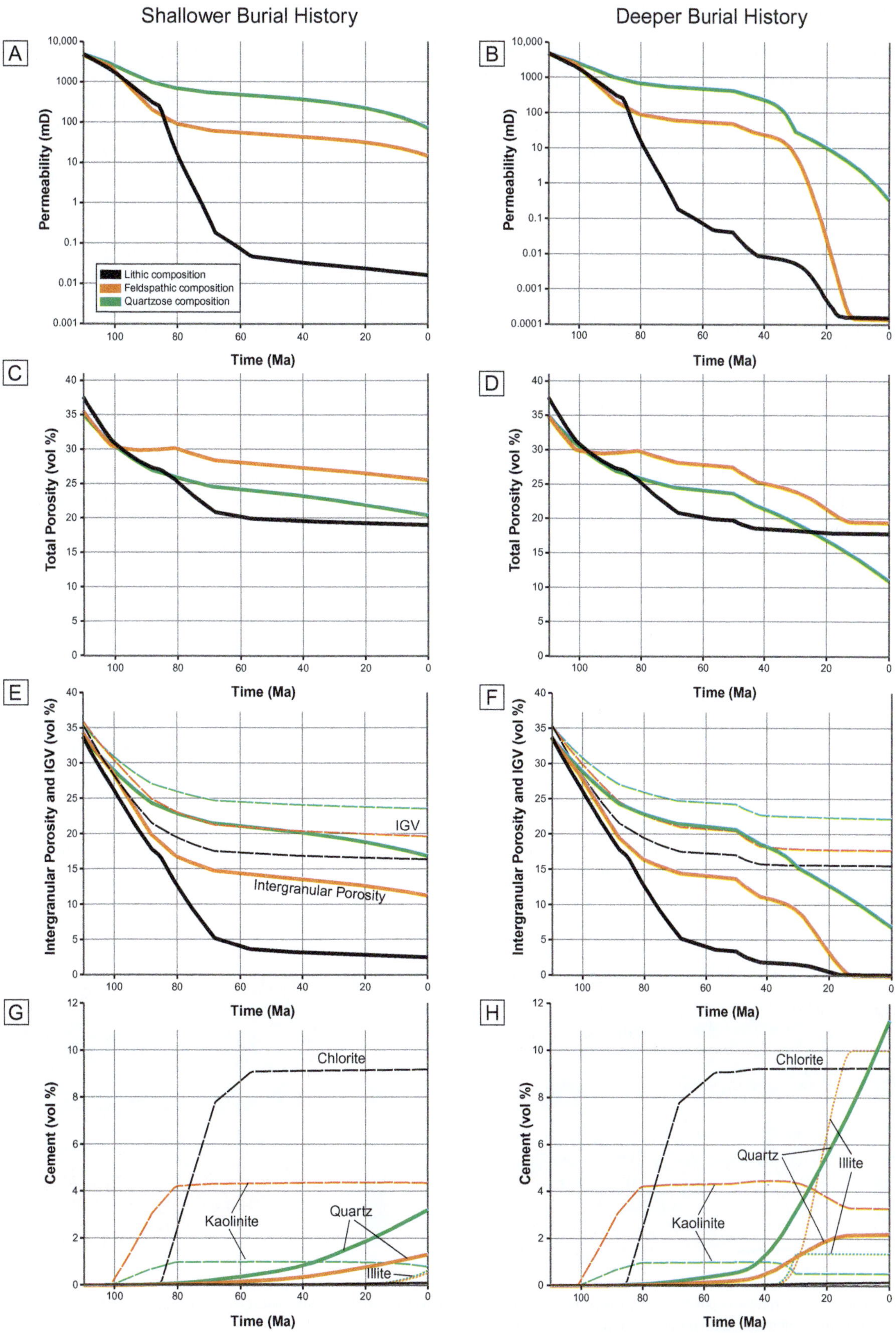


**FIG. 32** Touchstone simulation results for the input scenarios shown in Fig. 31 where plots in the left column correspond to the "*Shallower Burial History*" and those on the right use the "*Deeper Burial History*". (A and B) Permeability through time. (C and D) Total porosity through time. (E and F) Intergranular volume (IGV) with *dashed lines* and intergranular porosity with *solid lines* through time. (G and H) Volume percentages of various cement types through time. Line color codes: *green* for quartzose, *black* for lithic and *orange* for feldspathic framework grain compositions.

## Sensitivity to depositional environment

From the previous section, it is apparent that the reservoir quality varies substantially in response to framework grain composition and burial history, even for an ideal deepwater lithology. In this section, we continue to use the quartzose framework grain composition from the previous section, which had the best reservoir quality outcomes, together with the same two burial history cases (Fig. 33). We expand our analysis here, however, to consider reservoir quality variations among the facies types found in submarine fans. As discussed earlier and described in detail in Chapters 6–10, deepwater depositional systems exhibit considerable variations in depositional composition and texture over lateral scales of meters to kilometers.

In this analysis, we consider a set of scenarios consistent with facies spatial distributions within a submarine fan complex from Cretaceous deposits in the Jaca Basin in Northern Spain described by Bell et al. (2018). These workers analyzed well-exposed outcrops to map out channel axis, channel axis margin, lobe off-axis and lobe fringe deposits as illustrated in Fig. 33. We deviate from this dataset, however, by considering framework grain compositions consistent with the quartzose scenario in the previous section rather than using the more feldspathic compositions reported by Bell et al. (2018). We also assume mean grain sizes that are consistent with datasets that have maximum grain sizes on the finer end of the spectrum shown in Fig. 1.

For simplicity, these simulations assume a uniform depositional composition and texture for each depositional facies (Table 2 and Fig. 33). Mean grain size in these scenarios drops from upper fine for the channel axis, to lower fine at the channel margin, to very fine for the lobe off-axis and finally to coarse silt at the lobe fringe. Concurrently matrix increases from 0 to 6, 16 and 35 vol%, respectively, for each of these depositional facies. Grain sorting decreases from moderately well sorted for the channel to moderately sorted in the lobes and grain coatings are assumed to be poorly developed in all facies (30 area%). Finally, the abundances of feldspar grains and shale fragments increase slightly from channel axis to lobe fringe in response to hydrodynamic sorting effects as documented in other submarine fan deposits (Pyles et al., 2013; Stammer, 2014). All facies were then subject to the same "shallower" and "deeper" burial history cases used in the previous section (Fig. 31B).

The Touchstone simulation results indicate substantial variations in reservoir quality among submarine fan facies as well as through time for a given facies (Fig. 34). Permeability (Fig. 34A) at the time of deposition varies by over five orders of magnitude among the facies. As expected, the channel axis facies has the greatest permeability, whereas the lobe fringe facies has the lowest permeability. These results are particularly sensitive to the differences in matrix abundances among the facies while also reflecting mean grain size and grain sorting differences.

Permeability values also are strongly affected by diagenetic alteration, with decreases ranging from two to five orders of magnitude from the time of deposition to present-day for all facies types. The present-day reservoir qualities for the best *depositional* reservoir quality facies are more sensitive to burial history differences compared to poorer depositional facies. Thus, the range in simulated permeability among facies tends to become smaller with increasing thermal exposure. The present-day simulated permeability ranges among facies for the shallower and deeper burial cases are about six and three orders of magnitude, respectively. If thermal exposure were to exceed that of the deeper burial case, the permeability range among the facies scenarios would become even smaller as intergranular porosities in all facies types approach zero and absolute permeabilities drop to subeconomic values.

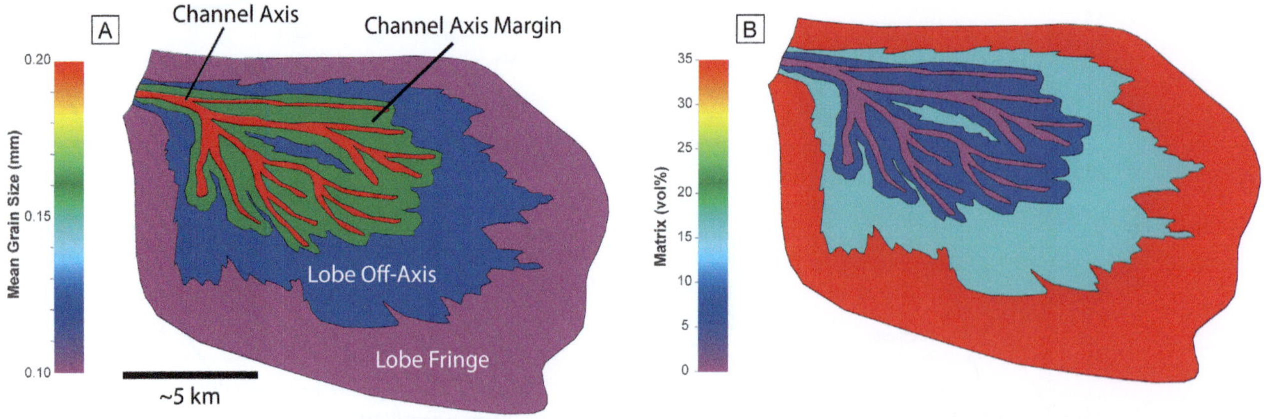


**FIG. 33** Pattern in interpreted depositional environments from the study of Bell et al. (2018) together with (A) mean grain size and (B) detrital clay matrix distributions used as input for reservoir quality simulations shown in Fig. 34.

**TABLE 2** Depositional composition and texture used as input for simulations of submarine fan diagenesis and reservoir quality shown in Figs. 32 and 33.

| Facies | Mean Grain Size, mm | Grain Sorting, std dev phi | Matrix, vol% | Q | F | R |
|---|---|---|---|---|---|---|
| Channel Axis | 0.20 | 0.50 | 0 | 89 | 7 | 4 |
| Channel Axis Margin | 0.16 | 0.63 | 6 | 85 | 8 | 7 |
| Lobe Off-axis | 0.12 | 0.69 | 16 | 80 | 11 | 9 |
| Lobe Fringe | 0.10 | 0.74 | 35 | 75 | 12 | 12 |

Q, F and R refer to the quartz, feldspar, rock fragment scheme of framework grain classification by Folk (1980).

As with the simulation results from the previous section, it is apparent that total porosity values (Fig. 34B) are a poor index of reservoir quality in deepwater sandstones. Results from the deeper burial scenario, for instance, show greater porosities in the lobe fringe facies compared to the other facies types despite generally having far lower permeabilities. This nonintuitive result reflects the fact that all present-day porosity in the lobe fringe facies simulations is associated with matrix micropores that are not readily occluded by cements. Intergranular porosity (Fig. 34C) provides a much better depiction of the relative reservoir quality variations among the facies types compared to total porosity.

The depositional intergranular volumes tend to be larger in the facies with finer mean grain sizes (Fig. 34D) despite somewhat poorer sorting (as discussed previously, poorer sorting decreases depositional IGV, whereas finer mean grain size leads to increased values, at least for sub medium grain sizes). The lobe fringe facies has a much higher IGV at the present-day compared to the other types because compaction is impeded once the entire intergranular volume is fully occupied by detrital matrix. For the remaining facies types, the present-day IGV values show only modest differences for a given burial scenario. In these examples, the slightly greater depositional IGV of the lobe off-axis facies is offset by a greater magnitude of compaction that results from its increased abundance of ductile framework grains compared to the channel facies types.

With the exception of the lobe fringe facies, differences in reservoir quality for a given facies type between the two burial history scenarios occur mainly in response to variable cement volumes. These variations involve a greater extent of reaction of kaolinite to form illite (Fig. 34E) as well as greater magnitudes of quartz cementation (Fig. 34F) in the deeper burial history case compared to the shallower one. Variations in the amount of quartz cement among the facies types for a given burial history reflect differences in detrital clay volumes and grain sizes. In particular, the lobe fringe and lobe off-axis facies have reduced extents of quartz cementation because their increased detrital clay matrix amounts reduce the intergranular pore volume that can be occluded by quartz overgrowths.

The variations in simulation results among facies lead to considerable spatial heterogeneity in reservoir properties over lateral distances on the order of meters to kilometers (Fig. 35, Videos S1–S8 in the online version at https://doi.org/10.1016/B978-0-323-91918-0.00011-6). The spatial heterogeneity in reservoir quality within an actual submarine fan complex of interest may deviate significantly compared to these scenarios depending on burial history differences as well as departures in the following attributes among facies compared to what we assume in this analysis: detrital matrix abundance, framework grain composition, mean grain size, grain sorting and grain coating. Additionally, there may be considerable heterogeneity in reservoir quality within a given facies type due to variable depositional characteristics. Nonetheless, the results demonstrate that while depositional reservoir quality is a fundamental control on the evolution of reservoir quality in submarine fan complexes, it alone is a poor basis for assessing the reservoir volume, spatial distribution and fluid-flow characteristics.

## Future directions

While the ability to quantitatively predict sandstone diagenesis and reservoir quality in many geologic settings has increased significantly over the past two decades, additional advances may be made by integrating existing models with predictive models from several related disciplines while also conducting research to develop additional process-oriented diagenesis models.

With respect to research topics, a priori models have yet to be developed that accurately predict volume abundances and textural characteristics resulting from number of important mineral reactions in reservoir sandstones such as those involving the formation of smectite, kaolin, many occurrences of chlorite, most zeolites and virtually all carbonate minerals. Additionally, to date no publications have demonstrated models capable of accurately predicting or reproducing the

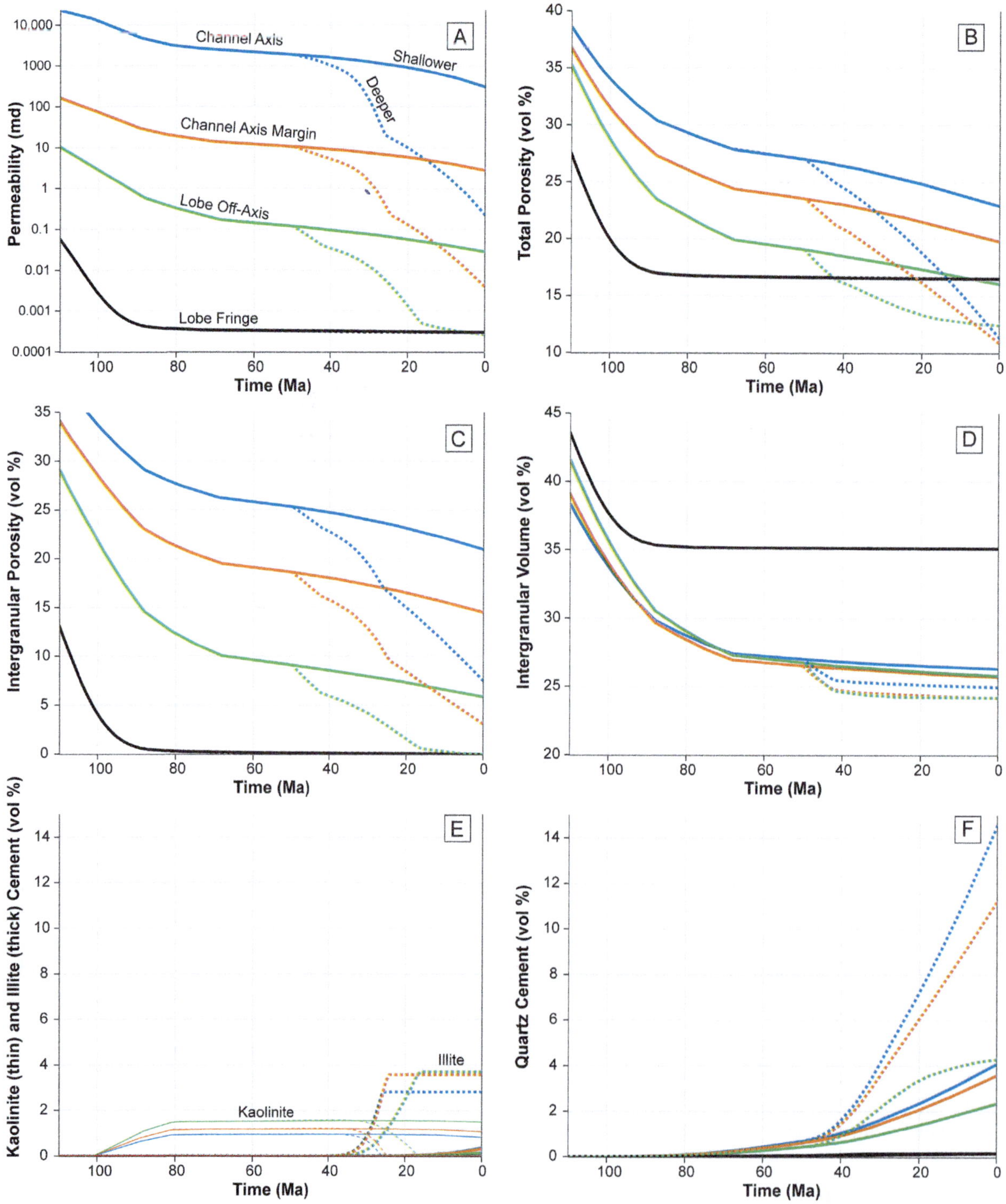


**FIG. 34** Touchstone simulation results for the four depositional environment scenarios described in the text. The solid lines in the plots use the "*Shallower Burial History*" case and the dotted lines the "*Deeper Burial History*" case shown in Fig. 31B. The plots show the following results as a function of geologic time: (A) permeability, (B) total porosity, (C) intergranular porosity, (D) intergranular volume (IGV), (E) kaolinite cement and illite cement and (F) quartz cement.

variations of intergranular volume within sandstone datasets that include samples that have experienced significant chemical compaction.

Existing and future process-oriented diagenesis models require input in the form of the depositional composition and texture of sands in addition to stresses, temperatures and fluid-fluxes and compositions through geologic time as well as

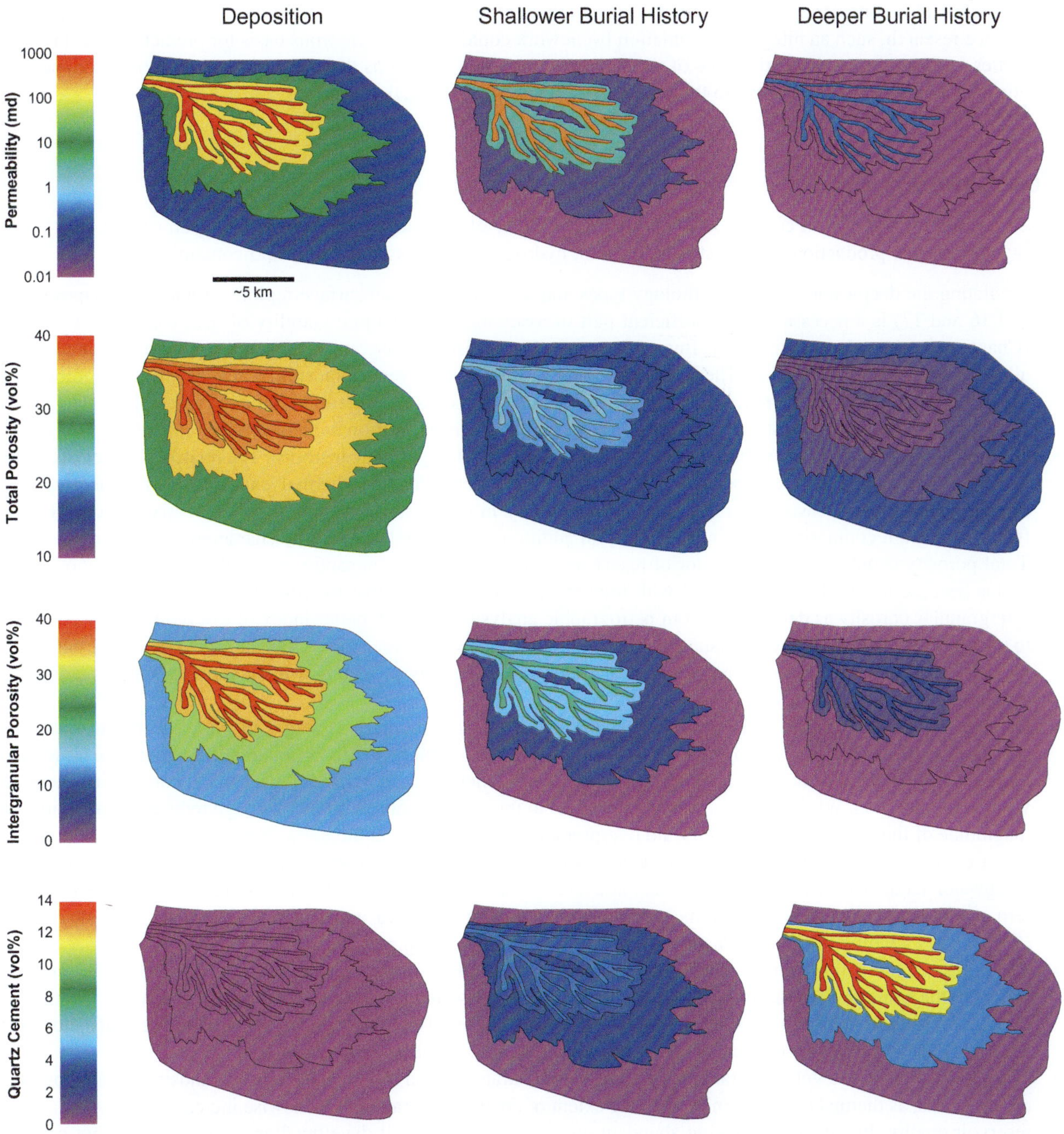


**FIG. 35** Touchstone simulation scenarios of spatial variations in reservoir quality within the submarine fan complex shown in Fig. 33 at deposition and at the present-day for the "*Shallower Burial History*" and "*Deeper Burial History*" cases (see Fig. 31B) for permeability, total porosity, intergranular porosity and quartz cement.

over engineering timescales. Thus, important advances also will come from the integration of diagenetic and rock property models with (1) quantitative models that predict the spatial and temporal variations in sediment depositional composition and texture, (2) 3D petroleum system/basin models capable of simulating stresses, temperatures and fluid-fluxes through geologic time as well as reservoir simulators that predict these properties over engineering timescales and (3) reactive transport models that use thermodynamics and kinetics to simulate geochemical reactions while also considering mass transport arising from diffusion and advection. An additional potential component of such a system is the integration of models of structural deformation capable of predicting the occurrence and properties of cataclastic zones and fracture systems while also considering diagenetic overprints on these structural features. In addition to its utility for fundamental geoscience research, such an integrated simulation framework could serve as a rigorous basis for predicting the flow-flow properties of reservoirs at the initial state of an engineering activity as well as through time as engineering induced diagenetic alteration occurs in response to the production or injection of reactive fluids.

## Conclusions

Reservoir quality is a critical control on the economic viability of sandstones associated with hydrocarbon, geothermal energy and lithium production and the storage of $CO_2$, hydrogen, compressed air and fluid contaminants.

- Predicting the deepwater sandstone lithology types and distributions in subsurface regions of interest (Chapters 6–11 and 16 and 17) is a necessary but insufficient part of assessing the economic viability of reservoirs prior to drilling (Chapter 20). Abundant detrital clay, fine or finer mean grain sizes, poor grain sorting and occurrence of laminations all act to reduce reservoir quality. There is no guarantee, however, that even the most favorable deepwater lithology type (Table 1) will have adequate reservoir quality because diagenesis has the potential to reduce reservoir quality to sub-economic levels for all sandstone lithologies, irrespective of their depositional characteristics.
- Sandstones undergo enormous changes in permeability through their geologic histories. The relative differences in permeability among different composition sands vary greatly depending on thermal exposure. Consequently, when selecting analogs as a basis for assessing reservoir quality in deepwater sandstones, it is essential to consider both framework grain composition and burial history in addition to other sedimentologic characteristics.
- Total porosity is not a reliable indicator of reservoir quality, particularly in sandstones with abundant clay or altered grains because much of it is associated with micropores that are ineffective for flow and that tend to be water-wet. Intergranular porosity as determined from petrographic analysis is a much better index of reservoir quality compared to total porosity in most deepwater sandstones. Petrographic data also have the important ability to reveal the nature and magnitude of compaction and cementation in analog samples thereby providing a foundation for understanding and predicting reservoir quality variations and patterns.
- Compaction typically is the main porosity loss mechanism during the earliest stages of burial. The significance of porosity loss from this process can be large in lithic-rich sandstones but generally is not large enough in quartzose sandstones to make reservoirs uneconomic. For feldspathic sandstones, the importance of mechanical compaction increases with grain alteration. While fluid overpressure can lead to improved reservoir quality through reduced compaction, the magnitude of this effect is larger when fluid overpressures develop early in the burial history and cementation is not a major control on reservoir quality. (See Chapter 18 for a discussion of simulation of fluid overpressure histories.)
- Deepwater sandstones that are subject to tectonic stresses are likely to compact more extensively compared to counterparts lacking such stresses. Tectonic stresses could be significant in a number of deepwater tectonic settings (Chapter 3) including fold belts (Chapter 4) and in the vicinity of diapirs (Chapter 5).
- The impact of thermal history on cementation is strongly dependent on the depositional composition of the sand. For instance, the lithic-rich scenario in the example simulations described above is relatively insensitive to the thermal exposure difference between the shallower and deeper burial cases. The feldspathic and quartzose scenarios, however, are highly sensitive to this difference. (See Chapter 18 for a discussion of thermal modeling.)
- The presence of well-developed grain coatings does not guarantee good reservoir quality. In sandstones with abundant quartz grains, as thermal exposure increases, the extent of coating coverage must likewise increase to preserve a given reservoir quality. In sandstones lacking abundant quartz grains, even well-developed grain coatings will not preserve reservoir quality given that quartz cementation is rarely volumetrically significant.
- A control on reservoir quality at the field scale is the significant variation in detrital clay, mean grain size, sorting and framework grain composition that can occur among depositional facies in response to hydrodynamic sorting effects and other factors (see Chapters 6, 7, 8 and 11). As illustrated in the submarine fan simulation scenarios described earlier (Fig. 35) and natural examples (e.g., Stammer, 2014; Marchand et al., 2015; Porten et al., 2019; Huang et al., 2020; and Bello et al., 2021), lithology-dependent variations can result in substantial reservoir quality differences in sandstones that have identical burial histories.
- Quantitative predictive models have been developed that accurately predict some, but not all, of the factors that control reservoir quality in sandstones. Such models use the following types of input information: the depositional composition and texture (see Chapters 6–11), temperature and effective stress histories (see Chapter 18) and model parameters that have been optimized using analog datasets (see Chapter 12 in Taylor et al., 2022).

## Acknowledgments

We are grateful for the useful suggestions for improvements to this chapter provided by Luca Caracciolo, Bill Heins, Jon Rotzien, Cindy Yeilding and Richard Sears.

## References

Ajdukiewicz, J.M., Nicholson, P.H., Esch, W.L., 2010. Prediction of deep reservoir quality using early diagenetic process models in the Jurassic Norphlet formation, Gulf of Mexico. AAPG Bull. 94, 1189–1227.

Alcantar, N., Israelachvili, J., Boles, J., 2003. Forces and ionic transport between mica surfaces: implications for pressure solution. Geochim. Cosmochim. Acta 67, 1289–1304.

Altaner, S.P., 1986. Comparison of rates of smectite illitization with rates of K-feldspar dissolution. Clays Clay Miner. 34, 608–611. https://doi.org/10.1346/CCMN.1986.0340517.

Al-Mufti, O.N., Arnott, R.W.C., 2020. The origin and significance of convolute lamination and pseudonodules in an ancient deep-marine turbidite system: from deposition to diagenesis. J. Sediment. Res. 90, 480–493.

Anjos, S.M.C., De Ros, L.R., De Souza, R.S., Silva, C.M.A., Sombra, C.L., 2000. Depositional and diagenetic controls on the reservoir quality of lower cretaceous pendência sandstones, Potiguar rift basin, Brazil. AAPG Bull. 84, 1719–1742.

Anjos, S.M.C., De Ros, L.R., Silva, C.M.A., 2003. Chlorite authigenesis and porosity preservation in the Upper Cretaceous marine sandstones of the Santos Basin, offshore eastern Brazil. In: Worden, R.H., Morad, S. (Eds.), Clay Mineral Cements in Sandstones. International Association of Sedimentologists, pp. 291–316. Special Publication 29.

Bahlis, A.B., De Ros, L.F., 2013. Origin and impact of authigenic chlorite in the upper cretaceous sandstone reservoirs of the Santos Basin, eastern Brazil. Pet. Geosci. 19, 185–199.

Baiyegunhi, T.L., Liu, K., Gwavava, O., Baiyegunhi, C., 2020. Impact of diagenesis on the reservoir properties of the cretaceous sandstones in the southern Bredasdorp Basin, Offshore South Africa. Minerals 10, 757. https://doi.org/10.3390/min10090757.

Baker, J.C., Havord, P.J., Martin, K.R., Ghori, K.A.R., 2000. Diagenesis and petrophysics of the early Permian Moogooloo sandstone, southern Carnarvon Basin, Western Australia. Am. Assoc. Pet. Geol. Bull. 84, 250–265.

Bazin, B., Brosse, E., Sommer, F., 1997a. Chemistry of oil-field brines in relation to diagenesis of reservoirs. 1. Use of mineral stability fields to reconstruct *in situ* water composition. Example of the Mahakam basin. Mar. Pet. Geol. 14, 481–495.

Bazin, B., Brosse, E., Sommer, F., 1997b. Chemistry of oil-field brines in relation to diagenesis of reservoirs. 2. Reconstruction of palaeo-water composition for modelling illite diagenesis in the greater Alwyn area (North Sea). Mar. Pet. Geol. 14, 497–511.

Beard, D.C., Weyl, P.K., 1973. Influence of texture on porosity and permeability of unconsolidated sand. Am. Assoc. Pet. Geol. Bull. 57, 345–369.

Beaufort, D., Cassagnabère, A., Petit, A., Lanson, B., Berger, G., Lacharpagne, J.C., Johansen, H., 1998. Kaolinite-to-dickite conversion series in sandstone reservoirs. Clay Miner. 33, 297–316.

Beaufort, D., Rigault, C., Billon, S., Billault, V., Inoue, A., Inoue, S., Patrier, P., 2015. Chlorite and chloritization processes through mixed-layer mineral series in low-temperature geological systems—a review. Clay Miner. 50, 497–523.

Bell, D., Kane, I.A., Pontén, A.S.M., Flint, S.S., Hodgson, D.M., 2018. Spatial variability in depositional reservoir quality of deep-water channel-fill and lobe deposits. Mar. Pet. Geol. 98, 97–115. https://doi.org/10.1016/j.marpetgeo.2018.07.023.

Bello, A.M., Jones, S., Gluyas, J., Acikalin, S., Cartigny, M., 2021. Role played by clay content in controlling reservoir quality of submarine fan system, forties sandstone member, Central Graben, North Sea. Mar. Pet. Geol. 128.

Berger, A., Gier, S., Krois, P., 2009. Porosity-preserving chlorite cements in shallow-marine volcaniclastic sandstones: evidence from cretaceous sandstones of the Sawan gas field, Pakistan. AAPG Bull. 93, 595–615.

Biot, M.A., 1941. General theory of three-dimensional consolidation. J. Appl. Phys. 12, 155–164.

Bjørklykke, K., Egeberg, P.K., 1993. Quartz cementation in sedimentary basins. AAPG Bull. 77, 1538–1548.

Bjørkum, P.A., 1996. How important is pressure in causing dissolution of quartz in sandstones? J. Sediment. Res. 66, 147–154.

Bjørkum, P.A., Walderhaug, O., 1990. Geometrical arrangement of calcite cementation within shallow marine sandstones. Earth Sci. Rev. 29, 145–161.

Bjørkum, P.A., Oelkers, E.H., Nadeau, P.H., Walderhaug, O., Murphy, W.M., 1998. Porosity prediction in quartzose sandstones as a function of time, temperature, depth, stylolites frequency, and hydrocarbon saturation. Am. Assoc. Pet. Geol. Bull. 82, 637–648.

Bjørlykke, K., Aagaard, P., 1992. Clay minerals in North Sea sandstones. In: Houseknecht, D.W., Pittman, E.D. (Eds.), Origin, Diagenesis, and Petrophysics of Clay Minerals in Sandstones. SEPM, pp. 65–80. Special Publication 47.

Bloch, S., Lander, R.H., Bonnell, L., 2002. Anomalously high porosity and permeability in deeply buried sandstone reservoirs: origin and predictability. Am. Assoc. Pet. Geol. Bull. 86, 301–328.

Boles, J.R., 1998. Carbonate cementation in Tertiary sandstones, San Joaquin Basin, California. In: Morad, S. (Ed.), Carbonate Cementation in Sandstones. vol. 26. International Association of Sedimentologists, pp. 261–283. Special Publication.

Boles, J.R., Ramseyer, K., 1987. Diagenetic carbonate in Miocene sandstone reservoir, San Joaquin basin, California. AAPG Bull. 71, 1475–1487.

Bonnell, L.M., Lander, R.H., Larese, R.E., 2014. Prediction of the formation of grain coating chlorite by the in situ alteration of volcanic rock fragments. (Abstract.). In: Reservoir Quality of Clastic and Carbonate Rocks: Analysis, Modelling and Prediction. Geological Society, London, pp. 112–113.

Browne, G.H., Slatt, R.M., 2002. Outcrop and behind-outcrop characterization of a late Miocene slope fan system, Mt. Messenger Formation, New Zealand. AAPG Bull. 86, 841–862.

Carvalho, M.V.F., De Ros, L.F., Gomes, N.S., 1995. Carbonate cementation patterns and diagenetic reservoir facies in the Campos Basin Cretaceous turbidites, offshore eastern Brazil. Mar. Pet. Geol. 12, 741–758. https://doi.org/10.1016/0264-8172(95)93599-Y.

Cavazza, W., Gandolfi, G., 1992. Diagenetic processes along a basin-wide marker bed as a function of burial depth. J. Sediment. Res. 62, 261–272.

Chan, M.A., 1985. Correlations of diagenesis with sedimentary facies in Eocene sandstones, western Oregon. J. Sediment. Res. 55, 322–333.

Chuhan, F.A., Bjørlykke, K., Lowrey, C.J., 2001. Closed-system burial diagenesis in reservoir sandstones: examples from the garn formation at Haltenbanken area, off-shore mid-Norway. J. Sediment. Res. 71, 15–26. https://doi.org/10.1306/041100710015.

De Boer, R.B., 1977. Pressure solution: theory and experiments. Tectonophysics 39, 287–301.

De Ros, L.F., 1998. Heterogeneous generation and evolution of diagenetic quartzarenites in the Silurian-Devonian Furnas formation of the Paraná Basin, southern Brazil. Sediment. Geol. 116, 99–128.

De Souza, R.S., De Assis Silva, C.M., 1998. Origin and timing of carbonate cementation of the Namorado sandstone (Cretaceous), Albacora field, Brazil: implications for oil recovery. In: Morad, S. (Ed.), Carbonate Cementation in Sandstones. vol. 26. International Association of Sedimentologists, pp. 309–325. Special Publication.

Dowey, P.J., Worden, R.H., Utley, J., Hodgson, D.M., 2017. Sedimentary controls on modern sand grain coat formation. Sediment. Geol. 353, 46–63.

Dutton, S.P., 2008. Calcite cement in Permian deep-water sandstones, Delaware Basin, West Texas: origin, distribution, and effect on reservoir properties. AAPG Bull. 92, 765–787.

Dutton, S.P., Loucks, R.G., 2010. Diagenetic controls on evolution of porosity and permeability in lower Tertiary Wilcox sandstones from shallow to ultradeep (200–6700 m) burial, Gulf of Mexico Basin, U.S.A. Mar. Pet. Geol. 27, 69–81.

Dutton, S.P., Loucks, R.G., Day-Stirrat, R.J., 2012. Impact of regional variation in detrital mineral composition on reservoir quality in deep to ultradeep lower Miocene sandstones, western Gulf of Mexico. Mar. Pet. Geol. 35, 139–153.

Ehrenberg, S.N., 1991. Kaolinized, potassium-leached zones at the contacts of the garn formation, Haltenbanken, mid-Norwegian continental shelf. Mar. Pet. Geol. 8, 250–269.

Ehrenberg, S.N., 1993. Preservation of anomalously high porosity in deep buried sandstones by grain-coating: example from the Norwegian Continental Shelf. Am. Assoc. Pet. Geol. Bull. 77, 1260–1286.

Ehrenberg, S.N., 1995. Measuring sandstone compaction from modal analyses of thin sections: how to do it and what the results mean. J. Sediment. Res. A65, 369–379.

Elias, B.P., Hajash, A., 1992. Changes in quartz solubility and porosity due to effective stress: an experimental investigation of pressure solution. Geology 20, 451–454.

Elliott, W.C., Matisoff, G., 1996. Evaluation of kinetic models for the smectite to illite transformation. Clays Clay Miner. 44, 77–87.

Ernst, W.G., Blatt, H., 1964. Experimental study of quartz overgrowths and synthetic quartzites. J. Geol. 72, 461–470.

Fetter, M., De Ros, L.F., Bruhn, C.H.L., 2009. Petrographic and seismic evidence for the depositional setting of giant turbidite reservoirs and the paleogeographic evolution of Campos Basin, offshore Brazil. Mar. Pet. Geol. 26, 824–853.

Folk, R.L., 1980. Petrology of Sedimentary Rocks. Austin. Hemphill Publishing Co., USA, p. 182.

Franks, S.G., Zwingmann, H., 2010. Origin and timing of late diagenetic illite in the Permian–Carboniferous Unayzah sandstone reservoirs of Saudi Arabia. AAPG Bull. 94, 1133–1159. https://doi.org/10.1306/04211009142.

Fu, M.-Y., Song, R.-C., Xie, Y.-H., Zhang, S.-N., Gluyas, J.G., Zhang, Y.-Z., Zhang, Y., 2016. Diagenesis and reservoir quality of overpressured deep-water sandstone following inorganic carbon dioxide accumulation: upper Miocene Huangliu formation, Yinggehai Basin, South China Sea. Mar. Pet. Geol. 77, 954–972. https://doi.org/10.1016/j.marpetgeo.2016.08.005.

Gaupp, R., Okkerman, J.A., 2011. Diagenesis and reservoir quality of Rotliegend sandstones in the northern Netherlands—a review. In: The Permian Rotliegend of the Netherlands. 98. SEPM, pp. 193–226. Special Publication.

Gaupp, R., Matter, A., Platt, J., Ramseyer, K., Walzebuck, J., 1993. Diagenesis and fluid evolution of deeply buried Permian (Rotliegende) gas reservoirs, Northwest Germany. AAPG Bull. 77, 1111–1128.

Giles, M.R., Indrelid, S.L., Beynon, G.V., Amthor, J., 2000. The origin of large-scale quartz cementation: evidence from large data sets and coupled heat-fluid mass transport modeling. In: Worden, R.H., Morad, S. (Eds.), Quartz Cementation in Sandstones. vol. 29. International Association of Sedimentologists, pp. 21–38. Special Publication.

Gradmann, S., Beaumont, C., 2012. Coupled fluid flow and sediment deformation in margin-scale salt-tectonic systems: 2. Layered sediment models and application to the northwestern Gulf of Mexico. Tectonics 31, TC4011. https://doi.org/10.1029/2011TC003035.

Gratier, J.P., Guiguet, R., 1986. Experimental pressure solution deposition on quartz grains – the crucial effect of the nature of the fluid. J. Struct. Geol. 8, 845–856.

Greene, G.W., Kristiansen, K., Meyer, E.E., Boles, J.R., Israelachvili, J.N., 2009. Role of electrochemical reactions in pressure solution. Geochim. Cosmochim. Acta 73, 2862–2872.

Griffiths, J., Worden, R.H., Utley, J.E.P., Brøstrom, C., Martinius, A.W., Lawan, A.Y., Al-Hajri, A.I., 2021. Origin and distribution of grain-coating and pore-filling chlorite in deltaic sandstones for reservoir quality assessment. Mar. Pet. Geol. https://doi.org/10.1016/j.marpetgeo.2021.105326.

Guilkey, J., Lander, R., Bonnell, L., 2021. A hybrid penalty and grid based contact method for the material point method. Comput. Methods Appl. Mech. Eng. 379, 113739. https://doi.org/10.1016/j.cma.2021.113739.

Hansen, H.N., Løvstad, K., Lageat, G., Clerc, S., Jahren, J., 2021. Chlorite coating patterns and reservoir quality in deep marine depositional systems—Example from the Cretaceous Agat Formation, Northern North Sea, Norway. Basin Res. https://doi.org/10.1111/bre.12581.

Hay, R.L., Sheppard, R.A., 2001. Occurrence of zeolites in sedimentary rocks: an overview. Rev. Mineral. Geochem. 45, 217–234.

Hayes, M.J., Boles, J.R., 1992. Volumetric relations between dissolved plagioclase and kaolinite in sandstones implications for aluminum mass transfer in the San Joaquin Basin California. In: Houseknecht, D.W., Pittman, E.D. (Eds.), Origin, Diagenesis, and Petrophysics of Clay Minerals in Sandstones. vol. 47. SEPM, pp. 111–123. Special Publication.

Heald, M.T., 1956. Cementation of Simpson and St. Peter sandstones in parts of Oklahoma, Arkansas, and Missouri. J. Geol. 64, 16–30.

Heald, M.T., Larese, R.E., 1974. Influence of grain coatings on quartz cementation. J. Sediment. Res. 44, 1269–1274.

Heins, W.A., Kairo, S., 2007. Predicting sand character with integrated genetic analysis. Geol. Soc. Am. Spec. Paper 420, 345–379. https://doi.org/10.1130/2006.2420(20).

Helmold, K.P., 1980. Diagenesis of Tertiary Arkoses, Santa Ynez Mountains, California. PhD dissertation, Stanford University, p. 225.

Helmold, K.P., Van De Kamp, P.C., 1984. Diagenetic mineralogy and controls on albitization and laumontite formation in Paleogene arkoses, Santa Ynez Mountains, California. In: McDonald, D.A., Surdam, R.C. (Eds.), Clastic Diagenesis. vol. 37. American Association of Petroleum Geologists Memoir, pp. 239–276.

Hendry, J.P., Trewin, N.H., Fallick, A., 1996. Low-mg calcite marine cement in cretaceous turbidites: origin, spatial distribution and relationship to seawater chemistry. Sedimentology 43, 877–900. https://doi.org/10.1111/j.1365-3091.1996.tb01508.x.

Higgs, K.E., Browne, G.H., Sahoo, T.R., 2019. Reservoir characterisation of syn-rift and post-rift sandstones in frontier basins: an example from the cretaceous of Canterbury and great south basins, New Zealand. Mar. Pet. Geol. 101, 1–29. https://doi.org/10.1016/j.marpetgeo.2018.11.030.

Houseknecht, D.W., 1987a. Assessing the relative importance of compaction processes and cementation to reduction of porosity in sandstones. Am. Assoc. Pet. Geol. Bull. 71, 1537–1564.

Houseknecht, D.W., 1987b. Assessing the relative importance of compaction processes and cementation to reduction of porosity in sandstones. Am. Assoc. Pet. Geol. Bull. 71, 633–642.

Houseknecht, D., 1988. Intergranular pressure solution in four quartzose sandstones. J. Sediment. Petrol. 58, 228–246.

Houseknecht, D.W., Hathon, L.A., 1987. Petrographic constraints on models of intergranular pressure solution in quartzose sandstones. Appl. Geochem. 2, 507–521.

Houseknecht, D.W., Ross Jr., L.M., 1992. Clay minerals in Atokan deep-water sandstone facies, Arkoma Basin: origins and influence on diagenesis and reservoir quality. In: Houseknecht, D.W., Pittman, E.D. (Eds.), Origin, Diagenesis, and Petrophysics of Clay Minerals in Sandstones. Vol. 47. SEPM, pp. 227–240. Special Publication.

Howard, J.J., 1992. Influence of authigenic-clay minerals on permeability. In: Houseknecht, D.W., Pittman, E.D. (Eds.), Origin, Diagenesis, and Petrophysics of Clay Minerals in Sandstones. 47. SEPM, pp. 257–264. Special Publication.

Hower, J., Eslinger, E.V., Hower, M.E., Perry, E.A., 1976. Mechanism of burial and metamorphism of argillaceous sediment: 1. Mineralogical and chemical evidence. Geol. Soc. Am. Bull. 87, 725–737.

Huang, W.-L., Longo, J.M., Pevear, D.R., 1993. An experimentally derived kinetic model for smectite-to-illite conversion and its use as a geothermometer. Clays Clay Miner. 41, 162–177.

Huang, Y., Kane, I.A., Zhao, Y., 2020. Effects of sedimentary processes and diagenesis on reservoir quality of submarine lobes of the Huangliu formation in the Yinggehai Basin, China. Mar. Pet. Geol. 120, 104526. https://doi.org/10.1016/j.marpetgeo.2020.104526.

Jackson, C.A.-L., Magee, C., Jacquemyn, C., 2020. Rift-related magmatism influences petroleum system development in the NE Irish Rockall Basin, offshore Ireland. Pet. Geosci. 26, 511–524.

Kantorowicz, J.D., 1984. The nature origin and distribution of authigenic clay minerals from Middle Jurassic Ravenscar and Brent Group sandstones. Clay Miner. 19, 359–375.

Kantorowicz, J.D., 1990. The influence of variations in illite morphology on the permeability of Middle Jurassic Brent Group sandstones, Cormorant Field, UK North Sea. Mar. Pet. Geol. 7, 66–74.

Ketzer, M.M., Morad, S., Amorosi, A., 2003. Predictive diagenetic clay-mineral distributions in siliciclastic rocks within a sequence stratigraphic framework. In: Worden, R.H., Morad, S. (Eds.), Clay Mineral Cements in Sandstones. vol. 29. International Association of Sedimentologists, pp. 43–61. Special Publication.

Kristiansen, K., Valtiner, M., Greene, G.W., Boles, J.R., Israelachvili, J.N., 2011. Pressure solution—the importance of electrochemical surface potentials. Geochim. Cosmochim. Acta 75, 6882–6892.

Land, L.S., 1997. Mass transfer during burial diagenesis in the Gulf of Mexico sedimentary basin: an overview. In: Montañez, I.P., Gregg, J.M., Shelton, K. L. (Eds.), Basin-Wide Diagenetic Patterns: Integrated Petrologic, Geochemical, and Hydrologic Considerations. vol. 57. SEPM, pp. 29–39. Special Publication.

Lander, R.H., Bonnell, L.M., 2010. A model for fibrous illite nucleation and growth in sandstones. AAPG Bull. 94, 1161–1187. https://doi.org/10.1306/04211009121.

Lander, R.H., Bonnell, L.M., 2018. Rock property prediction using process-oriented models and digital sedimentary petrology models. In: Extended abstract presented at 80th EAGE Conference and Exhibition, 11–14 June 2018, Copenhagen, Denmark., https://doi.org/10.3997/2214-4609.201801137.

Lander, R.H., Bonnell, L.M., 2020. Touchstone 8.2 User Guide. Geocosm LLC, Durango, Colorado, p. 335.

Lander, R.H., Walderhaug, O., 1999. Predicting porosity through simulating sandstone compaction and quartz cementation. Am. Assoc. Pet. Geol. Bull. 83, 433–449.

Lander, R.H., Bloch, S., Mehta, S., Atkinson, C.D., 1990. Burial diagenesis of paleosols in the Giant Yacheng gas field, People's Republic of China: bearing on illite reaction pathways. J. Sediment. Res. 61, 256–268.

Lander, R.H., Larese, R.E., Bonnell, L.M., 2008. Toward more accurate quartz cement models—the importance of euhedral vs none-euhedral growth rates. Am. Assoc. Pet. Geol. Bull. 92, 345–369.

Lanson, B., Beaufort, D., Berger, G., Baradat, J., Lacharpagne, J.C., 1996. Illitization of diagenetic kaolinite-to-dickite conversion series: late-stage diagenesis of the Lower Permian Rotliegend sandstone reservoir, offshore of the Netherlands. J. Sediment. Res. 66, 501–518.

Link, M.H., Welton, J.E., 1982. Sedimentology and reservoir potential of Matilija sandstone: an Eocene sand-rich deep-sea fan and shallow-marine complex, California. AAPG Bull. 66, 1514–1534.

Liu, M., Xiong, C., 2021. Diagenesis and reservoir quality of deep-lacustrine sandy-debris-flow tight sandstones in Upper Triassic Yanchang formation, Ordos Basin, China: implications for reservoir heterogeneity and hydrocarbon accumulation. J. Petrol. Sci. Eng. 202, 108548.

Liu, G., Hu, G., Shi, X., Ma, Y., Yin, X., Li, A., 2021. Carbonate cementation patterns and diagenetic reservoir facies of the Triassic Yanchang formation deep-water sandstone in the Huangling area of the Ordos Basin, Northwest China. J. Petrol. Sci. Eng. 203, 108608. https://doi.org/10.1016/j.petrol.2021.108608.

Lundegard, P.D., 1991. Sandstone porosity loss—a "big picture" view of the importance of compaction. J. Sediment. Petrol. 62, 250–260.

Macaulay, C.I., Fallick, A.E., Hasezeldine, R.S., 1992. Diagenetic pore waters stratified for at least 35 million years: Magnus oil field, North Sea: geologic note. AAPG Bull. 76, 1625–1634.

Macaulay, C.I., Fallick, A.E., Hasezeldine, R.S., 1993. Textural and isotopic variations in diagenetic kaolinite from the Magnus oilfield sandstones. Clay Miner. 28, 625–639.

Macaulay, C.I., Bryndzia, T., Inan, E., Diggs, T.N., 2012. Diagenesis of zeolite-cemented sands in the Frio Formation, Perdido Fold Belt, Western GoM. In: 2012 AAPG Annual Convention and Exhibition, April 22–25, 2012, Long Beach, California. AAPG Search and Discovery Article #90142.

Mansurbeg, H., El-Ghali, M.A.K., Morad, S., Plink-Björklund, P., 2006. The impact of meteoric water on the diagenetic alterations in deep-water, marine siliciclastic turbidites. J. Geochem. Explor. 89, 254–258.

Mansurbeg, H., Morad, S., Salem, S., Marfil, R., El-ghali, M.A.K., Nystuen, J.P., Caja, M.A., Amorosi, A., Garcia, D., La Iglesia, A., 2008. Diagenesis and reservoir quality evolution of Palaeocene deep-water, marine sandstones, the Shetland-Faroes Basin, British continental shelf. Mar. Pet. Geol. 25, 514–543.

Marchand, A.M.E., Apps, G., Li, W., Rotzien, J.R., 2015. Depositional processes and impact on reservoir quality in deepwater Paleogene reservoirs, US Gulf of Mexico. Am. Assoc. Pet. Geol. Bull. 99, 1635–1648.

Marfil, R., Hall, A., Garcia-Gil, S., Stamatakis, M.G., 1998. Petrology and geochemistry of diagenetically altered tuffaceous rocks from the middle Triassic of central Spain. J. Sediment. Res. 68, 391–403.

McHardy, W.J., Wilson, M.J., Tait, J.M., 1982. Electron microscope and x-ray diffraction studies of filamentous illitic clay from sandstones of the Magnus field. Clay Miner. 17, 23–29.

McKinley, J.M., Worden, R.H., Ruffell, A.H., 2003. Smectite in sandstones: a review of the controls on occurrence and behaviour during diagenesis. In: Worden, R.H., Morad, S. (Eds.), Clay Mineral Cements in Sandstones. 34, pp. 109–128. Int. Assoc. Sedimentol. Spec. Publ.

Morad, S. (Ed.), 1998. Carbonate Cementation in Sandstones, Special Publication 26 of the International Association of Sedimentologists. Blackwell Science, p. 511.

Morad, S., Al-Ramadan, K., Ketzer, J.M., De Ros, L.F., 2010. The impact of diagenesis on the heterogeneity of sandstone reservoirs: a review of the role of depositional facies and sequence stratigraphy. AAPG Bull. 94, 1267–1309.

Moraes, M.A.S., 1991. Diagenesis and microscopic heterogeneity of lacustrine deltaic and turbiditic sandstone reservoirs (lower cretaceous), Potiguar Basin, Brazil. AAPG Bull. 75, 1758–1771.

Niemeijer, A., Spiers, C.J., Bos, B., 2002. Compaction creep of quartz sand at 400–600 degrees C: experimental evidence for dissolution-controlled pressure solution. Earth Planet. Sci. Lett. 195, 261–275.

Noh, J.H., Boles, J.R., 1993. Origin of zeolite cements in the Miocene sandstones, North Tejon oil fields, California. J. Sediment. Petrol. 63, 248–260.

Oliveira, D.M., De Ros, L.F., 2018. Meteoric water and salt-dome-related diagenesis in tertiary Turbidite reservoirs from the Espírito Santo Basin, Brazil. J. Sediment. Res. 88, 1362–1380.

Parker, A., Lander, R., Bonnell, L.M., Gianniny, G., 2016. A Fresh Look at Controls on Sand Depositional Porosity, 68th Annual Rocky Mountain GSA Section Meeting., https://doi.org/10.1130/abs/2016RM-275871.

Paxton, S.T., Szabo, J.O., Ajdukiewicz, J.M., Klimentidis, R.E., 2002. Construction of an intergranular volume compaction curve for evaluating and predicting compaction and porosity loss in rigid-grain sandstone reservoirs. Am. Assoc. Pet. Geol. Bull. 86, 2047–2067.

Perez, R., Boles, A.R., 2005. An empirically derived kinetic model for albitization of detrital plagioclase. Am. J. Sci. 305, 312–343. https://doi.org/10.2475/ajs.305.4.312.

Pittman, E.D., Larese, R.E., 1991. Compaction of lithic sands: experimental results and applications. Am. Assoc. Pet. Geol. Bull. 75, 1279–1299.

Pittman, E.D., Larese, R.E., Heald, M.T., 1992. Clay coats: occurrence and relevance to preservation of porosity in sandstones. In: Houseknecht, D. W., Pittman, E.D. (Eds.), Origin, Diagenesis, and Petrophysics of Clay Minerals in Sandstones. vol. 47. SEPM, Special Publication, pp. 241–255.

Porten, K.W., Warchoł, M.J., Kane, I.A., 2019. Formation of detrital clay grain coats by dewatering of deep-water sands and significance for reservoir quality. J. Sediment. Res. 89, 1231–1249. https://doi.org/10.2110/jsr.2019.65.

Primmer, T.J., Cade, C.A., Evans, J., Gluyas, J., Hopkins, M.S., Oxtoby, N., Smalley, P.C., Warren, E.A., Worden, R.H., 1997. Global patterns in sandstone diagenesis: Their application to reservoir quality prediction for petroleum exploration. In: Kupecz, J.A., Gluyas, J., Bloch, S. (Eds.), Reservoir Quality Prediction in Sandstones and Carbonates. AAPG Memoir. American Association of Petroleum Geologists, Tulsa, pp. 61–78.

Pyles, D.R., Straub, K.M., Stammer, J.G., 2013. Spatial variations in the composition of turbidites due to hydrodynamic fractionation. Geophys. Res. Lett. 40, 3919–3923.

Pytte, A., 1982. The Kinetics of Smectite to Illite Reaction in Contact Metamorphic Shales. M.A. Thesis, Dartmouth College, Hanover, N.H, p. 78.

Pytte, A., Reynolds Jr., R.C., 1989. The kinetics of the smectite to illite reaction in contact metamorphic shales. In: Naeser, N.D., McCulloch, T.H. (Eds.), The Thermal History of Sedimentary Basins: Methods and Case Histories. Springer-Verlag, New York, pp. 133–140.

Ramseyer, K., Boles, J.R., 1986. Mixed-layer illite/smectite minerals in tertiary sandstones and shales, San Joaquin basin, California. Clays Clay Miner. 34, 115–124.

Rossi, C., Alaminos, A., 2014. Evaluating the mechanical compaction of quartzarenites: the importance of sorting (Llanos foreland basin, Colombia). Mar. Pet. Geol. 56, 222–238.

Ryan, P.C., Reynolds, R.C., 1996. The origin and diagenesis of grain-coating serpentine-chlorite in Tuscaloosa formation sandstone, US Gulf Coast. Am. Mineral. 81, 213–225.

Ryu, I.-C., Niem, A.R., 1999. Sandstone diagenesis, reservoir potential, and sequence stratigraphy of the Eocene Tyee Basin: Oregon. J. Sediment. Res. 69, 384–393.

Seyedolali, A., Dix, M., Triplehorn, D.M., 2002. Sand grain provenance controls style and amount of quartz cement. In: Geological Society of America Abstracts with Programs. vol. 34, p. 434. No. 6, September 2002.

Sibley, D.F., Blatt, H., 1976. Intergranular pressure solution and cementation of the Tuscarora orthoquartzite. J. Sediment. Petrol. 46, 881–896.

Sorby, H.C., 1908. On the application of quantitative methods to the study of the structure and history of rocks. Q. J. Geol. Soc. Lond. 64, 171–233.

Środoń, J., Eberl, D.D., 1984. Illite. In: Bailey, S.W. (Ed.), Mica: Reviews in Mineralogy 13. Mineralogical Society of America, pp. 495–539.

Stammer, J.G., 2014. Hydrodynamic Fractionation of Minerals and Textures in Submarine Fans: Quantitative Analysis from Outcrop, Experimental, and Subsurface Studies. Ph.D. Dissertation, Colorado School of Mines, p. 171.

Sullivan, M., Coombes, T., Imbert, P., Ahamdach-Demars, C., 1999. Reservoir quality and petrophysical evaluation of Paleocene sandstones in the west of Shetland area. In: Geological Society, London, Petroleum Geology Conference Series. vol. 5, pp. 627–633, https://doi.org/10.1144/0050627.

Tada, R., Siever, R., 1989. Pressure solution during diagenesis. Annu. Rev. Earth Planet. Sci. 17, 89–118.

Taylor, T.R., 1990. The influence of calcite dissolution on reservoir porosity in Miocene sandstones, Picaroon field, offshore Texas Gulf Coast. J. Sediment. Petrol. 60, 322–334.

Taylor, T.R., Land, L.S., 1996. Association of allochthonous waters and reservoir enhancement in deeply buried Miocene sandstones: picaroon field, corsair trend, offshore Texas. In: Crossey, L.J., Loucks, R., Totten, M.W. (Eds.), Siliciclastic Diagenesis and Fluid Flow: Concepts and Applications. vol. 55. SEPM, pp. 37–48. Special Publication.

Taylor, T.R., Stancliffe, R., Macaulay, C., Hathon, L., 2004. High temperature quartz cementation and the timing of hydrocarbon accumulation in the Jurassic Norphlet sandstone, offshore Gulf of Mexico, USA. In: Cubitt, J.M., England, W.A., Larter, S. (Eds.), Understanding Petroleum Reservoirs: Towards an Integrated Reservoir Engineering and Geochemical Approach. 237. geological society, London, pp. 257–278. Special Publications.

Taylor, T.R., Giles, M.R., Hathon, L.A., Diggs, T.N., Braunsdorf, N.R., Birbiglia, G.V., Kittridge, M.G., Macaulay, C.I., Espejo, I.S., 2010. Sandstone diagenesis and reservoir quality prediction: models, myths, and reality. AAPG Bull. 94, 1093–1132.

Taylor, T.R., Kittridge, M.G., Winefield, P., Bryndzia, L.T., Bonnell, L.M., 2015. Reservoir quality and rock properties modeling–Triassic and Jurassic sandstones, greater shearwater area, UK Central North Sea. Mar. Pet. Geol. 65, 1–21.

Taylor, T.R., Lander, R.H., Bonnell, L.M., 2022. Sandstone petrography, petrology, and modeling. In: SEPM Concepts in Sedimentology and Paleontology, vol. 13, pp. 1–356. https://doi.org/10.2110/sepmcsp.13.

Terzaghi, E., 1923. Die Berechnung der Duerchlässigkeitsziffer des Tones im Verlauf der hydrodynamischen Spannungserscheinungen. vol. 132 Szber Akademie Wissenschaft Vienna, Mathnaturwissenschaft Klasse IIa, pp. 125–138.

Tieh, T.T., Berg, R.R., Popp, R.K., Brasher, J.E., Pike, J.D., 1986. Deposition and diagenesis of upper miocene arkoses, Yowlumne and Rio Viejo fields, Kern County, California. AAPG Bull. 70, 953–969.

Tilley, B.J., Longstaffe, F.J., 1989. Diagenesis and isotopic evolution of porewaters in the Alberta Deep Basin: the Falher member and cadomin formation. Geochim. Cosmochim. Acta 53, 2529–2546.

Tobin, R.C., Schwarzer, D., 2014. Effects of sandstone provenance on reservoir quality preservation in the deep subsurface: experimental modelling of deepwater sand in the Gulf of Mexico. Special Publication 386. In: Scott, R.A., Smyth, H.R., Morton, A.C., Richardson, N. (Eds.), Sediment Provenance Studies in Hydrocarbon Exploration and Production. Geological Society of London, pp. 27–e47, https://doi.org/10.1144/SP386.17.

Tobin, R.C., McClain, T., Lieber, R.B., Ozkan, A., Banfield, L.A., Marchand, A.M.E., McRae, L.E., 2010. Reservoir quality modeling of tight-gas sands in Wamsutter field: integration of diagenesis, petroleum systems, and production data. Am. Assoc. Pet. Geol. Bull. 94, 1229–e1266. https://doi.org/10.1306/04211009140.

Toussaint, R., Aharonov, E., Koehn, D., Gratier, J.-P., Ebner, M., Baud, P., Rolland, A., Renard, F., 2018. Stylolites: a review. J. Struct. Geol. 114, 163–195. https://doi.org/10.1016/j.jsg.2018.05.003.

Velde, B., Vasseur, G., 1992. Estimation of the diagenetic smectite-to-illite transformation in time-temperature space. Am. Mineral. 77, 967–976.

Walderhaug, O., 1994a. Precipitation rates for quartz cements in sandstones determined by fluid inclusion microthermometry and temperature-history modelling. J. Sediment. Res. 64, 324–333.

Walderhaug, O., 1994b. Temperatures of quartz cementation in Jurassic sandstones from the Norwegian continental shelf evidence from fluid inclusions. J. Sediment. Petrol. 64, 311–323.

Walderhaug, O., 1996. Kinetic modeling of quartz cementation and porosity loss in deeply buried sandstone reservoirs. Am. Assoc. Pet. Geol. Bull. 80, 731–e745.

Walderhaug, O., 2000. Modeling quartz cementation and porosity in Middle Jurassic Brent Group sandstones of the Kvitebjørn field, northern North Sea. Am. Assoc. Pet. Geol. Bull. 84, 1325–1339.

Walderhaug, O., Bjørkum, P.A., 1998. Calcite cement in shallow marine sandstones: growth mechanisms and geometry. In: Morad, S. (Ed.), Carbonate Cementation in Sandstones. vol. 26. International Association of Sedimentologists, pp. 179–192. Special Publication.

Walderhaug, O., Bjørkum, P.A., 2003. The effect of stylolite spacing on quartz cementation in the Lower Jurassic Stø formation, southern Barents Sea. J. Sediment. Res. 73, 146–156. https://doi.org/10.2110/jsr.2006.015.

Walderhaug, O., Porten, K.W., 2007. Stability of detrital heavy minerals on the Norwegian continental shelf as a function of depth and temperature. J. Sediment. Res. 77, 992–1002.

Walderhaug, O., Lander, R.H., Bjørkum, P.A., Oelkers, E.H., Bjørlykke, K., Nadeau, P., 2000. Modelling quartz cementation and porosity in reservoir sandstones: examples from the Norwegian continental shelf. In: Worden, R.H., Morad, S. (Eds.), Quartz Cementation in Sandstones. vol. 29. International Association of Sedimentologists, pp. 39–49. Special publication.

Walderhaug, O., Oelkers, E.H., Bjørkum, P.A., 2004. An analysis of the roles of stress, temperature and pH in chemical compaction of sandstones – discussion. J. Sediment. Res. 74, 447–449.

Walderhaug, O., Bjørkum, P.A., Aase, N.E., 2006. Kaolin-coating of stylolites, effect on quartz cementation and general implications for dissolution at mineral interfaces. J. Sediment. Res. 76, 234–243.

Weyl, P.K., 1959. Pressure solution and the force of crystallization—a phenomenological theory. J. Geophys. Res. 64, 2001–2025.

Wilson, M.D., 1992. Inherited grain-rimming clays in sandstones from eolian and shelf environments: their origin and control on reservoir properties. In: Houseknecht, D.W., Pittman, E.D. (Eds.), Origin, Diagenesis, and Petrophysics of Clay Minerals in Sandstones. 47. SEPM, pp. 209–225. Special Publication.

Wooldridge, L.J., Worden, R.H., Griffiths, J., Utley, J.E.P., 2017. Clay-coated sand grains in petroleum reservoirs: understanding their distribution via a modern analogue. J. Sediment. Res. 87, 338–352.

Wooldridge, L.J., Worden, R.H., Griffiths, J., Utley, J.E.P., Thompson, A., 2018. The origin of clay-coated sand grains and sediment heterogeneity in tidal flats. Sediment. Geol. 373, 191–209.

Worden, R.H., Barclay, S.A., 2000. Internally-sourced quartz cement due to externally-derived $CO_2$ in sub-arkosic sandstones, North Sea. J. Geochem. Explor. 69–70, 645–649.

Worden, R., Armitage, P., Butcher, A., Churchill, J., Csoma, A., Hollis, C., Lander, R., Omma, J., 2018. Petroleum reservoir quality prediction: overview and contrasting approaches from sandstone and carbonate communities. In: Armitage, P.J., Butcher, A., Churchill, J., Csoma, A., Hollis, C., Lander, R. H., Omma, J., Worden, R.H. (Eds.), Reservoir Quality Prediction in Sandstones and Carbonates. 435, pp. 1–31. Geol. Soc. London spec. Publ.

Worden, R.H., Griffiths, J., Wooldridge, L.J., Utleya, E.P., Lawan, A.Y., Muhammeda, D.D., Simona, N., Armitage, P.J., 2020. Chlorite in sandstones. Earth Sci. Rev. 204, 103105. https://doi.org/10.1016/j.earscirev.2020.103105.

Zhang, Y., Pe-Piper, G., Piper, D.J.W., 2015. How sandstone porosity and permeability vary with diagenetic minerals in the Scotian Basin, offshore eastern Canada: implications for reservoir quality. Mar. Pet. Geol. 63, 28–45. https://doi.org/10.1016/j.marpetgeo.2015.02.007.

Chapter 14

# Applied paleontology in exploration and development

**Ryan Weber and David Jutson**
*Paleo-Data Inc., New Orleans, LA, United States*

## Introduction

Paleontology has likely been around since the pre-Christian era with Greek, Roman and Chinese philosophers recognizing stones embedded in rock resembled shellfish in the sea. Applications emerged with Nicholas Steno, Leonardo Da Vinci and William Smith who developed the understanding that specific types of fossils seemed to be associated with certain layers of rock. In fact, these naturalists began to relatively correlate rock layers, or strata, using the fossils contained within them. This gave rise to a method termed *biostratigraphy*. Although early biostratigraphers primarily utilized macrofossils to correlate, microfossils (i.e., those fossils that are examined with the aid of a microscope) were discovered to survive the destructive rigors of the rock drilling process through the collection of cuttings. Cuttings are the small pieces of rock or mud broken by the drill bit and transported to the surface by the fluid that lubricates the drill bit and removes the cut material from the bottom of the drill hole (Fig. 1). The fossilized remains of very small organisms remain preserved in these cuttings and have been utilized for correlation in boreholes since the mid-19th century in Polish Oil Fields (Payne et al., 1999).

Many different types of organisms and their fossilized remains can be utilized for biostratigraphy and paleoenvironmental reconstructions. The applicability and utility of a specific discipline of microfossil depends on the geological age and depositional environment of the strata penetrated. Due to the provenance of sediment that arrives in the deepwater, two assemblages can exist. The in-situ assemblage consists of those marine organisms that live on the sea surface or seafloor and leave fossilized remains syndepostionally with the sediments surrounding the fossils. The reworked assemblage consists of those marine organisms that lived in another environment or geological time that has been remobilized via sedimentation (i.e., turbidites) or other geological phenomena (i.e., slumping or salt movement). To understand the observed assemblage, which is often a mix of in-situ and reworked fossils, deepwater biostratigraphers often have to be versed in all types of fossils, even though it may take a lifetime to reach expert taxonomy of a specific discipline. *Secondarily*, the preservation of the original mineralization of the strata and its burial history plays a significant role in the utilization of microfossils (Fleisher and Lane, 1999). Therefore, the main microfossil groups utilized in deepwater settings are organized based on the original mineralization of the organism.

The first group to discuss is broadly termed calcareous microfossils. This includes any microscopic organism that utilized some form of calcium carbonate to produce their skeleton, test or shell. Calcareous nannofossils include nannoliths (discoasters, pentaliths, sphenoliths, etc.) and coccoliths/coccospheres, which are the remains of a single-celled Protist or phytoplankton termed Coccolithophores. Coccolithophore literally means "seed-stone bearing" in classical Greek. The organisms float and live in the photic zone of oceans, sometimes called the Planktic environment. Coccolithophores secret plates, or scales, of calcium carbonate-based minerals to form a sphere around its single cell. This cell can be 10–20 μm in diameter, so the skeletal plates are very small, possibly as small as 1 μm. Coccolithophores make up for their small size with their abundance. They bloom during peak conditions in so much quantity that these blooms can be easily visible from a geostationary satellite in space. Coccolithophores are the primary contributors to calcium carbonate mineral deposition on Earth and are often very abundant in marine mudrocks. Because coccolithophores are so well-preserved, widespread and rapidly evolving their fossils make for excellent biostratigraphic utility. Plus, they are easily prepared with a simple smear slide. However, nannofossils are limited to Triassic and younger rocks, as well as, marine rocks that were deposited above the carbonate compensation depth (the depth at which water depth at which the rate of supply of calcium carbonate from the surface is equal to the rate of dissolution—see chapter 6). Finally, because of their small size, nannofossils can be susceptible to reworking into newer sediment or contamination during the drilling or laboratory processes.

**Deepwater Sedimentary Systems.** https://doi.org/10.1016/B978-0-323-91918-0.00007-4

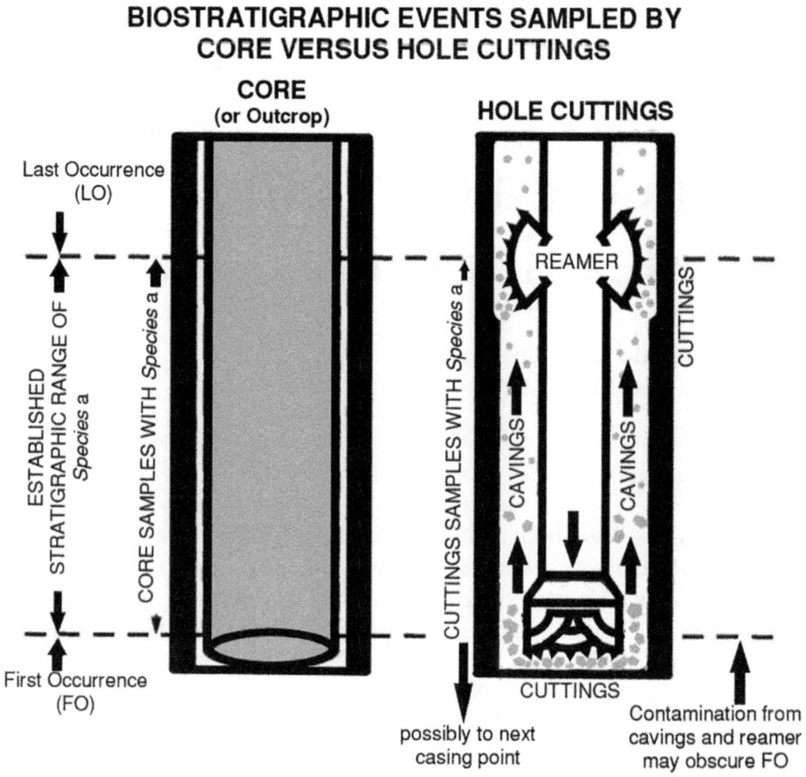


**FIG. 1** A comparison of bioevents observed in samples from a rock core compared to samples from drill cuttings. First occurrence (or origination) events can be obscured due to the nature of circulated cuttings in a wellbore compare to a rock core where the depth of the event is known. Some drill stems have a reamer on the same drill stem as the drill bit, and can further obfuscate bioevents. Comparing biostratigraphic data with other data, like gamma-ray logs, helps build confidence in sample integrity.

Foraminifera are also mostly calcareous including tests (or shells) from amoeboid protists and have survived since the Cambrian. There are three main groups of Foraminifera: the planktic, calcareous benthic and agglutinating benthic. Planktic foraminifera secrete their test from calcium carbonate and float and move around on the sea surface making them quite cosmopolitan much like nannofossils. Calcareous benthic foraminifera also secrete a calcareous test, and live on seafloor. Calcareous benthic foraminifera are environmentally sensitive occupying specific niches dependent on nutrient supply and salinity. These dependencies give fossilized calcareous benthic foraminifera a unique opportunity to recreate seafloor paleobathymetry and paleoenvironments. Agglutinating foraminifera also live on the seafloor and secrete glue to cement surrounding sediments to create their test. Agglutinating foraminifera are also environmentally sensitive, providing insight into the substrate at the time of deposition as well as grossly inferring the rate of deposition. Infaunal (burrowing) species tend to be elongated and smooth to travel into the substrate, whereas epifaunal (surface-dwelling) species tend to be more globular with flatter sides so as to allow easier motion over the substrate (Iaccarino et al., 2007). Laboratory preparation is relatively easy with disaggregating the mud for the silts/sands and sorting through sets of sieves down to 63-μm size fractionation. The relatively large size of the calcareous foraminifera (both Planktic and benthic) allows to specifically sample carbonate that records the ratio of stable isotopes (i.e., carbon and oxygen) in the geological past.

Another calcareous fossil that might prove some utility in the deepwater are the Ostracods, which are a bivalve-like shells of a shrimp-like crustacean. Ostracods are mostly benthic organisms thriving in a range of salinity settings from freshwater to fully marine. Marine ostracods exist, but not in the abundance of nannofossils or foraminifera. Ostracods have been found in sediments dating back to the Ordovician and are also environmentally sensitive to nutrient supply and salinity making them a good candidate for paleoenvironmetal studies. The significance of ostracods in the deepwater would be during times of high carbonate production and insight into sediment provenance.

Moving on to an allied discipline, palynology or the study of fossilized plant-produced organic material. This discipline focuses on a variety of organism groups such as dinoflagellates, pollen, spores, acritarchs and chitinozoans. Dinoflagellates are the cystic remains (dinocysts) of single-celled autotrophic or heterotrophic protists, which evolved in the Triassic. Interestingly, only 10% of living dinoflagellates have a cystic phase, which is the phase most likely to be preserved as microfossils. The majority of dinocysts tend to be deposited in the neritic or bathyal environments. Pollen and spores are the gametal remains from gymnosperm and angiosperm plants: fungi are first recorded from the basal Paleozoic, and bacteria and algae evolved in the Proterozoic. This group is largely nonmarine or lacustrine and can be deposited in deepwater settings in large abundances giving us some insight into sediment provenance. Acritarchs and Chitinozoans (flask-shaped) are organic remains of uncertain origins during the Proterozoic to Permian. Current research suggests that Acritarchs and Chitinozoans may be algal cystic remains. The advantage to palynology is that this group is preserved when calcareous fossils are not. These organisms are also environmentally sensitive, which gives an opportunity to infer paleoenvironment or paleoecology. Organic fossils due to their color sensitivity to heat can give an excellent insight into thermal maturity and burial history but can be preservationally challenged at high temperatures. Historically, palynology has had a lab-intense preparation using strong acids to dissolve many minerals in the rock samples to concentrate enough organic material to speed up the analysis (Fleisher and Lane, 1999). Recently, many new nonacid methods have emerged to reduce the laboratory risk (Halbritter et al., 2018).

Siliceous fossils include radiolarian and diatoms. Radiolaria are buoyancy-enhancing, intricate porcelanite skeletons of nonmotile protozoa zooplankton. Diatoms have a siliceous skeleton (called frustules) remains of photosynthesizing algae. Diatoms can be planktic and benthic, freshwater to marine, whereas radiolaria are marine and planktic. The size range varies from 20 to 200 μm. The siliceous group tends to outcompete calcareous groups in higher latitudes due to the oceanic solubility of silica and carbonate. The use of the siliceous group can be lab-intense to concentrate the fossils for analysis. Preservation can be a challenge due to the shallow burial depth of porcellanite to quartz. Remineralization above this depth helps the preservation.

Finally, Conodonts can provide some utility in Proterozoic and Paleozoic rocks. Conodonts are originally phosphatic tooth-like feeding apparatus of an extinct vertebrate, similar to modern-day lamprey. These organisms probably lived in the water column with high diversity and nutrient supply. Conodonts are usually 100–2000 μm, so are larger than the previously discussed fossils.

In all, the application of certain disciplines of fossils in deepwater sediments depends on the temporal and geographic distribution of the sediment provenance (summarized in Table 1). Preservation and burial history are also important considerations when applying micropaleontology. In all, most basins collect and analyze data from the calcareous groups and palynology due to their cosmopolitan distribution, good preservation in deepwater settings and more commercially friendly (see Fleisher and Lane, 1999 for additional detailed info on microfossil disciplines).

The fundamental unit that paleontologists and biostratigraphers use to communicate is the "fossil species." In biology, the definition of a "species" is any organism that can procreate. In the fossil record, paleontologists often do not have the luxury of knowing whether organisms could procreate. Therefore, paleontologists often use other criteria, like skeletal morphology or geological observations, to define a "fossil species." Fossil species are given names and references according to the same code (International Code of Botanical Nomenclature) as other scientific disciplines. Publishing, referencing and adhering to this code are essential to understanding the observations of the paleontologist.

## Logistics and methodologies

When it comes to commercial viability and value, logistics for sample collection, transport, laboratory preparation, analysis and reporting can be very challenging while exploring deepwater reservoirs. Many deepwater prospects and hydrocarbon fields are miles to hundreds of miles offshore and discovered by dynamically positioned vessels floating in the ocean. The same challenges exist when the target is beneath a land-based rig. The first challenge is collecting cuttings samples. As a borehole is drilled, a drill-bit cuts rock/sediments while injecting a fluid (often coined "drilling mud") to sweep the (now) open borehole from the material just cut by the bit, or cuttings. Ideally, the drilling fluid is a similar density as the formation so that the borehole remains open. That fluid is then circulated back up to the surface within the space outside of the drill stem and borehole wall. Once the fluid filled with cuttings, breaches the surface it is run through a set of screens meant to capture the cuttings so that the drilling fluid can be reused and circulated back down the drill stem. The screens (in shaking machines called "shakers") allow for the collection of samples. Utilizing pump stroke volumes, calculation of lag time to depth is possible. Lag times can be 4–6 h depending on the depth and hole size of the borehole. Therefore, some errors can occur in the depth of cuttings, however, mud loggers and engineers have greatly improved these calculations (Fig. 1).

Once samples are collected, some laboratory techniques are utilized to ease the analytical process. Transporting samples to a lab can be the largest source of time from shaker to report. Offshore vessels may not have daily transport available, so sometimes a portable lab and analysts are sent to the rig. Either way, samples are processed and examined under a microscope. Most modern analytical techniques capture abundance distribution data sample by sample. This is critically important to curate data as samples can often deteriorate, and curating samples through climate-controlled environments is often deemed noncommercially valuable. Therefore, collecting abundance data with up-to-date taxonomic frameworks is critically important to the long-term value of the data collected, so that the next generation with new context and a fresh perspective still finds value in the observations.

**TABLE 1** Summary of Microfossil types and factors to consider when choosing the utility of a specific fossil type.

| Microfossil type | | | Structural composition | Stratigraphic range | Environmental range | Main utility | Preparation |
|---|---|---|---|---|---|---|---|
| Foraminifera | Benthic | Calcareous | Calcium carbonate | Cambrian—recent | Brackish—fully marine—hypersaline | Local biostratigraphic markers paleoenvironmental indicators | Washing and boiling in hydrogen peroxide (6%) |
| | | Agglutinated | Clastic particles organic or carbonate wall | Cambrian—recent | Marsh—estuarine/ brackish—fully marine | Local biostratigraphic markers paleoenvironmental indicators | Washed through a 63 μ. sieve |
| | Planktic | | Calcium carbonate | M. Jurassic—recent | Fully marine planktic | Regional biostratigraphic markers paleodepth indicators with benthics | Dried and sieved into standard size fractions<br>In hard rocks—thin sections |
| Ostracods | | | Calcium carbonate | Cambrian/? Ordovician—recent | Freshwater—fully marine mainly benthic | Local biostratigraphic markers | |
| Calcareous nannofossils | | | Calcium carbonate | Late Triassic—recent | Fully marine planktic rare at depth shallower than middle shelf | Regional biostratigraphic markers | Smear slides samples |
| Radiolaria | | | Silica | Ltst Precambrian—recent | Fully marine planktic | Local and regional biostratigraphic markers | As for foraminifera but with a weak wash (10%) of hydrofluoric acid |
| Diatoms | | | Silica | Jurassic?/Early Cretaceous—recent | Freshwater to marine | Regional biostratigraphic markers | Smear of wet samples acid cleansing to remove calcium carbonate |
| Pollen | | | Sporopollenin | Angiosperm: Cret.—recent Gymnosperm: Lt. Dev.—recent | Terrestrial and "pseudoplanktic" | Paleoenvironmental and stratigraphic markers thermal maturation studies | Acid maceration |

| | | | | | |
|---|---|---|---|---|---|
| Spores | Sporopollenin polysaccharide | Fungal: Precamb.—Recent Plant: Lt. Silurian—recent | Terrestrial and "pseudoplanktic" | Paleoenvironmental markers thermal maturation studies | Acid maceration |
| Dinoflagellates | Organic wall related to Sporopollenin | Silur.?/Triassic—recent | Freshwater, mainly marine rare in open ocean | Regional biostratigraphic markers | Acid maceration |
| Achritarchs | Organic wall | Precambrian (mainly)—recent | ?Planktic | Regional biostratigraphic markers frequently the only fossil in sands | Acid maceration |
| Chitinozoans | Pseudochitinous material | Ordovician—Devonian | Fully marine | Regional biostratigraphic markers | Acid maceration |
| Conodonts | Calcium carbonate—Fluorapatite with organic matter | Precambrian—Lt. Triassic | Fully marine | Regional biostratigraphic markers | Acid dissolution of the matrix |

From the abundance distribution data, a biostratigrapher will utilize a "range" chart or a sample-by-sample listing to create observations of bioevents. Those bioevents are the main system to correlate across regional basins and sometimes globally. Many operators and consultancies put forth much effort in identifying these bioevents and determining the relative age in millions of years or mega annums (Ma). Nonetheless, bioevents are the main observation posted on a paleontological or biostratigraphically report. As these bioevents relate to many global geological cyclic events, geologists and geophysicists integrate observations from other data into the context of geological time. Hence, biostratigraphy is like one's watch or calendar. Biostratigraphy adds the context of time to the rocks/sediments which can be then be inferred to other geological events; either globally or regionally.

Biostratigraphic resolution is often dependent on sample collection. Samples are traditionally caught at thirty (30′) foot intervals. This is probably an artifact from catching samples per stand when "Kelly-Down" was the most accurate method for circulation-lag depth calculation. With recent advancements in automated pump-stroke volumetric calculations, samples can be accurately depth calculated within one (1′) foot. However, uncertainty at the shaker may cause ten (10′) samples the most reasonable maximum sample resolution (Fig. 1).

Biostratigraphic resolution is also dependent on data density, or the true and accurate subdivision of a stratigraphic sequence based on one or more fossils groups. This is largely depending on sedimentation rate. With a low sedimentation rate, fossil abundance may be high because there's less clastic sediment diluting the pelagic rain. Fine-scale biostratigraphy may be obscured by bioturbation and microfossils can dissolve on the sediment–water interface or primary diagenesis. In this scenario, bioevents appear compressed with respect to depth and can create apparent gaps in geological age. With a high sedimentation rate, fossil abundance is usually diluted by terrigenous clastic input which makes rare species even more difficult to observe. In this scenario, bioturbation is not as important. Delicate microfossils are rapidly buried prohibiting primary diagenesis. Bioevents appear expanded with respect to depth with many samples between bioevents. In age vs. depth plots, the slope of the line between bioevents represents accumulation rate and is a good proxy for expected sedimentary environments in clastic settings.

## Applied biostratigraphy

Even after over a century of application, biostratigraphy is still primarily a correlation tool. It provides the primary framework for regional stratigraphy in expanded passive margins, like the Gulf of Mexico or North Sea. Expanded passive margins tend to be a monotonous sequence of sand and shale with very little mineralogical difference. Bioevents or "Tops", are typically posted on logs and seismic as a basis for correlation and trend evaluation. Biostratigraphy is used as the primary tool to calibrate logs and seismic.

Once paleontological observations are collected and bioevents inferred, the context of geological time and other global or regional events can be understood. Within the space (or depth) between these bioevents can be further subdivided into a zone, because abundance distribution data is collected. There are many other observations that can be gleaned for abundances due to the environmental sensitivities of certain fossils/organisms. For example, the abundance of certain benthic foraminifera can infer paleobathymetry or depositional environment. The abundance of certain nannofossils, planktic foraminifera or dinocysts can infer surface water conditions or paleo-temperature. Since the data is usually collected with respect to depth, a record of geographical and depositional history is recorded.

When using bioevents from fossils collected in the muddy system of deepwater turbidites or pelagic input, it is important to remember that the data is unique to lithology. That's not to say lithology can affect abundance as many planktic fossils are not observed in coarse lithologies possibly due to sediment dilution, ablation during deposition or secondary diagenesis during fluid migration. Nonetheless, the principle of faunal succession in a regional sense is irrespective of localized lithological deposition. This means that if one coarse-grained package is a different geological age from another, the likelihood that they correlate or be connected is low.

Bioevents can be anything unique and repeatable in the faunal succession. As most boreholes are drilled from top-down and positive observations (e.g., a tangible appearance of something new) are quicker to identify, the extinction event of a particular fossil is often used. The extinction event has the appearance of a top, sometimes called the highest occurrence (HO), last appearance datum (LAD) or first-downhole occurrence (FDO). Positive observations can also include an increase in abundance, an Acme (peak abundance) and coiling changes. A coiling change event is some planktic foraminifera change their coiling direction (sinistrally or dextrally; clockwise or counter-clockwise) coincidentally with glacial cycles. It is thought foraminifera do this in response to temperature or salinity changes in sea surface conditions, but the true reason is still unknown. Negative observations (e.g., the lack of evidence) in a drilling borehole include the origination of a fossil species. These are sometimes called a base, first occurrence (FO), first appearance datum (FAD) or lowest down-hole occurrence (LDO). This is a bioevent in geological time where a new species first appears. Negative observations can be challenging to interpret as contamination is possible, especially when collecting cuttings from an open borehole. However, negative observations become important when trying to decipher depositional reworking from salt tectonics or borehole contamination.

A single species can have multiple bioevents during its' geological lifespan. The example used most often is a planktic foraminifera called *Globorotalia menardii*. This species has a useful top or HO, a highest persistent occurrence (HPO) where it is observed in almost every sample, an abundance Acme, a coiling direction change, it disappears for a while and comes back as a second HO. Finally, it also has an important persistent LO, and a rare LO. Now, paleontologists can observe 200–300 species in a typical borehole for a single fossil discipline. This means a borehole could have 600–900 columns of data to investigate bioevents with multiple fossil types (coccolithophore, foraminifera and dinocysts).

Correlation with biostratigraphy is usually in the form of horizons (Fig. 2). Two wells had been drilled close to each other in the northern North Sea. Both wells included sands with significant oil in the Eocene/Paleocene section. The production team planned to drill one well between two wells to produce oil as it was thought that through log correlation, the sands were connected (Fig. 2A). To make sure of this, they ordered a biostratigraphic study of the oil-bearing interval. The disciplines chosen were microfauna and palynology. The result (Fig. 2B) showed that by correlation points defined by biostratigraphic events, the two sands were not connected and would require a significantly different production design than had first been considered. The cost of the biostratigraphic study was a few thousand euros but without this study, the cost of the failed production without this study would have been many millions of euros.

Unfortunately, nature is not layer cake as geological events have some sort of localization to it, especially within a particular basin. Often, a set of bioevents converge within a geological horizon. For example, the Pliocene in the Gulf of Mexico onlaps onto the Miocene, sometimes nonconformably, which creates an apparent condensed section or unconformity. This means that the geologic horizon "Top of Miocene" will appear as a different age across the region. This is a critical phenomenon to address when integrating biostratigraphical reports.

Here is the recommended process to interpret a biostratigraphical report.

1. Identify the major age changes in the entire borehole
   a. Hypothesize the geological phenomenon that may cause these major age changes
   b. Integrate with other data to recognize changes (lithology in logs or onlapping in seismic)
2. Compare the age jumps (and corresponding data) to draw correlations between boreholes
3. Read qualifying comments on confidence and reliability of a bioevent/horizon
   a. Check the detailed abundance of the fossil species to gauge persistence and abundance
   b. Gauge supporting evidence from other fossil disciplines.
4. Only if the above steps have succeeded, then proceed to identify local or subregional events or assemblage changes.
   a. Sequence biostratigraphy (explained later)
   b. Faunal changes due to rise/fall of sea level (paleobathymetry) or deposition (agglutinating foraminifera)

Between the major horizons and bioevents, paleoecological information can be gleaned from fossil distribution data. Historically, the most used information is from benthic foraminifera, whose distribution is based on bottom water conditions. Variables like physical substrate, benthic water chemistry, salinity, nutrient supply and sedimentation rate can greatly affect the distribution of benthic foraminifera. These variables share a general relationship with water depth, due to the stratified nature of the world's oceans (Fig. 3). Biostratigraphers utilize the upper depth limit of certain benthic foraminifera to create an Ecological Zonation or Ecozones with 1 being the shallowest water depth and 6 or 7 being the deepest water depth. Various interpretational approaches exist between operators and consultancies, especially on the emphasis on certain taxonomic criteria. Some interpreters place emphasis on the deepest water indicator while others average assemblage based on abundance. In addition, a variable amount of down-dip transported fauna can give a shallower interpretation than the true actual water depth. In bathyal or abyssal environments, it can be useful to plot shelfal, neritic or littoral ecozones as a good estimation of the qualitative amount of down-dip transport and possibly identify mass-transport complexes or deposits (Fig. 6).

With a compendium of boreholes with interpreted ecozonations, paleobathymetry maps can be created to estimate shelf-edge locations in the geological past. Likewise, a comparison of ecozonations between bioevents can assess clastic progradation, aggradation or retrogradation. In complex basins, especially with mobile salt mini-basins, these ecozone maps can assess the geographical extent of down-dip (shelf or shallower) transport into the slope or bathyal environment. As an example, many foraminifera paleontologists observed bathyal assemblages along the Miocene Gulf of Mexico coastline. At the time, many biostratigraphers speculated a very large sea-level incursion was needed to explain these bathyal faunas. However, increased exploration illustrated that these bathyal assemblages were localized, not chronostratigraphically correlative, and distributed in wedge-like geography along the Louisiana Miocene coast. As exploration moved into deeper waters (see chapters 1, 19 and 20), expanded sections down-dip and chronostratigraphically correlative to these bathyal wedges were discovered with mixed bathyal or neritic assemblages. It is now hypothesized that these bathyal wedges are healed scars of failed shelves of sediments, meaning that a large sea-level incursion is no longer needed to explain the bathyal assemblage. This signifies that a large amount of the expanded sections in the deepwater has been remobilized and possibly redeposited, and making paleontological data more challenging to interpret and more insightful to the context of reservoir deposition and regional exploration.

Benthic foraminifera, especially agglutinating, also add the potential to grossly interpret lithological facies. Certain species prefer to live in/on a particular grain-size substrate or sediment accumulation rate. An index can be calculated between barrier-prone facies and reservoir-prone facies. Indices also exist with the planktic fossils/organisms. Nannofossils and dinocysts have indices for cold vs. warm, trophic levels and nearshore vs. offshore. Pollen and spores can be utilized to infer down-dip transport or fluvial input (Flores et al., 2005). If the geographical conditions allow, pollen and spores can

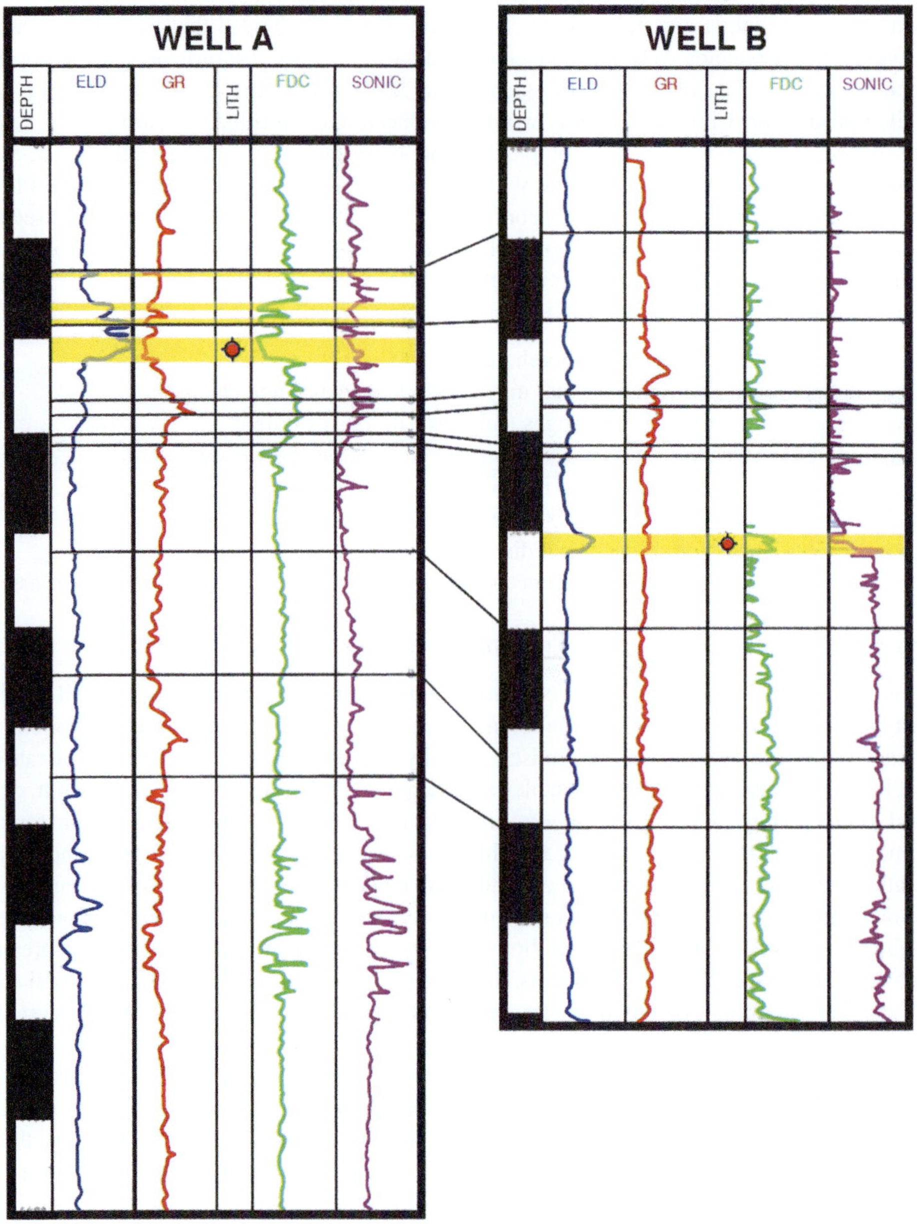


**FIG. 2** An example of well log correlations (A) pre-biostratigraphy and (B) post-biostratigraphy.

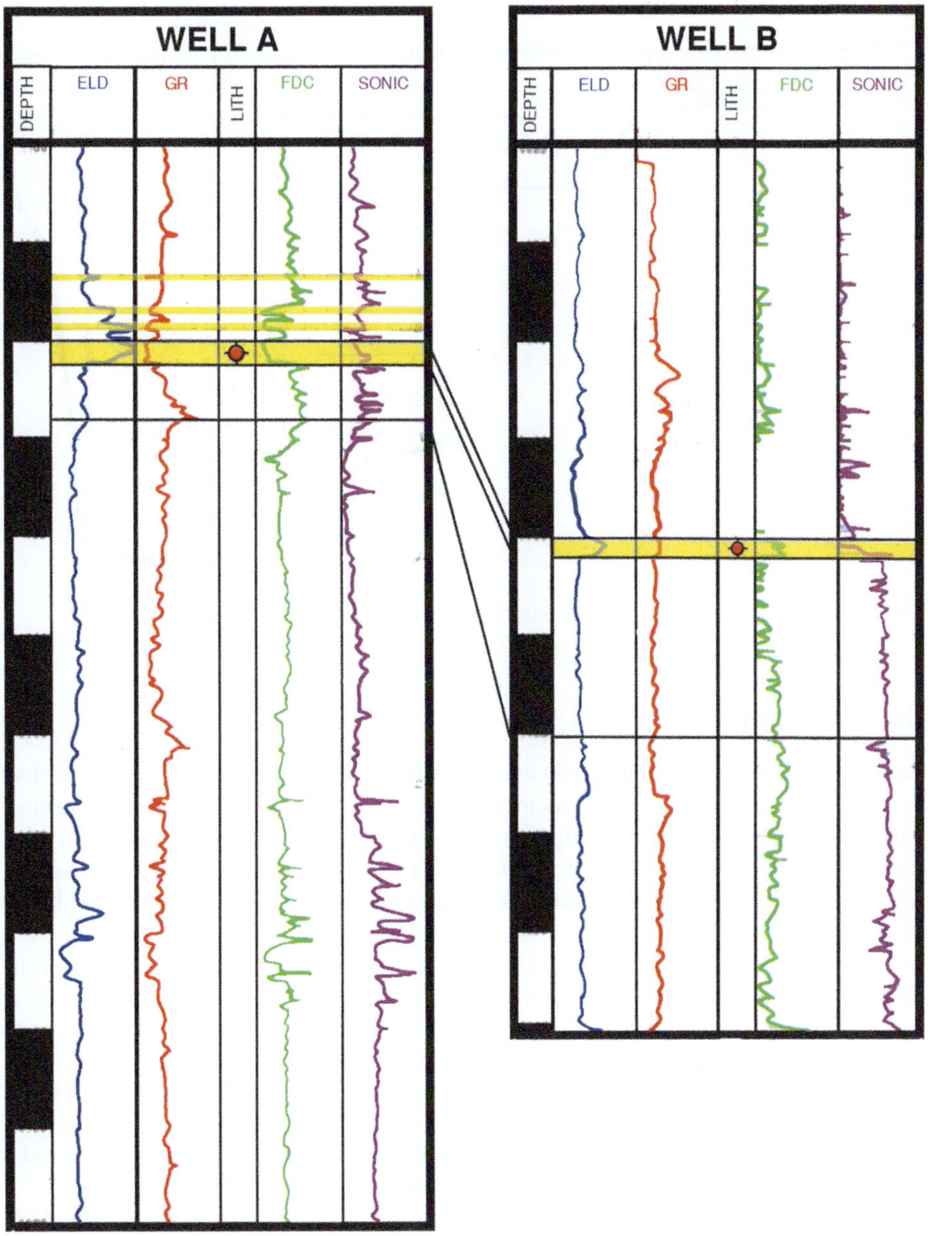


**FIG. 2, cont'd**

also be insightful to sediment transport. For example, Ponderosa Pine pollen is much more prevalent in the Eastern United States than the Juniper and Hickory dominance in the Western United States. These two areas also had independent river systems feeding sediment into the deepwater Gulf of Mexico and the provenance can be "unmixed" utilizing pollen analysis (Dickey and Yancey, 2010).

Similar to ecological facies, the ratio of stable isotopes is recorded in the sediments. Bulk sediment or bulk organic material has proven to provide a relative record of change in Carbon and Oxygen stable isotopes. Although it is better to analyze stable isotope ratios from the same mineralogy (i.e., foraminifera tests) between sample-to-sample; a record can still be obtained to create a much higher resolution of chronostratigraphy between bioevents and assemblages by comparing a borehole isotopic record with a world-wide standard isotopic record that has been calibrated to geological time. This has been extremely useful in Mesozoic records to correlate localized increases in total organic carbon relative to global Oceanic Anoxic Events (OAE—see chapter 2).

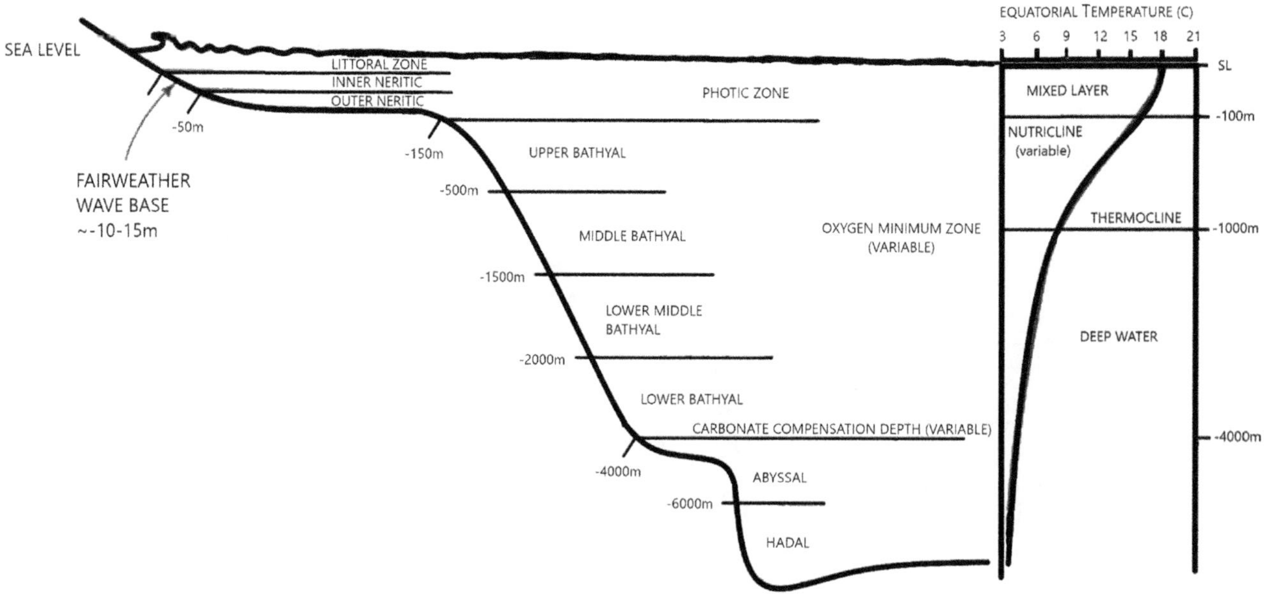


**FIG. 3** Example of oceanic microfauna environments and the relationship to water chemistry.

## Chronostratigraphy and geochronology

Major bioevents and horizons within a regional basin are summed up within a biostratigraphical age chart. Within the past few decades, many researchers have devoted careers to standardizing nomenclature and ages for geological events. Most notably is the International Commission on Stratigraphy, which is a collaboration of researchers who define precisely global units (systems, series and stages) that, in turn, are the basis for the units (periods, epochs and age) of the International Geological Time Scale; thus, setting global standards for the fundamental scale for expressing the history of the Earth. A subset of researchers expresses their case to place a golden spike to define a geologic boundary so that anyone in the world can collect data on that outcrop. From these golden spikes, geological time is calculated through numerous methods. Radiometric data, the idea of measuring parent-to-daughter ratio and radiometric decay to calculate origin timing is quite poplar. Magnetostratigraphy, the study of sea-floor spreading and Earth's magnetic reversals, has been very enlightening to geologic time scales. However, the science is now focused on relating any data (color, mineralogy, stable isotopes, etc.) to cycles related to orbital tuning or Milankovitch Cycles. Astronomers have now provided theoretical solutions to how the Earth's orbital has changed through time, which affects the insolation radiating on Earth, changing the climate on measurable scales. Many sections have been studied to discern cyclists and compared with the astronomical model to come up with geological age within 20–30,000 years of accuracy (Zalasiewicz et al., 2013).

Still, there are bioevents that are not observed globally. To calculate their relative age, biostratigraphers use mathematical processes, sometimes coined quantitative biostratigraphy. In industrial applications, the most common are graphic correlation and Ranking and Scaling. Graphic correlation is nonstatistical technique that establishes the coeval relationship between two sections by comparing the total stratigraphical ranges of taxa to easily compare accumulation rates and resolving correlative issues between two boreholes. Through graphic correlation on a significant number of boreholes, an optimum sequence or composite standard is obtained. Graphic correlation is an easy way to compare condensed sections, faults and unconformities among observed sections. The challenge becomes is that most of the geological time is not recorded in the rocks/sediments, so the optimal sequence becomes longer and longer as more boreholes are graphically correlated until the optimal sequence does not reflect reality. On the other hand, Ranking and Scaling is a statistical method that assumes depth as a reasonable proxy for geological time and scales the depths between bioevents from borehole to borehole. The result is a relative scale in depth for an optimum sequence and can be calibrated to time using known events from the previously discussed methods.

In exploration and production, having an accurate relative age (in Ma) becomes important when comparing local events with global phenomena, like sea level or relative sequences. Sequence stratigraphy has been a key to understanding shelf depositional processes, and many of the same biostratigraphical sequences observed on the shelf can be observed in the deepwater.

## Biostratigraphic sequences

Clastic deposition along a passive margin is often a monotonous sequence of sand and mud deposition as a result of deltaic and fluvial deposition changing with eustasy and accommodation. Sequence stratigraphy combines stratal architecture of sedimentary facies assumed to be the result of the repeated fill of accommodation (see chapter 12). Accommodation can be localized as opposed to the global change of eustasy. Organisms respond to these changes and are recorded in the rocks/ sediments. Utilizing abundances, indices and ecozones within a borehole can help to infer the vertical succession of systems tracts.

The most recognizable surface for biostratigraphers in a borehole is the Maximum Flooding Surface (MFS), or the depositional surface at the time sea level is at its highest point (Fig. 4). Planktic or pelagic fossils tend to be at the highest abundance while sea level is high, as the deepwater is starved of sediment. Planktic fossils have an inverse relationship with clastic supply, assuming that pelagic rain is relatively constant through geological time. In a typical borehole, look for maximum abundances of nannofossils, planktic foraminifera, dinocysts and radiolarians. Because bathyal sedimentation rate is low during higher sea levels, much geological time can be represented with little depth so the appearance of bioevents in relatively close depths may indicate condensed sections, which helps confirm an MFS. Likewise, a shift to deeper benthic foraminifera paleobathymetric ecozones would also be expected in a maximum flooding surface. High evolutionary turnover tends to occur in condensed sections (due to the apparent lack of time represented in the rock/sediment), so diversity indices (like Shannon or SHE functions) and similarity coefficients (like Cosine-Theta) can be useful to qualify surfaces. Peak numbers of diversity and low numbers in similarity (from the previous sample) usually denote a major oceanic change associated with maximum sea level.

The other end of the sequence spectrum is the Lowstand Systems Tract (LST). The biostratigraphic signal in an LST is marked by a low abundance of fossils or barren sand lithology (Fig. 5). Sometimes the LST may preserve a peak abundance of reworked material (i.e., older nannofossils, oxidized palynomorphs and neritic benthic foraminifera).

In upper and lower slope settings, the highstand systems tract (HST) is often not preserved or is very thin in the biostratigraphic. The subsequent Lowstand can overpower the HST signal or even erode it completely. The presence of clasts entrenched in coarse-grained material helps to confirm this hypothesis. The biostratigraphic signal of an HST should be an upward decrease in foraminifera and nannofossils, upward increase in epifaunal benthic foraminifera, reworked material (like older nannofossils, pollen, neritic foraminifera or detrital minerals). Finally, benthic foraminifera paleobathymetric ecozones should be shoal. In turbidite-dominant sedimentary environments, epifaunal agglutinating benthic foraminifera tend to increase in abundance near muddy-fan fringes, which would explain their abundance near the end of HSTs and at transgression initiation (Fig. 6).

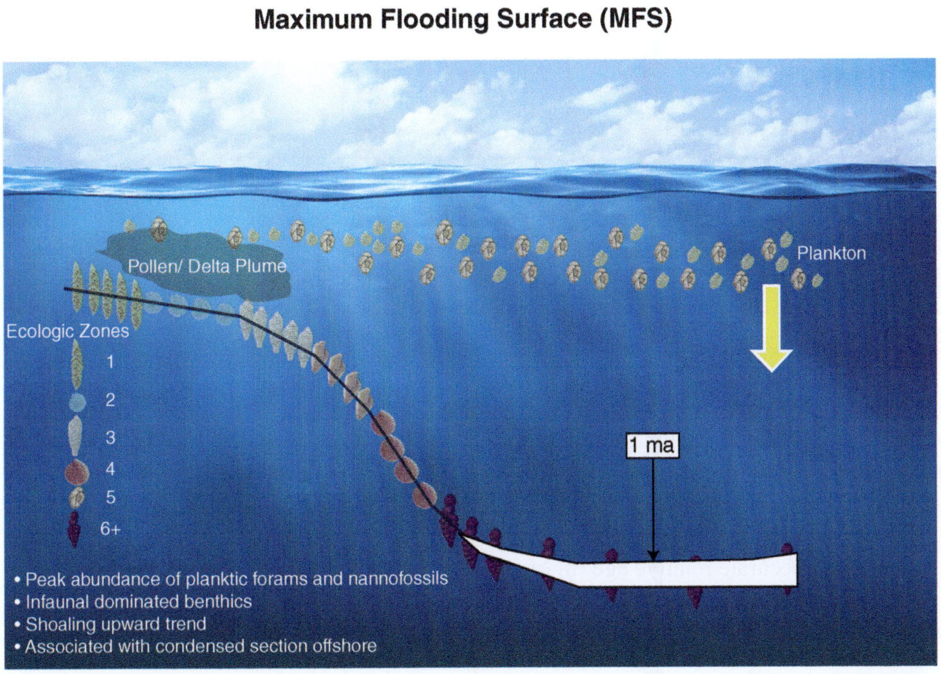


**FIG. 4** Hypothetical model of the deposition of microfossils during a maximum flooding surface. In this example, deepwater (bathyal to abyssal) sediments should be dominated by hemipelagics and fossils derived from planktic microfauna. A Mega-annum of sediment will be very thin as the blanket of hemipelagics fall to the seafloor.

FIG. 5 Hypothetical model of the deposition of microfossils during a lowstand period. In this example, deepwater (bathyal to abyssal) sediments will be diluted with deltaic and shelfal material. Fossils derived from shelfal calcareous benthic microfauna, pollen, spores and some dinoflagellates may dominate the muddy material. A Mega-annum of sediment will be thick as the sediments are deposited from shelf failures or some other process.

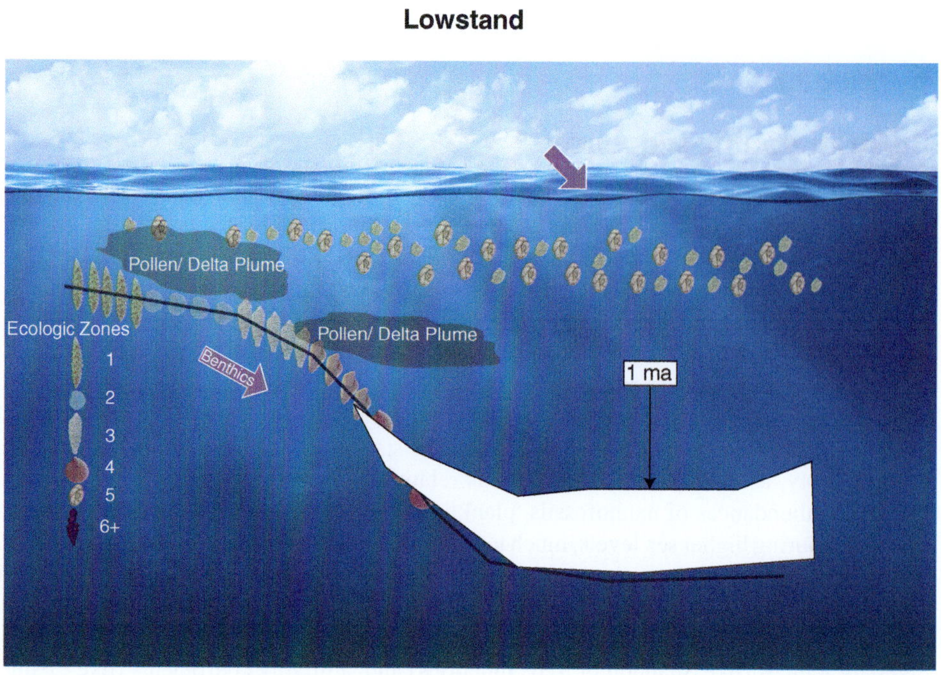


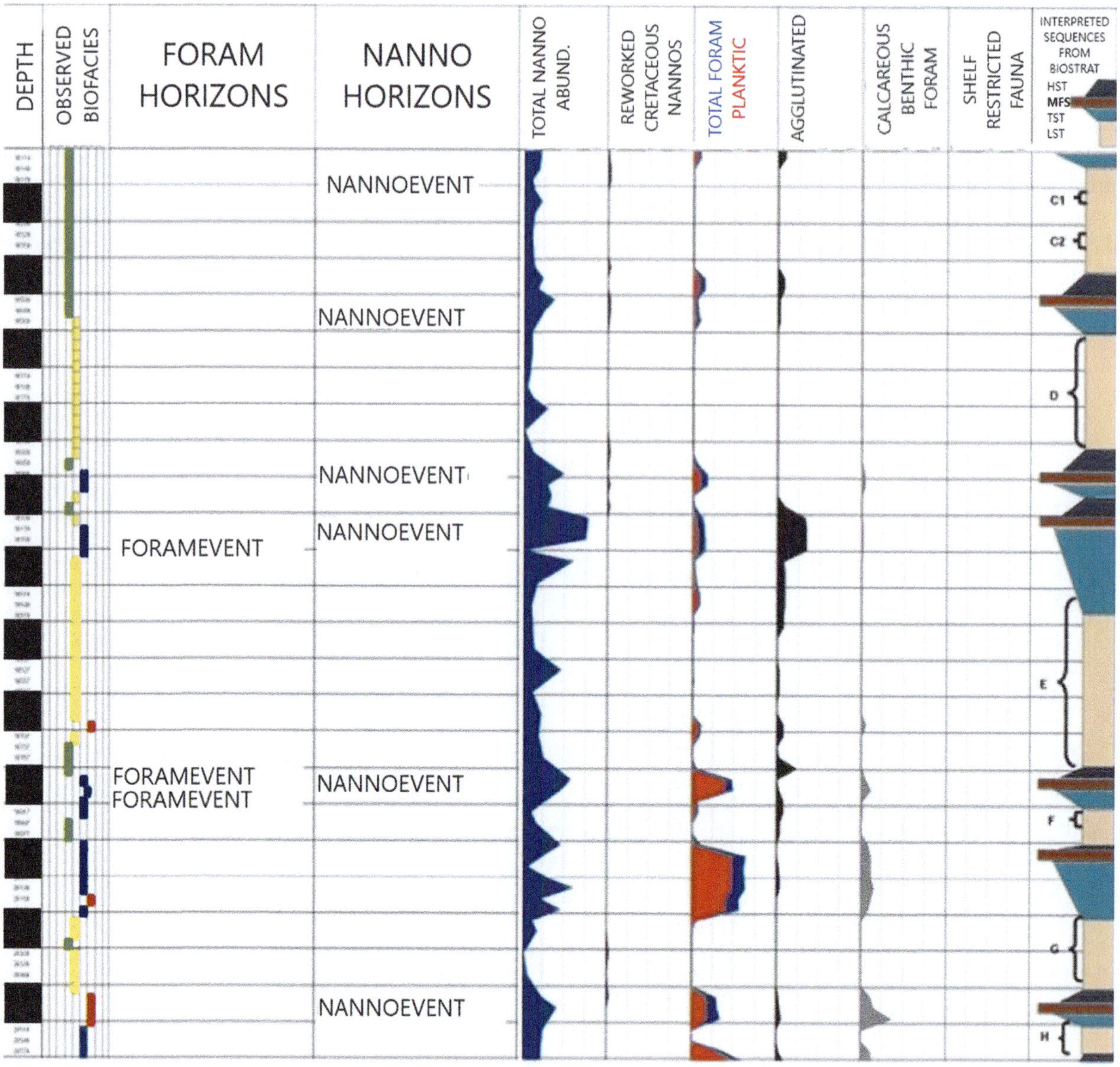


FIG. 6 An Gulf of Mexico example of sequence interpretation based on fossil abundances, observed biofacies or paleoenvironment and indices based on provenance (i.e., reworked cretaceous nannofossils and down-dip transported fauna).

The transgressive systems tract (TST), or the time of rising sea level, is seen in biostratigraphic data as an upward increase in planktic or pelagic fossils, infaunal benthic foraminifera and deeper benthic foraminifera paleobathymetric ecozones. Likewise, the TST should reflect an upward decrease in reworked material (older nannofossils, pollen, neritic foraminifera).

The challenge in bathyal and abyssal environments is that the main deposition for clastics is the interplay between pelagic rain and turbidites. Turbidites, by their nature, are remobilized sediment. Therefore, these signals can be muddled or mixed. For example, a mass transport deposit (MTD) could contain an MFS, which is often observed and creates a good hydrocarbon seal across a mini-basin. Turbidites and MTDs probably contain in situ faunas and floras deposited syn-depositionally with the generation of the MTD or turbidite (see chapters 6–11 for more information on these deposits). In major failures, older underlying sediment can be eroded and/or encapsulated with the in situ (syn-depositional) assemblage.

## Applications for deepwater exploration

Still, the primary tool for biostratigraphy is a correlation. Starting with mega-regional focus is to explore those mega sequences that have a history of reservoir prone lithologies. Age-depth plots utilizing biostratigraphic datums are a low-cost method to estimate accumulation rates (Figs. 7–9). Basins usually have some borehole information relative to age. A high slope in age vs. depth translates to large accumulation rates with a higher chance of reservoir-prone litho-facies while a low slope in age vs. depth translates to low accumulation rates with a higher chance of mud prone lithofacies. Flat slopes on age vs. depth usually denote nonconformable surfaces such as an unconformity or structural fault. The apparent age gaps and low slopes are good areas to evaluate seals and sealing capacity. The age-depth plot is also a cost-effective way to quickly compare numerous boreholes in one illustration to assess regional unconformities or seals (Fig. 8; Fillon, 2009; Crews et al., 2000). With recent technology, age-depth plots can be added to geophysical software to morph seismic data into a Wheeler diagram, which helps to visualize basin analyses and potential fluid migration (see chapters 2 and 18).

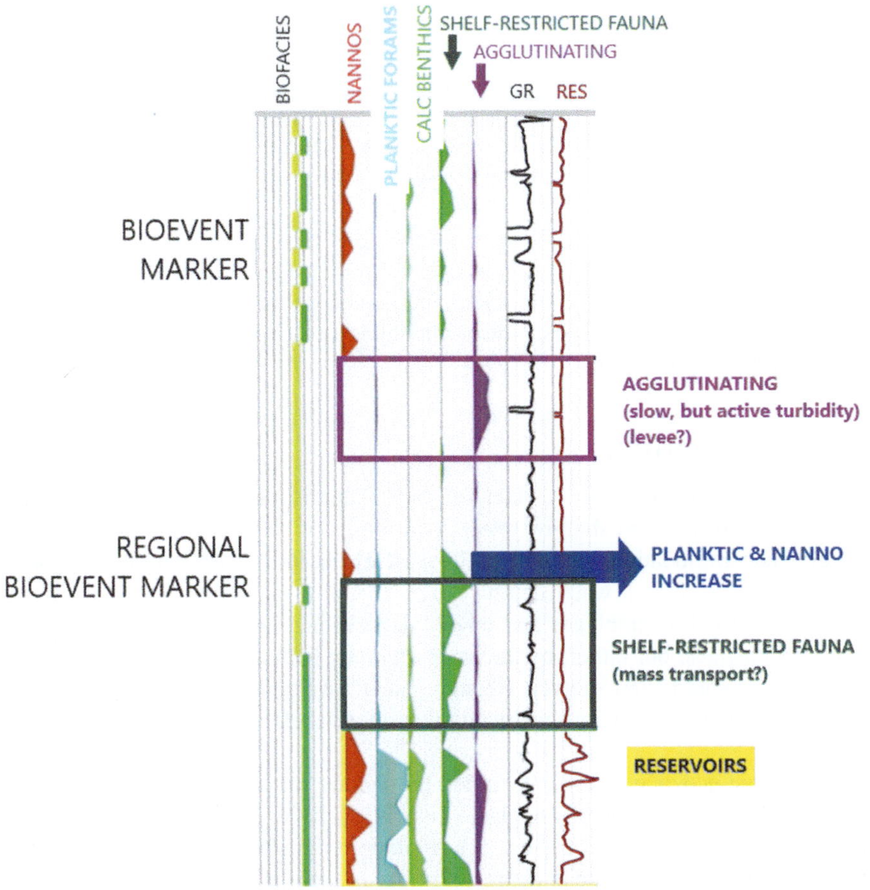


**FIG. 7** A deepwater Gulf of Mexico example of a hypothesized mass transport deposit (*green box*) based on fossil indices (shelf restricted fauna in *greencolumn*). In this example, this MTD likely remobilized a maximum flooding surface deposit as suggested by the high abundance of nannofossils (*red column*) and planktic foraminifera (*cyan column*) as well as a well-established chronostratigraphic bioevent. An increased abundance of agglutinating foraminifera above the hypothesized MTD (*violet box*) suggest slowed deposition (muddy or silty). A healed system (or another MTD) is suggested by the return of nannofossils and planktic forams near the top of this example (above *violet box*).

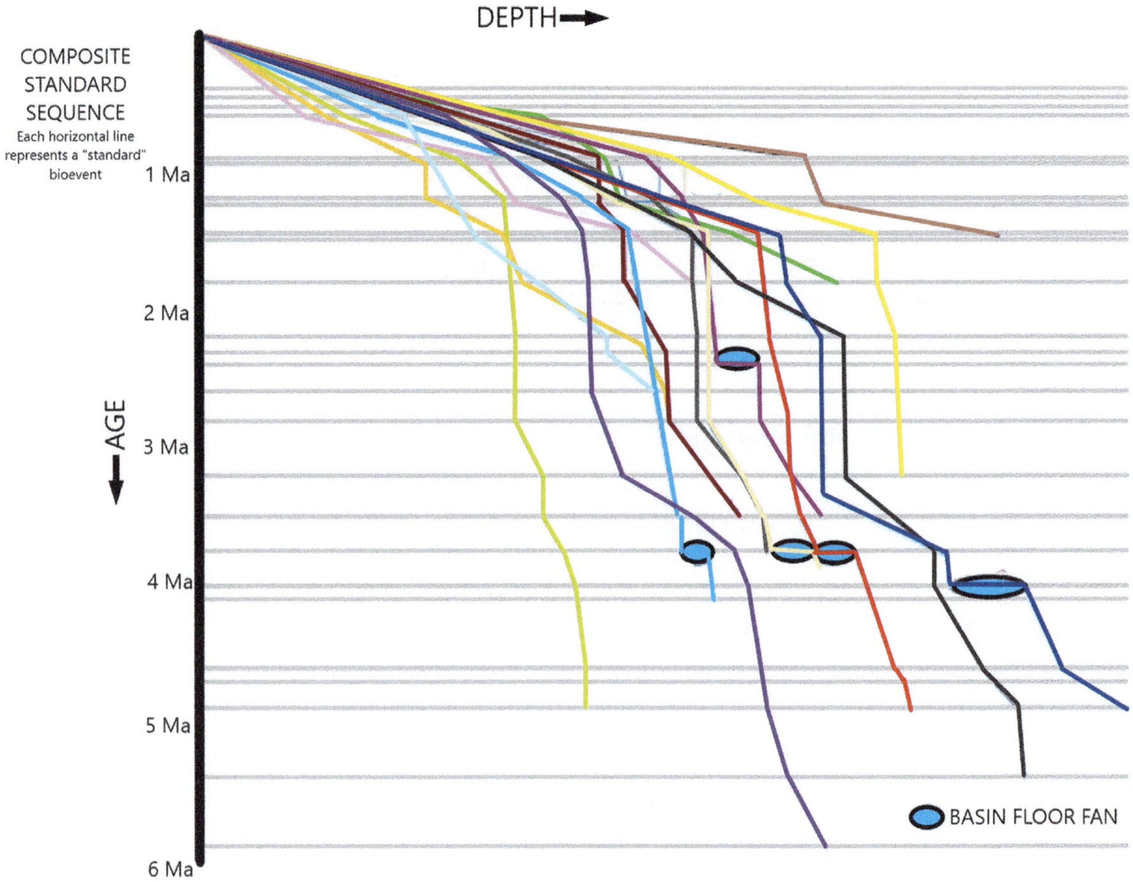


**FIG. 8** A nomograph of age-depth plots from the Garden Banks area in the Gulf of Mexico. Flat spots highlight an age gap along the wellbore path, suggestive of a non-conformity or a geologic structure (i.e., fault). Flat spots of a similar age across multiple wellbores suggest a regional non-conformity, whereas an isolated flat spot in one wellbore suggests a geologic structure (i.e., fault). *(Modified from Crews, J.R., Weimer, P., Pulham, A.J., Waterman, A.S., 2000. Integrated approach to condensed section identification in Intraslope basins, Pliocene-Pleistocene, Northern Gulf of Mexico. AAPG Bull. 84(10), 1519–1536.)*

Once prospects are identified, biostratigraphy is critical for basin analysis and assessing the risks to the petroleum system. Fossils can provide context to basin history, which inputs to reservoir presence, trap formation, burial history and seal deposition (see chapters 2 and 18). Even with basin analysis, biostratigraphy is a cost-effective method to quickly evaluate other prospects for correlation to other sandy packages.

## Deepwater ecology and depositional setting

Depositional environments on continental shelves shift in response to glacio-eustatic water depth fluctuations that are large compared with average water depths. The known or inferred shallow water depth limits of benthic foraminifera in shelf sediments therefore strongly reflect changing accommodation within a stratigraphic cycle. In marked contrast, deepwater continental slope and rise sediments are affected by cyclical water depth changes that are small compared with average water depths. The major changes in deepwater fossil assemblages and related depositional environments are therefore not driven primarily by local accommodation changes, but rather, by the more dramatic bathymetric and environmental changes occurring contemporaneously on nearby continental shelves. Changing shelf environments effectively controls sediment bypassing of the shelf edge and thus modulates sediment accumulation in linked shelf, slope and rise depocenters (see Chapters 3–5 and 8).

In deepwater sections, it is unlikely that large ecozone changes recorded by benthic foraminifera, e.g., from upper bathyal to middle bathyal or lower bathyal, can be attributed directly to glacio-eustatic cyclist of only ±120 m. Paleobathymetric changes on the slope reflect longer-term, noncyclical processes such as slope progradation, compaction, growth faulting, salt tectonism and thermal subsidence, which can elevate or depress portions of the seafloor from several hundred

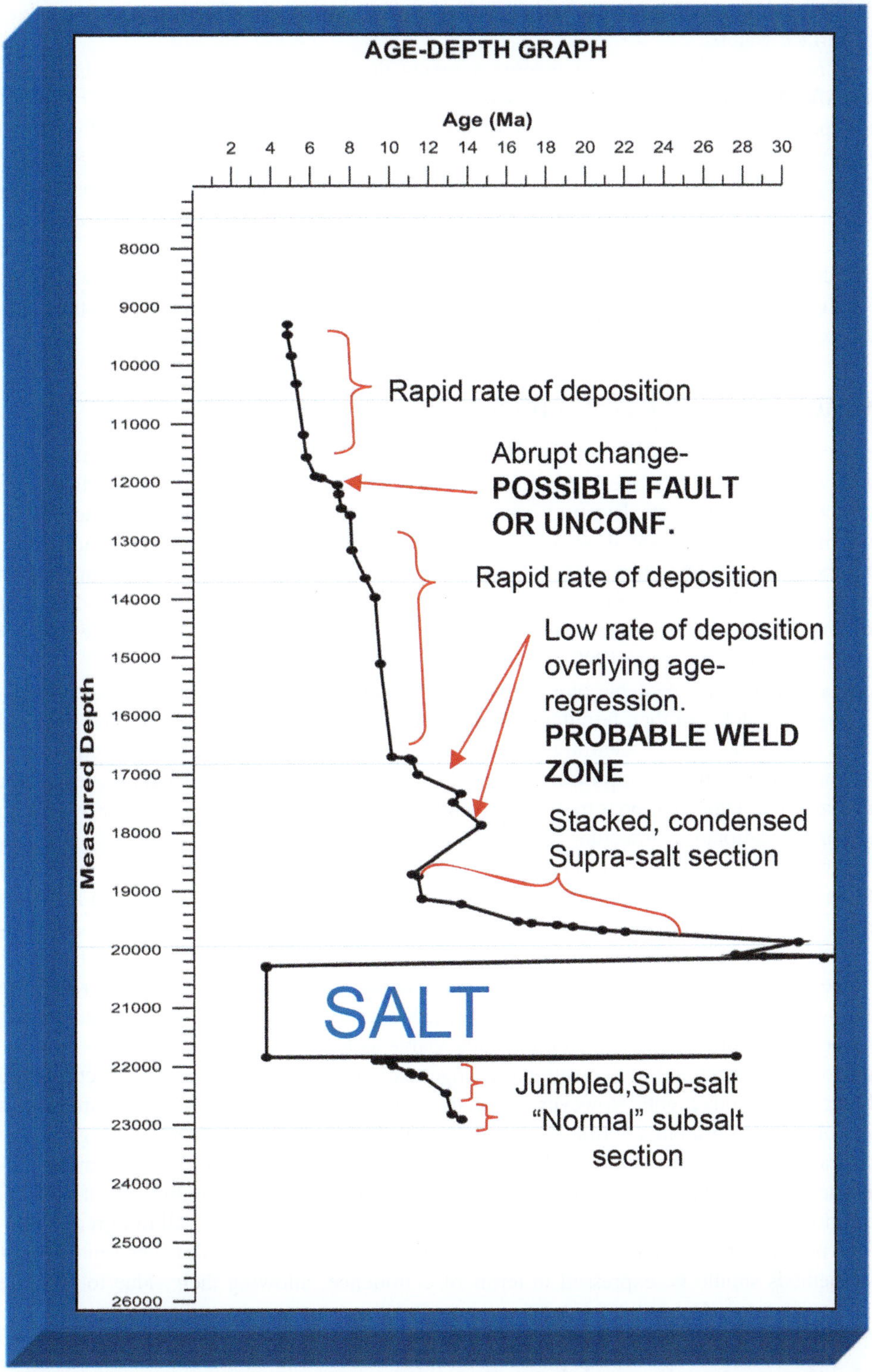


**FIG. 9** A salt-influence geo-history profile utilizing an age-depth plot from a wellbore in Green Canyon area of Gulf of Mexico that entered and exited an allochthonous salt body. *(Modified from Fillon, R. H., 2009. Linked Micropaleontology and Lithology of Cenozoic Deep Water Sediments: Application of Multiple Overlapping Foraminiferal Lithobiofacies to Cuttings Based Analysis of Reservoir Properties. In: Demchuk, T.D., Gary, A. (Eds.), SEPM Special Publication No. 93.)*

to several thousand meters. Foraminiferal assemblages in deepwater sediments contain: (1) in situ or indigenous benthic taxa that are representative of the combined bottom-water/bottom-sediment controlled benthic environment in a specific area (biotope); (2) planktonic taxa that, after death, are transported from their near-surface habitats into benthic biotopes; (3) allochthonous benthic taxa that are transported downslope, along or just above the seabed, into deepwater biotopes as empty tests. Foraminifera contained in deepwater biotopes convey information about processes acting on nearby continental shelves, as well as about the lithology and sedimentary processes taking place at the deepwater depositional site. Abundant transported deltaic taxa provide clear evidence of strong links between the shelf and down-slope deposystems while tests of planktonic foraminifera are but one of the many organic and inorganic constituents of "pelagic rain" and carry information about the overlying surface waters.

In turbidite and other expanded reservoir-prone deepwater subenvironments, distinctive foraminiferal assemblages are directly or indirectly related to specific physical–chemical parameters at the sediment–water interface, including (1) benthic energy level measured as current-speed and variability at the seabed over time intervals comparable with the life cycle of foraminifera (i.e., weeks to months); (2) physical nature of the seabed, including grain size, mineralogy, cohesiveness and firmness; (3) chemistry of surficial sediments and interstitial waters at and just below the seabed where many burrowing organisms live, i.e., pH, organic matter content, oxygen concentration and the dissolution potential of silica and carbonate; (4) stability of the seabed, i.e., rate and uniformity of sediment deposition, episodic sediment input, nondeposition and erosion. Specialized groups of benthic foraminifera have evolved to survive and indeed thrive, over a wide range of depositional variables. Benthic taxa indigenous to turbidite systems can therefore provide considerable insight into the depositional subenvironments of those systems such as proximal, distal and flank environments.

## Applications for deepwater reservoir development

Exploration emphasizes tying reservoirs into regional or semiregional biozonations. Although a regional biostratigraphic event needs to be a geographically widespread occurrence to be considered valid, the critical inferences can be made from the more aerially restricted, facies-controlled bioevents. Biostratigraphy has offerings at the reservoir scale and should play a major role in all phases of field life through appraisal, development and production, not just exploration. A more pragmatic, field-focused approach has been utilized allowing greater stratigraphic subdivision of the reservoir system beyond the control provided by regional bioevents. This have been utilized in mature provinces like the North Sea and Gulf of Mexico, with many fields now in the production or postplateau stage when a drive for better-integrated reservoir description is warranted to access remaining reserves (Payne et al., 1999).

Field-scale biostratigraphy aims to discover repeatable bioevents of potential field-wide chronostratigraphic significance to fingerprint muddy lithologies, which can then be integrated with petrophysical logs, and tied around the field to give log correlations constrained by biostratigraphy. Central to field-scale biostratigraphy is pushing data interpretation with integration and iteration with other geosciences disciplines. Although it may be possible to relate local bioevents back to a regional biozonation, the emphasis of a field zonation is to develop localized, perhaps facies-dependent observations to corroborate hypotheses on reservoir depositional architecture (O'Neill et al., 1999).

The brief period of geological time during which many reservoirs are deposited allows for taxonomic evolution to occur so regional bioevents will likely be too coarse resolution to provide value. Instead, the use of assemblages and local age events or flux of individual species or genera are utilized. These fluxes are biological responses to local dynamic and physiochemical changes in the paleowater mass. In the case of planktic or pelagic fossils typically reflect changes in the upper water mass within the photic zone while changes at the sediment–water interface affect benthic foraminifera. These changes enabled forms that could flourish under specific water mass niches to bloom for short periods of time while others might decline. This ephemeral aspect of blooms renders them potentially viable as local stratigraphic indices.

Relying on the biotic impact from environmental or depositional changes as opposed to evolutionary extinctions and originations could result in applying diachronous events. However, within the confines of a limited geographical area (like a field/prospect), and consideration of paleogeographic control, the assertion that these paleoenvironmental changes allow for a correlatable field-scale ecostratigraphy or facies stratigraphy. Some bioevents appear subtle at first and can be corroborated with increasing consistency with more boreholes. These studies usually involve a higher sample resolution (e.g., 10′ samples), especially if conventional or sidewall cores are available for analysis. The use of core allows greater integration with sedimentological and paleoenvironmental models; like ichnology (study of trace fossils). The reliability of bioevents created through these methods should be expressed in terms of confidence; allowing their value to be more clearly assessed (Fig. 1).

## Key messages on mud in the reservoir model

Biostratigraphy in reservoir development can have a profound impact on integrated reservoir characterization and description, from development planning to well placement. The objectives are to provide a framework for reservoir description by constraining log/seismic picks; provide inputs to reservoir architecture and heterogeneity through time-slice mapping, biofacies in depth and lateral heterogeneity; provide input to reservoir compartmentalization by inferring mud unit continuity as a baffle or barrier. Definitions of mudstone type through biofacies impact the understanding of stratigraphic reservoir compartmentalization to assess the relative lateral extent of mud units. Within turbidite reservoir packages, interturbiditic or channel muds comprise more proximal units and may be modeled as having a relatively lesser areal continuity, therefore acting more as a baffle to fluid flow. Whereas, hemipelagic muds (see chapters 6 and 7) may reflect more profound events in the basin and are likely to form really extensive blankets, which may act more as a barrier to fluid flow (if not disturbed by structural elements). Analyzing the relative assemblages of epifaunal and infaunal benthic foraminifera can provide a means to characterize mud units.

In a turbiditic environment, most thin-bedded intrareservoir muds are barren of microfossils. This is likely because the timing of the next turbidite (i.e., overlying package) buried the substrate before benthic communities could develop. A turbidite to a benthic community is much like a forest fire. A turbidite buries the underlying benthic community creating a prime niche for the opportunists to move in and gobble up the newly deposited nutrients. If allowed to continue without disaster, the major nutrients are used and more stable (oligotrophic) feeders move in to begin burrowing for buried nutrients. Therefore, benthic communities can be used as a relative proxy for reservoir deposition. Muds barren of benthic foraminifera indicate faster deposition, less laterally extensive, turbidites. Epifaunal benthic foraminifera, the opportunistic foraminifera feeding on nutrients on the seafloor, indicates a bit of a pause in sedimentation so suspension feeders can survive. The epifaunal feeders are sometimes termed the "doomed pioneers" in active sedimentation. Infaunal benthic foraminifera suggests that the pause in sedimentation occurred long enough so that the seafloor nutrients have been diminished that the burrowers move in to start utilizing buried sediments (Fleisher and Lane, 1999). Finally, with a long enough pause, hemipelagic or planktic fossils will begin to blanket the muds.

Understanding the lateral continuity of intrareservoir mud units may significantly impact the reservoir model, completion zones and perhaps well placement. Sometimes, the local bioevent scheme can help biosteer horizontal wells when a sandy package is bounded by unique bio-assemblages. To maximize value, work has to be focused and pragmatic but not utilized as a stand-alone geosciences discipline (Payne et al., 1999).

## Conclusions and future outlook

Amazingly, microfossils survive the drilling process and are critical to assisting drillers and geoscientist in correlations by providing chronostratigraphic context and clues to depositional environments. This cost-efficient process used for centuries provides context to deepwater sedimentary processes as the fossils can be multi-sourced: pelagic vs deltaic. When integrated with other data like logs and geophysical data, deepwater biostratigraphy is a very powerful tool.

The authors feel that the future research of biostratigraphy is steering toward integrating with stable isotope stratigraphy, sediment provenance and studying sediment-source materials, chronostratigraphic resolution via orbital cycles and documenting agglutinated benthic paleoenvironment and depositional environment. Currently, much research is based on the modern environment and how microfauna responds to changing environmental conditions, which have applications to paleoenvironment and environmental monitoring. Research into microfauna and their fossils will have multiple applications in the future.

## References

Crews, J.R., Weimer, P., Pulham, A.J., Waterman, A.S., 2000. Integrated approach to condensed section identification in intraslope basins, Pliocene-Pleistocene, Northern Gulf of Mexico. AAPG Bull. 84 (10), 1519–1536.

Dickey, R.L., Yancey, T.E., 2010. Palynological age control sediments bracketing the Paleocene-Eocene boundary, Bastrop, Texas. Gulf Coast Assoc. Geol. Soc. Trans. 60, 717–724.

Fillon, R.H., 2009. Linked micropaleontology and lithology of cenozoic deep-water sediments: application of multiple overlapping foraminiferal lithobiofacies to cuttings-based analysis of reservoir properties. In: Geologic Problem Solving with Microfossils: A Volume in Honor of Garry D. Jones, Thomas D. Demchuk, Anthony C. Gary. SEPM Special Publication No. 93.

Fleisher, R.L., Lane, H.R., 1999. Applied paleontology. In: Beaumont, E.A., Foster, N.H. (Eds.), Exploring for Oil and Gas Traps, Handbook No. 3, Treatise of Petroleim Geology. American Association of Petroleum Geologists, pp. 17-01–17-65.

Flores, J.-A., Sierro, F.J., Filippelli, G.M., Bárcena, M.Á., Pérez-Folgado, M., Vázquez, A., Utrilla, R., 2005. Surface water dynamics and phytoplankton communities during deposition of cyclic late Messinian sapropel sequences in the western Mediterranean. Mar. Micropaleontol. 56 (1–2), 50–79.

Halbritter, H., et al., 2018. Methods in palynology. In: Illustrated Pollen Terminology. Springer, Cham, https://doi.org/10.1007/978-3-319-71365-6_6.

Iaccarino, S., Premoli Silva, I., Biolzi, M., Foresi, L., Lirer, F., Turco, E., Petrizzo, M., 2007. Practical Manual of Neogene Planktonic Foraminifera. International School on Planktonic Foraminifera.

O'Neill, B.J., Duvernay, A.E., George, R.A., 1999. Applied palaeontology: a critical stratigraphic tool in Gulf of Mexico exploration and exploitation. In: Jones, R.W., Simmons, M.D. (Eds.), Biostratigraphy in Production and Development Geology. vol. 152. Geological Society Special Publication, pp. 303–308.

Payne, S.N.J., Ewen, D.F., Bowman, M.J., 1999. The role and value of "high-impact biostratigraphy" in reservoir appraisal and development. In: Jones, R. W., Simmons, M.D. (Eds.), Biostratigraphy in Production and Development Geology. vol. 152. Geological Society of London, London, pp. 5–22. Special publications.

Zalasiewicz, J., Cita, M.B., Hilgen, F., Pratt, B.R., Strasser, A., Thierry, J., Weissert, H., 2013. Chronostratigraphy and geochronology: a proposed realignment. GSA Today 23 (3).

Chapter 15

# Deepwater ichnology: New observations on contourites

**Francisco J. Rodríguez-Tovar, Olmo Miguez-Salas and Javier Dorador**
*Departamento de Estratigrafía y Paleontología, Universidad de Granada, Granada, Spain*

## Introduction: Deepwater ichnology

The deepwater environment is a complex setting in which numerous processes determine sediment deposition. The three major sedimentary processes in deepwater systems are the settling of pelagic and hemipelagic particles from the water column, sediment gravity flows and bottom-current controlled sedimentation (Hüneke and Mulder, 2011; Rebesco et al., 2014; de Castro et al., 2020a, b, 2021; also see Chapters 6–9). Differentiation and characterization of deepwater sediment can be quite complicated, and the establishment of valid diagnostic criteria remains a challenge.

Ichnology is the study of traces—both modern corresponding to neoichnology and fossils to paleoichnology—produced by organisms on or within a substrate (e.g., Bromley, 1996). The field of ichnology has experienced significant growth over the last decades. Trace refers to "A morphologically recurrent structure resulting from the life activity of an individual organism (or homotypic organisms) modifying the substrate, namely burrows, borings and etchings, tracks and trackways, coprolites, gastroliths, regurgitaliths, nests, woven cocoons, spider webs, leaf mines and bite and gnaw structures, but not eggs, embedment structures and plant reaction tissues" (Bertling et al., 2006). Traces, reflecting the behavior of producers, provide meaningful information on ecological and depositional parameters that affect the tracemaker community.

Two main paradigms, supporting the usefulness of the ichnological analysis, have been developed, the ichnofabric approach and the ichnofacies model. The ichnofabric approach embraces any textural or structural aspect resulting from bioturbation and bioerosion at any scale (Ekdale and Bromley, 1983; Rodríguez-Tovar and Dorador, 2015; Rodríguez-Tovar et al., 2020). It considers trace fossil assemblage and other ichnological features such as cross-cutting, tiering or degree of bioturbation (Buatois and Mángano, 2011; Ekdale et al., 2012; Knaust, 2017). The ichnofacies model (Seilacher, 1964, 1967) is based on the identification of certain features shared by ichnotaxa during a long time span that can be related with particular environmental conditions (Buatois and Mángano, 2011). Ichnofacies characterization is based on the integration of data from ichnological, sedimentological, stratigraphical and taphonomical information. Both the ichnofabric approach and the ichnofacies model determine the usefulness of ichnology in very different fields, including paleoecology, sedimentology, evolutionary paleoecology, paleoceanography and paleoclimatology (Mángano and Buatois, 2012). Paleoenvironmental conditions such as sedimentation rate, hydrodynamic energy, substrate consistency, bottom water oxygenation, salinity or food availability can be interpreted based on the trace fossil assemblage. Hence, ichnological content can be used to characterize deepwater processes and sedimentary intervals (Rodríguez-Tovar, 2022).

In recent years, ichnological information has been adopted to distinguish among pelagites, hemipelagites, contourites, turbidites and hyperpycnites (i.e., Mulder et al., 2002; Ponce et al., 2007; Wetzel et al., 2008; Rodríguez-Tovar and Hernández-Molina, 2018; Buatois et al., 2019; Rodríguez-Tovar, 2022; Fig. 1). This is well known for particular facies that have been profusely studied; for instance, pelagites, hemipelagites and turbidites show certain diagnostic ichnological features, including the ichnotaxa *Chondrites*, *Thalassinoides* and *Zoophycos* for pelagites and hemipelagites or graphoglyptids *Cosmorhaphe*, *Palaeodictyon* and *Spirorhaphe* for turbidites (Fig. 1; see reviews in Uchman and Wetzel, 2011, 2012; Wetzel and Uchman, 2012; Pickering and Hiscott, 2016). Thus, pelagites and hemipelagites are mainly associated with the *Zoophycos* ichnofacies, commonly related to low-energy subtidal settings, below the storm wave base, from the outer shelf to slope, while turbidites mainly correspond to the *Nereites* ichnofacies, common in basin-floor depositional environments, characterized by very slow yet continuous sedimentation of pelagic and hemipelagic materials, interrupted by periodic turbiditic deposition (MacEachern et al., 2007, 2012; Buatois and Mángano, 2011; Uchman and Wetzel, 2011, 2012; Wetzel and Uchman, 2012; Dorador and Rodríguez-Tovar, 2015). Facies such as hyperpycnites and contourites are

**Deepwater Sedimentary Systems.** https://doi.org/10.1016/B978-0-323-91918-0.00022-0

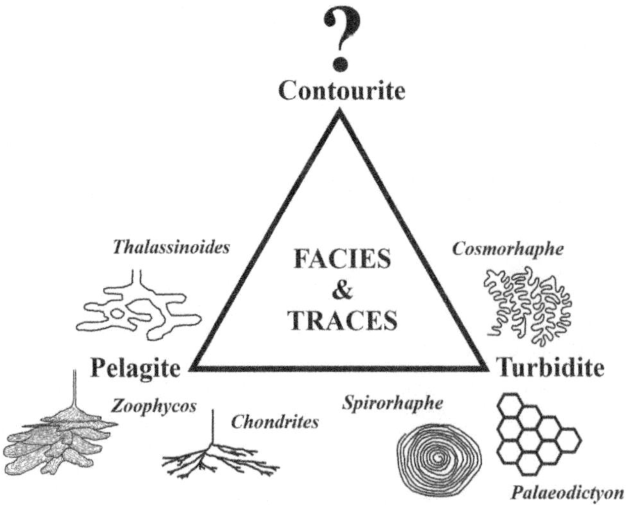


**FIG. 1** Main deepwater (marine) facies and some characteristic trace fossils.

comparatively understudied, and their conclusive characterization is difficult. Nevertheless, some authors consider bioturbation irrelevant for interpreting deepwater deposits such as contourites, turbidites and hyperpycnites, arguing that ichnological features have nothing to do with the fluid mechanics of the corresponding depositional processes (Shanmugan, 2018).

One example of this would be contourite deposits. Initial papers describing contourites and most subsequent research have treated bioturbation as a reliable diagnostic criterion when differentiating contourites from other deepwater deposits (Lowell and Stow, 1981; Gonthier et al., 1984; Chough and Hesse, 1985; Stow et al., 1986, 1998; Faugères and Stow, 1993; Mulder et al., 2001, 2002, 2013; Wetzel et al., 2008). Some studies interpret bioturbation as a secondary feature in contourites, however, and assign greater importance to other physical sedimentary features (Shanmugam et al., 1993; Shanmugam, 2000, 2002, 2017a; Martín-Chivelet et al., 2003, 2008). The review "Ichnological analysis of contourites: Past, present and future" (Rodríguez-Tovar and Hernández-Molina, 2018) demonstrated that ichnological research can improve criteria for differentiating between types of contourite facies, while also suggesting future research directions. In the wake of such publications, ichnological studies on contourites have increased significantly, involving different types of contourites in outcrop and in core examples.

According to the above, and based on the scientific and economic interest of contourites and deepwater sedimentation more generally (see all other chapters in this book), an updated report on the ichnological information from contourites is fundamental. This chapter therefore compiles available information on bioturbation features, ichnological studies and the resultant ichnofacies characterizations. Additionally, new techniques to advance in the ichnological analysis of contourites are introduced. In parallel, the relationship between trace fossil assemblages and paleoenvironmental conditions prevailing during contourite deposition is discussed. Finally, the influence of bioturbation on petrophysical properties is addressed, as these types of deepwater deposits hold economic interest for energy resource exploration and development, including oil and gas.

## Contourites and ichnology: Significant recent advances

The ichnological study of contouritic deposits is now advancing roughly at the same pace as turbidite or pelagic research, both of which made substantial progress in the past century (see Chapters 1, 6, 7). However, there is a disparity in the scientific community's knowledge about contourites relative to other deepwater deposits (Figs. 1 and 2). Although certain ichnotaxa are largely associated with pelagites (i.e., *Chondrites*, *Zoophycos*) and turbidites (i.e., *Cosmorhaphe*, *Palaeodictyon*, *Spirorhaphe*), particular trace fossils diagnostic of contourites have not been recognized to date (Fig. 1). A recent online search (December 2020) in Google Scholar for the term "contourites+bioturbation" yielded 2330 contributions, whereas a similar search for "turbidites+bioturbation" yielded 19,000 results, while "pelagic+bioturbation" harvested 28,900 (Fig. 2A). A more specific search about ichnology, "contourites+ichnology", yielded 299 items in Google Scholar, whereas "turbidites+ichnology" gave 3650 results and "pelagic+ichnology" just 2440, highlighting the same trend (Fig. 2B). Since the onset of the 21st century, and especially since 2010, the number of papers about *contourites + ichnology*

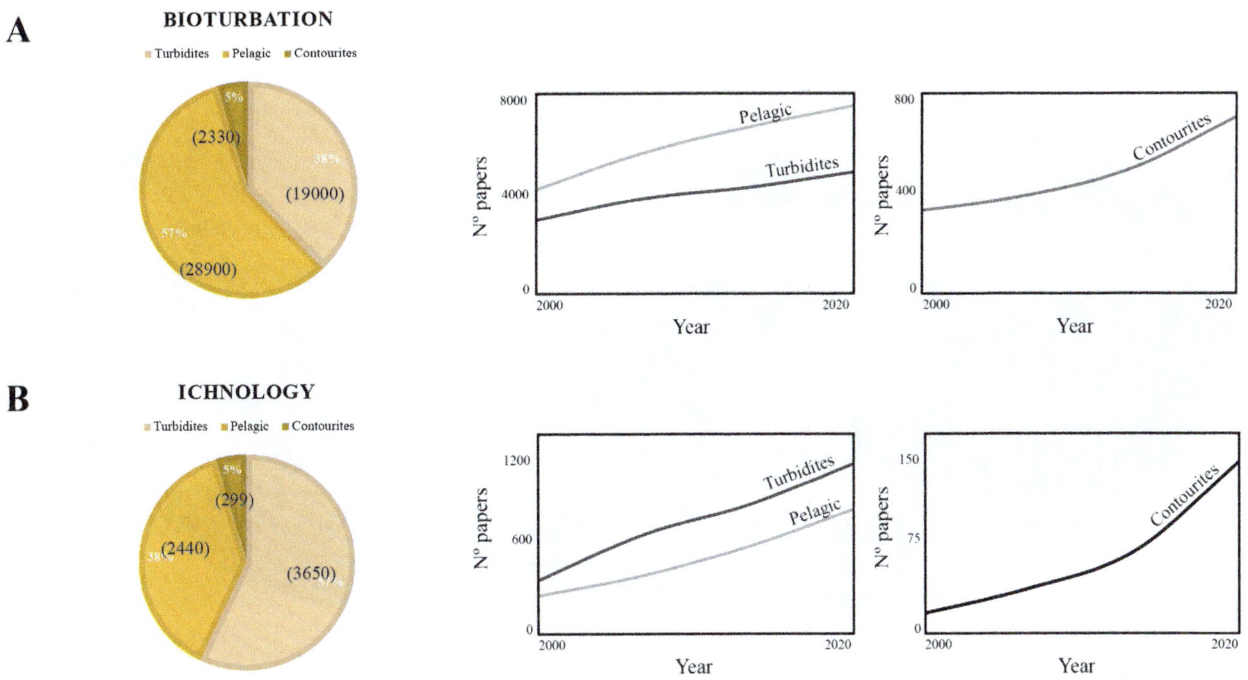


**FIG. 2** (A) Pie chart showing the results obtained with the keyword "bioturbation" together with the main deepwater (marine) facies from Google Scholar (on the right), and number of articles published during the last 20 years (on the left); (B) pie chart showing the results obtained with the keyword "ichnology" together with the main deepwater facies from Google Scholar (on the right), and number of articles published during the last 20 years (on the left). Note the significant increase in the number of published papers dealing with "contourites + bioturbation" or "contourites + ichnology" over the last 10 years in comparison with other deepwater facies.

or *contourites + bioturbation* has shown a significant increase (Fig. 2), even though contourites still represent less than 5% of deepwater ichnological research studies (note the number of papers for each type of deepwater deposit). Thus, despite significant progress, a long road of opportunity lies ahead.

The first step in trace fossil research is to address the ichnological attributes of trace fossil diversity and abundance. In building the foundations for an understanding of ichnology and contourites, it is important to consider these attributes in different contourite types. The sedimentological classification of contourites is diverse but mainly consists of clastic, calcareous and chemogenic types (see fig. 22 in Rebesco et al., 2014). The first two categories include muddy, silty, sandy and gravel contourites; the third consists of manganiferous and gravel-lag contourites (Stow and Faugères, 2008; Rebesco et al., 2014). During the early stages of contourite field development, most sedimentological papers described ichnological features using the terms "bioturbation," "burrows," and "mottled" (e.g., Faugères et al., 1984; Gonthier et al., 1984; Rodríguez-Tovar and Hernández-Molina, 2018 for a review), without conducting detailed ichnological analysis. As research developed, some ichnological papers began to build the basic pillars of contourite trace fossil diversity and abundance (e.g., Fu and Werner, 1994; Baldwin and McCave, 1999; Stow et al., 2002a), but without going into detail as to their relationship with contourite type. In the past decade, ichnological studies conducted on ancient contourite outcrops grew exponentially (Rasmussen and Surlyk, 2012; Miguez-Salas and Rodríguez-Tovar, 2019a, b; Reolid and Betzler, 2019; De Weger et al., 2021; Figs. 3 and 4). These recent studies have provided detailed ichnological information from contouritic outcrops in different areas (e.g., Cyprus or Morocco; Fig. 4) and of diverse ages. The inclusion of ancient outcrop studies has allowed the scientific community to gain a broader vision of diversity and abundance, with tridimensional information, and additional results from sediment core studies (Figs. 3 and 5).

Detailed ichnological studies on sediment cores have some persistent challenges: the narrow exposed surface, a two-dimensional view and the poor visibility of some biogenic structures, among other difficulties. Even so, the study of sediment cores provides, in most cases, a long and continuous record of deposits that are not exposed in outcrops, and their study has been essential for building current knowledge about the ichnological content of contourites (Rodríguez-Tovar and Hernández-Molina, 2018 for a review). For instance, studies involving cores drilled in the Gulf of Cadiz during IODP Expedition 339 have provided very interesting insights regarding the ichnological content of muddy and sandy contourites (e.g., Alonso et al., 2016; Jones, 2016; de Castro et al., 2020a, b; Fig. 5). Research into Quaternary deposits

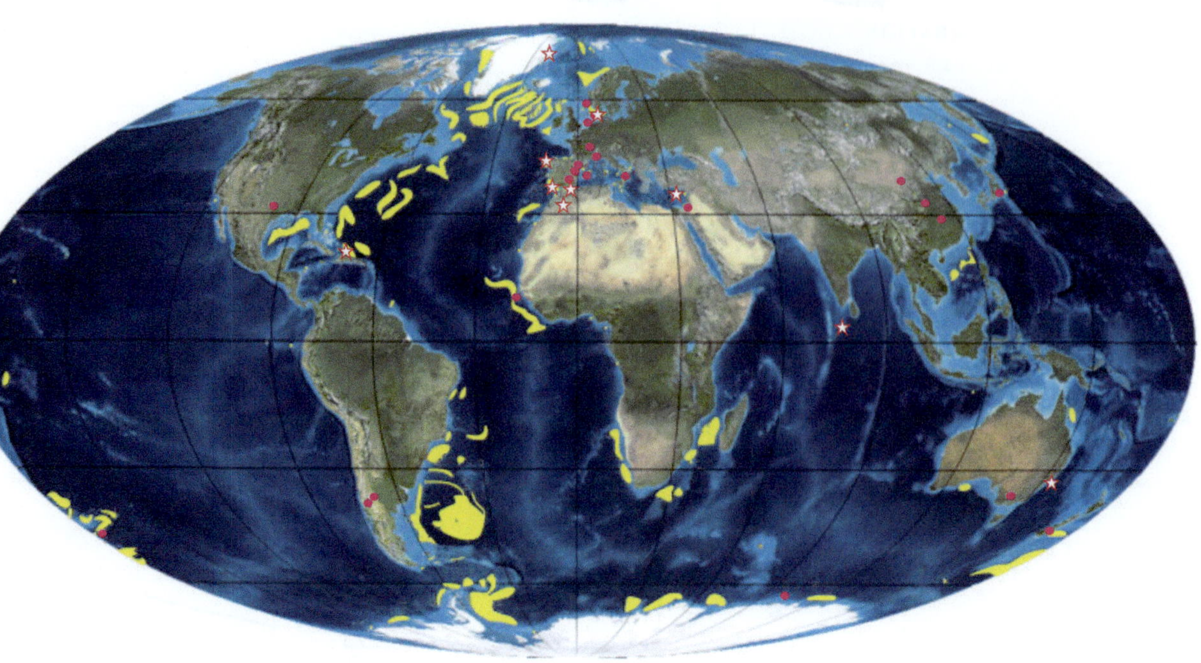


**FIG. 3** Global distribution of large contourite deposits in present (recent) ocean basins and in the ancient record as well as location of deposits where ichnological content has been studied in detail. *Modified from Rebesco, M., Hernández-Molina, F.J., Van Rooij, D., Wåhlin, A., 2014. Contourites and associated sediments controlled by deep-water circulation processes: state-of-the-art and future considerations. Mar. Geol. 352, 111–154.*

has been conducted along the NW Iberian margin (e.g., Dorador et al., 2019, 2021a) and in the Alboran Sea (Alonso et al., 2021; Fig. 5). Additionally, information has been obtained from older successions, for example, the upper cretaceous deposits of NE Greenland (Hovikoski et al., 2020). Reolid and Betzler (2019) studied Miocene to Pleistocene contourite cores from Maldives, providing some clues for comparison with other carbonate drifts, and de Castro et al. (2021) conducted a multiproxy analysis, including ichnological information, in a study of siliciclastic and bioclastic systems from Madagascar and Corsica. The integration of outcrop and core information now allows for a more detailed ichnofacies and ichnofabrics characterization of contourite deposits, contributing to the body of ichnological knowledge about contourites (Miguez-Salas and Rodríguez-Tovar, 2019a, b, 2021; Miguez-Salas et al., 2021a).

However, in spite of the significant increase in the number of ichnological contributions about contourites over the last decade (Fig. 3), additional studies are necessary to build a better understanding.

## Trace fossil assemblages in contourites

The recognition of trace fossils in contourites, as well as trace fossil assemblage diversity, has advanced parallel to the overall development of the field. Key turning points were the introduction of outcrop studies and the increased abundance of cores from ocean scientific expeditions (e.g., Bahr et al., 2014; Dorador et al., 2014a, b; Hernández-Molina et al., 2016 among others; see Chapter 9 in this book). Both sources have provided new data about bioturbation in contourites, permitting a differentiation between trace fossil assemblages of calcareous and clastic contourites. Meanwhile, in terms of diversity, outcrop research marks a breakthrough, as core diversity data tend to be limited and reiterative, especially when working with modern sediment cores. Outcrop studies, in which ichnotaxobase identification is much easier than in core data, provide tridimensional observations—even reaching assignation at the ichnospecies level (see Miguez-Salas and Rodríguez-Tovar, 2021). Despite such progress, the actual number of outcrops identified as ancient contourite deposits remains low: for example, Neogene Miura-Boso region, Japan, in Stow et al. (2002a); Eocene-Miocene Petra Tou Romiou, Cyprus, in Stow et al. (2002b); Devonian Eastern Moroccan Central Massif in Hüneke (2013); Late Ordovician Pingliang Formation, China, in Li et al. (2020); Late Miocene Rifian corridor, Morocco, in Capella et al. (2017). Thus, the true diversity of contourite deposits may be underestimated.

Muddy contourites, both calcareous and clastic, are typified by >50% clay, and composition may be dominated by siliciclastic or biogenic calcareous components (planktonic and benthic calcareous and siliceous organisms) (Rebesco et al., 2014). Additionally, Rebesco et al. (2014) indicate that sometimes “These deposits occur in thick units

**FIG. 4** General and detailed views of contourite outcrops from Cyprus (A–C) and Morocco (D–F): (A) Early Miocene calcareous contourite succession from Agios Konstantinos, (B) Early and Middle Miocene calcareous contourite succession from Petra Tou Romiou, (C) detailed view of muddy contourite facies from Petra Tou Romiou, (D) Late Miocene sandy clastic contourite facies at El Adergha and (E and F) Late Miocene sandy contourite deposits showing different orders of cross-stratification at Sidi Chahed. For additional information on the physical stratigraphy of these types of deposits, see Chapter 9.

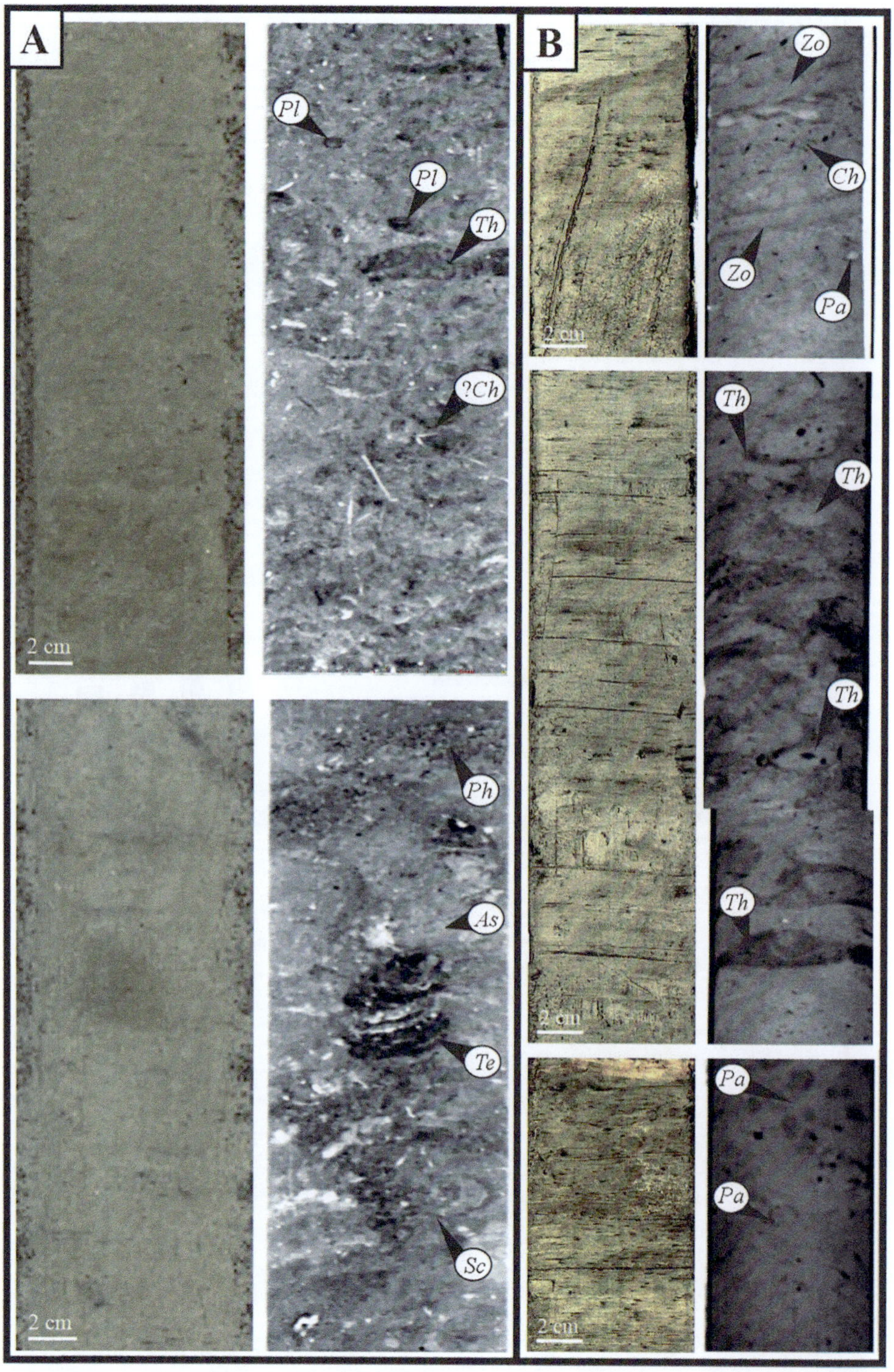


**FIG. 5** High-resolution (left) and CT images (right) of some core contouritic intervals from the Alboran Sea, offshore Spain (A) and the SW Iberian Margin, Spain and Portugal (B). *As*, *Asterosoma*; *Ch*, *Chondrites*; *Pa*, *Palaeophycus*; *pH*, *Phycosiphon*; *Pl*, *Planolites*; *Sc*, *Scolicia*; *Te*, *Teichichnus*; *Th*, *Thalassinoides*; *Zo*, *Zoophycos*.

and are rather difficult to differentiate from hemipelagic deposits." Throughout the literature, muddy contourites are usually characterized as highly bioturbated deposits with a common indistinct mottled appearance (Rodríguez-Tovar and Hernández-Molina, 2018). From an ichnological perspective, trace fossil diversity and abundance in muddy contourites should be comparable to those registered in pelagic and hemipelagic deposits. Ichnological analysis conducted on deposits has corroborated this. Accordingly, the trace fossil assemblages observed resemble those found in pelagic deposits. Trace

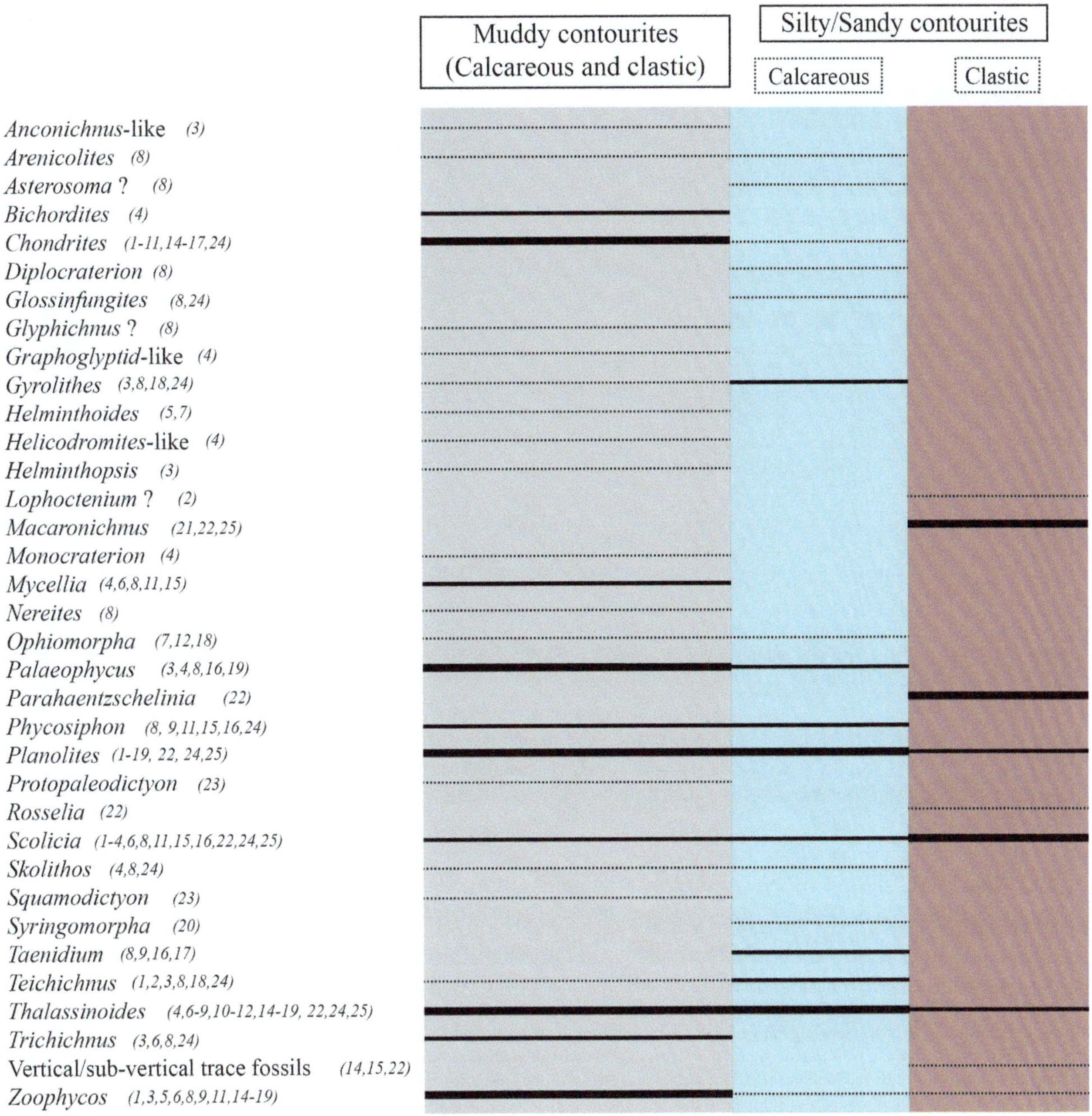


**FIG. 6** Ichnogenera differentiated in muddy (calcareous and clastic) and silty to sandy (calcareous and clastic) contourites. Numbers in the figure refer to (1) Faugères et al. (1984), Stow and Holbrook (1984); (2) Gonthier et al. (1984); (3) Fu and Werner (1994); (4) Baldwin and McCave (1999); (5) Stow et al. (2002b); (6) Löwemark et al. (2004); (7) Hüneke and Stow (2008); (8) Wetzel et al. (2008); (9) Rasmussen and Surlyk (2012); (10) Hüneke (2013); (11) Hanebuth et al. (2015); (12) Alonso et al. (2016); (13) Capella et al. (2017); (14) Brackenridge et al. (2018); (15) Petrovic et al. (2019); (16) Reolid and Betzler (2019); (17) Miguez-Salas and Rodríguez-Tovar (2019a); (18) Miguez-Salas and Rodríguez-Tovar (2019b); (19) Dorador et al. (2019); (20) Miguez-Salas et al. (2019); (21) Miguez-Salas et al. (2020a, b); (22) Miguez-Salas and Rodríguez-Tovar (2021); (23) Li et al. (2020); (24) Stow and Smillie (2020); (25) Miguez-Salas et al. (2021a, b).

fossil assemblage diversity in muddy contourites, whether in cores or outcrops, shows recurrent ichnogenera *Chondrites*, *Palaeophycus*, *Planolites*, *Thalassinoides* and *Zoophycos* (Figs. 6–8). Other frequent ichnogenera are Mycellia, *Phycosiphon*, *Scolicia* and *Taenidium* (Figs. 6–8).

Sandy contourites have a mixed siliciclastic to biogenic composition in which traction sedimentary structures—planar lamination, ripple cross-lamination and various types of cross-stratification— part of classic sedimentation units from

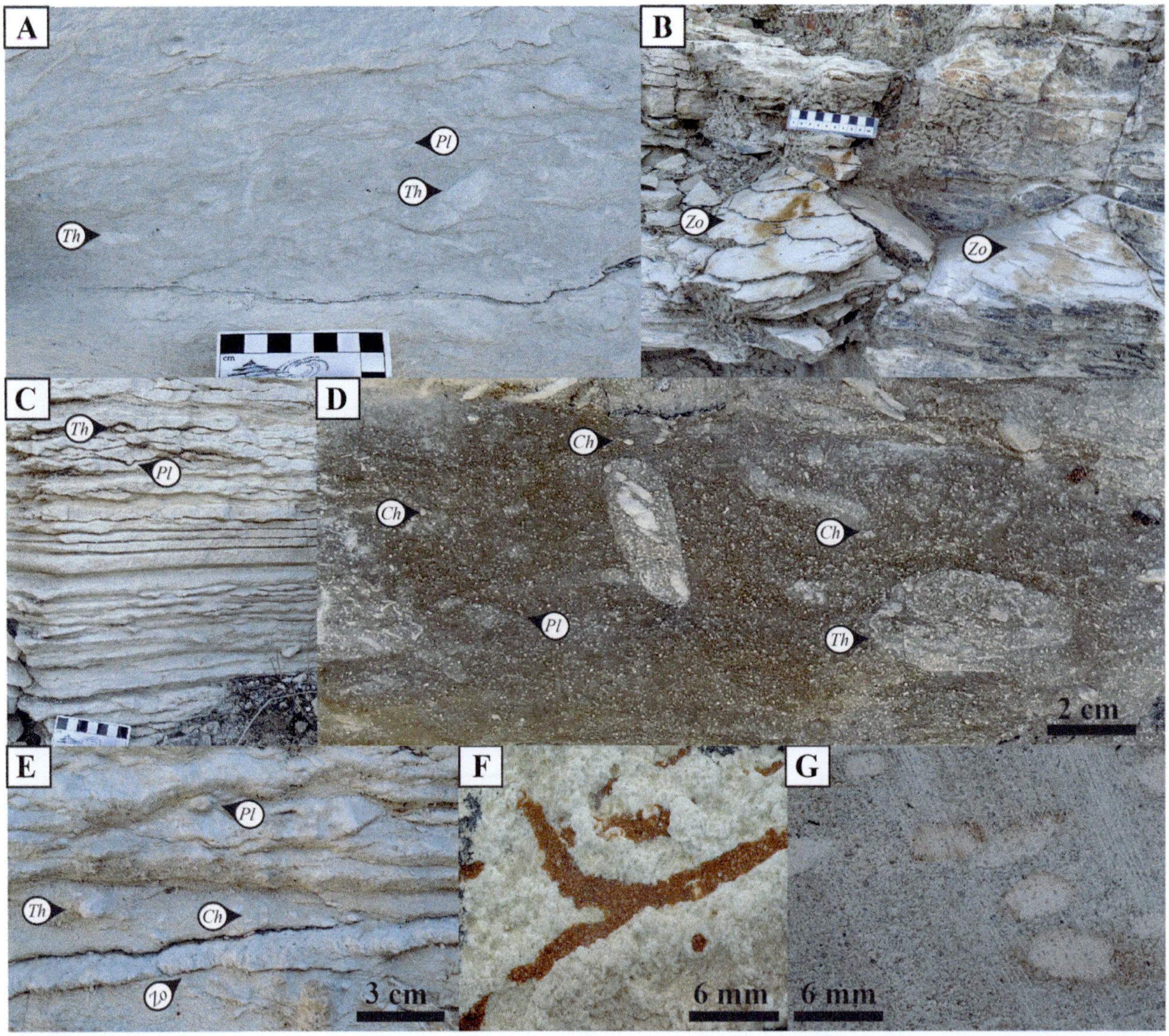


**FIG. 7** Trace fossils in Eocene to mid-Miocene calcareous contourites from Cyprus. (A) calcilutitic contourite facies with *Planolites* (*Pl*) *Thalassinoides* (*Th*) at Petra Tou Romiou outcrop; (B) calcilutitic contourite facies with *Zoophycos* (*Zo*) at Agios Konstantinos outcrop; (C) close-up outcrop examples of the calcarenitic contourite facies, showing *Planolites* (*Pl*) in the noncompacted layers and *Thalassinoides* (*Th*) at the base of the beds at Petra Tou Romiou outcrop; (D) muddy contourite with *Chondrites* (*Ch*), *Planolites* (*Pl*) and *Thalassinoides* (*Th*) at Petra Tou Romiou beach section; (E) calcareous sandy contourite with *Chondrites* (*Ch*), *Planolites* (*Pl*), *Thalassinoides* (*Th*) and *Zoophycos* (*Zo*) at Petra Tou Romiou outcrop; (F and G) detailed view of calcilutitic contourite facies with *Chondrites* from Kalavasos outcrop.

turbidites to contourites (detailed in Chapters 6 and 7) as described recently by de Castro et al. (2020a, b, 2021), become dominant (Rebesco et al., 2014). The sorting tends to be poor to moderate, partly owing to bioturbational mixing. These deposits can be bioturbated throughout, appearing as massive (structureless) at first sight. Such is the case of most core examples (i.e., Faro Drift, Gulf of Cadiz, southwestern Spain), where the differentiation of discrete trace fossils is hard to achieve (see Rodríguez-Tovar and Hernández-Molina, 2018). The trace fossils observed in cores comprise common *Thalassinoides*, *Planolites* and sub-vertical burrows. However, a restricted view of core sections can lead to underestimations of the real trace fossil assemblage diversity (Fig. 5). An ichnological study conducted on sandy siliciclastic contourite deposits in the Late Miocene Rifian Corridor outcrops (Morocco) has shown trace fossil assemblages of high abundance and low ichnodiversity (Miguez-Salas et al., 2021a). *Scolicia*, *Parahaentzschelinia* and *Macaronichnus* are common ichnogenera of these deposits (Miguez-Salas et al., 2020a; Miguez-Salas and Rodríguez-Tovar, 2021; Figs. 6–8).

Calcareous sandy contourites, such as the carbonaceous equivalent of siliciclastic sandy contourites, show common nondepositional surfaces, hardgrounds and burrowing (Rebesco et al., 2014). Primary sedimentary structures are

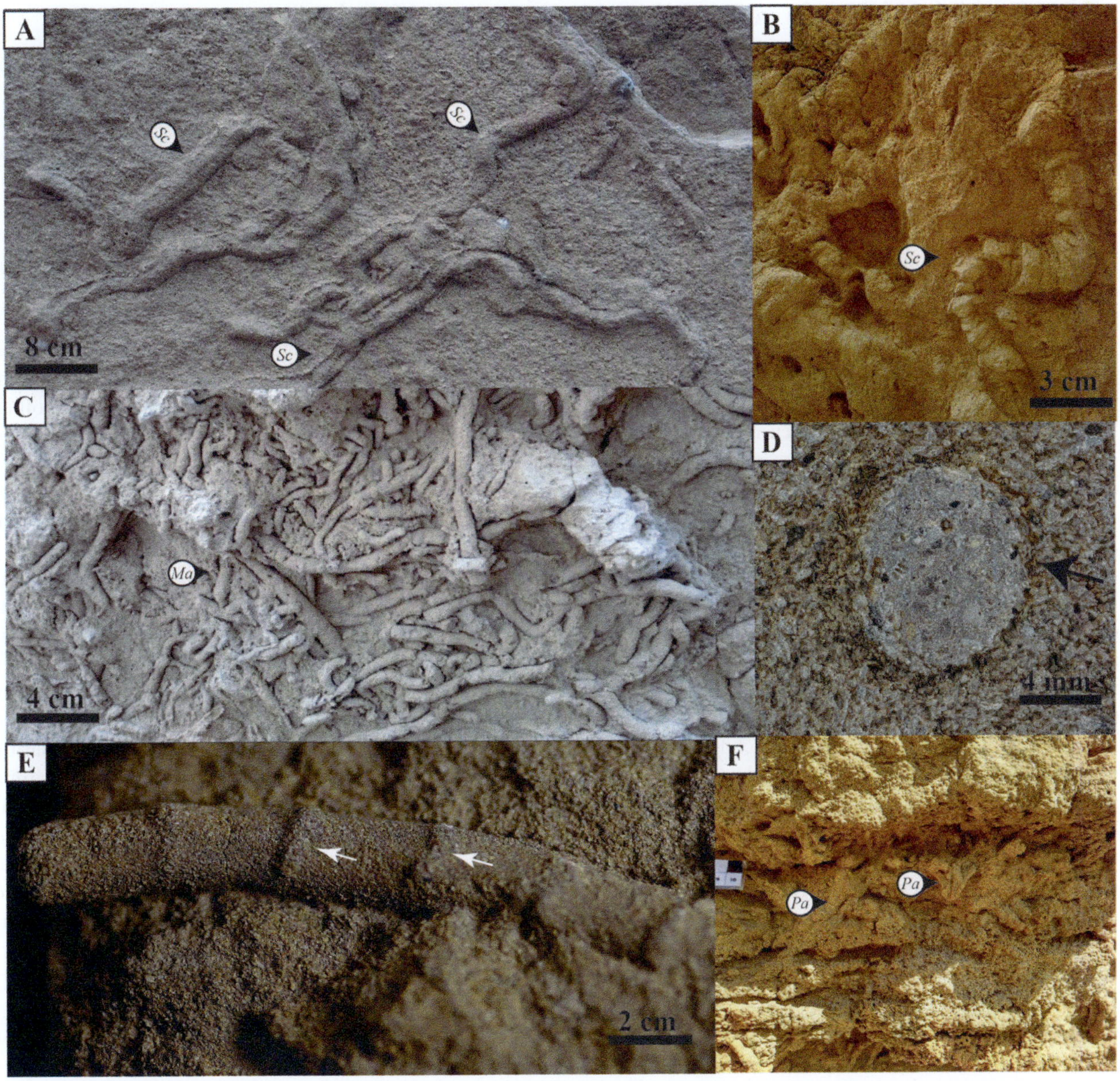


**FIG. 8** Trace fossils in Late Miocene sandy clastic contourites from Morocco. (A) *Scolicia* (*Sc*) at Sidi Chahed section; (B) *Scolicia* (*Sc*) at Kirmta section; (C) Dense *Macaronichnus* (*Ma*) assemblage at Kirmta section; (D and E) *Macaronichnus* (*Ma*) specimens, note the darker burrow rim (*black arrow* in D) and the lobes (*white arrows* in E) from El Adergha outcrop; (F) *Parahaentzschelinia* (*Pa*) at El Adergha outcrop.

commonly affected and disturbed by bioturbation. Ichnological analysis conducted on cores from carbonate drifts worldwide (e.g., Santaren Channel lining at the Great Bahama Bank, Danish Basin in Denmark) shows as recurrent ichnogenera *Palaeophycus*, *Phycosiphon*, *Planolites*, *Scolicia*, *Taenidium* and *Thalassinoides* (Rasmussen and Surlyk, 2012; Reolid and Betzler, 2019). The study of outcrop analogs has led to the discovery of other common ichnogenera (Figs. 6–8). Late Oligocene and early Miocene calcareous contourites in southern Cyprus (Petra Tou Romiou type outcrop and Agios Konstantinos section) record abundant *Planolites* and *Thalassinoides* together with common *Gyrolithes* and *Teichichnus*. The trace fossil assemblage diversity observed in carbonate drifts can be considered slightly higher than that of siliciclastic contourites. Nevertheless, compared to other marine deposits, biodiversity is clearly lower. Therefore, current observations raise two considerations: (i) high-energy contourite depositional settings would favor only specialized colonization, impoverishing trace fossil assemblage diversity or (ii) the current diversity is underestimated, and as contourite research progresses, the number of recognized ichnogenera will increase.

## Archetypal ichnofacies in contourites: A complex relationship

Ichnofacies are characterized by recurring ethological groupings of trace fossils reflecting specific combinations of organisms' responses to environmental conditions (Seilacher, 1967; Bromley, 1996; MacEachern et al., 2007, 2010, 2012). In marine environments, associations of trace fossils and ichnological features defining ichnofacies are narrowly related to ecological and depositional factors, which are mainly linked to the evolution from proximal to distal environments (Chapters 8 and 9) and changes in the relative sea level (Chapter 12) (MacEachern et al., 2007, 2010, 2012; Buatois and Mángano, 2011). Thus, ichnofacies are determined by the interplay of numerous parameters (e.g., hydrodynamic energy, sedimentation rate, substrate, oxygenation and food availability, among others). In this sense, the ichnofacies model stands as a valuable tool in sedimentary basin research, in some cases revealing a direct relationship between one particular ichnofacies and a concrete type of deposit, hence the processes involved. One example of this can be found in the *Nereites* ichnofacies, narrowly associated with turbidites, flysch and deepwater deposits.

In this context, the possibility of recognizing a typical trace fossil assemblage and then defining specific archetypal ichnofacies for contourites looms as a challenge. Highly detailed ichnological studies are necessary, involving different types of contourites, from both modern and ancient examples (Wetzel, 2008; Rodríguez-Tovar and Hernández-Molina, 2018). Yet ichnological analyses focusing on the ichnofacies assignation in contourites are scarce. An occasional detailed characterization of the trace fossil assemblages has been presented, as in the case of the upper Cretaceous chalk contourite drift complex, but assignation to archetypal ichnofacies is lacking (Rasmussen and Surlyk, 2012).

Recent years have witnessed more research involving ichnofacies in contourites from outcrops and cores. Ichnological analysis of the carbonate contourite drift at the Petra Tou Romiou outcrop (Cyprus) allowed for assignation of the trace fossil assemblages to the *Zoophycos* ichnofacies and the distal–archetypal–proximal *Cruziana* ichnofacies (Miguez-Salas and Rodríguez-Tovar, 2019b). *Zoophycos* ichnofacies was associated with chalky calcilutite sediment, interpreted as calcareous muddy contourites deposited during low-energy conditions (Hüneke et al., 2020). Increased energy in the environment led to a shift toward the first sandy contouritic facies and the establishment of the distal *Cruziana* ichnofacies with predominant horizontal trace fossils (Miguez-Salas and Rodríguez-Tovar, 2019b). The increase in ichnodiversity, particularly in vertical morphologies, was related to more energetic and proximal environments, as revealed by the transition from archetypal to proximal *Cruziana* ichnofacies (Miguez-Salas and Rodríguez-Tovar, 2019b). In the sandy contouritic drift of the Rifian Corridor, Morocco, ichnological analysis has been conducted in outcrops at the Sidi Chahed and Kirmta sections interpreted as high-energy slope contourite channels; however, the trace-fossil assemblages feature high abundance and low ichnodiversity, with a dominance of horizontal trace fossils including *Macaronichnus* and *Scolicia* (Miguez-Salas et al., 2020a, 2021a) produced by mobile deposit and detritus feeders, discarding a direct assignation to the proximal, high-energy, *Skolithos* ichnofacies. Accordingly, the trace-fossil assemblage could tentatively be assigned to an impoverished proximal *Cruziana* ichnofacies (MacEachern et al., 2012). In this case, paleoenvironmental conditions (e.g., nutrient supply; see below) other than hydrodynamic energy must be considered to explain the dominance of horizontal trace fossils and the absence of vertical biogenic structures. Occasionally, in association with strong bottom currents, in nondepositional horizons (stiff substrate to hardgrounds), a typical *Glossifungites* ichnofacies is developed (Wetzel et al., 2008). Core analysis of the carbonate drift from the Maldives, and its comparison to other carbonate drifts, suggests that *Cruziana* and *Zoophycos* may be the main ichnofacies from carbonate drifts (Reolid and Betzler, 2019).

According to the above, contouritic facies are related to several ichnofacies (i.e., mainly *Zoophycos* and *Cruziana* ichnofacies) instead of just one particular archetypal ichnofacies. Different types of contourites reflect a wide range of paleoenvironmental (ecological and depositional) conditions, and hydrodynamic energy in particular (see below) induces changes in the tracemaker community. Variable conditions furthermore affect the behavior of the tracemakers, extending to the trace fossil assemblage, impeding conclusive ichnofacies assignation.

## Paleoenvironmental conditions during bottom currents: The role of energy in the tracemaker community

In deepwater environments mainly dominated by pelagic to hemipelagic sedimentation, organic matter availability, sedimentation and oxygenation are the main environmental parameters affecting the benthic habitat and therefore determining ichnological features (Uchman and Wetzel, 2011; Wetzel and Uchman, 2012). Other parameters, such as temperature or salinity, play a comparatively minor role in the macrobenthic communities.

A different case entirely is that of deepwater settings affected by the incidence of bottom currents. The term contourites refer to "sediment deposited or substantially reworked by the persistent action of bottom currents" (Stow et al., 2002b; Rebesco et al., 2014). This term embraces a wide spectrum of sediment, conditioned by diverse current types. Numerous processes are linked to the circulation of deepwater masses and bottom currents (Rebesco et al., 2014). Primarily, deep ocean waters are formed in marginal seas or shallow shelf regions where the water becomes cool and dense or highly saline upon strong evaporation (Rebesco et al., 2014). The distribution of salt and heat in the deep ocean is strongly tied to these dense currents (Rebesco et al., 2014). For example, it is well known that the Mediterranean Outflow Water contributes to an increase in temperature and salinity at intermediate depths in the Atlantic Ocean (Khelifi et al., 2009; Rebesco et al., 2014; Rogerson et al., 2012). Because changes in salinity and temperature are narrowly linked to bottom currents, they would influence benthic communities, and then ichnological features, in contourite settings. However, as indicated by Rebesco et al. (2014), the paleontological content of contourites tends to resemble that of the surrounding pelagic and hemipelagic deposits. For this reason, changes in temperature and salinity associated with bottom currents could be discarded as major limiting factors, inducing significant changes in macrobenthic tracemaker communities. In turn, contour currents are typically associated with well-oxygenated water masses (Wetzel et al., 2008), meaning oxygenation would not be a limiting factor for benthic habitats affected by bottom currents.

Besides these three associated ecological features—salinity, temperature and oxygen—bottom currents are characterized by high-energy flows and the potential to produce extensive accumulations of sediment (Rebesco et al., 2014). Thus, hydrodynamic energy and rate of sedimentation could be major impact factors influencing the benthic habitat due to bottom-current activity. Changes in bottom-current velocity and their effects (e.g., erosion, nondeposition and variable rate of deposition) can cause changes in substrate features that might exert a major effect on the benthic habitat (Wetzel et al., 2008). The fact that benthic food is usually limited in the deepwater environment is also a determinant, as the abundance and evolution of benthic fauna are partially controlled by the amount of organic matter that reaches the sediment surface (Suess, 1980; Lampitt, 1985; Gooday and Turley, 1990; Wetzel and Uchman, 2018). Bottom currents may carry a considerable amount of particulate organic matter into deepwater environments (e.g., Thistle et al., 1985; Lucchi and Rebesco, 2007; Wetzel et al., 2008; Rebesco et al., 2014). It follows that a complex interaction between hydrodynamic energy, rates of sedimentation and erosion and benthic food flux would influence the macrobenthic tracemaker community in terms of diversity, abundance, size of organisms and behavior, therefore resulting in variations in ichnological features.

The variable incidence of bottom currents in terms of hydrodynamic energy and rate of sedimentation and benthic food on the benthic habitat has been clearly demonstrated in recent years, based on detailed ichnological analyses at different scales. At the regional scale, in terms of contourite drifts, bottom currents induced variations in composition and behavior of tracemakers, as revealed by ichnological features, affecting the complete trace fossil assemblage or particular ichnotaxa.

As examined earlier, the carbonate contourite drift at the Petra Tou Romiou outcrop (Cyprus) records an ichnofacies replacement, from *Zoophycos* ichnofacies to *Cruziana* ichnofacies, that suggests a relative sea level fall determining a change in the depositional context toward proximal and shallower settings and an increase in bottom-current flow velocity (Miguez-Salas and Rodríguez-Tovar, 2019b). The increase in energy associated to the *Cruziana* ichnofacies is reflected by several ichnological features. First, the presence of vertical burrows from the calcarenitic materials that penetrate the calcilutitic levels was tentatively related to the exhumation of omission suites during minor erosional episodes due to higher bottom-current flow velocity (Miguez-Salas and Rodríguez-Tovar, 2019b). Second, trace fossils (e.g., *Gyrolithes* and *Ophiomorpha*) produced by predominant filter-feeding organisms are absent in the calcarenite beds, but registered in the intercalated calcilutite beds, signaling worse conditions for colonization (Miguez-Salas and Rodríguez-Tovar, 2019b). The flow velocity and sedimentation rate were probably too high for these organisms; suspension feeding would have been impaired because the filter apparatus can easily be plugged by a high concentration of particles in suspension (Thistle et al., 1991). However, larger deposit feeders—those producing *Thalassinoides*, together with *Teichichnus*—are associated with the calcarenitic levels when the flow velocity is high (Miguez-Salas and Rodríguez-Tovar, 2019b). *Teichichnus* spreite structures may indicate an upward migration and, at least partly, equilibrium feedback of the tracemaker to an environment with considerable sedimentation rate (Knaust, 2018).

In the sandy contouritic drift of the Rifian Corridor, Morocco, contouritic outcrops show abundant, and in cases, exclusive presence of the trace fossil *Macaronichnus segregatis degiberti*, recorded for the first time in deepwater contourites deposited in a Late Miocene outer shelf to upper slope setting (Miguez-Salas et al., 2020a). *Macaronichnus* is typical of shallow-marine (mainly foreshore) environments, produced by deposit feeding opheliid polychaetes in high-to-intermediate latitudes. In the studied outcrops, *M. s. degiberti* density is greater near the sandstone units associated with contouritic channels (sides and axes), then decreases progressively away from the channels. The presence of *M. s. degiberti* only locally in the studied bottom-current deposits was interpreted to reveal the interplay between food supply, flow and seafloor heterogeneity; the main factor behind its occurrence would be greater food supply carried through the Mediterranean-Atlantic gateway within the bottom-current outflow. The variable interplay of those environmental conditions (food supply, flow and seafloor heterogeneity) favored the development of communities consisting of opportunistic smaller *Macaronichnus* tracemakers near the contouritic channel axis, and of larger *Macaronichnus* along the sandy drift, with a progressive decrease in density farther from the channels (Miguez-Salas et al., 2020a). The record of dense *Macaronichnus* ichnoasemblages in the contourites could be linked to a high nutrient supply provided by ancient bottom currents, as supported by the record of *Scolicia* (Miguez-Salas et al., 2021a). Indeed, *Scolicia* abundance and size usually increase along with a greater amount of benthic food (Wetzel et al., 2008). The strong paleo-Mediterranean Outflow Water bottom currents that dominated the slope, creating well-oxygenated and nutrient-rich benthic environments, would favor colonization by tracemakers able to exploit such accumulations of organic matter in the sediment. *Macaronichnus*, but also *Scolicia* producers, might develop an opportunistic behavior, creating rapid and extensive bioturbation, impeding colonization by other tracemakers—including suspension feeders—as revealed by the depauperate trace fossil assemblage (Miguez-Salas et al., 2021a).

Ichnological study of cores collected along the NW Iberian Margin shows systematic variation in ichnological content across proximal to distal depocenters within a large-scale elongated contourite drift (Dorador et al., 2019). Data on trace fossils from muddy contourites demonstrate that tracemakers' behavior depends on the relative distance to the core bottom current. Sedimentation rate and organic matter availability are higher in areas proximal to the core, where organic matter is rapidly buried. This prevents oxidation and makes organic matter available for shallow tier tracemakers, those living at or just below the seafloor (i.e., *Palaeophycus* producers). In distal settings, both sedimentation rate and organic matter availability are lower. Organic matter is rapidly oxidized at the surface, favoring the development of middle and deep tier tracemakers, whose activity takes place from a few to some tens of centimeters down within the substrate; hence, they transport organic matter to deeper layers of the sediment (i.e., *Zoophycos* producers). Ichnological features therefore shed light on the close relationship between tracemakers behavior, bottom-current intensity, sedimentation rate and organic matter availability, as well as variations depending on the distance to the core bottom current (see fig. 4 in Dorador et al., 2019).

At a shorter scale, ichnological analysis helps interpret variations in bottom-current features during contourite deposition (Rodríguez-Tovar et al., 2019). In view of the contourite facies model, the calcareous contourites from the Petra Tou Romiou-type section, southern Cyprus, were studied in detail as representative of the standard facies in a contourite bigradational sequence (Chapters 7 and 9). The bigradational sequences of sand-dominated contourites consist of whitish calcarenite beds with wavy layering in the middle portion of the sequence. There are gradual alternations of whitish and greenish calcilutites above and below these calcarenites. Ichnological features, such as the length, shape, diameter and orientation of individual burrow segments, the configuration of burrow systems and external features including fill material and taphonomy, were studied throughout the bigradational sequence. Ichnological and sedimentary results indicate a gradual transition from and toward the greenish and whitish calcilutites found above and below the calcarenite beds, showing a similar trace-fossil assemblage with flattened *Chondrites* isp. and *Planolites* isp. In the noncompacted layers, however, *Planolites* are nearly un-deformed and locally show longitudinal striae that allow assignation to *Planolites reinecki* Książiewicz (Rodríguez-Tovar et al., 2019). This finding was interpreted as indicative of variations in substrate consistency during deposition of the calcarenite beds, associated with intermittent bottom-current processes governing contourite deposition. This conclusion defies the traditional notion that sediment accumulation induced by the bottom current is an uninterrupted process, with a gradually changing flow velocity related to the bigradational sequence formation (e.g., the early models in Faugères et al., 1984; Gonthier et al., 1984; Stow and Holbrook, 1984; see Rebesco et al., 2014, and references therein). Discontinuous deposition and interruption in sedimentation during deposition of the bigradational contourite sequence have recently been confirmed through the ichnological analysis of cores from the Gulf of Cadiz contourite depositional system (de Castro et al., 2020a, b).

## Modern examples: A bridge to interpret the ancient contourite record

Ichnological analyses on contourites have mainly involved ancient examples; the study of modern bottom-current affected environments is very scarce. Neoichnology, the study of modern traces or *lebensspuren* based on the analysis of biogenic structures, is of special interest to evaluate the relationship between physicochemical stress and animal response (i.e., Dashtgard and Gingras, 2012). In modern marine environments, physical parameters (hydrodynamic energy, sediment grain size, sedimentation rate or substrate consistency) and chemical conditions (salinity and bottom or pore-water oxygenation) determine the behaviors of burrowing macrofauna and, consequently, neoichnological features. These conditions are especially stressful in shifting environmental settings, for example, those affected by bottom currents, inducing a variable response of the modern macrobenthic tracemaker community. Most neoichnological studies focus on marginal-marine and shallow-marine environments (i.e., Gingras et al., 2008; Dashtgard and Gingras, 2012; Zonneveld, 2016), whereas deepwater settings are less studied (Wetzel, 2008, 2010; Seike et al., 2012, 2020, and references therein). Still, the in-situ neoichnological analysis of modern environments affected by high-energy currents appears useful for evaluating the relationship between bottom-current dynamics and the deepwater benthic community.

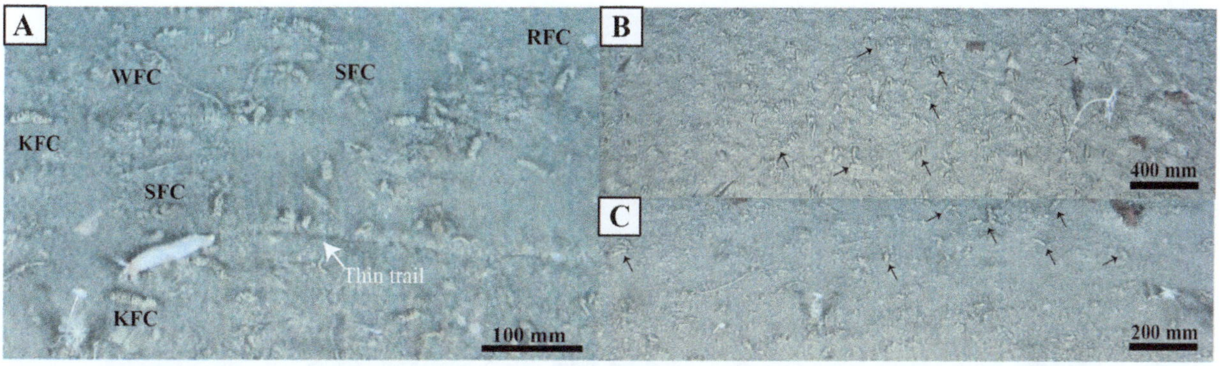


**FIG. 9** (A) Feeding *lebensspuren* morphotypes observed in storm-affected deepwater environments from the site Station M in the northeastern Pacific; (B and C) general view of Station M with dominated knotted fecal cast and smooth fecal cast assemblages (*black arrows*) after a high-energy event. Notes: *KFC*, knotted fecal casts; *RFC*, round fecal casts; *SFC*, smooth fecal casts; *WFC*, wavy fecal casts. Data and pictures of Station M were provided by the David and Lucile Packard Foundation (Monterey Bay Aquarium Research Institution, USA). *From Miguez-Salas, O., Huffard, C.L., Smith Jr., K.L., McGill, P.R., Rodríguez-Tovar, F.J., 2020b. Faunal assemblage changes, bioturbation and benthic storms at an abyssal station in the northeastern Pacific. Deep-Sea Res. I Oceanogr. Res. Pap. 160, 103277.*

With the development of new equipment, including camera systems (see Miguez-Salas et al., 2020b, and references therein for details), the activity of benthic communities in deep environments can be recorded, for instance those affected by episodic increases in bottom-current intensity. Miguez-Salas et al. (2020b) reported on the evolution of deepwater *lebensspuren* assemblages before, during and after high-energy episodes tied to storm events (Fig. 9). This study demonstrated variations in the diversity and abundance of *lebensspuren* assemblages with regard to changes in energy conditions affecting local-scale erosion, resuspension of unconsolidated seafloor sediment and reorganization of organic matter resources (Miguez-Salas et al., 2020b). Such results are of special significance not only for the field of neoichnology but also for paleoenvironmental reconstructions of bottom-current affected environments. Further development of this research theme will bring important advances in contourite analysis.

## New techniques to advance in the ichnological analysis of contourites

Ichnological information in contouritic deposits has traditionally been included as additional data, complementing facies analysis based on visual observations during sedimentary descriptions (e.g., Stow and Lovell, 1979; Faugères et al., 1984). However, comprehensive study of all ichnological properties and their subsequent interpretation are only possible when trace fossil characterization is viable (Knaust, 2012). This is a straightforward matter when ichnological analysis is conducted on contourite outcrops, but the approach becomes complicated in the case of core analysis. Over the last decade, this problem has been resolved by means of novel techniques applied to enhance the visibility of trace fossils in cores, especially in modern ones, facilitating the ichnological analysis of contouritic deposits (Fig. 10).

Digital image processing is an important tool for the ichnological analysis of deepwater deposits (see Dorador and Rodríguez-Tovar, 2018 for a review). When applied to contouritic deposits from cores, image treatment firmly supports ichnological analysis (e.g., Alonso et al., 2016, 2021; Jones, 2016; de Castro et al., 2020a, b, 2021). Fiji and ICY software (de Chaumont et al., 2012; Schindelin et al., 2012; see Miguez-Salas et al., 2019, for detailed description) have been successfully used to enhance trace fossil visibility and quantify bioturbation from contouritic samples both in outcrops and in cores (Miguez-Salas et al., 2019; Rodríguez-Tovar et al., 2020). Quantification of bioturbation allows correlation with the most traditionally used indices as the bioturbation index (Reineck, 1963, 1967; Taylor and Goldring, 1993) and the ichnofabric indices (Droser and Bottjer, 1986, 1989), both mainly based on the percentage of bioturbated surface. Penetrative techniques such as X-ray imaging or computed tomography (CT) have been also applied to this end. Some studies rely on X-ray radiographs to characterize bioturbation in contourites from cores with greater detail (e.g., Löwemark et al., 2004; Wetzel et al., 2008). CT is a relatively young technique that is growing in use in ichnological studies, as it allows one to obtain 3D information without destroying the rock sample or sediment core (e.g., Fu et al., 1994; Pemberton and Gingras, 2005; Baniak et al., 2013; Dorador and Rodríguez-Tovar, 2020). Its application in contouritic cores serves to image morphological features, to analyze the trace fossil assemblages and to quantify bioturbated rock volumes in subsurface oil and gas studies, among other purposes. CT data can also be used to study ichnoassemblages from contouritic intervals in

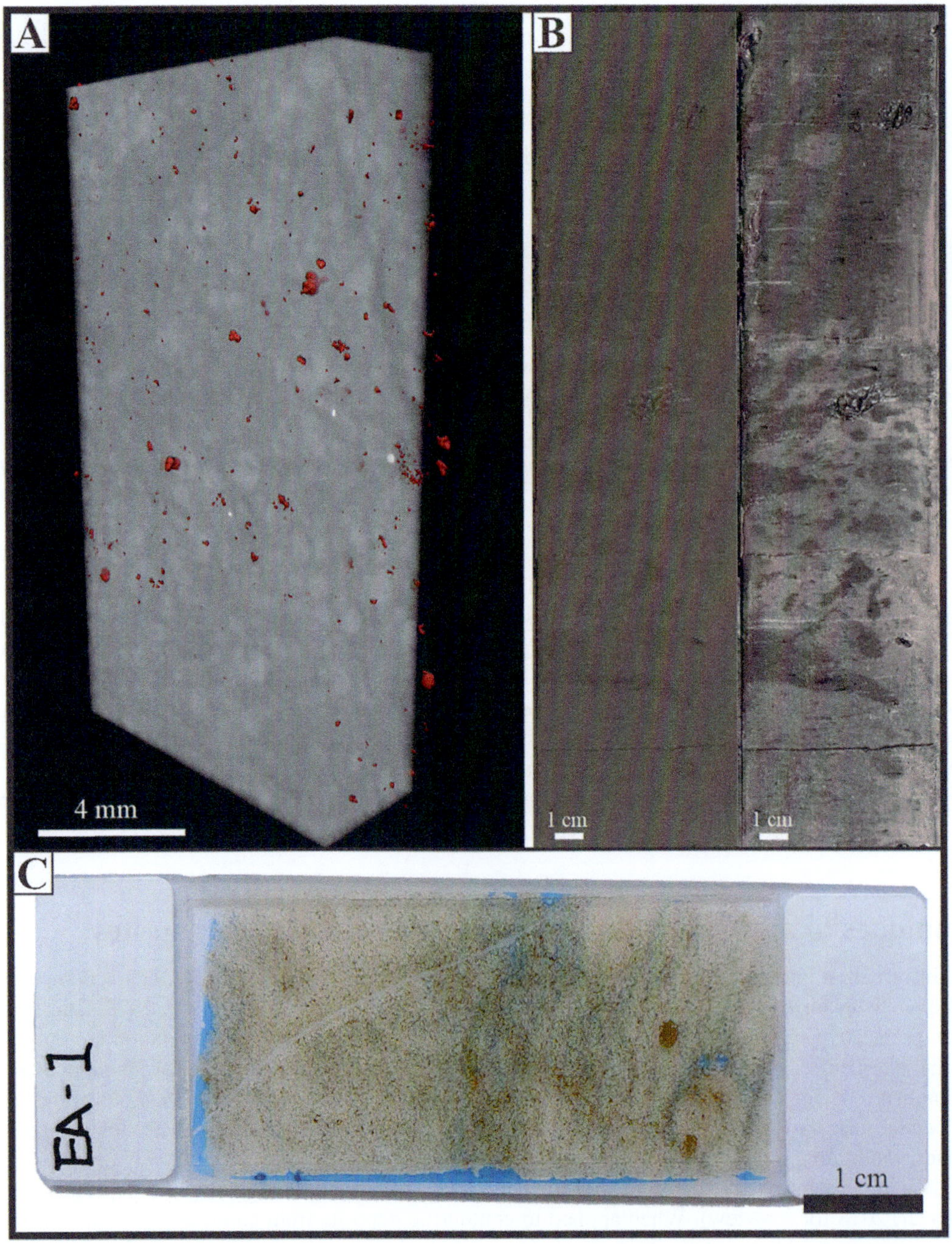


FIG. 10 Applied techniques in the ichnological analysis of contourites. (A) Micro-CT; (B) high-resolution image treatment; (C) thin sectioning with blue resin impregnation.

sediment cores. Integrative analyses of 2D slices and the 3D reconstruction of burrows proved helpful to characterize the trace fossil assemblages in sediment cores from the Galicia Interior Basin, offshore Portugal (Dorador et al., 2019, 2021a).

Thin-section analysis is a traditional method that is still commonly used in ichnology and recently applied to evaluate the ichnological content of contourite deposits (e.g., Dorador et al., 2021b). This technique aids the study of small morphological features, for instance the halo and inner core of *Macaronichnus* from Moroccoan contourite deposits (Dorador et al., 2021b). Moreover, some samples impregnated with blue resin facilitated study of the impact of bioturbation on petrophysical properties and the relative roles of *Macaronichnus* rim and tube cores affecting porosity (Dorador et al., 2021b; see detailed below). Mercury intrusion porosimetry has recently been applied to evaluate and compare the incidence on porosity of several ichnotaxa from different types of contourites (Rodríguez-Tovar et al., 2021).

## Bioturbation and petrophysical properties in contourites: Economic interest

The importance of contourites in the context of hydrocarbon exploitation has been addressed in scientific output, facies being of significant economic interest now and for the future (Rebesco et al., 2014; Shanmugam, 2017a, b and references therein). Contour currents affect hydrocarbon reservoirs in numerous forms, including reservoir geometry and quality and the distribution of reservoirs and sealing rocks (Rebesco et al., 2014). Particularly at the level of contourite facies, coarse-grained contourites can represent hydrocarbon reservoirs, whereas fine-grained contourites can represent seals (Rebesco et al., 2014). The porosity and permeability of particular types of contourites (i.e., bottom-current reworked sands) may be very high, suggesting even better reservoir quality than for turbidites (see Chapters 6, 7, 9; Shanmugam, 2017a, b).

One key aspect of contourites resides in bioturbation. It is largely considered a diagnostic feature, but its significance depends on the type of contourite (Rodríguez-Tovar and Hernández-Molina, 2018). As indicated by Rebesco et al. (2014), the activity of benthic organisms is common in contourites, destroying the primary sedimentary structures and fabric. Over the years bioturbation has come to be an important indicator for reservoir quality estimation in the oil and gas industry. Bioturbation potentially alters the fabric and texture of sedimentary rock, including primary sedimentary structures (i.e., lamination), redistribution of grains and—at a minor scale—compaction. It also prompts certain geochemical changes owing to the incorporation of organic matter, organism activity, feeding behavior and so on, determining significant alterations of the primary sedimentary matrix (Pemberton and Gingras, 2005; Gingras et al., 2012; Liu et al., 2019).

Biogenic activities can affect petrophysical properties such as porosity, permeability and connectivity (Pemberton and Gingras, 2005; Knaust, 2017; Eltom et al., 2019). It is generally believed that bioturbation diminishes permeability, since burrowing organisms reduce sorting and may favor sediment homogenization (see Gingras et al., 2012). Nevertheless, several examples have shown that bioturbation may enhance bulk permeability, modify pore structure and affect reservoir fluid flow (e.g., Pemberton and Gingras, 2005; Gordon et al., 2010; Cunningham et al., 2012; La Croix et al., 2013; Eltom et al., 2019; Liu et al., 2019; Quaye et al., 2019; Knaust et al., 2020). Thus, the influence of bioturbation on rock properties can be either positive or negative—that is, either increasing or decreasing porosity, respectively, with respect to the host rock. Detailed ichnological analyses should be performed on every bioturbated reservoir to evaluate the effect of bioturbation and estimate the reservoir quality (Gingras et al., 2012; Knaust, 2017). Traditionally, more attention is paid to particular characteristics of burrows that bear an impact on rock properties and reservoir quality: density, tiering, size, orientation, architecture, fill and lining (Gingras et al., 2012; Knaust, 2013). Pemberton and Gingras (2005) classified flow media based on the sedimentological, ichnological, diagenetic and stratigraphical context, recognizing several interrelated scenarios. In the wake of the paper, studies have explored a variety of depositional environments, sedimentary rocks and ichnotaxa, evaluating the influence of bioturbation on the distribution of porosity, permeability and implications for reservoir development (i.e., Gordon et al., 2010; Knaust, 2013, 2017; La Croix et al., 2013, 2017; Eltom et al., 2019; Liu et al., 2019; Knaust et al., 2020).

According to the above, one essential ichnological research theme to be developed in coming years within the framework of contourite analysis is the impact of ichnological features on the rock properties of contourites. This theme has been initiated through the analysis of several types of contourites and different ichnotaxa, applying diverse techniques (Dorador et al., 2021b; Miguez-Salas et al., 2021b; Rodríguez-Tovar et al., 2021). Several cases can be illustrated (Fig. 11).

### The role of *Macaronichnus*: Ichnological digital image analysis package and impregnated thin sections

Dorador et al. (2021b) analyzed the effect of the burrow lining on rock properties, as a particular characteristic of burrows scarcely studied to date. The study focused on the thickness and composition of the lining or mantle (rim) of *Macaronichnus* (Fig. 10A–C), which is known to be a relevant trace fossil regarding reservoir quality in bioturbated deposits, mainly in terms of its occurrence in dense concentrations (Gingras et al., 2002; Pemberton and Gingras, 2005; Knaust, 2009; Gordon et al., 2010; Quaye et al., 2019). Thus, thin sections of selected specimens of *Macaronichnus segregatis degiberti* from El Adergha and Kirmta contourite outcrops in northern Morocco were studied for petrographic characterization including impregnation with blue dye resin to highlight porosity. The host rock, tube core and rim were distinguished for these analyses. Additionally, porosity from all the thin sections was characterized by blue pixel count using Photoshop (Dorador and Rodríguez-Tovar, 2014, 2018; Dorador et al., 2014a, b), giving a representative mean value for the samples on a whole and from every single part (host rock, tube core, rim). Intensity of bioturbation was quantified using the Ichnological Digital Image Analysis Package (Dorador and Rodríguez-Tovar, 2014, 2018; Dorador et al., 2014a, b). Results reveal that *Macaronichnus* is a complex case, exhibiting dual behavior. Burrow rims have greater porosity than the host sediment and then a positive effect. Fillings in tube cores have lower porosity and then a negative effect. On this basis, the relative volume occupied by rims and tube cores should be determined to estimate the real effect of *Macaronichnus* on porosity.

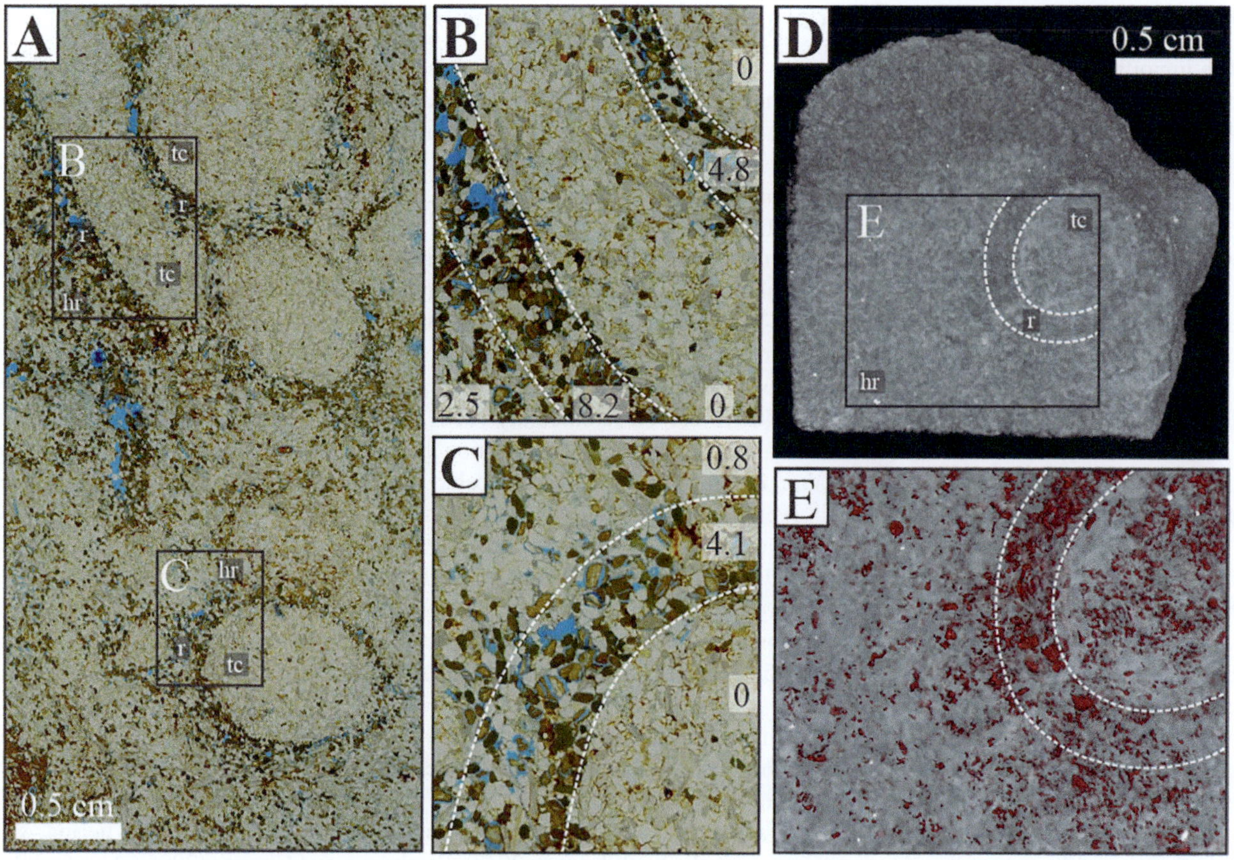


**FIG. 11** *Macaronichnus* and petrophysical properties in contourites. Samples with porosity zonation caused by the activity of *Macaronichnus* tracemaker observed in impregnated thin sections (A–C) and micro-CT analysis (D and E). Numbers in B and C show porosity values (%) of the host rock (hr) and in each zone of the *Macaronichnus* burrow as rim (r) and tube core (tc).

Any evaluation of the impact of *Macaronichnus* on petrophysical properties, hence on reservoir potential, based only on the intensity of bioturbation or the study of entire burrows would introduce significant misinterpretation.

## Clastic- vs calcareous-dominated contourites: X-ray micro-CT analysis

To improve characterization of the impact of bioturbation on petrophysical rock properties of contourites, Miguez-Salas et al. (2021b) applied a new methodology—X-ray micro-CT analysis (computed tomography)—a highly detailed procedure affording more precise porosity evaluation than traditional methods (Fig. 11D–E). Clastic- and calcareous-dominated contourite deposits were studied, focusing on the matrix and the main trace fossils. In the sandy clastic contourite deposits analyzed, *Macaronichnus* had the greatest impact on (modified) porosity distribution, but of variable influence; *M. segregatis* specimens show that porosity values are up to three times higher in the burrow rim than inside, generating a dual-porosity flow medium, while *M. s. degiberti* is indicative of burrow fill that is less porous than the host sediment, with the burrow rim alone recording high values. This variable influence on porosity distribution could be associated with different behaviors of the tracemaker during feeding activities, or even with different producers. In calcareous contourite deposits, *Chondrites* from muddy contourite chalk have a neutral effect on porosity, while *Thalassinoides* generate a minor effect on porosity, reaching slightly higher values in the burrow fill. Again, this minor effect is related to similarity between the passive infilling material and the host sediment, indicating the small influence of the tracemaker on the redistribution of grains. Results reveal that the application of precise techniques such as micro-CT analysis can advance the appraisal of bioturbation in contourites for detailed reservoir characterization studies.

## Trace fossils and interparticle/intraparticle porosity: Mercury intrusion porosimetry

The research study by Rodríguez-Tovar et al. (2021) applies, for the first time, the mercury intrusion porosimetry technique to evaluate the effects of various ichnotaxa on an array of contourites. Samples of dominant-calcareous contourites were selected from the Eocene-Early Miocene of Cyprus, with *Chondrites*, *Planolites* and *Thalassinoides*, and sandy clastic contourites from the Late Miocene Rifian Corridor of Morocco with *Macaronichnus*, *Parahaentzschelinia* and *Scolicia*, comparing the host rock with the infilling material. Total porosity shows similar values for the host sediment in clastic and calcarenitic contourites. However, significant differences in interparticle to intraparticle porosity are registered—interparticle porosity is almost absent in calcarenitic contourites, where intraparticle porosity is dominant, mainly related to the microfacies composition and the abundance of foraminifera chambers. Higher total porosity values were obtained for the infilling material of *Parahaentzschelinia*, *Scolicia* and *Macaronichnus*, possibly related to the tracemaker behavior and redistribution of grains during feeding activity. Thus, the role of bioturbation in rock-porosity modification within the studied contourites, hence its impact on reservoir quality, would be variable. For clastic contourites, bioturbation determines an increase in total porosity, especially in interparticle porosity, meaning a positive impact for reservoir quality; yet in calcarenitic and muddy chalk facies, bioturbation has a minor effect on porosity and therefore a neutral impact on the reservoir quality.

Altogether, advantages and disadvantages can be identified for the applied techniques, mainly according to the type of contourites, their ichnological features or particular objectives pursued. An integrative analysis is needed to determine the most appropriate technique in view of the particular situation in question (research in progress).

## Key messages

Detailed ichnological analysis of along-slope bottom-current deposits, scarcely applied to date, can be considered a key tool for an accurate characterization of the involved processes and the final results. Ichnological analysis must be based on detailed field and laboratory studies, involving outcrops and cores, and considering all the ichnological features.

Bioturbation can be considered a bridge between past and modern deepwater environments affected by bottom currents. Depositional and ecological conditions of ancient along-slope bottom currents can be evaluated on neoichnological studies of modern environments affected by high-energy currents. Thus, audiovisual techniques (i.e., photographs and videos) from deepwater areas should be analyzed in detail to study benthic organisms' activity during high-energy periods. Recent bioturbation structures produced on and within the sediment (i.e., *lebensspuren*) can be compared with trace fossils from ancient contourites allowing characterization and differentiation with respect to other deepwater facies as pelagites, hemipelagites and sediment gravity flow deposits. Integration of neo- and paleo-ichnological information provides a unique opportunity for understanding bottom-current processes and contouritic sediments.

Ichnological data must be considered when the economic interest of contourites and other deepwater deposits is evaluated, especially due to the influence of bioturbation on rocks' flow-media properties (e.g., porosity and permeability) and its impact on reservoir quality. The application of a multi-technique approach is revealed as the most effective way to provide an accurate and precise characterization.

## Conclusions

Ichnological study of contouritic deposits is showing great progress, demonstrating its relevance for the characterization and differentiation of these facies with respect to other deepwater sedimentary intervals.

To date, no trace fossil assemblage can be considered typical of contourites; they vary depending on the type of contourite.

The diversity of trace fossil assemblages observed in carbonate drifts thus far appears to be slightly higher than that of siliciclastic contourites.

Contouritic facies may be related to several ichnofacies (mainly *Zoophycos* and *Cruziana* ichnofacies), instead of just one archetypal ichnofacies.

The complex interaction between hydrodynamic energy, rates of sedimentation and erosion and benthic food flux associated with bottom currents is acknowledged as a major factor inducing changes in the macrobenthic tracemaker community and in ichnological features.

In situ neoichnological analysis of modern environments affected by high-energy currents stands as a useful tool for evaluating the relationship between bottom-current dynamics and the deepwater benthic community.

Detailed ichnological analyses of contourites and other deepwater deposits are critical for their discrimination and for understanding onshore and offshore rock properties, not only for oil and gas exploration and development, but also for any endeavor that requires careful characterization of reservoir to non-reservoir (i.e., seal, overburden) within a sedimentary basin.

## Acknowledgments

We would like to thank the Editors for their valuable comments and suggestions regarding this manuscript. Our research was funded by Projects CGL2015-66835-P and PID2019-104625RB-100 (Secretaría de Estado de I+D+I, Spain), B-RNM-072-UGR18 and A-RNM-368-UGR20 (FEDER Andalucía), and P18-460 RT-4074 (Junta de Andalucía) and Scientific Excellence Unit UCE-2016-05 (Universidad de Granada). The research of OM-S is funded through a pre-doctoral grant from the Ministerio de Educación, Cultura y Deporte (Gobierno de España). The research by JD was funded through the *Juan de la Cierva* Program (IJC2019-038866-I) of the *Ministerio de Ciencia e Innovación, Gobierno de España*.

## References

Alonso, B., Ercilla, G., Casas, D., Stow, D.A.V., Rodríguez-Tovar, F.J., Dorador, J., Hernández-Molina, F.J., 2016. Contourite vs gravity-flow deposits of the Pleistocene Faro drift (Gulf of Cadiz): sedimentological and mineralogical approaches. Mar. Geol. 377, 77–94.

Alonso, B., Juan, C., Ercilla, G., Cacho, I., López-González, N., Rodríguez-Tovar, F.J., Dorador, J., Francés, G., Casas, D., Vandorpe, T., Vázquez, J.T., 2021. Paleoceanographic and paleoclimatic variability in the Western Mediterranean during the last 25 cal. kyr BP. New insights from contourite drifts. Mar. Geol. 437, 106488.

Bahr, A., Jiménez-Espejo, F.J., Kolasinac, N., Grunert, P., Hernández-Molina, F.J., Röhl, U., Voelker, A.H., Escutia, C., Stow, D.A., Hodell, D., Alvarez-Zarikian, C.A., 2014. Deciphering bottom current velocity and paleoclimate signals from contourite deposits in the Gulf of Cádiz during the last 140 kyr: an inorganic geochemical approach. Geochem. Geophys. Geosyst. 15 (8), 3145–3160.

Baldwin, C.T., McCave, I.N., 1999. Bioturbation in an active deep-sea area: implications for models of trace fossil tiering. Palaios 14, 375–388.

Baniak, G.M., Gingras, M.K., Pemberton, S.G., 2013. Reservoir characterization of burrow-associated dolomites in the upper Devonian Wabamun Group, Pine Creek gas field, Central Alberta, Canada. Mar. Pet. Geol. 48, 275–292.

Bertling, M., Braddy, S.J., Bromley, R.G., Demathieu, G.R., Genise, J., Mikuláš, R., Nielsen, J.K., Nielsen, K.S.S., Rindsberg, A.K., Schlirf, M., Uchman, A., 2006. Names for trace fossils: a uniform approach. Lethaia 39, 265–286.

Brackenridge, R.E., Stow, D.A., Hernández-Molina, F.J., Jones, C., Mena, A., Alejo, I., Ducassou, E., Llave, E., Ercilla, G., Nombela, M.A., Perez-Arlucea, M., 2018. Textural characteristics and facies of sand-rich contourite depositional systems. Sedimentology 65 (7), 2223–2252.

Bromley, R.G., 1996. Trace Fossils. Biology, Taphonomy and Applications. Chapman & Hall, London.

Buatois, L.A., Mángano, M.G., 2011. Ichnology: Organism-Substrate Interactions in Space and Time. Cambridge University Press, Cambridge.

Buatois, L.A., Mángano, M.G., Pattison, S.A.J., 2019. Ichnology of prodeltaic hyperpycnite-turbidite channel complexes and lobes from the upper cretaceous Praire Canyon Member of the Mancos Shale, Book Cliffs, Utah, USA. Sedimentology 66, 1825–1860.

Capella, W., Hernández-Molina, F.J., Flecker, R., Hilgen, F.J., Hssain, M., Kouwenhoven, T.J., van Oorschot, M., Sierro, F.J., Stow, D.A.V., Trabucho-Alexandre, J., Tulbure, M.A., 2017. Sandy contourite drift in the late Miocene Rifian corridor (Morocco): reconstruction of depositional environments in a foreland-basin seaway. Sediment. Geol. 355, 31–57.

Chough, S.K., Hesse, R., 1985. Contourites from Eirik Ridge, south of Greenland. Sediment. Geol. 41, 185–199.

Cunningham, K.J., Sukop, M.C., Curran, H.A., 2012. Carbonate aquifers. In: Knaust, D., Bromley, R.G. (Eds.), Trace Fossils as Indicators of Sedimentary Environments. Developments in Sedimentology, vol. 64. Elsevier, Amsterdam, pp. 869–896.

Dashtgard, S.E., Gingras, M.K., 2012. Marine invertebrate neoichnology. In: Knaust, D., Bromley, R.G. (Eds.), Trace Fossils as Indicators of Sedimentary Environments. Developments in Sedimentology, vol. 64. Elsevier, Amsterdam, pp. 273–295.

de Castro, S., Hernández-Molina, F.J., Rodríguez-Tovar, F.J., Llave, E., Ng, L.Z., Nishida, N., Mena, A., 2020a. Contourites and bottom current reworked sands: bed facies model and implications. Mar. Geol. 428, 106267.

de Castro, S., Hernández-Molina, F.J., De Weger, W., Jiménez-Espejo, F.J., Rodríguez-Tovar, F.J., Mena, A., Llave, E., Sierro, F.J., 2020b. Contourite characterization and its discrimination from other deep-water deposits in the Gulf of Cadiz contourite depositional system. Sedimentology 68, 987–1027.

de Castro, S., Miramontes, E., Dorador, J., Jouet, G., Cattaneo, A., Rodríguez- Tovar, F.J., Hernández-Molina, F.J., 2021. Siliciclastic and bioclastic contouritic sands: textural and geochemical characterisation. Mar. Pet. Geol. 128, 105002.

de Chaumont, F., Dallongeville, S., Chenouard, N., Hervé, N., Pop, S., Provoost, T., Meas-Yedid, V., Pankajakshan, P., Lecomte, T., Montagner, Y., Lagache, T., Dufour, A., Olivo-Marin, J.-C., 2012. ICY: an open bioimage informatics platform for extended reproducible research. Nat. Methods 9, 690–696.

De Weger, W., Hernández-Molina, F.J., Miguez-Salas, O., de Castro, S., Bruno, M., Chiarella, D., Sierro, F.J., Blackbourn, G., Manar, M.A., 2021. Contourite depositional system after the exit of a strait: case study from the late Miocene Rifian Corridor, Morocco. Sedimentology. https://doi.org/10.1111/sed.12882.

Dorador, J., Rodríguez-Tovar, F.J., 2014. A novel application of digital image treatment by quantitative pixel analysis to trace fossil research in marine cores. Palaios 29, 533–538.

Dorador, J., Rodríguez-Tovar, F.J., 2015. Application of digital image treatment to the characterization and differentiation of deep-sea ichnofacies. Span. J. Paleontol. 30, 265–274.

Dorador, J., Rodríguez-Tovar, F.J., 2018. High-resolution image treatment in ichnological core analysis: initial steps, advances and prospects. Earth Sci. Rev. 177, 226–237.

Dorador, J., Rodríguez-Tovar, F.J., 2020. CroSSED sequence, a new tool for 3D processing in geosciences using the free software 3DSlicer. Sci. Data 7, 270.

Dorador, J., Rodríguez-Tovar, F.J., IODP Expedition 339 Scientists, 2014a. Digital image treatment applied to ichnological analysis of marine core sediments. Facies 60, 39–44.

Dorador, J., Rodríguez-Tovar, F.J., IODP Expedition 339 Scientists, 2014b. Quantitative estimation of bioturbation based on digital image analysis. Mar. Geol. 349, 55–60.

Dorador, J., Rodríguez-Tovar, F.J., Mena, A., Francés, G., 2019. Lateral variability of ichnological content in muddy contourites: weak bottom currents affecting organisms' behavior. Sci. Rep. 9 (1), 1–7.

Dorador, J., Rodríguez-Tovar, F.J., Mena, A., Francés, G., 2021a. Deep-sea bottom currents influencing tracemaker community: an ichnological study from the NW Iberian margin. Mar. Geol. 437, 106503.

Dorador, J., Rodríguez-Tovar, F.J., Miguez-Salas, O., 2021b. The complex case of *Macaronichnus* trace fossil affecting rock porosity. Sci. Rep. 11, 1975.

Droser, M.L., Bottjer, D.J., 1986. A semiquantitative field classification of ichnofabric. J. Sediment. Petrol. 56, 558–559.

Droser, M.L., Bottjer, D.J., 1989. Ichnofabric of sandstones deposited in high-energy nearshore environments: measurement and utilization. Palaios 4, 598–604.

Ekdale, A.A., Bromley, R.G., 1983. Trace fossils and ichnofabric in the Kjølby Gaard Marl, Upper Cretaceous, Denmark. Bull. Geol. Soc. Den. 31, 107–119.

Ekdale, A.A., Bromley, R., Knaust, D., 2012. The ichnofabric concept. In: Knaust, D., Bromley, R.G. (Eds.), Trace Fossils as Indicators of Sedimentary Environments. Developments in Sedimentology, vol. 64, pp. 139–155.

Eltom, H.A., Rankey, E.C., Hasiotis, S.T., Barai, R., 2019. Effect of bioturbation on petrophysical properties: insights from geostatistical and flow simulation modeling. Mar. Pet. Geol. 104, 259–269.

Faugères, J.-C., Stow, D.A.V., 1993. Bottom-current-controlled sedimentation: a synthesis of the contourite problem. Sediment. Geol. 82, 287–297.

Faugères, J.-C., Gonthier, E., Stow, D.A.V., 1984. Contourite drift molded by deep Mediterranean outflow. Geology 12, 296–300.

Fu, S., Werner, F., 1994. Distribution and composition of biogenic structures on the Iceland-Faeroe ridge: relation to different environments. Palaios 9, 92–101.

Fu, S., Werner, F., Brossmann, J., 1994. Computed tomography: application in studying biogenic structures in sediment cores. Palaios 9, 116–119.

Gingras, M.K., MacMillan, B., Balcom, B.J., Saunders, T., Pemberton, S.G., 2002. Using magnetic resonance imaging and petrographic techniques to understand the textural attributes and porosity distribution in *Macaronichnus*-burrowed sandstone. J. Sediment. Res. 72, 552–558.

Gingras, M.K., Dashtgard, S.E., MacEachern, J.A., Pemberton, S.G., 2008. Biology of shallow marine ichnology: a modern perspective. Aquat. Biol. 2, 255–268.

Gingras, M.K., Baniak, G., Gordon, J., Hovikoski, J., Konhauser, K.O., La Croix, A., Lemiski, R., Mendoza, C., Pemberton, S.G., Polo, C., Zonneveld, J.-P., 2012. Porosity and permeability in bioturbated sediments. In: Knaust, D., Bromley, R.G. (Eds.), Trace Fossils as Indicators of Sedimentary Environments. Developments in Sedimentology, vol. 64. Elsevier, Amsterdam, pp. 837–868.

Gonthier, E., Faugères, J.-C., Stow, D.A.V., 1984. Contourite facies of the Faro Drift, Gulf of Cadiz. In: Stow, D.A.V., Piper, D.J.W. (Eds.), Fine Grained Sediments, Deepwater Processes and Facies. vol. 15. Geol. Soc., London, Special Publication, pp. 275–291.

Gooday, A.J., Turley, C.M., 1990. Responses by benthic organisms to inputs of organic material to the ocean floor: a review. Philos. Trans. R. Soc. A Math. Phys. Eng. Sci. 331 (1616), 119–138.

Gordon, J., Pemberton, G., Gingras, M., Konhauser, K.O., 2010. Biogenically enhanced permeability: a petrographic analysis of *Macaronichnus segregatus* in the lower cretaceous Bluesky formation, Alberta, Canada. AAPG Bull. 52549 (94), 1779–1795.

Hanebuth, T.J., Zhang, W., Hofmann, A.L., Löwemark, A.L., Schwenk, T., 2015. Oceanic density fronts steering bottom-current induced sedimentation deduced from a 50 ka contourite-drift record and numerical modeling (off NW Spain). Quat. Sci. Rev. 112, 207–225.

Hernández-Molina, F.J., Sierro, F.J., Llave, E., Roque, C., Stow, D.A.V., Williams, T., Lofi, J., Van der Schee, M., Arnáiz, A., Ledesma, S., Rosales, C., 2016. Evolution of the gulf of Cadiz margin and Southwest Portugal contourite depositional system: tectonic, sedimentary and paleoceanographic implications from IODP expedition 339. Mar. Geol. 377, 7–39.

Hovikoski, J., Uchman, A., Weibel, R., Nøhr-Hansen, H., Sheldon, E., Ineson, J., Bjerager, M., Therkelsen, J., Olivarius, M., Larsen, M., Alsen, P., Bojesen-Koefoed, J., 2020. Upper cretaceous bottom current deposits, north-East Greenland. Sedimentology 67, 3619–3654.

Hüneke, H., 2013. Bioclastic contourites: depositional model for bottom-current redeposited pelagic carbonate ooze (Devonian, Moroccan Central Massif). Z. Dtsch. Ges. Geowiss. 164 (2), 253–277.

Hüneke, H., Mulder, T., 2011. Deep-sea sediments. In: Developments in Sedimentology. vol. 63. Elsevier, Amsterdam.

Hüneke, N.V., Stow, D.A.V., 2008. Identification of ancient contourites: problems and palaeoceanographic significance. In: Rebesco, M., Camerlenghi, A. (Eds.), Contourites. Developments in Sedimentology, vol. 60. Elsevier, pp. 323–344.

Hüneke, H., Hernández-Molina, F.J., Rodríguez-Tovar, F.J., Llave, E., Chiarella, D., Mena, A., Stow, D.A.V., 2020. Diagnostic criteria based on microfacies for calcareous contourites, turbidites and pelagites in the Eocene-Miocene slope succession, southern Cyprus. Sedimentology 68, 557–592.

Jones, C., 2016. Sandy Contourite Deposits in the Gulf of Cadiz: Sedimentological Aspects and its Evolutionary Implication with the Gibraltar Gateway. MSc Earth Science by Research Thesis, Royal Holloway University of London.

Khelifi, N., Sarnthein, M., Andersen, N., Blanz, T., Frank, M., Garbe-Schonberg, D., Haley, B.A., Stumpf, R., Weinelt, M., 2009. A major and long-term Pliocene intensification of the Mediterranean outflow, 3.5-3.3 Ma ago. Geology 37, 811–814.

Knaust, D., 2009. Ichnology as a tool in carbonate reservoir characterization: a case study from the Permian-Triassic Khuff Formation in the Middle East. GeoArabia 14, 17–38.

Knaust, D., 2012. Methodology and techniques. In: Knaust, D., Bromley, R.G. (Eds.), Trace Fossils as Indicators of Sedimentary Environments. Developments in Sedimentology, vol. 64. Elsevier, pp. 245–272.

Knaust, D., 2013. Bioturbation and Reservoir Quality: Towards a Genetic Approach. AAPG. Search and Discovery, Article #50900.

Knaust, D., 2017. Atlas of Trace Fossils in Well Core: Appearance, Taxonomy and Interpretation. Springer, Cham, Switzerland.

Knaust, D., 2018. The ichnogenus *Teichichnus* Seilacher, 1955. Earth Sci. Rev. 177, 386–403.

Knaust, D., Dorador, J., Rodríguez-Tovar, F.J., 2020. Burrowed matrix powering dual porosity systems—a case study from the Maastrichtian chalk of the Gullfaks Field, Norwegian North Sea. Mar. Pet. Geol. 113, 104158.

La Croix, A.D., Gingras, M.K., Pemberton, S.G., Mendoza, C.A., MacEachern, J.A., Lemiski, R.T., 2013. Biogenically enhanced reservoir properties in the Medicine Hat gas field, Alberta, Canada. Mar. Pet. Geol. 43, 464–477.

La Croix, A.D., MacEachern, J.A., Ayranci, K., Hsieh, A., Dashtgard, S.E., 2017. An ichnological-assemblage approach to reservoir heterogeneity assessment in bioturbated strata: insights from the lower cretaceous Viking Formation, Alberta, Canada. Mar. Pet. Geol. 86, 636–654.

Lampitt, R.S., 1985. Evidence for the seasonal deposition of detritus to the deep-sea floor and its subsequent resuspension. Deep-Sea Res. I 32 (8), 885–897.

Li, H., van Loon, A.J., He, Y., 2020. Cannibalism of contourites by gravity flows: explanation of the facies distribution of the Ordovician Pingliang formation along the southern margin of the Ordos Basin, China. Can. J. Earth Sci. 57 (3), 331–347.

Liu, H., Shi, K., Liu, B., Song, X., Guo, R., Li, Y., Wang, G., Wang, H., Shen, Y., 2019. Characterization and identification of bioturbation-associated high permeability zones in carbonate reservoirs of upper cretaceous Khasib Formation, AD oilfield, central Mesopotamian Basin, Iraq. Mar. Pet. Geol. 110, 747–767.

Lowell, J.P.B., Stow, D.A.V., 1981. Identification of ancient sandy contourites. Geology 9, 347–349.

Löwemark, L., Schönfeld, J., Werner, F., Schäfer, P., 2004. Trace fossils as a paleoceanographic tool: evidence from late quaternary sediments of the southwestern Iberian margin. Mar. Geol. 204, 27–41.

Lucchi, R.G., Rebesco, M., 2007. Glacial contourites on the Antarctic Peninsula margin: insight for palaeoenvironmental and palaeoclimatic conditions. In: Viana, A.R., Rebesco, M. (Eds.), Economic and Palaeoceanographic Significance of Contourite Deposits. vol. 276. Geological Society of London Special Publication, pp. 111–127.

MacEachern, J.A., Bann, K.L., Pemberton, S.G., Gingras, M.K., 2007. The ichnofacies paradigm: high-resolution paleoenvironmental interpretation of the rock record. In: MacEachern, J.A., Bann, K.L., Gingras, M.K., Pemberton, S.G. (Eds.), Applied Ichnology. Society for Sedimentary Geology Short Course Notes. SEPM, Society for Sedimentary Geology, pp. 27–64.

MacEachern, J.A., Pemberton, S.G., Gingras, M.K., Bann, K.L., 2010. Ichnology and facies models. In: James, N.P., Dalrymple, R.W. (Eds.), Facies Models 4. vol. 6. Geological Association of Canada, pp. 19–58.

MacEachern, J.A., Bann, K.L., Gingras, M.K., Zonneveld, J.-P., Dashtgard, S.E., Pemberton, S.G., 2012. The ichnofacies paradigm. In: Knaust, D., Bromley, R.G. (Eds.), Trace Fossils as Indicators of Sedimentary Environments. Developments in Sedimentology, vol. 64. Elsevier, pp. 103–138.

Mángano, M.G., Buatois, L., 2012. A multifaceted approach to ichnology. Ichnos 19, 121–126.

Martín-Chivelet, J., Fregenal-Martínez, M.A., Chacón, B., 2003. Mid-depth calcareous contourites in the latest cretaceous of Caravaca (Subbetic zone, SE Spain). Origin and paleohydrological significance. Sediment. Geol. 163, 131–146.

Martín-Chivelet, J., Fregenal Martínez, M.A., Chacón, B., 2008. Traction structures in contourites. In: Rebesco, M., Camerlenghi, A. (Eds.), Contourites. Developments in Sedimentology. vol. 60. Elsevier, pp. 159–181.

Miguez-Salas, O., Rodríguez-Tovar, F.J., 2019a. Stable deep-sea macrobenthic trace maker associations in disturbed environments from the Eocene Lefkara formation, Cyprus. Geobios 52, 37–45.

Miguez-Salas, O., Rodríguez-Tovar, F.J., 2019b. Ichnofacies distribution in the Eocene–early Miocene Petra Tou Romiou outcrop, Cyprus: sea level dynamics and palaeoenvironmental implications in a contourite environment. Int. J. Earth Sci. 108, 2531–2544.

Miguez-Salas, O., Rodríguez-Tovar, F.J., 2021. Trace fossil analysis of sandy clastic contouritic deposits in the late Miocene Rifian Corridor (Morocco): ichnotaxonomical and palaeoenvironmental insights. J. Afr. Earth Sci. 174, 104054.

Miguez-Salas, O., Dorador, J., Rodríguez-Tovar, F.J., 2019. Introducing Fiji and ICY image processing techniques in ichnological research as a tool for sedimentary basin analysis. Mar. Geol. 413, 1–9.

Miguez-Salas, O., Rodríguez-Tovar, F.J., De Weger, W., 2020a. *Macaronichnus* and contourite depositional settings: bottom currents and nutrients as coupling factors. Palaeogeogr. Palaeoclimatol. Palaeoecol. 545, 109639.

Miguez-Salas, O., Huffard, C.L., Smith Jr., K.L., McGill, P.R., Rodríguez-Tovar, F.J., 2020b. Faunal assemblage changes, bioturbation and benthic storms at an abyssal station in the northeastern Pacific. Deep-Sea Res. I Oceanogr. Res. Pap. 160, 103277.

Miguez-Salas, O., Rodríguez-Tovar, F.J., De Weger, W., 2021a. The Late Miocene Rifian corridor as a natural laboratory to explore a case of ichnofacies distribution in ancient gateways. Sci. Rep. 11, 4198.

Miguez-Salas, O., Dorador, J., Rodríguez-Tovar, F.J., Linares, F., 2021b. X-ray microtomography analysis to approach bioturbation's influence on minor-scale porosity distribution: a novel approach in contourite deposits. J. Petrol. Sci. Eng. 208, 109251.

Mulder, T., Migeon, S., Savoye, B., Faugères, J.-C., 2001. Inversely graded turbidite sequences in the deep Mediterranean. A record of deposits from flood-generated turbidity currents? Geo-Mar. Lett. 21, 86–93.

Mulder, T., Migeon, S., Savoye, B., Faugeres, J.-C., 2002. Reply to discussion by Shanmugam on Mulder et al. (2001, Geo-Marine Letters, 21, 86–93) inversely graded turbidite sequences in the deep Mediterranean. A record of deposits from flood generated turbidity currents? Geo-Mar. Lett. 22, 112–120.

Mulder, T., Hassan, R., Ducassou, E., Zaragosi, S., Gonthier, E., Hanquiez, V., Marchès, E., Toucanne, S., 2013. Contourites in the Gulf of Cadiz: a cautionary note on potentially ambiguous indicators of bottom current velocity. Geo-Mar. Lett. 33, 357–367.

Pemberton, S., Gingras, M., 2005. Classification and characterizations of biogenically enhanced permeability. AAPG Bull. 89, 1493–1517.

Petrovic, A., Lantzsch, H., Schwenk, T., Marquardt, J., Titschack, J., Hanebuth, T.J., 2019. Post-LGM upward shift of the Mediterranean outflow water recorded in a contourite drift off NW Spain. Mar. Geol. 407, 334–349.

Pickering, K.T., Hiscott, R.N., 2016. Deep Marine Systems: Processes, Deposits, Environments, Tectonics and Sedimentation. AGU and Wiley, Chichester.

Ponce, J.J., Olivero, E.B., Martinioni, D.R., López Cabrera, M.I., 2007. Sustained and episodic gravity flow deposits and related bioturbation patterns in Paleogene turbidites (Tierra del Fuego, Argentina). In: Bromley, R.G., Buatois, L.A., Mangano, M.G., Genise, J.F., Melchor, R.N. (Eds.), Sediment-Organism Interactions: A Multifaceted Ichnology. vol. 88. SEPM Special Publication, pp. 253–266.

Quaye, J.A., Jiang, Z., Zhou, X., 2019. Bioturbation influence on reservoir rock quality: a case study of Well Bian-5 from the second member Paleocene Funing formation in the Jinhusag, Subei Basin, China. J. Petrol. Sci. Eng. 172, 1165–1173.

Rasmussen, S.L., Surlyk, F., 2012. Facies and ichnology of an upper cretaceous chalk contourite drift complex, eastern Denmark, and the validity of contourite facies models. J. Geol. Soc. Lond. 169, 435–447.

Rebesco, M., Hernández-Molina, F.J., Van Rooij, D., Wåhlin, A., 2014. Contourites and associated sediments controlled by deep-water circulation processes: state-of-the-art and future considerations. Mar. Geol. 352, 111–154.

Reineck, H.E., 1963. Sedimentgefüge im Bereich der südliche Nordsee. Abhandlungen der Senckenbergischen Gesellschaft 505, 1–138.

Reineck, H.E., 1967. Layered sediments of tidal flat beaches, and shelf bottoms of the North Sea. In: Lauff, G.H. (Ed.), Estuaries. vol. 83. American Association for the Advancement of Science Special Publication, pp. 191–206.

Reolid, J., Betzler, C., 2019. The ichnology of carbonate drifts. Sedimentology 66, 1427–1448.

Rodríguez-Tovar, F.J., 2022. Ichnological analysis. A tool to characterize deep-marine processes and sediments. Earth Sci. Rev. 228, 104014.

Rodríguez-Tovar, F.J., Dorador, J., 2015. Ichnofabric characterization in cores: a method of digital image treatment. Ann. Soc. Geol. Pol. 85, 465–471.

Rodríguez-Tovar, F.J., Hernández-Molina, F.J., 2018. Ichnological analysis of contourites: past, present and future. Earth Sci. Rev. 182, 28–41.

Rodríguez-Tovar, F.J., Hernández-Molina, F.J., Hüneke, H., Llave, E., Stow, D., 2019. Contourite facies model: improving contourite characterization based on the ichnological analysis. Sediment. Geol. 384, 60–69.

Rodríguez-Tovar, F.J., Miguez-Salas, O., Dorador, J., 2020. Image processing techniques to improve characterization of composite ichnofabrics. Ichnos 27, 258–267.

Rodríguez-Tovar, F.J., Miguez-Salas, O., Dorador, J., 2021. Mercury intrusion porosimetry to evaluate the incidence of bioturbation on porosity of contourites. Riv. Ital. Paleontol. Stratigr. 127 (1), 149–161.

Rogerson, M., Rohling, E.J., Bigg, G.R., Ramirez, J., 2012. Paleoceanography of the Atlantic-Mediterranean exchange: overview and first quantitative assessment of climatic forcing. Rev. Geophys. 50, RG2003.

Schindelin, J., Arganda-Carreras, I., Frise, E., Kaynig, V., Longair, M., Pietzsch, T., Preibisch, S., Rueden, C., Saalfeld, S., Schmid, B., Tinevez, J.Y., 2012. Fiji: an open-source platform for biological-image analysis. Nat. Methods 9 (7), 676.

Seike, K., Jenkins, R.G., Watanabe, H., Nomaki, H., Sato, K., 2012. Novel use of burrow casting as a research tool in deep-sea ecology. Biol. Lett. 8, 648–651.

Seike, K., Shirai, K., Kubota, K., Ota, Y., Sassa, S., 2020. Does trace fossil size correspond with burrower population density? An example from the modern counterpart of the trace fossil *Bichordites*. Palaeogeogr. Palaeoclimatol. Palaeoecol. 557, 109946.

Seilacher, A., 1964. Biogenic sedimentary structures. In: Imbrie, J., Newell, N. (Eds.), Approaches to Palaeoecology. John Wiley & Sons, Chichester, UK, pp. 296–316.

Seilacher, A., 1967. Bathymetry of trace fossils. Mar. Geol. 5, 413–428.

Shanmugam, G., 2000. 50 years of the turbidite paradigm (1950s–1990s): deep-water processes and facies models—a critical perspective. Mar. Pet. Geol. 17, 285–342.

Shanmugam, G., 2002. Discussion on Mulder et al. 2001, Geo-Marine Letters 21, 86–93. Inversely graded turbidite sequences in the deep Mediterranean. A record of deposits from flood-generated turbidity currents? Geo-Mar. Lett. 22, 108–111.

Shanmugam, G., 2017a. The contourite problem. In: Mazumder, R. (Ed.), Sediment Provenance. Elsevier, pp. 183–254.

Shanmugam, G., 2017b. Contourites: physical oceanography, process sedimentology, and petroleum geology. Pet. Explor. Dev. 44 (2), 183–216.

Shanmugam, G., Spalding, T.D., Rofheart, D.H., 1993. Traction structures in deep-marine bottom current-reworked sands in the Pliocene and Pleistocene, Gulf of Mexico. Geology 21, 929–932.

Shanmugan, G., 2018. Bioturbation and trace fossils in deep-water contourites, turbidites, and hyperpycnites: a cautionary note. J. Indian Assoc. Sedimentol. 35, 13–32. Special Issue dedicated to George Devries Klein in celebrating his life and achievements.

Stow, D.A.V., Faugères, J.C., 2008. Contourite facies and the facies model. In: Rebesco, M., Camerlenghi, A. (Eds.), Contourites. Developments in Sedimentology. vol. 60. Elsevier, pp. 223–256.

Stow, D.A.V., Holbrook, J.A., 1984. North Atlantic contourites: an overview. In: Stow, D.A.V., Piper, D.J.W. (Eds.), Fine Grained Sediments, Deep-Water Processes and Facies. vol. 15. Geol. Soc., London, Special Publication, pp. 245–256.

Stow, D.A.V., Lovell, J.P.B., 1979. Contourites: their recognition in modern and ancient sediments. Earth Sci. Rev. 14, 251–291.

Stow, D.A.V., Smillie, Z., 2020. Distinguishing between deep-water sediment facies: turbidites, contourites and hemipelagites. Geosciences 10, 68.

Stow, D.A.V., Faugères, J.-C., Gonthier, E., 1986. Facies distribution and textural variation in Faro drift contourites: velocity fluctuation and drift growth. Mar. Geol. 72, 71–100.

Stow, D.A.V., Faugères, J.-C., Viana, A.R., Gonthier, E., 1998. Fossil contourites: a critical review. Sediment. Geol. 115, 3–31.

Stow, D.A.V., Ogawa, Y., Lee, I.T., Mitsuzawa, K., 2002a. Neogene contourites, Miura- Boso forearc basin, SE Japan. In: Stow, D.A.V., Pudsey, C.J., Howe, J.A., Faugères, J.-C., Viana, A.R. (Eds.), Deep-Water Contourite Systems: Modern Drifts and Ancient Series, Seismic and Sedimentary Characteristics. vol. 22. Geol. Soc., London, Memoirs, pp. 409–419.

Stow, D.A.V., Faugères, J.-C., Howe, J.A., Pudsey, C.J., Viana, A.R., 2002b. Bottom currents, contourites and dep-sea sediment drift: current state-of-the-art. In: Stow, D.A.V. (Ed.), Deep-Water Contourite Systems: Modern Drifts and Ancient Series, Seismic and Sedimentary Characteristics. vol. 22. Geol. Soc., London, Memoirs, pp. 7–20.

Suess, E., 1980. Particulate organic carbon flux in the oceans—surface productivity and oxygen utilization. Nature 288, 260–263.

Taylor, A.M., Goldring, R., 1993. Description and analysis of bioturbation and ichnofabric. J. Geol. Soc. London 150, 141–148.

Thistle, D., Yingst, J.Y., Fauchald, K., 1985. A deep-sea benthic community exposed to strong near bottom currents on the Scotian rise (western Atlantic). Mar. Geol. 66, 91–112.

Thistle, D., Ertman, S.C., Fauchald, K., 1991. The fauna of the HEBBLE site: patterns in standing stock and sediment-dynamic effects. Mar. Geol. 99, 413–422.

Uchman, A., Wetzel, A., 2011. Deep-sea ichnology: the relationships between depositional environment and endobenthic organisms. In: Hüneke, H., Mulder, T. (Eds.), Deep-Sea Sediments. Developments in Sedimentology, vol. 63. Elsevier, pp. 517–556.

Uchman, A., Wetzel, A., 2012. Deep-sea fans. In: Knaust, D., Bromley, R.G. (Eds.), Trace Fossils as Indicators of Sedimentary Environments. Developments in Sedimentology, vol. 64. Elsevier, pp. 643–671.

Wetzel, A., 2008. Recent bioturbation in the deep South China Sea: a uniformitarian ichnologic approach. Palaios 23, 601–615.

Wetzel, A., 2010. Deep-sea ichnology: observations in modern sediments to interpret fossil counterparts. Acta Geol. Pol. 60, 125–138.

Wetzel, A., Uchman, A., 2012. Hemipelagic and pelagic basin plains. In: Knaust, D., Bromley, R.G. (Eds.), Trace Fossils as Indicators of Sedimentary Environments. Developments in Sedimentology. vol. 64. Elsevier, pp. 673–701.

Wetzel, A., Uchman, A., 2018. The former presence of organic matter caused its later absence: burn-down of organic matter in oceanic red beds enhanced by bioturbation (Eocene Variegated Shale, Carpathians). Sedimentology 65 (5), 1504–1519.

Wetzel, A., Werner, F., Stow, D.A.V., 2008. Bioturbation and biogenic sedimentary structures in contourites. In: Rebesco, M., Camerlenghi, A. (Eds.), Contourites. Developments in Sedimentology. vol. 60. Elsevier, pp. 183–202.

Zonneveld, J.P., 2016. Applications of experimental neoichnology to paleobiological and evolutionary problems. Palaios, 275–279.

Chapter 16

# Imaging and interpretation: Seismic, rock physics and image log analysis workflows for deepwater systems

**Sumit Verma[a], Shuvajit Bhattacharya[b], Tom Fett[c], Per Avseth[d] and Ivan Lehocki[e]**

[a]*University of Texas Permian Basin, Odessa, TX, United States,* [b]*University of Texas at Austin, Austin, TX, United States,* [c]*Consultant, San Antonio, TX, United States,* [d]*Dig Science/NTNU, Oslo, Norway,* [e]*Dig Science/Lehocki Geospace, Oslo, Norway*

## Introduction

The advancement of seismic and well logging technology over the last two to three decades has provided the petroleum industry with a unique opportunity to conduct integrated reservoir characterization with improved certainty and efficiency. During the same period, hydrocarbon traps have been increasingly difficult to detect, the average discovery volumes have decreased and today reserve replacement is not the main goal for an industry in transition. Still, hydrocarbons will play an important role during a sustainable transition to a future dominated by renewable energy resources. Deepwater reservoirs, located in offshore Brazil, offshore Angola, the Gulf of Mexico and other areas will likely play a vital role during the transition period as there are still vast underexplored areas to be investigated.

In these settings, traps are often stratigraphic and subtle, with complex reservoir sand distributions. Hence, the detailed information that can be acquired from seismic amplitudes and wellbore imaging can be very useful. Seismic attributes can reveal geomorphology beyond conventional seismic resolution and amplitude vs offset (AVO) technology can potentially distinguish between anomalies caused by lithology effects and pore fluid effects. A great variety of workflows have been established to extract subsurface information from seismic and well log data, and in this chapter we can only scratch the surface of these (Moscardelli et al., 2006; Chopra and Marfurt, 2007; Kolla et al., 2007; Posamentier et al., 2007; Bhattacharya and Verma, 2020; Bhattacharya et al., 2020).

In the next Section "Seismic interpretation of deepwater depositional system on the North Slope, Alaska, USA", a case study from the Cretaceous Nanushuk-Torok sequence on the North Slope, Alaska, USA is discussed. Traditional seismic attributes and machine learning (ML) are used to map and interpret different geomorphological features commonly found in shelf-to-marine settings. Multiple seismic attributes, including amplitude, coherence, coherent energy, spectral decomposition and gray-level co-occurrence matrices, are used to identify shelf, channels, canyon, basin-floor fan, slump, submarine slide blocks and sediment wave deposits. Although many of these features are found across different areas in the world, not all of them are found together in one place. This is why the North Slope is selected as the study area. Each seismic attribute highlights some stratigraphic elements, including their internal and external heterogeneities, better than others. Therefore, different unsupervised ML approaches (self-organizing map or SOM and generative topographic map or GTM) are used to automate the process for seismic facies clustering and aid in geologic interpretation. Upon comparing results from two ML algorithms, GTM was better at providing details of seismic stratigraphic variations. In the next step, it is recommended to verify these geologic features with high-resolution image logs and core samples.

Following the North Slope case study, the Section "Quantitative interpretation of a North Sea turbidite system using rock physics", a case study from the Glitne Field, North Sea, Norway, is presented where rock physics and AVO feasibility modeling is applied to better understand the large variability in seismic amplitudes and AVO signatures observed across the Glitne submarine fan system. A key aspect of quantitative seismic interpretation is to understand what to expect at any given depth or location. For example: what is the expected seismic contrast between a cap-rock shale and a reservoir sandstone filled with hydrocarbons at a burial depth of 2 km (6560 feet)? And what is the expected difference between the seismic signature of a brine-filled scenario vs an oil-filled scenario? The answers to these questions are related to local geology. The key drivers to the contrast between sands and shales are compaction and diagenesis. As Brown (2011) pointed out: "Low

**Deepwater Sedimentary Systems.** https://doi.org/10.1016/B978-0-323-91918-0.00015-3

impedance rocks are normally more prospective than high impedance rocks." That is why bright spot technology was used as a screening tool in the 1970s and 1980s, and direct hydrocarbon indicators (DHIs) became a hot topic and the mapping of bright spots and flat spots became an exploration strategy. However, often the technology failed due to pit-fall scenarios. Over the years, the industry has realized that we need to understand when to expect bright amplitudes and when to expect dim amplitudes, and that dim spots can potentially be equally prospective as bright spots in certain areas or settings. This is where rock physics and AVO feasibility modeling comes in as an appropriate tool to better understand what to expect. In this Glitne Field case study, it is demonstrated that the interplay among compaction, facies and fluids can cause complex seismic amplitude responses even within the same field. A good integration between geological knowledge and rock physics understanding is needed to disentangle the various effects.

In the final case study, "Role of dipmeters and imaging devices in evaluating deepwater sedimentary intervals"; the use of borehole images for the interpretation of deepwater depositional systems utilizing several DWS examples is presented. The examples were chosen to represent various imaging applications and to illustrate some of the many presentations that are available. Other open hole logs, such as the gamma ray and calipers, are routinely run with the FMI which allows for a depth correlation to the other logs acquired from a well. The images provide a picture of the formation that is easily recognizable to most geologists and hopefully will be of use to both students and advanced users. These images help fill the gap between various studies conducted over a wide range of scales, from "mega" as with seismic to "micro" for the borehole imaging (BHI). The examples presented here are mainly from the USA Gulf Coast. They are both illustrative and are released for publication. These applications have proven useful world-wide, including the Far East and the Middle East, Europe , Africa and both of the Americas. Numerous references about this technology and open-hole logging in general can be found online at www.sprensky.com. Specific information about imaging can be found in Bouma (2000) and Fett et al. (1989). Also, most service companies have technical information about their imaging and logging tools on their websites. Schlumberger's Principles, though a bit dated, are still a useful basic logging reference (Schlumberger, 1990). The Atlas of Deep-Water Outcrops (AAPG Study 56) is a comprehensive and well-illustrated documentation of many DWS outcrops and models (Studlick et al., 2007).

Seismic and well log interpretation and technology is a large field of study and a single chapter can never capture every facet of such a complex topic. After reading Chapter 16, the reader will benefit greatly from pursuing additional literature on seismic and well log data imaging and interpretation workflows, as well as fundamental background theory and complementary methods. The use of seismic attributes for interpretation of the sub-surface is extensively captured by the books by Chopra and Marfurt (2007) and Barnes (2016), qualitative interpretation of seismic amplitudes from 3D seismic is comprehensively covered by Brown (2015), whereas Simm and Bacon (2014) addresses quantitative seismic amplitude interpretation. Quantitative seismic interpretation integrating rock physics and geology can be further studied in the books by Avseth et al. (2005) and Vernik (2016).

## Seismic interpretation of deepwater depositional system on the North Slope, Alaska, USA

### Brief background

This section briefly covers the fundamentals and application of 3D seismic interpretation techniques in shallow marine-to-deepwater settings. Multiple seismic attributes and ML approaches are shown to recognize and classify different stratigraphic features found in such geologic settings.

### Methods

#### *Seismic interpretation*

Processed and stacked 3D seismic amplitude sections are primary datasets to start initial stratigraphic interpretations. Seismic amplitude interpretation generally includes the interpretation of easy-to-identify geomorphological features (e.g., shelf, slope, slump, etc.) on seismic sections and the mapping of critical geologic horizons.

#### *Seismic attributes*

In the next step, one can compute several seismic attributes (discussed below) to unravel complex geologic features from the seismic data. Seismic attributes are quantities extracted from the seismic amplitude volume or section using various mathematical and statistical relations. Important seismic attributes are discussed below and are later shown in their application on the North Slope, Alaska.

## Dip magnitude and azimuth

Reflector dip attributes can reveal faults, channels, fractures, karsts, differential compaction and other stratigraphic features. It is also essential in computing other geometric attributes, such as coherence, coherent energy, curvature and aberrancy. The dip attribute consists of dip direction or dip azimuth and plunge or dip magnitude. Although there are different ways to compute dip attributes, one of the commonly used implementations described by Marfurt (2006) uses the reflector dip by a robust dip search.

## Coherence

The coherence attribute helps identify discontinuity in the reflector. The coherence algorithm is based on comparing the amplitude or waveform of adjacent traces. Chopra and Marfurt (2007) discussed the theory and applications of cross-correlation, semblance, variance, Eigen structure and Sobel filter similarity-based attributes. Eigen structure coherence is one of the more commonly used seismic coherences and is based on waveform similarity. Although in most cases, Eigen structure similarity works better than other similarity attributes in low signal to noise ratio data and excels at depicting geological discontinuities like faults, it fails to find discontinuities such as narrow channels when there is no change in waveform shape.

In contrast, the Sobel filter similarity detects changes in amplitude but is severely affected by the presence of acquisition footprints, which if not removed can impart systematic noise trends to the data. The Sobel filter is useful in detecting thin channels or the extent to which gas has seeped to shallower strata as both of these geological scenarios lead to amplitude changes. Fig. 1A shows a channel boundary based on the coherence attribute.

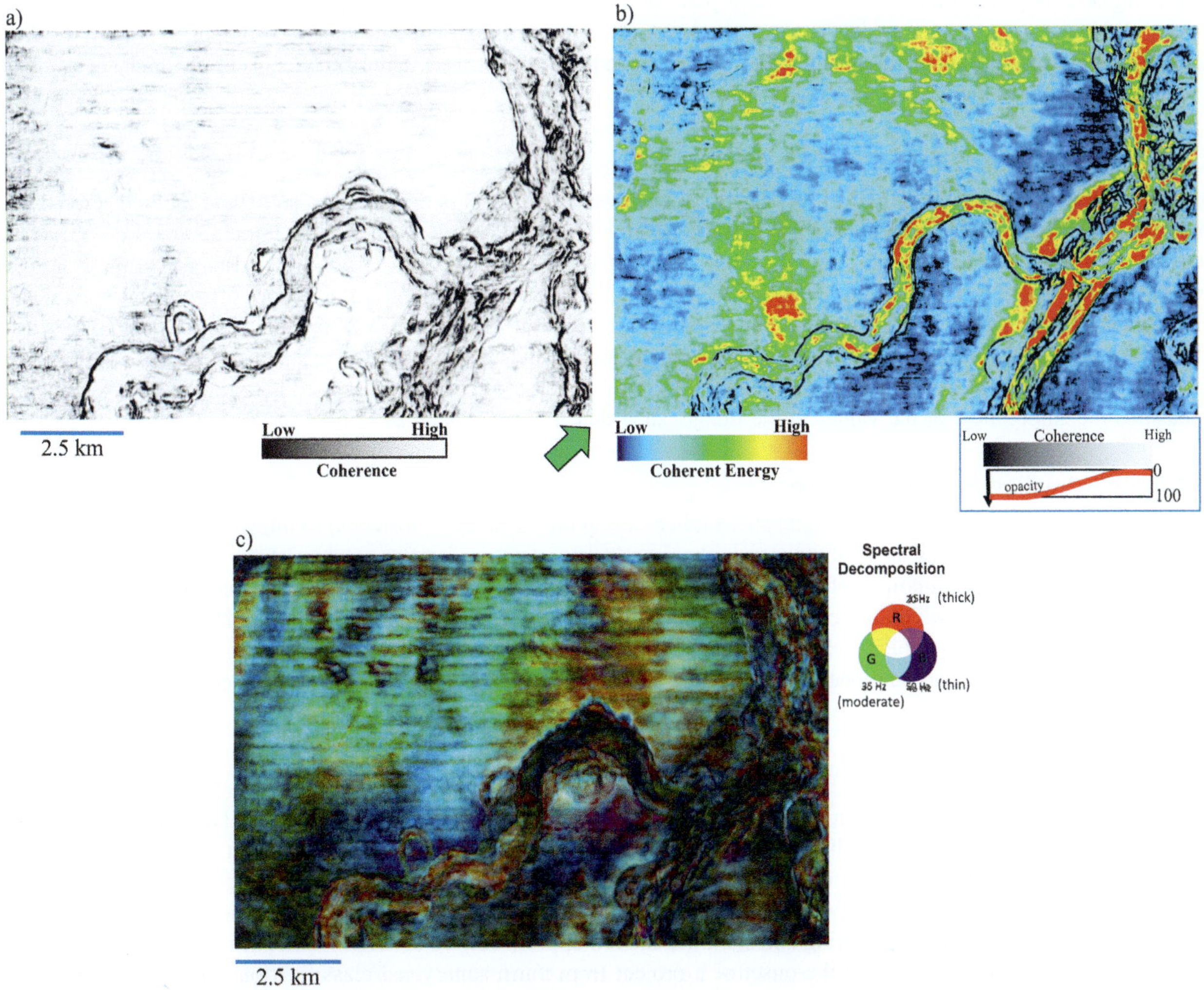


**FIG. 1** Channels observed in a 3D seismic survey (Tui 3D) from offshore New Zealand with, (A) coherence ~2.10 s, (B) coherent energy co-rendered with coherence at ~2.15 s, (C) spectral decomposition with RGB combination at ~2.10 s.

## Coherent energy

Energy or envelope is a commonly used seismic attribute, which is mathematically expressed as a square of the seismic amplitude. Marfurt (2006) applied singular value decomposition to minimize the noise component and obtained the first principal component of the data, which does not contain any steeply dipping noise. The energy (square of the amplitude) of this first principal component is the coherent energy. Coherent energy generally corresponds to the depositional environment. For example, a shale deposited in a transgressive sequence would show low coherent energy, as it corresponds to minimal reflection energy. In contrast, low-stand sand layered with the shales will lead to high coherent energy. Coherent energy is different than the commonly used attribute root-mean-squared (RMS) amplitude attribute. Unlike RMS amplitude, it is less susceptible to noise because the first few principal components of the seismic data contain the signal, whereas the latter components mostly contain noise (Chopra and Marfurt, 2007; Bhattacharya and Verma, 2020). Fig. 1B shows the external and internal heterogeneity of a channel based on a co-rendered coherent energy and coherence attribute.

## Spectral decomposition

Frequency-related attributes are very useful to quantify the dimension of complex geologic features. Seismic data generally consists of 10–100 Hz frequency range, which is not always able to resolve features of interest. Spectral decomposition of a seismic volume divides the data into different frequency bands, which contain different geological information. Different frequencies of seismic data have different tuning thicknesses, so we can interpret the relative thickness of channel sands by comparing spectral decomposition volumes of different frequencies (Widess, 1973). In general, three spectral decomposition volumes are co-rendered with RGB (red-green-blue) displays, where the red (lowest frequency) corresponds to the thicker parts of a channel, green (moderate frequency) corresponds to the medium thickness areas of a channel and the blue color (highest frequency) relates to the thinnest parts of a channel. When all the frequencies are present (in the thickest parts), the image will become white, whereas the black areas show the absence of all three components. Fig. 1C shows the multiple frequency content in a channel that we can use to infer its thickness. Thicker channels imply thick net sandstone for production or storage (Partyka et al., 1999).

## GLCM

In image processing science, GLCM (gray-level co-occurrence matrices) is defined as a statistical method of analyzing textural patterns. GLCMs are generally used on 2D seismic images. Gao (2003) defined computed voxel co-occurrence matrix (VCM) on 3D data. Regarding 3D seismic, currently GLCM, or texture attributes, is a more common term for VCM (Matos et al., 2003; Roy et al., 2013). GLCM geostatistically analyzes the sample values of a collection of seismic traces to effectively quantify seismic reflections' spatial organization. As such, GLCM reveals textures present in seismic data and homogeneity, energy, contrast, dissimilarity, entropy, mean and variance are the standard GLCM/seismic texture attributes. In the following case study, we used GLCM homogeneity for both unsupervised learning inputs. Homogeneity measures the smoothness of the values in the image, such that it highlights areas of invariant mean and variance.

## Attribute extraction along the surface

Experienced seismic interpreters use stratal slices (slices along the surfaces or horizons) to interpret stratigraphic features. The key reason for extracting an attribute along a horizon is that a horizon is often (almost always) mapped as a geologic timeline, a foundational component of seismic interpretation and sequence stratigraphy. Stratal slices remove the effects of structural dips and faults which are commonly due to post-depositional tectonics; removing such dipping effect shows deposits of the same stratigraphic levels. Although, it is a common practice for structural geologists to analyze the structure on the time slice; Verma and Scipione (2020), as well as Verma and Bhattacharya (2019), observed that attribute analysis along a stratal slice can reveal sub-seismic faults.

## Machine learning

Seismic attributes illuminate different geological features. Depending on the geologic question at hand, some of these features might be of interest, whereas others may not. Therefore, seismic interpreters prefer to look at more than one attribute to illuminate features of interest. Simultaneous analysis of different seismic attributes is challenging. Unsupervised classification or clustering, such as a SOM and generative topographic mapping (GTM) can help divide seismic data into different seismic facies, whereas clustering and semi-supervised approaches are beneficial since it is hard to have complete labels or interpretations available at the onset of a project to perform supervised classification (Zhao et al., 2015; Roy et al., 2013).

SOM is one of the widely used methods of unsupervised classification. SOM projects the data into a lower-dimensional space (2D) while preserving the topology. Generally, SOM outputs are two SOM projected volumes. We cross-plot these two volumes and apply an appropriate color scale to obtain our facies classification image (Roy et al., 2013). Similar to SOM, GTM generates two latent space volumes, which are cross-plotted to find different facies. The difference lies in the algorithm of transforming the higher dimensional data to 2D space; SOM searches for the prototype vector closest to the randomly chosen vector while GTM utilizes the probability-based expectation–maximization algorithm for model parameter estimation.

### *Seismic attribute interpretation pitfalls*

Most seismic attributes are not calibratable to well data, except impedance attributes. As such, attributes are often subject to misinterpretation due to various issues, including data acquisition, pre-stack and post-stack processing workflows. Often, we find acquisition footprints (e.g., shot-receiver locations) and surficial features (lakes) in shallow sections which create artifacts during the seismic imaging. Limitations in processing such as inaccurate velocity analysis and residual ground roll can also give rise to acquisition footprints (Marfurt, AASPI). In addition, velocity pull-up can occur due to the presence of high-velocity salt and carbonates. The first case study below shows an example of a velocity push-down effect near the contact of sand-rich submarine slide blocks and organic-rich mudrocks. If the geologic context is unknown, such features can be misconstrued as vertical faults. Also, these features create issues in seismic time-to-depth conversion. Structural leakage is another type of artifact introduced by not computing certain attributes (for example, coherence and frequency) along with local dip and azimuth. This phenomenon happens due to the steep dip of the reflectors so it is always recommended to compute dip and azimuth for the seismic volume first before computing any attributes on it.

## Case study: North Slope, Alaska, USA

An example of detailed 3D seismic attribute and ML-assisted interpretation from a shallow marine to deepwater reservoir from the North Slope, Alaska is shown (Fig. 2; Bird and Houseknecht, 2011). The North Slope has been producing hydrocarbons from the Prudhoe Bay, Kuprauk, Alpine and Milne Point fields for several decades. The primary oil-producing deep sedimentary formations include: the Paleozoic-aged Kekiktuk (sandstone) and Lisburne (limestone), the Triassic-aged Ivishak (sandstone) and Sag River (sandstone), the early Cretaceous-aged Kuparuk (sandstone) formations and Jurassic sandstones. Deep reservoirs are mostly fault-controlled, compared to shallow reservoirs. Since 2015, new exploration activities have focused on the Cretaceous and Cenozoic Nanushuk Formation and its basinward equivalent Torok. There have been several new discoveries in these two formations over the last few years, notably the Pikka and Willow

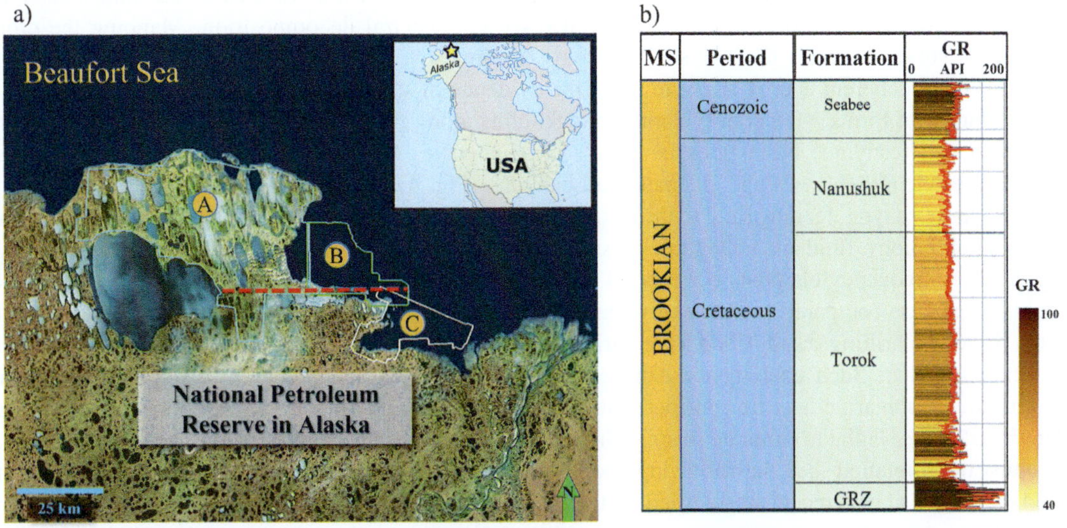


FIG. 2 (A) Study area in National Petroleum Reserve in Alaska (NPRA), North Slope Alaska. Note: Seismic survey A = NE-NPRA 3D, 2006; B = Harrison Bay 3D, 2006; C = Harrison Bay 3D, 2004. (B) Stratigraphic column on the North Slope Alaska, along with a gamma-ray log of a well across the Nanushuk-Torok, which is the focus of our study. *(Modified from Bird, K.J., Houseknecht, D.W., 2011. Geology and petroleum potential of the Arctic Alaska petroleum province. In: Spencer, A.M., Embry, A.F., Gautier, D.L., Stoupakova, A.V. (Eds.), Arctic Petroleum Geology, vol. 35. Geological Society, London, pp. 485–499. In press.)*

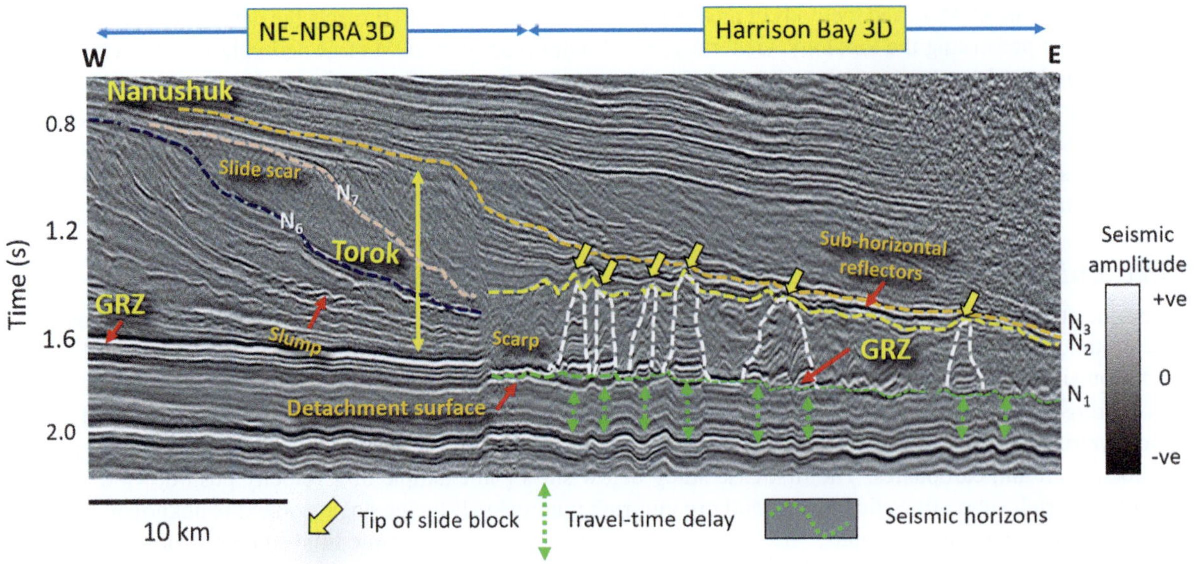


**FIG. 3** A seismic section along W-E in the study area, showing different geomorphological features identified on the seismic section (vertical exaggeration: 50). The white dash polygons indicate inferred submarine slide blocks. The yellow arrows show the tip of the inferred submarine slide blocks, whereas the green double-headed arrows indicate the velocity push-down effect below the slide blocks. Different horizons, such as N1 (detachment surface), N2 and N3 horizons are interpreted. N6 and N7 horizons on the left side cannot be traced in the right as the scarp affects these. The detachment surface, N1, is not traced to the left as it is not present there. Note: the red dash line in Fig. 2 indicates the location of the seismic section. (Modified after Bhattacharya, S., Verma, S., Rotzien, J., 2020. 3D seismic imaging of the submarine slide blocks on the North Slope, Alaska, Interpretation.)

Fields. Houseknecht et al. (2017) evaluated the Cretaceous aged Nanushuk and Torok formations for their hydrocarbon potential and estimated ~9 billion barrels of oil and 25 trillion cubic feet (TCF) of gas in place.

The Nanushuk and Torok formations are genetically related, and they form clinoforms together (Fig. 3). The Nanushuk Formation forms topsets with sand-prone coastal plain and shallow shelf marine sediments, whereas the Torok Formation is its chronological equivalent foresets (muddy slope and sandy turbidite) and bottom sets in the basin.

Three 3D post-stack seismic surveys are utilized in this case study. These 3D surveys are available from the Alaska Division of Geological and Geophysical Surveys (State of Alaska, 2017). Keeping the focus on stratigraphic analysis, the first step is to interpret seismic amplitude sections and compute an ensemble of seismic attributes, such as coherent energy, coherence, gray-level co-occurrence matrix (GLCM) and spectral decomposition. Mapping the clinoforms in all the 3D seismic surveys reveals that the shelf margins continue over tens of kilometers (Bhattacharya and Verma, 2020). Stratal slices of coherent energy attribute show the evidence of sand deposits (based on rock physics and drilling results) in two dominant forms: (1) laterally continuous features along the shelf edges and (2) sporadic presence on the basin-slope and basin (canyons, basin-floor fan and sediment wave) (Fig. 4). Co-rendering of coherent energy and coherence attributes clearly shows slope channels and canyons (Fig. 4). Slope channels have low sinuosity, and some of them meet the large canyon, feeding sand to the basin-floor fan. The channels have variable amplitude anomalies, indicative of changes in lithology, fluid and tuning thickness. Spectral decomposition helps to discern channels of varying thicknesses (Fig. 5). Low-frequency elements correspond to thick channels, and vice versa. Basin-floor fans have a stronger coherent energy response compared to the nearby features, indicating strong impedance contrasts, which is due to thick sand deposits (based on drilling results), and have variable dimensions; some have a single lobe, and others have multiple lobes coalescing together. Such coalescence can be visible on coherent energy or co-rendered coherent energy and coherence attributes. Submarine fans are considered to be one of the major hydrocarbon-producing elements in deepwater settings. However, it is essential to calibrate seismic-driven information with petrophysics and rock-physics data for a complete and robust interpretation (see the following section on rock physics and advanced petrophysics). Due to low oil mobility and dispersive clay, a recent well drilled on the submarine fan deposits in these formations resulted in a dry hole at multiple targets.

Apart from fan deposits, there are sediment wave deposits in the study area (Fig. 6A and B). Sediment wave deposits are parallel-to-subparallel elongate features on the basin floor. These features show distinctive, alternating, high and low amplitude responses and are parallel to the shoreline, indicating the possible influence of longshore current. Spectral

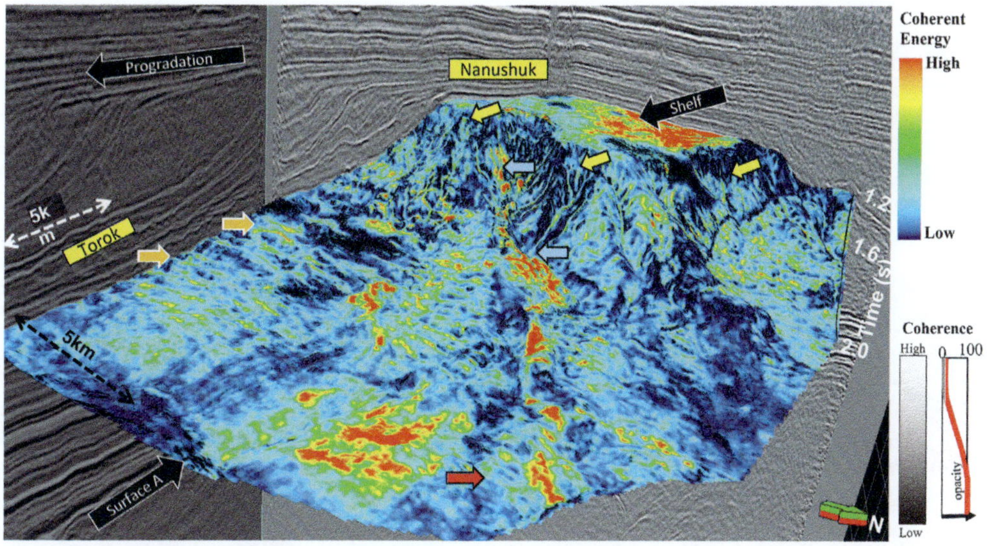


**FIG. 4** A 3D view of the Nanushuk-Torok succession with three seismic amplitude walls surrounding a coherent energy plot co-rendered with a multispectral coherence along surface A. The *orange arrows*: the sediment wave, *yellow arrows*: channel, *blue arrows*: canyon, *red arrow*: basin floor fan.

**FIG. 5** A 3D view of the Nanushuk-Torok succession with three seismic amplitude walls surrounding an RGB plot along a surface A. The spectral decomposition volumes correspond to three different frequencies, 20 Hz (*red*), 35 Hz (*green*) and 50 Hz (*blue*).

decomposition and coherent energy attributes are helpful to show the wavy pattern of these deposits. Different types of mass-transport deposits are present in the study area, including slumps, slide scars and slide blocks (Fig. 3). Based on 2D and 3D seismic data Weimer (1987), Homza (2004) and Bhattacharya et al. (2020) documented detailed characteristics of these submarine slide blocks, including the well-known Fish Creek Slide. These submarine slide blocks affect the lower portion of the Torok Formation. Most of these slide blocks overlie and downlap onto either the gamma-ray zone (GRZ) or Lower Cretaceous unconformity (Homza, 2004; Bhattacharya and Verma, 2020). The study revealed that the composition of the underlying units (i.e., GRZ, an organic-rich shale with low velocity) are also important as significant acoustic impedance contrasts created velocity push-down effect in the form of concave down seismic reflector shape.

Each seismic attribute highlights different stratigraphic elements better than others; an unsupervised ML approach is used to automatically generate clusters, corresponding to different stratigraphic features; SOM and GTM algorithms were

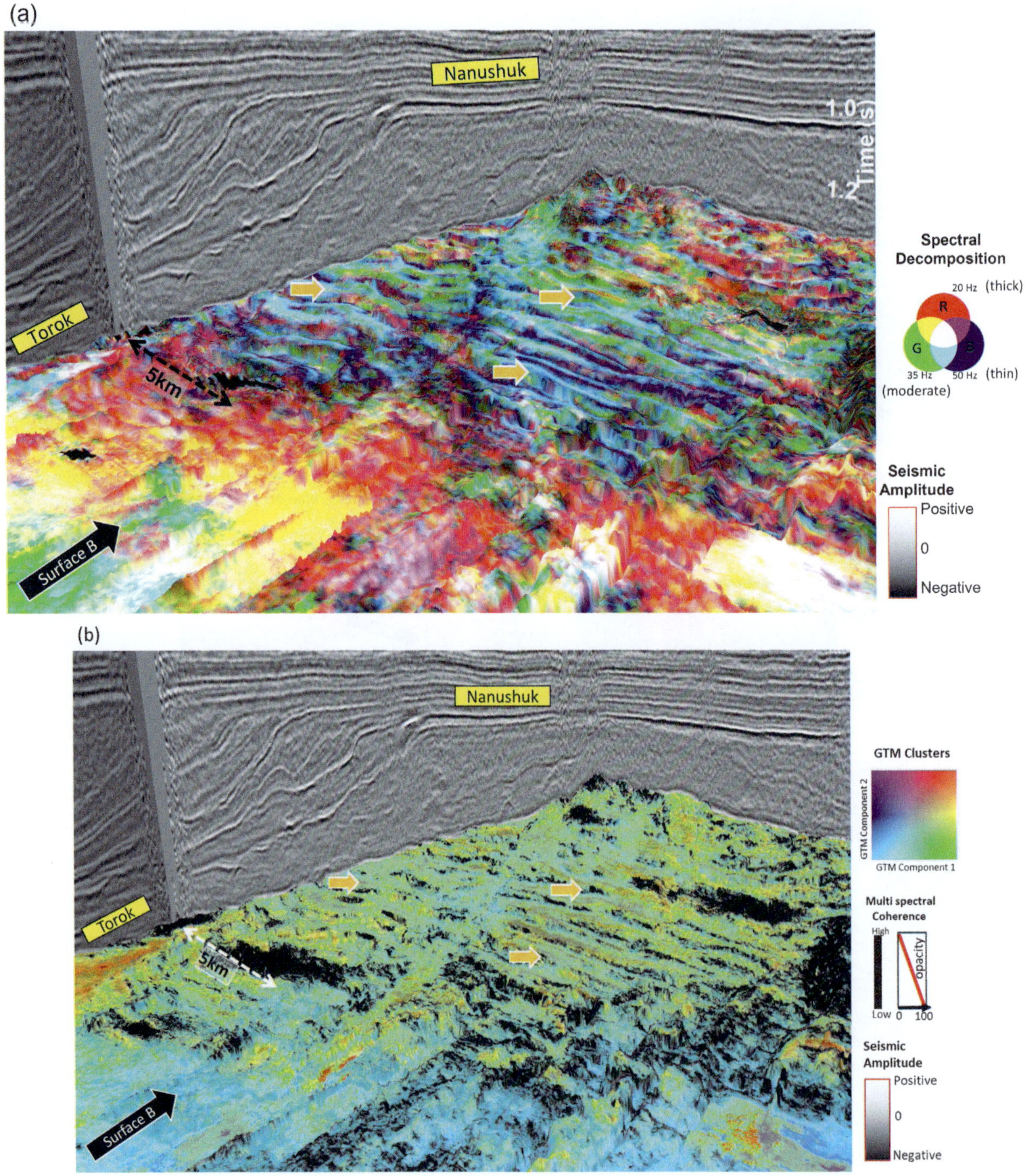


**FIG. 6** (A) A 3D view with three seismic amplitude walls surrounding a RGB plot along a surface B. The spectral decomposition volumes correspond to three different frequencies, 20 Hz (*red*), 35 Hz (*green*) and 50 Hz (*blue*). The *orange arrows*: the sediment wave.

used for clustering. In this study, four seismic attributes, such as coherent energy, GLCM homogeneity and spectral decomposition at 20 and 35 Hz frequencies were used as inputs to SOM and GTM models. A total of 256 clusters were generated as common commercial software allows a maximum of 256 colors. However, a human expert would not identify 256 geologic elements, the adjacent colors representing similarity in terms of computer vision, and similarity in seismic facies or geology.

GTM and SOM are particularly helpful to automatically map different geomorphological features when no borehole log and core data are available. Comparing SOM- and GTM-based results shows that GTM results provide more details than

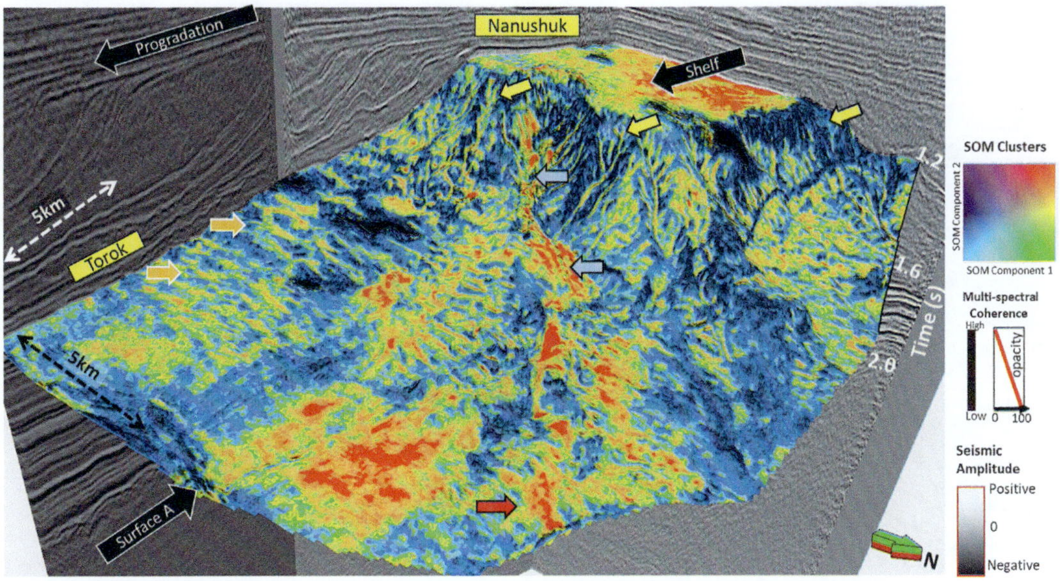


FIG. 7 A 3D view of the Nanushuk-Torok succession with three seismic amplitude walls surrounding a plot of multispectral coherence co-rendered with SOM-estimated facies along surface A.

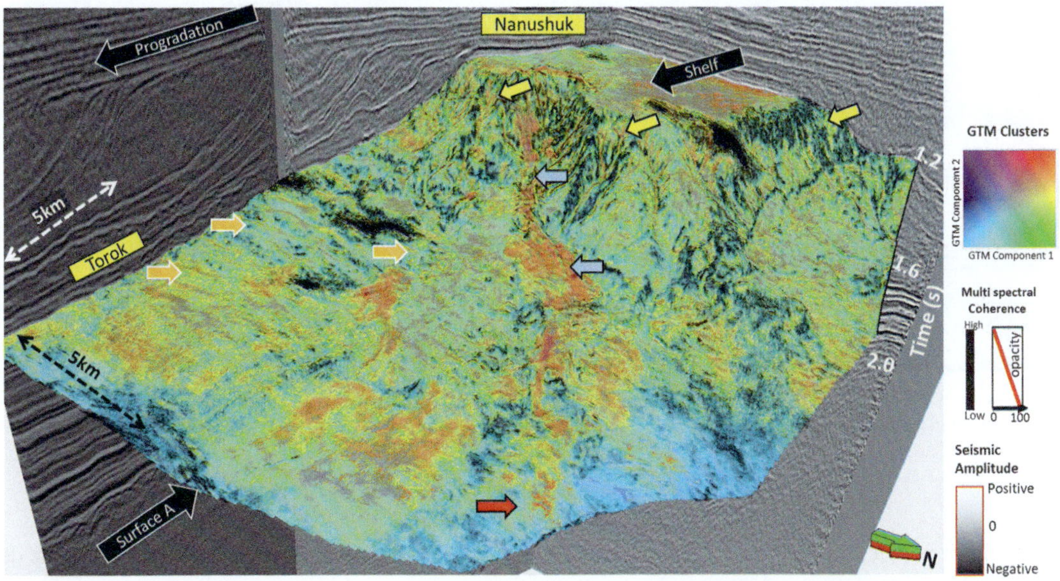


FIG. 8 A 3D view of the Nanushuk-Torok succession with three seismic amplitude walls surrounding a plot of multi-spectral coherence co-rendered with GTM-estimated facies along surface A.

SOM (Figs. 7 and 8), hence GTM was utilized for further stratigraphic interpretations. ML-based clusters indicate different geomorphological features, including sediment wave deposits, canyons, basin-floor fans and slope channels.

Further analysis of GTM-based cluster results at three different time-levels, along a prominent clinoform (level 1), 70ms (level 2) and 110ms (level 3) above level 1, indicated changes in seismic response which might be related to the changes in geologic processes over time (Figs. 8 and 9). Based on that, it is interpreted that sediment wave deposits continued after filling the deep canyon.

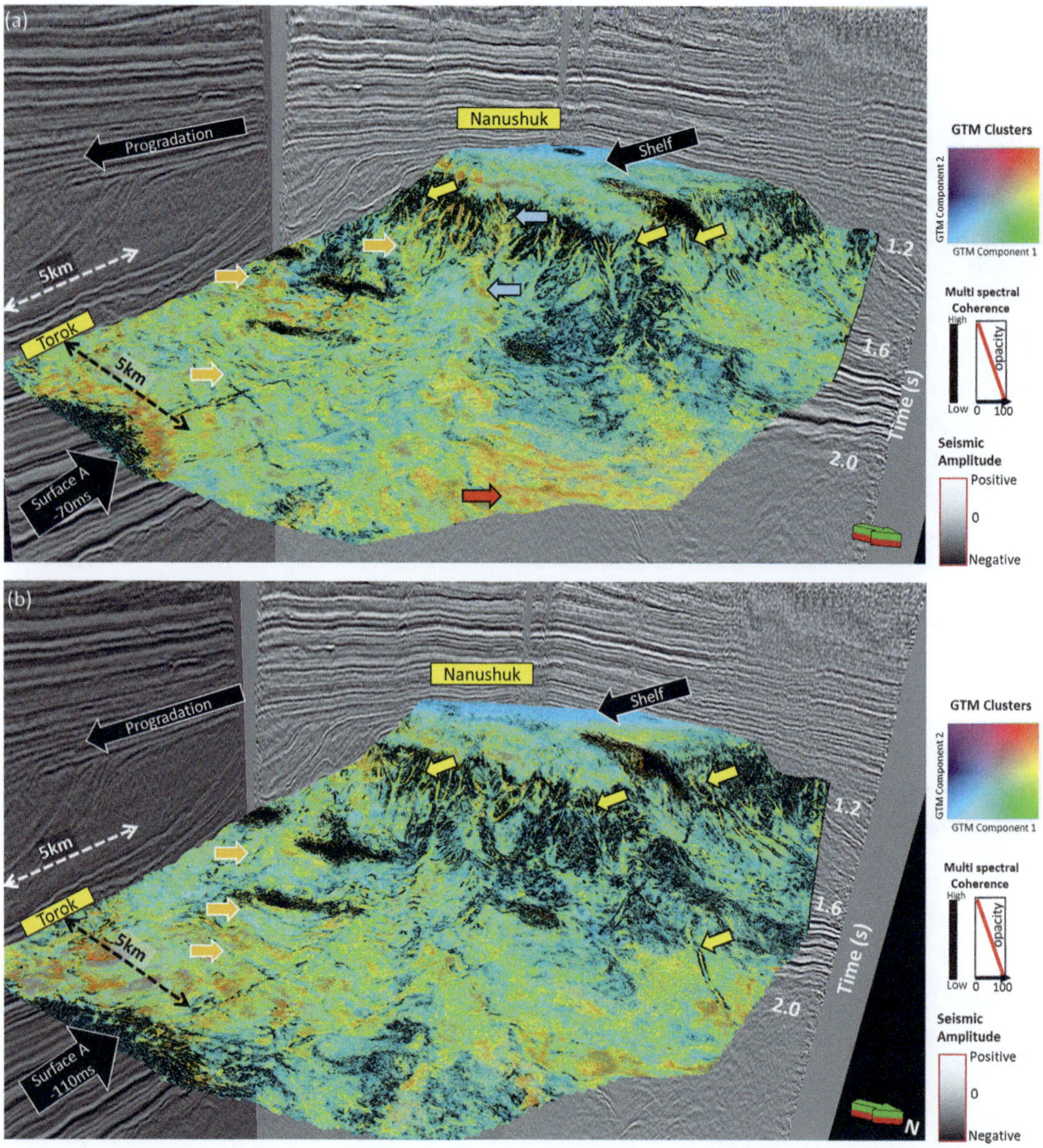


**FIG. 9** A 3D view with three seismic amplitude walls surrounding a plot of multispectral coherence co-rendered with GTM-estimated facies along (A) ~70 ms and, (B) ~110 ms above surface A.

## Summary

In this case study, it was found that multi-attribute analysis is an effective method in interpreting complex depositional features found in deepwater settings. Also, it is demonstrated that unsupervised ML approaches, such as GTM, are feasible for automated clustering and mapping of different geomorphological features (e.g., shelf, slope channel, basin-floor fan, canyon and sediment waves) based on their seismic response. Geologic expertise was used to interpret ML-derived features.

# Quantitative interpretation of a North Sea turbidite system using rock physics

The seismic interpretation workflow outlined in the previous case study is essentially qualitative interpretation of geomorphology and stratigraphic patterns in deepwater clastic systems. The seismic amplitudes, representing contrasts in elastic properties between individual layers, contain information about rock type, porosity, pore-fluid type and saturation—information that cannot be accurately obtained from qualitative seismic interpretation and attribute analysis. Quantitative seismic interpretation has emerged as a new technology over the last couple of decades, where rock physics tools are applied to extract more quantitative information about the subsurface rock properties from seismic amplitudes.

In this case study, quantitative seismic interpretation is demonstrated with an example from the North Sea. More specifically, rock physics and AVO feasibility modeling is applied to study the expected seismic amplitude response in a North Sea turbiditic oil field. The feasibility modeling and scenario testing are constrained by compaction trends for sandstones and shales, petrographic information from core analysis and local stratigraphy and facies associations. The observed AVO signatures in the studied Glitne Field show a large variability associated with diagenetic and depositional changes within the target formation (Heimdal Formation of Paleocene age). The reservoir sandstones are located at a burial depth of around 2 km below the seafloor, with limited tectonic influence. This is a depth (i.e., temperature) where quartz cementation has just started. However, the degree of cementation changes across the submarine fan system due to varying compaction history and clay content and heterogeneity of the reservoir sandstones (see also Lander et al., Chapter 13). Using combined diagenetic and rock physics modeling, one can predict the expected seismic properties along the top reservoir horizon (2D map view). Different scenarios can be tested (facies and fluids) and compared with observed real data, and it is found that the feasibility modeling mimics quite well the observed AVO responses across the studied submarine fan lobe complex when accounting for both compaction trends and depositional changes.

## Brief background and workflow

Quantitative seismic interpretation (Avseth et al., 2005; Simm and Bacon, 2014; Vernik, 2016) is an interdisciplinary field in geoscience that includes a variety of methodologies to improve the understanding of the earth subsurface from remote sensing data. Particularly, rock physics (Mavko et al., 2020) represents a link between qualitative geology and quantitative geophysical observables, and much research has been conducted on understanding both theoretical and empirical relationships between elastic properties and geological features (e.g., Han et al., 1986; Marion et al., 1992; Dvorkin and Nur, 1996; Vernik and Kachanov, 2010; Avseth et al., 2021). These relationships can be applied to predict reservoir properties from seismic amplitude data in what is referred to as quantitative seismic interpretation (Avseth et al., 2005). Quantitative seismic interpretation has proven to be an essential technology to detect, characterize and rank prospective deepwater siliciclastic reservoir sandstones (e.g., Avseth et al., 2001; Mukerji et al., 2001; Caers et al., 2001; Gonzales et al., 2016).

Fig. 10 shows a seismic amplitude map of Top Heimdal Formation reservoir sandstones in the Glitne Field, North Sea, where observed AVO anomalies at three different wells are juxtaposed (adapted from Avseth, 2000). The Glitne oilfield is located west of Stavanger, Norway, near the border with the UK continental shelf. Glitne was the smallest stand-alone field development on the Norwegian continental shelf during the production period from 2001 to 2013. The production of oil was from the upper part of the Heimdal complex representing Paleocene aged turbidite deposits. The stratigraphic evolution and facies associations of the Glitne fan system are found to be analogous to the Permian Tanqua Karoo deepwater fan system in South Africa (Keogh et al., 2008; Hodgetts et al., 2004) where a channel-levee complex fed a frontal lobe complex.

There are drastic changes in the AVO signatures associated with both geological trends and fluid changes. A "hard" class 1 AVO response (positive zero offset reflectivity and negative AVO gradient; see also Castagna et al., 1998) is observed both in the water-filled feeder-channel well (Well 1) and in the brine-filled lobe-margin well (Well 3). However, the oil-filled reservoir sands in the lobe-sands well (Well 2) show a class 3 AVO response (i.e., negative zero offset reflectivity and negative AVO gradient). Avseth (2000) showed the importance of honoring various seismic scale lithofacies when interpreting and classifying the seismic AVO anomalies across the Glitne turbidite system. The feeder-channel sandstones proved to be massive and quite cemented whereas the lobe sands are poorly consolidated to unconsolidated in the upper part. The lobe fringe on the northern side of the lobe complex (penetrated by Well 3) was characterized by more interbedded sandstones and shales. In this study, the importance of compaction trends and diagenetic variations across the Glitne field will also be demonstrated. The onset of quartz cement depends on temperature. Since the burial depth varies by several tens of meters across the Glitne Field, the present burial of the Top Heimdal Formation will control the degree of quartz cement. In addition, clay coating and reservoir heterogeneity can also affect the degree of cementation. Cementation is of particular interest because just a few percent of cement tends to stiffen the rock by gluing grains together. The cementation will therefore significantly reduce the fluid and stress sensitivities of the elastic properties. A forward modeling feasibility study is conducted, focusing on a selected sub-area of the lobe complex in the Glitne Field (marked by a white dashed rectangle in Fig. 10 and a black dashed rectangle in Fig. 13). Compaction trends for different facies (sandstones and shales) are used to constrain rock physics and AVO feasibility modeling. Eventually, it is demonstrated how the feasibility modeling can be used to guide the AVO quantitative interpretation by comparing observed AVO attributes with modeled AVO attributes.

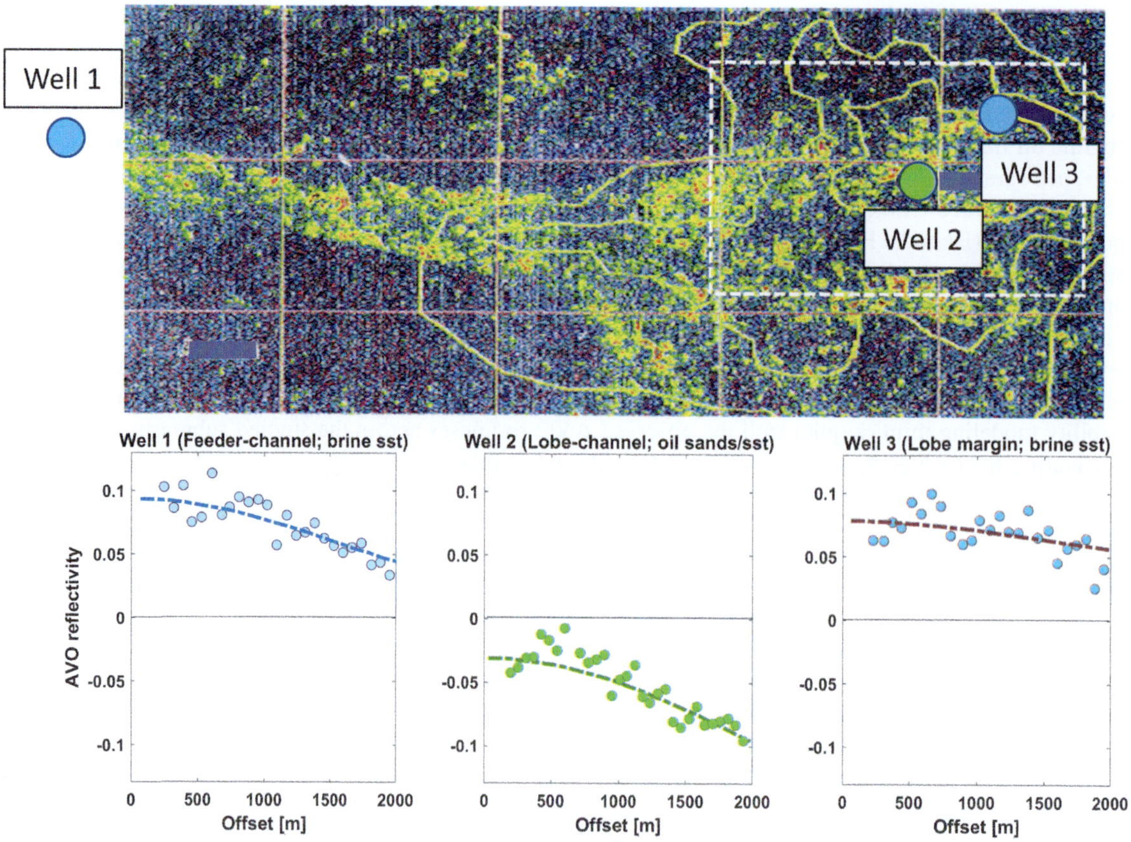


**FIG. 10** Seismic amplitude map from Top Heimdal horizon, Glitne Field (above). Well-calibrated AVO reflectivities picked from CDP gathers at three well locations are also shown. *(Adapted from Avseth, P., 2000. Combining Rock Physics and Sedimentology for Seismic Reservoir Characterization of North Sea Turbidite Systems. Unpublished PhD-thesis, Stanford University.)*

In general, the model-driven methodology presented in this study can be applied to any field or area and is particularly useful in areas with limited well control. It is important to account for burial history away from wells, especially in regions with complex tectonics, because the elastic properties will be affected by diagenetic processes that have happened after deposition (e.g., Lehocki and Avseth, 2021; Lehocki et al., 2020). Fig. 11 shows a flowchart of the applied methodology. Similar approaches have been presented by Mondal et al. (2020) applied to a case in the Mahanadi Basin, offshore East Coast of India, by Liu et al. (2021) applied to targets in the Campeche Basin in the Gulf of Mexico and by Went (2021) demonstrated on a stratigraphic trap in the basin margin offshore Mozambique. A complimentary methodology to forward feasibility modeling is to perform seismic inversion and quantitative lithology and fluids prediction from seismic data. This could be either model-driven or data-driven, or a hybrid of both, using ML. Examples of the latter are shown in Chapter 17.

## Data collection and geologic constraints

The first step in any quantitative seismic interpretation study is to collect and condition required input data (seismic data, well log data, core data, thin sections, etc.) and geologic constraints (stratigraphic information, burial history, facies maps, etc.). In this study, the data comprise a seismic horizon of the Top Heimdal reservoir in the Glitne Field, calibrated seismic AVO attributes (intercept and gradient), well log data from three wells, detailed petrography and core observations from one of the wells (Well 2) and helium porosity measurements from the same well. All the data are available from a public website (https://srb.stanford.edu/quantitative-seismic-interpretation), in association with case examples published by Avseth et al. (2005).

Fig. 12 shows well log data from well 15/5-5 (Well 2 in Fig. 10), located in a lobe-channel of the Glitne Field, where oil was encountered. The well log display includes petrophysical logs calibrated to core observations ($V_{clay}$ log from gamma-ray calibrated to $V_{clay}$ from XRD; Porosity from density log, calibrated to helium porosities), petrographic information

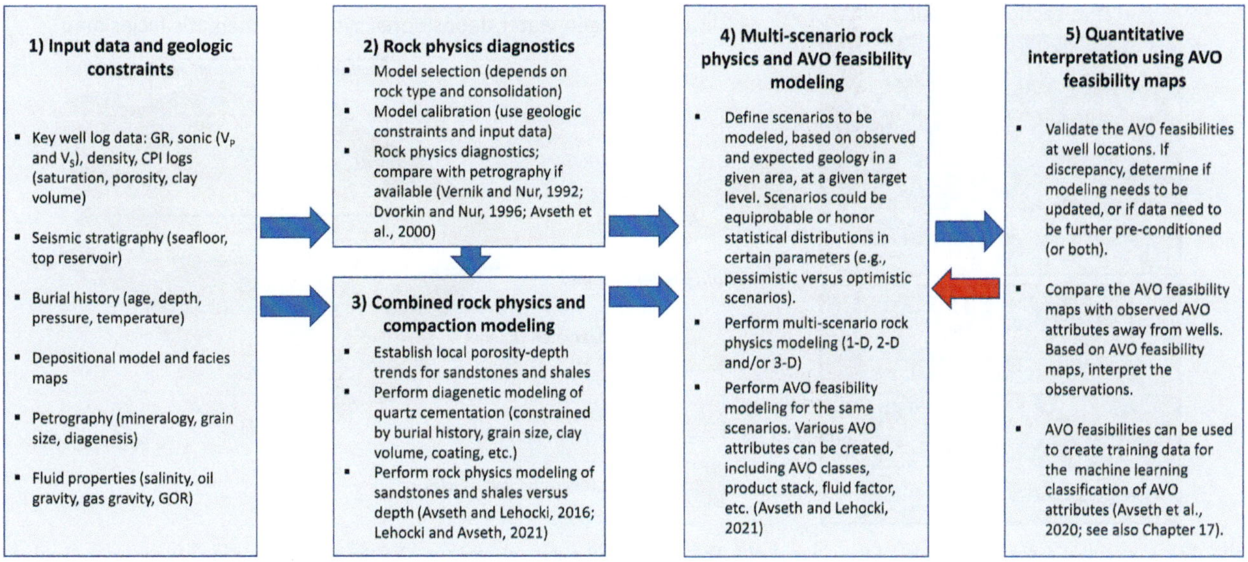


**FIG. 11** Flowchart of AVO feasibility modeling methodology presented in this study.

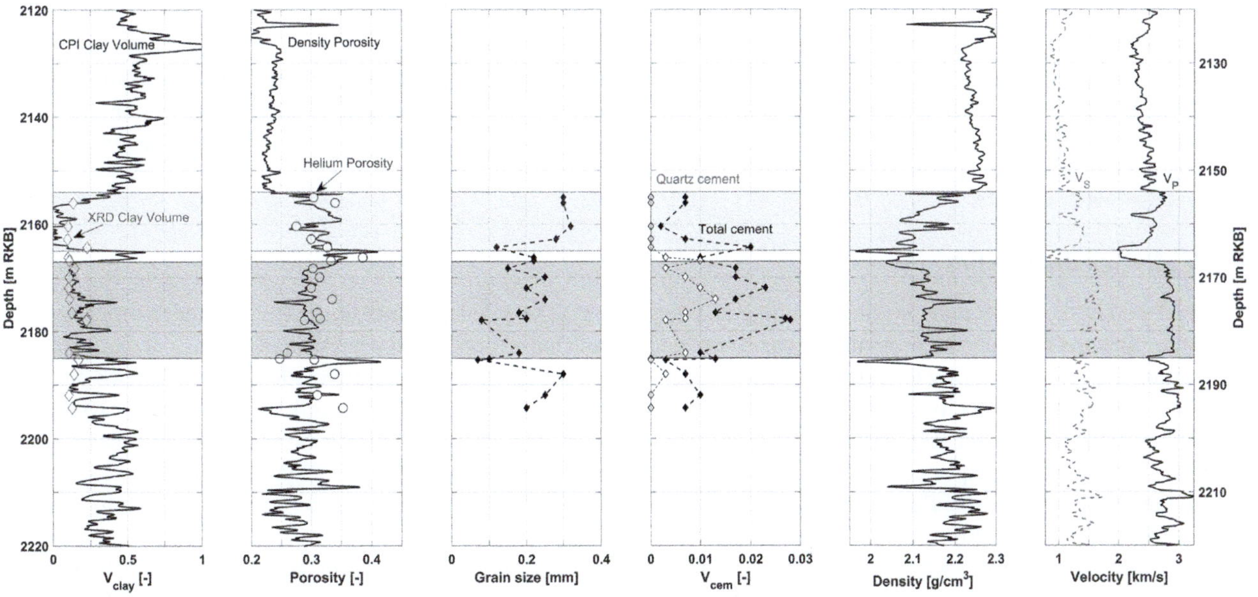


**FIG. 12** Petrophysical logs, petrographic data and elastic logs for well 2 Glitne Field, North Sea. The *gray*-shaded zones represent the key reservoir including Heimdal Formation unit 1 and unit 2.

(sorting and cement volume from thin-section modal analysis in the key reservoir sandstone units) and elastic logs (density, $V_P$, $V_S$; all brine saturated, i.e., fluid substituted in the oil zone from 2154 to 2187m (7065 to 7173 feet) RKB).

The reservoir represents a prograding deepwater lobe complex that was abruptly abandoned and overlaid by marine shales (Martinsen et al., 1996; Avseth et al., 2001). A conceptual depositional model of the Glitne Field is shown in Fig. 13. The depositional system is classified as a point-sourced sand/mud rich system where levees are poorly developed and sheet-sands are occurring in amalgamated channels (cf., Reading and Richards, 1994). Frontal splays are identified on seismic attribute maps, but not penetrated by wells (Avseth et al., 2001). Well 1 is penetrating the feeder channel complex, well 2 is located in the lobe-complex penetrating amalgamated channel sands, whereas well 3 is penetrating interbedded lobe fringe facies.

Despite the sand-rich nature of the lobe complex, the sandstones vary in texture. The reservoir comprises two major subunits of the Heimdal Formation indicated by different gray shadings in Fig. 12. Thin sections from these two units are

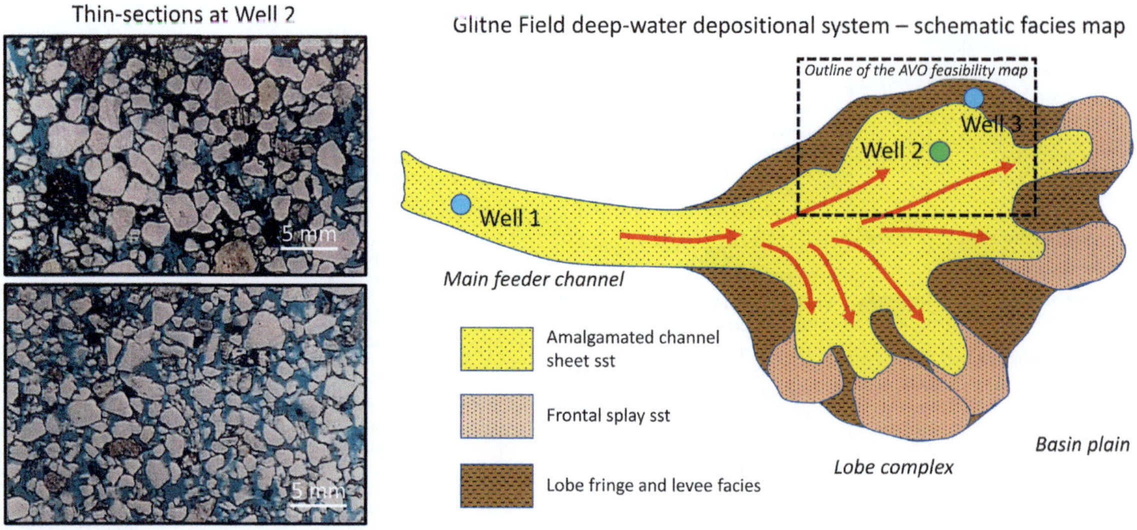


**FIG. 13** Conceptual depositional model of the Glitne Field. The outline of the area selected for AVO feasibility modeling is depicted in the dashed square. The two thin sections to the left are representative of the two major sandstone reservoir units encountered in Well 2 (Heimdal Formation units 1 and 2) with slightly different petrography (Fig. 12), yet significant differences in elastic properties (Fig. 15). *(Modified from Avseth, P. 2000. Combining Rock Physics and Sedimentology for Seismic Reservoir Characterization of North Sea Turbidite Systems. Unpublished PhD-thesis, Stanford University.)*

included in Fig. 13. An approximately 1-m thin debrite separates the two units. The total clay content in the reservoir sandstones is mostly in the range of 10 to 15%, predominantly illite (12% on average) and small amounts of smectite (1%) and kaolinite (1%), some of which are coating the uppermost part of the reservoir sands (see upper left thin-section in Fig. 13). The upper unit is friable with little or no quartz cement (0–0.5%) and a maximum of 1% of total cement. The lower unit has 2–3% quartz cement (Fig. 12, fourth subplot).

## Rock physics diagnostics

The next step is to conduct rock physics screening and diagnostics at selected well locations. This is done in order to understand the expected rock physics properties at the given location and burial depth, and what kind of geological parameters are controlling the local variability in elastic properties. One of the most common and useful methodologies in rock physics is to investigate porosity-velocity relationships (e.g., Han et al., 1986; Klimentos, 1991; Vernik and Nur, 1992; Dvorkin and Brevik, 1999; Marion et al., 1992; Smith et al., 2009). The objective is either to interpret borehole geophysical measurements or forward model expected seismic signatures as a function of controlling geological parameters, like clay content, rock texture, pressure or pore fluid saturations. A wide range of rock physics models can be utilized in rock physics diagnostics (Mavko et al., 2020).

The most common approach for high-porosity sands and sandstones is the Dvorkin and Nur (1996) approach, including the contact cement model and the friable sands model, further extended with the constant cement model by Avseth et al. (2000) (all shown in Fig. 14). These models are based on contact mechanics theory where elasticity is controlled by the characteristcis of grain contacts in granular media. These models have been proven particularly useful for high porosity sands and sandstones. The diagenetic trend comprises the Dvorkin-Nur (D—N) contact cement model at high porosities, and the modified upper Hashin-Shtrikman bound (MUHS) at intermediate-to-low porosities. The friable sand model is a combination of Hertz-Mindlin contact theory at high-porosity endmember and modified lower Hashin-Shtrikman bound (MLHS) for decreasing porosities. The Hashin-Shtrikman bounds have been proven to be useful heuristic "interpolators" from high porosity to low porosity domain for clean sandstones, where the lower bound has been found to mimic sorting variation, whereas the upper bound has been found to represent increasing cement volume (Avseth et al., 2010). The constant cement model is a combination of the friable sand model and the contact cement model. It is a useful model for a cemented sandstone reservoir at a given burial depth, assuming that the burial history is the same for the whole reservoir (cf., Walderhaug, 1996). In contrast, porosity will vary as a function of depositional porosity. Further explanations and the formulas for these models, or many alternative models, can be found in Mavko et al. (2020).

Fig. 15 shows the rock physics diagnostics of the two Heimdal Formation units indicated in the log display in Fig. 12. The P-wave velocity (fluid substituted to brine properties, to avoid ambiguities related to different pore fluids) is

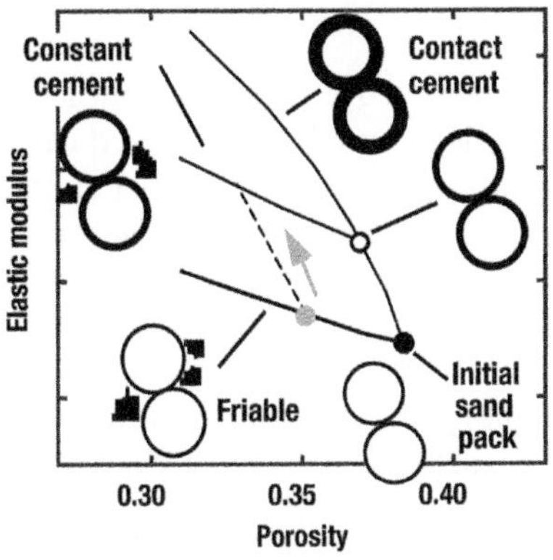


**FIG. 14** Rock physics diagnostic models for high porosity sandstones. *(Adapted from Avseth, P., Dvorkin, J., Mavko, G., Rykkje, J., 2000. Rock-physics diagnostic of North Sea sands: link between microstructure and seismic properties. Geophys. Res. Lett. 27, 2761–2764.)*

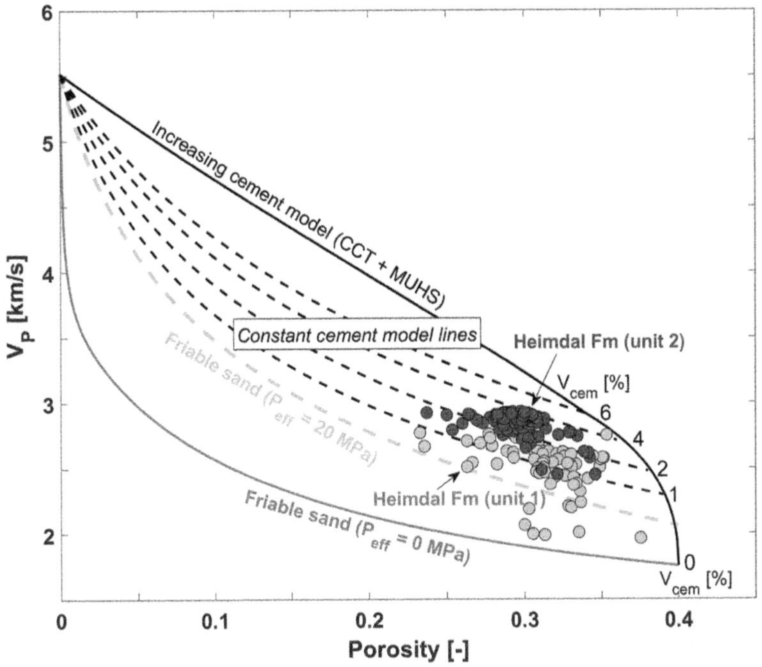


**FIG. 15** Rock physics diagnostics of two Heimdal Fm units indicated in Fig. 12, Glitne Field, North Sea. Note the larger cement volumes indicated in unit 2, in agreement with the petrographic observations in Fig. 3.

cross-plotted against total porosity (from density logs calibrated with helium porosities), and diagnostic rock physics models are superimposed. Several constant cement models are included to quantify the cement volume (cf., Avseth et al., 2010). These models have also been calibrated to the observed mineralogy in the reservoir sandstones, assuming 15% feldspar and 10% clays in addition to quartz as part of the solid matrix. The two different Heimdal Formation units are diagnosed with different cement volumes. The upper unit 1 is found to have very little (0–2%) cement, whereas the second unit below has slightly higher cement volume (2–4%). These cement volume estimates from the rock physics diagnostics match nicely with the thin-section observations (Fig. 12). The porosity values in the two units are overlapping, but average values are somewhat lower in the deeper unit. A significant velocity difference between the units can be explained by the difference in elastic stiffnesses associated with the different cement volumes (cf., Avseth et al., 2000). As seen in Fig. 12, the average grain sizes are smaller in the second unit. This could partly explain the differences in cement volume, as smaller grain size implies a larger specific surface area available for quartz overgrowth (Bjørlykke, 2015). However, the increasing cement volume from unit 1 to unit 2 could also be related to increasing depth (i.e., temperature) as the reservoirs have been exposed to continuous subsidence. The current reservoir temperature is around 75°C, slightly above the critical temperature for the onset of quartz cement (Walderhaug, 1996).

## Compaction modeling

Next, the rock-physics diagnostic approach shown above is extended by accounting for the complete burial history of the turbidite sandstones (i.e., from deposition to deep burial). More specifically, burial history modeling (mechanical and chemical compaction) is combined with rock-physics modeling (cf., Lehocki and Avseth, 2021). The sandstones in the Glitne Field have been exposed to continuous subsidence and limited or no *syn*- or post-depositional tectonic uplift (Baig et al., 2019). Hence, the burial history can be modeled with a close to linear subsidence rate since deposition in the Paleocene age. Contact theory is combined with empirical porosity depth trends and Walderhaug diagenetic modeling (Lander and Walderhaug, 1999; Avseth and Lehocki, 2016; see also Chapter 13, this book). The resulting velocity-depth trends for sands and sandstones are shown in Fig. 16 for the three wells penetrating the Glitne submarine fan at three different locations. During the mechanical compaction (down to 70°C at ca. 1.9km (6232 feet) burial depth), the P-wave velocities for unconsolidated sands are approaching 2.5km/s (8125 feet/s) (solid black lines in Fig. 16). Varying sorting implies varying depositional porosity and, therefore, slightly varying velocities with depth. The dashed black lines shown in Fig. 16 represent velocity-depth trends when depositional porosity varies between 0.35 and 0.45 (i.e., poorly sorted—well sorted), whereas the "expected" depth trend shown in solid black line represents moderately well-sorted sandstones with depositional porosity of 0.4. The well log data superimposed on the modeled depth trends are colored as a function of lithology, where the black points are sandstone intervals, and gray points are shaly intervals. The upper part of the Heimdal Formation representing the submarine fan shown in Fig. 1 is indicated by cyan colors. In Well 2, the cyan interval comprises both units (1 and 2) shown in Fig. 12. It is interesting to note that the Top Heimdal event is at slightly different depths in the different wells. The depth of Top Heimdal is several tens of meters deeper in Well 1 (main feeder-channel) compared with Well 2 (channel-lobe transition zone to proximal lobe setting, cf., Wynn et al., 2002). Finally, the depth of Top Heimdal is also deeper in Well 3 (lateral lobe fringe) than in Well 2. Therefore, it is imperative that the compaction degree and velocities are larger in the feeder-channel and lobe margin than in the lobe-channel due to varying diagenesis (i.e., quartz cementation). Assuming the temperature gradient is constant over the Glitne Field, the onset of chemical compaction occurs at around 1900m (6232 feet) below the mudline (water depth is around 110m, 360 feet) in all the wells. The rapidly increasing velocity below this depth is caused by the onset of chemical compaction. The Heimdal Formation reservoir sandstones in the three wells will be differently affected because they are at slightly different burial depths. This information is important to include in rock physics modeling and AVO analysis of the Glitne Field. In addition, the sand quality can also change between the wells due to facies variations. In particular, the net-to-gross and shaliness of the reservoir sands change from the relatively massive feeder and lobe channel sandstones in Wells 1 and 2 to the more interbedded sandstones in Well 3

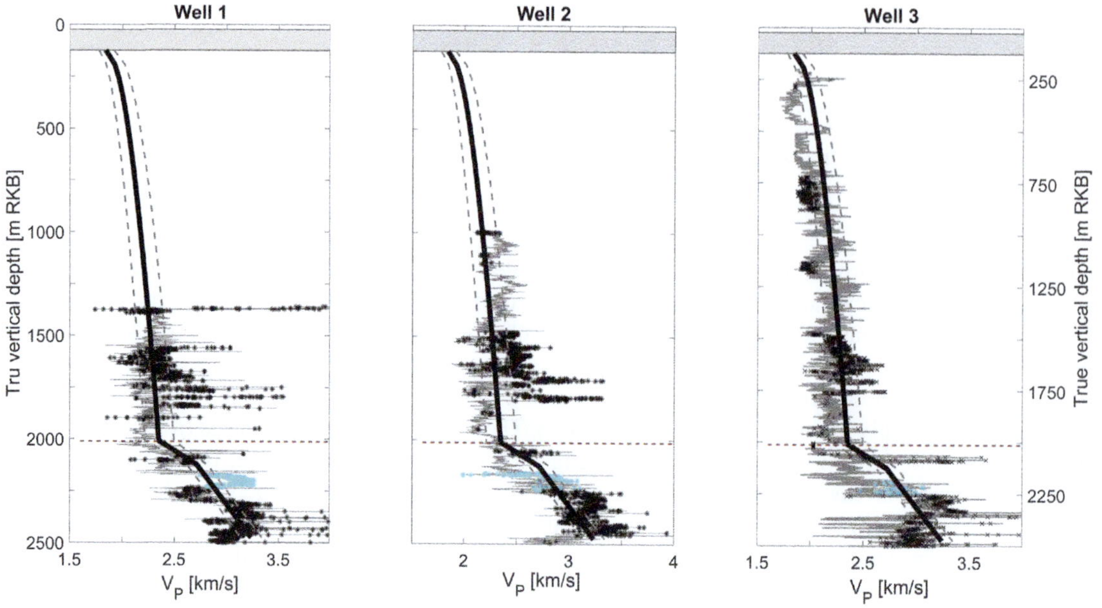


**FIG. 16** Combined rock physics and compaction modeling for sands and sandstones in the Glitne Field. Note the drastic increase in velocities when going from mechanical to chemical compaction domain (marked by *dashed horizontal lines*). The black dots are sandstone intervals, and *gray* colors are mainly embedding shales. Cyan represents the upper part of the Heimdal Formation submarine fan system investigated in this study.

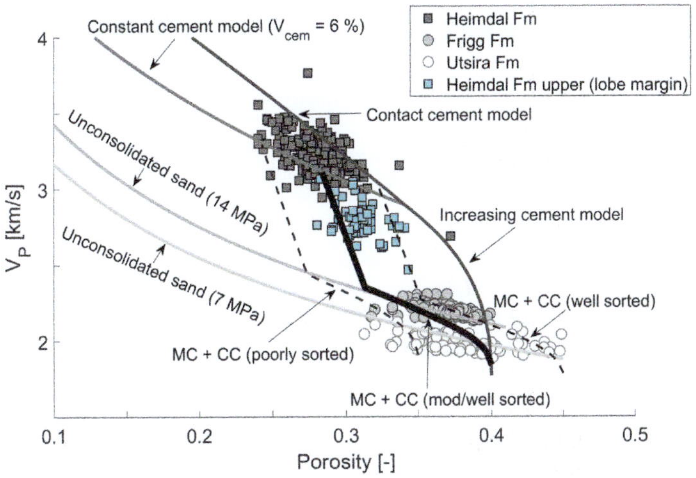


**FIG. 17** Crossplot of P-wave velocity ($V_P$) vs porosity for sands/sandstone data in Well 3. Models include both diagnostic models (cf., Fig. 2) and combined burial and rock physics models (*black solid* and *dashed lines*). Note the rapid increase in velocities from mechanical compaction to chemical compaction domain, corresponding to the transition indicated at around 1.9 km (6232 feet) burial depth in Fig. 7. The Heimdal Formation sandstones show velocities from around 2.8 to 3.5 km/s, whereas the shallower unconsolidated sands range from 2.0 to 2.4 km/s. The burial-constrained modeling captures the data nicely when depositional porosity is varied (i.e., associated with sorting variation) from 0.35–0.45.

(see also Avseth et al., 2001). The facies variation can explain the variability around the modeled depth trends. However, the overall trend in the data is captured nicely by the combined burial and rock physics modeling.

Fig. 17 shows a crossplot of P-wave velocity vs porosity for sands and sandstone data in Well 3, including data from different stratigraphic intervals in the overburden and the target level (i.e., Quaternary age Utsira Formation, Eocene age Frigg Formation and Paleocene age Heimdal Formation). The different intervals can be explained nicely by the conventional rock physics diagnostic approach (cf., Fig. 14), where the Utsira and Frigg formations are modeled using the friable sand model at their respective effective stress (7 and 14 MPa, respectively) and Heimdal Formation sandstone interval is modeled using the constant cement model. The cement volume will vary throughout the Heimdal Formation (due to depth variation and facies changes). The impact of chemical diagenesis on the sandstone properties is more realistically captured by the combined burial and rock physics modeling. The continuous change in the porosity-velocity relationship from deposition to present-day burial depth is modeled. The rapid increase in velocities can be clearly seen as one goes from the mechanical compaction domain (mean porosities reducing from 0.4 to ca. 0.3) to the chemical compaction domain (mean porosities decreasing beneath 0.3). The relatively large variability in velocities seen in the Heimdal Formation is nicely explained by the compaction trend (i.e., solid black lines with dashed lines representing variability in depositional porosity, i.e., varying sorting).

## AVO feasibility modeling and scenario testing

Next, the compaction models created above are used to constrain AVO modeling for various pore fluid scenarios in the Glitne Field to see if the observed variability in AVO signatures shown in Fig. 10 can be explained. The AVO feasibility methodology utilized for this purpose, was introduced by Avseth and Lehocki (2016) and further developed by Avseth et al. (2020) and Avseth and Lehocki (2021), first as 1D examples and followed by 2D examples (i.e., AVO feasibility maps).

Based on the combined burial and rock physics modeling presented in the previous section (cf., the flow chart in Fig. 11), corresponding depth trends of acoustic impedance and $V_P/V_S$ are estimated for sandstones for different pore fluid scenarios (i.e., oil- and brine-filled sandstones). Also, rock physics depth trends for shales (i.e., clay-supported, silty mudrocks) are obtained using calibrated models based on contact theory (see Avseth et al., 2008; Avseth and Lehocki, 2016). The diagenetic transformation from smectite-rich to illite-rich shales is accounted for, occurring at around the same temperature/depth as the commencement of quartz cementation (Bjørlykke, 2015). Alternatively, shale depth trends can be derived empirically from the well log data, as most of the overburden comprises shales (cf., Avseth et al., 2021; Avseth and Lehocki, 2021). The sandstone and shale elastic depth trends are superimposed on upscaled (5 m resolution) well log data from Well 2 in Fig. 18 for two different target depths. The temperature gradient in this well is 35°C/km, the average grain size in the target zone is 0.2–0.3 mm (fine-to-medium-grained sandstone) and the clay content in the reservoir sandstones is

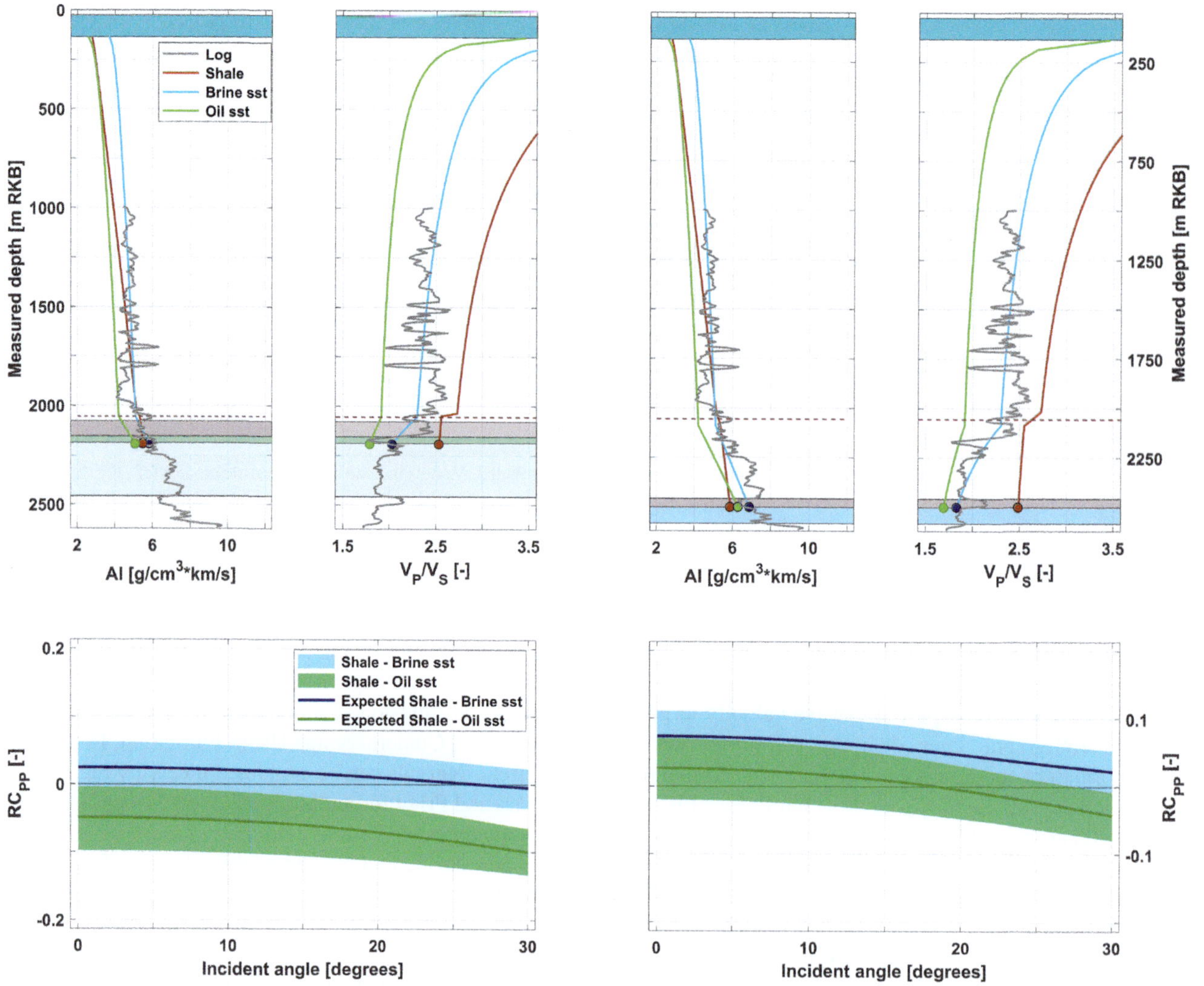


**FIG. 18** Combined burial and rock physics modeling for Paleocene sandstones in Well 15/5-5 in the Glitne Field, North Sea. Note the change in elastic properties from mechanical to chemical compaction at around 2000 m (6560 feet) burial depth (indicated by *dashed brown horizontal line*). The expected AVO signatures change drastically as we go from the Top Heimdal Formation horizon (left) to a slightly deeper target level (right). The expected oil response for the deeper example is similar to the expected brine response of the Top Heimdal Formation. This shows the importance of burial history as a key constraint in AVO analysis.

in the range of 10–15%, mostly detrital ductile rock fragments, illite and mica (Fig. 12). These are all important input parameters for the combined compaction and rock physics modeling (Lehocki and Avseth, 2021).

Based on the rock physics depth trends, one can simulate the expected AVO signatures at any given depth. Two examples are demonstrated in Fig. 18, one where the Top Heimdal Formation is modeled at observed burial depth, and one where the model is a slightly deeper sandstone with the same rock type. To honor uncertainties in the rock physics properties, 5% variability in $V_P$ and $V_S$, and 3% variability in density are added to the expected values from the depth trends, assuming normal distributions. These are values representative of the observed natural variability in mixed sand/mud-rich deepwater clastic systems (e.g., Avseth et al., 2003) that can be obtained from local well log data or statistics from analog areas. Also, realistic covariances between $V_P$ and $V_S$ are imposed (cf., Greenberg and Castagna, 1992; Avseth et al., 2005). The rock physics modeling at the Top Heimdal Formation (upper) fits accurately with the observed values in the well log data. An AVO class 2p-2 is expected for brine-saturated sandstones and a class 2–3 for oil-saturated sandstones. The top reservoir is located just beneath the onset of chemical compaction and the burial history is characterized by continuous subsidence (i.e., zero net erosion) since the deposition of the Tertiary age turbidite sandstones. For the slightly deeper modeling example, a slightly stiffer (i.e., more cemented) sandstone is expected with AVO class 1 for brine-saturated sandstones and class 2p for oil-saturated sandstones. This 1D AVO feasibility demonstration for the Glitne Field

**TABLE 1** Scenario input geological parameters for the 2-D AVO feasibility modeling.

| Scenario | Grain size | Clay volume | Clay coating |
|---|---|---|---|
| 1: Lobe channel/sheet sands; Heimdal Fm, unit 1 | 0.3 mm | 0.1 | 0.5 |
| 2: Lobe channel/sheet sands; Heimdal Fm, unit 2 | 0.2 mm | 0.05 | 0 |
| 3: Mixed scenario | 0.2–0.3 mm | 0.05–0.1 | 0–0.5 |

illustrates how important it is to account for the burial history when investigating expected AVO signatures, even in areas with continuous subsidence, small tectonic influence and relatively flat strata (see also Chapters 2 and 18, this book). Furthermore, it is interesting to observe how this modeling exercise can explain some of the large variability seen in the observed AVO signatures across the Glitne Field (Fig. 10).

Next, AVO feasibility modeling is performed in 2D, along the Top Heimdal Formation horizon. The focus is on a selected sub-area of the lobe complex within structure around Well 2 and Well 3 (see a rough outline in Fig. 13). Three different scenarios have been defined (see Table 1) to model. Scenario 1 is representative of what is observed in Heimdal Formation unit 1 at Well 2, where the reservoir sands are unconsolidated to poorly consolidated, likely due to a combination of the burial history and presence of clay coatings (see upper left thin-section in Fig. 13). The sands have just entered the chemical compaction zone where quartz cementation has commenced. However, the presence of clay coatings has potentially inhibited quartz cementation. In the modeling, the grain size is set to 0.3 (cf., thin-section observations), the clay content is set to 0.1 (cf., thin-section observations), whereas the degree of grain coating is assumed to be 0.5. According to Bello et al. (2021), a coating degree of more than 40–50% tends to arrest quartz cementation and preserve reservoir quality, based on a study of Paleocene age Forties Member turbidite sandstones in the Central North Sea. It is to be noted that the Glitne Field sandstones are classified in the transition between sub-arkose and sub-litharenite (Martinsen et al., 1996), whereas the sandstones studied by Bello et al. (2021) represent sub-arkose (Pettijohn et al., 1972; Folk, 1980). The shale depth trends used in the modeling are derived empirically from the well log data. The temperature gradient is assumed to be 35°C/km. The feasibility modeling of scenario 1 is shown in Fig. 19. An AVO class 2p response is expected for a brine-filled sandstone over the whole lobe structure (Fig. 19, upper left subplot). Fluid substituting to oil (using

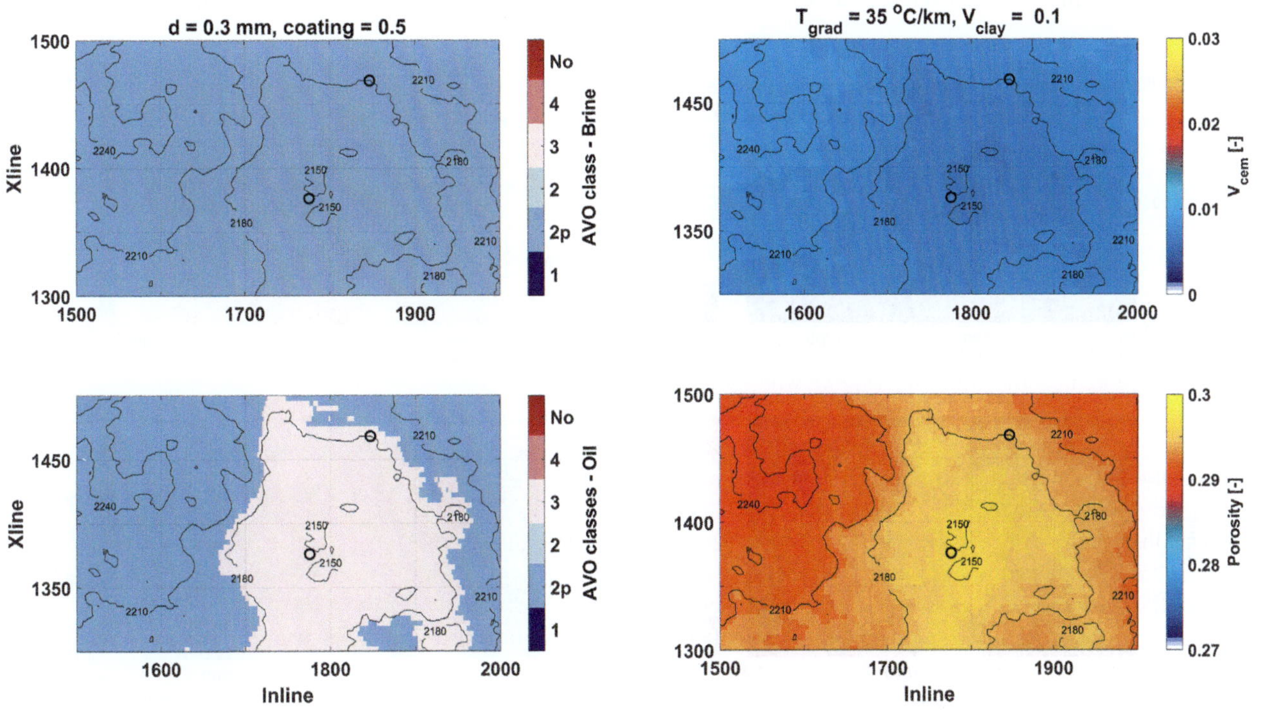


**FIG. 19** AVO and rock property feasibility maps for scenario 1, Top Heimdal Formation, Glitne lobe structure.

Gassmann theory; see Mavko et al., 2020) above the oil–water contact (2187m TVD RKB), predicts AVO class 3 for oil-filled sandstones. This matches with what is observed in the gathers at Well 2 location (Fig. 10) and with the 1D AVO feasibility modeling for Well 2 (Fig. 9, left). Also included are the 2D maps of predicted reservoir porosities and cement volumes within the Top Heimdal Formation (Fig. 10, right-hand subplots). These are derived from the Lander and Walderhaug (1999) modeling, given the input burial constraints (depositional porosity of 0.4, normal mechanical compaction). The predicted cement volume is around 1%, whereas expected porosities are around 0.3 in the upper part of the lobe, and slightly lower around the margins of the lobe. The cement volume and porosities shown in Fig. 10 (right) control the elastic properties used to estimate the AVO attributes shown in Fig. 10 (left). Note that porosity variations related to facies changes are not included in this scenario. The porosity variation shown in Fig. 19 is only associated with the compactional processes. Alternative scenarios could be run to test the effect of different facies, as depicted in the depositional model in Fig. 13. Many different scenarios could potentially be tested, both with regards to burial history and depositional features. For more detailed information about various input parameters to test in the scenario modeling, see Lehocki and Avseth (2021).

In scenario 2, a different type of Heimdal Formation sandstone is tested related to changes both in diagenesis and depositional facies. This time, a sandstone is modeled with slightly lower clay content and no clay coating. Also, the grain size is assumed to be slightly smaller than in scenario 1, assumed now to be 0.2. This scenario is more representative of the Heimdal Formation unit 2 (see the lower left thin-section in Fig. 13) and could also be more similar to the sandstones encountered in the feeder-channel (Well 1) and down the flank from Well 2 within the lobe complex. Due to the smaller grain size, lack of clay coating and slightly deeper burial, the sandstones will now be more cemented. This is indeed what is observed in Fig. 11. The cement volume over the lobe structure is now expected to be 2–3%. These results accurately fit the thin-section data of Heimdal Formation unit 2 (Fig. 12). Also, the porosities are slightly lower than for scenario 1. The resulting AVO feasibilities show an expected AVO class 1 for the brine-saturated sandstones and class 2p for oil-saturated sandstones (Fig. 20, left sub-plots).

Finally, scenario 3 is a mixture of scenarios 1 and 2, where scenario 1 is constrained to the lobe structure around well 2. In contrast, scenario 2 is confined to the area down-flank from the structural closure within the lobe complex. It would be optimal to have a facies map to constrain the spatial scenario mixing. For simplicity, it is assumed that the transition is depth consistent at 2166m TVD RKB (in agreement with stratigraphic observations in the wells and the fact that the layers are quite flat, see Avseth et al., 2005, fig. 4.30). The resulting AVO feasibility map is shown in Fig. 21. It is interesting to see the geological shape occurring, with AVO classes changing from class 1 in the brine-filled area down-flank the lobe structure, to class 3 in the oil-filled lobe channel. Cement volumes change from close to zero in the upper lobe sands to 3–4% down-flank, and porosities change from 0.29 to 0.30 in the lobe-channel sands to 0.27–0.28 in the down-flank sandstones. These

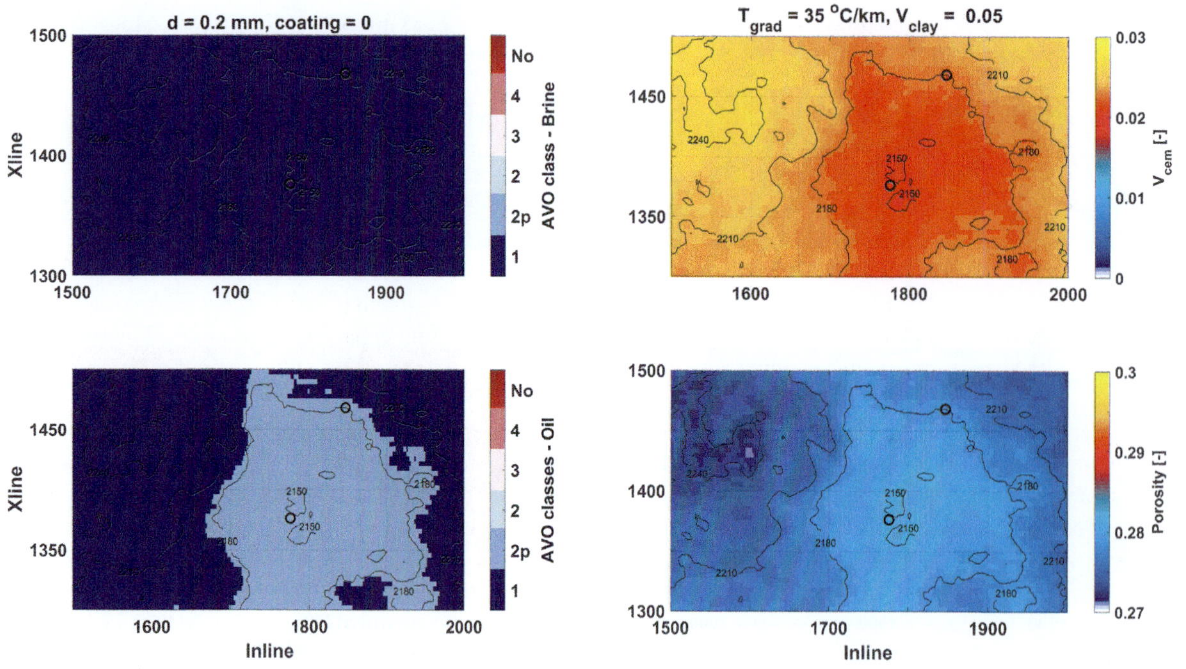


FIG. 20 AVO and rock property feasibility maps for scenario 2, Top Heimdal Formation, Glitne lobe structure.

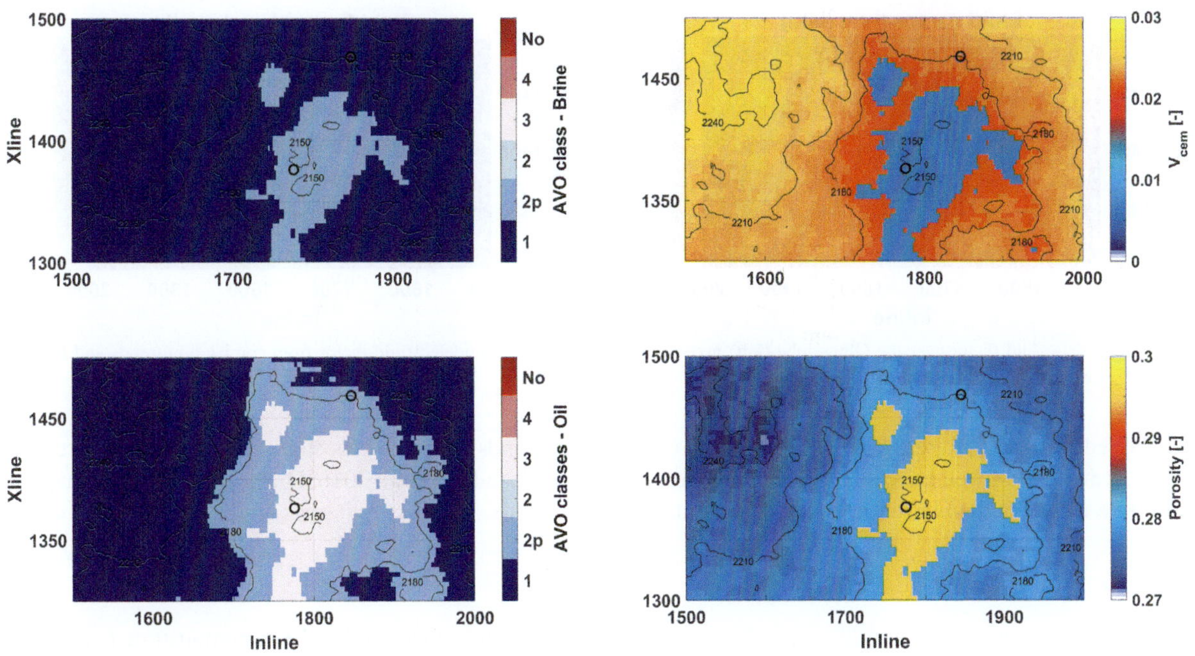


**FIG. 21** AVO and rock property feasibility maps for scenario 3, Top Heimdal Formation, Glitne lobe structure.

results derived from the combined compaction and rock physics modeling fit nicely with both petrographic observations (Fig. 12), rock physics diagnostics (Fig. 15) and seismic observations (Fig. 10). The next step is to demonstrate the use of the feasibility modeling results for quantitative seismic interpretation.

## Model-driven quantitative seismic interpretation

The AVO feasibility results above can be compared with real data observations, including pre-stack seismic gathers, angle-stacks and/or AVO attributes derived from angle-stacks. This is a workflow that can improve the risk assessment of a prospect during exploration, or of undrilled compartments in existing fields (i.e., during appraisal). The Glitne Field was discovered in 1995 and the production ended in 2012. However, the workflow can still be demonstrated on the Glitne AVO data. A great variety of AVO attributes can be created to boost the detectability of hydrocarbons from seismic data, also referred to as "direct hydrocarbon indicators" or DHIs (Avseth et al., 2005; Simm and Bacon, 2014). One of the most useful AVO attributes during derisking and ranking of hydrocarbon prospects is the fluid factor (FF) (Smith and Gidlow, 1987). The FF equation as a function of AVO reflectivities can be formulated as follows (see also Avseth et al., 2008):

$$FF = \frac{8}{5} \cdot R(0) - m \cdot [R(0) - G] \quad (1)$$

where $m$ is the product of the "mudrock line" slope (normally 1.16, cf., Greenberg and Castagna, 1992) and the background $V_S/V_P$ ratio (around 0.5). The FF will represent the deviation from the background trend in an AVO cross-plot of intercept ($R(0)$) and gradient ($G$), and it tends to cancel out lithology variations while enhancing the effect of fluids. Hence, it is comparable with the extended elastic impedance (Whitcombe et al., 2002) where fluid and lithology effects can be enhanced via optimal rotation angles of a background trend in the same AVO cross-plot (Hicks and Francis, 2006).

Fig. 22 shows a comparison of the observed FF derived from the real angle stack data in the Glitne Field compared with the modeled FF from the AVO feasibilities of oil-filled lobe sands (scenario 3, Fig. 21). The calibrated observed FF values and the modeled ones are quite similar (real data FF is based on calibrated amplitude values, whereas modeled FF is based on reflectivities). The same shape is recognized in the real data as in the feasibility model, likely reflecting good quality lobe-channel sands with trapped oil. Some other patches with relatively large negative FF values can also be seen in the real data. These could reflect local compartments of reservoir sands filled with oil, down-flank from the main structure, possibly with stratigraphic trapping. However, the down-flank smaller FF anomalies could also be ambiguities or noise in the data. This demonstration shows how rock physics and AVO feasibilities can be used semi-quantitatively to interpret observed anomalies and outline prospects or upside potential around existing fields. However, this also shows that it is important to

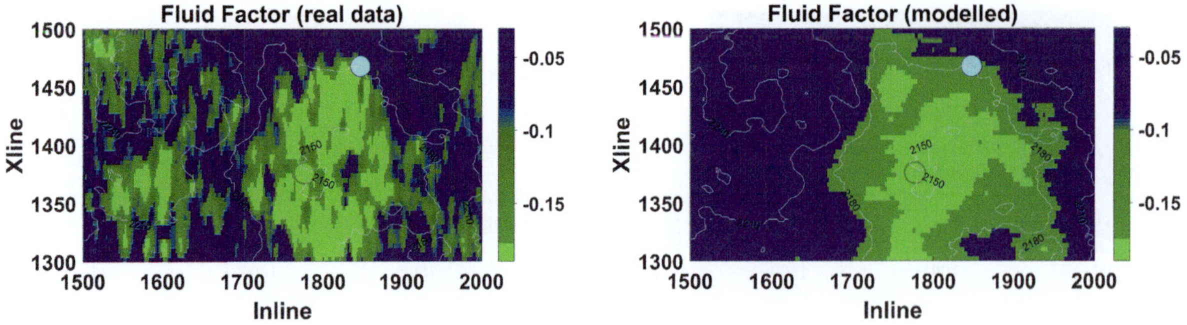


FIG. 22 Real data fluid factor over the Glitne lobe (left) juxtaposed with feasibility modeled fluid factor from compaction trends and lateral facies changes (cf., scenario 3).

honor the sedimentary features, facies variations and compactions trends when the expected seismic responses are modeled between or away from well locations (cf., Brevik et al., 2011; Avseth and Lehocki, 2016; Dræge et al., 2014). Chapters 7, 8 and 13 give more detailed introductions to sedimentary environments, facies associations and compaction trends.

### Summary

In this section, the use of 1D and 2D rock physics and AVO feasibility modeling has been demonstrated within a lobe complex of a submarine fan in the Glitne Field, North Sea. By combining diagenetic and rock physics modeling, constrained by petrography and facies interpretations, it is possible to create realistic scenarios of AVO feasibility maps that can be compared with real AVO data observations. The methodology demonstrated in this study can be integrated into a quantitative seismic interpretation workflow for improved reservoir mapping and prospect derisking from seismic AVO data and should be particularly useful for deepwater clastic systems with complex spatial distributions of lithofacies.

## Role of dipmeters and imaging devices in evaluating deepwater sedimentary intervals

### Introduction

This section presents and explains several borehole imaging (BHI) examples logged with the Formation MicroImager (FMI—a Mark of Schlumberger) in deepwater sediment (DWS). Borehole imaging devices (BHIs) have by far the highest resolution of the various wireline logs and these images are easily relatable to outcrops. They provide a vital link between other open hole logs and seismic. Borehole imaging devices (BHIs) are the culmination of the evolution of the dipmeter family of wireline tools. Schlumberger's Formation MicroImager (FMI) is the premier device in this family. It provides high-resolution pictures of the subsurface, and as with all dipmeters, the tools were designed to determine the dip of bedding planes and other features in boreholes. This requires a minimum of three points oriented in space. The depth of these features is determined from the logging cable with which they are lowered into the hole. In the case of the FMI, the orientation is provided by highly accurate accelerometers and magnetometers. Dips are computed to within 1 degree of accuracy for both the dip magnitude and direction. This accuracy is invaluable when used to help calibrate seismic for depth and inclinations. Micro-resistivity is a thinly focused measurement of resistivity, i.e., the resistance to the flow of an electrical current of a volume of rock. The FMI tool (see Fig. 23) has 192 micro-resistivity sensors that are recorded digitally every 0.1″ (0.254 cm). This information is processed to provide reliable and repeatable images of four sides of the borehole and can be used to identify lithologies and fractures as well as to delineate very thin (down to the centimeter range) beds. The data can be presented in many formats. Most of these presentations simulate looking at the formation and so the images are easily interpreted by geoscientists familiar with looking at rocks; they are analogous to looking at outcrops. The three-dimensional aspect of the information often requires special displays. These plots can be analyzed visually and statistically, and the lessons learned with BHI images aid in the evaluation of whole cores as seen in Figs. 25A and 31B.

### Applications of borehole imaging (BHIs)

All the examples shown in this section have been released on the condition that their exact location will not be disclosed. They were logged with Schlumberger's Formation MicroImager (see Fig. 23) and were recorded in feet as is customary for borehole logs in the areas. Most examples are from the USA Texas Gulf Coast and all are interpreted to have been deposited

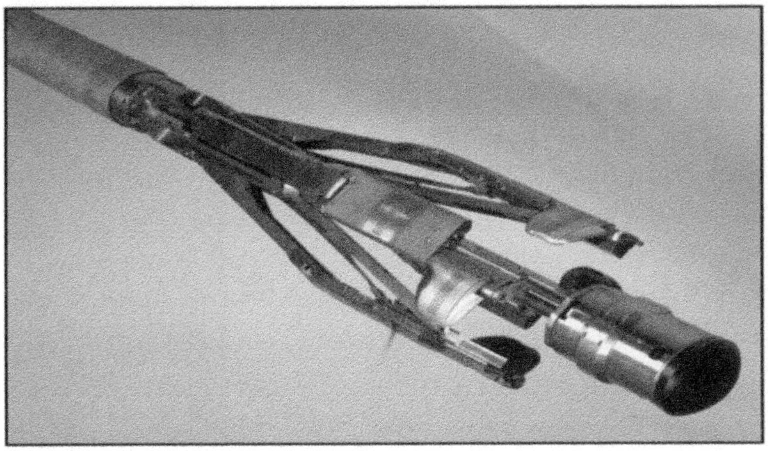

**FIG. 23** Picture of the Formation MicroImager (FMI) tool. It was introduced in 1991. The FMI is the state-of-the-art member of the "dipmeter" family of borehole imaging (BHI) wireline tools. It has four arms with one pad and one flapper on each arm. It has 192 Micro-Resistivity sensors and has 80% coverage in an 8″ hole. The FMI has a minimum hole size of 5 7/ 8″ and a maximum of 21′ It is rated to 325° F and 20,000psi.

in deepwater environments. A certain familiarity with the FMI tool characteristics and presentations is recommended. The references listed below will be useful in this regard.

## The usefulness of "3D to 2D" presentations in interpreting BHI, core and outcrop

The BorMap for-mat as seen on the left in Fig. 24 pro-vides a use-ful way to put the three-di-men-sional imag-ing in-for-ma-tion onto a two-di-men-sional pre-sen-ta-tion. It has be-come the stan-dard for BHIs. One con-se-quence of us-ing this for-mat is that dips ap-pear as si-nu-soids. The epoxy core pho-tos shown in Fig. 25B il-lus-trate how a core would look if dis-played in this BorMap like for-mat. Note the cross-beds and rip-ples. Both pho-tographs have res-o-lu-tions like those avail-able with the FMI. The photo in Fig. 25B sim-u-lates a BorMap-type for-mat and will help one vi-su-al-ize the fea-tures and ori-en-ta-tions of both cores and im-ages.

## Historical perspectives, acquisition and presentation of dipmeters (including BHIs)

The dipmeter family of tools began with the early three pad dipmeter (DM) introduced circa 1943 and evolved to the current state-of-the-art. Formation MicroImager (FMI) tool. This "family of tools" has grown steadily in sophistication, accuracy and versatility. The FMI tool measures Micro-resistivity with resolution down to 0.2″ (0.5 cm) and detection well below that (see Fig. 23). That is to say that features, such as thin beds, can be analyzed down to less than the centimeter-level; while very thin features, such as fractures, can be detected down to a millimeter or less. Additionally, precise orientation is provided by highly accurate magnetometers and accelerometers. Dipmeter interpretation has progressed from determining the structure and locating missing sections to sophisticated and oriented descriptions of reservoirs and their properties.

Meaningful stratigraphic interpretations are routinely produced. This ever-increasing amount of data required digital acquisition and special ways of presenting this information. The dipmeter tadpole plot (Fig. 24B) was developed to present dip vs depth. Interpretation techniques were developed by Schlumberger's Al Gilreath and others to interpret the dipmeter results. These relied on pattern recognition and client feedback.

The BorMap format as seen on the left in Fig. 24 provides a useful way to put the three-dimensional imaging information onto a two-dimensional presentation. It has become the standard for BHIs. One consequence of using this format is that dips appear as sinusoids. The epoxy core photos shown in Fig. 25B illustrate how a core would look if displayed in this BorMap-like format. Note the cross-beds and ripples. Both photographs have resolutions like those available with the FMI. The photo in Fig. 25B simulates a BorMap type format and will help one visualize the features and orientations of both cores and images.

The full dynamic range of data of each micro-resistivity measurement can be normalized and represented using various color schemes. In the most commonly used color scheme, white represents the highest resistivities (as for limestones). Black used to represent the lowest resistivities (i.e., the highest conductivities, as for shales) As resistivities decrease they go from orange to yellow, to tan, to brown and on to black for the lowest resistivities as shown in Figs. 26 and 27. The data can be normalized over a longer interval, as in the Static Images, which emphasizes lithology. Computations over shorter intervals, such as 2′ (0.61 m), highlight the details within that interval only. The Dynamic Images are normally computed

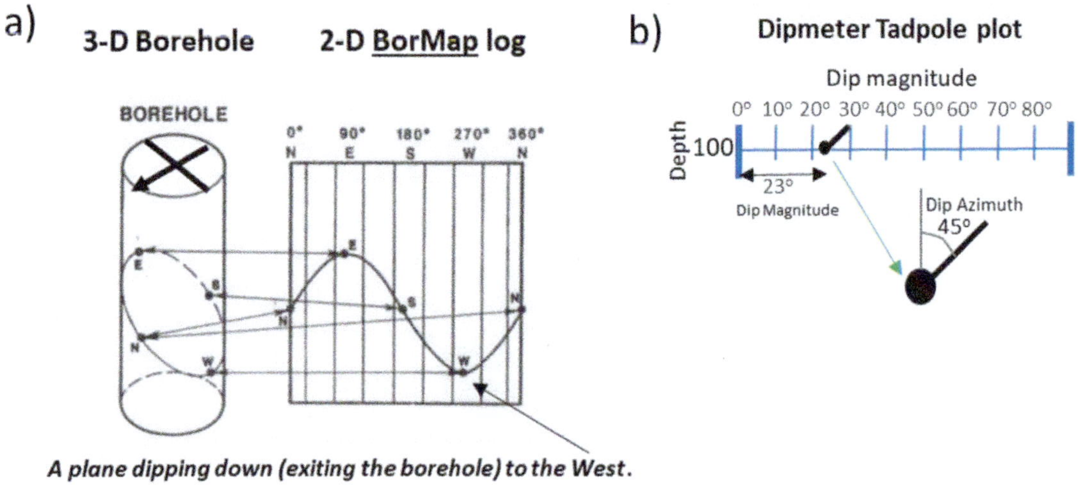


**FIG. 24** (A) Schematic diagram of a 3D borehole and the corresponding 2D *BorMap* image log obtained from borehole imaging (BHI) devices. (B) Traditional dipmeter *tadpole plot* log obtained from any dipmeter tool. Precise dip values can be derived from batch dipmeter computations and manually computed using workstations. Polar plots, such as Azimuth Frequency Plots (AFPs) are routinely produced. For horizontal holes, the "Top of Hole" is presented on the outer edges of the image. The "Bottom of the hole" is put in the middle. Note that in a horizontal hole, a pad pointing straight up has no azimuthal direction and is therefore undefined. The borehole orientation which is assumed to be the same as the tool is fully defined in three-dimensional space.

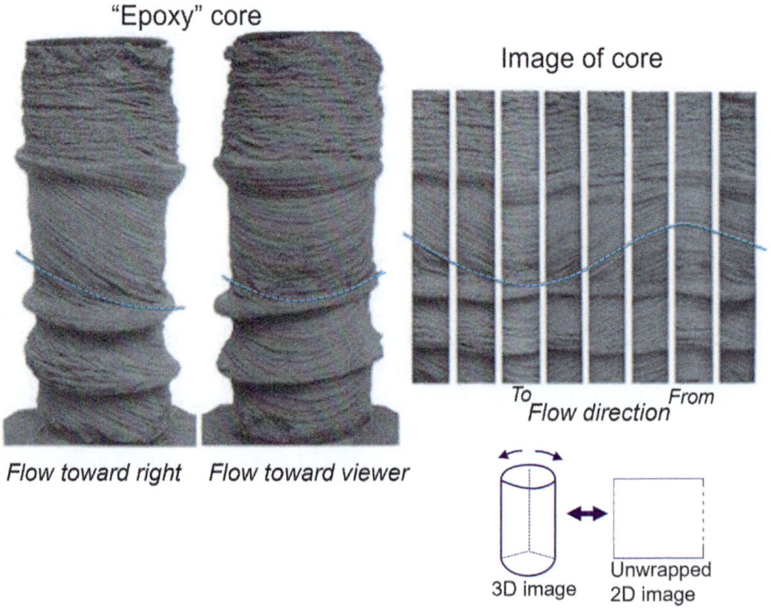


**FIG. 25** Photos of an epoxy core with cross-bedding and a "pseudo" BorMap of the core. Note the "Ripples" at the top of the core and the "trough cross-beds" below. Flow directions are indicated. The visual resolution is like that of FMI Images. The core was taken by Rufus Leblanc, Sr. on the Brazos River south of Houston, Texas, USA.

sequentially every 2′ (0.61 m). In the 1990s, analysis by Steve Hansen and Tom Fett illustrated the usefulness of comparing various normalizations to known lithologies, especially in sands and shales (see Fig. 27). The best results were obtained using "calibrated images", i.e., images that were adjusted to match the resistivities from other tools designed to measure true resistivity and the results were verified with core when available. This work led to several computer apps that presented formation bed thickness, bedding sequences, bed frequency, etc. (as in Fig. 29 Tracks 5 and 6). These and other details about BHI interpretation are discussed in AAPG Memoir 92 and in AAPG Memoir 72.

## Borehole imaging examples from DWSs

BHI information has proved invaluable for delineating lithology, contacts, high-resolution bed thicknesses, grain size, composition, texture, sedimentary features, etc. These parameters and their patterns are vital to the proper understanding of deepwater sediments. Several examples of BHI applications are discussed below.

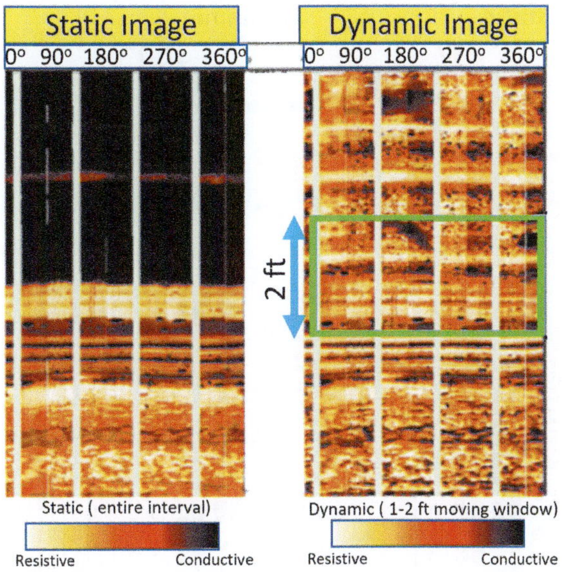


**FIG. 26** Data normalization: Static and Dynamic images from the same FMI measurements. The upper half of the zone is shaly. The lower half is sandy. The interval is interlaminated at the top and then bioturbated. The Static Images on the left are the better for determining lithology. Note the abrupt nature of contact in the middle of the static image. The Dynamic Images on the right are best for showing the internal details. in the two-foot (0.61 m) interval over which they were computed.

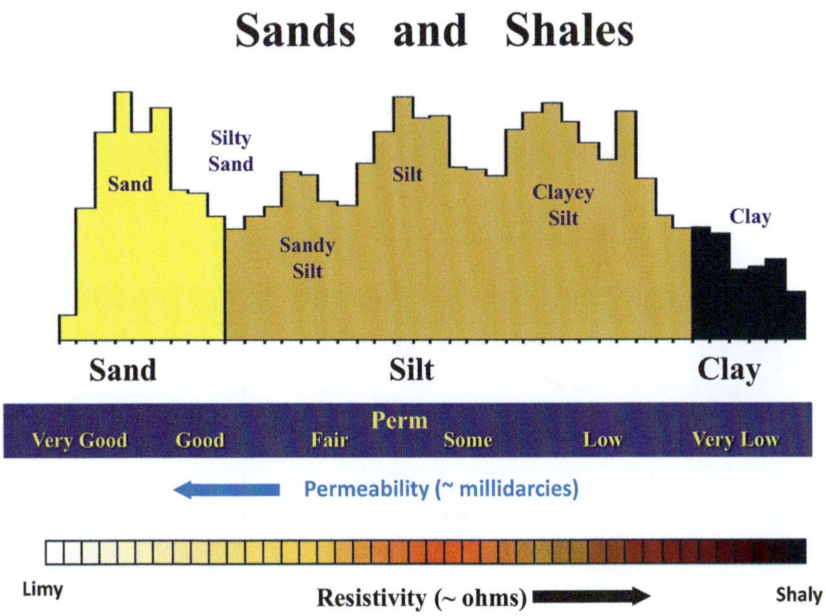


**FIG. 27** Determination of lithology from "calibrated" FMI images. The sands are shown as yellow and the shales as black. High resistivity items are represented by white. The best results are obtained when the images are "calibrated" with true resistivity devices. These analogies work best in sand and shale sequences and when computed zone-by-zone.

## *Log depth scaling and the value of dip information*

The images in the examples in Figs. 28–32 illustrate the effects of variable log scales on the appearance and perception of logs and images. Fig. 28 has both Static and Dynamic images They are from the USA Texas Gulf Coast. The images are presented on a 2′ (0.61 m) per division depth scale This log represents over 20′ (6.1 m) of the hole. A "thinning-upwards" sequence of beds is seen in the lower part of the log. These beds range in size down to an inch (2.54 cm) or less. The upper portion of the log is bioturbated and lacks planar surfaces, i.e., there are no sinusoidal surfaces. Knowledge of the geological setting would allow further inferences. The lower part of the log is a laminated bed (see similarly shaped sinusoids) and reflects low energy and hence a deepwater or other low energy environment. The darker layers of the images are clay-rich, and the lighter ones are sand-rich. The Static Images (Figs. 26A and 28A) provide the best indicators of lithology since the

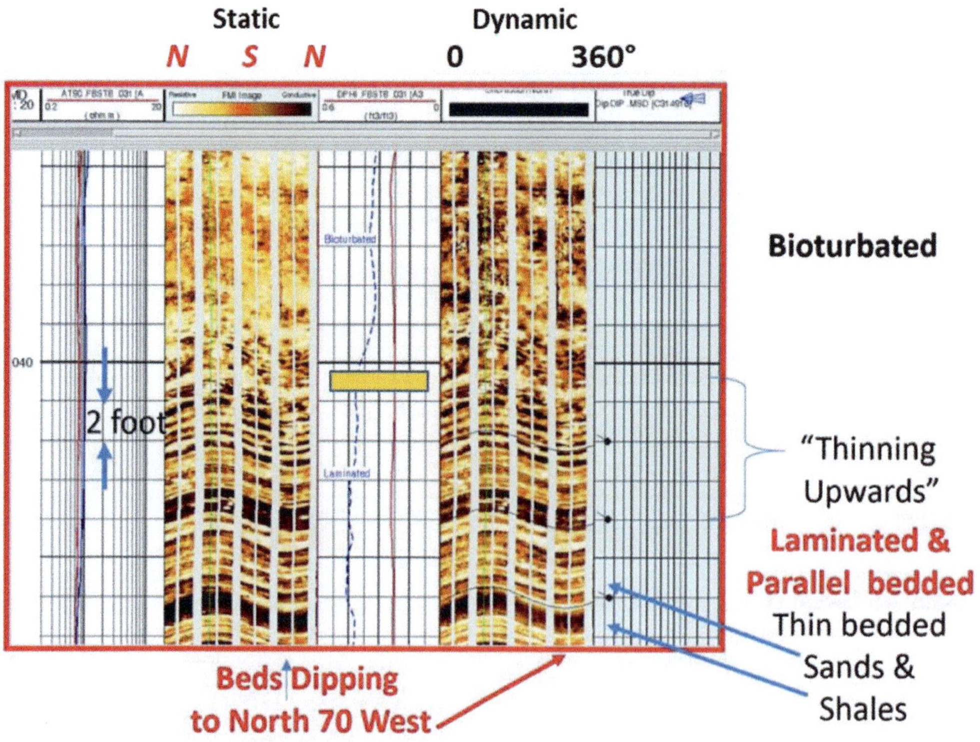


**FIG. 28** Static and Dynamic Images of DWSs from a vertical well in the USA Texas Gulf Coast. On the Static Image on the left, note the bioturbation in the upper half and the thin bedding in the lower. Utilize the Static Images to determine the lithology and the Dynamic Images for the finer details. The thinly bedded beds in the bottom are parallel bedded and have similar sinusoids. They represent Structural Dip and are dipping down at 7° to the North70°West. These dips were calculated manually by the analyst. Bulk batch dipmeter computations are also available.

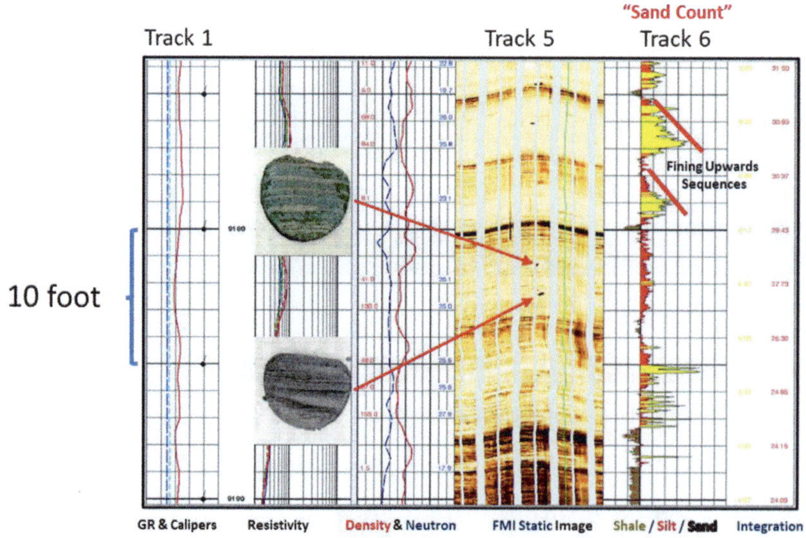


**FIG. 29** An FMI from a thin-bedded DWS in the USA Texas Gulf Coast onshore shows the location of sidewall cores as small black holes as indicated by the arrows. The photos of the ends of two sidewall cores correspond to the holes left on the borehole walls by the coring. Two fining-upwards sequences as seen on the images near the top. The **Sand Count** log (in Track 6) confirms this. These vertical sequences were not apparent on any of other borehole logs Note that the seemingly massive sand in the middle of the interval. has subtle layering as confirmed by the cores. The images are pre-sented on a 2 foot (0.61 m) per di-vi-sion depth scale. This log rep-re-sents over 20′ (6.1 m) of the hole. A "thin-ning-up-wards" sequence of beds is seen in the lower part of the log. These beds range in size down to an inch (2.54 cm) or less. The up-per por-tion of the log is bio-tur-bated and lacks pla-nar sur-faces, i.e., there are no si-nu-soidal sur-faces. Knowledge of the ge-o-log-i-cal set-ting would al-low fur-ther in-fer-ences. The lower part of the log is a lam-i-nated bed (see sim-i-larly shaped sinusoids).

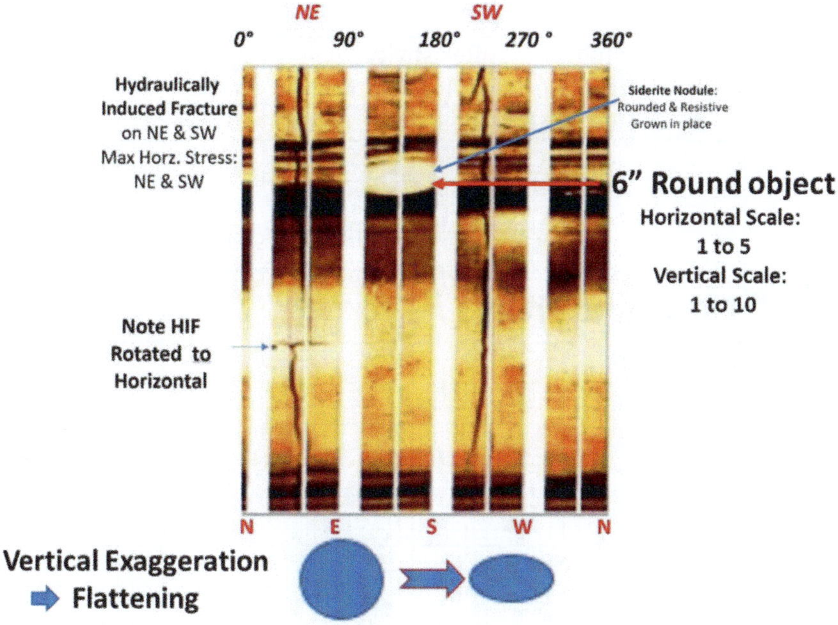


**FIG. 30** FMI image from USA Texas Gulf Coast showing hydraulically induced fractures (HIFs) and a siderite nodule. This image is presented on a horizontal scale of 1–5, where the image represents the 8″ (20 cm) borehole radius. This is the usual scale. The vertical (depth) scale is a "10 to 1" ratio. These vertical scales normally range from "1 to 5" to "1 to 1200". When the vertical and horizontal scales are not the same, there will be a vertical exaggeration and round objects will appear as ellipses, as illustrated. The two vertical lines are Hydraulically Induced Fractures (HIFs) that split the borehole on the NE and SW sides of the hole. The maximum horizontal stress is found to be Northeast—Southwest.

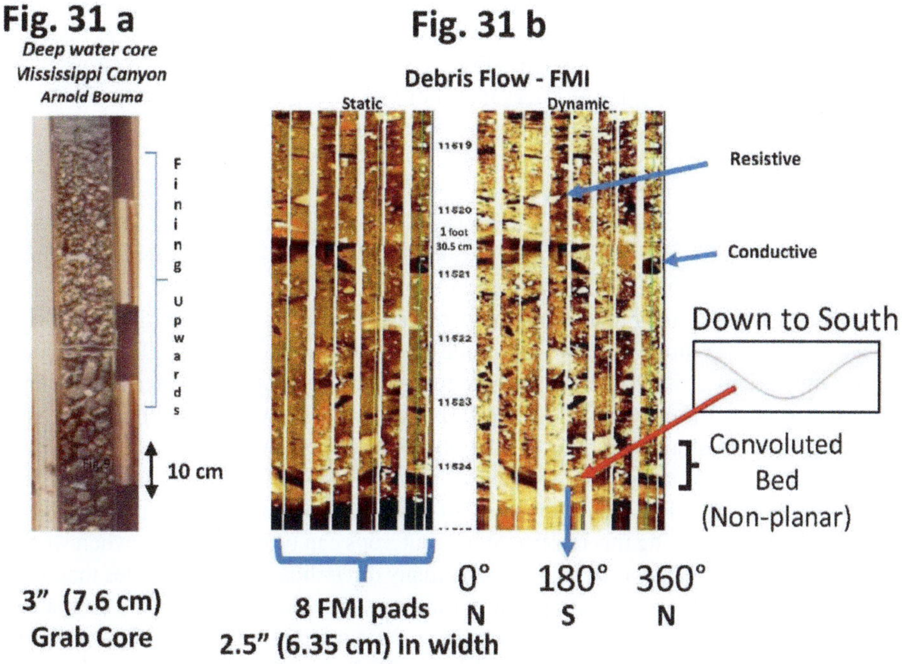


**FIG. 31** Deepwater "Core Grab" and example of a debrite. The photo on the left is a core grab obtained in deepwater several hundred miles offshore in the Gulf of Mexico. It was taken by Arnold Bouma. The gravelly sediment fine upwards. The static and dynamic images on the right are also from the Gulf of Mexico and are interpreted to be a debrite. The dark surface at the bottom does not fit a sinusoid and is not planar. See the sinusoid below images for comparison. This convoluted bed is roughly dipping south as is the sinusoid displayed next to it.

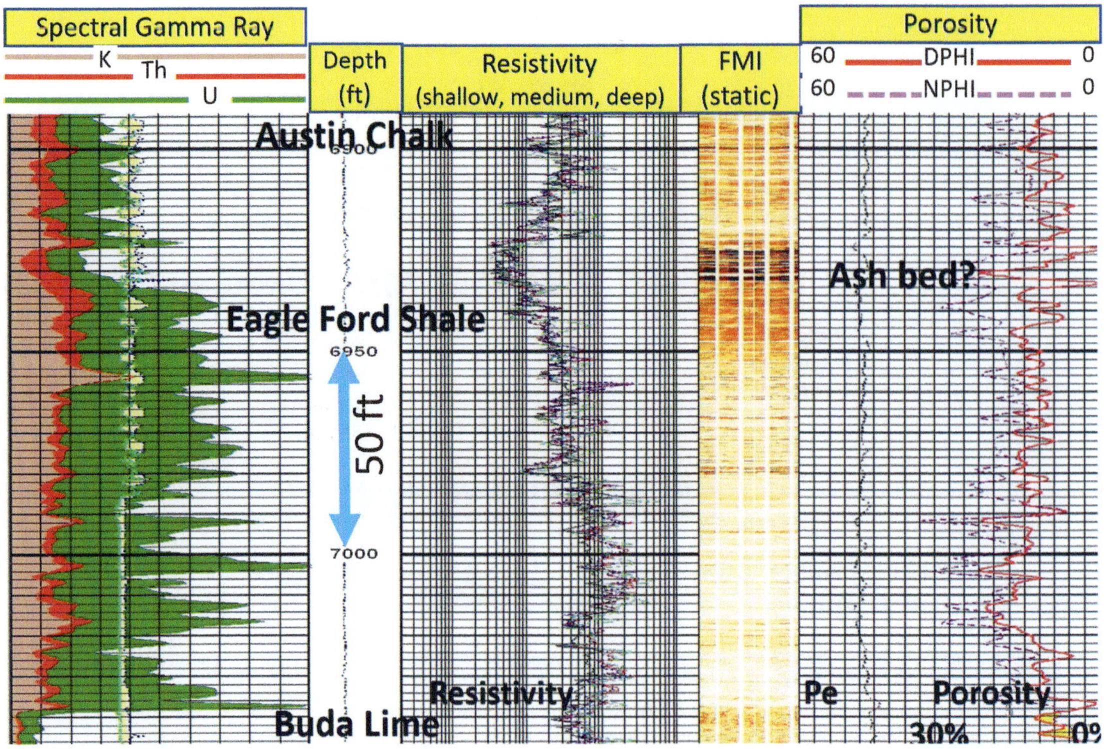


**FIG. 32** Merged log from South Texas with Triple Combo and FMI images. The Austin Chalk (AC) consists of alternating Lime and Marl layers. Note the lower resistivity ash (or kaolinite) bed at 6930′ (2113 m). The top of the Eagle Ford Shale (EFS) is normally picked at the uppermost high GR reading as at 6935′ (2114 m). This high radioactivity is due to the presence of Uranium which is shown on the Spectral GR as green. The Buda Lime at 7040′ (2146 m) shows some vugginess with darker spots on the Images and some porosity on the Density. Both the AC and EFS are naturally fractured, but the fractures do not show up at this depth scale. The bedding and lithology are delineated by the images down to a foot (30.5 cm) or so.

histogram used to match the resistivity to a color relationship is applied over the entire interval instead of over the short 2′ (0.61 m) intervals as are used in the Dynamic Images (Figs. 26B and 28B). The lower resistivities, as in the shales are consistently colored black. The beds in the lower interval part of Fig. 28 are dipping down to the WNW (counter-regional for the USA Texas Gulf Coast). This was due to the influence of normal listric growth faulting. Note that the sinusoids exit the borehole to the WNW. The dip magnitude scale is expanded for the lower dip magnitudes. This is done because a difference of one or two dips is far more important in the lower (structural) dip zones. At higher dip ranges a variation of a couple of degrees is usually not critical. Dipmeters utilize a hybrid scale with an expanded scale in the 0 to 10 degree range. As seen in the lower right of Fig. 28, the beds dip down to the WNW (N70°W) at 7 degrees magnitude.

An accurate value of "structural dip" is vital to proper mapping and any analysis of stratigraphic dips involves subtracting the structural dip. These dip magnitudes can help refine the seismic interpretation. Structural dip is picked on the dipmeter from the average dip direction and magnitude of an interval of interest. The computed dips are present-day true dips. The beds were laid down flat and subsequently tilted by structural changes. One needs to eliminate any changes in that original dip caused by cross-bedding, etc. Experience has shown that when picking structural dip, the best approach is to choose the lowermost average value of the observed dip magnitudes.

Dip direction and magnitude information are important to fully understand DWS deposits. The orientation of the various components can often be determined. The type and energy of various flows can be further evaluated using dip magnitude and dip relationships. Finding and defining this information in outcrops can be quite tedious when working in a single plane and can sometimes be ambiguous. BHI information identifies many dip indicators and provides their true dips. These values can then be analyzed statistically and presented graphically for improved understanding and better analysis. When dips are manually computed on a workstation, the accuracy of these dip values is a direct function of the skill of the interpreter. The true dip values are available from batch processing that is automatically computed based on various parameters. Manually computed dips can be determined by the interpretation of planar surfaces, as represented by sinusoids, using modern sophisticated software such as Schlumberger's TechLog. These interpretations are vital to the best interpretations but are also dependent on the skill of the analyst. Batch computations can conveniently provide many points that can then be analyzed statistically. Numerous references regarding dipmeter family tools, computations and presentations are listed below.

### Merging images with sidewall cores and "Integrated Sand Count" processing

Fig. 29 is another image of deepwater sediments from the Texas Gulf Coast. The cores were taken prior to the logging of the FMI. The locations of the core holes appear as black spots and are located on the south side of the hole or in the middle of the image. The hole azimuth tadpoles in the first log track (Track One) from the left indicate that the hole is inclined at 8° to the NNE. Photos of the ends of two sidewall cores confirm the lithology as indicated by the images. The Triple Combo curves (gamma-ray, resistivity, density and neutron porosity) are found in Tracks One, Three and Four, respectively. Track Five has the Static Images and Track Six contains a computed Sand Count log. For this log, brown represents Shale, red represents Silt and yellow represents Sand. Track Seven displays an integration of these lithologies. A closer inspection of the images in Fig. 29 reveals details within the various intervals themselves as noted. Note the "fining upwards" sequences above 9180′ (2799 m) and the laminated intervals at the bottom in both the sandy and shaly zones. The layer in the middle was not truly "massive" as confirmed by the core photos. The Dynamic Images (not displayed) showed thin-bedded layering. Expanding the depth scale would have revealed even smaller features. Locating core values at their exact location on the image demonstrates the synergy of multiple sources of information This is a good example of the power of using images to help "scale up and scale down" across various types of borehole information.

### Drilling induced fractures and present-day stress effects on BHIs

Fig. 30 is another FMI example from South Texas. The interval covers 2 ft (61 cm). For reference, note the azimuth scale in both degrees and the equivalent points of the compass as shown on the BorMap presentation. Several interesting features are worth noting. The dark vertical lines on two opposite sides (NE and SW) of the hole are Hydraulically Induced Fractures (HIF). HIFs are formed when the mud weight at least temporarily exceeds the strength of the rock and splits the borehole in two. It splits the hole by opening against the weaker present-day Minimum Horizontal Stress. The location of these HIFs on the Northeast and on the Southwest sides of the hole indicates that the orientation of the present-day Maximum Horizontal Stress is Northeast and Southwest. It is perpendicular to the Minimum Horizontal Stress. This is typical of the present-day stress orientations found in the South Texas area. The excess hydraulic pressure causing these HIFs is often not due to mud weight alone. It may be caused by temporary increases in the total hydrostatic pressure which can be attributed to hydraulic ram effects while running into the hole and by pump pressure surges. HIFs sometimes occur near faults due to their local stress perturbations.

Alternatively, a "weak borehole" with low hydrostatic pressure may result in Borehole Breakouts (BO). It tends to enlarge the hole in the direction of the present-day Minimum Stress or NW/SE in South Texas. Borehole breakout often produces an egg-shaped borehole. This elongation can be seen on the calipers. One caliper pair will be more elongated or washed out. It is not uncommon to see both HIFs and BOs in the same borehole due to temporary variations in mud and hole conditions. These HIFs and BOs usually occur in areas with significant differences between the Minimum and Maximum Present-Day Stresses. Both mechanically induced fractures are known as "drilling-induced" fractures, as compared to "naturally occurring" ones.

Fig. 30 also shows the slight rotation (wrapping) of the azimuthally oriented images. The images rotate from lower left to upper right. This is due to the wrap of the wireline cable and is normal for all borehole logging tools. The strong calipers on the FMI usually tend to reduce this wrapping. As it turns out, the FMI tool usually follows the same path for a while then quickly rotates 90° and then resumes a vertical path. This usually results in BHI "Repeat sections" that are nearly identical. This excellent repeatability is quite reassuring from a log quality perspective and is so commonplace that only short, "quality control" repeat logs are run for FMIs. It is preferable to run these repeats across the main zones of interest.

Unfortunately, this tendency of the BHI tools to follow the same path results in a lower chance of improving the coverage of the "gaps" between the pads. These stress-related effects can affect borehole stability and present-day stresses strongly influence the direction of hydraulic "fracking." This information is therefore important to both the drilling and completion engineers and it is vital for designing directional and horizontal drilling programs and completions. There is a large body of information regarding stress systems and the borehole imaging responses to them. (see Pailet, 1990). In Fig. 30, the egg-shaped white object near the top of the interval is interpreted as a siderite nodule. It is 6″ (15.24 cm) in diameter. While it appears to be egg-shaped, it is actually round. This is the result of the vertical exaggeration created by the different vertical and horizontal scales of the image. See the circular and elliptical shapes at the bottom of the page for an illustration of this effect.

The circle is flattened into an elliptical shape by the vertical exaggeration. In this log example, one can see that the object is resistive (white) and round (taking the vertical exaggeration into consideration) with smooth edges. It does not extend across the hole, but the images never-the-less show a good representation of its true size and shape.

## *Debrites from the US Gulf of Mexico*

The two images in Fig. 31 are from the US Gulf of Mexico. The photo on the left (Fig. 31A) is of a "grab core" taken on Leg 96 of the Deep-Sea Drilling Project in deepwater at over 200 feet (60 m), and hundreds of miles offshore in the US Gulf of Mexico. The core was acquired under the supervision of Arnold Bouma and Jim Coleman and proves the presence of pebble-sized sediment in an area that many previously thought would only contain fine-grained particles. Notice the size and shape of the pebbles in this ~3″ (7.6 cm) core.

Note the "fining-upward" nature of the deposit. This core should put an end to the debate over whether gravity-flow processes can transport reservoir-quality size rocks far out into the deepwater. The FMI images on the right are from offshore the USA Gulf of Mexico and have been interpreted as a Debris Flow. The various resistive or white and conductive or black objects seem to float in a brown matrix. There is no obvious grading. The resistivity, angularity and size of the objects can be easily seen. The FMI sensor array is 2.5″ (0.06 m) in width. The image depth scale is 1′ (0.3 m) per division. The images are from a depth of 11,500′ (3505 m) in a nearly vertical well. Note the non-planar, slightly convoluted bed near the bottom of the image. While the bed does not fit a plane, it does seem to have a subtle indication of dipping to the South. A south-dipping sinusoid is presented beside the Dynamic Image for reference. See Bouma (2000) for more BHI examples of slumping and other features.

Fig. 32 is a merged logging suite from the Upper Cretaceous of South Texas. This log covers over 100′ (30.5 m) and the FMI image information is still useful. Depth scales such as the "1 inch-to-100 feet" (i.e., a 1 to 1200 ratio), can be useful for getting the big picture and for comparing BHI data to other logs. Compare this BHI to the other examples to assess the pros and cons of differing log depth scales. This is greatly facilitated when viewing the images on a workstation. The upper part of the interval is across the Austin Chalk. The Spectral Gamma Ray in Track 1 indicates the amounts of Uranium (green), Thorium (red) and Potassium (brown) that are present. The Austin Chalk and the Eagle Ford Shale actually have similar properties other than their TOC (Total Organic Carbon) content and radioactivity. The top of the Eagle Ford Shale (EFS) is normally picked where the Gamma Ray radioactivity increases rapidly, as at 6934′ (2113 m) but the contact is usually gradational. The EFS interval is highly radioactive, mainly due to the presence of uranium. The Eagle Ford Shale also typically has relatively higher porosities and resistivities. On the other hand, ash beds (as seen at 6930′ in Fig. 32) usually have low resistivities and may have low or high radioactivity. The Eagle Ford Shale was deposited in a reducing environment which lends itself to the preservation of total organic carbon (TOC). Modern high-tech logs such as the LithoScanner can be used to calculate the amount of TOC. This is important in estimating the producibility of most unconventional resources. It is this preservation of TOC and the thickness of the Eagle Ford Shale (technically a marl) that have made the EFS a world-class unconventional resource play. This log bottoms in the Buda Lime which is a clean limestone. Below the Buda Lime lies the high gamma-ray, low resistivity Del Rio Marine Shale. This distinctive sequence of log profiles can be correlated across the USA South Texas. This merged log shows the usefulness of combining various high-tech logging tools. Having all the various devices, with their varying responses and vertical resolutions, at the same depths is obviously very synergistic. These measurements can then all be "scaled down" to match core and lab measurements or "scaled up" to match other logs as well as related geological and seismic studies. Routinely, the FMI and the Triple Combo Platform Express (PEX™) are acquired together in both vertical and horizontal wells.

## *Lower Permian Wolfcamp imaging example from West Texas with natural fracturing*

Fig. 33 is from the Wolfcamp Formation in the Permian Basin of West Texas, USA. These images show a series of thinly bedded sediments. The white layers are limestones which have been interpreted to be limy turbidites that flowed into this area from the west. Note that these limestones are not fractured, as can be seen on the images by their lack of vertical planar surfaces (i.e., high angle sinusoids). Fractures are only seen in the brown "silty" zones. The fractured areas are indicated on the log by the black arrows. Note that these fractures are "bed-contained" and they do not continue into the limestones. Only the "tops and bottoms" of the fracture sinusoids are seen (see the red arrows). It is fortuitous for this Wolfcamp production that the siltstones contain the great majority of fractures since they also contain the hydrocarbons. The contrast in mechanical properties between the lime and silt layers tends to increased fracturing. These fractures can then provide a primary means for getting the hydrocarbons to the well bore. Note that the 8 traces of the FMI are moving diagonally from left to right across the BorMap presentation. This is because due to torque from the wireline causing the pads to rotate clockwise. The BoreMap display shows the direction that the pads face at any depth. Pad 1 has a black line in the middle of the image for identification.

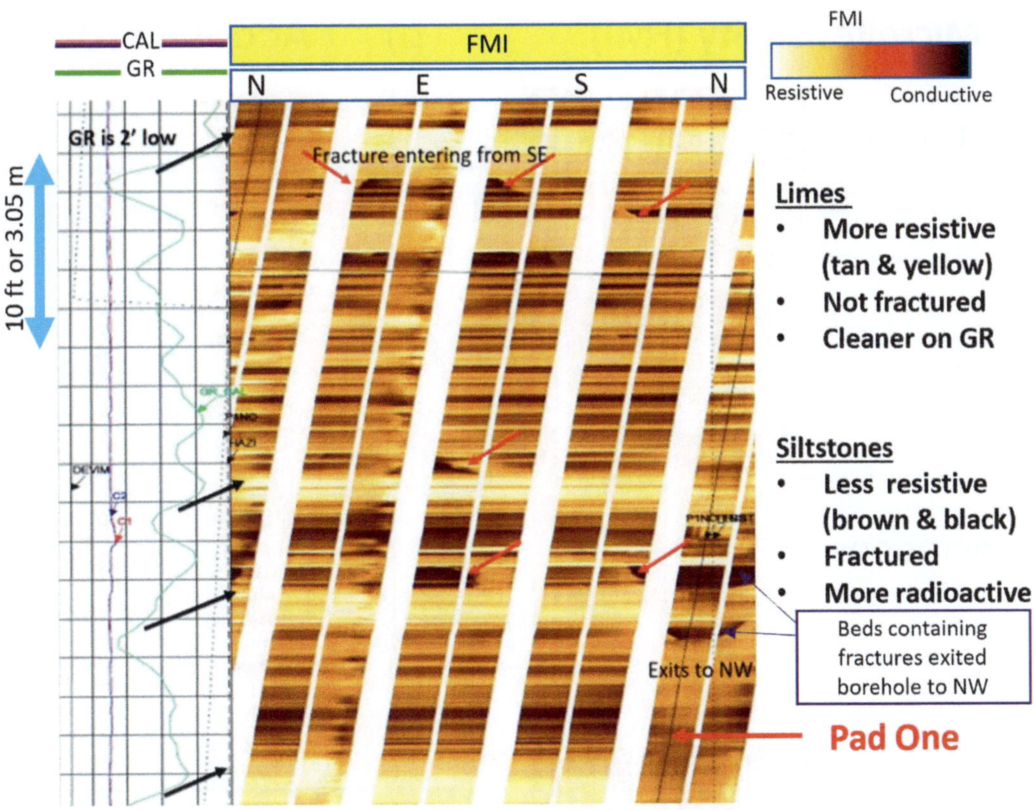


**FIG. 33** Fractured DWSs from the Wolfcamp of West Texas. The Siltstones (*brown*) are fractured. They also contain most of the hydrocarbons. The Limestones (tan) are not fractured. Note the absence of any part of a sinusoid in these layers. The Natural fractures are normally only found in the siltstones and are 'bed contained, i.e., the sinusoid does not extend out of the bed itself, but it can be used to calculate a dip. Note the tops of sinusoids enter the borehole from the SE while exiting to the NW. Note that the images rotate or "wrap" from left to right. There is a 2′ (61 cm) depth offset between the images and the gamma ray curve. That is, the gamma ray needs to be shifted up 2 feet (61 cm) to be on depth with the images.

### *Wolfcamp example with micro-resistivity (FMI) and acoustical (UBI) images*

The images in Fig. 34 are from an FMI logged in the Lower Permian Wolfcamp Formation of USA West Texas. The logs acquired simultaneously with the images are the Gamma Ray (in green—for correlation), four calipers (for borehole geometry) and borehole inclination (deviation and azimuth). This hole was nearly straight. There are two layers of concretions at 6480′ and 6486′ (1976 m and 1977 m). The small depth mismatch of the curves at 6374 ½′ (1943 m) was due to a small tool pull. Both images in Fig. 34B are from an acoustical imager, the Ultrasonic Borehole Imager (UBI). Both tools were logged over the same interval in the same well and are presented on depth with one another. The image on the left half of the acoustical images is the Acoustical Amplitude. It responds primarily to borehole rugosity. Usually, the Acoustical Amplitude image correlates best to the Micro-Resistivity images. The black spots on the image are pockmarks caused by roughness on the borehole surface. The acoustical image on the right is the Sonic Travel Time, also known as the Acoustical Radius. It is measured simultaneously with the Amplitude. It mainly measures multiple radii of the borehole. In this example, the darker vertical areas, as seen on the Northwest and Southeast sides of the Acoustical Radius, show Borehole Breakout (BO). Note that in contrast, neither the UBI Amplitude nor the FMI curves detected these drilling-induced features. The variation in the actual borehole radii is quite small, and yet the minimum stress direction can be determined. It is apparent in this case, as in most cases, that the Micro-Resistivity curves are best for delineating bedding planes and objects such as the concretion at 6384.5′ (1946 m). The Acoustical images often do a better job of detecting stress-related features.

In this West Texas field 1500′ (487 m) of whole core was obtained on three consecutive trips in the hole of the same well by utilizing the new "concentric sleeves" coring tools. The operator also analyzed the DNA of bacteria taken from samples in a nearby pilot hole. This data was then used while drilling horizontally to determine whether the laterals were staying in zone.

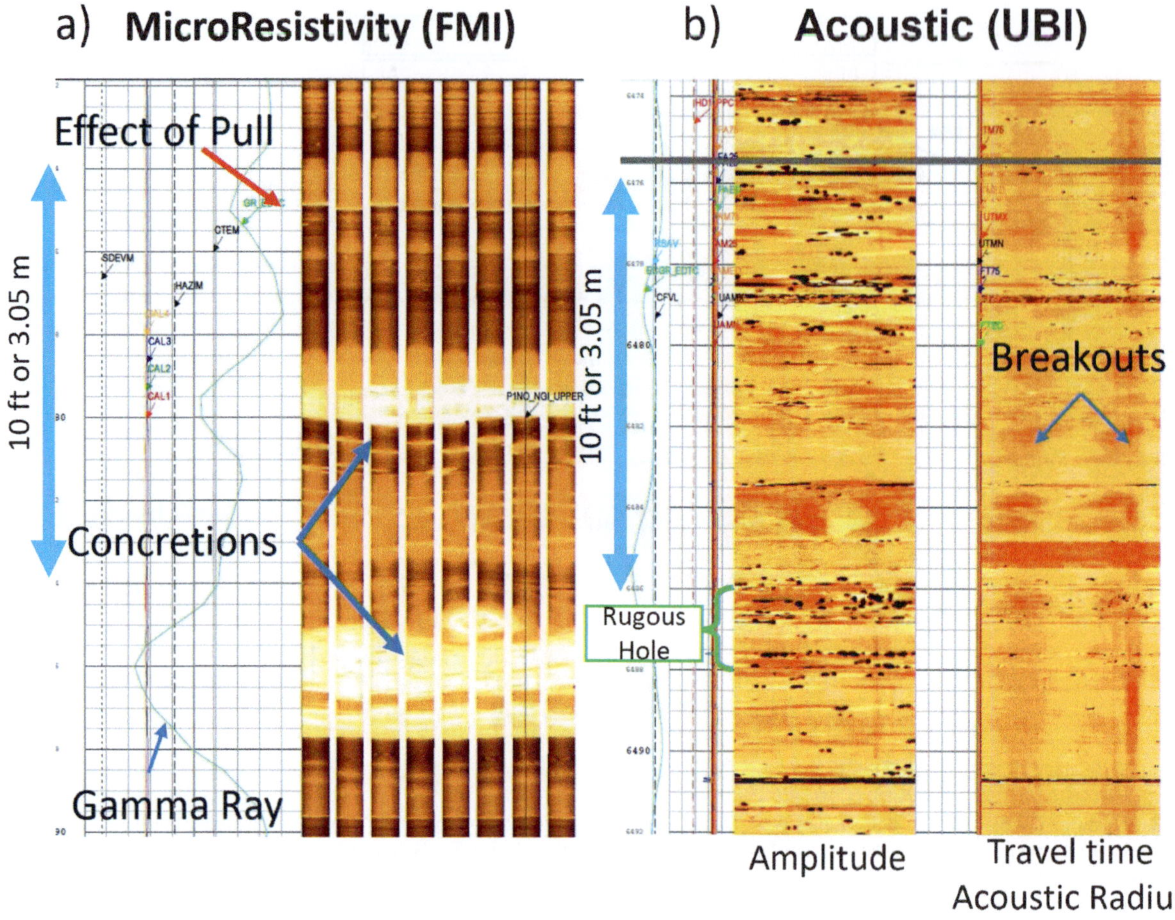


**FIG. 34** Microresistivity (FMI) and acoustical (UBI) Images in DWSs from the Wolfcamp in West Texas. Both tools were logged over the same interval. Note the excellent delineation of the bedding on the FMI. There is a small depth offset at 6481′ due to the recording of the pulls at different depths on the pads and on the flaps. Several concretions are imaged, but they are flattened due to the difference in the vertical and horizontal scales. On the UBI (on the right), the acoustical images have problems delineating the bedding and the concretions. The Acoustical Radius image shows Borehole Breakout as induced vertical "smudges" that the microresistivity images do not reflect.

### *Eocene Wilcox Lobo Formation—South Texas, USA*

Kim Henke, a student of Bob Berg at Texas A&M University, in her 1982 thesis analyzed the Eocene Wilcox Lobo Formation in South Texas and interpreted the sediments as DWSs (Henke, 1982). Dr. Berg was one of the early people who recognized the widespread distribution of turbidites in the USA Texas Coast subsurface. His approach involved the analysis of information from all available logs and cores. Lithology, quartz size, sedimentary features, texture, composition and cementation were plotted and evaluated. Berg also fully utilized any dipmeter information available. These detailed studies led him to believe that many of the formations, previously thought to have been deposited in coastal environments, were in fact deepwater sediments. This approach nicely illustrates the benefits of the in-depth analysis of all available data, including the grading of the beds. Some of these formations interpreted as DWSs included various Eocene Wilcox and Yegua age sediments. The interpretations were a bit controversial at the time as they contradicted previous interpretations. Many others have since recognized the presence of DWSs in the US onshore Gulf Coast. Growth faulting has proven to be a good way to mix coastal and DWS which no doubt lead to some of the confusion.

### *Miocene Messenger Formation—Taranaki Basin, North Island, New Zealand*

The classic 2002 Taranaki geological study illustrated in Browne and Slatt (Browne and Slatt, 2002) showcases the benefits of a coordinated study by academics oil and gas companies and service companies. It analyzes the Miocene Messenger Formation of the Taranaki Basin on the North Island of New Zealand. The drilling, coring and logging of two wells behind this outcrop were coordinated by a team headed by Roger Slatt. Two wells were drilled and fully cored. They were logged with the Triple Combo and the Formation MicroImager.

Dr. Slatt has also published on the deepwater Pennsylvanian Jackfork Formation in Southern Arkansas (Slatt et al., 2000). The outcrop at the De Gray Spillway is a classic example of thin-bedded turbidites. It is worth noting that similar thinly bedded pays zones have proven to be productive worldwide. Several highly productive, low resitivity Eocene Yegua Formation wells drilled south of Houston in the 1980s alerted the industry to the need for BHIs to valuate these thinly bedded pay sands. Many such beds are found in DWS environments. Ambrose et al. illustrate the characteristics of such sands (Ambrose et al., 2013). They show several funnel-shaped log sequences which were made up of "graded beds". These describe several deepwater depositions in the Eocene Wilcox and Yegua of the Texas Gulf Coast.

## Summary

The applications of borehole imaging devices (BHIs) have been demonstrated with several DWS examples. The various BHI formats have been explained with commentary on the figures and in the text. Their integration with seismic, other wireline logs, outcrops and cores was discussed.

## Additional resource material for utilizing legacy dipmeters and BHIs

The two dipmeter Guidelines included in the companion *Science Digest* summarize some basic rules of dipmeter interpretation developed in the late 1970s. They are still useful today. The Structural Guidelines summarize the rules that are applied for evaluating missing sections such as faults and unconformities. It also illustrates a method of estimating Water Depth from dipmeter tadpole plots. The Stratigraphic Guidelines show stratigraphic approaches that were used at the time. It illustrates a system of determining sand body geometry based on three generic sand body types. These included bars, channels and fans. This approach was based on experience in the passive margin of the Texas Gulf Coast. It became apparent that certain basic kinds of sand bodies, such as channels, fans and bars were found in both coastal and deepwater environments. The "Dip-oriented" Channels that feed Fans are well known in both environments. The "laterally oriented" sand bodies such as Barrier Bars, Offshore Bars, Crevasse Splays and Contourites were grouped in the generic Bar. These are usually "Strike- oriented". This "Channel, Fan and Bar" classification system conveniently side-stepped the question of "what was the depositional environment?" and what associated terminology to use. Primarily it provided a way to predict the direction of the "long axis of the sands".

Indications as to the shape, size and orientation of these sand bodies were routinely provided. It has become evident from the insights provided by 3D seismic and horizontal drilling that this simplistic approach had merit. Early efforts to determine the depositional environments led to research based primarily on the polling of several well-respected geologists. They were asked to estimate the probability of "finding" and "not finding" various sedimentary features that could be identified from logs in these six environments. These features included bed thicknesses and contacts, presence of graded beds, thickening upwards sequences, bioturbation, various types of cross-bedding, coals, rootlets, etc. The results were presented in an SPWLA paper (Fett, 1987). An appreciation of the variations in reservoir quality within the various sand bodies themselves is discussed by Bill Galloway (Galloway and Hobday, 1996). It shows the importance of these types of studies. It clearly demonstrates the need for better models and for further outcrop studies, especially in DWSs. The Atlas of Deep-Water Outcrops (AAPG Study 56) is a comprehensive and well-illustrated documentation of many DWS outcrops and models. The advent of BHI devices has made the identification of various sedimentary features much more exact. A single image log in a vertical well is the equivalent of having an 8‘’wide vertical window to examine an outcrop. Horizontal logs are like having an 8″ (20 cm) horizontal window. Modern seismic provides a framework to post the "log windows", all-be-it on a much different scale. Although these BHI windows are thin they can be quite detailed. The images from vertical wells provide useful vertical profiles. They also provide orientations and can be used to extrapolate information away from the borehole. Research about depositional systems by pioneers such as Bill Fischer, Bill Galloway, Jim Coleman and Rufus LeBlanc of Shell led the way in improving the understanding of the deposition of all sediments (LeBlanc, 1997). The Bureau of Economic Geology (BEG) continues to lead the way in this type of research which they generously share with the geological community. Other geologists, who focused primarily on the marine environments, like Arnold Bouma, Pete Vail, Tor Nielsen and Jon Rotzien have shown us how to better understand marine deposits. Ganapathy Shanmugam has pointed out that there is often more to DWSs than simple "turbidity flows" (Shanmugam, 2006).

Deepwater sediments are difficult to observe in real-time although flume experiments have provided insights. The element of timing has always been a major complicating variable in all stratigraphic studies. In the case of deepwater sediments, it can be especially vexing since almost all of the information is from ancient "flows" and can usually only be observed in outcrops, seismic and subsurface well logs. A true sense of perspective is often difficult to obtain especially in areas with "mass movements" such as Growth Faults. Dipmeters have been long utilized to delineate these Growth Faults, and borehole imaging devices such as the FMI provide even better information. Both provide a means to relate outcrop and seismic scale information to that provided by logs and cores. Major technological advances in the ability to study outcrops have greatly enhanced the ability to understand them. These include GPS, Lidar and Drones. Modern 3D Seismic is producing major improvements in the ability to understand the intricacies of DWSs. A host of new software is helping us visualize the bigger picture as never before. Subsurface data has become increasingly more sophisticated and Horizontal Logging and Logging While Drilling (LWD) have added new dimensions.

Continuous coring has become safer and more feasible and is especially important to the study. Deepwater sediments (DWS) are being analyzed and modeled as never before and the application of the Borehole Imaging devices, as discussed above, are just a sample of what is possible.

### Further reading

Several excellent references on deepwater sediments and borehole imaging devices are listed below. They have all been selected based on their generous use of illustrations. Geologic interpretation depends on visual analysis of samples, outcrops and subsurface data. This is a major reason that borehole images can be so useful to geologists. Since deepwater sedimentation is difficult to observe in real-time, sub-surface and outcrop examples of ancient rocks must be utilized. Three AAPG Texts (Memoir 72 and 92 and Study 56) illustrate the wealth of information that is available. The AAPG Memoirs on Depositional Environments of Sandstones (Mem.31) and of Carbonates (Mem.33) are visually rich resources and can provide important insights into all depositional environments. G. Shanmugam's *Deep-Water Processes and Facies* provides a wealth of "outside the box" perspectives. See the companion Science Digest for an example of his work. AAPG Memoir 72 Turbidite Systems Chapter 2 (edited by Arnold Bouma) specifically discusses borehole imaging devices. Schlumberger's 1989 and 2015 Tech Reviews and the AAPG Memoir 92 Dipmeter and BHI Technology are good sources of BHI-specific information. All service companies make tool brochures and case studies readily available. AAPG's Search and Discovery is an excellent geological search engine. The Gulf Coast Association of Geological Societies (GCAGS) Transactions have numerous studies from the Gulf Coast region. The SEPM provides great educational material. For further references, an excellent bibliography by Stephen Prensky of pertinent BHI papers is available online at www.ssprensky.com/bibliog.

### Key messages

In these case studies, multi-attribute analysis was shown to be an effective method in interpreting complex depositional features found in deepwater settings. Unsupervised Machine Learning approaches are feasible for automated clustering and mapping of different geomorphological features based on their seismic response that can then interpretated utilizing geologic expertise. In addition, rock physics and AVO feasibility modeling has been demonstrated within a lobe complex of a deepwater submarine fan and by combining diagenetic and rock physics modeling, constrained by petrography and facies interpretations, it is possible to create realistic scenarios of AVO feasibility maps that can be compared with real AVO data observations. Additionally, quantitative seismic interpretation workflows are useful in reservoir mapping and prospect derisking based on seismic AVO data and should be particularly useful for deepwater clastic systems with complex spatial distributions of lithofacies. The effective application of images from borehole imaging devices (BHI) can be integrated with seismic, other wireline logs, outcrops and cores in building a complete subsurface interpretation.

## Conclusion

An understanding of deepwater stratigraphy requires an integration of multiple datasets, including seismic, petrophysics and core data. This chapter discusses commonly used seismic and petrophysical tools and their fundamentals, scales of resolution, challenges, opportunities and example studies. 3D seismic imaging and seismic-attribute-guided analysis can provide a great visualization and understanding of the external heterogeneities of the stratigraphic features commonly found in deepwater systems. While conventional seismic attribute-based methods have become an integral part of the interpretation workflow in the commercial software, there is a need for quantitative seismic interpretation and calibration to underlying geology via seismic inversion, rock physics and petrophysical analysis to decipher the impact of lithology, pore space and fluid type on seismic anomalies. Petrophysical analysis based on high-resolution borehole imaging and other well logs highlights the internal heterogeneities, including depositional features, textural patterns, lithology, fractures and thin beds at a local scale. Utilizing all these tools together can provide a more detailed understanding of the system.

## Acknowledgments

S.V. and S.B. would like to thank the Alaska Department of Natural Resources, Division of Oil and Gas, for making the tax-credit 3D seismic data available to the University of Texas Permian Basin and the University of Alaska Anchorage (State of Alaska, 2017). They also thank Schlumberger and Eliis for donating Petrel™ and Paleoscan™ software to the University of Texas Permian Basin and the University of Texas at Austin, which were used for seismic interpretation. Authors thank the sponsors of the attribute-assisted seismic processing and interpretation industry consortium (AASPI) for funding. Authors used AASPI software for computing seismic attributes.

P.A. and I.L. would like to thank Equinor ASA (Norway) who provided the data used in the study shown in Section "Quantitative interpretation of a North Sea turbidite system using rock physics." The well log and seismic data shown in this section (Glitne Field) are publicly available on the following website: https://srb.stanford.edu/quantitative-seismic-interpretation. Thanks to Stanford Rock Physics & Borehole Geophysics Project for hosting the data website.

T.F. would like to thank the six experts who answered his survey regarding what to expect to find (and what not to find) in various depositional environments. They include Bob Berg, Jim Coleman, Bill Galloway, Rufus LeBlanc, Dick Selley and Oberto Serra. All have inspired his progression, starting in 1966, from Schlumberger Wireline Field Engineer to the Dipmeter Specialist for the Gulf Coast to the Division Geologist for Australasia and then in the Middle East.

The author of this chapter would like to thank Derek Adams and Rich Sears for helping with the edits.

## References

Ambrose, W.A., Loucks, R.G., Dutton, S.P., 2013. Depositional Systems and Controls on Reservoir Quality (Determined from Core Data) in Deeply Buried Tertiary Strata in the Texas-Louisiana Gulf of Mexico. Report of Investigations No. 278. The University of Texas at Austin – Bureau of Economic Geology, p. 80.

Avseth, P., 2000. Combining Rock Physics and Sedimentology for Seismic Reservoir Characterization of North Sea Turbidite Systems. Unpublished PhD-thesis, Stanford University.

Avseth, P., Lehocki, I., 2016. Combining burial history and rock-physics modelling to constrain AVO analysis during exploration. Lead. Edge 35, 528–534.

Avseth, P., Lehocki, I., 2021. 3D subsurface modeling of multi-scenario rock property and AVO feasibility cubes—an integrated workflow. Front. Earth Sci. 9, 1–13.

Avseth, P., Dvorkin, J., Mavko, G., Rykkje, J., 2000. Rock-physics diagnostic of North Sea sands: link between microstructure and seismic properties. Geophys. Res. Lett. 27, 2761–2764.

Avseth, P., Mukerji, T., Jørstad, A., Mavko, G., Veggeland, T., 2001. Seismic reservoir mapping from 3-D AVO in a North Sea turbidite system. Geophysics 66, 1157–1176.

Avseth, P., Flesche, H., van Wijngaarden, A.-J., 2003. AVO classification of lithology and pore fluids constrained by rock physics depth trends. Lead. Edge 22, 1004–1011.

Avseth, P., Mukerji, T., Mavko, G., 2005. Quantitative Seismic Interpretation. Cambridge University Press.

Avseth, P., Dræge, A., van Wijngaarden, A.-J., Johansen, T.A., Jørstad, A., 2008. Shale rock physics and implications for AVO analysis: a North Sea demonstration. Lead. Edge 27, 788–797.

Avseth, P., Mukerji, T., Mavko, G., Dvorkin, J., 2010. Rock-physics diagnostics of depositional texture, diagenetic alterations, and reservoir heterogeneity in high-porosity siliciclastic sediments and rocks—a review of selected models and suggested workflow. Geophysics 75, A31–A47.

Avseth, P., Lehocki, I., Feuilleaubois, L., Hansen, T.N., Angard, K., Reiser, C., 2020. Exploration workflow for real-time modelling of rock property and AVO feasibilities in areas with complex burial history—a Barents Sea demonstration. First Break 38, 51–56.

Avseth, P., Lehocki, I., Kjøsnes, Ø., Sandstad, O., 2021. Data-driven rock physics analysis of North Sea tertiary reservoir sands. Geophys. Prospect. 69, 608–621.

Baig, I., Faleide, J.I., Mondol, N.H., Jahren, J., 2019. Burial and exhumation history controls on shale compaction and thermal maturity along the Norwegian North Sea basin margin areas. Mar. Pet. Geol. 104, 61–85.

Barnes, A.E., 2016. Handbook of Poststack Seismic Attributes. Geophysical References Series No. 21. Society of Exploration Geophysicists.

Bello, A.M., Jones, S., Gluyas, J., Acikalin, S., Cartigny, M., 2021. Role played by clay content in controlling reservoir quality of submarine fan system, forties sandstone member, Central Graben, North Sea. Mar. Pet. Geol. 128, 105058.

Bhattacharya, S., Verma, S., 2020. Seismic attribute and petrophysics-assisted interpretation of the Nanushuk and Torok formations on the North Slope, Alaska. Interpretation 8 (2), SJ17–SJ34.

Bhattacharya, S., Tian, M., Rotzien, J., Verma, S., 2020. Application of seismic attributes and machine learning for imaging submarine slide blocks on the North Slope, Alaska. In: 90th Annual International Meeting, SEG, Expanded Abstracts.

Bird, K.J., Houseknecht, D.W., 2011. Geology and petroleum potential of the Arctic Alaska petroleum province. In: Spencer, A.M., Embry, A.F., Gautier, D.L., Stoupakova, A.V. (Eds.), Arctic Petroleum Geology. vol. 35. Geological Society, London, pp. 485–499. In press.

Bjørlykke, K., 2015. Petroleum Geoscience, from Sedimentary Environments to Rock Physics. Springer.

Bouma, A.H., 2000. Fine-Grained Turbidite Systems. vol. 72 AAPG Memoir.

Brevik, I., Callejon, A., Kahn, P., Janak, P., Ebron, D., 2011. Rock physicists step out of the well location, meet geophysicists and geologists to add value in exploration analysis. Lead. Edge 30, 1382–1391.

Brown, A.R., 2011. Interpretation of Three-Dimensional Seismic Data. American Association of Petroleum Geologists. https://doi.org/10.1306/M4271346.

Brown, J., 2015. Imaging: getting the picture downhole. Schlumberger Tech. Rev. 27 (2), 4–21.

Browne, G.H., Slatt, R., 2002. Outcrop and behind-outcrop characterization of a Late Miocene Slope fan system, Mt. Messenger Formation, New Zealand. AAPG Bull. 86 (5), 841–846.

Caers, J., Avseth, P., Mukerji, T., 2001. Geostatistical integration of rock physics, seismic amplitudes, and geologic models in North Sea turbidite systems. Lead. Edge 20, 308–312.

Castagna, J.G., Swan, H.W., Foster, D.J., 1998. Framework for AVO gradient and intercept interpretation. Geophysics 63, 948–965.

Chopra, S., Marfurt, K., 2007. Seismic Attributes for Prospect Identification and Reservoir Characterization. vol. 1 Society of Exploration Geophysicists, Tulsa, OK.

Dræge, A., Duffaut, K., Wiik, T., Hokstad, K., 2014. Linking rock physics and basin history—filling gaps between wells in frontier basins. Lead. Edge 33, 240–246.

Dvorkin, J., Brevik, I., 1999. Diagnosing high-porosity sandstones: strength and permeability from porosity and velocity. Geophysics 64, 795–799.

Dvorkin, J., Nur, A., 1996. Elasticity of high-porosity sandstones: theory for two North Sea data sets. Geophysics 61, 1363–1370.

Fett, T.H., 1987. A Practical Method for Identifying and Exploiting Clastic Depo. Environments. vol. 28 SPWLA.

Fett, T.H., et al., 1989. Using formation MicroScanner images. Schlumberger Tech. Rev., 16–40.

Folk, R.L., 1980. Petrology of Sedimentary Rocks. Hemphill Publishing Company, Austin, TX.

Galloway, W.E., Hobday, D.K., 1996. Terrigenous Clastic Depositional Systems. Springer-Verlag.

Gao, D., 2003. Volume texture extraction for 3D seismic visualization and interpretation. Geophysics. https://doi.org/10.1190/1.1598122.

Gonzales, E., Gesbert, S., Hofmann, R., 2016. Adding geological prior knowledge to Bayesian lithofluid facies estimation from seismic data. Interpretation 4, SL1–SL8.

Greenberg, M.L., Castagna, J.P., 1992. Shear-wave velocity estimation in porous rocks: theoretical formulation, preliminary verification and applications. Geophys. Prospect. 40, 195–209.

Han, D., Nur, A., Morgan, D., 1986. Effects of porosity and clay content on wave velocities in sandstones. Geophysics 51, 2093–2107.

Henke, K., 1982. Origin of the late Paleocene early Eocene Wilcox sandstones, Lobo Webb and Zapata Counties, Texas. Master's thesis. Texas A&M University.

Hicks, G.J., Francis, A.M., 2006. Extended elastic impedance and its relation to AVO crossplotting and VP/VS. In: EAGE Extended Abstract.

Hodgetts, D., Drinkwater, N.J., Hodgson, J., Kavanagh, J., Flint, S.S., Keogh, K., Howell, J.A., 2004. Three-dimensional geological models from outcrop data using digital data collection techniques: an example from the Tanqua Karoo depocenter, South Africa. 2004, In: Curtis, A., Wood, R. (Eds.), Geological Prior Information: Informing Science and Engineering. vol. 239. Geological Society, London, pp. 57–75. Special Publications.

Homza, T.X., 2004. A structural interpretation of the fish creek slide (lower cretaceous), northern Alaska. AAPG Bull. 88, 265–278.

Houseknecht, D.W., Lease, R.O., Schenk, C.J., Mercier, T.J., Rouse, W.A., Jarboe, P.J., Whidden, K.J., Garrity, C.P., Lewis, K.A., Heller, S., Craddock, W.H., Klett, T.R., et al., 2017. Assessment of Undiscovered Oil and Gas Resources in the Cretaceous Nanushuk and Torok Formations, Alaska North Slope, and Summary of Resource Potential of the National Petroleum Reserve in Alaska. U.S. Geological Survey Fact Sheet, pp. 2017–3088, https://doi.org/10.3133/fs20173088. 4 p.

Keogh, K., Berg, F.K., Petek, G., 2008. A method for quantifying geological uncertainties in assessing remaining oil targets: a case study from the Glitne Field, North Sea. In: Robinson, A., Griffiths, P., Price, S., Hegre, J., Muggeridge, A. (Eds.), The future of Geological Modelling in Hydrocarbon Development. vol. 309. The Geological Society, London, pp. 193–203. Special Publications.

Klimentos, T., 1991. The effects of porosity-permeability-clay content on the velocity of compressional waves. Geophysics 56, 1930–1939.

Kolla, V., Posamentier, H.W., Wood, L.J., 2007. Deep-water and fluvial sinuous channels—characteristics, similarities and dissimilarities, and modes of formation. Mar. Pet. Geol. https://doi.org/10.1016/j.marpetgeo.2007.01.007.

Lander, R., Walderhaug, O., 1999. Predicting porosity through simulating sandstone compaction and quartz cementation. AAPG Bull. 83, 433–449.

LeBlanc, R., 1997. Personal Communication.

Lehocki, I., Avseth, P., 2021. From cradle to grave: how burial history controls the rock-physics properties of quartzose sandstones. Geophys. Prospect. 69, 629–649.

Lehocki, I., Avseth, P., Mondol, N., 2020. Seismic methods for fluid discrimination in areas with complex geologic history—a case example from the Barents Sea. Interpretation 8, SA35–SA47.

Liu, Y., Casado, J.H., El-Toukhy, M., Tai, S., 2021. Basin-wide empirical rock-physics transform and its application in Campeche Basin. Lead. Edge 40, 178–185.

Marfurt, K., 2006. Robust estimates of 3D reflector dip and azimuth. Geophysics. https://doi.org/10.1190/1.2213049.

Marion, D., Nur, A., Yin, H., Han, D., 1992. Compressional velocity and porosity in sand-clay mixtures. Geophysics 57, 554–563.

Martinsen, O., Holmefjord, I.B., Midtbø, R.E., Torkildsen, G., Rykkje, J.M., 1996. Standard Core Description for Well 15/5–5. Norsk Hydro Internal Report.

Matos, M., Marfurt, K., Johann, P., 2003. Seismic color self-organizing maps. SBGf. https://doi.org/10.3997/2214-4609-PDB.195.1795_EVT_6YEAR_2009.

Mavko, G., Mukerji, T., Dvorkin, J., 2020. The Rock Physics Handbook, third ed. Cambridge University Press.

Mondal, S., Yadaw, A., Chatterjee, R., 2020. Rock-physics forward modelling to predict seismic behaviour: a case study for exploration target in Mahanadi basin, East Coast of India. Geophys. Prospect. 68, 2186–2194.

Moscardelli, L., Wood, L., Mann, P., 2006. Mass-transport complexes and associated processes in the offshore area of Trinidad and Venezuela. AAPG Bull. 90 (7), 1059–1088. https://doi.org/10.1306/02210605052.

Mukerji, T., Jørstad, A., Avseth, P., Mavko, G., Granli, J.G., 2001. Mapping lithofacies and pore-fluid probabilities in a North Sea reservoir: seismic inversions and statistical rock physics. Geophysics 66, 988–1001.

Pailet, F., 1990. Borehole Imaging. SPWLA. Reprint Volume.

Partyka, G., Gridley, J., Lopez, J., 1999. Interpretational applications of spectral decomposition in reservoir characterization. Lead. Edge. https://doi.org/10.1190/1.1438295.

Pettijohn, F.J., Potter, P.E., Siever, R., 1972. Petrographic Classification and Glossary: Sand and Sandstone. Springer, New York, NY.

Posamentier, H.W., Davies, R.J., Cartwright, J.A., Wood, L., 2007. Seismic geomorphology—an overview. In: Davies, R.J., Posamentier, H.W., Wood, L. J., Cartwright, J.A. (Eds.), Seismic Geomorphology: Applications to Hydrocarbon Exploration and Production.

Reading, H.G., Richards, M., 1994. Turbidite systems in deep-water basin margins classified by grain size and feeder system. AAPG Bull. 78, 792–822.

Roy, A., Dowdell, B.L., Marfurt, K.J., 2013. Characterizing a Mississippian tripolitic chert reservoir using 3D unsupervised and supervised multiattribute seismic facies analysis: an example from Osage County, Oklahoma. Interpretation 1 (2). https://doi.org/10.1190/INT-2013-0023.1.

Schlumberger, 1990. Log Interpretation Principles/Applications. Schlumberger.

Shanmugam, G., 2006. Deep-water processes and facies models: implications for sandstone petroleum reservoirs. In: Handbook of Petroleum Exploration and Production. vol. 5. Elsevier, Amsterdam.

Simm, R., Bacon, M., 2014. Seismic Amplitude—An Interpreter's Handbook. Cambridge University Press.

Slatt, R., Stone, C.G., Weimer, P., 2000. Characterization of slope and basin facies tracts, Jackfork Group, Arkansas, with applications WR deepwater (turbidite) reservoir management. In: Deep-Water Reservoirs of the World. vol. 20. GCSSEPM Foundation, https://doi.org/10.5724/gcs.00.15.0940.

Smith, G.C., Gidlow, P.M., 1987. Weighted stacking for rock property estimation and detection of gas. Geophys. Prospect. 35, 993–1014.

Smith, T., Sayers, C., Sondergeld, C., 2009. Rock properties in low-porosity/low-permeability sandstones. Lead. Edge 28, 48–59.

State of Alaska, Department of Natural Resources, Division of Oil & Gas, 2017. AS 43.55 Exploration Tax Credit Project. http://dog.dnr.alaska.gov/Information/GeologicalAndGeophysicalData.

Studlick, J., Shew, R., Nilsen, T., Steffens, G., 2007. Atlas of Deep-Water Outcrops. In: AAPG Studies in Geology, first ed. vol. 56. AAPG.

Verma, S., Bhattacharya, S., 2019. Delineation of complex fault network north slope, Alaska using seismic attributes. In: 89th Annual International Meeting, SEG, Expanded Abstracts, 1893-1897., https://doi.org/10.1190/segam2019-3214607.1.

Verma, S., Scipione, M., 2020. The early Paleozoic structures and its influence on the Permian strata, Midland Basin: insights from multi-attribute seismic analysis. J. Nat. Gas Sci. Eng. 82. https://doi.org/10.1016/j.jngse.2020.103521.

Vernik, L., 2016. Seismic Petrophysics in Quantitative Interpretation. Society of Exploration Geophysicists.

Vernik, L., Kachanov, M., 2010. Modeling elastic properties of siliciclastic rocks. Geophysics 75, E171–E182.

Vernik, L., Nur, A., 1992. Petrophysical classification of siliciclastics for lithology and porosity prediction from seismic velocities. AAPG Bull. 76, 1295–1309.

Walderhaug, O., 1996. Kinetic modelling of quartz cementation and porosity loss in deeply buried sandstone reservoirs. AAPG Bull. 80, 731–745.

Weimer, P., 1987. Seismic stratigraphy of three areas of lower slope failure, Torok Formation, northern Alaska. In: Tailleur, I., Weimer, P. (Eds.), Alaskan North Slope Geology. Pacific Section SEPM and the Alaska Geological Society, pp. 481–496.

Went, D.J., 2021. Practical application of global siliciclastic rock-property trends to AVA interpretation in frontier basins. Lead. Edge 40, 454–459.

Whitcombe, D.N., Connolly, P.A., Reagan, R.L., Redshaw, T.C., 2002. Extended elastic impedance for fluid and lithology prediction. Geophysics 67, 63–67.

Widess, M.B., 1973. How thin is a thin bed. Geophysics. https://doi.org/10.1190/1.1440403.

Wynn, R.B., Kenyon, N.H., Masson, D.G., Stow, D.A.V., Weaver, P.P.E., 2002. Characterization and recognition of deep-water channel-lobe transition zones. AAPG Bull. 86, 1441–1460.

Zhao, T., Jayaram, V., Roy, A., Marfurt, K.J., 2015. A comparison of classification techniques for seismic facies recognition. Interpretation. https://doi.org/10.1190/INT-2015-0044.1.

Chapter 17

# Seismic rock physics and machine learning for deepwater stratigraphic intervals

**Anshuman Pradhan[a], Per Avseth[b] and Tapan Mukerji[a]**
[a]*Stanford University, Stanford, CA, United States,* [b]*Dig Science/NTNU, Trondheim, Norway*

## Introduction

Quantitative seismic interpretation (Avseth et al., 2005; Simm and Bacon, 2014; Vernik, 2016) is a cross-disciplinary field in geoscience that combines a wide range of tools and methodologies with the goal of better understanding subsurface geology from seismic amplitude data. In particular, rock physics (Mavko et al., 2020) represents the bridge between qualitative geology and quantitative geophysical measurements.

Rock physics tools have become widely used in quantitative seismic interpretation, both for forward modeling (e.g., Bakke et al., 2013; Mondal et al., 2020; Avseth et al., 2021) and during seismic inversion studies (e.g., Mukerji et al., 2001a, b; Grana and Rossa, 2010; Bosch et al., 2010). In recent years, rock physics has also been successfully integrated with basin modeling and analysis of petroleum systems (e.g., AlKawai et al., 2018; Pradhan et al., 2020; Løseth et al., 2011). The application of rock physics tools has proven to be a valuable practice in particular for deepwater depositional settings, where distributions of sand bodies can be quite complex and not easily determined from more qualitative stratigraphic interpretations (e.g., Gonzales et al., 2016). In the last couple of decades, rock physics templates (Ødegaard and Avseth, 2004) and statistical rock physics (Mukerji et al., 2001a, b) have been commonly implemented in seismic inversion and classification workflows in classical two-step amplitude-vs-offset (AVO) inversion algorithms, where seismic data are first transformed to elastic parameters (e.g., Ma, 2001; Russell et al., 2006), and then elastic parameters are translated to reservoir/geologic properties via rock physics relationships (e.g., Doyen, 2007; Avseth et al., 2016). The rock physics templates can be used semiquantitatively to interpret clusters of data and to guide interactive classification of lithofacies and fluids from seismic amplitude data. In combination with statistical rock physics, the templates can also be used to extend observed data clusters and include modeled training data in the classification step. It is important to honor expected lithofacies and their distributions in a deepwater setting. Sandy facies can be both blocky and homogeneous, but also quite shaly and heterogeneous over short distances; for instance, in a channel-levee complex. In very recent years, machine learning and deep learning has become a very popular technology (Qadrouh et al., 2019) to boost both speed and performance (success rate) of seismic inversion (Das et al., 2019; Feng, 2020) and reservoir prediction and facies classification (e.g., Mosser et al., 2019; Downton et al., 2020). These algorithms do not need to operate in a low-dimensional physical domain (for instance acoustic impedance (AI) vs $V_P/V_S$), but in latent domains that project several dimensions of data, including other data and attributes like attenuation and travel time, into an optimal separation hyperplane (e.g., Bressan et al., 2020). This also allows for so-called one-step inversions where the seismic data are automatically transformed to reservoir geology without an apparent rock physics link (e.g., Purves et al., 2019; Cova et al., 2020). However, the need for geoscientific domain knowledge to both quality check the machine learning algorithms (e.g. training weight factors, hyperparameter tuning, etc.) and the classification results is still essential, especially when it comes to defining, labeling and augmenting the training data to be used in the classifications.

This chapter demonstrates how rock physics combined with geological processes is still a very important foundation for quantitative interpretations of deepwater clastic systems, whether one chooses to do a classical two-step (and commonly deterministic) AVO inversion (e.g., Avseth et al., 2016) or a more automated one-step (computational statistics/ML-based) AVO inversion (Rimstad et al., 2012; Fjellstad et al., 2021). First, fundamental rock physics and how rock physics properties are controlled by geological processes, both compactional and depositional, are elaborated on. This is demonstrated with examples from various deepwater clastic systems in the North Sea, including sandstones belonging to the Paleocene Heimdal Formation (introduced in Chapter 16) and the Cretaceous Agat Formation. Next, a brief summary of the concept of

**Deepwater Sedimentary Systems.** https://doi.org/10.1016/B978-0-323-91918-0.00006-2

seismic lithofacies (Avseth, 2000) is provided, followed by a recipe on how rock physics templates can aid in the quantitative interpretation of deepwater clastic systems. Finally, it is demonstrated how machine learning algorithms can be used in combination with statistical rock physics for improved quantitative seismic interpretation in these types of depositional environments, with an example from offshore Nile Delta (see also Chapter 11).

## The rock physics link between geology and geophysics

Rock physics can be defined as the link between geological features and geophysical observations (Fig. 1). However, for a given case there are often many more unknown geological parameters than independent geophysical observables. In seismic exploration, one rarely has more than two independent geophysical observables, for instance, two AVO attributes (intercept and gradient). Through seismic inversion, these reflectivity attributes can be converted to layer parameters, for instance, acoustic impedance and $V_P/V_S$, or elastic impedance at two orthogonal Chi-angles (Whitcombe et al., 2002). However, since the inverse problem is nonunique, given the many unknown geological parameters, several combinations of different elastic parameters can produce the same seismic signatures. The rock physics link seeks to establish some rules or constraints on how certain geological features are controlling elastic and seismic properties. This is possible through empirical, data-driven relations (e.g., Han et al., 1986; Avseth et al., 2021), or via theoretical or heuristic models that can be calibrated to some local observations (e.g., Dvorkin and Nur, 1996; Avseth et al., 2010; Vernik and Kachanov, 2010).

A multitude of research studies (e.g., Han et al., 1986; Marion et al., 1992; Jizba, 1991; Zimmer et al., 2007) have revealed the relationship between geological parameters like clay content, porosity, cement volume, sorting, net-to-gross, etc. and seismic properties like $V_P$, $V_S$ and density. Furthermore, anisotropy parameters can also be linked to geological features like micro-cracks and interbedding (Thomsen, 1986; Sayers, 1995) below the measurement resolution. By linking geological processes to physical parameters, rock physics models can be used in a predictive manner to say something about the expected properties within a given depositional environment, at a certain burial depth. However, it is important to disentangle the effect of burial from the effect of deposition, when interpreting seismic amplitude data. Therefore, rock physics models should honor the depositional and burial history. The rock physics diagnostic approach (Chapter 16) introduced by Dvorkin and Nur (1996) or the sandstone diagenesis model by Vernik and Kachanov (2010) are alternative ways of modeling the combined effect of depositional variability and compactional effects on the seismic properties of sandstones. Lehocki and Avseth (2021) extended the Dvorkin-Nur modeling approach and first modeled sorting and mechanical compaction, followed by chemical compaction. This is a more realistic modeling approach for normally compacted sandstone reservoirs and is demonstrated with Glitne Field data in Chapter 16.3.4.

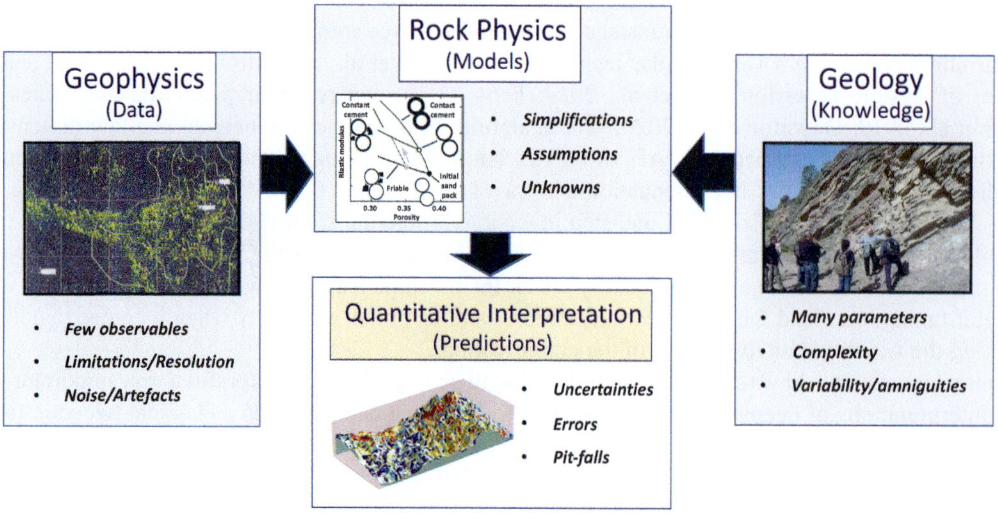


**FIG. 1** Seismic rock physics represents the link between geology and geophysics, or translation of elastic parameters to geological properties (or vice versa). Quantitative interpretation is the prediction of geological properties from geophysical observables using rock physics relationships.

## Effect of clay in reservoir sandstones

In deepwater settings, it is common to have great spatial variability in reservoir heterogeneity and clay content. The presence of clays in sandstone reservoirs can be quite complex and distributed in different ways (Khadem et al., 2021; see also Chapters 2 and 8). In deepwater turbidite sandstones, clays can be either pore-filling (dispersed) or laminating, or a combination of both (e.g., Nasser et al., 2011). Thin-section examples are shown in Fig. 2. One is from the Alvheim Field (Avseth et al., 2009) showing a few percent of pore-filling clays, whereas the other is from offshore Angola (Avseth et al., 2005b) showing laminated clays. Clay particles in sandstones can also be load-bearing when randomly distributed and are then referred to as structural clays (Khadem et al., 2021). These different distributions could reflect different depositional facies and subenvironments, where laminated clays will be more common in overbank and crevasse splay settings of relatively mud-rich deepwater systems, and pore-filling clays will be more common along channel/lobe axis of more sand-rich systems (e.g., Avseth et al., 2003; Gladstone et al., 2018). In proximal fans, amalgamated thick-bedded turbidite sand units typically have relatively low pore-filling clay volumes in the range of 5%–10% (Bello et al., 2021), whereas in more distal fan/lobe margin total pore-filling clay volume can range between 10% and 30% and have deleterious effect on reservoir quality. Pore-filling (i.e., pore-bridging, pore-lining or grain-coating) clays could also represent authigenic clays reflecting the interplay between detrital mineralogical composition and diagenetic alterations. Structural clays could occur due to remobilized sand-shale units or within more chaotic mass-flow units. Both porosities and seismic velocities will change significantly as a function of clay content and clay distribution pattern. The presence of clays will also affect the fluid and stress sensitivities in these shaly sandstones. Marion et al. (1992) came up with a topological model for seismic velocities in shaly sandstones and silty shales and, based on laboratory measurements, they found that the velocity-porosity trend in mixtures of sandstones and shales could sometimes create an overturned $V$-shape, where the stiffest rocks with the highest velocities were found when the sandstone porosities were filled up with clays. If the clays occur as lamina, they tend to weaken the rock frame significantly causing lower effective velocities and higher $V_P/V_S$ ratios, whereas porosities are not affected that much. Clays can also stiffen the rock frame by acting as coating cement (e.g., Dvorkin and Nur, 1996). On a larger scale, the interbedding characteristic (sand-shale ratio or net-to-gross) of deepwater systems will also affect the seismic properties within turbidite systems (e.g., Golikov et al., 2013), and can be associated with different depositional settings. Normally, the net-to-gross decreases basin-ward along a submarine fan system, and radially from a channel axis to the overbank area as discussed in Chapters 8 and 9 (see also Reading and Richards, 1994). Given the sand-rich nature of the turbidite systems in the Alvheim Field and Måløy Slope, we will assume net-to-gross of 1.0 in the rock physics modeling examples in the following section. Modeling of varying net-to-gross is beyond the scope of this study but could easily be handled using Backus averaging of interbedded sandstones and shales, see for instance Avseth et al. (2009) and Stovas et al. (2006).

## Rock physics modeling and diagnostics: Måløy Slope, North Sea

Next, we want to demonstrate the relationships between geological processes and rock physics properties, with examples from two different deepwater depositional systems in the North Sea; one representing an Early Cretaceous slope setting in the Måløy Slope area of the North Sea, offshore Norway, and the other representing a Paleocene basin floor setting in the Alvheim area of the North Sea. Fig. 3 shows well log data from three selected wells in the Måløy Slope area of the Northern North Sea, where we focus on the Cretaceous Agat Formation sandstones. These are deepwater lobe complex deposits shed

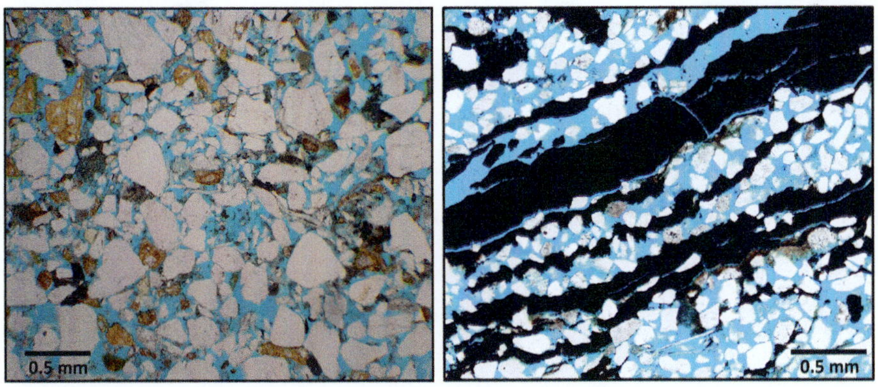


**FIG. 2** Thin-section images showing minor amounts of pore-filling clays (*dark brown* particles) to the left (Alvheim Field, see Avseth et al., 2009), and laminated clays (*dark brown*) to the right (Offshore West Africa example, see Avseth et al., 2005b). *Black bars* indicate 0.5 mm length.

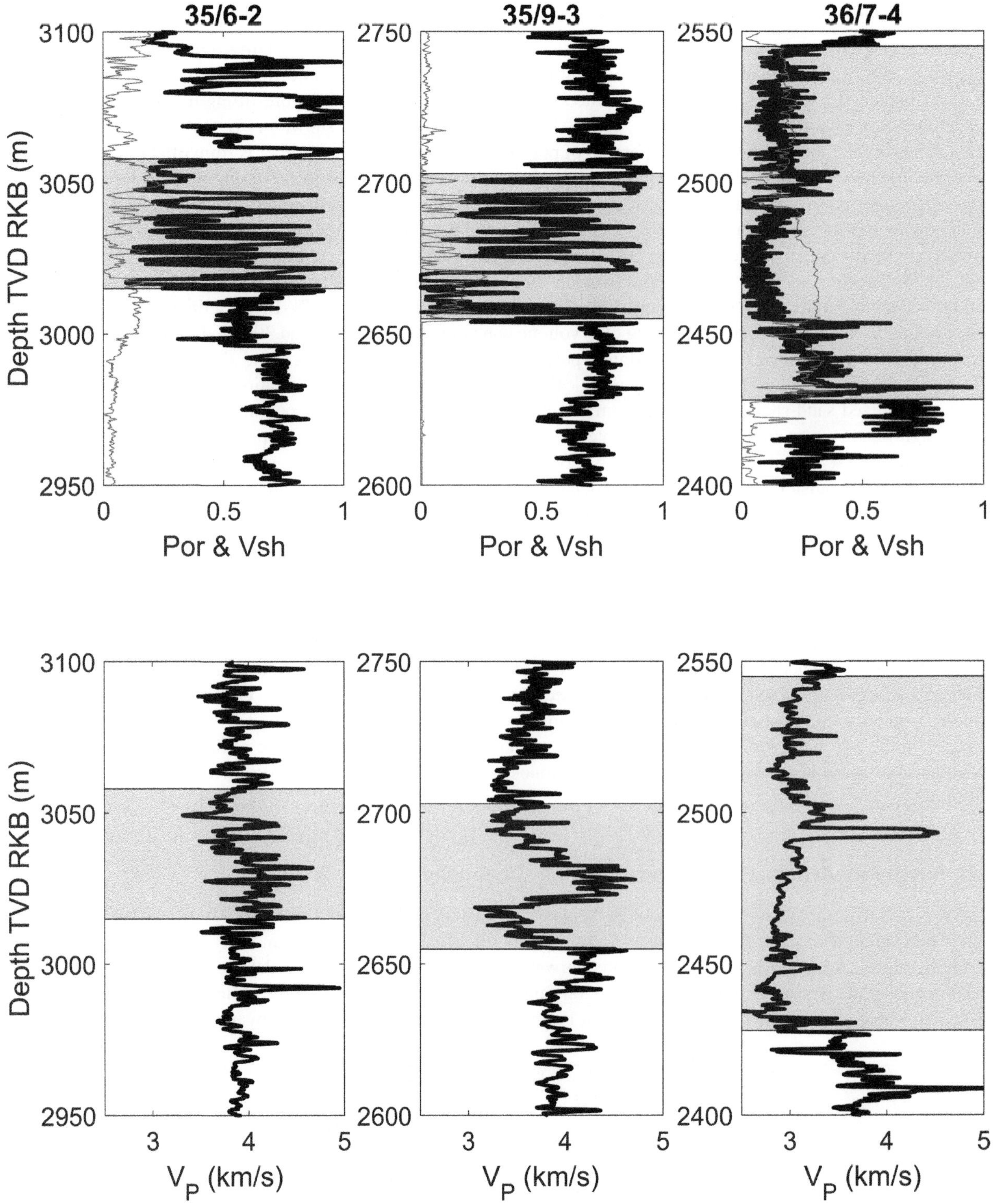


**FIG. 3** Well log data from three wells in the Måløy Slope area, Northern North Sea. The *gray zones* represent deepwater sandstones of the Cretaceous Agat Formation.

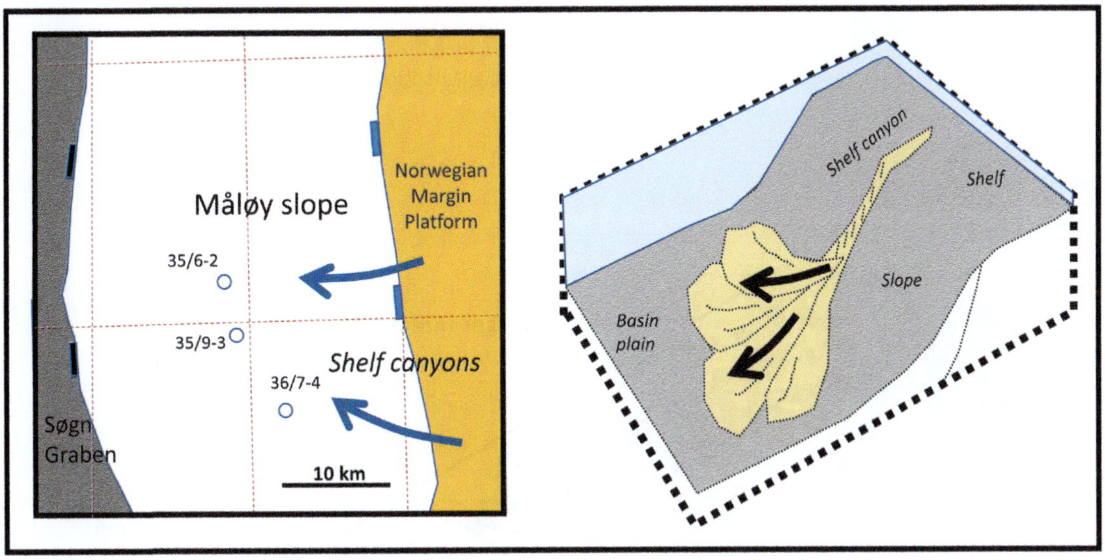


**FIG. 3—cont'd** The *lower left subplot* depicts the location of the wells relative to each other in a slope setting where sands have been transported via shelf canyons in the east, fed by rivers and deltas along the Norwegian coast.

off from canyons along the Norwegian coast. The sands were deposited in a slope-to-basin-floor setting by gravity flows, primarily turbidity currents, and debris flows (Bugge et al., 2001). Presence of water-escape dish structures in massive sandstones has confirmed the presence of turbidity sandstones (Lowe and LoPiccolo, 1974; Wahyuni et al., 2019), and iron oxides and dish-forming chlorite have affected the reservoir properties both by inhibiting quartz cementation and deteriorating the reservoir quality (Kombrink, 2021; Hurst and Buller, 1984). The wells penetrate the Agat Formation sandstone at different burial depths and in different depositional settings. Well 36/7-4 shows a blocky sandstone with homogenous character, with relatively high porosities, and the sandstones are buried at around 2425–2550m depth. This was a commercial oil and gas discovery (The Cara discovery; today the Duva Field; Olsen, 2017; Avseth et al., 2017). Well 35/9-3 is located more basinward and penetrated a more heterogeneous interval of Agat Formation sandstones (average net-to-gross around 0.6, see also Martinsen et al., 2005), with subcommercial oil and gas saturation in relatively intermediate porosity (~15%–25%) sandstones. Note that this well also penetrated the Agat Formation at a greater depth (ca. 2650–2700m RKB (Rotary Kelly Bushing)). Finally, well 35/6-2S is a dry well located even further basinward, and also here the Agat Formation interval showed a heterolithic $V_{sh}$ log pattern. The present depth is here around 3020–3060m RKB.

Next, seismic velocity changes between the wells are investigated. Fig. 3 also includes the brine-saturated P-wave velocity logs (fluid substitution from in situ hydrocarbon to brine has been performed using Gassmann theory, see Mavko et al., 2020 for more information) for the three wells. Fig. 4 shows a cross-plot of velocities vs porosities for the Agat Formation in the three wells. Diagnostic rock physics models, as described in Chapter 16.3, are also superimposed.

Geologic trends can be observed in the velocity-porosity cross-plot. The massive sandstones in the 36/7-4 well show porosities predominantly in the range between 23% and 32%, and corresponding P-wave velocities are in the range between 2.8 and 3.2km/s. According to the rock physics diagnostic models, calibrated to typical clean arenite mineralogy (mostly quartz, with minor amounts of feldspar and clays; cf. Vernik and Nur, 1992), cement volumes are estimated in the range of 2%–5% using the approach introduced by Avseth et al. (2009). The velocity-porosity trends parallel with the constant cement models observed for the Agat Formation in this well are likely associated with sorting variability (cf., Avseth et al., 2009). Sorting is a depositional parameter, and poorly sorted sand is normally also more clay-rich (i.e., silt and clay particles filling in the pore space between sand grains). The data from well 35/9-3 plots with overlapping and slightly lower porosities and higher velocities than what is observed for 36/7-4. This could be a combination of burial and depositional effects, as the Agat Formation is buried deeper, and located more basinward (well 35/9-3 is located around 9km northwest of well 36/7-4). The $V_{sh}$ log in Fig. 3 clearly shows a more interbedded nature, especially in the lower parts (around 2670–2700m). The data points are a bit more scattered when it comes to cement volume, which could reflect the interplay between depositional clay and diagenetic cement volumes. Overall, the data plot with lower porosities and higher velocities

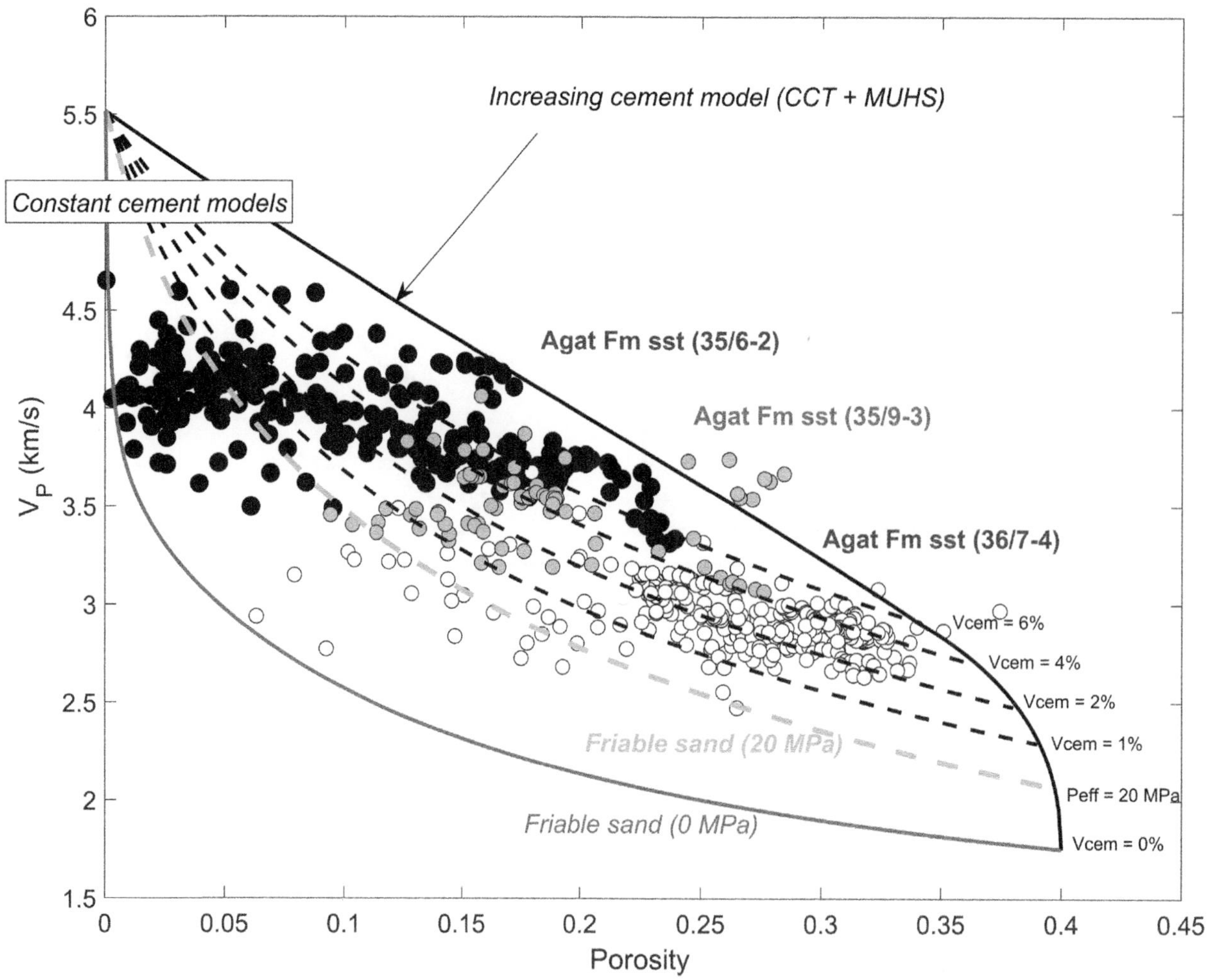


**FIG. 4** Velocity-porosity cross plot including data from the Agat Formation sandstones displayed as well logs in Fig. 3. Note the increasing velocity and decreasing porosity for the individual wells, but also from one well to another. As diagnosed from the rock physics models, the increase in velocities is associated with increasing cement volume (associated with increasing burial depth). However, for each well, the trends are associated with decrease in sorting and increase in pore-filling clay volume.

due to increased compaction relative to well 36/7-4. This is even more obvious for Well 35/9-2S. Here the porosities range from close to zero to 25% within the Agat Formation, whereas velocities range from 3.3 to 4.5 km/s. The cement volume according to the rock physics diagnostic models is above 6% for the cleaner, high porosity sandstones. However, the velocity-porosity trend with decreasing sorting observed in the data intersects the modeled constant cement lines, which is a pattern typically seen when clay content is gradually increasing (Avseth et al., 2005). The increasing pore-filling clay is causing a stiffer rock frame and increasing velocity, but the presence of clay is also inhibiting quartz cementation, meaning that the lower porosity (shaly) sandstones are likely less cemented than the higher porosity (clean) sandstones. The combination of diagenetic and depositional trends in these sandstones can be modeled with a more holistic and geologically consistent approach (Lehocki and Avseth, 2021; Avseth, 2020; see also Chapter 16.3), as seen in Fig. 5. First, we model the burial history (mechanical and chemical compaction) and associated porosity-velocity evolution for the cleanest sandstone (i.e., the high porosity end-member). Next, the Dvorkin and Gutierrez (2002) model constrained by this end-member, is used to model the sorting variability and increasing clay content. The silty shale leg is also modeled, between the critical clay content (when all sandstone porosity is filled up with clay) and the pure shale point, the latter determined from observed shale properties from embedding shale interval (i.e., the Rødby Formation cap-rock shale).

The modeling strategy presented here enables a better understanding of the complex trends and scatter observed in the well log data, and how they relate to geologic processes. It also enables extrapolation of models beyond the data observations, which is very useful in areas with limited well control. In this way, it will be easier to predict different lithology and fluid scenarios from seismic data during exploration (see also Chapter 16.3).

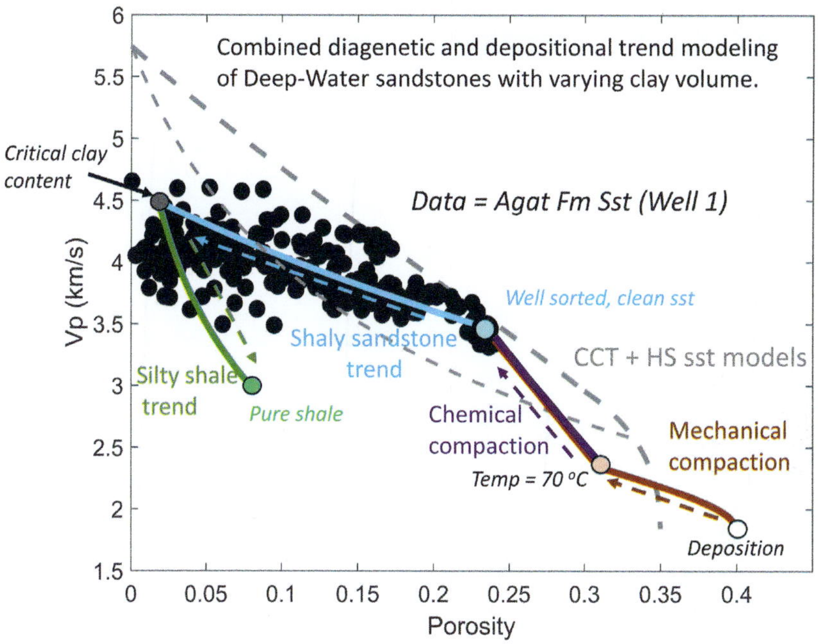


**FIG. 5** Combined burial and rock physics modeling (mechanical compaction + chemical compaction) and depositional trend modeling (shaly sandstone trend and silty shale trend) compared with the Agat Formation sandstone data in well 35/6-2S. The compaction modeling defines the high-porosity end-member of the cleanest Agat Formation sandstones at a given burial depth. For reference, diagnostic rock physics models are included, including contact cement theory (CCT) and Hashin-Shtrikman (HS) bounds (see Chapter 16 for more details).

## Rock physics modeling and diagnostics: Alvheim Field, North Sea

Next, analysis is done on the relationship between geological processes and rock physics properties in the Paleocene age Heimdal Formation sandstone reservoirs in the Alvheim area, North Sea. The rock physics properties and seismic signatures of these deepwater sandy deposits have been extensively studied in previous publications (e.g., Avseth et al., 2008, 2009, 2021; Lehocki and Avseth, 2010; Rimstad et al., 2012; Golikov et al., 2013). A depositional model for the area was presented by Fitzsimmons et al. (2005) who found that individual lobes were deposited along linear fairways confined to depositional topography created by earlier sequences, comprising both thick-bedded massive sandstone facies and heterolithic units. Amalgamation is common in the deeper parts of the Heimdal Formation, but less common in the later stages during fairway broadening that ended in an abandonment stage and shale draping of the deep-marine sands.

This study used six vertical wells that penetrate the Heimdal Formation sandstone reservoirs at around 2 km below the sea floor (Fig. 6). The focus is on the upper part of the Heimdal Formation sandstones, where excellent quality reservoir sandstones are capped by the marine Lista Formation shales. Individual sand lobes have structural closures filled with either oil or gas, and the filling history is quite complex due to stratigraphic compartmentalization (Hjellbakk et al., 2016) and local structural influence (tilting and faulting). The well log data in Fig. 6 include the in-situ P-wave sonic logs for each of the wells, and the upper reservoir interval is highlighted with colors depending on pore fluid (red = gas, green = oil, blue = brine). Rock physics compaction trends are superimposed, where the transition from mechanical to chemical compaction is honored, using the same workflow as presented in Chapter 16 for the Glitne Field. This area has likely experienced around 200–300 m uplift (Avseth and Lehocki, 2016), and the compaction trends have been calibrated to honor this uplift (indicated by the combined burial-velocity curve in subplot 1, using the approach introduced by Avseth and Lehocki, 2016). As demonstrated by Rimstad et al. (2012), this burial uplift associated with Miocene tilting of the area can have an impact on the seismic reservoir prediction in the Alvheim Field. The superimposed compaction depth trends match closely with the overall trends in the reservoir zone. Also, the modeled velocity trends fit nicely with the observed velocities in the overburden. However, note that the overburden mainly consists of shales that follow slightly different velocity-depth trends. Nevertheless, shales and sands in the mechanical compaction domain tend to have similar velocities (Avseth et al., 2001a, b). Shales also undergo chemical compaction and will stiffen in the chemical compaction domain due to smectite to illite transition and associated generation of microcrystalline quartz that will stiffen the rock frame of the illite-rich shales (Thyberg et al., 2009). Avseth et al. (2021) demonstrated that a transition from smectite rich to illite rich shales is occurring within the Lista Formation shales when temperatures are high enough to initiate the diagenetic reactions (ca. 60–70 °C), see also Bjørlykke and Aagaard (1992).

Fig. 7 shows two cross-plots of velocity vs porosity for the Heimdal reservoir zones in the six wells displayed in Fig. 6, where hydrocarbon zones have been fluid substituted to brine saturation for a more consistent comparison between the

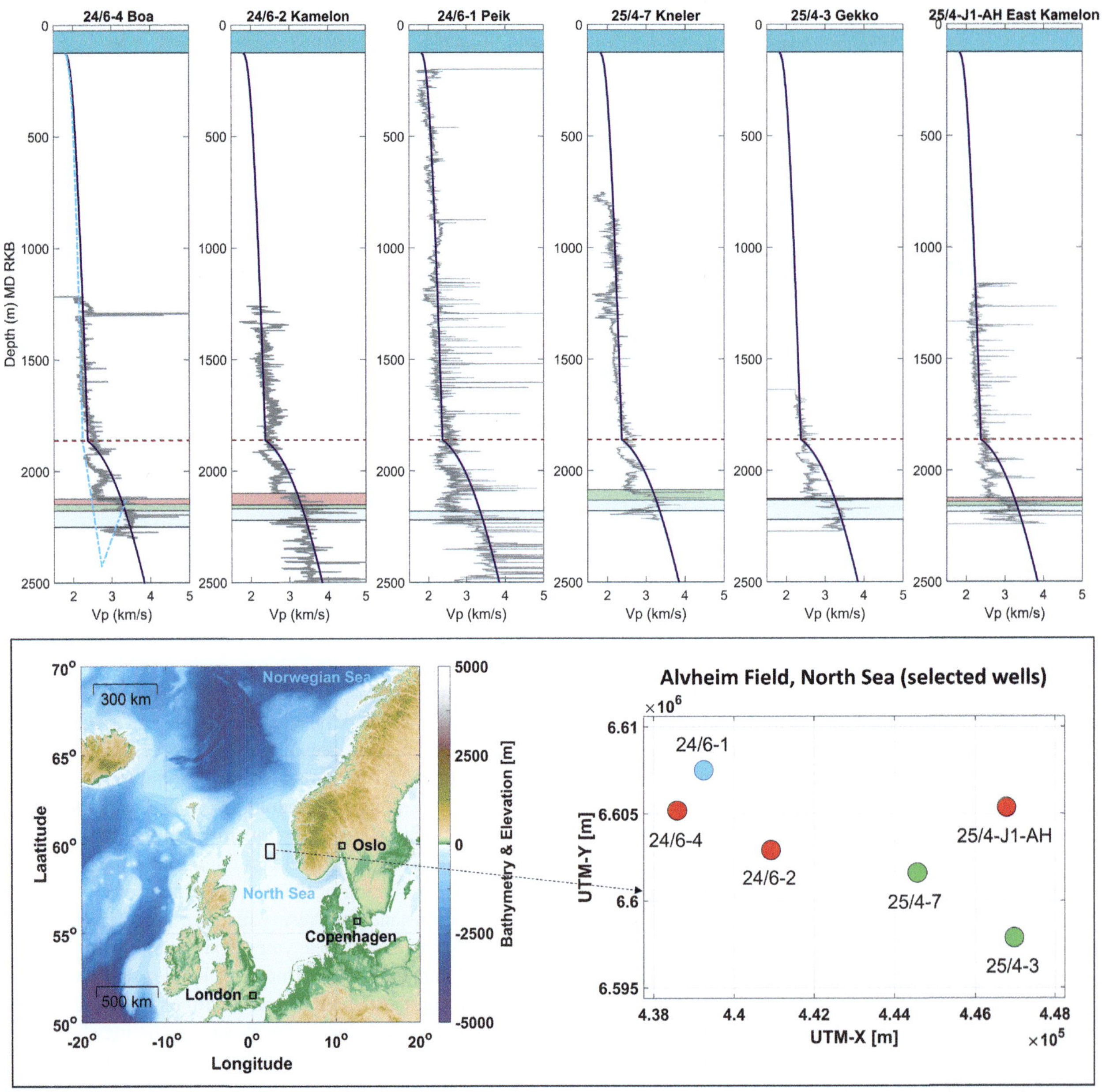


**FIG. 6** P-wave velocity logs and superimposed sandstone depth trends (mechanical + chemical compaction) for six wells in the Alvheim Field, North Sea. The *dark blue* trends are the present-day rock physics properties with depth, honoring the transition from mechanical to chemical compaction. The *cyan curve* in the left-most subplot illustrates the combined burial-rock physics evolution through geologic time, of a given point of present-day curve, showing that the sandstones have been buried a couple of 100 m deeper, then uplifted. Regional map and location of selected wells are depicted in the *lower subplot*.

wells. A selection of rock physics models (see Chapter 16.3, and Avseth et al., 2005; Mavko et al., 2020 for more details) have been superimposed on the scatter of data (left subplot) including the contact cement model and increasing cement model (with critical porosity of 0.4, and mineral moduli of relatively clean quartz-arenite with 10% clays) that constitute the upper stiff bound of the sandstones; a constant cement model (2%) that represents an average expected rock physics model for the studied sandstones at the given burial depth and the friable sand model at zero pressure, which represents a theoretical lower bound. A shale model is also superimposed, representative of the cap-rock Lista Formation shales (cf., Avseth et al., 2021), together with the Dvorkin-Gutierrez model calibrated to two different sandstone high-porosity end-members representing slightly different volumes of quartz cement. Finally, a laminated sand-shale model is also included, by mixing the same high-porosity sandstone points with a representative pure shale point. The data are scattered within the bounds, but some trends are apparent when comparing the data with the superimposed rock physics models. The cleanest

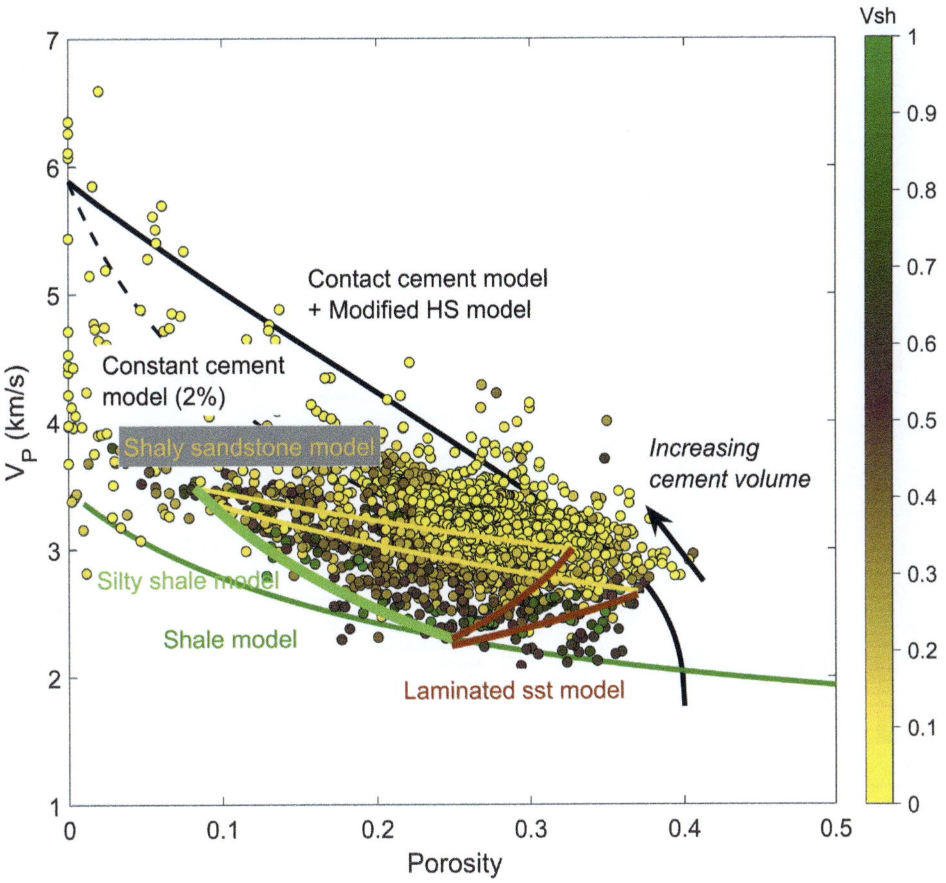


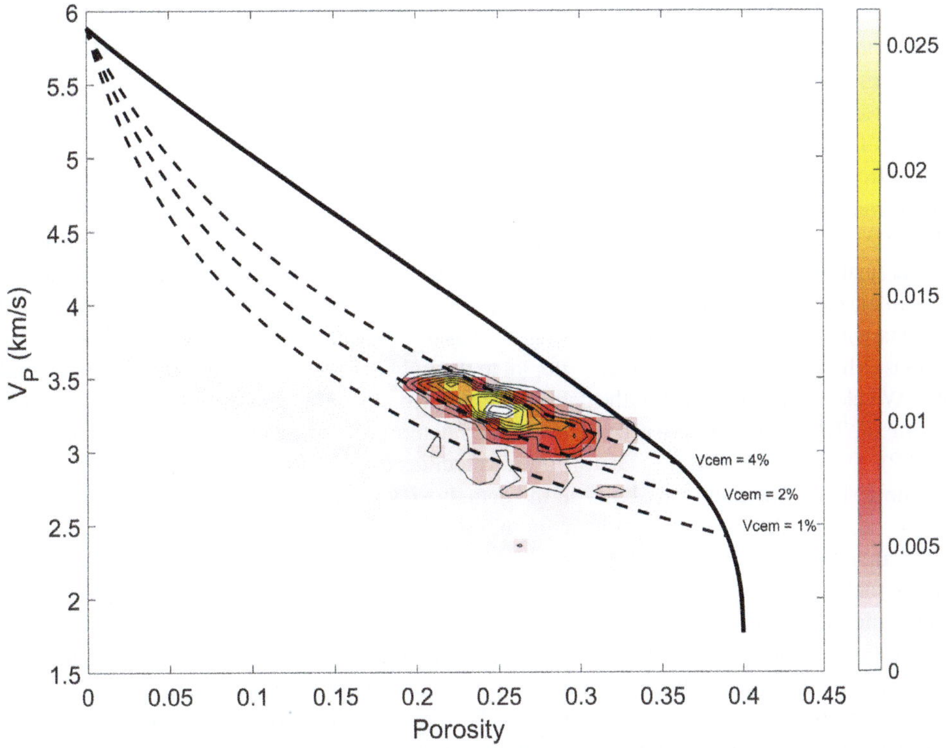


**FIG. 7** $V_P$ (brine saturated values) vs porosity (scatter data and statistical pdf) for six wells in the Alvheim Field, North Sea. The massive sandstones are the dominant facies, but laminated shales and heterolithics are also present within the Heimdal Formation reservoir zone.

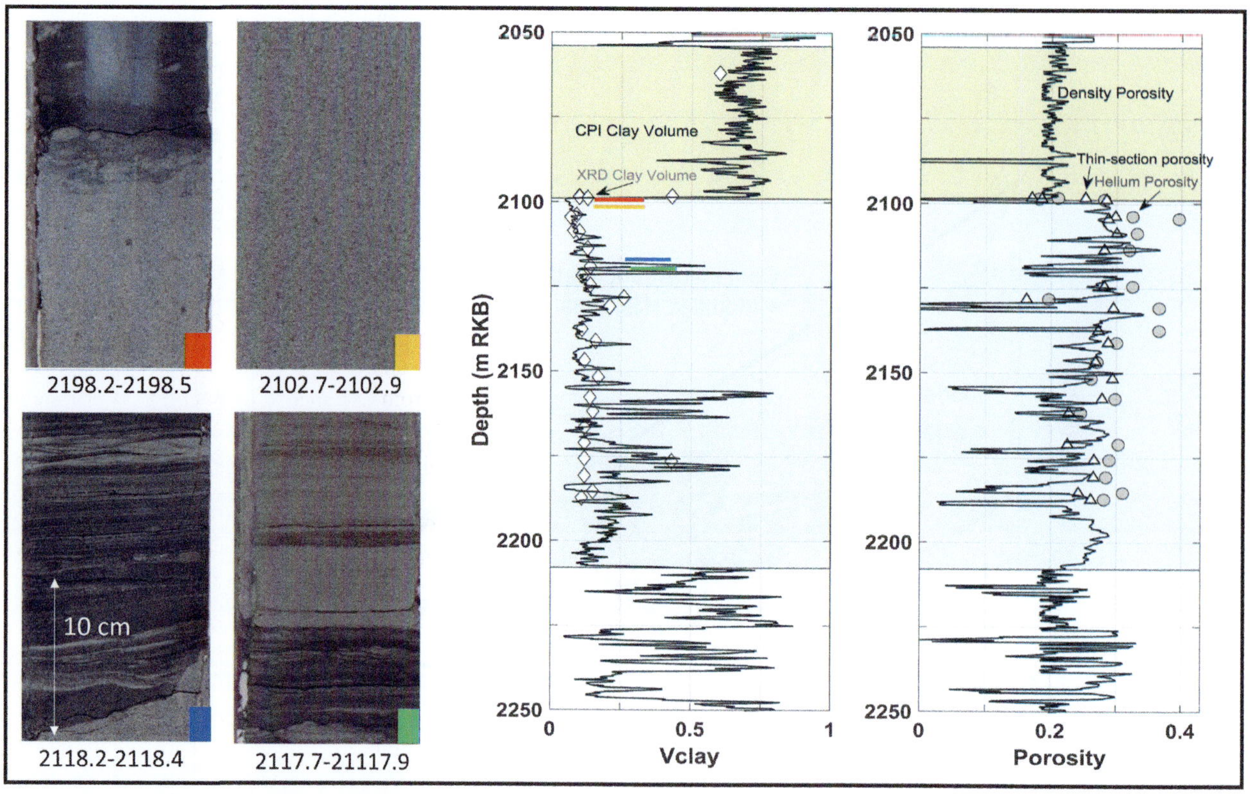


**FIG. 8** Core observations juxtaposed with petrophysical logs from well 24/6-2. Massive clean turbidite sandstones are observed in the *upper part*, and cm-scale clay laminated sandstones and thin-bedded turbidites are occurring within the reservoir, in zones where the clay volume log indicate meter-scale heterogeneity. *Blue zone* shows the upper Heimdal Formation reservoir zone, and *olive-green zone* shows the cap-rock Lista Formation.

sandstone data seem to follow the constant cement model, whereas the more shaly sandstones (brownish colored circles) follow either a dispersed trend (i.e., Dvorkin-Gutierrez shaly sandstone leg) or a more laminated trend, reflecting the fact that there are also heterolithic facies within the studied interval. Note that the clay laminated sandstones (at mm–cm scale) seem to be more abundant in the relatively heterolithic and interbedded zones (at meter scale), see comparison between core observations and petrophysical log data for well 24/6-2 in Fig. 8. Some of the scattered points could also be thin calcite stringers (i.e., high velocity-low porosity outliers), which are occasionally present in North Sea Paleocene sandstones (e.g., Jonk et al., 2005). The few data points falling outside the elastic bounds are likely noisy spikes in the data and could be filtered in further analysis (cf., Avseth et al., 2021). One way to filter away noise and highlight the dominant trends in the data, is to create statistical density plots or "heat-map" crossplots, where data are binned into cells, and the coloring represents the frequency of data within each cell. These can be regarded as normalized probability density functions (pdf) representing the data distributions (Mukerji et al., 2001a, b). The subplot to the right in Fig. 7 shows the pdf plot of the scattered data shown in the left subplot. We clearly see that most of the data are distributed along the constant cement models, mainly between 2% and 4% quartz cement. This agrees with the cement volumes determined from thin-section analysis in the Alvheim Field (Avseth et al., 2009). Still, there are also "tails" of lower velocities associated with heterolithic facies that tend to be less cemented and/or more shaly. The main variability in the data, however, is associated with varying sorting in the sandstone data, similar to what is observed for individual wells in the Måløy Slope area above.

In the Alvheim area, there appears to be little compactional variability, as the burial history is quite similar for all the wells. However, the fact that the burial depth of the top of the reservoir can vary by a few tens of meters, combined with the fact that the onset of quartz cement (ca. 70 °C, cf., Walderhaug, 1996) is very close to the top reservoir level, can cause some differences in quartz cementation in the different wells also in the Alvheim area. It can be difficult to separate the effect of temperature vs the effect of heterogeneity on the diagenesis (i.e., the velocity drop of the heterolithic "tails" in Fig. 7 could be due to increasing clay content or lower cement volume). The presence of hydrocarbons could also impact diagenesis (see Chapter 13). These ambiguities can make it extra challenging to separate fluid effects from lithology effects in the seismic data (cf., Fig. 1). Below, it is demonstrated how the power of $V_S$ information can help us to better discriminate between facies and fluids in these reservoirs, when we create rock physics templates of acoustic impedance vs $V_P/V_S$. Prior to that, however, the concept of seismic lithofacies is introduced, which is useful when separating and classifying different types of lithologies and fluid types from seismic amplitude data in these deepwater clastic systems.

## Seismic lithofacies in deepwater clastic systems

Seismic lithofacies were first defined by Avseth et al. (2000) as seismic scale sedimentary units with characteristic petrography, lithology and elastic properties. These were used as categories in AVO classification of lithology and fluids (e. g., Avseth et al., 2001a, b; Mukerji et al., 2001a, b), and should not be confused with traditional seismic facies introduced by Vail and Mitchum Jr. (1977), among others, to guide sequence stratigraphic interpretations from seismic data. The "Walther's Law" of facies (Walther, 1893-4) implies that lithofacies observed conformably in vertical sequences will also be found laterally next to each other in a given depositional system. Hence, the lithofacies allow us to connect vertical changes in geology to lateral changes in a predictive manner (i.e., facies associations). The seismic lithofacies can be represented by probability distribution functions (pdfs) based on data observations from several wells. They could also be augmented from observed data using rock physics models, including scenarios not observed at well locations, but still plausible in the area of interest. Various fluid scenarios not captured by well log observations could also be modeled and included as seismic lithofacies using Gassmann theory. The facies could be classified from well log data (Avseth and Mukerji, 2002), or from seismic amplitude data (Avseth et al., 2001a, b; Mukerji et al., 2001a, b). The defined facies should honor the scale effects, and for instance a homogeneous sandstone unit at the well log scale could be part of a heterolithic sandstone unit at the seismic scale (Takahashi, 2000).

For deepwater depositional systems, Avseth et al. (2000) defined the lithofacies listed in Table 1. The lithofacies are based on a descriptive scheme going from (high energy) thick-bedded units to (low-energy) thin-bedded units. The lithofacies chart is comparable and compatible with the chart provided in Table 2 in Chapter 7 in this book, yet with fewer categories. This is reasonable because the seismic scale and limited attributes (cf., Fig. 1) will not differentiate all the detailed facies in table 2 in Chapter 7. Subfacies are included, however, to honor important textural changes in the

**TABLE 1** Seismic lithofacies definitions and typical log motifs in deepwater clastic systems.

| Facies | Geological description of facies and subfacies | GR log motif |
|---|---|---|
| *Seismic lithofacies classification and description* | | |
| I Gravels and conglomerates | Gravels, conglomerates and pebbly sands. Sand-rich or mud-rich debris flow deposits | Complex. Can be blocky if "clean" |
| II Thick-bedded sandstone | **IIa:** Very clean, well sorted, massive sandstones with small amounts of quartz overgrowths. Water-escape structures are common. Clay content less than 10% | Usually blocky and smooth. Bell and funnel shapes can occur |
| | **IIb:** Clean, massive sandstones with clay coatings. Water-escape structures are prominent. Pore-filling clay content slightly higher than in facies $II_a$. (approximately 10%–15%) | |
| | **IIc:** Plan-laminated sandstone. Higher pore-filling clay content (10%–20%) and grain size in general smaller (fine to medium grained) than in facies IIa and IIb | Low, but increasing GR values, from IIa to IIc |
| | **IId:** Shaly sandstone (clay content between 20% and 40%) | Intermediate in IId |
| III Interbedded sandstone—shale | Interbedded sand-shale couplets, where sand units are relatively thin-bedded compared with Facies II types of sand (i.e., below seismic resolution) | Serrated |
| | | Intermediate GR values |
| IV Silty shales | Silly shales and thin-laminated silt shale couplets. (In Rock Physics often referred to as "sandy" shales) | Serrated high GR values |
| V Pure shales | Pure shales, often seen as thick, massive shale masses | Serrated/smooth very high GR values |
| VI Chaotic deposits | Syn-depositional deformation units, slide blocks, slump deposits, injection sands, shale diapirs, etc. | Serrated/complex |

Adapted from Avseth, P., Combining rock physics and sedimentology for seismic reservoir characterization of North Sea turbidite systems. Unpublished PhD thesis, Stanford University, 2000.

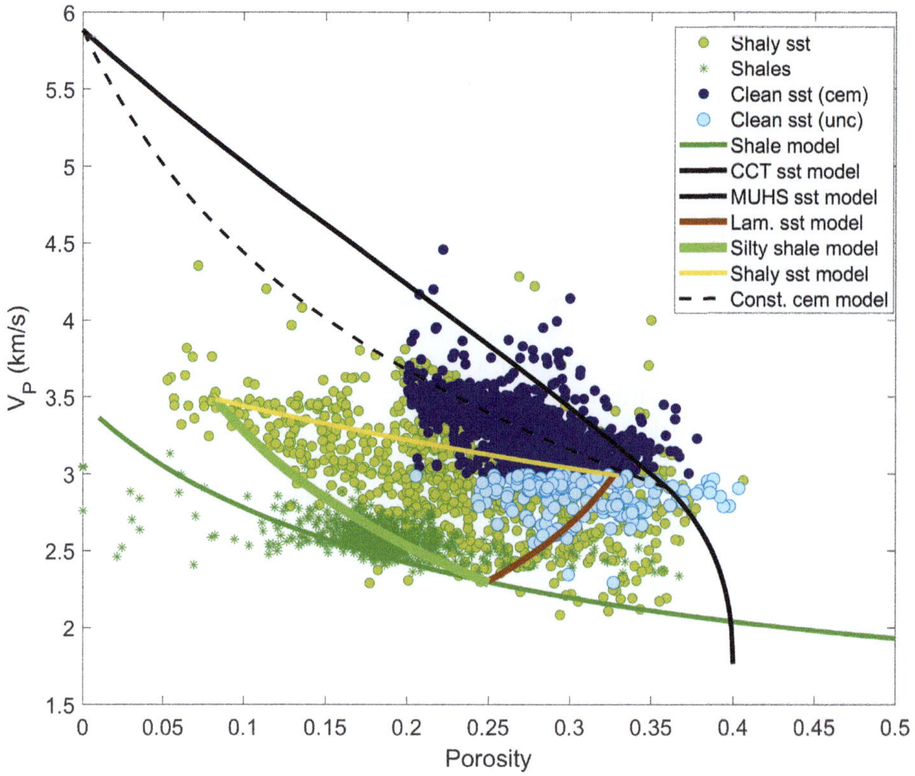


**FIG. 9** Seismic lithofacies clustering in $V_P$ vs porosity domain for the Alvheim wells. *Dark blue* is facies IIa according to the scheme of Avseth et al. (2000) and is the dominant facies type. Also less cemented facies IIb–IIc are present (*cyan colors*), and facies III (*light green*, heterolithics). The cap-rock Lista Formation is also included in the cross plot and represents facies IV–V (silty shales, *dark green*).

thick-bedded sandstone facies. Fig. 9 shows different facies for the Alvheim Field data, including clean sandstones with clay content less than 10% (Facies II according to Table 1), shaly sandstones (Facies III) with clay content larger than 20% and shales (Facies IV–V). The clean sandstones have been split up into cemented and unconsolidated populations (IIa vs IIb), guided by the rock physics diagnostic models (cf., Avseth et al., 2000). Again, it can be observed that the ambiguities in P-wave velocity for the shaly sandstones and the clean sandstones, can be resolved by utilizing the shear wave information and $V_P/V_S$ ratios (see following section).

The key take-away with the combined use of rock physics modeling and definitions of seismic lithofacies, is the benefit of understanding the wide scatter in the data, reducing the ambiguities associated with overlapping data and grouping the scatter into populations of data that can be associated with predictive geological patterns and trends in the seismic data. This will be beneficial both for the conventional two-step inversion and for one-step machine learning classification of seismic data, as shown in the following section.

## Rock physics templates for quantitative interpretation of deepwater clastic systems

In the following section, the use of rock physics templates will be demonstrated, with examples from the Måløy Slope and the Alvheim Field areas. Rock physics templates were introduced by Ødegaard and Avseth (2004) as site-specific rock physics models cross-plotted in terms of independent seismic observables, for instance, $V_P/V_S$ vs acoustic impedance (Buland et al., 1996), acoustic impedance vs elastic impedance (Connolly, 1999), near-angle impedance vs far-angle impedance (Mukerji et al., 1998) or lambda-rho vs mu-rho (Goodway et al., 1997). In the examples below, rock physics templates are created in terms of $V_P/V_S$ vs acoustic impedance.

### Rock physics templates and AVO analysis: Måløy Slope, North Sea

Fig. 10 shows the site-specific rock physics templates for the three wells studied in the Måløy Slope area. Note how the template models are updated from one well to the other. For the Cara discovery well (36/7-4), a constant cement model of 3% is used as determined from the rock physics diagnostics in Fig. 4. The fluid substituted models for oil and gas are also included. The oil in the Cara discovery is around 40 API (American Petroleum Institute) and hence a light oil. The measured (in situ values) well log data are superimposed on the templates, and a good match is observed between the data and the

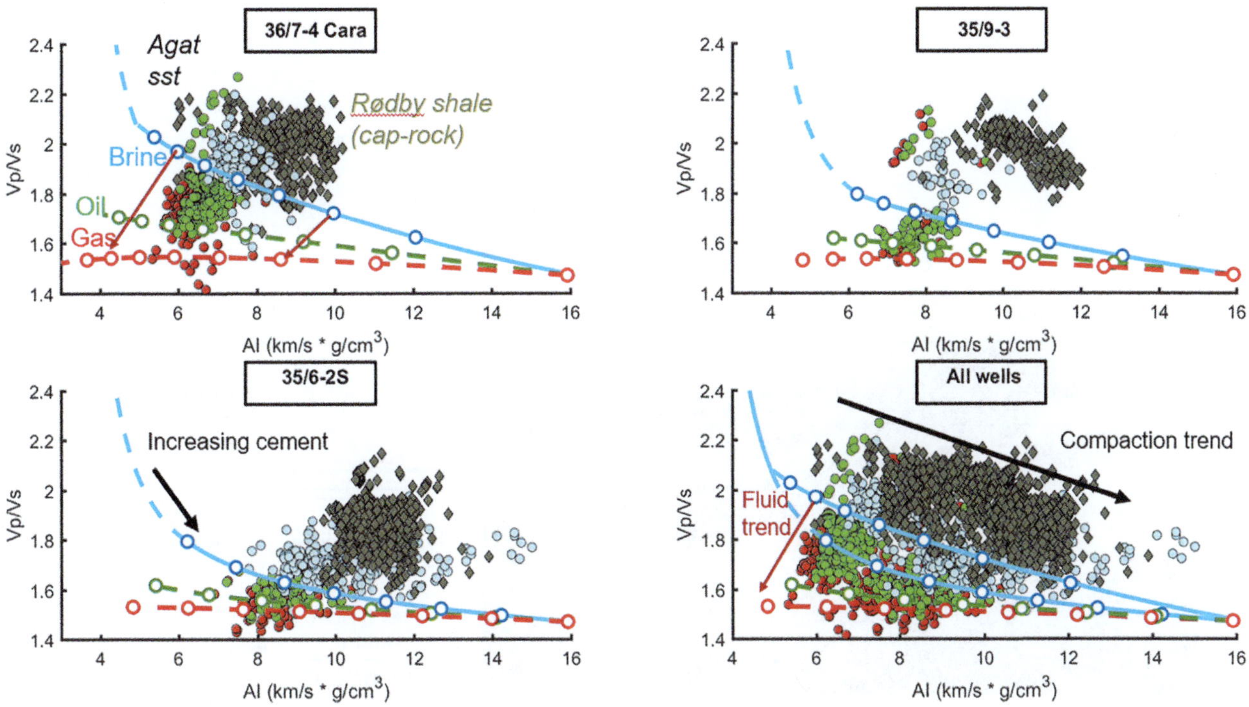


**FIG. 10** Rock physics templates for the different wells in the Måløy slope area (Agat Formation sst capped by Rødby Formation shales). Note the updated templates with different wells, as the reservoir is more deeply buried, and the rock physics templates include more cement. Also note that the brine sandstone data in wells 35/9-3 and 35/6-2S plots further away from the modeled sandstone trend, due to larger variability in sorting and clay content in the reservoir sandstones, than in well 36/7-4.

models. The Rødby cap-rock shale is also superimposed and shows acoustic impedances slightly higher than the brine-filled sandstones. The marly content of the cap-rock shale results in quite low $V_P/V_S$ ratios (around 2.0), and we do not expect large contrasts in the AVO gradients at the interface between the cap-rock shales and the hydrocarbon-filled reservoir.

For the two more basinward wells (35/9-3 and 35/6-2S), the template models are updated using an increasing cement volume model (Contact Cement Theory combined with modified upper bound Hashin-Shtrikman, CCT+MUHS, see Chapter 16 and Avseth et al., 2005). The updated template models fit accurately with the observed data in these two wells also (note that hydrocarbon values for 35/6-2S have been predicted using Gaussmann theory, as this is a dry well). Superimposing data from all three wells in the area, and the different template models (Fig. 8, lower right subplot), one can see that the data fall on an increasing compaction trend (from upper left to lower right in the template). Moreover, the shales follow a similar trend. Note that the oil sandstones in Well 36/7-4 fall between the constant cement model of 3% and the increasing cement model for brine. This observation highlights an inherent ambiguity between diagenetic trends and fluid trends in the seismic domain, where a combination of acoustic impedance and $V_P/V_S$ can represent two different scenarios, for example, a moderately cemented and well-sorted sandstone with brine, or a poorly cemented and moderately sorted sandstone with oil. This demonstrates the importance of having rock physics templates that honor the local burial history and can be updated from one location to the other. For instance, this could be the case if there is a significant structural dip in the stratigraphy, like in the Måløy Slope area, or if there has been laterally varying tectonics impacting the rock properties (see Barents Sea example by Lehocki et al., 2020).

The rock physics templates in Fig. 10 can be used to interpret and better understand observed trends in well log data, but also to say something about various "what if" scenarios. The template is a guide to say what happens when various geological parameters change, including soring, cementation, net-to-gross, fluid saturation, pressure, etc. Moreover, the rock physics templates can also be used as a guide or tool during the quantitative interpretation of AVO inversion data (Avseth et al., 2016). The rock physics templates can be used to generate various rock physics attributes that can be estimated directly from AVO inversion data, and then plotted as seismic attributes (Avseth et al., 2013; Avseth and Veggeland, 2015). An example from the Cara discovery is shown in Fig. 11 (Avseth et al., 2017), where the AVO inversion was performed predrill as a risk assessment tool. The fluid impedance is an attribute that varies with distance away from the constant cement model, along the Gassmann predicted fluid trend in a rock physics template. Fig. 11 shows how this attribute is separated between different fluids. Moreover, the shaly facies will have fluid impedance values higher than brine

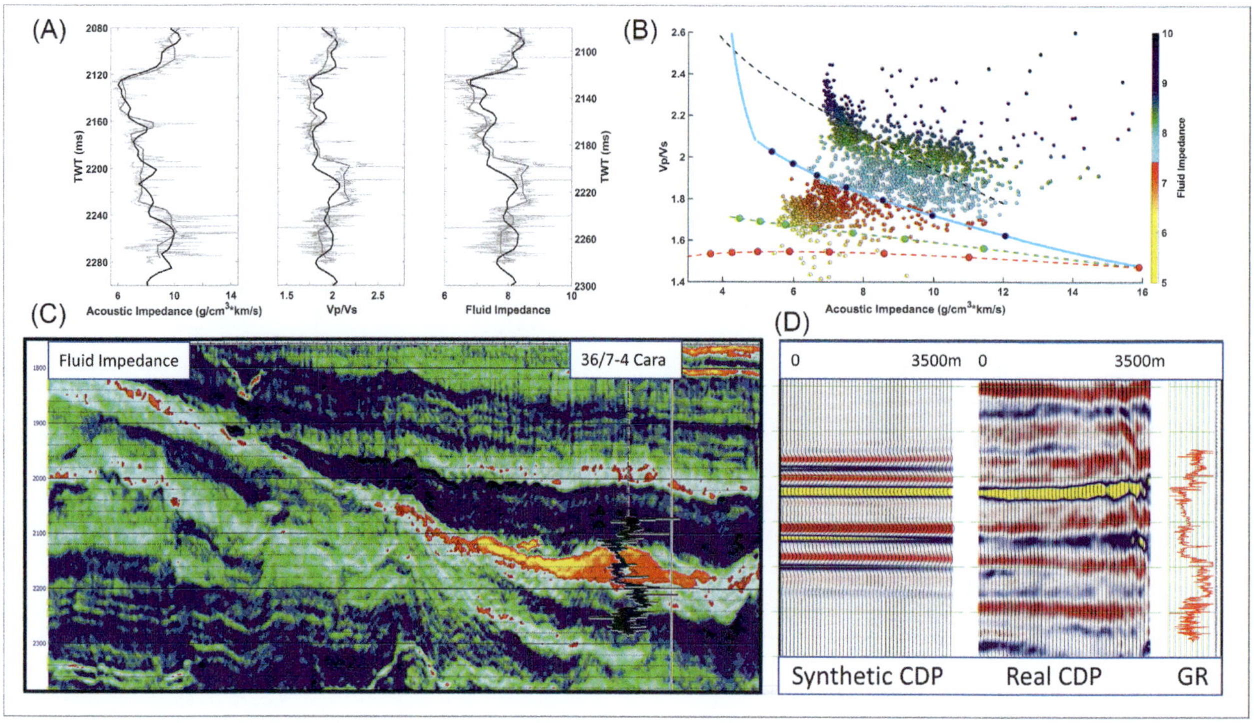


**FIG. 11** Applying rock physics templates (B) to guide the quantitative interpretation of AVO inversion data (A) along a seismic section (C) intersecting well 36/7-4 (The Cara discovery). The inversion and fluid impedance analysis (cf., Avseth and Veggeland, 2015) was done pre-drill, and the results from the RPT analysis were used to derisk the prospect (Avseth et al., 2017). AVO analysis of pre-stack gathers was also performed pre-drill (D), and the Top Agat Formation filled with gas (or light oil) was found to give a class 4 AVO anomaly with "flat" gradient (*yellow color* represents strong negative amplitude, *blue* is weaker negative and *red* is positive amplitude). The Cara discovery (now Duva field) represents a pinch-out stratigraphic deepwater clastic trap.

sandstones. Note that the template is fine-tuned to be valid for the Cara area, where the Agat Formation sandstones are expected to have around 3% cement volume. If the template was calibrated to a deeper area with higher cement volume, the oil zone could easily be interpreted to be brine-filled.

## Rock physics templates and AVO analysis: Alvheim Field, North Sea

Next, the use of rock physics templates and AVO classification for the Alvheim Field is demonstrated, using data from the wells in Fig. 6. Fig. 12 shows a rock physics template for the Alvheim Field with superimposed data from four of the six wells in Fig. 6 that have measured $V_S$ available. The data have been grouped into brine, oil and gas-filled sandstones. The Lista Formation shale properties are also superimposed, representing the cap-rock. As opposed to the Måløy Slope area, the shale properties show softer acoustic impedance than the brine-filled sandstones, whereas gas sandstones will have overlapping/similar acoustic impedances relative to the cap-rock shales. Furthermore, the $V_P/V_S$ contrasts are quite large between the sandstones and shales, as opposed to the Cretaceous case above. This is reflecting a relatively clay-rich, pelagic shale-type in the Alvheim Field, draping the basin-floor fans, whereas in the Måløy Slope the cap-rock shales are more marly, hemipelagic and likely more silty. From a seismic perspective, this difference illustrates the importance of also understanding the expected variability of the cap-rock shales, not only the reservoir sandstones.

Another interesting observation in Fig. 12 is that some of the oil sands have lower acoustic impedance than the gas sandstones and similar $V_P/V_S$ ratios as most of the brine sandstones. These represent the more heterogeneous reservoirs filled with oil (both in 26/4-2 oil zone, and the uppermost part of the oil zone in 25/4-7) and can be difficult to detect in seismic data. The heterogeneity can also induce anisotropic effects on the seismic response, and they can yield patchy saturation patterns at the seismic scale. Rock physics templates can be modified to honor both net-to-gross effects and anisotropy effects (Avseth et al., 2010; Golikov et al., 2013), but this is not done in this study.

Finally, an example of AVO classification is demonstrated on the Alvheim Field (Fig. 13). The rock physics templates and the statistical rock physics of different data populations can be used to group data clusters and perform AVO classification at the seismic scale. Based on calibrated AVO data (two left subplots in Fig. 13), combined with training data from

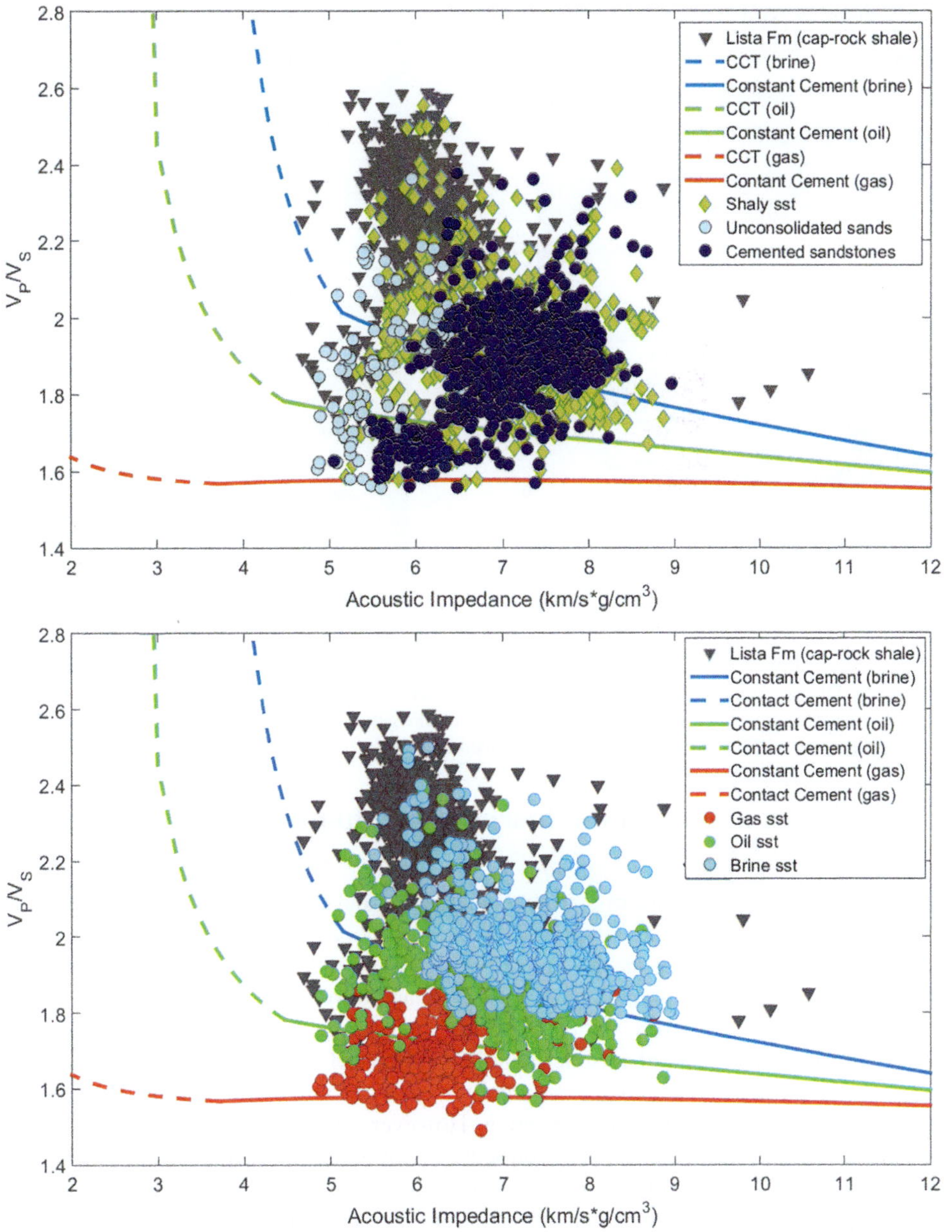


**FIG. 12** Rock physics template for Alvheim Field, including models for brine, oil and gas sandstones. (*Upper subplot*) Data colored by seismic litho-facies. (*Lower subplot*) Heimdal Formation sandstone data colored by pore fluids. The data shows a good separation between the different fluid scenarios. However, some of the oil sandstones represents heterolithic reservoirs and show properties falling close to or above the brine sandstone model (see also Avseth et al., 2009).

upscaled well log data, the AVO attributes can be converted into most likely facies/fluid class (Avseth et al., 2001a, b; Mukerji et al., 2001a, b; Avseth and Lehocki, 2016). The resulting facies and fluid prediction maps, using a quadratic distance classifier, are also shown in Fig. 13, where the location of the six wells is indicated. Overall, the classification results fit with what is observed in the wells in Fig. 6. For instance, most likely gas is predicted in Wells 24/6-4 and 24/6-2, and most likely oil is predicted in Well 25/4-7. However, as seen in the fluid map, gas is predicted in the lobe penetrated by well 25/4-7, just to the east of the well, yet this lobe has been proven to be only oil-filled. This shows the ambiguities in the seismic data. Rimstad et al. (2012) used rock physics depth trends and spatial statistics in a Bayesian inversion scheme and successfully predicted only oil in the lobe penetrated by 25/4-7. Simple linear methods will not easily resolve nonuniqueness and nonlinear behavior during the classification. This highlights the need for extensive use of nonlinear machine learning techniques during the classification of facies and fluids from seismic AVO data in deepwater clastic systems.

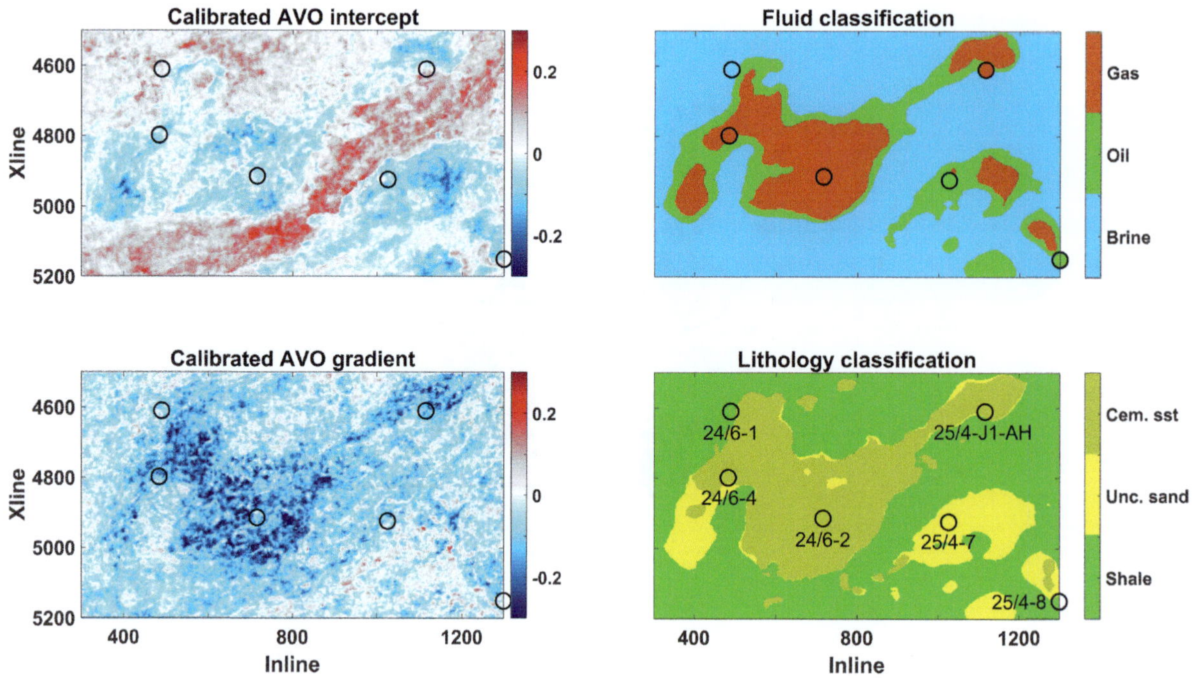


**FIG. 13** Rock physics-based QI analysis of the Alvheim Field. AVO classes for different facies and fluid scenarios (*right subplots*) are determined based on data observations (Fig. 10) and the most likely facies and fluids are predicted along the Top Heimdal horizon from calibrated AVO data (intercept and gradient, *left subplots*). The gas and oil-filled lobe structures are nicely identified in the data.

## Statistical rock physics and machine-learning for quantitative seismic interpretation

Rock physics models provide the link between geology and geophysics. They help us to quantify the trends of velocity-porosity as a function of geologic processes of sorting, cementation and diagenesis. But as any other model, these rock physics models are idealizations and approximations and are not perfect, as we can see from the actual data scatter around the theoretical model curves in the figures above (e.g., in Figs. 4 and 5). Different clusters of data (e.g., representing different pore fluid scenarios) have separation but there is also overlap between the clusters (e.g., Fig. 12). These are due to intrinsic heterogeneity of the rock properties, and these heterogeneities occur at different scales (see Chapter 1). Ignoring the scatter and overlap can lead to risks of misinterpretation. Instead of using deterministic physics-based (imperfect) models one could go to the other extreme and depend purely on data (imperfect) and multivariate statistical and machine learning techniques for quantitative seismic interpretation. However, the training data used in these machine learning methods may be incomplete and may not be fully representative of all of the scenarios one might expect to encounter in the targeted reservoir. We feel that the optimum strategy is to combine the best of each of these tools to generate better results than would be possible from purely statistical or purely deterministic techniques alone. One important role of physics-based models in machine learning workflows is to augment the training data to account for scenarios expected in the reservoir but not represented in the original training data. For example, we might have a well in the water leg of a sandstone reservoir, and we have the distributions and training data (velocities, impedances, etc.) only for the shales and the brine sands. We need to augment this data and derive the distributions for the oil sands using a physics-based model (in this case the well-known fluid substitution equation of Gassmann's model) to have a more complete representative training dataset that can then be used for machine learning. Derived distributions and augmentation of training data plays an important role in applications of machine learning for geosciences.

### Overview of some machine learning algorithms

Machine learning (ML) algorithms can generally be categorized into unsupervised and supervised learning algorithms. The former class of algorithms is applicable to datasets where there is no specific notion of separation between input and output variables. In machine learning parlance this is called an unlabeled dataset. In the absence of explicit output labels, the unsupervised ML algorithm extracts features identifying patterns or clusters that are intrinsically present in the input data. These features are subsequently used to make inferences about the geologic variable of interest such as seismic facies and stratigraphy (de Matos et al., 2007; Zhao et al., 2015; Bugge et al., 2020), direct hydrocarbon indicators and thin beds (Roden et al., 2015, 2017). Supervised learning algorithms, on the other hand operate in settings where there is the notion of input and output variables and the objective is to learn a functional relationship between the two. Let $\boldsymbol{x}$ denote the input or independent variable, and $\boldsymbol{y}$ denote the output or dependent variable. Variable $\boldsymbol{y}$ can represent discrete valued variables such as lithofacies or continuous valued variables such as porosity. Problems with categorical or continuous outputs are referred to as classification or regression problems, respectively. In both these learning problems, the objective is to learn the relationship

$$\boldsymbol{y} = \widehat{g}_w(\boldsymbol{x}) \tag{1}$$

using the training set $\{(\boldsymbol{x}_1, y_1), (\boldsymbol{x}_2, y_2), .., (\boldsymbol{x}_n, y_n)\}$ comprising of example input/output pairs of $\boldsymbol{x}$ and $\boldsymbol{y}$. The subscript $\boldsymbol{w}$ denotes that the function $\widehat{g}_w(.)$ is specified using parameters $\boldsymbol{w}$, which are unknown at the outset and are learned by minimizing a cost or penalty function calculated on the training examples. For regression problems, the least squares cost function

$$J(\boldsymbol{w}) = \frac{1}{2}\sum_{i=1}^{n}\left(\widehat{g}_w(\boldsymbol{x}_i) - y_i\right)^2 \tag{2}$$

is frequently employed. In other words, the goal is to find $\boldsymbol{w}$ such that the outputs predicted by the ML model $\widehat{g}_w(\boldsymbol{x}_i)$, corresponding to a training set example input $\boldsymbol{x}_i$, is as close as possible to the true output value $y_i$. In many classification problems, the machine learning model is designed to output the probability of outputs conditional to the inputs, denoted as $p_w(y\,|\,\boldsymbol{x})$. In these cases, parameters $\boldsymbol{w}$ are learned by maximizing the likelihood of the training examples

$$L(\boldsymbol{w}) = \prod_{i=1}^{n} p_w(y_i|\boldsymbol{x}_i) \tag{3}$$

where $L(\boldsymbol{w})$ is the likelihood function calculating the total probability of all the training examples under the ML model. Note that Eq. (3) assumes that each training example is independent of all others, and consequently the total probability can be expressed as a product over all the examples in the training set.

## Deep learning model architectures

Decades of research in the ML community have led to the development of several ML models, such as linear regression and classification models, kernel-based models and artificial neural networks, which may be trained to capture simple to complex relationships exhibited by the training data (Bishop, 2006). In recent years, the field of deep learning models in particular have witnessed tremendous amount of research interest and advances (Goodfellow et al., 2016) due to their ability to approximate complex non-linear relationships in high-dimensional spaces. The discussion of deep learning models in this chapter is limited to deep artificial neural networks (ANNs) and convolutional neural network layers (CNNs), which have been shown to be successful in a wide variety of geoscientific prediction tasks (Wu et al., 2019; Yang and Ma, 2019; Das et al., 2019; Pradhan et al., 2020).

In ANNs, the functional model $\widehat{g}_w(.)$ is specified by stacking together multiple hidden neural network layers. The basic building block of an ANN layer is the so-called neuron or perceptron, which applies the nonlinear transformation

$$y = \sigma(\boldsymbol{w}.\boldsymbol{x} + b) = \sigma\left(\sum_{j=1}^{m} w_j x_j + b\right), \tag{4}$$

to inputs $\boldsymbol{x}$. Here, $\boldsymbol{x}$ is the $m$-dimensional vector $[x_1, x_2, .. x_m]^T$. Learnable parameters of the neuron consist of the weight parameter vector $\boldsymbol{w} = [w_1, w_2, .. w_m]^T$ and bias term $b$. In the above, $\sigma(.)$ denotes a nonlinear function that in neural network terminology is called an activation function. Commonly employed activation functions include the sigmoid function, $\sigma(x) = 1/(1+e^{-x})$, the hyperbolic tangent function, $\sigma(x) = (e^x - e^{-x})/(e^x + e^{-x})$ and the rectified linear unit (ReLU), $\sigma(x) = \max(0, x)$. A single hidden ANN layer contains multiple neurons, each taking as inputs the outputs from the previous layer (Fig. 14). The number of hidden layers and the number of neurons in each layer are typically treated as hyper-parameters of the network and appropriate values may be assigned by a hyper-parameter tuning process during training. Although increasing the size of the hidden layers and depth (number of hidden layers) increases capability of the network to learn complex functions, it also makes the network prone to overfitting (Goodfellow et al., 2016) to noise and outliers present in the training dataset. Regularization techniques, described in detail later, may be employed to tackle overfitting during training of the network.

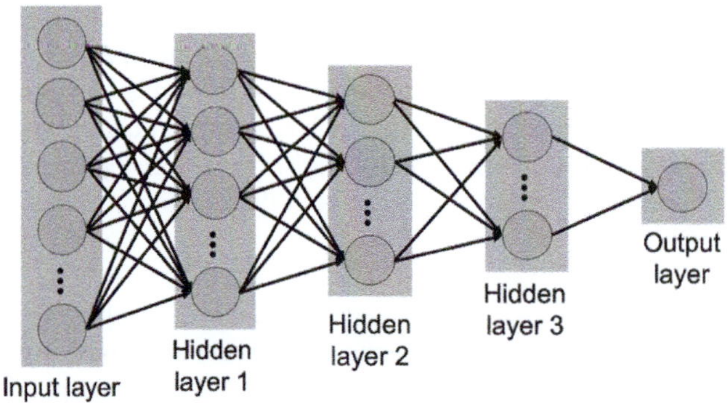


**FIG. 14** ANN architecture. Each hidden layer consists of multiple neurons (*circles*). Each neuron takes as input the outputs of all units in the previous layer.

A limitation associated with ANNs is that they require inputs in vectorized form. For example, making predictions with 3D seismic data requires the data to be flattened into a 1D vector, which then serves as input to the ANN. If the input 3D seismic cube has dimensions $100 \times 100 \times 100$, then a single neuron in the first hidden layer will have $10^6$ weight parameters. This requirement causes ANNs to demonstrate poor scalability to high-dimensional datasets. In addition, vectorizing the input makes it difficult to extract features capturing any spatial dependencies present in the input. Convolutional neural networks (CNNs) address these limitations by forcing the hidden layers, termed convolutional layers, to possess spatial dimensions such as width, height and depth. The relationship between the inputs and outputs is modeled by multiple hidden convolutional layers. Units in any specific hidden layer are not fully connected to all units of the previous layer as is the case in ANNs, rather a small subset of the units through convolutional filters or kernels. Fig. 15 shows pictorially how a convolutional filter operates on a windowed region of an input hidden layer. The filter performs a convolutional operation to obtain a single pixel of the subsequent layer

$$y^R = \sigma\left(\sum_{j=1}^{n_k} w_j x_j^R + b\right), \tag{5}$$

where $\{w_j; j=1,..,n_k\}$ denote the learnable weight parameters of the kernel, $n_k$ is the total number of parameters in the kernel, $b$ is the bias term and $\sigma(.)$ is the nonlinear activation function. Notation $x_j^R$ is used to denote $j$th pixel within the

**FIG. 15** Depiction of convolution operation in CNNs. Network layers are shown as *gray cubes* and convolutional filters are shown in *yellow and orange*. Outputs of the convolution operation are shown in *blue*. Output at a different pixel is obtained by spatially translating the convolutional kernel (*left and right plots*). Multiple kernels are used to obtain the channel dimension in the output layer (*top and bottom plots*).

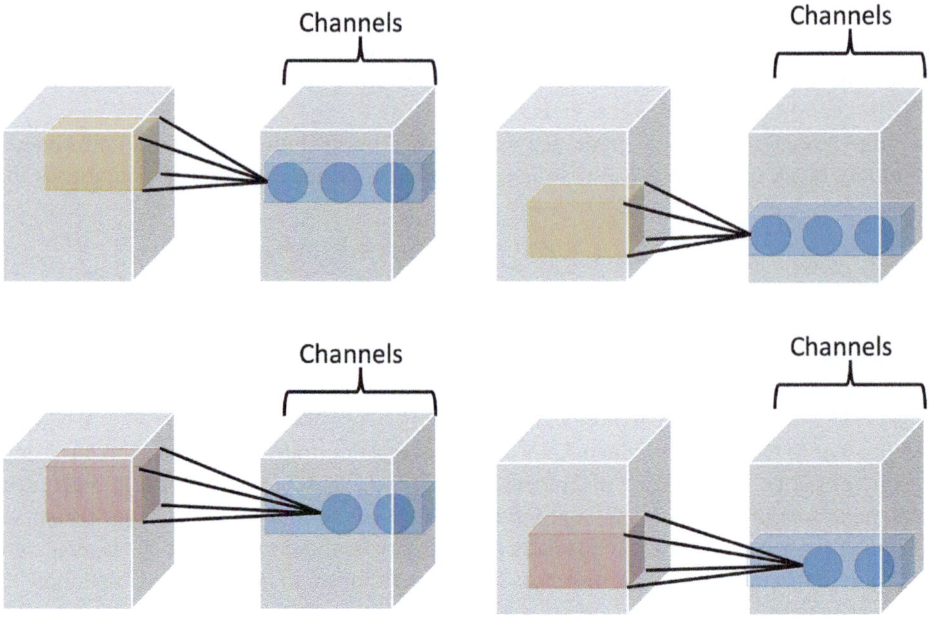

windowed region indexed by $R$. To obtain the output at another pixel, the filter is spatially translated, as shown in the right plot Fig. 13, and Eq. (5) is applied to the different region of the hidden layer. Every hidden layer can have multiple convolutional kernels, providing outputs along the channel dimension of the subsequent layer (see top and bottom plots of Fig. 15). Each kernel potentially activates according to distinct features present in the input layer. Stacking multiple convolutional layers and employing multiple kernels in each layer lends CNNs the capability to extract multiple complex feature representations from the input.

In addition to the number of convolutional layers and kernels, design of CNN architecture requires specification of various other parameters such as the kernel size, kernel stride and the type of convolution operation. Kernel size refers to the size of the windowed region over which the kernel operates while stride refers to the step length by which the kernel is spatially translated. Apart from convolutional layers, pooling layers, such as max and average pooling, and upsampling layers, such as transposed convolutional layers, are frequently employed in the CNN architectures. These layers can be used to reconcile any difference in dimensions of the input and output volumes. Dumoulin and Visin (2016) discuss the arithmetic of convolutional layers and pooling layers in detail. Krizhevsky et al. (2012) and Long et al. (2015) demonstrate how these layers can be employed to design popular deep CNN architectures that have been shown to be successful in a wide variety of image classification tasks.

## Training deep learning models

Training modern deep learning models is not a trivial task given the large number of learnable parameters and hyper-parameters involved that can lead to issues such as overfitting of the network to training data. A popular practice is to divide all of the training examples into three sets: training, validation and test sets. The training set is used to optimize the network parameters while the validation set is used for identifying overfitting and hyper-parameter tuning. The test set, kept completely blind during the entire training process, serves as a final check against overfitting and assessing prediction accuracy. Learning of the network parameters proceeds by optimization of the specified objective function (such as Eqs. 2 or 3) using the training set. The training procedure commences by assigning some initial values to all the parameters in the network. Glorot and Bengio (2010) discuss a popular method of parameter initialization in neural networks. Subsequently, the optimization is performed using gradient descent methods. Specifically, the parameter values are updated in the direction of the gradient of the objective function with respect to the parameters. Calculation of gradients in neural networks is fairly simple to accomplish by backpropagation (Bishop, 2006). Several gradient descent optimization algorithms effective for training deep learning models are available in literature (for the interested reader, the optimization algorithm used here is the Adam algorithm proposed by Kingma and Ba, 2014).

Parameter updates are performed iteratively, by making repeated passes through the training set. A single pass through the entire training set is termed an epoch. Training is generally required to be performed over multiple epochs until acceptable levels of prediction errors are obtained. In many cases, the network starts to overfit to the training set after a certain number of epochs. The validation set is employed to identify such phenomenon. Specifically, the model's performance on the validation set after every training epoch is evaluated. At the end of the training, network parameter values for the epoch with optimal validation set performance are retained. This procedure is referred to as early stopping of the training. Regularization techniques, such as $\ell_2$ or $\ell_1$ regularization and dropout (Srivastava et al., 2014) are also commonly employed to tackle overfitting in deep learning networks. These methods either add some form of penalty to the loss function or make architectural modifications such as to prevent the network from becoming overly complex during training. For instance, $\ell_2$ regularization penalizes the loss function if the network weight parameters become too large, restricting the network's ability to overfit data.

## Machine learning for deepwater reservoirs: Opportunities and challenges

The need for data-driven characterization of deepwater reservoirs with machine learning arises due to the various challenges associated with traditional physics-based inversion methodologies (Tarantola, 2005; Bosch et al., 2010). Given the large amount of uncertainty associated with geological understanding of deepwater systems, inversion methods are typically posed in a Bayesian formulation, where uncertainty on variables of interest is modeled with probability density functions $f(.)$. Let the target variable of interest be denoted by $\boldsymbol{y}$ and data variable by $\boldsymbol{x}$. Variable $\boldsymbol{y}$ could be reservoir litho-facies or porosity while $\boldsymbol{x}$ may represent seismic data. Bayesian inversion methods proceed by specifying prior geological beliefs on the target variable through prior distribution $f(\boldsymbol{y})$. Given data $\boldsymbol{x}_{obs}$ that is observed in the field, prior uncertainty on $\boldsymbol{y}$ can be updated by Bayes' rule to give the posterior uncertainty on $\boldsymbol{y}$:

$$f(\boldsymbol{y}|\boldsymbol{x}_{obs}) = \frac{f(\boldsymbol{x}_{obs}|\boldsymbol{y})f(\boldsymbol{y})}{f(\boldsymbol{x}_{obs})}. \tag{6}$$

Here, $f(\boldsymbol{y}\,|\,\boldsymbol{x}_{obs})$ and $f(\boldsymbol{x}_{obs}\,|\,\boldsymbol{y})$ denote the posterior and likelihood distributions respectively. The likelihood distribution expresses the probability of observing the data given any sample from the prior. The data likelihood may be computed with a physical forward model relating $\boldsymbol{x}$ and $\boldsymbol{y}$

$$\boldsymbol{x} = g_{physics}(\boldsymbol{y}). \tag{7}$$

The posterior distribution contains updated estimates of uncertainty with information contained in data $\boldsymbol{x}_{obs}$. Evaluating the posterior distribution in Eq. (6) is not trivial to accomplish in typical deepwater reservoir characterization applications. To realistically capture the complex stratigraphic architectures and heterogeneities encountered in deepwater reservoirs, it becomes necessary to specify prior $f(\boldsymbol{y})$ by geologically realistic geostatistical models (Linde et al., 2015) such as training-image based, object-based or surface-based models (Mariethoz and Caers, 2014; Pyrcz and Deutsch, 2014). These models do not have an analytical formulation for the underlying probability density function $f(\boldsymbol{y})$, thus requiring the inversion to be accomplished with Monte Carlo-based sampling methods (Tarantola, 2005). A limitation is that Monte Carlo methods become computationally demanding in high-dimensional problems. This issue is further exacerbated by the fact that for object-based and surface-based models, they are not easily conditioned to spatially dense seismic data (Pyrcz and Deutsch, 2014).

The advent of deep learning algorithms presents special opportunities to address the abovementioned shortcomings. Recent research advances have made it possible to train very deep neural networks without overfitting. As discussed previously, very deep networks can model highly complex relationships in high-dimensional spaces, thus making them well-suited to model complex deepwater reservoir architectures while honoring dense seismic data. In addition, training of deep learning models can be performed efficiently by gradient-based optimization methods. However, application of deep learning models to seismic problems comes with its own set of challenges. A common challenge relates to the lack of labeled training data, especially in 3D applications. Note that 3D seismic data is typically collected once in a field and the true labels of earth properties are available only at a sparse set of well locations in the domain of interest. The applicability of 4D seismic data will generally be limited since 4D monitor surveys are typically conducted a few times during the production cycle of a reservoir while the large-scale deep learning networks being considered would typically require few hundred examples for reliable training. Thus, it becomes difficult to form the training set containing repeated samples of the inputs and outputs. Several authors (Das et al., 2019; Pradhan et al., 2020, chapter 3) have proposed to address this issue by synthetic modeling of the training dataset. The general approach is to randomly sample the prior distribution $f(\boldsymbol{y})$ and model the corresponding $\boldsymbol{x}$ with a physics-based forward model (Eq. 7). In other words, the prior geological uncertainty space is explored, and the data are generated consistently with the underlying physical model. Deep learning model $\widehat{g}_w(.)$ trained with this dataset is used to make predictions with the real data $\boldsymbol{x}_{obs}$. To make reliable predictions with machine learning models trained on synthetic datasets, it is necessary to employ strategies such as prior falsification and error modeling. These are discussed in greater detail in the next section, where an application of reservoir facies estimation from 3D seismic data with deep CNNs from an offshore deltaic reservoir from Nile Delta is presented.

## Deep learning facies classification: Nile Delta case study

The goal of this case study is to demonstrate how deep CNNs may be used to make automatic predictions of reservoir facies models from seismic data. The target reservoir is a gas-producing reservoir in offshore Nile Delta located at a depth of 2100m. Presence of amalgamated, sinuous and leveed channels in the reservoir zone, as part of paleo slope-channel systems of Plio-Pleistocene age, along with thinly bedded sandstones and limestone, have been identified by previous seismic geomorphological analyses in the study area (Cross et al., 2009). The analysis is presented on an area with a spatial extent of 5km along $x$ and $y$ directions. The thickness of the reservoir zone varies spatially with an average thickness of 250m. The zone of interest was discretized into 100, 100 and 250 cells along $x$, $y$ and depth dimensions. Petrophysical logs were available at seven wells in the study area. Wells 1–4 were used in specifying the prior distribution of the model parameters and wells 5–7 are kept blind. Seismic data available in the zone of interest consist of 3D time migrated post stack, near stack and far stack volumes (Fig. 16). The data, with a dominant frequency of 18Hz, are sampled on a grid having 100, 100

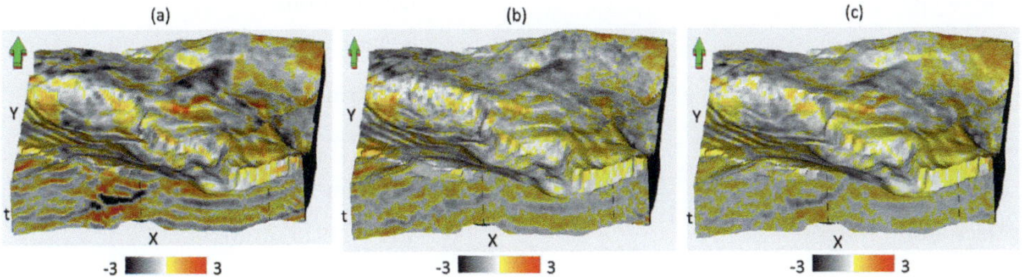


**FIG. 16** 3D (A) post-stack, (B) near stack and (C) far stack seismic volumes.

and 60 cells along $x$, $y$ and time dimensions. The objective is to predict the facies $\boldsymbol{y}$ from observed seismic data $\boldsymbol{x}_{obs}$ with CNNs.

## Creation of training dataset for machine learning

To realistically capture the complex channelized geometries of the reservoir elements, prior uncertainty on facies $f(\boldsymbol{y})$ is parameterized using an object-based geostatistical model. Geological understanding and prior geomorphological analyses of the reservoir architecture was used to guide the specification of the prior. Cross et al. (2009) have presented a detailed study of the depositional evolution, distribution and geometries of the stratigraphic elements of the reservoir system, based on analyses of 3D seismic data, core and well log data. Filling of the slope channel system can be broadly categorized into six stages. In the first stage, initiation of the reservoir system occurred primarily as a result of erosional downcutting of the Nile Delta slope through high density gravity flows and mass wasting along the channel margins. Deposition in the subsequent two stages was dominated by the formation of laterally amalgamated and aggradational sinuous channels. This was followed by the formation of channels with frontal splays and leveed channels, concentrated in the upper part of the reservoir. Presence of thinly bedded sands increasing the lateral connectivity of the channels was also identified. In the final stages, the channel system was sealed by the deposition of nonreservoir mudstones, siltstones and thin-bedded sands. Cross et al. (2009) identify the primary reservoir architectural elements likely to impact reservoir dynamic flow behavior to be laterally amalgamated channels, sinuous channels, channels with lateral and frontal splays, leveed channels and thinly bedded sands. Consequently, channel, splay, levee and thin bed facies along with background shale facies will be explicitly considered in this study. In the object-based modeling algorithm, leveed channels, splays and thin beds are modeled with sinusoid, lobe and ellipsoid objects, respectively. Table 2 shows the prior probability distributions on the various parameters controlling the geometry and shape of the facies geo-objects. Specification of these distributions was guided primarily by observations presented by Cross et al. (2009). For instance, channel bodies were identified to have a maximum thickness of 30m, consequently the prior distribution on the channel thickness parameter was assigned as a triangular distribution given by $\mathcal{T}(2\text{ m}, 15\text{ m}, 30\text{ m})$. Simulation rules were specified to generate models roughly mimicking the general reservoir architectural characteristics. For instance, individual sinusoidal objects were allowed to interact with each other such as to allow creation of laterally and vertically amalgamated channel complexes. In line with the observations of Cross et al. (2009), the spatial density of thin beds was specified to be greater toward the shallower part of the reservoir. Petrel commercial software was used to generate 2400 unconditional Monte Carlo simulations of object-based models representing the reservoir facies (Fig. 17). For channel objects, gas and brine fluid saturation scenarios are considered. The prior distribution for the depth of the gas-water contact (GWC) was modeled as a triangular distribution $\mathcal{T}(2415\text{ m}, 2430\text{ m}, 2450\text{ m})$ using information from water saturation logs available at the wells. For every facies realization, a sample of the GWC depth value is obtained from its prior and used to assign the fluid saturation scenarios in the channel objects.

To forward model the seismic angle stacks, realizations of P-wave velocity ($V_P$), S-wave velocity ($V_S$) and bulk density ($\rho_b$) are required. Some studies on rock physics modeling and probabilistic seismic petrophysical inversion have been conducted in this field (Aleardi and Ciabarri, 2017; Aleardi et al., 2018). Fig. 18 shows the global bivariate probability distributions for P-impedance and $V_P - V_S$ ratio as estimated from sonic, shear sonic, bulk density and interpreted petrophysical facies logs at the wells. The prior distributions on these properties are specified as multivariate Gaussian distributions for each facies. Variograms allow capturing the spatial correlation expected in the subsurface variability of the elastic properties (Avseth et al., 2005). Correlation lengths along vertical direction for $V_P$, $V_S$ and $\rho_b$ were determined for each individual facies by fitting the parametric variogram models to the experimental variogram calculated at wells. In the channel facies, $V_P$, $V_S$ and $\rho_b$ were found to have correlation length of approximately 10m while the corresponding

**TABLE 2** **Prior distributions on the parameters for facies objects.** $\mathcal{N}(\mu, \sigma)$: normal distribution with mean $\mu$ and standard deviation $\sigma$. $\mathcal{T}(a, c, b)$: triangular distribution with lower limit $a$, upper limit $b$ and mode $c$. $U(a, b)$: uniform distribution with lower limit $a$ and upper limit $b$.

| Geo-object | Parameters | Distributions |
|---|---|---|
| Channel | Global proportion | $\mathcal{N}(27\%, 6\%)$ |
| Channel | Amplitude | $\mathcal{T}(300\text{ m}, 400\text{ m}, 500\text{ m})$ |
| Channel | Wavelength | $\mathcal{T}(350\text{ m}, 600\text{ m}, 800\text{ m})$ |
| Channel | Width | $\mathcal{T}(100\text{ m}, 250\text{ m}, 450\text{ m})$ |
| Channel | Thickness | $\mathcal{T}(2\text{ m}, 15\text{ m}, 30\text{ m})$ |
| Channel | Orientation | $\mathcal{T}(240^0, 305^0, 310^0)$ |
| Levee | Width (fraction of channel width) | $\mathcal{T}(0.1, 0.3, 0.8)$ |
| Levee | Thickness (fraction of channel width) | $\mathcal{T}(0.3, 0.6, 0.9)$ |
| Splay | Global proportion (fraction of channel proportions) | $U(10\%, 50\%)$ |
| Splay | Minor width | $\mathcal{T}(180\text{ m}, 350\text{ m}, 700\text{ m})$ |
| Splay | Major to minor width ratio | $\mathcal{T}(0.7, 1.2, 2.2)$ |
| Splay | Thickness | $\mathcal{T}(2\text{ m}, 4\text{ m}, 9\text{ m})$ |
| Thin beds | Global proportion | $\mathcal{N}(12\%, 4\%)$ |
| Thin beds | Minor width | $\mathcal{T}(600\text{ m}, 800\text{ m}, 4000\text{ m})$ |
| Thin beds | Major to minor width ratio | $\mathcal{T}(0.8, 1, 1.2)$ |
| Thin beds | Thickness | $\mathcal{T}(1\text{ m}, 1.5\text{ m}, 2\text{ m})$ |
| Thin beds | Orientation | $\mathcal{T}(300^0, 305^0, 310^0)$ |

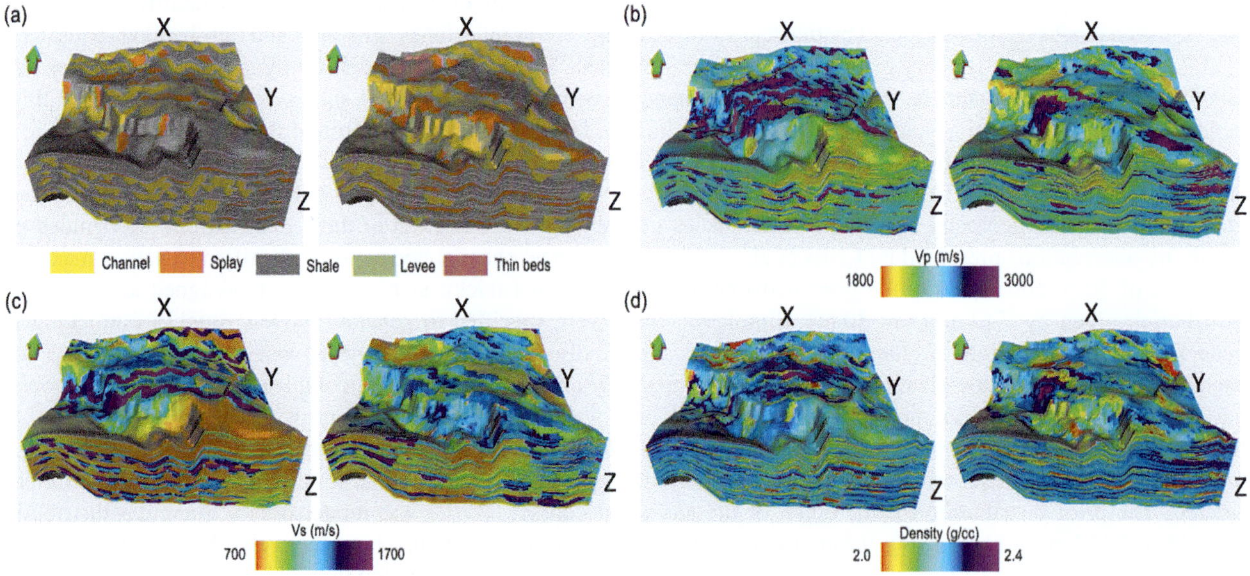


**FIG. 17** Prior realizations (2 out of 2400) of (A) facies, (B) $V_P$, (C) $V_S$ and (D) bulk density.

**FIG. 18** (A) Contour plots of the bivariate probability distributions for P-impedance and $V_P - V_S$ ratio for all facies. (B) Corresponding plots after lumping together all non-reservoir facies.

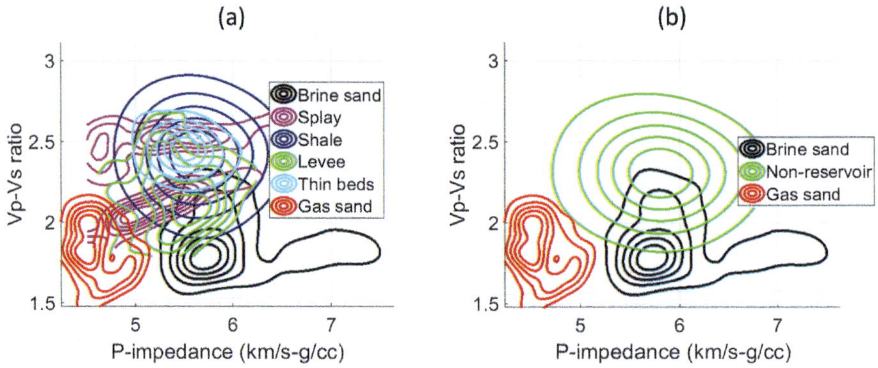

properties for shale facies had longer correlation lengths. Estimating the horizontal correlation length directly from well data is challenging due to absence of horizontal continuity in the data. Horizontal correlation ranges were assigned values between 1200 and 2500 m, using the ratio of the vertical to horizontal variogram range of the poststack seismic data as a rough guide. It was assumed that $V_S$ and $\rho_b$ of each facies are linearly correlated with $V_P$ as determined from well logs. Facies conditioned realizations of $V_P$ were generated by sequential Gaussian simulation (SGSIM; Deutsch and Journel, 1998). For each simulated $V_P$ realization, corresponding realizations of $V_S$ and $\rho_b$ (Fig. 17) were modeled by sequential Gaussian cosimulation (COSGSIM; Deutsch and Journel, 1998), honoring the facies-dependent interproperty correlations. Wavelets were extracted by seismic-well ties at well 1 and seismic angle gathers were generated using single-scattering forward modeling with the exact nonlinear Zoeppritz equation to generate poststack, near stack and far stack synthetic seismic data volumes (Fig. 19). The 2400 object-based facies realizations and their corresponding synthetic seismic angle-gathers comprise the training set for training the CNNs to perform facies classification at every pixel of the reservoir model grid with the observed seismic angle stack data. Note that the CNN is trained to predict three facies—gas sand, brine sand and nonreservoir facies—instead of the six facies modeled from the well data. Splay, levee, thin bed and shale facies

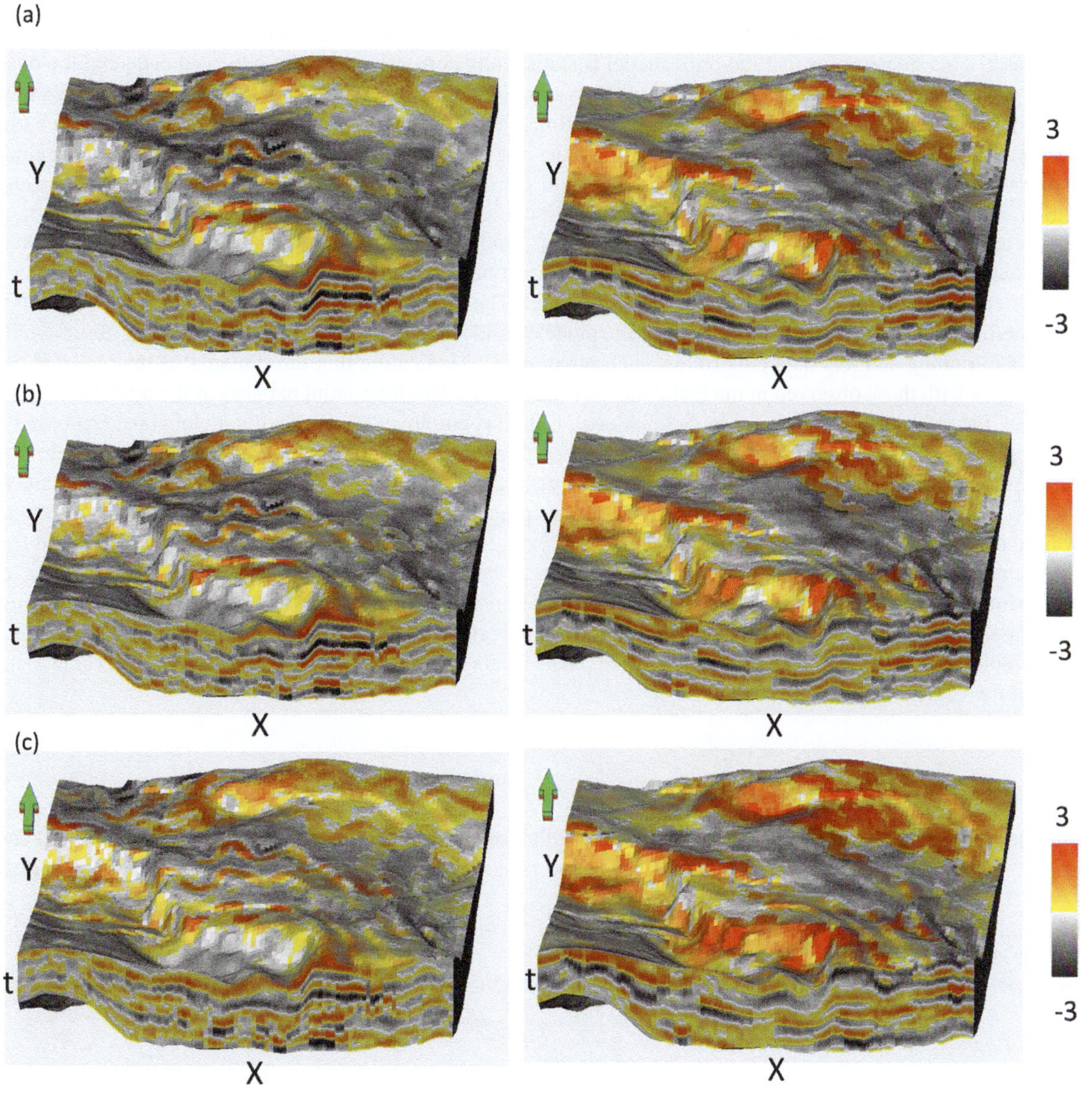


**FIG. 19** Forward modeled poststack (A), near-stack (B) and far-stack (C) synthetic seismic data for the two prior realizations shown in Fig. 15.

were lumped together into nonreservoir facies (Fig. 18). This was performed as well-log measurements showed a large amount of overlap between these facies in the P-impedance and $V_P - V_S$ domain, indicating that seismic data may not be able to individually distinguish the four nonreservoir facies.

The above approach facilitates generating as many realizations of the independent and dependent variables as required. While it may be possible to train the machine learning model with this synthetically generated dataset, the model may not provide reliable predictions with real data. This is especially due to the following factors: (1) imperfections arising out of simplifying assumptions in the geological and geophysical models, (2) noise present in the real data and (3) subjective decisions made during specification of the uncertainty model may be inconsistent with reality. The following steps are employed to ensure that the machine learning model performs reliably with real data: (1) modeling additive noise and (2) prior falsification. In the former, a probabilistic model for the noise arising due to modeling imperfections and noisy data is specified. This model is sampled to generate noise realizations, which are then added to the synthetic seismic forward model realizations. Intuitively, adding randomly sampled noise to the training dataset serves to confuse the network and prevents it from overfitting to the synthetic simulations. The residual traces or the mismatch between the synthetic and real traces, calculated from seismic-well tie at well 1, were used as a guide to specify the probability distribution for the noise variable. The noise is modeled as a spatially correlated multivariate Gaussian distribution and is added to the seismic synthetics.

Prior falsification (Scheidt et al., 2015, 2017) is the second step employed to account for the synthetic nature of the training dataset. This step ensures that the prior model is statistically consistent with the observed data, and the observed data are not outliers compared with the distribution of the synthetically generated data from the prior model. This analysis is necessary due to the subjectivity involved during specification of the prior distributions on the various uncertain variables. In the event that the prior distribution does not encompass the true subsurface model, the statistics of the prior realizations will deviate significantly from that of the true model. Since the true model is the variable being estimated and not available, falsification analysis is performed with the data variables. Specifically, the evaluation is made whether $\boldsymbol{x}_{obs}$ is an outlier with respect to relevant statistics of the data realizations $\{\boldsymbol{x}_i\}_{i=1}^n$ in the training dataset. In the event that the real data is an outlier, it would indicate that the prior distribution is inconsistent necessitating measures such as broadening the range of prior parameters or modifying the conceptual geological model. The first falsification analysis is made with regards to the global probability distributions of the amplitudes of the modeled and real angle stacks (Fig. 20). It can be observed that the contours of the training and real data pdfs follow each other closely, indicating that the statistics of the modeled seismic amplitudes agree with those observed in the real data. To establish whether the spatial patterns of the modeled geo-objects are consistent, the methodology proposed by Scheidt et al. (2015) is employed. Specifically, a global pattern-based quantitative comparison of the observed and training data is performed at multiple resolutions using 3D discrete wavelet transform (Mallat, 1989) and multidimensional scaling (Borg and Groenen, 1997). For every seismic volume, the above methodology extracts features capturing patterns at different frequencies and spatial directions. These features are subsequently used in the Mahalanobis outlier detection algorithm (Leys et al., 2018) to determine whether pattern-based features found in training examples conform with the patterns in the real data. Fig. 21 shows the results from the outlier detection algorithm. Modeled seismic volumes in the training set are represented by gray circles while the real seismic data is denoted by a red diamond. The outlier detection algorithm indicates that real data belongs to the same statistical population as the training examples, consequently the prior is not falsified. Hence, data modeled with the prior distributions (Table 2) may be used to train the CNN.

## Facies classification with CNNs

The simulated 2400 realizations from previous step were split into training, validation and test sets of 2000, 200 and 200 samples, respectively. The input to the network consists of poststack, near-stack and far-stack seismic volumes stacked as a

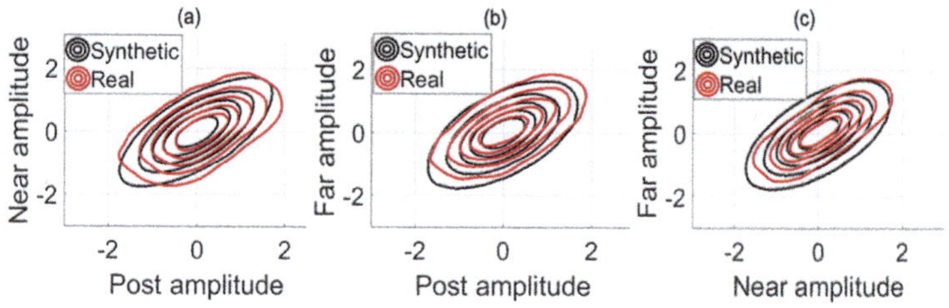


**FIG. 20** Kernel density estimates of the global bivariate distributions of (A) poststack vs near stack, (B) poststack vs far stack and (C) near stack vs far stack amplitudes.

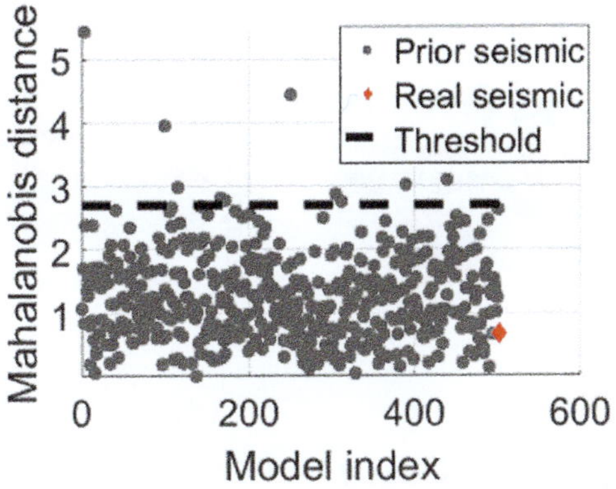


**FIG. 21** Results of prior falsification analysis showing real seismic data is consistent is synthetic seismic used to train the machine learning model. Mahalanobis distances of 500 samples of seismic data modeled from the prior, and real seismic data are shown on *y*-axis. Shown in dotted black line is the threshold value for outlier detection. Samples with Mahalanobis distance greater than threshold are deemed to be outliers.

4D tensor of dimensions $60 \times 100 \times 100 \times 3$. The architecture comprises of several 3D convolutional layers, interspersed with transposed convolutional layers. Transposed convolutional layers aid in upsampling the input volume, necessitated since the output is in depth and has differing dimensions than the input. The convolutional kernels are systematically dilated (Yu and Koltun, 2016) by expanding the kernel size with interleaved zero padding. This helps in aggregating features from large regions of the input without making the CNN too deep. The strategy of batch-normalization (Ioffe and Szegedy, 2015) of the network layers is also employed, which leads to robust feature extraction in deeper layers of the CNN by reducing the network's sensitivity to minute changes in the outputs of shallower layers. The softmax activation function is used in the last layer of the network which predicts the probability of the three facies categories (gas sand, brine sand and nonreservoir facies) at every pixel in the reservoir grid.

The network was trained for 40 epochs (~24 h run time) with the Adam optimization algorithm on a machine having 128 GB RAM and two 32 GB Tesla V100 GPUs. Pixelwise classification accuracy, averaged across all pixels in model grid, is used to evaluate the network's performance. The CNN makes predictions with classification accuracies of around 90% on the training, validation and test sets. The true facies for a test set model are compared with CNN predictions in Figs. 22 and 23. Results are presented along a cross-section and horizon slice from the model, where the poststack seismic data, true facies of the test model and CNN predicted gas sand probability are shown. In Fig. 22, the most-likely facies section is also plotted, which shows the facies with highest probability as predicted by the CNN at every pixel. It can be observed that the CNN almost perfectly predicts the curvilinear patterns of the channel objects in this synthetic test data. Spatial locations of the channels and the GWC is also predicted with high accuracy, indicating that the CNN has robustly learnt the functional mapping from seismic data to facies as encapsulated in the training set.

The trained network was subsequently employed to make predictions with real data. Fig. 24 shows the CNN predicted gas sand probability along a horizon. It can be seen that the spatial continuity of curvilinear channel-like features present in the data are preserved in the CNN predictions. Uncertainty in the CNN predictions, shown in the variance map, is mostly concentrated around the predicted channel-like features. CNN predictions are compared against petrophysical facies interpretations available at wells 2 and 6 (Fig. 25). At well 2, it can be seen that the network correctly assigns high probability at the location of the two gas sand beds. Similar results are obtained at well 6, which was kept completely blind during the prior specification procedure. The network correctly identifies the sand bed present in the deeper section of the well.

## Key messages

Rock physics models aid in interpreting and making inferences about geological parameters and processes, such as facies and fluid distributions, cementation and diagenesis, from geophysical measurements. However, as is the case with all models, these are idealizations of the true reservoir heterogeneities. To make reliable predictions and decisions, it is indispensable to reliably quantify the uncertainty arising due to the inability to model this intrinsic heterogeneity. Machine learning models are data-driven mathematical models that allow uncovering complex structures and make inferences with large-scale datasets such as seismic data. Recent advances in machine learning research have made it possible to make predictions and quantify uncertainties efficiently in high-dimensional settings. However, many machine learning models

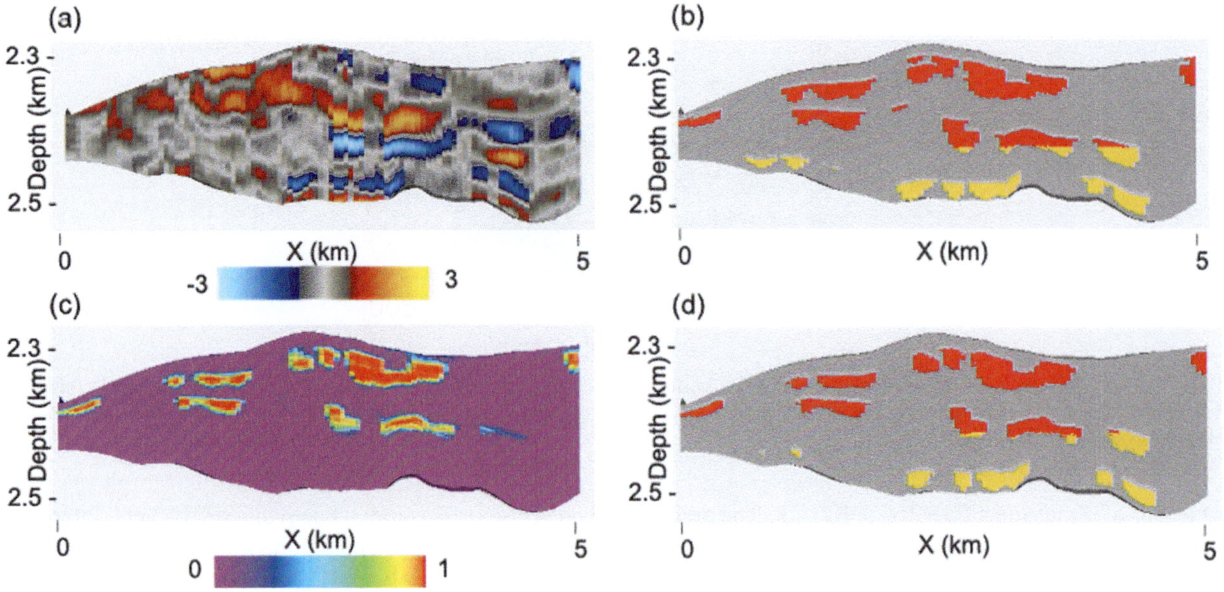


**FIG. 22** (A) Poststack seismic section, (B) true facies section, (C) CNN predicted gas sand probability and (D) most likely facies prediction. Gas sands, brine sand and nonreservoir facies are represented by *red*, *yellow* and *gray colors*, respectively.

**FIG. 23** Horizon slices of (A) poststack seismic, (B) true facies and (C) CNN predicted gas sand probability. Gas sands, brine sand and nonreservoir facies are represented by *red*, *yellow* and *gray colors*, respectively.

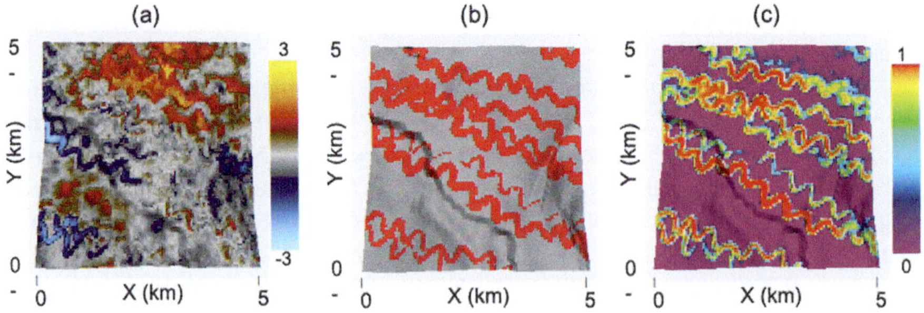


learn these relationships in a "black-box" fashion, making it challenging to attribute geological meaning to the models. This chapter advocated the integrated use of geological knowledge, rock physics and machine learning models to make reliable and efficient predictions about rock properties in a "geologically interpretable" manner. Fig. 26 presents a high-level overview of the proposed integrated workflow. The statistical rock physics and machine learning methodology described in this chapter may be broadly divided into four steps:

**(1)** *Defining the prior and checking for prior consistency*: Prior geological knowledge about the deepwater reservoir, in conjunction with well-log data, rock physics models and templates, is used to create a prior "geological" model space. Contained within this prior model are uncertainties on the conceptual geological and quantitative rock models. The different facies of interest, their styles of spatial distributions (e.g., sinuous channels, fans, etc.) and their rock physics relations (velocity-porosity, $V_P$–$V_S$, etc.) are defined in this step. However, since specification of the prior model requires making subjective modeling choices, a prior falsification step is needed, which establishes the consistency between the synthetic and real seismic data. The falsification analysis may be used to modify and tune the prior geological and rock physics models.

**(2)** *Monte Carlo simulations and generation of an augmented training dataset*: This prior model is used to generate multiple realizations of the subsurface (e.g., lithofacies distributions) and their corresponding seismic data, to be used for training the machine learning model. Generating the seismic data requires populating the facies models with the appropriate facies-dependent petrophysical (porosity, clay-content and saturations) and elastic properties using rock physics. A key feature in the use of rock physics models is to extend the training data using Monte Carlo simulations to account for scenarios that are of interest but were not encountered in the well data (e.g., different fluid saturations,

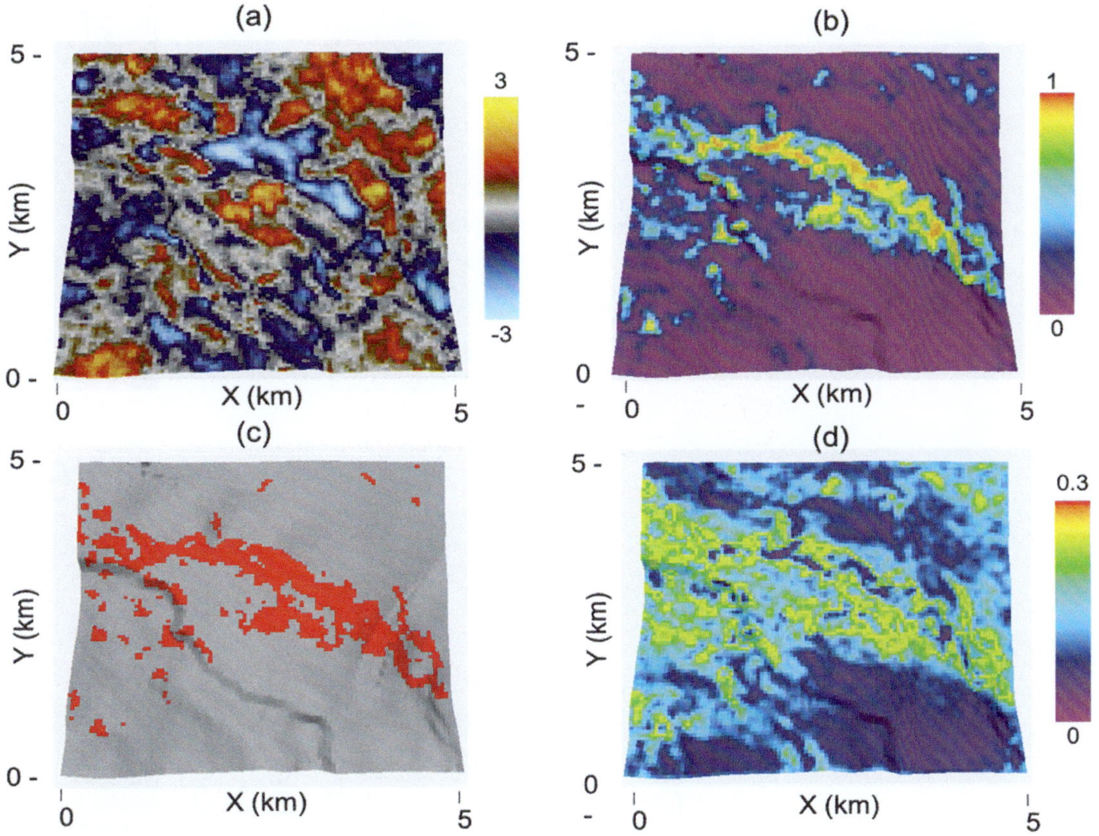


**FIG. 24** Horizon slices of (A) poststack seismic data, (B) CNN predicted gas sand probability, (C) most-likely facies and (D) prediction variance. Gas sands, brine sand and nonreservoir facies are represented by *red*, *yellow* and *gray colors*, respectively.

different sorting or diagenesis, etc.). The Monte Carlo simulations should honor spatial relations between facies as well as facies-specific correlations between petrophysical and elastic properties.

**(3)** *Training a machine learning model*: Once a consistent augmented training dataset is created, an appropriate machine learning model needs to trained. This requires selection of the seismic attribute that will be used as the predictor variables in the machine learning. Depending on the choice of the machine learning model this can involve either extracting the attributes from seismic data (e.g., via AVO analysis or impedance inversions) or using the full waveform as a high-dimensional predictor variable with deep neural networks. The training of any machine learning algorithm consists of dividing the augmented training dataset into a training set, a validation set and a testing set. Deep neural network architectures have hyperparameters that need to be tuned using the validation set and then tested on the test set. Modern deep learning training can involve significant computations requiring graphics processing units (GPUs). Various standard statistical classification accuracy metrics are evaluated on the test set to finally come up with a satisfactory trained machine learning model.

**(4)** *Applying the trained algorithm to the seismic data*: Finally, the trained machine learner or statistical classification algorithm is applied on the whole seismic dataset and used to make automatic and efficient predictions of desired reservoir properties. However, that is not the end of the process. It is imperative to cross-check the outputs of the machine-learning algorithms with geologic understanding and other subjective information of experienced interpreters encoded in their "natural neural networks," thus linking back and closing the loop with geologic domain knowledge.

## Conclusions

This chapter showed how seismic rock physics can be applied during quantitative seismic interpretation of deepwater clastic systems. Rock physics represents a quantitative link between geological processes and geophysical observables. This chapter demonstrates how this rock physics link can be used together with seismic AVO data and statistical techniques to improve the understanding of facies and fluid distributions in deepwater depositional systems, with two case examples

FIG. 25 Petrophysical interpretations of facies (*left*; sand: *light*, shale: *dark*), and CNN predicted gas sand probability (*right*) for (A) well 2 and (B) well 6.

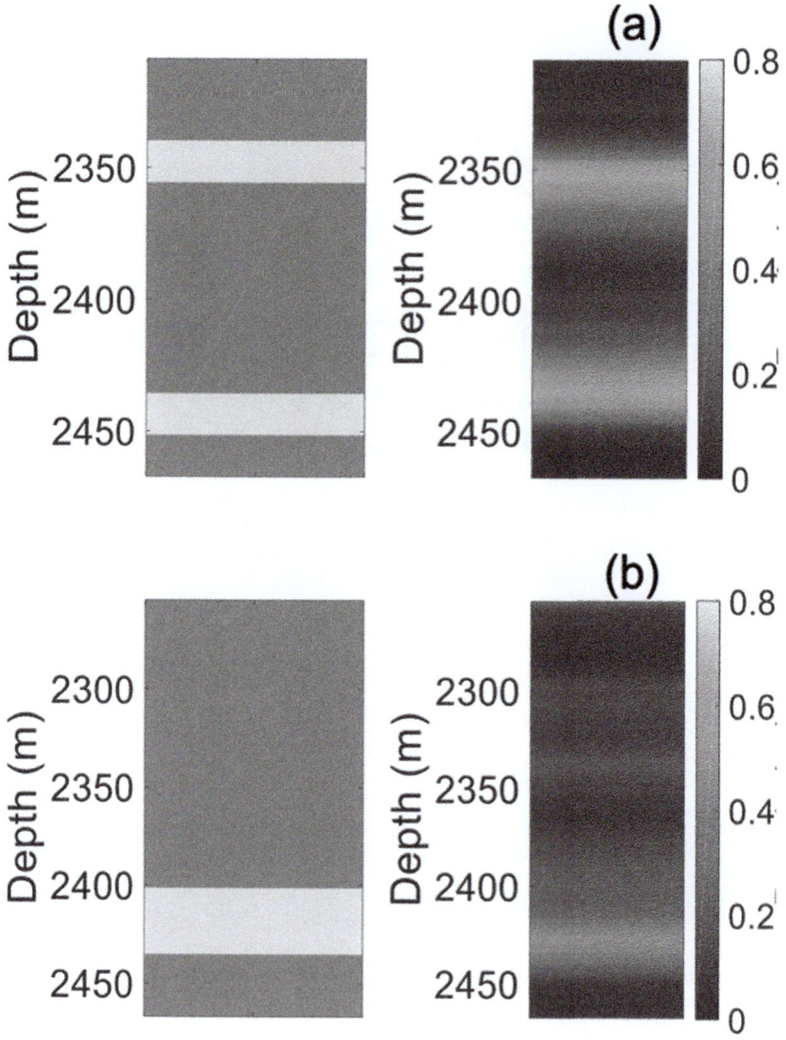


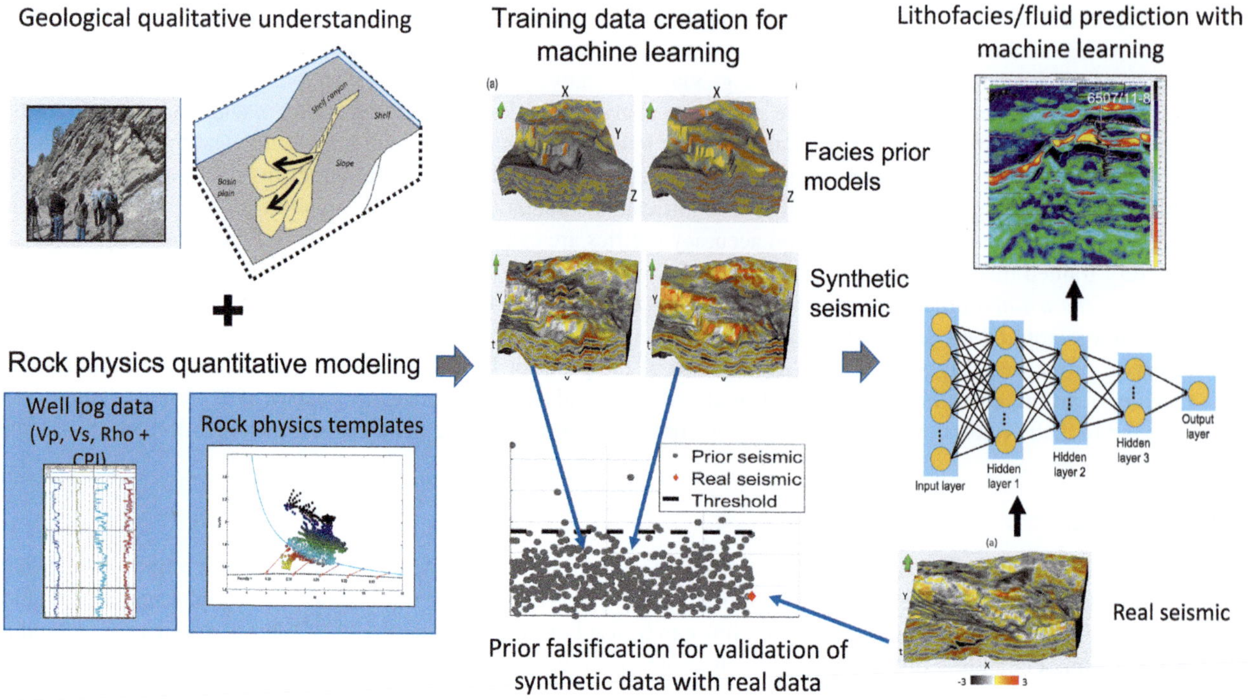


FIG. 26 High-level overview of geology and rock physics driven machine learning workflow for deepwater reservoir property predictions.

from Måløy Slope area and Alvheim Field, both in the North Sea. Furthermore, a case example from the Nile Delta demonstrates how machine learning and deep learning techniques can improve on the quantitative seismic interpretation results, in terms of classification success rate. The machine learning techniques can handle larger dimensions of data and solve nonlinear classification problems not easily solved via traditional quantitative seismic interpretation workflows. However, domain knowledge in geology and rock physics is important to ensure optimal results with machine learning based methods. Machine learning, especially modern deep learning algorithms require lots of training data examples. Labeled training examples are not always available in the quantities required for deep learning. Hence the available training data has to be augmented with synthetically generated training data, and this is where domain knowledge in geology and rock physics plays a critical role in ensuring that the synthetic training data is meaningful and representative of the system under study.

## Acknowledgments

We acknowledge operator AkerBP with partners (ConocoPhillips and Lundin Energy) of production license 203 (Alvheim Field), and operator Neptune Energy with partners (Idemitsu Petroleum, PGNiG Upstream and Sval Energi) of production license 636 (Duva Field) for seismic AVO data used in this study. Thanks to Q-eye Labs for contributions with the AVO inversion data used during the de-risking of the Cara discovery, and thanks to Ivan Lehocki (Dig Science) for contributions on the AVO classification in the Alvheim case example. We thank Edison E&P for providing the Nile Delta dataset used in this study. We offer special thanks to Giuseppe Bellentani, Francesco Siliprandi and Luca Di Cesare for their help and advice regarding the dataset. We also acknowledge the sponsors of the Stanford Center for Earth Resources Forecasting (SCERF), and thank Steve Graham, the Dean of the School of Earth, Energy, and Environmental Sciences at Stanford University for funding.

## References

Aleardi, M., Ciabarri, F., 2017. Assessment of different approaches to rock physics modeling: a case study from offshore Nile Delta. Geophysics 82 (1), MR15–MR25. https://doi.org/10.1190/geo2016-0194.1.

Aleardi, M., Ciabarri, F., Calabrò, R., 2018. Two-stage and single-stage seismic-petrophysical inversions applied in the Nile Delta. Lead. Edge 37 (7), 510–518. https://doi.org/10.1190/tle37070510.1.

AlKawai, W.H., Mukerji, T., Scheirer, A.H., Graham, S., 2018. Combining seismic reservoir characterization workflows with basin modeling in the Deepwater Gulf of Mexico Mississippi Canyon area. AAPG Bull. 102, 629–652.

Avseth, P., 2000. Combining rock physics and sedimentology for seismic reservoir characterization of North Sea turbidite systems. Unpublished PhD thesis, Stanford University.

Avseth, P., 2020. Rock Physics Diagnostic of Deepwater Clastics Honoring Burial and Depositional Trends. EAGE Extended Abstract.

Avseth, P., Lehocki, I., 2016. Combining burial history and rock-physics modelling to constrain AVO analysis during exploration. Lead. Edge 35, 528–534.

Avseth, P., Veggeland, T., 2015. Seismic screening of rock stiffness and fluid softening using rock-physics attributes. Interpretation 3, SAE85–SAE93.

Avseth, P., Dvorkin, J., Mavko, G., Rykkje, J., 2000. Rock-physics diagnostic of North Sea sands: link between microstructure and seismic properties. Geophys. Res. Lett. 27, 2761–2764.

Avseth, P., Mukerji, T., Jørstad, A., Mavko, G., Veggeland, T., 2001a. Seismic reservoir mapping from 3-D AVO in a North Sea turbidite system. Geophysics 66, 1157–1176.

Avseth, P., Mavko, G., Dvorkin, J., Mukerji, T., 2001b. Rock Physics and Seismic Properties of Sands and Shales as a Function of Burial Depth. SEG Extended Abstract.

Avseth, P., Mukerji, T., Mavko, G., 2005. Quantitative Seismic Interpretation. Cambridge University Press.

Avseth, P., van Wijngaarden, A.-J., Flesche, H., Fristad, T., Rykkje, J., Mavko, G., 2005b. Seismic Fluid Prediction in Poorly Consolidated and Clay Laminated Sands. EAGE Extended Abstract.

Avseth, P., Dræge, A., van Wijngaarden, A.-J., Johansen, T.A., Jørstad, A., 2008. Shale rock physics and implications for AVO analysis: a North Sea demonstration. Lead. Edge 27, 788–797.

Avseth, P., Jørstad, A., van Wijngaarden, A.-J., Mavko, G., 2009. Rock physics estimation of cement volume, sorting, and net-to-gross in North Sea sandstones. Lead. Edge 28, 98–108.

Avseth, P., Mukerji, T., Mavko, G., Dvorkin, J., 2010. Rock-physics diagnostics of depositional texture, diagenetic alterations, and reservoir heterogeneity in high-porosity siliciclastic sediments and rocks—a review of selected models and suggested workflow. Geophysics 75, A31–A47.

Avseth, P., Veggeland, T., Horn, F., 2013. Seismic screening for hydrocarbon prospects using rock-physics attributes. Lead. Edge 33, 266–274.

Avseth, P., Janke, A., Horn, F., 2016. AVO inversion in exploration—key learnings from a Norwegian Sea prospect. Lead. Edge 35, 405–414.

Avseth, P., Mukerji, T., 2002. Seismic lithofacies classification from well logs using statistical rock physics. Petrophysics 43, 70–81.

Avseth, P., Flesche, H., van Wijngaarden, A.-J., 2003. AVO classification of lithology and pore fluids constrained by rock physics depth trends. Lead. Edge 22, 1004–1011.

Avseth, P., Horn, F., Jansson, T., 2017. Zumba vs. Cara—From Failure to Success in Geophysical QI Risking. Extended abstract, NCS exploration 2017, Geonova.

Avseth, P., Lehocki, I., Kjøsnes, Ø., Sandstand, O., 2021. Data-driven rock physics analysis of North Sea tertiary reservoir sands. Geophys. Prospect. 69, 608–621.

Bakke, K., Kane, I., Martinsen, O.J., Petersen, S.A., Johansen, T.A., Hustoft, S., Hadler Jacobsen, F., Growth, A., 2013. Seismic modeling in the analysis of deepwater sandstone termination styles. AAPG Bull. 97, 1395–1419.

Bello, A.M., Jones, S., Gluyas, J., Acikalin, S., Cartigny, M., 2021. Role played by clay content in controlling reservoir quality of submarine fan systems, Forties Sandstone Member, Central Graben, North Sea. Mar. Pet. Geol. 128, 105058.

Bishop, C.M., 2006. Pattern Recognition and Machine Learning. Springer-Verlag, New York.

Bjørlykke, K., Aagaard, P., 1992. Clay minerals in North Sea sandstones. In: Houseknecht, D.W., Pittman, E.D. (Eds.), Origin, Diagensis and Petrophysics of Clay Minerals in Sandstones. 47. SEPM Special Publication, pp. 65–80.

Borg, I., Groenen, P., 1997. Modern Multidimensional Scaling: Theory and Applications. Springer.

Bosch, M., Mukerji, T., Gonzales, E., 2010. Seismic inversion for reservoir properties combining statistical rock physics and geostatistics: a review. Geophysics 75, 75A165–75A176.

Bressan, T.S., de Souza, M.K., Girelli, T.J., Junior, F.C., 2020. Evaluation of machine learning methods for lithology classification using geophysical data. Comput. Geosci. 139, 104475.

Bugge, T., Tveiten, B., Bäckström, S., 2001. The depositional history of the cretaceous in the northeastern North Sea. In: Martinsen, O.J., Dreyer, T. (Eds.), Sedimentary Environments Offshore Norway—Paleozoic to Recent. vol. 10. Norwegian Petroleum Society Special Publications, pp. 279–291.

Bugge, A.J., Lie, J.E., Evensen, A.K., Nilsen, E.H., Kolbjørnsen, O., Faleide, J.I., 2020. Data-driven identification of stratigraphic units in 3D seismic data using hierarchical density-based clustering. Geophysics 85, IM15-IM26.

Buland, A., Landrø, M., Andersen, M., Dahl, T., 1996. AVO inversion of troll field data. Geophysics 61, 1589–1602.

Connolly, P., 1999. Elastic impedance. Lead. Edge 18, 438–452.

Cova, R., Juhl Hansen, H., Jakobsen, A., Ross, R., Mutual, E., 2020. Facies Classification and Surface Interpretation Using Direct Probabilistic Inversion of AVO Data. SEG Extended Abstract.

Cross, N.E., Cunningham, A., Cook, R.J., Esmaie, A.T.E., Swidan, N.E., 2009. Three-dimensional seismic geomorphology of a Deepwater slope-channel system: the Sequoia field, offshore West Nile Delta, Egypt. AAPG Bull. 93 (8), 1063–1086. https://doi.org/10.1306/05040908101.

Das, V., Pollack, A., Wollner, U., Mukerji, T., 2019. Convolutional neural network for seismic impedance inversion. Geophysics, R869–R880.

de Matos, M.C., Osorio, P.L.M., Johann, P.R.S., 2007. Unsupervised seismic facies analysis using wavelet transform and self-organizing maps. Geophysics 72 (1), P9–P21. https://doi.org/10.1190/1.2392789.

Deutsch, C.V., Journel, A.G., 1998. GSLIB: Geostatistical Software Library and User's Guide, second ed. Oxford University Press, New York.

Downton, J.E., Collet, O., Hampson, D., Colwell, T., 2020. Theory-guided data science-based reservoir prediction of a North Sea oil field. Lead. Edge 39, 742–750.

Doyen, P.M., 2007. Seismic Reservoir Characterization: An Earth Modeling Perspective. European Association of Geoscientists and Engineers.

Dumoulin, V., Visin, F., 2016. A guide to convolution arithmetic for deep learning. arXiv. preprint arXiv:1603.07285.

Dvorkin, J., Gutierrez, M., 2002. Grain sorting, porosity, and elasticity. Petrophysics 43, 185–196.

Dvorkin, J., Nur, A., 1996. Elasticity of high-porosity sandstones: theory for two North Sea data sets. Geophysics 61, 1363–1370.

Feng, R., 2020. Unsupervised learning elastic rock properties from pre-stack seismic data. J. Petrol. Sci. Eng. 192, 107237.

Fitzsimmons, R., Veiberg, D., Kråkenes, T., 2005. Characterization of the Heimdal Sandstones within Alveim, Quads 24 and 25, Norwegian North Sea. In: Dore, A.G., Vining, B.A. (Eds.), Petroleum Geology: North-West Europe and Global Perspectives—Proceedings of the 6th Petroleum Geology Conference, pp. 123–131.

Fjellstad, T., Avseth, P., Omre, H., 2021. A one-step Bayesian inversion framework for 3D reservoir characterization based on a Gaussian mixture model—a Norwegian Sea demonstration. Geophysics 86, R221–R236.

Gladstone, C., McClelland, H.L.O., Woodcock, N.H., Pritchard, D., Hunt, J., 2018. The formation of convolute lamination in mud-rich turbidites. Sedimentology 65, 1800–1825.

Glorot, X., Bengio, Y., 2010. Understanding the difficulty of training deep feedforward neural networks. In: AISTATS.

Golikov, P., Avseth, P., Stovas, A., Bachrach, R., 2013. Rock physics interpretation of heterogeneous and anisotropic turbidite reservoirs. Geophys. Prospect. 61, 448–457.

Gonzales, E., Gesbert, S., Hofmann, R., 2016. Adding geological prior knowledge to Bayesian lithofluid facies estimation from seismic data. Interpretation 4, SL1-SL8.

Goodfellow, I., Bengio, Y., Courville, A., 2016. Deep Learning. The MIT Press.

Goodway, B., Chen, T., Downton, J., 1997. Improved AVO Fluid Detection and Lithology Discrimination Using Lame Petrophysical Parameters. SEG Expanded Abstracts, pp. 183–186.

Grana, D., Rossa, E.D., 2010. Probabilistic petrophysical-properties estimation integrating statistical rock physics with seismic inversion. Geophysics 75, O21–O37.

Han, D., Nur, A., Morgan, D., 1986. Effects of porosity and clay content on wave velocities in sandstones. Geophysics 51, 2093–2107.

Hjellbakk, A., Bang, A., Kotwicki, A., Langaas, K., Leonthin, M.A., Stuart, K.L., 2016. Stratigraphic Compartmentalization of the Alvheim Field. EAGE Extended Abstract.

Hurst, A., Buller, A.T., 1984. Dish structures in some Paleocene deep-sea sandstones (Norwegian Sector, North Sea); origin of the dish-forming clays and their effect on reservoir quality. J. Sediment. Res. 54, 1206–1211.

Ioffe, S., Szegedy, C., 2015. Batch normalization: Accelerating deep network training by reducing internal covariate shift. In: Bach, F., Blei, D. (Eds.), Proceedings of the 32nd International Conference on International Conference on Machine Learning—Volume 37 (ICML'15). vol. 37, pp. 448–456. JMLR.org.

Jizba, D., 1991. Mechanical and acoustical properties of sandstones and shales. PhD thesis, Stanford University.

Jonk, R., Hurst, A., Duranti, D., Parnell, J., Mazzini, A., Fallick, A.E., 2005. Origin and timing of sand injection, petroleum migration, and diagenesis in tertiary reservoirs, South Viking Graben, North Sea. AAPG Bull. 89, 329–357.

Khadem, B., Saberi, M.R., Avseth, P., 2021. Rock physics of sand-shale mixtures: classifications, theoretical formulations and study on real dataset. Mar. Pet. Geol. 134. https://doi.org/10.1016/j.marpetgeo.2021.105366, 105366.

Kingma, D., Ba, J., 2014. Adam: a method for stochastic optimization. arXiv. preprint arXiv:1412.6980.

Kombrink, H., 2021. A well's Headshot. GeoExpro. www.expronews.com.

Krizhevsky, A., Sutskever, I., Hinton, G., 2012. ImageNet classification with deep convolutional neural networks. In: Solla, S.A., Leen, T.K., Müller, K.R. (Eds.), Advances in Neural Information Processing Systems. The MIT Press, pp. 1097–1105.

Lehocki, I., Avseth, P., 2010. Rock Physics Quantification of Diagenetic Cement. EAGE Extended Abstract.

Lehocki, I., Avseth, P., 2021. From cradle to grave: how burial history controls the rock-physics properties of quartzose sandstones. Geophys. Prospect. 69, 629–649.

Lehocki, I., Avseth, P., Mondol, N., 2020. Seismic methods for fluid discrimination in areas with complex geologic history—a case example from the Barents Sea. Interpretation 8, SA35–SA47.

Leys, C., Klein, O., Dominicy, Y., Ley, C., 2018. Detecting multivariate outliers: use a robust variant of the Mahalanobis distance. J. Exp. Soc. Psychol. 74, 150–156. https://doi.org/10.1016/j.jesp.2017.09.011.

Linde, N., Renard, P., Mukerji, T., Caers, J., 2015. Geological realism in hydrogeological and geophysical inverse modeling: a review. Adv. Water Resour. 86, 86–101.

Long, J., Shelhamer, E., Darrell, T., 2015. Fully Convolutional Networks for Semantic Segmentation. CVPR.

Løseth, H., Wensaas, L., Gading, M., Duffaut, K., 2011. Can hydrocarbon source rocks be identified on seismic data? Geology 39, 1167–1170.

Lowe, D.R., LoPiccolo, R.D., 1974. The characteristics and origins of dish and pillar structures. J. Sediment. Res. 44, 484–501.

Ma, X.Q., 2001. Global joint inversion for the estimation of acoustic and shear impedance from AVO derived P- and S-wave reflectivity data. First Break 19, 557–566.

Mallat, S.G., 1989. A theory for multiresolution signal decomposition: the wavelet representation. IEEE Trans. Pattern Anal. Mach. Intell. 11 (7), 674–693. https://doi.org/10.1109/34.192463.

Mariethoz, G., Caers, J., 2014. Multiple Point Geostatistics: Stochastic Modeling With Training Images. Wiley Blackwell.

Marion, D., Nur, A., Yin, H., Han, D., 1992. Compressional velocity and porosity in sand-clay mixtures. Geophysics 57, 554–563.

Martinsen, O.J., Lien, T., Jackson, C., 2005. Cretaceous and Palaeogene turbidite systems in the North Sea and Norwegian Sea Basins: source, staging area and basin physiography controls on reservoir development. In: Geological Society, London, Petroleum Geology Conference series. 6, pp. 1147–1164.

Mavko, G., Mukerji, T., Dvorkin, J., 2020. The Rock Physics Handbook, third ed. Cambridge University Press.

Mondal, S., Yadav, A., Chatterjee, R., 2020. Rock-physics forward modelling to predict seismic behaviour: a case study for exploration target in Mahanadi basin, East Coast of India. Geophys. Prospect. 68, 2186–2194.

Mosser, L., Amato del Monte, A., Avseth, P., Dræge, A., MacGregor, L., 2019. Investigating the AVO Signature of a North Sea Oil-Field Using Pseudo-Wells and Unsupervised Deep Learning. EAGE Extended Abstract.

Mukerji, T., Jørstad, A., Mavko, G., Granli, J.R., 1998. Near and far offset impedances: seismic attributes for identifying lithofacies and pore fluids. Geophys. Res. Lett. 25, 4557–4560.

Mukerji, T., Jørstad, A., Avseth, P., Mavko, G., Granli, J.G., 2001a. Mapping lithofacies and pore-fluid probabilities in a North Sea reservoir: seismic inversions and statistical rock physics. Geohysics 66, 988–1001.

Mukerji, T., Avseth, P., Mavko, G., Takahashi, I., Gonzalez, E., 2001b. Statistical rock physics: combining rock physics, information theory, and geostatistics to reduce uncertainty in seismic reservoir characterization. Lead. Edge 20, 313–319.

Nasser, M., Mavko, G., Dvorkin, J., Ostroff, G., 2011. Rock physics analysis of deepwater sediments, West Africa. SEG Extended Abstract.

Ødegaard, E., Avseth, P., 2004. Well log and seismic data analysis using rock physics templates. First Break 22, 37–43.

Olsen, T.R., 2017. 36/7-4 Cara discovery – a stratigraphic success. NPF Exploration Revived Abstract.

Pradhan, A., Dutta, N.C., Le, H.Q., Biondi, B., Mukerji, T., 2020. Approximate Bayesian inference of seismic velocity and pore-pressure uncertainty with basin modeling, rock physics, and imaging constraints. Geophysics 85, ID19-ID34.

Purves, S., Alaei, B., Bonas, J.B., Aursand, P., 2019. Machine Learning From Core to Seismic; Porosity Prediction in the Norwegian North Sea. EAGE Extended Abstract.

Pyrcz, M.J., Deutsch, C.V., 2014. Geostatistical Reservoir Modeling. Oxford University Press, Oxford.

Qadrouh, A.N., Carcione, J.M., Alajmi, M., Alyousif, M.M., 2019. A tutorial on machine learning with geophysical applications. Boll. Geofis. Teor. Appl. 60, 375–402.

Reading, H.G., Richards, M., 1994. Turbidite systems in Deepwater basin margins classified by grain-size and feeder system. AAPG Bull. 78, 792–822.

Rimstad, K., Avseth, P., Omre, H., 2012. Hierarchical Bayesian lithology/fluid prediction: a North Sea case study. Geophysics 77, B69–B85.

Roden, R., Smith, T., Sacrey, D., 2015. Geologic pattern recognition from seismic attributes: principal component analysis and self-organizing maps. Interpretation 3 (4), SAE59–SAE83. https://doi.org/10.1190/INT-2015-0037.1.

Roden, R., Smith, T., Santogrossi, P., Sacrey, D., Jones, G., 2017. Seismic interpretation below tuning with multiattribute analysis. Lead. Edge 36, 330–339. https://doi.org/10.1190/tle36040330.1.

Russell, B., Hampson, D., Bankhead, B., 2006. An Inversion Primer. CSEG Recorder, pp. 96–106. Special Edition.

Sayers, C., 1995. Simplified anisotropy parameters for transversely isotropic sedimentary rocks. Geophysics 60, 1933–1935.

Scheidt, C., Jeong, C., Mukerji, T., Caers, J., 2015. Probabilistic falsification of prior geologic uncertainty with seismic amplitude data: application to a turbidite reservoir case. Geophysics 80. M89-M12.

Scheidt, C., Li, L., Caers, J., 2017. Quantifying Uncertainty in Subsurface Systems. Wiley-Blackwell, Wichester, UK.

Simm, R., Bacon, M., 2014. Seismic Amplitude—An Interpreter's Handbook. Cambridge University Press.

Srivastava, N., Hinton, G., Krizhevsky, A., Sutskever, I., Salakhutdinov, R., 2014. Dropout: a simple way to prevent neural networks from overfitting. J. Mach. Learn. Res. 15, 1929–1958.

Stovas, A., Landrø, M., Avseth, P., 2006. AVO attribute inversion for finely layered reservoirs. Geophysics, C25–C36.

Takahashi, I., 2000. Quantifying Information and Uncertainty of Rock Property Estimation from Seismic Data. PhD Thesis, Stanford University.

Tarantola, A., 2005. Inverse Problem Theory and Methods for Model Parameter Estimation. SIAM, p. 352.

Thomsen, L., 1986. Weak elastic anisotropy. Geophysics 51, 1954–1966.

Thyberg, B.I., Jahren, J., Winje, T., Bjørlykke, K., Faleide, J.I., 2009. From mud to shale: rock stiffening by micro-quartz cementation. First Break 27, 27–33.

Vail, P.R., Mitchum Jr., R.M., 1977. Seismic stratigraphy and global changes of sea level, part 1: overview. In: Payton, C.E. (Ed.), Seismic Stratigraphy—Applications to Hydrocarbon Exploration. vol. 26. Am. Assoc. Petrol. Geol. Memoir, pp. 51–52.

Vernik, L., 2016. Seismic Petrophysics in Quantitative Interpretation. Society of Exploration Geophysicists.

Vernik, L., Kachanov, M., 2010. Modeling elastic properties of siliciclastic rocks. Geophysics 75, E171–E182.

Vernik, L., Nur, A., 1992. Petrophysical classification of siliciclastics for lithology and porosity prediction from seismic velocities. AAPG Bull. 76, 1295–1309.

Wahyuni, F., Escalona, A., Augustsson, C., Marin, D., 2019. Early Cretaceous Tectonostratigraphic Evolution of the Sogn Graben and Måløy Slope, northern North Sea (Agat Fm.). EAGE Extended Abstract.

Walderhaug, O., 1996. Kinematic modeling of quartz cementation and porosity loss in deeply buried sandstone reservoirs. AAPG Bull. 80, 731–745.

Walther, J., 1893-4. Einleitung in die Geologie als historische Wissenschaft. 3 vols. Gustav Fischer, Jena.

Whitcombe, D., Connolly, P.A., Reagan, R.L., Redshaw, T.C., 2002. Extended elastic impedance for fluid and lithology prediction. Geophysics 67, 63–67.

Wu, X., Liang, L., Shi, Y., Fomel, S., 2019. FaultSeg3D: using synthetic data sets to train an end-to-end convolutional neural network for 3D seismic fault segmentation. Geophysics 84, IM35-IM45.

Yang, F., Ma, J., 2019. Deep-learning inversion: a next-generation seismic velocity model building method. Geophysics 84, R583–R599.

Yu, F., Koltun, V., 2016. Multi-scale context aggregation by dilated convolutions. In: Proc. Int. Conf. Learn. Representations.

Zhao, T., Jayaram, V., Roy, A., Marfurt, K.J., 2015. A comparison of classification techniques for seismic facies recognition. Interpretation 3 (4), SA29–SA58. https://doi.org/10.1190/INT-2015-0044.

Zimmer, M., Prasad, M., Mavko, G., Nur, A., 2007. Seismic velocities of unconsolidated sands: part 2—influence of sorting- and compaction-induced porosity variation. Geophysics 72, E15–E25.

## Further reading

Avseth, P., Lehocki, I., 2021. 3D subsurface modeling of multi-scenario rock property and AVO feasibility cubes—an integrated workflow. Front. Earth Sci. 9, 1–13.

Duchi, J., Hazan, E., Singer, Y., 2010. Adaptive subgradient methods for online learning and stochastic optimization. J. Mach. Learn. Res. 12, 2121–2159.

Pradhan, A., 2020. Statistical learning and inference of subsurface properties under complex geological uncertainty with seismic data. PhD thesis, Stanford University.

Pradhan, A., Mukerji, T., 2020. Seismic Inversion for Reservoir Facies Under Geologically Realistic Prior Uncertainty With 3D Convolutional Neural Networks. SEG Technical Program Expanded Abstracts, pp. 1516–1520.

Walden, A.T., White, R.E., 1998. Seismic wavelet estimation: a frequency domain solution to a geophysical noisy input-output problem. IEEE Trans. Geosci. Remote Sens. 36, 287–297.

Yi, B.Y., Lee, G.H., Kim, H., et al., 2013. Comparison of wavelet estimation methods. Geosci. J. 17, 55–63. https://doi.org/10.1007/s12303-013-0008-0.

Chapter 18

# Integrating forward stratigraphic modeling with basin and petroleum system modeling

Allegra Hosford Scheirer[a], Keyu Liu[b], Jianliang Liu[b] and Oliver Schenk[c]
[a]*Stanford University, Stanford, CA, United States,* [b]*China University of Petroleum, Qingdao, China,* [c]*Schlumberger, Aachen, Germany*

## Introduction

Physical processes that shape sedimentary basins vary over spatial and temporal scales that span many orders of magnitude. As discussed in previous chapters, centimeter-to-kilometer-sized ripples and dunes form over days to months to years, whereas ~1 km (~ .6 mi.) to ~1000 km (~6200 mi.)-sized deltas form on ~tens of thousands to million-year timescales. Tectonically, numerous geologic processes occur over similar spatial and timescales: volcanism, erosion, basin subsidence, mountain building, changes in water depth and many more. These orders of magnitude differences in spatial and temporal processes affect subsurface geology in varying ways. Thus, the principal challenge of geologic modeling is integrating at least a subset of these scales into a holistic quantitative representation of a sedimentary basin.

Further complicating the task of geologic modeling is that the rock record is incomplete on the spatial and temporal scales of interest. Even the most assiduous effort to meld together outcrop-scale observations, regional geophysical data, borehole measurements, physical analog experiments and an investigator's experience is fraught with uncertainty, raises more questions than answers and makes clear the insufficiency of purely observational approaches to understand the behavior and expression of geologic systems.

The goals of this chapter are to discuss in detail the modeling of sedimentary basins with two complementary techniques—forward stratigraphic modeling and basin and petroleum system modeling— and to demonstrate how integrating the two methodologies maximizes insight of geologic processes in sedimentary basins. The aims of this chapter are not to provide a comprehensive treatise on the mathematics of the underlying constitutive physics and chemistry—heat flow, fluid flow, compaction, kinetics and so forth—that comprise numerical modeling of sedimentary basins; numerous excellent publications already serve this purpose (Granjeon et al., 1999; Hantschel and Kauerauf, 2009; Kendall et al., 1991; Lerche, 1990a, b; Tetzlaff and Harbaugh, 1989). Rather, the chapter aims to demystify these two subdisciplines of basin-scale numerical modeling by providing readers with the working knowledge that is essential for maximizing the understanding of modeling depositional systems. By metaphorically drilling down into the core functionalities of stratigraphic modeling and basin and petroleum system modeling, it becomes clear how linking the two approaches complements each other.

The hands-on process of modeling a sedimentary basin begins before sitting behind a computer. Prior to starting to work with the software, the goals of forward stratigraphic and basin and petroleum system modeling should be identified clearly. Examples of modeling objectives for the forward stratigraphic approach include unraveling the interrelationship between accommodation, subsidence, sediment supply and relative sea level. Examples of objectives for basin and petroleum system modeling include determining the magnitude, timing and effects of source rock thermal maturation, the relative timing between petroleum migration and trap formation, the magnitude and timing of erosion events, and predictions of pore pressure.

This chapter begins with a history of the development of forward stratigraphic and basin and petroleum system modeling. Using as a guide the workflow shown in Fig. 1, subsequent sections describe in detail the "nuts and bolts" of constructing and analyzing each type of model to set the foundation necessary for the coupled forward stratigraphic-petroleum system models. Finally, two case studies illustrate how the techniques are combined and the results subsequently analyzed, providing a culmination of many of the technical aspects of the book.

Before continuing, it is essential to explicitly state two intentions of this work. First, although the discipline of basin and petroleum system modeling is (obviously) rooted in the simulation of petroleum generation, migration and accumulation (Chapter 2), basin modeling remains a significant technology to reconstruct the geologic evolution of the basin independent

**Deepwater Sedimentary Systems.** https://doi.org/10.1016/B978-0-323-91918-0.00002-5

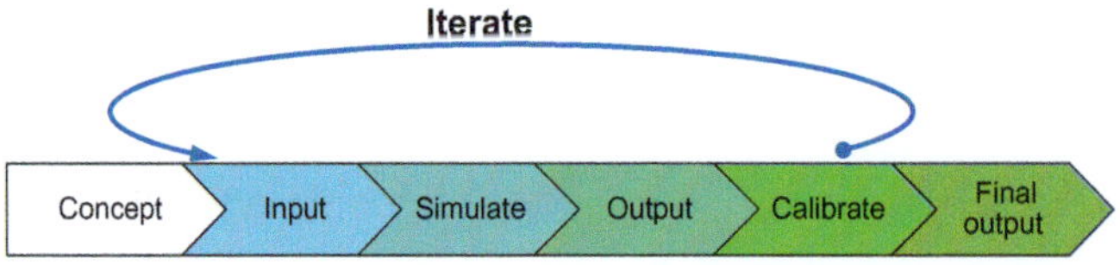


**FIG. 1** Modeling workflow. Schematic workflow for both forward stratigraphic and basin and petroleum system modeling.

of petroleum studies; parameters of interest may focus on pore pressure prediction, chemical diagenesis, deconvoluting tectonic loads from sediment and water loads, and rock mechanics, among many other subsurface processes described in previous chapters—even "just" the stratigraphic and structural evolution itself. Accordingly, in the remainder of this chapter, the phrase "basin model[ing]" is generally used interchangeably with "basin and petroleum system model [ing]." Context clarifies whether petroleum systems are involved. Second, this chapter comprises a deep dive into constructing and analyzing forward stratigraphic and basin and petroleum system models, both as separate disciplines and as a coupled technology. That said, the sheer breadth and depth of geological environments, their driving processes, their internal properties and so forth mean that some topics are left out or treated briefly. Even so, an extensive list of references from critical studies fills the gap.

## History of forward stratigraphic modeling and basin and petroleum system modeling

The study of linked processes that shape geologic systems is aimed at describing and explaining what is, and what cannot be, observed on the Earth. Investigations of the Earth's surface and subsurface originated in qualitative form; in the late 1800s, the father of modern geology Hutton (1788, 1795) authored two volumes totaling nearly a 1000 pages of prose to conceptualize uniformitarianism, the notion that in geology, the present is the key to the past.

In the mid-to-late 1800s, physical models were coupled with conceptual ones to study Earth processes in a hands-on manner. For example, a physical model constructed by Alphonse Favre examined the concept of orogenesis by layering clay onto a sheet of rubber, which was subsequently stretched and released with the goal of creating mountain structures (Bokulich and Oreskes, 2017). More than 200 years later, geoscientists continue to study geological processes in a controlled environment in which the variables are manipulated to evaluate various scenarios. However, as detailed by Bokulich and Oreskes (2017), such experiments suffer from laboratory artifacts—the limitations that the experimental apparatus physically impose on the replica of real-world systems—and scale effects: the inability to maintain the proper scaling relationship between variables. This latter problem regarding scale in tabletop models is relevant to petroleum geology in that M. King Hubbert (who remains widely known for predicting the year when oil production would peak and subsequently decline) recognized in 1945 the necessity for physical similarity, the concept that all physical quantities in a lab model must scale in constant ratios to each other to improve their applicability (Bokulich and Oreskes, 2017).

The advent of numerical modeling dates to the use of computer models in the late 1940s in the fields of meteorology and nuclear physics (Winsberg, 2019). Computer modeling expanded the geoscientists' toolkit; narrative descriptions of Earth processes combined with observations from physical experiments brought the modeling pathway full circle. From early programming languages—Cobol and Fortran in the late 1950s and BASIC (among others) in the mid-1960s—to Python and R today, computer-based modeling completes the powerful triad of (1) conceptual, (2) physical and (3) computational modeling approaches.

Computer modeling of sedimentary basins and petroleum systems dates at least as far back as the mid-late 1960s, when the Kansas Geological Survey initiated a new publication series, Computer Contributions, regarding the development and use of computer programs in the Earth sciences. This and succeeding publications in the series built upon computational advancements in the first half of the decade for solving partial differential equations, analyzing vector trends, calculating regression parameters, among many others. The inaugural paper in the series by Harbaugh (1966) documented collaborative research between Stanford University and the Kansas Geological Survey that aimed to advance the simulation of oil generation, migration and accumulation in sedimentary basins. The first stage of the research was to model depositional systems. Harbaugh (1966) developed a comprehensive mathematical model to simulate vertically-oriented sediment deposition and movement of the basement; deltaic sedimentation that resulted in realistic sedimentary sequences of topset, foreset and bottomset beds; winnowing of fine particles at the beach and the geographic distribution and relative vitality of organism communities on the seafloor—all in three dimensions.

Ojakangas (1967) followed up on these efforts with a program to simulate the development of structural and stratigraphic traps for the accumulation of petroleum. Demonstrated with a test case in the Wind River Basin, Wyoming,

USA, the program incorporated deposition, erosion, and compaction of shale. Because the algorithm lacked fluid migration, oil accumulations were deemed possible if reservoir rocks—sandstone and limestone—were present in structural highs (because they could potentially be filled with oil) and if oil had sufficient time to migrate into traps after their formation. Although the program lacked numerical formulations of fluid generation and migration, conclusions regarding potential oil accumulations on the basis of space, physical properties and geologic time nevertheless marked a step forward in petroleum system modeling.

A final example from this era makes clear the importance of basin-scale modeling to the petroleum industry. Citing the expense of acquiring data and its value for exploration management and planning, Grender et al. (1974) constructed a 3D model of the southern San Joaquin Basin, California, USA. Each cell in the model spanned 2 mi. × 2 mi. (3.2 km × 3.2 km) by 500 ft. (~13 m) from surface to basement; the model covered an area of ~1000 $mi.^2$ (~2600 $km^2$) and reached ~25,000 ft. (~7.6 km) in depth. Model parameters were limited to subsurface structure, four generalized lithologic types (coarse-grained marine, fine-grained marine, nonmarine and basement) and oil and exploration data. In the era of punch cards, Grender et al. (1974) detailed the arrangement of input parameters within data fields and grid cells, the subroutines that performed calculation tasks and the numerous options for graphical output. Circling back to the critical need to utilize existing exploration data, Grender et al. (1974) emphasized using model results to identify plays, the volume of oil within those plays and relatively untested areas that may indicate the remaining exploration potential of the area.

These 1960s and 1970s era publications and many more similar to them emphasized computational concepts—trial and error, probabilistic approaches, model validation, simulation time, irregular distribution of input data, grid discretization—and geologic concepts—fluid migration, faulting, salt tectonics, diagenesis—that persist in both forward stratigraphic and basin and petroleum system modeling today.

Forward dynamic modeling of sedimentary processes had its initial development in the middle 1960s to the early 1970s. The first modern stratigraphic model was proposed by Sloss (1962). The concept centered on the hypothesis that the stacking pattern of sedimentary rock is controlled by four variables: quantity of sediment supplied to a basin, rate of subsidence relative to base level, sediment flux distribution within the depositional system and sediment composition and texture. Although this model was conceptual rather than quantitative, it set the stage for a quantitative approach, which was formalized by Schwarzacher (1966) via a one-dimensional (1D) model that examined the relationship between sedimentation rate and water depth. This effort formed the basis of carbonate simulation modeling in the future.

One-dimensional modeling of a vertical stratigraphic succession is only the starting point for forward stratigraphic modeling. Largely due to the advancement in computing capability, the 1980s saw dozens of treatments in two and three dimensions (Merriam and Davis, 2001; Paola, 2000), such as modeling different sediment types (e.g., clastic, carbonate, organic) and the interaction between them. Simulation algorithms incorporated into early- and present-day software include geometric, diffusion, fuzzy logic, analytical and hydraulic models, yielding abundant idealized stratigraphic patterns.

Stratigraphic models that are constructed in two dimensions (2D) afford rapid and effective approaches for simulating depositional scenarios that are consistent with observations in reflection seismic, well log and outcrop data. Some representative 2D software packages include SEDPAK (Strobel et al., 1989), CARBONATE (Bosence and Waltham, 1990), 2D SedFlux (Syvitski and Hutton, 2001) and PHIL (Scheibner et al., 2003). However, 2D forward stratigraphic models fail to completely capture the physics controlling the processes of sediment transport, deposition and erosion. Accordingly, three dimensional (3D) approaches for forward stratigraphic modeling were developed to provide geologists more insight into the internal dynamics of depositional processes and the factors that influence and control stratal architectures.

Based on various modeling algorithms, numerous 3D modeling programs have been developed. These include Sedsim (Tetzlaff and Harbaugh, 1989), SedFlux 3D (Hutton and Syvitski, 2008) and GPM (Hill et al., 2009), which are based on hydraulic physics; Dionisos (Granjeon et al., 1999), which is based on diffusion; CARBONATE 3D (Warrlich et al., 2008), CarbSim (Duan et al., 2000) and CarboCAT (Burgess, 2013), which utilize analytical solutions; and FUZZIM (Nordlund, 1999) and Sedsim (for carbonate calculations), which take advantage of the relatively new field of fuzzy logic.

Several landmark books laid the foundation of petroleum geology, geochemistry and petroleum systems upon which basin modeling software is based. These include those by Hunt (1979), Tissot and Welte (1984) and Welte et al. (1997). The development sequence of commercial and proprietary software for basin and petroleum system modeling began in the 1980s. In contrast with forward stratigraphic modeling, which has its roots in the simulation of depositional systems in three dimensions (e.g., Harbaugh, 1966), basin and petroleum system modeling debuted in the mid-1980s as a tool for simulating sedimentary burial, thermal history and source rock maturity in one dimension. This development was driven by several factors; Hantschel et al. (2019) ascribe advancements during this time to laboratory experiments on source rocks, the use of Darcy flow in reservoir models and the advantages of flowpath physics to simulate petroleum migration, whereas Hermanrud (1993) cites the computational benefits of using Lagrangian rather than Eulerian coordinates and finite element rather than finite difference grids. During the 1980s, basin modeling software was developed by both commercial software companies and in-house oil companies (Hantschel et al., 2019). During these early days of basin modeling, models averaged about 100,000 grid cells in size (Hantschel et al., 2019).

In the 1990s, basin and petroleum system modeling began incorporating the petroleum system concept of Magoon and Dow (1994) as well as developing 2D models with grid cell sizes an order of magnitude larger than those used the decade before. The incorporation of invasion percolation, often abbreviated as IP in the literature, and combined migration schemes to basin modeling software rounded out the categories of petroleum flow physics used today (Hantschel et al., 2019). In addition, the incorporation of more components and sophisticated distributions of activation energies for kerogen kinetics (see below) improved predicted hydrocarbon classes that result from conversion of kerogen to oil, kerogen to gas, and oil to gas.

Finally, Hantschel et al. (2019) document numerous advancements of basin modeling in the 2000s: 3D models, parallelization of computer processors, local grid refinement (also called model-in-model and nested models), probabilistic algorithms for quantifying uncertainty, in-reservoir alteration of petroleum and capabilities for addressing additional petroleum systems such as gas hydrates and shale plays. Models commonly comprise tens to hundreds of millions of grid cells mainly driven by ever-increasing seismic resolution and computational advancements such as petrabytes of storage, thousands of compute nodes and CPU cores, and hundreds of graphics processing units (ubiquitously called GPUs).

In summary, widespread industry adoption of the technology combined with the rapid increase in computer processing speed and memory in the 21st century continue to drive new workflows, methodologies and applications of basin and petroleum system modeling. The critical importance of basin and petroleum system modeling across the petroleum industry, government and academic sectors is evidenced by the existence of models constructed for sedimentary provinces all over the world and their use for myriad applications.

## Forward stratigraphic and basin and petroleum system modeling

Publications about forward stratigraphic models and basin and petroleum system models commonly include three words to describe a typical modeling approach: deterministic, forward and iterative. *Deterministic* modeling produces the same results for a given set of inputs. In other words, given one set of input, a numerical model executes the same sequence of mathematical operations to obtain a single set of output, regardless of how many times it is run. *Forward* modeling, as opposed to inverse modeling, begins with a predetermined or known model, applies mathematical operations to it and calculates what the model structure will produce resulting from those calculations. Finally, *iterative* modeling means that the cycle of input-simulate-output-calibrate-final input (i.e., Fig. 1) is repeated; usually, iteration is performed to improve the reliability of model results but may also be aimed at other goals such as scenario testing.

In principle, a geologic model of a sedimentary basin is readily constructed by encoding information about the physical and chemical properties of geologic layers and their evolution through time. In practice, the input stage of forward stratigraphic modeling and basin and petroleum system modeling is dominated by the time and effort required to build the model. Prior to embarking on model construction, three overarching topics must be formalized: (1) stating project objectives; (2) scrutinizing basin stratigraphy to ensure that relevant model layers are chosen; (3) identifying required data, their quality and their spatial distribution in the study area. Table 1 compares typical input parameters for the two modeling approaches.

Data comprise the backbone of the model and typically come from a wide range of geoscience disciplines and sources; among others, these include surface geology, organic geochemistry, marine geophysics, petrophysics, remote sensing, seismic surveys, drilling, physical properties, scientific literature and global databases. Compiling these data is Herculean in scope. Key tasks include locating data; generating data with new physical analyses; aggregating, sorting and formatting data; evaluating data quality and upscaling and/or downscaling data. The upside of this prodigious effort is that data—much of it collected at great expense and over many years—are integrated and stored in a specific project or suite of projects. The principal downside is a significant investment of personnel hours.

### Forward stratigraphic modeling

#### Numerical algorithms and modeling workflow

There are four main categories of numerical algorithms for modeling sediment transport: (1) geometric models, (2) fluid-flow models, (3) diffusion-based models and (4) fuzzy logic models. Each category has various computational advantages and different practical applications.

**TABLE 1 Comparison of parameters for basin and petroleum system modeling and forward stratigraphic modeling.[a,b]**

| | Basin and petroleum system modeling | Forward stratigraphic modeling |
|---|---|---|
| Time span | Geologic existence of the basin | Specific time period of interest |
| Typical model depth | ~5 to ~40 km (~3 to ~25 mi.); larger endmember if a crustal model is used | Tens to hundreds of meters (or miles); not applicable in 1D |
| Typical model area | Hundreds to thousands of kilometers or miles (basin to subbasin scale) | Tens of meters (or feet) to hundreds of kilometers (or miles); not applicable in 1D |
| Typical grid cell size | 100 to 1000 m (~60 to 600 mi.) or more | Tens of meters (or feet) to hundreds of kilometers (or miles) |
| Resolution of paleowater depth | Relatively coarse spatial and temporal resolution | Relatively fine spatial and temporal resolution |

[a]*Neither basin modeling nor forward stratigraphic modeling require petroleum generation and flow modeling.*
[b]*Numeric ranges are rules of thumb.*

Geometric models are based on geometric algorithms that honor mass conservation: sediment in + sediment eroded = sediment in the fluid + sediment deposited + sediment out. These types of models are relatively simple to construct and effective for demonstrating concepts that are based on seismic and sequence stratigraphic analysis. Geometric models provide effective visualizations of changes in stratigraphic stacking patterns with sea level fluctuations. These models are generally constructed in 2D space and are straightforward to implement. Sedpak is a software package that uses geometric models; some examples of geometric models are reported by Jervey (1988), Strobel et al. (1989), Bosscher and Schlager (1992) and Bowman and Vail (1999).

Fluid-flow models use the physics of hydrodynamics to simulate sediment deposition, transport and erosion. An example software package that employs fluid-flow principles is Sedsim, which implements simplified Navier–Stokes equations for computational efficiency. In these types of models, both mass and energy are conserved, yielding simulated results that are more realistic than geometric models can produce. Fluid-flow models permit investigation of clastic sedimentation in fluvial-deltaic, shallow marine and deep marine depositional settings. Example use cases are found in Griffiths et al. (2001) and Tetzlaff and Harbaugh (1989).

Diffusion-based models are based on Fick's laws of diffusion. The diffusion approach attracts significant attention because of its simplicity and applicability in modeling different depositional systems. It is particularly effective in simulating carbonate deposition because diffusion physics is well suited to simulate water depth, energy level, temperature and clastic sediment dilution. On the other hand, diffusion-based physics fails to adequately capture the transport of sediments comprising multiple grain sizes. The software package DionisosFlow employs diffusion physics. Example studies are discussed by Granjeon (2014) and Hawie et al. (2019).

Finally, fuzzy logic algorithms simulate sedimentation according to empirical understanding and knowledge. The advantage of the fuzzy logic algorithm is that it essentially translates qualitative geological observations and descriptions into quantitative information for numerical treatment. It is particularly effective for simulating the deposition of carbonate and organic sediments because there are well-established databases on factors affecting carbonate and organic deposition based on extensive modern case studies in the Bahamas and the Great Barrier Reef (Davies et al., 1989; Eberli and Ginsburg, 1989). Demicco (2004a, b, c) synthesizes the concepts that underlie fuzzy logic and its application to earth science in general and forward stratigraphic modeling in particular. Some example studies performed with FUZZIM software, which was developed by Nordlund (1999), are given by El-Shahat et al. (2009) and Ghiasi-Freez et al. (2012).

The basic procedure of stratigraphic forward modeling encompasses the following: (1) constructing an initial depositional surface, (2) specifying erosion, sediment transport and deposition over a specified geological interval and (3) honoring mass balance and energy conservation (some of these processes are illustrated schematically in Fig. 2). Forward stratigraphic modeling can simulate stratal stacking and superposition of sediments during basin filling and can provide quantitative information of sediments at any given location during sedimentary basin evolution (Burgess and Wright, 2003; Griffiths et al., 2001; Kendall et al., 1991; Liu and Liu, 2021; Liu et al., 1998). A final product is a set of digital data volumes encompassing lithologies, sedimentary facies and chronostratigraphic information that are essential for basin modeling.

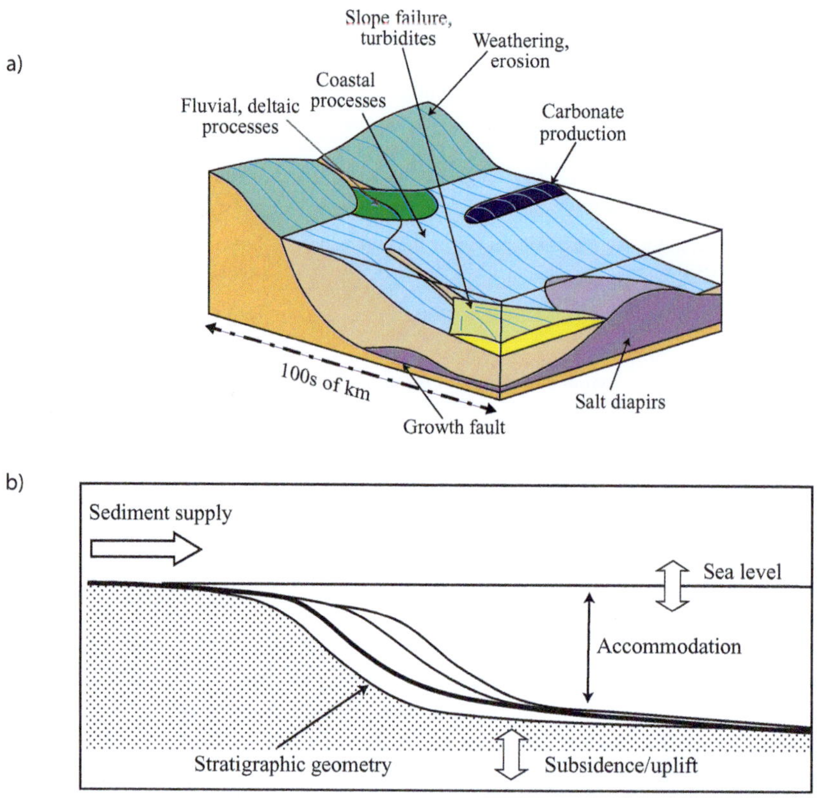


**FIG. 2** Examples of sedimentary processes considered in forward stratigraphic modeling. (A) Cartoon of surface and submarine processes that interact to drive sediment erosion, transport and deposition. (B) Three primary processes that control stratigraphic architecture are changing sea level, subsidence and uplift, and sediment supply; an objective of forward stratigraphic modeling is to study the interdependent role of each process. *(Panel A: Modified from Beicip-Franlab and Nalcor Energy (2020); panel B: Reprinted from Shafie and Madon (2008) with permission from Geological Society of Malaysia.)*

A key difference between forward stratigraphic modeling and basin modeling is that a subset of the geologic section—a particular stratigraphic interval of interest such as certain petroleum system elements—is modeled rather than the entire geological section, as is the case in basin modeling. The result is the stratal geometry and characteristics of the modeled stratal architecture on a scale much finer than a basin model can achieve. Moreover, modeling a subset of the geologic section with forward stratigraphic modeling allows the modeler to concentrate on rock units that are relevant to project goals but are not incorporated into the basin model for various reasons.

The modeling workflow is illustrated in Fig. 3. It starts with developing a conceptual geologic model of the stratigraphic interval of interest (Fig. 3A), which may be based on geological understanding, the stratigraphic interval from regional analysis of the target area, or modern analogs. At this step, the objectives of the simulation are set. Before starting the modeling process, it is essential to (1) understand the tectono-sedimentary evolution of the stratigraphic interval of interest, including the depositional processes, sedimentary facies, tectonic subsidence and periods of erosion or hiatus and (2) gather and prepare data required for modeling.

Forward stratigraphic modeling is capable of simulating deposition on spatial scales from a meter (e.g., flume tank) to hundreds of kilometers (several feet to hundreds of miles). Model grid cells can vary in size in both the horizontal and vertical dimensions, depending on the model objectives and required resolution. It can produce fine-scale heterogeneous sedimentary models. Typical grid cell dimensions for simulating reservoir heterogeneities range from tens to hundreds of meters. Most commercially available software packages such as Sedsim, DionisosFlow and GPM can generate 3D stratigraphic models.

Step 2 is to screen and set various input parameters used in the simulation (Fig. 3B). This is followed by running the simulation for the target time interval to generate a preliminary model (Fig. 3C). The simulation outputs are then compared with spatial and temporal data, such as from seismic sections and time slices, well logs or core data (Fig. 3D). If the discrepancy between the simulation results and the observed geological data is numerically inadequate or fails to address project objectives, steps 1 through 3 are repeated iteratively by adjusting input parameters until the modeled outputs and the known geological data adequately agree. The final result is a stratigraphic sedimentary model.

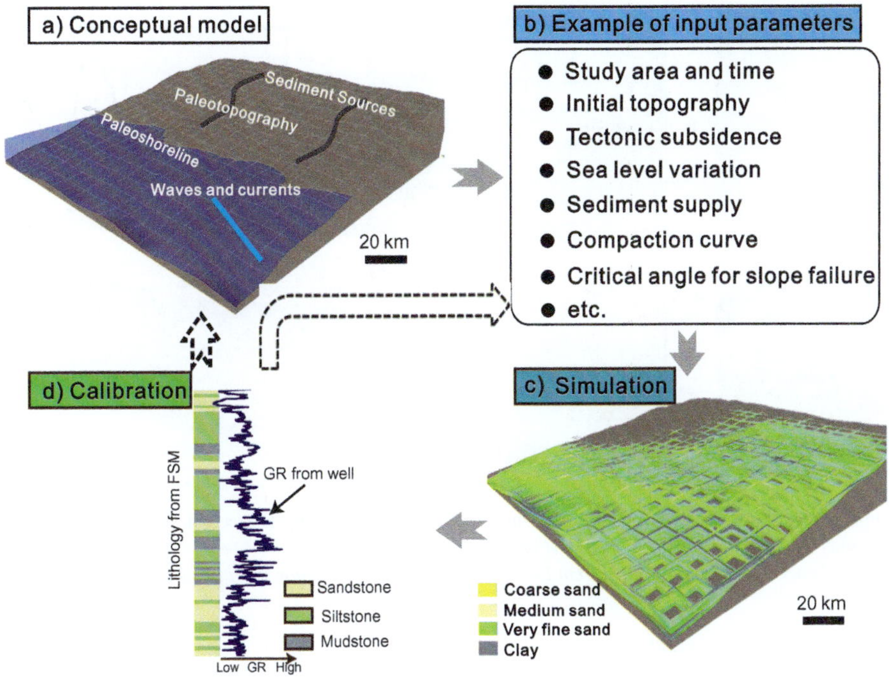


**FIG. 3** Schematic workflow for forward stratigraphic modeling. (A) Concept for model setup. *Brown shading*, basement surface; *purple shading*, sea surface; *gray lines*, sediment sources; *blue lines*, wave direction. Example grid geometry is shown. (B) Example of input parameters. (C) Example of model output illustrating predicted heterogeneity in sediment type. (D) Example of calibrating model output to observed data, which in this case is a comparison between modeled lithology and a gamma ray (GR) curve in a well. *(Panels A and C: Adapted from Liu et al. (2021) and reprinted with permission of the AAPG whose permission is required for further use. AAPG © 2021.)*

## *Input*

The main input parameters required in forward stratigraphic modeling include model setup, size of modeling area, the topography of the bottommost model layer, tectonic subsidence history, history of sea level change and aspects of sediment supply (e.g., some of the processes and features of sedimentary systems shown in Fig. 2). Sediment supply is specified by source locations, supply rates, the composition of clastic sediments and growth rates and erosion rates for carbonate and/or organic sediments. In addition, if a more intricate geologic model is needed, other input parameters for describing geologic processes such as wave, tide, contour current, gravity flow and eolian deposition may be specified. Sources of data for these modeling parameters may be the literature, geological surveys (especially 3D seismic data), drilling data, communications with experts and modern analogs, among many others. Some of the main input parameters are elaborated below.

### Time period and model area

In forward stratigraphic modeling, simulation duration refers to the time period over which the modeled rock units were deposited. This could range from hundreds of thousands to many millions of years. Sources of information for the age of subsurface units include published literature, paleontology and radiometric dating. Model variables are calculated for a time interval $\Delta t$. The value of $\Delta t$ depends on project objectives; for a given package of geologic units to be modeled, $\Delta t$ is set to a value equal to or less than the smallest time interval that is necessary to resolve each geologic unit of interest. If $\Delta t$ is larger than the unit deposited over the shortest amount of geologic time, that unit, along with all the sediments deposited during the time interval, are simulated together. The smaller the time interval, the higher the temporal resolution of the model.

Model setup also requires the geographic area of the basin (sometimes called the area of interest) to be defined. The spatial resolution of the model can be varied by increasing or decreasing the numbers of grid cells within the simulation area. Some modeling packages (e.g., Sedsim) have the capability of generating nested models by simultaneously producing models of different resolutions (e.g., a coarse-grid model and a fine-grid model nested within a regional one).

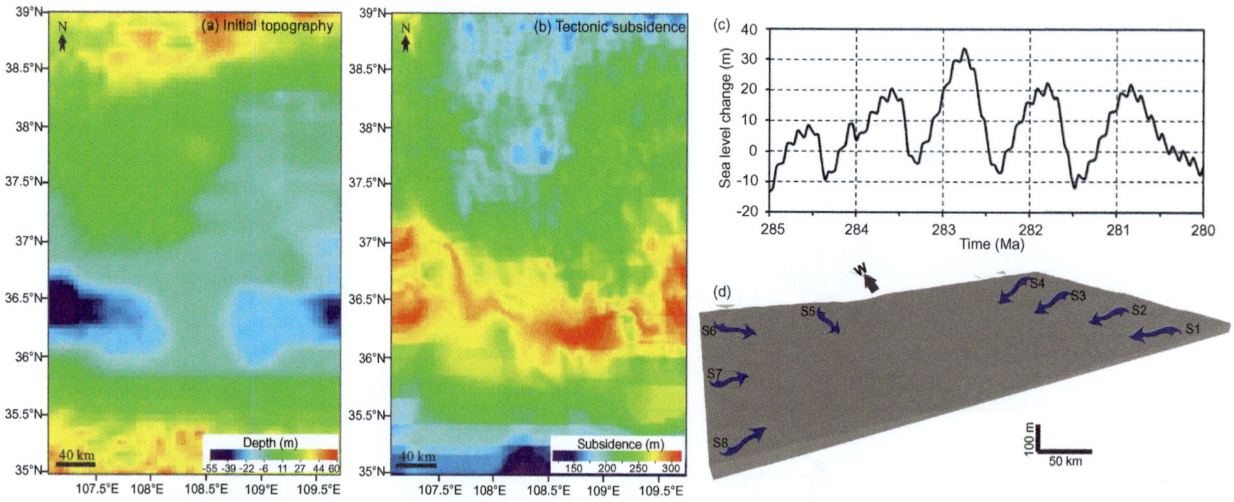


**FIG. 4** Examples of input parameters for a forward stratigraphic model. (A) Initial topography, (B) tectonic subsidence, (C) variations in water level and (D) sediment source locations (*blue arrows* labeled S1 to S8). *(Panels A and B span 350 km (217 mi.) east–west and 450 km (279 mi.) north-south with a 5 km (3.1 mi.) grid spacing. Panels C and D: Reprinted from Liu et al. (2016) with permission from Elsevier.)*

## Initial topography

The initial topography is the paleotopography of the layer upon which deposition is occurring; it is the foundation of the simulation (Figs. 3A and 4A). It is crucial to construct the initial topography accurately because it strongly influences computed depositional processes, facies and sedimentary stacking patterns. Several methods are recommended for reconstructing the initial topography, including flattening the top horizon of a deepwater facies on interpreted seismic data, backstripping, interpreting depositional facies and determining paleowater depths of the underlying formations upon which the initial base level (sea level or lake level) position is set.

## Tectonic subsidence

Accommodation is created by both eustatic sea level and tectonic subsidence. Tectonic subsidence (Fig. 4B) is the fundamental control on sediment accumulation and long-term stratal preservation (Burgess and Wright, 2003). Tectonic subsidence rates in most forward stratigraphic modeling programs can vary through time and space. Backstripping provides an effective method in computing the total subsidence of a stratal unit by separating the tectonic subsidence from the isostatic loading. Three-dimensional isostatic subsidence in Sedsim software is modeled by considering the loading of the sediments and the rock layers, the mantle density and the flexural rigidity of the crust. Most forward stratigraphic modeling programs can incorporate vertical tectonic movement in a simulation. A subsidence/uplift value is defined for each grid node of the initial topography, and the values can be changed at any time during the simulation.

## Sea level change

Sea level change is another factor controlling stratigraphic architecture, sediment stacking patterns and sedimentary facies. A marine depositional environment is influenced by two frequencies of sea level changes: low-frequency global eustatic sea level curves (e.g., Haq et al., 1987) and high-frequency eustatic oscillations that are associated with Milankovitch cycles (e.g., Forkner et al., 2010; Koerschner and Read, 1989). Composite sea level curves can be generated by modulating low-frequency sea level curves with a high-frequency one consistent with the Milankovitch cycles to produce an appropriate high-frequency composite sea level curve with amplitudes corresponding to the paleo-climate settings (e.g., greenhouse or icehouse; Liu et al., 2020) (Fig. 4C). This assumes that the components of orbital eccentricity, obliquity and precession extend beyond the currently observed limits of around six million years. The amplitude of sea level fluctuations associated with the frequency of change is difficult to determine without running trial experiments. However, the rate of movement of facies belts over bathymetric slopes can be used as a constraining criterion; such movement is a function mostly of residual slopes associated with sediment grain sizes.

## Sediment type

In addition to sediment source areas, modeling clastic sediments requires input parameters such as water discharge velocity and volume, and concentration and composition of the sediments. In some programs, clastic sediments composed of varying grain sizes and densities can be predefined; these properties are typically derived from the literature, published databases and/or modern analogs. For example, in the software package Sedsim, four predefined clastic and four predefined carbonate lithologies provide physical and chemical characteristics that are necessary for forward stratigraphic modeling.

Various algorithms have been successfully used in modeling the growth and erosion of carbonates. The key for accurately simulating carbonates is to establish the relationship between their growth or erosion and the controlling factors such as water depth, nutrient supply, wave energy, clastic input, temperature and salinity. Modern analogs and databases are often used to obtain these parameters and to define their interrelationships. For example, in Sedsim, carbonate deposition is simulated using rules of fuzzy sets, whereas a diffusion algorithm is adopted to simulate carbonate deposition in DionisosFlow.

As with carbonate modeling, the deposition of organic-rich sediments can be simulated by forward stratigraphic modeling via empirical relationships (diffusion or FUZZIM logic set). The deposition and spatial distribution of organic matter are controlled by sedimentary environments and facies, whereas their preservation depends on the redox conditions of the depositional setting and depositional rates. Type I and II organic matter are modeled by considering factors such as water depth, distance from the shoreline, salinity, temperature, climate setting (e.g., icehouse or greenhouse) and the depositional rate of sediment input. Type III organic matter is chiefly found in plant debris and is the dominant kerogen type in coal. Because this type of organic matter is typically deposited at the water table close to the coast or in the lower floodplain of fluvial-deltaic systems, depositional processes considered for modeling Type III organic matter include paleogeographic location, water depth and rates of terrestrial input.

Fig. 5 illustrates one example in which forward stratigraphic modeling was used to predict the original source rock quantity (total organic carbon, TOC, in weight %) and the original source organic quality (hydrogen index, HI, in mg hydrocarbon/g TOC) of petroleum source rocks. This study by Mann and Zweigel (2008) considered three essential conditions for organic matter modeling: the source of organic matter (marine and terrigenous), the preservation conditions of the organic matter during deposition and burial and the basin-fill stratigraphy—essentially, the parasequence stacking. The authors first demonstrate the method on a synthetic case and then deploy it to model basin fill and organic properties of the Late Jurassic Spekk Formation, a petroleum source rock of the mid-Norwegian Continental Shelf. Model output, which was constructed on a 2D seismic transect crossing the Froan Basin and calibrated to two wells coincident with the line, includes sand fraction, anoxia and total organic carbon (TOC) and hydrogen index (HI), the latter of which are shown in Fig. 5. Units I, II and II correspond to organic-rich petroleum source rocks in the lower part of the Spekk Formation, and Unit V is a source rock in the upper Spekk. Unit IV corresponds to the Rogn Sands. The patterns observed in the TOC and HI maps provide insight on the most important environmental conditions during deposition of the Spekk source rocks. Other studies that used forward stratigraphic modeling to predict organofacies include Bruneau et al. (2016) for the Lusitanian Basin, Bruneau et al. (2018) for the Paris Basin and Grohmann et al. (2021) for the Eastern Mediterranean Sea.

Although salt is less commonly encountered in forward stratigraphic modeling studies than siliciclastic and carbonate sediments, it can be an important feature to consider for many reasons including the effect of sediment routing around salt domes and canopies, the effect of salt kinematics on basin structure (i.e., salt-induced minibasins and paleobathymetry), sand draping on salt structures and its importance as an effective seal for petroleum accumulations. Hawie et al. (2019) discuss the incorporation of salt into forward stratigraphic modeling and show its influence on siliciclastic reservoir rocks in the Scotian Basin, offshore Canada. Salt can be modeled using rule-based or diffusion algorithms similar to that for simulating carbonate reefs and platforms.

## *Simulation and calibration*

Following the assignment of input parameters, simulation usually occurs in two stages. First, a coarse model is simulated—one with relatively large grid cells—to minimize computational time while also screening the suitability of the input parameters. For example, model outputs such as the depositional angle of repose and carbonate production rates are compared with observations from sediment isopach maps, seismic lines and so forth. If the discrepancy between the simulation results and observed data is large, the input parameters with relatively larger uncertainties should be iteratively modified until the modeled outputs and the known geological data are comparable. Next, the full-scale model is simulated.

Model calibration typically occurs on both a map basis—observed vs calculated output—and a well basis, in which predicted quantities are extracted from output maps and compared with well data. One type of map-based calibration of forward stratigraphic models is comparing modeled sediment distribution maps with gross depositional environment

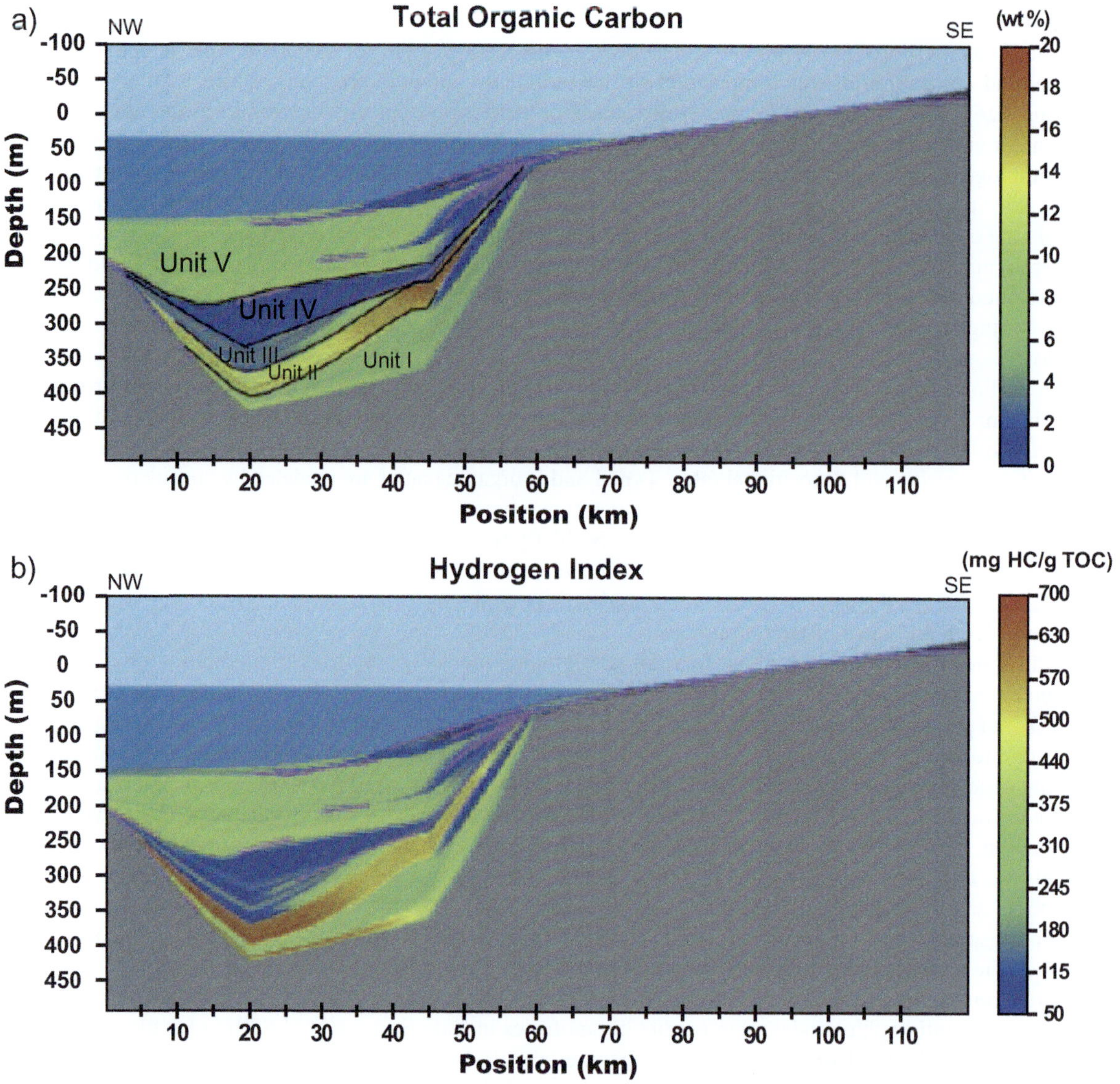


**FIG. 5** Example output from forward stratigraphic modeling. The model is of the Spekk Formation, mid-Norwegian Continental Shelf. Units I–III are organic-rich petroleum source rocks in the lower Spekk, and Unit V is a source rock in the upper Spekk. Unit IV corresponds to the Rogn Sands. (A) Total Organic Carbon (TOC) in wt% and (B) Hydrogen Index (HI) in mg HC/g TOC). Transect coincides with seismic line IKU-125-187 in the Froan Basin. *(Modified from Mann and Zweigel (2008). Reprinted with permission.)*

maps; although the comparison is qualitative, large-scale discrepancies are visible. Well data are essential for calibrating forward stratigraphic models because they are point locations where data can be compared with vertical extractions from calculated maps.

Fig. 6A highlights several ways in which simulation results were calibrated to observed data in the Ordos Basin, China (Liu et al., 2021). The study was designed to couple forward stratigraphic modeling with basin and petroleum system modeling, which is further discussed below. The stratigraphic modeling effort consisted of two separate components: one for the Permian tight gas and one for the Triassic tight oil petroleum systems. The calibration effort was robust: model output was compared to well data from 71 boreholes that were not used for model building. First, simulation results were compared to lithology, grain size and porosity data from available conventional core (Fig. 6A). Good agreement is observed within the corresponding modeled interval. The model was further calibrated to observed and modeled well logs after applying 2000 m (6562 ft.) of overburden to the simulated results so that the modeled section was appropriately compacted; because stratigraphic models typically comprise vertical sections on the scale of tens to hundreds of meters, the target interval is less compacted in the stratigraphic model than in nature. By using the gamma ray log as a proxy for sand–shale

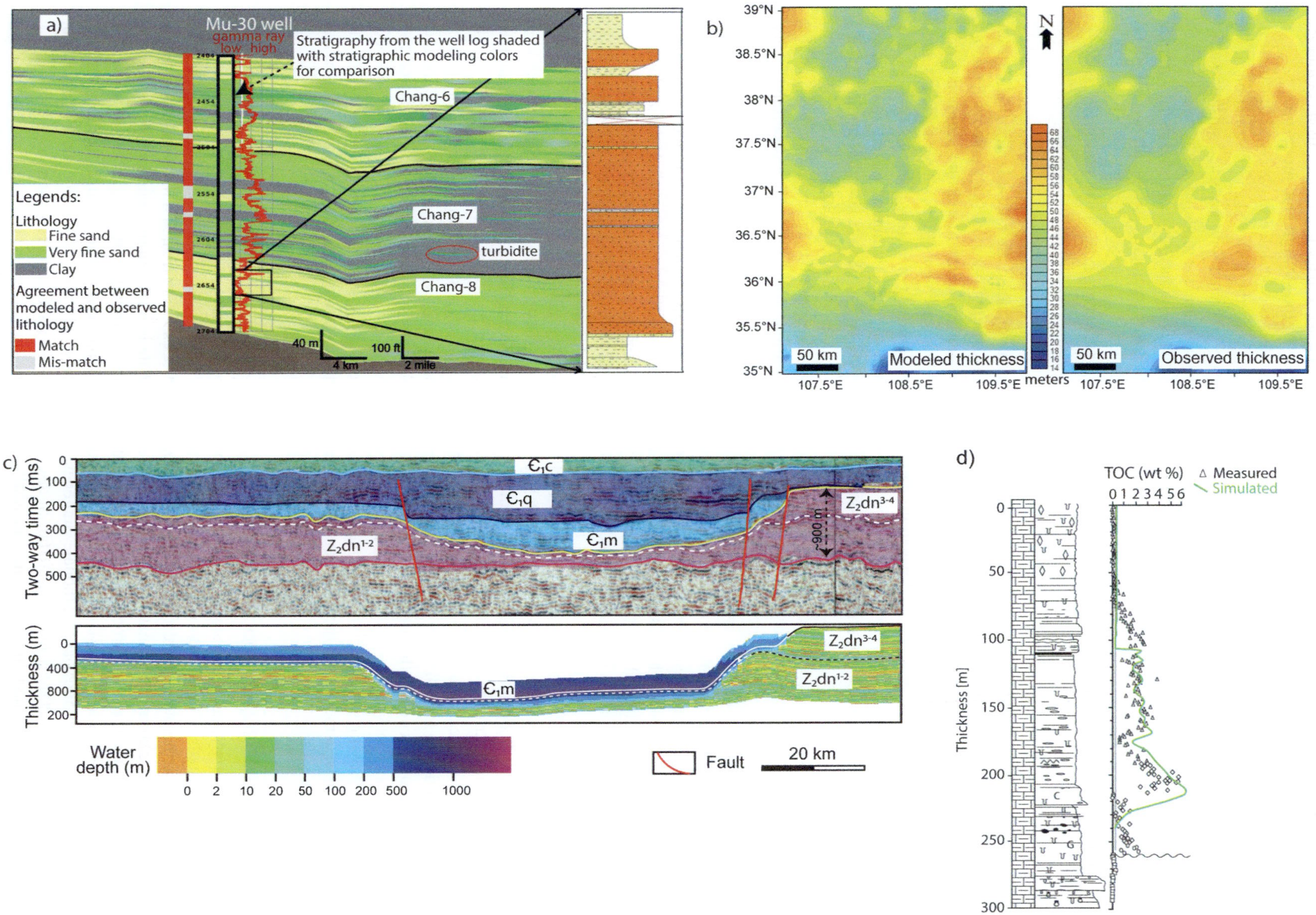


**FIG. 6** Example calibration of forward stratigraphic models. (A) Comparison of stratigraphic thickness and lithofacies between simulation results and data from well logs and cores from the Mu-30 well in the Ordos Basin, China. (B) Comparison of isopach maps between modeled and observed thicknesses of the lower Shanxi Formation in the Ordos Basin, China. (C) Comparison of depositional patterns observed on seismic section (top panel) with simulated facies (bottom panel). Note that the seismic section is in two-way travel time whereas the modeled section is in thickness. (D) Simulated (*green line*) and observed (*symbols*) TOC is from a forward stratigraphic modeling study in the Eastern Mediterranean Sea. *(Panel A: Adapted from Liu et al. (2021) and reprinted with permission of the AAPG whose permission is required for further use. AAPG © 2021; panel B: Adapted from Liu et al. (2016) with permission from Elsevier; panel C: Reprinted from Liu et al. (2020) with permission from Elsevier; panel D: Adapted from Grohmann et al. (2021) and used under Creative Commons Attribution License, https://creativecommons.org/licenses/by/3.0/us/legalcode.)*

ratio, variations of the modeled and observed lithology were found to agree by ~85% given the intrinsic vertical resolutions of each (~15 to 30 cm/~ 6 to ~ 12 inches for well logs and ~10s of m or feet for the stratigraphic model).

One of the most common model outputs used for calibrating stratigraphic models is calculated vs observed thickness of (compacted) model layers; Fig. 6B shows good agreement between isopach maps of the lower Shanxi Formation of the Ordos Basin, China (Liu et al., 2016a). A third calibration option is to compare predicted with observed seismic facies, as shown in Fig. 6C for the Sichuan Basin, China (Liu et al., 2020). Note that in this figure, the modeled stratigraphic section (lower panel) is a subset of that shown on the interpreted seismic line (top panel). This emphasizes that forward stratigraphic modeling focuses on a particular stratigraphic interval of interest rather than the entire geologic section. Finally, Fig. 6D is from a forward stratigraphic model of petroleum source rocks in the Eastern Mediterranean Sea (Grohmann et al., 2021). Despite the intrinsic difference in scale, Fig. 6D displays excellent agreement between modeled and observed TOC.

## Output

Forward stratigraphic modeling provides information on both uncompacted and compacted thicknesses of geologic strata, lithofacies types (e.g., grain sizes, compositions) and characteristics of sedimentary environments (e.g., paleowater depth, distance from the shoreline, sedimentation rate, primary productivity and oxygen concentration at the sediment–water interface). It predicts the spatial–temporal distribution of essential petroleum system elements (e.g., source rock, reservoir rock and seal rock) in basin locations that lack well penetrations. Forward stratigraphic modeling also fills in aspects of subsurface geology that cannot be imaged at typical seismic resolution. For example, for a dominant seismic frequency of 25 Hz, the detection limit (1/4 wavelength) of seismic reflection data deeper than 3000 m (9843 ft.) depth is approximately 30 m (98 ft.). This means that many of the fine-scale features of typical reservoir rocks cannot be imaged by current seismic reflection frequencies. Fine 3D stratigraphic models can contain sedimentary heterogeneities far more detailed than that imaged from seismic data. The modeling process itself can also help to test various scenarios, reduce uncertainties and allow for a better understanding of the geological model.

Fig. 7A illustrates the sand content of a forward stratigraphic model from the Campos Basin, Brazil (Acevedo et al., 2014), a petroleum province where turbidites host significant oil resources. The model was constructed by generating the initial topographic surface—the paleo basin floor—and defining the sediment sources, which were inferred from paleo-bathymetry. Input parameters such as sea level, transport coefficient and diffusion coefficient were varied within plausible ranges. The model was iteratively adjusted such that the amount and type of sediment carried by turbidity currents produced a reasonable match between seismic facies and modeled sequence architecture. Using sand content and geometry as proxies, model output clearly shows several episodes of margin incision and fill, shelf progradation from both sides of the basin and basin floor fans that formed as a result of the turbidity currents (Acevedo et al., 2014).

Fig. 7B illustrates output from a forward stratigraphic model by Hawie et al. (2019) for the Early Cretaceous (~130 to 101 Ma) section of the offshore Scotian Basin, Canada. The aim of the study was to examine the influence of salt canopies and domes on the shelf on the distribution of sandstone reservoir rocks in this gas-rich province. Two aspects of this study are unique with regard to the studies discussed to this point in the chapter. First, gross depositional maps are commonly used in forward stratigraphic modeling for calibration. Instead, these maps were used, along with seismic stratigraphy and bio-stratigraphy, for model input. Second, following model calibration, the authors performed a sensitivity analysis on two input variables—thickness and rock facies—by predefining the uncertainty range of each parameter, assuming independence between them and deploying standard computational methods. The water discharge map, a product of the model run prior to the sensitivity analysis, demonstrates the diversion of sedimentary pathways around the evolving salt canopies and domes within the modeled section.

## Strengths of forward stratigraphic modeling

One strength of forward stratigraphic modeling is the ability to simulate depositional processes from meter-scale (e.g., flume tank) to scales of hundreds of kilometers with variable horizontal and vertical resolutions depending on the desired number of geologic layers. Furthermore, a wide range of sediment types and processes are treated in forward stratigraphic modeling: sediments such as clastics, carbonates, organic matter, salt—and their interactions—and processes such as catchment erosion, eolian, fluvial, marine processes (wave, tide, storm and geostrophic currents), tectonic subsidence and uplift, base level (sea level or lake level), syn- and post-depositional compaction, isostasy, slope failure/slumping, density flows (debris flows to turbidites) and carbonate diagenesis (Griffiths et al., 2001; Griffiths and Paraschivoiu, 1998; Huang et al., 2012; Li et al., 2009; Liu et al., 2016b; Salles et al., 2010; Tetzlaff and Harbaugh, 1989).

As a result, forward stratigraphic modeling has been successfully applied to address various geological issues, including reconstructing sedimentary evolution in various depositional settings (Griffiths et al., 2001; Liu et al., 2016b), examining

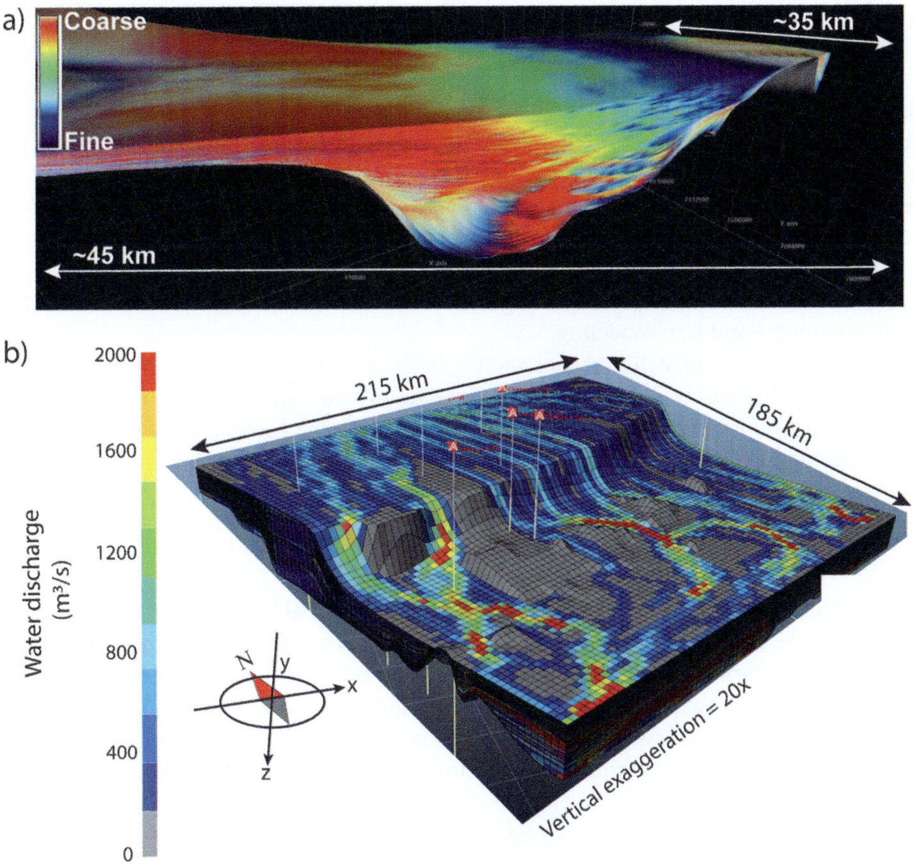


**FIG. 7** Examples of calibrated output from forward stratigraphic modeling. (A) Predicted sand fraction for a sequence in the Campos Basin, offshore Brazil (Acevedo et al., 2014). Basin floor fans, indicated by coarse sediment (*red shading*), were deposited by turbidity currents and fill the deep basin floor. The turbidity currents originated at shallow depths in the upper end of the basin (upper left of the figure). (B) Water discharge illustrating diversion of water flow and thus sediment transport around salt domes and canopies, Scotian Basin, offshore Canada. *(Panel A: From Acevedo et al. (2014), used with permission of Society of Petroleum Engineers; permission conveyed through Copyright Clearance Center, Inc.; Panel B: Adapted from Hawie et al. (2019) and reprinted with permission of the AAPG whose permission is required for further use. AAPG © 2019.)*

key mechanisms for depositional processes (Burgess, 2001), predicting distribution of organic-rich source rocks (Bruneau et al., 2016) and favorable stratigraphic traps in potential petroleum reservoir rocks (Warrlich et al., 2008). Forward stratigraphic modeling quantitatively predicts detailed depositional heterogeneities in time and space, thereby enhancing the understanding of sedimentary systems.

## Basin and petroleum system modeling

Basin and petroleum system modeling allows geoscientists to examine the dynamics of sedimentary basins and their associated fluids to evaluate subsurface processes such as the timing of petroleum generation, migration and accumulation in relation to trap formation. Other key characteristics that are predicted by basin modeling include pore pressure, elastic properties, chemical diagenesis and many more.

Geohistory analysis is a key aspect for investigating the dynamic evolution of sedimentary basins. It quantifies subsidence rates through geologic time to assess the vertical movement of each stratigraphic interval after deposition and indicates basin subsidence and uplift history. Geohistory analysis considers decompaction of each stratigraphic unit to its original thickness and corrections based on paleobathymetry and changes in eustasy, the latter of which may be neglected by some basin modeling software.

Decompaction begins by computationally removing from the present-day input geometry all but the bottommost model layer and metaphorically reinflating it to its depositional thickness. This is accomplished by (numerically) restoring the pore space that was lost due to the load of the overlying section. Next, the penultimate model layer is computationally placed on the bottommost model layer—which begins to compact because of the overlying load—and is decompacted to its original thickness. Then, the third layer from the bottom is deposited and decompacted, while also compacting the layers beneath it. The process continues until reaching the topmost model layer. The sequential process of decompacting and recompacting model layers are carried out during the first simulation assuming hydrostatic pressure conditions; the following simulation proceeds with pore pressure-controlled compaction. Model iterations are required if, for example, pore pressure or chemical diagenesis affect the evolution of porosity during the geologic time considered by the model.

Basin subsidence is caused by sediment and water loading as well as tectonic activity. Partitioning the magnitude of subsidence between sediment loading and tectonic forces is called backstripping, the output of which are paleogeometries that are used repeatedly to model sediment deposition and compaction. This process results in a predicted present-day geometry, which should match the input present-day geometry within a predetermined criterion. The geohistory-based approach described is termed event step modeling.

However, this straightforward approach to reconstructing basin geometry fails in complex tectonic environments such as in salt provinces and in fold-and-thrust belts; layer thicknesses are no longer predictable with simple relationships. A typical solution in this situation is to use a paleo step approach in which the geometry and position of each model layer is reconstructed at discrete times in the basin history. It is therefore critical that structural reconstruction include the time steps that are most important for addressing model objectives.

## *The petroleum system*

The petroleum system concept underpins the input, simulation, output and calibration steps of basin modeling. Codified by Magoon and Dow (1994), the petroleum system concept formalized for the first time disparate descriptions of rocks and fluids in petroliferous sedimentary basins. In so doing, the authors incorporated previous contributions while also introducing concepts and vocabulary that apply to all petroleum types and occurrences, from actively generating to spent source rocks, from biogenic to unconventional systems and from passive margin to strike-slip basins. As a mechanism for evaluating petroleum migration from source to trap, the petroleum system framework (1) organizes information, (2) promotes efficient investigations, (3) clarifies basin geology, (4) distinguishes basin analysis from play and prospect analysis, (5) provides an effective means of communication and (6) serves as a research tool.

The core of the petroleum system concept is captured by a single sentence in Magoon and Dow (1994): "A petroleum system is a *natural* system that encompasses a *pod of active source rock* and all the *related oil and gas*, which includes all the geologic *elements and processes* that are essential if a hydrocarbon accumulation is to exist." Breaking down this thesis into clauses illuminates the fundamentals behind the concept. *Natural* is self-explanatory: petroleum is generated from natural Earth material in the form of plant and animal matter. *Pod of active source rock* refers to the part of the source rock that actively generated, or is generating, oil and gas. The term *related oil and gas* establishes the key point that source rocks and petroleum are genetically linked on the basis of geologic and geochemical evidence.

Finally, *elements and processes* are the six constituents that must coincide in time and space for a petroleum system to exist (Fig. 8). *Elements* are the geologic units in the subsurface that together define the rock system: overburden rock + source rock necessary to generate petroleum, and reservoir rock + seal rock necessary to store petroleum. *Processes* marry the four rock components with two fluid components that depend on relative timing: the formation of traps and the generation–migration–accumulation of petroleum. If the timing of generation–migration–accumulation and the timing of trap formation occur in a favorable sequence, together these six interdependent constituents form the functional unit of a petroleum system.

### Setup

The modeling workflow for basin and petroleum system modeling is illustrated in Fig. 9. This schematic diagram forms the basis for describing each step of the basin modeling process.

### Geologic horizons

The core of a basin model is the set of geologic horizons that extends from the Earth's surface (the seabed, lake bottom, topographic surface, etc.) to the basement, which typically corresponds to a geologic surface (granite, continental crust, etc.) but may also be defined as an arbitrary surface considered to be the "economic basement"—the deepest geologic horizon below which economic deposits of Earth resources are unlikely to exist. The base of a petroleum system model may even extend below the sedimentary layers to the lithosphere–asthenosphere boundary; this is especially critical in rifted areas.

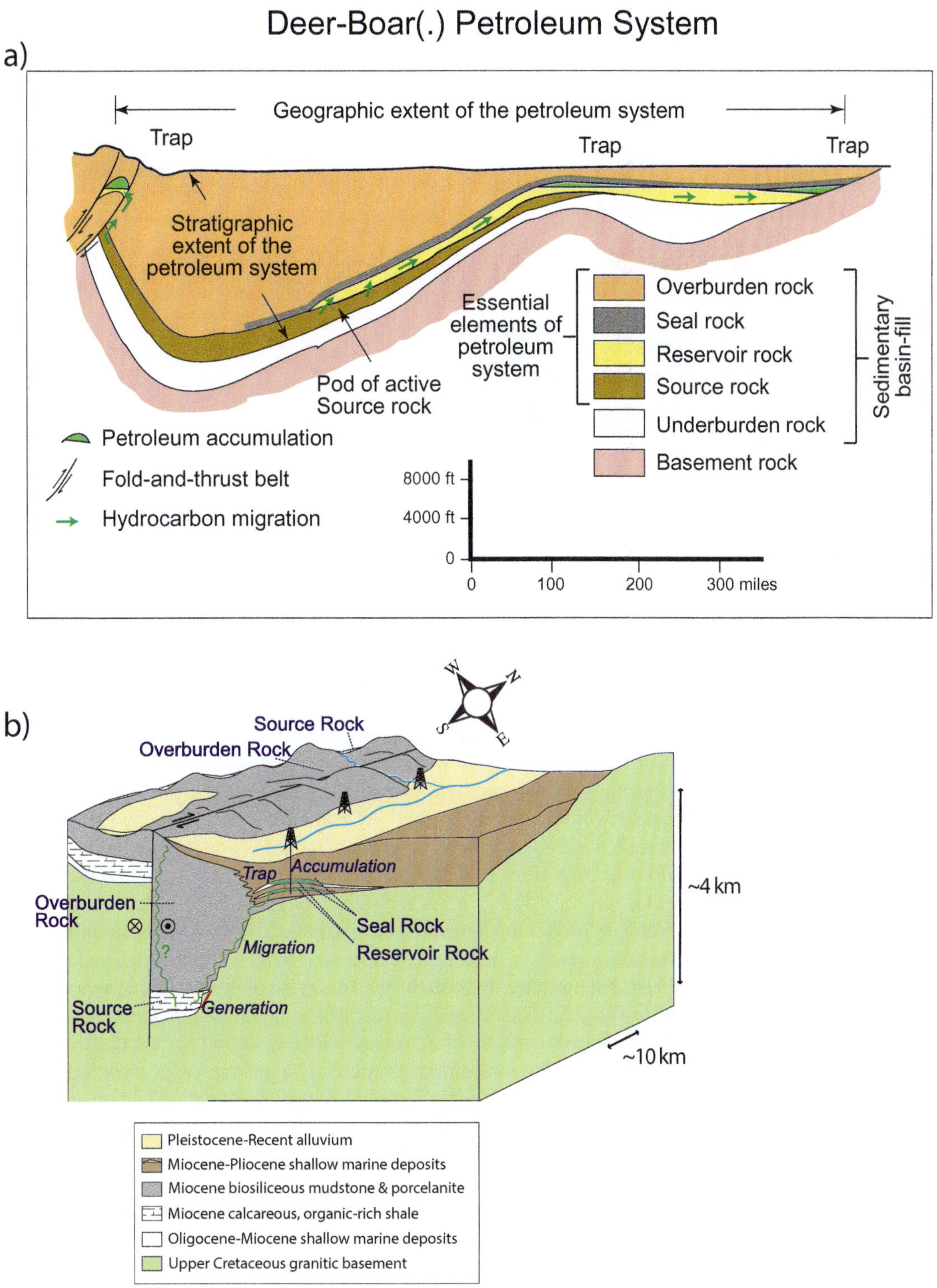


**FIG. 8** Schematic illustrations of the petroleum system concept. Two-dimensional and three-dimensional diagrams illustrate the petroleum system framework as defined by Magoon and Dow (1994). These comprise four essential elements of the rock framework—source rock, reservoir rock, seal rock and overburden rock—and two time-based processes: generation–migration–accumulation of petroleum and trap formation. (A) Cross-section of a fictitious petroleum system. (B) Three-dimensional view of a petroleum system based on the Salinas Basin, California, USA. *(Panel A: Modified from Magoon (1988); panel B: Reprinted from Menotti et al. (2018) and reprinted with permission of the AAPG whose permission is required for further use. AAPG © 2018.)*

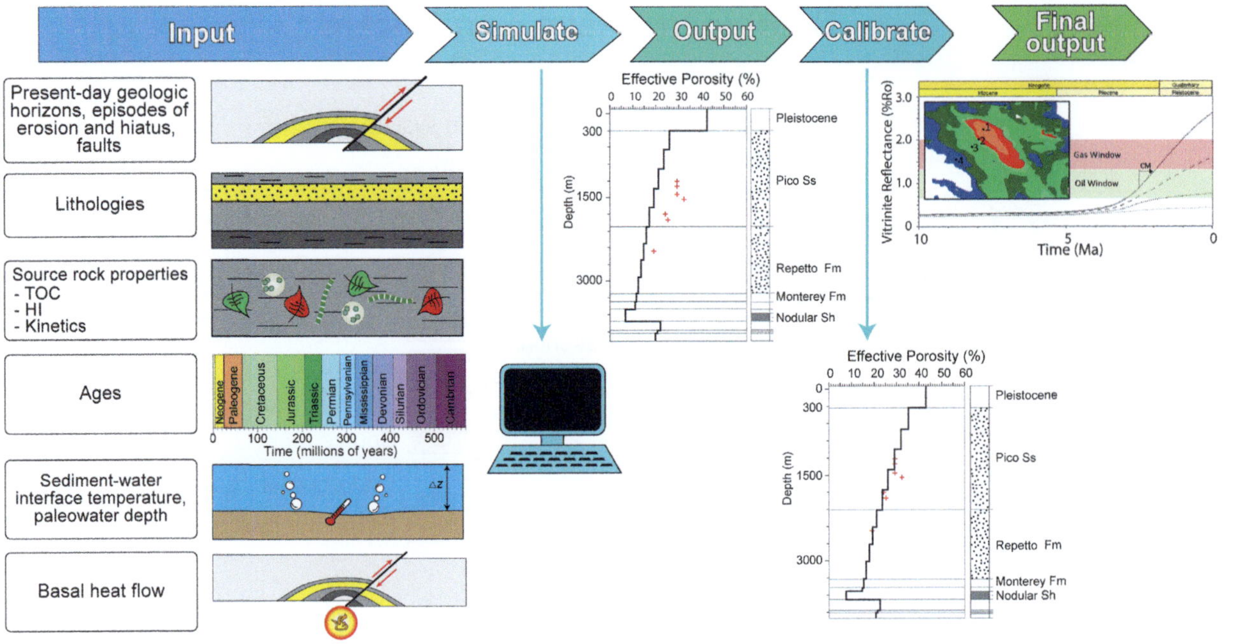


**FIG. 9** The five general modeling steps shown in Fig. 1 applied to basin and petroleum system modeling. The well shown in the Output and Calibrate steps is Bandini-1 in the Los Angeles Basin, California, USA.

Following the definition of the model space, the next step in basin modeling is to construct the present-day geometry of the basin. The model horizon that is the most straightforward to determine is the Earth's surface; topography and bathymetry data cover much of the earth, are readily available online, adhere to rigorous standards for quality control and are gridded over a range of scales. The remaining horizons in a basin model are commonly defined from a combination of data sources; these include wells, seismic surveys, geologic maps and cross-sections, published literature and even potential field data. In practice, horizons are often "pieced together" by combining data of many types.

An example of combining multiple sources of data to construct a basin model is detailed by Schultz (2016) for a 3D model of the Los Angeles Basin, California, USA (Fig. 10). In that model, the top and bottom surfaces originated from national data repositories: the National Geophysical Data Center for topography and bathymetry and the Southern California Earthquake Center Community Velocity Model supplemented by well picks along the basin margin for the basement. The remainder of the model layers were constructed by combining numerous sources of published and unpublished data, including (1) formation tops identified from a comprehensive unpublished database of 394 wells that was compiled by longtime expert on the Los Angeles Basin Tom Wright, (2) additional well control and guidance on interpolating between wells from published cross sections by the American Association of Petroleum Geologists and (3) unpublished structure maps also compiled by Tom Wright. The result is likely the most thorough, publicly available basin model of the petroliferous Los Angeles Basin in existence. Of note is the absence of seismic data in this heavily populated urban area. Of course, the adequacy of this hybrid approach depends on modeling objectives.

The initial basin model, that is, the model that is constructed for the present day, is parameterized in spatial coordinates (some combination of $x$, $y$ and $z$). The time coordinate for each horizon is its geologic age. This simple statement belies the most critical requirement of basin modeling, which is that each horizon in the model was deposited at a single time in the geologic history of the basin. In other words, basin models should be composed of chronostratigraphic units or time-equivalent surfaces. Model horizons that originate from seismic data are timelines by definition and thus "simply" require identification and depth conversion for this first step of model input.

In the absence of seismic data, model horizons are typically constructed using formation tops as identified in outcrop and subsurface wells. By definition, horizons that are compiled from geologic maps and wellbore lithology—either physical rock samples or log signature—are lithostratigraphic rather than chronostratigraphic in nature. Often, this is the only approach available to the basin modeler. However, a model horizon that is constructed based on correlated lithofacies commonly results in horizons that are time-transgressive. Unfortunately, this is an unavoidable pitfall of constructing horizons from formation tops. The absence of seismic data is especially detrimental in terms of geographic spacing of wells; exploration wells tend to cluster around subsurface structural highs, in and around discovered fields, and other areas where

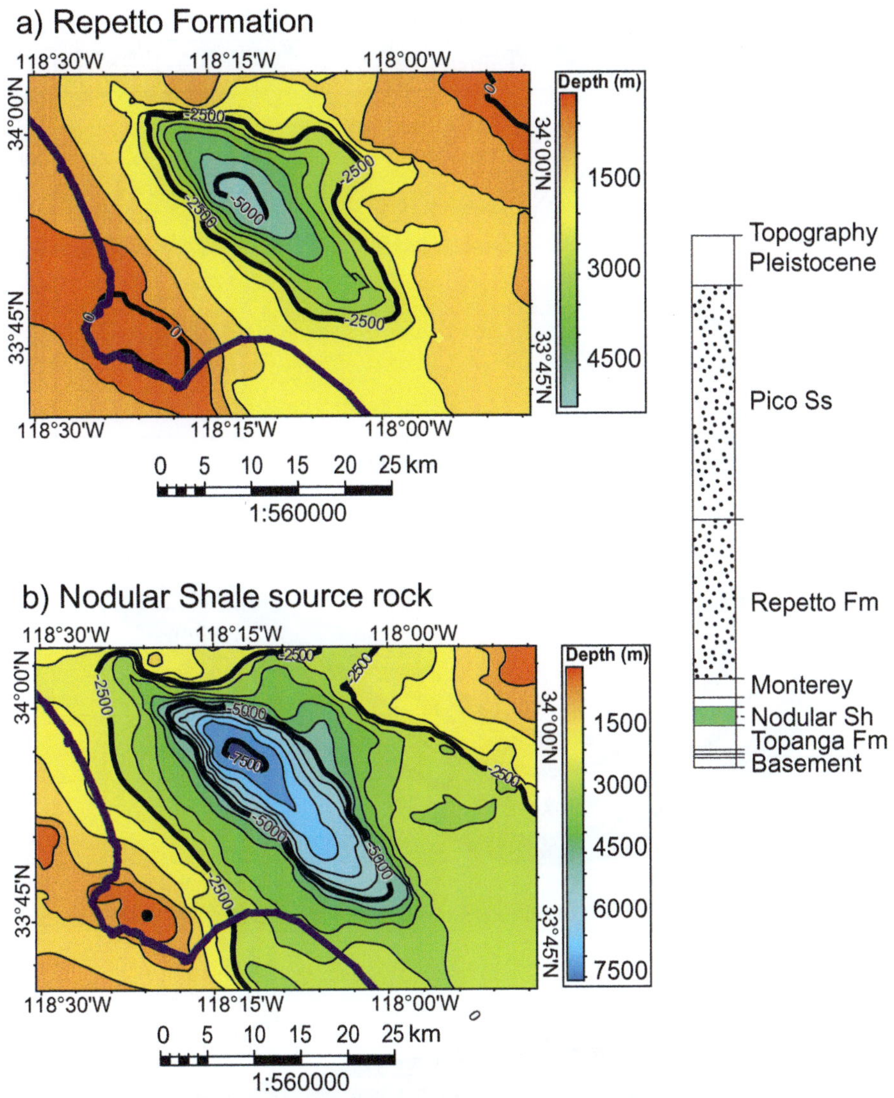


**FIG. 10** Examples of input horizons for a 3D basin and petroleum system model of the Los Angeles Basin, California, USA (Schultz, 2016). (A) The Repetto Formation is an important reservoir rock. (B) The Nodular Shale (*green shading* in adjacent rock column) is an organic-rich source rock. Ss, Sandstone; Fm, Formation, Sh, shale. *(Modified with permission from Schultz (2016).)*

the reservoir rock is present. Because 2D and 3D basin models are composed of continuous grids, horizon depths must therefore be interpolated between well locations.

Finally, the complexity of basin geology coupled with the limitations of numerical modeling sometimes requires the modeler to rely on intuition. For example, the basal part of a shale is often known or suspected to be the petroleum-generative portion of the unit but cannot be distinguished from the overlying, nongenerative shale section. In this case, the unit should be subdivided into two or more layers so that they are modeled separately with different litho- and organofacies. Synthetic horizons are also useful for constructing a hypothetical basin model, which is a valuable tool for testing scenarios or new workflows, in training courses and in frontier exploration provinces lacking data.

To this point, discussion of geologic units in a basin model has centered on geologic horizons utilizing structure maps (Fig. 10). The number of horizons in the model is $n$. From this point forward, the empty space between horizons will be considered as filled with solid rocks and the basin model is parameterized in terms of layers; the number of layers is $n - 1$. Layers are inherently parameterized in thickness; where model horizons are coincident because of truncation and onlapping, among other causes, the thickness between layers is zero. Rock layers define the length (1D), area (2D) and volume (3D) of the basin model depending on model dimensions. Fig. 11 illustrates models constructed in 1D, 2D and 3D.

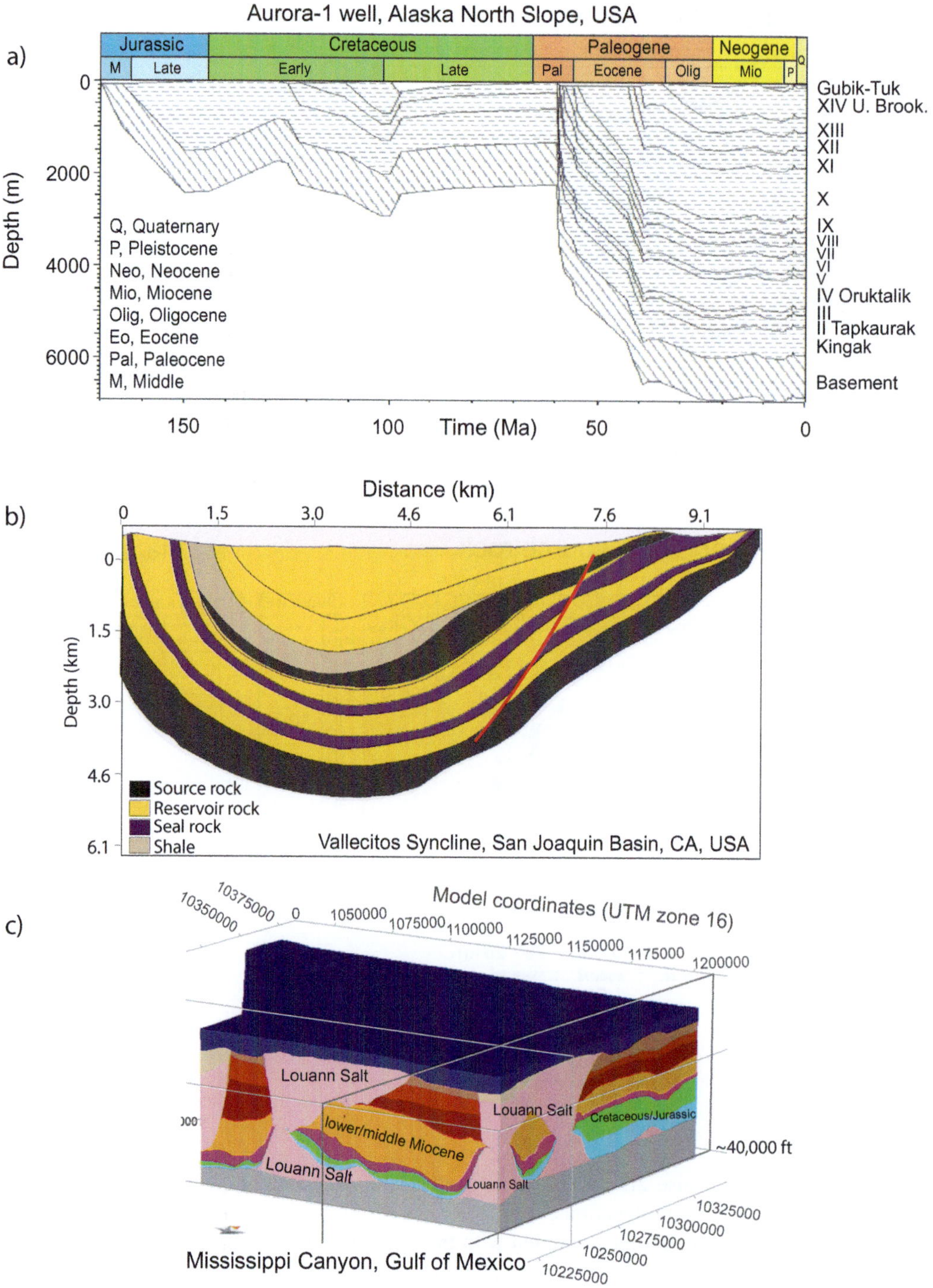


**FIG. 11** One-dimensional, 2D and 3D basin models. (A) Burial history plot of the Aurora-1 well, Alaska North Slope, USA. The model geometry at the present day (time = 0 Ma) is the original 1D model input. The model was built using data in a report by Banet (1993). The layer nomenclature is his. (B) Two-dimensional basin model highlighting the essential elements of the petroleum system of the Vallecitos Syncline, San Joaquin Basin, California, USA. (C) Regional-scale model constructed by Mosca et al. (2018) for the Mississippi Canyon area on the continental slope of the deepwater Gulf of Mexico. *(Panel B: Image incorporates information from He et al. (2014); panel C: Adapted from Mosca et al. (2018) and reprinted with permission of the AAPG whose permission is required for further use. AAPG © 2018.)*

The specifics of each model in Fig. 11 highlight similarities and differences in model construction, data sources and model objectives. The 1D model shown in Fig. 11A was created in modern software by using data tables from a 30-year-old report (Banet, 1993). At the time, the Aurora-1 well was one of a few publicly available boreholes drilled in northeastern Alaska. To further understand petroleum systems in the region, the author incorporated detailed knowledge of stratigraphy and abundant new organic geochemistry analyses to construct a 1D model comprising 38 layers, 11 episodes of hiatus and four erosional events. The 2D model, shown in Fig. 11B, complemented geochemical analyses for the Moreno and Kreyenhagen formations in the Vallecitos Syncline, San Joaquin Basin, California, USA. With biomarker geochemistry, He et al. (2014) distinguished the relative contributions of the two organic-rich source rocks, resulting in a new interpretation of the petroleum systems in the area. The shading in Fig. 11B highlights the essential elements of the petroleum systems. Finally, Fig. 11C illustrates a view of the 3D model by Mosca et al. (2018; discussed further below) of the Mississippi Canyon, Gulf of Mexico; the model spanned 3600 $km^2$ ($\sim$1390 $mi.^2$), was built on a 150 m $\times$ 150 m (492 $\times$ 492 ft.) grid and totaled 24 million grid cells. The model was developed to set the regional geologic and boundary conditions for a higher-resolution area-of-interest model that used to adjust for—in near-real time—changes in pore pressure while drilling.

Project objectives often require that horizontal and/or subhorizontal sublayers be defined within the main model layers. In general, introducing sublayers increases the vertical resolution of basin models. For example, structural geometries are improved when sublayers are inserted. Faults are difficult to implement within a finite element model because their relatively thin nature must follow grid cell boundaries, which are either vertical or horizontal, causing faults to resemble staircases rather than planes. Additional horizontal sublayers reduce the staggered appearance of faults (models containing thrust faults are treated as paleo-step models, as described above). A final example of the need for sublayers involves fluid migration; flowpath physics, described below, fails to charge sand lenses, channelized reservoir rocks and other multi-layer trapping geometries. Sublayers can better represent these scenarios and thus result in more realistic accumulations of petroleum (see Baur and Katz, 2018 for examples).

## Ages

The next step in basin and petroleum system modeling is associating each model horizon with an age, that is, its time of deposition. Geologic formations are dated with a variety of techniques, such as biostratigraphy, magnetostratigraphy, radiometric dating, detrital zircon, organic geochemical biomarkers and tephrochronology. Several dating methods can be used to determine the absolute age of a single rock unit. For example, investigators dated the Round Mountain Silt in the southeastern San Joaquin Basin, California, USA, using microfauna (foraminifera and diatoms), mollusks, large mammal bones and shark teeth, land mammal fossils, magnetic polarity and strontium isotopes. Within the resolution afforded by each individual technique, when taken together, the age indicators bracket the unit's deposition between 16.5 and 14 Ma (publication sources for each age determination are in Pyenson et al., 2009).

In contrast to the Round Mountain Silt example, horizon ages in the Gulf of Mexico are determined primarily with a single methodology—biostratigraphy. For a basin modeler, determining the ages of model horizons using micropaleontological information alone is daunting. Fortunately, a comprehensive study by Olson et al. (2015) developed a robust chronostratigraphic framework for Mesozoic strata in the Gulf of Mexico; this was founded on age-diagnostic biostratigraphic data in concert with seismic stratal correlations. The authors intended the framework to (1) apply equally well on the regional to the basin scale; (2) provide a means for correlating units between offshore and onshore sections; (3) establish the presence and age of petroleum source rocks and (4) aid non-specialists faced with micropaleontological information, among other reasons. Basin modelers grapple with the same issues during model construction.

The specificity of the ages determined with dating methods belies the common problems encountered in compiling ages for stratigraphic horizons. Each method suffers from various sources of uncertainty. For example, problems inherent to absolute age determination from biostratigraphic data include the following: (1) the lifespan of a particular species of microfauna may last for several million years, (2) a microfauna assemblage in a geologic unit of interest may be contaminated by younger species because of caving while drilling and (3) continual revision to the geologic timescale affects the absolute ages of age-diagnostic fossil horizons.

Despite the numerous dating methods available, often, the only information available for the age of a geologic horizon is a qualitative description such as "Late Cretaceous." Interpreting this vague statement is problematic: was the unit deposited during that epoch as defined in the geologic timescale (100.5 to 66.0 Ma; Walker et al., 2018) or during a narrower time interval that is geologically significant and well known within the basin? In cases such as these, the basin modeler uses their best judgment utilizing information from analog basins, stratigraphic relationships and/or trial and error.

Finally, and perhaps paradoxically, ages must be assigned to horizons that no longer exist in the basin, that is, they have been eroded away, and to hiatuses (time periods of no deposition). This underscores that basin models are based on events;

each geologic event that modified basin geometry and geology must be included if relevant to modeling objectives. The mechanics of incorporating erosion and non-deposition into the basin model framework differ by software package.

## Paleobathymetry

Paleobathymetry, also called paleowater depth, is a key input parameter in basin modeling. It is variously determined from microfauna assemblages, sedimentary facies, geochemical data and clinoform heights interpreted from seismic data and/or deduced from the tectonic environment. The paleowater depth at the time of deposition of each model layer and/or at other times in basin history is input by the user; the software restores the top of each layer to a depth below the sea level, thereby establishing a fixed vertical datum required for the geohistory analysis.

Paleobathymetry controls the paleogeometry of basin models. Although much lighter than sediments, water also contributes to the load exerted on the basin, thereby affecting its subsidence. Paleobathymetry also plays a significant role for thermal history analysis because together with estimated paleo-surface temperatures, it defines the upper thermal boundary condition of a basin model, the sediment–water interface temperature. See Immenhauser (2009) for a comprehensive compilation of paleobathymetry indicators in the rock record.

Fig. 12A illustrates the paleowater depth used for the 1D model of the Aurora-1 well, Alaska North Slope, USA, shown in Fig. 11A. The maximum water depth was its deepest (~250 m) about 100 Ma; several times, the basin had no water at all—it was subaerial. Banet (1993) based these values on sedimentary facies.

## Hiatus and erosion

Additional important input parameters are hiatus and erosion because they represent periods of nondeposition and removal of rock units, respectively. The amount of lost section and lithology can be critical for source-rock maturity and petroleum charge timing, reservoir rock properties and remigration of fluids caused by regional tilting.

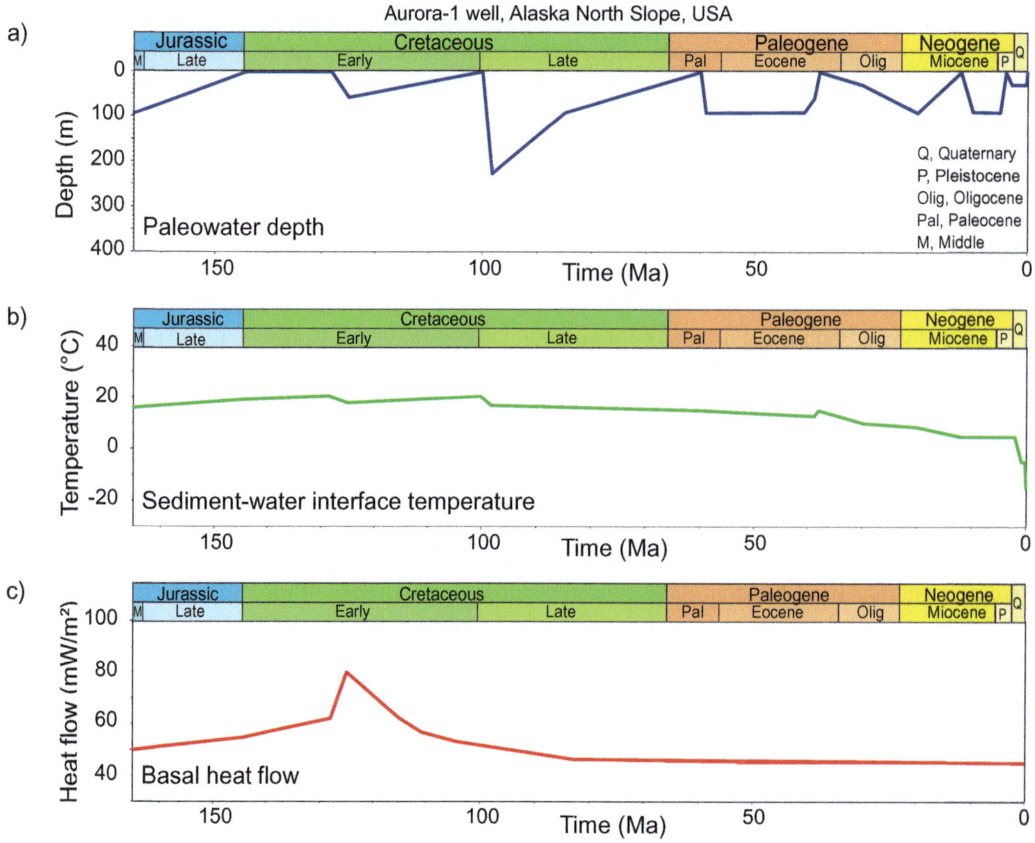


**FIG. 12** Boundary conditions for the 1D basin model of the Aurora-1 well, Alaska North Slope, USA. (A) Paleowater depth, (B) sediment–water interface temperature and (C) basal heat flow.

Well data such as vitrinite reflectance, fission-track and sonic-porosity logs can be used to estimate the amount and time periods of erosion that have occurred in a sedimentary basin. A noteworthy strength of 1D basin modeling is that scenarios of erosion can be quickly constructed, simulated and calibrated with these data. Burns et al. (2007) compared such methods for the Alaska North Slope, USA, as did Henriksen et al. (2011) for the Barents Sea. Unsurprisingly, both authors concluded that net erosion magnitudes differ by method. (Burns et al. (2007) favored the sonic-porosity log method, arguing that vitrinite reflectance and fission track data require knowledge of the thermal evolution of the basin; at the same time, he acknowledged that vitrinite reflectance may be more reliable in other basins. Henriksen et al. (2011) found that the standard deviation in net erosion values decreases when multiple methods are combined, thereby mitigating issues of data coverage and quality.

Other methods to constrain the timing, areal extent and amount of erosion include seismic interpretation. For example, seismic sections are flattened to the interpreted unconformity to estimate the depositional setting prior to erosion. The original (pre-erosion) thickness can then be reconstructed for 2D and 3D basin models within basin modeling software.

Multiple episodes of hiatus, uplift and erosion occurred at the location of the Aurora-1 well, Alaska North Slope, USA, and are evident in the burial history model shown in Fig. 11A. From 165 Ma to the present day, Banet (1993) documented eight periods of hiatus—all but one of them lasting 2 Myr or less—and six erosion events. The longest lasting erosion event occurred from 144 to 127 Ma, and its geological expression, termed the Lower Cretaceous Unconformity, is widespread across the basin both in outcrop and the subsurface.

## Lithofacies

The composition of each model layer is defined by its lithofacies. The choice of lithology affects everything from the seal properties of an oil or gas reservoir rock (e.g., capillary entry pressure) to geomechanical properties, to material properties such as thermal conductivity and heat capacity, to the initial porosity and compaction behavior with sedimentary burial. These properties and many more are needed to solve the numerous equations that govern basin evolution and petroleum generation such as heat flow, fluid flow, mechanical and chemical diagenesis, rock mechanics and so forth.

Information about the lithology of model layers comes from several sources, including cuttings and core from boreholes, outcrops, geologic maps, literature and assumed analogs. If wells are sparse and/or spaced at distances greater than the cell size of a basin model, interpolation is necessary, both between the wells and beyond them to the model limits, possibly creating unrealistic lithofacies. A similar problem ensues if wells are densely populated; in maturely explored oil and gas fields, boreholes may be spaced so closely together that interpolation between them may result in discontinuities or sharp contacts where lithology changes from well to well.

Although it varies by software package, lithofacies are typically specified in basin models as pure rock types (e.g., "sandstone," "shale," or "limestone"), mixtures of rock types and (rarely) combinations of minerals. The properties and compaction behavior of siliciclastic rock types are well understood from calibrated laboratory measurements. The situation is different for carbonates, mainly because of the diversity of texture, fabric, pore type and secondary processes that may occur during basin evolution, such as cementation and dolomitization. Salt represents a special lithology because it does not compact and acts as an impermeable seal rock (see below).

Mixtures of lithologies for a model layer can be specified by percentages of each rock type. For example, a unit called "sandy shale" could be specified in basin modeling software as a percentage of each lithology—30% sandstone and 70% shale for example. The computational manner in which rock properties are combined in mixed proportions differs both by physical property and by parameter. For example, transport parameters such as thermal conductivity are mixed according to layering of the grains and flow direction (Hantschel and Kauerauf, 2009). Parameters such as permeability and capillary entry pressure, however, depend on functions for accurate mixing (Hantschel and Kauerauf, 2009). Because of these complexities, a rule of thumb when mixing lithologies in basin modeling is to combine no more than two or three rock types.

Fig. 13 illustrates facies maps from deepwater settings, each of which was applied to model layers in 3D basin and petroleum system models. The image in Fig. 13A from Mosca et al. (2018) covers an area that is 1/10th the size of the regional model shown in Fig. 11C. This area-of-interest map aided engineers and geoscientists while drilling target Well A, as did calibration data from the offset Well B and the daily update of pressure conditions in the high-resolution model. The map in Fig. 13A is sand-shale percentage that was determined from the discontinuity seismic attribute, which is the inverse of seismic coherency. The method requires calibration to net-to-gross sand percentage data from nearby oil and gas fields but where possible, resolves sandstone layers that are 30 ft. (10 m) or larger.

In contrast to the seismic attribute approach of Mosca et al. (2018), Schulz's initial approach was more qualitative. It combined well observations, published stratigraphic columns and interpretations of submarine fan morphology and composition to derive submarine fan facies maps for key reservoir rock layers in a 3D basin and petroleum system model of the

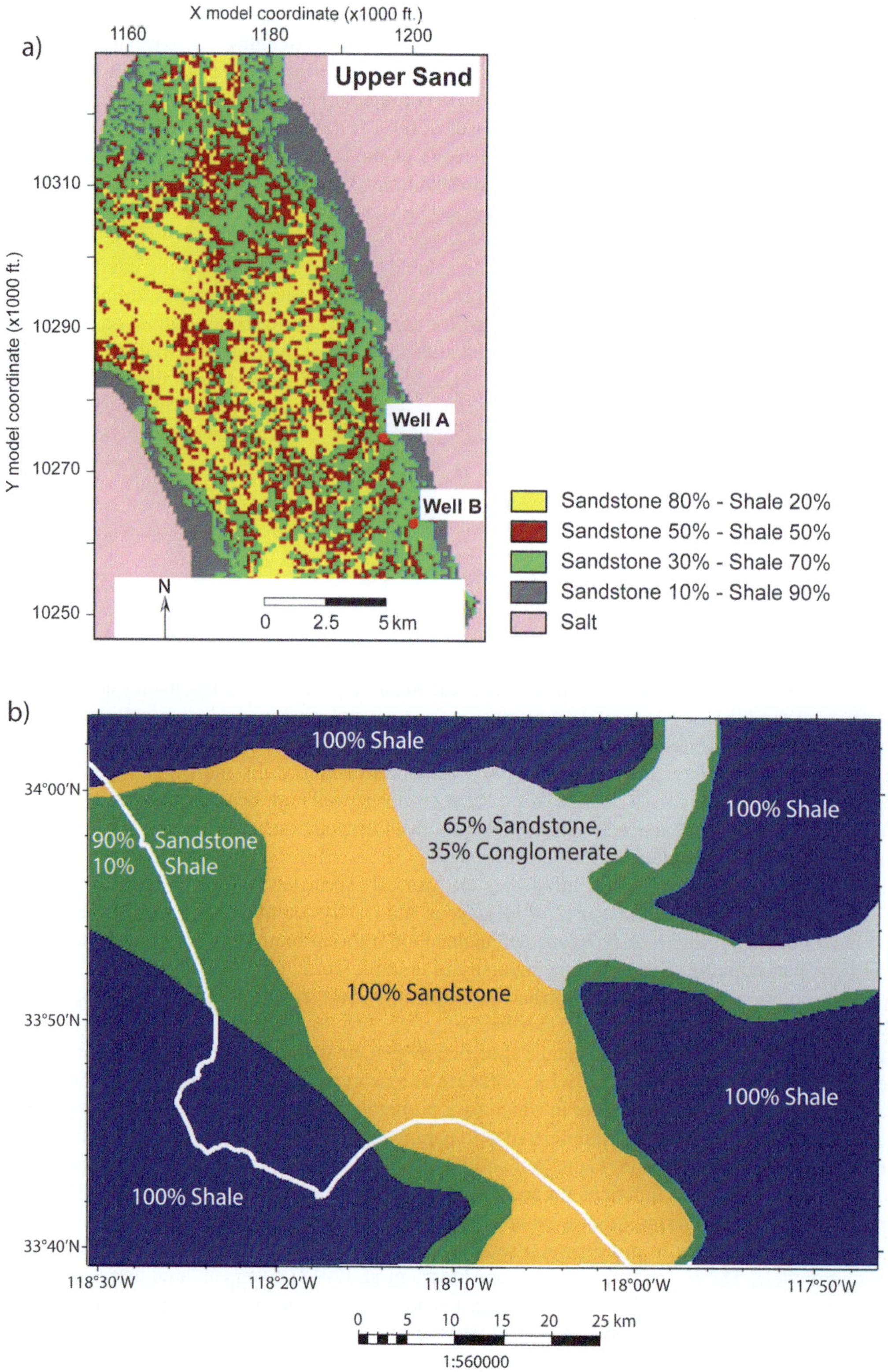


**FIG. 13** Examples of lithofacies used in basin and petroleum system modeling. (A) Lithofacies map for a high-resolution 3D basin model by Mosca et al. (2018) in the Mississippi Canyon, Gulf of Mexico. The map was created from the discontinuity seismic attributes, which is the inverse of seismic coherency. Wells A and B (filled cyan circles) were the target well and the reference (offset) well, respectively. Well A was drilled while guided by pore pressure predictions from the basin model, which was updated after each day of drilling. (B) Lithofacies map for the Monterey layer of the 3D basin and petroleum system model of the Los Angeles Basin, California, USA by Schultz (2016). *(Panel A: Adapted from* Mosca et al. (2018) *and reprinted with permission of the AAPG whose permission is required for further use. AAPG © 2018.; panel B: Modified with permission from Schultz (2016).)*

Los Angeles Basin, California, USA (selected horizon maps from that model are shown in Fig. 10). Fig. 13B shows that both pure and mixed lithologies were used to model the sediment types that resulted from turbidity, debris, fluidized sediment and grain flows during the late Miocene. Nearly all the oilfields in the Los Angeles Basin produce from reservoir rocks of this age (~5 to 7 Ma); thus, accurate lithofacies maps are essential for this unit and other reservoir rock layers.

Salt is treated as a layer in basin modeling. Typically, it is introduced into the model as a flat sheet, which then deforms with subsequent sedimentation. Salt geometry must be reconstructed in basin modeling because its geometry changes through time (as do all other model layers). These restorations can be accomplished as paleo-step models—as in a study by AlKawai et al. (2018) for allochthonous salt movement in the Thunder Horse minibasin in the Gulf of Mexico—or as event step models using built-in software tools as evidenced by Allwardt et al. (2009) in three unidentified minibasins in the Gulf of Mexico.

The 3D basin model by Mosca et al. (2018), of the Mississippi Canyon, Gulf of Mexico, shown in Fig. 11C, illustrates a model-building workflow in a salt basin and highlights its practical application. Following interpretation of syndepositional sediments in 3D seismic volumes, salt geometries were restored through time in an external software product. The dynamic geometry of the salt layer, which was originally deposited as a flat sheet at 170 Ma, developed by differential loading of the overlying sediments through geologic time. Its geometry was restored at eight discrete times and subsequently imported into basin modeling software. The salt restoration process was essential to achieve the purposes of this study, which was to predict pore pressure during drilling in near-real time. Thus, determining accurate salt geometries was essential because of their influence on effective stress and compartmentalization of minibasins that created closed or open pressure cells depending on their geometry, among other factors.

## Organofacies

A petroleum source rock is specified in basin and petroleum system modeling by its lithology (e.g., "organic-rich shale"), TOC in weight % and HI in mg hydrocarbon/g TOC. The values of TOC and HI, each of which can be specified as a constant value or can vary spatially—depending on data availability and software package—must be the original amount before the source rock generated petroleum. Original TOC and HI, denoted as $TOC_o$ and $HI_o$, respectively, can be determined empirically by calculating the fractional conversion of petroleum (e.g., Peters et al., 2005), by pyrolyzing samples of thermally immature rock samples from conventional cores, cuttings or outcrops, via forward stratigraphic modeling (e.g., Fig. 5), and/or from the literature. Good examples in the literature of determining and mapping the original organic quantity (TOC) and quality (HI) of source rocks for basin and petroleum system models are found in Peters et al. (2006) for the Alaska North Slope, USA, Peters et al. (2007) for the San Joaquin Basin, California, USA, and Zhang et al. (2019) for the Changling Depression, North East China.

In addition to $TOC_o$ and $HI_o$, modeling petroleum generation requires kinetics, equations that govern the thermal conversion of kerogen to petroleum, the conversion of petroleum to lighter products (secondary cracking) and the conversion of various compounds to other products (e.g., carbonate to $CO_2$), among many other reactions. Although the generation of petroleum from organic-rich source rocks is governed by a complex set of concurrent and consecutive reactions, each with unique kinetics, in basin and petroleum system modeling, the Arrhenius equation is assumed to adequately explain the set of sequential, independent and parallel first-order reactions that occur during the thermal alteration of organic matter. It is a first-order rate equation in which reaction rate is exponentially dependent on temperature: $k = Ae^{-E_a/RT}$, where $k$ is determined empirically and describes the speed of the reaction in units of inverse time (1/myr or 1/s); $A$ is the frequency factor, also in units of inverse time that describes the frequency at which molecules vibrate near each other and thus "encourage" reactions; $E_a$ is the activation energy in kcal/mole, a distribution function that expresses the energy threshold that must be surmounted by vibrating reactant molecules to initiate and continue reactions; $R$ is the universal gas constant ($8.314\,J\,K^{-1}\,mol^{-1}$) and $T$ is the temperature in Kelvin.

The choice of kerogen kinetics activates three critical decision points in basin and petroleum system modeling: (1) the assumed organofacies of the source rock, (2) the choice of software-provided algorithms (often called "default") or custom kinetics and (3) the number and types of petroleum products to be calculated. For example, the chosen kinetics might be appropriate to calculate a single (i.e., bulk) phase of petroleum, a vapor and liquid phase or multiple components that are controlled by the boiling point of hydrocarbons (Hantschel and Kauerauf, 2009).

When choosing kerogen kinetics, the first emphasis should be on obtaining custom values—those that are determined in the laboratory on thermally immature samples of the source rock being modeled. Often, this is not possible, so typically, the next approach is to use software-provided kinetics that are as similar as possible to the location, age and type of source rock as in the model. In the absence of this level of specificity, it is necessary to use generalized kinetics such as those by Pepper and Corvi (1995) and Baur (2019).

Source rock kinetics is a significant source of uncertainty in basin and petroleum system modeling. One geologic complication is that the organofacies of petroleum source rocks vary vertically (i.e., in a wellbore) and laterally across a sedimentary basin; often, samples on which kinetics are determined are unrepresentative of the source rock facies that generated petroleum in the subsurface. Moreover, choosing a single kinetics relationship in basin software requires the user to identify a single kerogen type (Types I, II, III), yet source rocks commonly contain macerals of all types. Finally, summarizing and building on almost three decades of research by prominent kineticists, Stainforth (2009) details numerous problematic assumptions regarding kerogen kinetics. One of the chief concerns is that some of the laboratory setups for determining them are suboptimal (Burnham, 2017; Peters et al., 2015; Schenk and Horsfield, 1993).

## Thermal boundary conditions

Basin and petroleum system models require thermal boundary conditions at the top and base of the model. The thermal boundary condition at the top of the model—usually called the sediment–water interface temperature—is the temperature at the seafloor or at the Earth's surface in subaerial settings (Fig. 12B). Determining the temperature at the sediment–water interface is difficult because detailed knowledge of ocean temperatures is available only for the sea surface rather than the seabed.

Accordingly, three pieces of information are required to reconstruct the sediment–water interface temperature: (1) temperature of the sea surface at the (2) location of the sedimentary basin (i.e., tectonic plate) through (3) geologic time. Paleolatitude is reconstructed from plate tectonic models and databases; these are widely available online. Paleosurface temperature is typically available in the climate literature or is built into some software packages (e.g., Wygrala, 1989). Once the sea surface temperature is reasonably determined, an empirical relationship is used to transform it to the sediment–water interface temperature; among others, Beardsmore and Cull (2001) and Hantschel and Kauerauf (2009) discuss such algorithms. In general, seawater temperature exhibits considerable variability from the surface to a few hundred meters depth, depending on location. At deeper depths, seawater temperatures converge to the Arctic bottom water temperature, which is 4 °C at present but was higher in the geologic past. In subaerial settings, the average present-day temperature is available from numerous sources and should be used as the value at time = 0 Ma in the model.

Finally, the ground temperature of an uplifted basin can differ substantially from global database values. Tong et al. (2016) demonstrate the effect of using standard database temperatures on the thermal maturity of the Cameo Coal in the Piceance Basin, Colorado, USA. Using a novel method (for basin modeling) to reconstruct paleotemperatures—paleobotanical data that depend in part on the morphology of fossil leaves—colder temperatures than predicted by default databases in this intermontane basin change both the timing and the degree of transformation of the source rock relative to those that result from using the standard database.

In basin modeling, the basal thermal boundary condition can be set either as the temperature at the base of the lithosphere (1333 °C) or as the basal heat flow (Fig. 12C), which can be estimated or calculated from models of the Earth's crust and mantle. Present-day heat flow data are commonly available in sedimentary basins. However, determining heat flow through geologic time is critical in basin and petroleum system modeling because it plays an outsized role in subsurface thermal conditions and thus maturation of organic-rich source rocks. One common way to estimate paleoheat flow is to characterize the basin type through geologic time (passive margin, rift, forearc basin, etc.) and associate each with average heat flow values. For example, the average basal heat flow in a forearc basin that is unassociated with magnetism is low; the average values are 20–50 $mW/m^2$. Higher values predominate in an active (syn-rift) rift or passive margin, averaging 65–110 $mW/m^2$ (e.g, Allen and Allen, 2005). The difficulty of this approach lies not so much in determining when and at which stage a sedimentary basin is in the Wilson Cycle, but instead, in choosing a single number at a single geologic time that adequately represents those conditions in the basin model. Consequently, the basal heat flow parameter in basin modeling is highly uncertain.

A crustal modeling approach is a robust method for reconstructing the heat that flows upward from the base of the lithosphere and into a sedimentary basin. Present-day lithospheric thicknesses are usually constrained by potential field modeling and seismic tomography. Lithospheric stretching and compression—described in canonical papers by McKenzie (1978) and White and McKenzie (1989)—result in thickness changes that affect the weight of the entire lithosphere and result in subsidence (with creation of accommodation) and uplift (combined with erosion). Crustal modeling then combines heat flow, lithospheric stretching, subsidence and water depth to predict regional basal heat flow values through geologic time (Hantschel and Kauerauf, 2009). A key advantage of such crustal models is the integration of the regional tectono-stratigraphic evolution on a crustal scale.

One example in which a crustal model was used in basin and petroleum system modeling is by Rüpke et al. (2010), who examined the structural, thermal and petroleum evolution of the Tano Basin, Ghana, Africa, a transform margin setting. Similarly, Tewari et al. (2019) used crustal modeling in the Kwanza Basin, Angola, Africa, presently a passive margin but formerly a margin in which rifting associated with the breakup of Gondwana migrated with time from proximal to distal settings. Modeling incorporated the multi-phase rifting process by subdividing the 275 km (~171 mi.) long section into five tectonic domains. Magmatic underplating can also be integrated in crustal models, such as those demonstrated by Gac et al. (2018) for the South West Barents Sea and by Schenk et al. (2019) for the Exmouth subbasin on the North West Shelf of Australia.

## *Simulation*

The simulation step of basin and petroleum system modeling is the bridge between the input phase of the workflow and the beginning of the *output-calibration-final output* phases. Myriad choices are offered within a simulator interface—and differ widely between software products—but a good first strategy is to minimize options for the initial simulation to allow the modeler to assess simulator performance. A critical geologic criterion of simulator performance is the accuracy with which the present-day geometry was recovered; simulation results comprise not the model layers that constituted the original basin model (unless enforced by the user), but rather the calculated result obtained after backstripping, decompacting and recompacting each geologic layer. Generally, a log file following simulation reports the accuracy of the recovered vs the present-day geometry and indicates if the user should adjust simulation parameters. Other key computational criteria, among many others, include the simulation time, which may govern if model objectives can be adequately met in a reasonable amount of time.

### Migration modeling

One of the two time-based processes of the petroleum system concept is generation–migration–accumulation of liquid and vapor (the other process being trap formation). Petroleum geologists are trained to think of the movement of petroleum in sequential terms: expulsion of petroleum from kerogen in organic-rich source rocks (primary migration), expulsion of petroleum from the source rock and its ensuing movement within a higher-permeability unit (secondary migration) and petroleum leakage from a trap (tertiary migration). However, basin modeling ignores this discrete categorization of petroleum flow and the relative time relationship that is implicit between the categories; in simple terms, basin and petroleum system modeling concerns the movement of petroleum in pore space.

Three main classes of driving forces are used in basin and petroleum system modeling to model the flow of petroleum in the subsurface: buoyancy (gravity and density contrast between water and petroleum), overpressure (viscosity and mobility) and capillary pressure (interfacial tension between water and petroleum that retards petroleum flow). Broadly speaking, in basin and petroleum system modeling, these forces are employed as flowpath, a time-independent approach that depends only on buoyancy; Darcy flow, which chiefly depends on permeability and fluid viscosity; and invasion percolation, which is parameterized as a function of buoyancy and capillary pressure. These methods are implemented in basin modeling software in 2D and 3D; in 1D, generation of petroleum is meaningful, but migration and accumulation of petroleum are not.

Flowpath migration is perhaps the most intuitive of the three main classes of migration physics because the driving force is gravity (buoyancy): $U = (\rho_w - \rho_{hc})gh$, where $U$ is the petroleum potential, $\rho_w$ is the density of water (~1000 kg/m$^3$), $\rho_{hc}$ is the density of petroleum in kg/m$^3$, $g$ is the acceleration of gravity (9.8 m/s$^2$) and $h$ is the column height. This simple formulation is independent of time, indicating that with the flowpath migration method, petroleum migrates and accumulates instantaneously.

The effect of the flowpath equation is that petroleum is expelled from a source rock, migrates vertically to the top of a high-permeability carrier bed and migrates along the steepest path along the base of a seal rock until a trap is reached; migration ends where the trap is overlain by a low-permeability seal rock and breaks through only if the buoyancy pressure of the petroleum is less than the capillary pressure of that seal (Hantschel and Kauerauf, 2009). In some modeling software, the overlying seal rock is parameterized by its permeability, either by the user or via a default value. Petroleum migrates upward by default unless downward charge is specified by the user.

The result of buoyant petroleum migration is that accumulations appear as realistic shapes that are independent of the model grid; that is, they appear similar in shape and size to oil and gas fields. Moreover, flowpath migration adequately captures fill and spill behavior to adjacent structural highs (Fig. 14A). The computational sequence of flowpath modeling provides information as to how the traps evolve at each model time step: (1) flowpaths are calculated in a manner similar to

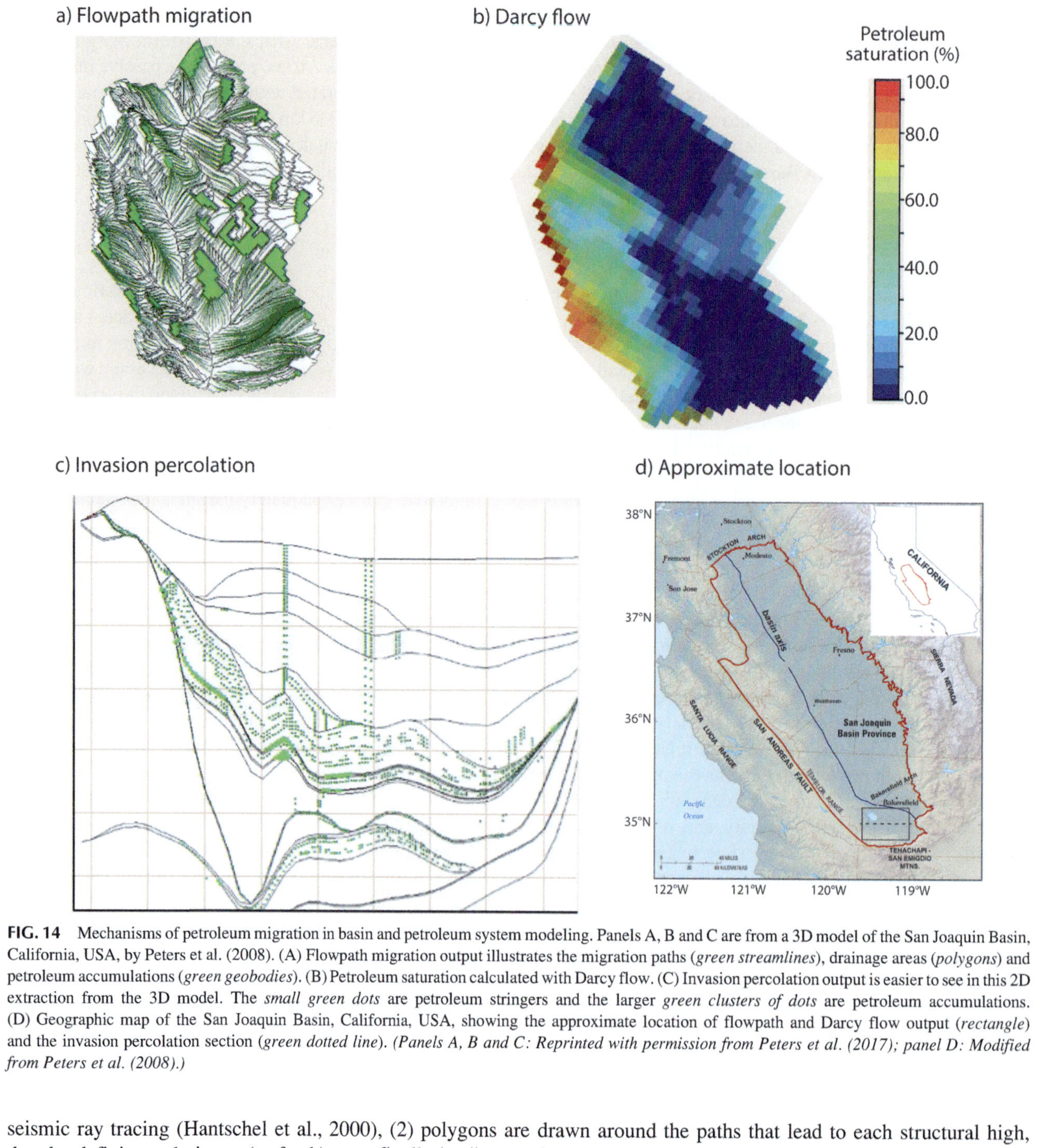


**FIG. 14** Mechanisms of petroleum migration in basin and petroleum system modeling. Panels A, B and C are from a 3D model of the San Joaquin Basin, California, USA, by Peters et al. (2008). (A) Flowpath migration output illustrates the migration paths (*green streamlines*), drainage areas (*polygons*) and petroleum accumulations (*green geobodies*). (B) Petroleum saturation calculated with Darcy flow. (C) Invasion percolation output is easier to see in this 2D extraction from the 3D model. The *small green dots* are petroleum stringers and the larger *green clusters of dots* are petroleum accumulations. (D) Geographic map of the San Joaquin Basin, California, USA, showing the approximate location of flowpath and Darcy flow output (*rectangle*) and the invasion percolation section (*green dotted line*). *(Panels A, B and C: Reprinted with permission from Peters et al. (2017); panel D: Modified from Peters et al. (2008).)*

seismic ray tracing (Hantschel et al., 2000), (2) polygons are drawn around the paths that lead to each structural high, thereby defining a drainage (or fetch) area, finally leading to (3) volumetric analysis of each petroleum accumulation.

Numerous advantages of flowpath migration make it attractive for modeling the movement of petroleum in the subsurface: simulation times are fast, thereby permitting the use of high-resolution grids such as those obtained from seismic surveys. Flowpath models also serve as sufficient first passes or approximations of petroleum volumes and potential reservoir rock behavior. Disadvantages include the geologically unrealistic assumption of instantaneous migration and accumulation, which in effect ignores lithologies between source and reservoir rocks. Other drawbacks include the inability to accumulate petroleum in low-permeability model layers and the inability to charge sandstone lenses and stacked sandstones, which lack the required carrier bed-seal rock geometry.

Darcy flow as incorporated into basin and petroleum system modeling software is often referred to as a full physics migration method because it accounts for all of the forces acting on the petroleum system: buoyancy, capillary pressure of the petroleum phase, pore pressure of the water phase and fluid viscosity. Both Sylta (2002) and Baur and Katz (2018), as do many other authors, provide equations that are used for two-phase Darcy flow in basin modeling software. Sylta (2002) also describes the physics that result in large losses of petroleum in low permeability layers. In contrast to the instantaneous nature of generation-migration-accumulation of petroleum that is a defining characteristic of flowpath physics, in Darcy flow, geologic time is explicit in the flow velocity term. Darcy flow is implemented using a relationship between saturation and the relative permeability curves of water and petroleum (Hantschel and Kauerauf, 2009). In layers where the modeled critical saturation is exceeded, petroleum moves to an adjacent cell. Because of these complexities, Darcy flow in basin modeling requires significant simulation time, and thus, it is common to coarsen the grid spacing to reduce computational expense.

Petroleum accumulations that result from Darcy flow modeling comprise saturations rather than discrete trap-filled bodies; the saturation geometries appear blocky because of the size of grid cells (Fig. 14B), a feature that also makes difficult the evaluation of hydrocarbon column heights (Baur and Katz, 2018). If deployed alone, Darcy flow is likely most appropriately used in basin modeling (1) for modeling unconventional petroleum systems, in which low-permeability source rocks behave also as reservoir rocks and seal rocks, (2) if the basin model might be integrated with reservoir models, which routinely employ Darcy physics, (3) for complex seals and (4) for nonequilibrium reservoirs.

The third type of petroleum migration incorporated into basin modeling software is invasion percolation, a mechanism that hinges on the concept of an invading fluid—petroleum—and a defending fluid—water. Buoyancy drives the invading petroleum upward along a path of least resistance, which is defined by the route of the smallest capillary entry pressures of the largest pore throats (Hantschel and Kauerauf, 2009). The initial migration pattern is random while petroleum "jumps" between pores, thereby establishing the least resistant path. Eventually, petroleum coalesces into discrete stringers along favorable pathways and builds up an oil column beneath barriers with higher capillary pressure (Fig. 14C). In the absence of heterogeneity, a single pathway is established and maintained; in this case, it is often necessary to introduce noise and anisotropy into the model layers to encourage a more realistic geometry of percolated cells (Hantschel and Kauerauf, 2009).

Similar to flowpath migration, invasion percolation also occurs instantaneously relative to geologic time; the rate of migration is limited only by expulsion from the source rock. However, in contrast to flowpath migration, invasion percolation depends not only on buoyancy but on capillary pressure as well. Accordingly, lithological heterogeneities within the model layers strongly influence oil migration. It is common in migration models run with invasion percolation to downscale the model cells into smaller subcells. Model output is thus of high resolution and results in accurate predictions of column heights. Invasion percolation is therefore often applied for modeling petroleum flow in regions with complex structures, such as in compressional tectonic regimes. Compared to flowpath migration and Darcy flow, invasion percolation shows its strength when modeling complex distribution of lithofacies such as those made possible by seismic attribute analyses.

Combinations of these migration physics are frequently used to model the flow of petroleum in rocks with different properties. Frequently termed domain decomposition, Darcy flow may be used for petroleum migration in a model in low-permeability model layers, whereas flowpath or invasion percolation is implemented in model layers of high permeability.

The brief descriptions of each method provided here barely scratch the surface of migration physics and their implementation in basin modeling software. The reader is directed to numerous references for more complete explanations (Baur and Katz, 2018; Hantschel and Kauerauf, 2009) and to individual studies cited in this chapter that use each method. Finally, commercial software packages differ in their implementation of petroleum migration. The recurring theme discussed throughout this chapter—that choices made in subsurface geologic modeling depend on project objectives—is equally true of the choice of migration physics.

## Output

Basin and petroleum system modeling software simulates a large quantity of output parameters. Among many others, these include temperature, pressure (hydrostatic, lithostatic, fracture pressure, etc.), thermal maturity of source rock (transformation ratio and vitrinite reflectance), rock stress, porosity, capillary entry pressure and, of course, petroleum saturation, masses and properties—all as a function of geologic time. The user can scroll forward and backward in time to investigate how each output parameter evolved in magnitude and areal extent in the basin. To minimize storage space and computation time, the user may set the simulator to compute a subset of all possible output parameters.

Depending on the dimensions of the model, results are analyzed in 1D, 2D and 3D. One-dimensional plots provide significant information at the locations of a well or pseudo-well. These include output as a function of depth (Fig. 9), time

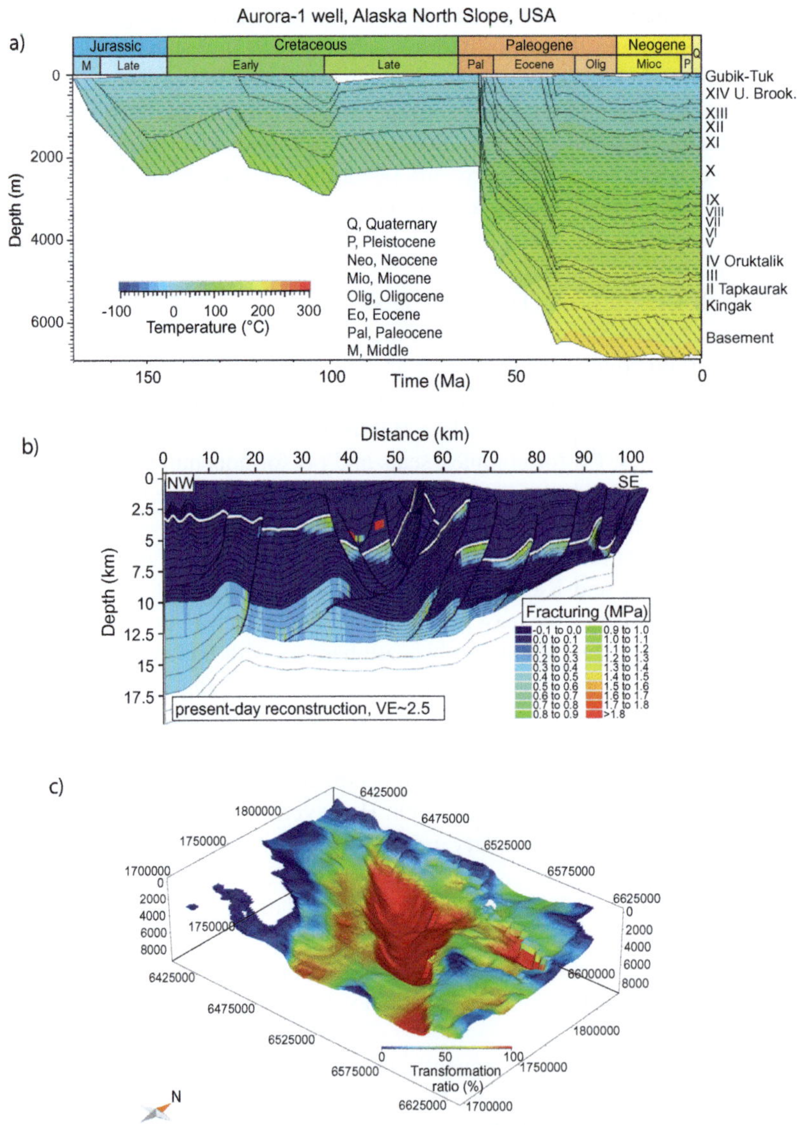


**FIG. 15** Examples of output from a basin and petroleum system model. (A) Burial history with predicted temperature overlay for a 1D basin model of the Aurora-1 well, Alaska North Slope, USA. The model was constructed using the data and nomenclature in a report by Banet (1993). Other elements of this model are shown in Figs. 11A, 12 and 18A. (B) Predicted fracturing gradient from a 2D model by Burgreen-Chan et al. (2015) from the East Coast Basin, offshore the North Island of New Zealand. The *heavy white line* denotes the top of the Wanstead Formation regional seal. (C) Transformation ratio overlain on a 3D structure map of the Nodular Shale source rock, Los Angeles Basin, California, USA. Three-dimensional basin model is by Schultz (2016). Other components of this model are shown in Figs. 9, 10 and 13B. *(Panel B: Reprinted with permission from* Burgreen-Chan et al. (2015)*, John Wiley & Sons Ltd, European Association of Geoscientists & Engineers and International Association of Sedimentologists; panel C: Modified with permission from* Schultz (2016).*)*

(Fig. 11A) and burial, which can be overlain with a predicted parameter such as temperature, pressure or thermal maturity (Fig. 15A). Plots such as these are generated either from 1D models at well or pseudo-well locations or extracted from simulated 2D or 3D basin models.

Results along 2D sections can be derived from 2D basin models or extracted from simulated 3D models. Cross-sections are useful for visualizing similar output parameters as for 1D models but additionally, 2D transects are valuable for examining trends along and across the structural axes of basins, through the pod of active source rock, across existing oil fields and other areas that address model objectives. In contrast to 1D models, 2D models permit the analysis of oil migration in relation to thermal maturation of the source rock and evolving stratigraphic and structural geometries, etc.

The model output shown in Fig. 15B is a 2D basin and petroleum system model constructed along a transect that is located offshore the North Island of New Zealand (Burgreen-Chan et al., 2015). The predicted model output—fracturing potential—is an overlay less frequently seen in the literature but is essential for addressing the objectives of the study, which centered on predicting overpressure. A second example, also less commonly encountered in the literature, is shown by Kroeger et al. (2015) for a different cross-section located offshore the North Island of New Zealand. This work aimed to predict gas hydrate saturations in shallow sediments, a relatively new capability in basin modeling software.

Simulated output from 3D basin and petroleum system models can be analyzed in 3D or in map view. Fig. 15C illustrates the present-day structure map of the Nodular Shale source rock overlain with its transformation ratio (Schultz, 2016). This figure derives from a 3D basin and petroleum system model of the Los Angeles Basin, California, USA, a prolific petroleum province. Such plots make possible the visualization and analysis of the dynamic evolution of the entire sedimentary basin. For example, one application is to examine the model prediction of the critical moment of petroleum generation to assess whether the trap has already formed by the time of petroleum generation-migration-accumulation or if the cap rock is predicted to be sufficiently sealing.

Quantitative outputs for the entire model or selected areas of interest can be further analyzed in mass balance tools, which provide important information on petroleum generation, adsorption, retention, expulsion, accumulation, secondary cracking and losses through time. Other functionality can be used to assess the properties of individual petroleum accumulations, including their filling history, the contribution of fluids generated from different source rocks through time, as well as spilling pathways and areas of seal breakthrough. Simulated subsurface accumulations can also be converted to pressure–temperature conditions at the surface to predict fluid properties such as American Petroleum Index gravity and gas–oil ratio.

Schenk et al. (2019) summarizes many types of output parameters for a 3D basin and petroleum system model of the Exmouth subbasin, North West Shelf, Australia (Figs. 16 and 17). From the designation of each model layer as a source rock, reservoir rock or seal rock (Fig. 16A) to a transformation map overlain on a structure map of the Upper Dingo Claystone source rock (Fig. 16C), to 1D extractions of transformation ratio and sealing efficiency through time in two locations in the basin (Fig. 16D), the diagram illustrates the types of outputs available for a thorough analysis of a petroleum source rock. For example, the model predicts significant Early Cretaceous burial for the Dingo Claystone source rock, petroleum generation over large areas of the Exmouth subbasin, and southward migration of generated and expelled petroleum from the basin center along Late Jurassic carrier beds. Both short-distance migration and long-distance migration are indicated. Taken together, these model predictions provide insight into the charge history of previously discovered accumulations and the possibility of additional discoveries.

## *Calibration*

The objective of calibrating a basin and petroleum system model is answering the question "do the model results make sense?" Calibration is the process of tuning input parameters so that model output is consistent with independent observations; significant variation between predictions and data requires that input parameters be adjusted, the model resimulated and the output parameters of interest re-evaluated. A typical workflow for model calibration is to examine model output at well locations to compare model predictions with observed data; Fig. 9 shows an example of an uncalibrated 1D model and a calibrated 1D model at the location of the Bandini-1 well in the Los Angeles Basin, California, USA.

A common way to begin calibrating a model is via trial and error; input parameters are changed according to expert knowledge gained from previous modeling projects and/or from understanding the geologic setting. One advantage of this approach is that several scenarios can be rapidly investigated. One drawback is that it is difficult to tease out with manual calibration the sensitivity of model results to input variables, especially if those variables are dependent. Hantschel and Kauerauf (2009) discuss more robust and systematic model schemes for calibration.

A recommended order for calibrating basin and petroleum system models is shown in Table 2: stratigraphic horizons, porosity, pressure, temperature, thermal maturity, and petroleum quantity and distribution. It may be counterintuitive to begin the calibration process with input horizons, but the present-day model geometry is the foundation of a basin model; errors in interpretation and/or horizon depths resulting from seismic time to depth conversion translate through to modeled paleogeography, compaction behavior and petroleum migration.

Porosity calibration follows because deposition and erosion are among the first computational steps. Correct estimation of the load of both the matrix (effective stress) and the pore fluid affects pressure and thermal conditions in later calculation steps. The following calibration step links porosity, pressure and permeability. Pore pressure data are generally limited to reservoir rocks, so these units are tuned first using porosity-effective stress pairs to modify the compaction curve. Next, the permeability of the seal rock and conduits are adjusted to match pressure data; permeability and capillary entry pressure data help to calibrate sealing behavior. Calibration is usually initially performed using 1D models at well or pseudo-well locations. Once calibrated, the adjusted porosity–effective stress and permeability–effective stress relationships are

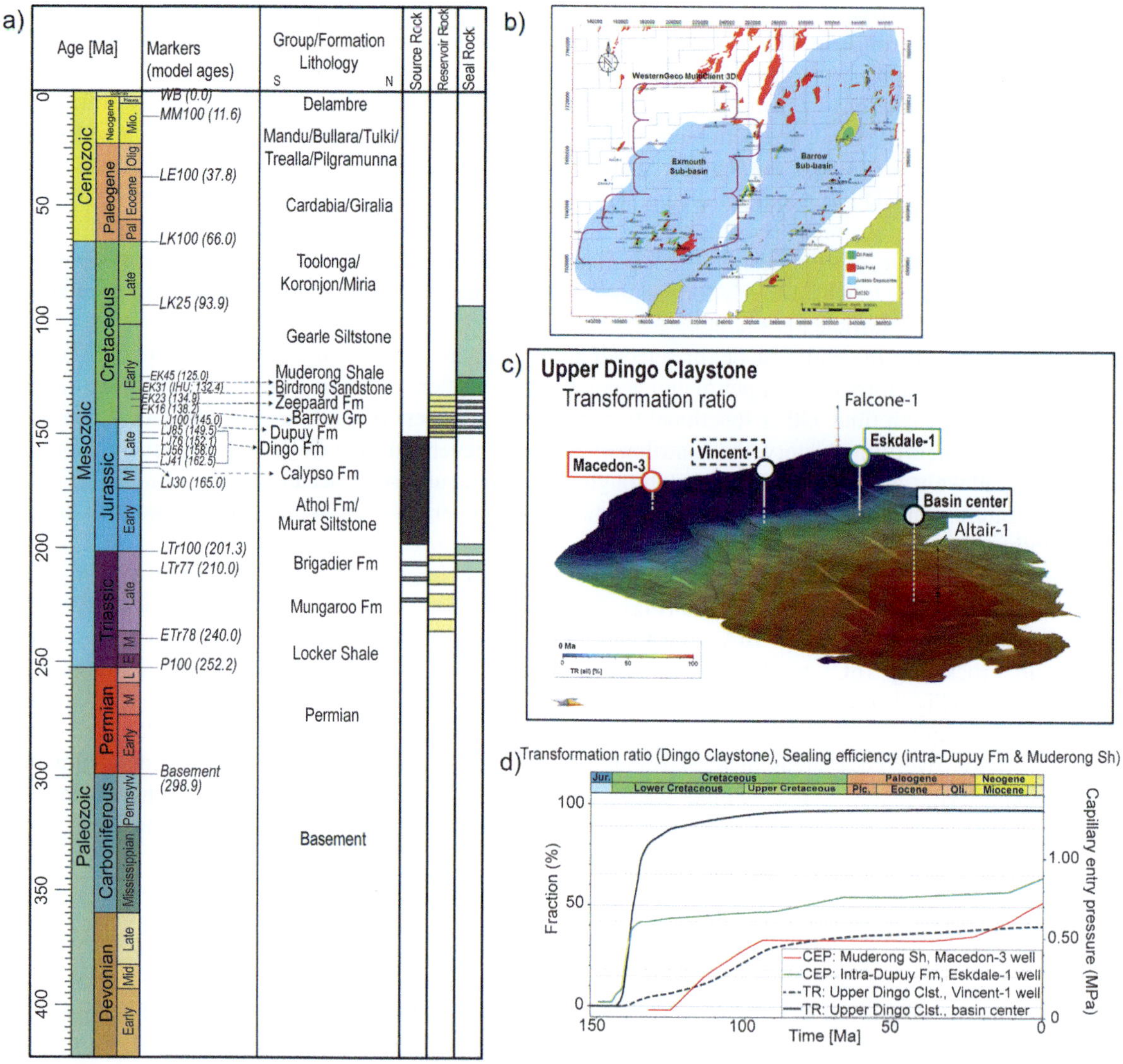


**FIG. 16** Comprehensive summary of output from a 3D basin and petroleum system model of the Exmouth subbasin, North West Shelf, Australia by Schenk et al. (2019). (A) Stratigraphic column highlights model horizons and ages. (B) Location map of the study area. (C) Thermal maturity expressed as transformation ratio on a 3D structure map of the Upper Dingo Claystone source rock. (D) Transformation rate through time of the of the Upper Dingo Claystone source rock at the basin center (*solid black line*) and at the basin margin (*black dashed line*). Sealing efficiency expressed as capillary pressure of two overlying geologic units overlying the source rock (*red and green lines*).

subsequently applied to the same model layers in 2D or 3D models and checked for a satisfactory fit. In practice, calibration steps are iterative, time intensive and require good bookkeeping as numerous scenarios are tested independently in the reservoir rocks and seal rocks.

Output analysis and calibration often focus on the generation–migration–accumulation of petroleum, asking the following: has petroleum been generated, where was petroleum generated, when was petroleum generated, could petroleum have migrated to a particular prospect and what are the properties of the petroleum? Thermal calibration is thus of particular importance. These output parameters are calibrated primarily by adjusting the two thermal boundary conditions (basal heat flow and sediment–interface temperature), but the magnitude and timing of erosion events also affect the thermal evolution of the basin.

Example calibration plots are shown in Fig. 18. Fig. 18A illustrates vitrinite reflectance data from the Aurora-1 well, Alaska North Slope, USA (Figs. 11A, 12 and 15A also pertain to Aurora-1). Vitrinite reflectance analyses tend to be rare; measurements are expensive and require more time to obtain relative to other proxies for thermal maturity. However, Fig. 18A belies this statement—dozens of measurements (green pluses) exist downhole. Calibration of these data (blue

**TABLE 2** A suggested workflow for calibrating basin and petroleum system models.

| Data type | Geologic process represented | Model input to examine | Model output to examine |
|---|---|---|---|
| Well tops | Seismic time-depth conversion errors | Compare depth surfaces to well tops | Well tops should match surfaces |
| Porosity | Compaction | Compaction curves | Porosity-depth curves and porosity data |
| Pressure | Compaction disequilibrium, pressure gradients, petroleum generation | Seal and reservoir porosity as functions of effective, compaction curves, permeability, reservoir connectivity | Pore pressure, hydrostatic pressure, lithostatic pressure, fracture pressure |
| Temperature | Heat conduction, convection, steady state vs transience | Temperature, sediment-water interface temperature, average geothermal gradient, basal heat flow, thermal conductivity | Temperature, surface and basal heat flow |
| Maturity | Source rock generation | Source rock TOC, HI and kinetics | Calculated vitrinite reflectance, transformation ratio, $T_{max}$, hydrogen index, generated products |
| Oil and gas quantity and distribution (3D or 2D) | Petroleum generation, expulsion, migration, accumulation, leakage | Source rock definition, primary and secondary kinetics, reservoir quality and connectivity, faults, seal capacity | Oil vs gas, distributions, composition, volumes, phases (in situ and surface), saturations, column heights, breakthrough, mass balances |

The workflow suggests both input and output parameters to examine during the calibration process. Modified from Noelle Schoellkopf (2011, personal communication).

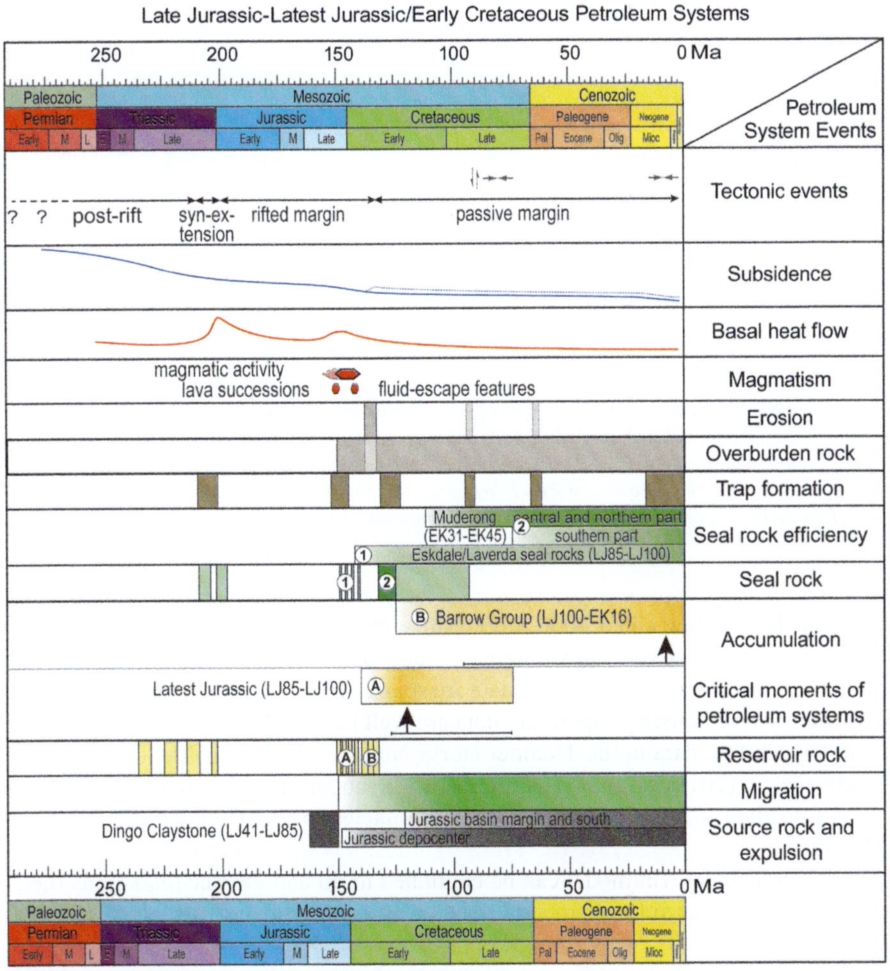


**FIG. 17** Comprehensive summary of the Late to Latest Jurassic and Early Cretaceous petroleum systems of the Exmouth subbasin, North West Shelf, Australia by Schenk et al. (2012).

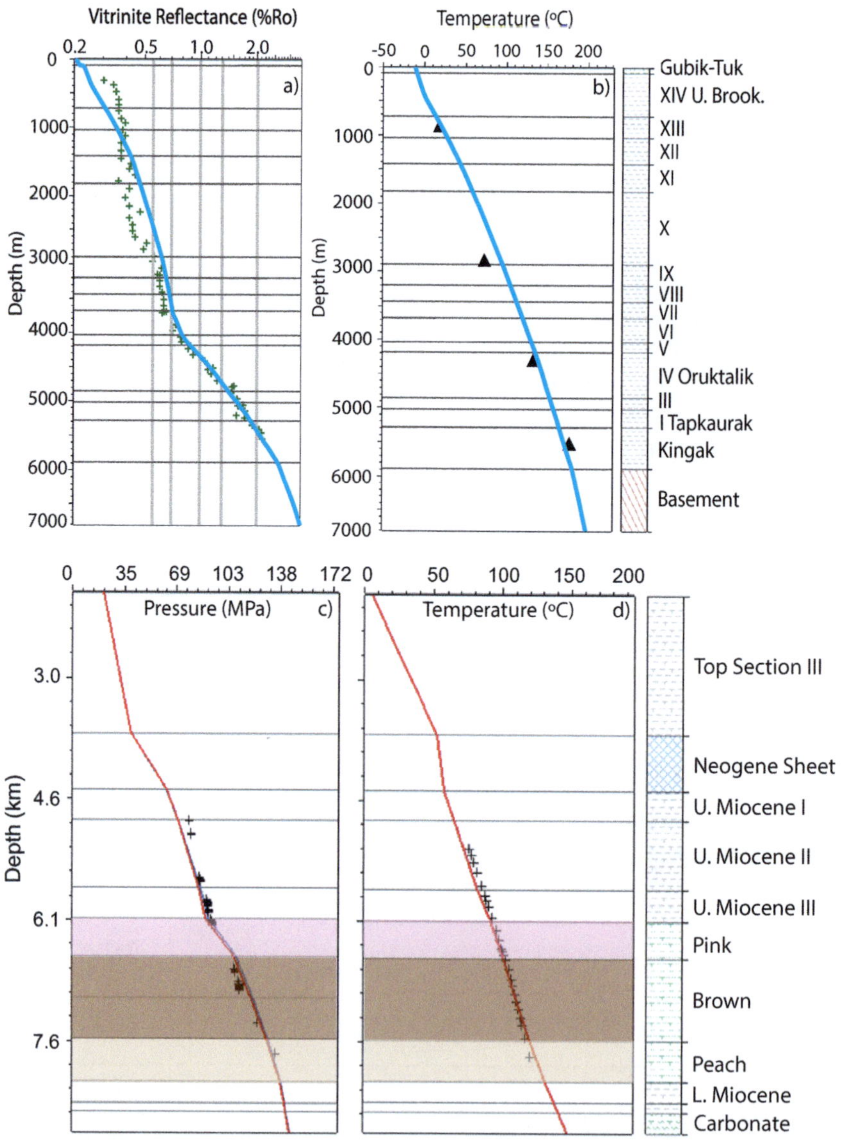


**FIG. 18** (A) Example of basin modeling prediction (*blue curve*) and observed data (*green pluses*) from a 1D basin model of the Aurora-1 well, Alaska North Slope, USA. The burial history of Aurora-1 is shown in Fig. 11A, and the predicted temperature overlain on the burial history is shown in Fig. 15C. (B) Calibrated pressure and temperature extracted from a 2D basin model at the location of the Thunder Horse North well in a minibasin located in the Mississippi Canyon, Gulf of Mexico. The *Pink*, *Brown* and *Peach layers* are reservoir intervals. *(Panel B: Adapted from AlKawai et al. (2018) and reprinted with permission of the AAPG whose permission is required for further use. AAPG ©2018.)*

line) is reasonable, particularly below about 4000 m (2.5 mi.) where the trend of the data forms a dogleg. Relatively poorer calibration upsection could derive from a number of factors; Peters et al. (2005) details potential causes of data irregularity in vitrinite reflectance data. Temperature data are sparse for the Aurora-1 well—just four measurements exist (black triangles, Fig. 18B). These limited temperature data are well fit. Fig. 18C and D illustrate excellent calibration of modeled pressure and temperature data in the Thunder Horse North well, Gulf of Mexico (AlKawai et al., 2018). In a novel approach, AlKawai et al. (2018) combined calibrated basin modeling output with a quantitative seismic interpretation workflow, which resulted in spatial distribution of lithofacies in the Thunder Horse minibasin of Mississippi Canyon, Gulf of Mexico.

Finally, a basin and petroleum system model can be calibrated to oil and gas accumulations (i.e., at known fields) to petroleum volume and saturation and to fluid properties such as gas–oil ratio.

### *Strengths of basin and petroleum system modeling*

Perhaps unrecognized by the wider geoscience community, basin and petroleum system modeling is a powerful tool for learning about and analyzing sedimentary basins irrespective of petroleum studies; the array of equations for chemical and physical processes makes several applications of basin modeling independent of petroleum possible. For example, basin models are well-suited to predict pore pressure within and surrounding a borehole. Calculated thermal histories can be used for understanding many processes in the subsurface, such as the smectite to illite transition and its effect on elastic properties and the effects of altering silica crystallography on porosity and permeability of the host rock.

Other applications of basin modeling other than the generation–migration–accumulation of petroleum are estimating erosion magnitudes, investigating rates of sedimentation and other aspects of basin burial history, feeding basin output variables to other applications such as reservoir modeling and to integrative workflows, applying rock physics templates to calculate elastic properties and many more.

It is clear from the previous discussion that the sheer quantity of parameters that are required for constructing a basin and petroleum system model is a daunting undertaking. Because the input, simulation and calibration stages of model building require enormous effort, project planning requires that sufficient time be allocated to thoroughly assess the model results. An upside of the investment of personnel resources—in gathering and assessing data quality in particular—is that data "rescue" benefits current and future colleagues who will be able to access compilations that did not exist prior to basin modeling. Moreover, the basin model becomes a living archive that can be updated as additional data are acquired and as additional concepts and interpretations come to light.

On the downside, calibrating basin models that are constructed and analyzed in a deterministic manner makes it difficult to assess the sensitivities of model results on any one variable—and even harder to determine the dependencies between variables.

It is critical to remember that basin and petroleum system models are large in scale; they span sedimentary basins that are tens to thousands of kilometers (or mi.) in width and/or length with the aim of investigating aspects of sedimentation and tectonics in the basin through time. As a result, the resolution of basin models is generally coarse—$1 \times 1$ km or larger (~half mi. to a few mi.) for regional models (although computational power and improvements of simulators can now address models of millions of cells in reasonable time). The downside of regional scale models is that individual grid cells often exceed the dimensions of many oil and gas fields and thus are inadequate for prospect-scale studies. Basin models are thus unsuitable for predicting accumulations hosted in small-offset, fault-bound compartments and for resolving petroleum accumulations in small-scale stratigraphic traps. Fortunately, finer resolution grids can be embedded within the basin-scale model; these so-called local grid-refinement or nested models provide a much higher resolution of an area or volume of interest (e.g., Mosca et al., 2018).

## Integrating basin modeling and forward stratigraphic modeling

The Earth's subsurface is heterogeneous on nearly every spatial and temporal scale; the chapters in this textbook describe rocks as tiny grains and as basin-filling turbidites and span depositional time from minutes in lab experiments to millions of years. The modeling techniques described in this chapter fit within those endmembers: in forward stratigraphic modeling, sediments ranging in size from one half to one thousandth of a millimeter (~hundredth of an inch to $10^{-5}$ in) are deposited over a few hundred thousand to several million years whereas lithified rock layers in basin models span meters to thousands of kilometers (~1 ft. to ~1000 mi. or more) in thickness with depositional times ranging from thousands to hundreds of millions of years.

The two case studies presented here—the Ordos Basin, China, and the Alaska North Slope, USA—originated as petroleum system studies. The facies representing petroleum system elements are heterogeneous. Integrated stratigraphic and basin modeling aim to complement each other in a number of ways (e.g., Arab et al., 2016; Liu and Liu, 2021; Liu et al., 2016a). Embedding fine-scale heterogeneities in sedimentary facies derived from a forward stratigraphic model into a basin model provides a holistic product that will provide a more robust understanding of, and predictions for, conventional and unconventional petroleum systems.

The coupling process is relatively straightforward. Stratigraphic model results for a particular set of model layers are exported and subsequently imported into basin modeling software, where they augment and/or replace equivalent layers. Three main output parameters from forward stratigraphic models can serve as critical initial input parameters for basin models (Fig. 19): (1) paleowater depth, (2) sedimentary facies and (3) horizon depths. Fig. 20 illustrates a real-world example of a coupled forward stratigraphic basin and petroleum system model.

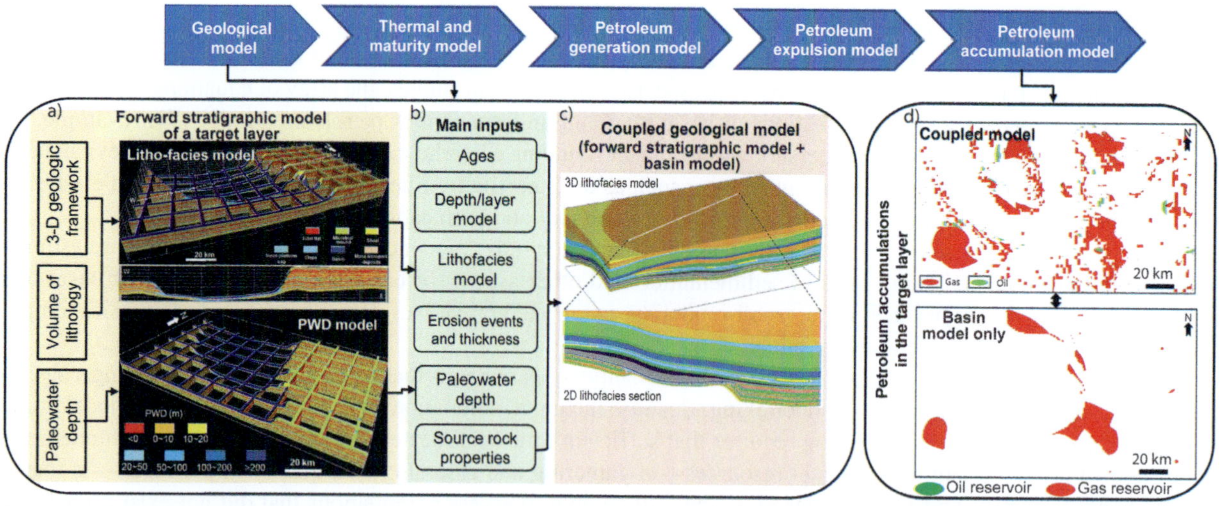


**FIG. 19** Workflow for coupling forward stratigraphic modeling with basin and petroleum system modeling. (A) The lithologic section of interest (also called the target layer) is defined within a geologic framework. (B) Lithology and paleowater depth are obtained from forward stratigraphic modeling, which are input, along with other variables, to a basin model. (C) Fine-scale heterogeneities within the target layer replace the lithology defined for the corresponding layer in the basin model. (D) Output parameters of interest are analyzed. Panel D demonstrates the use of the coupled approach for mapping and analyzing petroleum accumulations.

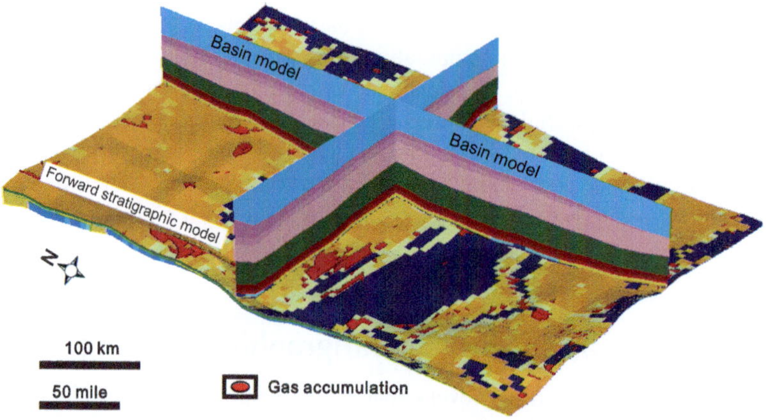


**FIG. 20** An example of the workflow shown in Fig. 19 applied to a geologic setting. The lithologically heterogeneous layers from the forward stratigraphic model are embedded in a "layer cake" geometry basin model. *(Adapted from Liu et al. (2021) and reprinted with permission of the AAPG whose permission is required for further use. AAPG ©2021.)*

Some aspects of coupling the two modeling approaches relate to the underlying model setup. For example, if simulated on a finer grid, as is often the case, the parameters to be exported from the forward stratigraphic model and subsequently imported to the basin model must be upscaled to the same resolution as the basin model. An additional complication is that forward stratigraphic models and basin models may not span the same geographic area; sediment sources are critical input to the stratigraphic model, but these terranes are rarely necessary to include in the geographic area of a basin model. This difference requires the output from forward stratigraphic modeling to be compressed or extended to fit into the area spanned by the basin model.

The following sections feature coupled forward stratigraphic and basin and petroleum system models that were constructed by the authors. Other studies available in the literature include the following: Grohmann et al. (2021), who estimated with forward stratigraphic modeling the lateral and vertical extent of organic matter to define in high resolution the organofacies of source rock layers in a basin model of the Eastern Mediterranean Sea, and Barabasch et al. (2019), who combined stratigraphic modeling with basin modeling to better understand the timing of source rock generation, the migration pathways and the effect of salt on petroleum systems in the Levant Basin.

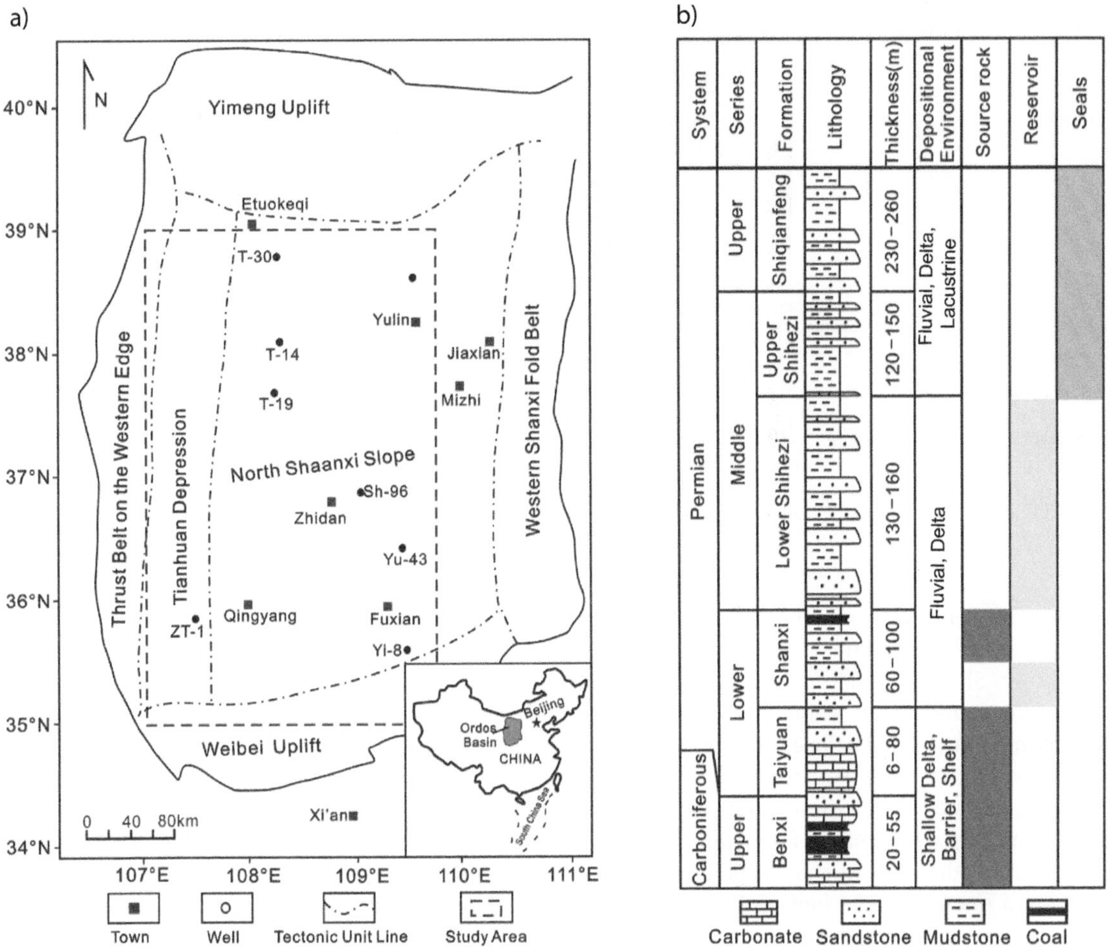


**FIG. 21** Geographic and stratigraphic context of coupled forward stratigraphic-basin and petroleum system model. (A) Location map of the Ordos Basin, China, showing the surrounding tectonic terranes (*dashed–dotted line*), model area (*red dashed rectangle*), town locations (*filled squares*) and locations of wells that were used for model calibration (*filled circles*). (B) Generalized stratigraphic column of the Upper Paleozoic, Ordos Basin, showing lithofacies, average stratal thicknesses and depositional environments.

## Case study 1: Ordos Basin, China

A coupled basin and petroleum system-forward stratigraphic model of the Upper Paleozoic tight gas petroleum system in the Ordos Basin, China, illustrates the workflow, results and advantages of integrating multiple modeling technologies. The basin is located in central China with an area of approximately 250 × 1000 km$^2$ (97 mi.$^2$ × 386 mi.$^2$; Fig. 21A). The basin has evolved through five tectono-sedimentary stages, including a mid–late Proterozoic aulacogen stage, an early Paleozoic shallow marine platform stage, a late Paleozoic littoral plain stage, a Mesozoic cratonic basin stage and a Cenozoic peripheral faulted basin stage. As a result of the Cenozoic deformation, six primary tectonic elements are present in the basin, including the Yimeng uplift to the north, the Weibei uplift to the south, the Western Edge thrust belt and the Tianhuan depression to the west, the North Shaanxi slope in the center and the Western Shanxi fold belt to the east. The study area covers the bulk of the central part of the basin.

Four major episodes of uplift and erosion shaped the basin in the Mesozoic: Late Triassic, late Middle Jurassic, Late Jurassic and Late Cretaceous. The geographic location and magnitude of erosion varied during each event; the final episode in the Late Cretaceous was particularly intense, removing more than 1600 m (5250 ft.) of sediment in the eastern part of the basin. These tectonic events were incorporated in the basin model and the magnitudes of eroded thicknesses were taken from the literature.

The Ordos Basin is the second largest onshore hydrocarbon-bearing basin in China, with numerous oil and gas discoveries. In 2010, the basin ranked first in gas production in China. By 2020, the combined annual oil and gas production in the basin amounted to more than 60 million tons (>400 million bbl) equivalent, ranking it the top hydrocarbon-producing basin in China. The upper Paleozoic tight gas system consists mainly of Upper Carboniferous and Permian strata, including the

**TABLE 3 Input data for the forward stratigraphic model of the lower Permian Shanxi Formation, Ordos Basin, China.**

| | |
|---|---|
| Time span | 285 to 280 Ma with a temporal resolution of 40 ka |
| Horizon grids | 92 grid cells north-south, 60 grid cells east-west, 5 km × 5 km (3.1 mi. × 3.1 mi.) km grid spacing |
| Initial basin topography | An unconformity surface with several basement highs on the northern and southern margins of the Ordos Basin, China |
| Subsidence | Subsidence ranging from 110 to 350 m (~360 ft. to ~1150 ft.) during deposition of the Lower Permian Shanxi Formation |
| Lake level curve | High-frequency lake-level oscillation curve modulated by a low-frequency one comprising the Milankovitch orbital frequencies |
| Sediment sources locations | Four on the northern basin margin and four on the southern margin |
| Sediment particle size | Coarse sand (0.55 mm; 0.02 in.), medium sand (0.28 mm; 0.01 in.), very fine sand (0.1 mm; 0.004 in.) and clay (0.001 mm; $0.4e^{-5}$ in.) |
| Sediment supply rate | Varies with the oscillation of the lake level over time |

Upper Carboniferous Benxi Formation, the Lower Permian Taiyuan and Shanxi formations, the Middle Permian Lower and Upper Shihezi formations and the Upper Permian Shiqianfeng Formation (Fig. 21B). Gas accumulates principally in the Shanxi and Lower Shihezi formations, which were deposited in fluvial and shallow deltaic environments. The essential elements of the unconventional petroleum system are the Upper Carboniferous and Lower Permian coal source rocks, the Lower-to-Middle Permian fluvial and deltaic reservoir rocks and the Upper Permian regional mudstone seal rock (Fig. 21B).

The objectives of the study were to produce a more geologically realistic model for the Paleozoic tight-gas system by taking advantage of a coupled forward stratigraphic basin modeling approach. A 3D stratigraphic model comprising 125 sublayers within the Lower Shanxi Formation was simulated by forward stratigraphic modeling. The simulation input parameters used for the forward stratigraphic modeling component are shown in Table 3. After numerous simulations by adjusting and fine-tuning the input parameters following the workflow shown in Fig. 3, a final sedimentary model for the Lower Permian Shanxi Formation was constructed. A 3D fence diagram of the model is shown in Fig. 22A. In general, sand developed in the northern and southern parts of the study area, whereas clay predominates in the basin center. Lithologies vary both laterally and vertically on the submeter scale. In general, thin fine-grained sand intervals develop between coarse and medium-grained sand intervals. Some fine sand intervals are intercalated with clay layers in the central part of the basin (Fig. 22B and C). Because of the limitations of computing resources, 20 of the simulated 125 sublayers were exported for basin and petroleum system modeling.

The stand-alone basin and petroleum system model was constructed with 15 stratigraphic layers, which included 9 upper Paleozoic units and 6 Mesozoic and Cenozoic units. Three of the units are source rocks. The lithologies for the basin model were defined using lithological ratios (e.g., sand/shale ratio) such as those illustrated in Fig. 13B, and the paleowater depth was determined similarly. For example, channel facies were considered primarily to be sandstone lithology, floodplain facies were regarded as shales with silt-rich lithology, and semi-deep (50–100 m) to deep (>100 m) lacustrine facies were considered a typical shale lithology. Similarly, channel facies were considered primarily to be sandstone lithology, floodplain facies were regarded as shales with silt-rich lithology, and semideep (50–100 m) to deep (>100 m) lacustrine facies were considered a typical shale lithology.

The upper thermal boundary condition, the sediment–seawater interface temperature, was calculated automatically using default software settings that account for basin location through time. The lower thermal boundary condition of basal heat flow was generally less than 50 $mW/m^2$ in the basin prior to the Jurassic. Because of a tectonically–driven thermal event in the Late Mesozoic, the basal heat flow dramatically increased to values of 81–95 $mW/m^2$ in the Late Jurassic and Early Cretaceous (140–100 Ma). Subsequently, the basal heat flow decreased gradually. At present, the basal heat flow values are in the range of 54–70 $mW/m^2$, with an average value of 61 $mW/m^2$.

The TOC and HI data of the source rock were obtained from measurements on core samples. Kinetics used for the coal source rock were obtained from custom pyrolysis experiments. Pepper and Corvi (1995) "TIIIH (DE)" default kinetics were assigned to the mudstone source rock, and "TII (B)" kinetics was assigned to the limestone source rock. Petroleum migration was simulated with a hybrid approach that used flowpath physics in high-permeability layers and Darcy flow in low-permeability layers.

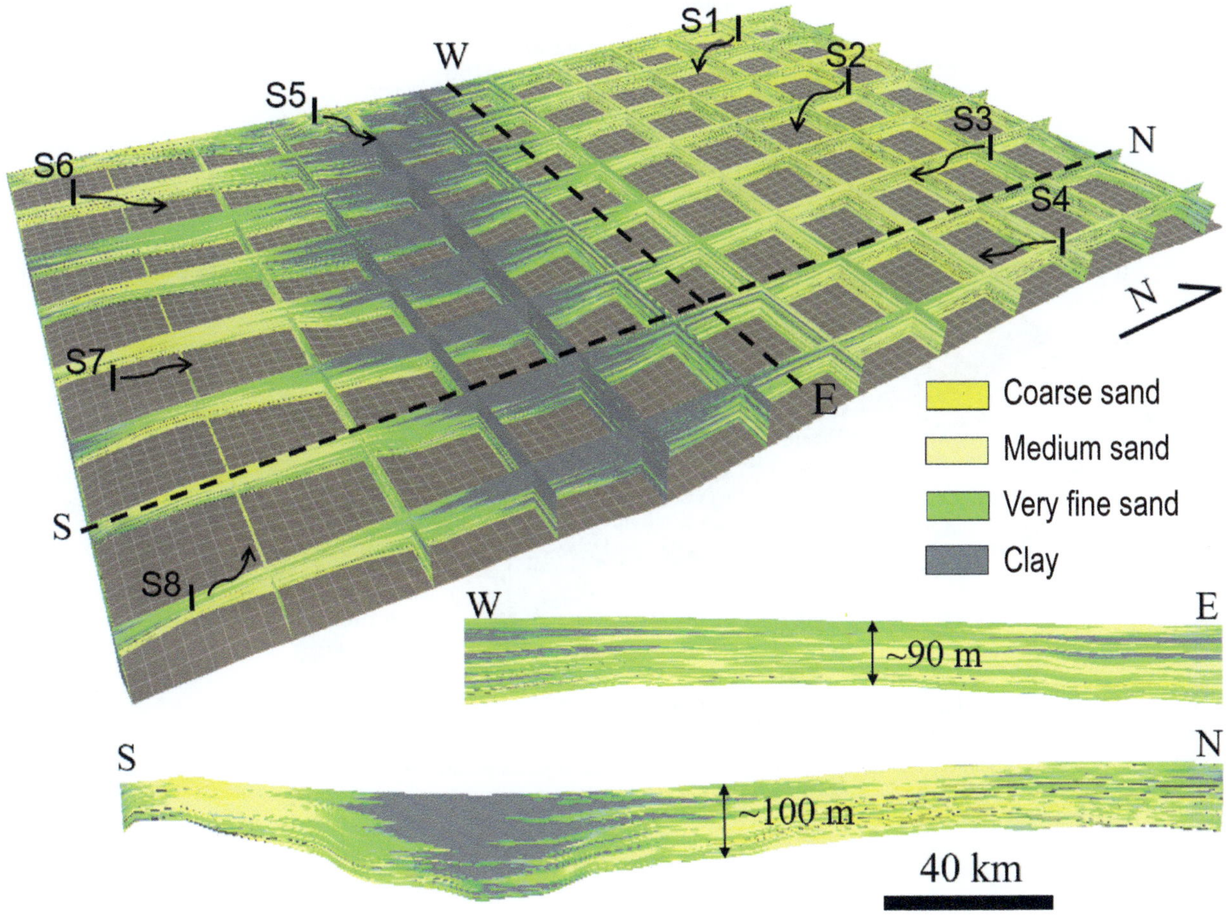


**FIG. 22** The lithofacies simulation output of the Lower Permian Shanxi Formation, Ordos Basin, China. (A) Basin-wide fence diagram of 3D forward stratigraphic model. The fences are spaced at 40 km (25 mi.); every eighth fence is displayed. Grid dimensions are 5 km × 5 km (3.1 × 3.1 mi.). Eight point sources (S1 to S8) supply sediment to the basin. (B) East–west-oriented cross-section. (C) North–south-oriented cross-section. The vertical exaggeration is 80×. *(Panels A, B, and C: Reprinted from Liu et al. (2016) with permission from Elsevier.)*

In the base case basin and petroleum system model, lithologies of both the Shanxi and the overlying Lower Shihezi formations were homogeneous in the vertical direction but heterogeneous laterally. Sandstone and siltstone were defined as reservoir rocks, whereas the mudstones were defined as seal rocks. Reservoir rocks in the Shanxi Formation are mainly distributed in the northern and to a lesser extent in the southern areas. Model results show only a few gas pools in the Shanxi Formation in the central and eastern parts (red dashed circles, Fig. 23). This is likely because of (1) the lack of regional seals immediately above the P1s layer and (2) the absence of lithologic heterogeneity for the Shanxi Formation.

Following construction of the conventional basin model, the Shanxi Formation layer was replaced with the 20 sublayers that were exported from the forward stratigraphic model. Each sublayer was assigned a facies map based on the forward stratigraphic model, thus significantly increasing the degree of lithologic heterogeneity of the Shanxi Formation basin model layer compared to the base case (i.e., the stand-alone basin model).

The volume and spatial distribution of gas accumulations combined from the 20 sublayers of the Shanxi Formation are shown in Fig. 24. Relative to the base case model, the integrated model resulted in many more gas accumulations. Gas chiefly accumulates near the depocenter, where alternating sandstone and mudstone layers that developed as a result of frequent oscillations of lake/sea level created favorable traps. Because of the lateral and vertical sedimentary heterogeneity incorporated in the Shanxi Formation from the forward stratigraphic model, gas migrated along multiple pathways, forming multiple small pools.

One output parameter that is useful for comparing the base case (basin model) with the coupled model (basin model+ forward stratigraphic) is the volume of predicted petroleum. Although both models generated about the same volume of

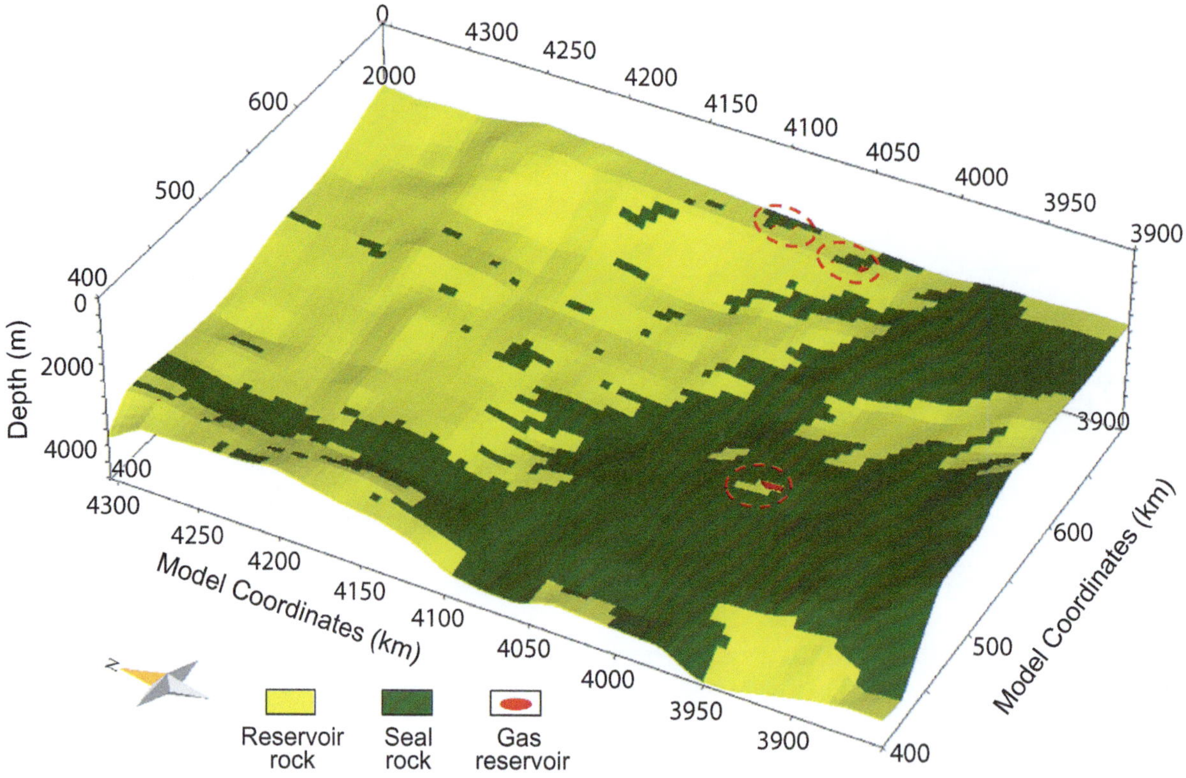


**FIG. 23** Base case for the traditional basin and petroleum system model. Modeled gas accumulations in the lower Permian Shanxi Formation correspond to the polygons encircled by red for emphasis. *(Reprinted from Liu et al. (2016) with permission from Elsevier.)*

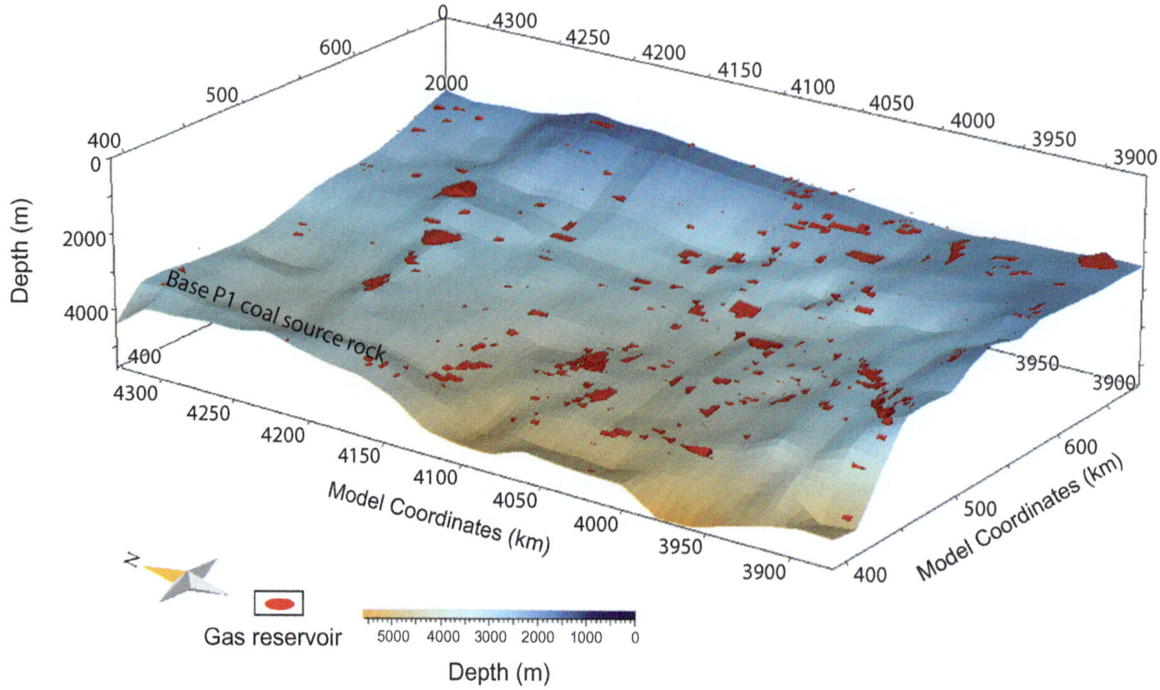


**FIG. 24** Spatial distribution in the Shanxi Formation of gas accumulations in all the sublayers of the coupled model. The results are plotted on the structural depth map of the base of the lower Permian coal. *(Reprinted from Liu et al. (2016) with permission from Elsevier.)*

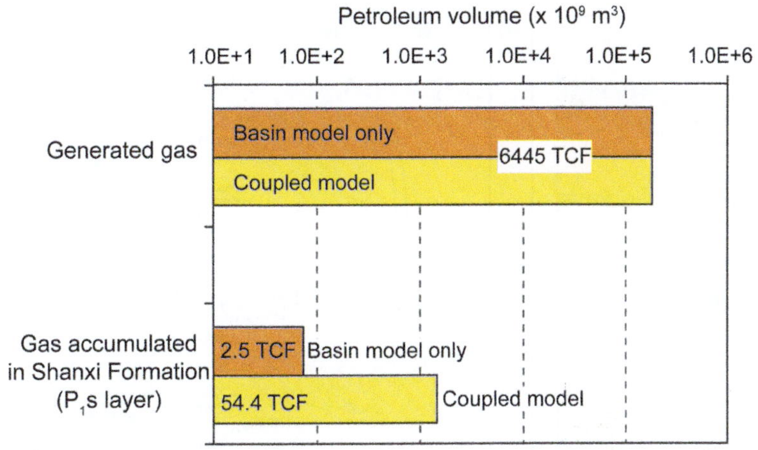


**FIG. 25** Comparison of gas volume in the Ordos Basin, China predicted in all reservoir rocks (top two bars) and in just the lower Permian Shanxi Formation ($P_1s$) (lower two bars). For each set of bars, the orange bar shows predicted volumes using basin modeling alone whereas the yellow bar is the volume predicted by the coupled forward stratigraphic-basin and petroleum system model. *TCF*, trillion cubic feet.

gas—6445 trillion cubic feet (TCF)—the Shanxi Formation accumulated 97% more gas (54.4 TCF) in the coupled model than in the base case (2.5 TCF; Fig. 25), almost certainly because of the highly heterogeneous nature of the Shanxi lithofacies that originated from the stratigraphic model. Of note, the gas volume in the coupled model is on the same order of magnitude as the volume obtained for the Shanxi Formation in an independent resource assessment (39.6 TCF; Yang et al., 2001). The integrated forward stratigraphic model–basin model appears to realistically approximate basin resources.

## Case study 2: Colville Basin (Alaska North Slope, USA)

The Alaska North Slope, USA, is an additional example that demonstrates the efficacy of integrating basin modeling with forward stratigraphic modeling. However, in contrast to the Ordos Basin case study, in which output from forward stratigraphic modeling was specifically generated for subsequent input to the basin model, the basin model of the Alaska North Slope preceded by several years the forward stratigraphic model. Moreover, the results of forward stratigraphic modeling—which were generated for two distinct intervals in the stratigraphic column and by two different experts—were then input into basin modeling for subsequent investigation.

Deemed a super basin by virtue of its petroleum production and reserves, systems and infrastructure (Sternbach, 2020), the Alaska North Slope (formally, the Colville Basin, Alaska, USA) is home to the two largest conventional oil fields in the United States, Prudhoe Bay and Kuparuk River (Energy Information Administration, 2015). The most prolific petroleum source rocks are the Triassic Shublik Formation, the basal part of the Jurassic Kingak Shale and the basal part of the Cretaceous Hue Shale (the Gamma Ray Zone; also called the Highly Radioactive Zone) (Fig. 26A). The Shublik and basal Kingak source rocks developed at the end stage of a passive margin and at the early stage of a rifting succession, respectively, whereas the Hue Shale developed in the basal Colville foreland basin succession known as the Brookian, so called because of sediment derivation from the nearby Brooks Range orogen (Bird, 2001).

The Colville Basin extends from the foothills of the Brooks Range to the edge of the continental shelf and buried Barrow Arch (offshore the Beaufort Sea coast) and extends from a subsurface platform near the US–Russian maritime boundary in the west to the US–Canada border to the east—a distance of ~1100 km (~680 mi.) (Fig. 26B). The basement surface of the basin shoals gradually from south to north, creating an asymmetric profile; the maximum preserved sediment thickness is 6 to 8 km (~3.7 to 5 mi.) foothills of the Brooks Range but just 1 to 3 km (~0.6 to 1.9 mi.) (Bird and Molenaar, 1992).

The Cretaceous-Tertiary sedimentary fill of the Colville Basin is dominantly marine to nonmarine siliciclastic, the erosional product of two orogenic episodes: one spanning Middle Jurassic to mid-Early Cretaceous and a second in the Cenozoic (Moore et al., 1994). These two tectonic events were separated by a Late Cretaceous period of relative tectonic quiescence and reduced sediment input. Brookian sediments began to fill the basin shortly after Brooks Range Uplift (Bird and Molenaar, 1992). West-to-east progradation of those sediments are revealed by ubiquitous clinoforms on seismic reflection data; the clinoforms record the ultimate shelf margin as sediment marched across the Colville Basin (Fig. 26B, C) (Houseknecht et al., 2009).

Significant sediment volumes associated with the Brookian progradation blanketed underlying units, thereby sufficiently thermally stressing source rocks to generate petroleum from the three major source rocks. Based on more than

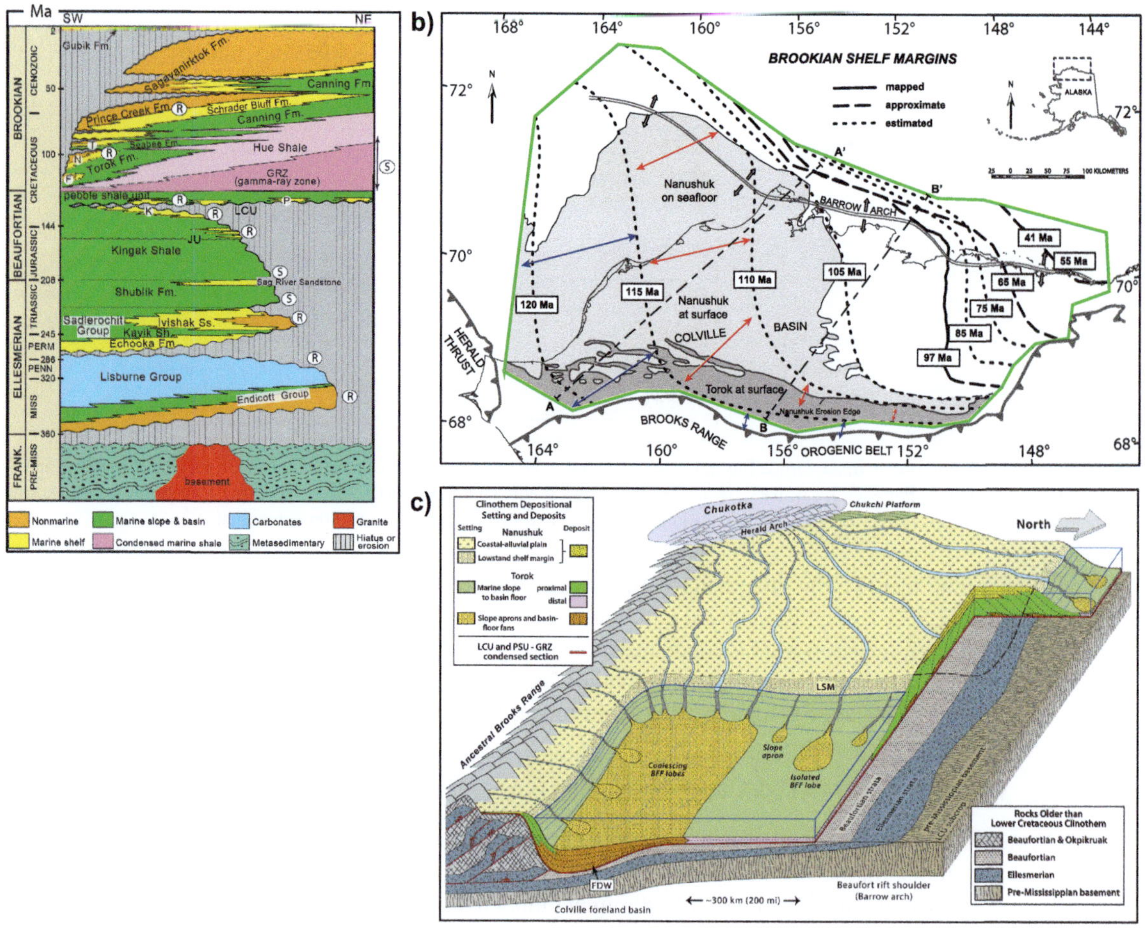


**FIG. 26** (A) Generalized chronostratigraphy of the onshore part of the Alaska North Slope. Tectono-stratigraphic sequences are in capital letters at the left. Oil-prone source rocks (*circled S*) and important reservoir rocks (*circled R*) are noted. F, Fortress Mountain Formation; GRZ, Gamma Ray Zone (this is the same as HRZ, Highly Radioactive Zone; both usages appear in the literature); JU, Jurassic unconformity; K, Kuparuk River Sandstone; LCU, Lower Cretaceous Unconformity; N, Nanushuk Formation; P, Put River Sandstone; T, Tuluvak Formation. (B) Base map of the Colville Foreland Basin illustrating major tectonic features and the geographic area of the 3D basin and petroleum system model (*green polygon*) of Schenk et al. (2012). Shelf margins of the Brookian Sequence prograde from west to east; the shelf margins for 120–105 Ma are estimates. Forward stratigraphic modeling focused on the sequences deposited between 122 and 115 Ma (modeled interval indicated by *blue arrows*), and the sequences deposited between 115 and 110 Ma (Zastepa and Schenk, 2017, pers. commun.; modeled interval indicated by *red arrows*). Sections A–A′ and B–B′ are shown in Fig. 27. (C) Generalized model for main provenance areas, sediment routing and deposition of the main architectural elements of the Lower Cretaceous clinothem. *(Panel A: From Hosford Scheirer and Bird (2020), used with permission of Society of Exploration Geophysicists; permission conveyed through Copyright Clearance Center, Inc.; panel C: Reprinted from Houseknecht (2019) with permission of the AAPG whose permission is required for further use. AAPG © 2019.)*

400 1D burial history models, Houseknecht et al. (2012) determined that the onset of petroleum generation was diachronous—essentially coincident with west-to-east progradation—beginning about ~110 Ma in the west and about 20 million years later in the east. A north–south trend in source rock maturation is superimposed on the east–west pattern because of the development of additional depocenters during the Cretaceous.

To evaluate in more detail the evolution of the region's multiple petroleum systems, a 3D basin and petroleum system model was constructed for the entire on- and offshore Colville Basin (Schenk et al., 2012). The geographic area spanned by the model is enormous: 275,000 km$^2$ (106,000 mi.$^2$) (green polygon in Fig. 26B). The model framework was constructed using ~48,000 km (~30,000 mi.) of interpreted 2D seismic lines and a database of more than 400 wells.

In the petroleum system framework of Magoon and Dow (1994), overburden rock—one of the four essential elements—is defined as the entire rock column above the source rock; the onset time of petroleum generation depends in part on its thickness. Achieving model objectives thus hinged on reconstructing the diachronous deposition and erosion of the Brookian Sequence. Because the thick Brookian sediments behave chiefly as overburden rocks in North Slope petroleum systems, the 3D model glossed over the stratigraphic and depositional details of this sequence (e.g., Fig. 27A, B).

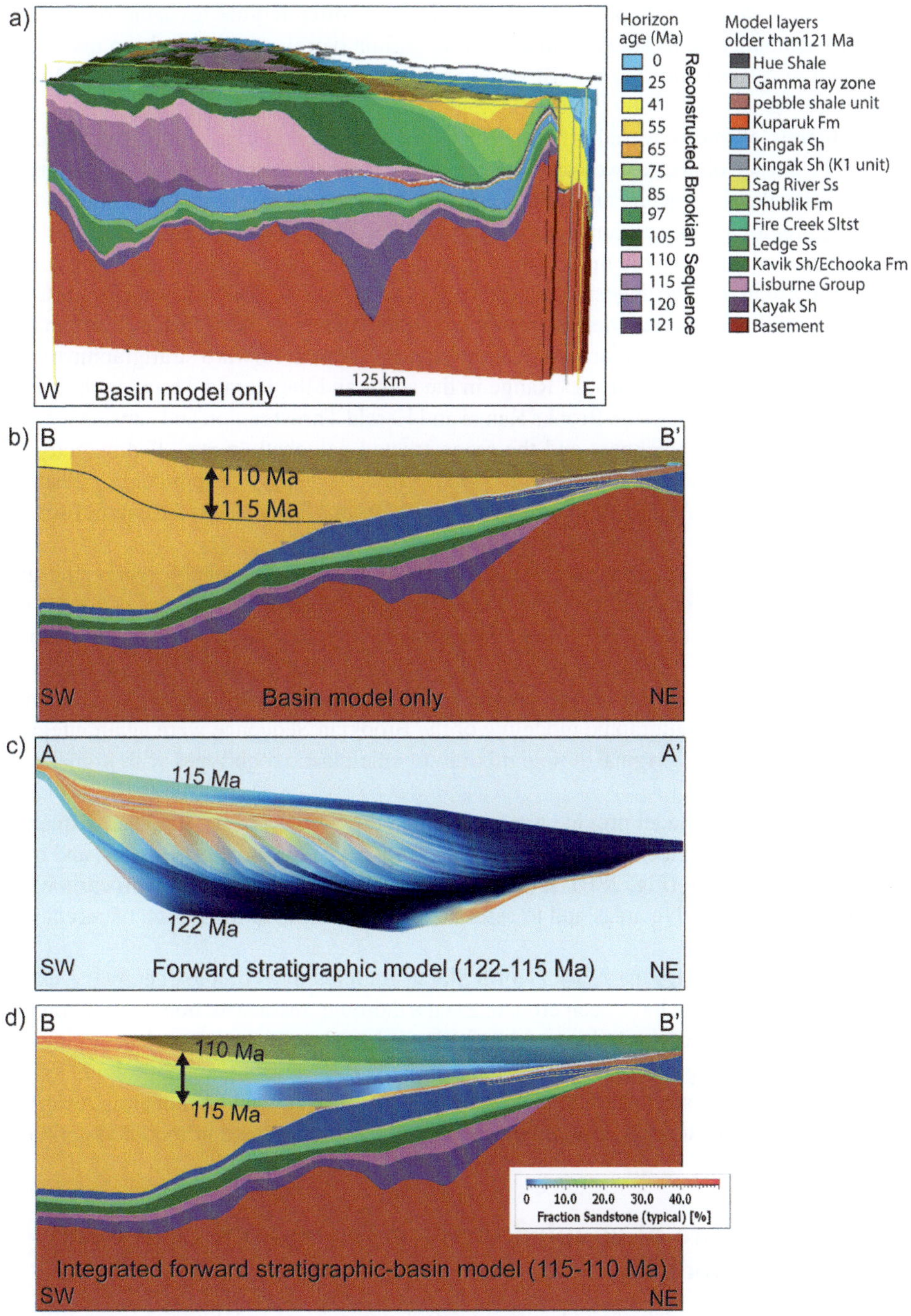


**FIG. 27** (A) Basin and petroleum system model of the Alaska North Slope, USA. The cut plane through the deepest part of the model highlights the prograding nature of the Brookian Sequence that was reconstructed in the 3D basin model. (B) Southwest–northeast-oriented cross-section B–B′ (location on Fig. 26B) from the 3D basin model illustrating simplified lithofacies (*layer colors*) for the Brookian Sequence at 115-110 Ma (*black arrow*). (C) Southwest-northeast-oriented cross-section A–A′ (location on Fig. 26B) illustrating the output of the forward stratigraphic model for the Brookian Sequence deposited between 122 and 115 Ma. *Blue shading* indicates shale-dominated facies and *orange shading* denotes sandstone-rich facies; (D) Southwest-northeast-oriented cross-section B–B′ (location on Fig. 26B) illustrating the integration of the forward stratigraphic model for the sequence deposited between 115 and 110 Ma within the 3D regional basin and petroleum system model. *(From Peters et al. (2019), used with permission.)*

As commonly occurs in petroleum-rich provinces, new discoveries often require revisions of numerical models. Since 2015, a spate of industry announcements suggests that at least 2 billion barrels of oil have been discovered in stratigraphic traps within the Nanushuk and Torok formations—Brookian strata deposited between ~125 and 104 Ma (mid-Early to early Late Cretaceous). An ensuing assessment by the US Geological Survey estimated that together, the two units may hold a mean of ~8.8 billion barrels of undiscovered oil and 39 TCF of undiscovered gas (Houseknecht and Nanushuk-Torok Assessment Team, 2017). These volumes as well as the predominantly stratigraphic traps where the petroleum accumulated emphasize the need to refine the architecture and composition of Brookian strata beyond what is already accomplished and possible within the mega-regional 3D petroleum system model of Schenk et al. (2012).

To that end, a pilot forward stratigraphic modeling study focused on the oldest Brookian Sequence deposited between 122 and 115 Ma, which approximates the depositional time of the lower Nanushuk Formation (Fig. 27C; Christ et al., 2016). Initial inputs to the stratigraphic forward models were derived from paleogeometries that were reconstructed from the 3D basin model (Schenk et al., 2012). The need for sediment provenance areas in forward stratigraphic modeling required that the model area be extended to include the Brooks Range in the south and the Herald Thrust in the west (sediment source locations are shown in Fig. 26B). Uplift in the Brooks Range and Herald Thrust region and subsidence variations within the basin were estimated from sediment thickness and the reconstructed paleobathymetry. Sedimentary source points were selected according to paleocurrent information from Houseknecht et al. (2009). The modeling results show prograding foresets building up from the Herald Arch in the southwest and the Brooks Range from the south, filling the Colville Basin and onlapping onto the Barrow Arch toward the north–northeast (Fig. 27C).

Continued software development and integration of applications spurred additional forward stratigraphic modeling of Brookian strata. This second effort focused on the Brookian Sequence deposited between 115 and 110 Ma (Zastepa and Schenk, pers. comm., 2017; Peters et al., 2019). The chosen model sequence approximates the depositional time of the upper Nanushuk, and thus, it is of key relevance to recent discoveries and stratigraphic traps. As with the prior stratigraphic modeling of Brookian strata, input to the forward model derived from output of the mega-regional basin model. The results from this second round of forward stratigraphic modeling of the Brookian Sequence were again integrated into the mega-regional basin model by replacing the original simple mixture of siliciclastic rock types with higher-resolution sandstone fraction maps (Fig. 27D).

The results of the coupled basin model provide a more detailed distribution of the modeled sequence compared to the original basin model, contribute to the identification of potential topset and turbidite geometries, and result in an improved understanding of the sealing potential (Fig. 28B). It also helps to identify more detailed hydrocarbon migration pathways and potential traps to predict prospective areas and to assess exploration risks in the region of recent giant oil discoveries (Fig. 28C).

Modeling challenges, particularly for the Alaska North Slope, are related to complex geometry, especially for younger intervals of the Brookian Sequence that have been eroded. Even a thorough reconstruction—evolution of paleobathymetry, diachronous deposition and erosion, regional tilting and so forth—is based on assumptions. Integrating basin and petroleum system modeling with forward stratigraphic modeling and eventually back to basin and petroleum system modeling compounds uncertainties. Understanding such uncertainties and evaluating their impact on modeling results are key. Therefore, calibrating the individual modeling steps is critical because it provides confidence in model assignments and results.

## Key messages on the future of coupled forward stratigraphic and basin and petroleum system models

The future of forward stratigraphic modeling and basin and petroleum system modeling lies principally with technological advances. Some of these relate to ever-increasing computing power and thus reduced simulation time. However, the majority of innovations that are on the horizon for these two approaches center on two main themes: (1) stochastic/probabilistic treatment and (2) automated workflows.

Both fields are beginning to deploy previously established techniques in uncertainty analysis. For example, Tong and Mukerji (2017) defined the distributions and standard deviation of nine input parameters in a 3D basin model of the Piceance Basin, Colorado, USA: HI, TOC, thickness of eroded strata, time period of erosion, basal heat flow, sediment–surface interface temperature (the basin was subaerial), source rock kinetics and the lithology of two model layers. Monte Carlo sampling of these parameters generated 500 basin models, the output of which was analyzed with two sensitivity analysis techniques. The impact of uncertain input parameters on model output was examined as a function of space and time to identify their impact on gas generation from the Cameo Coal in the Piceance Basin, Colorado, USA. An important conclusion is that the most sensitive parameter on model output changes through time. This result would be difficult, if not impossible, to identify with deterministic approaches alone.

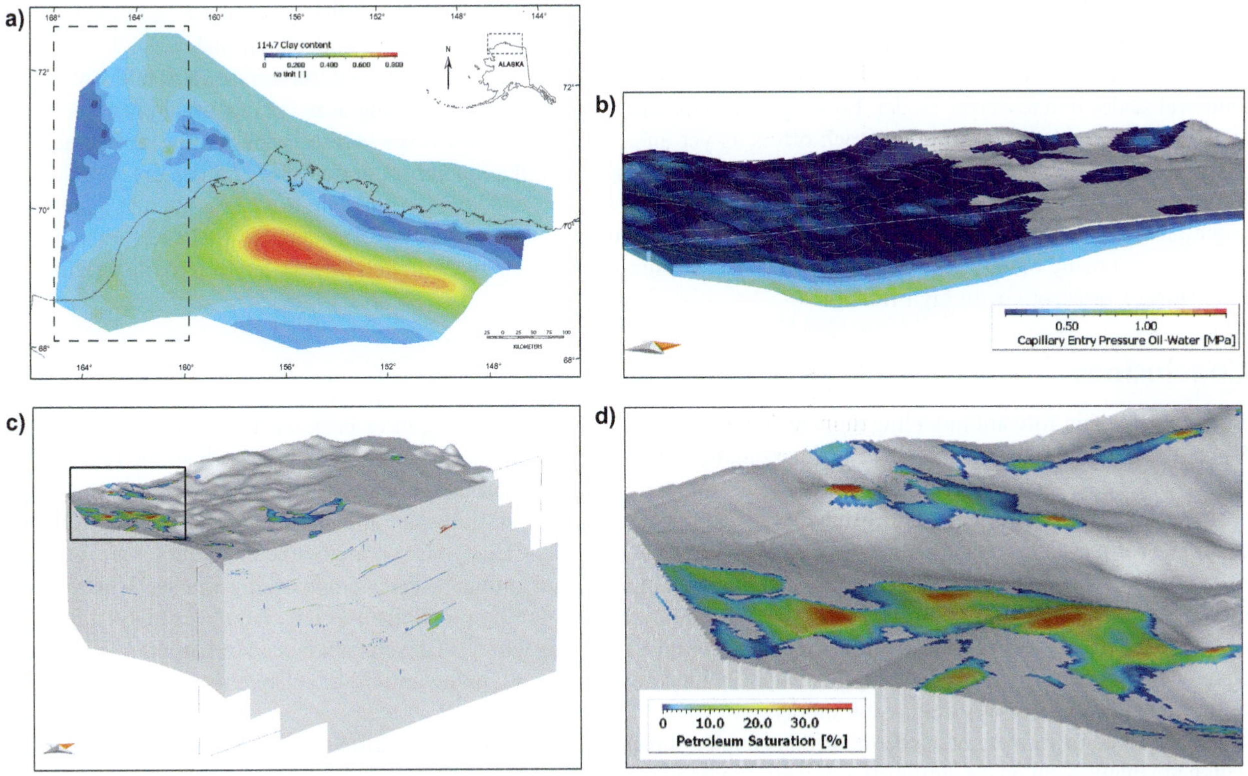


**FIG. 28** (A) Clay content predicted by the forward stratigraphic model of the Alaska North Slope, USA for the 114.7 Ma depositional surface was embedded in the 3D basin and petroleum system model to predict on a regional scale the (B) capillary entry pressure and (C, D) petroleum saturation. *(From Peters et al. (2019), used with permission.)*

A major advancement in basin and petroleum system modeling is the use of integrated approaches in which output from basin modeling constitutes the first step of a sequence of operations to achieve an outcome that would be difficult to accomplish with basin modeling alone. For example, AlKawai et al. (2018) developed a novel workflow that combined basin modeling predictions of temperature and pressure with rock physics models and well logs to invert seismic cubes into impedance. The impedance was then combined with probability distribution functions through a statistical classification scheme to determine the most likely lithofacies at both well locations and in areas away from wells. The probability of each predicted lithofacies was also obtained, thereby allowing more robust decision-making that could be used in future drilling. Fig. 18B illustrates the calibrated basin model upon which the workflow depended.

Uncertainty analysis is also developing in forward stratigraphic modeling. Aware of the computational expense necessary to test numerous scenarios, each with uncertain input and output parameters, Gervais et al. (2018) built functions to approximate the spatial outputs generated from forward stratigraphic modeling. Rather than use the simulator during the modeling sequence, the authors used a technique that is commonly used in reservoir modeling called metamodeling. The metamodel, also called a surrogate model, consisted of a synthetic test case that was constructed based on data from the Gulf of Mexico. The uncertain parameters treated in the metamodel are the three major processes that shape the sedimentary architecture of a basin (e.g., Fig. 2): accommodation, sediment and water supply, and sediment transport. The technique demonstrated numerous benefits: (1) the influence of each parameter was quantified throughout the basin, thereby highlighting the most significant geologic process occurring at each location; (2) based on those results, new interpretations of basin characteristics drove model calibration and (3) probability maps of sand proportion highlighted potential reservoir rocks. At the conclusion of their article, Gervais et al. (2018) advocate for integrating the resulting geological scenarios to initialize basin models.

Automated model building and calibration are perhaps the next frontier in the modeling world, as is interactive modeling in model building interfaces. A few of the benefits of interactive modeling include real-time updates of the thickness and lithology of the geologic section, real-time model calibration and real-time decisions of parameters for optimization. Machine learning is an intrinsic element of the examples described above as well as for automated approaches.

Notwithstanding the limitations of each modeling product, the value in combining the two approaches to elucidate sedimentological facies and geologic structure in the subsurface is evident. With the probabilistic approaches elaborated above, a single "final" model that has traditionally been the result of forward stratigraphic and basin and petroleum system modeling is replaced with a suite of models, along with quantitative determinations of the goodness of fit of each model. Basin-scale models are thusly accompanied by robustly determined probabilities to aid decision-making.

Building on the advantages of these integrated models, one logical extension is to step further down in both spatial and temporal scales to a reservoir model. Linking basin modeling with reservoir modeling is as yet incompletely achieved. In addition, the coupled modeling approach offers as yet unknown benefits for other investigations of the subsurface. For example, carbon capture and storage is growing as a technology for reducing atmospheric $CO_2$ (Chapter 21). Crucial geologic factors that can be addressed by coupling forward stratigraphic modeling with basin modeling include quantifying potential storage volume, evaluating seal integrity and determining the geometry of stacked, interbedded sandstone and shale, among many other factors. Following basin scale simulations, reservoir models will be key for characterizing specific storage reservoirs with high fidelity.

## Conclusions

Both stratigraphic forward modeling and basin and petroleum system modeling have been used to study subsurface geological processes and attendant paleoenvironmental conditions for many decades. For most of that time, the two technologies evolved separately, even though both disciplines required information about similar geologic characteristics (paleowater depth and compaction behavior, for example) and used similar data streams (seismic surveys, wellbores, etc.). The format of this chapter aimed to make that clear in three ways. First, the two modeling techniques were described using the same rubric—input, simulation, output, calibration, final output—while emphasizing that each step of the modeling process remain true to model objectives (Fig. 1). Moreover, the parallelism of the chapter format—a roadmap of sorts—purposely maintains a structure that enhances understanding when (1) the chapter is read by a nonmodeler, (2) the chapter is read by an expert in stratigraphic modeling but not in basin modeling and (3) the chapter is read by an expert in basin modeling but not in stratigraphic modeling.

Second, the organizational scheme of this chapter shines light not only on the differences between the techniques—which are many—but on the similarities as well. For example, by using the same format for the two modeling technologies, the reader can go back and forth through the text to compare and contrast how stratigraphic models and basin models are constructed and calibrated. That goal required a deeper dive into the nuts and bolts of each technology. Numerous ways to determine the age of a geologic horizon for a basin model or to determine the sediment source areas for a stratigraphic model—this is a level of specificity that is missing from typical journal article. The level of detail provided in this chapter stops just short of a how-to manual but gets close.

Finally, the detailed descriptions of how and why and when to build a forward stratigraphic model and a basin and petroleum system model laid the foundation for bringing them together via two case studies. More often the not, a synergistic approach is often lacking in academic, industry and government environments. Certainly the coupled approach espoused in this chapter requires that silos be eliminated in favor of collaboration between workgroups. One of many benefits of such collaboration is that experts in each discipline will significantly enhance their understanding of the geologic processes at work in the subsurface, not only in deepwater environments but in shelf and shallow environments as well.

It is clear that using integrated forward stratigraphic and basin and petroleum system models offers advantages that are unachievable when each method is implemented alone. Coupling the models significantly enhances the project life cycle so that the modeling team successfully accomplishes project objectives.

## Acknowledgments

Thank you to Ken Bird, Alan Burnham, Samer Bou Daher, Ken Peters, Noelle Schoellkopf, Agnieszka Zastepa, Alina-Berenice Christ, Dan Scheirer, past and present students and industrial affiliates of the Stanford University Basin and Petroleum System Modeling consortium, and the volume editors. Many thanks to Attila Bartha and Gail Bergen for speedy and thorough reviews that improved the manuscript.

## References

Acevedo, A., Madhoo, H.A., Khramtsov, A., Noomee, L., Tetzlaff, D.M., 2014. Techniques to understand reservoirs associated with deepwater sedimentological processes, from basin to field scale – a case study. In: International Petroleum Technology Conference (IPTC-18016-MS), pp. 1–6.

AlKawai, W.H., Mukerji, T., Hosford Scheirer, A., Graham, S.A., 2018. Combining seismic reservoir characterization workflows with basin modeling in the deepwater Gulf of Mexico Mississippi Canyon Area. AAPG Bull. 102, 629–652.

Allen, P.A., Allen, J.R., 2005. Basin Analysis—Principles and Applications. Blackwell Publishing, Malden, MA, p. 549.

Allwardt, J.R., Michael, G.E., Shearer, C.R., Heppard, P.D., Ge, H., 2009. 2D modeling of overpressure in a salt withdrawl basin, Gulf of Mexico, USA. Mar. Pet. Geol. 26, 464–473.

Arab, M., Belhai, D., Granjeon, D., Roure, F., Arbeaumont, A., Rabineau, M., Bracène, R., Lassal, A., Sulzer, C., Déverchère, J., 2016. Coupling stratigraphic and petroleum system modeling tools in complex tectonic domains: case study in the North Algerian Offshore. Arab. J. Geosci. 9, 1–32.

Banet Jr., A.C., 1993. A geochemical profile and burial history of Aurora 890 #1 OCS Y-0943 well offshore of the ANWR 1002 Area, Northeast Alaska: Bureau of Land Management Technical Report 16. pp. 1–51.

Barabasch, J., Ducros, M., Hawie, N., Bou Daher, S., Nader, F.H., Littke, R., 2019. Integrated 3D forward stratigraphic and petroleum system modeling of the Levant Basin, Eastern Mediterranean. Basin Res. 31, 228–252.

Baur, F., 2019. Predicting petroleum gravity with basin modeling: new kinetic models. AAPG Bull. 103, 1811–1837.

Baur, F., Katz, B., 2018. Some practical guidance for petroleum migration modeling. Mar. Pet. Geol. 93, 409–421.

Beardsmore, G.R., Cull, J.P., 2001. Crustal Heat Flow: A Guide to Measurement and Modeling. Cambridge University Press, Cambridge, UK, p. 324.

Beicip-Franlab and Nalcor Energy, 2020. Chapter 7—Stratigraphic modeling: seal capacity and integrity study. In: Offshore Newfoundland & Labrador Resource Assessment. Orphan Basin Area NL20-CFB01. https://oilandgas.nalcorenergy.com/wp-content/uploads/2020/09/2020-Orphan-Resource-Assessment-Public-Atlas-Final.pdf.

Bird, K.J., 2001. Alaska: a twenty-first-century petroleum province. In: Downey, M.W., Threet, J.C., Morgan, W.A. (Eds.), Petroleum Provinces of the Twenty-First Century: Tulsa, AAPG Memoir 74, pp. 137–165.

Bird, K.J., Molenaar, C.M., 1992. The North Slope Foreland Basin, Alaska. In: Macqueen, R.W., Leckie, D.A. (Eds.), Foreland Basins and Fold Belts. American Association of Petroleum Geologists Memoir 55, pp. 363–393.

Bokulich, A., Oreskes, N., 2017. Models in the geosciences. In: Magnani, L., Bertolotti, T.W. (Eds.), Springer Handbook of Model-Based Science. Springer, pp. 891–911.

Bosence, D, Waltham, D, 1990. Computer modeling the internal architecture of carbonate platforms. Geology 18 (1), 26–30.

Bosscher, H., Schlager, W., 1992. Computer simulation of reef growth. Sedimentology 39, 503–512.

Bowman, S.A., Vail, P.R., 1999. Interpreting the stratigraphy of the Baltimore Canyon section, offshore New Jersey with PHIL, a stratigraphic simulator. In: Harbaugh, J.W., Watney, W.L., Rankey, E.C., Slingerland, R., Goldstein, R.H., Franseen, E.K. (Eds.), Numerical Experiments in Stratigraphy: Recent Advances in Stratigraphic and Sedimentologic Computer Simulations. SEPM Special Publication 62, Tulsa, OK, pp. 117–138.

Bruneau, B., Chauveau, B., Duarte, L.V., Desaubliaux, G., Moretti, I., Baudin, F., 2016. 3D numerical modelling of marine organic matter distribution: example of the early Jurassic sequences of the Lusitanian Basin (Portugal). Basin Res. 30, 101–123.

Bruneau, B., Villié, M., Ducros, M., Chauveau, B., Baudin, F., Moretti, I., 2018. 3D numerical modelling and sensitivity analysis of the processes controlling organic matter distribution and heterogeneity—a case study from the Toarcian of the Paris Basin. Geosciences 8, 405.

Burgess, P.M., 2001. Modeling carbonate sequence development without relative sea-level oscillations. Geology 29, 1127–1130.

Burgess, P.M., 2013. CarboCAT: a cellular automata model of heterogeneous carbonate strata. Comput. Geosci. 53, 129–140.

Burgess, P.M., Wright, V.P., 2003. Numerical forward modeling of carbonate platform dynamics: an evaluation of complexity and completeness in carbonate strata. J. Sediment. Res. 73, 637–652.

Burgreen-Chan, B., Meisling, K.E., Graham, S., 2015. Basin and petroleum system modelling of the East Coast Basin, New Zealand: a test of overpressure scenarios in a convergent margin. Basin Res. 28, 536–567.

Burnham, A.K., 2017. Global Chemical Kinetics of Fossil Fuels: How to Model Maturation and Pyrolysis. Springer, Cham, Switzerland, p. 315.

Burns, W.M., Hayba, D.O., Rowan, E.L., Houseknecht, D.W., 2007. Estimating the amount of eroded section in a partially exhumed basin from geophysical well logs: an example from the North Slope. In: Galloway, J.P., Haeussler, P.J. (Eds.), Studies by the U.S. Geological Survey in Alaska, 2005: Reston, U.S. Geological Survey Professional Paper 1732-D, pp. 1–18.

Christ, A.-B., Schenk, O., Salomonsen, P., 2016. Using stratigraphic forward modeling to model the Brookian sequence of the Alaska North Slope. In: Raju, N.J. (Ed.), Geostatistical and Geospatial Approaches for the Characterization of Natural Resources in the Environment: Challenges, Processes and Strategies. Co-published by Springer International Publishing and Capital Publishing Company, Cham, Switzerland and New Delhi, India, pp. 623–626.

Davies, P.J., Symonds, P.A., Feary, D.A., Pigram, C.J., 1989. The evolution of the carbonate platforms of Northeast Australia. In: Crevello, P.D., Wilson, J. L., Sarg, J.F., Read, J.F. (Eds.), Controls on Carbonate Platforms and Basin Development. SEPM Society for Sedimentary Geology Special Publication 44, pp. 233–258.

Demicco, R.V., 2004a. Applications of fuzzy logic to stratigraphic modeling. In: Demicco, R.V., Klir, G.J. (Eds.), Fuzzy Logic in Geology. Elsevier, Inc., Burlington, pp. 121–156.

Demicco, R.V., 2004b. Fuzzy logic and earth science: an overview. In: Demicco, R.V., Klir, G.J. (Eds.), Fuzzy Logic in Geology: Burlington. Elsevier, Inc, pp. 63–104.

Demicco, R.V., 2004c. Fuzzy logic in geological sciences: a literature review. In: Demicco, R.V., Klir, G.J. (Eds.), Fuzzy Logic in Geology. Elsevier, Inc., Burlington, pp. 103–120.

Duan, T., Cross, T.A., Lessenger, M.A., 2000. 3D carbonate stratigraphic model based on energy and sediment flux. In: 2000 AAPG Annual Convention and Exhibition, New Orleans, Louisiana, May 19–22, 2000, v. Search and Discovery Article #90914.

Eberli, G.P., Ginsburg, R.N., 1989. Cenozoic progradation of northwestern Great Bahama Bank, a record of lateral platform growth and sea-level fluctuations. In: Crevello, P.D., Wilson, J.L., Sarg, J.F., Read, J.F. (Eds.), Controls on Carbonate Platforms and Basin Development. SEPM Society for Sedimentary Geology Special Publication 44, pp. 339–351.

El-Shahat, W., Villinski, J.C., El-Bakry, G., 2009. Hydrocarbon potentiality, burial history and thermal evolution for some source rocks in October oil field, northern Gulf of Suez, Egypt. J. Petrol. Sci. Eng. 68, 245–267.

Energy Information Administration, 2015. Top 100 U.S. OIl and Gas Fields. U.S. Department of Energy, Washington, DC, p. 12.

Forkner, R.M., Hinnov, L.A., Smart, P., 2010. Use of insolation as a proxy for high-frequency eustasy in forward modeling of platform carbonate cyclostratigraphy—a promising approach. Sediment. Geol. 231, 1–13.

Gac, S., Hansford, P.A., Faleide, J.I., 2018. Basin modelling of the SW Barents Sea. Mar. Pet. Geol. 95, 167–187.

Gervais, V., Ducros, M., Granjeon, D., 2018. Probability maps of reservoir presence and sensitivity analysis in stratigraphic forward modeling. AAPG Bull. 102, 613–628.

Ghiasi-Freez, J., Kadkhodaie-Ilkhchi, A., Ziaii, M., 2012. Improving the accuracy of flow units prediction through two committee machine models: an example from the South Pars Gas Field, Persian Gulf Basin, Iran. Comput. Geosci. 46, 10–23.

Granjeon, D., 2014. 3D Forward Modelling of the Impact of Sediment Transport and Base Level Cycles on Continental Margins and Incised Valleys. International Association of Sedimentology Special Publication 46, pp. 453–472.

Granjeon, D., Joseph, P., Harbaugh, J.W., Watney, W.L., Rankey, E.C., Slingerland, R., Goldstein, R.H., Franseen, E.K., 1999. Concepts and applications of a 3-D multiple lithology, diffusive model in stratigraphic modeling. In: Harbaugh, J.W., Watney, W.L., Rankey, E.C., Slingerland, R., Goldstein, R. H., Franseen, E.K. (Eds.), Numerical Experiments in Stratigraphy: Recent Advances in Stratigraphic and Sedimentologic Computer Simulations. Vol. 62. SEPM Society for Sedimentary Geology, Tulsa, OK, pp. 197–210.

Grender, G.C., Rapoport, L.A., Segers, R.G., 1974. Experiment in quantitative geologic modeling. Am. Assoc. Pet. Geol. Bull. 58, 488–498.

Griffiths, C.M., Paraschivoiu, E., 1998. Three-dimensional forward stratigraphic modeling of early cretaceous sedimentation on the Leveque and Yampi Shelves, Browse Basin. APPEA J. 38, 147–158.

Griffiths, C.M., Dyt, C.P., Paraschivoiu, E., Liu, K., 2001. Sedsim in hydrocarbon exploration. In: Merriam, D., Davis, J.C. (Eds.), Geologic Modeling and Simulation. Kluwer Academic, New York, pp. 71–97.

Grohmann, S., Fietz, W.S., Nader, F.H., Romero-Sarmiento, M.-F., Baudin, F., Littke, R., 2021. Characterization of Late Cretaceous to Miocene source rocks in the Eastern Mediterranean Sea: an integrated numerical approach of stratigraphic forward modelling and petroleum system modelling. Basin Res. 33, 846–874.

Hantschel, T., Kauerauf, A.I., 2009. Fundamentals of Basin and Petroleum Systems Modeling. Springer-Verlag, Berlin, p. 476.

Hantschel, T., Kauerauf, A.I., Wygrala, B., 2000. Finite element analysis and ray tracing modeling of petroleum migration. Mar. Pet. Geol. 17, 815–820.

Hantschel, T., Wygrala, B., Palmowski, D., 2019. Petroleum systems modeling on the move: scientific concepts, numerical models and exploration practices. In: AAPG Hedberg Conference, Houston, TX, 4–6 March 2019, pp. 1–28.

Haq, B.U., Hardenbol, J., Vail, P.R., 1987. Chronology of fluctuating sea levels since the Triassic. Science 235, 1156–1167.

Harbaugh, J.W., 1966. Mathematical simulation of marine sedimentation with IBM 7090/7094 computers. In: Merriam, D.F. (Ed.), Computer Contribution 22. State Geological Survey, The University of Kansas, Lawrence, KS, Lawrence, pp. 1–52.

Hawie, N., Marfisi, E., Saint-Ange, F., MacDonald, A.W.A., 2019. Statistical analysis of forward stratigraphic models in complex salt provinces: the central Scotian Basin case study. AAPG Bull. 103, 433–467.

He, M., Graham, S., Hosford Scheirer, A., Peters, K.E., 2014. A basin modeling and organic geochemistry study in the Vallecitos syncline, San Joaquin Basin, California. Mar. Pet. Geol. 49, 15–34.

Henriksen, E., Bjørnseth, H., Hals, T., Heide, T., Kiryukhina, T., Kløvjan, O., Larssen, G.B., Ryseth, A., Rønning, K., Sollid, K., 2011. Uplift and erosion of the greater Barents Sea: impact on prospectivity and petroleum systems. In: Spencer, A.M., Embry, A.F., Gautier, D.L., Stoupakova, A.V., Sørensen, K. (Eds.), Arctic Petroleum Geology. Geological Society Memoir 35, London, pp. 271–281.

Hermanrud, C., 1993. Basin modelling techniques—an overview. In: Dore, A.G., Augustson, J.H., Hermanrud, C., Stewart, D.J., Sylta, O. (Eds.), Basin Modelling: Advances and Applications: Amsterdam, Netherlands. Norwegian Petroleum Society (NPF) Special Publication 3, pp. 1–34.

Hill, J., Tetzlaff, D., Curtis, A., Wood, R., 2009. Modeling shallow marine carbonate depositional systems. Comput. Geosci. 35, 1862–1874.

Hosford Scheirer, A., Bird, K.J., 2020. A lateral well in the Shublik Formation, Alaska North Slope, with implications for development of unconventional resource potential. Interpretation 8, SJ36–SJ49.

Houseknecht, D.W., 2019. Evolution of the Arctic Alaska sedimentary basin. In: Miall, A.D. (Ed.), The Sedimentary Basins of the United States and Canada Amsterdam. Elsevier, pp. 719–745.

Houseknecht, D.W., 2017. Nanushuk–Torok Assessment Team. In: Assessment of undiscovered oil and gas resources in the Cretaceous Nanushuk and Torok formations, Alaska North Slope, and summary of resource potential of the National Petroleum Reserve in Alaska, 2017. U.S. Geological Survey Fact Sheet FS-2017-2088, pp. 1–4. Available at https://pubs.er.usgs.gov/publication/fs20173088.

Houseknecht, D.W., Bird, K.J., Schenk, C.J., 2009. Seismic analysis of clinoform depositional sequences and shelf-margin trajectories in Lower Cretaceous (Albian) strata, Alaska North Slope. Basin Res. 21, 644–654.

Houseknecht, D.W., Burns, W.M., Bird, K.J., 2012. Thermal maturation history of Arctic Alaska and the Southern Canada Basin. In: Harris, N.B., Peters, K.E. (Eds.), Analyzing the Thermal History of Sedimentary Basins: Methods and Case Histories. SEPM Special Publication No. 11, Tulsa, OK, pp. 199–219.

Huang, X., Dyt, C., Griffiths, C., Salles, T., 2012. Numerical forward modelling of fluxoturbidite flume experiments using Sedsim. Mar. Pet. Geol. 35, 190–200.

Hunt, J.M., 1979. Petroleum Geochemistry and Geology. W.H. Freeman and Company, San Francisco, p. 617.

Hutton, J., 1788. Theory of the Earth. vol. 1 Transactions of the Royal Society of Edinburgh, p. 278. Edinburgh.

Hutton, J., 1795. Theory of the Earth. vol. 2 Transactions of the Royal Society of Edinburgh, Edinburgh, p. 215.

Hutton, E.W.H., Syvitski, J.P.M., 2008. Sedflux 2.0: an advanced process-response model that generates three-dimensional stratigraphy. Comput. Geosci. 34, 1319–1337.

Immenhauser, A., 2009. Estimating palaeo-water depth from the physical rock record. Earth Sci. Rev. 96, 107–139.

Jervey, M.T., 1988. Quantitative geological modeling of siliciclastic rock sequences and their seismic expression. In: Wilgus, C.K., Hastings, B.S., Posamentier, H., Wagoner, J.V., Ross, C.A., Kendall, C.G.S.C. (Eds.), Sea-Level Changes—An Integrated Approach. vol. 42. The Society for Sedimentary Geology Special Publication 42, pp. 47–69.

Kendall, C.G.S.C., Strobel, J.S., Cannon, R., Bezdek, J., Biswas, G., 1991. The simulation of the sedimentary fill of basins. J. Geophys. Res. 96, 6911–6929.

Koerschner, W.F., Read, J.F., 1989. Field and modelling studies of Cambrian carbonate cycles, Virginia, Appalachians. J. Sediment. Res. 59, 654–687.

Kroeger, K.F., Plaza-Faverola, A., Barnes, P.M., Pecher, I.A., 2015. Thermal evolution of the New Zealand Hikurangi subduction margin: impact on natural gas generation and methane hydrate formation—a model study. Mar. Pet. Geol. 63, 97–114.

Lerche, I., 1990a. Basin Analysis—Quantitative Methods. Vol. 1 Academic Press, Inc., San Diego, CA, p. 564.

Lerche, I., 1990b. Basin Analysis—Quantitative Methods. Vol. 2 Academic Press, Inc., San Diego, CA, p. 570.

Li, F., Griffiths, C.M., Dyt, C.P., Weill, P., Feng, M., Salles, T., Jenkins, C., 2009. Multigrain seabed sediment transport modelling for the south-west Australian Shelf. Mar. Freshw. Res. 60, 774–785.

Liu, J., Liu, K., 2021. Estimating stratal completeness of carbonate deposition via process-based stratigraphic forward modeling. Sci. China Earth Sci. 64, 253–259.

Liu, K., Liang, T.C.K., Paterson, L., Kendall, C.G.S.C., 1998. Computer simulation of the influence of basin physiography on condensed section deposition and maximum flooding. Sediment. Geol. 122, 181–191.

Liu, J., Liu, K., Huang, X., 2016a. Effect of sedimentary heterogeneities on hydrocarbon accumulations in the Permian Shanxi Formation, Ordos Basin, China: insight from an integrated stratigraphic forward and petroleum system modelling. Mar. Pet. Geol. 76, 412–431.

Liu, Y., Dutta, N.C., Vigh, D., Kapoor, J., Hunter, C., Saragoussi, E., Jones, L., Yang, S., Eissa, M.A., 2016b. Basin-scale integrated earth-model building using rock-physics constraints. Lead. Edge 35, 141–145.

Liu, J., Liu, K., Li, C., Liu, W., 2020. Tectono-sedimentary evolution of the Late Ediacaran to early Cambrian trough in central Sichuan Basin, China: new insights from 3D stratigraphic forward modelling. Precambrian Res. 350, 105826.

Liu, K., Liu, J., Huang, X., 2021. Coupled stratigraphic and petroleum system modeling: examples from the Ordos Basin, China. AAPG Bull. 105, 1–28.

Magoon, L.B., 1988. The petroleum system-A classification scheme for research, exploration, and resource assessment. In: Magoon, L.B. (Ed.), Petroleum Systems of the United States. U.S. Geological Survey Bulletin 1870, pp. 2–15.

Magoon, L.B., Dow, W.G., 1994. The petroleum system. In: Magoon, L.B., Dow, W.G. (Eds.), The Petroleum System—From Source to Trap Memoir. vol. 60. American Association of Petroleum Geologists, Tulsa, OK, pp. 3–24.

Mann, U., Zweigel, J., 2008. Modelling source-rock distribution and quality variations: the organic facies modelling approach. In: de Boer, P., Postma, G., van der Zwan, K., Burgess, P., Kukla, P. (Eds.), Analogue and Numerical Modelling of Sedimentary Systems: From Understanding to Prediction. International Association of Sedimentologists Special Publication 40, Oxford, pp. 239–274.

McKenzie, D., 1978. Some remarks on the development of sedimentary basins. Earth Planet. Sci. Lett. v. 40, 25–32.

Menotti, T., Hosford Scheirer, A., Meisling, K., Graham, S.A., 2019. Integrating strike-slip tectonism with three-dimensional basin and petroleum system analysis of the Salinas Basin, California. AAPG Bull 103, 1443–1472.

Merriam, D.F., Davis, J.C. (Eds.), 2001. Geologic Modeling and Simulation: Sedimentary Systems. Kluwer Academic/Plenum Publishers, p. 352.

Moore, T.E., Wallace, W.K., Bird, K.J., Karl, S.M., Mull, C.G., Dillon, J.T., 1994. Geology of northern Alaska. In: Plafker, G., Berg, H.C. (Eds.), The Geology of Alaska, v. G-1, The Geology of North America. The Geological Society of America, Boulder, CO, pp. 49–140.

Mosca, F., Djordjevic, O., Hantschel, T., McCarthy, J., Krueger, A., Phelps, D., Akintokunbo, T., Joppen, T., Koster, K., Schupbach, M., Hampshire, K., MacGregor, A., 2018. Pore pressure prediction while drilling: three-dimensional earth model in the Gulf of Mexico. AAPG Bull. 102, 691–708.

Nordlund, U., 1999. FUZZIM: forward stratigraphic modeling made simple. Comput. Geosci. 25, 449–456.

Ojakangas, D.R., 1967. Mathematical Simulation of Oil Trap Development. PhD thesis, Stanford University, p. 150.

Olson, H., Snedden, J., Cunningham, R., 2015. Development and application of a robust chronostratigraphic framework in Gulf of Mexico Mesozoic exploration. Interpretation 3, SN39–SN58.

Paola, C., 2000. Quantitative models of sedimentary basin filling. Sedimentology 47, 121–178.

Pepper, A.S., Corvi, P.J., 1995. Simple kinetic models of petroleum formation. Part I—oil and gas generation from kerogen. Mar. Pet. Geol. 12, 291–319.

Peters, K.E., Walters, C.C., Moldowan, J.M., 2005. The Biomarker Guide: Cambridge. Cambridge University Press, UK, p. 1155.

Peters, K.E., Magoon, L.B., Bird, K.J., Valin, Z.C., Keller, M.A., 2006. North Slope, Alaska—source rock distribution, richness, thermal maturity, and petroleum charge. AAPG Bull. 90, 1–31.

Peters, K.E., Magoon, L.B., Valin, Z.C., Lillis, P.G., 2007. Source rock geochemistry of the San Joaquin Basin Province, California. In: Hosford Scheirer, A. (Ed.), Petroleum Systems and Geologic Assessment of Oil and Gas in the San Joaquin Basin Province, California, U.S. Geological Survey Professional Paper 1713, Chapter 11. U.S. Geological Survey, pp. 1–102. available at http://pubs.usgs.gov/pp/pp1713/.

Peters, K.E., Magoon, L.B., Lampe, C., Hosford Scheirer, A., Lillis, P.G., Gautier, D.L., 2008. A four-dimensional petroleum systems model for the San Joaquin Basin Province, California. In: Hosford Scheirer, A. (Ed.), Petroleum Systems and Geologic Assessment of Oil and Gas in the San Joaquin Basin Province, California, U.S. Geological Survey Professional Paper 1713, Chapter 12. U.S. Geological Survey, pp. 1–35. available at http://pubs.usgs.gov/pp/pp1713.

Peters, K.E., Burnham, A.K., Walters, C.C., 2015. Petroleum generation kinetics: single versus multiple heating-ramp open-system pyrolysis. AAPG Bull. 99, 591–616.

Peters, K.E., Schenk, O., Hosford Scheirer, A., Wygrala, B., Hantschel, T., 2017. Basin and petroleum system modeling of conventional and unconventional petroleum resources. In: Hsu, C.S., Robinson, P.R. (Eds.), Springer Handbook of Petroleum Technology. Springer, Heidelberg, pp. 381–417.

Peters, K.E., Magoon, L.B., Wygrala, B., 2019. Historical transformation of the petroleum system concept to computerized petroleum system models and linked technologies. In: 2019 AAPG Annual Convention and Exhibition, San Antonio, Texas, May 19-22, 2019, Search and Discovery Article No. 42385.

Pyenson, N.D., Irmis, R.B., Lipps, J.H., Barnes, L.G., Mitchell Jr., E.D., McLeod, S.A., 2009. Origin of a widespread marine bonebed deposited during the middle Miocene Climatic Optimum. Geology 37, 519–522.

Rüpke, L.H., Schmid, D.W., Hartz, E.H., Martinsen, B., 2010. Basin modelling of a transform margin setting: structural, thermal and hydrocarbon evolution of the Tano Basin, Ghana. Pet. Geosci. 16, 283–298.

Salles, T., Marchès, E., Dyt, C., Griffiths, C., Hanquiez, V., Mulder, T., 2010. Simulation of the interactions between gravity processes and contour currents on the Algarve Margin (South Portugal) using the stratigraphic forward model Sedsim. Sediment. Geol. 229, 95–109.

Scheibner, C., Kuss, H.-J., Speijer, R., 2003. Stratigraphic modelling of carbonate platform-to-basin sediments (Maastrichtian to Paleocene) in the Eastern Desert, Egypt. Palaeogeogr. Palaeoclimatol. Palaeoecol. 200, 163–185.

Schenk, H.J., Horsfield, B., 1993. Kinetics of petroleum generation by programmed-temperature closed- versus open-system pyrolysis. Geochim. Cosmochim. Acta 57, 623–630.

Schenk, O., Magoon, L.B., Bird, K.J., Peters, K.E., 2012. Petroleum system modeling of Northern Alaska. In: Peters, K.E., Curry, D.J., Kacewicz, M. (Eds.), Basin Modeling—New Horizons in Research and Applications. American Association of Petroleum Geologists Hedberg Series no. 4, Tulsa, OK, pp. 317–338.

Schenk, O., Tewari, S., Karvelas, A., 2019. Basin Study Report—Exmouth Sub-Basin: Geoscience Australia, v. July, 2019. pp. 1–357.

Schultz, L., 2016. Evaluating the Thermal History of the Los Angeles Basin, California Through 3-D Basin and Petroleum System Modeling. Masters of Science thesis, Stanford University, Stanford, CA, p. 71.

Schwarzacher, W., 1966. Sedimentation in Subsiding Basins. Nature 210, 1349–1350.

Shafie, K., Madon, M., 2008. A review of stratigraphic simulation techniques and their applications in sequence stratigraphy and basin analysis. Bull. Geol. Soc. Malaysia 54, 81–89.

Sloss, L.L., 1962. Stratigraphic models in exploration. J. Sediment. Res. 32, 415–422.

Stainforth, J.G., 2009. Practical kinetic modeling of petroleum generation and expulsion. Mar. Pet. Geol. 26, 552–572.

Sternbach, C.A., 2020. Super basin thinking: methods to explore and revitalize the world's greatest petroleum basins. AAPG Bull. 104, 2463–2506.

Strobel, J., Cannon, R., Kendall, C.G.S.C., Biswas, G., Bezdek, J., 1989. Interactive (SEDPAK) simulation of clastic and carbonate sediments in shelf to basin settings. Comput. Geosci. 15, 1279–1290.

Sylta, Ø., 2002. Quantifying secondary migration efficiencies. Geofluids 2, 285–298.

Syvitski, J.P.M., Hutton, E.W.H., 2001. 2D SEDFLUX 1.0C: an advanced process-response numerical model for the fill of marine sedimentary basins. Comput. Geosci. 27, 731–753.

Tetzlaff, D.M., Harbaugh, J.W., 1989. Simulating Clastic Sedimentation. Springer US, New York, p. 202.

Tewari, S., Rathee, D., Schenk, O., 2019. An insight into complex charge history and hydrocarbon distribution in Kwanza Basin. In: PESGB/HGS Africa Conference, 1st–2nd October 2019, London, UK v. poster.

Tissot, B.P., Welte, D.H., 1984. Petroleum Formation and Occurrence. Springer-Verlag, Berlin, p. 699.

Tong, Y., Mukerji, T., 2017. Generalized sensitivity analysis study in basin and petroleum system modeling, case study on Piceance Basin, Colorado. J. Petrol. Sci. Eng. 149, 772–781.

Tong, Y., Ibarra, D.E., Caves, J.K., Mukerji, T., Graham, S.A., 2016. Constraining basin thermal history and petroleum generation using paleoclimate data in the Piceance Basin, Colorado. Basin Res., 1–12.

Walker, J.D., Geissman, J.W., Bowring, S.A., Babcock, L.W., 2018. Geologic Time Scale v. 5.0. Geological Society of America, Boulder, CO, https://doi.org/10.1130/2018.CTS005R3C p.

Warrlich, G., Bosence, D., Waltham, D., Wood, C., Boylan, A., Badenas, B., 2008. 3D stratigraphic forward modelling for analysis and prediction of carbonate platform stratigraphies in exploration and production. Mar. Pet. Geol. 25, 35–58.

Welte, D.H., Horsfield, B., Baker, D.R. (Eds.), 1997. Petroleum and Basin Evolution: Insights From Petroleum Geochemistry, Geology, and Basin Modeling. Springer-Verlag, Berlin, p. 535.

White, R., McKenzie, D., 1989. Magmatism at rift zones: the generation of volcanic continental margins and flood basalts. J. Geophys. Res. 94, 7685–7729.

Winsberg, E., 2019. Computer simulations in science. In: Zalta, E.N. (Ed.), The Stanford Encyclopedia of Philosophy. Metaphysics Research Lab, Stanford University.

Wygrala, B.P., 1989. Integrated Study of an Oil Field in the Southern Po Basin, Northern Italy. PhD dissertation thesis, University of Cologne, West Germany, p. 217.

Yang, H., Xi, S.L., Liu, X.S., 2001. Generation, migration and accumulation of the Upper Paleozoic, Ordos Basin. In: Fu, C.D. (Ed.), Deep Basin Gas Research (in Chinese With English Abstract). Petroleum Industry Press, Beijing, China, pp. 88–103.

Zhang, J., Guo, J., Li, Y., Sun, Z., 2019. 3D-basin modeling of the changling depression, NE China: exploring petroleum evolution in deep tight sandstone reservoirs. Energies 12, 1043.

Chapter 19

# Technical (engineering) advancements enabling deepwater exploration and production

Lori Fremin[a], Richard A. Sears[b] and Charlie Williams[c]

[a]*Shell—Retired, Houston, TX, United States,* [b]*Energy Resources Engineering, Stanford University, Stanford, CA, United States,* [c]*Center for Offshore Safety, Executive Director—Retired, Houston, TX, United States*

## Introduction

Deepwater, and more generally offshore oil and gas production, traces its roots back to the turn of the 20th century when in Summerland, California, USA, an enterprising wildcatter constructed piers, extending from the beach, out into the Santa Barbara Channel on the Southern California coast. Derricks constructed on the piers, powered by generators on the beach produced oil over the next few years from sands at a depth of approximately 450ft (135 m) below sea level.

Fig. 1 is a photo, taken just after the turn of the 20th century of the coastline at Summerland, California, USA, just south of Santa Barbara.

Humble beginnings compared with today's deepwater developments. Of these early offshore wells, the average produced 2 barrels of oil per day[a] (Leffler, 2011). In contrast, a large modern deepwater structure such as the Olympus tension leg platform in approximately 3100ft (950 m) of water in the U.S. Gulf of Mexico, offshore Louisiana, is shown in Fig. 2. Olympus is the third tension leg platform that Shell installed in the greater Mars Basin in the U.S. Gulf of Mexico. The first of the three, Mars, was installed in 1996. Peak production from the Mars platform was in June 2000 at 208,000 barrels of oil and 217 million cubic feet of gas per day (Priest, 2019).

Despite these obvious differences, there are three things Summerland Field had in common with deepwater developments of today. First, there is the truism, "Oil is where you find it," and those who are developing oil fields will go where the oil is, be it jungles, deserts or deepwater. Second, throughout the history of offshore exploration and production, there has always been deepwater. In 1900, it was most any place beyond the beach, and it would be fair to say that the ultra-deepwater of the day was anything deeper than 35ft (11 m) where the piers ended. Third, moving into a new production play, such as deepwater, requires innovative engineering solutions to extend existing production technology into the new and challenging environment. Chapter 1 of this book details much of the history that brings us to today's deepwater, and this chapter will review some of the critical technical and engineering advancements that enable the large modern deepwater developments of today.

Everyone who has studied production of oil and natural gas from deepwater sedimentary systems knows well the magnificent, associated floating surface structures. These are triumphs of modern engineering that have made producing oil and gas from these deepwater systems possible. There is also a significant untold story of technologies and their development that is less known, without which, there would have been no deepwater development and centrally no economic justification to build these massive structures. An important aspect of this story is that like other pioneering plays in the oil and gas industry, the critical technologies were often poorly understood, or in many cases did not exist when deepwater discoveries were made. The industry had to develop the technologies with very challenging deadlines to enable major deepwater discoveries to become successful developments.

The case studies of this chapter seek to give an overview of a few of these key technologies. The purpose is to highlight that all world class developments of deepwater depositional systems are the result of excellent teamwork characterized by understanding, communication and collaboration between the subsurface experts, the surface engineering experts and the

[a]One barrel equals 42 US gallons.

Deepwater Sedimentary Systems. https://doi.org/10.1016/B978-0-323-91918-0.00019-0

FIG. 1 California, USA coastline at Summerland in the early 20th century. Oil production derricks were constructed on wooden piers built into the Santa Barbara Channel. The water depth at the ends of the piers is approximately 35 ft (11 m). Photographer: G.H. Eldridge.

FIG. 2 The Olympus Tension Leg Platform in the deepwater Gulf of Mexico, USA. Operated by Shell, Olympus, in approximately 3100 ft (950 m) of water, is the third tension leg platform installed in the prolific Mars Basin. *Source: Shell.*

acces experts in drilling and production engineering. Consideration of the technologies described here should offer a view of the significant complexity, excellent planning and extensive parallel technology successes that bring the depositional systems such as found in deepwater today, to economically successful world class developments.

Fig. 3 shows the breadth of the technical disciplines involved through the exploration and production lifecycle of a deepwater project. While some large multinational operating companies may have many, or most, of these discipline specialists on staff, the successful development of deepwater resources has been the product of unprecedented collaboration between project operators, coventure partners, contractors and governments.

Exploration and appraisal are described in some detail in Chapter 1 of this book as the Modern Exploration Common Process. Successful projects then move into development and involve the contributions of many engineering specialists. This process is also described in Chapter 1 as Engineering Principles for Deepwater Petroleum Exploration and Production.

The following case studies focus on examples from the U.S. Gulf of Mexico. As the most mature of deepwater environments with the longest and most robust history, the Gulf of Mexico has played a leading role in engineering technology advances that have enabled oil and natural gas developments in deepwater provinces worldwide. The global spread of technology and expertise from early experience in the U.S. Gulf of Mexico was critical in successfully opening deepwater plays elsewhere in the world, particularly in the West of Shetland province in the U.K. as well as offshore West and Central Africa. Early experience in Africa and Brazil with floating production systems was also an important contribution to the global deepwater knowledge base. Different aspects of the global development of deepwater knowledge and resources are detailed in other chapters of this book, particularly from a geological perspective in Chapters 3–5, and from a business perspective in Chapter 20.

**Technical Disciplines in Deepwater Exploration and Production Project Teams**

**Exploration and Appraisal**
- Geology
- Geophysics
- Petrophysics
- Reservoir Engineering
- Well Engineering
- Economics

**Development**
- Geology
- Geophysics
- Petrophysics
- Reservoir Engineering
- Drilling and Completions Engineering
- Economics
- Production Engineering and Technology
- Process and Facilities Engineering
- Materials and Corrosion Engineering
- Civil Engineering
- Marine Engineering
- Metocean
- Electrical Engineering
- Project Engineering and Scheduling
- Operational Readiness

**Production and Operations**
- Production Engineering and Technology
- Process and Facilities Engineering
- Rotating Equipment Engineering
- Pressure Equipment Engineers and Inspectors
- Control Systems
- Metocean
- Operational Readiness
- Chemical Engineering
- Materials and Corrosion Engineering
- Maintenance and Reliability Engineering
- Data and Analytics

**FIG. 3** The technical disciplines typically involved throughout the exploration and development value chain for deepwater assets.

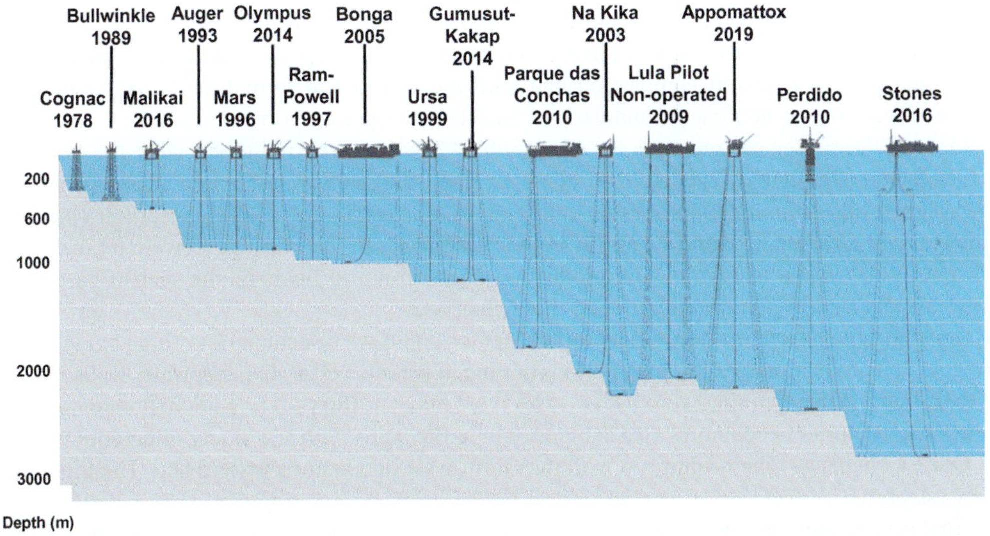


**FIG. 4** Deepwater development, water depth milestones established by Shell over a span of four decades, including fixed structures in the 1970s and 1980s in the U.S. Gulf of Mexico, Tension Leg Platforms in the U.S. Gulf of Mexico and Malaysia and various floating production systems in the Gulf of Mexico, Malaysia, Nigeria and Brazil. The deepest water fixed structure is the Bullwinkle Platform set in 1352 ft (412 m) of water while Shell's deepest water development is the Stones floating production facility in approximately 9500 ft (2900 m) of water. Both Bullwinkle and Stones are in the U.S. Gulf of Mexico. *Source: Shell.*

Fig. 4 shows a large number of water depth milestones set in deepwater, and it has often been the case that drilling and developing resources in deeper water is when new technology solutions are required. Much of the technology transfer to other deepwater basins in the world happened as major operators with deepwater Gulf of Mexico experience acquired licenses elsewhere and built exploration and development teams that included experienced professionals from the Gulf of Mexico. It's also worth noting that new technologies, particularly involving floating production systems, were undergoing development by Petrobras in deepwater offshore Brazil during the 1990s. That technology was used offshore Africa in the late 1990s and early 2000s, and then was available to be used in the 2000s when Floating Production Storage and Offloading vessels (FPSO) were approved for use in the U.S. Gulf of Mexico. In Chapter 20 of this book, Fig. 3 provides a timeline for discoveries and deepwater technology development globally.

Before discussing the engineering case studies, it's important to first reflect upon the advances in seismic technology that have been critical to the successful development of deepwater resources.

Imaging the subsurface utilizing seismic data is a complex technical process, initially developed for onshore exploration in the 1920s and enhanced since. The technology was adapted for the offshore environment as exploration offshore became more common place, and for a number of operational and physical reasons, offshore, and particularly deepwater seismic data is often the most effective in accurately imaging the subsurface.

The seismic method is well described in a number of texts and for background on seismic imaging and technology, the reader is referred to Hyne (2012).

Important aspects of seismic imaging in the context of deepwater systems are discussed and illustrated in Chapter 1 of this book. As a foundational element of the exploration and development process (Chapter 1; Fig. 13 and Table 4) seismic is used extensively to characterize the subsurface. The scale of investigation for reflection seismic data complements other exploration processes (Chapter 1; Figs. 10 and 11) and modern geophysical processing techniques make it possible to create detailed images of the subsurface (Chapter 1; Fig. 12A–C). Chapter 16 in this book provides insights into the use of seismic amplitude and other attributes to accurately characterize lithology as well as to distinguish between different pore filling fluids (e.g., gas, oil or water).

## Case study 1—High rate/high ultimate wells

The initial belief about U.S. Gulf of Mexico deepwater reservoirs was that they would be just like the shallow water Gulf. Shallow water wells and the formations being produced were characterized by low-productivity shaly sands with significant sand production issues. The wells required frequent well workovers, redrills and side-tracks because of a need to do maintenance or well replacement due to production tubing corrosion, inadequacies in sand control, safety valve repairs, production casing collapse from reservoir compaction and fault movement and formation damage and plugging. When a good well was producing it was at a rate of hundreds of barrels of oil per day (Priest, 2007).

Early drilling into deepwater sediments in the U.S. Gulf of Mexico occurred along what has been called the "Flex Trend". The Flex Trend is primarily characterized by upper slope sedimentary processes, including bypass, erosion and often complex channel fill. Sedimentary processes in this environment and the resulting reservoir characteristics are discussed in some detail in Chapters 6, 8 and 13 of this book. Discoveries in the Flex Trend were typically characterized by discontinuous, poorly developed reservoirs with production rates similar to fields on the shelf (Cossey, 2004) and many companies were discouraged by the complexities of these reservoirs.

It became apparent early on that the deepwater Gulf of Mexico could not be economic without being developed by fewer wells producing at high production rates for long periods of time. It was also clear that deepwater wells could cost more than $100 million to drill and if they required workovers, would cost many millions to re-enter and maintain—if they could be accessed at all. Thus, deepwater economics also demanded long life wells with few if any interventions and little required maintenance. Quite a challenge when compared with the Gulf of Mexico historic experience. The high rate/high ultimate well challenge was born.

Chapter 20 in this book provides more data on production rates from deepwater fields globally and the cash flow models presented can be used to quantify the benefits to project profitability resulting from the high production rates achieved in deepwater wells.

Job one for the technical experts was to develop a shared belief that high-rate production was possible in the deepwater. Within Shell Oil, this began with a series of technical workshops, field studies and forums that included international expert staff. The focus was on global high-rate producing wells that could demonstrate that it was possible and that high-rate technology could be developed. These studies were backed up by field tests in the Gulf of Mexico in the two fixed bottom producing structures that were in more than 1000 ft (305 m) of water (Cognac and Bullwinkle). These field tests initially faced great doubt but ultimately showed 5000 barrels of oil per day (BPD) per well could be achieved with no sand production. Despite all this work, the early consensus among most Gulf of Mexico operators was to use 3000 BPD per well for deepwater development planning purposes.

The impact of well rate, and therefore reservoir deliverability on surface system design was made clear from Shell's experience at Auger (see Figs. 4 and 5). Auger was Shell's first major deepwater Gulf of Mexico discovery and the tension leg platform was designed with 32 well slots based on early assumptions about well deliverability. In the end, the field was initially developed using only 14 wells, filling the initial production capacity. Over time, the facility's processing capacity has been expanded, allowing for further development of the Auger structure itself, as well as additional subsea wells exploiting nearby opportunities such as the Cardamom Field developed 25 years later (Shell, 2021).

**FIG. 5** The Auger tension leg platform, with its tendons and mooring system illustrated, is shown at scale projected over the city of New Orleans. Auger was installed in 1994 in 2720 ft (830 m) of water, approximately 214 miles (345 km) south of New Orleans in the U.S. Gulf of Mexico. *Source: Leffler, W. L., et al., 2011. Deepwater Petroleum Exploration & Production, A Nontechnical Guide. second ed. PennWell., Illustration used with permission: © 2011 PennWell Books. All rights reserved.*

It was the discovery and development of early deepwater fields such as Auger and Mars that significantly "reset" industry's expectations regarding deepwater reservoir development and sent the explorers into deepwater with new ideas, analogs and optimism, along with improved seismic technologies that were increasingly capable of imaging the reservoirs and complex structures encountered in the "Mini-Basin," "Fold Belt," and "Subsalt" plays developed during the late 1990s and into the 2000s. Figs. 6 and 7 show examples of modern seismic data in the major deepwater Gulf of Mexico plays that were initially exploited in the 1980s and 1990s. Details on the different deepwater plays, reservoirs and traps can be found in several of the other chapters in this book, including Chapters 4, 5, 7, 8 and 12.

Fig. 8 provides a more detailed illustration of the quality of 3D seismic that is available for the exploration and development of modern deepwater projects. In the right-hand panel in this example, the outline of a salt diapir can be interpreted from the 3D seismic and is shown in green. A portion of the seismic data set is shown in a semitransparent fashion and the presence of hydrocarbons can be interpreted as trapped up against the salt. In contrast to the 3D data set, the left-hand panel is a portion of a 2D seismic line acquired before the field was discovered. The 2D data was adequate for siting and derisking the initial exploration well but is inadequate to fully assess the distribution of hydrocarbons and to efficiently develop the field.

Another example of seismic data quality improvements that enabled deepwater exploration and development is shown in Fig. 9. Thunder Horse Field in the deepwater U.S. Gulf of Mexico is a complex dip and fault closure beneath overhanging salt. The setting and geometry of the field can be seen in the center panels of Fig. 7. Initial time domain processing produced the left-hand panel in Fig. 9 where the sediments beneath the overhanging salt are totally obscured. In contrast, the prestack processed, depth domain data in the right-hand panel shows improved imaging of the sediments beneath the overhanging salt and an improved image of the base of salt. Additionally, the depth processed 3D data also offered the ability to map a 4-way dip closed structure and to image the interpreted source rock interval and map the overall petroleum "fetch" into the structure. These improvements greatly reduced the uncertainty of the prospect, reduced overall risk and improved the positioning of the exploration well.

Modern techniques in subsurface imaging and interpretation in deepwater are detailed in this book in Chapter 16. For additional detail on seismic processing technology, the reader is again referred to Hyne (2012).

Fig. 4 summarizes four decades of deepwater technology development, showing one company's march into deepwater (Shell Oil). It began in 1978 with the installation of the Cognac platform in 1025 ft (313 m) of water and continues through

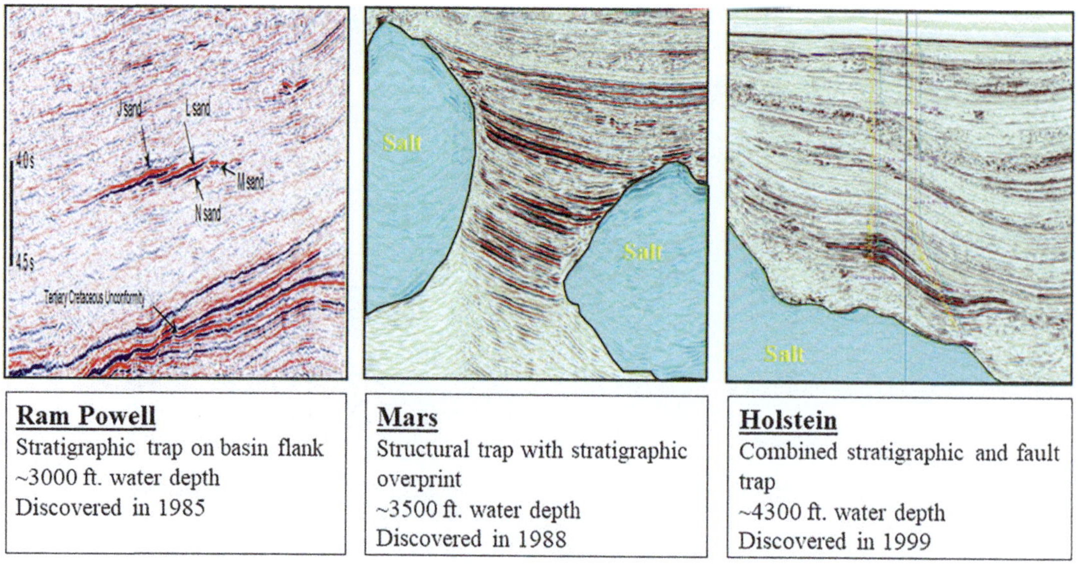


**FIG. 6** The first successful round of deepwater exploration in the U.S. Gulf of Mexico started in the mid-1980s as industry made a number of middle Miocene and younger discoveries. This play was primarily progressed using amplitude anomalies, or bright spots, on seismic data. The original discoveries were made using analysis of attributes on 2D seismic, but industry quickly evolved to acquiring subregional 3D seismic surveys to explore and develop these discoveries. Although industry continues to make discoveries in the play, the larger stand-alone fields were discovered by the mid-1990s after which the play was primarily focused on smaller accumulations as tie-backs to nearby infrastructure. The snapshots of three of these early discoveries are shown in this figure and are typical for this play (Yeilding, 2013). *Seismic Source: bp.*

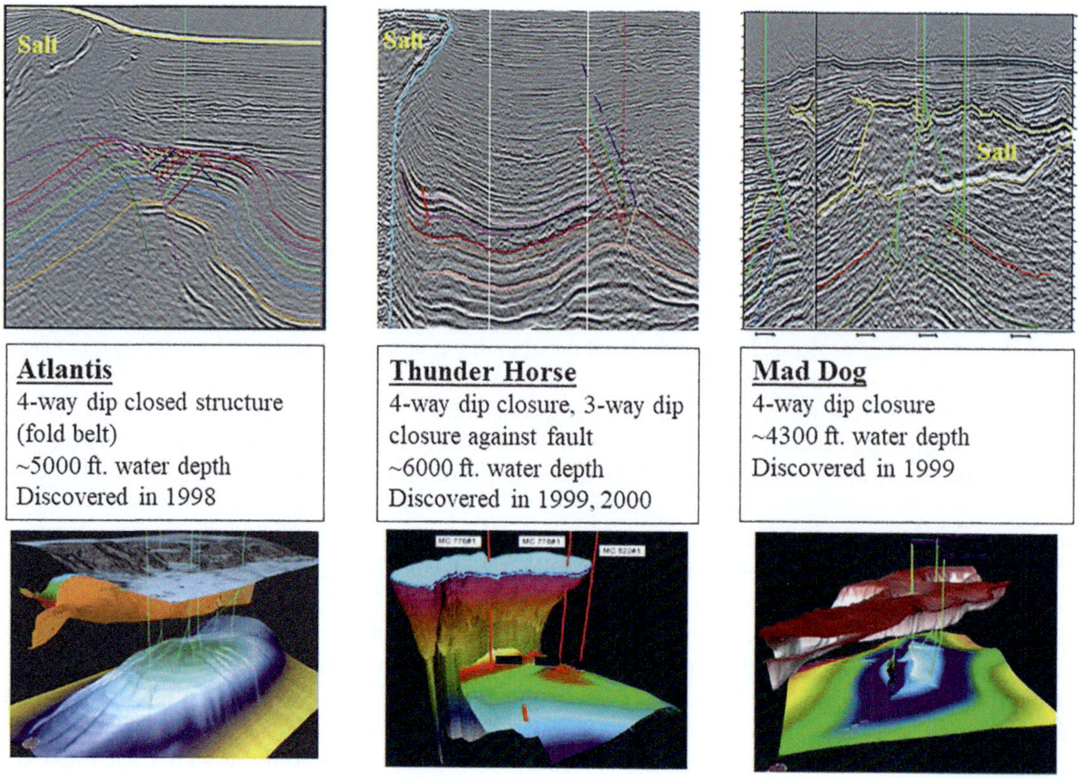


**FIG. 7** A second phase of exploration drilling in the deepwater U.S. Gulf of Mexico began in the mid-1990s, targeting older Miocene stratigraphy with robust structural culminations. Many of these prospects are partially or completely subsalt. These plays were focused on folds and turtle structures with middle and lower Miocene reservoirs of the central deepwater Gulf of Mexico. To properly describe prospectivity and position wells, 3D seismic and 3D prestack depth migration was required. To make this seismic cost-effective, industry participated in joint "group seismic shoots" followed by depth migration. Survey size increased from a few square miles to hundreds of square miles. Imaging was further enhanced by the larger oil companies using proprietary reprocessing focused on specific prospects. This figure shows three of the discoveries from this round of exploration: Mad Dog and Atlantis which are folds in the deepwater Atwater Fold Belt, and Thunder Horse which is an example of the middle and lower Miocene turtle play. Note the presence of near-surface salt, which creates significant complexity and uncertainty in exploring, appraising and developing these fields (Yeilding, 2013). *Seismic Source: bp.*

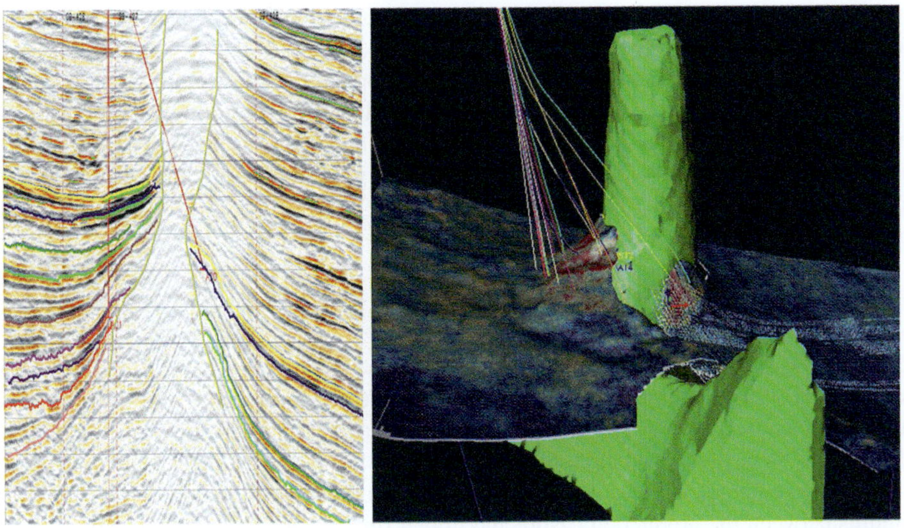

**FIG. 8** This figure illustrates the quality of 3D seismic that is available for the exploration and development of modern deepwater projects, as well as contrasting the imaging of a prospect with 3D vs. 2D seismic data. In the right-hand panel in this example, the outline of a salt diapir has been interpreted from the 3D seismic and is shown in *green*. A portion of the seismic data set is shown in a semitransparent fashion at one reservoir level and the presence of hydrocarbons can be interpreted as trapped up against the salt. In contrast to the 3D data set, the left-hand panel is a portion of a 2D seismic line acquired before the field was discovered. The 2D data was adequate for siting and derisking the initial exploration well but is inadequate to fully assess the distribution of hydrocarbons and to efficiently develop the field. *Source: Shell.*

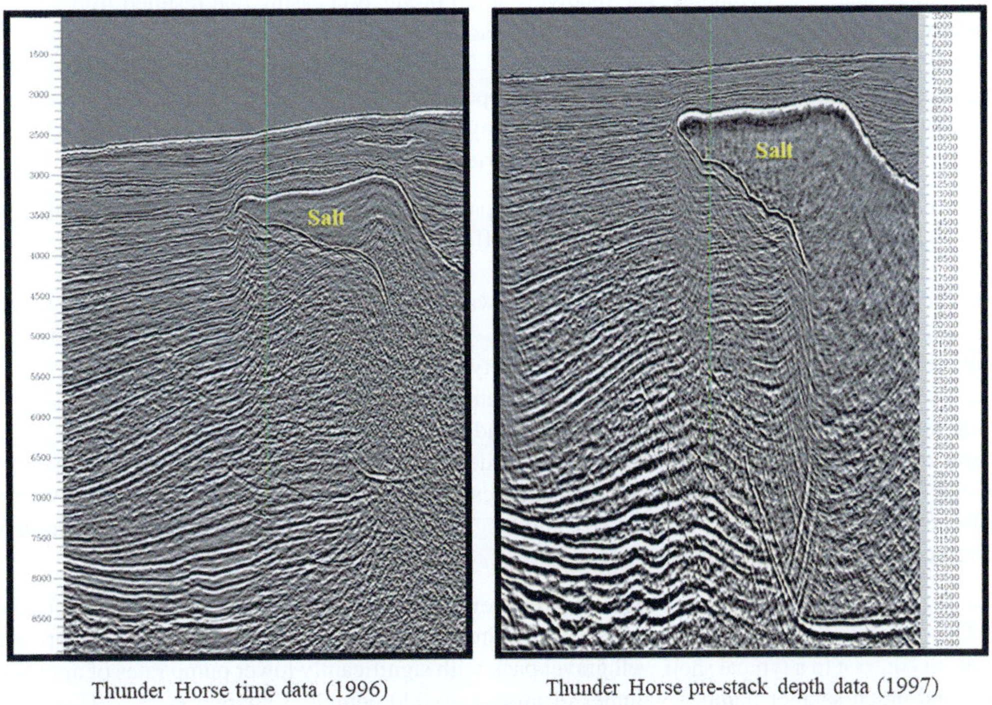


**FIG. 9** Another example of seismic data quality improvements that enabled deepwater exploration and development is shown here. Thunder Horse Field in the deepwater U.S. Gulf of Mexico is a complex dip and fault closure beneath overhanging salt. The setting and geometry of the field can be seen in the center panels of Fig. 7. Initial time domain processing produced the left-hand panel above, where the sediments beneath the overhanging salt are totally obscured. In contrast, the prestack processed, depth domain data in the right-hand panel shows improved imaging of the sediments beneath the overhanging salt and an improved image of the base of salt. Additionally, the depth processed 3D data also offered the ability to map a 4-way dip closed structure and to image the interpreted source rock interval and map the overall petroleum "fetch" into the structure. These improvements greatly reduced the uncertainty of the prospect, reduced overall risk and improved the positioning of the exploration well (Yeilding et al., 2005). *Seismic source: bp.*

today. Shell's deepest water depth facility is the Stones floating production facility installed in 2016 in the U.S. Gulf of Mexico at a water depth of approximately 9500 ft (2900 m) (Shell, 2021).

The necessary well engineering technology to enable the development of these significant discoveries is in three key areas: sand control, production tubing compatible with high-rate wells and production liners that would resist compaction failure. Concurrent advances in seismic imaging, as illustrated in this chapter (Figs. 6–9) as well as in Chapter 1 of this book (Fig. 12A–C), and the ability to calibrate seismic with well data to exploit seismic attribute analysis (Chapter 16 of this book) allowed for accurate reservoir imaging and characterization that was crucial in applying these development engineering technologies.

## Sand control

Sand control was one of the most daunting challenges in deepwater. In the early days of U.S. Gulf of Mexico oil and gas development, well maintenance resulting from sand production, well sand-ups and well equipment damage from sand cutting was not an uncommon occurrence. This was not the result of a lack of technology work. There had been continuing technology development to solve problems with sand production and formation damage. Solutions included formation consolidation and gravel packs utilizing high viscosity fluids with high gravel concentrations. The purpose of the gravel pack is to fill both the annulus between the production casing and gravel filtering screen and the perforation tunnels with gravel. Because of the thin pay zones typical of the Gulf of Mexico shelf and upper slope, the focus in gravel pack technology was to stay at low pumping pressure, pump relatively slowly and pump small high viscosity "slugs" of pack gravel and viscosifying carrier fluid.

The Gulf of Mexico deepwater reservoirs required significantly different technology to increase reliability, well production rate and well life. In the U.S. Gulf of Mexico, deepwater sands can be characterized as typically fine to very fine grained, dominantly quartz, compacted but with low cement, siliciclastic reservoirs. Deepwater reservoir quality and properties are described in detail in Chapter 13 of this book. In the context of production engineering, and particularly sand control, these reservoirs posed significant challenges. Gravel packing the longer pay zones in deepwater would require large pumped volumes and gravel volumes. The relatively small production casing that resulted from the multiple casing strings for deepwater drilling also created a small annular space between the production casing and gravel screen. A specialized version of gravel packing technology was developed. This concept was not widely used at the time, but this technique has the distinct advantage of creating short and wide openings into the reservoir filled with gravel in the formation outside the perforation "tunnels." This provides a bypass to any near wellbore limitations to flow. To achieve this successfully in deepwater over long intervals required work in a series of technology areas including:

1. Physical and mathematical models to achieve full gravel packing of the short openings in the formation and casing-to-screen annulus to ensure no voids. This not only ensured effective formation sand exclusion at high production rates but also protected the gravel screen from damage.
2. Optimizing the viscosifier concentration and type. This was to ensure the effective gravel transport and to minimize formation damage.
3. Optimizing the gravel type for reliability and permeability.
4. Optimizing the gravel screen designs and materials to enhance reliability.
5. Redesigning the installation tools for high pump rates and large gravel volumes.
6. Building gravel pack computer models to allow effective design, real-time monitoring and quality control. The models had to account for the formation and physical characteristics for each well. The model also had to provide good transport design, effective and complete packing and a size and length optimization for the openings in the formation that were specific for the particular well and pack.

All these technologies were brought together to result in the new gravel packing technology that dominates deepwater well design today. The packs are being pumped up to 50 barrels/minute and beyond with volumes of gravel up to and exceeding 250,000 pounds, in contrast to a typical shelf well gravel pack with significantly lower pump rates of approximately 10–15 barrels/minute, and much smaller pumped volumes of around 40,000 pounds (Lybarger et al., 1974). Also, contrary to established practice on the shelf, deepwater wells typically utilize low viscosity carrier fluids. A well designed and executed gravel pack can deliver long life and excellent well productivity never seen in previous sand control methods including the old gravel packs. Fig. 10 illustrates the design of a modern deepwater gravel pack.

Initially, it was unexpected that sand control would be required in deepwater reservoirs. There had been a long-standing rule in the Gulf of Mexico shelf that you did not need sand control below 10,000–12,000 ft (3000–3600 m) below sea level. However, the extensive work on reservoir formation properties in deepwater indicated it would be needed—even at

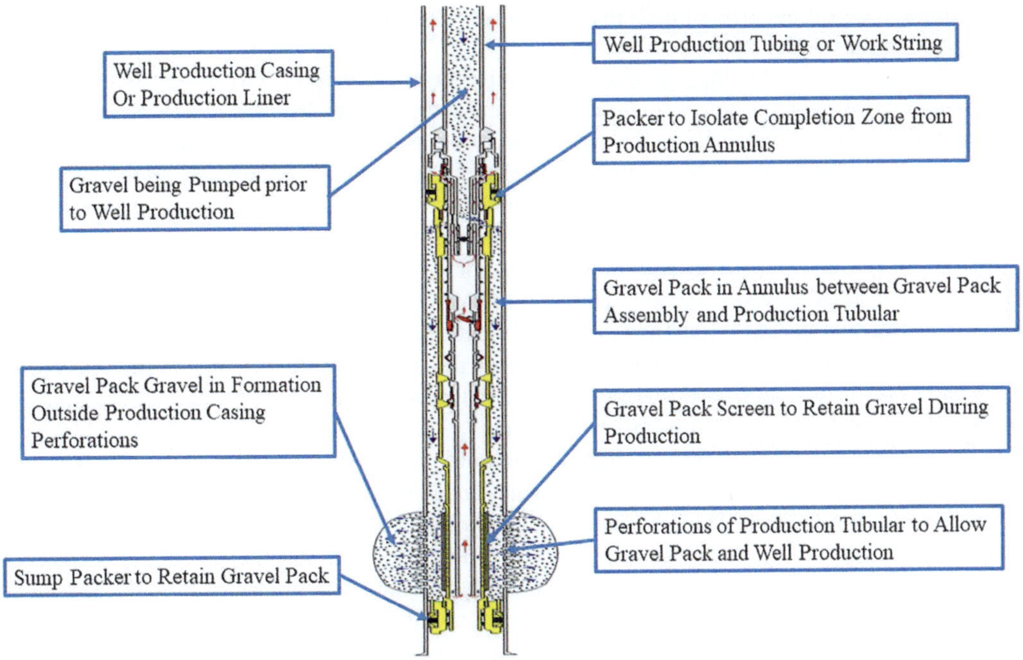


**FIG. 10** Schematic of a typical deepwater borehole gravel pack illustrating the major features and showing the complexity of the downhole configuration employed to control sand production in deepwater wells. The purpose of the gravel pack is to fill both the annulus between the production casing and gravel filtering screen and the perforation tunnels with gravel. *Illustration used with permission: ©2021 Baker Hughes. All rights reserved.*

measured depths of 20,000–30,000 ft (6000–9000 m). This was confirmed by deepwater well tests that were done to determine vertical connectivity of the many layers of deepwater turbidites. A detailed description of deepwater sedimentary deposits is provided in this book in Chapter 7, and a description of the deliverability of production from deepwater reservoirs is presented in Chapter 2.

## Production tubing design and long-term well integrity

Another important development in high-rate/high-ultimate deepwater wells was production tubing and production liners that would deliver long-term pressure integrity. In the majority of U.S. Gulf of Mexico shelf developments, carbon steel production tubing had been used with conventional threaded connections where the pipe thread provides the seal. It was not uncommon to have to workover wells to replace production tubing because of leaks from erosion and corrosion as well as leaks in the connections.

To solve the need for long-life and high integrity deepwater production tubing, corrosion resistant alloys were ultimately used with "premium" connections utilizing machined metal seals on both sides of the casing joint. If the correct corrosion resistant alloy is used, it has the added benefit of being resistant to special corrosion mechanisms if they exist due to the properties of the produced fluids.

There are a number of corrosion resistant alloy materials employed, including martensitic, austenitic, ferritic and duplex. Although corrosion resistant alloys had been available in the industry, they had not been widely used because of the expense; however, corrosion resistant alloys became cost effective when compared with the cost of frequently working over or replacing deepwater wells. There was focused research and development on optimizing the cost of the large volume of corrosion resistant material needed for deepwater, and it also became necessary to do technology development work on the alloying and process of manufacture.

Although corrosion resistant alloys have excellent corrosion and erosion resistance in general, they can have failures including cracking from certain environmental factors as well as due to processes in manufacture, handling, storage and alloying. They also have inspection challenges relative to conventional tubing. The corrosion resistant alloys are typically nonmagnetic, requiring the development of ultrasonic inspection methods. Ultimately both the production companies and the tubing manufacturers spent tens of millions of dollars on corrosion resistant alloy research and development for deepwater, which included work on quality control and inspection methods. The results of this research and technology development were highly successful. The corrosion resistant alloy tubing (and variants) in deepwater have proven to be a great success in achieving long-term integrity at high production rates. In fact, the tubing can often be reused multiple times which was unheard of with normal carbon steel tubing.

Another area of research and technology development for the corrosion resistant alloy deepwater production tubing was the threaded connections between the tubing sections. Technical solutions were needed in two areas, connections that prevented galling and had long-term leak resistance and connections with that would support the high vertical stress from extremely long deepwater tubing strings. Most corrosion resistant alloys tend to gall when used to make threaded connections, especially with metal seal surfaces. This galling is a problem for sealing integrity but also a severe problem for the tubing running and installation process. With the cost of deepwater drilling rigs and the cost of the corrosion resistant tubing itself, these installation failures would add substantially to the already expensive tubing and wells. In addition, deepwater wells could have extreme depths—up to 20,000–30,000 ft (6000–9000 m) measured depth which put high tension loads on the connections. Thus, the tubing connections had to accommodate these high loads while delivering a high integrity, long-life seal. Fortunately, high technology connection work had been progressing for high-temperature/high-pressure wells, but this technology had to be further developed to accommodate the requirements for deepwater connections and most especially for the use of the corrosion resistant metals themselves.

The last major technical challenge to deliver high-rate/high-ultimate deepwater well completions was work on the production liners that were installed across the producing intervals and in which the screens and gravel pack completions were installed. The deepwater U.S. Gulf of Mexico production intervals were highly compactable and the drive mechanism for production was largely compaction drive. The consequence of this with high rates of production would have been production liner failures. Research was done on the optimum grade and thickness of steel that could best resist compaction failures. This work was complicated by the fact that the deepwater Gulf of Mexico liners were relatively small in diameter because of the depth of the wells and the large number of casing strings and liners employed in drilling. Thus, the production liner thickness had to be balanced with the internal diameter needed to successfully install long gravel pack completions. In addition, work was done on rock mechanics to develop well path optimization models (van Oort et al., 2001).

Well path optimization minimized the challenges with mud weight management as well as reducing stress and potential failure on the production liners. All this work was to satisfy the need for long well life in deepwater developments. Water floods, although not originally considered in deepwater facilities design or field development plans, would be necessary to sustain well life via subsurface pressure maintenance to reduce the risk of formation compaction related casing collapse. Ultimately, it was pressure maintenance, combined with the technical work on production liners that would provide sufficient deepwater well life and integrity.

Common practice for some companies in early deepwater developments was to rely on seismic and petrophysical data, supported by reservoir models and global outcrop studies of deepwater depositional environments to support development plans and production rate estimates. Long-term production tests were typically not employed. This was an extension of their development practices on the shelf where engineers had learned that they could rely on available subsurface data to create reliable reservoir models and estimates of well deliverability. This also proved to be a significant differentiator for these companies with the resulting positive impact on development cost and schedule, compared with other companies that felt it was necessary to employ long-term production tests to predict reservoir deliverability.

## Case study 2—Deepwater Gulf of Mexico structures and infrastructure planning

The march to deepwater saw the final fixed bottom, or fixed leg, structure in the deepwater Gulf of Mexico with Shell's Bullwinkle platform in 1352 ft (412 m) of water. It was installed in 1988 in a single section launch. The structure's total weight was 77,000 tons including the decks with the jacket structure alone weighing 49,375 tons. This was an industry first and a remarkable engineering accomplishment. But was also an industry "last" as it clearly demonstrated that deeper water would need to be developed from floating and subsea infrastructure. It simply would not be practical or economic to continue to build large fixed structures in deeper and deeper water.

Fig. 11 shows the Bullwinkle production jacket being transported from the Corpus Christi, Texas shipyard where it was built, to its installation location in the Gulf of Mexico, 160 miles (258 km) southwest of New Orleans in 1352 ft (412 m) of water.

These fixed structures had dominated the world of offshore production up to this point. The projects were large-to-giant civil engineering projects, and during design there was some interaction between subsurface technologists and the civil and facility engineers relative to well count, but communication was sparse. Deepwater developments moved to an entirely different level of requirement in the necessity for the structural designers to fully consider and communicate with the subsurface technical experts and design for the subsurface development as well as custom design for the specific water depth and metocean conditions at the location. In this section, the importance of this interaction as well as other considerations is highlighted.

FIG. 11 The Bullwinkle production jacket being transported from the Ingleside, Texas shipyard where it was built, to its installation location in the U.S. deepwater Gulf of Mexico, approximately 160 miles (258 km) southwest of New Orleans in 1352 ft (412 m) of water. Bullwinkle, installed by Shell in 1988 is the deepest water fixed platform in the Gulf of Mexico and was the last fixed structure installed in the Gulf of Mexico deepwater. *Source: Shell.*

It is interesting to note that as development design commenced in the deepwater U.S. Gulf of Mexico, "ship-shaped" floating production systems that included storage were not allowed by regulation as there was concern over large volumes of oil being stored offshore in proximity to the busy shipping lanes of the Gulf. This put additional pressure on the whole system design to not only optimally match the subsurface with the surface facility design, but the design also had to include oil export requirements.

There are three decisions that the entire development team must make that are key to the full lifecycle design of the production facilities and to optimizing the economics of the field development. These are: the specific location and water depth of the development structures, whether direct well access would be required from the surface and whether development support should be provided for future subsea wells and/or nearby developments.

In the early days of the U.S. Gulf of Mexico deepwater, developments were quite remote and there were no nearby surface structures to provide production processing, production transfer and other necessary services. Thus, some number of surface floating structures needed to be built and installed to provide these capabilities. But the extreme cost of these floating structures made providing them for all future development wells, field extensions and nearby fields prohibitive. Thus "urban planning", "corridor planning" and "hub and spoke" development became important parts of the development planning process. It was no longer about just what was needed to be done for a particular field, but what should be done for economic and successful area development. How much floating surface structure capacity was needed and where and how many subsea well tie-ins and tie-ins to nearby fields, must be considered and provided?

In the U.S. Gulf of Mexico shelf, standardized fixed structural design was possible. In the deepwater Gulf of Mexico, the location, water depth, metocean specific conditions, seafloor conditions and potential for shallow hazards all impact the selection and design. Thus, the surface and subsurface teams must consider all these factors together. The general categories of structures are tension leg platforms (TLPs), spars, moored semisubmersibles that are floating production systems (FPSs) and floating production systems with storage and offloading (FPSO). Ship-shape floating systems with storage are now permitted and have been in use in the U.S. Gulf of Mexico since 2010 (Parshall, 2016). The different surface systems are illustrated in Figs. 4 and 12.

Important elements of surface system design that require input from the subsurface team are the areal distribution of the oil and gas resources, the number of development wells needed, the resource depths, pressures and temperatures and the estimated production per reservoir level and well. These are all key for facility design and the choice of surface structure. The three-dimensional distribution of resources and wells determine the number of structures necessary, the size of the structures and if there will be remote subsea well locations tied back to the producing floating infrastructure. Reservoir

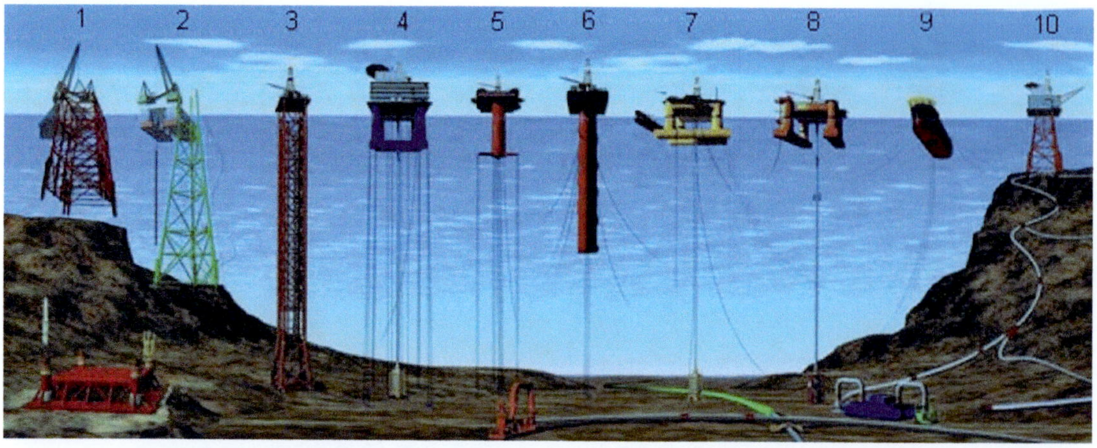


**FIG. 12** In the Gulf of Mexico shelf, standardized fixed structural design was possible. In the deepwater Gulf of Mexico, the location, water depth, metocean specific conditions, seafloor conditions and potential for shallow hazards all impact the selection and design. Thus, the surface and subsurface teams must consider all these factors together when planning and developing a project. The general categories of offshore structures are fixed platforms, #1 and 2, compliant towers, #3, tension leg platforms (TLP), #4 and 5, spars, #6, moored semisubmersibles that are floating production systems (FPS), #7 and 8, floating production systems with storage and offloading (FPSO), #9, and subsea completion and tieback to a host facility, #10. *Source: BOEM.*

deliverability dictates constraints on flow rates and required surface system capacity. It is also important for the subsurface team to advise of potential future nearby developments and their needs. The reader is referred to Chapters 7 and 8 in this book to consider the distribution and complexity of deepwater reservoirs and how that will affect the number and resulting complexity of wells drilled to exploit these resources, as well as Chapters 2 and 18 to understand controls on the environmental models of deepwater reservoirs.

Fig. 5 shows the Auger tension leg platform (TLP), with its tendons and mooring system, at scale projected over the city of New Orleans (Leffler et al., 2011). Auger was installed in 1994 in 2720 ft (830 m) of water, approximately 214 miles (345 km) south of New Orleans in the Gulf of Mexico. Based on experience with Auger, it was learned that for later TLP installations, the mooring system was not necessary as the tendons would serve to adequately stabilize the TLP. This offers advantages in cost and installation time, but most significantly is a major advantage for seabed access in areas of high resource richness such as the Mars Basin in the U.S. Gulf of Mexico where there are three tension leg platforms installed. Mooring lines, if necessary, typically extend several miles from the surface facility limiting the movement and use of mobile drilling rigs as well as complicating the installation of pipelines and other nearby producing facilities (Enze et al., 1994).

Another choice for the development team is whether there is a need for direct vertical access to the producing wells, driven by concerns over well life and the need for future well workovers. Initially TLP's were widely used because of this capability as the wellheads are at the surface on the TLP. Direct vertical access can be provided on spars and semisubmersible production facilities as well but is generally more technically challenging and the number of wells is limited. This also adds constraints on the deck loads and facility size. With the continuing advance in subsea tree and well design, as well as high levels of well life and reliability, there are an increasing number of developments without direct vertical access, particularly for field extensions and outsteps.

Fig. 13 illustrates many of the aspects of subsea and urban planning. Shown in this figure, a floating production system, like many installed in the deepwater U.S. Gulf of Mexico at fields such as Na Kika, Thunder Horse and the Independence Hub, gathers production from a number of wells on the seabed. The water depth of a development such as this could be several thousand feet with production wells distributed over tens of square miles. Subsea wellheads are typically clustered on the sea floor in groups surrounding a subsea manifold. The floating surface structure is connected to wells and manifolds by flowlines for the produced oil and gas as well as control lines from the host structure. Servicing of wells and subsea equipment is accomplished using deepwater drilling rigs, which can be very expensive, hence putting a premium on component and overall system reliability.

An example of this type of production system is the Na Kika development in the U.S. Gulf of Mexico, located about 144 miles (232 km) south of New Orleans, Louisiana, in water depths ranging from 5800 to 7600 ft (1768–2317 m). It started production in 2003 with subsea wells located from 5 to 12 miles (8–13 km) from the platform and tied back to the host through a total of 90 miles (145 km) of oil and gas flowlines (Hydrocarbons Technology, 2021).

Shell recognized early on the importance of high reliability subsea systems in deepwater environments. Fig. 14 illustrates Shell's early history with subsea developments in the Gulf, with production starting in 1996 and growing rapidly over

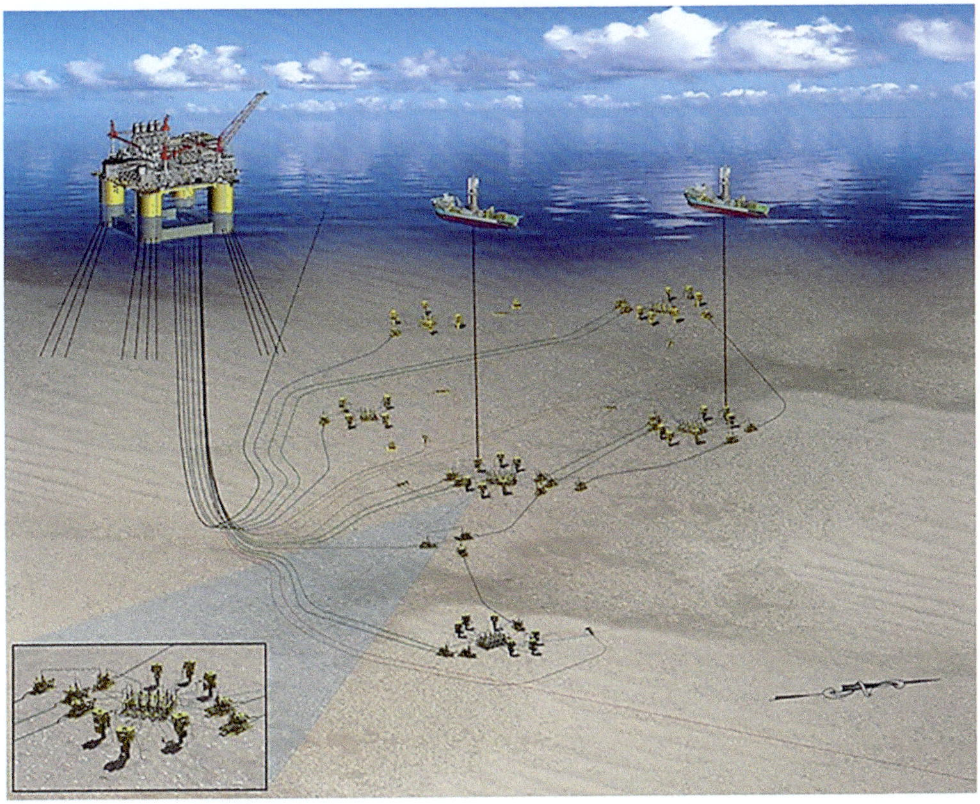

FIG. 13 Illustrated here are many aspects of subsea and urban planning. Shown in this figure, a floating production system, like many installed in the deepwater Gulf of Mexico at fields such as Na Kika, Thunder Horse and the Independence Hub, gathers production from a number of wells on the seabed. The water depth of a development such as this could be several thousand feet with production wells distributed over tens of square miles. Subsea wellheads are typically clustered on the sea floor in groups surrounding a subsea manifold. The floating surface structure is connected to wells and manifolds by flowlines for the produced oil and gas, as well as control lines from the host structure. Servicing of wells and subsea equipment is accomplished using deepwater drilling rigs, which can be very expensive, hence putting a premium on component and overall system reliability. *Source: Shell.*

the next several years.[b] As subsea production grew, it was important that subsea wells and systems were very reliable as servicing equipment on the seabed in deepwater requires a deepwater drilling rig. That work can then be very expensive (easily several hundreds of thousands of dollars per day), which would make deepwater developments uneconomic without very high well reliability and percentage of uptime.

Finally, the specific effects of metocean conditions must be considered and most especially Gulf of Mexico loop currents. Loop currents are areas of warm water fed from the Caribbean that travel into and through the Gulf of Mexico before exiting the Florida Strait and flowing into the Gulf Stream. Loop currents are temporally variable in that they sometimes barely enter the Gulf while at other times they can move well into the central Gulf of Mexico before heading back to the east. The resulting significant deepwater currents, which occur even at the seafloor must be carefully considered in the design elements for deepwater structures. This is particularly true for spars that are vulnerable to the currents and associated vortex eddies, which can result in significant structural vibration. Thus, spars have spiraling fins called strakes to shed the eddies caused by the currents, which unfortunately, adds to the spar's cross section causing additional loading on the structure and potentially complicating or limiting design choices necessary for optimum subsurface development (National Academy of Sciences, 2018).

Although not associated with production structure design, an important aspect of loop currents in the Gulf of Mexico is their impact on drilling and even construction operations. In particular, the loop currents can put loads on drilling risers and create drilling riser movements that halt drilling operations sometimes for a week or more. This has quite significant impact on the planning and cost of deepwater developments.

[b]Personal Communications, Schaff, Robert, Subsea Engineering Manager, Shell, and Maggio, Adam, Subsea Engineering Technical Authority, Shell.

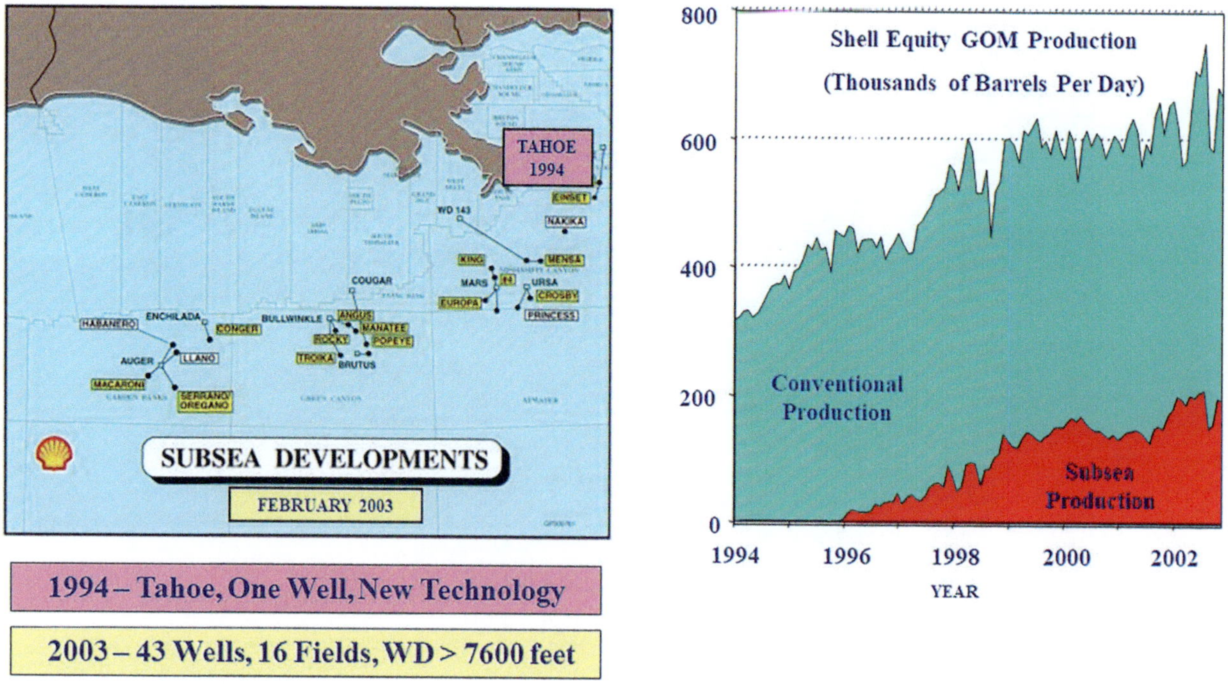


**FIG. 14** Shell experience with subsea completions in the deepwater U.S. Gulf of Mexico grew quickly during the 1990s and has continued to grow since. Subsea systems must be very reliable as servicing remote equipment on the sea floor in deepwater is very expensive. Success with subsea systems was enabled by effective technology development and deployment leading to high reliability subsea systems. *Source: Shell.*

## Case study 3—Flow assurance, pipelines, flowlines and risers

Some of the lesser known but significant challenges in deepwater development planning and decision making by the surface and subsurface teams are those in installing various pipelines and flowlines, assuring long-term structural integrity under high cyclical loads and assuring the flow inside the lines. These technical challenges are primarily driven by the extreme water depths, the currents, the cold sea water temperatures at the sea floor which are near freezing and the properties of the produced fluids.

Pipeline transfer of oil and gas to shore from the deepwater U.S. Gulf of Mexico was the only available export option up until the approval of the use of FPSOs in the early 2000s. This was economically sound given the extensive pipeline infrastructure that had been developed on the shelf over the preceding decades, along with the robust markets for both oil and gas landed along the U.S. Gulf Coast by pipeline. As noted earlier, floating production systems with storage that required offshore transfer to tankers were initially not allowed in the U.S. Gulf of Mexico which meant that the challenges associated with subsea pipelines in the deepwater environment had to be solved to enable the production of oil and gas resources.

Only a few aspects of these challenges are discussed here so they can be better considered by the development team. For reference, Chapter 2 in this book provides details on the fluid properties found in deepwater reservoirs.

Flow assurance in deepwater is likely the least widely known but one of the most important challenges to be overcome. All production streams contain some combination of natural gas, liquid hydrocarbons and water. At deepwater depths the ocean bottom temperatures are near freezing. All types of seafloor pipes are thus at very low ambient temperatures. Even the in-field flowlines and jumpers can be quite long and export pipelines are extremely long, up to hundreds of miles. Thus, any flow in these pipes can cool quickly and stay cold over very long distances. Looking again at Fig. 13 and considering the details of deepwater subsea developments, such as those provided earlier for Na Kika, these challenges are quite significant.

Wax, asphaltene and hydrate plug formation can occur in all types of flowlines and pipelines, leading to high remediation costs, and in some cases, they cannot be remediated at all. The realization of the potential severity of this in deepwater (a problem which did not exist in milder shelf conditions) drove a variety of technical developments that were necessary to ensure the success of deepwater projects.

These developments included modeling, heating and heat retention and chemical inhibition. Better modeling developed an understanding of the occurrence envelope for flow assurance problems, as well as potential remediation possibilities. This was supported by enhanced modeling and predictions for multiphase flow and slugging.

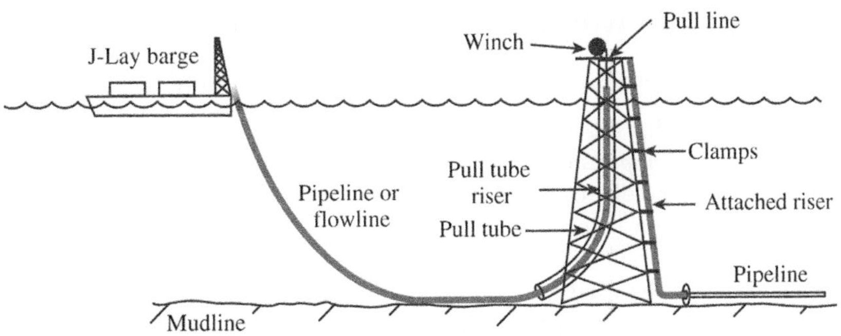


**FIG. 15** J-Lay method for deepwater pipeline installation. The pipe is deployed to the seabed from a J-Lay vessel. Also shown are common methods for attaching subsea pipelines to host platform facilities. Shown on the right side of the figure, risers can be pre-installed on the platform and connected to the pipeline at the seabed. Alternatively, the pipeline is dragged up through a pull tube that was pre-installed on the platform. *Source: Leffler, W.L., et al., 2011. Deepwater Petroleum Exploration & Production, A Nontechnical Guide. second ed. PennWell, Illustration used with permission: © 2011 PennWell Books. All rights reserved.*

Heating of pipelines included heat retention through insulation and double wall pipe and developing electrically heated pipe of various designs. These were all expensive and technically complex options.

Inhibition included the traditional methanol and glycol injection as well as specialized new types of inhibitors that could be used at lower rates of injection with the high rates of production in deepwater. Also hydrates, waxes and asphaltenes are not similar in structure, chemistry or formation occurrence. Thus, several types of inhibitors needed to be developed. Work continues today on lowering the risk of blocked pipelines in deepwater via better inhibition understanding and technology as well as performance modeling and active surveillance (Wasden, 2003).

Installing export pipelines in deepwater is challenging because of the extreme water depth. Laying and pulling in pipelines in deepwater requires long sections of pipe from the seafloor to the pipe-lay equipment on the surface. These long lengths result in very high loads to be supported and manipulated by the laying equipment. The equipment designed for shallow water and shelf installations was simply not capable of managing these loads. The result was a large technology development effort in much larger equipment as well as in the pipe-laying process itself. A common method today is "J-lay" in which the pipe is deployed almost vertically to the seafloor. The resultant shape of the pipeline during the laying process, resembling the letter J, allows it to tolerate the greater currents and lay-barge movements that occur in deepwater. Fig. 15 illustrates this common configuration for deepwater pipe laying (Leffler et al., 2011). Specialized methods of attaching jumper pipes and flowline pipes to subsea wells were required as well as advancements in flexible pipe technology (Perinet and Frazer, 2007).

The last special consideration for pipelines in deepwater is the attachment to the floating structure. In traditional fixed bottom structures, pipes (often referred to as risers) were attached to the outside of the structure or pulled into pull-tubes within the structure. Thus, the riser pipe was rigidly attached in the section that went from the seafloor to the production facility. With deepwater structures, the riser pipe connecting the seafloor pipe to the floating structure "hangs" from the structure and is subject to the movements of the floating structure and to the deepwater ocean currents. This includes vortex-induced vibrations (VIV), which potentially could destroy the riser (Allen, 1998).

Part of the solution to ensuring long term reliability of deepwater risers was the development of suspended steel catenary risers. The fatigue and stress on the riser are affected by the pipe design and material, the design of the connection to the floating structure at the surface, the pipe length and the tension on the pipe. Many technology advancements have occurred to accommodate the riser loads, deepwater currents and the motion of the floating production facilities.

Like other deepwater challenges, developing the optimal solution begins at the reservoir. The subsurface distribution of hydrocarbons (both in depth and areally) impacts the choice and design of surface facilities as well as the number and arrangement of wells, all factors that influence the flowline design including connections from seafloor to surface. In addition, the properties of the produced fluids will dictate what happens within deepwater flowlines and risers. So, in this—like other technical areas—project success depends on communication between the engineering and geoscience specialists during project design.

A particularly challenging deepwater development was the Mensa Field in the U.S. Gulf of Mexico. Discovered in 1987 in approximately 5300 ft (1616 m) of water, Mensa started producing its gas a decade later through flowlines running some 68 miles (109 km) to an existing host platform on the shelf. It was through the Mensa development that Shell learned how to conquer the complex flow assurance challenges in deepwater (Razi and Bilinski, 2012; Denney, 2013).

## Case study 4—"Doing More with Less" stepping into the data and information age

### (From daily and manual surveillance, to continuous surveillance, to proactive technical monitoring)

This case study explores the journey that the oil and gas industry has taken to apply 21st century technologies to the ever-present challenges of optimizing development, monitoring production and increasing reliability.

The industry has long used remotely operated and semiautonomous underwater vehicles for construction and inspection. On platforms, drones are now being used for inspecting hard to access areas and the development of long-range heavy lift drones is set to change how equipment and tools are delivered to rigs. In addition, 3D printing on offshore facilities can minimize or even eliminate the need for parts and tools to be sent from shore.

Beyond these physical revolutions, the use of information technology is enabling the development of models in areas such as predictive maintenance, and combined with fiberoptic and satellite communications, the offshore industry is able to use "big data" to enhance remote monitoring and intervention. Offshore oil and gas is entering an era where digitalization is enabling a more proactive approach to reliability management, enabling the industry to "do more with less."

After the boom and growth in deepwater that was exaggerated by the large upswing in oil prices in the early 2000s, the industry experienced the inevitable trough in response to dropping prices. For several years beginning in 2015, an oversupply of oil due to the increase in onshore light tight oil and natural gas suppressed prices, leading to a challenging economic environment where deepwater operators, contractors, vendors and suppliers have had to learn to "do more with less" to be successful operationally and economically. This case study looks at the increased availability of operational and technical data over the last few years, and the use of that data for Exception Based Surveillance (EBS), descriptive and predictive analytics and the operation of remote control rooms (RCRs) in an effort to increase operational efficiency and decrease costs in a safe and sustainable manner.

Over the years, deepwater operators and engineers have shifted from "simple" and standard operator rounds where they would walk the deck to adding in some manual monitoring of real-time operational data gathered from the facilities, wells and pipelines. Now the output of operator rounds is being captured using digital tools. Over time the manual monitoring of data has shifted to continuous monitoring with alerts, alarms and automated notifications. Although the offshore control systems were often very simple and "crude," the data from these systems were an untapped and very powerful resource. Offshore operating companies have started to pay attention to pioneers, such as Aera Energy (a California based onshore operator) who hired external big data experts, process expertise from Toyota and architectural expertise from academia and nontraditional consultant houses to change how they work and improve their business with the use of readily available operational data. Getting the concept of enterprise architecture right was critical for this small joint venture company and its success was highlighted in several conferences in 2008 and 2009 (Ross et al., 2006; Allan et al., 2013).

Aera Energy LLC was formed as a Joint Venture in 1997. At the time of formation, the company established a vision to operate an energy company in a different way. Rather than pattern their company after other oil and gas companies, Aera sought to operate more like Toyota, Southwest Airlines, FedEx and digital age companies like Amazon. The company achieved this vision and has been a standard setter in the energy industry for over two decades.

The Aera Way, as it has become known, is founded on a key playbook including:

- Leveraging "Enterprise Architecture" as a strategy for the organization,
- Adopting a manufacturing mindset and applying concepts such as Lean, Business Process Management, Six Sigma and the Toyota Production System,
- Investing in people and developing a learning organization and a management team that can lead change and
- Integrating people, process, technology and data to leverage digital age thinking and strategies.

Although none of these concepts are new, there is something unique about Aera. First, Aera recognized the need to apply all these concepts as an integrated system to achieve its vision. Second, and more importantly, Aera succeeded at implementing these ideas. Many aspects of the Aera Playbook are well documented in bestselling textbooks and taught in colleges and universities. Yet most companies struggle to effectively implement them into the DNA of the company and the ways of working. Aera continues to successfully apply these concepts to create unsurpassed value creation and achieve its vision through a workforce that is willing to learn new ways of working and a leadership team that knows how to lead change (Cox et al., 2001).[c]

Similar to Aera, "thinking outside of the box" would need to become the norm to help break the confines of the traditional upstream ways of working in the offshore industry. As the offshore communication infrastructure started to build

[c]Personal Communication, Bob Palermo: Leadership and Management Consultant and former Vice President, Aera Energy Center for Excellence.

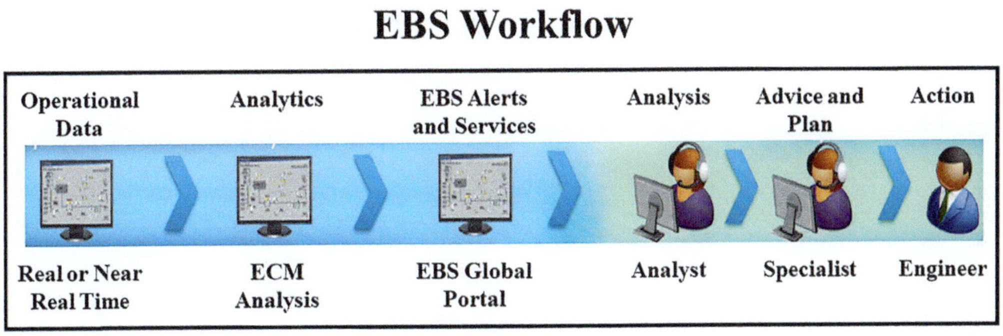


**FIG. 16** Illustration of the Exception Based Surveillance (EBS) workflow. Real-time, or near real-time data is gathered from operational systems. This time-series data is subjected to statistical analysis where it is compared with long running trends. The deviation of the real-time data from these established trends is defined as the error and thresholds can be set to alert engineers and/or operators of the deviation. This Error Correction Model (ECM) is available to operational and remote staff through a global portal where analysts and technical specialists can advise engineering and operational staff on recommended actions. *Source: Shell.*

out across the Gulf of Mexico and companies began working together across assets, the data and communication with onshore offices became more and more reliable. Operational and control data became accessible at remote locations with very little latency, which shifted the industry from manual to semiautomatic alerts to improve the standard daily surveillance with less effort. Once the data was available, data analytics became extremely powerful, allowing engineers to see trends over time, identify developing operational problems and optimize production.

The significance of this is that early experience in deepwater facilities was that a significant number of predictive analytic alerts were related to wells and well data. The analytics were flagging potential issues related to wells and well performance, which ties directly back to the subsurface, well deliverability and the subsurface development plan.

Initially, simple descriptive analytics were useful to the production operators and engineers to understand platform production operations (Gartner, 2021). Over time, concepts such as "rate of change" and "contingent rate of change" monitoring become more dependable, and relationships based on real-time operational and control data are then set as business rules from which notifications, alerts and/or control actions can be taken with or without manual intervention. Business rules have also allowed offshore operators to standardize some equipment monitoring plans and learnings and have been used to improve data quality, which, in turn, reduces the amount of time people must spend checking the integrity of the data. This was essentially a defining moment for Exception Based Surveillance (EBS). Fig. 16 shows an example workflow for EBS.

EBS is where the engineers or the operators are now working problems where the data shows a break in the pattern or an "abnormality." With this surveillance the engineer is no longer spending time "looking at all of the data for a problem." Instead, a potential problem is signaled to the engineer or operator as an "alert." Fig. 17 shows an example calculation of simple analytics created for a deepwater production facility. Here, the differential pressure across the crude oil cooling unit is measured and compared with the theoretical pressure drop calculated for the unit. The difference between the theoretical and measured pressures results from paraffin build-up within the cooler. The paraffin content is related to the composition and chemistry of the produced oil and an alert here indicates that production operations, specifically well performance, will be impacted by the relationship between crude oil composition and the well and surface facilities.[d]

Over the past few years, predictive analytics have become more common place (Gartner, 2021). This is where the data and the data relationships feed algorithms that are based on recognized historical patterns. Alerts are automatically generated when the algorithm detects a pattern that has historically caused an unwanted event or incident. In some companies, these alerts are manually handled and in others there are acceptance or escalation criteria used to automatically mitigate the situation with simple notifications sent to the engineers or operators. Early recognition of production issues results in improved production performance and can be used to update a field development plan.

In addition, many operators have fully replicated the offshore data and control systems in remote onshore centers. These RCRs have a range of capabilities depending on the company and the regulatory requirements for the asset. Full replication of the control system enables control of the subsea wells, direct vertical access wells, pipelines, production equipment and marine systems from onshore. Fig. 18 illustrates a model for remote monitoring and control along with some of the benefits that can be realized with Remote Assisted Operations.

[d]Personal Communications, Graham McVinnie, Proactive Technical Monitoring and Process Engineering Manager, Shell, and Danyelle Kennedy, Bridge Operations and Hydrocarbon Allocation Manager, Shell.

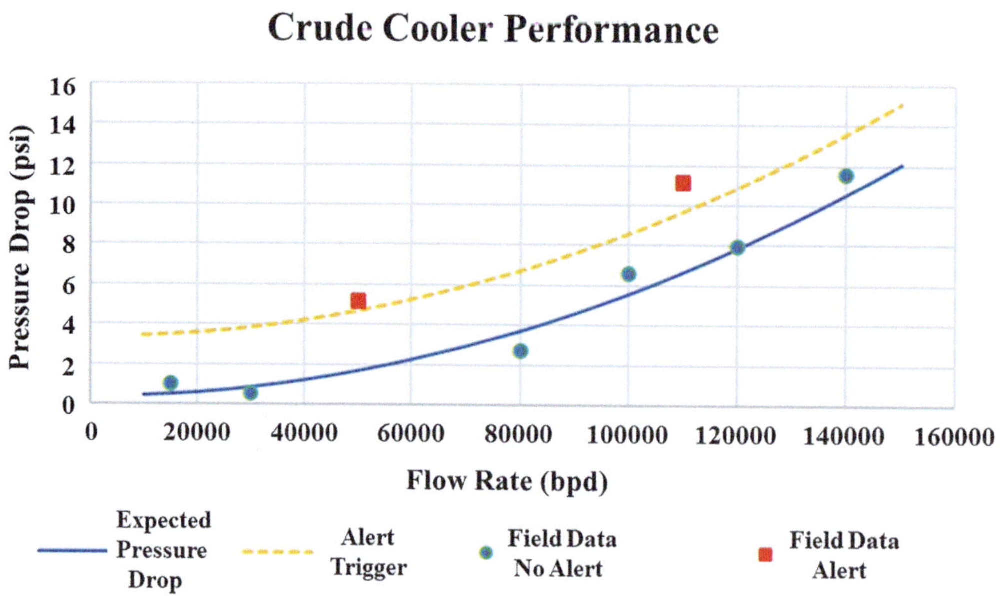


**FIG. 17** Following on from the EBS workflow shown in Fig. 16, we see here an example calculation of the analytics that are created from ECM analysis for a deepwater production facility. Here, the differential pressure across the crude oil cooling unit is measured (*red squares* and *blue dots*), which is compared with the theoretical pressure drop calculated for the unit (blue line). Pressure drop in pounds per square inch (psi) is plotted relative to flow rate measured in barrels per day (bpd). The difference between the theoretical and measured pressures results from paraffin build up within the cooler. The paraffin content is related to the composition and chemistry of the produced oil. Facility engineers set an error level for an alert to be triggered (*yellow dashed line*) notifying operational and/or engineering staff of data outside of accepted limits. An alert here indicates that production operations, specifically well performance, will be impacted by the relationship between crude oil composition and the well and surface facilities. *Source: Shell, Personal Communications, Graham McVinnie, Proactive Technical Monitoring and Process Engineering Manager, Shell, and Danyelle Kennedy, Bridge Operations and Hydrocarbon Allocation Manager, Shell.*

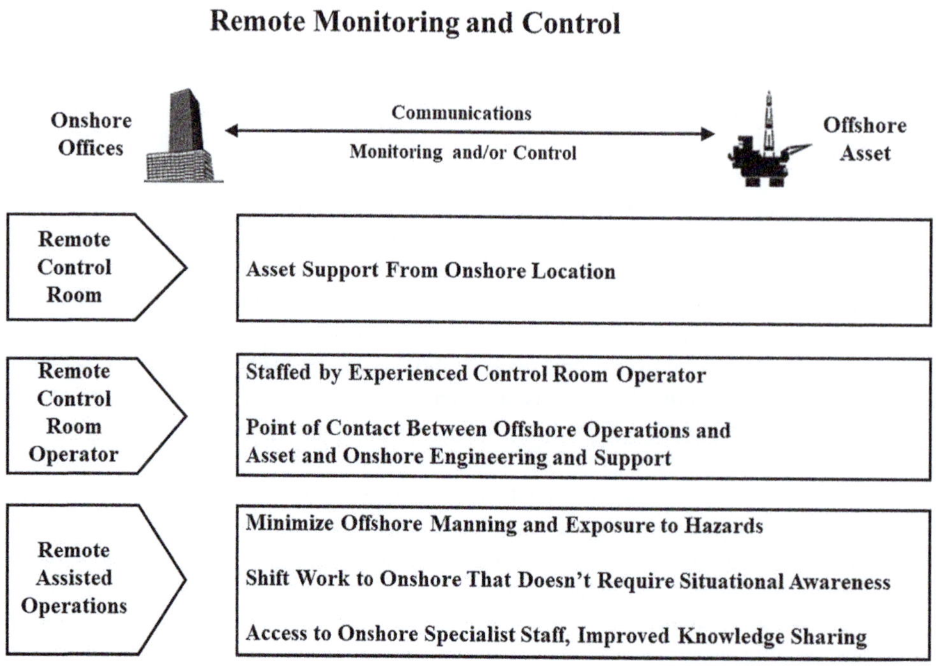


**FIG. 18** Shown is a model for remote monitoring and control along with some of the benefits that can be realized with Remote Assisted Operations. Offshore assets are connected to an onshore support facility with often dedicated communications and data links. The onshore control room is staffed with experienced operational and engineering staff who have immediate access to engineering and specialist staff in the onshore offices. This allows onshore staff to access real-time operational information and provide operational input to offshore staff without the time, expense and safety exposure of having to travel offshore. It also allows for some tasks that do not require offshore situational awareness to be completely controlled from the onshore facility. *Personal Communication, Riley Parker, Instrumentation and Control Engineering Discipline Lead, Shell.*

Through the 2020 COVID-19 pandemic, remote capabilities were used to enable the reduction of personnel offshore to reduce potential exposure. Each company must decide the extent of remote control they will employ after considering the health, safety, security and environment risks associated with the asset, but regardless of where any one deepwater operating company sits in this journey, the industry continues to shift toward more and more progressive analytics and processing schemes that enable use of more and more of the currently available data on deepwater production facilities. The net result is improved facility performance, enabling operators to do more work with less resources and ultimately resulting in greater and more efficient production of deepwater resources. It is worth noting that the availability of data continues to increase through the use of additional sensors, more advanced equipment monitoring technology and improved communication methods that provide secure and automated data access.

## Continuously evolving and emerging technology

Over the past 15 years, lean or continuous improvement methods have been put in place to improve problem solving skills and create standard processes to improve the efficiency of teams around standardized and centrally available data. Teams began learning that thinking through the overall "system" is required before making further improvements. In essence, a critical product of the last decade for success in deepwater has been the application of "systems thinking," where the system necessarily includes the reservoir itself, including consideration of the reservoir fluid properties, the wells and flow lines to bring production to the surface, the production and processing facilities and the export system. Success over the full lifecycle of deepwater projects requires this level of purposeful communication and planning. Deepwater operators have accelerated the shift in ways of working by reducing the number of personnel onboard production facilities and increasing the use of operational data systems and where possible, remote monitoring and control systems.

Emerging technologies, including concepts such as digitally modeling platform operations and systems and the further development of predictive analytics that feed artificial intelligence will continue to improve the ties between the subsurface challenges of reservoirs and fluids and the well and surface systems designed to bring those fluids to market safely and efficiently. Operators are now starting to build digital twins of offshore facilities to enhance operator training and to provide new insights into their operation and performance (Baker, 2020).

## Key messages

Deepwater exploration and production has been a component of the oil and gas industry for well over 100 years. What has changed is the specific water depth regarded as deepwater. As the industry has moved into deeper waters offshore, new technologies have been developed and applied across the entire upstream (exploration and production) industry.

In some cases, particularly seismic imaging, technology developments were not exclusive to the deepwater environment. In drilling and production however, the new technologies were not only initially developed for operating in deepwater, but were required for the industry to operate in this challenging environment and in particular, to economically explore for and develop deepwater hydrocarbon reserves.

The first three case studies described in this chapter explore these unique challenges and solutions. High rate/high ultimate wells considered the characteristics of deepwater reservoirs and fluids as well as production flow rates, and how well design and completion methods enable successful developments. Due to the cost and complexity of the deepwater operating environment, offshore producing structure design and infrastructure planning is critically important and must be considered earlier and in a more multidisciplinary fashion than earlier shallow water developments. Unique to deepwater environments are the environmental characteristics in which infrastructure is built. This, along with the characteristics of the produced fluids puts great pressure on the design of pipelines, flowlines and risers and the management of production through them to assure trouble free flow.

The fourth case study, moving into the digital and information age, explores solutions being employed in modern deepwater developments to improve operations and ensure success in today's challenging economic environment.

## Conclusion

Through more than 50 years of drilling and development of deepwater oil and gas resources, technology has always been a key enabler of success, both technically and economically. This chapter has endeavored to show a few of the areas where technology development, often occurring at the same time as the development work on the resources themselves, has been key to this success.

As impressive as these technologies are, along with the production systems and structures that they support, there should be little doubt that the continued efficient exploitation of discovered resources and the development of new discoveries will require even more impressive systems and technologies. Along with the physical systems, significant and continuing evolution of the data and management systems which will support these technologies and ensure their effective application are expected (Yeilding et al., 2005).

Looking farther ahead, many of the systems developed for deepwater oil and gas exploration and development will no doubt be critical enablers for the future offshore development of alternative energy and resources.

## References


Allan, M.E., Gold, D.K., Reese, D.W., 2013. In: Application of Toyota's Principles & Lean Processes to Reservoir Management: More Tools to Overload the Toolbox or a Step Change in Our Business? Aera Energy LLC. Presented at the SPE Western Regional & AAPG Pacific Section Meeting, 2013 Joint Technical Conference, Monterey, California, USA. SPE 165331-MS.

Allen, D.W., 1998. In: Vortex-Induced Vibration of Deepwater Risers. Paper presented at the Offshore Technology Conference, Houston, Texas, 1998. Paper Number: OTC-8703-MS https://doi.org/10.4043/8703-MS.

Baker, J., 2020. Digital twins are propelling the oil and gas industry into the future of asset optimization. NS Energy. https://www.nsenergybusiness.com/features/digital-twins-oil-gas/.

Cossey, S.P.J., 2004. Exploration and production in the Deepwater Gulf of Mexico. Offshore Magazine. https://www.offshore-mag.com/home/article/16756707/exploration-and-production-in-the-deepwater-gulf-of-mexico.

Cox, G.A., Johnston, R.M., Palermo, R.M., 2001. In: Implementing Enterprise Architecture—Putting Quality Information in the Hands of Oil and Gas Knowledge Workers. Presentation at the SPE Western Regional Meeting, Bakersfield, California, USA. Aera Energy LLC. SPE 68794.

Denney, D., 2013. Mensa field: deepwater gulf of Mexico case study. Journal of Petroleum Technology.

Enze, et al., 1994. Auger TLP Design, Fabrication and Installation Overview, Offshore Technology Conference.

Gartner, 2021. 4 Types of Analytics. https://www.gartner.com/en/newsroom/press-releases/2014-10-21-gartner-says-advanced-analytics-is-a-top-business-priority.

Hydrocarbons Technology, 2021. https://www.hydrocarbons-technology.com/projects/nakika/.

Hyne, N.J., 2012. Nontechnical Guide to Petroleum Geology, Exploration, Drilling & Production. PennWell Corporation.

Leffler, W.L., et al., 2011. Deepwater Petroleum Exploration & Production, A Nontechnical Guide, second ed. PennWell.

Lybarger, J.H., Scheuerman, R.F., Willard, R.O., 1974. In: Water-Base, Viscous Gravel Pack System Results in High Productivity in Gulf Coast Completions. Paper presented at the SPE Symposium on Formation Damage Control, New Orleans, Louisiana. Paper Number: SPE-4774-MS. https://doi.org/10.2118/4774-MS.

National Academy of Sciences, 2018. Consensus Study, Understanding and Predicting the Gulf of Mexico Loop Current: Critical Gaps and Recommendations. https://www.nationalacademies.org/our-work/advancing-understanding-of-gulf-of-mexico-loop-current-dynamics–.

Parshall, J., 2016. Twenty-year effort brings only two FPSO vessels to US GOM. Journal of Petroleum Technology.

Perinet, D., Frazer, I., 2007. In: J-Lay and Steep S-Lay: Complementary Tools for Ultradeep Water. Paper presented at the Offshore Technology Conference, Houston, Texas, U.S.A., April 2007. Paper Number: OTC-18669-MS https://doi.org/10.4043/18669-MS.

Priest, T., 2007. The Offshore Imperative. Texas A&M University Press.

Priest, T., 2019. Shell Oil's Deepwater Mission to Mars. https://dsps.lib.uiowa.edu/mars-deepwater-gulf/wp-content/uploads/sites/66/2019/09/Priest-Shell-Oils-Deepwater-Mission-to-Mars_June-2019.pdf.

Razi, M., Bilinski, P., 2012. In: Mensa field, Deepwater Gulf Of Mexico—Case Study, Shell Exploration & Production, SPE 159741. SPE Annual Technical Conference and Exhibition, San Antonio, Texas.

Ross, J.W., Weill, P., Robertson, D.C., 2006. Enterprise Architecture as Strategy. Harvard Business School Press.

Shell, 2021. Auger: From Deepwater Pioneer to New Energy Giant. https://www.shell.com/about-us/major-projects/cardamom/auger-from-deep-water-pioneer-to-new-energy-giant.html. Shell project details are available at https://www.shell.com/about-us/major-projects.html.

van Oort, E., Nicholson, J., D'Agostino, J., 2001. In: Integrated Borehole Stability Studies: Key to Drilling at the Technical Limit and Trouble Cost Reduction. Paper Presented at the SPE/IADC Drilling Conference, Amsterdam, Netherlands, February 2001. Paper Number: SPE-67763-MS https://doi.org/10.2118/67763-MS.

Wasden, F.K., 2003. In: Flow Assurance in Deepwater Flowlines and Pipelines—Challenges Met, Challenges Remaining. Paper presented at the Offshore Technology Conference, Houston, Texas, May 2003. Paper Number: OTC-15184-MS https://doi.org/10.4043/15184-MS.

Yeilding, C.A., 2013. What a Difference a Few Decades Makes: Exploration History of the US Gulf of Mexico Deepwater. University of Houston, Department of Earth and Atmospheric Sciences, Annual Sheriff Lecture.

Yeilding, C.A., Yilmaz, B.L., Rainey, D.I., Pfau, G.E., Boyce, R.L., Wendt, W.A., Judson, M.H., Ray, A.K., Chen, R., Peacock, S.G., Duppenbecker, S.D., Bowman, M.J., Sternbach, C.A., 2005. The history of a new play: Thunder horse discovery, deepwater gulf of Mexico. In: Downey, M.W., Friedman, G.M. (Eds.), Discoverers of the 20th Century: Perfecting the Search. AAPG Special Publication.

Chapter 20

# Economic considerations and market condition effects in deepwater

Cheryl R. Collarini[a], Henry S. Pettingill[b,c] and Joe L. Stires[d]
[a]Collarini Energy Experts, Covington, LA, United States, [b]Rose and Associates, Houston, TX, United States, [c]Geo Ventures International, Ainsa, Spain, [d]Noble Energy, Retired, Houston, TX, United States

## Introduction

The business of exploring for oil and gas never has been for the faint of heart, and that is especially true in frontier areas. As undiscovered resources continue to reside in more difficult environments, particularly in deepwater, the costs, risks and complexity of the decisions to pursue their discovery and development continue to increase. Evaluating whether to move a floating rig 200 or more miles away from the sight of land (1 million feet) to drill a well in 10,000 ft. (3000+ m) of water then another 25,000 ft. (7600+ m) below the seabed, in order possibly to find one or two juicy reservoirs that are 500 ft. (150 m) thick at a cost of $150 million before knowing whether investing another $5 billion to develop them is sensible takes a lot of trust in the people and the technology. Perhaps this thinking is similar to the faith in technology that led Neil Armstrong, Buzz Aldrin and Michael Collins to agree to fly to the Moon in 1969. Interestingly, some of that same "space age" technology has influenced the deepwater technology in use today. There is no doubt that an amalgamation of talent, knowledge, equipment and technology is required to justify the decisions to commit large amounts of money to complete major deepwater development projects. While the earlier chapters of this book focus on the geoscience and engineering technology of deepwater sedimentary systems, this chapter covers the myriad factors important to financial success, including the existence and density of reserves, the rate of recovery, market conditions influencing product price and cost of recovery, and political and fiscal issues that affect profitability. Finally, it offers a glimpse into the future possibilities that deepwater development holds for energy beyond hydrocarbon production.

## Overview—Key economic drivers, market conditions, risk vs reward

Market conditions influencing the decisions to invest in oil and gas, and the success of those investments, exist at the global, regional, national and local levels. These conditions include the **physical environment, the economic environment, the political and regulatory environments** and the **limits of technology**. While investment decisions are occasionally made based on humanitarian reasons, this chapter addresses the more common situation: that those decisions are governed by the expectation that financial benefit will accrue to the investor. Even prior to deciding to drill the initial (wildcat) well, some preliminary economic analysis is conducted that indicates a reasonable chance of profitability of a new project. After, most companies have adopted a four-part evaluation method that begins with a feasibility study, followed by the pre-Front-End Engineering Design (pre-FEED), the FEED and the Final Investment Decision (FID). Each of these "stage gates" leading to the final decision requires an evaluation of profitability that includes the estimated volumes of recoverable hydrocarbons, the associated revenues and costs, and the risks, physical, political and financial. At each gate, if the evaluation indicates adequate profitability and acceptable risk (usually compared to other opportunities the investor is considering), the project proceeds to the next phase.

The economics of deepwater are extremely challenging. The analysis leading to the investment decision begins with estimates of the resource that is hydrocarbon volumes in place. From that, recoverable volumes are estimated, which depend primarily upon the reservoir rock, drive mechanism and fluid properties but also on the mechanics of the connection, or route, from the reservoir to the sales point. The forecast sales volumes and product prices define the expected revenue, or cash inflow. From that revenue the cash outflows must be subtracted: operating expenses to produce the product, capital investment to effect the project, the share taken via contract or law by the mineral owners, whether private

Deepwater Sedimentary Systems. https://doi.org/10.1016/B978-0-323-91918-0.00005-0

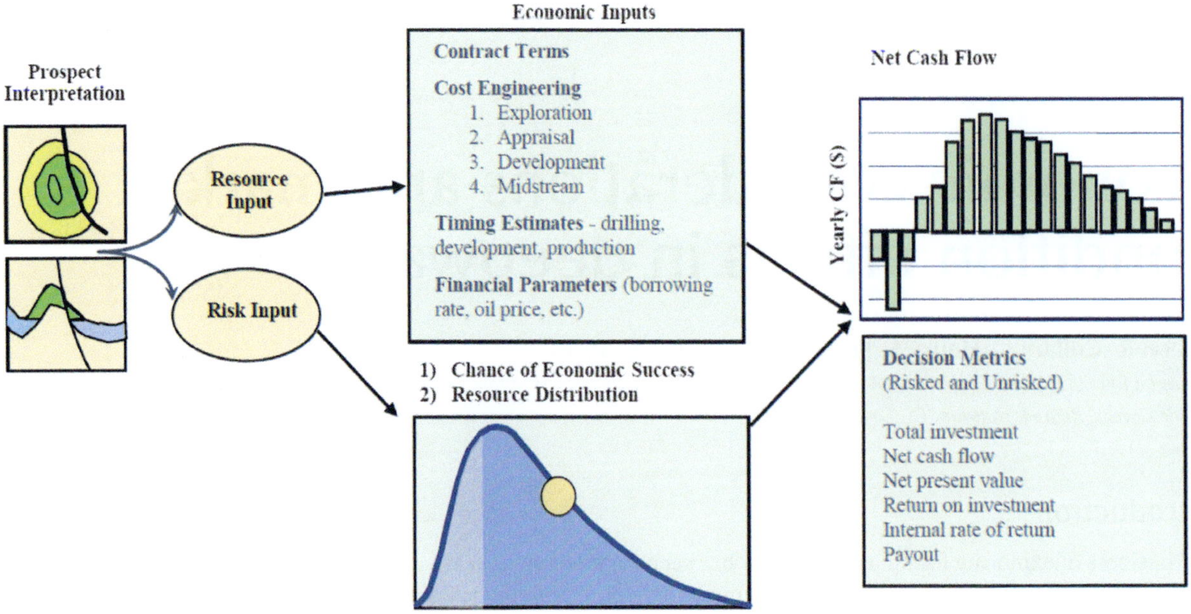


**FIG. 1** Exploration and development valuation workflow. Determination of resources technically in place and recoverable is but one aspect of the decision to pursue a project. Costs, contract terms, financial parameters (including product prices and markets), timing of the investments and production, and finally the risk of finding and the distribution of volumes all feed into the economics that determine whether to invest. Many different experts are required to contribute to each project team.

or public, and any taxes levied by governments. No different from any private investment in any industry, the return on that investment is calculated by comparing the amount of net profit to the investment amount. Fig. 1 illustrates the work process from concept to final investment decision.

Of note, the rate of recovery is as important as the total, and that is also determined by the reservoir properties and drive mechanism, the development program, including the number of wells and the timing of their drilling, completion designs, facility type and size, and the capacity of pipelines to the sales points (see Chapter 19 for more discussion on these topics). Finally, the time value of money issues also can be affected by many other factors, including the speed of government and regulatory approvals, access to funding and more. Each variable that affects profitability is determined, or at least influenced, by global and local conditions. While these conditions affect all oil and gas investments, deepwater projects are more sensitive to most due to both the high cost of exploration and development and long project life; they require robust product prices and favorable cost and government takes compared to many other areas. Each of these variables is examined in more detail in the following sections.

## The physical environment

### Deepwater defined

The definition of deepwater as it relates to offshore drilling has evolved over time (Fig. 2). At one time, it was considered to be 600ft. (183 m), the approximate depth at the edge of the outer continental shelf of the Gulf of Mexico, which was actively developing at the time, the early1970s. Later that same decade it extended to approximately 1500ft. (457 m), which still today appears to be the approximate technical, economic limit for the use of fixed offshore platforms, those pinned to or resting on the sea floor without additional anchors or other support (Petronius, a compliant tower design in the US Gulf of Mexico, reached 1754 ft. water depth (535 m) in 2000). As the industry has advanced to deeper and deeper water, some references call depths up to 5000 ft. (1524 m) "deep" or "mid-water" and ultra-deepwater beyond that. Because the distances between the continental shelf edges and dry land vary from continent to continent, and indeed from country to country within each continent, the maturity of basin development varies as well. Those with larger shelves, such as the Gulf of Mexico, Northwest Australia and the North Sea, had been developing for decades before technology for deeper water became commercially available (see Chapters 1 and 19 for additional information).

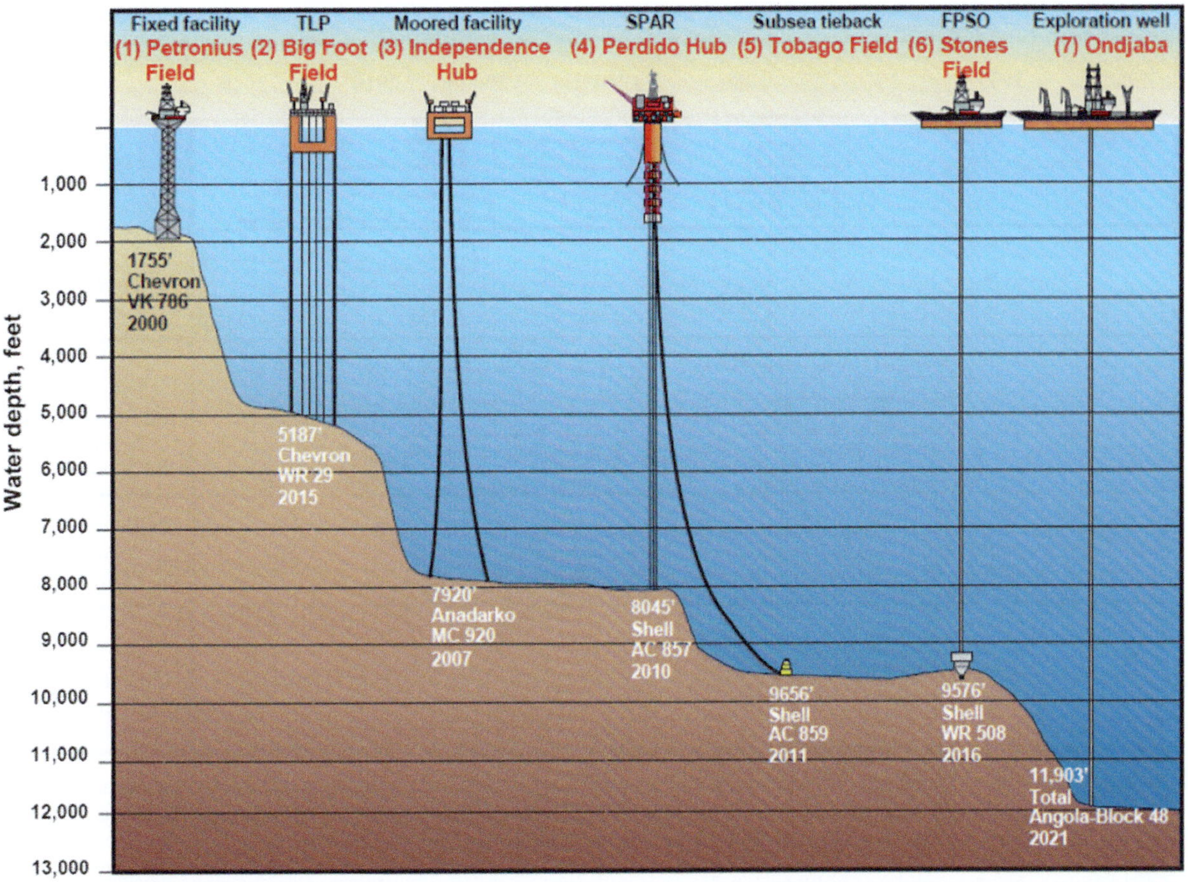


**FIG. 2** World-record water depths for drilling and development facilities and years set. This figure shows the progression of facility design from the fixed offshore platform installed by Chevron (Texaco) to the deepest exploration well drilled using a drillship. *TLP*, tension leg platform; *SPAR*, single point anchor reservoir; *FPSO*, floating production and storage offloading. *(From Weimer, P., Pettengill, H.S., Roden, R., 2022. The history of deepwater exploration, part 3. AAPG Explorer (January), 16–19.)*

The initial discoveries in new deepwater basins started slowly, but their number has increased considerably during the last decade. An overall progression of milestones is depicted in Fig. 3 for water depths greater than 1500 ft. (457 m). The first discovery in deepwater was in Northwest Australia in 1978; two new basins, the Gulf of Mexico and Brazil, were added in the 1980s; 10 new basins in the 1990s in West Africa; six new basins in the 2000s in Asia, Europe and the Middle East; and 14 new basins in North and South America since 2010. In a mature basin, a smaller percentage of its original hydrocarbons remains to be recovered because some of the fields will have been discovered and developed already. This reduces the exploration opportunities, as a percentage, compared to frontier basins. Areas where deepwater begins very close to shore will not have had the opportunity to develop as much, since deepwater technologies are more recent. Some of these technologies include advanced subsea systems that can be remotely installed using robotic vehicles and operated, including real-time data recordation via complex umbilical controls that allow continuous performance monitoring. In addition, mature basins have existing infrastructure that makes smaller oil and gas accumulations more commercially attractive and likely to be exploited. This physical reality influences where companies desire to invest in deepwater opportunities.

## Geography of deepwater

While some offshore drilling occurred late in the 19th century, the offshore industry began to emerge in the 1950s and 1960s, supported by the Offshore Technology Conference, first held in 1969. The two most mature offshore basins are the Gulf of Mexico and the North Sea, which were extremely active beginning in the 1940s and 1970s, respectively. These have extensive shallow water shelves which allowed pipelines, intermediate and host facilities and other supporting infrastructure to develop over time as deepwater technology also advanced. In addition, the proximity to ready markets in the United States and Europe nurtured early development. The intermediate facilities allowed multiple stages of gas compression and liquids pumping, pipeline sharing, and commercial use of otherwise fallow platforms and their related equipment which would be ready for abandonment.

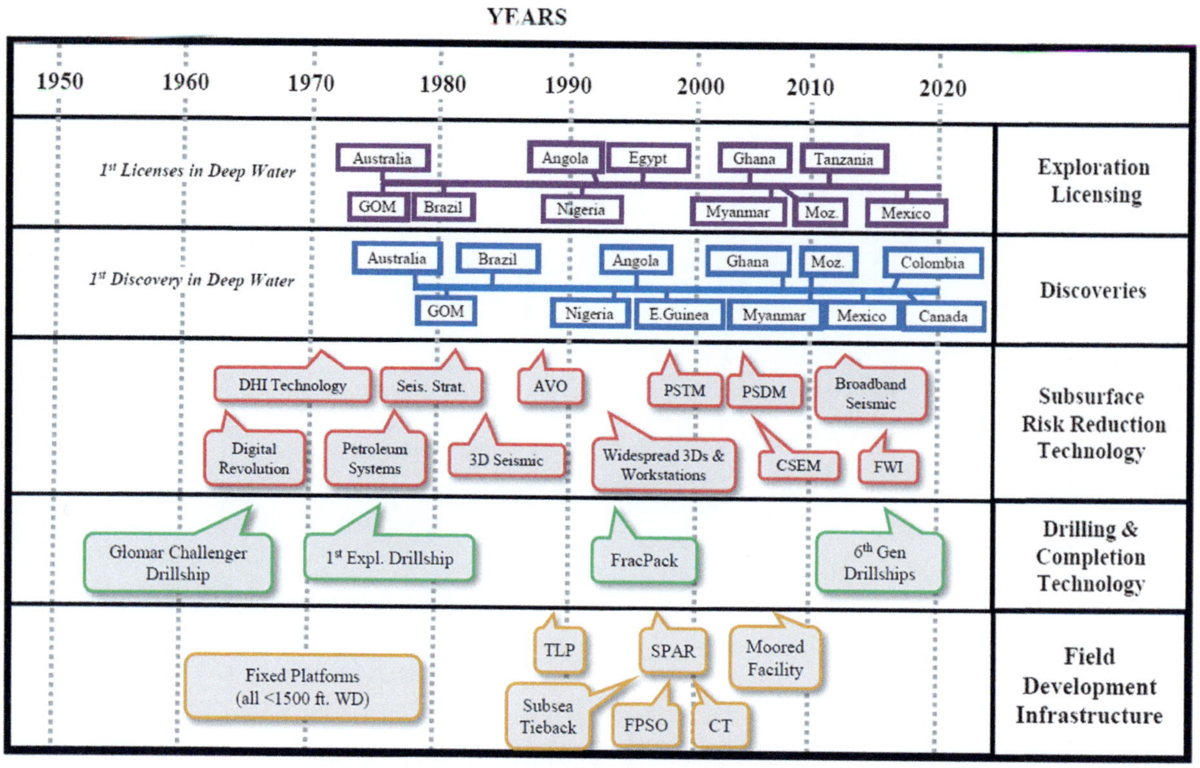


**FIG. 3** Deepwater timeline indicating key Exploration and Development technical events. Milestone events are located according to the year of first occurrence or deployment. For subsurface risk reduction technology: *DHI*, Direct Hydrocarbon Indicators and includes amplitude anomalies, gas wipeouts and other gas flags; *Seis. Strat.*, seismic stratigraphy; *AVO*, amplitude versus offset (a type of DHI characteristic); *PSTM*, pre-stack time migration; *PSDM*, pre-stack depth migration; *CSEM*, controlled-source electromagnetism; *FWI*, full waveform inversion. For field development infrastructure types: *FPSO*, floating production storage and offloading; *TLP*, tension leg platform; *SPAR*, single point anchor reservoir; *CT*, compliant tower; Moored facilities include semi-submersibles. *(Modified from Weimer, P., Pettengill, H.S., Roden, R., 2022. The history of deepwater exploration, part 3. AAPG Explorer (January), 16–19.)*

Exploration companies seek favorable geographic, political and safety environments in keeping with established and foreseeable developing technologies. Areas with a more abrupt drop from land to deepwater have been slower to develop, such as the Eastern Mediterranean. Additionally, some places have had delayed development due to political issues, such as Timor Leste, where gas was discovered in 1974 but only recently is able to be developed.

More than 100 oil-producing countries in the world share the many deepwater basins, as illustrated in Table 1. Each has its own fiscal regime as it pertains to oil and gas. These regimes are the methodologies by which governments are compensated for the removal of their non-renewable natural resources. They may comprise royalty and income tax, production sharing contracts, service contracts or combinations of them; each country determines how it will conduct its business with investor companies. Adjusting these terms can encourage or discourage investment in their country (and sometimes states or other divisions within the country). Because the terms affect the economics of an investment directly, understanding them is essential for a credible estimate of the size of resource needed to make a project profitable and direct where exploration should be focused. More explanation of these regimes follows later in this chapter.

## Discovery size

Determining the amount of product available to sell begins with the assessment of the actual hydrocarbons in place in the targeted reservoirs. "*Total petroleum initially-in-place* is that quantity of petroleum that is estimated to exist originally in naturally occurring accumulations. It includes that quantity of petroleum that is estimated, as of a given date, to be contained in known accumulations prior to production plus those estimated quantities in accumulations yet to be discovered (equivalent to 'total resources')" (SPE, 2007).

No matter the location, oil- or gas-in-place depends on the quality of the rock, the depth (pressure), and the size of the accumulation (see Chapters 2, 6–12 for more on the topics of reservoirs and rock types). Using all the scientific data at hand, a range of rock volumes can be estimated for each potential reservoir in the prospect area. From that, a

**TABLE 1** Deepwater basins with discoveries.

| Map key | Basin | Discovery | Year |
|---|---|---|---|
| 1 | Eastern Canada Bay | Bay du Nord | 2015 |
| 2 | Northern Gulf of Mexico | Jolliet | 1981 |
| 3 | Southern Gulf of Mexico | Lakach | 2004 |
| 4 | Colombia | Kronos | 2015 |
| 5 | Trinidad & Tobago | LeClerc | 2016 |
| 6 | Guyana | Liza | 2015 |
| 7 | Potiguar | Piti | 2014 |
| 8 | Sergipe Alagoas | Moita Bonita | 2012 |
| 9 | Espirito Santo | Tanganika | 2014 |
| 10 | Campos | Albacora | 1984 |
| 11 | Santos | Tambuatá | 1999 |
| 12 | North Falklands | Sea Lion | 2010 |
| 13 | Outeniqua | Brulpadda | 2019 |
| 14 | Angola | Girassol | 1996 |
| 15 | Cabinda | Cabinda | 1997 |
| 16 | Gabon | Gabon | 1998 |
| 17 | Equatorial Guinea | Zaifro | 1995 |
| 18 | Nigeria | Bonga | 1995 |
| 19 | Ghana-Cote D'Ivoire | Jubilee | 2007 |
| 20 | Senegal-Mauritania | Chinguetti | 2001 |
| 21 | Nile | Rosetta | 1997 |
| 22 | Levant | Mari-B | 2000 |
| 23 | Tanzania | Pweza | 2010 |
| 24 | Mozambique | Windjammer | 2010 |
| 25 | Shetlands | Foinaven | 1993 |
| 26 | Central Norwegian Shelf | Ormen Lange | 1997 |
| 27 | Black Sea-Romania | Domino | 2012 |
| 28 | Krishna-Godvari | R-1: Annapurna | 2011 |
| 29 | Myanmar | Shwe Yee Htun | 2016 |
| 30 | Brunei-Sabah | Kikeh | 2002 |
| 31 | Mahakam | West Seno | 1998 |
| 32 | West Philippine Sea | Malampaya | 1993 |
| 33 | NW shelf Australia (3 sub-basins) | Scarborough | 1978 |

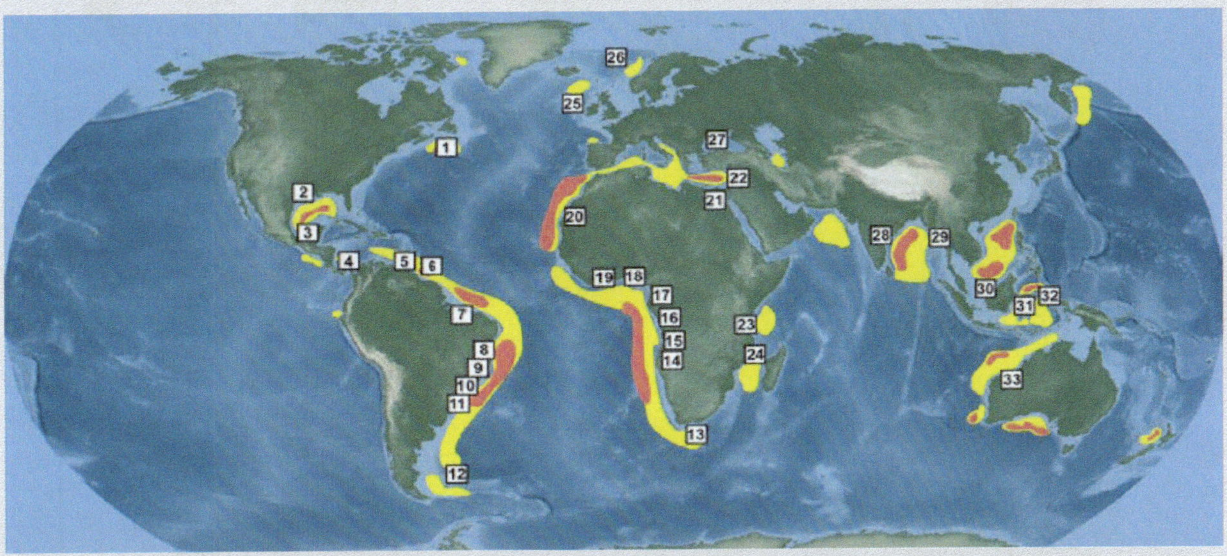


*Figure 1. Global map of all basins in more than 1500-feet water depth with fields or economic discoveries. For each basin, the first discovery and its year are listed in parentheses 1. Eastern Canada (Bay du Nord, 2015), 2. Northern Gulf of Mexico (Jolliet, 1981), 3. Southern Gulf of Mexico (Lakach, 2004), 4. Colombia (Kronos, 2015), 5. Trinidad & Tobago (LeClerc, 2016), 6. Guyana (Liza, 2015), 7. Potiguar (Piti, 2014), 8. Sergipe Alagoas (Moita Bonita, 2012), 9. Espirito Santo (Tanganika, 2014), 10. Campos (Albacora, 1984), 11. Santos (Tambuatá, 1999), 12. North Falklands (Sea Lion, 2010), 13. Outeniqua (Brulpadda, 2019), 14. Angola (Girassol, 1996), 15. Cabinda (1997), 16. Gabon (1998), 17. Equatorial Guinea (Zafiro, 1995), 18. Nigeria (Bonga, 1995), 19. Ghana-Cote D'Ivoire (Jubilee, 2007), 20. Senegal-Mauritania (Chinguetti, 2001), 21. Nile (Rosetta, 1997), 22. Levant (Mari-B, 2000), 23. Tanzania (Pweza, 2010), 24. Mozambique (Windjammer, 2010), 25 Shetlands (Foinaven, 1993), 26. Central Norwegian shelf (Ormen Lange, 1997), 27. Black Sea-Romania (Domino, 2012), 28. Krishna-Godivari (R-1: Annapurna, 2001), 29. Myanmar (Shwe Yee Htun, 2016), 30. Brunei-Sabah (Kikeh, 2002), 31. Mahakam (West Seno, 1998), 32. West Philippine Sea (Malampaya, 1993). 33. NW shelf- Australia (three sub-basins) (Scarborough, 1978). Areas with good exploration potential are shown in yellow (deepwater, 1500–6000 ft.) and orange (ultra-deepwater, deeper than 6000 ft.).*

Based on Weimer, P., Pettingill, H.S., 2020. The History of Deepwater Exploration, Part 2, AAPG Explorer, December.

corresponding range of original-oil-in-place (OOIP) is calculated in stock tank barrels of oil (Imperial units): $OOIP = 7758 * A * h * \phi * (1 - S_{wi})/B_{oi}$, where $A$ is area in acres, $h$ is net thickness in feet ($Ah$ is net rock volume), $\phi$ is porosity of the rock, $S_{wi}$ is the initial water saturation of the rock and $B_{oi}$ is the formation volume factor that converts reservoir barrels of oil to stock tank barrels, which are what is sold. In standard units, OOIP in $m^3 = A * h * \phi * (1 - S_{wi})/B_{oi}$ where $A * h$ is net pay volume in $m^3$. For gas reservoirs, the formula for OGIP is the same, except that the denominator is $B_{gi}$, which converts reservoir cubic feet or meters to standard conditions of atmospheric pressure and specified temperature for sales.

## Net pay volume

Before drilling a wildcat, the $A$ and $h$ components of the aforementioned volumetrics equation are usually the variables with largest range of uncertainty, and hence the major control on in-place resources. $A*h$, also known as Net Rock Volume (NRV), can be calculated two ways:

**(1)** Area (acres or km$^2$) * Net Thickness (feet or m) * Geometric Factor (%)
**(2)** GRV (acre-feet or m$^3$) * Net/Gross (%)

where GRV is gross rock volume; Net/Gross is the percentage of the reservoir interval that is porosity-bearing, and Geometric Factor is a necessary correction for "feather acreage" along the periphery of the trap where the hydrocarbon-water contact reduces $h$.

Early in the global deepwater play, it was found that many large discoveries were driven by either a large trap area, a large net pay column or both (Fig. 4A and B). However, those with a large net pay column relative to their trap area were

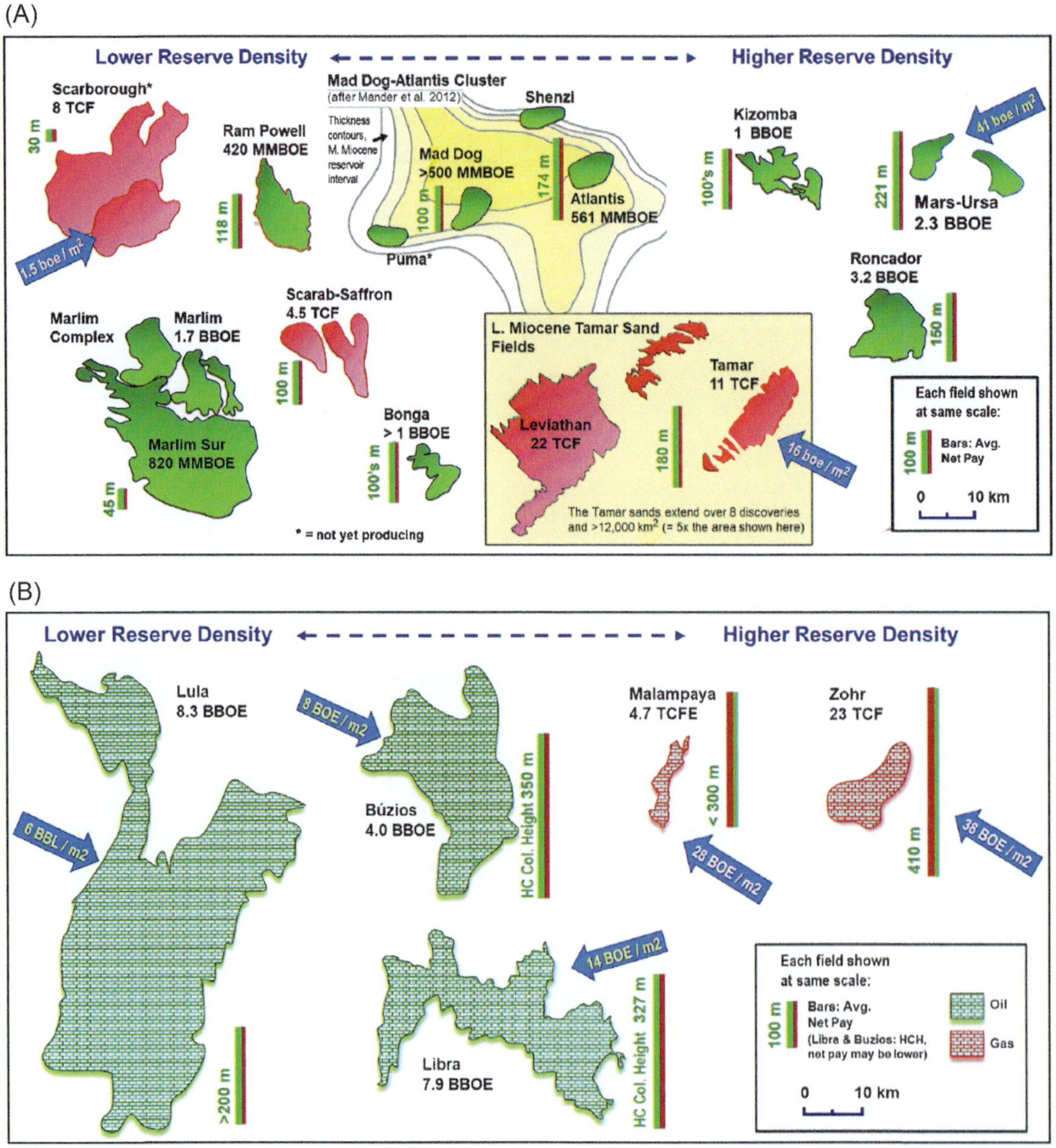


**FIG. 4** (A) Net rock volume (NRV) and reserve density—sandstone reservoirs. Fields with higher reserve density contain more hydrocarbon volume per area, dictated by net pay thickness and rock properties including porosity and water saturation. Higher reserve density areas are more likely to be profitable, because they require fewer wells and more closely contained facilities. (B) Net rock volume (NRV) and reserve density—carbonate reservoirs. Carbonate field examples ranked from lower to higher reserve density. *(From Pettingill, H.S., 2020a. Energy and Power Density: The Deepwater Advantage. Manuscript posted on LinkedIn (updated from April 2020 AAPG Explorer Article by Heather Saucier).)*

generally brought onstream earlier, since the development plan required fewer wells per millions of barrels of oil equivalent (MMBOE), and the well ultimates were much higher, making the economics more attractive. An extreme example of this is the Mars discovery in the US deepwater Gulf of Mexico (Fig. 4A), which owes its large net pay per area to turbidite sediment ponding within highly confined salt minibasins. In contrast, the 1979 Scarborough gas discovery (offshore Australia), for which the reservoir was deposited over a vast area of an unconfined turbidite system, remains undeveloped over 40 years after its discovery. For more information on turbidites, see Chapters 6–8.

## Reservoir properties

Porosity is a major factor in the oil-in-place calculation. Very clean sandstones with coarse-grained sand of similar size tend to be highly porous and permeable. They also tend to have a lower water saturation than do sands with a smaller grain size. Poorly sorted sands have a mix of grain sizes, including silts, allowing the smaller grains to fit in between the larger ones. This not only reduces the overall porosity but also allows for a higher water saturation due to the larger surface area of the grains present. These dirtier sands begin with a comparatively reduced OOIP. An exception to this rule is highly quartzose sands (> 90% quartz) such as Tamar and Leviathan in the Eastern Mediterranean, where poorly sorted sands have almost 30% porosity and permeability exceeding one Darcy. Carbonate reservoirs tend to be constructed from a much more diverse array of grain sizes and can be very porous with little to no interstitial water. Their pore compartments range from small to cavernous ("vuggy") and may contain very large OOIP relative to sandstone reservoirs.

The percentage of oil-in-place that is recoverable (recovery efficiency or RE) depends upon a number of factors including the rock quality (porosity, permeability), reservoir shape, reservoir recovery mechanism, aquifer size and fluid properties, namely density and viscosity. Primary reservoir recovery mechanisms are water drive, depletion drive (gas expansion) and gravity drainage. Most reservoirs exhibit a combination of two or more of these mechanisms. Since the product must be recovered to be sold, evaluating the likely drive mechanism, therefore the RE, is an important factor in the evaluation.

Permeability is the measure of flow capacity of the rock, and it is determined by how well the pore spaces are connected. Porosity is also an important factor in recovery, primarily because highly porous rock also tends to be more permeable. This is not always the case, however. Sometimes the cementation between grains can block flow paths and cause the connectivity to be very low. When permeability is low, recovery tends to be low, because hydrocarbons either can be blocked from flowing at all or will not be able to flow to the wells fast enough to be commercial unless treated by stimulating, such as fracking. Local permeability can be measured accurately in core samples but overall reservoir permeability and continuity at the reservoir scale are best measured using pressure transient analysis. This method relies on a reasonable flow period followed by a shut-in period and results in a mathematical calculation of both permeability and near wellbore damage (skin) within the radius of investigation, which depends upon the length of the test and quality of the rock. The intelligent well technology developed for deepwater completions allows pressures to be measured in real time at the formation, so obtaining the data is not a special effort once development wells are placed on production. This methodology also can detect nearby reservoir boundaries, such as faults, contacts or stratigraphic barriers that may or may not be discernable using seismic; this is helpful in planning subsequent development. See Chapters 2, 6–13 for more on these topics.

## Recovery efficiency

Calculating recovery efficiency is straightforward. The same oil or gas in place formulas can be used to calculate the remaining oil in place throughout the life of the field. At the time of field abandonment, the water saturation and formation volume factors at abandonment are used, and then, the ultimate recovery expected is the difference between the OOIP and the remaining oil in place. Complicating that calculation is that the saturation values must be estimated, both at original conditions and at abandonment. Primary considerations affecting those values are gas-oil ratio (GOR) for oil reservoirs, condensate yield for gas reservoirs and hydrocarbon density, which are the important factors in determining the formation volume factors for oil and gas at both original and abandonment conditions. Also, there are reservoirs where compression of the rock is a significant factor that can reduce the reservoir porosity as pressure declines during production.

Reservoir architecture, as discussed in Chapters 8 and 9, is a major factor in determining the percentage of oil and gas in place that can be swept from the reservoir. Understanding it as much as possible before development allows wells to be located in the optimum positions to maximize recovery. How the reserves are swept into the producing wells depends upon the reservoir architecture, which includes the overall structure and shape of the reservoir, how it is compartmented due to faulting and other discontinuities and the continuity and connectivity of the oil or gas reservoirs. These elements influence how many wells are needed, how the wells are to be drilled, either as straight holes or directionally, and even what size hole will be needed. Both rock quality and reservoir continuity are major factors in designing the completions, including the number, size and location of perforations, whether and how sand control should be used and flexibility in opening and closing perforated segments. All of these actions affect the amount of oil and gas that can be recovered and the rates at which they can be produced. It is particularly important in deepwater wells because intervention to correct well failures or improve performance can be very expensive and may not be commercially justifiable.

The reservoir drive mechanism is a primary factor in the determination of recovery efficiency. As mentioned in the previous section of this chapter, reservoirs are characterized by their drive mechanisms, water drive, depletion drive or gravity drainage. Water drive occurs when the aquifer is very large compared to the oil accumulation, and the rock is permeable enough to allow the water to readily expand and flow into and through the oil reservoir. Even though water is generally considered to be incompressible (hydraulic controls, for example) in very large quantities its expansion can be significant. Water drive is the most efficient primary recovery drive mechanism for oil reservoirs, as the water sweeps the oil in front of and above it into the production wells. Recovery efficiency can be 40% or more, with some very clean reservoirs reporting recovery of 65% of OOIP. Depletion drive occurs when an aquifer does not exist or is small compared to the size of the hydrocarbon accumulation. In this type of reservoir, the pressure quickly declines as the reservoir fluids are produced. In an oil reservoir, these energy associated with reservoir pressure dissipates as gas in the reservoir expands and is produced, leaving much of the oil behind, and by itself, depletion drive can result in very low recovery efficiency, sometimes 10% or less. In a gas reservoir, however, this mechanism is most efficient, resulting in 80% or greater recovery efficiency in some instances, because the gas remaining at depletion is at very low pressure. In a water-driven gas reservoir, water sweeps much of the gas, but what is left behind is at higher pressures, while in lower pressure gas fields, water production can be operationally problematic. Water drive gas reservoirs still can recover 50% or more of the gas originally in place, depending upon how high the wells are placed structurally in the reservoir. In a depletion drive gas reservoir, most of the time the well position does not have an impact on recovery efficiency.

As discussed in Chapter 19, the operator can increase recovery efficiency beyond primary expectations by implementing improved (secondary or tertiary) recovery methods. Injecting water into an oil reservoir to maintain pressure and to help sweep the oil is commonly done onshore and has been employed in some deepwater developments, such as in the Mars Field in the US Gulf of Mexico. Injecting produced gas back into an oil reservoir serves the same purpose in preserving reservoir pressure longer and has the added benefit of helping to reduce carbon emissions by avoiding gas flaring, particularly in frontier basins that lack ready access to natural gas markets. Adding compression to produce from lower pressure gas reservoirs is almost always economically justifiable. Also, in many gas reservoirs that produce significant condensate, re-injection of produced gas is done for much of the life to keep the reservoir pressure high enough to prevent liquids from falling out in the reservoir, which would have a significant negative effect on recovery efficiency. Tertiary or enhanced recovery methods, such as lean gas or carbon dioxide ($CO_2$) injection or microbe introduction are also possibilities that are proven onshore and may be commercial in deepwater applications. $CO_2$ injection has the potential benefit of capturing and injecting anthropogenic carbon dioxide that might otherwise be released into the atmosphere (Chapter 21). The commerciality of any enhanced recovery method depends upon project-specific parameters, such as reserve size, local infrastructure, cost to implement and the ability to place wells in the most effective positions in the reservoir.

Recovery estimates in deepwater fields are typically more complex than in onshore and shallow water fields because the produced fluids must travel a significantly longer distance to reach the sales point. It is not unusual for producing deepwater wells to be 10 miles (16 km) or more from the host processing facility, whereas onshore or in shallow water, that distance is usually 15,000 ft. (460 m) or less. Moving petroleum that extra distance requires energy. Utilization of seafloor and/or downhole boosting equipment is a big part of the design equation that affects the recovery efficiency for deepwater reservoirs, which in turn determines the volume of product that ultimately will be sold. In addition, the ultimate recovery of each reservoir is dependent upon the detailed facility design, because pumps, compression and related equipment will determine the final reservoir pressures that can be achieved as the field is depleted.

Reservoir simulation models are an important tool for estimating the ultimate recovery and production rates over time, and once the geologic architecture and rock properties are set up, it is relatively simple to vary the number of wells and their placement in the reservoir to optimize recovery. These models are kept updated over the producing life through a process called history matching where the model parameters are adjusted so that the model output matches the actual production of the wells and field. This allows the operator to update the estimate of reserves remaining as well as to evaluate re-development alternatives. For more information on recovery efficiency, drive mechanisms or reservoir modeling, see Hyne (2012).

## The importance of reserve density, high rate and high recovery

The extremely large investment needed to explore and develop in deepwater demands that successful prospects produce at high rates (HR), to allow rapid recovery of the investment, as well as high per-well ultimate recoveries (HU), so that overall profitability is achieved. Critical parameters of HRHU reservoirs, are the same as discussed in Chapter 19 and above. For hydrocarbon recovery, they affect OOIP, RE and initial rate and include reservoir quality, reservoir architecture, porosity and permeability, reservoir pressure and temperature, reservoir drive mechanism and fluid characteristics.

Each well must return enough profit not only to justify its own investment but also to support the overlying infrastructure required to get its production to market including flowlines, platforms, pipelines, and surface facilities and the expenses to keep them operating. A primary measure of the per-well ultimate recovery is the reserve density of a field. Reserve density (also known as energy density) is expressed in recoverable hydrocarbons per unit area, as shown in Fig. 4. Higher reserve densities allow higher recovery per capital invested because fewer development wells are required to produce the same reserves.

Quantitative thresholds for HRHU designation have varied over time and by company. In the deepwater arena, greater than 12,000 barrels of oil equivalent per day and 25 million barrels per well represents the top quartile (Fig. 5). HRHU reservoirs are necessary to justify the high exploration and development costs of many deepwater and ultra-deepwater discoveries.

Initially, the best deepwater reservoirs with high reserve densities were recognized in deep marine sandstones, comprising depositional lobes (amalgamated sheet sands as discussed in Chapter 8) in the deepwater northern Gulf of Mexico (United States) and amalgamated channel-fill in offshore West Africa (Fig. 4A). These reservoirs have excellent connectivity and continuity, porosity and permeability, reservoir drive, and oil quality. Importantly, during the last decade the pre-salt fields of the Santos Basin, offshore Brazil (lacustrine microbialites) and the Zohr Field, offshore Egypt (Miocene buildup) clearly show that carbonate strata may also be HRHU reservoirs, with large reserve densities (Figs. 4B and 5).

Wells can be either high-rate, high-ultimate or both (HRHU), hence there is a low degree of correlation between the two parameters as shown in Fig. 5. This scatter is because there are several controls on each parameter, some of which are operator-dependent, like well rates that are controlled by completion design or facility limitations. For example, some high-rate wells in the US Gulf of Mexico are single-well subsea tieback completions of fields with fewer than five MMBOE of reserves, so they have low well ultimates. Reservoir architecture can influence both rate and well ultimate recovery, although in different ways. Complicated channelized reservoirs may have baffles, barriers and reduced bed continuity that limit well recovery (sometimes due to sweep/aquifer strength issues but also sometimes due to preferential water breakthrough) regardless of the initial sustained flow rate. More homogenous sheet-like reservoirs can have high rates and recovery if the reservoir area (trap area) is large with a strong drive mechanism, although they can have limited per-well ultimate recovery if the trap is small. Faulting is important to ultimate recovery as it is a direct determinant of reservoir structure; sometimes faults are not sealing or are partial flow inhibitors, but usually they form reservoir boundaries that can

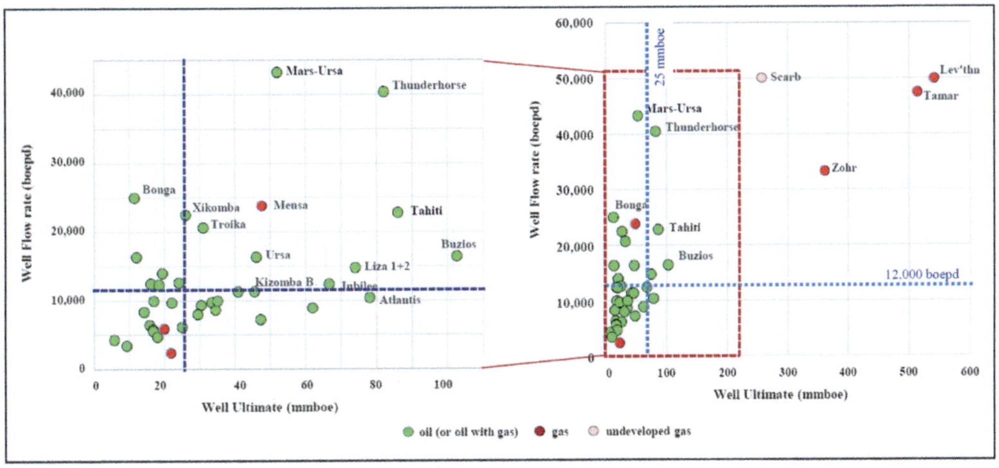


**FIG. 5** High-rate, high-ultimate wells. Well flow rates versus per-well ultimate recovery shown for representative giant deepwater fields. It is important to note that there is not a significant correlation between well ultimate and flow rate due to varying reservoir sizes and characteristics as discussed in the text. The figure on the left is the same as the outlined area of the figure on the right at an expanded scale. *(From Weimer, P., Pettengill, H.S., Roden, R., 2022. The history of deepwater exploration, part 3. AAPG Explorer (January), 16–19.)*

limit recovery area. HRHU wells demand both relatively high reserve density and reservoir continuity. Sometimes this cannot be detected from seismic alone and is only discovered from production performance.

Reserve density depends primarily upon net porosity thickness (*-h), and production rate is generally correlative with permeability times thickness; both are additionally dependent on reservoir drive and reservoir architecture (which affect recovery efficiency). A related metric is *power density* (Bryce, 2010; Pettingill, 2020a), which is a measure of the power output (in btu) per area of an energy source. Successful deepwater oil and gas fields are known for having among the highest power densities of energy sources (Fig. 6), which is a large driver to their economic attractiveness.

While advantageous, HRHU reservoirs are not always necessary for the economic success of deepwater reservoirs. Many sedimentary basins of the world are mature basins with well-developed infrastructure (either onshore or offshore). As a consequence, lower rates of production can be economic in specific cases. For further reading, see Chapter 1.

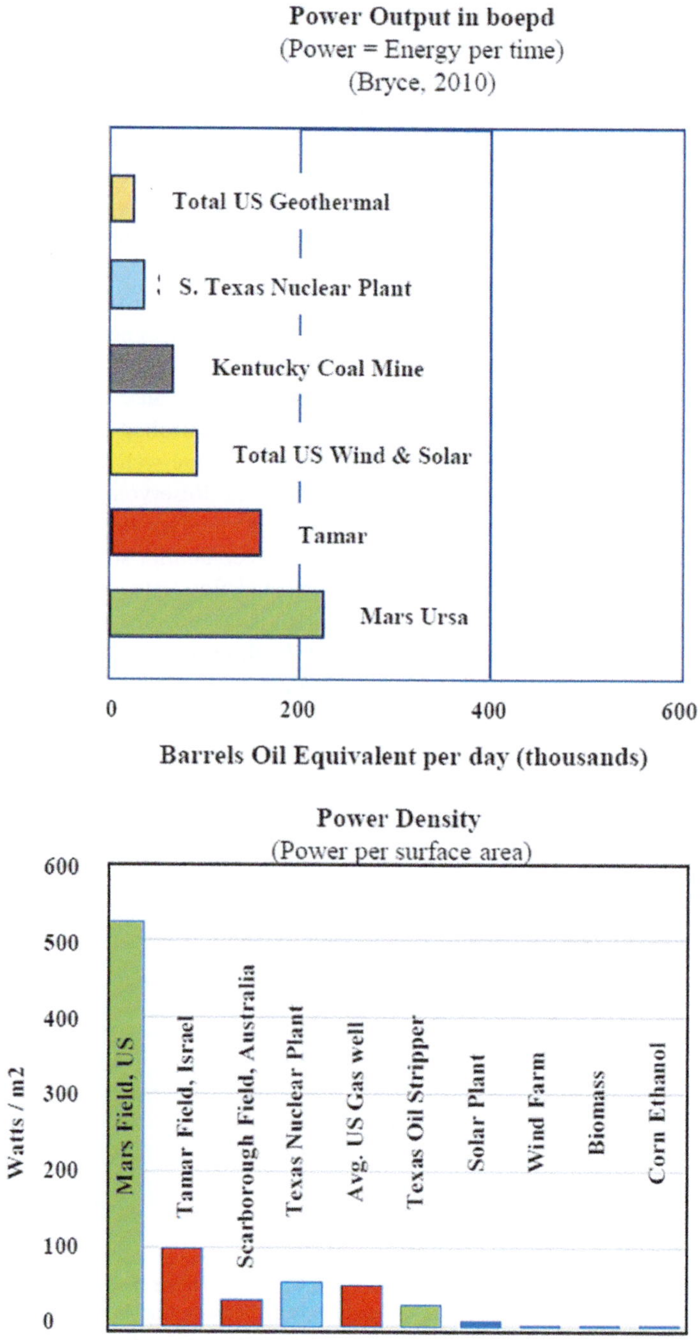


**FIG. 6** Power output and power density. Power output and power density comparison for deepwater fields and other energy sources illustrates the importance of deepwater oil and gas to the total global energy resource. *(From Bryce, R., 2010. Power hungry: the myths of "green" energy and the real fuels of the future. Public Affairs. 414 pp. ISBN:978-1-58648-789-8; Pettingill, H.S., 2020a. Energy and Power Density: The Deepwater Advantage. Manuscript posted on LinkedIn (updated from April 2020 AAPG Explorer Article by Heather Saucier).)*

### Forecasting the rates of production

In evaluating an investment opportunity, many alternative development scenarios may be examined. The number and depth of wells that will be drilled and how (subsea or platform-supported "dry" tree for example), the kind of completion and platform and processing facility type and size are all considered. This depends on the number of reservoirs expected and how wells can be deployed to execute a sensible depletion plan. This may require the consideration of multiple scenarios of completion type (single vs multiple zone or complex "intelligent" completion options) and of timing of drilling wells to effect efficient field depletion. All of this depends upon the expected distribution in space of reservoir accumulations determined using seismic and other exploration techniques.

Because of the strong initial pressures and the thickness of the reservoirs, deepwater wells are able to produce initially at high rates that allow attractive payouts of the capital investments. The initial rates can be estimated using engineering techniques that consider reservoir thickness, permeability, pressure and fluid properties as well as completion type, including casing and tubing size, perforation density and other aspects of completion design. The two most important of these parameters to determining rate are permeability and thickness. In addition to flow from the formation to the wellhead, oil and gas must also move through the flowline to the facility. In deepwater developments these flowlines can be several miles long on the seabed in addition to the vertical depth of the water, which might also be 2 miles (3 km). This post-wellhead flow is through low-temperature water, which can have a marked impact on flow, especially when more viscous and/or low gas-oil-ratio fluids are being transported. The unique issues and challenges to efficient flow have resulted in a new specialty field in petroleum engineering called flow assurance engineering, discussed in detail in Chapter 19.

## The economic environment

Every investment decision relies upon a comprehensive analysis of its expected profitability, which is calculated using financial fundamentals. Cash profit is the difference between cash coming in and cash going out. This is also called net cash flow. These amounts, forecast to occur over time, determine the profitability of an investment, summarized in key indicators, such as internal rate of return (IRR), return on investment (ROI) and net present value (NPV). The words *profit* and *income* are also used by accountants in standardized accounting reports and tax calculations and in government share calculations defined by contracts or law. They often are not calculated in the same way as the net cash flows used by the exploration and development team and by management or investors to make the FID.

### Economic analysis fundamentals

In the oil industry, cash in (**revenue**) is the product of hydrocarbon sales volumes and their respective prices. In some circumstances, there are other sources of revenue such as agreements to process or transport other producers' production. And occasionally there are non-hydrocarbon saleable products, such as $CO_2$ and sulfur, though sometimes those same components are seen as contaminants and reduce value. Usually, these other revenue sources are not major contributors, but in some situations, they can be important to the investment decision.

There are many categories of cash outlays, and they are called by different names in different locations. There are the costs for ongoing operations, like salaries, fuel, transportation, chemicals, materials and so on to keep production going, normally called **operating expenses**. There are monies paid to people or entities that own the minerals (**entitlements**). These may take the form of royalties, a percentage of the production as it is produced or production sharing or other contractual methods. Entitled parties are landowners in countries that allow private ownership of minerals (United States and Canada) or countries and/or localities depending upon their laws. Also included in entitlements are monies paid to governments in the form of taxes, production, fees or net profit sharing. Finally, there is the **investment** itself, which comprises exploration, appraisal, development, redevelopment (if applicable) and abandonment expenditures. The exploration and appraisal investments generally are spent in advance of knowing whether any production will ever result.

Companies considering investments forecast the cash flows that are expected to occur over the entire life of the project; they calculate measures of profitability to allow them to compare opportunities within their portfolios and make the choices that will maximize the value of their organizations. Since they are deciding where to invest, these measures are expressed in terms of the efficiency of the investment, such as ROI, IRR and NPV. ROI is the multiple of the investment expected to be recovered from the net cash before the investment is subtracted. IRR is the equivalent interest rate on the investment; it can be compared directly to the interest rate of any other investment, such as a bond or a bank. NPV is the interest-adjusted time value of money that is received over time; it allows a fair comparison of two or more series of cash flows realized irregularly over different periods of time. Most profit-making companies desire to maximize their NPV at a reasonable interest rate, such as what they could otherwise make at a bank. To do that, they often calculate the ratio of NPV to the time-adjusted investment and rank their inventory of opportunities. In that way, when they have limited investment funds, they are able to ration funds in a manner that selects the projects that will deliver the most NPV per dollar invested.

These calculations all begin with detailed best estimates of every element affecting cash flows. Most of those elements depend on various market conditions in local to global environments. While these are common to all oil and gas investments, deepwater exploration and development requires extremely large expenditures, and proportional returns, so they are more sensitive to market conditions.

## Product prices are market-determined

The only other variable in revenue determination besides sales volume is product price. While both oil and natural gas are global commodities, oil is more transportable than gas in its standard state; natural gas prices are more likely to be determined by local markets. The United States, Europe and Canada have extensive distribution systems for natural gas from industrial to residential end users. Worldwide, natural gas is displacing coal for electricity generation since its emissions are far less harmful to the environment. In addition, in places where natural gas is plentiful, such as Australia, Trinidad, Qatar, Indonesia and the United States, operators are able to liquify the gas, known as LNG, and export it by tanker to gas-poor markets, like Japan, where the use of international pipelines is not feasible.

The oil and gas price markets follow basic supply and demand theory, which states that when the supply of a product exceeds the demand, sellers will drop the price until they create additional demand. Conversely, when there is a shortage of a product, sellers will increase the price to make more money until people stop buying. It is a cycle. Recently, oil supply increased in North America and was at an all-time high in 2019 as a result of tremendous drilling activity early in the decade spurred by high oil prices. On the other hand, the COVID-19 pandemic, which reduced demand tremendously in 2020, had a devastating effect on prices until restrictions began being lifted late in the year. Some semblance of balance seems to have been reached with oil prices in the $50 to $80 range late in 2021, however, supply and demand principles are the cornerstone of economics and global politics and natural occurrences can also have significant effects on the short-term price fluctuations of oil and natural gas.

Crude oil prices of the many crude oil streams produced globally tend to move together, although there are persistent differentials among grades depending on the quality. The realized prices are also reduced by the cost to get the product to market. Both crude oil and petroleum product prices can be affected by events that have the potential to disrupt the flow to market, including geopolitical and weather-related developments. These events may lead to actual disruptions or create uncertainty about future supply or demand, which can lead to higher volatility in prices. Much of the world's crude oil is located in regions that have had their oil production disrupted due to political events. Several major oil price shocks have resulted from supply disruptions triggered by political events, most notably the Arab Oil Embargo in 1973–74, the Iran-Iraq war in the late 1970s and the Persian Gulf War in 1990. Given the history of oil supply disruptions emanating from political events, market participants are always assessing the possibility of future disruptions and their potential impacts.

The Organization of Petroleum Exporting Countries (OPEC) is a cartel formed to establish oil-exporting policies and set prices. The original founding members, Iran, Iraq, Kuwait, Saudi Arabia and Venezuela in 1960 had control of a significant percentage of world oil production. "OPEC's objective is to co-ordinate and unify petroleum policies among Member Countries, in order to secure fair and stable prices for petroleum producers; an efficient, economic and regular supply of petroleum to consuming nations; and a fair return on capital to those investing in the industry" (OPEC). Today, there are 13 OPEC member countries that produce about 40% of the world's production and purport to control almost 80% of the world's oil reserves. Their actions in terms of controlling global supply have a definite impact upon the price.

Weather can also disrupt oil supply. Hurricanes in 2005, for example, shut down oil and natural gas production in the Gulf of Mexico as well as nearby refineries resulting in petroleum price increases in the region and throughout the distribution system as supplies dropped. Severely cold weather also can strain markets as producers attempt to supply enough of the product, such as heating oil and natural gas, to consumers in a short amount of time, resulting in higher prices. Other events, such as refinery outages or pipeline problems can restrict the flow of oil and products, driving up prices. However, the influence of these types of factors on oil prices tends to be relatively short-lived. Once the problem subsides and oil and product flows return to normal, prices usually return to previous levels.

The US Energy Information Administration produces an enormous amount of timely analytical and educational information regarding all forms of energy. Fig. 7 shows the historical spot prices of West Texas Intermediate grade (WTI) crude oil in nominal and real terms (inflation adjusted to April 2021 dollars).

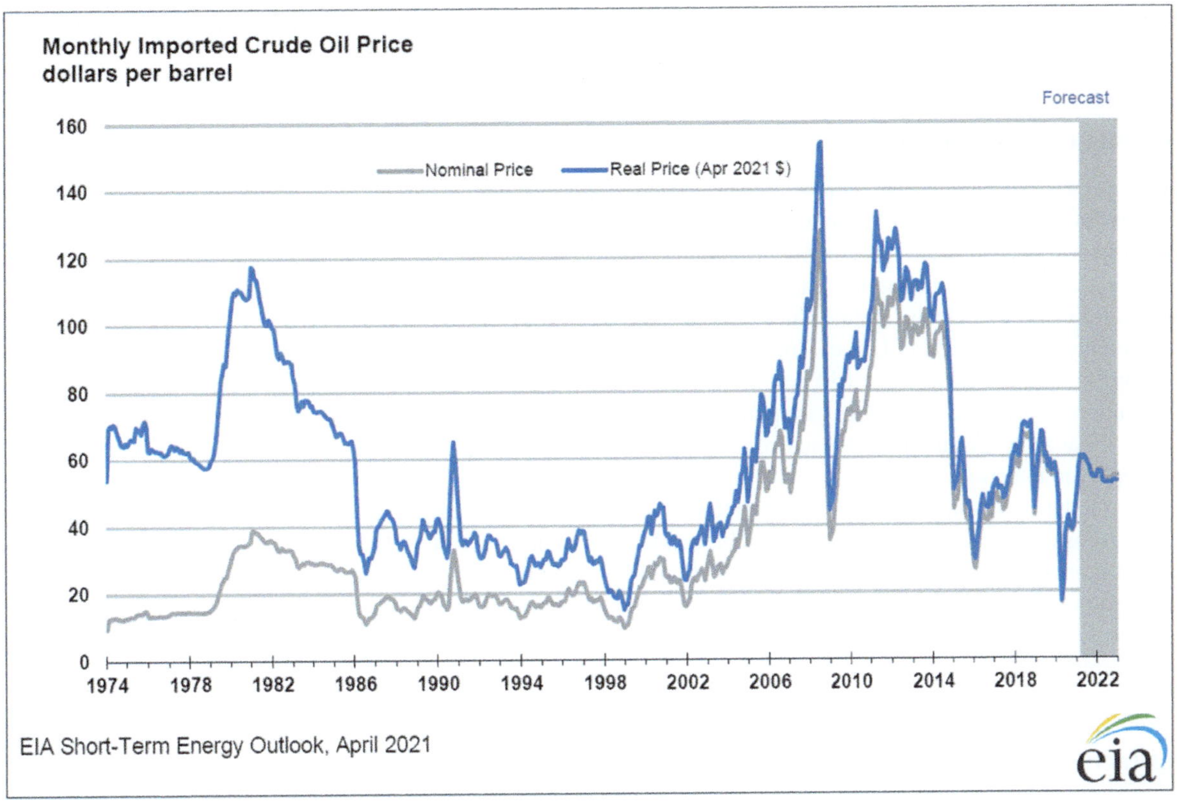


**FIG. 7** West Texas Intermediate crude oil price versus time. The Energy Information Agency publishes historical oil prices as both nominal (US dollars of the day) and real (adjusted for inflation to equivalent April 2021 prices). West Texas Intermediate is a reference price for crude oil trading. *(US Energy Information Administration (EIA), April 2021. What Drives Crude Oil Prices: Spot Prices.)*

In evaluating the potential economic benefit of a deepwater investment, not only must the volumes of oil and gas be predicted over time (production forecast), but the prices expected to be received also must be estimated. Noting the volatility of the historical prices in Fig. 7, it is difficult to see a predictable pattern. Since 1983, crude oils and since 1990, natural gas have been traded as commodities on the New York Mercantile Exchange (NYMEX). Contracts for futures are bought and sold, and give some indication of what the market believes these products will sell for years into the future. There are several crude oil futures traded, the most common in the United States being WTI, sold per stock tank barrel; for natural gas the most common is Henry Hub, sold per MMBtu. Brent crude, which originates in the North Sea, is also a benchmark traded on NYMEX. Both WTI and Brent are desirable light, sweet crudes; Brent generally trades at a slightly higher price than WTI, because it is easier to get to market, therefore less expensive to transport. These are considered benchmarks to which each specific project's expectations may be compared to estimate price differentials.

To estimate a reasonable forecast of product price to be received, the expected market must be assessed. In the United States, the crude market is well established and fairly predictable. In other more remote locations, there may be many market options to consider, and that is an assessment in itself. In deepwater, there are often issues of oil quality, which comprise viscosity compared to the benchmark and impurity content, such as sulfur, heavy metals and asphaltenes, which must be remediated before or during the refining process. These generally deduct from the price received. For natural gas, there may also be impurities, such as nitrogen and carbon dioxide, but there may also be enhancements, specifically higher Btu content that will increase the per-cubic-foot price. For both, transportation may be a major cost, and it is generally taken to be a deduction from the actual price received. Pipeline tariffs are levied by distance traveled and some deepwater projects are far from the market. Depending upon the proximity of existing infrastructure to the project, transportation cost may dictate the type of facility needed to optimize profitability. For example, the use of floating production systems may be chosen over laying long pipelines. These development options will be discussed later.

Revenue, or cash inflow, can be calculated by month or year by just multiplying the estimated sales volumes by their respective net product prices. Recognizing uncertainty is important in price forecasting as well. It is often assessed using a sensitivity analysis, which is calculating "what if" scenarios by varying prices and re-calculating profitability. Most companies use "floor" or expected minimum prices to see if a worst-case pricing scenario will still result in an acceptable profit.

But directional accuracy is more important than precision, because the objective is to compare opportunities in a portfolio at the same point in time and these will be done using the same global price assumptions.

## Operating costs support the continuing production operation

Operating costs are ongoing cash outlays needed to keep production going, so they are normally not assumed to be incurred before production begins. Prior to production, cash outlays are usually considered to be an investment. Ongoing expenses, while an important part of the investment decision, are not generally the largest consideration. These costs include labor, fuel, insurances, boat and helicopter transportation, which is a big fixed cost in deepwater even in a mature basin, living expenses for crew, chemicals, maintenance and periodic well work. In addition, any overhead caused by the project that would not otherwise be incurred, a new shore base for example, should be included as a cost in the evaluation. In the case of extensions to mature basins, there will probably be little overhead addition. However, in new areas, there could be significant additional costs associated with transportation and living expenses for rotating personnel from the home country to the location, for accounting and international contract specialists and perhaps even for in-country training for development of local employee expertise.

Many modern deepwater projects incorporate wells and facilities that can be operated remotely for increased efficiency and lower cost. These automated systems include extensive monitoring of wells and equipment, which promotes lower costs and more efficient and safer operations. It should be noted that while operating costs have a minor impact on cash flows in the early years of production, since they are small compared to revenue when production is high, they are ultimately the determining factor in commerciality in the late years of field life and usually determine when fields are abandoned or sold.

## Investments cause the project to begin (and continue)

The biggest part of the cash outlays for any project, especially in deepwater, is the capital investment. In addition to their size, the timing of the expenditures increases their impact relative to the revenue stream, because most of the investments occur before any product sales result in revenue. As such, the cycle time from discovery to first production has a large economic impact on profitability, especially on the rate of return and the NPV. Because this cycle time tends to be long for deepwater projects, often 5 to 10 years or more, many elements of the forecast made to support the investment decision are subject to change, including the investment and the production forecast, but especially the price of oil. As technology matures, standardization of components and their designs continues to have an important impact on cycle time.

Data to analyze the time from discovery to first production for existing developments is not readily available globally but is compiled for the Gulf of Mexico by the US Bureau of Ocean Energy Management (BOEM). The average cycle time is 7 years, with a high of 25 years and a low of 1 year for all 180 named deepwater fields as shown in Fig. 8A. It can be seen that there is no discernable trend when correlated to discovery year or to water depth, though logically that might be expected. Refining the data to only those fields where platforms were installed (eliminating the fields that exist entirely subsea) is a little more enlightening (Fig. 8B). There is a trend downward from the early 1980s, where cycle time was close to 10 years, to 2014, when the cycle time is less than 5 years. This might result from a technology learning curve extending through the 35-year period where improved designs "on the shelf" may have shortened both the design time and time to the FID. Or it could reflect the uptick in prices that occurred in the few years before 2014. On that same graph of cycle time vs discovery year, there are a number of high-cycle-time outliers in the mid-1980s that almost assuredly result from the plummet in oil prices during that time. This would logically have postponed moving a discovery to development due to poor economics. Also of note in that same Fig. 8B is the lack of correlation between cycle time and water depth, especially if the obvious outliers are eliminated. Logically, this reflects the fact that floating platforms, which are necessary beyond about 1500 ft. (457 m) of water, do not take measurably more time, because the vast majority of the effort in design and construction relate to the topsides and not to the mooring systems. While it was also of interest to examine whether the size of the operator affected the time to first production, the data available did not preserve the original operator's name, and many of the fields have changed hands since production began, so such an analysis was not meaningful.

Projects often benefit from good timing versus the cost and oil price cycles, whereas unlucky timing of the same can deteriorate a project's profitability. Ironically, the optimal entry point may be during price downturns, when costs are low and most companies reduce their capital budgets. On the flip side, relatively high oil or gas prices during the production phase have a great positive impact, but investing in that environment means higher costs as well. More often than not in deepwater, a field produces over a timespan that catches multiple price cycles. The long life of these projects requires a

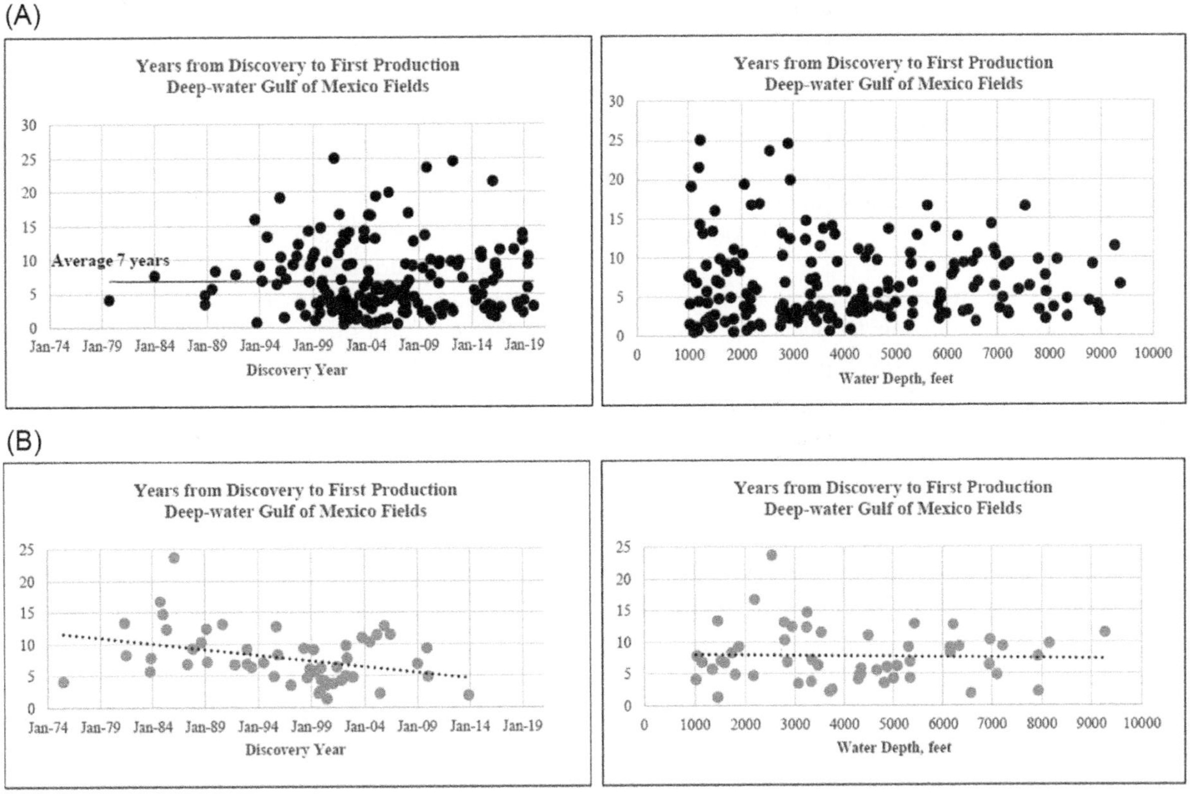


**FIG. 8** (A) Years from discovery well to first production, all Gulf of Mexico deepwater fields. Years from discovery well to first production, all Gulf of Mexico deepwater fields correlated against discovery year and water depth shows no relationship to either factor. Years from discovery to first production, Gulf of Mexico deepwater fields with platforms. Years from discovery well to first production, all Gulf of Mexico deepwater fields with platforms correlated against discovery year and water depth shows no relationship to water depth but a trend toward fewer years between discovery and first production. This may result from a learning curve, improved technology and growing infrastructure allowing pipeline connectivity. *(U.S. Bureau of Ocean Energy Management Data, 2021.)*

long-term investment philosophy. Fortunately, the investment decision process is taken in steps, so that the entry costs, while not small, are a manageable percentage of the cost of an entire development.

## Pre-wildcat exploration

Before even drilling a wildcat well, the rights to drill and the reason to drill must be in hand. In mature areas, such as the Gulf of Mexico and the North Sea, seismic data is probably already available for licensing or the rights to acquire it can be obtained easily. While the cost of licensing, processing and (perhaps) re-processing are not insignificant, they are small compared to the cost of drilling, especially drilling in the wrong place. In this situation, the first investment would be in a seismic program, possibly area-wide and not even attributed to a specific prospect but to general exploration overhead. Therefore, the initial investment in the project would be the acquisition of the lease or license in a competitive bidding round. In the United States federal offshore Outer Continental Shelf (OCS), these are called lease sales where certain areas are offered to companies under sealed bids with universal terms that include royalties (a percentage of gross production) payable to the government statutorily. The amount of money paid to the government upon leasing is called the *signature bonus*, which for federal offshore leases can be as low as $25 per acre and as high as the market will bear, sometimes in the hundreds of millions of dollars for a 5000-acre standard block. In other areas that use competitive bidding, the initial payment may be called a *licensing* fee. In many cases, there also may be annual rentals to maintain the rights, which also encourage timely action on the part of the lessee.

In some countries, especially in less mature and frontier areas where technical information may not be as available, a work program may be negotiated where the bidding or negotiating party proposes a pre-defined work program. This program may comprise a commitment to acquire seismic, complete a technical report and drill exploratory wells over an agreed period of time. In this situation, the pre-wildcat investment would comprise seismic acquisition and processing over the agreed time frame. In any case, the desire to reduce risk by employing seismic analysis can add years to the evaluation process, reducing the NPV of the discovery and the project's IRR.

## Deciding to drill the wildcat

Before deciding to spend the money on the initial well, an economic evaluation must be done to justify the investment. Why is the evaluation important? Because the successful wildcat can lead the way to billions of dollars in appraisal and development investment (see Chapter 1 for more information on the exploration process). Each well can cost over $100 million, more in very deep formations and much more if trouble is encountered while drilling. Why drill it if there is not enough potential to justify the go-forward plan? Before drilling, there are two possible outcomes associated with the decision to drill: (1) the wildcat is a dry hole technically or commercially and (2) the wildcat finds an accumulation greater than the threshold required for commercial development. The economic benefit inclusive of risk must exceed the minimum criteria of the investor in order to proceed. That entails a credible evaluation of each of those alternative outcomes: (1) the loss of the dry hole cost and (2) the expected value of the success including all subsequent appraisal, development and production cash flows until field abandonment. Practically, because there is limited information, those estimates are more uncertain than they will be later in the process but there still can be a realistic estimate of what will happen given success.

## Project chance of success: Exploration risk mitigation

An exploration program is risky, only about 25% of deepwater wildcats have been successful in recent years (Rystad Energy, 2021). And most new projects require an appraisal program to determine commercial viability prior to project sanction (FID). The exceptions to this are infrastructure-led exploration (ILX) wildcats, which are smaller prospects offset to an existing production facility. Sometimes, the wildcat itself can be completed and connected directly to that facility in a relatively short amount of time.

Reducing technical risk is essential to making sound economic decisions. Typically, exploration technical risk (the chance of technical success) for a prospect is determined by estimating individual "geologic chance factors," which are multiplied together to get an overall chance of success. These chance factors are the elements that must occur geologically to result in the presence of producible hydrocarbons, namely source rock presence, maturity, migration, trap, seal and timing. Various companies use different names for these elements and some include other components. It is very important to be internally consistent when making investment decisions among the opportunities available to an organization; peer review is commonly used to achieve this. The general process is the same in deepwater as other exploration arenas, however, several technology advancements have had a dramatic impact on the quality of risk assessment in deepwater.

The term "uncertainty" generally refers to the range of possible outcomes of resource size, flow rates and even economic parameters such as price. If risk means the likelihood of hydrocarbons existing in a specific location, uncertainty means the assessment of the range of the volume of those hydrocarbons in that location. Advancements in geophysical technology have played a key role in reducing both risk and uncertainty in deepwater exploration (see Fig. 3). The digital revolution in seismic recording and processing of the 1960s preceded the drilling and development in deepwater by more than a decade. In the early 1970s, geophysicists became aware of Direct Hydrocarbon Indicators (DHIs) and how seismic amplitude information was facilitating the direct detection of hydrocarbons. Because DHI technology zeroes in on fluid prediction, they were able to greatly de-risk charge (the result of having mature source rock, with hydrocarbon generation and migration occurring once a trap is in place). Because the seismic amplitude response is partially a result of reservoir porosity, DHIs can often help reduce the risk associated with reservoir quality. For exploration wells with DHIs, the global success rate is about 50% (Rudolph, 2001; Roden et al., 2012), about double the global success rate for deepwater overall. It is important to develop a calibrated DHI if possible, because false DHIs exist. The main four failure causes are: (1) a thick, wet sand, (2) low-saturation gas (LSG), (3) exotic lithologies and (4) non-hydrocarbon gases (Forrest et al., 2010; Simm and Bacon, 2014).

In addition, the publication of the seminal AAPG Memoir 26 in 1977 revealed the principles of seismic stratigraphy and how depositional sequences could be interpreted from seismic data. At about the same time, geoscientists learned to link seismic signatures ("seismic facies") to some lithologies and stratigraphic architectures. For deepwater exploration, this led to much better understanding of the chance of reservoir and seal presence (and hence the associated risks).

The biggest impact on modern geophysics, specifically on deepwater exploration, has been the evolution of 3D seismic technology including acquisition, processing and interpretation, including the development of interpretation workstations (see Chapters 1–5, 9, 16–19 for more on this topic). Even though initial 3D survey acquisition began in the 1970s, 3D surveys did not become practical until the mid-1980s when computer workstations enabled the efficient interpretation of large amounts of data. Within 10 years of the development of computer interpretation workstations, 3D surveys became the tool of choice used for exploration and development around the world. There have been many improvements in 3D seismic acquisition with the constant drive to acquire more offset angles, higher density of receivers, more channels, more frequencies (broadband) and more azimuths with greater efficiency (Monk, 2020). Early in the deepwater play history, 3D seismic acquisition improved from longer cable lengths (offset) and tighter line spacing. This improvement in 3D imaging made seismic maps much more reliable, which in turn had the effect of greatly reducing the trap geometry risk, including faulting and juxtaposition issues that affected both trap and seal risk. It also resulted in better prediction of reservoir and seal presence, hence lowering risk.

The 1980s also saw the introduction of Amplitude versus Offset (AVO) technology, where changes in amplitude with offset on pre-stack seismic data provided insights on fluid effects (see Chapters 1 and 16). In conjunction with acquisition and the evolution of AVO technology have been improvements in seismic processing as a result of growth in computer power. Because AVO technology constitutes a significant component of modern DHI workflows, it contributes to the risk reduction provided by the presence of DHIs (Simm and Bacon, 2014; Roden et al., 2012—see Chapter 16).

In the latter half of the 1990s, pre-stack time migration algorithms evolved, followed a few years later by pre-stack depth migration. These advanced migration approaches have greatly improved seismic imaging with pre-stack depth migration necessary in subsalt settings. Full waveform inversion (FWI) has enabled more accurate delineation of velocity wavefields, and broadband processing techniques have increased the bandwidth to improve resolution. Today's computer infrastructure not only allows interpreters to handle vast amounts of data in a short timeframe, but also enables the implementation of advanced algorithms only dreamed about a few decades ago.

A major challenge in seismic imaging occurs in basins with extensive salt features, both autochthonous and allochthonous (read more in Chapter 5). For example, shallow, thick allochthonous salt structures with irregular tops and bases in the northern Gulf of Mexico present a challenge for seismic imaging. Once some sub-salt discoveries were made, there was renewed interest in being able to accurately image below the salt. Routine post-stack migration processing was routinely accompanied by pre-stack depth migration (see the seismic data examples in Chapter 19). Later, seismic acquisition improved significantly with the development of Wide Azimuth (WAZ) techniques, and more recently, the implementation of Ocean Bottom Node (OBN) seismic, both of which have produced dramatic improvements in seismic imaging, particularly below salt. WAZ is a seismic acquisition technique that employs a wide distribution of source-receive azimuths, usually involving multiple source boats. OBN consists of placing an array of seismic receivers on the ocean bottom, which offers not only significant data improvement, but also the ability to re-acquire the 3D seismic at a later time to monitor reservoir changes in seismic response of the reservoir through time as it depletes ("4D"). A major improvement in imaging geologically complex areas such as sub-salt was achieved with Full-waveform Inversion (FWI), an advanced seismic processing methodology that determines high-resolution, high-fidelity velocity models of the subsurface. These techniques have resulted in improvements in signal-to-noise ratio (particularly sub-salt), reservoir illumination, seismic migration processing and seismic velocity modeling for structural imaging (see Chapter 1 for more information).

These advances in seismic acquisition and processing further improved the aforementioned understanding of trap risk from 3D seismic and, in some cases, more accurate assessment of DHIs. In addition to the impact on the chance of geologic or technical success for deepwater wells, these advances improved the ability to assess prospect volumetric estimates by giving more insight to reservoir architecture and geometry. In addition to improving success-case economics, they help provide a better understanding of the chance of economic success, since deepwater development success is dependent on achieving minimum threshold volumes (minimum economic field size).

## Prospect maturation and associated costs

In light of the economic drivers related to the cost of exploration drilling, the associated dry hole risk and the uncertainty in prospect size, the ability to reduce both the risk and uncertainty is critical in arriving at a "drill or drop" decision. This generally involves the collection and analysis of geological, geophysical, petrophysical and reservoir engineering data. Much of this is described in the prior section on exploration risk reduction. The main goals of prospect technical maturation are to assure a reasonable level of confidence in the assessment of chance of success such that the risked economics are acceptable to the corporation and to achieve confidence in the range of success-case volumetric outcomes expected. For this latter objective, it is critical that there is a high chance that a discovery size will exceed the minimum required for economic development, while at the same time determining that the expected volumes provide an acceptable reward for the risks being taken ("risk/reward"). In addition, the "upside" case and its associated probability of occurring is often taken into consideration, such as unusually large discovery sizes that could be "company-makers" and/or could be stand-alone successes that could establish production and infrastructure in an unproven frontier play, thus making future exploration in the area much more attractive.

The capital and staffs costs associated with this technical maturation must be considered in the pre-drill economics, as must the time required to perform it. For instance, seismic acquisition and processing required for technical maturation of some exploration prospects can exceed $20 million and take up to 2 years to complete. Geoscience studies and the associated data may incur costs exceeding 1 million dollars and may increase the time required to reach a decision. In addition to these costs in technical maturation, there may be acreage rentals and other ongoing overhead costs, also called general and administrative (G&A) costs, and these must be considered in the pre-drill economics that lead to the decision to drill or drop.

## Drilling technology and associated costs

The methods used to drill and develop fields in deepwater are extremely involved and represent some of the most complex construction technology developed in any industry (for more information, see Chapters 1 and 19). Being able to drill safely is fraught with many engineering challenges. Rigs capable of drilling in deepwater are collectively termed "floaters" and are kept in position by mooring lines to the ocean floor or by dynamic positioning (using thrust from an array of engines). The first drill ship used in deepwater drilling was developed in association with the Deep Sea Drilling Project launched in 1968, and the first exploration drilling ship was developed in the mid-1970s. Since then, six generations of drill ships have been developed by the drilling industry, with each new generation offering advancements from the prior generation. The deepest water depth achieved as of 2021 for an exploration well is 3628 m (11,900 ft.) offshore Angola (Fig. 2).

The cost of drilling wells with these rigs is a major factor in deepwater economics. Daily costs comprise the rental rate of the rig itself and all the services associated with the drilling operation. Day rates are largely driven by market conditions (supply vs demand) and have ranged from less than $100,000 to more than $500,000 (2013–14). The "spread rate" is the total cost to operate the rig per day, which includes drilling mud and management, downhole equipment, services, logging, cementing, standby vessels and associated staffing costs. Typically, in deepwater the spread rate is double the day rate. It is not uncommon for drilling to take 6 months or more per well, with total costs sometimes exceeding $200 million for both exploration and production wells. For development drilling from a platform, a large platform rig utilizing dry (surface) wellheads and flow control valves (customarily called Christmas trees) may be used at considerably less cost.

Onshore and in shallow water, it is common to preserve a successful wildcat for future re-entry and production; in deepwater, they are usually expendable wells because of the much larger cost to drill a well capable of being used as a producer. Such a well needs larger casing and a bigger hole, which means larger bits, and so on, including more mud, more rig time and more services. The wildcat, even when successful, sometimes will be treated as a dry hole, because the cost will have been spent in order to gain information of great value to the future of the project. It is designed to gain the most information for the least cost ("fail cheap"), because there is usually a high probability it physically will be a dry hole. This low-cost design may mean a smaller hole than would be needed for a producing completion and/or a shallower hole than needed to evaluate all possible horizons if the main pay needed to justify development is not that deep. These design choices would mean fewer bits, fewer rig days and smaller pipe, therefore lower well cost. A common exception to this is ILX, where the probability of success of a potential tieback well to an existing facility is high and the extra costs may be worth the higher chance of saving both time and money in the event of success. Some people call these wells "develocats."

## The appraisal decision

Given a successful wildcat, which confirms the existence of hydrocarbons, there is generally a plan in place to drill one or more additional wells to appraise and calibrate the extent and quality of the discovery. This confirmation is important to the evaluation of future costs and production that will help justify making the FID or project sanction. The actual investment type and amount will depend upon this information, including the number and clustering method of wells and completions, the size of platform and processing equipment and the size of flowlines and pipelines or alternative transportation methods. It will also allow a more refined estimate of production over time, which will indicate the revenues that are expected to be realized. It is usual to construct a decision model that includes values and discrete probabilities of occurrence for various outcomes. Decision analysis, developed at Stanford University in the 1960s, is a field of study all its own. A decision tree is one of the tools often used to organize the model, and a simple example is shown in Fig. 9. Implied after the second appraisal well is a continuing sequence until enough information is known to make the FID. Actual complex models have been known to cover walls before the final decision is made.

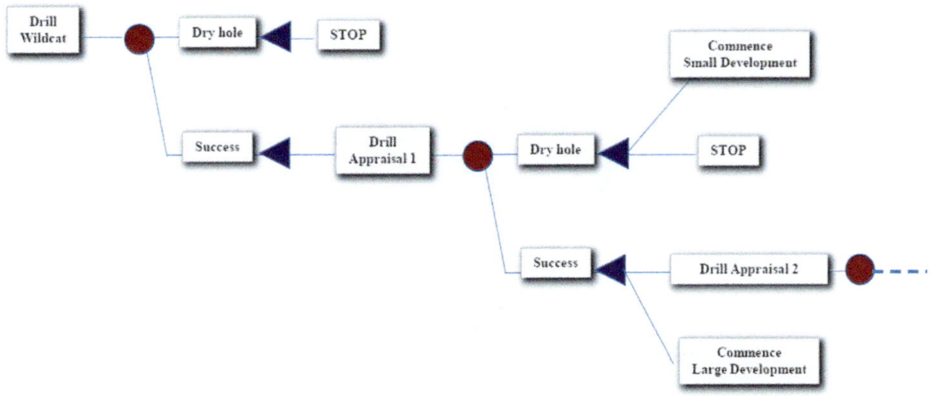


**FIG. 9** Example appraisal and development decision tree. Evaluating the possible outcomes of a major deepwater project can be very complex, requiring dry hole and discovery probability estimates and ranges of outcomes of hydrocarbon volumes at each step of the way. A decision tree is an important tool used by economic analysts to keep the evaluation organized. At each decision point (triangles), a full economic analysis must be done, including revenues from sale of oil and gas, costs of operating and transporting, and the investment required. The economic benefits are measured against other investment opportunities before the final investment decision (FID) is made.

After each well, wildcat, appraisal or development, the economic evaluation of the prospect will need to be refined using the information gained. Volumes, costs, investments, even prices may all change with each piece of data and when it is acquired. And usually, the risks will be better defined as more knowledge reduces uncertainty. The appraisal program itself needs thoughtful design, because gathering information about the size of the prospect is very expensive. Each well drilled should provide not only useful information but also reduce the uncertainty about key characteristics that define the extent of a possible development. It does take a team of many members with collective industry knowledge and experience to contribute to the evaluation.

## The final investment decision: Development

Even while appraisal drilling is ongoing, the concept definition or pre-FEED phase of the evaluation may begin. In the subsurface, reservoir models, both static and dynamic, integrate the new information, moving to a higher level of complexity. Above the mudline (or seabed), the new insights begin to drive toward an optimal solution. The choice of development concept can vary dramatically depending on the nature of the project. Regardless of scale, the activities in this phase can include process scheme optimization, studies of particular aspects of design (technology selection), finalizing concepts of field architecture, preliminary piping and instrumentation diagrams (P&ID) that describe the production process, enquiries for long-lead items, cost estimating and economic assessment, project schedule and development of risk mitigation plans. Concurrently, the number and depths of wells and types of completions and their forecast producing rates over time are incorporated into the FEED design (Genesis, 2020).

*The major categories of options at this time include the following:*

## Well spacing and field layout

Well layout begins with the reservoir configuration, which dictates where the well drainage or "take" points must be to optimize oil and gas recovery. Usually, this requires at least one proposed completion from each commercial reservoir that is identified or expected. Sometimes there will be anticipated water or gas injection wells used to maintain pressure and improve recovery. From those bottomhole locations, various schemes will be evaluated to work back toward the best surface (or seabed) starting points. The project architecture may include surface drilled wells from a platform, individual subsea well locations, subsea well clusters connected together and then tied back to a platform or combinations of the above. The wet wellheads and trees (valves) are at the seafloor; the wells are drilled by floating rigs and completed using subsea trees. This allows one strategically placed production facility to service many wells over a reasonably large field area, such as the Na Kika development in the Gulf of Mexico (see Chapter 19). Subsea systems comprise the well system (includes the downhole completion system and the subsea tree), the production system (includes protective structures, manifolds, templates, intervention systems and subsea processing systems) and the pipeline system (includes tie-ins, umbilicals, risers, injection pipelines and production pipelines).

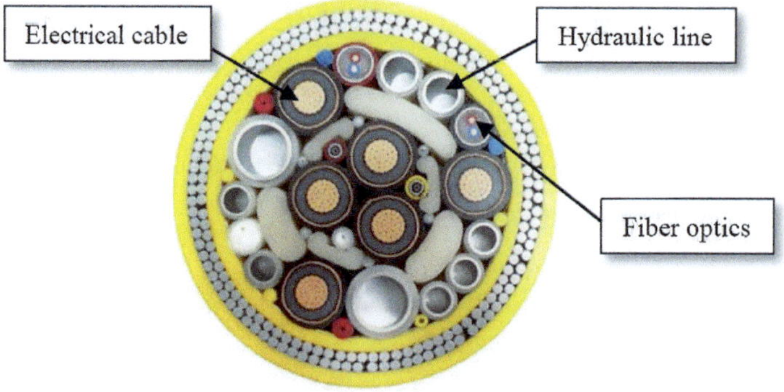


**FIG. 10** Umbilical bundle. Cross section of the Jack/St. Malo umbilical for Chevron; 2 umbilicals, totaling 42 km, 450 km south of New Orleans, Louisiana, at 2100 m water depth. *(Image courtesy of Nexans. Vittone, M., 2015. Monster Cables: Connecting the Subsea Factory. gCaptain.com.)*

Subsea developments have been made possible by technologies such as subsea trees, risers and umbilical lines. An umbilical is an assembly of hydraulic hoses, electrical cables and optical fibers used to remotely control subsea equipment at both the wellhead and downhole. It is an essential part of every subsea production system, without which sustained economical subsea petroleum production is not possible. The various connections via umbilical control the opening and closing of wellhead valves, transmit pressure, temperature and other information from the bottom of wells, and can even switch completion intervals. The umbilical is often run concurrent with, but is separate from, the flowlines that transport the produced fluids from the well to the production equipment. Fig. 10 shows a cross-section of an umbilical bundle (Vittone, 2015).

The production equipment may be located on the seafloor rather than on a platform. Subsea processing, originally conceived as a way to overcome the challenges of extremely deepwater, also has become a viable solution for fields located in harsh conditions where processing equipment on the surface might be at risk. Additionally, subsea processing is an emergent application to increase production from mature or marginal fields by providing pressure boosting equipment nearer to the bottom of the wells.

The deployment of subsea equipment requires robotic equipment, which must be supported by specialized, and expensive, vessels. Repairs or intervention with installed subsea equipment is therefore very expensive. This type of expense can result in late-life economic failure of a well or even the entire subsea development.

## Completion technology

Further to the discussion in Chapter 19, another critical technology and cost driver is the completion technology used to develop the reservoirs. During the 1980s and 1990s, deepwater discoveries were primarily in Oligocene and younger reservoirs in the Gulf of Mexico, Brazil and West Africa. These reservoirs had relatively low temperatures, with little to no diagenesis, and had limited reservoir consolidation. Developing these reservoirs was an engineering challenge, because formation sand would collapse and flow into the borehole even at low production rates, causing plugged perforations and the reservoirs to stop producing pre-maturely. Although gravel pack technology was commonly used in shallower sands in the 1970s, by the mid-1990s, the technology of combined hydraulic fracturing and gravel packing (frack-packs) was developed. This allowed these reservoirs to be produced at much higher rates while maintaining completion integrity. As a direct consequence, the high-rate, high-ultimate-recovery wells can be produced at rates up to 50,000 to 60,000 barrels per day from these very thick and permeable unconsolidated sands (see Fig. 5). A well-placed, high-rate completion in a reservoir can sometimes eliminate the need for multiple wells to deplete the same reserves over a similar period of time.

Intelligent completions are another advance in deepwater technology. They incorporate permanent downhole sensors and surface-controlled downhole flow control valves to enable the monitoring, evaluation and active management of production (or injection) in real time without mobilizing a rig. Data is transmitted to surface for local or remote monitoring in a digital well platform via the umbilical connections discussed above (Schlumberger, 2021). The information provided allows management of rates for optimum reservoir recovery. Initially used in deepwater wells where rig intervention is expensive and high-risk, intelligent completions since have proven their value in managing production from multilateral wells, horizontal wells with multiple zones, wells in heterogeneous reservoirs and wells in mature reservoirs.

Completion design is a major part of the development planning for the FID. This overall plan feeds into the production forecast, which in turn influences the design of the facility for processing production as well as the platform that supports the facility and the transportation mechanism for getting oil and gas to market.

## Surface platform and facility

Moving closer to market from the subsea architecture, the production must flow to market. Usually, the development is far enough from shore to require a platform to support the required production on site. Stepping out from the limits of the fixed platform, there are various alternatives (Fig. 2). The tension-leg platform began being researched in the early 1970s and came into practical use by Conoco in the North Sea in the 1980s. Moored floating platforms, including the massive Thunder Horse semi-submersible in the US Gulf of Mexico, are active options today as are Single Point Anchor Reservoir (SPAR) designs, which some consider to be the most stable; resembling a bobbing fishing cork. Finally, a floating production storage and offloading (FPSO) unit is a floating vessel used by the offshore oil and gas industry for the production and processing of hydrocarbons and for the storage of oil. Rather than requiring a pipeline outlet to get to market, the FPSO is designed to receive hydrocarbons, process them and store oil until it can be offloaded onto a tanker for transporting to market. The first FPSO was built in 1977 from a converted oil tanker and used by Shell in the Spanish Mediterranean Sea. The progression of the technology is shown in Fig. 2. Any one of these options is applicable today and should be considered in the pre-FEED. The choices to consider depend upon the water depth, the size of the discovery and well productivity (which will determine the size of the platform and equipment) and the distance from and terrain separating the project from market (which could determine whether a pipeline system or FPSO is more effective).

The FPSO is by far the most common approach outside the US Gulf of Mexico, where using the established pipeline network has been more economically attractive. One reason to favor the FPSO is that ample deck space allows high throughputs that cannot be handled by most platforms. In addition, produced oil never leaves custody of the producer, reducing the possibility of spills, amid other issues. The first US FPSO, in the Cascade/Chinook field, began producing in 2012 after overcoming a series of regulatory and economic hurdles. When using an FPSO, the natural gas produced while processing the oil must be addressed as part of the development plan. Depending upon the volume of gas, the economics and the regulatory environment, options include a gas pipeline to shore, flaring, re-injection or liquefying the gas for transportation to other markets.

Since the platforms and equipment are usually the largest part of the overall investment, a considerable amount of time and study are expended to evaluate alternatives, their costs and their benefits. Because an enormous amount of investment is involved even in smaller deepwater developments, the time spent in evaluating tens of alternative schemes involving a varying number of wells and completions, platform types, facility sizes, pipeline alternatives, even joint ventures and more will result in the most effective investment. Innovative business models have been pursued to lower the costs, such as third-party financing of facilities including leased FPSOs, shared pipelines and so on.

None of these investments is without risk. Many platforms have been damaged or lost due to hurricanes and cyclones, and some sabotaged politically. The loss is great in terms of both direct investment and lost or delayed production. During the devastating hurricane season in the Gulf of Mexico in 2005, the Typhoon platform was completely upended, and the Thunder Horse platform was tilted and endangered, though ultimately saved and began producing in 2008. The good news from that season and subsequent ones is that more platforms in deepwater are now equipped to measure wind speed and wave heights. The bad news was that wave heights and wind strengths were much higher than ever expected and designed for. However, deepwater platforms have proven to be very robust in their design.

Finally, floating platforms are moveable; while it has not yet been done except with FPSOs (more in Chapters 1 and 19), they may be able to be reconfigured for use at alternative locations so may have value even after field depletion. This salvage or reinvented value should be included in the FID evaluation, but because it will occur far into the future, its present value is generally not significant and may in practice be ignored.

## Finally, payment for the rights

Globally, the minerals that underlie the surface of the land or water are considered to belong to someone. In a few places, namely Canada and the United States, these rights belong to people who own the land even though they may be able to sell those rights separately from the surface rights. In the rest of the world, and indeed under waters or lands claimed by the state, provincial and federal governments in the United States and Canada, the minerals are considered to belong to the people, and as such are controlled and regulated by their governments. Consequently, when seeking the right to drill for oil and gas, the host government agencies are the other parties to any agreements consummated.

The general approaches to deepwater acreage acquisition have evolved in different countries, but there are two main methods: direct negotiation and competitive bidding. Four general kinds of bidding types have been employed.

**Sealed Bids**, for which there are various parameters that may be bid on, used individually or in combination:

**Signature bonus**, which is the cash payment immediately upon signing the agreement (lease), also has additional terms and laws which have associated costs. In some cases, the signature bonus is the sole criterion; there is no work program, and the highest acceptable bid from a qualified party wins the opportunity.

**Work program**, where the bidding party tenders a defined work program that includes acquiring and processing seismic, drilling a certain number of wells, etc. This type is common with Production Sharing Contracts, discussed below.

**Contract terms**, where the biddable parameters are set and comprise a combination of contract variables (royalty rate, production share, cost recovery, signature bonus, etc.).

**Open gazettal**, which may be used in defined bidding rounds or as an open process with no bidding rounds. Bids may be submitted in accordance with a defined combination of biddable parameters (just as described above for sealed bids), but the bid is then published for 30 to 90 days, during which other companies can submit competing bids.

Regardless of the lease acquisition method, any initial payment, such as the signature bonus, is also considered an investment and should be part of the economic evaluation prior to signing.

Once the rights are acquired, the producer (or producer group) is subject to the fiscal terms associated with the lease or contract. Each oil producing nation has its own set of terms under which it operates, collectively called its "fiscal regime." These contract terms vary widely and can change over time. Fig. 11 shows the basic types of petroleum contracts that determine fiscal aspects. Tax rates, government take and fixed product prices are just some of the variables in the contract terms, and for production sharing contracts, cost recovery and profit split are additional drivers. The fiscal regimes of nations fall into three general categories: royalty-income tax, production sharing agreements and service contracts. In each of these, both parties' interests are usually aligned, so that the more efficient the operation and the more value is extracted, the more benefit both realize.

**Royalty-income tax** is the system used by many mature oil-producing nations, including the United States, Canada, Australia, the United Kingdom and Norway. Royalty is a percentage of the gross production removed and sold from the contract area. It may be paid in kind or in value as sold. Income taxes are paid according to the respective laws of the nations themselves. Even though most of this text is about exploration and development in deepwater itself, deepwater depositional systems also occur where dry land is today. Especially in the United States, onshore royalty and statutory taxes additionally exist, so a short explanation of that is included here.

Using the United States onshore as an example, the initial lease is negotiated with the mineral owner, usually an individual but sometimes a company or other organization, a state or local government (usually under waterways) or the federal government where national forests or other reserved lands exist. Both the signature bonus, usually expressed in dollars-per-acre, and royalty percentage are negotiable, as are the other terms of the lease, but there are standard forms that most people use, most

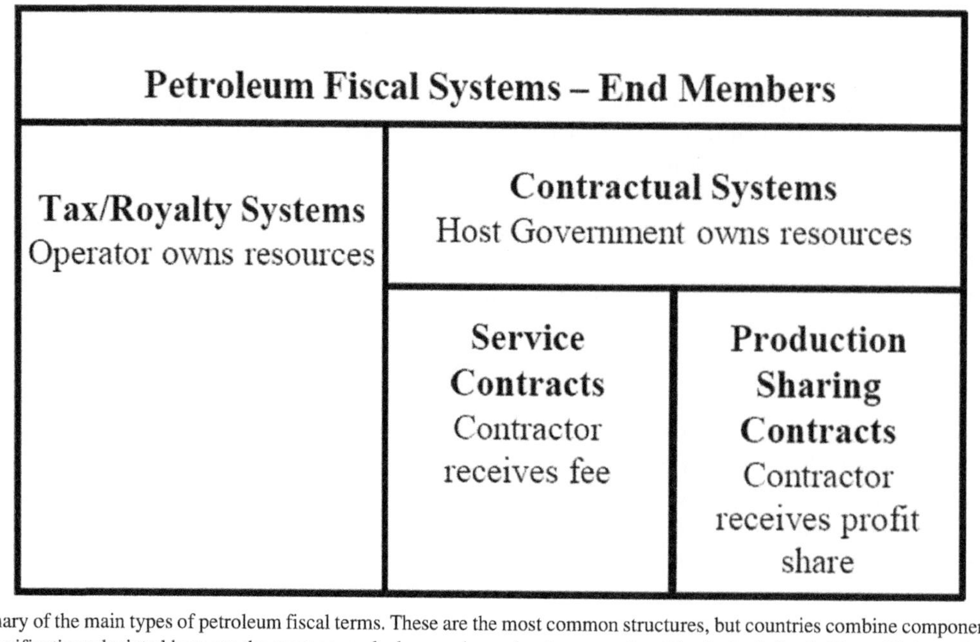


**FIG. 11** Summary of the main types of petroleum fiscal terms. These are the most common structures, but countries combine components of the various systems. The classifications depicted here are the most general where end-member types can be objectively defined. *(From Weimer, P., Pettengill, H.S., Roden, R., 2022. The history of deepwater exploration, part 3. AAPG Explorer (January), 16–19.)*

frequently those published by the American Association of Professional Landmen. In addition to the federal income taxes that are established by law for all entities in the country, there may be significant state and local taxes that must be included in the evaluation for investment decisions. Each state's laws are unique, but there are certain common elements. Severance taxes are levied by the states for removing the common resources of the people. These taxes are usually assessed as a percentage of the gross sales, but are sometimes assessed as a fixed amount per barrel of oil or MMBtu of gas. It is also common practice for the royalty owner to pay their share of those taxes. Also, counties and localities may assess property taxes, called ad valorem taxes. These are assessed as a percentage of assets, in some places physical assets, in some places reserves. And, some states also levy income taxes. Any of these may change over time with changes in the laws, and in some way, all must be accounted for in an evaluation. State and local taxes are not assessed on federal offshore leases in the United States.

In the United States onshore, it is customary to refer to the gross, or total, amount of production, reserves and even costs as the "8/8ths" amount. This is also the total working interest of all owners who are lessees of the mineral rights. This phraseology probably evolved to distinguish the royalty amount (historic custom of 1/8 of production) from the total produced. The net revenue interest is equal to the total minus the royalty, which would be 7/8ths or 87.5% on those vintage leases. In the United States federal offshore, there are competitive lease sales by offshore area usually held at least once a year, though that schedule changes with changing government policies and industry interest. The only element of the bidding is the signature bonus; the bids are sealed and opened publicly, and the highest qualified bid wins unless the bids are deemed too low when compared to the assessed value of the lease. The Bureau of Offshore Energy Management (BOEM) is the government agency that determines the minimum acceptable bonus for each tract, most of which are usually 3 miles by 3 miles or about 5000 acres. The terms of the leases and the royalty rates are set by the government, in 2021, royalties were 18.75% for deepwater tracts. The same income tax laws apply, which generally allow deductions from revenues for expenses when spent and for investments over time according to the rules for depreciation, depletion and amortization. Under a royalty-income tax legal framework, reserves still in the ground belong to the company and are an asset on its books.

**Production sharing contracts** are used by many nations, including Malaysia, Indonesia, Nigeria, Qatar and Vietnam. Production sharing agreements were first used in Bolivia in the early 1950s, although the first implementation similar to today's structure occurred in Indonesia in the 1960s. In production sharing agreements, the host government awards the execution of exploration and production activities to an oil company. The oil company bears the financial risk of the initiative and explores, develops and ultimately produces the field. When successful, the company is permitted to use the money from produced oil to recover capital and operating expenditures, known as "cost oil." The remaining money is known as "profit oil," and is split between the government and the company according to agreed percentages. The key concept is that the host government owns the reserves and reimburses the operator, so national assets are never owned by the foreign entity. Further, the government reimburses the capital costs needed for development and ultimately owns the fixed assets deployed, including the wells. In most production sharing agreements, changes in global oil prices affect the company's share of production, which adds to company risk. Production sharing agreements are beneficial to countries that lack the expertise and/or capital to develop their resources and wish to attract foreign companies to do so. They can be very profitable agreements for the oil companies involved, but they involve considerable risk, because of the additional capital spent by the contractor before production commences.

There may be other provisions that affect the economic evaluation of the opportunity, and some of the production sharing agreements has complicating factors, such as a product set-aside for local use, special petroleum taxes and so on.

**Service contracts** are agreements whereby the oil company, also referred to as the contractor, agrees to bear all capital and services risk for the exploration and development. In return, if exploration efforts are successful and oil is produced, the government allows the contractor to recover costs through the sale of the oil or gas. In addition, the contractor is paid a fee per barrel of oil produced thereafter for their risk and expertise, though it also may have provisions permitting the company to buy back an amount of crude at international prices from established production. The fee is often subject to taxes.

A National Oil Company (NOC) or other government entity is by law the sole title holder of the area under the agreement, so all petroleum deposits and oil and/or gas produced are their property at the well head. The commercial oil company acts as a contractor for the government entity, and as such carries out on their behalf all operations necessary for the exploration and development of oil deposits. The contractor is solely responsible for providing all necessary funds at their own risk. This means that unless oil is found in commercial quantities, the contractor will not be reimbursed for the expenses incurred in the unsuccessful search for oil. However, if oil is found in commercial quantities, the contractor puts the discovery on production. Thereafter, it may be operated either by the government entity or by the contractor. The cost incurred up to the time of discovery is considered as a loan to be debited to the government entity's account. This loan is repaid with interest and risk fees (Bansal, 2017). In this situation, there may be a legal question as to whether the investing companies are able to book associated reserves for public reporting depending upon the language used in the contract.

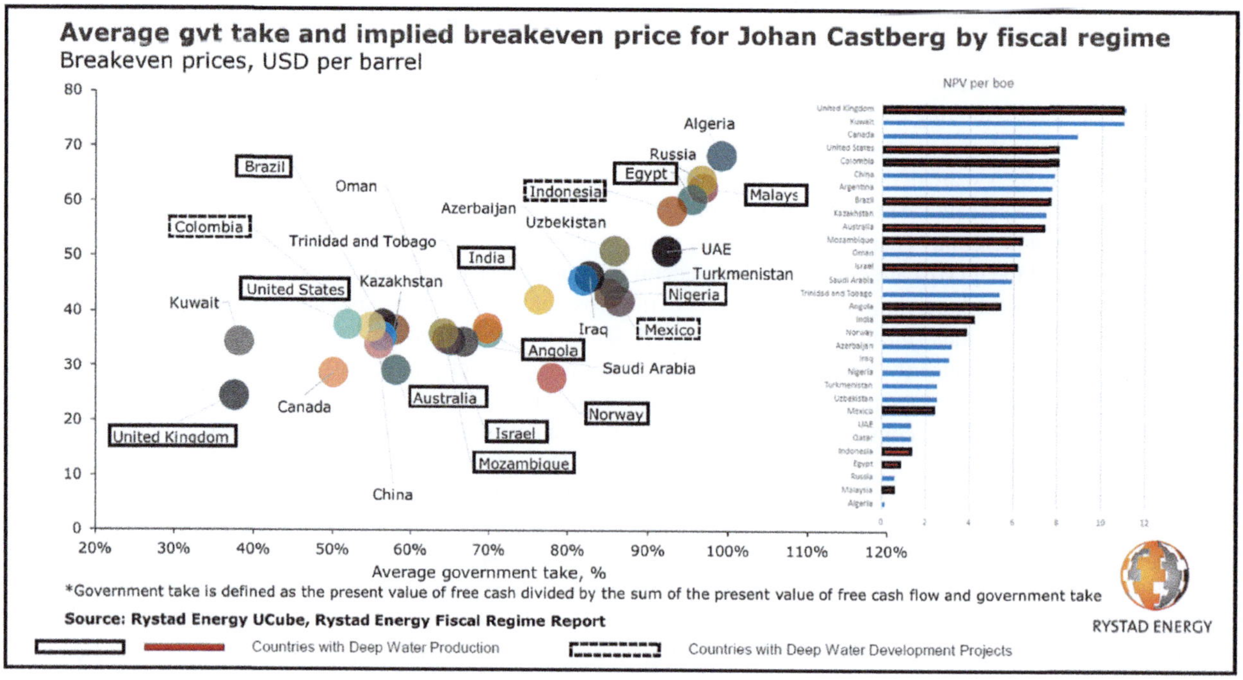


**FIG. 12** Government take, breakeven prices and NPV/boe by country. **Government take, breakeven prices and NPV/boe for offshore areas in various countries. The comparisons are calculated using a shallow water reference field from the Barents Sea, Norway and show how the same project would fare in various locations under those countries' fiscal terms as they currently exist.**Government take, breakeven prices and NPV/boe for offshore areas in various countries. The comparisons are calculated using a shallow water reference field from the Barents Sea, Norway and show how the same project would fare in various locations under those countries' fiscal terms as they currently exist. *(Modified from Rystad Energy, March 2021. Wildcat Safari Running Dry: Onshore Success Rate Dips to Lowest on Record, Falls for Fourth Year in a Row, www.rystadenergy.com/newsevents/news/press-releases.)*

In the end, the fiscal regime has a significant effect on the potential economic viability of a project. Sometimes, this can be a reason for negotiating terms, especially in the event of a marginal discovery that is identified between the exploration and production phases. Countries adjust their fiscal regimes over time in order to attract investors, and investors respond by exploring in areas that not only have resource potential but also have attractive potential returns. For example, "Guyana initiated a new top-quartile fiscal regime to encourage exploration. After commercial production begins, the Licensee is allowed to recover all capital and operating costs from 'cost oil,' which for the first three years is up to 75% of production and thereafter up to 65% of production. The Licensee's share of the remaining production or 'profit oil,' for the first five years is 50% of the first 40,000 barrels of oil/day and 47% for additional production; and thereafter 45% in full satisfaction of all income taxes and royalties. Using full-cycle economics, including recovery of capital and operating costs, the Licensee's share of gross revenue is estimated to be 63%" (CGX, 2021).

When a country does not seek financial infusion and wishes to conserve its natural resources, it will make its fiscal terms tougher. Fig. 12 shows a comparison of terms among oil producing nations.

## Case study: Different regimes

To summarize the economic environment, two examples are shown below for a moderate project developed under two fiscal regimes, the United States, which is a royalty-income tax regime, and a generic production sharing contract example. This will give an idea of the differences in key profit indicators in two different regimes for the same size project.

The **Alpha Prospect** is under consideration for acquisition by a hypothetical company, Newco. No money has been spent to date, and the possibility of paying a signature bonus is being evaluated. To share the risk, Newco is partnering with another major company, each proposing to invest 50% of the working interest. In the United States Gulf of Mexico, the royalty rate for deepwater leases is 18.75%. The anticipated signature bonus to be paid is $25 million, although the economic analysis would be conducted using different bid amounts before the actual bid is submitted. The technical review has resulted in an expected ultimate recovery of 200 million barrels of oil and 250 billion cubic feet of gas, and the probability of economic success is estimated to be 23.8%. The prospect is in a mature area of the basin with pipelines in the vicinity; the water depth is 2500 ft. (762 m). The bonus will be paid in Year 1, and there will be a 3D seismic study in Year 2, costing $20 million which may include licensing, additional acquisition and reprocessing, and evaluation to refine a wildcat location.

It will cost $100 million to drill the expendable wildcat in Year 3, and $90 million each for two appraisal wells in Year 4; these will be saved for production and completed later. Meanwhile, the FEED study will be undertaken at a cost of $25 million. Based on those results (which could change later depending on what they find), a 100,000 barrels of oil per day (BOPD) capacity floating production facility and subsea equipment will be manufactured at a total cost of $1850 million in Years 5–7. Also in Year 6, an oil pipeline will be installed and ready for tie in, and in Year 7 a gas pipeline will be installed; each will be 25 miles long and tie into an existing facility and cost $25 million apiece. A rig will be put on location in Year 7, and three more wells will be drilled; all five (including the appraisal wells) will be completed at a total cost of $440 million. First production will occur in Year 8 at 100,000 BOPD, the capacity of the facility. Phase 2 drilling will occur in Year 10, which will allow production to return to near facility capacity. Five wells will be drilled then at a total cost of $600 million.

For purposes of the financial analysis in what Newco calls the base case, the prices are forecast to be $65 per barrel for oil and $2.75/MMBtu for gas. Both prices and costs are estimated to escalate at 3% per year. Operating costs of $60 million per year cover all transportation, fuel, labor, maintenance and other expenses. The production is forecast by well to stay constant for two years, then to decline at 25%–35% per year based upon the decline patterns exhibited by similar deepwater reservoirs. The total productive life of the field is 14 years, relatively short compared to average field lives. The allowable deductions for calculating income tax assume current (2021) United States tax law, and the income tax rate used is 28%, an estimate of what will be in effect under proposed tax law changes at the time of this writing. All of the above costs and production figures are expressed as gross numbers. Newco's net costs are to be their 50% share, and their net production will be half of the 81.25% of production (40.63%) after payment of the government royalty share.

The annual cash flow projections are shown in Fig. 13, and the summary of economic results is included as Fig. 14, including the key profit indicators. The after-tax, risked rate of return is 21.3%, which is acceptable by many companies' standards for projects of this type. The project also indicates it will return 3.5 times the total investment on an unrisked, before-tax basis.

The **Zulu Prospect** is under consideration for acquisition by Newco as well. No money has been spent to date, and Newco has the same 50% working-interest partner. In this country, the terms presented are a production sharing contract. There is a 5% royalty, and there is no income tax. The Country will own a 10% working interest in the project but will bear none of the

**Alpha Prospect**
**After-Tax Cashflows**

| Year | 1 | 2 | 3 | 4 | 5 | 6 | 7 | 8 | 9 | 10 | 11 | 12 | 13 | 14 | 15 | 16 | 17 | 18 | 19 | 20 | 21 |
|---|---|---|---|---|---|---|---|---|---|---|---|---|---|---|---|---|---|---|---|---|---|
| Net MMBO | - | - | - | - | - | - | - | 7.02 | 14.09 | 11.48 | 14.09 | 10.06 | 7.21 | 5.18 | 3.74 | 2.71 | 1.97 | 1.43 | 1.05 | 0.77 | 0.46 |
| Net BCF | - | - | - | - | - | - | - | 8.78 | 17.61 | 14.35 | 17.61 | 12.57 | 9.01 | 6.48 | 4.68 | 3.39 | 2.46 | 1.79 | 1.31 | 0.96 | 0.57 |
| Average Price, $/BOE | - | - | - | - | - | - | - | 69.74 | 71.84 | 73.99 | 76.21 | 78.50 | 80.85 | 83.28 | 85.78 | 88.35 | 91.00 | 93.73 | 96.54 | 99.44 | 102.42 |
| Net Revenues, $MM | - | - | - | - | - | - | - | 591.95 | 1,222.78 | 1,026.36 | 1,297.24 | 953.92 | 704.07 | 521.56 | 387.72 | 289.20 | 216.41 | 162.44 | 122.28 | 92.30 | 56.82 |
| Net Opex, $MM | - | - | - | - | - | - | - | 36.90 | 38.00 | 39.14 | 40.32 | 41.53 | 42.77 | 44.06 | 45.38 | 46.74 | 48.14 | 49.59 | 51.07 | 52.61 | 54.18 |
| Operating Cashflows, $MM | - | - | - | - | - | - | - | 555.06 | 1,184.77 | 987.22 | 1,256.93 | 912.39 | 661.30 | 477.50 | 342.34 | 242.46 | 168.27 | 112.85 | 71.20 | 39.69 | 2.64 |
| Net Capex, $MM | 12.50 | 10.30 | 53.05 | 112.00 | 89.10 | 287.89 | 509.06 | 306.94 | 388.27 | 208.89 | - | - | - | - | - | - | - | - | - | - | - |
| **UNRISKED** | | | | | | | | | | | | | | | | | | | | | |
| Net Cashflow, pre-tax, $MM | (12.50) | (10.30) | (53.05) | (112.00) | (89.10) | (287.89) | (509.06) | 248.12 | 796.51 | 778.32 | 1,256.93 | 912.39 | 661.30 | 477.50 | 342.34 | 242.46 | 168.27 | 112.85 | 71.20 | 39.69 | 2.64 |
| Net Income Taxes, $MM | - | - | - | - | - | - | - | 99.93 | 260.70 | 197.02 | 272.54 | 176.07 | 106.47 | 55.00 | 71.95 | 60.07 | 47.12 | 31.60 | 19.94 | 11.11 | 0.74 |
| Net ATAX Cashflow, $MM | (12.50) | (10.30) | (53.05) | (112.00) | (89.10) | (287.89) | (509.06) | 148.19 | 535.81 | 581.30 | 984.38 | 736.32 | 554.84 | 422.50 | 270.40 | 182.39 | 121.15 | 81.25 | 51.27 | 28.58 | 1.90 |
| **RISKED** 23.8% | | | | | | | | | | | | | | | | | | | | | |
| Net Revenues | - | - | - | - | - | - | - | 140.59 | 290.41 | 243.76 | 308.10 | 226.56 | 167.22 | 123.87 | 92.08 | 68.68 | 51.40 | 38.58 | 29.04 | 21.92 | 13.49 |
| Net Opex | - | - | - | - | - | - | - | 8.76 | 9.03 | 9.30 | 9.58 | 9.86 | 10.16 | 10.46 | 10.78 | 11.10 | 11.43 | 11.78 | 12.13 | 12.49 | 12.87 |
| Net Capex | 12.50 | 10.30 | 53.05 | 26.60 | 21.16 | 68.37 | 120.90 | 72.90 | 92.21 | 49.61 | - | - | - | - | - | - | - | - | - | - | - |
| Net Cashflow, pre-tax | (12.50) | (10.30) | (53.05) | (26.60) | (21.16) | (68.37) | (120.90) | 58.93 | 189.17 | 184.85 | 298.52 | 216.69 | 157.06 | 113.41 | 81.31 | 57.58 | 39.96 | 26.80 | 16.91 | 9.43 | 0.63 |
| Net Income Taxes, $MM | - | - | - | - | - | - | - | 23.73 | 61.92 | 46.79 | 64.73 | 41.82 | 25.29 | 13.06 | 17.09 | 14.27 | 11.19 | 7.50 | 4.74 | 2.64 | 0.18 |
| Net ATAX Cashflow, $MM | (12.50) | (10.30) | (53.05) | (26.60) | (21.16) | (68.37) | (120.90) | 35.20 | 127.25 | 138.06 | 233.79 | 174.88 | 131.77 | 100.34 | 64.22 | 43.32 | 28.77 | 19.30 | 12.18 | 6.79 | 0.45 |

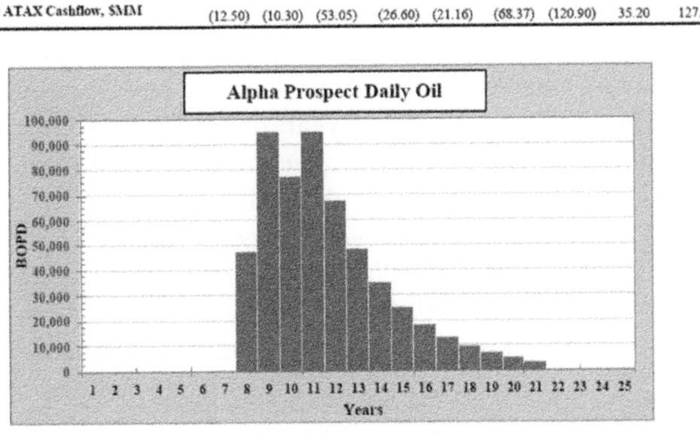


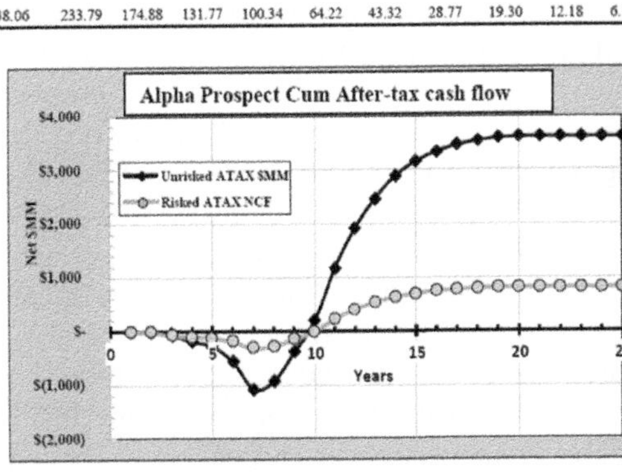


**FIG. 13** Cash flow model, Newco Alpha project, royalty-income tax terms.

| **Alpha Prospect Economic Summary** | | | | | |
|---|---|---|---|---|---|
| **Company Working Interest** | | 50.00% | Pe = 0.2375 | | |
| **Company Revenue Interest** | | 40.63% | | | |
| **Net Reserves** | **Unrisked** | **Risked** | **Net Capex** | **Unrisked** | **Risked** |
| MMBO | 81.251 | 19.297 | Exploration | 174.19 | 99.2 |
| BCF | 101.564 | 24.121 | Development | 1,803.81 | 428.4 |
| MMBOE | 98.178 | 23.317 | Total | 1,978.00 | 527.61 |

| **IRR** | **Unrisked** | | **Risked** | |
|---|---|---|---|---|
| | **BTAX** | **ATAX** | **BTAX** | **ATAX** |
| Company Net | 36.70% | 29.30% | 26.90% | 21.30% |
| **DCF** | **Unrisked** | | **Risked** | |
| **Discounted at:** | **BTAX** | **ATAX** | **BTAX** | **ATAX** |
| 0% | $5,036.60 | $3,626.40 | $1,138.40 | $803.40 |
| 5% | $2,628.50 | $1,822.10 | $573.10 | $381.60 |
| 10% | $1,393.10 | $912.70 | $285.30 | $171.20 |
| 15% | $738.20 | $441.80 | $134.50 | $64.10 |
| 20% | $382.20 | $193.80 | $54.00 | $9.20 |

| **Capital Efficiency** | | **Unrisked ATAX** | **Risked ATAX** |
|---|---|---|---|
| CF&D | | $20.15 | $22.63 |
| NPV/I | Exploration | $6.70 | $2.20 |
| | Total Project | $0.91 | $0.60 |

**FIG. 14** Economic summary, Alpha project income tax-royalty terms. This table summarizes the cash flow analysis presented in Fig. 13 for review by decisionmakers. BTAX refers to the before-income-tax cash flows. ATAX refers to the after-income-tax cash flows. The DCF or discounted cash flow is the total of all the yearly cash flows, brought back to a common point in time, usually the time of the investment being considered, at the indicated interest rates. These and the other indicators are more fully explained in the text.

investment costs. There will be a signature bonus in the amount of $20 million. The technical parameters are the same as the Alpha Prospect including ultimate recovery, probability of success, water depth and timing of investments. However, the oil will be taken directly from the floating production facility, in this case an FPSO, by tanker, and there will be a $50 million dollar expenditure in Year 7 for a gas pipeline to shore for local use. No gas will be sold by the contractor group.

In addition to the costs to operate the project, there will be an annual rental payment to the Country of $1,000,000 and a requirement to spend $250,000 to train local talent for employment in the industry. Instead of income tax, the Country shares profits after the contractor group recovers its costs out of production. During the first 36 months of production, 75% of the oil sales can be taken by the contractor group to recover its investments; the remaining production is split under the negotiated terms of the contract between the working interest owners and the Country. After that time, any investment left to be recovered may be taken from 65% of the oil sales. The allowable split in the first 5 years of production is 50% to the working interest owners for the first 40,000 BOPD and 60% to the Country of any production above that level. After 5 years, the Country takes 55% of the first 40,000 BOPD and 65% of any production above that level.

All of the above costs and production figures are expressed as gross numbers. Newco's net costs are to be their 45% share after accounting for the Country participation and that of its partner. The annual cash flow projections and economic summary are shown in Figs. 15 and 16, respectively. The risked rate of return is 19.6%, which is still acceptable by many companies. The project also indicates it will return 2.4 times the total investment on an unrisked basis. It is not uncommon for contracts such as this to have more complicating elements, which make the analysis and negotiating complex and the bailiwick of financial analysis experts.

| Zulu Prospect After-Tax Cashflows | 1 | 2 | 3 | 4 | 5 | 6 | 7 | 8 | 9 | 10 | 11 | 12 | 13 | 14 | 15 | 16 | 17 | 18 | 19 | 20 | 21 |
|---|---|---|---|---|---|---|---|---|---|---|---|---|---|---|---|---|---|---|---|---|---|
| **Company Net Production Share** | | | | | | | | | | | | | | | | | | | | | |
| MMBO | - | - | - | - | - | - | - | 7.186 | 14.412 | 10.078 | 7.567 | 5.683 | 4.149 | 2.956 | 2.197 | 1.654 | 1.264 | 0.983 | 0.780 | 0.633 | 0.470 |
| BCF | - | - | - | - | - | - | - | - | - | - | - | - | - | - | - | - | - | - | - | - | - |
| Average Price, $/BOE | | | | | | | | $ 79.94 | $ 82.34 | $ 84.81 | $ 87.35 | $ 89.98 | $ 92.67 | $ 95.45 | $ 98.32 | $ 101.27 | $ 104.31 | $ 107.44 | $ 110.66 | $ 113.98 | $ 117.40 |
| Net Revenues, $MM | $ - | $ - | $ - | $ - | $ - | $ - | $ - | $ 574.47 | $ 1,186.67 | $ 854.71 | $ 661.05 | $ 511.31 | $ 384.55 | $ 282.12 | $ 215.97 | $ 167.47 | $ 131.84 | $ 105.62 | $ 86.31 | $ 72.10 | $ 55.18 |
| Net Opex, $MM | $ - | $ - | $ - | $ - | $ - | $ - | $ - | $ 33.21 | $ 34.20 | $ 35.23 | $ 36.29 | $ 37.37 | $ 38.50 | $ 39.65 | $ 40.84 | $ 42.07 | $ 43.33 | $ 44.63 | $ 45.97 | $ 47.34 | $ 48.77 |
| Operating Cashflows, $MM | $ - | $ - | $ - | $ - | $ - | $ - | $ - | $ 541.27 | $ 1,152.47 | $ 819.49 | $ 624.77 | $ 473.93 | $ 346.05 | $ 242.47 | $ 175.13 | $ 125.41 | $ 88.51 | $ 60.99 | $ 40.34 | $ 24.76 | $ 6.41 |
| Net Capex Incl NOC Carry, $MM | $ 10.00 | $ 10.93 | $ 53.67 | $ 112.63 | $ 89.73 | $ 287.89 | $ 509.06 | $ 306.94 | $ 388.27 | $ 208.89 | $ - | $ - | $ - | $ - | $ - | $ - | $ - | $ - | $ - | $ - | $ - |
| **UNRISKED** | | | | | | | | | | | | | | | | | | | | | |
| Net Cashflow, pre-tax, $MM | ($10.00) | ($10.93) | ($53.67) | ($112.63) | ($89.73) | ($287.89) | ($509.06) | $234.33 | $764.20 | $610.59 | $624.77 | $473.93 | $346.05 | $242.47 | $175.13 | $125.41 | $88.51 | $60.99 | $40.34 | $24.76 | $6.41 |
| Net Income Taxes, $MM | $0 | $0 | $0 | $0 | $0 | $0 | $0 | $0 | $0 | $0 | $0 | $0 | $0 | $0 | $0 | $0 | $0 | $0 | $0 | $0 | $0.00 |
| Net ATAX Cashflow, $MM | ($10.00) | ($10.93) | ($53.67) | ($112.63) | ($89.73) | ($287.89) | ($509.06) | $234.33 | $764.20 | $610.59 | $624.77 | $473.93 | $346.05 | $242.47 | $175.13 | $125.41 | $88.51 | $60.99 | $40.34 | $24.76 | $6.41 |
| Cumulative ATAX NCF, $MM: | ($10.00) | ($20.93) | ($74.60) | ($187.22) | ($276.95) | ($564.84) | ($1,073.90) | ($839.57) | ($75.37) | $535.22 | $1,159.99 | $1,633.92 | $1,979.97 | $2,222.44 | $2,397.57 | $2,522.98 | $2,611.50 | $2,672.49 | $2,712.83 | $2,737.59 | $2,744.00 |
| **RISKED** 23.8 | | | | | | | | | | | | | | | | | | | | | |
| Net Production MMBO | - | - | - | - | - | - | - | 1.71 | 3.42 | 2.39 | 1.80 | 1.35 | 0.99 | 0.70. | 0.52 | 0.39 | 0.30 | 0.23 | 0.18 | 0.15 | 0.11 |
| Net Revenues | $0 | $0 | $0 | $0 | $0 | $0 | $0 | $136.44 | $281.83 | $202.99 | $157.00 | $121.44 | $91.33 | $67.00 | $51.29 | $39.78 | $31.31 | $25.08 | $20.50 | $17.12 | $13.11 |
| Net Opex | $0 | $0 | $0 | $0 | $0 | $0 | $0 | $7.89 | $8.12 | $8.37 | $8.62 | $8.88 | $9.14 | $9.42 | $9.70 | $9.99 | $10.29 | $10.60 | $10.92 | $11.24 | $11.58 |
| Net Capex | $10.00 | $10.93 | $53.67 | $26.75 | $21.31 | $68.37 | $120.90 | $72.90 | $92.21 | $49.61 | $0.00 | $0.00 | $0.00 | $0.00 | $0.00 | $0.00 | $0.00 | $0.00 | $0.00 | $0.00 | $0.00 |
| Net Cashflow, pre-tax | ($10.00) | ($10.93) | ($53.67) | ($26.75) | ($21.31) | ($68.37) | ($120.90) | $55.65 | $181.50 | $145.02 | $148.38 | $112.56 | $82.19 | $57.59 | $41.59 | $29.78 | $21.02 | $14.49 | $9.58 | $5.88 | $1.52 |
| Net Income Taxes, $MM | - | - | - | - | - | - | - | - | - | - | - | - | - | - | - | - | - | - | - | - | - |
| Net ATAX Cashflow, $MM | ($10.00) | ($10.93) | ($53.67) | ($26.75) | ($21.31) | ($68.37) | ($120.90) | $55.65 | $181.50 | $145.02 | $148.38 | $112.56 | $82.19 | $57.59 | $41.59 | $29.78 | $21.02 | $14.49 | $9.59 | $.88 | $1.52 |
| Cumulative ATAX NCF, $MM: | ($10.00) | ($20.93) | ($74.60) | ($101.34) | ($122.65) | ($191.03) | ($311.93) | ($256.28) | ($74.78) | $70.24 | $218.62 | $331.18 | $413.37 | $470.95 | $512.55 | $542.33 | $563.35 | $577.84 | $587.42 | $593.30 | $594.82 |

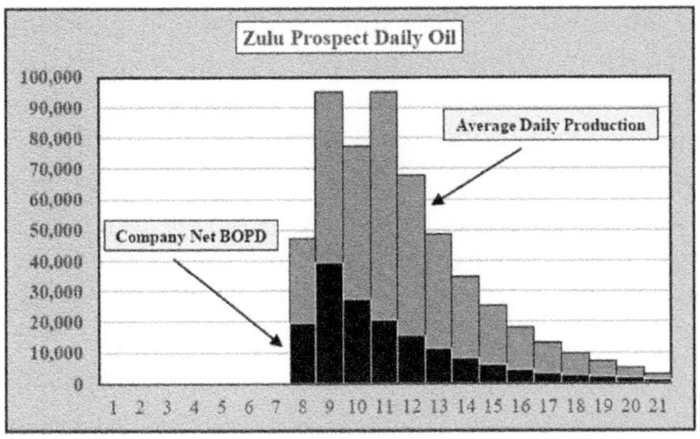


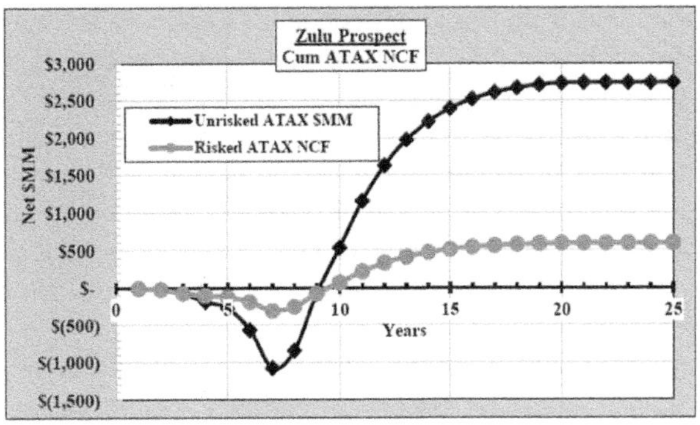


**FIG. 15** Cash flow model, Newco Zulu project, Production Sharing Contract (PSC) terms.

**Zulu Prospect**
**Economic Summary**

| | | |
|---|---|---|
| Company Working Interest | 45.000% | Pe =0.2375 |
| Company Paying Interest | 50.000% | |
| Effective Revenue Interest | 30.005% | |

| Net Reserves | Unrisked | Risked | Net Capex | Unrisked | Risked |
|---|---|---|---|---|---|
| MMBO | 60.011 | 14.253 | Exploration | $174.42 | $98.42 |
| BCF | - | - | Development | $1,803.58 | $428.23 |
| MMBOE | 60.011 | 14.253 | Total | $1,978.00 | $526.65 |

| IRR | Unrisked BTAX | Unrisked ATAX | Risked BTAX | Risked ATAX |
|---|---|---|---|---|
| Company Net | 28.0% | 28.0% | 19.6% | 19.6% |

| DCF Discounted at: | Unrisked BTAX | Unrisked ATAX | Risked BTAX | Risked ATAX |
|---|---|---|---|---|
| 0% | $2,744.0 | $2,744.0 | $594.8 | $594.8 |
| 5% | $1,389.1 | $1,389.1 | $279.7 | $279.7 |
| 10% | $696.922 | $696.9 | $121.0 | $121.0 |
| 15% | $333.8 | $333.8 | $39.4 | $39.4 |
| 20% | $140.3 | $140.3 | ($2.5) | ($2.5) |

| Capital Efficiency | | Unrisked ATAX | Risked ATAX |
|---|---|---|---|
| CF&D | | $32.96 | $36.95 |
| NPV/I | Exploration | $5.15 | $1.58 |
| | Total Project | $0.69 | $0.43 |

**FIG. 16** Economic summary, Zulu project, Production Sharing Contract (PSC) terms. This table summarizes the cash flow analysis presented in Fig. 15 for review by decisionmakers. The physical project is the same as the Alpha Prospect, but the government is different, and the fiscal terms are not the same. BTAX and ATAX in this example are labeled as before and the numbers are the same, because the country takes its entitlement in production sharing rather than as tax.

## The political and regulatory environment

Key above ground risks include not just product prices but also political risk in the host country. Political events may have local, regional or global impacts. Such events may include conflicts (or even wars), peace treaties, alliances and un-alliances, changes in the parties in power in an influential power and so on. Country officials are expected to act in the best interests of their citizens, and usually they do, but sometimes they do not. Because regulations are promulgated by governments, these two are often related. Political change may affect costs or even revenues, and regulatory change almost always increases cost. Political risk is not quantifiable, but it is real, and there are steps taken to mitigate it.

Political issues play a major role in the determination of whether to invest in a petroleum or other energy venture in a particular location. Today, there is a process defined as "creeping nationalization" under which a host government, by legislation or regulation, may unilaterally change the terms and conditions of a petroleum agreement to the detriment of the oil company. To protect against this, companies will negotiate "stabilization or equilibrium" clauses in their agreements that provide that if the fiscal or economic benefits of the agreement are adversely affected by legislation or regulation, the parties shall adjust the fiscal and contractual benefits of the agreement to give the companies the same benefits as under the government terms when the agreement was signed. The company should also provide for international arbitration in the event of a dispute or disagreement with the government under their agreement. With respect to the risk of the applicable law being the law of the host country, a company should provide for international law, regulations, traditions and customs applicable to international oil and gas ventures between a company and a government to also be applicable (Boulos, 2017). Nevertheless, any political disruption can cause financial harm, if not in actual cash flow, then in delays in production and consequent receipts. Depending on the size of the company, it may have little recourse in a dispute with the host government.

An important event that affected global regulation of deepwater drilling was the Macondo drilling accident of April 2010, also known as Deepwater Horizon. The response of many countries was to freeze all drilling activity until the accident was studied and the issues resolved, such that learning from the disaster could be applied to their own basins. The effect on global deepwater drilling was to stop activity for about 1 year. In the Gulf of Mexico, no new leases were approved until 2011, and the first well drilled was 9 months after the incident. These actions had a significant impact on the profitability of projects already in development, and on the companies involved, and delayed those investments in exploration that were pending.

Safety and environmental compliance are part of the regulatory equation, and much of this is becoming a global consensus. Aging infrastructure, including pipelines in mature basins, older platforms used as hosts or intermediary stops and aging and less regulated oil and LNG tankers are all concerns in producing from deepwater. Any of these issues can have a cost impact on a project. On the other hand, producing in deepwater has advantages. The high reserve density reduces the areal footprint on the earth's surface that is required per Btu of energy produced. This includes all aspects of construction, transportation, staff required, etc., since a single well in the deepwater can produce what tens or hundreds of onshore wells can deliver. In addition, deepwater projects are located far from populated areas, and even if a spill or other polluting incident were to occur, its impact will be somewhat diminished compared to a similar onshore or near-shore incident. Notwithstanding that, spill incidents can impact the businesses and lives of many along the adjacent shorelines, especially including those involved in fishing and tourism.

Government policies have a direct economic impact on energy investment. Most of the 2020s' issues are environmental, related to carbon emissions and climate change. While some of these issues are foretelling the future energy environment, others are being pursued synergistically with deepwater oil projects. Examples are: inclusion of as much solar powered equipment as can reasonably fit on an offshore facility; consideration of wind power as an adjunct to deepwater development; using carbon dioxide ($CO_2$) injection as carbon storage or even enhanced oil recovery (EOR) use. $CO_2$ emissions and flaring are generally monitored and minimized in deepwater operations, greatly reducing the carbon footprint, especially in comparison to the many offshore wells' production replaced by each HRHU well in deepwater (Chapter 21). Many governments reward environmental activities that they endorse, and these can be real financial incentives that should be included in the economic evaluations. Enough incentive will cause investors to move to the area encouraging the activity, because these monetary amounts will be included in the FID analysis.

## Key messages

The 1960s and ongoing space programs have been credited with developing technologies that spurred many industrial advances. And, it is no secret that medical computer technology and seismic 3D technology use similar mathematical theory and physics technology. Enormous technical advances have occurred in oil and gas related to deepwater that enable functionality that was unheard of 10–20 years ago. Data acquisition is far advanced, and the ability to assimilate it, while far advanced, is still being developed, as mentioned in Chapter 19 as it affects production. Satellite data transmittal has been an enormous advance that has added to safety and environmental responsibility. The need to physically transport personnel has been reduced due to the improved ability to operate remotely. All of this in general has helped the industry move rapidly into deeper and deeper areas of exploration and will serve to underpin the emerging offshore renewables industry as well.

While the technologies used in deepwater seem to have occurred nearly overnight, they have been developing for well over 50 years, accelerating more recently with general technological advances such as computer speed and availability. Future technologic advances in deepwater may include multiple lateral drilling to longer lengths, deeper drilling, more environmentally reliable subsea processing and perhaps better exploration methods than the current seismic techniques provide.

## Conclusion

Commercial deepwater exploration is a long-term game. Given this, a deepwater discovery and subsequent producing life will experience several oil and gas price cycles. Deepwater projects require extremely large investments, often in the billions of dollars, and the decisions require complex evaluations of many parameters. Some of these are within the investor's control, such as the size of a facility, but many are unknowns dictated by nature, by changing governments and by market conditions that affect supply and demand and in turn, product pricing. The proper assessment of each opportunity requires the contribution of many professional and technical experts in a number of fields comprising business, geoscience, legal, engineering and more. This same fundamental, collective expertise will assuredly apply to advancing current and alternative energy technologies into the future.

## Appendix: Supplementary material

Supplementary material related to this chapter can be found on the accompanying CD or online at https://doi.org/10.1016/B978-0-323-91918-0.00005-0.

## Acknowledgments

Many thanks to Dr. Paul Weimer for the use of his figures and to his content contributions on technical topics.

## References

Bansal, A.K., 2017. International Petroleum Fiscal Systems—Service Contracts. www.linkedin.com/pulse/international-petroleum-fiscal-systems-service-contracts-bansal. (Chapter 6).

Boulos, A.J., 2017. Assessing Political Risk., p. 6. www.ipaa.org/wp-content/uploads/2017.

Bryce, R., 2010. Power hungry: the myths of "green" energy and the real fuels of the future. In: Public Affairs., ISBN: 978-1-58648-789-8. 414 pp.

CGX Energy Inc., 2021. Guyana Fiscal Regime. cgxenergy.com/fiscal-regime.

Forrest, M., Roden, R., Holeywell, R., 2010. Risking seismic amplitude anomaly prospects based on database trends. Lead. Edge 29, 570–574.

Genesis Oil and Gas Consultants Ltd, 2020. Concept Definition and Pre-FEED. www.genesisenergies.com/capbilities/concept-definition-and-pre-feed.

Hyne, N.J., 2012. Nontechnical Guide to Petroleum Geology, Exploration, Drilling & Production, third ed. PennWell, Tulsa, Oklahoma, USA. In press.

Monk, D., 2020. Survey design and seismic acquisition for land, marine, and in-between. In: Society of Exploration Geophysicists, Distinguished Instructor Series 23., ISBN: 9781560803706. 214 pp.

Pettingill, H.S., 2020a. Energy and Power Density: The Deepwater Advantage. Manuscript posted on LinkedIn (updated from April 2020 AAPG Explorer Article by Heather Saucier).

Roden, R., Forrest, M., Holeywell, R., 2012. Risking seismic amplitude anomaly prospects calibrated to an industry-wide database. In: Abstract, 74th EAGE Conference & Exhibition Workshop, Evaluation and Risk Assessment of Amplitude Supported Prospects—A Practical Exercise, June 8.

Rudolph, K.W., 2001. DHI/AVO Analysis Best Practices: A Worldwide Analysis. AAPG-SEG Distinguished Lecture. https://archives.aapg.org/education/dist_lect/slides/2001_02/rudolph.pdf.

Rystad Energy, March 2021. Wildcat Safari Running Dry: Onshore Success Rate Dips to Lowest on Record, Falls for Fourth Year in a Row. www.rystadenergy.com/newsevents/news/press-releases.

Schlumberger, 2021. Intelligent Completions, Monitor Your Reservoir and Control Production in Real Time. www.slb.com/completions/well-completions/intelligent-completions.

Simm, Bacon, 2014. Seismic Amplitudes. Cambridge University Press. 271 pp.

Society of Petroleum Engineers (SPE), 2007. Petroleum Resources Management System; Prepared by the Oil and Gas Reserves Committee., p. 3.

Vittone, M., 2015. Monster Cables: Connecting the Subsea Factory. gCaptain.com (2015).

## Further reading

Hess, 2019. Investor presentation. In: Scotia Howard Weil Energy Conference, March 26, 2019. (slides from Hess website).

Inikori, S.O., Spring, L.Y., Ageh, E., Van der Bok, J.W., 2007. The development of world-class high rates, high ultimate recovery wells in deepwater turbidites–Bonga field, offshore Nigeria. In: Paper Presented at the SPE Annual Technical Conference and Exhibition, Anaheim, California, USA, November 2007., https://doi.org/10.2118/110360-MS. Paper Number: SPE-110360-MS.

Lowry-Yilmaz, 2020. A balanced portfolio of super basins for exploration and exploitation. In: Luncheon Keynote, AAPG Super Basins Conference, February 13, 2020, Sugarland TX.

Meckel, L.D., 2020a. High rate high ultimate wells. In: Conference Proceedings, Second HGS and EAGE Conference on Latin America, December 2020. vol. 2020, pp. 1–5, https://doi.org/10.3997/2214-4609.202086008.

Meckel, L.D., 2020b. Benchmarking Deepwater Well Rates and Ultimate Recoveries. LinkedIn Post. August 2020 https://www.linkedin.com/pulse/benchmarking-deepwater-well-rates-ultimate-recoveries-trey-meckel/?articleId=6696093540969263104.

Oil and Gas Reserves Committee of the Society of Petroleum Engineers, 2007. Petroleum Resources Management System.

Organization of the Petroleum Exporting Countries, Brief History. www.opec.org/opec_web/en/about_us/24.htm.

Pettingill, H.S., 2020b. Oral Presentation, AAPG Superbasins Conference.

Pettingill, H.S., Weimer, P.W., 2020. Deep water reservoirs from giant fields: energizing the globe, past, present and future. In: AAPG Super Basins Conference, Feb. 13, 2020, Sugarland TX.

Peyton, C.E. (Ed.), 1977. Seismic Stratigraphy—Applications to Hydrocarbon Exploration. AAPG Memoir 26, ISBN: 0-89181-302-0.

Priest, T., 2019. Shell Oil's Deepwater Mission to Mars. University of Iowa Press. 109 pp.

Rudolph, K.W., Goulding, F.J., 2017. Benchmarking exploration predictions and performance using 20+ years of drilling results: one company's experience. AAPG Bull. January 2016.

Saucier, H., 2020. Power density shows deepwater fields as most environmentally friendly energy source. In: Interview with Henry S. Pettingill, AAPG Superbasins Conference. AAPG Explorer, vol. 41(4).

Smil, V., 2015. Power Density: A Key to Understanding Energy Sources and Uses. The MIT Press. 306 pp.

US Energy Information Administration (EIA), April 2021. What Drives Crude Oil Prices: Spot Prices.

Weimer, P., 2017. Deeper Waters: how science and technology pushed exploration to greater depths. AAPG Explorer (January).

Weimer, P., Pettingill, H.S., 2004. Global overview of deepwater exploration and production. In: The Petroleum Systems of Deepwater Settings, Distinguished Instructor Course Notes, SEG/EAGE Short Course, pp. 219–258 (Chapter 2).

Weimer, P., Pettingill, H.S., 2020. The History of Deepwater Exploration, Part 2. AAPG Explorer. December.

Weimer, P., Slatt, R.M., Dromgoole, P., Bowman, M., Leonard, A., 2000. Developing and managing turbidite reservoirs: case histories and experiences: results of the 1998 EAGE/AAPG research conference. AAPG Bull. 84 (4), 453–465.

Weimer, P., Pettengill, H.S., Roden, R., 2022. The history of deepwater exploration, part 3. AAPG Explorer (January), 16–19.

Chapter 21

# Deepwater sedimentary systems: The next 100 years of deepwater

Cindy A. Yeilding[a], Richard A. Sears[b], Zackary M. Donovan[c] and F. Javier Hernández-Molina[d]
[a]bp America (retired), Houston, TX, United States, [b]Energy Resources Engineering, Stanford University, Stanford, CA, United States, [c]Episcopal High School, Bellaire, TX, United States, [d]Department of Earth Sciences, Royal Holloway, University of London, Egham, Surrey, United Kingdom

## Introduction and objectives

The objective of this chapter is to reflect on future utilization of the deepwater environment and explore how the geoscience principles described in this textbook could support the safe and sustainable development of some or all of these ideas. A number of possible "uses" in the deepwater environment are outlined in this chapter; however, the objective is to identify, not advocate, for these uses.

While this book explores aspects of the past and present associated with deepwater science and engineering, this chapter is focused on envisioning potential future deepwater uses and highlighting opportunities for geoscience technology and research. Using a "crystal ball," also known as imagination informed by both fresh eyes and decades of experience in the exploration of new concepts, ideas, technologies and business development in deepwater sedimentary systems, this chapter will examine:

- context and drivers on human activities in deepwater,
- future energy demand scenarios,
- how geologic processes could impact deepwater developments and activities,
- current and potential future uses and industries in deepwater and
- societal and policy drivers in deepwater.

Potential areas of geoscience research and technology development (Appendix), as well as a vignette from the US deepwater Gulf of Mexico (Box 1) are also included.

Numerous future applications for the offshore are envisioned in this chapter: some are underway now, others are currently in planning and development and others are still "mind candy" yet to be clearly defined. Some will prove too difficult to implement and execute safely and cost-effectively in the near term, and others will progress logically into deeper and deeper water. To prepare for the opportunities, challenges and risks associated with continued development of the offshore, particularly deepwater, the geological processes detailed in the previous 20 chapters of this book need to be examined and analyzed, and potential impacts on infrastructure and operations understood before moving forward with novel projects and deepwater developments.

## Context

### A new frontier: Ocean research and understanding of deepwater geologic processes

More than 70% of Earth's surface is covered by water, and much of this (~63%) is beyond the shelf break at 120 to 200 meters (393 to 650 feet) or greater of water depth (Kennett, 1982; Seibold and Berger, 2010) and generally characterized as *deepwater*. The offshore environment is a vast frontier which is predominantly underexplored, and in many cases, unexplored. To quote NASA oceanographer Dr. Gene Feldman in 2009, "*We have better maps of the surface of Mars and the moon than we do of the bottom of the ocean*" (NASA, 2021). For example, from the ocean's surface through the seabed, scientists have identified almost a quarter of a billion species, but some researchers estimate that this represents only about 10% of the ocean's species (National Geographic, 2021). Keeping these aspects in mind, in spite of decades of research and in particular the development of oil and gas resources, deepwater is truly a frontier.

Deepwater Sedimentary Systems. https://doi.org/10.1016/B978-0-323-91918-0.00017-7

**Box 1 Deepwater Gulf of Mexico.**

*Looking back to "see" ahead*

The most important concept in looking ahead at the future of deepwater is to remember that *"we just don't know what we don't know."* As late as the 1970s, oil and gas industry dogma was that there was little, if any, opportunity in deepwater. Many wells drilled on the U.S. Gulf of Mexico shelf had been drilled deeper to penetrate older deepwater facies where very little sand had been encountered, leading to the general conclusion that deepwater environments were starved for reservoir quality sediment. Geoscientists, however, observed features in seismic data in deepwater areas that looked very much like oil and gas discoveries on the shelf (see for example, see Fig. 8, Chapter 19 of this book) and continued to progress exploration concepts further offshore.

Even with early major discoveries in the late 1980s and early 1990s (see Fig. 2, Chapter 1 and Table 1, Chapter 20 of this book), the recognition of the scale of the deepwater opportunity took time to develop. Fig. 1 illustrates this evolution of oil and gas discoveries in the U.S. Gulf of Mexico deepwater. In the mid-1980s, there were under 10 well penetrations in the deeper waters of the Gulf, with the first big discovery in > 600 meters (2000 feet) of water, Ram Powell, in 1985. At this time, there were significant dogmas about the basin, including but not limited to: (1) there would be no reservoir, (2) sand could be present but it would not be an effective reservoir, (3) no effective source rock was predicted for this part of the basin and (4) even if hydrocarbons were discovered, developments would never be economic. These dogmas were disproved through the thorough application of the first principles of geoscience and exploration, technology advances, supportive policies and economic factors. There were some fabulous discoveries and many stupendous failures but ultimately the basin has matured to delivering over 1 mm barrels of oil equivalent (boe)/day, \$2 to 7.5 billion dollars in annual royalties, over \$30 billion into the U.S. economy in annually and over 300,000 annual domestic jobs (Fig. 1; Natural Resources Revenue Data, 2022).

Economics, necessity, policy and innovation will determine which of the opportunities described in this chapter could become reality and when they might be practical. As industrialization of the world's oceans is considered, a complete understanding of deepwater systems and processes through the water column, at the seabed and in the subsurface will be necessary to make safe and sound choices. Those that progress from concept to application will need to consider the impacts of depositional processes and their geologic setting as part of the development plan. The work in this book represents one of the most comprehensive compilations of the current understanding of deepwater systems and processes but there is still a huge amount of research to be carried out, data to be acquired and paradigms to be broken.

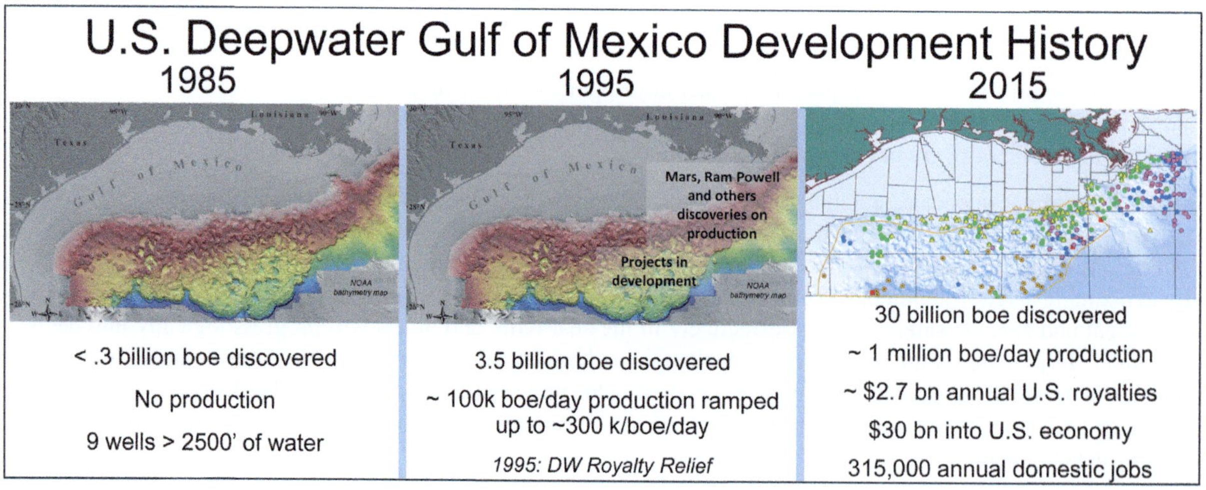


**FIG. 1** Example of the timing and evolution of a deepwater hydrocarbon province from the U.S. deepwater Gulf of Mexico. The first decade of deepwater exploration resulted in significant hydrocarbon volumes discovered as exploration concepts were tested, followed by two decades of additional order of magnitude growth in discovered volumes and production (Yeilding, 2021). Discoveries were enabled through new exploration concepts, improved seismic imaging, improved drilling rigs and capabilities and supportive policies.

At the seabed, real-time records of deepwater processes are exceedingly rare. Deepwater systems were essentially unknown until November 18, 1929, when a magnitude 7.2 earthquake off the coast of Newfoundland, Canada triggered a submarine landslide that raced down the Continental Slope and snapped 12 telegraph cables on the Atlantic Ocean floor (Derouin, 2017). This particular submarine landslide initiated the study of downslope processes, including debris flows, turbidity currents and mass transport, all key sediment transport processes described in the preceding chapters of this book (see Chapters 6–10). A more recent example was recorded in January 2020. A turbidity flow originating from the mouth of the sediment-swollen African Congo River traveled more than 1100 km (683 miles) through the Congo Canyon into the deep Atlantic Ocean. Similar to the 1929 event, two submarine communication cables were broken by the flow, resulting in a significant disruption of internet traffic from and between African countries (Talling et al., 2021).

While insights into the Earth's oceans are growing rapidly, to date only about 20% of the ocean floor has been mapped using detailed (high-resolution) multi-beam bathymetric technology. This represents more than triple the information available in 2017 (when it is estimated that about 6% of the ocean floor had been mapped), however, much of the seabed remains uncharacterized, rendering the ocean as a vast frontier for exploration and offering the opportunity for many new scientific discoveries (Burton, 2020). Begun in 2017, the Nippon Foundation-GEBCO Seabed 2030 Project aims to complete the mapping of the global ocean floor by 2030 (Nippon Foundation-GEBCO, 2021) and will provide a significantly improved dataset for the offshore.

Knowledge of deepwater sedimentary systems has developed over the last several decades largely through the study of marine geology. Analysis of marine outcrops on land, including deepwater sedimentary deposits, provided much of the foundation for the understanding of modern-day deepwater processes. A significant portion of recent knowledge of deepwater systems comes from academic research—from offshore research boreholes (including the oceanic drilling programs, DSDP, ODP and IODP[a]), as well as military activities and data acquired and characterized by the oil and gas industry. The boom in deepwater oil and gas exploration following the energy crisis of the 1970s, as well as a resurgence in the 1990s and 2000s, verified the significant opportunities in deepwater basins around the world. This history is laid out in detail in Chapter 1 of this book.

## Future resource scenarios

Earth's human population is currently about 7.9 billion people, 3.2 times more than the population at the beginning of the 1960s. This growth in population will likely drive increases in energy, resource and food demand. With this increased demand, it is likely that human development will continue to depend upon and expand farther into marine systems, driving further evaluation of the deep ocean for a variety of activities, including energy production, aquaculture, shipping, mining and military uses. This movement to offshore is logical: it reduces the impact on livable land, removes infrastructure from line of sight and from populated areas, and in many situations the economic, societal and technological opportunities are more robust offshore than onshore.

Fossil hydrocarbon resources have dominated global energy systems since the late 19th century when coal use surpassed wood as the major source of primary energy. About a century later, oil became the dominant fuel and it remains so today, supplying 31% of global primary energy demand in 2020 (Nakicenovic, 1996; bp, 2021). Natural gas provided an additional 25% of global primary energy in 2020. Together, oil and natural gas have provided more than 50% of the world's primary energy since the middle of the 20th century (Nakicenovic, 1996).

Several major energy companies, consultancies and government agencies create projections of global energy supply and demand for the next 3–4 decades.[b] The scenarios they create are based on their own assumptions about technology, politics and economics and so they often differ in the detailed makeup of future energy supply and demand. Despite the detailed differences the scenarios generally have one thing in common. Global energy use is enormous and because of that scale, and the fact that similarly enormous capital has been invested in our current energy sources and infrastructure, it can be said that global energy is a stable, slowly evolving system (Sears, 2020).

To date, deepwater development has been dominated by the oil and gas industry. Without the economic incentive provided by hydrocarbon production, it is unlikely that scientists would have developed the detailed understanding of deepwater seabed processes and of the subsurface. The extent to which the world continues to depend on deepwater oil and gas resources for fuel and on the products derived from petroleum (Ranken Energy Corporation, 2021) will frame the future. Over the next few decades, continued deepwater hydrocarbon production will have an impact on the development of other deepwater uses and sources. For this reason, it is important to consider several future energy scenarios in assessing the future of deepwater.

### Energy resources: Oil and gas, renewables and the impact on deepwater

A representative future scenario is the Reference Case for the U.S. Energy Information Administration International Energy Outlook (EIA IEO). Completed each year, Fig. 2 is taken from the EIA IEO 2021 Reference Case. Comparisons can be

[a] The Deep Sea Drilling Program (DSDP, later known as International Ocean Drilling Program or IODP; Warme et al., 1981) and other shipboard expeditions (Stow et al., 1985) made numerous discoveries about seafloor morphology and the near-sub-surface around the world (Chapter 1 in this book).

[b] For details on future energy scenarios, many scenarios plus supporting details are published on the websites of energy related organizations, such as Shell, bp, ExxonMobil, Total Energies, Equinor, The US Energy Information Agency (EIA), the International Energy Agency (IEA), Rystad Energy and DNV.

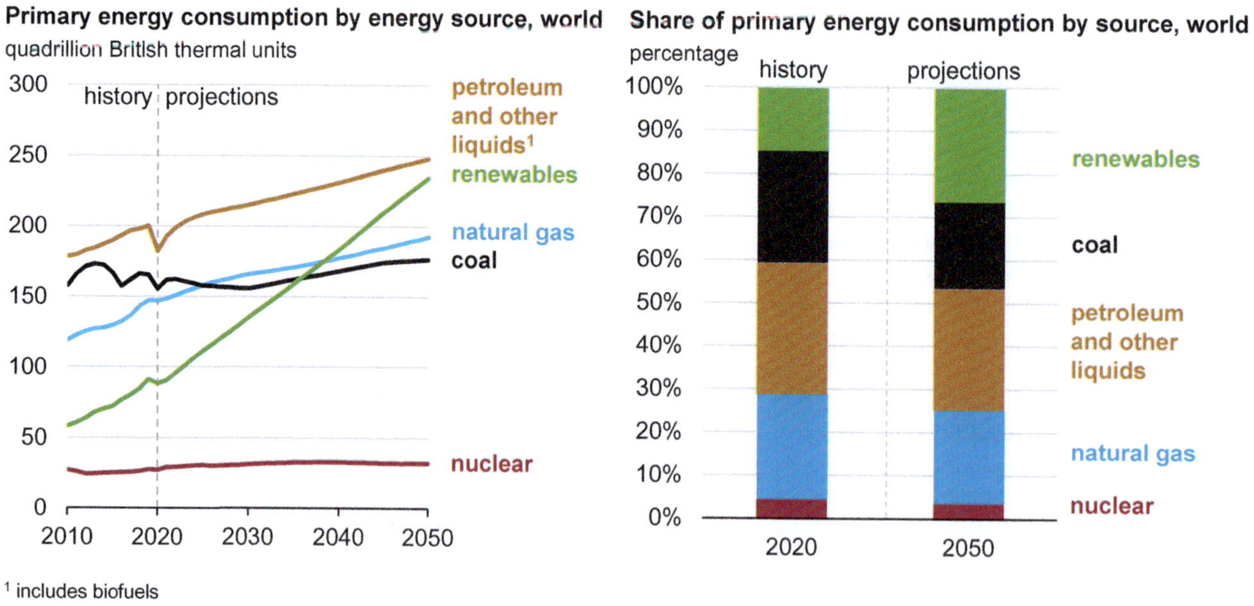


**FIG. 2** World primary energy consumption, 2010–50, 2021 EIA reference case scenario. The left-hand panel shows actual (2010–20) and projected (2021–50) consumption by fuel. This scenario shows an increase in total primary energy demand of 47% from 2020 to 2050. Consumption of all energy sources increases from today with renewables growing significantly faster than all other sources. The right-hand panel shows the percentage share of energy consumption by source for 2020 and 2050. By percentage share, all sources of primary energy decrease between 2020 and 2050 except renewables which approximately double their share of energy consumption. *Data from the US Energy Information Agency, 2021 International Energy Outlook (EIA IEO, 2021).*

made between different future scenarios (see, for example, Sears, 2020; Wood Mackenzie, 2021) and there are features common to many of them.[c]

Many published scenarios project a significant increase in global energy demand, typically 30%–50% by the middle of the 21st century. For example, EIA IEO posits a 47% increase in demand by 2050. This large increase in overall energy demand puts additional pressure on the necessary rate of growth in renewables for there to be a significant decrease in demand for fossil fuels.

It is future demand for fossil fuels that will have a significant impact on activities in deepwater in the coming decades. In 2019, global oil and natural gas production totaled approximately 150 million barrels oil equivalent per day (BOE/D) (bp, 2021). Deepwater production topped 10 million BOE/D for the first time in 2019 and based on existing projects and those planned for development over the next few years, deepwater was poised to grow to more than 14.5 million BOE/D by 2024 ( JPT, 2019). While oil and gas production both declined slightly in 2020 due to the impacts of COVID-19, production is up sharply in 2021 (EIA STEO, 2021) as demand rebounds and increases.

Not only does deepwater represent a significant fraction of global production, it is also among the most efficient production in the world (Offshore Energy, 2020a, b; Rystad Energy, 2020), which suggests that as long as there is significant demand for oil and natural gas, deepwater will be an attractive environment for the energy industry.

Looking again at EIA IEO (2021) and other future scenarios, another general feature is that as energy demand grows over the coming decades, renewable energy grows faster than any other energy source, but not fast enough to fully meet growing demand in the near to mid-term. The result is significant demand growth is predicted for both oil and natural gas to meet this demand over the coming few decades.

Of the published scenarios, one that differs significantly from EIA IEO (2021) is the Shell "Sky" scenario (Shell, 2021). In this scenario, energy demand still grows significantly over the coming decades, approximately doubling by late in the 21st century. In this scenario, global energy systems experience unprecedented improvements in technology, particularly significant gains in the efficiency of energy services. As a result, demand for oil increases slightly over the current decade,

---

[c] Comparing different scenarios is not always straight-forward. Scenarios, by definition, are based on assumptions about the future. Different assumptions, particularly on technology and public policy can have significant impacts on how the scenarios play out. The IEA NZE scenario is a good example of this. Also, one must pay attention to the time frame as published energy scenarios today look forward to 2040, 2050 and even 2070. Finally, different scenarios count differently. For example, the EIA IEO (2021) includes hydroelectricity in its total for renewable energy. Many other scenarios count hydroelectricity separately.

peaking around 2030 and then falling significantly through the rest of this century. In "Sky," oil demand in 2050 is slightly less than today and by 2070, it has dropped to 50–60 million barrels/day, a little more than half of today's level. While these are significant decreases, oil and natural gas are still required to supply 10% and 6% of global primary energy respectively in the Shell "Sky" scenario, down from 32% and 22% today.

One additional scenario to be discussed briefly is from the International Energy Agency (IEA). In mid-2021, the IEA published their "Net Zero by 2050" scenario (IEA NZE, 2021). The aim is to achieve global net zero emissions, or a balance of greenhouse gases produced and removed, by 2050. Similar to, but more aggressive than Shell's Sky scenario, IEA NZE relies on significant technology advances to grow the use of renewables, limit demand for fossil fuels through electrification of transportation and widespread recycling of plastics and in managing emissions. As a result, in 2050 oil demand is 70% lower than in 2020 at 24 million barrels/day, and annual natural gas demand falls by 55% to 1750 billion cubic meters. Total energy demand increases by 30% by 2050, somewhat less than other scenarios.

Across the range of future scenarios currently imagined, there is one constant; that oil and natural gas will be a significant part of the global energy system for at least the next few decades, and possibly much longer. In some scenarios it is significantly diminished from today; in others, demand for hydrocarbons grows. In either case, the impact on deepwater is that the oil and gas industry will be active there for decades to come.

Much of the current deepwater infrastructure will therefore still be in use, including offshore platforms, pipelines and onshore support bases. The active presence of the industry could improve prospects for the future utilization and potential repurposing of deepwater infrastructure, or just as easily delay or complicate the development of new uses. It depends very much on the time frame involved and potential operational, political or economic synergies.

There is no doubt though that in order to operate efficiently, the oil and gas industry will continue to develop a better understanding of offshore processes within the water column and at the seabed, as well as a more detailed understanding of the subsurface. In an environment such as deepwater where it is unclear what the future has in store increased knowledge will likely be very beneficial to all forms of use.

## Vision for future deepwater and the impact of marine geological processes

As humanity and human activities develop over the next 100 years, there will be many compelling reasons to consider expanding activities into the deepwater environment. The objective of this section is to envision a range of potential uses and activities that could take place in deepwater offshore areas. Compared to onshore, offshore activities have less impact on livable land and in many situations the opportunity is more robust offshore than onshore. Deepwater typically has the advantage of moving operations well away from sensitive coastal ecosystems and heavily trafficked coastal areas. This also creates an opportunity to move infrastructure outside of the shoreline "line of sight" (often 32–40 km or 20–25 miles). Established deepwater industries include oil and gas exploration and production, fish farming, military uses, shipping and communications infrastructure. Many other industries, including wind and solar energy production and geologic storage, have begun commercial developments offshore and are planning future projects. There are a myriad of additional economic ventures and infrastructure development uses under research and consideration, including mining (where exploratory permits have already been issued) and human habitats. While this chapter emphasizes the need to research the impacts of geologic processes on offshore developments, a broad range of physical, environmental, economic, social and political challenges, issues and impacts need to be researched prior to major industrial progression from onshore to offshore. Most of these are outside the scope of this text.

Potential geologic impacts must be studied and considered as part of the analysis of potential safety and environmental impact and risk assessment for a deepwater activity. As more industries move offshore and into deepwater, it becomes even more critical to understand the impacts of the physical geologic, oceanographic and environmental processes from the subsurface, at the seabed and the water column (Fig. 3). Any forces or activity at the sea surface, within the water column or at the seabed, such as a major storm, sediment input from updip or a submarine slide, can impact offshore developments. For example, landslides as documented in Chapter 10 (the Storegga Slide) can displace enormous volumes of sediment and damage or destroy infrastructure.

As humanity considers increased industrialization and uses of the deepwater environment, it is critical to understand the geological and hydrological setting and the impacts of geological processes on these activities. The potential developments and applications described in this chapter impact one or more of these settings (Table 1).

Table 1 and Fig. 4 are generalized schematics that illustrate many, but not all, of the potential areas that can be impacted by marine sedimentary or subsurface processes. A key message of this chapter is that the impacts of these processes on industrial, environmental, agricultural and societal activities as well as supporting infrastructure in the ocean must be considered prior to moving forward with development. Before a range of deepwater development concepts are introduced, a brief synopsis of deepwater geologic processes is discussed.

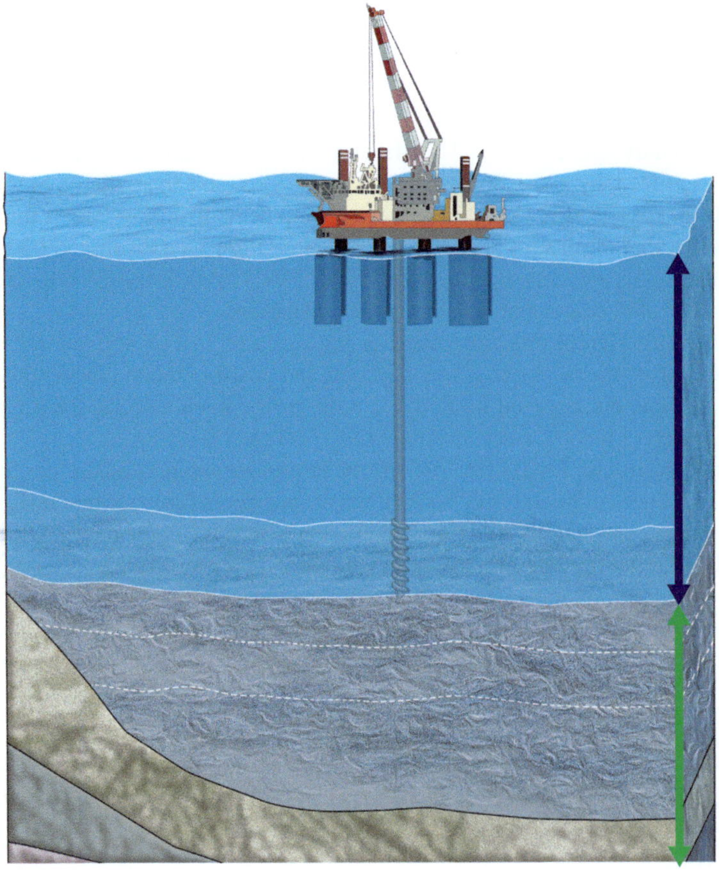


**FIG. 3** Schematic illustrating zones of interest for geologic processes: the ocean surface, the water column, the seabed and the subsurface below the seabed.

Understanding the impacts of water mass circulation (Chapter 9) and the associated oceanographic processes (internal waves, eddies, etc.) and chemical and physical processes in deepwater are critical to the risk assessment, opportunity analysis and project planning for future deepwater opportunities. Continued development of the subsurface in deepwater requires advancement of concepts and technologies for oil and gas as well as building knowledge and expertise in how sediments, pressures and fluids will be impacted through future uses such as carbon storage and mineralization.

The ocean water column typically contains variable amounts of clay-sized sediment and is the primary medium through which coarser grained sediment travels before being deposited in the deepwater. Although most deepwater sediment is pelagic or hemipelagic in nature—derived from the settling of the clay-sized particles—sedimentary processes also include turbidity currents, debris flows, transitional flows and other mass movements including creeps, slumps and slides (see Chapter 6). Marine sediment can also be transported and reworked through bottom currents (see Chapter 9). As previously mentioned, these currents and flows can cause significant damage to infrastructure in the water column and at the sea floor, and possibly trigger other sedimentary processes (as mass transport) and natural disasters such as tsunamis (Laberg and Camerlenghi, 2008; Leon et al., 2017; Miramontes et al., 2018; Brackenridge et al., 2020). Prior to deepwater drilling in the oil and gas industry today, geohazards teams are responsible for assessing the risk that sediment gravity flows, mass-transport events and shallow gas accumulations pose to drilling or subsea installation and operations. The ocean floor, or seabed, is the site of bottom current activity, sediment transport and reworking, and extensive biological activities. Processes like slumping, sliding and bottom currents are also active at the seabed. Chemical processes also occur (mineralization and mineral deposits) at or near the seabed. These processes are discussed in previous chapters in this book

**TABLE 1** Examples of deepwater activities and location of infrastructure.

| Use | Sea level | Water Column | Seabed | Subsurface |
|---|---|---|---|---|
| **Energy Resources, including:** | | | | |
| *Petroleum* | ■ | ■ | ■ | ■ |
| *Wind* | ■ | ■ | ■ | □ |
| *Solar* | ■ | ■ | ■ | □ |
| *Tidal* | ■ | ■ | ■ | □ |
| *Wave* | ■ | ■ | ■ | □ |
| *Geothermal* | ■ | ■ | ■ | ■ |
| *Nuclear* | ■ | ■ | ■ | ■ |
| *Physiochemical Conversion* | ■ | ■ | ■ | □ |
| **Subsurface Storage including:** | | | | |
| *Carbon Dioxide ($CO_2$)* | □ | □ | ■ | ■ |
| *Hydrogen ($H_2$)* | □ | □ | ■ | ■ |
| **Environmental Resources including:** | | | | |
| *Minerals* | □ | ■ | ■ | □ |
| *Hydrates* | □ | □ | ■ | ■ |
| **Aquaculture (Mariculture) including:** | | | | |
| *Plants including Seaweed, Algae* | ■ | ■ | ■ | □ |
| *Animals including Fish, Shellfish* | ■ | ■ | ■ | □ |
| **Infrastructure including:** | | | | |
| *Human Communities. Seasteads* | ■ | ■ | ■ | ■ |
| *Pipelines* | □ | ■ | ■ | □ |
| *Cables* | □ | ■ | ■ | □ |
| *Energy Storage* | □ | ■ | ■ | ■ |
| *Recycled Structures* | ■ | ■ | ■ | ■ |
| **Monitoring:** | | | | |
| | **Activity benchmarked and monitored to assure safety and integrity** | | | |

■ Zone impacted through activity, location and impact of structure

□ Zone to be analyzed prior to and during infrastructure installation

This table provides a generalized assessment of the zones described in Fig. 3 and illustrates the zones where equipment, operational activity and infrastructure associated with different types of offshore activities are likely to be present. Floating or fixed platforms, mooring lines, production and monitoring equipment, drones and remotely operated vehicles, geophysical arrays, fiber optic cables, pumps, manifolds and wells are examples of structures associated with offshore development. Understanding the geologic processes impacting these environments are critical to evaluating the viability of these activities in deepwater and to conducting safe operations. For equipment at or near sea level, the impacts of storm winds, currents, marine pollution should be considered in operations. Within the water column and at seabed, potential impacts of turbidity flows, mass transport, hemipelagic and pelagic sedimentation and ocean currents could impact operations and structural components of an offshore facility. In the subsurface, active faulting, mass movement and earthquakes are examples of geologic processes that could impact infrastructure.

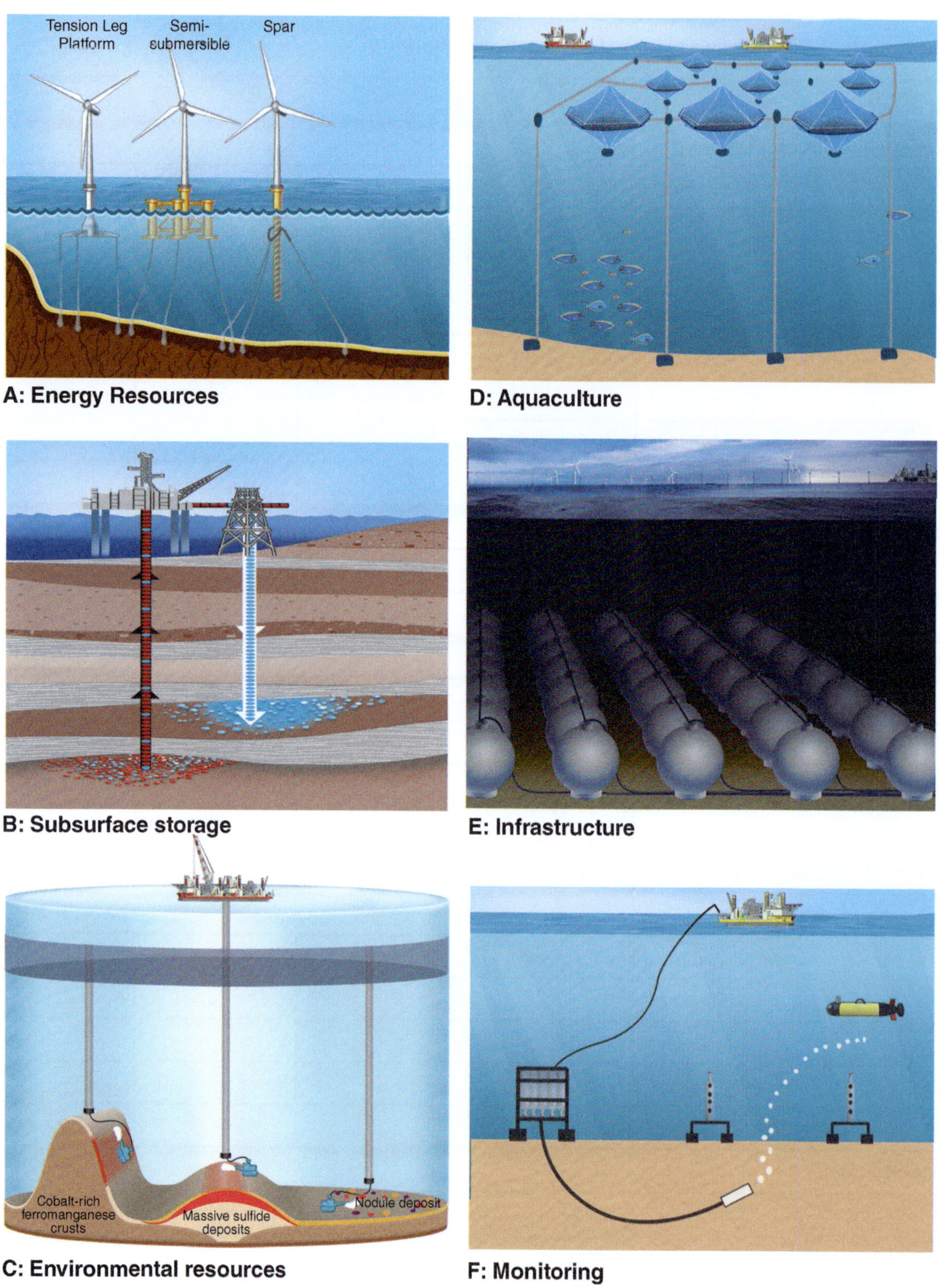


**FIG. 4** Examples of conceptual structures demonstrating types of infrastructure potential in the offshore including: (A) Energy resources: an example of floating wind structures including turbines, platform, hull and mooring lines, (B) subsurface $CO_2$ storage example including rigs, platforms and wells, (C) environmental resources mining example including floating production facility, wells/riser and seabed equipment, (D) aquaculture example including boats, cages and mooring lines, (E) infrastructure example including storage structures, cables, shallow water fixed platform and cables and (F) monitoring example including boats, cables, seabed monitors, autonomous underwater vessels and wells. It is critical to understand the potential impacts of geologic events on infrastructure at/near sea level, in the water column, on the seabed and in the subsurface. Examples of infrastructure are derived from illustrations provided by BSEE (Bureau of Safety and Environmental Enforcement, 2022), StEnSea (drawing from HOCHTIEF; Fraunhofer IEE, 2022), IEA Offshore CCUS and Monitoring, drawing from K. Davies (IEA, 2020a) and Pew Trust (PEW Charitable Trusts, 2022).

and the impacts of these processes, and many others, should be assessed prior to moving forward with the types of developments highlighted in the following sections.

## *Energy resources*

### General context

The ocean has always been a source of sustenance and commerce for humanity, primarily through fishing and harvesting. Local energy has been provided for centuries through the gathering of driftwood and harvesting of whale oil and seaweed. Energy from tides has also been used to grind grain (Tide Mill Institute, 2021). In the mid-20th century, oil and gas exploration and development moved into the offshore (Chapters 1, 19 and 20 in this book) and eventually into deepwater. In the initial decades of the 21st century, renewable energy projects such as wind and solar developments began being deployed offshore. As infrastructure grows and technology improves, these developments are likely to move into deeper water. Other forms of offshore renewable energy resources including tidal, wave and geothermal power could become integral parts of the energy portfolio. The high-level descriptions below are provided to frame a range of existing and potential energy resources that could be implemented in deepwater. Also, the impacts of geologic processes on activities at the sea surface, in the water column, at the seabed and in the subsurface are introduced.

### Oil and gas

Offshore oil and gas exploration has been active since the early 20th century with the first wells in a few feet of water (Chapter 19 in this book). In the past 100 years, oil and gas supply from the offshore has increased significantly, representing approximately 30% of global petroleum production in 2015 (EIA, 2016). The first deepwater discoveries were made in the mid to late 1980s in the U.S. Gulf of Mexico and Brazil, and subsequent exploration opened up new frontiers in offshore West and sub-Saharan Africa, Europe and other basins. In just over 3 decades, deepwater production has grown to 10 mmboe/day, representing 10% of global oil production ( JPT, 2019; IEA, 2019).

Many economic models indicate an approximately 30%–50% increase in global primary energy demand by 2050 (EIA IEO, 2021; IEA NZE 2021; Shell 2021). Established deepwater basins have proven petroleum systems (Chapters 2 and 18 in this book) with significant investment in infrastructure and access to global markets. Although major new discoveries may be less frequent in these mature provinces, smaller prospects proximal to existing infrastructure often have low exploration risk and less environmental impact with attractive economics when successful (Chapter 20 in this book). Proven emerging deepwater regions with significant petroleum resources like offshore Guyana, offshore northwest Africa, northeastern Canada and offshore India host growing infrastructure and will also supply petroleum for the next several decades (for a timeline of oil and gas activity, please see Fig. 3, Chapter 20 in this book; for a description, timeline and table of major deepwater discoveries, see Chapter 1). How energy supply and demand will evolve over the coming decades is uncertain, but many scenarios suggest that deepwater petroleum production will continue to be a part of the energy mix for at least the next 20 and perhaps for more than 50 years (Rystad Energy, 2020).

In addition, as society demands lower carbon energy sources (see Chapter 1 in this book for additional reading on this topic), offshore oil and gas production in proven basins could remain a part of the energy mix for longer than many other sources of petroleum because of their lower environmental impact and attractive economics. Offshore production, and particularly deepwater production, typically has smaller carbon and environmental footprints (Fig. 5, so offshore exploration and production could prove to be preferred in the future (Oberstoetter and Usoro, 2021). Secondary and tertiary recovery methods from existing deepwater fields could also produce lower carbon petroleum, with the potential for $CO_2$ enhanced oil recovery (EOR) to sequester carbon dioxide as well.

### Offshore renewable energy, general

Low carbon offshore renewable energy resources are of great interest to governments, communities and investors. Wind and solar energy are increasingly moving offshore, with deepwater projects in the planning and installation stages (Chalier and Justus, 1993; Yang and Copping, 2017). Marine renewable energy can be grouped into three categories: (1) energy resources based on the dynamic characteristics of the ocean, including energy conversion from currents, waves and tides; (2) energy resources based on the physicochemical characteristics of the ocean, which include Ocean Thermal Energy Conversion (OTEC), conversion from the salinity gradient and biomass energy (bioconversion); (3) others including wind, solar and geothermal energy (Chalier and Justus, 1993; Williams and Taylor, 2012; Shields and Payne, 2014; Shadman et al., 2019; Samad et al., 2022). Research and development of marine energy from waves, currents, tides, OTEC, salinity gradients and geothermal energy are underway and all of these are technically feasible if not yet economic (Chalier and Justus, 1993; Shadman et al., 2019). Comprehensive studies led by government and research institutions offer both regional assessments of

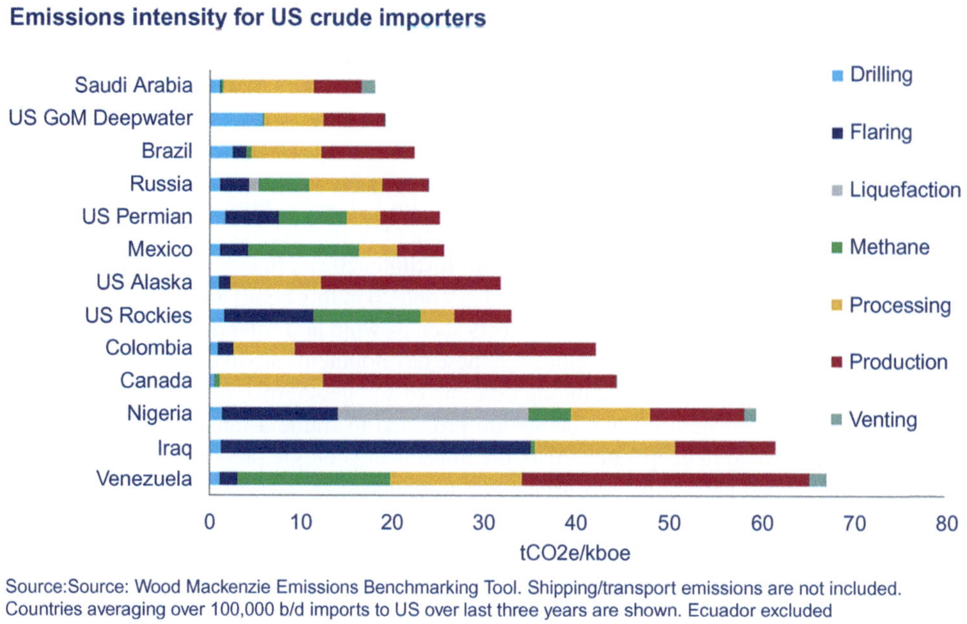


**FIG. 5** Carbon emissions intensity for U.S. crude oil importers. Note that the U.S. Gulf of Mexico deepwater and Brazil (which is dominated by deepwater production) have lower carbon intensities than all other significant crude oil sources except Saudi Arabia. *From Oberstoetter and Usoro (2021).*

renewable energy potential (see Musial et al., 2017 for an assessment of the U.S. Gulf of Mexico) and specific assessments of renewable energy potential by type (see ESMAP, 2019 for assessment of global offshore wind technical potential).

There are multiple advantages to the development of offshore renewables. Potential energy resources are plentiful in the water column as well as at or near the ocean's surface, and if proven safe and economic, would offer a wide range of potential low carbon resources to meet societal energy demands. Deepwater could be an attractive location given that most areas will be beyond the onshore "line of sight," not interfering with coastal views. With less seabed infrastructure and therefore potentially lower cost, floating facilities to capture energy from wind, solar and tides are possibly less obstructive to sea life and more practical to maintain. Some facilities may also be mobile, offering the ability to relocate as needed. Given the focus on reducing $CO_2$, methane and other greenhouse gases, there is growing funding for research and potentially increasing financial subsidies, and offshore renewable energy projects are attracting significant interest and potential capital investment.

Several of the significant technical challenges for deepwater renewables are energy transmission and storage. This could be addressed by establishing a cable infrastructure to transport energy. Alternatively, fuel cells or portable batteries could be charged and physically transported via ship to transfer electricity and energy or energy can be stored in ocean or seabed containers. Where available, offshore subsurface salt caverns can be utilized for energy storage (see geologic storage, this chapter).

One novel approach being pursued today is that of an Offshore Energy Island. Constructed as a hub for gathering and transporting offshore energy to shore, energy islands can help to minimize development costs and the environmental footprint associated with offshore energy development. Currently, offshore energy islands are being planned as grounded artificial islands in the shallow water of the North Sea (Danish Energy Agency, 2021) to gather and move renewable wind energy to shore. The concept has also been employed in the past by the oil and gas industry in the exploration and development of offshore and remote resources. Four artificial islands (THUMs islands) in San Pedro Bay, off the coast of Southern California, U.S., near the city of Long Beach serve as operations and processing hubs for oil and gas production from the Wilmington Field (see Chapter 1). Artificial Islands have also been built offshore the north coast of Alaska, U.S., and in the Persian Gulf (Karam, 2015).

Deepwater also possesses risks and uncertainties. The impact of ocean industrialization on the environment, particularly in deepwater, is not yet fully understood. Increased environmental, biological and geological research and pilot studies should be undertaken before a major scale-up of these industries occurs. Baseline studies to accurately characterize the deepwater environment are needed to assess the impact of future development.

## Solar energy

Solar is one of the most widely available renewable resources and with 70% of the Earth's surface covered by ocean, there is a lot of potential for offshore solar development (Chalier and Justus, 1993). The first commercial project, Zon-op-Zee, offshore Netherlands (North Sea) was piloted in 2019 (Offshore Energy, 2020a) with 56 solar panels. The solar installation has an initial capacity to generate 17 kilowatts of power. with potential to be expanded to 50–100 kilowatts in the near-term. To date (2022), Asia dominates offshore solar with significant projects installed or planned for China (263 gigawatts), India (78 gigawatts) and Japan (35 gigawatts) over the next decade (Radowitz, 2021) and there is wide interest in continued development of offshore solar projects globally. Solar panels can be mounted on fixed or floating platforms. Floating photovoltaic systems are also promising, with several pilot projects under development globally.

## Wind energy

Offshore wind energy is attractive for multiple reasons. In general, winds are stronger and more consistent offshore than onshore. There is less visual impact from shore if the facility is positioned beyond the line of sight. Larger wind turbines deployed offshore can lead to lower cost and greater production efficiency per unit (Chalier and Justus, 1993). In 2020, offshore turbines generated roughly three times as much energy as onshore wind turbines (Global Citizen, 2020). Like other forms of offshore energy, wind energy can be produced using fixed seabed platforms or floating facilities. The first offshore wind farm, Vindeby, consisted of 11 turbines and was commissioned in Denmark by Ørsted in 1991 (Ørsted, 2021). Vindeby was rather modest in scale, generating a total of 5 megawatts (MW) of power. By 2021, there were 169 offshore wind installations globally, with multiple future projects under consideration. The largest individual offshore wind turbines being installed today produce up to 14 MW and the largest offshore wind farm today, Hornsea 2, in the UK North Sea, has a planned capacity of 1300 MW. For comparison, a typical nuclear power plant in service in the United States today produces about 1000 MW.

By 2021, there were an additional 24 offshore wind farms under construction globally, with over 50% of that capacity being built in China. Total offshore global wind energy by end 2021 is predicted to reach over 34 gigawatts (GW). The majority of offshore wind installations are in Europe (primarily the North Sea) and China, with smaller operations in another 15 countries around the world (World Forum Offshore Wind, 2021).

In 2020, global offshore floating wind capacity was over 26.5 GW with floating wind turbines currently deployed in offshore Europe, USA, UK, Japan and China (Musial et al., 2021). As solutions to manage long distance energy transmission are developed, wind facilities can be pushed farther offshore into deeper water. Floating offshore wind farms, supported by floating or subsea substations to collect and transmit power, offer great promise for deepwater and are currently being studied and evaluated (Maslin, 2021b).

## Wave, tidal and current energy

Waves, tides and ocean currents produce a tremendous amount of kinetic energy (Chalier and Justus, 1993). Harnessing this energy and converting it to electricity is feasible but has not been applied at scale due to a combination of environmental, economic and technological challenges. Interest in development of renewable energy is creating a resurgence in research, technology development and demonstration projects for these resources.

Wave energy is generated by wind movement across a body of water (Pizzo et al., 2021). Despite the development of multiple demonstration and pilot projects, wave energy has yet to see widespread commercial success, and most applications for wave energy are still in the development and testing phase. In addition to mechanical devices such as wave energy converters, energy can be harnessed from waves using piezoelectric materials like quartz ( Jbaily and Yeung, 2015) which convert physical stress directly into electrical energy.

Tidal energy is generated as the gravitational pull of the moon and sun combine with the Earth's rotation to create ocean tides (National Geographic, 2021a, b). Unless someone steals the moon (please see the 2010 movie "*Despicable Me*"), there will always be tides that are predictable and ubiquitous. The tidal power industry is still relatively immature but offers promise as a dependable and renewable energy resource. Energy created from tidal movement can be used to generate electricity primarily along coastlines; however, the environmental implications need to be researched and well understood before the technology is scaled up. Alternatively, turbines could be developed for installation in deeper water where ocean currents could be harnessed to create hydropower.

Deepwater has advantages for power generation from the tides. The deep marine environment is more calm compared with the sea surface, and not typically subjected to the same extreme events. For example, wind turbines near the sea surface might be subjected to hurricane force winds and must be engineered appropriately. The tides are also predictable and repeatable which has certain advantages in supplying a power grid. There are many areas globally with the potential for harnessing tidal energy for electricity generation (Offshore Energy, 2015) where there is significant movement of a large volume of water due to tidal forces. Small-scale tidal power projects have been successfully operating for decades; the La Rance tidal power plant in France has been in operation since 1966 (Evans, 2019). There are significant engineering and environmental challenges and there are no industrial scale tidal power stations in operation today, but the potential is there, and deepwater might prove to be an attractive environment for harnessing that potential.

### Geothermal

An area of growing interest is offshore geothermal energy production (Chalier and Justus, 1993). To date, there are no active offshore geothermal projects; however, offshore geothermal could be a viable energy resource. Geothermal energy could be generated through a range of approaches and technologies (Hanson, 2020), including:

- an ocean platform-based power plant where geothermal steam moves through a pipeline from the seabed to the surface to power turbines,
- a land-based power plant where the two-phase geothermal fluid is separated at the seabed, borrowing multiphase fluid processing technology from the offshore oil and gas industry, and then directing the pure steam onto land via pipeline,
- an underwater power plant producing electricity and transporting it to land,
- a binary power plant on land which uses a heat exchanger located on the seabed, heating the circulating working fluid with geothermal fluids, and piping the working fluid to shore and
- a transmission line connected to a thermoelectric device using the temperature difference between the geothermal fluid and the ocean. This temperature difference could be quite significant, with geothermal fluids at temperatures exceeding 200 degrees Celsius and the surrounding ocean at the seabed in deepwater is typically near 0 degrees.

Countries with active volcanism and thin continental crust like Iceland and Norway are investigating the feasibility of offshore geothermal energy production (Richter, 2019).

Another area of investigation is the conversion of existing offshore oil wells into geothermal energy producers. This re-utilization of offshore wells and platforms could represent an effective way to utilize existing infrastructure to provide local renewable power and geothermal energy. Energy production from decommissioned oil and gas infrastructure is feasible and is being studied in areas like the North Sea (Maslin, 2021a).

### Nuclear

In the mid-1970s, floating nuclear power plants were investigated as an offshore alternative to land-based nuclear power generation. These plants offer advantages but also carry significant risks. More recently, a team from MIT is investigating the concept of siting a conventional nuclear reactor on an offshore floating platform (Stauffer, 2015). These facilities could utilize many of the technologies matured by the oil and gas industry and with nuclear-powered ocean vessels. The current appetite for nuclear facilities onshore or offshore is low, but new risk reduction technologies, the growing need to develop non-$CO_2$ emitting energy resources and a shift in public sentiment could enable the development of a floating nuclear power plant industry. Indeed, the world's first floating nuclear power plant will be deployed in Russia (Foro Nuclear, 2021) in the coming decade.

## *Offshore storage*

### General

The offshore subsurface offers multiple opportunities to store gasses and fluids in either a permanent or temporary capacity. Carbon dioxide ($CO_2$) can be stored permanently underground in saline aquifers, through enhanced oil recovery (EOR) and via mineralization. There is also the potential to store $CO_2$ in offshore coal seams. Storing hydrogen, ammonia or electricity temporarily underground until needed is another potential option to store energy.

### Carbon dioxide storage

One of the promising offshore opportunities is carbon dioxide storage. Carbon capture, use and storage (CCUS) combines processes and technologies to capture carbon dioxide ($CO_2$) from industrial emissions or directly to the air and then compress and transport the $CO_2$ to be converted into products or stored permanently underground (Fig. 6; NPC, 2019). There are numerous options for storing $CO_2$ in geological formations. At this time, the most easily accessible and scalable solutions are subsurface storage via EOR or injection into a saline reservoir, and both of these solutions are viable for offshore.

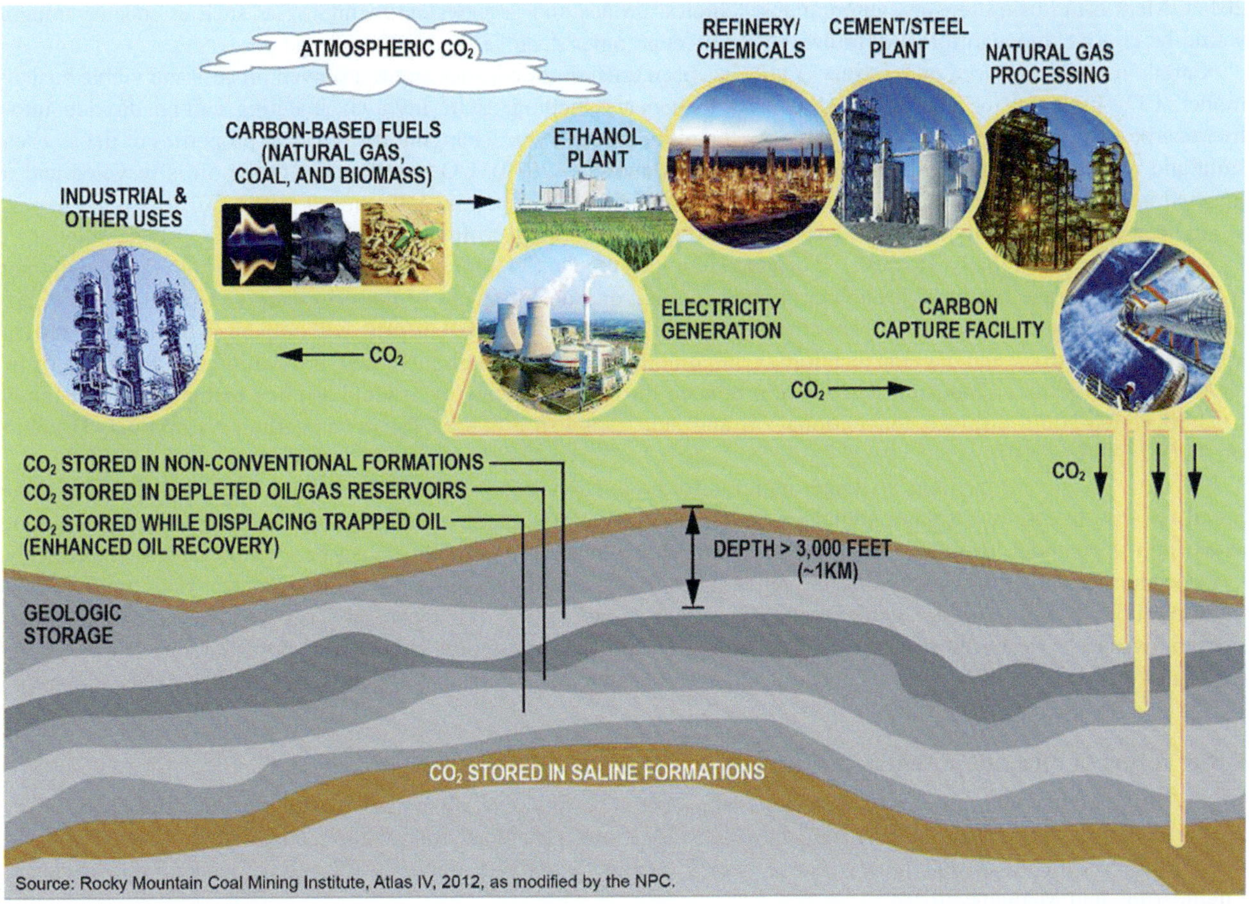


**FIG. 6** The technology chain and phases of carbon capture, use and storage (NPC, 2019). *Figure courtesy of the National Petroleum Council, as modified from the Rocky Mountain Coal Mining Institute original illustration.*

$CO_2$ can also be stored through mineralization or in depleted oil and gas fields (IPCC, 2005) and in deep, unminable coal seams.

Globally, there are a number of $CO_2$ capture and storage projects operating and in development, both onshore and offshore, that are validating critical technologies (Global CCS Institute, 2021). CCUS is at a tipping point globally and is an industry very much in its infancy, and offshore storage could play a significant role in its future. Onshore carbon capture projects can also link to offshore storage. Some projects which employ offshore $CO_2$ storage include Sleipner (Norway, North Sea), Snohvit (Norway, Barents Sea), Gorgon (Australia), Santos Basin EOR (Brazil, Santos Basin) and the planned Net Zero Teesside industrial hub (UK, North Sea; NPC 2019; Global CCS Institute, 2021).

## Offshore $CO_2$ storage methods: EOR, saline storage, abandoned oil fields and recycling platforms

In saline storage, $CO_2$ is injected into saltwater-bearing sedimentary saline aquifers. Captured $CO_2$ is compressed into a super-critical fluid, transported and injected into a saline or "wet" formation. Global storage volumes available in saline reservoirs do not appear to be a constraint and potential offshore carbon storage sites are plentiful in many of the sedimentary basins around the world. For example, saline reservoir storage potential estimated for the western U.S. Gulf of Mexico (offshore Texas and Louisiana) from five calibrated geologic intervals alone is assessed at over 550 billion metric tons (Carr, et al., 2011). For calibration, the USA produces 5.1 billion metric tons of $CO_2$ per year.

Offshore CCUS, though still economically and regulatory challenged, is a realistic option. Offshore provinces are often managed by a government, which offers one landowner and a common regulatory framework (NPC, 2019). Offshore is distanced from populated areas so there is little infrastructure nearby and there are no populated areas or potable aquifers to be threatened. Because of the physical properties of supercritical $CO_2$ (the most likely phase to be injected in underground storage operations) the overlying water column would likely help to control any potential subsea leak (NPC,

2019). Monitoring using sensors, autonomous vehicles, drones and geophysical technologies such as seismic imaging would be relatively straightforward, utilizing many technologies deployed today for oil and gas.

Carbon dioxide ehnanced oil recovery (EOR) has been utilized for decades and is a proven process and viable business model. $CO_2$ EOR can be employed as a *tertiary recovery* method. EOR involves injecting carbon dioxide into a hydrocarbon-bearing reservoir. $CO_2$ dissolves in the oil in the reservoir, changing the physical properties of the reservoir fluid and resulting in a much higher recovery factor (Mawdsley, 2019). $CO_2$ produced with the oil is recycled and re-injected into the reservoir, where ultimately, up to 99% of the $CO_2$ is permanently stored (NPC, 2019).

EOR, both onshore and offshore, does not likely offer enough volume to address the significant amounts of $CO_2$ storage needed globally. However, it offers a stepping stone to $CO_2$ storage at scale and will further the development of the carbon storage business through continued funding and development of the necessary $CO_2$ infrastructure, both onshore and offshore (NPC, 2019). The oil and gas industry, through its EOR efforts, will help build expertise in moving and storing $CO_2$, drive and develop technology and monitoring capabilities, encourage innovation and provide jobs in a developing $CO_2$ storage industry. Petroleum produced through EOR may even yield a lower carbon or even negative $CO_2$ oil which could displace conventional oil and help establish a market for lower carbon oil products (Denbury Inc., 2022; MIT, 2010).

Offshore oil and gas facilities no longer required for production (e.g., platforms, wells, pipelines, etc.) could be repurposed for $CO_2$ injection, but there are technical, commercial and regulatory factors which need to be considered when assessing the potential for re-use. Trapping mechanisms and reservoir suitability must be assessed and analyzed. Other key questions include the evaluation of the potential to re-use existing wells as injection wells, or existing pipelines for transportation, or will projects require completely new infrastructure? These technical aspects as well as economic considerations must be evaluated on a case-by-case basis. Offshore platforms could also serve as potential hosts for $CO_2$ Direct Air Capture (DAC) facilities, offering the potential to store captured $CO_2$ locally in the underlying subsurface stratigraphy.

## Other forms of offshore potential carbon capture: Mineralization

Another form of geologic carbon dioxide storage which could be employed offshore is in situ mineralization (NPC, 2019). This is caused by accelerated "weathering," in which $CO_2$ forms a chemical bond with reactive minerals found in mantle peridotites, basaltic lavas and other reactive rocks, creating carbonate minerals (National Academies of Sciences, Engineering, and Medicine 2019).

These include potential opportunities for sub-surface mineralization offshore where offshore ultramafic rocks are shallow enough to serve as injection host rocks (Keleman and Matter, 2008).

## Other forms of offshore potential carbon capture: Ocean carbon storage

The ocean naturally sequesters $CO_2$ but the process of "intentional" storage of $CO_2$ in the deep ocean could be applied at an industrial scale to store carbon. Because of the unique stability characteristics of supercritical $CO_2$, deepwater has also been considered for potential carbon storage in seawater at or just above the seabed. $CO_2$ could be stored in deepwater at the seabed as a carbon dioxide hydrate, a liquid, a gas or in an aqueous phase (IPCC, 2005; House et al., 2006).

## Salt canopies and salt domes

Salt domes and salt canopies provide another subsurface storage option in the offshore. Salt domes have long been an onshore storage site for petroleum (for example, the U.S. Strategic Petroleum Reserve) and the use of salt domes for storage is being investigated for offshore energy and hydrogen storage. Many deepwater basins have salt domes and canopies which could developed as energy storage sites, including the USA, Gulf of Mexico, offshore Angola, UK and Norwegian North Sea, and Brazil (see Chapter 5 for more on deepwater salt tectonics).

There are many potential options for storage in excavated caverns within salt, and in porous reservoirs around and under salt (where the impermeable salt serves as the trapping element). In addition to providing sites for permanent $CO_2$ sequestration in surrounding sediments, salt caverns offer opportunities for storing green hydrogen produced offshore, ammonia (as a dense state fluid for energy storage) produced from hydrogen (Patonia and Poudineh, 2020) or compressed air (Ali, 2019). Hazardous waste could also be stored in offshore sedimentary intervals or salt structures.

## Other forms of offshore energy storage: Batteries and bags

In addition to subsurface storage, there are other options to store energy offshore. One technology being investigated is the use of batteries on ships or platforms to store excess energy produced by renewable resources. Offshore energy storage "bags"

(Energy Digital, 2020) are another emerging technology which could be used to store energy as compressed air in the water column.

The European Marine Energy Center is producing hydrogen gas as a means to store unused renewable energy produced from tidal and wind energy (EMEC, 2017). Hydrogen gas is produced in the outer Orkney islands, off the northeast coast of Scotland, by 500–1000 kW solid oxide fuel cells—or *electrolyzers*, for short—that runs in regenerative mode to achieve electrolysis of fresh water and produce both hydrogen and oxygen.

## *Deep sea mining (environmental resources)*

The increased need for rare earth elements (REE) and other metals for batteries, computers, cell phones and other electronics has raised the potential for mining minerals from the ocean (rare earth elements were introduced in Chapter 10 of this book as a way to link source to sink). This includes minerals collected from the water column, seabed or near surface. Minerals and methane are two potential products which could attract commercial attention in deepwater environments and there is growing interest in offshore seabed mineral mining (Miller et al., 2018). Mineral mining directly from seawater is also technically feasible and is being actively researched.

### Hydrates

Gas hydrate is a solid crystalline material composed of gases (including methane, $CH_4$; ethane, $C_2H_6$ and carbon dioxide, $CO_2$) and water (2021, USGS, gas hydrate page). Gas hydrates are observed in some deepwater sedimentary intervals, present at the seabed and/or in the near-surface sediment. Hydrates have been studied as a possible source of methane and some subsea hydrate development and permits have been issued. Demonstration projects to produce gas hydrates have been carried out in Alaska, Canada and Japan (Collet et al., 2009; Green Car Congress, 2019).

Carbon dioxide storage could be integrated into hydrate mining where carbon dioxide is injected into the hydrate to replace the methane being released. The $CO_2$ forms a solid hydrate, which then restabilizes the sediment, enabling methane extraction while storing $CO_2$ at or near the seabed (World Ocean Review, 2014). There is much to be learned about the potential development of hydrates, and any activities associated with seabed or shallow sediment hydrate production should proceed with caution: not only is methane a potent greenhouse gas but hydrate accumulations host many sensitive biological communities which could be harmed by development activities.

### Minerals mining

The technologies utilized in energy transition activities such as magnets and batteries for electric vehicle motors, wind turbines and electricity generation and transmission will require significant amounts of critical minerals, and their supply emerges as a growth area in mining (IEA, 2020b; Musial et al., 2021; IEA NZE, 2021). The total market size of critical minerals like copper, cobalt, manganese and various rare earthelements (REE) is projected to grow almost five-fold between 2020 and 2050 in the IEA net zero pathway scenario (Fig. 7; IEA NZE, 2021). Many of these critical minerals are available in the deepwater environment, either on the ocean floor or dispersed within the water column (Hernández-Molina, 1996), and these minerals also occur as sediment-hosted minerals in onshore deepwater deposits (Maestro-González, 2021).

Mineral deposits in the ocean have been studied since the 1860s. Beginning in the 1960s, the United Nations began to develop a comprehensive framework for ocean governance including ocean minerals, resulting in the formation of the International Seabed Authority in 1982, to help regulate exploration and exploitation of seabed minerals. There is growing interest in offshore seabed mineral mining and at this time (2022), exploratory permits for seabed mining have been issued by the International Seabed Authority to 29 contractors around the world (Ocean Foundation, 2021).

Of primary commercial interest are three types of seabed marine mineral deposits (Lodge, 2021). The first are polymetallic nodules rich in elements like manganese, iron, copper, nickel, cobalt, lead, zinc and other minerals hosting REEs. These are found at the seabed, either exposed or partially buried in seabed sediment. The most heavily researched deposits are in the Clarion-Clipperton Zone of the eastern Pacific Ocean at depths of 3500–5500m (~11,483–18,045 feet). Additional deposits have been identified in the Indian Ocean and other parts of the Pacific Ocean (Lodge, 2021).

The second type of marine mineral desposits, polymetallic sulfides or sea floor massive sulfides, are associated with hydrothermal activity and can occur along plate tectonic boundaries including volcanic arcs. Deposits generally form in approximately 2000m (~6562 feet) of water and are rich in copper, iron, zinc, silver and gold (Lodge, 2021).

The third area of commercial interest is cobalt crusts that form through seawater precipitation of minerals in seawater and include iron, manganese, nickel, cobalt, copper and other rare metals. These are typically observed between

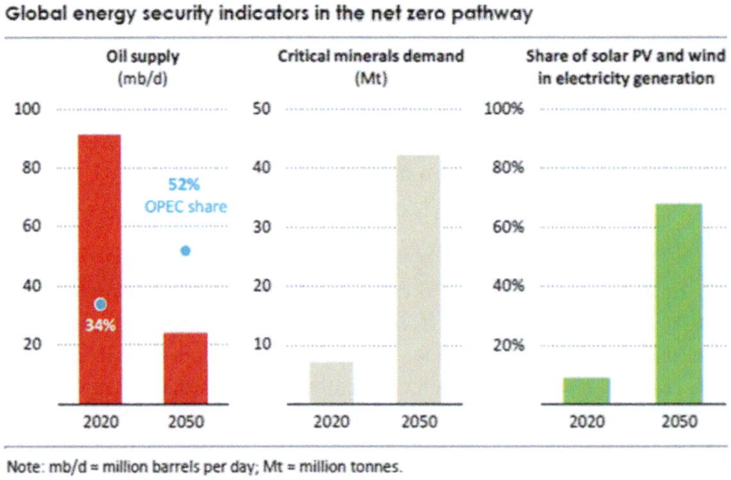


**FIG. 7** Potential demand in critical minerals projected by 2050 in the IEA Net Zero Pathways scenario (IEA NZE, 2021). This scenario invokes almost a five-fold increase in global critical minerals demand by 2050.

400 and 7000m (~1312 and 2297 feet) of water, associated with the flanks and tops of marine seamounts (Lodge, 2021; Fig. 8).

Especially important is understanding the potential impacts of mining on the biological communities associated with these geographies and the impact on these sensitive biological communities. There is also evidence that large-scale seabed mining processes could release carbon stored in deep sea sediment (Gronewold, 2019).

Critical mineral extraction from seawater is also feasible as a mining technique. Seawater contains large amounts of minerals, dissolved gases and organic molecules (e.g., Loganathan et al., 2017). Some of the most valuable minerals include the 17 REEs, precious metals, lithium and uranium. These minerals can be extracted using passive adsorption or electrochemical processes. These methods require power to operate, and most technologies are still in the basic research and development stage. Electricity generated through a local, ideally renewable source is likely the most cost-effective source of power (LiVecchi et al., 2019). Gases like carbon dioxide, oxygen and hydrogen can also be extracted from seawater. Salts, such as sodium chloride (common table salt), can easily be mined from seawater offshore; however, these are likely more attractive as a byproduct of desalinization required for other reasons rather than being attractive as a stand-alone industry.

## *Aquaculture*

Aquaculture is the process of farming in water: raising fish, shellfish, algae and crops which can be used to produce food, fuels and other products. Ocean farming (aka "mariculture") is already an established industry with nearshore regulations in some regions and vast potential for operations to move farther offshore as demand increases. In 2012, global output for fish farming was over 70 million tons—exceeding beef production for the first time. Nearly half of all fish and shellfish consumed are now farmed. In 2021, the commercial market for the farmed fish industry was over $300 billion and is projected to reach almost $450 billion by 2027, creating a significant industry while contributing to global food security and nutrition (National Marine Fisheries Service, 2021; Maximize Market Research, 2019; O'Shea et al., 2019). Fish farms located in deeper waters experience stronger ocean currents than coastal waters, which can mitigate some of the issues related to nearshore farming such as fecal matter waste and pollution (Gunther, 2018).

Marine agriculture is also an established and growing business. Kelp and other seaweed for human and animal consumption have been harvested from the sea for generations, and microalgae production is a rapidly expanding industry, with many co-benefits including ecosystem services like improved water quality and carbon sequestration (Ocean Stewardship Coalition, 2021). The currently $6 billion seaweed industry is expected to double in the next few years (Degnarain, 2020).

Algae (both macro and micro) has been studied for decades as a feedstock for biofuels. The economic and technology challenges to commercial production of biofuel from algae are significant, however research continues along a number of potential pathways to biofuel production from algae (NREL, 2021). Recent research also highlights the potential for seaweeds to play a role in hydrogen production (Zhang et al., 2020). Challenges still remain to wide-scale expansion of

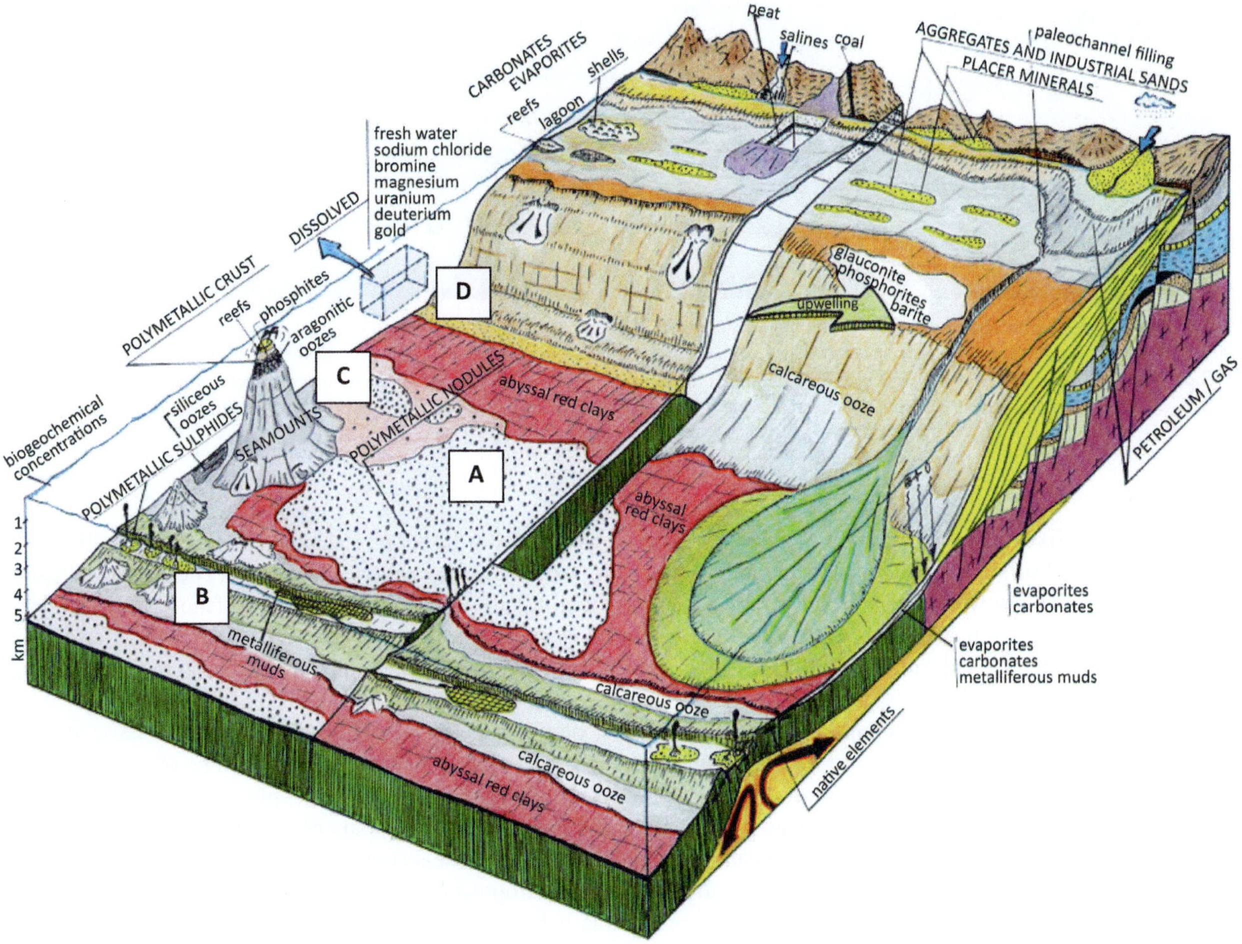


**FIG. 8** Locations and types of the main mineral resources on the seabed in deepwater environments. A = polymetallic nodules, B = polymetallic sulfides, C = polymetallic (cobalt) crusts, D = minerals dissolved in seawater. *Figure courtesy of Hernández-Molina (1996).*

mariculture in deepwater settings including safety-at-sea, potential for conflict with other ocean users (e.g., maritime shipping), operational costs and lack of regulatory frameworks and security (particularly in Areas Beyond National Jurisdiction, ABNJ). However, recent enhancements in ocean technologies and the increased focus on climate solutions (including ocean-based carbon dioxide removal, CDR) have the potential to overcome many of the challenges and reach the full potential of ocean-based productivity.

## Infrastructure

Infrastructure development, including communications systems, energy transmission and housing, will be required to continue the expansion of industrial applications from the oceans. This includes development for energy resources, mining, farming and other uses. Many of these technologies and applications require significant investment, as the ocean environment presents challenges and infrastructure must be resilient to harsh conditions and actively monitored. While new infrastructure will certainly be required for many new industrial activities and applications, there are likely many opportunities to reuse and redevelop existing industry infrastructure such as oil and gas platforms and pipelines, potentially avoiding the environmental and economic impacts of decommissioning infrastructure and accelerating new offshore developments.

### Offshore cities (seasteads)

Humanity has been building offshore for various reasons for over a millennium. For example, Venice (now Italy) was initially populated in the lagoon of the Adriatic Sea to protect locals from the Huns (Merali, 2002). The Netherlands has been integrating floating infrastructure into some coastal communities for hundreds of years. More recently, humans

**FIG. 9** A tryptic of examples of past, present and future offshore living communities. (A) Venice, which has existed as an offshore community for over 1000 years, (B) the Prosafe Safe Concordia, owned by Prosafe, an offshore accommodation vessel or "floatel," which houses staff working on an offshore gas (bp Cassia C) platform and (C) an artistic rendition of a future offshore city ("seastead"). *Panel (A): Image courtesy of Tyler Donovan; Panel (B): Image courtesy of Andee Nelson, used with permission from McDermott and Prosafe; Panel (C): Image courtesy of OCEANIX/BIG Bjarke Ingels Group.*

have constructed fixed offshore islands for leisure and tourism such as Palm Jumeirah in Dubai (Geology.com, 2022). Currently, commercial offshore habitation developments are being explored, for example: in Dubai (The World Islands), a development in shallow water exists in the Arabian Gulf. A floating city in the Indian Ocean, offshore of the Maldives is under development, with the first of 20,000 inhabitants moving in 2024 (Lewis, 2022). Offshore cities offer romance and uniqueness but could also create a pragmatic solution to onshore overcrowding, high land costs and mitigation against rising sea level.

Deepwater municipal development is also feasible. Ships, platforms and "floatels" (living quarters associated with development and servicing of offshore infrastructure) offer a 2022 version of deepwater offshore cities (restaurants, catering, rooming and power). For example, there are currently approximately 35,000 people working offshore in the U.S. Gulf of Mexico on oil and gas platforms (NAS, 2021) and hundreds of thousands of people habituate the ocean at any one time on cruise ships, fishing boats, oceanographic and military vessels (Fig. 9).

Advantages of offshore cities include the ability to utilize local hydropower and develop mariculture for fish and marine vegetation. These could be developed as completely renewable habitations with appropriate access to energy, food and water and other societal basics. These could offer a living solution adjacent to highly populated areas. Offshore cities could be developed around a central industry: minerals extraction, farming, energy production, then possibly diversified into more multi-functional regional centers through time. Alternatively, offshore cities could develop as needed as alternative residences adjacent to onshore communities. Excess energy and food production could be exported back to shore.

Offshore renewables could be combined with activities such as farming to form multi-use facilities and optimize logistics and ocean footprint.

Another opportunity is to integrate technologies on or near these facilities to help clean up ocean pollution, including plastics and other debris. This application could be piloted and then scaled up as technologies mature and as needed for environmental protection.

## Pipelines, cables, fiber and transmission

Infrastructure for energy transmission as well as pipelines for liquid and gas transportation will also continue to be developed as hubs that serve as energy storage sites. Many existing hubs have been built and are operated by the oil and gas industry, but offshore sites are under development associated with renewable energy infrastructure such as the Danish Energy Island (Danish Energy Agency, 2021).

Communication infrastructure and hardware such as fiber optic and internet cable are also installed in many parts of the ocean on the seabed (Fig. 10). Currently, 99% of the telecommunications data traffic across oceans is carried by undersea cables and this infrastructure continues to expand across the oceans.

There are currently >1,500,000 km (932,000 miles) of cables on the seabed (Alcatel, 2021). Deepwater research has assisted in identifying risk zones, including passages through straits and areas vulnerable to deep sea currents and as pinch points due to sea floor irregularities. When submarine cables are damaged or broken, repairs frequently cost at least $1 mm each. In addition, there can be huge commercial and financial losses and increased business risk for the companies which use the cables, particularly banks and other financial institutions.

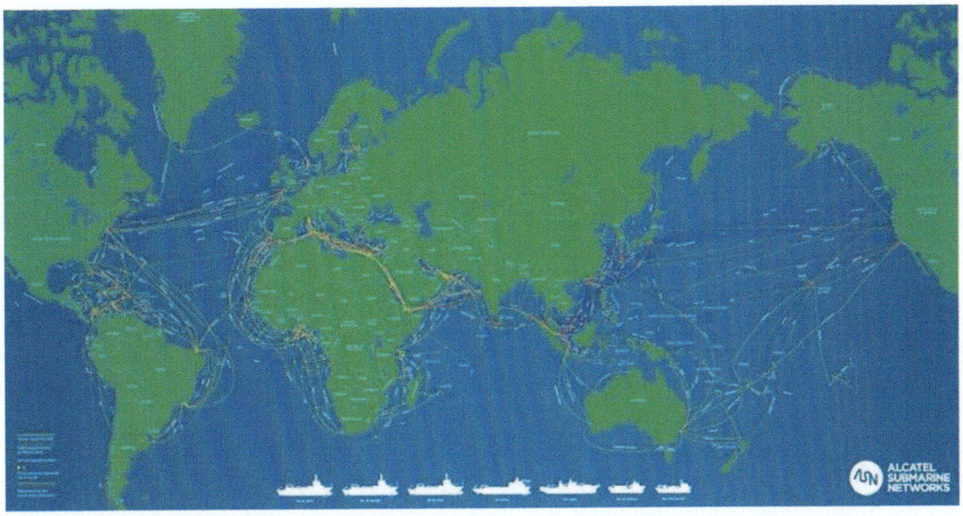


**FIG. 10** Map of global offshore telecommunication cables (Alcatel, 2021). *Image with permission and courtesy of ASN, Alcatel Submarine Networks.*

## Repurposed oil and gas infrastructure

Decommissioned oil and gas platforms offer significant opportunities to re-utilize existing offshore infrastructure. Decommissioning of offshore oil and gas infrastructure been considered a difficult and expensive task for the past several decades, but interest in repurposing these types of facilities has helped change a threat into an opportunity. Studies have been done to consider installation of solar and/or wind energy generation on or near existing platforms and there are proposals to create multi-purpose facilities.

An example of reuse of oil and gas infrastructure is the conversion of oil and gas infrastructure into artificial reefs for the tourism and recreational fishing industries (Bureau of Safety and Environmental Enforcement, 2021), but there many additional opportunities (Fig. 11) including:

- repurposing of oil and gas structures as wind, solar or tidal energy farms,
- offshore generation of green hydrogen and transportation to shore in existing or new pipeline networks,
- use of offshore platforms as hosts for carbon storage projects,
- redevelopment of offshore structures for sea water desalination and utilizing existing pipelines for transportation to shore,
- use of existing and future infrastructure to further develop aquaculture,
- potential use as offshore living quarters, either as a temporary living for worker or leisure or a more permanent offshore seastead (Fig. 11).

As a novel example of repurposing, in late 2021, Saudi Arabia announced their intention to redevelop an offshore oil rig complex into a theme park as a tourist destination (CNN, 2021).

**FIG. 11** Examples of reuse of decommissioned oil and gas infrastructure. (A) Image from U.S. BSEE "Rigs to Reef" program, which submerges and utilized decommissioned oil and gas infrastructure to form artificial "reefs"; (B) model for reuse of an offshore oil and gas platform as a wind and/or solar energy generation facility; (C) an artistic rendition of a repurposed oil and gas platform as a seastead, including local wind power and on-site agriculture. *Panel (A) courtesy of U.S. Bureau of Safety and Environmental Enforcement; Panel (B) image courtesy of SD Wind Energy; Panel (C) Image courtesy of X_LAND by XTU Architects.*

Are these opportunities feasible? The answer for most is "probably." It depends upon many factors, including the condition and ratings of the existing equipment, the costs to repurpose vs new construction, safety and environmental assessments, local ocean conditions and certainly economics and demand for the service provided.

### Mitigation and protection

Humanity will also design and consider installing barriers to mitigate the impacts of extreme weather and rising sea levels. Modifications to the ocean could be made to help protect onshore populations and resources from storm damage or tsunamis (some of these are caused by submarine landslides as described in Chapter 10). The Dutch have been engineering the sea for centuries but there are modern fortifications under consideration as well. For example, the Ike Dike, offshore Texas, US, is a coastal barrier to protect Houston and other parts of the Texas Gulf Coast from hurricane storm surge (Texas A&M Galveston, 2021) or the MoSE system designed to protect the city of Venice from intermittent flooding (Merali, 2002). These are both examples of shallow water mitigation tools but as industry and infrastructure progress into deepwater, it is certainly conceivable that protective barriers might be needed.

The environmental responses, especially the geological and biological impacts, to large-scale barriers and blockades needs to be carefully analyzed prior to installation of these types of developments. A potential opportunity when constructing barriers and fortifications of this magnitude is to use $CO_2$ sequestering materials like carbon sequestering concrete or reinforced wood to construct these features, creating a carbon sequestration opportunity as well as mitigation infrastructure.

## *Other uses*

There are numerous other industries and uses for deepwater including shipping, leisure and tourism via commercial cruises, private boats and fishing. One example is the tourism at Antarctica which has grown from a few hundred at the beginning of the tourist industry in the 1950s; 38,478 visitors in 2015–16; 56,000 tourists during the 2018–19 to a total of 73,991 in the 2019–20 season (IAATO, 2020).

There is certainly a vast knowledge of the ocean within various military institutions around the world who will continue to build that understanding and look to the deep oceans in pursuit of their role in protecting national sovereignty.

## *Enabling technologies*

Enabling technologies such as continuous fluid and acoustic monitoring, automation, digital and big data analysis and higher resolution imaging will continue to evolve and will likely be a significant part of ensuring that existing deepwater oil and gas operations remain both safe and economic and that new energy and other types of developments are feasible. Remote operation centers are in use today and in the not-too-distant future, some new deepwater developments will be completely unmanned, following on from recent shallow water developments that are fully controlled from onshore operations centers (Baker Hughes, 2021). Future offshore resource developments will likely be sparsely or unpopulated, utilizing onshore remote operation centers and remotely operated intervention and maintenance.

Robots, drones, remotely operated vehicles (ROVs) and autonomous underwater vehicles (AUVs) are utilized today and will continue to conduct inspections, deliver equipment and reduce or eliminate required personnel onsite and the need to transport personnel offshore by helicopter. Drones currently under development can carry cargos of several hundred pounds over distances of a few hundred miles (The Verge, 2019). In the offshore, this capability will revolutionize maintenance and equipment delivery, reduce the need for personnel on board and result in significant improvements in economy, safety and efficiency.

Monitoring equipment can be installed at or near the seabed and airborne detection technologies like satellite imaging and drone-based detection can aid in on-site detection. Research instruments to monitor beneath or adjacent to facilities and on the seafloor utilizing geophysical arrays, drones and satellites are technologies that will continue to be matured for deepwater operations. This monitoring and verification of activities will be important, as well as the enhanced ability to detect offshore seismic activity which could impact operations or damage offshore infrastructure.

Continued development of resilient materials capable of functioning in remote and often harsh deepwater environments such as more efficient photovoltaic materials is critical. For example, novel materials will enable the construction of smaller more lightweight wind turbines for installation on offshore structures, while at the same time will lead to much larger and more efficient turbines that will make up future deepwater floating wind farms. The ability to 3D print tools and parts as needed on site will boost efficiencies and reduce costs and risk associated with procuring and shipping, waiting on a part or sending a specialized technician offshore for repairs (DNV, 2021).

Technologies such as these being applied today in the oil and gas industry, and others under development for the broader range of potential activities will be key to the successful future of many industries with the potential for remote operations in deepwater.

## *Societal impacts of deepwater development*

The deepwater offers significant opportunities for humanity but at what potential risks, uncertainties and costs? The impact on life, ocean environmental disruptions, marine noise and waste disposal are just a few of the potential issues. The ocean is under-explored and to apply the earlier quote, "*we don't know what we don't know*" about the ecosystem and the hydrosphere. There are currently, and will continue to be, challenges of collaboration and international governance. The risks, opportunities and potential policy actions described in the following text include some of the societal points of debate and policy considerations that are and will continue to be important in informing and driving intelligent moves into deepwater or in informing decisions to avoid actions. The significance of environmental, social and governance (ESG) factors is growing and will likely play an increasing role in decisions made by corporations and governments regarding deepwater development.

## *Risks*

### Marine pollution

*Noise*: Marine life uses hearing to hunt, communicate and navigate, and marine noise can disrupt and stress mammals and fish. Mitigation of marine sound is vital and technologies that reduce or eliminate disruption to the marine habitat created through human activities are important to continued progress in eliminating marine noise pollution. Robust environmental impact assessments are required for most oceanographic and geophysical cruises. These studies are required by governments before any acoustic or seismic acquisition is permitted. Marine observers are often a part of the crew for these studies to ensure that marine fauna are not impacted by noise. Researchers are also developing lower noise geophysical technologies to reduce the risk to marine mammals and fish.

*Marine pollution:* Marine pollution, primarily chemicals and plastic, is a significant and growing issue for the deepwater. These pollutants must be addressed, cleaned up and avoided, and opportunities to reduce existing marine pollution such as onshore chemical runoff, marine exhaust and fuels, waterborne and seabed plastics should be exploited.

### Climate change

The impacts of severe weather and rising sea level due to climate change will continue to increase the difficulty of operating developments and maintaining infrastructure offshore. In addition, nearly 2.4 billion people (40% of the world's population) live within 100 km (~ 62 mi.) of the coastline, and 1 billion people (~ 12.8% of the world's population) live in coastal areas <10 m (~ 33 feet) above sea level (Kulp and Strauss, 2019) which can result in displacement of, societal disruption and increased poverty and health issues as societies are displaced. Sea level rise will also impact shorebases and ports which provide services and support for offshore developments. As discussed previously in this chapter, development of offshore communities and municipalities could become viable and practical in response to rising sea level.

### Stranded assets

Assuming a long-term decrease in petroleum demand due to geopolitical and environmental pressures, substitution of renewable fuels and/or economic challenges, some offshore resources that are technically viable will likely never be developed, resulting in "stranded assets." This is not a new phenomenon; many environmentally sensitive areas with petroleum potential, including much of the offshore U.S. and Europe, are already protected by policies banning exploration and production activities. This protection will likely increase with the growing importance of ESG factors (Environmental, Social and Governance) which drives engagement with a broader range of stakeholders in the decision to offer offshore acreage for petroleum exploration and production. However, because of the factors stated above, it is unlikely that this will have a significant impact on existing deepwater production or new major discoveries, unless the decrease in petroleum demand is on a scale similar to the IEA Net Zero scenario (IEA NZE, 2021).

## *Key messages on opportunities*

### Economic impacts

There are several key opportunities associated with further development of the Earth's ocean resources. The potential economic aspects are appealing: jobs, new industries, revenue and access to many types of critical resources such as energy and minerals. New shorebases and supply chains will be needed to support offshore renewable developments such as wind and solar farms. Re-use of existing oil and gas infrastructure could help existing shorebases and communities thrive and continue. Providing food and energy for a growing global population is critical and the offshore offers potential to develop these resources. Policies, practices, regulations and experience from century of operations in oil and gas offer expertise, lessons learned and methodologies that can be transferred to other industries and further developed. Finally, new technologies, methods and materials will continue to be developed to increase safety, reduce costs, increase efficiency, reliability and resilience of offshore infrastructure.

### Ocean pollution

As noted above, Earth's oceans have several daunting pollution issues. 4.8 to 12.7 million tons of plastic enter the world's oceans every year (Geyer et al., 2017). Marine plastic, including pelagic debris and microplastics (<5 mm or .2 inches), is estimated to have reached approximately 5.25 trillion particles, around 250,000 tons (van Sebille et al., 2015) afloat but, unfortunately, most ocean-borne plastic is missing, probably is stored in deepwater submarine canyons, channels, ocean trenches and contouritic drift deposits (Pham et al., 2014; Woodall et al., 2014; Koelmans et al., 2017; Kane and Fildani, 2021; Kane and Clare, 2019; Kane et al., 2020).

Could new ocean infrastructure offer the opportunity to reverse this by collecting macro- and micro-plastics and other marine trash? Key areas currently being researched include the behaviors and potential mitigation of garbage patches and the microplastic distribution in submarine fans (Pohl et al., 2020) due to plastic-rich sediment gravity flows emanating from canyons linked to major urban centers. An understanding of ocean currents, bottom currents (Chapter 9), gravity flows and mass flows (Chapters 8 and 10) is helping researchers tackle these pollution challenges. One potential benefit of increased deepwater development and infrastructure is to create symbiotic processes to capture plastics in the ocean and to help mitigate future pollution.

### Data acquisition, sharing and collaboration

For over 50 years, the oil and gas industry has collected detailed datasets of offshore and deepwater metocean conditions. Academic, military and shipping organizations have collected detailed data as well. The knowledge gained from these extensive datasets can and should be available for consideration of future deepwater developments, and opportunities are being exploited today to leverage existing infrastructure to build on the oil and gas industry's experience and knowledge base (for example, National Academies of Sciences, Engineering, and Medicine, 2018). The oceans are a shared resource and unprecedented collaboration between key stakeholders will be vital to making informed and intelligent decisions about future activities.

## *Policies and governance*

The world's oceans are common to all of the planet and the health of the oceans is critical to life on Earth. Methods must be developed to understand and apply the hard lessons learned from atmosphere and climate science and assure that they are fully integrated into ocean governance. As the oceans are a globally shared resource, decisions made by individual governments must continue to be considered holistically.

### Economic zones

In 1982, the United Nations adopted the Law of the Sea establishing territorial waters for coastal countries and the Exclusive Economic Zone (EEZ) extending 200 nautical miles (~ 370 km) from shore (Ablos, 2006; OECD, 2021). The 200-mile EEZ was timely in that by the early 1980s oil and gas exploration was moving into deeper waters, particularly in the U.S. Gulf of Mexico where, after decades of exploration and development activity, exploration of the shallow water shelf was mature with limited opportunity for new major finds. As the industry progressed into deeper water and made important new discoveries (see, for example, Chapter 19, Fig. 5), the foundation was laid for building the current understanding of the deepwater processes detailed throughout the chapters of this book. At this time, additional areas beyond the current EEZ are under consideration for inclusion in coastal state sovereign rights (Fig. 12).

### Life cycle analysis

The development and application of a full life cycle analysis (LCA) approach to actions impacting the ocean is needed, utilizing consistent standards and measurements with common definitions and metrics. New policies, standards and regulations will need to be developed. These should include undertaking environmental impact studies and risk assessments,

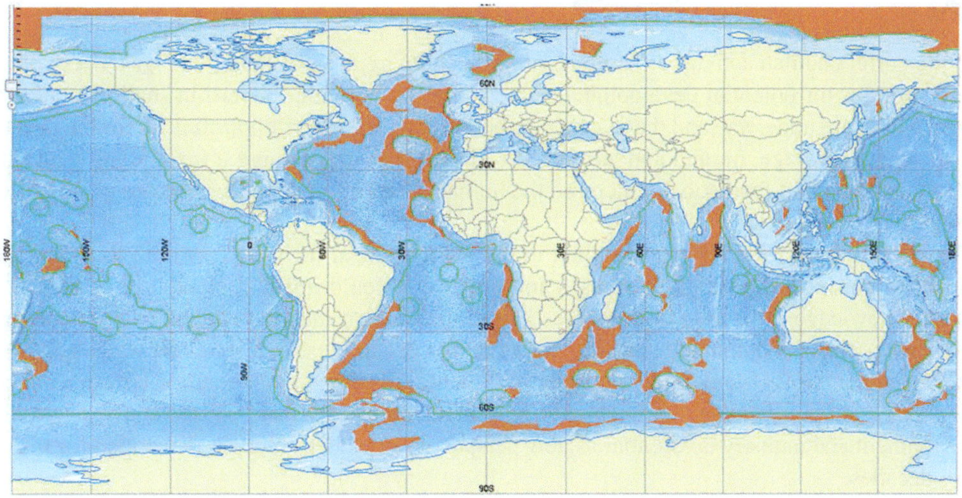


**FIG. 12** Global map indicating (in red) the requested areas beyond the 200 nautical miles (~ 370 km) that would allow additional coastal state sovereign rights for exploring and exploiting, conserving and managing natural living or non-living resources of their deepwaters superjacent (Madon, 2017).

both of which should be informed by the learnings and insights from this volume, previous industrial experiences and other technical analysis. A regulatory framework and global collaboration that recognizes the potential opportunities in deepwater while protecting this unique and complex environment is required.

## *Geoscience research and technology future considerations*

Since the 1950s, understanding of deepwater sedimentary systems at modern day and during the late Quaternary has matured, which has enabled a greater understanding of the ancient deposits of the fossil record (Chapter 1). Current knowledge of deepwater processes allows for the identification and characterization of their sedimentary deposits (e.g., Pickering et al., 1989; Talling et al., 2012, Pickering and Hiscott, 2016, Stow and Smillie, 2020, de Castro et al., 2020—see Chapters 6–10). These deposits are derived mainly from sediment gravity flow, mass-transport, bottom current, pelagic or authigenic processes. But the conceptual sedimentary models that have been applied for these ancient deposits are derived mainly from the interpretation of the sedimentary facies associations (e.g., Bouma, 1962; Gonthier et al., 1984), and to a lesser extent, where flows have been visualized and/or quantified in natural settings as well as laboratories (e.g., Azpiroz-Zabala et al., 2017).

Research opportunities for the next few decades includes the monitoring and quantification of deepwater sedimentary processes through permanent monitoring and better calibration of numerical models. High-resolution seabed mapping is currently underway (Nippon Foundation-GEBCO, 2021) which will provide a more detailed framework for the modern sea floor and a better understanding of the link between processes and local seafloor irregularities. -Observation and measurement of a greater number of gravity flows and bottom currents is essential to fully understand and quantify the dynamics of sediment laden water and the associated secondary oceanographic processes (eddies, cores, deep tides, internal waves, etc.). From this, a new revolution in the knowledge of the deep ocean will be possible, providing data on the nature and variability of the processes and enabling the development of new models for deepwater sediments and their lithological and petrophysical characteristics.

Knowledge and understanding of deepwater sedimentary systems have increased significantly during the past few decades. Academia, industry and governments have advanced exploration into the deep ocean, and deepwater geoscience will likely play a vital role during the energy transition, as there are still vast underexplored areas to be investigated for a number of uses, as detailed in this chapter. New technological advances and the ability to integrate and understand better data from outcrop, core, log and seismic scale (see Chapters 1–18) will allow researchers to develop better geologic models (see Chapters 8, 17–18) for deepwater sedimentary systems. It is essential to continue to observe and quantify sedimentary processes in real time in the modern ocean with the help of numerical modeling and flume experiments. To achieve this, collaboration between academic, government, military, non-governmental organizations and industry partners is more needed than ever, and consequently, a symbiosis of science, environment and industry should be developed, enabling new perspectives for coming decades. For insights on future research needed for the concepts described in the first 20 chapters of this textbook, please see Appendix.

## Conclusions

While most of the current human activity offshore is military, shipping, leisure and oil and gas, many other industries are looking to the deepwater for potential development. Renewables are moving forward in shallow water, offshore carbon dioxide storage projects are underway and there are a number of potential future opportunities for the deepwater, including energy resources, energy storage, mining and aquaculture. These activities require a more significant infrastructure and technology advances and could dramatically increase the human footprint in the oceans, especially as facilities and infrastructure progress into deepwater.

Whether facilities and infrastructure are installed in the subsurface, at the seabed, within the water column or at sea level, it is critical to understand geologic processes and how these could impact potential development and operations in the deepwater. Knowledge of deepwater processes and sedimentation must be applied when planning, installing and operating deepwater projects. Continued technological advances of, and research into deepwater geologic processes are also required to develop improved models for deepwater sedimentary systems. Collaboration between industrial, academic, environmental groups and governmental partners will be critical for success, and consequently, a symbiosis of science, policy, the environment and industry is essential as new perspectives on deepwater are developed for the next generations and for coming decades.

## Appendix

As populations increase and the demand for energy grows over the coming decades, the opportunities for development of the deepwater will become more viable and offer potentially attractive options to provide resources and infrastructure to support these needs. This chapter muses upon a future potential industrial and societal uses for deepwater. A number, but certainly not all, of developments and uses are described to envision the types of opportunities possible. Again, the objective is not to advocate for any specific technology but to envision a number of potential opportunities as well as highlight societal and policy risks and opportunities associated with these types of developments. In all cases, the geoscience principles described in this textbook should be applied while planning and developing future offshore infrastructure and the potential risks associated with potential industrialization of this new frontier.

The need for continued research, establishing baselines, monitoring and verification and data gathering from the offshore, particularly in deepwater environments, is clear and needs to be foremost in research and development by academia, government, environmental groups and industry. This will require broad collaboration across stakeholders, including more open sharing of research, data and technologies. The understanding of the geology and impact of geological processes must be a critical part of environmental analysis and risk analysis of any deepwater project.

### Considerations for future applications and geoscience research based on the framework provided by this volume

#### *Chapter 1: Introduction to deepwater sedimentary systems*

The philosophies and workflows for deepwater petroleum exploration have remained relatively unchanged since the 1980s. Looking forward, the creative process of exploration, coupled with higher resolution visualization of the seafloor and subsurface and continuing advances in the understanding of deepwater sedimentary systems, will enhance the interpretation and understanding of deepwater systems.

Continued advancements in geophysical imaging and the development of artificial intelligence (AI) methodologies for all forms of exploration (including but not limited to oil and gas, minerals and saline reservoirs) will bring on a new wave of exploration for energy geoscience and subsurface storage contributing to a reduction of risk and uncertainty for all deepwater activities.

The fundamentals of the exploration process: risk assessment, multiple models and economic analysis, can be applied to decision making in other offshore industries.

#### *Chapter 2: Source rocks and fluid systems in deepwater plays*

Petroleum Systems assessment will benefit from greater detail and geologic resolution, as well as additional measurement and calibration of the chemistry of organic matter and petroleum and in the physics of heat and fluid flow and petroleum fluids. Forward numerical basin models with geochemical inversion will continue to develop as a robust workflow that allows for replication, calibration and ultimately grater predictability of petroleum systems.

Continued improvements in computing capacity will create opportunities for larger and more complex subsurface models, eventually reaching the basin scale with the fidelity of current field-based models. Detailed petroleum systems models can and will be critical for assessing the viability of subsurface storage for $CO_2$ and hydrogen, and for modeling pore pressure and potential fluid flow in these and other subsurface applications.

## Chapter 3: Tectonic and crustal models

The proliferation of deep (10–15s TWT) and high-quality 2D and 3D seismic surveys along hydrocarbon-bearing continental margins has allowed for more accurate plate tectonic reconstruction, timing of plate movements and basin creation to final stages of filling. In turn, this has allowed the oil and gas industry to characterize, interpret and model petroleum risk elements with lower uncertainty, leading to higher quality exploration, and the opening of new frontier basins in offshore areas such as the Atlantic Conjugate Margins, Gulf of Mexico, eastern Africa and the Arctic. Similarities among global continental margin, marginal rift and sedimentary basin fill indicate that while there are nuances in continental margin basin creation and sedimentation, the recipe for these prolific petroleum-bearing basins is more ordered and simplistic than originally envisaged. Further geophysical data and core calibration will enable researchers to continue to develop these structural characterizations and improve predictive models.

## Chapter 4: Deepwater passive margin fold belts

There is increasing interest in salt and shale mobile substrates in deepwater systems, especially in passive margin fold belts because of their potential for creating complex structural traps. The current and future availability of higher resolution wide-azimuth 2D and 3D seismic data should improve the descriptions of opportunities within these settings. Advances in deepwater drilling technology will allow for more focused exploitation of hydrocarbon accumulations in the near-to-medium term, and of other subsurface opportunities (e.g., storage) in the longer term.

Future research should focus on fluid pressures, pore pressure prediction and characterization and sediment compaction, as well as the onset and evolution of compressional structures. These features could form structural traps for $CO_2$ storage, energy and hydrogen storage and hydrocarbon exploration targets.

## Chapter 5: Salt tectonics in deepwater settings

With recent improvements in geophysical imaging and geologic calibration, there are robust models for salt tectonics in deepwater settings. Future research should include continued imaging improvements, better understanding about how deepwater passive margins evolve and further development of insights into how salt structures are influenced by crustal separation as well as the original depositional and structural configuration. This research will greatly improve the capacity for structural and stratigraphic prediction adjacent to and beneath salt features.

Improved geophysical imaging technologies will enhance the ability to image geology adjacent to and below salt features. The mechanics of salt movement and the evaluation of the salt sediment interface will be integral parts of the assessment of salt domes and canopies for offshore storage of $CO_2$, hydrogen and other forms of energy. Sulfur and other mineral mining could also be associated with salt domes.

## Chapter 6: Deepwater sedimentary processes

Gravity flows, water mass circulation (and associated bottom currents) and pelagic and hemipelagic sedimentation are the dominant deepwater sedimentary processes. These processes have been found to be more sophisticated that initially thought, especially when they interact with each other, leading to complex and interbedded stratigraphic successions that can be challenging to characterize and to interpret. Deepwater sedimentation preserves more of Earth's history and environmental evolution than sedimentation found in shallow marine, coastal or terrestrial environments. Greater analytical accuracy and models need to be developed to gain a more robust interpretation of Earth's past events to predict its future. Future research is needed to develop a better understanding of their interaction through modern seafloor monitoring, improved numerical models and more robust tank experiments.

Protection against the forces of submarine sediment gravity flows will be a critical part of any structure tethered or anchored to the seabed as well as infrastructure at or near the seafloor. This includes subsea production and injection systems, wellheads and seabed cables. Improved prediction and monitoring systems to help predict and detect updip slumping and sliding due to increased sediment input or tectonic activity or high velocity local bottoms current by eddies or internal waves would help operators or municipalities respond to gravity flow or bottom current incidents and protect infrastructure.

## Chapter 7: Deepwater sedimentation units

There are a great variety of deepwater sedimentary deposits because of the wide range of sediment transport and depositional processes, but it is necessary to establish a better understanding of the dynamics between hydrodynamic processes and related deposits as well as in which physiographic domains they form. Current sedimentary models are under revision and new more complete models should be developed in the coming decades, especially for those sediments being redeposited, winnowed or reworked by more than one processes. As with the submarine processes described in Chapter 6, protection against the forces of submarine sediment gravity flows or bottom currents will be a critical part of any structure tethered or anchored to the seabed as well as infrastructure at or near the seafloor.

It is surprising how limited facies models are for deepwater sedimentary intervals. Chapter 7 presents new models that are relatively uncalibrated compared to the turbidite paradigm established in the mid-20th century. Future research will more firmly establish the true range and variability of deepwater sedimentation unit models as researchers continue to probe new outcrops, seafloors and sub-surface environments.

## Chapter 8: Deepwater sedimentary environments

Future exploration of deepwater sedimentary environments requires stronger integration of deepwater sedimentary facies and facies association between outcrop, core sample, well log and seismic scales. Special focus should be applied to mixed depositional systems or between different sub-environments in any one particular system. This should permit better reconstruction of ancient deepwater environments and better characterize their role in greater oceanographic, biologic and sedimentary processes and their economic-related resources. Being able to deconstruct these environments at a bed- to sub-bed-level and identify sediment transport patterns will help reduce the uncertainty in interpreting the evolution of seafloors around the world and better understand the impacts of human, autogenic and allogenic inputs. An improved characterization of the range, variability and uncertainty in deepwater sedimentary facies in these environments will be important for assessing the viability for future subsurface exploration as well as the potential for trapping and storing $CO_2$ and other potential gasses or fluids.

## Chapter 9: Contourite and mixed deepwater sedimentary systems

Contourites and mixed depositional systems represent a major part of continental margin and adjacent abyssal plain sedimentation in many of the world's oceans. The growing interest incontourite and mixed depositional systems demonstrates that these systems represent significant deep-marine sedimentary environments. Questions regarding how bottom currents have shaped the deepwater stratigraphic record and past ocean basins could and should be used to guide a vast body of future research. Direct measurement of current directions, velocities and sediment densities will be important for generating new sedimentary models for deepwater sediments and for developing of offshore infrastructure, aquaculture and mining.

## Chapter 10: Mass-transport, injections and other forms of sediment mobility

Mass transport, injectites, mud volcanoes and other styles of sediment remobilization are important processes in deepwater marine systems, especially with regard to geohazards, climatic change, petroleum plays and offshore infrastructure. Future studies of these deposits at different scales will help address their characteristics and relationships with other deposits, and enable improved identification of potential impacts in areas such as basin analysis and energy geosciences. The stability of the subsurface and seabed where these processes can occur should be thoroughly evaluated prior to the installation of seabed infrastructure. To improve the understanding of catastrophic events, researchers will be able to evaluate the recurrence interval and overall risk of large submarine landslides that can cause tsunamis and other natural disasters.

## Chapter 11: Source-to-sink analysis of deepwater systems

The relatively new and rapidly growing practice of source-to-sink analysis integrates many avenues of research with multi- and interdisciplinary workflows that link orogenic events, paleoenvironmental change and basin formation and filling. Source-to-sink analysis reduces geological uncertainty to guide business decisions in the economic sectors that exploit deepwater deposits from petroleum and critical mineral exploration to carbon capture and sequestration. Future deepwater studies should include expanded research scopes to consider geologic factors that divert sediments away from gravity flows to be deposited along the slope (or even in proximal basins) and should publish their findings as open-access geological databases and software for rigorous statistical data interrogation—a trend that has already begun and is welcomed by the source-to-sink community. A strong understanding of sediment provenance and feedback processes within the broader sediment routing system will be important in refining downdip mineralogic and lithologic predictions.

## Chapter 12: Sequence stratigraphy of deepwater systems

Sequence stratigraphy is one of the single most powerful predictors of lithology in coastal and shallow marine environments, and sequence stratigraphic characterization can be further employed to unravel the more complex sedimentation patterns occurring in deepwater. Applications of sequence stratigraphy to deepwater environments have become better calibrated in recent years, but non-diagnostic variability in the sedimentological makeup of systems tracts reflects the unique tectonic and depositional settings of each sedimentary basin and needs to be rationalized on a case-by-case basis. It is necessary to better understand the shifts of deepwater depositional elements to discriminate between regionalstratigraphic trends vs. local sedimentological variations in terms of timing and frequency of sedimentation cycles.

The relative position of bottom current deposits in relation with turbidites, MTDs and pelagic and hemipelagic sediments must be clarified and a better understanding of the link between relative sea level changes and oceanographic processes is required. Calibration to lithological data (well cuttings, cores) and biostratigraphic information will document continued examples of calibrated cycles. Higher resolution seismic and improved imaging methods will enable more detailed geobody modeling within a sequence stratigraphic framework.

## Chapter 13: Reservoir quality and diagenesis of deepwater sandstones

Understanding reservoir quality and diagenesis of deepwater sandstones is critical to the economic viability of projects associated with hydrocarbons, geothermal energy and lithium or other mineral production from reservoir fluids, as well as the storage of $CO_2$, hydrogen, compressed air and fluid waste. The impact that the diagenetic factors can have on reservoir quality needs to be understood through simulation scenarios and future research integrating existing diagenetic quantitative models with predictive models from structural geology, flow-flow properties of reservoirs, geochemistry and other disciplines.

## Chapter 14: Applied paleontology in oil and gas exploration and development

Applied paleontology in exploration, development and production ties reservoirs into regional or semi-regional biozonations. Improved stratigraphic resolution of the reservoir system, beyond the control provided by regional bio events, is required and must be based on data integration and iteration with other geoscience disciplines. Detailed data such, as eco-stratigraphy, facies-stratigraphy, relative and absolute age dating and chemical composition to help calibrate climactic and other changes is required.

## Chapter 15: Ichnology of deepwater contourite systems

Ichnological content can be used to characterize deepwater processes and sediment, since they can be diagnostic criteria to discriminate between turbidites, pelagic and hemipelagic deposits and contourites. But more studies are necessary for a better understanding of their specific trace fossil assemblage and new and better techniques would enhance the visibility of trace fossils in sediments. The influence of bioturbation on rock properties should be studies further and applied in future reservoir evaluation.

## Chapter 16: Subsurface imaging and interpretation

Subsurface data (2D and 3D seismic data and well log imaging technology) over the decades have provided a unique opportunity to conduct reservoir characterization with improved certainty and efficiency. Seismic attributes and pre-stack amplitude data need to be better calibrated to well data and further assessed with rock physics modeling to avoid misinterpretation and to enhance the interpretation of structural features, sand body geometries and facies and pore fluid parameters. Advanced geophysical data acquisition and imaging, including improvements driven by increases in computational capacity and the use of artificial intelligence methodologies, will continue to improve subsurface descriptions.

## Chapter 17: Seismic rock physics and machine learning for deepwater stratigraphic intervals

Seismic rock physics together with seismic AVO data and statistical techniques improve the understanding of stratigraphic intervals, sedimentary facies and fluid distributions in deepwater depositional systems. Improved rock physics models are needed, including better datasets for deep learning algorithms that can be used for quantitative seismic interpretation and the prediction of reservoir properties. Improved imaging and rock physics will eventually provide detailed geoform maps, enabling detailed seabed and subsurface geologic mapping needed for exploration, subsurface storage and overburden evaluation.

### *Chapter 18: Integrating forward stratigraphic modeling with basin and petroleum systems modeling*

Basin modeling and forward stratigraphic modeling need to be integrated and interpreted to maximize the understanding of a sedimentary basin. The ability to construct models in three dimensions, with time serving as the fourth dimension, would clarify understanding of the physical processes that occur in the subsurface and would better elucidate sedimentological facies, reservoirs and geologic structure of the subsurface. A collaborative effort from a number of disciplines, including organic geochemists , sedimentary geologists, biostratigraphers, geodynamicists and geophysicists will be necessary to create improved models for basins and petroleum systems.

This will improve the efficiency of hydrocarbon exploitation, CCUS and geothermal development and in addition will be a valuable skill set in any future endeavor involving the subsurface.

### *Chapter 19: Technical (engineering) advancements enabling deepwater exploration and production*

The engineering advancements that have enabled deepwater exploration and production have been the result of excellent teamwork characterized by understanding, communication and collaboration between the sub-surface experts, the surface engineering experts and the "access" experts in drilling and production engineering. New technologies will continue to be developed, along with the production systems and structures that they support for the efficient exploitation of discovered resources and the development of new hydrocarbon discoveries. The future development of deepwater energy and resources will benefit from those technologies developed to date as well as future technological advances.

Future work in novel materials, process technology, predictive and descriptive analytics and computing systems, as well as developments in artificial intelligence and big data applications and associated modeling will all be critical to expanding deepwater infrastructure into new geographic areas as well as new applications.

### *Chapter 20: Economic considerations and market condition effects in deepwater*

Deepwater projects require extremely large investments, often in excess of abillion dollars. Economic considerations and market conditions also vary with geo-political factors related to the location of the resources being pursued. Economic factors include product prices and costs of development and operations, but also other factors such as the size, location and uncertainty of the resources as well as the impact of the national and local regulations that will govern the specific project. The oil and gas industry is moving rapidly into deeper and deeper water, and the industry's accumulated knowledge and expertise will serve to underpin the emerging offshore renewables industry as well.

The proper assessment of any new opportunity in deepwater requires the contribution of many professional and technical experts in a number of fields comprising business, geoscience, legal, engineering and more.

### Potential glossary contributions

Floatels: living quarters associated with development and servicing of offshore infrastructure.

ESG: Environmental, social and governance.

Stranded assets.

Secondary recovery.

Tertiary recovery.

Electrolysis.

Video for Science Direct site: https://youtu.be/Lwq1j3nOODA.

## Acknowledgments

This chapter benefited from a robust technical edit from Jon Rotzien and the authors are grateful for his insights. Advice from Nigel Jenvey, Julie Wilson, Sam Walker, Howard Herzog, Dan Schrag, John Guy, Matt Boisne, Miki O'Farrell, Ellen King and Lori Fremin is greatly appreciated. Finally, many thanks to Elsevier for their expertise, patience and guidance in creating a cohesive reference book from almost 60 authors.

## References

Ablos, 2006. A Manual on Technical Aspects of the United Nations Convention on the Law of the Sea—1982, Prepared by the IHO, IAG, IOC Advisory Board on Law of the Sea (ABLOS), fourth ed. International Hydrographic Bureau MONACO. Special Publication, No. 51. (March).

Alcatel, 2021. Animated Maps: Submarine Cables. https://www.youtube.com/watch?v=6dkiqJ_IZGw.

Ali, U., 2019. How salt caverns could transform renewable energy storage for the U.S. Power Technol. https://www.power-technology.com/features/how-salt-caverns-could-transform-renewable-energy-storage-for-the-us/.

Azpiroz-Zabala, M., Cartigny, M.J., Talling, P.J., Parsons, D.R., Sumner, E.J., Clare, M.A., Simmons, S.M., Cooper, C., Pope, E.L., 2017. Newly recognized turbidity current structure can explain prolonged flushing of submarine canyons. Sci. Adv. 3 (10). https://doi.org/10.1126/sciadv.1700200.

Baker Hughes, 2021. Pass the Remote: Bringing Equinor's Offshore Roles in to Land. https://www.bakerhughes.com/company/energy-forward/pass-remote-bringing-equinors-offshore-roles-land.

Bouma, A.H., 1962. Sedimentology of Some Flysch Deposits. A Graphic Approach to Facies Interpretation. Elsevier, p. 168.

bp, 2021. Statistical Review of World Energy. https://www.bp.com/en/global/corporate/energy-economics/statistical-review-of-world-energy.html.

Brackenridge, R.E., Nicholson, U., Sapiie, B., Stow, D.A.V., Tappin, D.R., 2020. Indonesian Throughflow as a preconditioning mechanism for submarine landslides in the Makassar Strait. Geol. Soc. Lond. Spec. Publ. 500, 195–217. 1 April 2020 https://doi.org/10.1144/SP500-2019-171.

Bureau of Safety and Environmental Enforcement, 2021. Rigs to Reefs. https://www.bsee.gov/what-we-do/environmental-compliance/environmental-programs/rigs-to-reefs.

Bureau of Safety and Environmental Enforcement, 2022. BSEE Renewable Energy Policy. What We Do | Bureau of Safety and Environmental Enforcement.

Burton, K., 2020. One Fifth of the Ocean Floor Now Mapped. Geographical. https://geographical.co.uk/nature/oceans/item/3735-one-fifth-of-the-ocean-floor-now-mapped-81-per-cent-to-go.

Carr, D.L., Trevino, R., Meckel, T., Brenton, C., Yang, C., Miller, E., 2011. Executive Summary: Task 15—NATCARB Atlas Update—$CO_2$ Sequestration Capacity, Offshore Western Gulf of Mexico. GCCC Digital Publication Series #11-24, University of Texas at Austin, Bureau of Economic Geology, p. 2.

Chalier, R.H., Justus, J.R., 1993. Ocean Energies: Environmental, Economic and Technological Aspects of Alternative Power Sources. Elsevier Oceanography Series, 56, Elsevier, p. 554.

CNN, 2021. Saudi Arabia to Launch Enormous Oil Themed 'Extreme Park'. https://www.cnn.com/travel/article/saudi-arabia-oil-rig-theme-park/index.html. October 2021.

Collet, T., Johnson, A., Knapp, C., Boswell, R., 2009. AAPG Memoir 89, Natural Gas Hydrates—Energy Resource Potential and Associated Geologic Hazards. AAPG.

Danish Energy Agency, 2021. https://ens.dk/en/our-responsibilities/wind-power/energy-islands/denmarks-energy-islands.

de Castro, S., Hernández-Molina, F.J., Rodriguez-Tovar, F.J., Llave, E., Ng, Z.L., Nishida, N., Mena, A., 2020. Contourites and bottom current reworked sands: bed facies model and implications. Mar. Geol. 428, 106267. https://doi.org/10.1016/j.margeo.2020.106267.

Degnarain, N., 2020. Ocean Crops: Is This the Next Frontier for Agriculture? Forbes. https://www.forbes.com/sites/nishandegnarain/2020/07/29/ocean-crops-is-this-the-next-frontier-for-agriculture/?sh=c9226965c955.

Denbury Inc., 2022. https://www.denbury.com/operations/CCUS-Operations/Denbury-Carbon-Solutions/default.aspx.

Derouin, S., 2017. Benchmarks: November 18, 1929, Turbidity Currents Snap Trans-Atlantic Cables. https://www.earthmagazine.org/article/benchmarks-november-18-1929-turbidity-currents-snap-trans-atlantic-cables.

DNV, 2021. 3D Printed Parts Could Benefit Oil and Gas, Offshore and Maritime Supply Chains. https://www.dnv.com/oilgas/laboratories-test-sites/article/3D-printed-parts-could-benefit-oil-gas-offshore-and-maritime-supply-chains.html.

EIA, 2016. Offshore Production Nearly 30% of Global Crude Oil Output in 2015. October https://www.eia.gov/todayinenergy/detail.php?id=28492.

EIA IEO, 2021. International Energy Outlook. https://www.eia.gov/outlooks/ieo/.

EIA STEO, 2021. Short-Term Energy Outlook. https://www.eia.gov/outlooks/steo/report/global_oil.php.

EMEC, 2017. European Marine Energy Centre Hydrogen R&D Facilities. https://www.emec.org.uk/facilities/hydrogen/.

Energy Digital, 2020. "Energy Bag" Offshore Wind Energy Storage. https://energydigital.com/smart-energy/energy-bag-offshore-wind-energy-storage.

ESMAP, 2019. Going Global: Expanding Offshore Wind to Emerging Markets. World Bank, Washington, DC.

Evans, S., 2019. La Rance: Learning From the World's Oldest Tidal Project. Power Technology. October 2019 https://www.power-technology.com/features/la-rance-learning-from-the-worlds-oldest-tidal-project/.

Foro Nuclear, 2021. What is a Floating Nuclear Power Plant? https://www.foronuclear.org/en/updates/in-depth/what-is-a-floating-nuclear-power-plant/.

Fraunhofer IEE, 2022. StEnSea – Stored Energy in the Sea. (Accessed 24 June 2022).

Geology.com, 2022. The Artificial Islands of Dubai. Geology.com. https://geology.com/satellite/artificial-islands-of-dubai/.

Geyer, R., Jambeck, J.R., Law, K.L., 2017. Production, use, and fate of all plastics ever made. Sci. Adv. 3, e1700782. https://doi.org/10.1126/sciadv.1700782.

Global CCS Institute, 2021. Global Status of CCS 2021. Global CCS Institute, Australia.

Global Citizen, 2020. https://www.globalcitizen.org/en/content/offshore-wind-rise-and-future/.

Gonthier, E., Faugères, J.-C., Stow, D.A.V., 1984. Contourite facies of the Faro Drift, Gulf of Cadiz. In: Stow, D.A.V., Piper, D.J.W. (Eds.), Fine Grained Sediments, Deepwater Processes and Facies. vol. 15. Geological Society, London, pp. 275–291. Special Publication.

Green Car Congress, 2019. DOE and International Partners Successfully Gas Hydrate Test Well on Alaska North Slope. https://www.greencarcongress.com/2019/01/20190124-hydrates.html.

Gronewold, N., 2019. Seabed-Mining Foes Press U.N. to Weigh Climate Impacts: Minerals Used in Electronics are Found on the Seafloor, but Disturbing Them Could Release Carbon. Scientific American. https://www.scientificamerican.com/article/seabed-mining-foes-press-u-n-to-weigh-climate-impacts/.

Gunther, M., 2018. Can Deepwater Aquaculture Avoid the Pitfalls of Coastal Fish Farms? Yale Environment 360. https://e360.yale.edu/features/can-deepwater-aquaculture-avoid-the-pitfalls-of-coastal-fish-farms.

Hanson, P., 2020. https://www.thinkgeoenergy.com/green-electricity-from-heat-under-the-sea-offshore-geothermal-energy/.

Hernández-Molina, F.J., 1996. Deep Marine Sedimentary Systems. University of Cadiz, p.50.

House, K.D., Schrag, D.P., Harvey, C.F., Lackner, K.S., 2006. Permanent carbon dioxide storage in deep-sea sediments. Proc. Natl. Acad. Sci. 103, 12291–12295. https://doi.org/10.1073/pnas.0605318103.

IAATO, 2020. Antarctic Visitor Figures 2019–2020. https://iaato.org/wp-content/uploads/2020/06/IAATO-on-Antarctic-visitor-figures-2019-20-FINAL.pdf.

IEA, 2019. World Energy Outlook 2019. IEA, Paris. https://www.iea.org/reports/world-energy-outlook-2019.

IEA, 2020a. Environmental Monitoring of Offshore Carbon Capture and Storage. IEA, Paris. In press https://www.iea.org/articles/environmental-monitoring-of-offshore-carbon-capture-and-storage.

IEA, 2020b. The Role of Critical Minerals in Clean Energy Transition: World Energy Outlook Special Report. https://www.iea.org/reports/the-role-of-critical-minerals-in-clean-energy-transitions/executive-summary.

IEA NZE, 2021. Net-Zero by 2050. IEA. https://www.iea.org/reports/net-zero-by-2050.

IPCC, 2005. In: Metz, B., Davidson, O., de Coninck, H.C., Loos, M., Meyer, L.A. (Eds.), IPCC Special Report on Carbon Dioxide Capture and Storage. Prepared by Working Group III of the Intergovernmental Panel on Climate Change. Cambridge University Press, Cambridge, United Kingdom and New York, NY, USA, p. 442.

Jbaily, A., Yeung, R.W., 2015. Piezoelectric devices for ocean energy: a brief survey. J. Ocean Eng. Mar. Energy 1, 101–118. https://doi.org/10.1007/s40722-014-0008-9.

Journal of Petroleum Technology (JPT), 2019. Global Deepwater Production Hits 10 Million BOE/D. https://jpt.spe.org/global-deepwater-production-hits-10-million-boed-milestone.

Kane, I.A., Clare, M.A., 2019. Dispersion, accumulation, and the ultimate fate of microplastics in deep-marine environments: a review and future directions. Front. Earth Sci. 7, 80. https://doi.org/10.3389/feart.2019.00080.

Kane, I.A., Fildani, A., 2021. Anthropogenic pollution in deep-marine sedimentary systems—a geological perspective on the plastic problem. Geology 49 (5), 607–608. https://doi.org/10.1130/focus052021.1. 2021.

Kane, I.A., Clare, M.A., Miramontes, E., Wogelius, R., Rothwell, J.J., Garreau, P., Pohl, F., 2020. Seafloor microplastic hotspots controlled by deep-sea circulation. Science 368, 1140–1145.

Karam, Q., 2015. Artificial Islands Construction in Offshore Areas for Oil and Gas Exploration Industry: An Environmental Perspective. Kuwait Institute for Scientific Research. 2015 https://www.researchgate.net/publication/283894321_Artificial_Islands_Construction_in_Offshore_Areas_for_Oil_and_Gas_Exploration_Industry_An_Environmental_Perspective.

Keleman, P.B., Matter, J., 2008. In situ carbonation of peridotite for $CO_2$ storage. Proc. Natl. Acad. Sci. U. S. A. 105, 17295–17300.

Kennett, J., 1982. Marine Geology. Prentice-Hall, p. 813.

Koelmans, A.A., Kooi, M., Law, K.L., van Sebille, E., 2017. All is not lost: deriving a top-down mass budget of plastic at sea. Environ. Res. Lett. 12 (11), 1–9.

Kulp, S.A., Strauss, B.H., 2019. New elevation data triple estimates of global vulnerability to sea-level rise and coastal flooding. Nat. Commun. 10, 4844. https://doi.org/10.1038/s41467-019-12808-z.

Laberg, J.S., Camerlenghi, A., 2008. The significance of contourites for submarine slope stability. In: Rebesco, M., Camerlenghi, A. (Eds.), Contourites. Developments in Sedimentology, vol. 60. Elsevier, Amsterdam, pp. 537–556.

Leon, R., Somoza, L., Urgeles, R., Medialdea, T., Ferrer, M., Biain, A., Garcia-Crespo, J., Mediato, J.F., Galindo, I., Yepes, J., Gonzalez, F.J., Gimenez-Moreno, J., 2017. Multi-event oceanic island landslides: new onshore-offshore insights from El Hierro Island, Canary Archipelago. Mar. Geol. 393, 156–175.

Lewis, N., 2022. A Floating City in the Maldives Begins to take Shape. CNN. https://www.cnn.com/style/article/maldives-floating-city-spc-intl/index.html.

LiVecchi, A., Copping, A., Jenne, D., Gorton, A., Preus, R., Gill, G., Robichaud, R., Green, R., Geerlofs, S., Gore, S., Hume, D., McShane, W., Schmaus, C., Spence, H., 2019. Powering the Blue Economy: Exploring Opportunities for Marine Renewable Energy in Maritime Markets. Report by US Department of Energy (DOE). https://www.energy.gov/sites/prod/files/2019/03/f61/Chapter%206.pdf.

Lodge, M., 2021. The International Seabed Authority and Deep Seabed Mining. https://www.un.org/en/chronicle/article/international-seabed-authority-and-deep-seabed-mining. 2021.

Loganathan, P., Naidu, G., Vigneswaran, S., 2017. Mining valuable minerals from seawater: a critical review. Environ. Sci. Water Res. Technol. 3, 37–53. https://doi.org/10.1039/C6EW00268D.

Madon, M., 2017. The Continental Shelf – five decades of progress (1966-2016). Bull. Geol. Soc. Malays. 63, 145–156.

Maestro-González, A., 2021. In: The future of mining: the exploitation of marine geological resources as global commons. In: Security in the Global Commons and Beyond: Advanced Sciences and Technologies for Security Applications. Springer, https://doi.org/10.1007/978-3-030-67973-6_4.

Maslin, E., 2021a. Geothermal Energy: A New Life for Old Offshore Oil Wells? Offshore Engineer. February 2021 https://www.oedigital.com/news/485619-geothermal-energy-a-new-life-for-old-offshore-oil-wells.

Maslin, E., 2021b. Floating Offshore Wind: Attention Turns Subsea for Power Transport. Offshore Engineer. December 2021 https://www.oedigital.com/news/493140-floating-offshore-wind-attention-turns-subsea-for-power-transport?utm_source=AOGDigital-ENews-2021-12-28&utm_medium=email&utm_campaign=OEDigital-ENews.

Mawdsley, M., 2019. An Introduction to Miscible Flooding for Enhanced Oil Recovery. September 2019 https://www.gljpc.com/blog/introduction-miscible-flooding-enhanced-oil-recovery.

Maximize Market Research, 2019. Fish Farming Market – Global Industry Analysis and Forecast (2021-2027) by Environment, Fish Type, and Region. https://www.maximizemarketresearch.com/market-report/global-fish-farming-market/30238/#.

Merali, Z., 2002. Saving Venice. Scientific American. https://www.scientificamerican.com/article/saving-venice/.

Miller, K.A., Thompson, K.F., Johnston, P., Santillo, D., 2018. An overview of seabed mining including the current state of development, environmental impacts and knowledge gaps. Front. Mar. Sci. 4.

Miramontes, E., Garziglia, S., Sultan, N., Jouet, G., Cattaneo, A., 2018. Morphological control of slope instability in contourites: a geotechnical approach. Landslides 15 (6), 1085–1095. https://doi.org/10.1007/s10346-018-0956-6.

MIT, 2010. Role of Enhanced Oil Recovery in Accelerating the Deployment of Carbon Capture and Sequestration. MIT Energy Initiative. July 2010.

Musial, W., Spitsen, P., Beiter, P., Marquis, M., Cooperman, A., Hammond, R., Shields, M., 2021. Offshore Wind Energy Market Report: 2021 Edition. U.S. Department of Energy, p. 119. https://www.energy.gov/sites/default/files/2021-08/Offshore%20Wind%20Market%20Report%202021%20Edition_Final.pdf

Musial, W., Tegen, S., Driscoll, R., Roberts, O., Kilcher, L., Scott, G., Beiter, P., 2017. Survey and Assessment of the Ocean Renewable Resources in the U.S. Gulf of Mexico, OCS Study BOEM 2020-017. https://espis.boem.gov/final%20reports/BOEM_2020-017.pdf.

Nakicenovic, N., 1996. Freeing energy from carbon. Daedalus 125 (3).

NAS, 2021. Committee on Progress and Opportunities Toward Decreasing the Risk of Offshore Energy Operations. i*n preparation*, National Academy of Sciences.

NASA, 2021. Who are NASA's Earth Explorers. Dr. Gene Feldman. https://www.nasa.gov/audience/forstudents/5-8/features/oceans-the-great-unknown-58.html.

National Geographic, 2021a. https://www.nationalgeographic.org/encyclopedia/ocean/.

National Geographic, 2021b. https://www.nationalgeographic.org/encyclopedia/ocean/.

National Marine Fisheries Service, 2021. Fisheries of the United States, 2019. U.S. Department of Commerce, NOAA Current Fishery Statistics No. 2019. U.S. Department of Commerce., p. 167. https://www.fisheries.noaa.gov/national/sustainable-fisheries/fisheries-united-states. Submitted for publication.

National Petroleum Council (NPC), 2019. Meeting the Dual Challenge: A Roadmap to At-Scale Deployment of Carbon Capture, Use, and Storage. https://dualchallenge.npc.org/downloads.php.

National Academies of Sciences, Engineering, and Medicine, 2018. National Academies' Gulf Research Program to Collaborate With Shell on Gulf of Mexico Ocean Observation Effort. https://www.nationalacademies.org/news/2018/11/national-academies-gulf-research-program-to-collaborate-with-shell-on-gulf-of-mexico-ocean-observation-effort.

National Academies of Sciences, Engineering, and Medicine, 2019. Negative Emissions Technologies and Reliable Sequestration: A Research Agenda. The National Academies Press, Washington, DC, https://doi.org/10.17226/25259.

Natural Resources Revenue Data, 2022. https://revenuedata.doi.gov/downloads/revenue/. (Accessed 4 April 2018).

Nippon Foundation-GEBCO, 2021. The Nippon Foundation-GEBCO Seabed 2030 Project: 100% of the Ocean Floor Mapped by 2030. https://seabed2030.org.

NREL, 2021. https://www.nrel.gov/bioenergy/algal-biofuels.html.

Oberstoetter, M., Usoro, M., 2021. Could Restricting Oil Production in the US Gulf of Mexico Lead to Carbon Leakage? https://www.woodmac.com/news/opinion/could-restricting-oil-production-in-the-us-gulf-of-mexico-lead-to-carbon-leakage/.

Ocean Foundation, 2021. Seabed Mining. https://oceanfdn.org/seabed-mining/.

Ocean Stewardship Coalition, 2021. Seaweed as a Nature-based Climate Colution: Vision Statement. UN Global Compact, p.18. file:///C:/Users/art.donovan/AppData/Local/Microsoft/Windows/INetCache/Content.Outlook/EFIZLSEP/Seaweed%20as%20a%20Nature-Based%20Climate%20Solution.pdf.

OECD, 2021. Protected Areas. https://data.oecd.org/biodiver/protected-areas.htm.

Offshore Energy, 2015. Estimate of Global Potential Tidal Resources. https://www.offshore-energy.biz/estimate-of-global-potential-tidal-resources/.

Offshore Energy, 2020a. Oceans of Energy Has Doubled the Capacity of Its First Offshore Floating Solar Farm in the Dutch North Sea. https://www.offshore-energy.biz/oceans-of-energy-beefs-up-its-dutch-north-sea-solar-farm/.

Offshore Energy, 2020b. Offshore Deepwater Oil Production One of the Cheapest Sources of New Supply as Costs Reach New Low. https://www.offshore-energy.biz/offshore-deepwater-oil-production-one-of-cheapest-sources-of-new-supply-as-costs-reach-new-low-rystad-says/.

O'Shea, T., Jones, R., Markham, A., Norell, E., Scott, J., Theuerkauf, S., Waters, T., 2019. Towards a Blue Revolution: Catalyzing Private Investment in Sustainable Aquaculture Production Systems. The Nature Conservancy and Encourage Capital, Arlington, VA.

Ørsted, 2021. 1991-2001, The First Offshore Wind Farm. https://orsted.com/en/about-us/whitepapers/making-green-energy-affordable/1991-to-2001-the-first-offshore-wind-farms.

Patonia, A., Poudineh, R., 2020. Ammonia as a Storage Solution for Future Decarbonized Energy Systems. The Oxford Institute for Energy Studies. https://www.oxfordenergy.org/wpcms/wp-content/uploads/2020/11/Ammonia-as-a-storage-solution-for-future-decarbonized-systems-EL-42.pdf.

PEW Charitable Trusts, 2022. Deep Sea Mining: The Basics. https://www.pewtrusts.org/en/research-and-analysis/fact-sheets/2017/02/deep-sea-mining-the-basics.

Pham, C.K., Ramirez-Llodra, E., Alt, C.H.S., Amaro, T., Bergmann, M., Canals, M., 2014. Marine litter density and distribution in European seas, from shelves to deep basins. PLoS One 9, e95839. https://doi.org/10.1371/journal.pone.0095839.

Pickering, K.T., Hiscott, R.N., 2016. Deep Marine Systems. Processes, Deposits, Environments, Tectonics and Sedimentation. AGU & Wiley, p. 657.

Pickering, K.T., Hiscott, R.N., Hein, F.J., 1989. Deep Marine Environments. Clastic Sedimentation and Tectonics. Unwin Hyman Ltd, London.

Pizzo, N., Deike, L., Ayet, A., 2021. How does the wind generate waves? Phys. Today 74 (11). November.

Pohl, F., Eggenhuisen, J.T., Kane, I.A., Clare, M.A., 2020. Transport and burial of microplastics in deep-marine sediments by turbidity currents. Environ. Sci. Technol. 54 (7), 4180–4189. https://doi.org/10.1021/acs.est.9b07527. 2020.

Radowitz, B., 2021. Solar and Offshore Wind to Help Renewables Double in Asia this Decade: Fitch Solutions. Recharge: Global News and Intelligence for the Energy Transition. https://www.rechargenews.com/transition/solar-and-offshore-wind-to-help-renewables-double-in-asia-this-decade-fitch-solutions/2-1-938872.

Ranken Energy Corporation, 2021. https://www.ranken, 2021energy.com/index.php/products-made-from-petroleum.

Richter, A., 2019. Scientists From Iceland and Oil Nation Norway Want to Help the World to Generate Electricity From Offshore Geothermal Heat Sources—A Process That Causes No $CO_2$ Emissions or Conflicts About Landscape Destruction. ThinkGeoenergy. https://www.thinkgeoenergy.com/green-electricity-from-heat-under-the-sea-offshore-geothermal-energy/.

Rystad Energy, 2020. Oil Production Costs Reach New Lows, Making Deepwater One of the Cheapest Sources of Novel Supply. https://www.rystadenergy.com/newsevents/news/press-releases/oil-production-costs-reach-new-lows-making-deepwater-one-of-the-cheapest-sources-of-novel-supply/.

Samad, A., Sannasiraj, S.A., SundarParesh, V., Halder, P., 2022. Ocean Wave Energy Systems. Hydrodynamics, Power Takeoff and Control Systems. Springer, p. 578.

Sears, R.A., 2020. Energy futures: oil and the oil industry. Brown J. World Aff. XXVI (II). Spring/Summer.

Seibold, E., Berger, W.H., 2010. The Sea Floor. An Introduction to Marine Geology, third ed. Springer Verlag, p. 369.

Shadman, M., Silva, C., Faller, D., Wu, Z., de Freitas Assad, L.P., Landau, L., Levi, C., Estefen, S.F., 2019. Ocean renewable energy potential, technology, and deployments: a case study of Brazil. Energies 12, 36–58. https://doi.org/10.3390/en12193658.

Shell, 2021. Sky Scenario. https://www.shell.com/promos/business-customers-promos/download-latest-scenario-sky/_jcr_content.stream/1530643931055/eca19f7fc0d20adbe830d3b0b27bcc9ef72198f5/shell-scenario-sky.pdf.

Shields, M.A., Payne, A.I.L., 2014. Marine Renewable Energy Technology and Environmental Interactions. Springer, p. 176.

Stauffer, N., 2015. A New Look for Nuclear Power. https://news.mit.edu/2015/new-look-floating-nuclear-power-0624.

Stow, D.A.V., Smillie, Z., 2020. Distinguishing between deep-water sediment facies: turbidites, contourites and hemipelagites. Geoscience 10, 68.

Stow, D.A.V., Cremer, M., Droz, L., Normark, W.R., O'Connell, S., Pickering, K.T., Stelting, C.E., Meyer-Wright, A.A., DSDP Shipboard Scientists, 1985. Mississippi fan sedimentary facies, composition, and texture. In: Bouma, A.H., Normark, W.R., Barnes, N.E. (Eds.), Submarine Fans and Related Turbidite Systems. Springer-Verlag, New York, pp. 259–266, https://doi.org/10.1007/978-1-4612-5114-9_38.

Talling, P.J., Masson, D.G., Sumner, E.J., Malgesini, G., 2012. Subaqueous sediment density flows: depositional processes and deposit types. Sedimentology 59, 1937–2003.

Talling, P., et al., 2021. Novel Sensor Array Helps to Understand Submarine Cable Faults Off West Africa. EarthArXiv.

Texas A&M Galveston, 2021. Ike Dike: A Coastal Barrier Protecting the Houston/Galveston Region From Hurricane Storm Surge. https://www.tamug.edu/ikedike/.

The Verge, June 2019. Giant Cargo Drones Will Deliver Packages Farther and Faster. https://www.theverge.com/2019/6/10/18657150/autonomous-cargo-drones-delivery-boeing-aircraft-faa-regulation.

Tide Mill Institute, 2021. https://www.tidemillinstitute.org/definition-of-a-tide-mill/Institute.

van Sebille, E., Wilcox, C., Lebreton, L., Maximenko, N., Hardesty, B.D., van Franeker, J.A., Eriksen, M., Siegel, D., Galgani, F., Law, K.L., 2015. A global inventory of small floating plastic debris. Environ. Res. Lett. 10 (12), 124006.

Warme, J.E., Douglas, R.G., Winterer, E.L. (Eds.), 1981. The Deep Sea Drilling Project: A Decade of Progress. vol. 32. SEPM Special Publication, https://doi.org/10.2110/pec81.

Williams, C., Taylor, C., 2012. Ocean Energy: Technologies and Potential Environmental Effects (Energy Science, Engineering and Technology). Nova Science Pub Inc, UK, p. 180.

Wood Mackenzie, 2021. What Different Scenarios Tell Us About the Future of Oil and Gas. https://www.woodmac.com/news/the-edge/what-different-scenarios-tell-us-about-the-future-of-oil-and-gas/.

Woodall, L.C., Sanchez-Vidal, A., Canals, M., Paterson, G.L.J., Coppock, R., Sleight, V., 2014. The deep sea is a major sink for microplastic debris. R. Soc. Open Sci. 1, 140317. https://doi.org/10.1098/rsos.140317.

World Forum Offshore Wind, 2021. https://w2m8n2q3.rocketcdn.me/wp-content/uploads/2021/08/WFO_Global-Offshore-Wind-Report-HY1_2021.pdf.

World Ocean Review, 2014. Methane Hydrate. https://worldoceanreview.com/en/wor-3/methane-hydrate/mining-impacts/.

Yang, Z., Copping, A., 2017. Marine Renewable Energy. Resource Characterization and Physical Effects. Springer, p. 387.

Yeilding, C.A., 2021. AAPG CCUS Conference, Keynote Address: Put the yoU in CCUS: Let's Explore!. March 23–24.

Zhang, K., Kim, W., Park, A., 2020. Alkaline thermal treatment of seaweed for high-purity hydrogen production with carbon capture and storage potential. Nat. Commun. 11, 3783. https://doi.org/10.1038/s41467-020-17627-1.

# Glossary

**Above grade** Above-grade slopes (Prather et al., 2016) are defined as being steeper than the equilibrium gradient and may be caused by rapid sea-level rise, tectonic over-steepening or alternating clastic and carbonate deposition.

**Abyssal plain** Flat ocean floor far from sedimentary input points typically greater than 3000 m water depth.

**Accommodation** Space made available for potential sediment accumulation by sea-level rise, subsidence or a combination of these two processes (Jervey, 1988).

**Acoustic impedance** Product of density times the sound wave velocity in a material.

**Active margin** Basin margin defined by extensional, compressional, transform or a combination of the tectonic processes.

**Ad valorem taxes** Ad valorem taxes are assessed based upon the value of the property (assets). These property taxes are paid to states and local taxation districts in each area of operation and are based on a percentage value of the physical assets or reserves. For forecasting purposes, they are commonly estimated as a percentage of revenue.

**Adam optimization algorithm** A stochastic optimization algorithm used to estimate the optimal values of neural network weight parameters from training data.

**Airgun** Marine apparatus designed to send acoustic pulses into the water column and subsurface for seismic acquisition.

**Angle-stacks** Seismic data stacked over a limited range of angle of incidence to be utilized in AVO screening or inversion.

**Anisotropy** Variation of velocity with the direction of wave propagation.

**Antidune** Bedform formed by supercritical flow in which the surface of flow is in phase with the bed.

**Appraisal** The business of assessing the size and characteristics of an exploration discovery prior to commercial production.

**Architectural element** 1-, 2- or 3-dimensional geobody.

**Architectural element analysis** The practice of characterizing geobodies based on scale and genetic relationship.

**Argillaceous sandstone** Sandstone interval enriched in clay-sized particles, typically greater than 10%–15% of the rock volume.

**Arrhenius equation** A formula that describes the temperature dependence of chemical reactions.

**Artificial neural network** Deep learning network consisting of multiple interconnected hidden layers, with each layer consisting of a set of multiple units called neurons.

**Attenuation** Decrease in wave energy as the wave propagates.

**Authigenic** Solid matter that forms in a sediment or rock after the time of deposition.

**AVO** Amplitude Variation with Offset. Variation of the amplitude of sound waves reflected from subsurface horizons as a function of the angle of reflection.

**AVO attributes** Intercept and gradient attributes extracted from seismic gathers and/or calculated from forward modeling used as principal means of investigation both in qualitative and quantitative seismic interpretation.

**AVO background trend** Diagonal trend line in the intercept-gradient crossplot near which the bulk of the data(points) reside, commonly interpreted as shale overlain by shale or brine-filled sand(stone).

**AVO classes (1, 2p, 2, 3 and 4)** Division of AVO data based on the intercept-gradient quadrant they plot in. Class 1: high impedance gas sands overlain by low impedance shales. Class 2: near-zero impedance contrast (top shales and reservoir shales share similar elastic properties). Class 3: high impedance shales above low impedance sands. Class 4: a subclass of Class 3, where the low impedance sands are overlain by high impedance cap rocks.

**AVO crossplot** A qualitative technique for visual inspection of AVO gradient vs. intercept points.

**Back stripping** Numerical technique to remove the effects of the sediment load in order to isolate the effect of tectonics on the basin. Requires corrections of compaction, paleobathymetry and paleo-water depth.

**Barrels of oil equivalent** To compare oil and gas fields, the concept of barrels of oil equivalent (BOE) attempts to equate gas value to oil value. Originally on a Btu basis, 6000 cubic feet of gas is approximately one equivalent barrel of oil. However, there can be wide disparities in price equivalence between the products, so the Btu conversion is academic and not necessarily an indicator of value.

**Base level** A surface of equilibrium that sedimentary processes strive to attain, at which neither erosion nor deposition takes place (Barrell, 1917).

**Base of slope** Zone of intermediate gradient between continental slope and basin floor.

**Basin floor** Zone of very low gradient basinward of the base of the slope.

**Basin modeling** Basin modeling utilizes mathematical algorithms to reconstruct the geohistory and thermal evolution of the subsurface geology; petroleum system modeling uses linked algorithms designed to reconstruct the evolution of petroleum.

**Bathymetric profiling** The practice of characterizing the depth of water.

**Bathymetry** Measurement of the depth of water.

**Bayes' rule** A mathematical equation that allows calculating the probability of an event based on the prior knowledge about its occurrence.

**Bayesian** Relating to Bayes' rule.

**Bed fabric** Aggregate grain arrangement and pattern of lithologies within an interval of rock.

**Bed termination** End of a sedimentation unit that can be caused by lack of sediment, truncation, onlap or other forms of pinch-out.

**Bed thickness** Length measurement from base to top of the sedimentation unit.

***Bi*-gradational contourite sequences** Standard sequence for dominantly muddy to sandy contourite deposits in drifts. Muddy and fine sandy contourite facies are typically arranged in a coarsening-up to fining-upward cycle (inverse to normal grading with divisions C1-C2-C3-C4-C5, where C3 represents the coarser interval) related to long-term variations in bottom-current velocity, or due to temporal variations in local sediment supply. Partial basal or top sequence types are also possible.

**Biochemical sedimentation** Sedimentation that occurs due to organisms and water chemistry to form in a lake, ocean or river.

**Biogenic structures** Any evidence of organism activity other than the production of body parts (Buatois and Mángano, 2011).

**Bioturbation** Process by which the primary consistency and structure of a sediment are modified by the activity of organisms (Bromley, 1996).

***Bi*-partite** Referring to a sedimentation unit or bed that has two principal first-order architectural elements.

**Blowout preventer (BOP)** Specialized valve or mechanical device used to seal, control and monitor oil and gas flow from the subsurface.

**Borehole imaging devices (BHI)** Wireline or logging devices that are designed to determine the dip of planar and near planar surfaces in a borehole and to provide images of these surfaces.

**Bottom current** Semipermanent water-mass flow capable of eroding, transporting and depositing sediments on the seafloor.

**Bottom current reworked sands (BCRS)** Sandy deposits originally derived from a gravity (turbiditic) flow and partially or completely reworked by bottom currents.

**Bouma Sequence** A sedimentation unit that has five principal subdivisions Ta, Tb, Tc, Td and Te representing suspension, traction and suspension sedimentation stages, respectively, due to a waning (unsteady) turbidity current.

**Braided channel system** Multiple simultaneously active channels facilitating sediment transport.

**Brent** Brent is the oil benchmark (or index or reference) price generally used outside the United States. It is a light, sweet crude and other oils are priced at sale based on differences in gravity and contaminants.

**Brownfield** Oilfield near the end of its production life.

**Burial history** A set of data representing, at minimum, the burial depth, effective stress and temperature through geologic time for a point of interest in a basin.

**Burrow** Space within sediment occupied and maintained by an animal (Bromley, 1996).

**Burrow fill** Material that fills a burrow, may be active (backfill) or passive (gravitational) (Bromley, 1996).

**Burrow rim** Characteristic mantle concentration surrounding the tube core in *Macaronichnus*.

**Calibration** The process of adjusting input parameters to achieve model output that is consistent with the observations.

**Canyon** A relatively narrow pathway with steep walls that generally acts to collect and funnel sediment from the shelf to deepwater.

**Capital investment** A capital investment is money spent on assets that will produce income over their useful lives.

**Carbonate** Class of rock formed principally by chemical or biochemical processes with $CO_3^{2-}$.

**Carbon capture, utilization and storage (CCUS)** A process that captures carbon dioxide emissions and reuses or stores in geologic formations (National Petroleum Council, 2019 CCUS Report).

**Cement** Authigenic matter that forms within what was previously an intergranular pore. Most cements are made up of minerals, although cements may also be made up of organic matter such as bitumen or other noncrystalline phases such as opal-CT.

**Chance node** A chance node is a point on a decision tree that indicates there is more than one possible outcome of an event.

**Chance of economic success** The chance of economic success is the probability of finding hydrocarbons in commercial quantities; that is, finding enough oil to make the project profitable.

**Channel** Deep and narrow conduits for sediment transport and deposition whose confinement can be generated due to (i) construction of aggrading levees, (ii) erosion and incision from sediment transport processes (Cronin et al., 2000; Wynn et al., 2002, 2007) or even (iii) tectonic confinement (Anderson, 1998).

**Chaotic deposit** Generic term for any number of poorly sorted or remobilized deposits that commonly feature a random arrangement of grains, structures or bedding.

**Chemical compaction** The most common definition for chemical compaction in sedimentary petrology is compaction that arises due to the dissolution of mass at grain contacts. It is also commonly referred to as "pressure solution" and "contact dissolution." Unfortunately, some publications, particularly in the rock physics literature, use the term "chemical compaction" when they are in fact referring to quartz overgrowth cementation.

**Chondritic Uniform Reservoir (CHUR)** The mean chemical composition of the part of the Solar Nebula from which chondrites are formed during the formation of the Solar System. It is often referred to in the context of Nd isotope evolution (e.g., Bouvier et al., 2008).

**Chronostratigraphic (chronostratigraphy)** The subdivision of the stratigraphic record into time-based rock units as opposed to rock units (lithology). Methods such as biostratigraphy, radiometric dating and astrochronology can help to reconstruct the absolute ages of rock units.

**Clast support** Grains transported and supported by other grains (lacking matrix).

**Clastic** Class of rock formed principally by quartz, feldspar and lithic (rock fragment) grains.

**Climbing-ripple cross-lamination** Sedimentary structure formed from preservation of most, if not all, of the ripple bedform.

**Cohesive flow** Flow that is generally laminar, in which support is provided by a muddy matrix.

**Compaction** Bulk volume reduction. Note that compaction generally is not the sole cause of porosity loss in sandstones given that dissolution, cementation and replacement may also affect porosity.

**Compliant tower** A compliant tower, like a fixed platform, employs a steel tubular jacket to support the surface facilities. Unlike fixed platforms, compliant towers yield to the water and wind movements in a manner similar to floating structures. Like fixed platforms, they are secured to the seafloor with piles, but they have smaller dimensions, and the stresses on the steel are mitigated using buoyant sections. These may be guyed with mooring lines or self-standing. Each is designed according to the water depth, weather conditions and soil strength.

**Compressional domain** A zone of deformation in a passive margin fold belt system dominated by thrusting, folding, squeezing of diapirs and salt nappe extrusion.

**Concordia** When concordant $^{206}Pb$ and $^{238}U$ data are plotted against each other, they fall along a concordia curve (Wetherill, 1956).

**Condensate yield** The condensate yield is the amount of liquid hydrocarbon that exists in the reservoir as a gas but is separated as a liquid at standard conditions. It is usually expressed as barrels per million cubic feet of gas and sells for about the same price as a barrel of light sweet oil.

**Confined** Sedimentary environment in a relatively narrow fairway.

**Conglomerate** Clastic rock type defined by most grains exceeding very coarse-grained sand (2 mm in diameter).

**Constant cement model** Rock physics model assuming that sands of varying porosity/sorting have the same/constant amount of cement.

**Constructional channel** Channel that is net aggradational (vs. incisional).

**Contact cement model** Rock physics model that captures/quantifies the substantial increase of elastic properties of sands due to the initial precipitation of quartz cement to intergranular space.

**Contact dissolution** Dissolution at grain contacts that leads to compaction. It is also commonly referred to as "pressure solution" and "chemical compaction."

**Contourite Depositional Complex (CDC)** Association of different CDSs connected laterally (and vertically) and related with the same water mass in the same or adjacent basins.

**Contourite Depositional Systems (CDSs)** An association of depositional (drifts) and related erosional features related to bottom currents belonging to one particular water mass.

**Contourite terraces** Sub-horizontal morphologic elements identified in deep marine settings, both at the present sea floor and in the sedimentary record, which have developed over time in constructional and erosional phases. They develop on the top and proximal domain of plastered drifts.

**Contourites** Sediments deposited or substantially reworked by the persistent action of bottom currents

**Contouritic drifts (or just "drifts").** Large to very large accumulations of contourite deposits typically identified based on their morphology and long-term sedimentary architecture in 2D or 3D seismic data (or very occasionally in outcrops).

**Convolutional neural networks** A class of artificial neural networks used to learn relationships from spatial data with spatial filters called kernels.

**Core** In the ichnological sense, the central part of a zone fill (Bromley, 1996).

**Cosimulation** Geostatistical simulation of multiple variables by honoring the mutual correlations.

**Cosmogenic radionuclides (CRN)** Rare isotopes created when a high-energy cosmic ray collides with the nucleus of an atom. CRN geochronometers occur across a wide elemental range with variable half-life periods and they may be measured in soil, rocks, groundwater and the atmosphere to date the age of a wide variety of geological processes.

**Counterregional fault** Syn-depositional growth faults dipping landwards, i.e., the opposite direction of the regional normal faulting.

**Creaming curve** As exploration has continued to explore into deeper waters and more remote environments, petroleum reserves have continued to increase without a period of significant plateau, highlighting the remaining potential and importance of these reservoirs to help meet future energy needs.

**Critical taper angle** The taper angle at which the wedge propagates along the basal detachment. The critical taper angle is a function of the physical properties of the wedge material, pore fluid pressure, incoming sediment thickness, etc.

**Cross stratification** Broad class of sedimentary structures derived from the dune bedform.

**Cumulative frequency distribution** A cumulative frequency distribution is a tabular or graphical summary of a collection of data, showing the probability of observations occurring that are less than or equal to a specified value. The distribution may also be constructed to show the probability of values that are greater than or equal to a specified value.

**Debris flow** Type of sediment gravity flow defined by laminar, cohesive flow with matrix support or frictional grain collisions.

**Debrite** Deposit formed by a debris flow.

**Decision node** A decision node is a point on a decision tree that indicates when a decision must be made.

**Decision tree** A decision tree is a diagram or chart depicting the possible decisions, probabilities of success and outcomes of these decisions and their relationships. It is a sum of products exercise that should capture the full range of outcomes.

**Deepwater** Depth of water that is below wave base (also deepwater, which is commonly used as a modifying term).

**Deepwater sediment** Sediment deposited in deep or quiet water due to turbiditic and related processes, generally below the influence of tidal and fluvial currents.

**Density current** Current that is formed due to two or more masses of fluid, each with a different density.

**Denudation** A general term for the chemical and physical processes of erosion, leaching, stripping and reducing that wear away the terrestrial surface, transporting earth material from areas of higher elevation to areas of lower elevation.

**Depletion drive** Depletion drive occurs when there is no aquifer or a small one compared to the size of the oil or gas accumulation. As oil or gas is produced, the reservoir pressure depletes and gas is preferentially produced with liquids left behind.

**Deposit feeder** Animal that obtains nutriment from particulate organic matter and microbes incorporated within the sediment (Bromley, 1996).

**Depositional architecture** Geometric characterization and genesis of a geobody.

**Depositional contourite features** Morphological bottom current (contouritic) features generated by dominant depositional processes (e.g., drifts, sedimentary waves).

**Depositional element** Group of beds that forms a 3D body of sediment (geobody).

**Depositional model** Representation of a body of sediment or sedimentary rock.

**Depositional sequence** Interval of genetically related rock bounded at its top and base by erosional unconformities and their correlative conformities.

**Depositional system** Group of genetically related depositional environments linked by similar sedimentary processes.

**Derisk** A popular term among explorationists for minimizing the risk of a given prospect using geoscientific methods and/or geophysical technology.

**Detachment layer** Typically salt or overpressured shale layer along which basinward translation of the younger rocks occur.

**Deterministic** A process by which a particular set of input parameters will always produce the same set of outputs when a model is simulated.

**Detrital** Solid matter that was present at the time of deposition.

**Detritus feeder** Deposit feeder that feeds on the nutrient-rich topmost layer of the substrate (Bromley, 1996).

**Development** The business of producing petroleum by drilling additional wells prior to full-scale production and after the appraisal phase.

**DHI** Direct Hydrocarbon Indicator: an anomalous seismic attribute value that can be directly linked to the presence of hydrocarbons in a reservoir.

**Diagenesis** The physical, chemical and biological processes that transform sediment into rock during burial.

**Dip** Is the acute angle that a planar surface makes with a horizontal plane. This angle includes both direction and magnitude and is normally designated relative to true dip.

**Dipmeters** Are a family of wireline or logging devices that are designed to determine the dip of planar and near planar surfaces in a borehole. The family includes both legacy "dipmeter" tools and BHI's.

**Direct hydrocarbon indicators** Direct hydrocarbon indicators (DHIs) are anomalous seismic amplitudes that occur due to a change in pore fluids, which cause a change in the bulk rock's elastic properties and may indicate the presence of hydrocarbons.

**Discount rate** The discount rate is the interest rate used in calculating the present value of future cash flows. Either the greater of the incremental cost of capital or the opportunity rate is used, depending on corporate policy.

**Discounted return on investment (DROI)** The discounted return on investment is calculated as the present value of the net cash flows plus investment divided by the present value of the investment. This returns the multiples of investments, adjusted for time value of money. It is used in capital budgeting for ranking opportunities under a limited capital budget. It is not the same as the Internal Rate of Return. While DROI may also be calculated in exactly the same way as NPVI, there are several different calculations given this same name.

**Dish structure** Sedimentary structure formed by water elutriation, in a concave up shape often on the scale of centimeters, but can be meters.

**Distributary channel** Channel that undergoes bifurcation and division to spread flows laterally.

**Downslope sedimentation trends** The systematic organization of sedimentary deposits from upslope to downslope according to first principles.

**Drape** Sedimentary deposit of equal thickness, typically hemipelagic to pelagic, but can be derived from dilute flows.

**Dynamic model** Representation of a body of sediment or sedimentary rock used for simulation of fluid movement.

**Dynamic positioning** Drill ships use GPS and the computer controlled thrust from a number of engines to keep them in position over the well continuously. This method is called dynamic positioning.

**Economic limit** The economic limit is the rate of oil or gas production at which an entity no longer generates enough cash flow, after taxes, to cover the operating expenses for that period.

**Effective date** The effective date of an economic evaluation is time zero for determining present value. It is the date of the first investment in a prospect evaluation or the first day of the budget year for budget economic cases. Note that expenditures prior to the effective date are sunk costs and should not be included in cash flow calculations, but they may impact tax calculations in succeeding periods and those effects must be accounted for in the final analysis.

**Effective pressure** The pressure that each grain size "sees or feels" at any given depth (= lithostatic pressure – pore pressure).

**Effective stress** The stress that is transmitted through the solid portion of the rock or sediment. Note that the effective stress may exceed the vertical effective stress in compressional regimes.

**Elastic bounds** Theoretical bounds on elastic properties when mixing different minerals and/or fluids.

**Enhanced recovery** Enhanced recovery, also known as tertiary recovery, is the result of actions taken to increase the recovery efficiency of the reservoir beyond the natural mechanisms and beyond improved or secondary recovery, which includes natural gas or water injection. Enhanced recovery comprises injecting carbon dioxide, detergents, other chemicals or microbes to move more oil from the reservoir to the market.

**Entitlements** In the context of global oil, gas and minerals, the entitled parties are those who own or control the minerals and therefore have the rights via lease or other contracts to allow companies to explore and produce them in exchange for compensation.

**Epeiric** Broad shallow (shelf) marine environment responsible for thick carbonate successions globally; common in Paleozoic and Mesozoic time, yet not in the present day.

**Erosion** The loss of solid rock volume by sedimentary or tectonic processes.

**Erosional contourite features** Morphological bottom current (contouritic) features generated by dominant erosional processes (e.g., channels, furrows).

**Escalation** Escalation is the forecast increase of prices and costs over time due to inflation, according to a prediction of future values or trends.
**Eustatic sea level** Also, global sea level, which is the distance from the center of the Earth to the surface of the sea.
**Exhumation** Displacement of rocks with respect to the Earth's surface.
**Exploration** The business of locating new petroleum resources by generating prospects and drilling to test a concept.
**Exploration common process** The step-by-step method of gathering, generating and integrating surface and subsurface data to define leads and drill prospects.
**Expulsion rollover** Expulsion of underlying salt by prograding strata that thicken basinward.
**Extended elastic impedance** Effective impedance with shear information dependent on the coordinate rotation angle in the AVO intercept-gradient domain.
**Extensional domain** A zone of deformation in a passive margin foldbelt system dominated by listric and planar normal faults.
**Extrudites** Venting features resulting from extrusion of sand to the surfaces (such as the seafloor) and connected to and fed by underlying sand dikes.
**Facies** Interval of a particular data type that can be grouped based on similar traits (seismic, log, litho-, bio-, ichno-, magneto-, etc.).
**Facies model** Approximation or representation of a certain flow or rock type used to predict deposits in a depositional system or environment.
**Fall** Loose debris is removed from a steep slope and accumulates at the base of the slope as talus.
**Falling-stage systems tract** A systems tract defined by a forced regressive stacking pattern, which accumulates during relative sea-level fall (Catuneanu, 2019a).
**Fan** Accumulation of sediment with its apex at a sedimentary point source that widens in a cone-shaped, broad dispersal pattern.
**Fan apron** Multiple cone-shaped sediment accumulations that coalesce and overlap, with apexes along a line source.
**Fan valley** Accumulation of sediment with its apex at a sedimentary point source that widens in a broad dispersal pattern in a topographic low often generated by the prior evacuation of sediment, such as a mass-transport scar.
**Feasibility modeling** Modeling performed to evaluate the potential detectability of a given scenario and/or the applicability of a given technology.
**FEED and pre-FEED** FEED is the acronym for Front-End Engineering Design. It is the phase of evaluating a deepwater project that entails various studies to determine technical issues and estimate the rough investment cost of the production facility including platforms, gathering systems and pipelines. Pre-FEED occurs beforehand and is a less comprehensive evaluation to select the general concept of systems to be used if the project is successful.
**Feeder system** Depositional environment responsible for delivery of sediment down system or to another environment of deposition, a sediment input zone.
**FID** FID is the acronym for final investment decision, the sanctioning of a project by the investor's (company's) governing body or management. This decision allocates the funding and allows the project to begin.
**Filter feeding** Subdivision of suspension feeding, in which an organic filtration mechanism is involved in food acquisition (Buatois and Mángano, 2011).
**Fine-grained contourite** Defined and interpreted by De Castro et al. (2020) as having characteristic basal turbidite deposits overlain by along-slope-deposited sediment in one sedimentation unit.
**Fiscal regime** A fiscal regime is defined by the set of laws, rules and contracts used by a specific country to govern the way companies can explore for and produce oil and gas. It specifies how costs and revenues (or profits) are allocated between the country and the company.
**Floating production storage and offloading (FPSO)** A Floating Production Storage and Offloading (FPSO) platform is typically tied to the sea bottom by chains and anchors. The hull of the vessel is tethered to a turret and allowed to rotate depending on wind and waves. Storage capacity varies, but can be 1.5 MMBO, or more. Whether it is a purpose-built hull or converted tanker, there is a large amount of deck space and robust capacity, allowing large amounts of production equipment. As a result, production rates can range up to 150,000 BOPD.
**Flow** Movement of a mixture of sediment and medium, such as water.
**Flow bulking** Process of increasing the size and density of the sediment gravity flow.
**Flow capacity** Amount of sediment able to be carried in a flow, such as a turbidity flow.
**Flow competency** Coarsest clast size able to be carried in a flow, such as a turbidity flow.
**Flow partitioning** Segmentation of a flow into different parts based on rheology and process.
**Flow rate** Amount of fluid delivered over a given time period.
**Flow steadiness** Trait of flow that describes the rate determined by distance traveled over a given length, and if that rate varies from one time to the next.

**Flow uniformity** Trait of flow that describes its direction of movement, and if that direction changes from one time to the next.

**Fluid escape feature** Phenomenon related to the passing or leaking of fluids in a system, such as a gas chimney or some pockmarks.

**Fluid factor** A measure indicating the distance from the background trend in the AVO crosslot. Points lying away from the background trend can be directly interpreted as hydrocarbon-bearing.

**Fluid overpressure** Where the fluid pressure is greater than expected from the hydrostatic gradient. In such cases the effective stress transmitted through the solid portion of the rock is less than it would be otherwise.

**Fluidization** Process of suspension of grains by drag forces imparted by upward flow, which tends to drive pore-fluid upward and initiates localized sand fluidization.

**Flume** Tank usually filled with water and sediment to simulate physical submarine processes, such as sediment gravity flows, turbidity currents, debris flows, transitional flows and bottom currents.

**Formation MicroScanner (FMS) and the Formation MicroImager (FMI)** The primary BHI devices presented in this discussion are Schlumberger's FMS and FMI.

**Formation volume factor** Formation volume factors convert oil or gas from reservoir conditions to the stock tank or standard conditions. They are determined in a laboratory using actual fluid samples from well tests.

**Forward stratigraphic modeling** The process of modeling the spatial and temporal distribution of sediments related to depositional processes and changing sea level, among other drivers of change related to basin fill. Also called geologic process modeling.

**FPSO** Floating Production Storage and Offloading vessel first used offshore in Brazil, now globally, to produce petroleum in frontier areas.

**Friable sand model** Rock physics model describing the change of elastic properties of unconsolidated sands as a function of porosity, grain contact characteristics and effective pressure.

**Fringe** Distal margin of a depositional element, such as a splay.

**Frontier exploration** The business of locating petroleum resources that are considered higher risk; frontier exploration involves relatively untested plays or in regions far from infrastructure (i.e., the Arctic).

**Full-waveform inversion (FWI)** Full-waveform inversion is an advanced seismic processing methodology that determines high-resolution, high-fidelity velocity models of the subsurface through iterations that minimize the difference between observed data and modeled seismic waveforms.

**Gas hydrates** Naturally occurring crystalline solid formed of water and gas that contains huge amounts of methane and have the potential to provide an immense resource of natural gas from the marine sediments and permafrost.

**Gas-oil ratio** The gas-oil ratio is the amount of gas dissolved in oil in the reservoir, each measured at standard conditions and expressed in cubic feet per barrel.

**Gassmann's theory** Elasticity model used to compute effects of different fluids on elastic moduli of porous rocks.

**Gateways** Gateways (sometimes referred to as seaways, passages, straits, corridors and channels) are relatively narrow marine passages, connecting ocean basins, of varying depth that control surface, intermediate or deepwater circulation.

**Geochronology** The science of dating rocks, minerals, fossils and events using radioactive isotopes.

**Geologic chance factors** When determining the chance of technical success, many companies assess individual elements, or geologic chance factors, that must occur for hydrocarbons to exist. These include source, maturity, migration, trap and seal, which are multiplied together to estimate the probability of geologic success.

**Graded** Slopes (also in-grade) that have a smoothed concave-up theoretical reference profile and are efficient at transferring sediment to the basin floor.

**Graded bed** Sedimentation unit in which grains are sorted from coarsest at the base to finest at the top.

**Gradient** Angle of the sloping terrain or bathymetry.

**Grain** Particle present at the time of deposition with a long axis length in the range from silt (0.0625 mm) to granule (4 mm).

**Grain coating** A thin (typically $<30\,\mu m$) coating of material over the surface of a grain. Grain coatings may inhibit the nucleation of overgrowth cements such as quartz and reduce permeability by narrowing or closing pore throats. Note that coatings may be made up of a broad range of both detrital and authigenic materials. Common coating types that are effective for inhibiting quartz cementation include detrital matrix, infiltrated clay, pore-lining chlorite and microcrystalline quartz.

**Grain population** Group of clasts in the same size range (diameter), typically three in most deepwater systems—gravels, sands and muds.

**Gravelly contourite** Defined and interpreted by De Weger et al. (2020) as having interleaved gravelly to sandy deposits in one sedimentation unit formed by along-slope processes (contour currents).

**Gravitational failure** Gravity-driven deformation, typically along a salt or overpressured shale detachment layer.

**Gravity drainage** Gravity drainage is a very efficient recovery mechanism where oil is generally pumped from a reservoir as the weight of the oil allows it to flow to the bottom where the wells. Recovery efficiency can be very high, but the time for recovery is extremely slow, making its commerciality difficult.

**Gravity gliding** Sliding in the direction of downdip along a detachment layer due to the gravitational force of the overburden.

**Gravity spreading** Vertical collapse and lateral spreading of rock under its own weight. The presence of a dipping detachment layer allows for downslope movement.

**Grid** A mesh of regular or irregularly shaped cells that is used to partition continuous geometric space into discretized space for numerically solving differential equations.

**Gross working interest** Gross working interest, or working interest, is the ownership in a given lease, field or other reporting entity. Each working interest owner is liable for its share in all investments and expenses in the amount of its portion of the working interest percentage (net working interest).

**Growth strata** Growth or *syn*-tectonic strata are stratigraphic intervals that were deposited during deformation, identified by stratal thinning toward the fold axis. Therefore, the ages of growth strata typically define the timing of deformation.

**Halokinetic** To do with salt and salt diapirism and salt tectonics.

**Healed (slope) accommodation** Healed slopes and slopes that have been readjusted over time (Ross et al., 1994) often feature erosional truncation, sediment bypass and marine onlap of apron complexes.

**Heat flow** Heat transfer achieved by processes of conduction (transfer of thermal energy by contact; dominant in the shallow lithosphere), convection (transfer of heat by movement of solid or fluid; dominant within the mantle) and radiation (e.g., the Sun, important mainly for the Earth's surface). The basal heat flow represents the lower thermal boundary condition of basin models and can be predicted by crustal models that calculate the tectonic stretching and relaxation of the lithospheric mantle and the crust.

**Hemipelagic** Fine-grained sedimentary deposit formed by vertical settling of particles under the influence of gravity.

**Hemipelagic wedge** Tapering unit of sediment deposited by hemipelagic processes.

**Hemipelagite** Deposit consisting of hemipelagic sediment.

**Henry Hub** Henry Hub is the benchmark (or index or reference) price for natural gas traded in the United States. This refers to the point in South Louisiana where at one time most gas produced offshore entered the interstate market.

**Hertz-Mindlin contact theory** A granular medium model for calculation of effective elastic properties of two identical spheres/grains: used to describe the properties of precompacted granular rocks.

**High rate-high ultimate wells** Also called HRHU, high rate, high ultimate wells are needed to produce large volumes of oil quickly and in ultimate recoverable volumes over their lives. This category is an indicator of the kinds of wells that are usually needed to make a deepwater project commercial.

**High-density turbidity current** Flow in which grains constitute >10% of the flow and can exceed 30% sediment volume.

**High-energy flow** Robust and energetic flow capable of traveling at high speeds, carrying a high sediment load and able to interact with and scour the bed.

**Highstand systems tract** A systems tract defined by a normal regressive stacking pattern, which follows a transgression of equal hierarchical rank (Catuneanu, 2019a).

**Homopycnal** Dense fluvial flow that is more buoyant than the ambient water mass it enters.

**Hybrid depositional systems** Hybrid depositional systems are built mostly by synchronous interactions between along-slope bottom currents and downslope turbidity flows, which form bottom current reworked sands, reworked turbidite deposits and transitional facies between contourite and turbidite deposits.

**Hybrid event** Sediment gravity flow that mixes sandy and muddy constituents, either due to the combination of fluid turbulence (turbidity) and matrix strength (cohesion), or processes such as delamination that incorporate muddy substrate into an originally sandy flow.

**Hybrid system** Depositional system formed by sediment gravity flow, mass-transport and bottom current processes.

**Hydrostatic gradient** The fluid pressure gradient that exists in a freely connected column of water in the absence of significant fluid movement.

**Hyperparameter tuning** A procedure to estimate the optimal values of neural network hyperparameters such as number of hidden layers and convolutional kernels.

**Hyperpycnal flow** Dense fluvial flow that is less buoyant than the ambient water mass it enters.

**Hypopycnal** Flow of a lower river water density entering a higher density standing water density in the basin.

**Ichnoassemblage or trace fossil assemblage** Group of trace fossils preserved in a rock unit or sedimentary facies, with no assumption with regard to time of emplacement or recurrence in the stratigraphic record (Buatois and Mángano, 2011).

**Ichnocoenosis** An ecologically pure assemblage of trace fossils, i.e., deriving from the work of a single endobenthic community (Bromley, 1996).

**Ichnodiversity** Number of ichnotaxa in an ichnocoenosis (Bromley, 1996).

**Ichnofabric** Any aspect of the texture and internal structure of a substrate resulting from bioturbation and bioerosion at any scale (Buatois and Mángano, 2011).

**Ichnofacies (archetypal ichnofacies)** Association of trace fossils that is recurrent in time and space, and that directly reflects environmental conditions such as bathymetry, salinity and substrate character (Bromley, 1996).

**Ichnology** The study of traces produced by organisms (both animals and plants) on or within a substrate, and includes all issues related to bioturbation, bioerosion and biodeposition (Buatois and Mángano, 2011).

**Ichnotaxobase** Morphological feature of a trace fossil that is considered a valid basis for ichnotaxonomy (Bromley, 1996).

**IGV** Abbreviation of intergranular volume.

**Illitization** The reaction that results in the formation of the clay mineral illite. This reaction may involve kaolinite (or its dickite polymorph), smectite or mixed layer illite/smectite reactants. For reactions involving smectite, the reaction may progress through a series of intermediate phases with variable proportions of smectite interlayers that display varying degrees of stacking order.

**In grade** See graded slope.

**Incised channel** Conduit for sediment transport that is kept open by erosional processes occurring in the channel.

**Income tax** Income taxes are payable to the state and federal governments. They are calculated according to the individual laws, generally as a percentage of net income as defined by applicable law.

**Infrastructure-led exploration (ILX)** Infrastructure-led exploration refers to the search for oil and gas where a platform, subsea cluster, pipeline and/or other infrastructure already exists that would allow smaller accumulations to be commercial and support the operations already underway. This may occur when a company is the operator or when it is reasonable to believe an agreement can be made with the operator to process new production through the existing facility.

**Injectites** Intrusions and remobilization of sediment, commonly sand, that exhibit a range of geometries, including dikes, sills, columnar and irregular intrusions, and may be associated with sand extrusions. The development of sand injectites is related to elevation of pore pressure, hydrofracture of sealing strata, liquefaction, fluidization and injection of sand. It may be triggered by different mechanisms, such as seismicity, overpressuring caused by rapid loading, thermal pressurization and fluid migration.

**Instraslope basin** Type of basin located on the continental slope.

**Intelligent completions** Intelligent completions incorporate permanent downhole sensors and control valves to enable remote monitoring of data and management of production in real time.

**Interfaces** Interfaces represent sharp or transitional boundaries (pycnoclines) between water masses. These interfaces influence deep marine sedimentation through secondary oceanographic processes (tides, internal waves, etc.).

**Intergranular porosity** Pores between grains that are larger than ~30 μm in diameter (pores below this threshold are considered to be part of the microporosity).

**Intergranular volume** The volume between grains and commonly abbreviated as "IGV." This parameter is an important measure of the packing density of grains at the time of deposition and is also used to assess the compaction state of sandstones. IGV is determined as the sum of the intergranular porosity and all matrix and cement types that are not associated with transgranular fractures.

**Internal rate of return** The internal rate of return is the equivalent bank rate of interest that results from an investment. It is calculated by finding the discount rate that causes the net present value of a series of cash flows to equal zero. It is used to screen projects against the company's cost of money, whether debt or equity or a combination of both.

**Inversion** Transformation of seismic data to interpret and describe reservoir properties.

**Kernel** Filters with tunable weight parameters that perform convolution operations on a hidden layer of convolutional neural network.

**Kernel density estimation (KDE)** A function that stacks a Gaussian "bell curve" on top of each measurement, whose standard deviation is determined by the local probability density (Vermeesch, 2012).

**Laminar flow** Flow lines that are parallel in a moving body of fluid and do not cross (not turbulent).

**Laminar sublayer** Roughly centimeter-thick layer near the bed beneath a turbulent flow that is characterized by parallel flow lines (laminar flow).

**Levee** Positive topographic feature that forms a natural lateral boundary to channel environments.

**Linear normal fault** Fault angle is planar, i.e., has a high dip angle.

**Lining** Material applied to the burrow wall by the occupant. Passively accumulated material may be included, which adheres to the wall while the burrow is open (Bromley, 1996).
**Liquefaction** Process by which grains are no longer supported by intergranular friction but are momentarily suspended by the ambient pore fluid.
**Listric normal fault** A curved normal fault, generally concave upward, whose dip decreases with depth.
**Lithofacies** Rock or lithology properties assigned to a basin model. Important properties for basin models include thermal conductivity, heat capacity, radiogenic heat production, permeability, compressibility and capillary entry pressure. Also, mappable rock unit consisting of beds with similar bed thickness, typically with similar aggregate grain fabrics and grain size.
**Lithostatic pressure** The stress on a body of rock imposed by the weight of overlying matter.
**Lithostratigraphic (lithostratigraphy)** Subdivision of rock record according to lithology. The fundamental lithostratigraphic subdivision of the rock record is a formation, which is identified by its lithological characteristics and stratigraphic position on a wider scale.
**Lobe** See splay.
**Low-density turbidity current** Flow in which grains constitute $<10\%$ of the flow.
**Lowe Sequence** A sedimentation unit that has six principal subdivisions R1, R2, R3, S1, S2, S3 representing traction, traction carpet and suspension sedimentation stages, respectively, due to a waning (unsteady) high-density turbidity current.
**Low-energy flow** Relatively tranquil flow traveling at low speeds, carrying a low sediment load, typically not associated with erosion.
**Lowstand systems tract** A systems tract defined by a normal regressive stacking pattern, which follows a forced regression of equal hierarchical rank (Catuneanu, 2019a).
**Machine learning algorithms** Algorithms that automatically learn to make predictions from data.
**Macroporosity** Pores that are clearly distinguished using a petrographic microscope ($\sim>30\,\mu m$ in diameter); includes both primary and secondary pores.
**Mahalanobis distance** A measure of mathematical distance used to quantify relative distances between samples in a dataset.
**Mahalanobis outlier detection algorithm** An algorithm that detects outliers in a given set of data samples using Mahalanobis distances.
**Mantle** Outer zone of a zoned backfill (Bromley, 1996).
**Mass movement** Movement of material due to gravity acting on the solid particles in the mass.
**Mass-transport complex** Depositional and architectural complex comprised of multiple individual mass-transport deposits, which are coeval or closely chronologically and genetically related.
**Mass-transport deposit** Deposit resulting from a single depositional event of submarine mass flows with laminar behavior when moving downslope, including slides, slumps and debris flow.
**Mass-transport system** Depositional system characterized by a high proportion of mass-transport deposits.
**Matrix** Clay to fine silt sized matter that is deposited in the sediment (mud). Matrix, unlike pseudomatrix, is considered part of the intergranular volume.
**Matrix support** Grains are transported and supported by a muddy, cohesive matrix.
**Mean grain size** The volume weighted average of the grain size for a sample. See Chapter 6 in Taylor et al. (in press) for recommended measurement techniques.
**Mechanical compaction** Compaction that arises from some combination of grain rearrangement, ductile deformation or brittle failure.
**Megasequence** Group of depositional sequences defined at its base and top by tectonic drivers, such as significant mountain building events.
**Microporosity** Pores less than $\sim 30\,\mu m$ in diameter. Such pores are difficult to clearly resolve in a petrographic microscope and are often associated with clays or partially dissolved solids.
**Migration** Lateral or longitudinal movement of a depositional element, often referring to a channel.
**Mixed contourite features** Morphological bottom current (contouritic) features generated by a combination of depositional and erosional processes (e.g., contourite terraces, sub-circular depressions).
**Mixed depositional systems** Mixed features built by the interactions between along-slope currents and other transport or depositional processes. Mixed (turbidite-contourite) depositional systems are built by numerous, frequent interactions between along-slope bottom currents and downslope turbidity flows, which may be synchronous or asynchronous and

form intercalations of bottom current reworked sands, reworked turbidite deposits, turbidites, contourites and other deepwater deposits.

**Mobile substrate** Sedimentary accumulation that is not well lithified and prone to soft sediment deformation at a large scale, often affecting seafloor bathymetry and subsurface sedimentary layering to multiple kilometers below mudline.

**Modal category** A classification used to define the volume fraction of a rock component. Broadly, these categories are related to various types of pores as well as the following types of solid materials: grains, matrix, cements or replacements.

**Modified Upper/Lower Hashin-Shtrikman bounds** Heuristic bounds giving the narrowest possible range of elastic moduli as a function of porosity for any mineral-fluid mixture. Used to interpolate elastic properties between a high porosity end member and pure solid.

**Mottled (appearance)** Homogenized appearance of a texture, where primary sedimentary structures are not preserved, due to intense bioturbation.

**Mud volcanism** Geothermal phenomenon similar to mud diapirism, but with distinctive geological features and forms of landform development and quite different evolution processes that may develop both onshore and offshore.

**Mud volcanoes** Fluid venting structures formed by the extrusion to the surface or seafloor of a three-phase mixture of solids and fluids composed of sediments (mainly mud) and at least two fluid phases, water and gas (and occasionally liquid hydrocarbons), as a result of a combination of gas overpressure (high pore fluid pressure at great depth), gravitative instability of shales and fluid migration through faults or fissures.

**Mudclast** Grain of mud of variable size (generally sand to gravel sized).

**Muddy contourite** Muddy deposits (size $<0.063$ mm) deposited or substantially reworked by the action of bottom currents. These are homogeneous and highly bioturbated deposits, often with an indistinct mottled appearance, and can also show distinct burrows of varied ichnofacies. Bioclastic and/or carbonate components can be present.

**Mudstone** Rock that is predominantly composed of silt and/or clay.

**Multichannel streamer** Several-km-long cables with geophones used for seismic acquisition.

**Multi-dimensional scaling (MDS)** MDS can be used to visualize similarities and patterns within a dataset.

**Neoichnology** The study of modern traces or lebensspuren (Buatois and Mángano, 2011).

**Net cash flow** Net cash flow is the total cash generated by the project over a specified period, usually its commercial life. It is calculated as all cash in minus all cash out, or as revenues less expenses, investments and taxes. It differs from similar accounting definitions in that it does not include noncash items such as depreciation or amortization.

**Net gas** Net gas is the company's share of gas available for sale. For a tax-royalty contract, it is calculated as the total produced gas, less fuel and lease use, times the company's net revenue interest in the gas production stream. For a production-sharing contract, it is calculated as the gas production after fuel usage, shrinkage, other owners' shares, royalty interests and profit- or production-sharing interests are deducted.

**Net oil** Net oil is the company's share of liquid hydrocarbons available for sale. For a tax royalty contract, it is the total produced oil times the company's net revenue interest in the produced liquids. Under a production-sharing contract, it is the liquid production after other owners' shares, royalty interests and profit- or production-sharing interests are deducted.

**Net present value** The net present value of an investment is the net monetary gain or loss predicted to result from a proposed investment, discounted to the present at a specified interest rate. It is calculated using all the cash flows, including the initial investment, and discounting to the effective date using the discount rate.

**Net present value index** Net present value index is calculated as the net present value divided by the present value of the investment. It shows the value created by each dollar of investment and is used in capital budgeting for ranking investment opportunities under a limited budget.

**Net revenue interest** The net revenue interest (NRI), usually expressed for a lease in a royalty-income tax regime, is the percentage of the revenue that is owned by the working interest owners after all the royalties are paid.

**Net working interest** The net working interest in a project is the amount a company or partner owns in the project and the obligation it has toward expenditures

**Neuron** Elementary unit of artificial neural networks that enables learning input-output functional relationships from data with tunable weight parameters.

**New ventures** The business of finding areas that may eventually be used for exploratory drilling activity.

**Nonreservoir** Interval of rock that is of poor porosity and permeability, that is unable to hold and release a fluid, such as petroleum, during production.

**NYMEX** The New York Mercantile Exchange (NYMEX) has been trading futures in oil since 1983 and natural gas since 1990. This allows a market benchmark for forecasting future prices, although like the stock market in general, it fluctuates over time.

**Ocean bottom node seismic acquisition (OBN)** Ocean bottom node seismic acquisition is a type of ocean bottom acquisition where the seismic receivers are free-standing nodes fixed on the ocean floor, rather than towed on a streamer cable at the ocean surface. Ocean bottom receivers can also be connected to a cable laid down by a boat; however, this is limited to shallow water (less than 1 km), whereas ocean bottom nodes can be applied below thousands of meters of water depth. The advantages of OBN are that it acquires shear wave data, can provide wider azimuth range and can reacquire data over time ("4D").

**Ocean currents** An ocean current is a continuous, directional movement of seawater generated by forces acting upon the water including wind, the Coriolis Effect, waves, cabbeling or other thermobarometric and salinity differences.

**Oil revenue** Oil revenue is the company's revenue from sales of oil after other owners' shares and royalties are paid. For a tax and royalty contract, oil revenue is the product of net oil times oil price, before all other deductions. For a production-sharing contract, net oil revenue is the total of cost recovery and profit share, after royalties, taxes and all other deductions.

**OPEC** The Organization of Petroleum Exporting Countries (OPEC) is a cartel formed to establish oil-exporting policies and set prices in order to secure fair and stable prices for petroleum producers; an efficient, economic and regular supply of petroleum to consuming nations and a fair return on capital to those investing in the industry.

**Operating expense** Operating expenses are the cash outlays expended in the daily operation and maintenance of production from a given lease, field or other area of operation. Direct operating expenses include, but are not limited to, boats, helicopters, processing fees, fuel, salaries of field personnel, chemicals, equipment rental and so on. Indirect expenses usually include only overheads that are chargeable to that operation, such as a shore base.

**Opportunistic strategy** Population strategy used by an organism that can rapidly colonize vacant niches, e.g., a habitat depopulated by a temporary environmental change (Bromley, 1996).

**Organofacies** Mappable rock unit containing distinctive assemblage of organic matter without regard to the mineralogy.

**Outcrop analog** A group of rocks that can be seen with the naked eye, generally in situ, that provides a useful comparison to subsurface rocks of similar nature.

**Out-of-grade** See above grade.

**Overbank** Areas outside and away from canyons, channels, levees and splays in areas prone to passive accumulation of sediment known as background sedimentation including dilute flows and hemipelagic to pelagic sedimentation.

**Overburden** Interval of rock that is stratigraphically above intervals of source, reservoir and seal.

**Overgrowth** An authigenic mineral that grows after nucleating on a detrital material. Common overgrowth types include quartz, K-feldspar, albite, calcite and dolomite.

**Overhead expense** Overhead expenses are cash outlays for office salaries, equipment rentals, engineering salaries, supplies and so on, which are not chargeable directly to a single location.

**Paleo-climate** Either a model output or climatic conditions derived from known living habits of flora and fauna found in the rock record.

**Paleocurrent** Direction of travel derived from a sedimentary structure, allowing interpretation of sediment transport and depositional patterns at the age of the sedimentary deposit.

**Paleo-drainage reconstruction** A description of fluvial systems that were present over the AOI during the target time interval.

**Paleo-geographic model** Description of the erosional and depositional environments in paleo-space.

**Paleo-water depth** The depth of the water (sea or lake) at the time a rock is deposited.

**Partial annealing zone (PAZ)** In fission track thermochronology, the temperature range over which fission tracks in thermochronometric minerals anneal; the PAZ range is different for different thermochronometers, e.g., fission tracks in apatite anneal over a range of temperatures from ~60°C to ~110°C, dependent on apatite chemistry.

**Partial retention zone (PRZ)** In (U-Th)/He thermochronology, the temperature range over which the daughter product starts to be retained in the system; the PRZ varies for different thermochronometers; for example, $^4$He is fully retained in apatite below ~40°C, partially retained between ~40°C and 70°C and not retained above ~70°C (Farley, 2000, 2002).

**Passive margin** A continental margin formed originally by rifting, largely devoid of active tectonics and separated by an ocean.

**Passive margin foldbelt** A zone of deformation formed above salt or shale, usually caused by a combination of gravity sliding above a basinward-dipping detachment and gravity spreading of a sedimentary wedge with a seaward-dipping bathymetric surface.

**Pelagic** Vertical settling of particles under the influence of gravity; continuous process in the open ocean with very low rates of accumulation.

**Perched** Referring to sedimentary accumulation such as an apron that is above the base-of-slope and has filled the ponded accommodation.

**Permeability** Permeability is the flow capacity of rock, measured in darcies or millidarcies, determined by how well the pore spaces are connected.

**Petrography** Characterization of the composition and texture of sediments and rocks.

**Petroleum system** The petroleum system is a unifying concept that encompasses the four essential elements (source rock, in which oil and gas were formed; reservoir rock, in which oil and gas accumulate; seal rock, which prevents oil and gas from escaping the reservoir rock and overburden rock, which thermally stresses the source rock) and two time-based processes (trap formation and the generation-migration-accumulation of petroleum).

**Petrology** The study of the composition and texture of sandstones and how they evolve through geologic time.

**Phi (grain size)** A measure of grain size that is defined as the negative log base two of the grain size in mm: $f = -\log_2$ (mm).

**Plate model and paleo-geological model** Representation in the map view of where crustal fragments were during the time period of interest. Geological information such as emergent hinterland areas and the composition of those rocks.

**Plumites** Thin-bedded sand-mud couplets resulting from hypopycnal flow.

**Polymorph** Minerals that share the same chemical composition but have different crystal structures. Examples include calcite and aragonite as well as kaolinite and dickite.

**Ponded** Salt withdrawal creates ponded areas in topographic lows, commonly ranging from ~15 to 50 km in diameter. Ponded slopes can be healed (*healed-slope morphology*) by filling of the halokinetic basins and deposition of a perched apron after the ponded mini-basin is filled. High-relief ponded aprons in contrast tend to have higher sand content, particularly in the proximal portions, where intense channelization and scouring probably impart a relatively high connectivity to the various sand units.

**Power density** Related to reserve density, power density expresses the power output per area of an energy source and can be used to compare hydrocarbon resources to other energy source footprints.

**Presalt** Sedimentary layer that is both stratigraphically lower and older than the evaporite deposits overlying it.

**Pressure solution** Dissolution that occurs at contacts between solid objects including grains. Many clastic sedimentary petrologists avoid using this term because it implies that pressure is the primary driving force for dissolution, which contrasts with some natural observations and experimental data that indicate a number of other factors may be of comparable or greater significance. "Contact dissolution" is a better term to use given that it does not imply a specific driving force for the phenomenon.

**Pressure transient analysis** Pressure transient analysis (PTA) is a method for analyzing the reservoir permeability and wellbore damage (skin) by flowing a well and analyzing how its pressure builds up over time. It may also be used to detect reservoir boundaries.

**Primary porosity** Pores that were present at the time of deposition. Although the term is often used interchangeably with intergranular porosity, these terms strictly speaking are not synonymous in as much as there may be primary intragranular pores in grains at the time of deposition (for instance, in diatom tests).

**Prior falsification** A protocol for validation of the consistency of prior knowledge with data for Bayesian applications.

**Probability** Probability is the chance that a specified outcome will occur. It is the ratio of the number of times the specified outcome occurs to the total number of outcomes.

**Probability density plot (PDP)** A function that stacks a Gaussian "bell curve" on top of each measurement whose standard deviation is determined by the analytical precision (Vermeesch, 2012).

**Probability distribution function** A mathematical function that, for given input value, gives the relative likelihood of observing it among all possible values.

**Production** The business of full-scale extraction of a resource after a successful exploration and development phase.

**Production-sharing contract** In a production-sharing contract, the host government allows an oil company to explore and develop. The oil company pays for the investments and, when successful, is permitted to recover capital and operating expenditures from production. The remaining money is split between the government and the company according to agreed percentages.

**Pseudomatrix** Clay-rich regions within a sandstone that are associated with deformed grains. Such regions are sometimes misidentified as detrital clay matrix.

**Pycnocline** A pycnocline is the boundary separating two liquid layers of different densities. In the oceans, the formation of a pycnocline may result from changes in salinity, temperature or both.

**QEMSCAN** Quantitative evaluation of minerals by scanning electron microscopy.

**QFR** A framework grain classification scheme (quartz, feldspar, rock fragments) based on the scheme of Folk (1980).

**Quadratic distance classifier** A statistical classifier that uses a quadratic decision surface to assign classification categories to data samples.

**Quality adjustment** Price differential against benchmark crude prices assessed for gravity, viscosity, sulfur and other contaminants, including heavy metals.

**Quantitative seismic interpretation** The process of interpreting seismic data in a quantitative manner.

**Recovery efficiency** Recovery efficiency is the percentage of original oil in place that can be commercially produced. It depends upon the reservoir architecture and drives mechanism.

**Relative sea level** Sea level relative to a subsurface reference horizon, which measures changes in accommodation independently of sedimentation (Jervey, 1988).

**Replacement** Replacement refers to the solid material that formed within what was previously a pore that is associated with the dissolution of a solid modal category. Like cements, most replacements are made up of minerals, although organic matter and inorganic but noncrystalline phases such as opal-CT may also occur. A given phase such as kaolinite may have formed simultaneously as a cement and replacement within a sample. Furthermore, these related modal categories may be virtually identical with respect to their crystal habits and compositions. The distinction between replacement and cement, however, is critically important for accurately determining the sample IGV.

**Reserve density** Reserve density is the amount of recoverable hydrocarbons per unit area. Higher reserve density areas are more likely to be commercial in deepwater.

**Reservoir** Rock interval with high porosity and permeability (conventional), or rock type with high porosity, that can be stimulated (i.e., hydraulically fractured) to enable the production of fluids.

**Reservoir barrels** One reservoir barrel of oil is equivalent to 42 US gallons at reservoir temperature and pressure, including the dissolved gas that will be separated at the surface. The formation volume factor converts reservoir barrels to stock tank barrels.

**Reservoir deliverability** Geologic risk that assesses rock quality, oil quality and the ease of fluid flow through a reservoir.

**Reservoir modeling** The scientific and business practice of representing and characterizing a porous and permeable interval of rock using sophisticated software programs.

**Reservoir quality** Reservoir quality describes the ability of the rock pore system to store and produce significant volumes of fluids.

**Return on investment** The return on investment is the total gain from an investment divided by the capital cost over the life, expressed as a percentage. It is calculated as the undiscounted net cash flow plus the investment divided by the undiscounted investment. It is a measure of the efficiency of an investment opportunity in generating profits and is a popular way to rank investments in capital budgeting, though it does not recognize the timing of the cash flows.

**Revenues** Revenues are all inflows of cash or other enhancements that result from the company's ongoing operations.

**Reworked sedimentation unit** Sedimentation unit, such as a turbidite, that has been modified by a secondary process, such as a bottom current, to change the deposit.

**Reynolds number** A measure of inertial over viscous forces that describes the turbulence of a fluid.

**Rheology** Deformation and flow of matter, especially the non-Newtonian flow of liquids and the plastic flow of solids, determined by sediment type, distribution and amount of ambient fluid.

**Rock physics diagnostics** A methodology where rock physics models are used to infer information about rock texture if elastic moduli and porosity are known.

**Royalty** Royalty is the interest a lessor (landowner) retains in the production of minerals from a lease by virtue of the lease contract, concession agreement, profit-sharing agreement or regulation. It can be taken in kind or as a share of the sales revenue depending on the agreement.

**Sandstone** Rock that is predominantly composed of grains of sand (grains that range between 62 μm and 2 mm in diameter).

**Sandy contourite** Sandy contourites (fine- and coarse-grained) are sand deposits (size 0.063–2 mm) that are fully reworked (redeposited) by bottom currents. They can be characterized by as mostly poor to moderate (0.8–2 $\Phi$), but occasionally by well-sorted deposits with laminations and/or tabular-to-wedge bedding, from a few centimeters to several meters in thickness. Bioturbation is frequent, but less significant compared to muddy contourites. Sediment has a mixed siliciclastic/biogenic composition.

**Seal** Interval of rock that neighbors a reservoir rock, yet prevents fluid flow from the reservoir into another neighboring body of rock.

**Seal failure** Seal rock that caps hydrocarbon accumulation can fail due to high reservoir capillary pressure or late-stage deformation.

**Secondary porosity** Pores that form in response to the dissolution of solid materials such as grains, matrix or cements with diameters greater than 30 μm (large enough to be clearly discerned in an optical petrographic microscope). Intragranular pores and moldic pores are types of secondary pores.

**Sediment concentration** Amount of sediment, in % volume, in a flow.

**Sediment flux** Defines a volume of sedimentary material passing through a point in space over a specific period of time; for example, the sediment flux of siliciclastic material exiting the outfall point of a continental-scale system may be measured in megatons/year (MT/yr).

**Sediment gravity flow** Mixture of sediment and ambient fluid that flows downslope due to gravity acting on the solid particles in the flow.

**Sediment source terrane** Regions of the Earth's surface that denude and produce derivative sediment that enters source-to-sink systems; e.g., plutonic regions within a craton or exhumed sedimentary strata.

**Sediment supply** The volume of sediment delivered to a certain area.

**Sedimentary system** Group of depositional environments in which sediment is created and/or transported and may be deposited or carried through the environments without deposition.

**Sedimentation** Process of sediment accumulation due to sediment supply in excess of the energy of the sediment-transport agent (Catuneanu, 2019a).

**Sedimentation mechanics** The study of how sediment is entrained, transported and deposited according to the laws of physics.

**Sedimentation unit (bed)** The sedimentary deposit of one event, for instance, a single turbidity current will deposit a single sedimentation unit recording the event.

**Sediment-water interface temperature** The temperature at the seabed. It is the upper thermal boundary condition for basin models. It can be determined from estimated paleo-mean surface temperatures corrected by the paleo-water depth.

**Seismic attribute** Mathematically derived quantity from seismic data, that can enhance the certain information, which might be less obvious in traditional seismic amplitude image.

**Seismic geomorphology** It is the study of subsurface sedimentary strata produced by sedimentation and erosion with the help of seismic.

**Seismic inversion** Quantitative estimation of rock properties such as acoustic impedance from seismic reflection data.

**Seismic lithofacies** Rock units at seismic scale with characteristic lithology, bedding configuration, petrography and elastic properties.

**Sensitivity analysis** Sensitivity analysis is a risk analysis technique. It evaluates the changes in a project's profitability as a result of possible changes in the assumptions used, in order to assess the economic impact and the risks of key uncertainties. Variables often subject to evaluation are product price, reserve size, production rates, operating cost, capital investment and timing.

**Service contract** Service contracts are agreements between a country and an oil company, also referred to as a contractor. The company agrees to bear all capital and expense for the exploration and development. In return, if oil is produced, the government allows the contractor to recover costs through the sale of the oil and/or gas. In addition, the contractor is paid a fee per barrel of oil produced for taking the risk and providing expertise.

**Severance taxes** Severance taxes are assessed by most states in the United States as compensation for the removal of their natural resources. Each state has different rates for different minerals. Most are taxed as a percentage of revenue, but some are taxed at a fixed dollar amount per unit-of-production.

**Shelf** Flat or gently sloping sedimentary environment located basinward of the shoreface and landward of the top of the continental slope.

**Shortening** Reduction of the original horizontal bed length due to compression.

**Signal propagation** The signal of external environmental forcings on an S2S system is often recorded as geomorphic and stratigraphic products; this term refers to the down-system transfer of a quantifiable record of these earth-surface processes within an S2S system (see Romans et al., 2016).

**Signature bonus** Signature bonus is the amount of money initially paid by the contractor to the owner of the mineral rights for a mineral lease or contract area that will give the lessee the right to explore and develop the underlying minerals. It may also refer to a lump sum payment stipulated by contract and triggered by some event, such as a discovery, production test or a certain production rate, being reached. May also be called a licensing fee.

**Siliciclastic** Of or pertaining to sedimentary environments dominated by sediment characterized by quartz, feldspar and lithic grains.

**Simulation** The process by which a computer solves numerically, rather than analytically, a set of equations.

**Single point anchor reservoir (SPAR)** The SPAR is a platform anchored to the sea bottom using multiple chains and anchors. This design allows wellheads at the surface (dry trees), simplifying completion and workover requirements. The upright design, like a "tin can" or "bobbing cork," gives favorable sea response and handles robust loads, allowing large amounts of production equipment.

**Sink-to-source analysis** A suite of analytical techniques and workflows that examine the effects of the catchment (source) and basin (sink) dynamics on the generation, transport, composition, distribution and depositional timing of clastic sediments in modern and ancient sedimentary systems.

**Slide** A coherent mass of sediment that moves along a planar glide plane.

**Slope** Sedimentary environment characterized by a shallow gradient sloping toward the ocean basin, located basinward of the shelf and landward of the basin floor.

**Slump** Coherent mass of sediment that moves on a concave-up glide plane and undergoes rotational movements causing significant internal deformation and formation of compressional features (folds and thrusts).

**Soft sediment deformation** Alteration of primary sedimentary fabric prior to lithification from the consolidation of sediment, typically due to differential compaction and water escape of different types of sediment (mud vs. sand).

**Softmax activation function** A mathematical function used in the last layer of a neural network to normalize network outputs to probabilities over classification categories.

**Sorting** A measure of the distribution in the relative sizes of silt and larger sized grains. Sorting values are commonly defined in Trask units or as the standard deviation in grain size defined in phi units.

**Specific storage** A term from hydrology that refers to the unit volume of fluid that is released from a reservoir in response to a unit change in fluid pressure (hydraulic head).

**Splay** Body of sediment deposited in an area of relative unconfinement, originating from an area of relative confinement.

**Spreite** Laminated biogenic structure composed of closely spaced successive tunnel walls produced as a burrow is shifted laterally (broadside-on) through the sediment (Bromley, 1996).

**Standard deviation in phi** A grain sorting metric that is derived by converting grain size measurements to phi units and then calculating the standard deviation. This sorting unit is equivalent to "Folk sorting" (Folk, 1980) when the grain size distribution in phi units is Gaussian.

**Stochastic** A probabilistic approach to modeling in which input parameters are randomly sampled from a pre-defined probability distribution function.

**Stock tank barrels** One stock tank barrel of oil is equivalent to 42 US gallons at 60°F and atmospheric pressure.

**Strain partitioning** A deformation process in which the total strain experienced on rock is heterogeneously distributed in terms of the strain intensity.

**Stratigraphic hierarchy** A system of classification of stratigraphic cycles on the basis of their absolute or relative scales (Catuneanu, 2019b).

**Stratigraphic sequence** A cycle of change in stratal stacking patterns, defined by the recurrence of the same type of sequence stratigraphic surface in the sedimentary record (Catuneanu, 2019a).

**Structureless sandstone** Sandy interval lacking sedimentary structures (also known as *massive*).

**Stylolite** A surface where the significant dissolution of solid material has occurred. Such surfaces may include a concentration of insoluble materials and may nucleate on clay laminae.

**Subcritical flow** If the Froude Number is less than 1, in which gravitational forces exceed the viscous forces, the flow is called a *subcritical flow*, otherwise known as *slow and tranquil* flow state.

**Subsidence** Sinking or gradual downward settling of the Earth's surface with little or no horizontal motion.

**Super basin** A basin with 5 BBOE produced and the equivalent amount remaining, several pay zones and play types, and substantial existing infrastructure (Fryklund and Stark, 2016).

**Supercritical flow** If the value is greater than 1, in which viscous forces exceed the gravitational forces, the flow is termed a *supercritical flow*, also known as a *fast and rapid* flow state.

**Supply and demand theory** The law of supply and demand is a theory that describes the interaction between sellers and buyers of a resource. The theory defines the relationship between the price of the product and the willingness of people to either buy or sell it. Generally, as price increases, people are willing to supply more and demand less, and vice versa when the price falls, until the market settles on a mutual volume and price.

**Suspension** Sedimentation process formed by vertical settling of grains; can range from sand to gravel and silt to clay.

**Suspension feeder** Animal that traps and feeds on seston (Bromley, 1996).

**Sweeping cross stratification** Low-angle (5–20 degrees) surfaces formed by dune migration and preserved in the deposit.

**Systems tract** A linkage of contemporaneous depositional systems forming the subdivision of a sequence (Brown and Fisher, 1977).

**Taper angle** The taper angle of a wedge is the angle between its free surface and basal detachment.

**Tension leg platform (TLP)** A TLP is a floating platform anchored to the sea bottom using numerous pipes or tendons. This design allows wellheads at the surface (dry trees), simplifying completion and workover requirements. Weight limitations are an issue, which tends to limit production equipment and rates.

**Terrigenous** Derived from or having to do with a continental setting.

**Thermochronology** The quantitative study of the thermal histories of rocks using temperature-sensitive radiometric dating methods (e.g., $^{40}Ar/^{39}Ar$, K-Ar, fission-track and (U-Th)/He).

**Thermochronometer** The pairing of a thermochronology age dating method with a host mineral or bulk rock, commonly apatite, zircon, magnetite, rutile and titanite.

**Thin-skinned deformation** Style of deformation that only involves overburden sedimentary rocks and not deeper basement rocks.

**Tier (shallow, middle, deep)** Location within the substrate (Bromley, 1996).

**Toe-of-slope** The transitional zone at the lower reach of a continental slope where the maximum gradient turns from steep (>1–2°) to shallow (<1°). At toe-of-slope, depositional processes become dominant over erosional processes.

**Topological** A topological rock physics model assumes the constituent parts are interrelated or arranged together, like the clay-particles filling the pore space of sand grains.

**Touchstone™** A software system developed by Geocosm LLC for analysis, simulation and prediction of sandstone and siltstone diagenesis and reservoir quality.

**Trace fossil** Fossilized structure produced in unlithified sediment, sedimentary rock or other substrate by the activity or growth of organisms (Bromley, 1996).

**Tracemaker** Animal responsible for the trace fossil (Bromley, 1996).

**Traction** Process of sedimentation defined by transport of grain aggregates along the bed, often forming ripples, plane beds or dunes.

**Traction carpet** Interval in a high-density turbidity current deposit that is inversely graded and formed due to frictional freezing (Lowe R2, S2 divisions).

**Transgressive systems tract** A systems tract defined by a transgressive stacking pattern (Posamentier et al., 1988).

**Transitional flow** Sediment gravity-flow that forms due to fluid turbulence (turbidity) and matrix strength (cohesion).

**Translational domain** A zone in a passive margin foldbelt system where lateral translation along a detachment layer occurs and is a neutral zone of shortening. Commonly recognized by the presence of stacked ramp syncline basins.

**Transportation fees** Cost assessed to move each barrel from the owners' production point to the end purchaser.

**Transpression** Type of strike-slip deformation that is oblique, creating local compression.

**Transtension** Type of strike-slip deformation that is oblique, creating local extension.

**Trapping configuration** Disposition of strata that permits focusing of petroleum in a pooled area.

**Trask sorting** A grain sorting metric that is determined as the square root of the ratio of the 75th and 25th percentile sizes in mm.

**Turbidite** Deposit from a turbidity current.

**Turbulent flow** Flow in which there is extensive mixing of fluid vortices.

**Type section** Interval or area of rock that is described as the most complete or best of its type.

**Ultimate recovery** The ultimate recovery of a reservoir is the total amount of oil produced or expected to be produced over its entire producing life. It is the original oil or gas in place times their respective recovery efficiencies.

**Umbilical lines** Umbilical lines are bundles of hydraulic tubes, electrical cables and optic fibers that allow control of subsea wellheads from a remote or surface facility. They are essential for subsea wellheads.

**Unconfined** Sedimentary environment offering a wide zone for available sedimentation.

**Uplift** Displacement of Earth's surface with respect to the geoid (see England and Molnar, 1990; "displacement in the direction opposite to the gravity vector").

**Vertical effective stress** The portion of the stress is borne by the solid portion of the rock that is due to vertical sediment loading. Note that sandstone compaction responds to overall effective stress, which may exceed the vertical effective stress due to tectonic forces.

**Vp** P-wave velocity.

**Vs** S-wave velocity.

**Water drive** Water drive is the mechanism that occurs when a lateral or underlying aquifer moves into the reservoir, displacing petroleum until the aquifer is depleted. Water production eventually exceeds petroleum production, and the well or field produces too much water to be economically viable.

**Water elutriation fabric** Aggregates of grains in a sedimentary rock that show evidence of water escape during initial stages of diagenesis, during compaction; dish and pipe structures are common types of water elutriation fabrics in turbidites, as well as flame structures.

**Water masses** A water mass is an identifiable body of water with a common formational history, whose physical properties differ from that of the surrounding water and which originated in a particular region of the ocean.

**Water saturation** Water saturation is the percentage of the pore space between the grains of rock that is occupied by water. In an oil reservoir, the oil saturation is equal to 1.0-water saturation.

**Wellhead price** Price received for the sale of gas or crude oil. For oil, it is typically the benchmark price less transportation fees and less quality adjustment. For gas, it is typically based on Btu content at the index price, less a transportation fee.

**West Texas Intermediate (WTI)** West Texas Intermediate is the oil benchmark (index or reference) price generally used in the United States. It is a light, sweet crude and other oils are priced at sale based on differences in gravity and contaminants.

**Wide-azimuth seismic acquisition (WAZ)** WAZ is a seismic data acquisition technique that employs a wide distribution of source-receiver azimuths. This is typically achieved by employing multiple source boats. The advantage of having a range of azimuths is that the data can image complex geology better than traditional narrow-azimuth seismic (NAZ).

**Yield strength** The maximum stress that can be applied without changing the material permanently; debris flows with higher yield strength behave more as plastics than fluid mixtures.

**Zoeppritz equation** A mathematical equation that calculates the partitioning of seismic wave energy at a subsurface horizon.

**Zone fill** Burrow fill that consists of two or more discrete sediment types arranged concentrically; a mantle surrounding a core (Bromley, 1996).

# Index

Note: Page numbers followed by *f* indicate figures, *t* indicate tables, and *b* indicate boxes.

## Q

## T

CPI Antony Rowe
Eastbourne, UK
August 24, 2023